ENGINEERING

DRAWING & DESIGN

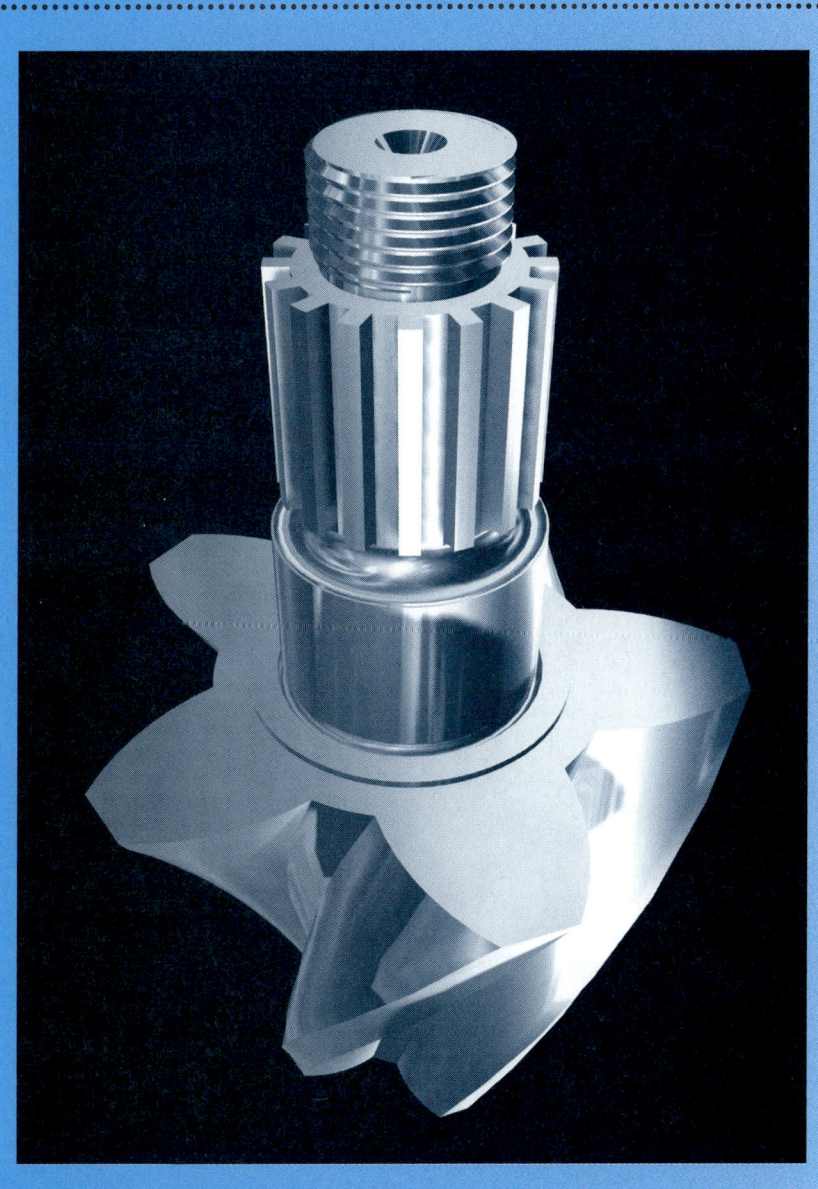

ENGINEERING

4th Edition

DAVID A. MADSEN
Faculty Emeritus, Former Department Chair Drafting Technology, Autodesk Premier Training Center
Clackamas Community College, Oregon City, Oregon
Autodesk Developer Network Member
Director Emeritus, American Design Drafting Association

DAVID P. MADSEN
Instructor, Program Coordinator Drafting Technology
Northwest College, Powell, Wyoming
Autodesk Developer Network Member
SolidWorks Research Associate
American Design Drafting Association Member

J. LEE TURPIN
Former Drafting Technology Instructor, Department Chair, Vocational Counselor
Clackamas Community College, Oregon City, Oregon

DRAWING & DESIGN

THOMSON

DELMAR LEARNING

Australia Brazil Canada Mexico Singapore Spain United Kingdom United States

Engineering Drawing and Design, Fourth Edition
David A. Madsen, David P. Madsen, J. Lee Turpin

Vice President, Technology and Trades SBU:
David Garza

Director of Learning Solutions:
Sandy Clark

Senior Acquisitions Editor:
James DeVoe

Development:
Mary Clyne

Marketing Director:
Deborah S. Yarnell

Marketing Coordinator:
Mark Pierro

Production Director:
Patty Stephan

Senior Production Manager:
Larry Main

Senior Content Project Manager:
Stacy Masucci

Content Project Manager:
Mary Beth Vought

Technology Project Manager:
Kevin Smith

Technology Project Specialist:
Linda Verde

Editorial Assistant:
Tom Best

Library of Congress Cataloging-in-Publication Data
Library of Congress Cataloging-in-Publication Data
Madsen, David A.
   Engineering drawing & design / David A. Madsen, David P. Madsen, J. Lee Turpin.—5th ed.
       p. cm.
   Includes index.
   1. Mechanical drawing. I. Title: Engineering drawing and design. II. Madsen, David P. III. Turpin, J. Lee. IV. Title.
   T353.M1965 2007
   604.2—dc22
                                2005056945

ISBN-13: 978-1-4180-2987-6
ISBN-10: 1-4180-2987-4

NOTICE TO THE READER

CONTENTS

Chapter 20—Belt and Chain Drives 654

Chapter 21—Welding Processes and Representations 672

SECTION 6
SPECIALTY DRAFTING AND DESIGN 701

Chapter 22—Industrial Process Piping 702

Chapter 23—Structural Drafting 743

CD Appendices: There is a CD available at the back of *Engineering Drawing and Design*, 4e. The CD contains a variety of valuable features for you to use as you learn engineering drawing and design. The CD icon, placed throughout this textbook, directs you to features found on the CD.

CD Appendix A—American National Standards of Interest to
    Designers, Architects, and Drafters
CD Appendix B—ASME Standard Line Types
CD Appendix C—Dimensioning and Tolerancing Symbols and
    ASME Dimensioning Rules
CD Appendix D—Designation of Welding and Allied
    Processes by Letters
CD Appendix E—Symbols for Pipe Fittings and Valves
CD Appendix F—Computer Terminology and Hardware
CD Appendix G—ASME/ANSI Exercises

CD Instructions: Access the CD to view the CD chapters, appendices, chapter tests, and selected chapter problems:

■ Place the CD in your CD drive.

■ The CD should open (start) automatically.

■ If the CD does not start automatically, pick the Start button in the lower left corner of your screen, and select Run . . ., followed by accessing the CD drive on your computer.

■ Pick the desired chapter, appendix, test, or problem application button on the left side of the CD window.

PREFACE

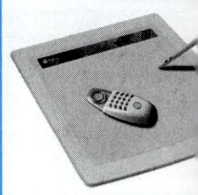

For more than twenty years, students and instructors have relied on *Engineering Drawing and Design* to provide A-to-Z drafting and design instruction that keeps pace with industry's most important developments. Throughout its history of substantive revisions, *Engineering Drawing and Design's* authors have led the field by responding to user requests with timely coverage and relevant applications. The fourth edition continues the tradition with these exciting additions:

- 3-D CADD, solid modeling, and parametric applications highlighted in an expanded, full-color chapter.
- Updated U.S. National CADD standards.
- Updated ANSI, ASME, and related standards.
- An expanded and updated chapter on civil drafting returned to the main text.
- All-new material in Chapter 27 following the design process from start to finish, and providing an inside view of an actual product design from concept through manufacturing at Milwaukee Electric Tool Corporation.
- Dozens of new relevant CADD applications.
- Hundreds of new drafting and design problems, including team projects.

In addition, *Engineering Drawing and Design*, fourth edition, continues the proven success of its previous editions:

- Engineering Design Applications.
- CADD Applications throughout.
- Manual drafting content to support this application as needed.
- Numerous illustrative examples supporting text content.
- Actual industry drawing examples to pull chapter content together.
- Professional Perspectives.
- Math Applications throughout and a design drafting related math chapter on CD.
- Step-by-step layout methods.

- Engineering layout techniques.
- Practical and useful appendices.
- Real industry problems throughout.
- Chapter tests for examination or review on CD.
- Updated Website Research options throughout.
- ASME identification problems.

Engineering Drawing and Design includes the American National Standards Institute (ANSI) and the American Society of Mechanical Engineers (ASME) standards and common alternates, such as the American Welding Society (AWS), the American Institute for Steel Construction (AISC), and the Construction Specifications Institute (CSI). Also presented, when appropriate, are standards and codes for specific engineering fields. One important foundation to engineering drawing and design, and the implementation of a common approach to graphics nationwide, is the standardization in all levels of drawing and design instruction. When you become a professional, this text will serve as a valuable desk reference.

AMERICAN DESIGN DRAFTING ASSOCIATION (ADDA) APPROVED PUBLICATION

The content of this text is considered a fundamental component of the design drafting profession by the American Design Drafting Association. This publication covers topics and related material, as stated in the ADDA Curriculum Certification Standards and the ADDA Certified Drafter Examination Review Guide. Although this publication is not conclusive with respect to ADDA standards, it should be considered a key reference tool in pursuit of a professional design-drafting career.

ENGINEERING DRAWING AND DESIGN CURRICULUM OPTIONS

This conversational, easy-to-use text is comprehensive enough to satisfy your entire curriculum. The chapters can be used as presented or rearranged to fit any of the following courses:

- The Engineering Design Process
- Engineering Graphics
- Computer-Aided Design and Drafting (CADD)
- Mechanical Drafting
- Descriptive Geometry
- Manufacturing Materials and Processes
- Welding Processes and Weldment Drawings
- Geometric Dimensioning and Tolerancing (GD&T)
- Tool Design and Drafting Fundamentals
- Mechanisms: Linkages, Cams, Gears, and Bearings
- Belt and Chain Drives
- Pictorial Drawings
- 3-D CADD, Solid Modeling, Animation, and Virtual Reality
- Structural Drafting
- Civil Drafting
- Industrial Pipe Drafting
- Heating, Ventilating, and Air-Conditioning (HVAC)
- Pattern Development
- Precision Sheet Metal Drafting
- Fluid Power
- Engineering Charts and Graphs
- Electrical Drafting
- Electronic Schematic Drafting

CHAPTER FORMAT

Each chapter provides realistic examples, illustrations, and related tests and problems. The examples illustrate recommended drafting and design presentation based on ASME/ANSI standards and other related national standards and codes, with actual industry drawings used for reinforcement. The correlated text explains drawing techniques and provides professional tips for skill development. Step-by-step examples provide a logical approach to setting up and completing the drawing problems. Each chapter has these special features:

Engineering Design Application

The Engineering Design Application leads the content of every chapter. This section gives you an early understanding of the type of engineering project that is found in the specific design and drafting area discussed in the chapter.

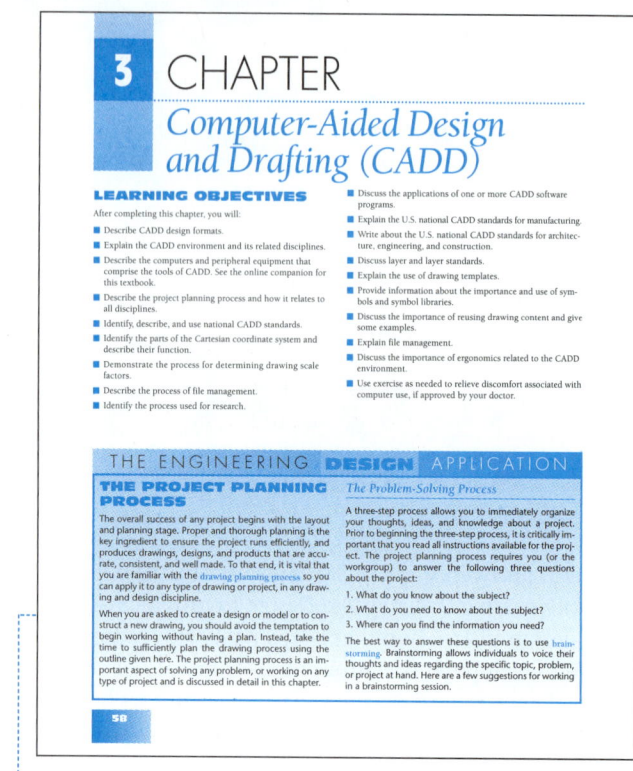

THE ENGINEERING DESIGN APPLICATION

CADD Applications

CADD Applications are provided in each chapter to illustrate how the use of CADD is streamlining the design and drafting process.

CADD APPLICATIONS

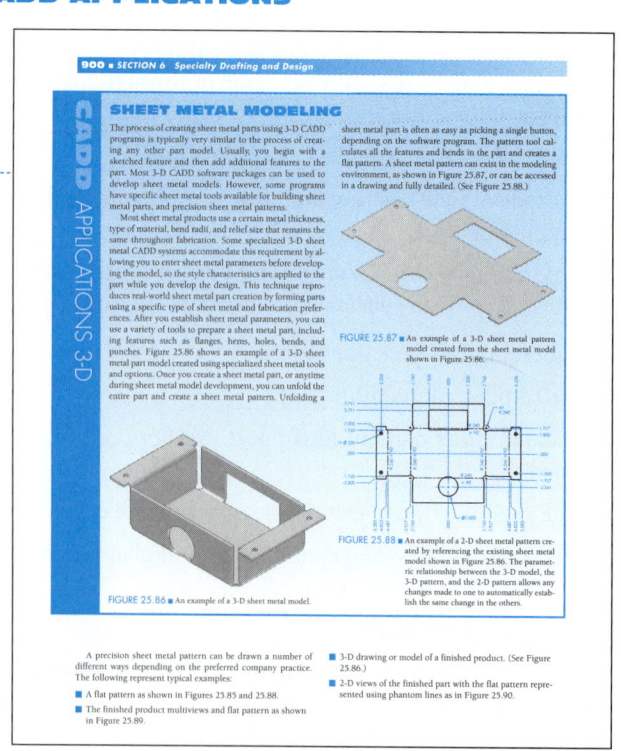

Step-By-Step Drafting Procedures And Techniques

Each chapter has step-by-step instructions for applying manual and computer-aided drafting techniques.

Math Applications

Math Applications, math problems, and practical drafting problems appear in every chapter. These elements provide examples and instruction on how math is used in the specific discipline. CD Chapter 3, Engineering Drawing and Design Math Applications, is a comprehensive math chapter covering math instruction used in the drafting and design industry. The CD Chapter 3 Math instruction coorelates with the Math Applications and Problems found at the end of most chapters.

MATH APPLICATIONS

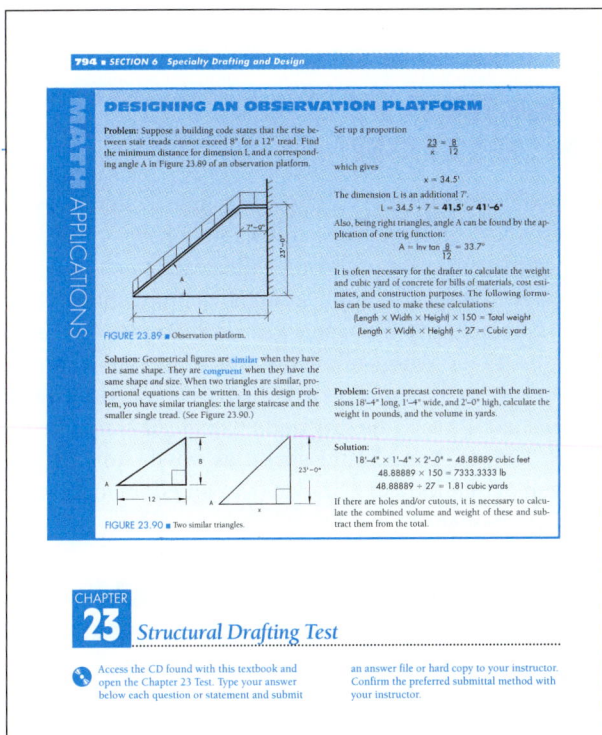

Metric Applications

Values presented in the content are in inches and are identified with (in.). Related metric values are given in parentheses using the appropriate equivalencies. Specific metric applications are given throughout this textbook. Metric applications are a very important part of the drafting and design industry. Chapter ending problems are also given for inch and metric applications and are labeled accordingly with (inch) or (metric). Illustrative examples provided throughout the text are displayed using metric values in millimeters, unless otherwise specified.

Related Tests

Related chapter tests appear on the CD for examination or review. You will see a CD icon at the end of each chapter referring you to the CD. The tests ask for short answers or drawings to confirm your understanding of chapter content.

Problems

Each chapter ends with numerous real-world drafting and design problems for you to practice what you have learned. This book contains over 1,000 problems, varying from basic to complex. Problems are presented as real-world engineering sketches, pictorial engineering layouts, and actual industry projects. You can use manual or computer-aided drafting depending on your course objectives. Advanced problems are given for challenging applications or for use as team projects. Several chapters will refer students to the CD for additional problems.

RELATED TESTS

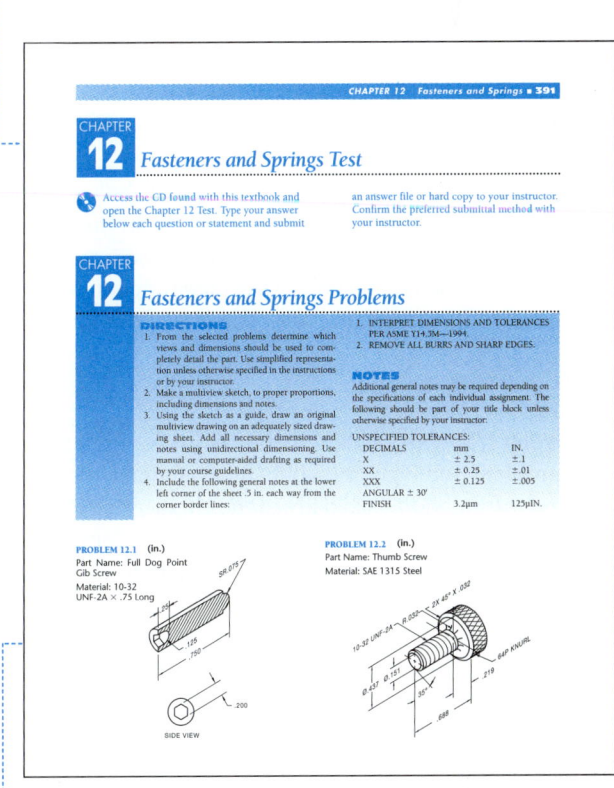

PROBLEMS

Professional Perspective

The Professional Perspective is a boxed article at the end of most chapters explaining how to apply the skills and knowledge discussed in the chapter to a real-world, job-related setting. Professional perspectives give you an opportunity to hear what actual engineers, designers, and drafters have to say about the design drafting industry and what you can expect as a drafter. Some contain representative examples of a drawing related to chapter content.

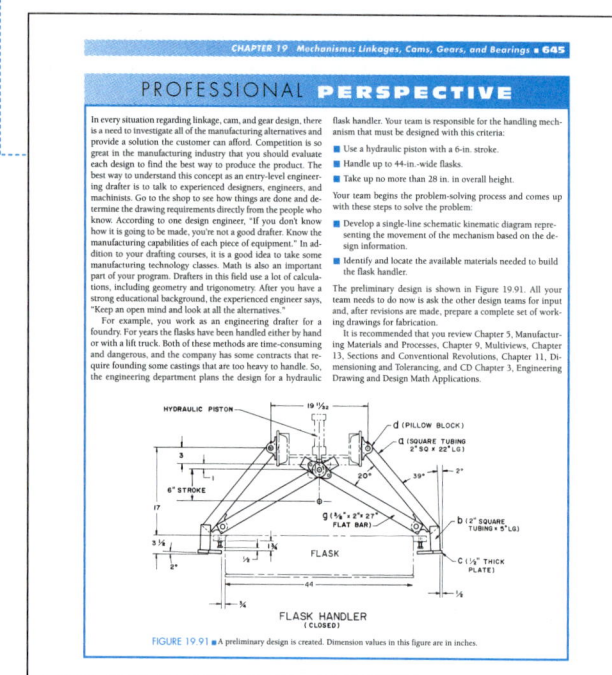

Website Research

Website Research is a feature that is placed at the end of most chapters, providing key websites where you can do additional research, find standards, and seek manufacturing information or vendor specifications related to the chapter content. Notice: websites and links may change after the date of this publication.

Related Appendices

Each chapter refers to the key appendices for your reference and use in problem solving. The appendices contain the types of charts and information used daily in the engineering design and drafting environment. Appendices can be copied for use as desk references as needed. These appendices include common fastener types and data, fits and tolerances, metric conversion charts, tap drill charts, and other manufacturing information used on engineering drawings. There is a complete list of abbreviations and a comprehensive glossary. It is also recommended that you learn to use other resources, such as the *Machinery's Handbook*, ASME/ANSI standards, other specific industry standards related to chapter content, and appropriate vendor's catalogues.

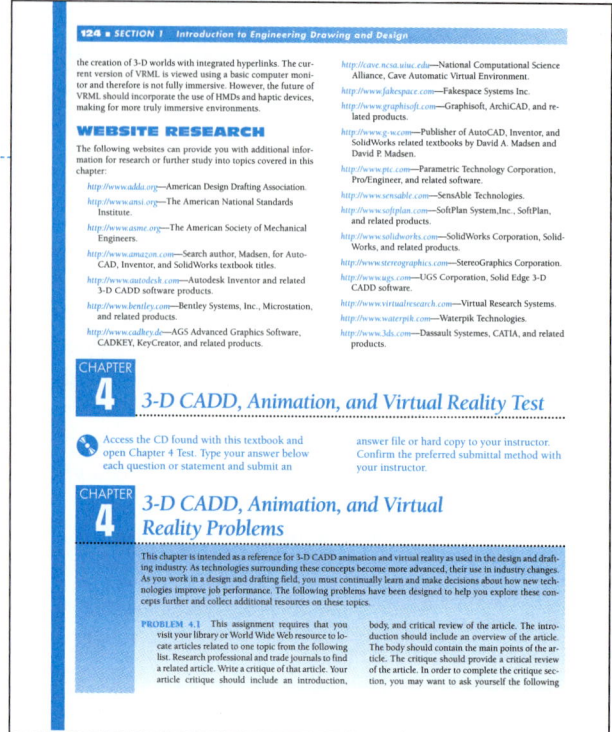

PULLING IT ALL TOGETHER

Looking at actual industry drawings is an excellent way to pull together what you have learned. Chapters have actual industry drawings placed along with other illustrative examples so you can see how the specific application is accomplished in the "real world". Chapters also often end with an example of a fairly complex industry drawing, allowing you to see how previously learned content is applied in industry. These drawings have been selected for their quality and characteristics related to the chapter content. You should spend as much time as possible looking at actual industry drawings in an effort to help you visualize the thought that goes into creating drawings.

CD ACCOMPANYING ENGINEERING DRAWING AND DESIGN

The CD at the back of this textbook contains a variety of valuable features to aid your study of engineering drafting and design. Watch for the CD icons placed throughout the textbook, directing you to one of the following CD features:

■ Specialty drafting and design chapters:
  ■ Fluid Power
  ■ Engineering Charts and Graphs
  ■ Engineering Drawing and Design Math Applications
■ Chapter Tests
■ A variety of extra Chapter Problems

- Additional Appendices
- ASME/ANSI exercises

CD INSTRUCTIONS

Access the CD to view the CD chapters, appendices, chapter tests, and selected chapter problems using these instructions:

- Place the CD in your CD drive.
- The CD should open (start) automatically.
- If the CD does not start automatically, pick the Start button in the lower left corner of your screen, and select Run . . . , followed by accessing the CD drive on your computer.
- Pick the desired chapter, appendix, test, or problem application button on the left side of the CD window.

INDUSTRY APPROACH TO PROBLEM SOLVING

Your responsibility as a drafter is to convert engineering sketches or instructions to formal drawings. This text explains how to prepare drawings from engineering sketches by providing you with the basic guides for layout and arrangement in a knowledge-building format. One concept is presented before the next is introduced. Problem assignments are presented in order of difficulty within each chapter and throughout the text. The concepts and skills learned in one chapter are built upon and used in following chapters, so that by the end of the text, you have the ability to solve problems using a multitude of previously learned discussions, examples, and activities. The problems are presented as pictorial or actual industrial layouts in a manner that is consistent with the engineering environment. Early problems provide suggested layout sketches. It is not enough for you to duplicate drawings from given assignments; you must be able to think through the process of drafting development. The goals and objectives of each problem assignment are consistent with recommended evaluation criteria and based on the progression of learning activities.

COMPUTER-AIDED DESIGN AND DRAFTING

Computer-aided design and drafting (CADD) is presented throughout this text. CADD topics include:

- CADD hardware appendix.
- CADD software used in drafting and design.
- U.S. National CADD Standards.
- Recommended CADD layers.
- CADD material requirements.
- Specific CADD applications.

- CADD menus and symbol libraries for specific engineering drafting applications.
- Increased productivity with CADD.
- 3-D CADD and solid modeling as used in industry.
- Parametric CADD applications.
- The CADD environment in industry.

DESIGN PROJECTS

Chapters contain projects that let you practice your design knowledge and skills. These are advanced problems requiring you to systematically determine a desired solution. There are challenges in manufacturing knowledge, tolerances, accuracy, and other issues related to the specific discipline.

TEAM PROJECTS

Some projects are designed to be solved in a team approach. Groups of students can develop their own team organization and establish the best course of action to create the desired solutions.

COURSE PLAN

Section 1—Introduction to Engineering Drawing and Design

Introduction to Engineering Drawing and Design

Chapter 1 provides a detailed look at drafting as a profession and includes a brief history, occupations, professional organizations, occupational levels, opportunities, career requirements, seeking employment, CADD issues, workplace ethics, copyrights, patents, and trademarks.

Manual Drafting Equipment, Media, and Reproduction Methods

Chapter 2 covers the study of traditional manual drafting equipment, materials, and reproduction methods for drafting, with specific instruction on how to use tools and equipment.

Computer-Aided Design and Drafting (CADD)

Chapter 3 provides an in-depth introduction to CADD, including U.S. National Standards, developing CADD skills, drafting with the computer, using menus, CADD in industry, file management, reusing drawing content, CADD symbols and symbol libraries, layers and layer management, creating and using drawing templates, 2-D and 3-D coordinate systems, introduction to 3-D modeling and animation, ergonomics and CADD workplace exercises, the CADD environment, the job market, CADD drawing materials, and CADD software manufacturers. CD Appendix F provides basic coverage of computer terminology and computer drafting equipment if you need to learn the fundamentals or have a refresher.

3-D CADD, Animation, and Virtual Reality

Following comprehensive CADD coverage, Chapter 4 provides the modern use of solid modeling with actual industry applications. The different types of CADD models are explained, along with illustrative examples. You will learn the function of parametric solid model parts and assemblies and see how to use automation to create 2-D drawings from models. You will learn how models can be animated to function as in a real world situation. Virtual reality is explained and demonstrated with use in actual industry and architectural environments.

Manufacturing Materials and Processes

Chapter 5 gives you the most complete coverage of manufacturing materials and processes found in any textbook of this type. The chapter includes product development, manufacturing materials, material numbering systems, hardness and testing, casting and forging methods, metal stamping, manufacturing and design of plastics, design and drafting related to manufacturing processes, complete machining processes, computer numerical control, computer-integrated manufacturing, machine features and drawing representations, surface texture, design of machine features, and statistical process control. A solid understanding of materials and processes will prove to be a valuable asset as you continue your education of drafting and design using this textbook.

Section 2—Fundamental Applications

Sketching, Lines, Lettering, and Geometric Construction

Sketching, lines, lettering, and geometric constructions are explained and extensively detailed in Chapters 6, 7, and 8. These chapters show you how to properly use manual and computer-aided drafting in accordance with ASME/ANSI standards. The sketching techniques you learn will continue to be used throughout your drafting and design education and into the profession. Sketching skills are important in preparation for creating manual and CADD drawings and for communication with others in the industry. Lines and lettering make up the foundation of engineering drawings. Proper application of lines and lettering based on correct national standards is important for you to learn now and use throughout your career. Geometric construction is the basis of all geometric shapes and drafting applications, whether you are using manual or computer-aided drafting. These fundamentals will be used throughout your drafting and design education and into the profession.

Section 3—Drafting Views and Annotations

Multiviews and Auxiliary Views

Chapters 9 and 10 provide a complete study of multiviews and auxiliary views, in accordance with ASME/ANSI standards, with accurate and detailed instruction on topics such as view selection and placement, first and third-angle projection, and viewing techniques. These chapters provide step-by-step examples on how to lay out multiview and auxiliary view drawings.

Dimensioning and Tolerancing

Chapter 11 is developed in accordance with ASME Y14.5M and provides complete coverage on dimensioning systems, rules, specific and general notes, tolerances, symbols, inch and metric limits and fits, drawing specifications, dimensioning castings, forgings, plastic drawings, and dimensioning for computer-aided design (CAD) and computer-aided manufacturing (CAM). You are provided with a step-by-step example showing how to lay out a fully dimensioned multiview drawing. Additional coverage includes the basics of tool design and drafting applications.

Fasteners and Springs

Chapter 12 presents fastener and spring terminology. The complete range of fastening devices is covered, including screw threads, thread cutting, thread forms, thread representations and notes, washers, dowels, pins, rings, keys and keyseats, and rivets. Types of springs, spring ends, and spring applications are included. Fasteners and springs are covered after dimensioning practices so you can apply dimensions to your spring drawings, and then sectioning techniques follows so you can apply fastening systems to future drawings.

Sections, Revolutions, and Conventional Breaks

Chapter 13 explains every type of sectioning practice available to the mechanical engineering drafter. This chapter also includes proper representation of sectioned features that should remain unsectioned in a sectional view, conventional revolutions, and conventional breaks. Section views can be used as needed in drafting problems throughout the rest of this textbook.

Geometric Dimensioning and Tolerancing

In Chapter 14, geometric tolerancing symbols, terms, and applications are presented in accordance with ASME Y14.5M. This chapter features datums, feature control, basic dimensions, geometric tolerances, material condition, position tolerance, virtual condition, geometric tolerancing applications, and CADD usage. Geometric dimensioning and tolerancing education follows dimensioning and sectioning so you can apply these concepts as needed throughout your continued drafting and design education.

Pictorial Drawings and Technical Illustrations

Chapter 15 provides a complete review of three-dimensional drafting techniques used in manual and computer-aided drafting. The chapter includes extensive discussion of isometric, diametric, trimetric, perspective, exploded drawings, and shading methods. The content of this chapter can be introduced here or earlier in your education, depending on course objectives.

Section 4—Descriptive Geometry

Descriptive Geometry I and II

Chapters 16 and 17 provide the most complete and easy-to-understand descriptive geometry coverage available in any engineering drawing and design textbook. Topics include step-by-step explanation and drafting of true length of lines, true size and shape of planes, angles between lines and planes, slope, bearing, percent of grade, edge view of planes, true distance between a line and plane, true distance between a point and plane, true distance between two skew lines, finding the piercing point of a line into a plane, the true angle between intersecting planes, and graphically solving vector force problems. Problems are presented as actual engineering situations. Descriptive geometry can be taught as a separate course, along with multiviews and auxiliary views, prior to multiviews and auxiliary views, or as presented in this format.

Section 5—Working Drawings

Working Drawings

Working drawings, introduced in Chapter 18, pull together everything you have learned, including multiviews, auxiliary views, sectioning practices, dimensioning techniques, and fasteners, allowing you to draw a complete product with all of its individual parts, an assembly of the product, and a parts list that includes all of the parts listed, identified, and correlated with the detail drawings. Working drawings include details, assemblies, and parts lists, along with the most extensive discussion on engineering changes available. Also provided is a complete analysis of how to prepare a set of working drawings from concept to final product, and how to implement an engineering change. A large variety of problem projects are based on actual products found in a variety of applications in the real world. The working drawing section continues with assemblies created with linkages, cams, gears, bearings, and weldments.

Mechanisms: Linkages, Cams, Gears, Bearings, Belt and Chain Drives

Chapters 19 and 20 provide you with extensive coverage of linkage mechanisms, cams, gears, bearings, and belt and chain drives. The content includes how to design and draw these features as details and in assembly. Special attention is given to design calculations for gears and cams. Detailed information is given on the selection of bearings and lubricants. The use of vendors' catalog information is stressed in the design of belt and chain drive systems. Actual mechanical engineering design problems are provided for gear train and cam plate design.

Welding Processes and Representations

Welding coverage in Chapter 21 provides an in-depth introduction to processes, welding drawings and symbols, weld types, symbol usage, weld characteristics, weld testing, welding specifications, and prequalified welded joints.

Section 6—Specialty Drafting and Design

Engineering Drafting and Design Fields of Study

Chapters 22 through 26 provide you with instruction in specific engineering drawing and design fields. These chapters can be used individually for complete courses or combined with previous chapters as needed to fit your curriculum objectives. These chapters contain comprehensive instruction, problems, and tests. These fields include:

Industrial Process Piping — Chapter 22 provides explanations and detailed examples of pipe and fittings, piping symbols, valves and instrumentation, pumps, tanks and equipment, flow diagrams, piping plans and elevations, piping isometrics, and piping spools.

Structural Drafting — Chapter 23 covers reinforced and precast concrete, truss and panelized framing, timbers, laminated beams, steel joists and studs, prefabricated systems, structural steel, structural welding, and structural details. You have the option of completing one of several sets of complete structural drawings for light commercial buildings.

Civil Drafting — Chapter 24 discusses the complete discipline of mapping, including legal descriptions, survey terminology, and site plans. Civil drafting includes road layout, cuts and fills, and plan and profile drawings. You will go through the step-by-step process for drafting a road layout, road cuts and fills, plan and profile drawings, site plan, topography contours, and a site grading plan.

Heating, Ventilating, and Air-Conditioning (HVAC), Pattern Development, and Precision Sheet Metal Drafting — Chapter 25 provides complete coverage of HVAC systems, heat exchangers, HVAC symbols, single- and double-line ducted systems, working from an engineer's sketch, sections and details, common sheet metal pattern developments with step-by-step layout procedures, sheet metal intersections, bend allowances calculations, and precision sheet metal drawings.

Electrical Drafting and Power Substation Design — Chapter 26 provides the only coverage of its type in electrical power transmission. The content is specific to electrical drafting of wiring diagrams, cable assemblies, one-line and elementary diagrams, electrical power system symbols, plot plans, bus layouts, ground layouts, conduit layouts, electrical floor plan symbols, power supply plans, and schematics.

Electronic Schematic Drafting — Chapter 26 continues with content exclusively related to the electronics industry. This chapter includes topics such as block diagrams, electronic components and symbols, engineer's sketches, component numbering, operational amplifer schematics, integrated circuit systems, logic diagrams, large-scale integration schematics, surface mount technology, printed circuit technology, electronics artwork, layers, marking, drilling, assembly drawings, photo drafting, and pictorial diagrams.

Section 7—Engineering Design

The Engineering Design Process

The engineering design process chapter is placed at the end of the textbook specifically to allow you an opportunity to learn a vast amount about drafting and design theory and skills before proceeding with your own designs or the designs of your school or company. Chapter 27 provides an introduction to several engineering design systems and detailed coverage with step-by-step use of the phase gate design process. You will see how all aspects of the design process fit together. You will take the design of a product from engineering sketches through production, implementing CAD/CAM, parametric design, rapid prototyping, computer-aided engineering (CAE), concurrent engineering, collaborative engineering, reverse engineering, team projects, and other innovative topics. Creativity, collaboration, and the design process are emphasized in this chapter. Much of the discussion derives from information obtained from Milwaukee Electric Tool Corporation. Paralleling the general design process information is the tracking of an actual new product design from concept through full-production manufacturing. The product used for this design sequence example is the Milwaukee Electric Tool Corporation's V28 Lithium-Ion Sawzall® reciprocating saw.

Section 8—Specialty Drafting and Design on the Companion CD

Fluid Power—CD Chapter 1 gives complete coverage on forces, pressure, work, power, hydraulics, pneumatics, diagrams, symbols, equipment, systems, and operations. This chapter is located on the CD accompanying *Engineering Drawing and Design*, fourth edition.

Engineering Charts and Graphs—CD Chapter 2 provides the most complete coverage on rectilinear, log and semilog scales, surface, column, pie, concurrency, alignment, organizational, statistical process control (SPC), distribution, and pictorial charts. The chapter includes an in-depth analysis of chart design and step-by-step chart drafting techniques. This chapter is located on the CD accompanying *Engineering Drawing and Design*, fourth edition.

Engineering Drawing and Design Math Applications — CD Chapter 3 provides comprehensive math instruction for engineering drafting, and related fields. The content parallels the math applications and problems found in chapters throughout the textbook. The content is presented with numerous examples in a manner that is easy to use and understand.

CHAPTER LENGTH

Chapters are presented in individual learning segments that begin with basic concepts and build until each chapter provides complete coverage of each topic. The content of each chapter generally should be divided into logical teaching segments, pro-

viding you with an opportunity to create basic drawing as you progress into more complex problems.

Applications

This text contains CADD discussion and examples, engineering layout techniques, working from engineer's sketches, professional practices, and actual industry examples. The problem assignments are based on actual real-world products and designs. Special emphasis has been placed on providing realistic problems. Problems are presented as 3-D drawings, engineering sketches, and layouts in a manner that is consistent with industry practices. Many of the problems have been supplied by industry. Each problem solution is based on the step-by-step layout procedures provided in the chapter discussions. Problems are given in order of complexity so you gain exposure to a variety of engineering experiences. Early problems recommend the layout to help you save time. Advanced problems require you to go through the same thought process that a professional faces daily, including drawing scale, sheet size selection, view selection and layout, dimension placement, sectioning placement, and many other activities. Problems can be solved using manual or computer-aided drafting as determined by the individual course guidelines. All problems should be solved in accordance with recommended ANSI, ASME, or other related industry practices.

A Special Notice About Problems

You should always approach a problem with critical analysis based on view selection and layout and dimension placement when dimensions are used. Do not assume that problem information is presented exactly as the intended solution. Problems can be deliberately presented in a less-than-optimal arrangement. Beginning problems provide given layout examples allowing you to solidify your knowledge before advancing on your own. Advanced problems often require you to evaluate the accuracy of provided information. Never take the given information or accuracy of the engineering sketch for granted. This is also true as you progress into industry.

Problems are found in the textbook following each chapter, and additional chapter problems are located on the CD accompanying this textbook. A wide range of problems are available in both locations. In some cases, basic problems are placed on the CD, allowing you to copy the problems for solution. In other cases, advanced problems are found on the CD if you want an extra challenge, and for instructors to provide a variety of practical opportunities for their students. Watch for the CD icon identifying problems found on the CD. Instructions, such as the following, direct you to the CD problems:

 PROBLEMS. Access the CD found with this textbook and open the problem of your choice, or as assigned by your instructor. Solve the problems using the instructions

provided with this chapter or on the CD, unless otherwise specified by your instructor.

 MATH PROBLEMS. Access the CD found with this textbook and open the math problem of your choice, or as assigned by your instructor. Solve the problem or problems using the instructions provided.

Using Chapter Tests

 Chapter tests provide complete coverage of each chapter and can be used for instructional evaluation, as study questions, or for review. The chapter tests are located on the CD accompanying this textbook. Watch for the CD icon. Instructions, such as the following, direct you to the CD test:

 Access the CD found with this textbook to view the Chapter Test.

APPENDICES

The appendices contain the types of charts and information used daily in the engineering design and drafting environment. In addition to using the appendices found in this book, it is recommended that you learn to use other resources, such as the *Machinery's Handbook*, ASME/ANSI standards, other specific industry standards related to chapter content, and appropriate vendors' catalogues. The textbook appendices include:

- Abbreviations
- Conversion Charts
- Mathematical Rules Related to the Circle
- General Applications of SAE Steels
- Surface Roughness Produced by Common Production Methods
- Wire Gages (Inches)
- Sheet Metal Gages (Inch)
- Sheet Metal Thicknesses (Millimeters)
- Standard Allowances, Tolerances, and Fits
- Unified Screw Thread Variations
- Metric Screw Thread Variations
- ASTM and SAE Grade Markings for Steel Bolts and Screws
- Cap Screw Specifications

- Machine Screw Specifications
- Set Screw Specifications
- Hex Nut Specifications
- Key and Keyseat Specifications
- Tap Drill Sizes
- Concrete Reinforcing Bar (Rebar) Specifications
- Common Welded Wire Reinforcement Specifications
- ASTM A500 Square and Rectangular Structural Tubing Specifications
- Structural Metal Shape Designations
- Corrosion-Resistant Pipe Fittings
- Valve Specifications
- PVC Pipe Dimensions in Inches
- Rectangular and Round HVAC Duct Sizes
- Spur and Helical Gear Data
- CADD Drawing Sheet Sizes, Settings, and Scale Factors

CD Appendices

Access the CD accompanying *Engineering Drawing And Design*, fourth edition, for the following appendices:

- American National Standards of Interest to Designers, Architects, and Drafters
- ASME Standard Line Types
- Dimensioning and Tolerancing Symbols and ASME Dimensioning Rules
- Designation of Welding and Allied Processes by Letters
- Symbols for Pipe Fittings and Valves
- Computer Terminology and Hardware

Special CD Appendix— ASME/ANSI Exercises

The ASME/ANSI Exercises found in this appendix are actual industry drawing files containing *intentional ASME/ANSI errors*. You can correct the drawing files using CADD or redline prints to conform to accepted ASME/ANSI standards. This provides a valuable supplement for learning ASME/ANSI standards. Searching actual industry drawings in an effort to find errors helps you form a keen eye for correct drafting presentation and compliance with national standards. This activity is the function of a drafting checker in industry. Possessing this skill allows you to become more familiar with proper ASME standards, correct drawing layout, and proper dimension placement when creating your own drawings and when correcting drawings created by others.

SUPPLEMENTS

e.resource

This is an educational resource that creates a truly electronic classroom. It is a CD-ROM containing tools and instructional resources that enrich your classroom experience and make the instructor's preparation time shorter. The elements of *e.resource* link directly to the text and tie together to provide a unified instructional system. With *e.resource*, your instructor can spend time teaching, not preparing to teach.

Features contained on the *e.resource* include:

Syllabus

Lesson plans created by chapter. You have the option of using these lesson plans with your own course information.

Chapter Hints

Objectives and teaching hints that provide the basis for a lecture outline that helps instructors to present concepts and material. Key points and concepts can be graphically highlighted for student retention.

PowerPoint® Presentations

These slides provide the basis for a lecture outline that helps the instructor present concepts and material. Key points and concepts can be graphically highlighted for student retention.

Exam View Computerized Test Bank

Over 800 questions of varying levels of difficulty are provided in true/false and multiple-choice formats so instructors can assess student comprehension.

Drawing Files

Files for many of the drawings in the text are listed in chapter folders. Instructors and students can modify these files to create new illustrations or chapter problems, or instructors can make them available to students who desire to work within AutoCAD.

ASME Exercises

The same ASME/ANSI Exercises found on the student CD are also located on the *e.resource,* with accompanying solution drawings.

Video and Animation Resources

These .avi files graphically show the execution of key concepts and commands in drafting, design, and AutoCAD and let instructors bring multimedia presentations into the classroom.

Online Companion

Access the online companion for *Engineering Drawing and Design*, fourth edition, to view the Web Chapters.

Go to: *www.delmar.com* and follow these steps:

1. In the Technical Learning Solutions box, pick technology learning solutions or courseware.

2. From the Technical Learning Solutions list, select Online Companions.

3. From the Select Discipline pull-down list, pick CAD and Drafting.

4. In the Online Companions list, select Engineering Drawing and Design by David A. Madsen.

5. Select Release 4 Downloads.

6. Pick features as desired.

Alternately Go to:

http://www.delmarlearning.com/companions/index.asp?isbn= 1418029874 and select Release 4 Downloads, then pick the desired Web Chapter.

Workbook

While hundreds of problems are found throughout the core textbook, a workbook is developed to correlate with *Engineering Drawing and Design*, 4e. This workbook follows the main text and contains "survival information" related to the topics covered. The problem assignments take you from the basics of line and lettering techniques to drawing projects. Problems can be done on CADD or manually. ISBN 1-4180-2988-2.

Solutions Manual

A solutions manual is available with answers to end-of-chapter review questions and solutions to end-of-chapter problems. Solutions are also provided for the workbook problems. ISBN 1-4180-2990-4.

Videos

Two video sets are available, containing four 20-minute videos each. The videos correspond to the topics addressed in the text.

Set #1, ISBN 0-7668-3903-6. Set #2, ISBN 0-7668-3908-7. Video sets are also available on Interactive Video CD-ROM. Set #1, ISBN 0-7668-3914-1. Set #2, ISBN 0-7668-3915-X.

ACKNOWLEDGMENTS

Co-Authors

I would like to express a special thanks to my co-authors:

David P. Madsen—My son has always had a special interest in drafting and began working toward a career in drafting and education from an early age. Dave has over 10 years of experience in the drafting industry. His wide range of practical experience includes solid modeling and mechanical, commercial, and residential architectural, structural, civil, electrical, electronic, and technical illustration drafting. Dave holds an Associate of Science degree in General Studies, a Bachelor of Science degree in Technology Education, and a Master of Science degree in Educational Policy, Foundations, and Administrative Studies with specialization in Postsecondary, Adult, and Continuing Education. Teaching, education, and learning, especially as it applies to drafting technology and computer-aided design and drafting, have always been Dave's passion. His knowledge of drafting and related industries and enthusiasm for education is evident in the classroom and in his writing. Dave is currently the Instructor of Drafting Technology at Northwest College in Powell, Wyoming. He is an Autodesk Developer Network member, SolidWorks Research Associate, and American Design Drafting Association member. Dave is also the author of *Autodesk Inventor and Its Applications*, and *SolidWorks and Its Applications*, published by the Goodheart-Willcox Publisher. I am proud to have Dave as a co-author of *Engineering Drawing and Design*, fourth edition. His contribution is the rewriting of Chapter 4, 3-D CADD, Animation, and Virtual Reality, all new CADD Applications, and a comprehensive review and edit of chapters throughout the book.

J. Lee Turpin—Lee Turpin is a former Drafting Technology Instructor/Department Chair, and Vocational Counselor at Clackamas Community College, Oregon City, Oregon. Lee has over 30 years of experience, including industry, teaching, and counseling. Lee is my longtime friend and mentor. Without Lee's influence, my drafting and design educational career may not have blossomed. Lee is truly the kind of person everyone wants as a friend. He is one of the kindest people I have ever known. As a teacher, Lee cared about student success, and later he continued as a vocational counselor, directing many students into rewarding careers. I want to thank Lee for his experience in presenting descriptive geometry to students in a manner that is easy to follow and understand. Lee is the author of Chapters 16 and 17, covering the full range of descriptive geometry concepts.

Steve Brown—Steve is the author of CD Chapter 3, Engineering Drawing and Design Math Applications and most math applications and math problems. The math instruction is given in an innovative and practical manner that immediately addresses the real-world issues. Steve has a great ability to make math interesting, easy, and enjoyable to learn. Steve is a math instructor at Clackamas Community College, Oregon City, Oregon.

Co-Authors of Previous Editions

Dr. James Folkestad, the CADD modeling, animation, and virtual reality coverage found in *Engineering Drawing and Design*, third edition.

Karen Schertz, contributing the Engineering Design Process chapter found in *Engineering Drawing and Design*, third edition.

Terence M. Shumaker, original implementation of the introduction to CADD and related topics, industrial pipe drafting, and pictorial drawing chapters in *Engineering Drawing and Design*, first and second editions.

Catherine Stark, coverage of fluid power and civil drafting in *Engineering Drawing and Design*, first and second editions.

Reviewers and Industry Professionals

I would like to give special thanks and acknowledgment to the many professionals who helped formulate this work. Also, thanks to those who reviewed the manuscript in an effort to help publish the best possible text:

David Umfress

Marguerite Newton, Niagara County Community College

Rob Wright, Sherando High School

David W. Smith, Cincinnati State

Joseph I. Greenfield, State University of New York at Delhi

Patricia Godin Healey, Wake Technical Community College, Raleigh, North Carolina

Pam Koehler, Dunwoody College of Technology

Adam Wills, Tennessee Technology Center at Nashville

Robert Lurker, Central Missouri State University

Jeffrey Jahnke, Century College

Jerold Sowls, Fox Valley Technical College

In addition, special thanks go to Jerry Griess for his detailed and insightful technical edit of the entire text and solutions manual.

Many figures appearing in Chapter 23 and 25 were reprinted from Jefferis and Madsen, *Architectural Drafting and Design*, fourth and fifth editions, by Thomson Delmar Learning.

The quality of this text is enhanced by the support and contributions from industry and vendors. The list of contributors is extensive, and acknowledgment is given at each figure illustration. The following individuals and their companies gave an extraordinary amount of support toward the development of new content and illustrations for this fourth edition:

Milwaukee Electric Tool Corporation for helping with the specific design process information and the tracking of an actual product design from concept through manufacturing. The product used for this design sequence example is the Milwaukee Electric Tool Corporation's V28 Lithium-Ion Sawzall® reciprocating saw.

Special thanks is given to Tom James, Senior Vice President, Engineering for the Techtronic Industries Global Portable Power Tools Group, for supporting the use of a Milwaukee Electric Tool Corporation product design from conception through production. Additional special thanks is given to the following Milwaukee Electric Tool Corporation professionals who gave extensive time and support to the design sequence content:

David P. Serdynski, Senior Design Engineer

Troy Thorson, Design Engineer

Design content review:
Scott Bublitz, Industrial Designer—R&D Concept Team
Richard S. Peterson, Brand Media Manager
David Selby, Director of Engineering

PTC for numerous 3-D CADD models and related drawings; Michael M. Campbell, Vice President of Product Management, PTC.

The following people and their companies provided significant support with illustrative examples and actual industry drawing carried over to the fourth edition from previous editions:

Keith McDonald, FLIR Systems, Inc., Portland, Oregon

Vin Mehta, Hunter Fan Company, Memphis, Tennessee

Tom Pearce, Stanley Hydraulic Tools, A Division of the Stanley Works, Portland, Oregon

Dick Button, Fastener content, Fisher Controls International, Inc., Marshalltown, Iowa

Randee R. Buckle, Wendy's International, Inc., Dublin, Ohio

Martin Soll, Tool Design content, GD&T Project

Additional Industry and Professional Support

Accugraph Corporation, Ontario, Canada

Aerojet TechSystems Company, Sacramento, California

American National Standards Institute, New York, New York

Berol USA RapiDesign, Burbank, California

Bishop Graphics, Inc., Westlake Village, California

CALCOMP, Anaheim, California

Chartpak, Northampton, Massachusetts

Computervision Corporation, Bedford, Massachusetts

Consul and Mutoh, Ltd., Anaheim, California

Hyster Company, Portland, Oregon

Parker-Hannifin, Cleveland, Ohio

Jean K. Shirkofl, consultant, Portland, Oregon

T & W Systems, Huntington Beach, California

Ricardo Wilkins, Michael Spiegel, Fluid Air Components, Portland, Oregon

American Design Drafting Association (ADDA), Newbern, Tennessee

Olen Parker, Executive Director and Corporate Operations Officer. Special thanks to the Board of Directors and Board of Governors for their commitment, extra contribution, and support for the advancement of the design drafting profession. For a list and information about these special people, go to *www.adda.org*, select the Leadership button on the left navigation bar, followed by picking Board of Directors or Board of Governors. In addition to their professional support, I thank many of these people for their friendship.

Professional Consultants

Doug Major, Mark Hartman, John Boertjens, CAD Technology Corp., Franklin, North Carolina

Kasey Cassell, The Cork Screw Project

Bill Curtis, Curtis and Associates, Portland, Oregon

Tom Durston, Applications Engineer/EDA, Consultant

Cindy Easton, Technical Communications Manager, OrCAD, Inc., Beaverton, Oregon

Sabine Gossart, SolidWorks Corporation, Concord, Massachusetts

David P. Hammer, Hammer Inc., Medford, New Jersey

Chris Herford, Gary Roberts, Ball Valve Project

Richard Hertel, The Oil Pump Project

Daniel L. Hodgin, KH2A Engineering, Inc., Portland, Oregon

Kim Lockwood, The Flashlight Project

John Melloy, Fly Tying Vice Project

Karl Mayhew, Interface Engineering, Portland, Oregon

Mike McIntyre, The Landing Gear Project

John Melloy, Light Fixture Project

Jack Pitcher, The Table Vice Project

Roy H. Reiterman, P. E., Roy H. Reiterman & Associates, Consulting Engineers, WRI Technical Consultant

Everett Russ, PAE Consulting Engineers, Inc., Portland, Oregon

Leonard Sosnovske, Nevada Power, Las Vegas, Nevada

Timothy Y. Taylor, Matthew Bell, The Motor Project

Connie Willmon, Ankrom Moisan Architects, Portland, Oregon

Deborah Parker Wong, The Medialink Group, Kentfield, California

Ricardo Wilkins, Michael Spiegel, Fluid Air Components, Portland, Oregon

Rod Rawls, Reel Project, Drill Pump Project, Pipe Wrench Project, Auxiliary View, and Sectioning Projects

The publisher wishes to acknowledge the following contributors to the supplements package:

Kevin Standiford, Workbook and *e-resource*

Patricia Godin Healey, PowerPoint® Presentations

Adam Wills, Computerized Test Bank

From the Author

Professionally speaking, Design Drafting is my life! I was forever hooked after my first high school drafting experience. My education includes a Bachelor of Science degree and Master of Education degree from Oregon State University with a focus on anything "drafting". I have many years of practical experience in industry, as a design drafter in a wide variety of technical disciplines. I have also owned my own architectural design and construction business. My academic career started with my first teaching job at Centennial High School, in Gresham, Oregon. After three years, and with the help of my great friend and mentor, Lee Turpin, I seized the opportunity to join Clackamas Community College. I retired as Department Chair in 2001 after over 30 years of teaching. During my tenure, I was proud to be a member of the ADDA and its Board of Directors. The special honor of Director Emeritus was granted to me by the ADDA in 2005. My first publishing experience came in 1976 with *Geometric Dimensioning and Tolerancing*. Titles authored in architectural, civil engineering, and an AutoCAD series followed. I am proud that my son, David P., has also chosen to be a drafting educator and textbook author. I love my farm, fishing, hunting, Oregon, and Hawaii. I give special thanks to my wife, Judy, for her patience and support during the revision of this textbook.

To students and educators using *Engineering Drawing and Design*, fourth edition:

I have tried very hard to make this edition the best ever and for it to be the best book available for teaching and learning drafting and design. The communication style is intended to help make learning engineering drawing and design as interesting as possible. Numerous illustrations are included in an effort to keep reading to a minimum and to accurately demonstrate proper standards. Actual industry applications are a key element so you can relate the learning content with real-world use. While every effort has been made to comply with national standards, and eliminate errors, some may have been missed due to the magnitude of content found in this textbook. In an effort to make future improvements, please notify the publisher of any errors or missing content.

Sincerely,

David A. Madsen

Introduction to Engineering Drawing and Design

1 CHAPTER

Introduction to Engineering Drawing and Design

LEARNING OBJECTIVES

After completing this chapter, you will:

- Explain topics related to the history of engineering drafting.
- Identify categories and disciplines related to drafting.
- Define *drafter* and identify current terminology.
- Identify the professional organization that is dedicated to the advancement of design and drafting.
- Discuss the requirements for becoming a drafter.
- List and explain points to consider when seeking employment.
- Discuss issues related to computer-aided design and drafting (CADD).
- Explain workplace ethics and related issues.
- Identify topics related to copyrights, patents, and trademarks.

A SHORT HISTORY OF ENGINEERING DRAWING

By J. H. Oakey

We are all aware of the marvelous drawings and the inventive genius of Leonardo da Vinci. It is naturally assumed that he was the originator of drafting and that after his work, all inventors and engineers carefully duplicated his designs in some form in their drawings. This is not so. History is not well documented in the realm of inventors and engineers. The only source for the detailed history of Leonardo's work is his own careful representation. His drawings were those of an artist. They were three-dimensional (3-D) and they generally were without dimensional notations, as shown in Figure 1.1. Craftsmen worked from the 3-D representation, and each machine or device was one-of-a-kind; parts were not interchangeable. In fact, interchangeability was not realized except in special demonstrations until the development of the micrometer in the late 1800s, and even then it was not easily achieved.

Without the concept of interchangeability, accurate drawings were not necessary. Inventors, engineers, and builders worked on each product on a one-of-a-kind basis, and parts were manufactured from hand sketches or hand drawings on blackboards. Coleman Sellers, in the manufacture of fire engines, had blackboards with full-size drawings of parts. Blacksmiths formed parts and compared them to the shapes on the blackboards. Coleman Seller's son, George, recalls lying on his belly using his arms as a radius for curves as his father stood over him directing changes in the sketches until the drawings were satisfactory. Most designs used through the 1800s were accomplished by first completing a hand sketch of the object to be built. These were then converted into wooden models (3-D modeling) from which patterns were constructed. This practice was followed well into the twentieth century by some. Most of us are familiar with the stories of Henry Ford and his famous blackboards. What is new is that these were also the Henry Ford "drafting tables." Henry would sketch cars and parts three-dimensionally and have patternmakers construct full-size models in wood.

An effort to create a program to standardize drawing came when the Franklin Institute was founded in Philadelphia in 1824. It was founded "to advance the general interests" of mechanics and entrepreneurs "by extending a knowledge of mechanical science." One of the goals of the Institute was to establish a mechanical drawing school. Unfortunately, the academic program was never realized because two factions argued over whether to base it on classical academics (Latin and Greek) or on science and practical courses. Eventually, the founders abandoned the academic purposes of the Institute.

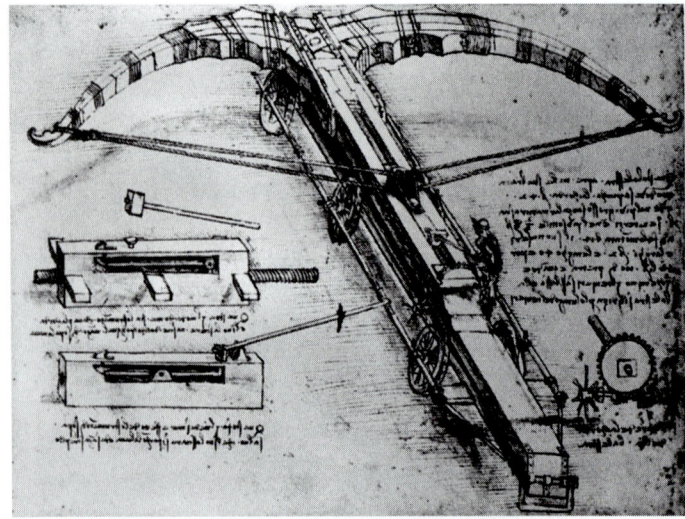

FIGURE 1.1 ■ The drawing of a design by Leonardo da Vinci.

FIGURE 1.2 ■ 3-D modeling is replacing board drafting. *Courtesy Solid Works Corporation.*

Also in 1824, in upstate New York, Amos Eaton founded a school "for the purpose of instructing persons . . . in the application of science to the common purpose of life." This school has grown to be known today as Rensselaer Polytechnic Institute.

In 1862, with the coming of federal government assistance in the form of the Morrill Act, more technical schools emerged, and we can assume that mechanical drawing *slowly* became a part of the intellectual tool kits of trained mechanics, engineers, and inventors.

This can, in part, put to rest the common thought, as we look at pictures of pyramids, steam engines, and other engineering examples, that "it all started with a drawing." Certainly there were freehand sketches, cartoons, and other forms of graphic models. It is probably accurate to say that before most things were built, they were tested with a 3-D model. Just as mechanical drafting slowly took the place of hand sketches and wooden models, today, computer design and drawing and 3-D computer modeling are replacing board drafting. An example is shown in Figure 1.2. The eventual utility and order of computer modeling and design remain to be fully established as part of the developing graphic process.

DRAFTING TECHNOLOGY

According to the *Dictionary of Occupational Titles,* published by the U.S. Department of Labor, drafting is grouped with professional, technical, and managerial occupations. This category includes occupations concerned with the theoretical and practical aspects of such fields of human endeavor as architecture; engineering; mathematics; physical sciences; social sciences; medicine and health; education; museum, library, and archival sciences; law; theology; the arts; recreation; administrative specialties; and management. Also included are occupations in support of scientists and engineers and other specialized activities such as piloting aircraft, operating radios, and directing the course of ships. Most of these occupations require substantial

educational preparation, usually at the college, junior college, community college, or technical institute level.

Men and women employed in the drafting profession often are referred to as drafters. A general definition of drafter, as prepared by the Career Information System, follows:

Drafters translate data and sketches of engineers, architects, and scientists into detailed drawings that are used in manufacturing and construction. Their duties may include interpreting directions given to them, making sketches, preparing drawings to scale, and specifying details. Drafters may also calculate the strength, quality, quantity, and cost of materials. They utilize various drafting tools and computer equipment, engineering practices, and math to complete drawings.

According to the *Occupational Outlook Handbook,* published by the U.S. Department of Labor and Bureau of Labor Statistics, the definition of drafter closely parallels that of the Career Information System, with the following additional information:

Most drafters use computer-aided drafting (CAD) systems to prepare drawings. These systems employ computer workstations which create a drawing on a video screen. The drawings are stored electronically so that revisions can be made easily. These systems also permit drafters to easily and quickly prepare variations of a design. Although this equipment has become easier to operate, CAD is only a tool. Persons who produce technical drawings using CAD still function as a drafter, and need most of the knowledge of traditional drafters—relating to drafting skills and standards—as well as CAD skills.

Due to advancement in the technology used to create drawings, drafters are becoming technicians. The term drafter is commonly used, but often replaced by drafting technician, CAD technician, CADD (computer-aided design and drafting) technician, and CAD/CAM (computer-aided manufacturing) technician.

OCCUPATIONS IN ARCHITECTURE AND ENGINEERING DRAFTING FIELDS

There are several types of drafting technology occupations. While drafting in general has one basic description, specific drafting areas have unique conceptual and skill characteristics. The types of drafting occupations fall into three general professional areas: architecture, engineering, and manufacturing. The following are specific drafting areas as defined by the *Dictionary of Occupational Titles.* The terminology introduced throughout the following occupational definitions is explained in detail in related chapters.

FIGURE 1.3a ■ Computer-generated elevations. *Courtesy Piercy & Barclay Designers, Inc.*

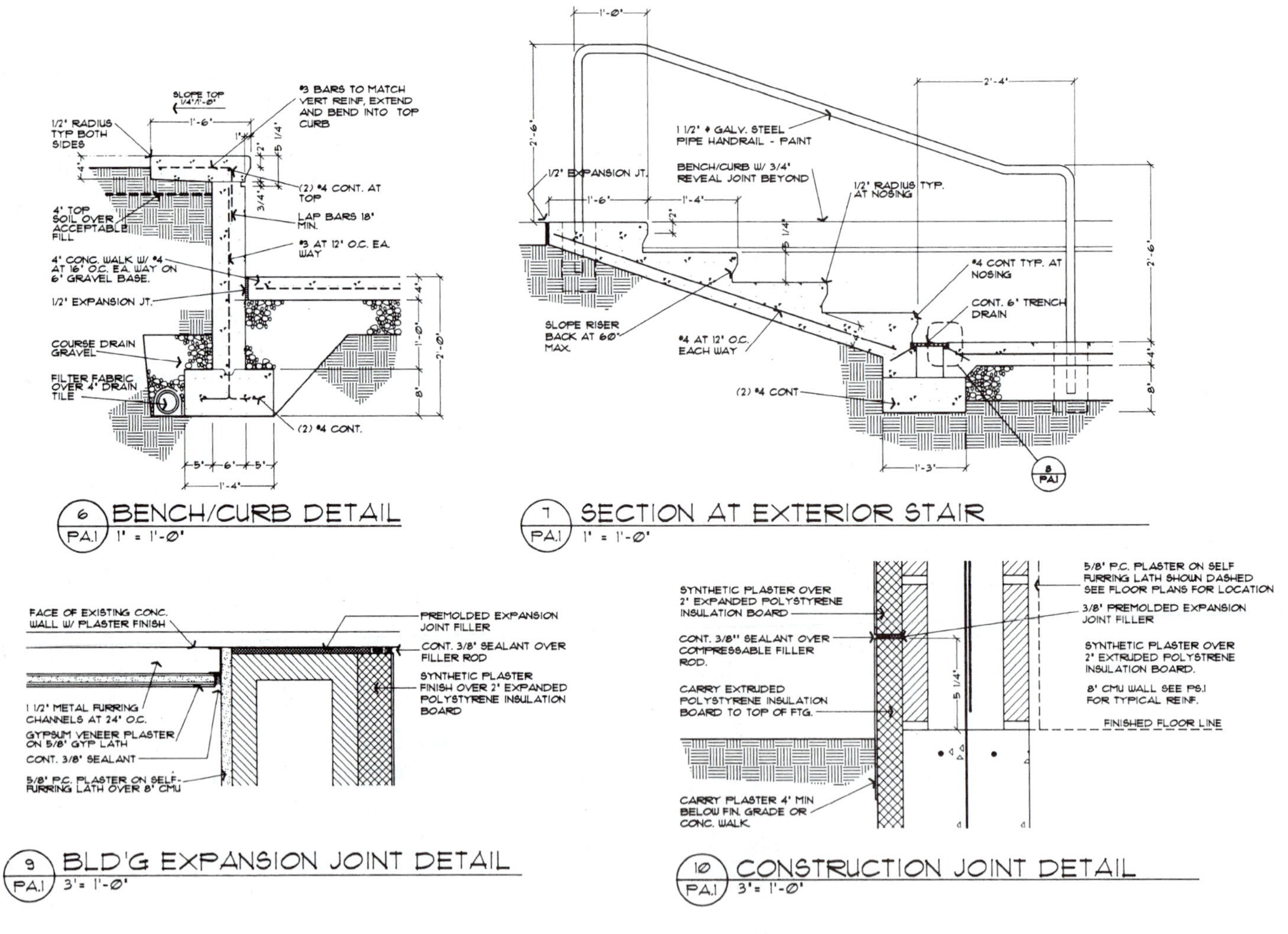

FIGURE 1.3b ■ Computer-generated architectural details. *Courtesy Soderstrom Architects PC.*

Architectural Drafter

Draws artistic architectural and structural features of any class of buildings and like structures; designs and details; confirms compliance with building codes; may specialize in planning architectural details according to structural materials used. (See Figure 1.3a and Figure 1.3b.)

Landscape Drafter

Prepares detailed scale drawings from rough sketches or other data provided by landscape architect; may prepare separate detailed site plan, grading and drainage plan, lighting plan, paving plan, irrigation plan, planting plan, and drawings and detail of garden structures; may build models of proposed landscape construction and prepare colored drawings for presentation to client. (See Figure 1.4.)

Electrical Drafter

Prepares electrical-equipment working drawings and wiring diagrams used by construction and repair crews who erect, install, and repair electrical equipment and wiring in communications centers, power plants, industrial establishments, commercial or domestic buildings, or electrical distribution systems, performing duties described under drafter. (See Figure 1.5.)

Aeronautical Drafter

Specializes in preparing engineering drawings of developmental or production airplanes and missiles and ancillary equipment, including launch mechanisms and scale models of prototype aircraft, as planned by aeronautical engineers.

Electronic Drafter

Drafts wiring diagrams, schematics, and layout drawings used in manufacture, assembly, installation, and repair of electronic equipment such as television cameras, radio transmitters and receivers, audio-amplifiers, computers, and radiation detectors, performing duties as described under Drafter; drafts layout and detail drawings of racks, panels, and enclosures; may conduct service and interference studies and prepare maps and charts related to radio and television surveys; may by designated according to equipment drafted. (See Figure 1.6.)

Civil Drafter

This category is also known by the following titles: drafter, civil engineering; drafter, construction; and drafter, engineering. A Civil Drafter prepares detailed construction drawings, topographic profiles, and related maps and specification sheets used in planning and construction of highways, river and harbor improvements, flood control, drainage, and other civil engineering projects, performing duties as described under drafter; plots maps and charts showing profiles and cross sections indicating relation of topographical contours and elevations to buildings, retaining walls, tunnels, overhead power lines, and other structures; drafts detailed drawings of structures and installations such as roads, culverts, fresh water supply and sewage disposal systems, dikes, wharfs, and breakwaters; computes volume of excavations and fills and prepares graphs and hauling diagrams used in earth-moving operations; may accompany survey crew in field to locate grading markers or to collect data required for revision of construction drawings; may be designated according to type of construction. (See Figure 1.7, page 8.)

Structural Drafter

Performs duties of drafter by drawing plans and details for structures consisting of reinforced steel, concrete, masonry, wood, and other structural materials; produces plans and details of foundations, building frame, floor and roof framing, and other structural elements. (See Figure 1.8, page 8.)

Castings Drafter

Drafts detailed drawings for castings, which require special knowledge and attention to shrinkage allowance and such factors as minimum radii of fillets and rounds.

Patent Drafter

Drafts clear and accurate drawings of mechanical devices for use by a patent lawyer in obtaining patent rights.

Tool Design Drafter

Drafts detailed drawing plans for manufacture of tools, usually following designs and specifications indicated by a Tool Designer.

Mechanical Drafter

Drafts detailed working drawings of machinery and mechanical devices, indicating: dimensions and tolerances, fasteners, joining requirements, and other engineering data. Drafts multiple-view assembly and subassembly drawings as required for manufacture and repair of mechanisms; performs other duties as described under drafter. Mechanical drafting, in general, is the core of the engineering drafting industry. (See Figure 1.9, page 9.)

Directional Survey Drafter

Plots oil- or gas-well boreholes from photographic subsurface survey recordings and other data; computes and represents diameter, depth degree, direction of inclination, location of equipment, and other dimensions and characteristics of boreholes.

Geological Drafter

Draws maps, diagrams, profiles, cross sections, directional surveys, and subsurface formations to represent geological or geophysical stratigraphy location and oil deposits; performs duties described under drafter, correlates and interprets data obtained from topographical surveys, well logs, or geophysical prospecting reports, utilizing special symbols to denote geological physical formations or oil field installations; may finish drawings in mediums and according to specifications required for reproduction by blueprinting, photographing, or other duplication methods.

Geophysical Drafter

Draws subsurface contours in rock formations from data obtained by geophysical prospecting party; plots maps and diagrams from computations based on recordings of seismograph gravity meter, magnetometer, and other petroleum prospecting instruments and from prospecting and surveying field notes.

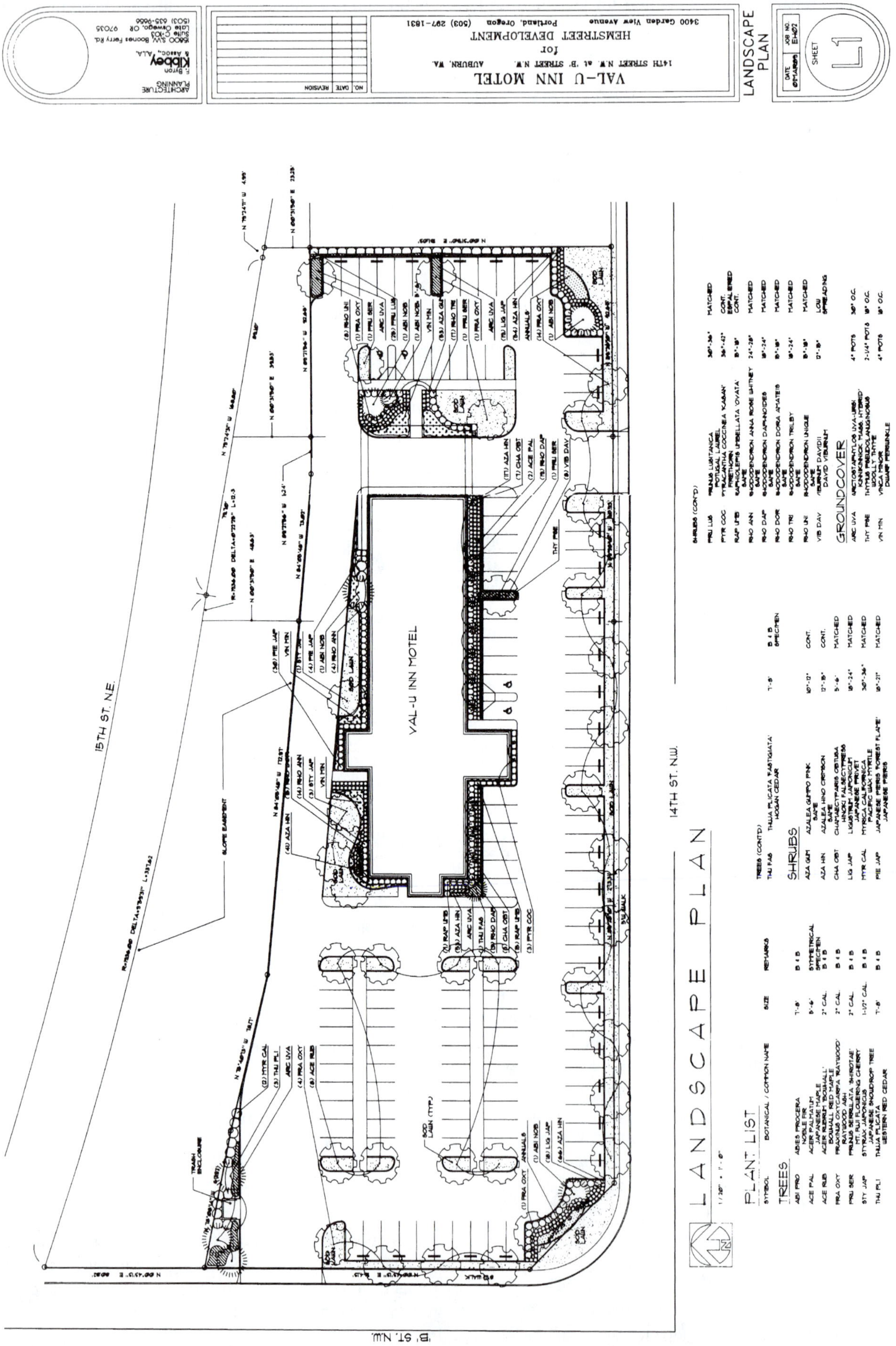

FIGURE 1.4 ■ Landscape plan. *Courtesy OTAK, Inc., for landscaping, and Kibbey & Associates, for site plan.*

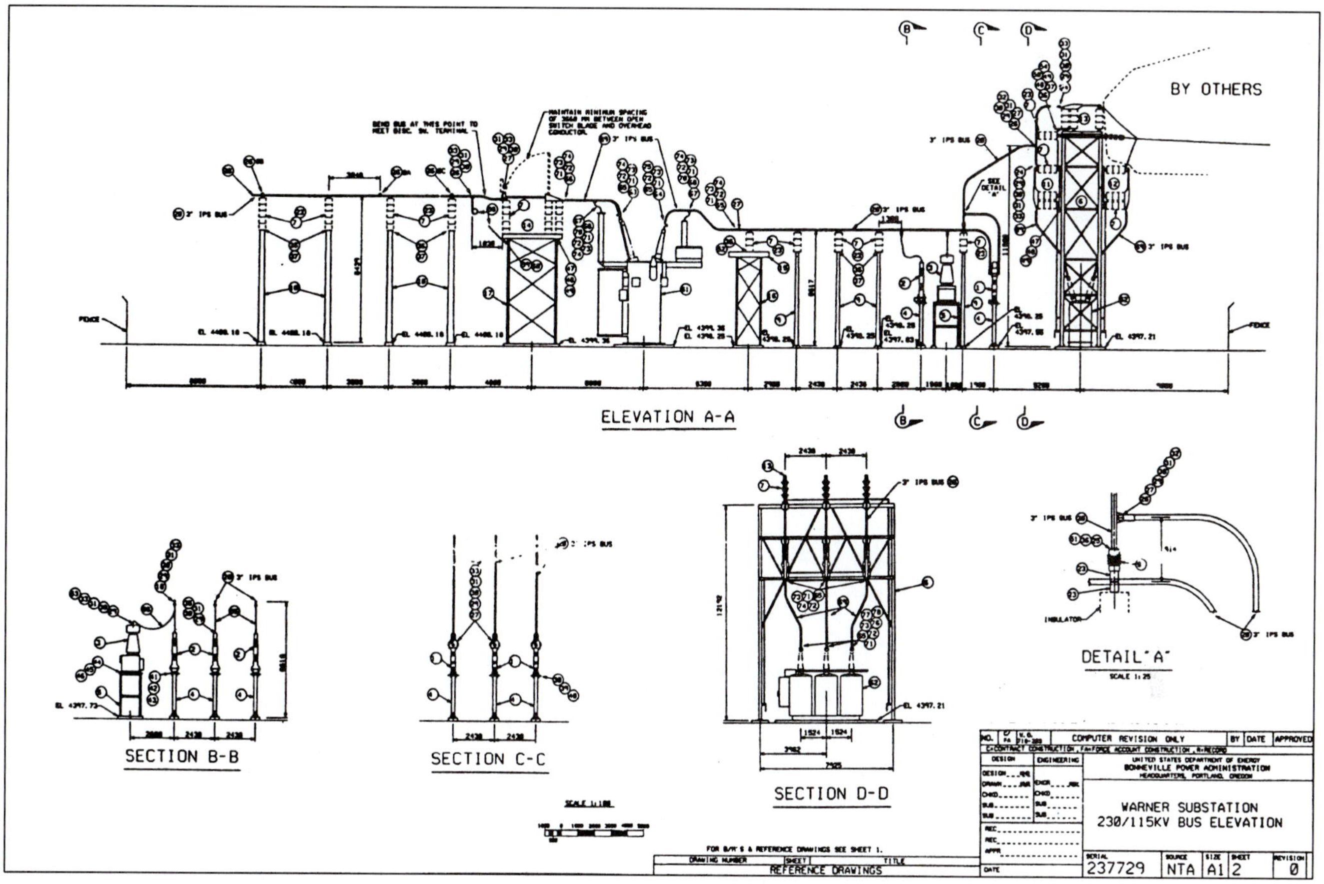

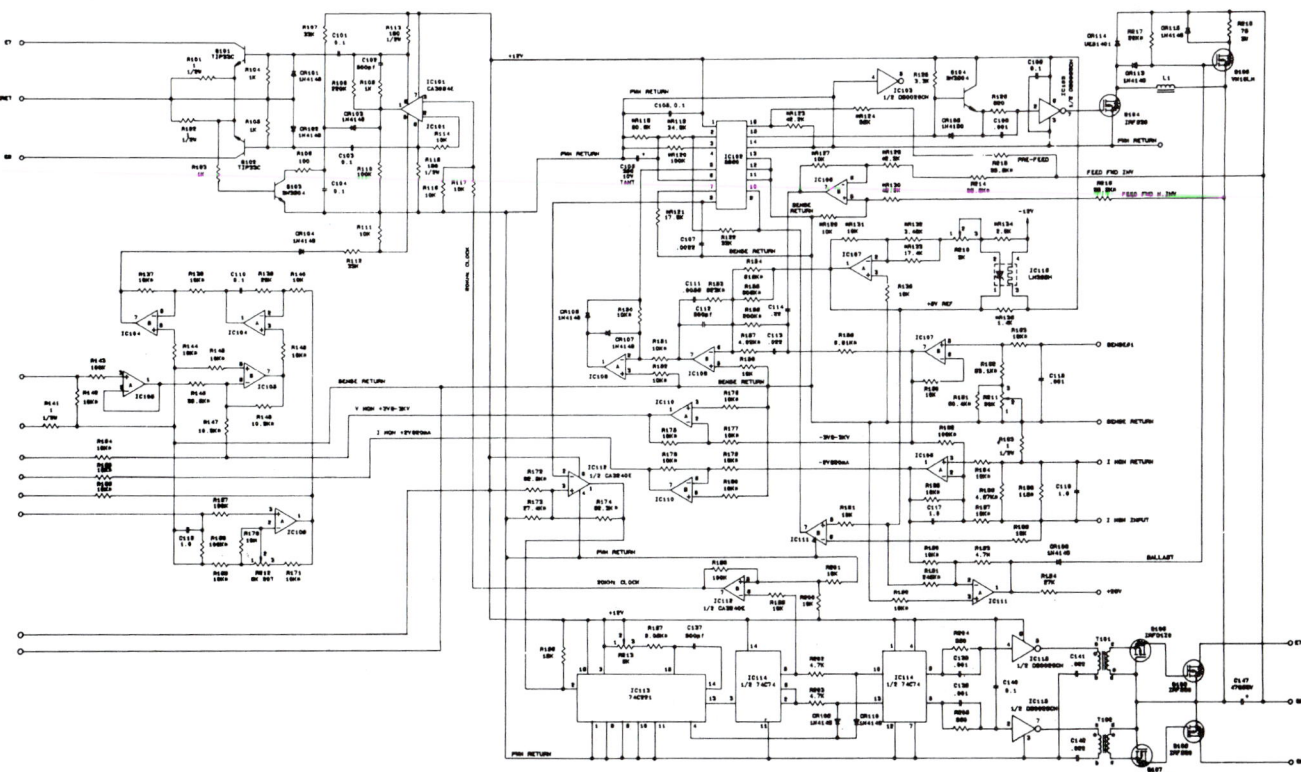

FIGURE 1.5 ■ Electrical drafting plan and elevation. *Courtesy Bonneville Power Administration.*

FIGURE 1.6 ■ Computer-generated electronics schematic. *Courtesy Versacad Corporation.*

FIGURE 1.7 ■ Civil drafting, computer-generated drafting subdivision plat. *Courtesy Glads Project.*

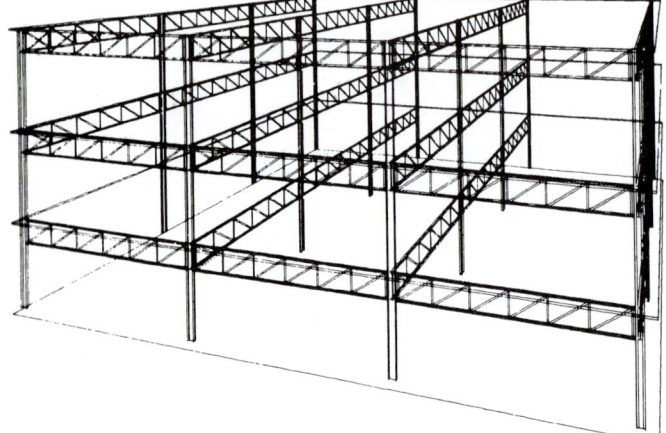

FIGURE 1.8 ■ Computer-generated structural perspective. *Courtesy Computervision Corporation.*

Heating and Ventilating Drafter

Also known as heating ventilating and air-conditioning (HVAC) Drafter; specializes in drawing plans for installation of heating, air-conditioning, and ventilating equipment; may calculate heat loss and heat gain for buildings for use in determining equipment specifications, following standardized procedures; may specialize in drawing plans for installation of refrigeration equipment. (See Figure 1.10.)

Plumbing Drafter

Also known as piping drafter, specializes in drafting plans for installation of plumbing and piping equipment for residential, commercial, and industrial installations. Commercial and industrial piping are closely related to industrial pipe drafting. (See Figure 1.11.)

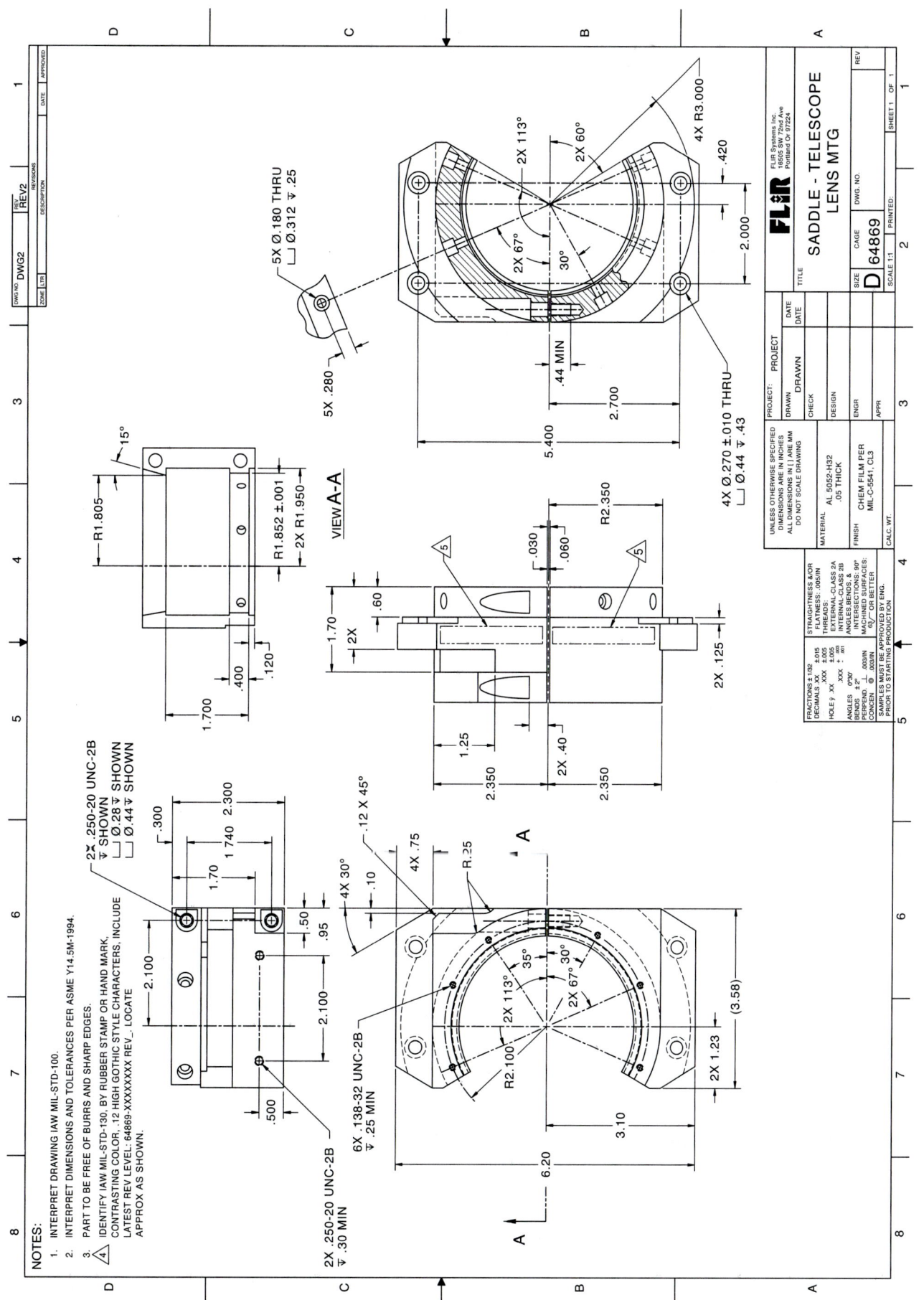

FIGURE 1.9 ■ Computer-aided mechanical drafting. *Courtesy of FLIR Systems, Inc.*

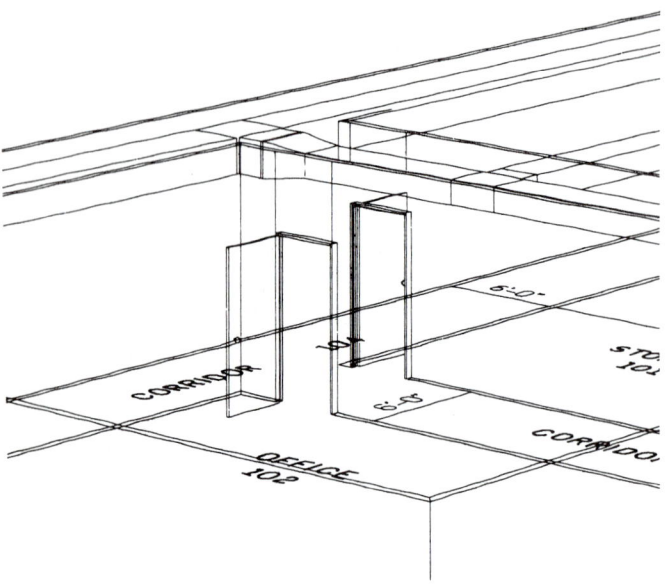

FIGURE 1.10 ■ CADD computer-generated HVAC pictorial. *Courtesy Computervision Corporation.*

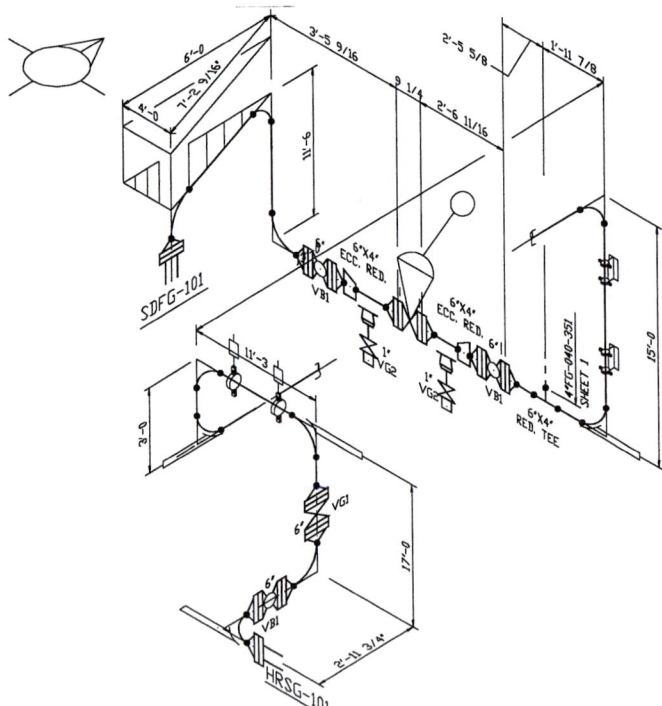

FIGURE 1.11 ■ CADD isometric piping layout. *Courtesy Autodesk, Inc.*

Automotive Design Drafter

Designs and drafts working layouts and master drawings of automotive vehicle components, assemblies, and systems from specifications, sketches, models, prototype and verbal instructions, applying knowledge of automotive vehicle design, engineering principles, manufacturing processes and limitations, and drafting techniques and procedures, using drafting instruments and work aids; analyzes specifications, sketches, engineering drawings, ideas, and related design data to determine critical factors affecting design of components based on knowledge of previous designs and manufacturing processes and limitations; draws rough sketches and performs mathematical computations to develop, design, and work out detailed specifications of components; applies knowledge of mathematical formulas and physical laws; performs preliminary and advanced work in development of working layouts and final master drawings adequate for detailing parts and units of design; makes revisions to size, shape, and arrangement of parts to create practical design; confers with automotive engineer and others on staff to resolve design problems; specializes in design of specific type of body or chassis components, assemblies or systems such as door panels, chassis frame and supports, or braking system.

Industrial Pipe Drafter

Also known as an oil and gas drafter; drafts plans and drawings for layout, construction, and operation of oil fields, refineries, and pipeline systems from field notes, rough or detailed sketches, and specifications; develops detail drawings for construction of equipment and structures, such as drilling derricks, compressor stations, gasoline plants, frame, steel, and masonry buildings, piping manifolds and pipeline systems, and for manufacture, fabrication, and assembly of machines and machine parts. (See Figure 1.12.)

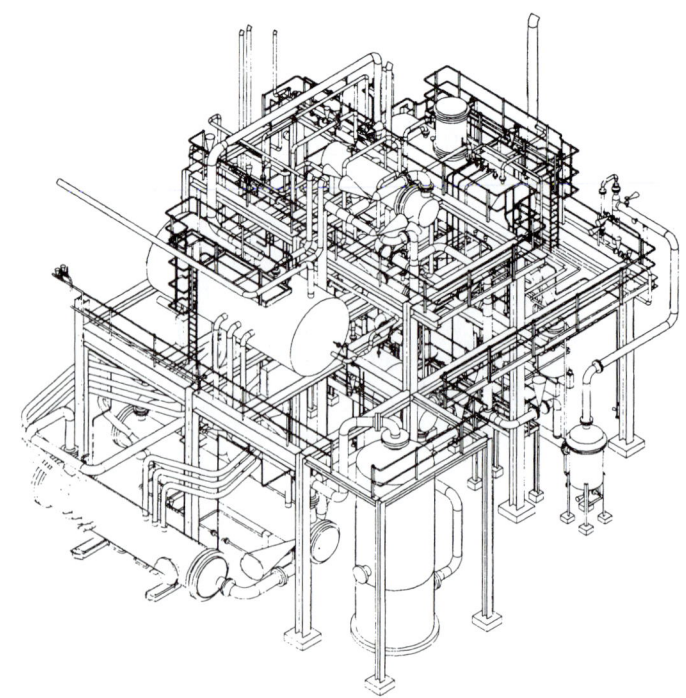

FIGURE 1.12 ■ CADD piping pictorial. *Courtesy Computervision Corporation.*

Technical Illustrator

Lays out and draws illustrations for reproduction in reference works, brochures, and technical manuals dealing with assembly, installation, operation, maintenance, and repair of machines, tools, and equipment; prepares drawings from blueprints, designs mockups, and photographs by methods and techniques suited to specified reproduction process or final use, such as diazo, photo-offset, and projection transparencies, using drafting and optical equipment; lays out and draws schematic, perspective, axonometric, orthographic, or oblique-angle views to depict function, relationship, and assembly sequence of parts and assemblies, such as gears, engines, and instruments; shades or colors drawing to emphasize details or to eliminate undesired background using ink, crayon, airbrush, and overlays; pastes instructions and comments in position on drawing; may draw cartoons and caricatures to illustrate operation, maintenance, and safety manuals and posters. (See Figure 1.13.)

Cartographic Drafter

Also known as a **cartographer**, this person draws maps of geographical areas to show natural and construction features, political boundaries, and other features; analyzes survey data, survey maps and photographs, computer- or automated-mapping products, and other records to determine locations and names of features; studies records to establish boundaries of properties and local, national, and international areas of political, economic, social, or other significance. Geological and topographical maps are drawn by a cartographic drafter.

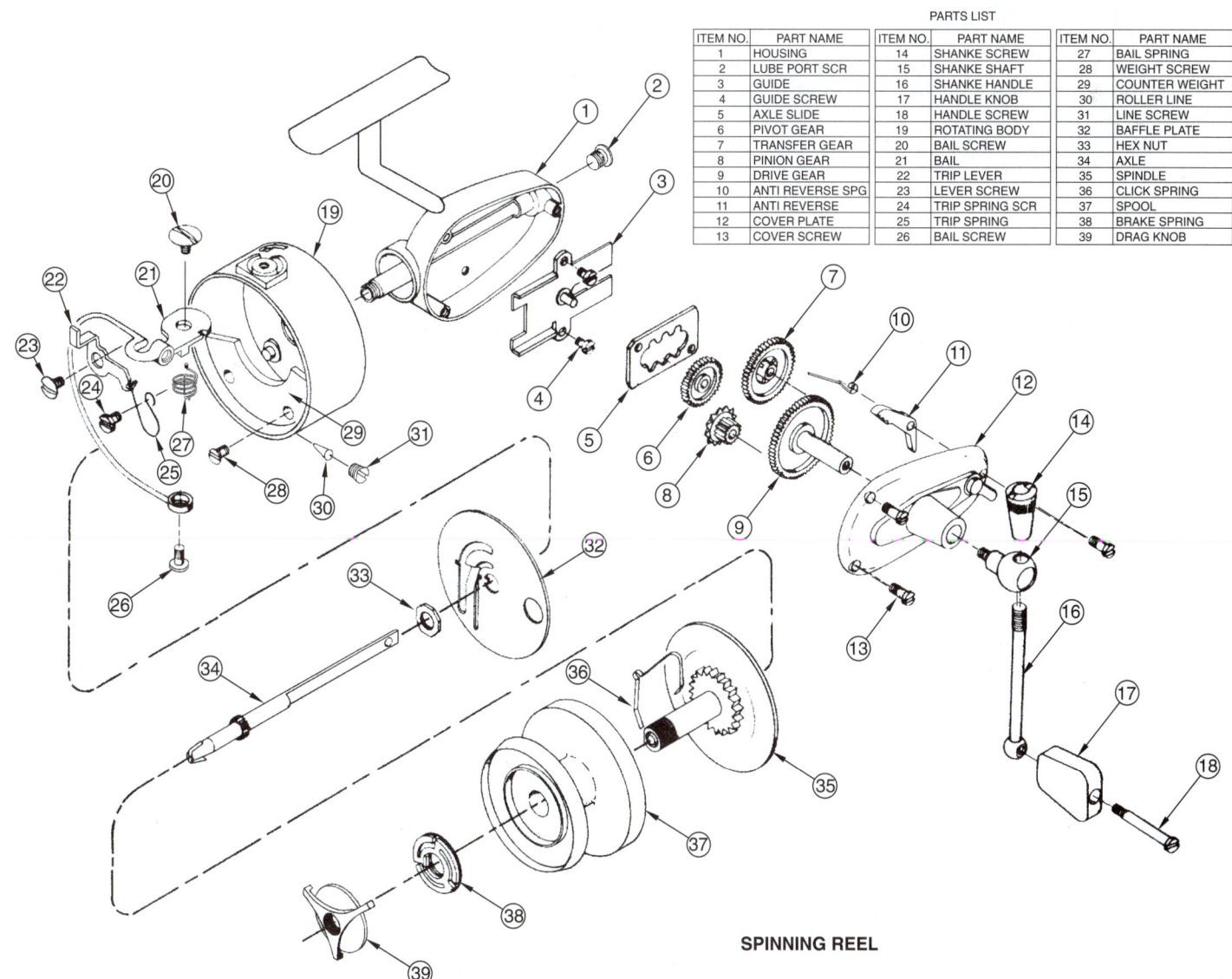

ITEM NO.	PART NAME	ITEM NO.	PART NAME	ITEM NO.	PART NAME
1	HOUSING	14	SHANKE SCREW	27	BAIL SPRING
2	LUBE PORT SCR	15	SHANKE SHAFT	28	WEIGHT SCREW
3	GUIDE	16	SHANKE HANDLE	29	COUNTER WEIGHT
4	GUIDE SCREW	17	HANDLE KNOB	30	ROLLER LINE
5	AXLE SLIDE	18	HANDLE SCREW	31	LINE SCREW
6	PIVOT GEAR	19	ROTATING BODY	32	BAFFLE PLATE
7	TRANSFER GEAR	20	BAIL SCREW	33	HEX NUT
8	PINION GEAR	21	BAIL	34	AXLE
9	DRIVE GEAR	22	TRIP LEVER	35	SPINDLE
10	ANTI REVERSE SPG	23	LEVER SCREW	36	CLICK SPRING
11	ANTI REVERSE	24	TRIP SPRING SCR	37	SPOOL
12	COVER PLATE	25	TRIP SPRING	38	BRAKE SPRING
13	COVER SCREW	26	BAIL SCREW	39	DRAG KNOB

PARTS LIST

SPINNING REEL

FIGURE 1.13 ■ Technical illustration, exploded isometric assembly.

Photogrammetrist

Analyzes source data and prepares mosaic prints, contour-map profile sheets, and related cartographic materials requiring technical mastery of photogrammetric techniques and principles; prepares original maps, charts, and drawings from aerial photographs and survey data and applies standard mathematical formulas and photogrammetric techniques to identify, scale, and orient geodetic points, estimations, and other planimetric or topographic features and cartographic detail; graphically represents aerial photographic detail, such as contour points, hydrography, topography, and cultural features, using precision stereoplotting apparatus or drafting instruments; revises existing maps and charts and corrects maps in various states of compilation; prepares rubber, plastic, or plaster three-dimensional relief models.

PROFESSIONAL ORGANIZATION

The American Design Drafting Association (ADDA) is a nonprofit, professional organization dedicated to the advancement of design and drafting.

The ADDA sponsors the following programs and activities for the design drafting profession:

■ Leadership

■ Drafter Certification

■ Curriculum Certification

■ Student Chapters

■ National Design Drafting Contest

■ *Drafting Examination Review Guides*

■ U.S. Department of Labor Occupational Outlook Handbook—Free Download

■ Employment Center

■ Annual Conference, and Additional Member Resources

The basic knowledge and skill required to do board drafting is also helpful in preparing input data for and/or operating computerized drafting systems such as digitizers, plotters, and photo composition devices.

Whatever means are used to generate drawings, persons skilled in drafting techniques are in demand. Therefore, drafting can be a rewarding and challenging career to those who can visualize and portray ideas graphically.

HOW TO BECOME A DRAFTER

Persons considering drafting as a career should be able to visualize what is to be drawn and be mechanically minded. Potential drafters should be willing to work in cross-functional teams and learn new concepts, and be able to adjust to varying working conditions. Above all, they should be neat and accurate in their work and be detail oriented.

Today, most industries, to fill their drafting job openings, are employing only graduates who have received specialized training in design and drafting from technical institutes, junior or community colleges, and vocational schools. The specialized training includes basic and advanced drawing, mathematics (algebra, geometry, and trigonometry), physics or chemistry, English, humanities, technology courses, and courses in specific fields such as aeronautical, architectural, automotive, electrical, electronics, illustrative, mapping, mechanical, piping, structural, and sheet metal drafting. Computer-aided drafting is essential training for today's employment market. Mechanical and visual aptitude are also important.

DRAFTING OCCUPATIONAL LEVELS

There are several levels of advancement for drafters, based on educational background and practical experience. For advancement, most companies require career employees to hold a two-year college or trade-school degree; however, persons with an equivalent or greater amount of practical experience in a specific job-related area also can obtain advanced occupational levels.

Occupational levels differ slightly from one industry to another, although the general classifications are accurate nationwide.

According to the *Occupational Outlook Handbook*, published by the U.S. Department of Labor and Bureau of Labor Statistics, entry-level or junior drafters do routine work under close supervision. After gaining experience, intermediate-level drafters progress to more difficult work with less supervision. They may be required to exercise more judgment and perform calculations when preparing and modifying drawings. Drafters may advance to senior drafter, designer, or supervisor. Many employers pay for continuing education, and with appropriate college degrees or experience, drafters may go on to become engineering technicians, engineers, or architects.

DRAFTING JOB OPPORTUNITIES

Drafting job opportunities, which include all possible drafting employers, fluctuate with national and local economies. This is no different from most other employment in industry or construction. Drafting is tied closely to manufacturing and construction so that a slowdown or speedup in these industries nationally affects the number of drafting jobs available. This same effect upon drafting opportunities may be experienced at the local level or with specific industries. For example, construction may be strong in one part of the country, and slow in another, so the demand for drafters in those localities is strong or slow accordingly. Other examples are demonstrated when automobile manufactures experience poor sales, and drafting opportunities decline, or when hi-tech industries expand, and more drafters are needed. Public and private indicators suggest that the demand for drafters will continue to be strong for graduates of two-year postsecondary drafting curriculums. The emphasis should be placed on drafting skills, computer-aided drafting, math, English, and oral and written communication skills.

The types of drafting jobs available are also controlled by local demands. A predominance of mechanical drafting jobs is found in metropolitan areas where manufacturing is strong, while in outlying areas there may be more civil or structural drafting jobs. Each local area has a need for more of one type of drafting skill than another. Also, drafting curriculums in different geographical areas usually specialize in the fields of drafting that help fill local employment needs. A broader range of opportunities exist in many local areas due to the flexibility of electronic data transfer, making it possible to do any task worldwide. Some drafting programs offer a broad-based education so graduates can have versatile employment opportunities. When selecting a school, look into curriculum, placement potential, and local demand. Talk to representatives of local industries for an evaluation of the drafting curriculum.

Opportunities for advancement for drafters are excellent, although dependent on the advancement possibilities of the specific employer. Advancement also depends on an individual's initiative, ability, product knowledge, and willingness to continue to be educated. Additional education for advancement usually includes increased levels of mathematics, pre-engineering, engineering, software, and advanced drafting. Drafting has traditionally been an excellent stepping-stone to designing, engineering, and management.

SEARCHING FOR A DRAFTING POSITION

Entry-level drafting positions require you to be prepared to meet the needs and demands of industry. The previous discussions explained the recommended education level and type of training required for most job-entry opportunities. Entry into this career marketplace depends on your training and ability and on the market demand. Some geographic areas around the country may have a greater demand than others. The amount of demand often is based on the economy of the region, the state, and the country. The job market usually changes with the economy. More jobs are commonly available when the economy is doing well. During these positive economic times, it may be easier to find a job than when the economy is down. In a poor economic environment, your training, skills, and personal presentation may make the difference in finding an employment opportunity. A two-year postsecondary degree in a college program such as Drafting Technology, Computer-Aided Drafting, or Engineering Technology can provide a big advantage when seeking a position in this industry. Programs of this type normally have a quality cross section of training in design and drafting, math, and communication skills. These programs also typically have job preparation and placement services to aid their graduates. Many of these schools have direct industry contacts that help promote hiring opportunities. Training programs also often have cooperative work experience (CWE) or internships where their students work in industry for a designated period of time while completing the degree requirements. These positions allow a company to determine

if the student is a possible candidate for full-time employment and provide the student with valuable on-the-job experience that can be included on a resume. Even if you do not go to work at the company where you do the CWE or internship, you can get a letter of recommendation for your portfolio.

When the local economy is doing extremely well and drafting job opportunities are plentiful, it may be possible to find a job with less than a two-year college degree. If you want to find entry-level employment in a job market of this type, you can take intensive training in computer-aided drafting and drafting practices. The actual amount of training required depends on how well you do and whether you can match an employer who is willing to hire with your level of training. Many people have entered the industry in this manner, although you would be well advised to continue schooling toward a degree while you are working.

The following are some points to consider when you are ready to seek employment:

- ■ Get your resume in order. Take a resume preparation course or get some help from your instructors or a career counselor. Your resume must be a quality and professional representation of you. When an employer has many resumes, the best stand out. A possible resource is the book *Resumes for Dummies.*

- ■ Write an application or cover letter. You can get help on this from the same people who help with your resume. The application letter should be professionally and clearly written, short and to the point, and should give reasons why you would be an asset to the company. A possible resource is the book *Cover Letters for Dummies.*

- ■ Prepare a portfolio. The portfolio should contain examples of school and industry drawings that you have completed. The drawings should be neatly organized and of the type that helps you target the specific industry discipline that you are seeking. For example, model your portfolio with mechanical drawings if you are interviewing with a manufacturing industry. Display architectural drawings if you are interviewing with an architect or building designer. Include your letters of recommendation from employers and instructors.

- ■ Register with the department, school, and state employment service. Watch the employment ads in local newspapers, and check out Internet employment sites, which are discussed later.

- ■ Make a realistic decision about the type of place where you want to work and the salary and benefits you realistically think you should get. Base these decisions on sound judgment. Your instructors should have this information for the local job market. Do not make salary your first issue when seeking a career position. The starting salary is often just the beginning at many companies. Consider advancement potential. A drafting technology position often is a stepping-stone to many opportunities, such as design, engineering, and management.

■ Research prospective companies to learn about their business or products. The Internet is a good place to seek information, because most companies have a website. This type of research also can help you during an interview.

■ Be prepared when you get an interview. First impressions are critical. You must look your best and present yourself well. Figure 1.14 shows a job candidate making the first introduction for a possible employment opportunity. Always be on time or early. Relax as much as you can. Answer questions clearly and to the point, but with enough detail to demonstrate that you known what you are talking about. It is often unwise to talk too much. Show off your portfolio. Be prepared to take a CAD test or demonstrate your skills. A possible resource is the book *Interviewing for Dummies*.

■ Ask intelligent questions about the company during an interview because you need to decide if you want to work there. For example, you may not want to work for a company where they have no standards, poor working conditions, and pirated software. You might prefer to work for a company where they have professional standards, layering systems, a pleasant work environment, and advancement possibilities.

■ Respond fast to job leads. The employment marketplace is often very competitive. You need to be prepared and move quickly. Follow whatever instructions are given for you to apply. Sometimes employers want you to go in person to fill out an application, and sometimes they want you to fax or mail a resume. Either way, you can include your application letter and resume. Sometimes they want you to call for a pre-interview screening.

■ In an active economy, it is common to get more than one offer. If you get an offer from a company, take it if you have no doubts. However, if you are uncertain, ask for 24 or 48 hours to make a decision. If you get more than one offer, weigh the options carefully. There are advantages and disadvantages with every possibility. Make a list of the advantages and disadvantages with each company and carefully consider them.

■ Once you have made a decision, you need to feel good about it and move on with enthusiasm. Figure 1.15 shows a new employee working at a CADD position.

■ Send a very professional thank-you letter to the companies that were considering you. Phone the companies and send a follow-up letter where you had other offers. You never know when you might need to apply at these companies in the future, so this is an important step.

Employment Opportunities on the Internet

The Internet has become a valuable place to seek employment. There are hundreds of websites that are available to help you prepare for and find a job. Many websites allow you to post your resume for possible employers and apply for jobs. Some employers screen applicants over the Internet. Figure 1.16 shows a person looking for a job opportunity on the Internet. The only caution is that your personal information displayed through the Internet is available for anyone to read. However, some websites, such as *http://www.Monster.com* provide a safe place to post your resume for only employers to review. You should always confirm that the terms of agreement provide you with a safe place to do a job search.

FIGURE 1.14 ■ A candidate for an employment opportunity making the first introduction. You must present yourself well. First impressions are very important.

FIGURE 1.15 ■ A new employee, working at a CADD position. *Courtesy Hewlett-Packard Company.*

FIGURE 1.16 ▪ A person looking for a job opportunity on the Internet.

DRAFTING SALARIES AND WORKING CONDITIONS

Salaries in drafting professions are comparable to salaries of other professions with equal educational requirements. Employment benefits vary according to each individual employer. However, most employers offer vacation and health insurance coverage, while others include dental, life, and disability insurance.

COMPUTERS IN DESIGN AND DRAFTING

The acronym CAD means computer-aided design, although many people refer to it as computer-aided drafting. CADD is computer-aided design and drafting. The correctness of this is not extremely important, but what is important is that CAD or CADD has revolutionized the way business is done in design and drafting for the engineering and architecture industries. The use of CADD, has made the design and drafting process more accurate and faster. This has allowed for added creativity among engineers, architects, designers, and senior drafters. During the early years of CADD development, many people felt that the computer reduced the individual creativity and artistic impression of the designer and drafter. CADD has now enhanced the ability for a person to be creative by providing many new tools such as solid modeling, animation, and virtual reality. The CADD system allows the drafter to produce drawings that are accurate, very neat and legible, and matched to industry standards. Even architectural drawings, which have always had an artistic flair with lettering and line styles, can now be produced on CADD to match the quality of the finest handwork available. Plus these drawings are consistent from one person or company to the next.

The following discussions relate to topics that are discussed in detail where they apply throughout this textbook. Design coordination is comprehensively covered in Chapters 4 and 27; 3-D modeling, parametric design, animation, and virtual reality is explained in Chapter 4; and the national CAD standards are discussed in Chapter 3.

Design Coordination

Software is the instructions stored in your computer and determine what you can do with the computer. CADD software programs offer a wide variety of applications that make the design and drafting process efficient. Parametric solid modeling allows the computer program to create objects and automatically update designs based on the information you provide. Parametric design enables you to enter certain values or information, which the software uses to create or change drawings based on this input. An example is the design of a gear, where the computer prompts you for variables such as specific diameters, type and number of teeth, and hub and keyway specifications. The program then automatically tells you if the information is complete and accurate, provides additional data, and creates a 3-D model or a detail drawing of the gear.

With regard to computer usage, the term virtual refers to something that appears to have the properties of being real or actual. A virtual design stores your work in a file where the information remains integrated, up-to-date, and easy to manage. Working from the virtual design, you can create and modify designs in 2-D (two-dimensional) and 3-D (three-dimensional) views and automatically create sections, elevations, details, and other working drawings easily. The virtual design integrates every aspect of the design, so changes are automatically updated in all drawings, saving time and reducing the risk of errors along the way. Figure 1.17 shows the coordination taking place through the entire engineering process of design, documentation, communication, and collaboration. Engineers, architects, and designers can share their work directly across a local network or through the Intranet and Internet. The Intranet links communication between computers within a company or organization, while the Internet is a worldwide network of communication between computers.

Animation and Virtual Reality

Virtual reality (VR) can be used to demonstrate products or to display material on a website. Virtual reality or VR refers to a world that appears to be a real world, having many of the properties of a real or actual world. The VR world often appears and feels so real that it is almost as if it is real. This is where the computer is used to simulate environments, including inside and outside of the product or building, and including sound and touch. Walk-through can be characterized as a camera in a computer program that is set up like a person walking through a building, around a product or building, or through a landscape. Fly-through is similar, but the camera is like a helicopter flying through the area. Fly-through is generally not used to describe a tour through a building. Walk-through or fly-through is the effect of a computer-generated movie where

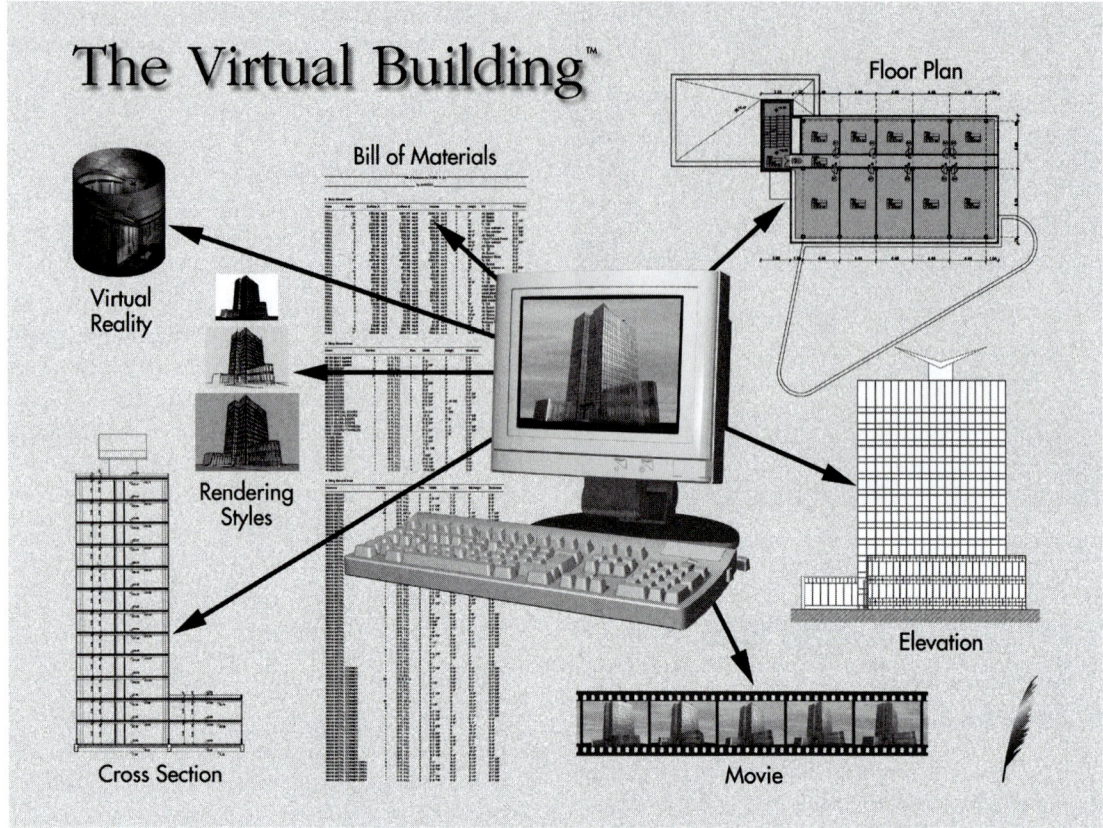

FIGURE 1.17 ■ The coordination taking place through the entire engineering process of design, documentation, communication, and collaboration. *ArchiCAD® image courtesy of Graphisoft®.*

the computer images represent the real architecture, or VR presentation where the computer images turn or move as you turn your head in the desired direction. Realistic renderings, animations, and VR are excellent tools to show the client how the building will look inside and out, or how a product operates. Design ideas can be created and changes made easily at this stage.

Rapid Prototyping (RP)

Rapid prototyping is a manufacturing process by which a solid physical model of a part is made directly from 3-D CADD model data, without any special tooling. An RP model is a physical 3-D model that can be created far more quickly than by using standard manufacturing processes. Examples of RP are stereolithography and fused deposition modeling, or 3-D printing. Rapid prototyping equipment accepts 3-D CADD files, slices the data into thin cross sections, and constructs layers from the bottom up, bonding one on top of the other, to produce physical prototypes. Computer-aided design software such as Pro/Engineer allows you to export an RP file from a solid model in the form of an .stl file.

A computer using post-processing software slices the 3-D CADD data into .005–.013-in. thick cross-sectional planes. Each slice or "layer" is composed of closely spaced lines resembling a honeycomb. The slice is shaped like the cross section of the part. The cross sections are sent from the computer to the rapid prototyping machine, which builds the part one layer at a

time. The stereolithography machine has a "vat" that contains a photosensitive liquid epoxy plastic, and a flat platform or starting base resting just below the surface of the liquid. (See Figure 1.18.) A laser, controlled with bidirectional motors, is positioned above the vat perpendicular to the surface of the polymer. The first layer is bonded to the platform by the heat of a thin laser beam that traces the lines of the layer onto the surface of the liquid polymer. When the first layer is completed, the platform is lowered the thickness of a layer. Additional layers are bonded on top of the first in the same manner, according to the shape of their respective cross section. This process is repeated until the prototype part is complete.

A second type of rapid prototyping called solid object 3-D printing uses an approach similar to inkjet printing. During the build process, a print head with hundreds of jets builds models by dispensing a thermoplastic material in layers. The printer can be networked to any CADD workstation and operates with the push of a few buttons. (See Figure 1.19.) Additional RP coverage is provided in Chapter 27.

National CADD Standards

There are two sets of national CADD standards developed by different groups for special purposes. One set of CADD standards is developed for the manufacturing industry and the other is created for the architecture, engineering, and construction industry. While the creation of these standards is dissimilar, their use can be applied across all design and drafting fields. The application, content, and purpose of these standards are discussed in detail in Chapter 3.

The National Council for Advanced Manufacturing (NACFAM) publishes the *Computer-Aided Drafting and Design (CADD) Skill Standards.* This publication provides the skills needed for beginner CADD users.

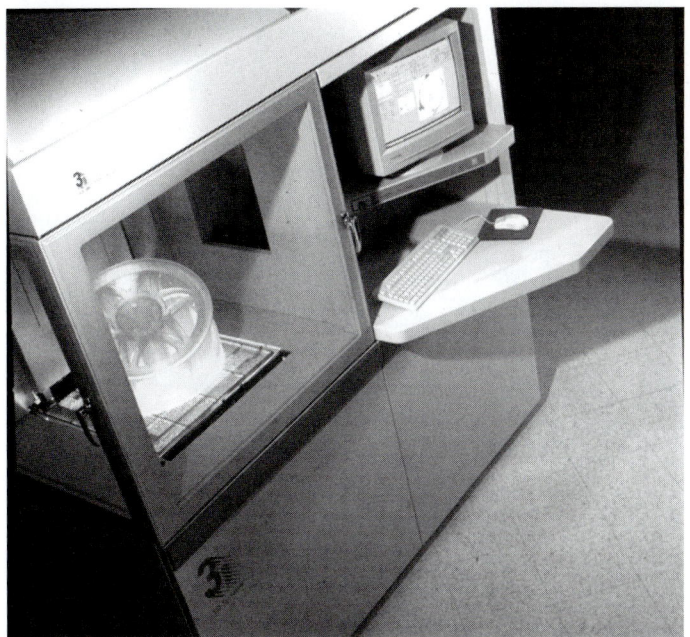

FIGURE 1.18 ■ This stereolithography system builds a physical prototype, layer by layer, using liquid epoxy plastic that is hardened by a laser to precise specification. *Courtesy 3D Systems.*

(b)

(a)

(c)

FIGURE 1.19 ■ (a) The 3-D printer employs a print head with hundreds of jets to build models by dispensing a thermoplastic material in layers. *Courtesy 3D Systems.* (b) A designer removes a model built on a solid object printer. *Courtesy 3D Systems.* (c) Engineers built this prototype wheel from a CAD file using a solid object printer. *Courtesy 3D Systems.*

FIGURE 1.20 ■ An office with computer workstations and a manual drafting table. *Courtesy Hewlett-Packard Company.*

The *United States National CAD Standard* was created by the CADD/GIS Technology Center, the American Institute of Architects (AIA), the Construction Specifications Institute (CSI), the United States Coast Guard, the Sheet Metal and Air Conditioning Contractors National Association (SMACNA), the General Services Administration (GSA), and the National Institute of Building Sciences' (NIBS) Facilities Information Council. These standards are discussed in Chapter 3.

The Computer

The personal computer or PC has become the primary tool in most engineering and architectural firms for the creation of drawings and calculation of engineering data. The PC that is set up for work applications is commonly referred to as a workstation. An engineering or architectural office might have a workstation set up for each person and one or more manual drafting tables for layout work as needed. Figure 1.20 shows an office with computer workstations and a manual drafting table.

The PC belongs to a class of computers called microcomputers. The most commonly used microcomputer systems for CADD applications are IBM systems and computers made by other manufacturers, which are referred to as IBM compatibles or clones. The input devices such as the mouse and digitizer, and the output devices such as the printer or plotter, are referred to as peripherals. Figure 1.21 shows a computer system being used in an architectural office. The drawing on the screen is being printed on a plotter.

WORKPLACE ETHICS

Ethics are rules and principles that define right and wrong conduct. A code of ethics is a formal document that states an organization's values and the rules and principles that employees are

FIGURE 1.21 ■ A computer system being used in an architectural office. *Courtesy Hewlett-Packard Company.*

expected to follow. In general, codes of ethics contain these main elements: be dependable, obey the laws, and be good to customers. An example of a company with a corporate code of ethics is the Lockheed Martin Corporation, recipient of the American Business Ethics Award. According to information found in the Lockheed Martin website, Lockheed Martin aims to "set the standard" for ethical business conduct, to be achieved through six virtues:

1. *Honesty:* to be truthful in all our endeavors; to be honest and forthright with one another and with our customers, communities, suppliers, and shareholders.

2. *Integrity:* to say what we mean, to deliver what we promise, and to stand for what is right.

3. *Respect:* to treat one another with dignity and fairness, appreciating the diversity of our workforce and the uniqueness of each employee.

4. *Trust:* to build confidence through teamwork and open, candid communication.

5. *Responsibility:* to speak up, without fear of retribution, and report concerns in the workplace, including violations of

ONCE IS ALWAYS ENOUGH WITH CADD

by Karen Miller
Graphic Specialist with Tektronix, Inc.

Reusability is one of the important advantages of CADD. With CADD it is never necessary to draw anything more than once. This advantage is enhanced further by developing a CADD symbols library. Building a parts library for reusability has increased productivity, decreased development costs, and set the highest standards for quality at the Test and Measurement Documentation Group at Tektronix, Inc.

The parts library began by reusing 3-D isometric parts created by the AutoCAD illustrator and saved as blocks (symbols) in a Parts Library Directory. By pathing the named blocks back to this library directory, each block becomes accessible to any directory and drawing file. This allows the CADD illustrator to insert library symbols into any drawing simply by typing the block name.

The illustrator adds new parts to the library as a product is disassembled and illustrated. Each part is given the next available number as its block name in the library, as shown in Figure 1.22.

Originally, the AutoCAD drawings were combined with text (written using Microsoft Word) in Ventura Publisher to create technical publications. Now, the AutoCAD drawings are added to document files in Interleaf technical publishing software. The entire parts library is available to both AutoCAD and Interleaf users (Figure 1.23).

From the point of view of cost management, the parts library has saved hundreds of hours of work. From the illustrator's view, the parts library helps improve productivity and frees time for new or complex projects.

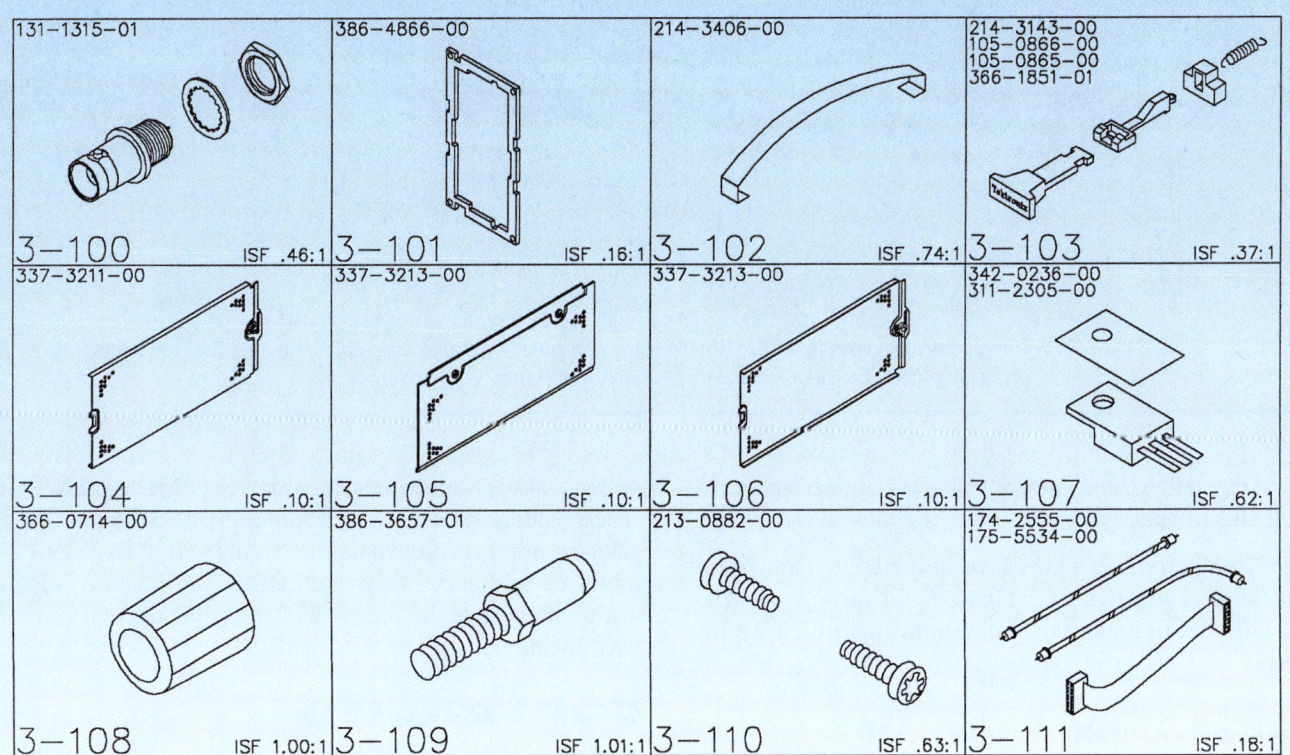

FIGURE 1.22 ■ CADD parts library. *Courtesy Karen Miller, Graphic Specialist, Tektronix, Inc.*

(Continued)

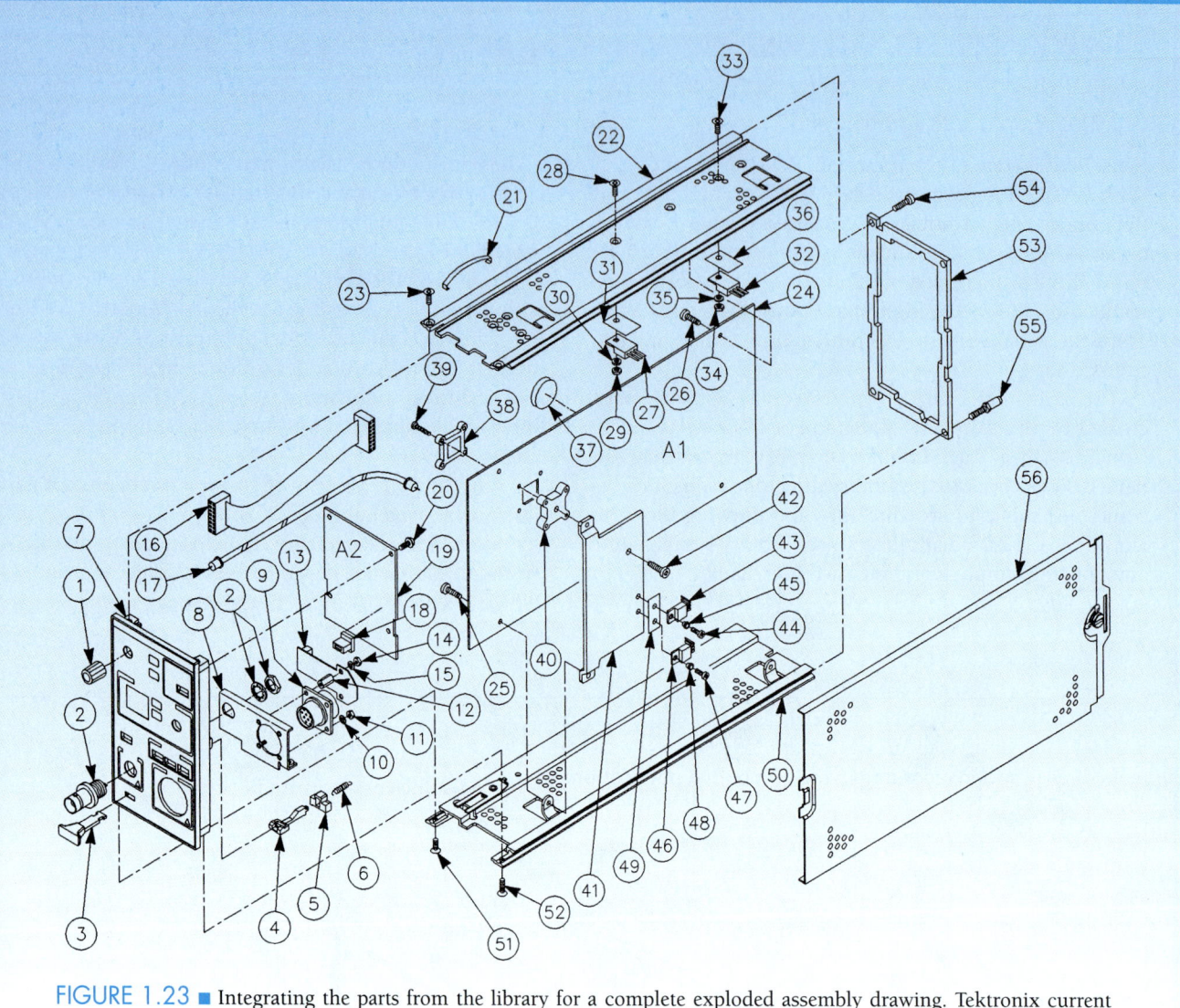

FIGURE 1.23 ■ Integrating the parts from the library for a complete exploded assembly drawing. Tektronix current probe amplifier. *Courtesy Karen Miler, Graphic Specialist, Tektronix, Inc.*

laws, regulations, and company policies, and seek clarification and guidance whenever there is doubt.

6. *Citizenship:* to obey all the laws of the United States and the other countries in which we do business and to do our part to make the communities in which we live better.

Intellectual Property Rights

The success of a company often relies on the integrity of its employees. Products are normally the result of years of research, engineering, and development. This is referred to as the intellectual property of the company. Protection of intellectual property can be critical to the success of the company in a competitive industrial economy. This is why it is very important for employees to help protect design ideas and trade

secrets. Many companies manufacture their products in a strict, secure, and secret environment. You will often find proprietary notes on drawings that inform employees and communicate to the outside world that the information contained in the drawing is the property of the company and cannot be used by others.

COPYRIGHTS

A copyright is the legal rights given to authors of "original works of authorship." Copyright and patent law is established in the U.S. Constitution, article 1, section 8, which empowers the U.S. Congress "to promote the progress of science and useful arts, by securing for limited times to authors and inventors the exclusive right to their respective writings and discoveries." Copyrights control exclusively the reproduction and distribution of the work by others. In the United States, published or unpublished works that are typically copyrightable include:

■ Literary works, including computer programs and compilations.

■ Musical works, including any accompanying words.

■ Dramatic works, including any accompanying music.

■ Pantomimes and choreographic works.

■ Pictorial, graphic, and sculptural works.

■ Motion pictures and other audiovisual works.

■ Sound recordings.

■ Architectural works, and certain other intellectual works.

Copyright protection exists from the time the work is created in fixed form. Fixed form may not be directly observable as long as it can be communicated with the aid of a machine or device. The copyright in the work of authorship immediately becomes the property of the author who created the work. Copyright is secured automatically when the work is created, and the work is created when it is fixed in a copy or phonorecorded for the first time. Copies are material objects from which the work can be read or visually perceived either directly or with the aid of a machine or device. A copyright notice can be placed on visually perceptible copies. The copyright notice should have the word "Copyright," the abbreviation "Copr.," or the symbol © (® for phonorecords of sound recordings); the year of first publication; and the name of the owner of the copyright. Copyright registration is a legal formality intended to make a public record, but it is not a condition of copyright protection. A copyright claim is registered by sending a copy of unpublished work or two copies of the work as published, with a registration application and fee, to the Library of Congress, Copyright Office, Washington D.C. The application forms are available from the Copyright Office upon request. The previous discussion provides some basic general guidelines. You should confirm specific details and requirements with the Copyright Office.

PATENTS

A patent for an invention is the grant of a property right to the inventor, issued by the Patent and Trademark Office (PTO). The term of a new patent is 20 years from the date on which the application for the patent was filed in the United States or, in special cases, from the date an earlier related application was filed, subject to the payment of maintenance fees. United States patent grants are effective only within the United States, U.S. territories, and U.S. possessions. The patent law states, in part, that any person who "invents or discovers any new and useful process, machine, manufacture, or composition of matter, or any new and useful improvement thereof, may obtain a patent," subject to the conditions and requirements of the law.

The patent law specifies that the subject matter must be "useful." The term useful refers to the condition that the subject matter has a useful purpose and must operate. Laws of nature,

physical phenomena, and abstract ideas cannot be patented. A complete description of the actual machine or other subject matter is required to obtain a patent.

Application for a Patent

There are two types of patent applications, one is nonprovisional and the other is provisional. A nonprovisional patent application is for the full patent, which lasts 20 years. The provisional patent application is for a temporary patent that lasts for one year.

Nonprovisional Application for a Patent

According to the PTO, a nonprovisional application for a patent is made to the Assistant Commissioner for Patents and includes:

1. A written document that has a specification, and an oath or declaration.

2. A drawing in those cases in which a drawing is necessary.

3. The filing fee.

All application papers must be in the English language, or a translation into the English language is required. All application papers must be legibly written on only one side either by a typewriter or mechanical printer in permanent dark ink or its equivalent in portrait orientation on flexible, strong, smooth, nonshiny, durable, white paper. The papers must be presented in a form having sufficient clarity and contrast between the paper and the writing to permit electronic reproduction. The application papers must all be the same size—either 21.0 cm by 29.7 cm (DIN size A4) or 21.6 cm by 27.9 cm (8 1/2 × 11 in.), with a top margin of at least 2.0 cm (3/4 in.), a left-side margin of at least 2.5 cm (1 in.), a rightside margin of at least 2.0 cm (3/4 in.), and a bottom margin of at least 2.0 cm (3/4 in.) with no holes made in the submitted papers. It is also required that the spacing on all papers be 1 1/2 or double spaced, and the application papers must be numbered consecutively (centrally located above or below the text) starting with page one.

All required parts of the application must be complete before the application is sent, and it is best to send all of the parts together. All applications received in the PTO are numbered in serial order, and the applicant will be informed of the application serial number and filing date by a filing receipt.

Provisional Application for a Patent

Since June 8, 1995, the PTO has offered inventors the option of filing a provisional application for patent that was designed to provide a lower-cost first patent filing in the United States and to give U.S. applicants equality with foreign applicants. Claims and oath or declaration are not required for a provisional application. Provisional application provides the means to establish an early effective filing date in a patent application and permits the term "Patent Pending" to be applied in connection with the invention. Provisional applications may not be filed for design inventions. The filing date of a provisional application is the date

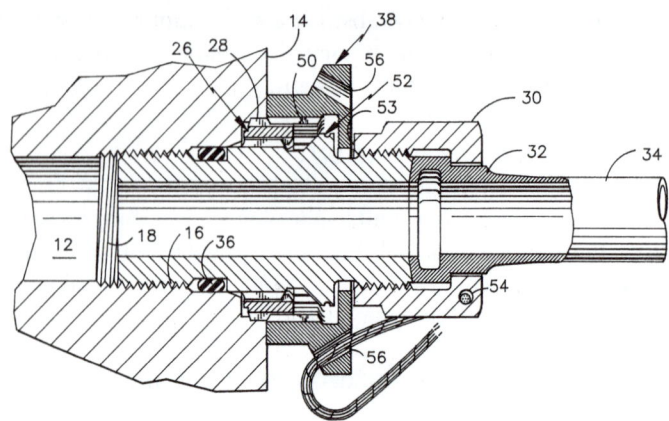

FIGURE 1.24 ■ A proper patent drawing that is used as an example in the Patent and Trademark Office publication, *Guide for the Preparation of Patent Drawings.*

on which a written description of the invention, drawings if necessary, and the name of the inventor(s) are received in the PTO.

Patent Drawings

According to the *Guide for the Preparation of Patent Drawings,* published by the U.S. Department of Commerce, Patent and Trademark Office, drawings form an integral part of a patent application. The drawing must show every feature of the invention specified. Figure 1.24 shows a proper patent drawing that is used as an example in a PTO publication. There are specific requirements for the size of the sheet on which the drawing is made, the type of paper, the margins, and other details relating to creating the drawing. The reason for specifying the standards in detail is that the drawings are printed and published in a uniform style when the patent issues, and the drawings also must be understood by persons using the patent descriptions.

Drawings must be created using solid black ink lines on white media. Color drawings or photographs are accepted on rare occasions, but there is an additional petition and other specific requirements for submitting a color drawing or photograph.

The previous discussion provides some basic general guidelines. You should confirm specific details and requirements with the Patent and Trademark Office. Additional information about the preparation of patent drawings is found in Chapter 27 of this text.

TRADEMARKS

According to the PTO publication *Basic Facts about Registering a Trademark,* a **trademark** is either a word, phrase, symbol or design, or combination of words, phrases, symbols or designs, that identifies and distinguishes the source of the goods or services of one party from those of others. A **service mark** is the same as a trademark except that it identifies and distinguishes the source of a service rather than a product. Normally, a mark for goods appears on the product or on its packaging, while a service mark appears in advertising for the services. A trademark is

different from a copyright or a patent; as previously explained, a copyright protects an original artistic or literary work, while a patent protects an invention.

Trademark rights start from the actual use of the mark, or the filing of a proper application to register a mark in the Patent and Trademark Office stating that the applicant has a genuine intention to use the mark in commerce regulated by the U.S. Congress. Federal registration is not required to establish rights in a mark, nor is it required to begin use of a mark. However, federal registration can secure benefits beyond the rights acquired by just using a mark. For example, the owner of a federal registration is presumed to be the owner of the mark for the goods and services specified in the registration, and to be entitled to use the mark nationwide. Generally, the first party who either uses a mark in commerce or files an application in the PTO has the ultimate right to register that mark. The PTO's authority is limited to determining the right to register. The right to use a mark can be more complicated to determine, particularly when two parties have begun use of the same or similar marks without knowledge of one another and neither has a federal registration. Only a court can make a decision about the right to use. A federal registration can provide significant advantages to a party involved in a court proceeding. The PTO cannot provide advice concerning rights in a mark. Only a private attorney can provide such advice.

Trademark rights can last indefinitely if the owner continues to use the mark to identify its goods or services. The term of a federal trademark registration is ten years, with ten-year renewal terms. However, between the fifth and sixth year after the date of initial registration, the registrant must file an official paper giving certain information to keep the registration alive. The registration is canceled if this is not done.

The previous discussion provides some basic general information. You should confirm specific details and requirements with the Patent and Trademark Office.

WEBSITE RESEARCH
Professional Design Drafting Organization

The American Design Drafting Association (ADDA) website at *http://www.adda.org* has valuable information and publications related to the design drafting profession.

Employment in Drafting

To find resources on the Internet, you can do a search using related words such as employment, employment services, career services, human resources, job hunt, job fair, job listings, job seekers, and resume services. Your school or state employment service probably also has a website for you to use when seeking employment. This textbook does not endorse or recommend any specific job search option. If you are seeking national employment as well as local opportunities, the following is a list of a few websites where you can find such information as on-line career resources, job updates, employer profiles, discussion groups, classified ads, job fairs, entry-level positions for college graduates, resume writing, resume posting, career guidance, salary guides, and professional organizations:

http://www.ajb.dni.us—America's Job Bank

http://www.careermag.com—Career Magazine

http://www.careerpath.com, http://www.careermosaic.com, and *http://www.careerbuilder.com*—Career Builder

http://www.careers.org—Careers

http://www.espan.net—E-Span's Interactive Employment Network

http://www.monster.com—Find jobs, post a resume, network, and get advice.

http://www.dbm.com—The Riley Guide: Employment Opportunities and Job Resources on the Internet

http://www.jobweb.com—Jobweb

http://www.usjobnet.com, http://www.nbew.com—US Job Net

http://www.bestjobsusa.com—Best Jobs USA

Free services specifically for members of the design and drafting industry:

http://www.adda.org—American Design Drafting Association

National CADD Standards

The National Council for Advanced Manufacturing (NACFAM) publishes the *Computer-Aided Drafting and Design (CADD) Skill Standards*. The NACFAM website: *http://www.nacfam.org*

The *United States National CAD Standard* information is found on the U.S. National CAD Standard website: *http://www.nationalcadstandard.org*

Copyrights, Patents, and Trademarks

Detailed information, available publications, and application procedures and fees can be found on the U.S. Copyright Office website: *http://www.lcweb.loc.gov*

Detailed information, available publications, and application procedures and fees can be found on the Patent and Trademark Office website: *http://www.uspto.gov*

PROFESSIONAL PERSPECTIVE

The design and drafting profession can provide a rewarding career for people who enjoy detailed work and have a mechanical aptitude that gives the ability to visualize. Math and communication skills are also important. The fundamental standards involved in the industry are found in mechanical drafting. This is the type of drafting done for manufacturing. Recommended drafting standards for this industry are controlled by the American National Standards Institute (ANSI) and the American Society of Mechanical Engineers (ASME). There are a variety of different design drafting disciplines that include the general categories of architectural; structural; civil; industrial pipe; heating, ventilating, and air-conditioning; sheet metal; electrical; electronic; and technical illustration. Some educational programs provide training in specific disciplines, while others provide diversified training in several areas. The opportunity to experience more than one discipline allows you to find an industry that you prefer. An entry-level position may not be in your chosen field, but you should be able to find employment in your desired area with experience and an open job market.

There are a variety of opportunities that allow people to expand career potential into related areas such as tool design or cartography. Many people who enter the drafting industry begin to move up quickly into design, checking, purchasing, estimating, and management. The ability to advance in this field depends on your skill, product knowledge, attitude, ability to communicate, continued education, and enthusiasm.

CHAPTER 1 *Introduction to Engineering Drawing and Design Test*

 Access the CD found with this textbook to view the Chapter 1 Test. Confirm the preferred submittal method with your instructor.

CHAPTER 1

Introduction to Engineering Drawing and Design Problems

DIRECTIONS

Select one or more of the following topic areas as determined by your instructor or course guidelines, and write a 300 to 500-word report on the selected topic or topics. Prepare each report using a word processor. Use double-spacing, proper grammar and spelling, and illustrative examples where appropriate. Use, but do not copy, the information found in this chapter and additional research information.

- History of drafting
- One or more drafting fields of your choice
- Why it is never necessary to draw anything more than once with CADD
- American Design Drafting Association (ADDA)
- Searching for a drafting position
- Employment opportunities on the Internet
- Requirements for becoming a drafter
- Computers in design and drafting
- Parametric design
- Virtual design

- VR
- Rapid prototyping
- National CADD standards
- Workplace ethics
- Intellectual property rights
- Copyrights
- Patents
- Trademarks
- Professional perspective
- Your own selected topic that relates to the content of this chapter

Manual Drafting Equipment, Media, and Reproduction Methods

LEARNING OBJECTIVES

After completing this chapter, you will:

- Describe and demonstrate the use of various manual drafting tools and equipment.
- Read engineer's, architect's, and metric scales, and drafting machine verniers.
- Discuss and use drafting media, sheet sizes, and title block information.
- Explain common reproduction methods.

INTRODUCTION

This chapter discusses and demonstrates the type of equipment that is used for manual drafting. Manual drafting is the type of drafting that is done by hand with pencil or ink on a medium such as paper or polyester film using drafting instruments and equipment. Manual drafting is also typically called hand drafting. Manual drafting has been replaced by computer-aided design and drafting (CADD) in most of the drafting industry. Some companies use CADD and have manual drafting tables available for use, as shown in Figure 2.1. CADD equipment and the reasons for the evolution of drafting from manual to CADD are discussed in detail in Chapter 3 of this textbook. Some people say that CADD will completely take the place of manual drafting someday, while others say that manual drafting may always be used for some applications, such as updating old drawings. Only the future will tell the true story; however, it is true that CADD has revolutionized the way design and drafting is done, and there are constantly new advances made in this industry. It does not matter if you are using manual drafting or CADD, you still need to know the basics of drafting. These basics are founded on national and industry-specific standards.

DRAFTING EQUIPMENT

Drafting tools and equipment are available from a number of vendors selling professional drafting supplies. For accuracy and long life, always purchase high-quality equipment. Local ven-

FIGURE 2.1 ■ An architectural office with computer workstations and a manual drafting table in use. *Courtesy Hewlett-Packard Company.*

dors can be found by looking in the yellow pages of your telephone book under headings such as: Drafting Room Equipment & Supplies, Blue-printing, Architects' Supplies, Engineering Equipment & Supplies, or Artists' Materials & Supplies.

Drafting supplies and equipment can be purchased in a kit, or items can be bought individually. Whether equipment is purchased in a kit or by the individual tool, the items that are normally needed include the following:

- One 0.3 mm automatic drafting pencil with 4H, 2H, and H leads.
- One 0.5 mm automatic drafting pencil with 4H, 2H, H, and F leads.
- One 0.7 mm automatic drafting pencil with 2H, H, and F leads.
- One 0.9 mm automatic drafting pencil with H, F, and HB leads. Drafters may elect to purchase two or more pencils and use a different grade of lead in each. Doing so reduces the need to constantly change leads. Some drafters use a light blue lead for layout work. A combination of 0.5, 0.7, and 0.9 mm pencils and leads is good for sketching and

other related activities if you are doing computer-aided drafting.

- 6-in. bow compass.
- Dividers.
- Eraser. Select an eraser that is recommended for drafting with pencil on paper.
- Erasing shield.
- 8-in. 30°–60° triangle.
- 8-in. 45° triangle.
- Irregular curve.
- Scales:
  - Triangular architect's scale.
  - Triangular civil engineer's scale.
  - Triangular metric scale.
- Drafting tape.
- Circle template (small holes).
- Circle template (large holes).
- Lettering guide (optional).
- Arrowhead template (optional).
- Sandpaper sharpening pad.
- Dusting brush.

DRAFTING FURNITURE

Tables

There is a large variety of drafting tables available, ranging from economical models to complete professional workstations. Drafting tables are generally sized by the dimensions of their tops. Standard tabletop sizes range from 24 × 36 in. (600 × 900 mm) to 42 × 84 in. (1000 × 2000 mm).

The features to look for in a good-quality, professional table include:

- One-hand tilt control.
- One-hand or foot height control.
- The ability to position the board vertically for use with the track drafting machine.
- An electrical outlet.
- A drawer for tools and/or drawings.

Some manufacturers ship tables with tops that are ready to draw on. Others ship tables with tops of steel or basswood that need to be covered. Basswood can be a drawing surface; however, most offices commonly cover drafting-table tops with smooth, specially designed surfaces. The material, usually vinyl, provides the proper density for effective use under normal drafting conditions. After compass or divider points have pierced the surface, the small holes close to provide a smooth surface for continued use. Drafting tape is commonly used to adhere drawings to the tabletop, although some drafting tables have magnetized tops and use magnetic strips to attach drawings.

Chairs

Better drafting chairs have the following characteristics:

- Padded or contoured seat design.
- Height adjustment.
- Foot rest.
- Fabric that allows air to circulate.
- Sturdy construction.

DRAFTING PENCILS, LEADS, AND SHARPENERS

Mechanical Pencils or Lead Holders

The term mechanical pencil, also referred to as lead holder, is applied to a pencil that requires a piece of lead to be manually inserted; by some physical action such as twist, push, or pull, the lead is mechanically, or semiautomatically, advanced to the tip. These have been replaced by automatic pencils for most use.

Leads for mechanical pencils are bought separately. Leads are graded by hardness and designated by a number and letter, or one or two letters. The designation is usually found along or on one end of the lead. Always sharpen the end opposite the designation. Figure 2.2a shows a mechanical pencil.

Automatic Pencils

The term automatic pencil refers to a pencil with a lead chamber that, at the push of a button or tab, advances the lead from the chamber to the writing tip and, when a new piece of lead is needed, advances the new piece to the tip. Automatic pencils are designed to hold leads of one width so you do not need to sharpen the lead. These pencils are available in several different lead sizes. Drafters have several automatic pencils. Each pencil has a different grade of lead hardness and is used for a specific technique. (See Figure 2.2b.)

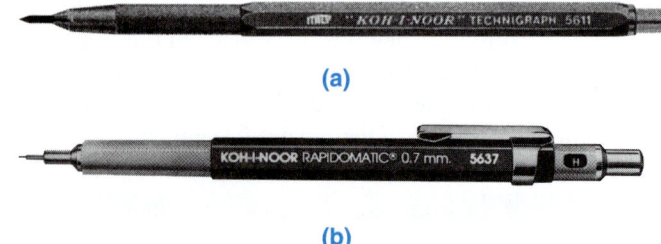

(a)

(b)

FIGURE 2.2 ■ (a) Mechanical pencil, also known as a lead holder. *Courtesy Koh-I-Noor, Inc.* (b) An automatic pencil. Lead widths are 0.3, 0.5, 0.7, and 0.9 mm. *Courtesy Koh-I-Noor, Inc.*

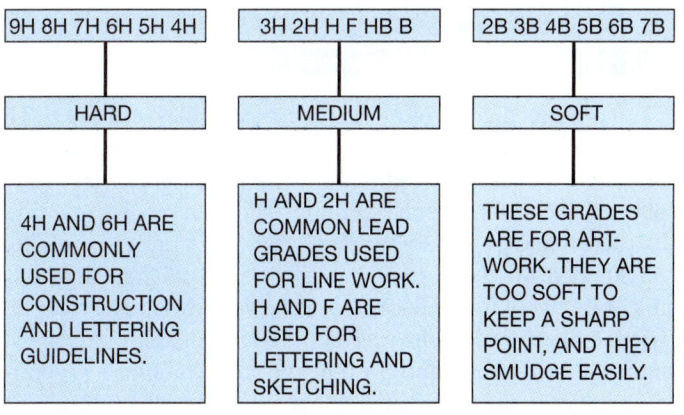

9H 8H 7H 6H 5H 4H	3H 2H H F HB B	2B 3B 4B 5B 6B 7B
HARD	MEDIUM	SOFT
4H AND 6H ARE COMMONLY USED FOR CONSTRUCTION AND LETTERING GUIDELINES.	H AND 2H ARE COMMON LEAD GRADES USED FOR LINE WORK. H AND F ARE USED FOR LETTERING AND SKETCHING.	THESE GRADES ARE FOR ART-WORK. THEY ARE TOO SOFT TO KEEP A SHARP POINT, AND THEY SMUDGE EASILY.

FIGURE 2.3 ■ The range of lead grades.

Lead Grades

The leads you select for line work and lettering depend on the amount of pressure you apply and other technique factors. Experiment until you identify the leads that give the best line quality. Leads commonly used for thick lines range from 2H to F, while leads for thin lines range from 4H to H, depending on individual preference. Construction lines for layout and guidelines are very lightly drawn with a 6H or 4H lead. Figure 2.3 shows the different lead grades.

Polyester and Special Leads

Polyester leads, also known as plastic leads, are for drawing on polyester drafting film, often called by its trade name, Mylar®. Plastic leads come in grades equivalent to F, 2H, 4H, and 5H and are usually labeled with a prefix and number, for example, P1 and P2. Some companies make a combination lead for use on both vellum and polyester film.

Colored leads have special uses. Red lead is commonly used for corrections. Red prints as a black line. Blue leads can be used on the original drawing by the supervisor to tell the drafter the corrections that need to be made. Blue lead does not show on the print when the original drawing with blue lines on it is run through a diazo machine. Some drafters use light blue lead for all layout work and guidelines because it does not reproduce on a diazo printer; however, it may show on a photocopier.

Basic Pencil Technique

Automatic pencils do not require rotation, although some drafters feel rotating the pencil makes darker lines. The automatic pencil should be held near vertical. The full surface of the lead is used when the pencil is in a vertical position. Provide enough pressure and go over each line enough times to make a line dark and crisp. Take care not to make it too thick. Figure 2.4 shows some basic pencil motions.

Sanding Block

Especially useful for sharpening compass lead, this is one of the simplest devices used for sharpening. Sandpaper stapled to a

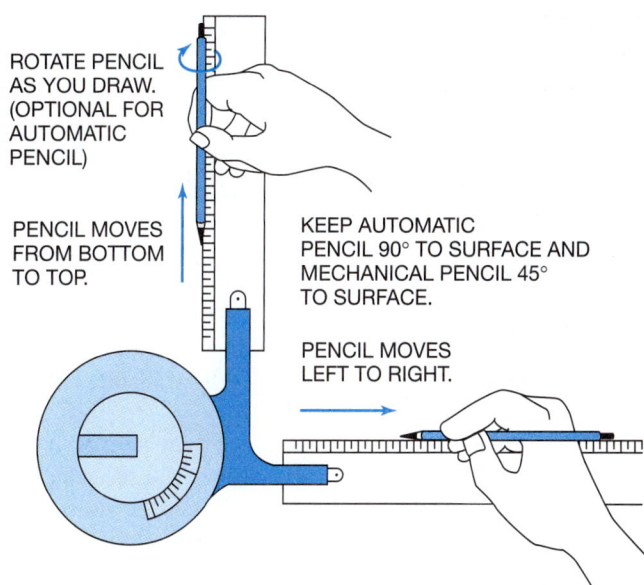

FIGURE 2.4 ■ Basic pencil motions. Move the pencil in the opposite direction if you are left-handed.

wooden paddle is called a sandpaper block, or sanding block. Plastic lead fills sandpaper rapidly so several sheets are needed as compared to graphite lead. To avoid smudging your drawing, use the sanding block away from your drawing table and dispose of the graphite carefully.

Pocket Pointer

A portable sharpener for mechanical pencils is the pocket pointer. This pointer contains blades that sharpen lead to a conical point. The pocket pointer works with either graphite or plastic lead.

Cutting Wheel Pointer

The best type of conical-point sharpener for mechanical pencil holders is a mechanical lead pointer with a tool-steel cutting wheel. Use the slots provided in the top to expose the right length of lead to get a sharp or slightly dull point. The slightly dull point is used for lettering. This sharpener works on graphite or polyester leads equally well.

TECHNICAL PENS AND ACCESSORIES
Technical Pens

Also known as technical fountain pens, technical pens have improved in quality and ability to produce excellent inked lines. These pens function on a capillary action where a needle acts as a valve to allow ink to flow from a storage cylinder through a small tube, which is designed to meter the ink so a specific line width is created. Technical pens may be purchased individually or in sets. The different tip sizes used to make various line widths range from a narrow number 6 × 0 (.005 in./0.13 mm)

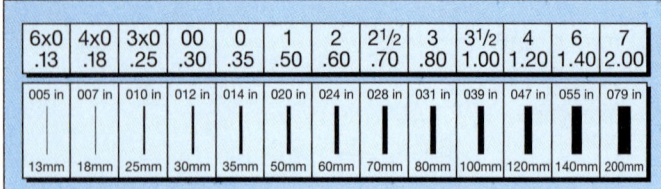

6x0	4x0	3x0	00	0	1	2	2½	3	3½	4	6	7
.13	.18	.25	.30	.35	.50	.60	.70	.80	1.00	1.20	1.40	2.00
005 in	007 in	010 in	012 in	014 in	020 in	024 in	028 in	031 in	039 in	047 in	055 in	079 in
13mm	18mm	25mm	30mm	35mm	50mm	60mm	70mm	80mm	100mm	120mm	140mm	200mm

FIGURE 2.5 ■ Technical pen line widths. *Courtesy Koh-I-Noor, Inc.*

to a wide number 7 (.079 in./2 mm). Figure 2.5 shows a comparison of some of the different line widths available with technical pens.

In addition to having the advantage of a constant line width, technical pens have a reservoir that allows you to make inked lines for a long period of time before ink must be added. Technical pens can be used with templates to make circles, arcs, and symbols. Compass adapters to hold technical pens are also available. Technical pen tips are designed to fit into scribers for use with lettering guides.

Pen Cleaning

Read the cleaning instructions that come with the brand of pen that you purchase. Some pens require disassembly for cleaning while others should not be taken apart. Pens should be cleaned before each filling or before being stored for a long period of time. Clean the technical pen nib, cartridge, and body separately in warm water or in a special cleaning solution.

Ultrasonic pen cleaners are available to clean points. Pens are placed in a tank where millions of energized microscopic bubbles, generated by ultrasonic action, carry cleaning solution into the smallest openings of the drawing point to scrub the tube inside and out.

Specially formulated pen cleaner should be used for best results. Pen cleaner also can be used to soak pen points for cleaning by hand.

Ink

Drafting inks should be opaque, or have a matte or semiflat black finish that will not reflect light. The ink should reproduce without hot spots or line variation. Drafting ink should have excellent adhesion properties for use on paper or film. Certain inks are recommended for use on film in order to avoid peeling, chipping, or cracking. Inks recommended for use in technical pens also have nonclogging characteristics. This property is especially important for use in high-speed computer-graphics plotters.

When selecting an ink, be sure to purchase one for the job you want done. First, determine how the ink will be applied, that is, from a technical pen, computer plotter, airbrush, fountain pen, calligraphy pen, or with a brush. Second, determine the surface the ink will be used on, such as paper (vellum or bond), polyester film, or acetate. Third, determine if your use requires the ink to be opaque, fast-drying, waterproof, or erasable.

ERASERS AND ACCESSORIES

The common shapes of erasers are rectangular and stick. The stick eraser works best in small areas. There are three basic types of erasers: pencil, ink, and plastic. The plastic eraser is used for plastic lead or ink on polyester film. These erasers are identified by their white or translucent color. Select an eraser that is recommended for the particular material used. When used with ink, apply very light pressure, and be very careful, because the friction developed by the speed of erasure can easily damage the drafting film surface and prevent the adhesion of ink when redrawing over the erased area. Moistening the eraser during use on polyester film helps to reduce any damage to the drawing surface, but do not do this on paper.

Erasing Tips

When erasing, the idea is to remove an unwanted line or letter. You do not want to remove the surface of the paper or polyester film. Erase only hard enough to remove the unwanted line. However, you must bear down hard enough to eliminate the line completely. If all the line does not disappear, ghosting results. A ghost is a line that seems to have been eliminated but still shows on a print. Lines that have been drawn so hard as to make a groove in the drawing sheet also can cause a ghost. To remove ink from vellum, use a pink or green eraser, or an electric eraser. Work the area slowly. Do not apply too much pressure or erase in one spot too long or you can go through the paper. On polyester film, use a vinyl eraser and/or a moist cotton swab. The inked line usually comes off easily, but use caution. If you destroy the matte surface of either a vellum or polyester sheet, you are not able to redraw over the erased area.

Lead may be picked up from the drawing board surface and transmitted to the back of the drawing surface. When this happens, the drawing must be turned over and the graphite removed from the back surface of the drawing.

Electric Erasers

Professional manual drafters use electric erasers. Those with cords that plug in are best, but cordless, rechargeable units are also available. When working with an electric eraser, you do not need to use much pressure because the eraser operates at high speed. The purpose of the electric eraser is to remove unwanted lines quickly. *Use caution:* these erasers can also remove paper quickly!

Erasing Shield

Erasing shields are thin metal or plastic sheets with a number of differently shaped holes. They are used to erase small, unwanted lines or areas. For example, if you have a corner overrun, place one of the slots of the erasing shield over the area to be removed while covering the good area.

Eradicating Fluid

Eradicating fluid is primarily used with ink on film or for removal of lines from sepia (brown) prints. The eradicating fluid is most often applied with a brush or cotton swab. If in doubt about its application, follow the manufacturer's instructions; however, application is usually done by lightly moistening the area to be corrected. Then the solution is wiped with a tissue, being careful to remove all residue. Eradicating fluid is especially effective for removing aged ink lines and for erasure of large areas.

Cleaning Agents

Special eraser particles are available to help reduce smudging and to keep the drawing and your equipment clean. The particles also help float triangles, straightedges, and other drafting equipment to reduce line smudging. Use this material sparingly, since too much can cause your lines to become fuzzy. Cleaning powders are not recommended for use on ink drawings or on polyester film.

Dusting Brush

Use a dusting brush to remove eraser particles from your drawing. Doing so helps reduce the possibility of smudges. Avoid using your hand to brush away eraser particles; the hand tends to cause smudges, which reduces drawing neatness.

  The brush should be cleaned regularly with soap and water because it picks up graphite particles and casts a slight film over the drawing.

DRAFTING INSTRUMENTS
Kinds of Compasses

Compasses are used to draw circles and arcs. However, using a compass can be time-consuming. Use a template, whenever possible, to make circles or arcs more quickly. A compass is especially useful for large circles.

  There are several basic types of compasses; however, a bow compass, shown in Figure 2.6, is used for most drawing applications. Beam compasses consist of a bar with an adjustable needle, and a pencil or pen attachment for swinging large arcs or circles. Also available is a beam that is adaptable to the bow compass. Such an adapter works only on bow compasses that have a removable leg.

Use of Compass

Keep both the compass needle point and lead point sharp. The points are removable for easy replacement. The better compass needle points have a shoulder on them. The shoulder helps keep the point from penetrating the medium more than necessary. Compare the needle points in Figure 2.7.

  The compass lead should, in most cases, be one grade softer than the lead you use for straight lines, because less

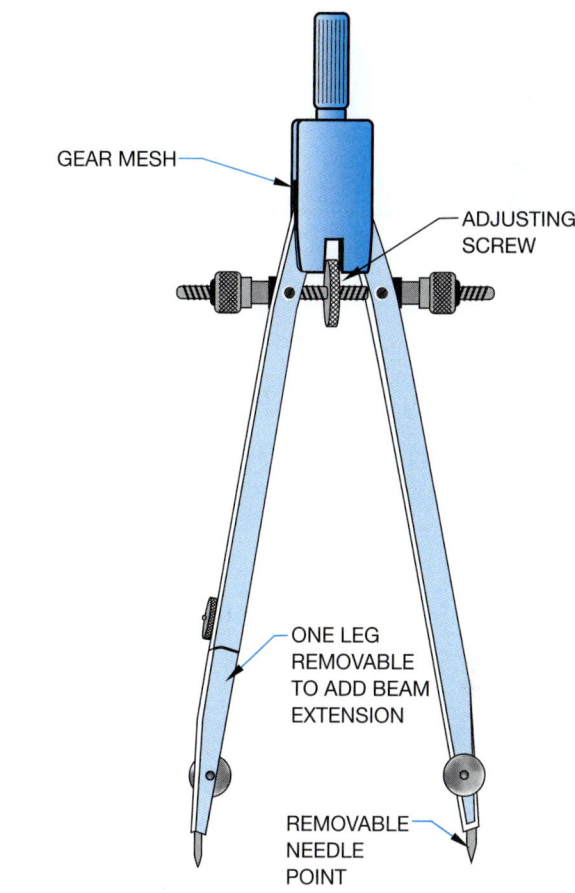

FIGURE 2.6 ■ Bow compass.

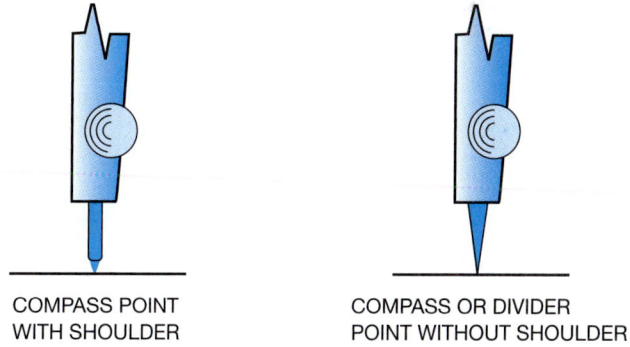

FIGURE 2.7 ■ Compass points.

pressure is used on a compass than on a pencil. Keep the compass lead sharp. An elliptical point is commonly used with the bevel side away from the needle leg. Keep the lead and the point equal in length. Figure 2.8 shows properly aligned and sharpened points on a compass.

  Use a sandpaper block to sharpen the elliptical point. Be careful to keep the graphite residue away from your drawing and off your hands. Remove excess graphite from the point with a tissue or cloth after sharpening. Sharpen the lead often.

  Some drafters prefer to use a conical point in their compass.

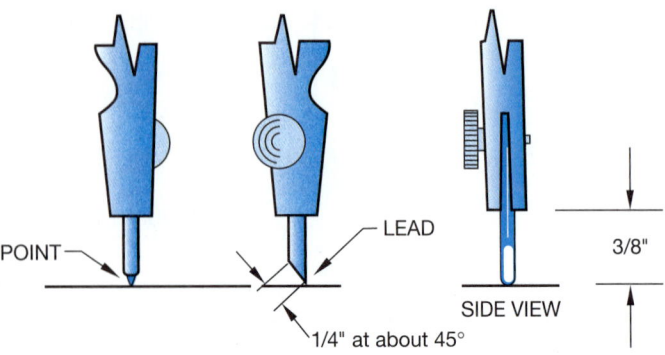

FIGURE 2.8 ■ Properly sharpened and aligned elliptical compass point.

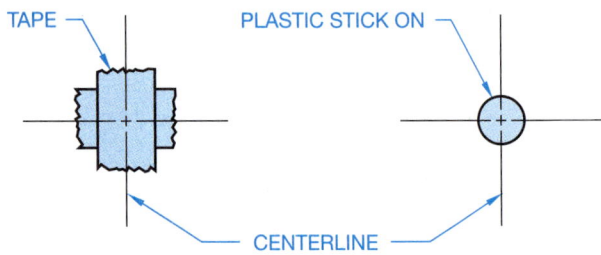

FIGURE 2.9 ■ Drawing sheet protection from compass point.

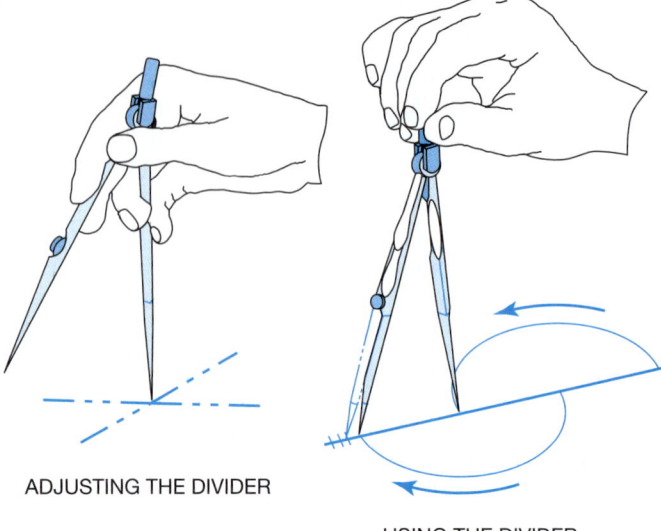

FIGURE 2.10 ■ Using a divider.

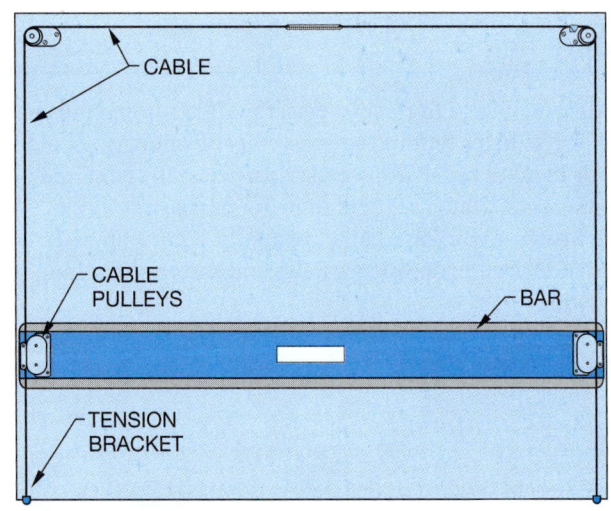

FIGURE 2.11 ■ Parallel bar.

Protecting the Sheet During Compass Use

If you are drawing a number of circles from the same center, you will find that the compass point causes an ugly hole in your drawing sheet. Reduce the chance of making such a hole by placing a couple of pieces of drafting tape at the center point for protection. There are small plastic circles available for this purpose. Place one at the center point, then pierce the plastic with your compass. (See Figure 2.9.)

Dividers

Dividers are used to transfer dimensions or to divide a distance into a number of equal parts.

Note: Do not try to use dividers as a compass.

Some drafters prefer to use bow dividers because the center wheel provides the ability to make fine adjustments easily. Also, the setting remains more stable than with standard friction dividers.

A good divider should not be too loose or tight. It should be easily adjustable with one hand. In fact, you should control a divider with one hand as you lay out equal increments or transfer dimensions from one feature to another. Figure 2.10 shows how the divider should be handled when used.

Proportional Dividers

Proportional dividers are used to reduce or enlarge an object without the need of mathematical calculations or scale manipulations. The center point of the divider is set at the correct point

for the proportion you want. Then you measure the original size line with one side of the proportional divider and the other side automatically determines the new reduced or enlarged size.

Parallel Bar

The **parallel bar** slides up and down the board to allow you to draw horizontal lines. (See Figure 2.11.) Vertical lines and angles are made with triangles in conjunction with the parallel bar. The parallel bar is commonly found in architectural drafting offices because architectural drawings are frequently very large. Architects often need to draw straight lines the full length of their boards, and the parallel bar is ideal for such lines.

Triangles

There are two standard triangles. One has angles of 30°–60°–90° and is known as the **30°–60° triangle**. The other has angles of

45°–45°–90° and is known as the 45° triangle. Figure 2.12 shows these popular triangles.

Some drafters prefer to use triangles in place of a vertical drafting machine scale as shown in Figure 2.13. The machine protractor or the triangle can be used to make angled lines. Drafters who use parallel bars rather than drafting machines also use triangles to make vertical and angled lines.

Triangles can also be used as straightedges to connect points for drawing lines without the aid of a parallel bar or machine scale. Triangles are used individually or in combination to draw angled lines in 15° increments. (See Figure 2.14.) Also available

are adjustable triangles with built-in protractors that are used to make angles of any degree up to a 45° angle.

Templates

Circle Templates

Circle templates are available with circles in a range of sizes beginning with 1/16 in. (1.5 mm). The circles on the template are marked with their diameters and are available in fractions, decimals, or millimeters. The parts of a circle are shown in Figure 2.15. Sample circle templates are shown in Figure 2.16. A popular template is one that has circles, hexagons, squares, and triangles.

Always use a circle template rather than a compass. Circle templates save time and are very accurate. For best results when making circles, try to keep your pencil or pen perpendicular to the paper. To obtain proper width lines with a pencil, use a 0.9 mm automatic pencil.

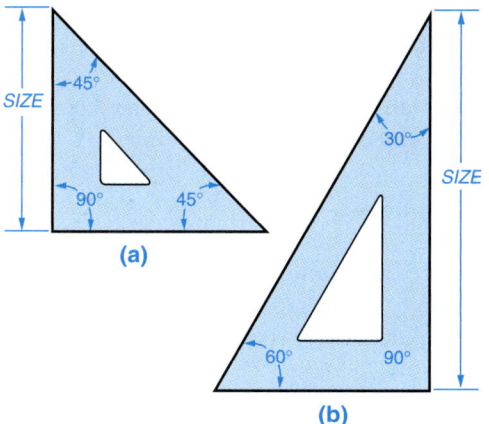

FIGURE 2.12 ■ 45° and 30°–60° triangles.

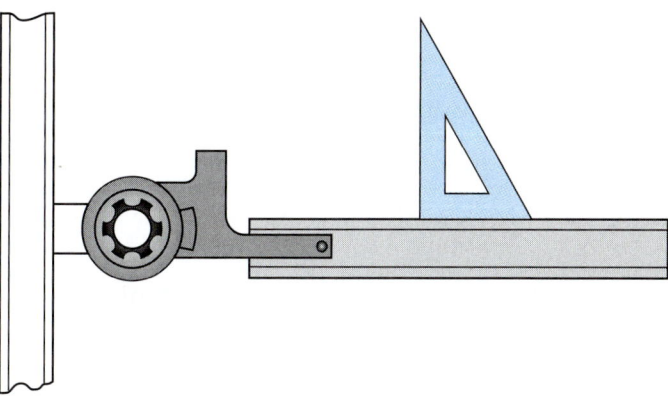

FIGURE 2.13 ■ Using a triangle with a drafting machine.

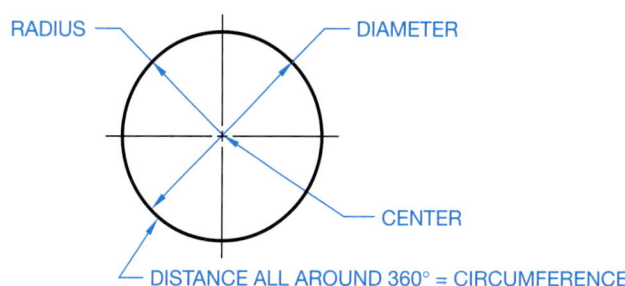

FIGURE 2.15 ■ Parts of a circle.

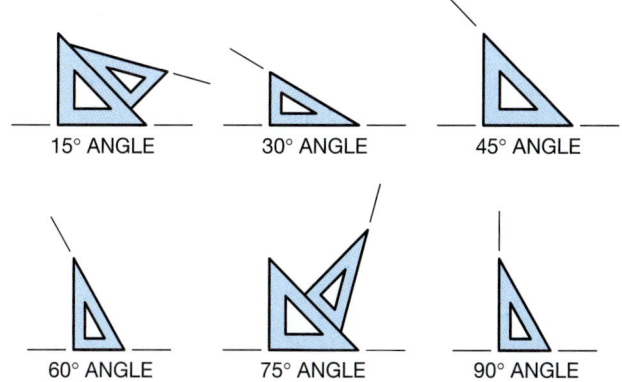

FIGURE 2.14 ■ Angles that may be made with the 30°–60° and 45° triangles individually or in combination.

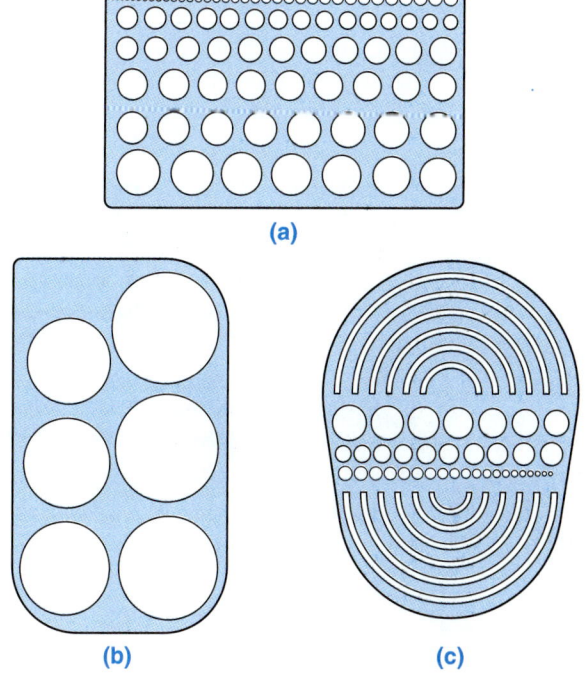

FIGURE 2.16 ■ (a) Small circles. (b) Large, full circles. (c) Large, half circles.

To use a circle template properly, first draw the centerlines of your circle. Then exactly align the dashes on the template with the centerlines as shown in Figure 2.17. Proceed to trace the outline of the circle.

To draw arcs with a circle template, use one of two methods. One method is to draw the centerlines of the arc, align the template with the proper diameter, and draw the arc. Keep in mind that the template circles are marked in diameter while the size of an arc is given in radius. Remember to divide the template size in half to find the proper arc radius. The other method of drawing arcs is to lightly draw outside construction lines, then fill in the arc to the points of tangency. This method is demonstrated in Chapter 8, Geometric Construction. Be sure the connection between the arc and the straight line is smooth.

When using a circle template and a technical pen, keep the pen perpendicular to the paper. Some templates have risers built in to keep the template above the drawing sheet. Without this feature there is a risk of ink running under a template that is flat against the drawing. If your template does not have risers, purchase and add template lifters, use a few layers of tape placed on the underside of the template (although tape does not always work well), or place a second template with a larger circle under the template you are using. (See Figure 2.18.)

Ellipse Templates

Ellipses are circles seen at an angle. The parts of an ellipse are shown in Figure 2.19.

The type of pictorial drawing known as isometric projects the sides of objects at a 30° angle in each direction away from the horizontal. Isometric circles are ellipses aligned with the horizontal right or left planes of an isometric box, as in Figure 2.20. Isometric ellipse templates automatically position the ellipse at the proper angle of 35° 16'. (See Figure 2.21.)

Isometric sketching is covered in Chapter 6 and Isometric drawing is discussed in detail in Chapter 15.

Use a Template

Never use a compass if you can use a template. Templates increase drafting speed and are very accurate.

Irregular Curves

Irregular curves are commonly called **French curves**. These curves have no constant radii. An irregular curve is shown in Figure 2.22. A **radius curve** is composed of a radius and tangent.

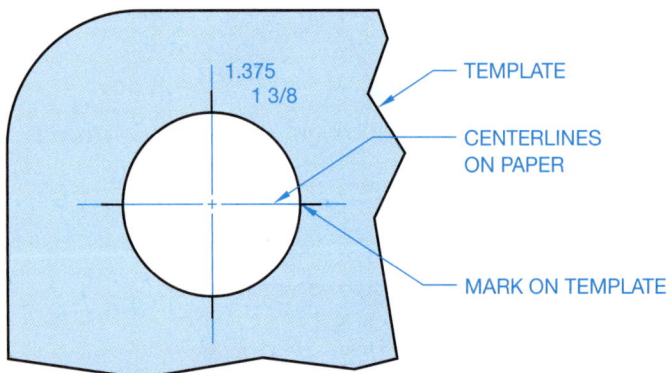

FIGURE 2.17 ■ Using a circle template to draw a circle.

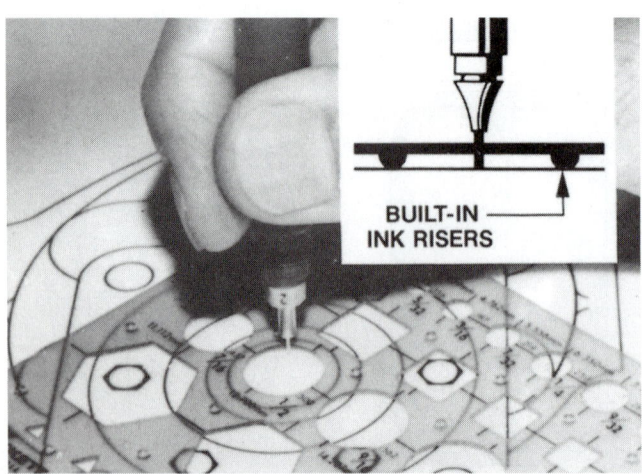

FIGURE 2.18 ■ A template with built-in risers for inking. *Courtesy Chartpak.*

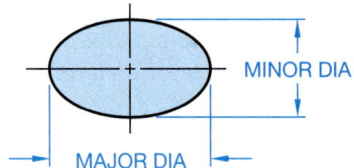

FIGURE 2.19 ■ Parts of an ellipse.

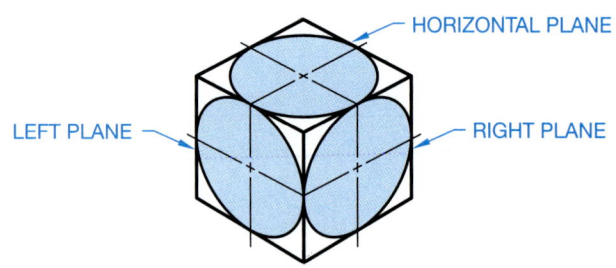

FIGURE 2.20 ■ Ellipses in isometric planes.

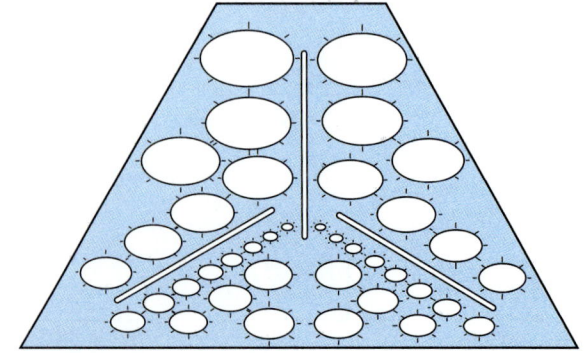

FIGURE 2.21 ■ Isometric ellipse template.

FIGURE 2.22 ■ Irregular or French curves.

The radius on these curves is constant and ranges from 3 ft to 200 ft. (900–60,000 mm). These curves are commonly used in highway drafting. In addition to these two kinds of curves, there are available ship's curves. The curves in a set of ship's curves become progressively larger and, like French curves, have no constant radii. They are used for layout and development of ships' hulls. Flexible curves are also available that allow you to adjust to a desired curve.

In order to draw an irregular curve, points on the curve are developed or plotted. The points are then connected with a construction line. Then a flexible curve is bent to fit, or a portion of an irregular curve is matched to the construction line to include at least three plotted points. Care must be taken to make the curve flow smoothly. This smooth flow of the line is accomplished by never drawing the full length of the curve, but by overlapping each successive setting of the irregular curve. (See Figure 2.23.)

DRAFTING MACHINES

Drafting machines have a vernier head allowing you to measure angles accurately to 5' (minutes). Drafting machines, for the most part, take the place of triangles and parallel bars. The draft-ing machine maintains a horizontal and vertical relationship between scales, which also serve as straightedges. A protractor allows the scales to be set quickly at any angle. There are two types of drafting machines, arm and track. Although both types are excellent tools, the track machine has generally replaced the arm machine in industry. A major advantage of the track machine allows the drafter to work with a board in the vertical position. A vertical drafting surface position is generally more comfortable to use than a horizontal table. For school or home use, the arm machine can be a good economical choice.

Arm Drafting Machine

The arm drafting machine is compact and less expensive than a track machine. The arm machine clamps to a table and through an elbowlike arrangement of supports, allows you to position the protractor head and scales anywhere on the board. An arm drafting machine is shown in Figure 2.24.

Track Drafting Machine

A track drafting machine has a traversing arm that moves left and right across the table and a head unit that moves up and down the traversing arm. There is a locking device for both the

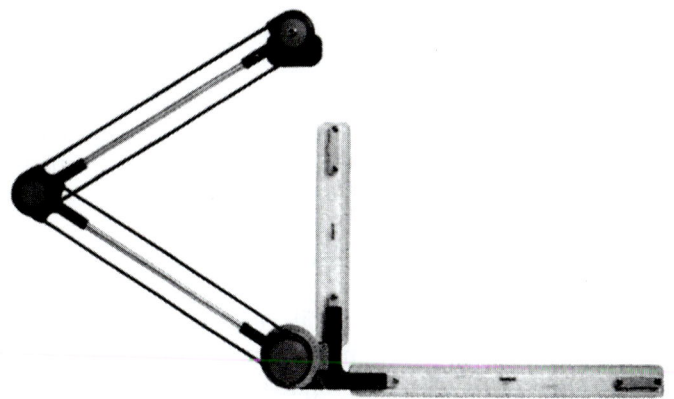

FIGURE 2.24 ■ Arm drafting machine. *Courtesy Vemco America, Inc.*

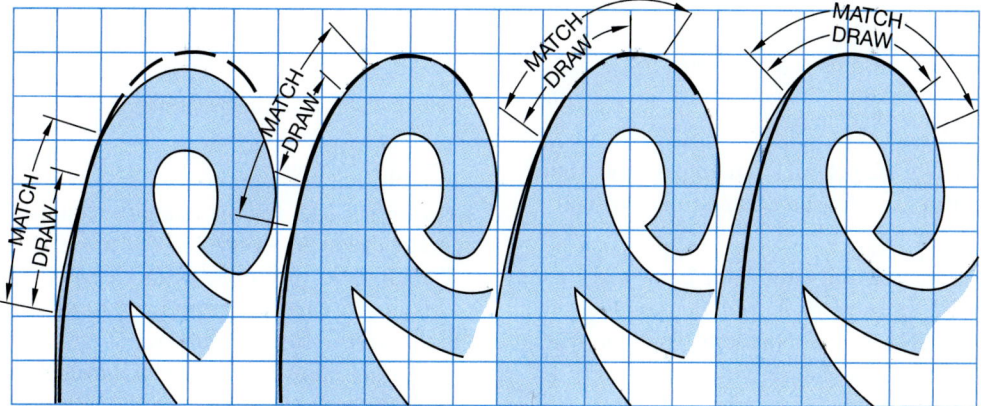

FIGURE 2.23 ■ Irregular curve development.

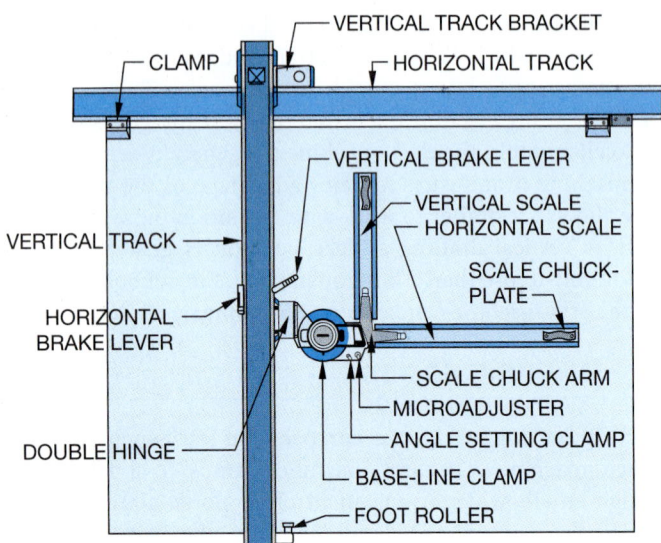

FIGURE 2.25 ■ Track drafting machine and its parts. *Courtesy Mutoh America, Inc.*

head and the traversing arm. The shape and placement of the controls of a track machine vary with the manufacturer, although most brands have the same operating features and procedures. Figure 2.25 shows the component parts of a track drafting machine.

Sizes of Drafting Machine

When ordering a drafting machine, the specifications should relate to the size of the drafting board on which it is mounted. For example, a 37 1/2 × 60 in. (950–1500 mm) machine properly fits a table of same size.

Controls and Machine Head Operation

Drafting machine heads contain the controls for horizontal, vertical, and angular movement. Although each brand of machine contains similar features, controls may be found in different places on different brands. (See Figure 2.26.) Most machines have the following controls:

■ Baseline adjustment—releases the scales so they can move but the protractor is not affected.

■ Index control—permits automatic stops every 15°. It can also be pushed in and locked to let you adjust the machine to any angle.

■ Indexing clamp—locks the protractor at angles other than 15° increments so you can draw an accurate line without the protractor moving.

To operate the drafting machine protractor head, place your hand on the handle and, using your thumb, depress the index thumbpiece. Doing so allows the head to rotate. Each increment marked on the protractor is one degree with a label every 10°. As the vernier plate (the small scale numbered from 0 to 60) moves past the protractor, the zero on the vernier aligns with

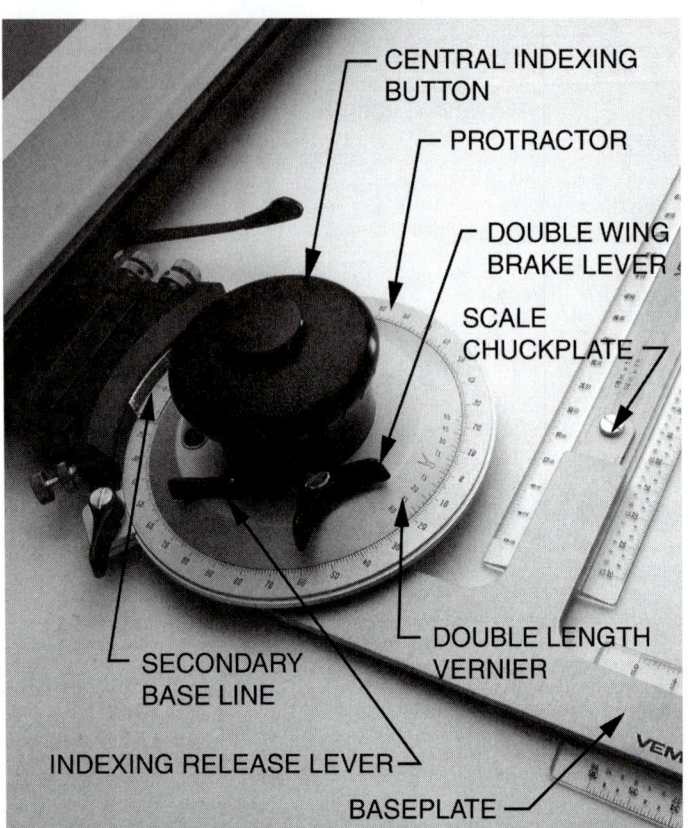

FIGURE 2.26 ■ Drafting machine head controls and parts. *Courtesy Vemco Corporation.*

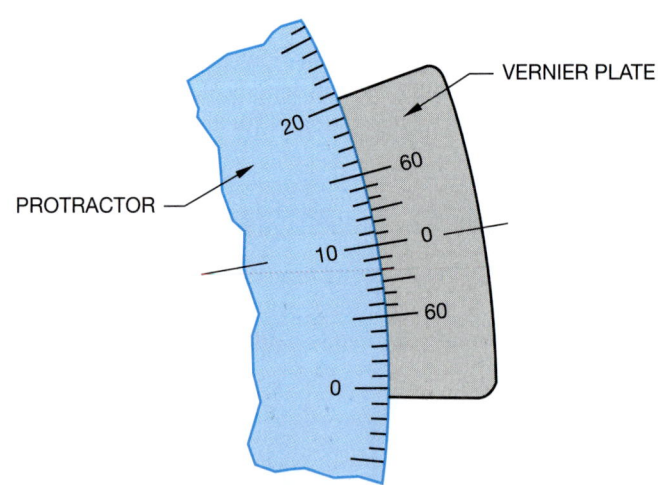

FIGURE 2.27 ■ Vernier plate and protractor showing a reading of 10°.

the angle that you want to read. For example, Figure 2.27 shows a reading of 10°. As you rotate the handle, notice the head automatically locks every 15°. To move the protractor past the 15° increment, you must again depress the index thumbpiece.

Having rotated the protractor head 40° clockwise, the machine is in the position shown in Figure 2.28. The vernier plate at the protractor reads 40°, which means that both the horizontal and vertical scale have moved 40° from their original position at 0° and 90°, respectively. The horizontal scale reads directly

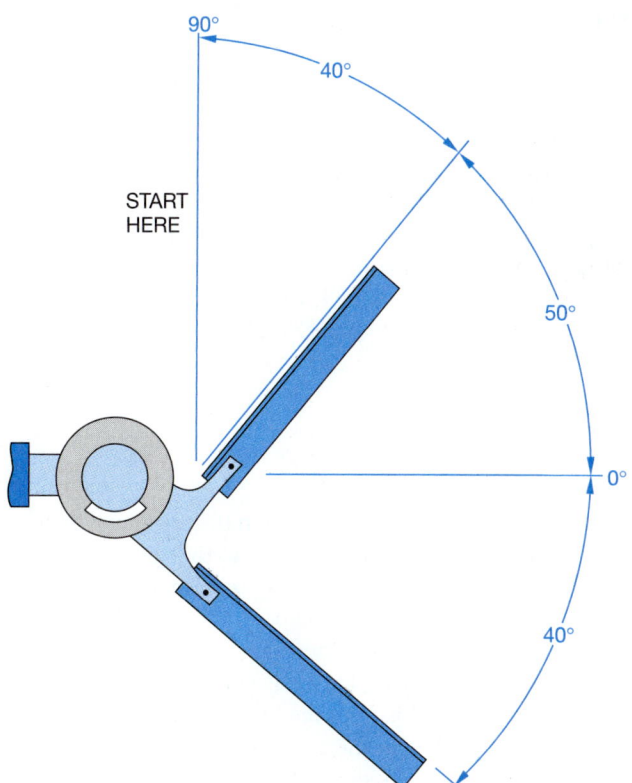

FIGURE 2.28 ■ Angle measurement with the drafting machine.

from the protractor starting from 0°. The vertical scale reading begins from the 90° position. The key to measuring angles is to determine if the angle is to be measured from the horizontal or vertical starting point. See the examples in Figure 2.29.

Measuring full degree increments is easy since you simply match the zero mark on the vernier plate with a full degree mark on the protractor. See the reading of 12° in Figure 2.30. The vernier scale allows you to measure angles as accurately as 5' (minutes). Remember, 1 degree equals 60 minutes (1° = 60') and 1 minute equals 60 seconds (1' = 60").

Reading and Setting Angles with the Vernier

To read an angle other than a full degree, assume the vernier scale is set at a **positive angle** as shown in Figure 2.31. Each mark on the vernier scale represents 5'. First see that the angle to be read is between 7° and 8°. Then find the 5' mark on the **upper half** of the vernier (the direction in which the scale has been turned) that is most closely aligned with a full degree on the protractor. In this example, it is the 40' mark. Add the minutes to the degree just passed. The correct reading, then, is 7°40'. The procedure for reading **negative angles** is the same, except you read the minute marks on the **lower half** of the vernier.

Suppose you want to set the angle 7°40' as shown in Figure 2.31. First release the protractor brake and disengage the indexing mechanism with the thumb control. Rotate the protractor arm counterclockwise until the zero of the vernier is at 7°. Then slowly continue the rotation until the 40' mark on

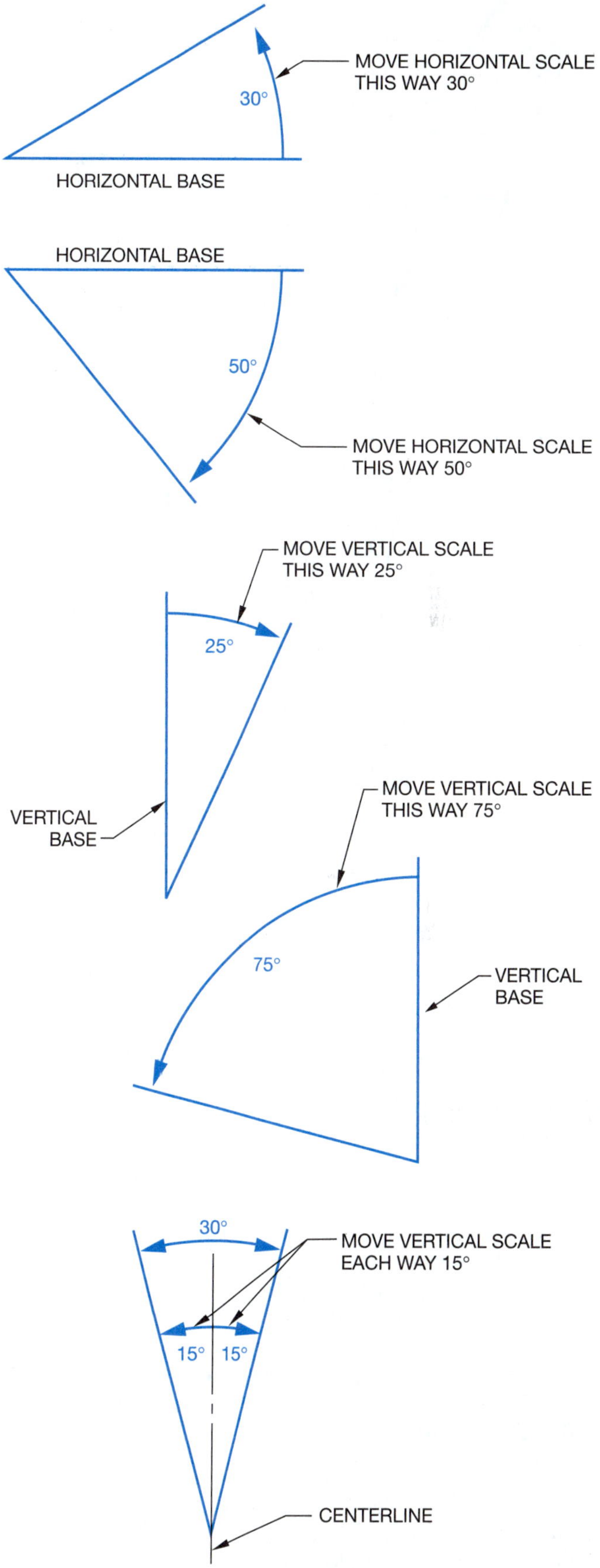

FIGURE 2.29 ■ Angle measurements from either a horizontal or vertical reference line.

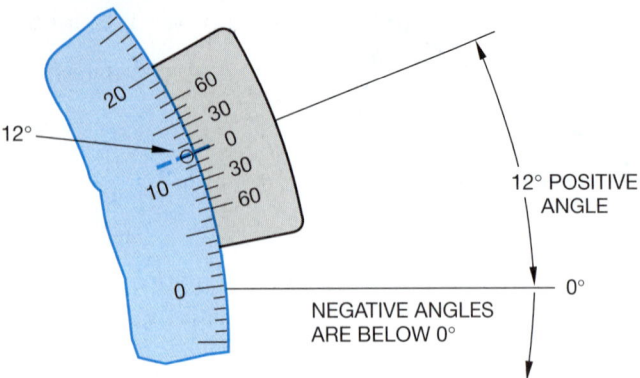

FIGURE 2.30 ■ Measuring full degrees.

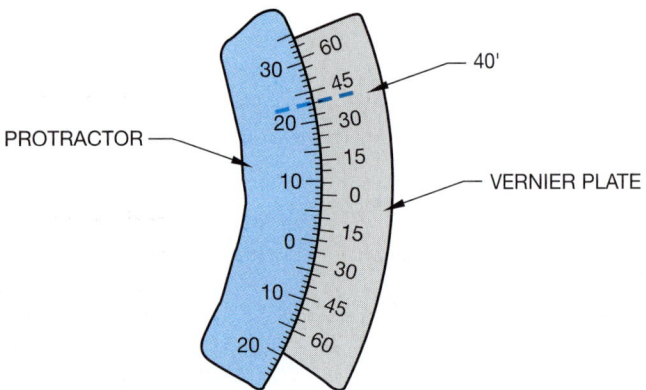

FIGURE 2.31 ■ Reading positive angles with the vernier.

the upper half of the vernier aligns with the nearest degree mark on the protractor. Lock the protractor brake and draw the line. The procedure for setting negative angles is essentially the same except for turning the protractor head in a clockwise direction and reading the lower half of the vernier.

Machine Setup

To insert a scale in the baseplate chuck, place the scale flat on the board and align the scale chuckplate with the baseplate chuck on the protractor head. Firmly press, but do not drive, the scale chuckplate into the baseplate chuck. To remove a scale, use the scale wrench provided with the machine. Removing a scale by hand without the aid of a key could result in damage to the scale and/or machine. Refer to the operation manual provided with your drafting machine.

Scale Alignment

Before drawing with any drafting machine, the scales should be checked for alignment and, if needed, be adjusted at right angles to each other. For best results with a track drafting machine, the scales should also be aligned with respect to the horizontal track. Both operations can be accomplished by following the instructions provided with your operation manual.

SCALES

Scale Shapes

There are four basic **scale shapes**, as shown in Figure 2.32. The **two-bevel scale** is also available with chuckplates for use with standard arm or track drafting machines. These machine scales have typical calibrations, and some have no scale reading for use as a straightedge alone. Drafting machine scales are purchased by designating the length needed, 12, 18, or 24 in., and the scale calibration such as metric, engineer's full scale in tenths and half scale in twentieths, or architect's scale 1/4" = 1'−0". Many other scales are available.

Scale Notation

The **scale of a drawing** is usually noted in the title block or below the view of an object that differs in scale to that given in the title block. Drawings are scaled so the object represented can be illustrated clearly on standard sizes of paper. It would be difficult, for example, to make a full-size drawing of a Boeing 747; thus, a scale that reduces the size of such a large object must be used. Machine parts are often drawn full size or even two, four, or ten times larger than full size, depending on the actual size of the part.

The scale selected, then, depends on:

■ The actual size of the part.

■ The amount of detail to be shown.

■ The media size selected.

■ The amount of dimensioning and notes required on the part.

Commonly used scales should be selected. Avoid scales that are not identified in the following discussions.

The following scales and their notations are frequently used on mechanical drawings:

INCH SCALES
Full scale	=	FULL or 1:1
Half scale	=	HALF or 1:2
Quarter scale	=	QUARTER or 1:4
Twice scale	=	DOUBLE or 2:1
Four times scale	=	4:1
Ten times scale	=	10:1

METRIC SCALES
Full scale	=	1:1
Half scale	=	1:2

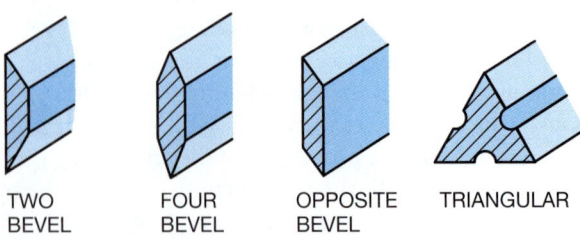

FIGURE 2.32 ■ Scale shapes.

One fifth scale	=	1:5
One twenty-fifth scale	=	1:25
One thirty-three and one-third scale	=	1:33 1/3
One seventy-five scale	=	1:75

Some scales used on architectural drawings are noted as follows:

1/8"	=	1'−0"	1"	=	1'−0"
1/4"	=	1'−0"	1 1/2"	=	1'−0"
1/2"	=	1'−0"	3"	=	1'−0"

Some scales used in civil drafting are noted as follows:

1"	=	10'	1"	=	50'
1"	=	20'	1"	=	60'
1"	=	30'	1"	=	100'

Metric Scale

ANSI/ASME According to the American National Standards Institute (ANSI) and the American Society of Mechanical Engineers (ASME):

The commonly used SI (International System of Units) linear unit used on engineering drawings is the **millimeter**. The commonly used U.S. customary linear unit used on engineering drawings is the **decimal inch**. On drawings where all dimensions are either in inches or millimeters, individual identification of units is not required. However, the drawing shall contain a note stating: UNLESS OTHERWISE SPECIFIED, ALL DIMENSIONS ARE IN INCHES (or MILLIMETERS as applicable). Where some millimeters are shown on an inch-dimensioned drawing, the millimeter value should be followed by the abbreviation *mm*. Where some inches are shown on a millimeter-dimensioned drawing, the inch value should be followed by the abbreviation *IN*.

Metric symbols are as follows:

millimeter	−	ııııı
centimeter	=	cm
decimeter	=	dm
meter	=	m
dekameter	=	dam
hectometer	=	hm
kilometer	=	km

Some metric-to-metric equivalents are the following:

10 millimeters	=	1 centimeter
10 centimeters	=	1 decimeter
10 decimeters	=	1 meter
10 meters	=	1 dekameter
10 dekameters	=	1 kilometer

Some metric-to-U.S. customary equivalents are the following:

1 millimeter	=	.03937 inch
1 centimeter	=	.3937 inch
1 meter	=	39.37 inches
1 kilometer	=	.6214 mile

Some U.S. customary-to-metric equivalents are the following:

1 mile	=	1.6093 kilometers	=	1609.3 meters
1 yard	=	914.4 millimeters	=	.9144 meter
1 foot	=	304.8 millimeters	=	.3048 meter
1 inch	=	25.4 millimeters	=	0.254 meter

To convert inches to millimeters, multiply inches by 25.4.

Figure 2.33 shows the common scale calibrations found on the triangular metric scale. One advantage of the metric scales is that any scale is a multiple of ten; therefore, any reductions or enlargements are easily performed. No mathematical calculations should be required when using a metric scale. Always select a direct reading sale. To avoid the possibility of error, avoid multiplying or dividing metric scales by anything but multiples of ten.

Civil Engineer's Scale

The triangular civil engineer's scale contains six scales, two on each of its sides. The civil engineer's scales are calibrated in multiples of ten. The scale margin displays the scale represented on a particular edge. The following table shows some of the many scale options available when using the civil engineer's scale. Keep in mind that *any* multiple of ten is available with this scale.

Civil Engineer's Scale						
Divisions	Ratio	Scales Used with This Division				
10	1:1	1" = 1"	1" = 1'	1" = 10'	1" = 100'
20	1:2	1" = 2"	1" = 2'	1" = 20'	1" = 200'
30	1:3	1" = 3"	1" = 3'	1" = 30'	1" = 300'
40	1:4	1" = 4"	1" = 4'	1" = 40'	1" = 400'
50	1:5	1" = 5"	1" = 5'	1" = 50'	1" = 500'
60	1:6	1" = 6"	1" = 6'	1" = 60'	1" = 600"

The ten scale is often used in mechanical drafting as a full, decimal-inch scale, shown in Figure 2.34. Increments of 1/10 (.1) in. can easily be read on the 10 scale. Readings of less than .1 in. require you to approximate the desired amount, as shown in Figure 2.34. Some scales are available that refine the increments to 1/50th of an inch. The ten scale is also used in civil drafting for scales of 1" = 10' or 1" = 100' and so on (see Figure 2.35).

The 20 scale is commonly used in mechanical drawing to represent dimensions on a drawing at half scale (1:2). Figure 2.36 shows examples of half-scale decimal dimensions. The 20 scale is also used for scales of 1" = 2', 1" = 20', 1" = 200', as shown in Figure 2.37.

The remaining scales on the engineer's scale can be used in a similar fashion. For example, 1" = 5', 1" = 50', and so on. Figure 2.37 shows dimensions on each of the civil engineer's scales. The 50 scale is popular in civil drafting for drawing plats of subdivisions, plot plans, and site plans.

Architect's Scale

The triangular architect's scale contains 11 different scales. On ten of them, each inch represents a foot and is subdivided into

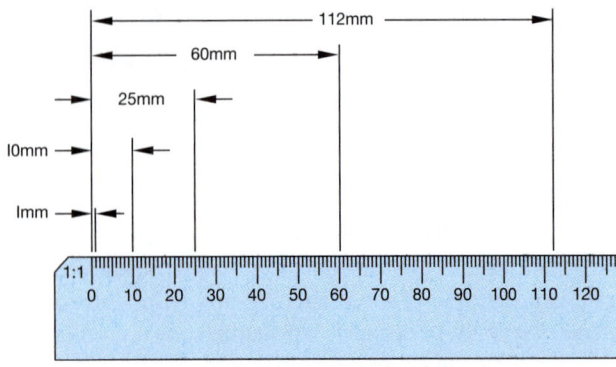

FULL SCALE = 1:1

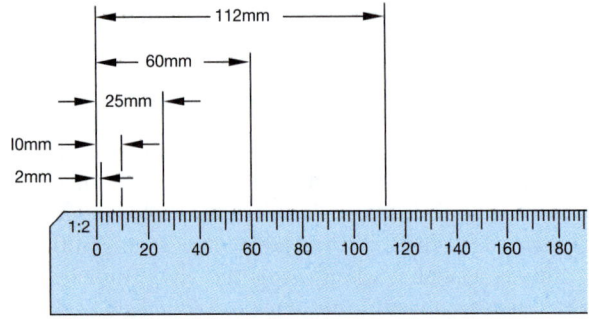

HALF SCALE = 1:2

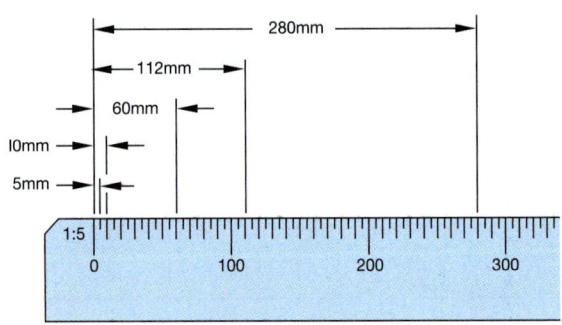

ONE FIFTH SCALE = 1:5

FIGURE 2.33 ■ Metric scale calibrations.

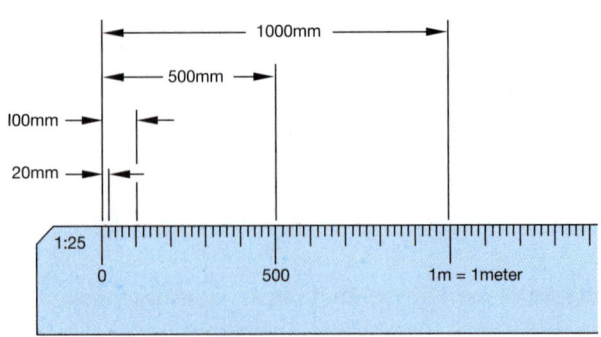

ONE TWENTY FIFTH SCALE = 1:25

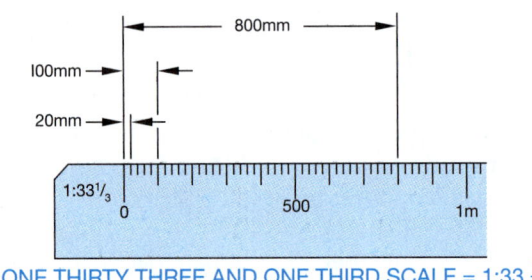

ONE THIRTY THREE AND ONE THIRD SCALE = 1:33 $\frac{1}{3}$

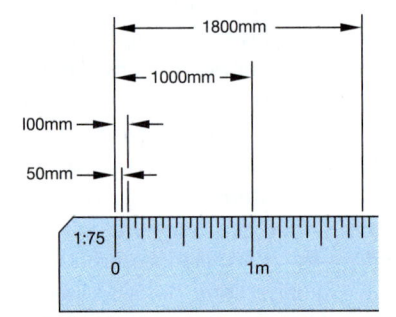

ONE SEVENTY FIFTH SCALE = 1:75

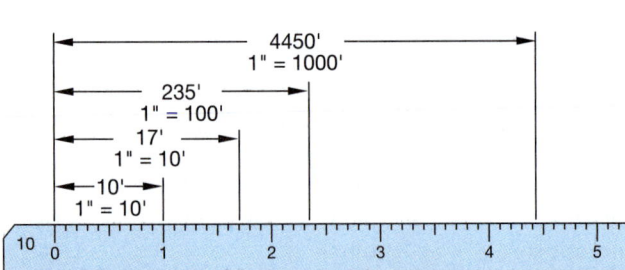

FIGURE 2.35 ■ Civil engineer's scale, units of 10.

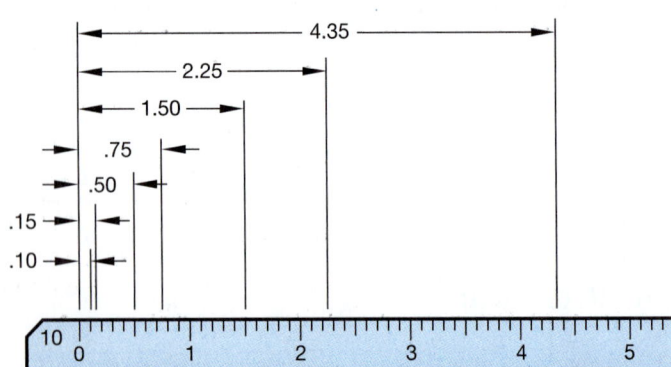

FIGURE 2.34 ■ Full engineer's decimal scale (1:1).

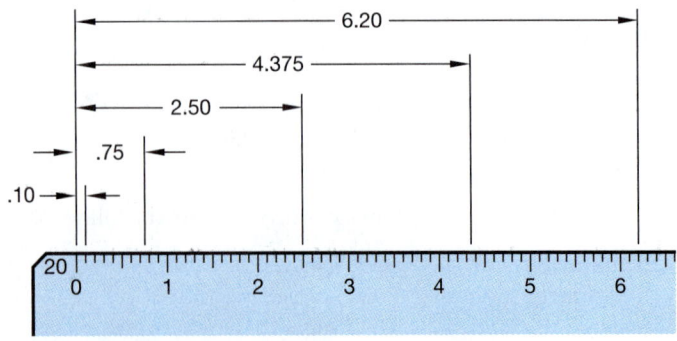

FIGURE 2.36 ■ Half scale on the engineer's scale (1:2).

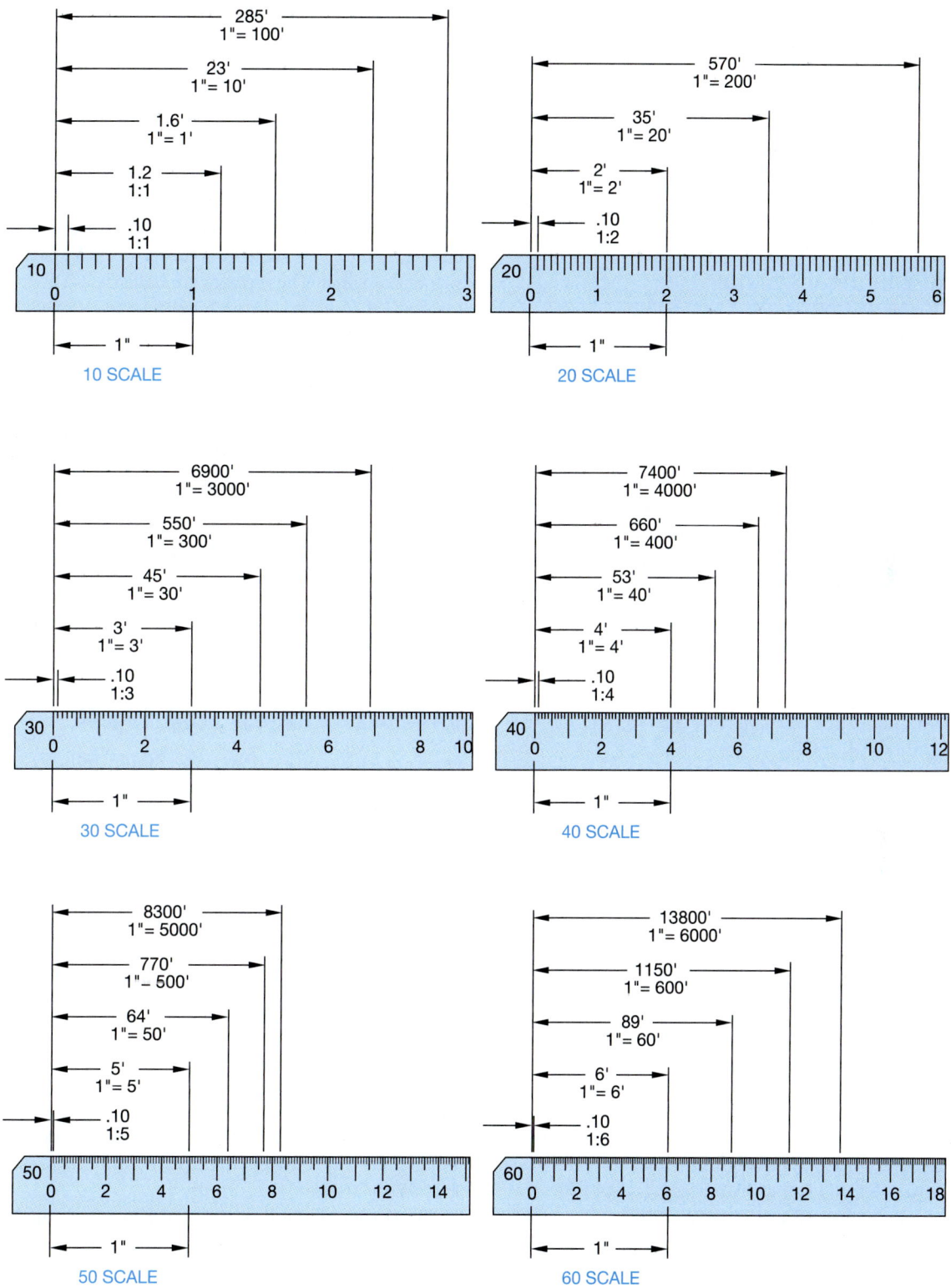

FIGURE 2.37 ■ Civil engineer's scales.

multiples of 12 parts to represent inches and fractions of an inch. The eleventh scale is the full scale with a 16 in the margin. The 16 means that each inch is divided into 16 parts and each part is equal to 1/16 of an inch. Look at Figure 2.38 for a comparison between the 10 engineer's scale and the 16 architect's

scale. Figure 2.39 shows an example of the full architect's scale, while Figure 2.40 shows the fraction calibrations.

Look at the architect's scale examples in Figure 2.41. Note the form in which scales are expressed on a drawing. The scale notation can take the form of a word (full, half, double), or a ratio (1:1, 1:2, 2:1), or an equation of the drawing size in inches or fractions of an inch to one foot (1" = 1'–0", 3" = 1'–0", 1/4" = 1'–0"). The architect's scale commonly has scales running in both directions along an edge. Be careful when reading a scale from left to right. Do not confuse its calibrations with the scale that reads from right to left. Full feet values are read from zero toward the middle of the scale and inch values are read at the calibrations provided between zero and the margin of each scale.

Mechanical Engineer's Scale

The **mechanical engineer's scale** is commonly used for mechanical drafting when drawings are in fractional or decimal inches. The mechanical engineer's scale typically has full-scale divisions

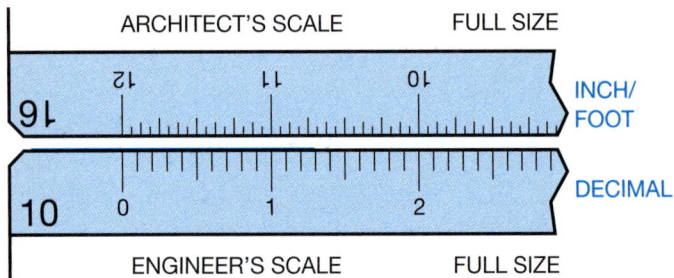

FIGURE 2.38 ▪ Comparison of architect's scale (16) and full engineer's scale (10).

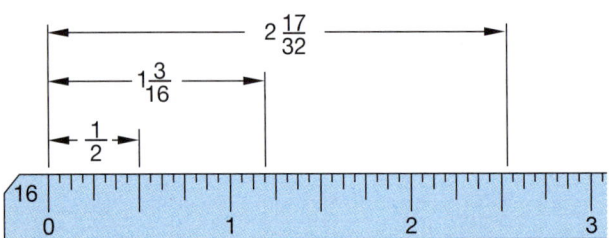

FIGURE 2.39 ▪ Full (1:1), or 12" = 1'–0", architect's scale.

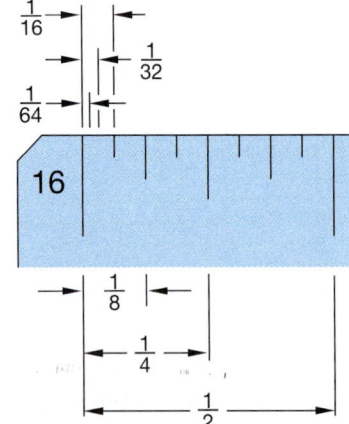

FIGURE 2.40 ▪ Enlarged view of architect's (16) scale.

that are divided into 1/16, 10, and 50. The 1/16 divisions are just like the 16 architect's scale where there are 12 inches and each inch is divided into 1/16 in. increments, or sometimes 1/32 in. divisions. The 10 scale is the same as the 10 civil engineer's scale, where each inch is divided into 10 parts, with each division being .10 inch. The 50 scale is for scaling dimensions that require additional accuracy, because each inch has 50 divisions. This makes each increment 1/30 in. or .02 in. (1 ÷ 50 = .02). Figure 2.42 shows a comparison between these scales. The mechanical engineer's scale also has options for scaling down the size of drawings. These scales are half size 1:2 (1/2" = 1"), quarter size 1:4 (1/4" = 1"), and eighth size 1:8 (1/8" = 1"). These scales are shown in Figure 2.42. A drawing that is represented at full scale (1:1), half scale (1:2), and quarter scale (1:4) is shown in Figure 2.43 for your comparison.

CLEANING DRAFTING EQUIPMENT

Cleaning the drafting equipment each day helps keep your drawing clean and free of smudges. This is easily done with mild soap and water or with a soft rag or tissue. Avoid using harsh cleansers or products that are not recommended for use on plastic. It is also a good practice to clean your hands periodically to remove graphite and oil. Also, keep your hands dry.

DRAFTING MEDIA

Several factors, other than cost, should influence the purchase and use of drafting media. The considerations include durability, smoothness, erasability, dimensional stability, and transparency.

Durability should be considered if the original drawing will have a great deal of use. Originals may tear or wrinkle and the images become difficult to see if the drawings are used often.

Smoothness relates to how the medium accepts line work and lettering. The material should be easy to draw on so the image is dark and sharp without a great deal of effort on your part.

Erasability is important because errors need to be corrected and changes frequently made. When images are erased, ghosting should be keep to a minimum. **Ghosting** is the residue that remains when lines are difficult to remove. The unsightly ghost images are reproduced in a print. Materials that have good erasability are easy to clean up.

Dimensional stability is the quality of the media to not alter size due to the effects of atmospheric conditions such as heat, cold, and humidity. Some materials are more dimensionally stable than others.

Transparency is one of the most important characteristics of drawing media. The diazo reproduction method requires light to pass through the material. The final goal of a drawing is good reproduction, so the more transparent the material the better the reproduction, assuming that the image drawn is professional quality. Transparency is not important when using the photocopy process.

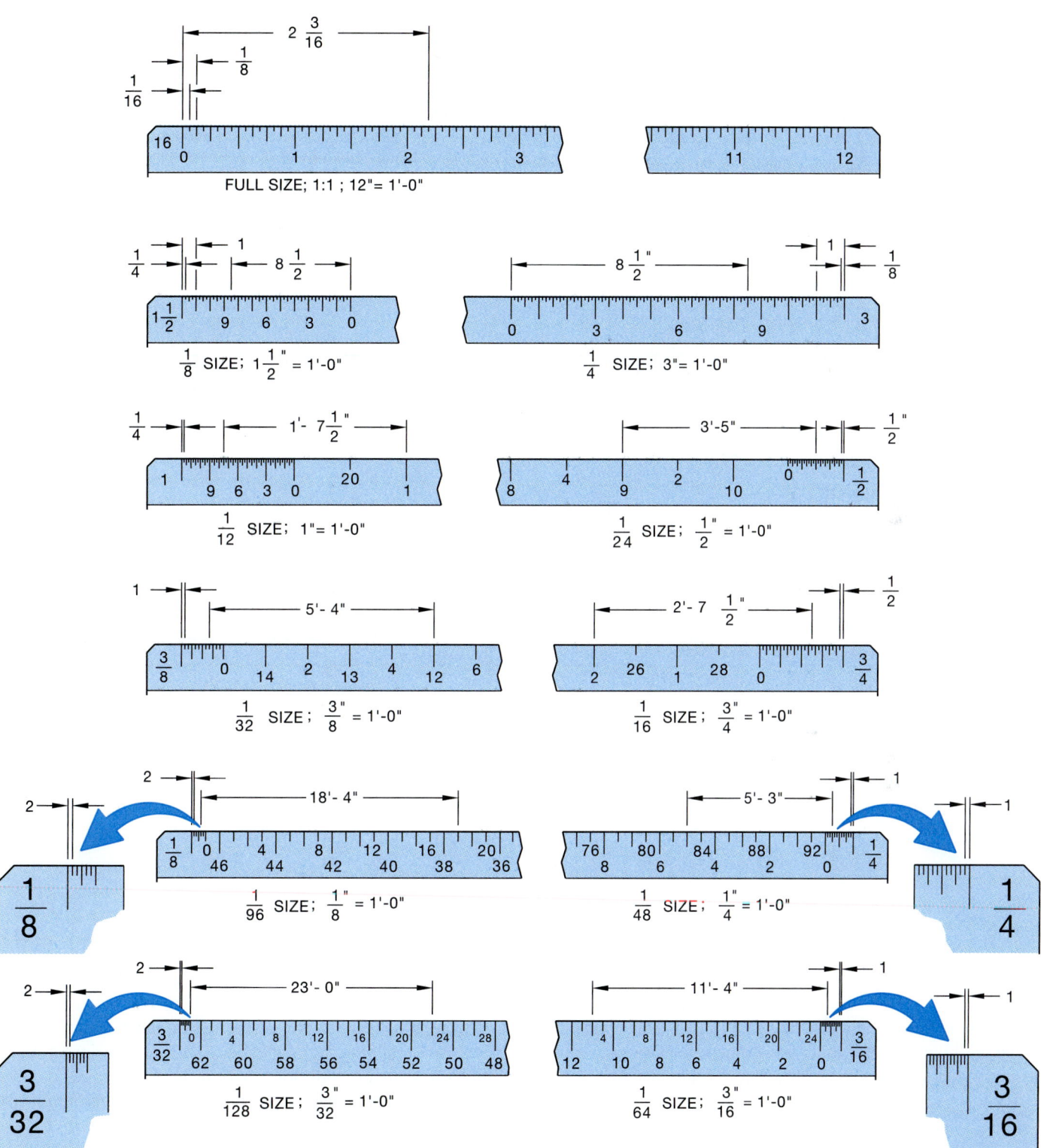

FIGURE 2.41 ■ Architect's scale examples.

PAPERS AND FILMS

Vellum

Vellum is drafting paper that is specially designed to accept pencil or ink. Lead on vellum is the most common combination used for manual drafting. Many vendors manufacture quality vellum for drafting purposes. Each claims to have specific qualities that you should consider in the selection. Vellum is the least expensive material having good smoothness and transparency. Use vellum originals with care. Drawings made on vellum that require a great deal of use could deteriorate, as vellum is not as durable as

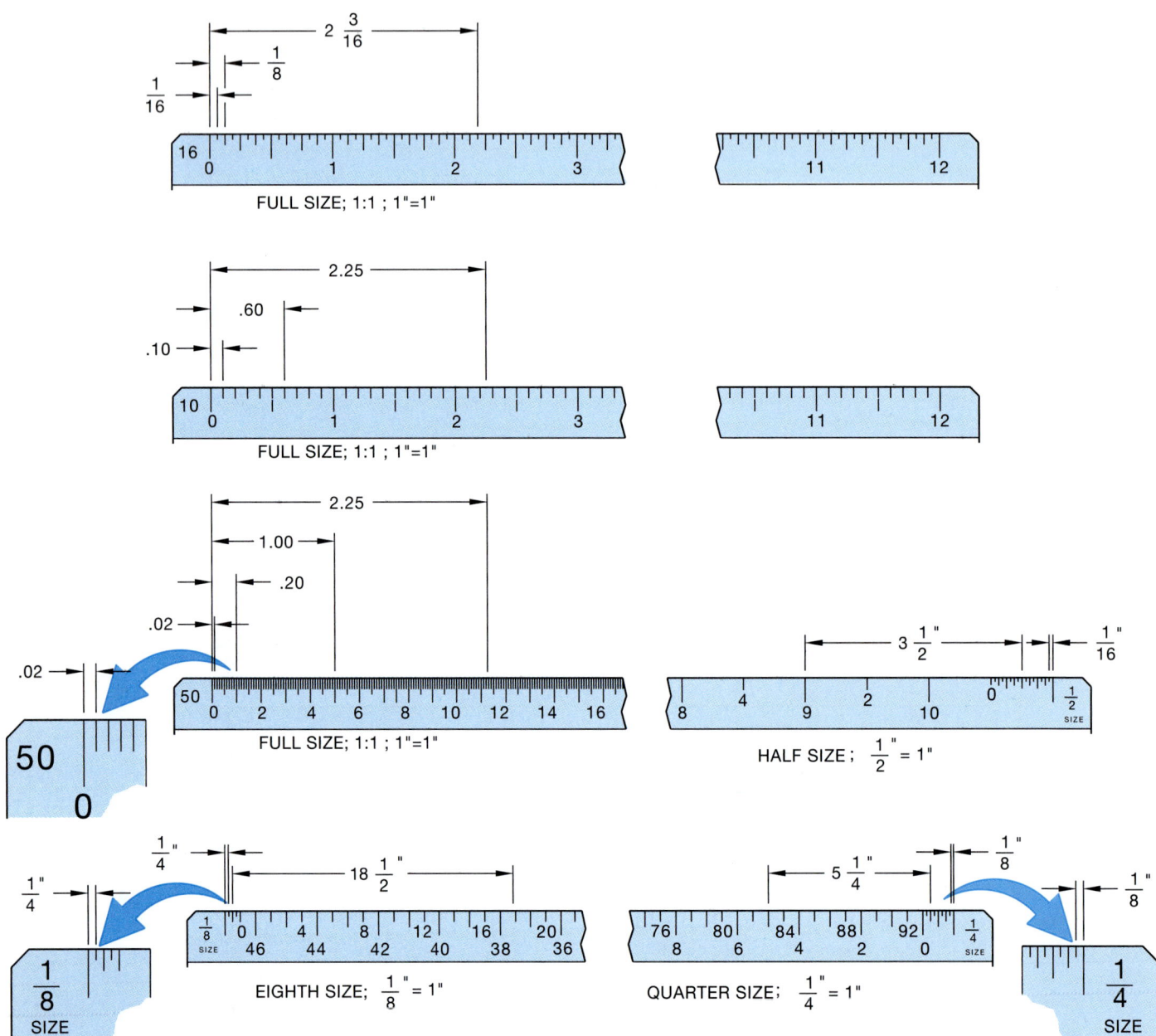

FIGURE 2.42 ■ A comparison of three types of scales.

other material. Some brands are better erased than others. Affected by humidity and other atmospheric conditions, vellum generally is not as dimensionally stable as other materials.

Polyester Film

Polyester film, also known by its brand name, Mylar®, is a plastic material offering excellent dimensional stability, erasability, transparency, and durability. Drawing on Mylar is best accomplished using ink or special polyester leads. Do not use regular graphite leads as they smear easily. Drawing techniques that drafters use with polyester leads are similar to graphite leads except the polyester leads are softer and feel like a crayon when used.

Mylar is available with a single or double matte surface. Matte is a nonglossy, slightly textured surface. The double matte film has texture on both sides so drawing can be done on either side, if necessary. Single matte film is the most common is use with the non-drawing side having a slick surface.

Protect the Polyester Film Surface

When using polyester film, you must be very careful not to damage the matte by erasing. Erase at right angles to the direction of your lines and do not use too much pressure. Doing so helps minimize damage to the matte surface. Once the matte is destroyed and removed, the surface will not accept ink or pencil. Also, be cautious about getting moisture on the polyester film

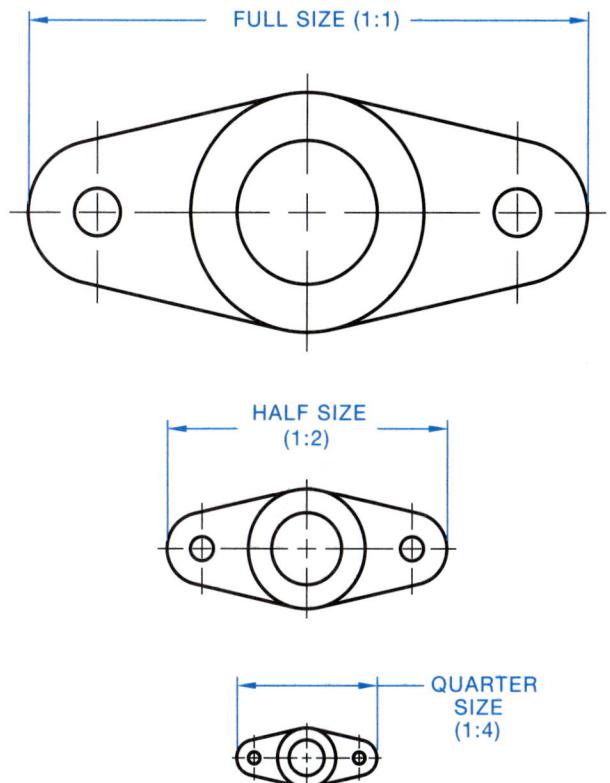

FIGURE 2.43 ■ A sample drawing represented at full scale, half scale, and quarter scale.

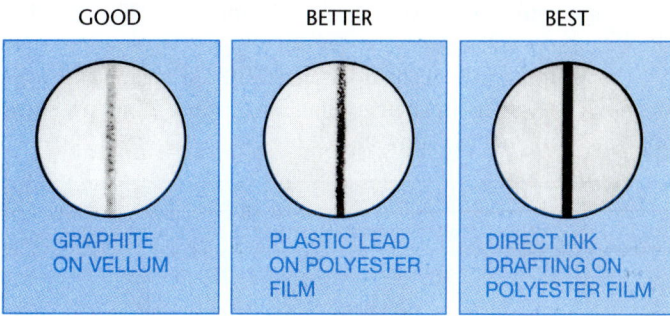

FIGURE 2.44 ■ A magnified comparison of graphite on vellum, plastic lead on polyester film, and ink on polyester film. *Courtesy Koh-I-Norr, Inc.*

surface. Oil from your hands can cause your pen to skip across the material.

Normal handling of drawing film is bound to soil it. Inked lines applied over soiled areas do not adhere well and in time will chip off. It is always good practice to keep the film clean. Soiled areas can be cleaned effectively with special film cleaner.

Polyester film is much more expensive than vellum; however, it should be considered when excellent reproductions, durability, dimensional stability, and erasability are required of original drawings.

Reproduction

The one thing most designers, engineers, architects, and drafters have in common is their finished drawings are intended for reproduction. The goal of every professional is to produce drawings of the highest quality that give the best possible prints when they are reproduced.

Many of the factors that influence the selection of media for drafting have been discussed; however, the most important factor is reproduction. The primary combination that achieves the best reproduction is the blackest, most opaque lines or images on the most transparent base or material. Each of the materials mentioned makes good prints if the drawing is well done. If the only concern is the quality of the reproduction, ink on polyester film is the best choice. Some products have better characteristics than others. Some individuals prefer certain products. It is

up to the individual or company to determine the combination that works best for their needs and budget. The question of reproduction is especially important when sepias must be made. Sepias are second- or third-generation originals. See Sepias in this chapter.

Look at Figure 2.44 for a magnified view of graphite on vellum, plastic lead on polyester film, and ink on polyester film. Judge for yourself which material and application provides the best reproduction. As you can see from Figure 2.44, the best reproduction is achieved with a crisp, opaque image on transparent material. If your original drawing is not good quality, it does not get better on the print.

SHEET SIZES, TITLE BLOCKS, AND BORDERS

ASME/ANSI standard sheet sizes and format are specified in the documents ASME Y14.1, *Drawing Sheet Size and Format,* and ASME Y14.1M, *Metric Drawing Sheet Size and Format.*

All professional drawings have title blocks. Standards have been developed for the information put into the title block and on the surrounding sheet next to the border so the drawing is easier to read and file than drawings that do not follow a standard format.

Sheet Sizes

ANSI Y14.1 specifies sheet size specifications in inches as follows:

Size Designation	Size in Inches	
	Vertical	Horizontal
A	8 1/2	11 (horizontal format)
	11	8 1/2 (vertical format)
B	11	17
C	17	22
D	22	34
E	34	44
F	28	40

There are four additional size designations (G, H, J, and K); these apply to roll sizes.

The M in the title of the document Y14.1M means all specifications are given in metric. Standard metric drawing sheet sizes are designed as follows:

Size Designation	Size in Millimeters	
	Vertical	Horizontal
A0	841	1189
A1	594	841
A2	420	594
A3	297	420
A4	210	297

Longer lengths are referred to as elongated and extra-elongated drawing sizes. These are available in multiples of the short side of the sheet size.

Standard inch sheet sizes are shown in Figure 2.45a, and metric sheet sizes are shown in Figure 2.45b.

Zoning

Some companies use a system of numbers along the top and bottom margins and letters along the left and right margins called zoning. Notice in Figure 2.45a that numbered and lettered zoning begins on C-size drawing sheets, and in Figure 2.45b zoning is used on A3 sheet sizes and larger. Zoning allows the drawing to read like a road map. For example, you can refer to the location of a specific item as D4, which means that the item can be found at or near the intersection of D across and 4 up or down.

Title Blocks and Borders

Companies generally have title blocks and borders preprinted on drawing sheet to reduce drafting time and cost. Drawing sheet sizes and sheet format items such as borders, title blocks, zoning, revision columns, and general note locations have been standardized so that the same general relationship exists between engineering drawings internationally. The ASME Y14.1 and ASME Y14.1M documents specify the exact size and location for each item found on the drawing sheet. It is recommended that standard sheet sizes and format be followed to improve readability, handling, filing, and reproduction. Each company can use a slightly different design, although the following basic information is located in approximately the same place on most engineering drawings:

1. Title block. Lower right corner.
   a. Company name.
   b. Confidential statement.
   c. Unspecified dimensions and tolerances.
   d. Sheet size.
   e. Drawing number.
   f. Part name.
   g. Material.
   h. Scale.
   i. Drafter's name or initials.
   j. Checker's name or initials.
   k. Engineer's name or initials.
   l. Revision level.
2. Revision column. Upper right corner, over or next to the title block.
   a. Revision symbol, number or letter.
   b. Description.
   c. Drafter.
   d. Date.
3. Border line.
   a. With or without zoning.

Figure 2.46 shows a sample title block.

Title Block Definitions

■ **Tolerances** are discussed in detail in Chapter 11, although it is important to know that a tolerance is a given amount of acceptable variation in a size or location dimension. All dimensions have a tolerance.

■ **Millimeters** and **inches**. All dimensions are in millimeters (mm) or inches (in.) unless otherwise specified.

■ **Unless otherwise specified** means that, in general, all of the features or dimensions on a drawing have the relationship or specifications given in the title block unless a specific note or dimensional tolerance is provided at a particular location in the drawing.

■ **Unspecified tolerances** refers to any dimension on the drawing that does not have a tolerance specified. This is when the dimensional tolerance required is the same as the general tolerance shown in the title block.

■ **Revisions**. When parts are redesigned or altered for any reason, the drawing is changed. All drawing changes are commonly documented and filed for future reference. When this happens, the documentation should be referenced on the drawing so that users can identify that a change has been made. Before any revision can be made, the drawing must be released for manufacturing.

Title and Revision Block Instructions

The title block displayed in Figure 2.47a shows most of the common elements found in industry title blocks. The revision block, Figure 2.47b, found in this format is located in the upper right corner of the drawing sheet. Some companies use other placement.

The large numbers shown on Figure 2.47 refer to the following instructions for completing the title and revision blocks. All

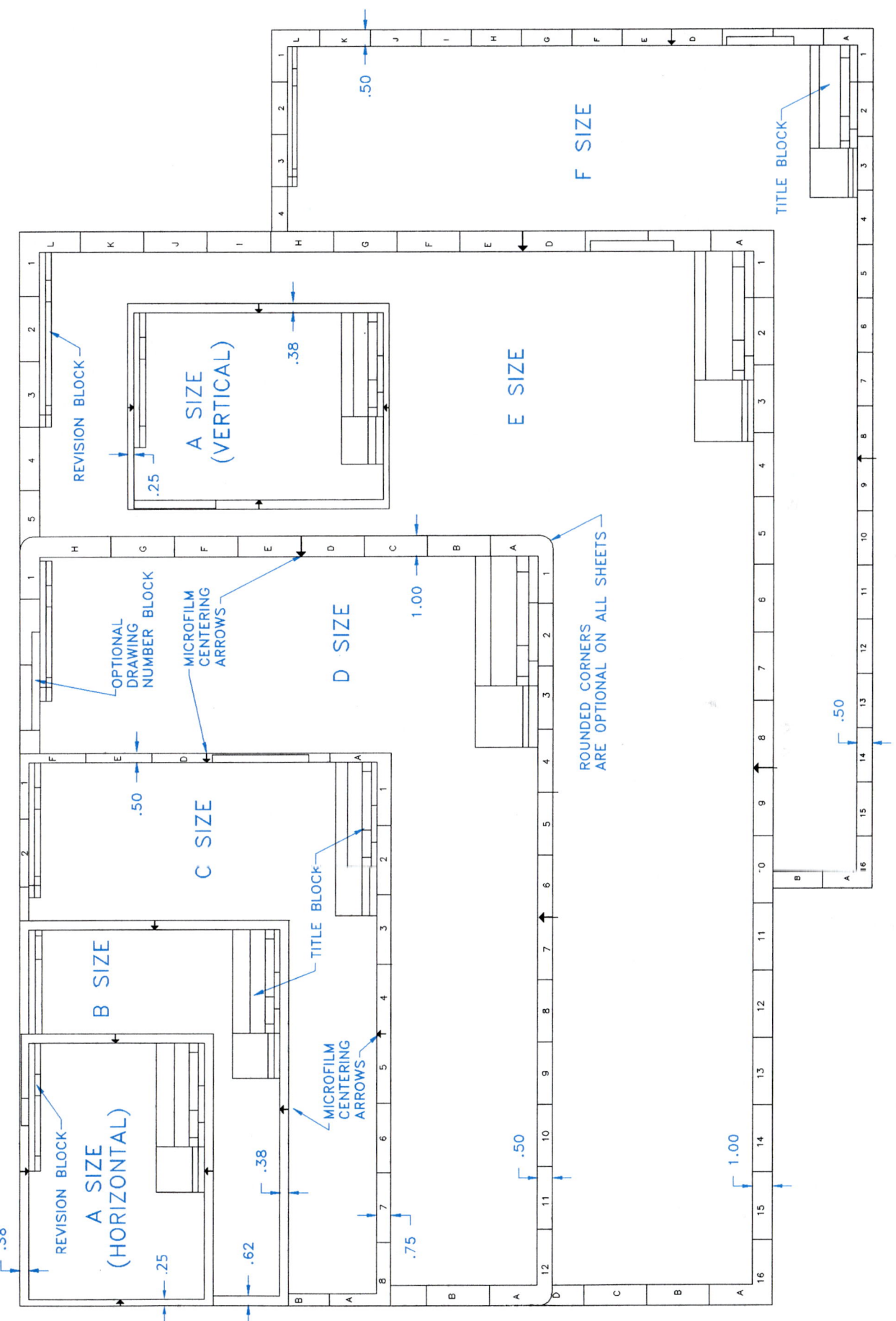

FIGURE 2.45a ■ Standard inch drawing sheet sizes.

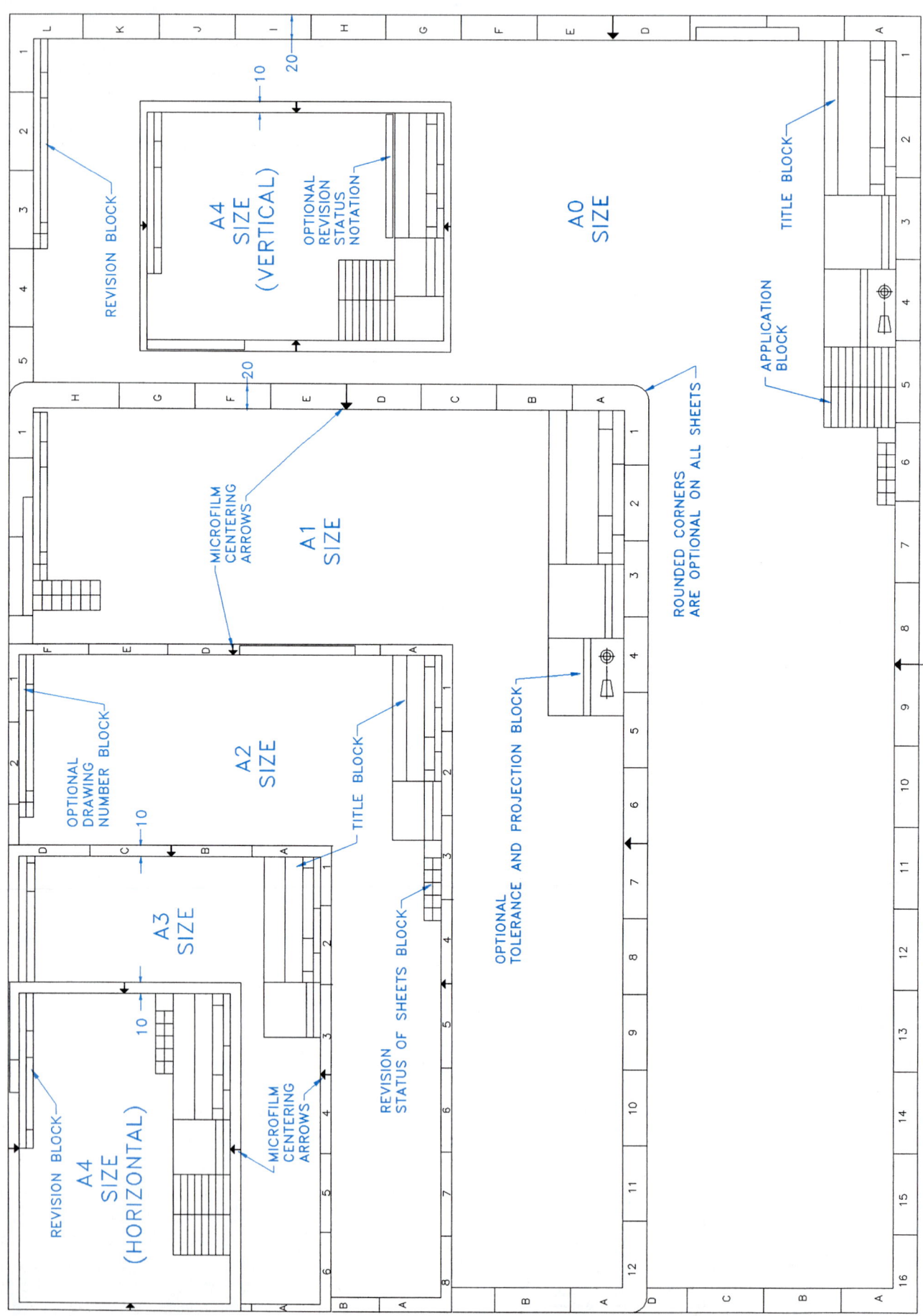

FIGURE 2.45b ■ Standard metric drawing sheet sizes.

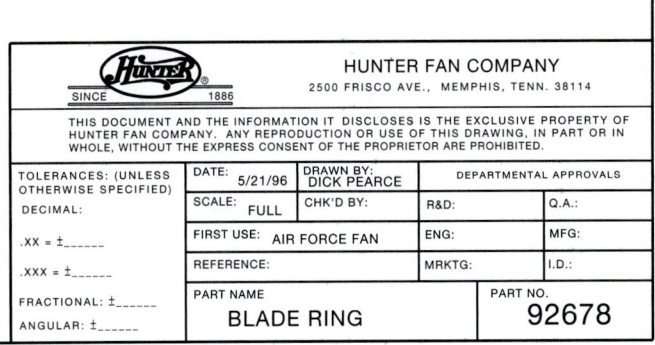

FIGURE 2.46 ■ Sample title block. *Courtesy Hunter Fan Company, Memphis, TN.*

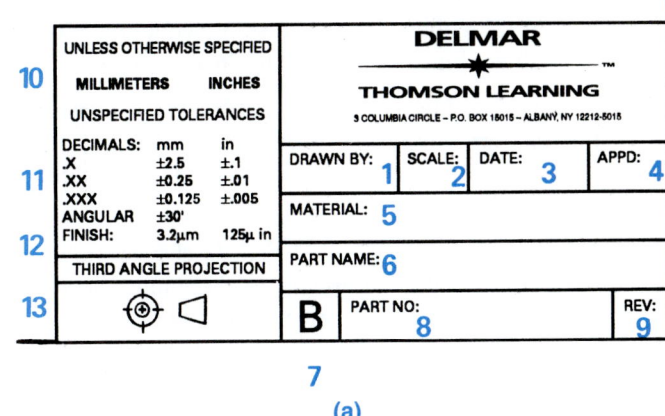

(a)

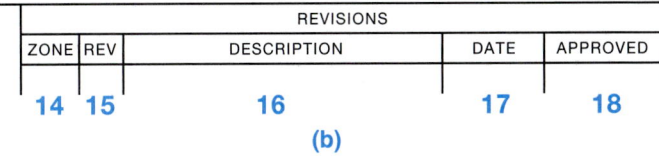

(b)

FIGURE 2.47 ■ (a) Title block elements. (b) Revision block elements.

lettering should be between .10 in. to .24 in. (2.5–6mm) high as indicated in the following:

Drawing Elements	Minimum Letter Heights INCH	Letter Drawing Sizes INCH	Letter Heights METRIC (mm)	Drawing Sizes METRIC
Title, drawing size	.24	D, E, F, H, J, K	6	A0, A1
CAGE Code, drawing number, and revision letter in the title block	.12	A, B, C, G	3	A2, A3, A4
Zone letters and numbers in borders	.24	All sizes	6	All sizes
Drawing block headings	.10	All sizes	2.5	All sizes
All other characters	.12	All sizes	3	All sizes

The lettering style should be vertical uppercase Gothic freehand, or mechanical lettering as specified by the instructor or company standards. Computer graphic systems are to use vertical uppercase letters of a ROMAN font unless otherwise specified.

1. DRAWN BY: Identify yourself by using all your initials such as DRC, DAM, JLT, unless otherwise specified.

2. SCALE: Examples of scales to fill in this block are FULL or 1:1, HALF or 1:2, DBL or 2:1, QTR or 1:4, NONE.

3. DATE: Fill in this block using the following order: day, month, and year such as 18 NOV 06, or numerically with month, day, and year such as 11/18/06.

4. APPD: This block is to be used by the instructor or checker who initials his or her approval of the drawing.

5. MATERIAL: Describe here the material used to make the part, for example, BRONZE, CAST IRON, SAE 4320.

6. PART NAME: Insert here the name of the part, such as COVER, HOUSING.

7. B: Gives sheet sizes shown in Figure 2.45. When preprinted borders are used, this sheet size designation is usually printed.

8. PART NO: Fill in the drawing number of the part. Most companies have their own part numbering system. While numbering systems differ, they are often keyed to categories such as the disposition of the drawing (casting, machining, assembly), materials used, related department within the company, or a numerical classification of the part.

9. REV: Fill in the revision letter of the part or drawing. A new or original drawing is — (dash) or 0 (zero). The first time a drawing is revised, the — or 0 changes to an A, for the second drawing change a B is placed here, and so on. The letters, *I, O, S, X,* and *Z* are not used. When all of the available letters *A* through *Y* have been used, double letters such as *AA* and *AB*, or *BA* and *BB* are used. Some companies use revision numbers rather than letters.

10. UNLESS OTHERWISE SPECIFIED: Unless otherwise specified, drawing dimensions will be given in millimeters or inches. When the predominant dimensions are in MILIMETERS, neatly blacken out or remove INCHES, when in INCHES, blacken out or remove MILLIMETERS.

11. UNSPECIFIED TOLERANCES: One, two, and three place decimals are established as MILLIMETERS or INCHES in a manner the same as item 10. For example, if the drawing dimensions are predominately MILLIMETERS, then INCH-related tolerances are blackened out or removed. Angular tolerances for unspecified angular dimensions are ±30'. These applications depend on company practice.

12. FINISH: In this block fill in the unspecified surface finish, for surfaces that are identified for finishing without a specific callout. Surface finish is discussed in Chapters 5 and 11.

13. THIRD-ANGLE PROJECTION: The drawing assignments in this text are done using third-angle projection unless otherwise specified. A complete discussion of this topic is found in Chapter 9.

14. ZONE: This is the drawing zone where the change is located. Remember from an earlier discussion, the zoning allows the drawing to read like a road map; numbers along the top and bottom, and letters along the sides, direct you to a specific location on the drawing. If the location of the change is D4, then D4 is placed in the ZONE column for this change. The ZONE column is used only if the drawing has zoning.

15. REV: Fill in the revision letter or number in this block, such as A, B, C, etc. Succeeding letters are to be used for each Engineering Change Notice (ECN) or group of related ECNs regardless of quantity of changes on an ECN. The REV: block in the title block (9) must be changed to agree with the last REV letter in the revision block. Some companies use revision numbers.

16. DESCRIPTION: In this block, fill in the Engineering Change Notice (ECN) letter covering the Engineering Change Request (ECR) or group of ECRs that requires the drawing to be revised. Another option is to give a short description of the change. An **Engineering Change Notice** is

a numbered document that provides information about the change. ECNs are discussed in Chapter 21.

17. DATE: Fill in the day, month, and year on which the ECN package is ready for release to production, such as 6 Apr 06, or use month, day, and year numbers such as 4/6/06.

18. APPROVED: This column is for the initials of the person approving the change, and optional date.

Drafting revisions are completed in chronological order by adding horizontal columns and extending the vertical column lines.

The previous example showed you a typical title block and revision block. Each company has its own design and format that is similar, and most CADD programs have predefined sheet layout options with borders and title blocks. These are normally based on the ASME standard title blocks and revision blocks and can usually be customized for your own applications. The document ASME Y14.1 gives the recommended title block and revision block dimensions shown in Figure 2.48. ASME Y14.1 states the dimensions shown on the sample title blocks and revision blocks are recommended and may be varied to accommodate your company requirements and to fit with the lettering requirements of ASME Y14.2M. The title and revision blocks shown in Figure 2.48a is the recommendation for A-, B-, C-, and G-size inch sheets. The recommended title block for D-, E-, F-, H-, J-, and K-size sheets is bigger, measuring 7.62 in. long and 2.50 in., high. The title and revision blocks shown in Figure 2.48b is the recommen-

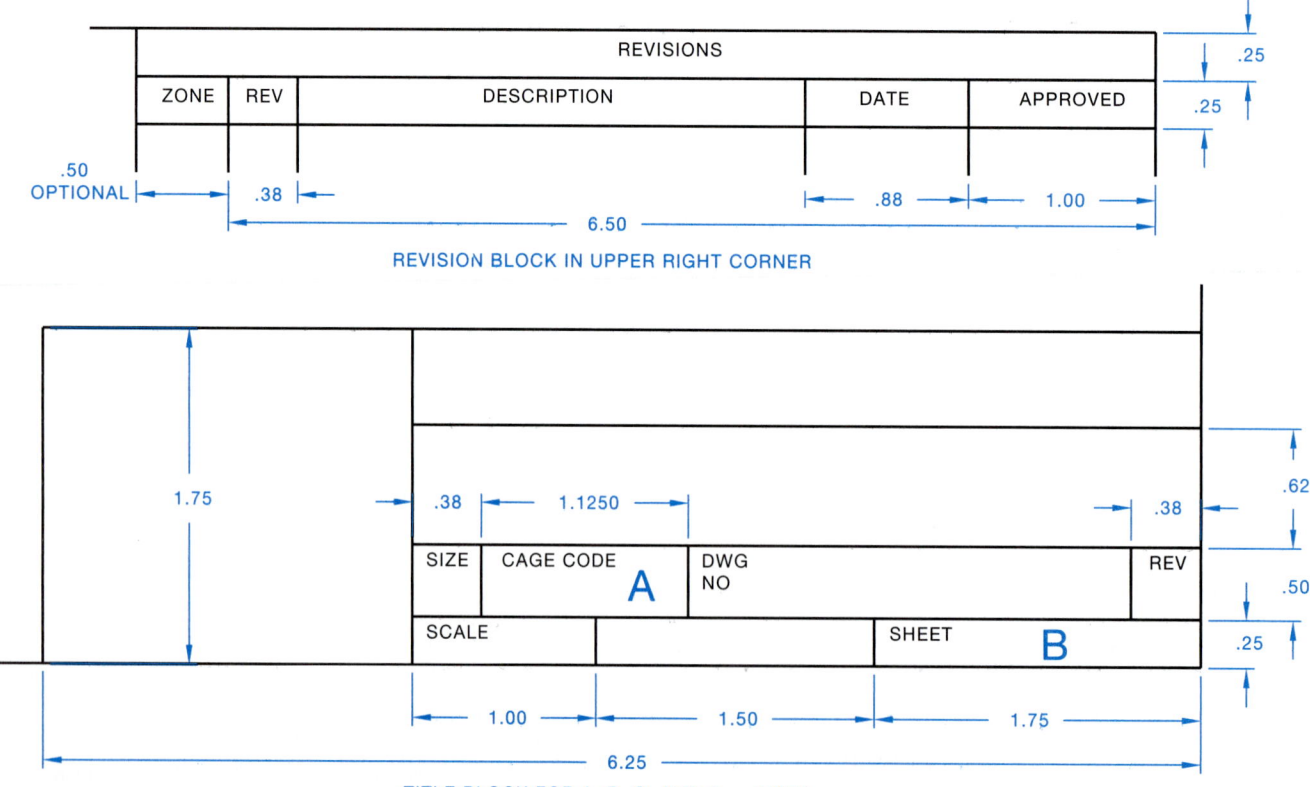

FIGURE 2.48a ▪ Title block and revision block inch dimensions recommended by ASME Y14.1.

dation for A2, A3, and A4 metric sheets. The recommended title block for A0 and A1 sizes is bigger, measuring 135–149 mm long and 20 mm high.

The revision block is recommended for the upper right corner of the drawing border, but some companies place it in other locations, such as over the title block or to the left of the title block. Look again at Figure 2.48. Notice the large letters *A* and *B*. These are items found in the ASME Y14.1 examples not displayed in Figure 2.47. These are described as follows:

The **CAGE Code** is a five number code assigned by the Defense Logistic Service Center (DLSC) to all Department of Defense contractors. CAGE stands for *Commercial And Government Entity*. The CAGE Information report also contains contact information for the vendor.

A An optional **FSCM** number can also go here. *FSCM* stands for *Federal Supply Code for Manufacturers*. The FSCM number is a five-digit numerical code used on all drawings that produce items used by the federal government.

B The **SHEET** compartment is for the identification of sheet numbers. This is used by companies that anticipate having the drawing for a particular application occupy more than one sheet. 1 OF 1 is placed here if the drawing is on one sheet. If the drawing is on two sheets, the first has 1 OF 2 and the second has 2 OF 2 in this compartment.

Notice that the ASME title block in Figure 2.48 has DWG NO rather than PART NO, as in Figure 2.47. Use of PART NO is a company preference.

TAPING DOWN YOUR DRAWING

Before you tape down your drawing, be sure your hands and equipment are clean. You should always use specially designed "drafting tape" because it can be removed without tearing the drafting paper. If you are using a drafting machine, first be sure that the scales are properly adjusted. Refer to the manufacturer's instructions. Never fold the drafting original, and try not to roll it. Follow these steps:

■ Start taping down the drafting paper or film by aligning one edge with your horizontal scale or parallel bar.

■ Use short pieces of tape that are approximately 1 in. (25 mm) long.

■ Tape one corner, normally the top left or right corner.

■ Move your hand across the sheet toward the opposite corner, forcing the paper tight to the table, and then tape the opposite corner.

■ Tape the other top corner and then force the paper down toward the opposite corner and tape it.

The drafting paper or film should be very tightly taped to your drafting board or table. If there are any edges curling up, you can add some tape to the edges to keep them down. Be very careful when you move the drafting equipment across the table, because catching an edge of the paper can easily tear the sheet and ruin a professional drawing.

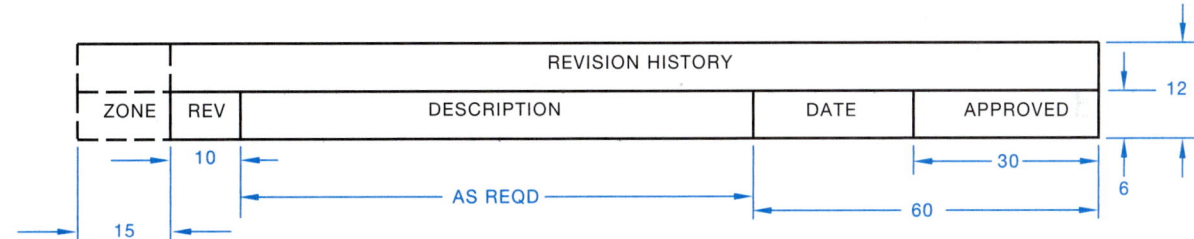

REVISION BLOCK IN UPPER RIGHT CORNER

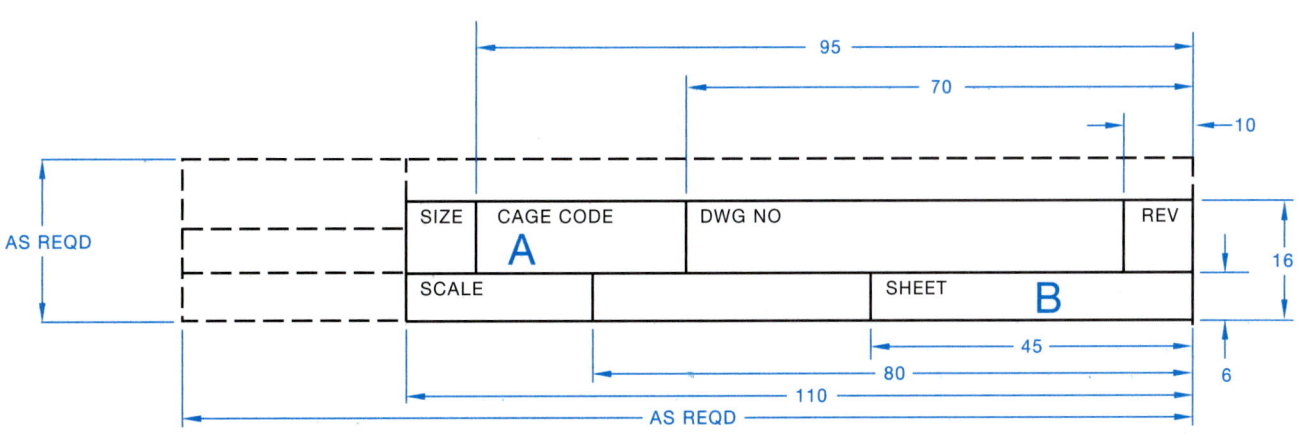

TITLE BLOCK FOR A2, A3, AND A4 SIZES

FIGURE 2.48b ■ Title block and revision block metric dimensions recommended by ASME Y14.1M.

DIAZO REPRODUCTION

Diazo prints, also known as ozalid dry prints or blue-line prints, are made with a printing process that uses an ultraviolet light that passes through a translucent original drawing to expose a chemically coated paper or print material underneath. The light does not go through the dense, black lines on the original drawing, thus the chemical coating on the paper beneath the lines remains. The print material is then exposed to ammonia vapor which activates the remaining chemical coating to produce blue, black, or brown lines on a white or clear background. The print that results is a diazo, or blue-line print, not a blueprint. The term blueprint is now a generic term that is used to refer to diazo prints even though they are not true blueprints.

Making a Diazo Print

Before you run a print using a complete sheet, run a test strip to determine the proper speed setting.

To make a diazo print, place the diazo material on the feedboard, coated (yellow) side up. Position the original drawing, image side up, on top of the diazo material, making sure to align the leading edges. Use diazo material that corresponds in size with your original.

Using fingertip pressure from both hands, push the original and diazo material into the machine. The light passes through the original and exposes the sensitized diazo material except where images (lines and lettering) exist on the original. When the exposed diazo material and the original drawing emerge from the light section, remove the original and carefully feed the diazo material into the developer section, as shown in Figure 2.49.

In the developer section, the sensitized material that remains on the diazo paper activates with ammonia vapor and blue lines form, although some diazo materials make black or brown lines. Special materials are also available to make other colored lines, or to make transfer sheets and transparencies. Figure 2.49 shows the diazo print process.

Storage of Diazo Materials

Diazo materials are light-sensitive so they should always be kept in a dark place until ready for use. Long periods of exposure to room light cause the diazo chemicals to deteriorate and reduce the quality of your print. If you notice brown or blue edges around the unexposed diazo material, it is getting old. Always keep it tightly stored in the shipping package and in a dark place such as a drawer. Some people prefer to keep all diazo material in a dark refrigerator. Doing so preserves the chemical for a long time.

Sepias

Sepias are diazo materials that are used to make secondary originals. A secondary or second-generation original is actually a print of an original drawing that can be used as an original. The term generation refers to the number of times a copy of an original drawing is reproduced and used to make other copies. Changes can be made on the secondary original while the original drawing remains intact. The diazo process is performed with material called sepia. Sepia prints normally form dark brown lines on a clear (translucent) background; therefore, they are sometimes called brown lines. In addition to being used as a secondary original to make alterations to a drawing without changing the original, sepias also are used when originals are required at more than one company location.

Sepias are normally made with the original face down on the sensitized sepia material. The balance of the print process is the same as just described, except that the sepia should be fed into the developer coated-side down. Doing so allows the coating on the sepia material to come into direct exposure with the ammonia in the development chamber. Experiment both ways and observe the quality of your reproductions. Standard diazo copies may reproduce better with the coated-side down; it depends on your machine and materials.

Sepias are made in reverse (reverse sepias) so that corrections or changes can be made on the matte side (face side) and erasures can be made on the sensitized side. Sepia materials are available in paper or polyester. The resulting paper sepia is similar to vellum; and the polyester sepia is polyester film. Drawing changes can be made on sepia paper with pencil or ink, and on sepia polyester film with polyester lead or ink. Figure 2.50 shows an illustration of the reverse sepia print process. The use of sepia reproduction has mostly been replaced by computer-aided drafting.

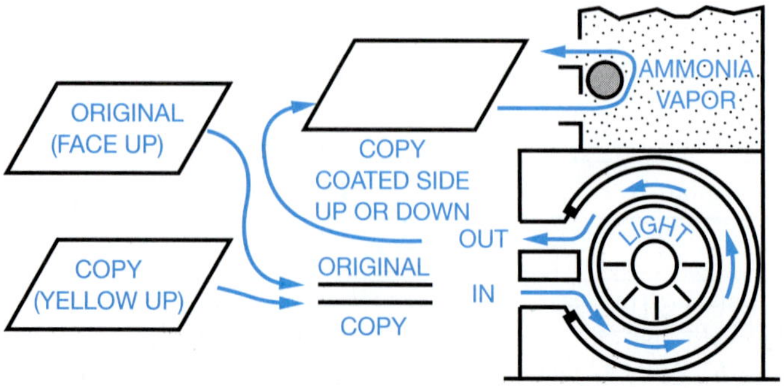

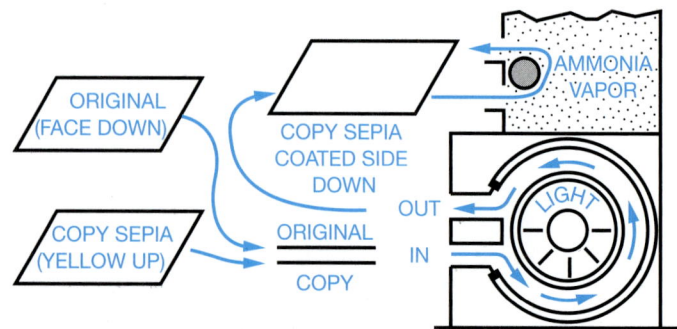

FIGURE 2.49 ■ The diazo print process.

FIGURE 2.50 ■ The reverse sepia print process.

Evolution in Reproduction Technology

The photocopy process and CADD printers and plotters are rapidly replacing the diazo process. The photocopy process, which is explained later, has many advantages over the diazo process. The advantages include quality reproduction in many sizes, use of most common materials, and no hazardous ammonia. The printers and plotters used with the CADD workstation create drawings on paper or other media, which are referred to as hard copy. Printers such as the laser printer or the ink-jet printer allow the company to produce quality hard copy drawings very quickly. The hard copy drawings can be printed on vellum material for further reproduction using the diazo or photocopy process. Chapter 3 of this text covers the options available with a CADD system.

SAFETY PRECAUTIONS FOR DIAZO PRINTERS

Ammonia has a strong, unmistakable odor. It may be detected, at times, while operating your diazo printer. Diazo printers are designed and built to provide safe operation from ammonia exposure. Some machines have ammonia filters while others require outside exhaust fans. When print quality begins to deteriorate and it has been determined that the print paper is in good condition, the ammonia may be old. Ammonia bottles should be changed periodically, as determined by the quality of prints. When the ammonia bottle is changed, it is generally time to change the filter also.

When handling bottles of ammonia or when replacing a bottle supplying a diazo printer, be sure to follow the manufacturer's recommendations. Specific procedures should be provided with your machine that explain safety issues, such as eye protection, ammonia contact, inhaling ammonia, ammonia exposure first aid, and ultraviolet light exposure when running the diazo machine. Keep in mind that ammonia is a hazardous waste and must be disposed of properly. Refer to your local, state, or federal Environmental Protection Agency (EPA) guidelines for proper disposal of ammonia. Several governmental agencies including the Occupational Health and Safety Agency [OSHA] and the Environmental Protection Agency [EPA] provide guidelines pertinent to identifying, classifying and disposing of hazardous materials and waste by-products. Manufacturing facilities can also generate numerous hazardous materials including paints, solvents, oils and fluids that need to be safely handled, stored and disposed of properly according to established guidelines.

PHOTOCOPY REPRODUCTION

Photocopy printers are also known as engineering copiers. Prints can be made on bond paper, vellum, polyester film, colored paper, or other translucent materials. The reproduction capabilities also include instant print sizes ranging from 45 to 141 percent of the original size. Larger or smaller sizes are possible by enlarging or reducing in two or more steps, that is by making a second print from a print.

Almost any large original can be converted into a smaller-sized reproducible print, and then the secondary original can be used to generate diazo or photocopy prints for distribution, inclusion in manuals, or for more convenient handling. Also, a random collection of mixed-scale drawings can be enlarged or reduced and converted to one standard scale and format. Reproduction clarity is so good that halftone illustrations (photographs) and solid or fine line work have excellent resolution and density.

Also, engineering copiers do not use ammonia, which is an important consideration when purchasing a print machine.

HOW TO PROPERLY FOLD PRINTS

Prints come in a variety of sizes ranging from small 8 1/2 × 11 in. to 34 × 44 in. or larger. It is easy to file the 8 1/2 × 11 in. size prints because standard file cabinets are designed to hold this size. There are file cabinets available called flat files that can be used to store full-size unfolded prints. However, many companies use standard file cabinets. Larger prints must be properly folded before they can be filed in a standard file cabinet. It is also important to properly fold a print if it is to be mailed.

Folding large prints is much like folding a road map. Folding is done in a pattern of bends that results in the title block and sheet identification ending up on the front. This is desirable for easy identification in the file cabinet. The proper method used to fold prints also aids in unfolding or refolding prints. Look at Figure 2.51 to see how large prints are properly folded.

MICROFILM

Microfilm is photographic reproduction on film of a drawing or other document that is highly reduced for ease in storage and sending from one place to another. When needed, equipment is available for enlargement of the microfilm to a printed copy. Special care must be taken to make an original drawing of the best possible quality. The reason for this is that during each "generation," the process of "blowback" makes the lines narrower in width and less opaque than the original. The term generation refers to the number of times a copy of an original drawing is reproduced and used to make other copies. For example, if an original drawing is reproduced on sepia or microfilm, and the sepia or microfilm is used to make other copies, this is referred to as a second generation or secondary original. When this process has been done four times, the drawing is called a fourth generation. The term blowback means bringing the drawing from the microfilm back to a printed copy. A true test of the original drawing's quality is the ability to maintain good reproductions through the fourth generation of reproduction.

In many companies, original drawings are filed in drawers by drawing number. When a drawing is needed, the drafter finds

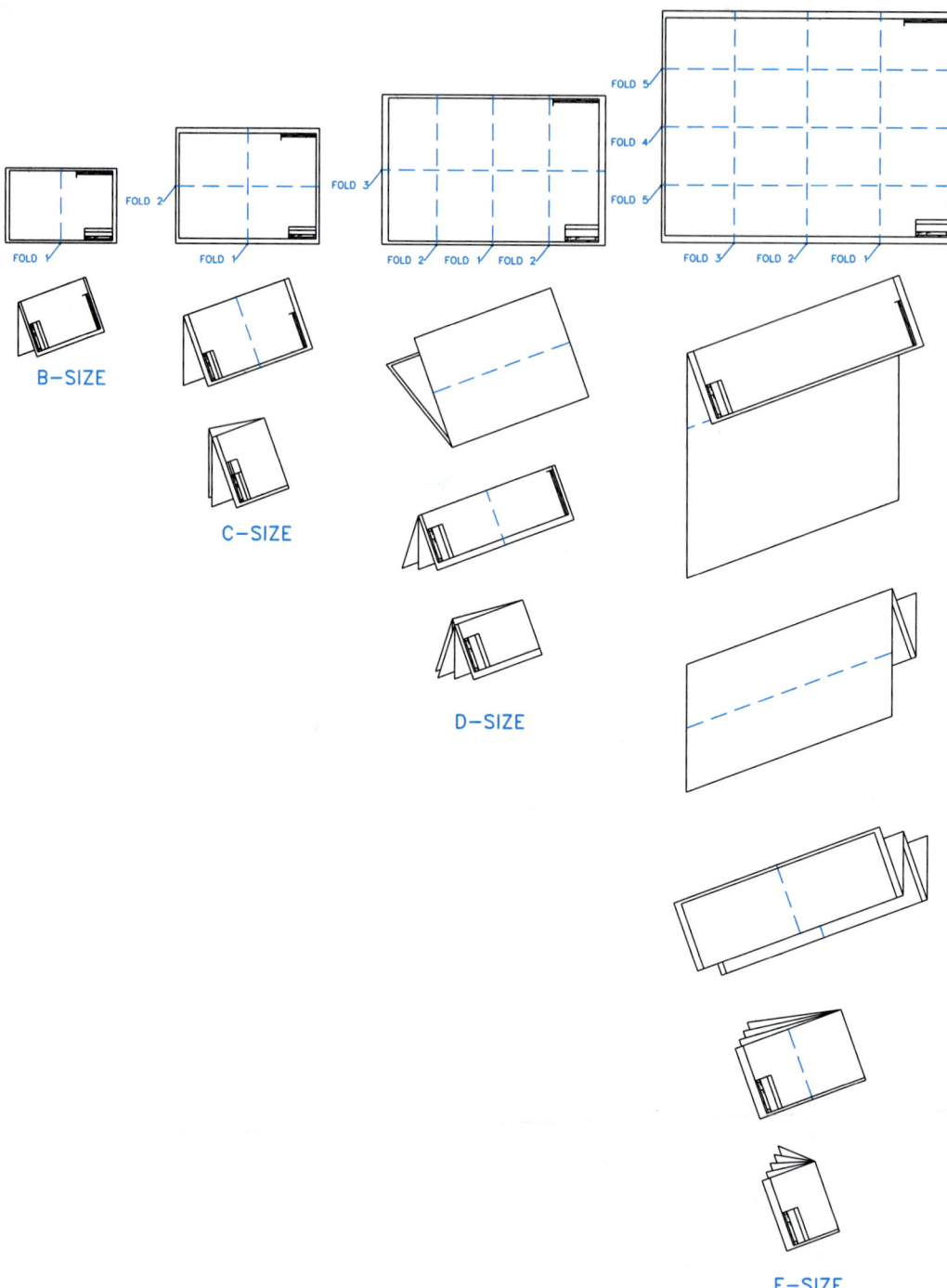

FIGURE 2.51 ■ How to properly fold B-size, C-size, D-size, and E-size prints.

the original, removes it, and makes a copy. This process works well although, depending on the company's size or the number of drawings generated, drawing storage often becomes a problem. Sometimes an entire room is needed for drawing storage cabinets. Another problem occurs when originals are used over and over. They often become worn and damaged, and old vellum becomes yellowed and brittle. Also, in case of a fire or other kind of destruction, originals can be lost and endless hours of drafting vanish. For these and other reasons, microfilm has been used for storage and reproduction of original drawings. ASME Y14.5M provides recommended microfilm reduction factors for different sheet sizes. These help ensure standardization.

While the microfilm storage of old drawings still exists in some companies, CADD files have replaced the use of microfilm for most modern applications. The CADD Application explains the reasons behind this evolution.

DIGITIZING EXISTING DRAWINGS

Although not used much today, digitizing is a method of transferring drawing information from a digitizer into the computer. A digitizing tablet can be used to take existing drawings and digitize the information into the computer. Digitizer tables range in size from 6 × 6 in. to 44 × 60 in. In most cases, companies do not have the time to convert existing drawings to CADD, because the CADD systems are heavily used for new product drawings. Therefore, some businesses digitize existing drawings for other companies.

While the digitizing of existing drawings still has some applications, scanning of existing drawings into CADD has become widely used because of improvements in this technology. Scanning of existing drawings is discussed next. Digitizer tablets are still used by some CADD systems; however, CADD software has rapidly evolved for use with a mouse to access commands from pull-down menus, tool bar buttons, icon menus, and dialog boxes. Mouse technology also has advanced to provide maximum flexibility when working at a CADD station.

OPTICAL SCANNING OF EXISTING DRAWINGS

Although not used as much today, scanning is a method of reproducing existing drawings in the form of computer drawings. Scanners work much the same way as taking a photograph of the drawing. One advantage of scanning over digitizing existing drawings is that the entire drawing, including dimensions, symbols, and text, is transferred to the computer. A disadvantage is that some drawings, when scanned, require a lot of editing to make them presentable. The scanning process picks up visual graphic information from the drawing. What appears to be a dimension, for example, is only a graphic representation of the dimension picked up from the original drawing. What you get is an exact duplicate of the existing drawing. However, if you want the dimensional information to be technically accurate, the dimensions must be redrawn and edited on the computer. There are software packages available that make this editing process quick and easy.

After the drawing is scanned, the image is sent to a raster converter that translates information to digital or vector format. A raster is an electron beam that generates a matrix of pixels. Pixels make up the drawing image on the computer. A raster editor is then used to display the image for changes.

In the final analysis, companies involved in scanning can, in many cases, reproduce existing drawings more efficiently than companies that digitize existing drawings.

When an existing drawing has been transferred to the computer, it becomes a CADD drawing file that may be edited as necessary. Optical scanning accessories can automatically scan drawings from paper, vellum, film, or blue line and convert the copy image into a computer file.

CADD VS. MICROFILM

Microfilm had become an industry standard for storage and access of drawings. Large international companies especially relied on the microfilm network to ensure that all worldwide subsidiaries had the ability to reproduce needed drawings and related documents. One big advantage of microfilm was the ability to archive drawings. Archive means to store something permanently for safekeeping. The use of CADD in the engineering and construction industries has made it possible to create and store drawings electronically on a computer or optical disk. This has made it possible to quickly retrieve drawings that have been stored. A big advantage of this advancement involves use of these drawings. When CADD-generated drawings are retrieved, they are of the same quality as when they were originally drawn. CADD drawings can be used to make multiple copies, or to efficiently redesign a product. In addition to the quality of the stored CADD drawing being the same as the original, the drawing file can be sent anywhere in the word over the Internet or within your company through the Intranet. The Internet is a worldwide network of computers. The Intranet is a network of computers within your company.

CAD/CAM

The optimum efficiency of design and manufacturing methods is achieved without producing a single paper copy of a drawing of a part. Computer networks can directly link the engineering and manufacturing departments through the integration of computer-aided design (CAD) and computer-aided manufacturing/machining (CAM) software. This integration is referred to as CADD/CAM. The drafter or designer creates a 3-D model or 2-D engineering drawing of a part using CADD software. CAM software is then used to convert the geometry to computer numerical control (CNC) data which is read by the numerically controlled machine tool. Often, the CAD/CAM system is electronically connected to the machine tool. This electronic connection is called networking. This direct link is referred to as direct numerical control (DNC), and requires no additional media, such as paper, disks, CDs, or tape, to transfer information from engineering to manufacturing.

MATH APPLICATIONS

ANGLE MEASUREMENT IN RADIANS

Although drafting machines, surveying instruments, and many modern machine tools work with degrees and minutes, many computer and calculator applications require angles to be measured in decimal fractions of a degree. It is important to be able to apply math to convert from one type of angle measurement to another because not all machines and computers do this automatically.

To convert from minutes of a degree to decimal degrees, divide the number of minutes by 60. For example, 7° 40' is the same as 7.666666°, or rounding, 7.67° because 40 ÷ 60 = .666666. To convert from decimal degrees to minutes, multiply the decimal portion of the angle by 60. For example, 18.75° is the same as 18° 45' because .75 × 60 = 45.

WEBSITE RESEARCH

The following websites can provide you with information about drafting media, supplies, and equipment:

http://www.koh-I-noor.com

http://www.staedtler.com

http://www.chartpak.com

The following websites provide information about a specific brand of drafting machine and options:

http://www.mutoh.com

http://www.allinonesupply.com

http://www.asme.org—American Society of Mechanical Engineers (ASME)

http://www.ansi.org—American National Standards Institute (ANSI)

http://www.govsupport.com—A source for CAGE Code listed supplier parts and inventory

http://www.neco.navy.mil—Navy Electronic Commerce Online

CHAPTER 2

Manual Drafting Equipment, Media, and Reproduction Methods Test

 Access the CD found with this textbook to view the Chapter 2 Test. Confirm the preferred submittal method with your instructor.

READING A TITLE BLOCK

Given the following title block, describe the numbered elements 1 through 13. Label your answers as shown below.

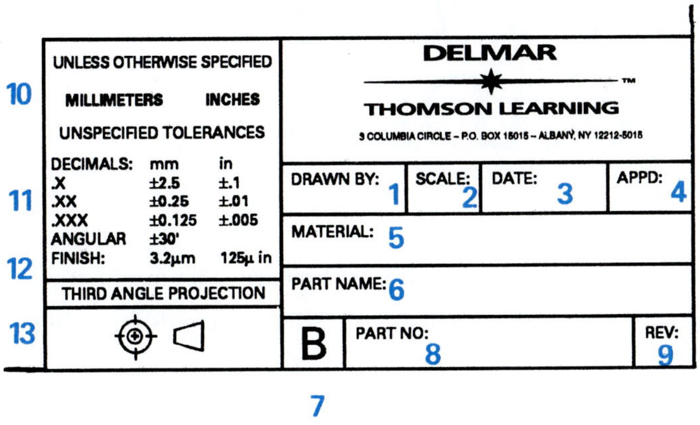

UNLESS OTHERWISE SPECIFIED			DELMAR				
10 MILLIMETERS	INCHES		THOMSON LEARNING ™				
UNSPECIFIED TOLERANCES			3 COLUMBIA CIRCLE – P.O. BOX 15015 – ALBANY, NY 12212-5015				

	DECIMALS:	mm	in	DRAWN BY: **1**	SCALE: **2**	DATE: **3**	APPD: **4**
11	.X	±2.5	±.1				
	.XX	±0.25	±.01	MATERIAL: **5**			
	.XXX	±0.125	±.005				
	ANGULAR	±30'		PART NAME: **6**			
12	FINISH:	3.2μm	125μ in				

THIRD ANGLE PROJECTION					
13 ⊕ ◁		**B**	PART NO: **8**		REV: **9**

7

PROBLEM 2.11	(1) _____		PROBLEM 2.18	(8) _____
PROBLEM 2.12	(2) _____		PROBLEM 2.19	(9) _____
PROBLEM 2.13	(3) _____		PROBLEM 2.20	(10) _____
PROBLEM 2.14	(4) _____		PROBLEM 2.21	(11) _____
PROBLEM 2.15	(5) _____		PROBLEM 2.22	(12) _____
PROBLEM 2.16	(6) _____		PROBLEM 2.23	(13) _____
PROBLEM 2.17	(7) _____			

MATH PROBLEMS

Convert the following angle measurements to decimal degrees.

PROBLEM 2.24	15'
PROBLEM 2.25	7° 30'
PROBLEM 2.26	18° 5'
PROBLEM 2.27	200° 18'
PROBLEM 2.28	−13° 42'

Convert the following angle measurements to degrees and minutes.

PROBLEM 2.29	60.4°
PROBLEM 2.30	9.5°
PROBLEM 2.31	.27°
PROBLEM 2.32	177.8°
PROBLEM 2.33	−45.1°

3 CHAPTER

Computer-Aided Design and Drafting (CADD)

LEARNING OBJECTIVES

After completing this chapter, you will:

- Describe CADD design formats.
- Explain the CADD environment and its related disciplines.
- Describe the computers and peripheral equipment that comprise the tools of CADD. See the online companion for this textbook.
- Describe the project planning process and how it relates to all disciplines.
- Identify, describe, and use national CADD standards.
- Identify the parts of the Cartesian coordinate system and describe their function.
- Demonstrate the process for determining drawing scale factors.
- Describe the process of file management.
- Identify the process used for research.
- Discuss the applications of one or more CADD software programs.

- Explain the U.S. national CADD standards for manufacturing.
- Write about the U.S. national CADD standards for architecture, engineering, and construction.
- Discuss layer and layer standards.
- Explain the use of drawing templates.
- Provide information about the importance and use of symbols and symbol libraries.
- Discuss the importance of reusing drawing content and give some examples.
- Explain file management.
- Discuss the importance of ergonomics related to the CADD environment.
- Use exercise as needed to relieve discomfort associated with computer use, if approved by your doctor.

THE ENGINEERING DESIGN APPLICATION

THE PROJECT PLANNING PROCESS

The overall success of any project begins with the layout and planning stage. Proper and thorough planning is the key ingredient to ensure the project runs efficiently, and produces drawings, designs, and products that are accurate, consistent, and well made. To that end, it is vital that you are familiar with the drawing planning process so you can apply it to any type of drawing or project, in any drawing and design discipline.

When you are asked to create a design or model or to construct a new drawing, you should avoid the temptation to begin working without having a plan. Instead, take the time to sufficiently plan the drawing process using the outline given here. The project planning process is an important aspect of solving any problem, or working on any type of project and is discussed in detail in this chapter.

The Problem-Solving Process

A three-step process allows you to immediately organize your thoughts, ideas, and knowledge about a project. Prior to beginning the three-step process, it is critically important that you read all instructions available for the project. The project planning process requires you (or the workgroup) to answer the following three questions about the project:

1. What do you know about the subject?
2. What do you need to know about the subject?
3. Where can you find the information you need?

The best way to answer these questions is to use brainstorming. Brainstorming allows individuals to voice their thoughts and ideas regarding the specific topic, problem, or project at hand. Here are a few suggestions for working in a brainstorming session.

THE ENGINEERING DESIGN APPLICATION *(continued)*

- One person records all statements.
- Place a reasonable time limit on the session, or agree to let it run until all ideas are exhausted.
- Every statement, idea, or suggestion is a good one.
- Do not discuss or evaluate any statement or suggestion.

Return to the brainstorming list later to evaluate the items. Throw out items that the group agrees are not valid. Rank the remaining ideas in order of importance or validity. Now you are ready to conduct research and gather information required to begin and complete the project.

Research Techniques

Even though the Internet can be an inviting method of conducting research, it can also be time-consuming and nonproductive if it is not used in a logical manner. It is still important to know how to use traditional resources such as libraries, print media, and professional experts in the area of your study. Any research should be conducted using a simple process to find information efficiently.

- Define and list your topic, project, or problem.
- Identify key words related to the topic.
- Identify all resources with which you are familiar that can provide information.
- Using the Internet, conduct quick keyword searches on your topic.
- Check with libraries for lists of professional indexes, periodicals and journals, specialized trade journals, and reference books related to specialized topics.
- Contact schools, companies, and organizations in your local area to find persons who are knowledgeable in the field of your research.
- Create a list of all potential resources.
- Prioritize the list and focus first on the most likely resource to provide the information you need.
- Begin your detailed study of the prioritized list.
- Ask questions.

Preparing a Drawing

After planning the drawing project as previously outlined, establish a method for starting and completing the actual drawing. Always begin any drawing or design project by creating freehand sketches. A basic procedure to use for this drawing construction process is as follows:

- Create a freehand sketch of the intended drawing layout showing all views, sections, details, and pictorials. Apply all required dimensions and notes. Using numbered items or colored pencils, label or list the components in the order they are to be drawn.
- Try to determine if special views are required.
- Note the geometric shapes that must be drawn from scratch.
- Determine if existing shapes can be edited to create new features.
- Find predrawn objects that can be inserted or referenced in the drawing.
- Determine the locations and type of dimensions to be used.
- Locate all local notes, general notes, and view titles.
- Try to decide if using object values such as size, location, and areas can assist you in drawing additional features, or provide you with useful information required for the project.
- Determine the types of prints and plots or electronic images that are required for the project.
- Begin work on the project.

It may seem to the beginning drafter that these project planning procedures are an unnecessary amount of work just to plan a drawing or project. But keep in mind that if you do not take the time, at the beginning of the project, to plan and gather the information you need, you still have to do so at some time during the life of the project. Therefore, it is better to begin a project from a solid foundation of plans and information than to interrupt your work at many stages of the project to find the information you need.

If you make the project planning process a regular part of your work habits, any type of engineering design process you encounter is easier.

CADD DEFINED

The term **computer-aided design and drafting (CADD)** refers to the entire spectrum of design and drafting with the aid of a computer, from straight lines to color animation. CADD is a process in which you use a computer program for design and drafting applications. CADD is used in all industry design and drafting fields covered throughout this textbook, and is demonstrated in two-dimensional (2-D) and three-dimensional (3-D) representations. CADD is used to create information in graphic form, and is also used to design and store other forms of information such

FIGURE 3.1 ■ The screen display of this piping system was generated with a 3-D modeling software package. *Courtesy Applications Development, Inc.*

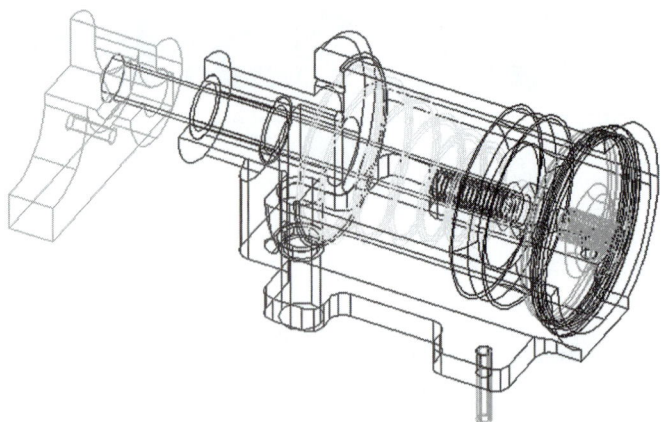

FIGURE 3.2 ■ Lines defining the edges of a model appear as wires, hence the name *wireframe*.

as engineering analyses, cost calculations, and material lists. A range of artistic capabilities are found under the heading of CADD, and drafting is just one of them. Three-dimensional industrial modeling and analysis is one of the specialized areas that has developed as a result of CADD. (See Figure 3.1.) **Computer-aided design (CAD)** describes the use of computer programs to create design and engineering applications. CAD refers to computer-aided design, but it is also a common reference to **computer-aided drafting**.

CADD Design Formats

Productivity gains realized by the use of CADD tools are directly related to the proper use of those tools. In any discipline of drafting and design, the final format, detail, and accuracy of the drawings and design are directly related to the function of the part, model, or layout. For example, a simple part that is to be mass manufactured without additional testing and analysis can be drawn quickly as a 2-D multiview drawing (see Chapter 9). However, a prototype design of a new component that must be analyzed and tested might best be developed as a 3-D solid model. The following sections provide a brief introduction to these and other design formats that the CADD drafter must choose in the planning stages of any project.

2-D Drawings

The abbreviation 2-D refers to **two-dimensional**: having only length and width (or width and height) dimensions. The views of the object appear in flat form and are normally rotated 90° from each other. This form of drafting is best used for manufactured objects rather than presentation purposes. Two-dimensional drawings are most often dimensioned and contain notes and text that describe features and details on the part, map, or plan. Two-dimensional drawings created using CADD are identical to drawings created using manual drafting techniques. The only difference is that you use CADD software tools to create objects

instead of a pencil and eraser. Chapters 9 and 10 cover creating 2-D drawings and Chapter 11 explains how to dimension these drawings.

3-D Wireframe Models

Objects initially drawn using 3-D techniques are often composed of lines connecting corners of the object. These individual points on the object are referred to as **vectors**. Technically a vector is a quantity having magnitude (length) and direction, but this can be simplified to say that a vector is a line defined by two end points. Therefore, a 3-D model must be composed of numerical values for each corner of the object. Lines connecting these corners appear as wires, hence the name **wireframe**. (See Figure 3.2.)

Three-dimensional wireframes are themselves not very useful because they are difficult to visualize. Wireframes can be the basis for creating other types of 3-D models. Three-dimensional surface and solid models are often displayed as wireframes for the purpose of redisplaying (regenerating) the model quickly.

Three-dimensional modeling is an integral part of the design, manufacturing, and construction industry and contributes to increased productivity in all aspects of a project. Drafters should develop a good working understanding of this skill because of the many and varied job opportunities that are available.

3-D Surface Models

A 3-D **surface model** is a "hollow" object on which flat planes connect all corners of the object, and the geometry is described by its surfaces. For example, automobile body panels require the construction of accurate surface shapes. High-end surface modeling software allows the user to create surface models that can be edited, analyzed, and tested. These files can then be used in the manufacture of the part.

Basic surface modeling software is used to create realistic **presentation** models. Basic surface models appear to be solid but contain no mass property data. Surface characteristics such as color, texture, light, and shadows can be applied to the model in order to create a more realistic presentation. Surface models are also used for situations in which the presentation of the object is important, such as the shaded and rendered objects used in architectural presentations, television commercials, movies, and computer animations. In addition, this type of model is used in 3-D worlds, which are composite models designed for geographical information systems (GIS), video games, and virtual reality applications.

3-D Solid Models

3-D solid modeling is a 3-D modeling method that accurately describes both the exterior and interior of a part or assembly. Solid models are constructed differently depending on the software program. Some solid model programs allow you to produce basic solid models using tools including solid primitives, which are objects such as boxes, cones, spheres, and cylinders that are combined, subtracted, and edited to produce a final model. In contrast, many parametric solid modeling programs construct solid models beginning with a 2-D sketch, followed by a sketched feature, such as an extrusion or revolution created from the sketch. Additional features can then add solid material to or subtract solid material from the initial base feature to generate a final model. Materials can be specified for the model, which can then be submitted to testing and analysis. In addition, the solid model can be assigned surface properties and lighting for use in accurate presentations. An additional aspect of solid models that enhances productivity and usefulness is the ability to cut through any part of the model to create sections. (See Chapter 4.) This allows designers and engineers to view and edit the model while looking at interior features. (See Figure 3.3.)

The solid model can also be used to quickly create a prototype composed of plastic. This discipline is called rapid prototyping and is discussed later in this section.

CADD Software Products

The following provides general information about a variety of CADD software manufacturer products. Much of the information is taken from the related websites. The information provided is intended as an introduction only. For more information, refer to the related websites referenced at the end of this chapter. This

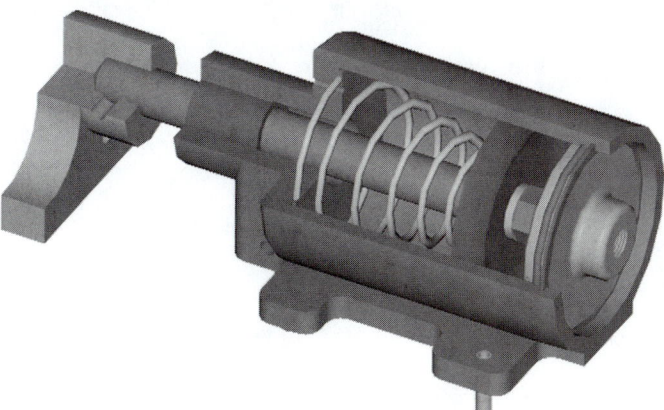

FIGURE 3.3 ■ Solid models can enhance productivity because of the ability to cut through any part of the model to create sections.

discussion is not intended to promote or endorse any of the products represented. Other CADD software products are available. Do an Internet search using keywords such as computer-aided drafting, computer-aided design, computer-aided design and drafting, CADD, and CADD software to find additional products.

■ **AGS Advanced Graphics Software**
AGS Advanced Graphics Software is the manufacturer of KeyCreator, formerly CADKEY software. KeyCreator has tools for designers in the manufacturing industry used for design, modeling, and rendering. KeyCreator also computes numerical control data for milling complex and free-form surfaces.

■ **Autodesk, Inc.**
Autodesk's core product is AutoCAD®. AutoCAD is a 2-D and 3-D drafting and design software product. AutoCAD LT® software is a 2-D drafting product. Autodesk products are available for industries such as building, infrastructure, manufacturing, media and entertainment, and wireless data. Autodesk has a variety of CADD software programs created for specific disciplines such as architectural, civil, electrical, and mechanical. The following provides a small sample of some of the additional products:

Autodesk Inventor® Series is a comprehensive, integrated 3-D and 2-D design solution.

Autodesk® Civil 3D™ software is used for civil engineering applications using a dynamic engineering model.

Autodesk® Architectural Desktop provides the efficiency of real-world building objects, design, and documentation productivity for architects and architectural designers.

Autodesk® Revit® is used for building information modeling.

■ **Bentley Systems, Inc.**
Bentley Systems provides software for building plant, civil, and geospatial, architecture, engineering, construction, and operations. MicroStation is Bentley's key product for the design, construction, and operation applications.

■ **Dassault Systemes**
Dassault Systemes is the manufacturer of the CATIA software for 3-D representation for managing product information, automation, and validation of design and manufacturing data. Mechanical design products address sheet metal and mold requirements through applications improving productivity and reducing time-to-market.

■ **Graphisoft**
Graphisoft is the manufacturer of the ArchiCAD Design/Building Series, which is a set of tools for builders and residential designers. Graphisoft packages are based on open standards, allowing you to create data without recreation. A Graphisoft product called Virtual Building™ manages the full information cycle of buildings from concept through occupancy. The program contains information about building materials and characteristics. Virtual

Building™ is a 3-D digital database that tracks all elements that make up a building, allowing the designer to use items such as surface area and volume, thermal properties, room descriptions, costs, product information, and window, door, and finish schedules.

■ **Parametric Technology Corporation**
Parametric Technology Corporation is the manufacturer of Pro/Engineer, also known in industry as Pro/E. Pro/Engineer provides 3-D solutions that detail the form, fit, and function of products. Pro/Engineer is a parametric solid modeling program for CADD and offers automated computer-aided manufacturing (CAM) applications. In the computer-aided engineering (CAE) environment, Pro/Engineer tests and optimizes products for structural, thermal, and dynamic performance.

■ **SoftPlan Systems, Inc.**
SoftPlan Architectural Design Software is a residential and light commercial CAD software package. SoftPlan allows you to create floor plans, cross-sections, elevations, framing plans, detail drawings, and site plans. Drawings are created assembled with features such as walls, windows, doors, and beams.

SoftView takes the drawing created in SoftPlan and generates a three-dimensional rendering of the model. You can also create photorealistic interior and exterior 3-D renderings from any view.

■ **SolidWorks Corporation**
The SolidWorks® 3-D mechanical design software provides manufacturers with 3-D design and 2-D drafting capabilities. SolidWorks design validation tools allow you to simulate real-world conditions and to test multiple options.

■ **UGS, The PLM Company**
UGS produces Solid Edge, a 3-D CADD software with design management capabilities allowing modeling and process workflows.

INDUSTRY AND CADD

Computer-aided drafting systems are small enough to occupy just a few square feet of desk space. Individual contractors, designers, and architects can have one in their home or office. CADD is used in all aspects of drafting, design, and engineering, and in the disciplines of architecture, HVAC, structural, civil, process piping, landscape design, pattern development, machine design, and solid modeling.

Computer-Aided Manufacturing (CAM)

Computer-aided manufacturing (CAM) requires the use of computers to assist in the creation or modification of manufacturing control data, plans, or operations. Computers are integral to the manufacturing process. Computerized welding machines, machining centers, punch-press machines, and laser-cutting machines are commonplace. Many firms are en-

gaged in **computer-aided design/computer-aided manufacturing (CAD/CAM)**. In a CAD/CAM system, a part is designed on the computer and transmitted directly to the computer-driven machine tools that manufacture the part. Within that process there are other computerized steps along the way, including the following:

STEP 1 The CAD program is used to create the product geometry. This can be in the form of 2-D dimensioned multiview drawings, or as 3-D models.

STEP 2 The drawing geometry is used in the CAM program to generate instructions for the computer numerical control (CNC) machine tools. This is commonly referred to as CAD/CAM integration.

STEP 3 The CAM program uses a series of commands to instruct CNC machine tools by setting up tool paths. The tool path includes the selection of specific tools to accomplish the desired operation.

STEP 4 The CAM programmer establishes the desired tool and tool path. The final CNC program is generated when the postprocessor is run. A **postprocessor** is an integral piece of software that converts a generic, CAM system tool path into usable CNC machine code (G-code). The **CNC program** is a sequential list of machining operations in the form of a code that is used to machine the part as needed. Figure 3.4 illustrates the CADD 3-D model, the tool path, and the G-code for the part.

STEP 5 The CNC program is verified using the software's simulator.

STEP 6 The CNC code is created.

STEP 7 The program is run on the CNC machine tool to manufacture the desired number of parts.

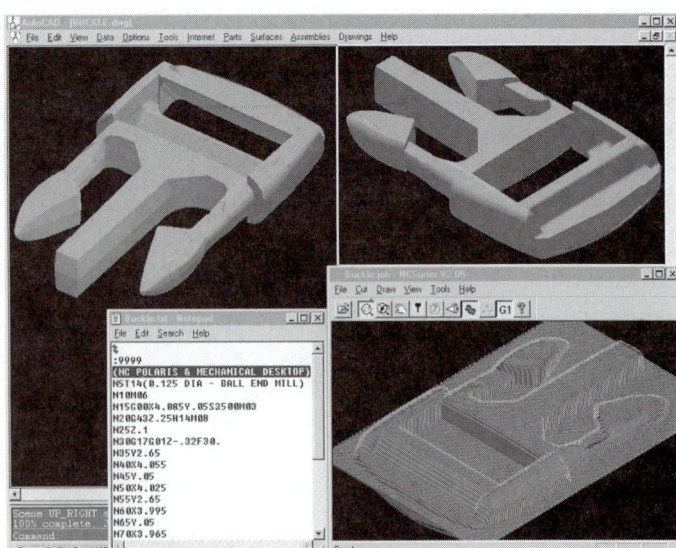

FIGURE 3.4 ■ This screen display illustrates the CADD 3-D model, the tool path, and the G-code for the part. *Courtesy NC Micro.*

Computer Numerical Control (CNC)

Computer numerical control (CNC) is also known as numerical control (NC). CNC is just one critical aspect of CAM in which a computerized controller uses motors to drive each axis of a machine such as a mill to manufacture parts in a production environment. The motors of the machine rotate based on the direction, speed, and length of time that is specified in the CNC program file. This file is created by a programmer and contains programming language such as G-codes, which are preparatory functions such as tool moves, and M-codes, which are miscellaneous functions such as tool changes and coolant settings.

Computer numerical control was a major innovation in manufacturing. It has lead to increased productivity because the consistency of the process has lowered manufacturing costs, increased product quality, and led to the development of new techniques. Persons possessing skills in CADD and CNC can find a variety of opportunities in manufacturing industries.

Rapid Prototyping (RP)

Rapid prototyping is a manufacturing process by which a solid physical model of a part is made directly from 3-D CAD model data, without any special tooling. An RP model is a physical 3-D model that can be created far more quickly than by using standard manufacturing processes. This discipline is also referred to as stereolithography, or 3-D printing. Rapid prototyping equipment accepts 3-D CAD files, slices the data into thin cross sections, and constructs layers from the bottom up, bonding one on top of the other, to produce physical prototypes. Computer-aided design and drafting software such as AutoCAD, Autodesk Inventor, and SolidWorks allow you to export a file from a solid model in a form usable by the rapid prototyping machine.

A computer using post-processing software, slices the 3-D CAD data into very thin, often .005 in. thick, cross-sectional planes. Each slice or "layer" is composed of closely spaced lines resembling a honeycomb. The slice is shaped like the cross section of the part. The cross sections are sent from the computer to the rapid prototyping machine, which builds the part one layer at a time. The rapid prototyping machine has a "vat" that contains a photosensitive liquid epoxy plastic and a flat platform or starting base resting just below the surface of the liquid. (See Figure 3.5.) A laser, controlled with bidirectional motors, is positioned above the vat perpendicular to the surface of the polymer. The first layer is bonded to the platform by the heat of a thin laser beam that traces the lines of the layer onto the surface of the liquid polymer. When the first layer is completed, the platform is lowered the thickness of a layer. Additional layers are bonded on top of the first in the same manner, according to the shape of their respective cross section. This process is repeated until the prototype part is complete.

A second type of rapid prototyping, called solid object 3-D printing, uses an approach similar to inkjet printing. During the

FIGURE 3.5 ■ This stereolithography system builds a physical prototype, layer by layer, using liquid epoxy plastic that is hardened by laser to precise specifications. *Courtesy 3D Systems.*

build process, a print head with hundreds of jets builds models by dispensing a thermoplastic material in layers. The printer can be networked to any CAD workstation and operates with the push of a few buttons. (See Figure 3.6.)

Rapid prototyping has helped to revolutionize product design and manufacture. The development of physical models can be accomplished in significantly less time when compared to traditional machining processes. Changes to a part can be made on the CAD 3-D model then sent to the RP equipment for quick reproduction. Engineers can use these models for design verification, sales presentations, investment casting, tooling, and other manufacturing functions. Additionally, medical imaging, CAD, and RP have made it possible to quickly develop medical models such as replacement teeth.

Computer-Integrated Manufacturing (CIM)

This is the concept of an automated factory integrated by a CAD/CAM system using robotics and computer-controlled machinery, from delivery of the raw material to completion and shipping of the finished product. The field of CIM incorporates the disciplines of CAD, CAM, robotics, electronics, hydraulics, pneumatics, computer programming, and process control. Computer-integrated manufacturing enables all persons within a company to access and utilize the same database that would normally be used by designers and engineers.

(a)

(b)

(c)

FIGURE 3.6 ■ (a) The 3-D printer employs a print head with hundreds of jets to build models by dispensing a thermoplastic material in layers. *Courtesy 3D Systems.* (b) A designer removes a model built on a solid object printer. *Courtesy 3D Systems.* (c) Engineers built this prototype wheel from a CAD file using a solid object printer. *Courtesy 3D Systems.*

thought of as the umbrella discipline that includes CADD, CAD/CAM, CNC, CIM, and rapid prototyping. It includes, but is not limited to, the following topics:

■ Mechanical design and product development automation.

■ Surface and solid modeling.

■ Simulation, analysis, testing, and optimization of mechanical structures and systems.

■ Product data management (PDM) of engineering data and documents.

■ Using the Internet and other technologies to collaborate on projects.

■ Manufacturing using NC programming and RP systems and services.

■ Utilizing computers including personal computers (PCs), workstations, and networking.

Computer-Aided Engineering (CAE)

The use of computers in design, analysis, and manufacturing of a product, process, or project is called computer-aided engineering (CAE). Computer-aided engineering can also be

Within computer-aided engineering, the computer and its software controls most, if not all, portions of the manufacturing. A basic CIM system can include transporting the stock material from a holding area to the machining center at which several machining functions are performed. From there, the part can be moved automatically to another station at which additional pieces are attached, then on to an inspection station, and from there to shipping or packaging.

THE CADD WORKSTATION

The electronic tools used with CADD are referred to as **hardware**. An individual CADD workstation relies on a computer for data processing, calculations, and communications with other pieces of **peripheral equipment**. Peripheral equipment is any additional hardware item that uses the computer's files and performs functions that the computer cannot handle. A typical CADD workstation is shown in Figure 3.7.

The additional functions and services provided by the peripheral equipment fall into three categories: input, output, and storage. **Input** means to put information inside the computer to be acted on in some way. Input can come from the keyboard, a mouse (or other similar input device), or a digitizer. **Output** refers to information that is sent from the computer to a receiving point such as a monitor, a plotter, or a printer. **Storage** refers to disks and drives that allow the operator to store programs, drawing files, symbols, and data.

For a wide range of information related to the CADD workstation and other CADD equipment, refer to the CD provided with this book. This additional content includes information covering:

■ Computers

■ Monitors

■ Input devices

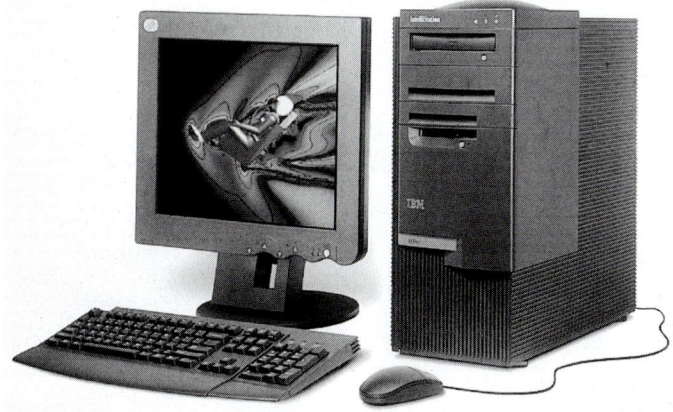

FIGURE 3.7 ■ This CADD workstation has a computer, keyboard, mouse, and flat-panel monitor. *Courtesy of International Business Machines Corporation.*

■ Data storage and media devices

■ Plotters and printers

NATIONAL CADD STANDARDS

There are parallel and unique sets of national CADD standards developed by two important sectors of United States business and industry. One national CADD standard is created by the National Council for Advanced Manufacturing (NACFAM) and publishes the *Computer-Aided Drafting and Design, CADD Skill Standards, National Occupational Skill Standards.* This publication provides the skills needed for beginner CADD users. The other national CADD standard is the *U.S. National CAD Standard* created by the CADD/GIS Technology Center, the American Institute of Architects (AIA), the Construction Specifications Institute (CSI), the United States Coast Guard, the Sheet Metal and Air Conditioning Contractors National Association (SMACNA), the General Services Administration (GSA), and the National Institute of Building Sciences' (NIBS) Facilities Information Council. This standard establishes a specific system for organizing and presenting building design documents and information.

While the content of the two national standards is different, there are elements of each that can be used when developing the CADD standards for any industry, company, or school. The following outlines the features of each national CADD standard.

CADD Skill Standards

This national CADD standard is set forth in the document titled, *Computer-Aided Drafting and Design, CADD Skill Standards, National Occupational Skill Standards,* published by the National Council for Advanced Manufacturing (*http//:www.nacfam.org*). This standards publication represents skills that are core to all CADD disciplines, generic to all software, and are for entry-level applications. The CADD Skill Standards are created using a format that identifies the specific skill, the objective for using the skill, and the standard from which the skill is established. These are examples of the format used in this standard:

Example 1

Skill	Create freehand technical sketches.
Objectives	Identify the purpose of a freehand sketch.
Standard	Create proportionate and legible technical freehand, orthographic, pictorial, schematic, and diagram sketches.

Example 2

Skill	Create a new drawing.
Objectives	Demonstrate the ability to open a drawing file and create a drawing.
Standard	Open a drawing file and create a drawing.

The CADD Skill Standards are helpful in uniting education and industry by providing needed learning objectives for use by students and employees to achieve the highest-quality skill applications.

The CADD Skill Standards are divided into the following skill areas, practices, and objectives:

Fundamental Drafting Skills

Drafting Skills

- Media and materials
- Measurement systems
- Annotation placement (notes and symbols)
- Line styles and weights
- Title blocks and drawing formats
- Standard symbols
- Drawing reproduction
- Sketching

Orthographic Projections

- Identify, create, and place orthographic views
- Identify, create, and place auxiliary views
- Identify, create, and place section views

Pictorial Drawings

- Identify and create axonometric views
- Identify and create oblique views
- Identify perspective drawings

Dimensioning

- Correctly apply dimensioning rules
- Dimension objects
- Dimension complex shapes
- Dimension features from a centerline
- Dimension a theoretical point of intersection
- Use dual dimensioning standards
- Use size and location dimensioning practices
- Use dimensioning styles
- Place tolerances and geometric dimensioning and tolerancing (GD&T)

Hardware

- Demonstrate proper equipment care
- Operate and adjust input devices
- Operate and adjust output devices
- Use correct handling and operation of storage media
- Start and shut down a workstation
- Adjust monitor controls for maximum comfort and usability
- Recognize availability of information services

Physical and Safety Needs

- Demonstrate an understanding of ergonomic considerations and personal safety

Operating Systems

- Start and exit software programs and applications
- Demonstrate proper file management techniques
- Format disks
- Identify, create, and use directory structures and change paths
- Demonstrate proper file maintenance and backup procedures
- Translate, import, and export data files between formats
- Use help files
- Save drawings to storage devices

Basic CADD Skills

Create

- Create a new drawing
- Perform drawing setup
- Construct geometric shapes
- Annotate drawings using appropriate text style and size
- Use and control accuracy enhancement tools
- Identify, create, store, and use appropriate symbols and symbol libraries
- Create wireframe drawings and solid models
- Create objects using primitives
- Create 2-D drawings from 3-D models
- Create 3-D wireframe models from 2-D geometry

Edit

- Use geometric and non-geometric editing commands
- Control coordinates and display scale
- Control object properties
- Use viewing commands
- Use display commands
- Use standard parts and symbol libraries
- Plot drawings using correct layout and scale

Manipulate

- Use layering techniques
- Use grouping methods
- Minimize file size

Analyze

- Use query commands

Dimensioning

- Properly use associative dimensioning

Advanced CADD Skills

Create

- Create a wireframe and solid model
- Create non-analytic surfaces using modeling
- Create analytic surfaces using modeling with planes and analytic curves
- Create offset surfaces
- Find the intersection between two surfaces
- Create joined surfaces
- Create a fillet or blend between two surfaces
- Create feature-based geometry, such as holes, slots, fillets, and rounds
- Create cut sections
- Construct and label exploded assembly drawings

Edit

- Trim, extend, and manipulate surfaces
- Edit control points
- Move, copy, and scale primitives

Manipulate

- Perform axis view clipping
- Extract wireframe data from surface and solid models
- Shade and render an object

Analyze

- Extract geometric data
- Extract attribute data
- Identify gaps in non-intersecting surfaces
- Get surface properties
- Obtain mass properties data, such as moments of inertia

Productivity and Work Habits

- Customize to improve performance
- Manipulate associated non-graphical data
- Use template and library files to establish drawing standard presets
- Develop geometry using parametric programs

U.S. National CAD Standard

In 1995, a group of agencies including the CADD/GIS Technology Center (CGTC), the American Institute of Architects (AIA), the Construction Specifications Institute (CSI), the U.S. Coast Guard, the Sheet Metal and Air Conditioning Contractors National Association (SMACNA), and the National Institute of Building Sciences (NIBS) came together to develop a single CADD standard for the United States, now referred to as the U.S. National CAD Standard (NCS). Standards are very important in

drafting. The purpose of the standard is to allow consistent and streamlined communication among owners, architects and designers, and construction teams. Use of the standard can result in reduced costs for developing and maintaining office standards and greater efficiency in the transfer of building design information from design to construction. The standard is not intended to be applied with any specific CAD software program, but AutoCAD and MicroStation were selected as the primary applications because of their popularity.

Major Elements of the U.S. National CAD Standard

Uniform Drawing System

The Construction Specifications Institute (CSI) created the Uniform Drawing System™, which is an eight-module system for organizing and presenting building design information. The Uniform Drawing System was adopted by the U.S. National CAD Standard. The basic content found in each module is as follows, per the CSI website:

- Module 1—Drawing Set Organization establishes set content and order, sheet identification, and file naming for a set of construction drawings.
- Module 2—Sheet Organization provides format for sheets. Includes drawing, title block, and production reference areas and their content. Also includes a coordinate-based location system and preferred sheet sizes. The U.S. National CAD Standard CD has ready-to-use sheet formats.
- Module 3—Schedules sets consistency in format, terminology, and content. Additional guidelines include how to "build" a project-specific schedule and an organizational system for identifying and filing schedules. The U.S. National CAD Standard CD has ready-to-use schedule formats.
- Module 4—Drafting Conventions addresses standard conventions used in drawings: drawing orientation, layout, symbols, material indications, line types, dimensions, drawing scale, diagrams, notation, and cross-referencing.
- Module 5—Terms & Abbreviations provides standard terms and standard abbreviations used in construction documents and specifications. It provides consistent spelling and terminology, standardizes abbreviations, and notes common usage.
- Module 6—Symbols addresses commonly used standard symbols, classifications, graphic representation, and organization in creating, understanding and fulfilling the intent of construction documents. The CD includes a search index for selecting and viewing the symbols and a copy function to bring the selected symbol into CAD applications. This is a product of a joint effort with the CADD/GIS Technology Center and CSI.
- Module 7—Notations provides guidelines for notation classification, format, components, and location; use of notes; terminology; and linking to specifications.

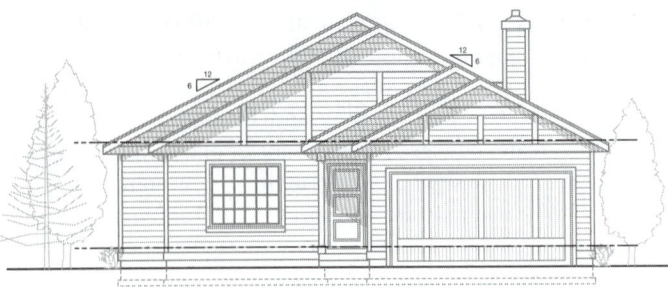

FRONT ELEVATION

FIGURE 3.8 ■ The model file contains individual elements that make up the final drawing, such as walls, doors, windows, and the various drawing features.

Discipline	Designator
Structural	S
Architectural	A
Interiors	I
Equipment	Q
Fire protection	F
Plumbing	P
Process	D
Mechanical	M
Electrical	E
Telecommunications	T
Resources	R
Other disciplines	X
Contractor/Shop drawings	Z
Operations	O

The **model file type** characterizes the type of drawing. There is a long list of possible model file types. The following are specifically related to the general and architectural discipline designators:

Discipline	Code	Definition
General		
	BS	Border sheet
	KP	Keyplan
Architectural		
	3D	Isometric/3D
	AC	Area calculations/Occupancy plan
	CP	Reflected ceiling plan
	DT	Detail
	EL	Elevation
	FP	Floor plan
	QP	Equipment plan
	RP	Roof plan
	SC	Section
	SH	Schedule
	XD	Existing/Demolition plan

■ Module 8—Code Conventions identifies types of general regulatory information that should appear on drawings, locates code-related information in a set of drawings, and provides standard graphic conventions. Can be a tool to expedite code review by designers and plan review authorities.

Drawing Set Organization

Drawing set organization covers drawing units, file naming, and sheet identification.

Drawing Units. CADD systems use real-world units for drafting with feet and inches, feet and tenths of feet, and meters and millimeters as appropriate for the specific drawing.

File Naming. There are two file types identified in the standard: model files and sheet files. The **model files** contain the individual elements that make up the final drawing, such as the walls, doors, windows, dimensions, and the various drawing features on their own layers, as shown in Figure 3.8. CADD layers are discussed later. The **sheet files** bring the model files together to create the composite drawing, and include the border and title block. (See Figure 3.9.) CADD layers are covered in detail later in this chapter.

File names are established in a standard format based on four elements—the project code, the discipline designator, the model file type, and the user definable code. The **project code** is optional and is determined by the designer or architect. The project code can have up to twenty characters. The **discipline designator** has two characters, with the second character being a hyphen (-). The first character of the discipline designator is one of the following:

Discipline	Designator
General	G
Hazardous materials	H
Survey/Mapping	V
Geotechnical	B
Civil works	W
Civil	C
Landscape	L

The last four characters are **user definable** such as *F1* for "floor one." If all four characters are not used, the remaining spaces are filled with *Xs,* for example F1XX.

A complete file name might be PR2004A-FPF1XX.dwg, where:

■ PR2004 is the project code.

■ A- is the discipline designator.

■ FP is the model file type.

■ F1XX is the user definable code.

■ .dwg is the AutoCAD file extension (.dgn for a MicroStation file).

Sheet Identification. Sheet names are similar to file names, with the first characters being the project code, and the next two characters the discipline designator with a level 2 designator, which adds one more character to the discipline for specific applications, such as *AD* for "Architectural Demolition." This is followed by a sheet-type designator, followed by a two-character

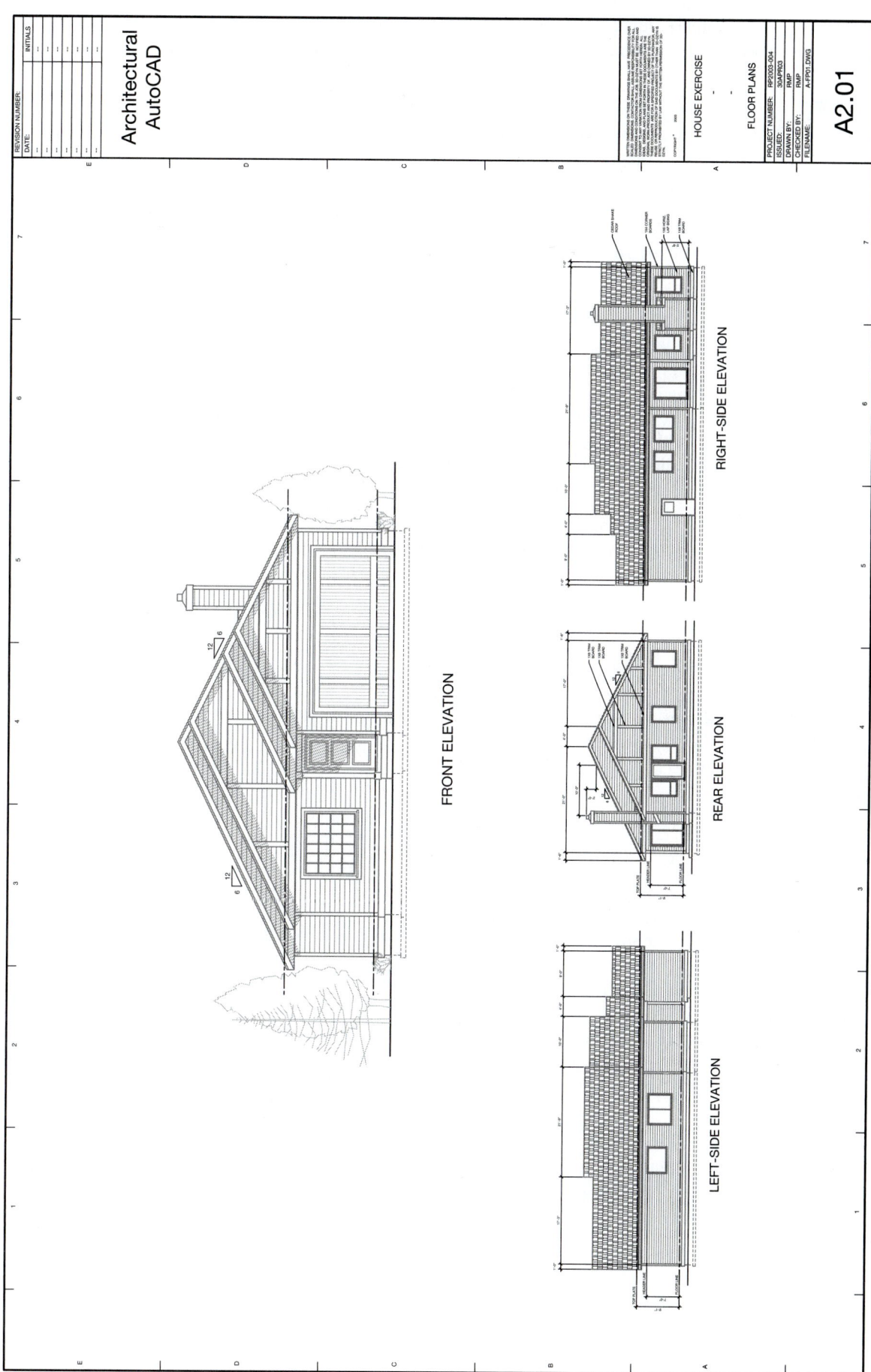

FIGURE 3.9 ■ The sheet file brings the model files together to create the composite drawing and includes the border and title block.

sheet sequence number—for example, 02-12. Three more characters can be added for user definable content. The sheet-type designators are:

Sheet Type	Designator
General	0
Plans	1
Elevations	2
Sections	3
Large scale views	4
Details	5
Schedules and diagrams	6
User defined	7 and 8
3D	9

A sample sheet name might be PR2004AD102XXX.dwg, where:

■ PR2004 is the project code.

■ AD is the discipline designator with level 2, Architectural Demolition.

■ 1 is the sheet type.

■ 02XXX is the sheet sequence number and user definable characters.

■ .dwg is the AutoCAD file extension (.dgn for a Micro-Station file).

The sheet identifier is located in the title block and in any reference to the sheet on a drawing. Coordinating between the sheet name and the sheet identifier are the discipline designator, sheet-type designator, and the sheet sequence number. For example, in the sheet reference number A-124:

■ A- represents the architectural discipline.

■ 124 indicates that 12 is the total number of sheets and 4 is the specific sheet (such as sheet 4 of 12).

Organizing Drawing Files. Drawing files must be organized so you or anybody else in the office can easily find a drawing when needed. When organizing files, create a project folder that contains subfolders for each of the projects. The subfolders can be named for the project, client, or job number. The files in each project subfolder can include drawing files, word processing documents, and spreadsheets as needed. This helps keep the entire project together, making it easier to manage. Figure 3.10 shows a file structure for organizing drawing files.

CADD Graphics

As you will learn in Chapter 7, drafting is a graphic language made up of features such as lines, symbols, and text. The standard deals with these drawing elements.

Line Weights. A variety of line weights, or thicknesses, help improve drawing clarity. Certain features can be drawn with thicker lines in an effort to make them stand out. Thus, line thickness and line weight are used interchangeably in CADD. These recommended line weights are different than those identified in the

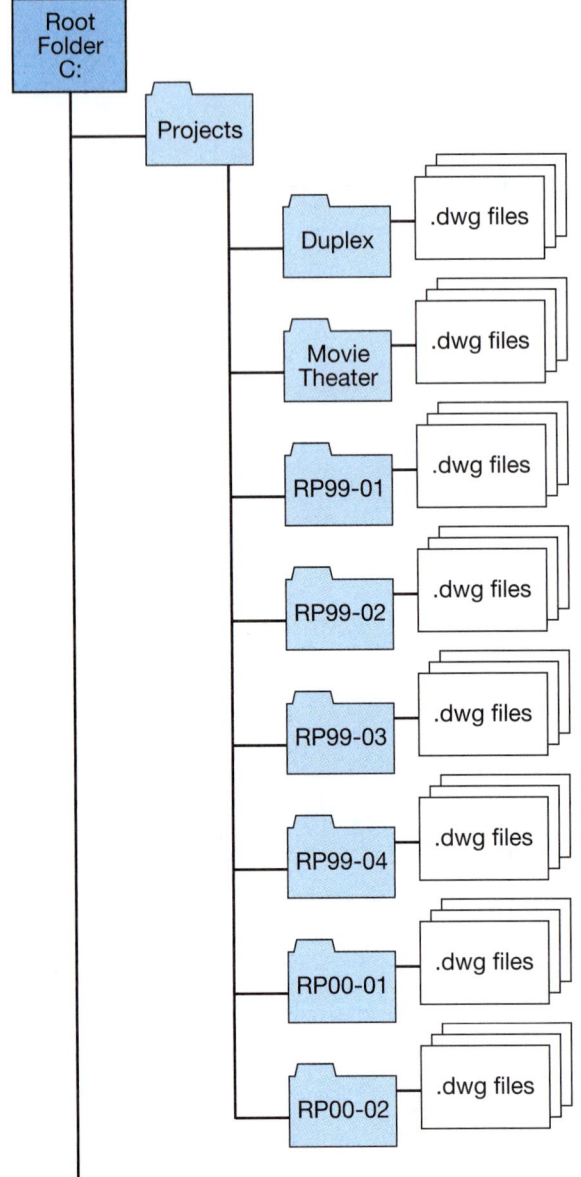

FIGURE 3.10 ■ A file structure for organizing drawing files.

ASME standard used in this textbook. Confirm the preference with your instructor or supervisor. While a large variety of line weights are possible, the standard acknowledges the following:

Line Thickness	Inches	Millimeters	Use
Fine	.007	0.18	Use sparingly for features such as patterns.
Thin	.010	0.25	Dimensioning features, phantom lines, hidden lines, centerlines, long break lines, schedule grid lines, and background objects.

Line Thickness	Inches	Millimeters	Use
Medium	.014	0.35	Minor object lines, dimension text, text for notes and in schedules.
Wide	.020	0.50	Major object lines, cutting-plane lines, short break lines, and title text.
Extra wide	.028	0.70	Minor title underlines, schedule outlines, large titles, special emphasis object lines, and elevation and section grade lines.
Option 1	.040	1.00	Major title underlines, and separating portions of a drawing.
Option 2	.055	1.40	Border lines, cover sheet lines, and artwork.
Option 3	.079	2.00	Same as Option 2.

Line Types. A variety of line types are included in the standard, as shown in Figure 3.11. The commonly used line types are introduced and displayed in Chapter 7, with examples of their applications.

Line Colors. The use of color in a CADD drawing and a plotted drawing aids in clarity and visualization. Although most CADD systems can display up to 256 colors, the standard recommends white (black), red, green, blue, gray, yellow, magenta, and cyan. However, colors such as yellow, magenta, and cyan can be difficult to see.

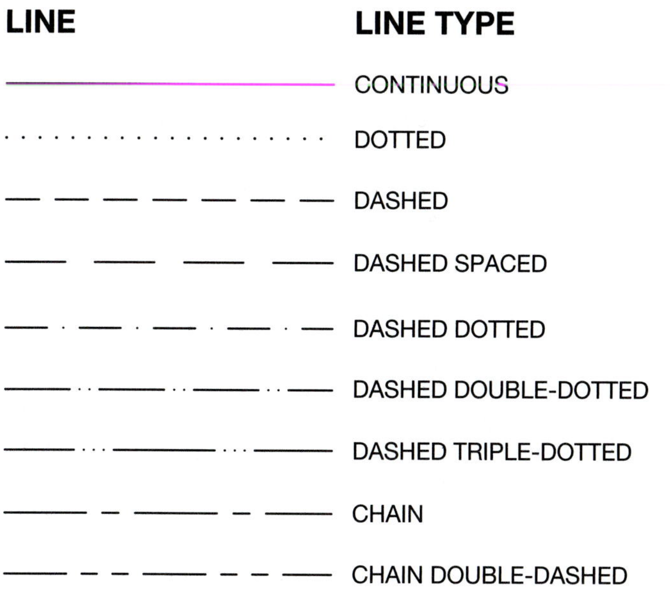

LINE / LINE TYPE

LINE	LINE TYPE
————————	CONTINUOUS
· · · · · · · · · · ·	DOTTED
— — — — — —	DASHED
— — — —	DASHED SPACED
— · — · — · —	DASHED DOTTED
— ·· — ·· —	DASHED DOUBLE-DOTTED
— ··· — ··· —	DASHED TRIPLE-DOTTED
—— — —— — ——	CHAIN
—— — — —— — —	CHAIN DOUBLE-DASHED

FIGURE 3.11 ■ The U.S. National CAD Standard recommended line types.

ABCDE	12345	MONOTEXT FONT
ABCDE	12345	PROPORTIONAL FONT
ABCDE	*12345*	*SLANTED FONT*
ABCDE	**12345**	**FILLED FONT**
ABCDE	12345	OUTLINE FONT

FIGURE 3.12 ■ The U.S. National CAD Standard recommended text styles.

Text Styles and Fonts. Text and text fonts are introduced and discussed in Chapter 7. The text fonts recommended by the standard are the basic monotext font and a variety of forms of the Roman font. See Figure 3.12. These fonts are recommended by the standard, but they do not allow for the creativity often desired in an architectural drawing. CADD programs have fonts available that closely duplicate the artistic style found in manual drawings, as shown in Chapter 7. Recommended text heights are also covered in Chapter 7.

Terms and Abbreviations

The standard provides a complete listing of terms and abbreviations used in the construction industry. The proper terms and abbreviations are provided throughout this text where they apply to specific content. Abbreviations should be used to save time and drawing space.

Notes on Drawings. Drawing notes are covered in Chapter 11. The types of notes recommended by the standard are:

- General notes
- General discipline notes
- General sheet notes
- Keynotes

CADD Symbols

CADD symbols are introduced in this chapter and are discussed and shown where they apply to specific content throughout this book. The standard provides a complete range of recommended symbols for all disciplines. The following is a list of the classification of symbols by type:

- Reference symbols
- Line symbols
- Identity symbols
- Object symbols
- Material symbols
- Text symbols

Sheet Sizes, Borders, and Title Blocks

Sheet sizes, borders, and title blocks were discussed and shown in Chapter 2. The standard recommends that title blocks be placed vertically along the right side of the sheet and contain the following compartments:

■ Designer identification block—the logo or name of the agency or company.

■ Issue block—revision area or addenda.

■ Management block—information about the designer, checker, and drafter.

■ Project identification block—project name and location.

■ Sheet identification block—the sheet identifier.

Placement of the title block along the right side of the drawing sheet is common in architecture, engineering, and construction-related industries, such as structural, industrial pipe, and civil drafting fields discussed in this textbook. The title block, discussed in Chapter 2 and other chapters throughout this text, is commonly placed in the lower right corner of the drawing sheet as recommended by the ASME standard.

Sheet Sizes. Recommended ANSI/ASME and ISO sheet sizes were discussed in Chapter 2. The ANSI/ASME standard sheet sizes are generally used for inch drawings and the ISO sheet sizes are used for metric drawings. Although the NCS does not recommend specific sheet sizes, U.S. government agencies often require ANSI/ASME sheet sizes.

Drawing Scales

Drawing scales were introduced in Chapter 2 and apply to this content. There is also additional CADD-related coverage later in this chapter. Recommended drawing scales are also discussed in chapters where they apply to the discipline and can differ from the standard. The drawing scales presented by the U.S. National CAD Standard are primarily used on architectural, engineering, and construction-related drawings. Scales used for mechanical drawings are different from architectual scales and are discussed in Chapter 2 and throughout this text where applied to mechanical drawings for manufacturing. The following combines the standard recommended scales for a variety of drawing types in inches and metric:

Drawing Type	Inch	Metric
Site plans	1" = 20'	1:200
	1" = 30'	1:400
	1" = 40'	1:500
	1" = 50'	1:600
	1" = 60'	1:700
	1" = 100'	1:1000
	1" = 200'	1:2000
	1" = 400'	1:5000
	1" = 500'	1:6000
	1" = 1000'	1:10000
	1" = 2000'	1:20000
Floor plans	1/4" = 1'-0"	1:50
	1/8" = 1'-0"	1:100
	1/16" = 1'-0"	1:200
Roof plan	1/16" = 1'-0"	1:200
Exterior elevations	1/8" = 1'-0"	1:100
	1/16" = 1'-0"	1:200

Drawing Type	Inch	Metric
Interior elevations	1/4" = 1'-0"	1:50
	1/8" = 1'-0"	1:100
Full sections	1/4" = 1'-0"	1:50
	1/8" = 1'-0"	1:100
	1/16" = 1'-0"	1:200
Wall sections	1/2" = 1'-0"	1:20
	3/4" = 1'-0"	
Stair details	1" = 1'-0"	1:10
	1 1/2" = 1'-0"	
Details	1" = 1'-0"	1:10
	1 1/2" = 1'-0"	
	3" = 1'-0"	1:5

Dimensioning

The dimensioning standards and practices for other applications are covered in the chapters where they correlate. Both feet and inches and metric units are covered and relate to this standard.

GETTING STARTED WITH A CADD PROJECT

The most important and productive time you can spend working on any project or drawing is the time you use to plan it. This point cannot be overemphasized. Always plan your work carefully before you begin to use the tools required to create the drawing.

When you plan a project and map out the directions required to complete it, you are implementing a "process" by which you can do your work. This process can be applied to any project. But when you begin the actual work of building the drawing or model, you use tools, skills, techniques, formulae, materials, and standards to complete the job. These items are referred to as the "content" of the discipline. If you begin a project by using content before process, your task is much more difficult, takes longer to complete, and can contain more mistakes. Therefore, take a few minutes to study the following procedures for planning your work. Although this was introduced at the beginning of this chapter, the following gives you more information to consider.

Using the Planning Process—A Review

A three-step process lets you immediately organize your thoughts, ideas, and knowledge about the specific project. You can use this process at the beginning of the project to map the overall goals, and also use it at any point during the project to guide you through smaller tasks and goals. The planning process requires you or your team to answer the following three questions about the project:

1. What do you know about the subject?

2. What do you need to know about the subject?

3. Where can you find the information you need?

The best way to answer these questions is to use brainstorming. In this process, individuals voice their thoughts and ideas regarding the specific topic, problem, or project at hand. All ideas are listed by a recorder and are not discussed or criticized. This technique is a proven method to tap the knowledge, creativity, and energy of individuals and groups. Here are a few suggestions for working in a brainstorming session.

■ One person records all statements.

■ Place a reasonable time limit on the session, or agree to let it run until all ideas are exhausted.

■ Every statement, idea, or suggestion that is made is a good one.

■ Do not discuss or evaluate any statement or suggestion.

After the session, take a break or discuss other items. When you return to the brainstorming list, it is time to evaluate the items. Throw out any items that the group agrees are not valid. Rank the remaining ideas in order of importance or validity.

Prior to beginning this brainstorming process, it is critically important that you read all instructions available for the project. Use different colored highlighters to mark the following:

■ Project details.

■ Items needing further study, or those you do not understand.

■ References to other resources or information about the project.

Now look at the three-step planning process questions again.

What Do You Know About the Subject?

List everything that you are sure about regarding the topic. For example, if you are working on a new drawing of a mechanical part, write down all the things you know about the object, such as the material, machining features, and tolerances. Do you know the type of drawing (2-D, isometric, or 3-D) and the standards to be used? How is the object to be manufactured? Do you know basic items such as paper size and scale of the drawing, text heights and font styles, number of views, sections, and details required? You can even list specific CADD software commands and techniques that you know can be used to complete this drawing.

What Do You Need to Know About the Subject?

List anything you are not sure about. This includes things you think you know, but about which you are not sure. Do you have a question about a detail listed in the instructions? List it here. List any CADD commands or procedures that you think you may need but that you are not confident in using.

Where Can You Find the Information You Need?

Try to think of all additional resources that could assist you in completing the project. This can include textbooks; written instructions; vendor brochures and catalogs; published standards; experts in the discipline such as instructors, engineers, architects, product representatives; and other resources such as trade journals, websites, academic papers, government publications, and even conferences and conventions.

As stated earlier, after answering these questions using the brainstorming technique, return to your lists. Eliminate unnecessary items and prioritize the remainder. If you make these procedures a regular part of your planning process, you find any project you work on flows smoothly, takes less time, and contains fewer errors.

Choosing the Drawing Format

Perhaps the most critical question to ask at the beginning of any project is, "What is the function of this project?" This question is important for engineers, architects, and designers. Then a similar question must be asked by the drafter, "What is this drawing used for?" You can break this down into discipline-specific questions as follows:

■ *Mechanical design*—What is the function of the part? Will this drawing be used for manufacturing, inspection and testing, or presentation?

■ *Architectural/structural*—What type of structure is required, and what is its intended use? Is the drawing to be used by contractors, for the purpose of acquiring building permits, or for presentation to the client?

■ *Civil*—Who will use the map and what for? If the map is a property plat survey or large land survey, what is the required accuracy?

■ *Commercial mechanical (piping, HVAC, electrical)*—What is the purpose of the building, how many people occupy it, and how much and what types of equipment are required? Is the drawing used for preliminary design and layout, estimating, construction, or client presentation?

It should be evident that when a drafter or designer is preparing to begin a new drawing in any of the varied disciplines, it is vital to determine the purpose of the drawing. Although engineers and designers are responsible for questions related to the overall design and materials, drafters must focus on the uses of each drawing created. A discussion of drawing formats was presented early in this chapter. Now, look at some questions that assist the drafter in deciding on the proper format to use.

■ Do machinists use traditional methods to manufacture the part from the drawing?

■ Are CNC machines to be used to manufacture the part?

■ Is the drawing to be used for 2-D presentation purposes only?

■ Is the drawing to be reduced and printed in a book for use by nontechnical persons, or in a company parts catalog or instruction manual?

■ Is the drawing used for client presentation in order to win a contract?

■ Must the drawing be flexible so that the design can be used for presentation purposes, but also used in the testing, analysis, and manufacturing processes?

Once the purpose and function of the drawing has been decided, you can then begin the actual process of creating the drawing. Following a planned procedure for beginning a new drawing is a sure step to completing a successful project.

Drawing Setup Steps

Throughout every step of a project, you should always apply the three-step problem-solving process. Stopping periodically to answer the three questions always keeps you aware of what information you need and where you need to get it.

After you plan the overall project and determine the number and type of drawings required, you should also have a plan indicating the progression, or chronology, of drawings. Just as every drawing begins with a single line, every project begins with a single drawing. Once you have determined the type of drawing to be created, the following drawing setup procedure is a basic outline provided to assist you in starting any type of drawing. This process can be modified to suit specific CADD software needs and applications.

1. Determine the width (X axis) and height (Y axis) of the real-world drawing area. This is called the drawing limits. For example, if you are drawing three views of a part and you estimate the size of views, dimensions, and extra space to be 36" × 28", then your drawing limits are approximately 36" × 28".

2. Determine the sheet size required for the drawing, for example, C-size or 22" × 17".

3. Determine the scale of the final plotted or printed drawing. Calculate the scale factor if it is unknown; for example, the drawing limits of 36" × 28" occupies a C-size using a HALF scale which is 1:2 or .5" = 1". The scale factor is 1 ÷ .5 = 2. For a detailed discussion of calculating scale factors, see the next section, Scaling and Scale Factors.

4. Open the appropriate template or base drawing. Remember that the template drawing should contain objects, values, and settings that are common to that type of drawing. This includes border and title block, title block text, general notes, text styles, dimension styles, layers, and display settings.

5. Establish layers, line types, and line weights if they are not part of the template drawing. Always use national and company standards when specifying line types and line weights. See Chapter 7 for information on lines.

6. Create text styles using the scale factor if they are not already saved in the template drawing. To calculate the proper CAD text height for plotted text of .125", multiply .125" × 2. The scale factor gives you a text height of .25".

7. Create a proper dimension style using the text style with a text height. Set all dimension sizes such as text height, arrows, ticks, and extension line offsets to their actual plot-

ted measurements. Then set the overall dimension scale to the scale factor of the drawing.

Following these basic steps helps you always begin drawings in a consistent and accurate manner. Remember, it is most efficient to save the text and dimension styles in a template drawing so you do not have to create them each time you begin a drawing.

SCALING AND SCALE FACTORS

Drawings created using manual techniques are laid out using a scale, whereas drawings created with CADD are drawn full scale and plotted to a specific scale. Mechanical drawings use scales such as 1:2, 1:4, 1:10, 2:1, 5:1, and 10:1. Architectural drawings are representations of buildings and use scales such as 1/2" = 1', 1/4" = 1', and 1/8" = 1'. Some of the smallest scale drawings are in the civil engineering field. A small-scale drawing shows a large area, while a large-scale drawing shows a small area. These drawings represent areas of land measured in feet or miles, and some of the scales used are 1" = 10', 1" = 20', and 1" = 50'.

The most important point to remember when using CAD is you always draw full size and plot at the proper scale. This also means you must plan ahead and know the scale of the final plotted drawing. This is discussed in detail later. Drawing in CAD at full scale means that if a part measures 4.35", you draw it 4.35" long. Even if the final drawing is plotted on a B-size sheet at 1:4, it is still drawn at full size in CADD. A house that measures 48'-6" along one wall must be drawn that exact size in CADD. You will soon realize that it is critically important to plan your work carefully so you set up the correct drawing limits within which to work. Establishing the correct drawing limits allows you to draw anything full size within those limits—from a microscopic computer circuit to a map of the solar system.

The planning that should occur prior to beginning the drawing involves, but is not limited to, the following items related to scaling:

■ Size of paper for final plot or print.

■ Size of the object to be drawn.

■ Height of plotted text.

■ Type of dimensioning to be used.

Planning the scale at which a drawing is to be plotted is important because the plotted scale dramatically affects the size of some drawing objects such as text. First, determine what you have to draw and the size of paper for the plot. Next, determine the scale at which the drawing should be plotted. For example, a drawing of a mechanical part must be plotted at a scale of 1:4. This means when you plot in CADD and use the proper scale, all objects in the drawing are reduced to one quarter of their original size. What would happen to text in this case if it were drawn in CADD .125" high? It would be reduced to one quarter (.25) of its size, which means the plotted text is .25 × .125 = .03125" high! This is too small. That is why you must plan your drawing carefully, and take the final plotted text height into consideration.

Text height and dimension text heights are directly affected by the plotting scale, and must be considered early in the planning of your drawing. Use the following procedure to create a text style and dimension style that work together to create properly sized dimension text heights.

1. Create a new text style using the font of your choice, such as Roman for mechanical or Stylus BT for architectural. Set the text height.

2. Create a new dimension style. Give it a name that reflects the type of dimensioning to be used, for example, "Mechanical" or "Architectural."

3. Choose the text style you created in step 1.

4. Select the proper text height, for example, .125.

5. Set all values for dimension line, extension line, and arrowhead sizes and offsets.

6. Set the scale factor according to the drawing scale. The scale factor is explained in detail in the following math but is the number of times smaller or larger the drawing display is from the original part or location. For example, a mechanical drawing scale of 1:4 means the drawing is reduced four times, or to one quarter of the original size. Therefore, the scale factor is 4.

7. Save the dimension style and set it current so it can be used. Whenever you begin a new drawing, always remember to set the appropriate dimension style current.

You can easily find the scale factor of any drawing scale using the following math:

Mechanical

1:4
↓
.25" = 1"
↓
1/.25 = 4

The number 4 represents the scale factor of the scale 1:4. You know the part is reduced four times when it is plotted. Therefore, you must set the dimension scale to the value of 4. Now, whenever you place a dimension on your drawing, all of its size components such as text, arrowheads, and offsets are multiplied by the dimension scale value of 4. This makes them four times their actual size, but they look correct on your monitor when you draw the features of the part at their actual size.

The process for identifying the scale factor of architectural and civil drawings is often more involved than with mechanical drawings, though the principle is exactly the same. In contrast to mechanical drawing scales, architectural or civil drawing scales typically contain different units, such as 1/4" = 1'−0" or 1" = 200'. For these applications, you must first convert both numbers to the same unit format. Once you make the conver-

sion, you can then use the math previously shown to identify the scale factor. For example:

Architectural

1/4" = 1'−0"
↓
1/4" = 12"
↓
.25" = 12"
↓
12/.25 = 48

The scale factor for a drawing scale of 1/4" = 1'−0" is 48.

Civil

1" = 200'
↓
1" = 2400"
↓
2400/1 = 2400

The scale factor for a drawing scale of 1" = 200' is 2400.

The second important aspect of using the scale factor of the drawing is the text height. Always multiply the intended plotted text height by the scale factor to find the CADD text height. For example, if you want to have text displayed .125" high on a drawing with a scale of 1:4, first multiply this by the scale factor, in this case 4.

$$.125 \times 4 = .5$$

The value of .5 is the scaled text height. Therefore, you should set the text height to .5 when drawing in CADD. It appears correct on the screen. When you plot the drawing at the scale of 1:4, the text height is multiplied by the scale to produce the plotted text height; thus, $.5 \div 1/4 = 2$, which is the proper text height.

This same technique is used for architectural, civil, and any other scaled drawings. Calculations for a .125" text height for the architectural and civil scales shown above are given here:

Architectural	**Civil**
.125 × 48 = 6	.125 × 2,400 = 300

Therefore, a CADD text height of 6 inches produces .125" text when the architectural drawing is plotted correctly. Likewise, text height of 300" in the civil drawing produces the proper .125" text height when the drawing is plotted.

As you can see, planning your drawing regarding scales and scale factors is a critically important task and one that should become an integral part of your drawing habits. Recommended ASME/ANSI text heights are discussed in Chapter 7.

In Chapter 7 you will discover the general text height is .12 in. (3 mm) as recommended by the American Society of Mechanical Engineers (ASME). The previous examples use .125" as the desired text height. As shown, if the desired drawing text height is .125 and the scale factor is 4, then the scaled text height is .5 (.125 × 4 = .5). If you use the ASME recommended drawing text height of .12, then the scaled text height is calculated as .12 × 4 = .48. Using this value, .48 is entered for scaled text height values. This is fine if the practice matches your company or school standards, but many use the .125 drawing text heights for convenience. You should confirm the preferred practice with your supervisor or instructor.

USING LAYERS

In CADD terminology, a layer is a group of objects, drawing elements, or components that share one or more common characteristics. When layers are "stacked" on top of each other, they are registered perfectly. For example, an architectural drawing can contain a layer for walls, windows, plumbing, and electrical. A multiview mechanical drawing can have layers for object lines, hidden lines, dimensions, and section lines. When all layers of a drawing are visible, the drawing is complete and all components are located accurately. (See Figure 3.13.)

The benefits of layers are many, and their use can increase drawing flexibility, productivity, and clarity. The following information explains a few of the reasons layers are used:

■ Related components and information can be placed on individual layers.

■ Drawing clarity is improved by assigning different colors, line types, and line weights to layers.

■ Layers can be turned on or off as needed to place or remove information from the screen.

■ Individual layers can be turned off for plotting and printing purposes. For example, the floor plan and heating, ventilation, and air-conditioning (HVAC) plan can be plotted together and used by the HVAC contractor for estimating and bidding purposes.

Layers Used in Mechanical Drafting for Manufacturing

There are a variety of layer use options found in mechanical drafting for manufacturing. Companies commonly establish their own layer standards. In mechanical drafting, each different style of line and text, such as object lines, centerlines, hid-

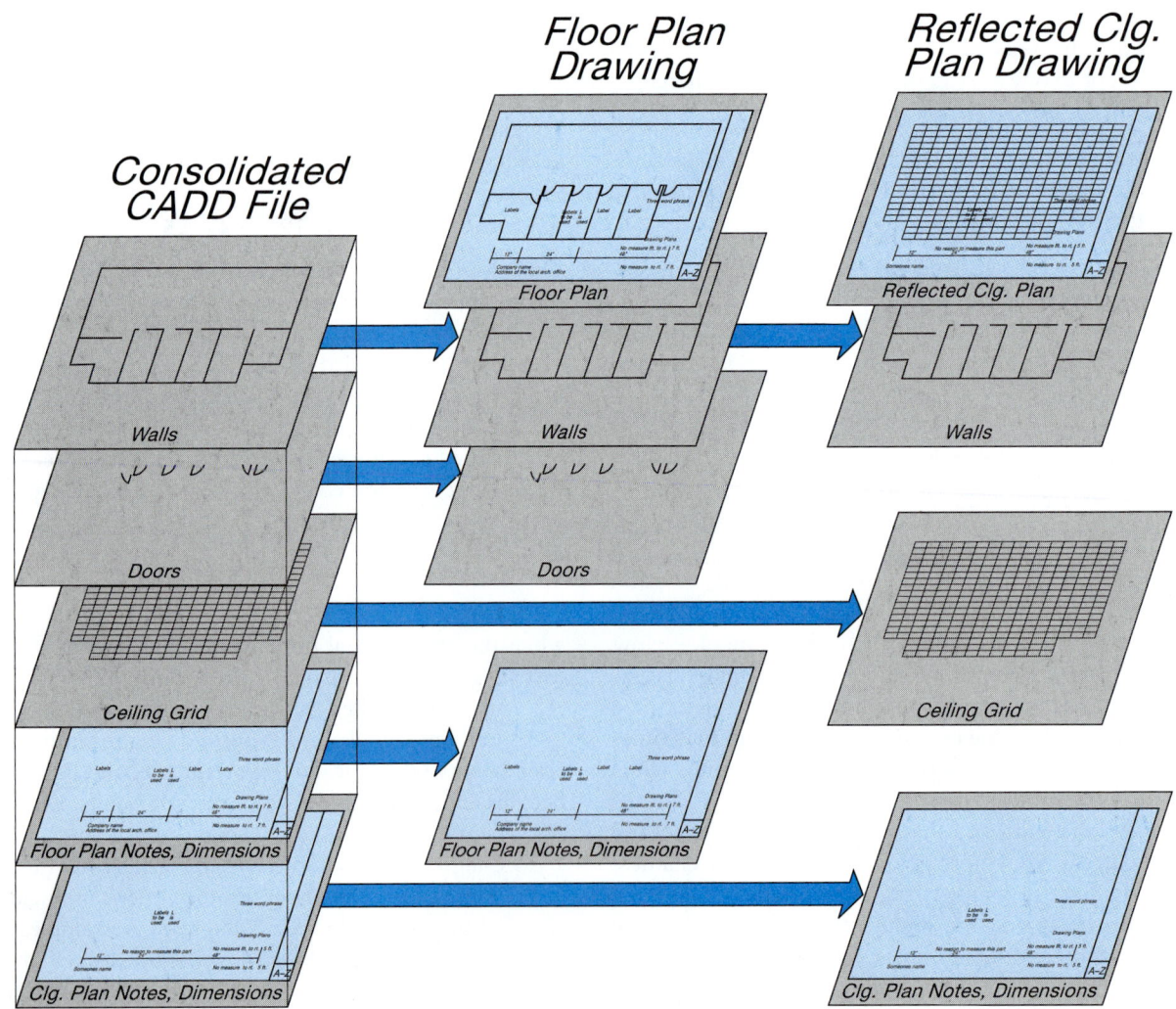

FIGURE 3.13 ■ An example of using layers to share drawing information.

den features, dimensions, notes, sections, and symbols might be placed on a separate layer controlling the visibility of views. Layer names including line type, line weight (line thickness), and color can be something like this:

Layer Name	Line Type	Line Weight	Color
Object	Continuous	.032 in./0.6 mm	Black
Hidden	Hidden	.016 in./0.3 mm	Blue
Center	Centerline	.016 in./0.3 mm	Green
Dimension	Continuous	.016 in./0.3 mm	Red
Construction	Continuous	.016 in./0.3 mm	Yellow
Border	Continuous	.032 in./0.6 mm	Black
Phantom	Phantom	.016 in./0.3 mm	Magenta
Section	Continuous	.016 in./0.3 mm	Brown

Layers can also be given more complex names. The name can include the drawing number, color code, and layer content. For example, the layer name DWG100-2-DIMEN can refer to drawing DWG 100, color 2 which is yellow, and DIMEN because the layer is used for dimensions.

Layer names can generally have up to 255 characters and can include letters, numbers, and certain special characters, including spaces. Confirm the specific requirements with your CADD software instructions. While it is good practice to keep layer names as short as possible, this makes it possible to establish layer names based on the product. For example, the layer name can include the assembly, subassembly, part name or number, line type, line weight, and color. Confirm the CADD layer practice with your company or school before establishing the layers for your CADD template.

Layers Used in the Architectural, Engineering, and Construction Industry

The architectural, engineering, and construction (AEC) industry has specific CADD layer standards titled *CAD Layer Guidelines*, published by The American Institute of Architects (AIA) Press. These layer guidelines were created by four professional organizations and three government agencies and provide consistent guidelines for organizing drawings in the building design professions. These standards are used for the practices covered in the structural, industrial pipe, and civil drafting chapters of this textbook.

Some CADD systems automatically set drawing elements on separate layers; others require that you create your own layering system. These layers are part of the drawing and can be turned on or off as needed. In addition to the layer identification system, many companies set up other layer systems based on their own project needs. For example, layers can be identified using a system in which elements are divided into three groups: major group, minor group, and basic group. The major group contains the type of drawing, such as "Floor Plan." The minor group identifies the elements on the drawing, such as "Electrical." The basic group labels the type of information, such as "Lines" or "Text." Here is an example of this system of layer identification: FL1ELECL, which means first floor (FL1),

electrical (ELEC), and lines (L). This is a typical layout of the three groups:

Major Group	Minor Group	Basic Group
BSMT	HVAC	LINE
FL1	ELEC	TXT
FL2	PLUMB	DIM
GAR	WALLS	INSERT
SITE	DOORS	HATCH
SCHED	WINDOWS	
NOTES	APPL	
FDN		
ELEV		
ROOF		
REFLECT		

Layer names should be kept short, such as FDN for "foundation." Layer identification can also contain numbers specifying the pen width used to produce the plot. For example, 1 = 0.25 mm, 2 = 0.35 mm, 3 = 0.5 mm, and 4 = 0.7 mm pen width.

The American Institute of Architects (AIA) recommends the use of layer names that contain two to four parts. Refer to the AIA standard, *CAD Layer Guidelines*, for complete information. The AIA-recommended layering system has a discipline code, a major group, a minor group, and a status group of information in the layer names. The discipline code is identified by a letter, followed by a hyphen, and describes the project element, as follows:

A   Architecture, interiors, and facilities
S   Structural
M   Mechanical
P   Plumbing
F   Fire protection
E   Electrical
C   Civil engineering and site work
L   Landscape architecture

The major groups reduce the layer names to type of information found in the discipline code such as doors, windows, walls, ceilings, furniture, or equipment. A minor group can also be added to the layer name for further definition. Layers might have minor groups such as NOTE, SYMB (for "symbols"), DIMS (for "dimensions"), or TITLB (for "title block information"). In the recommended format, the major group is followed by a dash, and the minor group is used. This format also provides for status characters that are entered after the minor group. These status entries are for special project requirements that relate to new and existing construction in remodeling projects. An example of a simple layer name with only the discipline code and major group identified is shown in Figure 3.14a, and a major and minor group layer name is shown in Figure 3.14b. The AIA layer name system with major group and status field is shown in Figure 3.15, and the layer name system with major group, minor group, and status field is shown in Figure 3.15.

Each chapter provides recommended AIA layer names that are related to the layers used in the topics covered. The complete

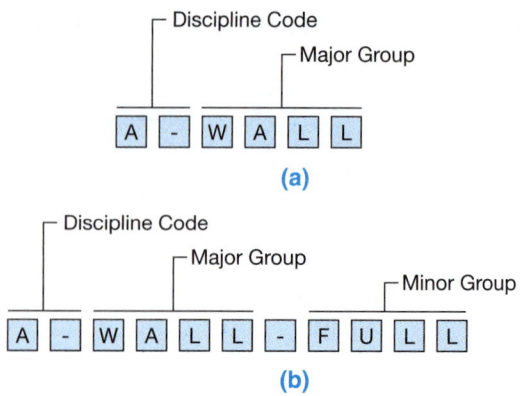

(a)

(b)

FIGURE 3.14 ■ Examples of AIA-recommended format for simple layer names showing the use of (a) major group only and (b) major and minor group.

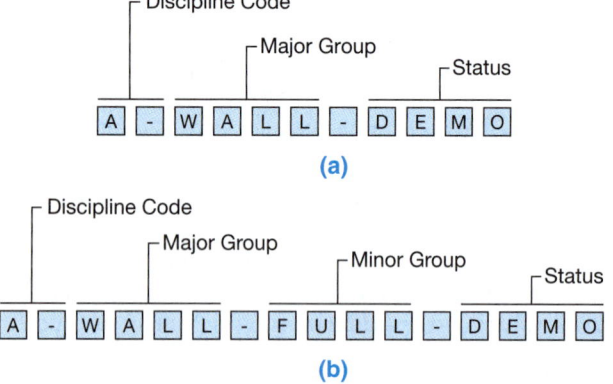

(a)

(b)

FIGURE 3.15 ■ Examples of AIA-recommended format for layer names showing use of (a) major group and status field and (b) major group, minor group, and status field.

layer naming system recommended by the AIA is displayed in the *CAD Layer Guidelines*.

Architectural Layer Systems in Use

A big advantage of CADD is the ability to send drawing files on disk or through the Internet or an Intranet to different subcontractors involved in a project. This is important in commercial applications because the architect often creates the building design and has others design the electrical, plumbing, and mechanical systems. The mechanical system encompasses heating, ventilating, and air-conditioning (HVAC). To make this system work effectively, the architect prepares a base drawing, which represents the floor plan and its bearing walls and partitions (Figure 3.16). A disk copy, Internet, or network communication of this drawing is sent to the interior design department to create the layers containing the nonbearing facility layout, which includes the room layout with fixtures and appliances (Figure 3.17). The CADD drawing files then go to the plumbing, electrical, and HVAC contractors for design of their related elements. Figure 3.18 shows the plumbing layer for the previous building. A composite drawing of all layers is shown in Figure 3.19. Each layer is generally set up in its own color on the computer screen.

CADD layers provide these advantages:

■ Reduced drafting time.

■ Improved project coordination; this is achieved by sharing information that is common to different elements of the project or common to several floors in a multistory building.

■ Ease of preparing design alternatives; the same drawing can be shared by turning layers on or off as needed for display.

THE IMPORTANCE OF TEMPLATE DRAWINGS

A template is a pattern of a standard or commonly used part that is created once, then used on future drawings. A collection of symbols in a symbol library could be called a template. If you create a base drawing that contains standard components, values, and settings, it is referred to as a template drawing.

Most schools and companies use template drawings. These template drawings save time and help produce a certain amount of consistency in the drawing process. First look at what a template drawing is composed of, and then at how it can be used.

Template Drawing Contents

Template drawings should be stored in a common location that is accessible to everyone who needs them. This is often on a network computer. If you maintain your own templates, be sure they are stored in at least two different locations. Template drawings can and should be updated and added to; therefore, it is important that you replace all old copies with the updated versions. Keep on file a variety of template drawings that contain settings for different drawing disciplines and scales. Template drawings can contain, but are not limited to, the following items:

■ Border and title block. Standard title block text information can be filled in.

■ One or more model and layout tabs.

■ Several named text styles with heights to match different scale drawings.

■ A text style with text height for dimensioning purposes.

■ Named dimension styles with values and settings for specific drawing scales.

■ Named layers containing colors, line types, and line weights.

■ Display settings for point styles, multiline styles, and line weights.

■ User profiles containing display screen menu layouts, colors, fonts, and configuration.

■ Drafting settings such as object snaps, grid, snap, units, and limits.

■ Section patterns (hatch) and scales.

■ Plot styles and settings.

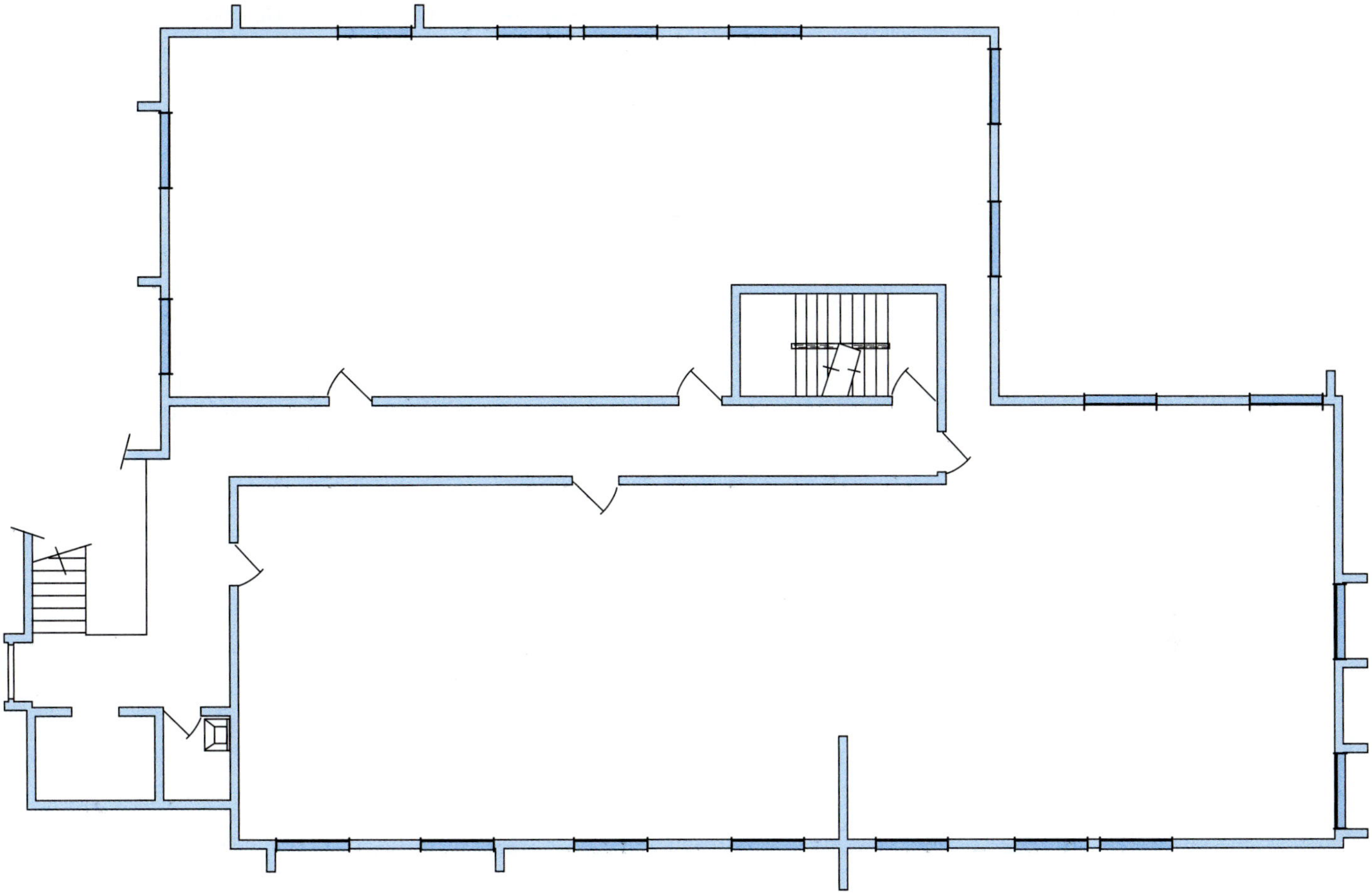

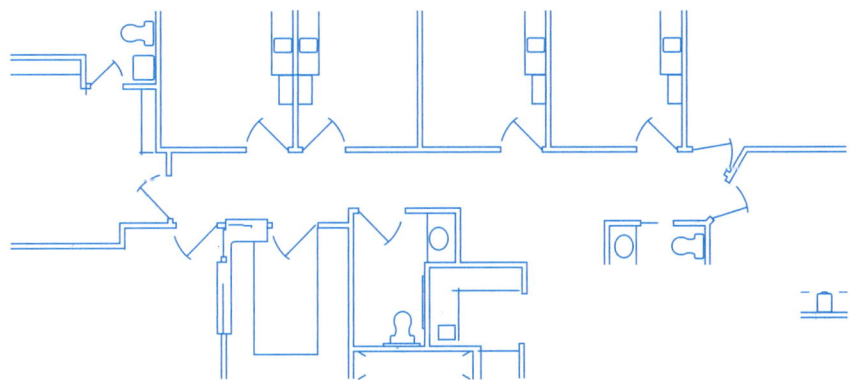

FIGURE 3.16 ■ Base layer sheet for exterior and support structure is unchanged through the project. *Courtesy Structureform.*

FIGURE 3.17 ■ Interior design of nonbearing partitions may be changed as desired using layers. *Courtesy Structureform.*

Using Template Drawings

Template drawings help create a productive drawing session and ensure drawing consistency. Be sure your template drawings are stored in an easily accessible location such as the local hard disk or the network server, and keep backup copies on other disks or optical media. Use the following steps to work with template drawings whenever you begin a new drawing project.

1. Open the template file.

2. Immediately use the SAVE or SAVE AS command, depending on your CADD system, to save the new drawing with a name that is different from the template name. Be sure to save the new drawing in a different location than that of the templates, such as a specific project folder.

3. Begin work on the new drawing project.

—s— = EXIST. 4"ϕ C.I. SEWER LINE
– – – – – = NEW C.I. SEWER LINE
—w— = EXIST. 1"ϕ WATER LINE
- - - - - - - - = NEW WATER LINE
= NEW 10 GALLON ELECTRIC WATER HEATER OVER RESTROOM ON PLATFORM. WATER HEATER TO BE AMERICAN APPLIANCE 4 GSM 10 OR EQUAL AS LISTED IN THE DIRECTORY OF CERTIFIED WATER HEATERS BY THE CALIFORNIA ENERGY COMMISSION (NOTE: PROVIDE 3/4" + COPPER P+T VALVE DISCHARGE LINE TO OUTSIDE FOR EACH WATER HEATER).
10 GAL ELEC W.H.
2"vent

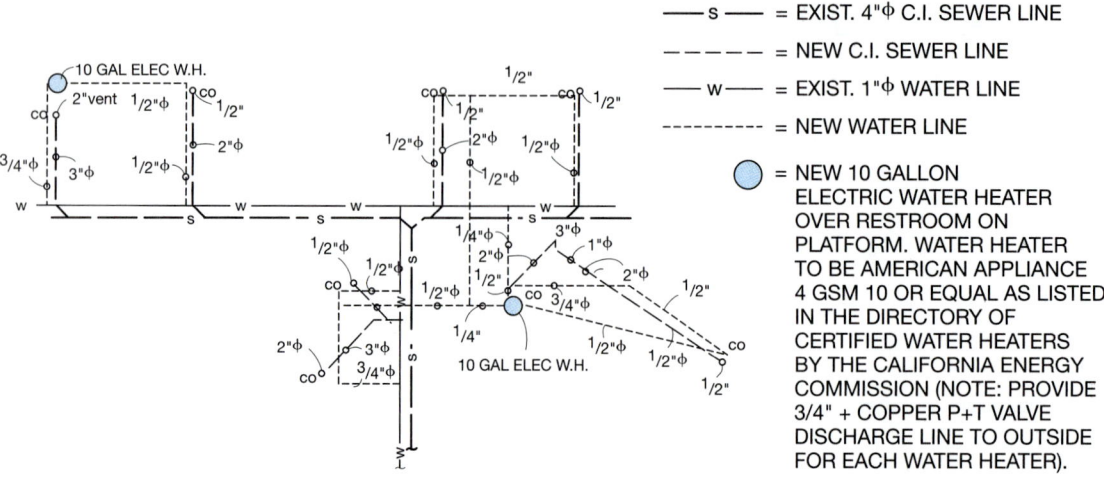

FIGURE 3.18 ■ Plumbing layer. *Courtesy Structureform.*

—s— = EXIST. 4"ϕ C.I. SEWER LINE
– – – – = NEW C.I. SEWER LINE
—w— = EXIST. 1"ϕ WATER LINE
- - - - - - - = NEW WATER LINE
= NEW 10 GALLON ELECTRIC WATER HEATER OVER RESTROOM ON PLATFORM. WATER HEATER TO BE AMERICAN APPLIANCE 4 GSM 10 OR EQUAL AS LISTED IN THE DIRECTORY OF CERTIFIED WATER HEATERS BY THE CALIFORNIA ENERGY COMMISSION (NOTE: PROVIDE 3/4" + COPPER P+T VALVE DISCHARGE LINE TO OUTSIDE FOR EACH WATER HEATER).
10 GAL ELEC W.H.
2"vent

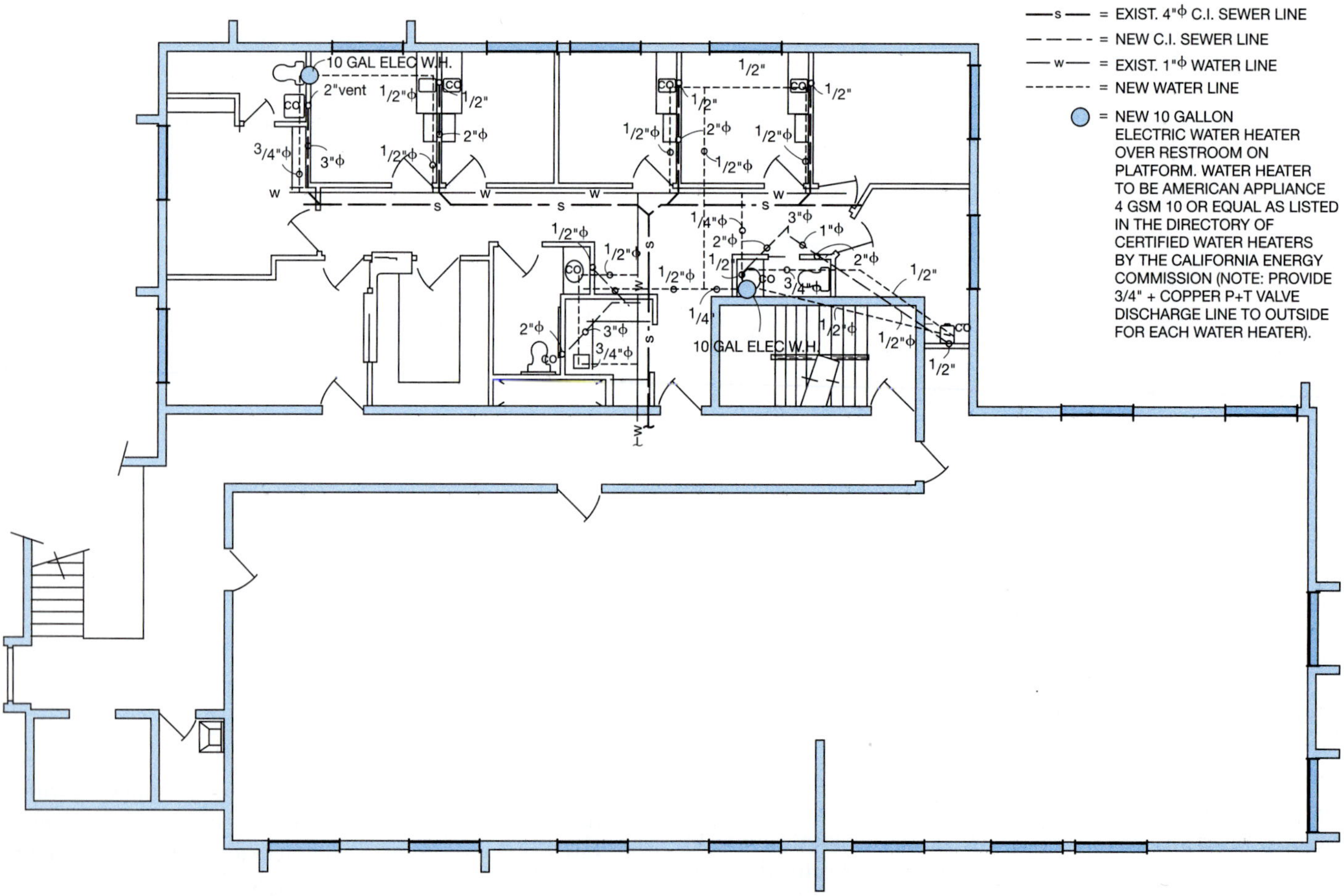

FIGURE 3.19 ■ Composite of base layer, interior design layer, and overlays. *Courtesy Structureform.*

Now you have a base drawing that contains many of the standard items required for any drawing. As you work, you may discover additional items that should be in your template drawings. Add them as you think of them, because this leads to greater productivity in future work.

BASIC DRAWING FUNCTIONS

Most CADD software is structured along similar drawing processes. The following discusses general categories of drawing functions. You should always consult your software HELP files for specific information on commands and functions.

After you step through the procedure of setting up a new drawing, there are at least seven additional areas used to complete the drawing. They are basic geometry construction, object editing, using predrawn objects and symbols, text and dimension placement and editing, drawing display, drawing queries, and printing. Each of these areas are discussed in general terms to provide you with a framework for working efficiently on any project.

Preparing a Drawing

After planning the overall scope of the drawing project as outlined previously, you should focus on a method for starting and completing the actual drawing. The least stressful and most productive method to use for preparing your drawing plan is to create a freehand sketch. A basic procedure to use for this drawing construction process is as follows:

1. Create a freehand sketch of the intended drawing layout. This does not have to be to scale. Sketch all views, sections, details, and pictorials that are required. Apply all required dimensions and notes. Using numbered items or colored pencils, label or list the components in the order they are to be drawn. Use a color such as red to provide notes on the sketch regarding specific commands or functions to use on certain features. Let the sketch be your roadmap for the project.

2. Determine if special views are required. Also decide if you need named views or symbols that can be restored quickly for working in detailed areas of the drawing.

3. Note the geometric shapes that must be drawn from scratch.

4. Carefully study the sketch and determine what features can be created by editing existing shapes.

5. Determine if there are any objects or components that have already been drawn, such as objects in a symbol library, that can be inserted or referenced into the drawing.

6. Determine the locations and type of dimensions to be used.

7. Locate all local notes, general notes, and view titles.

8. Specific commands can provide you with information about the drawing and its objects. Try to decide if using object values such as size, location, and areas can assist you in drawing additional features or provide you with useful information required for the project. Commands such as these are called inquiry commands.

9. Determine the types of prints and plots or electronic images that are required for the project. This should include paper sizes, scales, line thickness, textures, and colors. Also include plots that require specific layers or features to be printed in bold or grayscale.

Basic Geometric Construction

In its most basic form, all art is created with just two geometric shapes: the arc and the straight line. CADD software contains a variety of commands that let you draw an assortment of geometric shapes such as circles, arcs, polygons, rectangles, multiple line objects, complex curves, and ellipses.

One of your first tasks when planning the drawing with a freehand sketch is to determine which shapes can be constructed using basic geometry commands. Use your sketch to note object values such as radii, diameters, angles, and lengths. Also note the command names on your sketch if you are just learning the software. An additional step you can use is to number the basic shapes in the order you plan to draw them.

Object Editing

A good rule to use when constructing a CADD drawing is "draw as little as possible." If you carefully plan your work, you begin to see with any type of drawing, that shapes and features are repeated, used in different sizes, or even appear rotated or mirrored from one another. Try to locate features on the freehand sketch that fall into these categories.

After you have drawn the basic objects in the drawing, see how many of the original features can be copied, rotated, stretched, compressed, or mirrored in some manner to create more objects. In this way, you actually draw just a few objects, and then use those to create new geometry. This process is referred to as editing and can become a very productive and efficient method to use in any project. Productive editing commands include MIRROR, SCALE, ROTATE, STRETCH, TRIM, MOVE, and COPY.

MIRROR

MIRROR is one of a variety of editing commands for use in creating efficient and accurate drawings. The MIRROR command produces reverse copies of an object or symbol by specifying an axis on which to reflect the object. The image can be duplicated at any distance from the original or appear to be attached to it, such as the object in Figure 3.20.

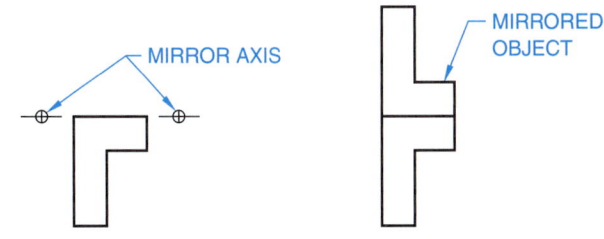

FIGURE 3.20 ■ The MIRROR command can be used to reflect a feature.

Using Predrawn Objects

A productive feature of CADD is the ability to draw an object or component once, then save it for future use. These objects can be referred to as symbols, cells, blocks, and reference files. Regardless of the terminology, the value of the objects is they can be used over and over on any drawing. This saves time in drawing creation and leads to an increase in productivity.

A most important thing for you to remember is to always be aware of the kinds of predrawn symbols that are used by your school or company. Before you begin work on any drawing, always consult available company and project standards for information on the use of symbols. CADD programs allow you to use predrawn symbols and entire drawings.

Using Symbols

When an object is saved using a specific name, it is always available for insertion in the current drawing and other drawings.

Any objects or drawings can be saved as a single object. The selected objects are saved as a drawing file with an insertion point. This drawing file can then be inserted in any drawing. If you think a drawing you create can be useful in other situations, save it as a symbol for later use. When you insert the symbol into any other drawing, it is placed using the specified insertion point.

Reference Drawings

When it is necessary to insert an entire drawing or part of a drawing into the current drawing, it can be referenced. This means the reference drawing is not actually inserted in the current drawing; its name is just attached. Then, when you open the drawing containing reference files, the CADD program finds the reference files and displays them on the screen. Referenced drawings generally do not become part of the current drawing. Therefore, it is always important to keep referenced files in the same location. This process is a little more complex than using symbols, so it is best to consult your CADD textbook or Help files for detailed instructions.

This topic is discussed in greater detail in the section titled Creating and Using Symbols.

Text and Dimensions

Dimensions should always be placed on a drawing after the geometry is completed, followed by local notes and text. When using CADD software, it is important to establish dimension styles and text styles before you begin a drawing. It is best to have dimension styles and text styles saved in template drawings so they are available for use when the drawing is opened.

Most drawings created for production purposes rely on accurate and readable dimensions. Therefore, it is important to plan the style and location of dimensions before you begin the drawing. It is critical that you become familiar with your school, company, or national standards regarding dimension and text styles.

Drawing Display

Throughout the life of a drawing, you typically change the display hundreds of times. This means that you zoom in on an object to get closer, or pan around the drawing to see something that is off the screen.

ZOOM

A command such as ZOOM contains several options allowing you to increase and decrease the displayed size of the drawing. For example, you can place a zoom window around an area that needs to be enlarged. After the zoom window is created, the entire screen is filled with the area that was inside the window. This command is useful when working in a detailed area of a drawing. Once a new window is displayed, the command can be used again and again to get even closer to the object or feature. Figure 3.21 illustrates how ZOOM works. Do not confuse zoom or any other view tool with the commands used to edit a drawing. View tools allow you to look at a drawing in a variety of ways but do not actually modify the size, location, or shape of drawing objects.

3-D Views

Display options for 3-D models are endless. If you work with 3-D models, it is important to learn the different methods for 3-D display. One benefit of a 3-D model is it can be displayed in a variety of formats. For example, a single drawing can contain standard dimensioned 2-D views, plus one or more 3-D views of the model at different scales using different shading and coloring techniques. Several different 3-D views can be displayed at different scales on a single drawing.

For example, notice the two views of the truck shown in Figure 3.22. The left side view is a plan or top view, and a 3-D view is on the right side.

Drawing Queries

A query is a question. For example, typical drawing queries involve finding the distance between two points, the area of an object, the length of a line or object, or the diameter of a circle. These functions are often referred to as inquiry commands. An

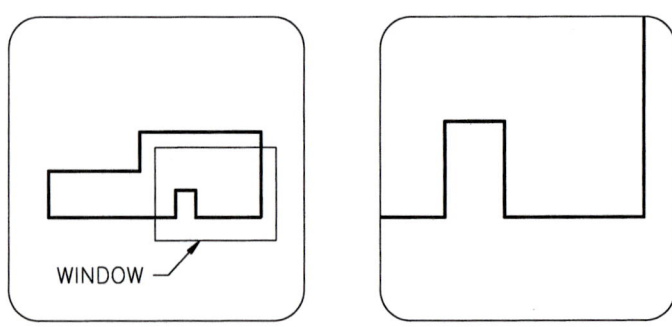

WINDOW

FIGURE 3.21 ■ The ZOOM command brings the object, or a feature of the object, closer.

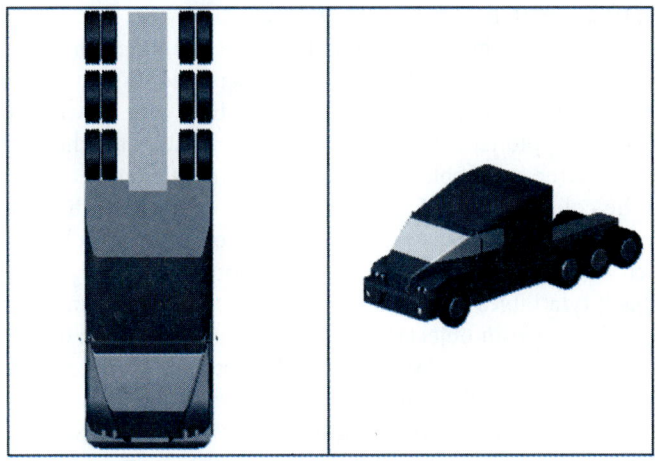

FIGURE 3.22 ■ The left-side view is a plan, and the coordinate location of the viewer is X=0, Y=0, Z=1. The coordinate location of the viewer is X=4, Y=4, Z=2 in the 3-D view on the right side.

example of an inquiry command is finding the area of an object, such as a parking lot.

If you know the project you are working on requires the calculation of areas, it is best to create the objects in such a manner that the area can be calculated quickly. Additionally, inquiry commands can be used to check your work for accuracy. It is good practice to learn how to check your drawings for accuracy. Less time required for error correction later in the project leads to greater efficiency and productivity.

CADD Plotting Guidelines

A drawing created in CADD exists in two forms: soft copy and hard copy. Soft copy is the electronic version of the drawing that you see on the computer screen, or the actual data file. Hard copy is a physical drawing produced on paper or some other media by a printer or plotter.

Scale Factors

A scale factor is a numerical value used in the proper scaling of text, dimension objects—such as dimension text and arrowheads or slashes—and the size of the model limits. When drawing in CADD, the objects are always full scale. For example, a 50'–0" × 30'–0" (15240 = 9144 mm) building is drawn 50 × 30' After the building is drawn, the notes and other text are added. If you use standard text height, then the text needs to be plotted 1/8" (3 mm) high on the drawing. If 1/8" text is placed into the 50 × 30' building, the text on the computer screen and on the plot is too small to read. This is where the scale factor becomes important. Scale factors were introduced earlier in this chapter.

The scale factor of a drawing is something that you need to plan in advance and is part of your CADD template. As discussed earlier, a template is a file containing standard settings such as layer names, dimension and text styles, and line types applied to the drawing. To make the text readable on the plotted paper, the desired plotted text height of 1/8" is multiplied by the scale fac-

tor to determine the height of the text when in the drawing. The scale factor is also used to scale dimension features.

The scale factor can be determined by first deciding the drawing scale for the plotted drawing. The scale factor is then determined by dividing the plotted units into the drawing units. The following formula demonstrates this concept:

Plotted Units = Drawing Units

Plotted Units *Drawing Units* = Scale Factor

A mechanical drawing to be plotted at a scale of 1:2, which is HALF scale, has a scale factor calculated as follows:

1:2 scale is 1/2" = 1"

.5" = 1"

1 ÷ .5 = 2

The scale factor is 2.

A mechanical drawing to be plotted at a scale of 2:1, which is DOUBLE scale, has a scale factor calculated as follows:

2:1 scale is 2" = 1"

.2" = 1"

1 ÷ 2 = .5

The scale factor is .5.

An architectural drawing to be plotted at 1/4" = 1'−0" has a scale factor calculated like this:

1/4" = 1'−0"

.25" = 12"

12/.25 = 48

The scale factor is 48.

Once the scale factor has been determined, calculate the actual text height to be used in the CADD program. If the desired text height is .125", multiply the plotted text height by the scale factor. For a mechanical drawing plotted at a scale of 1:2, the text height is:

1:2 scale is 1/2" = 1"

.125" × 2 = .25"

.125" high text on the plotted drawing is drawn .25" high.

For the 1/4" = 1'−0" architectural scale drawing the text height is:

.125" × 48" = 6"

.125" high text on the plotted drawing is drawn at 6" high

Common scale factors are provided in Appendix CC.

The Model and Sheet Layout

CADD files are created as models that are placed on sheets. AutoCAD, for example, refers to the place where drawing objects are created as model space, and the sheet is called layout space or paper space. Layout space is an area used to lay out the sheet of paper to be plotted. Everything you draw in CADD is drawn full scale in model space. When you enter layout space, a real-size sheet of paper is placed over the top of the model space drawing. Border line, title blocks, and notes are also found on the real-size paper layout space. The model space drawings and the layout space sheet are put together for plotting. In Figure 3.23, the drawing in model space is combined with the layout space sheet. When setting up the drawing to be plotted, multiple models can be arranged on the layout space sheet. For example, a foundation sheet can have the foundation plan at one scale and separate details at another scale as shown in Figure 3.24. Each CADD program has its own model file and sheet layout format. Many of these are covered throughout this text. Refer to your CADD Help files or related textbook for specific application.

Getting Ready to Plot

When the layout is prepared as desired, the sheet can be set up for plotting. During this process you can select plotter settings, pen settings, and paper size. Each layout can have a unique sheet setup. For example, you can create two layouts for a floor plan. One layout can be assigned a 36 × 24" C-size sheet and another can be assigned an 8 1/2 × 11" A-size sheet. This allows you to maximize the plotting possibilities with the same drawing. Your CADD program generally has a sheet or page setup dialog box where plotter configurations can be selected. If you use the Microsoft Windows operating system where printers and plotters are configured, they also show up on the list.

After the plotter is selected, you can specify the desired pen settings. To do this, a plot style table, for example, can be assigned to the layout. A **plot style table** is a pen configuration that allows you to have complete control over how your drawing appears when plotted. Two common plot style tables are the color table and the style table. **Color tables** assign different pen values to the colors in the CADD program. For example, if the color Red is assigned a heavyweight pen that plots with a black color, then any objects with a red color are plotted with heavy black lines. **Style tables** are names with pen settings that can be assigned to layers or objects. For example, a style table can include a named style called Walls, which has been assigned a heavy black line. The Walls pen style can then be assigned to the Wall layer in the drawing. Anything that is drawn on the Wall layer then plots with a heavy black line. After the desired plot style table is set up, you are ready to send the drawing to the plotter. Standard applications of line weights and colors were discussed earlier in this chapter. Additionally, line weights are covered in Chapter 7 and throughout this book where they apply to specific drawing features. There are likely more settings required for your specific CADD program, such as AutoCAD, MicroStation, or Pro/Engineer. Carefully review the Help files, or an in-depth textbook for your CADD software for detailed information.

It is also possible to plot your drawing to an electronic file. This is often done to allow your drawing to be published to the Internet and viewed through a Web browser. Each CADD program has a format or utility for sharing and viewing files.

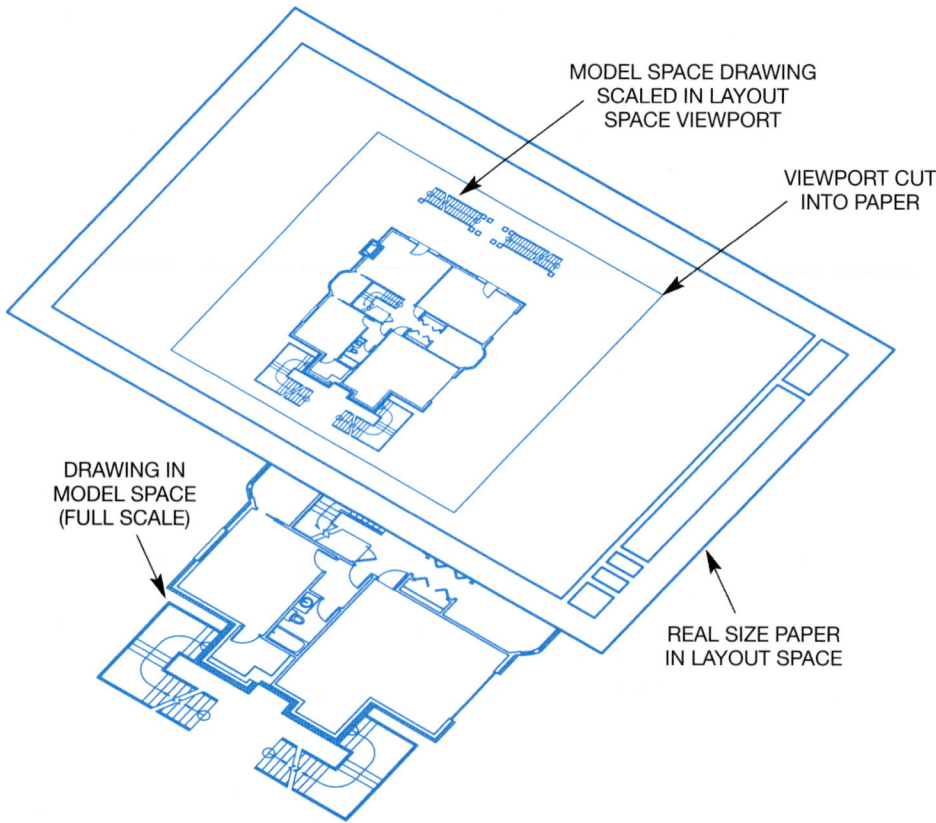

FIGURE 3.23 ■ The drawing in model space being combined with the layout space sheet.

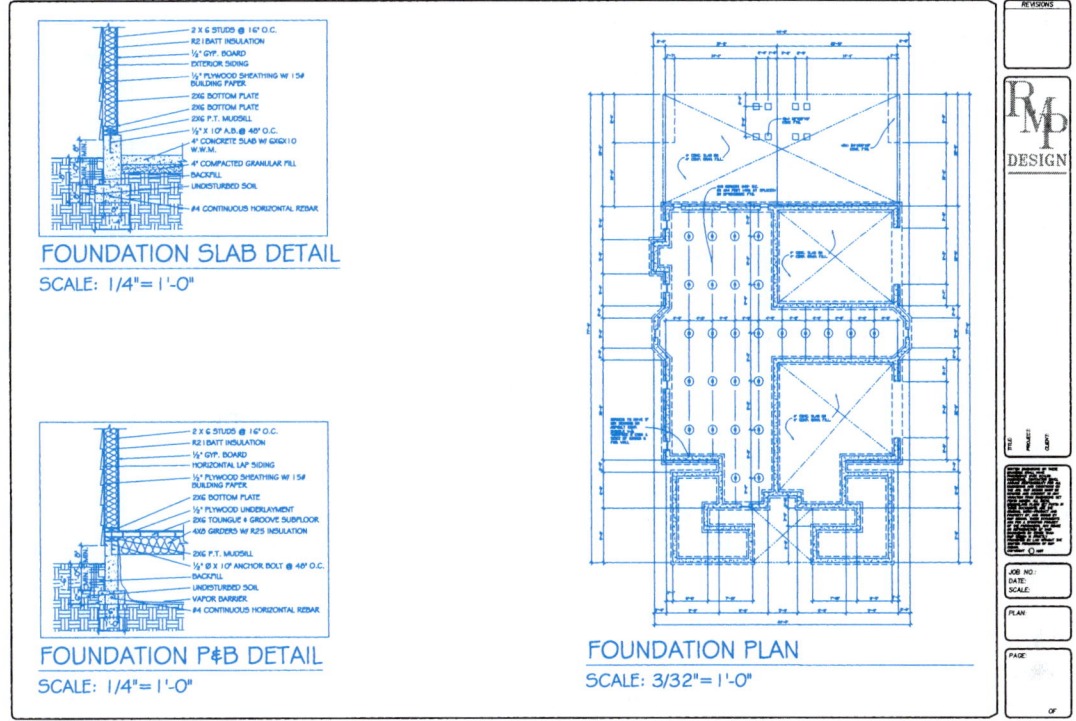

FIGURE 3.24 ■ A foundation sheet with the foundation plan at one scale and separate details at another scale ready to be plotted.

DRAWING WITH THE COMPUTER

Creating good drawings with a CADD system involves your ability to learn and remember, hand-to-eye coordination between the input device and the screen, your ability to understand and visualize, and your typing ability, although this varies among CADD systems.

In this section you will learn how you use these skills to create drawings. The intent is to describe the general concepts, techniques, and commands used when working with any computer drafting system. A familiarity with the basics of CADD allows you to more readily understand the application of any system. Slightly different coordinate entry procedures are applied depending on the CADD program you are using. The following discussion provides general information and specific applications where needed for demonstration purposes. Confirm the point entry input format used for your CADD program by referring to the related Help files or textbook.

The Cartesian Coordinate System

Using Absolute Coordinates

The Cartesian coordinate system is a rectangular coordinate system that locates a point by its distances from intersecting, perpendicular planes. Drawings done with a computer are based on the Cartesian (rectangular) coordinate system, so it is critical for you to develop a solid understanding of its features.

The two-dimensional rectangular coordinate system is illustrated in Figure 3.25. The point of intersection of the dark ver-

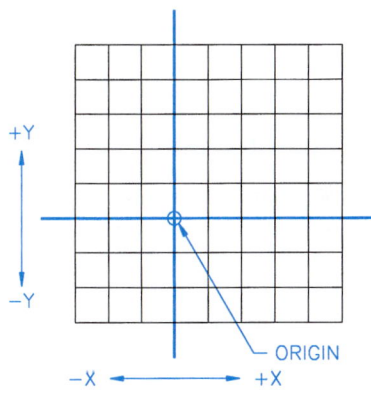

FIGURE 3.25 ■ Two-dimensional Cartesian (rectangular) coordinate system.

tical and horizontal lines (planes) is called the origin. The origin has a value of zero. Values along the horizontal increase as you move to the right of the thick vertical line. This is called the X axis. The Y axis is the vertical line and values increase as you move upward from the origin. The values just mentioned are positive. Note in Figure 3.25 that each axis can also have negative values. Negative X values are to the left of the vertical plane, and negative Y values are below the horizontal plane. Study this figure carefully before continuing.

The best way to understand the rectangular coordinate system is to work with a piece of ten-lines-to-the-inch graph paper. Locate the intersection of any two heavy perpendicular grid lines on the graph paper. Mark this point as the origin. Both X and Y have values of zero at the origin. At each inch mark on the horizontal line to the right of the origin, write the values 1, 2, 3, 4, and so on. Do the same for the inch lines above the origin along the vertical

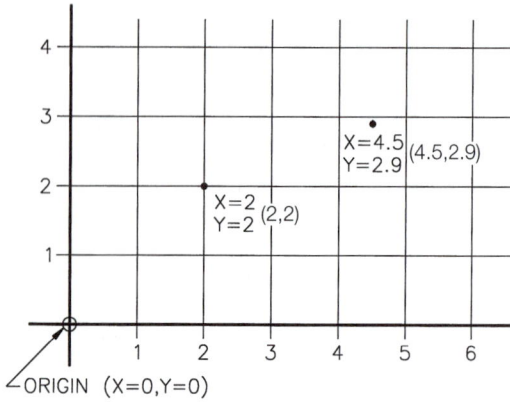

FIGURE 3.26 ■ Points located using absolute coordinates.

axis line. (See Figure 3.26.) Now locate the point of X=2, Y=2. First, count two inches to the right on the X axis, then count two inches straight up on the Y axis. This point has the value of X = 2, Y = 2. Only one point in this coordinate system can have that value. Coordinate locations are always listed with the X value given first; for example, the previous point is written 2,2.

Not all points have even inch values. Locate the point X=4.5, Y=2.9 on your grid paper. This point is located directly from the origin and not from any other point. This point is written as 4.5, 2.9. Count the grid squares along the X and Y axes to find this point, or use a scale and measure. Remember that every point must have at least two coordinate values. A point of just 4.5 can be any location 4.5 in. from the X plane.

Using Relative Coordinates

Objects can also be drawn using a moving or relative origin. In this method, each point becomes the origin and the next point is located in relation (relative) to the previous point. This method is called relative coordinates. If each point becomes the origin for the next point, then numbers to the left and below the origin are negative. See Figure 3.27 for an example of this method.

The following gives the coordinates used to locate the points A, B, C, and D in Figure 3.27:

Point	Coordinate System		Coordinates
A	Absolute Coordinates	1,1	1 unit on X axis, 1 unit on Y axis from Origin.
B	Relative Coordinates	@2,0	2 units on X axis, 0 units on Y axis from Point A.
C	Relative Coordinates	@2,2	2 units on X axis, 2 units on Y axis from Point B.
D	Relative Coordinates	@–3,–1	–3 units on X axis, –1 unit on Y axis from Point C.

The relative coordinate method is useful when using the dimensions of each line to lay out an object. It is important that you

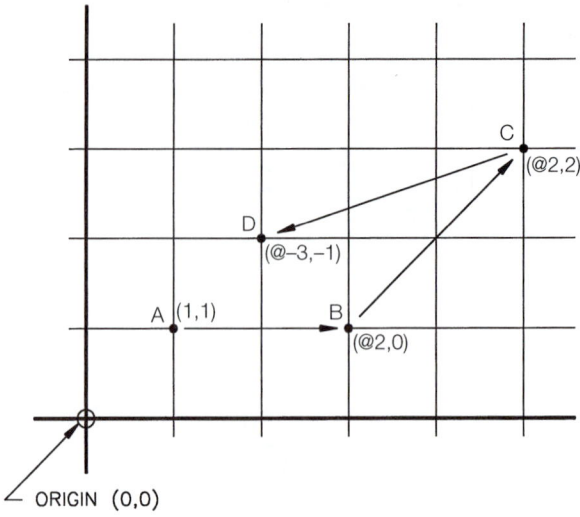

FIGURE 3.27 ■ Points located using relative coordinates.

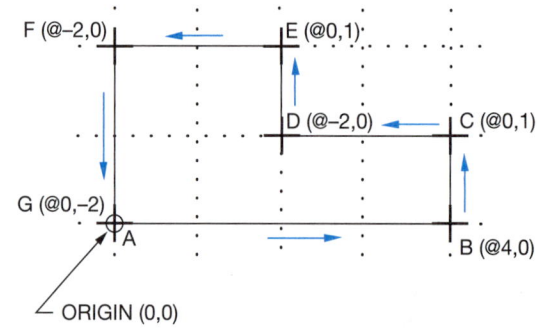

FIGURE 3.28 ■ Object drawn using relative coordinates.

keep in mind the positive and negative aspects of this method, because each point becomes the origin of a coordinate system. For example, moving to the left of a point gives the next point a negative X value. The object in Figure 3.28 shows the relationship of points on an object constructed with relative coordinates.

The following gives the coordinates used to locate the points A, B, C, D, E, F, and G in Figure 3.28:

Point	Coordinate System		Coordinates
A	Origin	0,0	
B	Relative Coordinates	@4,0	4 units on X axis, 0 unit on Y axis from Point A, the Origin.
C	Relative Coordinates	@0,1	0 units on X axis, 1 unit on Y axis from Point B.
D	Relative Coordinates	@–2,0	–2 units on X axis, 0 units on Y axis from Point C.
E	Relative Coordinates	@0,1	0 units on X axis, 1 unit on Y axis from Point D.
F	Relative Coordinates	@–2,0	–2 units on X axis, 0 units on Y axis from Point E.
G	Relative Coordinates	@0,–2	0 units on X axis, –2 units on Y axis from Point F.

For example, when using AutoCAD, relative coordinate values must be preceded with the @ symbol and entered at the keyboard as follows:

@2,2

This entry locates a point two units on the X axis and two units on the Y axis from the previous point.

Using Polar Coordinates

Not all points can be located properly with a rectangular coordinate system. For example, a point can be at a 45° angle from horizontal at a radius distance of 1.5 in. from a reference point. Such a location defines a point by polar coordinates. Notice in Figure 3.29 that the horizontal line has a value of 0°. Angles are measured in a counterclockwise direction from horizontal. The radius distance R is measured from a specified reference, or center point.

A point located using polar coordinates is based on the distance from a fixed point at a given angle and can be absolute polar coordinates measured from the Origin, or relative polar coordinates measured from the previous point. When preceded by the @ symbol, a polar coordinate is measured from the previous point. If the @ symbol is not included, the coordinate is located relative to the origin. When entering a polar coordinate in AutoCAD, for example, the < symbol separates the two values and establishes that a polar or angular increment follows. Polar coordinates are entered at the keyboard like this:

Absolute polar coordinate: 1.5<45

Relative polar coordinate: @1.5<45

Figure 3.30 is the same object drawn in Figure 3.28, but the values given are polar coordinates. The first number at each corner of the object is the radius, and the second number is the angle. Study this object until you are familiar with the concept of polar coordinates.

The following gives the coordinates used to locate the points A, B, C, D, E, F, G, and H in Figure 3.30:

Point	Coordinate System	Coordinates	
A	Origin	0,0	
B	Absolute Polar Coordinates	2<45	2 units from 0,0 at 45 degrees.
C	Relative Polar Coordinates	@4<0	4 units from B at 0 degrees.
D	Relative Polar Coordinates	@1<90	1 unit from C at 90 degrees.
E	Relative Polar Coordinates	@2<180	2 units from D at 180 degrees.
F	Relative Polar Coordinates	@1<90	1 unit from E at 90 degrees.
G	Relative Polar Coordinates	@2<180	2 units from F at 180 degrees.
H	Relative Polar Coordinates	@2<270	2 units from G at 270 degrees.

Using Three-Dimensional (3-D) Coordinates

To construct a three-dimensional (3-D) coordinate system, you need to add a third dimension, the Z axis. The third axis, Z, projects perpendicular from the drawing surface. This means a positive Z value projects toward you from the surface of the display screen when viewing the object from the top or plan view. (See Figure 3.31.) The intersection of these three axes forms the origin, which has the numerical value of X = 0, Y = 0, and Z = 0 (0,0,0). Any point behind the origin on the Z axis has a negative value.

The object shown in Figure 3.32 is in wireframe form, and the X, Y, Z coordinate values of each point are indicated. The first number of each set is X, the second Y, and the third Z. Study the object in Figure 3.32 until you understand the technique to establish points in three dimensions (3-D). Figure 3.33 shows how to locate a specific 3-D coordinate point.

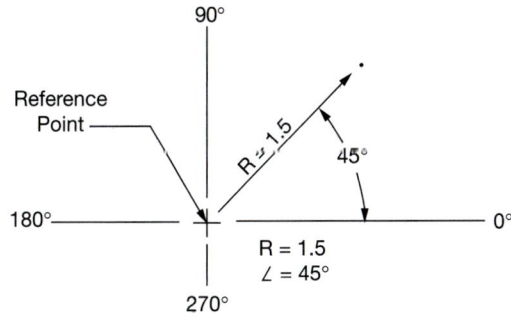

FIGURE 3.29 ■ Locating a point using polar coordinates.

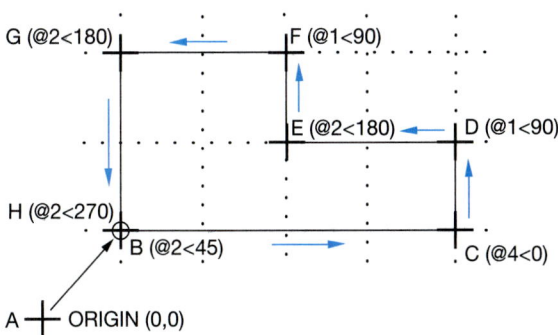

FIGURE 3.30 ■ Object drawn using polar coordinates.

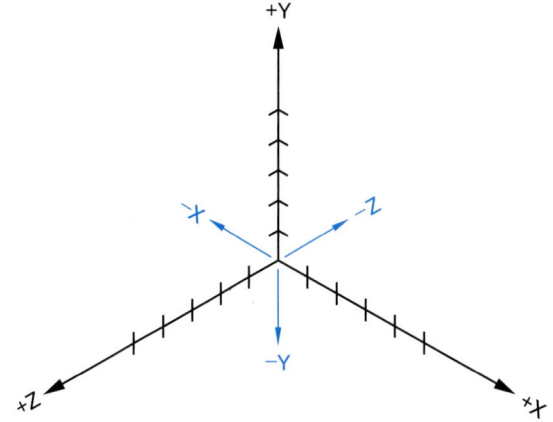

FIGURE 3.31 ■ Three-dimensional coordinate system.

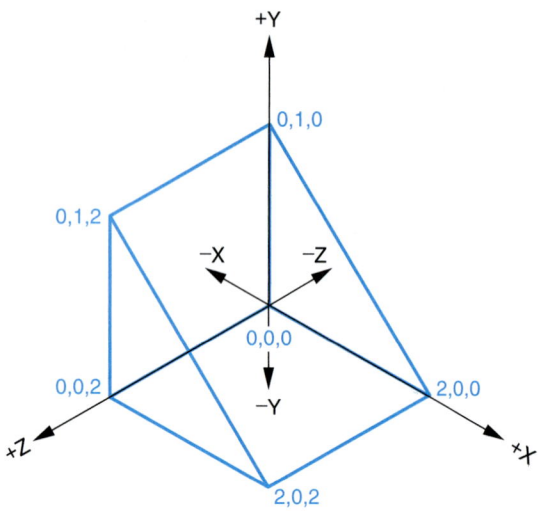

FIGURE 3.32 ■ Object plotted using 3-D coordinate system.

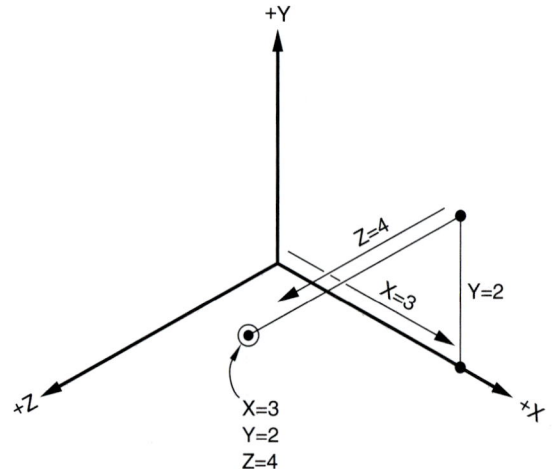

FIGURE 3.33 ■ Locating a point using 3-D coordinates.

Notice in Figure 3.32 that the coordinate system can accommodate negative coordinates. An object can be drawn in any or all of the four quadrants. An object can just as easily be drawn using all negative coordinates as positive ones. Most drafters work with positive coordinates unless it is more convenient to locate the origin in the center of the object or a feature.

Three-Dimensional and Parametric Solid Modeling

The previous discussion covered the basics of 3-D coordinates used to draw basic three-dimensional objects. Creating a 3-D representation of a design can help overcome visualization problems and produce a realistic, testable product model. There are three different 3-D modeling techniques including wireframe, surface, and solid modeling. These types of models and their applications are discussed in detail with numerous examples in Chapter 4. The following provides a brief introduction. Wireframe models are the most basic 3-D CADD models, and contain only information about object edges and the intersection of edges. Like wireframe models, surface models contain

information about object edges and the intersection of those edges, but go a step further to include information about object surfaces. Similar to surface models, solid models contain information about object edges, the intersection of those edges and surfaces, and also include data about object volume and mass.

Parametric Solid Modeling

Parametric solid modeling involves the basic idea of developing solid models that contain parameters, which are controls, limits, and checks that allow you to easily and effectively make changes and updates. This means that when you describe the size, shape, and location of model geometry using specific parameters, you can easily modify those specifications to explore alternative design options.

The parametric concept is also referred to as intelligence, because parametric modeling occurs as a result of the software program's ability to store and manage model information. This information includes knowledge of every model characteristic such as calculations, sketches, features, dimensions, geometric parameters, when each piece of the model was created, and all other model history and properties.

There are several parametric solid modeling CADD software programs available to the designer and drafter. Autodesk Inventor, Pro/Engineer, and SolidWorks are three of these applications. Many parametric solid modeling programs are surprisingly similar in the way they function. In fact, once you learn the basic process of creating a model using specific software, such as Autodesk Inventor, you can usually effectively transition to different parametric solid modeling software, such as SolidWorks.

Although some designs can contain a single part, such as a bracket, most products can consist of at least two parts, such as a bracket and a bolt, while other products can contain hundreds of parts and subassemblies in an assembly. An assembly is a grouping of one or more design components. Components can include part models and subassemblies. Subassemblies are groups of components that are often standard product items and can be used several times in various assemblies. For example, subassemblies can include items like wire harnesses or switches, which are composed of two or more parts, and may be used more than once in the final assembled product.

In a parametric solid assembly mode, multiple parts and subassemblies are brought together and mated to form a complete assembly. Mates are used in the assembly environment, in which they link individual components together to create the assembly. Once the assembly of a mechanism is created, you can observe the movement of components in an assembly. A crank shaft and piston movement is shown at three different positions in Figure 3.34. You can drive a model to show the amount of movement between components, pause movement, and detect collisions between components. A good example of a driven model can be seen with the assembly shown in this model.

CREATING AND USING SYMBOLS

One of the great time-saving features of CADD technology is the ability to draw a feature once, save it, and then use it over and

FIGURE 3.34 ■ A crank shaft and piston movement shown at three different positions. You can drive a model to show the amount of movement between components, pause movement, and detect collisions between components. A good example of a driven model can be seen with the assembly shown in this model. *Courtesy of PTC.*

over without ever having to draw it again. That is why one of the first things a company does after purchasing a CADD system is to begin developing a symbol library. A symbol library is a collection of symbols that can be used on any drawing.

Drawing a Symbol

The process of creating a symbol is easy. To create a symbol, begin a drawing with the name of the symbol as the file name. When you start the actual drawing of the symbol, begin by placing the origin of your coordinate system at a spot on the symbol that would be convenient to use as a point of reference or location on a drawing. Look at the symbols in Figure 3.35. These symbols have small circles at their location points. This is the spot you point to on your drawing to locate the symbol. This point is called the insertion point.

When the symbol has been drawn, it must then be stored for future use. Applications can differ with each CADD program, but symbol, feature, part, and subassembly libraries are possible options. Confirm the specific command or terminology with your CADD program Help files or textbook.

Creating a Symbol Library

After all of the required symbols are drawn, they can be used in a variety of ways. It is important that all the symbols be located in a common area, such as a directory on the hard drive, or a

drive or folder on the network server. If you have the ability to customize the software's menus, the symbols can be added to pull-down menus, a toolbar, or a tool palette for easy access. These methods of locating the symbols in menus allow them to be picked and then placed in the drawing, eliminating the need to type file names.

Using Symbols

Most of the editing features used to revise drawings can be used to edit symbols. In addition, there are several commands that allow you to work with symbols.

Symbols are fluid just like the rest of the drawing and can be changed, moved, renamed, and copied. Commands such as ROTATE, COPY, MIRROR, and MOVE all can be used to work with symbols.

REUSING DRAWING CONTENT

In virtually every drafting discipline, individual drawings created as part of a given project are likely to share a number of common elements. All of the drawings within a specific drafting project generally hold the same set of standards. Drawing features such as the text size and font used for text, standardized dimensioning methods and appearances, layer names and properties, drafting symbols, drawing layouts, and even typical drawing details are often duplicated in many different drawings. These and other components of CADD drawings are referred to as drawing content. One of the most fundamental advantages of computer-aided drafting systems is the ease with which content can be shared between drawings. Once a frequently used drawing feature has been defined, it can be used again as needed, in any number of drawing applications.

Drawing templates represent one way to reuse drawing content that has already been defined. Creating your own customized

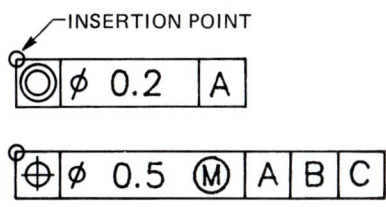

FIGURE 3.35 ■ Location or insertion points on symbols.

drawing template files provides an effective way to start each new drawing using standard settings. However, drawing templates provide only a starting point. During the course of a drawing project, you often need to add content to the current drawing that has been defined previously in another drawing. Some drawing projects require you revise an existing drawing rather than start a completely new drawing. For other projects, you may need to duplicate the standards used in a drawing supplied by a client. With most CADD programs, content that has already been defined in previous drawings can be copied and reused in other drawings with a simple **drag-and-drop** operation. Drag-and-drop is a feature that allows you to perform tasks by moving your cursor over the top of an icon in a folder, or menu source. The icon represents a document, drawing, folder, or other application. Next, press and hold the pick button on your pointing device, followed by dragging the cursor to the desired folder or drawing and release the pick button. The selected content is added to your current drawing or chosen file.

Several categories of drawing content, including symbols, dimension styles, layers, layouts, line types, text styles, and externally referenced drawings, can be applied in this manner.

An example of a CADD program feature used to share drawing content is the AutoCAD DesignCenter. Figure 3.36 shows the DesignCenter with the variety of folders containing drawing content for different drafting disciplines. In this example, the Fasteners folder is selected on the left and its contents are displayed on the right side.

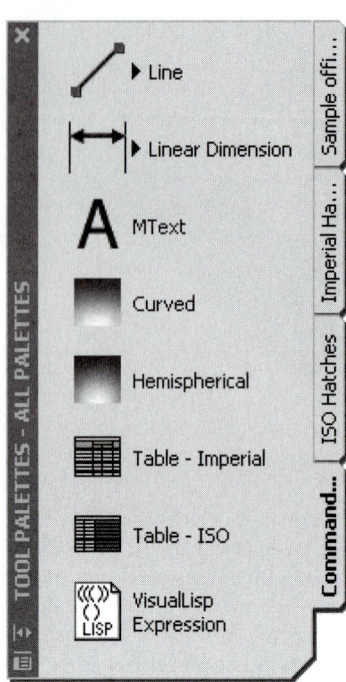

FIGURE 3.37 ■ Sample tool palettes are included with AutoCAD.

CADD programs also have features such as tool palettes and toolbars that contain often-used block symbols and patterns for use in a drawing. **Toolbars** contain various buttons that activate commands. Tool palettes are customizable with symbols and patterns to suit your own specific needs.

These different options provide access to commonly used commands, symbols, and patterns. These palettes can be customized with your own symbols and patterns, and new palettes can also be created. An example is the AutoCAD tool palette shown in Figure 3.37.

STORING A DRAWING

The storage area for the computer drafting system is the hard drive or other storage device. Permanent storage is best done using optical disks or other media. The disk drive does the actual storing of your drawing on the disk. It can also read the data from the disk and send it to the computer.

The act of storing a drawing that is on the screen is handled by menu commands such as SAVE and SAVE AS. After selecting the proper command, you are given a chance to change the file name. After that, pressing the ENTER key saves the drawing to the folder where you want the drawing stored.

Drawings should be saved often. This could be a matter of personal preference or company policy. Practice varies, but it is a good habit to save your work every 10 to 15 minutes, just in case something happens to the computer. It is better to lose 10 minutes of work rather than the work you have done all day.

FILE MANAGEMENT

The term **file management** refers to the methods and processes of creating a management system, then storing, retrieving, and

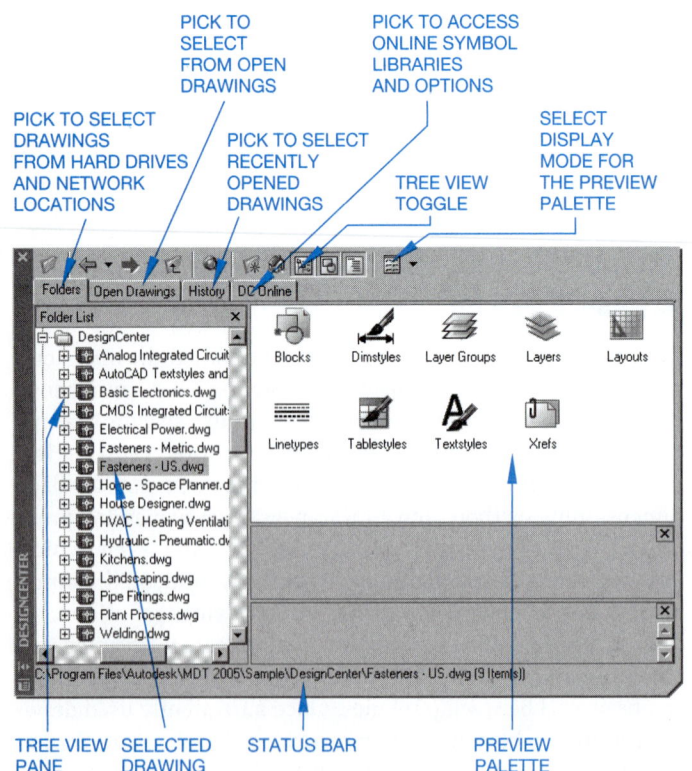

FIGURE 3.36 ■ DesignCenter is used to copy content from one drawing to another.

maintaining the files within the guidelines of the computer system. From the CADD drafter's viewpoint, this can be as simple as opening a drawing file and saving it in the same location. The drafter may also be responsible for creating a new hard disk directory tree made up of folders and subfolders for new projects, and ensuring that a variety of file types are stored in their proper locations.

Be aware that a wide range of file management systems is used in industry. The important thing for you to remember is to become familiar with the system used in your school or company. One of your goals should be to handle your work in the most efficient and productive manner possible. Proper file management procedures can help you reach this goal regardless of the project.

Rather than describe a specific type of file management system, it is more important for you to be familiar with some of the aspects of file management and maintenance procedures. Proper file management should include the following:

- Save all working files on a regular basis. When you save a file, save it to at least two different media, such as the local hard disk and a flexible or optical disk. In addition, establish automatic safeguards for saving your work on a regular basis. Most CADD programs have an automatic save function to save your drawing periodically.

- Create storage media directories and folders for all files.

- Never save data files such as CADD drawings, text documents, and spreadsheets to folders that contain application programs.

- Create separate folders for each project, and subfolders for different types of files in the project.

- Template file originals should be stored in a protected location that allows only selected individuals to edit and save these important files.

- Separate drawing files from other types of files such as text documents, spreadsheets, and database files.

- Maintain off-site locations for file storage and archiving for safekeeping.

- Establish a regular schedule for deleting all backup, temporary, and unnecessary files from your storage media.

- If you have hard disk maintenance software on your computer, set it to defragment your hard disk on a regular basis. When application software is used, and files are opened and saved, portions of the files are saved in different physical locations on the hard disk and storage media. As more of this defragmentation occurs, it can begin to slow the hard drive's access to the files. Disk maintenance software can rearrange software and data on the hard disk and reclaim space occupied by deleted files.

- Install only licensed software on your computer system. Always report any unlicensed software to your systems manager.

- Never install software without approval from your instructor, supervisor, or systems manager.

Network Systems

The network **server** is the computer that acts as the manager, or control center, for the network. Drives, directories, and folders for software applications and data files must be maintained on the server. Schools and companies using computer networks are faced with a much more complex file management task. In addition to maintaining all of the software and data files that are required on the local computers, a variety of applications and data backup locations must be managed on the network server. In addition, the server allows many users to be connected to peripheral equipment such as plotters, scanners, and other reproduction facilities. (See Figure 3.38.)

A network that connects the computers within an office or company is called a **local area network (LAN)**. A network that connects computers in widely scattered locations is called a **wide area network (WAN)**. Within these systems, there are also electronic communication services such as e-mail and local office intranet systems, Internet connections, file security systems, and computer virus detection applications. Therefore, even though you may not be responsible for the management and maintenance of these systems, it is important that you know how they function. This allows you and others who use the network to work more efficiently and also makes the network administrator's job easier.

THIS IS WHAT THE FUTURE LOOKS LIKE: WEB-BASED COLLABORATION

The design and drafting industry is beginning to see the development of technologies that allow teamwork over the World Wide Web (WWW). This technology has major implications for the evolution of the practice. Design and drafting is a cooperative process involving multiple disciplines. One of the most time-consuming and risky aspects of the practice is document management. Hundreds of drawings and documents can be issued during the design process. All of these documents can be in various states of completion. The coordination of this information is a big task. Web-based coordination uses the power of sophisticated database management systems in conjunction with the Internet to simplify and streamline document management and team communications. These products allow designers, drafters, product suppliers, manufacturers, contractors, and owners to communicate and coordinate throughout the entire project, regardless of where each participant is located.

This changing technology also has implications for the way design projects are produced. Web-based coordination allows for the increased use of outsourcing. **Outsourcing** means sending parts of a project out to subcontractors for completion. Outsourcing to multiple resources requires consistency in procedure and similarity in tools. This interactive teamwork of the future requires clear measurement of progress. The collaboration tools now being developed give designers a higher level of communication and control over the production process. E-mail allows tracking throughout the project.

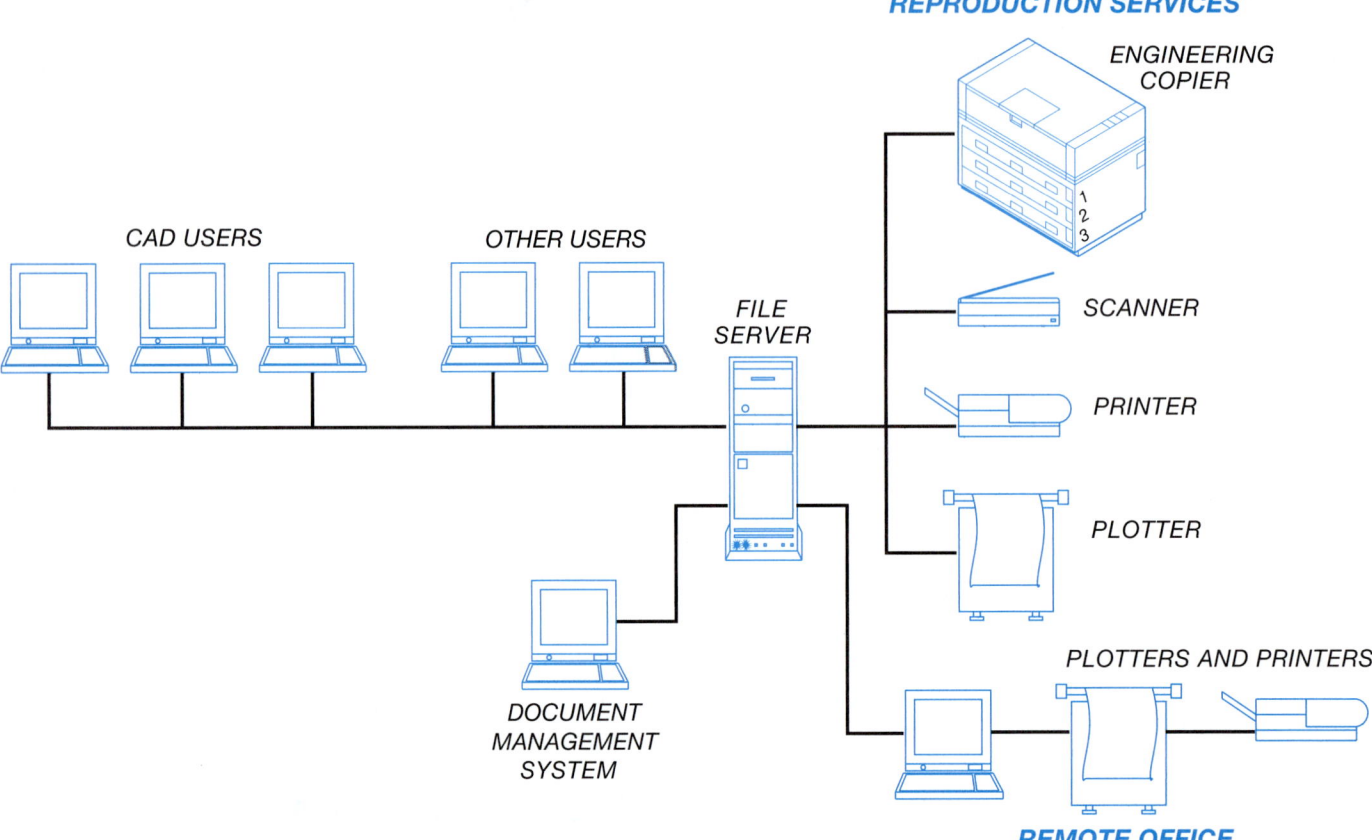

FIGURE 3.38 ■ A typical network system in an architectural office.

The outsourcing model of the future is based on defined tasks rather than hours worked. These discrete tasks are based on interrelated logical steps rather than drawing sheets. Drawings move forward on the basis of their relationship to other drawings and the available information. The development of drawings and documents is treated as stepped tasks or slices of time. Outsource personnel are independent contractors responsible for managing their own time rather than punching a clock. These professional subcontractors are found all over the world, and they are connected through the WWW. Through this huge network of potential workers, it is possible to find professionals who produce high-quality drawings on schedule and economically.

Web-based collaboration also affects the relationship of the designer and the client. Because the work product is visible on the Internet, the owner can view its progress at any time. Progress on the documents is obvious, and the owner can involve more people in reviewing the work. This model of practice enhances the owner's involvement in the process and improves the sense of overall teamwork.

Research Techniques

Within the framework of new Web-based CADD, the ability to conduct thorough and detailed research is enhanced. Having access to resources throughout the world, information on any topic is often just a few minutes away. Therefore, it is suggested that you take time to study the techniques of conducting good Internet research. Several website addresses are provided at the end of this chapter to assist you in this effort. Do not forget that it is still important to know how to use traditional resources such as libraries, print media, and professional experts in the area of your study.

Even though the Web can be an inviting method of conducting research, it can also be time-consuming and nonproductive if it is not used in a logical manner. Any research should be conducted using a simple process assists you in finding information quickly and efficiently. Keep these items in mind when conducting research:

■ Define and list your topic, project, or problem.

■ Identify key words of the topic.

■ Identify all resources with which you are familiar that can provide information.

■ Using the Internet, conduct quick keyword searches on your topic. Write down the most likely websites and links to your topic, or copy the site locations to a text file in your word processing software. Avoid exploring sites in detail at this stage. Library websites often have excellent links for specific subject areas. Keep in mind that most periodical and professional journal documents on websites have been archived since only about the mid-1980s.

■ Check with libraries for lists of professional indexes, periodicals and journals, specialized trade journals, and reference books related to specialized topics. A wealth of information is still available in book and magazine form. Older magazine and journal articles not found on the web are often in microfiche form in libraries.

■ Contact schools, companies, and organizations in your local area to find persons who are knowledgeable in the field of your research.

■ Create a list of all potential resources. Be sure to list enough information about the resource so you are able to find it again quickly.

■ Prioritize the list by putting the most likely resources at the top. Focus first on the most likely resource to provide the information you need.

■ Begin your detailed study of the prioritized list.

Any project, whether a simple drawing or an entire set of construction plans, requires study, research, and testing. Regardless of the type of work you are doing, never forget about the rich resources and information locked within the human brain—information that is the result of many hours of study, personal experiences, professional achievements, successes, and failures. Often the knowledge to be gained from consulting professionals is far more valuable than days of Internet and library research. Always remember the most basic and simple form of research: ask questions.

ERGONOMICS

Ergonomics is the study of a worker's relationship to physical and psychological environments. There is concern about the effects of the CADD working environment on the individual worker. Some studies have found that people should not work at a computer workstation for longer than about four hours without a break. Physical problems can develop when someone is working at a poorly designed CADD workstation; these problems can range from injury to eyestrain. The most common type of injury is referred to as repetitive strain injury (RSI), but it also has other names, which include repetitive movement injury (RMI), cumulative trauma disorder (CTD), and occupational overuse syndrome (OOS). One of the most common effects is carpal tunnel syndrome (CTS), which develops when inflamed muscles trap the nerves that run through your wrists. Symptoms of CTS include tingling, numbness, and possibly loss of sensation. An RSI usually develops slowly and can affect many parts of your body. You should be concerned if you experience different degrees and types of discomfort that seem to be aggravated by computer usage.

Drafters who are required to do tedious, repetitive tasks on the computer could develop symptoms of fatigue. Consideration should be given to allowing CADD employees to do a variety of jobs or take periodic breaks.

Many factors can help reduce the possibility of RSI. These factors are outlined in the following sections.

Ergonomic Workstations

Figure 3.39 shows an ergonomically designed workstation. In general, a workstation should be designed so you sit with your feet flat on the floor, with your calves perpendicular to the floor and your thighs parallel to the floor. Your back should be straight, your forearms should be parallel to the floor, and your wrists should be straight. For some people, the keyboard should be either adjustable or separate from the computer to provide

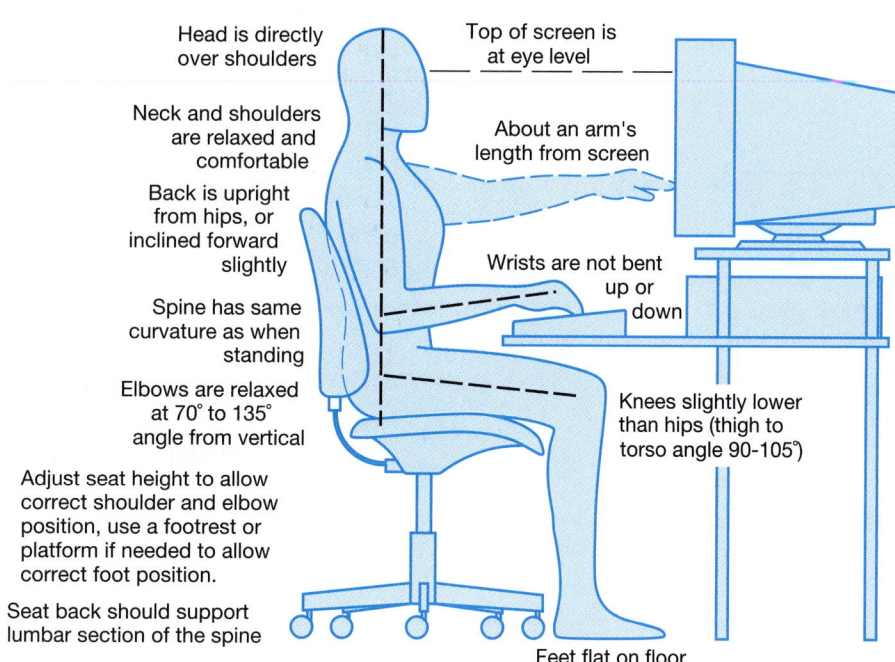

Head is directly over shoulders

Top of screen is at eye level

Neck and shoulders are relaxed and comfortable

About an arm's length from screen

Back is upright from hips, or inclined forward slightly

Spine has same curvature as when standing

Wrists are not bent up or down

Elbows are relaxed at 70° to 135° angle from vertical

Knees slightly lower than hips (thigh to torso angle 90-105°)

Adjust seat height to allow correct shoulder and elbow position, use a footrest or platform if needed to allow correct foot position.

Seat back should support lumbar section of the spine

Feet flat on floor

FIGURE 3.39 ■ An ergonomically designed workstation.

more flexibility. The keyboard should be positioned, and arm or wrist supports can also be used, to reduce elbow and wrist tension. Also, when the keys are depressed, a slight sound should be heard to assure the key has made contact. Ergonomically designed keyboards are available.

The monitor is a concern. It should be 18" to 28", or approximately one arm's length, away from your head. The screen should be adjusted to 15° to 30° below your horizontal line of sight. Eyestrain and headache can be a problem with extended use. If the position of the monitor is adjustable, you can tilt or turn the screen to reduce glare from overhead or adjacent lighting. Some users have found that a small amount of background light is helpful. Monitor manufacturers offer large, flat, nonglare screens that help reduce eyestrain. Some CADD users have suggested changing screen background and text colors weekly to give variety and reduce eyestrain.

The chair should be designed for easy adjustments, to give you optimum comfort. It should be comfortably padded. Your back should be straight or up to 10° back, your feet should be flat on the floor, and your elbow-to-hand movement should be horizontal when you are using the keyboard, mouse, or digitizer. The mouse or digitizer puck should be close to the monitor so movement is not strained and equipment use is flexible. You should not have to move a great deal to look directly over the cursor to activate commands.

Positive Work Habits

In addition to an ergonomically designed workstation, your own personal work habits can contribute to a healthy environment. Try to concentrate on good posture until it becomes second nature. Keeping your feet flat on the floor helps improve posture. Try to keep your stress level low, because increased stress can contribute to tension, which can aggravate physical problems. Take breaks periodically to help reduce muscle fatigue and tension. You should consult with your doctor for further advice and recommendations.

Exercise

If you feel the pain and discomfort that can be associated with computer use, some stretching exercises can help. Figure 3.40 shows some exercises that can help reduce workstation-related problems. Some people have also had success with yoga, biofeedback, and massage. Consult with your doctor for advice and recommendations before starting an exercise program.

The Plotter

The plotter makes some noise and is best located in a separate room next to the workstation. Some companies put the plotter in a central room, with small office workstations around it. Others prefer to have plotters near the individual workstations, which can be surrounded by acoustical partition walls or partial walls.

Other Factors

The CADD working environment can be different from traditional drafting rooms in that air-conditioning and ventilation should be designed to accommodate the computers and equipment. Carpets should be antistatic. Noise should be kept to a minimum.

WEBSITE RESEARCH

The following list of websites provides access to information on many of the CADD-related topics discussed in this chapter. Useful websites that you locate should be bookmarked in your web browser. In addition, you should create a text document that contains a list of useful websites with links to those sites. Separate the site names by categories. This list provides you with a resource that is a backup to your browser's bookmarks, and also with a file that you can add to at any time you are not working on the Internet.

http://www.nacfam.org—National Coalition for Advanced Manufacturing (NACFAM), publisher of the CADD Skill Standards.

http://www.tsc.wes.army.mil—A/E/C CAD Standard from CAD/GIS Technology Center. Order a CD of the CAD Standard.

http://www.nationalcadstandard.org—U.S. National CAD Standard.

http://www.aia.org—American Institute of Architects.

http://www.csinet.org—Construction Specifications Institute.

http://www.nahb.org—National Association of Home Builders.

http://www.adda.org—American Design Drafting Association.

http://www.autodesk.com—Autodesk, Inc., AutoCAD, and related products.

http://www.bentley.com—Bentley Systems, Inc., Microstation, and related products.

http://www.cadkey.de—AGS Advanced Graphics Software, CADKEY, KeyCreator, and related products.

http://www.graphisoft.com—Graphisoft, ArchiCAD, and related products.

http://www.ptc.com—Parametric Technology Corporation, Pro/Engineer, and related software.

http://www.softplan.com—SoftPlan System, Inc., SoftPlan, and related products.

http://www.solidworks.com—SolidWorks Corporation, SolidWorks, and related products.

http://www.3ds.com—Dassault Systemes, CATIA, and related products.

http://www.nsf.gov—National Science Foundation's engineering website.

http://www.albany.edu—University at Albany, New York State Universities. Conducting research on the Internet.

http://www.uiuc.edu—University of Illinois at Urbana–Champaign. Resources for conducting research. Pick the IRIS link for additional links to other research sources.

http://www.cad-forum.com—CAD Forum. List of professional organizations for architecture, engineering, design, CAD, and construction. Excellent links to information on every discipline of CADD.

http://www.acec.org—American Consulting Engineers Council. An organization with approximately 6,000 member firms.

Sitting at a computer workstation for several hours in a day can produce a great deal of muscle tension and physical discomfort. You should do these stretches throughout the day, whenever you are feeling tension (mental or physical).

As you stretch, you should breathe easily—inhale through your nose, exhale through your mouth. Do not force any stretch, do not bounce, and stop if the stretch becomes painful. The most benefit is realized if you relax, stretch slowly, and *feel* the stretch. The stretches shown here take about 2½ minutes to complete.

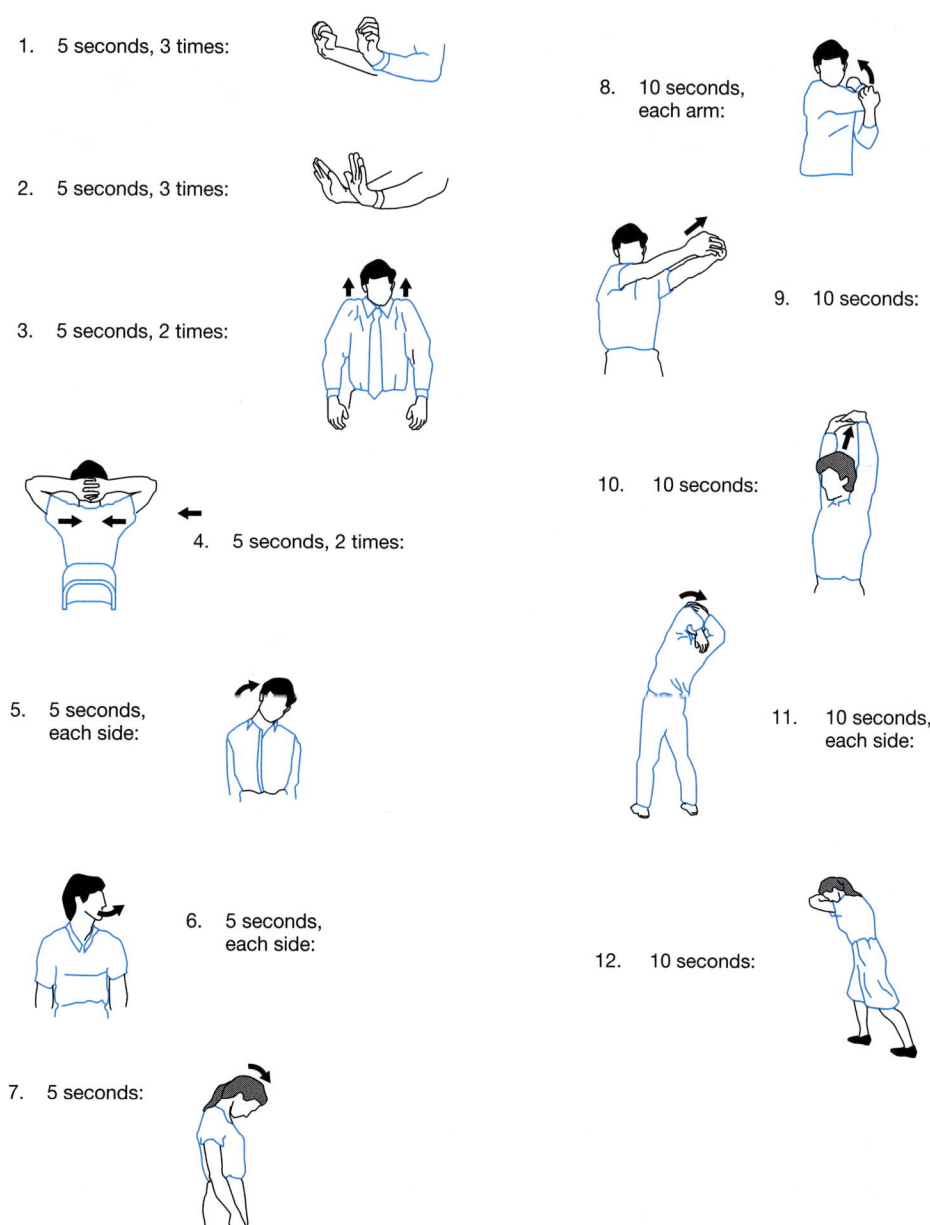

FIGURE 3.40 ■ Stretches that can be used at a computer workstation to help avoid repetitive strain injury.

http://www.buildernet.com—BuilderNet. Products and suppliers for architects, engineers, and contractors.

http://www.hoovers.com—Hoover's Online, The Business Network. "List of Lists." Links to a wide variety of companies and websites.

http://www.thomasregister.com—Thomas Register list of American manufacturers.

http://www.siriuslink.com—Sirius Digital Archiving. CAD Internet resources.

http://www.caenet.com—*Computer-Aided Engineering* magazine.

http://www.tii-tech.com—TII Technical Education Systems. CIM product information and useful links for industry, training, and educational sites.

http://www.nanozine.com—*Nano Technology* magazine.

http://www.thevrsource.com—The VR Source Newswire. Product reviews, new product listings, monthly specials, and more.

http://www.hitl.washington.edu—Human Interface Technology Labs, University of Washington, Seattle.

PROFESSIONAL PERSPECTIVE

A wide variety of jobs are available for qualified CADD professionals. Keep in mind that the kinds of tasks may not always be traditional drafting functions. In addition to creating drawings, you can be responsible for working in some of the following areas:

- Preparing freehand sketches on the shop floor or at a job site, then converting the sketch to a finished CADD drawing.

- Digital image creation and editing.

- Text documents such as reports, proposals, and studies.

- Incorporation of CADD drawings and images into text documents.

- Conducting research for job proposals, feasibility studies, or purchasing specifications.

- Evaluating and testing new software.

- Training staff members in the use of new software or procedures.

- Collecting vendor product information for new projects.

- Speaking on the phone and dealing personally with vendors, clients, contractors, and engineers.

- Checking drawings and designs created by others for accuracy.

- Researching computer equipment and preparing bid specifications for purchase.

Human resource directors who most often hire employees agree that persons who possess a set of good general skills usually become good employees. The best jobs are found by those students who have developed a good working understanding of the project planning process and can apply it to any situation. The foundation on which this process is based rests on the person's ability to communicate well verbally, apply solid math skills (through trigonometry), write clearly, exhibit good problem-solving skills, and know how to use resources to conduct research and find information.

These general qualifications also serve as the foundation for the more specific skills of your area of study. These include a good working knowledge of drawing layout and construction techniques based on applicable standards, and a good grasp of CADD software used to create drawings and models. In addition, those students who possess the skills needed to customize the CADD software to suit their specific needs may be in demand.

What is most important for the prospective drafter to remember is the difference between content and process. This was discussed previously in this chapter, but deserves a quick review. **Content** applies to the details of an object, procedure, or situation. Given enough time, you can find all of the pieces of information needed to complete a task, such as creating a drawing or designing a model. **Process** refers to a method of doing something, usually involving a number of steps. By learning a useful process, you will find it easier to complete any task and find all of the information (content) you need. It is beneficial to learn a good process for problem solving and project planning that can be used in any situation. By using the process for any task, it becomes easier to determine what content is needed.

For these reasons, it is strongly recommended that you focus your efforts on learning and establishing good problem-solving and project-planning habits. These skills make the task of locating the content you need for any project easier and contribute to making all aspects of your life more efficient, productive, and relaxing.

http://www.ai.about.com—About.com. Artificial intelligence topics and links.

http://websearch.about.com—About.com. Web searching techniques.

http://www.g-w.com—Publisher of AutoCAD related textbooks by David A. Madsen and David P. Madsen. Look for the following titles:

AutoCAD and its Applications, Architectural Drafting Using AutoCAD, Architectural Desktop and Its Applications, In-

ventor and Its Applications, and *SolidWorks and Its Applications.*

http://www.amazon.com—Search author, Madsen, for the following CAD textbook titles:

AutoCAD and Its Applications, Architectural Drafting Using AutoCAD, Architectural Desktop and Its Applications, Inventor and Its Applications, and *SolidWorks and Its Applications.*

RECTANGULAR COORDINATE SYSTEM

One of the most important aspects of learning to operate a CADD system is the rectangular coordinate system. The polar system is one form of rectangular coordinates that uses a distance and angle for measurement. Understanding angles and how they relate to each other is useful in the solution of a variety of problems. For example, you may be given a flat sheet metal part or a property plat to draw, and the sketch contains only included angles. It is up to you to convert these included angles into polar coordinates in order to draw the object. (See Figure 3.41.)

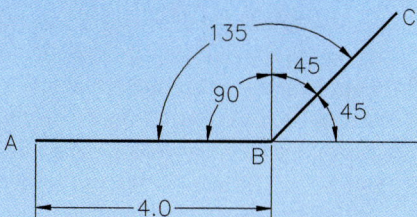

FIGURE 3.42 ■ Calculating angles for polar coordinate entry.

In order to draw to point D you must remember the rule that intersecting lines create equal opposite angles. By drawing a horizontal line through point C and projecting the previous line past the object, you see that one half of the 90° angle is above the horizontal. Therefore, the angle of the line to point D is the result of subtracting 45° from 180°. To draw a line from point C to point D, enter a relative length of 3" and an angle of 135°.

Try to complete the remainder of the object (see Figure 3.43) using the previous examples as a guide.

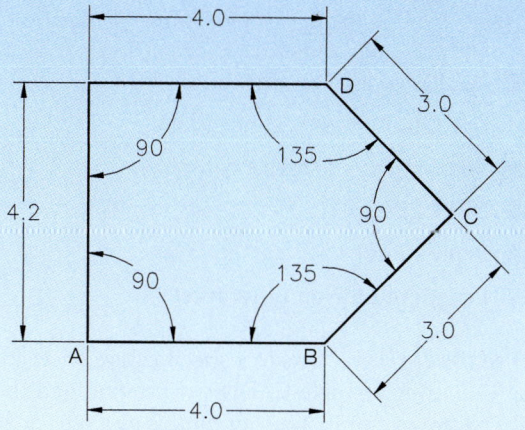

FIGURE 3.41 ■ Convert the included angles into polar coordinates.

The first line can be drawn easily by entering a length, 4", and an angle of 0°. The angle of the second line is determined by subtracting 135° from 180° (the angle of a straight line). This gives the angle of 45° measured counterclockwise from horizontal. Thus, to draw to point C, enter a relative length of 3" and an angle of 45°. (See Figure 3.42.)

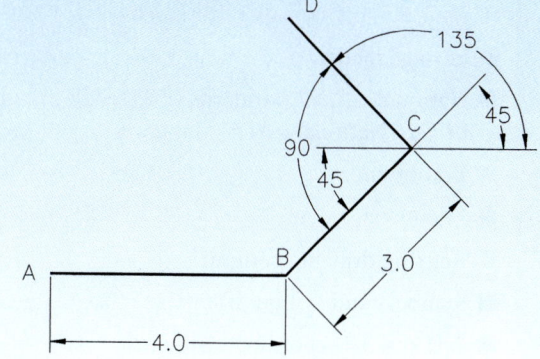

FIGURE 3.43 ■ Calculating angles for polar coordinate entry.

MATH APPLICATIONS

CHAPTER 3

Computer-Aided Design and Drafting (CADD) Test

 Access the CD found with this textbook to view the Chapter 3 Test. Confirm the

preferred submittal method with your instructor.

CHAPTER 3

Computer-Aided Design and Drafting (CADD) Problems

PROBLEM 3.1   This problem involves a written report. Choose one of the following topics listed and write a report on that subject. Be specific, especially if the manufacturer makes more than one model of CADD system. Check with your instructor about any specifics relating to your report. Research materials can be the Internet, trade magazines and journals, textbooks, company brochures and instruction manuals, telephone interviews, personal interviews, field trips, attendance at seminars, workshops, conferences, conventions, and the CD with this book.

- CADD system (choose a specific system)
- Computer
- Input devices
- Digitizer, graphics tablet
- Output devices: plotter, printer, display screen (monitor or terminal)
- Storage devices: disk drive, optical disk drive, tape drive
- Drafting media: paper, film, pens
- Storage media
- National CADD Standards (CAD Skill Standards) or U.S. National CADD Standard
- Ergonomics
- File management
- Reusing drawing content
- Symbols and symbol libraries
- 2-D and 3-D coordinate systems
- Layers and layer standards (mechanical or architectural)
- Drawing templates

PROBLEM 3.2   Write a short report that discusses the differences among pen plotters, inkjet plotters, and laser printers. You may use any sources that can provide the needed information, such as the Internet, textbooks, magazines, advertising brochures, demonstrations, interviews, and personal experiences. Research and information on the CD with this book can be used. Be sure to discuss the following points in your report:

- Type of media required such as paper and pens
- Media size
- Price
- Quality of final product
- Usefulness in your application
- Warranties and service contracts
- Durability
- Reputation
- Length of time on the market

PROBLEM 3.3   Create a specification sheet for purchasing a single CADD workstation and the software to run it. You can do this as a report, a form, or a list. Research and information found on the CD with this textbook can be used. The assignment should include the following pieces of equipment and software:

- Desktop computer—specifications, speeds, and capacities change rapidly, so when researching, look for the most current specifications in the following components:
  - Computer memory
  - Processor (CPU) speed

- Hard drive
- Portable storage media device such as floppy disk, optical disk, and high-capacity disk
- Graphics card with memory
- Ports for input devices, printers, speakers, etc.
- Color monitor—CRT or flat-panel
- Input device such as mouse, trackball, digitizer
- A/B-size plotter (pen or inkjet), or D-size plotter
- Computer operating system software
- CADD software—current version

Keep the following things in mind when creating your specification list.

- Contact at least three computer hardware vendors for prices.
- Provide each vendor with the same specifications.
- Do not specify brand names of equipment unless instructed to do so.
- Determine the brand of CADD software you want before obtaining prices.
- You may have to go to different vendors for the CADD software.

PROBLEM 3.4 Using existing equipment or vendor's catalogs, measure the equipment required for a CADD workstation (computer, monitor, keyboard, and pointing device), and design a workstation table. Sketch the table first, then draw it on your CADD system.

After drawing the table, create a computer lab arrangement of from 12 to 20 CADD workstations using your table design. Consider the following when designing the lab:

- Marker board location
- Glare from windows
- Aisles and access
- Power connections
- Location of pen plotters or printers
- Storage for paper, plotter pens, and supplies
- Instructor podium or workstation
- Projection system for instructor workstation (use overhead projector)

Remember to draw as little as possible. Once you have drawn one workstation, it can be saved as an object in a symbol library, or just copied as many times as needed in the drawing. If your drawing is used for actual CADD lab layout purposes, keep in mind that additional information for the workstation furniture and computer components can be added to the drawing later. As you continue your CADD studies, always be aware of items that can be added to the drawing and its components to enhance the value of the drawing. If you are using AutoCAD, you can add *attributes* to symbols to give the drawing "intelligence." Consult your textbook or user's manual for detailed instructions. If you are not yet using CADD, you can use sketches to your best ability prior to learning sketching in Chapter 6.

PROBLEM 3.5 This problem involves the creation of template drawings. These drawings should be constructed early in your CADD studies and then used and added to as you progress. Refer to the discussion of template drawings in this chapter before you begin. Create a minimum of one template drawing for each sheet size and drafting discipline in which you will be working. For example, if you will be creating mechanical and civil drawings using B, C, and D sheet sizes for each, you should create six template drawings. The following items should be included in your template drawings:

- Border and title block
- Title block text
- Standard layers
- Text styles for 1/8" text and 1/4" high text (titles)
- Dimension style
- Named views of the title block and the entire drawing

PROBLEM 3.6 For this problem you need to have access to the Internet. Use the search function of your Web browser to locate information on the following topics:

- CADD design formats
- The CADD environment
- Computer hardware (refer to the CD found with this textbook)
- Research techniques
- The project planning process
- The drawing planning process
- One or more CADD software programs
- 2-D and 3-D coordinate systems
- U.S. National CADD Standards for Manufacturing
- U.S. National CADD Standards for Architecture, Engineering, and Construction
- Layer and layer standards
- Drawing templates
- Symbols and symbol libraries
- Reusing drawing content
- File management
- Ergonomics
- Rapid prototyping
- Solid modeling
- Computer-integrated manufacturing (CIM)

When you locate resources and links for each topic, try to get information in the following areas:

■ Schools and training facilities offering courses

■ Research papers, theses, and white papers

■ Research organizations

■ Vendors of equipment, software, and resources

Create a text file of the list given above and include the website addresses below each topic in the format of active links if your software supports this feature.

PROBLEM 3.7   Before working on this problem, you must complete problem 3.6. After you have located information for each topic in problem 3.6, choose the one that interests you the most. Write a short, informative report on the topic using word processing software. Include at least two illustrations in the report. Try to copy images from the Internet and paste them in your document if they are relevant. If your word processing software supports it, include at least two active links in the report to a website related to the topic. One of the links should be embedded in one of the images.

When you have completed the report, send it as an e-mail attachment to your instructor. Request that the instructor test your website links to be sure they function properly.

PROBLEM 3.8   Use the Internet or your library's resources to find professional trade journals and magazines. Try to find at least one publication in the disciplines of CADD, mechanical engineering, architectural design, structural engineering, solid modeling, rapid prototyping, civil engineering and surveying, geographical information systems (GIS), numerical control, 3-D design, virtual reality, and nanotechnology. Keep a list of the publications that you find in each field. Add journals of other disciplines to the list as you find them.

Throughout the course of your studies, try to read one article a week from one of the journals. Choose a different journal each week.

MATH PROBLEMS

PROBLEM 3.9   Draw the following sheet metal part using CADD software. Use the entry method of your choice. Space is provided in which you can write the coordinate values needed to draw a line to each point or use a separate sheet.

A to B_____

B to C_____

C to D_____

D to E_____

E to F_____

F to G_____

G to A_____

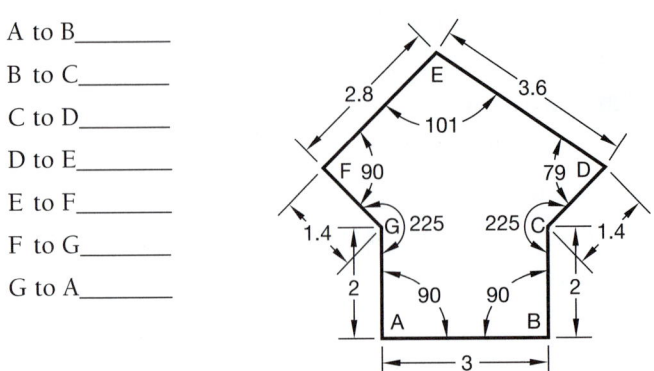

PROBLEM 3.10   The following property plat has been given to you in the form of a sketch with included angles. You are to draw the property plat full size using the dimensions given. Do not use decimal values for angles or distances. Space is provided in which you can write the coordinate values needed to draw a line to each point. Write your answers using a polar coordinate notation. For example, a line 87'−3" long, at an angle of 65° is written @ 87'3"<65.

A to B_____

B to C_____

C to D_____

D to E_____

E to F_____

F to A_____

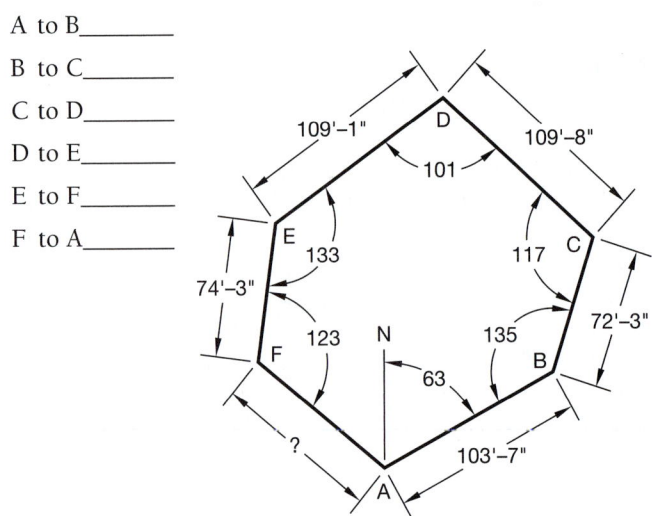

3-D CADD, Animation, and Virtual Reality

LEARNING OBJECTIVES

After completing this chapter, you will:

- Name the ASME standard for models and drawings in digital format.

- Explain the difference between wireframe, surface, and solid modeling.

- Discuss the difference between and importance of the solid modeling kernels.

- Discuss the function of parametric solid modeling.

- Define constraints and relations.

- Explain how to work with parts, features, subassemblies, and assemblies.

- Discuss sketches and sketched features.

- Provide information about placed features, reference features, pattern features, and catalog features.

- Discuss the creation of 2-D drawings from solid models and their parametric relationship.

- Explain the use of animation in the design drafting profession.

- Discuss virtual reality in a variety of design drafting fields.

THE ENGINEERING DESIGN APPLICATION

Today's business world is extremely competitive in nature. Teams must work to produce high-quality products that are brought to the marketplace at the correct time. If a product is introduced to the market too late, the competition may have already captured the majority of the customers, leaving only a minority behind for the latecomer. The latecomer in this case has a major hurdle to overcome, that of winning customers over from the competition. Research clearly shows that once customers have selected a product, they are reluctant to change. Therefore, it is critical to gain early entry into the marketplace and to capture a customer base. Furthermore, early entry into the marketplace means an extended sales life for the product. Simply put, getting to the marketplace first is a major strategy for success.

Design teams are one business strategy that is being used by business and industry to accelerate the design process and thus get product to the market. Traditionally, the design process happened in a linear fashion. For example, the design engineer would complete a design and then pass it to manufacturing. When manufacturing received the design, often they would have to make changes, changes that required time. The design process would get longer and longer as manufacturing and the engineer passed the design back and forth requesting changes.

Traditionally these changes were very difficult to make. Some of this can be contributed to the fact that communication methods were not very flexible. Drawing documentation was done by hand, and often nonengineers had a difficult time understanding the drawings. Additionally, changes required laborous erasing and sometimes redrawing of prints that were stamped with approvals and passed back to manufacturing. Lack of flexibility is greatly improved using conventional CADD software, which allows you to make changes quickly and print a new drawing.

Today, teams consisting of individuals from marketing, design, engineering, and manufacturing work together using integrated CADD solid modeling software. This software allows for making changes in a more fluid manner. Changes can be made by marketing and design while the engineers' requirements are protected through geometric and drawing constraints. Also parts, assemblies, subassemblies, and drawings can be related to one another, so if one component changes, the associated components and drawings also update.

This teamwork and the ability to quickly modify product design is critical in today's competitive marketplace. Reaching the marketplace with innovation before the

(Continued)

THE ENGINEERING DESIGN
APPLICATION *(continued)*

competition has several critical benefits for the early-to-market company, including:

■ The product's sales life is lengthened.

■ The company gains a pricing advantage.

■ The company can start product development later and therefore can use more up-to-date technology to develop products.

■ The company has a marketplace advantage by gaining customers who potentially become loyal customers.

As a drafter it is part of your job to make things happen and make them happen quickly so that your company can gain these advantages and win in today's marketplace. Drafters are very critical members of the engineering design process.

INTRODUCTION TO 3-D CADD

CADD initially become accepted in the drafting and design industry in the late 1970s and early 1980s. Since that time, significant changes in computer hardware and CADD software have occurred. In the constantly changing CADD world, you must be prepared to learn new drafting tools and techniques and be open to attending classes, seminars, and workshops on a regular basis.

CADD software continues to improve in a variety of ways. CADD programs are easier to use than ever before and contain multiple tools and options, allowing you to produce better quality and more accurate drawings in less time. As CADD applications improve, the traditional requirements of a drafter often become less important, while the ability to use new CADD software and application specific tools increases. For many designers and drafters, the future of CADD has arrived in the form of three-dimensional (3-D) modeling, animation, and virtual reality.

ASME/ANSI The standard related to a model, and a model and drawing in digital format, is ASME Y14.41, *Digital Product Definition Data Practices.* The standard establishes requirements for the preparation of digital product definition data, called data sets. A data set is a complete product definition, including a design model, related annotation or text, and supporting documentation. The standard is also intended to provide a guide for computer-aided design and drafting (CADD) software programs to develop improved modeling and annotation practices for CADD and engineering disciplines.

3-D CADD MODEL FORMATS

Traditional two-dimensional (2-D) drawings, such as the drawing shown in Figure 4.1, are very useful for documenting engineering design requirements. However, it can be difficult for a person to understand and visualize a product if they have not been trained to interpret 2-D drawings. Additionally, 2-D drawings are limited in their ability to help designers and engineers analyze and

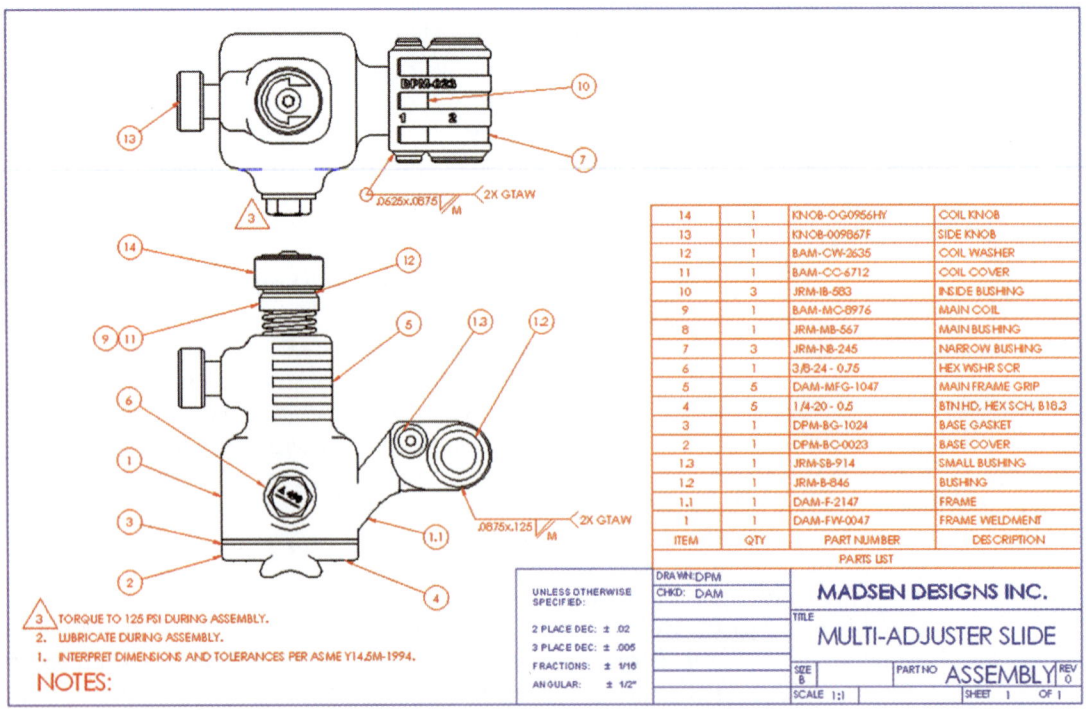

FIGURE 4.1 ■ An example of a traditional 2-D drawing.

FIGURE 4.2 ■ An example of a 3-D model. This is the 3-D model of the 2-D drawing shown in Figure 4.1.

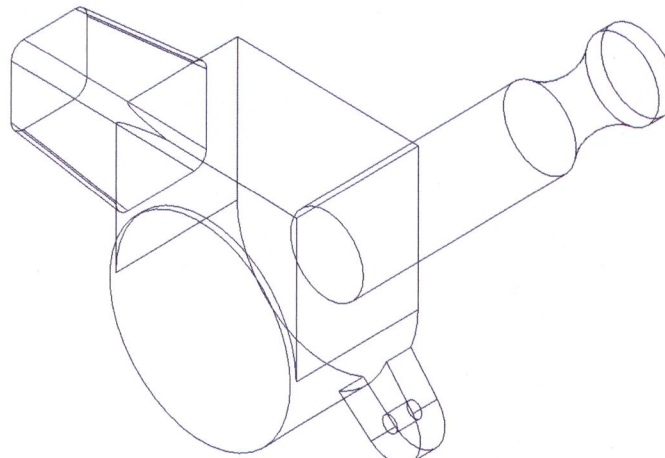

FIGURE 4.3 ■ A wireframe model consists of edges and intersections of edges.

test products, and to present ideas. Creating a 3-D representation of a design, such as the model shown in Figure 4.2, can help overcome visualization problems and produce a realistic, testable product model. There are three different 3-D modeling techniques including wireframe, surface, and solid modeling.

Wireframe Modeling

Wireframe models are the most basic 3-D CADD models, and contain only information about object edges and the intersection of edges. As the name implies, a wireframe model appears as if it is constructed from wires, as shown in Figure 4.3. Although wireframe models provide an outline of the 3-D object, they can be difficult to read and interpret. This can be improved on some models by hiding the lines that fall behind object features, but sometimes even a hidden line representation can be difficult to visualize.

The primary advantage to using wireframe models is that wireframe model CADD files are very small, because they only contain information about object edges and the intersection of edges. However, a wireframe model does not contain information about object surfaces. As a result, complex surfaces, which are found on many products, cannot be represented. Also, wireframe models cannot define object volume, which means that wireframe models are not useful for calculating mass or generating machining code.

Surface Modeling

Similar to wireframe models, surface models contain information about object edges and the intersection of those edges, but they go a step further to include information about object surfaces. (See Figure 4.4.) Surface models can provide realistic design representations and allow designers to create complex curves and forms. As a result, surface models are used by designers who are primarily concerned with external product shape and appearance. Figure 4.5 provides an example of a surface model with complex shapes that are common on many home appliances, such as the deep fryer shown.

Surface models displaying realistic model surfaces provide a means of communicating design intent to those who have a difficult time visualizing traditional 2-D drawings. Surface models

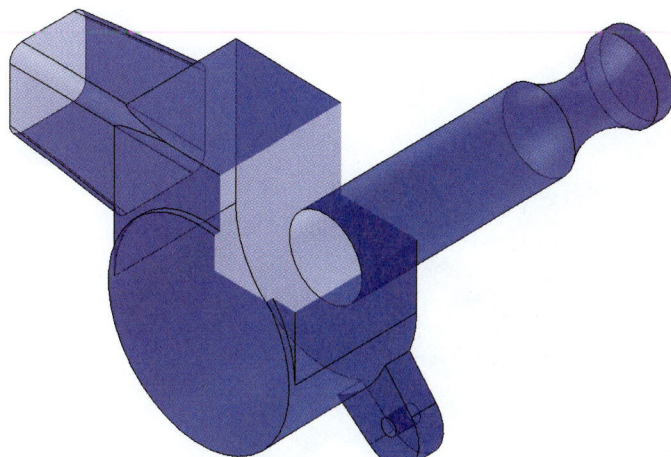

FIGURE 4.4 ■ A surface model describes object surfaces. This model displays a translucent surface color and has been sectioned to reveal the hollow interior. This is a surface model of the same object shown in wireframe in Figure 4.3.

FIGURE 4.5 ■ Product design with complex forms and shapes. *Courtesy Unigraphics Solutions Inc.*

are also useful for creating complex shapes. Despite these advantages, hollow surface models are limited because they do not contain mass property data.

Solid Modeling

Like surface models, **solid models** contain information about object edges, the intersection of those edges and surfaces, and also include data about object volume and mass. (See Figure 4.6.) Solid models are the most realistic 3-D models and allow you to analyze exterior and interior object characteristics. Designers and

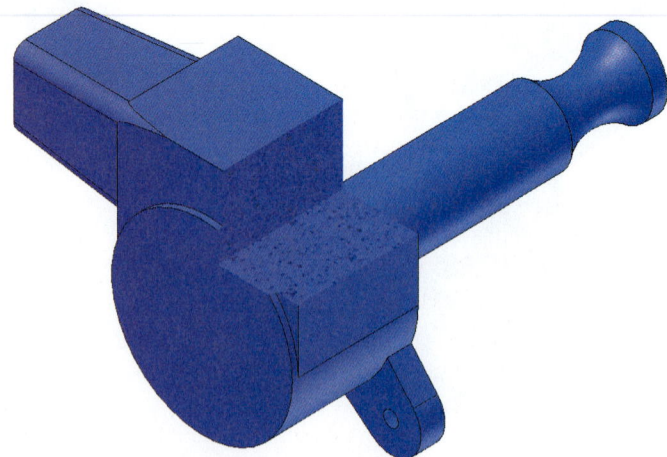

FIGURE 4.6 ■ A solid model describes object surfaces and volume. This model has been sectioned to reveal the solid interior. This is a solid model of the same object shown as a surface model in Figure 4.4.

Characteristic	Basic Model	Parametric Model	Hybrid Model
Analysis	No	Yes–add on software	Yes
Animation	No	Yes–add on software	Yes
Feature based	No	Yes	Yes
History	No	Yes	Yes
Intelligence	No	Yes	Yes
Surface modeling	No	Limited	Yes
Updateable	No	Yes	Yes
Volumetric Information	Yes	Yes	Yes

FIGURE 4.7 ■ Characteristics of basic, parametric, and hybrid solid modeling programs.

engineers use solid models to perform interference and collision checks, mass calculations, and simulations, and to generate machining code.

Solid models contain a great deal of information compared to surface models and significantly more data than wireframe models. However, certain solid models have more functionality than others. Some solid model software programs are used to create very basic solids, while other programs can be used to generate fully parametric models, in addition to wireframe and surface models. The following information describes the basic, parametric, and hybrid model formats. The table shown in Figure 4.7 describes these solid model programs and the characteristics of each.

Basic Solid Models

Many solid modeling programs allow you to generate only **basic solids** or **dumb solids**. Dumb solids are created using basic solid modeling methods and contain very little, if any, information about specific parameters, dimensions, constraints, part history, or features. Only fundamental volume information, such as length, width, and height, are stored in dumb solids. As shown in Figure 4.8, most solid modeling programs allow you to join, subtract, and intersect solid shapes. However, with basic solid models, information about the joining, subtracting, and intersecting of shapes is not stored. As a result, many issues can arise during the design process due to the fact that the model cannot be easily modified and updated. For example, if you cut a hole through a basic model using a subtraction operation then realize the hole is no longer needed, you must eliminate the hole using a repair or join operation, because the hole data is no longer available. If the hole cannot be repaired, you may need to totally recreate the model.

Basic solid modeling is a step above non-parametric wireframe and surface modeling. However, the lack of information stored in basic solids and the inability to effectively manipulate basic solid models to easily explore alternative design options is a tremendous disadvantage to using basic solid modeling programs. Designers and drafters using dumb solid software feel this disadvantage because they are often unable to respond to the rapid pace of the product development process.

INITIAL PART MODEL SKETCH

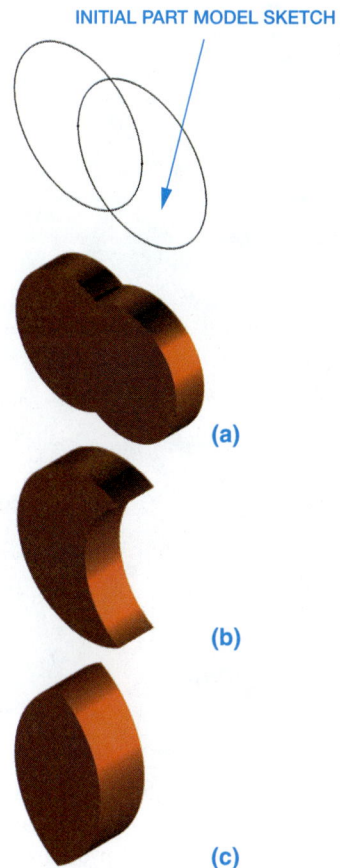

(a)

(b)

(c)

FIGURE 4.8 ■ An example of a join (a), subtraction (b), and intersection (c) solid operation.

Parametric Solid Models

Parametric solid modeling CADD programs eliminate the disadvantages associated with basic solid models. **Parametric solid modeling** involves the development of solid models containing parameters, which are controls, constraints, and checks that allow you to easily and effectively make changes and updates. This means that when you describe the size, shape, and location of model geometry using specific parameters, you can easily modify those specifications to explore alternative design options. In addition, if the modifications or new parameters conflict with existing geometry, the model cannot be built without an alert, like the error shown in Figure 4.9.

The parametric concept is also referred to as intelligence, because parametric modeling occurs as a result of the software programs' ability to store and manage model information. In contrast to dumb solids, everything associated with a parametric solid model is stored in a database. This information includes knowledge of every model characteristic such as calculations, sketches, features, dimensions, geometric parameters, when each piece of the model was created, and all other model history and properties.

The best way to understand the basic principle of parametric association is to study the size of a parametric geometric shape. For example, the cylinder shown in Figure 4.10a was drawn one

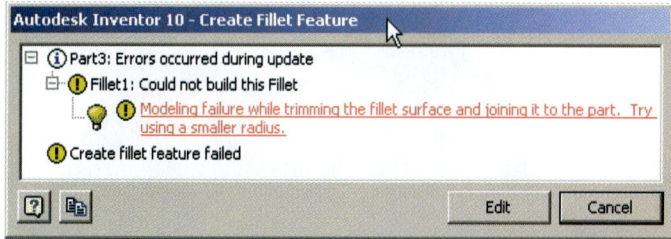

FIGURE 4.9 ■ An example of a modeling error that must be resolved before the model can be built correctly.

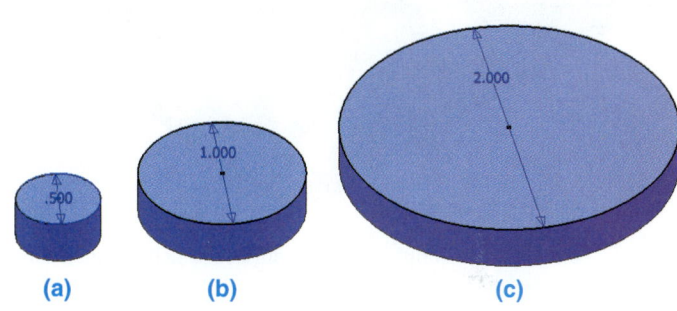

(a)   (b)   (c)

FIGURE 4.10 ■ An example of a basic parametric relationship defining the diameter of a cylinder.

half inch in diameter. When the dimension shown in Figure 4.10b is added to the cylinder, that defines the cylinder as one inch in diameter, the size of the cylinder automatically changes according to the one-inch diameter dimension. When a new dimension of two inches in diameter is applied to the cylinder, as shown in Figure 4.10c, the size of the cylinder automatically changes from one to two inches in diameter. This example shows how the size of the cylinder is associated with the cylinder dimension. The cylinder only increases in size when the parametric value of the dimension is edited.

Hybrid Models

Hybrid CADD modeling software programs incorporate the functionality of wireframe, surface, and basic and parametric solid modeling. Hybrid modeling systems often are described as separate CADD applications. However, most often, hybrid modelers take the form of parametric solid modeling programs that contain wireframe, surface, and basic solid tools, which have the ability to convert these models to parametric solids.

Working with a variety of model formats can prove very beneficial. For example, some complex shapes can only be developed using surface models, while solid models provide object volume and mass information. Figure 4.11 shows a surface that was used to help create a complex solid model. Bringing the advantages of each model format together ensures product design occurs as effectively as possible. Hybrid modeling programs can provide the flexibility for a designer to create complex surface model shapes, while allowing an engineer to convert those shapes to a solid model that can be analyzed.

FIGURE 4.11 ■ An example of using a surface to help create a complex solid model.

PARAMETRIC SOLID MODELS

There are several parametric solid modeling software programs available to the designer and drafter. Autodesk Inventor, Pro/Engineer, and SolidWorks are three of these applications. Many parametric solid modeling programs are surprisingly similar in the way they function. In fact, once you learn the basic process of creating a model using specific software, such as Autodesk Inventor, you can usually effectively transition to different parametric solid modeling software, such as SolidWorks.

The following information includes general concepts associated with many parametric solid modeling software programs. The work environments, model elements, and model creation techniques used are very similar throughout parametric solid modeling programs. Additionally, you should become familiar with parameters, which are critical to correct and effectively use parametric solid models.

The Parametric Solid Model Work Environments

Every parametric solid modeling software program contains a variety of work environments that allow you to generate parametric solid models. Most parametric solid modeling applications function using part and assembly work environments, or file types. In some cases you work with a single part file, while other projects require you to produce several or even hundreds of parts and assemblies.

Part files are used to create single parts and parts that are used in assemblies. Parts can be developed separately in individual part files or can be created inside an assembly file in relation to other parts. An example of a part is shown in Figure 4.12.

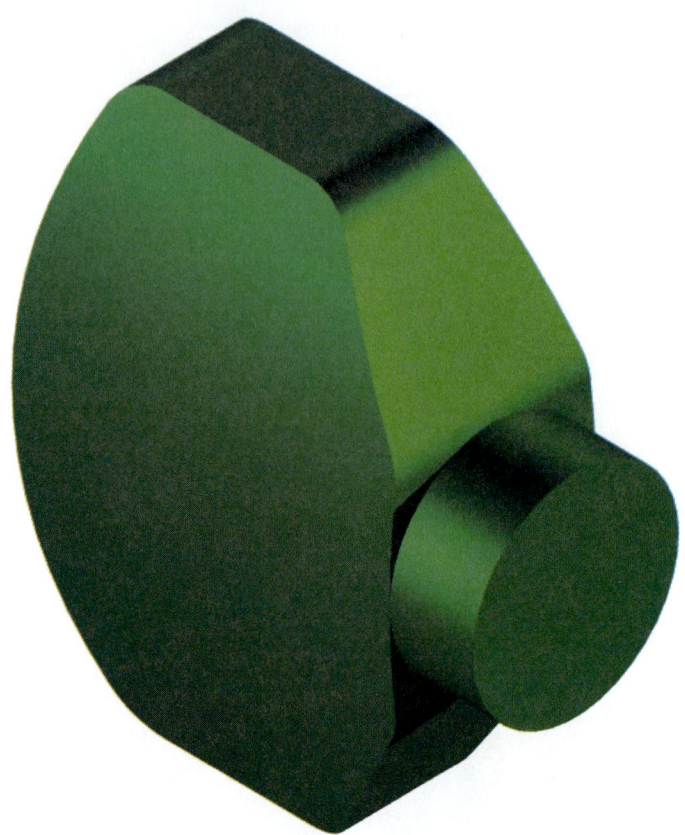

FIGURE 4.12 ■ An example of a parametric solid model part that is in the assembly shown in Figure 4.13.

Assembly files are used to create assemblies and subassemblies, and reference multiple part and subassembly files. An **assembly** is a grouping of components, a **component** is a term used to describe the individual parts and subassemblies used to create an assembly, and a **subassembly** is an assembly that is added to another assembly. An example of an assembly is shown in Figure 4.13.

In addition to part and assembly work environments, most parametric solid modeling systems include additional work environments and file types. The tools and options required for specific modeling applications, such as creating wireframe models or basic solids, are contained in separate work environments. In other cases, totally unique files are used to develop additional model or documentation requirements. For example, Autodesk Inventor uses a presentation file to create exploded, animated, and stylized assembly views. An example of a presentation is shown in Figure 4.14.

Parametric Solid Part Model Elements

Parametric solid models include assemblies and parts. However, an assembly is just a grouping of part and subassembly models. As a result, you must understand the basic elements of a part model before advancing to assemblies. Every part model contains at least one sketch and at least one sketched feature. A more complex model can include placed, reference, pattern, and catalog features. The following describes each of these elements:

- **Sketches** represent the first step in the creation of a model. Sketches provide the profile, or foundation and guide, used for developing models containing sketched features. An example of a sketch is shown in Figure 4.15.

- **Sketched features** are products of sketches. Every parametric solid modeling system contains several sketch feature tools. Sketched features are typically created as extrusions, revolutions, sweeps, lofts, coils, and other shapes, depending on the design, sketch geometry, and available CADD software tools. Most of the time, the initial feature, known as the **base feature**, is produced as a sketched feature. An example of a sketched feature is shown in Figure 4.16.

- **Placed features** are, as the name implies, placed on to an existing feature without using a sketch. Placed features are also known by a few other names, including **built-in**, **added**, or **automated features**. Shells, fillets, chamfers, threads, and face drafts are considered placed features. Generating placed features requires defining the size dimensions and geometric characteristics in a dialog box and selecting a location, such as a point or an edge for the feature.

  Placed features are often created toward the end of the design process, but they can be added anytime if at least one sketched feature is present. Figure 4.17 shows an example of a placed feature and the dialog box used to create the feature. Notice how all of the geometric size information is described using the dialog box.

- **Reference features**, also commonly known as **work features** or **reference geometry**, are used to direct the location and arrangement of other features. Reference features are used primarily for construction purposes in order to

FIGURE 4.13 ■ An example of a parametric solid model assembly.

FIGURE 4.14 ■ An example of an Autodesk Inventor presentation file. This is the exploded presentation drawing of the solid model assembly shown in Figure 4.13.

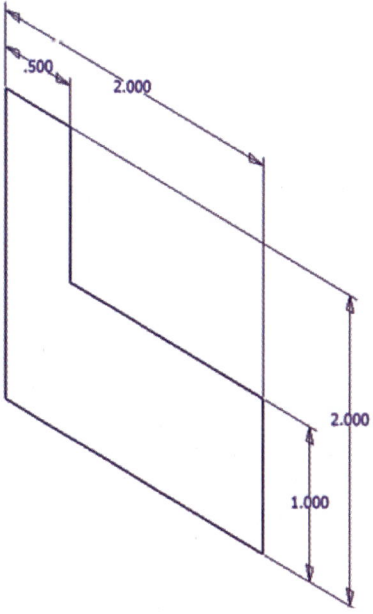

FIGURE 4.15 ■ An example of a parametric solid model sketch used to create a model.

FIGURE 4.16 ■ An example of a parametric solid model sketched feature created using a sketch. This particular feature is an extrusion created from the sketch in Figure 4.15.

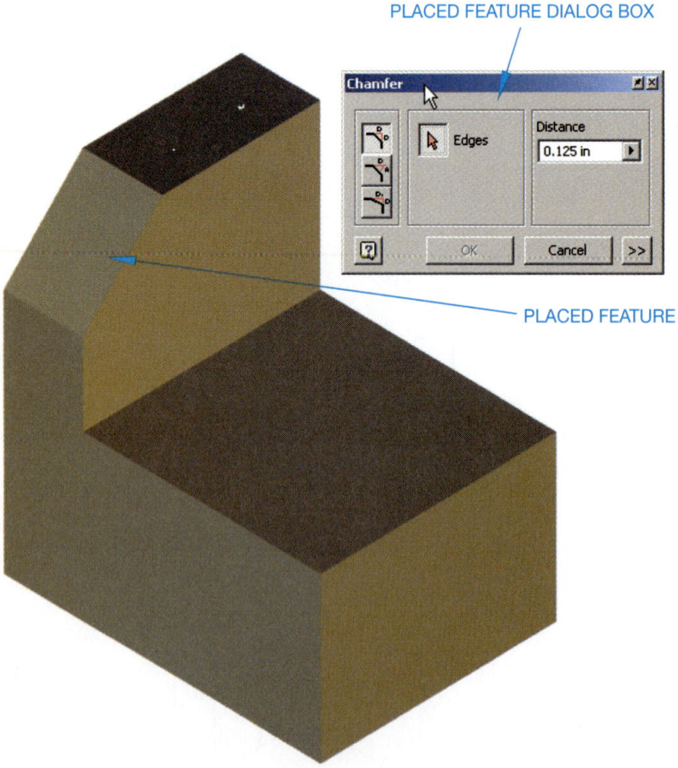

PLACED FEATURE DIALOG BOX

PLACED FEATURE

FIGURE 4.17 ■ An example of a parametric solid model placed feature, and the dialog box used to create the feature. This particular feature is a chamfer.

develop additional sketches and features in areas where no geometry is available. Reference planes are used to create construction planes, reference axes are used to create construction lines, or axes, and reference points are used to create construction points anywhere on a feature or in three-dimensional space. Each of these features can be parametrically associated with a model and change when a feature or sketch is modified. In the same respect, if a reference feature is modified, features attached to the reference feature also reflect the changes.

■ Catalog features, also known as library features, are similar to placed features, because you can place a catalog feature onto an existing feature or part. However, catalog features are often far more complex than a standard shell, fillet, chamfer, or face draft. (See Figure 4.18.) A catalog feature is a copy of an existing feature or set of features you create and then save in a library to be used in other models.

Some CADD programs also allow you to generate derived components which are very similar to catalog features, but are more complicated and can contain a complete model consisting of several features, or even multiple parts. Derived components are often used as base features, and can be inserted into existing models.

■ Pattern features are developed using an existing feature or group of features, such as extrusions or holes, in order to make a pattern of the feature or features. Different from sketched or placed features, pattern features copy an existing feature or features, generating occurrences of the feature or features. Pattern features do not actually create completely new features. Most parametric solid modeling systems allow you to build three types of pattern features, including rectangular patterns, also commonly known as linear patterns, circular patterns, and mirrored features. A rectangular pattern, shown in Figure 4.19, copies and organizes a feature into a designated number of rows and columns and places the features a specified distance apart from each other. A circular pattern, shown in Figure 4.20, copies and organizes a

CATALOG FEATURE

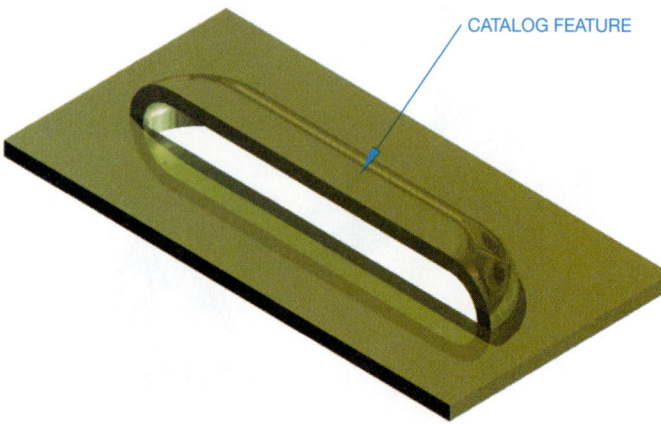

FIGURE 4.18 ■ An example of a catalog feature. This particular catalog feature is a sheet metal louver containing a sketched feature and a fillet and round created using a placed feature.

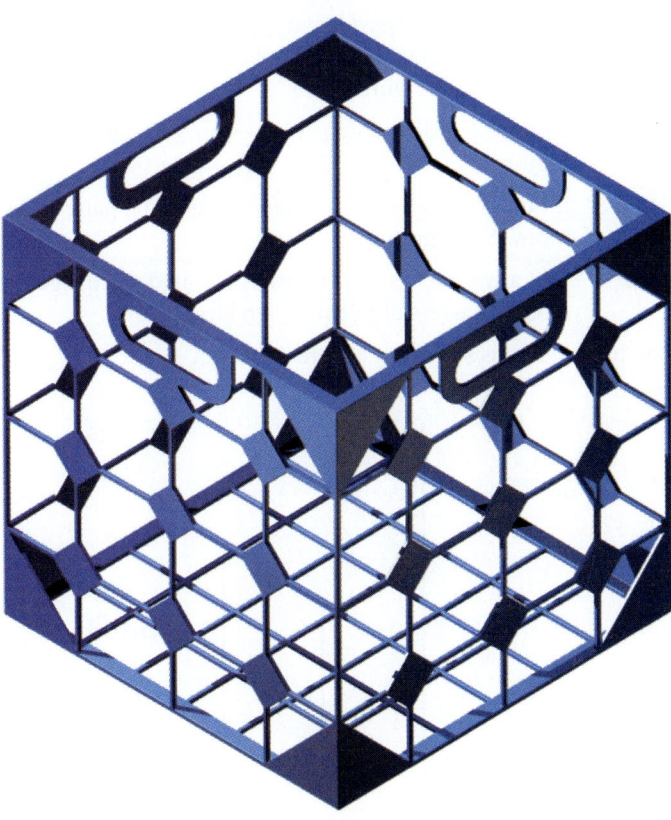

FIGURE 4.19 ■ An example of using rectangular feature patterns.

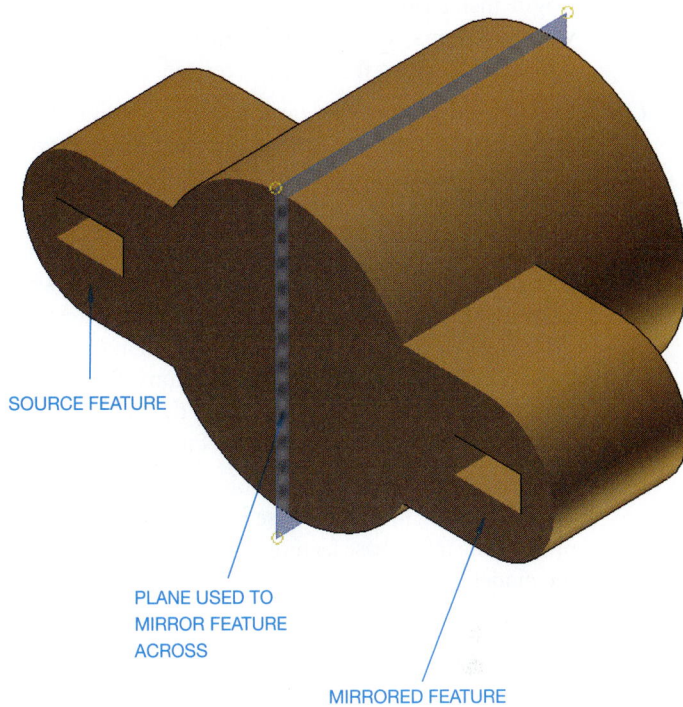

SOURCE FEATURE

PLANE USED TO
MIRROR FEATURE
ACROSS

MIRRORED FEATURE

FIGURE 4.21 ■ An example of a mirrored feature. The source feature is the original feature, and the mirrored feature is a mirror image of the source feature across the mirror plane.

feature around an imaginary circle a designated number of times and places each feature a specified distance from the other. A mirrored feature, shown in Figure 4.21, mirrors an existing feature across a plane.

Creating a Parametric Solid Part Model

Several basic steps are required to develop a 3-D parametric part model using most parametric solid modeling programs. Again, the basic steps required to build a parametric solid part model are very similar from one parametric solid modeling system to another. Only a portion of the steps required to build a part model may be necessary to construct a very simple model, while additional steps are usually needed to build a complex model. Although you can develop your own specific method of creating a model, the general process of model development occurs in the following steps. These steps assume you choose to use default work environment settings, have already set up custom work environment preferences, or use a template with predefined settings:

STEP 1 Define work and sketch environment settings specific to the model.

STEP 2 Open a sketch on the default XY, XZ, or YZ plane, or use the default sketch plane that may automatically be opened due to the current application configuration.

STEP 3 Begin your sketch at the origin, or 0,0,0-X,Y,Z point, allowing the sketch to become fully defined at the

FIGURE 4.20 ■ An example of a circular feature pattern.

origin instead of floating in space. With some software programs, you may need to project the origin onto the sketch so it can be referenced.

STEP 4 Draft sketch geometry to create a general outline of the initial base feature. You should plan and develop your sketch in terms of the 3-D model it creates. (See Figure 4.22.)

STEP 5 Add parameters to further define the sketch. **Parameters** are parametric associations that define the size, shape, and location of model geometry. Parameters are defined using geometric characteristics, often known as **constraints** or **relations**, and dimensions. Parameters are held constant throughout model development but can be altered when necessary. (See Figure 4.23.)

STEP 6 Option 1: Use sketch feature tools, such as EXTRUDE and REVOLVE, to create a base feature from the sketch profile. The initial base feature is the basic foundation of the model. (See Figure 4.24.)

Option 2: Insert a derived component to use as a base feature.

STEP 7 Option 1: Draft sketch geometry on the base feature, as shown in Figure 4.25a, and use feature tools to create additional features that join, subtract, or intersect the base feature. (See Figure 4.25b.)

Option 2: Use placed feature tools such as SHELL, FILLET, and CHAMFER to add placed features to sketched features. (See Figure 4.26.)

Option 3: Add catalog features to your model. (See Figure 4.27.)

STEP 8 If necessary, create additional objects using pattern features. (See Figure 4.28.)

STEP 9 Edit any sketches or features as required to change the model or assess different design ideas.

The final model is shown in Figure 4.29.

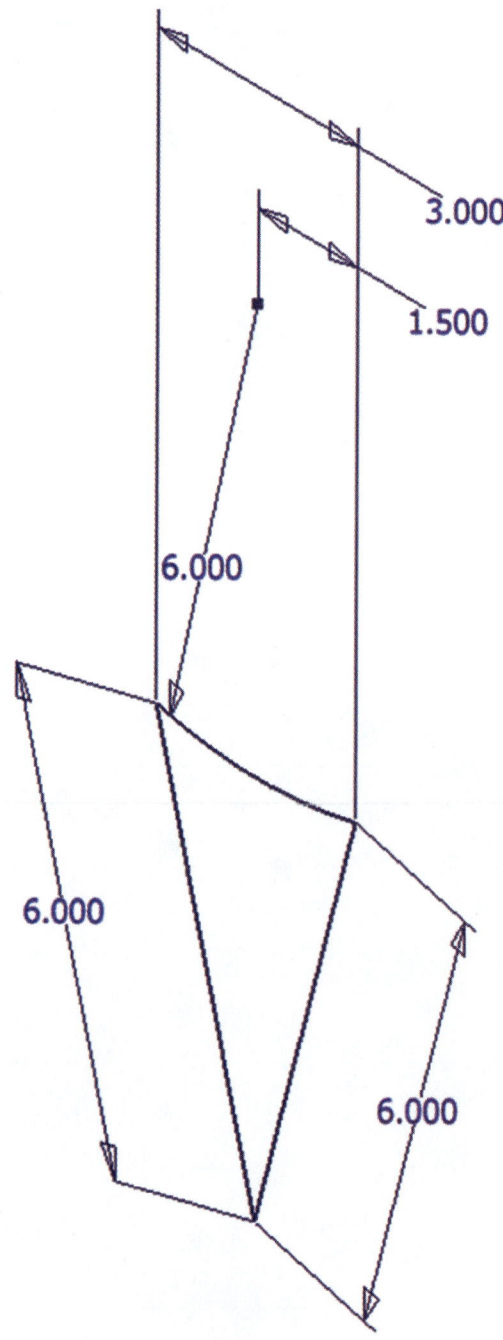

FIGURE 4.23 ■ Adding geometric characteristics and dimensions.

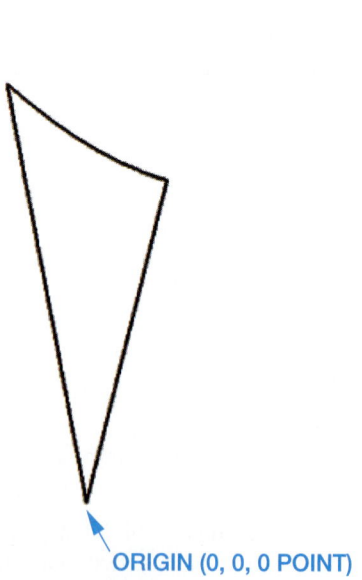

ORIGIN (0, 0, 0 POINT)

FIGURE 4.22 ■ Initial sketch shapes fixed to the origin, or 0,0,0 point.

FIGURE 4.24 ■ Developing a feature, in this example, an extrusion, using initial sketch geometry.

FIGURE 4.25b ■ Creating an additional model feature, in this example, an extrusion that joins the base feature.

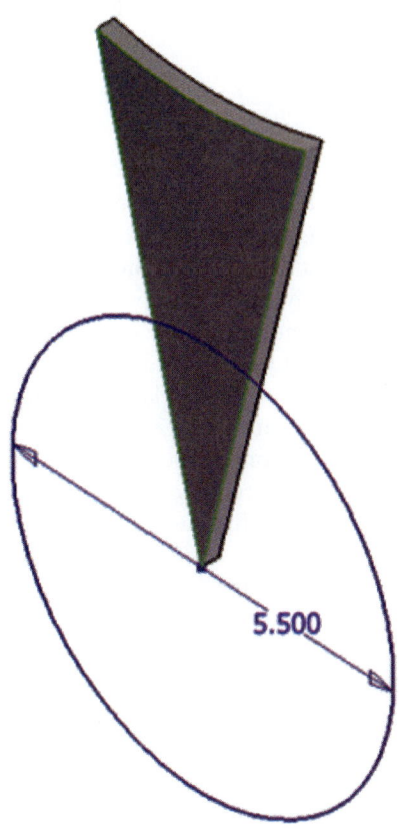

FIGURE 4.25a ■ Sketching on a base feature.

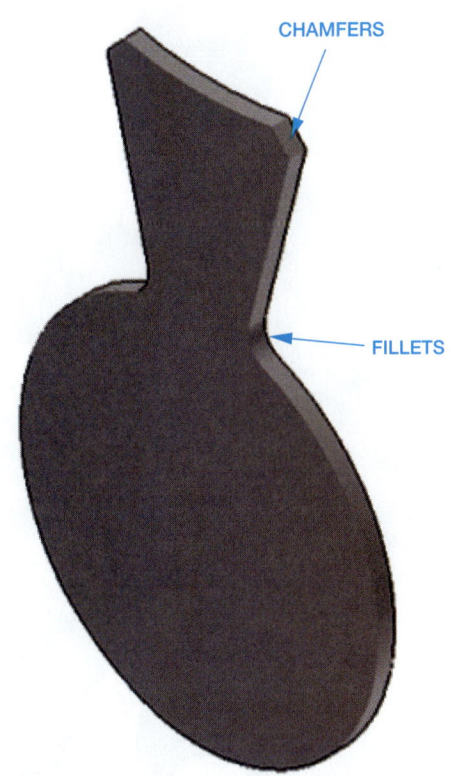

FIGURE 4.26 ■ Adding placed features.

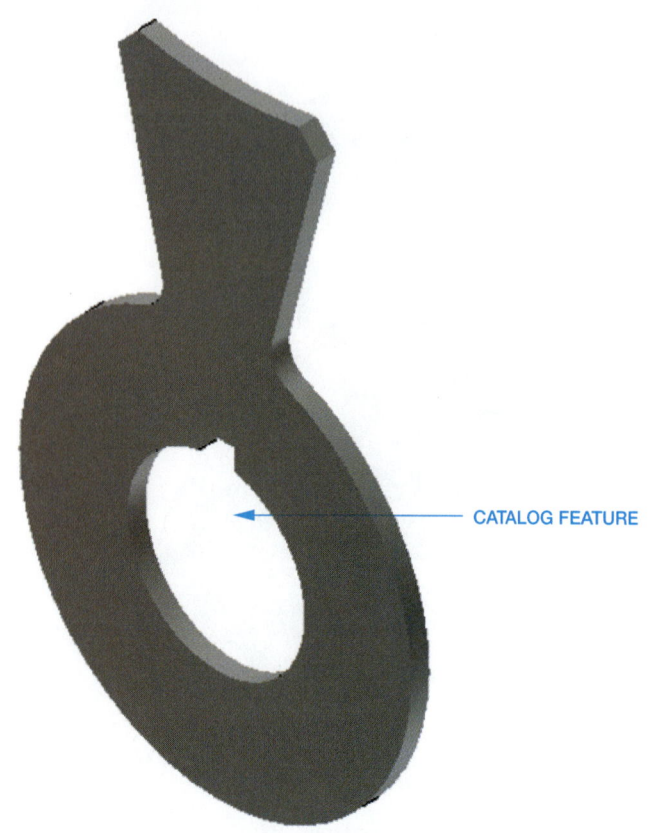

CATALOG FEATURE

FIGURE 4.27 ■ Adding a catalog feature.

FIGURE 4.28 ■ Adding pattern features to the model.

FIGURE 4.29 ■ The finished model with additional features.

Parameters

As previously described, the design process typically begins with a sketch. Usually base feature sketches and shapes are easy to create and are meant as the initial parametric part model concept. Often the first sketch is a crude representation of an idea, but the sketch soon becomes accurately defined using parameters. Parameters include dimensions and geometric characteristics, often known as constraints or relations. Constraints and relations describe the relationship of geometric shapes. For example, a constraint or relation can define if the length of one line is equal to another, if the center of a small circle is concentric to a larger circle, or if two lines are perpendicular.

The concept of using parameters is a very important principle in parametric solid modeling. Parameters are required for design development and are used throughout the design process, from the initial sketch to fully defined features. You must use parameters to constrain a part as accurately as possible to preserve your design intent. For example, if the center of a 20 mm diameter hole must be 32 mm away from an edge, you must specify and control this information using dimensions and geometric characteristics. Parametric associations allow you to:

■ Define your part using dimensions and geometric characteristic tools.

■ Allow design elements to remain constant, protecting your model and design ideas. To accomplish this, many parametric modeling systems tell you if a sketch is over-constrained or if a modeling failure occurs. An over-constrained sketch has too many dimensions or geometric controls and may not allow the software to recognize or resolve the sketch geometry. (See Figure 4.30a.) A modeling failure is the result of dimensions or geometric controls that are impossible to apply to the model. Over-constrained and modeling failure situations must be resolved before you can create a feature and a fully parametric model. A sketch or model also cannot contain enough parameters. An under-constrained sketch or model has elements that are unclear, can be changed or moved, and remain undefined and unstable. (See Figure 4.30b.) As you progress through the design process, you must define your sketches and features so they become fully-constrained, as shown in Figure 4.30c, except in some situations that require the model to adapt to changes made to other shapes. This ensures that your model is parametric and that your design intentions are not violated.

■ Make any necessary changes to the design of a model, allowing yourself to almost immediately assess design alternatives by changing, adding, or deleting sketches, features, dimensions, and geometric controls.

■ Develop additional files that reference existing files. Assemblies and 2-D drawings are created almost instantaneously using existing models because most dimensions and geometric associations have already been defined in the part, subassembly, or assembly.

■ Develop parameter-driven assemblies that allow changes made to individual parts to be reproduced automatically as changes in the assembly and assembly drawing.

■ Use adaptive parts in assemblies. Adaptive parts are extremely effective when you may not know the exact dimensions of a part or you may not fully understand the relationship between assembly components. Adaptive parts modify automatically if another part changes.

■ Develop equations that drive your models, allowing a few dimensions to define the entire model, or even to create a family of related parts.

Creating a Parametric Solid Assembly Model

An assembly is a grouping of one or more design components. Figure 4.31 shows an assembly consisting of subassemblies and

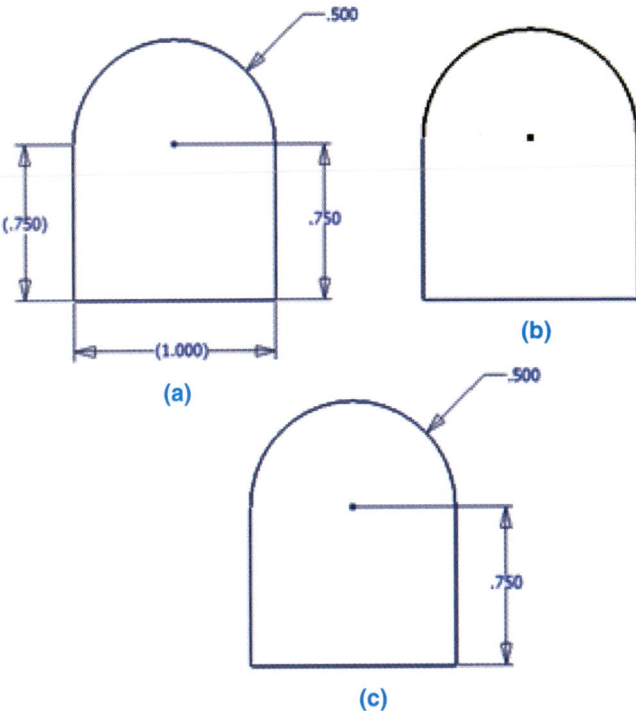

FIGURE 4.30 ■ (a) A sketch in over-constrained, (b) under-constrained, and (c) fully constrained condition.

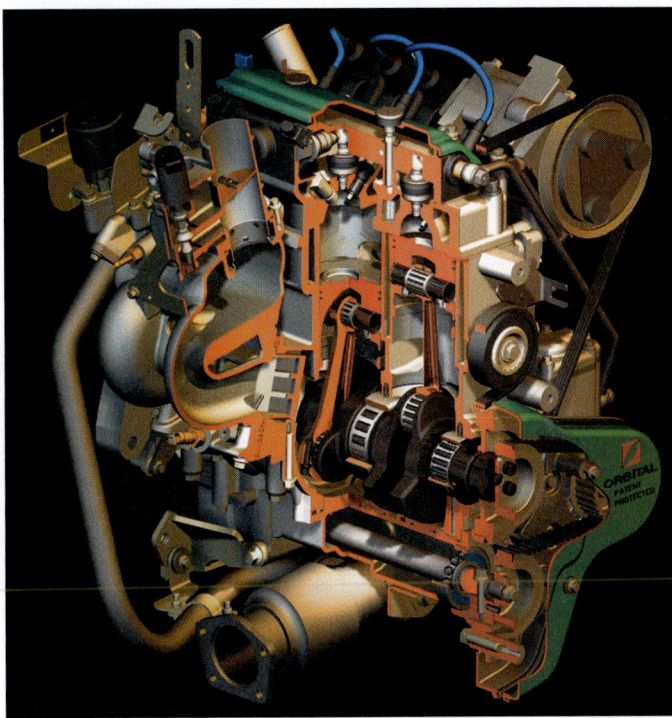

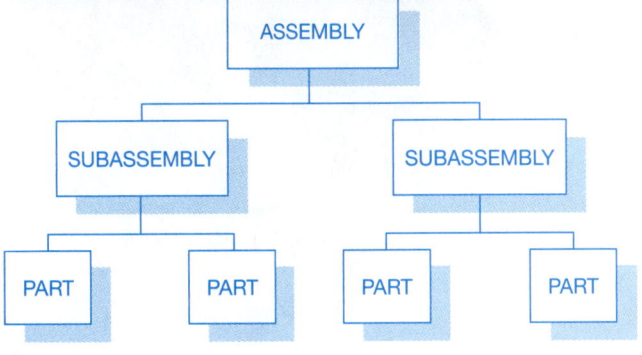

FIGURE 4.31 ■ An engine assembly consisting of parts and subassemblies, along with a flow chart demonstrating the chain from the assembly to subassemblies and to parts. *Courtesy Unigraphics Solutions Inc.*

parts, along with a flow chart demonstrating the chain from the assembly to subassemblies, and to parts.

Components can include part models and subassemblies. Subassemblies are groups of components that are often standard product items and may be used several times in various assemblies. For example, subassemblies can include items like wire harnesses or switches, which are composed of two or more parts, and can be used more than once in the final assembled product. Another example of a subassembly is the alternator (see Figure 4.32) found on the engine assembly shown in Figure 4.33. When you look at the engine assembly in Figure 4.33, the alternator subassembly can be seen on the right rear side. Although some designs contain a single part, such as a bracket, many products consist of at least two parts, such as a bracket and a bolt. Other products can contain hundreds of parts and subassemblies.

In a parametric solid assembly model file, multiple parts and subassemblies are brought together and constrained, or mated, to form a complete assembly. As described earlier, parameters define relationships between sketch geometry when creating parts. Constraints, or mates, are used similarly in the assembly environment, in which they link individual components together to create the assembly. Figure 4.34 shows an assembly constraint that mates the axis of an eye bolt with the axis of a nut, lock washer, and washer.

There are two basic options for developing assemblies and subassemblies using most parametric solid modeling systems. One method involves inserting existing parts and subassemblies into an assembly file, followed by constraining, or mating, the components. This is referred to as bottom-up assembly. The second technique involves building individual components in an assembly file without inserting existing components. This option is commonly referred to as creating components in-place. This is referred to as top-down assembly. For this method, you may find it helpful to think of an assembly file as a part or subassembly file that also contains assembly tools. Although inserting components and mating them together is effective, it is often faster, easier, and more productive to develop components within an assembly file.

Inserting existing components and building components in-place represent fundamental approaches to assembly creation. However, usually both assembly creation methods are used to develop more complex assemblies. Often standard base parts or subassemblies are individually built and then inserted into an assembly file. Other times, entire, usually less complex, assemblies are generated by developing all components inside a single assembly file. The advantage to creating components in-place is that you can quickly develop an assembly and analyze design intent as you model a product. In addition, when you build a com-

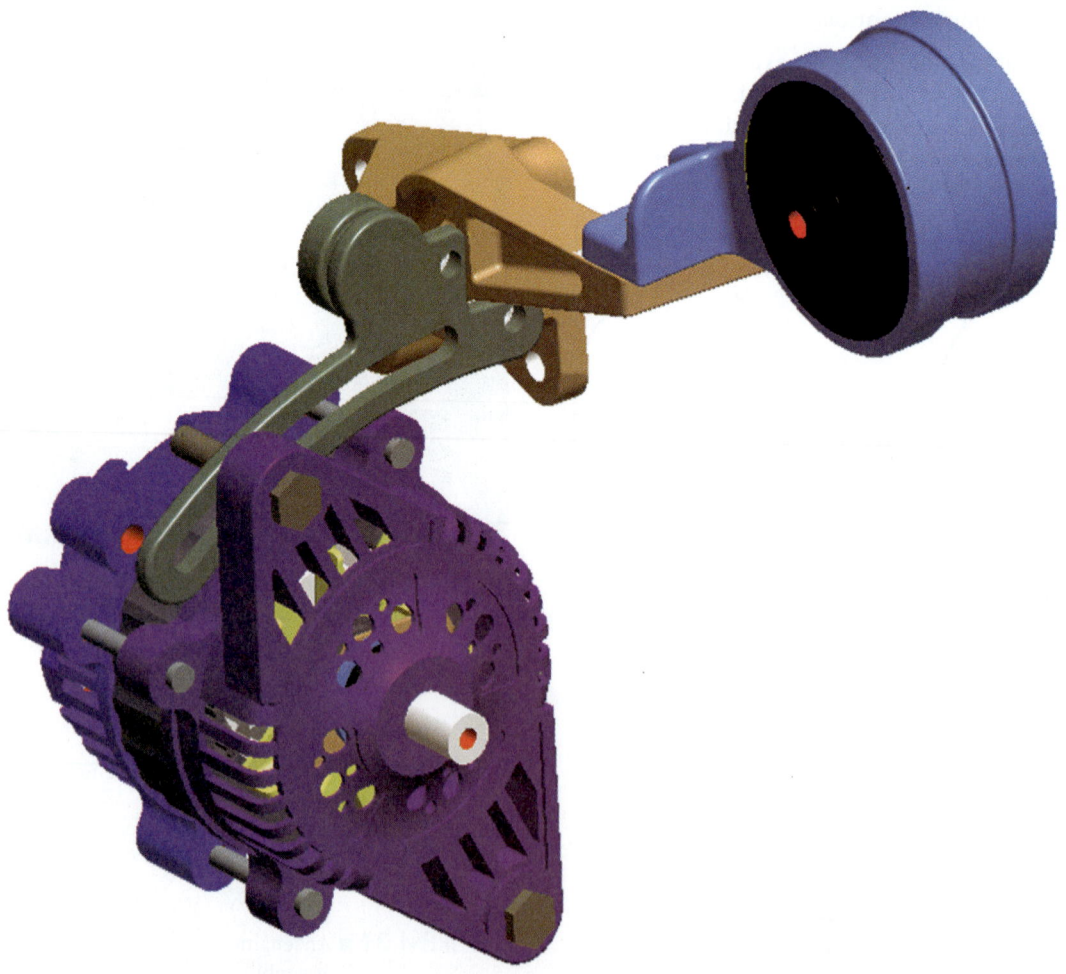

FIGURE 4.32 ■ Alternator subassembly for the assembly shown in Figure 4.33. *Courtesy PTC.*

FIGURE 4.33 ■ Engine assembly. The alternator subassembly shown in Figure 4.32 can be seen on the right rear side of the engine assembly. *Courtesy PTC.*

FIGURE 4.34 ■ Using assembly constraints, or mates, to create a parametric assembly model.

ponent inside an assembly file, a separate, fully parametric, individual part or subassembly file is also generated.

Editing Parametric Solid Models

The parametric nature of parametric solid modeling software allows you to edit model parameters anytime during the design process. Sketches, features, parameters, parts, and assemblies all can be manipulated to explore alternative design options or to adjust a model according to new or different information, because the original data used to build the model is stored in the model. Often modifying a single parameter is all that is required to revise a model. Other times, a completely different product design is built by editing several existing model parameters. The example in Figure 4.35 shows how changing a few model parameters can significantly alter a product design in very little time. Another important aspect of parametric solid modeling is that geometry is tied to the constraints and the constraints are tied to the geometry. Because of this, one automatically changes to reflect a change in the other. Also, the assembly and 2-D drawing automatically update as the models are changed, and the model changes as the others are edited.

In most cases, the tools and options used to edit models are similar or identical to the tools used to initially create the model. For example, the diameter of the hole shown in Figure 4.36 can be modified easily using the same dialog box used to initially create the hole.

CREATING 2-D PARAMETRIC DRAWINGS FROM PARAMETRIC SOLID MODELS

Some parametric solid modeling CADD systems, such as Autodesk Inventor, Pro/Engineer, and SolidWorks, have extended the functionality of parametric solid models by including parametric 2-D drawing capabilities. This ability allows you to use 3-D models to quickly and easily generate traditional 2-D part and assembly drawings. Figure 4.37 illustrates this concept by

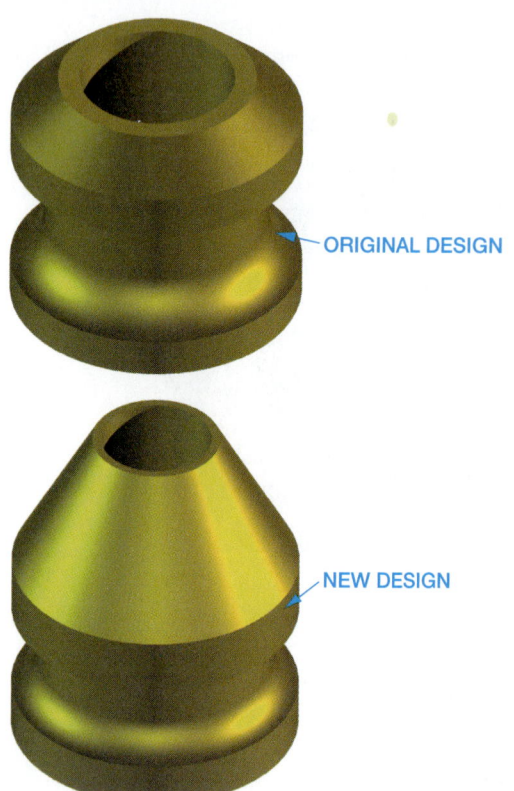

FIGURE 4.35 ■ Significantly modifying a design by quickly editing a few model parameters.

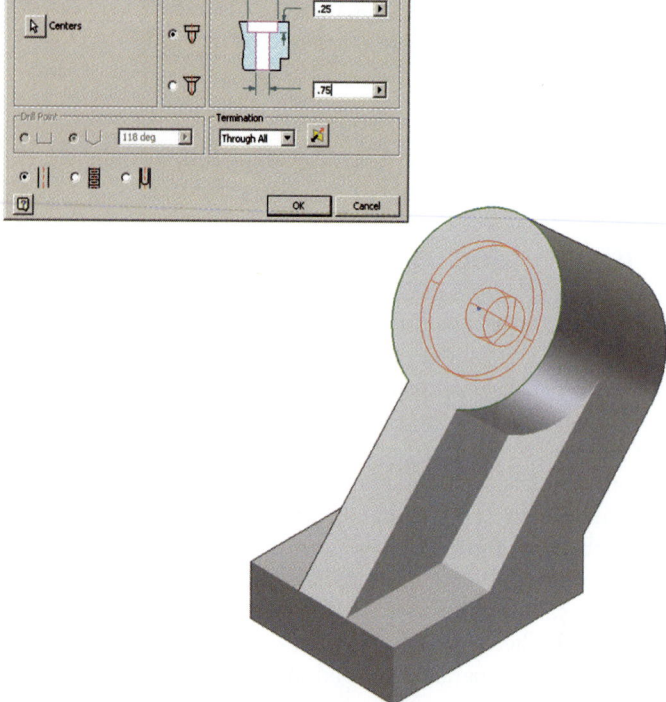

FIGURE 4.36 ■ Editing a parametric solid model hole using the Holes dialog box.

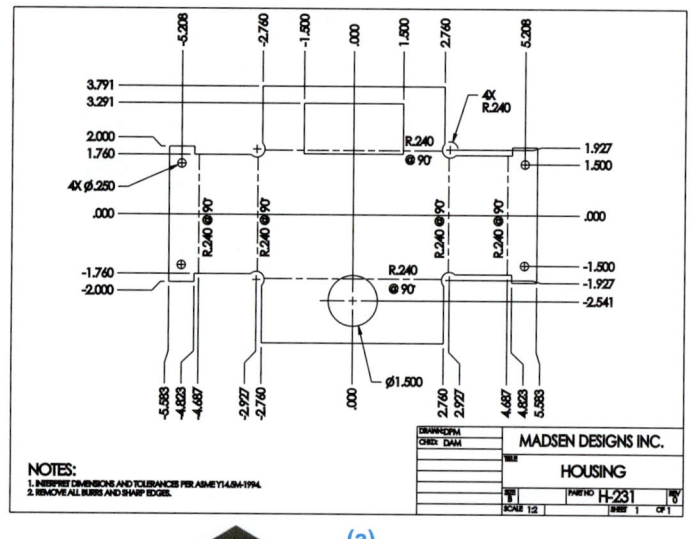

(a)

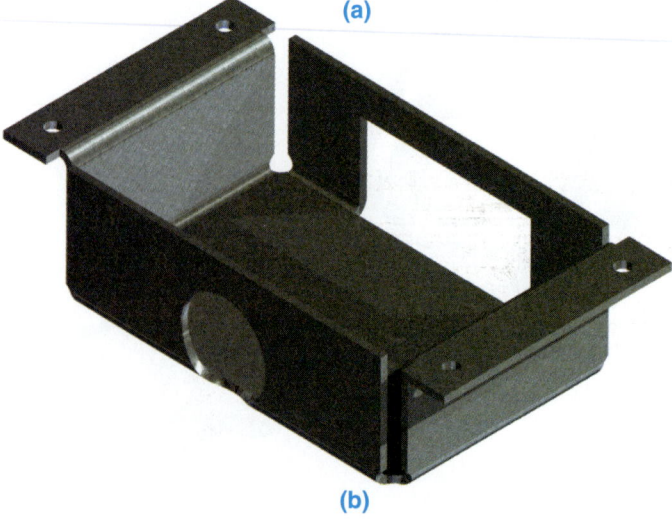

(b)

FIGURE 4.37 ■ (a) A 2-D drawing created by referencing (b) a 3-D model.

showing an example of a 2-D drawing created using information stored in a 3-D model.

Models and drawings are parametric, which means that when you edit the model, the corresponding drawing is updated to reflect the new design. Similarly, you can edit a model by modifying the parametric model dimensions inside the drawing file. A 3-D parametric solid modeling program that has 2-D drawing capabilities is truly a design and drafting system that allows you to start with very basic product design ideas and end with a virtual prototype and complete set of working drawings.

Using existing part model geometry, you can create a monodetail drawing, which is a drawing of a single part on one sheet, or a multidetail drawing, which is a drawing of several parts on one sheet. You can also reference existing assembly models to create assembly and subassembly drawings. Because models already contain all the geometry needed to produce 2-D drawings, the process of creating a 2-D drawing by referencing a model takes much less time than traditional 2-D CAD drafting. Most of the work is done in the part and assembly model environments, and it is just a matter of extracting the model information in the drawing environment to generate the 2-D drawing. Chapter 18 covers 2-D part and assembly drawings in detail.

ANIMATION

Animation is the process of making drawings or models move and change according to a sequence of predefined images. Computer animations are made by defining, or recording, a series of still images in various positions of incremental movement that, when played back, no longer appear individually as static images, but appear to be unbroken motion. Though a true animation cannot be shown in this book, Figure 4.38 provides an example of three images taken from an animation of an assembly process. Try to imagine what the complete animation looks like based on the still images shown. This animation shows the parts coming together to build the assembly. An animation of a dynamic moving assembly is shown in Figure 4.39, where a crank shaft is turned, displaying three different positions of the cylinders.

Animations are used in many ways in the design and documentation process, or they can be created for totally different purposes. The function of an animation depends on the type of application and requirement. The following information describes a few of these applications:

■ Design—CADD animations effectively describe product design. For example, moving or dragging parts and subassemblies is a very effective way to explore the motion and relationship of assembly components by showing how components in an assembly move and function, and for interference detection between parts. Another example is creating a full-length video used to show how products are assembled and disassembled. Animations help people understand and visualize designs in ways that 2-D drawings and even motionless 3-D models cannot provide. Companies often use animations to effectively communicate design ideas to customers, analyze product function, and explore alternative design concepts.

■ Film—Computer animations are used heavily in the movie and television industries to add spectacular visual effects. In fact, some animated movies and television programs are created entirely using computer animation techniques.

■ Games—Animations provide the foundation for developing computer and video games. The increasing complexity of computer animation is resulting in video games that are more realistic and exciting than ever before.

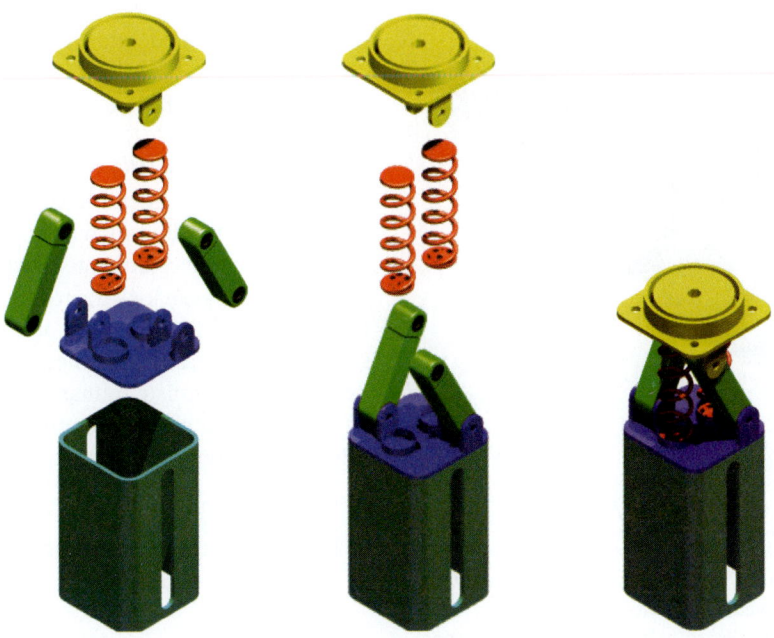

FIGURE 4.38 ■ An example of three images taken from an animation of an assembly process.

■ Education—Computer animations are a great tool for educators. Teachers and trainers create animations that can be used as an additional learning tool in the classroom or even as a complete online or distance learning presentation.

As previously described, animations can range from very simple movement of solid model components in an assembly to large scale videos or presentations complete with dialogue, music, and a variety of graphics. Many CADD programs, especially parametric solid modeling software, contain tools and options allowing you to generate basic animations. Other systems, like 3ds Max or VIZ from Autodesk, contain advanced animation tools that let you render solid models into very realistic 3-D motion simulations.

Storyboarding

It is always a good idea to do some preproduction work before you record an animation. Storyboarding is a process by which you sketch out the key events of the animation. These sketches help ensure the key scenes are included to complete the story or demonstration. The process of storyboarding is used by video producers to preplan their production before costly studio editing time is incurred because professional video editing suites can run upward of $300 an hour. As discussed earlier, rendering can take days to complete even on a high-speed computer. If scenes are left out of the animation, the animation has to be redone, causing you to lose significant time and money. Renderings, like video productions, are different from live-action film productions where improvising takes place. Improvising does not occur while rendering animations, and therefore they must be precisely preplanned.

When storyboarding an animation, keep the focus on your audience. This focus should include the overall length of the animation, key points that must be demonstrated, and how these key points are to be best illustrated. Storyboarding is a simple process that can be done on note cards or plain paper. Include sketches of the key scenes that show how these key events should be illustrated, and the time allotted for each.

Inverse Kinematics

Inverse kinematics (IK) is a method used to control how solid objects move in an assembly. IK joins solid objects together using natural links or joints such as that illustrated in the sequence of frames of the universal joint shown in Figure 4.40. For example, IK relationships can lock the rotation of an object around one particular axis. Adding this type of information allows the solid assembly to move as the finished product would move. IK is used extensively to animate human and mechanical joint movements.

Building and simulating an IK model involves a number of steps. These steps include:

■ Building a solid model of each jointed component.

■ Linking the solid model together by defining the joints.

■ Defining the joint behavior at each point, such as direction of rotation.

■ Animating the IK assembly, which can be accomplished through an animation sequence.

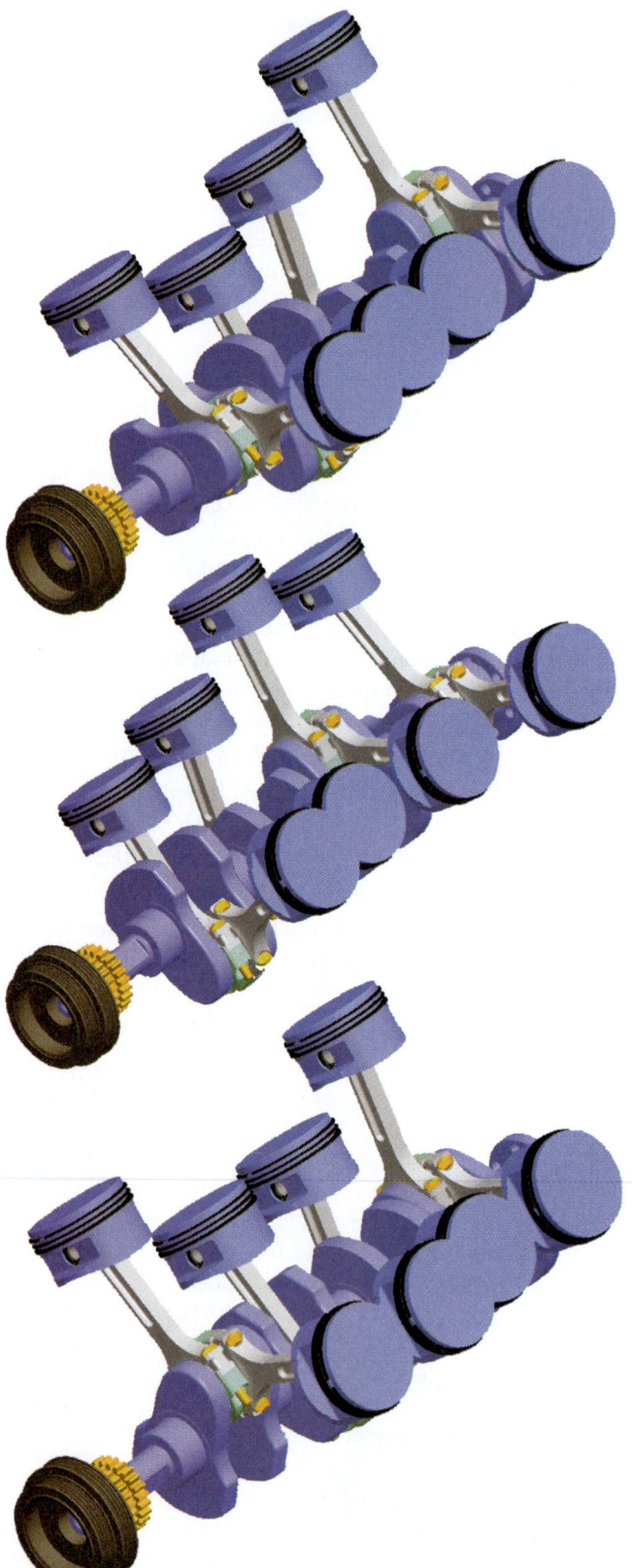

FIGURE 4.39 ■ A crank shaft and piston movement shown at three different positions. You can drive a model to show the amount of movement between components, pause movement, and detect collisions between components. A good example of a driven model can be seen with the assembly shown in this model. *Courtesy PTC.*

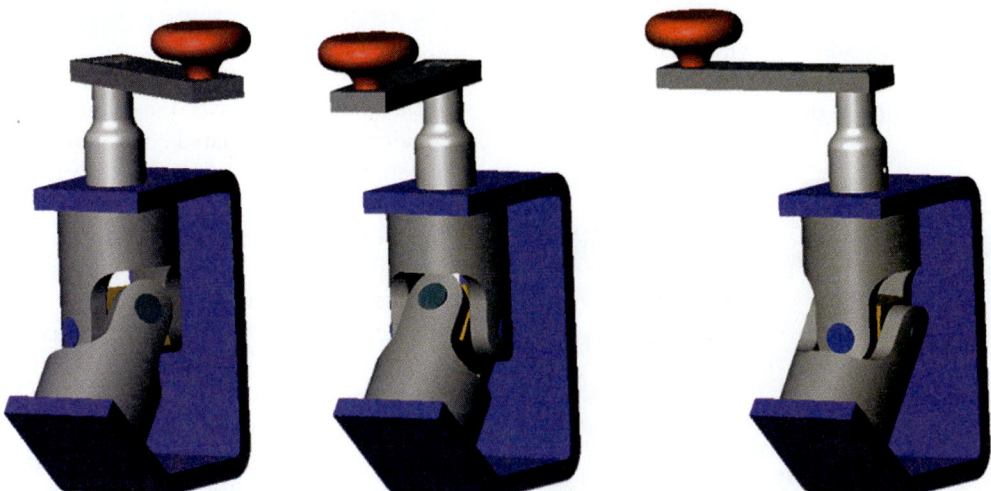

FIGURE 4.40 ■ An example of inverse kinematics (IK) is used to animate a simple part. This simple universal joint has been constrained around its axis. When the crank handle is turned, the constrained assembly rotates appropriately.

Animation can be applied to anything; it can be used in conjunction with any product to aid its explanation. For example, a robotic arm can be more easily understood if all the parts are moving. An animation that showed the parts of an assembly actually coming together in sequence is an impressive way to train assembly workers. Animation can be used purely to illustrate a product in an environment, making for an unforgettable presentation that retains people's attention. The options for animation are limited only by creativity and imagination.

Exporting Your Animations

Most rendering software allows you to preview the animation sequences before rendering is executed. This feature is a good way to verify that an animation meets your expectations. When finished, you select a rendering output file format and instruct the software to render your animation to a file. Animation software renders to a number of different file formats that allow for convenient playback. Common files formats are AVI, Quicktime, and MPEG.

INTRODUCTION TO VIRTUAL REALITY

Virtual reality (VR) is a situation in which what is seen and experienced appears real, but is not real. A popular image of VR is a person who wears a helmet or goggles containing two small flat-panel screens, one for each eye. On one hand is worn an instrumented glove that senses finger movement. The virtual reality user actually seems to be inside the model and can see his or her hand in the screen. The user can move around inside the model, and he or she can also interact with the model.

Virtual reality technology is a logical step in the design process. In the past, there has not been the ability to place a person inside the models being created and see them from a variety of viewpoints before the models are built. A VR system gives you the capability to interact with a model of any size from molecu-

lar to astronomical. Surgeons can learn on virtual patients and practice a real operation on a virtual body constructed from scanned images of the human patient. Home designers can walk around inside the house, stretching, moving, and copying shapes in order to create the finished product.

Buildings can be designed and placed on virtual building sites. Clients can take "walk-through" tours of a building before it is built and make changes as they "walk." Scientists can conduct experiments on a molecular level by placing themselves inside a model of chemical compounds. Using telerobotics, a person can "see" through a robot's eyes while in a safe virtual environment, in order to guide a robot into a hazardous situation. These applications are possible using a variety of equipment.

A common basic application VR is called through-the-window VR, also referred to as "passive" VR. It is the manipulating of a 3-D model with input from a mouse, trackball, or 3-D motion control device. This allows more than one person to see and experience the 3-D world. A variation on this is a flat-panel display with handles for movement. The window VR unit in Figure 4.41 is designed to allow natural interaction with the virtual environment. Museum and showroom visitors can walk up, grab the handles, and instantly begin interacting. Observers can follow the action by moving beside the primary user. A variety of handle-mounted buttons imitate keyboard keystrokes, joystick buttons, or 3-D motion control device buttons.

A second type of through-the-window VR consists of a stereoscopic viewing panel that mounts directly over the monitor screen. The viewer wears lightweight, passive, polarized eyewear. The viewing panel "shutters" the images directly on the screen to generate 3-D images by users wearing the special glasses. (See Figure 4.42.) This technology also allows several persons to view the same image on the screen.

A field of tremendous opportunities is developing around the creation of virtual worlds. These worlds are detailed 3-D models of a wide variety of subjects. Virtual worlds need to be constructed for many different applications. Persons who can construct realistic 3-D models can be in great demand. The fields of VR geographic information systems (GIS) are combined to create "intelligent" worlds from which data can be obtained while occupying the virtual world. In the future, many cities will have a virtual model on their website.

FIGURE 4.41 ■ This unit is designed to allow natural interaction with the virtual environment. Museum and showroom visitors can walk up, grab the handles, and instantly begin interacting. Observers can follow the action by moving beside the primary user. *Courtesy Virtual Research Systems, Inc.*

FIGURE 4.42 ■ This type of through-the-window VR consists of a stereoscopic viewing panel, which mounts directly over the monitor screen. The viewer wears lightweight, passive, polarized eyewear. *Courtesy Stereo-Graphics Corporation.*

USING VIRTUAL REALITY

As previously introduced, virtual reality is a term used to describe a system that allows one or more people to move and react in a computer-simulated environment. In this environment, you manipulate virtual objects using various types of devices as though they are real objects. This simulated world gives you a feeling of being immersed in a real world.

Virtual reality requires special interface devices that transmit the sights, sounds, and sensations of the simulated world. In return, these devices record your speech and movement and transmit them back to the simulation software program.

To interact visually with the simulated world, you wear a head-mounted display (HMD), which directs computer images at each eye (see Figure 4.43). The HMD tracks your head movements, including the direction in which you are looking. Using this movement information, the HMD receives updated images from the computer system, which is continually recalculating the virtual world based on your head motions. The computer generates new views at a fast rate, which prevents the view from appearing halting and jerky and from lagging behind your movements. The HMD can also deliver sounds to your earphones. The tracking feature of the HMD also can be used to update the audio signal to simulate surround sound effects.

HMD technology is moving toward smaller, less cumbersome units. The use of lightweight eyewear, a special monitor, and sensing devices allow you to see in 3-D, as graphically demonstrated in Figure 4.44.

Two alternative concepts have been introduced to overcome uncomfortable head-mounted display gear: the BOOM and the CAVE. The Binocular Omni-Orientation Monitor (BOOM) from Fakespace is a head-coupled stereoscopic display device (see Figure 4.45). The heavy display, instead of being worn, is attached to a multilink arm system that is counterbalanced. The person can guide the counterbalanced display while looking into it like binoculars. The system is guided with tracking that is attached to the counterbalanced arms. The Cave Automatic Virtual Environment (CAVE) projects stereo images on the

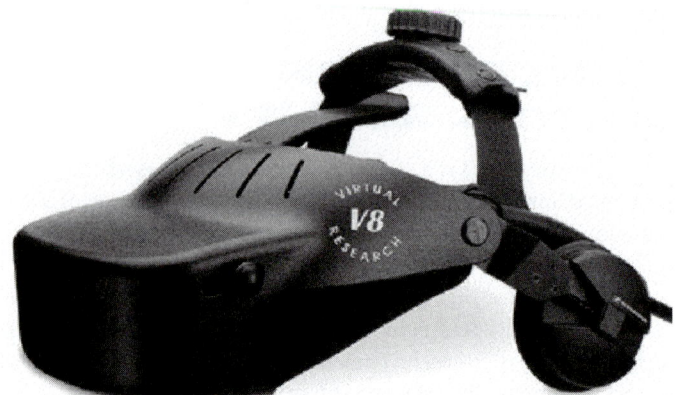

FIGURE 4.43 ■ This head-mounted display incorporates both stereo viewing with two small monitors and stereo sound using the attached earphones; it is the principal component of *immersive* VR. *Courtesy Virtual Research Systems, Inc.*

FIGURE 4.44 ■ This lightweight eyewear, a special monitor, and sensing devices generate alternating images, allowing you to see in 3-D. *Courtesy StereoGraphics Corporation.*

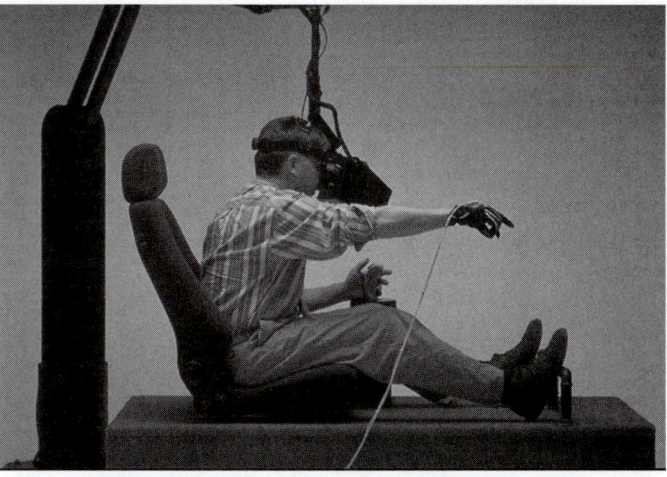

FIGURE 4.45b ■ This *suspended display system* is designed for applications such as vehicle simulation and cockpit modeling. The operator can stand or sit, with both hands free to manipulate real or virtual controls and input devices. *Courtesy Fakespace Systems Inc.*

FIGURE 4.45a ■ Binocular Omni-Orientation Monitor (BOOM) allows viewers to view 3-D images without wearing a head-mounted display. *Courtesy Fakespace Systems Inc.*

walls and floor of a room (see Figure 4.46). CAVE was developed at the University of Illinois at Chicago to allow users to wear lightweight stereo glasses and to walk around freely inside the virtual environment. Several persons can participate within the CAVE environment as the tracking system follows the lead viewer's position and movements.

The most challenging physical sensation to simulate in a virtual world is the sense of touch. A haptic interface is a device that relays the sense of touch and other physical sensations. In this environment, your hand and finger movements can be tracked, allowing you to reach into the virtual world and handle objects. Haptic interfaces of this type hold great potential for design engineers, allowing various team members to manipulate a product design in a virtual environment in a natural way.

Although you can handle an object, it is difficult to generate sensations associated with human touch, for example, the sensations that are felt when a person touches a soft surface, picks up a heavy object, or runs a finger across a bumpy surface. To simulate these sensations, very accurate and fast computer-controlled motors generate force by pushing against the user. These haptic devices are synchronized with HMD sight and sound, and the motors must be small enough to be worn without interfering with your natural movement. A simple haptic device is the desktop stylus (see Figure 4.47). This device can apply a small force, through a mechanical linkage, to a stylus held in the user's hand. When the stylus encounters a virtual object, the user is provided feedback that simulates the interaction. In addition, if the stylus is dragged across a textured surface, it responds with the proper vibration.

In the future, engineers may use VR to increase productivity in a variety of areas, including virtual mock-up, assembly, and design reviews. These applications may include the realistic simulation of human factors, such as snap-fits, key component function, and experiencing of virtual forms. Virtual assemblies may include fit evaluation, maintenance path planning, manufacturability analysis, and assembly training.

FIGURE 4.46 ■ Cave Automatic Virtual Environment (CAVE) allows viewers to view 3-D images projected on the walls of a room. *Courtesy Fakespace Systems Inc.*

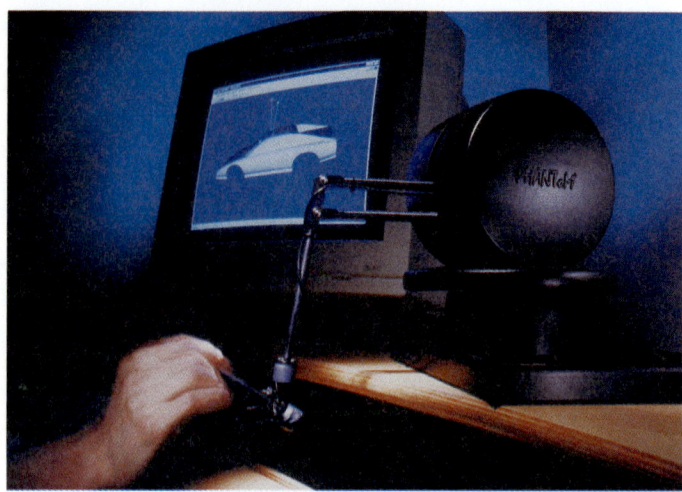

FIGURE 4.47 ■ Phantom haptic device provides haptic feedback to the user. *Courtesy SensAble Technologies.*

An emerging area in the world of virtual reality is Web-enabled **virtual reality modeling language (VRML)**. VRML is a formatting language that is used to publish virtual 3-D settings, called "worlds," on the World Wide Web. Once the developer has placed the world on the Internet, the user can view it using a Web browser plug-in. This web-browser plug-in contains controls that allow the user to move around in the virtual world, as the user would like to experience it. Currently VRML is a standard authoring language that provides authoring tools for

PROFESSIONAL PERSPECTIVE

CADD solid modeling has significantly changed the process by which products at Waterpik® Technologies, Inc., are developed. Waterpik Technologies, Inc. uses Pro/Engineer to develop its products. This high-end solid modeler has allowed the product development team to reduce the product lifecycle, which is the time it takes a product to get to the market and has allowed them to remain competitive in today's ultra-aggressive marketplace.

Waterpik Technologies began using Pro/Engineer about 15 years ago. Within this short period of time, the number and diversity of products introduced has changed significantly. Many more new products are now introduced annually than in the past. Furthermore, the new products are being developed using a smaller number of employees than in previous years. Simply stated, a smaller number of people have been able to produce a significantly larger number of products. These gains in product development have been made possible because of the advent of CADD solid modeling.

CADD solid models have provided an important tool for improving communications. At Waterpik, the CADD solid model has been used for demonstrating future products. Before solid models, industrial engineers relied on basic multiview drawings or handmade models. Multiview 2-D drawings are difficult for the untrained person to read and interpret. Therefore, a communication gap was created between those who have difficulty reading the drawings such as marketing and sales, and those who could read the 2-D drawings such as engineering and manufacturing. The CADD solid model has given marketing, sales, and engineering a way to visualize the design before any physical models are created. This improvement in communications has shortened the time needed to make a product and has improved the product's design.

Maybe even more significant is the impact that solid modeling has had on the appearance of the products at Waterpik. According to Tim Hanson, Waterpik Technologies, "Solid modeling has significantly changed the complexity of our products. Before we had to design things that were easy to machine. This meant products with straight lines or minimal curves. Today our products do not have a straight line on them." These changes can be seen by comparing Waterpik's designs from before and after the introduction of solid modeling. The toothbrush shown in Figure 4.48 was produced in 1989, before solid modeling. Notice that the product's shape is based on simple forms and straight

PROFESSIONAL PERSPECTIVE

lines. In contrast, the more modern toothbrush shown in Figure 4.49 was produced using solid modeling. This product has very few simple forms and is comprised of a number of complex surfaces. Furthermore, the Flosser shown in Figure 4.50 has an extremely complex form that incorporates a variety of shapes and complex curves. These complex shapes and forms simply were not possible before CADD solid modeling was introduced.

All of these advances in speed, communication, and product design occurred at Waterpik because of solid modeling. "Without the solid-modeling tool we would not be able to produce the number of high-quality products that we produce today," according to Tim Hanson, Waterpik Technologies.

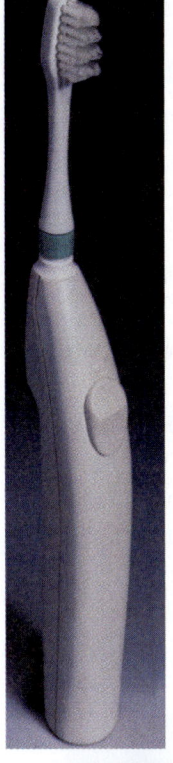

FIGURE 4.49 ■ An example of a toothbrush design after the introduction of solid modeling. *Courtesy Waterpik Technologies, Inc.*

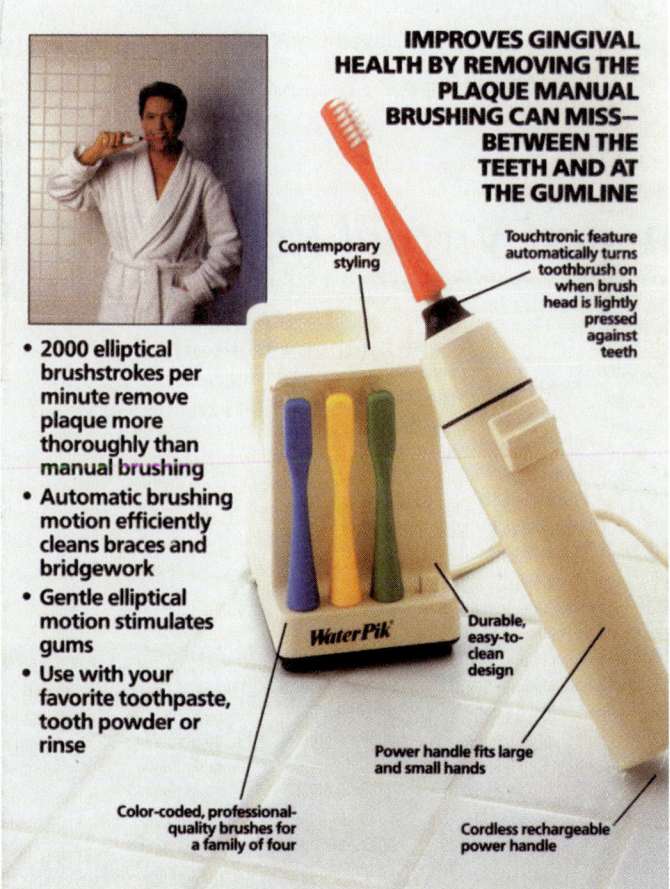

IMPROVES GINGIVAL HEALTH BY REMOVING THE PLAQUE MANUAL BRUSHING CAN MISS— BETWEEN THE TEETH AND AT THE GUMLINE

Contemporary styling

Touchtronic feature automatically turns toothbrush on when brush head is lightly pressed against teeth

- 2000 elliptical brushstrokes per minute remove plaque more thoroughly than manual brushing
- Automatic brushing motion efficiently cleans braces and bridgework
- Gentle elliptical motion stimulates gums
- Use with your favorite toothpaste, tooth powder or rinse

WaterPik

Durable, easy-to-clean design

Color-coded, professional-quality brushes for a family of four

Power handle fits large and small hands

Cordless rechargeable power handle

FIGURE 4.48 ■ An example of a toothbrush design before the introduction of solid modeling. *Courtesy Waterpik Technologies, Inc.*

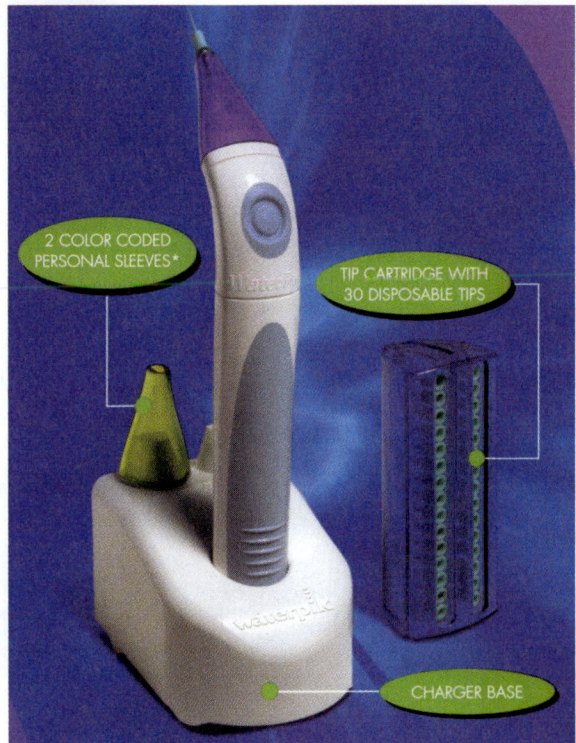

2 COLOR CODED PERSONAL SLEEVES*

TIP CARTRIDGE WITH 30 DISPOSABLE TIPS

CHARGER BASE

FIGURE 4.50 ■ An example of a freeform design of the Teledyne Flosser. *Courtesy Waterpik Technologies, Inc.*

the creation of 3-D worlds with integrated hyperlinks. The current version of VRML is viewed using a basic computer monitor and therefore is not fully immersive. However, the future of VRML should incorporate the use of HMDs and haptic devices, making for more truly immersive environments.

WEBSITE RESEARCH

The following websites can provide you with additional information for research or further study into topics covered in this chapter:

http://www.adda.org—American Design Drafting Association.

http://www.ansi.org—The American National Standards Institute.

http://www.asme.org—The American Society of Mechanical Engineers.

http://www.amazon.com—Search author, Madsen, for Auto-CAD, Inventor, and SolidWorks textbook titles.

http://www.autodesk.com—Autodesk Inventor and related 3-D CADD software products.

http://www.bentley.com—Bentley Systems, Inc., Microstation, and related products.

http://www.cadkey.de—AGS Advanced Graphics Software, CADKEY, KeyCreator, and related products.

http://cave.ncsa.uiuc.edu—National Computational Science Alliance, Cave Automatic Virtual Environment.

http://www.fakespace.com—Fakespace Systems Inc.

http://www.graphisoft.com—Graphisoft, ArchiCAD, and related products.

http://www.g-w.com—Publisher of AutoCAD, Inventor, and SolidWorks related textbooks by David A. Madsen and David P. Madsen.

http://www.ptc.com—Parametric Technology Corporation, Pro/Engineer, and related software.

http://www.sensable.com—SensAble Technologies.

http://www.softplan.com—SoftPlan System,Inc., SoftPlan, and related products.

http://www.solidworks.com—SolidWorks Corporation, SolidWorks, and related products.

http://www.stereographics.com—StereoGraphics Corporation.

http://www.ugs.com—UGS Corporation, Solid Edge 3-D CADD software.

http://www.virtualresearch.com—Virtual Research Systems.

http://www.waterpik.com—Waterpik Technologies.

http://www.3ds.com—Dassault Systemes, CATIA, and related products.

CHAPTER 4

3-D CADD, Animation, and Virtual Reality Test

 Access the CD found with this textbook to view the Chapter 4 Test. Confirm the preferred submittal method with your instructor.

CHAPTER 4

3-D CADD, Animation, and Virtual Reality Problems

This chapter is intended as a reference for 3-D CADD animation and virtual reality as used in the design and drafting industry. As technologies surrounding these concepts become more advanced, their use in industry changes. As you work in a design and drafting field, you must continually learn and make decisions about how new technologies improve job performance. The following problems have been designed to help you explore these concepts further and collect additional resources on these topics.

PROBLEM 4.1   This assignment requires that you visit your library or World Wide Web resource to locate articles related to one topic from the following list. Research professional and trade journals to find a related article. Write a critique of that article. Your article critique should include an introduction, body, and critical review of the article. The introduction should include an overview of the article. The body should contain the main points of the article. The critique should provide a critical review of the article. In order to complete the critique section, you may want to ask yourself the following questions: For whom is the book or article written? Are the author's viewpoints consistent with other authorities? Are the author's viewpoints supported with valid reasons? Does the author adequately cover the subject? Where do you agree or disagree with the author? Why? How might the article be strengthened? What implications do you find for your field of work?

- 3-D CADD
- Parametric modeling
- Animation
- Virtual reality

PROBLEM 4.2   This assignment requires that you visit the World Wide Web to find additional information on the use of 3-D CADD, animation, or virtual reality. Select one or more of the following industries and locate two companies providing related services. Develop a written and/or oral presentation on each company's services, depending on your course objectives. Include examples of the technologies each company uses to provide these services.

- Household consumer products
- Portable electric tools
- Aeronautical engineering
- Aerospace engineering
- Automotive engineering
- Light equipment engineering
- Heavy equipment engineering
- Consumer electronics
- Residential architecture
- Commercial architecture
- Automation and robotics
- Medical device design and manufacturing
- Medical animation
- Design intent

PROBLEM 4.3   This assignment requires that you conduct research on 3-D CADD applications. Visit two solid modeling software companies via the World Wide Web. Develop a report about those companies. Provide information about the following items related to each company.

- Functionality of software, such as analysis tools and surface modeling
- Major customers/users of software
- How each company makes revisions or updates to its software
- Other pertinent information

PROBLEM 4.4   Conduct research and write a report of 200 words or more in length, using illustrative examples, covering on one or more of the following topics related to 3-D CADD:

- A comparison between wireframe modeling, surface modeling, and solid modeling

- Parametric solid models
- The parametric solid model work environment
- Parametric solid modeling elements
- Parametric solid assembly models
- Creating 2-D parametric drawings from parametric solid models

PROBLEM 4.5   Conduct research and write a report of 200 words or more in length, using illustrative examples, covering animation as related to one or more of the following fields:

- Design
- Film
- Games
- Education
- Architecture

PROBLEM 4.6   Conduct research and write a report of 200 words or more in length, using illustrative examples, covering on one or more of the following topics related to animation:

- Storyboarding
- Inverse kinematics

PROBLEM 4.7   Conduct research and write a report of 200 words or more in length, using illustrative examples, covering innovative animation topics not covered in this chapter.

PROBLEM 4.8   Conduct research and write a report of 200 words or more in length, using illustrative examples, covering one or more of the following topics related to virtual reality:

- Head-mounted displays
- Walk through tours
- Passive VR
- Stereoscopic viewing of VR
- Virtual reality and geographic information systems
- Binocular Omni-Orientation Monitor
- Cave Automatic Virtual Environment

PROBLEM 4.9   Conduct research and write a report of 200 words or more in length, using illustrative examples, covering innovative virtual reality topics not covered in this chapter.

PROBLEM 4.10   Conduct research and write a report of 200 words or more in length, using illustrative examples, covering at least two hardware or software programs related to each of the fields of animation and virtual reality.

Manufacturing Materials and Processes

LEARNING OBJECTIVES

After completing this chapter, you will:

■ Define and describe various manufacturing materials, material terminology, numbering systems, and material treatment.

■ Discuss casting processes and terminology.

■ Explain the forging process and terminology.

■ Describe manufacturing processes.

■ Define and draw the representation of various machined features.

■ Explain tool design and drafting practices.

■ Draw a basic machine tool.

■ Discuss the statistical process quality control assurance system.

■ Evaluate the results of an engineering and manufacturing problem.

■ Explain the use of computer-aided manufacturing (CAM) in today's industry.

■ Discuss robotics in industry.

■ Identify a variety of manufacturing processes used to create plastic products.

THE ENGINEERING DESIGN APPLICATION

Your company produces die cast aluminum parts. A customer has explained that any delay in parts shipment requires them to shut down a production line. It is very costly. It is critical to your customer that certain features on their castings fall within specification limits, and the engineering department has decided to develop an early warning system in an attempt to prevent production delays.

You are asked to develop new quality control charts with two levels of control. The maximum allowable deviation is indicated by the upper and lower specification limits as specified by the dimensional tolerances on the part drawing. A mean value is established between the nominal dimension and dimensional limits, and these are set up as upper and lower control limits. The chart instructs the inspector to immediately notify the supervisor if the feature falls outside of control limits, and to halt production if it is outside of specification limits. This provides a means of addressing problems before they can interfere with production, shifting the emphasis from revision to prevention. A sample chart with two levels of control is shown in Figure 5.1.

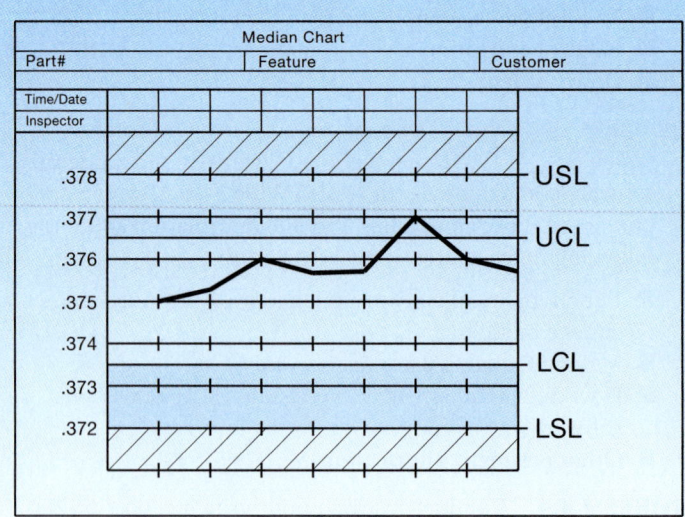

FIGURE 5.1 ■ Sample quality control chart with both control and specification limits.

INTRODUCTION

A number of factors influence product manufacturing. Beginning with research and development (R & D), a product should be designed to meet a market demand, have good quality, and be economically produced. The sequence of product development begins with an idea and results in a marketable commodity, as shown in Figure 5.2.

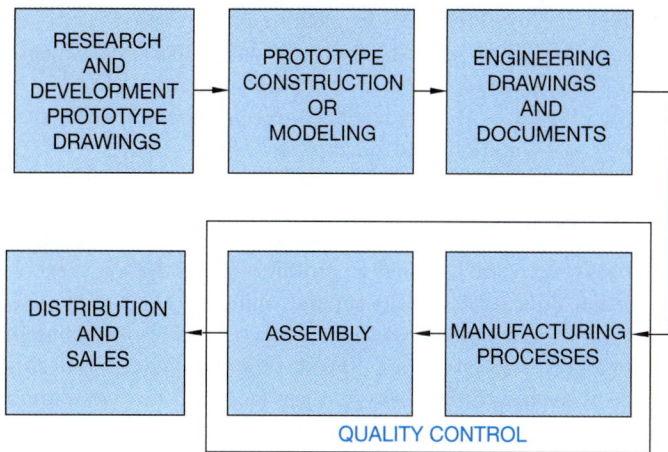

FIGURE 5.21 ■ Sequence of product development.

MANUFACTURING MATERIALS

There is a wide variety of materials available for product manufacturing that fall into three general categories: metal, plastic, and inorganic materials. Metals are classified as ferrous, nonferrous, and alloys. Ferrous metals contain iron, such as cast iron and steel. Nonferrous metals do not have iron content; for example, copper and aluminum. Alloys are a mixture of two or more metals.

Plastics or polymers have two types of structure: thermoplastic and thermoset. Thermoplastic material can be heated and formed by pressure and, upon reheating, the shape can be changed. Thermoset plastics are formed into a permanent shape by heat and pressure and cannot be altered by heating after curing. Plastics are molded into shape and only require machining for tight tolerance situations or when holes or other features are required that would be impractical to produce in a mold. It is common practice to machine some plastics for parts, such as gears and pinions.

Inorganic materials include carbon, ceramics, and composites. Carbon and graphite are classified together and have properties that allow molding by pressure. These materials have low tensile strength (ability to be stretched) and high compressive strength, with increasing strength at increased temperatures. Ceramics are clay, glass, refractory, and inorganic cements. Ceramics are very hard, brittle materials that are resistant to heat, chemicals, and corrosion. Clay and glass materials have an amorphous structure, while refractories must be bonded together by applying temperatures. Due to their great heat resistance, refractories are used for high-temperature applications, such as furnace liners. Composites are two or more materials that are bonded together by adhesion. Adhesion is a force that holds together the molecules of unlike substances when the surfaces come in contact. These materials generally require carbide cutting tools or special methods for machining.

Ferrous Metals

The two main types of ferrous metals are cast iron and steel. These are the metals that contain iron. There are many classes of cast iron and steel that are created for specific types of application based on their composition.

Cast Iron

There are several classes of cast iron, including gray, white, chilled, alloy, malleable, and nodular cast iron. Cast iron is primarily an alloy of iron and 1.7% to 4.5% of carbon, with varying amounts of silicon, manganese, phosphorus, and sulfur.

Gray Cast Iron

Gray iron is a popular casting material for automotive cylinder blocks, machine tools, agricultural implements, and cast iron pipe. Gray cast iron is easily cast and machined. It contains 1.7% to 4.5% carbon and 1% to 3% silicon.

> **ANSI/ASTM** The American National Standards Institute (ANSI) and American Society for Testing Materials (ASTM) specifications A48–76 group gray cast iron into two classes: easy to manufacture (20A, 20B, 20C, 25A, 25B, 25C, 30A, 30B, 30C, 35A, 35B, 35C); and more difficult to manufacture (40B, 40C, 45B, 45C, 50B, 50C, 60B, 60C). The prefix denotes the minimum tensile strength in thousand pounds per square inch.

White Cast Iron

White cast iron is extremely hard, brittle, and has almost no ductility. Ductility is the ability to be stretched, drawn, or hammered thin without breaking. Caution should be exercised when using this material, because thin sections and sharp corners may be weak and the material is less resistant to impact uses. This cast iron is suited for products with more compressive strength requirements than gray cast iron (compare over 200,000 pounds per square inch [psi] with 65,000 to 160,000 psi). White cast iron is used where high wear resistance is required.

Chilled Cast Iron

When gray iron castings are chilled rapidly, an outer surface of white cast iron results. This material has the internal characteristics of gray cast iron and the surface advantage of white cast iron.

Alloy Cast Iron

Elements such as nickel, chromium, molybdenum, copper, or manganese can be alloyed with cast iron to increase the properties of strength, wear resistance, corrosion resistance, or heat resistance. Alloy iron castings are commonly used for such items as pistons, crankcases, brake drums, and crushing machinery.

Malleable Cast Iron

The term malleable means the ability to be hammered or pressed into shape without breaking. Malleable cast iron is produced by heat-treating white cast iron. The result is a stronger, more ductile, and shock-resistant cast iron that is easily machinable.

ANSI/ASTM The specifications for malleable cast iron are found in ANSI/ASTM A47–77.

Nodular Cast Iron

Special processing procedures, along with the addition of magnesium- or cerium-bearing alloys, result in a cast iron with spherical-shaped graphite rather than flakes, as in gray cast iron. The results are iron castings with greater strength and ductility. Nodular cast iron can be chilled to form a wear-resistant surface ideal for use in crankshafts, anvils, wrenches, or heavy-use levers.

Steel

Steel is an alloy of iron containing 0.8% to 1.5% carbon. Steel is a readily available material that can be worked in either a heated or cooled state. The properties of steel can be changed by altering the carbon content and heat treating. Mild steel (MS) is low in carbon (less than 0.3%), and is commonly used for forged and machined parts, but cannot be hardened. Medium carbon steel (0.3% to 0.6% carbon) is harder than mild steel yet remains easy to forge and machine. High carbon steel (0.6% to 1.50% carbon) can be hardened by heat treating, but is difficult to forge, machine, or weld.

Hot-rolled steel (HRS) characterizes steel that is formed into shape by pressure between rollers or by forging when in a red-hot state. When in this hot condition, the steel is easier to form than when it is cold. An added advantage of hot forming is a consistency in the grain structure of the steel, which results in a stronger, more ductile metal. The surface of hot-rolled steel is rough, with a blue-black oxide buildup. The term cold-rolled steel (CRS) implies the additional forming of steel after initial hot rolling. The cold-rolling process is used to clean up hot-formed steel, provide a smooth, clean surface, ensure dimensional accuracy, and increase the tensile strength of the finished product.

Steel alloys are used to increase such properties as hardness, strength, corrosion resistance, heat resistance, and wear resistance. Chromium steel is the basis for stainless steel and is used where corrosion and wear resistance is required. Manganese alloyed with steel is a purifying element that adds strength for parts that must be shock and wear resistant. Molybdenum is added to steel when the product must retain strength and wear resistance at high temperatures. When tungsten is added to steel, the result is a material that is very hard and ideal for use in cutting tools. Tool steels are high in carbon and/or alloy content so that the steel will hold an edge when cutting other materials. When the cutting tool requires deep cutting at high speed, then the alloy and hardness characteristics are improved for a classification known as high-speed steel. Vanadium alloy is used when a tough, strong, nonbrittle material is required.

Steel castings are used for machine parts where the use requires heavy loads and the ability to withstand shock. These castings are generally stronger and tougher than cast iron. Steel castings have uses as turbine wheels, forging presses, gears, machinery parts, and railroad car frames.

Stainless Steel

Stainless steels are high-alloy chromium steels that have excellent corrosion resistance. In general, stainless steels contain at least 10.5% chromium, with some classifications of stainless steel having between 4% and 30% chromium. In addition to corrosion resistance properties, stainless steel can have oxidation and heat resistance and can have very high strength. Stainless steel is commonly used for restaurant and hospital equipment and for architectural and marine applications. The high strength-to-weight ratio also makes stainless steel a good material for some aircraft applications.

Stainless steel is known for its natural luster and shine that makes it look like chrome. It is also possible to add color to stainless steel, which is used for architectural products such as roofing, hardware, furniture, kitchen, and bathroom fixtures.

The American Iron and Steel Institute (AISI) identifies stainless steels with a system of 200, 300, or 400 series numbers. The 200 series stainless steels contain chromium, nickel, and manganese; the 300 series steels have chromium and nickel; and the 400 series steels are straight-chromium stainless steels.

Steel Numbering Systems

AISI/SAE. The American Iron and Steel Institute (AISI) and the Society of Automotive Engineers (SAE) provide similar steel numbering systems. Steels are identified by four numbers, except for some chromium steels, which have five numbers. For a steel with the identification SAE 1020, the first two numbers (10) identify the type of steel and the last two numbers (20) specify the approximate amount of carbon in hundredths of a percent (0.20% carbon). The letter L or B may be placed between the first and second pair of numbers. When this is done, the L means that lead is added to improve machinability, and the B identifies a boron steel. The prefix E means that the steel is made using the electric furnace method. The prefix H indicates that the steel is produced to hardenability limits. Steel that is degassed and deoxidized before solidification is referred to as killed steel and is used for forging, heat treating, and difficult stampings. Steel that is cast with little or no degasification is known as rimmed steel, and has applications where sheets, strips, rods, and wires with excellent surface finish or drawing requirements are needed. General applications of SAE steels are shown in Appendix D. For a more in-depth analysis of steel and other metals, refer to the *Machinery's Handbook.** Additional information about numbering systems is provided later in this chapter.

---

*Erick Oberg, Franklin D. Jones, and Holbrook L. Horton, *Machinery's Handbook,* 27th ed. (New York. Industrial Press Inc., 2004).

Hardening of Steel

The properties of steel can be altered by heat treating. Heat treating is a process of heating and cooling steel using specific controlled conditions and techniques. When steel is initially formed and allowed to cool naturally, it is fairly soft. Normalizing is a process of heating the steel to a specified temperature and then allowing the material to cool slowly by air, which brings the steel to a normal state. In order to harden the steel, the metal is first heated to a specified temperature, which varies with different steels. Next the steel is quenched, which means to cool suddenly by plunging into water, oil, or other liquid. Steel can also be case hardened using a process known as carburization. Case hardening refers to the hardening of the surface layer of the metal. Carburization is a process where carbon is introduced into the metal by heating to a specified temperature range while in contact with a solid, liquid, or gas material consisting of carbon. This process is often followed by quenching to enhance the hardening process. Tempering is a process of reheating a normalized or hardened steel through a controlled process of heating the metal to a specified temperature, followed by cooling at a predetermined rate to achieve certain hardening characteristics. For example, the tip of a tool may be hardened while the rest of the tool remains unchanged.

Under certain heating and cooling conditions and techniques, steel can also be softened using a process known as annealing.

Hardness Testing

There are several methods of checking material hardness. The techniques have common characteristics based on the depth of penetration of a measuring device or other mechanical systems that evaluate hardness. The Brinell and Rockwell hardness tests are popular. The Brinell test is performed by placing a known load, using a ball of a specified diameter, in contact with the material surface. The diameter of the resulting impression in the material is measured and the Brinell Hardness Number (BHN) is then calculated. The Rockwell hardness test is performed using a machine that measures hardness by determining the depth of penetration of a spherical-shaped device under controlled conditions. There are several Rockwell hardness scales depending on the type of material, the type of penetrator, and the load applied to the device. A general or specific note on a drawing that requires a hardness specification may read: CASE HARDEN 58 PER ROCKWELL "C" SCALE. For additional information, refer to the *Machinery's Handbook*.

Nonferrous Metals

Nonferrous metals do not contain iron and have properties that are better suited for certain applications where steel may not be appropriate.

Aluminum

Aluminum is corrosion resistant, lightweight, easily cast, conductive of heat and electricity, may be easily extruded, and is very malleable. Extruding is shaping the metal by forcing it through a die. Pure aluminum is seldom used, but alloying it with other elements provides materials that have an extensive variety of applications. Some aluminum alloys lose strength at temperatures generally above 250°F (121°C), while they gain strength at cold temperatures. There are a variety of aluminum alloy numerical designations used. A two- or three-digit number is used, and the first digit indicates the alloy type, as follows: 1 = 99% pure, 2 = copper, 3 = manganese, 4 = silicon, 5 = magnesium, 6 = magnesium and silicon, 7 = zinc, 8 = other. For designations above 99%, the last two digits of the code are the amount over 99%. For example, 1030 means 99.30% aluminum. The second digit is any number between 0 and 9 where 0 means no control of specific impurities and numbers 1 through 9 identify control of individual impurities.

Copper Alloys

Copper is easily rolled and drawn into wire, has excellent corrosion resistance, is a great electrical conductor, and has better ductility than any metal except for silver and gold. Copper is alloyed with many different metals for specific advantages, which include improved hardness, casting ability, machinability, corrosion resistance, elastic properties, and lower cost.

Brass

Brass is a widely used alloy of copper and zinc. Its properties include corrosion resistance, strength, and ductility. For most commercial applications, brass has about a 90% copper and 10% zinc content. Brass can be manufactured by any number of processes including casting, forging, stamping, or drawing. Its uses include valves, plumbing pipe and fittings, and radiator cores. Brass with greater zinc content can be used for applications requiring greater ductility, such as cartridge cases, sheet metal, or tubing.

Bronze

Bronze is an alloy of copper and tin. Tin in small quantities adds hardness and increases wear resistance. Tin content in coins and medallions, for example, ranges from 4% to 8%. Increasing amounts of tin also improves the hardness and wear resistance of the material but causes brittleness. Phosphorus added to bronze (phosphor bronze) increases its casting ability and aids in the production of more solid castings, which is important for thin shapes. Other materials, such as lead, aluminum, iron, and nickel, can be added to copper for specific applications.

Precious and Other Specialty Metals

Precious metals include gold, silver, and platinum. These metals are valuable because they are rare, costly to produce, and have specific properties that influence use in certain applications.

Gold

Gold for coins and jewelry is commonly hardened by adding copper. Gold coins, for example, are 90% gold and 10% copper.

The term carat is used to refer to the purity of gold, where 1/24 gold is one carat. Therefore, $24 \times 1/24$ or 24 carats is pure gold. Fourteen-carat gold, for example, is 14/24 gold. Gold is extremely malleable, corrosion resistant, and is the best conductor of electricity. In addition to use in jewelry and coins, gold is used as a conductor in some electronic circuitry applications. Gold is also used in applications where resistance to chemical corrosion is required.

Silver

Silver is alloyed with 8% to 10% copper for use in jewelry and coins. Sterling silver for use in such items as eating utensils and other household items is 925/1000 silver. Silver is easy to shape, cast, or form, and the finished product can be polished to a high-luster finish. Silver has uses similar to gold because of corrosion resistance and ability to conduct electricity.

Platinum

Platinum is rarer and more expensive than gold. Industrial uses include applications where corrosion resistance and a high melting point are required. Platinum is used in catalytic converters because it has the unique ability to react with and reduce carbon monoxide and other harmful exhaust emissions in automobiles. The high melting point of platinum makes it desirable in certain aerospace applications.

Columbium

Columbium is used in nuclear reactors because it has a very high melting point, 4380°F (2403°C), and is resistant to radiation.

Titanium

Titanium has many uses in the aerospace and jet aircraft industries because it has the strength of steel, the approximate weight of aluminum, and is resistant to corrosion and temperatures up to 800°F (427°C).

Tungsten

Tungsten has been used extensively as the filament in light bulbs because of its ability to be drawn into very fine wire and its high melting point. Tungsten, carbon, and cobalt are formed together under heat and pressure to create tungsten carbide, the hardest manmade material. Tungsten carbide is used to make cutting tools for any type of manufacturing application. Tungsten carbide saw blade inserts are used in saws for carpentry so the cutting edge will last longer. Such blades make a finer and faster cut than plain steel saw blades.

A Unified Numbering System for Metals and Alloys

Many numbering systems have been developed for the identification of metals. The organizations that developed these numbering systems include the American Iron and Steel Institute (AISI), Society of Automotive Engineers (SAE), American Society for Testing Materials (ASTM), American National Standards Institute (ANSI), Steel Founders Society of America, American Society of Mechanical Engineers (ASME), American Welding Society (AWS), Aluminum Association, and Copper Development Association, as well as military specifications by the U.S. Department of Defense and federal specifications by the General Accounting Office.

A combined numbering system created by the ASTM and the SAE was established in an effort to coordinate all the different numbering systems into one system. This system avoids the possibility that the same number might be used for two different metals. This combined system is the Unified Numbering System (UNS). The UNS is an identification numbering system for commercial metals and alloys; it does not provide metal and alloy specifications. The UNS system is divided into the following categories.

UNS Series	Metal
Nonferrous Metals and Alloys	
A00001 to A99999	Aluminum and aluminum alloys
C00001 to C99999	Copper and copper alloys
E00001 to E99999	Rare-earth and rare-earth-like metals and alloys
L00001 to L99999	Low-melting metals and alloys
M00001 to M99999	Miscellaneous metals and alloys
P00001 to P99999	Precious metals and alloys
R00001 to R99999	Reactive and refractory metals and alloys
Z00001 to Z99999	Zinc and zinc alloys
Ferrous Metals and Alloys	
D00001 to D99999	Specified mechanical properties steels
F00001 to F99999	Cast irons
G00001 to G99999	AISI and SAE carbon and alloy steels, excluding tool steels
H00001 to H99999	AISI H-classification steels
J00001 to J99999	Cast steels, excluding tool steels
K00001 to K99999	Miscellaneous steels and ferrous alloys
S00001 to S99999	Stainless steels
T00001 to T99999	Tool steels

The prefix letters of the UNS system often match the type of metal being identified. For example, A for aluminum, C for copper, and T for tool steels. Elements of the UNS numbers typically match numbers provided by other systems; for example, SAE1030 is G10300 in the UNS system.

PLASTICS AND POLYMERS

A general definition of plastic is any complex, organic, polymerized compound capable of being formed into a desired shape by molding, casting, or spinning. Plastic retains its shape under ordinary conditions of temperature. Polymerization is a process of joining two or more molecules to form a more complex molecule with physical properties that are different from the original molecules. The terms plastic and polymer are often used to mean the same thing. Plastics can range in any state from liquid to solid. The main elements of plastic are generally common petroleum products, crude oil, and natural gas.

Rubber is an elastic hydrocarbon polymer which occurs as a milky emulsion, known as latex, in the sap of a number of plants but can also be produced synthetically. Over half of the rubber used today is synthetic, but several million tons of natural rubber are produced annually and is necessary, for example, for the automotive industry and the military. Additional discussion about rubber is provided where used for specific applications.

There are many types of plastics that are available for use in the design and manufacture of a product. These plastics generally fall into two main categories: these are thermoplastics and thermosets. The thermoplastics can be heated and formed by pressure, and the shape can change when reheated. Thermosets are formed into shape by heat and pressure and cannot be reheated and changed into a different shape after curing. Most plastic products are made with thermoplastics, because they are easy to make into shapes by heating, forming, and cooling. Thermoset plastics are the choice when the end product is used in an application where heat exists, such as the distributor cap and other plastic parts found on or near the engine of your car. In addition to thermoplastics and thermosets, there are elastomers. Elastomers are polymer-based materials that have elastic qualities not found in the two types of plastics previously defined. Elastomers are generally able to be stretched at least equal to their original length and return to their original length after stretching.

Thermoplastics

Although there are thousands of different thermoplastic combinations, the following gives you some of the more commonly used alternatives. You may recognize some of them by their acronyms, such as PVC.

Acetal—Acetal is a rigid thermoplastic that has good corrosion resistance and machinability. While it will burn, it is good in applications where friction, fatigue, toughness, and tensile strength are factors. Some applications include gears, bushings, bearings, and products that come in contact with chemicals or petroleum.

Acrylic—Acrylics are used when a transparent plastic is needed in clear or a variety of colors. Acrylics are commonly used in products that you see through or through which light passes because, in addition to transparency, they are scratch and abrasion resistant, and they hold up well in the weather. Examples of use include windows, light fixtures, and lenses.

Acrylic-styrene-acrylonitrile (ASA)—ASA has very good weatherability for use as siding, pools and spas, exterior car and marine parts, outdoor furniture, and garden equipment.

Acrylonitrile-butadiene-styrene (ABS)—ABS is one of the most commonly used plastics because of its excellent impact strength, reasonable cost, and ease of processing. ABS also has good dimensional stability, temperature resistance above 212°F (100°C), and chemical and electrical resistance. ABS is commonly used in products that you see daily, such as electronics enclosures, knobs, handles, and appliance parts.

Cellulose—There are five versions of cellulose: nitrate, acetate, butyrate, propionate, and ethyl cellulose.

■ *Nitrate* is tough but very flammable, explosive, and difficult to process. Nitrate is commonly used to make products such as photo film, combs, brushes, and buttons.

■ *Acetate* is not explosive but it is slightly flammable, is not solvent resistant, and becomes brittle with age. Acetate has the advantage of being transparent, can be made in bright colors, is tough, and is easy to process. It is often used for transparency film, magnetic tape, knobs, and sunglass frames.

■ *Butyrate* is similar to acetate but is used in applications where moisture resistance is needed. Butyrate is used for exterior light fixtures, handles, film, and outside products.

■ *Propionate* has good resistance to weathering, is tough and impact resistant, and has reduced brittleness with age. Propionate is used for flashlights, automotive parts, small electronics cases, and pens.

■ *Ethyl cellulose* has very high shock resistance and durability at low temperatures but has poor weatherability.

Fluoroplastics—There are four types of fluoroplastics that have similar characteristics. These plastics are very resistant to chemicals, friction, and moisture. They have excellent dimensional stability for use as wire coating and insulation, nonstick surfaces, chemical containers, O-rings, and tubing.

Ionomers—Ionomers are very tough; resistant to abrasion, stress, cold, and electricity; and very transparent. They are commonly used for cold food containers, other packaging, and film.

Liquid crystal polymers—This plastic can be made very thin with excellent temperature, chemical, and electrical resistance. Uses include cookware and electrical products.

Methyl pentenes—Methyl pentenes have excellent heat and electrical resistance and are very transparent. These plastics are used for items such as medical containers and products, cooking and cosmetic containers that require transparency, and pipe and tubing.

Polyallomers—This plastic is rigid with very high impact and stress fracture resistance at temperatures between −40° to 210°F (−40° to 99°C). Polyallomers are used when constant bending is required in the function of the material design.

Polyamide (nylon)—Commonly called nylon, this plastic is tough; abrasion, heat, and friction resistant; and strong. Nylon is corrosion resistant to most chemicals, but is not as dimensionally stable as other plastics. Nylon is used for combs, brushes, tubing, gears, cams, and stocks.

Polyarylate—This material is impact, weather, electrical, and extremely fire resistant. Typical uses include electrical insulators, cookware, and other options where heat is an issue.

Polycarbonates—Excellent heat resistance, impact strength, dimensional stability, and transparency are positive characteristics of this plastic. Additionally, this material does not stain or corrode but has moderate chemical resistance. Common uses include food containers, power tool housings, outside light fixtures, and appliance and cooking parts.

Polyetheretherketone (PEEK)—PEEK has excellent heat, fire, abrasion, and fatigue resistance qualities. Typical uses include high electrical components, aircraft parts, engine parts, and medical products.

Polyethylene—This plastic has excellent chemical resistance and has properties that make it good to use for slippery or nonstick surfaces. Polyethylene is a common plastic that is used for chemical, petroleum, and food containers; plastic bags; pipe fittings; and wire insulation.

Polyimides—This plastic has excellent impact strength, wear resistance, and very high heat resistance, but it is difficult to produce. Products made from this plastic include bearings, bushings, gears, piston rings, and valves.

Polyphenylene oxide (PPO)—PPO has an extensive range of temperature use from $-275°$ to $375°F$ ($-170°$ to $191°C$), and is fire retardant and chemical resistant. Applications include containers that require super-heated steam, pipe and fittings, and electrical insulators.

Polyphenylene sulfide (PPS)—PPS has the same characteristics as PPO, but it is easier to manufacture.

Polypropylene—This is an inexpensive plastic to produce and has many desirable properties, including heat, chemical, scratch, and moisture resistance. It is also resistant to continuous bending applications. Products include appliance parts, hinges, cabinets, and storage containers.

Polystyrene—This plastic is inexpensive and easy to manufacture, has excellent transparency, and is very rigid. However, it can be brittle and has poor impact, weather, and chemical resistance. Products include model kits, plastic glass, lenses, eating utensils, and containers.

Polysulfones—This material resists electricity and some chemicals, but it can be damaged by certain hydrocarbons. While somewhat difficult to manufacture, it does have good structural applications at high temperatures and can be made in several colors. Common applications are hot water products, pump impellers, and engine parts.

Polyvinylchloride (PVC)—PVC is one of the most common products found for use as plastic pipes and vinyl house siding because of its ability to resist chemicals and the weather.

Thermoplastic polyesters—There are two types of this plastic that exhibit strength and good electrical, stress, and chemical resistance. Common uses include electrical insulators, packaging, automobile parts, and cooking and chemical use products.

Thermoplastic rubbers (TPR)—This resilient material has uses where tough, chemical-resistant plastic is needed. Uses include tires, toys, gaskets, and sports products.

Thermosets

Thermoset plastics make up only about 15% of the plastics used because they are more expensive to produce; they are generally more brittle than thermoplastics; and once they are molded into shape, they cannot be remelted. However, their use is important in products that require a rigid and harder plastic than thermoplastic materials, and in applications where heat could melt thermoplastics. The following list provides information about common thermoset plastics.

Alkyds—These plastics can be used in molding processes, but they are generally used as paint bases.

Melamine formaldehyde—This is a rigid thermoset plastic that is easily molded, economical, nontoxic, tough, and abrasion and temperature resistant. Common uses include electrical devices, surface laminates, plastic dishes, cookware, and containers.

Phenolics—The use of this material dates back to the late 1800s. This plastic is hard and rigid, has good compression strength, is tough, and does not absorb moisture; but it is brittle. Phenolic plastics are commonly used for the manufacture of electrical switches and insulators, electronics circuit boards, distributor caps, and binding material and adhesive.

Unsaturated polyesters—It is common to use this plastic for reinforced composites, also known as reinforced thermoset plastics (RTP). Typical uses are for boat and recreational vehicle construction, automobiles, fishing rods, tanks, and other structural products.

Urea formaldehyde—These plastics are used for many of the same applications as the previously described plastics, but they do not hold up to sunlight exposure. Common uses are for construction adhesives.

Elastomers

The two main types of plastics have been introduced as thermoplastics and thermosets. However, there are elastomers, which are types of polymers that are elastic, much like rubber. Elastomers can be referred to as synthetic rubbers. Synthetic rubbers produce almost twice as many products as natural rubber. **Natural rubber** is a material that starts as the sap from some trees. Natural rubber and many synthetic rubbers are processed by combining with adhesives and using a process called vulcanization. **Vulcanization** is basically the heating of the material in a steel mold that forms the desired shape. The following provides you with information about the most commonly used elastomers.

Butyl rubber—This material has a low air penetration ability and very good resistance to ozone and aging but has poor petroleum resistance. Common uses include tire tubes and puncture-proof tire liners.

Chloroprene rubber (Neoprene)—The trade name Neoprene was the first commercial synthetic rubber. This material has better weather, sunlight, and petroleum resistance than natural rubber. It is also very flame resistant, but does not resist electricity. Common uses include automotive hoses and other products where heat is found, gaskets, seals, and conveyor belts.

Chlorosulfonated polyethylene (CSM)—CSM has excellent chemical, weather, heat, electrical, and abrasion resistance. It is typically used in chemical tank liners and electrical resistors.

Epichlorohydrin rubber (ECO)—This material has great petroleum resistance at very low temperatures. For this reason, it is used in cold weather applications such as snow-handling equipment and vehicles.

Ethylene proplene rubber (EPM) and *ethylene proplene diene monomer (EPDM)*—This is a family of materials that have excellent weather, electrical, aging, and good heat resistance. These materials are used for weatherstripping, wire insulation, conveyor belts, and many outdoor products.

Fluoroelastomers (FPM)—FPM materials have excellent chemical and solvent resistance up to 400°F (204°C). FPM is expensive to produce, so it is used only when its positive characteristics are needed.

Nitrile rubber—This material resists swelling when immersed in petroleum. Nitrile rubber is used for any application involving fuels and hydraulic fluid, such as hoses, gaskets, O-rings, and shoe soles.

Polyacrylic rubber (ABR)—This material is able to resist hot oils and solvents. ABR is commonly used in situations such as transmission seals, where it is submerged in oil.

Polybutadiene—This material has qualities that are similar to those of natural rubber. It is commonly mixed with other rubbers to improve tear resistance.

Polyisoprene—This material was developed during World War II to help with a shortage of natural rubber, and it has the same chemical structure as natural rubber. However, this synthetic is more expensive to produce.

Polysulfide rubber—The advantage of this rubber is that it is petroleum, solvent, gas, moisture, weather, and age resistant. The disadvantage is that it is low in tensile strength, resilience, and tear resistance. Applications include caulking and putty, sealants, and castings.

Polyurethane—This material has the ability to act like rubber or hard plastic. Due to the combined rubber and hard characteristics, products include rollers, wear pads, furniture, and springs. In addition to these products, polyurethene is used to make foam insulation and floor coverings.

Silicones—This material has a wide range of makeup, from liquid to solid. Liquid and semiliquid forms are used for lubricants. Harder forms are used where nonstick surfaces are required.

Styrene butadiene rubber (SBR)—SBR is very economical to produce and is heavily used for tires, hoses, belts, and mounts.

Thermoplastic elastomers (TPE)—The other elastomers typically require the fairly expensive vulcanization process for production. TPE, however, is able to be processed with injection molding just like other thermoplastics, and any scraps can be reused. There are different TPEs, and in general, they are less flexible than other types of rubber.

MANUFACTURING PROCESSES

Casting, forging, and machining processes are used extensively in the manufacturing industry. It is a good idea for the entry-level drafter or pre-engineer to be generally familiar with types of casting, forging, and machining processes, and to know how to prepare related drawings. Any number of process methods may be used by industry. For this reason, it is best for the beginning drafting technician to remain flexible and adapt to the standards and techniques used by the specific company. As the drafting technician gains knowledge of company products, processes, and design goals, he or she may begin to produce designs. It is common for a drafter to become a designer after three years of practical experience.

Castings

Castings are the end result of a process called founding. Founding, or casting, as the process is commonly called, is the pouring of molten metal into a hollow or wax-filled mold. The mold is made in the shape of the desired casting. There are several casting methods used in industry. The results of some of the processes are castings that are made to very close tolerances and with smooth finished surfaces. In the simplest terms, castings are made in three separate steps:

1. A pattern is constructed that is the same shape as the desired finished product.

2. Using the pattern as a guide, a mold is made by packing sand or other material around the pattern.

3. When the pattern is removed from the mold, molten metal is poured into the hollow cavity. After the molten metal solidifies, the surrounding material is removed and the casting is ready for cleanup or machining operations.

Sand Casting

Sand casting is the most commonly used method of making castings. There are two general types of sand castings: green sand and dry sand molding. Green sand is a specially refined sand that is mixed with specific moisture, clay, and resin, which work as binding agents during the molding and pouring procedures. New sand is light brown in color: the term "green sand" refers to the moisture content. In the **dry sand** molding process, the sand does not have any moisture content. The sand is bonded together with specially formulated resins. The end result of the green sand or the dry sand molds is the same.

Sand castings are made by pounding or pressing the sand around a **split pattern**. The first or lower half of the pattern is placed upside down on a molding board, then sand is pounded or compressed around the pattern in a box called a **drag**. The drag is then turned over, and the second or upper half of the pattern is formed when another box, called a **cope**, is packed with sand and joined to the drag. A fine powder is used as a parting agent between the cope and drag at the parting line. The parting line is the separating joint between the two parts of the pattern or mold. The entire box, made up of the cope and drag, is referred to as a **flask**. (See Figure 5.3.)

Before the molten metal can be poured into the cavity, a passageway for the metal must be made. The passageway is called a **runner** and **sprue**. The location and design of the sprue and runner are important to allow for a rapid and continuous flow of metal. Additionally, **vent holes** are established to allow for gases, impurities, and metal to escape from the cavity. Finally, a riser (or group of risers) is used, depending on the size of the casting, to allow for the excess metal to evacuate from the mold and, more importantly, to help reduce shrinking and incomplete filling of the casting. (See Figure 5.4.) After the casting has solidified and cooled, the filled risers, vent holes, and runners are removed.

Cores

In many situations, a hole or cavity is desired in the casting to help reduce the amount of material removal later or to establish a wall thickness. When this is necessary, a **core** is used. Cores are made from either clean sand mixed with binders, such as resin, and baked in an oven for hardening, or ceramic products when a more refined surface finish is required. When the pattern is made, a place for positioning the core in the mold is established; this is referred to as the **core print**. After the mold is made, the core is then placed in position in the mold at the core print. The molten metal, when poured into the mold, flows around the core. After the metal has cooled, the casting is removed from the flask and the core is cleaned out, usually by shaking or tumbling. Figure 5.5 shows cores in place. Cored features help reduce casting weight and save on machining costs. Cores used in sand casting should generally be more than one inch (25.4 mm) in cross section. Cores used in precision casting methods can have much closer tolerances and fine detail. Certain considerations must be taken for supporting very large or long cores when placed in the mold. Usually in sand casting, cores require extra support when they are three times longer than the cross-sectional dimension. Depending on the casting method, the material, the machining required, and the quality, the core holes should be a specified dimension smaller than the

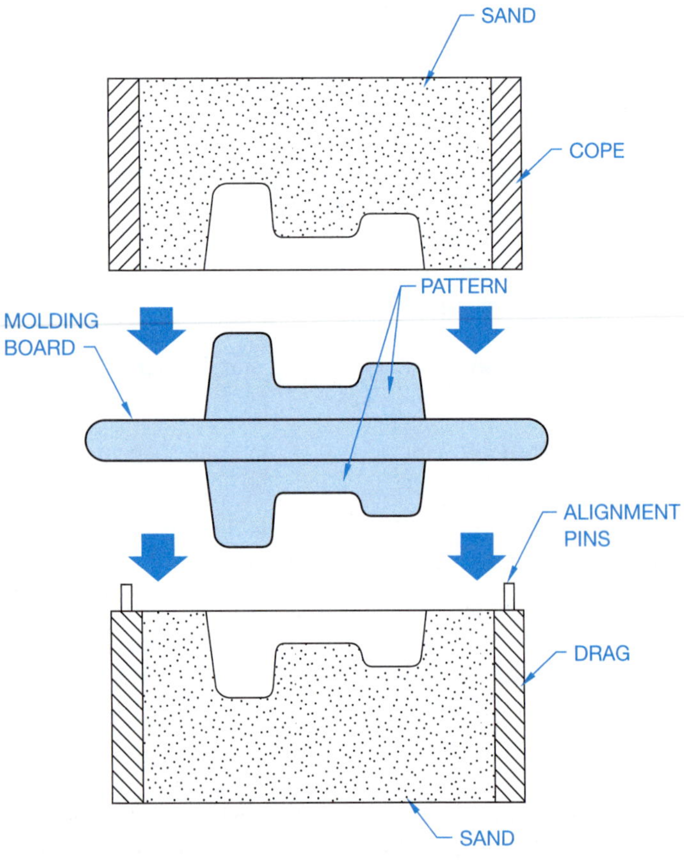

FIGURE 5.3 ■ Components of sand casting process.

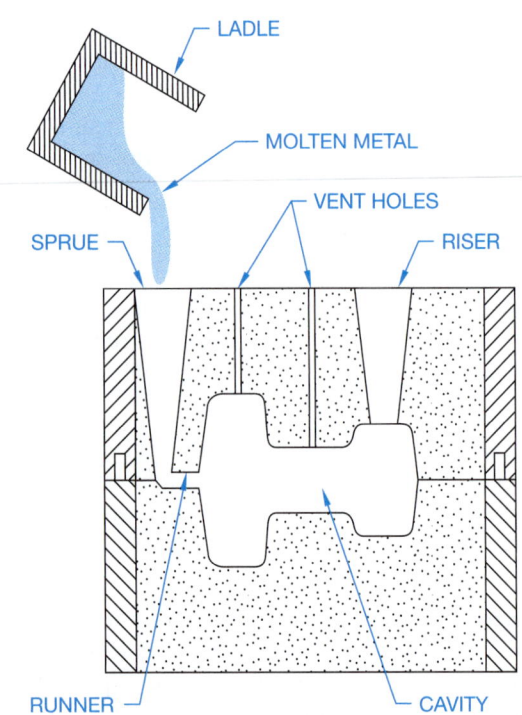

FIGURE 5.4 ■ Pouring molten metal into a sand casting mold.

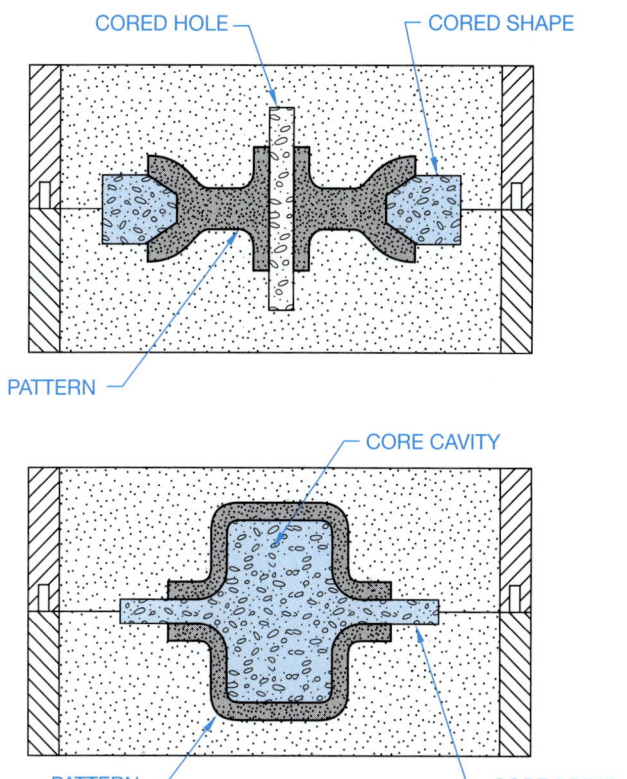

FIGURE 5.5 ■ Cores in place.

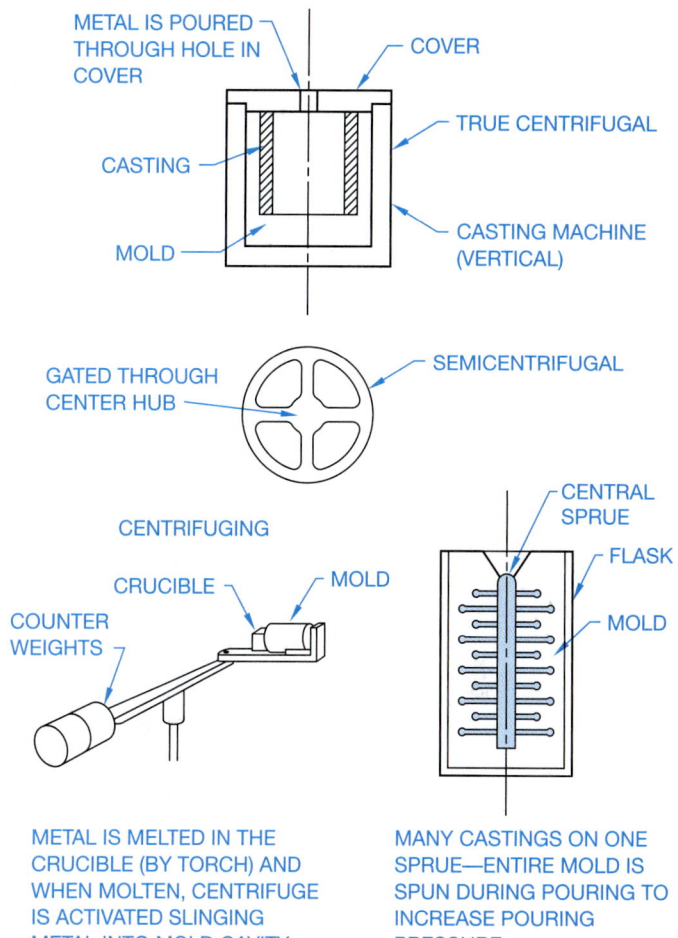

FIGURE 5.6 ■ Centrifugal casting.

desired end product if the hole is to be machined to its final dimension. Cores in sand castings should be between .125 to .5 in. (3.2 to 12.7 mm) smaller than the finished size.

Centrifugal Casting

Objects with circular or cylindrical shapes lend themselves to centrifugal casting. In this casting process, a mold is revolved very rapidly while molten metal is poured into the cavity. The molten metal is forced outward into the mold cavity by centrifugal forces. No cores are needed because the fast revolution holds the metal against the surface of the mold. This casting method is especially useful for casting cylindrical shapes such as tubing, pipes, or wheels. (See Figure 5.6.)

Die Casting

Some nonferrous metal castings are made using the die casting process. Zinc alloy metals are the most common, although brass, bronze, aluminum, and other nonferrous products are also made using this process. Die casting is the injection of molten metal into a steel or cast iron die under high pressure. The advantage of die casting over other methods, such as sand casting, is that castings can be produced quickly and economically on automated production equipment. When multiple dies are used, a number of parts can be cast in one operation. Another advantage of die casting is that high-quality precision parts can be cast with fine detail and a very smooth finish.

Permanent Casting

Permanent casting refers to a process in which the mold can be used many times. This type of casting is similar to sand casting in that molten metal is poured into a mold. It is also similar to die casting because the mold is made of cast iron or steel. The result of permanent casting is a product that has better finished qualities than can be gained by sand casting.

Investment Casting

Investment casting is one of the oldest casting methods. It was originally used in France for the production of ornamental figures. The process used today is a result of the cire perdue, or lost-wax, casting technique that was originally used. The reason that investment casting is called lost-wax casting is that the pattern is made of wax. This wax pattern allows for the development of very close tolerances, fine detail, and precision castings. The wax pattern is coated with a ceramic paste. The shell is allowed to dry and then is baked in an oven to allow the wax to melt and flow out; thus "lost" as the name implies. The empty ceramic mold has a cavity that is the same shape as the precision wax pattern. This cavity is then filled with molten metal. When the metal solidifies, the shell is removed and the

HOW IT WORKS

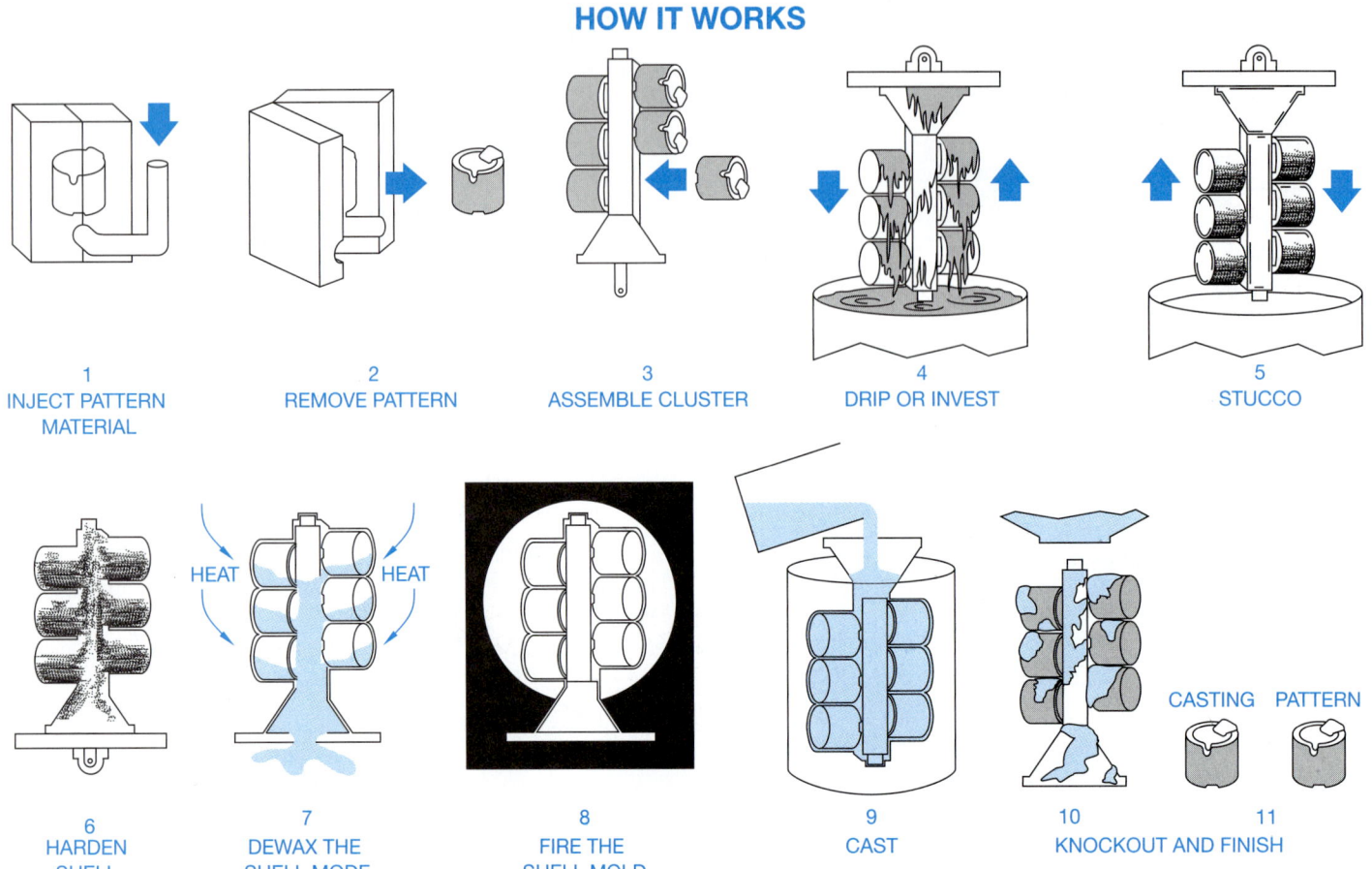

1
INJECT PATTERN
MATERIAL

2
REMOVE PATTERN

3
ASSEMBLE CLUSTER

4
DRIP OR INVEST

5
STUCCO

6
HARDEN
SHELL

7
DEWAX THE
SHELL MODE

8
FIRE THE
SHELL MOLD

9
CAST

10
KNOCKOUT AND FINISH

11

CASTING   PATTERN

FIGURE 5.7 ■ Investment casting. *Courtesy Precision Castparts Corporation.*

casting is complete. Generally very little cleanup or finishing is required on investment castings. (See Figure 5.7.)

Forgings

Forging is a process of shaping malleable metals by hammering or pressing between dies that duplicate the desired shape. The forging process is shown in Figure 5.8. Forging can be accomplished on hot or cold materials. Cold forging is possible on certain materials or material thicknesses where hole punching or bending is the required result. Some soft, nonferrous materials can be forged into shape while cold. Ferrous materials such as iron and steel must be heated to a temperature that results in an orange-red or yellow color. This color is usually achieved between 1800°–1950°F (982° to 1066°C). Forging is used for a large variety of products and purposes. The advantage of forging over casting or machining operations is that the material is shaped into the desired form, and in the process, it retains its original grain structure. Forged metal is generally stronger and more ductile than cast metal and exhibits a greater resistance to fatigue and shock than machined parts. Notice in Figure 5.9 that the grain structure of the forged material remains parallel to the contour of the part, while the machined part cuts through the cross section of the material grain.

Hand Forging

Hand forging is an ancient method of forming metals into desired shapes. The method of heating metal to a red color and then beating it into shape is called smithing or, more commonly, blacksmithing. Blacksmithing is used only in industry for finish work but is still used for horseshoeing and the manufacture of specialty ornamental products.

Machine Forging

Type of machine forging include upset, swaging, bending, punching, cutting, and welding. Upset forging is a process of forming metal by pressing along the longitudinal dimension to decrease the length while increasing the width. For example, bar stock is upset forged by pressing dies together from the ends of the stock to establish the desired shape. Swaging is the forming of metal by using concave tools or dies that result in a reduction in material thickness. Bending is accomplished by forming metal between dies, changing it from flat stock to a desired contour. Bending sheet metal is a cold forging process in which the metal is bent in a machine called a break. Punching and cutting are performed when the die penetrates the material to create a hole of any desired shape and depth or to remove material by cutting away. In forge welding, metals are joined to-

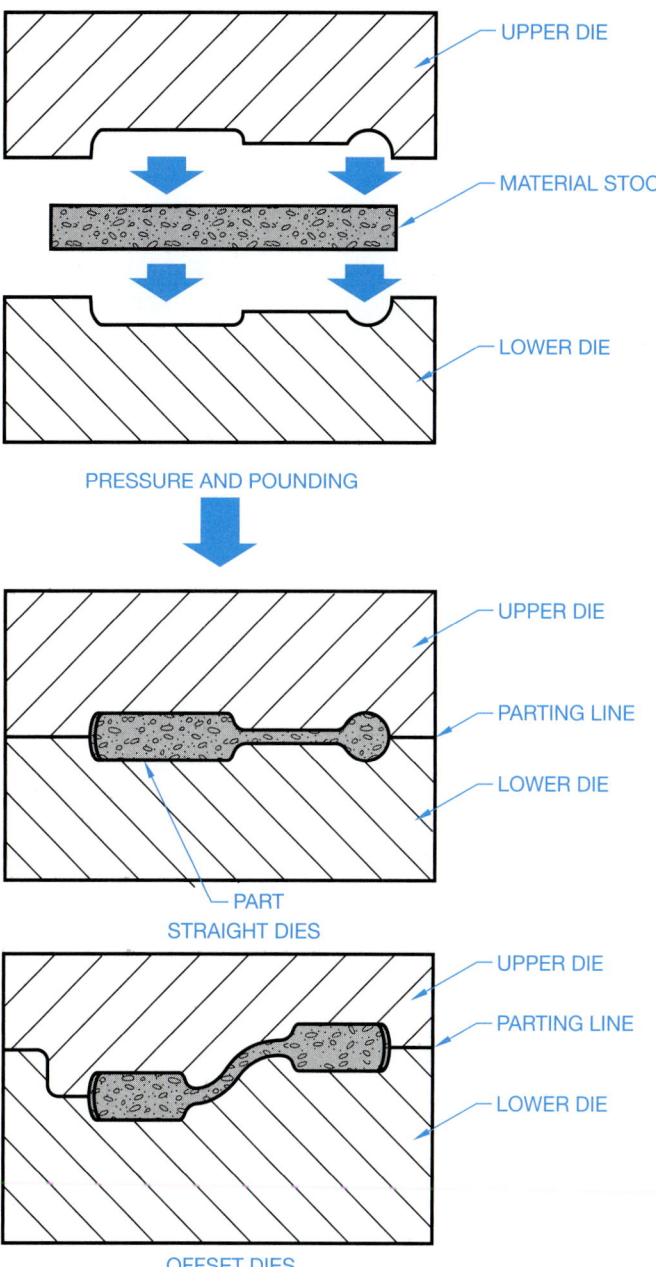

FIGURE 5.8 ▪ The forging process.

gether under extreme pressure. Material that is welded in this manner is very strong. The resulting weld takes on the same characteristics as the metal before joining.

Mass-production forging methods allow for the rapid production of the high-quality products shown in Figure 5.10. In machine forging, the dies are arranged in sequence so that the finished forging is done in a series of steps. Complete shaping can take place after the material has been moved through several stages. Additional advantages of machine forging include:

▪ The part is formed uniformly throughout the length and width.

▪ The greater the pressure exerted on the material, the greater the improvement of the metallic properties.

▪ Fine grain structure is maintained to help increase the part's resistance to shock.

▪ A group of dies can be placed in the same press.

Metal Stamping

Stamping is a process that produces sheet metal parts by the quick downward stroke of a ram die that is in the desired shape. The machine is called a punch press. The punch press can "punch" holes of different sizes and shapes, cut metal, or form a variety of shapes. Automobile parts such as fenders and other body panels are often produced using stamping. If the punch press is used to create holes, then the ram has a die in the shape of the desired hole or holes that pushes through the sheet metal. The punch press also can produce a detailed shape by pressing sheet metal between a die set, where the shape of the desired part is created between the die on the bed of the machine and the matching die on the ram. Stamping is generally done on cold sheet metal, as compared with forging that is done on hot metal that is often much thicker. The stamping process is useful in producing a large number of parts. The process can often fabricate thousands of parts per hour. This is a very important consideration for mass production.

Powder Metallurgy (PM)

The powder metallurgy (PM) process takes metal-alloyed powders and feeds them into a die where they are compacted under pressure to form the desired shape. The compacted metal is then removed from the die and heated at temperatures below the melting point of the metal. This heating process is referred to as sintering, which forms a bond between the metal powder particles, and is discussed later in this chapter.

Powder metallurgy processes can be a cost-effective alternative to casting, forging, and stamping. Powder metallurgy manufacturing can produce quality, precision parts at the rate of thousands per hour.

Metal injection molding (MIM) is another powder metallurgy process that can produce very complex parts. This process injects a mixture of powder metal and a binder into a mold under pressure. The molded product is then sintered to create properties in the metal particles that are close to a casting.

Powder forging (PF) is another powder metallurgy process that places the formed metal particles in a closed die where pressure and heat are applied. This is similar to the forging process. This process produces precision products that have good impact resistance and fatigue strength.

MACHINE PROCESSES

The concepts covered in this chapter serve as a basis for many of the dimensioning practices presented in Chapter 11. A general understanding of machining processes and the drawing representations of these processes is a necessary prerequisite to dimensioning practices. Problem applications are provided in the following chapters because a knowledge of dimensioning practice

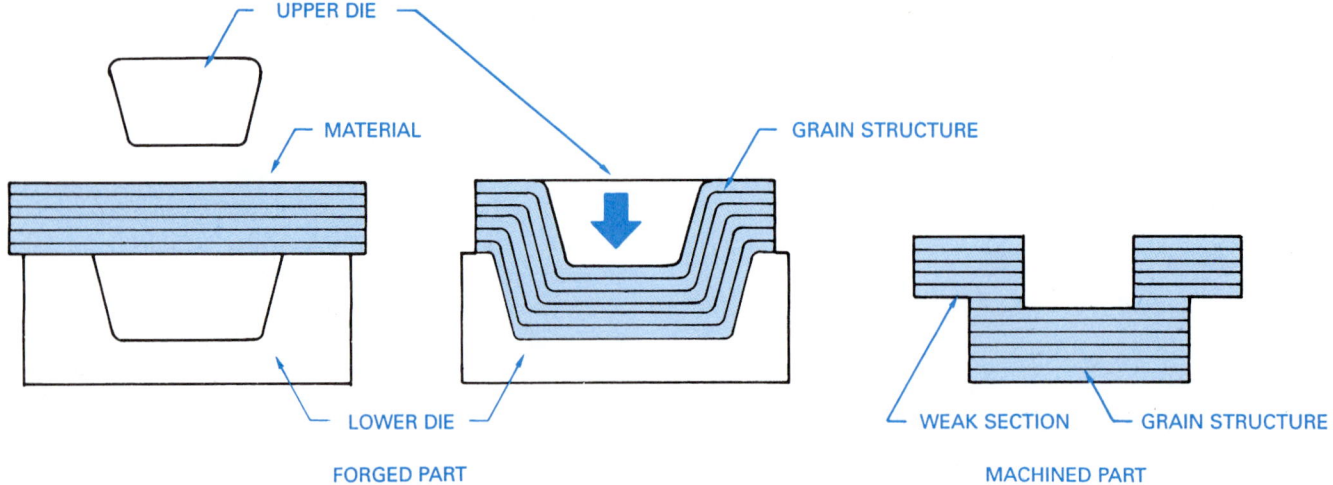

FIGURE 5.9 ■ Forging compared to machining.

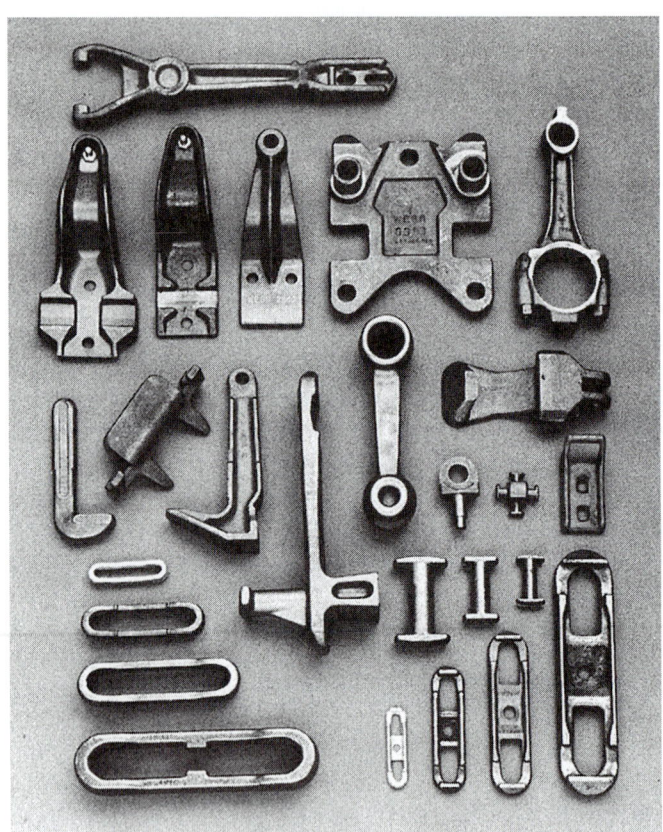

FIGURE 5.10 ■ Forged products. *Courtesy Jarvis B. Webb Co.*

is important and necessary to be able to complete manufacturing drawings. Machining is a general term used to define the process of removing excess or unwanted material with machine tools for rough or finish turning, boring, drilling, milling, or other processes. Machining is used for the manufacture of almost all metal products, and is also used to create or complete some plastic products. Machine tools are power-driven tools, such as a lathe, drill, mill, or other tools discussed in the following content. A machinist is a person who specializes in machining.

Machine Tools

Drilling Machine

The drilling machine, often referred to as a drill press (Figure 5.11), is commonly used to machine-drill holes. Drilling machines are also used to perform other operations, such as reaming, boring, countersinking, counterboring, and tapping. During the drilling procedure, the material is held on a table while the drill or other tool is held in a revolving spindle over the material. When drilling begins, a power or hand-feed mechanism is used to bring the rotating drill in contact with the material. Mass-production drilling machines are designed with multiple spindles. Automatic drilling procedures are available on turret drills. These turret drills allow for automatic tool selection and spindle speed. Several operations can be performed in one setup—for example, drilling a hole to a given depth and tapping the hole with a specified thread.

Grinding Machine

A grinding machine uses a rotating abrasive wheel, rather than a cutting tool, for the purpose of removing material. (See Figure 5.12.) The grinding process is generally used when a smooth, accurate surface finish is required. Extremely smooth surface finishes can be achieved by honing or lapping. Honing is a fine abrasive process often used to establish a smooth finish inside cylinders. Lapping is the process of creating a very smooth surface finish using a soft metal impregnated with fine abrasives, or fine abrasives mixed in a coolant that floods over the part during the lapping process.

Lathe

One of the earliest machine tools, the lathe (Figure 5.13), is used to cut material by turning cylindrically shaped objects. The material to be turned is held between two rigid supports, called centers, or in a holding device called a chuck or collet, as shown in Figure 5.14. The material is rotated on a spindle while a cutting tool is brought into contact with the material. The cutting tool is

FIGURE 5.11 ■ Drilling machine. *Courtesy Delta International Machining Corporation.*

FIGURE 5.12 ■ Grinding machine. *Courtesy Litton Industrial Automation.*

FIGURE 5.13 ■ Lathe. *Courtesy Hardinge Brothers, Inc.*

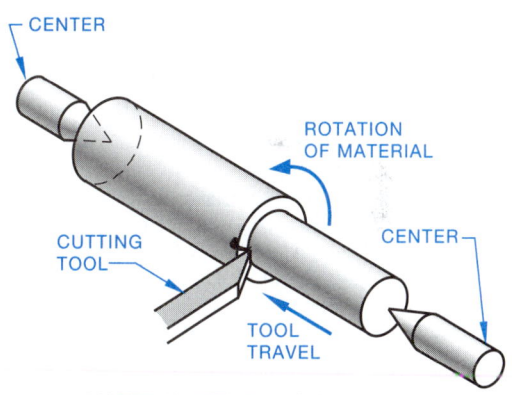

MATERIAL BETWEEN CENTERS

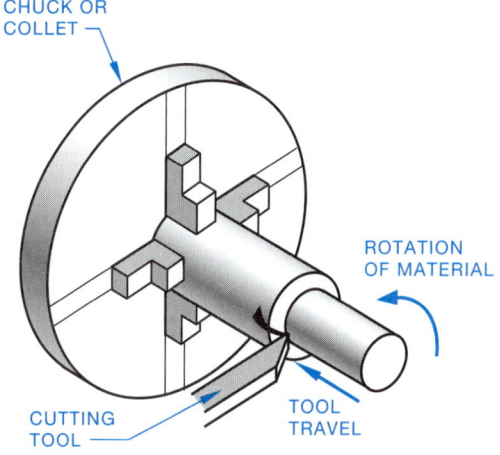

MATERIAL IN CHUCK

FIGURE 5.14 ■ Holding material in a lathe.

supported by a tool holder on a carriage that slides along a bed as the lathe operation continues. The turret is used in mass-production manufacturing where one machine setup must perform several operations. A turret lathe is designed to carry several cutting tools in place of the lathe tailstock or on the lathe carriage. The operation of the turret provides the operator with an automatic selection of cutting tools at preestablished fabrication stages. Figure 5.15 shows an example of eight turret stations and the tooling used.

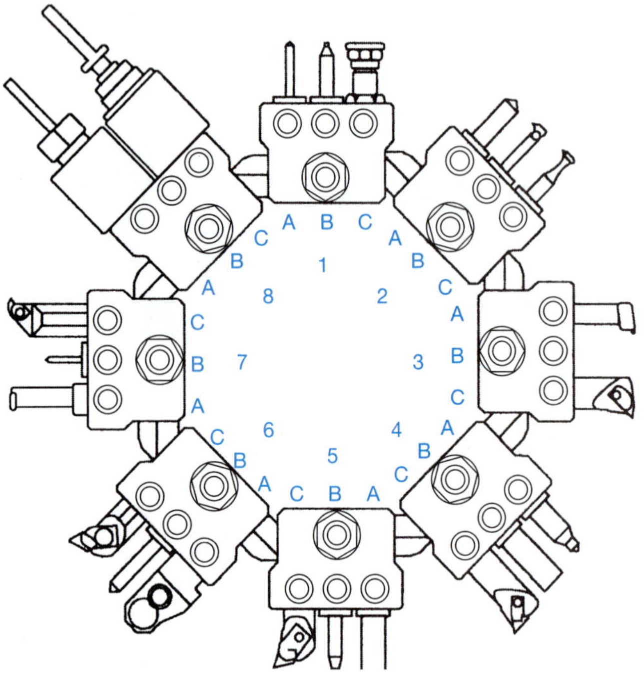

Milling Machine

The milling machine (Figure 5.16) is one of the most versatile machine tools. The milling machine uses a rotary cutting tool to remove material from the work. The two general types of milling machines are horizontal and vertical mills. The difference is in the position of the cutting tool, which can be mounted on either a horizontal or vertical spindle. In the operation, the work is fastened to a table that is mechanically fed into the cutting tool, as shown in Figure 5.17. There are a large variety of milling cutters available that influence the flexibility of operations and shapes that can be performed using the milling machine. Figure 5.18 shows a few of the milling cutters available. Figure 5.19 shows a series of milling cutters grouped together to perform a milling

STATION		TOOLING
	Model	Description
1A		Drill with Bushing
1B		Center Drill with Bushing
1C	T20 -5/8	Adjustable Revolving Stock Stop
2A		Drill with Bushing
2B		Boring Bar with Bushing
2C		Threading Tool with Bushing
3A		Grooving Tool
3C		Insert Turning Tool
4A		Center Drill with Bushing
4B		Flat Bottom Drill with Bushing
4C		Insert Turning Tool
5A		Drill with Bushing
5B		Step Drill with Bushing
5C		Insert Turning Tool
6A	T8 -5/8	Knurling Tool
6B		Drill with Bushing
6C		Insert Turning Tool
7A		Grooving Tool
7B		Drill with Bushing
7C		Insert Threading Tool
8A	TT -5/8	"Collet Type" Releasing Tap Holder
		Tap Collet
		Tap
8C	TE -5/8	Tool Holder Extension
	T19 -5/8	Floating Reamer Holder
		Reamer with Bushing

FIGURE 5.15 ■ A turret with eight tooling stations and the tools used at each station. *Courtesy Toyoda Machinery USA, Inc.*

FIGURE 5.16 ■ Close-up of a horizontal milling cutter.

FIGURE 5.17 ■ Material removal with a vertical machine cutter. *Courtesy The Cleveland Twist Drill Company.*

operation. **End milling cutters**, as shown in Figure 5.20, are designed to cut on the end and the sides of the cutting tool. Milling machines that are commonly used in high-production manufacturing often have two or more cutting heads that are available to perform multiple operations. The machine tables of standard horizontal or vertical milling machines move from left to right (x-axis), forward and backward (y-axis), and up and down (z-axis), as shown in Figure 5.21.

The Universal Milling Machine

Another type of milling machine, known as the **universal milling machine**, has table action that includes x-, y-, and z-axis movement plus angular rotation. The universal milling machine looks much the same as other milling machines, but has the advantage of additional angular table movement, as shown in Figure 5.22.

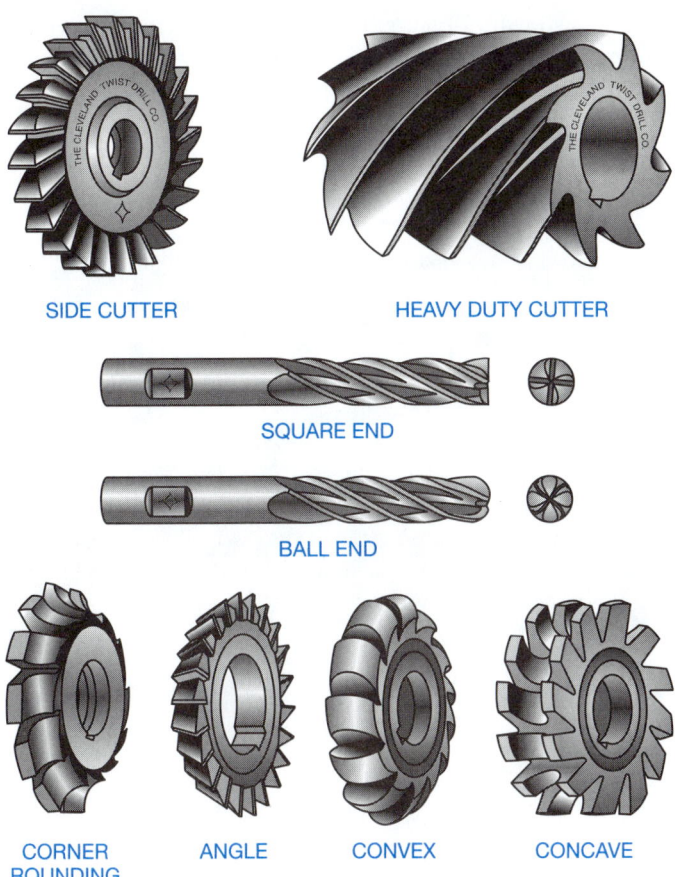

SIDE CUTTER   HEAVY DUTY CUTTER

SQUARE END

BALL END

CORNER ROUNDING   ANGLE   CONVEX   CONCAVE

FIGURE 5.18 ■ Milling cutters. *Courtesy The Cleveland Twist Drill Company.*

FIGURE 5.20 ■ End mill operation. *Courtesy The Cleveland Twist Drill Company.*

FIGURE 5.19 ■ Grouping horizontal milling cutters for a specific machining operation. *Courtesy The Cleveland Twist Drill Company.*

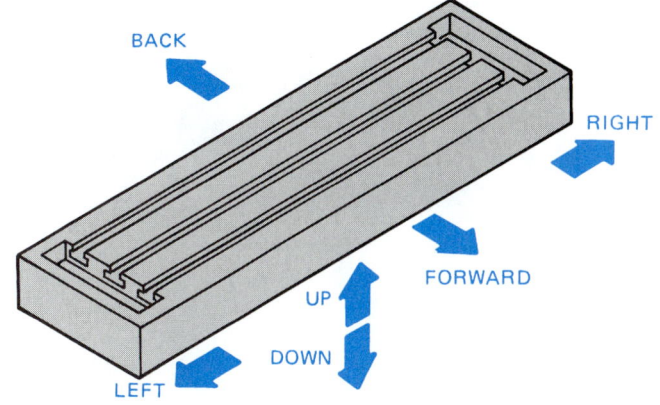

TABLE MOVEMENTS OF THE PLAIN MILLING MACHINE

FIGURE 5.21 ■ Table movements on a standard milling machine.

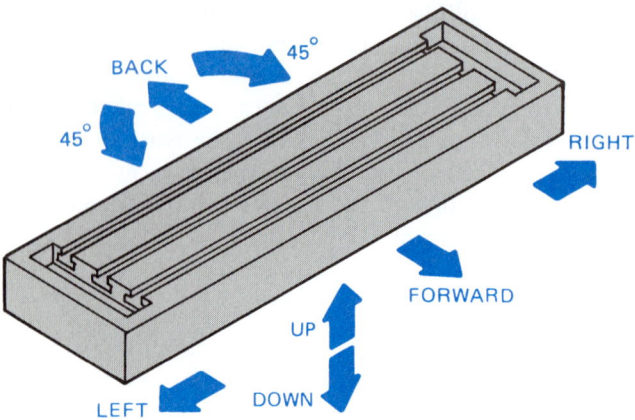

TABLE MOVEMENTS OF THE UNIVERSAL MILLING MACHINE

FIGURE 5.22 ■ Table movements on a universal milling machine.

This additional table movement allows the universal milling machine to produce machined features, such as spirals, not possible on conventional machines.

Saw Machines

Saw machines can be used as cutoff tools to establish the length of material for further machining, or saw cutters can be used to perform certain machining operations such as cutting a narrow slot, called a kerf.

Two types of machines that function as cutoff units only are the power hacksaw and the band saw. These saws are used to cut a wide variety of materials. The hacksaw, as shown in Figure 5.23, operates using a back-and-forth motion. The fixed blade in the power hacksaw cuts material on the forward motion. The metal-cutting band saw is available in a vertical or horizontal

design, as shown in Figure 5.24. This type of cutoff saw has a continuous band that runs either vertically or horizontally around turning wheels. Vertical band saws are also used to cut out irregular shapes.

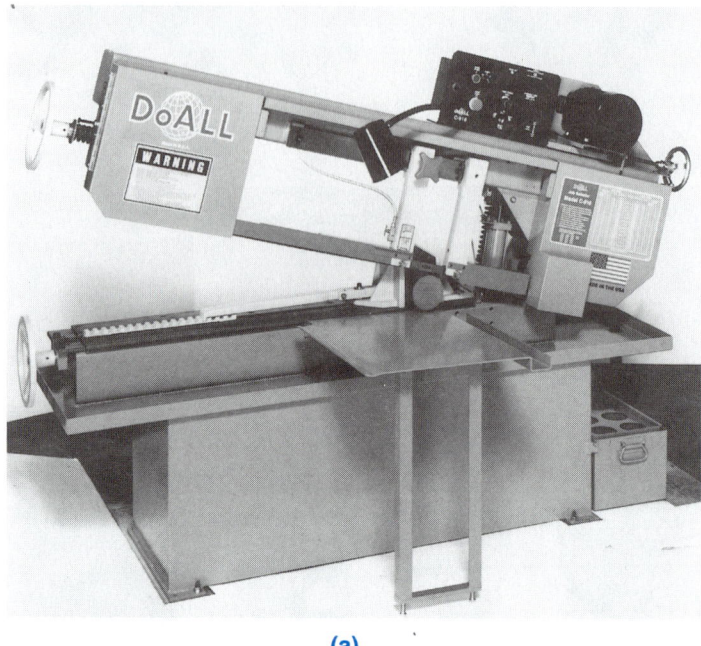

(a)

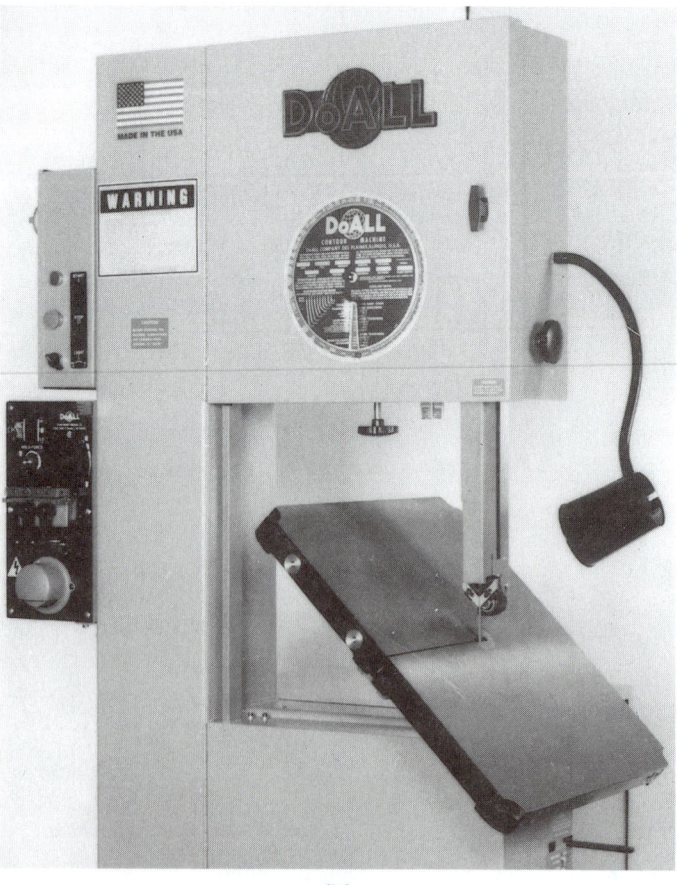

(b)

FIGURE 5.24 ■ (a) Horizontal band saw; (b) vertical band saw. *Courtesy DoAll Company.*

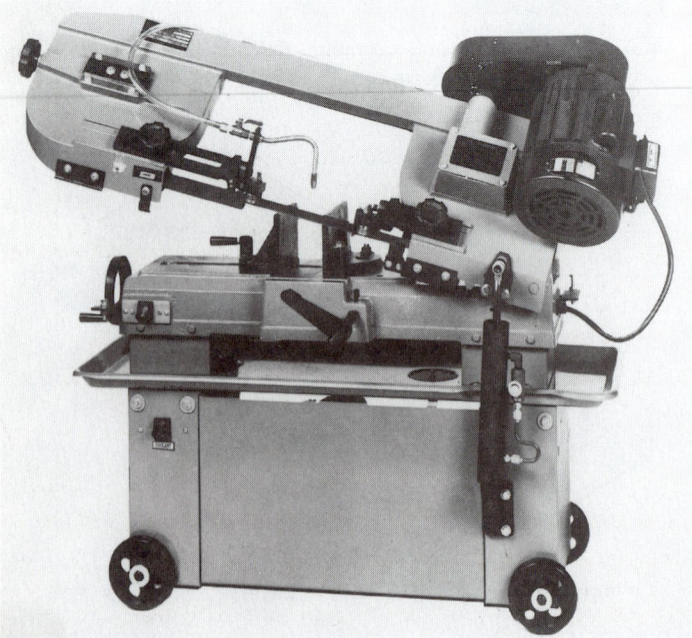

FIGURE 5.23 ■ Power hacksaw. *Courtesy JET Equipment and Tool.*

FIGURE 5.25 ■ Circular blade saw. *Courtesy The Cleveland Twist Drill Company.*

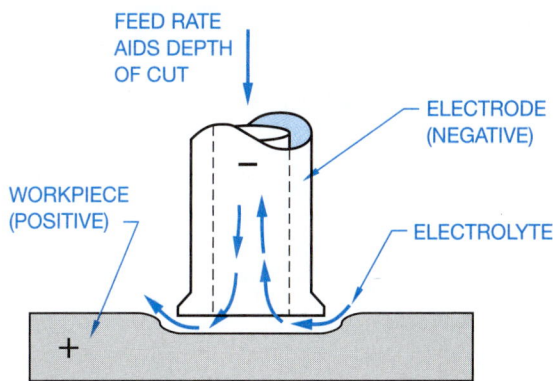

FIGURE 5.26 ■ Electrochemical machining (ECM).

Saw machines are also made with circular abrasive or metal cutting wheels. The abrasive saw can be used for high-speed cutting where a narrow saw kerf is desirable or when very hard materials must be cut. One advantage of the abrasive saw is its ability to cut a variety of materials—from soft aluminum to case-hardened steels. Cutting a variety of metals on the band or the power hacksaw requires blade and speed changes. A disadvantage of the abrasive saw is the expense of abrasive discs. Many companies use this saw only when versatility is needed. The abrasive saw is usually found in the grinding room where abrasive particles can be contained but may also be used in the shop for general-purpose cutting. Metal-cutting saws with teeth, also known as cold saws, are used for precision cutoff operations, cutting saw kerfs, slitting metal, and other manufacturing uses. Figure 5.25 shows a circular saw blade.

Water-jet cutting is used on composite materials and thin metal with a computer-controlled 55,000 pounds per square inch (psi) water jet. Cuts are made with holding tolerances of .0008 in. (0.020 mm) and without generating heat.

Shaper

The shaper is used primarily for production of horizontal, vertical, or angular flat surfaces. Shapers are generally becoming out of date and are rapidly being replaced by milling machines. A big problem with the shaper in mass-production industry is that it is very slow and cuts only in one direction. One of the main advantages of the shaper is its ability to cut irregular shapes that cannot be conveniently reproduced on a milling machine or other machine tools. However, other more advanced multiaxis machine tools are now available that quickly and accurately cut irregular contours.

Chemical Machining

Chemical machining uses chemicals to remove material accurately. The chemicals are placed on the material to be removed while other areas are protected. The amount of time the chemi-

cal remains on the surface determines the extent of material removal. This process, also known as chemical milling, is generally used in situations where conventional machining operations are difficult. A similar method, referred to as chemical blanking, is used on thin material to remove unwanted thickness in certain areas while maintaining "foil" thin material at the machined area. Material can be machined to within .00008 in. (0.002 mm) using this technique.

Electrochemical Machining

Electrochemical machining (ECM) is a process in which a direct current is passed through an electrolyte solution between an electrode and the workpiece. Chemical reaction, caused by the current in the electrolyte, dissolves the metal, as shown in Figure 5.26.

Electrodischarge Machining

In electrodischarge machining (EDM), the material to be machined and an electrode are submerged in a dielectric fluid that is a non-conductor, forming a barrier between the part and the electrode. A very small gap of about .001 inch (0.025 mm) is maintained between the electrode and the material. An arc occurs when the voltage across the gap causes the dielectric to break down. These arcs occur about 25,000 times per second, removing material with each arc. The compatibility of the material and the electrode is important for proper material removal. The advantages of EDM over conventional machining methods include its success in machining intricate parts and shapes that otherwise cannot be econmomically machined, and its use on materials that are difficult or impossible to work with, such as stainless steel, hardened steels, carbides, and titanium.

Electron Beam (EB) Cutting and Machining

In electron beam (EB) machining, an electron beam generated by a heated tungsten filament is used to cut or "machine" very accurate features into a part. This process can be used to machine holes as small as .0002 in. (0.005 mm) or contour irregular shapes with tolerances of .0005 in. (0.013 mm). Electron beam cutting techniques are versatile and can be used to cut or machine any metal or nonmetal.

Image labels (Figure 5.26): FEED RATE AIDS DEPTH OF CUT; ELECTRODE (NEGATIVE); WORKPIECE (POSITIVE); ELECTROLYTE

COMPUTER NUMERICAL CONTROL (CNC) MACHINE TOOLS

Computer-aided design and drafting (CADD) has a direct link to computer-aided manufacturing (CAM) in the form of computer numerical control (CNC) machine tools. CAM provides a direct link between CADD and CNC machine tools. A flowchart for the CNC process is shown in Figure 5.27. Figure 5.28 shows a CNC machine. In most cases, the drawing is generated in a computer and this information is sent directly to the machine tool for production.

Many CNC systems offer a microcomputer base that incorporates a monitor display and full alphanumeric keyboard, making programming and on-line editing easy and fast. For example, data input for certain types of machining may result in the programming of one of several identical features while the other features are oriented and programmed automatically, such as the five equally spaced blades of the centrifugal fan shown in Figure 5.29. Among the advantages of CNC machining are increased productivity, reduction of production costs, and manufacturing versatility.

There is a special challenge when preparing drawings for CNC machining. The drawing method must coordinate with a system of controlling a machine tool by instructions in the form of numbers. The types of dimensioning systems that relate directly to CNC applications are tabular, arrowless, datum, and related coordinate systems. The main emphasis is to coordinate the dimensioning system with the movement of the machine tools. This can be best accomplished with datum dimensioning systems in which each dimension originates from a common beginning point. Programming the CNC system requires that cutter compensation be provided for contouring. This task is automatically computer-calculated in some CNC machines, as shown in Figure 5.30. Definitions and examples of the dimensioning systems are discussed in Chapter 11, Dimensioning and Tolerancing.

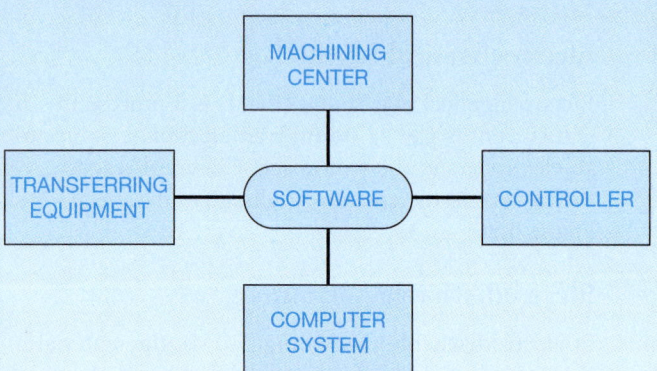

FIGURE 5.27 ■ The computer numerical control (CNC) process.

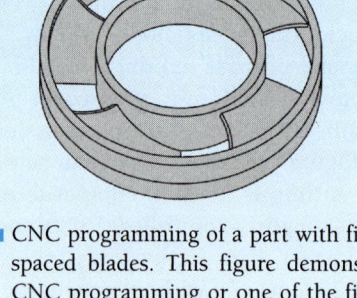

FIGURE 5.29 ■ CNC programming of a part with five equally spaced blades. This figure demonstrates the CNC programming or one of the five equally spaced blades and the automatic orientation and programming of the other four blades. *Courtesy Boston Digital Corporation.*

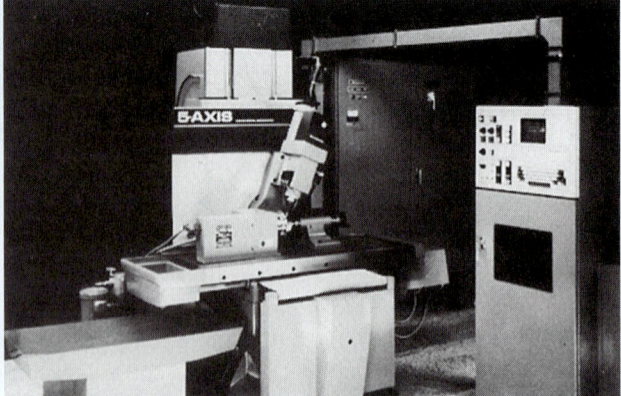

FIGURE 5.28 ■ A CNC machine. *Courtesy Boston Digital Corporation.*

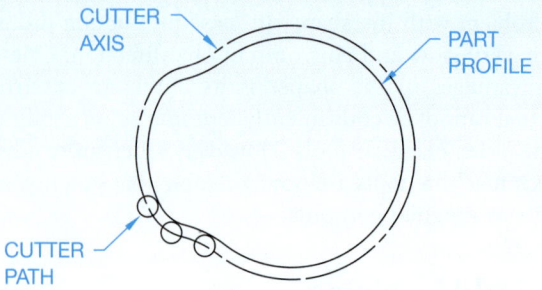

FIGURE 5.30 ■ Automatic cutter compensation for profile machining. *Courtesy Boston Digital Corporation.*

Ultrasonic Machining

Ultrasonic machining, also known as impact grinding, is a process in which a high-frequency mechanical vibration is maintained in a tool designed to a specific shape. The tool and material to be machined are suspended in an abrasive fluid. The combination of vibration and abrasion causes material removal.

Laser Machining

The laser is a device that amplifies focused light waves and concentrates them in a narrow, very intense beam. The term LASER comes from the first letters of the words "Light Amplification by Stimulated Emission of Radiation." Using this process, materials are cut or machined by instant temperatures up to 75,000°F (41,649°C). Laser machining can be used on any type of material and produces smooth surfaces without burrs or rough edges.

Machined Features and Drawing Representations

The following discussion provides a brief definition of the common manufacturing-related terms. The figures that accompany each definition show an example of the tool, a pictorial of the feature, and the drawing representation. The terms are organized in categories of related features, rather than in alphabetical order.

Drill

A drill is used to machine new holes or enlarge existing holes in material. The drilled hole can go through the part, in which case the note THRU can be added to the diameter dimension. When the views of the hole clearly show that the hole goes through the part, then the note THRU can be omitted. When the hole does not go through, the depth must be specified. This is referred to as a blind hole. The drill depth is the total usable depth to where the drill point begins to taper. A drill is a conical-shaped tool with cutting edges, normally used in a drill press. The drawing representation of a drill point is a 120° total angle. (See Figure 5.31.)

Ream

The tool used in the reaming process is called a reamer. The reamer is used to enlarge or finish a hole that has been drilled, bored, or cored. A cored hole is cast in place, as previously discussed. A reamer removes only a small amount of material: for example, .005 to 0.16 in. (0.1–1.6 mm) depending on the size of a hole. The intent of a reamed hole is to provide a smooth surface finish and a closer tolerance than is available with the existing hole. A reamer is a conical-shaped tool with cutting edges similar to a drill; however, a reamer does not create a hole as with a drill. Reamers can be used on a drill press, lathe, or mill. (See Figure 5.32.)

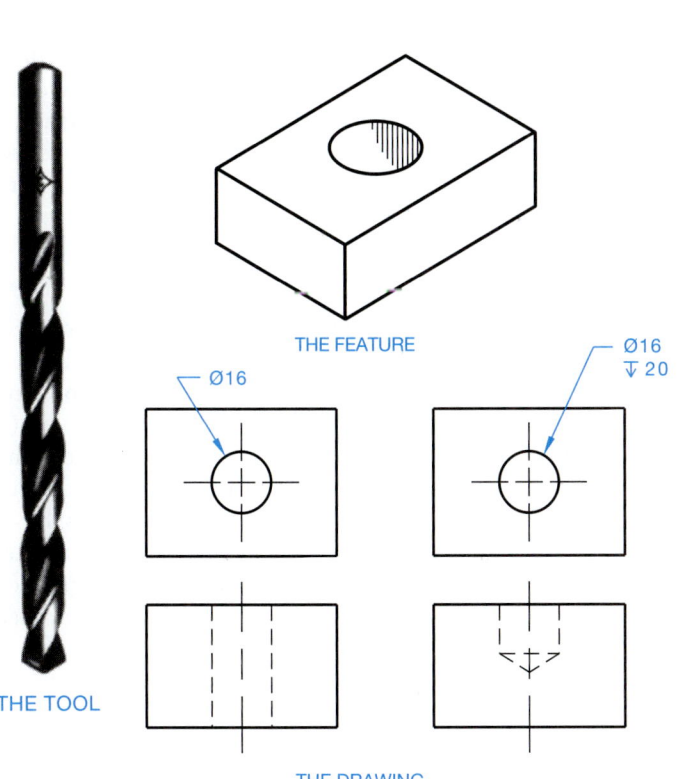

FIGURE 5.31 ■ Drill. *Tool photo courtesy The Cleveland Twist Drill Company.*

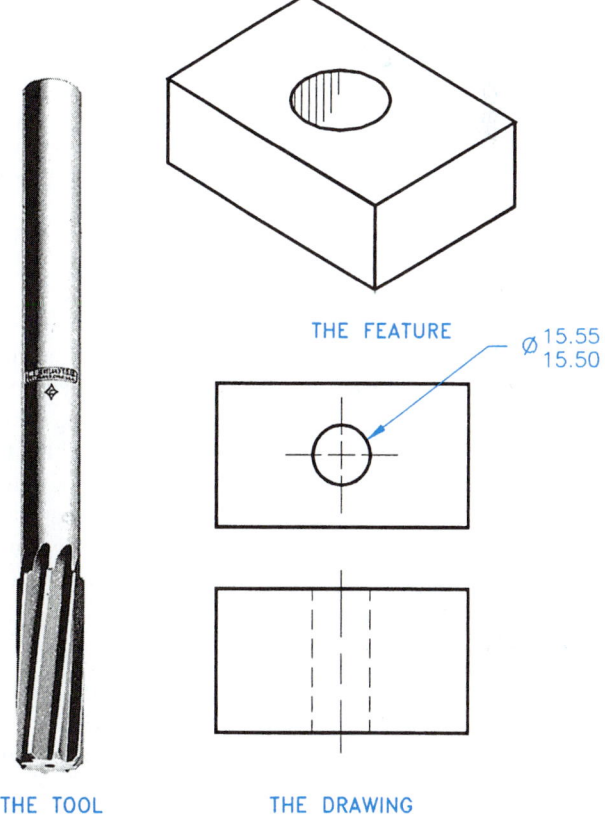

FIGURE 5.32 ■ Reamer. *Tool photo courtesy The Cleveland Twist Drill Company.*

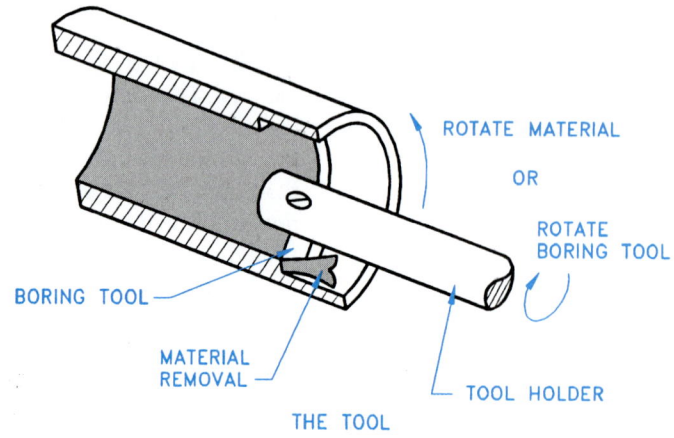

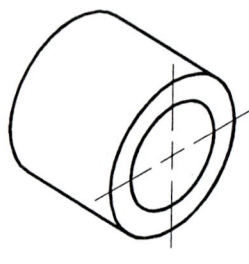

THE FEATURE

FIGURE 5.33 ■ Bore.

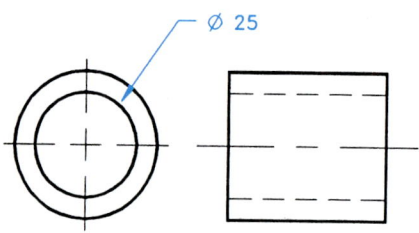

THE DRAWING

Bore

Boring is the process of enlarging an existing hole. The purpose may be to make a drilled or cored hole in a cylinder or part concentric with or perpendicular to other features of the part. A boring tool is used on machines such as a lathe, milling machine, or vertical bore mill for removing internal material. (See Figure 5.33.)

Counterbore

The **counterbore** is used to enlarge the end(s) of a machined hole to a specified diameter and depth. The machined hole is made first, and then the counterbore is aligned during the machining process by means of a pilot shaft at the end of the tool. Counterbores are usually made to recess the head of a fastener below the surface of the object. You should be sure that the diameter and depth of the counterbore are adequate to accommodate the fastener head and fastening tools. (See Figure 5.34.)

Countersink

A **countersink** is a conical feature in the end of a machined hole. Countersinks are used to recess the conically shaped head of a fastener, such as a flathead machine screw. The drafter should specify the countersink note so the fastener head is recessed slightly below the surface. The total countersink angle should match the desired screw head. This angle is generally 80° to 82° or 99° to 101° (See Figure 5.35.)

Counterdrill

A **counterdrill** is a combination of two drilled features. The first machined feature may go through the part, while the second feature is drilled, to a given depth, into one end of the first. The result is a machined hole that looks similar to a countersink-counterbore combination. The angle at the bottom of the counterdrill is a total of 120°, as shown in Figure 5.36.

Spotface

A **spotface** is a machined, round surface on a casting, forging, or machined part on which a bolt head or washer can be seated. Spotfaces are similar in characteristics to counterbores, except that a spotface is generally only about .08 in. (2 mm) or less in depth. Rather than a depth specification, the dimension from the spotface surface to the opposite side of the part can be given. This is also true for counterbores; however, the depth dimension is commonly provided in the note. When no spotface depth is given, the machinist will spotface to a depth that establishes a smooth cylindrical surface. (See Figure 5.37.)

Boss

A **boss** is a circular pad on forgings or castings that projects out from the body of the part. While more closely related to castings and forgings, the surface of the boss is often machined smooth for a bolt head or washer surface to seat on. Also, the boss com-

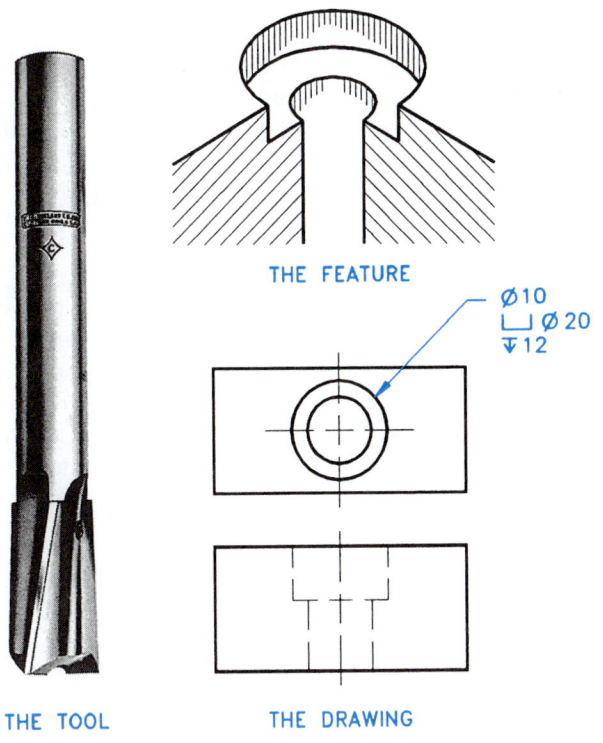

THE FEATURE

Ø10
⌴ Ø 20
↧ 12

THE TOOL   THE DRAWING

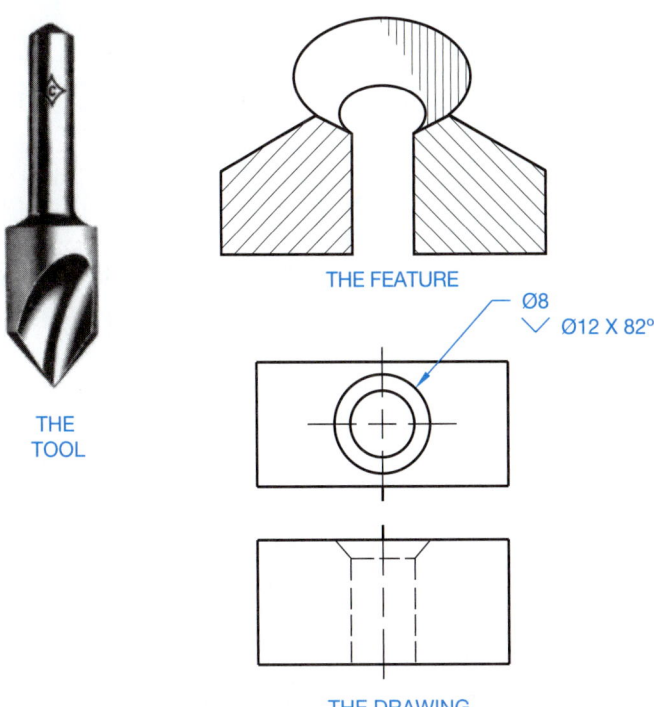

THE FEATURE

Ø8
⌵ Ø12 X 82°

THE TOOL

THE DRAWING

FIGURE 5.35 ■ Countersink. *Tool photo courtesy The Cleveland Twist Drill Company.*

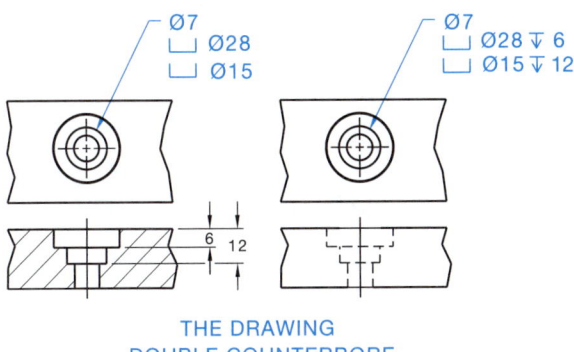

Ø7
⌴ Ø28
⌴ Ø15

Ø7
⌴ Ø28 ↧ 6
⌴ Ø15 ↧ 12

6   12

THE DRAWING
DOUBLE COUNTERBORE

FIGURE 5.34 ■ Counterbore. *Tool photo courtesy The Cleveland Twist Drill Company.*

monly has a hole machined through it to accommodate the fastener's shank. (See Figure 5.38.)

Lug

Generally cast or forged into place, a lug is a feature projecting out from the body of a part, usually rectangular in cross section. Lugs are used as mounting brackets or function as holding devices for machining operations. Lugs are commonly machined with a drilled hole and a spotface to accommodate a bolt or other fastener. (See Figure 5.39.)

Pad

A pad is a slightly raised surface projecting out from the body of a part. The pad surface can be any size or shape. The pad may be cast, forged, or machined into place. The surface is often ma-

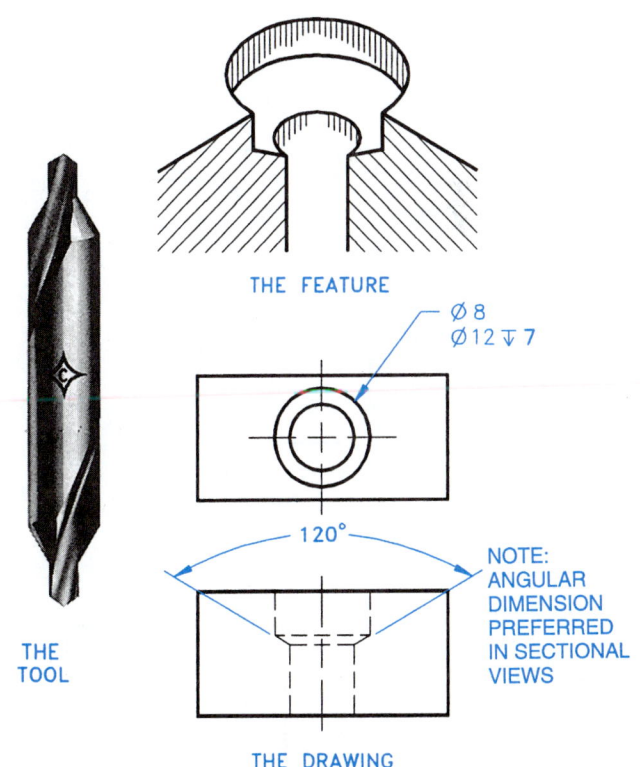

THE FEATURE

Ø 8
Ø12 ↧ 7

120°

NOTE:
ANGULAR
DIMENSION
PREFERRED
IN SECTIONAL
VIEWS

THE TOOL

THE DRAWING

FIGURE 5.36 ■ Counterdrill. *Tool photo courtesy The Cleveland Twist Drill Company.*

chined to accommodate the mounting of an adjacent part. A boss is a type of pad, although the boss is always cylindrical in shape. (See Figure 5.40.)

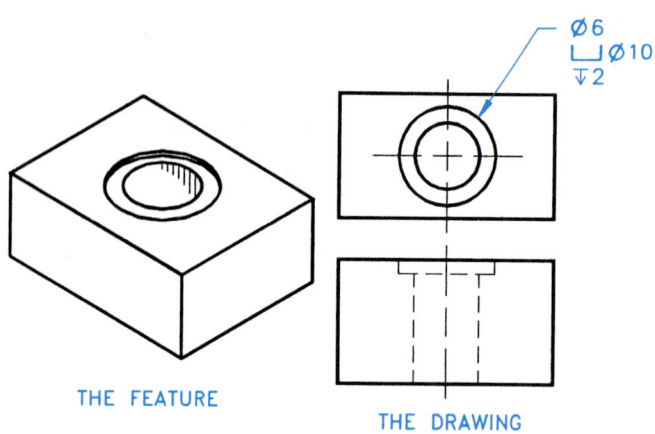

THE FEATURE

THE DRAWING

FIGURE 5.37 ■ Spotface.

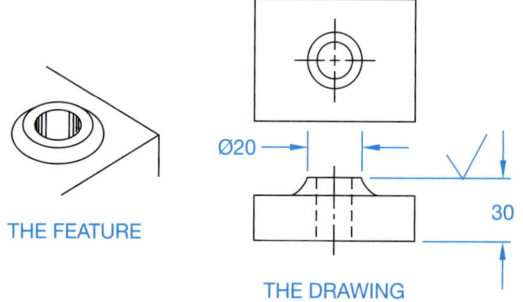

THE FEATURE

THE DRAWING

FIGURE 5.38 ■ Boss.

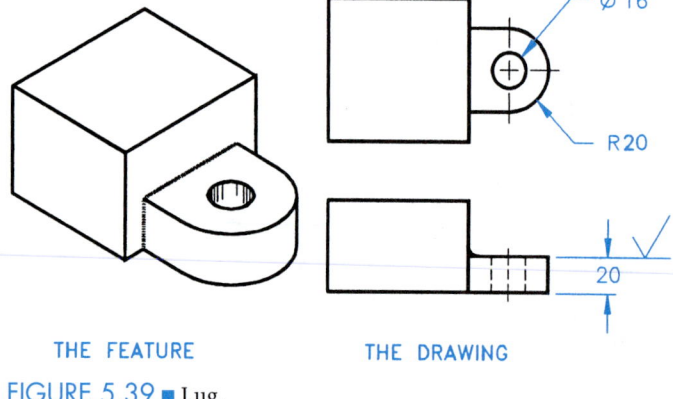

THE FEATURE

THE DRAWING

FIGURE 5.39 ■ Lug.

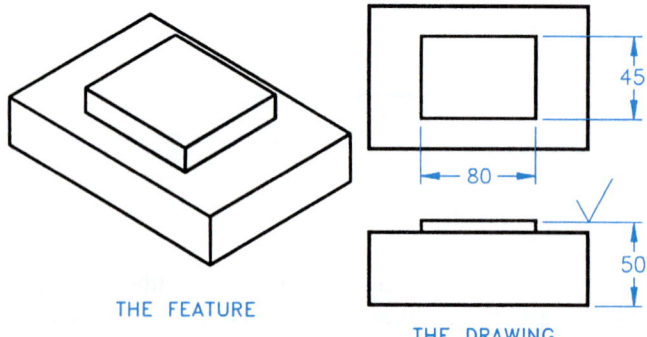

THE FEATURE

THE DRAWING

FIGURE 5.40 ■ Pad.

Chamfer

A **chamfer** is the cutting away of the sharp external or internal corner of an edge. Chamfers can be used as a slight angle to relieve a sharp edge, or to assist the entry of a pin or thread into the mating feature. (See Figure 5.41.) Verify alternate methods of dimensioning chamfers in Chapter 11.

Fillet

A **fillet** is a small radius formed between the inside angle of two surfaces. Fillets are often used to help reduce stress and strengthen an inside corner. Fillets are common on the inside corners of castings and forgings to strengthen corners and reduce stress during molding. Fillets are also used to help a casting or forging release a mold or die. Fillets are arcs given as radius dimensions. The fillet size depends on the function of the part and the manufacturing process used to make the fillet. (See Figure 5.42.)

Round

A **round** is a small-radius outside corner formed between two surfaces. Rounds are used to refine sharp corners, as shown in

EXTERNAL
CHAMFER

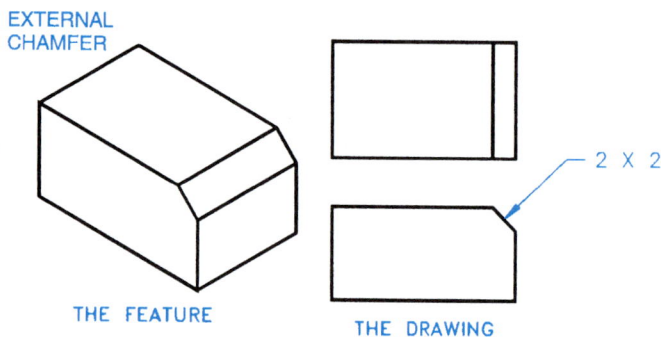

THE FEATURE

THE DRAWING

CYLINDRICAL CHAMFER

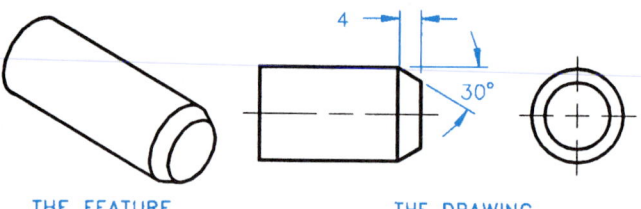

THE FEATURE

THE DRAWING

INTERNAL
CHAMFER

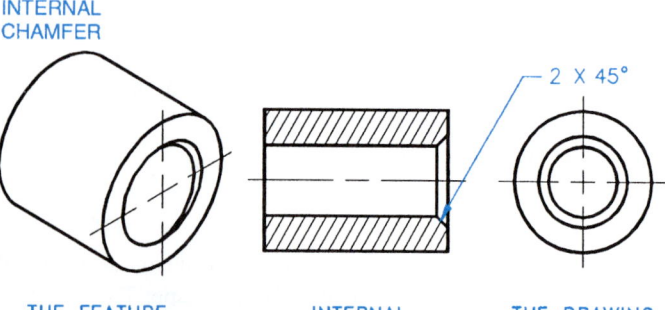

THE FEATURE   INTERNAL   THE DRAWING

FIGURE 5.41 ■ Chamfers.

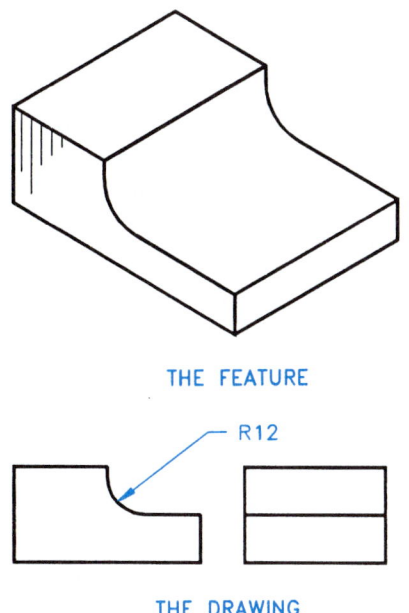

THE FEATURE

R12

THE DRAWING

FIGURE 5.42 ■ Fillet.

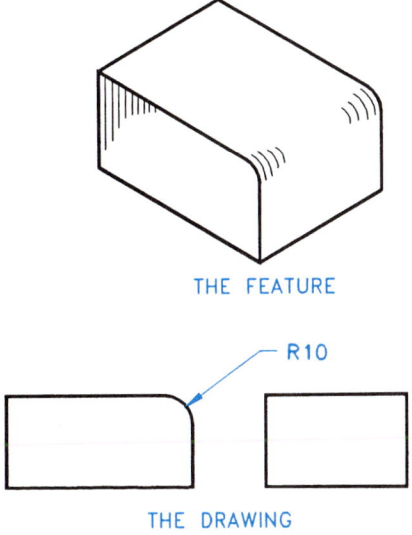

THE FEATURE

R10

THE DRAWING

FIGURE 5.43 ■ Round.

Figure 5.43. In some situations where a sharp corner must be relieved and a round is not required, a slight corner relief can be used, referred to as a **break corner**. The note BREAK CORNER can be used on the drawing. Another option is to provide a note that specifies REMOVE ALL BURRS AND SHARP EDGES. **Burrs** are machining fragments that are often left on a part after machining.

Dovetail

A **dovetail** is a slot with angled sides that can be machined at any depth and width. Dovetails are commonly used as a sliding mechanism between two mating parts. (See Figure 5.44.)

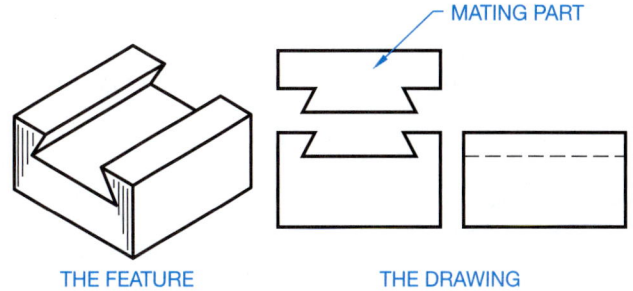

MATING PART

THE FEATURE   THE DRAWING

FIGURE 5.44 ■ Dovetail.

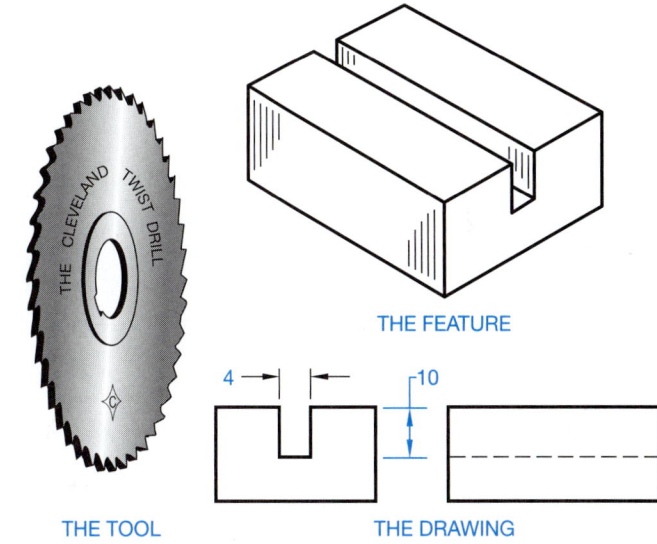

THE TOOL   THE DRAWING

FIGURE 5.45 ■ Kerf. *Tool photo courtesy The Cleveland Twist Drill Company.*

Kerf

A **kerf** is a narrow slot formed by removing material while sawing or using some other machining operation. (See Figure 5.45.)

Key, Keyseat, Keyway

A **key** is a machine part that is used as a positive connection for transmitting torque between a shaft and a hub, pulley, or wheel. The key is placed in position in a **keyseat**, which is a groove or channel cut in a shaft. The shaft and key are then inserted into a hub, wheel, or pulley where the key mates with a groove, called a **keyway**. There are several different types of keys. The key size is often determined by the shaft size. (See Figure 5.46.) Types of keys and key sizes are discussed in Chapter 12, Fasteners and Springs.

Neck

A **neck** is the result of a machining operation that establishes a narrow groove on a cylindrical part or object. There are several different types of neck grooves, as shown in Figure 5.47. Dimensioning necks is clearly explained in Chapter 11.

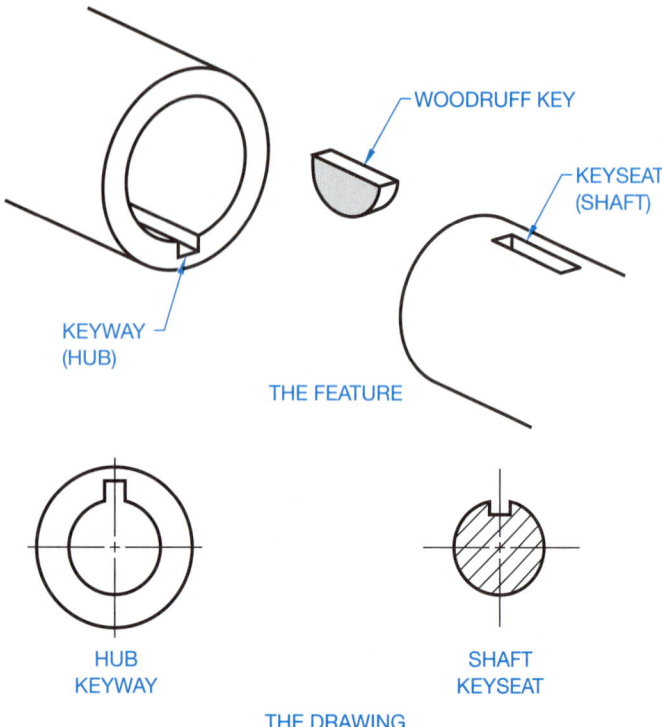

FIGURE 5.46 ■ Key, keyseat, keyway.

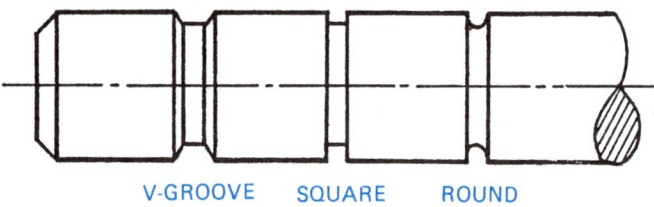

FIGURE 5.47 ■ Neck.

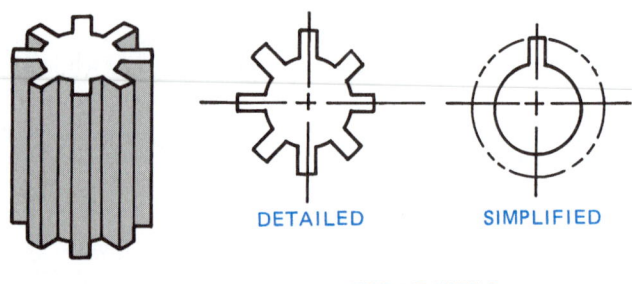

FIGURE 5.48 ■ Spline.

Spline

A spline is a gearlike, serrated surface on a shaft and in a mating hub. Splines are used to transmit torque and allow for lateral sliding or movement between two shafts or mating parts. A spline can be used to take the place of a key when more torque strength is required or when the parts must have lateral movement. (See Figure 5.48.)

Threads

There are many different forms of threads available that are used as fasteners to hold parts together, to adjust parts in alignment with each other, or to transmit power. Threads that are used as fasteners are commonly referred to as screw threads. External threads are thread forms on an external feature, such as a bolt or shaft. The machine tool used to make external threads is commonly called a die. (See Figure 5.49.) Threads can be machined on a lathe using a thread-cutting tool. The external thread can be machined on a fastener, such as the hexagon head bolt shown in Figure 5.49, or on a shaft without a head. This is referred to as a threaded rod. An external thread can also be machined on a shaft of one diameter that meets a shaft of a larger diameter. Sometimes a neck is machined where the external thread meets the fastener head or the larger-diameter shaft. This neck is referred to as an undercut, as shown in the right example in Figure 5.49. The undercut is used to eliminate the possibility of a machine radius at the head and to allow for a tight fit between the thread and the head when assembled. Internal threads are threaded features on the inside of a hole. The machine tool that is commonly used to cut internal threads is called a tap. (See Figure 5.50.)

T-slot

A T-slot is a slot of any dimension that is cut to resemble a "T." The T-slot may be used as a sliding mechanism between two mating parts. (See Figure 5.51.)

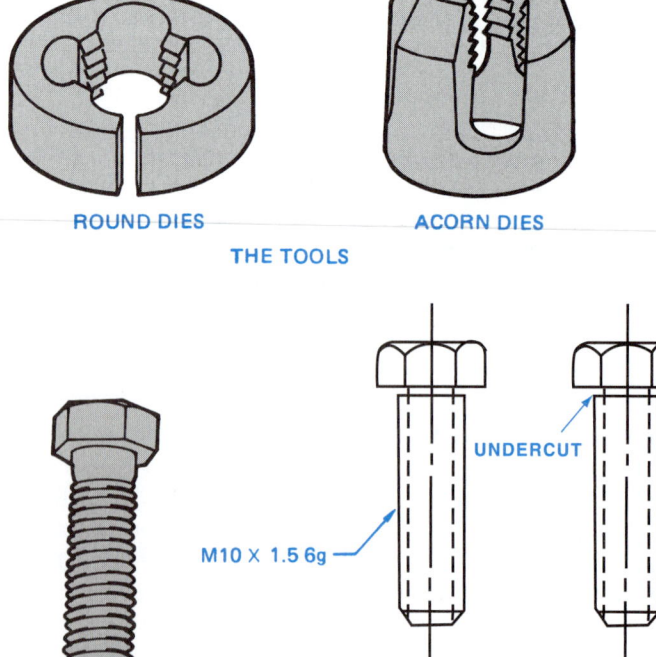

FIGURE 5.49 ■ External thread.

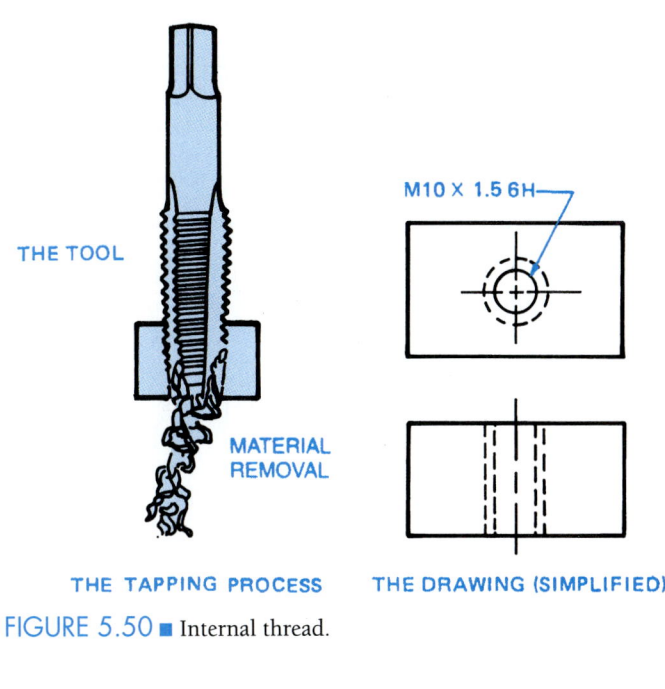

FIGURE 5.50 ■ Internal thread.

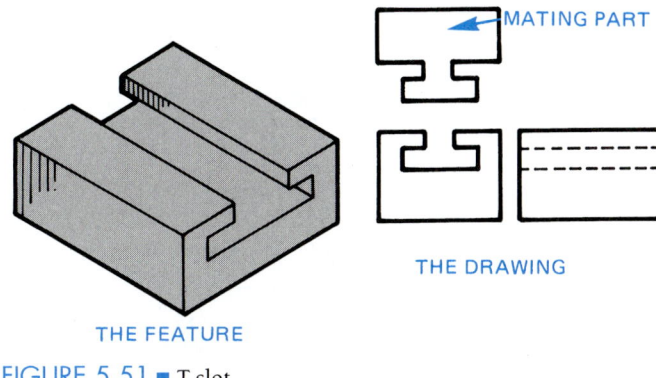

FIGURE 5.51 ■ T-slot.

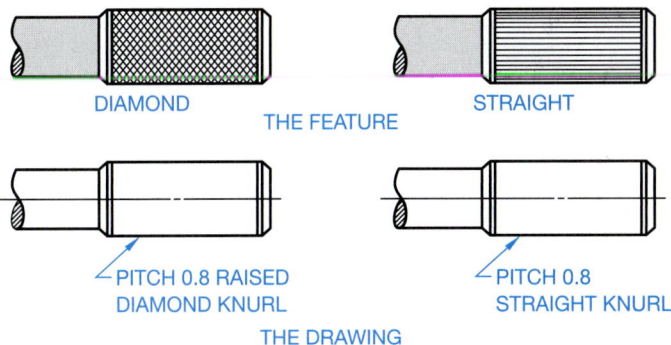

FIGURE 5.52 ■ Knurl.

Knurl

Knurling is a cold forming process used to uniformly roughen a cylindrical or flat surface with a diamond or straight pattern. Knurls are often used on handles or other gripping surfaces. Knurls may also be used to establish an interference (press) fit between two mating parts. The actual knurl texture is not displayed on the drawing. (See Figure 5.52.)

Surface Texture

Surface texture, or **surface finish**, is the intended condition of the material surface after manufacturing processes have been implemented. Surface texture includes such characteristics as roughness, waviness, lay, and flaws. Surface roughness is one of the most common characteristics of surface finish, which consists of the finer irregularities of the surface texture due, in part, to the manufacturing process. The surface roughness is measured in micrometers (μm) or microinches (μin.) **Micro (μ)** means millionth. The roughness averages for different manufacturing processes are shown in Appendix E. Surface finish is considered a dimensioning specification and therefore is discussed in detail in Chapter 11.

Design and Drafting of Machined Features

Drafters should gain a working knowledge of the machining processes and capabilities of their companies. Drawings should be prepared that allow machining within the capabilities of the available machine tools. If the local machine tools do not produce parts that meet functional requirements, then the machining may need to be performed elsewhere, or the company may have to purchase the necessary equipment. However, the first consideration should be looking for the least-expensive method to get the desired result. Avoid overmachining. Machining processes are expensive, so drawing requirements that are not necessary for the function of the part should be avoided. For example, surface finishes become more expensive to machine as the roughness height decreases. So, if a surface roughness of 125 microinches is adequate, then do not use a 32-microinch specification just because you like smooth surfaces. For example, notice the difference between 63- and 32-microinch finishes. A 63-microinch finish is a good machine finish that can be performed using sharp tools at high speeds with extra fine feeds and cuts. The 32-microinch callout, on the other hand, requires extremely fine feeds and cuts on a lathe or milling machine and in many cases requires grinding. The 32-microinch finish is more expensive to perform.

In a manufacturing environment where cost and competition are critical considerations, a great deal of thought must be given to meeting the functional and appearance requirements of a product for the least possible cost. It generally does not take very long for an entry-level drafter to pick up these design considerations by communicating with the engineering and manufacturing departments. Many drafters become designers, checkers, or engineers within a company by learning the product and being able to implement designs based on the company's manufacturing capabilities.

Manufacturing Plastic Products

Earlier discussions provided some general information about the different types of plastics and synthetic rubbers, and many of the products that are commonly made from these materials. The traditional plastic product manufacturing processes include injection molding, extrusion, blow molding, compression molding, transfer molding, and thermoforming. These processes are explained in the following discussion.

The Injection Molding Process

Injection molding is the most commonly used process for creating thermoplastic products. The process involves injecting molten plastic material into a mold that is in the form of the desired part or product. The mold is in two parts that are pressed together during the molding process. The mold is then allowed to cool so the plastic can solidify. When the plastic has cooled and solidified, the press is opened and the part is removed from the mold. The injection molding machine has a hopper where either powder or granular material is placed. The material is then heated to a melting temperature. The molten plastic is then fed into the mold by an injection nozzle or a screw injection system. The injection system is similar to a large hypodermic needle and plunger that pushes the molten plastic material into the mold. The most commonly used injection system is the screw machine. This machine has a screw design that transports the material to the injection nozzle. While the material moves toward the mold, the screw also mixes the plastic to a uniform consistency. This mixing process is an important advantage over the plunger system. Mixing is especially important when color is added or when using recycled material. The process of creating a product and the parts of the screw injection and injection nozzle systems are shown in Figure 5.53.

The Extrusion Process

The **extrusion process** is used to make continuous shapes such as moldings, tubing, bars, angles, hose, weatherstripping, films, and any product that has a constant cross-sectional shape. This process creates the desired continuous shape by forcing molten plastic through a metal die. The extrusion process typically uses the same type of injection nozzle or a screw injection system that is used in the injection molding process. The contour of the die establishes the shape of the extruded plastic. Figure 5.54 shows the extrusion process in action.

The Blow Molding Process

The **blow molding process** is commonly used to produce hollow products such as bottles, containers, receptacles, and

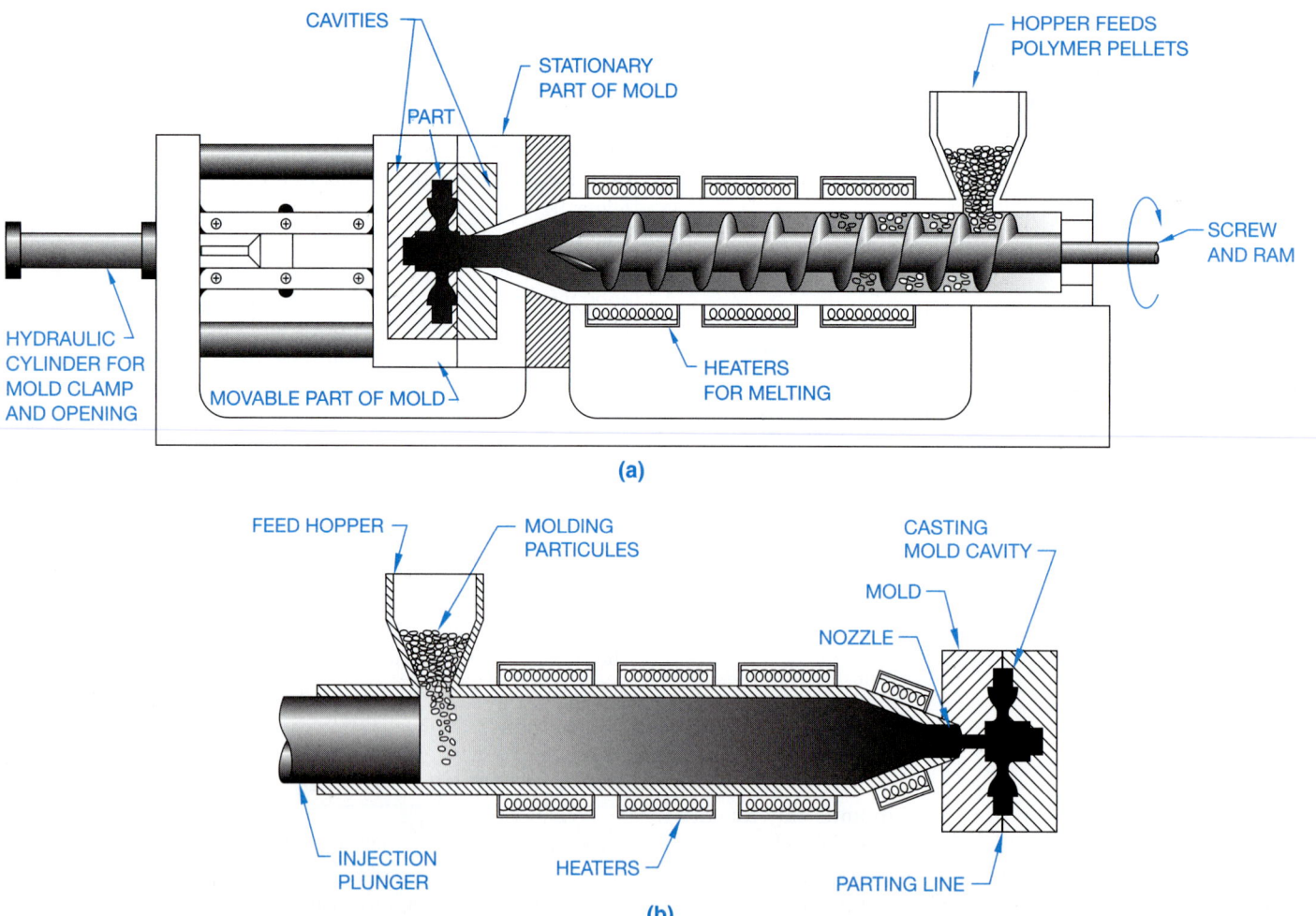

FIGURE 5.53 ■ (a) The process of creating a product and the parts of the screw injection system. (b) The process of creating a product and the parts of the nozzle injection system.

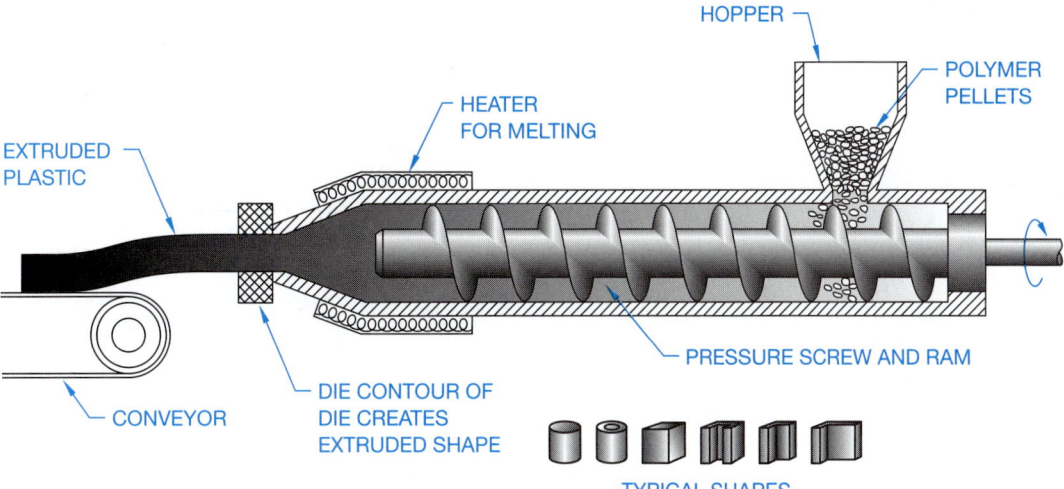

FIGURE 5.54 ■ The extrusion process in action.

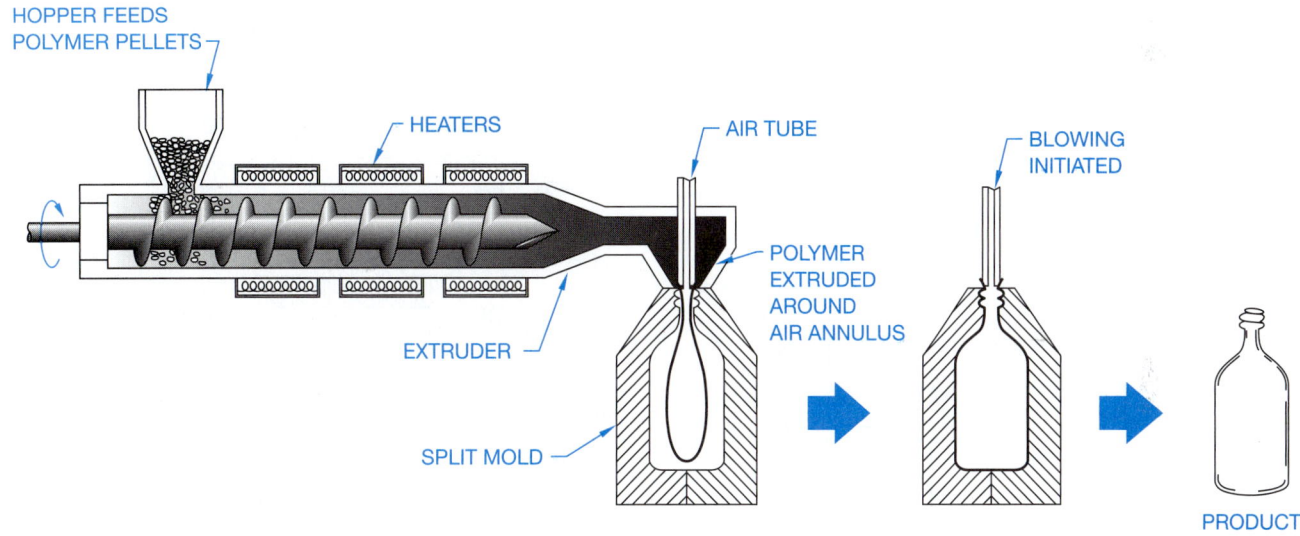

FIGURE 5.55 ■ The blow molding process.

boxes. This process works by blowing hot polymer against the internal surfaces of a hollow mold. The molten plastic enters around a tube that also forces air inside the material, which forces it against the interior surface of the mold. The polymer expands to a uniform thickness against the mold. The mold is formed in two halves, so when the plastic cools, the mold is split to remove the product. The blow molding process is shown in Figure 5.55.

The Calendering Process

The calendering process is generally used to create products such as vinyl flooring, gaskets, and other sheet products. This process fabricates sheet or film thermoplastic or thermoset plastics by passing the material through a series of heated rollers. The space between the sets of rollers gets progressively smaller until the distance between the last set of rollers establishes the desired material thickness. Figure 5.56 illustrates the calendering process used to produce sheet plastic products.

The Rotational Molding Process

The rotational molding process is typically used to produce large containers such as tanks, hollow objects such as floats, and other similar types of large, hollow products. This process works by placing a specific amount of polymer pellets into a metal mold. The mold is then heated as it is rotated. This forces the molten material to form a thin coating against the sides of the mold. When the mold is cooled, the product is removed. The rotational molding process is shown in Figure 5.57.

The Solid Phase Forming Process

The solid phase forming process can be used to make a variety of objects including containers, electrical housings, automotive

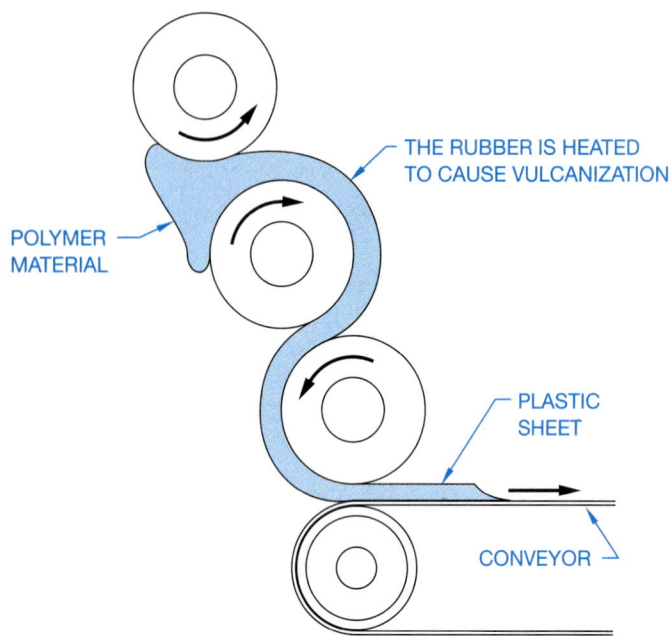

FIGURE 5.56 ■ The calendering process used to produce sheet plastic products.

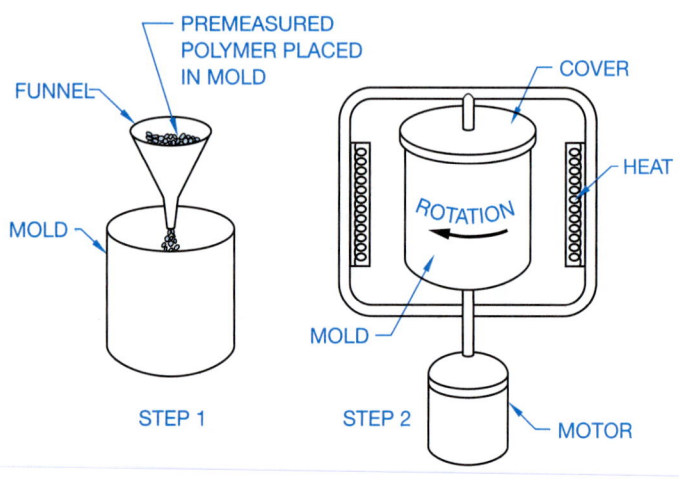

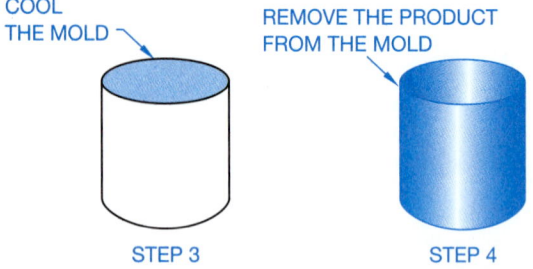

FIGURE 5.57 ■ The rotational molding process.

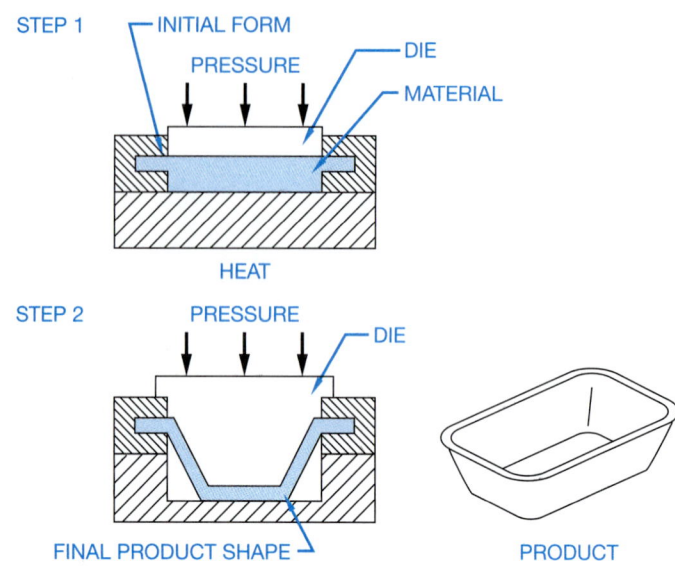

FIGURE 5.58 ■ The solid phase forming process.

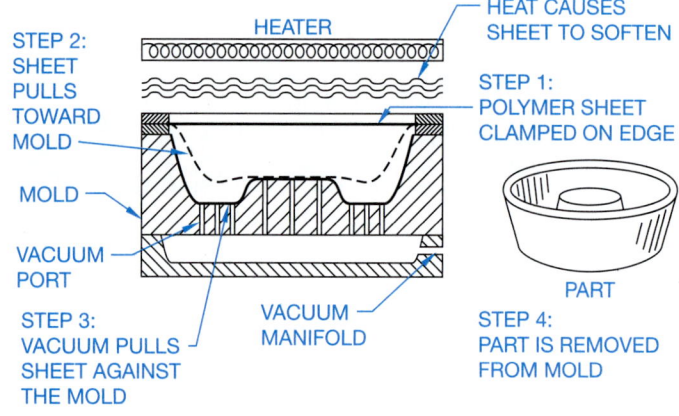

FIGURE 5.59 ■ Thermoforming of plastic. Vacuum pressure is commonly used to suck the hot material down against the mold.

Thermoforming of Plastic

Thermoforming of plastic is similar to solid phase forming, but can be done without a die. This process is used to make all types of thin-walled plastic shapes such as containers, guards, fenders, and other similar products. The process works by taking a sheet of material and heating it until it softens and sinks down by its own weight into a mold that conforms to the desired final shape. Vacuum pressure is commonly used to suck the hot material down against the mold, as shown in Figure 5.59.

Free-Form Fabrication of Plastic

The **free-form fabrication process** uses a computer model that is traced in thin cross sections to control a laser that deposits layers of liquid resin or molten particles of plastic material to form the desired shape. Using the laser to fuse several thin coatings of powder polymer to form the desired shape is also an option for

parts, and anything that has detailed shapes. Products can be stronger using this process, because the polymer is heated but not melted. The solid phase forming process works by placing material into an initial hot die where it takes the preliminary shape. As the material cools, a die that matches the shape of the desired product forms the final shape. This process is similar to metal forging. Figure 5.58 shows the solid phase forming process.

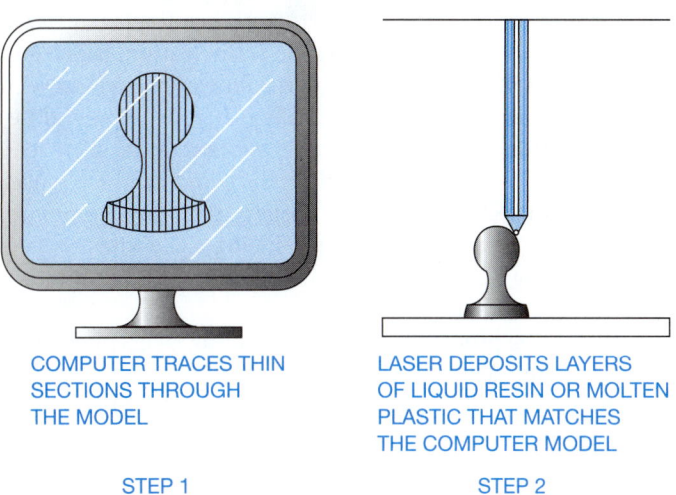

COMPUTER TRACES THIN
SECTIONS THROUGH
THE MODEL

LASER DEPOSITS LAYERS
OF LIQUID RESIN OR MOLTEN
PLASTIC THAT MATCHES
THE COMPUTER MODEL

STEP 1   STEP 2

FIGURE 5.60 ■ The free-form fabrication process.

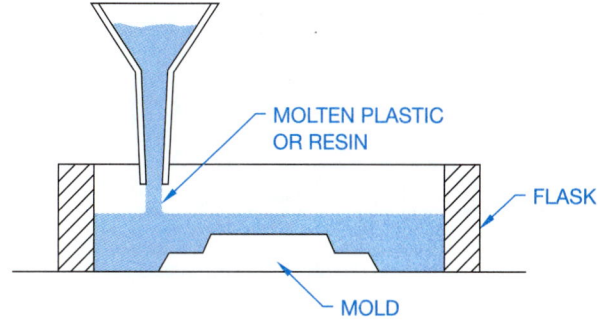

FIGURE 5.61 ■ Casting thermoplastics.

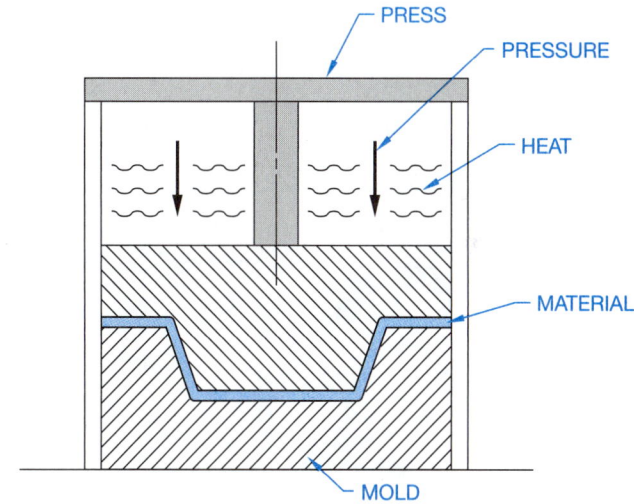

FIGURE 5.62 ■ The compression molding process.

this process. This process is often referred to by the trade name Stereolithography. Figure 5.60 shows the free-form process.

Thermoset Plastic Fabrication Processes

The manufacture of thermoset products can be more difficult than thermoplastic fabrication, because thermosets cannot be remelted once they have been melted and formed the first time. This characteristic makes it necessary to keep the manufacturing equipment operational for as long of a period as possible. When process equipment is shut down, it must be thoroughly cleaned before the thermoset material solidifies. The injection molding process discussed earlier can be used with thermoset plastic, but extreme care must be taken to ensure that the equipment is kept clean. For this reason, other methods are commonly used for manufacturing thermoset products. Thermoset plastic production generally takes longer than thermoplastic production, because thermoset plastics require a longer cure time. The most common production practices are casting, compression molding, foam molding, reaction injection molding, transfer molding, sintering, and vulcanization.

Casting Thermoset Plastics

The method used to cast plastics is very similar to the permanent casting of metals that was explained earlier in this chapter. When casting plastic, the molten polymer or resin is poured into a metal flask with a mold that forms the desired shape of the product. The plastic product is removed when it has solidified. The casting process is shown in Figure 5.61.

Compression Molding and Transfer Molding

Compression molding and transfer molding are common fabrication processes for thermosets. The compression molding process uses a specific amount of material that is heated and placed in a closed mold where additional heat and pressure are applied until the material takes the desired shape. The material

is then cured and removed from the mold. The compression molding process is shown in Figure 5.62.

Transfer molding is similar to compression molding for thermoset plastic products. In this process, the material is heated and then forced under pressure into the mold.

Foam Molding and Reaction Injection Molding

Foam molding is similar to casting, but this process uses a foam material that expands during the cure to fill the desired mold. The foam molding process can be used to make products of any desired shape, or sheets of foam products. The foam molding process is used to make a duck decoy in Figure 5.63.

The reaction injection molding process is similar to the foam process, only because the material that is used expands to fill the mold. This process is often used to fabricate large parts such as automobile dashboards and fenders. Polymer chemicals are mixed together under pressure and then poured into the mold where they react and expand to fill the mold.

The Sintering Process

Sintering is a process that takes powdered particles of material and produces detailed products under heat, pressure, and chemical reaction. The powder particles do not melt, but they do join

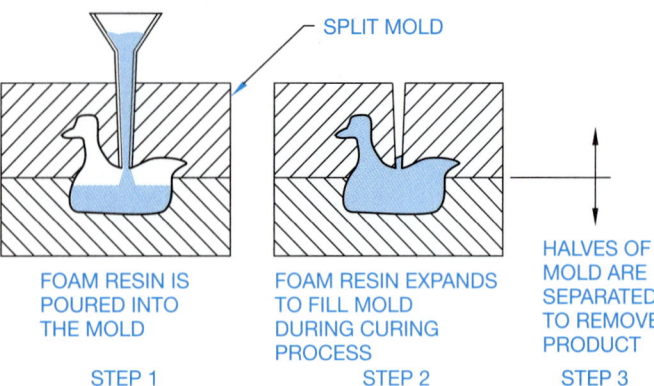

FIGURE 5.63 ■ The foam molding process.

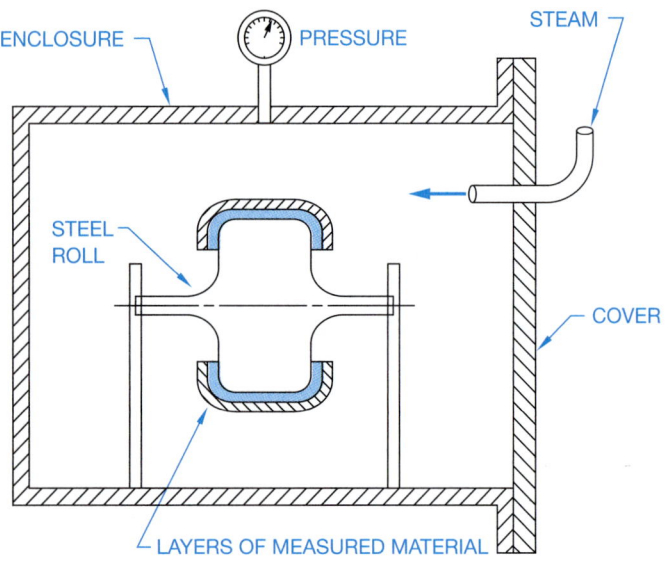

FIGURE 5.64 ■ The vulcanization process.

together into a dense and solid structure. This process is used when creating products for high-temperature applications.

The Vulcanization Process

Vulcanization is used to make rubber products such as tires and other circular and cylindrical shapes, although other shapes are commonly produced. This process works by wrapping measured layers of the polymer around a steel roll in the form of the desired product. The steel roll and material is placed inside an enclosure where steam is introduced to create heat and pressure. This process forces the molecules of the material to form a solid rubber coating on the roll. While the vulcanization process is a chemical reaction that cannot be displayed, a representation of the process is shown in Figure 5.64.

Manufacturing Composites

Composites are also referred to as reinforced plastics. Earlier in this chapter, there was a discussion about composite materials.

These materials combine polymers with reinforcing material such as glass, graphite, thermoplastic fibers, cotton, paper, and metal. The result is very strong material for use in products such as boats, planes, automobile parts, fishing rods, electrical devices, corrosion-resistant containers, and structural members. Composites are less expensive to produce and can be stronger and lighter in weight than many metals, including aluminum, steel, and titanium. The basics of producing composite products are the layering of polymer and reinforcing material in alternate coats or in combination.

The layering process combines alternating layers of polymer resin with reinforcing material such as glass. The number of layers determines the desired thickness. During the curing process, the resin saturates the reinforcing material, creating a unified composite. Rolling or spraying over the reinforcing material can apply layering in the resin. A similar method uses a spray of resin combined with pieces of reinforcing material. This method is called chopped fiber spraying.

Another process uses machines to wind resin-saturated reinforcement fibers around a shaft. That process is referred to as filament winding.

A process called compression molding is similar to the compression molding process that was explained earlier for thermoset plastic products. The difference is that mixing the polymer with reinforcing fibers creates composites. A similar process takes continuous reinforcing strands through a resin bath and then through a forming die. The die forms the desired shape. This method is often used to create continuous shapes of uniform cross section.

A process called resin transfer molding is used to make quality composite products with a smooth surface on both sides. This method places reinforcing material into a mold and then pumps resin into the mold.

A process referred to as vacuum bag forming uses vacuum pressure to force a thin layer of sheet-reinforced polymer around a mold. Figure 5.65 shows a variety of processes used to make composite products.

TOOL DESIGN

In most production machining operations, special tools are required to either hold the workpiece or guide the machine tool. Tool design involves knowledge of kinematics (study of mechanisms) in Chapter 19, machining operations, machine tool function, material handling, material characteristics. Tool design is also known as jig and fixture design. In mass-production industries, jigs and fixtures are essential to ensure that each part is produced quickly and accurately within the dimensional specifications. These tools are used to hold the workpiece so that machining operations are performed in the required positions. Application examples are shown in Figure 5.66. Jigs are either fixed or moving devices that are used to hold the workpiece in position and guide the cutting tool. Fixtures do not guide the cutting tool but are used in a fixed position to hold the workpiece. Fixtures are often used in the inspection of parts to ensure that the part is held in the same position each time a dimensional or other type of inspection is made.

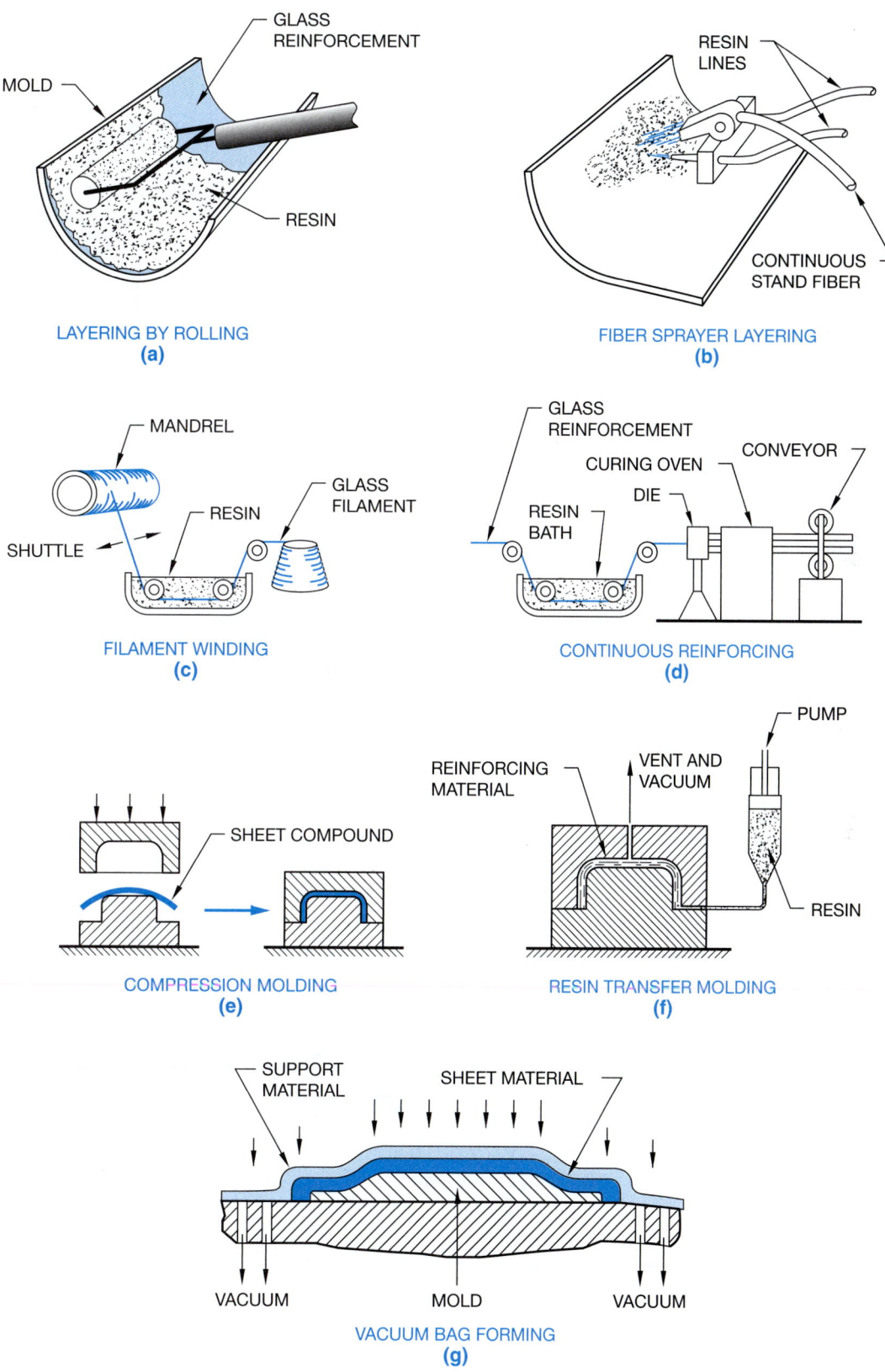

FIGURE 5.65 ■ Common processes used to make composite products. (a) The layering process using rolling. (b) The layering process using spraying. (c) Filament winding. (d) The continuous reinforcing strands process. (e) Compression molding. (f) Resin transfer molding. (g) Vacuum bag forming.

APPLICATION EXAMPLES

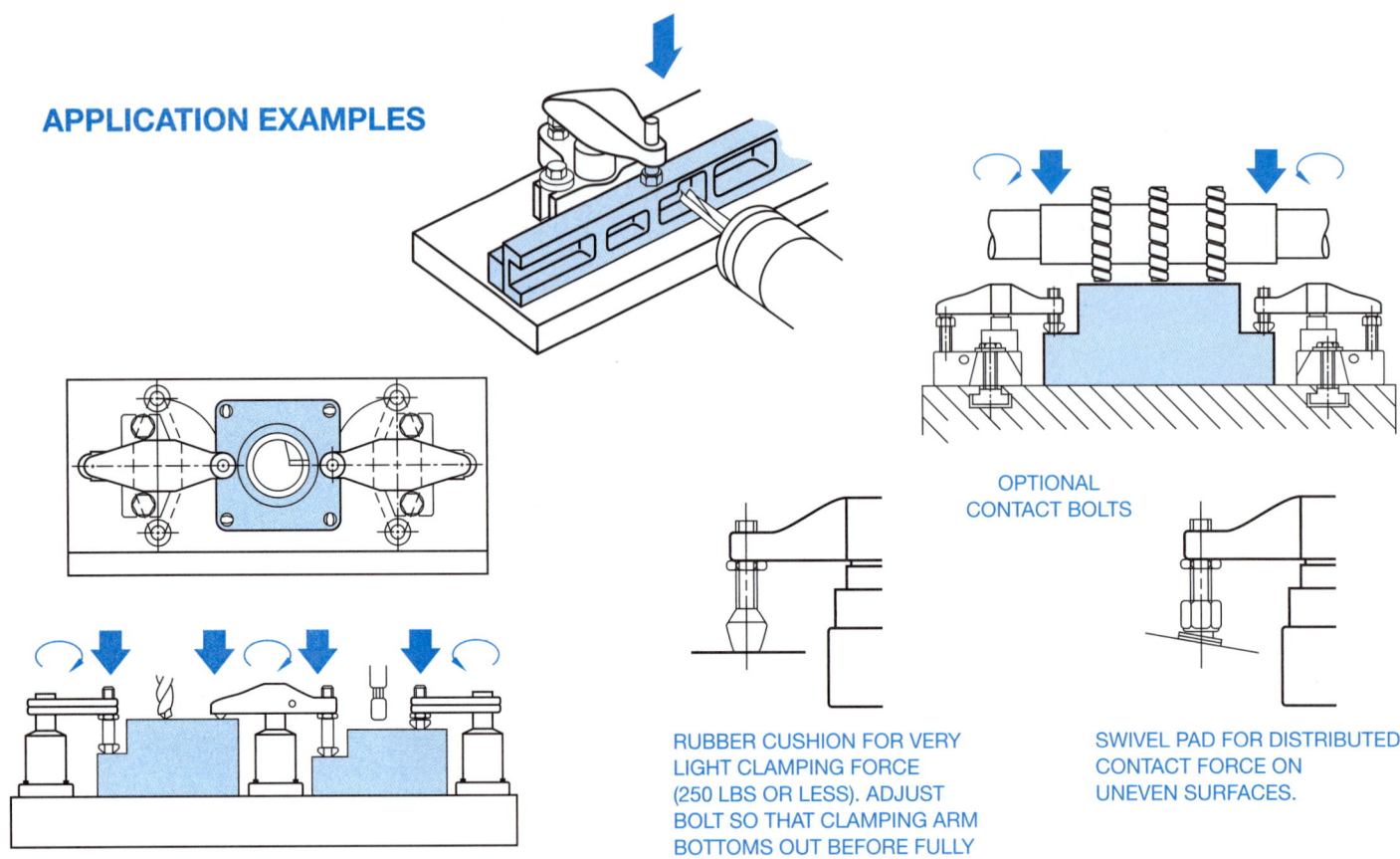

OPTIONAL
CONTACT BOLTS

RUBBER CUSHION FOR VERY
LIGHT CLAMPING FORCE
(250 LBS OR LESS). ADJUST
BOLT SO THAT CLAMPING ARM
BOTTOMS OUT BEFORE FULLY
COMPRESSING THE CUSHION.

SWIVEL PAD FOR DISTRIBUTED
CONTACT FORCE ON
UNEVEN SURFACES.

FIGURE 5.66 ■ Fixture application examples. *Courtesy Carr Lane Manufacturing Company.*

Jig and fixture drawings are prepared as an assembly drawing where all of the components of the tool are shown as if they are assembled and ready for use, as shown in Figure 5.67. Assembly drawings are discussed in Chapter 18. Components of a jig or fixture often include such items as fast-acting clamps, spring-loaded positioners, clamp straps, quick-release locating pins, handles, knobs, and screw clamps. Normally the part or workpiece is drawn in position using phantom lines, in a color such as red, or a combination of phantom lines and color. Additional coverage is provided in Chapter 11.

FIGURE 5.67 ■ Fixture assembly drawing. *Courtesy Carr Lane Manufacturing Company.*

COMPUTER-INTEGRATED MANUFACTURING (CIM)

Completely automated manufacturing systems combine computer-aided design and drafting (CADD), computer-aided engineering (CAE), and computer-aided manufacturing (CAM) into a controlled system known as **computer-integrated manufacturing (CIM)**. In Figure 5.68, CIM brings together all the technologies in a management system, coordinating CADD, CAM, CNC, robotics, and material handling from the beginning of the design process through the packaging and shipment of the product. The computer system is used to control and monitor all the elements of the manufacturing system. Before a more complete discussion of the CIM systems, it is a good idea to define some of the individual elements. Additional discussion is provided throughout this textbook where applied to specific content.

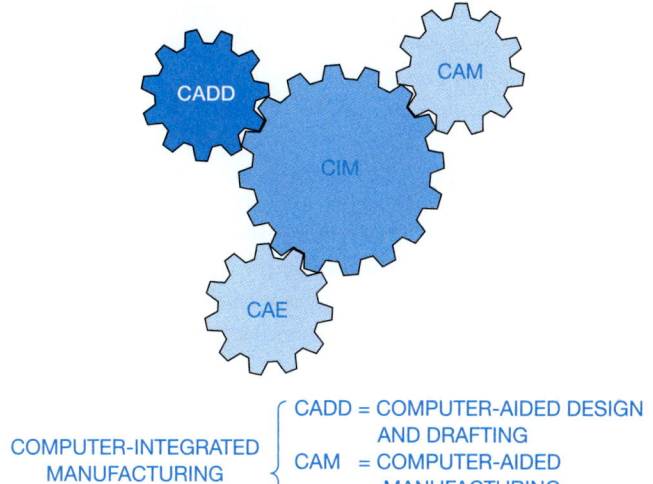

COMPUTER-INTEGRATED MANUFACTURING (CIM)
{
CADD = COMPUTER-AIDED DESIGN AND DRAFTING
CAM = COMPUTER-AIDED MANUFACTURING
CAE = COMPUTER-AIDED ENGINEERING

FIGURE 5.68 ■ Computer-integrated manufacturing (CIM) is the bringing together of the technologies in a management system.

Computer-Aided Design and Drafting (CADD)

Engineers and designers use the computer as a flexible design tool. Designs can be created graphically using 3-D models. The effects of changes can be quickly seen and analyzed. This allows the engineer to perform experimentation, perform stress analysis, and make calculations right on the computer. In this manner, engineers can be more creative and improve the quality of the product at less cost. Computer-aided drafting is a partner to the design process. Accurate quality 2-D drawings are created from the designs. Drafting with the computer has increased productivity from two to twenty times over manual techniques depending on the project and skill of the drafter. The computer has also revolutionized the storage of drawings. There is no longer a need for rooms full of drawing file cabinets. Complete CADD coverage is found in Chapters 3 and 4 of this textbook.

Computer-Aided Engineering (CAE)

Computer-aided engineering (CAE) uses three-dimensional models of the product for finite element analysis. This is done in a program in which the computer breaks the model up into finite elements, which are small rectangular or triangular shapes. Then the computer is able to analyze each element and determine how it will act under given conditions. It also evaluates how each element acts with the other elements and with the entire model. CAE also allows the engineer to simulate function and motion of the product without the need to build a real prototype. In this manner, the product can be tested to see if it works as it should. This process has also been referred to as predictive engineering, in which a computer software prototype, rather than a physical prototype, is made to test the function

and performance of the product. In this manner, design changes may be made right on the computer screen. Key dimensions are placed on the computer model, and by changing a dimension, you can automatically change the design or you can change the values of variables in mathematical engineering equations to automatically alter the design. Additional CAE and finite element analysis is found in Chapter 27.

Computer Numerical Control (CNC)

Computer numerical control (CNC) is the use of a computer to write, store, and edit numerical control programs to operate a machine tool. Numerical control (NC) is a method of controlling a machine tool using coded computer language. This was discussed in the CADD Applications earlier in this chapter.

Computer-Aided Manufacturing (CAM)

Computer-aided manufacturing (CAM) is a concept that surrounds use of computers to aid in any manufacturing process. CAD and CAM work best when product programming is automatically performed from the model geometry created during the CAD process. Complex 3-D geometry can be quickly and easily programmed for machining.

Computer-Aided Quality Control (CAQC)

Information on the manufacturing process and quality control is collected by automatic means while parts are being manufactured, and is the function of computer-aided quality control (CAQC). This information is fed back into the system and compared to the design specifications or model tolerances. With this type of monitoring of the mass-production process, it can be ensured that the highest product quality is maintained.

Robotics

According to the Society of Manufacturing Engineers (SME) Robot Institute of America, a robot is a reprogrammable multifunctional manipulator designed to move material, parts, tools, or specialized devices through variable programmed motions for the performance of a variety of tasks. This process is called robotics. Reprogrammable means the robot's operating program can be changed to alter the motion of the arm or tooling. Multifunctional means the robot is able to perform a variety of operations based on the program and tooling it uses. A typical robot is shown in Figure 5.69.

To make a complete manufacturing cell, the robot is added to all of the other elements of the manufacturing process. The other elements of the cell include the computer controller, robotic program, tooling, associated machine tools, and material-handling equipment.

Closed-loop (servo) and open-loop (nonservo) are the two types of control used to position the robot tooling. The robot is constantly monitored by position sensors in the closed-loop system. The movement of the robot arm must always conform

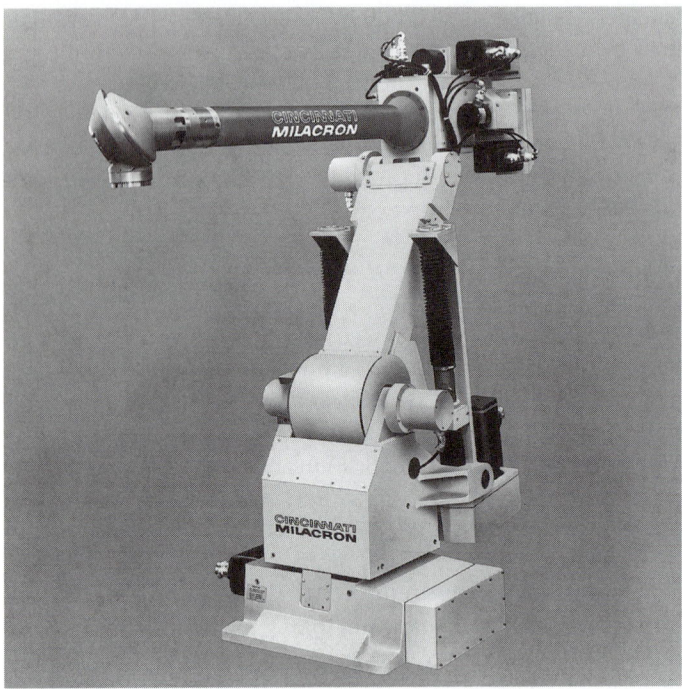

FIGURE 5.69 ■ A typical robot. *Courtesy Cincinnati Milacron Marketing Company.*

to the desired path and speed. Open-loop robotic systems do not constantly monitor the position of the tool while the robot arm is moving. The control happens at the end of travel, where limit controls position the accuracy of the tool at the desired place.

The Human Factor

The properly operating elements of the CIM system provide the best manufacturing automation available. What has been described in this discussion might lead you to believe that the system can be set up, turned on, and the people can go home. Not so. An important part of CIM is the human element. People are needed to handle the many situations that happen in the manufacturing cell. The operators constantly observe the machine operation to ensure that material is feeding properly and quality is maintained. In addition to monitoring the system, people load raw material into the system, change machine tools when needed, maintain and repair the system as needed, and handle the completed work when the product is finished.

INTEGRATION OF COMPUTER-AIDED DESIGN AND COMPUTER-AIDED MANUFACTURING (CAD/CAM)

Computer-aided design (CAD) and computer-aided manufacturing (CAM) can be set up to create a direct link between the design and manufacture of a product. The CAD program is used to create the product geometry. This can be in the form of 2-D

multiview drawings as discussed in Chapter 9, Multiviews, or as 3-D models as explained in Chapter 4, 3-D CADD, Animation, and Virtual Reality. The drawing geometry is then used in the CAM program to generate instructions for the computer numerical control (CNC) machine tools employing stamping, cutting, burning, bending, and other types of operations discussed throughout this chapter. This is commonly referred to as **CAD/CAM integration**. The CAD and CAM operations can be performed on the same computer, or the CAD work can be done at one location and the CAM program can be created at another location. If both the design and manufacturing are done within the company at one location, then the computers can be linked through the local area network. If the design is done at one location and the manufacturing done at another, then the computers can be linked through the Internet, or files can be transferred on disk or CD.

CAD/CAM is commonly used in modern manufacturing because it increases productivity over conventional manufacturing methods. The CAD geometry or model is created during the design and drafting process and is then used directly in the CAM process for the development of the CNC programming. The coordination can also continue into computerized quality control.

The CAD/CAM integration process allows the CAM program to import data from the CAD software. The CAM program then uses a series of commands to instruct CNC machine tools by setting up tool paths. The tool path includes the selection of specific tools to accomplish the desired operation. This can also include specifying tool feed rates and speeds, selecting tool paths and cutting methods, activating tool jigs and fixtures, and selecting coolants for material removal. Some CAM programs automatically calculate the tool offset based on the drawing geometry. An example of tool offset is shown in Figure 5.70. CAM programs such as SurfCAM and MasterCAM directly integrate the CAD drawing geometry from programs such as AutoCAD as a reference. The CAM programmer then just establishes the desired tool and tool path. The final CNC program is generated when the postprocessor is run. A **postprocessor** is an integral piece of software that converts a generic, CAM-system tool path into usable CNC machine code (G-code).

The **CNC program** is a sequential list of machining operations in the form of a code that is used to machine the part as needed. Typically known as G & M code, these codes invoke preparatory functions and control tool and machine movement, spindle speed and direction, and other operations such as clamping, part manipulation, and on-off switching. The machine programmer is trained to select the proper machine tools. A separate tool is used for each operation. These tools can include milling, drilling, turning, threading, and grinding. The CNC programmer orders the machining sequence, selects the tool for the specific operation, and determines the tool feed rate and cutting speed depending on the material.

CAM software programs such as SurfCAM and MasterCAM increase productivity by assisting the CNC programmer in creating the needed CNC code. CAD drawings from programs such as AutoCAD or solid models from software such as SolidWorks

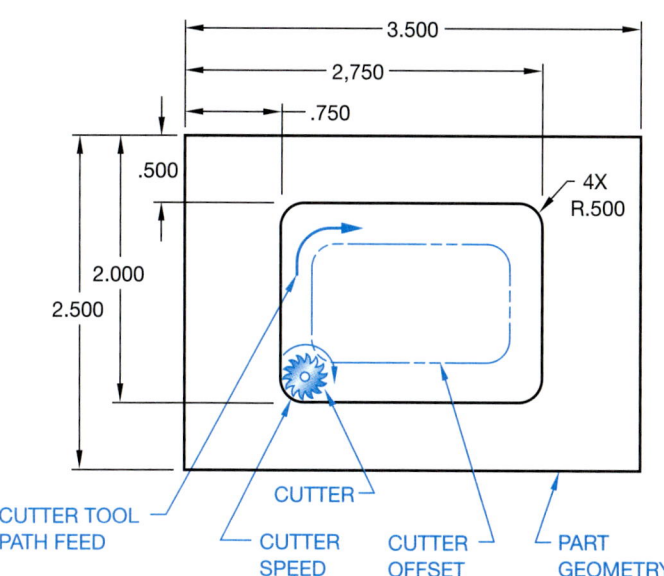

FIGURE 5.70 ■ The machine tool cutter, tool path, cutter speed, and cutter offset.

can be transferred directly to the CAM program. The following sequence of activities is commonly used to prepare the CAD/CAM integration:

1. Create the part drawing or model using programs such as AutoCAD, Autodesk Inventor, or Solid Works. However, programmers commonly originate the model in the CAD/CAM software, such as MasterCAM or SurfCAM.

2. Open the CAD file in the CAM program.

3. Run the CAM software program such as SurfCAM or MasterCAM to establish the following:

   ■ Choose the machine or machines needed to manufacture the part.

   ■ Select the required tooling.

   ■ Determine the machining sequence.

   ■ Calculate the machine tool feed rates and speeds based on the type of material.

   ■ Verify the CNC program using the software's simulator.

   ■ Create the CNC code.

4. Prove-out the program on the CNC machine tool.

5. Run the program to manufacture the desired number of parts.

STATISTICAL PROCESS CONTROL (SPC)

A system of quality improvement is helpful to anyone who turns out a product or is engaged in a service and who wants to improve the quality of work and at the same time increase the out-

put, all with less labor and at reduced cost. Competition is here to stay, regardless of the nature of the business. Improved quality means less waste and less rework, resulting in increased profits and an improved market position. Customarily, many managers see quality as a drag on profits. Quality is often placed after cost and delivery, because some managers believe that high quality can be achieved only through costly and slow inspection processes. Many managers are now seeing the influence of quality on sales because high quality has become an important criterion in their customers' purchase decisions. In addition, poor quality is expensive. It is estimated that between 15% and 40% of the American manufacturer's product cost is a result of unacceptable output. Regardless of the goods or service produced, it is always less costly to do it right the first time. Improved quality improves productivity, increases sales, reduces cost, and improves profitability. The net result is continued business success.

Traditionally, a type of quality control/detection system has been used in most organizations in the United States. This system comprises customer demand for a product, which is then manufactured in a process made up of a series of steps or procedures. Input to the process includes machines, materials, workforce, methods, and environment, as shown in Figure 5.71. Once the product or service is produced, it goes on to an inspection operation where decisions are made to ship, scrap, rework, or otherwise correct any defects when discovered (if discovered). In actuality, if nonconforming products are being produced, then some are being shipped. Even the best inspection process screens out only a portion of the defective goods. Problems inherent in this system are that it does not work very well and it is costly. American businesses have become accustomed to accepting these limitations as the "costs of doing business."

The most effective way to improve quality is to alter the production process, the system, rather than the inspection process. This entails a major shift in the entire organization from the detection system to a prevention mode of operation. In this system, the elements (inputs, process, product or service, customer) remain the same, but the inspection method is significantly altered or eliminated. A primary difference between the two systems is that in the prevention system, statistical

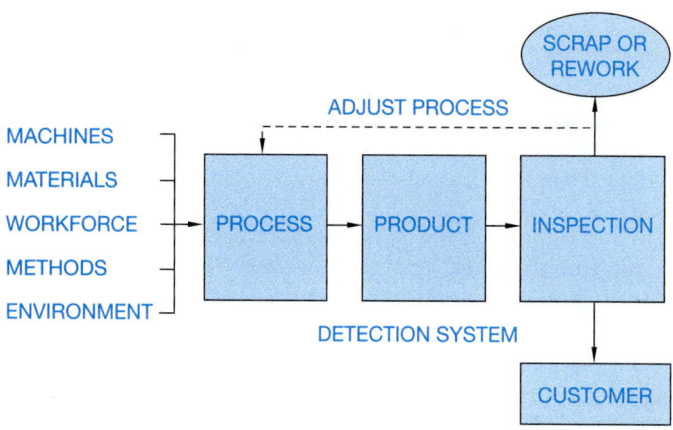

FIGURE 5.71 ■ Quality control/detection system.

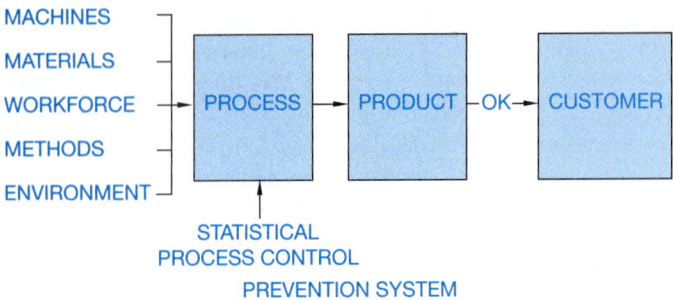

FIGURE 5.72 ■ Quality control/prevention system.

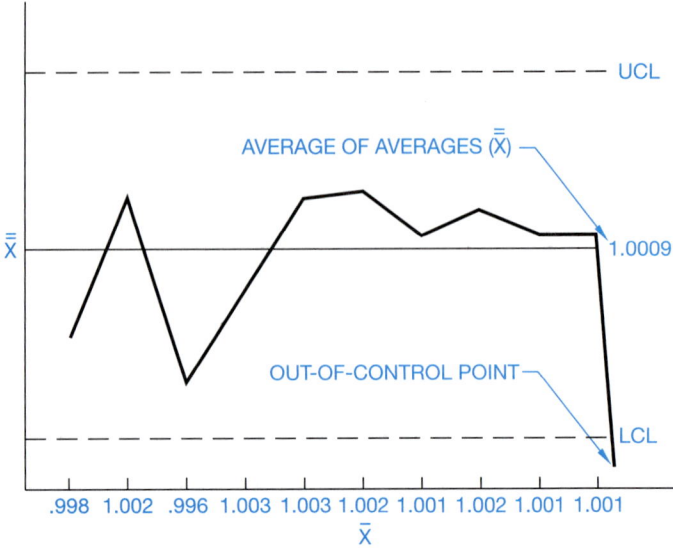

FIGURE 5.73 ■ Quality control chart.

techniques and problem-solving tools are used to monitor, evaluate, and provide guidance for adjusting the process to improve quality. **Statistical process control (SPC)** is a method of monitoring a process quantitatively and using statistical signals to either leave the process alone, or change it. (See Figure 5.72.) It involves several fundamental elements:

1. The process, product, or service must be measured. It can be measured using either **variables** (a value that varies), or **attributes** (a property or characteristic) from the data collected. The data should be collected as close to the process as possible. If you are collecting data on a particular dimension of a manufactured part, it should be collected by the machinist who is responsible for holding that dimension.

2. The data can be analyzed using control charting techniques. **Control charting techniques** use the natural variation of a process, determining how much the process can be expected to vary if the process is operationally consistent. The control charts are used to evaluate whether the process is operating as designed, or if something has changed.

3. Action is taken based on signals from the control chart. If the chart indicates that the process is **in control** (operating consistently), then the process is left alone at this point. On the other hand, if the process is found to be **out of control** (changing more than its normal variability allows), then action is taken to bring it back into control. It is also important to determine how well the process meets specifications and how well it accomplishes the task. If a process is not in control, then its ability to meet specifications is constantly changing. Process capability cannot be evaluated unless the process is in control. Process improvement generally involves changes to the process system that will improve quality or productivity or both. Unless the process is consistent over time, any actions to improve it may be ineffective.

Manufacturing quality control often uses computerized monitoring of dimensional inspections. When this is done, a chart is developed that shows feature dimensions obtained at inspection intervals. The chart shows the expected limits of sample averages as two dashed parallel horizontal lines, as shown in Figure 5.73. It is important not to confuse control limits with tolerances—they are not related to each other. The control limits come from the manufacturing process as it is operating. For

example, if you set out to make a part $1.000 \pm .005$ in., and you start the run and periodically take five samples and plot the averages ($\bar{x}$) of the samples, the sample averages will vary less than the individual parts. The control limits represent the expected variation of the sample averages if the process is stable. If the process shifts or a problem occurs, the control limits signal that change. Notice in Figure 5.73 that the $\bar{x}$ values represent the average of each five samples; $\bar{\bar{x}}$ is the average of averages over a period of sample taking. The **upper control limit (UCL)** and the **lower control limit (LCL)** represent the expected variation of the sample averages. A sample average may be "out of control," yet remain within tolerance. During this part of the monitoring process, the out-of-control point represents an individual situation that may not be a problem; however, if samples continue to be measured out of control limits, then the process is out of control (no longer predictable) and therefore may be producing parts outside the specification. Action must then be taken to bring the process back into statistical control, or 100% inspection must be resumed. The SPC process only works when a minimum of twenty-five sample means are in control. When this process is used in manufacturing, part dimensions remain within tolerance limits and parts are guaranteed to have the quality designed.

WEBSITE RESEARCH

The following websites can assist you in doing additional research on subjects such as manufacturing materials, manufacturing practices, tooling, and related areas:

http://www.allamericanproducts.com—All American Products: tooling components.

http://www.ab.com—Allen-Bradley: industrial electrical controls.

http://www.aluminum.org—Aluminum Association.

(Continued on page 164)

MANUFACTURING MATERIALS

Many parametric solid modeling software programs allow you to assign a material to your model that closely replicates the material used to manufacture the product. For example, if a part is made from mild steel, you are able to select a mild steel material from a list of available materials, or create a custom mild steel material to apply to the model. The material characteristics and amount of information stored in a model varies depending on the software program. Most solid modeling applications allow you to apply materials that use specific colors and textures that closely match the actual look of a material, such as mild steel. You may also have the option of adjusting color, lighting, and texture characteristics to further imitate the true look of the selected material. Some solid modeling programs incorporate tools and options that can be used to calculate physical properties associated with the model according to the selected material. In addition, often when you select a material in the modeling environment, the material name and corresponding properties are parametrically available for use in other applications such as parts lists and bills of materials.

MANUFACTURING PROCESSES

Parametric solid modeling software programs can be used to simulate actual manufacturing processes. The three-dimensional, step-by-step procedures used to generate a model closely resemble the virtual manufacturing of a product, from the initial stages of manufacturing to the finished product. The entire process can be modeled before any machines are used to manufacture parts and assemblies. In addition, many modeling programs and add-in applications contain specialized tools and options for replicating specific manufacturing tasks. The following information provides a few examples of how manufacturing processes can be simulated using parametric solid modeling software:

■ **Casting and forging.** Standard model feature tools can be used to produce patterns, molds, dies, and complete parts, such as the pattern and molds shown in Figure 5.74. However, many programs contain tools for defining specific casting and forging elements such as parting lines and draft angles. Other applications incorporate specialized casting and forging commands that greatly enhance the ability to quickly and accurately model patterns, molds, cores, and dies.

■ **Metal stamping.** In Chapter 25 you will learn how CADD can be used to model sheet metal products. The process of stamping metal can also be modeled using CADD software. Extrusions that cut through features

FIGURE 5.74 ■ Standard model feature tools can be used to produce patterns, molds, dies, and complete parts, such as pattern and molds. This is an example of a basic pair of molds and a pattern created using solid modeling software.

are available with most modeling programs, and can be used to produce stamped metal parts. Some software also includes dedicated sheet metal stamping, or punching, tools that closely replicate the actual process of stamping metal. An example of a stamped metal part model is shown in Figure 5.75.

FIGURE 5.75 ■ An example of a sheet metal-part model with two metal stamps.

(Continued)

CADD APPLICATIONS 3-D

■ **Machine processes.** Multiple machining processes can be modeled using parametric solid modeling applications. For example, the part shown in Figure 5.76 was modeled using a revolved feature tool. Revolutions are ideal for simulating the use of a lathe to manufacture a product. Other examples include adding basic features such as rounds and chamfers to replicate using a grinding or milling machine, and drilling holes using modeling tools that allow you to define the exact characteristics of a drilling operation, such as the counterbored holes shown in Figure 5.77.

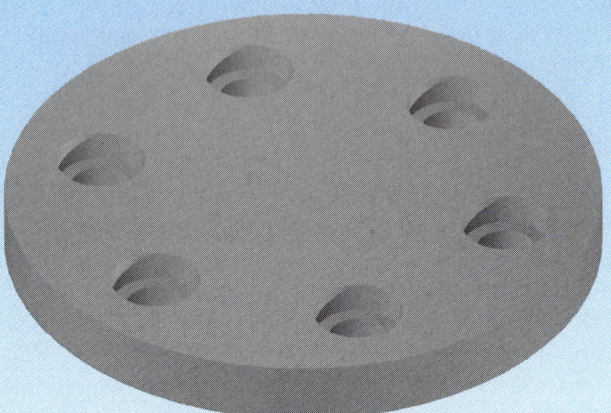

FIGURE 5.77 ■ Other examples of providing machine process in model creation include adding basic features such as rounds and chamfers to replicate using a grinding or milling machine, and drilling holes using modeling tools that allow you to define the exact characteristics of a drilling operation, such as counterbored holes. The counterbored holes in this part simulate several machining operations.

FIGURE 5.76 ■ An example of a part, modeled using a revolved feature tool. Revolutions are ideal for simulating the use of a lathe to manufacture a product.

http://www.asme.org—American Society of Mechanical Engineers.

http://www.aise.org—American Institute of Steel Construction

http://www.ides.com—Plastic materials directory.

http://www.carrlane.com—Carr Lane Manufacturing: tooling components.

http://www.copper.org—Copper Development Association.

http://www.destaco.com—De-Sta-Co Industries: toggle clamps.

http://www.emtec.org—Edison Materials Technology Center: manufacturing materials research.

http://www.idesinc.com—IDES Inc.: plastic materials information.

http://www.industrialpress.com—*Machinery's Handbook:* manufacturing materials, processes, specifications, and data.

http://www.jergensinc.com—Jergens Inc.: tooling components.

http://www.mitutoyo.com—Mitutoyo: precision manufacturing devices.

http://www.mjvail.com—M. J. Vail Co., Inc.: wholesale distributor.

http://www.azom.com—The A to Z of materials.

http://www.reidtool.com—Reid Tool Supply Co.: tooling components.

http://www.stratasys.com—Stratasys Inc.: rapid prototyping.

http://www.wolverinetool.com—Wolverine Tool Co.: heavy-duty toggle clamps.

http://www.worldsteel.org—International Iron and Steel Institute.

PROFESSIONAL PERSPECTIVE

Mechanical drafting and manufacturing are very closely allied. The mechanical drafter should have a general knowledge of manufacturing methods, including machining processes. It is common for the drafter to consult with engineers and machinists regarding the best methods to implement on a drawing from the design to the manufacturing processes. Part of design problem solving is to create a design that is functional, and one that can also be manufactured using available technology and at a cost that justifies the end product. The solutions to these types of concerns depend on how familiar the drafter and designer are with the manufacturing capabilities of the company. For example, if the designer over-tolerances a part or feature, the result can be a rejection of the part by manufacturing or an expensive machining operation. The drafter should also know something about how machining processes operate so that drawing specifications do not call out something that is not feasible to manufacture. The drafter should be familiar with the processes that are used in machine operations so the notes that are placed on a drawing conform to the proper machining techniques. Notes for machining processes are given on a drawing in the same manner they are performed in the shop. For example, the note for a counterbore is given as the diameter and depth of the hole (first process) and then the diameter and depth of the counterbore (second process). Also, the drafter should know the specification given will, in fact, yield the desired result. One interesting aspect of the design drafter's job is the opportunity to communicate with people in manufacturing and come up with a design that enables the product to be manufactured easily.

FINDING MEANS FOR SPC

Here is a set of data consisting of weekly samplings of a 2-in. bolt making machine. The process is considered "out of control" if sample averages exceed a UCL of 2.020 in. or are less than an LCL of 1.980 in. In preparation of a chart, find $\bar{\bar{x}}$ for each week, and find $\bar{x}$ for the set of four weeks. Was the process "out of control" during any week?

Week 1	Week 2	Week 3	Week 4
1.970	2.022	1.980	2.003
2.013	2.014	2.001	1.989
2.051	2.011	1.898	1.993
1.993	2.001	1.888	2.019
1.992	2.009	1.979	2.003

Solution: In the world of math statistics, a bar over a letter denotes an average (mean). The formula for finding the average of $\bar{x}$ values is usually written as $\bar{x} = \Sigma x/n$, where $\bar{x}$ is an individual data value and n is the total number of data values. The Σ is the capital Greek letter sigma, which stands for repeated addition. The formula is the mathematical way of saying to add up all the data and then divide by the total number of data. Applying this idea to each week separately:

$\bar{x}$ for Week 1 = (1.970 + 2.013 + 2.051 + 1.993 + 1.992) ÷ 5
$\qquad = (10.019) \div 5 = 2.0038$ or **2.004**

Similarly,
$\quad \bar{x}$ for Week 2 = 2.0114 or **2.011**
$\quad \bar{x}$ for Week 3 = 1.9492 or **1.949**
$\quad \bar{x}$ for Week 4 = 2.0014 or **2.001**

$\bar{\bar{x}}$ is the mean *of these means*. It is found by the same method:

$\bar{\bar{x}}$ = (2.0038 + 2.0114 + 1.9492 + 2.0014) ÷ 4
$\qquad = 1.99145$ or **1.991**

The mean for Week 3 is lower than the LCL, so the process was "out of control," and corrective measures should be taken at the end of that week.

MATH APPLICATIONS

CHAPTER 5 *Manufacturing Materials and Processes Test*

Access the CD found with this textbook to view the Chapter 5 Test. Confirm the preferred submittal method with your instructor.

CHAPTER 5

Manufacturing Materials and Processes Problems

This chapter is intended as a reference for manufacturing materials and processes. The concepts discussed serve as a basis for further study in the following chapters. A thorough understanding of dimensioning practices is necessary before complete manufactured products can be drawn. Problem assignments ranging from basic to complex manufacturing drawings are assigned in the following chapters.

PROBLEMS 5.1 THROUGH 5.14   Provide a brief but complete definition of each of the following terms. Use your own words to describe the appearance of the actual feature and the related drawing used to represent the feature.

PROBLEM 5.1   Counterbore

PROBLEM 5.2   Chamfer

PROBLEM 5.3   Countersink

PROBLEM 5.4   Counterdrill

PROBLEM 5.5   Drill (not through the material)

PROBLEM 5.6   Fillet

PROBLEM 5.7   Round

PROBLEM 5.8   Spotface

PROBLEM 5.9   Dovetail

PROBLEM 5.10   Kerf

PROBLEM 5.11   Keyseat

PROBLEM 5.12   Keyway

PROBLEM 5.13   T-slot

PROBLEM 5.14   Knurl

PROBLEMS 5.15 THROUGH 5.48 The following topics require research and/or industrial visitations. It is recommended that you research current professional magazines, visit local industries, or interview professionals in the related field. Your reports should emphasize the following:

■ Product

■ The process

■ Special manufacturing considerations

■ The link between manufacturing and engineering

■ Current technological advances

Select one or more of the topics listed below or as assigned by your instructor and write a report of approximately 500 words for each.

PROBLEM 5.15   Casting

PROBLEM 5.16   Forging

PROBLEM 5.17   Conventional machine shop

PROBLEM 5.18   Computer numerical control (CNC) machining

PROBLEM 5.19   Surface roughness

PROBLEM 5.20   Tool design

PROBLEM 5.21   Chemical machining

PROBLEM 5.22   Electrochemical machining (ECM)

PROBLEM 5.23   Electrodischarge machining (EDM)

PROBLEM 5.24   Electron beam (EB) machining

PROBLEM 5.25   Ultrasonic machining

PROBLEM 5.26   Laser machining

PROBLEM 5.27   Statistical process control (SPC)

PROBLEM 5.28   Thermoplastics

PROBLEM 5.29   Thermosets

PROBLEM 5.30   Elastomers

PROBLEM 5.31   Metal stamping

PROBLEM 5.32   Powder metallurgy

PROBLEM 5.33   Manufacturing thermoplastic products

PROBLEM 5.34   Manufacturing thermoset plastic products

PROBLEM 5.35   Manufacturing composites

PROBLEM 5.36   Cast iron

PROBLEM 5.37   Steel

PROBLEM 5.38   Aluminum

PROBLEM 5.39   Copper alloys

PROBLEM 5.40   Precious and other specialty metals

PROBLEM 5.41   Computer-aided manufacturing (CAM)

PROBLEM 5.42   Computer-aided engineering (CAE)

PROBLEM 5.43   Computer-integrated manufacturing (CIM)

PROBLEM 5.44   Robotics

PROBLEM 5.45   Computer-aided design

PROBLEM 5.46   Computer-aided drafting

PROBLEM 5.47   Manufacturing cell

PROBLEM 5.48   CAD/CAM integration

MATH PROBLEMS

Here is a set of six monthly samples from a process. The UCL is 5.05 and the LCL is 4.95.

Jan	Feb	Mar	Apr	May	Jun
5.06	5.03	5.04	5.04	4.95	4.95
4.94	5.04	5.04	5.07	4.95	4.97
4.99	5.05	5.05	5.05	4.93	4.98
4.99	4.99	5.06	5.02	4.92	5.01
5.00	4.98	5.06	5.00	4.95	5.00

PROBLEM 5.49   Find $\bar{x}$ for each month.

PROBLEM 5.50   Find $\bar{\bar{x}}$ for the six-month period.

PROBLEM 5.51   Conclude whether the process was "out of control" for any of the months.

Fundamental Applications

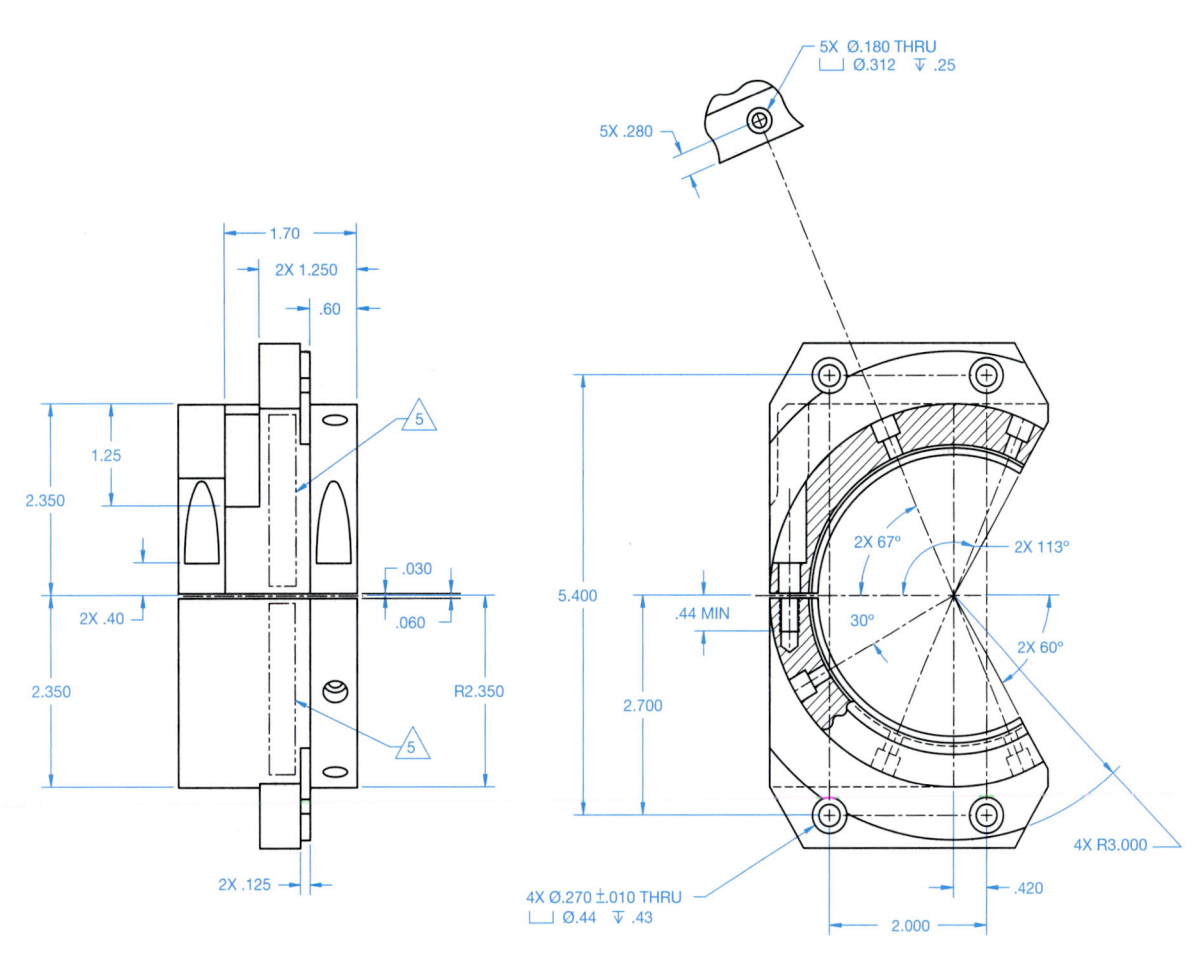

Courtesy Flir Systems, Inc.

6 CHAPTER

Sketching

LEARNING OBJECTIVES

After completing this chapter, you will:

- Sketch lines, circles, arcs, and other geometric shapes.
- Sketch multiviews.
- Sketch isometric views.
- Answer questions related to sketching.

THE ENGINEERING DESIGN APPLICATION

The engineering drafter often works from sketches or written information provided by the engineer. When you receive an engineer's sketch and are asked to prepare a formal drawing, you should follow these steps:

STEP 1 Whether you are using manual drafting or CADD, prepare your own sketch the way you think it should look on the final drawing, taking into account correct drafting standards.

STEP 2 Evaluate the size of the object so you can determine the scale and sheet size for your final manual drawing or the screen limits for your final computer-aided drawing.

STEP 3 Lay out the drawing either very lightly using construction lines or light blue lead. The construction lines can be easily erased if you make a mistake. If you are using a computer, begin the layout on the screen and if you make an error, just erase it and try again. Editing your CADD work is fast and easy.

STEP 4 Complete the final manual drawing by darkening all construction lines to proper ASME standard line weights and run a copy for checking. After completing the computer drawing, a check plot may be made on the plotter or printer. This gives you a chance to check your work on paper. Sometimes it is easier to check a drawing on paper than it is to check it on screen.

The example in Figure 6.1 shows a comparison between the engineer's rough sketch and the finished drawing.

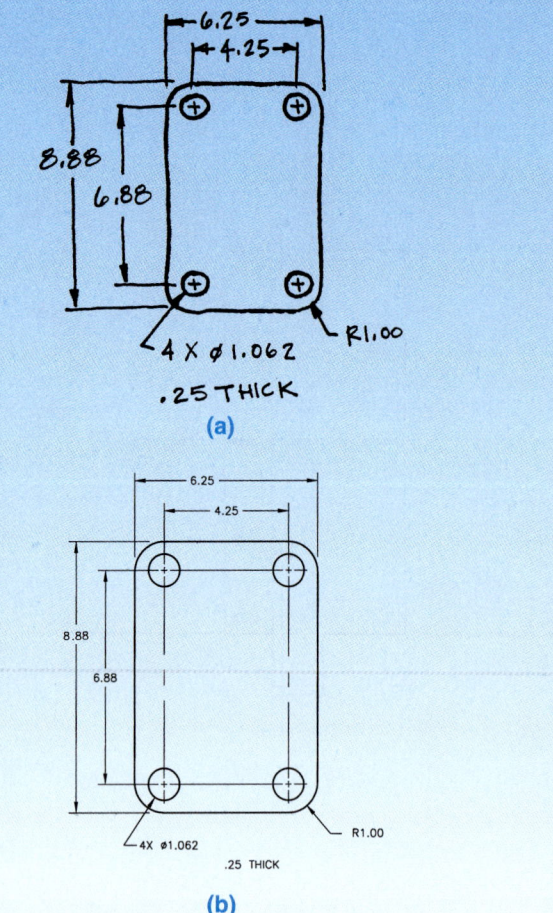

FIGURE 6.1 ■ A comparison between (a) an engineer's rough sketch and (b) the finished drawing.

SKETCHING

Sketching is freehand drawing; that is, drawing; without the aid of drafting equipment. Sketching is convenient since all that is needed is paper, pencil, and an eraser. There are a number of advantages and uses for freehand sketching. Sketching is fast visual communication. The ability to make an accurate sketch quickly often can be an asset when communicating with people at work or at home. Especially when technical concepts are the topic of discussion, a sketch can be the best form of communication. Most drafters prepare a preliminary sketch to help organize thoughts and minimize errors on the final drawing. The computer drafter usually prepares a sketch on graph paper to help establish the coordinates for drawing components. Some drafters use sketches to help record the stages of progress when designing, until a final design is ready for implementation into formal drawings. A sketch can be a useful form of illustration in technical reports. Sketching is also used in job shops where one-of-a-kind products are made. In the job shop, the sketch is often used as a formal production drawing. When the drafter's assignment is to prepare working drawings for existing parts or products, the best method to gather shape and size descriptions about the project is to make a sketch. The sketch can be used to quickly lay out dimensions of features for later transfer to a formal drawing.

The quality of a sketch depends on the intended purpose. Normally a sketch does not have to be very good quality as long as it adequately represents what you want to display. *Speed is a big key to sketching.* You normally want to prepare the sketch as fast as possible while making it as easy and clear to read as possible. Sometimes a sketch does need to have the quality of a formal presentation. A sketch can be used as an artistic impression of a product, or as a one-time detail drawing for manufacturing purposes. However, the sketch is normally used in preliminary planning or to relate a design idea to someone very quickly. The quality of your classroom sketches depends on your course objectives. Your instructor may want quality sketches or very quick sketches that help you establish a plan for further formal drafting. You should confirm this in advance. In the professional world, your own judgment determines the nature and desired quality of the sketch.

TOOLS AND MATERIALS

Sketching equipment is not very complicated. As mentioned, all you need is paper, pencil, and an eraser. The pencil should have a soft lead; a common number 2 pencil works fine or an automatic 0.7- or 0.9-mm pencil with F or HB lead is also good. The pencil lead should not be sharp. A dull, slightly rounded pencil point is best. Different thicknesses of line, if needed, can be drawn by changing the amount of pressure you apply to the pencil. The quality of the paper is not critical either. A good sketching paper is newsprint, although almost any kind works. Actually, paper with a surface that is not too smooth is best. Many engineering designs have been created on a napkin around a lunch table. Sketching paper should not be taped down to the table. The best sketches are made when you are able to move the paper to the most comfortable drawing position. Some people make horizontal lines better than vertical lines. If this is your situation, then move the paper so that vertical lines become horizontal. Such movement of the paper may not always be possible, so it does not hurt to keep practicing all forms of lines for best results. Graph paper is also good to use for sketching because it has grid lines that can be used as a guide for your sketch lines.

SKETCHING STRAIGHT LINES

Lines should be sketched in short, light, connected segments as shown in Figure 6.2. If you sketch one long stroke in one continuous movement, your arm tends to make the line curved

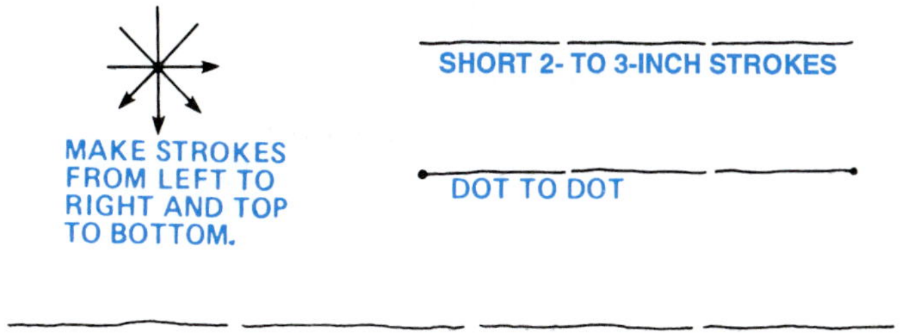

FIGURE 6.2 ■ Sketching short line segments.

FIGURE 6.3 ■ Long movements tend to cause a line to curve.

A•   •B

FIGURE 6.4 ■ Step 1, use dots to identify both ends of a line.

rather than straight, as shown in Figure 6.3. Also, if you make a dark line, you may have to erase if you make an error, whereas if you draw a light line there often is no need to erase.

Following is the procedure used to sketch a horizontal straight line with the dot-to-dot method:

STEP 1 Mark the starting and ending positions, as in Figure 6.4. The letters A and B are only for instruction. All you need are the points.

STEP 2 Without actually touching the paper with the pencil point, make a few trial motions between the marked points to adjust the eye and hand to the anticipated line.

STEP 3 Sketch very light lines between the points by drawing short light strokes (2-to 3-in. long). Keep one eye directed toward the end point while keeping the other eye directed on the pencil point. With each stroke, an attempt should be made to correct the most obvious defects of the preceding stroke so the finished light lines are relatively straight. (See Figure 6.5.)

STEP 4 Darken the finished line with a dark, distinct, uniform line directly on top of the light line. Usually the darkness can be obtained by pressing on the pencil. (See Figure 6.6.)

Very long straight lines can often be sketched by using the edge of the paper or the edge of a table as a guide. To do this, position the paper in a comfortable position with your hand

FIGURE 6.5 ■ Step 3, use short light strokes.

A•————————————————————————•B

FIGURE 6.6 ■ Step 4, darken to finish the line.

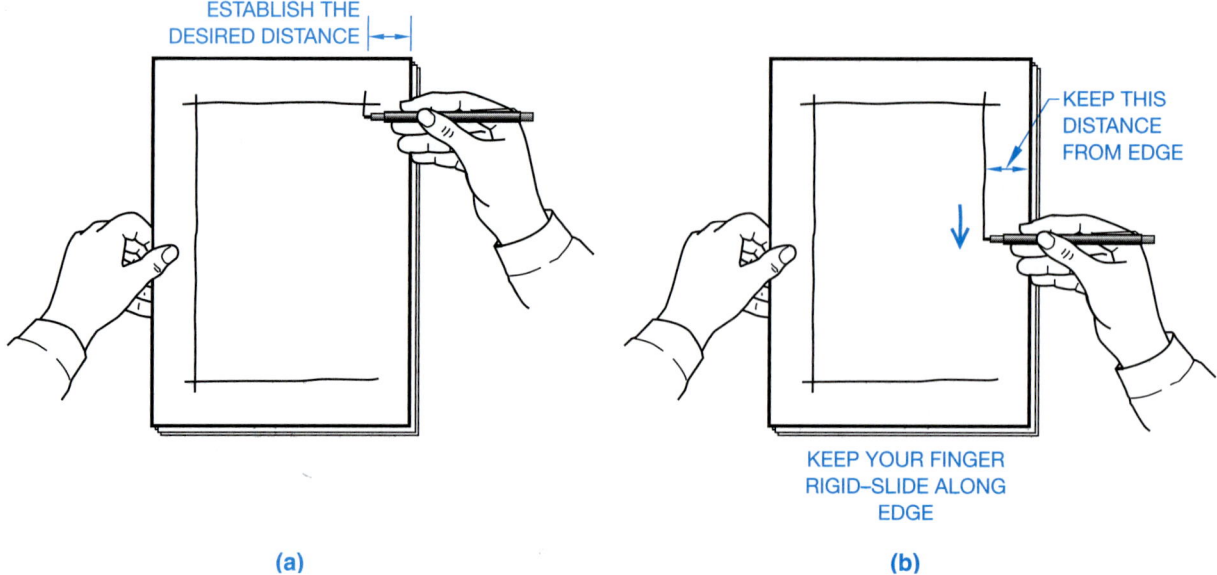

(a)   **(b)**

FIGURE 6.7 ■ Sketching very long straight lines using the edge of the sheet as a guide. (a) Place your hand along the edge as a guide. (b) Move your hand and the pencil along the edge of the paper using your finger or palm as a guide to keep the pencil a constant distance from the edge.

placed along the edge as shown in Figure 6.7a. Extend the pencil point out to the location of the line. Next, place one of your fingers or the palm of your hand along the edge of the paper as a guide. Now, move your hand and the pencil continuously along the edge of the paper as shown in Figure 6.7b. A problem with this method is that it works best if the line is fairly close to the edge of the paper. A sketch does not have to be perfect anyway, so a little practice should be good enough.

SKETCHING CIRCULAR LINES

Figure 6.8 shows the parts of a circle. There are several sketching techniques to use when making a circle; this text explains the quick freehand method for small circles, the box method, the centerline method, the hand-compass method, and the trammel method for very large circles.

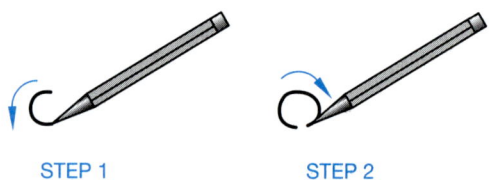

STEP 1   STEP 2

FIGURE 6.9 ■ Sketching a small circle just like drawing the letter o.

Sketching Quick Small Circles

Small circles are easy to sketch if you treat them just like drawing the letter o. You should be able to do this in two strokes by sketching a half circle on each side as shown in Figure 6.9.

Using the Box Method

It is always faster to sketch a circle without first creating other construction guides, but it can be difficult. The **box method** can help you by providing a square that contains the desired circle. Start this method by very lightly sketching a square box that is equal in size to the diameter of the proposed circle as shown in Figure 6.10. The very light lines are called **construction lines**. Next, sketch diagonals across the square. This establishes the center and allows you to mark the radius of the circle on the diagonals as shown in Figure 6.11. Use the sides of the square and the marks on the diagonals as a guide to sketch the circle. Create the circle by drawing arcs that are tangent to the sides of the square and go through the marks on the diagonals as shown in Figure 6.12. If you have trouble sketching the circle as dark and

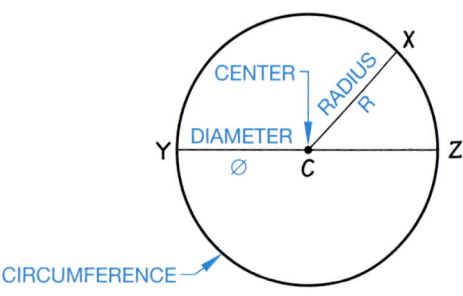

FIGURE 6.8 ■ The parts of a circle.

thick as desired, sketch it very lightly first and then go back over it to make it dark. You can easily correct very lightly sketched lines, but it is difficult to correct very dark lines. Your construction lines do not have to be erased if they are sketched very lightly.

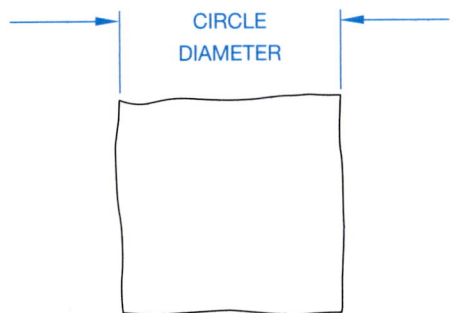

FIGURE 6.10 ■ Very lightly sketch a square box that is equal in size to the diameter of the proposed circle.

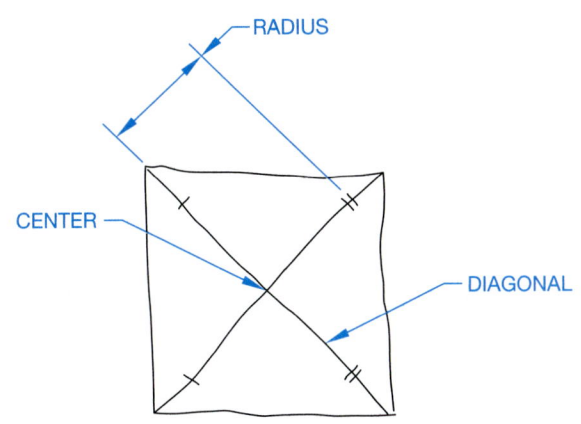

FIGURE 6.11 ■ Sketch light diagonal lines across the square, and mark the radius on the diagonals.

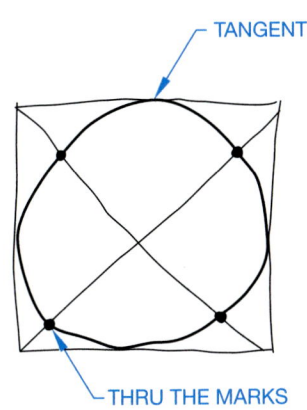

FIGURE 6.12 ■ Create the circle by sketching arcs that are tangent to the sides of the square and go through the marks on the diagonals.

Using the Centerline Method

The centerline method is similar to the box method, but without a box. This method uses very lightly sketched horizontal, vertical, and two 45° diagonal lines as shown in Figure 6.13. Next, mark the approximate radius of the circle on the centerlines as shown in Figure 6.14. Create the circle by drawing arcs that go through the marks on the centerlines as shown in Figure 6.15.

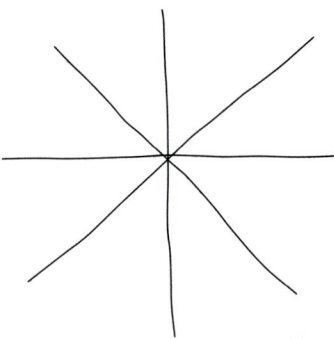

FIGURE 6.13 ■ Sketch very light horizontal, vertical, and 45° lines that meet at the center of the proposed circle.

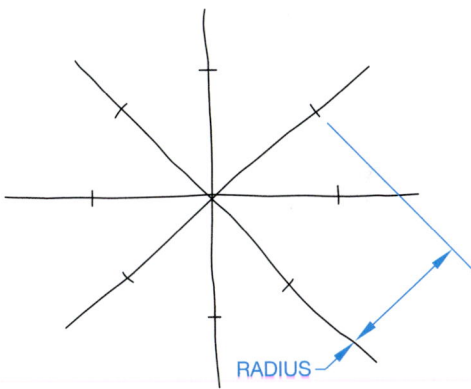

FIGURE 6.14 ■ Mark the approximate radius of the circle on the centerlines created in Figure 6.13.

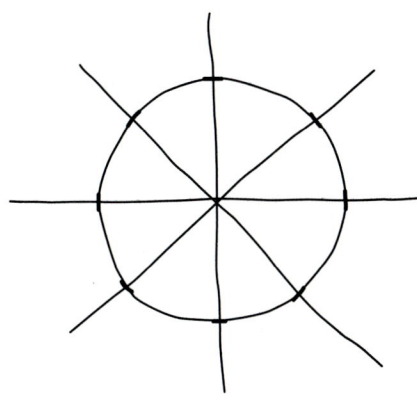

FIGURE 6.15 ■ Create the circle by sketching arcs that go through the marks on the centerlines.

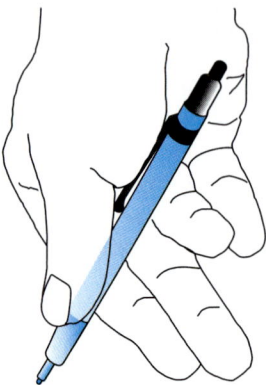

FIGURE 6.16 ■ Step 2, holding the pencil in the hand compass.

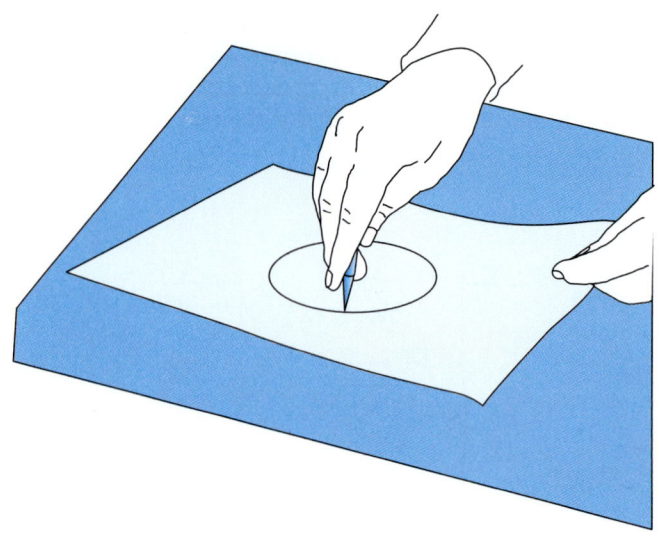

Using the Hand-Compass Method

The hand-compass method is a quick and fairly accurate method of sketching circles, although it is a method that takes some practice.

STEP 1 Be sure that your paper is free to rotate completely around 360°. Remove anything from the table that might stop such a rotation.

STEP 2 To use your hand and a pencil as a compass, place the pencil in your hand between your thumb and the upper part of your index finger so your index finger becomes the compass point and the pencil becomes the compass lead. The other end of the pencil rests in your palm as shown in Figure 6.16.

STEP 3 Determine the circle radius by adjusting the distance between your index finger and the pencil point. Now, with the desired approximate radius established, place your index finger on the paper at the proposed center of the circle.

STEP 4 With the desired radius established, keep your hand and pencil point in one place while rotating the paper with your other hand. Try to keep the radius steady as you rotate the paper. (See Figure 6.17.)

STEP 5 You can perform step 4 very lightly and then go back and darken the circle or, if you have had a lot of practice, you may be able to draw a dark circle as you go.

Trammel Method

The trammel method should be avoided if you are creating a quick sketch, because it takes extra time and materials to set up this technique. Also, the trammel method is intended for large to very large circles that are difficult to draw when using the other methods. The following examples demonstrate the trammel method to create a small circle. This is done to save space. Use the same principles to draw a large circle.

STEP 1 Make a trammel to sketch a 6-in. (150 mm) diameter circle. Cut or tear a strip of paper approximately 1 in.

FIGURE 6.17 ■ Step 4, rotate the paper under your finger center point.

(25 mm) wide and longer than the radius, 3 in. (75 mm). On the strip of paper, mark an approximate 3-in. radius with tick marks, such as the A and B in Figure 6.18.

STEP 2 Sketch a straight line representing the circle radius at the place where the circle is to be located. On the sketched line, locate with a mark the center of the circle to be sketched. Use the marks on the trammel to mark the other end of the radius line as shown in Figure 6.19. With the trammel next to the sketched line, be sure point B on the trammel is aligned with the center of the circle you are about to sketch.

STEP 3 Pivot the trammel at point B, making tick marks at point A as you go, as shown in Figure 6.20, until the circle is complete.

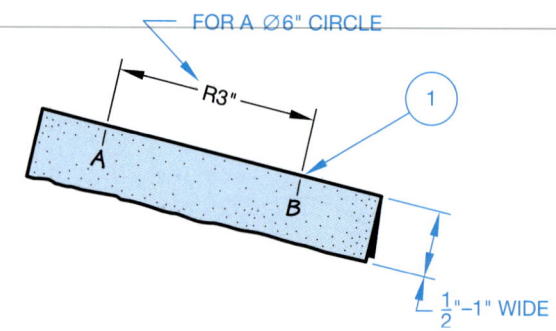

FIGURE 6.18 ■ Step 1, make a trammel.

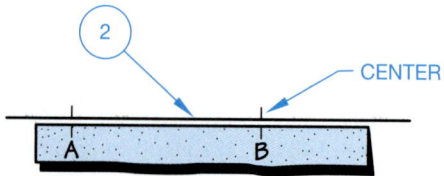

FIGURE 6.19 ■ Step 2, locate the center of the circle.

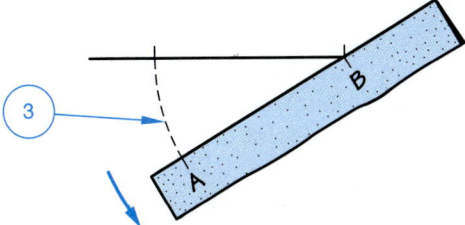

FIGURE 6.20 ■ Step 3, begin the circle construction.

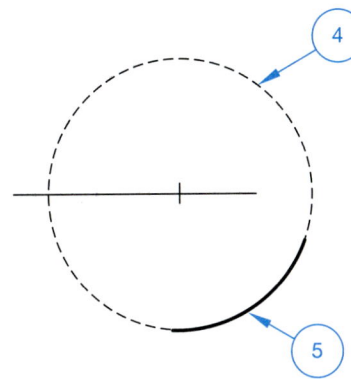

FIGURE 6.21 ■ Steps 4 and 5, darken the circle.

STEP 4   Lightly sketch the circumference over the tick marks to complete the circle, then darken it (step 5), as shown in Figure 6.21.

Another similar trammel method, generally used to sketch very large circles, is to tie a string between a pencil and a pin. The distance between the pencil and pin is the radius of the circle. Use this method when a large circle is to be sketched, since the other methods may not work as well. Workers at a construction site sometimes use this method by tying the string to a nail and driving the nail at the center location.

SKETCHING ARCS

Sketching arcs is similar to sketching circles. An **arc** is part of a circle as you can see in Figure 6.22. An arc is commonly used as a rounded corner or at the end of a slot. When an arc is a rounded corner, the ends of the arc are typically tangent to adjacent lines. In short, **tangent** means that the arc touches the line at only one point and does not cross over the line as shown in Figure 6.22. An arc is generally drawn with a radius. The most comfortable way to sketch an arc is to move the paper so your hand faces the inside of the arc. Having the paper free to move helps with this practice.

One way to sketch an arc is to create a box at the corner. The box establishes the arc center and radius as shown in Figure 6.23. You can also sketch a 45° construction line from the center to the outside corner of the box, and mark the radius on the 45° line. (See Figure 6.23.) Now, sketch the arc by using the tangent points and the mark as a guide as shown in Figure 6.24. You should generally connect the straight lines to the arc after the arc is created, because it is usually easier to sketch straight lines than it is to sketch arcs.

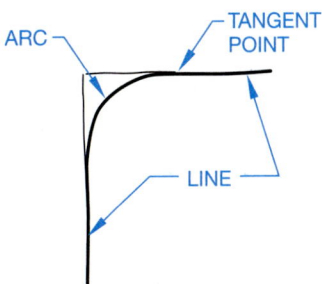

FIGURE 6.22 ■ An arc is part of a circle. This arc is used to create a rounded corner. Notice that the arc creates a smooth connection at the point of tangency with the straight lines.

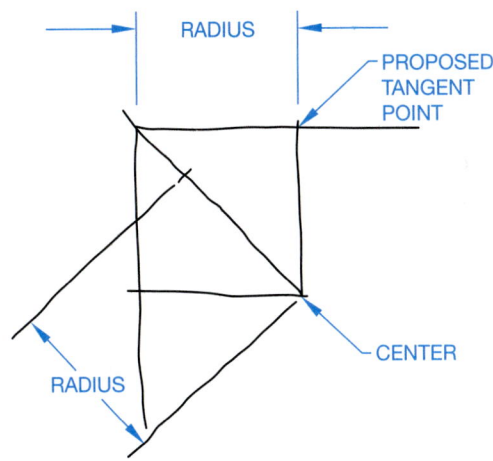

FIGURE 6.23 ■ The box establishes the center and radius of the arc. The 45° diagonal helps establish the radius.

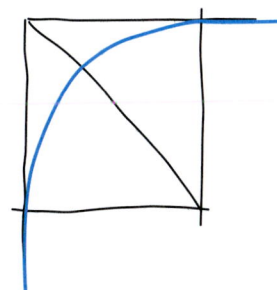

FIGURE 6.24 ■ Sketch the arc using the tangent points and mark on the diagonal as a guide for the radius.

The same technique can be used to sketch any arc. For example, a full radius arc is sketched in Figure 6.25. This arc is half of a circle, so using half of the box method or centerline method works well.

SKETCHING ELLIPSES

If you look directly at a coin, it represents a circle. As you rotate the coin, it takes the shape of an **ellipse**. Figure 6.26 shows the

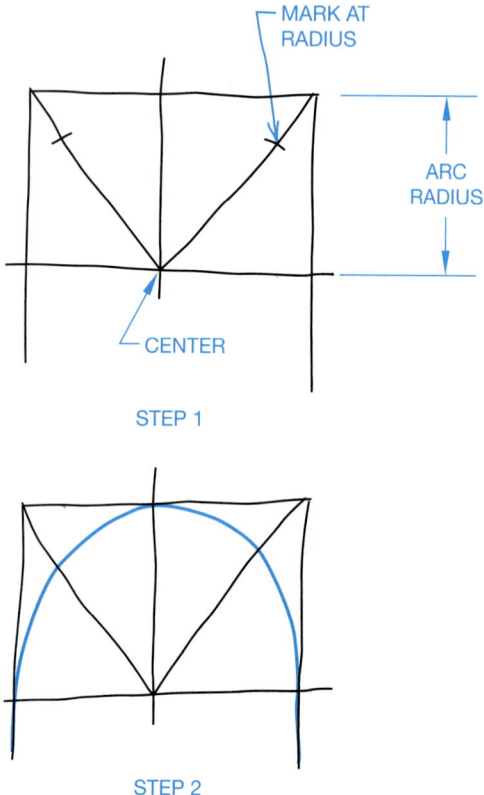

STEP 1

STEP 2

FIGURE 6.25 ■ Sketching a full radius arc uses the same method as sketching any arc or circle.

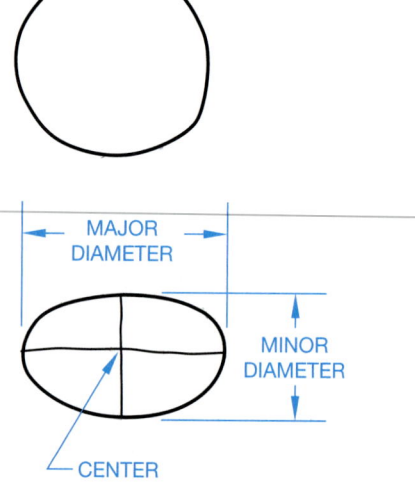

FIGURE 6.26 ■ The relationship between an ellipse and a circle.

Figure 6.27a. Next, sketch crossing lines from the corners of the minor diameter to the midpoint of the major diameter sides as in Figure 6.27a. Now, using the point where the lines cross as the center, sketch the major diameter arcs. (See Figure 6.27b.) Use the midpoint of the minor diameter sides as the center to sketch the minor diameter arcs as shown in Figure 6.27b. Finally, blend in connecting arcs to fill the gaps as shown in Figure 6.27c.

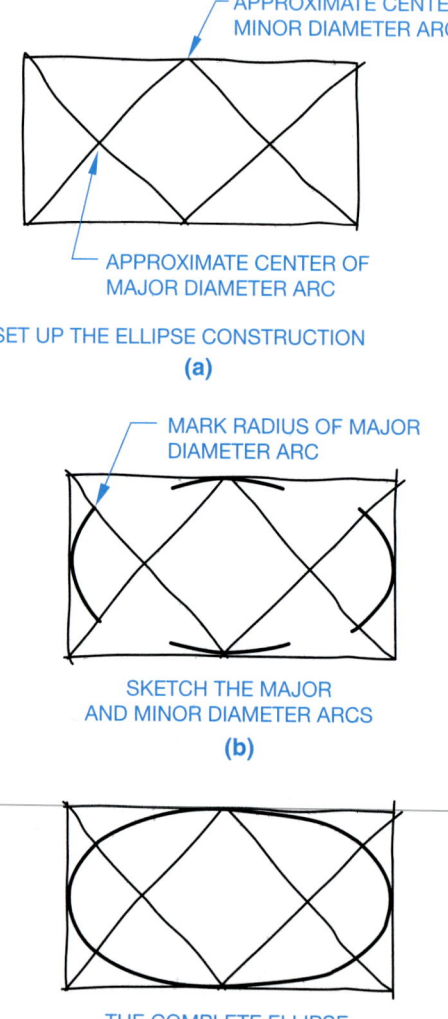

SET UP THE ELLIPSE CONSTRUCTION
(a)

SKETCH THE MAJOR
AND MINOR DIAMETER ARCS
(b)

THE COMPLETE ELLIPSE
(c)

FIGURE 6.27 ■ Sketching an ellipse. (a) Sketch a light rectangle equal in length and width to the major and minor diameters of the desired ellipse. Sketch crossing lines from the corners of the minor diameter to the midpoint of the major diameter sides. (b) Use the point where the lines cross as the center to sketch the major diameter arcs. Use the midpoint of the minor diameter sides as the center to sketch the minor diameter arcs. (c) Blend in connecting arcs to fill the gaps.

relationship between a circle and an ellipse and shows the parts of an ellipse.

If you can fairly accurately sketch an ellipse without a construction line, then do it. If you need help, an ellipse also can be sketched using a box method. To start this technique, sketch a light rectangle equal in length and width to the major and minor diameters of the desired ellipse as shown in

MEASUREMENT LINES AND PROPORTIONS

When sketching objects, all the lines that make up the object are related to each other by size and direction. In order for a sketch to communicate accurately and completely, it must be drawn in the same proportion as the object. The actual size of the sketch depends on the paper size and how large you want the sketch to look. The sketch should be large enough to be clear, but the proportions of the features are more important than the size of the sketch.

Look at the lines in Figure 6.28. How long is line 1? How long is line 2? Answer these questions without measuring either line, but instead relate each line to the other. For example, line 1 could be stated as being half as long as line 2, or line 2 called twice as long as line 1. Now you know how long each line is in relationship to the other (proportion), but you do not know how long either line is in relationship to a measured scale. No scale is used for sketching, so this is not a concern. Whatever line you decide to sketch first determines the scale of the drawing. This first line sketched is called the **measurement line**. Relate all the other lines in the sketch to that first line. This is one of the secrets of making a sketch look like the object being sketched.

The second thing you must know about the relationship of the two lines in the above example is their direction and position relative to each other. For example, do they touch each other, are they parallel, perpendicular, or at some angle to each other? When you look at a line, ask yourself the following questions (for this example use the two lines given in Figure 6.29):

1. How long is the second line?
   a. same length as the first line?
   b. shorter than the first line? How much shorter?
   c. longer than the first line? How much longer?
2. In what direction and position is the second line related to the first line?

Typical answers to these questions for the lines in Figure 6.29 are as follows:

1. The second line is about three times as long as the first line.
2. Line two touches the lower end of the first line with about a 90° angle between them.

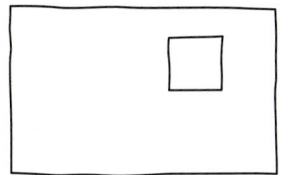
FIGURE 6.30 ■ Space proportions.

Carrying this concept a step further, a third line can relate to the first line or the second line and so forth. Again, the first line drawn (measurement line) sets the scale for the whole sketch.

This idea of relationship can also apply to spaces. In Figure 6.30, the location of the square can be determined by space proportions. A typical verbal location for the square in this block might be as follows: the square is located about one-half square width from the top of the object or about two square widths from the bottom, and about one square width from the right side or about three square widths from the left side of the object. All the parts must be related to the whole object.

INTRODUCTION TO THE BLOCK TECHNIQUE

Any illustration of an object can be surrounded with some sort of an overall rectangle, as shown in Figure 6.31. Before starting a sketch, visualize in your mind the object to be sketched inside a rectangle. Then use the measurement-line technique with the rectangle, or block, to help you determine the shape and proportion of your sketch.

Procedures In Sketching

STEP 1 When starting to sketch an object, visualize the object surrounded with an overall rectangle. Sketch this rectangle first with very light lines. Sketch the proper proportion with the measurement-line technique, as shown in Figure 6.32.

STEP 2 Cut out or cut away sections using proper proportions as measured by eye, using light lines, as in Figure 6.33.

STEP 3 Finish the sketch by darkening in the desired outlines for the finished sketch. (See Figure 6.34.)

LINE 1 ————————
LINE 2 ———————————————

FIGURE 6.28 ■ Measurement lines.

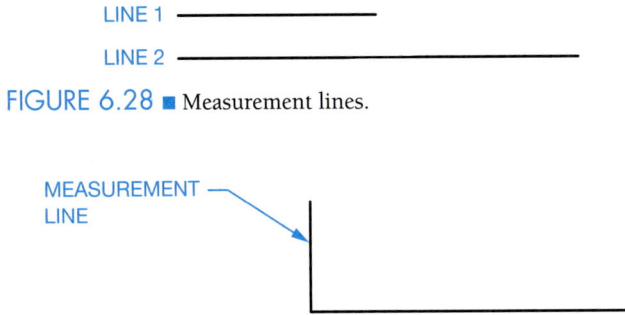

FIGURE 6.29 ■ Measurement line.

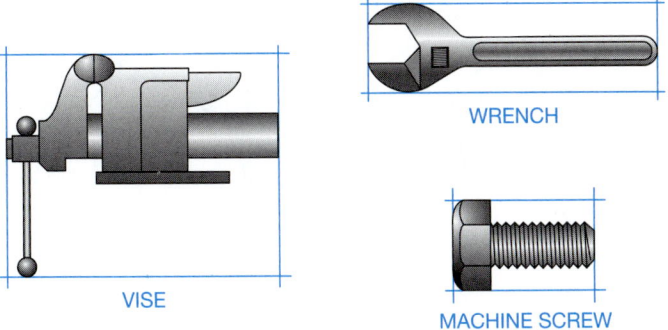
FIGURE 6.31 ■ Block technique.

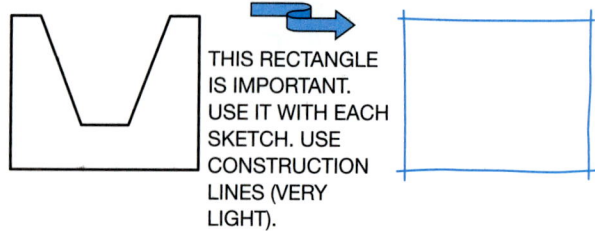

FIGURE 6.32 ■ Step 1, outline the drawing area with a block.

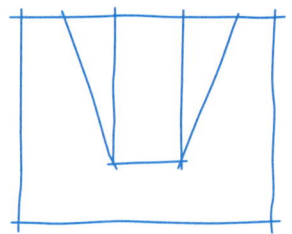

FIGURE 6.33 ■ Step 2, draw features to proper proportions.

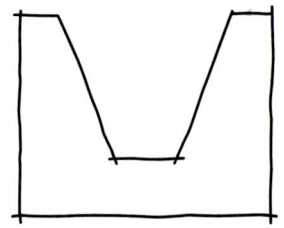

FIGURE 6.34 ■ Step 3, darken the object lines.

Sketching Irregular Shapes

By using a frame of reference or an extension of the block method, irregular shapes can be sketched easily to their correct proportions. Follow these steps to sketch the cam shown in Figure 6.35.

STEP 1 Place the object in a lightly constructed box. (See Figure 6.36.)

STEP 2 Draw several equally spaced horizontal and vertical lines as shown in Figure 6.37. If you are sketching an object already drawn, just draw your reference lines on top of the object's lines to establish a frame of reference. If you are sketching an object directly, you have to visualize these reference lines on the object you sketch.

STEP 3 On your sketch, correctly locate a proportioned box similar to the one established on the original drawing or object, as shown in Figure 6.38.

STEP 4 Using the drawn box as a frame of reference, include the grid lines in correct proportion, as seen in Figure 6.39.

STEP 5 Then, using the grid, sketch the small irregular arcs and lines that match the lines of the original, as in Figure 6.40.

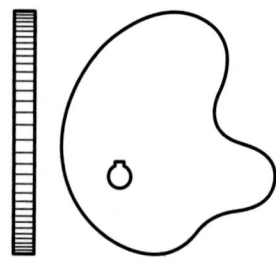

FIGURE 6.35 ■ Cam.

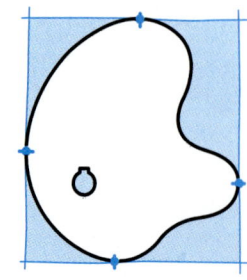

FIGURE 6.36 ■ Step 1, box the object.

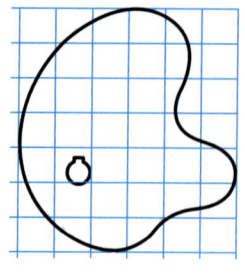

FIGURE 6.37 ■ Step 2, evenly spaced grid.

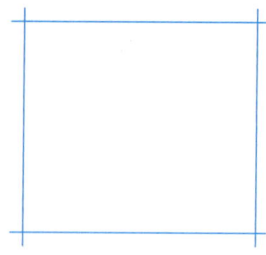

FIGURE 6.38 ■ Step 3, proportioned box.

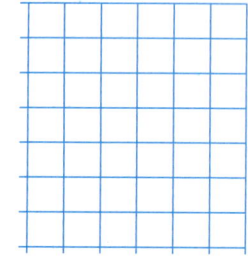

FIGURE 6.39 ■ Step 4, regular grid.

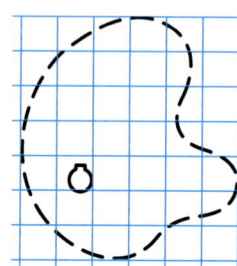

FIGURE 6.40 ■ Step 5, sketched shape using the regular grid.

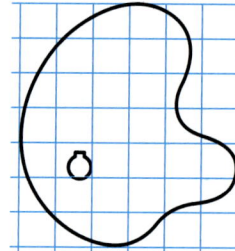

FIGURE 6.41 ■ Step 6, completely darken the outline of the object.

STEP 6 Darken the outline for a complete proportioned sketch, as shown in Figure 6.41.

CREATING MULTIVIEW SKETCHES

Multiview projection is also known as orthographic projection. Multiviews are two-dimensional views of an object that are established by a line of sight that is perpendicular (90°) to the surface of the object. When making multiview sketches, a systematic order should be followed. Most drawings are in the multiview form. Learning to sketch multiview drawings will save you time when making a formal drawing. The pictoral view shows the object in a 3-D (three-dimensional) picture, while the multiview shows the object in a 2-D representation. Figure 6.42 shows an object in 3-D and 2-D. Chapter 9 provides complete information about multiviews.

Multiview Alignment

To keep your drawing in a common form, sketch the front view in the lower left portion of the paper, the top view directly above

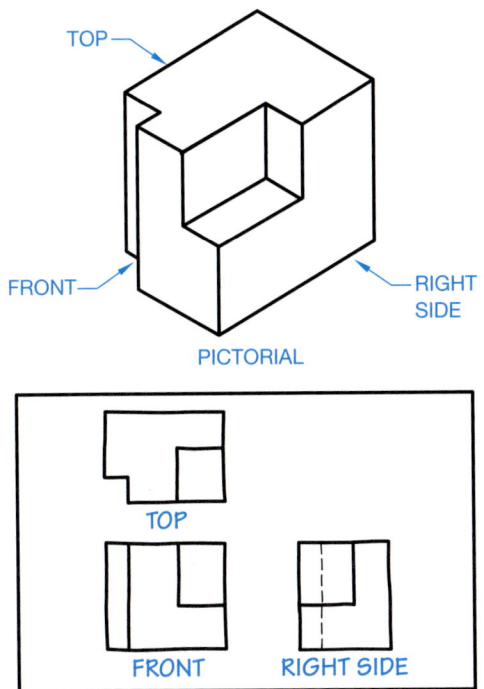

FIGURE 6.42 ■ Views of objects shown in pictorial view and multiview.

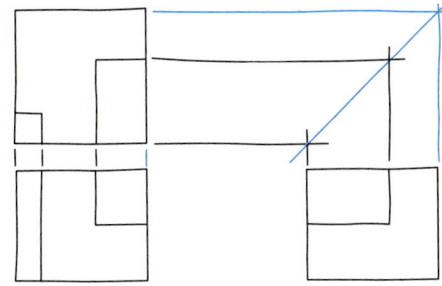

FIGURE 6.44 ■ Step 2, block out shapes.

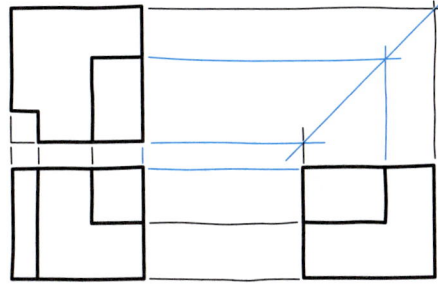

FIGURE 6.45 ■ Step 3, darken all object lines.

the front view, and the right-side view to the right side of the front view. (See Figure 6.42.) The views needed may differ depending on the object. Your ability to visualize between 3-D objects and 2-D views is very important in understanding how to lay out a multiview sketch. Multiview arrangement is explained in detail in Chapter 9.

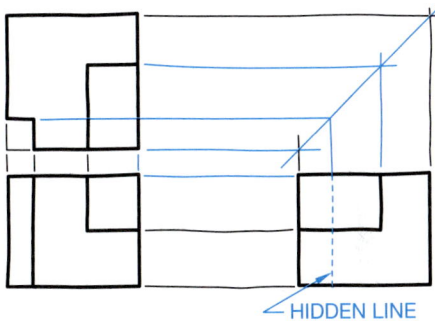

FIGURE 6.46 ■ Step 4, draw hidden features.

Multiview Sketching Technique

Steps in sketching multiviews:

STEP 1 Sketch and align the proportional rectangles for the front, top, and right-side views of the object given in Figure 6.42. Sketch a 45° line to help transfer width dimensions. The 45° line is established by projecting the width from the top view across and the width from the right-side view up until the lines intersect as shown in Figure 6.43. This 45° line is often called a **mitre line**.

STEP 2 Complete the shapes by cutting out the rectangles, as shown in Figure 6.44.

STEP 3 Darken the lines of the object as in Figure 6.45. Remember, keep the views aligned for ease of sketching and understanding.

STEP 4 In the views where some of the features are hidden, show those features with hidden lines, which are dashed lines as shown in Figure 6.46. Start the practice of sketching thick object lines and thin hidden lines.

CREATING ISOMETRIC SKETCHES

Isometric sketches provide a three-dimensional pictorial representation of an object. Isometric sketches are easy to create and make a very realistic exhibit of the object. The surface features or the axes of the objects are drawn at equal angles from horizontal. Isometric sketches tend to represent the objects as they appear to the eye. Isometric sketches help in the visualization of an object,

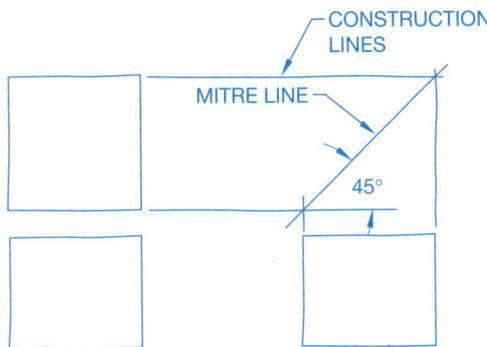

FIGURE 6.43 ■ Step 1, block out views.

because three sides of the object are sketched in a single three-dimensional view. Chapter 15 covers isometric drawings in detail.

Establishing Isometric Axes

In setting up an isometric axis, you need four beginning lines: a horizontal reference line, two 30° angular lines, and one vertical line. Draw them as very light construction lines. (See Figure 6.47.)

STEP 1 Sketch a horizontal reference line (consider this the ground-level line.)

STEP 2 Sketch a vertical line perpendicular to the ground line and somewhere near its center. The vertical line is used to measure height.

STEP 3 Sketch two 30° angular lines, each starting at the intersection of the first two lines as shown in Figure 6.47.

Making an Isometric Sketch

The steps in making an isometric sketch are as follows:

STEP 1 Select an appropriate view of the object.

STEP 2 Determine the best position in which to show the object.

STEP 3 Begin your sketch by setting up the isometric axes. (See Figure 6.48.)

STEP 4 By using the measurement-line technique, draw a rectangular box, using correct proportion, which could surround the object to be drawn. Use the object shown in Figure 6.49 for this example. Imagine the rectangular

box in your mind. Begin to sketch the box by marking off the width at any convenient length as in Figure 6.50. This is your measurement line. Next, estimate and mark the length and height as related to the measurement line. (See Figure 6.51.) Sketch the three-dimensional box by using lines parallel to the original axis lines. (See Figure 6.52.) Sketching the box is the most critical part of the construction. It must be done correctly; otherwise your sketch will be out of proportion. All lines drawn in the same direction must be parallel.

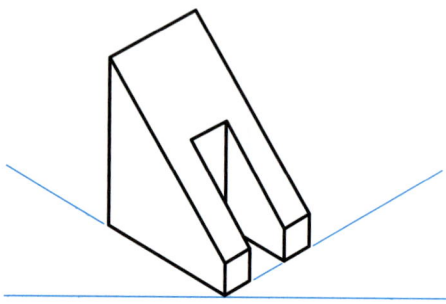

FIGURE 6.49 ■ Given object.

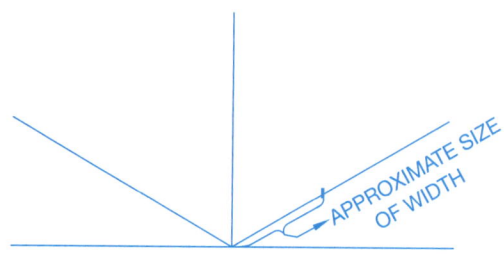

FIGURE 6.50 ■ Step 4, lay out the width.

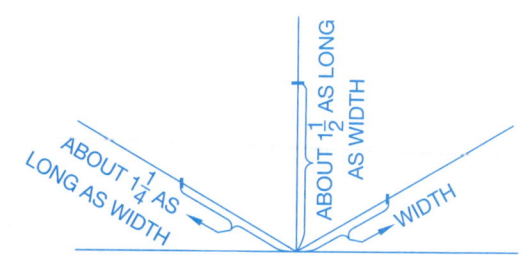

FIGURE 6.51 ■ Step 4, lay out the length and height.

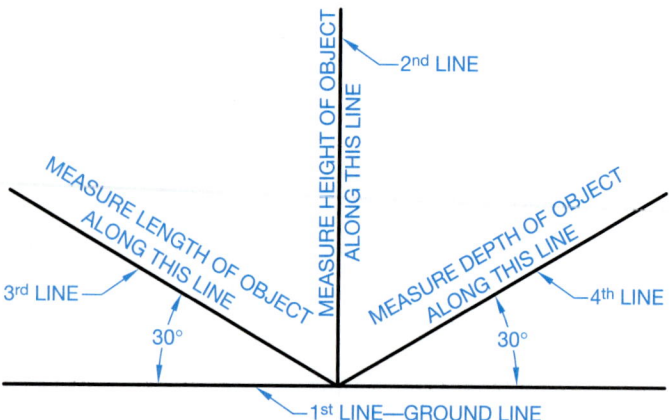

FIGURE 6.47 ■ Isometric axis.

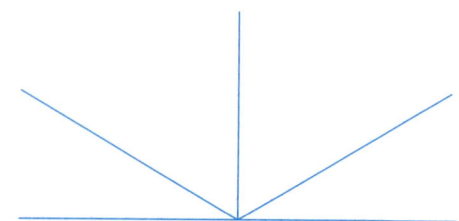

FIGURE 6.48 ■ Step 3, sketch the isometric axis.

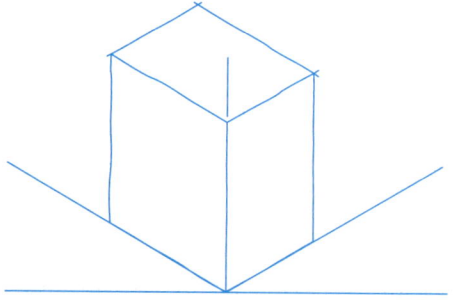

FIGURE 6.52 ■ Step 4, sketch the 3-D box.

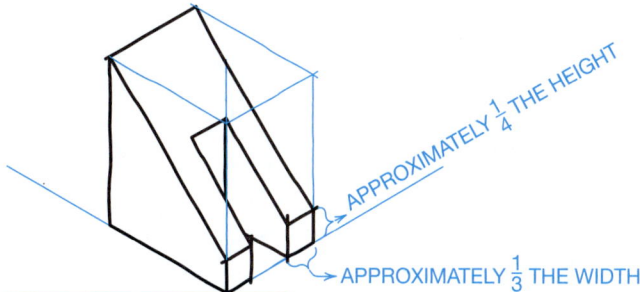

FIGURE 6.53 ■ Step 5, sketch the features.

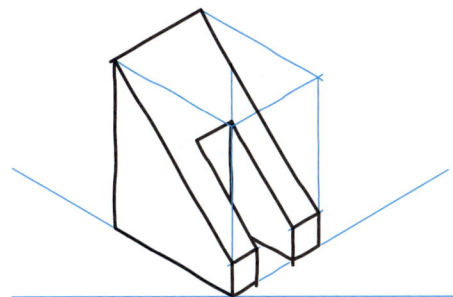

FIGURE 6.54 ■ Step 6, darken the outline.

STEP 5 Lightly sketch in the slots, insets, and other features that define the details of the object. By estimating distances on the rectangular box, the features of the object are easier to sketch in correct proportion than trying to draw them without the box. (See Figure 6.53.)

STEP 6 To finish the sketch, darken all the object lines (outlines) as in Figure 6.54. For clarity, do not show any hidden lines.

Nonisometric Lines

Isometric lines are lines that are on or parallel to the three original isometric axes lines. All other lines are nonisometric lines. Isometric lines can be measured in true length. **Nonisometric lines** appear either longer or shorter than they actually are. (See Figure 6.55.) You can measure and draw nonisometric lines by connecting their end points. You can find the end points of the nonisometric lines by measuring along isometric lines. To locate where nonisometric lines should be placed, you have to relate

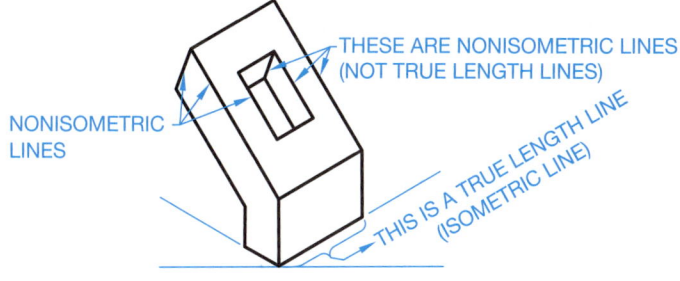

FIGURE 6.55 ■ Nonisometric lines.

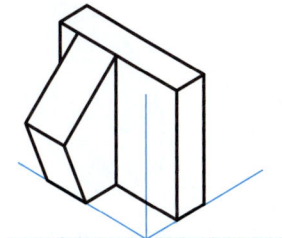

FIGURE 6.56 ■ Guide.

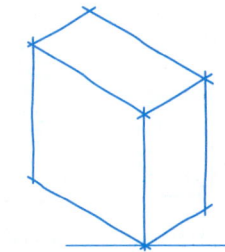

FIGURE 6.57 ■ Step 1, sketch the box.

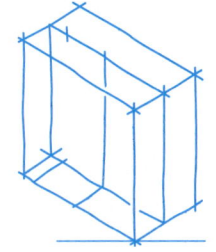

FIGURE 6.58 ■ Step 2, sketch isometric lines.

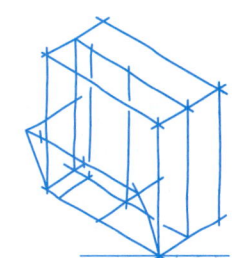

FIGURE 6.59 ■ Step 3, locate nonisometric line end points.

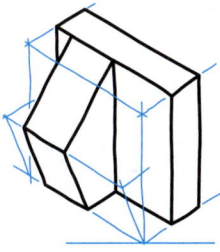

FIGURE 6.60 ■ Step 4, complete the sketch and darken all outlines.

to an isometric line. Follow through these steps, using the object in Figure 6.56 as an example.

STEP 1 Develop a proportional box, as in Figure 6.57.

STEP 2 Sketch in all isometric lines, as shown in Figure 6.58.

STEP 3 Locate the starting and end points for the nonisometric lines. (See Figure 6.59.)

STEP 4 Sketch the nonisometric lines, as shown in Figure 6.60, by connecting the points established in step 3. Also darken all outlines.

Sketching Isometric Circles

Circles and arcs appear as ellipses in isometric views. To sketch isometric circles and arcs correctly, you need to know the relationship between circles and the faces, or planes, of an isometric cube. Depending on which face the circle is to appear, isometric circles look like one of the ellipses shown in Figure 6.61. The angle the ellipse (isometric circle) slants is determined by the surface on which the circle is to be sketched.

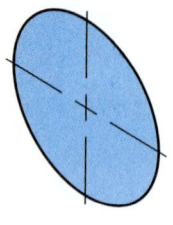 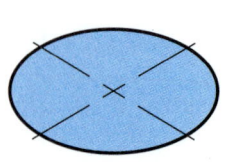 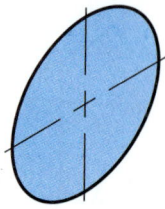

LEFT PLANE   HORIZONTAL PLANE   RIGHT PLANE

FIGURE 6.61 ■ Isometric circles.

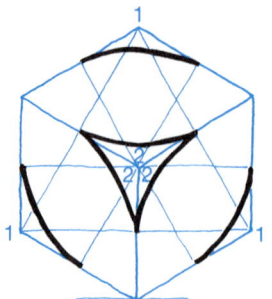

FIGURE 6.64 ■ Step 3, sketch arcs from points 1 and 2 as centers.

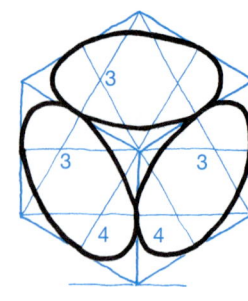

FIGURE 6.65 ■ Step 4, sketch arcs from points 3 and 4 as centers.

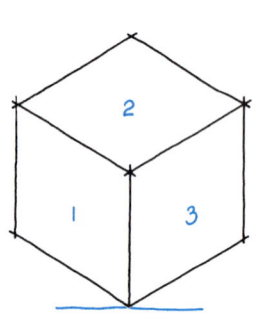 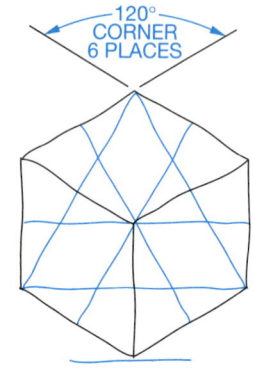

FIGURE 6.62 ■ Step 1, draw an isometric cube.

FIGURE 6.63 ■ Step 2, four-center isometric ellipse construction.

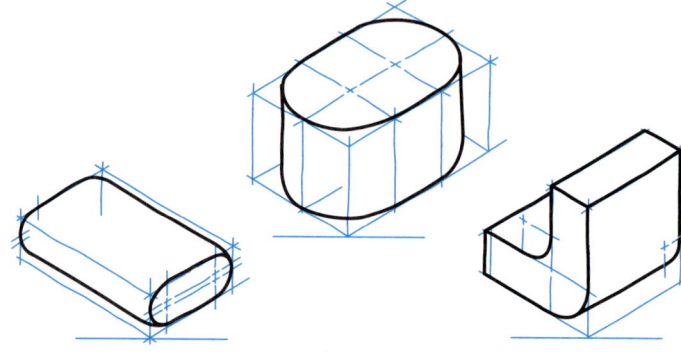

FIGURE 6.66 ■ Sketching isometric arcs.

To practice sketching isometric circles, you need isometric surfaces to put them on. The surfaces can be found by first sketching a cube in isometric. A cube is a box with six equal sides. Notice, as shown in Figure 6.62, that only three of the sides can be seen in an isometric drawing.

Four-Center Method

The four-center method of sketching an isometric ellipse is easier to perform, but care must be taken to form the ellipse arcs properly so the ellipse does not look distorted.

STEP 1 Draw an isometric cube similar to Figure 6.62.

STEP 2 On each surface of the cube, draw line segments that connect the 120° corners to the centers of the opposite sides. (See Figure 6.63.)

STEP 3 With points 1 and 2 as the centers, sketch arcs that begin and end at the centers of the opposite sides on each isometric surface. (See Figure 6.64.)

STEP 4 On each isometric surface, with points 3 and 4 as the centers, complete the isometric ellipses by sketching arcs that meet the arcs sketched in step 3. (See Figure 6.65.)

Sketching Isometric Arcs

Sketching isometric arcs is similar to sketching isometric circles. First, block out the overall configuration of the object, then establish the centers of the arcs. Finally, sketch the arc shapes as shown in Figure 6.66. Remember that isometric arcs, just like isometric circles, must lie in the proper plane and have the correct shape.

PROFESSIONAL PERSPECTIVE

Sketching

Freehand sketching is an important skill if you are a manual or a CADD drafter. Preparing a sketch before you begin a formal drawing may save many hours of work. The sketch assists you in the layout process because it allows you to:

■ Decide how the drawing should appear when finished.

■ Determine the size of the drawing.

■ Determine the sheet size for manual drafting or the screen limits for CADD.

■ Establish the coordinate points for the computer drawing.

A little time spent sketching and planning your work saves a lot of time in the final drafting process. Sketches are also a quick form of communication in any professional environment. You can often get your point across or communicate more effectively with a sketch.

CADD SKETCHING

Sketches are very important engineering drafting and design tools. The very first drawings were sketches, and even after hundreds of years, sketches continue to be widely used to effectively communicate ideas and information in a variety of ways. In fact, sketches are an essential piece of modern three-dimensional modeling software programs. Sketching is done using a computer, and it is required when producing CADD models. Part sketches typically represent the first step in the creation of a model and provide the initial geometry, pattern, and profile used to develop models.

Just as with sketches drawn by hand, model sketches generally are easy to create and should not require very much time to complete. Initial sketch geometry is usually approximate, and contains very few dimensions or other geometric relationships. The typical approach is to draw a basic shape, or part outline, quickly and easily, and then add dimensions and geometric information. Often only two steps are required to develop a basic feature sketch. The first step involves drawing simple shapes using sketch tools such as LINE, CIRCLE, and ARC. (See Figure 6.67.) The second step requires defining the size and shape of the sketch using dimensions and geometric relationships as shown in Figure 6.68. The sketch is then used to build a model. (See Figure 6.69.)

Keep the following information in mind when developing model sketches:

■ Fully close a sketch loop when developing a profile. If a gap or opened loop exists, you are not able to create some sketch features.

■ Develop simple, often incomplete sketch geometry, and create as many objects as possible using feature tools. For example, place a round on an existing feature using feature tools, instead of drawing and dimensioning an arc in the sketch environment, using sketch tools.

■ Fully define your sketch if applicable, but do not allow the sketch to become overdefined.

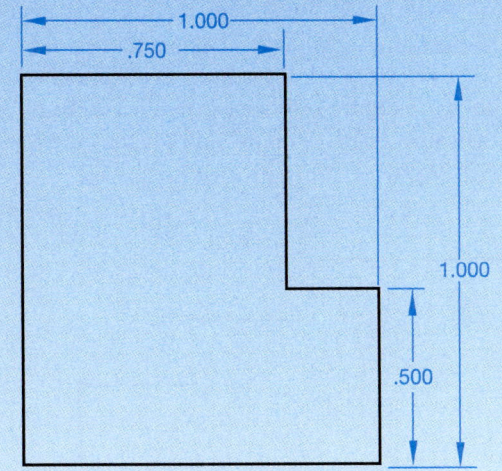

FIGURE 6.68 ■ Defining a sketch.

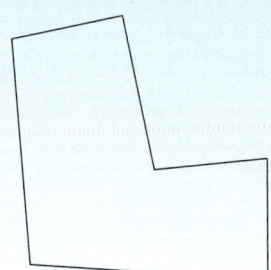

FIGURE 6.67 ■ A basic sketch created using sketching tools.

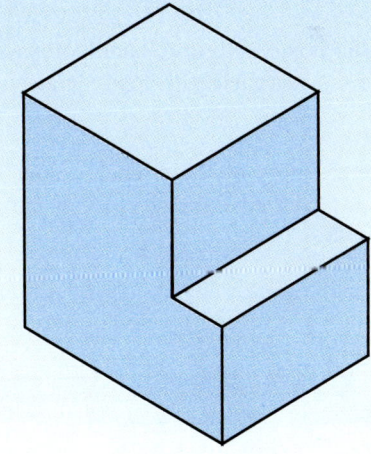

FIGURE 6.69 ■ A simple model created by extruding a sketch.

CHAPTER 6
Sketching Test

 Access the CD found with this textbook to view the Chapter 6 Test. Confirm the preferred submittal method with your instructor.

CHAPTER
6
Sketching Problems

Use proper sketching materials and techniques to solve the following sketching problems, on 8 1/2 × 11 in. bond paper or newsprint, unless otherwise specified by your instructor. Use very lightly sketched construction lines for all layout work. Darken the finished lines, but do not erase the layout lines, unless otherwise specified by your instructor.

PROBLEM 6.1   List on a separate sheet of paper the length, direction, and position of each line shown in the drawing. Remember, do not measure the lines with a scale. Example: Line 2 is the same length as line 1 and touches the top of line 1 at a 90° angle.

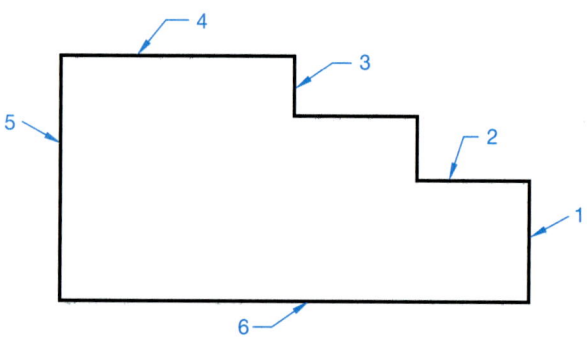

PROBLEM 6.2   Use the box, centerline, hand-compass, and trammel methods to sketch a circle with approximately a 4-in. diameter.

PROBLEM 6.3   Make a sketch of the wrench below. Use a frame of reference to make your sketch twice as big as the given sketch.

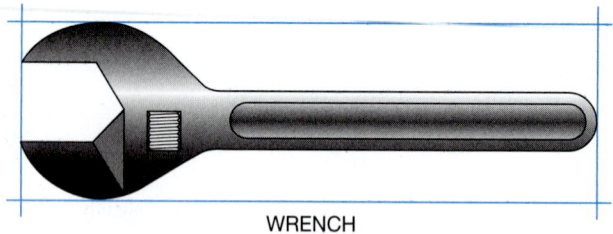

WRENCH

PROBLEM 6.4   Make a sketch of the machine screw below. Use a frame of reference to make your sketch twice as big as the given sketch.

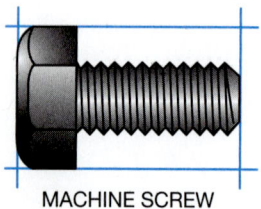

MACHINE SCREW

PROBLEM 6.5   Make a sketch of the vice below. Use a frame of reference to make your sketch twice as big as the given sketch.

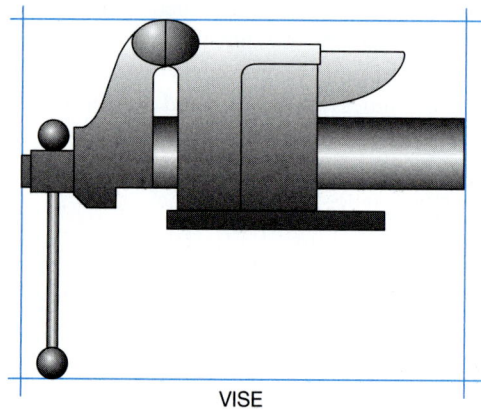

VISE

PROBLEM 6.6   Make a sketch of the patio, swimming pool, and spa below. Use a frame of reference to make your sketch twice as big as the given sketch.

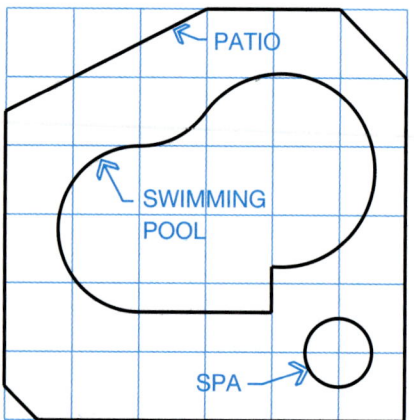

PROBLEMS 6.7 through 6.12: Access the CD found with this textbook and open the problem of your choice, or as assigned by your instructor. Solve the problem or problems using the same instructions provided for other problems in this chapter, unless otherwise specified.

CHAPTER 7

Lines and Lettering

LEARNING OBJECTIVES

After completing this chapter, you will:

- Identify the lines found on a given industry drawing.
- Draw ASME standard lines using manual drafting and computer-aided drafting.
- Solve engineering problems using manual and computer-aided drafting.
- Do freehand lettering and use lettering equipment.
- Use a CADD system to create text.
- Answer questions related to lines and lettering.

THE ENGINEERING DESIGN APPLICATION

The Engineering Design Application provided in Chapter 6 discussed sketches used during the design process and the importance of converting engineering sketches to formal drawings that properly convey national drafting standards. The wide variety of drafting and design disciplines are covered throughout this textbook. These disciplines include the major categories of manufacturing, structural, civil, sheet metal, and electrical drafting applications. There is a vast amount of information for you to learn, and each discipline has its own techniques and practices. Your challenge, as a professional drafter, is to properly convert engineering sketches and other given information into formal drawings that effectively communicate the design intent. If you take your profession seriously, your desire will be to create drawings that provide complete and accurate information and represent the best possible quality.

The content in this textbook guides you to this objective. Drafting standards, practices, and applications are provided, but it is your responsibility to convert this information into the completion of formal drawings that meet different challenges with each drawing you create. The example engineering sketch and resulting drawing shown in

Figure 6.1 represents the basic idea, but you will learn what is needed to solve problems and produce drawings that are much more complex. Your discovery in this chapter includes lines and letters used in a specific manner for you to communicate an engineering design concept to the people who manufacture or construct the product or structure. The drawings you create can be considered a technical work of art. The lines and letters are combined in a manner that clearly and skillfully demonstrates your communication. Actual industry drawing examples are given throughout this book to help reinforce and pull together what you have learned in each chapter. While these drawings represent only one possible solution to the specific problem, they are intended to show what you can accomplish and to demonstrate the quality you should aspire to achieve.

Take some time now to look at the drawing found in Figure 7.55. This drawing is complex, but it shows what you will be able to accomplish. Carefully study and practice the content covered in this textbook. Take every opportunity to look at real-world industry drawings provided and available at companies near your home or school.

LINES

Drafting is a graphic language using lines, symbols, and notes to describe objects to be manufactured or built. Lines on drawings must be of a quality that reproduce easily. All lines are dark, crisp, sharp, and of the correct thickness when properly drawn. There is no variation in darkness, only a variation in thickness, known as **line contrast**. Certain lines are drawn thick so they stand out clearly from other information on the drawing. Other lines are drawn thin. Thin lines are not necessarily less impor-

tant than thick lines, but they are subordinate for identification purposes.

ASME The American Society of Mechanical Engineers (ASME) recommends two-line thickness with bold lines twice as thick as thin lines. This line standard relates to both manual and computer-aided drafting. Standard line thicknesses are 0.6 mm for thick lines and 0.3 mm for thin lines. The actual width of lines may be more or less than the recommended thickness depending on the size

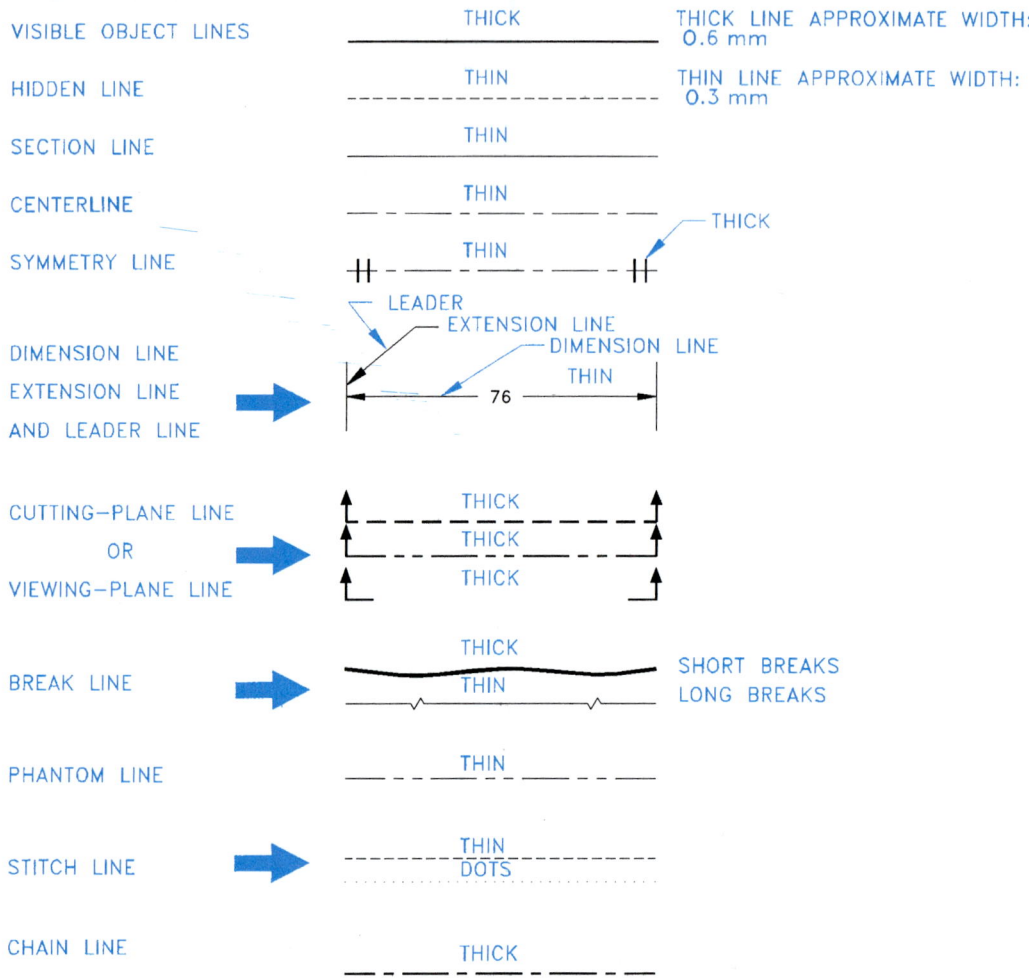

FIGURE 7.1 ■ Line conventions, width, and type of lines.

of the drawing or the size of the final reproduction. Drawings that meet military documentation standards require three thicknesses of lines: thick, medium, and thin. Figure 7.1 shows widths and types of lines as taken from ASME standard, *Line Conventions and Lettering,* ASME Y14.2M. Figure 7.2 shows a sample drawing using the various kinds of lines. The ASME Y14.2M specifies the recommended line widths for thick and thin lines as 0.6 and 0.3 mm respectively. The approximate inch conversion is thick lines .02 in. (0.6 mm) and thin lines .01 in. (0.3 mm).

TYPES OF LINES
Introduction to Lines

The following discussion is an introduction to the lines that are commonly used on engineering drawings. You will use only a few of these lines as you work on the problems for this chapter. You will use additional lines as you go on to learn about specific applications throughout this text. For example, the lines used in

dimensioning are covered in detail in Chapter 11, Dimensioning and Tolerancing, and sectioning practices are covered in Chapter 13, Sections, Revolutions, and Conventional Breaks. You will put the use of lines to practice in every chapter where specific engineering drafting applications are fully explained.

Construction and Guidelines

Construction lines are used for laying out a drawing. Construction lines are drawn very lightly so they do not reproduce and are not mistaken for any other line on the drawing. Construction lines are drawn with a 4H to 6H pencil and, if drawn properly, do not need to be erased. Use construction lines for all preliminary work.

Construction lines are also commonly used in CADD for layout work. These lines are usually drawn using a unique format called **layers**, which can be named something such as CONSTRUCTION and assigned a specific color. If necessary, review the layers discussion in Chapter 3. Construction lines created using a separate format can be removed easily or set so they do not display or plot when the layout work is finished.

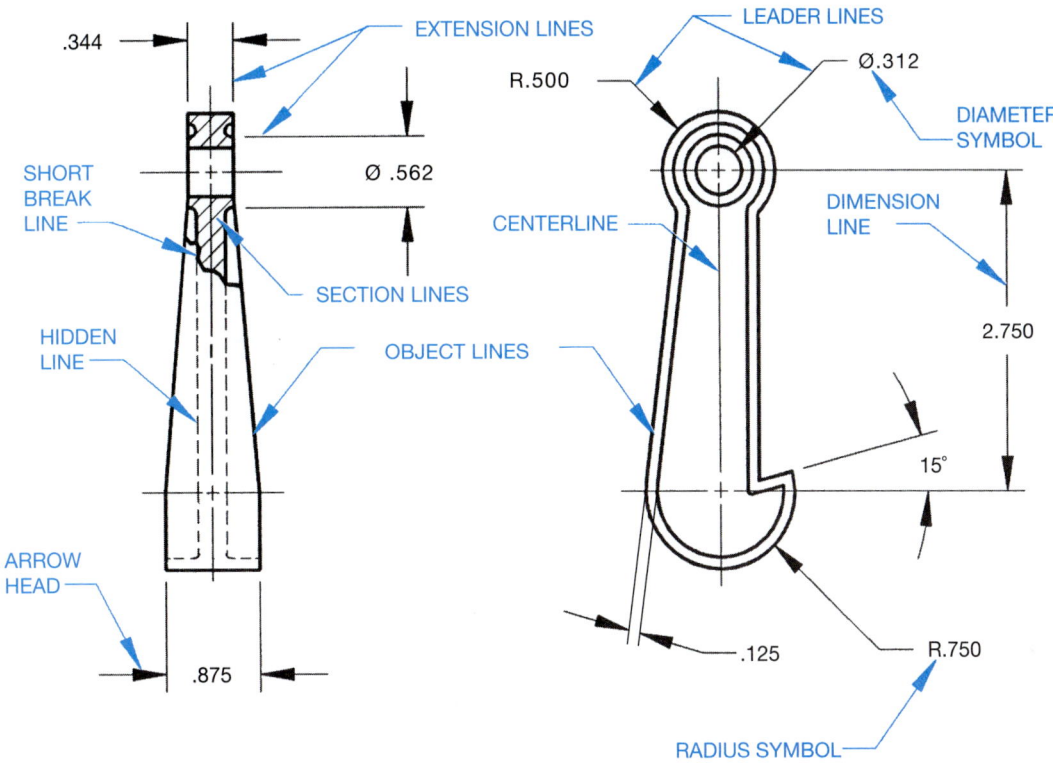

FIGURE 7.2 ■ Sample drawing with a variety of lines displayed.

Guidelines are drawn the same as construction lines and do not reproduce when properly drawn. Guidelines are used to keep lines of lettering in perfect alignment and of a constant height. For example, if lettering on a drawing is to be .12 in. (3 mm) high, then guidelines are drawn very lightly .12 in. (3 mm) apart. Guidelines must always be used for lettering. Some drafters prefer to use a light-blue lead for all guidelines and layout work. Light-blue lead does not reproduce and may be cleaner than graphite lead.

Object Lines

Object lines, also called visible lines or outlines, describe the visible surface or edge of the object. They are drawn as thick lines as shown in Figure 7.3. Thick lines, remember, are usually drawn .02 in. (0.6 mm) wide. Drafters usually use a soft lead, H or F, to

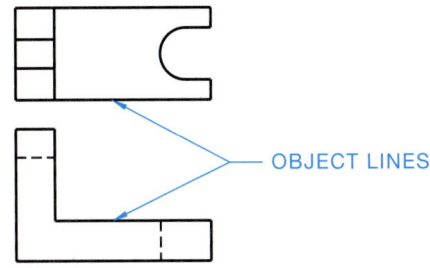

FIGURE 7.3 ■ Object lines.

draw object lines properly. You need to experiment to see which lead gives you the best line quality. For inked drawings, object lines can be made with a number 2 (0.6 mm) pen.

When using CADD, object lines are usually drawn using a unique format referred to as layers, which can be named something such as OBJECT and assigned a specific color, the object line style, and the recommended thickness of .02 in. (0.6 mm). Depending on the CADD system used, the line thickness can be displayed on the screen while drawing, or it may only be represented in the final print or plot of the drawing. Some CADD programs refer to line thickness as line weight.

Object Line Quality

To determine if your object lines are dark and crisp enough, turn the paper over and hold it up to the light. The line should have a dark, consistent density. A light table, if available, also works very well to check darkness.

You can also make a diazo or photocopy reproduction of your drawing to observe the quality. Properly drawn lines and lettering make quality prints. If the print lines are fuzzy, your original probably needs more work.

Most manual drafters use automatic pencils with a 0.7-mm lead for object lines.

Line quality is easy to achieve when using CADD. This is because the CADD program controls the way your lines look on the screen and the printer or plotter controls the way the lines

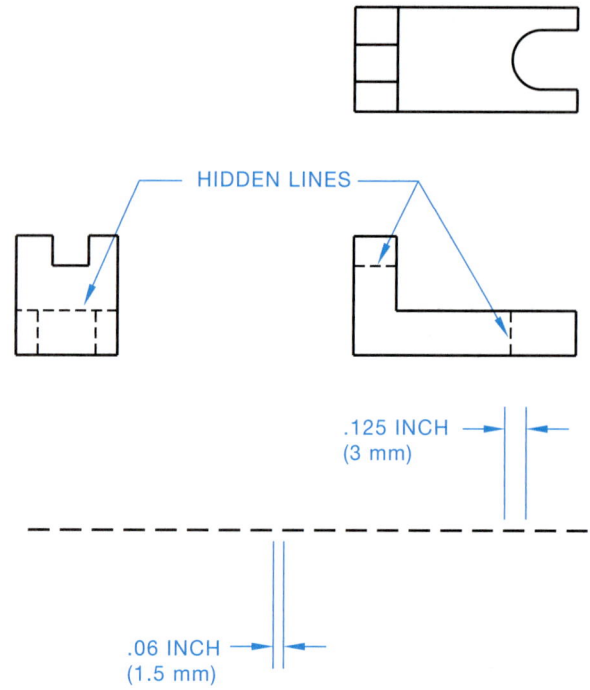

FIGURE 7.4 ■ ■ Hidden line representation.

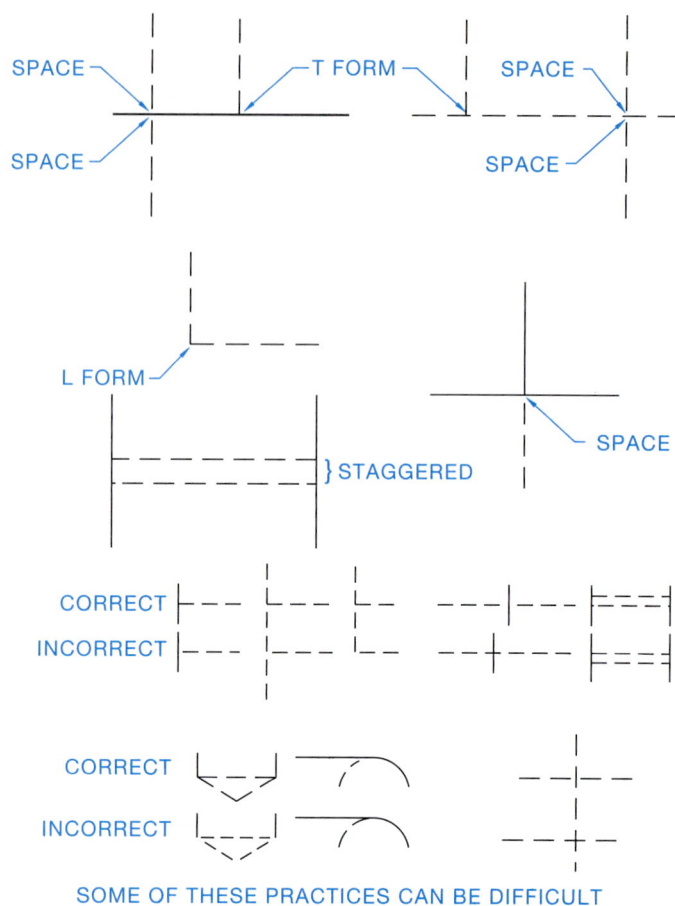

SOME OF THESE PRACTICES CAN BE DIFFICULT TO DO WITH CADD.

FIGURE 7.5 ■ ■ Hidden line rules.

look on paper. Modern engineering printers and plotters generally produce high-quality drawings.

Hidden Lines

A hidden line represents an invisible edge on an object. Hidden lines are thin lines, half as thick as object lines for contrast. Figure 7.4 shows hidden lines properly drawn with .125-in. (3-mm) dashes spaced .06 in. (1.5 mm) apart. Hidden lines, as all thin lines, can be drawn effectively with a 0.3-mm to 0.5-mm automatic drafting pencil, or a number 00 (0.3-mm) technical pen tip when inking. The best way to draw hidden lines is to draw dash lengths and spaces by eye. This takes some practice, but it is the fastest way.

Hidden Line Rules

The drawings in Figure 7.5 show situations where hidden lines meet or cross object lines and other hidden lines. These situations represent rules that should be followed when possible.

Hidden lines also should be represented as previously discussed when using CADD. The dash length and spacing can be adjusted to match the desired standard when using most CADD programs. Some initial experimentation may be needed to get the correct line representation depending on the size of your drawing and other scale factors within the CADD program. You may also have some difficulty applying all of the hidden line rules, but most CADD systems should give you the desired T and L form shown in Figure 7.5. Hidden lines are usually drawn using a unique format called layers, which can be named some-

thing such as HIDDEN. A hidden line format contains information such as the hidden line style, a contrasting color that helps distinguish hidden lines from other lines, and the recommended thickness of .01 in. (0.3 mm).

Centerlines

Centerlines are used to show and locate the centers of circles and arcs, and are used to represent the center axis of a circular or symmetrical form. Centerlines are thin lines on a drawing. They should be about half as thick as an object line. The long dash is about .75 to 1.50 in. (19-35 mm). The spaces between dashes are about .062 in. (1.5 mm) and the short dash about .125 in. (3 mm) long. The length of long lines varies for the situation and size of the drawing. Try to keep the long lengths uniform throughout the centerline. (See Figure 7.6.) Centerlines should start and end with long dashes. Centerlines should extend uniformly a short distance, such as .125 in. (3 mm) or .25 in. (6 mm), past an object. The distance the centerline goes past the object depends on the size of the drawing and company standards. The uniform distance past an object is maintained unless a longer distance is needed for dimensioning purposes.

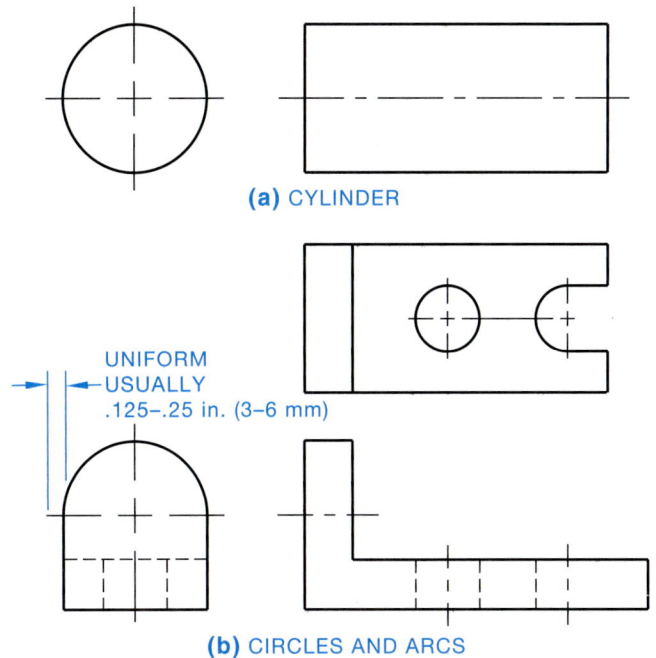

(a) CYLINDER

UNIFORM
USUALLY
.125–.25 in. (3–6 mm)

(b) CIRCLES AND ARCS

FIGURE 7.6 ∎ Centerline representation and examples.

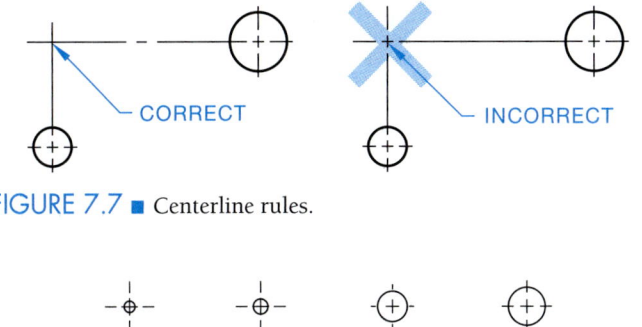

CORRECT   INCORRECT

FIGURE 7.7 ∎ Centerline rules.

FIGURE 7.8 ∎ Centerlines for small circles.

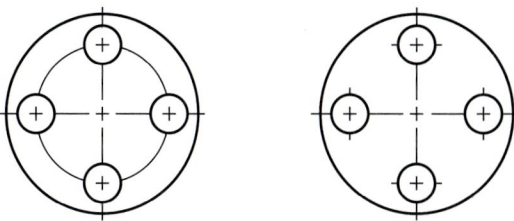

FIGURE 7.9 ∎ Bolt circle centerline options.

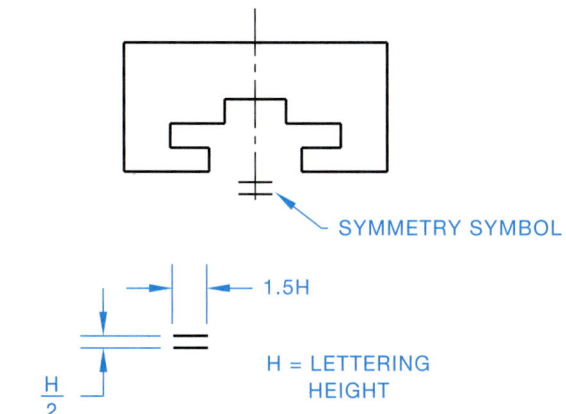

SYMMETRY SYMBOL

1.5H

H = LETTERING
HEIGHT

H/2

FIGURE 7.10 ∎ Symmetry symbol.

(See Figure 7.6.) When a centerline is used to establish a dimension, the centerline continues as an extension line without a gap where the centerline ends and the extension line begins. This is discussed more in Chapter 11. Small centerline dashes should cross only at the center of a circle or arc. (See Figure 7.7.) Small circles should have centerlines as shown in Figure 7.8. Centerlines do not extend between views of the drawing, except as discussed with auxiliary views in Chapter 10. Very short centerlines can be continuous without the standard long dash–short dash configuration, but this should be avoided if possible and there is no confusion with other lines.

Centerlines for holes in a bolt circle can be drawn either of two ways depending on how the holes are located, as shown in Figure 7.9. A **bolt circle** is a pattern of holes arranged in a circle.

When a centerline represents symmetry, as in the centerplane of an object, the symmetry symbol, shown in Figure 7.10, can be used if needed for clarity.

The centerline is commonly drawn with a 0.3- to 0.5-mm automatic pencil. A 0.3-mm pencil can provide best results with a soft lead and a 0.5-mm pencil with a harder lead. Remember the results should be dark, crisp, and sharp lines that are half as thick as object lines. When inking, a number 00 (0.3 mm) technical pen gives the best results.

Centerlines should also be represented as previously discussed when using CADD. The dash length and spacing can be adjusted to match the desired standard when using most CADD programs. Some initial experimentation may be needed to get the correct line representation depending on the size of your drawing and other scale factors within the CADD program. You may also have some difficulty applying all of the centerline rules, but most CADD systems should give you the desired results with some variation. Centerlines are usually drawn using a unique format called layers, which can be named something such as CENTER. A centerline format contains information such as the centerline style, a contrasting color that helps distinguish centerlines from other lines, and the recommended thickness of .01 in. (0.3 mm).

Extension Lines

Extension lines are thin lines used to establish the extent of a dimension. Extension lines begin with a .06 in. (1.5 mm) space from the object and extend to about .125 in. (3 mm) beyond the last dimension, as shown in Figure 7.11. Extension lines may cross object lines, centerlines, hidden lines, and other extension

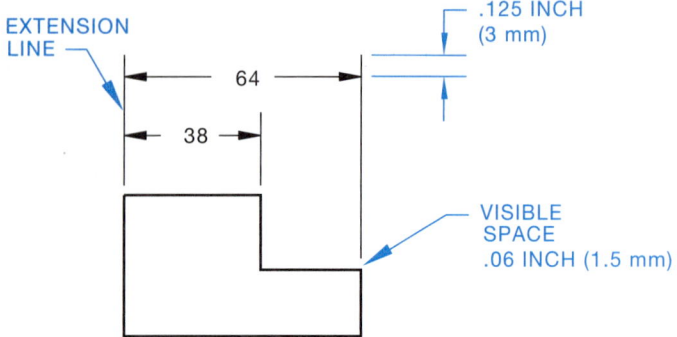

FIGURE 7.11 ■ Extension lines.

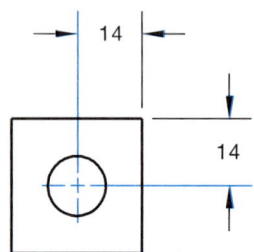

FIGURE 7.12 ■ The centerline becomes an extension line when used for dimensioning.

lines, but they may not cross dimension lines. Circular features, such as holes, are located by their centers in the view where they appear as circles. In this practice, centerlines become extension lines as shown in Figure 7.12.

Dimension Lines and Leader Lines

Dimension lines are thin lines capped on the ends with arrowheads and broken along their length to provide a space for the dimension numeral. Dimension lines indicate the length of the dimension. (See Figure 7.13.)

Leaders, or leader lines, are thin lines used to connect a specific note to a feature as shown in Figure 7.14. Leaders may be drawn at any angle, but 45°, 30°, or 60° lines are most common. Slopes greater than 75° or less than 15° from horizontal should be avoided. The leader has a .25-in. (6-mm) shoulder at one end that begins at the center of the vertical height of the lettering and an arrowhead at the other end pointing to the feature. If the leader were to continue from the point where the arrowhead touches the circle, it would intersect the center. (See Figure 7.15.)

Arrowheads

Arrowheads are used to terminate dimension lines and leaders. Properly drawn arrowheads should be three times as long as they are wide. All arrowheads on a drawing should be the same size. Do not use small arrowheads in small spaces. (Limited space dimensioning is covered in Chapter 11.) Some companies

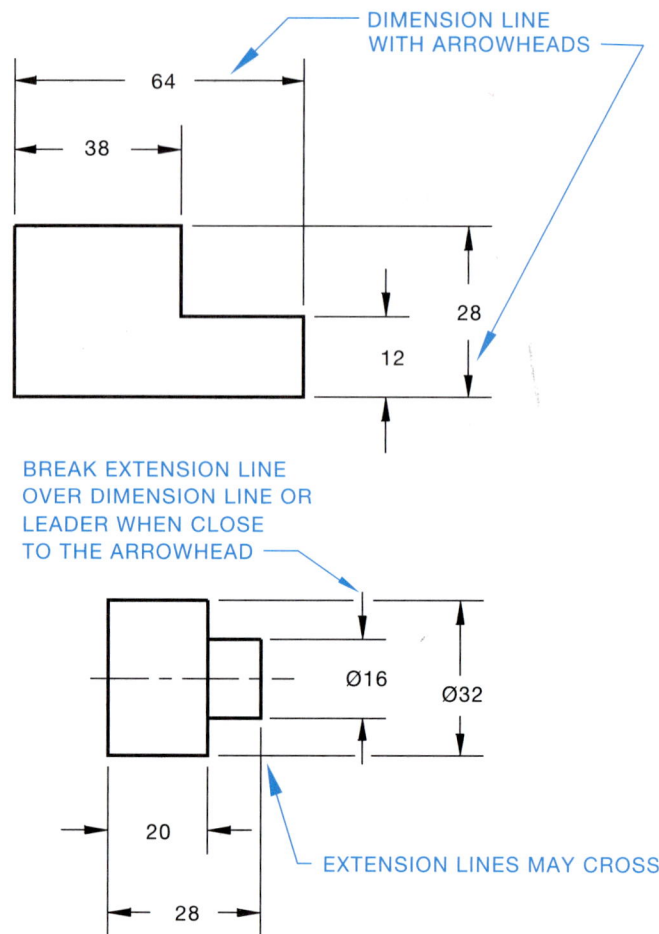

FIGURE 7.13 ■ Dimension lines.

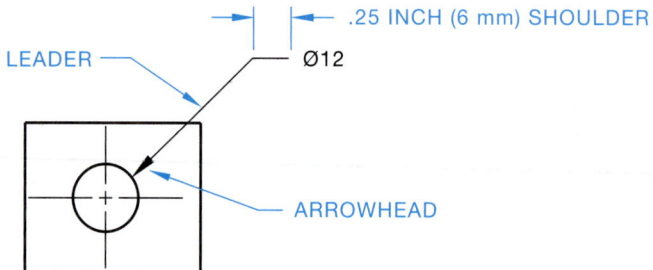

FIGURE 7.14 ■ Leader line.

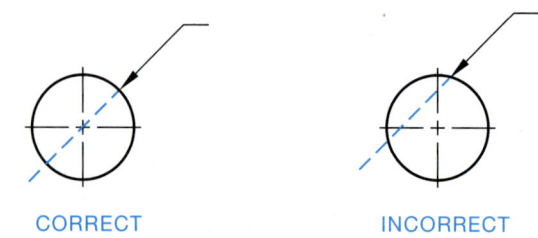

FIGURE 7.15 ■ Circle to leader line relationship. The path of the leader should pass through the center.

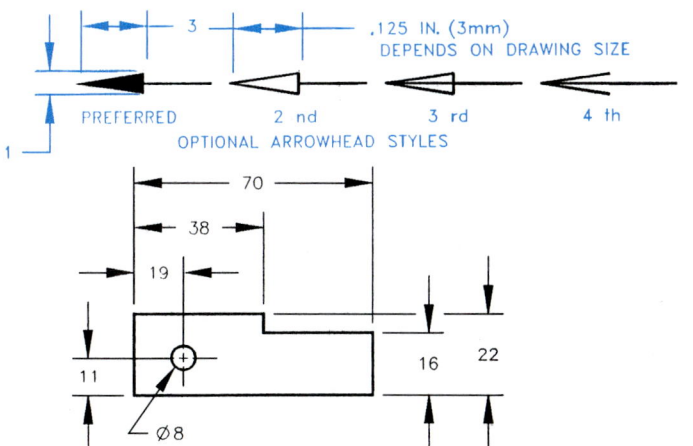

FIGURE 7.16 ■ Arrowheads.

require that arrowheads be drawn with an arrow template, while others accept properly drawn freehand arrowheads. (See Figure 7.16.) Individual company preference dictates whether arrowheads are filled in or left open as shown.

CADD programs normally allow you to select from a variety of available arrowhead options. You should be able to match the desired ASME standard or the standard used by your company or drafting application. Most CADD systems allow you to control the display of every dimension element including the size of arrowheads, and the location of arrowheads in relationship to the dimension text placement. The commonly preferred arrowhead for engineering drawings is solid-filled, but this standard can be different between offices. The solid-filled-arrowhead stands out clearly on the drawing and helps identify the dimension location. Dimensions are usually drawn using a unique format called layers, which can be named DIM or DIMENSIONS, for example. A dimension format contains information such as the line style, text placement, arrowhead style and size, a contrasting color that helps distinguish dimensions from other lines and text, and the recommended thickness of .01 in. (0.3 mm). Chapter 11 provides additional information and detailed standards for dimensioning practice and applications.

Cutting-Plane and Viewing-Plane Lines

Cutting-plane lines are thick lines used to identify where a sectional view is taken. Viewing-plane lines are also thick and are used to identify where a view is taken for view enlargements or for partial views. Cutting-plane and viewing-plane lines are properly drawn in either of the two ways. The approximate dash and space sizes are shown in Figure 7.17. The cutting-plane line takes precedence over the centerline when used in the place of a centerline.

The scale of the view can be increased or remain the same as the view from the viewing plane, depending on the clarity of information presented. When the location of the cutting plane or viewing plane is easily understood or if the clarity of the draw-

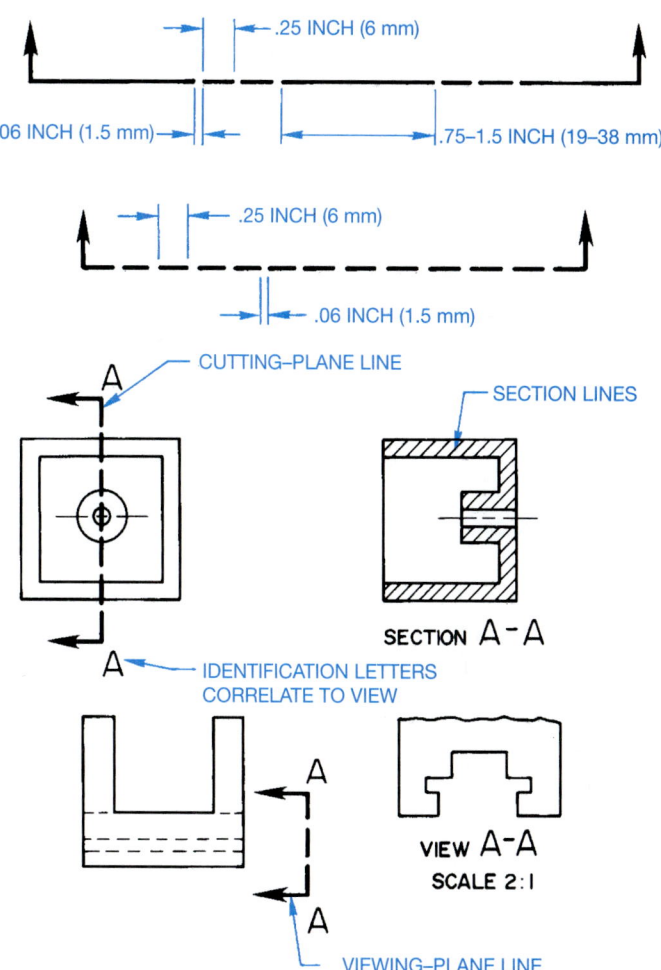

FIGURE 7.17 ■ Cutting- and viewing-plane lines.

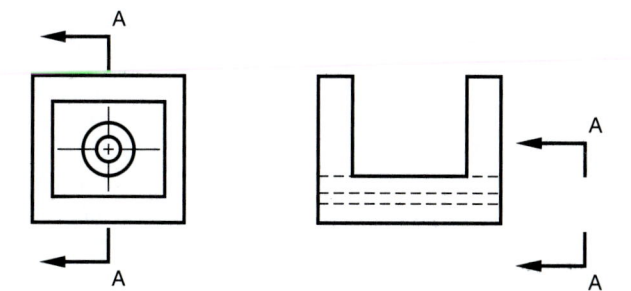

FIGURE 7.18 ■ Simplified cutting- and view-plane lines.

ing is improved, the portion of the line between the arrowheads can be omitted as shown in Figure 7.18.

Section Lines

Section lines are thin lines used in the view of a section to show where the cutting-plane line has cut through material.

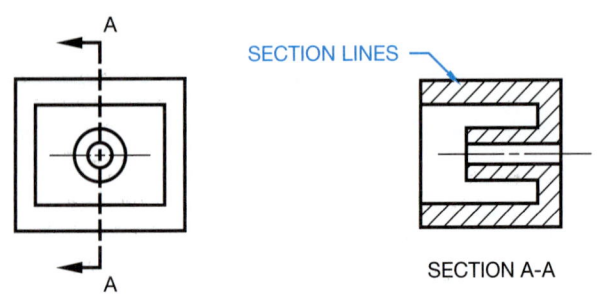

FIGURE 7.19 ■ Section lines.

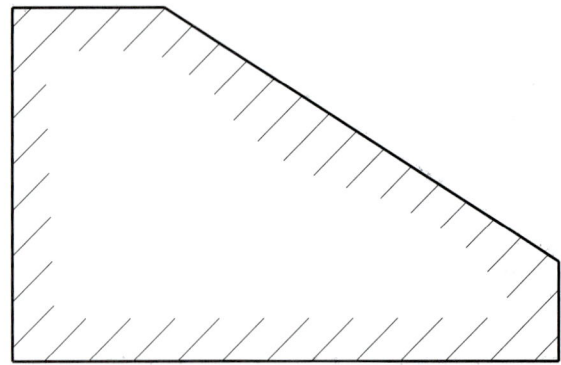

FIGURE 7.22 ■ Outline section lines.

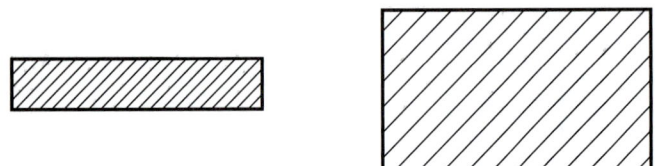

FIGURE 7.20 ■ Space between section lines.

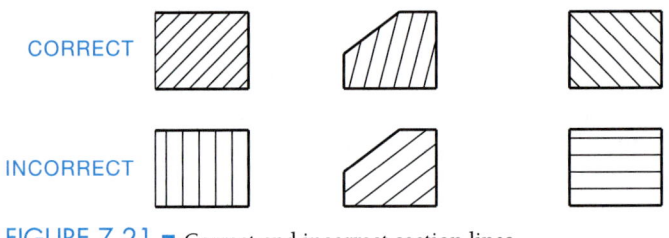

FIGURE 7.21 ■ Correct and incorrect section lines.

(See Figure 7.19.) Section lines are drawn equally spaced at 45° but can not be parallel or perpendicular to any line of the object. Any convenient angle may be used to avoid placing section lines parallel or perpendicular to other lines of the object; 30° and 60° are common. Section lines that are more than 75° or less than 15° from horizontal should be avoided. Section lines should be drawn in opposite directions on adjacent parts. (See Figure 7.25.) For additional adjacent parts, any suitable angle can be used to make the parts appear clearly separate. The space between section lines can vary depending on the size of the object, but the minimum space recommended by the ASME standard is .06 in. (1.5 mm). (See Figure 7.20.) Figure 7.21 shows correct and incorrect applications of section lines. When a very large area requires section lining, you can use outline section lining as shown in Figure 7.22.

The section lines shown in Figures 7.19 through 7.22 were all drawn as general section-line symbols. General section lines can be used for any material and are specifically used for cast or malleable iron. Coded section-line symbols, as shown in Figure 7.23, are not commonly used on detail drawings, as they are more difficult to draw and the drawing title block usually identifies the type of material in the part. Coded section lines can be used when the material must be clearly rep-

resented, as in the section through an assembly of parts made of different materials. (See Figure 7.24.) Very thin parts, less than .016-in. (4-mm) thick, can be shown without section lining; only the outline is shown. This option is often used for a gasket as shown in Figure 7.25.

CADD programs normally allow you to select from a variety of available section line options. You should be able to match the desired ASME standard section line symbol. The CADD system allows you to set the section line type, spacing, and angle. Section lines are normally set up on a specific layer and with a specific color so they stand out clearly on the drawing. Section lines are very easy to draw with CADD, because all you have to do is select the object to be sectioned, or pick a point inside of the area to be sectioned, or in some cases, section lines are automatically placed when a section view is drawn. Normally, the object to be sectioned must be a closed geometric shape without any gaps in the perimeter. Some CADD systems also allow you to place section lines in an area that is not defined by an existing geometric shape. Section lines are often drawn using graphic pattern tools such as AutoCAD's HATCH command.

Break Lines

There are two types of break lines: the short break and long break lines. The thick, short break is very common on detail drawings, although the thin, long break can be used for breaks of long distances at your choice. (See Figure 7.26.) Short break lines are generally drawn freehand, and long break lines are drawn with instruments or a template. Other conventional breaks can be used for cylindrical features as in Figure 7.27.

Phantom Lines

Phantom lines are thin lines made of one long and two short dashes alternately spaced. Phantom lines are used to identify alternate positions of moving parts, adjacent positions of related parts, repetitive details, or the contour of filleted and rounded corners. (See Figure 7.28.)

CAST OR MALLEABLE
IRON AND GENERAL
USE FOR ALL MATERIALS

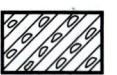
CORK, FELT, FABRIC,
LEATHER, AND FIBER

MARBLE, SLATE,
GLASS, PORCELAIN

STEEL

SOUND INSULATION

EARTH

BRONZE, BRASS, COPPER,
AND COMPOSITIONS

THERMAL INSULATION

ROCK

WHITE METAL, ZINC,
LEAD, BABBITT, AND
ALLOYS

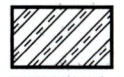
TITANIUM AND
REFRACTORY
MATERIAL

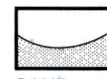
SAND

MAGNESIUM, ALUMINUM,
AND ALUMINUM ALLOYS

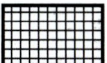
ELECTRIC WINDINGS,
ELECTROMAGNETS,
RESISTANCE, ETC.

WATER AND OTHER
LIQUIDS

RUBBER, PLASTIC, AND
ELECTRICAL INSULATION

CONCRETE

ACROSS GRAIN
WITH GRAIN } WOOD

FIGURE 7.23 ■ Coded section lines.

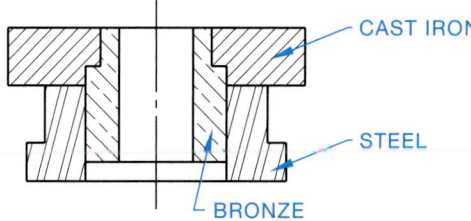

CAST IRON

STEEL

BRONZE

FIGURE 7.24 ■ Coded section lines in assembly.

VERY THIN MATERIAL
IN SECTION

FIGURE 7.25 ■ Very thin material in section is drawn without section lines.

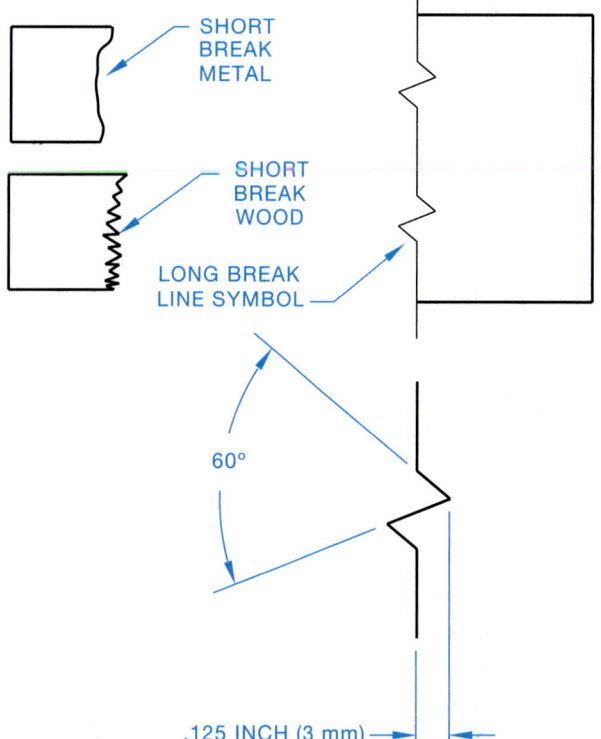

SHORT
BREAK
METAL

SHORT
BREAK
WOOD

LONG BREAK
LINE SYMBOL

60°

.125 INCH (3 mm)

FIGURE 7.26 ■ Long and short break lines.

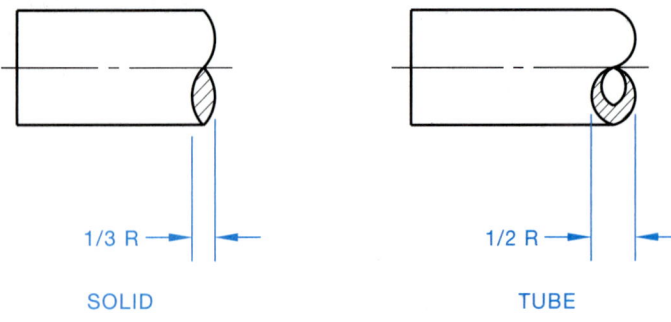

1/3 R — SOLID

1/2 R — TUBE

FIGURE 7.27 ■ Cylindrical conventional breaks.

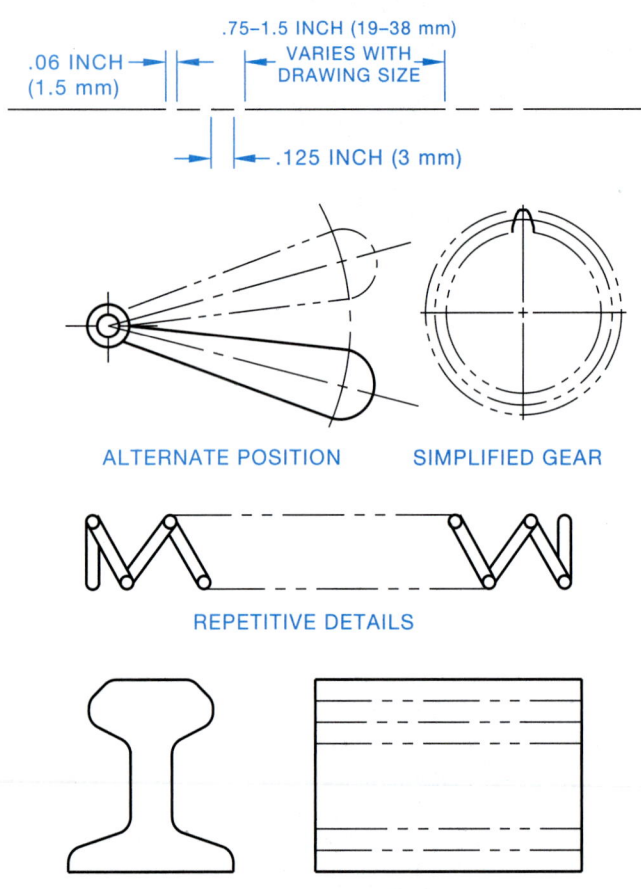

.06 INCH (1.5 mm)

.75–1.5 INCH (19–38 mm) VARIES WITH DRAWING SIZE

.125 INCH (3 mm)

ALTERNATE POSITION   SIMPLIFIED GEAR

REPETITIVE DETAILS

SHOWING FILLETED AND ROUNDED CORNERS

FIGURE 7.28 ■ Phantom line representation and examples.

Chain Lines

Chain lines are thick lines of alternately spaced long and short dashes used to indicate that the portion of the surface, next to the chain, line receives some specified treatment. (See Figure 7.29.)

Stitch Lines

There are two types of acceptable **stitch lines**. One is drawn as thin, short dashes, the other as .01-in. (0.3-mm) diameter dots

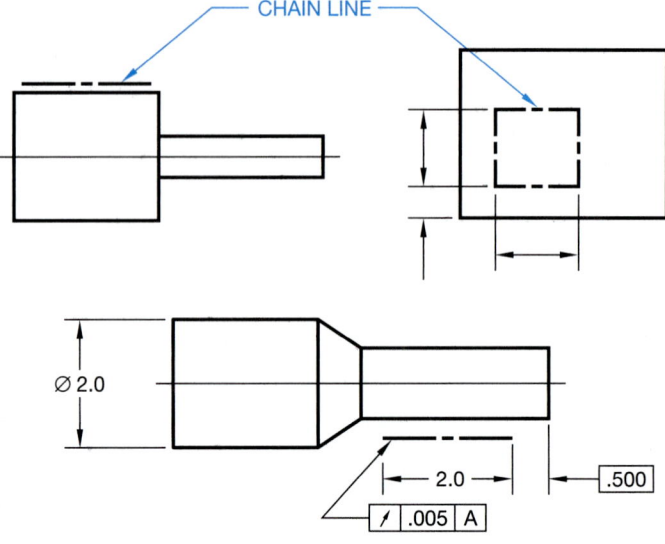

CHAIN LINE

FIGURE 7.29 ■ Chain lines.

Ø 2.0   2.0   .500

/ .005 A

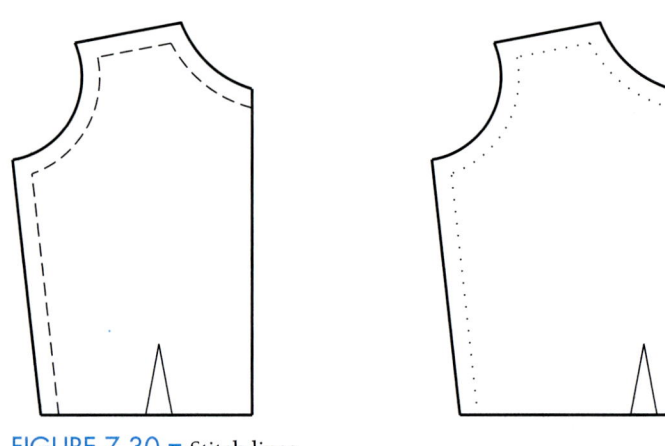

FIGURE 7.30 ■ Stitch lines.

spaced .12 in. (3 mm) apart. They are used to indicate the location of a stitching or sewing process as shown in Figure 7.30.

PENCIL AND INK LINE TECHNIQUES

Pencil Techniques

There are four basic properties of correctly drawn lines: uniformity, contrast, darkness, and sharpness. **Line uniformity** means that all lines are drawn their proper thickness without variation. For example, all thick lines (object, cutting-plane) are the same degree of thickness. All thin lines (center, hidden, dimension, extension) are the same thinness. **Line contrast** is the variation that exists between different types of lines, thick and thin.

ASME/MIL According to ASME standards, thick and thin are the two line thicknesses used. Military (MIL) standards recommend three line thicknesses: thick (cutting and viewing-plane, short break, and object), medium

MAKE ARCS FIRST — CONNECT ARCS —

FIGURE 7.31 ■ Draw circles and arcs first.

(hidden and phantom), and thin (center, dimension, extension, leader, long break, and section). The content and examples used in this chapter are based on the ASME standard.

All lines should be the same darkness with no variation. Lines must be drawn opaque so light will not pass through. Lines are sharp when edges are clear and crisp. Fuzzy edges on lines result from the texture of the paper, or too-soft pencil lead. When using an automatic pencil, do not let the lead out too far.

Wash your hands and equipment frequently to keep the drawing clean. Try not to handle your pencil leads.

Before darkening any lines, lay out your entire drawing with construction lines. A combination of pressure and pencil lead hardness makes the best lines. If you are right-handed, it is generally best to draw horizontal lines from left to right and vertical lines from bottom to top. The opposite may be true for left-handed drafters. Do not use excessive pressure. You do not have to dig trenches in the drawing board. The more pressure you use, the more difficult it is to erase. If your hand cramps while drawing or lettering, you are forcing yourself into an unnatural stroking technique. Relax! Try different methods until you find a suitable system of strokes.

Make thick lines with a 0.7- or 0.9-mm automatic pencil held vertically to establish full contact between the lead and the paper. In order to obtain lines that are dark, crisp, and sharp, you may need to go over each line several times. If light passes through your lines, they are drawn too lightly. Hold your drawing up to a light or place it on a light table to see if light passes through.

Always draw radii first, then connect the straight lines. This is also a good technique for ink drawings. (See Figure 7.31.) If you make the straight lines first, you can have trouble aligning the arcs.

Line Layout

First, draw thin horizontal lines beginning from the top of the sheet and work downward. Draw vertical thin lines next, starting on the left side of the sheet and working toward the right side for righthanded drafters. Next, draw all circles and arcs. Last, draw all horizontal and vertical thick lines in the same direction as thin lines. Do all lettering when the line work is complete. Try to avoid going back over lines or lettering that has already been drawn in order to help avoid smudging.

Inking Techniques

Ink or graphite only adheres cleanly to one side of vellum. If the ink beads or skips while making a line, you are working on the

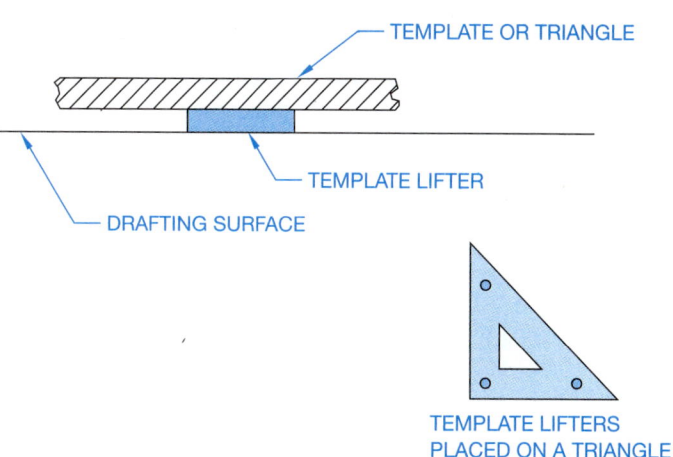

FIGURE 7.32 ■ Template lifters.

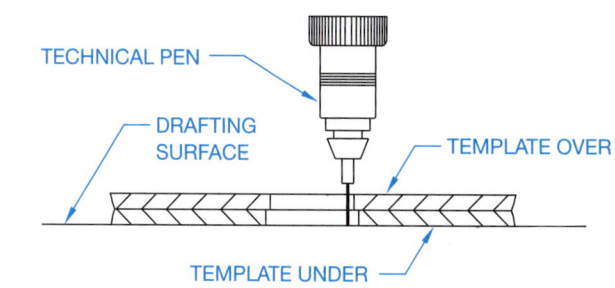

FIGURE 7.33 ■ Second template as a spacer.

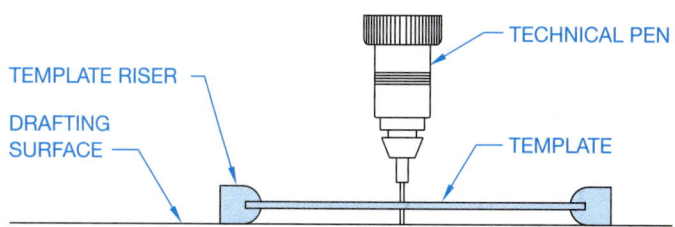

FIGURE 7.34 ■ Template risers.

wrong side of the paper. Paper with preprinted borders and title blocks have the correct side predetermined. Some manufacturers have a watermark that is visible when the paper is held up to the light. When the watermark is readable, the paper is right side up. For polyester film, there is either a single or double matte surface. **Matte** refers to the textured surface as opposed to a glossy surface. Double matte surface can be drawn on either side. Single matte surface requires you to draw on the matte side only, not the shiny, or glossy, side.

Use template lifters or templates and other equipment designed for inking that keep the edge of the tool slightly off the paper. Template lifters are plastic shapes that can be attached to the back of a template or other instrument to help keep the instrument off the drafting surface, as seen in Figure 7.32. Another option is to place a template under the one being used. (See Figure 7.33.) There are also template risers available that are long plastic strips that fit on the edges of the template to help keep it off of the drawing surface. Figure 7.34

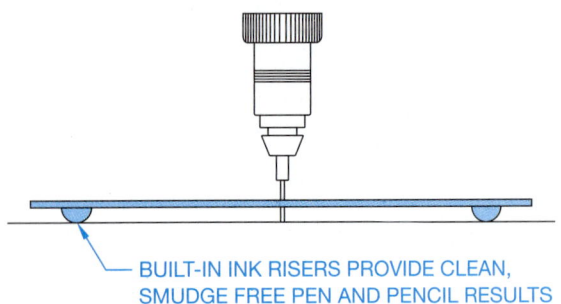

BUILT-IN INK RISERS PROVIDE CLEAN,
SMUDGE FREE PEN AND PENCIL RESULTS

FIGURE 7.35 ■ Built-in ink risers. *Courtesy Chartpak.*

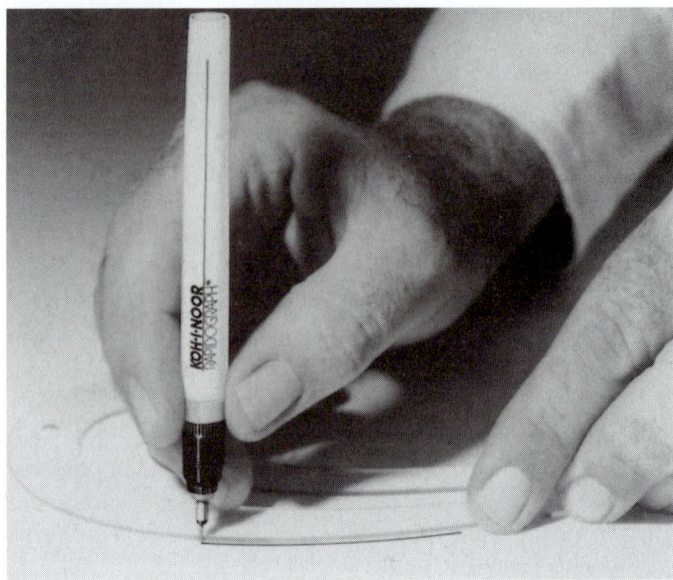

FIGURE 7.36 ■ Proper technical pen use. *Courtesy Koh-I-Noor, Inc.*

shows how the template riser functions when attached to a template. Drafting machine scales have long been manufactured with edge relief for inking. Now, templates, triangles, and other equipment are being made with ink risers built in, as shown in Figure 7.35. If you draw with ink without taking these precautions, the ink could easily flow under the template and cause a mess.

Periodically check your technical fountain pens for leaks around the tip to prevent your hands from smearing your drawing. Also check the tip for a drop of ink before you begin a line. Have a piece of cloth or a tissue available to help keep the tip free of ink drops. Keep your pens clean and loaded with fresh ink to help keep proper ink flow for trouble free use. Also, keep a piece of paper handy for scratching your pen to start the ink flowing.

Ink your lines beginning with thin horizontal lines, and work from top to bottom and from left to right if right-handed. The technique of drawing horizontal lines top to bottom and vertical lines left to right allows inked lines to dry as you go along and saves valuable drafting time. Inked lines are dry when they do not appear glossy. Glossy lines are wet; keep equipment away from them. Next, ink all circles and arcs. Ink all thick horizontal and vertical lines in the same manner as thin horizontal and vertical lines. Finally, do all lettering.

When using technical pens, hold the pen perpendicular (90°) to the paper for the best results. Do not apply any pressure to the pen. Allow the pen to flow easily over the vellum or polyester film. Figure 7.36 shows proper technical pen use. If the technical pen is not held 90° to the surface, the resulting line may be fuzzy or rough on the edge. Move your technical pen at a constant speed. Do not go too fast and do not slow at the ends of lines. Following these hints will help keep your line consistent in width and keep pen point from skipping.

Erasing Ink

Erasing ink from vellum is possible in small areas, but doing so may destroy its surface. A smooth surface cannot be inked over again satisfactorily. An electric eraser is best used here, but overuse burns a hole through the paper. Electric erasers are excellent tools, but be careful to touch the paper surface lightly.

To erase on polyester film, it is best to apply a little water with a felt tip or a clean cotton swab. Remove any excess water with a blotter. Allow the area to dry thoroughly. The ink remover recommended by the ink manufacturer can also be used. A polyester eraser also removes ink. Use an eraser with care so you do not destroy the matte surface, which makes further drafting in that area difficult.

Polyester Lead on Film

When pencils are used on polyester film, the recommended lead is made of polyester. The quality of line on polyester film depends on how well the matte finish is maintained. If you use single matte film, do not try to draw on the glossy side; use polyester lead on the matte side.

Pencil Skills with Polyester Lead

■ Draw with a single line, in one direction. Tracing a line in both directions deposits a double line and causes smearing and damage to the matte.

■ Draft with a light touch. Drafting films require up to 40% less pressure than other media. Smearing and embossing can be reduced with less pressure. It takes practice to use polyester lead after drawing with graphite lead.

■ The surface of your drafting board is also a factor in line quality. Use a recommended backing material on your table. Check with your local vendor.

■ Erase with a vinyl eraser. If an electric eraser is used, be very careful not to destroy the matte.

CADD APPLICATIONS

DRAWING LINES

The basics of drawing lines with CADD are applicable to all systems, whether they use a puck, stylus light pen, thumbwheels, joystick, or mouse. The term pointers refers to the pointing device used to select commands and digitize points. The term digitize or pick means to press the stylus to the tablet or press the proper mouse or keyboard button to select, or draw, a point. Several different commands are given for one function in the following discussions, to give you an idea of the terms used in different systems.

Once you are in the drawing mode, your location on the screen is indicated by a cursor. The drawing area cursor is often displayed as a crosshair, but it can be shown as a wand or point, depending on the CADD program. When you move the pointer, the cursor moves. Unless you press a button or click the pointer, nothing is drawn on the screen. Some systems may continually display the coordinates of the cursor location on the screen. These values change as you move the pointer. When constructing your drawing, it is best to look at the video display as you draw. You will then be able to see if your drawing looks right.

Begin drawing a line by first selecting the proper command such as LINE. The LINE command is used to create straight lines by selecting locations in the drawing area. Pick a point, and the coordinates of that point are displayed on the screen. Select another point and a straight line is drawn between the two. This shows as line 1 in Figure 7.37. Each selection of the pointer after that draws a line. The steps required to draw an object are shown in Figure 7.37a-d. There is no need to select LINE between each selected point. To exit the command and move to another feature, press the [Enter] key and access the LINE command again. This allows you to locate a new coordinate point without drawing an unwanted line.

Some systems use what is called a rubber band line. After you pick one point, a line is attached to that point and to the cursor and moves with the cursor. Selecting another point draws the first line and another is now attached to the cursor.

In most situations, unique line formats are used when drawing lines with CADD. Many systems, such as AutoCAD's layers, allow you to define the name, color, line type, thickness (line weight), and the display characteristics for each type of line. Drafters often use many of these predefined line formats when creating drawings. For example, a format that is used to draw visible object lines might be named OBJECT, be black in color, use a solid line type, and have a thickness of .02 in. (0.6 mm). A format used to draw hidden lines, for example, might be named HIDDEN, be green in color for contrast, use a hidden line type, and have a thickness of .01 in. (0.3 mm). These different line configurations can be indicated either before or after lines are drawn, though it is good practice to set the line format before you begin drafting. Any line drawn after you enter a specific format has the characteristics that you have chosen, and you must select a different line format each time you draw a different form of line. Still, you can modify the properties of a line you have already drawn using a variety of editing tools such as the CHANGE or PROPERTIES commands.

Drawing lines using tools such as the LINE command previously discussed, represents the fundamental approach to creating CADD drawings. After all, a drawing is just a grouping of lines and shapes. As you learn more about CADD and specific programs, you will become familiar with a multitude of commands that allow you to create and modify objects, dimensions, and other complex shapes without using several line tools.

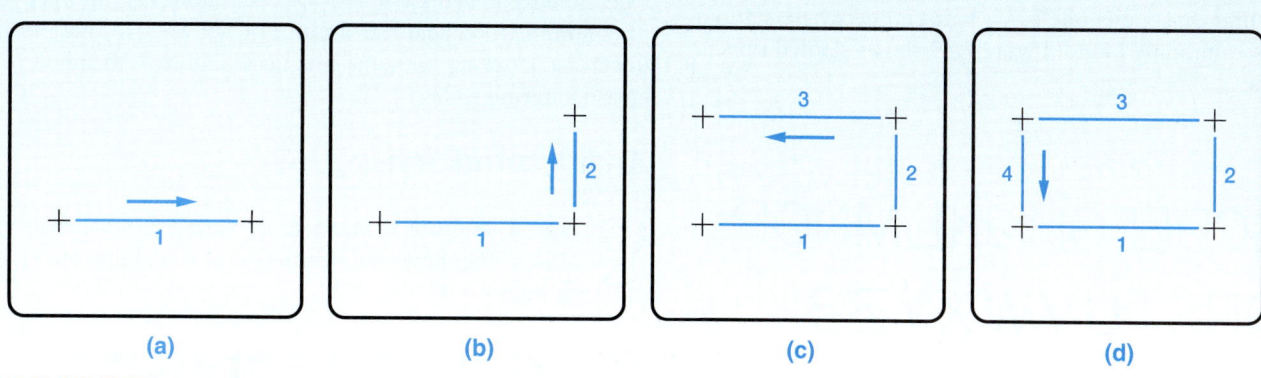

(a)   (b)   (c)   (d)

FIGURE 7.37 ■ Using a CADD program to draw an object with lines one point at a time.

LETTERING

Information on drawings that cannot be represented graphically by lines can be presented by lettered dimensions notes, and titles. It is extremely important for these lettered items to be exact, reliable, and entirely legible in order for the reader to have confidence in them and have no doubt as to their meaning. This is especially important when using reproduction techniques that require a drawing to be reduced in size using photocopy or microfilm. Poor lettering ruins an otherwise good drawing.

SINGLE-STROKE GOTHIC LETTERING

ASME The standard for lettering was established in 1935 by the American National Standards Institute (ANSI). This standard is now conveyed by the American Society of Mechanical Engineers document ASME Y14.2M, *Line Conventions and Lettering.*

The standardized lettering format was developed as a modified form of the Gothic letter font. The term **font** refers to a complete assortment of any one size and style of letters. The simplification of the Gothic letters resulted in elements for each letter that became known as single-stroke Gothic lettering. The name sounds complex but it is not. The term **single stroke** comes from the fact that each letter is made up of a single straight or curved line element that makes it easy to draw and clear to read. There are upper- and lowercase, vertical, and inclined Gothic letters, but industry has become accustomed to using vertical uppercase letters as the standard. (See Figure 7.38.)

OTHER LETTERING STYLES

Inclined Lettering

Some companies prefer inclined lettering. The general slant of inclined letters is 68°. One edge of the Ames Lettering Guide has a 68° slant, which may be used to help maintain the proper angle. Structural drafting is one field or discipline where slanted lettering is commonly found. Figure 7.39 shows slanted uppercase letters.

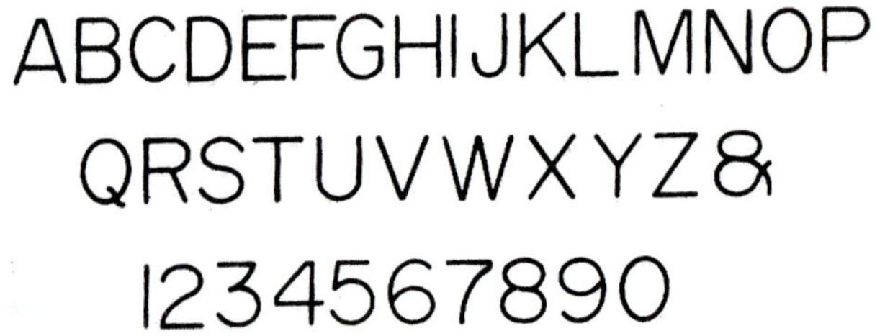

ABCDEFGHIJKLMNOP
QRSTUVWXYZ&
1234567890

FIGURE 7.38 ■ Vertical uppercase single-stroke Gothic letters and numbers.

ABCDEFGHIJKLMNOP
QRSTUVWXYZ&
1234567890
5
2

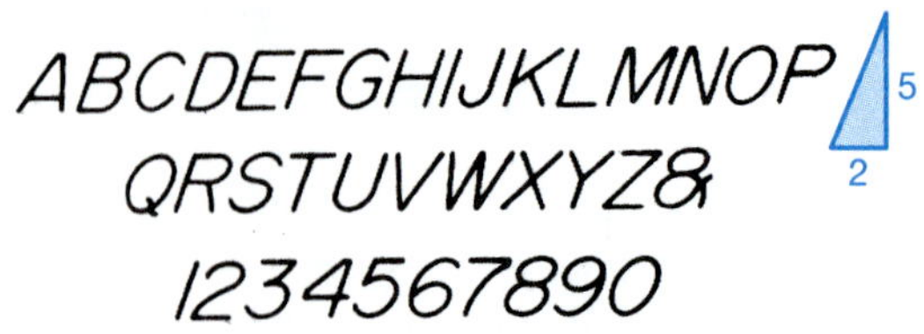

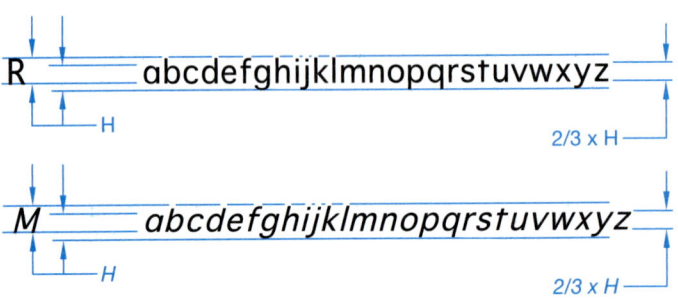

FIGURE 7.39 ■ Uppercase inclined letters and numbers.

R
abcdefghijklmnopqrstuvwxyz
H
2/3 x H
M
abcdefghijklmnopqrstuvwxyz
H
2/3 x H

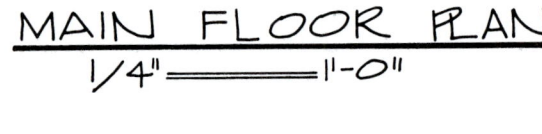

FIGURE 7.40 ■ Lowercase lettering.

MAIN FLOOR PLAN
1/4" = 1'-0"
LIVING
ABCDEFGHIJKLM
NOPQRSTUVWXYZ
1 2 3 4 5 6 7 8 9 10

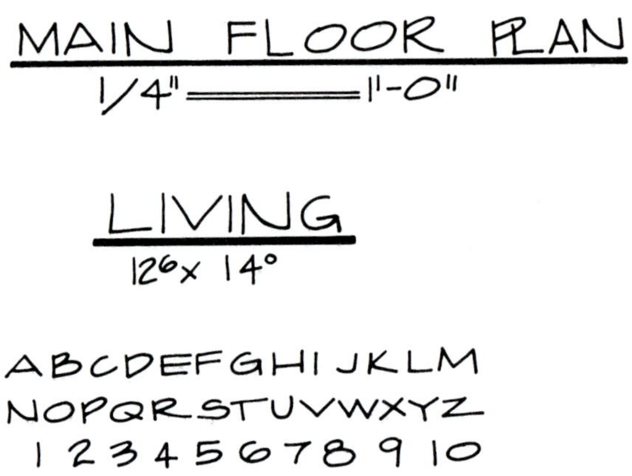

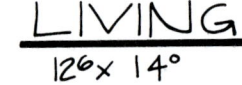

FIGURE 7.41 ■ Architectural lettering.

Lowercase Lettering

Occasionally, lowercase letters are used; however, they are very uncommon in mechanical drafting. Civil or map drafters use lowercase lettering for some practices. Figure 7.40 shows lowercase lettering styles.

Architectural Styles

Architectural lettering is much more varied in style than mechanical lettering; however, neatness and readability are essential. (See Figure 7.41.)

LETTERING LEGIBILITY

ASME The minimum recommended lettering size on engineering drawings is .12 in. (3 mm). All dimension numerals, notes, and other lettered information should be the same height except for titles, drawing numbers, section and view letters, and other captions, which are .24 in. (6 mm) high.

The following table gives the recommended minimum lettering heights for different elements of the drawing based on drawing sheet sizes:

Drawing Elements	Minimum Letter Heights INCH	Drawing Sizes INCH	Letter Heights METRIC (mm)	Letter Drawing Sizes METRIC
Title, drawing size	.24	D, E, F, H, I, J, K	6	A0, A1
CAGE code, drawing number, and revision letter in the title block	.12	A, B, C, G	3	A2, A3, A4
Section and view letters	.24	All sizes	6	All sizes
Zone letters and numbers in borders	.24	All sizes	6	All sizes
Drawing block headings	.10	All sizes	2.5	All sizes
All other characters	.12	All sizes	3	All sizes

Either vertical or inclined lettering can be used on a drawing depending on company preference. However, only one style of lettering should be used on a drawing. Lettering must be dark, crisp, and opaque for the best possible reproducibility.

The composition or spacing of letters in words and between words in sentences should make the individual letters uniformly spaced with approximately equal background areas. This usually requires that letters such as I, N, or S be spaced slightly farther from their adjacent letters than L, A, or W. The space between words in a note or sentence should be about the same as the height of the letters. The space between two numerals with a decimal point between them is a minimum of two-thirds of the letter height.

Notes should be lettered horizontally on the sheet. When lettering notes, sentences, or dimensions requires more than one line, the vertical space between the lines should be a minimum of one-half the height of letters. The maximum recommended space between lines of lettering is equal to the height of the letters. Some companies prefer to use the minimum space to help conserve space, while other companies prefer the maximum space for clarity. (See Figure 7.42.)

Additional specific information about notes is provided in Chapter 11, Dimensioning and Tolerancing.

VERTICAL FREEHAND LETTERING

Vertical freehand lettering is the standard for mechanical drafting. The ability to perform good-quality lettering quickly is important. A common comment among employers hiring entry-level drafters is the ability to do quality lettering and line work. Although standard, not all companies require freehand lettering. Some companies allow drafters the flexibility of freehand lettering or using a template. As most companies are now using computer-aided drafting, traditional lettering skills may become obsolete.

Always use lightly drawn horizontal guidelines that are spaced equal to the height of the letters. Some people need vertical guidelines to help keep their letters vertical. The ability to perform quality freehand lettering requires a great deal of practice for most people.

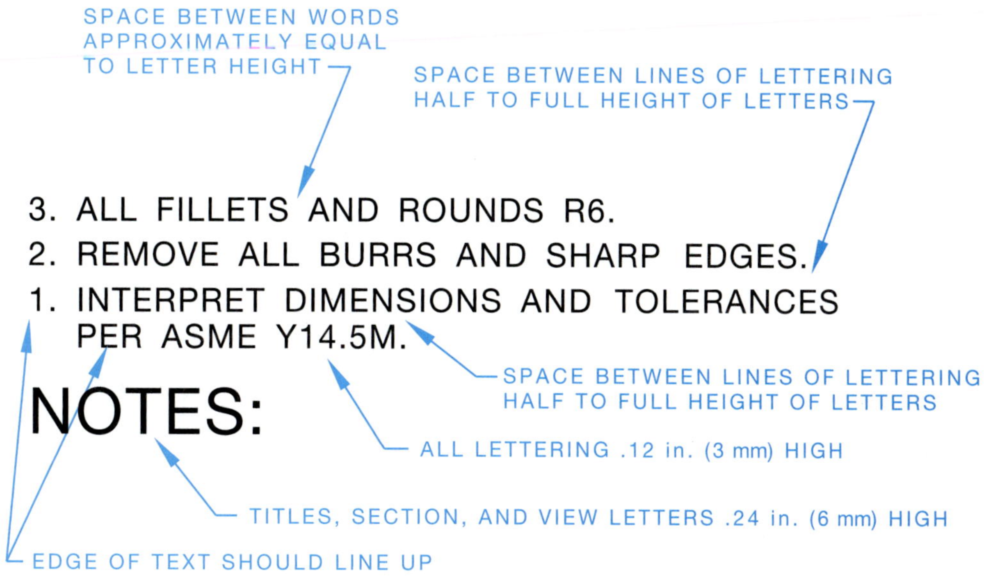

SPACE BETWEEN WORDS APPROXIMATELY EQUAL TO LETTER HEIGHT

SPACE BETWEEN LINES OF LETTERING HALF TO FULL HEIGHT OF LETTERS

3. ALL FILLETS AND ROUNDS R6.
2. REMOVE ALL BURRS AND SHARP EDGES.
1. INTERPRET DIMENSIONS AND TOLERANCES PER ASME Y14.5M.

NOTES:

SPACE BETWEEN LINES OF LETTERING HALF TO FULL HEIGHT OF LETTERS

ALL LETTERING .12 in. (3 mm) HIGH

TITLES, SECTION, AND VIEW LETTERS .24 in. (6 mm) HIGH

EDGE OF TEXT SHOULD LINE UP

FIGURE 7.42 ■ Spacing of letters, words, and notes.

Vertical Capital Letters

Straight Elements

Use a 0.5-mm automatic pencil for lettering. This kind of pencil does not need sharpening and H, F, of HB leads are usually easy to control for effective lettering. Remember, you need to experiment with leads to determine which gives you the best results. Letters should be dark and crisp. Fuzzy letters print as poorly as fuzzy lines.

Letters composed of straight lines are shown in Figure 7.43. You should become familiar with these letter forms and the strokes needed to make them. The arrows in Figure 7.43 indicate the direction of the stroke used to form the letters. However, if these strokes are uncomfortable, you should develop your own procedure. Try the recommended strokes first. Because these letters are made of single-stroke elements, try not to let the strokes combine, as the result can be curves where straight elements should be used.

Horizontal guidelines should be used for all lettering at all times. If you have difficulty keeping your letters vertical, use vertical guidelines. Guidelines should be very light 6H or 4H pencil lines. Some drafters prefer to use a light-blue lead rather than a graphite lead for guidelines. Light blue does not reproduce in the diazo or photocopy processes. When lettering, protect your drawing by resting your hand on a clean protective sheet placed over your drawing. This prevents smearing and smudging, as shown in Figure 7.44.

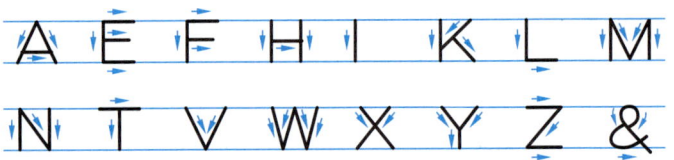

FIGURE 7.43 ■ Recommended strokes for vertical, uppercase Gothic letters with straight elements.

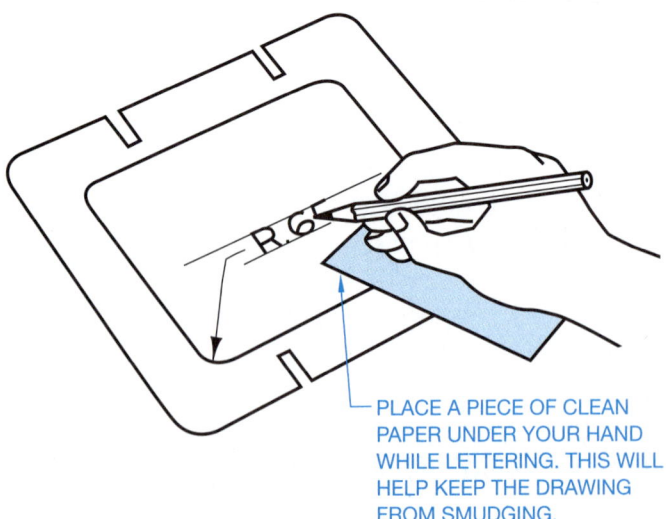

FIGURE 7.44 ■ Place clean paper under your hand when lettering.

Look at Figure 7.42 and notice the title NOTES at the bottom and the related notes numbered 1, 2, and 3 in increasing order above the title. This is common practice with manual drafting, because it allows you to add new notes above as they become necessary. This practice is not necessary with CADD, because the entire group of notes can be moved as needed to accommodate new notes. The placement of notes with CADD is generally with the title NOTES first and the numbered notes 1, 2, and 3 below. Additional detailed note and note placement coverage is provided in Chapter 11.

Curved Elements

Letters containing arcs are shown in Figure 7.45. Notice the difference in the sizes of arcs and circles used for different letters. The recommended letter elements, as with straight elements, are made up of a series of suggested strokes. These strokes, when used as shown, provide the best lettering results. Vertical guidelines are often as asset for the best lettering results.

Vertical Numerals and Fractions

Vertical numerals, as seen in Figure 7.46, are also made up of recommended strokes. Numerals are the same height as capital letters.

Fractions

Fractions are not as commonly used on engineering drawings as decimal inches or millimeters but are common on architectural and structural drawings. When fractions are used on a drawings, the fraction numerals should be the same size as other numerals on the drawing. The fraction bar should be drawn in line with the direction of the dimension. For example, if all dimensions read

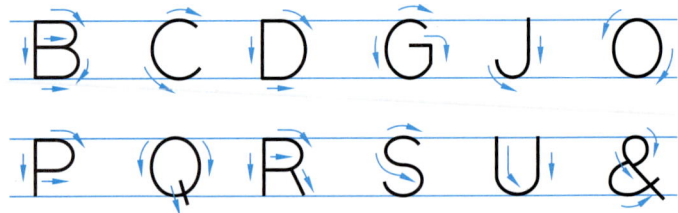

FIGURE 7.45 ■ Recommended strokes for vertical, uppercase Gothic letters with curved strokes.

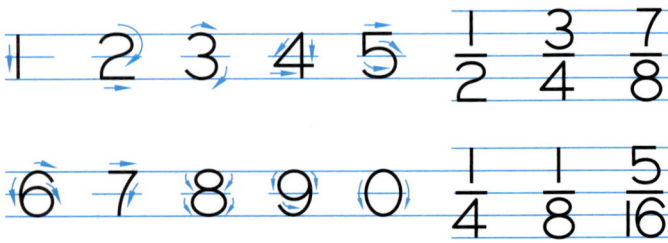

FIGURE 7.46 ■ Vertical numerals and fractions.

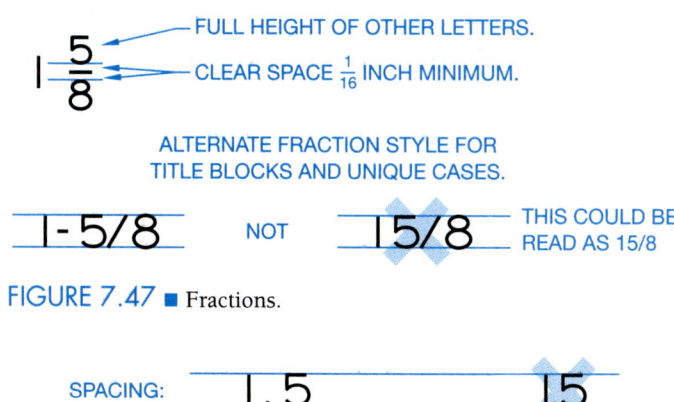

FIGURE 7.47 ■ Fractions.

FIGURE 7.48 ■ Spacing of decimal point in numerals.

horizontally, all fraction bars are horizontal. The fraction's numerals should not touch the fraction bar. A space of 0.6 in. (1.5 mm) between the numeral and bar is suggested. The fraction bar can be diagonal in certain situations such as when used in a general note, in a drawing title, or when using CADD. (See Figure 7.47.)

Decimal Points

The placement of the decimal point in a decimal dimension is critical. If the decimal point is crowded or drawn too lightly, it can be lost and the result is an unclear dimension. Always make the decimal point dark and bold. Also space the numerals far enough to clearly provide room for the decimal point. Two-thirds the height of letters is recommended. (See Figure 7.48.)

LETTERING TECHNIQUES

Always use guidelines. Straight, even letters of consistent height look better than letters of varying heights. Even when using guidelines, be sure to extend each letter directly to the guidelines. Letters that periodically extend beyond or fall short of the guidelines tend to make the words or notes irregular.

Try an H or F pencil for lettering if you have light touch or, if your touch is heavy, try a 2H pencil. Use a 0.5-mm automatic pencil. Lines should be black, crisp, and sharp. All vertical lines are made perpendicular, starting at the top for each stroke; all horizontal lines are made from left to right. Balance angles for the letters A, V, W, X, and Y about a vertical guideline. Use a round form for curved letters. Be careful to not allow any space to show between the letter and the guideline.

COMPOSITION

The term **composition**, used in this application, refers to the process of placing letters within words and words within paragraphs in an artistic and unified manner.

As a rule of thumb, curved letters can be placed close together; straight letters should be placed further apart. Good lettering composition is evident when all letters in a word look as if they have the same amount of space between them. To achieve this ap-

pearance it is often necessary to shorten the horizontal strokes of open letters such as L and J. When strokes are parallel and next to each other, as the WA in WALL and both Ns and I in PLANNING, they should be placed a little farther apart.

If you letters are wiggly or if you are nervous, try pressing hard to make your lines straighter. If you are pressing too hard, try to relax the pressure. Also try making each letter as rapidly as possible. This tends to eliminate wiggly letters. If your lead is too hard, wiggly letters can result; try a softer lead.

MAKING GUIDELINES

Guidelines are very lightly drawn lines equal to the height of letters in distance apart. As previously mentioned, some drafters prefer to use a light blue lead so guidelines do not reproduce.

Ames Lettering Guide

A commonly used device for making guidelines is the versatile Ames Lettering Guide shown in Figure 7.49. With an Ames Lettering Guide it is possible to draw guidelines and sloped lines for lettering from 1/16 to 2 in. (1.5–50 mm) in height. Instructions for use are normally found with the lettering guide when purchased. The lettering guide may also be used for parallel lines, needed for such purposes as section lines, including brick, tile, and concrete block, or a music staff.

OTHER LETTERING AIDS AND GUIDELINE METHODS

Other guideline lettering aids for equidistant spacing of lines have parallel slots ranging in width from 1/16 to 1/4 in. (1.5–6 mm). These lettering guideline aids are not as complex as the Ames Lettering Guide, but they are also not as flexible.

Another method of making guidelines used by a few drafters is to place an 1/8-in. (3-mm) grid paper under the drawing. The lines show through the drawing sheet and guidelines need not

FIGURE 7.49 ■ Ames Lettering Guide. *Courtesy Olson Manufacturing Company, Inc.*

be drawn. Be careful with this method, because lines of lettering may not be as straight as with conventional guidelines.

BASIC LETTERING CONSIDERATIONS

The following are basic guidelines that can be used to help you create quality manual lettering:

■ Always use two guidelines. Three are helpful for beginners. The center guideline is for the letter crossbars.

■ To adjust eye and hand coordination, first letter highly, then darken your lettering. Doing so allows you to correct mistakes. This technique should not be necessary after you get some lettering experience.

■ Make all letters and numbers on the drawing at least .12 in. (3 mm) high, except for titles, which are .24 in. (6 mm) high. All letters should be the same height. Be consistent.

■ Make all lettering dark. You may have to press on the pencil to get dark letters.

■ When practicing lettering, practice no more than 15 minutes per session. Otherwise your hand can cramp and any further practice may be of less value. Practice every day to gain speed and neatness as you letter.

LETTERING GUIDE TEMPLATES

Some companies prefer that drafters use lettering guides so uniformity is maintained. Standard lettering guide templates are available with vertical Gothic letters and numerals ranging in height from 3/32 to 3/8 in. (2–9.5 mm). (See Figure 7.50.) Lettering guides are also available in many other lettering styles, including slanted Gothic, and Microfont letters in either upper- or lowercase.

MECHNICAL LETTERING EQUIPMENT

Mechanical lettering equipment has typically been used for government projects and in civil drafting. Mechanical lettering equipment is available in kits with templates for letters and numerals in a wide range of sizes. A complete lettering equipment kit includes a scriber, templates, tracing pins, and lettering pens. Figure 7.51 shows the component parts of a lettering equipment set. Complete instructions are normally found with the lettering equipment when purchased.

PULLING IT ALL TOGETHER

The drawing shown in Figure 7.55 provides an example of line and lettering work that is commonly used in engineering drawing. The line types are labeled for your reference.

WEBSITE RESEARCH

The following websites can provide you additional information for research or further study into topics covered in this chapter:

http://www.asme.org—Find information and publications related to the American Society of Mechanical Engineers.

http://www.ansi.org—The American National Standards Institute. Information about national and international drafting standards.

http://www.adda.org—American Design Drafting Association—Drafting Reference Guide.

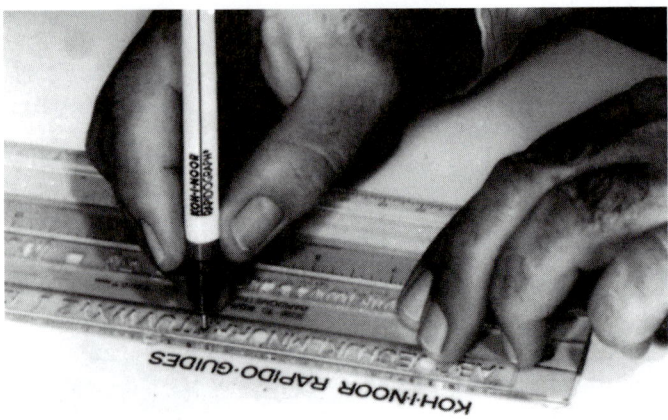

FIGURE 7.50 ■ Lettering guide template in use. *Courtesy Koh-I-Knoor, Inc.*

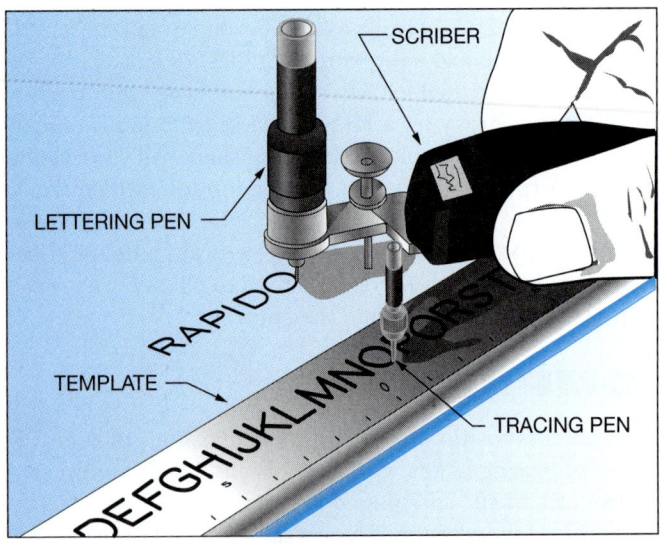

FIGURE 7.51 ■ Components of a lettering equipment set, and using the mechanical lettering equipment. *Courtesy Koh-I-Noor, Inc.*

LETTERING

Lettering is called text or attributes when using CADD. Adding text to a CADD drawing is one of the most efficient and effective tasks associated with computer-aided drafting, because the usually tedious task of freehand lettering is eliminated. When correct drafting practices and text settings are used, text created using CADD offers the following benefits:

■ Characters (words and numbers) are drawn at constant shapes and sizes.

■ Symbols containing complex shapes are quickly added to lines of text.

■ Information can be displayed using a number of text styles.

■ Increased speed, accuracy, and legibility.

■ Easy reproduction and modification.

■ Specialized tools and options such as spell-checking.

The process of placing text varies depending on the CADD system, the tools and commands used to generate the text, and the purpose of the text. However, typically, adding text is just a matter of defining and selecting the desired text style, picking the text location, and typing characters on the keyboard.

Lettering Styles

Most CADD programs contain standard, or default, text styles. The term default refers to any value that is maintained by the computer for a command or function that has variable parameters. For example, the default text height might be .12 inch. Though the default text style can be used to add drawing text, typically a new or modified style is more appropriate. CADD systems have multiple text options that allow you to create a variety of text styles based on the nature of the drawing and the specified drafting standards. Font, height, width, and slant angle are just a few of the characteristics that can be applied to text. You can change the style of existing text, or create new text with a modified style, by picking a menu command, symbol, or by typing a command at the keyboard. Figure 7.52 shows some character styles used in CADD.

Locating Text

The process of locating text is very much the same with CADD as it is with manual drafting. It is your responsibility as the drafter to correctly place notes, dimensions, and other drawing information according to appropriate drafting standards. Though some CADD systems automatically

ABCDEFGHIJKLMNOPQRSTUVWXYZ
abcdefghijklmnopqrstuvwxyz
1234567890

ABCDEFGHIJKLMNOPQRSTUVWXYZ
abcdefghijklmnopqrstuvwxyz
1234567890

ABCDEFGHIJKLMNOPQRSTUVWXYZ
abcdefghijklmnopqrstuvwxyz
1234567890

FIGURE 7.52 ■ Samples of CADD character font styles.

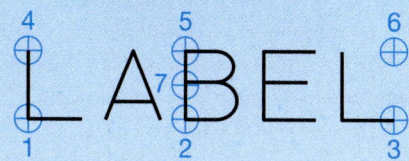

FIGURE 7.53 ■ Text can be placed on a drawing by choosing one of several location points shown here.

place text in certain predefined locations, such as in the cells of parts lists and title blocks, in most situations you must decide exactly where text is located, or at least set up the options required for an automated text placement system to function properly. The drafter still needs to decide where to locate dimensions, notes, and parts lists, but with CADD the process of placing notes is a little more technical. Most CADD systems provide for keyboard entry of location and size coordinates, thus allowing the drafter to accurately locate text. But using the keyboard to input text location coordinates takes time. A quicker way to select the text location is to move the screen cursor to the desired location and pick the point with the mouse.

Text is often further positioned by selecting one of several places on the lettering. Figure 7.53 shows an example of several points on the text that can be used for positioning purposes. Positioning and aligning text is referred to as justification.

Positioning and justification options vary depending on the CADD system. For example, some programs allow you to locate text between two points, which calculates the size of each letter so the text properly fits in the selected space.

Once you identify the location of text, you have the ability to change certain text style characteristics, such as the text height, and define other properties, such as the angle of rotation, also known as the direction, rotation angle,

(Continued)

or text path. The angle of rotation is the angle from horizontal that the text is on. An example of text rotation angle is shown in Figure 7.54. Text located on a horizontal line has a direction of 0° and text that reads from the bottom up vertically has a direction of 90°.

With the style, size, and rotation angle decided, pick a location on the drawing and type the text at the keyboard. This process can occur in a couple of ways. You could be asked to first pick the text location using a pointing input device and then type the text. The second method is the reverse of the first. Type the text as it appears on the screen with the cursor at the point of location that you previously specified. Then *drag* the text to the proper location, and press a button on the mouse, puck, or keyboard, to place the text. A nice thing about locating notes and labels on a drawing with a CADD system is the ability to move them around instantaneously as often as you want.

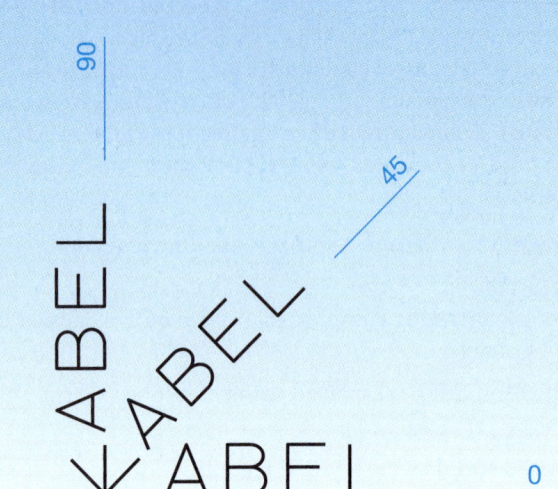

FIGURE 7.54 ■ Examples of text rotation angles.

PROFESSIONAL PERSPECTIVE

LINES

Quality line work is dependent on the opaqueness of the lines. The diazo reproduction process creates an excellent print if the light is unable to pass through the lines. If the line does not make an opaque image, you will not get a good print. Do not blame it on the print machine. You cannot make poor lines look good if they are not opaque. The job is difficult when drawing with pencil on vellum. You need the right combination of pressure, lead hardness, and skill to produce properly executed lines. This combination is different for each individual. Some people draw acceptable lines with an F lead while others are successful with a 2H lead. Also, one lead may not work well for all lines. You may need to have a few automatic pencils with different lead grades. If you are inking on polyester film, you will get good-quality lines, but you may have trouble getting used to working with the pitfalls of ink, such as smearing and error cleanup. Either manual technique takes a lot of practice.

Drafting with the computer is a different story, because all you need to do is make lines on the screen and then it is up to the plotting or printing process to reproduce a quality drawing on paper. CADD programs allow you to set line type, line thickness, and color as desired. You can also plot in different colors if you want to emphasize a part of the drawing, if a color plotter or printer is available.

Lettering

The appearance of manual drawings can be greatly enhanced by quality freehand lettering. An otherwise good drawing looks unprofessional with poor lettering. However, good freehand lettering does not come easily for some people; it takes a lot of practice. The only substitute for practice is a natural talent. For some people, lettering comes easily.

CADD lettering is a different story. Lettering on a computer is as easy as typing, and the lettering comes out fast and is perfect every time. Another exciting aspect of CADD lettering is the many styles available. There are even lettering styles available that duplicate the artistic appearance of the best freehand architectural lettering. Freehand lettering may not be important when the computer age totally takes over the drafting industry. For now, however, lettering on manual drawings is very important.

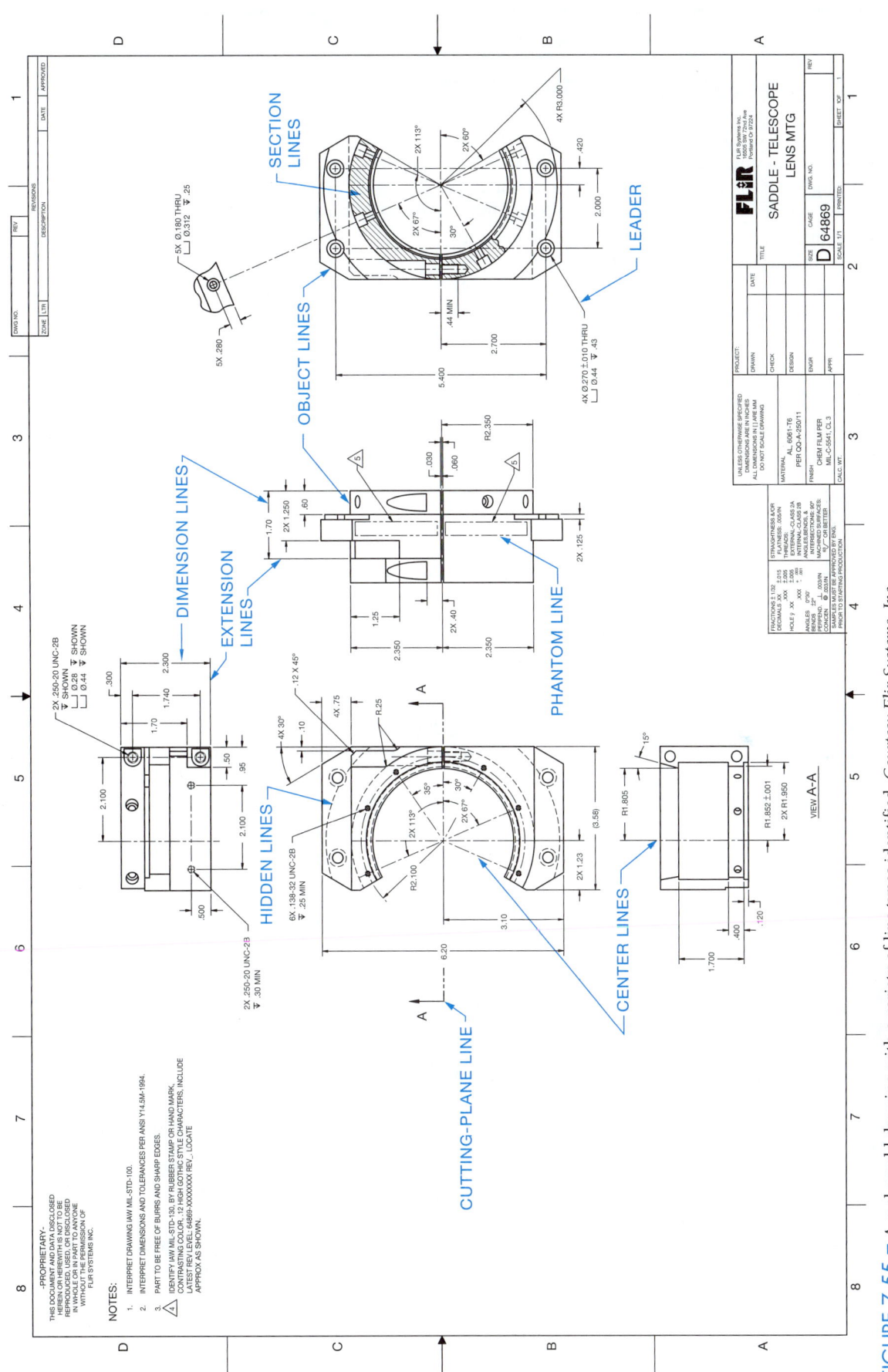

FIGURE 7.55 ■ A real-world drawing with a variety of line types identified. *Courtesy Flir Systems, Inc.*

MATH APPLICATIONS

FRACTIONAL ARITHMETIC

The United States is one of the few countries of the world that does not commonly use a metric system of measurement. It is important for a drafter to be able to handle the arithmetic of both common fractions and decimal fractions. Determine the overall height of Figure 7.56 and the overall width of fifteen of these pieces laid side to side.

The solution to overall height involves an addition of mixed numbers. Notice the use of a common denominator (16) to add the fractions:

$$2\frac{1}{16} + 1\frac{1}{4} + 2\frac{3}{8} = 2\frac{1}{16} + 1\frac{4}{16} + 2\frac{6}{16} = 5\frac{11}{16}$$

The solution to the width of fifteen pieces requires multiplication of one width × 15:

$$5\frac{7}{8} \times 15 = \frac{47}{8} \times \frac{15}{1} = \frac{705}{8} = 88\frac{1}{8}" = 7\text{-}4\frac{1}{8}"$$

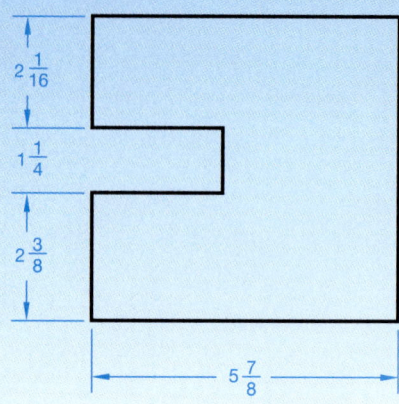

FIGURE 7.56 ■ Fractionally dimensioned part.

CHAPTER

7 *Lines and Lettering Test*

Access the CD found with this textbook to view the Chapter 7 Test. Confirm the preferred submittal method with your instructor.

CHAPTER

7 *Lines and Lettering Problems*

PART 1 • LINE PROBLEMS
DIRECTIONS

1. Using the selected engineer's layout as a guide only, make an original drawing using manual or CADD techniques as required by your course objectives. Select an appropriate scale. Draw *only* the object lines, centerlines, hidden lines, and phantom lines as appropriate for each problem. Do not dimension. Keep in mind that the engineer's sketches are rough and not meant for tracing.

2. Use construction lines (very lightly drawn) to prepare the entire drawing. When satisfied with the product, darken the drawing using proper line technique.

3. Complete the title block.

   a. The title of the drawing is given.
   b. The material the part is made of is given.
   c. The drawing number is the same as the problem number.
   d. Specify the scale and other unspecified information.

PROBLEM 7.1 **Object lines (in.)**

Part Name: Plate

Material: .25-in.-thick Mild Steel

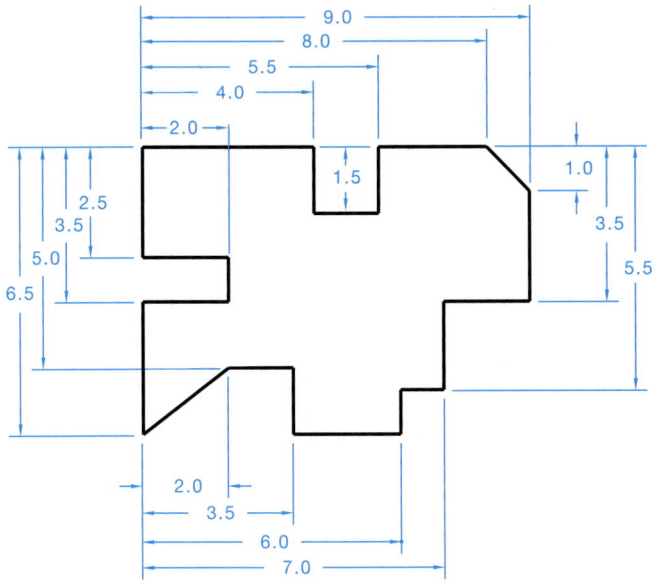

PROBLEM 7.2 **Straight object lines only (in.)**

Part Name: Milk Stencil

Material: .015-in.-thick wax-coated cardboard

Used as a stencil to spray paint identification on crates of milk.

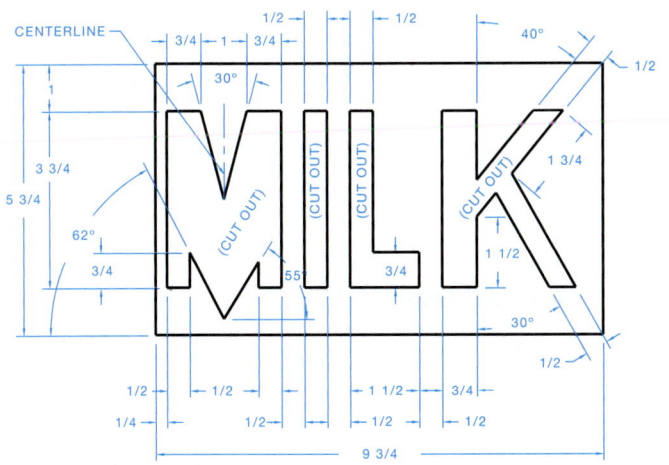

PROBLEM 7.3 **Arcs, object lines, and centerlines (in.)**

Part Name: Latch

Material: .25-in.-thick Mild Steel

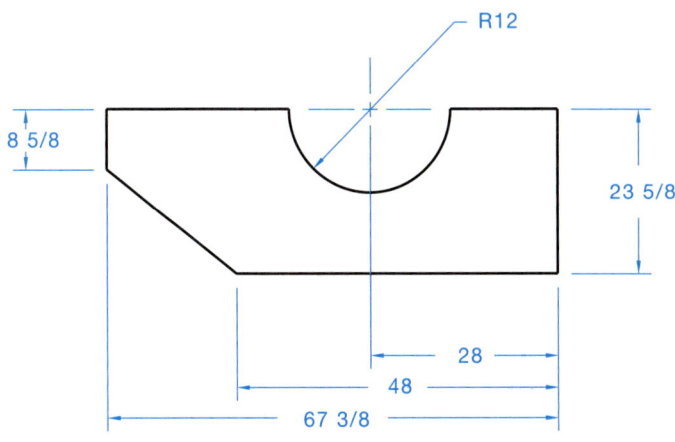

PROBLEM 7.4 **Circle, object lines, and centerlines (in.)**

Part Name: Stove Back

Material: .25-in.-thick Mild Steel

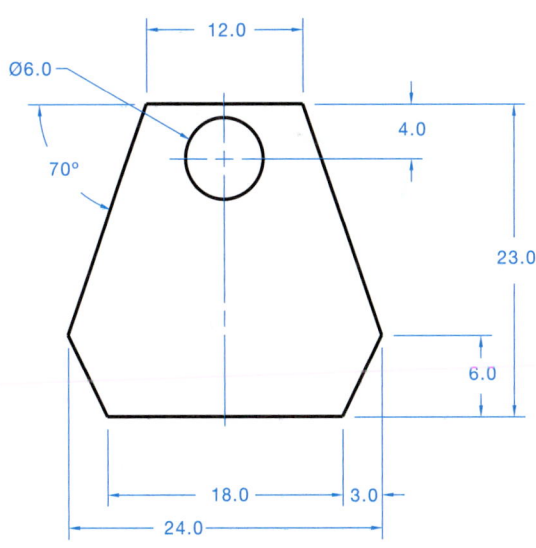

PROBLEM 7.5   **Circle and arc object lines and center-lines (in.)**

Part Name: Bogie Lock

Material: .25-in.-thick Mild Steel

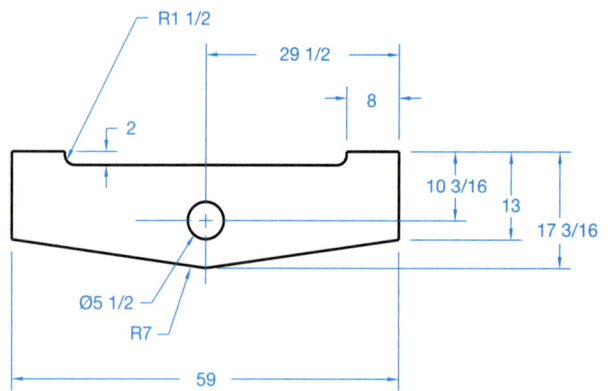

 PART 2, PROBLEMS 7.6 through 7.14: Access the CD found with this textbook and open the problem of your choice, or as assigned by your instructor. Solve the problem or problems using the same instructions provided for other problems in this chapter, unless otherwise specified.

PART 3 • LETTERING PROBLEMS DIRECTIONS

Use vertical freehand Gothic lettering, or computer-aided drafting as required by the specific instructions for each assignment. Do all freehand lettering on an A-size drawing sheet with .12-in. (3 mm) letters. Space lines of lettering .12 in. apart unless otherwise specified by your instructor. Use guidelines for all freehand lettering.

PROBLEM 7.15   Use vertical Gothic freehand lettering to letter the following statement:

THE QUALITY OF THE FREEHAND LETTERING GREATLY AFFECTS THE APPEARANCE OF THE ENTIRE DRAWING. DRAFTERS LETTER IN PENCIL OR INK. PROPER FREEHAND LETTERING IS DONE WITH A SOFT, SLIGHTLY ROUNDED POINT IN A MECHANICAL DRAFTING PENCIL, OR A 0.5 MM AUTOMATIC PENCIL WITH 2H, H, OR F GRADE LEAD DEPENDING UPON INDIVIDUAL PRESSURE. LETTERING IS DONE BETWEEN VERY LIGHTLY DRAWN GUIDELINES. THESE GUIDELINES ARE DRAWN PARALLEL AND SPACED EQUAL TO THE HEIGHTS OF THE LETTERS. GUIDELINES HELP TO KEEP YOUR LETTERING UNIFORM IN HEIGHT. LETTERING STYLES MAY VARY BETWEEN COMPANIES SOME COMPANIES REQUIRE THE USE OF LETTERING DEVICES.

PROBLEM 7.16   Use vertical Gothic freehand lettering to letter the following statement:

MOST MECHANICAL DRAFTING THAT DOES NOT USE CAD DOES USE VERTICAL FREEHAND LETTERING. THE QUALITY OF THE FREEHAND LETTERING GREATLY AFFECTS THE APPEARANCE OF THE ENTIRE DRAWING. MANY MECHANICAL DRAFTING TECHNICIANS USE FREEHAND LETTERING WITH PENCIL ON VELLUM OR WITH POLYESTER LEAD ON MYLAR. LETTERING IS COMMONLY DONE WITH A SOFT, SLIGHTLY ROUNDED LEAD IN A MECHANICAL PENCIL OR A 0.5 MM LEAD IN AN AUTOMATIC PENCIL. LETTERING IS DONE BETWEEN VERY LIGHTLY DRAWN GUIDELINES. GUIDELINES ARE SPACED PARALLEL AT A DISTANCE EQUAL TO THE HEIGHT OF THE LETTERS. GUIDELINES ARE REQUIRED TO HELP KEEP ALL LETTERS THE SAME UNIFORM HEIGHT.

PROBLEM 7.17   Use vertical Gothic freehand lettering and a CADD system to letter the following notes:

1. INTERPRET DIMENSIONS AND TOLERANCES PER ASME Y14.5M-1994.

2. UNLESS OTHERWISE SPECIFIED, ALL DIMENSIONS ARE IN MILLIMETERS.

3. REMOVE ALL BURRS AND SHARP EDGES.

4. ALL FILLETS AND ROUNDS R6.

5. CASEHARDEN 62 ROCKWELL C SCALE.

6. AREAS WHERE MATERIAL HAS BEEN REMOVED SHALL HAVE SMOOTH TRANSITIONS AND BE FREE OF SCRATCHES, GRIND MARKS, AND BURRS.

7. FINISH BLACK OXIDE.

8. PART TO BE CLEAN AND FREE OF FOREIGN DEBRIS.

Compare the difference between freehand and CADD lettering as to speed and appearance.

MATH PROBLEMS

 PROBLEMS 7.18 through 7.22: Access the CD found with this textbook and open the math problem of your choice, or as assigned by your instructor. Solve the problem or problems using the instructions provided.

CHAPTER 8

Geometric Construction

LEARNING OBJECTIVES

After completing this chapter, you will:

- Draw parallel and perpendicular lines.
- Construct bisectors and divide lines and spaces into equal parts.
- Draw polygons.
- Draw tangencies.
- Draw ellipses.
- Solve an engineering problem by making a formal drawing with geometric constructions from an engineer's sketch or layout.

THE ENGINEERING DESIGN APPLICATION

You should always approach an engineering drafting problem in a systematic manner. Plan how you propose to solve the problem. As an entry-level drafter you need to do this planning with sketches and written notes. These sketches and notes help you decide:

- The scale to use to effectively display the drawing. If the scale is too small, complex detail can be lost, but if the scale is too large, the drawing can take up too much space and cause you to spend too much time creating the drawing.

- What paper size or CADD drawing area should be established? This depends on the final plotted scale of the drawing, the amount of detail, and the specifications required.

Warning! Engineering sketches are often difficult to read. This is typical because engineers normally do not have the time or the skills to prepare a very neat and accurate sketch. Actual engineering sketches may be out of proportion and may be missing information. The dimensions frequently do not comply with ASME or other related standards. It is your responsibility to convert the engineer's communication into a formal drawing that is accurate and drawn to proper standards.

After the initial decisions about scale and paper size are made, you need to proceed with the problem in a logical manner. Refer to the engineer's sketch shown in Figure 8.1 as you follow these layout steps for either manual drafting or CADD:

STEP 1 Do all preliminary manual work using construction lines so if you make an error, it is easy to erase. Begin by establishing the centers of the ⌀57 and ⌀25.5 circles and then draw the circles.

STEP 2 Draw the concentric circles of ⌀71.5 and ⌀40.

STEP 3 Locate and draw the 6X R7.5 arcs and then using tangencies blend the R3 radii with the outside arcs.

STEP 4 Draw the 2X R7 arcs tangent to the large inside arcs.

STEP 5 Draw the centerlines for the 6X ⌀7 circles and then draw the circles in place.

STEP 6 Darken all thin lines and then darken all object lines. If there are arcs and straight lines, draw the arcs first and then blend the straight lines into the tangencies. The finished drawing is shown in Figure 8.2.

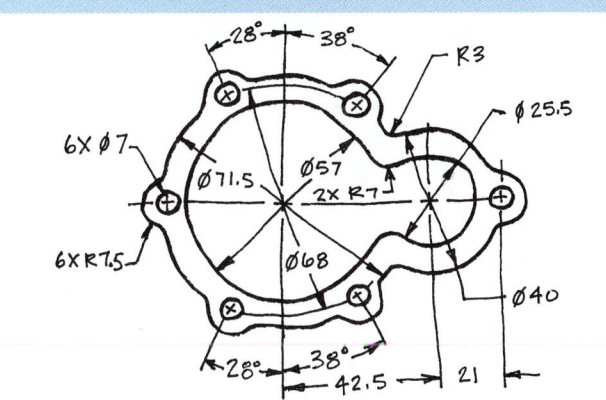

FIGURE 8.1 ■ Engineer's sketch.

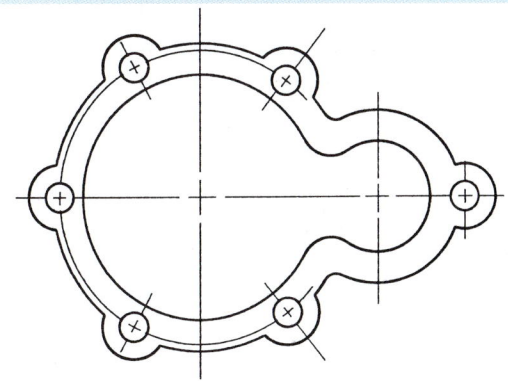

FIGURE 8.2 ■ The complete drawing (without dimensions) for engineer's sketch shown in Figure 8.1.

CADD APPLICATIONS

PREPARING A DRAWING FROM A SKETCH

Sometimes preparing a CADD drawing from an engineering sketch can seem overwhelming. Other times you may find that you could have used other tools, techniques, commands, or procedures to create a drawing in less time. If you spend some time planning and laying out your drawing in advance, it can help to reduce these issues.

If the engineering sketch is good quality, you can use it to plan your CADD layout and create a drawing according to correct drafting standards. However, if the engineering sketch is rough and does not represent your plans for making a professional-looking drawing, then you may want to prepare a new sketch before using CADD. Your sketch can be as simple or detailed as necessary and might include dimensions drawn according to ASME or other related standards.

Another step you may want to consider is the development of a project outline. Like your sketch, an outline can be as general or complete as you feel necessary. An outline, like the example of a solid model part development shown in Figure 8.3, can act as a setup plan and progress checklist. Planning with a sketch or an outline (or both) can save you time in completing the project, because you have a good idea of what to do before you start working with CADD and what to do during the drawing process.

Project: DF24-0004-07
File: C:/DF24 WING/FLAP SUPPORT/SHAFT
Date: 08-20-06
Scale: 2:1
Sheet: C

Outline

1. Open or create a model project.
2. Open a part file template, and create a new part file.
3. Define work and sketch environment settings.
4. Project the origin onto the sketch plane.
5. Draft sketch geometry (initial base feature outline).
6. Add geometric constraints and dimensions.
7. EXTRUDE an initial feature from the sketch profile.
8. Draft sketch geometry on the base feature, and REVOLVE the sketch.
9. Add HOLE.
10. Place CHAMFER and FILLETS.
11. Resave.

FIGURE 8.3 ■ A CADD project outline.

GEOMETRIC CONSTRUCTION

Machine parts and other manufactured products are made up of different geometric shapes ranging from squares and cylinders to complex irregular curves. Geometric constructions are methods that can be used to draw various geometric shapes or to perform drafting tasks related to the geometry of product representation and design. The proper use of geometric constructions requires a basic understanding of plane geometry. When using geometric construction techniques, extreme accuracy and the proper use of drafting instruments are important. Pencil or compass leads should be sharp and instruments should be in good shape. Always use very lightly drawn construction lines for all preliminary work.

When computers are used in drafting, the task of creating most geometric construction related drawings becomes much easier, although the theory behind the layout of geometric shapes and related constructions remains the same.

CHARACTERISTICS OF LINES

A straight line segment is a line of any given length, such as lines A–B shown in Figure 8.4.

A curved line may be in the form of an arc with a given center and radius or an irregular curve without a defined radius as shown in Figure 8.5.

Two or more lines may intersect at a point as in Figure 8.6. The opposite angles of two intersecting lines are equal; a = a, b = b.

Parallel lines are lines equidistant throughout their length, and if they were to extend indefinitely they would never cross. (See Figure 8.7.)

FIGURE 8.4 ■ Horizontal and vertical lines.

FIGURE 8.5 ■ Arc and irregular curve.

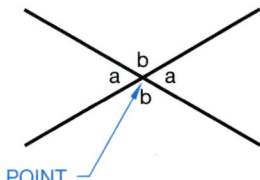

FIGURE 8.6 ■ Intersecting lines.

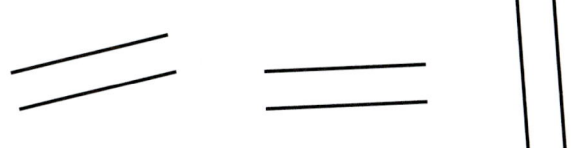

FIGURE 8.7 ■ Parallel lines.

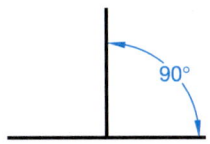

FIGURE 8.8 ■ Perpendicular lines.

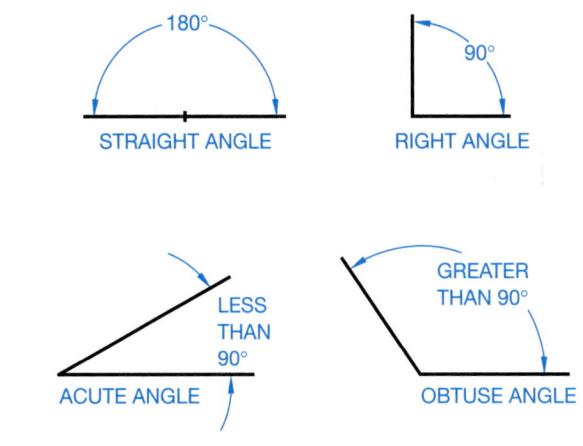

FIGURE 8.9 ■ Types of angles.

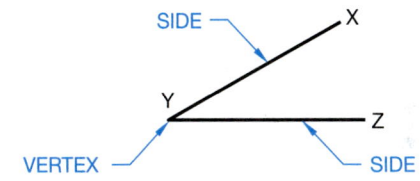

FIGURE 8.10 ■ Parts of an angle.

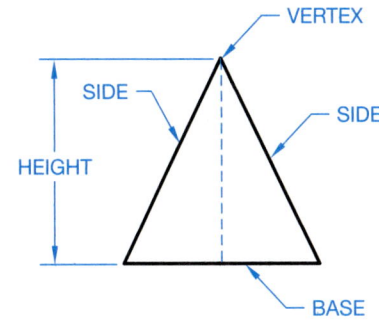

FIGURE 8.11 ■ Parts of a triangle.

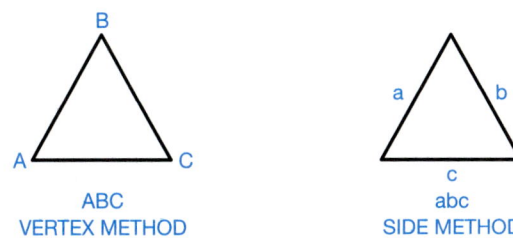

FIGURE 8.12 ■ Labeling triangles.

Perpendicular lines intersect at a 90° angle as shown in Figure 8.8.

GEOMETRIC SHAPES

Angles

Angles are formed by the intersection of two lines. Angles are sized in **degrees** (°). Components of a degree are **minutes** (') and **seconds** ("). There are 60 minutes (') in one degree and there are 60 seconds (") in one minute. 1° = 60', 1' = 60". The four basic types of angles are shown in Figure 8.9. A straight angle equals 180°, while a right angle equals 90°. An acute angle contains more than 0° but less than 90°, and an obtuse angle contains more than 90° but less than 180°.

The parts of an angle are shown in Figure 8.10. An angle is labeled by giving the letters defining the line ends and vertex, with the vertex always between the ends. An example is angle XYZ where Y is the vertex.

Triangles

A **triangle** is a geometric figure formed by three intersecting lines creating three angles. The sum of the interior angles always equals 180°. The parts of a triangle are shown in Figure 8.11. Triangles are labeled by lettering the vertex of each angle, such as ABC, or by labeling the sides abc as shown in Figure 8.12.

There are three kinds of triangles: acute, obtuse, and right, as shown in Figure 8.13. Two special acute triangles are the equi-lateral, which has equal sides and angles, and the isosceles, which has two equal sides and angles. An isosceles triangle may also be obtuse. In a scalene triangle, no sides or angles are equal.

Right triangles have certain unique geometric characteristics. Two internal angles equal 90° when added. The side opposite the 90° angle is called the **hypotenuse**, as shown in Figure 8.14. A

EQUILATERAL
TRIANGLE—
ALL SIDES AND
ANGLES EQUAL

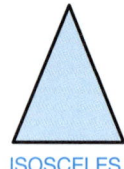
ISOSCELES
TRIANGLE—
TWO SIDES
AND TWO
ANGLES EQUAL

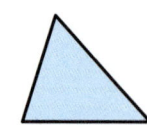
ACUTE SCALENE
TRIANGLE—
NO EQUAL SIDES
OR ANGLES

ACUTE TRIANGLES
(NO INTERIOR ANGLE IS GREATER THAN 90°)

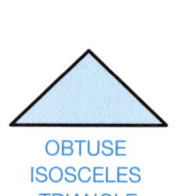
OBTUSE
ISOSCELES
TRIANGLE

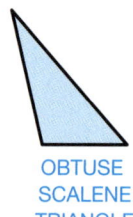
OBTUSE
SCALENE
TRIANGLE

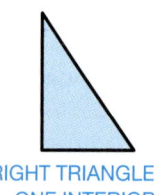
RIGHT TRIANGLE—
ONE INTERIOR
90° ANGLE

OBTUSE TRIANGLES
(ONE INTERIOR ANGLE GREATER THAN 90°)

FIGURE 8.13 ■ Types of triangles.

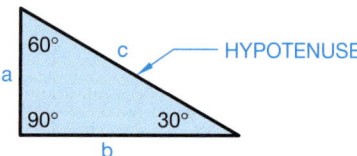

FIGURE 8.14 ■ Right triangle.

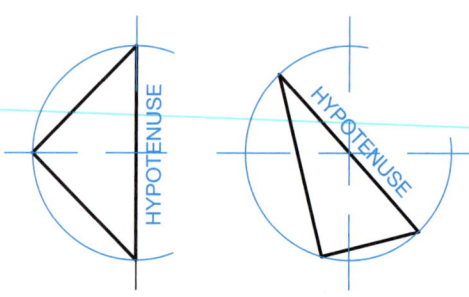

FIGURE 8.15 ■ A semicircle is formed when an arc is drawn through the vertices of a right triangle.

semicircle (half circle) is always formed when an arc is drawn through the vertices of a right triangle, as shown in Figure 8.15.

Quadrilaterals

Quadrilaterals are four-sided polygons that can have equal or unequal sides or interior angles. The sum of the interior angles is 360°. Quadrilaterals with parallel sides are called **parallelograms**. (See Figure 8.16.)

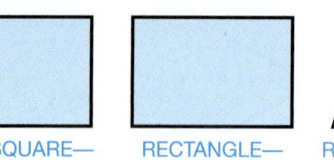
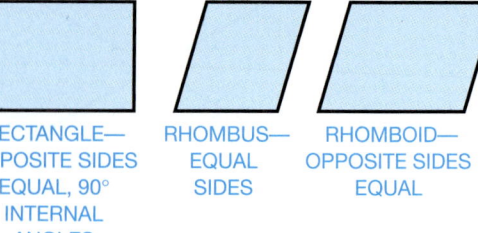

SQUARE—
EQUAL
SIDES, 90°
INTERNAL
ANGLES

RECTANGLE—
OPPOSITE SIDES
EQUAL, 90°
INTERNAL
ANGLES

RHOMBUS—
EQUAL
SIDES

RHOMBOID—
OPPOSITE SIDES
EQUAL

PARALLELOGRAMS

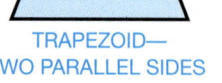

TRAPEZOID—
TWO PARALLEL SIDES

TRAPEZIUM—
NO PARALLEL SIDES

FIGURE 8.16 ■ Quadrilaterals.

Regular Polygons

Some of the most commonly drawn geometric shapes are regular polygons. **Regular polygons** have equal sides and equal internal angles. Polygons are closed figures with at least three sides and up to any number of sides. The relationship of a circle to a regular polygon is that a circle may be drawn to touch the corners (circumscribed) or the sides (inscribed), known as flats, of a regular polygon. (See Figure 8.17.) This relationship is an advantage in constructing regular polygons. Some common regular polygons are shown in Figure 8.18.

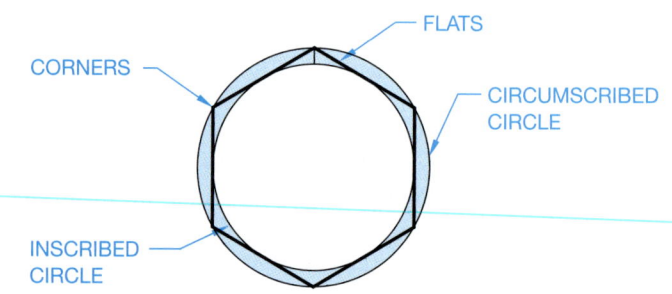

FIGURE 8.17 ■ Regular polygon.

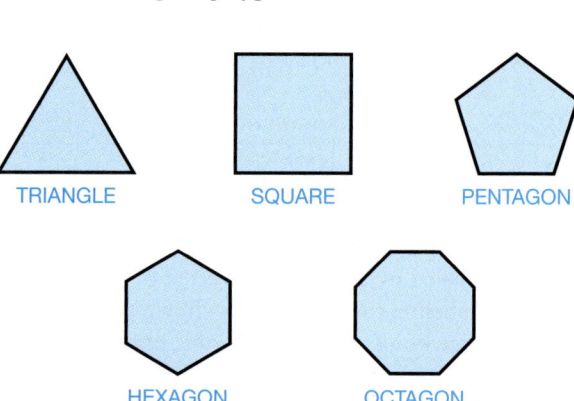

FIGURE 8.18 ■ Common regular polygons.

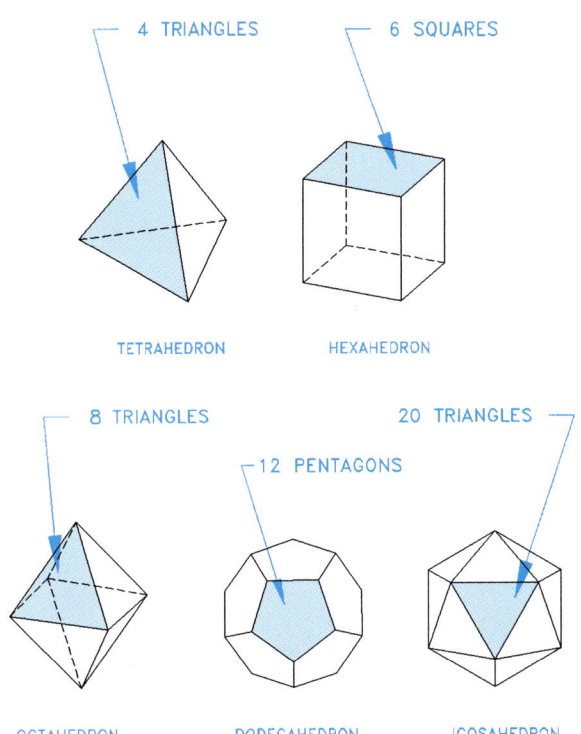

FIGURE 8.19 ■ Regular solids.

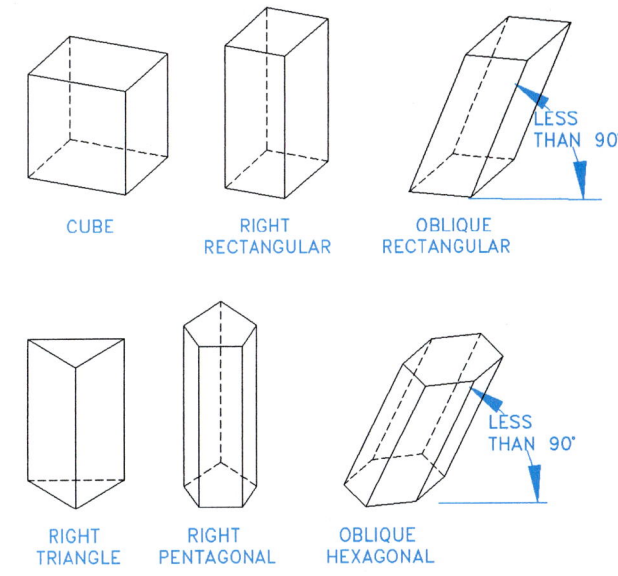

FIGURE 8.20 ■ Common prisms.

Regular Solids

Solid objects constructed of regular polygon surfaces are called regular polyhedrons. (See Figure 8.19.) A polyhedron is a solid formed by plane surfaces. The surfaces are referred to as faces.

Prisms

A prism is a geometric solid object with ends that are the same size and shaped polygons and sides that connect the same corresponding corners of the ends. Figure 8.20 shows a few common examples. A right prism has sides that meet 90° with the ends. An oblique prism has sides that are at an angle to the ends.

Pyramid Prisms

A pyramid prism has a regular polygon-shaped base and sides that meet at one point called a vertex as shown in Figure 8.21. A geometric solid is truncated when a portion is removed and a plane surface is exposed. The axis is an imaginary line that connects the vertex to the midpoint of the base. It can also be referred to as an imaginary line around which parts are regularly arranged.

Circles

A circle is a closed curve with all points along the curve at an equal distance from a point called the center. The circle has a total of 360°.

The following describes the primary parts of a circle and two circle relationships. These parts and relationships are also shown in Figure 8.22:

■ Center is located in the exact middle of the circle.

■ Circumference is the distance around the circle, or the circle edge.

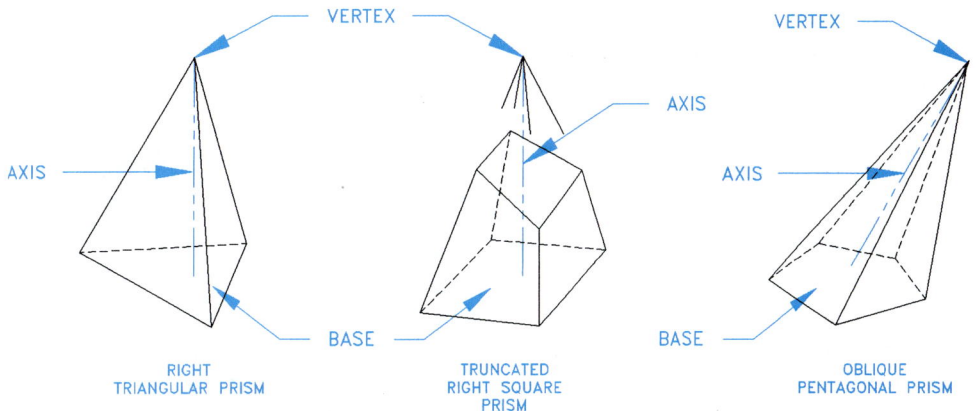

FIGURE 8.21 ■ Common pyramid prisms.

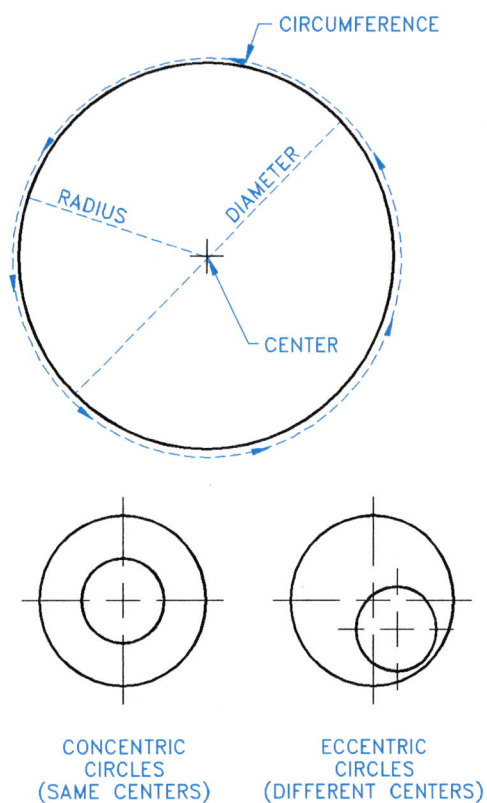

FIGURE 8.22 ■ Parts of a circle and circle characteristics.

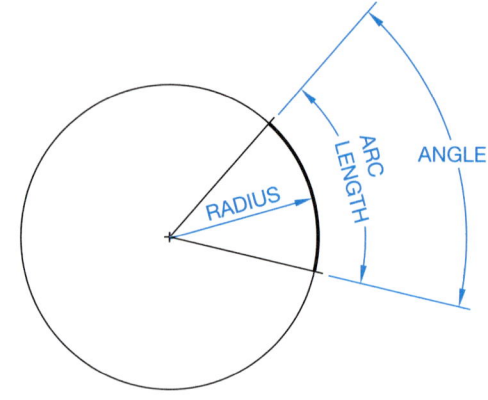

FIGURE 8.23 ■ Arc.

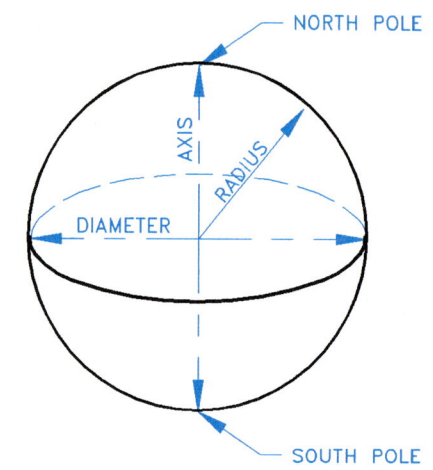

FIGURE 8.24 ■ Sphere.

■ **Diameter** is the distance across the circle, from one side to another, through the center. Diameter is identified using the diameter symbol: $\varnothing$.

■ **Radius**, identified using an uppercase **R**, is the distance from the center of the circle to the circumference. A circle's radius is always half of the diameter.

■ **Concentric circles** share the same center.

■ **Eccentric circles** have different centers.

An **arc** is part of the circumference of a circle. The arc can be identified by a radius, an angle, or a length. (See figure 8.23.)

Spheres

A **sphere** is three-dimensional and is the shape of a ball. Every point on the surface of a sphere is equidistant from the center. If you think of the earth as a sphere, the North and South poles are at the end of its axis as shown in Figure 8.24.

Tangents

Straight or curved lines are **tangent** to a circle or arc when the line touches the circle or arc at only one point. If a line were to connect the center of the circle or arc to the point of tangency, the tangent line and the line from the center would form a 90° angle. (See Figure 8.25.)

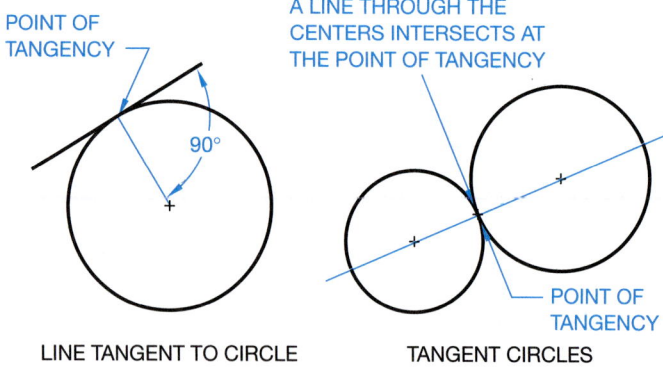

FIGURE 8.25 ■ Tangency.

COMMON GEOMETRIC CONSTRUCTIONS

Parallel Lines

Parallel lines are lines evenly spaced at all points along their length and will not intersect even when extended. Figure 8.26 shows an example of parallel lines. The space between parallel lines can be any distance.

FIGURE 8.26 ■ Parallel lines.

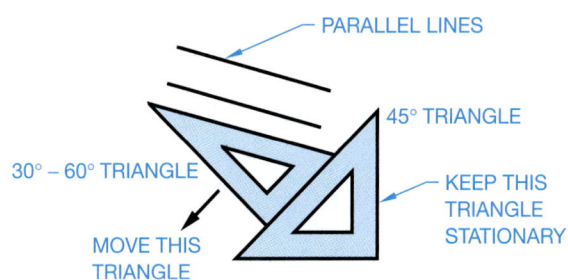

FIGURE 8.27 ■ Constructing parallel lines with triangles.

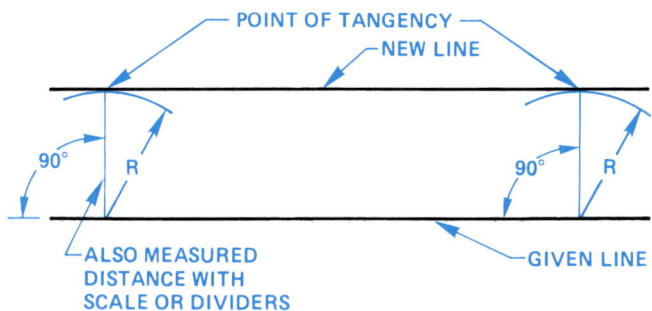

FIGURE 8.28 ■ Constructing parallel lines with a straightedge and compass.

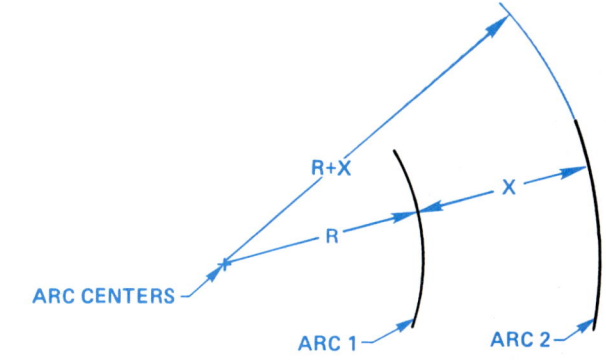

FIGURE 8.29 ■ Parallel arcs.

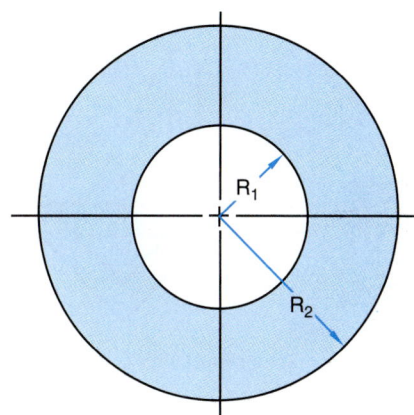

FIGURE 8.30 ■ Concentric circles.

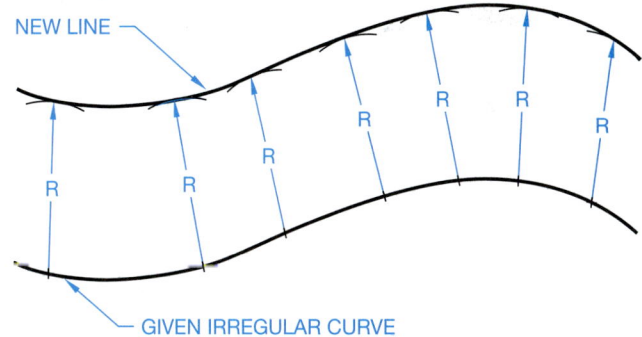

FIGURE 8.31 ■ Constructing parallel irregular curves.

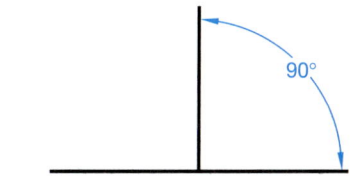

FIGURE 8.32 ■ Perpendicular lines.

Parallel lines can be drawn horizontally, vertically, or at any angle using a drafting machine. When a drafting machine is not available, parallel lines can be drawn with the aid of standard triangles as shown in Figure 8.27. Parallel lines can be drawn with a straightedge and compass when the distance between lines is established. Use the compass radius to draw arcs near the ends of the given line. The parallel line is then drawn at the points of tangency of the arcs as shown in Figure 8.28. Draw very light construction lines for all construction layout.

Parallel or concentric arcs can be drawn where the distance between arcs is equal. Establish parallel arcs by adding the radius of Arc 1 (R) to the distance between the arcs (X), as shown in Figure 8.29.

Concentric circles are parallel circles drawn from the same center. The distance between circles is equal, as shown in Figure 8.30.

Parallel irregular curves are drawn any given distance (R) apart by using a compass set at the given distance. A series of arcs is lightly drawn along the given curve at the required distance. The line to be drawn parallel is then constructed with an irregular curve at the points of tangency of the construction arcs. (See Figure 8.31.)

Perpendicular Lines

Perpendicular lines intersect at 90°, or a right angle, as seen in Figure 8.32. Perpendicular lines are drawn with the horizontal

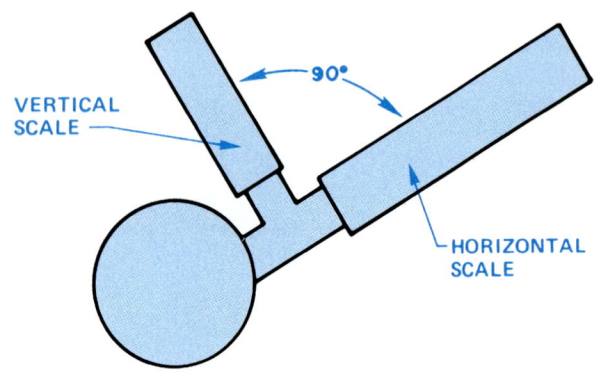

FIGURE 8.33 ■ Perpendicular drafting machine scales.

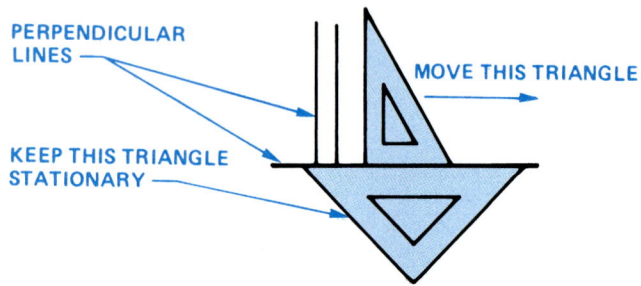

FIGURE 8.34 ■ Perpendicular lines with triangles.

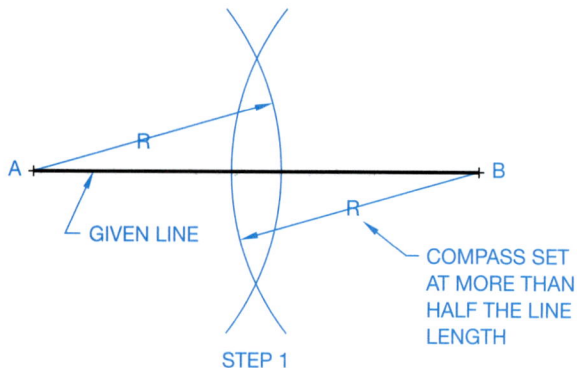

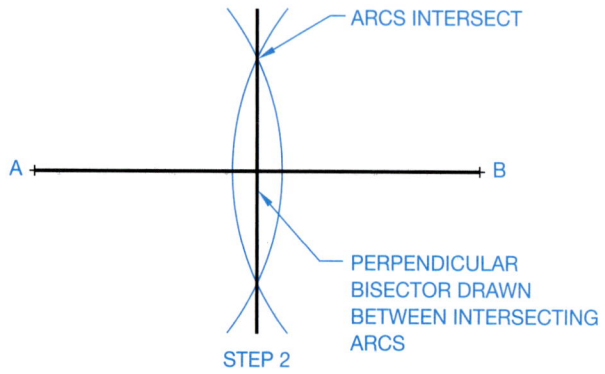

FIGURE 8.35 ■ Constructing a perpendicular bisector.

and vertical scales of a drafting machine, even if the scales are set at an angle away from horizontal, as seen in Figure 8.33. Perpendicular lines can also be drawn with a straightedge and triangle or two triangles as shown in Figure 8.34.

Perpendicular Bisector

The perpendicular bisector of a line can be found using a straightedge and a compass. A bisector is a line that divides an object into two equal parts.

STEP 1   Set the compass radius more than halfway across the given line segment and draw two intersecting arcs from the line ends.

STEP 2   Connect the points where the arcs intersect, as in Figure 8.35.

Bisecting an Angle

Any given angle can be divided into two equal angles using a compass and straightedge.

STEP 1   Adjust the compass to any radius (R₁) and draw an arc as shown in Figure 8.36.

STEP 2   At the points where the arc intersects the sides of the angle, draw two arcs of equal radius (R₂).

STEP 3   Connect a straight line from the vertex of the angle to the point of intersection of the two arcs.

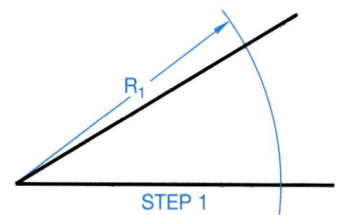

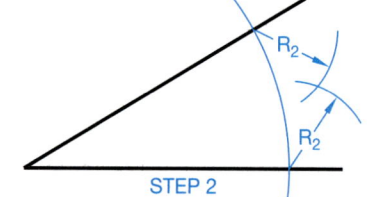

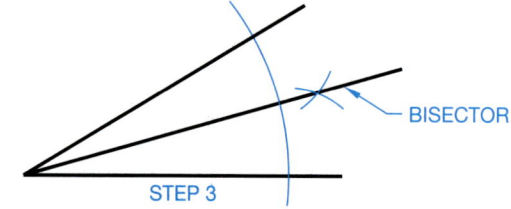

FIGURE 8.36 ■ Bisecting an angle.

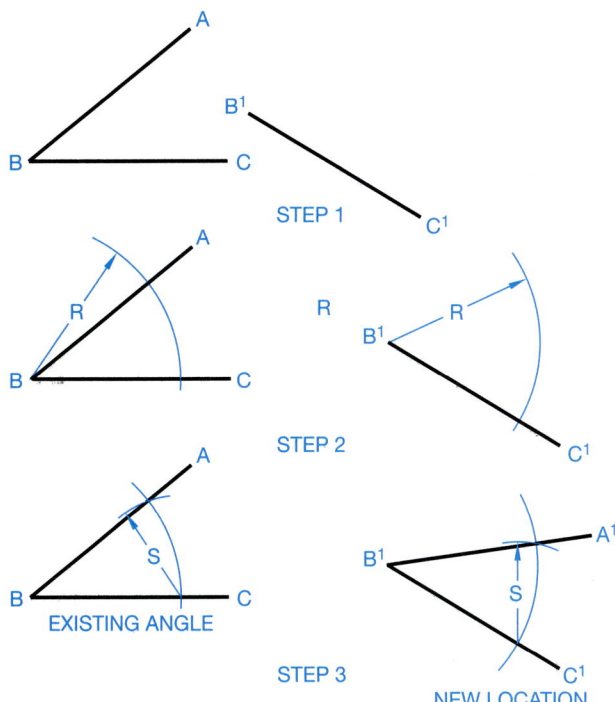

FIGURE 8.37 ■ Transferring a given angle to a new location.

Transferring an Angle

A given angle can be transferred to a new location as shown in Figure 8.37. This method also works well for transferring a triangle.

STEP 1 Given angle ABC, draw line B^1C^1 in the new location and position.

STEP 2 Draw any arc, R, with its center at B on the given angle. Transfer and draw arc R with its center at point B^1.

STEP 3 Using the intersection of one side of the given angle and arc R as center, set a compass distance to the opposite intersection of arc R. Then transfer the new radius, S, to the new position using the intersection of B^1C^1 and arc R as center. Connect a line from B^1 to the intersection of the two arcs, R and S, as shown in Figure 8.37.

Dividing a Line into Equal Parts

Any given line can be divided into any number of equal parts. Divide given line AB, shown in Figure 8.38, into eight equal parts.

STEP 1 Draw a construction line at any angle from either end of line AB as shown in Figure 8.39.

STEP 2 Use a scale or dividers to divide the angled construction line into eight equal parts. Use any size divisions extending down the angled line to approximately the

FIGURE 8.38 ■ Given line AB to divide into equal parts.

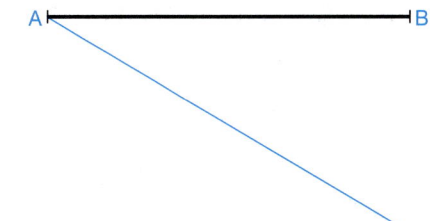

FIGURE 8.39 ■ Step 1, angled construction line.

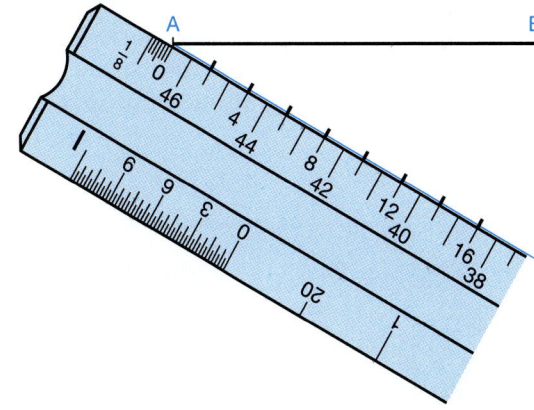

FIGURE 8.40 ■ Step 2, divide angled line into required number of equal parts. A divider is also commonly used to divide the angled line into equal parts.

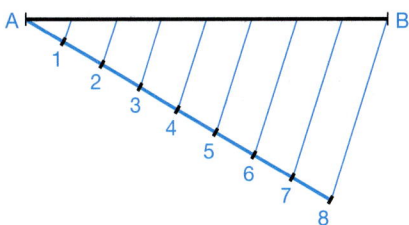

FIGURE 8.41 ■ Step 3, connect parallel line segments.

length of the given line. (See Figure 8.40.) Using a divider is generally more accurate, but using a scale can be accurate if you are careful.

STEP 3 Connect the last point (8) of the construction line to point B of the given line. Then draw lines parallel to line 8B from each of the numbered points on the construction line to the given line AB. (See Figure 8.41.) You now have the given line, AB, divided into eight equal parts. This same process can be used for any number of equal parts.

Dividing a Space into Equal Parts

You can divide any given space into any number of equal parts. Given the space shown in Figure 8.42, divide it into 12 equal parts.

STEP 1 Between the lines that establish the given space, place a scale with the required number of increments (12) so that

GIVEN SPACE

FIGURE 8.42 ■ Dividing a given space into equal parts.

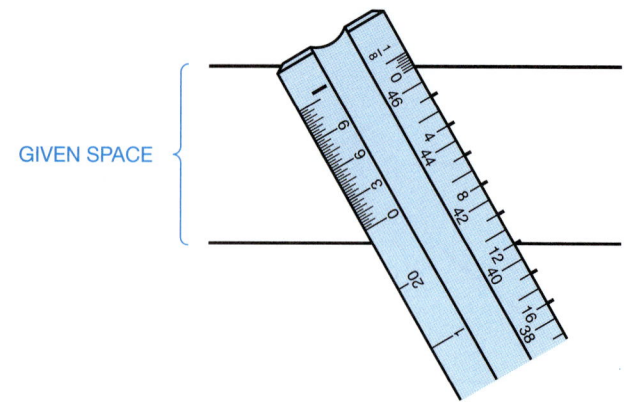

GIVEN SPACE

the zero on the scale is on either line and the 12 is on the other line. Mark each increment. (See Figure 8.43.)

STEP 2   Remove the scale. Each increment you have marked has divided the given space into 12 equal spaces. Figure 8.44 shows parallel lines drawn at each mark to complete the space division.

FIGURE 8.43 ■ Step 1, place a scale with required number of increments between the lines.

COMMON GEOMETRIC CONSTRUCTIONS

CADD APPLICATIONS

Understanding plane geometry and geometric construction is critical to the development of both manual and CADD drawings. However, the process of applying geometric constructions is typically much easier and more accurate with CADD because of the multitude of geometric construction tools available with most systems. Still, the more you know about geometric characteristics and shapes, the more effectively and efficiently you can prepare quality drawings.

The tools and commands available to complete geometric constructions differ with each CADD program, but the techniques are often very similar. The following information describes how CADD can be used to apply the common geometric constructions previously defined in this chapter.

■ Parallel lines can be created using options such as: PARALLEL object snap, OFFSET command, and PARALLEL constraint or relation. These tools typically require you to select an existing line or shape and define an offset distance, pick a line or object and enter the parallel line length, or pick two non-parallel lines.

■ Perpendicular lines are drawn with options including: PERPENDICULAR object snap, PERPENDICULAR object snap tracking, and PERPENDICULAR constraint or relation. These tools usually require you to select an existing line or shape and specify the length of the perpendicular line, or pick two non-perpendicular lines.

■ Perpendicular bisectors can be drawn with a number of different tools such as a combination of MIDPOINT object snap and PERPENDICULAR object snap tracking, or MIDPOINT and PERPENDICULAR constraints or relations. These options typically re-

quire you to select the midpoint of an existing line or shape and define the length of the perpendicular line.

■ Bisecting an angle is most effectively accomplished using specific bisecting tools such as the BISECT option of the XLINE command. However, a variety of other CADD tools exist that allow you to bisect an angle including ANGULAR dimensions and SYMMETRIC constraints.

■ Transferring an angle is quickly and easily done using CADD commands such as: ROTATE, a combination of ROTATE and MOVE, or ALIGN.

■ Dividing a line into equal parts is accomplished with tools such as the DIVIDE command. The DIVIDE command allows you to pick a line and enter the number of segments you want the line to be divided into. Reference points are then placed at each segment, but the line is not actually cut into pieces.

■ Dividing a space into equal parts can be done using a number of tools, depending on the application and given information. In many cases, several commands are required to accomplish this task. Typical tools used for dividing a space into equal parts include the DIVIDE, OFFSET, or COPY commands, and the ARRAY or RECTANGULAR PATTERN tools available with some CADD systems.

The CADD geometric construction tools and techniques described here can save a significant amount of time and help you generate very accurate drawings. However, if a geometric construction tool or command is not available for a specific application, you can always rely on the techniques identified in this chapter. Remember, a drawing is nothing more than shapes and text, whether it was produced using manual drafting or CADD.

GIVEN SPACE

FIGURE 8.44 ■ Step 2, given space divided into equal parts.

CONSTRUCTING POLYGONS

Polygons

A polygon is any closed plane geometric figure with three or more sides or angles. Pentagons, hexagons, and octagons are polygons and are described next. You can draw any regular polygon if you know the distance across the flats and the number of sides.

There are 360° in a circle; so, if you need to draw a 12-sided polygon, divide 360° by 12 (360 ÷ 12 = 30°) to determine the central angle of each side. Since there are 30° between each side of a 12-sided polygon, divide a circle with a diameter equal to the distance across the flats into equal 30° parts as shown in Figure 8.45. Then connect the 12 radial lines with line segments that are tangent to each arc segment, as shown in Figure 8.45.

Drawing a Triangle Given Three Sides

The following technique is known as triangulation. Using given triangle sides x, y, and z in Figure 8.46, draw a triangle as shown in the following steps:

STEP 1 Lay out one of the given sides. Select a side for the base. z has been chosen as shown in Figure 8.47. From one end of line z, strike an arc equal in length to one of the other lines, x, for example, as in Figure 8.47.

STEP 2 From the other end of line z, strike an arc with a radius equal to the remaining line, y. Allow this arc to intersect the previous arc. Where the two arcs cross, draw lines to the ends of the base line, as in Figure 8.48, to complete the triangle. This method is called triangulation. Triangulation is commonly used in many disciplines and especially useful in sheet metal pattern development, in Chapter 25.

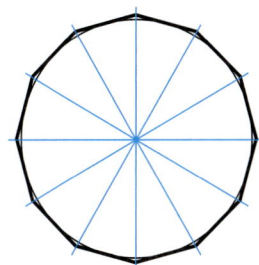

FIGURE 8.45 ■ Step 1, divide the circle into 12 parts; step 2, connect the 12 radial lines.

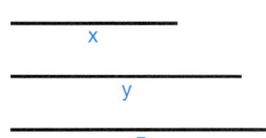

FIGURE 8.46 ■ Construct a triangle given three sides.

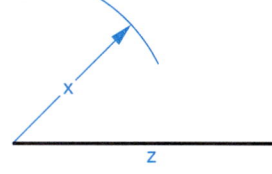

FIGURE 8.47 ■ Step 1, lay-out line z and swing arc x.

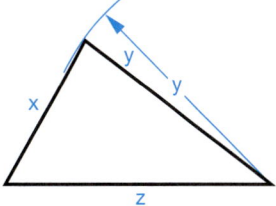

FIGURE 8.48 ■ Step 2, swing arc y.

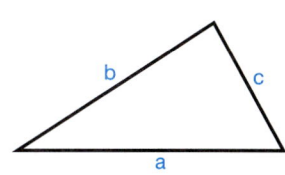

FIGURE 8.49 ■ Transfer a given triangle to a new location.

Transferring a Triangle

Any given polygon may be constructed or transferred to a new location using triangulation. Given the triangle, abc, in Figure 8.49, transfer it to a new location as shown in the following steps:

STEP 1 Transfer one of the sides of the triangle to the new location by measuring its length or using dividers. (See Figure 8.50.)

STEP 2 From one end of line a in the new location, draw an arc equal in length to line b. From the other end of line a, draw an intersecting arc equal in length to line c. (See Figure 8.51a.)

STEP 3 Connect both ends of line a to the intersection of the arcs to form the triangle in its new position as shown in Figure 8.51b.

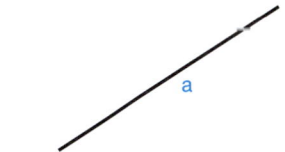

FIGURE 8.50 ■ Step 1, establish new location.

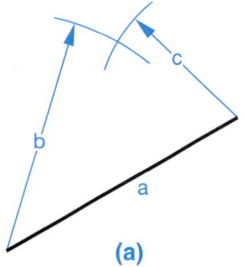

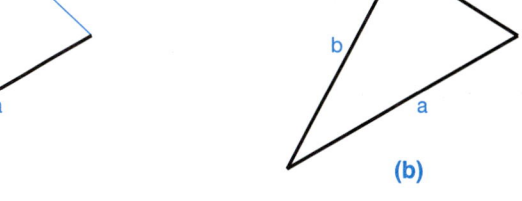

(a) **(b)**

FIGURE 8.51 ■ (a) Step 2, swing arcs b and c. (b) Step 3, connect the ends of line a to intersection of arcs.

Constructing Right Triangles

A right triangle can be drawn when the length of the two sides next to the 90° angle are given, as shown in Figure 8.52.

STEP 1 Draw line a perpendicular to line b. (See Figure 8.53.)

STEP 2 Connect a line from the end of line a to the end of line b to establish the right triangle. (See Figure 8.54.)

Draw a right triangle given the length of the hypotenuse and one side as in Figure 8.55.

STEP 1 Draw line c and establish its center. You can use the perpendicular bisector method to find the center. (See Figure 8.56.)

STEP 2 From the center of line c, draw a 180° arc with the compass point at the line center and the compass lead at the line end. (See Figure 8.57.)

STEP 3 Set the compass with a radius equal in length to the other given line, a. From one end of line c draw an arc intersecting the previous arc as shown in Figure 8.58.

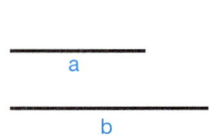

FIGURE 8.52 ■ Construct a right triangle given two sides.

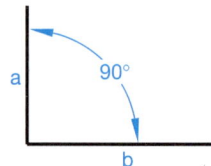

FIGURE 8.53 ■ Step 1, draw line a perpendicular to line b.

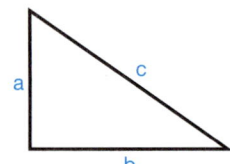

FIGURE 8.54 ■ Step 2, connect hypotenuse.

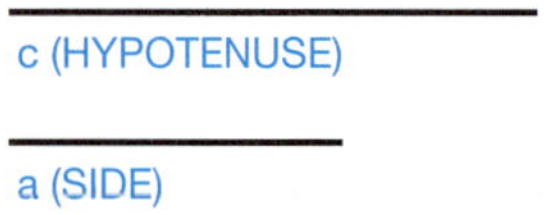

c (HYPOTENUSE)

a (SIDE)

FIGURE 8.55 ■ Construct a right triangle given a side and hypotenuse.

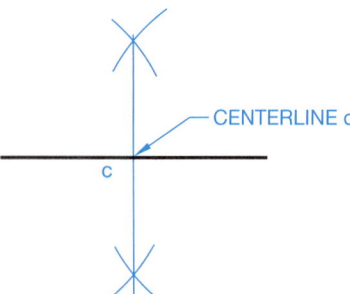

FIGURE 8.56 ■ Step 1, draw the hypotenuse and establish its center.

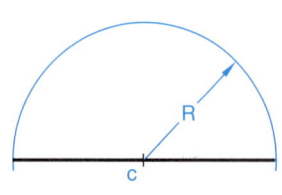

FIGURE 8.57 ■ Step 2, draw a 180° arc from the center.

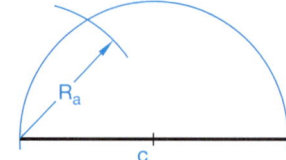

FIGURE 8.58 ■ Step 3, use radius R_a, to establish the end of line a.

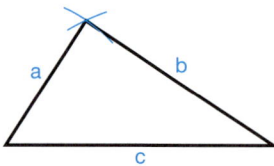

FIGURE 8.59 ■ Step 4, complete the right triangle.

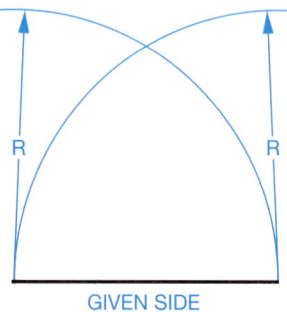

GIVEN SIDE

FIGURE 8.60 ■ Step 1, to construct an equilateral triangle given one side, swing equal arcs r.

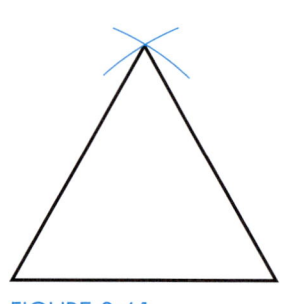

FIGURE 8.61 ■ Step 2, complete the equilateral triangle.

STEP 4 From the intersection of the two arcs in step 3, draw two lines connecting the ends of line c to complete the required right triangle. (See Figure 8.59.)

Constructing an Equilateral Triangle

An equilateral triangle can be drawn, given the length of one side, as demonstrated in the following steps.

STEP 1 Draw the given side length. Then set a compass with a radius equal to the length of the given side. Use this compass setting to draw intersecting lines from the ends of the given side as in Figure 8.60.

STEP 2 Connect the point of intersection of the two arcs to the ends of the given side to complete the equilateral triangle as shown in Figure 8.61.

Constructing Squares

Square-head bolts or square nuts are sometimes drawn in conjunction with manufactured parts. Use geometric constructions when necessary; however, when practical, use square templates or special square-head bolt and nut templates. There are several methods that can be used to draw a square, each of which relates to the characteristics of the square shown in Figure 8.62.

Draw a square with the length of one side given as follows:

STEP 1 Draw the length of the given side. Then with the drafting machine, or other appropriate method of creating a 90° angle, draw lines from each end equal in length to the given side. (See Figure 8.63.)

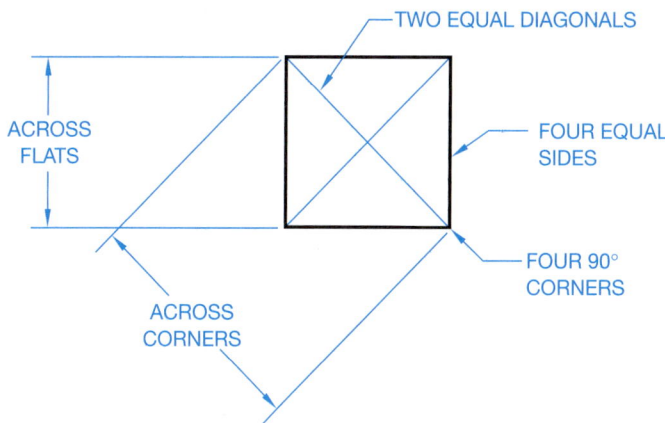

FIGURE 8.62 ■ Elements of a square.

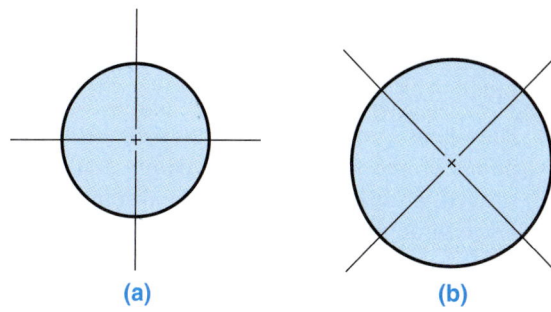

FIGURE 8.65 ■ Step 1, draw a circle to construct a square with flats or corners given.

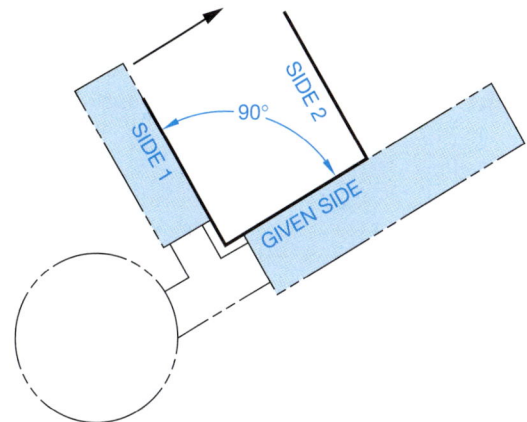

FIGURE 8.63 ■ Step 1, use the drafting machine to establish 90° angles and slides.

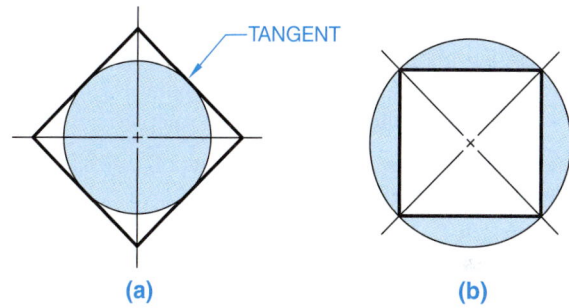

FIGURE 8.66 ■ Step 2, a circumscribed square at (a) and an inscribed square at (b).

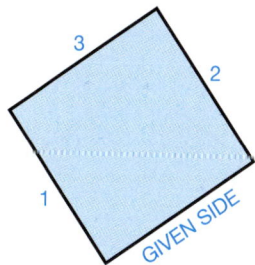

FIGURE 8.64 ■ Step 2, complete the square by drawing side 3.

STEP 2 Connect the end of sides 1 and 2 drawn in step 1 to complete the square with side 3, shown in Figure 8.64.

Draw a square with the distance across the flats or across the corners given.

STEP 1 Draw a circle equal in diameter to the distance across the flats, Figure 8.65a, and a circle with a diameter equal to the distance across the corners, Figure 8.65b. Notice that the position of the centers dictates the position of the squares.

STEP 2 Draw 45° lines tangent to the circle and extending to the centerlines as seen in Figure 8.66a. Draw lines inside the

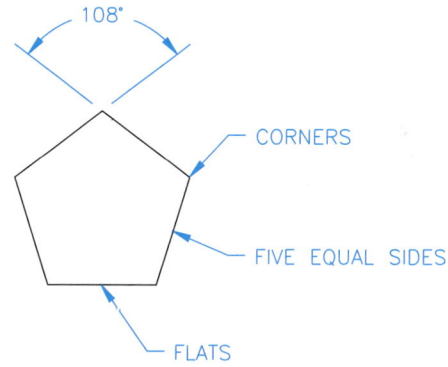

FIGURE 8.67 ■ Elements of a regular pentagon. Each side of a regular pentagon is equal and its interior angles are 108°.

circle connecting the intersections of the centerlines and the circle as seen in Figure 8.66b.

Constructing a Regular Pentagon

Pentagons are five-sided polygons. Each side of a regular pentagon is equal and its interior angles are 108° as shown in Figure 8.67. Pentagons can be tricky to draw, primarily because the angles used are not as convenient as those used with other types of polygons. Given the diameter of a circle, use the following steps to construct a pentagon:

Method 1:

STEP 1 Lightly draw a circle using the given diameter. (See Figure 8.68.)

STEP 2 Draw line 1 to define the circle's diameter. The end points of line 1 are labeled as points A and B. As described, a circle's diameter is the distance across the circle through the center. Line 1 in Figure 8.68 is horizontal but can be drawn at any angle depending on the desired angle of the pentagon.

STEP 3 If needed, bisect line 1, using the techniques previously discussed, to determine the center of the circle. The center of the circle is identified as point C in Figure 8.68.

STEP 4 Draw line 2 perpendicular to line 1, from point C to the circumference of the circle. The distance of line 2 is the radius of the circle. The intersection of line 2 and the circle's circumference is labeled as point D in Figure 8.68 and defines the first pentagon corner.

STEP 5 Bisect the distance between B and C to find point E as shown in Figure 8.68.

STEP 6 Place your compass needle at point E and draw arc 1 from point D to line 1. The intersection of the arc and line 1 is identified as point F in Figure 8.69.

STEP 7 Place your compass needle at point D and draw arc 2 from point F to the circumference of the circle. This intersection is labeled as point G in Figure 8.69 and defines another pentagon corner.

STEP 8 Set your compass to the distance between points D and G, and with the compass needle at point D, draw arc 3, which intersects the circumference of the circle at point H as shown in Figure 8.70. Continue the process of locating the pentagon corners by placing the compass needle at point H and drawing arc 4, which intersects the circle's circumference at point I. Next, place the compass needle at point I, and draw arc 5, which intersects the circle's circumference at point J. (See Figure 8.70.)

STEP 9 Connect points G and D, D and H, H and I, I and J, and J and G to complete the pentagon as shown in Figure 8.71.

Method 2:

STEP 1 Lightly draw a circle using the given diameter. (See Figure 8.72.)

STEP 2 Divide the circle into five equal parts with light lines from the circle's center past the circumference as shown in Figure 8.73. Each part is 72° because 360° ÷ 5 = 72°.

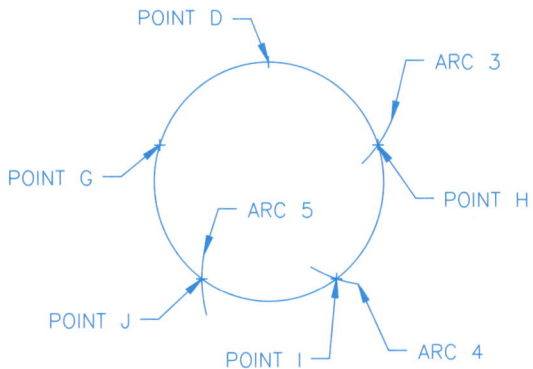

FIGURE 8.70 ■ Constructing a pentagon, Method 1, STEP 8.

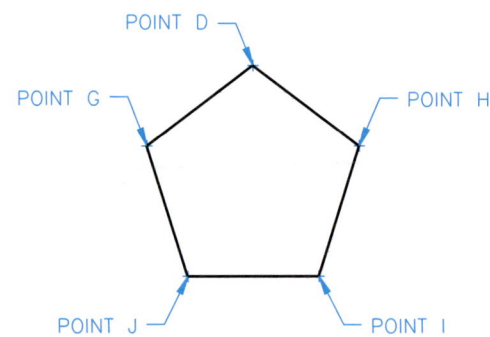

FIGURE 8.71 ■ Constructing a pentagon, Method 1, STEP 9.

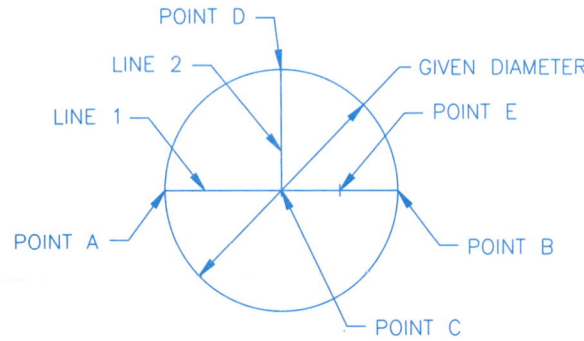

FIGURE 8.68 ■ Constructing a pentagon, Method 1, Steps 1 through 5.

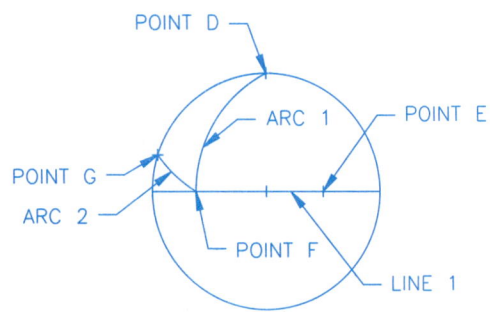

FIGURE 8.69 ■ Constructing a pentagon, Method 1, Steps 6 and 7.

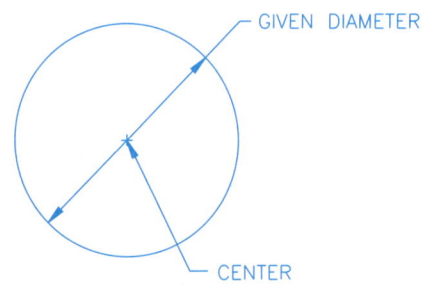

FIGURE 8.72 ■ Constructing a pentagon, Method 2, STEP 1.

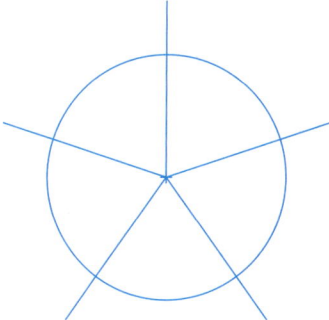

FIGURE 8.73 ■ Constructing a pentagon, Method 2, STEP 2.

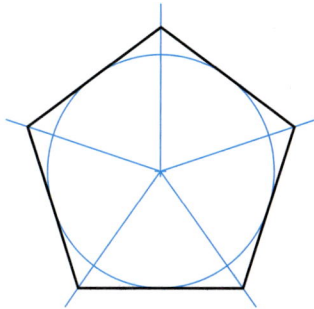

FIGURE 8.74 ■ Constructing a pentagon, Method 2, Step 3, circumscribed.

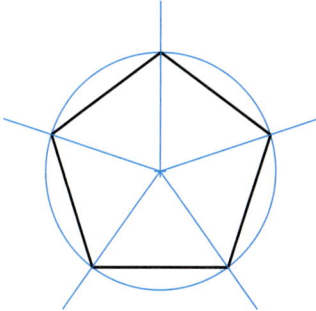

FIGURE 8.75 ■ Constructing a pentagon, Method 2, Step 3, inscribed.

STEP 3 Connect the points of tangency with the dividing lines for a pentagon that is outside of the circle, as shown in Figure 8.74, or connect where the lines intersect the circle for a pentagon inside the circle. (See Figure 8.75.)

Constructing a Regular Hexagon

Hexagons are six-sided polygons. Each side of a regular hexagon is equal and its interior angles are 120° as shown in Figure 8.76. Hexagons are commonly used as the shape for the heads of bolts and for nuts. Generally, the dimension given for the size of a hexagon is the distance across the flats. The distances across the flats and across the corners are shown in Figure 8.76.

There are two easy geometric construction methods used to draw a hexagon. One gives the distance across the flats; the

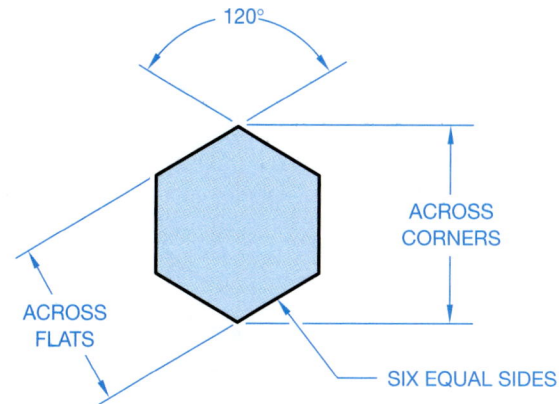

FIGURE 8.76 ■ Elements of a regular hexagon.

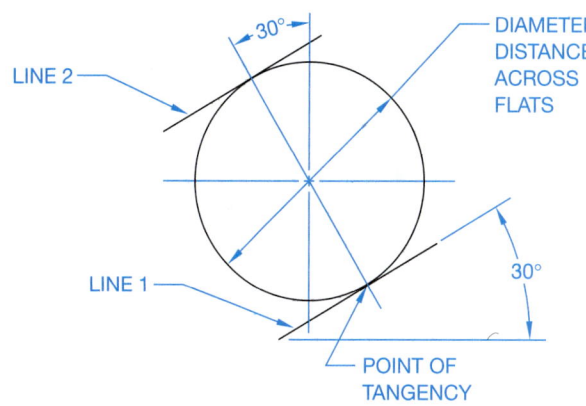

FIGURE 8.77 ■ Steps 1 and 2, constructing a hexagon.

other gives the distance across the corners. When practical, use a hexagon template or a special hex-head bolt, screw, or hex nut template.

Across the Flats

Given the distance across the flats, do the following to construct a hexagon:

STEP 1 Lightly draw a circle with the given distance as the diameter. (See Figure 8.77.)

STEP 2 Set your drafting machine at 30° from horizontal or use a 30–60 triangle. Then lightly draw lines 1 and 2 tangent to the circle as seen in Figure 8.77.

STEP 3 Now, set the angle at 30° the other way from horizontal and draw lines 3 and 4 tangent to the circle as shown in Figure 8.78.

STEP 4 Draw lines 5 and 6 vertical and tangent to the circle and darken the six lines to complete the required hexagon as in Figure 8.79.

Across the Corners

Given the distance across the corners of a hexagon, use the following construction steps:

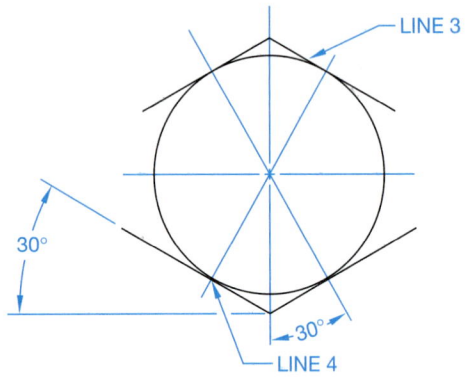

FIGURE 8.78 ■ STEP 3.

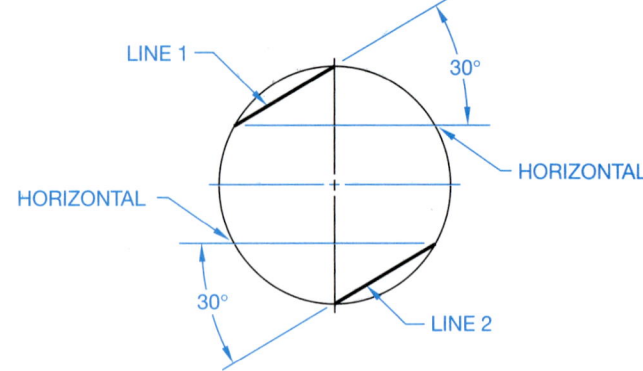

FIGURE 8.81 ■ STEP 2.

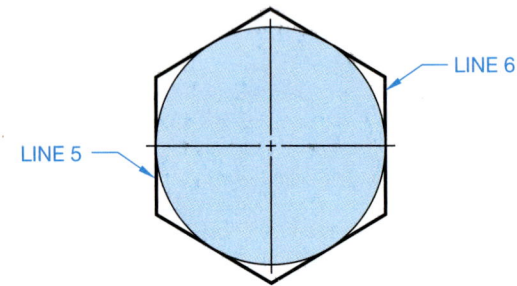

FIGURE 8.79 ■ Step 4, complete the hexagon.

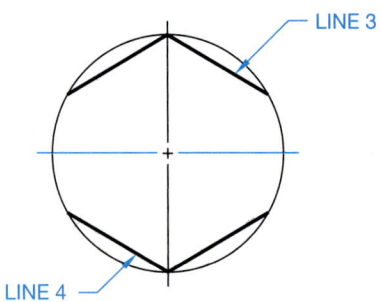

FIGURE 8.82 ■ STEP 3.

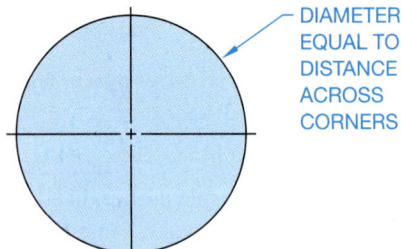

FIGURE 8.80 ■ Step 1, construct a hexagon given the distance across the corners.

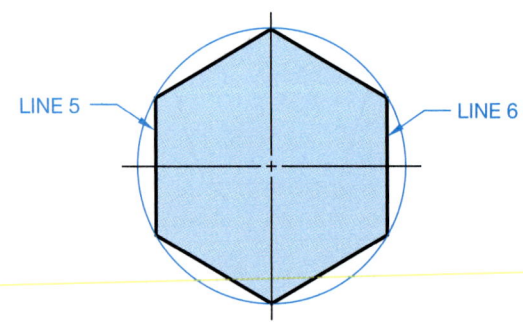

FIGURE 8.83 ■ Step 4, complete the hexagon.

STEP 1   Lightly draw centerlines in the desired location of the hexagon center. Then lightly draw a circle with the diameter equal to the given distance across the corners. Use construction lines as shown in Figure 8.80.

STEP 2   From the location where the vertical centerline touches the circle, draw two lines, 30° from horizontal, inside the circle as shown in Figure 8.81.

STEP 3   Draw lines 3 and 4 at 30° from horizontal in the other direction, as seen in Figure 8.82.

STEP 4   Draw vertical lines 5 and 6 and darken the object lines to create the required hexagon shown in Figure 8.83.

The position of a hexagon can be rotated by establishing the six sides in an alternate relationship to the circle centerlines, as shown in Figure 8.84. Be sure that the included angles are 120°

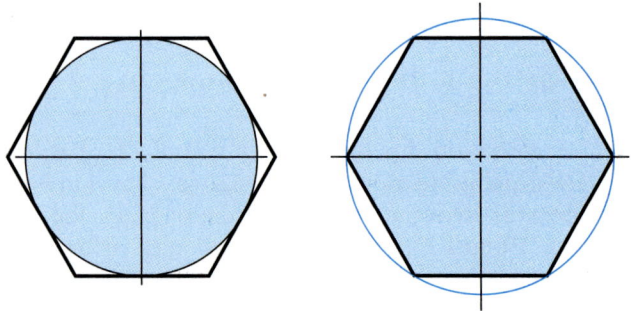

FIGURE 8.84 ■ Alternate hexagon positions.

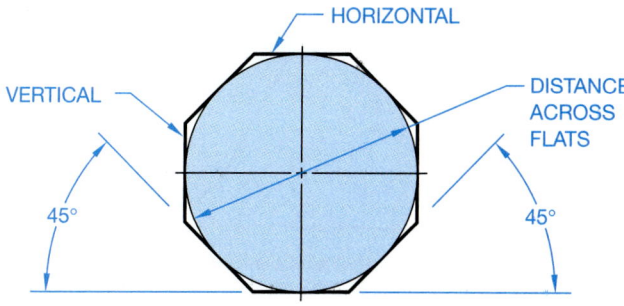

FIGURE 8.85 ■ Elements of an octagon.

and each side is equal in length. When you can, use a template to draw hexagons and other shapes. Templates save time. Many different geometric templates are available for your use.

Constructing a Regular Octagon

The octagon can be drawn using the same principle applied to the construction of a hexagon. A regular octagon has eight equal sides. If you use a 45° line as shown in Figure 8.85, you can easily draw an octagon about the given circle with a diameter equal to the distance across the flats using 45°, horizontal, and vertical lines.

CONSTRUCTING CIRCLES AND TANGENCIES

To be tangent to a circle or arc, a line must touch the circle or arc at only one point, and a line drawn from the center of the circle or arc must be perpendicular to the tangent line at the point of tangency. (See Figure 8.86.) Lines drawn tangent to cir-

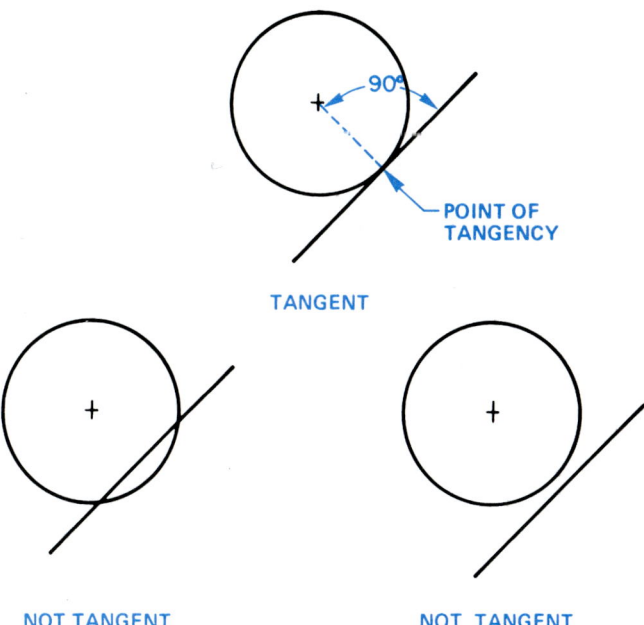

FIGURE 8.86 ■ Tangency.

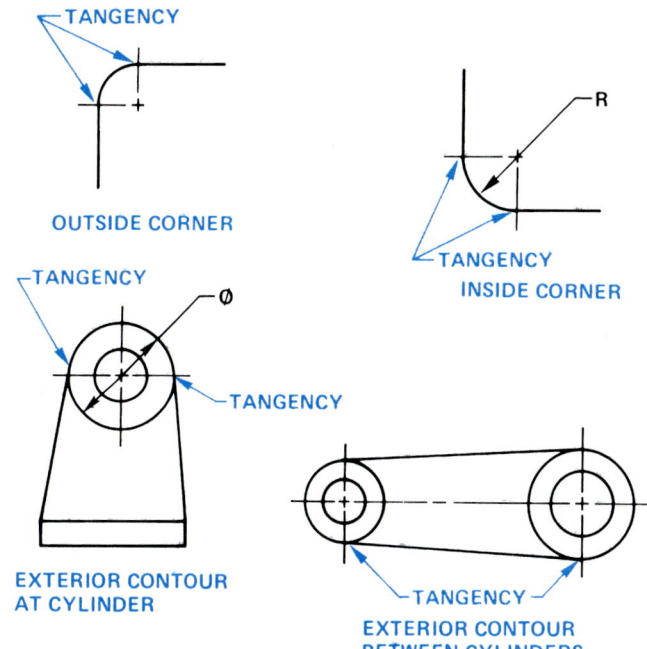

FIGURE 8.87 ■ Common tangencies.

cles or arcs are common in engineering drafting. It is important that tangent lines be drawn with a smooth transition at the point of tangency. Figure 8.87 shows some common tangency examples.

Drawing Arcs and Tangencies

When drafting tangent features, it is usually best to lightly draw the centerlines, then draw the arcs or circles, and finally draw the tangent lines. It is easier to draw a line tangent to a circle or arc than it is to draw a circle or arc tangent to a line. (See Figure 8.88.) When possible, use a template to draw the arcs and circles. Keep in mind that template circles are calibrated in diameters while arcs have radius dimensions. So, for example, a .50-in. (12.7 mm) radius requires using a 1.00-in. (25.4 mm) diameter circle. The combination of templates and a drafting machine provides the easiest and fastest results. Use a compass only if necessary. Figure 8.89 shows the relationship between a well-drawn and a poorly drawn tangency.

Draw an arc tangent to a given acute or obtuse angle in much the same way as previously shown for lines and circles. Given

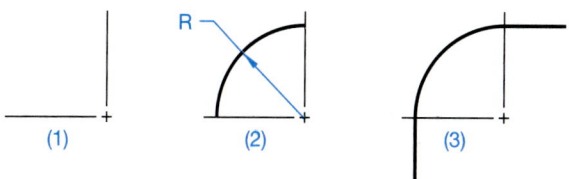

FIGURE 8.88 ■ Drawing lines tangent to an arc and an arc at a 90° corner.

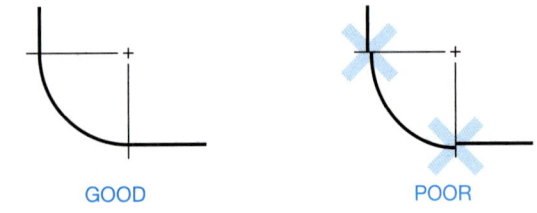

GOOD POOR

FIGURE 8.89 ■ Good and poor tangency examples.

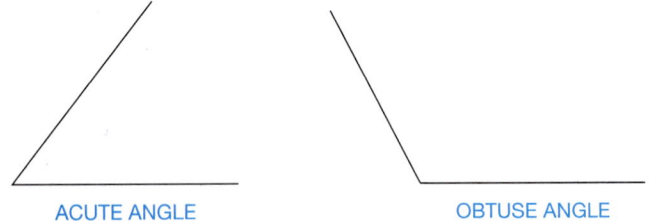

ACUTE ANGLE OBTUSE ANGLE

FIGURE 8.90 ■ Construct an arc tangent to given angles.

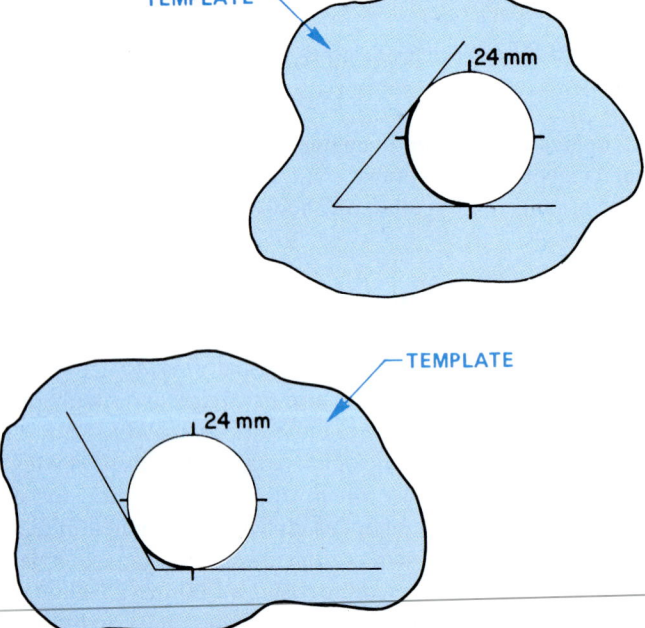

TEMPLATE

24 mm

TEMPLATE

24 mm

FIGURE 8.91 ■ Step 1, draw a tangent arc with a template.

the acute and obtuse angles in Figure 8.90, draw an arc with a 12-mm radius tangent to the sides of each angle.

STEP 1 Select a circle template with a 24-mm diameter circle. Move the template into position at each angle, leaving a small space for the lead or pen point thickness, and draw the arc as shown in Figure 8.91.

STEP 2 Remove the template and draw the sides of each angle to the points of tangency of the arcs. (See Figure 8.92.)

If the arc center is required for dimensioning purposes, use the following procedure:

STEP 1 Establish the arc centers by lightly drawing lines parallel to the given sides at a distance from the sides equal to

FIGURE 8.92 ■ Step 2, complete tangent arcs.

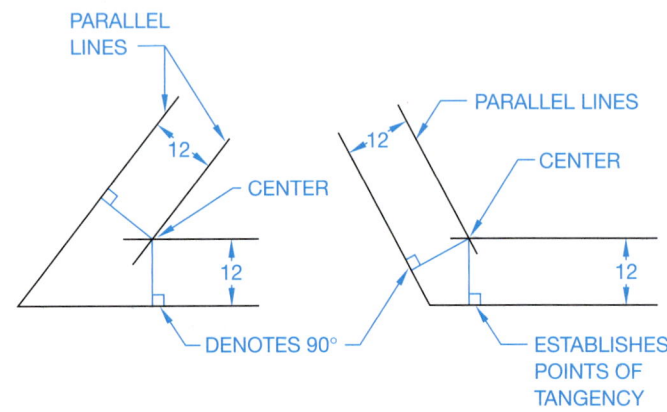

PARALLEL LINES

12

CENTER

12

DENOTES 90°

PARALLEL LINES

12

CENTER

12

ESTABLISHES POINTS OF TANGENCY

FIGURE 8.93 ■ Step 1, drawing tangent arcs with a compass. This procedure also works for arcs at 90° corners.

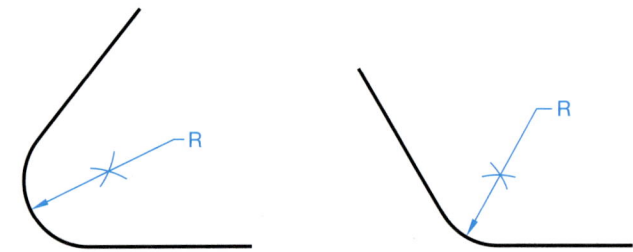

R R

FIGURE 8.94 ■ Step 2, complete tangent arcs.

the radius of the arc. In this example the radius is 12 mm. (See Figure 8.93.)

STEP 2 Set the compass at the given radius, 12 mm, and draw two arcs tangent to the given sides. Connect the sides to the points of the tangency of the arcs as shown in Figure 8.94.

Drawing an Arc Tangent to a Given Line and a Circle

The important point to remember is that the distance from the point of tangency to the center of the tangent arc is equal to the radius of the tangent arc. Given the machine part in Figure 8.95, draw an arc tangent to the cylinder and base with a 24-mm radius.

STEP 1 Draw a line parallel to the base at a distance equal to the radius of the arc, 24 mm. Then draw an arc that intersects the line, drawn with a radius equal to the given arc plus the radius of the cylinder (24 + 11 = 35 mm). (See Figure 8.96.)

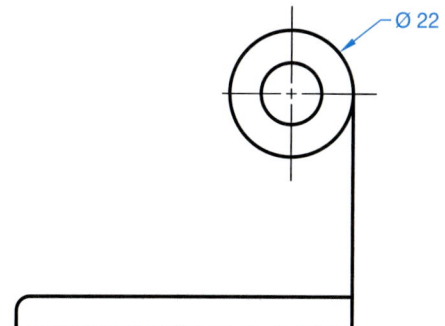

FIGURE 8.95 ■ Construct an arc tangent to a given line and circle.

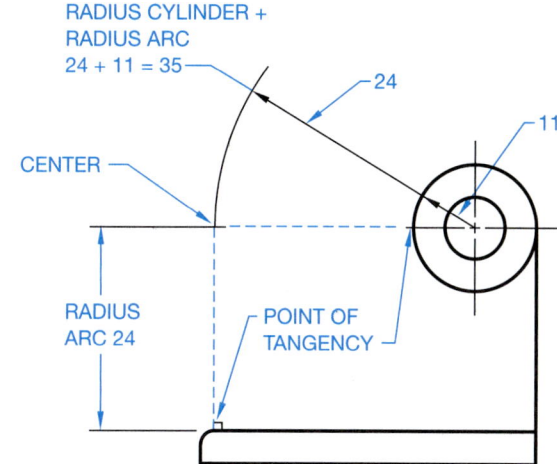

FIGURE 8.96 ■ STEP 1.

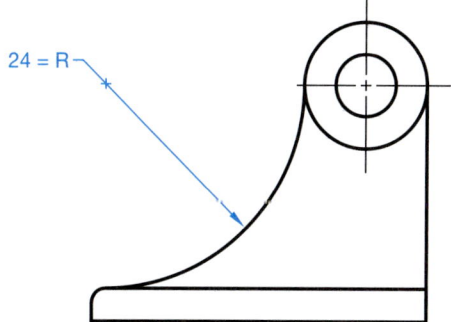

FIGURE 8.97 ■ STEP 2.

STEP 2   Use a compass or a template to draw the required radius from the center point established in STEP 1. (See Figure 8.97.)

The previous example demonstrated an internal arc tangent to a circle or arc and a line. A similar procedure is used for an external arc tangent to a given circle and line as shown in Figure 8.98. In this case, the radius of the cylinder (circle) is subtracted from the arc radius to get the center point.

The following examples, Figures 8.99 through 8.101, are situations that commonly occur on machine parts where an arc can be tangent to given cylindrical or arc shapes. Keep in mind that

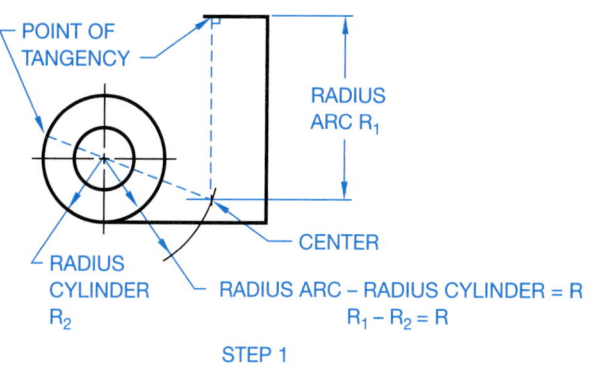

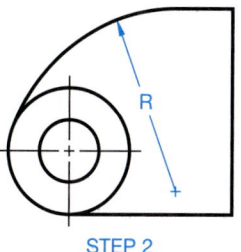

FIGURE 8.98 ■ Construct an external arc tangent to a line and cylinder.

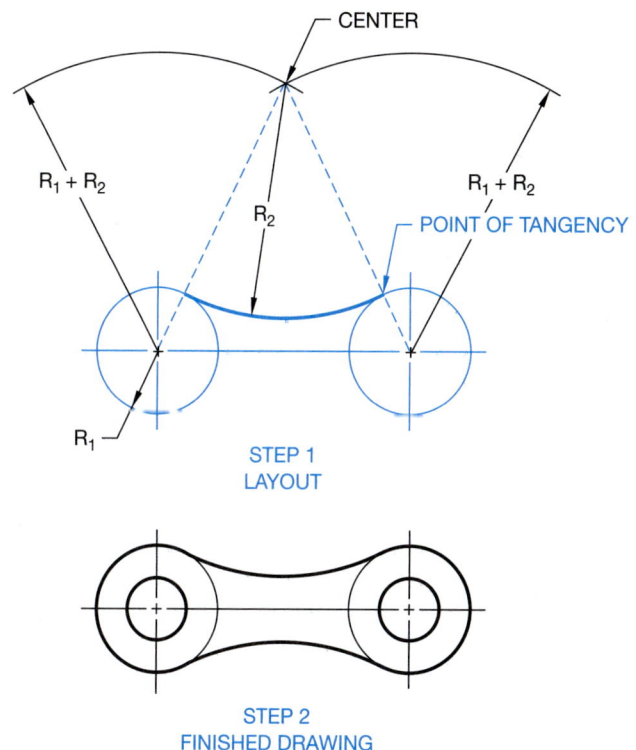

FIGURE 8.99 ■ Chain link.

the methods used to draw the tangents are the same as just described. The key is that the center of the required arc is always placed at a distance equal to its radius from the points of tangencies. To achieve this, you will have to either add or subtract the

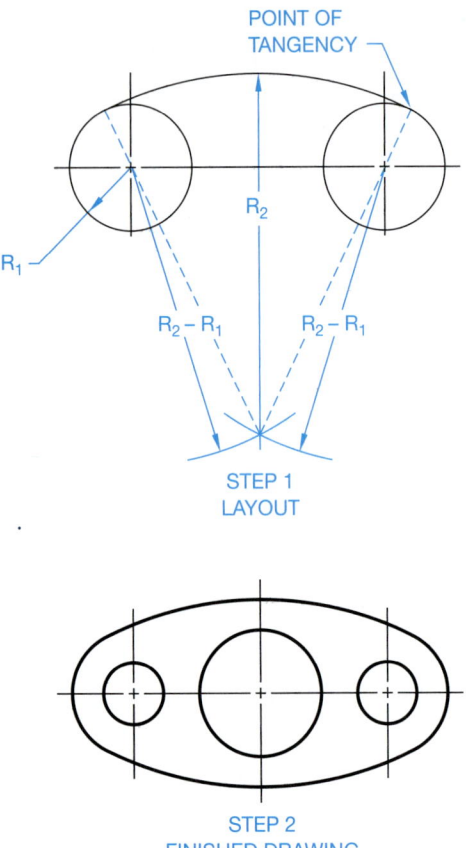

FIGURE 8.100 ■ Gasket.

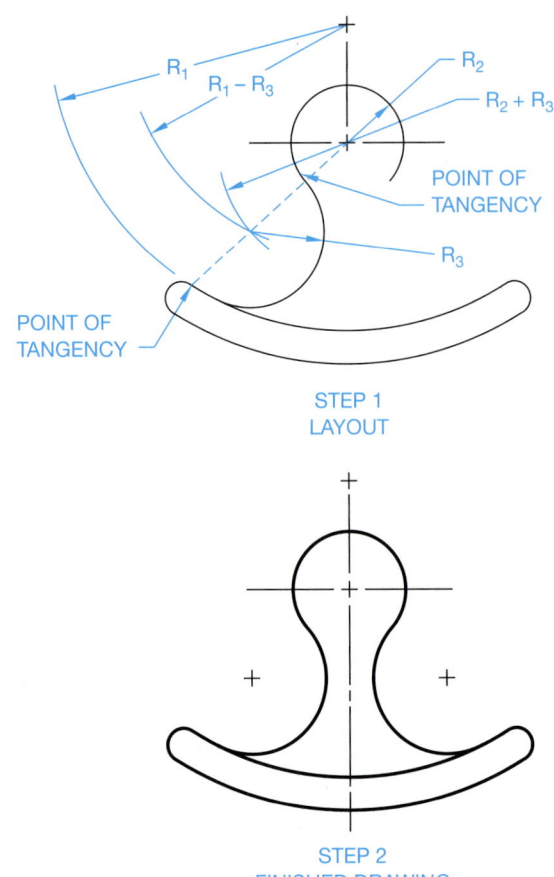

FIGURE 8.101 ■ Hammer head.

given radius and the required radius, depending on the situation. When possible, use a template to draw circles and arcs.

Drawing an Ogee Curve

An S curve, commonly called an ogee curve, occurs in situations where a smooth contour is needed between two offset features. The following steps show how to draw an S curve with equal radii.

STEP 1 Given the offset points A and B in Figure 8.102, draw a line between points A and B and bisect that line to find point C.

STEP 2 Draw the perpendicular bisector of lines AC and CB. From point A, draw a line perpendicular to the lower line that intersects the AC bisector at X. From point B draw a line perpendicular to the upper line that intersects the CB bisector at Y. (See Figure 8.103.)

STEP 3 With X and Y as the centers, draw a radius from A to C and a radius from B to C as shown in Figure 8.104.

STEP 4 If the S curve line, AB, represents the centerline of the part, then develop the width of the part parallel to the centerline using concentric arcs to complete the drawing. (See Figure 8.105.)

When an S curve has unequal radii, the procedure is similar to the previous example, as shown in Figure 8.106.

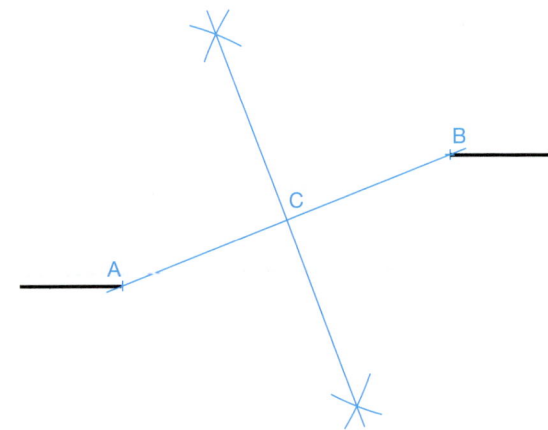

FIGURE 8.102 ■ Step 1, construct an S curve.

CONSTRUCTING AN ELLIPSE

When a circle is viewed at an angle, an ellipse is observed. The angular relationship of an ellipse to a circle is shown in Figure 8.107. If a surface with a through hole is inclined 45°, the representation is a 45° ellipse, as shown in Figure 8.108. An ellipse has a major diameter and a minor diameter. In Figure 8.108 the

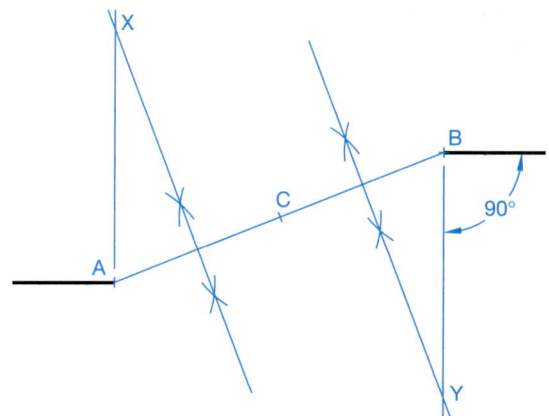

FIGURE 8.103 ■ Step 2, construct an S curve.

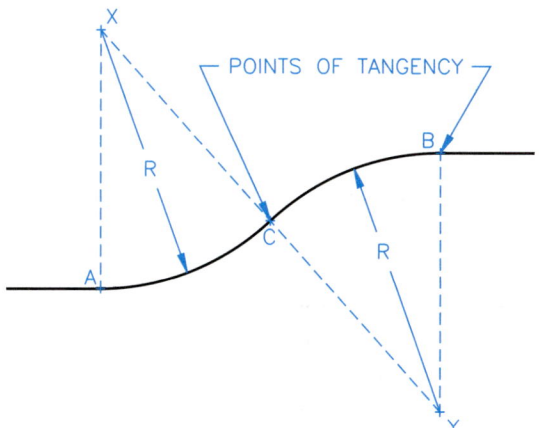

FIGURE 8.104 ■ Step 3, construct an S curve.

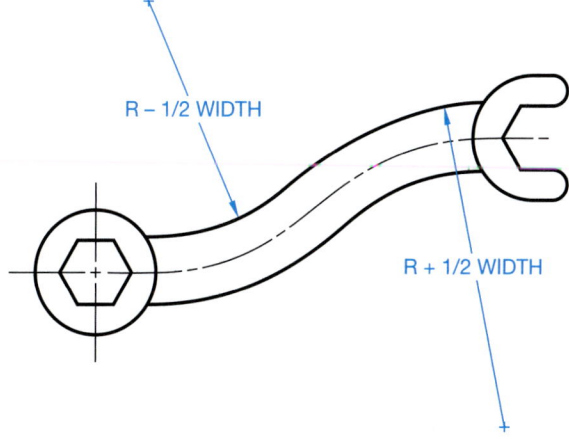

FIGURE 8.105 ■ Step 4, a wrench with a complete S curve.

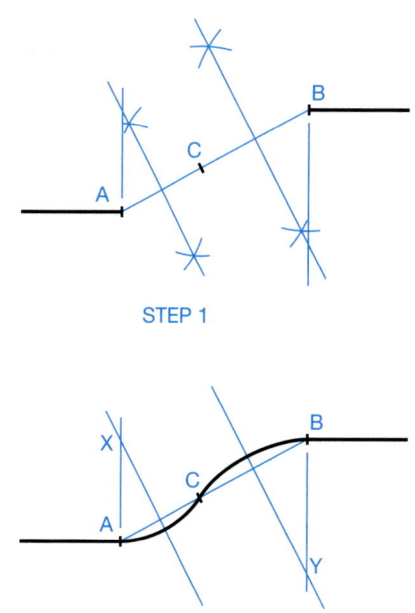

STEP 1

STEP 2

FIGURE 8.106 ■ Constructing an S curve with unequal radii.

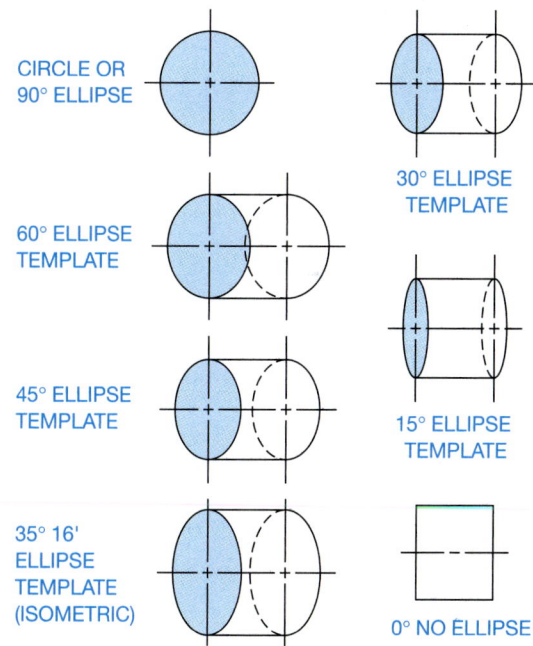

FIGURE 8.107 ■ Ellipses are established by their relationship to a circle turned at various angles.

minor diameter is established by projecting the top and bottom edge from the slanted surface. The major diameter, in this case, is equal to the diameter of the circle.

If possible, always use a template to draw elliptical shapes. Ellipse templates are available from 10° to 85°. The ellipse

shown in Figure 8.108 could have been drawn by projecting the centerlines and then lining up the center marks of a 45° ellipse template. (See Figure 8.109.)

Approximating the Ellipse

If the elliptical shape does not exactly fit the available template sizes, use one that is close. How close depends on your company

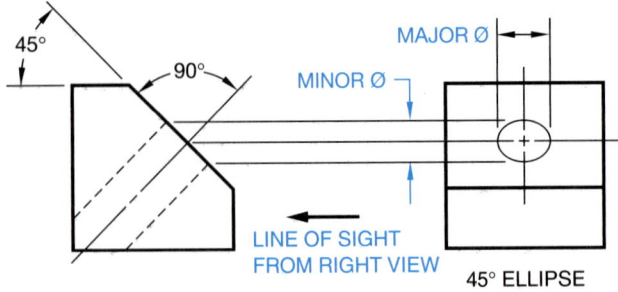

FIGURE 8.108 ■ Elliptical view.

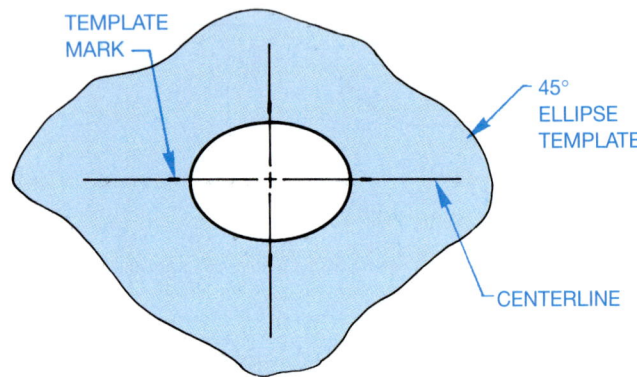

FIGURE 8.109 ■ Drawing an ellipse with a template.

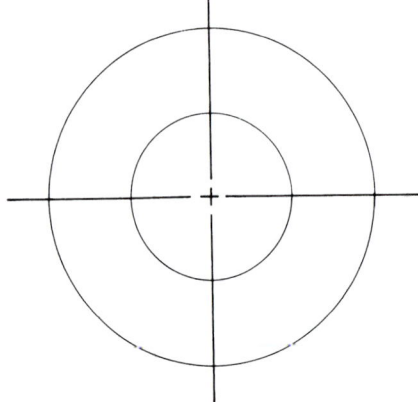

FIGURE 8.110 ■ Step 1, constructing an ellipse, concentric circle method.

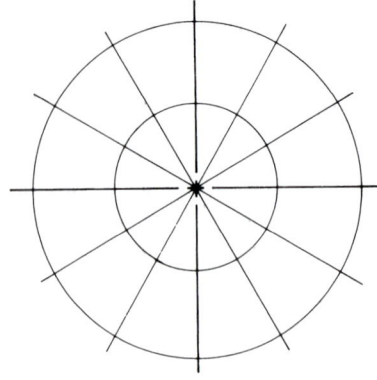

FIGURE 8.111 ■ Step 2, concentric circle method.

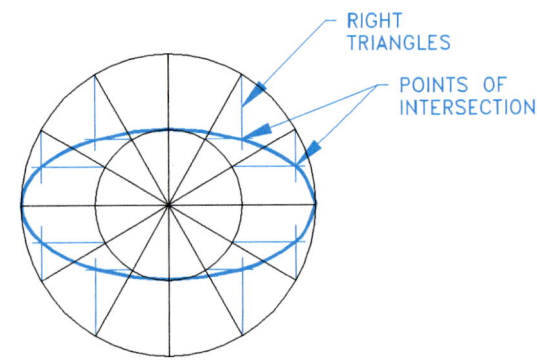

FIGURE 8.112 ■ Steps 3 and 4, complete ellipse.

standards. Most companies would rather have you draw a close representation than take the time to lay out an ellipse geometrically.

Using the Concentric Circle Construction Method

When templates are not available or are not close enough for the ellipse that must be drawn, several methods can be used to draw elliptical shapes. One of the most practical is the concentric circle method described in the following steps:

STEP 1 Use construction lines to draw two concentric circles, one with a diameter equal to the minor ellipse diameter and the other with a diameter equal to the major ellipse diameter. (See Figure 8.110.)

STEP 2 Use the drafting machine or triangles to divide the circles into at least 12 equal parts; 30° and 60° each way from horizontal. More parts give better accuracy but also consume more time. (See Figure 8.111.)

STEP 3 From the points of intersection of each line at the circumference of the circles, draw horizontal lines from the minor diameter and vertical intersecting lines from the major diameter, essentially creating a series of right triangles, although the intersection points only are needed. (See Figure 8.112.)

STEP 4 Using an irregular curve carefully connect the points created in step 3 to complete the ellipse as shown in Figure 8.112.

DRAWING GEOMETRIC SHAPES

The process used for drawing geometric shapes with CADD normally involves three steps:

1. Select the command.

2. Locate the feature.

3. Provide the needed information to draw the feature.

Rectangle

A rectangle is one of the easiest shapes to draw. Only two points are needed. The first is one corner of the rectangle, which can be any of the four corners. Once you pick this corner, the direction location of the second point has already been determined. The second point is the corner of the rectangle opposite the first corner that was picked. The computer then connects these two points with the straight lines needed to form the rectangle. Figure 8.113 illustrates this process. Rectangles and boxes can be drawn using the LINE command, but it takes longer.

Polygon

Any desired regular polygon is easy to draw with CADD. The command is usually POLYGON. The command prompts often begin by asking for the number of sides of the polygon. Then the command continues by asking for the center of the polygon. After you select the center, you must specify if the polygon is inscribed within or circumscribed about a circle. This is the same decision you make when drawing a polygon manually. As previously mentioned, an inscribed polygon is measured from the polygon corners, while a circumscribed polygon is measured from the polygon flats. Next, enter the radius or, in some cases, the diameter of the circle, and the polygon is automatically drawn. (See Figure 8.114.) POLYGON commands may also have an option that allows you to draw the desired polygon by picking the end points of an edge as shown in Figure 8.115.

Circle

A circle, for example, can be drawn several different ways. After selecting the CIRCLE command, the information that you have about the circle determines the method used to draw it. Four methods are shown in Figure 8.116, (a) through (d). The first method (a) requires that the center (1) and one point on the circle (2) be picked before the circle can be drawn. The second, example (b), requires that two opposite points (1 and 2) on the circumference (the diameter) be selected. The third example (c), requires that three points be picked on the circumference. The final method (d) requires that the center point be selected and the radius entered as a number selected from a menu, or typed on the keyboard.

Arc

Arcs are typically constructed in a similar manner as circles. But arcs are only portions of circles and therefore the two end points of the arc must be established. Common methods of constructing an arc are shown in Figure 8.117,

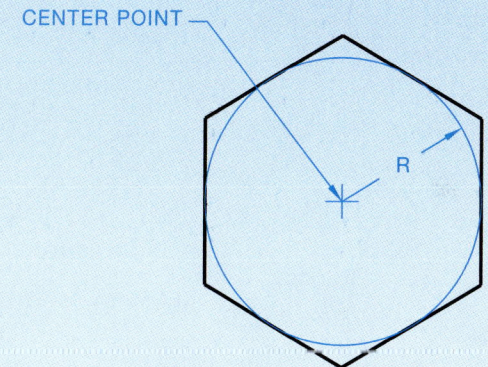

FIGURE 8.114 ■ Drawing a polygon by picking the center and entering the radius.

FIGURE 8.113 ■ Rectangle construction using opposite corners.

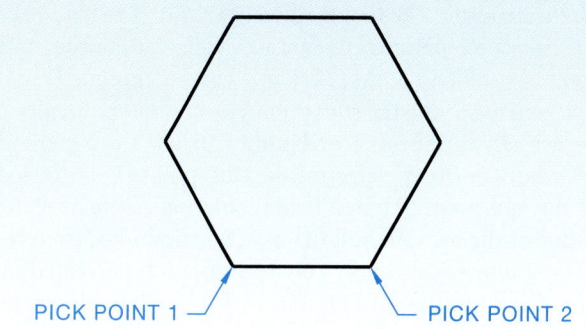

FIGURE 8.115 ■ Drawing a polygon automatically by picking the end points of a side.

(Continued)

CADD APPLICATIONS

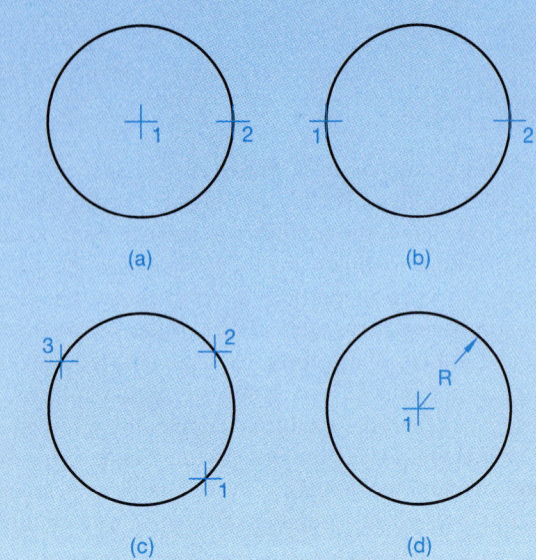

FIGURE 8.116 ■ Four ways to draw a circle.

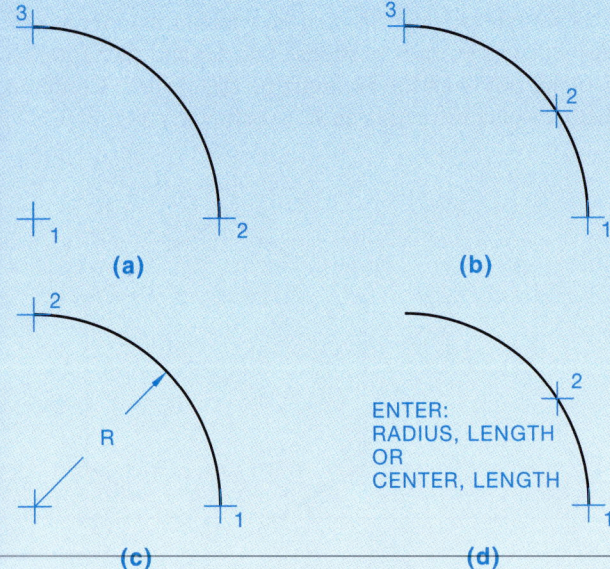

FIGURE 8.117 ■ Four ways to draw an arc.

(a) through (d). The first method (a) requires that you enter the center point of the arc (1), followed by the start point (2) and end point (3). The second technique (b) allows you to specify the start point (1), another point along the arc (2), and the arc's end point (3). The third method (c) creates an arc by selecting the start point (1), followed by the end point (2), and finally entering a value for the radius of the arc. Method (d) has two variations, each requiring numerical input. The first variation is created by selecting a start point (1), followed by a point along the arc (2), and then entering the arc radius and the length of the arc. The second variation requires the same two points

(1 and 2) but instead of entering the radius and length, you specify the arc's center point and length.

Tangencies

Tangencies are easy to draw with the computer. Most CADD programs have a command or option such as TANGENT, which allows you to pick a circle or arc and automatically draw a line or another circle or arc tangent to the selected object.

Drawing objects with tangencies using manual drafting techniques is often tricky. The feature must touch at exactly the point of tangency. While the methods discussed in this chapter allow you to do this with consistent success, the use of CADD systems makes this task automatic.

Most CADD programs have commands and options such as TANGENT object snap, or TANGENT constraint or relation, that allow you to automatically define a point of tangency between combinations of lines, arcs, circles, and ellipses. Another example is AutoCAD's TTR (Tangent, Tangent, Radius) option of the CIRCLE command. This allows you to automatically draw a circle tangent to lines, circles, or arcs simply by picking the objects and specifying the desired radius of the circle as shown in Figure 8.118.

Ellipse

Ellipses are quick and easy to draw on a CADD system. Most programs have a command such as ELLIPSE. In this command you usually select or type points to define the major and minor diameters or enter their values and the computer automatically draws an ellipse based on the information.

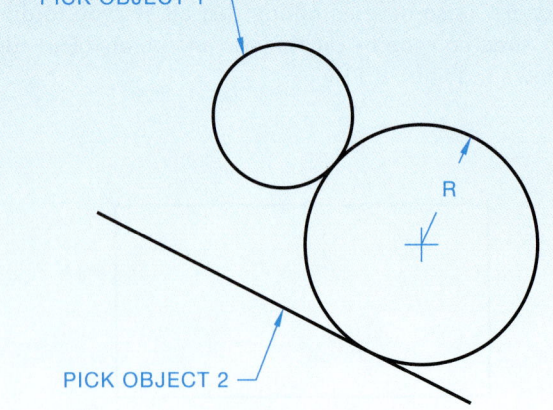

FIGURE 8.118 ■ Automatically drawing tangencies with CADD.

WEBSITE RESEARCH

The following websites can provide you with additional information:

http://www.industrialpress.com—Industrial Press. Online trigonometry tables.

http://www.g-w.com—AutoCAD texbooks by Madsen.

PROFESSIONAL PERSPECTIVE

Geometric constructions are found in nearly every engineering drafting assignment. Some geometric constructions are as simple as a radius corner while others are very complex. As a professional drafter, you should be able to quickly identify the type of geometric construction involved in the problem and solve it with one of the techniques you have learned. When you solve these problems, accuracy and carefully drawn construction lines are critical. Use the best tools available. For example, a professional drafter hardly ever draws an ellipse using construction methods; he or she uses the large variety of el-lipse templates available. Always use a template if you can, although, if all else fails, you do know how to construct the ellipse. Even though the geometric construction methods are presented for the manual drafter, the principles are the same for the CADD drafter. You should always refer to these principles and techniques to ensure that your CADD application has performed the task correctly. You will find with CADD that some of the very time-consuming constructions, such as dividing a space into equal parts or drawing a pentagon, are not only extremely fast and are almost perfectly accurate.

INTERIOR ANGLES OF REGULAR POLYGONS

For a regular polygon with *n* sides, the formula for calculating interior angles is: $\theta = (n - 2/n)180°$. (See the interior angle θ in Figure 8.119.)

The Greek letter **theta (θ)** is commonly used in geometry to stand for the measurement of an angle. As an example of using this formula, the interior angle of the hexagon in Figure 8.119 is:

$$\theta = (6 - 2/6)180° = (4/6)180° = 120°$$

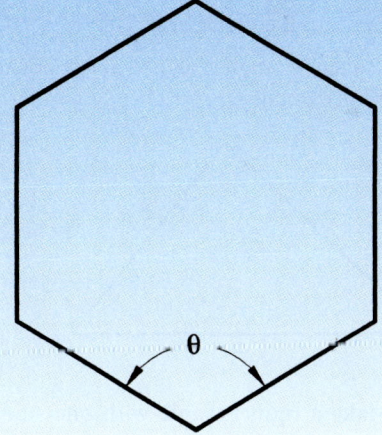

FIGURE 8.119 ■ Regular hexagon.

MATH APPLICATIONS

CHAPTER 8 *Geometric Construction Test*

 Access the CD found with this textbook to view the Chapter 8 Test. Confirm the preferred submittal method with your instructor.

CHAPTER
8
Geometric Construction Problems

DIRECTIONS

1. Problems can be completed using manual or computer-aided drafting, depending on your course guidelines.
2. Use very lightly drawn lines (construction lines) for all preliminary work and darken in only the completed product. *Do not* erase the construction lines when you complete the problems. This helps your instructor observe your drawing technique.
3. Geometric construction problems are presented as written instructions, drawings, or a combination of both. If the problems are presented as drawings without dimensions, transfer the drawing from the text, using dividers and scales, to your drawing sheet. Draw dimensioned problems full scale (1:1) unless otherwise specified. Draw all object, hidden, and centerlines. Do not draw dimensions.

PROBLEM 8.1   Draw two tangent circles with their centers on a horizontal line. Circle 1 has a 64-mm diameter and circle 2 has a 50-mm diameter.

PROBLEM 8.2   Make a perpendicular bisector of a horizontal line that is 79 mm long.

PROBLEM 8.3   Make an angle 48° with one side vertical. Bisect the angle.

PROBLEM 8.4   Divide a 96-mm line into 7 equal spaces.

PROBLEM 8.5   Draw two parallel horizontal lines each 50 mm long with one 44 mm above the other. Divide the space between the lines into 8 equal spaces.

PROBLEM 8.6   Transfer triangle a, b, c to a new position with side c 45° from horizontal.

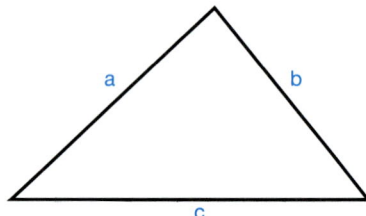

PROBLEM 8.7   Make a right triangle with one side 35 mm long and the other side 50 mm long.

PROBLEM 8.8   Make a right triangle with the hypotenuse 75 mm long and one side 50 mm long.

PROBLEM 8.9   Make an equilateral triangle with 65-mm sides.

PROBLEM 8.10   Make a square with a distance of 50 mm across the flats.

PROBLEM 8.11   Draw a hexagon with a distance of 2.5 in. across the flats.

PROBLEM 8.12   Draw a hexagon with a distance of 2.5 in. across the corners.

PROBLEM 8.13   Draw an octagon with a distance of 2.5 in. across the flats.

PROBLEM 8.14   Draw a rectangle 50 mm × 75 mm with 12-mm radius tangent corners.

PROBLEM 8.15   Draw two separate angles, one a 30° acute angle and the other a 120° obtuse angle. Then draw a .5-in. radius arc tangent to the sides of each angle.

PROBLEM 8.16   Given the following incomplete part, draw an inside arc with a 2.5-in. radius tangent to ∅A and line B.

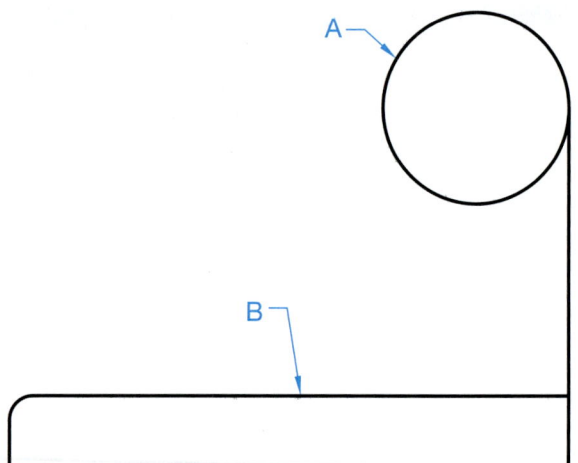

PROBLEM 8.17   Draw two circles with 25 mm diameters and 38 mm between centers on horizontal. On the upper side between the two circles, draw an inside radius of 44 mm. On the lower side, draw an outside radius of 76 mm.

PROBLEM 8.18   Draw an S curve with equal radii between points A and B given below:

+
B

A
+

PROBLEM 8.19 Draw two lines, one on each side of the ogee curve drawn in problem 8.18 and .5 in. away.

PROBLEM 8.20 Make an ellipse with a 38 mm minor diameter and a 50 mm major diameter.

 PROBLEMS 8.21 through 8.44: Access the CD found with this textbook and open the problem of your choice, or as assigned by your instructor. Solve the problem or problems using the same instructions provided for other problems in this chapter, unless otherwise specified.

MATH PROBLEMS

 PROBLEMS 8.45 through 8.50: Access the CD found with this textbook and open the math problem of your choice, or as assigned by your instructor. Solve the problem or problems using the same instructions provided.

Drafting Views and Annotations

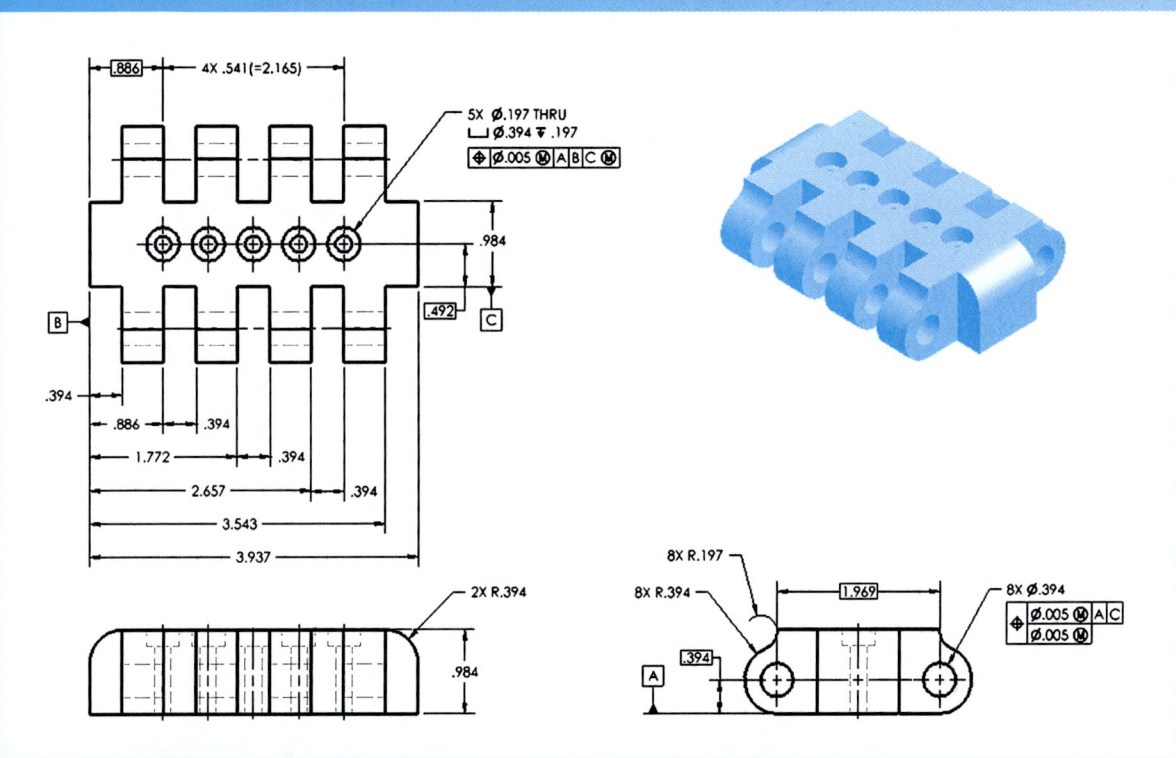

CHAPTER

Multiviews

LEARNING OBJECTIVES

After completing this chapter, you will:

■ Prepare single- and multiview drawings.

■ Select appropriate views for presentation.

■ Draw view enlargements.

■ Establish runouts.

■ Explain the difference between first- and third-angle projection.

■ Prepare formal multiview drawings from an engineer's sketch and actual industry layouts.

THE ENGINEERING DESIGN APPLICATION

The engineer has just handed you a sketch of a new part design. (See Figure 9.1.) The engineer explains that a multiview drawing is needed on the shop floor as soon as possible so a prototype can be manufactured and tested. As the engineering drafter, it is your responsibility to create a drawing that shows all necessary manufacturing data. Based on drafting guidelines, you first decide which view should be the front view. Having established the front view, you must now determine what other views are required in order to show all the features of the part. Using visualization techniques based on the glass box helps you to decide which views are needed. (See Figure 9.2.) Unfolding the box puts all of the views in their proper positions, so sketch the layout and complete the drawing. (See Figure 9.3.)

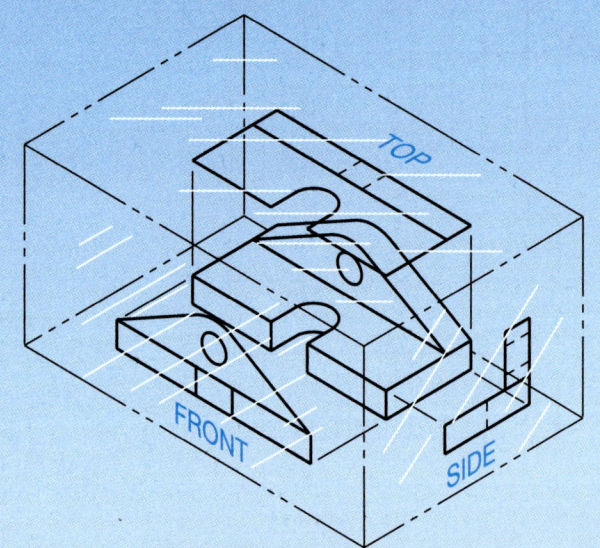

FIGURE 9.2 ■ Using the glass box principle to visualize the needed views.

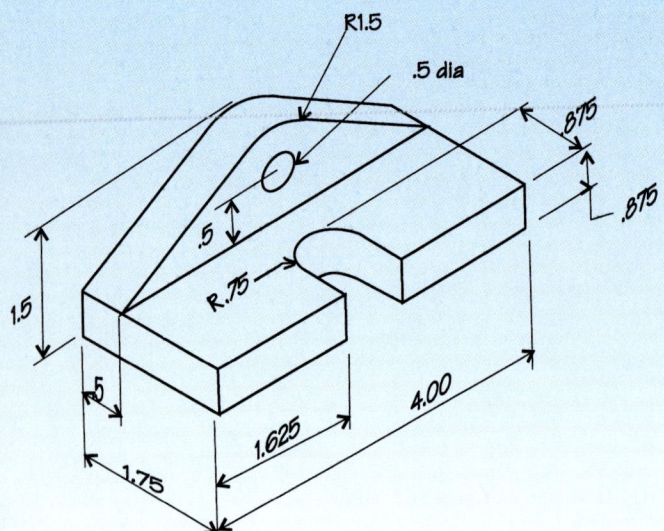

FIGURE 9.1 ■ Engineer's rough sketch.

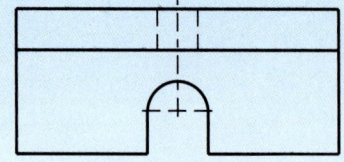
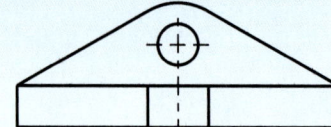
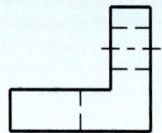

FIGURE 9.3 ■ A multiview drawing of the part without dimensions.

ASME/ANSI This chapter is developed in accordance with the ASME/ANSI standard for multiview presentation, titled *Multi and Sectional View Drawings*, ANSI Y14.3, published by The American Society of Mechanical Engineers (ASME). This standard is available from the American National Standards Institute, 1430 Broadway, New York, NY 10018, or The American Society of Mechanical Engineers, 345 E 47th Street, New York, NY 10017. The content of this discussion provides an in-depth analysis of the techniques and methods of multiview presentation.

Actual objects such as your pencil, computer monitor, or this textbook are typically easy to recognize and visualize because they are three-dimensional physical items. However, complex three-dimensional items are difficult to describe and can be even more complicated to draw and dimension because there is too much depth and so many sides. As a result, orthographic projection is used to change physical objects and three-dimensional ideas into two-dimensional drawings that effectively describe the design and features of an object, so the object can be documented and manufactured.

Orthographic projection is any projection of the features of an object onto an imaginary plane called a plane of projection. The projection of the features of the object is made by lines of sight that are perpendicular to the plane of projection. When a surface of the object is parallel to the plane of projection, the surface appears in its true size and shape on the plane of projection. In Figure 9.4, the plane of projection is parallel to the surface of the object. The line of sight (projection from the object) is perpendicular to the plane of projection. Notice also that the object appears three-dimensional (width, height, and depth) while the view on the plane of projection has only two dimensions (width and height). In situations where the plane of projection is not parallel to the surface of the object, the resulting orthographic view is foreshortened, or shorter than true length. (See Figure 9.5.)

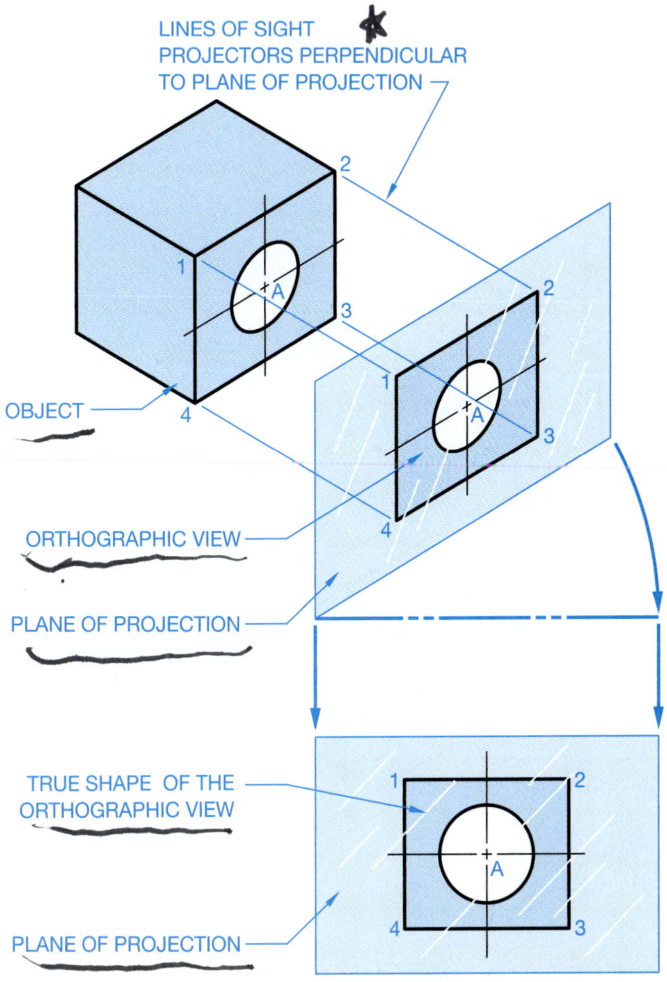

LINES OF SIGHT PROJECTORS PERPENDICULAR TO PLANE OF PROJECTION

OBJECT

ORTHOGRAPHIC VIEW

PLANE OF PROJECTION

TRUE SHAPE OF THE ORTHOGRAPHIC VIEW

PLANE OF PROJECTION

FIGURE 9.4 ■ Orthographic projection to form orthographic view.

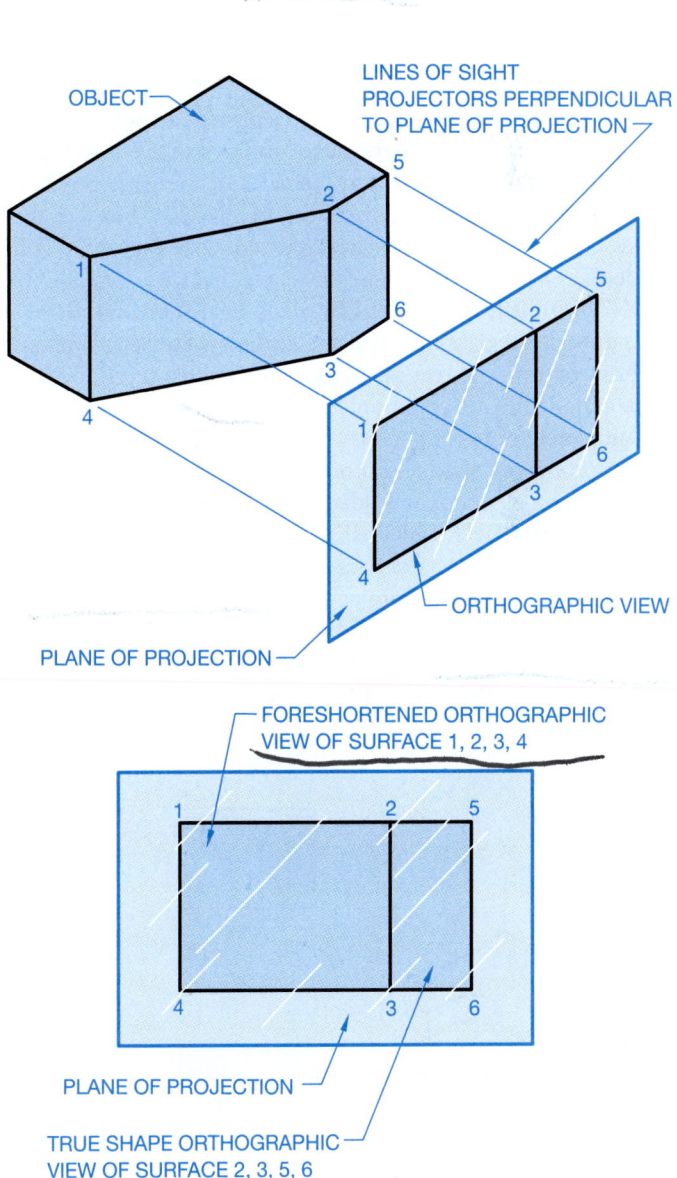

LINES OF SIGHT PROJECTORS PERPENDICULAR TO PLANE OF PROJECTION

OBJECT

ORTHOGRAPHIC VIEW

PLANE OF PROJECTION

FORESHORTENED ORTHOGRAPHIC VIEW OF SURFACE 1, 2, 3, 4

PLANE OF PROJECTION

TRUE SHAPE ORTHOGRAPHIC VIEW OF SURFACE 2, 3, 5, 6

FIGURE 9.5 ■ Projection of a foreshortened orthographic surface.

MULTIVIEWS

Multiview projection establishes views of an object projected upon two or more planes of projection by using orthographic projection techniques. The result of multiview projection is a multiview drawing. A multiview drawing represents the shape of an object using two or more views. Consideration should be given to the choice and number of views used so, when possible, the surfaces of the object are shown in their true size and shape.

It is easier for an individual to visualize a three-dimensional picture of an object than it is to visualize a two-dimensional drawing. In mechanical drafting, however, the common practice is to prepare completely dimensioned detail drawings using two-dimensional views, known as multiviews. Figure 9.6 shows an object represented by a three-dimensional drawing, also called a pictorial, and three two-dimensional views, or multiviews, also known as orthographic projection. The multiview method of drafting represents the shape description of the object.

Glass Box

If the object in Figure 9.6 is placed in a glass box so the sides of the glass box are parallel to the major surfaces of the object, you can project those surfaces onto the sides of the glass box and create multiviews. Imagine the sides of the glass box are the planes of projection previously discussed. (See Figure 9.7.) If you look at all sides of the glass box, you have six total views: FRONT, TOP, RIGHT-SIDE, LEFT-SIDE, BOTTOM, and REAR. Now, unfold the glass box as if the corners were hinged about the front (except the rear view) as demonstrated in Figure 9.8. These hinge lines are commonly called fold lines.

Completely unfold the glass box onto a flat surface and you have the six views of an object represented in multiview. Figure 9.9 shows the glass box unfolded. Notice the views are labeled FRONT, TOP, RIGHT, LEFT, REAR, and BOTTOM. This is the arrangement that the views are always found in when using multiviews in third-angle projection. Third-angle projection is the principal multiview projection used by U. S. industries. First-angle projection is commonly used in European countries. Additional third- and first-angle projection coverage is provided later in this chapter.

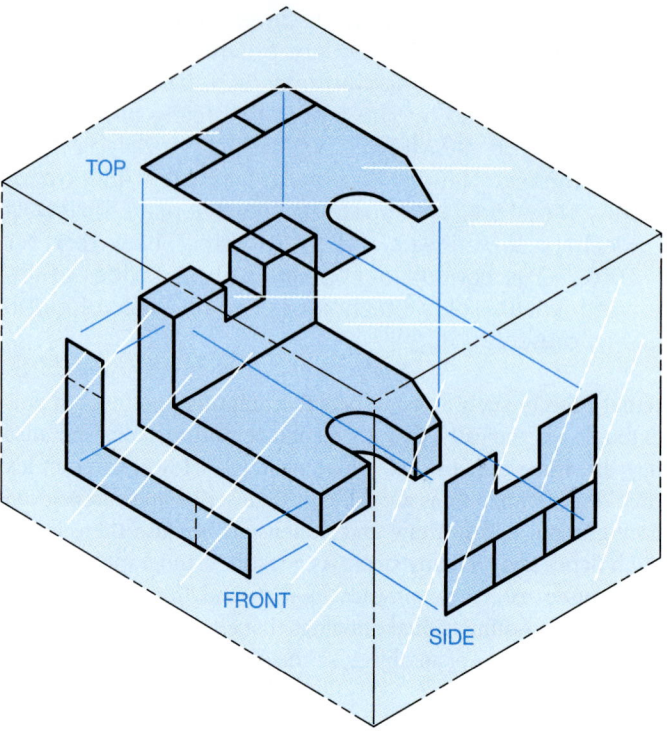

FIGURE 9.7 ■ The glass box principle.

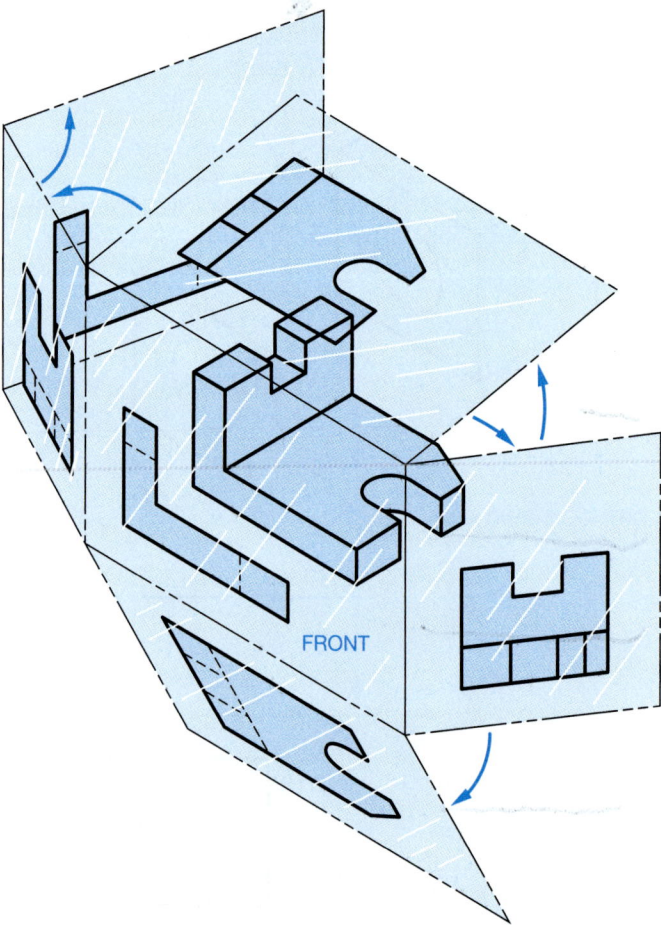

FIGURE 9.8 ■ Unfolding the glass box at hinge lines, also called fold lines.

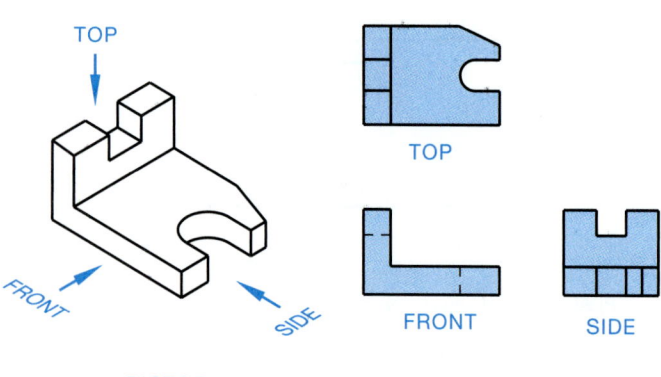

PICTORIAL   MULTIVIEWS

FIGURE 9.6 ■ Pictorial versus multiviews.

Look at Figure 9.9 in more detail so you can see the common items between views. Knowing how to identify features of the object common between views aids you in the visualization of multiviews. Notice in Figure 9.10 that the views are aligned. The top view is directly above, and the bottom view is directly below, the front view. The left-side view is directly to the left, while the right-side view is directly to the right, of the front view. This format allows you to project points directly from one view to the next to help establish related features on each view.

Now, look closely at the relationship among the front, top, and right-side views. A similar relationship exists with the left-side view. Figure 9.11 shows a 45° line projected from the corner of the fold, or reference line (hinge), between the front, top, and side views. The 45° line is often called a **mitre line.** This 45° line is used as an aid in projecting views. All of the features established on the top view can be projected to the 45° line and then down onto the side view. This projection works because the depth dimension is the same between the top and side views. The reverse is also true. Features from the side view may be projected to the 45° line and then over to the top view.

The same concept of projection achieved in Figure 9.11 using the 45° line also works by using a compass at the intersection of the horizontal and vertical fold lines. The compass establishes the common relationship between the top and side views as shown in Figure 9.12. Another method commonly used to transfer the size of features from one view to the next is to use dividers to transfer distances from the fold line of the top view to the fold line of the side view. The relationships between the fold line and these two views are the same, as shown in Figure 9.13.

The front view is usually the most important view and the one from which the other views are established. There is always one dimension common between adjacent views. For example, the width is common between the front and top views and the height between the front and side views. Knowing this allows you to relate information from one view to another. Take one more look at the relationship among the six views shown in Figure 9.14.

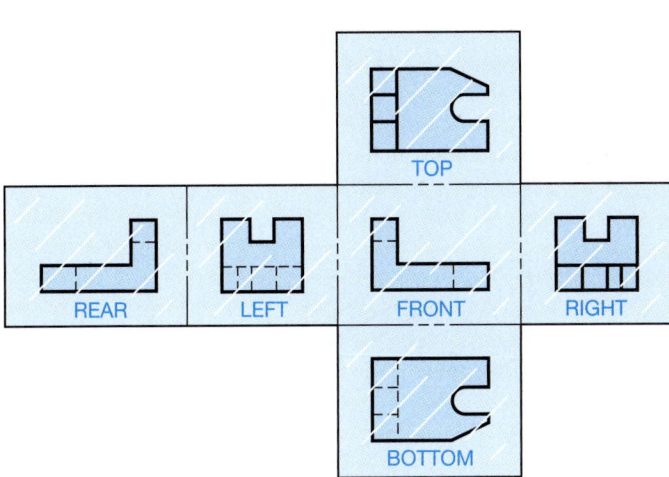

FIGURE 9.9 ■ Glass box unfolded.

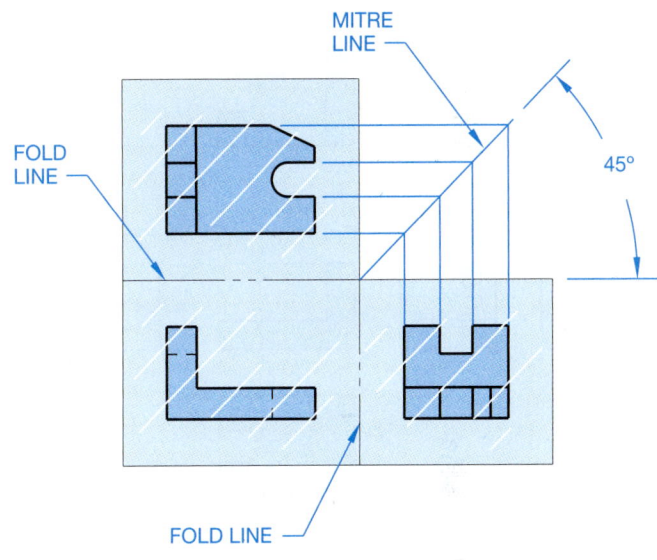

FIGURE 9.11 ■ 45° projection line.

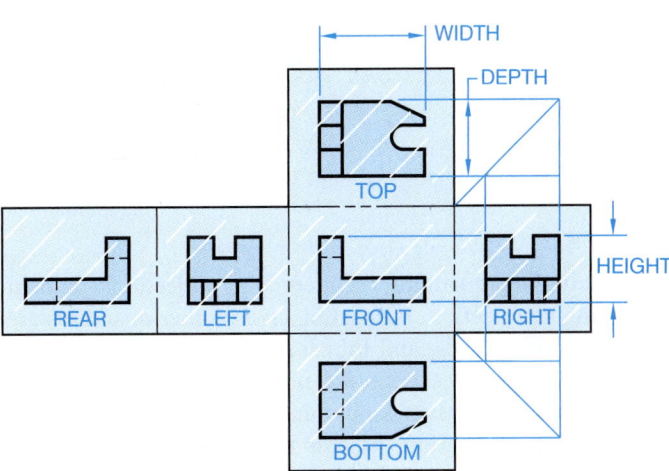

FIGURE 9.10 ■ View alignment.

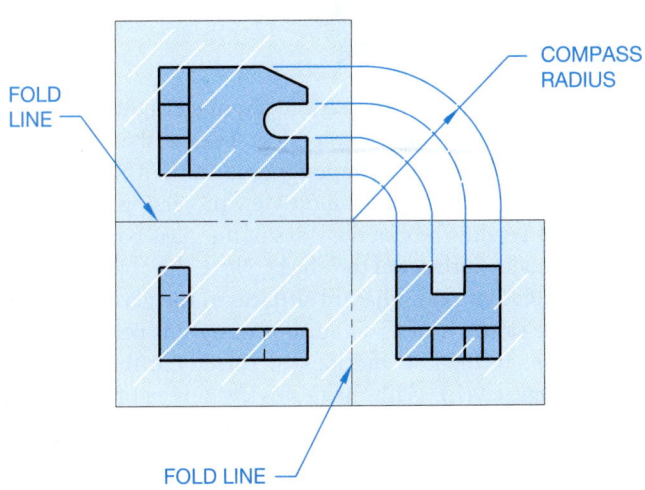

FIGURE 9.12 ■ Projection with a compass.

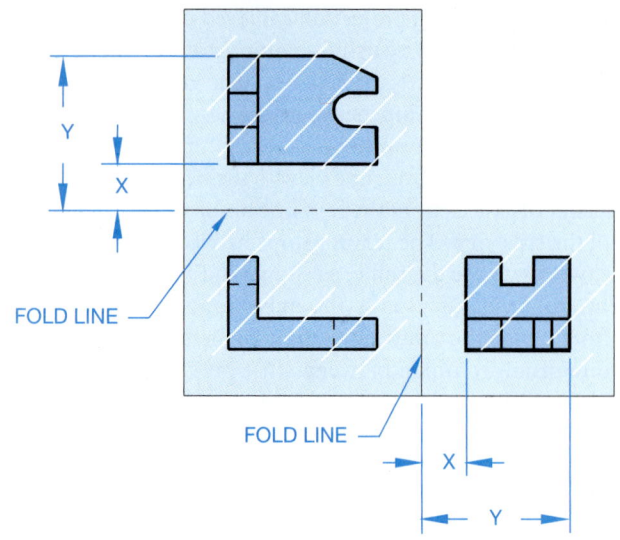

FIGURE 9.13 ■ Using dividers to transfer view projections.

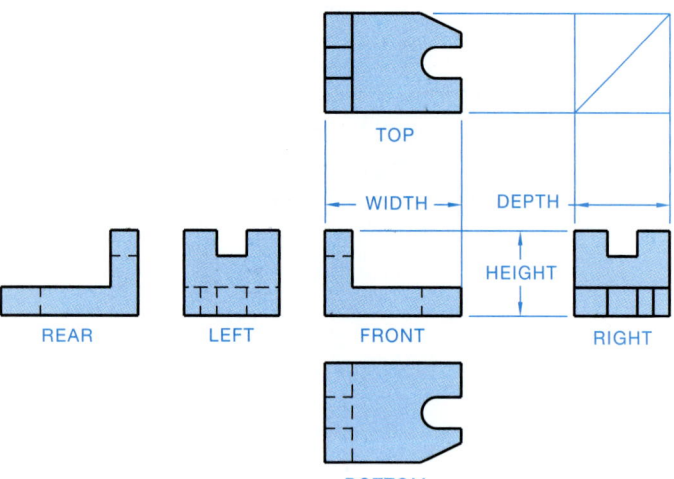

FIGURE 9.14 ■ Multiview orientation.

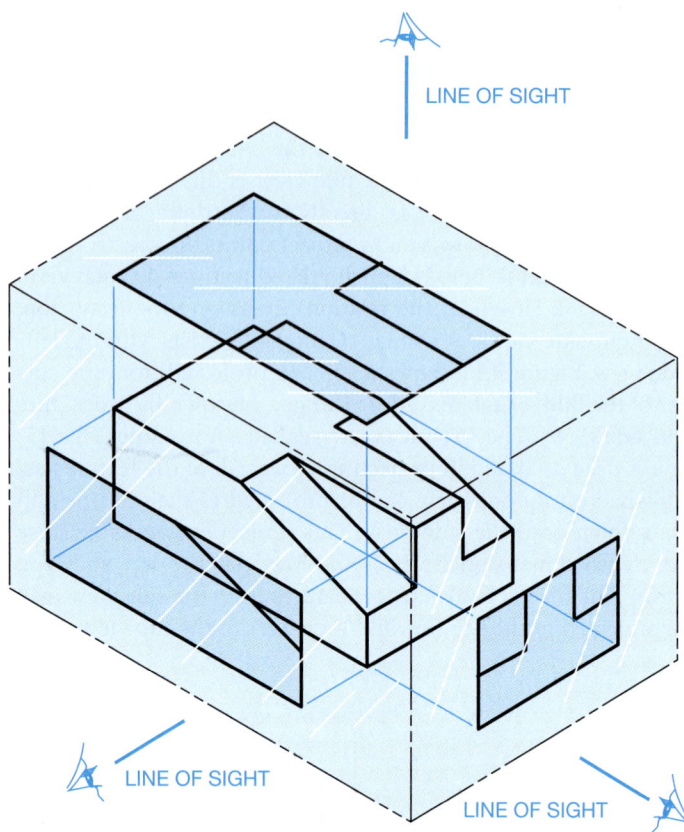

FIGURE 9.15 ■ Glass box in third-angle projection.

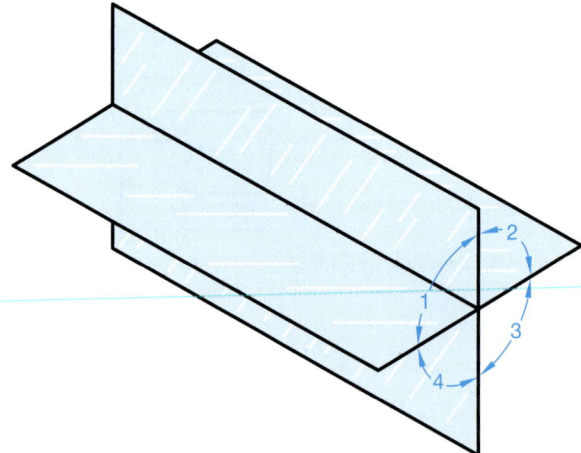

FIGURE 9.16 ■ Quadrants of spatial visualization.

THIRD-ANGLE PROJECTION

The method of multiview projection described in this chapter is also known as third-angle projection. This is the method of view arrangement that is commonly used in the United States. In the previous discussion on multiview projection, the object was placed in a glass box so the sides of the glass box were parallel to the major surfaces of the object. Next, the object surfaces were projected onto the adjacent surfaces of the glass box. This achieved the same effect as if your line of sight is perpendicular to the surface of the box and looking directly at the object, as shown in Figure 9.15. With the multiview concept in mind, assume an area of space is divided into four quadrants, as shown in Figure 9.16.

If the object is placed in any of these quadrants, the surfaces of the object are projected onto the adjacent planes. When placed in the first quadrant, the method of projection is known as first-angle projection. Projections in the other quadrants are termed second-, third-, and fourth-angle projections. Second- and fourth-angle projections are not used, though first- and third-angle projections are common.

Third-angle projection, as commonly used in the United States, is achieved when you take the glass box from Figure 9.15 and place it in quadrant three from Figure 9.16. Figure 9.17

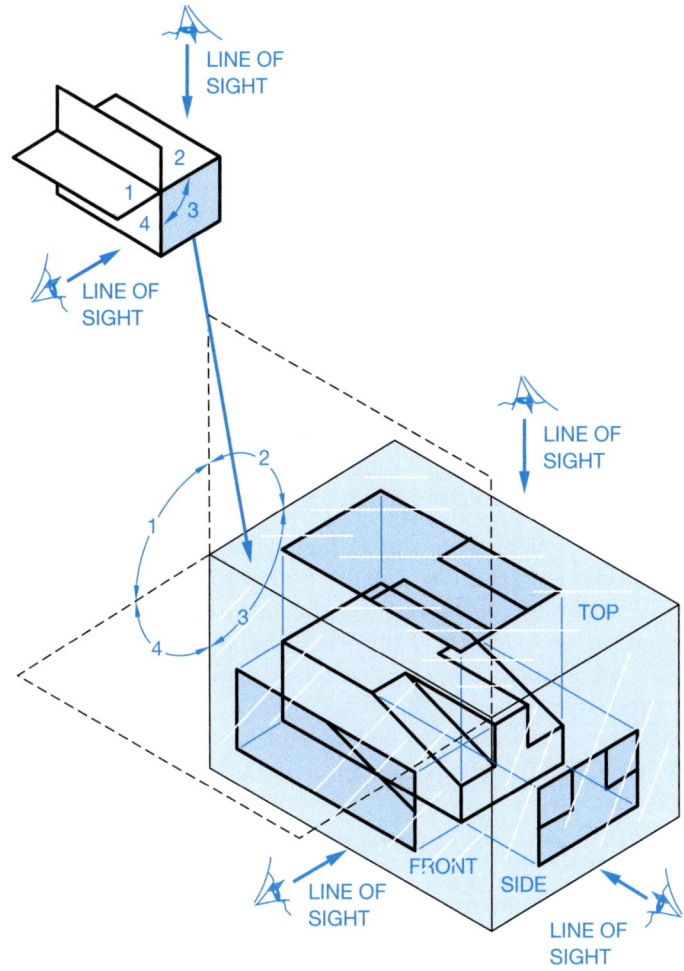

FIGURE 9.17 ■ Glass box placed in the third-quadrant for third-angle projection.

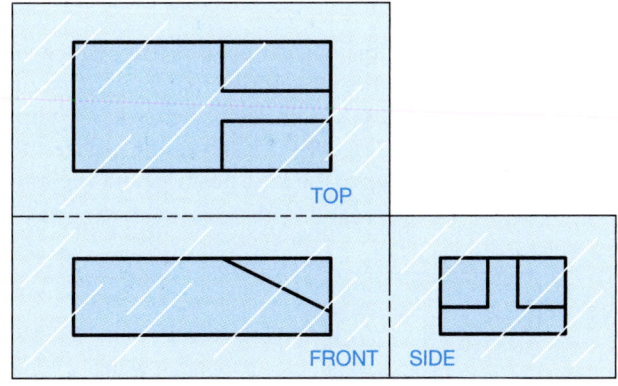

FIGURE 9.18 ■ Third-angle projection.

shows the relationship of the glass box to the projection planes in the third-angle projection. In this quadrant, the projection plane is between your line of sight and the object. When the glass box, in the third-angle projection quadrant, is unfolded, the result is the multiview arrangement previously discussed and shown in Figure 9.18.

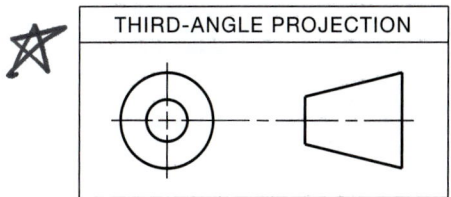

FIGURE 9.19 ■ Third-angle projection symbol.

A third-angle projection drawing is accompanied by a symbol on or next to the drawing title block. The standard third-angle projection symbol is shown in Figure 9.19.

FIRST-ANGLE PROJECTION

First-angle projection is commonly used in Europe and other countries of the world. This method of projection places the glass box in the first quadrant. Views are established by projecting surfaces of the object onto the surface of the glass box. In this projection arrangement, however, the object is between your line of sight and the projection plane, as you can see in Figure 9.20. When the glass box in the first-angle projection quadrant is unfolded, the result is the multiview arrangement shown in Figure 9.21.

A first-angle projection drawing is accompanied by a symbol on or adjacent to the drawing title block. The standard

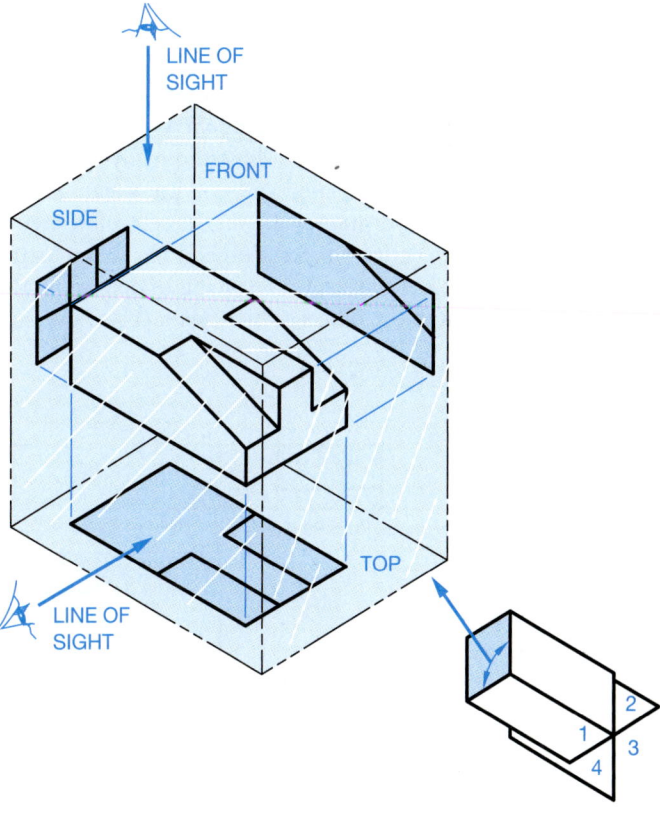

FIGURE 9.20 ■ Glass box in first-angle projection.

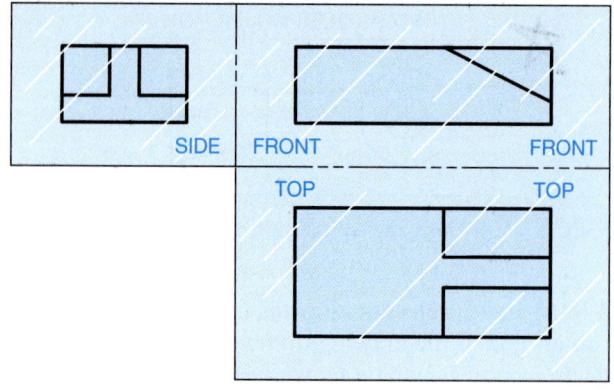

FIGURE 9.21 ■ First-angle projection.

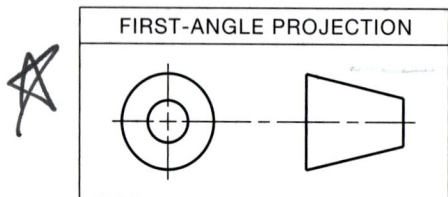

FIGURE 9.22 ■ First-angle projection symbol.

first-angle projection symbol is shown in Figure 9.22. Figure 9.23 shows a comparison of the same object in first- and third-angle projections.

VIEW SELECTION

Although there are six primary views that you can select to completely describe an object, it is seldom necessary to use all six. As a drafter, you must decide how many views are needed to properly represent the object. If you draw too many views you make the drawing too complicated and you are wasting time, which costs your employer money. If you draw too few views, you have not completely described the object. The manufacturing department then has to waste time trying to determine the complete description of the object, which again costs your employer money.

Selecting the Front View

Usually, you should select the front view first. The front view is generally the most important view and, as you learned from the glass box description, the front view is the origin of all other views. There is no exact way for everyone to always select the same front view, but there are some guidelines to follow. The front view should:

■ Represent the most natural position of use.

■ Provide the best shape description or most characteristic contours.

■ Have the longest dimension.

■ Have the fewest hidden features.

■ Be the most stable and natural position.

Take a look at the pictorial drawing in Figure 9.24. Notice the front view selection. This front view selection violates the best shape description and fewest hidden features guidelines. However, the selection of any other view as the front would violate other rules, so in this case there is possibly no absolutely correct answer. Given the pictorial drawings in Figure 9.25, identify the view that you believe would be the best front view for each. Possible answers are given in the caption.

Other View Selection

Use the same rules when selecting other views needed as you do when selecting the front view:

■ Most contours.

■ Longest side.

■ Least hidden features.

■ Best balance or position.

Given the six views of the object in Figure 9.26, which views would you select to completely describe the object? If your selection is the front, top, and right side, then you are correct. Now, take a closer look. Figure 9.27 shows the selected three views. The front view shows the best position and the longest

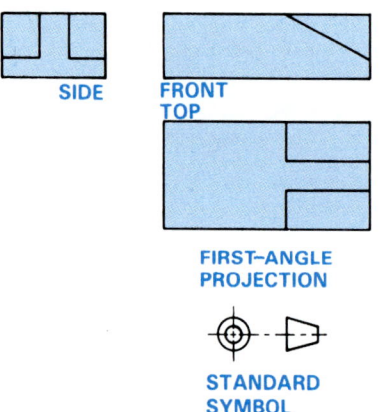

FIRST–ANGLE PROJECTION

STANDARD SYMBOL

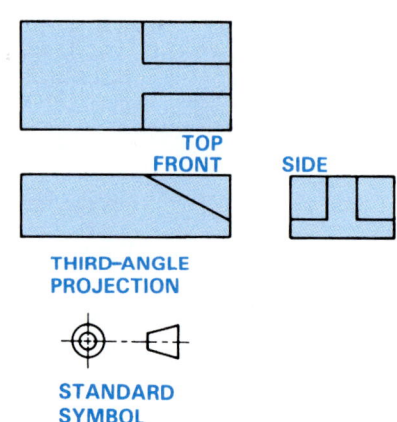

THIRD–ANGLE PROJECTION

STANDARD SYMBOL

FIGURE 9.23 ■ First-angle and third-angle projection.

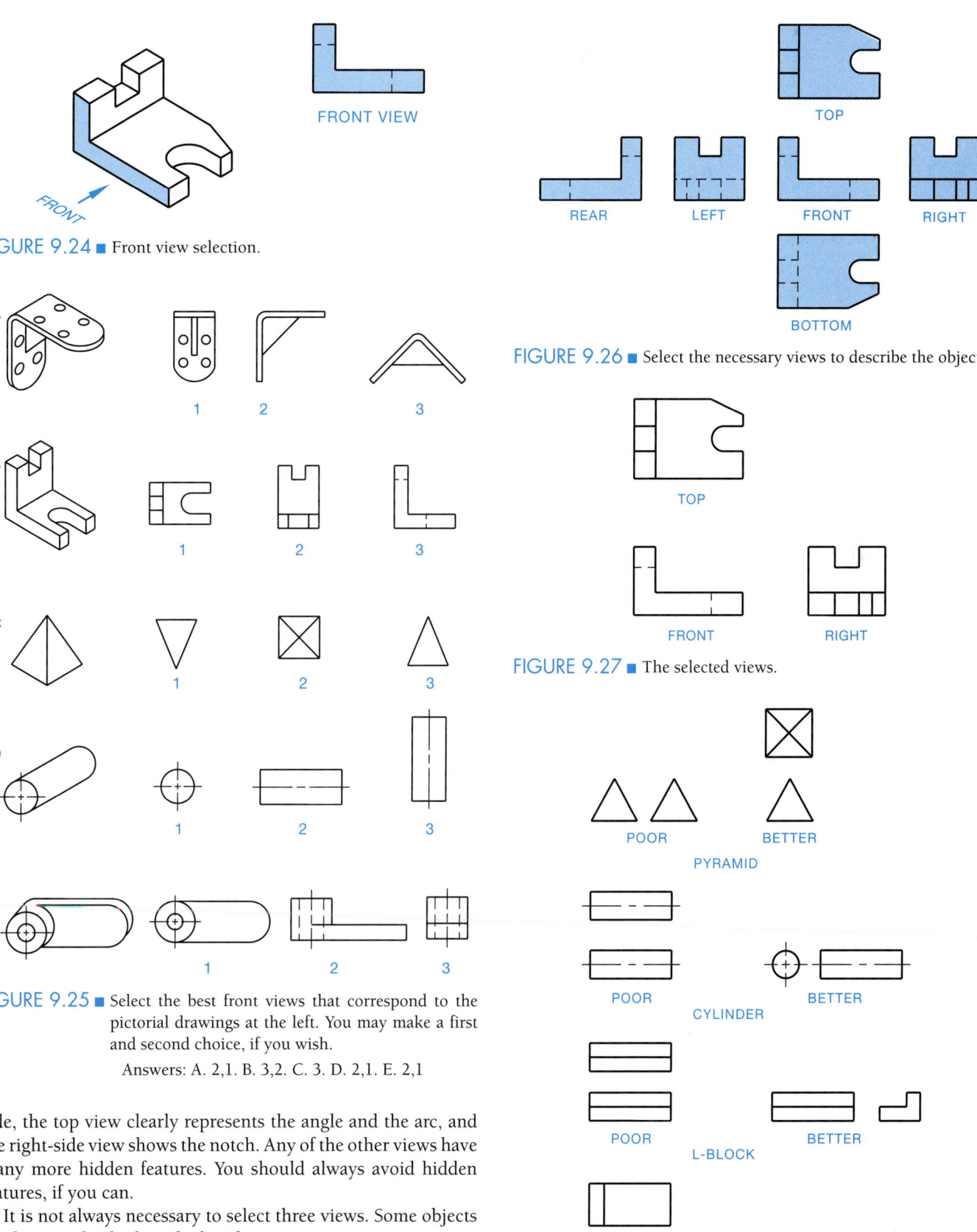

FIGURE 9.24 ■ Front view selection.

FIGURE 9.25 ■ Select the best front views that correspond to the pictorial drawings at the left. You may make a first and second choice, if you wish.

Answers: A. 2,1. B. 3,2. C. 3. D. 2,1. E. 2,1

FIGURE 9.26 ■ Select the necessary views to describe the object.

FIGURE 9.27 ■ The selected views.

FIGURE 9.28 ■ Selecting two views.

side, the top view clearly represents the angle and the arc, and the right-side view shows the notch. Any of the other views have many more hidden features. You should always avoid hidden features, if you can.

It is not always necessary to select three views. Some objects can be completely described with two views or even one view. When selecting fewer than three views to describe an object, you must be careful which view, other than the front view, you select. (See Figure 9.28.) Whenever possible, spend some time looking at actual industry drawings. Doing this gives you a better understanding of how the views of an object are arranged.

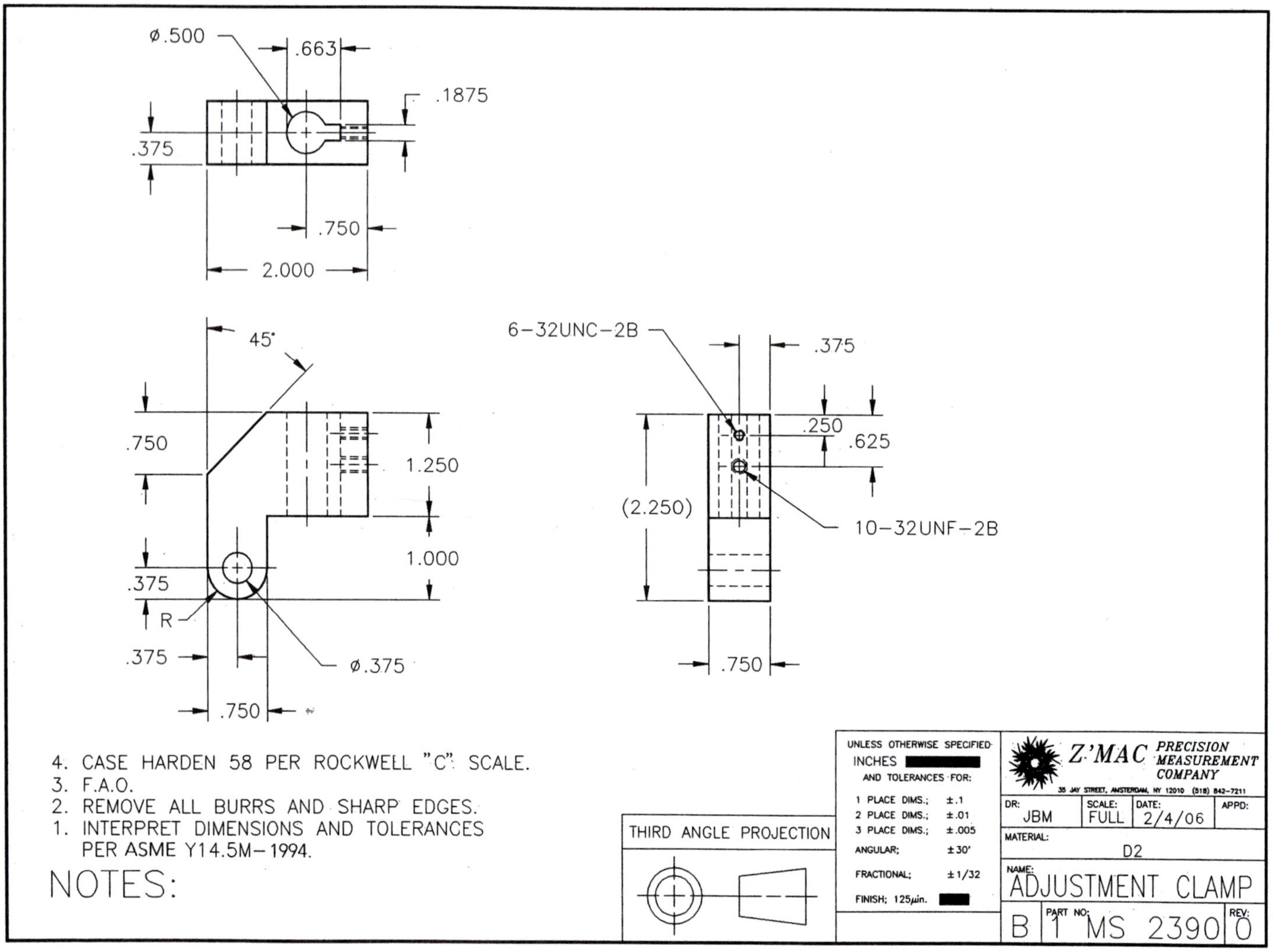

FIGURE 9.29 ■ An actual industry drawing with three views used to describe the part. *Courtesy Jim B. MacDonald.*

Figure 9.29 is an industry drawing showing an object displayed using three views.

One-View Drawings

One-view drawings are also often practical. When an object has a uniform shape and thickness, more than one view is unnecessary. Figure 9.30 shows a gasket drawing where the thickness of the part is identified in the materials specifications of the title block. The types of parts that fit into this category include gaskets, washers, spacers, and similar features.

Although two-view drawings are generally considered the industry's minimum recommended views for a part, other objects that are clearly identified by dimensional shape can be drawn with one view, as in Figure 9.31.

In Figure 9.31, the shape of the pin is clearly identified by the 25-mm diameter. In this case, the second view would be a circle

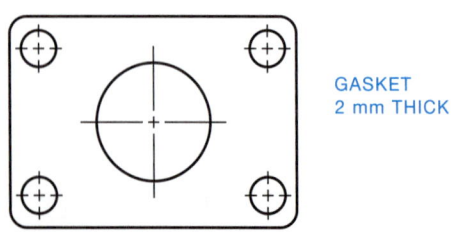

FIGURE 9.30 ■ One-view drawing.

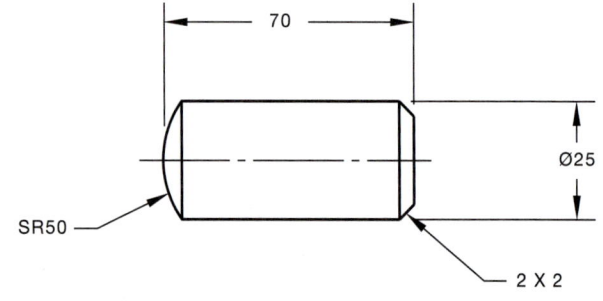

FIGURE 9.31 ■ One-view drawing.

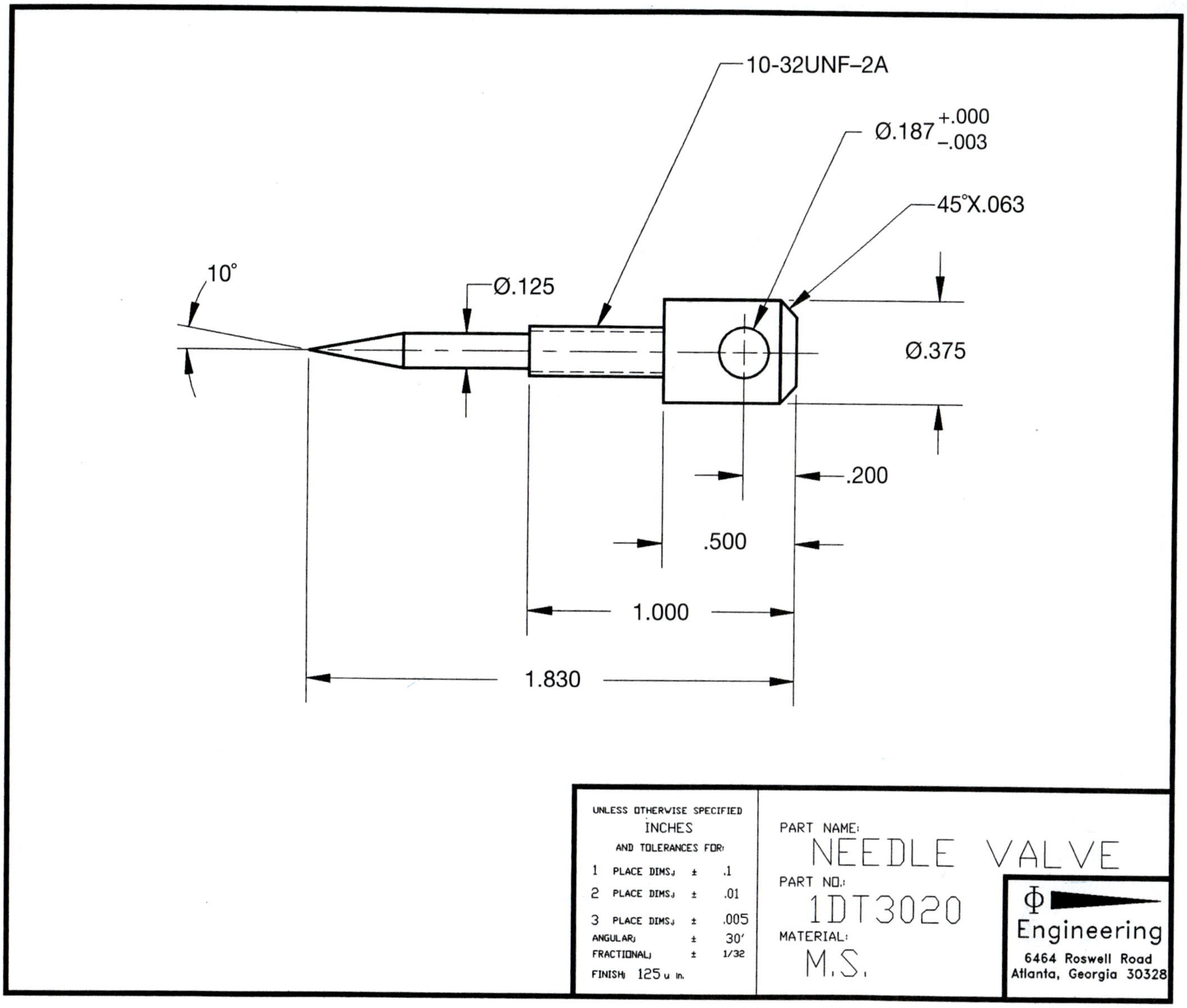

10-32UNF–2A

$\varnothing.187^{+.000}_{-.003}$

45°X.063

10°

$\varnothing.125$

$\varnothing.375$

.200

.500

1.000

1.830

UNLESS OTHERWISE SPECIFIED INCHES AND TOLERANCES FOR:		
1 PLACE DIMS.	±	.1
2 PLACE DIMS.	±	.01
3 PLACE DIMS.	±	.005
ANGULAR.	±	30'
FRACTIONAL.	±	1/32
FINISH: 125 u in.		

PART NAME:
NEEDLE VALVE
PART NO.:
1DT3020
MATERIAL:
M.S.

Φ► Engineering
6464 Roswell Road
Atlanta, Georgia 30328

FIGURE 9.32 ■ An actual industry drawing with one view used to describe the part. *Courtesy IO Engineering.*

and would not necessarily add any more valuable information to the drawing. The primary question to keep in mind is, can the part be easily manufactured from the drawing without confusion? If there is any doubt, then the adjacent view should probably be drawn. As an entry-level drafter, you should ask your drafting supervisor to clarify the company policy regarding the number of views to be drawn. Figure 9.32 provides an actual industry drawing of an object with one view.

Partial Views

When symmetrical objects are drawn in limited space or when there is a desire to save valuable drafting time, partial views can be used. The top view in Figure 9.33 is a partial view. Notice how the short break line is used to show that a portion of the view is omitted. Caution should be exercised when using partial

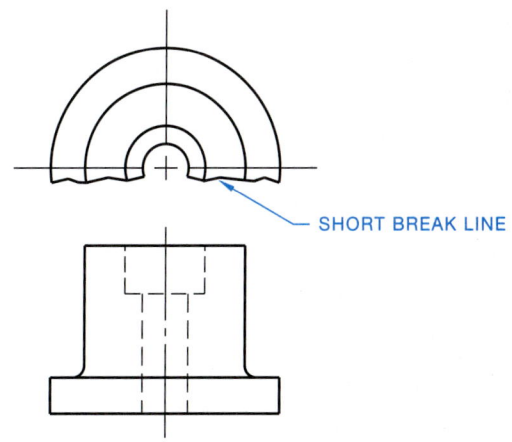

SHORT BREAK LINE

FIGURE 9.33 ■ Partial view.

views, as confusion could result in some situations. If the partial view reduces clarity, then draw the entire view. When using CADD, it is easier to draw the full view.

View Enlargement

When part of a view has detail that cannot be clearly dimensioned due to the drawing scale or complexity, a view enlargement can be used. To establish a view enlargement, a thick phantom line circle is placed around the area to be enlarged. This circle is broken at a convenient location and an identification letter is centered in the break. The height of this letter is generally .24 in. (6 mm) minimum so it stands out from other lettering. Arrowheads are then placed on the line next to the identification letter, as shown in Figure 9.34. An enlarged view of the detailed area is then shown in any convenient location in the field of the drawing. Common enlargement scales are 2:1, 4:1, and 10:1. The view identification and scale is placed below the enlarged view, such as DETAIL A. The lettering height is placed as a title, which is .24 in. (6 mm). The word DETAIL can be the same height or less than the identification letter depending on company preference. (See Figure 9.34.) An actual industry drawing with three principal views and a view enlargement is provided in Figure 9.35.

Alternate View Placement

It is always best to place views in the alignment based on the six principal views previously discussed. This helps ensure that the drawing is read based on traditional view-projection methods. However, alternate view placement can be used, if absolutely necessary. Using an alternate position for a view is possible when

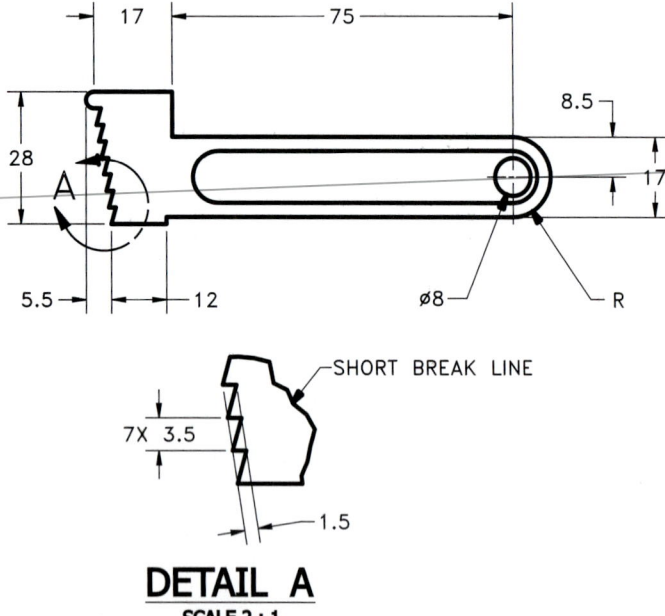

FIGURE 9.34 ■ View enlargement example.

space is a problem. For example, the rear view can be placed aligned with and to the right of the right side view. Any use of this practice should be confirmed with your company standards.

Removed Views

In some cases it is necessary to place a view out of normal arrangement with the other views. The situations that make this necessary can be available sheet space or when you want to enlarge the view. This is done by placing viewing plane lines identifying where the view is taken. At each end of the viewing plane, lines are labeled with A for the first removed view and B for the second removed view. Consecutive letters are used for additional removed views. The view is then moved to a desired location on the drawing and labeled to correspond with its viewing plane line, such as VIEW A-A. The removed view is kept in the same alignment as its normal arrangement. The removed view can be kept at the same scale as the other views or it can be enlarged. When the view is enlarged, the scale is placed under the view name, such as SCALE 2:1. This topic was briefly discussed in Chapter 7, and an example was shown in Figure 7.17. It is preferable to keep the removed view on the same sheet from where the view was taken. If a view is placed on a different sheet from where its viewing plane line is located, the sheet number and the zone of the cross-reference location is given with the view title, such as SEE SHEET 1 ZONE B4. Any use of this practice should be confirmed with your company standards.

Rotated Views

Rotating views from their normal alignment with the six principal views should be avoided, but it is possible if necessary. Situations such as limited sheet space, view size, or an effort to keep all views on one sheet can contribute to this possibility. If this is done, you need to give the angle and direction of rotation under the view title, such as ROTATED 90° CW. CW is the abbreviation for clockwise, and CCW is the abbreviation for counterclockwise. Any use of this practice should be confirmed with your company standards.

PROJECTION OF CONTOURS, CIRCLES, AND ARCS

Some views do not clearly identify the shape of certain contours. You must then draw the adjacent views to visualize the contour. (See Figure 9.36.)

The edge view of a surface shows that surface as a line. The true contour of the slanted surface is seen as an edge in the front view of Figure 9.36 and describes the surface as an angle, while the surface is foreshortened, slanting away from your line of sight, in the right-side view. In Figure 9.37, select the front view that properly describes the given right-side view. All three front views in Figure 9.37 could be correct. The side view does

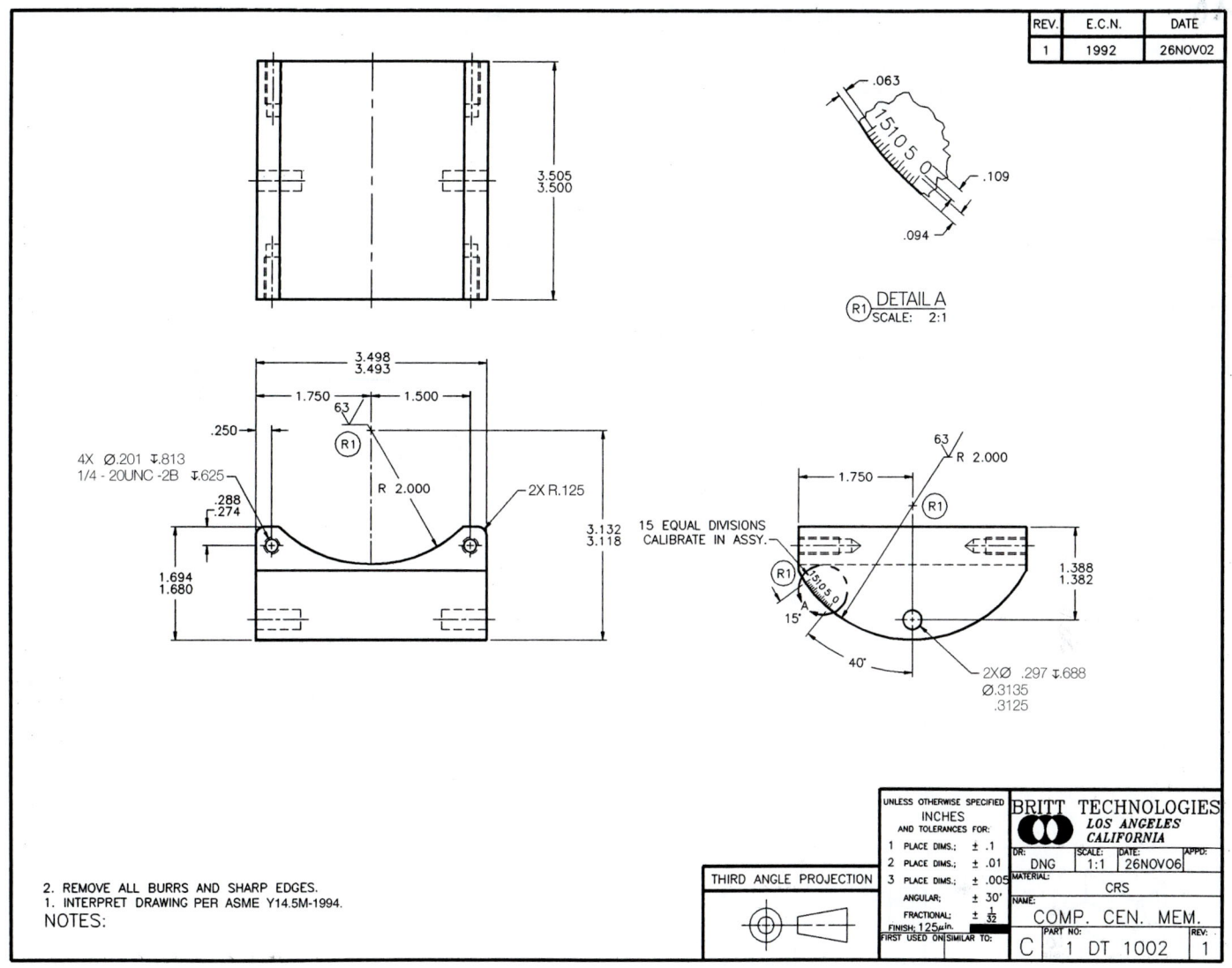

FIGURE 9.35 ■ An actual industry drawing with three principal views and a view enlargement used to describe the part. *Courtesy BRITT Technologies.*

not help the shape description of the front view contour. (See Figure 9.38.)

Cylindrical shapes appear round in one view and rectangular in another view, as seen in Figure 9.39. Both views in Figure 9.39 may be necessary as one shows the diameter shape and the other shows the length. The ability to visualize from one view to the next is a critical skill for a drafter. You may have to train yourself to look at two-dimensional objects and picture three-dimensional shapes. You can also use some of the techniques discussed here to visualize features from one view to another.

Chamfers

A chamfer is the cutting away of the sharp external or internal corner of an edge. Chamfers are used as a slight angle to relieve a sharp edge, or to assist the entry of a pin or thread into the mating feature. A chamfer is an inclined surface and is shown as a slanted edge or a line, depending on the view. The size and shape of chamfers is projected onto views. Figure 9.40 shows three examples of chamfered features. Notice how the true contour of the slanted surfaces is seen as edges in the front views, which describe the chamfers as angles. The surfaces shown in the top view of example (a) and the right-side views of examples (b) and (c) are foreshortened and slant away from your line of sight.

Circles on Inclined Planes

When the line of sight, in multiviews, is perpendicular to a circular feature such as a hole, the feature appears round, as shown in Figure 9.41. When a circle is projected onto an inclined surface, its view is elliptical in shape, as shown in Figure 9.42. The

β148

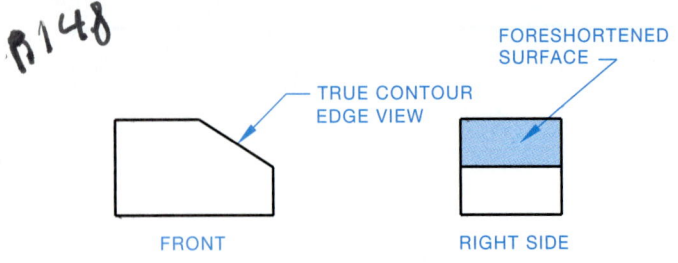

FIGURE 9.36 ■ Contour representation.

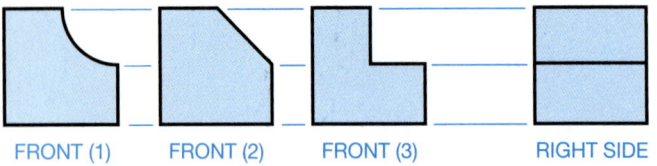

FIGURE 9.37 ■ Select the front view that properly goes with the given right-side view.

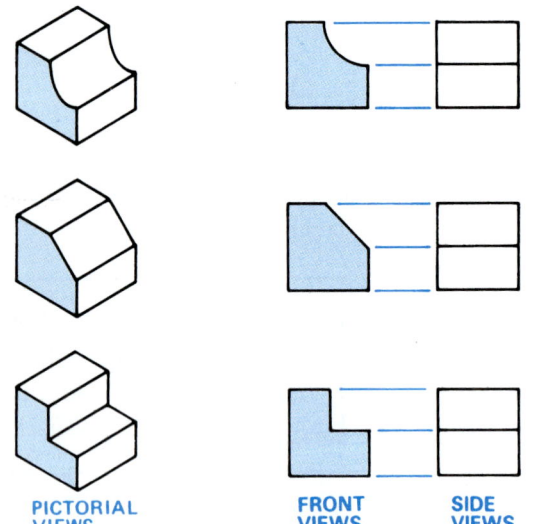

FIGURE 9.38 ■ The importance of a view that clearly shows the contour of a surface.

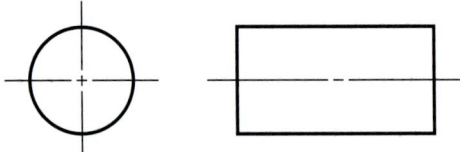

FIGURE 9.39 ■ Cylindrical shape representation.

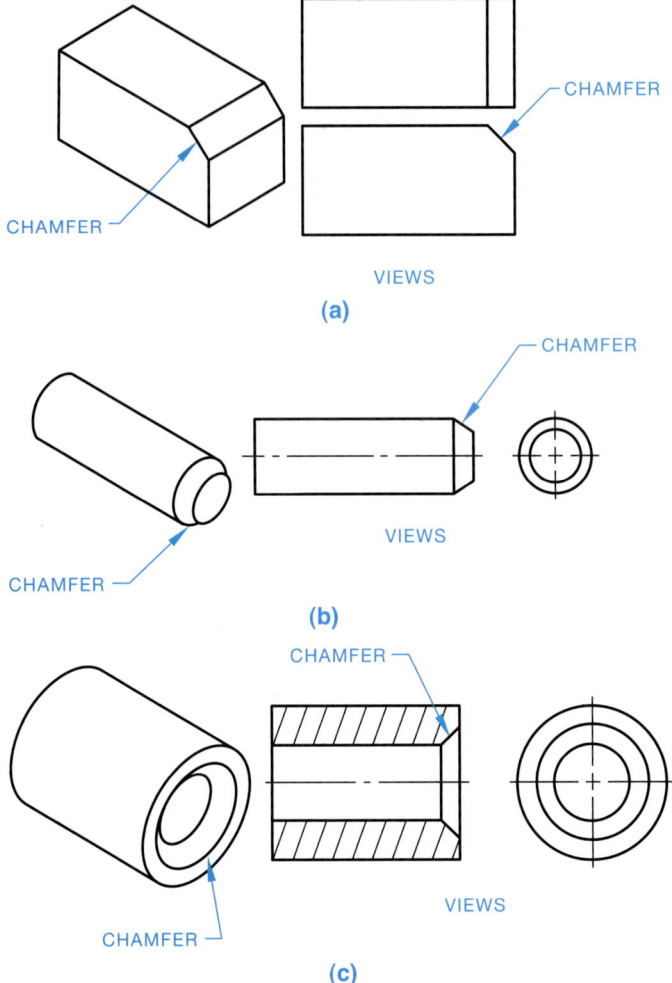

FIGURE 9.40 ■ Three examples of chamfered features.

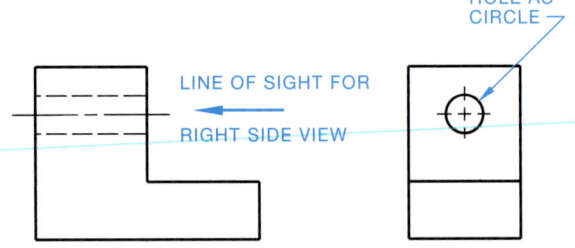

FIGURE 9.41 ■ View of a hole projected as a circle and its hidden view through the part.

ellipse shown in the top and right-side views of Figure 9.42 was established by projecting the major diameter from the top to the side view and the minor diameter to both views from the front view, as shown in Figure 9.43. The major diameter in this example is the hole diameter.

The rectangular areas projected from the two diameters in the top and right-side views of Figure 9.43 provide the bound-aries of the ellipse to be drawn in these views. The easiest method of drawing the ellipse is to find an ellipse template that has a shape that fits within the area.

Ellipse templates, as discussed in Chapters 2 and 6, desig-nate elliptical shapes by the angle of the circle in relationship to the line of sight. If the slanted surface is 30°, then a 30° ellipse template is used to draw the elliptical shape. Keep in mind that the other view requires a 60° ellipse. When the angle is 45°, then both views are the same. The ellipse template selected should have a shape that fits into the rectangle constructed, as shown in Figure 9.44. If you cannot find a representative ellipse that fits

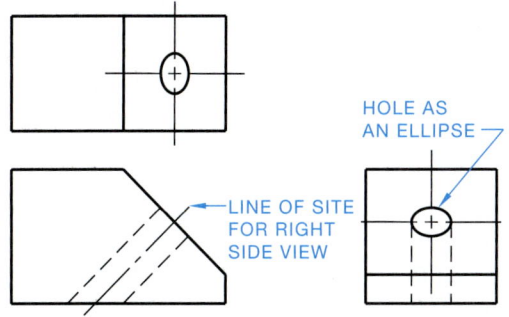

FIGURE 9.42 ■ Hole represented on an inclined surface as an ellipse.

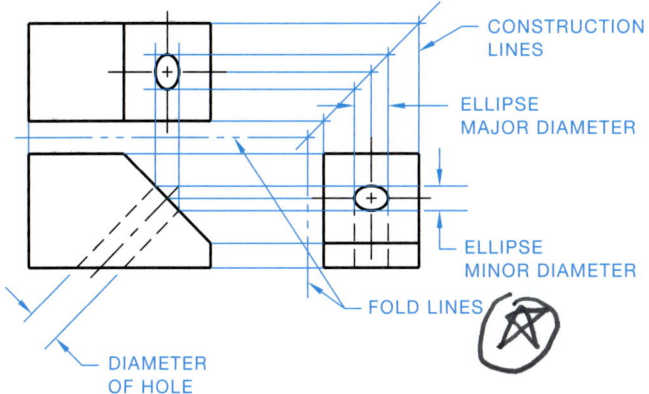

FIGURE 9.43 ■ Establish an ellipse in the inclined surface.

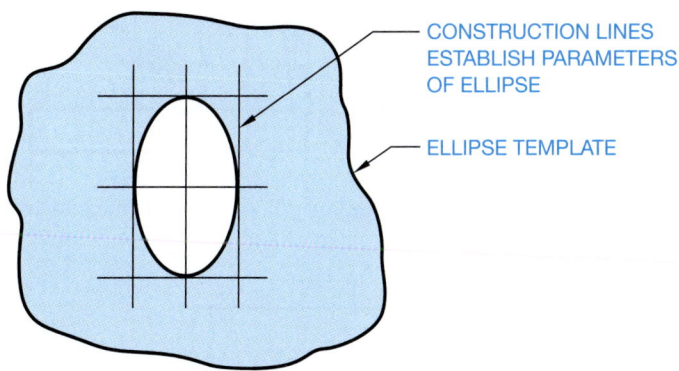

FIGURE 9.44 ■ Ellipse template.

exactly, then use the closest size possible, or verify the drawing procedure with your supervisor. Establishing an elliptical shape with geometric construction is much more time-consuming but may be necessary when templates are not available. The elliptical shapes are a representation and may not have to be perfect, depending on company policy. Also, notice that the hidden lines for the hole have been omitted in the top view of Figure 9.43. These lines would require a great amount of extra work without adding significantly to the function of the drawing. Such a practice of omitting lines for an accurate representation can be used with caution and if it complies with company practice. Remem-

ber that the drawing must easily communicate the manufacture of the part. The use of CADD makes it easy to draw object lines or hidden ellipses.

Arcs on Inclined Planes

When a curved surface from an inclined plane must be drawn in multiview, a series of points on the curve establishes the contour. Begin by selecting a series of points on the curved contour as shown in the right-side view of Figure 9.45. Project these points from the right side to the inclined front view. From the point of intersection on the inclined surface in the front view, project lines to the top view. Then project corresponding points from the right side to the 45° projection line and onto the top view. Corresponding lines create a pattern of points, as shown in the top view of Figure 9.45. You can also use a compass or dividers to transfer measurements from the fold lines, as previously discussed. After the series of points is located in relationship to the front and top views, connect the points with an irregular curve. See the completed curve in Figure 9.46.

Fillets and Rounds

Fillets are slightly rounded inside curves at corners, generally used to ease the machining of inside corners or to allow patterns to release more easily from castings and forgings. Fillets can also be designed into a part to allow additional material on inside corners for stress relief. (See Figure 9.47.)

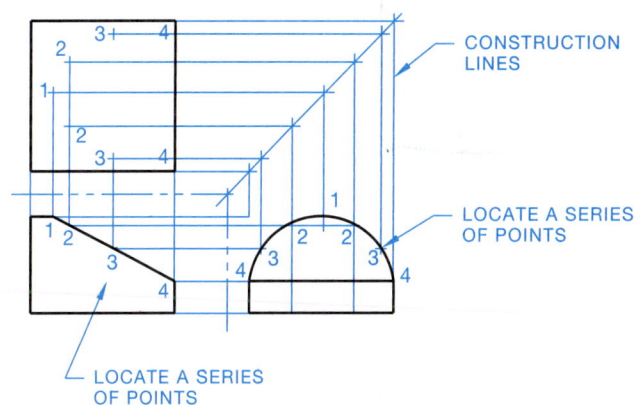

FIGURE 9.45 ■ Locating an inclined curve in multiview.

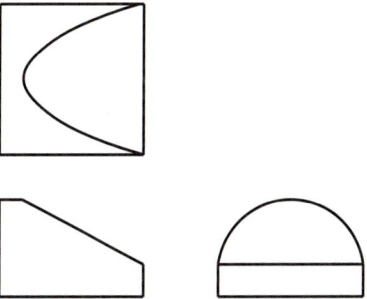

FIGURE 9.46 ■ Completed curve.

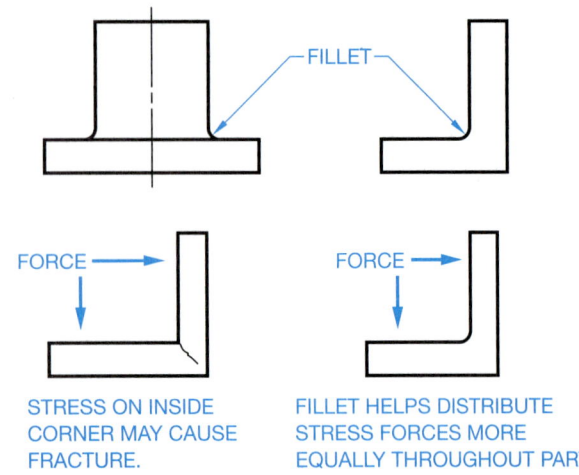

FILLET

FORCE →   FORCE →

STRESS ON INSIDE CORNER MAY CAUSE FRACTURE.   FILLET HELPS DISTRIBUTE STRESS FORCES MORE EQUALLY THROUGHOUT PART.

FIGURE 9.47 ■ Fillets.

Other than the concern of stress factors on parts, certain casting methods require that inside corners have fillets. The size of the fillet often depends on the precision of the casting method. For example, very precise casting methods can have smaller fillets than green-sand casting, where the exactness of the pattern requires large inside corners. Fillets are also common on machined parts, because it is difficult to make sharp inside corners.

Rounds are rounded outside corners that are used to relieve sharp exterior edges. Rounds are also necessary in the casting and forging process for the same reasons as fillets. Figure 9.48 shows rounds represented in views.

A machined edge causes sharp corners, which can be desired in some situations. However, if these sharp corners are to be rounded, the extent of roundness depends on the function of the part. When a sharp corner has only a slight relief, it is referred to as a break corner, as shown in Figure 9.49.

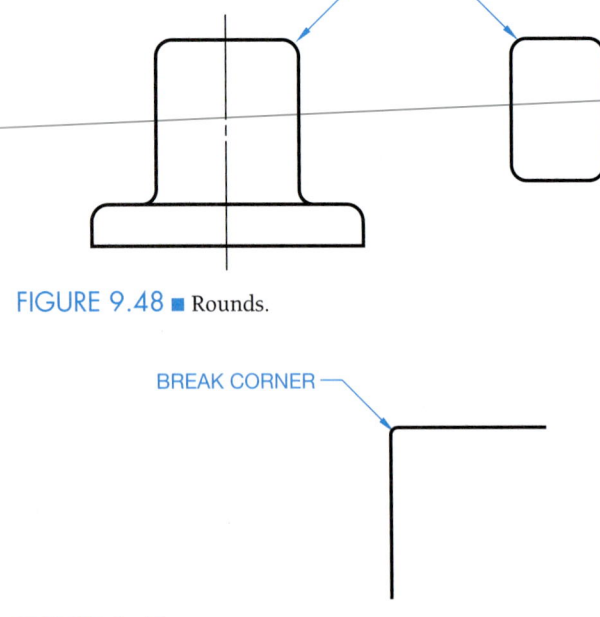

ROUNDS

FIGURE 9.48 ■ Rounds.

BREAK CORNER

FIGURE 9.49 ■ Break corner.

Rounded Corners in Multiview

An outside or inside slightly rounded corner of an object is represented in multiview as a contour only. The extent of the round or fillet is not projected into the view as shown in Figure 9.50. Cylindrical shapes may be represented with a front and top view, where the front identifies the height and the top shows the diameter. Figure 9.51 shows how these cylindrical shapes should be represented in multiview. Figure 9.52 shows the representation of the contour of an object as typically displayed in multiview using phantom lines. This is done to clearly accent the rounded feature.

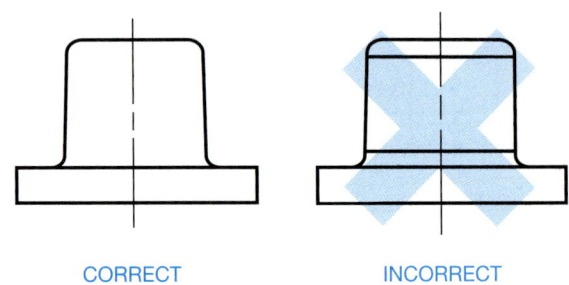

CORRECT   INCORRECT

FIGURE 9.50 ■ Rounds and fillets in multiview.

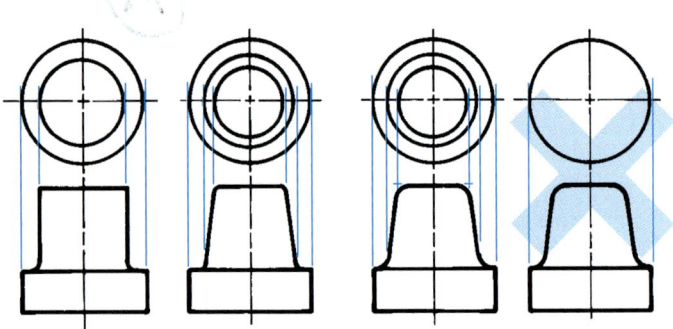

FIGURE 9.51 ■ Rounded curves and cylindrical shapes in multiview.

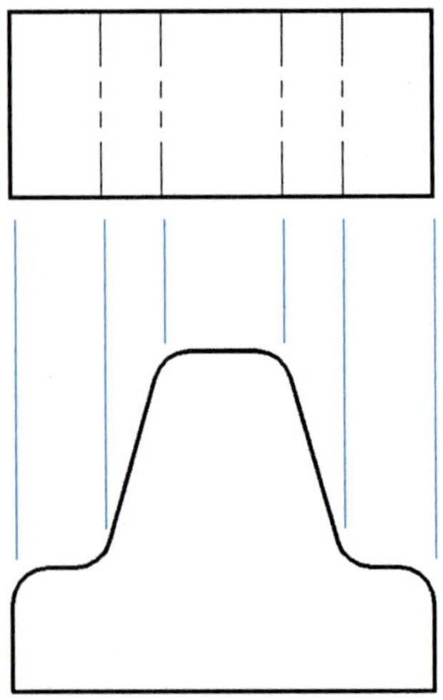

FIGURE 9.52 ■ Contour in multiview.

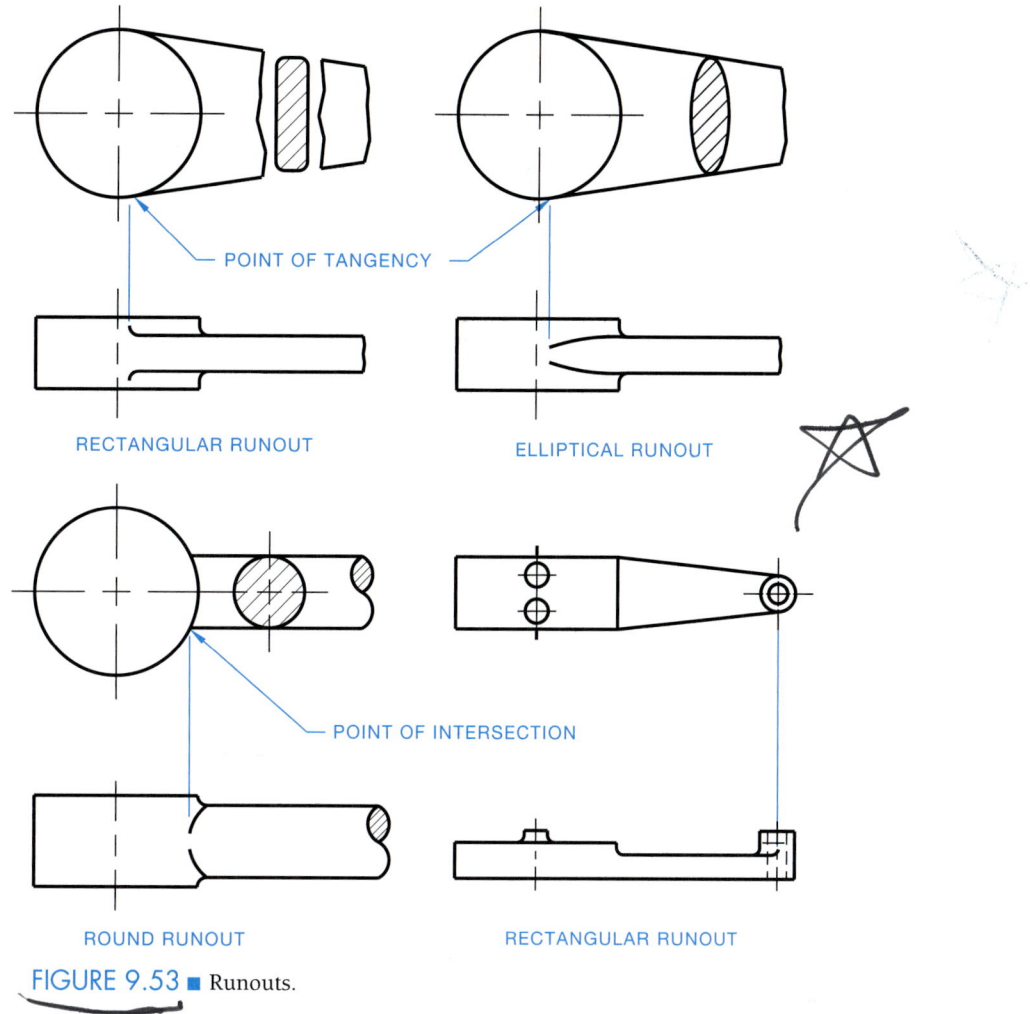

POINT OF TANGENCY

RECTANGULAR RUNOUT   ELLIPTICAL RUNOUT

POINT OF INTERSECTION

ROUND RUNOUT   RECTANGULAR RUNOUT

FIGURE 9.53 ■ Runouts.

Runouts

The intersections of features with circular objects are projected in multiview to the extent where one shape runs into the other. The characteristics of the intersecting features are known as <u>runouts</u>. The runout of features intersecting cylindrical shapes is projected from the point of tangency of the intersecting feature, as shown in Figure 9.53. Notice also that the shape of the runout varies when drawn at the cylinder depending on the shape of the intersecting feature. Rectangular-shaped features have a fillet at the runout, while curved (elliptical or round) features contour toward the centerline at the runout. Runouts may also exist when a feature such as a web intersects another feature, as shown in Figure 9.54.

LINE PRECEDENCE

When drawing multiviews, it is common for one type of line to fall in line with a different line type. As a drafter, you need to decide which line to draw. This is known as <u>line precedence</u>. You draw the line that is most important based on the following rules, and as shown in Figure 9.55:

- Object lines take precedence over hidden lines and center-lines.

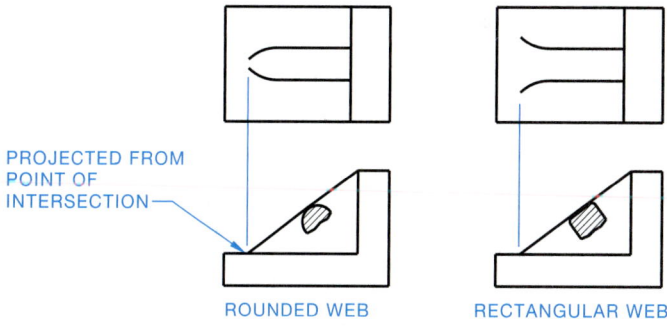

PROJECTED FROM POINT OF INTERSECTION

ROUNDED WEB   RECTANGULAR WEB

FIGURE 9.54 ■ Other types of runouts.

- Hidden lines take precedence over centerlines.

- In sectioning (covered in Chapter 13), cutting-plane lines take precedence over centerlines.

When an object line is drawn over a centerline, the ends or tails of the centerline are drawn slightly beyond the outside of the view or feature. Use a uniform distance, such as .125 in. (3 mm) or .25 in. (6 mm), past an object. The distance the centerline goes past the object depends on the size of the drawing and company standards. The uniform distance past an object is maintained unless a longer distance is needed for dimensioning purposes.

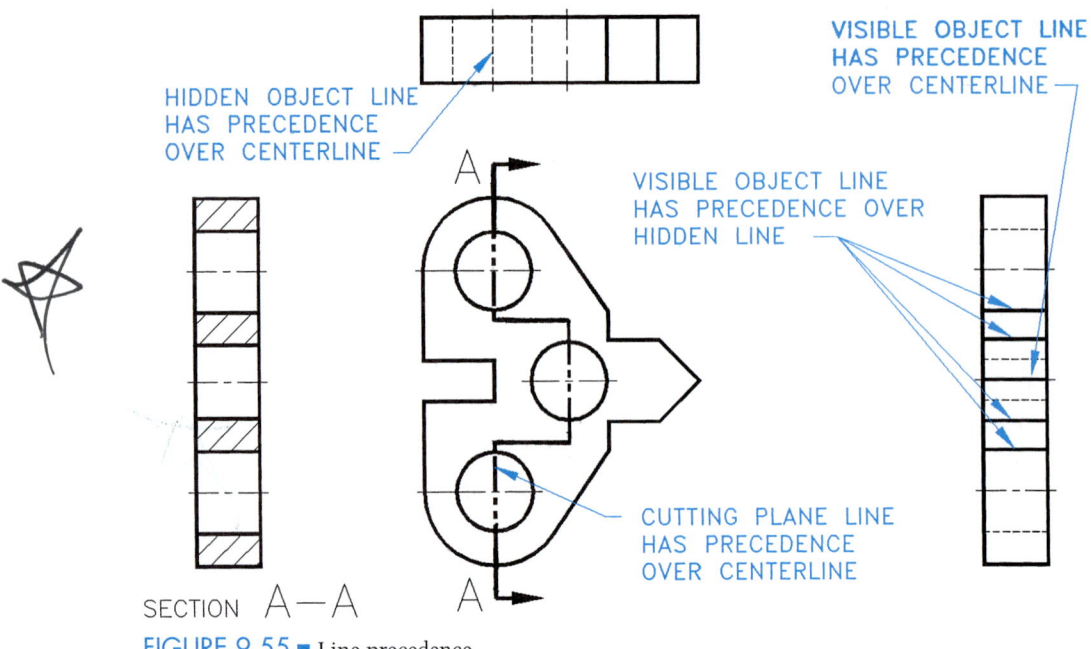

HIDDEN OBJECT LINE
HAS PRECEDENCE
OVER CENTERLINE

VISIBLE OBJECT LINE
HAS PRECEDENCE
OVER CENTERLINE

VISIBLE OBJECT LINE
HAS PRECEDENCE OVER
HIDDEN LINE

CUTTING PLANE LINE
HAS PRECEDENCE
OVER CENTERLINE

SECTION A—A

FIGURE 9.55 ■ Line precedence.

When a centerline is used to establish a dimension, the center-line continues as an extension line without a gap where the centerline ends and the extension line begins. This is discussed more in Chapter 11. Centerlines do not extend between views of the drawing, except as discussed with auxiliary views in Chapter 10. Figure 9.55 shows examples of line precedence.

MULTIVIEW ANALYSIS AND REVIEW

The following is a review covering the features and importance of using multiviews and orthographic projection:

■ The multiview system of orthographic projection is covered in the ASME Y14.3M standard.

■ The two internationally accepted view projection systems are third-angle and first-angle projection. Both systems are discussed in this chapter, and third-angle projection is used throughout this textbook.

■ Multiview drawings represent the shape of an object using two or more views; however, one view can be used when a third dimension such as depth, thickness, or diameter can be given in a note or as a dimension on the object.

■ There are six principal views: front, top, right side, left side, bottom, and rear.

■ All views are aligned in either third-angle projection or first-angle projection, and the appropriate third- or first-angle projection symbol should be placed on the drawing.

■ Adjacent views must be aligned, unless otherwise specified. Adjacent views are two adjoining views aligned by projection.

■ Related views must be aligned, unless otherwise specified. Related views are two views adjacent to the same intermediate view.

■ There is always one dimension that is the same between views.

■ The front view is normally considered the most important view from which other views are established. Specific view selection guidelines are provided in this chapter.

■ The number of required views can vary depending on the complexity of the object and must be selected to completely describe the shape of the object and features on the object.

■ A line in a view is in true length when the line is parallel to the projection plane.

■ A surface of a view is in true size and shape when the surface is parallel to the projection plane.

■ A line in a view is foreshortened when the line is not parallel to the projection plane.

■ A surface of a view is foreshortened when the surface is not parallel to the projection plane.

■ Additional view placement options are discussed in this chapter and include: removed views, rotated views, and cross-referenced views. Use of these options should be carefully confirmed with your company practices.

RECOMMENDED REVIEW

It is recommended that you review the following parts of Chapter 7 before you begin working on multiview drawings.

(Continued on page 257)

MULTIVIEWS

Most CADD programs contain a variety of tools and options that allow you to develop multiview drawings very efficiently and accurately. Options such as GRID, SNAP, ORTHO, and coordinate entry methods allow you to easily create new objects in a given drawing area and to organize different views. Another very effective multiview creation and modification technique involves the use of construction lines.

An earlier discussion explained and showed you how to set up multiviews by projecting features from one view to another using construction lines. Construction lines are used for layout purposes only and can be erased or left on the drawing if they are very light and do not reproduce. Construction lines are also widely used in CADD for layout work. These lines are usually drawn using a unique format, commonly called layers, that can be named something such as CONSTRUCTION and assigned a specific color. Once construction lines have been used to serve their purpose, such as establishing views, they can be removed with a DELETE or ERASE command, or their layer can be frozen or turned off.

CADD construction lines can be drawn using basic drafting commands such as LINE, but there are also specific tools that improve the usefulness of construction lines. One example is AutoCAD's XLINE command. The XLINE command allows you to create construction lines of infinite length that can be placed horizontally, vertically, at an angle, offset from an existing line, or even used to bisect an angle. Figure 9.56 shows how construction lines can be used in CADD to project features between views.

VIEW ENLARGEMENT

CADD software greatly enhances your ability to create precise view enlargements or details. There are several possible techniques for creating enlarged views, depending on the CADD application. Combinations of general editing commands such as COPY and SCALE allow you to copy existing geometry to a new location and enlarge the items to any size. Another very effective technique is to use drawing display settings such as AutoCAD's Model Space and Paper Space (Layout). With this example, you can increase the displayed size of a particular feature of an existing view using a scaled viewport. Working with drawing display options is extremely effective because you do not have to copy and enlarge or otherwise recreate a view. Instead, you just show a piece of the view using different display characteristics.

CHAMFERS

Though chamfers can be added to a drawing using a basic LINE command, many CADD programs allow you to draw a chamfer between any two lines using a specific chamfering tool. For example, AutoCAD has a command called CHAMFER that can be used to create a specific size chamfer. If the lines intersect, they are automatically trimmed and the chamfer is drawn. If the lines do not meet, they are automatically extended and the chamfer is drawn. The usual process for adding chamfers to a drawing is to access the CHAMFER command, specify the required chamfer distances or a distance and an angle, define any other special options, and then pick two lines. Figure 9.57 shows some before and after examples of using the CHAMFER command.

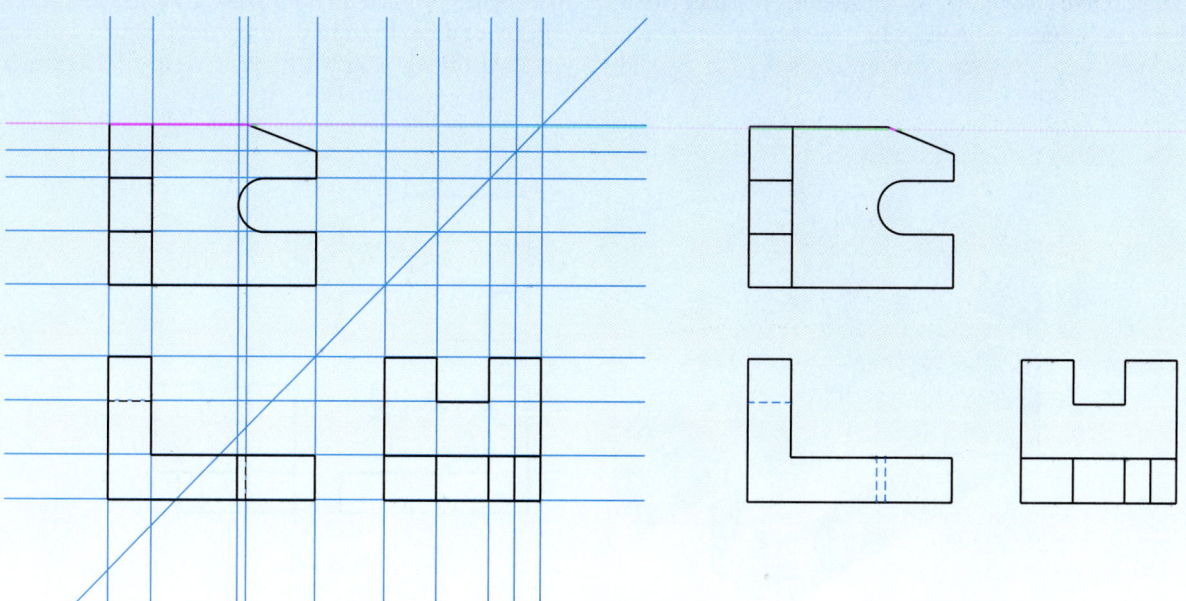

FIGURE 9.56 ■ Construction lines used in CADD to project features between views. The views on the right have the construction lines turned off.

(Continued)

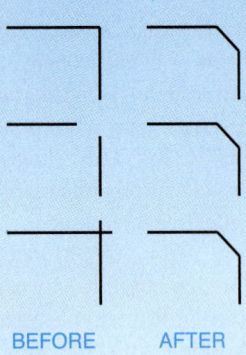

BEFORE   AFTER

FIGURE 9.57 ■ Before and after examples of using the CHAM-FER command.

Fillets and Rounds

CADD programs allow you to draw a tangent fillet or round between any two lines or a line and an arc. For example, the AutoCAD command for this operation is called FILLET, but it draws any radius exterior or interior cor-

ner. Much like placing chamfers as previously described, if the lines intersect, they are automatically trimmed and the fillet is drawn, and if the lines do not meet, they are automatically extended and the fillet is drawn. The typical process for adding fillets and rounds to a drawing is to access the FILLET command, specify the required arc radius and any other special options, and then pick two lines or a line and an arc. Figure 9.58 shows some before and after examples of using the FILLET command.

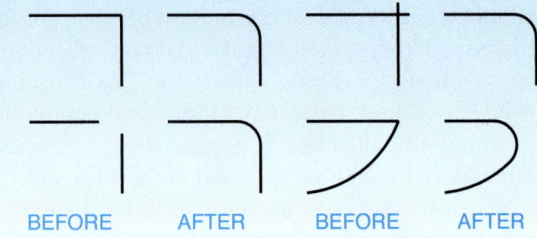

BEFORE   AFTER   BEFORE   AFTER

FIGURE 9.58 ■ Before and after examples of using the FILLET command.

MULTIVIEWS

Some three-dimensional solid modeling programs combine solid modeling with parametric two-dimensional drawing capabilities. Drafters new to solid modeling and the two-dimensional drawing capabilities of some solid modeling programs are often amazed by the speed at which entire multiview drawings can be created from a solid

model of the object. When using solid modeling programs, you do not actually draw multiview geometry in the drawing environment, because view information already exists in a model you create. Multiview drawings are generated quickly and easily from existing three-dimensional models. (See Figure 9.59.) All you have to do is extract the

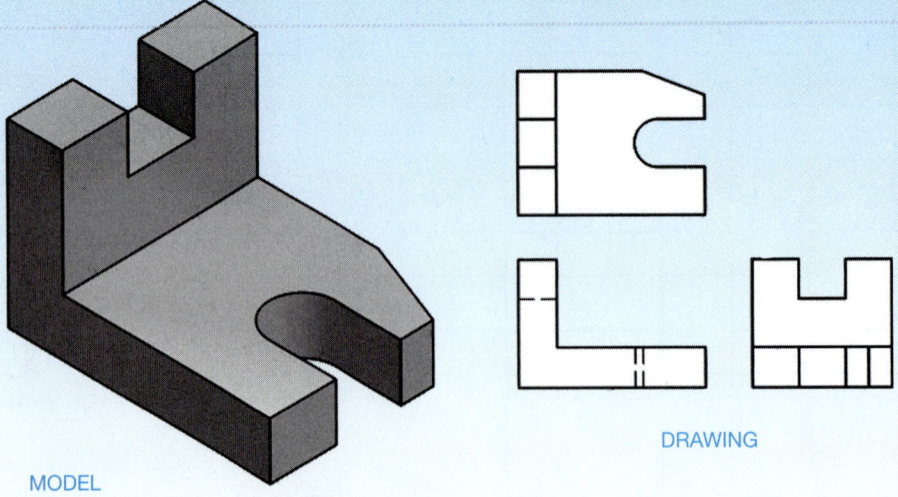

MODEL

DRAWING

FIGURE 9.59 ■ Multiview drawings are generated quickly and easily from existing three-dimensional models.

model information in the drawing environment to generate the multiviews. For example, assuming the model is correct and you want to create a front view, all you have to do is select FRONT from a menu and pick a location for the view on the sheet. In addition, the parametric nature of solid modeling programs ensures that when you edit a model, the corresponding drawing updates to reflect the new design. Similarly, you can change a model by modifying the parametric model dimensions inside the two-dimensional drawing.

There are two basic options for developing a multiview drawing by referencing an existing model. One technique is to insert a single view, such as a front view, using tools like BASE VIEW or MODEL VIEW. Then you can project additional views, like the top and right side views, using the PROJECTED VIEW tool. The second option involves using predefined view layouts to insert multiple drawing views at a single time. Options like Autodesk Inventor's SHEET FORMAT and SolidWorks' STANDARD 3 VIEW can be used to almost instantly create a number of drawing views.

VIEW ENLARGEMENT

The ease of producing enlarged views is greatly enhanced with some solid modeling programs that contain two-dimensional drawing options. Tools such as DETAIL VIEW allow you to create a view enlargement by referencing an existing drawing view, which, as previously described, parametrically associates a view with the model. (See Figure 9.60.) Using tools like DETAIL VIEW typically involves creating a circle, or in some cases a rectangle, around a specific feature on an existing drawing view, such as the front view. Then it is just a matter of picking a location for the view enlargement. The required phantom line circle, arrowheads, identification letter, and view title are automatically added to the drawing.

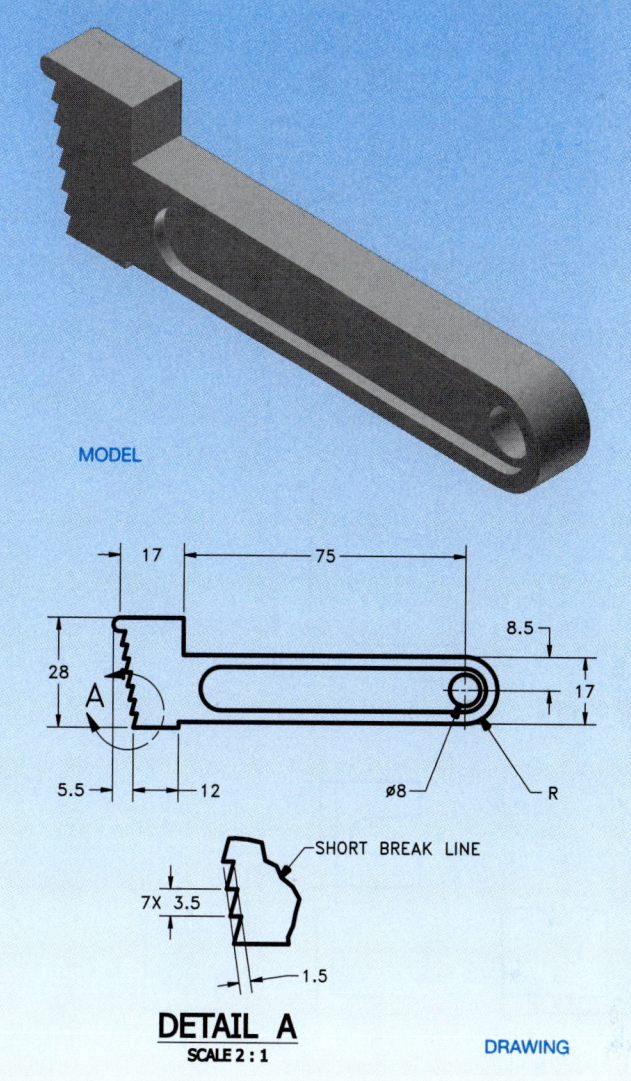

FIGURE 9.60 ■ Tools such as DETAIL VIEW allow you to create a view enlargement by referencing an existing drawing view, which parametrically associates all views with the model.

This will refresh your memory about how related lines are properly drawn.

- Object Lines.
- Viewing- and Cutting-Planes.
- Hidden Lines.
- Break Lines.
- Centerlines.
- Phantom Lines.

MULTIVIEW LAYOUT

Many factors influence the drawing layout. Your prime goal should be a clear and easy-to-read interpretation of selected views with related information. Although this chapter deals with multiview presentation, it is not totally realistic to consider view layout without thinking about the effects of dimension placement on the total drawing. Chapter 11 correlates the multiview drawings of *shape description* with dimensioning, known as **size description**.

The initial steps in view layout should be performed using rough sketches. By using rough sketches, you can analyze which views you need before you begin formal drafting. Sketches do not have to be perfect. Try to sketch as fast as you can to save time.

Sketching the Layout

Consider the engineering sketch in Figure 9.61 as you evaluate the proper view layout.

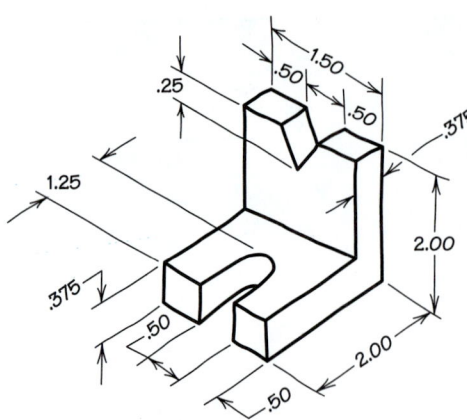

FIGURE 9.61 ■ Engineering sketch.

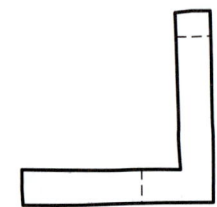

FIGURE 9.62 ■ Sketch the front view.

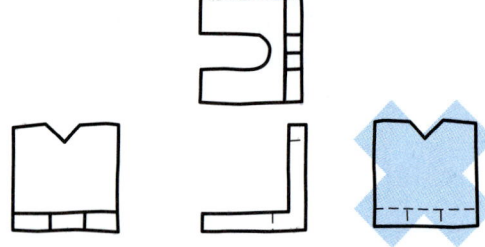

FIGURE 9.63 ■ Sketch the required views.

STEP 1 Select the front view using the rules discussed in this chapter. Sketch the front view that you have picked. Try to keep your sketch proportional to the actual object, as in Figure 9.62. However, keep in mind that a sketch does not have to be perfect. It should be done quickly to save time, while helping you lay out the drawing.

STEP 2 Select the other views needed to completely describe the shape of the V-BLOCK MOUNT, as shown in Figure 9.63.

The front, top, and left-side views clearly define the shape of the V-BLOCK MOUNT. Now lay out the formal drawing using the sketch as a guide. Several factors must be considered before you begin:

1. Size of drawing sheet.

2. Scale of the drawing.

3. Number and size of views.

4. Amount of blank space required for future revisions.

5. Dimensions and notes (not drawn at this time).

Drawing the Layout

STEP 1 Use an A3-size (297 × 420 mm) or B-size (11 × 17 in.) drawing sheet. The recommended working area is shown in Figure 9.64. The amount of blank area on a drawing depends on company standards. Some companies want the drawing to be easy to read with no crowding; others may want as much information as possible on a sheet. Generally, .50 in. (12.7 mm) should be the minimum space between the drawing and border line. More space is preferred. An area for future revision generally (but not always) should be left between the title block and upper right corner. An area for general notes should be available to the left of the title block for ASME/ANSI standard layout or in the upper left corner for military standard layout. The remaining area is the space available for drawing. On Figure 9.64, this area is about 10 × 6 in. (250 × 15 mm).

STEP 2 After determining the approximate working area, use your rough sketch as a guide to establish the actual size of the drawing by adding the overall dimensions and the space between views. (See Figure 9.64.) The amount of space selected will not crowd the views. The amount to select is an arbitrary decision and one where the drafter must use good judgment. Keep in mind that, for now, you will not consider dimensions when evaluating space requirements. The effect of dimensions on drawing layouts is discussed in Chapter 11.

STEP 3 Now, in each direction, subtract the total drawing size from the total space available and divide by two. This

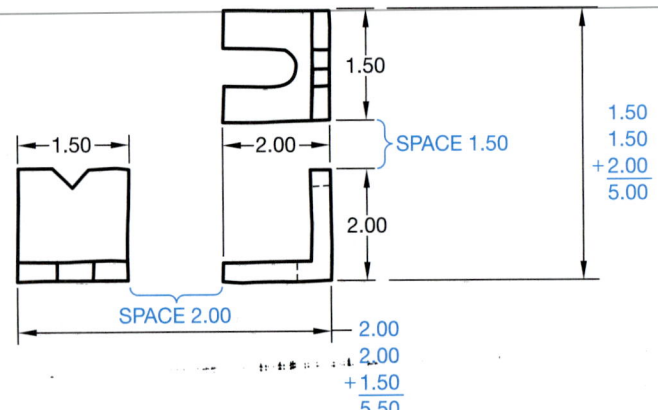

FIGURE 9.64 ■ Rough sketch with overall dimensions and selected space between views.

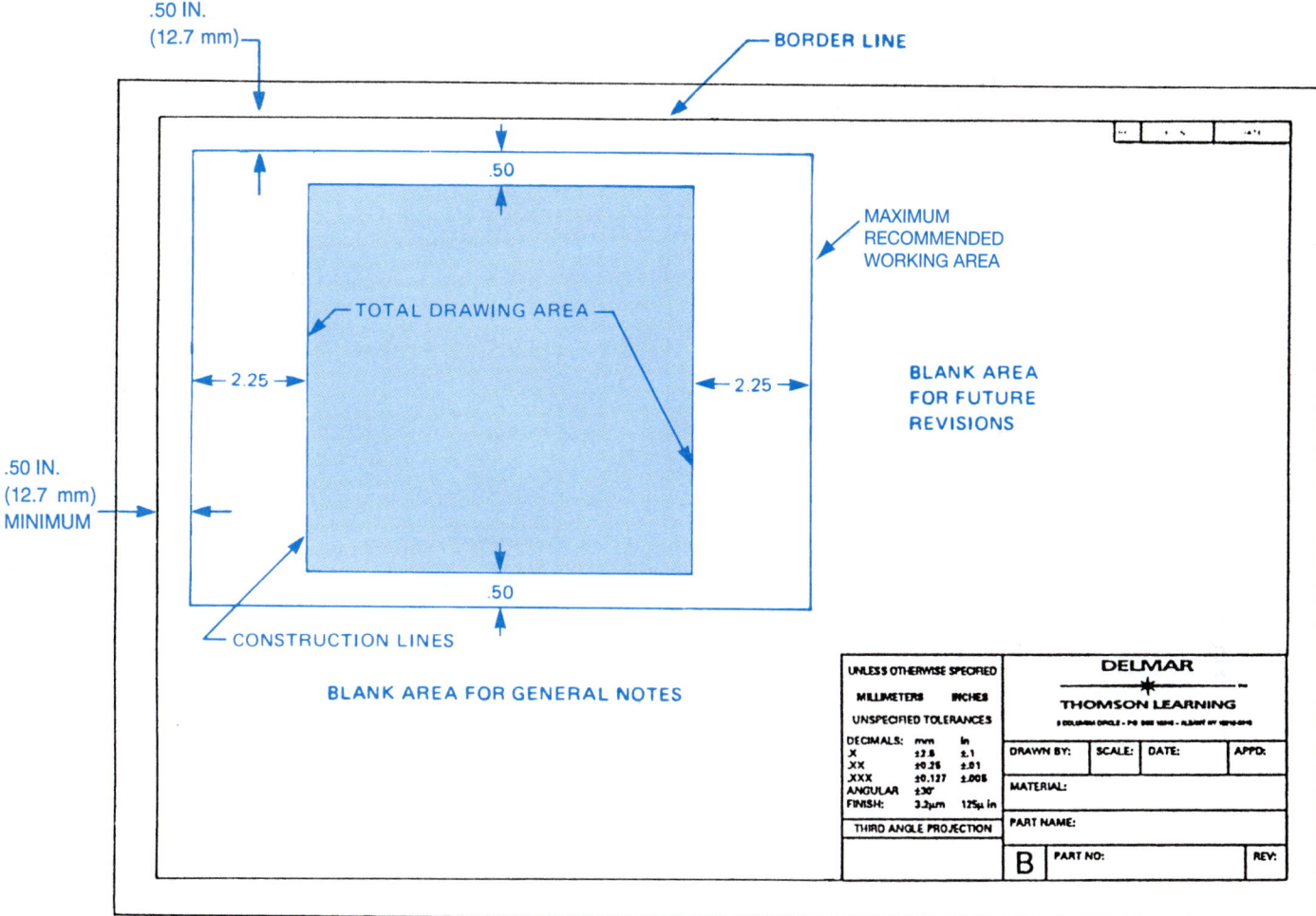

FIGURE 9.65 ∎ The recommended working area and layout of the total drawing area.

gives you the boundaries of the drawing area. Use construction lines to block out the total drawing areas that you have selected. (See Figure 9.65.)

Calculations:

Total space height	6.00
Total drawing height	−5.00
	1.00 ÷ 2 = .50
Total space width	10.00
Total drawing width	−5.50
	4.50 ÷ 2 = 2.25

STEP 4 Within the area established, use construction lines to block out the views that you selected in the rough sketch. Use multiview projection as discussed in this chapter, beginning with the front view. (See Figure 9.66.)

The construction lines, if properly drawn, will not have to be erased. Remember that construction lines are drawn very lightly with a 6H, 4H, or light blue lead. In any case, if properly drawn, the construction lines should not reproduce in the diazo or photocopy process.

STEP 5 Complete your drawing by using proper techniques to draw the finish lines. To help keep the final drawing neat, remember to:

a. Work from top to bottom.
b. Work from left to right if right-handed or from right to left if left-handed.
c. Draw thin lines first.
d. Draw circles and arcs next.
e. Draw centerlines.
f. Draw object lines.
g. Do all lettering and place a clean paper under your hand while lettering.
h. Avoid moving equipment or your hands over completed areas.
i. Keep your equipment and hands clean.
j. Cover large completed areas with blank paper to help avoid smudging.

Figure 9.67 shows the completed drawing.

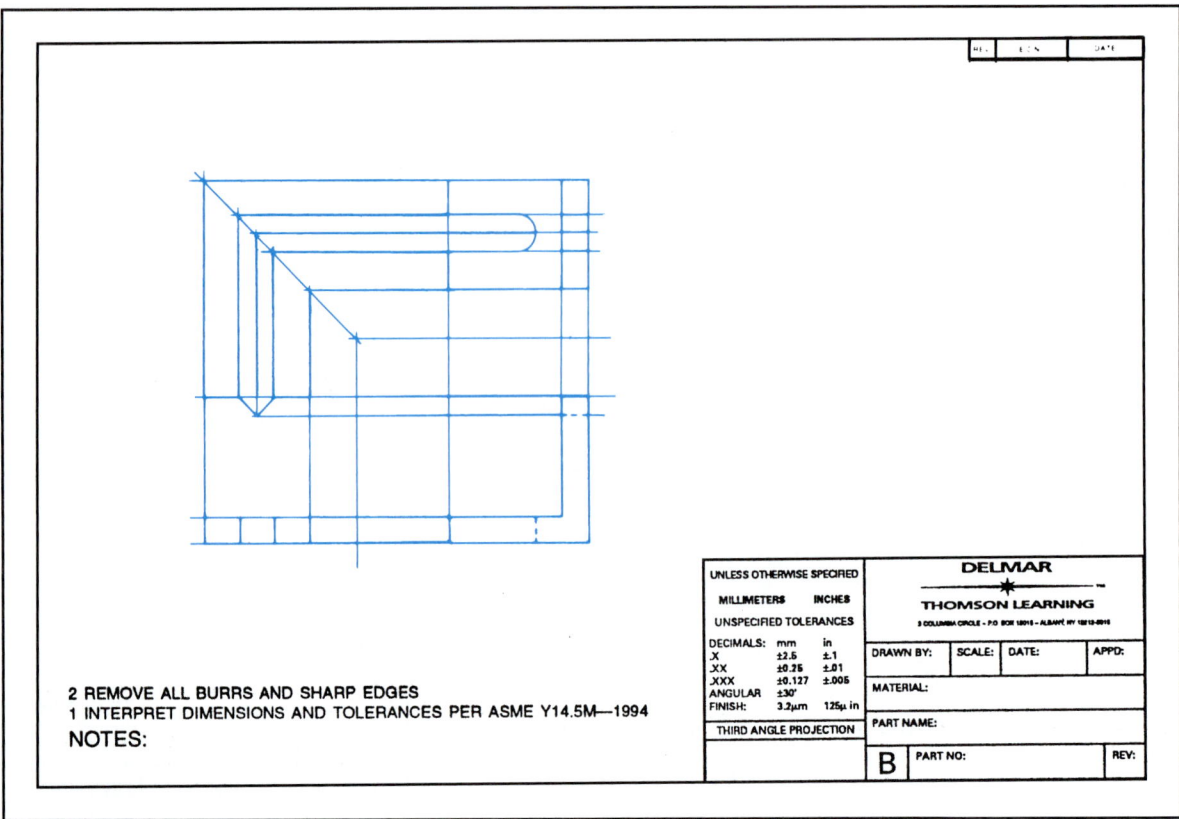

FIGURE 9.66 ■ Lay out the views using construction lines.

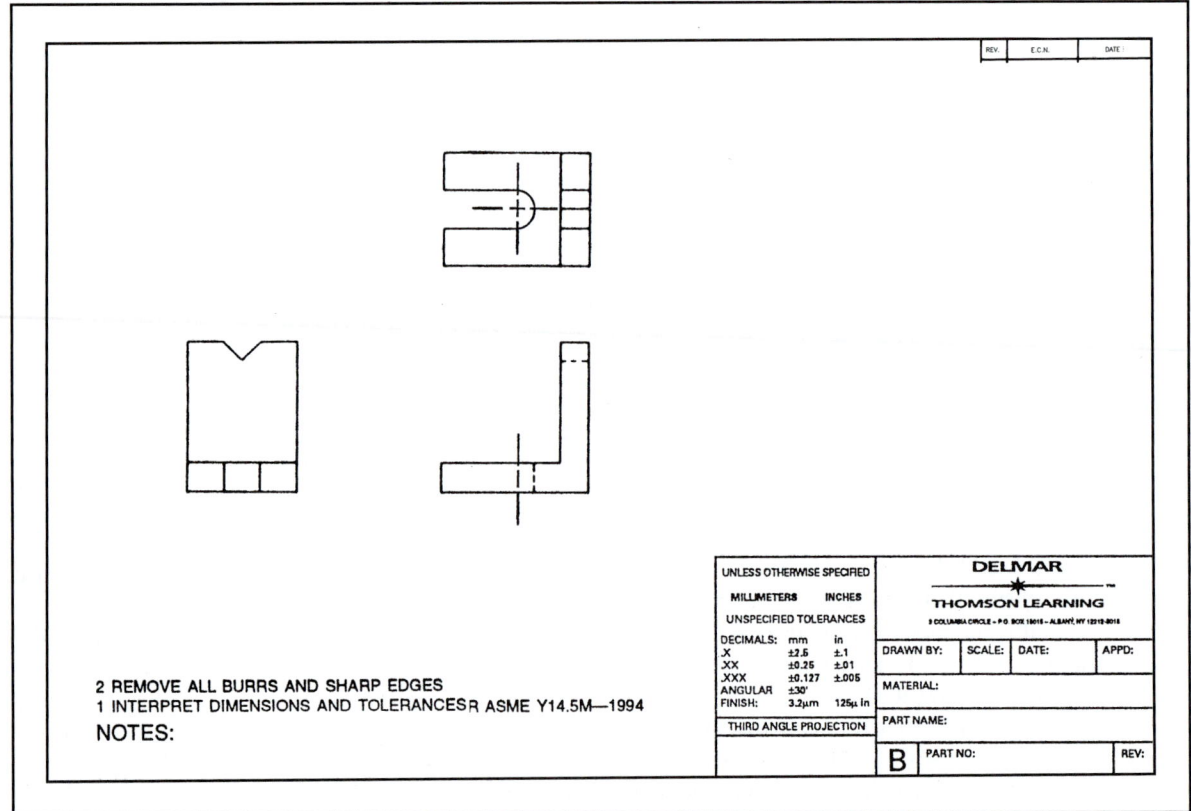

FIGURE 9.67 ■ Completed multiview drawing without dimensions.

DRAWING LAYOUT

When setting up a CADD drawing, many of the same considerations are made as when setting up a manual drawing. Often, a sketch is used to determine the total drawing area just as with manual drafting. The sheet size can then be determined and established with a Startup Wizard that guides you through the setup process and helps you establish the drawing layout, or you can use individual commands including LIMITS found in AutoCAD. The LIMITS command asks you to specify the length and height of the drawing area by establishing the lower left corner and the upper-right corner. One nice thing about using CADD is that if you make a mistake in calculating the right sheet size, you can change it anytime.

Another option is to use a template. A template in CADD is a file that contains standard settings that are applied to the new drawing. These settings can contain predefined drawing layouts, borders, title blocks, and other common drafting components, features, and standards. Templates are normally available in all of the standard inch and metric sheet sizes. Figure 9.68 shows an ANSI A title block template from AutoCAD.

When using CADD, the views are generally drawn full size and then scaled as needed to fit into a paper layout before plotting or printing.

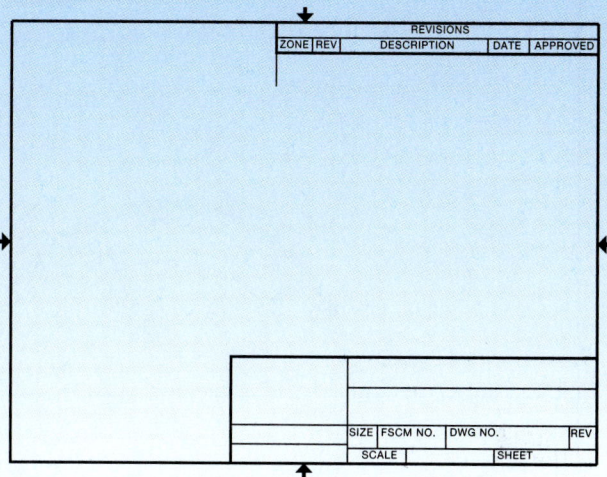

FIGURE 9.68 ■ A CADD ANSI A title block template.

WEBSITE RESEARCH

The following websites can provide you additional information for research or further study into topics covered in this chapter:

http://www.asme.org—Find information and publications related to the American Society of Mechanical Engineers.

http://www.ansi.org—The American National Standards Institute. Information about national and international drafting standards.

http://www.adda.org—American Design Drafting Association. Information related to the drafting and design profession.

PROFESSIONAL PERSPECTIVE

If you follow the view selection guidelines discussed in this chapter, you should be able to handle any drafting project. Establishing and laying out the necessary views of a part can be one of the most challenging aspects of drafting. The views that you select and the way you lay out the drawing can make the difference between having a drawing that is easy to read and understand or a drawing that is confusing and difficult to read. The practice of view projection that you learned in this chapter provides you with the foundation for effectively selecting and laying out multiviews. Look at the drawing in Figure 9.69. This is a complex part that is laid out with a number of multiviews that help the reader see every surface of the part. Become familiar with practices that are used in the industry, and spend as much time as you can looking at drawings that have been created by professional drafters.

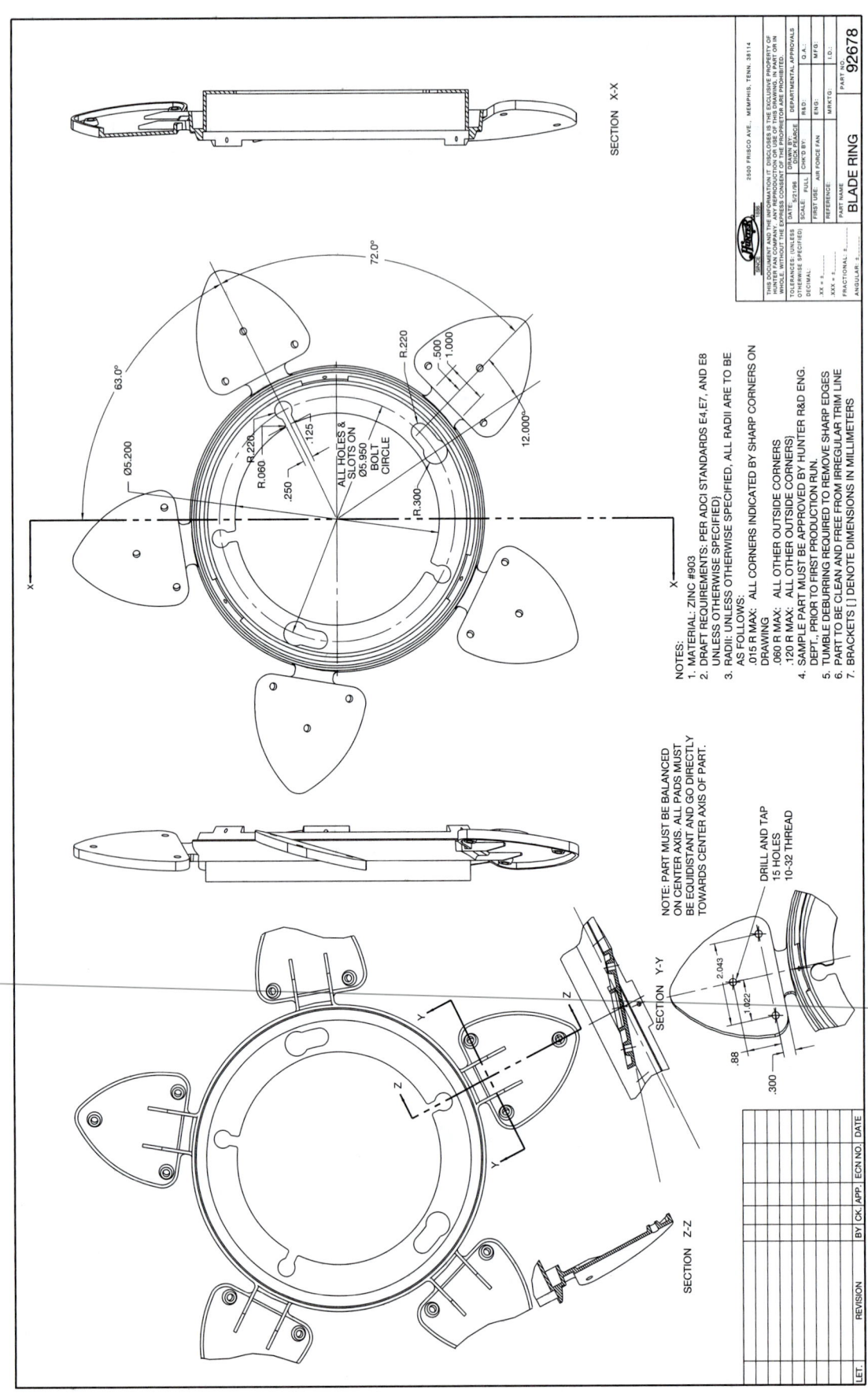

FIGURE 9.69 ■ A complex part displayed on the drawing using several carefully selected views. *Courtesy of Hunter Fan Company.*

WEIGHT OF AN ELLIPTICAL PLATE

Problem: A large steel plate in the shape of an ellipse is to be moved. The plate has the dimensions shown in Figure 9.70.

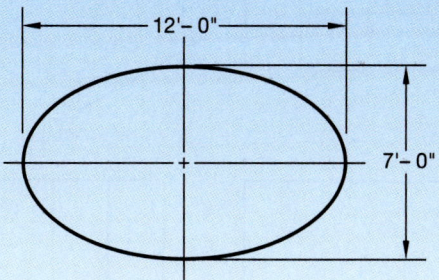

FIGURE 9.70 ■ The major and minor diameters of the elliptical plate.

The steel is known by its thickness to weigh 19 pounds per square foot. What is the weight of this plate?

Solution: The area of an ellipse, as well as other figures, is found in the CD Chapter 3, Engineering Drawing and Design Math Applications. It is given by the formula, Area = $\pi x y$, as defined by Figure 9.71.

This formula says to multiply the constant π (which is about 3.14159) by dimension x, then to multiply that result by dimension y. For the steel plate, $x = 6'$ and $y = 3.5'$, so the area is (3.14159) (6) (3.5) or 65.98 square feet (ft²). Now, the weight may be found:

$$65.98 \text{ ft}^2 \times 19 \frac{\text{lb}}{\text{ft}^2} = 1253 \text{ lb.}$$

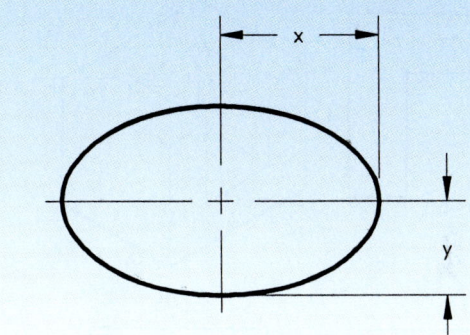

FIGURE 9.71 ■ Ellipse.

CHAPTER 9 *Multiviews Test*

Access the CD found with this textbook and open the Chapter 9 Test. Type your answer below each question or statement and submit an answer file or hard copy to your instructor. Confirm the preferred submittal method with your instructor.

CHAPTER 9 *Multiviews Problems*

PROBLEMS 9.1 through 9.45, PARTS 1 through 4: Access the CD found with this textbook and open the problem of your choice, or as assigned by your instructor. Solve the problems using the instructions provided on the CD, unless otherwise specified by your instructor.

DIRECTIONS

These problems provide you with views that contain missing lines or missing views. Draw the missing lines or missing views as appropriate. Pictorial views are provided to aid in visualization. You do not need to draw the pictorial view. Measure the given views and transfer the measurements to your manual or CADD drawing. Set up your drawings with a properly sized border and title block. Properly complete the information in the title block.

PROBLEM 9.46   Pocket Block

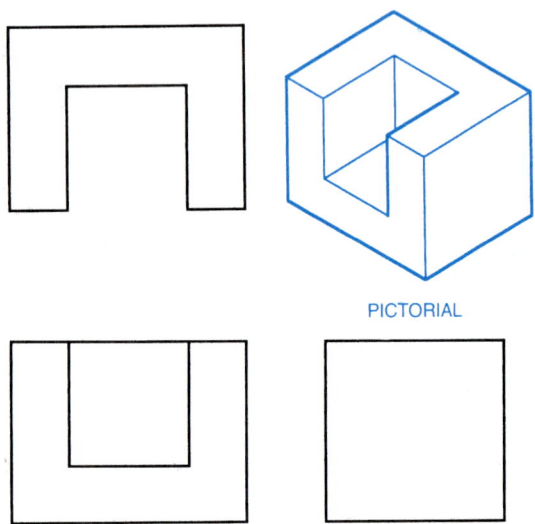

PICTORIAL

PROBLEM 9.47   Angle Gage

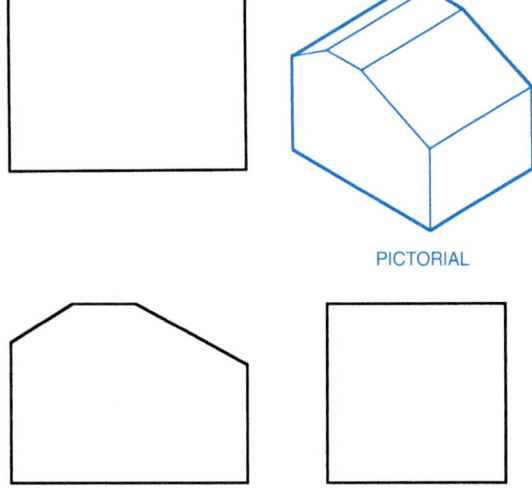

PICTORIAL

PROBLEM 9.48   Base

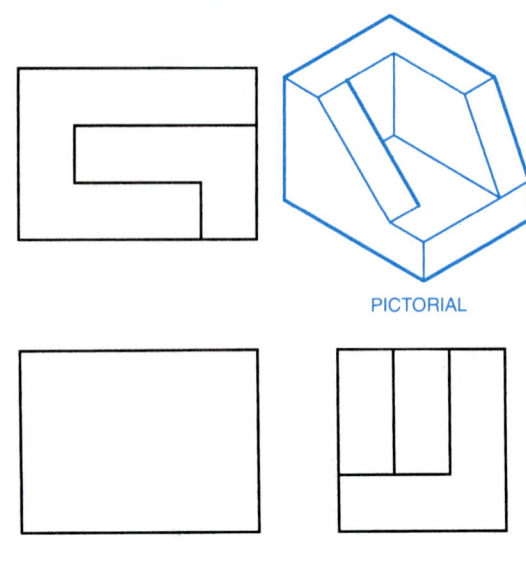

PICTORIAL

PROBLEM 9.49   Corner Block

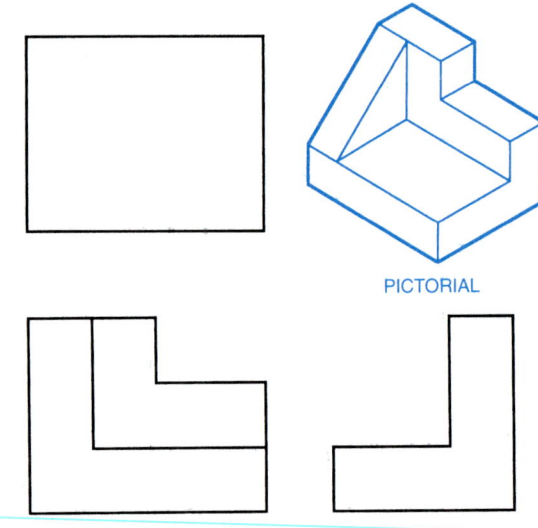

PICTORIAL

PROBLEM 9.50   Cylinder Block

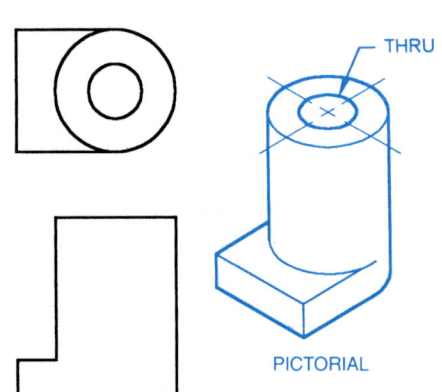

PICTORIAL

PROBLEM 9.51 Shaft Block

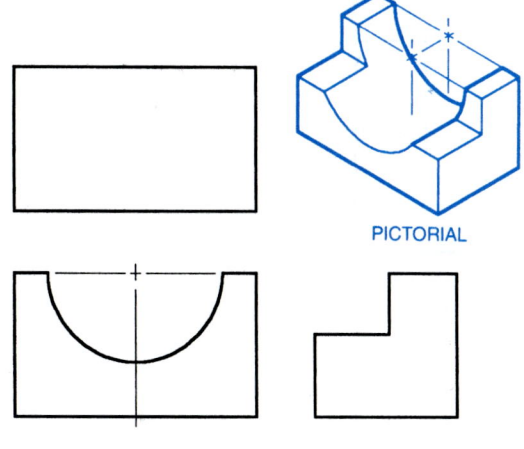

PICTORIAL

PROBLEM 9.52 Gib

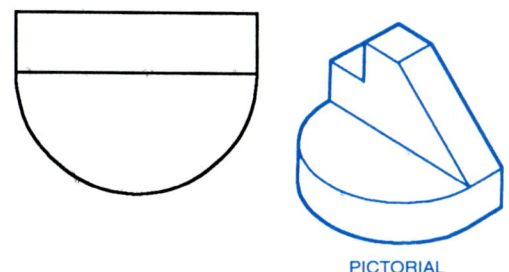

PICTORIAL

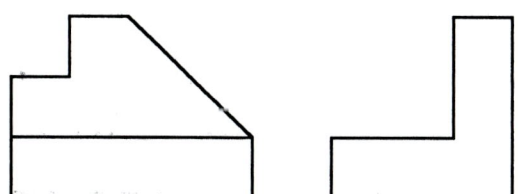

PROBLEM 9.53 Eccentric

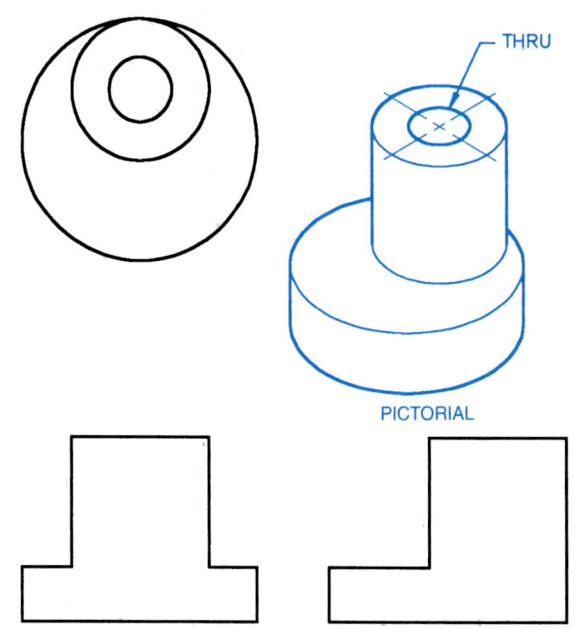

THRU

PICTORIAL

PROBLEM 9.54 Guide Block

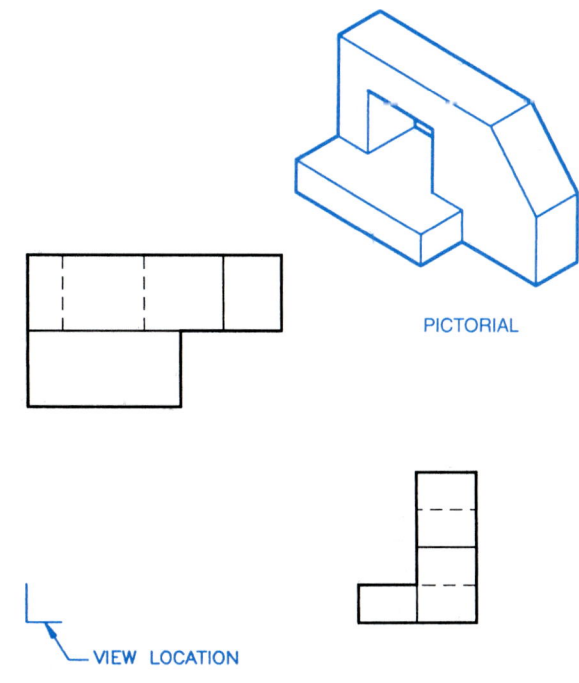

PICTORIAL

VIEW LOCATION

PROBLEM 9.55 Key Slide

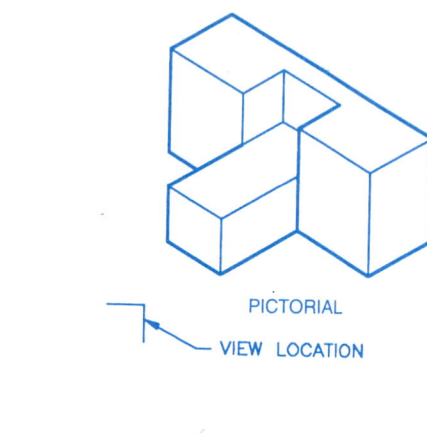

PICTORIAL

VIEW LOCATION

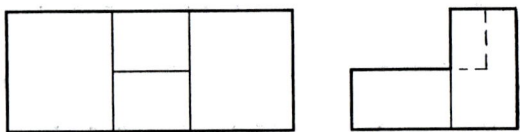

PROBLEM 9.57 Clevis

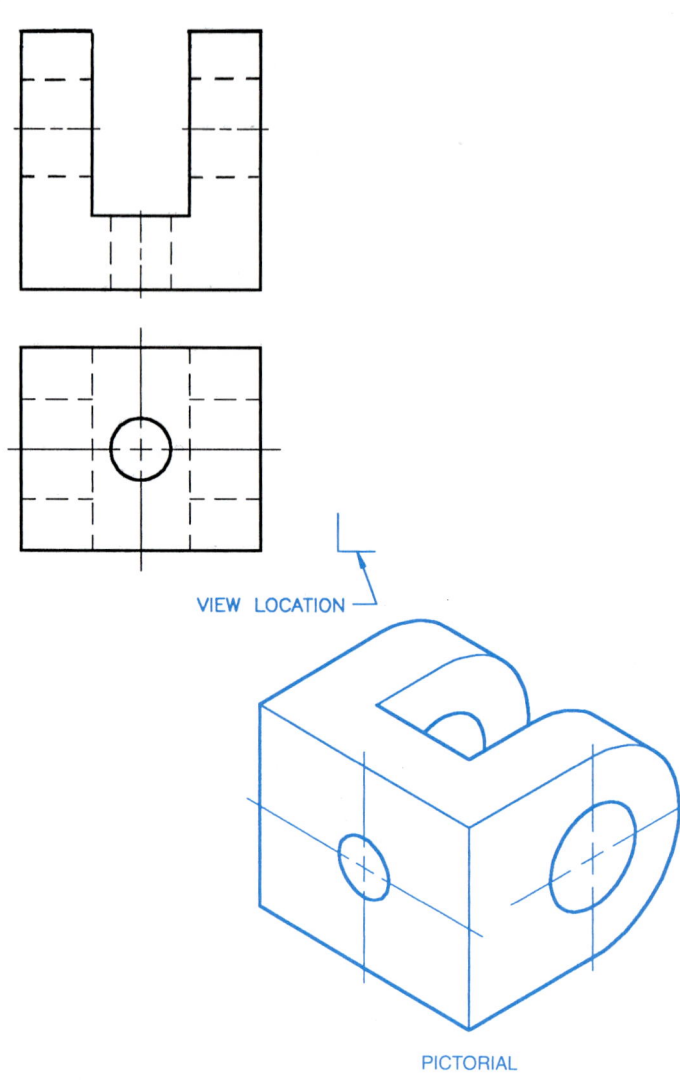

VIEW LOCATION

PICTORIAL

PROBLEM 9.56 Angle Bracket

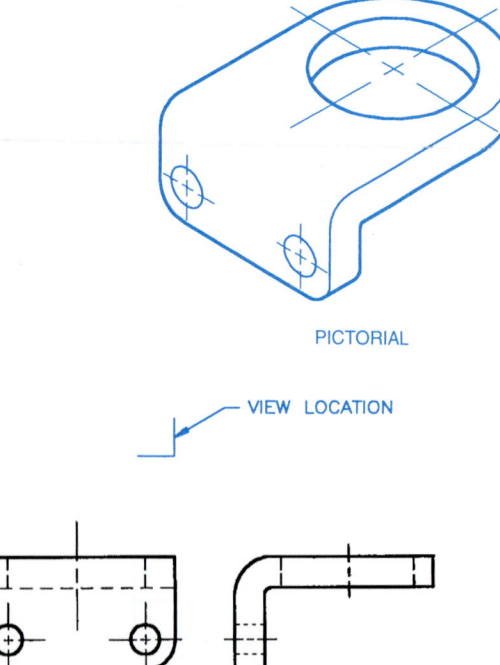

PICTORIAL

VIEW LOCATION

DIRECTIONS

These next problems, Problem 9.58 and Problem 9.59, provide you with pictorial views that contain dimensions. A suggested view layout is provided for your reference. Draw the views. The pictorial views are provided to aid in visualization. You do not need to draw the pictorial view. Use the given dimensions to create your manual or CADD drawing. Set up your drawings with a properly sized border and title block. Properly complete the information in the title block. Do not draw the dimensions.

PROBLEM 9.58 **V-block (metric)**

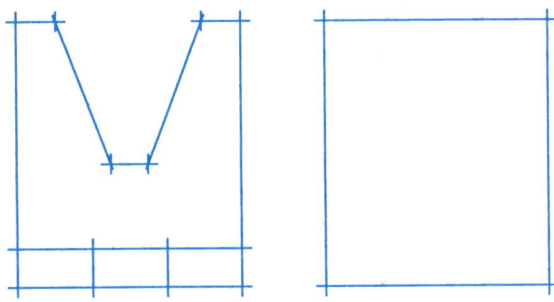

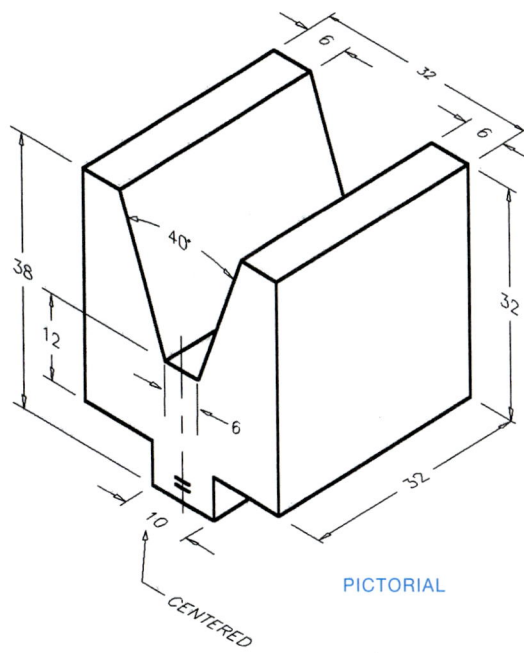

PICTORIAL

PROBLEM 9.59 **Guide Base (in.)**

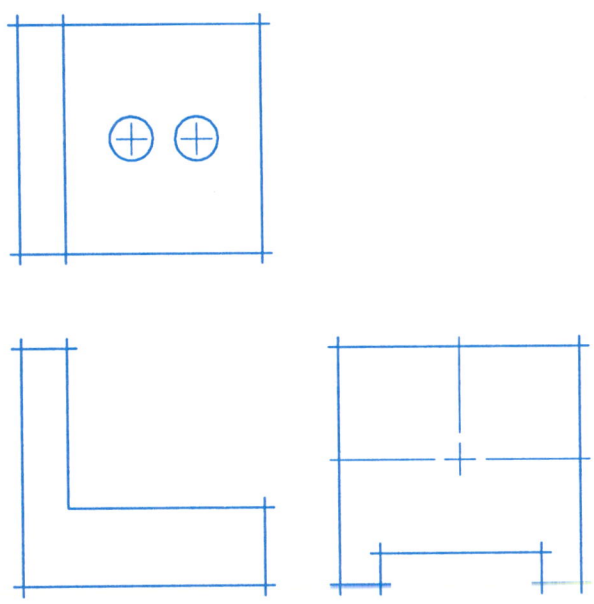

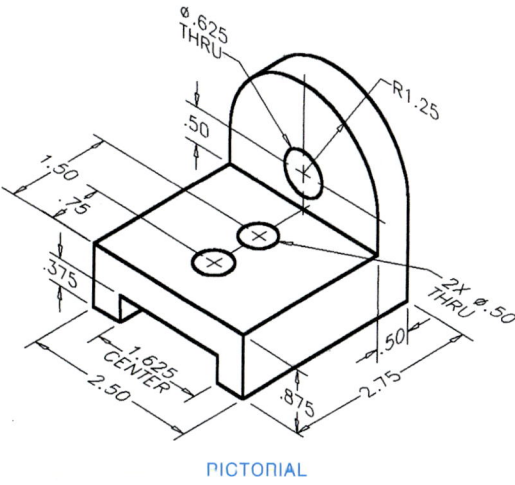

PICTORIAL

DIRECTIONS

These next problems, Problem 9.60 through Problem 9.89, provide you with a pictorial view or multiview layout that contains dimensions. Use the given information to select and draw the necessary multiviews using manual drafting or CADD. In some cases, the suggested number of views is given. Do not draw the pictorial view. Set up your drawings with a properly sized border and title block. Properly complete the information in the title block. Do not draw the dimensions.

PROBLEM 9.60 **One-view cylindrical object (in.)**

Part Name: Sleeve Bearing

Material: Phosphor Bronze

Problem based on original art courtesy Production Plastics.

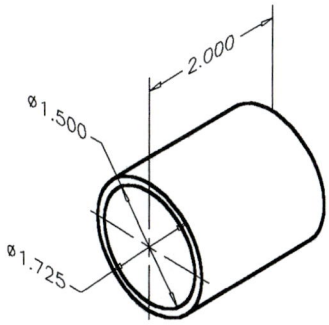

PROBLEM 9.61 **One-view cylindrical object (in.)**
Part Name: Pin
Material: SAE1035

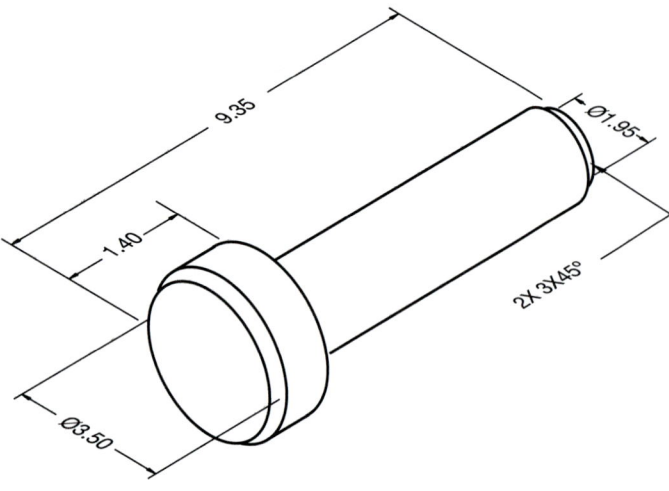

PROBLEM 9.63 **One-view arcs, circles, and centerlines (in.)**
Part Name: Gasket
Material: .050-thick Brass

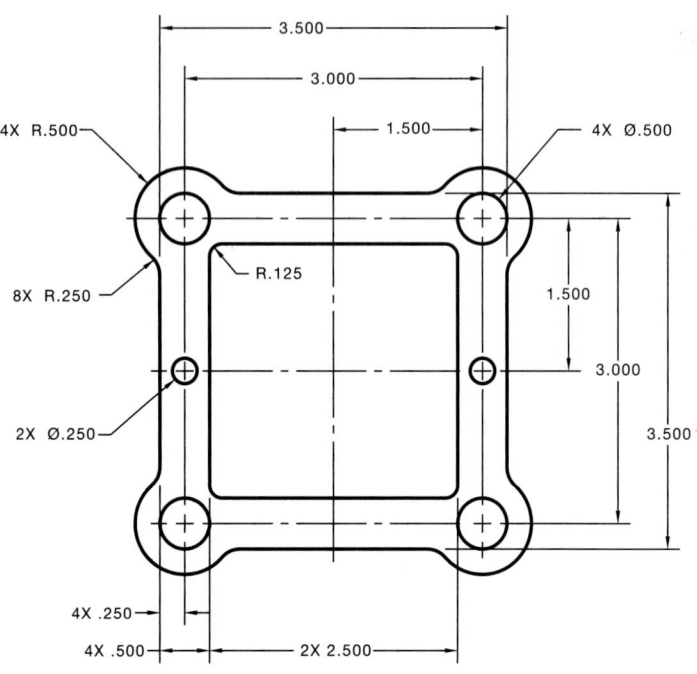

PROBLEM 9.62 **One-view arc and hole (in.)**
Part Name: Door Lock
Material: Mild Steel

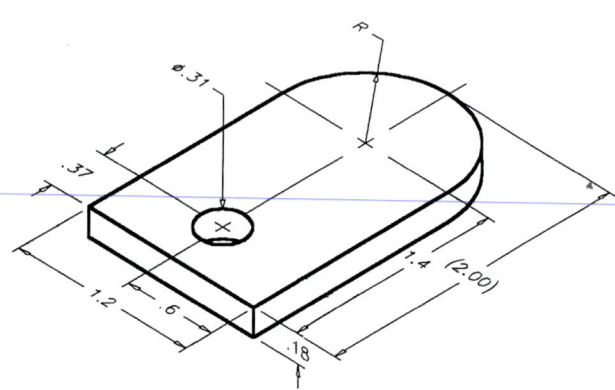

PROBLEM 9.64 **One-view arcs and centerlines (in.)**
Part Name: Clip
Material: .025-thick SAE 3140

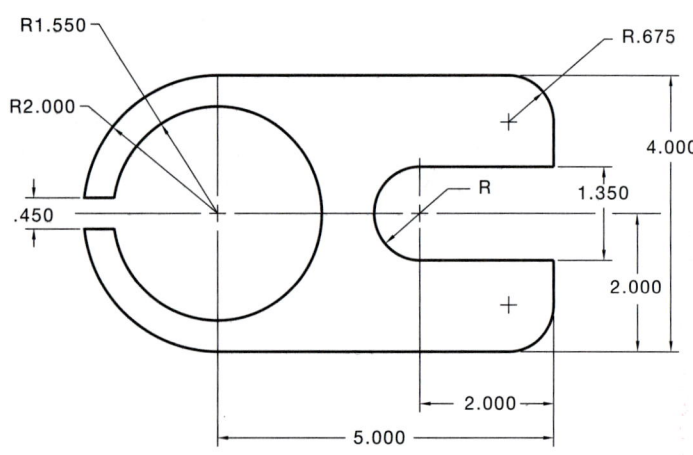

PROBLEM 9.65 One-view circle and arc object lines and centerlines (in.)

Part Name: Gasket

Material: .062-in.-thick Cork

Gasket for hydraulic pump.

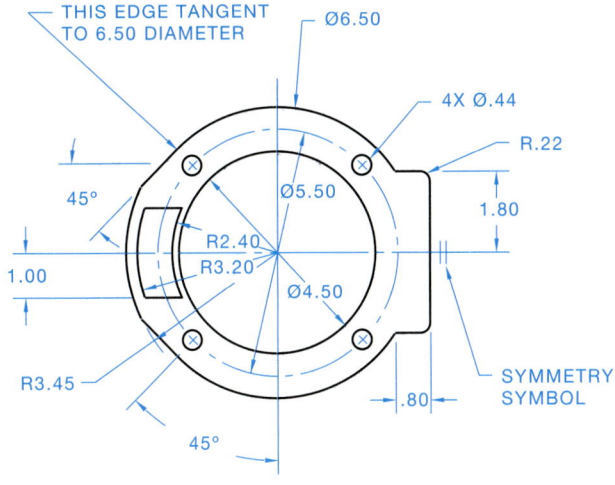

PROBLEM 9.67 Two views nontangent arcs (in.)

Part Name: Washer

Material: SAE 1060

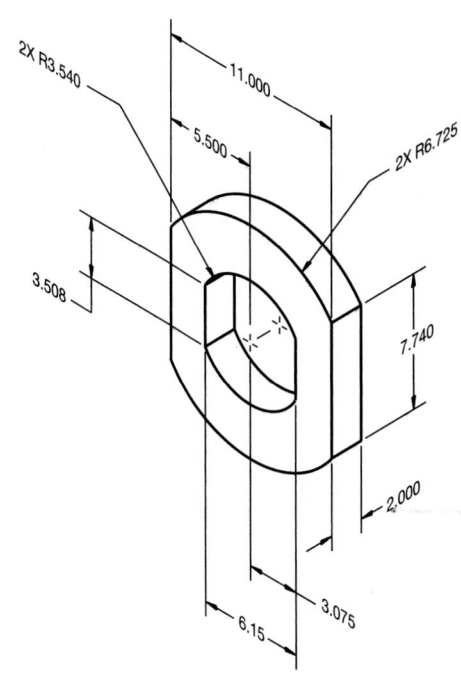

PROBLEM 9.66 Arc object lines, centerlines, phantom lines, and leader lines (in.)

Part Name: Bogie Lock

Material: .25-in.-thick Mild Steel

SPECIFIC INSTRUCTIONS: Connect the leader lines and place the notes on the drawing.

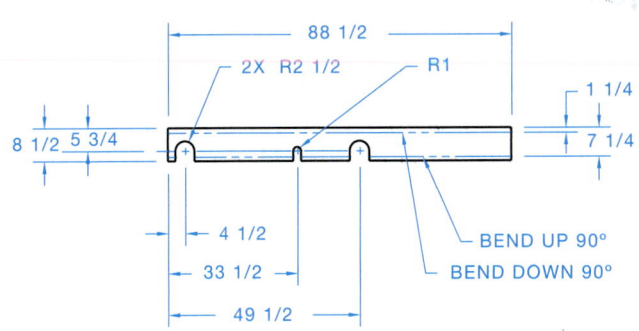

PROBLEM 9.68 Planes and slot (in.)

Part Name: Step Block

Material: Mild Steel

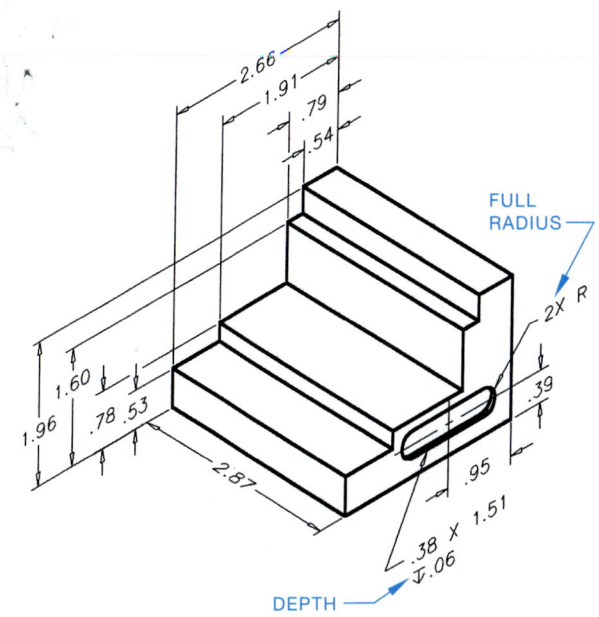

PROBLEM 9.69   **Cylinders and circles (in.)**

Part Name: Roll End Bearing

Material: Phosphor Bronze

Problem based on original art courtesy Production Plastics.

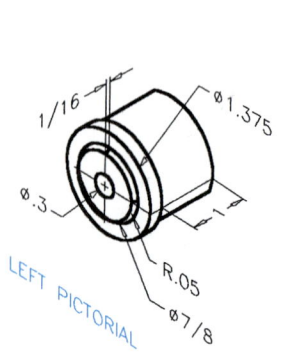

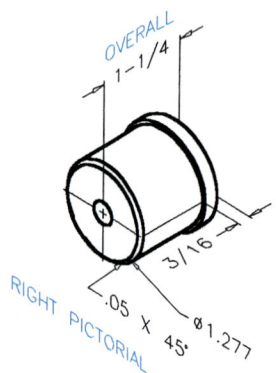

PROBLEM 9.70   **Cylinders and circles (in.)**

Part Name: Roll End Bearing

Material: Phosphor Bronze

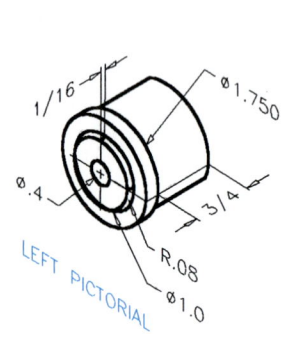

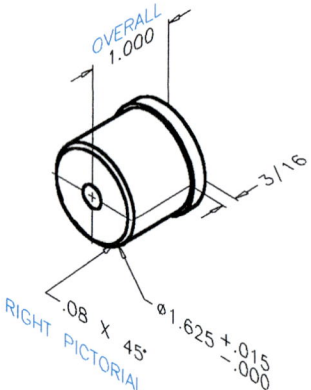

PROBLEM 9.71   **Two views with holes (in.)**

Part Name: Pivot Bracket

Material: SAE 3135

multi-view
9/9/08
PDF

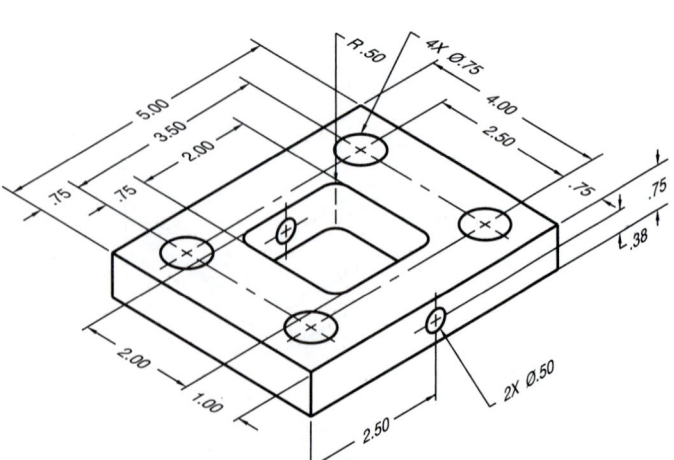

PROBLEM 9.72   **Three views with holes and slot (in.)**

Part Name: Sliding Bracket

Material: Phosphor Bronze

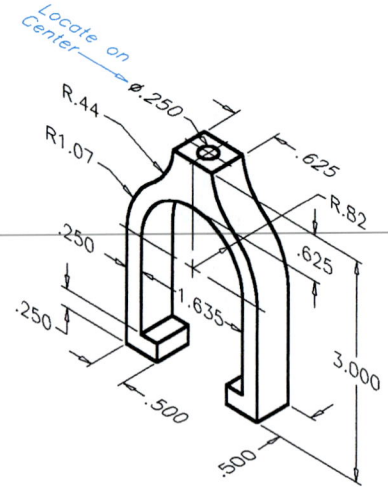

PROBLEM 9.73   **Circle arcs and planes (in.)**

Part Name: V-block Clamp

Material: SAE 1080

PROBLEM 9.74   **View enlargement (in.)**
Part Name: Drill Gauge
Material: 16-GA Mild Steel

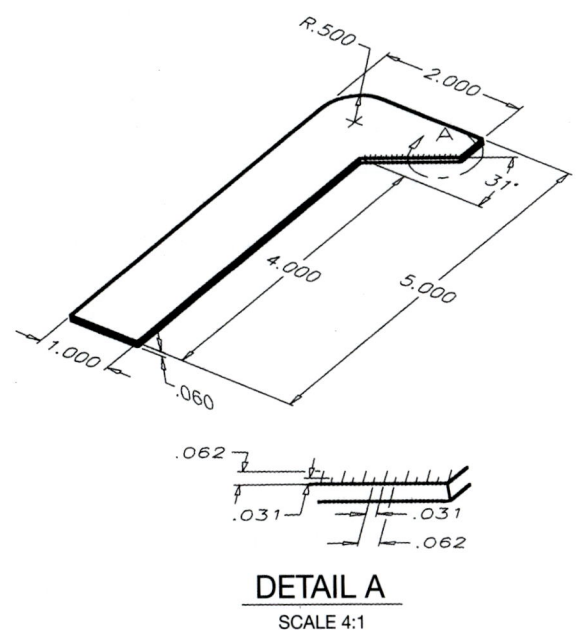

DETAIL A
SCALE 4:1

PROBLEM 9.75   **Two views with angled surface, countersinks (in.)**
Part Name: Angle Bracket
Material: SAE 1145

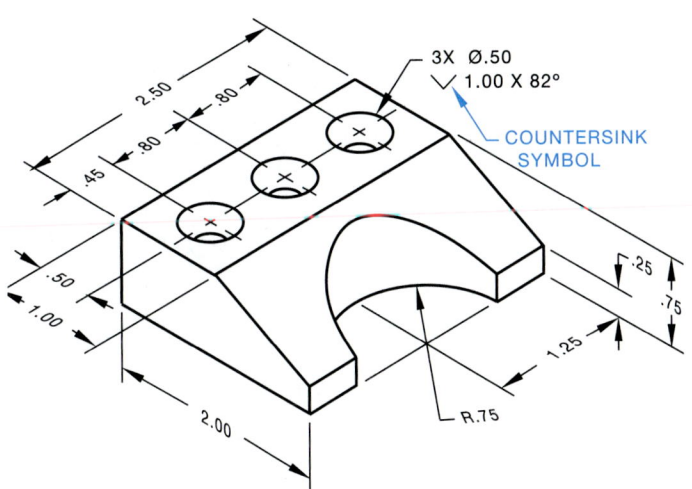

PROBLEM 9.76   **Multiple features (in.)**
Part Name: Lock Ring
Material: SAE 1020

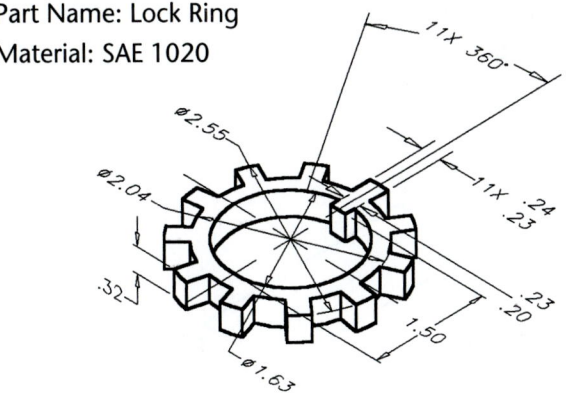

PROBLEM 9.77   **Angles and holes (in.)**
Part Name: Bracket
Material: SAE 1020

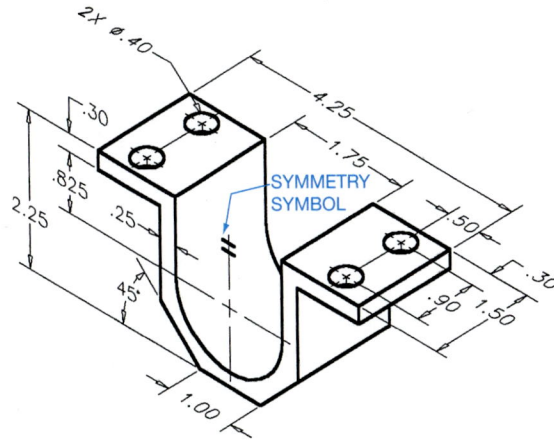

PROBLEM 9.78   **Multiviews (metric)**
Part Name: Guide Rail
Material: SAE 4310

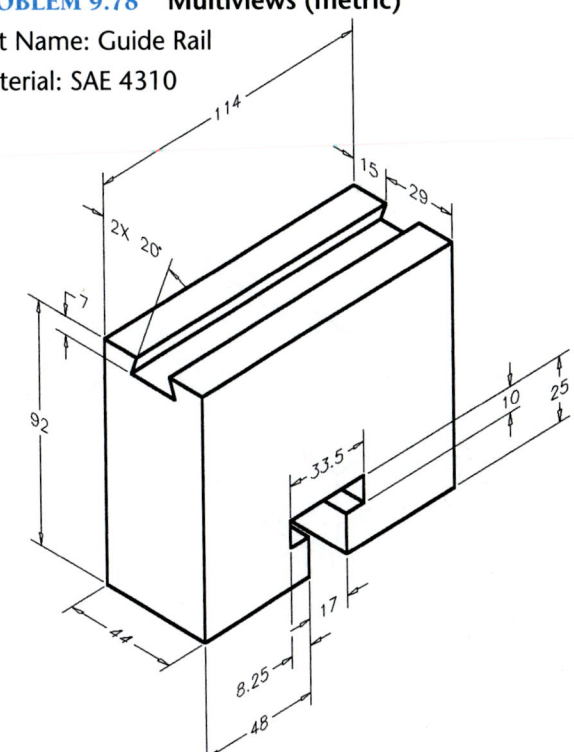

PROBLEM 9.79   Multiviews (in.)
Part Name: Support Bracket
Material: Plastic

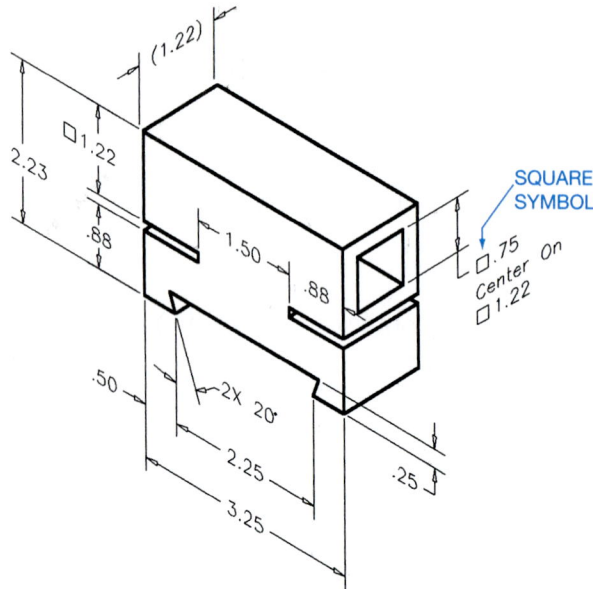

PROBLEM 9.81   Multiviews, angles, and holes (metric)
Part Name: Angle Bracket
Material: Mild Steel
SPECIFIC INSTRUCTIONS: Center 2× Ø15 holes and provide location dimension.

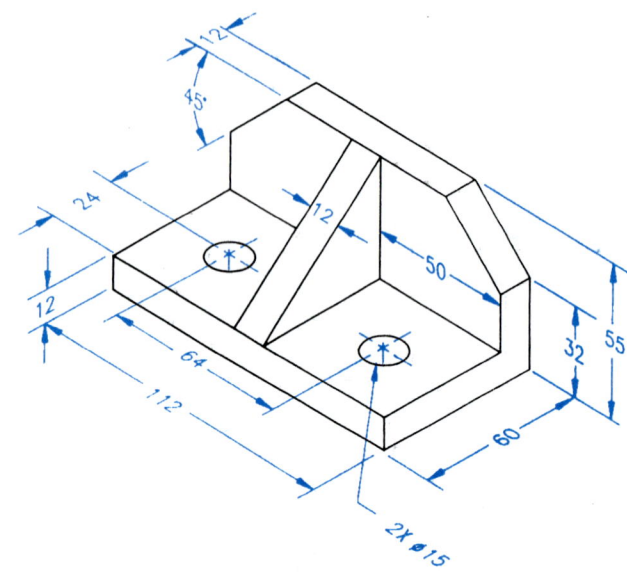

PROBLEM 9.82   Multiviews (in.)
Part Name: Support
Material: Mild Steel

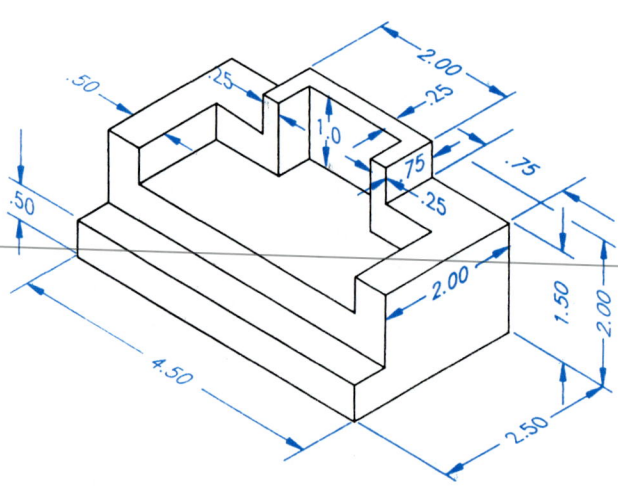

PROBLEM 9.80   Multiviews (in.)
Part Name: V-block
Material: A-Steel

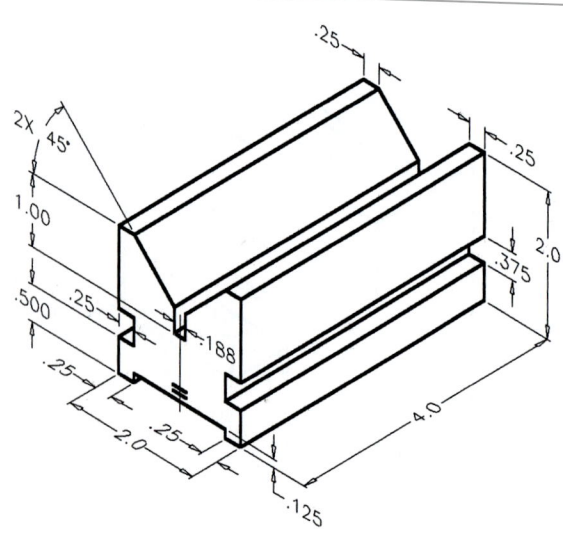

PROBLEM 9.83 Multiviews, circles, and arcs (in.)
Part Name: Chain Link
Material: SAE 4320

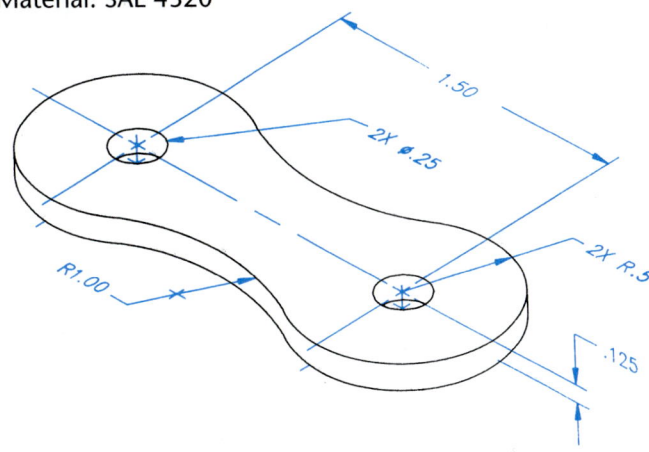

PROBLEM 9.84 Multiviews, circles, and arcs (in.)
Part Name: Pivot Bracket
Material: Cold-Rolled Steel

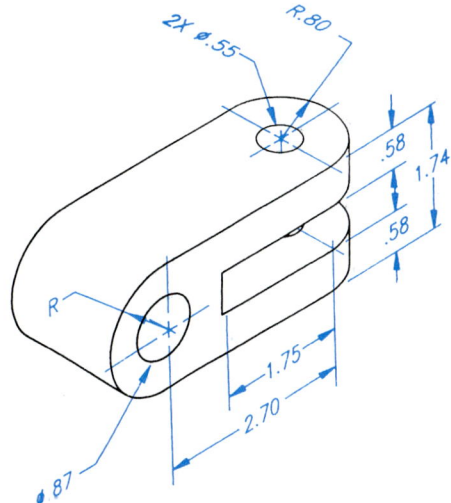

PROBLEM 9.85 Multiview arcs and circles (metric)
Part Name: Hinge Bracket
Material: Cast Aluminum

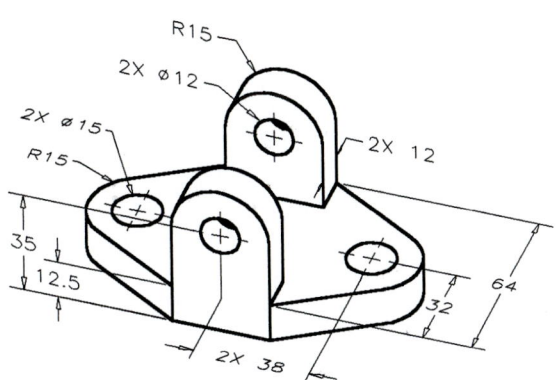

PROBLEM 9.86 Multiview circles and arcs (in.)
Part Name: Bearing Support
Material: SAE 1040

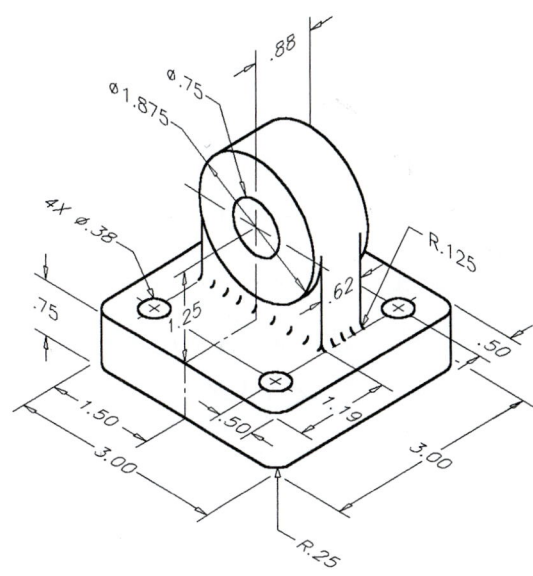

PROBLEM 9.87 Multiviews (metric)
Part Name: Gate Latch Base
Material: Aluminum

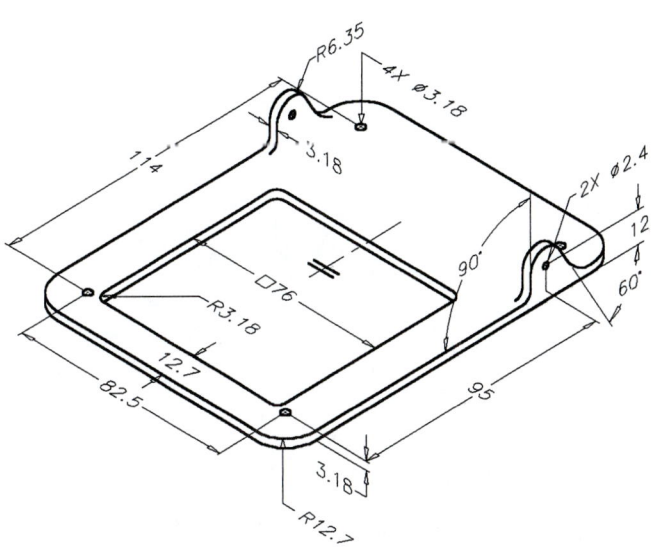

Part Name: Mounting Bracket

Material: SAE 1020

All fillets: R.13

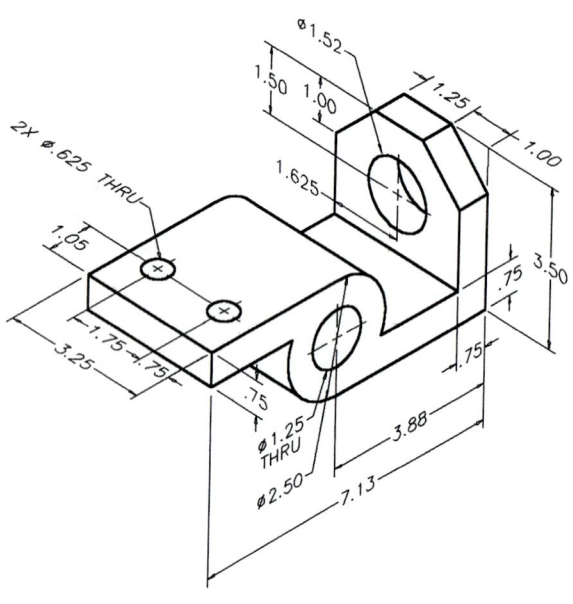

MATH PROBLEMS

 PROBLEMS 9.90 through 9.100: Access the CD found with this textbook and open the math problem of your choice, or as assigned by your instructor. Solve the problem or problems using the instructions provided.

PROBLEM 9.89 **Multiviews (in.)**

Part Name: Support Base

Material: SAE 1040

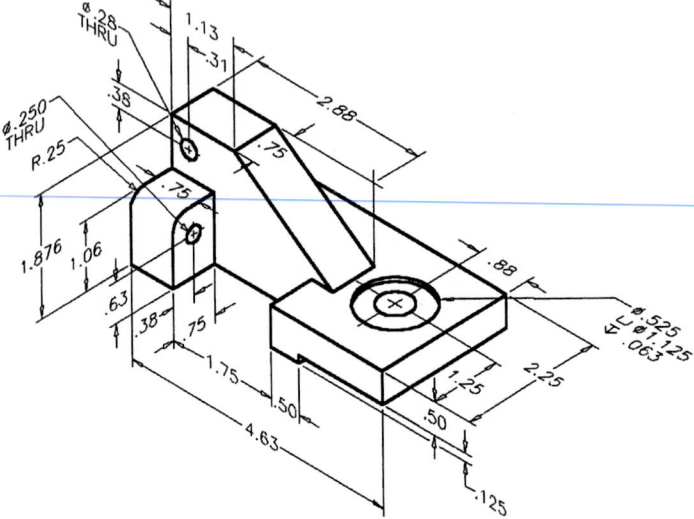

CHAPTER 10

Auxiliary Views

LEARNING OBJECTIVES

After completing this chapter you will:

- Describe the purpose of an auxiliary view.
- Explain how an auxiliary view is projected.
- Discuss and draw viewing-plane lines related to auxiliary views.
- Draw primary and secondary auxiliary views along with the related multiviews from given engineering problems.

THE ENGINEERING DESIGN APPLICATION

You have been asked to create a multiview detail drawing for a guide bracket, and the engineer has provided a sketch of the part. (See Figure 10.1.) As you study the sketch, you discover that one of the surfaces on the bracket is not parallel to any of the six principal viewing planes. You determine that your multiview drawing will need an auxiliary view in order to show the angled face in its true size and shape. Through sketching your layout ideas, you conclude that a complete auxiliary view would not add clarity to the drawing and decide instead on a partial auxiliary view. The final layout is shown in Figure 10.2.

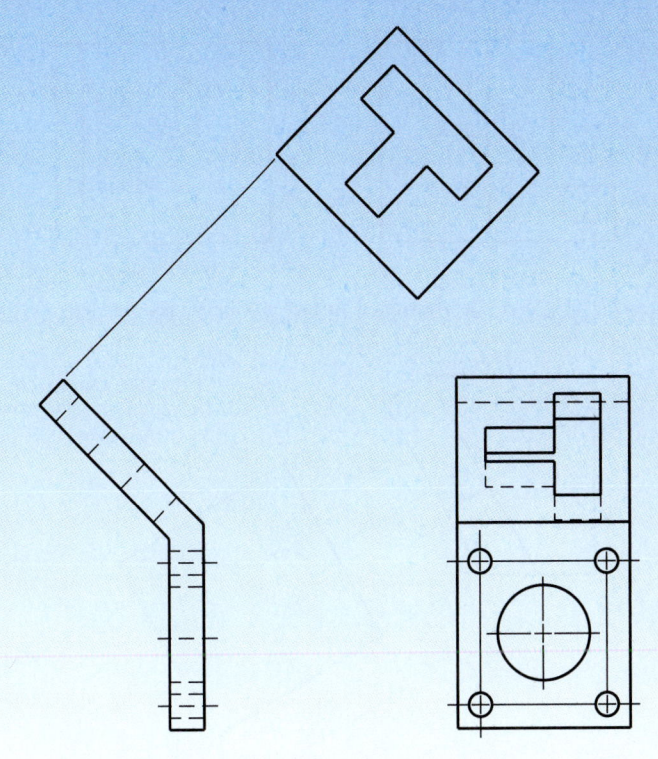

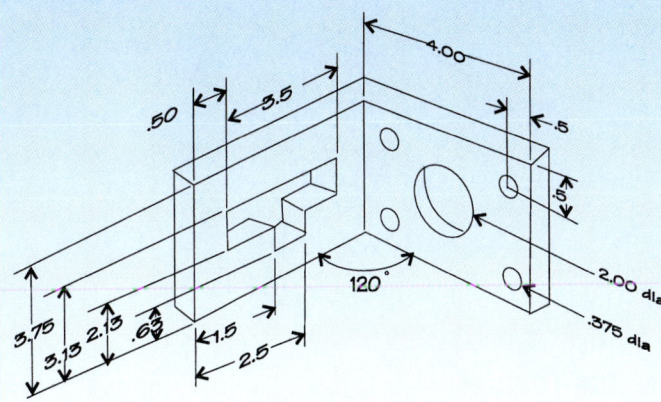

FIGURE 10.1 ■ Engineer's rough sketch.

FIGURE 10.2 ■ Final layout with partial auxiliary view (dimensions not included).

AUXILIARY VIEWS

ANSI/ASME The standard for auxiliary view presentation is found in ANSI Y14.3, titled *Multi and Sectional View Drawings,* American National Standards Institute (ANSI), published by the American Society of Mechanical Engineers (ASME).

Auxiliary views are used to show the true size and shape of a surface that is not parallel to any of the six principal views.

When a surface feature is not perpendicular to the line of sight, the feature is said to be foreshortened, or shorter than true length. These foreshortened views do not give a clear or accurate representation of the feature. It is not proper to place dimensions on foreshortened views of objects. Figure 10.3 shows three views of an object with an inclined foreshortened surface.

An auxiliary view allows you to look directly at the inclined surface in Figure 10.3 so you can view the surface and locate the hole in its true size and shape. An auxiliary view is projected from the inclined surface in the view where that surface appears

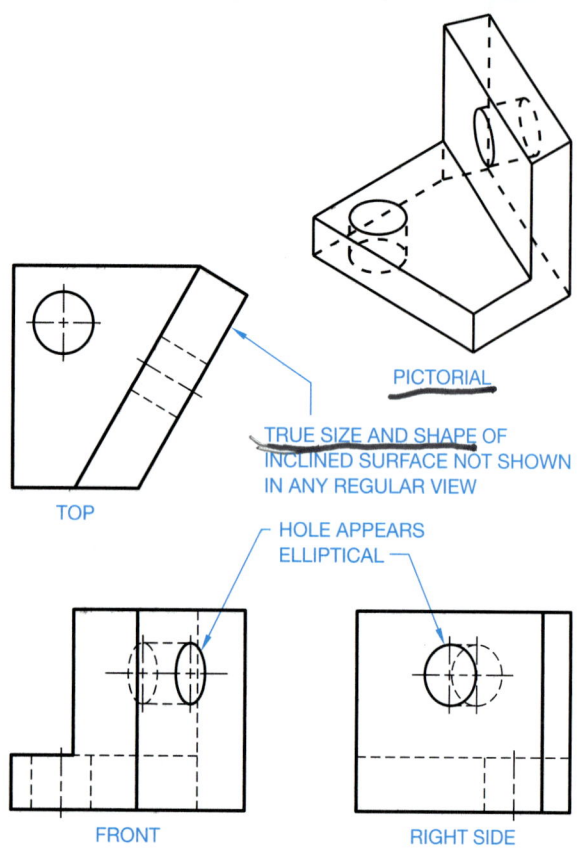

PICTORIAL

TRUE SIZE AND SHAPE OF
INCLINED SURFACE NOT SHOWN
IN ANY REGULAR VIEW

TOP

HOLE APPEARS
ELLIPTICAL

FRONT   RIGHT SIDE

FIGURE 10.3 ■ Foreshortened surface auxiliary view needed.

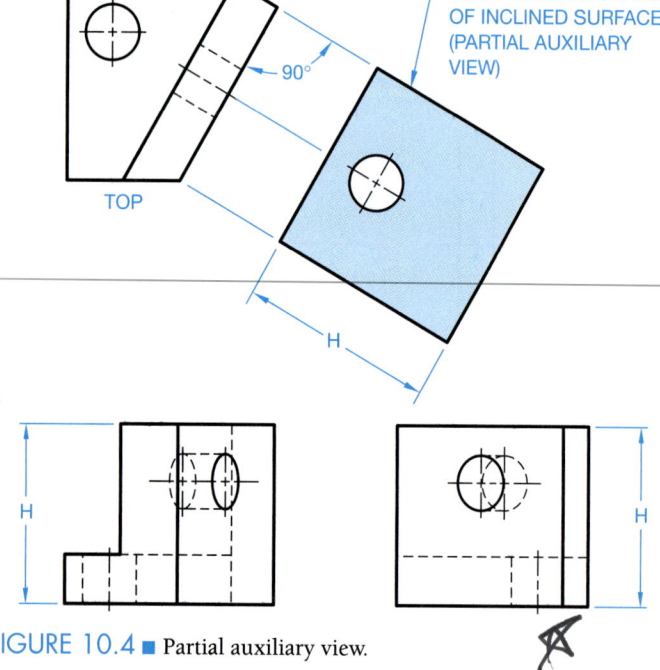

TRUE SIZE AND SHAPE
OF INCLINED SURFACE
(PARTIAL AUXILIARY
VIEW)

90°

TOP

H

H   H

FIGURE 10.4 ■ Partial auxiliary view.

as a line. The projection is at a 90° angle. (See Figure 10.4.) The height dimension, *H*, is taken from the view that shows the height in its true length.

Notice in Figure 10.4 that the auxiliary view shows only the true size and shape of the inclined surface. This is known as a

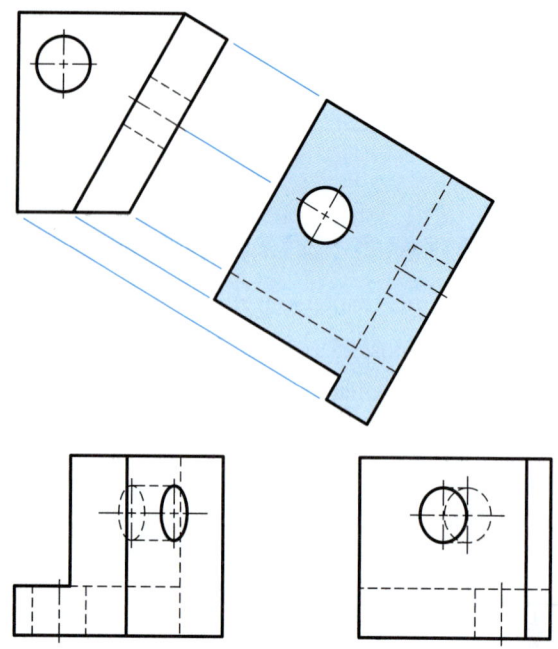

FIGURE 10.5 ■ Complete auxiliary view.

THE AUXILIARY PLANE
IS PARALLEL TO THE
INCLINED SURFACE

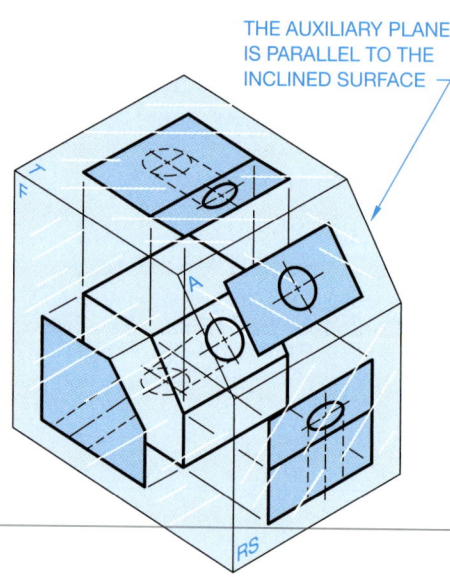

FIGURE 10.6 ■ The object in a glass box.

partial auxiliary view. A full auxiliary view, Figure 10.5, also shows all the other features of the object projected onto the auxiliary plane. Normally the information needed from the auxiliary view is the inclined surface only, and the other areas do not usually add clarity to the view. However, each object must be considered separately. Usually, you do not need to draw those areas that do not add clarity to the drawing.

Look at the glass box principle that was discussed earlier, as it applies to auxiliary views, as shown in Figure 10.6. Figure 10.7 shows the glass box unfolded. Notice that the fold line between the front view and the auxiliary view is parallel to the edge view of the slanted surface.

The auxiliary view is projected from the edge view of the inclined surface, which establishes true length. The true width of

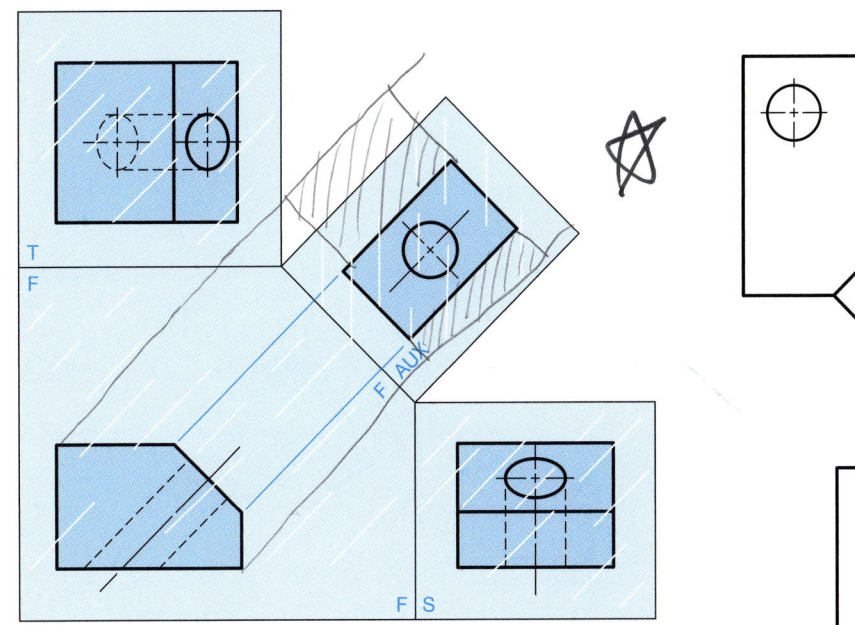

FIGURE 10.7 ■ The glass box unfolded.

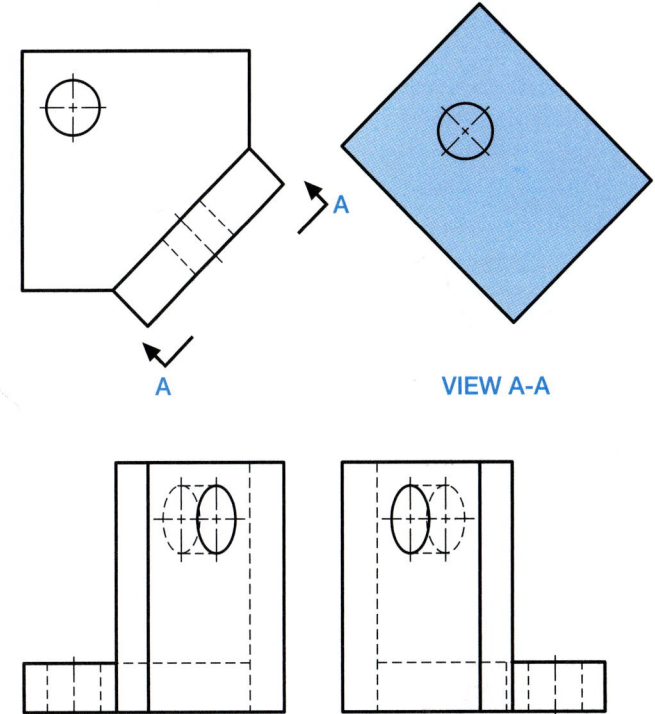

FIGURE 10.9 ■ Establishing auxiliary view with viewing plane.

the inclined surface may be transferred from the fold lines of the top or side views as shown in Figure 10.8. Projection lines are drawn as very light construction lines and should not reproduce. Hidden lines are generally not shown on auxiliary views unless the use of hidden lines helps clarify certain features. The auxiliary view can be projected directly from the inclined surface, as in Figure 10.7. When this is done, a centerline or projection line may continue between the views to indicate alignment and view relationship. This centerline or projection line is often used as an extension line for dimensioning purposes. If view alignment is

clearly obvious, the projection line may be omitted. When it is not possible to align the auxiliary view directly from the inclined surface, then a viewing-plane line can be used and the view placed in a convenient location on the drawing (Figure 10.9).

The viewing-plane line is labeled with letters so the view can be clearly identified. This is especially necessary when several

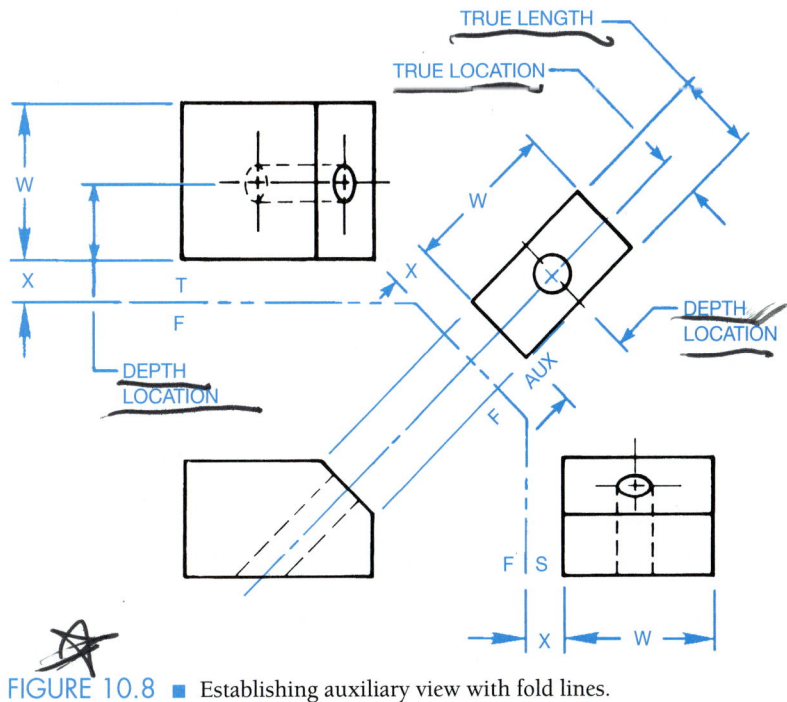

FIGURE 10.8 ■ Establishing auxiliary view with fold lines.

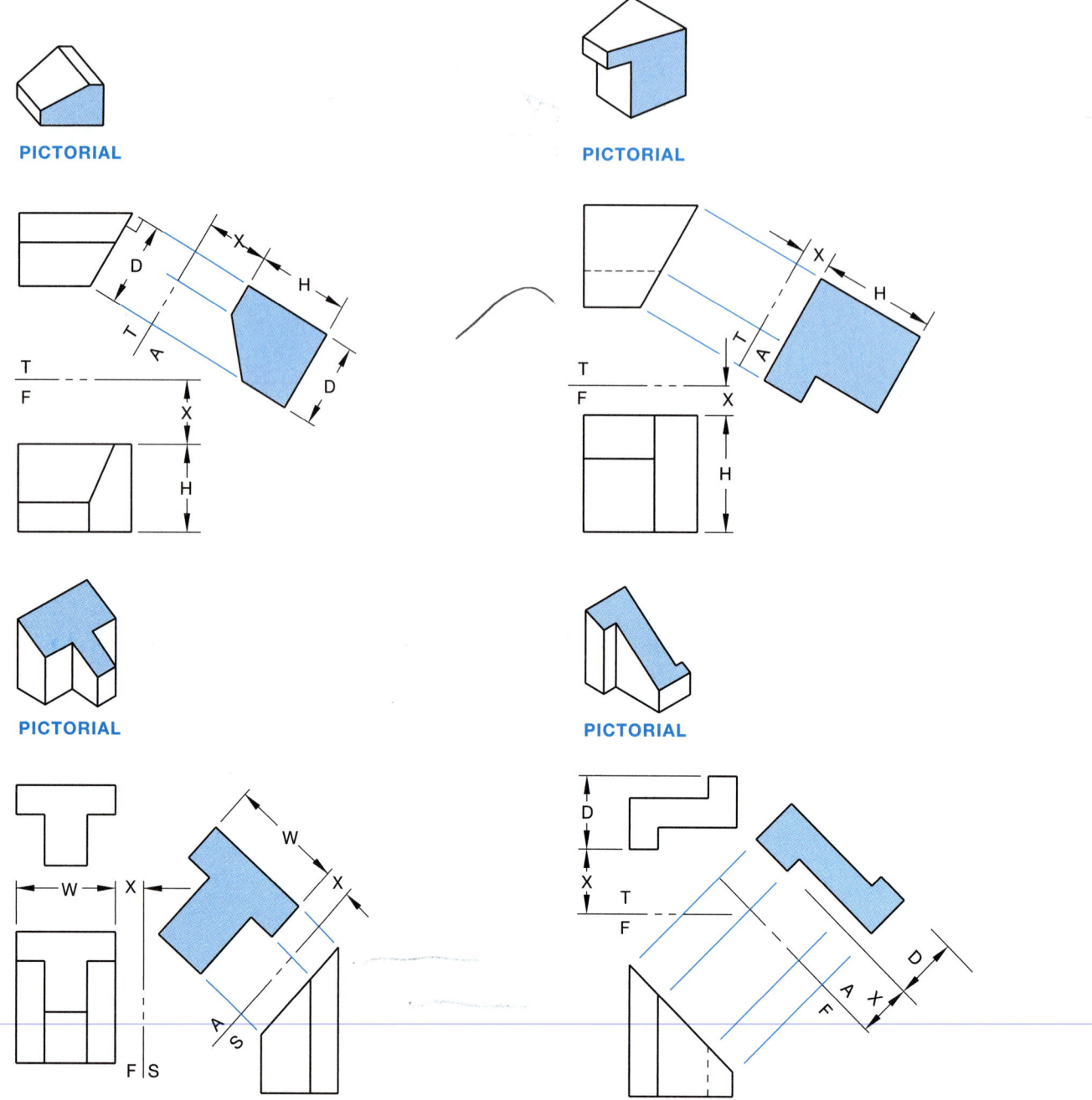

FIGURE 10.10 ■ Partial auxiliary view examples.

viewing-plane lines are used to label different auxiliary views. Place views in the same relationship as the viewing-plane lines indicate. Do not rotate the auxiliary, but leave it in the same relationship as if it were projected from the slanted surface. Where multiple views are used, orient views from left to right and from top to bottom. The views are labeled to correlate with their viewing plane, such as VIEW A-A. The word VIEW and A-A are title height of .24 in. (6 mm), or the word VIEW can be smaller than the A-A depending on company preference.

Figure 10.10 shows some other examples of partial auxiliary views in use. It is important to visualize the relationship of the slanted surfaces, edge view, and auxiliary view. If you have trou-

ble with visualization, it is possible to establish the auxiliary view through the mechanics of view projection. Use the following steps:

STEP 1 Number each corner of the inclined view so the numbers coincide from one view to the other, as shown in Figure 10.11. Carefully project one point at a time from view to view. Some points have two numbers depending on the view.

STEP 2 With the corresponding points numbered in each view, draw an auxiliary fold line parallel to the edge view of the slanted surface. The auxiliary fold line may be any

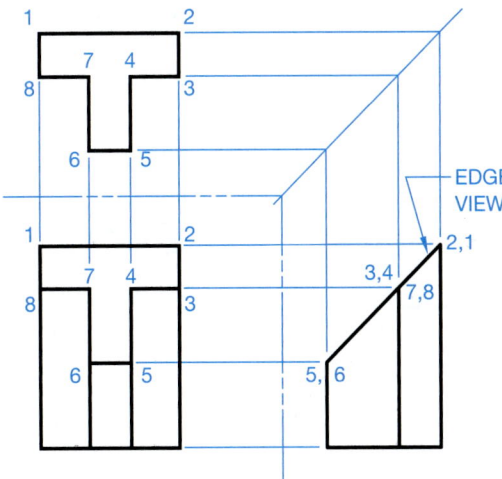

FIGURE 10.11 ■ Step 1, multiview layout.

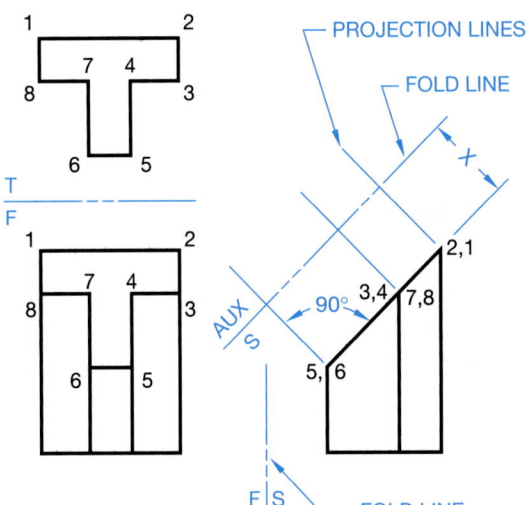

FIGURE 10.12 ■ Step 2, establish the auxiliary fold line.

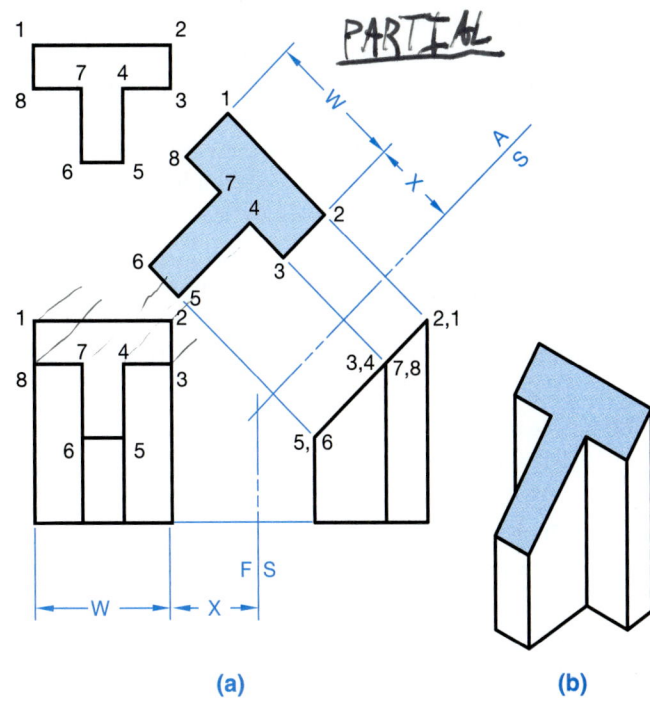

FIGURE 10.13 ■ (a) Step 3, lay out auxiliary view. (b) Pictorial to help visualize object.

convenient distance from the edge view. Project the points on the edge view perpendicular (90°) across the auxiliary fold line. (See Figure 10.12.)

STEP 3 Establish the distance to each point from the adjacent view and transfer the distance to the auxiliary view as shown in Figure 10.13a. Sometimes it is helpful to sketch a small pictorial to assist in visualization. (See Figure 10.13b.)

AUXILIARY VIEW REVIEW AND STEPS USED TO CREATE AN AUXILIARY VIEW

This chapter has discussed the purpose of auxiliary views and demonstrated how auxiliary views are created in relation to the principal multiviews. The following reviews the reasons for drawing auxiliary views and gives you steps that can be used when creating these views.

Reasons for Auxiliary Views

- To find the true length of an oblique line.
- To find the true size and shape of an inclined surface.
- To find the edge view of an oblique surface if there is a true length line that is a boundary of that oblique surface.
- To find the point view of an inclined line.
- To look at the object in a different plane that is not one of the principal planes, so the object can be viewed differently or to start successive auxiliary views.

Based on the previous reasons for creating an auxiliary view, pick a line or edge view of sight to look at the object. For example, if you are trying to find the true size and shape of a surface:

STEP 1 Find the edge view of the foreshortened surface. This is where the foreshortened surface appears as a line or edge view as shown in Figure 10.11.

STEP 2 Draw a fold line (reference plane) parallel to the edge view of the foreshortened surface as shown in Figure 10.12.

STEP 3 Your line of sight for the auxiliary view is perpendicular to the edge view of the foreshortened surface. Draw projection lines perpendicular to the fold line to begin creating the auxiliary view. (See Figure 10.12.)

STEP 4 Determine the type of auxiliary view you are creating, such as a height, width, or depth auxiliary view depending on which view you are projecting from to create the new auxiliary view. In Figure 10.12, you are projecting the auxiliary view from the right side view. The right side view does not have width so you are creating a width auxiliary view.

STEP 5 In this example, you get the width measurements from either the front or top views. The front view is used to get the width measurements in Figure 10.13. Draw a vertical fold line (reference plane) between the front and right side views as shown in Figure 10.12.

STEP 6 Measure the distance X from the fold line to the front view and the width W in the front view as shown in Figure 10.13.

STEP 7 Transfer the measurements X and W to the auxiliary view as shown in Figure 10.13. Be sure the measurements are made and transferred along projection lines that are perpendicular to the fold lines.

STEP 8 Connect the points to create the true size and shape of the foreshortened surface, which is the desired auxiliary view you are trying to establish. (See Figure 10.13.)

DRAWING CURVES IN AUXILIARY VIEWS

Curves are drawn in auxiliary views in the same manner as those shapes described previously. When corners exist, they are used to lay out the extent of the auxiliary surface. An auxiliary surface with irregular curved contours requires that the curve be divided into elements so the element points can be transferred from one view to the next, as shown in Figure 10.14.

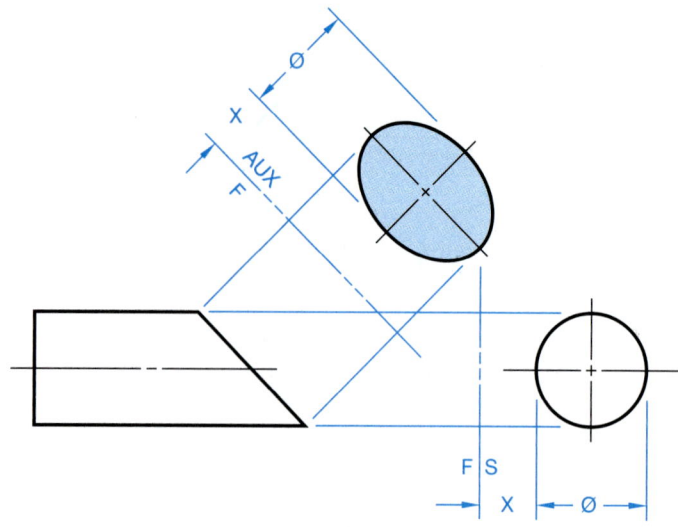

FIGURE 10.15 ■ Elliptical auxiliary view.

The contour of elliptical shapes may be plotted, or if the shape coincides with an available elliptical template, a template should always be used to save time. (See Figure 10.15.)

VIEW ENLARGEMENTS

In some situations, you may need to enlarge an auxiliary view so a small detail can be shown more clearly. Figure 10.16 shows an object with a foreshortened surface. The two principal views are clearly dimensioned at a 1:1 scale. The 1:1 scale is too small to clarify the shape and size of the slot through the part. A viewing-plane line is placed to show the relationship of the auxiliary view. The auxiliary view can then

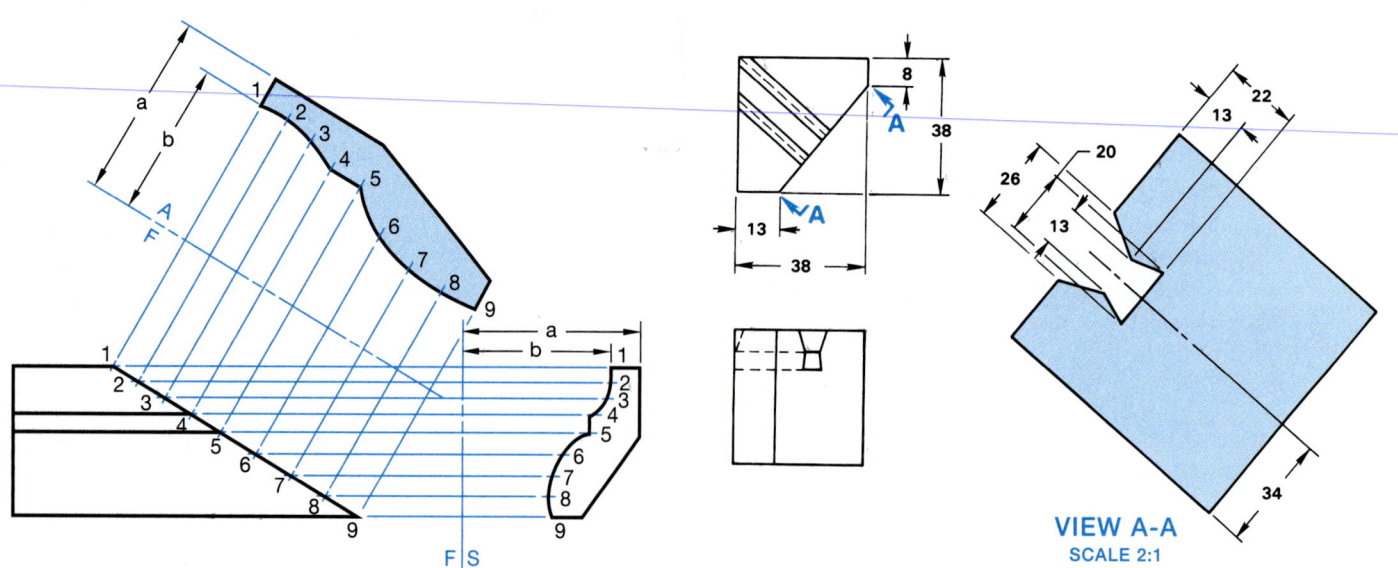

FIGURE 10.14 ■ Plotting curves in auxiliary view.

FIGURE 10.16 ■ Auxiliary view enlargement.

be drawn in any convenient location at any desired scale; in this case a 2:1 scale is used.

SECONDARY AUXILIARY VIEWS

There are some situations when a feature of an object is in an oblique position in relationship to the principal planes of projection. These inclined, or slanting, surfaces do not provide an edge view in any of the six possible multiviews. The inclined surface in Figure 10.17 is foreshortened in each view, and an edge view also does not exist. From the discussion on primary auxiliary views, you realize that projection must be from an edge view to establish the auxiliary view.

In order to obtain the true size and shape of the inclined surface in Figure 10.17, a secondary auxiliary view is needed. The following steps may be used to prepare a secondary auxiliary:

STEP 1 Only two principal views are necessary to work from, as the third view does not add additional information. Establish an element in one view that is in true length, as shown in Figure 10.18. Label the corners of the inclined surface and draw a fold line perpendicular to one of the true-length elements.

STEP 2 The purpose of this step is to establish a primary auxiliary view that displays the slanted surface as an edge. Project the slanted surface onto the primary auxiliary

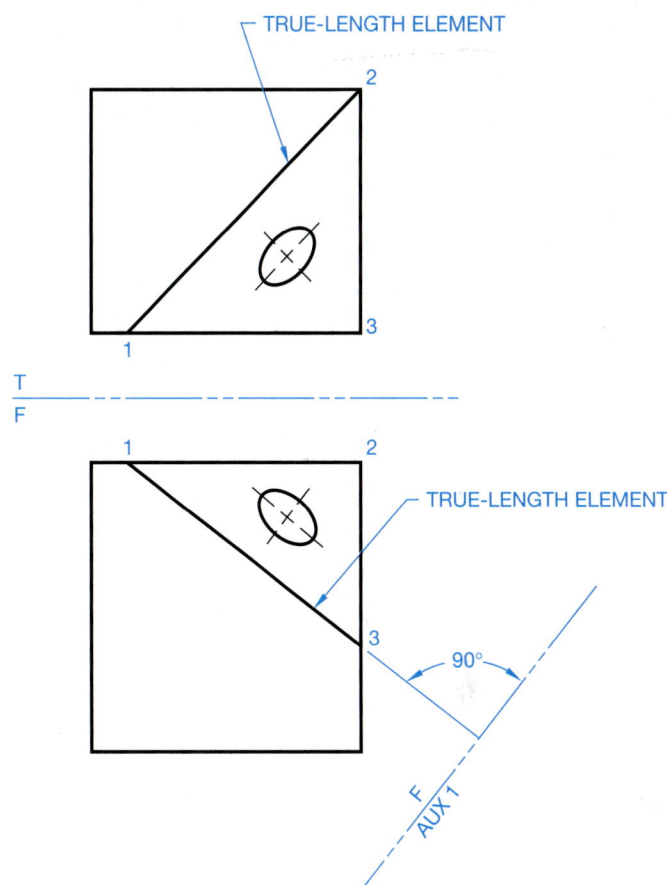

FIGURE 10.18 ■ Step 1, draw a fold line perpendicular to the true-length element.

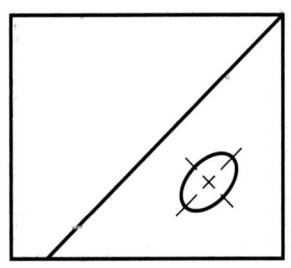

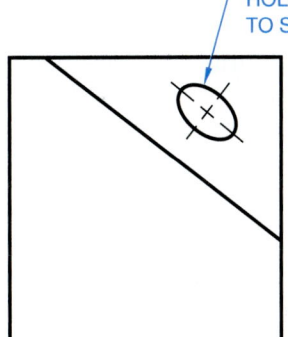

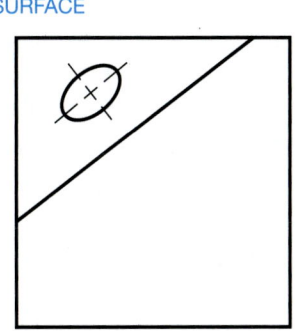

FIGURE 10.17 ■ Oblique surface. There is no edge view in the principal views.

plane, as shown in Figure 10.19. Doing so results in the inclined surface appearing as an edge view or line.

STEP 3 Now with an edge view established, the next step is the same as the normal auxiliary view procedure. Draw a fold line parallel to the edge view. Project points from the edge view perpendicular to the secondary fold line to establish points for the secondary auxiliary view, as shown in Figure 10.20.

In Figure 10.20, the primary auxiliary view established all corners of the inclined surface in a line, or edge view. This edge view is necessary so the perpendicular line of projection (sight) for the secondary auxiliary assists in establishing the true size and shape of the surface. In many situations, both the primary and secondary auxiliary views are used to establish the relationship between features of the object. (See Figure 10.21, page 284.) The primary auxiliary view shows the relationship of the inclined feature to the rest of the part, and the secondary auxiliary view shows the true size and shape of the inclined features plus the true location of the holes.

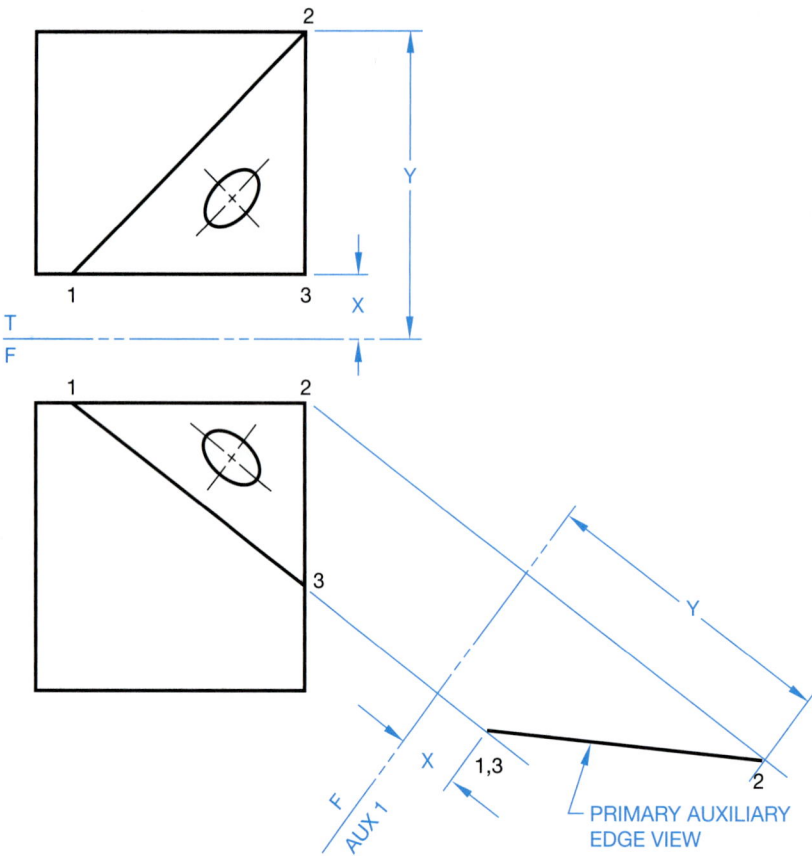

FIGURE 10.19 ■ Step 2, primary auxiliary edge view of oblique surface.

AUXILIARY VIEW ANALYSIS AND REVIEW

The following is a review covering the features and importance of using auxiliary views, multiviews, and orthographic projection:

■ The multiview system, including auxiliary views, is covered in the ASME Y14.3M standard.

■ The two internationally accepted view projection systems are third-angle and first-angle projection. Both systems are discussed in this chapter, and third-angle projection is used throughout this textbook.

■ Multiview drawings represent the shape of an object using two or more views, and there are six principal views: front, top, right side, left side, bottom, and rear.

■ All views are aligned in either third-angle projection or first-angle projection, and the appropriate third- or first-angle projection symbol should be placed on the drawing.

■ Adjacent views must be aligned, unless otherwise specified.

■ Related views must be aligned, unless otherwise specified.

■ There is always one dimension that is the same between adjacent views.

■ The front view is normally considered the most important view from which other views are established.

■ The number of required views can vary depending on the complexity of the object and must be selected to completely describe the shape of the object and features on the object.

■ A line in a view is in true length when the line is parallel to the projection plane.

■ A surface of a view is in true size and shape when the surface is parallel to the projection plane.

■ A line in a view is foreshortened when the line is not parallel to the projection plane.

■ A surface of a view is foreshortened when the surface is not parallel to the projection plane.

■ Auxiliary views are used to show the true size, shape, and relationship of features of a surface that is not parallel to any of the principal planes of projection.

■ A primary auxiliary view is one that is adjacent to and aligned with a principal view.

■ A secondary auxiliary view is one that is adjacent to and aligned with a primary auxiliary view or with another secondary auxiliary view.

■ Additional view placement options are discussed in this chapter and in Chapter 9 and include: removed views, rotated views, and cross-referenced views. Use of these options should be carefully confirmed with your company practices.

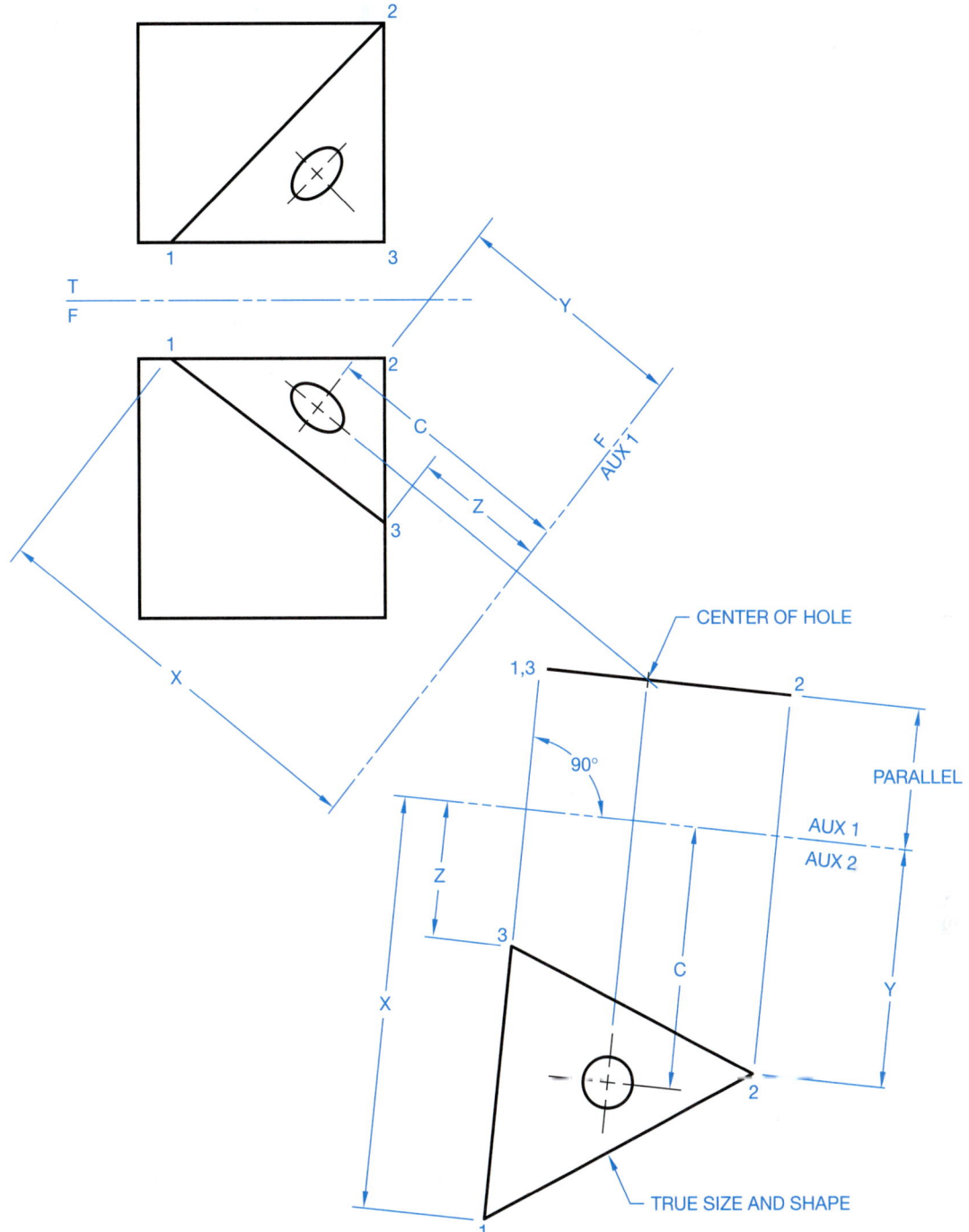

FIGURE 10.20 ■ Step 3, secondary auxiliary projected from the edge view to get the true size and shape of the oblique surface.

DESCRIPTIVE GEOMETRY

Descriptive geometry, covered in detail in Chapters 16 and 17, is used for determining issues such as finding the true lengths of lines, the end view of a line, the true size and shape of a surface or plane, the angle between two lines, the angle between two planes, the intersection between two planes, and the intersection between a cone or a cylinder and a plane. Descriptive geometry problems are solved graphically by projecting points onto se-

lected adjacent projection planes in an imaginary projection system. The multiview and auxiliary view concepts you have learned about in Chapter 9 and in this chapter serve as a basis for further study of descriptive geometry concepts. What you have learned in this chapter and in Chapter 9, coupled with what you will learn in the descriptive geometry chapters, will then help you solve additional drafting problems as you study advanced chapters covering specialty drafting fields and problems you encounter in industry.

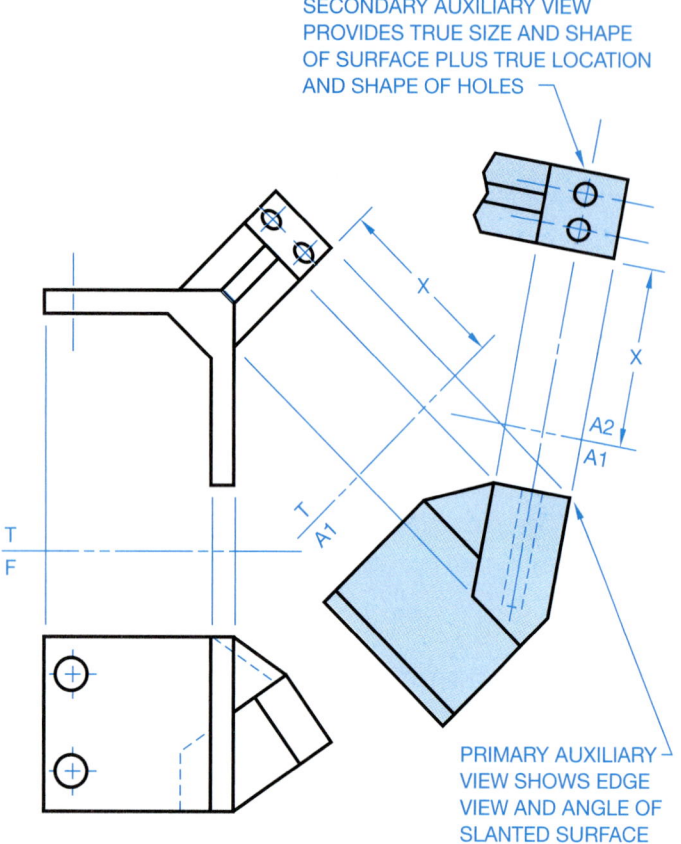

SECONDARY AUXILIARY VIEW PROVIDES TRUE SIZE AND SHAPE OF SURFACE PLUS TRUE LOCATION AND SHAPE OF HOLES

PRIMARY AUXILIARY VIEW SHOWS EDGE VIEW AND ANGLE OF SLANTED SURFACE

FIGURE 10.21 ■ The primary and secondary auxiliary views.

AUXILIARY VIEW LAYOUT

Working with auxiliary views is a little more complex than working with the normal multiviews. The auxiliary view is often difficult to place in relationship to the other views. Space requirements many times cause the auxiliary view to interfere with other views. Whenever possible, it is best to project the auxiliary view directly from the inclined surface with one projection line connecting the inclined surface to the auxiliary view. This arrangement makes it easier for the reader to correctly interpret the relationship of the views. This method is shown in the layout steps provided in Figures 10.11, 10.12, and 10.13.

When drawing space is limited, it is possible to use the viewing-plane method to display the auxiliary view. This technique is shown in Figure 10.9. The advantage of the viewing-plane method is that it allows you to place the auxiliary view in any convenient place on the drawing. Multiple auxiliary views should be placed in a group arranged from right to left and from top to bottom on the drawing. One unorthodox method that is often used by entry-level drafters is to place viewing planes all over the drawing and shift views from their normal position. This practice should be avoided, because the normal multiview projection is always best. You should follow the same layout procedure outlined for multiviews in Figures 9.61 through 9.67.

The steps summarized here work the same for manual drafting or CADD:

■ Select and sketch the front view.

CADD APPLICATIONS 2-D

DRAWING VIEWS

So far in your study you have learned that multiview and auxiliary view drawings are the orthographic projection of two or more planes at right angles to each other. One important aspect of this technology is accuracy. As a manual drafter, accuracy depends on your sharp pencil and a keen eye for technique. Even solutions in the best manual drafting have a degree of error due to line widths or the slight tilt of the pencil in relation to the paper. Multiview and auxiliary view drawings produced with CADD are perfect, assuming correct drafting standards and CADD techniques are used throughout the drawing process. The reason for this perfection has to do with the capabilities of the hardware and software, and the fact that the computer is actually displaying the mathematical counterpart of the graphic analysis. In other words, drawing views are a representation of the mathematical coordinates of the problem being presented. Therefore, the drawing accuracy is only controlled by the accuracy potential of the computer. Figure 10.22 shows a multiview drawing on a CADD monitor.

Auxiliary Views

Most CADD systems provide the drafter with "drawing aids" such as a screen grid similar to the light blue lines preprinted on vellum to assist the manual drafter. The screen grid is displayed as dots. These grid dot patterns may be displayed in any convenient increment to assist in accurately placing the views. The drafter may also establish designated increments for the screen cursor crosshairs to snap. This allows the drafter to accurately determine distances on the drawing without guesswork. Another asset of the CADD system is the display of drawing distance and coordinates. For example, when you draw a line on the screen, the computer displays the exact line length. Drawing multiviews with these kinds of drawing aids makes the job easy and accurate. When drawing auxiliary views, the CADD drawing aids are flexible and may be rotated at the angle of the auxiliary view to assist in accurately laying out the view as shown in Figure 10.23.

A CADD system also allows you to work with colored layers that provide for a clearer picture of the problem and

CADD APPLICATIONS 2-D

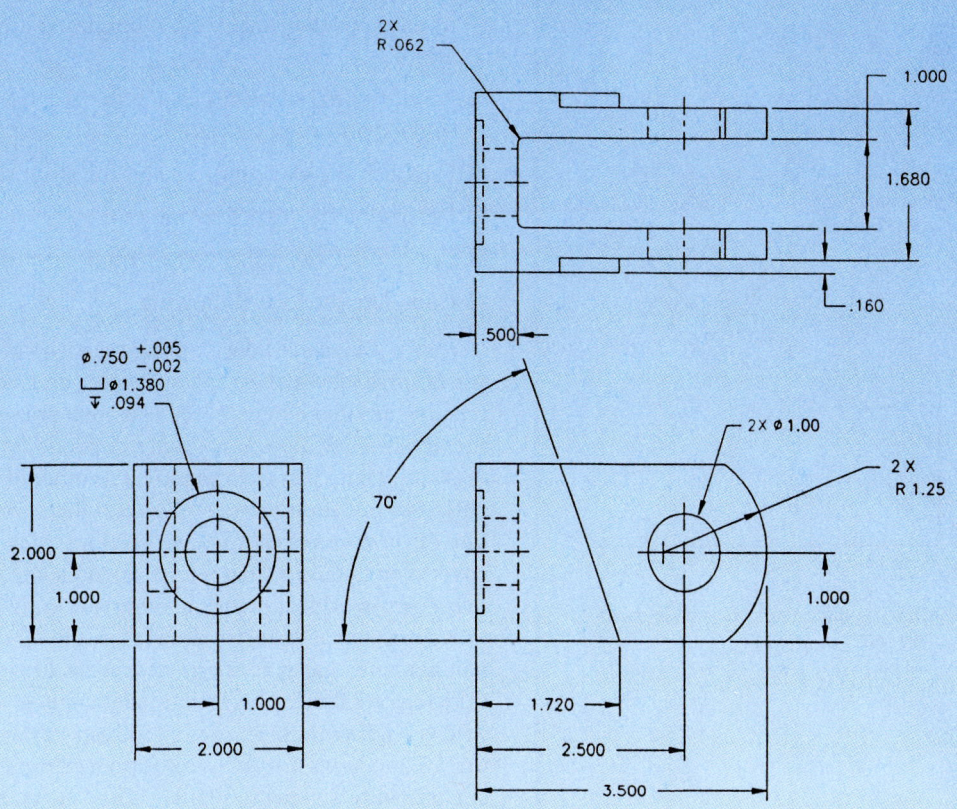

FIGURE 10.22 ■ Using CADD to create multiviews.

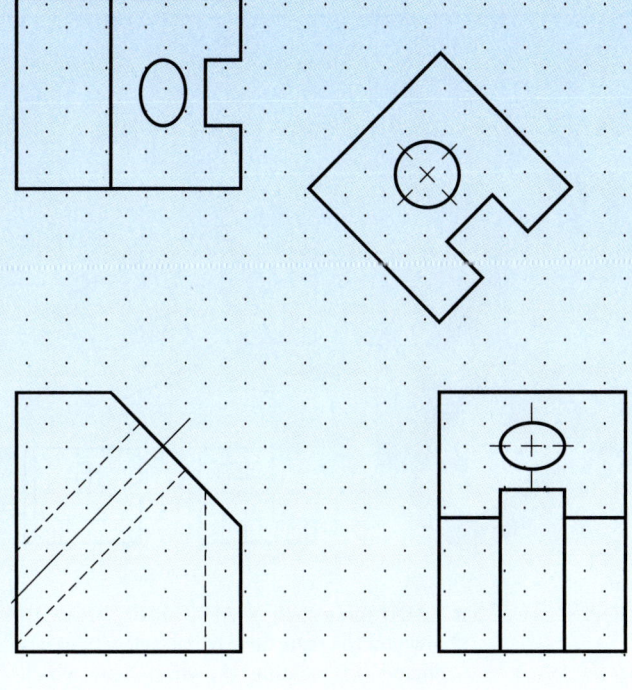

FIGURE 10.23 ■ Using the rotated grid points in CADD to help draw the auxiliary view.

solution. For example, you can set a layer for projection lines shown in any color, such as red. This helps you visualize the entire drawing and allows you to accurately project a feature from one view to the next. The layers may be turned on or off at your convenience. When you complete the drawing and you are ready to make a plot, all you have to do is turn off the projection line layer and the drawing is ready to plot.

The lines that you use to project between views are construction lines. Construction lines are only used for layout and construction and are not part of the drawing when they are turned off. AutoCAD's XLINE command, for example, creates construction lines of infinite length. They may be drawn horizontally, vertically, at an angle, or offset a given distance from another line or object. The XLINE can also be used to bisect an angle. The XLINE command ANGLE option is effective for laying out auxiliary views. This technique allows you to project from an inclined surface to establish the correct size and shape, and location of the auxiliary view as shown in Figure 10.24.

(Continued)

CADD APPLICATIONS 2-D

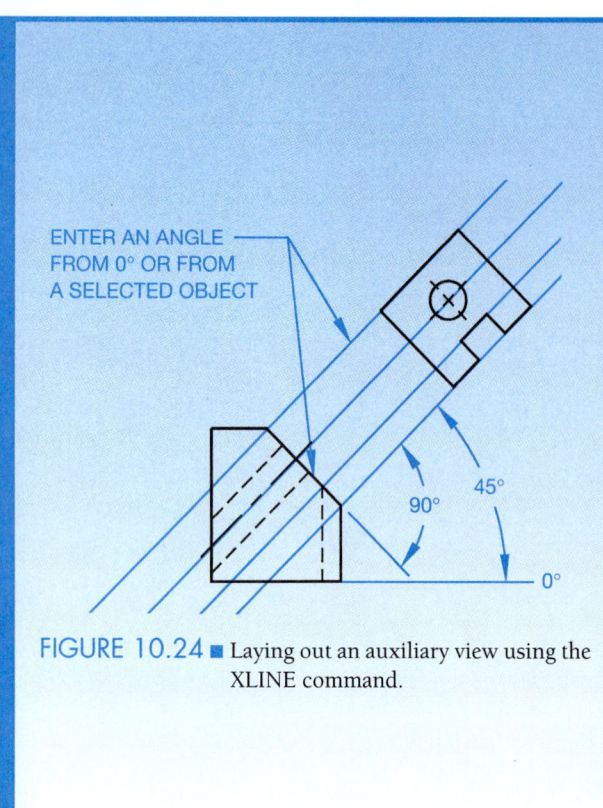

FIGURE 10.24 ■ Laying out an auxiliary view using the XLINE command.

ENTER AN ANGLE FROM 0° OR FROM A SELECTED OBJECT

90°
45°
0°

■ Select and sketch the other principal views in proper relationship to the front view, eliminating any unnecessary views.

■ An additional step is to sketch the auxiliary view or views in their proper positions.

■ Establish the working area and the sheet size to be used.

■ Determine the total drawing area.

■ Lay out the views using construction lines.

■ Complete the formal drawing.

Instruction on the development of views often focuses on the use of three views, such as a front, top, and side view, or a front, top, and auxiliary view. While this view orientation is common, the actual view selection totally depends on the part you are drawing. Many parts can require a number of views that can include several multiviews and an auxiliary view. The combination of views can even include a sectional view. Section views are covered in detail in Chapter 13. Spend some time looking at as many real-world drawings as you can find. You may discover that some actual industry drawing could be done in a different manner, and that you might have ideas for improving the way drawings are done. This is your challenge as you begin to create your own drawings. The actual industry drawing shown in Figure 10.26 shows a front view, top view, right-side view, a view created with a viewing plane (VIEW A-A), a broken out sectional view, and a partial auxiliary view projected from the sectional view. This example demonstrates the complexity of some drawings.

CADD APPLICATIONS 3-D

AUXILIARY VIEWS

The parametric two-dimensional drawing capabilities of some three-dimensional solid modeling programs allow you to easily produce a variety of drawing views, including auxiliary views. Tools such as AUXILIARY VIEW allow you to create an auxiliary view by referencing an existing drawing view, which parametrically associates a view with the model. (See Figure 10.25.) Using tools like AUXILIARY VIEW typically involves selecting the edge of the inclined surface from an existing drawing view, followed by picking a location for the auxiliary view. You can also choose to add auxiliary view information, such as a view label, by selecting the appropriate options. Usually when you create an auxiliary view, all possible object geometry is projected, which means that you cannot automatically create a partial auxiliary view. However, you can have the ability to hide or turn off the visibility of edges, resulting in a partial auxiliary view.

MODEL   DRAWING

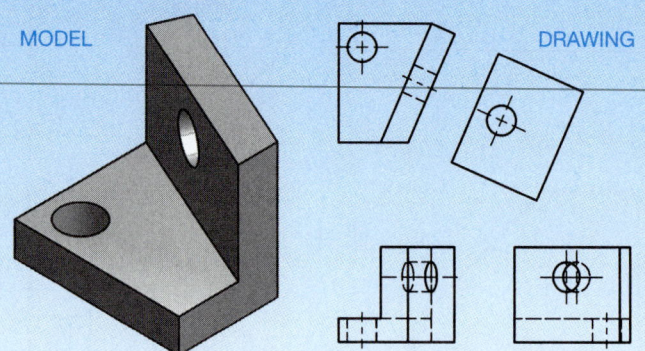

FIGURE 10.25 ■ CADD tools such as AUXILIARY VIEW allow you to create an auxiliary view by referencing an existing drawing view, which parametrically associates a view with the model.

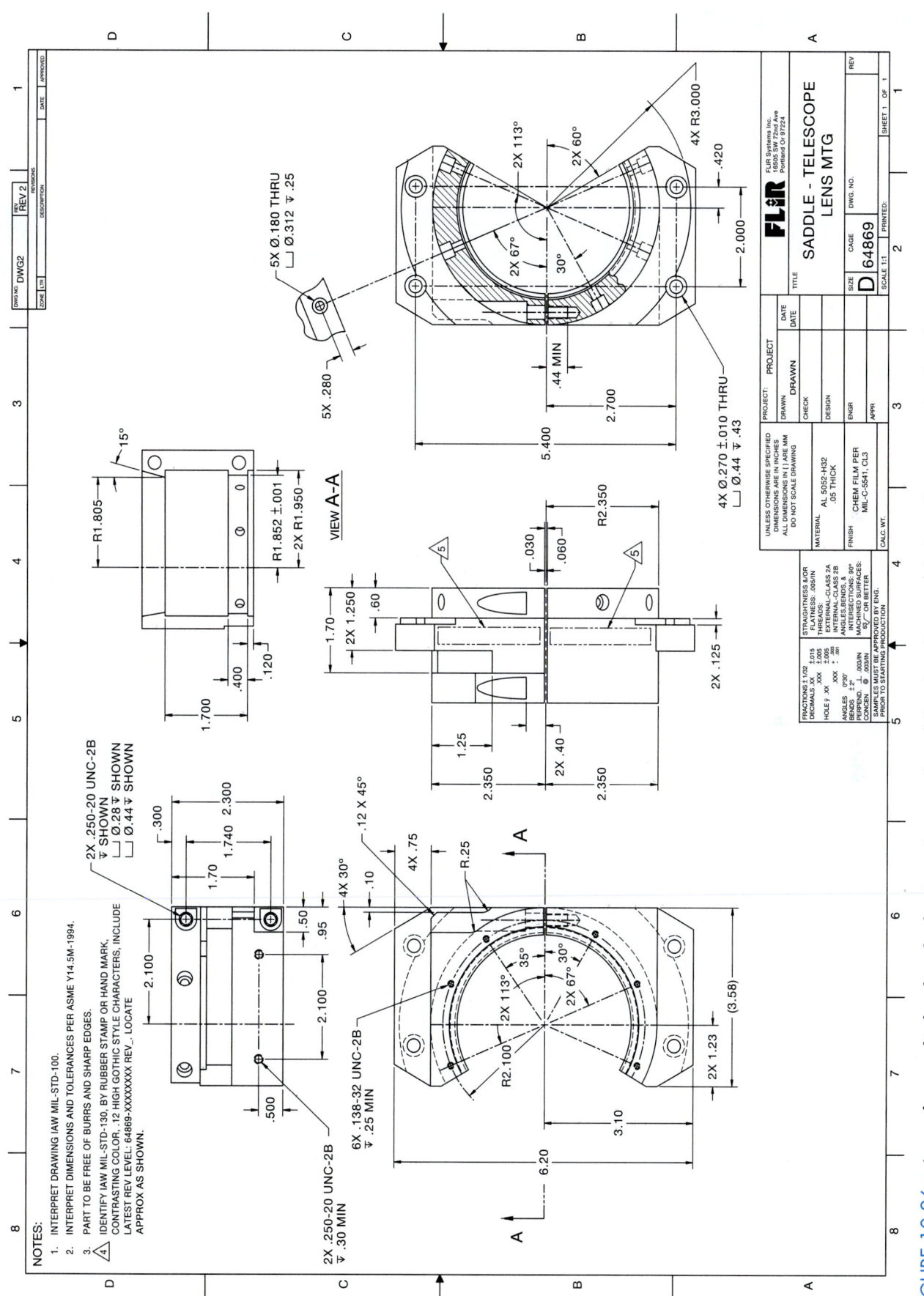

FIGURE 10.26 ■ A complex part displayed on the drawing using several carefully selected views including a front, top, and right-side view; a view created from a viewing plane; a sectional view; and an auxiliary view. *Courtesy Flir Systems.*

WEBSITE RESEARCH

The following websites can provide you additional information for research or further study into topics covered in this chapter:

http://www.asme.org—Find information and publications related to the American Society of Mechanical Engineers.

http://www.ansi.org—The American National Standards Institute. Information about national and international drafting standards.

http://www.adda.org—American Design Drafting Association. Information related to the drafting and design profession.

PROFESSIONAL PERSPECTIVE

For many people, one of the most difficult aspects of engineering drafting is the need to visualize two-dimensional views. This "special" ability is natural for some, while others must carefully analyze every part of a multiview and auxiliary view drawing in order to fully visualize the product. Ideally, you should be able to look at the views of an object and readily formulate a pictorial representation in your mind. The principal reason that multiviews are aligned is to assist in the visualization and interpretation process. The front view is the most important view, and that is why it contains the most significant features. As you look at the front view, you should be able to gain a lot of information about the object then visually project from the front view to other views to gain an understanding of the entire product.

It is up to the drafter to select the best views to completely describe the object. Too few views make the object difficult or impossible to interpret, while too many views make the drawing too complex and waste valuable drafting time. If you follow the guidelines set up in this chapter, you should be able to successfully complete the task. Remember, always begin with a sketch. With the sketch, you can quickly determine view arrangement and spacing. Without the preliminary sketch, you might use a lot of drafting time and end up discovering that the layout does not work as you expected.

MATH APPLICATIONS

PROJECTED AREA

Problem: Suppose the steel plate in the shape of an ellipse with an area of 65.98 ft^2 (from the Math Application in Chapter 9) is set at a 60° angle on the ground. With the sun overhead, what is the area of the *shadow* of the plate on the ground? (See Figure 10.27.)

Solution: The shadow's area is called the *projected area* in math. It is found by multiplying the true area by the *cosine* of the inclined angle. Using a calculator, cos 60° = .5, so the area of the shadow is 65.98 × .5 = 32.99 or 33 ft^2. The principle of multiplying true area by cosine to obtain projected area applies to all 2-D shapes. Math Applications in later chapters further explore right triangles and trigonometry functions, such as cosine.

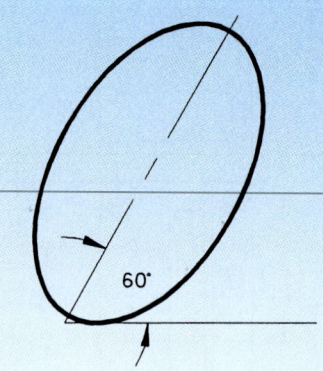

FIGURE 10.27 ■ Tilted ellipse.

CHAPTER 10 *Auxiliary Views Test*

 Access the CD found with this textbook to view the Chapter 10 Test. Confirm the preferred submittal method with your instructor.

CHAPTER 10 *Auxiliary Views Problems*

 PARTS 1 through 3, PROBLEMS 10.1 through 10.14: Access the CD found with this textbook and open the problem of your choice, or as assigned by your instructor.

Solve the problems using the instructions provided on the CD, unless otherwise specified by your instructor.

PROBLEM 10.15 **Primary auxiliary view (in.)**

Title: Angle V-block
Material: SAE 4320

DIRECTIONS

These problems provide you with dimensioned views and a proposed auxiliary view location, or pictorial views that contain dimensions and a recommended view layout. Draw the views. The pictorial views are provided to aid in visualization and to display the dimensions. You do not need to draw the pictorial view. Use the given dimensions to create your manual or CADD drawing. Set up your drawings with a properly sized border and title block. Properly complete the information in the title block. Do not draw the dimensions.

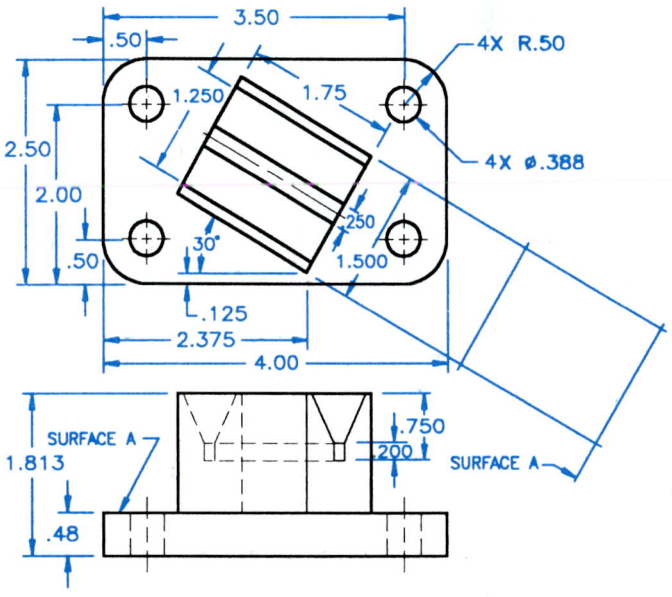

PROBLEM 10.16 **Primary auxiliary view (in.)**

Title: Cylinder Support

Material: Cast Iron

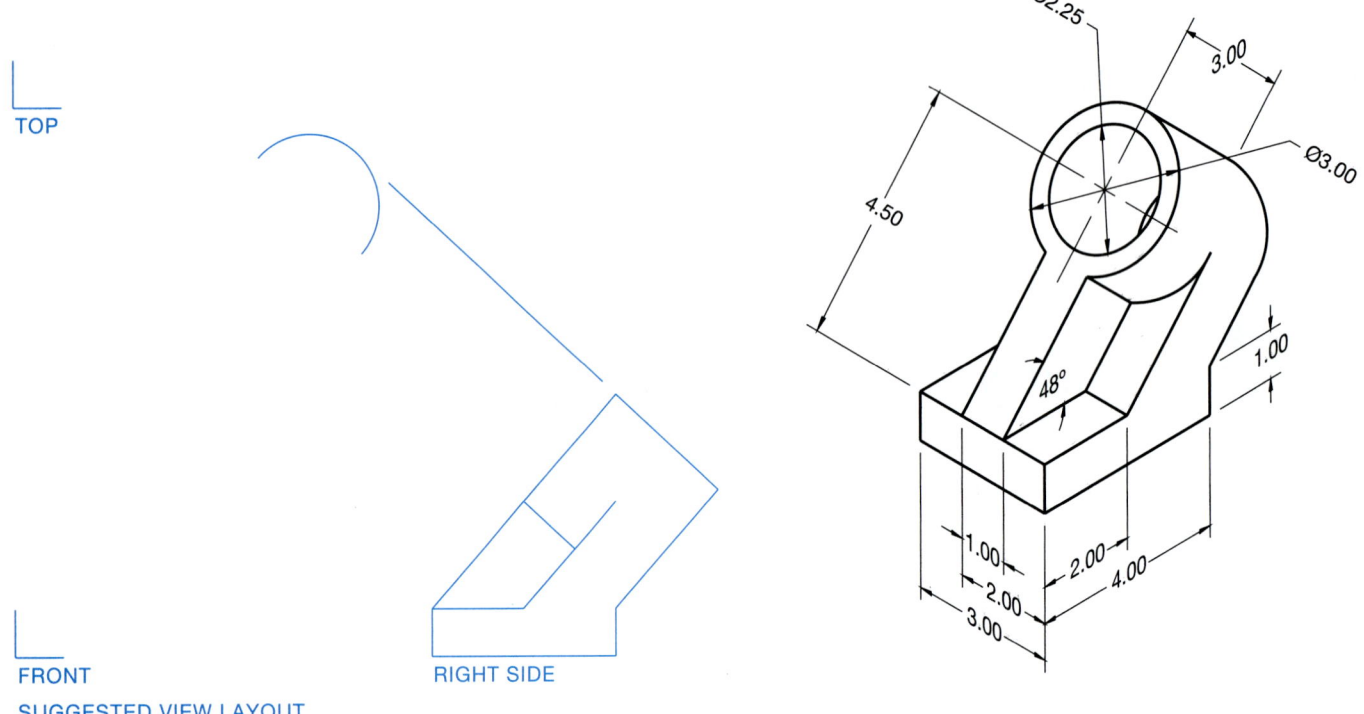

TOP

FRONT

RIGHT SIDE

SUGGESTED VIEW LAYOUT

PROBLEM 10.17 **Primary auxiliary view (in.)**

Title: 135 Bracket

Material: Aluminum

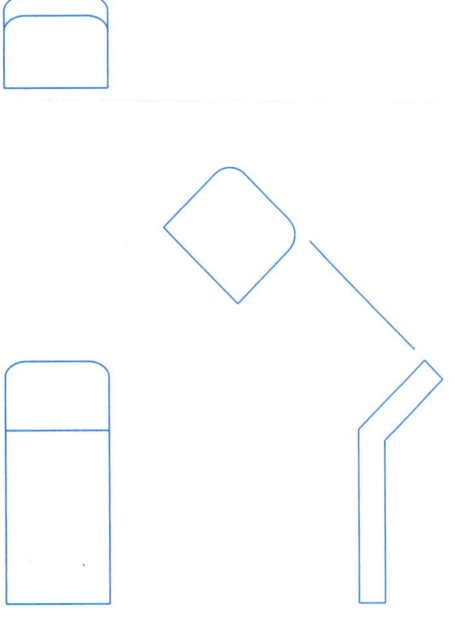

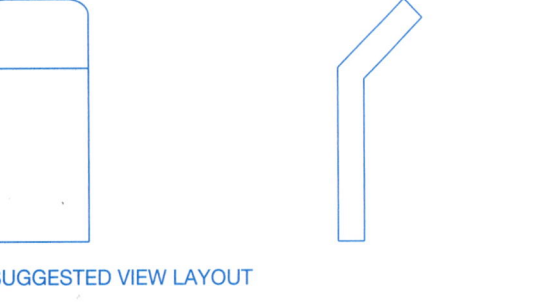

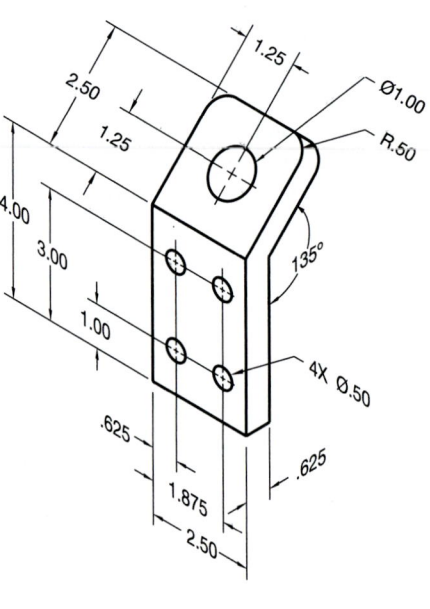

SUGGESTED VIEW LAYOUT

PROBLEM 10.18   Primary auxiliary view (in.)

Title: 135-1 Bracket
Material: Aluminum

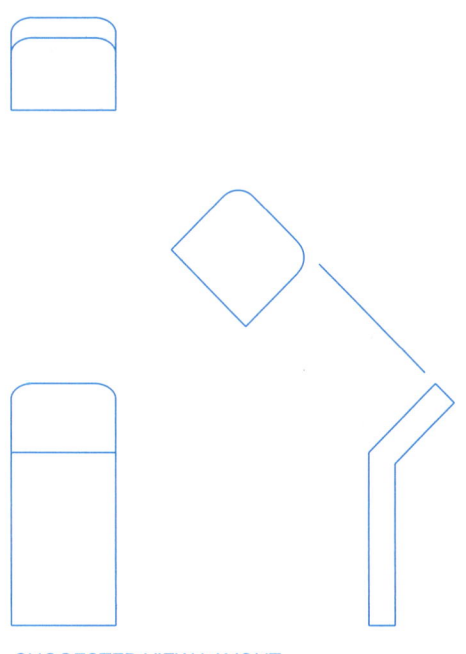

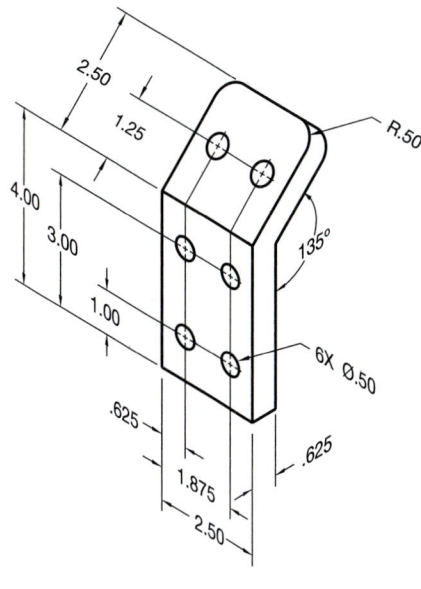

SUGGESTED VIEW LAYOUT

PROBLEM 10.19   Primary auxiliary view (in.)

Title: Support Base
Material: Cast Aluminum

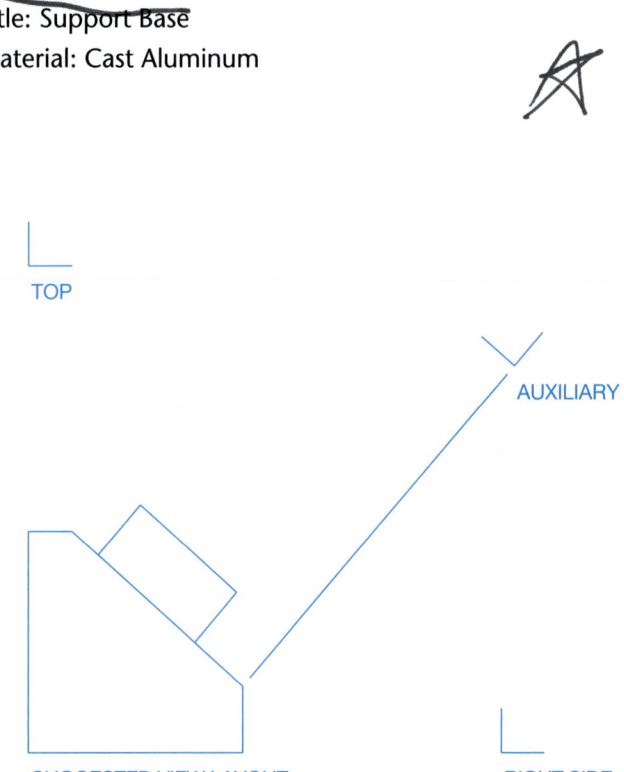

TOP

AUXILIARY

SUGGESTED VIEW LAYOUT

RIGHT SIDE

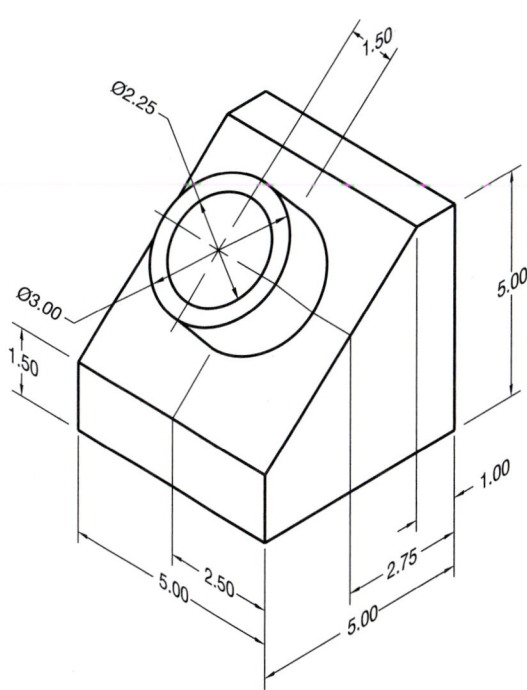

PROBLEM 10.20 **Primary auxiliary view (in.)**

Title: Shaft Support

Material: SAE 1020

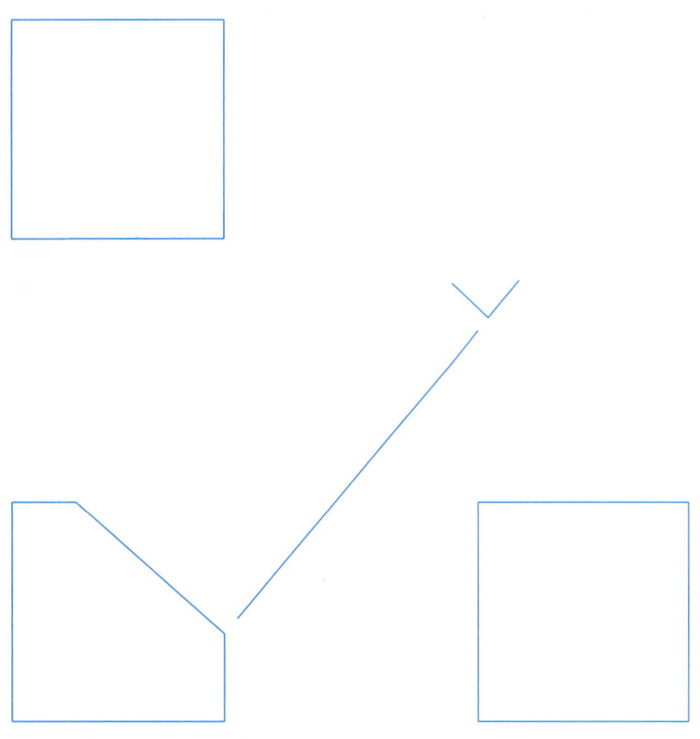

SUGGESTED VIEW LAYOUT

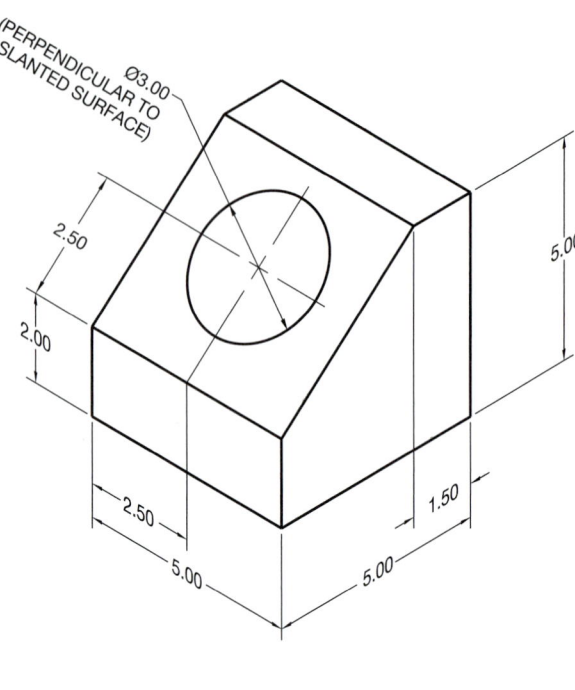

PROBLEM 10.21 **Primary auxiliary view (in.)**

Title: Spacer

Material: Cast Iron

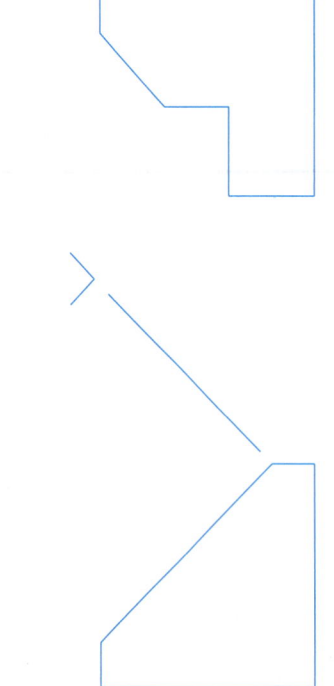

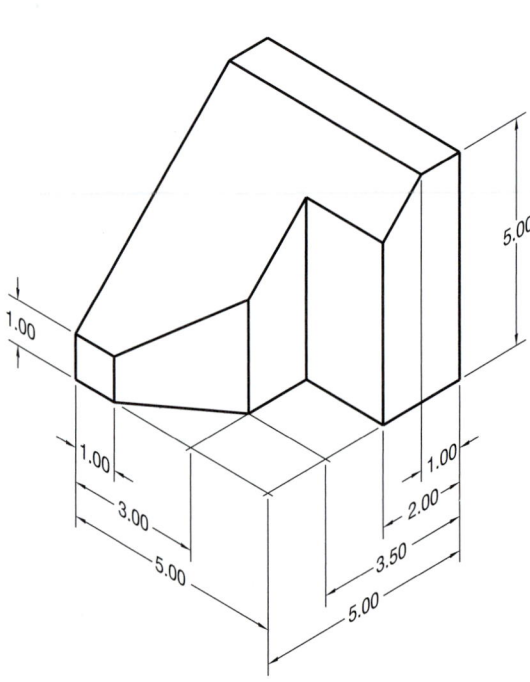

SUGGESTED VIEW LAYOUT

PROBLEM 10.22   **Primary auxiliary view (in.)**

Title: T-Block

Material: SAE 4320

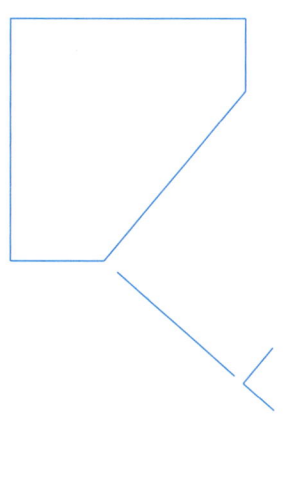

SUGGESTED VIEW LAYOUT

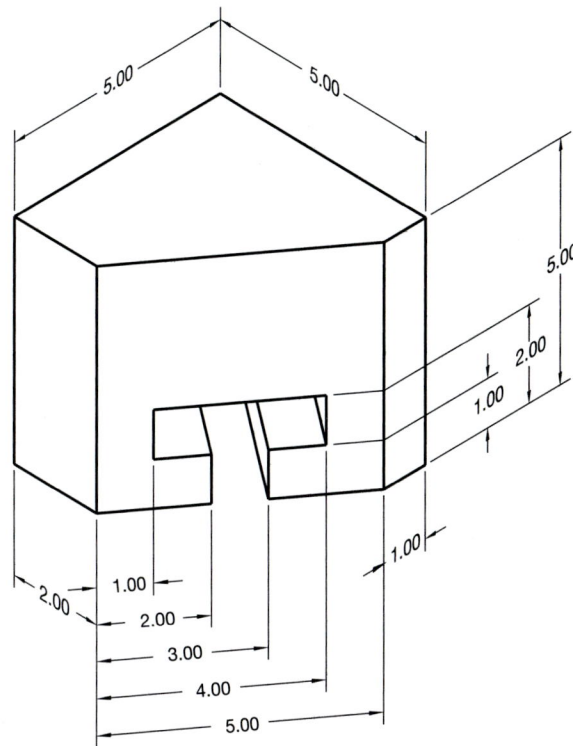

PROBLEM 10.23   **Primary auxiliary view (in.)**

Title: T-Wedge

Material: Mild Steel

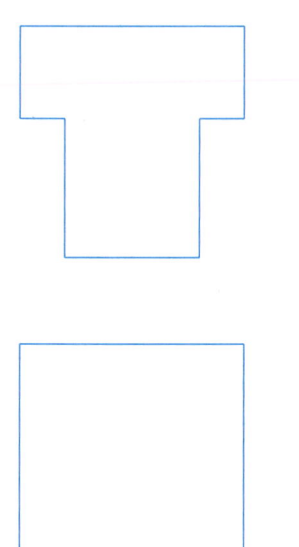

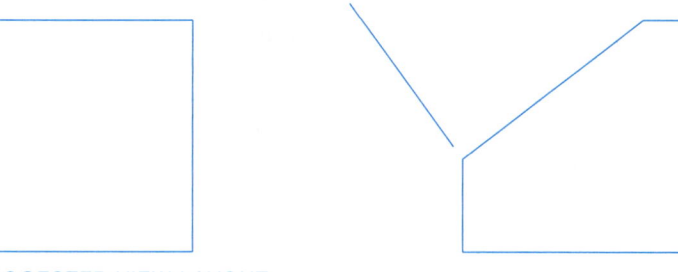

SUGGESTED VIEW LAYOUT

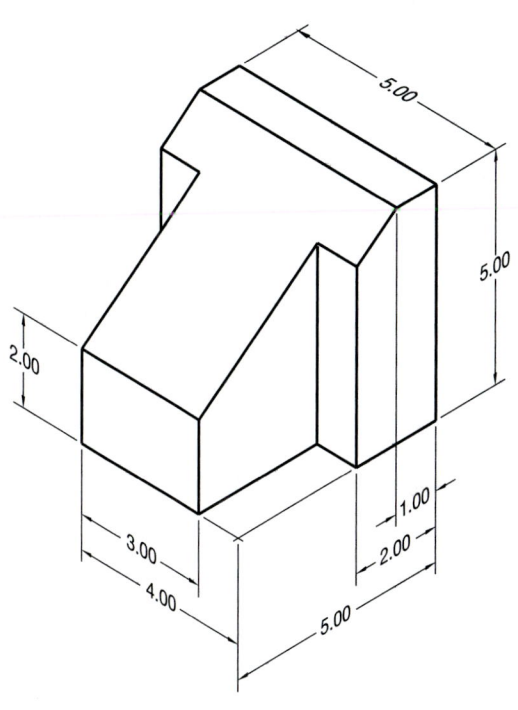

PROBLEM 10.24   Primary auxiliary view (in.)

Title: Brace

Material: Cast Iron

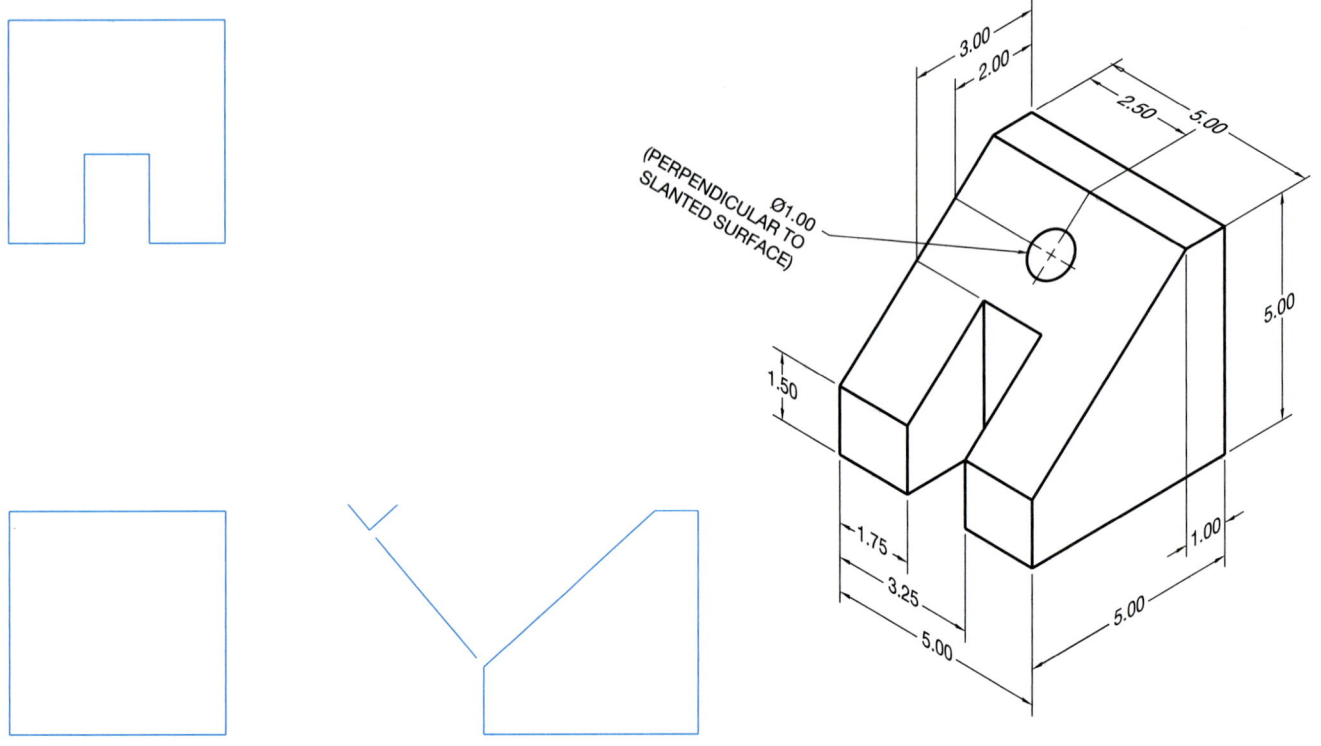

SUGGESTED VIEW LAYOUT

 PART 5, PROBLEMS 10.25 through 10.37: Access the CD found with this textbook and open the problem of your choice, or as assigned by your instructor. Solve the problems using the instructions provided on the CD, unless otherwise specified by your instructor.

MATH PROBLEMS

 PROBLEMS 10.38 through 10.47: Access the CD found with this textbook and open the math problem of your choice, or as assigned by your instructor. Solve the problem or problems using the instructions provided.

Dimensioning and Tolerancing

LEARNING OBJECTIVES

After completing this chapter, you will:

- Identify and use common dimensioning systems.
- Explain and apply dimensioning standards based on ASME Y14.5M.
- Apply proper specific notes for manufacturing features.
- Place proper general notes and delta notes on a drawing.
- Interpret and use correct tolerancing techniques.
- Prepare completely dimensioned multiview drawings from engineering sketches and industrial drawings.

- Apply draft angles as needed to a drawing.
- Dimension CAD/CAM machine tool drawings.
- Prepare casting and forging drawings.
- Provide surface finish symbols on drawings.
- Solve tolerance problems including limits and fits.
- Use an engineering problem as the basis for your layout techniques.
- Describe the purpose of ISO 9000 Quality Systems Standard and related standards.

THE ENGINEERING DESIGN APPLICATION

As mentioned later in this chapter, a complete detail drawing is made up of multiviews and dimensions. The layout techniques of a detail drawing must include an analysis of how the views and dimensions go together to create the finished drawing. The most effective way to form preliminary layout ideas is through the use of rough sketches as described in Chapter 6. The following gives you a preliminary look at the steps used in the creation of a completely dimensioned multiview drawing from an engineer's sketch.

While this is only a brief introduction to what you will learn throughout this chapter, it gives you an idea of the process a drafter goes through when given this assignment.

Consider the engineering sketch in Figure 11.1 as you evaluate how to prepare a complete detail drawing. The procedure is the same with CADD or manual drafting.

STEP 1   Select and make a rough sketch of the proper multiviews. Leave plenty of space between views so dimensions can be added. The engineer's sketch is not always accurate. It is your responsibility to convert the engineering ideas from the rough stage to a formal drawing. The information may be correct; however, the organization of information and proper layout are your duties. (See Figure 11.2.)

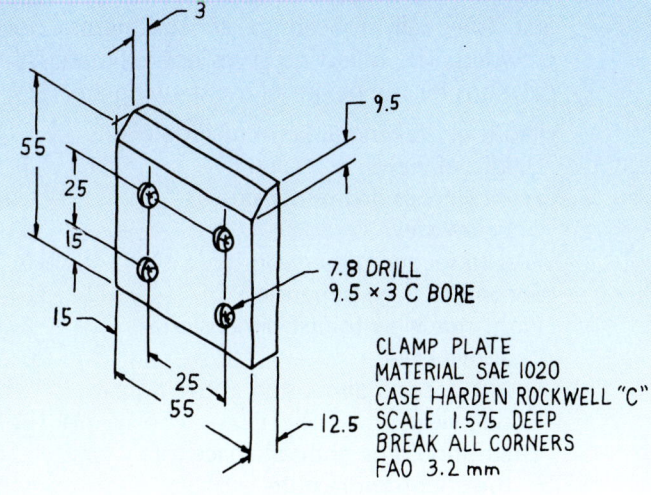

3

9.5

55

25

15

15

7.8 DRILL
9.5 × 3 C BORE

CLAMP PLATE
MATERIAL SAE 1020
CASE HARDEN ROCKWELL "C"
SCALE 1.575 DEEP
BREAK ALL CORNERS
FAO 3.2 mm

25
55

12.5

FIGURE 11.1 ■ Engineering sketch.

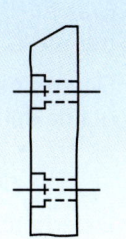

FIGURE 11.2 ■ Rough sketch of selected multiviews.

(Continued)

THE ENGINEERING DESIGN APPLICATION *(continued)*

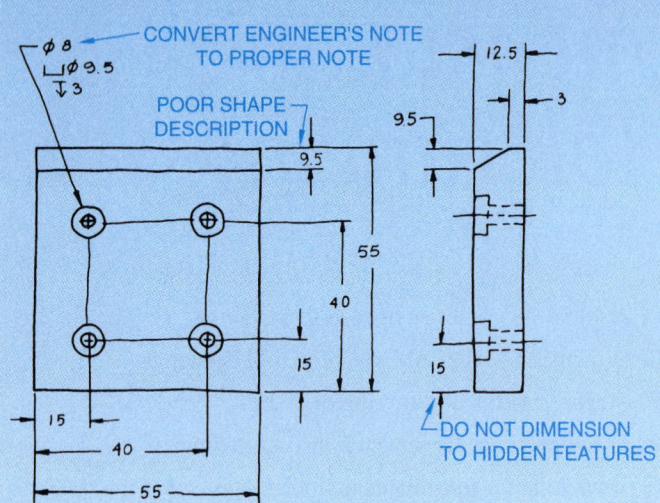

FIGURE 11.3 ■ Rough sketch with dimensions and notes placed. In this rough sketch, as in the real world, the dimensions and features are often out of proportion.

STEP 2 Place dimensions and notes on your multiview sketch. (See Figure 11.3.) Keep the following dimensioning rules in mind as you select and place the dimensions:

■ Begin with smallest dimension closest to the object and place dimensions that progressively increase in size further away from the object.

■ Do not crowd dimensions.

■ Dimension between views where possible.

■ Dimension to views that show the best shape of features.

■ Group dimensions when possible.

■ Do not dimension size or location dimensions to hidden features.

■ Stagger adjacent dimension numerals.

■ Use leaders to properly label specific notes.

■ Convert all information to proper drafting standards.

Evaluate these basic rules as you decide where dimensions should be placed.

STEP 3 Determine the scale to use based on the amount of fine detail to be shown and the total size of the object.

The clamp plate can easily be drawn full scale.

STEP 4 Determine the sheet size based on the drawing scale to be used, the total drawing area, including views and dimensions, and the amount of clear area needed for general notes and future revisions. Refer to Chapter 7.

Look at Figure 11.3 as you follow the calculations to determine the width and height of the drawing area needed for the clamp plate. The engineer's sketch shows dimension values in millimeters, which is the commonly used metric value. You should do the preliminary layout work using metric values if the drawing is in metric. Use inch values if the drawing is dimensioned in inches. The spacing suggestions are presented as guidelines and can vary depending on your specific drawing, and school or company practices. The spacing calculations are also approximate and are used as a general guide for setting up your drawing. When using CADD, these values can be adjusted easily while working on your drawing and can be changed to fit any desired sheet size during the final layout process. Additional information is provided throughout this chapter.

Length of drawing area (in millimeters)	
Front view width	55
Space between views:	
Front view to first dimension	20
First to second dimension	10
Second to third dimension	10
Open space	20
Right-side view to dimension	20
Right-side view depth	+ 12.5
Drawing area total width	147.5
Round off drawing area to	148

The space from the view to the first dimension and the spaces between dimensions are a drafting decision. The first space is often, but not always, larger than distances to additional dimension lines. Each dimension line, after the first, is equally spaced so dimensions are not crowded. The following gives approximate calculations for the height of the drawing area:

Height of drawing area (in millimeters)	
Height of views	55
Front view to first dimension below view	20
First to second dimension	10
Second to third dimension	10
Right-side view to first dimension above view	20
First to second dimension above right-side view	+ 10
There should be enough space for the counterbore note.	
Drawing area total height	125

The total width and height of the drawing area is 148 × 125. Keep in mind that these cal-

THE ENGINEERING DESIGN APPLICATION *(continued)*

culations are approximate. There should be adequate space on your drawing for some flexibility. Select a B (A3 metric) size drafting sheet for this drawing. Figure 11.4 shows the 148 × 125 drawing area blocked out on a B-size sheet. Before you proceed, be sure your equipment, table, and hands are clean when using manual drafting.

STEP 5 Use construction lines to draw the selected multiviews.

STEP 6 Use construction lines to place hidden, extension, center, dimension, and leader lines, using selected spacing and placement from rough sketch. Also, use guidelines to prepare for the placement of lettering. This drafter chose to change the hole locations to datum dimensioning. This method is defined and advantages given later in this chapter. (See Figure 11.5.)

STEP 7 Complete the detail drawing using proper drafting technique in the following sequence:

1. Darken all horizontal thin lines from top to bottom and all vertical thin lines from left to right if right-handed or right to left if left-handed when using manual drafting.

2. Draw all circles, arcs, and object lines.

3. Draw all straight object lines in the same order as thin lines.

4. Do all lettering from the top to the bottom of the sheet. Letter all specific and general notes and fill in the title block. To help keep your drawing clean, place a blank sheet of paper under your hand while lettering.

5. Draw all arrowheads.

6. Check your line quality by placing the original over a light table. If light passes through the lines or lettering, the work is not dark enough. Another way to check your work is to run a test-strip diazo copy. If the quality of the print is not adequate, you may need to work more on the drawing. A print does not improve the quality of a drawing.

The process of laying out a detail drawing involves similar steps with CADD when compared to manual drafting. Some of the similarities were

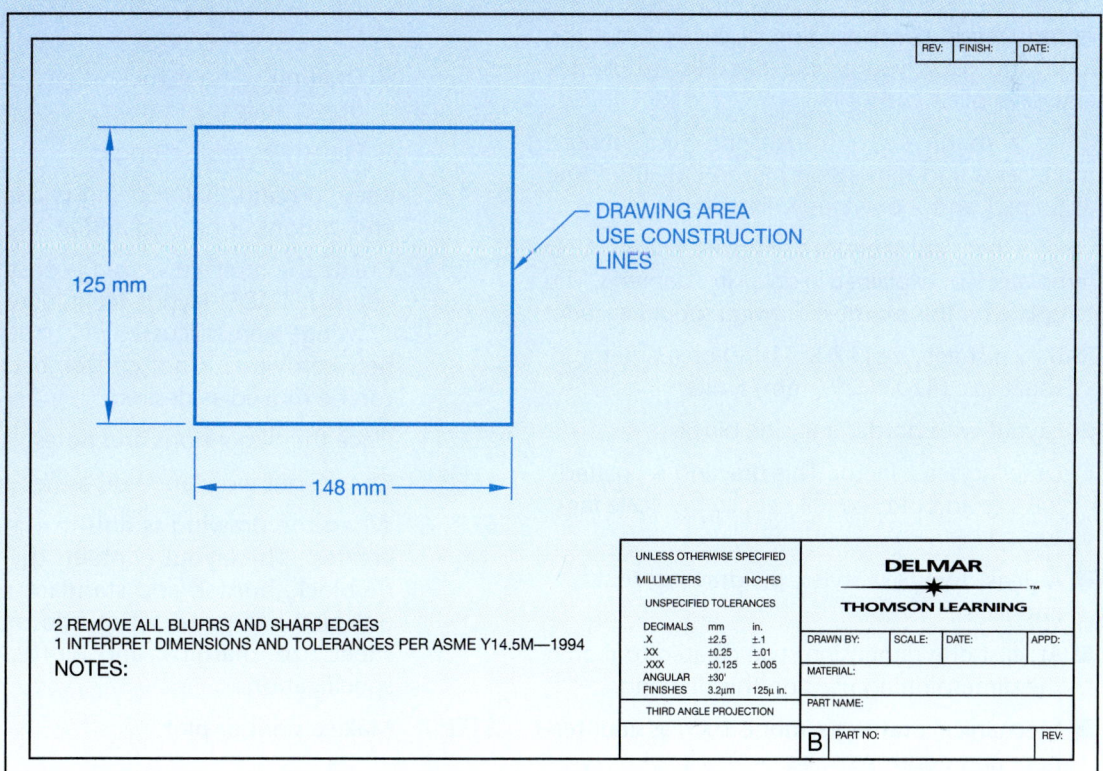

FIGURE 11.4 ■ Establish the approximate drawing area.

(Continued)

THE ENGINEERING DESIGN APPLICATION *(continued)*

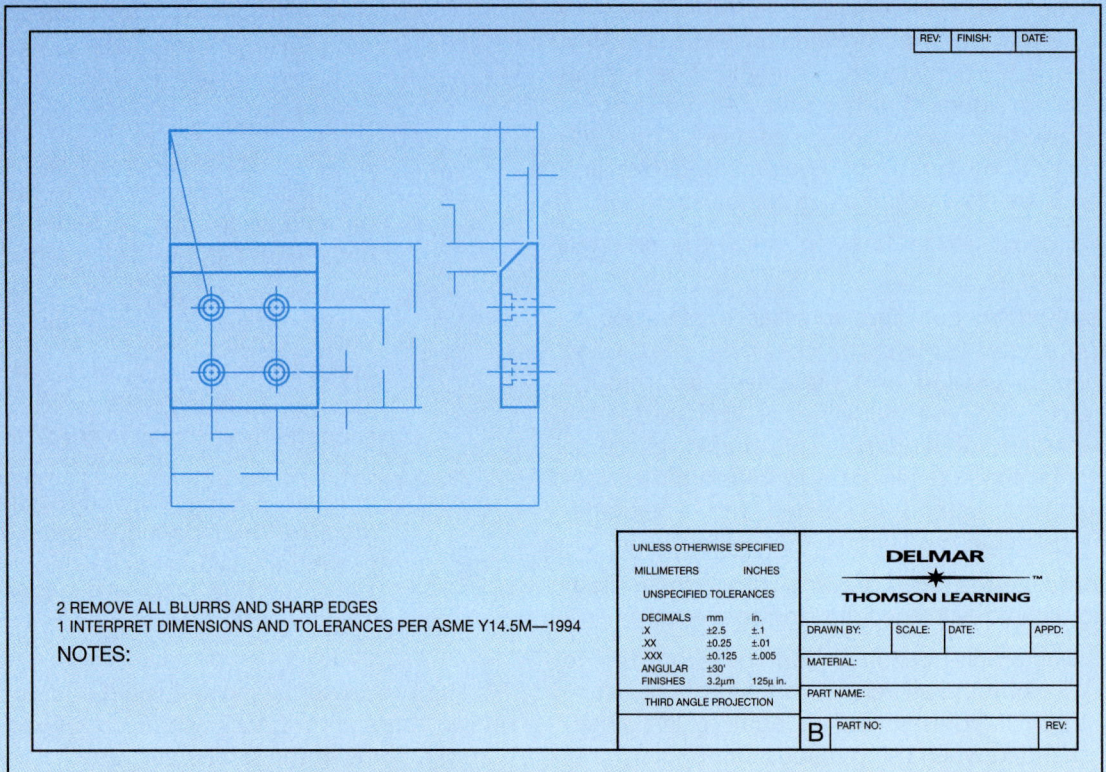

2 REMOVE ALL BLURRS AND SHARP EDGES
1 INTERPRET DIMENSIONS AND TOLERANCES PER ASME Y14.5M—1994
NOTES:

FIGURE 11.5 ■ Lay out the multiview's object lines, hidden lines, centerlines, extension and dimension lines, and leader lines; establish guidelines for lettering.

introduced in the preceding discussion, but the following gives you a brief checklist to use for CADD applications:

STEP 1 Make a rough sketch organizing your desired multiviews and dimension placement, the same as Steps 1 and 2 previously discussed.

STEP 2 Open a template or create a template. Using CADD templates was explained in detail in Chapter 3. The template for this example drawing should include:

- B inch sheet size (17 × 11 in.) or A3 metric sheet size (420 × 297 mm) limits.
- Layout with border and title block.
- Drawing scale factor. This drawing is created full size and plotted full size, so the scale factor is 1.
- At least two text styles, for drawing text and titles.
- At least one dimension style created to match the dimensioning used on this drawing.
- Mechanical drafting symbols such as counterbore and depth symbols.
- Named layers containing colors, line types, and line weights.

- Desired menu options.
- Drafting settings such as grid, snap spacing, object snaps, and units.
- Plot styles and settings.

Keep in mind that you can change these values and options as needed and at any time.

STEP 3 Create the multiviews using construction lines as needed. CADD layout techniques for multiview drawings were discussed in Chapter 9. The space between views is not critical, because the views can be moved as desired.

STEP 4 Place the dimensions and notes.

STEP 5 Check your work and edit as needed.

STEP 6 When the drawing is finished, a layout can be created. The layout contains the drawing, a title block, border, and standard notes. The layout also has page setup information, such as sheet size, margins, and plotter configuration specifications.

STEP 7 Make a print or plot.

Figure 11.6 shows the completed detail drawing of the clamp plate.

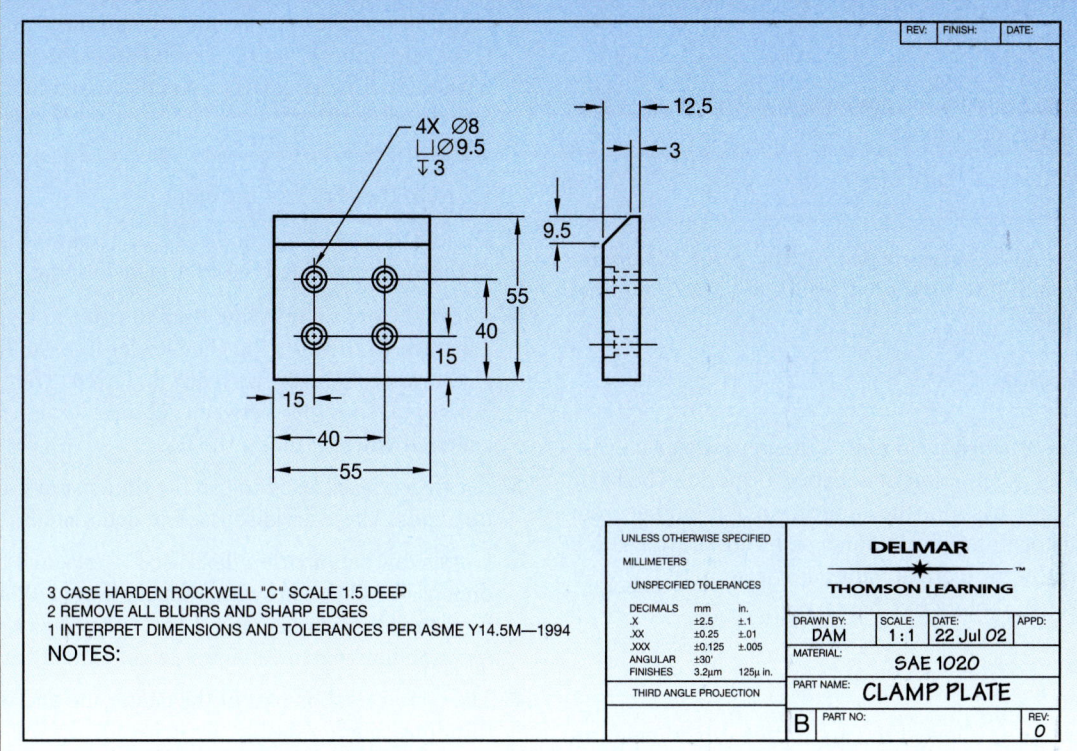

FIGURE 11.6 ■ Complete the drawing and add all lettering.

INTRODUCTION TO DIMENSIONING

ANSI/ASME The standard adopted by the American National Standards Institute (ANSI) and published by the American Society of Mechanical Engineers (ASME) is titled *Dimensioning and Tolerancing,* ASME Y14.5M. This standard is available through the ASME at 345 East 47th Street, New York, NY 10017. The standard that controls general dimensional tolerances found in the title block or in general notes is ASME Y14.1M, *Metric Drawing Sheet Size and Format.* Refer to the CD found at the back of the book, Appendix C, for ASME Dimensioning Rules.

The dimension numerals are displayed using millimeters (mm) in dimensions on example drawings throughout this chapter, unless otherwise specified. Where values are given within the text content, inches are followed by millimeters in parentheses, for example, .24 in. (6 mm). If the given measurement is a specification for a standard feature such as text height, the inch standard is followed by the metric standard, even if the exact metric conversion is not accurate; for example, the standard text height is .12 in. (3 mm). In some applications, an inch value is given followed by a rounded metric conversion in parentheses. The exact metric conversion is not given in most cases in an effort to keep the information in whole millimeters.

A complete detail drawing includes multiviews and dimensions, which provide both shape and size description. There are two classifications of dimensions: size and location. Size dimensions are placed directly on a feature to identify a specific size or may be connected to a feature in the form of a note. The relationship of features of an object is defined with location dimensions.

Notes are a type of dimension that generally identify the size of a feature or features with more than a numerical specification. For example, a note for a counterbore gives size and identification of the machine process used in manufacturing. There are basically two types of notes: local notes and general notes. Local notes are connected to specific features on the views of the drawing. Local notes are also commonly called specific notes, because they are specific to a feature. General notes are placed separate from the views and relate to the entire drawing.

It is important for you to effectively combine shape and size descriptions so the drawing is easy to read and understand. There are many techniques that help implement this goal. You should carefully evaluate the dimensioning rules while preparing detail drawings and should keep in mind, at all times, never to crowd information on a drawing. It is better to use larger paper than to crowd the drawing, unless otherwise indicated by your company or instructor.

DIMENSIONING CHARACTERISTICS AND DEFINITIONS

Actual Size

The part size as measured after production is known as actual size, also known as produced size.

Allowance

The tightest possible fit between two mating parts. Maximum material condition (MMC) external feature − MMC internal feature = allowance.

Basic Dimension

A basic dimension is considered to be a theoretically exact size, location, profile, or orientation of a feature or point. The basic dimension provides a basis for the application of tolerance from other dimensions or notes. Basic dimensions are drawn with a rectangle around the numerical value. For example: .625 or 30°. This is discussed more in Chapter 14.

Bilateral Tolerance

A bilateral tolerance is allowed to vary in two directions from the specified dimension.

$$\text{Inch examples are } .250 \, {}^{+.002}_{-.005} \text{ and } .500 \pm .005.$$

$$\text{Metric examples are } 12 \, {}^{+\,0.1}_{-\,0.2} \text{ and } 12 \pm 0.2.$$

Datum

A datum is considered to be a theoretically exact surface, plane, axis, center plane, or point from which dimensions for related features are established.

Datum Feature

The datum feature is the actual feature of the part that is used to establish a datum.

Dimension

A dimension is a numerical value used on a drawing to describe size, shape, location, geometric characteristic, or surface texture.

Dimensioning Components

The types of lines and the text used in dimensioning were introduced in Chapter 7. The following provides a brief review of these and gives additional information related to dimensioning components. Look at Figure 11.7 as you review the following. The terminology used in Figure 11.7 also correlates with dimensioning terminology used in AutoCAD:

■ Dimension lines indicate the length of the dimension. They are thin lines capped on the ends with arrowheads and broken along the length, providing a space for the dimension numeral. The gap between the dimension line and the dimension numeral varies, but is commonly .06 in. (1.5 mm). Dimension lines for architectural and structural drafting have different characteristics as discussed in Chapter 23, Structural Drafting.

■ An angular dimension line is an arc with the center of the arc from the vertex of the angle.

■ Dimension text is normally .12 in. (3 mm) high, centered in the space provided in the dimension line.

■ A leader line is a thin line used to connect a specific note to a feature on the drawing. The leader line can be at any angle between 15°–75°, with 45° preferred. There is a short horizontal shoulder between .12 and .24 in. (3–6 mm), centered where it meets the note.

■ Arrowheads are used to cap the dimension line and leader line ends. These are discussed in detail later.

■ Extension lines are thin lines used to establish the extent of a dimension. They start with a small offset of .06 in. (1.5 mm) from the object and extend .12 in. (3 mm) past the last dimension line. Extension lines do not extend between views.

■ The center dash is part of the centerline and is drawn as thin lines. Center dashes are generally drawn .12 in. (3 mm). Additional center dash information is provided later in this chapter. The center dashes in Figure 11.7 cross at the center of the circle and are used to dimension the center location. Centerlines become extension lines as they extend out to the dimension. Centerlines do not extend between views, unless as applied to an auxiliary view as discussed in Chapter 10.

■ The centerline space is commonly .06 in. (1.5 mm). This is the space between the short and long dashes of the centerline. The centerline offset is the part of the centerline extending beyond the circle or other related feature. The centerline extension is generally .12–.24 in. (3–6 mm).

Feature

A feature is any physical portion of an object, such as a surface or hole.

Geometric Tolerance

Geometric tolerance is the general term applied to the category of tolerances used to control form, profile, orientation, location, and runout. See Chapter 14.

Least Material Condition (LMC)

Least material condition (LMC) is the opposite of maximum material condition. The LMC is the lower limit for an external feature and the upper limit for an internal feature.

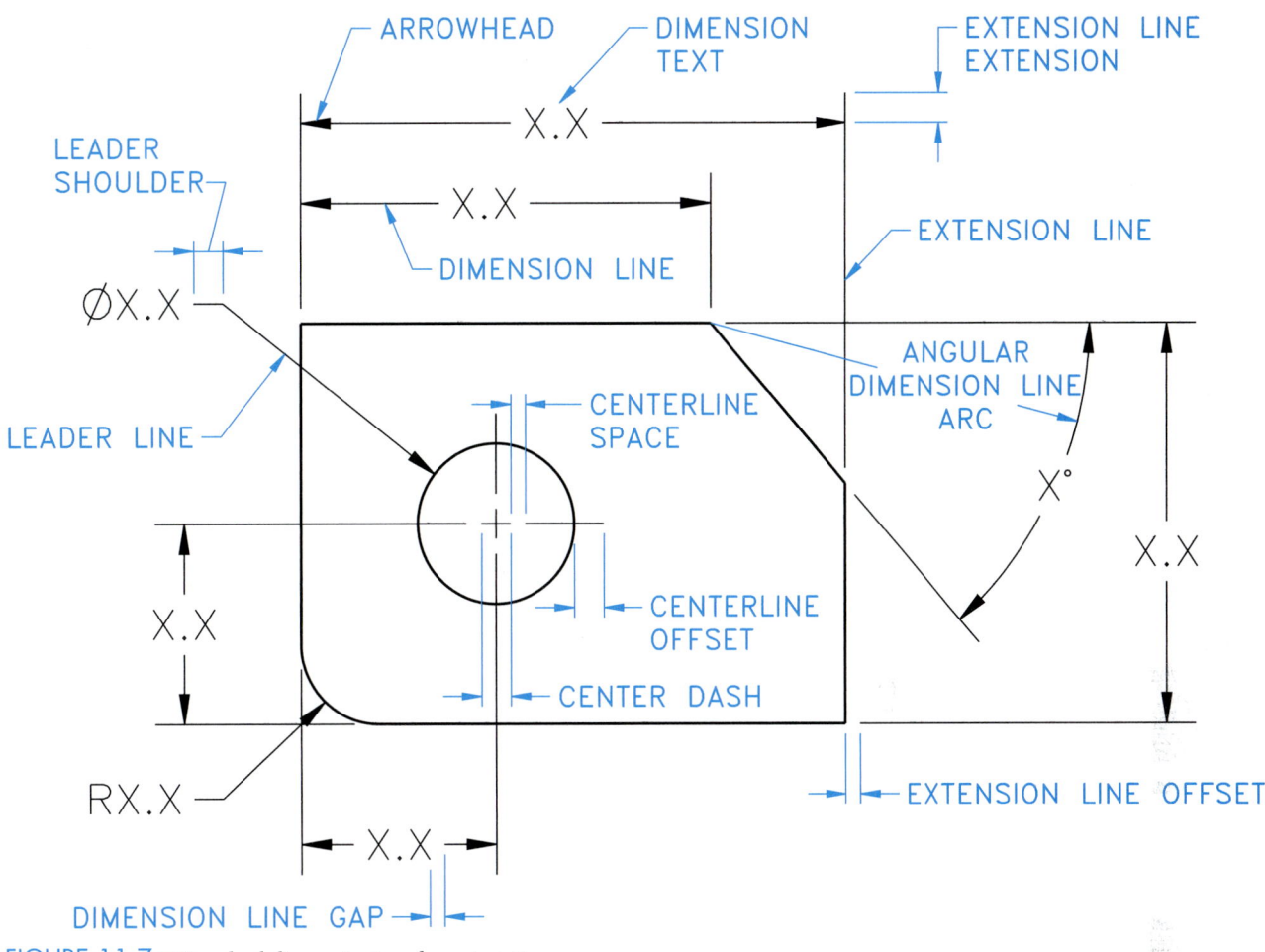

FIGURE 11.7 ■ Standard dimensioning characteristics.

Limits of Dimension

The limits of a dimension are the largest and smallest possible boundary to which a feature can be made as related to the tolerance of the dimension. Consider the following inch dimension and tolerance: .750 ± .005. The limits of this dimension are calculated as follows: .750 ± .005 = .755 upper limit and .750 − .005 = .745 lower limit. For the metric dimension 19.00 ± 0.15, 19.00 + 0.15 = 19.15 is the upper limit, and 19.00 − 0.15 = 18.85 is the lower limit.

Maximum Material Condition (MMC)

The maximum material condition (MMC), given the limits of the dimension, is the situation where a feature contains the most material possible. MMC is the largest limit for an external feature and the smallest limit for an internal feature.

Nominal Size

A dimension used for general identification such as stock size or thread diameter is referred to as nominal.

Reference Dimension

A reference dimension is a dimension usually without tolerance, used for information purposes only. A reference is a repeat of a given dimension or established from other values shown on the drawing. It does not govern production or inspection. Reference dimensions are enclosed in parentheses on the drawing. See Figure 11.8 and Figure 11.9.

Specified Dimension

The specified dimension is the part of the dimension from which the limits are calculated. For example, the specified dimension of .625 ± .001 in. is .625. In the following metric example, 15 ± 0.1 mm, 15 is the specified dimension.

Tolerance

The tolerance of a dimension is the total permissible variation in size or location. Tolerance is the difference of the lower limit from the upper limit. For example, the limits of .500 ± .005 in. are .505 and .495, making the tolerance equal to .010 in. These

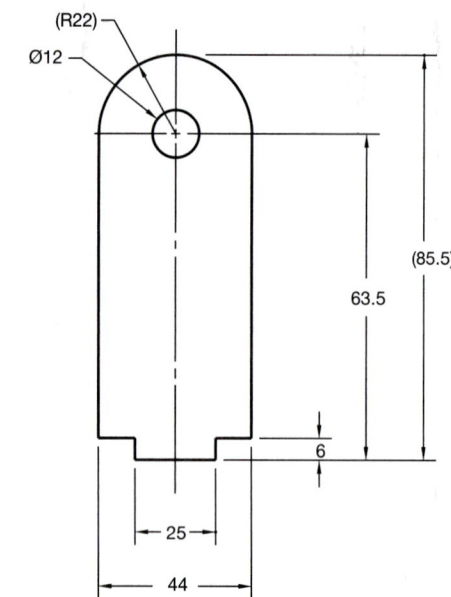

FIGURE 11.8 ■ Unidirectional dimensioning.

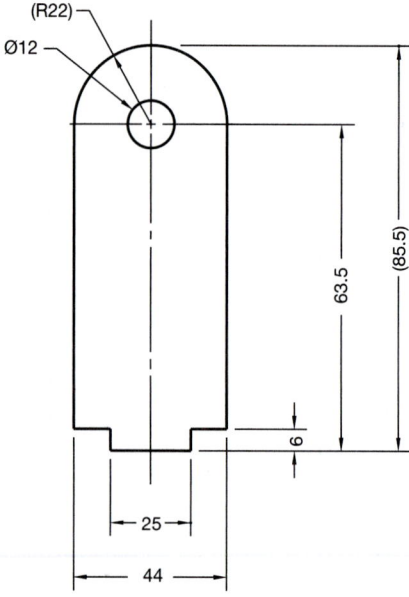

FIGURE 11.9 ■ Aligned dimensioning.

are the tolerance calculations: .500 + .005 = .505 upper limit. .500 − .005 = .495 lower limit. .505 − .495 = .010 tolerance. The tolerance for this metric example, 12 ± 0.1 mm, is 0.2 mm. This is the tolerance calculation: 12 + 0.1 = 12.1 upper limit. 12 − 0.1 = 11.9 lower limit. 12.1 − 11.9 = 0.2 tolerance.

Unilateral Tolerance

A unilateral tolerance is a tolerance that has a variation in only one direction from the specified dimension, as in $.875 \begin{smallmatrix} +.000 \\ -.002 \end{smallmatrix}$ in. or $22 \begin{smallmatrix} 0 \\ -0.2 \end{smallmatrix}$, $22 \begin{smallmatrix} +0.2 \\ 0 \end{smallmatrix}$ mm.

DIMENSIONING SYMBOLS

ASME/ANSI standards specify symbols to be used on drawings. With the influence of computer-aided drafting, the use of symbols is easy. When symbols are drawn manually, a template should be used because many of the symbols are difficult to draw properly freehand, and the template drawing is generally neater and more uniform. Any attempt to draw symbols using drafting tools other than specially designed templates takes too much time. Individual symbols are represented and detailed throughout this chapter each time they are introduced.

Symbols aid in clarity and ease of drawing presentation, and they save time. The standard dimension symbols are shown in Figure 11.10.

DIMENSIONING SYSTEMS

Unidirectional Dimensioning

Unidirectional dimensioning is commonly used in mechanical drafting. It requires that all numerals, figures, and notes be lettered horizontally and be read from the bottom of the drawing sheet. Figure 11.8 shows unidirectional dimensioning in use. The dimension in parentheses (80) is a reference dimension, which is discussed later in this chapter.

Aligned Dimensioning

Aligned dimensioning requires that all numerals, figures, and notes be aligned with the dimension lines so they can be read from the bottom for horizontal dimensions and from the right side for vertical dimensions. This method of dimensioning is commonly used in architectural and structural drafting. (See Figure 11.9.)

Tabular Dimensioning

Tabular dimensioning is a system in which size and location dimensions from datums or coordinates (x, y, z axes) are given in a table identifying features on the drawing. Figure 11.11 shows a method of tabular dimensioning.

Arrowless Dimensioning

Arrowless dimensioning is similar to tabular dimensioning in that features are identified with letters and keyed to a table. Location dimensions are established with extension lines as coordinates from determined datums. Arrowless dimensioning is also called ordinate dimensioning and is often referred to as dimensioning without dimension lines. (See Figure 11.12.)

Chart Drawing

Chart drawings are used when a particular part or assembly has one or more dimensions that change depending on the specific

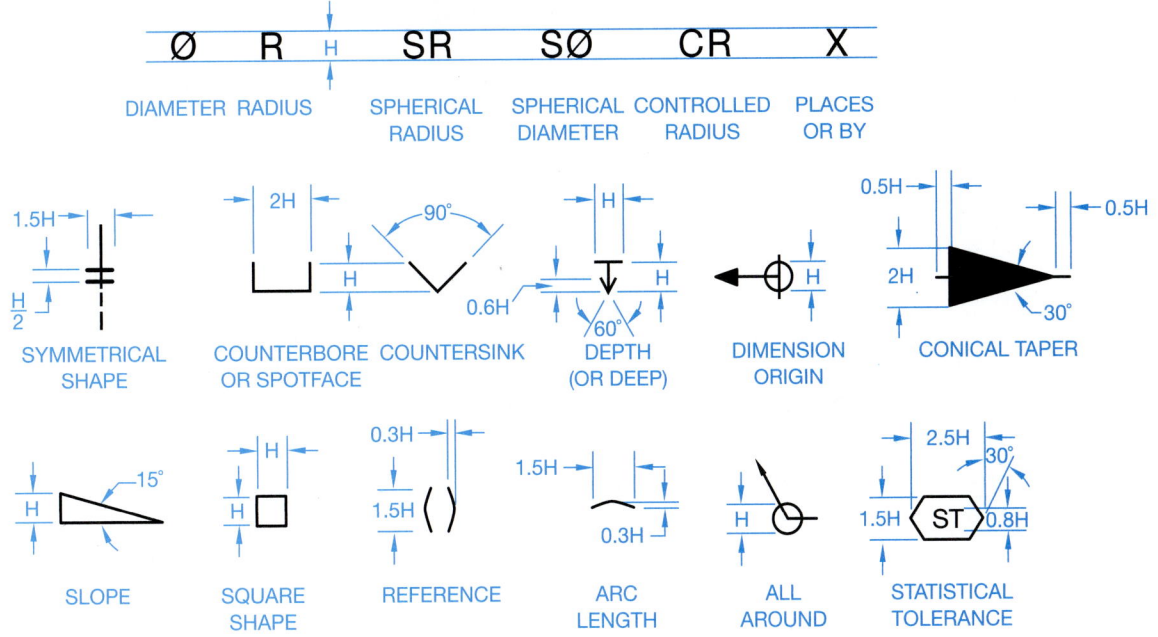

H = LETTER HEIGHT

FIGURE 11.10 ■ Standard ASME recommended dimensioning symbols.

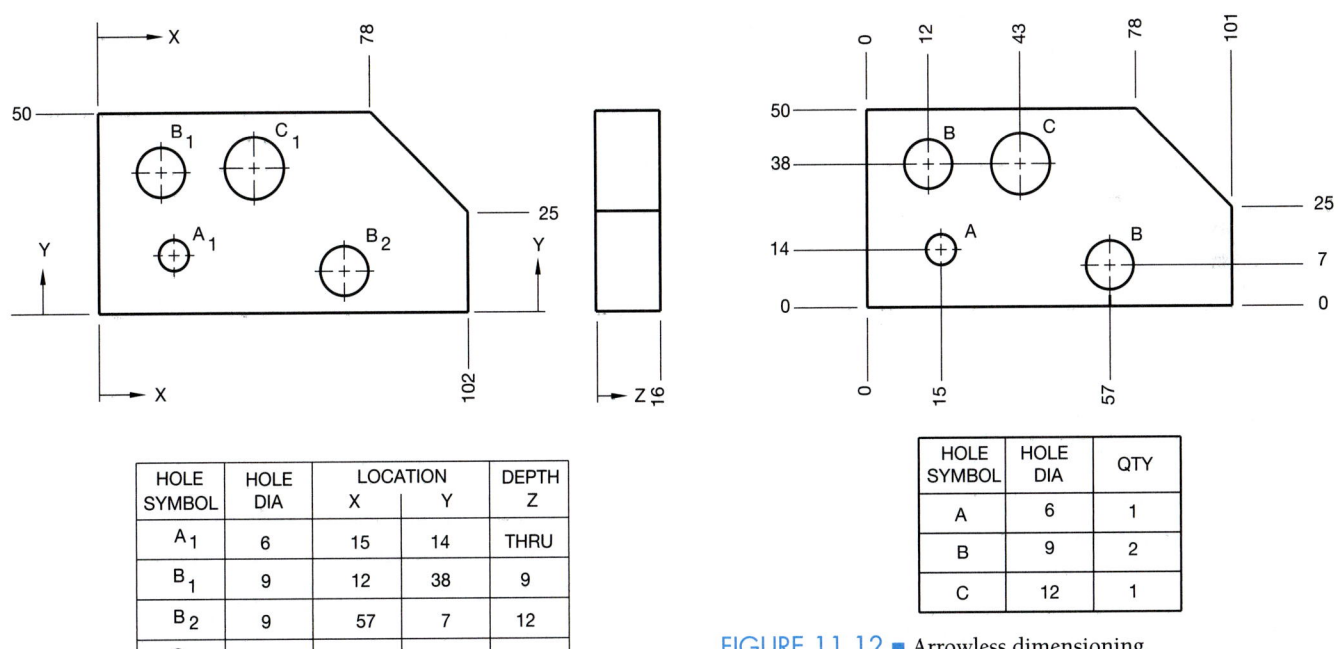

HOLE SYMBOL	HOLE DIA	LOCATION		DEPTH
		X	Y	Z
A_1	6	15	14	THRU
B_1	9	12	38	9
B_2	9	57	7	12
C_1	12	43	38	THRU

FIGURE 11.11 ■ Tabular dimensioning.

HOLE SYMBOL	HOLE DIA	QTY
A	6	1
B	9	2
C	12	1

FIGURE 11.12 ■ Arrowless dimensioning.

application. For example, the diameter of a part and the lengths have alternate dimensions required for different purposes.

The variable dimension is usually labeled on the drawing with a letter in the place of the dimension. The letter is then placed in a chart where the changing values are identified. Figure 11.13 is a chart drawing showing dimensions having alternate sizes. The view drawn represents a typical part, and the dimensions are labeled A and B. The correlated chart identifies the various (A) lengths available at given (B) diameters. The chart in this example also shows purchase part numbers for each specific item. This method of dimensioning is commonly used in vendor or specification catalogues for alternate part identification.

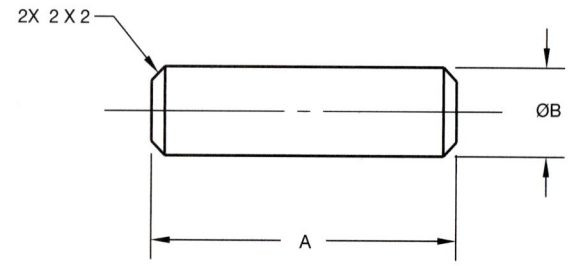

LENGTH A	B=20.3	B=38.1	B=50.8	B=57.2
	PART NO.	PART NO.	PART NO.	PART NO.
76	DP20.3-76.2	DP38.1-76.2	DP50.8-76.2	DP57.2-76.2
101	DP20.3-101.6	DP38.1-101.6	DP50.8-101.6	DP57.2-101.6
127	DP20.3-127	DP38.1-127	DP50.8-127	DP57.2-127
152	DP20.3-152.4	DP38.1-152.4	DP50.8-152.4	DP57.2-152.4

FIGURE 11.13 ■ Chart drawing.

DIMENSIONING FUNDAMENTALS

Dimension layout standards are established to help ensure dimensions are represented the same on all drawings. There are a variety of variables that make up the dimensioning layout as shown in Figure 11.14. The variables displayed in this figure represent the recommended standard.

Dimensioning Units

The metric International System of Units (SI) is featured predominately in this text because SI units (millimeters) supersede United States (U.S.) customary units specified on engineering drawings.

Metric units expressed in millimeters or U.S. customary units expressed in decimal inches are considered the standard units of linear measurement on engineering documents and drawings. The selection of millimeters or inches depends on the needs of the individual company. When all dimensions are either in millimeters or inches, the general note, UNLESS OTHERWISE SPECIFIED, ALL DIMENSIONS ARE IN MILLIMETERS (or INCHES), should be placed on the drawing. Inch dimensions should be followed by *IN.* on predominantly millimeter drawings, and *mm* should follow millimeters on predominantly inch-dimensioned drawings.

While **not** an ASME standard, some companies use a method of placing inch and metric equivalents together on each dimension. This is known as dual dimensioning. The inch dimension can be followed by millimeters in brackets or separated by a slash; for example, 1.00 [25.4] or 1.00/25.4. A general note should accompany the drawing to identify this practice: DIMENSIONS IN [ ] ARE MILLIMETERS (or INCHES), or INCH/MILLIMETERS or MILLIMETERS/INCHES.

Decimal Points

Dimension numerals containing decimal points should be allowed adequate space at the decimal so there is no crowding with the numerals. A minimum of two-thirds the height of the letters is recommended. The decimal should be clear and bold and in line with the bottom of the numerals; for example, 1.750.

A specified dimension in inches is expressed to the same number of decimal places as its tolerance, and zeros are added to the right of the decimal point if needed; for example, .500 ±.002.

Where millimeter dimensions are less than one, a zero precedes the decimal, as in 0.8. Decimal inch dimensions do not have a zero before the decimal, as in .75. Proper spacing of numerals with a decimal is shown in Figure 11.15.

Neither the decimal point nor a zero is shown when the metric dimension is a whole number; for example, 24. Also, when the metric dimension is greater than a whole number by a fraction of a millimeter, the last digit to the right of the decimal point is not followed by a zero, as in 24.5. This is true unless tolerance numerals are involved. (Refer to Figure 11.18.) Both the plus and minus values of a metric or inch tolerance have the same number of decimal places and zeros are added to fill in where needed.

Fractions

Fractions are used on engineering drawings, but they are not as common as decimal inches or millimeters. Fractions are commonly used on architectural, structural, and other construction related drawings. Fraction dimensions generally mean a larger tolerance than decimal numerals. When fractions are used on a drawing, the fraction numerals should be the same size as other numerals on the drawing. The fraction bar should be drawn in line with the direction the dimension reads. For unidirectional dimensioning, the fraction bars are all horizontal. For aligned dimensioning, the fraction bars are horizontal for dimensions that read from the bottom of the sheet and vertical for dimensions that read from the right. The fraction numerals should not be allowed to touch the fraction bar. A space of .06 in. (1.5 mm) is recommended between the bar and the fraction numerals. In a few situations—for example, when a fraction is part of a general note, material specification, or title—the fraction bar can be placed diagonally, as shown in Figure 11.16. This is also a common practice using CADD, although fractions can be placed as previously described.

Arrowheads

Arrowheads are used to terminate dimension lines and leaders. Properly drawn arrowheads should be drawn three times as long as they are high. All arrowheads on a drawing should be the same size. Do not use small arrowheads in small spaces. Limited-space dimensioning practice is covered in this chapter (see Figure 11.19).

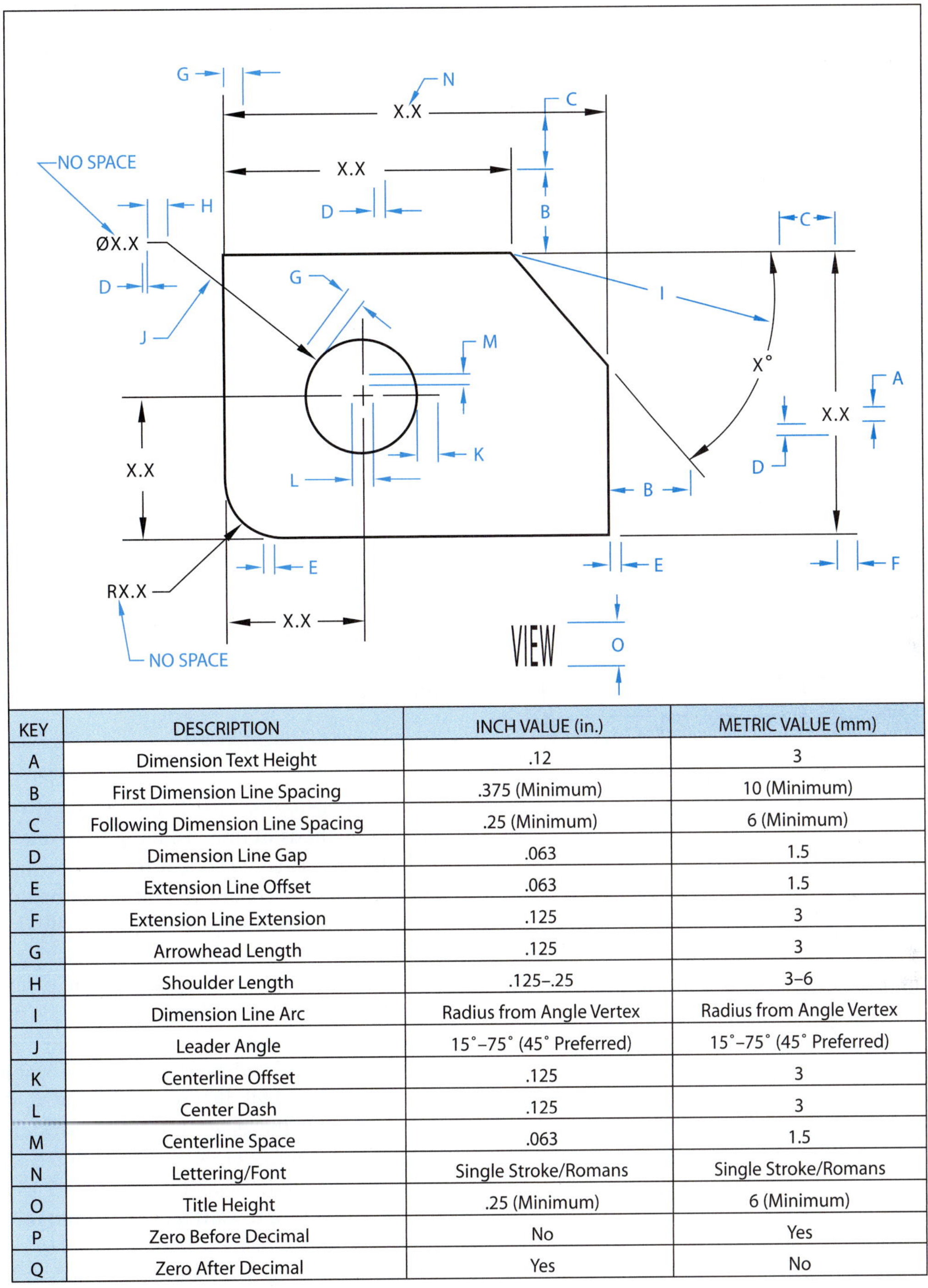

FIGURE 11.14 ■ Standard dimensioning layout standards and specifications.

KEY	DESCRIPTION	INCH VALUE (in.)	METRIC VALUE (mm)
A	Dimension Text Height	.12	3
B	First Dimension Line Spacing	.375 (Minimum)	10 (Minimum)
C	Following Dimension Line Spacing	.25 (Minimum)	6 (Minimum)
D	Dimension Line Gap	.063	1.5
E	Extension Line Offset	.063	1.5
F	Extension Line Extension	.125	3
G	Arrowhead Length	.125	3
H	Shoulder Length	.125–.25	3–6
I	Dimension Line Arc	Radius from Angle Vertex	Radius from Angle Vertex
J	Leader Angle	15°–75° (45° Preferred)	15°–75° (45° Preferred)
K	Centerline Offset	.125	3
L	Center Dash	.125	3
M	Centerline Space	.063	1.5
N	Lettering/Font	Single Stroke/Romans	Single Stroke/Romans
O	Title Height	.25 (Minimum)	6 (Minimum)
P	Zero Before Decimal	No	Yes
Q	Zero After Decimal	Yes	No

GOOD   POOR

FIGURE 11.15 ■ Spacing of numerals with a decimal.

Some companies require that arrowheads be drawn with an arrow template, while others accept properly drawn freehand arrowheads. (See Figure 11.17.) Individual company preference dictates if arrowheads are filled in solid or left open as shown. CADD users sometimes prefer the open style, which helps increase regeneration and plotting speed. However, most prefer the appearance of the filled-in arrowhead. Reduced regeneration and plotting speed is less of a concern with current technology. The filled-in arrowheads look better and make the dimension easier to read.

FOR TITLE BLOCKS, MATERIAL STOCK SIZES, AND
GENERAL NOTES

THIS 1-5/8   NOT
                   THIS 15/8

THIS COULD BE READ AS 15/8.

FIGURE 11.16 ■ Numerals in fractions.

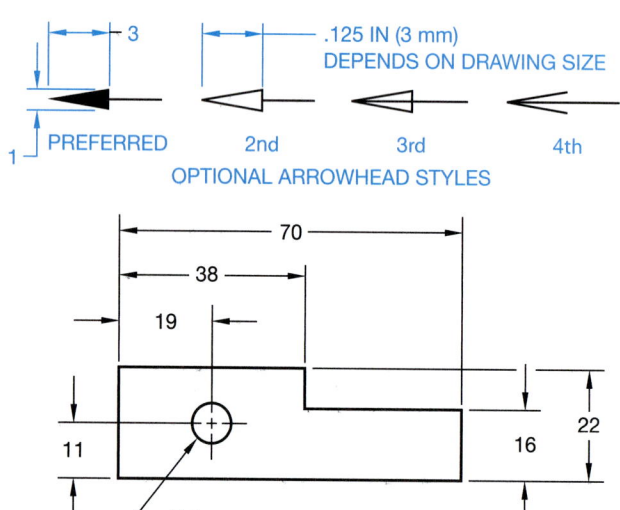

FIGURE 11.17 ■ Arrowheads.

Dimension Line Spacing

Dimension lines should be placed at a uniform distance from the object, and all succeeding dimension lines should be equally spaced. Figure 11.18 shows the minimum acceptable distances for spacing dimension lines. In actual practice the minimum distance is too crowded. Judgment should be used based on

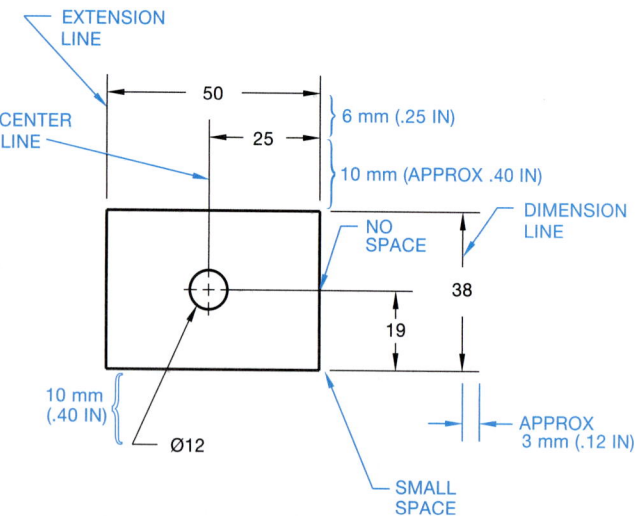

FIGURE 11.18 ■ Minimum dimension line spacing.

space available and information presented. Never crowd dimensions, if at all possible. Always place the smallest dimensions closest to the object and progressively larger dimensions outward from the object. Group dimensions and dimensions between views when possible (See Figure 11.18.)

Relationship of Dimension Lines to Numerals

Dimension numerals are centered on the dimension line, unless another placement is necessary. Alternate options are discussed later. Numerals are commonly all the same height and are lettered horizontally (unidirectional). A space equal to at least half the height of lettering should be provided between numerals in a tolerance. The numeral, dimension line, and arrowheads should be placed between extension lines when space allows. When space is limited, other options should be used. Figure 11.19 shows several dimensioning options. Evaluate each of the examples carefully as you dimension your own drawing assignments. Figure 11.20 shows some correct and incorrect dimensioning practices. Keep in mind that some computer-aided drafting programs do not necessarily acknowledge all of the rules or accepted examples. Some flexibility on your part is needed to become accustomed to some of the potential differences.

Chain Dimensioning

Chain dimensioning also known as point-to-point dimensioning, is a methods of dimensioning from one feature to the next. Each dimension is dependent on the previous dimension or dimensions. This is a common practice, although caution should be used as the tolerance of each dimension builds on the next, which is known as tolerance buildup, or stacking. Figure 11.21 also shows the common mechanical drafting practice of providing an overall dimension while leaving one of the intermediate dimensions blank. The overall dimension is often a critical dimension that should stand independent in relationship to the other dimensions. Also, if all dimensions are given, then the actual size may not equal the given overall dimension due to tolerance buildup. An example of tolerance buildup is when three chain dimensions have individual tolerances of ±.15 and each feature is manufactured at or toward the +.15 limit: the potential tolerance buildup is three times +.15, for a total of .45. The overall dimension would have to carry a tolerance of ± 0.45 to accommodate this buildup. If the overall dimension is critical, such an amount may not be possible. Thus, either one intermediate dimension should be omitted or the overall dimension omitted. The exception to this rule is when a dimension is given only as reference. A reference dimension is enclosed in parentheses, as in Figure 11.22a. Figure 11.22b shows the overall dimension of an object as a reference. Chain dimensioning is commonly used in architectural drafting and related construction industries.

Datum Dimensioning

Datum dimensioning is a common method of dimensioning machine parts whereby each feature dimension originates from

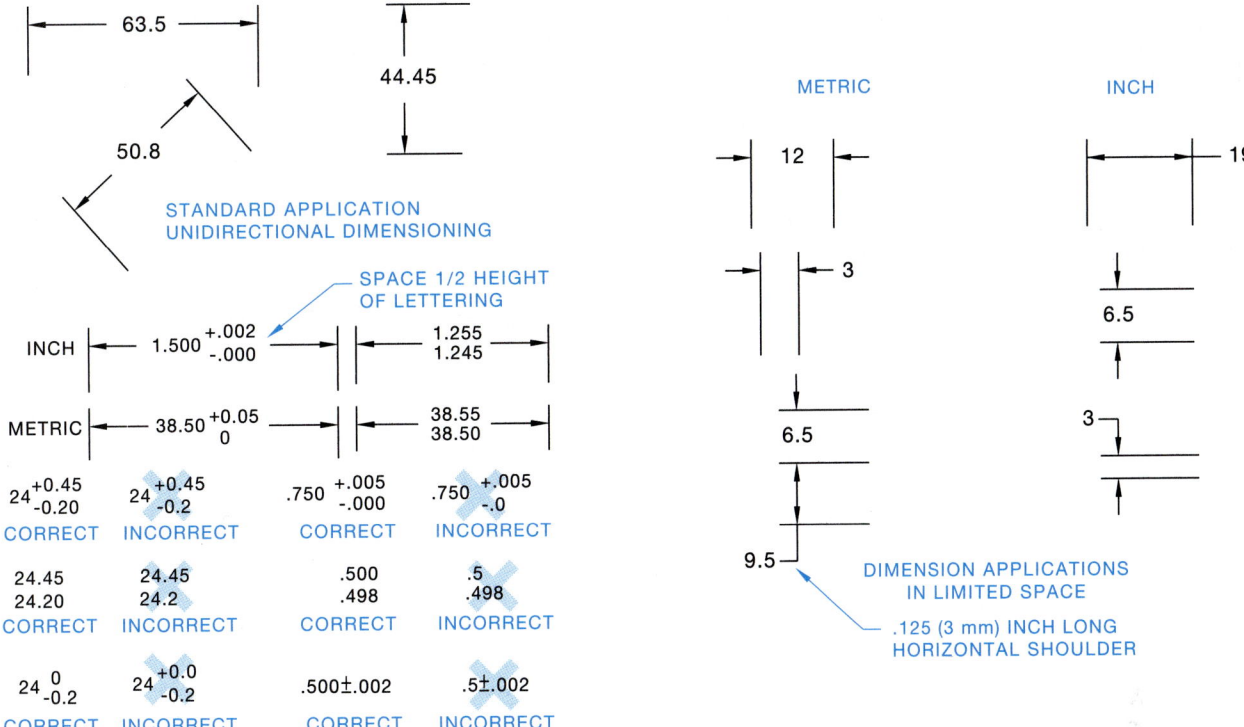

FIGURE 11.19 ■ Dimensioning applications to limited spaces. Metric values are used unless otherwise specified.

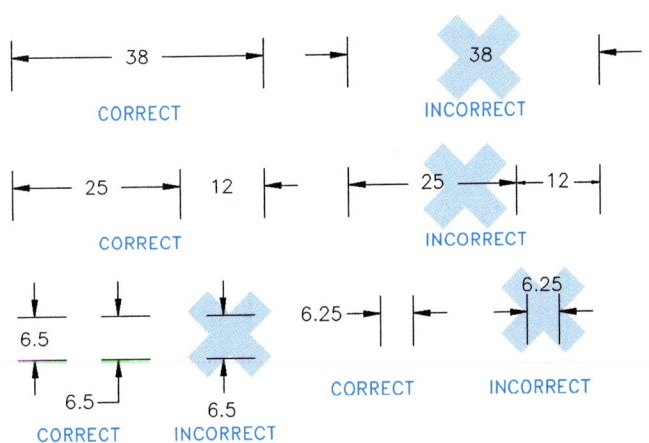

FIGURE 11.20 ■ Correct and incorrect dimensioning practices.

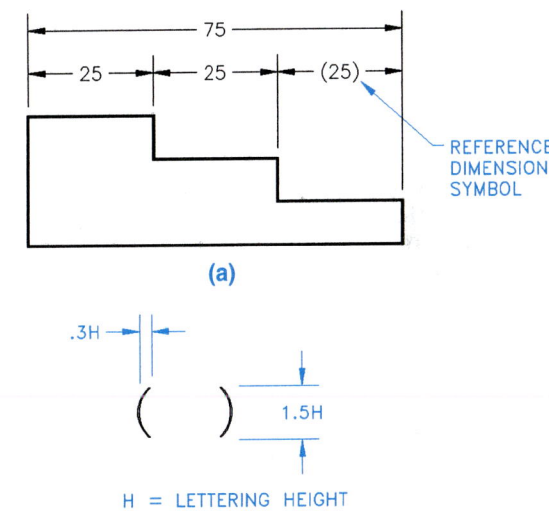

FIGURE 11.22 ■ Reference dimensions examples and reference dimension symbol.

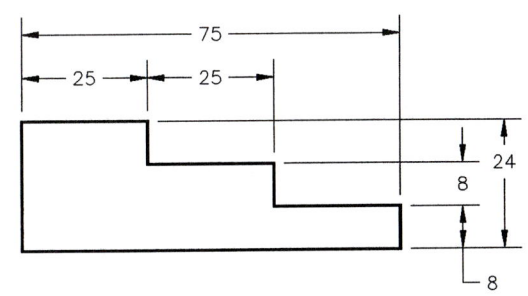

FIGURE 11.21 ■ Chain dimensioning.

a common surface, axis, or center plane. (See Figure 11.23.) Each dimension in datum dimensioning is independent so there is no tolerance buildup. Figure 11.24 shows how dimensions can be symmetrical about a center plane used as a datum.

Stagger Adjacent Dimensions

Notice in Figure 11.24 how dimension numerals are staggered rather than being stacked directly above one another. Always **stagger adjacent dimensions** when possible. Doing so helps clarity and reduces crowding. To accomplish this, the dimen-

sion numeral of some dimensions need to be offset from the center of the dimension line.

Dimensioning Symmetrical Objects, Cylinders and Square Features

Figure 11.24 also shows use of the symmetrical symbol. Both halves of the object are the same. Part of the right side is removed, and a short break line is used to save space. Use this practice only if necessary, because it can cause confusion if improperly read. If in doubt and if space permits, draw the entire object.

Dimension cylindrical shapes in the view where the cylinders appear rectangular. The diameters are identified by the diameter symbol, and the circular view can be omitted. (See Figure 11.25.) Square features are dimensioned in a similar manner using the square symbol shown in Figure 11.26.

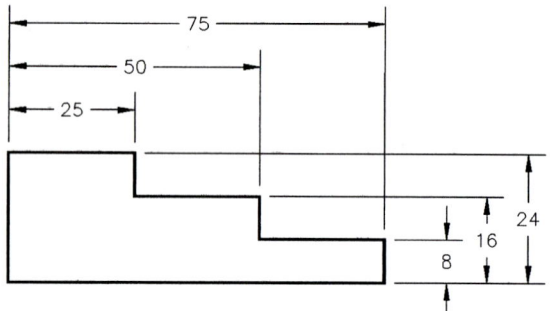

FIGURE 11.23 ■ Datum dimensioning from a common surface.

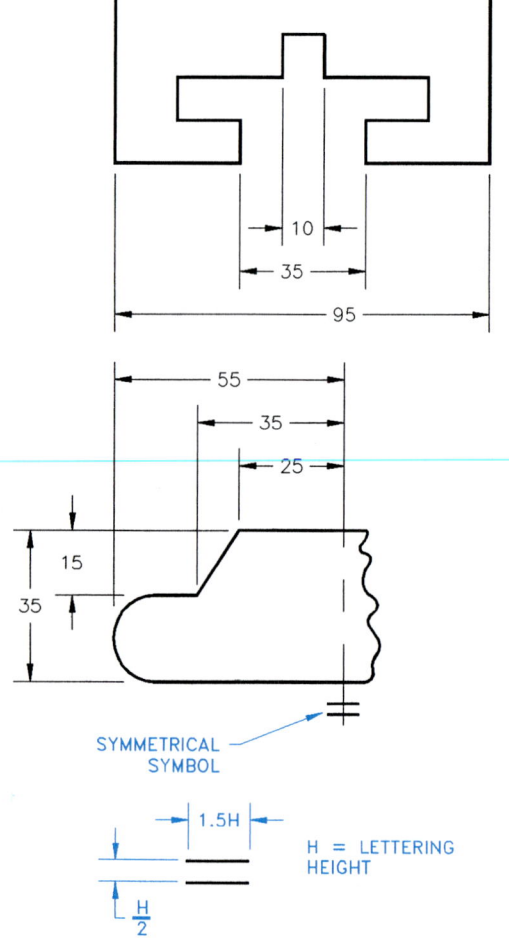

FIGURE 11.24 ■ Datum dimensioning from center-plane datums and symmetrical symbol.

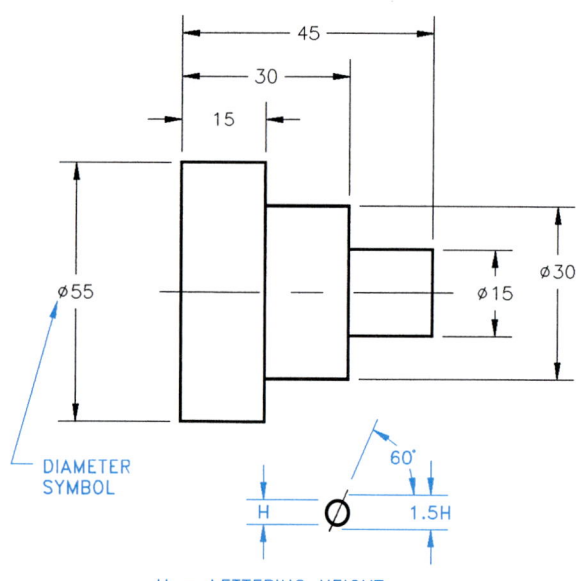

FIGURE 11.25 ■ Dimensioning cylindrical shapes and diameter symbol. Note: Some CADD programs do not automatically provide the properly sized diameter symbol, as shown in this example.

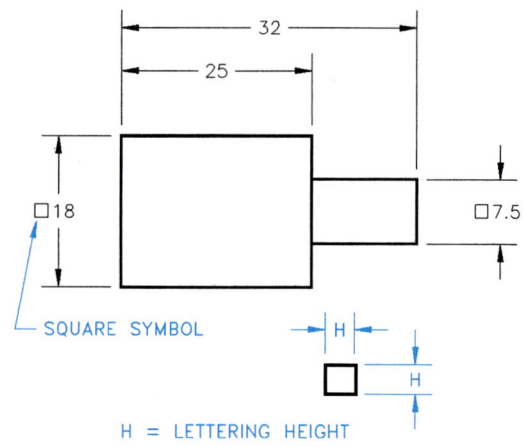

FIGURE 11.26 ■ Dimensioning square features and square symbol.

PREFERRED DIMENSIONING PRACTICES

The drawings in Figure 11.27 show both correct and incorrect dimensioning techniques. These are mostly suggested options for correct and incorrect dimensioning. Good judgment should be used with each case.

Always try, when possible, to:

■ Avoid crossing extension lines, but **do not** break them when they do cross.

■ **Never** cross extension lines over dimension lines, or break the extension line over the dimension line, when there is **absolutely no other solution.**

■ Break extension lines when they cross **over or near** an arrowhead.

■ **Avoid** dimensioning over or through the object.

■ **Avoid** dimensioning to hidden features.

■ **Avoid** unnecessary long extension lines.

■ **Avoid** using any line of the object as an extension line.

■ Dimension **between views** when possible.

■ **Group** adjacent dimensions.

■ Dimension to views that provide the **best shape description.**

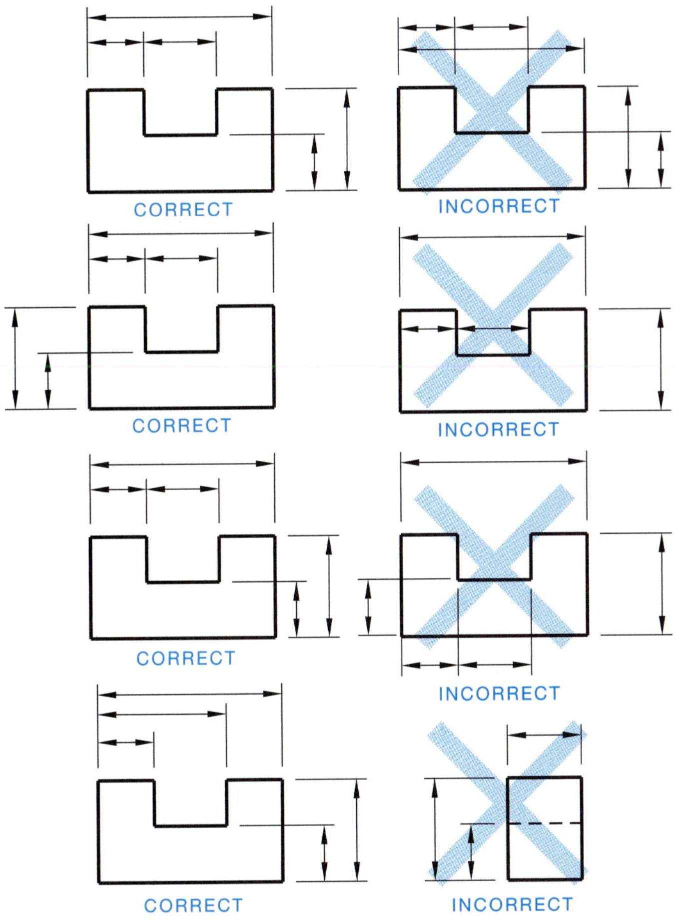

Dimensioning Angles

Angular surfaces can be dimensioned as coordinates, as angles in degrees, or as a flat taper. (See Figure 11.28.) Angles are calibrated in degrees, symbol °. There are 360° in a circle. Each degree contains 60 minutes, symbol ′. Each minute has 60 seconds, symbol ″. 1° = 60′; 1′ = 60″. Notice in the angular method in Figure 11.28, the dimension line for the 45° angle is drawn as an arc. The radius of this arc is centered at the vertex of the angle.

Dimensioning Chamfers

A **chamfer** is a slight surface angle used to relieve a sharp corner. Chamfers of 45° are dimensioned with a note, while other chamfers require an angle and size dimension, or two size dimensions as seen in Figure 11.29. A note is used on 45° chamfers because both sides of a 45° angle are equal. When placing the 45° chamfer note, the size is followed by the (X) symbol and then the 45° angle without spaces.

Dimensioning Conical Shapes

Conical shapes should be dimensioned when possible in the view where the cone appears as a triangle, as in Figure 11.30. A conical taper can be treated in one of three possible ways, as shown in Figure 11.31.

FIGURE 11.27 ■ Correct and incorrect dimensioning examples.

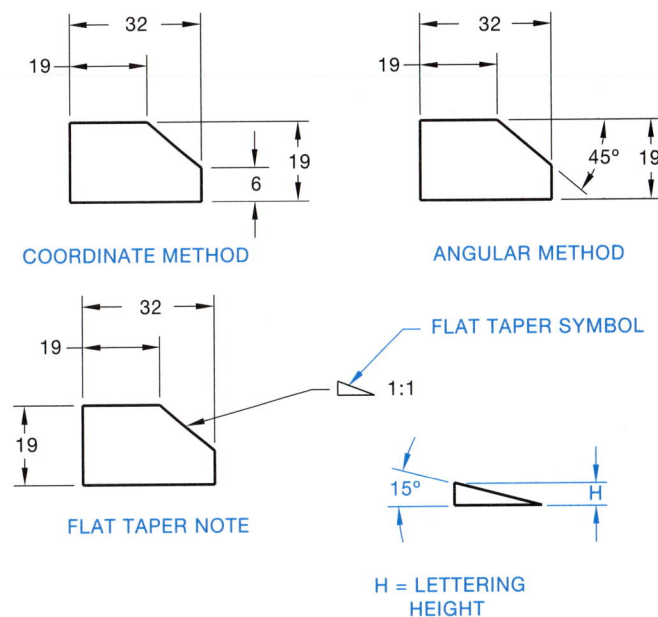

COORDINATE METHOD

ANGULAR METHOD

FLAT TAPER NOTE

FLAT TAPER SYMBOL

1:1

15° H

H = LETTERING HEIGHT

FIGURE 11.28 ■ Dimensioning angular surfaces and flat taper symbol.

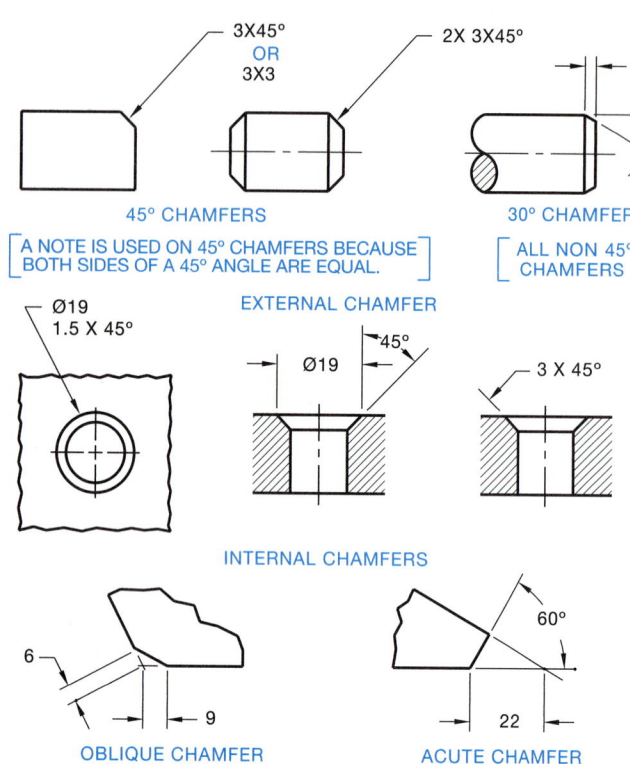

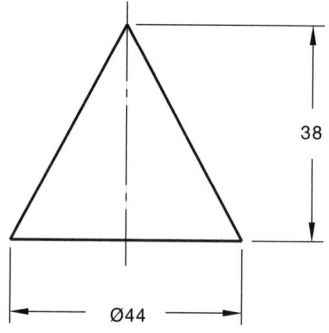

FIGURE 11.29 ■ Dimensioning chamfers.

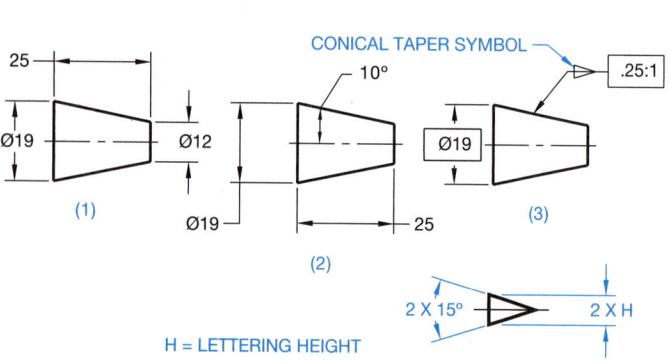

FIGURE 11.30 ■ Dimensioning conical shapes.

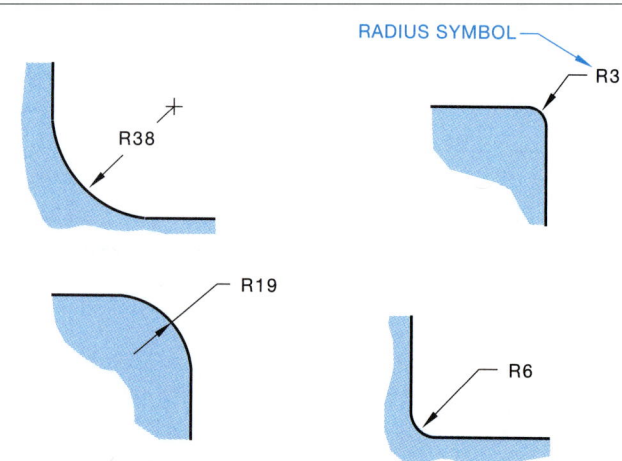

FIGURE 11.31 ■ Dimensioning conical tapers and the conical taper symbol.

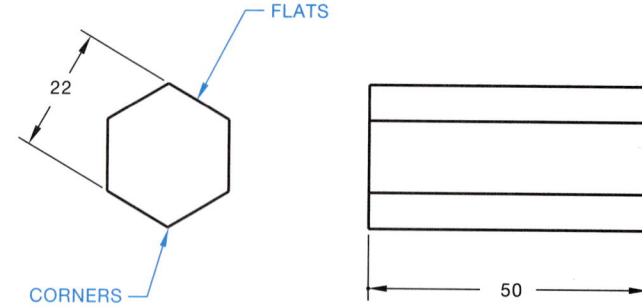

FIGURE 11.32 ■ Dimensioning hexagons.

Dimensioning Hexagons and Other Polygons

Dimension hexagons and other polygons across the flats in the views where the true shape is shown. Provide a length dimension in the adjacent view, as shown in Figure 11.32.

Dimensioning Arcs

Arcs are dimensioned with leaders and radius dimensions in the views where they are shown as arcs. The leader can extend from the center to the arc or point to the arc, as shown in Figures 11.33 and 11.34. The letter R precedes all radius dimensions. Depending on the situation, arcs can be dimensioned with or without their centers located. Figure 11.31 shows a very large arc with the center moved closer to the object. To save space, a break line is used in the leader and the shortened locating dimension. The length of an arc can also be dimensioned one of three ways, as shown in Figure 11.35.

In a situation where an arc lies on an inclined plane and the true representation is not shown, the note TRUE R is used to specify the actual radius. However, dimensioning the arc in an auxiliary view of the inclined surface is better, if possible. (See Figure 11.36.)

The symbol CR refers to controlled radius. Controlled radius means the limits of the radius tolerance zone must be tangent to the adjacent surfaces, and there can be no reversals in the contour.

FIGURE 11.33 ■ Dimensioning arcs—no center located, and the radius symbol.

The *CR* control is more restrictive than use of the *R* radius symbol where reversals in the contour of the radius are permitted.

A spherical radius is dimensioned with the abbreviation *SR* preceding the numerical value, as shown in Figure 11.37.

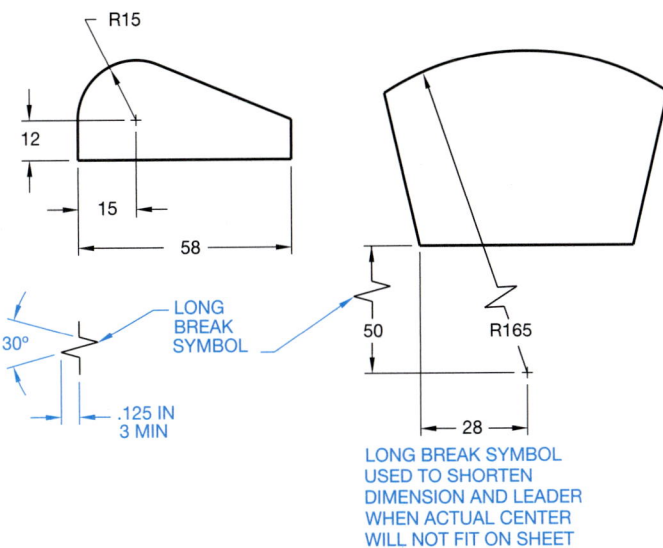

FIGURE 11.34 ■ Dimensioning arcs—with centers located, and the long break symbol.

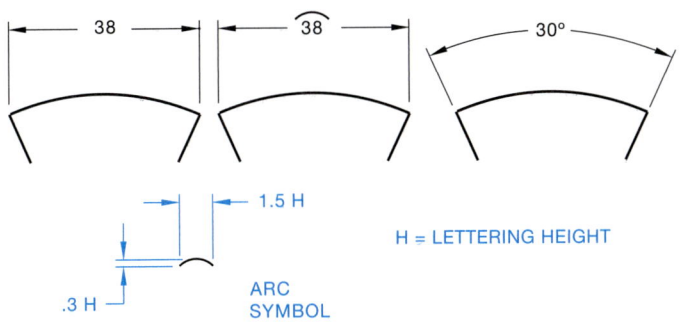

FIGURE 11.35 ■ Dimensioning arc length and the arc symbol.

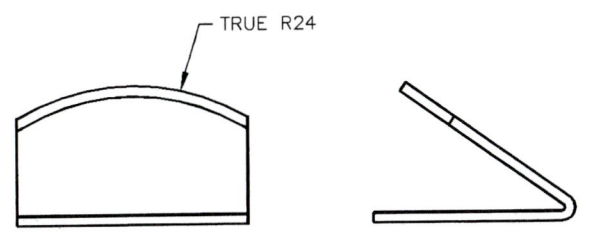

FIGURE 11.36 ■ Dimensioning a true radius in an inclined plane.

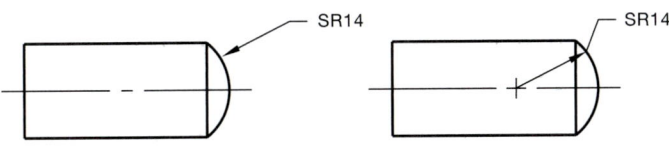

FIGURE 11.37 ■ Dimensioning a spherical radius.

Dimensioning Contours Not Defined as Arcs

Coordinates or points along the contour are located from common surfaces or datums as shown in Figure 11.38. Figure 11.39 shows a curved contour dimensioned using oblique extension

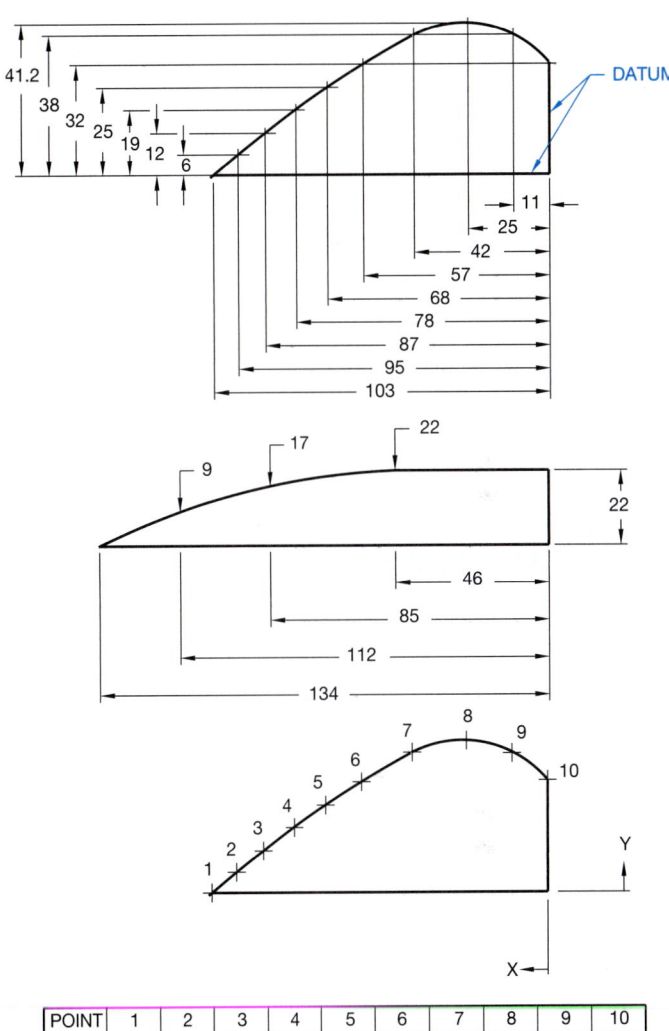

POINT	1	2	3	4	5	6	7	8	9	10
X	-103	-95	-87	-78	-68	-57	-42	-25	-11	0
Y	0	6	12	19	25	32	38	41	38	32

FIGURE 11.38 ■ Dimensioning contours not defined as arcs.

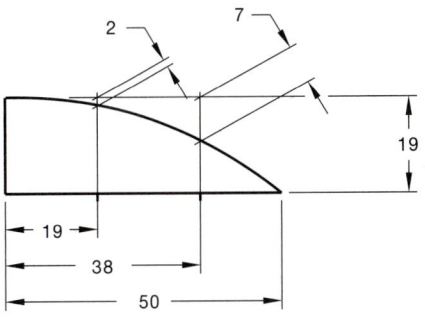

FIGURE 11.39 ■ Dimensioning a curved contour using oblique extension lines.

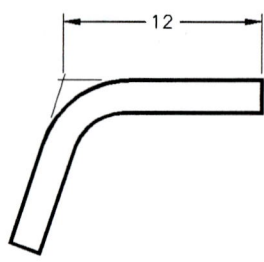

FIGURE 11.40 ■ Locating a point established by extension lines.

lines. While this technique can be used, it is not as common as the previous method.

Locating a Point Established by Extension Lines

If the sides of the object in Figure 11.40 are extended beyond the bend, they would meet at the intersection of the extension lines. This imaginary point is where the dimension often originates in this type of situation.

NOTES FOR SIZE FEATURES

Holes

Hole sizes are dimensioned with leaders to the view where they appear as circles, or dimensioned in a sectional view. When leaders are used to establish notes for holes, the shoulder should be centered on the beginning or the end of the note. When a leader begins at the left side of a note, it should originate at the beginning of the note. When a leader begins at the right side of a note, it should originate at the end of the note (See Figure 11.41.)

Figure 11.42 shows the leader touching the side of the circle and, if it were to continue, would intersect the center of the cir-

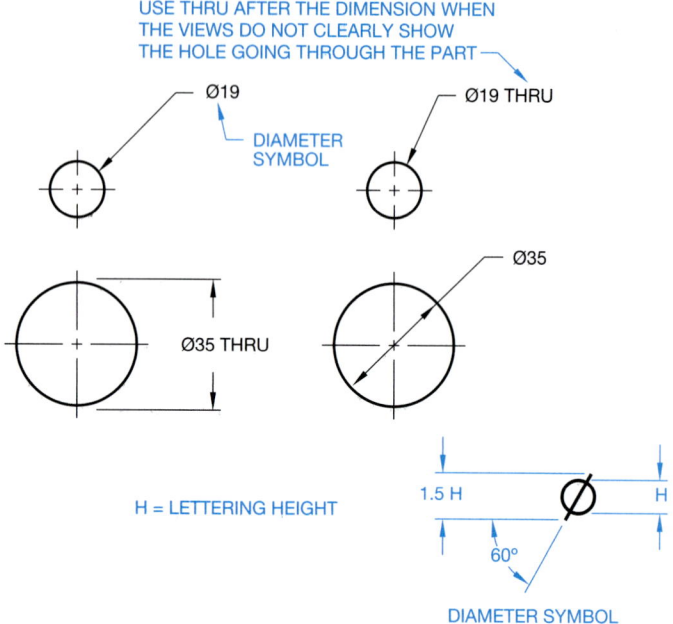

FIGURE 11.43 ■ Dimensioning hole diameters and the diameter symbol.

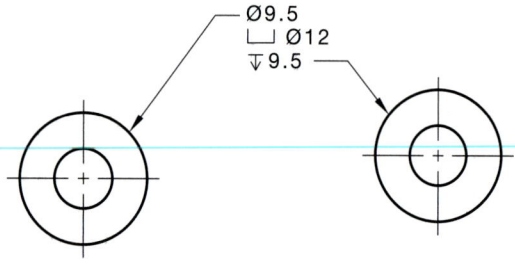

FIGURE 11.41 ■ Leader orientation to the note. Center leader shoulder at beginning or end of note.

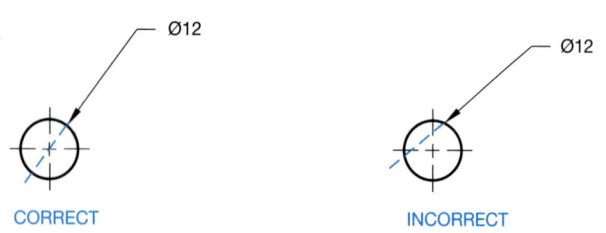

CORRECT   INCORRECT

FIGURE 11.42 ■ Leader orientation to the circle. Leader arrow should *point* to center.

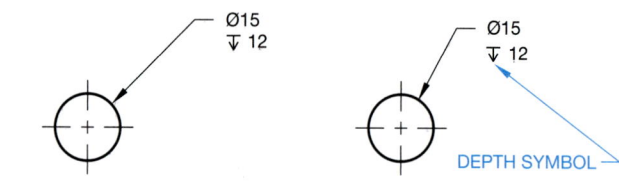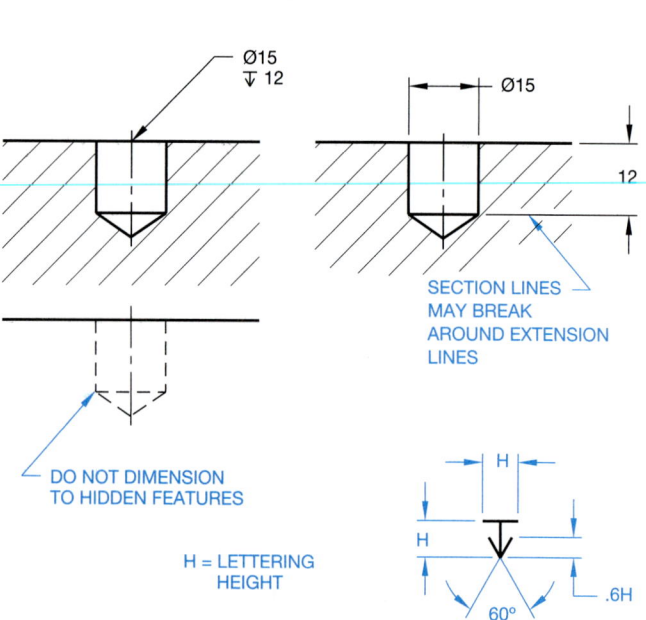

FIGURE 11.44 ■ Dimensioning hole diameters and depths, and the depth symbol. The machine process such as DRILL is not given in the note.

cle. Leaders can be drawn at any angle, preferably 45°, but between be 15° and 75° from horizontal is acceptable. Do not draw horizontal or vertical leaders.

A hole through a part can be noted with the word THRU following the numeral, if not obvious. The diameter symbol precedes the diameter numeral, as dimensioned in Figure 11.43. If a hole does not go through the part, the depth must be noted in the circular view or in section, as shown in Figure 11.44.

Counterbore

A counterbore is often used to machine a diameter below the surface of a part so a bolt head or other fastener can be recessed. Counterbore and other similar notes are given in the order of machine operations with a leader in the view where they appear as circles. (See Figure 11.45a.)

Multiple counterbores are dimensioned in a similar manner as shown in Figure 11.45d.

ASME/ANSI The ASME/ANSI standard recommends that the elements of each note shown in Figure 11.45 through 11.48 be aligned as shown. However, due to individual preference or drawing space constraints, the elements of the note can be confined to fewer lines, as shown in Figure 11.45b. **Never** separate individual note components, as shown in Figure 11.45c.

Countersink or Counterdrill

A countersink, or counterdrill, is also often used to recess the head of a fastener below the surface of a part. (See Figure 11.46.)

Spotface

A spotface is used to provide a flat bearing surface for a washer face or bolthead. (See Figure 11.47.) Follow the counterbore guidelines when lettering the spotface note.

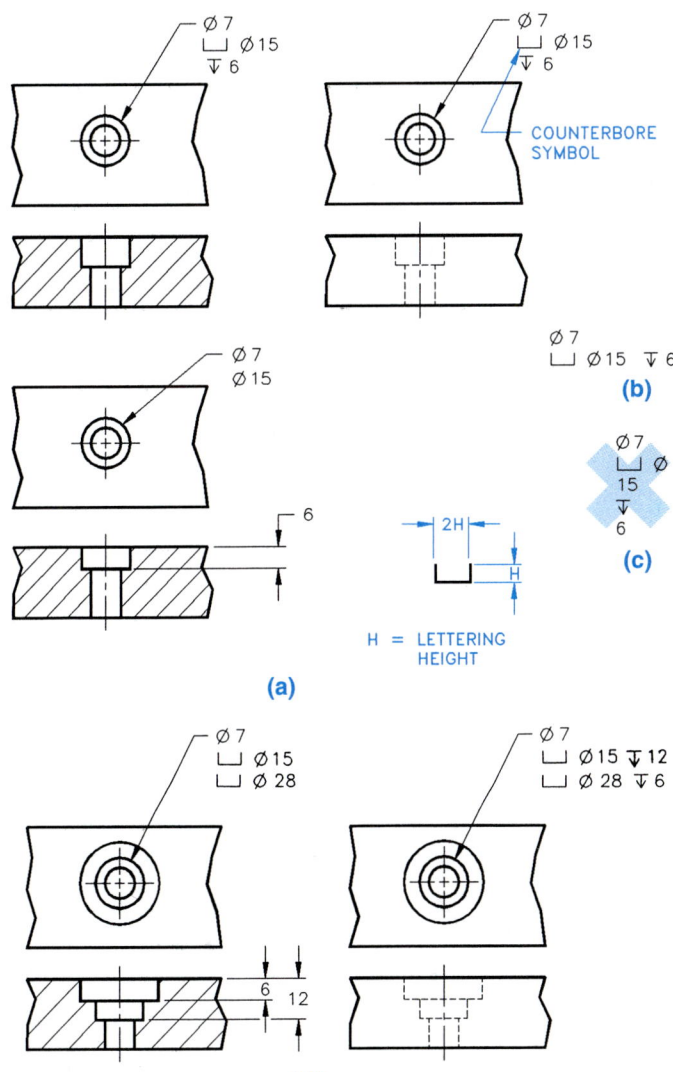

(a)

(b)

(c)

H = LETTERING HEIGHT

(d)

FIGURE 11.45 ■ (a) Counterbore note; (b) alternate note with elements of note grouped on one line; (c) never split individual note elements. The proper counterbore/spotface symbol is also displayed.

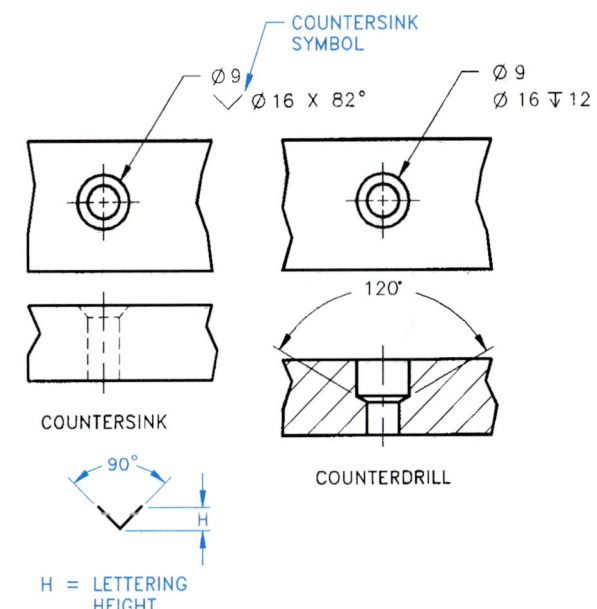

H = LETTERING HEIGHT

FIGURE 11.46 ■ Countersink, counterdrill notes, and dimensions. The properly drawn countersink symbol is also shown.

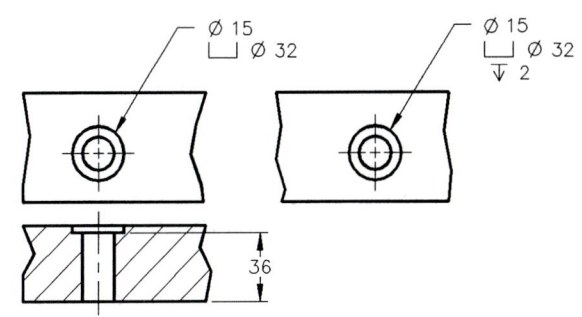

FIGURE 11.47 ■ Spotface note and dimensions.

Multiple Features

When a part has more than one feature of the same size, The features are dimensioned with a note specifying the number of like features, as shown in Figure 11.48. If a part contains several features all the same size, the methods shown in Figure 11.50 is used.

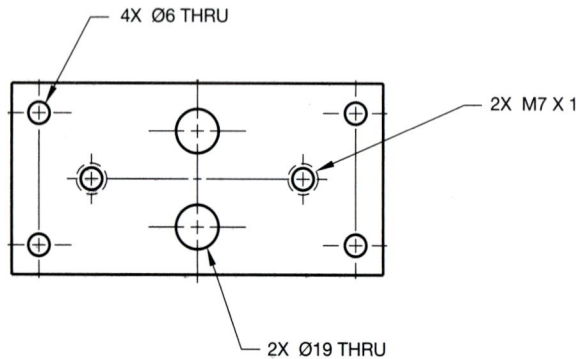

FIGURE 11.48 ■ Dimension notes for multiple features. The number of like features is placed first, followed with the by (X) symbol.

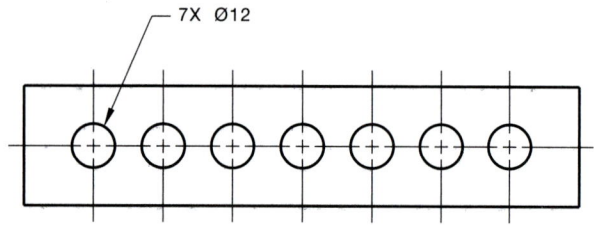

FIGURE 11.49 ■ Dimension notes for multiple features all of common size.

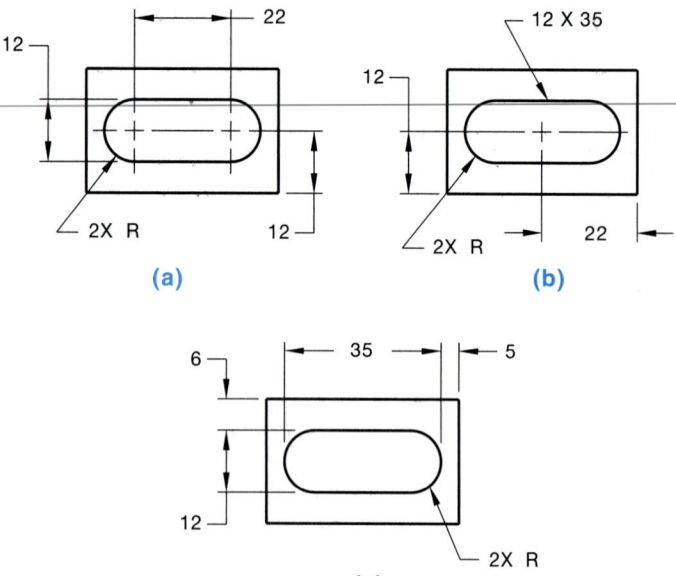

FIGURE 11.50 ■ Dimensioning slotted holes with full radius ends.

Slots

Slotted holes can be dimensioned in one of three ways, as shown in Figure 11.50. The slot dimensioning practice shown in Figure 11.50a is generally used when milling is the desired machining process. Figure 11.50b displays the normal dimensioning practice when using a punching process to make slotted holes. The preferred slot dimensioning practice for the laser cutting process is shown in Figure 11.50c. These methods are used only when the ends are fully rounded and tangent to the sides. In Figure 11.50, notice the note dimensioning the slot ends reading: 2X R. The 2X means there are two radii, one on each end, followed by a space and the letter R. R is the radius note, and no size is given because a full radius is assumed. When the ends of a slot or external feature have a radius greater than the width of the feature, then the size of the radius must be given, as shown in Figure 11.51.

Keyseats

Keyseats are dimensioned in the view that clearly shows their shape by width, depth, length, and location, as in Figure 11.52.

Knurls

Knurls are dimensioned with notes and leaders pointing to the knurl in the rectangular view, as shown in Figure 11.53a. ASME/ANSI **does not** recommend showing a knurl representation on the view as in Figure 11.53b, although some companies prefer this practice.

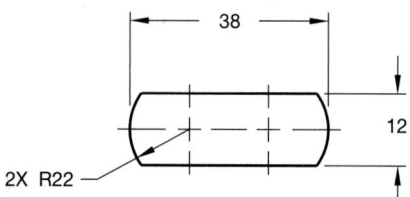

FIGURE 11.51 ■ Dimensioning slot or external feature with end radius larger than feature width.

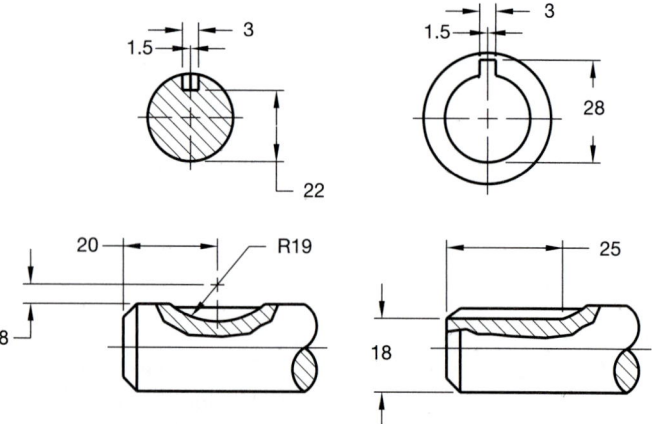

FIGURE 11.52 ■ Dimensioning keyseats.

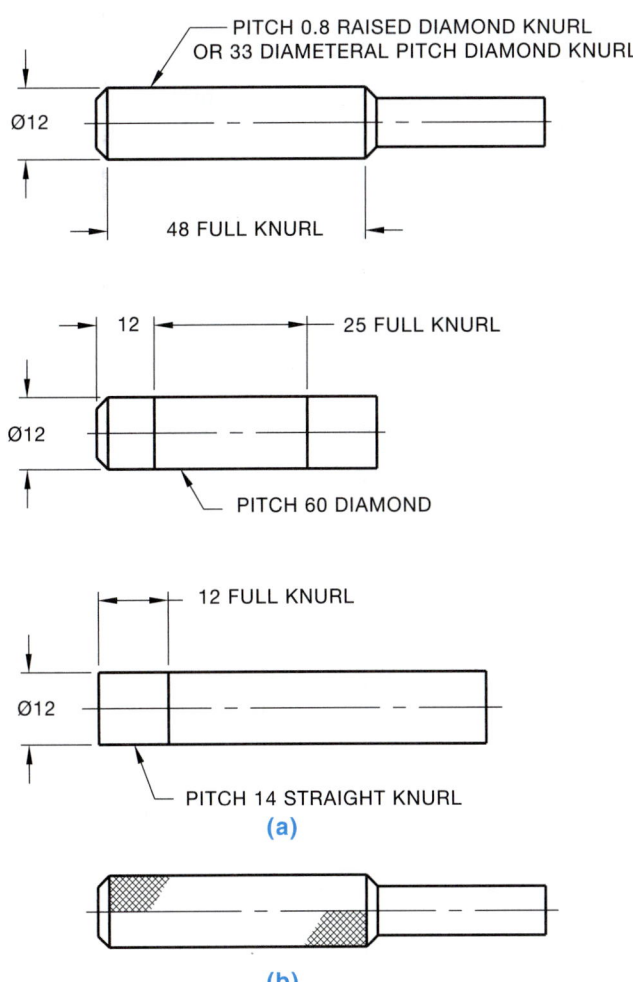

FIGURE 11.53 ■ (a) Dimensioning knurls. (b) Optional knurl representation.

Necks and Grooves

Necks and grooves are dimensioned as shown in Figure 11.54. This example shows multiple groove characteristics being dimensioned with GOOD, BETTER, and BEST options. In some of the examples, notice the dimensions placed within the objects. While this method should be avoided, these cases can require this practice. If the extension lines are not too long, it is possible to extend these dimensions through the object to the outside. The use of any of these options and practices should be confirmed with your company or school standards.

LOCATION DIMENSIONS

In general, dimensions identify either size or location. The previous dimensioning discussions focused on size dimensions. Location dimensions to cylindrical features, such as holes, are given to the center of the feature in the view where they appear as a circle, or in a sectional view. Rectangular shapes are located to their sides, and symmetrical features can be located to their centerline or center plane. Some location dimensions also control size.

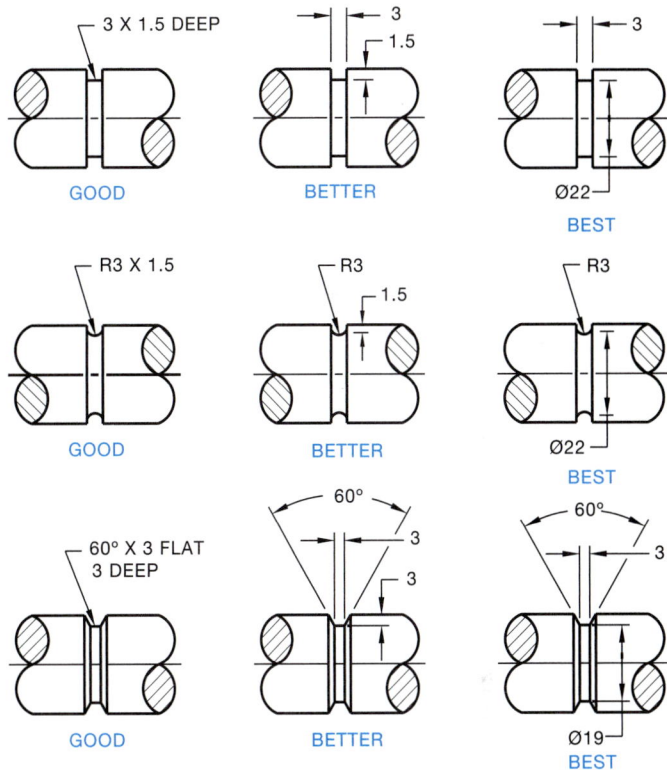

FIGURE 11.54 ■ Dimensions for necks and grooves.

Locating Holes

Locate a hole center in the view where the hole appears as a circle, as shown in Figure 11.55.

Rectangular Coordinates

Linear dimensions are used to locate features from planes or centerlines, as shown in Figure 11.56. This is referred to as rectangular coordinate dimensioning.

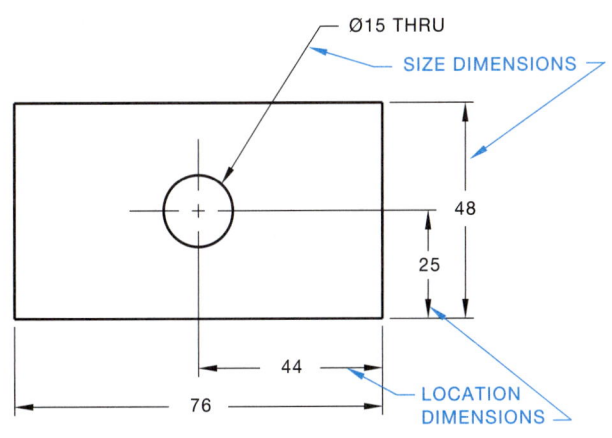

FIGURE 11.55 ■ Size and location dimensions.

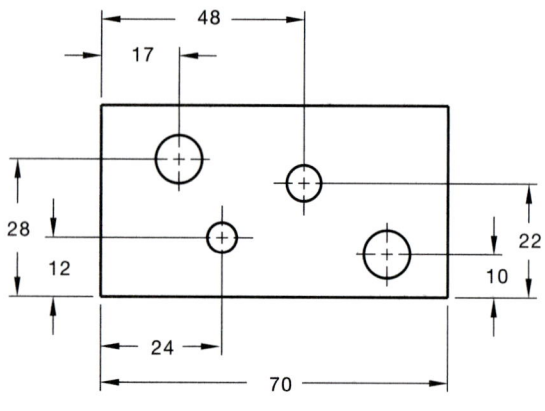

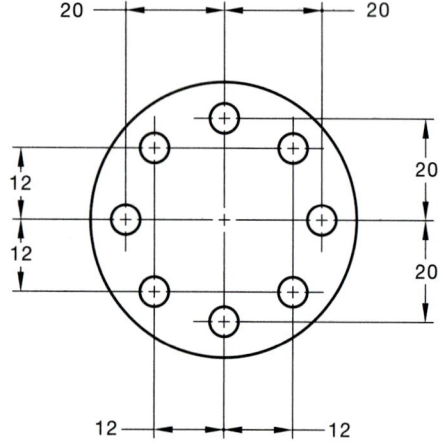

FIGURE 11.56 ■ Rectangular coordinate location dimensions.

Polar Coordinates

Angular dimensions locate features from planes or centerlines, as shown in Figure 11.57. This is called polar coordinate dimensioning.

Repetitive Features

Repetitive features are located by noting the number of times a dimension is repeated, and giving one typical dimension and the total length as reference. This method is acceptable for chain dimensioning. (See Figure 11.58.) If chain dimensioning is not acceptable, due to possible tolerance stacking, datum dimensioning should be used to locate the multiple features. With datum dimensioning, each feature (hole) is dimensioned independently from a common datum and tolerance stacking does not occur. Locating multiple tabs is shown in Figure 11.59. This method also works for slots.

When repetitive features on an object are nearly the same size, they can be shown with an identification letter, such as Y. (See Figure 11.60.)

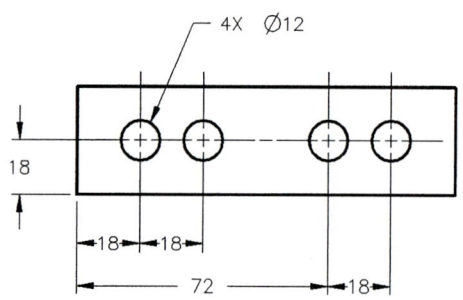

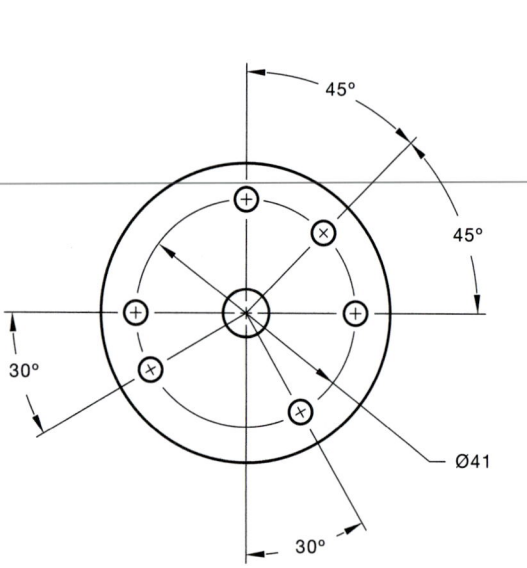

FIGURE 11.57 ■ Polar coordinate dimensions. If all features are equally spaced, a note such as 6X 60° can be used in one location.

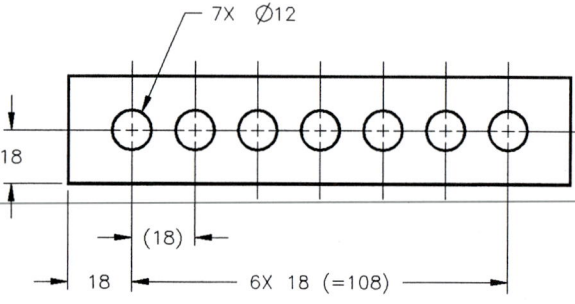

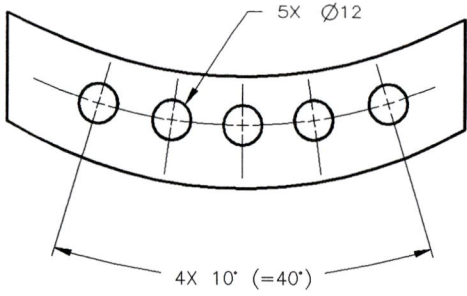

FIGURE 11.58 ■ Dimensioning repetitive features.

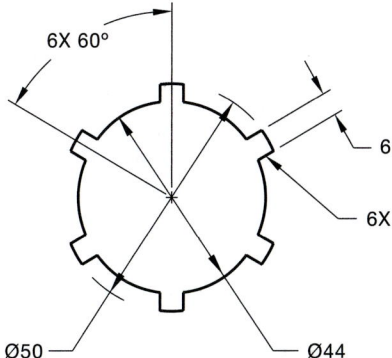

FIGURE 11.59 ■ Locating multiple tabs.

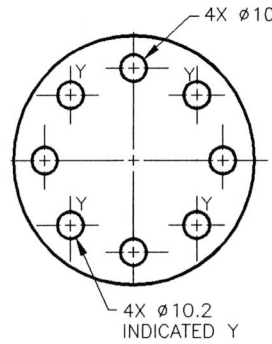

FIGURE 11.60 ■ Dimensioning similar-sized multiple features.

DIMENSION ORIGIN

When the dimension between two features must clearly identify from which feature the dimension originates, the dimension origin symbol is used. This method of dimensioning means the origin feature must be established first and the related feature is then dimensioned from the origin. (See Figure 11.61.)

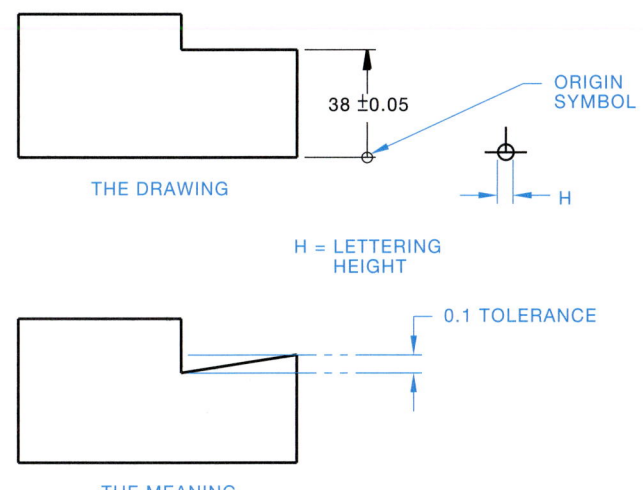

FIGURE 11.61 ■ Dimension origin and the dimension origin symbol.

DIMENSIONING AUXILIARY VIEWS

Dimensions should be placed on views that provide the best size and shape description of an object. In many instances the surfaces of a part are foreshortened and require auxiliary views to completely describe true size, shape, and the location of features. When foreshortened views occur, dimensions should be placed on the auxiliary view for clarity. In unidirectional dimensioning, the dimension numerals are placed horizontally so they read from the bottom of the sheet. When aligned dimensioning is used, the dimension numerals are placed in alignment with the dimension lines. (See Figure 11.62.) Keep in mind that unidirectional dimensioning is the preferred ASME standard for numeral placement.

GENERAL NOTES

ASME/ANSI General notes are located next to the title block for ASME/ANSI drawings. The exact location of the general notes depends on specific company or school standards. A common location for general notes is the lower-left corner of the drawing, usually .5 in. each way from the border line. Military standards specify general notes be located in the upper-left corner of the drawing.

The notes previously discussed are classified as specific notes because they refer to specific features of an object. General

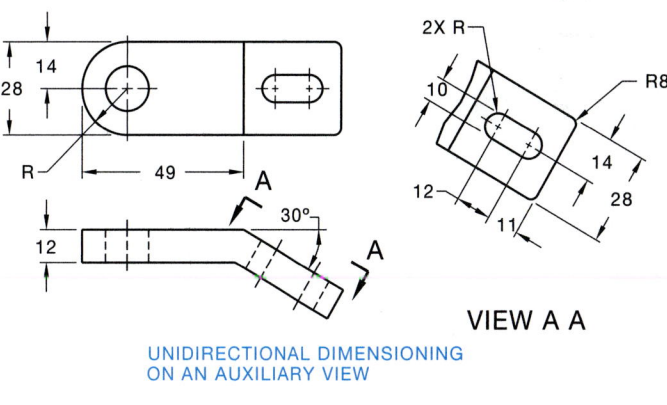

UNIDIRECTIONAL DIMENSIONING
ON AN AUXILIARY VIEW

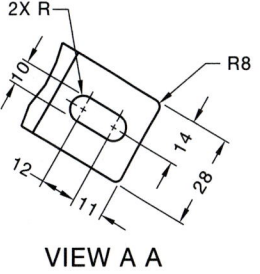

VIEW A A

ALIGNED DIMENSIONING
ON AN AUXILIARY VIEW

FIGURE 11.62 ■ Dimensioning auxiliary views.

(*Continued on page 321*)

DIMENSIONING WITH CADD

The proper placement and use of dimensions is one of the most difficult aspects of drafting. It requires thought and planning. The drafter must first determine the type of dimension that suits the application then place it on the drawing. The dimensioning options for the CADD drafter are many, and the proper usage of these options should be learned early and practiced whenever possible. A variety of dimensioning options is available with most CADD systems. These tools and commands can be selected using a menu, toolbar, or other input method. The process for adding dimensions is similar regardless of the CADD system or dimension tool you use and typically involves selecting existing drawing view items followed by picking a location for the dimension.

AutoCAD, for example, has the DIMLINEAR (DIMLIN) command that allows you to automatically place any straight dimension whether it is horizontal, vertical, or at an angle. All you do is pick the extension line origins and the dimension line location. To place a horizontal dimension, first select the DIMLIN command. You may then be prompted to select the first extension line origin by placing the screen cursor so that the crosshair is on the first corner, as shown in step 1 of Figure 11.63. Pick this point and then locate the crosshairs on the opposite end of the

feature and pick that point, as shown in step 2 of Figure 11.63. You have now established the length of the feature. The dimension is calculated by the computer. Next, you need to pick the distance away from the feature where the dimension line should be located. After this step is done, the dimension is automatically drawn on the screen. (See Figure 11.63.)

Another way to place dimensions on a drawing is to pick the extension line origins on the object. The CADD dimensioning then automatically calculates the extension line offset and places the extension lines, dimension lines, text, and arrowheads. This can be done in either a datum or chain dimensioning format, as shown in Figure 11.64. When you are dimensioning circles, the dimension line and text numeral are automatically placed inside the circle by picking the circle, or you have the option of dimensioning with a leader. Figure 11.65 illustrates some of the options. Dimensioning arcs is also easy. All you have to do is pick the arc to be dimensioned and enter the dimension text; the leader with text numeral is placed on the drawing automatically, as shown in Figure 11.66.

Some CADD programs also allow you to automatically dimension a group of objects just by selecting them. All you do is select a group of objects followed by picking the

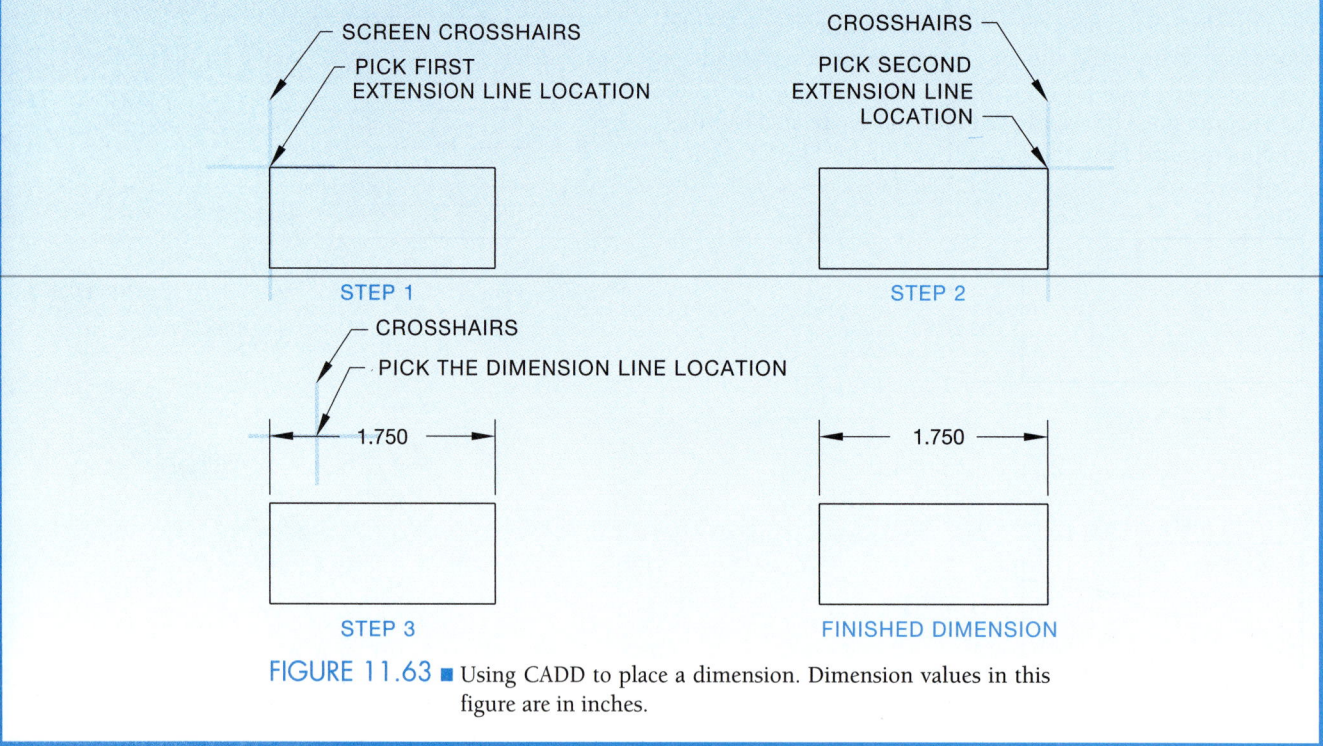

FIGURE 11.63 ■ Using CADD to place a dimension. Dimension values in this figure are in inches.

dimension line location, and the CADD system automatically draws the extension lines, dimension lines, and arrowheads, and places the dimension text. The AutoCAD command that performs this operation is QDIM.

As previously described, the computer calculates dimension values, which means that dimensions are associated with drawing geometry. For example, if you draw a one-inch line and then dimension the line, the dimension text value displays 1. You do not enter, or otherwise modify, the dimension value of 1, because it is already set by the one-inch line length you specified when the line was drawn. Under most circumstances, dimensions associated with objects represent a basic parametric relationship that should not be overridden. This is important because when you change the size or shape of a drawing feature, the dimensions associated with the drawing automatically update to reflect the new design. Using the previous example, if you change the length of the line from 1 to 2 inches, the extension lines, dimension lines, and arrowheads adapt to the new line size, and the dimension text value is replaced with a 2.

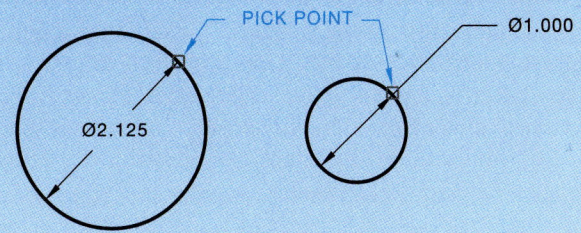

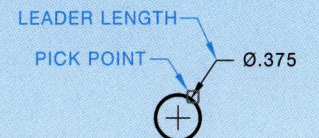

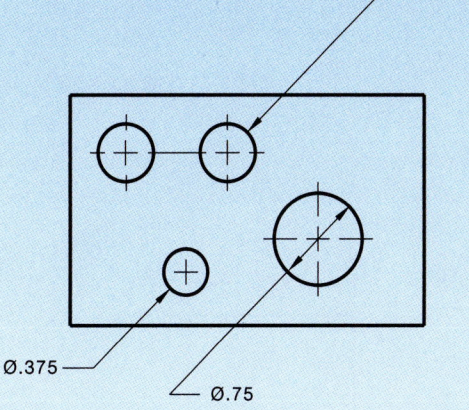

FIGURE 11.65 ■ Dimensioning circles with CADD. Dimension values in this figure are in inches.

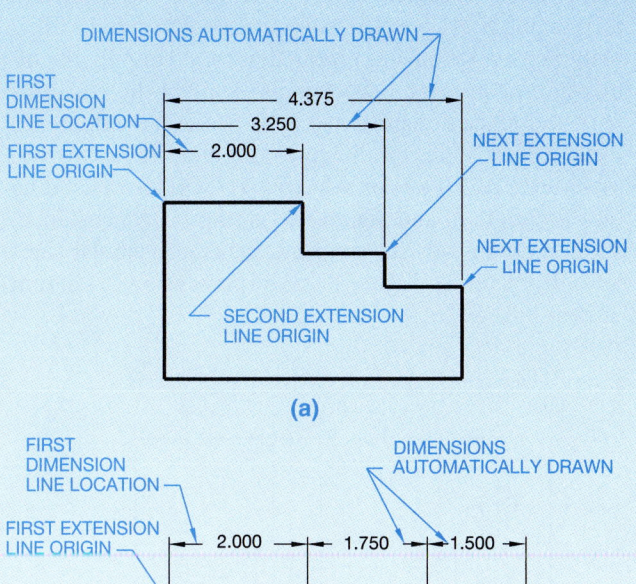

FIGURE 11.64 ■ Locating dimensions at extension line origins for (a) datums and (b) chain dimensioning. Dimension values in this figure are in inches.

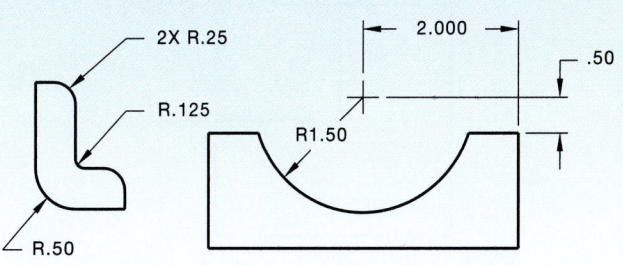

FIGURE 11.66 ■ Dimensioning arcs with CADD. Dimension values in this figure are in inches.

CADD APPLICATIONS

USING DIMENSION STYLES

CADD systems allow you to draw dimensions for any desired application, such as mechanical, architectural, structural, or civil drafting projects. The standards or variables you use to set up the way dimensions appear on your drawing often are stored using specific dimension styles. Default dimension style values can be set to the appropriate standards, such as ASME for mechanical drawings, but you have the flexibility to change default values and create new dimension styles for specific applications. Figure 11.67 shows some of the flexibility available with changing predefined default values.

There are a wide variety of variables you can set when creating a dimension style. For example, if you want to dimension a mechanical drawing, set the dimension style variables to the values described throughout this text. It is good practice to define dimension styles before you place dimensions, so dimensions are displayed correctly as you draw. However, most CADD systems allow you to modify dimension style variables at any time and update existing dimensions according to new display characteristics. Dimensioning for drafting applications such as structural, industrial pipe, and civil drafting are discussed where they apply in this textbook.

In addition to the major CADD software packages, there are hundreds of third-party software packages. The term third-party refers to programs that support or may be used in conjunction with the main package. In third-party software for dimensioning, the principal emphasis seems to be on datum, tabular, arrowless, and geometric tolerancing packages. These are the types of software that help increase speed and productivity and simplify the dimensioning process. For example, some of the programs have automatic dimensioning, which allows the drafter to pick all of the items for dimensioning and, when finished, press the ENTER key or select AUTOMATIC, depending on the program, and watch everything be dimensioned. Some packages also automatically generate a tabular chart for hole reference. Figure 11.68 displays arrowless datum dimensioning.

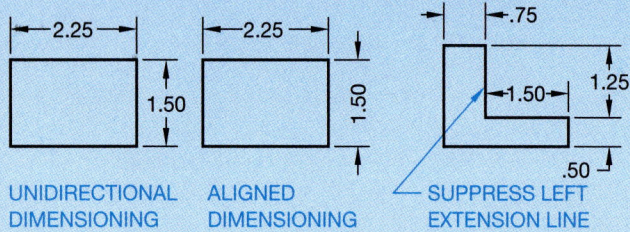

UNIDIRECTIONAL
DIMENSIONING

ALIGNED
DIMENSIONING

SUPPRESS LEFT
EXTENSION LINE

SUPPRESS EXTENSION LINE,
UNIDIRECTIONAL DIMENSIONS,
TEXT OUTSIDE OF EXTENSION LINES.

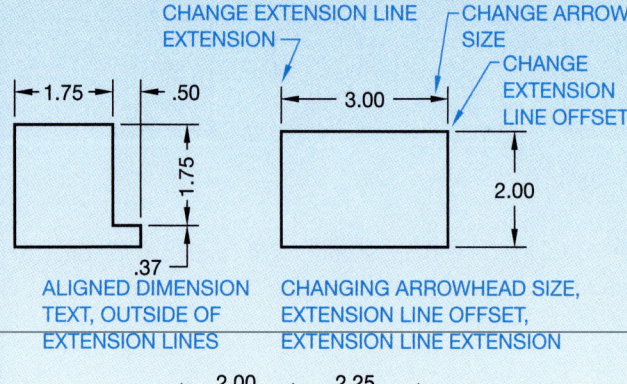

ALIGNED DIMENSION
TEXT, OUTSIDE OF
EXTENSION LINES

CHANGING ARROWHEAD SIZE,
EXTENSION LINE OFFSET,
EXTENSION LINE EXTENSION

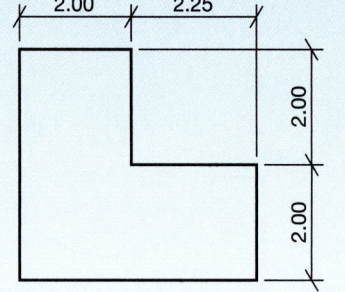

ARCHITECTURAL APPLICATION OF DIMENSION
TEXT ABOVE DIMENSION LINE, TICK MARKS

FIGURE 11.67 ■ Changing dimension default values for specific applications. Dimension values in this figure are in inches.

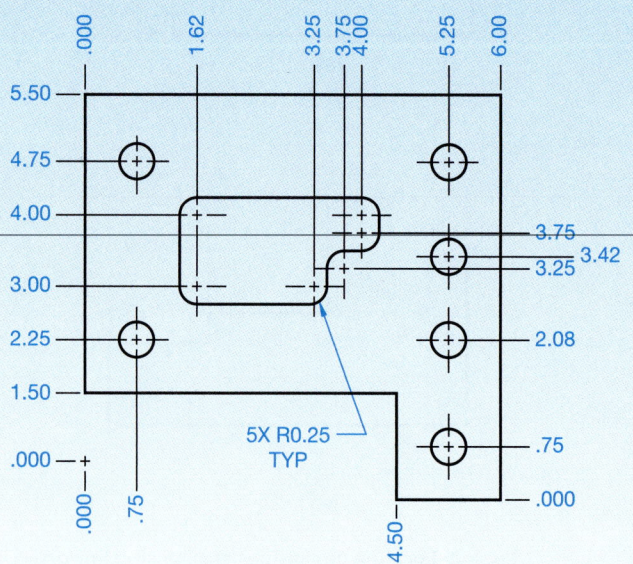

FIGURE 11.68 ■ Datum dimensioning with a third-party software package, Auto-DATUM. Dimension values in this figure are in inches. *Courtesy CADMASTER, Inc.*

DIMENSION LAYERS

A time-tested method of creating several drawings containing different information on the same outline is known as overlay drafting. A base drawing is generated, then an overlay is placed over it and a specific type of information is drawn on the overlay. For example, an architectural floor plan is drawn as the base, and then overlays are drawn for the plumbing plan, electrical plan, and so on. The base can be combined with any individual overlay or combinations thereof to create a print of just the information required. Also, several drafters can be given copies of the base and each can create an overlay containing specific information. This enables several drafters to essentially work on the same drawing at the same time; a great time-saver. The final drawing is then composed of several layers or levels.

The concept of layering is a prime component of most CADD systems. Layers set CADD drawing apart from manual drafting by building intelligence into the drawing, allowing you to organize information on the drawing into a wide variety of objects and in different colors, such as various line types, text for notes, dimensions and their components, and other features. This gives you control over the visibility of these objects. For example, you can turn off or freeze the dimensions so they are not displayed. Figure 12.65 shows a simple example of the layering concept.

Figure 11.69a shows drawing geometry created with visible object lines and hidden lines both drawn on their own layers. In Figure 11.69b you can clearly see how dimensions are drawn using a separate dimension layer. The complete drawing with views and dimensions combined is shown in Figure 11.69c with all layers visible.

Color is a valuable tool when working with layers. Many color systems enable the user to assign any color desired to any layer. Some systems have a limited number of colors, while others possess the ability to create thousands of hues. Color gives even more meaning to the layering concept, because several layers displayed on the screen at the same time can be distinguished from each other. The colors on the screen can be transferred to the plotter by assigning specific layer or plot characteristics, such as line weight, color, and Plot or No Plot configuration.

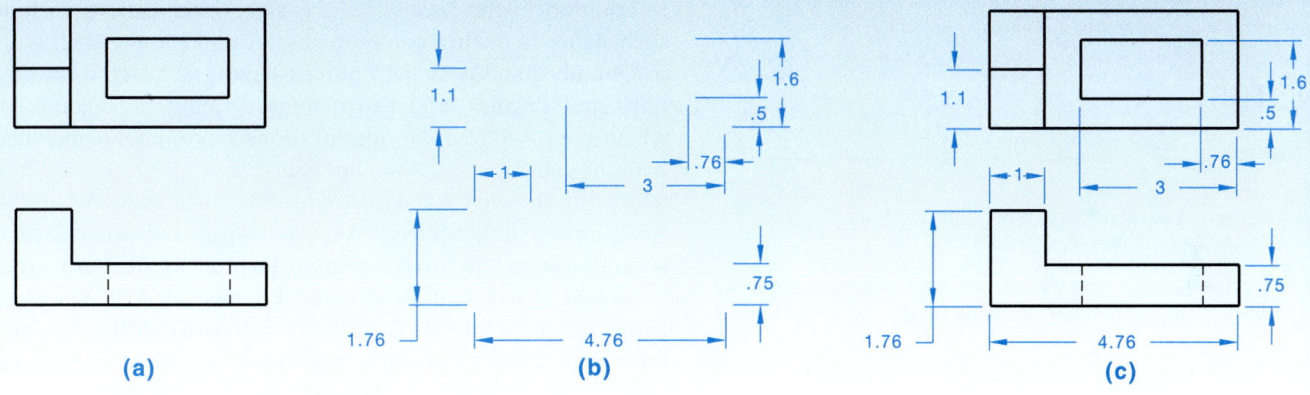

(a)   **(b)**   **(c)**

FIGURE 11.69 ■ Layering enables the drafter to focus on one aspect of the drawing at a time. Dimension values in this figure are in inches.

notes relate to the entire drawing. Each drawing contains a certain number of general notes either in or near the drawing title block. (See Figure 11.70.) General notes in the title block are concerned with such items as the following:

■ Material specifications.

■ Dimensions: inches or millimeters.

■ General tolerances.

■ Confidential note, copyrights, or patents.

■ Name of drafter/drawing creator.

■ Scale.

■ Date.

■ Part name.

■ Drawing size.

■ Part number.

■ Number of revisions.

■ First-angle or third-angle projection symbol.

■ ANSI/ASME, MIL (military), or other standard reference.

■ General machining, finish, or paint specifications.

■ Identification of features that are the same throughout the drawing.

Figure 11.71 shows some general notes commonly included on ASME/ANSI standard format drawings. Other common locations for general notes are the lower-right corner of the drawing directly above the title block, just to the left of the title block, or in the upper-left corner of the drawing.

General notes placed outside the title block can include some of the information previously listed for inside the title block.

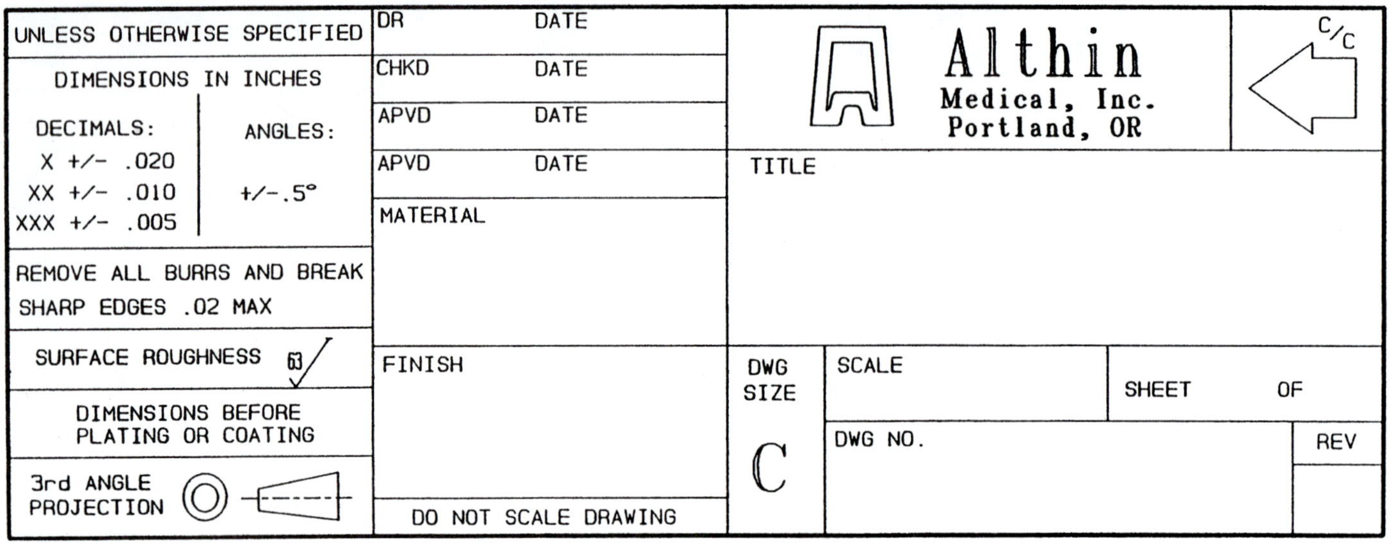

UNLESS OTHERWISE SPECIFIED	DR	DATE			Althin		C/C
DIMENSIONS IN INCHES	CHKD	DATE			Medical, Inc.		
	APVD	DATE			Portland, OR		
DECIMALS: ANGLES:	APVD	DATE		TITLE			
X +/− .020	MATERIAL						
XX +/− .010 +/−.5°							
XXX +/− .005							
REMOVE ALL BURRS AND BREAK SHARP EDGES .02 MAX							
SURFACE ROUGHNESS	FINISH		DWG SIZE	SCALE		SHEET OF	
DIMENSIONS BEFORE PLATING OR COATING			C	DWG NO.			REV
3rd ANGLE PROJECTION	DO NOT SCALE DRAWING						

FIGURE 11.70 ■ General title block information. *Courtesy Althin Medical, Inc.*

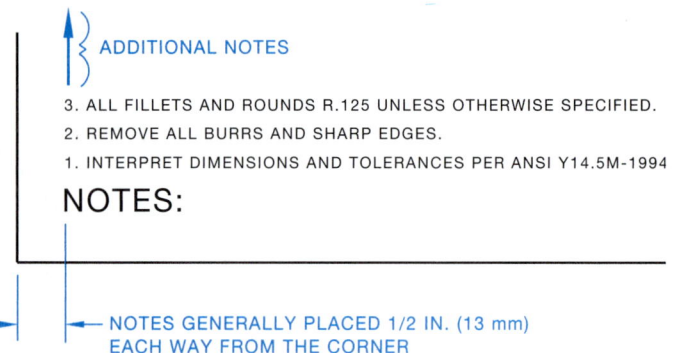

ADDITIONAL NOTES

3. ALL FILLETS AND ROUNDS R.125 UNLESS OTHERWISE SPECIFIED.
2. REMOVE ALL BURRS AND SHARP EDGES.
1. INTERPRET DIMENSIONS AND TOLERANCES PER ANSI Y14.5M-1994

NOTES:

NOTES GENERALLY PLACED 1/2 IN. (13 mm) EACH WAY FROM THE CORNER

FIGURE 11.71 ■ General notes located in the lower-left corner or upper-right corner of the sheet.

This depends on company or school standards. General notes placed outside the title block commonly contain this type of information:

■ A reference to the standard or standards used on the drawing. Some companies list a variety of applicable standards.

■ Machining practices used on the entire part, such as removal of all burrs and sharp edges. A **burr** is a rough edge left by a cutting tool or other operation.

■ Dimensions common throughout the drawing, such as all fillets and rounds of the same radius.

■ Other specifications related to the entire drawing, such as common surface finish, painting, or other treatments.

The dimension, machining practice, or other specifications relating to specific features are shown on the drawing with dimensioning practices and using specific notes as discussed throughout this chapter.

Specifications, further defined, is any written information or instructions included on the drawing or with a set of drawings, giving all necessary information not shown in the drawing field. In addition to the previous discussion, specifications include such items as quality requirements; manufacturer name, type, part number, and details for purchase parts or material applications and finishes; and instructions defining the manner in which work is to be done. Specifications are commonly included with the general notes when possible. These specifications often refer to an industry standard application, such as ASME, ANSI, American Welding Society (AWS), Aerospace Industries Association of America (AIAA), American Institute of Steel Construction (AISC), Society of Automotive Engineers (SAE), American Petroleum Institute (API), Military Standards (MILSTD), and many others. Reference can be made to details found in these and other nationally and internationally accepted standards by code or content, because their meaning is clear and consistent in the industry. Examples of these types of specifications are similar to the following general notes found on drawings:

■ DRAFT REQUIREMENTS PER AIAA STANDARDS E4, E7, AND E8, UNLESS OTHERWISE SPECIFIED.

■ HEAT MANIFOLD ITEM 11 TO 200°F.

■ CHILL UPPER BEARING ITEM 21 TO -50°F.

■ HEAT TREAT PER MIL-H-6875 TO H1100 CONDITION.

■ PENETRATE INSPECT FINISHED PART PER MIL-STD-271, GROUP III.

The words *UNLESS OTHERWISE SPECIFIED* are often placed with notes to remind the reader that the given specification generally applies, but can be modified by other information provided on the drawing or in other documents.

Notice in Figure 11.71 that the word NOTES is lettered first, followed by the first, second, and additional notes. Depending on company standards, the word NOTES is lettered .24 in. (6 mm) high, but some companies prefer to omit the word NOTES as an unnecessary preface to obvious general notes. The space between notes is from one-half to full height of the lettering, and the notes are normally lettered .12 in. (3 mm) in height. The first note, INTERPRET DIMENIONS AND TOLERANCES PER ASME Y14.5M—1994, should be included on all new drawings. Additional notes depend on the information required to support the drawing.

The technique of placing general notes reading from the bottom up as in Figure 11.71 has been done for the ease of manual drafting. The idea here is additional notes can be added easily without the need to erase everything and start over. With the use of CADD, the need to place notes in this manner is up to the preference of the company or school. Placing notes with CADD is actually easier when the notes are entered from top to bottom. If additional notes are needed at a later time, you can easily move the existing notes up and add the new notes below. Figure 11.72 shows general notes placed with CADD so they read from top to bottom. These notes are placed in the lower left corner of the drawing. When notes are placed to conform to military standards, then the notes are always placed in the upper left corner and they read from top to bottom.

Delta Notes

The term **delta** refers to a triangle placed on the drawing for reference. The triangle is commonly placed next to a dimension such as 2.625 ⟁ , or other location where it applies to a feature or item. This is used to refer the reader to a general note that relates to this item. This method is often used when it applies to a specific feature, but placing the note directly on the drawing is very difficult because the note is too long, or the note applies

NOTES:

1. INTERPRET DRAWING IAW MIL STD 100. CLASSIFICATION PER MIL-T-31000, PARA 3.6.4.

2. INTERPRET DIMENSIONS AND TOLERANCES PER ASME Y14.5M-1994.

3. PART TO BE FREE OF BURRS AND SHARP EDGES.

4. BAG ITEM AND IDENTIFY IAW MIL-STD-130, INCLUDE CURRENT REV LEVEL: 64869-0956356 REV 42375.

⟁5⟁ DIMENSION APPLIES BEFORE PLATING.

— DELTA

ADDITIONAL NOTES

FIGURE 11.72 ■ Notes can be conveniently placed to read from the first note downward with CADD. This makes it easy to continue from one note to the next. Editing notes is easy with CADD, because all you have to do is move the group of notes up to place additional notes when needed.

to several dimensions or features. Look at the note ⟁ in Figure 11.72. This note relates to the same delta item located somewhere on the drawing. Some companies use symbols other than a triangle. Hexagons and circles also can be used, but the triangle is common.

TOLERANCING
Definitions

As previously mentioned, a tolerance is the total permissible variation in a size or location dimension. Dimensions are placed on a drawing to ensure the parts fit together in an assembly and that all parts function as intended in that assembly. The following discusses the related tolerance terminology and the variety of fits used for different applications.

A **specified dimension** is that part of the dimension from which the limits are calculated. For example, 15.8 is the specified dimension of 15.8 ± 0.2.

A **bilateral tolerance** is allowed to vary in two directions from the specified dimension, as in $6.5 \, ^{+0.1}_{-0.3}$ or 6.5 ± 0.2.

The $6.5 \, ^{+0.1}_{-0.3}$ is an **unequal bilateral tolerance**, and the 6.5 ± 0.2 is called an **equal bilateral tolerance**.

A **unilateral tolerance** varies in only one direction from the specified dimension: $22 \, ^{0}_{-0.2}$ or $19.5 \, ^{+0.5}_{0}$.

Limits are the largest and smallest possible sizes a feature can be as related to the tolerance of the dimension.

Example 1:	19.0 ± 0.1
Upper limit:	$19.0 + 0.1 = 19.1$
Lower limit:	$19.0 - 0.1 = 18.9$
Example 2:	$9.5 \, ^{0}_{-0.5}$
Upper limit:	$9.5 + 0 = 9.5$
Lower limit:	$9.5 - 0.5 = 9.0$

The tolerance, being the total permissible variation in the dimension, is easily calculated by subtracting the lower limit from the upper limit.

Example 3:	22.0 ± 0.1
Upper limit:	22.1
Lower limit:	-21.9
Tolerance	0.2
Example 4:	$31.75 \, ^{+0.10}_{0}$
Upper limit:	31.85
Lower limit:	-31.75
Tolerance	0.10

All dimensions on a drawing have a tolerance except reference, maximum, minimum, or stock size dimensions. Dimensions on a drawing can read as in Figure 11.73, with **general tolerances** specified in the title block of the drawing. General tolerance specifications as given in a typical industry title block are shown in Figure 11.74, where x refers to one-place decimal dimensions, xx is for two-place decimals, and xxx refers to the tolerance for three-place decimal dimensions. Using this title

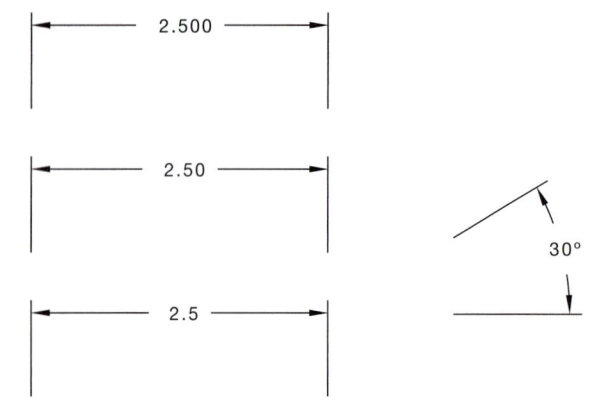

FIGURE 11.73 ■ Typical drawing dimensions. Dimension values in this figure are in inches.

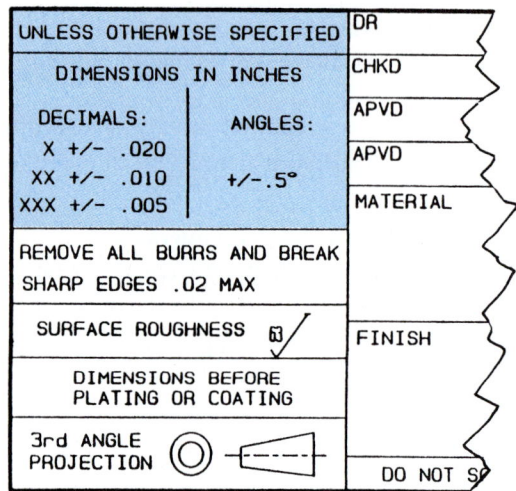

FIGURE 11.74 ■ General tolerances from a company title block. Dimension values in this figure are in inches. *Courtesy Althin Medical, Inc.*

block as an example, the tolerances of the dimensions in Figure 11.73 are as follows (given in inches):

■ 2.500 has .xxx ± .005 applied; 2.500 ± .005, tolerance equals .010.

■ 2.50 has .xx ± .010 applied; 2.50 ± .010, tolerance equals .020.

■ 2.5 has .x ± .020 applied; 2.5 ± .020, tolerance equals .040.

■ 30° has Angles ± 0.5° applied; 30° ± 0.5°, tolerance equals 1°.

Dimensions requiring tolerances different from the general tolerances given in the title block must be specified in the dimension on the drawing. These are referred to as **specific tolerance** dimensions and are represented with **plus/minus dimensioning** or **limits dimensioning** as in Figure 11.75.

Statistical Tolerancing

Statistical tolerancing is the assigning of tolerances to related dimensions in an assembly based on the requirements of statistical process control (SPC). SPC is discussed in Chapters 5 and

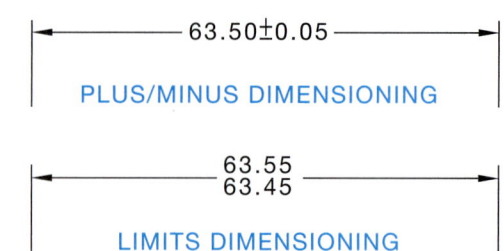

FIGURE 11.75 ■ Specific tolerance dimensions.

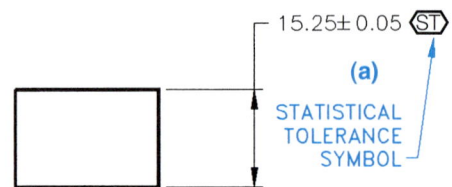

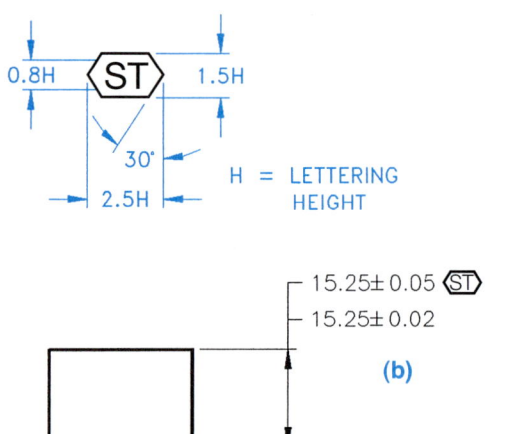

FIGURE 11.76 ■ Statistical tolerancing application and notes, and the statistical tolerancing symbol.

14. Statistical tolerancing is displayed in dimensioning as shown in Figure 11.76a. When the feature could be manufactured using SPC or conventional means, it is necessary to show both the statistical tolerance and the conventional tolerance as in Figure 11.76b. The appropriate general note should also accompany the drawing as shown in Figure 11.76.

Maximum and Least Material Conditions

Maximum material condition, abbreviated **MMC**, is the condition of a part or feature when it contains the most amount of material within the stated limits. The key is *most material*. The MMC of an external feature is the upper limit. (See Figure 11.77.) The MMC of an internal feature is the lower limit. (See Figure 11.78.)

The **least material condition** (**LMC**) is the opposite of MMC. LMC is the least amount of material possible in the size of a feature within the stated limits. The LMC of an external fea-

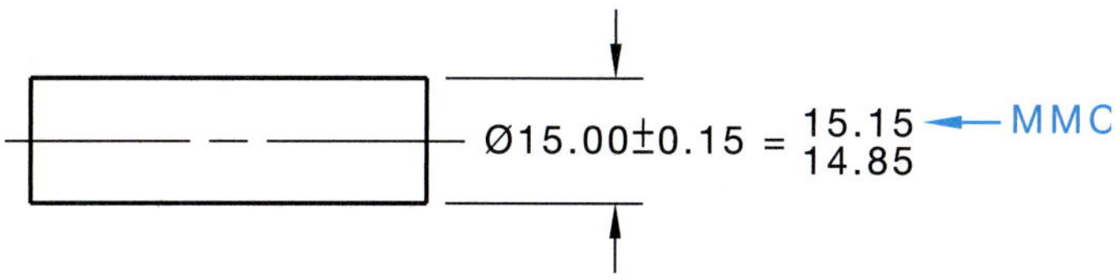

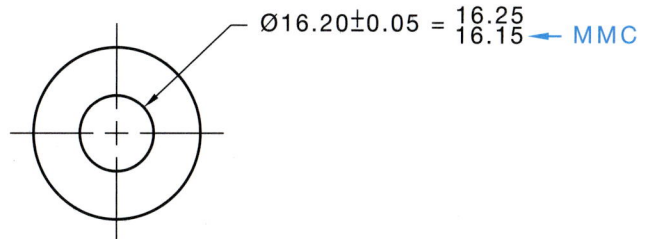

FIGURE 11.77 ■ Maximum material condition (MMC) of an external feature. Plus/minus dimension shown.

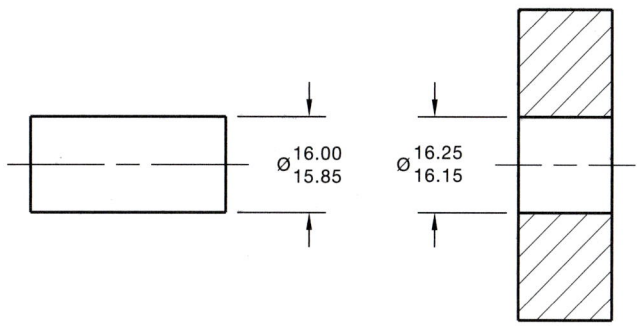

FIGURE 11.78 ■ Maximum material condition (MMC) of an internal feature. Plus/minus dimension shown.

FIGURE 11.79 ■ A clearance fit between two mating parts. Limits dimensions shown.

ture is its lower limit. The LMC of an internal feature is its upper limit.

Clearance Fit

A clearance fit is a condition when, due to the limits of dimensions, there is always a clearance between mating parts. The features in Figure 11.79 have a clearance fit. Notice the largest limit of the shaft is smaller than the smallest hole limit.

Allowance

The allowance of a clearance fit between mating parts is the tightest possible fit between the parts. The allowance is calculated with the formula:

MMC Internal Feature
−MMC External Feature
Allowance

The allowance of the parts in Figure 11.79 is:

MMC Internal Feature	16.15
−MMC External Feature	−16.00
Allowance	0.15

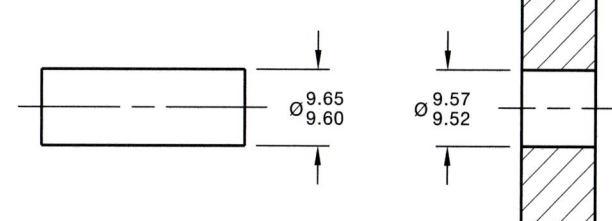

FIGURE 11.80 ■ An interference fit between two mating parts. Limits dimensions shown.

Interference Fit

An interference fit also known as force or shrink fit is the condition that exists when, due to the limits of the dimensions, mating parts must be pressed together. Interference fits are used, for example, when a bushing must be pressed onto a housing or when a pin is pressed into a hole. (See Figure 11.80.)

Types of Fits

Selection of Fits

In selecting the limits of size for any application, the type of fit is determined first based on the use or service required from the equipment being designed; then the limits of size of the mating parts are established to ensure the desired fit is produced. The number of standard fits described here cover most applications.

Designation of Standard ANSI Fits

Standard fits are designated by means of the following symbols, which facilitate reference to classes of fit for educational purposes. The symbols are not intended to be shown on manufacturing drawings; instead, sizes should be specified on drawings. The letter symbols used are as follows:

RC   Running, or Sliding, Clearance Fit

LC   Locational Clearance Fit

LT   Transition Clearance, or Interference, Fit

LN   Locational Interference Fit

FN   Force, or Shink, Fit

These letter symbols are used in conjunction with numbers representing the class of fit: thus FN 4 represents a class 4 force fit.

Description of Standard ANSI Fits

The classes of fits are arranged in three general groups known as running and sliding fits, locational fits, and force fits. Standard fit tables are given in Appendix I.

Running and Sliding Fits (RFC). Running and sliding fits are intended to provide a similar running performance with suitable lubrication allowance throughout their range of sizes. The clearances for the first two classes, used chiefly as sliding fits, increase more slowly with the diameter than do the clearances for the other classes, so accurate location is maintained even at the expense of free relative motion. These fits are described as follows:

- ■ RC1—Close sliding fits are intended for the accurate location of parts that must assemble without perceptible play.

- ■ RC2—Sliding fits are intended for accurate location, but with greater maximum clearance than class RC1. Parts made to this fit move and turn easily but are not intended to run freely, and in the larger sizes may seize with small temperature changes.

- ■ RC3—Precision running fits are about the closest fits that can be expected to run freely and are intended for precision work at slow speeds and light journal pressures. However, these are not suitable where appreciable temperature differences are likely to be encountered.

- ■ RC4—Close running fits are intended chiefly for running fits on accurate machinery with moderate surface speeds and journal pressures, where accurate location and minimum play is desired.

- ■ RC5 and RC6—Medium running fits are intended for high running speeds, or heavy journal pressures, or both.

- ■ RC7—Free running fits are intended for use where accuracy is not essential, or where large temperature variations are likely to be encountered, or under both these conditions.

- ■ RC8 and RC9—Loose running fits are intended for use where wide commercial tolerances may be necessary, together with an allowance, on the external member.

Locational Fits (LC, LT, and LN). Locational fits are fits intended to determine only the location of the mating parts; they provide rigid or accurate location, as with interference fits, or provide some freedom of location, as with clearance fits. Accordingly, they are divided into three groups: clearance fits (LC), transition fits (LT), and interference fits (LN). These fits are described as follows:

- ■ LC—Locational clearance fits are intended for parts that are normally stationary, but which can be freely assembled or disassembled. They range from snug fits for parts requiring accuracy of location, through medium clearance fits for parts such as spigots, to looser fastener fits where freedom of assembly is of prime importance.

- ■ LT—Locational transition fits are a compromise between clearance and interference fits. They are for applications where accuracy of location is important but either a small amount of clearance or interference is permissible.

- ■ LN—Locational interference fits are used where accuracy of location is of prime importance, and for parts requiring rigidity and alignment with no special requirements for bore pressure. Such fits are not intended for parts designed to transmit frictional loads from one part to another by

virtue of the tightness of fit. Such conditions are covered by force fits.

Force Fits (FN). Force, or shrink, fits constitute a special type of interference fit normally characterized by maintenance of constant bore pressures throughout its range of sizes. The interference, therefore, varies almost directly with diameter, and the difference between its minimum and maximum values is small so as to maintain the resulting pressures within reasonable limits. These fits are described as follows:

- ■ FN1—Light drive fits are those requiring light assembly pressures and producing more or less permanent assemblies. They are suitable for thin sections or long fits, or in external cast iron members.

- ■ FN2—Medium drive fits are suitable for ordinary steel parts, or for shrink fits on light sections. They are about the tightest fits that can be used with high-grade cast iron external members.

- ■ FN3—Heavy drive fits are suitable for heavy steel parts or for shrink fits in medium sections.

- ■ FN4 and FN5—Force fits are suitable for parts that can be highly stressed, or for shrink fits where the heavy pressing forces required are impractical.

Establishing Dimensions for Standard ANSI Fits

The fit used in a specific situation is determined by the operating requirements of the machine. When the type of fit has been established, the engineering drafter refers to tables that show the standard hole and shaft tolerances for the specified fit. One source of these tables is the *Machinery's Handbook*. Tolerances are based on the type of fit and nominal size ranges, such as $0-.12$, $.12-.24$, $.24-.40$, $.40-.71$, $.71-1.19$, and $1.19-1.97$ in. So, if you have a 1-in. nominal shaft diameter and an RC4 fit, refer to Figure 11.81 to determine the shaft and hole limits. The hole and shaft limits for a 1-in. nominal diameter are:

- ■ Upper hole limit $= 1.000 + .0012 = 1.0012$

- ■ Lower hole limit $= 1.000 + 0 = 1.000$

- ■ Upper shaft limit $= 1.000 - .0008 = .9992$

- ■ Lower shaft limit $= 1.000 - .0016 = .9984$

- ■ You then dimension the hole as $\varnothing 1.0012 - 1.0000$, and the shaft as $\varnothing .9992 - .9984$.

An actual drawing showing the use of standard ANSI fits is given in Figure 11.82.

Standard ANSI/ISO Metric Limits and Fits

The standard for the control of metric limits and fits is governed by the document ANSI B4.2, *Preferred Metric Limits and Fits*. The system is based on symbols and numbers relating to the internal or external application and the type of fit. The specifica-

NOMINAL SIZE RANGE IN INCHES	RC4 STANDARD TOLERANCE LIMITS	
	HOLE	SHAFT
0–.12	+.0006	–.0003
	0	–.0007
.12–.24	+.0007	–.0004
	0	–.0009
.24–.40	+.0009	–.0005
	0	–.0011
.40–.71	+.0010	–.0006
	0	–.0013
.71–1.19	+.0012	–.0008
	0	–.0016
1.19–1.97	+.0016	–.0010
	0	–.0020

FIGURE 11.81 ■ Standard RC4 fits for nominal sizes ranging from 0 to 1.97 inches. Standard fit tables are given in Appendix J.

TYPE OF FIT	ISO SYMBOL		DESCRIPTION OF FIT
	HOLE	SHAFT	
CLEARANCE FIT	H11/c11	C11/h11	Loose running
	H9/d9	D9/h9	Free running
	H8/f7	F8/h7	Close running
	H7/g6	G7/h6	Sliding
	H7/h6	H7/h6	Locational clearance
TRANSITION FIT	H7/k6	K7/h6	Locational transition
	H7/n6	N7/h6	Locational transition
INTERFERENCE FIT	H7/p6[1]	P7/h6	Locational interference
	H7/s6	S7/h6	Medium drive
	H7/u6	U7/h6	Force

FIGURE 11.83 ■ Description of metric fits. Standard fit tables are given in Appendix J.

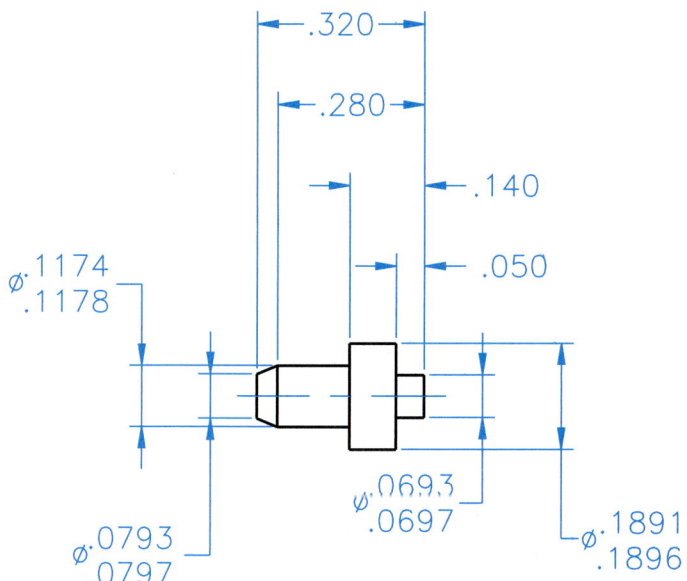

FIGURE 11.82 ■ An actual drawing showing the use of standard ANSI fits. The diameter dimension limits were calculated using the RC4 fits found in Figure 11.81.

tions and terminology for fits are slightly different from the ANSI standard fits previously described. The metric limits and fits are divided into three general categories: clearance fits, transition fits, and interference fits:

■ Clearance fits are generally the same as the running and sliding fits explained earlier. With clearance fits, a clearance always occurs between the mating parts under all tolerance conditions.

■ With transition fits, a clearance or interference can result due to the range of limits of the mating parts.

■ When interference fits are specified, a press or force situation exists under all tolerance conditions.

Refer to Figure 11.83 for the ISO symbol and descriptions of the different types of metric fits.

The metric limits and fits are designated in a dimension one of three ways. The method used depends on individual company or school standards and the extent of use of the ISO system. When most companies begin using this system, the tolerance limits are calculated and shown on the drawing followed by the tolerance symbol in parentheses; for example, 25.000 − 24.979 (25 h7). The symbol in parentheses represents the basic size, 25, and the shaft tolerance code, h7. The term basic size is the dimension from which the limits are calculated, just like the term *specified dimension* that was previously introduced.

When the feature is external as in the previous h7 shaft specification, a lowercase letter is used in the symbol. Uppercase letters are used in the symbol for internal features such as a hole. An example of the symbol for an internal feature is H7.

When companies become accustomed to using the system, they represent dimensions with the code followed by the limits in parentheses, as follows: 25 h7 (25.000 − 24.979).

Finally, when a company has used the system long enough for interpreters to understand the designations, the code is placed alone on the drawing, like this: 25 h7.

A metric fit is specified by providing the basic size common to the internal and external mating features, followed by a symbol corresponding to each feature. In this application, the internal feature symbol is first, followed by a backward slash, and then followed by the external features symbol, for example 25H8/f7.

The ISO symbols for the different types of metric fits are listed as follows:

ISO Symbol		
Basic Internal	Basic External	Type of Fit
H11/c11	C11/h11	Loose running fit.
H9/d9	F8/h7	Close running fit.
H8/f7	F8/h7	Sliding fit.
H7/g6	G7/h6	Sliding fit.
H7/h6	H7/h6	Locational clearance fit.
H7/k6	K7/h6	Locational transition fit.
H7/n6	N7/h6	Additional accuracy locational transition fit.
H7/p6	P7/h6	Locational interference fit.
H7/s6	S7/h6	Medium drive fit.
H7/u6	U7/h6	Force fit.

When it is necessary to determine the dimension limits from code dimensions, use the charts in Appendix I of this textbook, ANSI B4.2, or the *Machinery's Handbook*. For example, if you want to determine the limits of the mating parts with a basic size of 30 and a close running fit, refer to the chart shown in Figure 11.84. The hole limits for the 30-mm basic size are ∅30.033−30.000 (30 h8), and the shaft dimension limits are ∅29.980−29.959 (30 f7).

A drawing showing the use of the previously discussed standard ANSI/ISO metric limit and fit applications is given in Figure 11.85. Figure 11.85a shows the use of the ISO system with the limits followed by the tolerance symbol in parentheses. Figure 11.85b shows the use of the ISO system with the tolerance symbol followed by the limits in parentheses. Figure 11.85c shows using only the ISO system tolerance symbol on the dimensions where metric limits and fits are applied.

BASIC SIZE	CLOSE RUNNING FIT	
	HOLE (h8)	SHAFT (f7)
20	20.033	19.993
	20.000	19.959
25	25.033	24.980
	25.000	24.959
30	30.033	29.980
	30.000	29.959
40	40.039	39.975
	40.000	39.950
50	50.039	49.975
	50.000	49.950

FIGURE 11.84 ■ Tolerances of close running fits for basic sizes ranging from 20 to 50 mm. Standard fit tables are given in Appendix I.

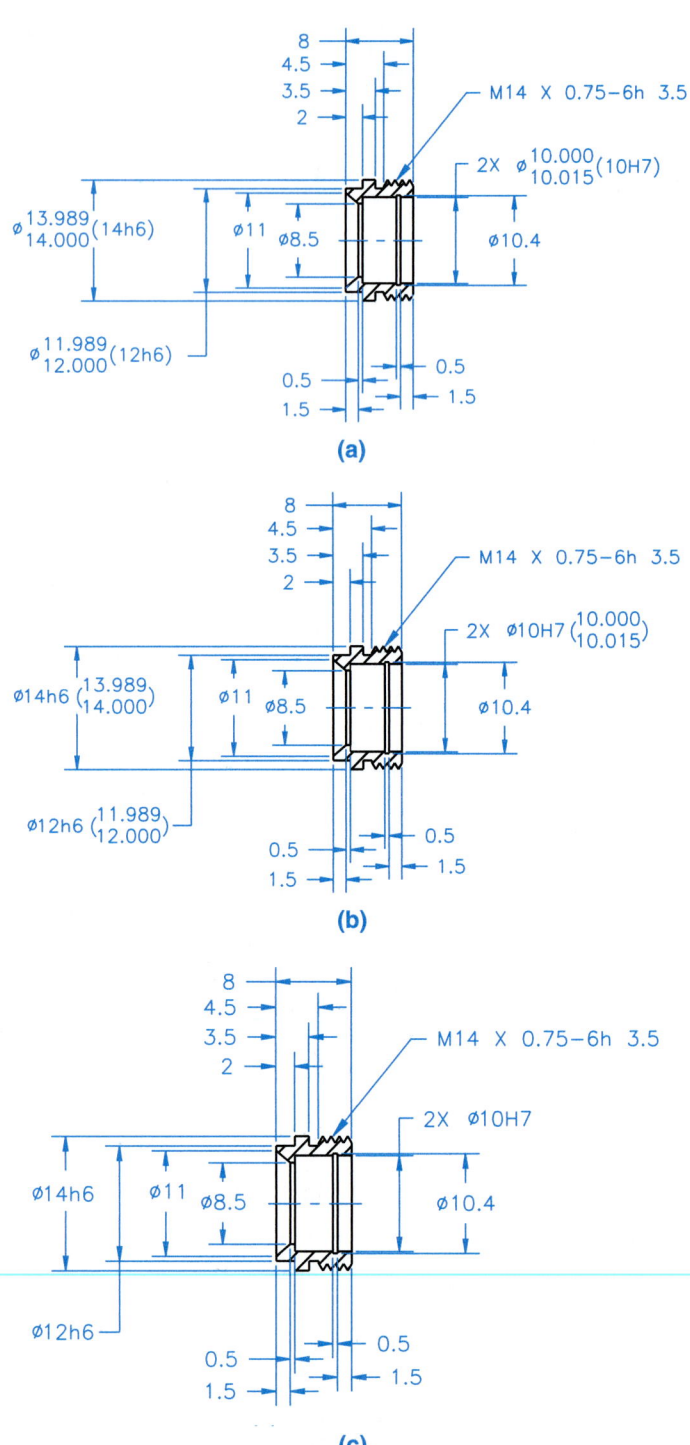

(a)

(b)

(c)

FIGURE 11.85 ■ A drawing showing the use of standard ANSI/ISO metric limits and fits.

DIMENSIONS APPLIED TO PLATINGS AND COATINGS

When platings such as chromium, copper, and brass, or coatings such as galvanizing, polyurethane, and silicone, are applied to a part or feature, the specified dimensions should be defined in re-

DIMENSIONING AND TOLERANCING

In contrast to standard two-dimensional CADD systems, solid model drawing dimensions and tolerances exist in two forms: model and drawing. Model dimensions and tolerances are added when you create a solid model. Every time you create a feature, a model dimension is stored in the system. The depth of an extrusion, diameter of a cylinder, and location of a hole are all examples of model dimensions. You may also have the option of adding tolerances to model size and location dimensions.

Model dimensions and tolerances, or parameters, are available and can be used when you dimension drawing views. In some cases, model dimensions and tolerances are automatically displayed when you insert drawing views. If this is an option, you can potentially create a fully dimensioned multiview drawing in just a matter of seconds. However, dimensions associated with model geometry are rarely displayed perfectly when initially added to drawing views, because they generally require some modification. Another important concept regarding model dimensions

and tolerances available with some CADD systems is they parametrically control the model. This means that you can actually modify the size and shape of a model by adjusting model dimensions from inside the drawing environment.

Drawing dimensions and tolerances are placed in addition to, or as a substitute for, model dimensions, and they function much like dimensions added using traditional CADD programs. If model dimensions are not appropriate, or if you need to add more dimensions, a number of drawing dimension tools are available to fully dimension drawing views. Drawing dimensions are parametrically associated with model geometry. For example, if you create a counterbore hole in a model, you can use tools such as HOLE CALLOUT or HOLE/THREAD NOTES to add a complete note. (See Figure 11.86.) Additionally, 2-D drawing dimensions automatically update when changes are made to the size or shape of the corresponding 3-D model. You can redefine a model by changing model dimensions, but you cannot redefine a model by editing drawing dimensions.

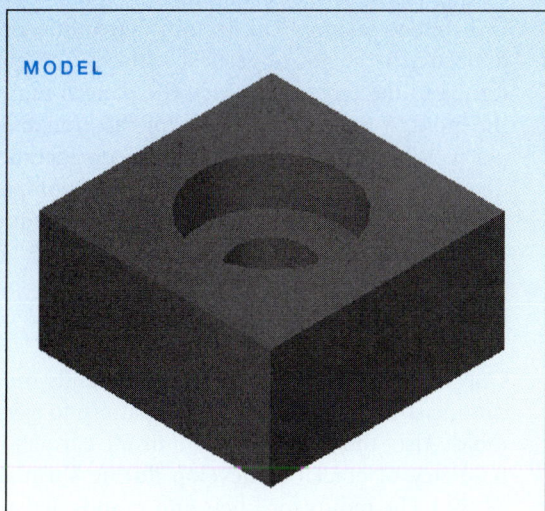
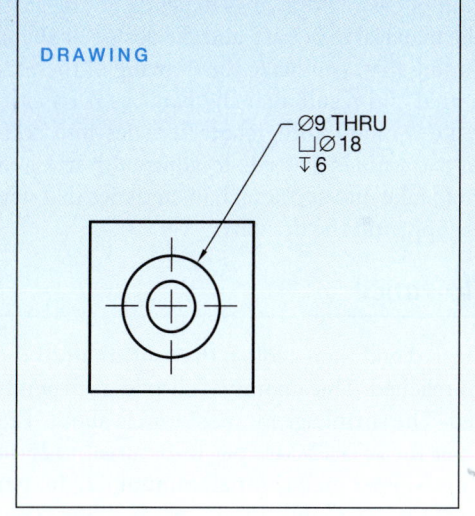

MODEL

DRAWING

Ø9 THRU
⌴Ø18
↧6

FIGURE 11.86 ■ Creating a counterbore hole in a model, using tools such as HOLE CALLOUT or HOLE/THREAD NOTES to add a complete note.

lation to the coating or plating process. A general note that indicates the dimensions apply before or after plating or coating is commonly used and specifies the desired variables; for example, DIMENSIONAL LIMITS APPLY BEFORE (AFTER) PLATING (COATING). A leader connecting a specific note to a surface can also be used for specific applications. Notice the dot replaces the arrowhead when the leader points to the surface in Figure 11.87. The dot is .06 in. (1.5 mm) minimum in diameter.

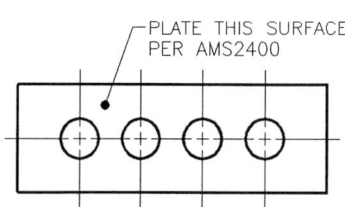

PLATE THIS SURFACE
PER AMS2400

FIGURE 11.87 ■ A dot replaces the arrowhead on a leader connecting a specific note to a surface.

MAXIMUM AND MINIMUM DIMENSIONS

In some situations, a dimension with an unspecified tolerance requires that the general tolerance be applied in one direction only from the specified dimension. For example, when a 12-mm radius shall not exceed 12 mm, the dimension reads R12 MAX. Therefore, when it is desirable to establish a maximum or minimum dimension, the abbreviations MAX or MIN are applied to the dimension. A dimension with a specified tolerance reads as previously discussed, for example:

$$R12\,{}^{\,0}_{-0.05}, \text{ or } R12\,{}^{+0.05}_{\,0}.$$

CASTING DRAWING AND DESIGN

ASME/ANSI Standards for the drafting of castings and forgings are recommended in ASME Y14.8M, *Castings and Forgings.*

A review of Chapter 5, *Manufacturing Materials and Processes,* is recommended before proceeding to the next few sections.

The end result of a casting drawing is the fabrication of a pattern. The preparation of casting drawings depends on the casting process used, the material to be cast, and the design or shape of the part. When doing this, you make the drawing of the part the same as the desired end result after the part has been cast. You also need to take certain casting characteristics into consideration, and the patternmaker needs to adjust the size and shape of the pattern to take into account characteristics that you do not intentionally apply on the drawing.

Shrinkage Allowance

When metals are heated and then cooled, they shrink until the final temperature is reached. The amount of shrinkage depends on the material used. The shrinkage for most iron is about .125 in. per ft. (.4 mm per meter), .250 in. per ft. for steel, .125 to .156 in. per ft. (.4 mm–.5 per meter) for aluminum, .22 in. per ft. (.7 mm per meter) for brass, and .156 in. per ft. (.5 mm per meter) for bronze. Values for shrinkage allowance are approximate, since the exact allowance depends on the size and shape of the casting and the contraction of the casting during cooling. You normally do not need to take shrinkage into consideration because the patternmaker applies shrink rules that use expanded scales to take into account the shrinkage of various materials.

Draft

Draft is the taper allowance on all vertical surfaces of a pattern, which is necessary to facilitate the removal of the pattern from the mold. Draft is not necessary on horizontal surfaces because the pattern easily separates from these surfaces without sticking. Draft angles begin at the parting line and taper

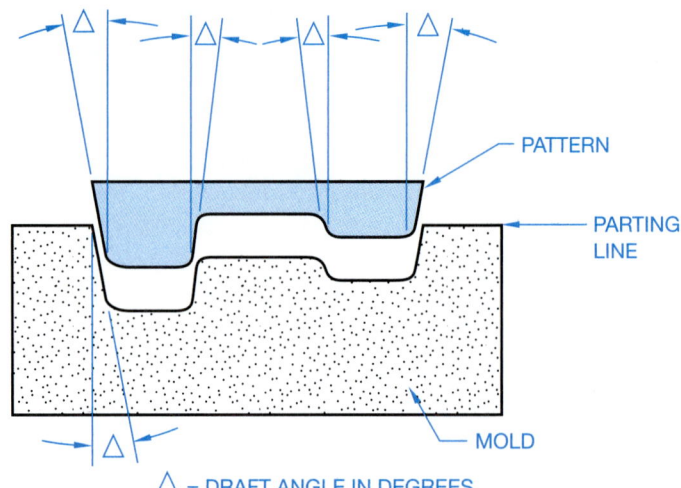

FIGURE 11.88 ■ Draft angles for castings.

away from the molding material. (See Figure 11.88.) The draft is added to the minimum design sizes of the product. Draft varies with different materials, size and shape of the part, and casting methods. For example, little if any draft is necessary in investment casting. The factors that influence the amount of draft are the height of vertical surfaces, the quality of the pattern, and the ease with which the pattern must be drawn from the mold. A typical draft angle for cast iron and steel is .125 in. per ft. Whether you take draft into consideration on a drawing depends on company standards. Some companies leave draft angles to the patternmaker, while others require you to place draft angles on the drawing.

Fillets and Rounds in Casting

One of the purposes of fillets and rounds on a pattern is the same as draft angles: to allow the pattern to eject freely from the mold. Also, the use of fillets on inside corners helps reduce the tendency of cracks to develop during shrinkage. (See Figure 11.89.) The radius for fillets and rounds depends on the material to be cast, the casting method, and the thickness of the part. The recommended radii for fillets and rounds used in sand casting is determined by part thickness as shown in Figure 11.90.

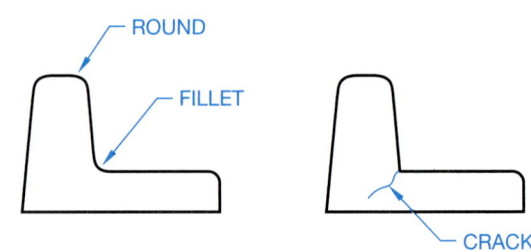

FIGURE 11.89 ■ Fillets and rounds for castings.

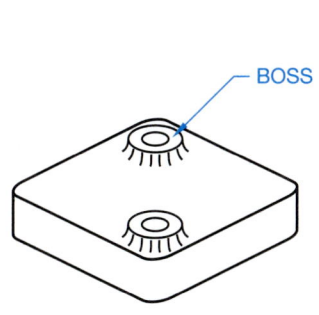

FIGURE 11.90 ■ Recommended fillet and round radii for sand castings.

The radii shown are $R = \frac{t}{4}$, $R = t$, and $R = \frac{T+t}{2}$.

MACHINING ALLOWANCE

Extra material must be left on the casting for any surface to be machined. As with other casting design characteristics, the machining allowance depends on the casting process, material, size, and shape of the casting, and the machining process to be used for finishing. The standard finish allowance for iron and steel is .125 in. (3 mm), and for nonferrous metals such as brass, bronze, and aluminum, .062 in. (1.5 mm). In some situations the finish allowance can be as much as .5 to .75 in. (12 to 20 mm) for castings that are very large or have a tendency to warp.

Other machining allowances are the addition of lugs, hubs, or bosses on castings that are otherwise hard to hold. You can add these items to the drawing or the patternmaker can add them to the pattern. These features may not be added for product function, but they serve as aids for chucking or clamping the casting in a machine. (See Figure 11.91.)

CASTING DRAWINGS

There are several methods used to prepare drawings for casting and machining operations. The method used depends on company standards. A commonly used technique is to prepare two drawings, one a casting drawing and the other a machining drawing.

Casting Drawing

The casting drawing, as shown in Figure 11.92, shows the part as a casting. Only dimensions necessary to make the casting are shown in this drawing.

Machining Drawing

The other drawing is of the same part, but this time only machining information and dimensions are given. The actual casting goes to the machine shop along with the machining drawing so the features can be machined as specified. The machining drawing is shown in Figure 11.93.

Combined Casting and Machining Drawing

Another method of preparing casting and machining drawings is to show both casting and machining information together on one drawing. This technique requires the patternmaker to add machining allowances. The drawing can have draft angles specified in the form of a note. The patternmaker must add the draft angles to the finished sizes given. With draft angles and finish allowances omitted, you need to consult with the patternmaker to ensure the casting is properly made. A combination casting/machining drawing is shown in Figure 11.94, page 334.

Drawing Phantom Lines to Show Machining Allowance and Draft Angles

Another technique is to draw the part as a machining drawing and then use phantom lines to show the extra material for machining allowance and draft angles, as shown in Figure 11.95, page 335.

FORGING, DESIGN, AND DRAWING

Draft for forgings serves much the same purpose as draft for castings. The draft associated with forging is found in the dies.

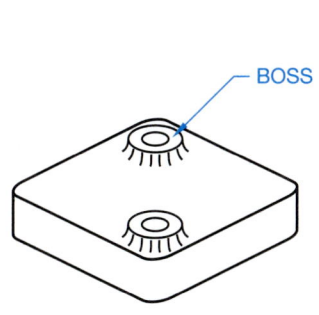
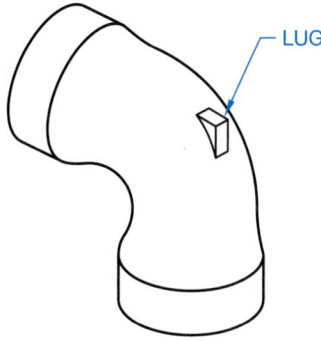
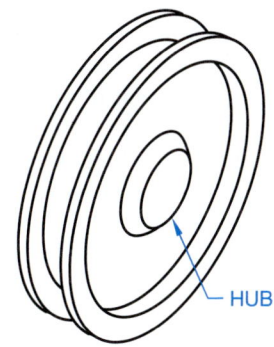

MOUNTING BOSSES FOR
BASE PLATE. ONLY REQUIRES
MACHINING BOSS SURFACES.

HOLDING LUG FOR
SURFACE MACHINING ELBOW

MACHINING HUB
EXTENSION FOR PULLEY

FIGURE 11.91 ■ Cast features added for machining.

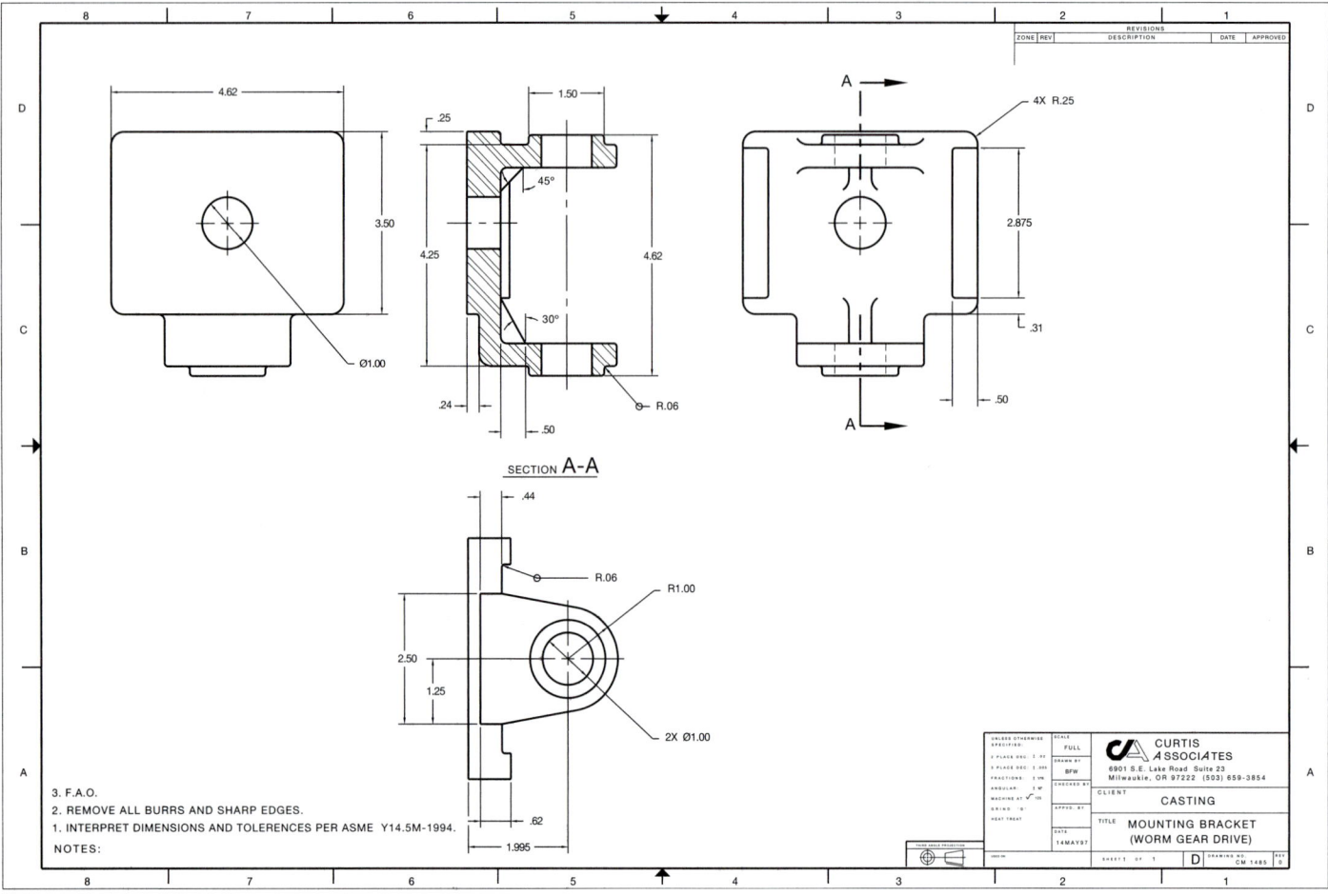

FIGURE 11.92 ■ Casting drawing. Dimension values in this figure are in inches. *Courtesy Curtis Associates.*

The sides of the dies must be angled to facilitate the release of the metal during the forging process. If the vertical sides of the dies do not have draft angle, then the metal sticks in the die. Internal and external draft angles can be specified differently because the internal drafts in some materials have to be greater to help reduce the tendency of the part to stick in the die. While draft angles can change slightly with different materials, the common exterior draft angle recommended is 7°. The internal draft angles for most soft materials is also 7°, but the recommended interior draft for iron and steel is 10°.

The application of fillets and rounds to forging dies is to improve the ejection of the metal from the die. Another reason, similar to that for casting, is increased inside corner strength. One factor that applies to forgings as different from castings is that fillets and rounds that are too small in forging dies can substantially reduce the life of the dies. Recommended fillet and round radius dimensions are shown in Figure 11.96, page 335.

Forging Drawings

A number of methods can be used in the preparation of forging drawings. One technique used in forging drawings that is clearly different from the preparation of casting drawings is the addition of draft angles. Casting drawings usually do not show draft angles. Forging detail drawings usually do show draft.

Before a forging can be made, the dimensions of the stock material to be used for the forging must be determined. Some companies leave this information to the forging shop to determine. Other companies have their engineering department make these calculations. After the stock size is determined, a drawing showing size and shape of the stock material is prepared. The blank material is dimensioned, and the outline of the end product is drawn inside the stock view using phantom lines. (See Figure 11.97, page 336.)

Forgings are made with extra material added to surfaces that must be machined. Forging detail drawings are made to show the desired end product with the outline of the forging shown in phantom lines at areas that require machining. (See Figure 11.98, page 336.) Notice the double line around the perimeter showing draft angle. Another option used by some companies is to make two separate drawings, one a forging drawing and the other a machining drawing. The forging drawing shows all of the views, dimensions, and specifications that relate only to the production of the forging, as shown in Figure 11.99, page 337. The machining drawing gives views, dimensions, and specifications related to the machining processes, as shown in Figure 11.100, page 337.

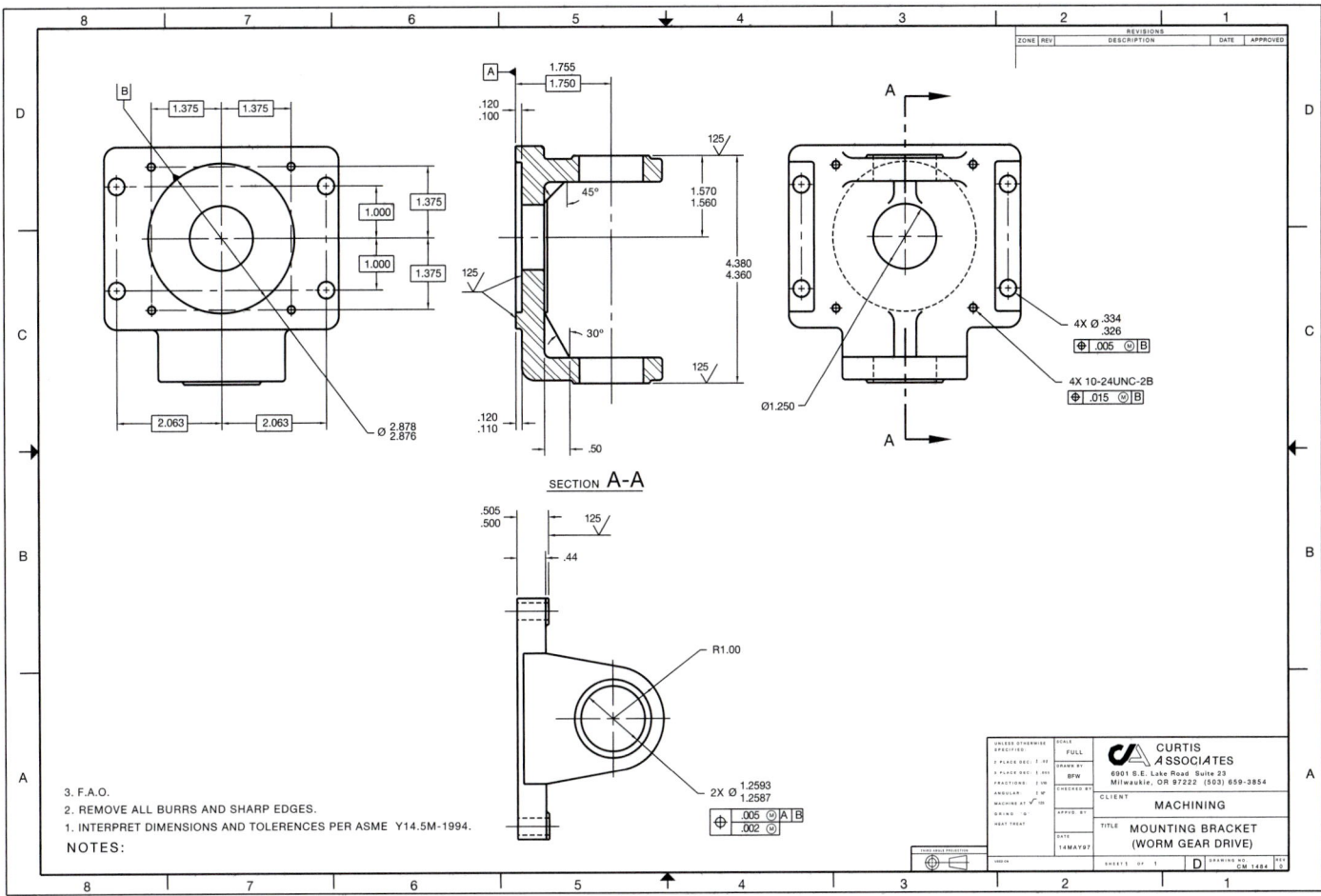

FIGURE 11.93 ■ Machining drawing. Dimension values in this figure are in inches. *Courtesy Curtis Associates.*

DRAWINGS FOR PLASTIC PART MANUFACTURING

As discussed in Chapter 5, there is a wide variety of plastic materials and manufacturing processes for creating plastic parts. Much like castings and forgings of metal parts, plastic manufacturing often requires draft angles to be applied to the design. This depends on the type of plastic and the manufacturing method used. Draft angles allow the finished plastic part to be ejected or removed from the mold without difficulty. Also associated with plastic parts, as with metal castings and forgings, is a parting line. The parting line is the location on the part where it has been separated from the mold. When labeled on the drawing, the abbreviation PL is used. Figure 11.101, page 337, shows some examples of parting lines.

Draft angles can be specified in a note, such as .010 MAX DRAFT ANGLE. In this case, the patternmaker uses this amount of draft as a guide and produces a pattern that has draft angles that are less than .010 on each side where needed for the manufacturing process. These draft angles are generally established within the specified part dimensions rather than added on to the part dimensions. A draft angle tolerance can also be specified as a zone. This can be shown on the drawing as in Figure 11.102, page 337, or it can be specified as a tolerance in a general note or in the title block. A general note might read: ALL DRAFT ANGLES .010, or ALL DRAFT ANGLES 6°. The engineer or the mold maker determines the amount of draft angle.

Another method of specifying draft is the plus draft and minus draft methods. This is abbreviated as +DFT or −DFT and is placed with the feature dimensions on the part. In the +DFT application, the draft is added to the dimension for external dimensions and removed from internal dimensions as shown in Figure 11.103a, page 338. In the −DFT method, the draft is removed from the external dimension and added to the internal dimension as shown in Figure 11.103b, page 338. Both +DFT and −DFT can be combined on a drawing as shown in Figure 11.103C, page 338.

MACHINED SURFACES

As you learned in Chapter 5, *Manufacturing Materials and Processes*, there are a wide variety of machine tools available to produce parts in manufacturing. A review of the machine tools and machining processes discussed in Chapter 5 is recommended.

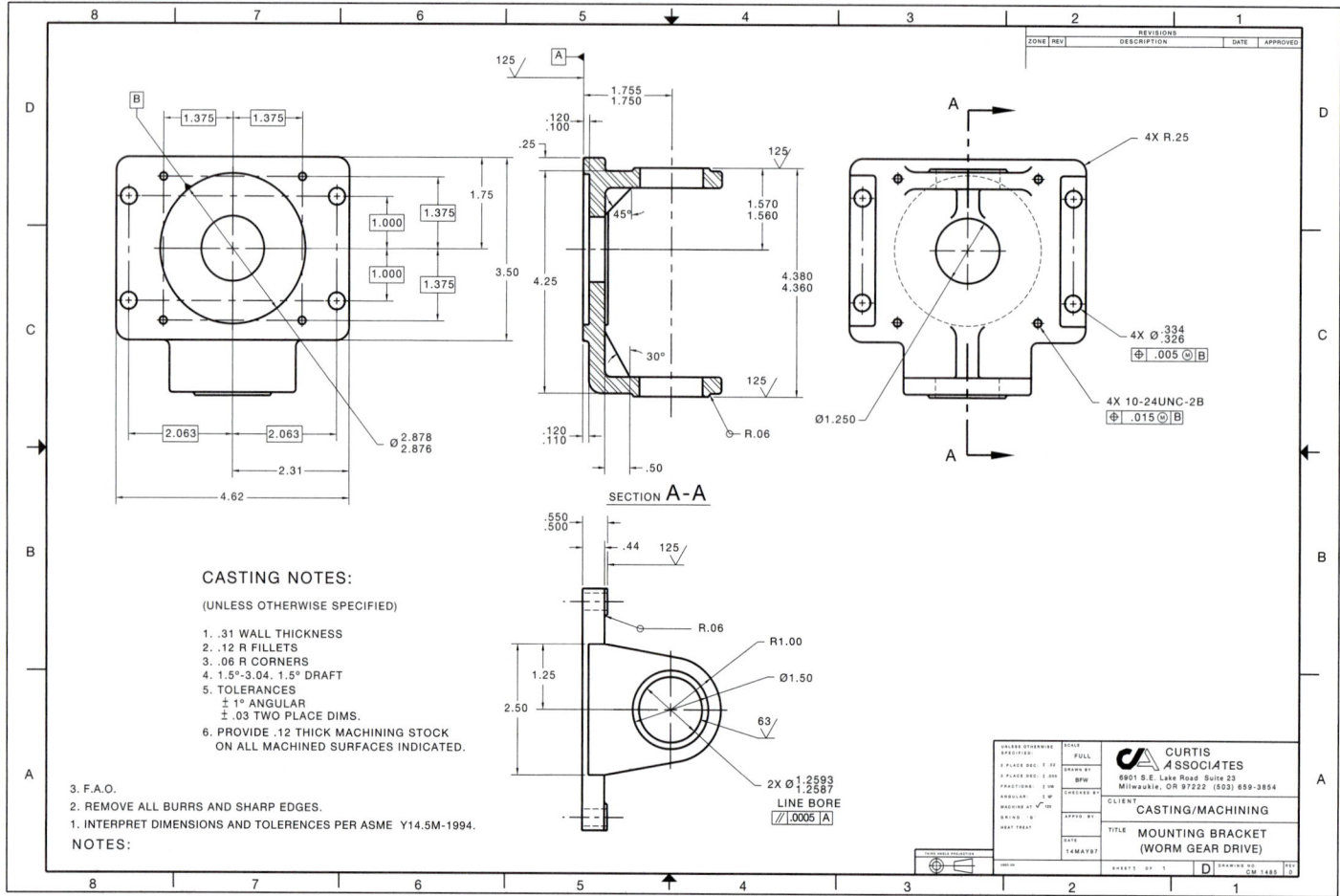

FIGURE 11.94 ■ Drawing with casting and machining information. Dimension values in this figure are in inches. *Courtesy Curtis Associates.*

Surface Finish Definitions

Surface finish or **surface texture** is the intended condition of the material surface after the machining processes have been completed. **Surface texture** includes such characteristics as roughness, waviness, lay, and flaws. Carefully look at Figures 11.104 through 11.111 (pages 338 through 340) as you study the following terminology.

Surface Finish

Surface finish refers to the roughness, waviness, lay, and flaws of a surface. Surface finish is the specified smoothness required on the finished surface of a part that is obtained by machining, grinding, honing, or lapping. The drawing symbol associated with surface finish is shown in Figure 11.104.

Surface Roughness

Surface roughness refers to fine irregularities in the surface finish and is a result of the manufacturing process used. Roughness height is measured in micrometers, μm (millionths of a meter), or in microinches, μin (millionths of an inch). (See Figure 11.108.)

Surface Waviness

Surface waviness is the often widely spaced condition of surface texture usually caused by such factors as machine chatter, vibrations, work deflection, warpage, or heat treatment. Waviness is rated in millimeters or inches. (See Figure 11.108.)

Lay

Lay is the term used to describe the direction or configuration of the predominant surface pattern. The lay symbol is used if considered essential to a particular surface finish. The characteristic lay symbol can be attached to the surface finish symbol, as shown in Figure 11.105.

Surface Finish Symbol

Some of the surfaces of an object are machined to certain specifications. When this is done, a surface finish symbol is placed on the view where the surface or surfaces appear as lines (edge view). (See Figure 11.106.) The finish symbol on a machine drawing alerts the machinist that the surface must be machined to the

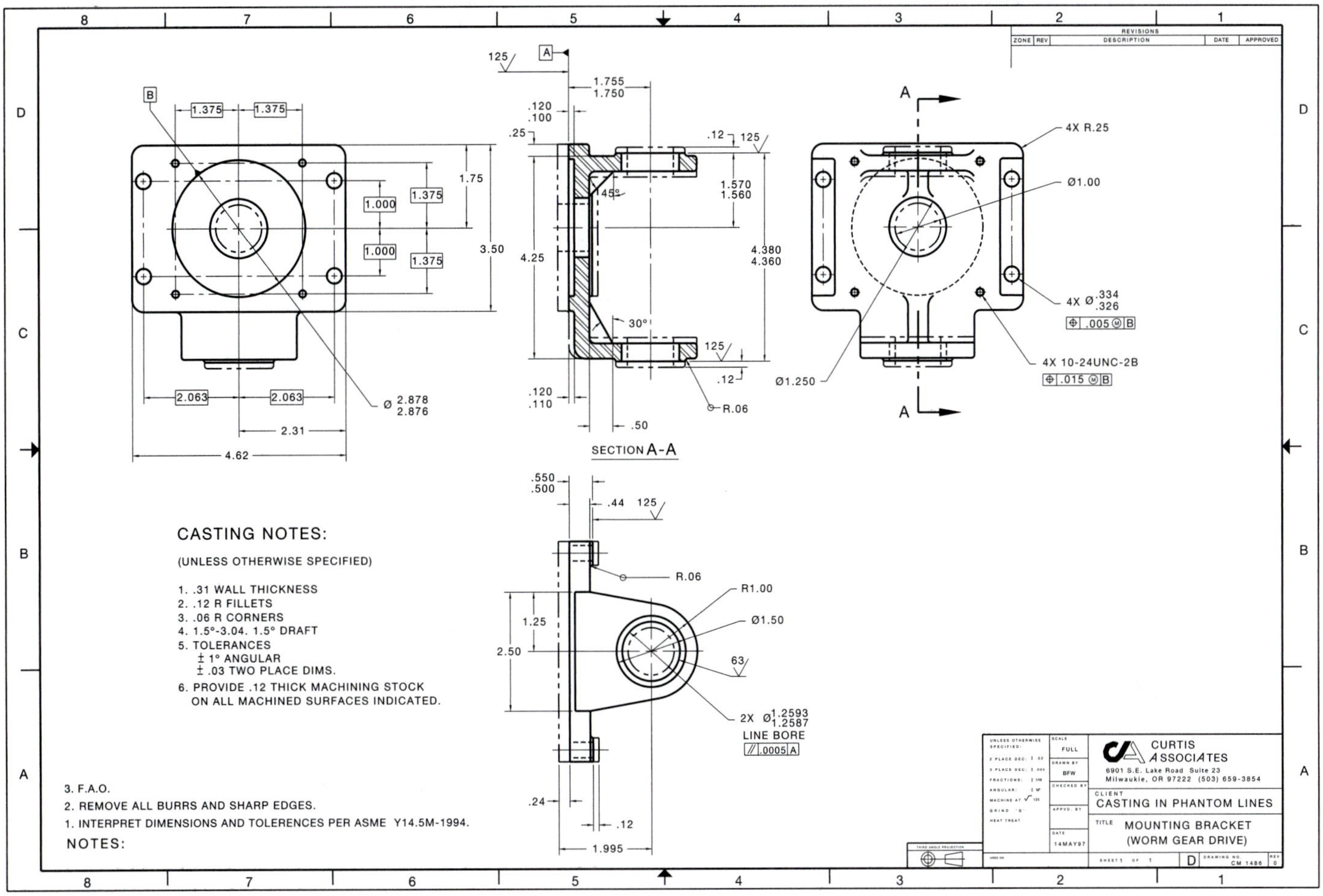

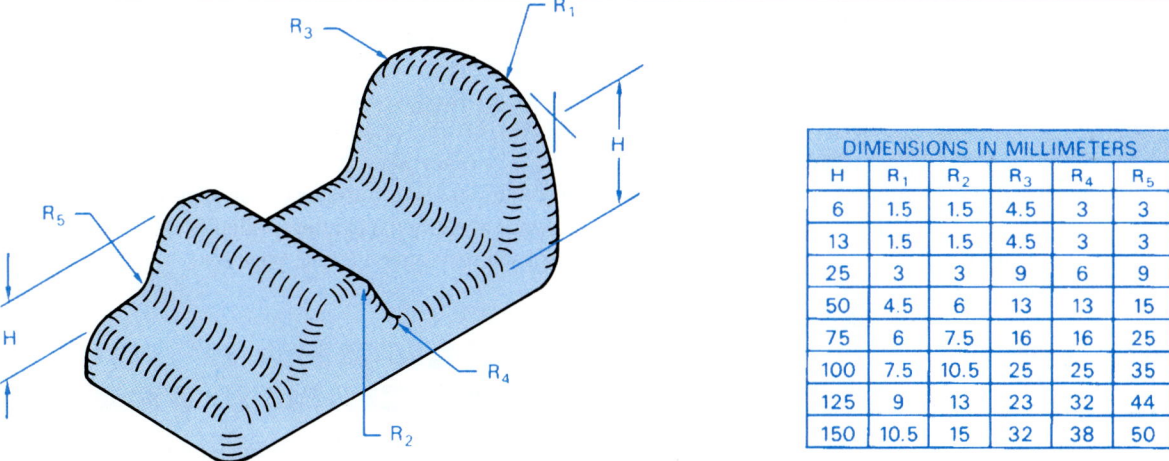

CASTING NOTES:

(UNLESS OTHERWISE SPECIFIED)

1. .31 WALL THICKNESS
2. .12 R FILLETS
3. .06 R CORNERS
4. 1.5°-3.04. 1.5° DRAFT
5. TOLERANCES
   ± 1° ANGULAR
   ± .03 TWO PLACE DIMS.
6. PROVIDE .12 THICK MACHINING STOCK
   ON ALL MACHINED SURFACES INDICATED.

SECTION A-A

3. F.A.O.
2. REMOVE ALL BURRS AND SHARP EDGES.
1. INTERPRET DIMENSIONS AND TOLERENCES PER ASME Y14.5M-1994.

NOTES:

CURTIS
ASSOCIATES
6901 S.E. Lake Road Suite 23
Milwaukie, OR 97222 (503) 659-3854

CLIENT
CASTING IN PHANTOM LINES

TITLE
MOUNTING BRACKET
(WORM GEAR DRIVE)

FIGURE 11.95 ■ Phantom lines used to show machining allowances. Dimension values in this figure are in inches. *Courtesy Curtis Associates.*

DIMENSIONS IN MILLIMETERS					
H	R₁	R₂	R₃	R₄	R₅
6	1.5	1.5	4.5	3	3
13	1.5	1.5	4.5	3	3
25	3	3	9	6	9
50	4.5	6	13	13	15
75	6	7.5	16	16	25
100	7.5	10.5	25	25	35
125	9	13	23	32	44
150	10.5	15	32	38	50

FIGURE 11.96 ■ Recommended fillets and rounds for forgings.

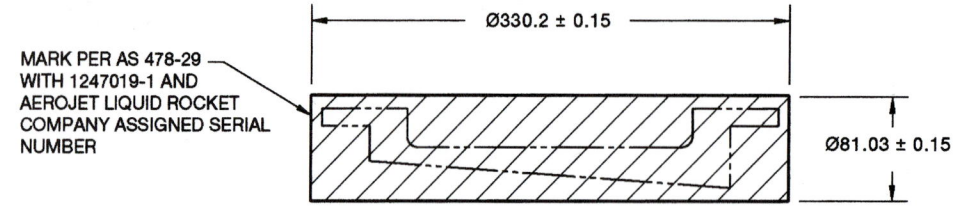

FIGURE 11.97 ■ Blank material for forging process. *Courtesy Aerojet Propulsion Division.*

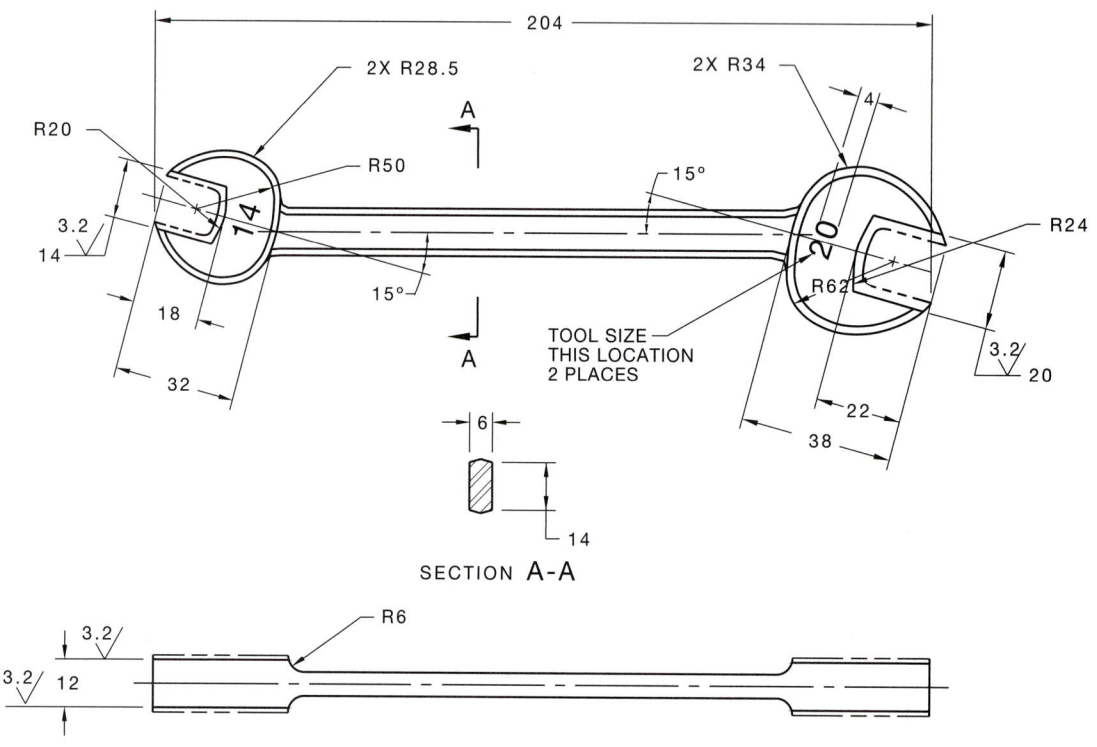

FIGURE 11.98 ■ Phantom lines used to show machining allowance on a forging drawing.

given specification. The finish symbol also tells the pattern maker or die maker that extra material is required in a casting or forging.

The surface finish symbol is properly drawn using a thin line as detailed in Figure 11.107. The numerals or letters associated with the surface finish symbol should be the same height as the lettering used on the drawing dimensions and notes.

Often only the surface roughness height is used with the surface finish symbol. For example, 3.2 means 3.2 micrometers. When other characteristics of a surface texture are specified, they are shown in the format represented in Figure 11.108. For example, roughness width cutoff is a numerical value establishing the maximum width of surface irregularities to be included in the roughness height measurement. Standard roughness width cutoff values for inch specifications are .003, .010, .030, .100, .300, and 1.000; .030 is implied when no specification is given.

Figure 11.109 is a magnified pictorial representation of the characteristics of a surface finish symbol. Figure 11.110 shows some common roughness height values in micrometers and microinches, a description of the resulting surface, and the process by which the surface is produced. When a maximum and mini-

mum limit is specified, the average roughness height must lie within the two limits.

When a standard or general surface finish is specified in the drawing title block or in a general note, a surface finish symbol without roughness height specified is used on all surfaces that are the same as the general specification. When a part is completely finished to a given specification, the general note FINISH ALL OVER, or abbreviation FAO, or FAO 125 μIN, can be used. The placement of surface finish symbols on a drawing can be accomplished a number of ways, as shown in Figure 11.111. Additional elements can be applied to the surface finish symbol, as shown in Figure 11.112, page 340.

DESIGN AND DRAFTING OF MACHINED FEATURES

You should gain a working knowledge of machining processes and the machining capabilities of the company for which they are used. Drawings should be prepared that allow machining

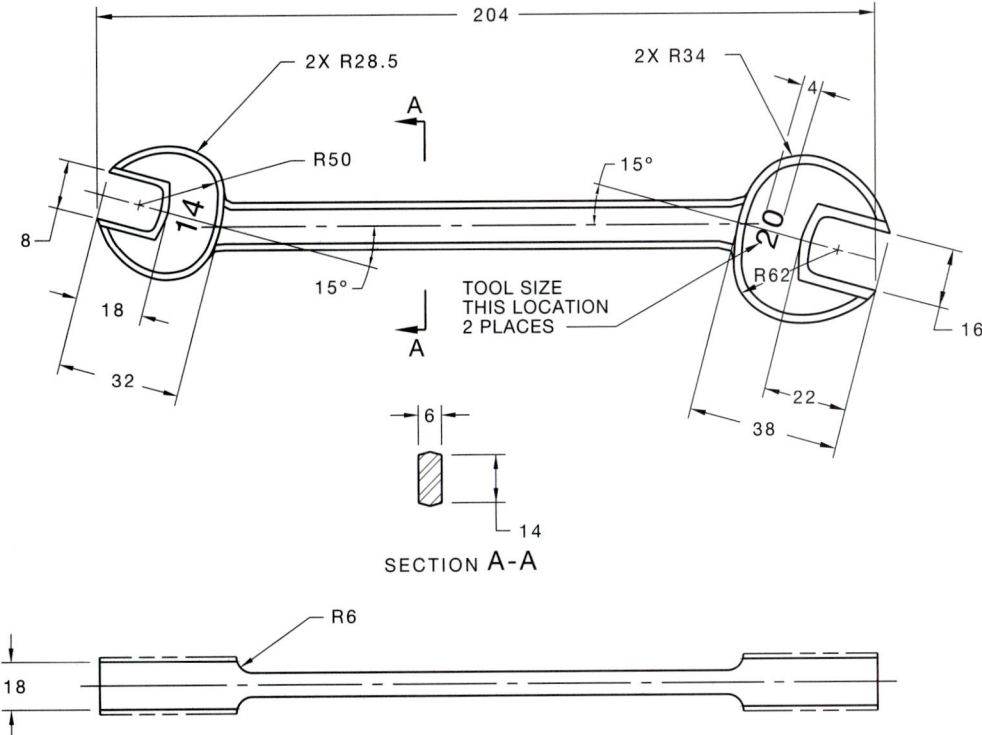

FIGURE 11.99 ■ Forging drawing with forging dimensions only.

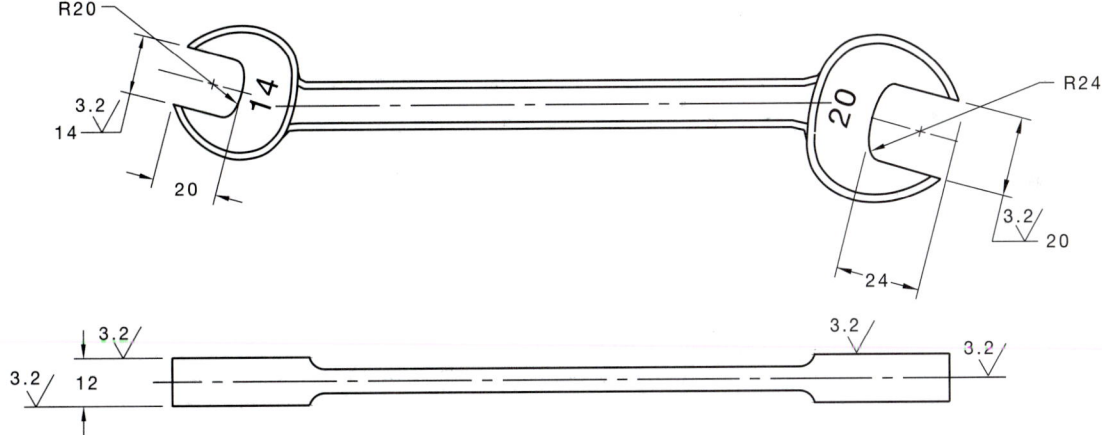

FIGURE 11.100 ■ Machining drawing with machining dimensions only.

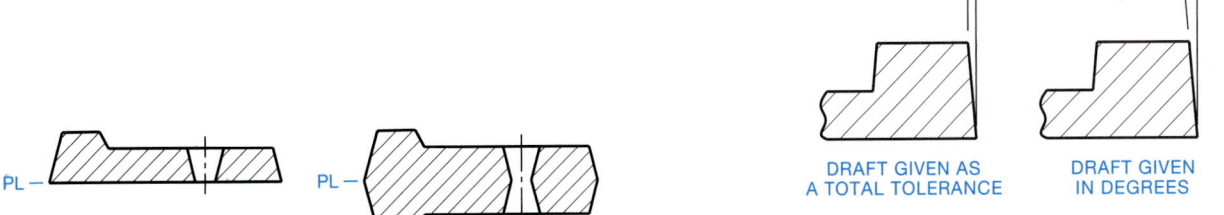

FIGURE 11.101 ■ Parting lines labeled on a drawing.

FIGURE 11.102 ■ Draft angle tolerances shown on a drawing. Dimension values in this figure are in inches.

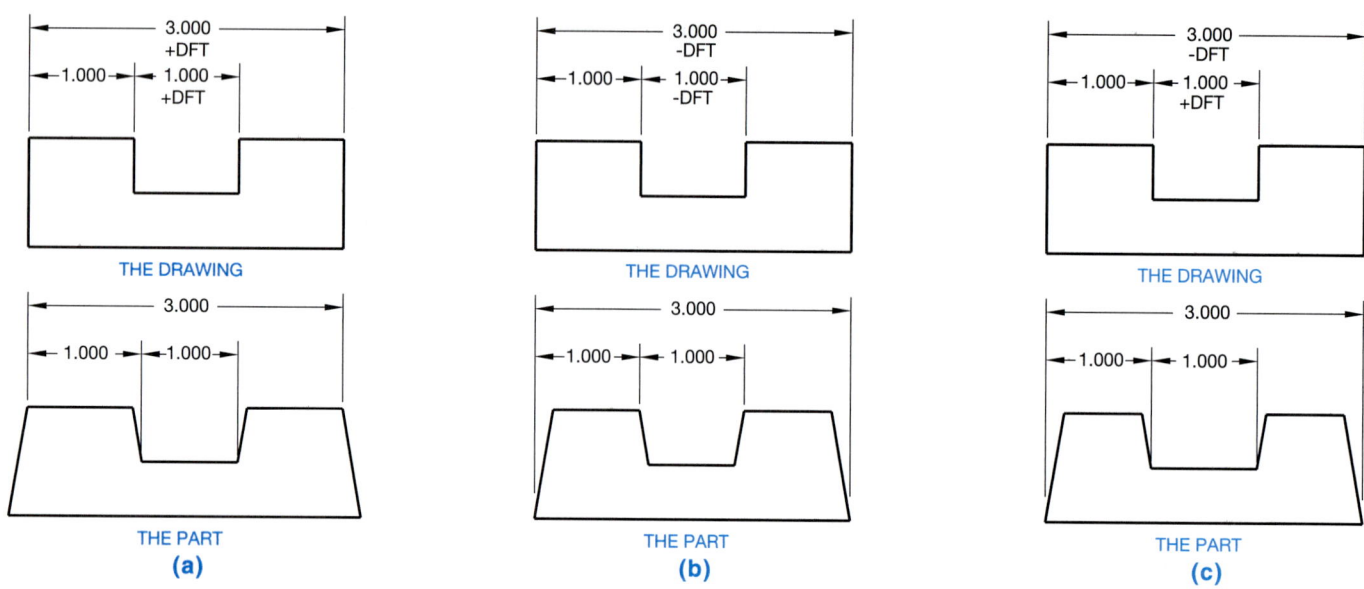

FIGURE 11.103 ■ Draft can be specified on a drawing with (a) the plus draft method, (b) the minus draft method, or (c) a combination of both methods. Dimension values in this figure are in inches.

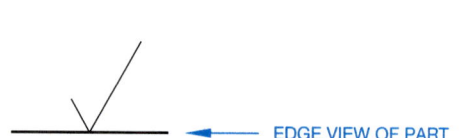

FIGURE 11.104 ■ Surface finish symbol.

within the capabilities of the machinery available. The first consideration should be the least-expensive method to get the desired result. Avoid overmachining. Machining processes are expensive, so do not call for requirements on a drawing that are not necessary for the function of the part. For example, as the roughness height decreases, surface finishes become more expensive to machine. So if a surface roughness of 125 microinches is adequate, then do not use a 32-microinch specification just because you like smooth surfaces. Another example is demonstrated by the difference between the 63- and 32-microinch finish. A 63-microinch finish is a good machine finish that can be performed using sharp tools at high speeds with extra-fine feeds and cuts. The 32-microinch callout requires extremely fine feeds and cuts on a lathe or milling machine and in many cases requires grinding. The 32 finish is more expensive to perform. In a manufacturing environment where cost and competition are critical considerations, a great deal of thought must be given to meeting the functional and appearance requirements of a product at the least possible cost. It generally does not take very long for an entry-level drafter to pick up these design considerations by communicating with engineering and manufacturing department personnel. Many drafters become designers, checkers, or engineers within a company by learning the product and being able to implement designs based on manufacturing capabilities.

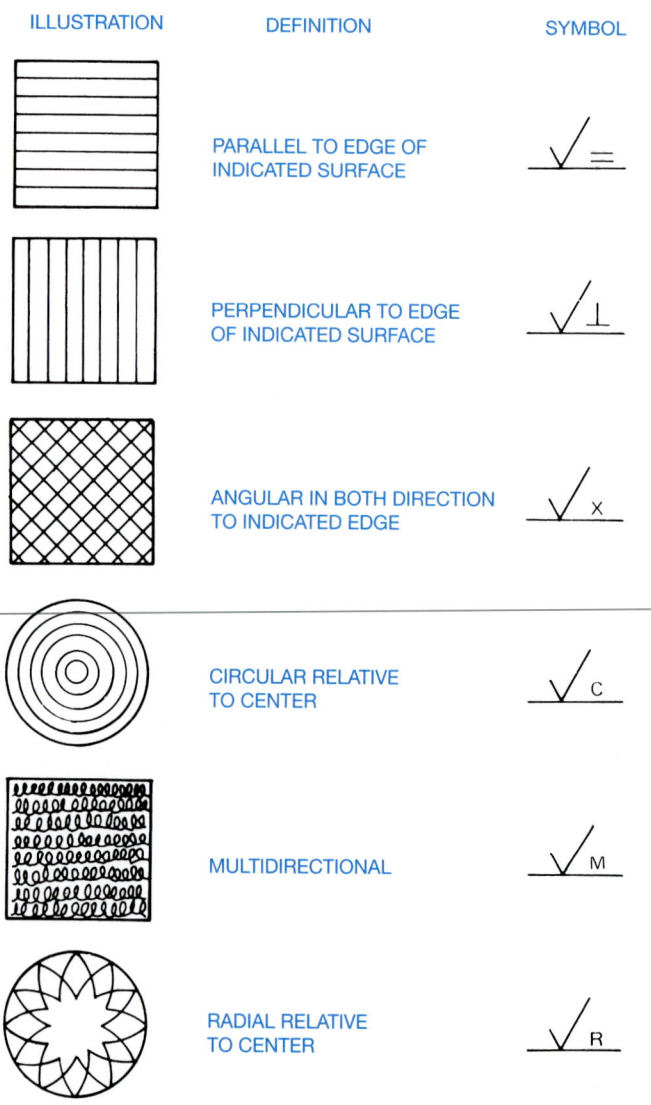

FIGURE 11.105 ■ Characteristic lay added to the surface finish symbol.

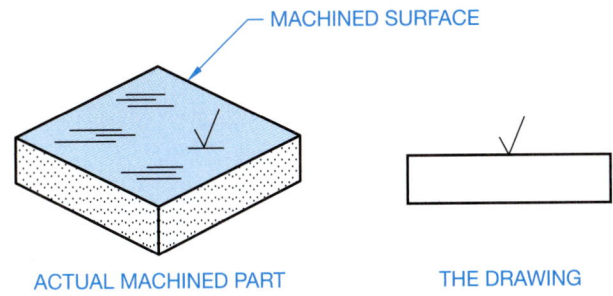

FIGURE 11.106 ■ Standard surface finish symbol placed on the edge view. The symbol should always be placed horizontally when unidirectional dimensioning is used.

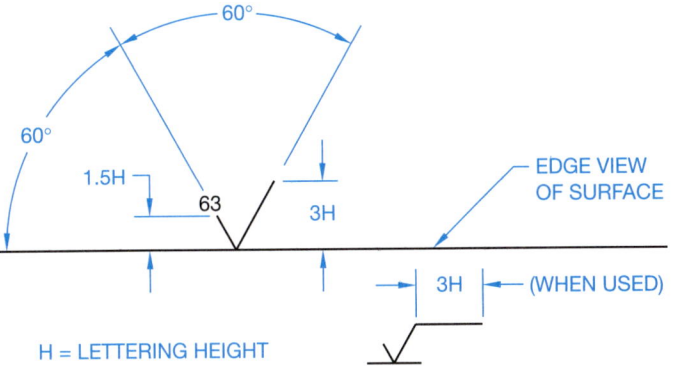

FIGURE 11.107 ■ Properly drawn surface finish symbol.

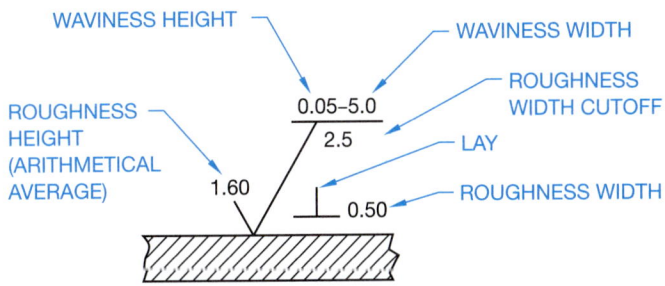

FIGURE 11.108 ■ Elements of a complete surface finish symbol.

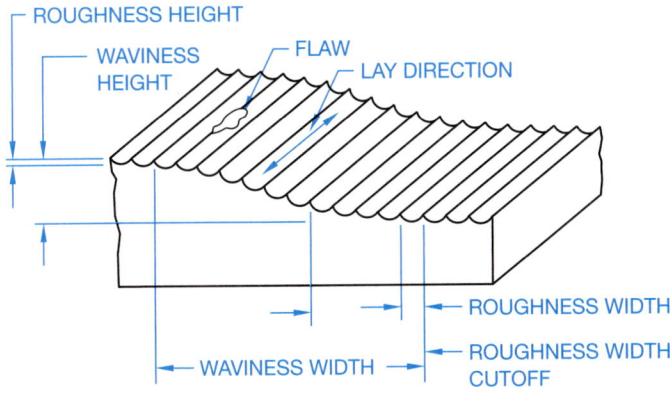

FIGURE 11.109 ■ Surface finish characteristics magnified.

TOOL DESIGN

In most production machining operations, special tools are required to either hold the workpiece or guide the machine tool. Tool design involves knowledge of kinematics (study of mechanisms) covered in Chapter 19 of this text, machining operations, machine tool function, material handling, and material characteristics. **Tool design** is also known as **jig and fixture design**. In mass-production industries, jigs and fixtures are essential to ensure each part is produced quickly and accurately within the dimensional specifications. These tools are used to hold the workpiece so machining operations are performed in the required positions. Application examples are shown in Figure 11.113. Jigs are either fixed or moving devices used to hold the workpiece in position and guide the cutting tool. Fixtures do not guide the cutting tool, but are used in a fixed position to hold the workpiece. Fixtures are often used in the inspection of parts to ensure that the part is held in the same position each time a dimensional or other type of inspection is made.

Jig and fixture drawings are prepared as an assembly drawing where all of the components of the tool are shown as if they are assembled and ready for use, as shown in Figure 11.114. Assembly drawings are discussed in detail in Chapter 18. Components of a jig or fixture often include such items as fast-acting clamps, spring-loaded positioners, clamp straps, quick-release locating pins, handles, knobs, and screw clamps, as shown in Figure 11.115, page 342. Normally the part or workpiece is drawn in position using phantom lines, in a color such as red, or a combination of phantom lines and color.

*The Tool Design Process**

Hammers, pliers, screwdrivers, wrenches, and other related items are tools, but they are not the kind of tools that are created by **tool designers**. **Tools**, as referred to in this discussion, are specially designed and built manufacturing aids, normally in production, that are used to assist operators in the manufacture of specific parts. There are many different kinds of tools that fit this definition. These include machining fixtures, welding fixtures, drill fixtures, drill jigs, inspection fixtures, progressive dies, injection molds, and many others. Tools can be very small and simple, to very large and complicated. Some tools are mechanisms that resemble machinery, but there is a clear difference between tools and machines. Tools are dedicated to a specific part, family of parts, or product, while machines are for general usage across many parts and products. This differentiation is made more for accounting and tax purposes than it is for technical reasons. Machines are capital investments, while tools are expense items related to a particular part or product. In many cases, the tool is so complex that from a technical perspective, it is a machine, but from an accounting perspective, it is a tool. The following are some very simplified definitions of the typical types of tools used in manufacturing.

■ **Drill jigs** are used in drilling operations using hand drills, drill presses, or radial drills, rather than milling machines. The drill jig registers the part relative to the critical datums.

*Courtesy of Martin Soll.

MICROMETERS	ROUGHNESS HEIGHT RATING MICRO INCHES	SURFACE DESCRIPTION	PROCESS
25	1000	VERY ROUGH	SAW AND TORCH CUTTING, FORGING, OR SAND CASTING.
12.5	500	ROUGH MACHINING	HEAVY CUTS AND COARSE FEEDS IN TURNING, MILLING, AND BORING.
6.3	250	COARSE	VERY COARSE SURFACE GRIND, RAPID FEEDS IN TURNING, PLANING, MILLING, BORING, AND FILING.
3.2	125	MEDIUM	MACHINING OPERATIONS WITH SHARP TOOLS, HIGH SPEEDS, FINE FEEDS, AND LIGHT CUTS.
1.6	63	GOOD MACHINE FINISH	SHARP TOOLS, HIGH SPEEDS, EXTRA-FINE FEEDS AND CUTS.
0.80 / 0.40	32 / 16	HIGH-GRADE MACHINE FINISH	EXTREMELY FINE FEEDS AND CUTS ON LATHE, MILL, AND SHAPERS REQUIRED. EASILY PRODUCED BY CENTERLESS, CYLINDRICAL, AND SURFACE GRINDING.
0.20	8	VERY FINE MACHINE FINISH	FINE HONING AND LAPPING OF SURFACE.
0.050 0.100	2 – 4	EXTREMELY SMOOTH MACHINE FINISH	EXTRA-FINE HONING AND LAPPING OF SURFACE. MIRROR FINISH.
0.025	1	SUPER FINISH	DIAMOND ABRASIVES.

FIGURE 11.110 ■ Common roughness height values with a surface description and associated machining process.

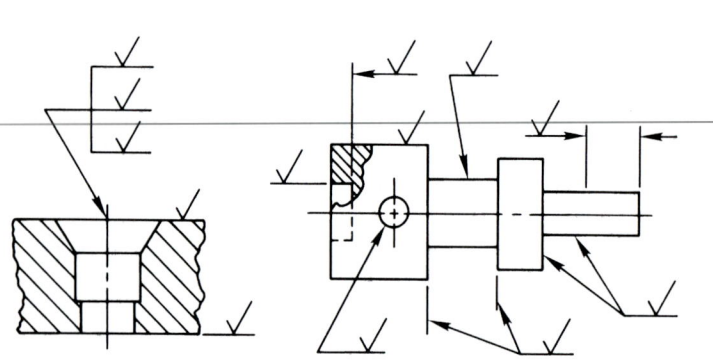

FIGURE 11.111 ■ Proper placement of surface finish symbols.

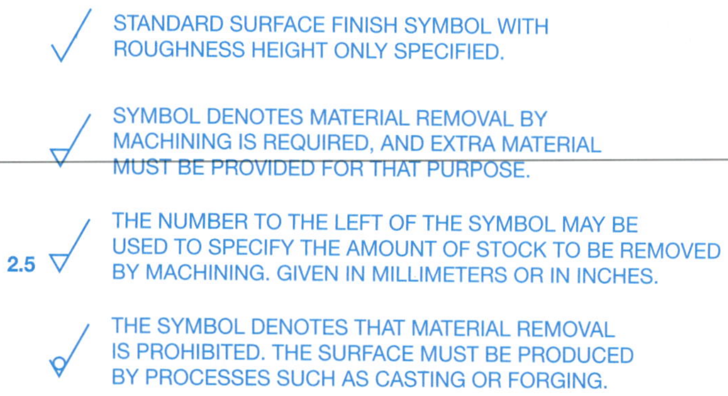

STANDARD SURFACE FINISH SYMBOL WITH ROUGHNESS HEIGHT ONLY SPECIFIED.

SYMBOL DENOTES MATERIAL REMOVAL BY MACHINING IS REQUIRED, AND EXTRA MATERIAL MUST BE PROVIDED FOR THAT PURPOSE.

2.5 THE NUMBER TO THE LEFT OF THE SYMBOL MAY BE USED TO SPECIFY THE AMOUNT OF STOCK TO BE REMOVED BY MACHINING. GIVEN IN MILLIMETERS OR IN INCHES.

THE SYMBOL DENOTES THAT MATERIAL REMOVAL IS PROHIBITED. THE SURFACE MUST BE PRODUCED BY PROCESSES SUCH AS CASTING OR FORGING.

FIGURE 11.112 ■ Material removal elements added to the surface finish symbol.

Datums are important points, axes, surfaces, or planes from which features are established and dimensioned Datums are discussed in detail earlier in this chapter and in Chapter 14. The operator then drills through a drill bushing or bushings in the jig, to locate the hole or holes in the part.

■ **Drill fixtures** are sometimes referred to as **holding fixtures**. Drill fixtures are used for drilling operations using milling machines, either manual or computer numerical control (CNC). The drill fixture registers the part relative to the critical datums, but does not have drill bushings. The machine axes are used to locate the hole or holes.

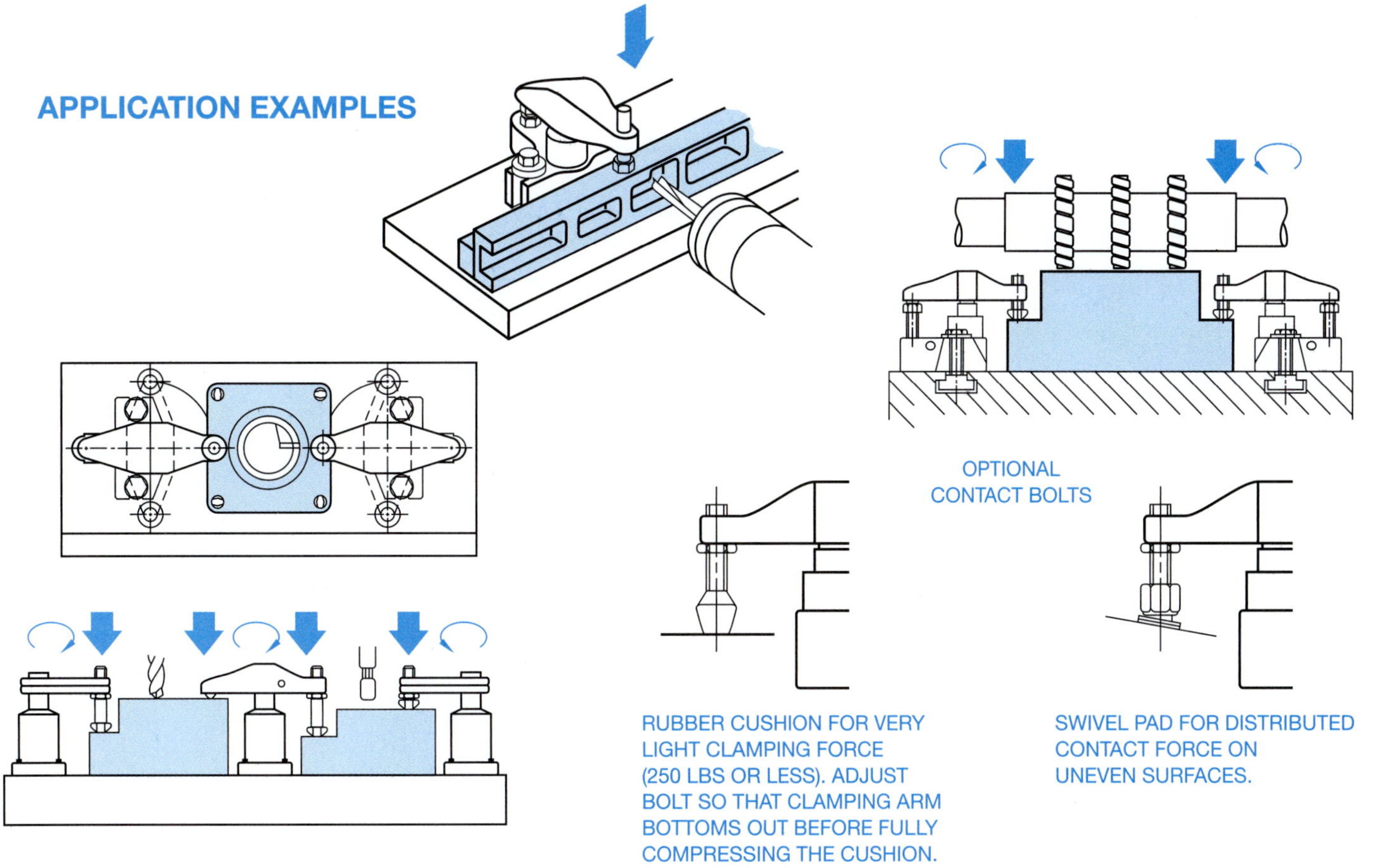

APPLICATION EXAMPLES

OPTIONAL
CONTACT BOLTS

RUBBER CUSHION FOR VERY
LIGHT CLAMPING FORCE
(250 LBS OR LESS). ADJUST
BOLT SO THAT CLAMPING ARM
BOTTOMS OUT BEFORE FULLY
COMPRESSING THE CUSHION.

SWIVEL PAD FOR DISTRIBUTED
CONTACT FORCE ON
UNEVEN SURFACES.

FIGURE 11.113 ■ Fixture application examples. *Courtesy Carr Lane Manufacturing Company.*

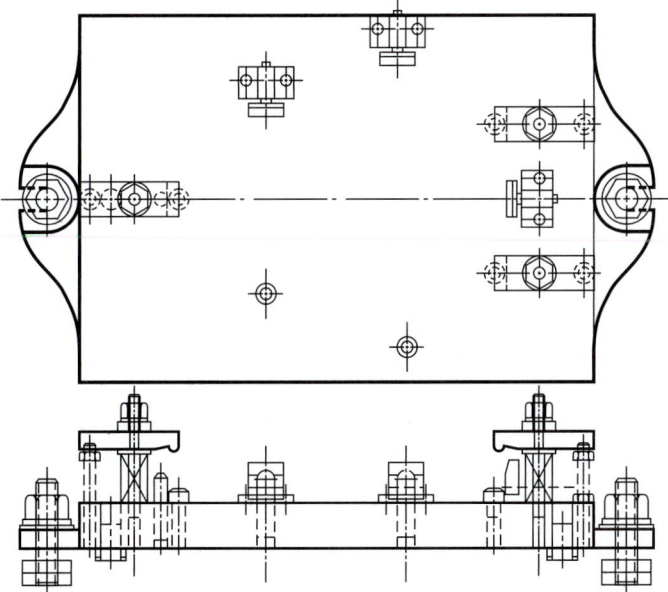

FIGURE 11.114 ■ Fixture assembly drawing. *Courtesy Carr Lane Manufacturing Company.*

Drill fixtures must resist the force of the drilling operation. Typically, drilling forces are relatively low and are only in one direction.

■ **Machining fixtures** are used for machining operations using milling machines that are either manual or CNC. The machining fixture registers the part relative to the critical datums. The machine axes are used to locate the feature or features to be machined. Machining fixtures must resist the force of the machining operation; typically, machining forces are relatively high and can be in any direction.

■ **Welding fixtures** are more accurately termed **welding jigs**, but the names have become synonymous in the industry. Welding fixtures are used to hold two or more pieces in the proper position and orientation so the pieces can be welded together.

■ **Inspection fixtures** are very common in the casting industry. They are used to hold a part, registering it on the critical datums, while an inspector checks critical feature sizes and/or locations.

■ **Progressive dies** are the tooling used in punch press operations. Continuous stock of relatively thin material is fed into the punch press from coils. As the punch press cycles, the material is cut to appropriate shape, holes are punched, or other operations are performed to make a finished part. Cookie cutters, scissors, and paper hole punches are comparable to progressive dies.

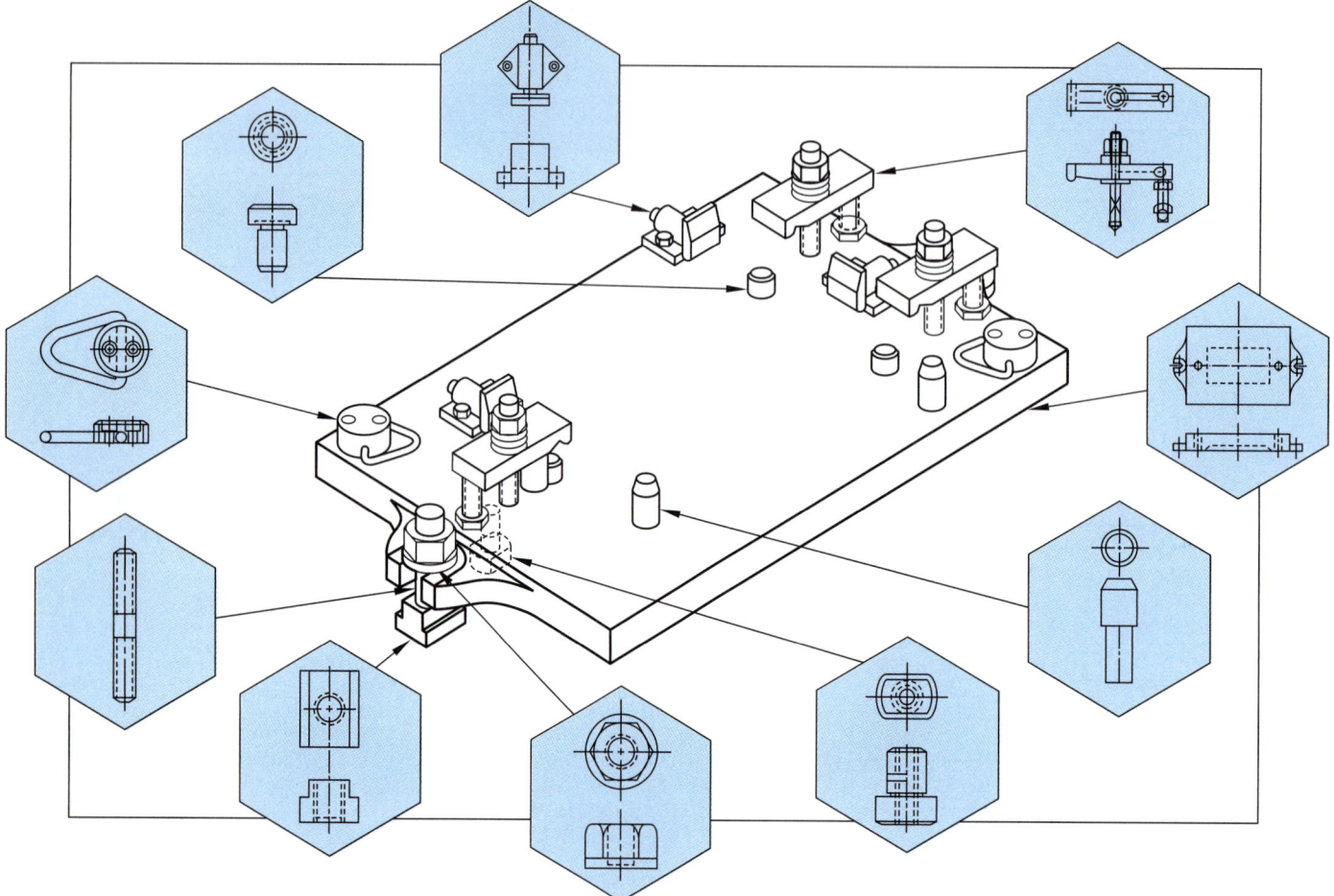

FIGURE 11.115 ■ Fixture components. *Courtesy Carr Lane Manufacturing Company.*

CADD APPLICATIONS

JIG AND FIXTURE DESIGN

Computer-aided design and drafting (CADD) jig and fixture design programs make it possible for the designer or engineer to electronically construct the tooling for a specific application directly on the computer terminal. As with other CADD programs, tooling component libraries, consisting of fully detailed and accurate jig and fixture components, are available. The speed of tooling design sharply increases as compared to manual methods. Items such as clamps, locator pins, rests, or fixture bases can be retrieved from the library and inserted into your design quickly by using the specific functions provided by the software.

Figure 11.116 shows a very simple drill jig tool. This drill jig is designed and manufactured to assist a shop employee in performing a specific task on a production part. The part is very simple, and the employee could measure and drill the hole without any tool; but for production work, the tool makes the location of the hole much faster, and the accuracy of the hole location is less dependent on the employee's abilities. This type of drill jig is referred to as a pickoff jig; *pickoff* because it sits on the part, rather than the part being put into the fixture, and *jig* because it actually locates a feature, which is the .25-in. diameter hole in this example.

Now that you have an idea of what makes up a tool, the following gives you the three basic elements of tool design:

1. Tool design is the art of visualizing how shop personnel will accomplish a specific task.

2. Conceptualizing hardware to assist in the accomplishment of that task.

3. Creating drawings so the hardware can be manufactured.

Visualizing and *conceptualizing* are not to be confused with inventing. Tool designers are not inventors. In fact, a good tool designer may not consider him- or herself to be creative. While

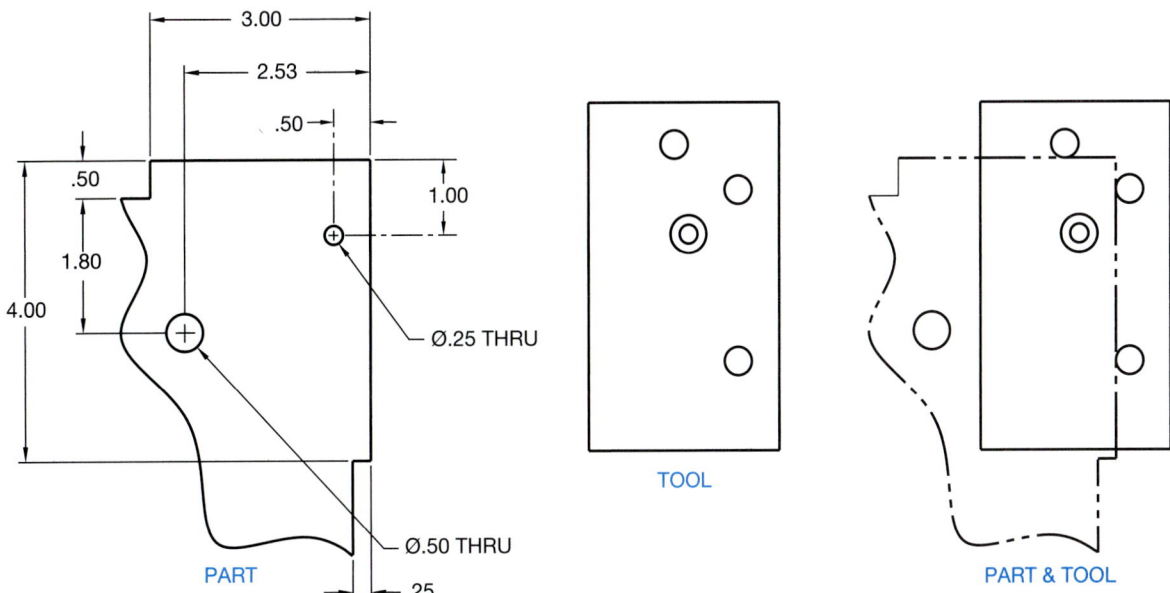

FIGURE 11.116 ■ A simple example of a PART and a TOOL. The task of the employee is to drill a .25-in. diameter hole in a specific location. It is very efficient for the employee to use the TOOL to locate the hole. He or she places the TOOL on top of the PART, holds it with one hand, and drills the hole through the PART, using the hole in the TOOL to locate the hole in the PART. Dimension values in this figure are in inches.

many tool designers are creative, tool designers use existing products and items whenever possible. Many times, an existing design can be modified to satisfy a new requirement, and good tool designers often save their employer time and money by using existing designs. A tool designer *must* be familiar with shop practices, and *must* be able to visualize shop personnel accomplishing specific tasks within the shop environment. This visualization leads to a concept of tools that the shop personnel use to assist in the accomplishment of the task. The tool designer *must* also be very good at print reading.

The part print contains all the information the tool designer needs. The tool designer must be able to find the important information. Looking back to the example in Figure 11.116, you are shown only a small corner of the part. Even this small corner has much more information than is required for this drill jig. Print reading is the first step in the tool design process. The information necessary to design the drill jig must be found on the part print. Some dimensions are directly related to the drill jig, some are incidental to the tool design, and some are totally unrelated to this drill jig. Looking at Figure 11.116, it is relatively easy to *sort out the applicable information*. Dimensions defining the feature size and location are directly related to the required tool. Dimensions that define the size and locations of other features that impact the tool are of incidental interest. Dimensions that define the size and locations of features not impacting the design of the tool are not relevant to the tool designer. More complicated part prints require in-depth study to find the relevant information.

Tool designers often receive their assignments from a manufacturing engineer (ME). The ME defines the specific task and how it is to be accomplished. Consider the example in Figure 11.116 again. The tool designer receives a tooling design request (TDR) from the ME. The TDR has an engineering drawing of the part attached, and might contain the following information:

PART NAME:---------PART

PART NO:-------------XXXX-XX

OPERATION:---------DRILL .25" DIA HOLE

MACHINE:-----------USE HAND DRILL

FIXTURE:------------PICKOFF JIG LOCATING
THE HOLE .50" AND 1.00"
FROM THE EDGES OF THE PLATE

The decision to use a pickoff jig and hand drill rather than a radial drill or even a CNC machine is made prior to the tool designer's involvement in the project. If the ME had determined that the best way to accomplish this task is to put this part on a CNC machine to drill the hole, then the TDR would be written accordingly. The tool designer would design a drill fixture that would be mounted on the CNC machine table. The fixture would hold and clamp the PART. The PART would be located with accurate registration to the two critical edges that define the location of the .25-in. diameter hole.

The Tool Designer's Tools

A tool designer uses manual drafting or CADD practices to design a fixture. He or she must capture the concept on paper or computer screen in order for the fixture to be built. The tool designer must also be familiar with standard tooling components that are available from numerous manufacturers. Various components, such as rest pads, clamps, pins, and drill bushings, to name a few, are available in a wide variety of sizes and shapes. Remember, tool designers do not create something that already exists, but they use standard components whenever possible.

Tools, in general, must possess some of the following qualities:

■ Reliability.

■ Repeatability.

■ Ease of use.

■ Ease of manufacture.

■ Ease of maintenance and repair.

Figure 11.117 shows the design that would go to the tool room for a machinist to use to build the pickoff drill jig previously discussed and shown in Figure 11.116. The drawing shown is a combination assembly drawing and detail drawing. More complex fixtures require the fixture assembly to be drawn on one sheet, and the component details to be drawn on separate sheets. Detail and assembly drawings are covered in Chapter 18, Working Drawings.

Notice that the PART in Figure 11.116 is dimensioned using conventional dimension and extension lines, and the DRILL JIG drawing in Figure 11.117 is dimensioned using arrowless dimensioning. Either type of dimensioning practice can be used on both drawings.

The PLATE (Figure 11.117, item 1) is the only piece that must be made in the tool room. The *tool room* is the shop where tools are manufactured. The DOWEL PIN (Figure 11.117, item 2) and the DRILL BUSHING (Figure 11.117, item 3) are purchased components. The following evaluates if this jig would meet the list of quality requirements for tools:

■ *Reliability*—Yes, this jig is simple, and very little could break or cause problems.

■ *Repeatability*—Yes, again the simplicity makes repeatability inherent. The drill jig registers directly on the datum surfaces from which the hole location is defined.

■ *Ease of use*—Yes, the jig offers the operator a simple means to locate the hole.

■ *Ease of manufacture*—Yes, with only one manufactured part, and four purchased parts, it will be very easy to make.

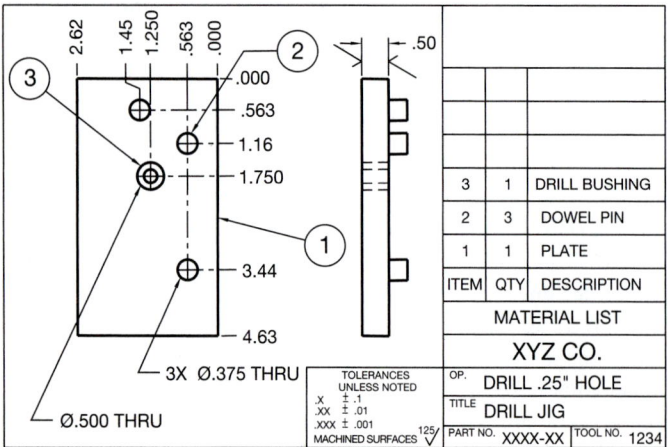

FIGURE 11.117 ■ A simplified example of a tool design drawing.

■ *Ease of maintenance and repair*—Yes, should something break or wear out, the component parts can be easily replaced.

In summary, you have learned about tools, tool design, and tool designers. Tools are shop aids, sometimes simple, and sometimes very complex. Tool design is the process of turning a concept of a tool into drawings so that the fixture can be manufactured. Tool designers are the people who imagine the concepts and turn them into drawings.

AN INTRODUCTION TO ISO 9000

The ISO 9000 Quality Systems Standard was established to encourage the development of standards, testing, quality control, and the certification of companies, organizations, and institutions where these practices are implemented. ISO stands for the International Organization for Standardization. The ISO is an international organization that is made up of nearly 100 countries. The United States is a member country that is represented by the American National Standards Institute (ANSI). ISO 9000 can certify companies, organizations, and institutions when their engineering, drafting standards, manufacturing, and quality control meet the requirements of a model quality management system established by the ISO 9000 organization. A certification is obtained by passing an inspection by an independent representative of the ISO. ISO 9000 certification is also referred to as registration. There are a number of reasons why a company may want to become ISO 9000 certified. The reasons can include:

■ In order to do business with customers that require ISO 9000 certification. This includes agencies of the U.S. government.

■ To compete in a competitive market that requires strict attention to quality control.

■ To lower cost and improve profits.

■ To improve and maintain customer confidence.

■ A desire to improve product quality and place emphasis on customer relations.

■ To manufacture and sell products in the European Union markets where this certification is required.

■ To require that subcontractors meet the same expectations for quality control.

■ The ISO 9000 certification also satisfies requirements established by other local and national organizations.

■ To improve company standards.

The ISO 9000 Quality Systems Standard is made up of a series of five international standards that provide leadership in the development and completion of a successful quality management system. These five standards are briefly described as follows:

ISO 9000-1 This standard provides direction and definitions that describe what each standard contains and assists companies in the selection and use of the appropriate ISO standard for the desired results.

ISO 9001 This is the model that can be used by any organization for designing, documenting, and implementing ISO standards. The model takes a product through the process of design, drafting, manufacturing, quality control, installation, and service.

ISO 9002 This standard is the same as ISO 9001, except that it does not contain the requirement of documenting the design and development process.

ISO 9003 This standard is for companies or organizations that only need to demonstrate through inspection and testing methods that they are providing the desired product or service.

ISO 9004-1 This is a set of guidelines that can be used to assist organizations in the development and implementation of a quality management system.

In addition to the ISO 9000 series are the QS 9000 and AS 9000 standards. The QS 9000 is the Automotive Requirements, and the AS 9000 is the Aerospace Standard. These standards contain all of ISO 9001 plus requirements beyond ISO 9001. These standards were developed by the U.S. automotive and aerospace industries for specific applications and needs that customize the standard for their industries.

PROFESSIONAL PERSPECTIVE

The proper placement and use of dimensions is one of the most difficult aspects of drafting. It requires careful thought and planning. First, you must determine the type of dimension that fits the application, then place it on the drawing. While the CADD system makes the actual placement of dimensions quick and easy, it does not make the planning process any easier. Making preliminary sketches is very important before beginning a drawing. The preliminary sketch allows you to select and place the views and then place the dimensions. When you were drawing only the views the space requirements were not as critical, but a drawing without dimensions is unrealistic.

The problem with dimensions is they take up a lot of space. One of the big issues an entry-level drafter faces is determining the space requirements for dimensions. In many cases these requirements are underestimated. As a rule of thumb, if you think the drawing is crowded on a particular size sheet, then play it safe and use the next larger size. Most companies want the drawing information spread out and easy to read, but be careful, because some companies want the drawing crowded with as much information as you can get on a sheet. This text advocates a clean and easy-to-read drawing that is not crowded. Out on the job, however, you must do what is required by your employer. If you are drawing an uncrowded drawing, you should consider leaving about one-quarter of the drawing space clear of view and dimensional information. Usually this space is above and to the left of the title block. This space provides adequate room for general notes and engineering changes. Engineering changes are covered in Chapter 18.

Follow the dimensioning rules, guidelines, and examples discussed in this chapter and use proper dimensioning standards. Try hard to avoid breaking dimensioning standards. Following are some of the pitfalls to watch out for as a beginning drafter:

■ Do not crowd dimensions. Keep your dimension line spacing equal and far enough apart to clearly separate the dimension numerals.

■ Do not dimension to hidden features. This also means do not dimension to the centerline of a hole in the hidden view. Always dimension to the view where the feature appears as an object line.

■ Dimension to the view that shows the most characteristic shape of an object or feature. For example, dimension shapes where you see the contour, dimension holes to the circular view, and dimension cylindrical shapes in the rectangular view.

■ Do not stack adjacent dimension numerals; stagger them so they are easier to read.

■ Group dimensions as much as you can. It is better to keep dimensions concentrated to one location or one side of a feature rather than spread around the drawing. This makes the drawing easier to read.

■ If you are using manual drafting, use a template to draw symbols. If you use computer-aided drafting, make a standard symbol library or use software with symbols available. Symbols speed up the drafting process, and they clearly identify the feature. For example, if you draw a single view of a cylinder, the diameter ∅ dimension is identified in the rectangular view and drawing the circular view may not be necessary.

■ Look carefully at the figures in this text and use them as examples as you prepare your drawings.

Try to put yourself in the place of the person who has to read and interpret your drawing. Make the drawing as easy to understand as possible. Keep the drawing as uncluttered and as simple as possible, yet still complete. Figure 11.118 shows an industry drawing of a complex part. Notice how the drafter carefully selected views and dimension placement to make the drawing as easy to read as possible.

Now, go back to the beginning of this chapter and read *The Engineering Design Application* segment again. The steps used to lay out a dimensioned multiview drawing were presented to give you a general idea about the process a drafter goes through when converting an engineering sketch into a formal drawing. Read the steps again and look at Figures 11.1 through 11.6. This review, along with spending some time looking at the complex actual industry drawing in Figure 11.118, will help you pull together the dimensioning basics that you learned throughout this chapter.

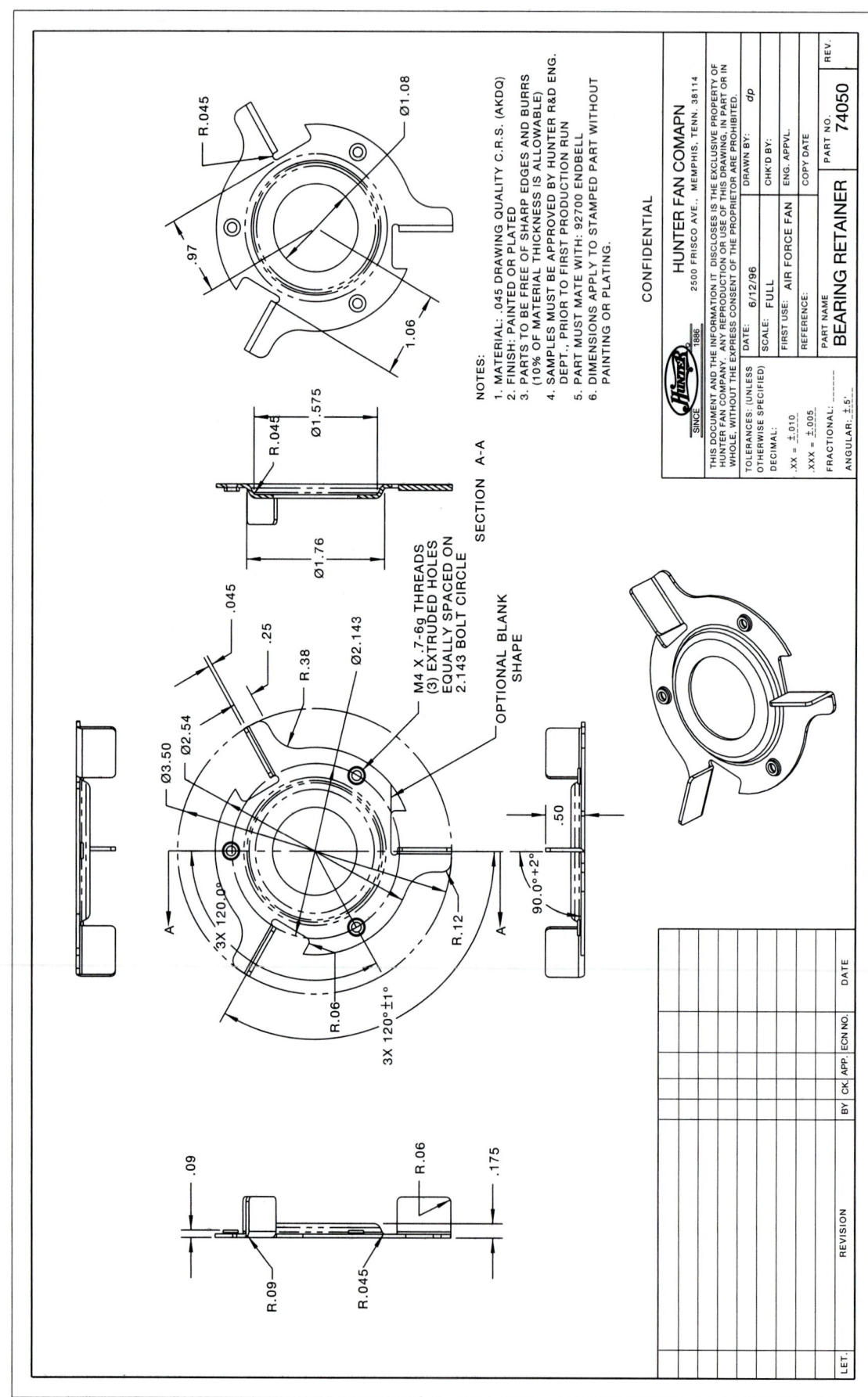

FIGURE 11.118 ■ A complex drawing with dimensions placed in the proper locations to aid in reading the drawing. Notice the pictorial drawing in the bottom center. This is shown to help you visualize the part. Dimension values in this figure are in inches. *Courtesy Hunter Fan Company, Memphis, TN.*

CADD APPLICATIONS

DIMENSIONING FOR CADD/CAM

The implementation of computer-aided design and drafting (CADD) and computer-aided manufacturing (CAM) in industry is best accomplished when common control of the computer exists between engineering and manufacturing. The success of this automation is, in part, relative to the standardization of operating and documentation procedures. CADD can be accomplished through the same coordinate dimensioning systems previously described in this chapter. Standard dimensioning systems are used to establish a geometric model of the part, which in turn is displayed at a CADD workstation. The data retrieved from this model is the mathematical description of the part to be produced. The drafter must dimension the part completely and accurately so that each contour or geometric shape of the part is continuous. The dimensioning systems that locate features or points on a feature in relation to x, y, and z axes derived from a common origin are most effective, such as datum, tabular, arrowless, and polar coordinate dimensioning. The x, y, and z axes originate from three mutually perpendicular planes that are generally the geometric counterpart of the sides of the part when the surfaces are at right angles as seen in Figure 11.119. If the part is cylindrical, two of the planes intersect at right angles to establish the axis of the cylinder and the third is perpendicular to the intersecting planes as shown in Figure 11.120. The x, y, and z coordinates that are used to establish features on a drawing are converted to x, y, and z axes that correspond to the linear and rotary motions that occur in CAM.

The position of the part in relation to mathematical quadrants determines whether the x, y, and z values are positive or negative. The preferred position is the mathematical quadrant that allows for programming positive commands for the machine tool. Notice in Figure 11.121 that the positive x and y values occur in quadrant 1.

In CAD/CAM and computer-integrated manufacturing (CIM) programs, the drawing is made on the computer screen and sent directly to the computer numerical control machine tool without generating a hard copy of the drawing. In this situation it becomes important for the drafter to understand the machine tool operation. Figure 11.122a shows a drawing created for CAD/CAM, and Figure 11.122b shows the same drawing as represented on the computer screen prior to generating the machine tool program. Notice the dimensions are not displayed on the drawing in Figure 11.122b. The data used to input the information from the original design drawing is used by the computer to establish the machine tool paths. The CAD/CAM program also allows the operator to show the tool path, as shown in Figure 11.123. The tool path display allows the operator to determine if the machine tool will perform the assigned machining operation.

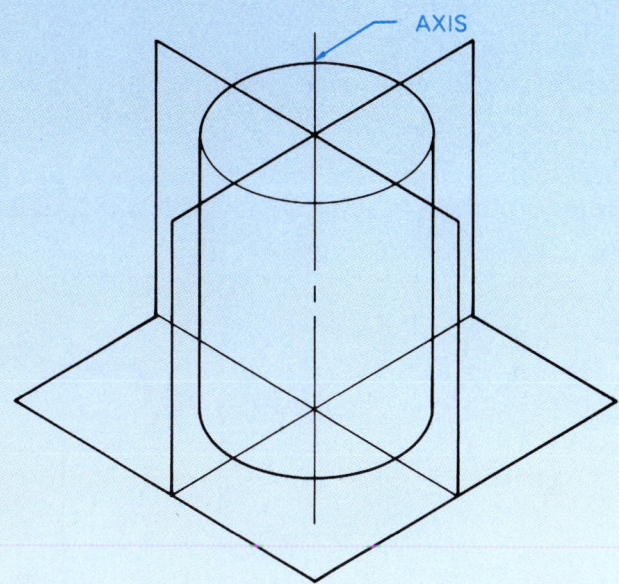

FIGURE 11.120 ■ The planes related to the axis of a cylindrical feature.

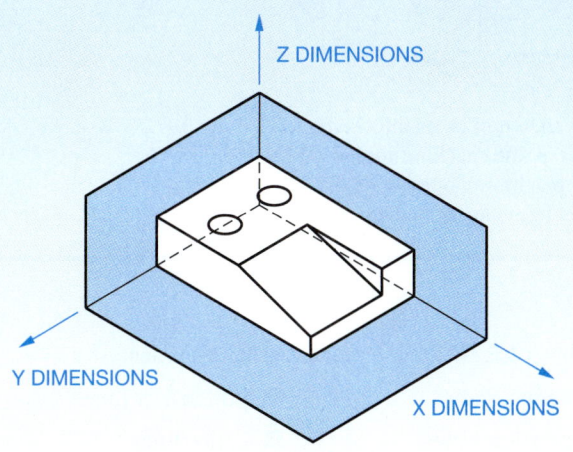

FIGURE 11.119 ■ Features of a part dimensioned in relation to X, Y, and Z axes.

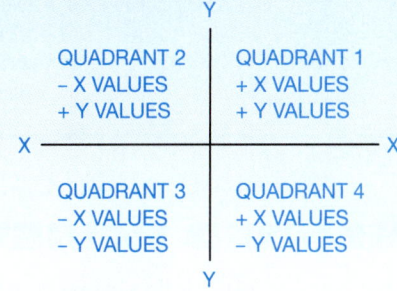

FIGURE 11.121 ■ Determination of X and Y values in related quadrants.

(Continued)

CADD APPLICATIONS

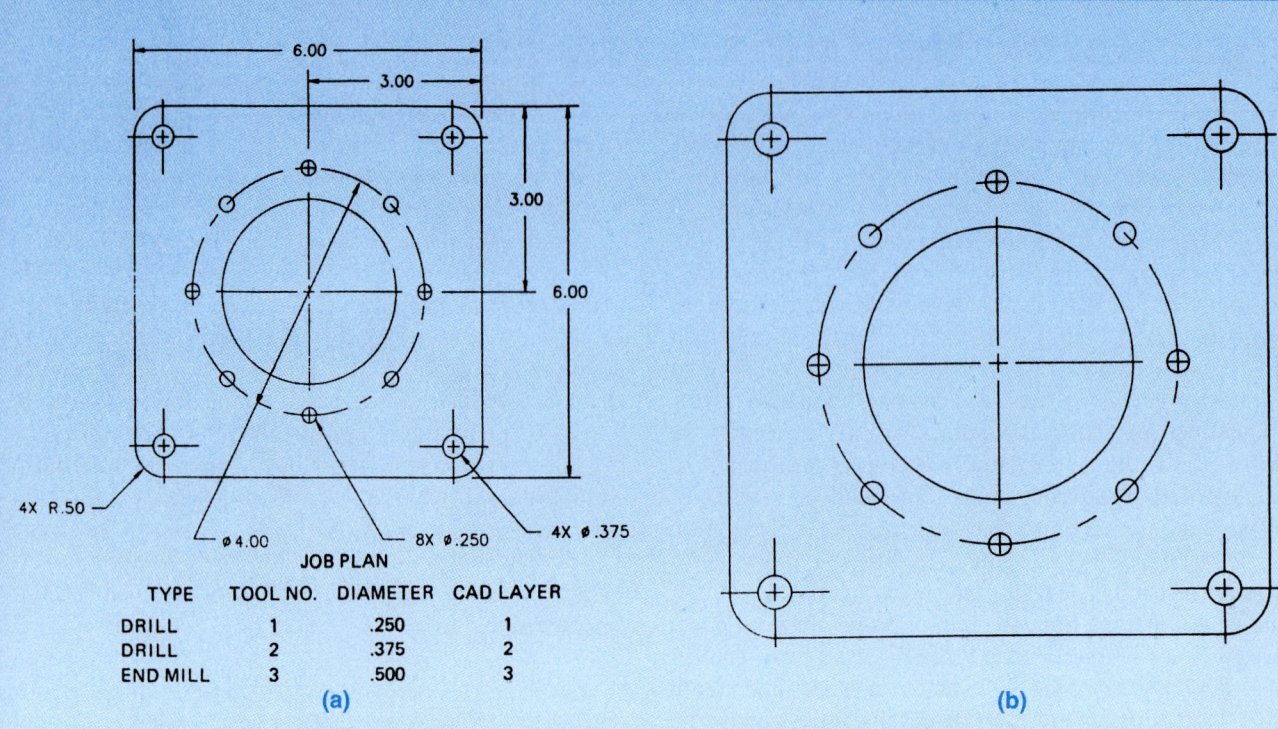

JOB PLAN

TYPE	TOOL NO.	DIAMETER	CAD LAYER
DRILL	1	.250	1
DRILL	2	.375	2
END MILL	3	.500	3

(a)

(b)

FIGURE 11.122 ■ (a) A drawing, created for CAD/CAM. Dimension values in this figure are in inches. (b) Drawing from (a) represented on the computer screen prior to generating the machine tool program.

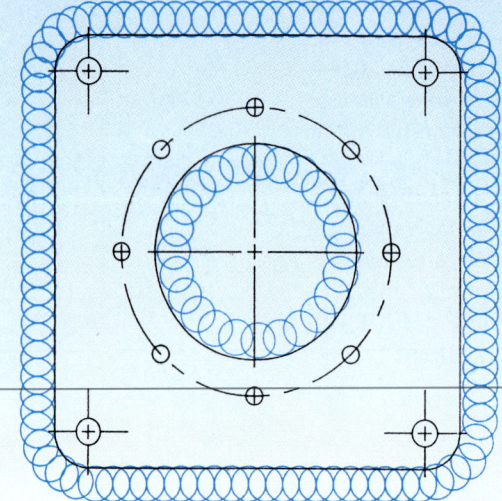

FIGURE 11.123 ■ The tool path display, shown in color, allows the operator to determine if the machine tool will perform the assigned machining operations. Dimension values in this figure are in inches.

RECOMMENDED REVIEW

It is recommended that you review the following parts of Chapter 7 before you begin working on fully dimensioned multiview drawings. This will refresh your memory about how related lines are properly drawn.

- Object Lines.
- Viewing Planes.
- Extension Lines.
- Hidden Lines.
- Break Lines.
- Dimension Lines.
- Centerlines.
- Phantom Lines.

■ Leader Lines.

■ Lettering Standards and Practices.

Also review Chapters 9 and 10 covering multiview and auxiliary view drawings, and Chapter 5 discussing manufacturing materials and processes.

WEBSITE RESEARCH

The following websites can provide you additional information for research or further study into topics covered in this chapter.

http://www.asme.org—Find information and publications related to the American Society of Mechanical Engineers, including ASME Y14.5M, *Dimensioning and Tolerancing.*

http://www.ansi.org—The American National Standards Institute. Information about national and international drafting standards, including ASME Y14.5M, *Dimensioning and Tolerancing.*

http://www.adda.org—American Design Drafting Association, *Drafting Reference Guide.*

http://www.industrialpress.com—Information about the *Machinery's Handbook.* This is a valuable resource for manufacturing standards, sizes, tolerances, fits, materials, and anything else you can think of for design and drafting.

http://www.industrialpress.com—On line trigonometry tables.

http://www.globalspec.com—The Engineering Web™ makes it faster and easier for you to research topics, products, and services by limiting your search to technical and engineering-related websites.

http://www.g-w.com—Select "Technical and Trades Technology," followed by "CAD/Animation/Drafting," and look for the latest, or your desired, edition of *AutoCAD and Its Applications, Basic.* This provides in-depth coverage on the use of AutoCAD to create dimensions based on ASME and other accepted standards.

http://www.amazon.com—Search Books for *AutoCAD and Its Applications.*

FINDING DIAGONALS

Suppose you need the distance from one point on a drawing to another, such as between points A and B in Figure 11.124.

Find the distance from C to D in Figure 11.125.

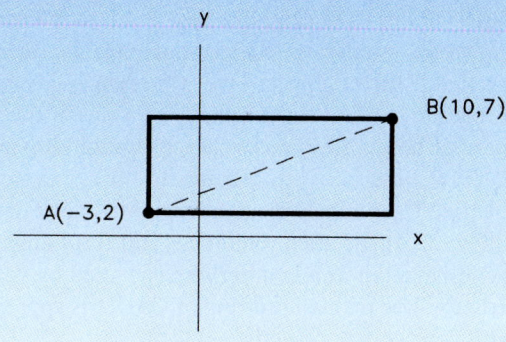

FIGURE 11.124 ■ Finding the diagonal of a rectangle.

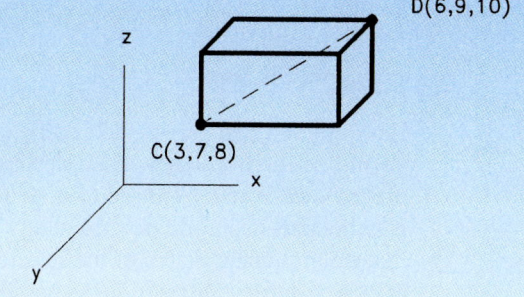

FIGURE 11.125 ■ Finding the diagonal of a cube.

Solution:
Sides: height (h) = 5, width (w) = 13
Diagonal (d) = $\sqrt{h^2 + w^2}$
d = 13.9

Solution:
Sides: height (h) = 2, width (w) = 2
      length (L) = 3
Diagonal (d) = $\sqrt{h^2 + w^2 + L^2}$
d = 4.12

MATH APPLICATIONS

CHAPTER 11

Dimensioning and Tolerancing Test

 Access the CD found with this textbook to view the Chapter 11 Test. Confirm the preferred submittal method with your instructor.

CHAPTER 11

Dimensioning and Tolerancing Problems

PROBLEMS CONTINUED FROM PREVIOUS CHAPTERS

In addition to the problems found in this chapter, you can go back to the following previous chapters and complete the drawings by adding the given dimensions. If manual drafting was used, dimensions can be placed if enough space was made available. Otherwise, the drawings need to be redrawn. If CADD was used, open the existing drawing and edit views by moving to accommodate dimension placement.

Chapter 7   Lines and Lettering: Problems 7.1 through 7.11.

Chapter 8   Geometric Construction: Problems 8.21 through 8.43.

Chapter 9   Multiviews: Problems 9.58 through 9.89.

Chapter 10   Auxiliary Views: Problems 10.15 through 10.36.

DIRECTIONS

1. From the selected sketch, determine which view should be the front view. Then determine which other views, if any, you need to draw to fully display the part in a multiview drawing.
2. Make a multiview sketch of the selected problem as close to correct proportions as possible. Be sure to indicate where you intend to place the dimension lines, extension lines, arrowheads, and hidden features, to help you determine the spacing for your final drawing.
3. Using the sketch you have just developed as a guide, make an original multiview drawing on an adequate size drawing sheet and at an appropriate scale. Include all dimensions needed using unidi-rectional dimensioning. Use computer-aided drafting if specified by your course outline. Include the following general notes at the lower left corner, .5 in. each direction from the borderline using .12-in.-high lettering. Letter the word NOTES with .24-in.-high lettering.

1. INTERPRET DIMENSIONS AND TOLER-ANCES PER ASME Y14.5M—1994.
2. REMOVE ALL BURRS AND SHARP EDGES.

Notes

Additional notes may be required depending on the specifications of each individual assignment. Remember that each problem assignment is given as an engineer's layout to help simulate actual drafting conditions. Initial problems provide a suggested layout and problems become more complex as you continue.

Dimensions and views on engineers' layouts may not be placed in accordance with acceptable standards. You need to carefully review the chapter material when preparing the layout sketch. In some problems, the engineer's layout "assumes" certain information, such as the symmetry of a part or the alignment of holes. You need to place enough dimensions or draw lines between features to "fully" dimension the part. Use a mechanical drafting title block with the following unspecified tolerances and third-angle projection symbol unless otherwise specified by your instructor. Use manual drafting or CADD as appropriate with your course objectives.

UNSPECIFIED TOLERANCES:

DECIMALS	mm	IN.
X	± 2.5	± .1
XX	± 0.25	± .01
XXX	± 0.125	± .005
ANGULAR ± 30′		
FINISH	3.2μm	125μIN.

PROBLEM 11.1 **Basic practice (metric)**
Part Name: Step Block
Material: SAE 1020

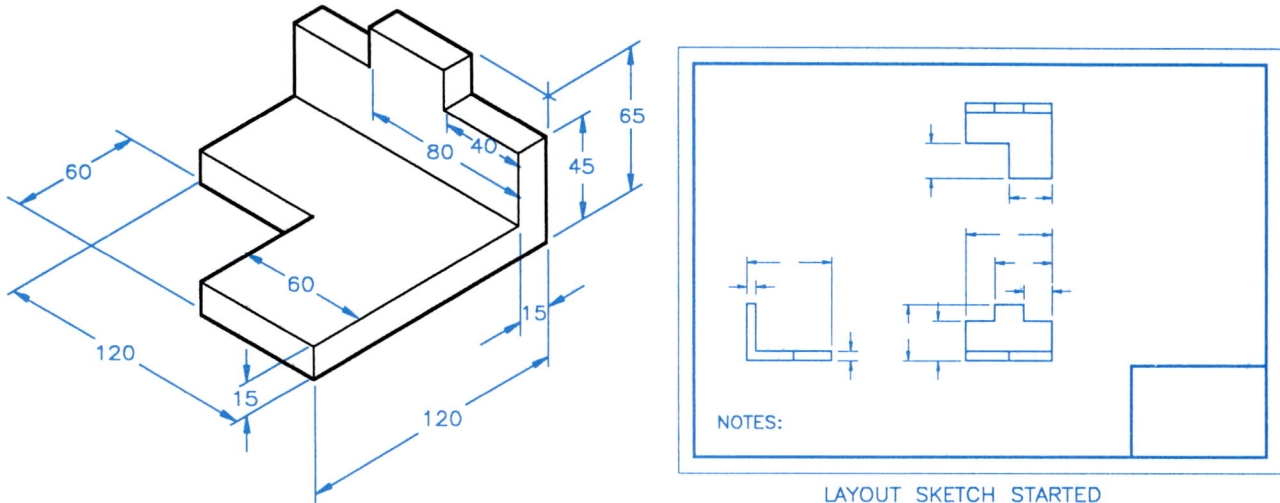

LAYOUT SKETCH STARTED

PROBLEM 11.2 **Basic practice (metric)**
Part Name: Machine Tool Wedge Plate
Material: SAE 4320

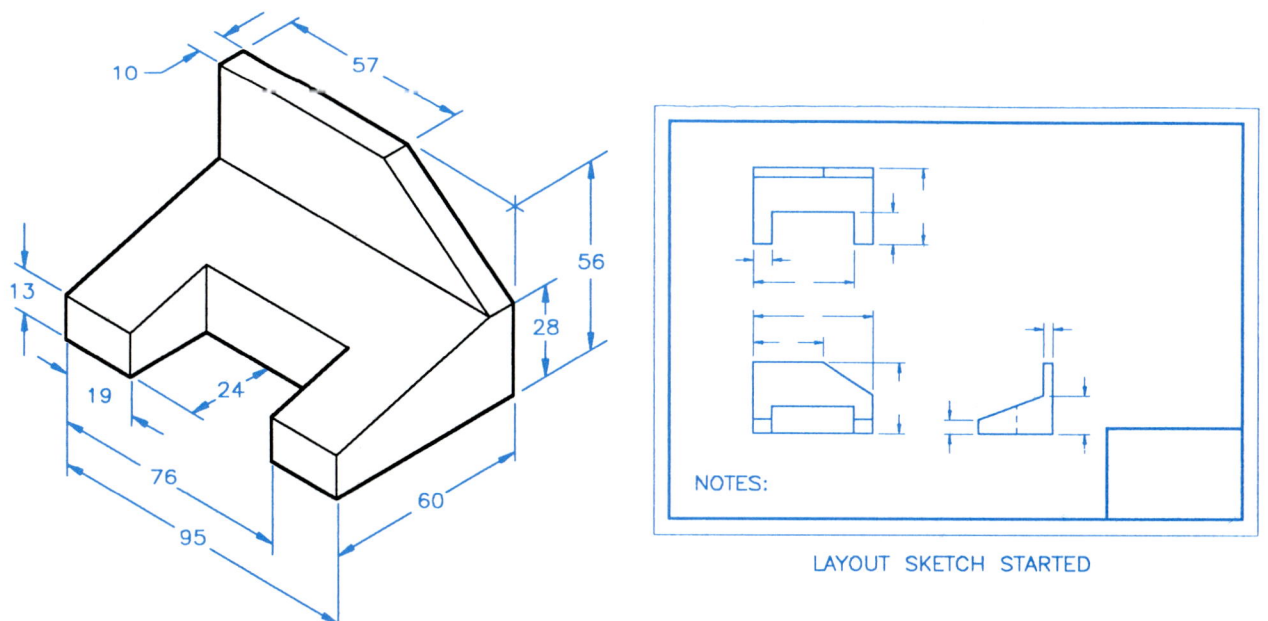

LAYOUT SKETCH STARTED

PROBLEM 11.3   Dimensioning basic practice (in.)

Part Name: V-guide

Material: SAE 4320

Fractions: ± 1/32

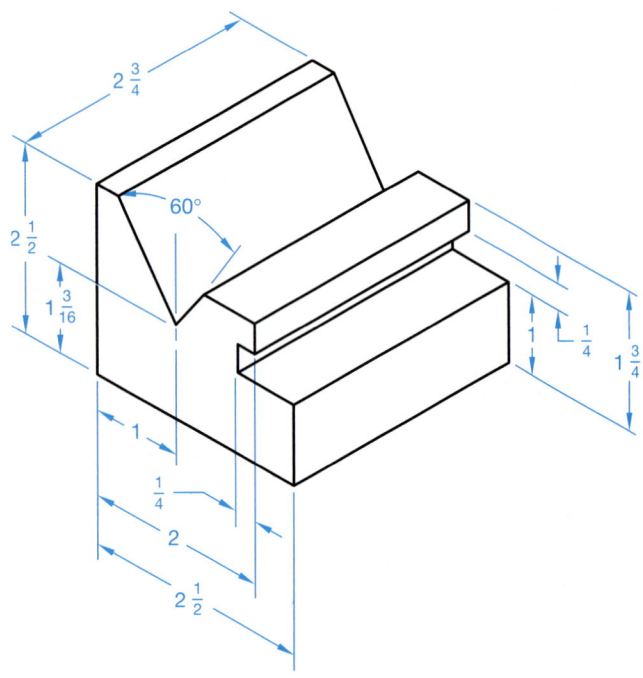

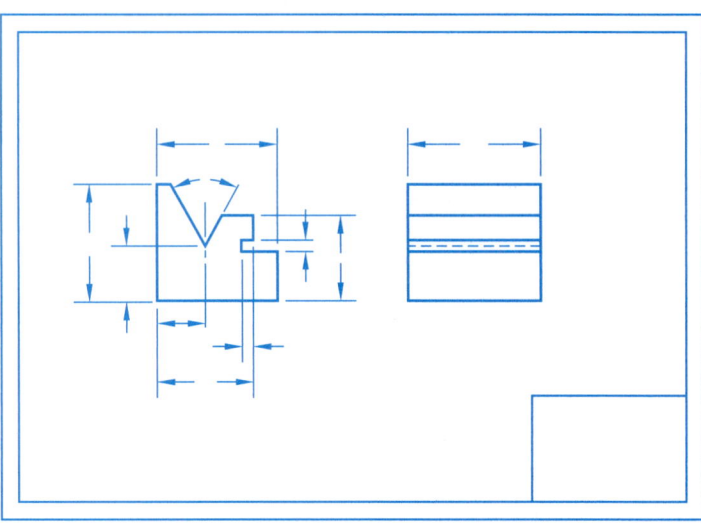

LAYOUT SKETCH
OPTIONAL SINGLE VIEW WITH LENGTH
GIVEN IN GENERAL NOTE OR TITLE BLOCK

PROBLEM 11.4   Circles and arcs. (in.)

Part Name: Rest Pad

Material: SAE 1040

Fillets: R.125

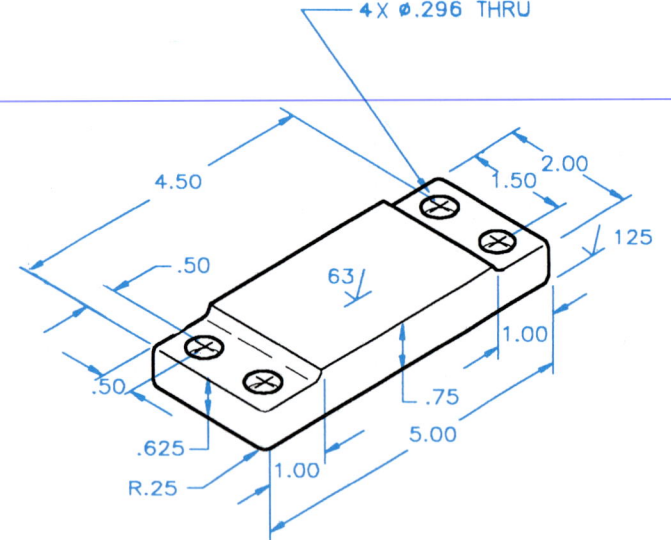

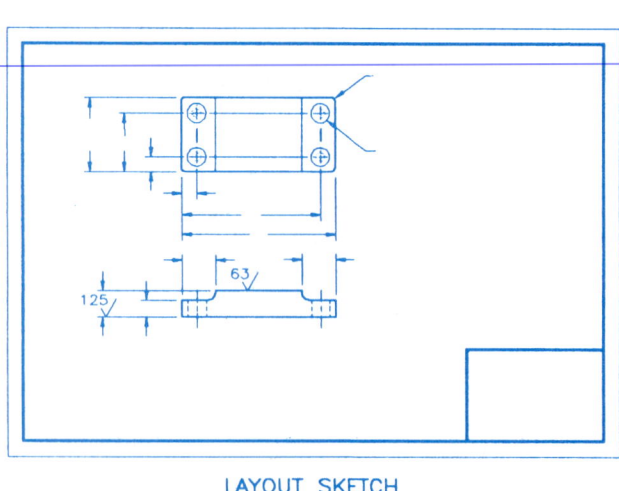

LAYOUT SKETCH

PROBLEM 11.5 **Limited space (metric)**
Part Name: Angle Bracket
Material: Mild Steel (MS)

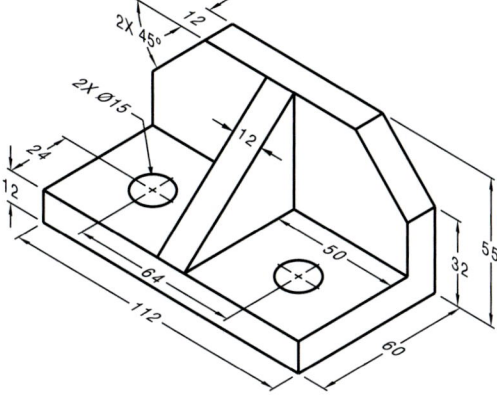

PROBLEM 11.7 **Limited spaces (metric)**
Part Name: Selector Slide Kicker
Material: Al 1510

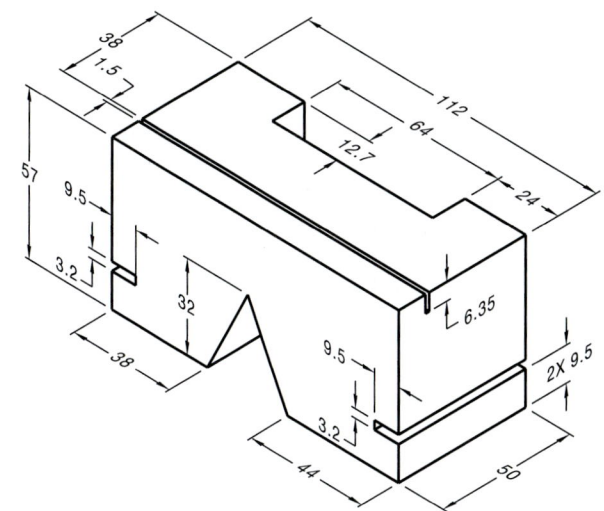

PROBLEM 11.6 **Dimensioning contours and limited spaces (in.)**
Part Name: Support
Material: Aluminum

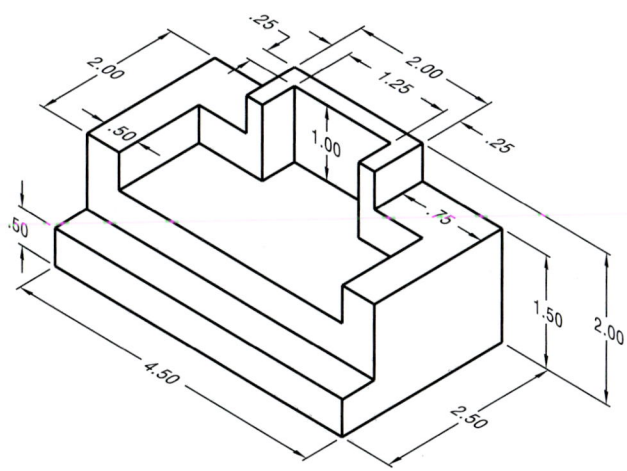

PROBLEM 11.8 **Circles and arcs (in.)**
Part Name: Chain Link
Material: SAE 4320

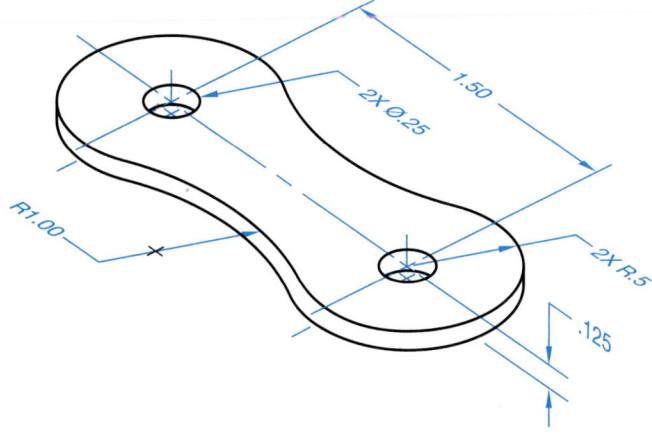

PROBLEM 11.9 Holes and limited space (in.)

Part Name: Pivot Bracket

Material: SAE 1040

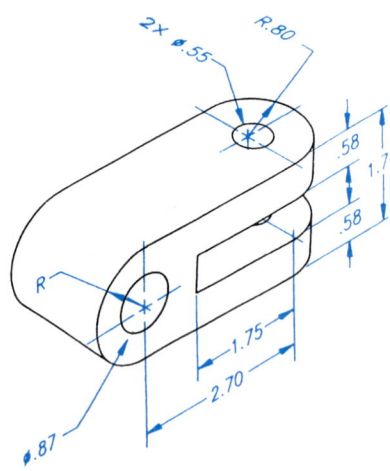

PROBLEM 11.10 Holes, angles, and arcs (in.)

Part Name: Journal Bracket

Material: Cast Iron (CI)

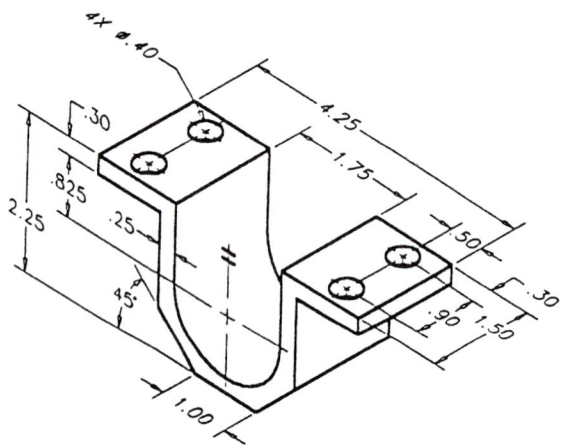

PROBLEM 11.11 Dimensioning multiple features (in.)

Part Name: Lock Ring

Material: SAE 1020

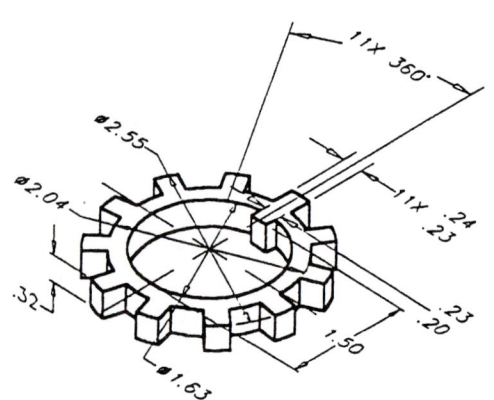

PROBLEM 11.12 Single view (in.)

Part Name: Idler Gear Shaft

Material: MIL-S-7720

Problem based on original art courtesy Aerojet TechSystems Co.

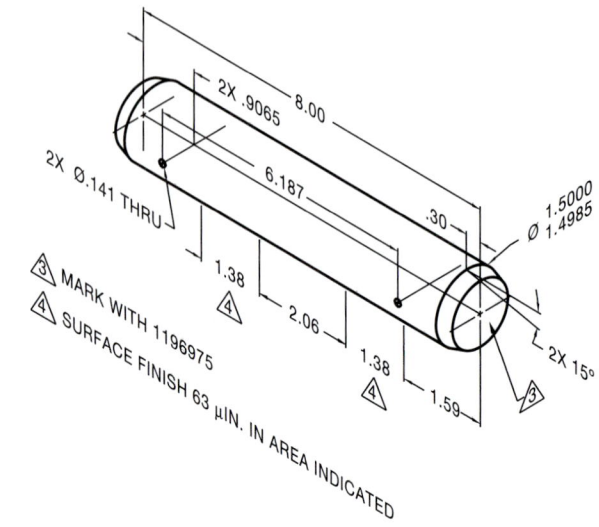

PROBLEM 11.13 Circles and arcs (metric)

Part Name: Bearing Support

Material: SAE 1040

Fillets: R6

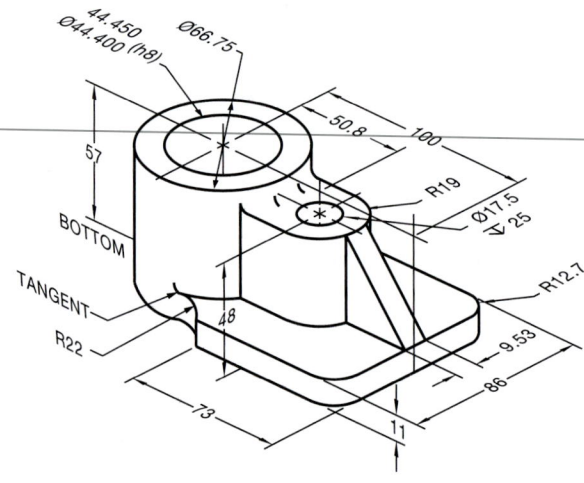

PROBLEM 11.14 Circles and arcs (metric)

Part Name: Hinge Bracket

Material: Cast Aluminum

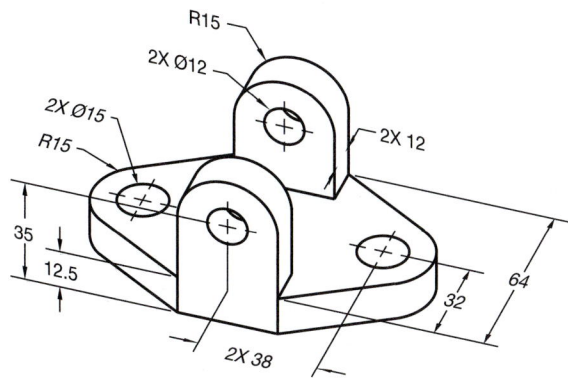

PROBLEM 11.15 Machine features (in.)

Part Name: Spacer

Material: SAE 1030

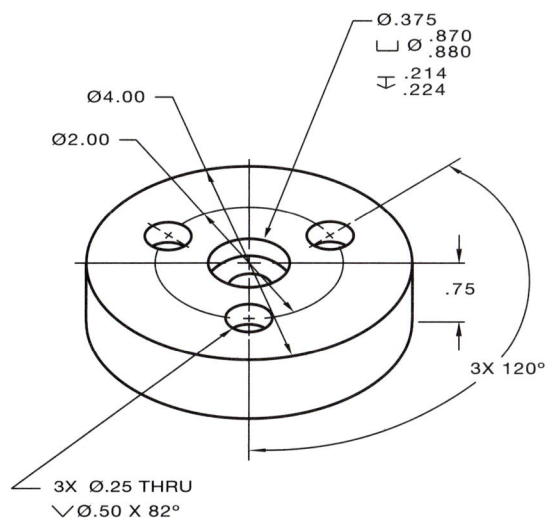

PROBLEM 11.16 Polar coordinate dimensioning (in.)

Part Name: Spacer

Material: Plastic

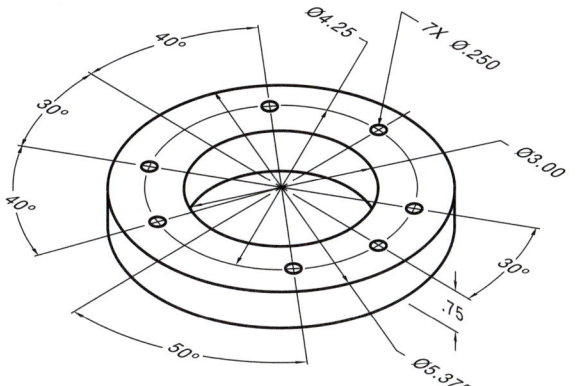

PROBLEM 11.17 Repetitive features (in.)

Part Name: Slot Plate

Material: Aluminum

SPECIFIC INSRUCTIONS: Using the RC4 standard tolerance limits found in Figure 11.81, Appendices, or the *Machinery's Handbook,* calculate and apply limits values to the ∅1.00 dimension.

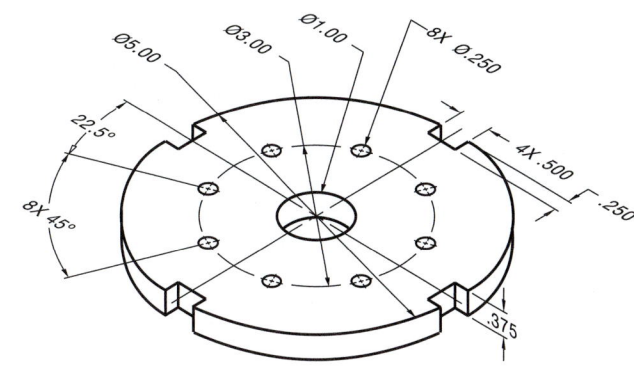

PROBLEM 11.18 Tabular dimensioning (metric)

Part Name: Mounting Base

Material: Stainless Steel

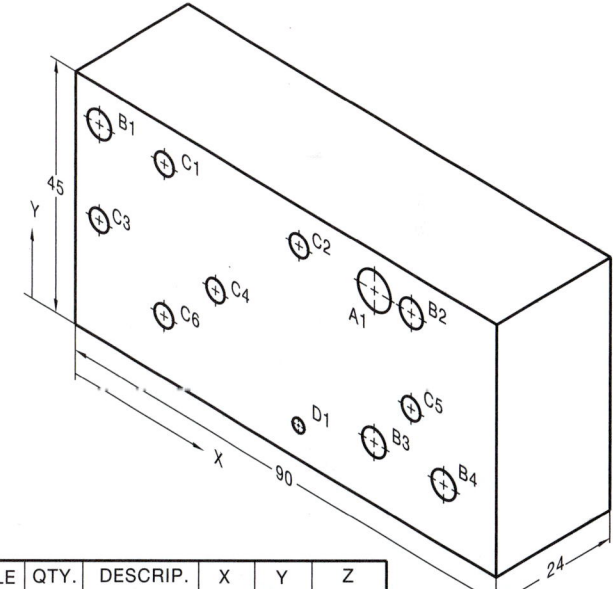

HOLE	QTY.	DESCRIP.	X	Y	Z
A1	1	Ø7	64	38	18
B1	1	Ø5	5	38	THRU
B2	1	Ø5	72	38	THRU
B3	1	Ø5	64	11	THRU
B4	1	Ø5	79	11	THRU
C1	1	Ø4	19	38	THRU
C2	1	Ø4	48	38	THRU
C3	1	Ø4	5	21	THRU
C4	1	Ø4	30	21	THRU
C5	1	Ø4	72	21	THRU
C6	1	Ø4	19	11	THRU
D1	1	Ø2.5	48	6	THRU

PROBLEM 11.19 Dimensioning circles, arcs, and slots (in.)

Part Name: Top Pipe Support Bracket

Material: SAE 1020

Fillets: R12

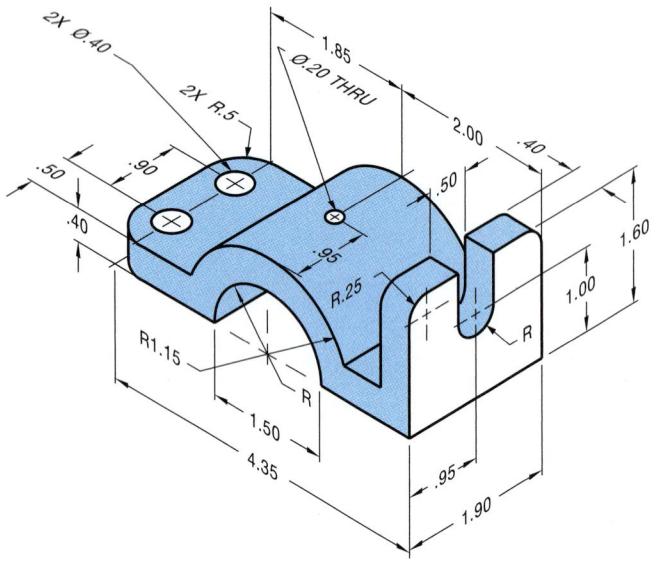

 PROBLEMS 11.21 through 11.32: Access the CD found with this textbook and open the problem of your choice, or as assigned by your instructor. Solve the problems using the instructions provided with this chapter or on the CD, unless otherwise specified by your instructor.

MATH PROBLEMS

 PROBLEMS 11.33 through 11.42: Access the CD found with this textbook and open the math problem of your choice, or as assigned by your instructor. Solve the problem or problems using the instructions provided.

PROBLEM 11.20 Chain dimensioning (metric)

Part Name: Control Housing Cover

Material: Cast Iron

Note: Do not draw a sectional view. Sections are covered in Chapter 13. Consider a bottom view to show wall thickness.

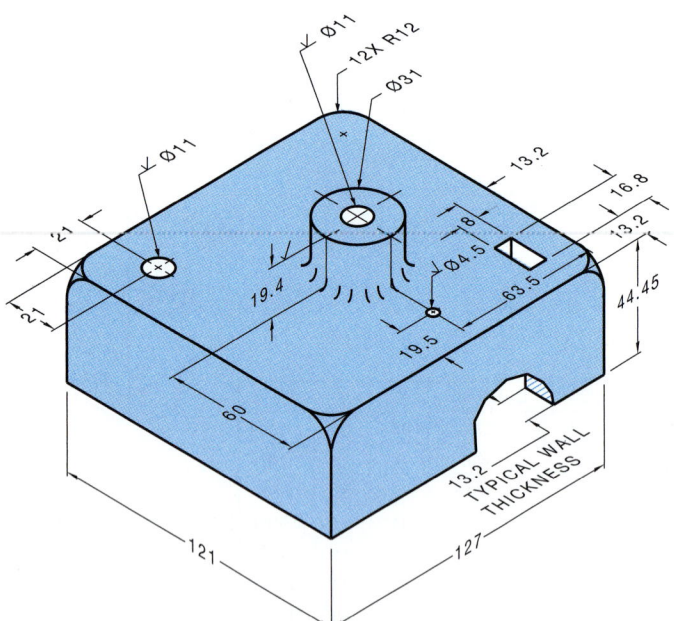

Fasteners and Springs

LEARNING OBJECTIVES

After completing this chapter, you will:

- Draw screw thread representations and provide correct thread notes.

- Read, write, and interpret written fastener specifications.
- Prepare drawings for fastening devices.
- Draw completely dimensioned spring representations.
- Prepare formal drawings from engineers' sketches and actual industrial layouts.

THE ENGINEERING DESIGN APPLICATION

You are asked to make a drawing from an engineer's notes for a fastening device that is needed for one of the products the engineer is designing. The engineer's notes look like Figure 12.1. You may be thinking:

- A 10-gage shaft threaded with 32–UNF–2A. Maybe you can use a 10–32UNF–2A threaded rod and cut it to .375 lengths.

- Then you need to chamfer both ends with .015 × 35° and provide a slot in one end that is .030 wide and .047 deep.

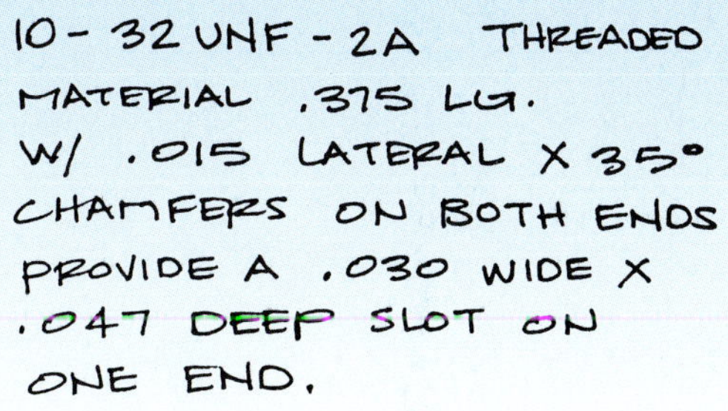

```
10 - 32 UNF - 2A   THREADED
MATERIAL .375 LG.
W/ .015 LATERAL X 35°
CHAMFERS ON BOTH ENDS
PROVIDE A .030 WIDE X
.047 DEEP SLOT ON
ONE END.
```

FIGURE 12.1 ■ Engineer's notes.

So, you go to work preparing a drawing like the one shown in Figure 12.2. The point here is that the engineer provided a lot of information in the thread note. In fact, standard screws and other fasteners may be completely described without a drawing. A written specification can be used to completely describe an object. For example, 1/2–13UNC–2 × 1.5 LG FILLISTER HEAD MACHINE SCREW.

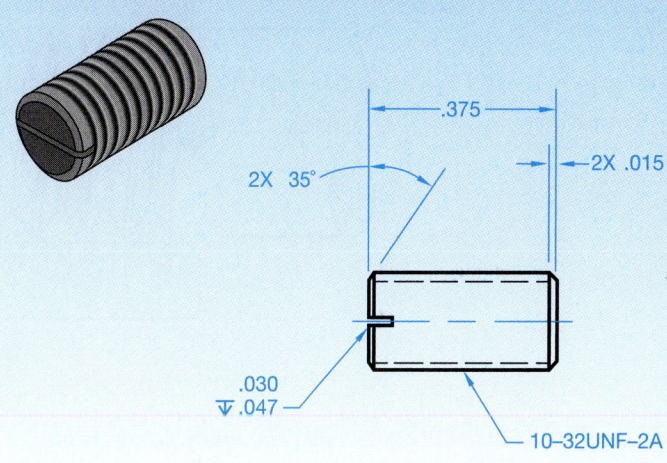

FIGURE 12.2 ■ CADD model and drawing from engineer's notes.

ASME/ANSI This chapter introduces you to the methods of specifying and drafting fasteners and springs. Fasteners include screw threads, keys, pins, rivets, and weldments. There are two types of springs, helical and flat. The American National Standards Institute documents that govern the standards for fasteners and springs are *Screw Thread Representation*, ANSI Y14.6; *Screw Thread Representation (Metric Supplement)*, ANSI Y14.6aM; *Symbols for Welding and Nondestructive Testing Including Brazing*, ANSI/AWS A2.4; and *Mechanical Spring Representation*, ANSI Y14.13M.

SCREW THREAD FASTENERS

Screw threads are a helix or conical spiral formed on the external surface of a shaft or internal surface of a cylindrical hole, as shown in Figures 12.3 and 12.4. Screw threads are used for an unlimited number of services, such as for holding parts together as fasteners, for leveling and adjusting objects, and for transmitting power from one object or feature to another.

The standardization of screw threads was achieved among the United States, United Kingdom, and Canada in 1949. A need for interchangeability of screw thread fasteners was the purpose of this standardization and resulted in the Unified Thread Series. The Unified Thread Series is now the American standard for screw threads. Prior to 1949 the United States standard was the American National screw threads. The unification standard occurred as a result of combining some of the characteristics of the American National screw threads with the United Kingdom's long-accepted Whitworth screw threads. Screw thread systems were revised again in 1974 for metric application. The modifications were minor and based primarily on metric translation. In order to emphasize that the Unified screw threads evolve from inch calibrations, the term Unified Inch Screw Threads is used while the term Unified Screw Threads Metric Translation is used for the metric conversion.

Screw Thread Terminology

Refer to Figure 12.3 and Figure 12.4 as a reference for the following definitions related to external and internal threads.

Axis

The thread axis is the centerline of the cylindrical thread shape.

Body

That portion of a screw shaft that is left unthreaded.

Chamfer

An angular relief at the last thread to help allow the thread to more easily engage with a mating part.

Classes of Threads

A designation of the amount of tolerance and allowance specified for a thread.

Crest

The top of external and the bottom of internal threads.

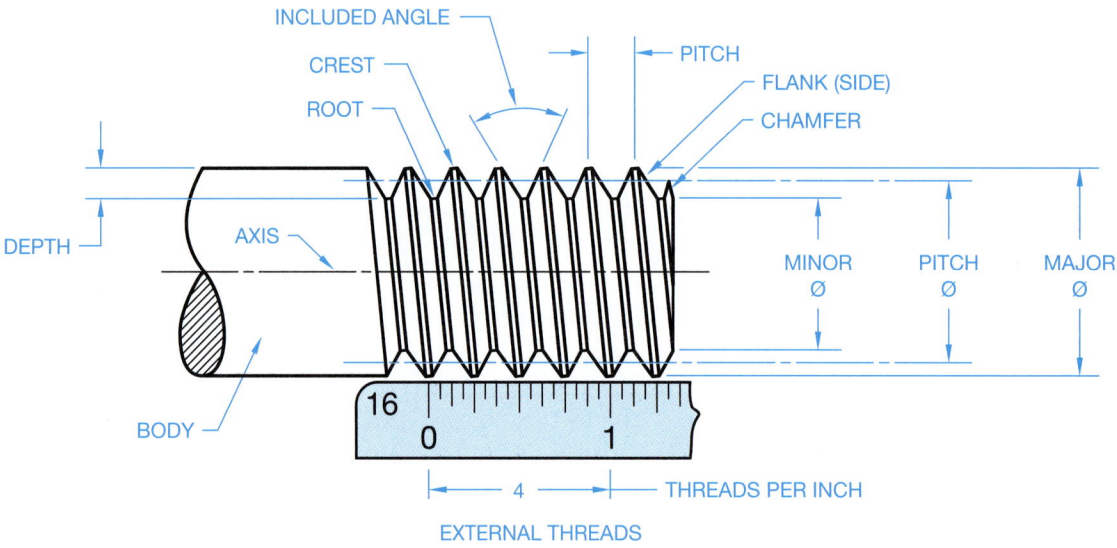

FIGURE 12.3 ■ External screw thread components.

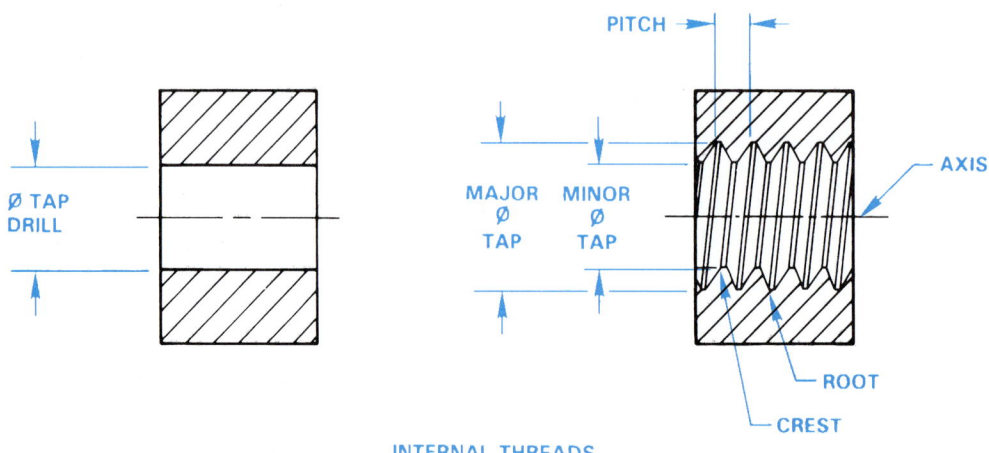

FIGURE 12.4 ■ Internal screw thread components.

Depth of Thread

Depth is the distance between the crest and the root of a thread, measured perpendicular to the axis.

Die

A machine tool used for cutting external threads.

Fit

Identifies a range of thread tightness or looseness.

Included Angle

The angle between the flanks (sides) of the thread.

Lead

The lateral distance a thread travels during one complete rotation.

Left-Hand Thread

A thread that engages with a mating thread by rotating counterclockwise, or with a turn to the left when viewed toward the mating thread.

Major Diameter

The distance on an external thread from crest to crest through the axis. For an internal thread the major diameter is measured from root to root across the axis.

Minor Diameter

The dimension from root to root through the axis on an external thread and measured across the crests through the center for an internal thread.

Pitch

The distance measured parallel to the axis from a point on one thread to the corresponding point on the adjacent thread.

Pitch Diameter

A diameter measured from a point halfway between the major and minor diameter through the axis to a corresponding point on the opposite side.

Right-Hand Thread

A thread that engages with a mating thread by rotating clockwise, or with a turn to the right when viewed toward the mating thread.

Root

The bottom of external and the top of internal threads.

Tap

A tap is the machine tool used to form an interior thread. Tapping is the process of making an internal thread.

Tap Drill

A tap drill is used to make a hole in material before tapping.

Thread

The part of a screw thread represented by one pitch.

Thread Form

The design of a thread determined by its profile.

Thread Series

Groups of common major diameter and pitch characteristics determined by the number of threads per inch.

Threads per Inch

The number of threads measured in one inch. The reciprocal of the pitch in inches.

THREAD-CUTTING TOOLS

A tap is a machine tool used to form an internal thread as shown in Figure 12.5. A die is a machine tool used to form external threads. (See Figure 12.6.)

A tap set is made up of a taper tap, a plug tap, and a bottoming tap as shown in Figure 12.7. The taper tap is generally used for starting a thread. The threads are tapered to within ten threads from the end. The tap is tapered so the tool more evenly distributes the cutting edges through the depth of the hole. The plug tap has the threads tapered to within five threads from the

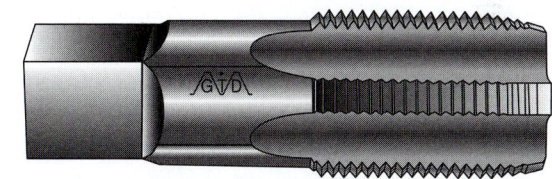

FIGURE 12.5 ■ Tap. *Courtesy Greenfield Tap & Die, Division of TRW, Inc.*

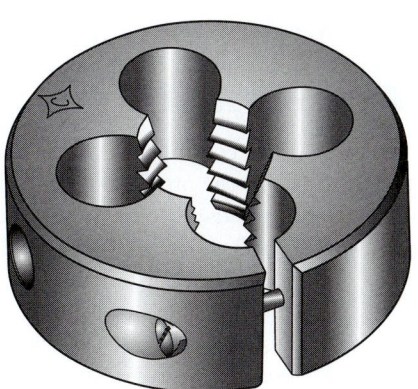

FIGURE 12.6 ■ Die. *Courtesy Cleveland Twist Drill, an Acme-Cleveland Company.*

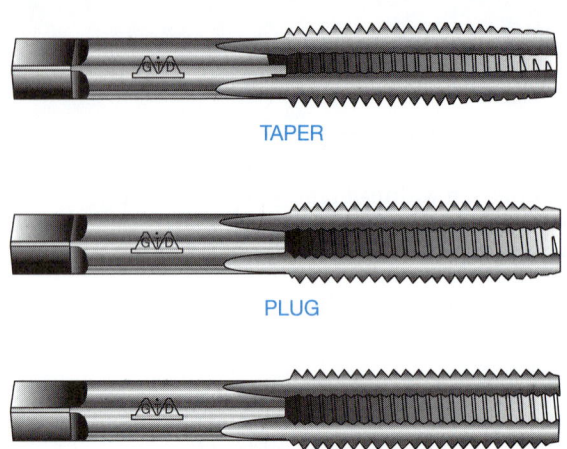

TAPER

PLUG

BOTTOMING

FIGURE 12.7 ■ Tap set includes taper, plug, and bottoming taps. *Courtesy Greenfield Tap & Die, Division of TRW, Inc.*

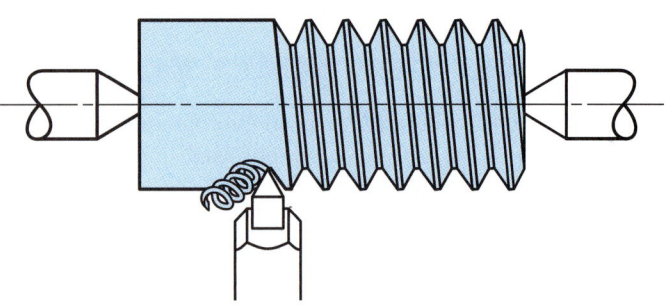

FIGURE 12.8 ■ Thread cutting on a lathe.

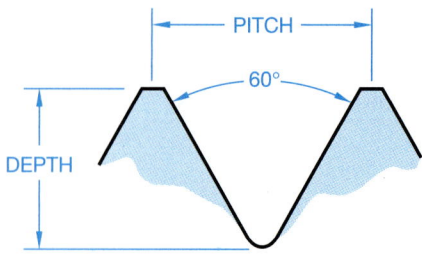

FIGURE 12.9 ■ Unified thread form. The unified thread form is the most commonly used thread. See Appendix J.

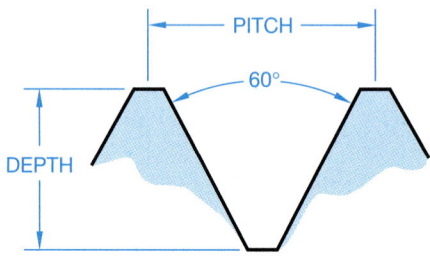

FIGURE 12.10 ■ American National thread form.

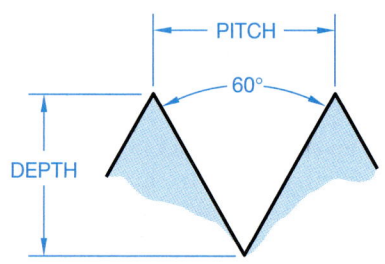

FIGURE 12.11 ■ Sharp-V thread form.

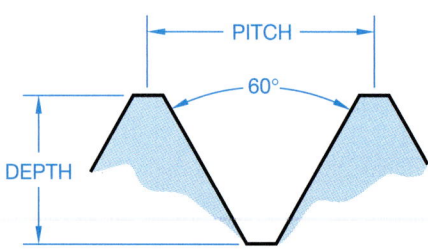

FIGURE 12.12 ■ Metric thread form. See Appendix K.

end. The plug tap can be used to completely thread through material or thread a **blind note** (a hole that does not go through the material) if full threads are not required all the way to the bottom. The **bottoming tap** is used when threads are needed to the bottom of a blind hole.

The die is a machine tool used to cut external threads. Thread-cutting dies are available for standard thread sizes and designations.

External and internal threads can also be cut on a lathe. A lathe is a machine that holds a piece of material between two centers or in a chucking device. The material is rotated as a cutting tool removes material while traversing along a carriage that slides along a bed. Figure 12.8 shows how a cutting tool can make an external thread.

THREAD FORMS

Unified threads are the most common threads used on threaded fasteners. Figure 12.9 shows the profile of a Unified thread.

American National threads, shown in profile in Figure 12.10, are similar to the Unified thread but have a flat root. Still in use today, the American National thread has generally replaced the sharp-V thread form.

The **sharp-V thread**, although not commonly used, is a thread that will fit and seal tightly. It is difficult to manufacture because the sharp crests and roots of the threads are easily damaged. (See Figure 12.11.) The sharp-V thread was the original United States standard thread form.

Metric thread forms vary slightly from one European country to the next. The International Organization for Standardization (ISO) was established to standardize metric screw threads. The ISO thread specifications are similar to the Unified thread form. (See Figure 12.12.)

Whitworth threads are the original British standard thread forms developed in 1841. These threads have been referred to as parallel screw threads. The Whitworth thread forms are primarily used for replacement parts. (See Figure 12.13.)

Square thread forms, shown in Figure 12.14, have a longer pitch than Unified threads. Square threads were developed as threads that would effectively transmit power; however, they are difficult to manufacture because of their perpendicular sides. There are modified square threads with 10° sides. The square thread is generally replaced by Acme threads.

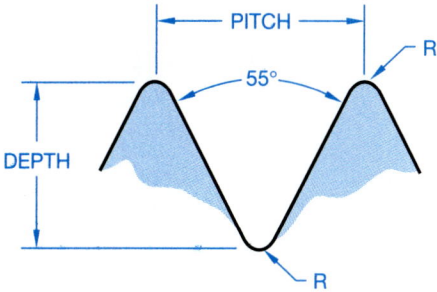

FIGURE 12.13 ■ Whitworth thread form.

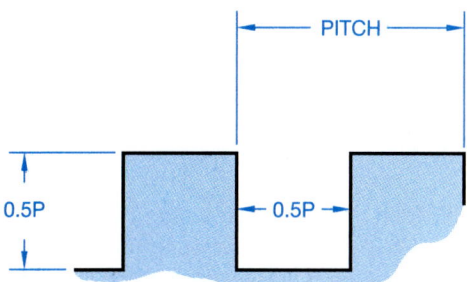

FIGURE 12.14 ■ Square thread form.

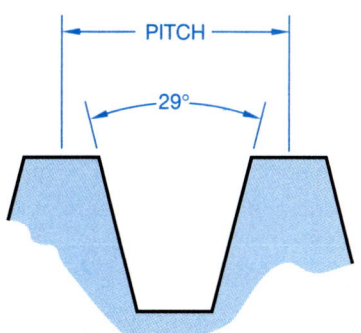

FIGURE 12.15 ■ Acme thread form.

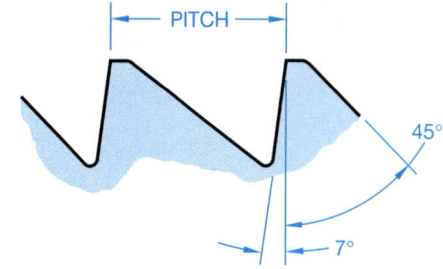

FIGURE 12.16 ■ Buttress thread form.

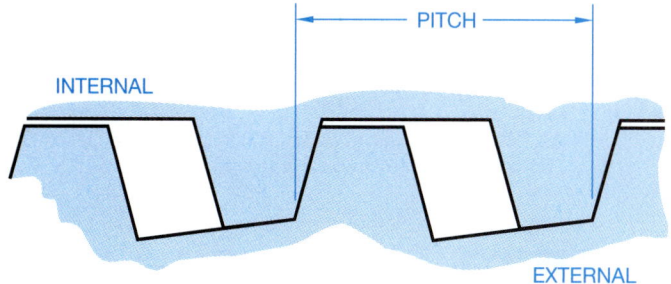

FIGURE 12.17 ■ Dardelet self-locking thread form.

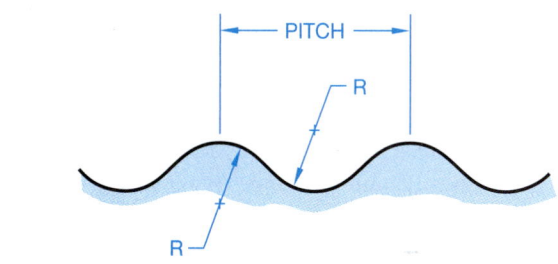

FIGURE 12.18 ■ Rolled thread form.

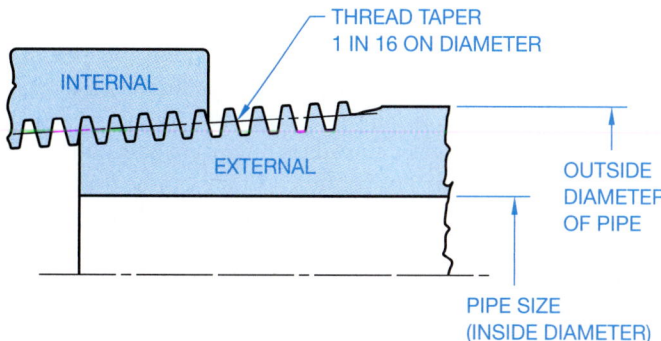

FIGURE 12.19 ■ American National Standard taper pipe thread form.

Acme thread forms are commonly used when rapid traversing movement is a design requirement. Acme threads are popular on such designs as screw jacks, vice screws, and other equipment and machinery that requires rapid screw action. A profile of the Acme thread form is shown in Figure 12.15.

Buttress threads are designed for applications where high stress occurs in one direction along the thread axis. The thread flank or side that distributes the thrust or force is within 7° of perpendicularity to the axis. This helps reduce the radial component of the thrust. The buttress thread is commonly used in situations where tubular features are screwed together and lateral forces are exerted in one direction. (See Figure 12.16.)

Dardelet thread forms are primarily used in situations where a self-locking thread is required. These threads resist vibrations and remain tight without auxiliary locking devices. (See Figure 12.17.)

Rolled thread forms are used for screw shells of electric sockets and lamp bases. (See Figure 12.18.)

American National Standard taper pipe threads are the standard threads used on pipes and pipe fittings. These threads are designed to provide pressure-tight joints or not, depending on the intended function and materials used. American pipe threads are measured by the **nominal pipe size**, which is the inside pipe diameter. For example, a 1/2-in. pipe size has an outside pipe diameter of .840 in. (See Figure 12.19.) Pipe thread design considerations and drafting practices are discussed later in this text.

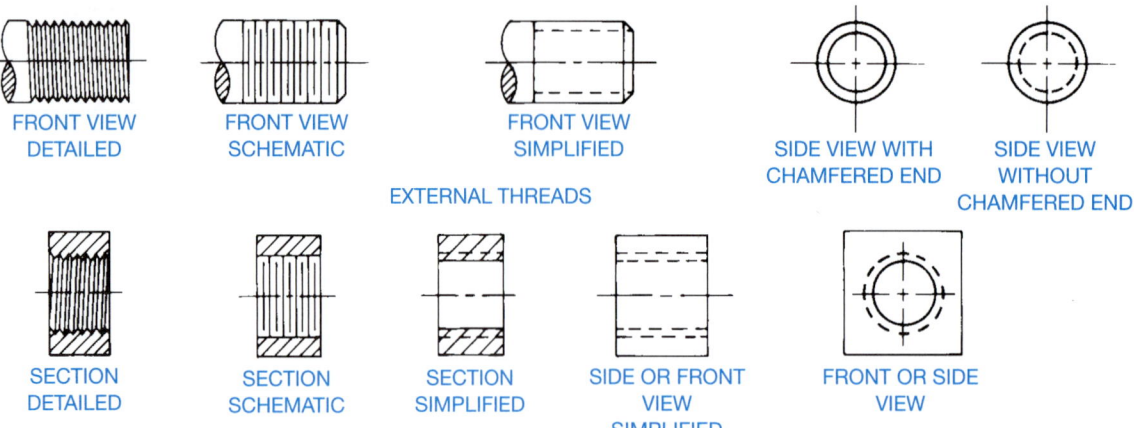

FRONT VIEW
DETAILED

FRONT VIEW
SCHEMATIC

FRONT VIEW
SIMPLIFIED

SIDE VIEW WITH
CHAMFERED END

SIDE VIEW
WITHOUT
CHAMFERED END

EXTERNAL THREADS

SECTION
DETAILED

SECTION
SCHEMATIC

SECTION
SIMPLIFIED

SIDE OR FRONT
VIEW
SIMPLIFIED

FRONT OR SIDE
VIEW

INTERNAL THREADS

FIGURE 12.20 ■ Thread representations.

THREAD REPRESENTATIONS

There are three methods of thread representation in use: detailed, schematic, and simplified, as shown in Figure 12.20. The detailed representation is used in special situations that require a pictorial display of threads such as in a sales catalog or a display drawing. Detailed thread representations are not commonly used on most manufacturing drawings because they are much too time-consuming to draw. Schematic representations are also not commonly used in industry. Although they do not take the time of detailed symbols, they do require extra time to draw. Some companies prefer to use the schematic thread representation.

The actual use and purpose of the drawing helps determine which thread symbol to use. It is possible to mix representations on a particular drawing if clarity is improved, but this practice is generally avoided. The simplified representation is the most common method of drawing thread symbols. Simplified representations clearly describe threads and they are easy and quick to draw. They are also very versatile as they can be used in all situations, while the other representations cannot be used in all situations. Figure 12.21 shows simplified threads in different applications.

Also, notice how the use of a thread chamfer slightly changes the appearance of the thread. Chamfers are commonly applied to the first thread to help start a thread in its mating part.

When an internal screw thread does not go through the part, it is common to drill deeper than the depth of the required thread when possible. This process saves time and reduces the chance of breaking a tap. The thread can go to the bottom of a hole, but to produce it requires an extra process using a bottoming tap and some clearance is required. Figure 12.22 shows a simplified representation of a thread that does not go through. The bolt should be shorter than the depth of thread so the bolt does not hit bottom. Notice in Figure 12.22 the hidden lines representing the major and minor thread diameters are spaced far enough apart to be clearly separate. This spacing is important because on some threads the difference between the major and minor diameters is very small and, if drawn as they actually appear, the lines would run together. The hidden line dashes are also drawn staggered for clarity. Figure 12.23 shows a bolt fastener as it appears drawn in assembly with two parts using simplified thread representation.

Drawing Simplified Threads

Simplified representations are the easiest thread symbols to draw and are the most commonly used in industry. The following steps show how to draw simplified threads.

STEP 1 Draw the major thread diameter as an object line for external threads and hidden line for internal threads as in Figure 12.24.

STEP 2 Draw the minor thread diameter, which is about equal to the tap drill size found in a tap drill chart. If the minor diameter and major diameter are too close together, then slightly exaggerate the space. The minor diameter is a hidden line for the external thread and a hidden line staggered with the major diameter lines for the internal thread. The minor diameter is an object line for the internal thread in section. (See Figure 12.25.)

The simplified thread representation can be drawn with the same steps when CADD is used. For example, when using AutoCAD, the object lines are drawn with a continuous line type and the hidden lines are drawn with a hidden line type. The line weights can be set to display the proper line contrast between object lines and hidden lines. The lines can also be put on their own layers as needed. Simplified thread symbols can also be inserted as blocks in AutoCAD and scaled as needed upon insertion.

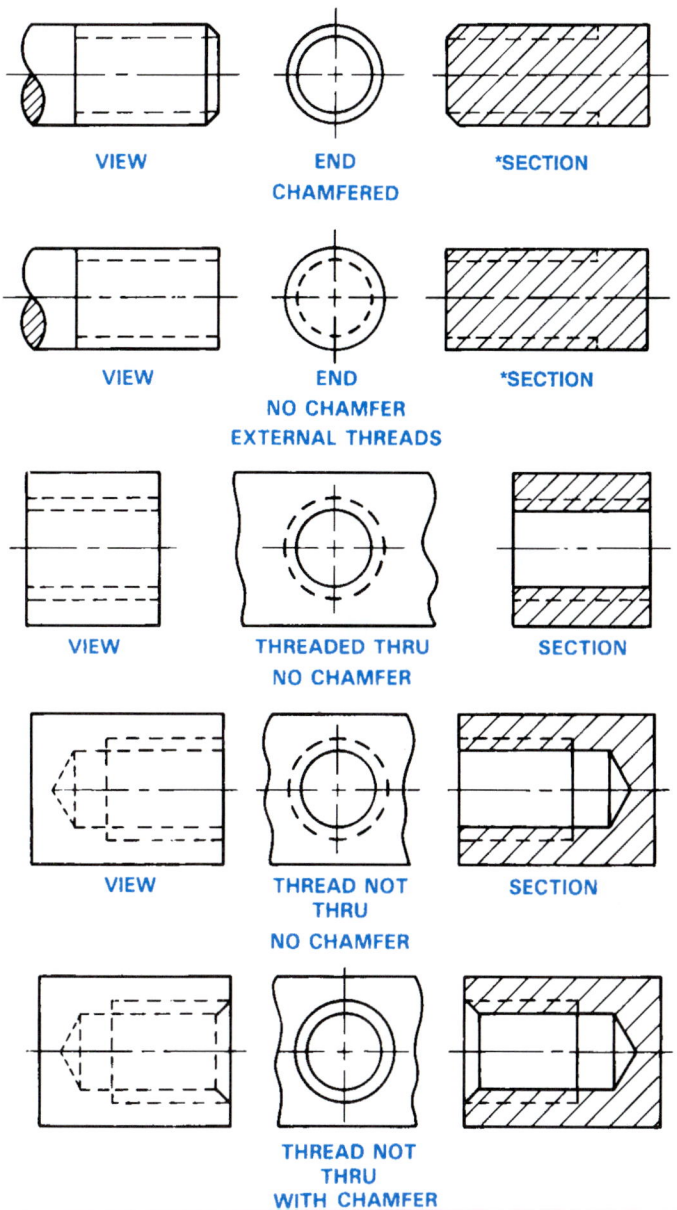

FIGURE 12.21 ■ Simplified thread representations. *Threaded shafts area not sectioned unless there is a need to expose an internal feature.

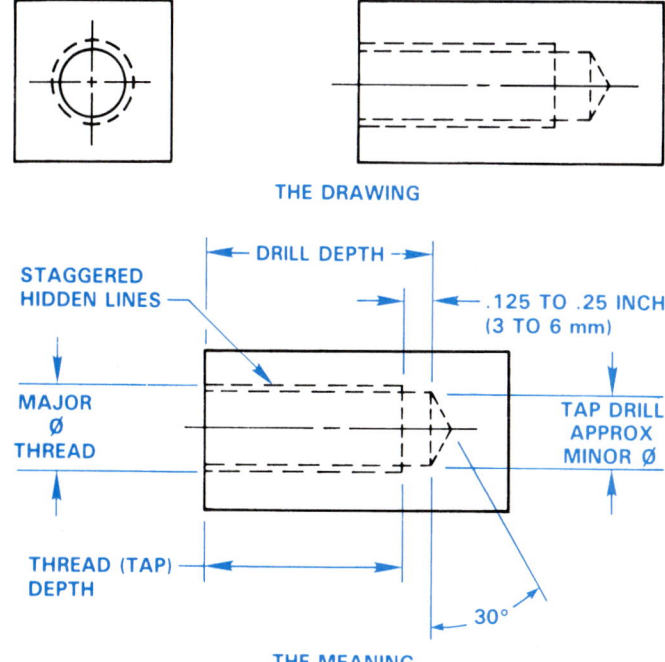

FIGURE 12.22 ■ The simplified internal thread that does not go through the part.

Drawing Schematic Threads

Schematic thread representations are drawn to approximate the appearance of threads by spacing lines equal to the pitch of the thread. When there are too many threads per inch to draw easily, then the distance can be exaggerated for clarity. The following steps show how to draw schematic thread representations:

STEP 1 Draw the major diameter of the thread. Schematic symbols can only be drawn in section for internal threads. Then lay out the number of threads per inch (if conve-

nient and space is available) using a thin line at each space. Figure 12.26 uses eight threads per inch.

STEP 2 Draw a thick line equal in length to the minor diameter between each pair of thin lines drawn in step 1. (See Figure 12.27.)

The schematic thread representation can be drawn with the same steps when using CADD. For example, when using Auto-CAD, all of the lines are drawn with a continuous line type. The line weights can be set to display the proper line contrast. The shorter lines representing the minor diameter are drawn the same thickness as the outside object lines, and the longer lines representing the major diameter are drawn as thin lines that are the same thickness as other thin lines in your drawing. The lines can be put on their own layers as needed. The lines representing the schematic threads can also be copied as needed with a command such as ARRAY that makes multiple copies of an object. Schematic thread symbols can be inserted as blocks in AutoCAD and scaled as needed upon insertion.

Drawing Detailed Threads

Detailed thread representations are the most difficult and time-consuming thread symbols to draw. They may be necessary for some applications as they most closely approximate the actual thread. Detailed external and sectioned threads can be drawn, but detailed internal threads may not be drawn in multiview. Detailed internal threads can be drawn only in section. The following steps show how to draw detailed thread representations for external and internal threads in section.

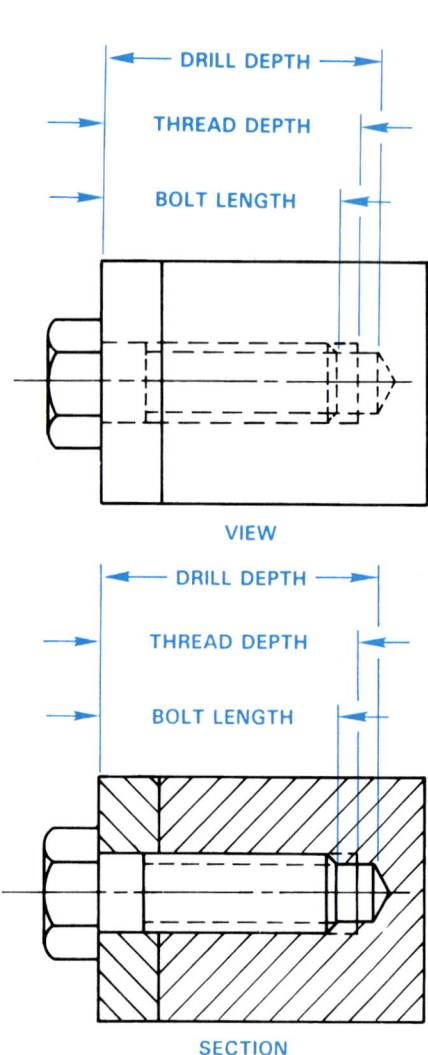

VIEW

SECTION

FIGURE 12.23 ■ The simplified external thread in a threaded assembly where the thread does not go through the part.

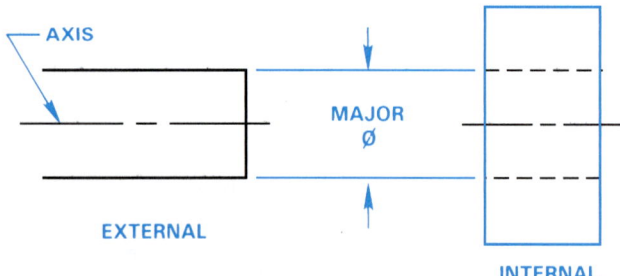

EXTERNAL

INTERNAL

FIGURE 12.24 ■ Step 1, establishing the simplified major thread diameter.

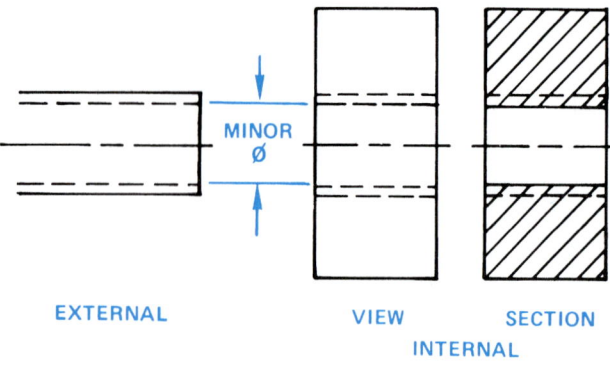

EXTERNAL

VIEW
INTERNAL

SECTION

FIGURE 12.25 ■ Step 2, establishing the simplified minor diameter.

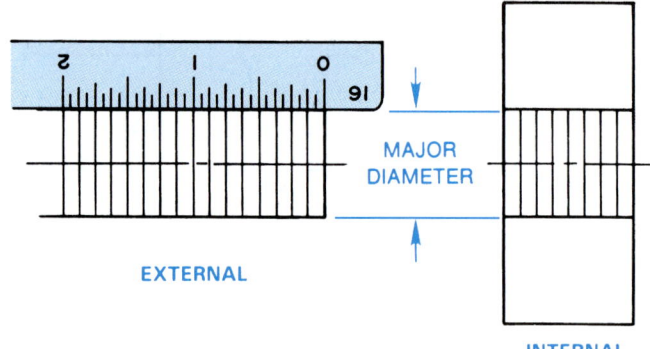

EXTERNAL

INTERNAL

FIGURE 12.26 ■ Step 1, establishing the major diameter and the number of threads per inch on the schematic thread.

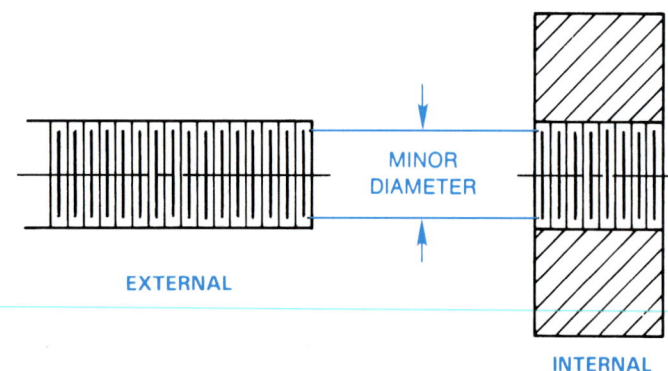

EXTERNAL

INTERNAL

FIGURE 12.27 ■ Step 2, establishing the minor diameter and completing the schematic thread.

STEP 1   Use construction lines to lightly draw the major and minor diameters of the thread as shown in Figure 12.28.

STEP 2   Divide one edge of the thread into equal parts; in this case, eight threads per inch so the pitch is .125 in., as shown in Figure 12.29. If the pitch is too small (much less than .125 in.), then exaggerate the distance. Remember, these symbols are representations and while they should have an appearance close to the actual thread, they do not have to be exact if the result is too difficult and time-consuming to draw, or unclear due to crowding.

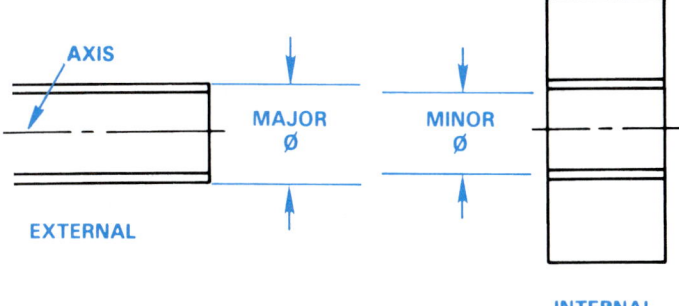

FIGURE 12.28 ■ Step 1, establishing the major and minor diameters for the detailed thread.

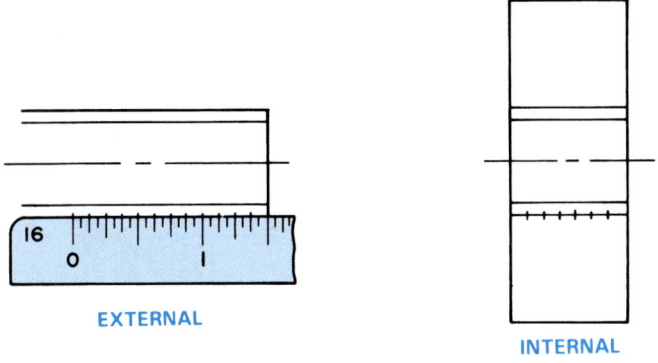

FIGURE 12.29 ■ Step 2, establishing the number of threads per inch on the detailed thread.

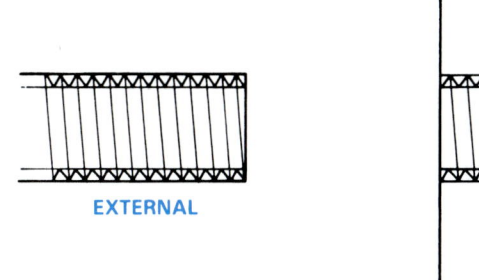

FIGURE 12.31 ■ Step 4, drawing the thread form on the detailed thread.

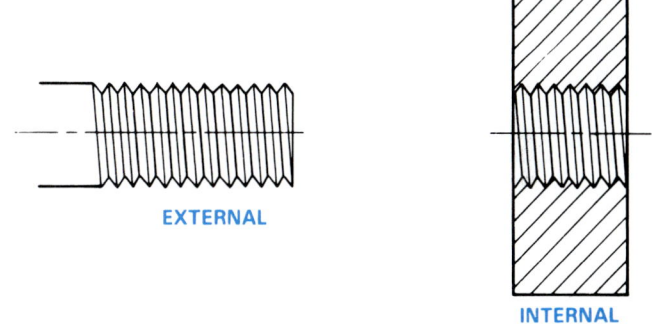

FIGURE 12.32 ■ Step 5, the complete detailed thread representation.

STEP 3 Stagger the opposite side one-half pitch and draw parallel thin lines equal to the spaces established in step 2. (See Figure 12.30.)

STEP 4 Draw Vs at 60° to form the root and crest of each thread. (See Figure 12.31.)

STEP 5 Complete the detailed thread representation by connecting the roots of opposite threads by drawing parallel lines as shown in Figure 12.32. Accuracy is very important if detailed threads are to turn out satisfactorily.

The detailed thread representation can be drawn with the same steps when CADD is used. For example, when using AutoCAD,

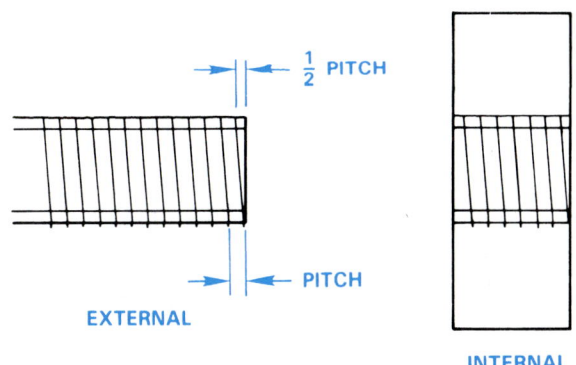

FIGURE 12.30 ■ Step 3, constructing the thread pitch on the detailed thread.

the object lines are drawn with a continuous line type. You can draw one V thread form and then use the ARRAY command to conveniently copy the desired number of threads onto one side of the fastener. Then use the MIRROR command to copy the threads to the other side, and the MOVE command to move them over one-half pitch. Then draw one set of major diameter and minor diameter lines followed by using the ARRAY command to duplicate them across the entire thread. Use the TRIM command as needed to finish the thread detail. The thread representation can be put on its own layer as needed. Detailed thread symbols can also be created as a block and then scaled as needed upon insertion.

Detailed thread representations can be used to draw any thread form using the same steps previously shown. The difference occurs in drawing the profile of the thread. For example, Vs are used to draw Unified, sharp-V, American Standard, and metric threads. The profile changes for other threads. An industry drawing of a fastener using a detail thread representation is shown in Figure 12.33.

The detailed thread drawings for American National Standard taper pipe threads are the same as for Unified threads except that the major and minor diameters taper at a rate of .0625 in. per inch. Figure 12.34 shows that pipe threads may be drawn tapered or straight depending on company preference. The thread note clearly defines the type of thread. Additional information about pipe thread design and drafting is provided later in this chapter.

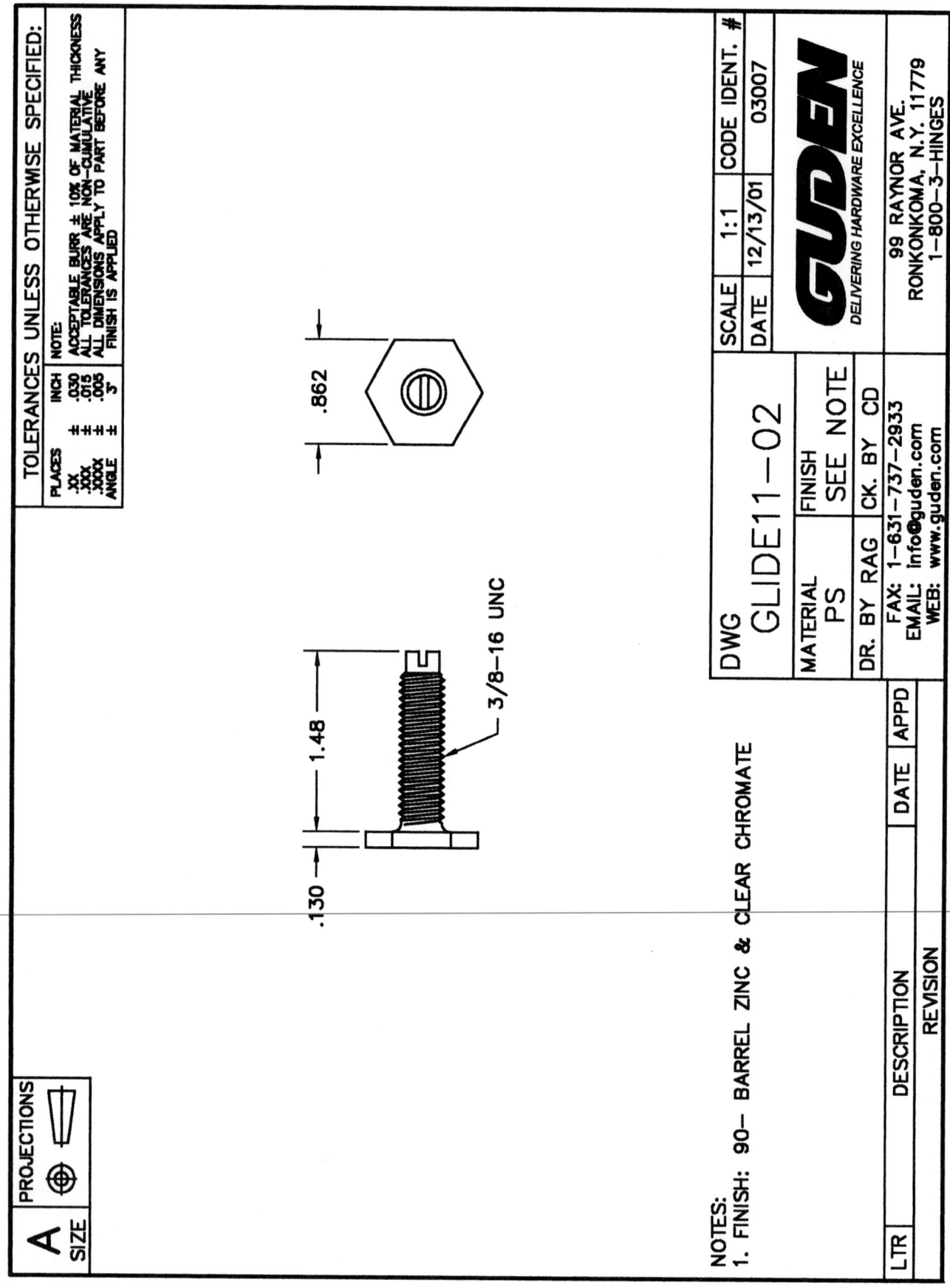

FIGURE 12.33 ■ An industry drawing using the detailed thread representation. *Courtesy H.A. Gluden Co., Inc.*

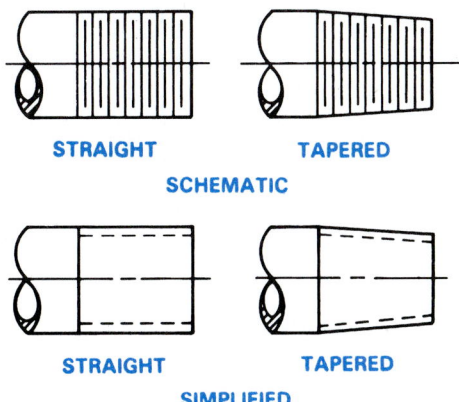

STRAIGHT   **TAPERED**

SCHEMATIC

STRAIGHT   **TAPERED**

SIMPLIFIED

FIGURE 12.34 ■ Straight and tapered pipe thread representations.

THREAD NOTES

Simplified, schematic, and detailed thread representations clearly show where threads are displayed on a drawing. However, the representations alone do not give the full information about the thread. As the term representation implies, the symbols are not meant to be exact but they are meant to describe the location of a thread when used. The information that clearly and completely identifies the thread being used is the thread note. The thread note must always be in the same order and be accurate; otherwise the thread may be manufactured incorrectly.

Unified and American National Threads

The thread note shall always be drawn in the order shown. The components of the note are described as follows:

1/2–13 UNC–2A
(A) (B)(C)   (D)(E)(F)(G)   (H)
(A)   The major diameter of the thread in inches followed by a dash (–). The major diameter is generally given as a fractional value.
(B)   Number of threads per inch.
(C)   Series of threads are classified by the number of threads per inch as applied to specific diameters and thread forms, such as coarse or fine threads. UNC (in the example) means Unified National Coarse. Others include UNF for Unified National Fine, UNEF for Unified National Extra Fine, or UNS for Unified National Special. The UNEF and UNS thread designations are for special combinations of diameter, pitch, and length of engagement. American National screw threads are identified with UN for external and internal threads, or UNR, a thread designed to improve fatigue strength of external threads only. The series designation is followed by a dash (–).
(D)   Class of fit is the amount of tolerance. 1 means a large tolerance, 2 is a general-purpose moderate tolerance, and 3 is for applications requiring a close tolerance.

(E)   A means an external thread (shown in the example) while B means an internal thread. (B replaces A in this location.) The A or B may be omitted if the thread is clearly external or internal, as shown on the drawing.
(F)   A blank space at (F) means a right-hand thread. A right-hand thread is assumed. LH in this space identifies a left-handed thread.
(G)   A blank space at (G) identifies a thread with a single lead, that is, a thread that engages one pitch when rotated 360°. If a double or triple lead is required, then the word DOUBLE or TRIPLE must be lettered here.
(H)   This location is for internal thread depth or external thread length in inches. When the drawing clearly shows that the thread goes through, this space is left blank. If clarification is needed, then the word THRU is lettered here.

Metric Threads

The metric thread notes shown below are the recommended standard as specified by the International Organization for Standardization (ISO). The note components are described as follows:

M 10 × 1.5–6H
(A)(B)   (C) (D) (E) (F)
(A)   M is the symbol for ISO metric threads.
(B)   The nominal major diameter in millimeters, followed by the symbol X, meaning *by*.
(C)   The thread pitch in millimeters, followed by a dash (–).
(D)   The number can be a 3, 4, 5, 6, 7, 8, or 9, which identifies the grade of tolerance from fine to coarse. The larger the number, the larger the tolerance. Grades 3 through 5 are fine, and 7 through 9 are coarse. Grade 3 is very fine and grade 9 is very coarse. Grade 6 is the most commonly used and is the medium tolerance metric thread. The grade 6 metric thread is comparable to the class 2 Unified screw thread. A letter placed after the number gives the thread tolerance class of the internal or external thread. Internal threads are designated by uppercase letters such as *G* or *H*, where *G* means a tight allowance and *H* identifies an internal thread with no allowance. The term allowance refers to the tightness of fit between the mating parts. External threads are defined with lowercase letters such as *e*, *g*, or *h*. For external threads, *e* denotes a large allowance, *g* is a tight allowance, and *h* establishes no allowance. Grades and tolerances below 5 are intended for tight fits with mating parts; those above 7 are a free class of fit intended for quick and easy assembly. When the grade and allowance are the same for both the major diameter and the pitch diameter of the metric thread, then the designation is given as shown, 6H. In some situations where precise tolerances and allowances are critical between the major and pitch diameters, separate specifications could be

used, for example, 4H 5H, or 4g 5g, where the first group (4g) refers to the grade and allowance of the pitch diameter, and the second group (5g) refers to the grade and tolerance of the major diameter. A fit between a pair of threads in indicated in the same thread note by specifying the internal thread followed by the external thread specifications separated by a slash, for example, 6H/6g.

(E) A blank space at (E) denotes a right-hand thread, a thread that engages when turned to the right. A right-hand thread is assumed unless an LH is lettered in this space. LH, which describes a left-hand thread, must be specified for a thread that engages when rotated to the left.

(F) The depth of internal threads or the length of external threads in millimeters is provided at the end of the note. When the thread goes through the part, this space is left blank, although some companies prefer to letter the description THRU.

Other Thread Forms

Other thread forms, such as Acme, are noted on a drawing using the same format. For example, 5/8–8 ACME–2 describes an Acme thread with a 5/8-in. major diameter, 8 threads per inch, and a general purpose class 2 thread fit. For a complete analysis of threads and thread forms, refer to the *Machinery's Handbook* published by Industrial Press, Inc.

American National Standard taper pipe threads are noted in the same manner with the letters NPT (National pipe thread) used to designate the thread form. A typical note may read 3/4–14 NPT. Additional information about pipe thread design is provided later in this chapter.

Thread Notes on a Drawing

The thread note is usually applied to a drawing with a leader in the view where the thread appears as a circle for internal threads as shown in Figure 12.35. External threads can be dimensioned with a leader as shown in Figure 12.36, with the thread length given as a dimension or at the end of the note. An internal thread that does not go through the part can be dimensioned as in Figure 12.37. Some companies require the drafter to indicate the complete process required to machine a thread including noting the tap drill size, tap drill depth if through, the thread note and thread depth if not through. (See Figure 12.38.) A thread chamfer can also be specified in the note as shown in Figure 12.39.

Many companies require only the thread note and depth. The complete process is determined in manufacturing.

MEASURING SCREW THREADS

When measuring features from prototypes or existing parts, the screw thread size can be determined on a fastener or threaded part by measurement. Measure the major diameter

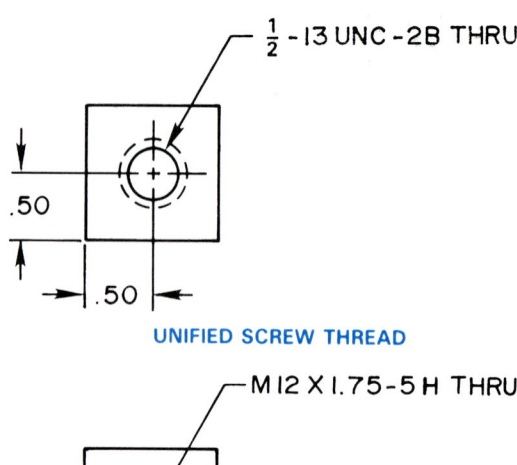
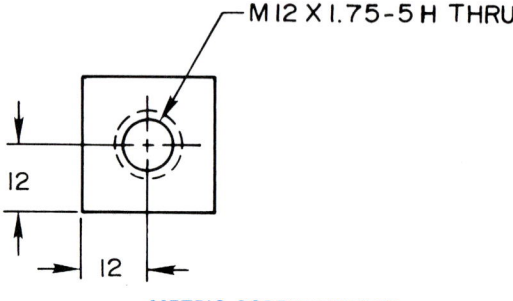

FIGURE 12.35 ■ Drawing and noting internal screw threads (simplified representation).

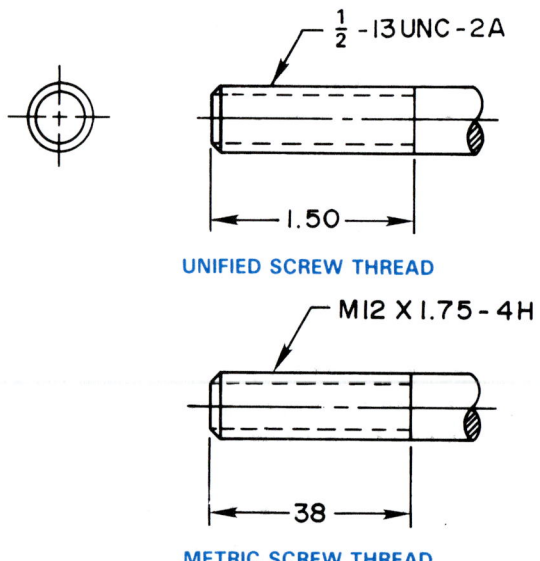

FIGURE 12.36 ■ Drawing and noting external screw threads.

with a vernier caliper or micrometer. Determine the number of threads per inch when a rule or scale is the only available tool by counting the number of threads between inch graduations. The quickest and easiest way to determine the thread specification is with a screw pitch gage, which is a set of thin leaves with teeth on the edge of each leaf that correspond to standard thread sections. Each leaf is stamped to show the

(Continued on page 371)

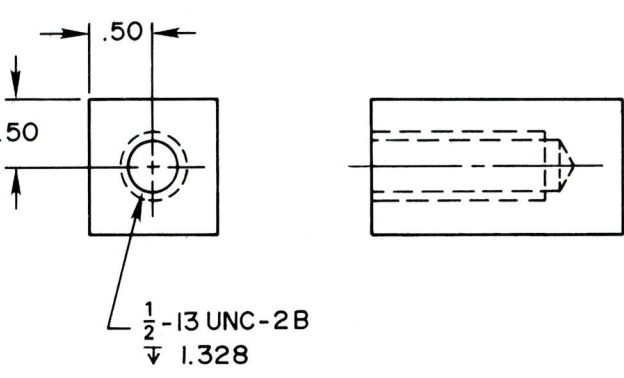

UNIFIED SCREW THREAD

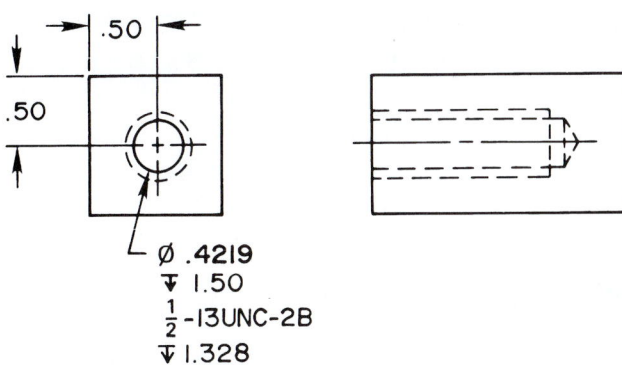

FIGURE 12.38 ■ Showing tap drill depth and thread depth.

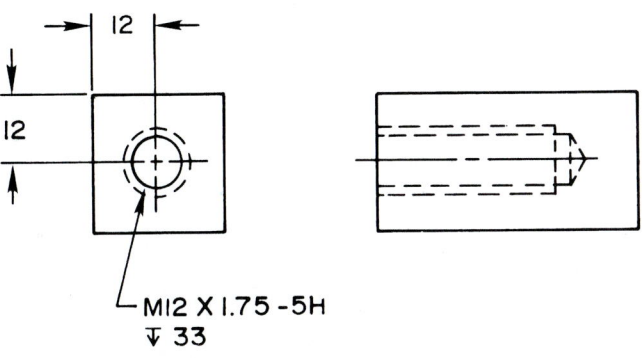

METRIC SCREW THREAD

FIGURE 12.37 ■ Drawing and noting internal screw threads with a given depth.

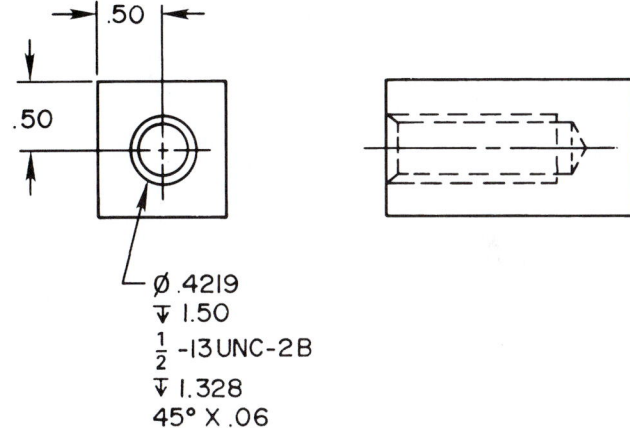

FIGURE 12.39 ■ Showing tap drill and thread depth with a chamfer.

THREAD NOTES

Thread information is quickly and accurately added to drawings using CADD. As described throughout this chapter, threads are drawn using CADD tips and techniques that are similar to manual drafting. However, CADD tools such as Layers, and specific commands including COPY, MOVE, OFFSET, ARRAY, and TRIM greatly aid in the process of adding any style of thread representation. In addition to the tools and commands used to draw thread representations, most CADD systems contain specific dimension or annotation options that are very effective for adding thread notes.

A thread note is typically attached to a leader that is capped with an arrowhead and the arrowhead and leader points to the thread representation. Though thread notes can be drawn using basic CADD drafting commands such as LINE and TEXT, there are specific tools that quickly and efficiently allow you to produce thread notes complete with a leader. One example is AutoCAD's LEADER or QLEADER command. The LEADER command allows you to use predefined dimension style settings to place a leader and type the required thread note using your keyboard. If the thread note is added to an internal feature, a variety of symbols can be used to fully describe the hole and desired thread. Thread notes are a type of dimension, and as a result, are typically drawn using the dimension style and layer. Figure 12.40 shows how a leader can be used in CADD to quickly create a thread note.

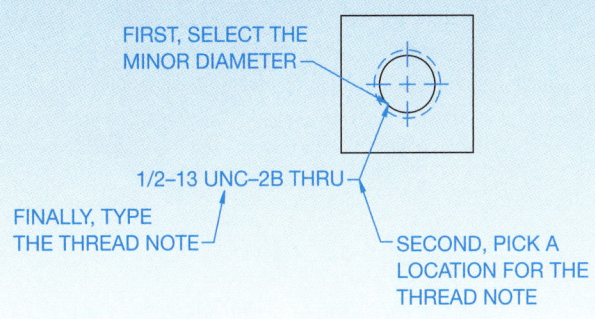

FIGURE 12.40 ■ Adding a thread note using the LEADER command.

CADD APPLICATIONS 2-D

THREAD REPRESENTATIONS

Typically, you can create threaded model features one of two ways, depending on the CADD system. The first option is to draw an actual thread profile, then generate the threaded feature using tools such as COIL or HELIX. This option creates a very realistic 3-D thread representation for applications when a true threaded feature is required. (See Figure 12.41.)

The second option available in some CADD systems involves defining threads by selecting or entering thread specifications, but not actually creating true thread geometry. This option involves using a tool such as HOLE to add internal thread information, or THREADS to describe external threads, again depending on the software. Figure 12.42 shows the Autodesk Inventor dialog box used to add threads to the hole shown. Notice how all of the thread specifications are entered in the dialog box. Though a real model of threads is not created, the thread characteristics are stored in the file, and you may have the option of displaying a bitmap image of threads for representation and visualization purposes, such as the image shown in Figure 12.42. A bitmap is an image of any kind, such as a picture,

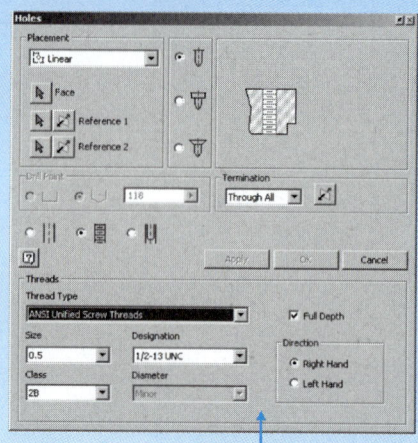

THREAD SPECIFICATIONS

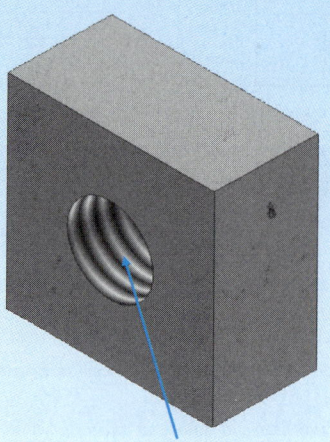

THREAD REPRESENTATION
(BITMAP IMAGE)

FIGURE 12.42 ■ An example of creating threads using the HOLES tool.

drawing, text character, or photo, composed of a collection of tiny individual dots.

THREAD NOTES

As described throughout this book, one of the benefits of using a 3-D CADD program that combines 2-D drawing capabilities is the power to dimension a drawing by referencing existing model information. In contrast to standard 2-D CADD systems, parametric solid model thread specifications are added to the model using tools such as HOLE and THREADS. Thread notes can then easily be added to a 2-D drawing using commands such as HOLE CALLOUT or HOLE/THREAD NOTES. For example, if you create a 1/2-13 UNC-2B hole in a model, the thread specifications

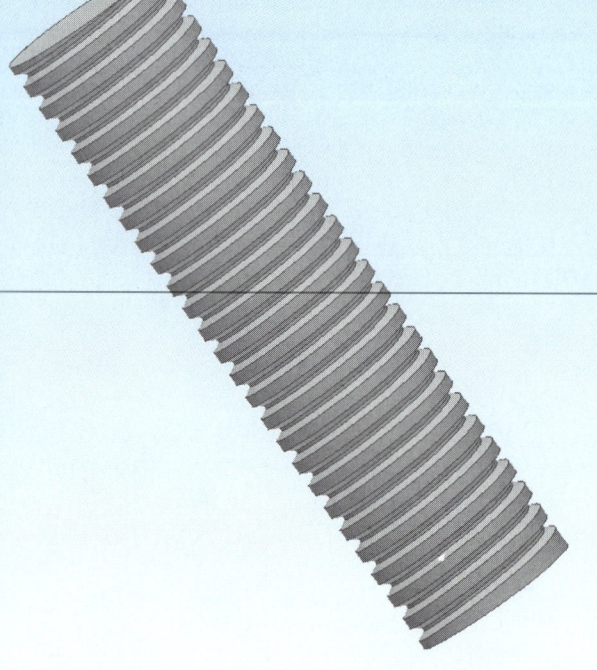

FIGURE 12.41 ■ An example of threads created using a COIL or HELIX tool.

are stored in the model, and it is just a matter of referencing the model thread parameters in the drawing to add a complete thread note. (See Figure 12.43.) Additionally, parametrically drawn thread notes automatically update when changes are made to the specifications of the corresponding model.

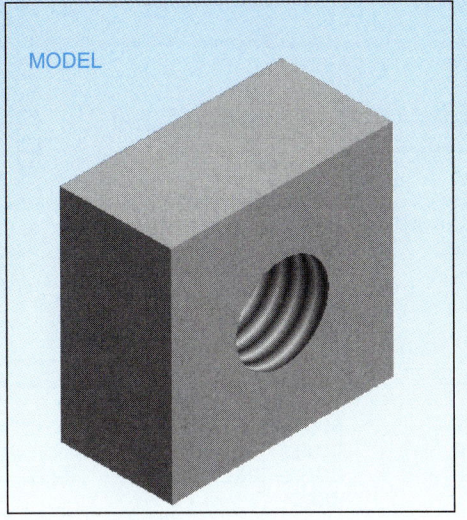 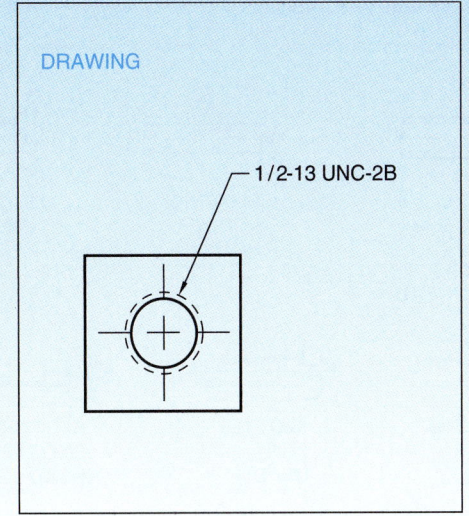

FIGURE 12.43 ■ Adding a thread note to a drawing by referencing existing model parameters.

number of threads per inch. Therefore, if the major diameter measures .625 in. and the number of teeth per inch is 18, then by looking at a thread variation chart you find that you have a 5/8–18 UNF thread.

THREADED FASTENERS

Bolts and Nuts

A bolt is a threaded fastener with a head on one end and is designed to hold two or more parts together with a nut or threaded feature. The nut is tightened upon the bolt or the bolt head can be tightened into a threaded feature. Bolts can be tightened or released by torque applied to the head or to the nut. Bolts are identified by a thread note, length, and head type; for example, 5/8–11 UNC–2 × 1 1/2 LONG HEXAGON HEAD BOLT. Figure 12.44 shows various types of bolt heads. Figure 12.45 shows common types of nuts. Nuts are classified by thread specifications and type. Nuts are available with a flat base or a washer face.

Machine Screws

Machine screws are thread fasteners used for general assembly of machine parts. Machine screws are available in coarse (UNC) and fine (UNF) threads, in diameters ranging from .060 in. to .5 in., and in lengths from 1/8 in. to 3 in. Machine

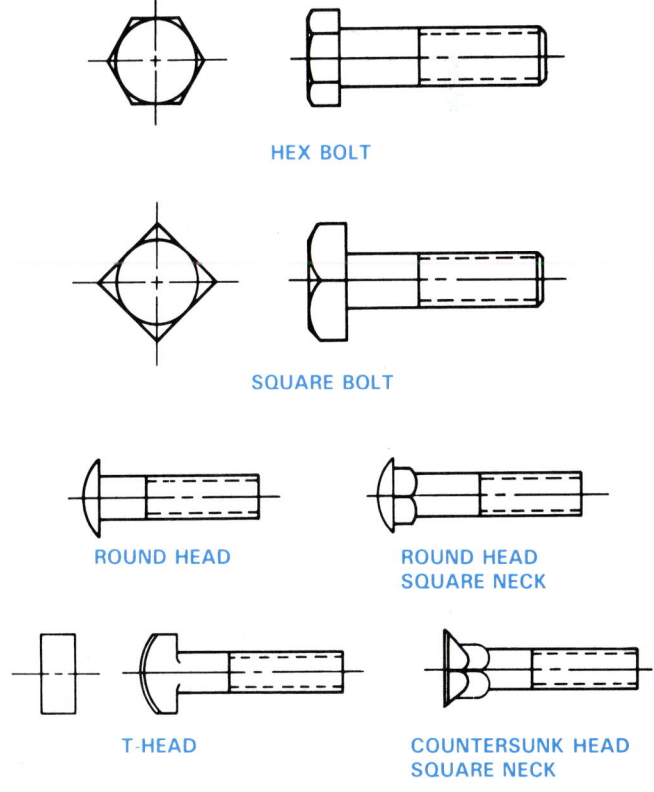

FIGURE 12.44 ■ Bolt head types.

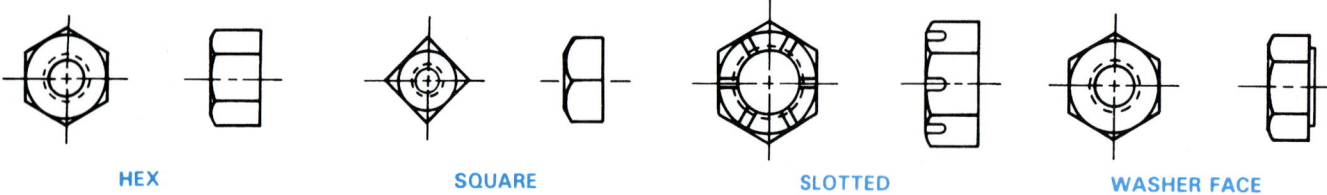

FIGURE 12.45 ■ Types of nuts. See Appendix P.

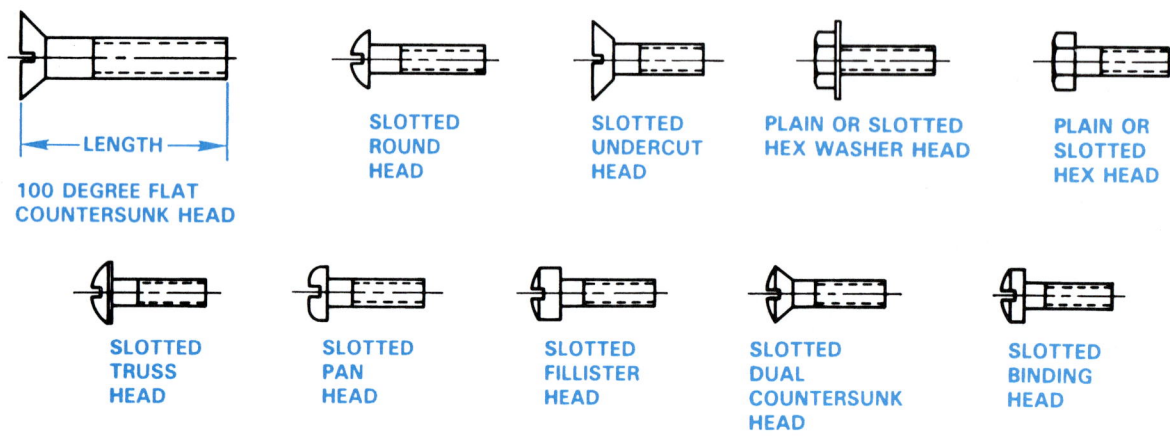

FIGURE 12.46 ■ Types of machine screw heads. See Appendix N.

screws are specified by thread, length, and head type. Machine screws have no chamfer. There are several types of heads available for machine design flexibility. (See Figure 12.46.)

Cap Screws

Cap screws are fine-finished machine screws that are generally used without a nut. Mating parts are fastened where one feature is threaded. Cap screws have a variety of head types and range in diameter from .060 in. to 4 in., with a large range of lengths. Lengths vary with diameter; for example, lengths increase in 1/16-in. increments for diameters up to 1 in. For diameters larger than 1 in., lengths increase in increments of 1/8 in. or 1/4 in.; the other extreme is a 2-in. increment for lengths over 10 in. Cap screws have a chamfer to the depth of the first thread. Standard cap screw head types are shown in Figure 12.47.

Set Screws

Set screws are used to help prevent rotary motion and to transmit power between two parts such as a pulley and shaft. Purchased with or without a head, set screws are ordered by specifying thread, length, head or headless, and type of point. Headless set screws are available in slotted and with hex or spline sockets. The shape of a set screw head is usually square. Standard square-head set screws have cup points, although other points are available. Figure 12.48 shows optional types of set screw point styles.

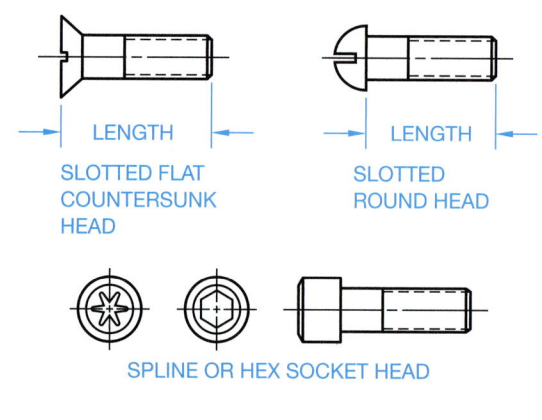

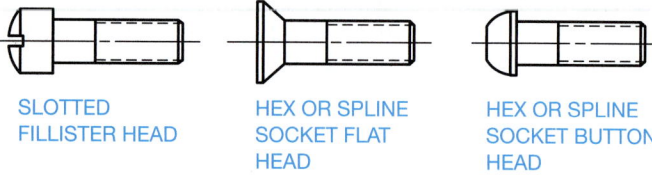

FIGURE 12.47 ■ Cap screw head styles. See Appendix M.

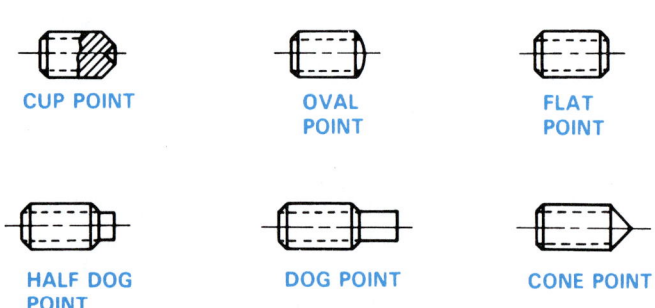

FIGURE 12.48 ■ Set screw point styles. See Appendix O.

THREAD DESIGN GUIDELINES*

The ASME Y14 series of standards includes the ASME Y14.6 and Y14.6aM, *Screw Thread Representation* inch and metric series. While these standards provide the guidelines for representing threads on the drawing, they do not include design criteria.

The focus of this discussion is to provide design guidelines for threaded features that better accommodate manufacturing practices and tooling. Standard thread-forming tools generally have lead-in chamfers that produce two or more incomplete threads on the leading edge of the tool as shown in Figure 12.49. These incomplete threads are called **runout**. Designing features that do not provide enough allowance for runout or sufficient room for the tool affect tool life and reduce the probability of a good thread.

Designing Threaded Holes

Holes threaded to receive a fastener are produced as either a through hole or a blind hole. Through holes are preferred over blind holes from a manufacturing standpoint. This eliminates consideration of incomplete threads, facilitates chip disposal, and allows use of most effective production methods. Unless the thickness of the material to be threaded is considerably greater than the required thread length, or unless design requirements prevent it, the pilot hole (tap drill) for the thread should be drilled through the section. For example, if the total material thickness is equal to or less than the thread depth plus two times

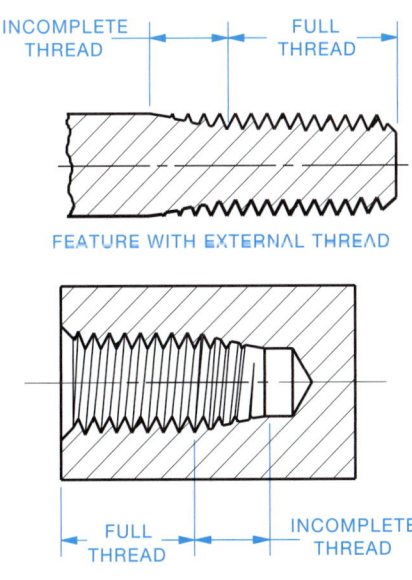

FIGURE 12.49 ■ Standard thread-forming tools generally have lead-in chamfers that produce two or more incomplete threads on the leading edge of the tool.

*Courtesy of Dick Button, Standards Coordinator
Fisher Controls International, Inc.,
Marshalltown, IA.*

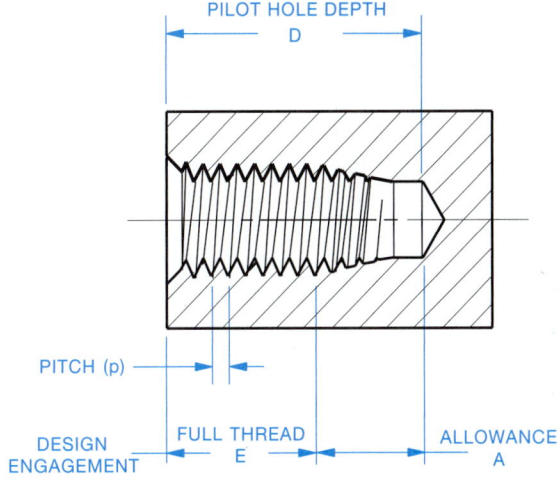

$$D = E + A$$
$$A = 5p + ADDITIONAL$$

ADDITIONAL TO BE ESTABLISHED LOCALLY BASED ON EXPERIENCE WITH PARTICULAR MANUFACTURING ENVIRONMENT. CONSIDER 1 TO 2 EXTRA PITCHES.

FIGURE 12.50 ■ Specifications for a threading for a blind hole.

the allowance A shown in Figure 12.50, then make the feature a through hole.

A **blind hole** is a hole that does not go through. If a blind hole is required, the part design should allow the pilot hole to be considerably deeper than the required thread length to maximize machining procedures. There must be an allowance for the chamfer (tap lead) that is greater than the typical lead for the style of threading tool expected to be used. The most common method of producing internal threads is with taps. ASME/ANSI B94.9 covers designs of ten different types of taps, furnished in taper, plug, or bottoming chamfers. This was introduced earlier in this chapter.

Taper taps generally have lead-in chamfers equal to or greater than 10 pitches. They are typically used with through holes in high-speed or high-quantity production applications.

Bottoming taps have lead-in chamfers equal to 2 pitches. They are used for special applications where through holes are not acceptable and the space is limited for pilot hole depth allowance. These should be avoided if at all possible by considering a different design approach. It is difficult to obtain a good start for the tapping process because there is such a short chamfer. Also, depending on the type of material being machined, bottoming taps may not produce "clean" threads.

Plug taps have lead in chamfers of 5 to 7 pitches and are the most commonly used production tool for thread forming. Typical is the cutting tap, but gaining popularity is the cold forming tap. Because cold forming does not remove material and cold work hardens some materials. This method, when done correctly, usually produces a stronger thread.

When determining pilot hole depth, this formula should be a minimum:

Pilot hole depth = Required thread depth + [5p]

(p = pitch, distance between two thread crests, as shown in Figures 12.3 and 12.4). Tap drill sizes are given in Appendix R.

Additional allowance for chips from the machining is beneficial. According to ASME/ANSI Y14.6, the dimension specified on the drawing for thread depth is to be the minimum measured distance to the last full thread, which means crowding the allowance for chips, tool chamfer, and general machining tolerances could jeopardize the quality of the thread.

A thread chamfer (or countersink) in the pilot hole before tapping helps prevent burrs from the tapping process. A reasonable countersink is a 90° inclusive, 0.01 to 0.02 in. (0.2 to 0.5 mm) larger than the major diameter of the thread.

Designing External Threads

Externally threaded features are used with a variety of applications that require considerations for manufacture as well as design function. This discussion deals with the standard Unified or metric (M) profile V thread. These are used primarily for fastening and joining.

Just as with the internal thread production, there must be allowance for incomplete threads due to tool runout. Generally, an allowance of three pitch lengths minimum should be used for design of tool runout. **Tool runout** is the distance a tool can go beyond the required full thread length as shown in Figure 12.51. Whenever design permits, this allowance should be increased. If design requirements do not permit the full allowance for standard tools, other possibilities must be investigated with your manufacturing source. For long-term cost-effectiveness of a product, it is better to design around standard tooling.

There are three physical possibilities to consider when designing external threaded features:

1. Apply the thread to a straight rod sized to the major diameter of the thread. This includes continuously threaded studs and rods as shown in Figure 12.52a.

2. Apply the thread to a feature diameter that is reduced in size from its adjacent feature, creating a shoulder as shown in Figure 12.52b.

3. Apply the thread to a feature diameter that is larger in size than the adjacent feature. (See Figure 12.52c.)

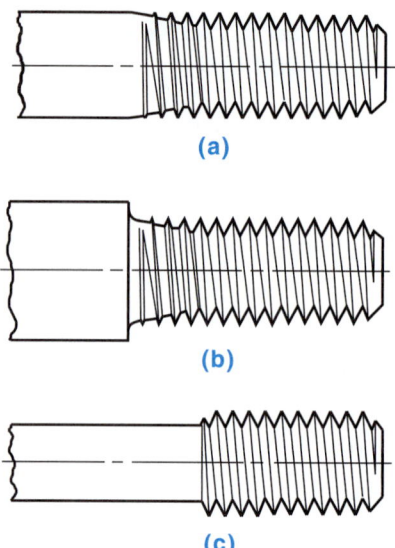

FIGURE 12.52 ■ There are three physical properties to consider when designing external threads: (a) Apply the thread to a straight rod sized to the major diameter of the thread. (b) Apply the thread to a feature diameter that is reduced in size from its adjacent feature, creating a shoulder. (c) Apply the thread to a feature diameter that is larger in size than the adjacent feature.

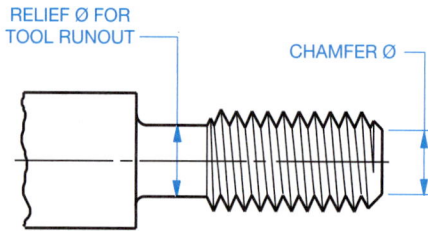

FIGURE 12.53 ■ If a relief is used to provide room for tool runout, the applied relief diameter should not exceed the minor diameter of the threads.

External threads should have a lead-in chamfer for tooling. The applied chamfer diameter should not exceed the minor diameter of the thread as indicated in ASME B1.1, *Unified Inch Screw Threads,* or ASME B1.1M, *Metric Screw Thread—M Profile.* If a relief is used to provide room for tool runout, the applied relief diameter should not exceed the minor diameter of the thread as shown in Figure 12.53. Chamfer and relief calculations and tool runout lengths for drawing application can be made available to designers and drafters by preparing a chart of allowance values to be used for commonly used thread sizes as displayed in Table 12.1. Each design and manufacturing group must determine an acceptable allowance for their environment and product. The design that is applied to large assemblies may not be applicable to small precision products.

When designing a threaded joint with a shoulder, there must be the provision of either a counterbore on the internal threaded feature, as shown in Figure 12.54a, or a relief on the external

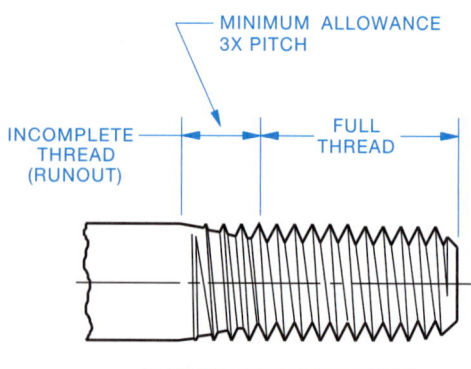

FIGURE 12.51 ■ Tool runout is the incomplete threads that a tool can go beyond the required full thread length.

TABLE 12.1 EXAMPLE OF CHART FOR RELIEF OR CHAMFER ALLOWANCE FOR INCH THREADS (Use with Figure 12.53)	
Threads per Inch	For Relief ∅ or Chamfer ∅, Subtract from Major ∅
20	.08
18	.08
16	.09
14	.10
13	.11
12	.12
11	.13
10	.14
8	.17

The allowances for this table were created by determing the difference between the maximum major and minor ∅, then adding .02.

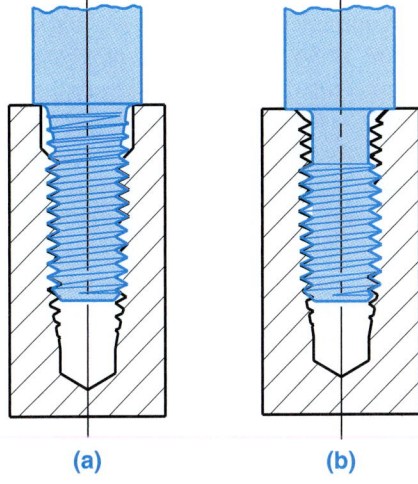

FIGURE 12.54 ■ When designing a threaded joint with a shoulder, provide (a) a counterbore on the internal threaded feature, or (b) a relief on the external threaded feature.

threaded feature, as shown in Figure 12.54b, to accommodate the tool runout.

Designs using threads with greater than normal length of engagement require special designations and gaging tolerances to properly verify the length during the manufacturing process. Typical applications are an internal thread in soft material such as plastic or aluminum, where longer engagement is required to increase sheer strength, or a continuously threaded rod used in an assembly for an adjusting mechanism. In either case, binding of the assembly may occur if the special length of thread is not properly noted. For example, if a standard thread gage is used to verify a long external threaded feature, the gage typically covers

only 15 pitches. However, if the thread is assembled into a feature that covers greater than 15 pitches, the gage area will not have verified the amount of thread relationship equal to that of the assembly. Normal length of engagement and gaging lengths are defined in ASME B1.1 and B1.13M, section 5, *Allowances and Tolerances*. Normal length of engagement and full thread length of a threaded feature are not necessarily the same.

Designing Pipe Threads

Pipe threads appear to be such a simple concept, and yet, depending on the desired function, they can consume many hours of engineering time attempting to define them adequately and manufacturing time trying to get the expected results.

There are three major functions of pipe threads:

1. An assembly that creates a pressure-tight joint, either with a sealer or without a sealer.

2. An assembly with a free/loose-fitting mechanical joint that is not pressure-tight.

3. An assembly with a rigid mechanical joint that is not pressure-tight.

The assembly with a rigid mechanical joint that is not pressure-tight is used for structural purposes such as railings or racks.

The assembly with a free/loose-fitting mechanical joint that is not pressure-tight can be used for fixture joints, mechanical joints with locknuts, and hose couplings that have a gasket to seal the joint.

The assembly that creates a pressure-tight joint, either with a sealer or without a sealer for pressure-tight joints, can be obtained by these two separate pipe thread types:

■ Taper pipe threads for general use, NPT.

■ Dryseal pipe threads, NPTF.

The dryseal pipe thread form is based on the NPT thread, but with some modifications and greater accuracy of manufacture, it can make a pressure-tight joint without a sealer. General use NPT threads must have a sealing compound applied to get a pressure-tight joint.

ANSI/ASME inch series NPT has wide use throughout the world due to the U.S. influence and increase of piping products and construction. There is also the DIN (German) metric pipe thread and the ISO pipe thread that is based on the British Whitworth pipe thread. The ISO pipe thread uses a designation that is similar to the NPT designation and a few of the sizes appear to assemble with the NPT thread, so be aware of this when dealing with products or manufacturing from other countries.

Taper pipe threads, NPT, are the focus of this discussion. ANSI/ASME B1.20.1, *Pipe Threads, General Purpose (Inch)*, is the standard for manufacture and gaging of NPT threads. ANSI/ASME Y14.6, *Screw Thread Representation*, is the standard for presentation and designation on drawings. Figure 12.55

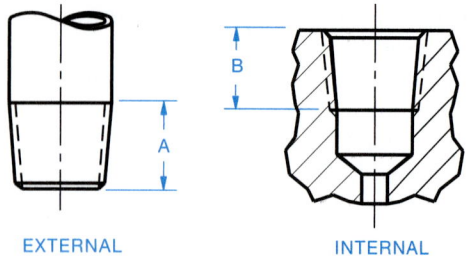

EXTERNAL   INTERNAL

FIGURE 12.55 ■ Recommended methods of representing NPT threads on detail drawings. The dimensions A and B are for reference to Table 12.2 and are not shown on the drawing.

shows recommended methods of representing NPT threads on detail drawings. An introduction to this was provided earlier in this chapter.

The drawing angle for representing the taper of a pipe thread should be 1 1/2° to 2° from the axis of the thread, or a 3 1/2° included angle. The values A and B in Table 12.2 refer to Figure 12.55, and provide a suggested distance on the drawing to represent where the thread ends, or the runout of imperfect threads. These dimensions are for drawing representation purposes only and must not be used for manufacture. *Do not specify a dimension on the drawing.*

When defining the criteria on the drawing for internal threads, two major factors for producing a good thread are the drill diameter and the depth of the pilot hole. There are many published drill guides for pipe thread tapping and most list a single drill size. The *Machinery's Handbook* is a good reference. However, various materials require adjustments to the pilot hole size due to differing machining characteristics. A better approach is to specify the pilot hole size on the drawing as a maximum size, then let manufacturing put in a pilot hole size that suits the process. When specifying the depth of the pilot hole, provide clearance for the tap lead that permits cutting into the hole far enough to produce thread for full gaging. Most standard pipe thread taps require at least a 7-pitch lead.

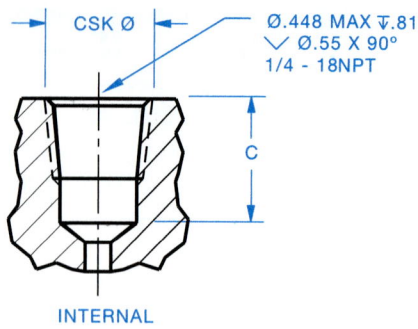

INTERNAL

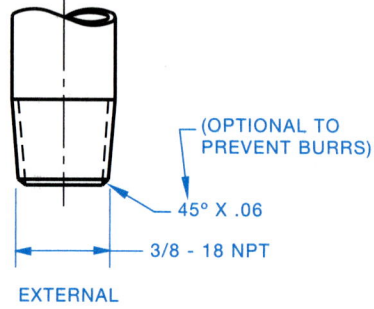

EXTERNAL

FIGURE 12.56 ■ Methods for showing NPT thread specifications on a drawing.

Specify the requirements on the drawing according to the examples in Figure 12.56. The dimensions given in Table 12.3 provide a guideline for values that can be used in dimensioning the pipe thread. These values have been derived from calculations and experience over many years of manufacturing NPT threads on products. These figures and values should not be taken as absolute for every application and process. The production of NPT threads has always been a bit of an art form, and it takes cooperation between engineering and manufacturing at each factory to develop the best criteria.

External NPT threads applied to a feature that has a shoulder similar to the example in Figure 12.56 must have room for the tooling. The part should be designed to provide as much room as possible. Dimension D from Figure 12.57 must be greater than dimension A in Figure 12.55. See the values from Table 12.2 for dimension A.

TABLE 12.2		
NPT Size	A	B
1/16 – 27	.38	.31
1/8 – 27	.38	.31
1/4 – 18	.58	.50
3/8 – 18	.61	.50
1/2 – 14	.78	.62
3/4 – 14	.78	.62
1 – 11 1/2	.97	.75
1 1/4 – 11 1/2	1.00	.75
1 1/2 – 11 1/2	1.03	.78
2 – 11 1/2	1.06	.78

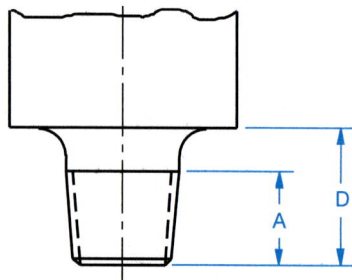

FIGURE 12.57 ■ Applying an NPT thread to a feature that has a shoulder. Dimension D must be greater than dimension A, which is found in Table 12.2. These dimensions are for reference only and are not displayed on the drawing.

TABLE 12.3 MAXIMUM HOLE SIZES FOR NPT THREADS				
NPT Thread		Produced Hole	Countersink ∅	Dimension C
Size	Pitch	Max ∅	(Tol ± 0.02)	Min
1/16	27	.253	.34	.56
1/8	27	.345	.44	.56
1/4	18	.448	.56	.81
3/8	18	.583	.69	.81
1/2	14	.721	.86	1.06
3/4	14	.931	1.06	1.06
1	11 1/2	1.168	1.34	1.25
1 1/4	11 1/2	1.513	1.69	1.31
1 1/2	11 1/2	1.752	1.91	1.31
2	11 1/2	2.226	2.38	1.31

In the process of manufacturing taper pipe threads, several considerations for producing a good joint are:

■ Pilot holes that are round.

■ Minimal waviness on produced thread, both internal and external (watch tool alignment).

■ Full crests for at least the first three threads, internal.

■ L1 gaging that goes deep instead of shallow.

■ Proper installation torque. Too much torque can be damaging.

LAG SCREWS AND WOOD SCREWS

Lag screws are designed to attach metal to wood or wood to wood. Before assembly with a lag screw, a pilot hole is cut into the wood. The threads of the lag screw then form their own mating thread in the wood. Lag screws are sized by diameter and length. Wood screws are similar in function to lag screws and are available in a wide variety of sizes, head styles, and materials.

SELF-TAPPING SCREWS

Self-tapping screws are designed for use in situations where the mating thread is created by the fastener. These screws are used to hold two or more mating parts when one of the parts becomes a fastening device. A clearance fit is required through the first series of features or parts while the last feature receives a pilot hole similar to a tap drill for unified threads. The self-tapping screw then forms its own threads by cutting or displacing material as it enters the pilot hole. There are several different types of self-tapping screws with head variations similar to cap screws. The specific function of the screw is important as these screws may be designed for applications ranging from sheet metal to hard metal fastening.

THREAD INSERTS

Screw thread inserts are helically formed coils of diamond-shaped wire made of stainless steel or phosphor bronze. The inserts are used by being screwed into a threaded hole to form a mating internal thread for a threaded fastener. Inserts are used to repair worn or damaged internal threads and to provide a strong thread surface in soft materials. Some screw thread inserts are designed to provide a secure mating of fasteners in situations where vibration or movement could cause parts to loosen. Figure 12.58 shows the relationship among the fastener, thread insert, and tapped hole.

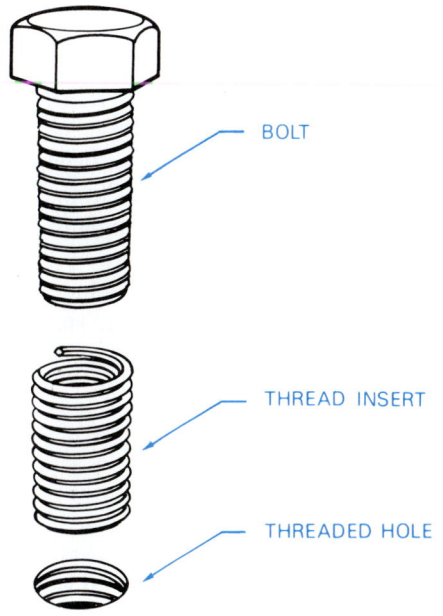

BOLT

THREAD INSERT

THREADED HOLE

FIGURE 12.58 ■ Thread insert.

SELF-CLINCHING FASTENERS

A self-clinching fastener is any device, usually threaded, that displaces the material around a mounting hole when pressed into a properly sized drilled or punched hole. This pressing or squeezing process causes the displaced sheet material to cold flow into a specially designed annular recess in the shank or pilot of the fastener. A serrated clinching ring, knurl, ribs, or hex head prevents the fastener from rotating in the metal when tightening torque is applied to the mating screw or nut. When properly installed, self-clinching fasteners become a permanent and integral part of the panel, chassis, bracket, printed circuit board, or other item in which they are installed. They meet high performance standards and can allow for easier disassembly of components for repair or service.

Self-clinching fasteners generally take less space and require fewer assembly operations than caged and other types of locking nuts. They also have greater reusability and more holding power than sheet metal screws. They are used mainly where good pullout and torque loads are required in sheet metal that is too thin to provide secure fastening by any other method.

Self-clinching fasteners traditionally fall into the categories of nuts, spacers and standoffs, and studs.

Self-Clinching Nuts

Self-clinching nuts feature thread strengths greater than those of mild screws and are commonly used wherever strong internal threads are needed for component attachment or fabrication assembly. During installation, a clinching ring locks the displaced metal behind the fastener's tapered shank, resulting in a high push-out resistance. High torque-out resistance is achieved when the knurled platform is embedded in the sheet metal. The clinching action of self-clinching nuts occurs on the fastener side of the thin sheet, with the reverse side remaining flush. A self-clinching nut is shown in Figure 12.59a.

Self-Clinching Spacers and Standoffs

Self-clinching spacers and standoffs are used where it is necessary to space or stack components away from a panel. Thru-threaded or blind types are generally standard, but variations have been developed to meet emerging applications, primarily in the electronics industry. These types include standoffs with concealed heads, others that allow boards to snap into place for easier assembly and removal, and those designed specifically for use in printed circuit boards. Figure 12.59b shows an example of a self-clinching standoff.

Self-Clinching Studs

Self-clinching studs are externally threaded self-clinching fasteners that are used where the attachment must be positioned before being fastened. Flush-head studs are normally specified,

FIGURE 12.59 ■ Self-clinching fasteners: (a) a self-clinching nut; (b) a self-clinching standoff; (c) a self-clinching stud. *Courtesy of Penn Engineering & Manufacturing Corp. and Hammer Inc. Advertising & Public Relations.*

but variations are available for desired high torque, thin sheet metal, or electrical applications. Figure 12.59c shows a self-clinching stud.

HOW TO DRAW VARIOUS TYPES OF SCREW HEADS

As you have found from the previous discussion, screw fasteners are classified, in part, by head type, and there is a large variety of head styles. A valuable drafting reference is the *Machinery's Handbook*, which clearly lists specifications for all common head types.

Hexagon head fasteners are generally drawn with the hexagon positioned across the corners vertically in the front view. When projected across the flats, the hex head looks like a square head. The following steps show how to draw a hex head bolt. Hex nuts are drawn in the same manner.

STEP 1 Use construction lines to draw the end view of the hexagon with the distance across the flats. A 3/4-in. nominal thread measures 1 1/8 in. across the flats. Position the bolt head so a front-view projection is across the corners as shown in Figure 12.60.

STEP 2 Block out the major diameter and length, and bolt head height. The same 3/4-in. hex bolt has a head height, H, of 11/32 in. (See Figure 12.61.)

STEP 3 Project the hexagon corners to the front view. Then establish radii B, centers with 60° angles as shown in Figure 12.62. Draw the radii.

STEP 4 Draw a 30° chamfer tangent on each side to the small radius arcs. Use object lines to complete both views and draw the thread representation. (See Figure 12.63.)

Square-head bolts are drawn in the same manner as hex-head bolts. Refer to Chapter 8, Geometric Construction, for instructions and examples on drawing square and hexagonal shapes. Use the *Machinery's Handbook* for specific dimensions related to the fastener head type you are drawing.

A few common cap screw heads are shown with approximate layout dimensions in Figure 12.64.

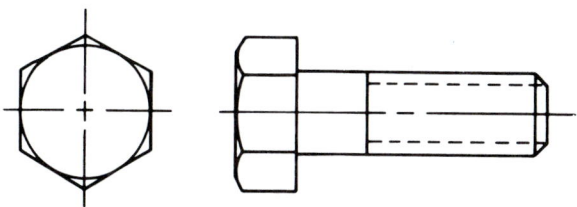

FIGURE 12.63 ■ Step 4, completed drawing.

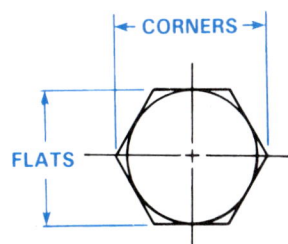

FIGURE 12.60 ■ Step 1, the flats and corner of a hexagon.

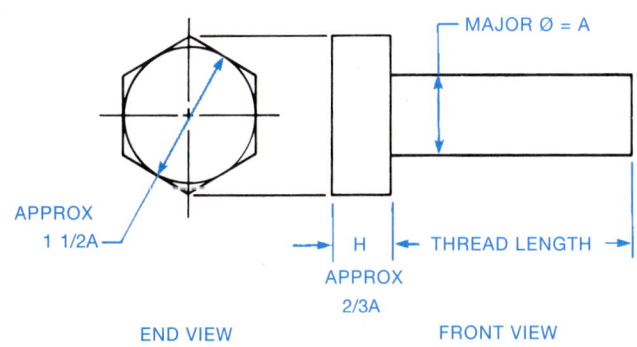

FIGURE 12.61 ■ Step 2, layout with construction lines.

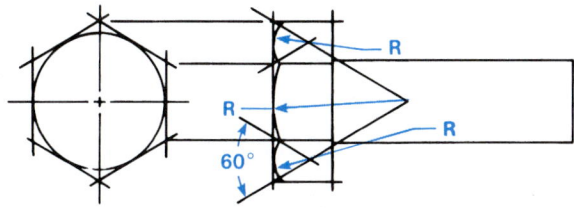

FIGURE 12.62 ■ Step 3, hex head layout.

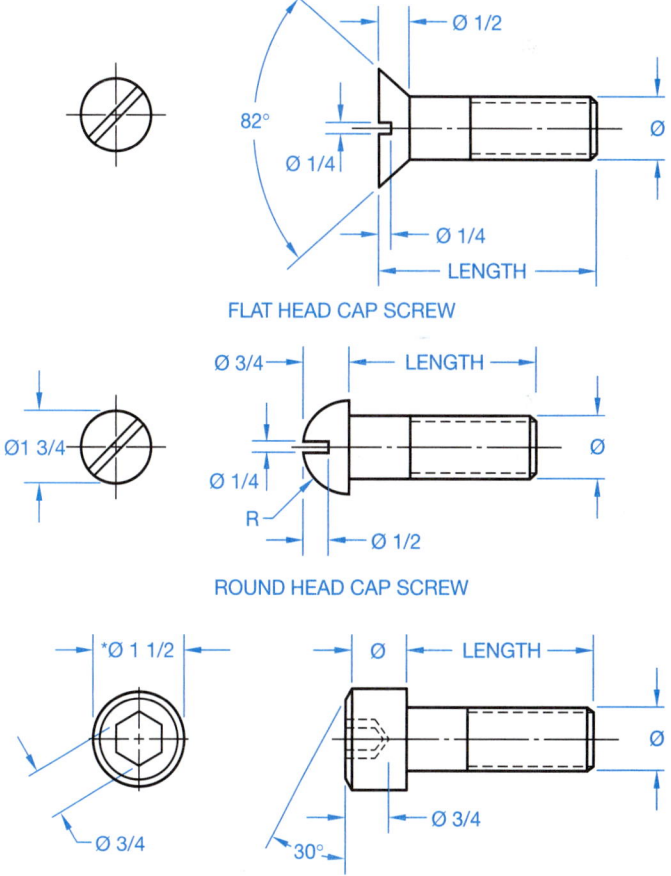

FLAT HEAD CAP SCREW

ROUND HEAD CAP SCREW

HEX SOCKET HEAD CAP SCREW

FIGURE 12.64 ■ Layout specification for common cap screws. *For metric screw sizes, the ratio of screw diameter to head diameter is 1.5 for 12 mm and above, but for sizes 3 mm to 10 mm it is 1.5 times plus 1 mm. Consult the *Machinery's Handbook* for sizes 2 mm and smaller.

Using Templates to Draw Screw Heads

When bolt heads, screw heads, or nuts are drawn only occasionally, then it can be satisfactory to use the layout techniques. However, whether you draw bolt or screw heads a few times or many times each day, it is always better to use templates. There is a large variety of templates designed to allow you to quickly and easily draw bolt or screw heads and detailed screw threads.

Using CADD to Draw Screw Heads

A CADD system can be used to very quickly and accurately draw bolt heads, screw heads, and nuts. The basic commands that are commonly used include LINE, ARC, CIRCLE, and POLYGON. The fastener heads are commonly constructed from a series of lines, arcs, circles, and polygons. The most common polygon is the hexagon, which creates the "hex" head bolt or screw, and the nut. The different types of fasteners can be created as blocks and then inserted and scaled into the drawing as needed. This practice saves a great deal of time. Special applications software is also available for inserting fasteners into the drawing. This was discussed earlier in this chapter.

Drawing your own CADD symbols for insertion into drawings is one option. There are also a wide variety of CADD libraries available for use on your drawings. Symbols are often found within your CADD software program as discussed in Chapter 3, or available from third-party sources. Many of these symbol libraries are free. An example of free CADD symbols can be found at *www.thomasnet.com*. After entering this website, pick CAD Drawings at the top of the screen. The next screen is CAD Register.com. Pick View CADBlocks Drawings in the CADBlocks area. Follow the on-screen instructions to access a wide variety of free CAD blocks.

DRAWING NUTS

A nut is used as a fastening device in combination with a bolt to hold two or more pieces of material together. The nut thread must match the bolt thread for acceptable mating. Figure 12.65

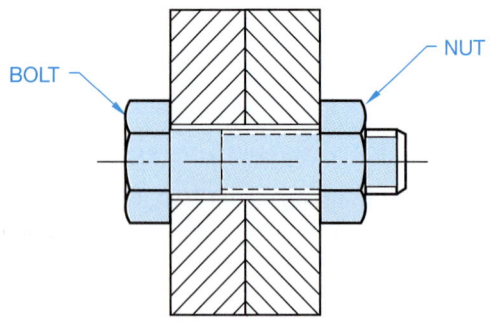

FIGURE 12.65 ■ Nut and bolt relationship known as a floating fastener.

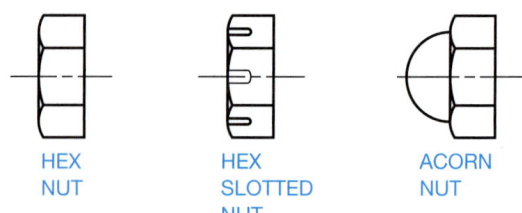

HEX NUT   HEX SLOTTED NUT   ACORN NUT

FIGURE 12.66 ■ Common nuts.

shows the nut and bolt relationship. The hole in the parts must be drilled larger than the bolt for clearance.

There are a variety of nuts in hexagon or square shapes. Nuts are also designed slotted to allow them to be secured with a pin or key. Acorn nuts are capped for appearance so the bolt thread is concealed. Self-locking nuts are available with neoprene gaskets that help keep the nut tight when movement or vibration is a factor. Figure 12.66 shows some common nuts.

Hex nuts are manually drawn in a manner similar to the method used to draw hex head fasteners shown in Figures 12.60 through 12.63. Use the *Machinery's Handbook* to get specific dimensions and information for the specific hex nut you are drawing. Square nuts and other types of fasteners are also detailed in the *Machinery's Handbook* for your reference.

DRAWING WASHERS

Washers are flat, disc-shaped objects with a center hole to allow a fastener to pass through. Washers are made of metal, plastic, or other materials for use under a nut or bolt head, or at other machinery wear points, to serve as a cushion or a bearing surface, to prevent leakage, to relieve friction, or as a locking device. (See Figure 12.67.) Washer thickness varies from .016 in. to .633 in. Refer to the *Machinery's Handbook* for washer tables with sizes and specifications.

DRAWING DOWEL PINS

Dowel pins used in machine fabrication are metal cylindrical fasteners that retain parts in a fixed position or keep parts aligned. Generally, depending on the function of the parts, one or two dowel pins are sufficient for holding adjacent parts. Dowel pins must generally be pressed into a hole with an interference tolerance of between .0002 in. to .001 in., depending on the material and the function of the parts. Figure 12.68 shows

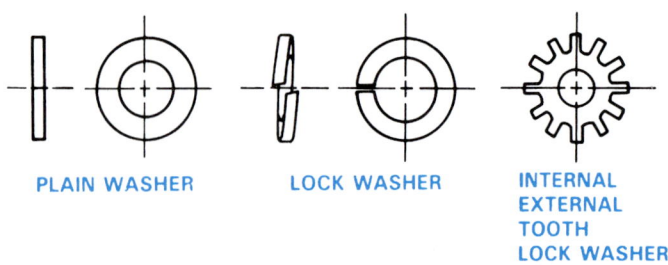

PLAIN WASHER   LOCK WASHER   INTERNAL EXTERNAL TOOTH LOCK WASHER

FIGURE 12.67 ■ Types of washers.

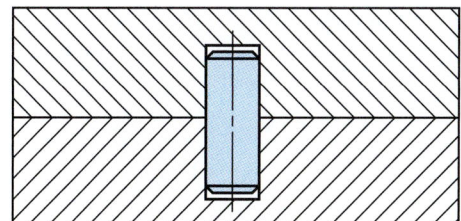

FIGURE 12.68 ■ Dowel pin in place, sectional view.

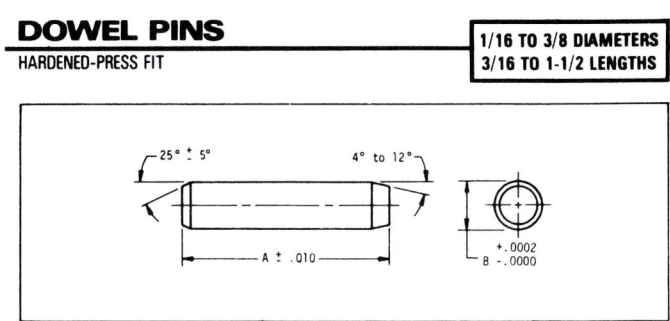

DOWEL PINS
HARDENED-PRESS FIT

| 1/16 TO 3/8 DIAMETERS |
| 3/16 TO 1-1/2 LENGTHS |

Material: 416 Stainless Steel (Clear Passivate) Hardened to: Rockwell C 36-40
Conforms to specification MS-16555
For 303 stainless pins, please see Cat. No. EPS-A1 and EPS-B1

	1/16 DIA. PIN B = .0626		3/32 DIA. PIN B = .0938		1/8 DIA. PIN B = .1251		5/32 DIA. PIN B = .1563	
A	Cat. No.	Price	Cat. No.	Price	Cat. No.	Price	Cat. No.	Price
3/16	EPS-D1-1	$.09	*EPS-D2-1	$.14	-		-	
1/4	EPS-D1-2	.09	*EPS-D2-2	.15	*EPS-D3-2	$.14	-	
5/16	EPS-D1-3	.10	EPS-D2-3	.11	*EPS-D3-3	.15	-	
3/8	EPS-D1-4	.10	EPS-D2-4	.11	EPS-D3-4	.13	*EPS-D4-4	$.17
7/16	EPS-D1-5	.10	EPS-D2-5	.11	EPS-D3-5	.13	*EPS-D4-5	.17
1/2	EPS-D1-6	.11	EPS-D2-6	.11	EPS-D3-6	.13	EPS-D4-6	.17

FIGURE 12.69 ■ Dowel pin chart drawing. *Courtesy NORDEX, Inc.*

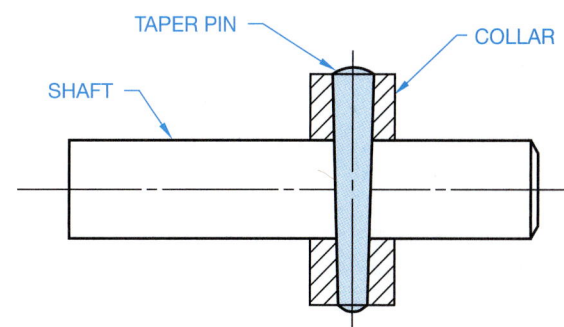

FIGURE 12.70 ■ Taper pin in assembly, sectional view.

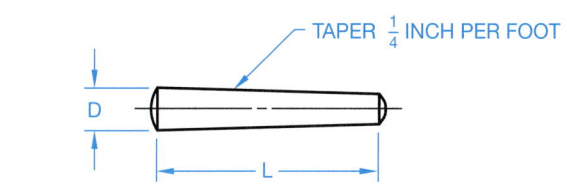

FIGURE 12.71 ■ Taper pin.

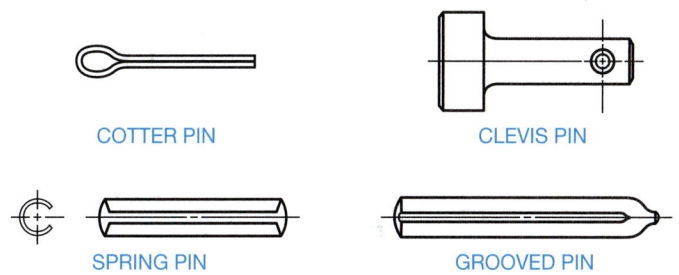

FIGURE 12.72 ■ Other common pins.

the section of two adjacent parts and a dowel pin. Figure 12.69 is a chart drawing of some standard dowel pins.

TAPER AND OTHER PINS

For applications that require perfect alignment of accurately constructed parts tapered dowel pins may be better than straight dowel pins. Taper pins are also used for parts that have to be taken apart frequently or where removal of straight dowel pins may cause excess hole wear. Figure 12.70 shows an example of a taper pin assembly. Taper pins, shown in Figure 12.71, range in diameter, *D,* from 7/0, which is .0625 in., to .875 in., and lengths, *L,* vary from .375 in. to 8 in.

Other types of pins serve functions similar to taper pins, such as holding parts together, aligning parts, locking parts, and transmitting power from one feature to another. Other common pins are shown in Figure 12.72.

RETAINING RINGS

Internal and external retaining rings are available as fasteners to provide a stop or shoulder for holding bearings or other parts on a shaft. They are also used internally to hold a cylindrical feature in a housing. Common retaining rings require a groove in the shaft or housing for mounting with a special plier tool. Also available are self-locking retaining rings for certain applications. (See Figure 12.73.)

KEYS, KEYWAYS, AND KEYSEATS

Standards for keys were established to control the relationship among key sizes, shaft sizes, and tolerances for key applications. A key is an important machine element, which is employed to provide a positive connection for transmitting torque between a shaft and hub, pulley, or wheels. The key is placed in position in a keyseat, which is a groove or channel cut in a shaft. The shaft and key are then inserted into the hub, wheel, or pulley, where

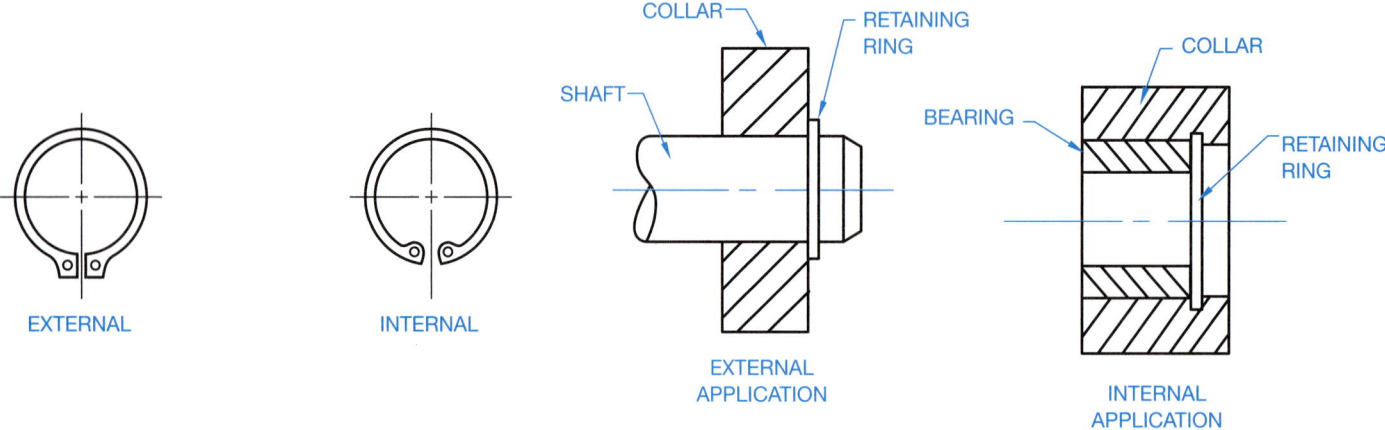

FIGURE 12.73 ■ Retaining rings.

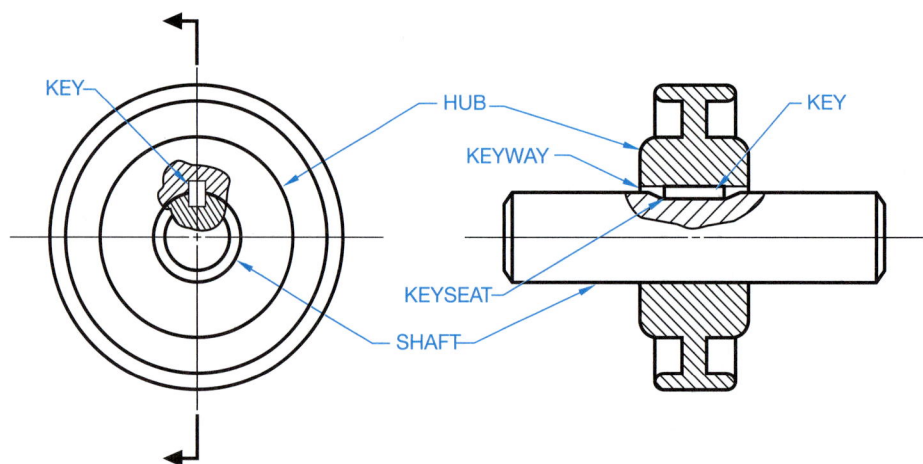

FIGURE 12.74 ■ Relationship between the key, keyseat, keyway, shaft, and hub.

the key mates with a groove called a **keyway**. Figure 12.74 shows the relationship among the key, keyseat, keyway, shaft, and hub.

Standard key sizes are determined by shaft diameter. For example, shaft diameters ranging from 7/8 in. to 1 1/4 in. would require a 1/4-in. nominal width key. Keyseat depth dimensions are established in relationship to the shaft diameter. Figure 12.75 shows the standard dimensions for the related features. For a shaft with a 3/4-in. diameter, the recommended shaft dimension, S, is .676 in. and the hub dimension, T, is .806 in. when using a rectangular key. Types of keys are shown in Figure 12.76. Appendix Q provides key and keyseat specifications.

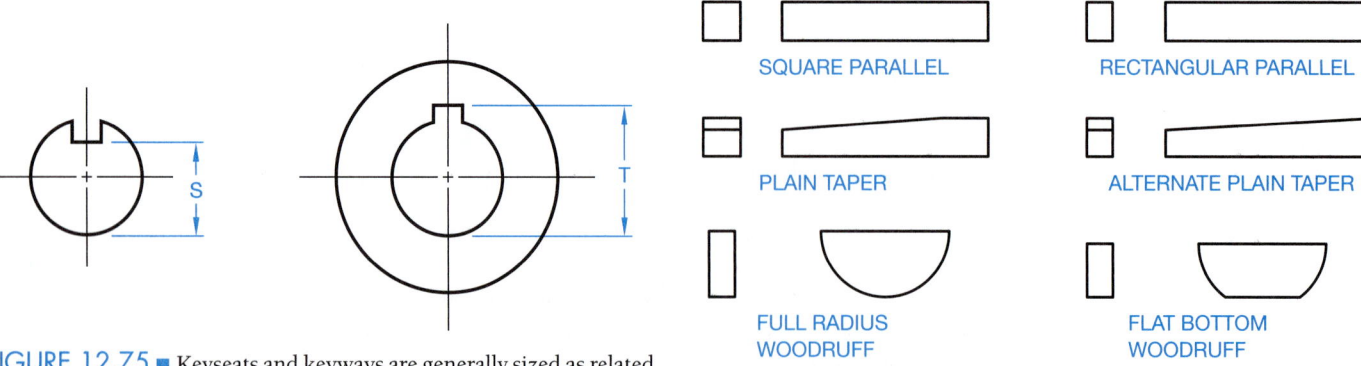

FIGURE 12.75 ■ Keyseats and keyways are generally sized as related to shaft size.

FIGURE 12.76 ■ Types of keys.

RIVETS

A **rivet** is a metal pin with a head used to fasten two or more materials together. The rivet is placed through holes in mating parts and the end without a head extends through the parts to be headed-over (formed into a head) by hammering, pressing, or forging. The end with the head is held in place with a solid steel bar known as a dolly while a head is formed on the other end. Rivets are classified by body diameter, length, and head type. (See Figure 12.77.)

DESIGNING AND DRAWING SPRINGS

A **spring** is a mechanical device, often in the form of a helical coil, that yields by expansion or contraction due to pressure, force, or stress applied. Springs are made to return to their normal form when the force or stress is removed. Springs are designed to store energy for the purpose of pushing or pulling machine parts by reflex action into certain desired positions. Improved spring technology provides springs with the ability to function for a long time under high stresses. The effective use of springs in machine design depends on five basic criteria, including: material, application, functional stresses, use, and tolerances.

Continued research and development of spring materials have helped to improve spring technology. The spring materials most commonly used include high-carbon spring steels, alloy

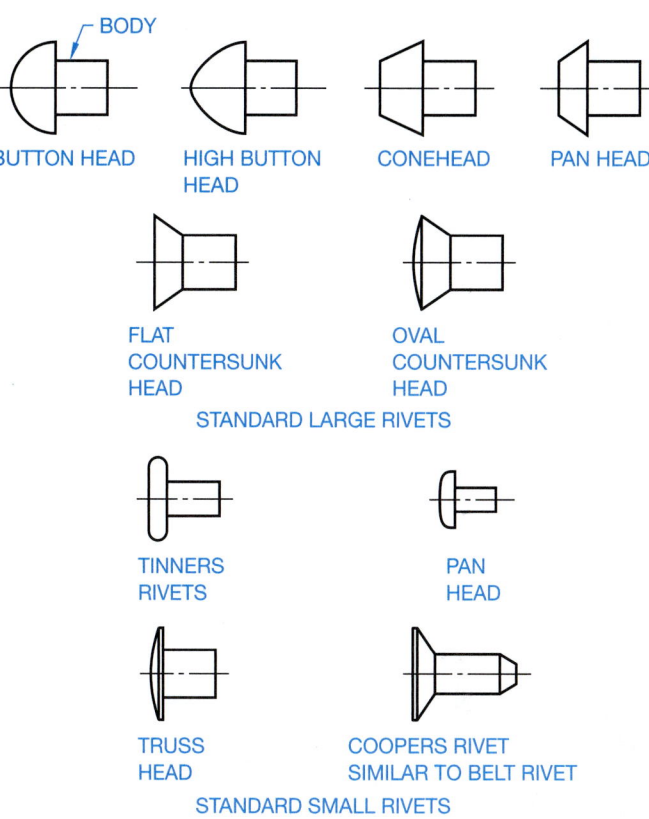

FIGURE 12.77 ■ Common rivets.

FASTENER SYMBOL LIBRARIES

One of the most powerful features of CADD software is the ability to reuse drawing and model content. This is especially true when designing and drafting products that contain fasteners. Often fasteners are standard parts that are used multiple times for several different projects. As a result, creating or purchasing a library of fasteners can prove very beneficial. As shown in this chapter, many fasteners contain complex shapes, especially if drawing threaded fasteners that require the use of detailed thread representations. Reusing existing fasteners and using fastener symbol libraries can save a significant amount of time and allow designers and drafters to focus on the design and documentation of a product instead of the task of creating individual fasteners using basic drawing tools.

An example of a 2-D fastener library is shown in Figure 12.78. This particular library contains several commonly used fasteners that can be added to a specific drawing view in the form of a symbol. Once the base symbol is inserted, you have the option of creating a unique fastener by editing the size and shape of the symbol. For example, you may want to adjust the length of a bolt. Many of these CADD library applications allow you to dynamically adjust the diameter, length, and thread specification and add the proper thread note.

FASTENER TOOLS

Some CADD programs and specialized fastener software systems make the process of adding fasteners to drawings and models extremely quick and easy. Instead of drawing and creating fasteners, or even picking fasteners from a symbol library, there are some CADD systems that allow you to select fastener variables, and automatically generate a fastener based on those specifications. CADD fastener software allows you to select the following variables, depending on the particular program:

■ Type of bolt or screw.

■ Type of head, if any.

■ Type and location of washers, if used.

■ Threading of holes or studs.

(Continued)

CADD APPLICATIONS

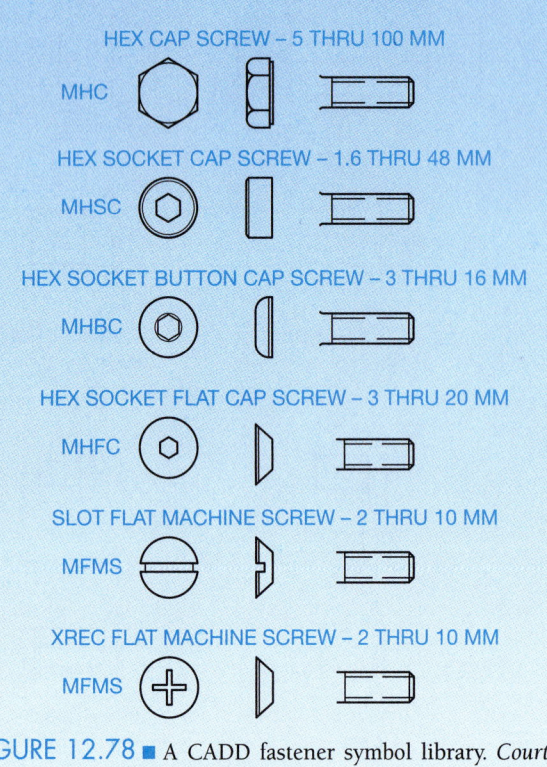

HEX CAP SCREW – 5 THRU 100 MM

MHC

HEX SOCKET CAP SCREW – 1.6 THRU 48 MM

MHSC

HEX SOCKET BUTTON CAP SCREW – 3 THRU 16 MM

MHBC

HEX SOCKET FLAT CAP SCREW – 3 THRU 20 MM

MHFC

SLOT FLAT MACHINE SCREW – 2 THRU 10 MM

MFMS

XREC FLAT MACHINE SCREW – 2 THRU 10 MM

MFMS

FIGURE 12.78 ■ A CADD fastener symbol library. *Courtesy CAD Tech Corporation.*

■ Type of thread representation.

■ Type of nut, if used.

After selecting these variables, typically you must choose a location to place the fastener. Then using parametrics based on your specifications, a drawing or model of the fastener is automatically created. An example of a

fastener generated using SolidWorks' SMART FASTENER tool is shown in Figure 12.79. In this example, a hex bolt, flat washer, lock washer, and hex nut are created by selecting an assembly hole and defining fastener specifications.

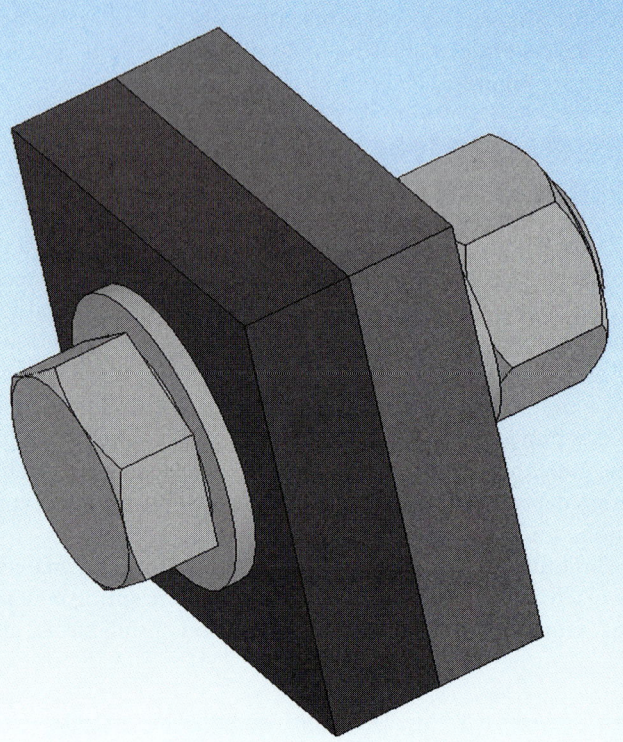

FIGURE 12.79 ■ An example of using a specialized CADD software fastener tool to create a complete fastener.

spring steels, stainless spring steels, music wire, oil-tempered steel, copper-based alloys, and nickel-based alloys. Spring materials, depending on use, may have to withstand high operating temperatures and high stresses under repeated loading.

Spring design criteria are generally based on material gage, kind of material, spring index, direction of the helix, type of ends, and function. Spring wire gages are available from several different sources ranging in diameter from number 7/0 (.490 in.) to number 80 (.013 in.). The most commonly used spring gages range from 4/0 to 40. There are a variety of spring materials available in round or square stock for use depending on spring function and design stresses.

The spring index is a ratio of the average coil diameter to the wire diameter. The index is a factor in determining spring stress, deflection, and the evaluation of the number of coils needed and the spring diameter. Recommended index ratios range between 7 and 9, although other ratios commonly used range from 4 to 16. The direction of the helix is a design factor when springs must operate in conjunction with threads or with one spring inside of another. In such situations the helix

of one feature should be in the opposite direction of the helix for the other feature.

Compression springs are available with ground or unground ends. Unground, or rough, ends are less expensive than ground ends. If the spring is required to rest flat on its end, then ground ends should be used. Spring function depends on one of two basic factors, compression or extension. Compression springs release their energy and return to their normal form when compressed. Extension springs release their energy and return to the normal form when extended. (See Figure 12.80.)

Spring Terminology

The springs given in Figure 12.81 show common characteristics that are identified throughout the following discussion.

Spring Ends

Compression springs have four general types of ends: open or closed ground ends and open or closed unground ends, shown

FIGURE 12.80 ■ Compression and extension springs.

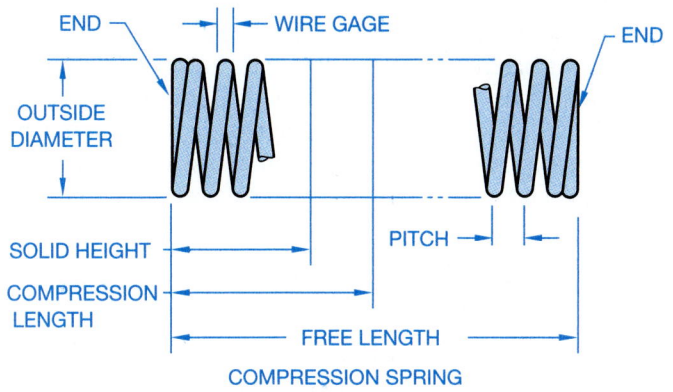

FIGURE 12.81 ■ Spring characteristics.

in Figure 12.82. Extension springs have a large variety of optional ends, a few of which are shown in Figure 12.83.

Helix Direction

The helix direction can be specified as right-hand or left-hand. This is the direction of the twist used to create the spring. (See Figure 12.82.)

Free Length

The length of the spring when there is no pressure or stress to affect compression or extension. (See Figure 12.81.)

Compression Length

The compression length is the maximum recommended design length for the spring when compressed. (See Figure 12.81.)

Solid Height

The solid height is the maximum compression possible. The design function of the spring should not allow the spring to reach

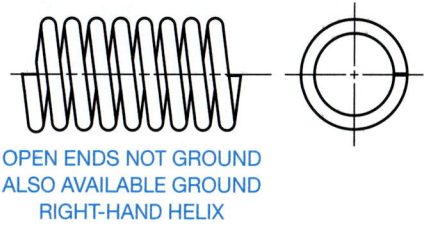

OPEN ENDS NOT GROUND
ALSO AVAILABLE GROUND
RIGHT-HAND HELIX

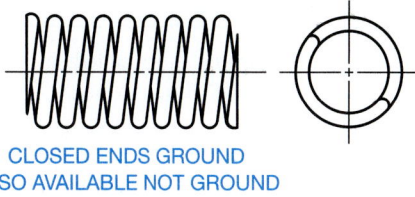

CLOSED ENDS GROUND
ALSO AVAILABLE NOT GROUND
LEFT-HAND HELIX

FIGURE 12.82 ■ Helix direction and compression spring end types.

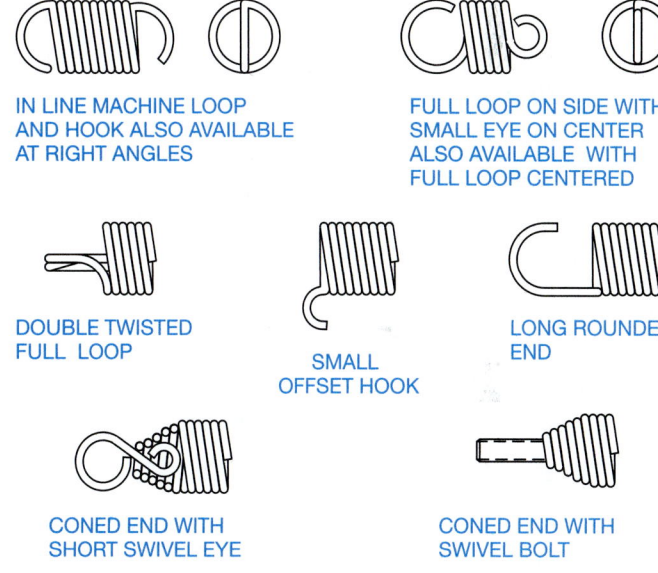

IN LINE MACHINE LOOP
AND HOOK ALSO AVAILABLE
AT RIGHT ANGLES

FULL LOOP ON SIDE WITH
SMALL EYE ON CENTER
ALSO AVAILABLE WITH
FULL LOOP CENTERED

DOUBLE TWISTED
FULL LOOP

SMALL
OFFSET HOOK

LONG ROUNDED
END

CONED END WITH
SHORT SWIVEL EYE

CONED END WITH
SWIVEL BOLT

MANY OTHER COMBINATIONS ARE AVAILABLE

FIGURE 12.83 ■ Extension spring end types.

solid height when in operation unless this factor is a function of the machinery. (See Figure 12.81.)

Loading Extension

The extended distance to which an extension spring is designed to operate. (See Figure 12.81.)

Pitch

The pitch is one complete helical revolution, or the distance from a point on one coil to the same corresponding point on the next coil. (See Figure 12.81.)

Material: 302 Stainless Steel (Spring Tamper) Passivated

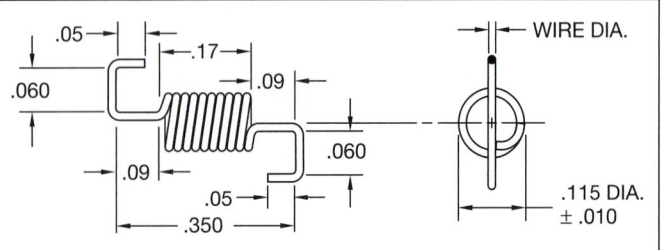

FIGURE 12.84 ■ Torsion spring, also called antibacklash spring. *Courtesy NORDEX, Inc.*

Torsion Springs

Torsion springs are designed to transmit energy by a turning or twisting action. Torsion is defined as a twisting action that tends to turn one part or end around a longitudinal axis while the other part or end remains fixed. Torsion springs are often designed as antibacklash devices or as self-closing or self-reversing units. (See Figure 12.84.)

Flat Springs

Flat springs are arched or bent flat-metal shapes designed so when placed in machinery they cause tension on adjacent parts. The tension can be used to level parts, provide a cushion, or position the relative movement of one part to another. One of the most common examples of flat springs is leaf springs on an automobile.

SPRING REPRESENTATIONS AND SPECIFICATIONS

There are three types of spring representations: detailed, schematic, and simplified, as seen in Figure 12.85. Detailed spring drawings are used in situations that require a realistic representation, such as vendor's catalogs, assembly instructions, or detailed assemblies. Schematic spring representations are commonly used on drawings. The single-line schematic symbols are easy to draw and clearly represent springs without taking the additional time required to draw a detailed spring. The use of simplified spring drawings is limited to situations where the clear resemblance of a spring is not necessary. While very easy to draw, the simplified spring symbol must be accompanied by

clearly written spring specifications. The simplified spring representation is not very useful in assembly drawings or other situations that require a visual comparison of features.

Spring Specifications

No matter which representation is used, several important specifications must accompany the spring symbol. Spring information is generally lettered in the form of a specific or general note.

Spring specifications include outside or inside diameter, wire gage, kind of material, type of ends, surface finish, free and compressed length, and number of coils. Other information, when required, may include spring design criteria and heat treatment specifications. The information is often provided on a drawing as shown in Figure 12.86. The material note is usually found in the title block.

Drawing Detailed Spring Representations

Detailed spring drawing requires a great degree of accuracy. First, determine the outside diameter, number of coils, wire diameter, and free length or compressed length.

The following steps show how to draw a detailed spring representation with the following specifications:

Material: 2.5-mm diameter high-carbon spring steel

Outside Diameter: 16 mm

Free Length: 50 mm

Number of Coils: 6

STEP 1 Using construction lines, draw a rectangle equal to the outside diameter (16 mm) wide and the free length (50 mm) long. (See Figure 12.87.)

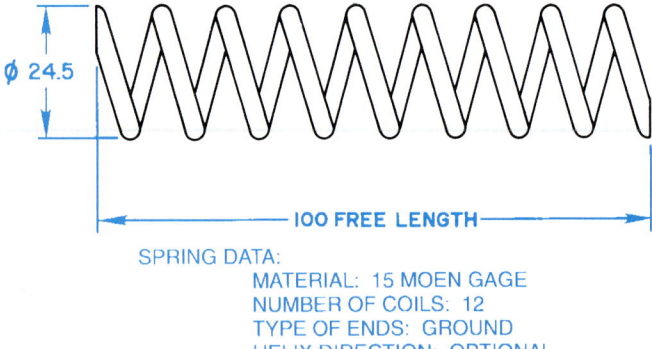

SPRING DATA:
    MATERIAL: 15 MOEN GAGE
    NUMBER OF COILS: 12
    TYPE OF ENDS: GROUND
    HELIX DIRECTION: OPTIONAL

FIGURE 12.86 ■ Detailed spring drawing with spring data chart.

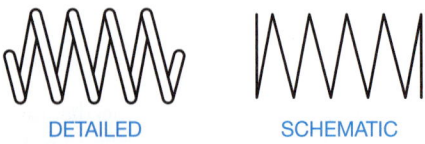

DETAILED   SCHEMATIC   SIMPLIFIED

FIGURE 12.85 ■ Spring representations.

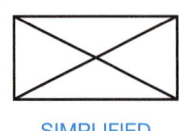

FIGURE 12.87 ■ Step 1, preliminary spring layout.

STEP 2 Along one inside edge of the length of the rectangle, lay out seven equally spaced full circles with a 2.5-mm diameter (wire size). The layout may be done by dividing the length into six equal spaces. (See Figure 12.88.) On the other inside edge, lay out a half circle at each end. Beginning at a distance of 1/2 P away from one end, draw the first of six full (2.5 mm) circles with equal spaces between them.

STEP 3 Connect the circles drawn in Step 2 to make the coils. Draw lines from a point of tangency on one circle to a corresponding point on a circle on the other side. Draw the last element on each side down the edge of the rectangle for ground ends. To draw unground ends, make the last element terminate at the axis of the spring. (See Figure 12.89.)

For detailed coils in a longitudinal sectional view, leave the circles from Step 2 and draw that part of the spring that appears as if the front half were removed as shown in Figure 12.90. Then fill in or section-line the circles.

Drawing Schematic Spring Representations

Schematic spring symbols are much easier to draw than detailed representations, while clearly resembling a spring. The follow-

ing steps show how to draw a schematic spring symbol for the same spring previously drawn.

STEP 1 Use construction lines to draw a rectangle equal in size to the outside diameter by the free length as seen in Figure 12.87.

STEP 2 Establish six equal spaces at P distance along one edge of the rectangle. Along the opposite edge begin and end with a space equal to 1/2 P. Establish five equal spaces between the 1/2 P ends. (See Figure 12.91.)

STEP 3 Beginning on one side, draw the elements of each spring coil as shown in Figure 12.92.

Drawing a Simplified Spring Representation

Simplified spring representations are drawn by making a rectangle equal in size to the major diameter and free length and then drawing diagonal lines as shown in Figure 12.93.

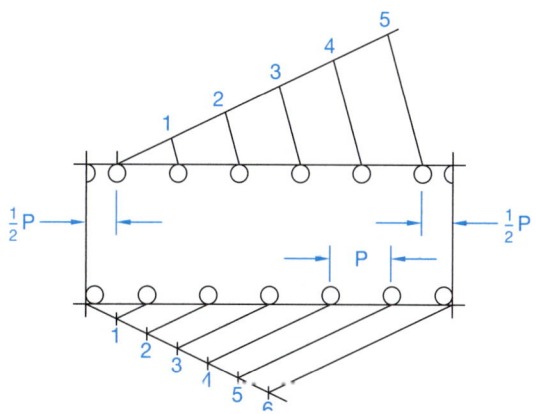

FIGURE 12.88 ■ Step 2, spacing the coils.

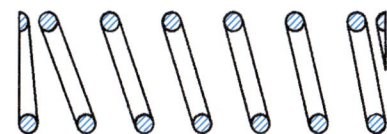

FIGURE 12.89 ■ Step 3, connect the coils to complete the detailed spring representation.

FIGURE 12.90 ■ Detailed spring representation in section.

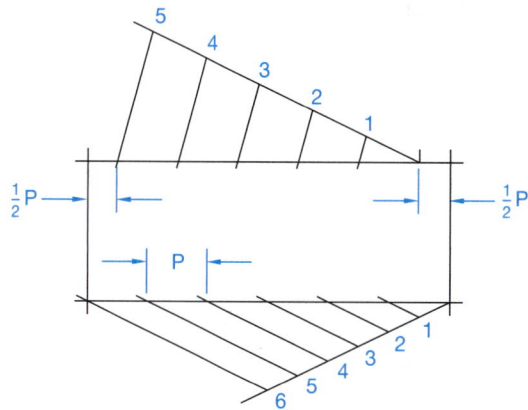

FIGURE 12.91 ■ Step 2, spacing the coils.

FIGURE 12.92 ■ Step 3, complete the schematic representation.

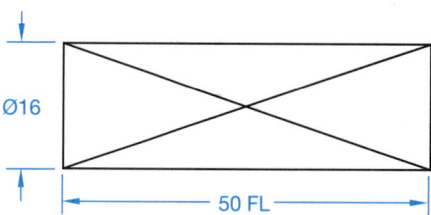

FIGURE 12.93 ■ Simplified spring representation.

CADD APPLICATIONS 2-D

DRAWING SPRINGS

Springs are drawn using CADD techniques that are similar to manual drafting. Many of the manual drafting geometric construction methods described in this chapter are effective for producing detailed, schematic, and simplified spring representations. However, CADD tools such as layers, and specific commands including COPY, MOVE, OFFSET, ARRAY, TRIM, and DIVIDE greatly aid in the process of drawing any style of spring. Additionally, like fasteners, springs are often standard parts that are used multiple times within a product or for several different projects. Reusing existing springs and using spring symbol libraries can save a significant amount of time by allowing designers and drafters to insert commonly used springs, or modify a basic spring to produce a unique drawing. 2-D spring software is also available that automatically generates springs based on specific characteristics.

RECOMMENDED REVIEW

It is recommended that you review the following parts of Chapter 7 before you begin working on fully dimensioned fastener and spring drawings. This will refresh your memory about how related lines are properly drawn.

- Object Lines.
- Viewing Planes.
- Extension Lines.
- Hidden Lines.
- Break Lines.
- Dimension Lines.
- Centerlines.
- Phantom Lines.
- Leader Lines.

Also review Chapters 9 and 10 covering multiview and auxiliary view drawings, and Chapter 11 covering dimensioning practices.

WEBSITE RESEARCH

The following websites can provide you with additional information for research or further study into topics covered in this chapter.

http://www.asme.org—Find information and publications related to the American Society of Mechanical Engineers.

CADD APPLICATIONS 3-D

CREATING SPRINGS

Many 3-D modeling programs contain a tool, such as COIL or HELIX for example, that allows you to generate 3-D springs. Springs are created using these tools by revolving a sketch profile, usually of a circle, around and up or down an axis. (See Figure 12.94.) Many 3-D CADD commands used to develop springs contain options allowing you to define a spring based on specific variables including:

- Outside diameter.
- Wire gage.
- Solid height.
- Compression length.
- Free length.
- Right or left-hand helix.
- Number of revolutions.
- Pitch.
- Extension or compression type.
- Open or closed ends.
- Ground or unground ends.
- Special ends.

FIGURE 12.94 ■ An example of a 3-D solid model spring.

Some CADD programs and specialized spring software systems make the process of adding springs to models very fast and easy. Instead of creating springs using feature tools, there are some CADD systems that allow you to select spring variables and automatically generate a spring based on those specifications. CADD spring software allows you to select the following options, depending on the particular program:

■ Wire diameter.

■ Number of coils.

■ Coil direction.

■ Outside diameter.

■ Free length.

■ Compressed length.

■ End type.

After selecting these variables, typically you must choose a location to place the spring. Then, using parametrics based on your specifications, a model of the spring is automatically created.

PROFESSIONAL PERSPECTIVE

There are three basic methods for drawing screw thread and spring representations: using construction techniques, using a template, or using CADD. The manual construction techniques should be avoided in most cases because they are very time-consuming. However, in some special applications these methods may be used effectively. The manual drafter should use a template when at all possible to draw screw threads, fastener heads, and spring representations. For the CADD user, there are many software programs available that allow drawing of screw threads in simplified, schematic, or detailed representations. A complete variety of head types can be drawn in addition to springs in simplified, schematic, or detailed representations. This type of CADD software makes an otherwise complex task very simple and saves a lot of drafting time.

The entry-level drafter often spends unnecessary time trying to make a screw thread representation look exactly like the real thing. It is important to have an accurate drawing, but with regard to screw threads, the thread note tells the reader the exact thread specifications. This is why threads are shown on a drawing as a representation rather than a duplication of the real thing. The simplified representation is the most commonly used technique, because it is fast and easy to draw.

Your goal as a professional drafter is to make the drawing communicate so that there is no question about what is intended. In simple terms, make the drawing clear and accurate using the easiest method, and make sure the thread note and specifications are complete and correct as shown in the real world example found in Figure 12.96.

http://www.ansi.org—The American National Standards Institute. Information about national and international drafting standards.

http://www.adda.org—American Design Drafting Association.

http://www.industrialpress.com—Information about the *Machinery's Handbook*. This is a valuable resource for all types of fasteners and fastener specifications.

http://www.cadtechcorp.com—CADD fasteners software applications.

http://www.thomasnet.com—CAD symbols.

http://www.avdel.textron.com—Technical information on fasteners.

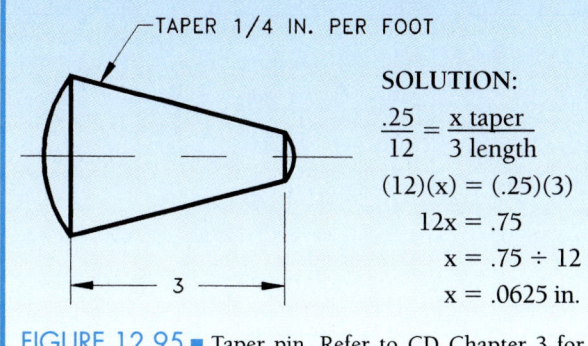

RATIO AND PROPORTION OF A TAPER PIN

PROBLEM: The taper of a pin is .25:12. What is the taper for the 3-in. long pin shown in Figure 12.95?

TAPER 1/4 IN. PER FOOT

SOLUTION:

$$\frac{.25}{12} = \frac{x \text{ taper}}{3 \text{ length}}$$

$$(12)(x) = (.25)(3)$$

$$12x = .75$$

$$x = .75 \div 12$$

$$x = .0625 \text{ in.}$$

FIGURE 12.95 ■ Taper pin. Refer to CD Chapter 3 for additional math instructions.

MATH APPLICATIONS

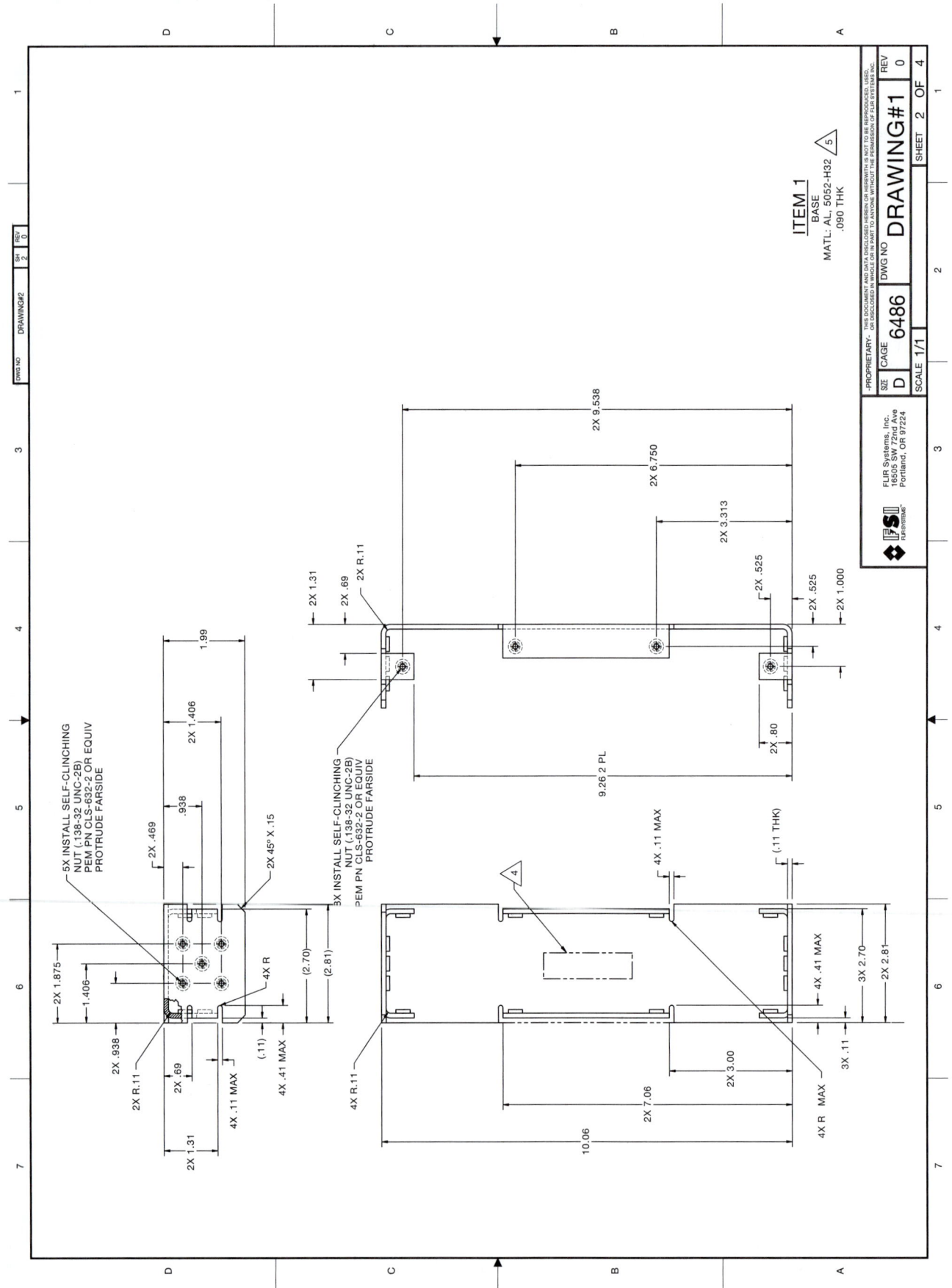

FIGURE 12.96 ■ An actual industry drawing with fasteners displayed and specified. *Courtesy Flir Systems, Inc.*

CHAPTER 12

Fasteners and Springs Test

Access the CD found with this textbook to view the Chapter 12 Test. Confirm the preferred submittal method with your instructor.

CHAPTER 12

Fasteners and Springs Problems

DIRECTIONS

1. From the selected problems determine which views and dimensions should be used to completely detail the part. Use simplified representation unless otherwise specified in the instructions or by your instructor.
2. Make a multiview sketch, to proper proportions, including dimensions and notes.
3. Using the sketch as a guide, draw an original multiview drawing on an adequately sized drawing sheet. Add all necessary dimensions and notes using unidirectional dimensioning. Use manual or computer-aided drafting as required by your course guidelines.
4. Include the following general notes at the lower left corner of the sheet .5 in. each way from the corner border lines:

1. INTERPRET DIMENSIONS AND TOLERANCES PER ASME Y14.5M—1994.
2. REMOVE ALL BURRS AND SHARP EDGES.

NOTES

Additional general notes may be required depending on the specifications of each individual assignment. The following should be part of your title block unless otherwise specified by your instructor:

UNSPECIFIED TOLERANCES:

DECIMALS	mm	IN.
X	± 2.5	±.1
XX	± 0.25	±.01
XXX	± 0.125	±.005
ANGULAR ± 30'		
FINISH	3.2μm	125μIN.

PROBLEM 12.1 (in.)

Part Name: Full Dog Point Gib Screw

Material: 10-32 UNF-2A × .75 Long

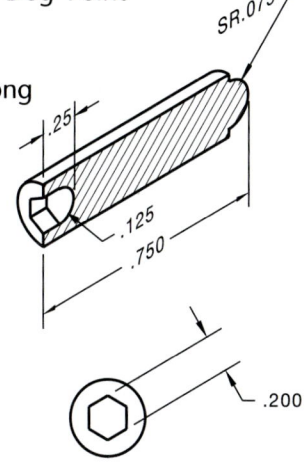

SIDE VIEW

SR.075
.25
.125
.750
.200

PROBLEM 12.2 (in.)

Part Name: Thumb Screw

Material: SAE 1315 Steel

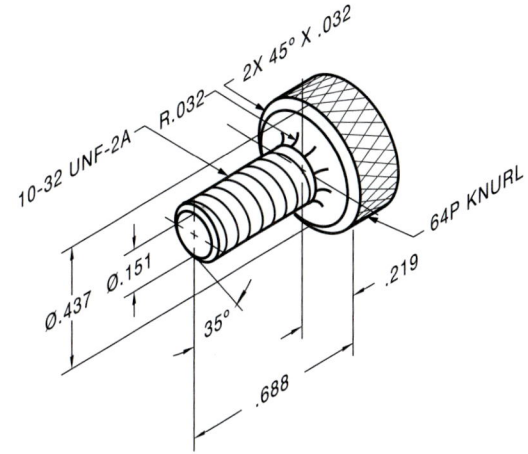

2X 45° X .032
R.032
10-32 UNF-2A
64P KNURL
Ø.437
Ø.151
35°
.219
.688

PROBLEM 12.3 (in.)

Part Name: H-Step Threading Screw

Material: SAE 3130

SPECIFIC INSTRUCTIONS: Draw each of the three threads using the representation specified below each thread note.

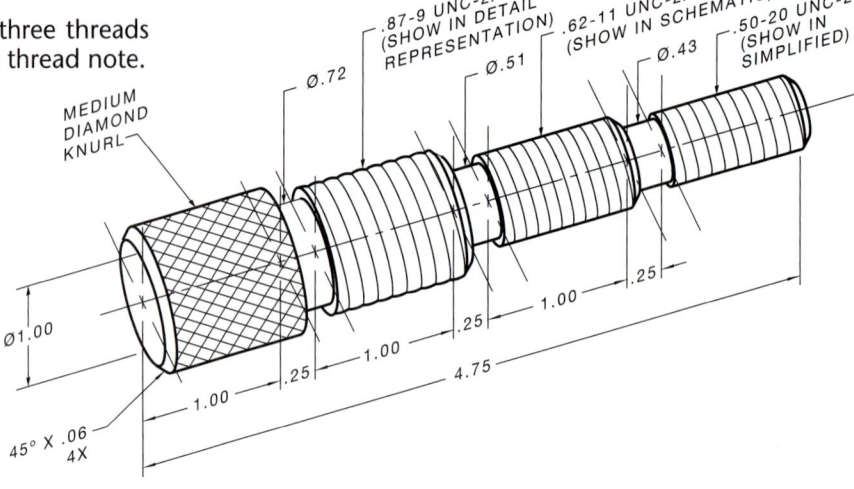

PROBLEM 12.4 (metric)

Part Name: Knurled Hex Soc Hd Step Screw

Material: SAE 1040

Case Harden: 1.6-mm deep per Rockwell C Scale

Finish: 2 μm; Black Oxide

PROBLEM 12.5 (in.)

Part Name: Machine Screw

Material: Stainless Steel

Finish All Over: 2μm

PROBLEM 12.6 (metric)

Part Name: Lathe Dog

Material: Cast Iron

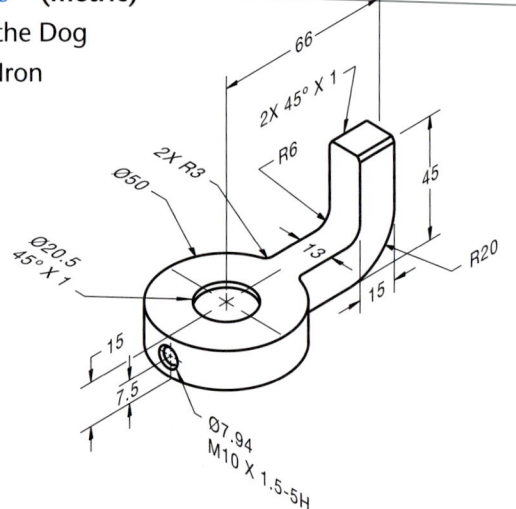

PROBLEM 12.7 (in.)

Part Name: Threaded Step Shaft
Material: SAE 1030

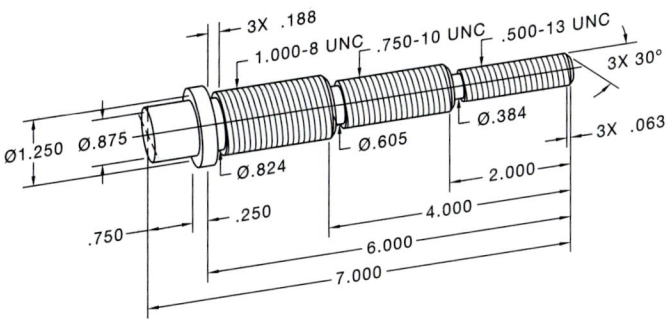

PROBLEM 12.8 (in.)

Part Name: Washer Face Nut
Material: SAE 1330 Steel

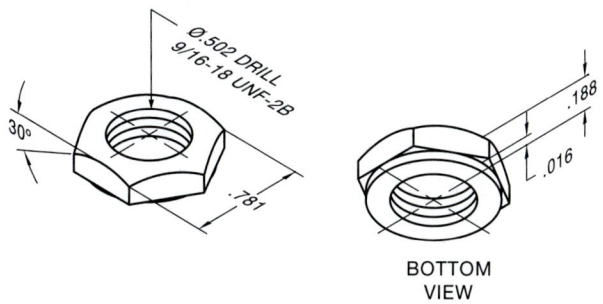

BOTTOM
VIEW

PROBLEM 12.9 (in.)

Part Name: Shoulder Screw
Material: SAE 4320 Steel

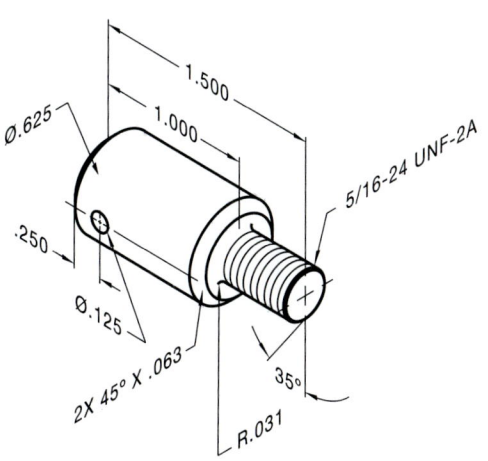

PROBLEM 12.10 (in.)

Part Name: Stop Screw
Material: SAE 4320
Hex Depth: .175

Note: Medium Diamond Knurl at Head

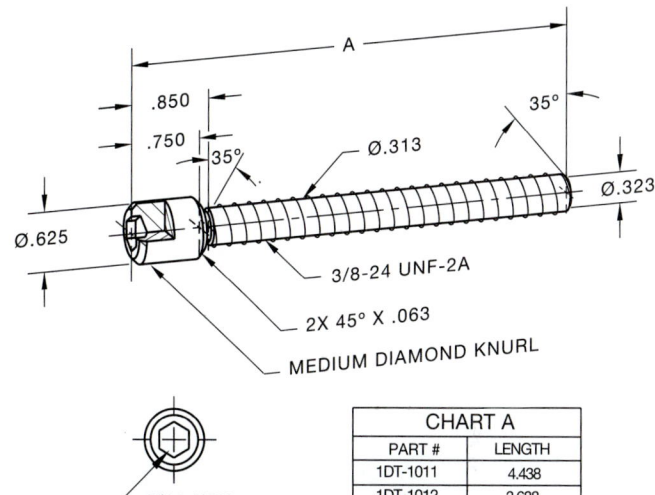

5/16 HEX

CHART A	
PART #	LENGTH
1DT-1011	4.438
1DT-1012	3.688

PROBLEMS 12.11 through 12.22: Access the CD found with this textbook and open the problem of your choice, or as assigned by your instructor. Solve the problems using the instructions provided with this chapter or on the CD, unless otherwise specified by your instructor.

MATH PROBLEMS

PROBLEMS 12.23 through 12.29: Access the CD found with this textbook and open the math problem of your choice, or as assigned by your instructor. Solve the problem or problems using the instructions provided.

Sections, Revolutions, and Conventional Breaks

LEARNING OBJECTIVES

After completing this chapter, you will:

- Draw proper cutting-plane line representations.
- Draw sectional views, including full, half, aligned, broken-out, auxiliary, revolved, and removed sections.
- Identify features that should remain unsectioned in a sectional view.

- Prepare drawings with conventional revolutions and conventional breaks.
- Modify the standard sectioning techniques as applied to specific situations.
- Make sectional drawings from given engineers' sketches and actual industrial layouts.

THE ENGINEERING DESIGN APPLICATION

You are working with an engineer on a special project and she gives you a rough sketch of a part from which you are to make a formal drawing. At first glance, you think it is an easy part to draw, but with further study you realize that it is more difficult than it looks. The rough sketch, shown in Figure 13.1, has a lot of hidden features. So your first thought is, "What do I use for the front view?" You decide that a front exterior view would show only the diameter and length. So instead of drawing an outside view, you decide to use a full section to expose all of the interior features and, at the same time, give the overall length and

diameter. "This is great!" you say. Then you realize there are 6 holes spaced 60° apart that remain as hidden features in the left-side view. This is undesirable, because you do not want to section all of the holes. So you decide to use a broken-out section to expose only two holes. This allows you to dimension the $6 \times 60°$ and the $6 \times Ø6^{+0.2}_{0}$ holes all at the same time. Now, with all this thinking and planning out of the way, it only takes you three hours to completely draw and dimension the part, and you are ready to give the formal drawing, shown in Figure 13.2, to the checker.

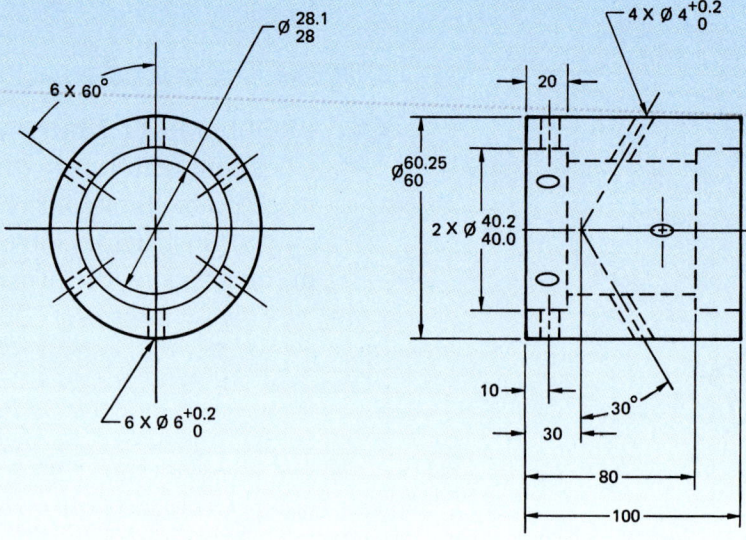

FIGURE 13.1 ■ Engineer's layout drawing.

THE ENGINEERING DESIGN APPLICATION *(continued)*

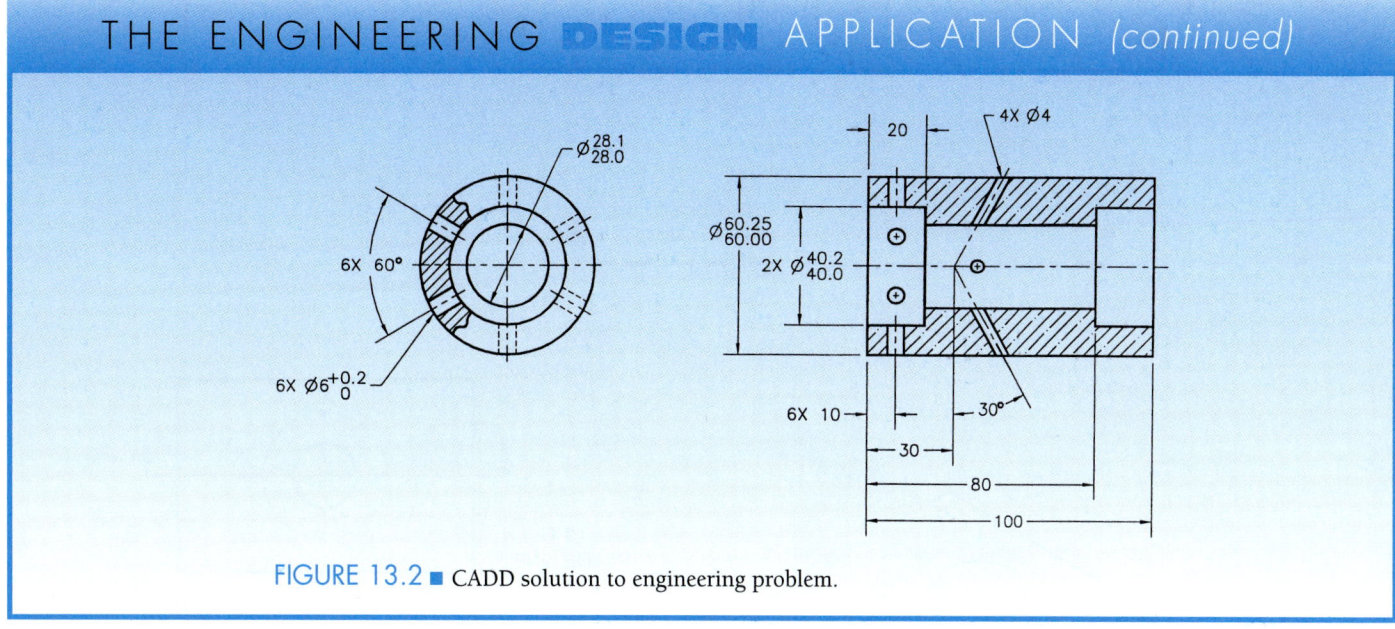

FIGURE 13.2 ■ CADD solution to engineering problem.

ANSI The American National Standards Institute document that governs sectioning techniques is titled *Multi and Sectional View Drawings,* ANSI Y14.3. The engineering standard *Line Conventions and Lettering,* ASME Y14.2M, covers the principles of drawing recommended section lines. The content of this chapter is based on the ANSI standard and provides an in-depth analysis of the techniques and methods of sectional view presentation.

SECTIONING

Sections, or *sectional views* are used to describe the interior portions of an object that are otherwise difficult to visualize. Interior features that are described using hidden lines are not as clear as if they are exposed for viewing as visible features. It is also a poor practice to dimension to hidden features. The sectional view allows you to expose the hidden features for dimensioning. Figure 13.3 shows an object in conventional multiview representation and using a sectional view. Notice how the hidden features are clarified in the sectional view. There are a variety of sectioning methods for different applications discussed throughout this chapter.

CUTTING-PLANE LINES

The sectional view is created by placing an imaginary cutting plane through the object cutting away the area to be exposed. The adjacent view becomes the sectional view by removing the portion of the object between the viewer and the cutting plane. (See Figure 13.4).

The sectional view should be projected from the view that has the cutting plane as you normally project a view in multiview. The cutting-plane line is a thick line representing the cutting plane as shown in Figure 13.4. The cutting plane line can be drawn using

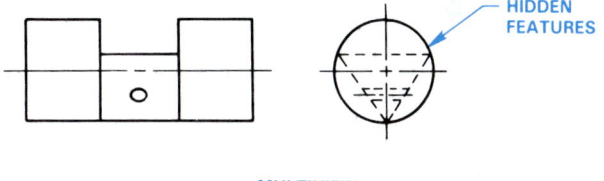

MULTIVIEW

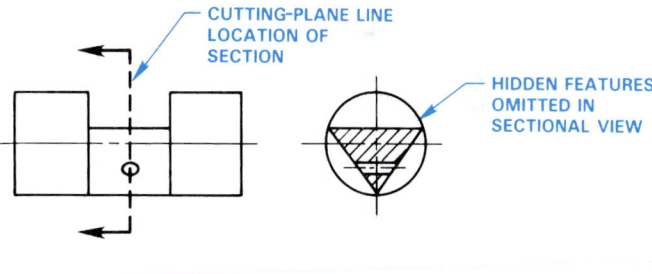

SECTIONAL VIEW

FIGURE 13.3 ■ Conventional multiview compared to a sectional view.

alternating long and two short dashes, or evenly spaced dashes. The long dashes can vary in length depending on the size of the drawing, but the short dashes are generally .25 in. (6 mm) in length. Refer to Figure 7.17 in Chapter 7 to see an example of the cutting-plane line styles. The cutting-plane line is capped on the ends with arrowheads showing the direction of sight of the sectional view. The cutting-plane line arrowheads maintain the same 3:1 length-to-width ratio as dimension-line arrowheads. Cutting-plane line arrowheads can be the same size as dimension-line arrowheads, but they are commonly drawn larger to allow them to show up better on the drawing. While there are no specific requirements for cutting-plane line arrowhead sizes, they are typically twice the size of dimension-line arrowheads, for example:

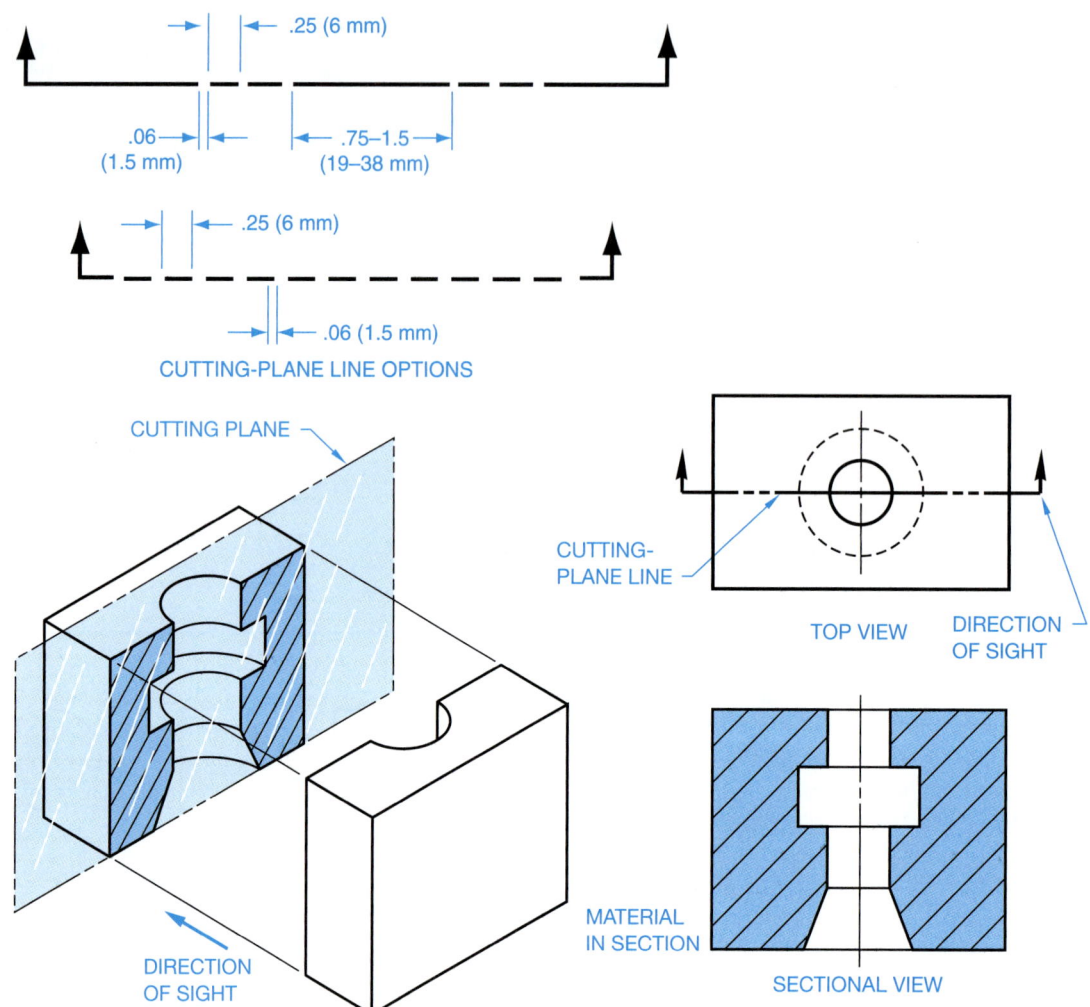

FIGURE 13.4 ■ Cutting-plane line.

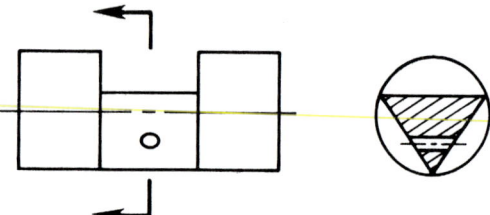

FIGURE 13.5 ■ Simplified cutting-plane line.

.25 in. (6 mm) long. This depends on the size of the drawing and your school or company standards. When the extent of the cutting plane is obvious, only the ends of the cutting-plane line can be used as shown in Figure 13.5. Such treatment of the cutting plane also helps keep the view clear of excess lines.

If lack of space restricts the normal placement of a sectional view, the view can be placed in an alternate location. When this is done, the sectional view should not be rotated but should remain in the same orientation as if it is a direct projection from the cutting plane. The cutting planes and related sectional views

should be labeled with letters beginning with AA, as shown in Figure 13.6. The text height for cutting-plane line labels and the correlated view identification is generally the same text height used for drawing titles, which is typically .25 in. (6 mm).

The cutting-plane line can be completely omitted when the location of the cutting plane is clearly obvious as shown in Figure 13.2 and Figure 13.7. When in doubt, use the cutting-plane line.

SECTION LINES

Section lines are thin lines used in the view of the section to show where the cutting-plane line has cut through material. (See Figure 13.8.) Also refer to Chapter 7. Section lines usually are drawn equally spaced at 45° but cannot be parallel or perpendicular to any line of the object. Any convenient angle can be used to avoid placing section lines parallel or perpendicular to other lines of the object. Angles of 30° and 60° are common. Section lines that are more than 75° or less than 15° from horizontal should be avoided. Section lines must never be drawn either horizontally or vertically. Figure 13.9 shows some common errors in drawing sec-

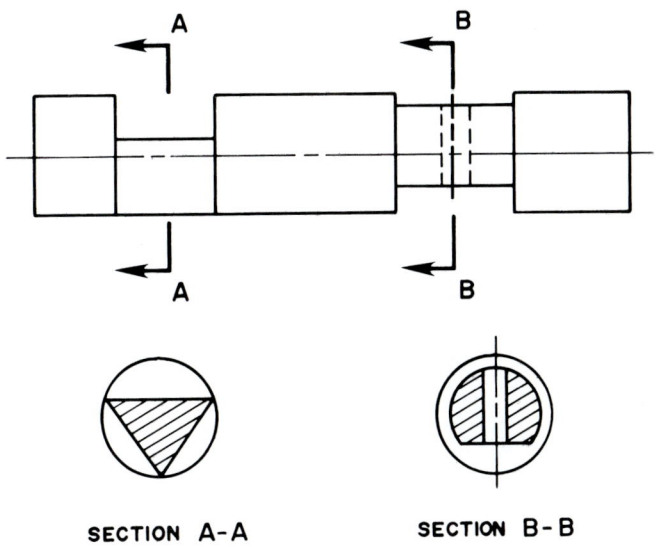

FIGURE 13.6 ■ Labeled cutting-plane lines and related sectional views.

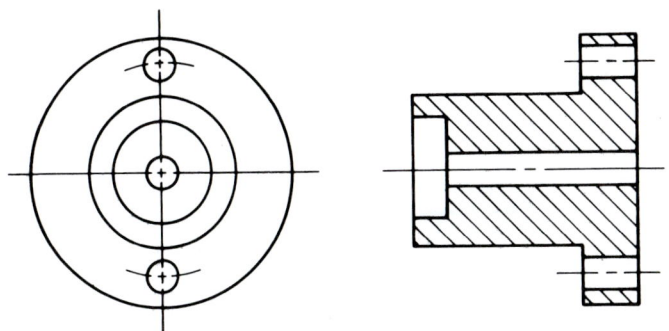

FIGURE 13.7 ■ An obvious cutting-plane line may be omitted.

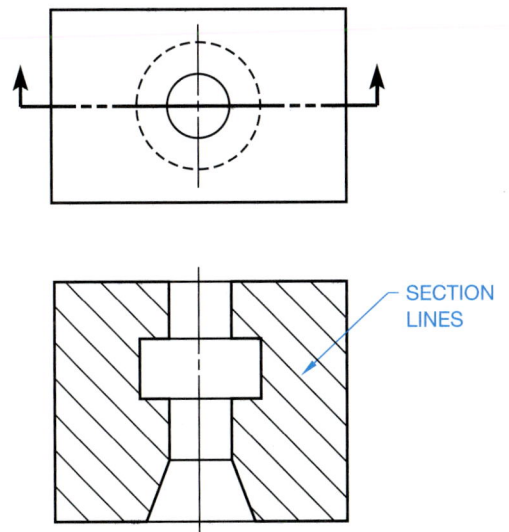

FIGURE 13.8 ■ Section lines.

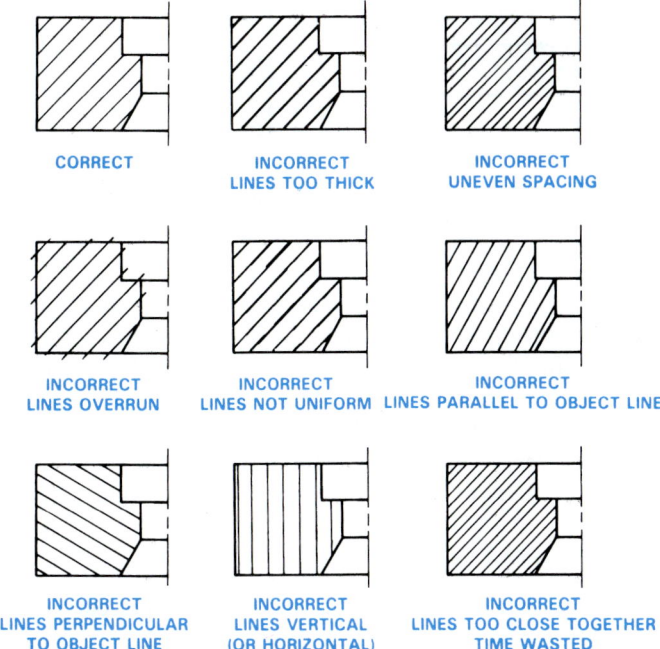

FIGURE 13.9 ■ Common section line errors.

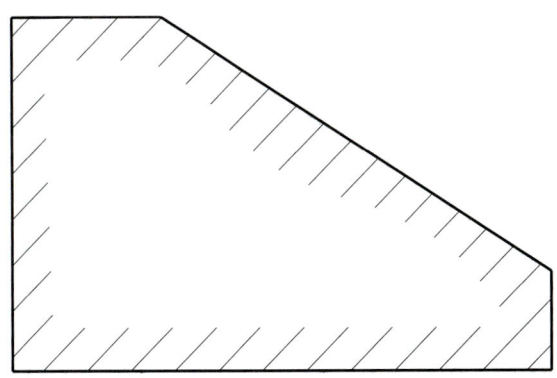

FIGURE 13.10 ■ Outline section lines.

tion lines. Section lines should be drawn in opposite directions on adjacent parts, and when several parts are adjacent, any suitable angle can be used to make the parts appear clearly separate. When a very large area requires section lining, you can elect to use outline section lining as shown in Figure 13.10. Confirm the approval of this practice with your company or school before using this option.

Equally spaced section lines specify either a general material designation or cast iron and malleable iron. This method of drawing section lines is quick and easy with the actual material identification located in the drawing title block.

General section lines are evenly spaced. The amount of space between lines depends on the size of the part. Very large parts have larger spacing than very small parts. (See Figure 13.11.) Use your own judgment. The key is to clearly represent section lines without an unnecessary expenditure of time.

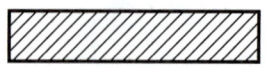 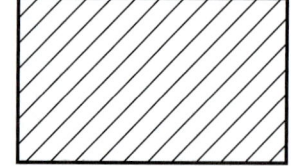

FIGURE 13.11 ■ Space between section lines.

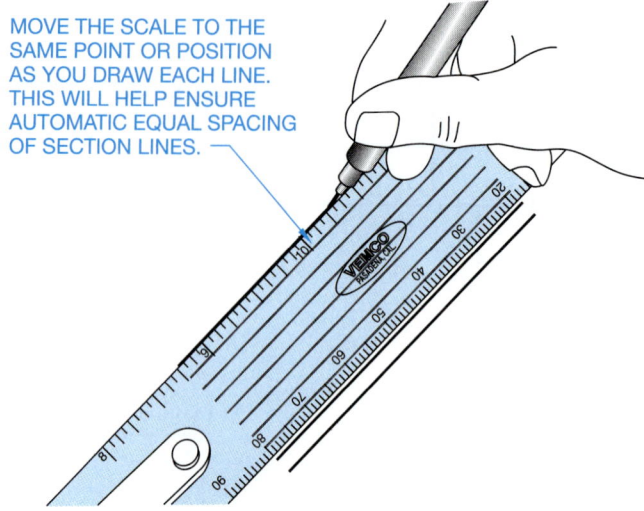

MOVE THE SCALE TO THE SAME POINT OR POSITION AS YOU DRAW EACH LINE. THIS WILL HELP ENSURE AUTOMATIC EQUAL SPACING OF SECTION LINES.

There are a number of ways to space section lines equally without measuring each space. One good way is to calibrate the space using the same point or mark on the drafting machine scale as you draw each line. This method works with clear plastic scales as shown in Figure 13.12. Another technique is to use the *Ames Lettering Guide* for drawing equally spaced section lines. Refer to Chapter 7 to review the use of this tool.

Using Coded Section Lines

Coded section lines are used if you want to represent specific material section line symbols in the sectional view. Coded section line symbols are shown in Figure 13.13. While the ASME Y14.3M standard covering sectioning applications recommends omitting section lines in thin sections, some companies prefer the practice of using solid fill on very thin sections.

FIGURE 13.12 ■ Equally spacing section lines with the drafting machine scale.

Coded section lining is more time-consuming to draw than general section lining when manual drafting is used. Coded section lining can be used effectively when a section is taken through an assembly of adjacent parts of different materials as seen in Figure 13.14.

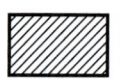
CAST OR MALLEABLE IRON AND GENERAL USE FOR ALL MATERIALS

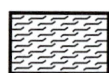
CORK, FELT, FABRIC, LEATHER, AND FIBER

MARBLE, SLATE, GLASS, PORCELAIN

STEEL

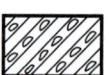
SOUND INSULATION

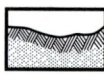
EARTH

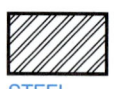
BRONZE, BRASS, COPPER, AND COMPOSITIONS

THERMAL INSULATION

ROCK

WHITE METAL, ZINC, LEAD, BABBITT, AND ALLOYS

TITANIUM AND REFRACTORY MATERIAL

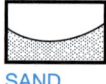
SAND

MAGNESIUM, ALUMINUM, AND ALUMINUM ALLOYS

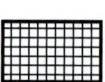
ELECTRIC WINDINGS, ELECTROMAGNETS, RESISTANCE, ETC.

WATER AND OTHER LIQUIDS

RUBBER, PLASTIC, AND ELECTRICAL INSULATION

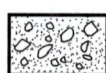
CONCRETE

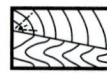
ACROSS GRAIN } WOOD
WITH GRAIN

FIGURE 13.13 ■ Coded section lines.

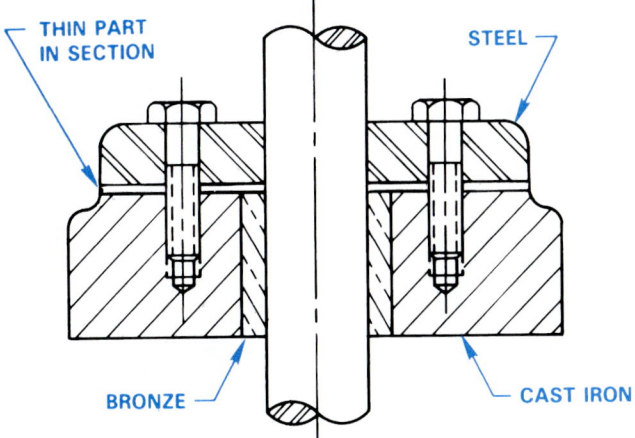

FIGURE 13.14 ■ Assembly section, coded section lines.

Very Thin Features in Section

Very thin features, less than .15 in. (4 mm) thick, can be drawn without section lines. When using this practice, only the outline is drawn as shown in Figure 13.14. This option is often used for features such as a gasket, as shown in the Figure 13.14 assembly.

FULL SECTIONS

A full section is drawn when the cutting plane extends completely through the object, usually along a center plane as shown in Figure 13.15. The object shown in Figure 13.15 could have used two full sections to further clarify hidden features. In such a case the cutting planes and related views are labeled. (See Figure 13.16.)

HALF SECTIONS

A half section is used when a symmetrical object requires sectioning. The cutting-plane line of a half section actually removes one quarter of the object. The advantage of a half section is the sectional view shows half of the object in section and the other half of the object as it normally appears. Thus the name half section. (See Figure 13.17.) Notice that a centerline is used in the sectional view to separate the sectioned portion from the unsectioned portion. Hidden lines are generally omitted from sectional views unless their use improves clarity.

OFFSET SECTIONS

Staggered interior features of an object are sectioned by allowing the cutting-plane line to offset through the features as shown in Figure 13.18. Notice in Figure 13.18 that there is no line in the sectional view indicating a change in direction of the cutting-plane line. Normally the cutting-plane line in an offset section extends completely through the object to clearly display the location of the section.

Section lines can be omitted in the sectional view when drawing clarity is not affected. Confirm this practice with your

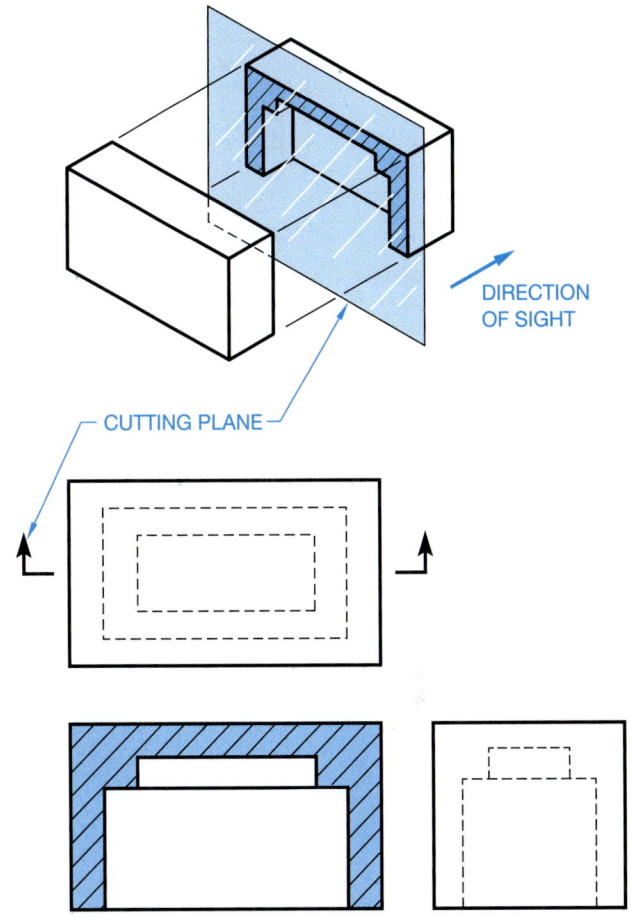

FIGURE 13.15 ■ Full section.

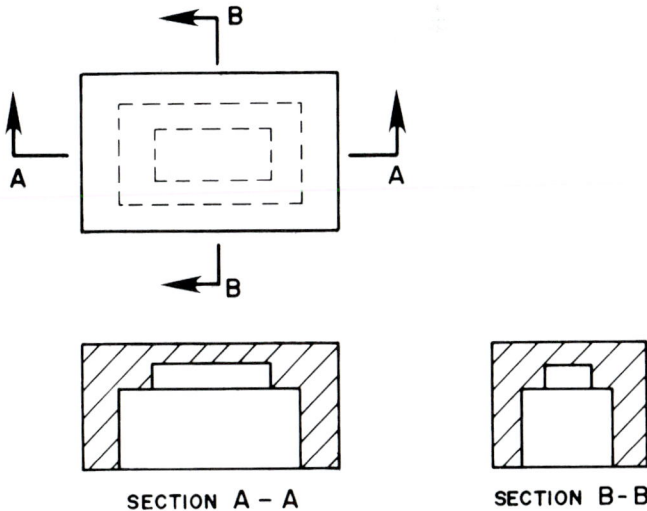

SECTION A – A   SECTION B-B

FIGURE 13.16 ■ Full sections.

supervisor or instructor. You should be cautious when omitting section lines, because doing so can cause confusion.

As previously mentioned, there is no line in the sectional view of a 2-D drawing indicating the change in direction of the

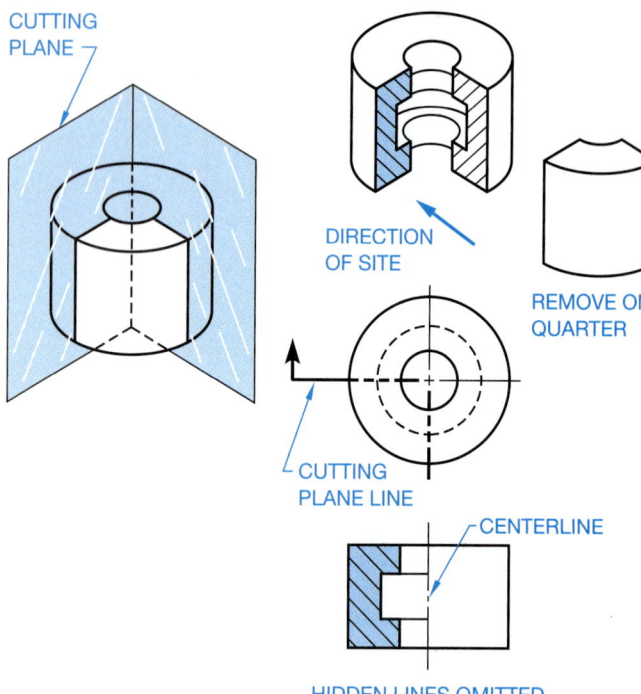

FIGURE 13.17 ■ Half section.

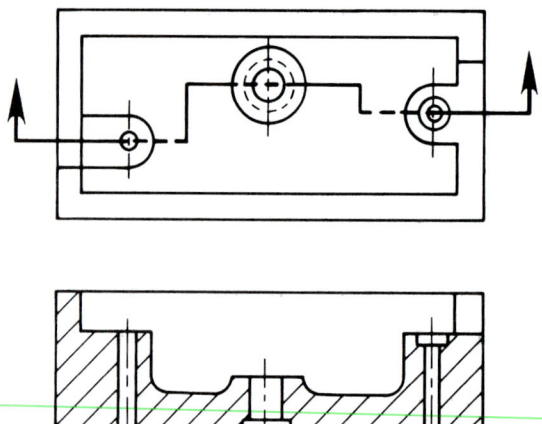

FIGURE 13.18 ■ Offset section.

cutting-plane line. This practice is different when creating 3-D drawings or parametric models using CADD. In this application, the CADD program automatically creates a line representing the edge where the cutting plane changes direction in the 3-D sectional view.

ALIGNED SECTIONS

Similar to the offset section, the aligned section cutting-plane line also staggers to pass through offset features of an object. Normally the change in direction of the cutting-plane line is less than 90° in an aligned section. When this section is taken, the

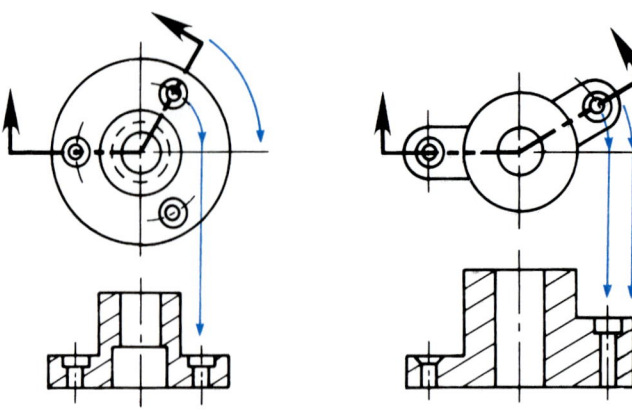

FIGURE 13.19 ■ Aligned section.

sectional view is drawn as if the cutting plane is rotated to a plane perpendicular to the line of sight as shown in Figure 13.19

UNSECTIONED FEATURES

Specific features of an object are commonly left unsectioned in a sectional view if the cutting-plane line passes through the feature and parallel to it. The types of features that are left unsectioned for clarity are bolts, nuts, rivets, screws, shafts, ribs, webs, spokes, bearings, gear teeth, pins, and keys. (See Figure 13.20.) When the cutting-plane line passes through the previously described features perpendicular to their axes, then section lines are shown as seen in Figure 13.21.

While it is most common to draw the outline of the previously discussed features and display them without section lines, a less used method called alternate section lines can be used. The normally unsectioned feature is drawn using hidden lines and every other section line is drawn through the feature, as shown in Figure 13.22.

INTERSECTIONS IN SECTION

When a section is drawn through a small intersecting shape, the true projection can be ignored when using manual drafting, because of the fine detail. The detail is too complex to represent. (See Figure 13.23a and b.) However, larger intersecting features are drawn as their true representation. (See Figure 13.23c and d.) The professional decision is up to you. When using 2-D CADD or 3-D modeling, the true projection is automatically displayed by the CADD program and an accurate representation is provided.

CONVENTIONAL REVOLUTIONS

When the true projection of a feature results in foreshortening, the feature should be revolved onto a plane perpendicular to the

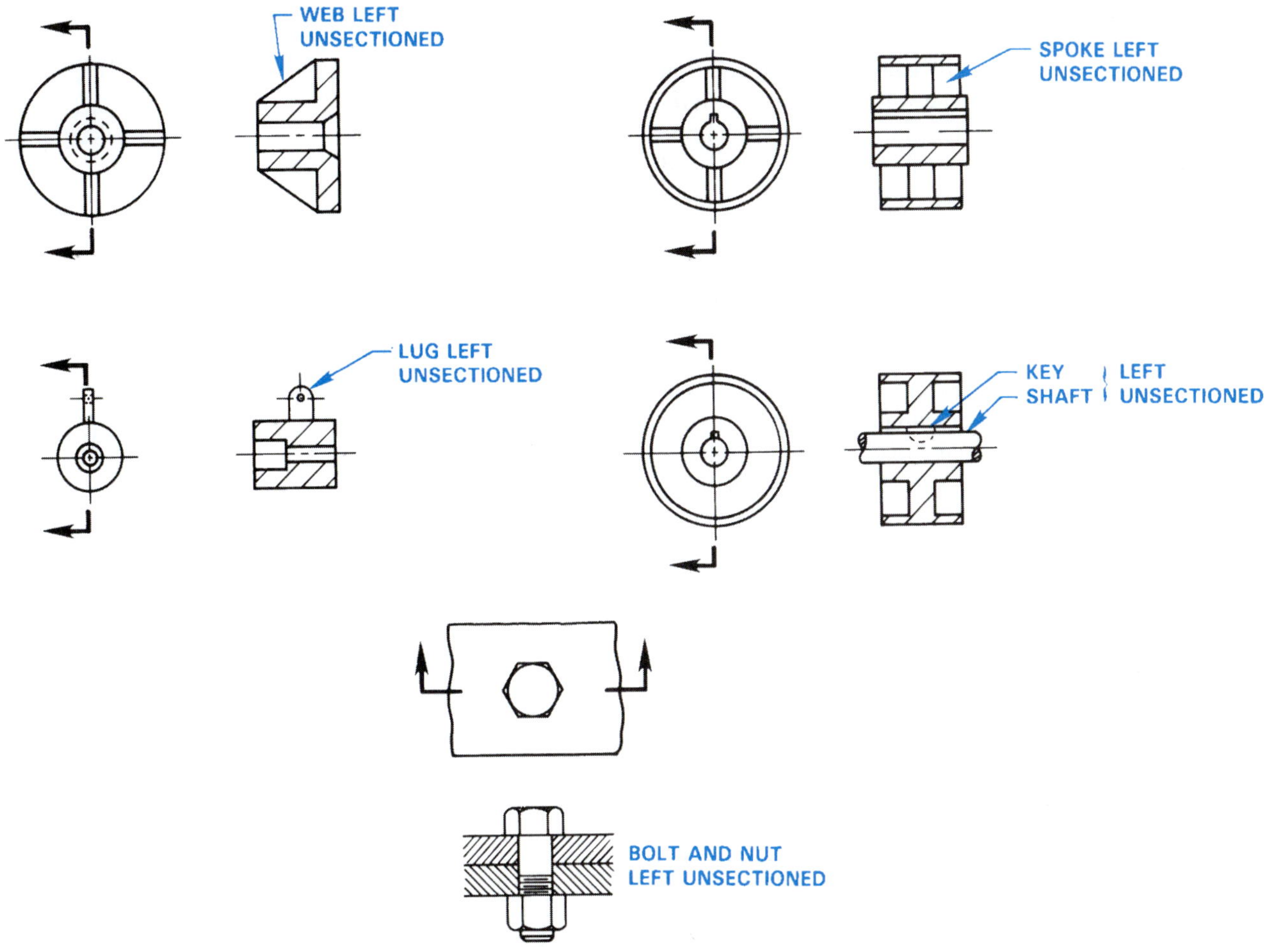

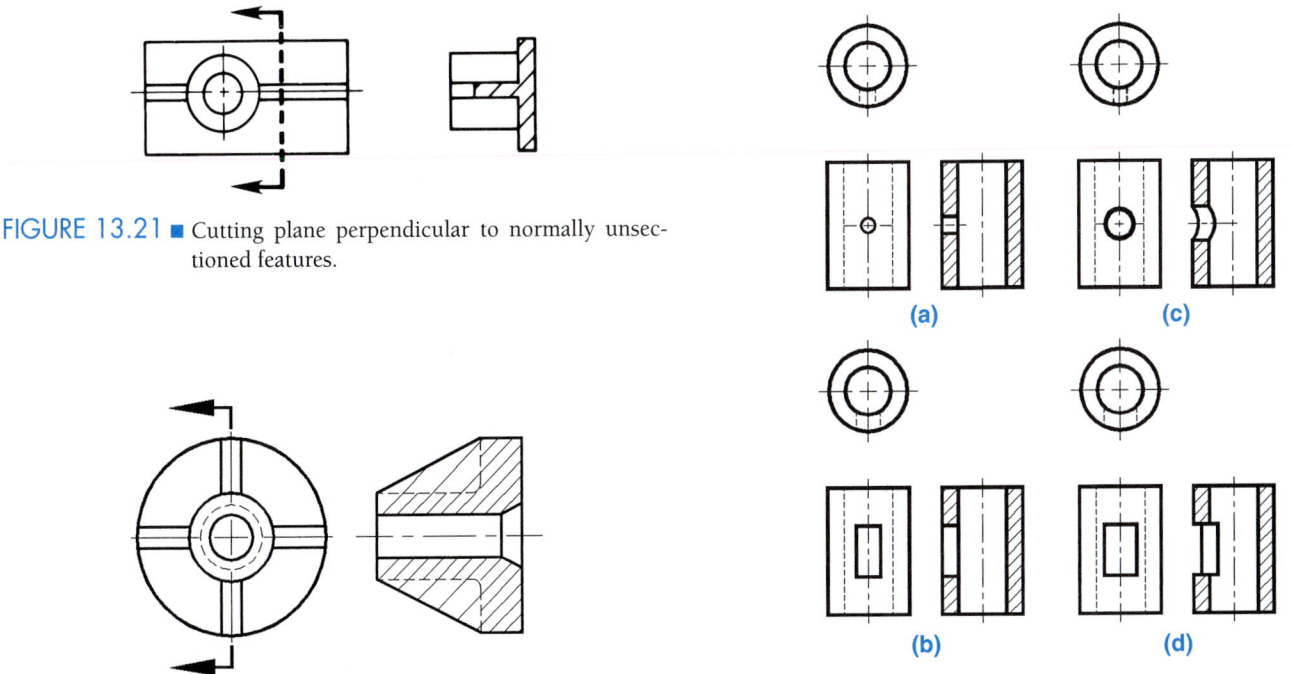

FIGURE 13.20 ■ Certain features are not sectioned when a cutting-plane line passes parallel to their axes.

FIGURE 13.21 ■ Cutting plane perpendicular to normally unsectioned features.

FIGURE 13.22 ■ Alternate section line method.

FIGURE 13.23 ■ Intersections in section.

line of sight as in Figure 13.24. The revolved spoke shown gives a clear representation with a minimum of drafting time. Figure 13.25 shows another illustration of conventional revolution compared to true projection. Notice how the true projection results in a distorted and foreshortened representation of the spoke. The revolved spoke in the preferred view is clear and easy to draw. The practice illustrated here also applies to features of unsectioned objects in multiview as shown in Figure 13.26. When using 2-D CADD or 3-D modeling, the true projection is automatically displayed by the CADD program and an exact representation of the object is provided. While the true projection is not in compliance with ASME Y14.3M recommendations of providing conventional revolutions, it is difficult for the CADD system to create a drawing other than true projection. The 2-D drawing can be edited to comply with conventional revolution standards by using commands such as ROTATE. However, caution should be used, because most CADD programs establish a parametric relationship between the 3-D model and the 2-D drawing. This parametric relationship means that any changes made to the 3-D model or the 2-D drawing automatically affects the other. Confirm the preferred application with your instructor or company.

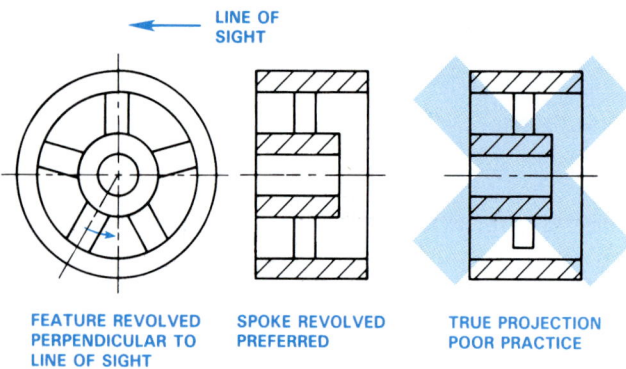

FIGURE 13.24 ■ Conventional revolution.

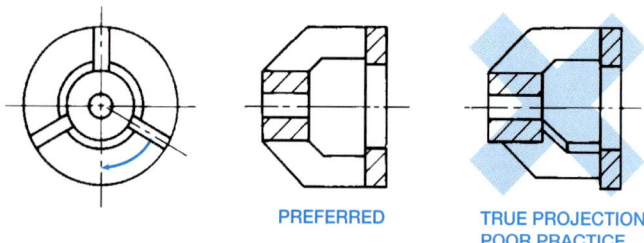

FIGURE 13.25 ■ Conventional revolution in section.

BROKEN-OUT SECTIONS

Often a small portion of a part may be broken away to expose and clarify an interior feature. This technique is called a broken-out section. There is no cutting-plane line used, as you can see in Figure 13.27. A short break line is generally used with a broken-out section.

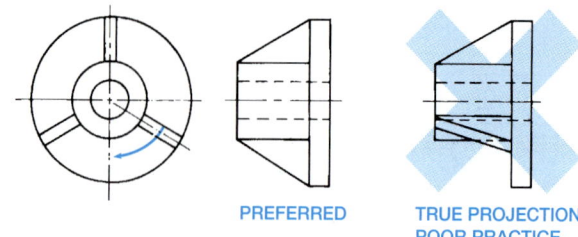

FIGURE 13.26 ■ Conventional revolution in multiview.

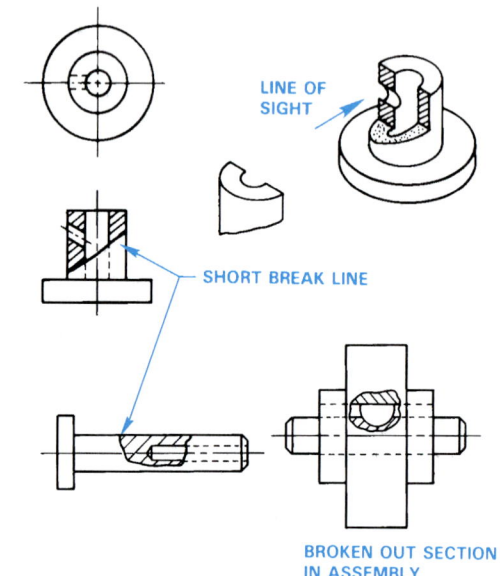

FIGURE 13.27 ■ Broken-out sections.

AUXILIARY SECTIONS

A section that appears in an auxiliary view is known as an auxiliary section. Auxiliary sections are generally projected directly from the view of the cutting plane. If these sections must be moved to other locations on the drawing sheet, they should remain in the same relationship (not rotated) as if taken directly from the view of the cutting plane. (See Figure 13.28.)

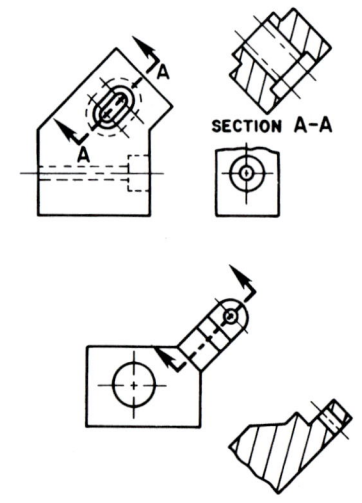

FIGURE 13.28 ■ Auxiliary sections.

SECTIONING

Typically, two primary steps are needed to create a sectional view using CADD software. The first step involves developing a drawing view that will act as the sectioned view. This process is identical to creating an external multiview or auxiliary view, depending on the application. The only difference is that instead of using hidden lines, you use visible object lines. If you already have a drawing view with hidden lines, you can easily transform the view into a sectional view by changing the hidden lines from a hidden line format, or layer, to a visible object layer. Once you develop sectional view geometry, the next step is to add section lines, which are applied with CADD as a graphic pattern.

The power of CADD is obvious when creating section lines. Drawing section lines using manual drafting practices can be a tedious, lengthy task, especially when working on large or complex sectional views. It is also often difficult to draw each line uniformly and equally spaced. However, with CADD, section lines can be added in a matter of seconds instead of what often feels like hours using manual drafting, and the task of producing uniformly and equally spaced section lines is automatic. Additionally, most CADD drafting packages have a great number of available section line types that can be used to create even the most complex assembly sections requiring multiple section line styles. There are standard material section lines for use on mechanical drawings, and numerous other patterns for architectural, structural, civil, and other engineering drafting disciplines.

As previously mentioned, CADD section lines are created as a graphic pattern. When using a program such as AutoCAD for example, section lines are drawn using the HATCH command. When you access the HATCH command, the Hatch and Gradient dialog box is displayed. The Hatch and Gradient dialog box allows you to select a specific graphic pattern, modify the pattern as needed, and add the pattern to a specific area, known as a boundary.

AutoCAD's graphic patterns are called Hatch patterns, and can be accessed using a few different techniques. You can select a pattern by name using the Pattern: drop-down list, or you can pick the ellipsis (. . .) or Swatch button to visually select the desired pattern using the Hatch Pattern Palette shown in Figure 13.29. Once you choose the desired Hatch pattern, you have the option of adjusting the display of the pattern as needed by changing the angle and scale of the pattern symbols. An example of changing Hatch pattern scale is shown in Figure 13.30. This example shows how you can modify the section line scale according to the scale of the drawing but also create a completely different pattern if necessary.

When using CADD to section an area, you must specify where the section lines are to be placed so you do not end up with lines in an unwanted area. If you want to place section lines inside a circle or square, all you need to do is pick the circle or square. (See Figure 13.31.) Some CADD pro-

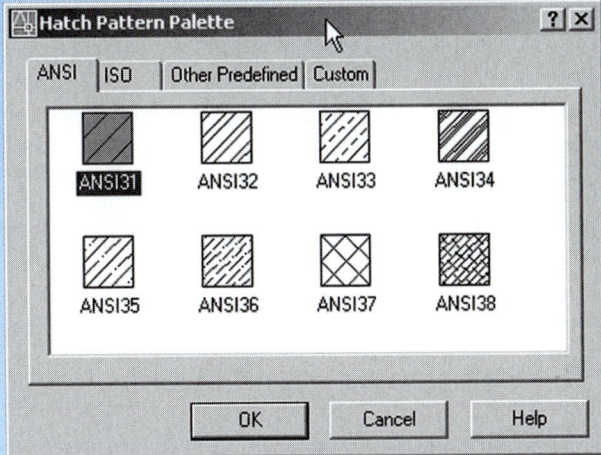

FIGURE 13.29 ■ AutoCAD Hatch Pattern Palette where many different section line symbols and patterns are available for use in your drawings. The section line symbols shown here are the ANSI patterns, which are only a few of the options.

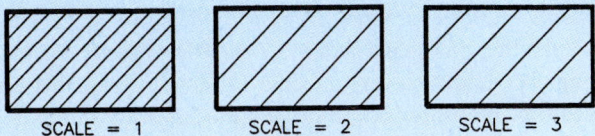

FIGURE 13.30 ■ Three different section line scale factors for CADD applications.

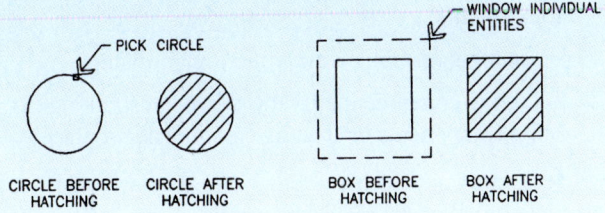

FIGURE 13.31 ■ Selecting objects for sectioning with CADD.

grams allow you to draw section lines in an enclosed area by picking a point within the area. In AutoCAD, for example, the enclosed area is called a boundary. You get an error message, "valid boundary not found," if there is a gap in the

(Continued)

CADD APPLICATIONS 2-D *(continued)*

boundary, or if there is another issue with the selected boundary. A common boundary gap might be a example, the enclosed area is called a boundary. You get an error message, "valid boundary not found," if there is a gap in the boundary, or if there is another issue with the selected boundary. A common boundary gap might be a corner where lines do not meet. You may have to "ZOOM" in on the corner to actually see a very small gap. If this happens, correct the problem and try to section the area again as shown in Figure 13.32.

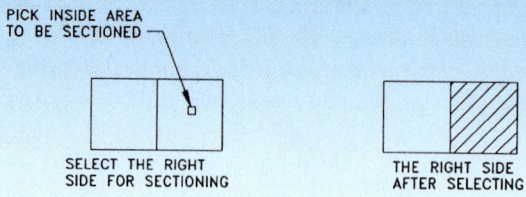

FIGURE 13.32 ■ Recommended drawing sequence for section lining adjacent areas with CADD.

CADD APPLICATIONS 3-D

MODEL SECTIONS

Many 3-D CADD software programs provide the ability to section a model. Sectioning a model is similar to sectioning a 2-D drawing view, in that both attempt to describe the interior portions of an object that are otherwise difficult to visualize. Both part and assembly models can be sectioned depending on the particular 3-D CADD software. Sectioning part models is very effective for creating sketches and features that intersect or originate from other model features. Sectioning an assembly is highly useful for a number of applications including the following:

- Clearly displaying a design.
- Creating stylized model views.
- Working with complex assemblies with multiple components.
- Visualizing assembly component relationships.
- Accessing component elements when constraining existing components or creating components in-place.

Sectioning a model is a temporary display option and can be altered or turned off at any time. An example of a sectioned model is shown in Figure 13.33.

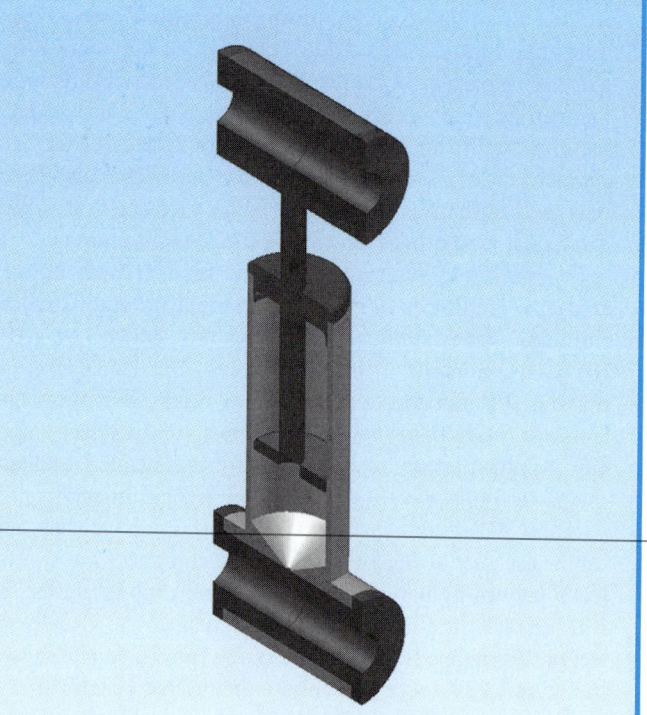

FIGURE 13.33 ■ An example of a sectioned 3-D model.

DRAWING SECTIONS

3-D CADD packages that combine 2-D drawing capabilities usually contain commands allowing you to produce a variety of sectional views. Tools such as SECTION VIEW allow you to create a sectional view by referencing an existing drawing view that is parametrically associated with the model. (See Figure 13.34.) Using tools like SECTION VIEW typically involves creating a cutting-plane line through a feature on an existing drawing view such as the front view. Then it is just a matter of picking a location for the sectional view. The new view is automatically sectioned, a cutting-plane line with arrowheads and identification letters is added, section lines are applied, and you often have the option of displaying a view label and scale, if appropriate.

CADD APPLICATIONS 3-D (continued)

MODEL

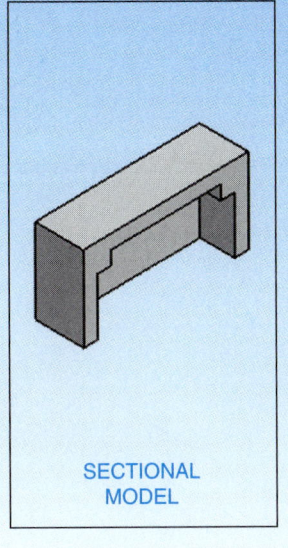

SECTIONAL MODEL

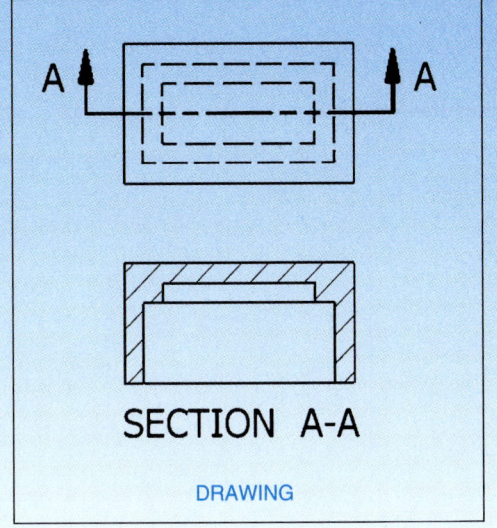

SECTION A-A

DRAWING

FIGURE 13.34 ■ Creating a sectional drawing view by referencing an existing drawing view that is parametrically associated with the model.

CONVENTIONAL BREAKS

When a long object of constant shape throughout its length requires shortening, conventional breaks can be used. These breaks can be used effectively to save time, paper, or space or to increase the scale of an otherwise very long part. Figure 13.35 shows typical conventional breaks. Used on metal shapes, the short break line is drawn thick, freehand, and slightly irregular. Notice that the actual length can be given with a long break line used as a dimension line. For wood shapes, the short break line is drawn freehand as a thick, very irregular line.

The break line for solid round shapes may be drawn freehand or with an irregular curve or template. The shape widths should be approximately 1/3 radius and should be symmetrical about the horizontal centerline and the vertical guidelines as shown.

The break lines for tubular round shapes may be drawn freehand or with an irregular curve or template. The total shape widths should be approximately 1/2 radius and should be symmetrical about the horizontal centerline and vertical guidelines as shown.

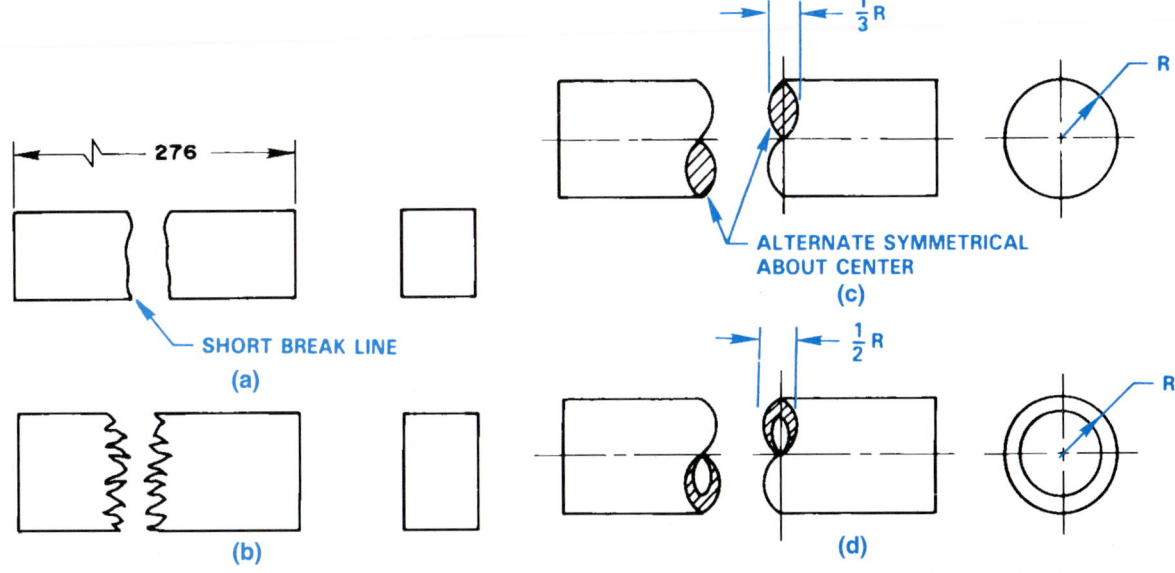

FIGURE 13.35 ■ Conventional breaks for various shapes are (a) metal shapes, (b) wood shapes, (c) cylindrical solid shapes, and (d) cylindrical tubular shapes.

CONVENTIONAL BREAKS

Conventional breaks can be easily created in the 2-D drawing environment available with some 3-D CADD software programs. Tools such as BROKEN VIEW and BREAK allow you to create a conventional break by referencing an existing drawing view that is parametrically associated with the

model. (See Figure 13.36.) Using conventional break tools usually involves defining how much of an existing drawing view, such as the front view, you want to remove, and selecting the type of break lines to use. The new view is automatically broken and break lines are added.

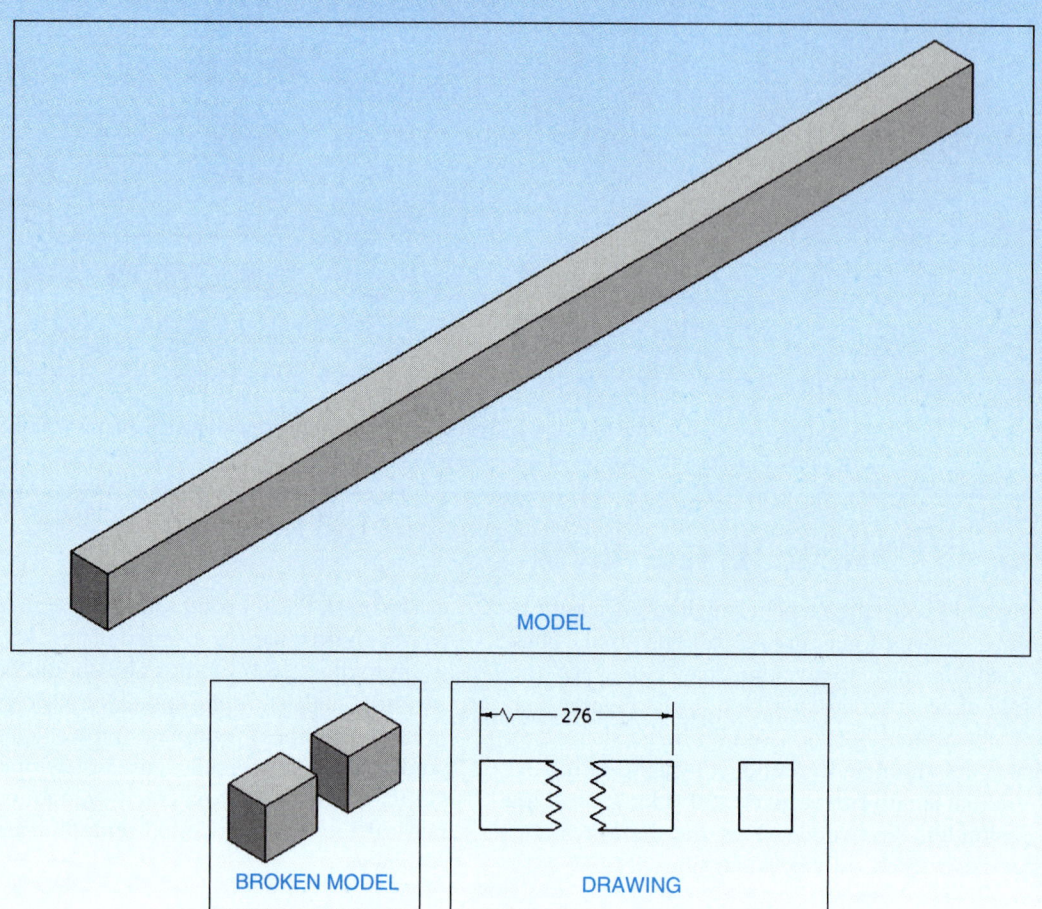

FIGURE 13.36 ■ Creating a conventional break by referencing an existing drawing view, that is parametrically associated with the model.

REVOLVED SECTIONS

When a feature has a constant shape throughout the length that cannot be shown in an external view, a revolved section may be used. The desired section is revolved 90° onto a plane perpendicular to the line of sight as shown in Figure 13.37. Revolved sections can be represented on a drawing one of two ways as shown in Figure 13.38. In Figure 13.38a the revolved section is drawn on the part without breaking a portion of the part away. The revolved section can also be broken away as seen in Figure 13.38b. The surrounding space can be used for dimensions as shown in Figure 13.39.

Notice in Figure 13.39 that very thin parts, less than 4 mm thick, in sections can be left unsectioned. This practice is also common when sectioning a gasket or similar feature.

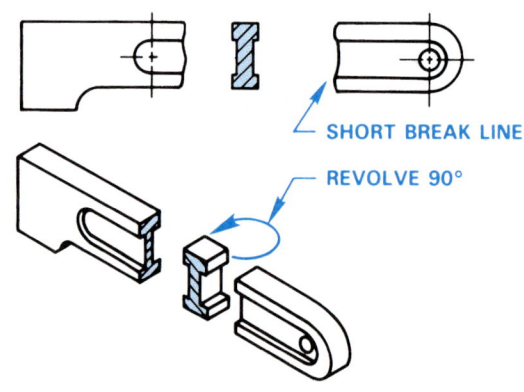

FIGURE 13.37 ■ Revolved section.

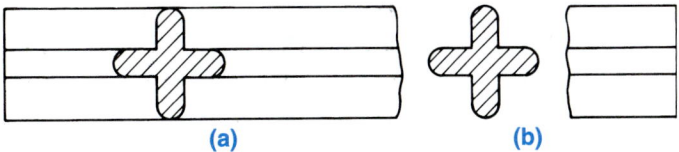

FIGURE 13.38 ■ (a) Revolved section not broken away; (b) revolved section broken away.

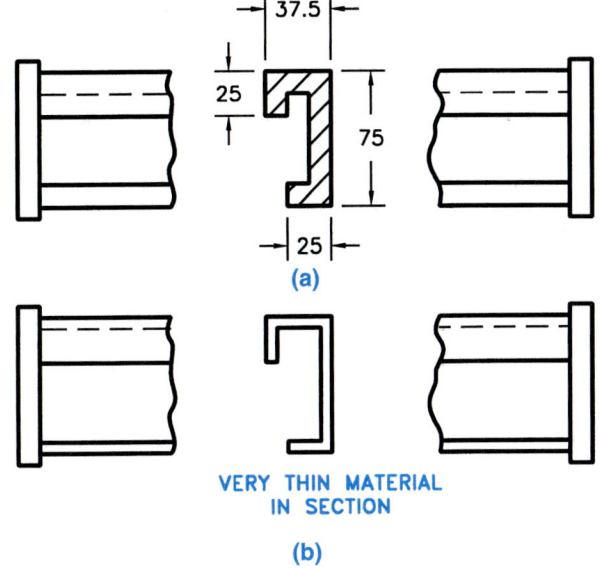

FIGURE 13.39 ■ (a) Dimensioning a broken-away revolved section; (b) a revolved section through thin material. Section lines are omitted when thin material less than 4 mm thick is sectioned.

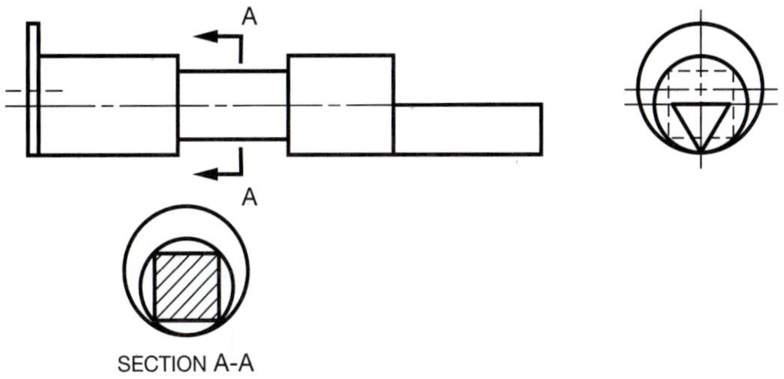

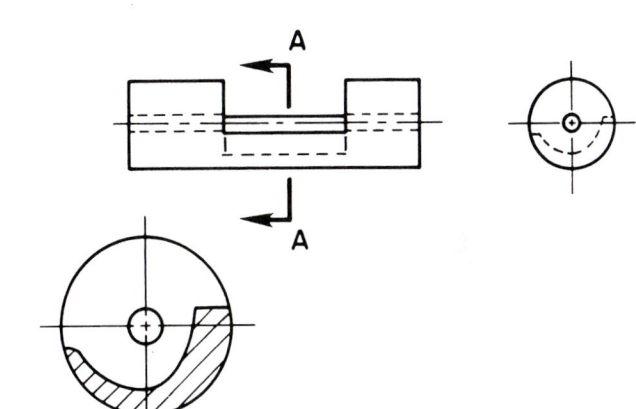

SECTION A-A

FIGURE 13.40 ■ Removed section.

SECTION A- A
SCALE: 2 : I

FIGURE 13.41 ■ Enlarged removed section.

While the ASME Y14.3M standard covering sectioning applications recommends omitting section lines in thin sections, some companies prefer the practice of using solid fill on very thin sections.

Using conventional breaks was introduced before revolved sections because conventional breaks are commonly used in combination with revolved sections as previously discussed and shown in Figures 13.38 and 13.39.

REMOVED SECTIONS

Removed sections are similar to revolved sections except they are removed from the view. A cutting-plane line is placed through the object where the section is taken. Removed sections are not generally placed in direct alignment with the cutting-plane line but are placed in a surrounding area as shown in Figure 13.40.

Removed sections may be preferred when a great deal of detail makes it difficult to effectively use a revolved section. An additional advantage of the removed section is that it can be drawn to a larger scale so close detail can be more clearly identified as shown in Figure 13.41. The sectional view should be labeled, as shown, with SECTION A-A and the revised scale which in this example is 2:1. The predominant scale of the principal views is shown in the title block.

Multiple removed sections are generally arranged on the

sheet in alphabetical order from left to right and top to bottom. (See Figure 13.42.) The cutting planes and related sections are labeled alphabetically excluding the letters *I*, *O*, and *Q*, as they can be mistaken for numbers. When the entire alphabet has been used, label your sections with double letters beginning with *AA*, *BB*, and so on.

Another method of drawing a removed section is to extend a centerline adjacent to a symmetrical feature and revolve the section of the centerline as shown in Figure 13.43. The removed section may be drawn at the same scale or enlarged as necessary to clarify detail.

RECOMMENDED REVIEW

It is recommended that you review the following parts of Chapter 7 before you begin working on fully dimensioned sectioning drawings. This will refresh your memory about how related lines are properly drawn.

- Object Lines.
- Viewing Planes.
- Extension Lines.
- Hidden Lines.
- Break Lines.
- Dimension Lines.
- Centerlines.
- Phantom Lines.
- Leader Lines.

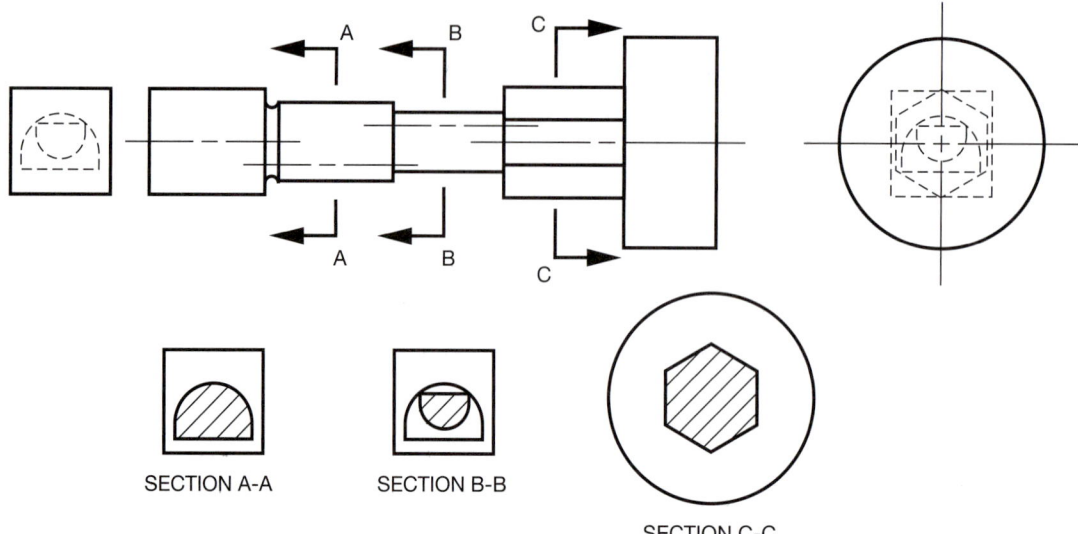

FIGURE 13.42 ■ Multiple removed sections. Hidden lines in the profile views help illustrate the importance of sectional views.

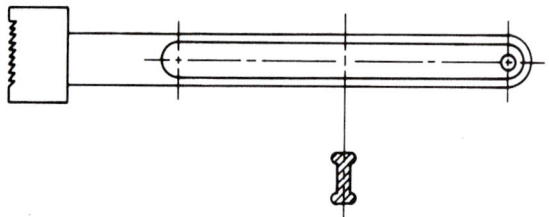

FIGURE 13.43 ■ Alternate removed section method.

Study these topics more carefully as related directly to sectioning practices:

■ Cutting Planes.

■ Coded Section Lines.

■ Section Lines.

Also review Chapters 9 and 10 covering multiview and auxiliary view drawings and Chapter 11 for recommended dimensioning practices.

PROFESSIONAL PERSPECTIVE

Your role as an engineering drafter is communication. You are the link between the engineer and manufacturing, fabrication, or construction. Up to this point you have learned a lot about how to perform this communication task, but the hard part comes when you have to apply what you have learned. Your goal is to make every drawing clear, complete, and easy to interpret. There are a lot of factors to consider, including:

■ Selection of the front view and related views.

■ Deciding if sections are to be used.

■ Placement of dimensions in a logical format so that the size description and located features are complete.

This is a difficult job, but you are motivated to do it. And just when you have mastered how to select and dimension multiviews, you are faced with preparing sectional drawings, which means a dozen main options and endless applications. But don't worry because you have made it this far! Here are some key points to consider:

■ First you need to completely show and describe the outside of the part.

■ If there are a lot of hidden lines for internal features, you know immediately that a section is needed, because you cannot dimension to hidden lines.

■ Analyze the hidden lines and decide on the type of section that will work best.

■ Review each type of sectioning technique until you find one that looks best for the application.

■ Be sure you clearly label the cutting-plane lines and related multiple sections so they correlate.

■ Half sections are good for showing both the outside and inside of the object at the same time, but be cautious about their use, because sometimes they confuse the reader.

You are not alone in this new venture. The problem assignments in this chapter recommend a specific sectioning technique during this learning process.

Figure 13.44 shows a carefully created drawing of a complex part. Along with front, left-side, and rear views, there is a full section and auxiliary section.

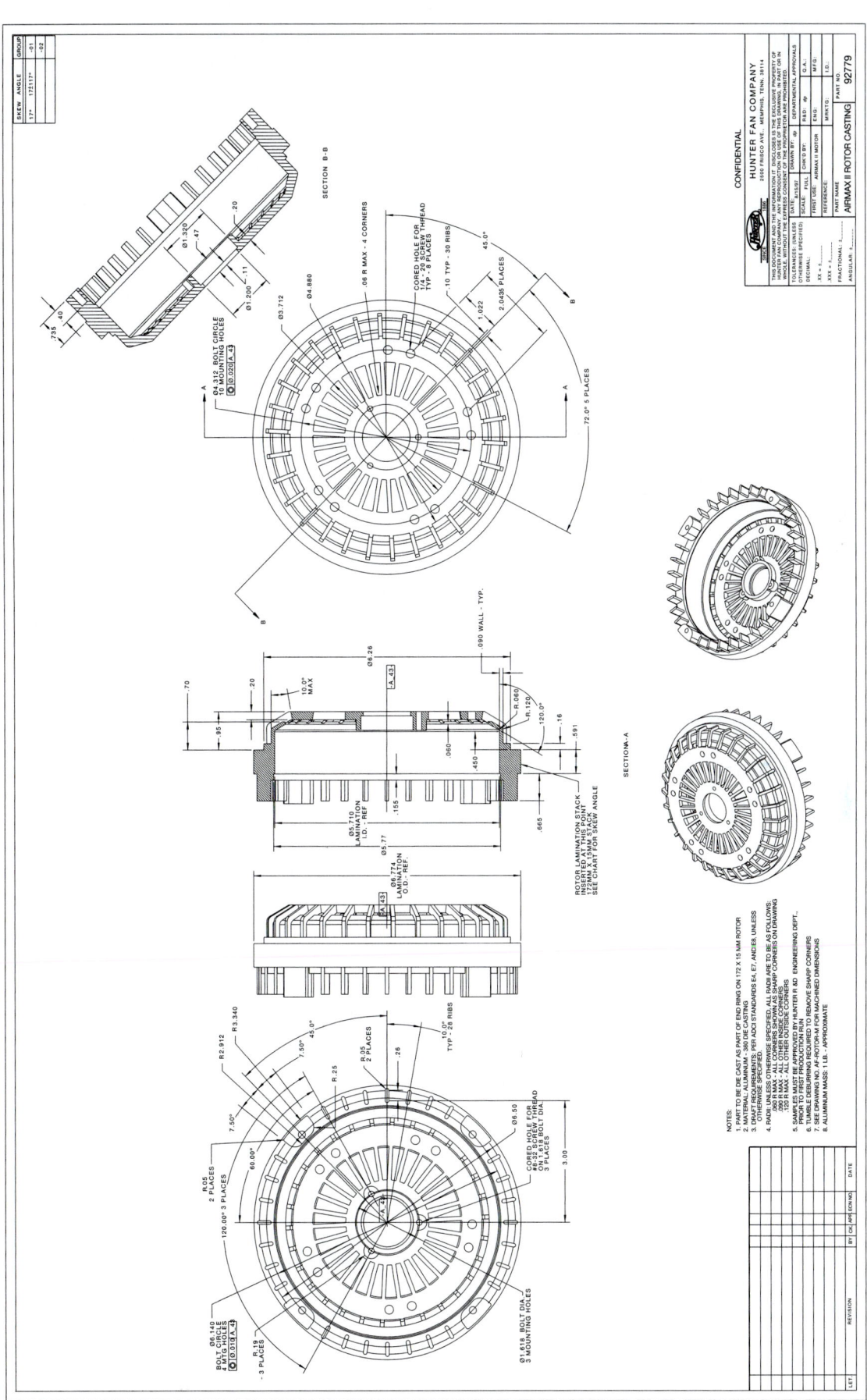

FIGURE 13.44 ■ An actual industry drawing of a complex part, using front, left-side, and rear views along with a full and auxiliary section. *Courtesy Hunter Fan Company, Memphis, TN.*

MATH APPLICATIONS

DISTANCE BETWEEN HOLES ON A BOLT CIRCLE

Problem: Find the center-to-center distance between adjacent holes on the bolt circle shown in Figure 13.45.

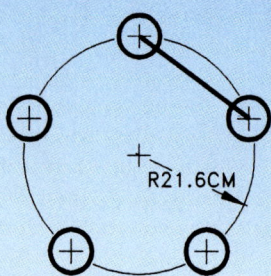

FIGURE 13.45 ■ Bolt circle.

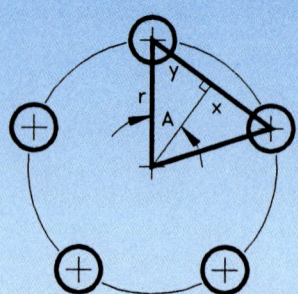

FIGURE 13.46 ■ Bolt circle with a right triangle drawn.

Solution: Problems like this involve finding the side of a right triangle. The first step is to construct a right triangle onto the drawing, as shown in Figure 13.46. From the definition of *sine*. (See Math Instruction Appendix in the Online Companion.)

From the definition of *sine*. (See CD Chapter 3.)

$$\sin A = \frac{y}{r}$$

Because there are five holes on the circle, angle A must be (half of 360°) ÷ 5, or 36°. Hypotenuse r is equal to the radius of 21.6 cm, so you now have $\sin 36° = \frac{y}{21.6}$. Using a calculator to find the sine of 36° gives $.5878 = \frac{y}{21.6}$.

Multiplying both sides of this equation by 21.6 gives the length of side y of the triangle: y = (21.6) (.5878) = 12.696. Finally, side y is half of the required center-to-center distance, so d = 2(12.696) = **25.4 cm.**

WEBSITE RESEARCH

http://www.asme.org—American Society of Mechanical Engineers standards.

CHAPTER 13 *Sectioning Test*

Access the CD found with this textbook to view the Chapter 13 Test. Confirm the preferred submittal method with your instructor.

CHAPTER 13 *Sectioning Problems*

PROBLEMS 13.1 through 13.4, PART 1: Access the CD found with this textbook and open the problem of your choice, or as assigned by your instructor. Solve the problems using the instructions provided on the CD, unless otherwise specified by your instructor.

PART 2

1. From the selected engineer's sketch or layout, determine the views and sections that are needed.
2. Make a sketch of the selected views and sections as close to correct proportions as possible. Do not spend a lot of time, as the sketch is only a guide. Indicate where the cutting-plane lines and dimensions are to be placed.
3. Using the sketch you have developed as a guide, draw an original sectioned multiview drawing on an adequately sized drawing sheet. Select a scale that properly details the part on the selected sheet size. Use unidirectional dimensioning. Use manual or computer-aided drafting as required by your course guidelines.
4. Include the following general notes at the lower left corner of the sheet .5 in. each way from the corner border lines:

    1. INTERPRET DIMENSIONS AND TOLERANCES PER ASME Y14.5M—1994.

NOTES

Additional notes may be required depending on the specifications of each individual assignment. A tolerance block is recommended as shown in problems for Chapter 11 unless otherwise specified.

PROBLEM 13.5   Full Section (in.)

Part Name: Fitting

Material: Bronze

Finish All Over: 63μin.

Note: Refer to Chapter 11 for proper dimensioning practices. Engineering sketches may not display correct practices. This problem may require a front view, side view showing the hexagon, and a full section to expose the interior features for dimensioning.

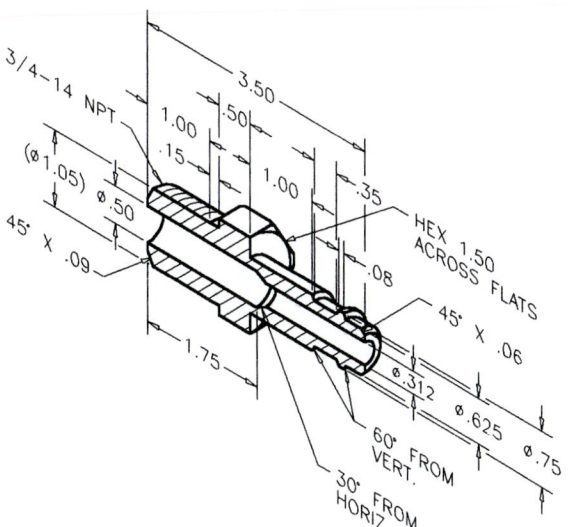

2. REMOVE ALL BURRS AND SHARP EDGES.
UNSPECIFIED TOLERANCES:

DECIMALS	mm	IN.
X	±2.5	±.1
XX	±0.25	±.01
XXX	±0.125	±.005
ANGULAR ± 30′		
FINISH	3.2μm	125μIN.

5. The engineering layouts may not be dimensioned properly. Verify the correct practice before placing dimensions; for example, the diameter symbol should precede the diameter dimension, and leaders should not cross over dimension lines. Check other line and dimensioning techniques for proper standards. Actual industrial drawings are provided as advanced problems throughout.

PROBLEM 13.6   Full section and view enlargement (in.)

Part Name: Spring

Material: SAE 1060

Problem based on original art courtesy Stanley Hydraulic Tools, a Division of the Stanley Works.

Heat Treat:

1. Austenitize at 1475° F.
2. Direct quench in agitated oil.
3. Temper to R_cC 44–46.

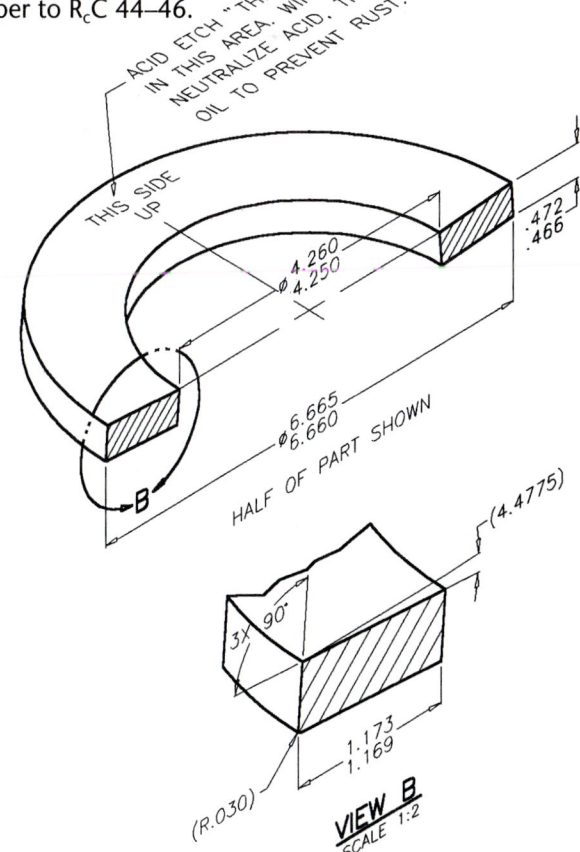

PROBLEM 13.7 Full section (metric)
Part Name: Hydraulic Valve Cylinder
Material: Phosphor Bronze
All Fillets and Rounds: R.1

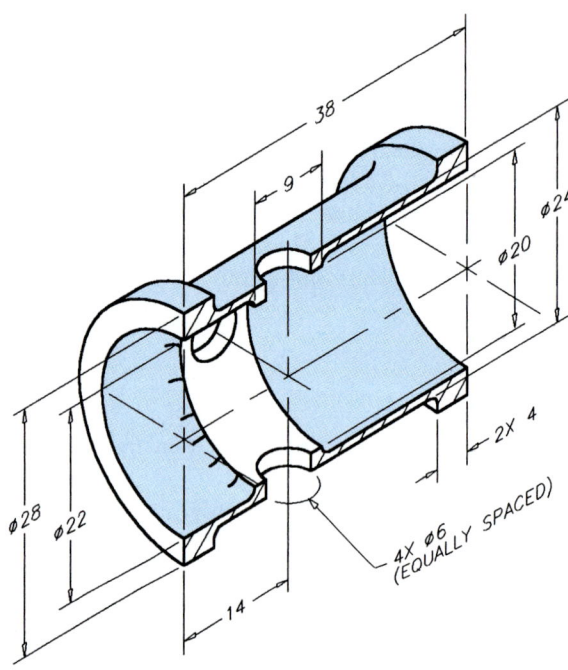

PROBLEM 13.8 Full section (in.)
Part Name: Machine Plate
Material: 6160 T6 Steel

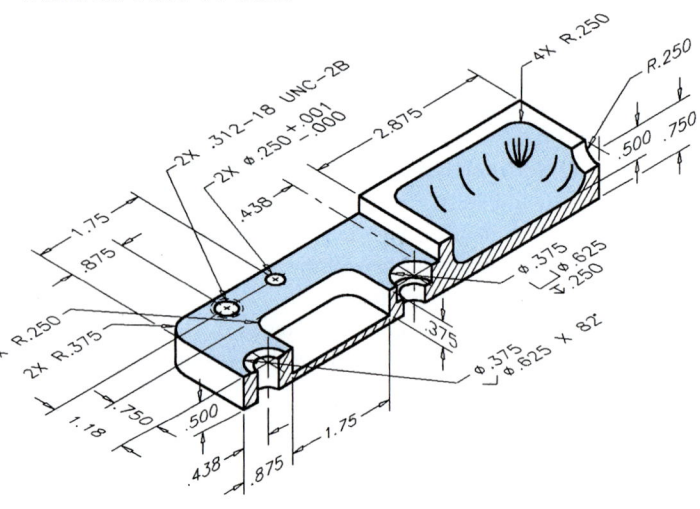

PROBLEM 13.9 Full section (in.)
Part Name: Face Plate
Material: 6160 T6 Steel

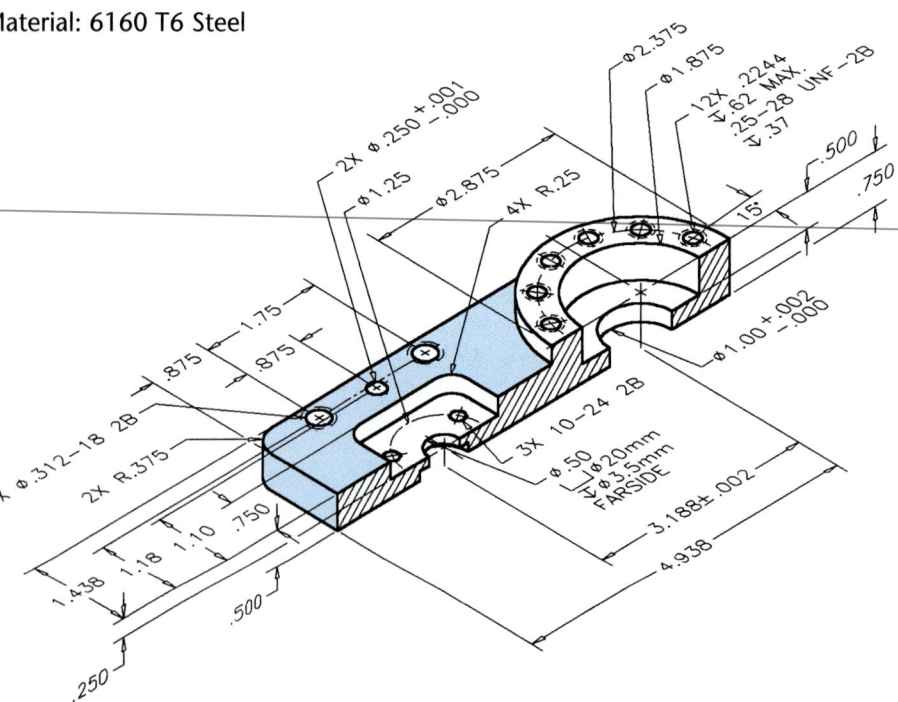

PROBLEM 13.10 **Full section, dimensioning cylindrical shapes (in.)**

Part Name: Plug

Material: Phosphor Bronze

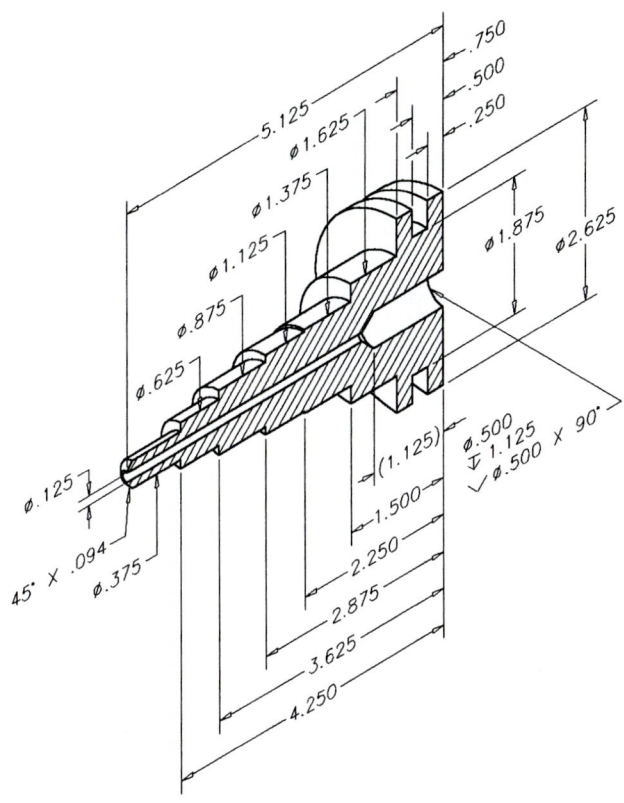

PROBLEM 13.12 **Full and broken-out section (metric)**

Part Name: Hydraulic Valve Cylinder

Material: Phosphor Bronze

Note: Proposed sections and section lines are not given in engineer's layout. Refer to The Engineering Design Application at the beginning of this chapter.

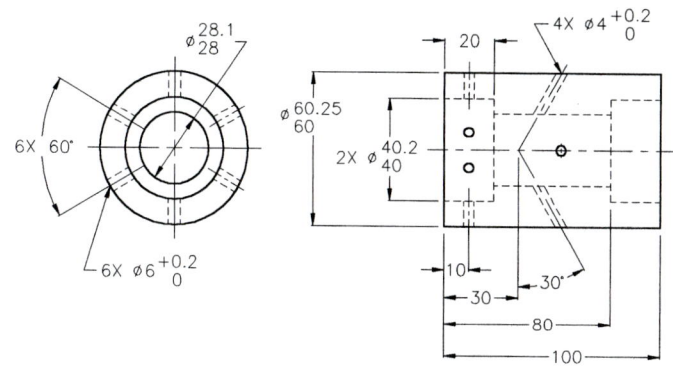

PROBLEM 13.13 **Full section (in.)**

Part Name: Hanger

Material: SAE 1030

Fillets and Rounds: R.062

PROBLEM 13.11 **Full section or half section (in.)**

Part Name: Hub

Material: Cast Iron

SPECIFIC INSTRUCTIONS: Convert the broken-out section in the given drawing to a full section.

Note: Type of section used affects view selection.

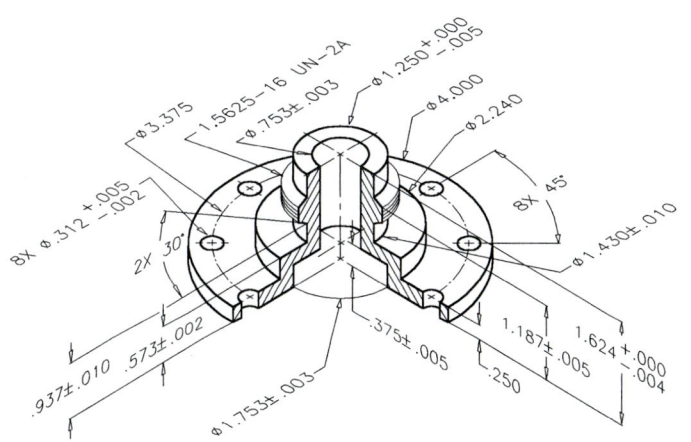

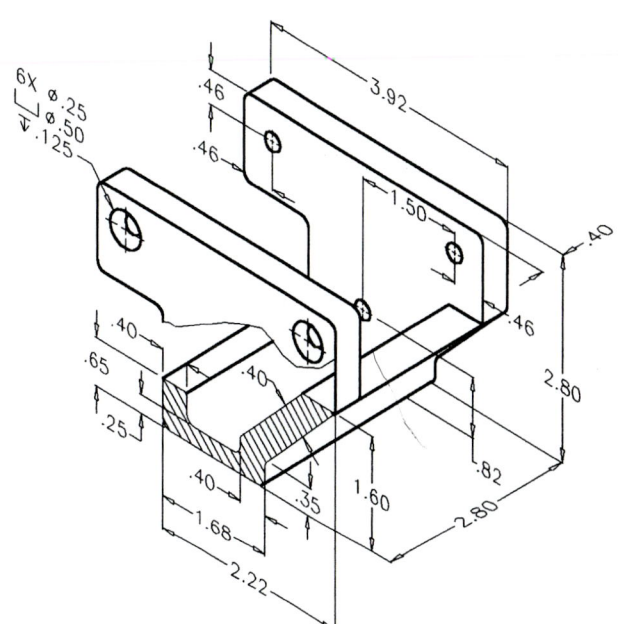

PROBLEM 13.14: Access the CD found with this textbook and open this problem if assigned by your instructor. Solve the problem using the previous instructions and instructions found on the CD, unless otherwise specified by your instructor.

PROBLEM 13.15 **Half section (in.)**

Part Name: Dial

Material: Bronze

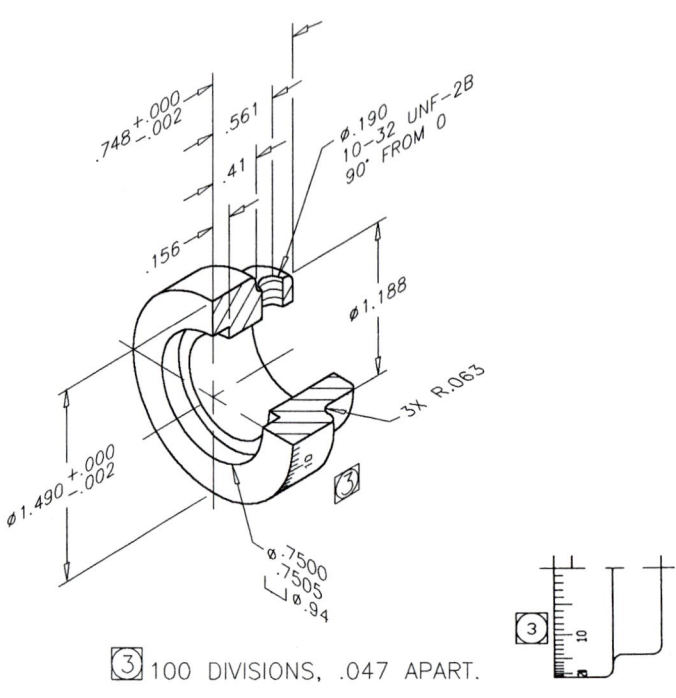

③ 100 DIVISIONS, .047 APART.

PROBLEM 13.16 **Half section (in.)**

Part Name: Rod Support

Material: 6061–T6 Aluminum

Fillets: R.03 unless otherwise specified.

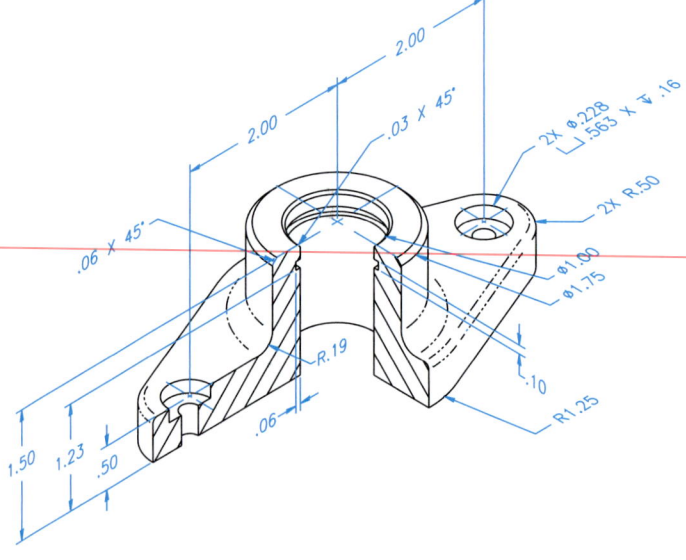

PROBLEMS 13.17 through 13.19: Access the CD found with this textbook and open the problem of your choice, or as assigned by your instructor. Solve the problems using the instructions provided with this chapter or on the CD, unless otherwise specified by your instructor.

PROBLEM 13.20 **Offset section (in.)**

Part Name: Slide Bracket

Material: AISI 1020

Fillets: R.25 unless otherwise specified.

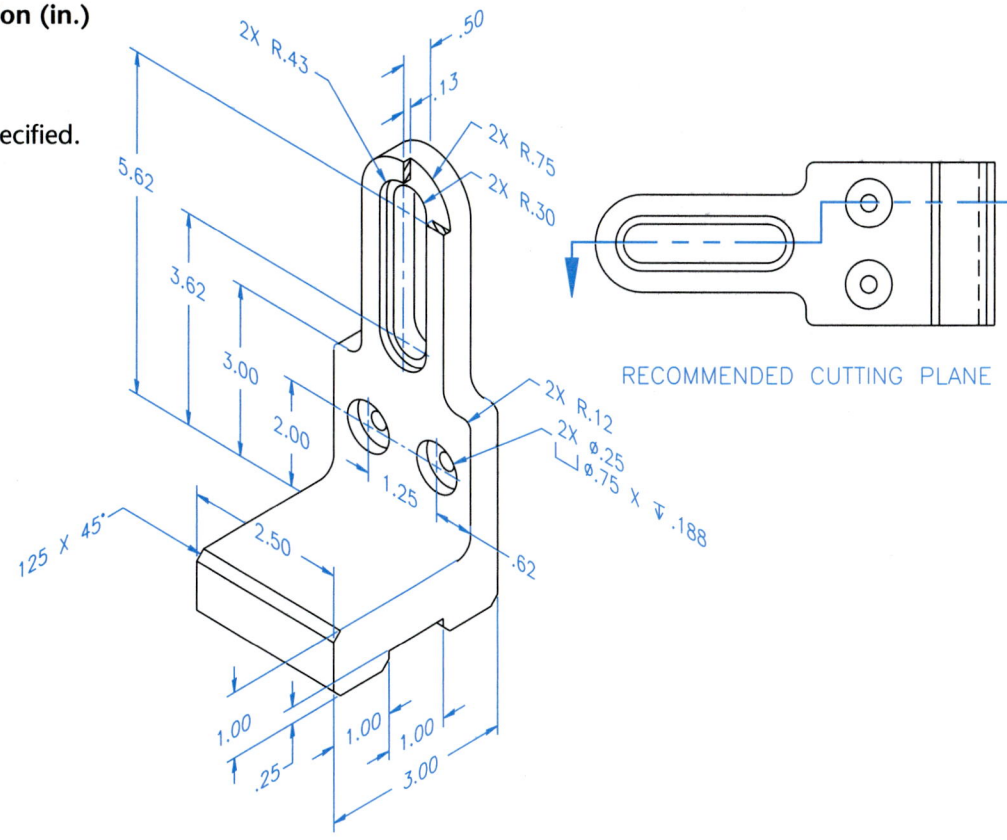

RECOMMENDED CUTTING PLANE

PROBLEM 13.21 **Offset section (in.)**

Part Name: Mounting Plate

Material: AISI 1020

Fillets: R.03 unless otherwise specified.

RECOMMENDED CUTTING PLANE

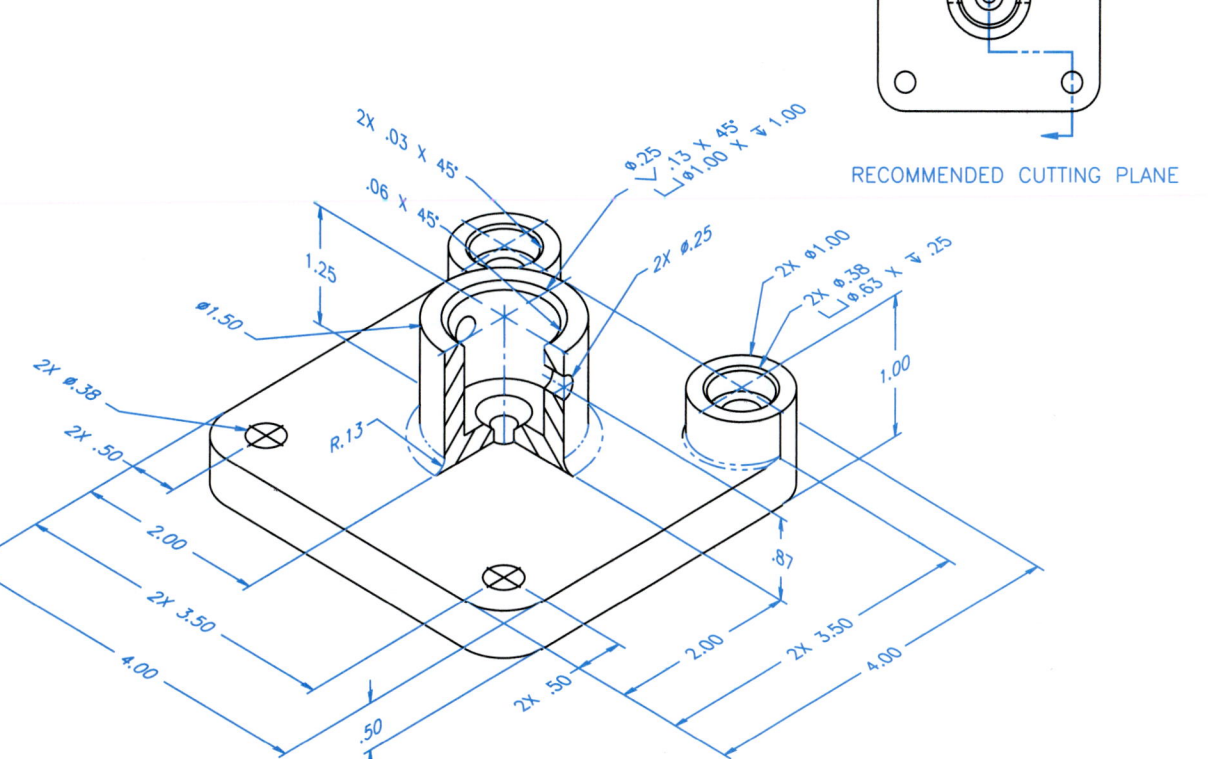

PROBLEM 13.22 Offset section (in.)

Part Name: Die Casting

Material: SAE 6150

SPECIFIC INSTRUCTIONS: Convert all dimensions to ANSI Y14.5M standards as shown and discussed in this text.

Courtesy of Kris Altmiller.

RECOMMENDED CUTTING PLANE

PROBLEM 13.23 Offset section (in.)

Part Name: Drill Plate

Material: SAE 1020

Case Harden: 55 Rockwell C Scale

Fillets and Rounds: R.12

FAO: 63μin.

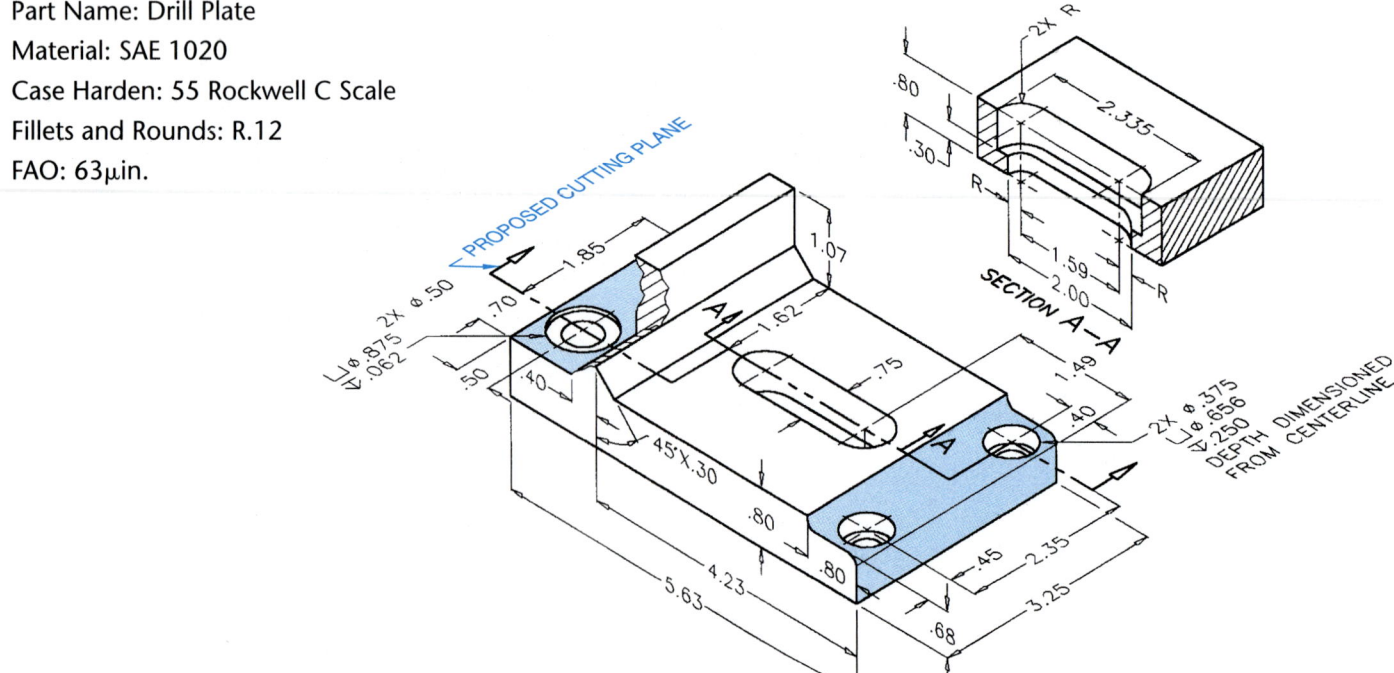

PROBLEM 13.24 **Aligned section (in.)**

Part Name: Shaft Retainer

Material: SAE 4340

Fillets and Rounds: R.06 unless otherwise specified.

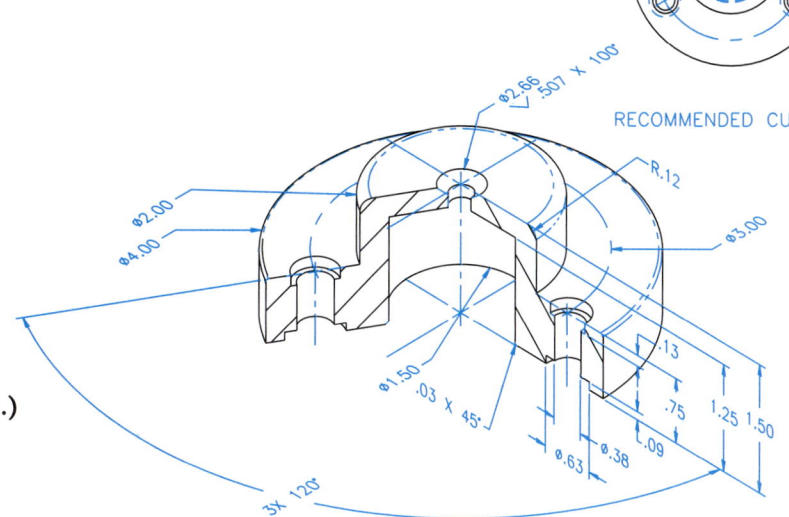

RECOMMENDED CUTTING PLANE
ø2.66
.507 X 100°
R.12
ø2.00
ø4.00
ø3.00
ø1.50
.03 X 45°
.13
.75
1.25
1.50
.09
ø.63
ø.38
3X 120°

PROBLEM 13.25 **Aligned section (in.)**

Part Name: Hub

Material: SAE 3145

Fillets and Rounds: R.125

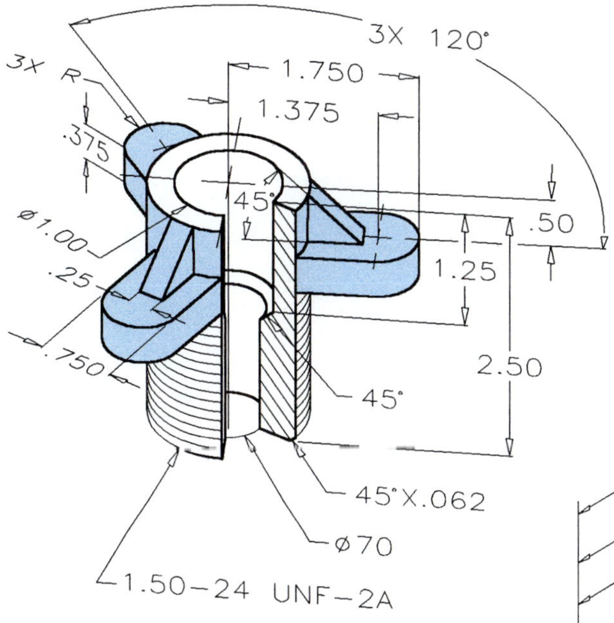

3X 120°
3X R
1.750
1.375
.375
45°
.50
ø1.00
1.25
.25
.750
2.50
45°
45°X.062
ø70
1.50–24 UNF–2A

PROBLEM 13.26 **Broken-out section (in.)**

Part Name: Taper Shaft

Material: SAE 4320

FAO: 16μin.

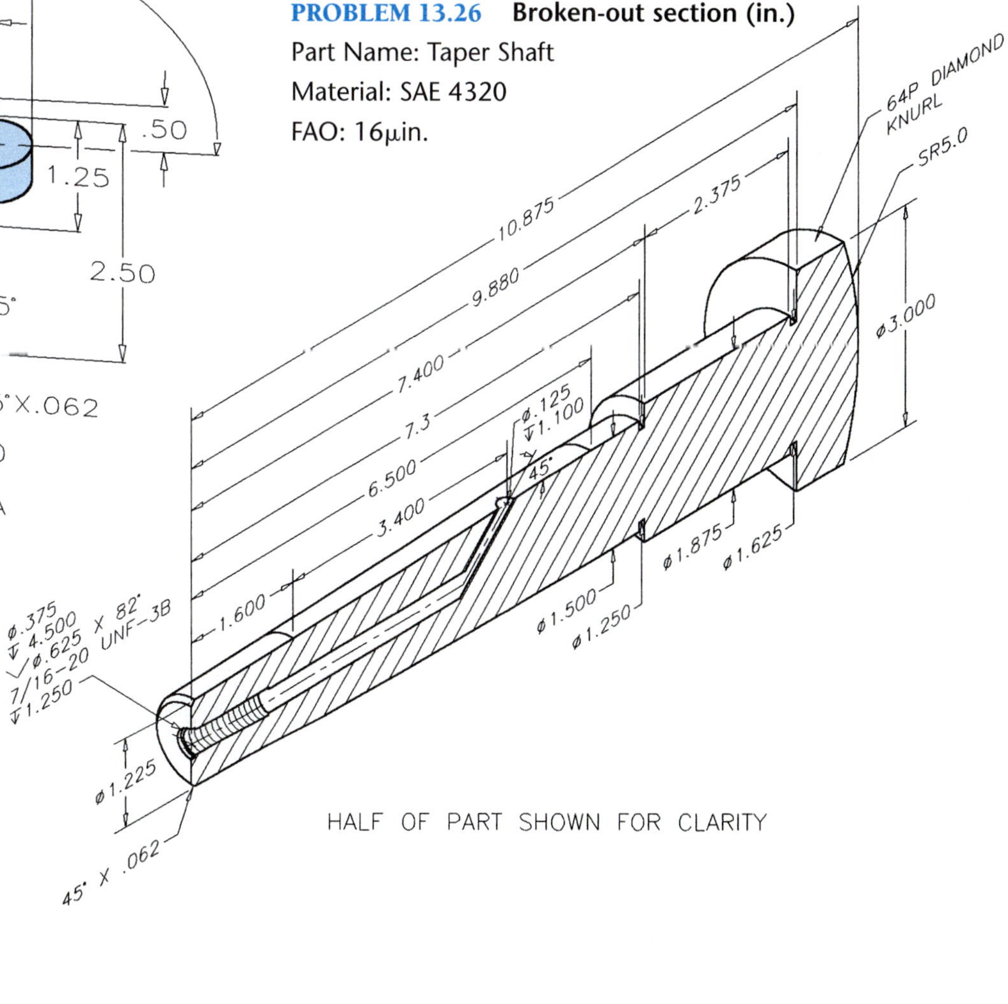

64P DIAMOND KNURL
SR5.0
10.875
2.375
9.880
ø3.000
7.400
ø.125
↧1.100
7.3
45°
6.500
3.400
ø1.875
ø1.625
1.600
ø1.500
ø1.250
ø.375
↧ 4.500
⌵ø.625 X 82°
7/16–20 UNF–3B
↧1.250
ø1.225
45° X .062
HALF OF PART SHOWN FOR CLARITY

PROBLEM 13.27 Auxiliary section (in.)

Part Name: Angle Bracket

Material: 6061-T6 Aluminum

Fillets: R.13 unless otherwise specified.

Rounds: R.38 unless otherwise specified.

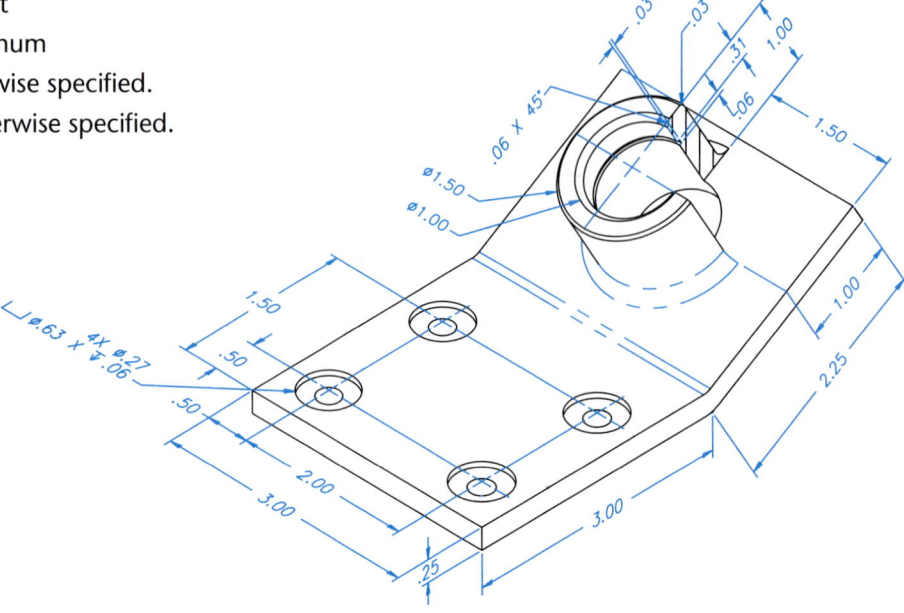

PROBLEM 13.28 Auxiliary section (metric)

Part Name: Gear Base

Material: SAE 2340

Finish All Over: 0.8μm

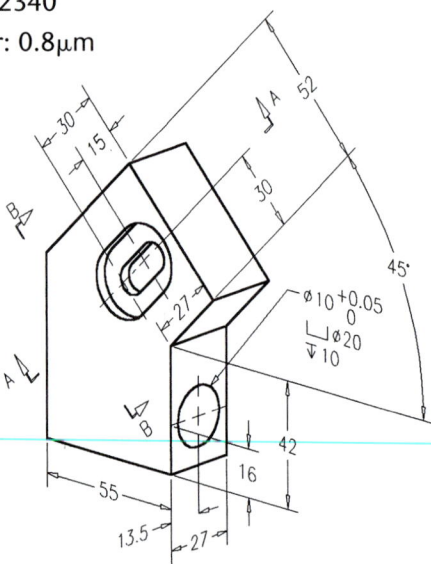

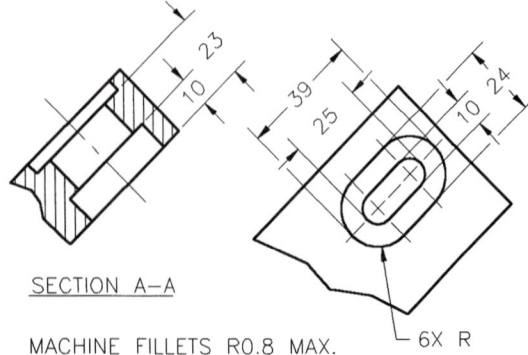

SECTION A–A

MACHINE FILLETS R0.8 MAX.

VIEW B–B

PROBLEM 13.29 Broken-out section, view enlargement (in.)

Part Name: Clamp Cap

Material: Cast Aluminum

Fillets and Rounds: R.06

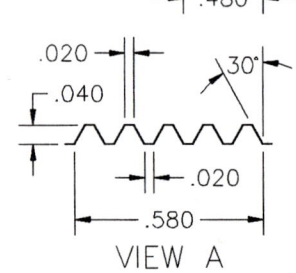

VIEW A

PROBLEM 13.30 **Broken-out section (metric)**

Part Name: 50 mm 45° Elbow

Material: Cast Iron

Fillets and Rounds: R4mm

SPECIFIC INSTRUCTIONS: Consider bottom view or auxiliary view for hole pattern dimensions.

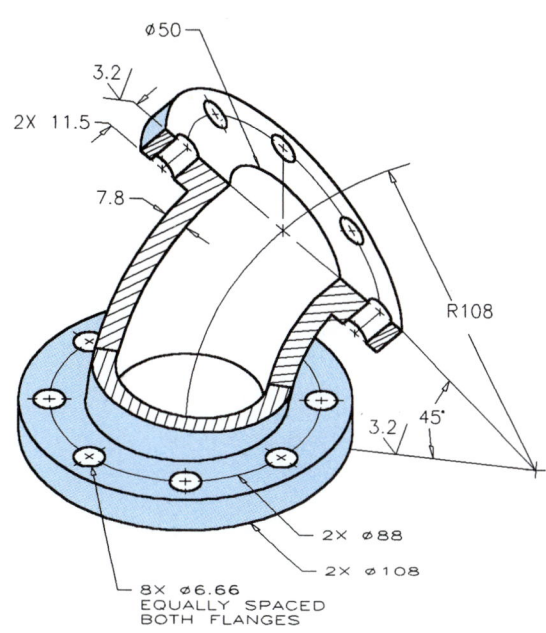

PROBLEM 13.31 **Revolved section (in.)**

Part Name: End Loading Arm

Material: SAE 2310

Fillets and Rounds: R.25

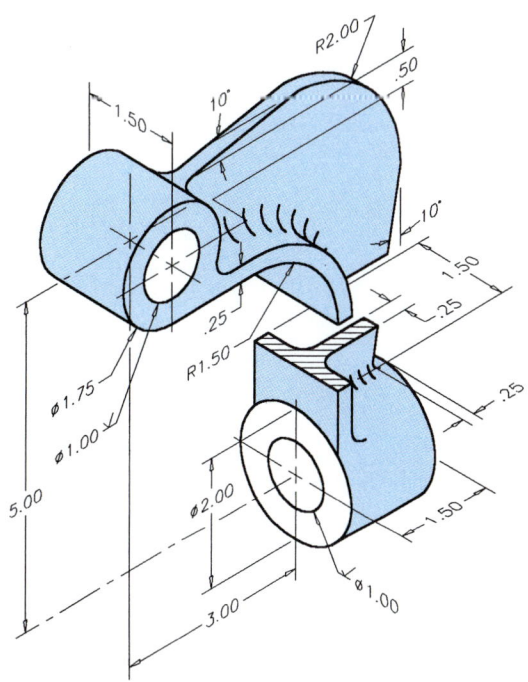

PROBLEM 13.32 **Revolved section (in.)**

Part Name: Offset Handwheel

Material: Bronze

All Fillets and Rounds: R.12

Finishes: 125μin.

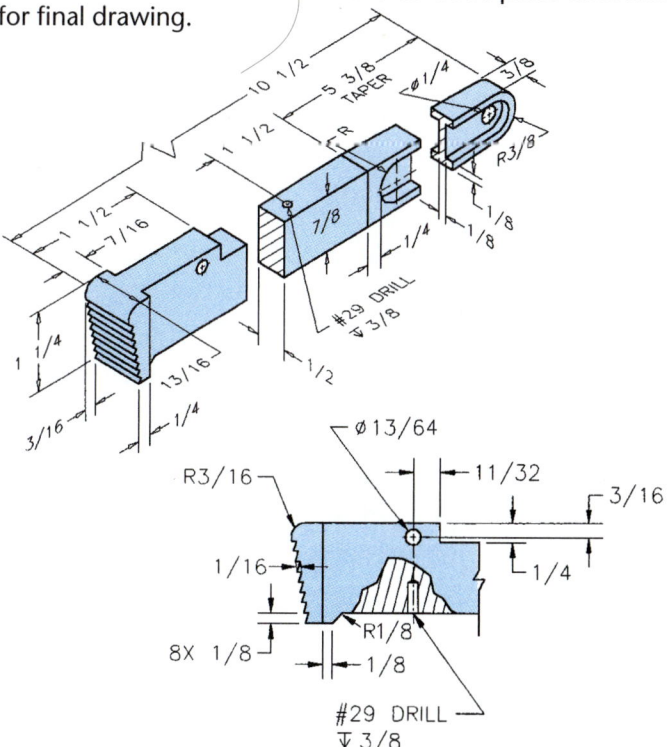

PROBLEM 13.33 **Broken-out, revolved section, view enlargement (in.)**

Part Name: Pipe Wrench Handle

Material: SAE 5120

Fillets and Rounds: R.06

SPECIFIC INSTRUCTIONS: Engineering sketch is given in fractional inches, convert all size dimensions to two-place decimals and location dimensions to three-place decimals for final drawing.

PROBLEM 13.34   **Removed section (metric)**

Part Name: Valve Stem

Material: Phosphor Bronze

Finish All Over: 1 μm

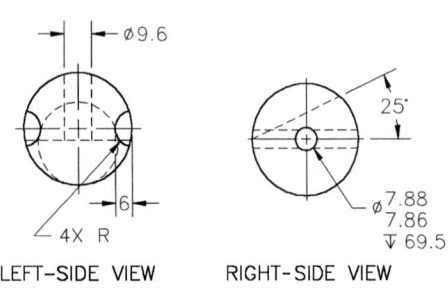

LEFT–SIDE VIEW   RIGHT–SIDE VIEW

NOTE: Some hidden lines omitted for clarity.

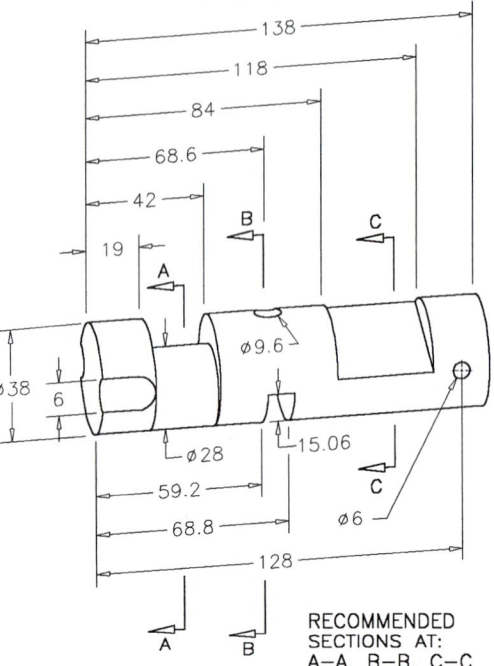

RECOMMENDED
SECTIONS AT:
A–A, B–B, C–C

PROBLEM 13.35   **Offset section (in.)**

Part Name: Die Plate Casting

Material: SAE 3120

Courtesy of Kris Altmiller.

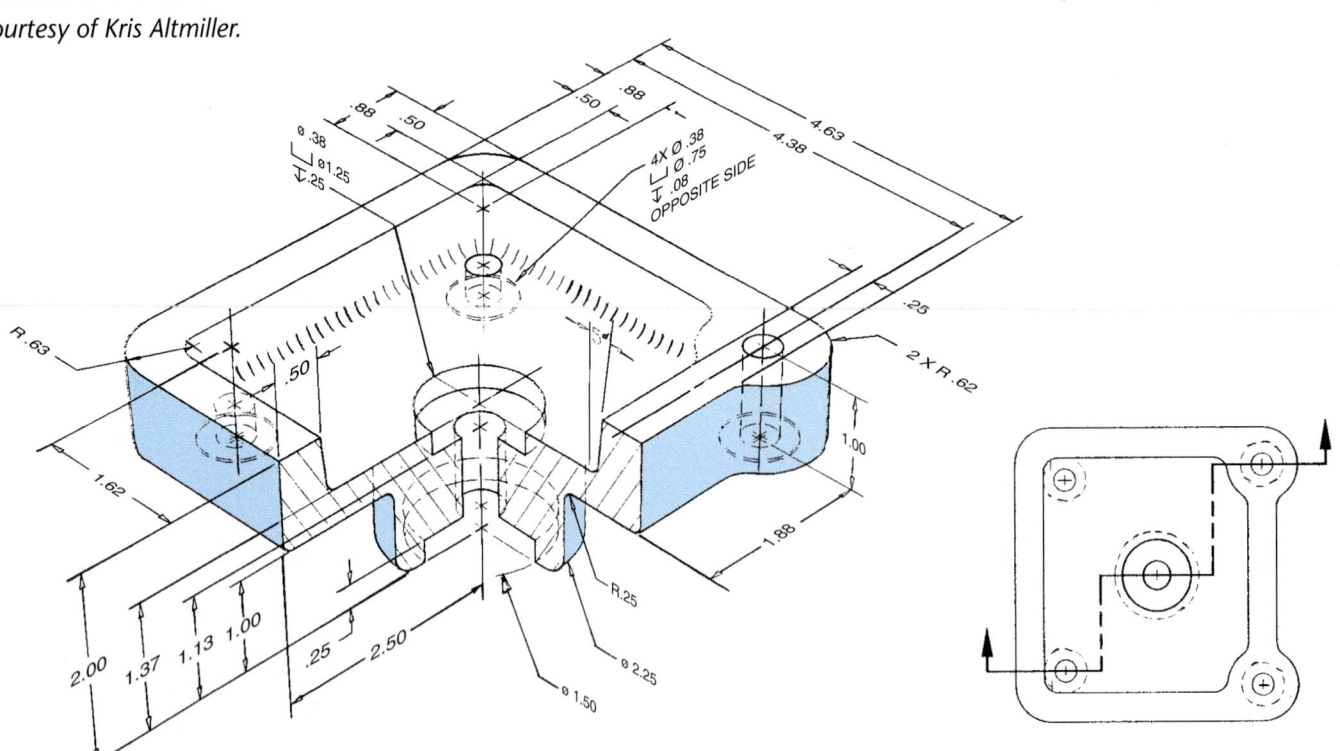

RECOMMENDED CUTTING PLANE

ALL FILLETS AND ROUNDS R.25 UNLESS OTHERWISE SPECIFIED

PROBLEM 13.36   **Broken-out section (in.)**

Part Name: Slide Bar Connector

Material: SAE 4120

SPECIFIC INSTRUCTIONS: Convert chain dimensioning to datum dimensioning.

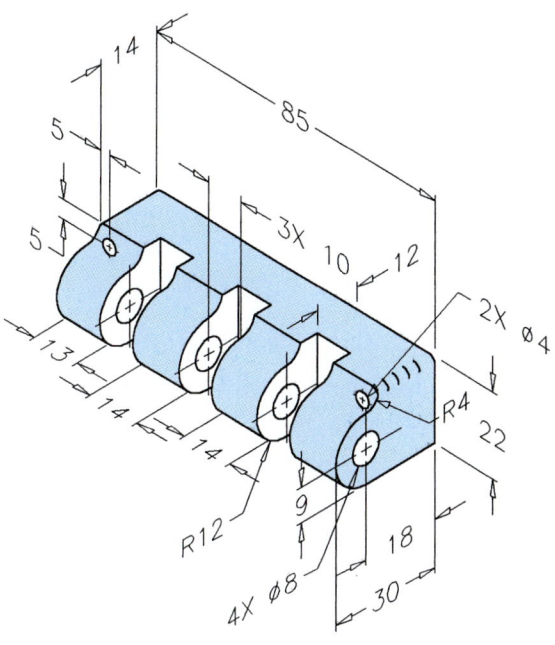

PROBLEM 13.38   **Conventional break (in.)**

Part Name: Leg

Material: Ø2.00 Schedule 40 A120

Problem based on original art courtesy TEMCO.

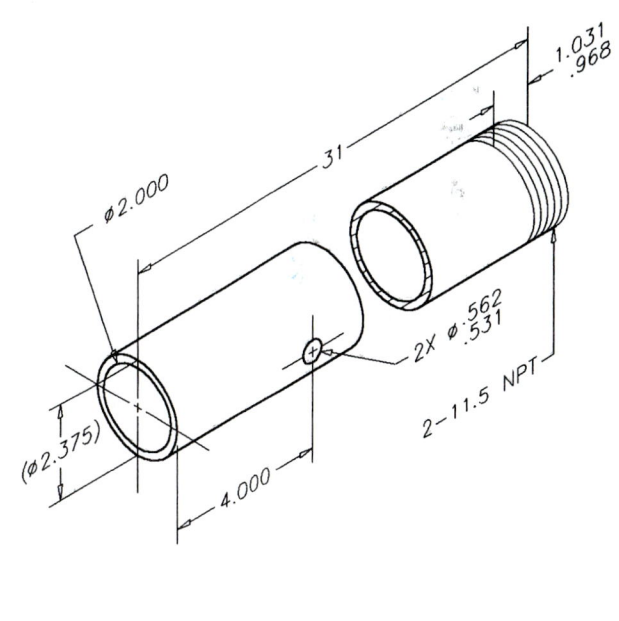

PROBLEM 13.37   **Broken-out section (in.)**

Part Name: Drain Tube

Material: Ø1.00 × .065

Problem based on original art courtesy TEMCO.

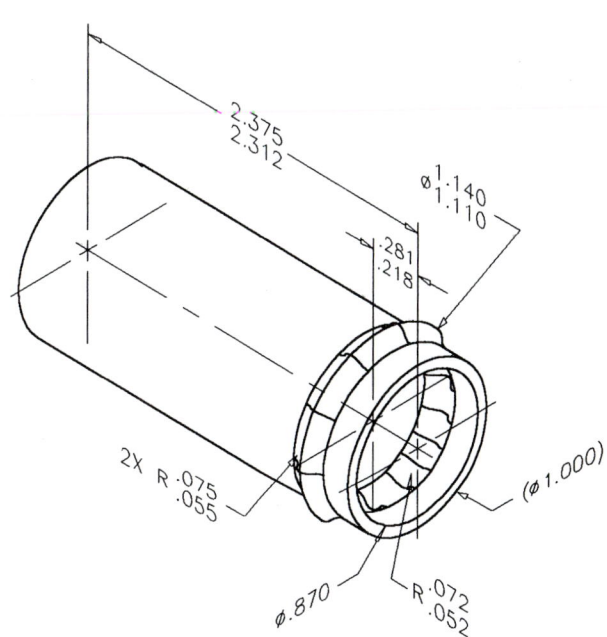

PROBLEM 13.39   **Offset section (in.)**

Part Name: Cover

Material: CI

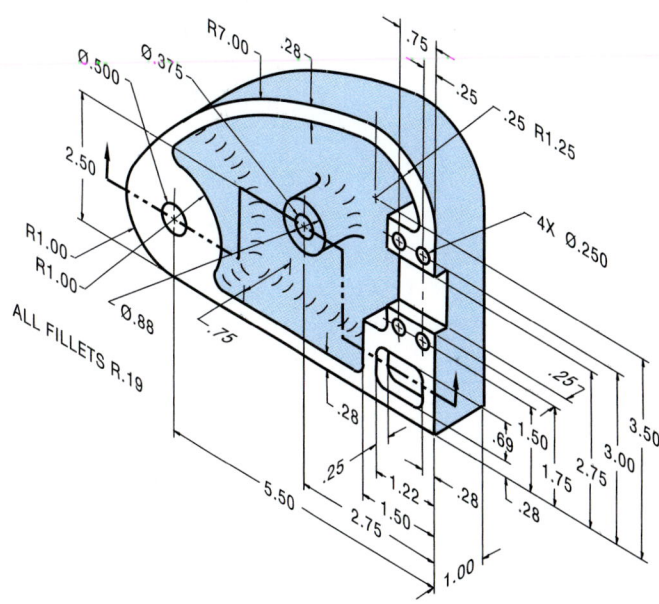

PROBLEMS 13.40 through 13.51: Access the CD found with this textbook and open the problem of your choice, or as assigned by your instructor. Solve the problems using the instructions provided with this chapter or on the CD, unless otherwise specified by your instructor.

MATH PROBLEMS

 PROBLEMS 13.52 through 13.59: Access the CD found with this textbook and open the math problem of your choice, or as assigned by your instructor. Solve the problems using the instructions provided.

Geometric Dimensioning and Tolerancing

LEARNING OBJECTIVES

After completing this chapter, you will:

- Label datum features on a drawing.

- Establish basic dimensions where appropriate.

- Place proper feature control frames on drawings, establishing form, profile, orientation, runout, and location geometric tolerances.

- Use and interpret material condition symbols.

- Determine the virtual condition of features.

- Use geometric tolerancing to completely dimension objects from an engineer's sketch and actual industrial layouts.

- Solve an engineering problem from engineer's notes.

THE ENGINEERING DESIGN APPLICATION

There have been some problems in manufacturing one of the parts due to the extreme tolerance variations possible. The engineer requests that you revise the drawing shown in Figure 14.1 and gives you these written instructions:

- Establish datum A with three equally spaced datum target points at each end of the ∅28.1–28.0 cylinder.

- Establish datum B at the left end surface.

- Make the bottom surfaces of the 2X ∅40.2–40.0 features perpendicular to datum A by 0.06.

- Provide a cylindricity tolerance of 0.3 to the outside of the part.

- Make the 2X ∅40.2–40.0 features concentric to datum A by 0.1.

- Locate the 6X ∅6 + 0.2 holes with reference to datum A at MMC and datum B with a position tolerance of 0.05 at MMC.

- Locate 4X ∅4 holes with reference to datum A at MMC and datum B with a position tolerance of 0.04 at MMC.

This is an easy job because you did the original drawing on CADD and you have a custom geometric tolerancing package. You go back to the workstation and within one hour you have the check plot shown in Figure 14.2 ready for evaluation.

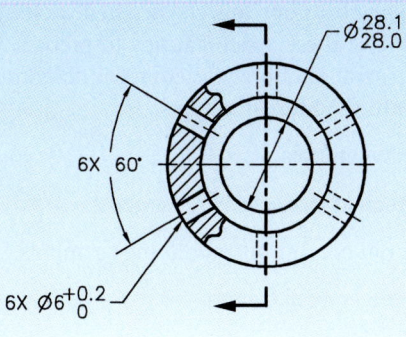
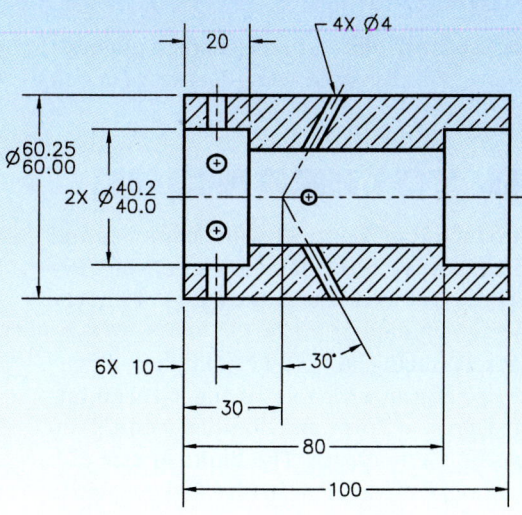

FIGURE 14.1 ■ The original drawing to be revised with geometric tolerancing added from engineer's notes.

(Continued)

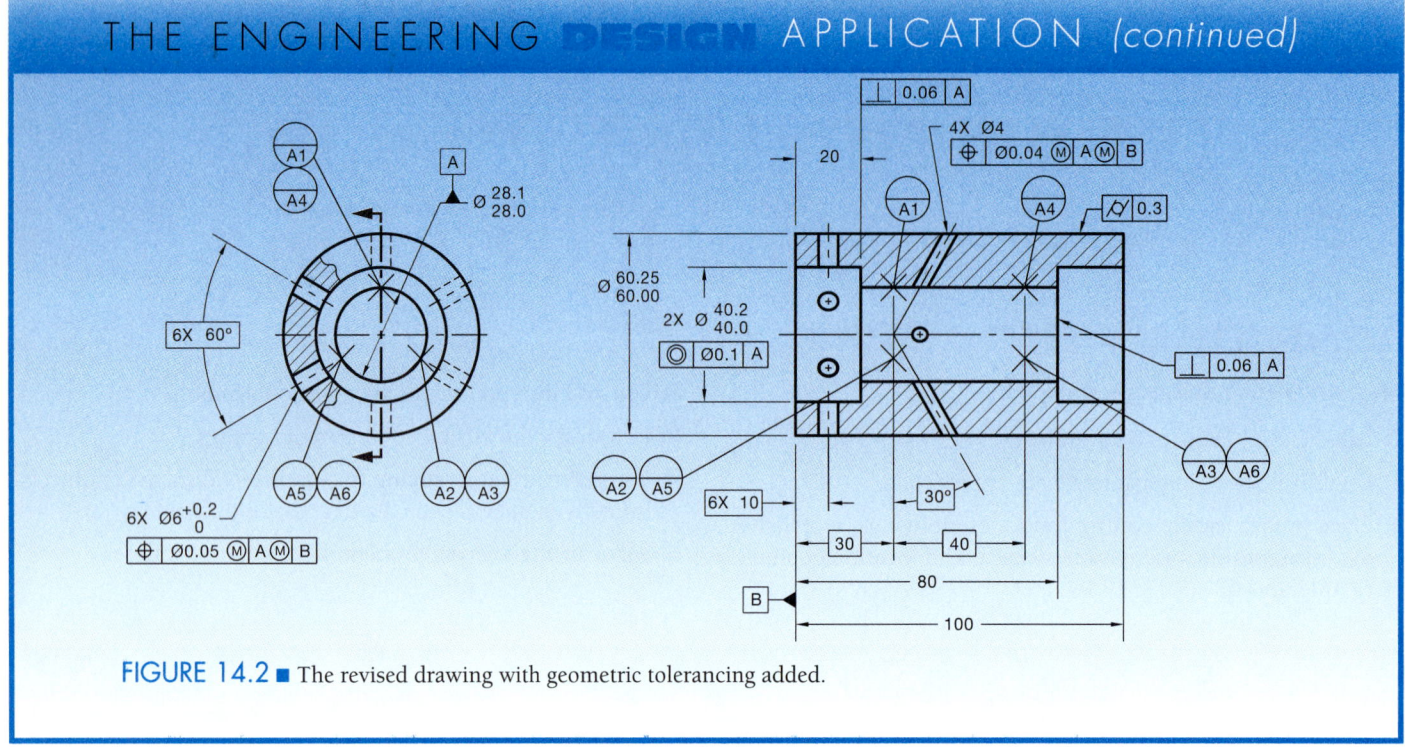

FIGURE 14.2 ■ The revised drawing with geometric tolerancing added.

Geometric tolerancing (GT) is the dimensioning and tolerancing of individual features of a part where the permissible variations relate to characteristics of form, profile, orientation, runout, or the relationship between features. This subject is commonly referred to as **geometric dimensioning and tolerancing (GD & T)**.

> **ANSI** The standards for geometric dimensioning and tolerancing (GD & T) are governed by the American National Standards Institute document ASME Y14.5M titled *Dimensioning and Tolerancing,* published by the American Society of Mechanical Engineers.

Unless otherwise specified, illustrative examples provided throughout this chapter are displayed using metric values in millimeters.

GENERAL TOLERANCING

Tolerancing was introduced in Chapter 11 with definitions and examples of how dimensions are presented on a drawing. Review Chapter 11 to establish a solid understanding of dimensioning and tolerancing terminology.

The term **general tolerancing**, as used here, implies dimensioning without the use of geometric tolerancing. General tolerancing applies a degree of form and location control by increasing or decreasing the tolerance. The **limits of size** of a feature control the amount of variation in size and geometric form. This is the boundary between maximum material condition (MMC) and least material condition (LMC). MMC and

LMC are defined later in this chapter. The form of the feature may vary between the upper limit and lower limit of a size dimension. This is known as **extreme form variation**. (See Figure 14.3.) The addition of GT is necessary to control form but will still permit a relatively large size tolerance.

SYMBOLOGY

Symbols on drawings represent specific information that would otherwise be difficult and time-consuming to duplicate in note form. Symbols must be clearly drawn to required size and shape so they communicate the desired meaning uniformly. The exact drafting specifications are given when each symbol is first introduced. Use these specifications to prepare your own CADD blocks or symbols library. Geometric tolerancing symbols are divided into the following types:

1. Datum symbols.

2. Geometric characteristic symbols.

3. Material condition (modifying) symbols.

4. Feature control frame.

5. Supplementary symbols.

A variety of professional templates and CADD programs are available to help save time when preparing GT symbols. It is recommended that all dimensioning symbols be drawn with a template or CADD program for good representation, accuracy, and speed.

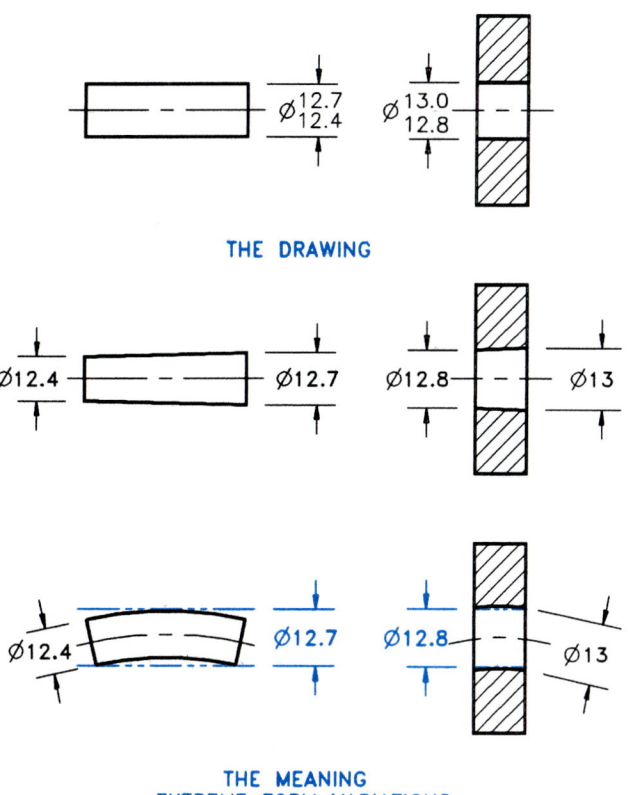

THE DRAWING

THE MEANING
EXTREME FORM VARIATIONS

FIGURE 14.3 ■ Extreme form variations of given tolerances.

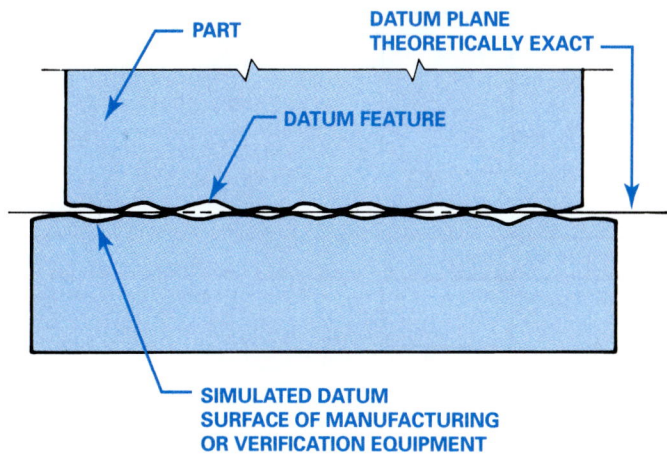

FIGURE 14.4 ■ Magnified representation of a datum feature.

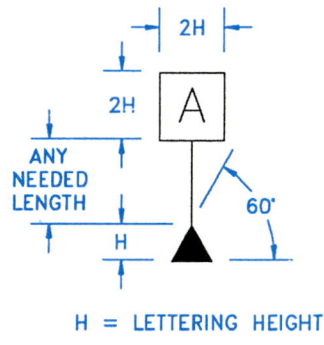

H = LETTERING HEIGHT

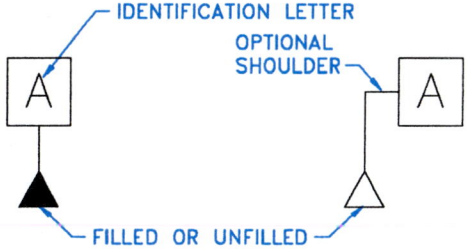

FIGURE 14.5 ■ Datum feature symbol.

DATUM FEATURE SYMBOLS

Datums are considered to be theoretically perfect surfaces, planes, points, or axes that are established from the true geometric counterpart of the datum feature. The datum feature, as shown in Figure 14.4, is the actual feature of the part that is used to establish the datum. Datums are used to originate size and location dimensions. Without the use of GT, datums are often implied. When dimensions originate at a common surface, for example, that surface is assumed to be the datum. The only problem with this method is that the manufacturing operation may not read the datum in the same way as the engineering department, or the implied datum may be ignored all together. With the use of defined datums, each part is made with dimensions that begin from the same origin; there is no variation.

Datum feature symbols are commonly drawn using thin lines with the symbol size related to the size of the lettering on a drawing, as shown in Figure 14.5. The triangular base may be filled or unfilled but should be consistently applied in a drawing. However, the filled triangle helps the symbol show up better on a drawing.

Datum feature symbols are placed in the view that shows the edge of a surface or are attached to a diameter or symmetrical dimension when associated with a centerline or center plane. (See Figure 14.6.)

DATUM REFERENCE FRAME

Datum referencing is used to relate features of a part to an appropriate datum or datum reference frame. A datum indicates the origin of a dimensional relationship between a toleranced feature and a designated feature or features on a part. The datum identification symbol on a drawing relates to

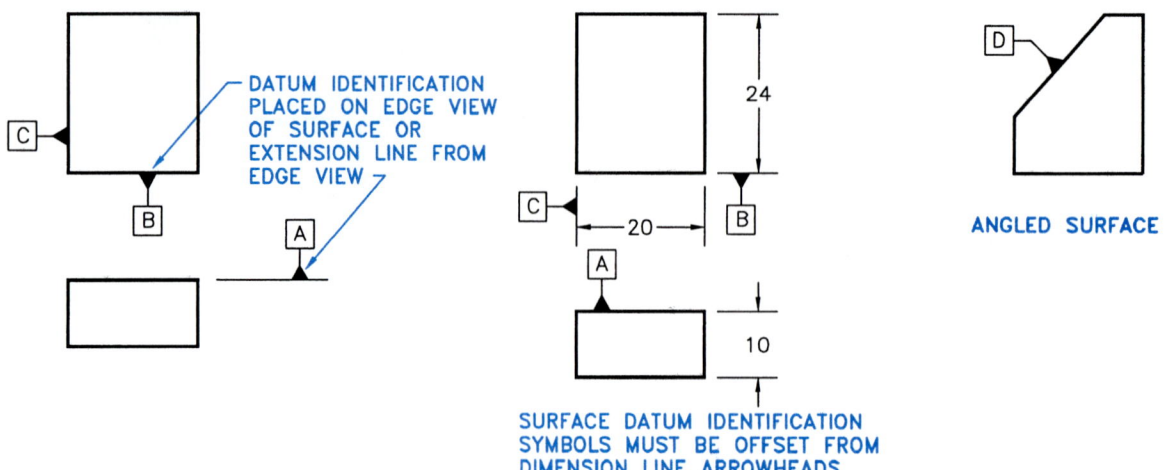

DATUM IDENTIFICATION PLACED ON EDGE VIEW OF SURFACE OR EXTENSION LINE FROM EDGE VIEW

SURFACE DATUM IDENTIFICATION SYMBOLS MUST BE OFFSET FROM DIMENSION LINE ARROWHEADS

ANGLED SURFACE

SURFACE DATUMS

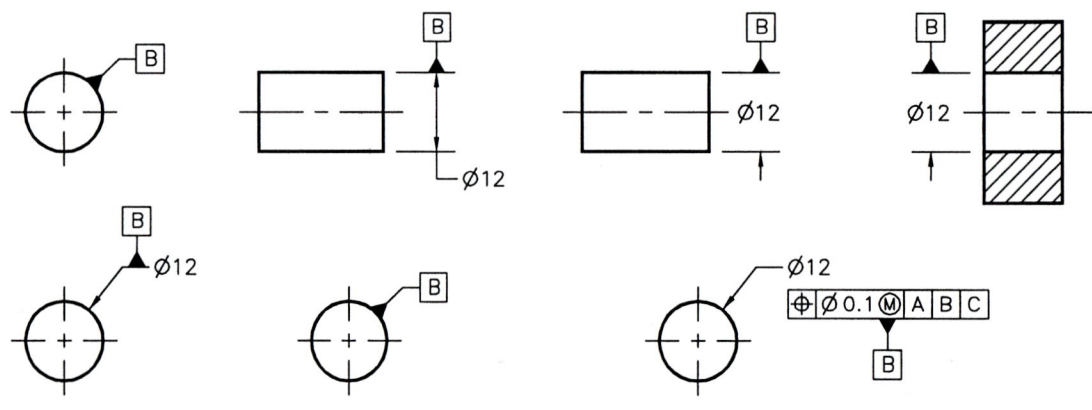

AXIS DATUMS

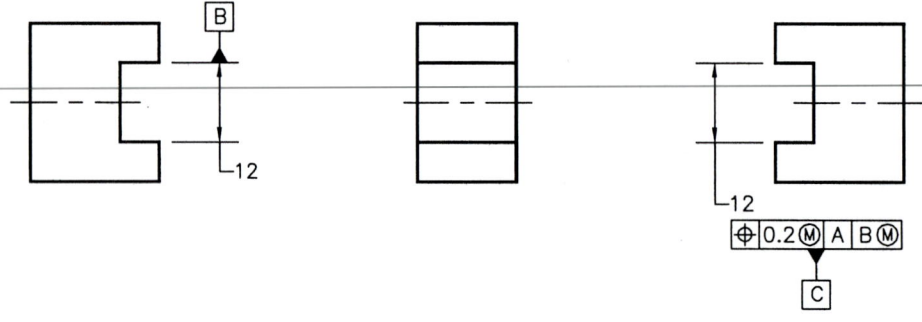

AXIS AND CENTER PLANE DATUM IDENTIFICATION SYMBOLS MUST ALIGN WITH OR REPLACE THE DIMENSION LINE ARROWHEAD OR BE PLACED ON THE FEATURE, LEADER SHOULDER, OR FEATURE CONTROL FRAME.

CENTER PLANE DATUMS

FIGURE 14.6 ■ Datum identification.

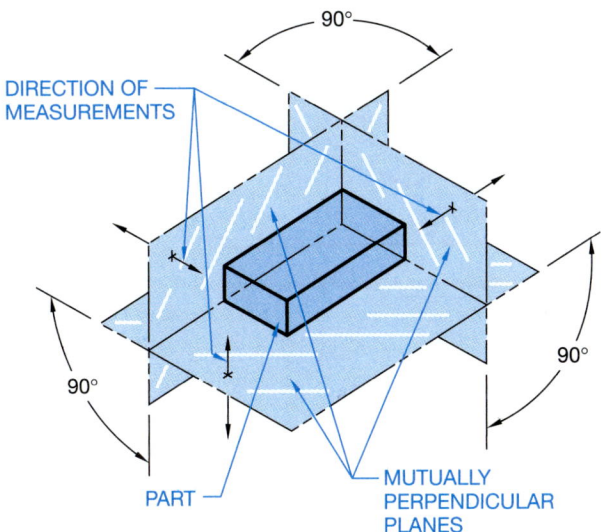

FIGURE 14.7 ■ Datum reference frame.

an actual feature on the part known as the **datum feature**. The part is then placed on the surface of an inspection table or manufacturing verification equipment. This inspection table or equipment is referred to as the **simulated datum**. The **datum** is the theoretically exact plane, axis, or point that is established by the true geometric counterpart of the specified datum feature. Measurements cannot be made from the true geometric counterpart, because it is theoretical or assumed to exist. The machine tables, surface plates, or inspection tables are of such high quality that they are used to simulate the datums from which measurements are taken and dimensions are verified. This way, each dimension always originates from the same reliable location.

Sufficient datum features, those most important to the design of a part, are chosen to position the part in relation to a set of three mutually perpendicular planes, jointly called a **datum reference frame.** This reference frame exists in theory only and not on the part. Therefore, it is necessary to establish a method for simulating the theoretical reference frame from the actual features of the part. This simulation is accomplished by positioning the part on appropriate datum features to adequately relate the part to the reference frame and to restrict the motion of the part in relation to it. (See Figure 14.7.) These planes are simulated in a mutually perpendicular relationship to provide direction as well as the origin for related dimensions and measurements. Thus, when a part is positioned on the datum reference frame, dimensions related to the datum reference frame by a feature control frame or note are thereby mutually perpendicular. This theoretical reference frame constitutes the three-plane dimensioning system used for datum referencing.

DATUM FEATURES

A datum feature is selected on the basis of its geometric relationship to the toleranced feature and the requirements of the

design. To ensure proper part interface and assembly, corresponding features of mating parts are also selected as a datum feature where practical. Datum features must be readily recognizable on the part. Therefore, in the case of symmetrical parts or parts with identical features, physical identification of the datum feature on the part may be necessary. A datum feature should be accessible on the part and be of sufficient size to permit processing operations.

Unless otherwise specified, the three reference planes are mutually perpendicular. Planes in the datum reference frame are placed in order of importance. For example, the plane that originates most size and location dimensions, or has the most functional importance, is the **primary**, or first, datum plane. The next datum is the **secondary**, or second, datum plane and the third element of the frame is called the **tertiary**, or third, datum plane. The desired **order of precedence** is indicated by entering the appropriate datum reference letters from left to right in the feature control frame. For instructional purposes, it may be convenient to label datums as A, B, and C, although in industry other letters are also used to identify datums such as D, E, and F, or X, Y, and Z. Letters that should be avoided are O, Q, and I. Figure 14.8a shows a part and the planes that are chosen as datum features. Notice in Figure 14.8b how the order of precedence of datum features relates the part to the datum reference frame. The datum features are identified as surfaces A, B, and C. These surfaces are most important to the design and function of the part. Surfaces A, B, and C are the primary, secondary, and tertiary datum features respectively, since they appear in that order in the feature control frame.

Multiple datum frames may be established for some parts depending on the complexity and function of the part. The relationship between datum frames is often controlled by a representative angle. (See Figure 14.9.)

Partial Surface Datums

In some situations, it can be more realistic to apply a datum feature symbol to a portion of a surface rather than the entire surface. For example, when a long part has related features located in one or more concentrated places, then the datum may be located next to the important features. When this is done, a chain line is used to identify the extent of the datum feature. The location and length of the chain line must be dimensioned with basic dimensions. (See Figure 14.10.)

Coplanar Surface Datums

Coplanar surfaces are two or more surfaces that are in the same plane. The relationship of coplanar surface datums may characterize the surfaces as one plane or datum in correlated geometric tolerance callouts as shown in Figure 14.11.

DATUM TARGET SYMBOLS

In many situations, it is not possible to establish an entire surface or surfaces as datums, or it may not be practical to coordinate a

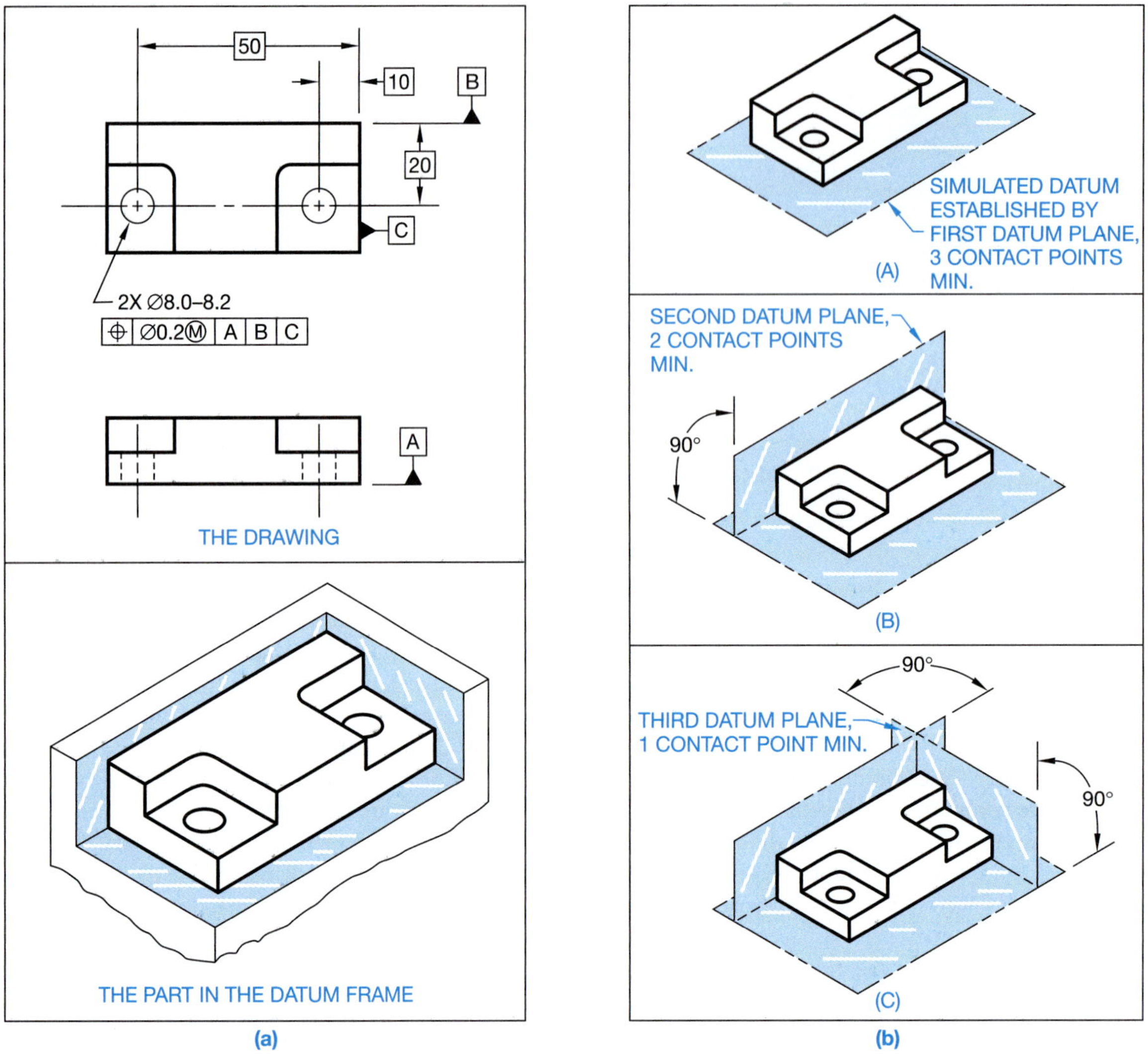

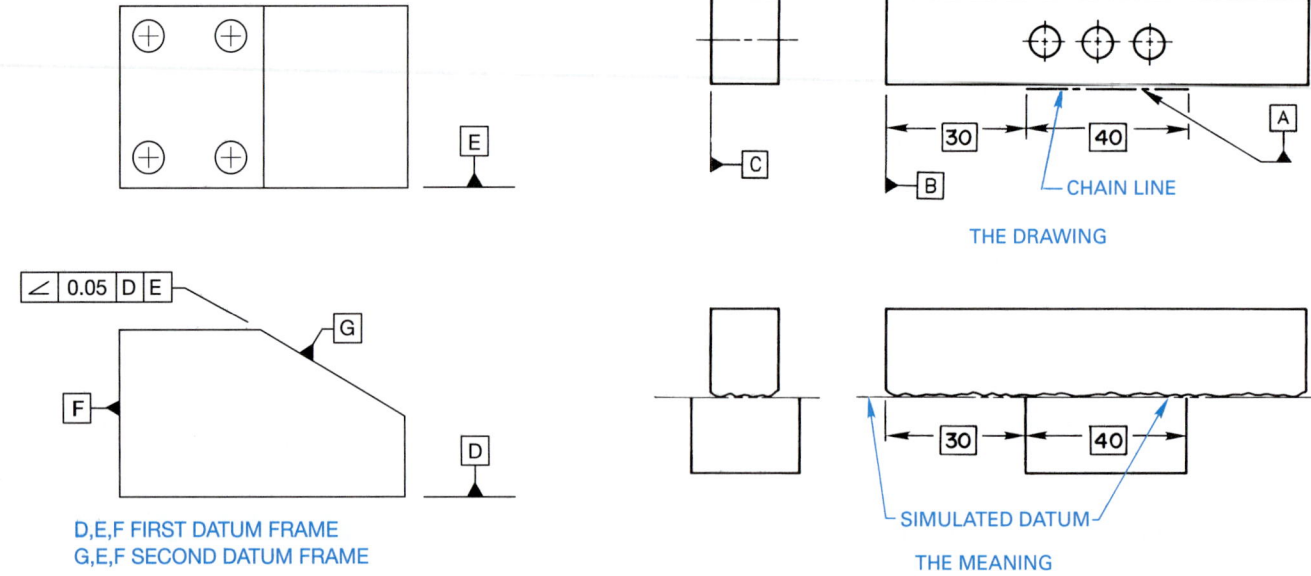

FIGURE 14.8 ■ (a) Part in which datum features are plane surfaces; (b) sequence of datum features relates part to datum reference frame.

FIGURE 14.9 ■ Multiple datum frames.

FIGURE 14.10 ■ Partial surface datums.

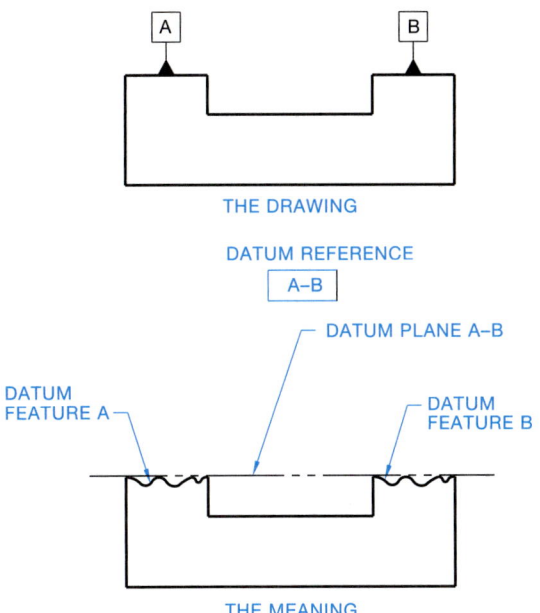

FIGURE 14.11 ■ Coplanar surface datums.

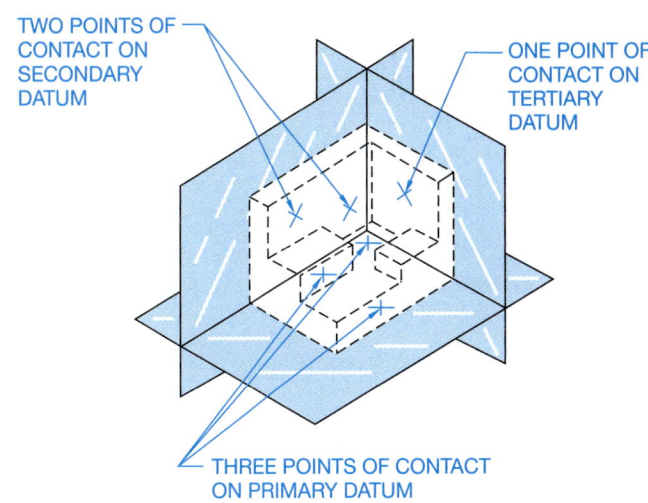

FIGURE 14.12 ■ Datum frame established by datum target symbols.

partial datum surface with the part features. When this happens due to the size or shape of the part, then datum targets may be used to establish datum planes. Datum targets are especially useful for parts with surface or contour irregularities, such as sand castings or forgings, for some sheet metal parts subject to bowing or warpage, or for weldments where heat can induce warpage. Datum targets are designated points, lines, or surface areas that are used to establish the datum reference frame. The primary datum is established by three points or contact locations. The three locations are placed on the part so they form a stable, well-spaced pattern and are not in a line. The secondary datum is created by two points or locations. The tertiary datum is established by one point or location. Remember, the purpose of the datum frame is to prepare a stable position for the part dimensions to be established in relationship to corresponding datums. For this reason, the three points on the primary datum are used to provide stability similar to a three-legged stool. The two points on the secondary datum provide the needed amount of stability when the object is placed against the secondary datum. Finally, the tertiary datum requires only one point of contact to complete the stability between the three datum frame elements. (See Figure 14.12.)

The datum target symbol is drawn as a circle using thin lines and is connected with a leader that points to a target point, line, or surface area. The datum target symbol is divided into two halves by a horizontal line. The top half of the symbol is reserved for the diameter of a circular datum target area, when used. The bottom half of the symbol is used for datum target identification. For example, if there are three datum target points on a datum, the first point may be labeled A1, the second A2, and the third A3. The datum target point is located on the surface or edge view from adjacent datums with basic dimensions. (See Figure

14.13.) Chain dimensioning is commonly used to locate datum target points and target areas. The location dimensions must originate from datums. Datum target areas are located to their centers. (See Figure 14.14.) When the surface of a part has an irregular contour or different levels, the contact points, lines, or areas may lie on the same plane and different lengths of locating pins may be used. The pins used to make point contact are usually rounded. The pins for target area contact have a flat surface equal in diameter to the specified target area.

DATUM AXIS

The datum frame established by a cylindrical object is the base of the part. The two theoretical planes, represented in Figure 14.15 by the X and Y center planes, cross to establish the datum axis. The actual secondary datum is the cylindrical surface. The center planes located at X and Y are used to indicate the direction of dimensions that originate from the datum axis. (See Figure 14.16.)

Datum target points, lines, or surface areas also may be used to establish a datum axis. A primary datum axis may be established using two sets of three equally spaced targets: a set at one end of the cylinder and the other set near the other end as shown in Figure 14.17, page 432. When two cylindrical features of different diameter are used to establish a datum axis, then the datum target points are identified in correspondence to the adjacent cylindrical datum features as shown in Figure 14.18, page 432. Cylindrical datum target areas and circular datum target lines also may be used to establish the datum axis of cylindrically shaped parts as shown in Figure 14.19, page 432. A secondary datum axis is established by placing three equally spaced targets on the cylindrical surface. (See Figure 14.20, page 433.)

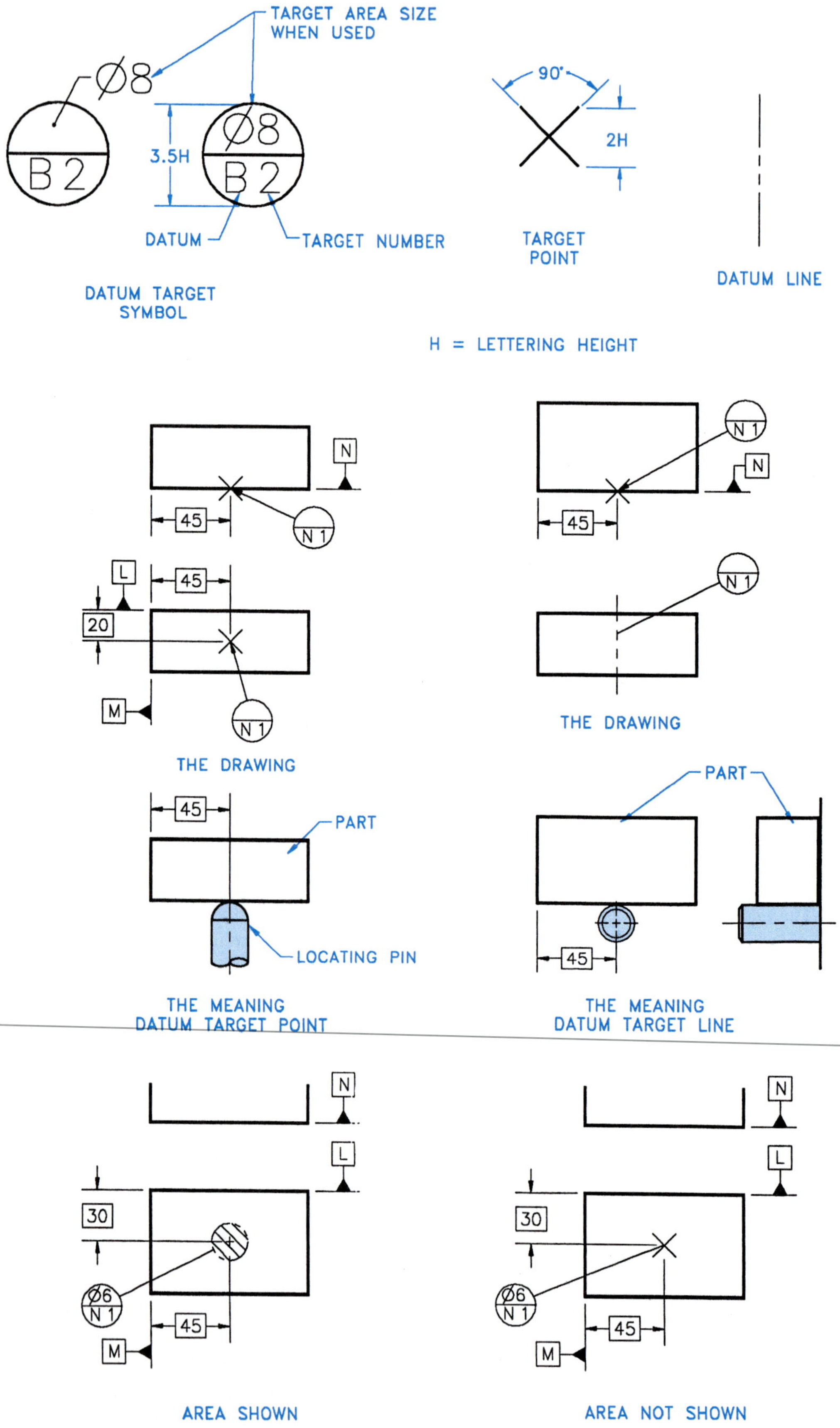

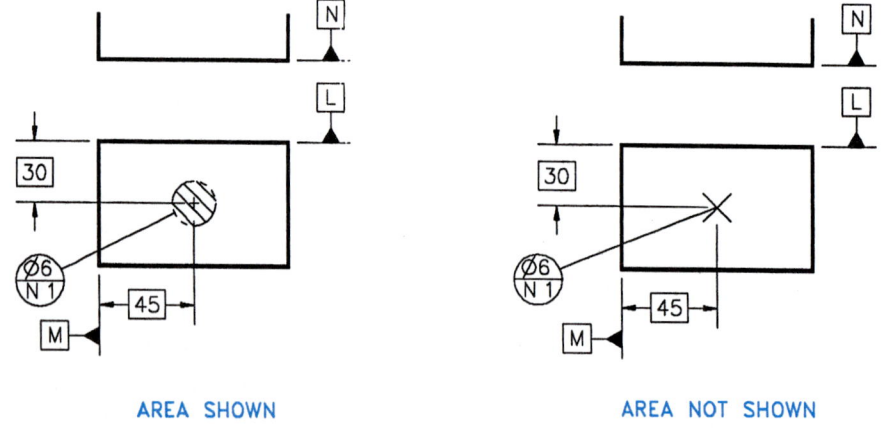

FIGURE 14.13 ■ Datum target symbol, target point, datum line, and target area.

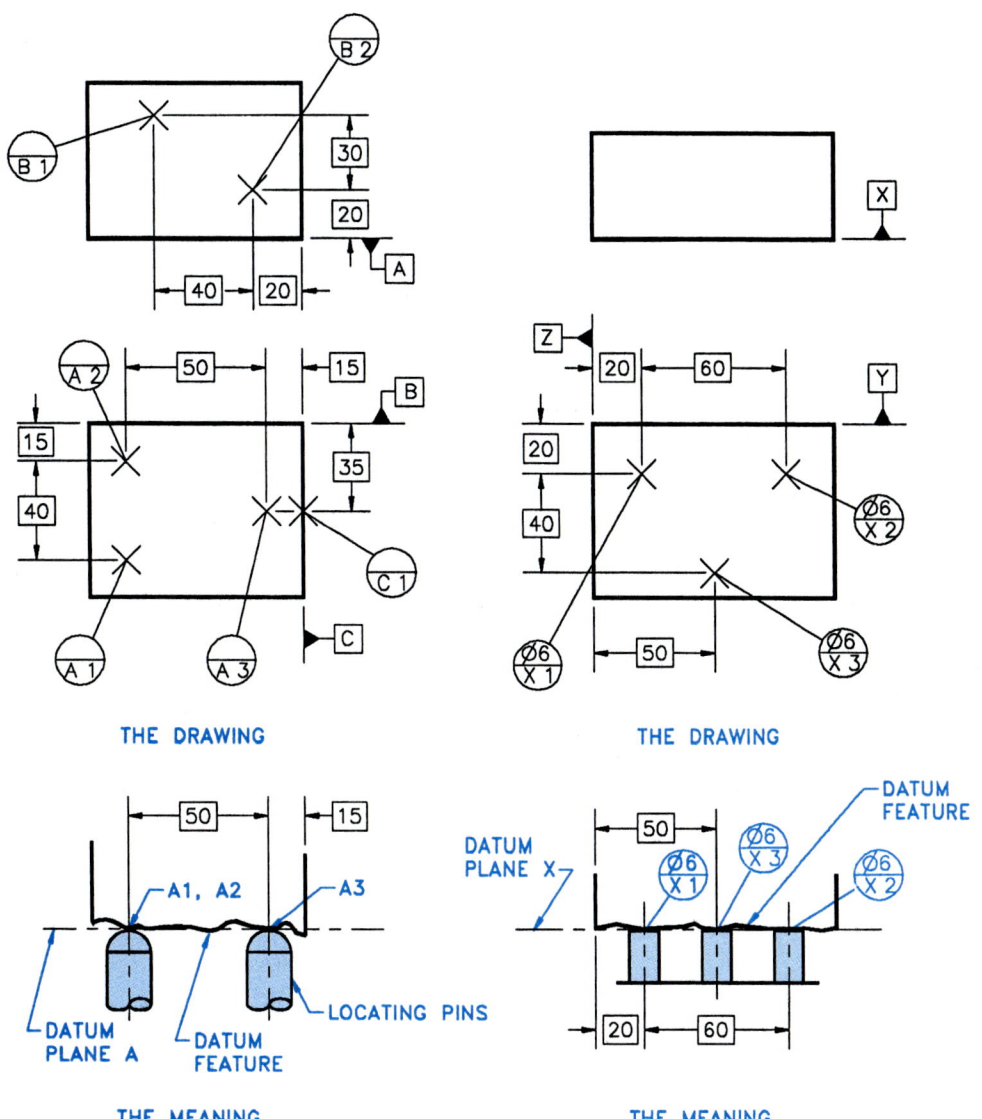

FIGURE 14.14 ■ Locating datum targets.

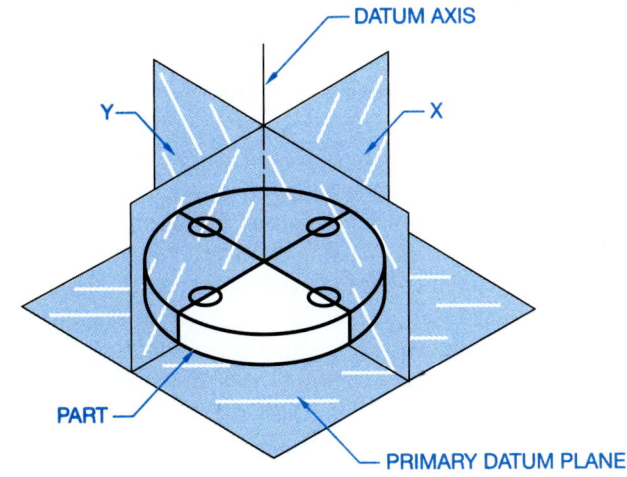

FIGURE 14.15 ■ Datum frame for cylindrical object.

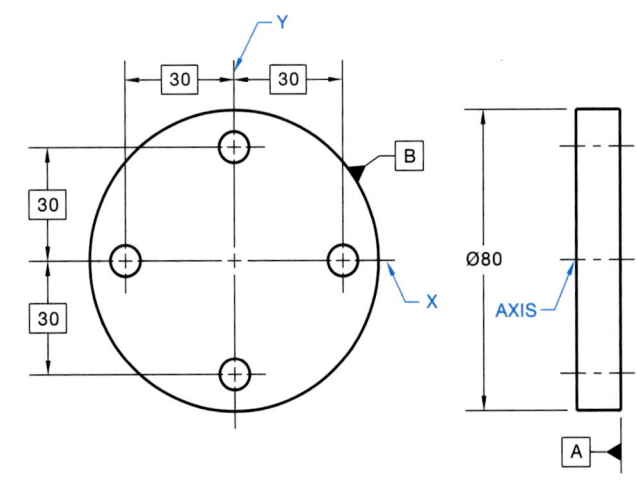

FIGURE 14.16 ■ Datum axis.

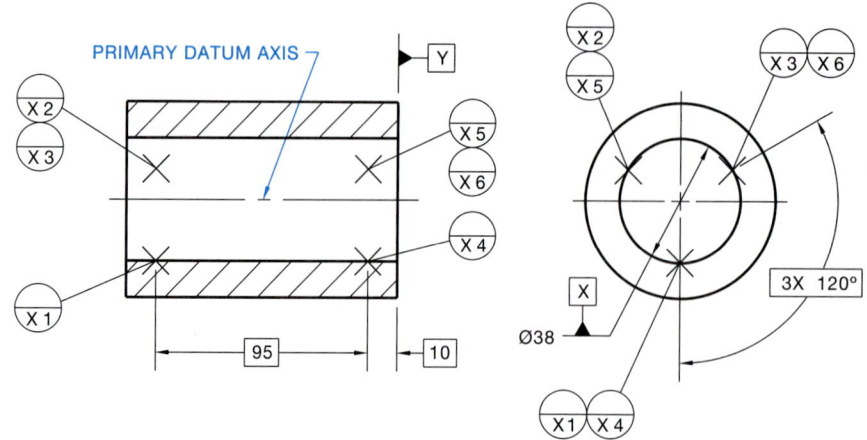

FIGURE 14.17 ■ Establishing a primary datum axis with two sets of three equally spaced targets.

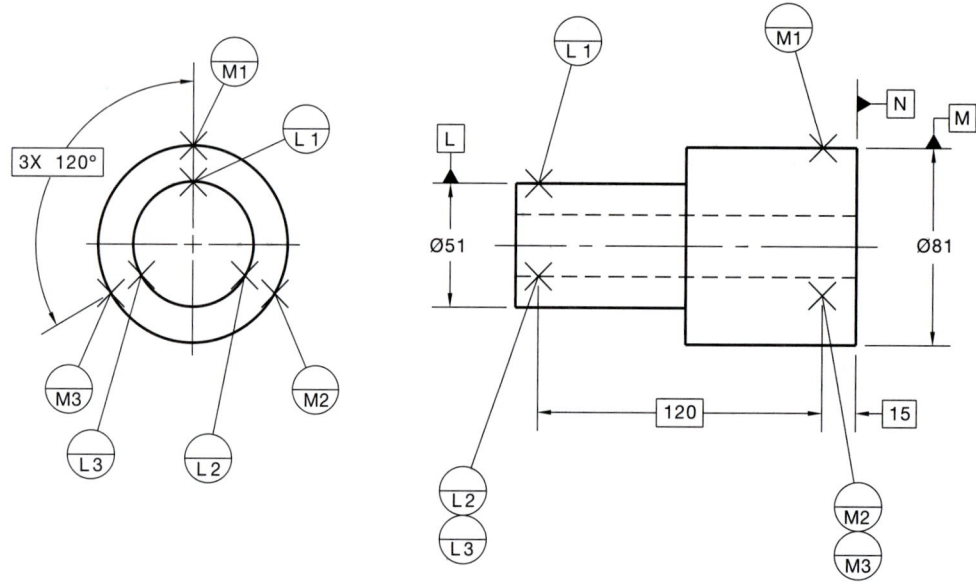

FIGURE 14.18 ■ Datum targets identified on adjacent cylindrical features.

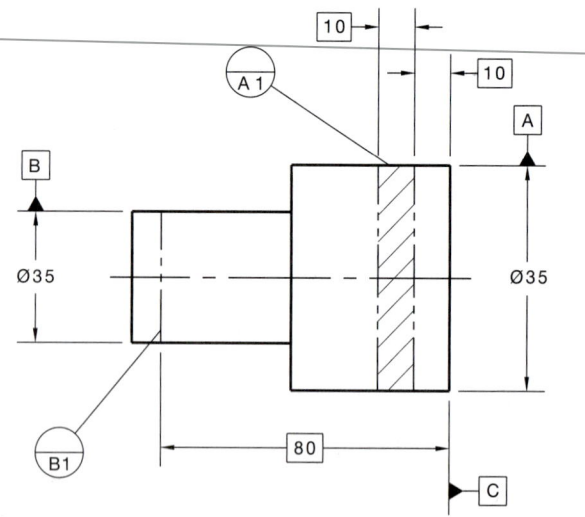

FIGURE 14.19 ■ Cylindrical datum target areas and circular datum target lines.

FEATURE CONTROL FRAME

The **feature control frame** is used to relate a geometric tolerance to a part feature. The elements in a feature control frame must always be in the same order. The most basic format is when a geometric characteristic and related tolerance are applied to an individual feature as shown in Figure 14.21. The next expanded format is when a geometric characteristic, tolerance zone descriptor, and material condition symbol are used in the feature control frame. (See Figure 14.22.) One, two, or three datum references may be included in a feature control frame as shown in Figure 14.23. The material condition symbol that applies to the feature tolerance is always placed after the tolerance. When a material condition symbol is applied to a datum reference, it is placed after the datum reference and in the same compartment as shown in Figure 14.24.

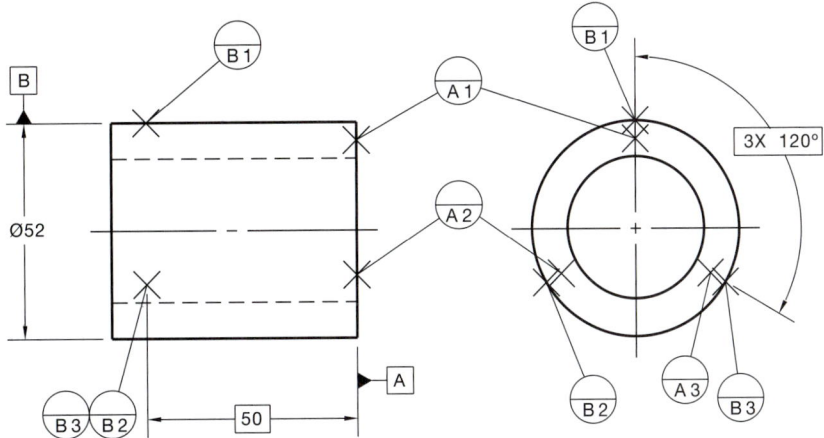

FIGURE 14.20 ◼ Establishing a secondary datum on a cylindrical object with three equally spaced targets.

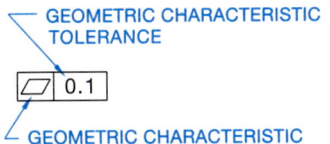

FIGURE 14.21 ◼ Feature control frame with geometric characteristics and related tolerance.

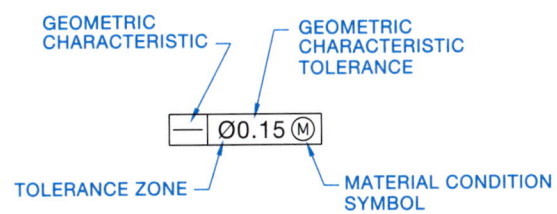

FIGURE 14.22 ◼ Feature control frame with geometric characteristic, tolerance zone descriptor, and material condition symbol.

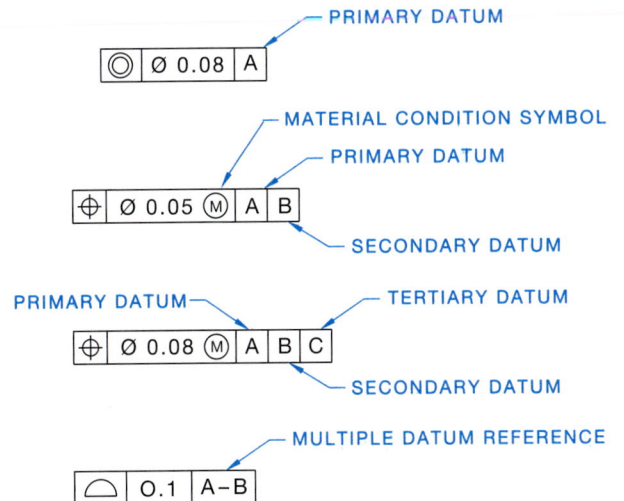

FIGURE 14.23 ◼ Applying one, two, or three datum references to the feature control frame.

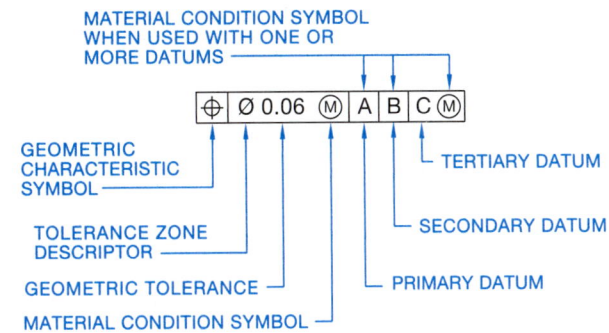

FIGURE 14.24 ◼ Elements of a feature control frame.

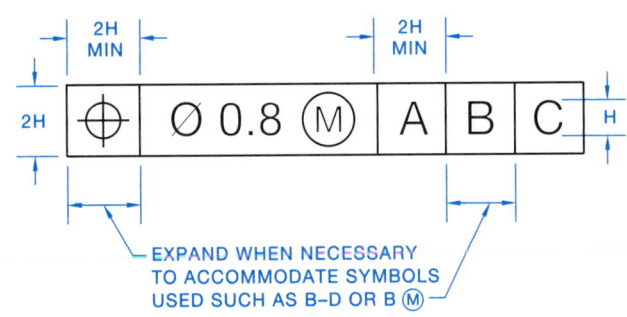

H = LETTERING HEIGHT

FIGURE 14.25 ◼ Detailed feature control frame and application.

The feature control frame is drawn with thin lines and connected to the feature with a leader or extension line. (See Figure 14.25.) The feature control frame may be combined with a datum feature symbol when the feature controlled by the geometric tolerance also serves as a datum. The datum feature symbol and the datum reference in the feature control frame are considered separately. When a datum feature symbol and a feature control frame are combined, the feature control frame should be shown first. The datum feature symbol is centered on the base of the feature control frame. (See Figure 14.26.)

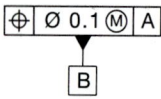

FIGURE 14.26 ■ Combination datum feature symbol and feature control frame.

BASIC DIMENSIONS

A basic dimension is defined as any size or location dimension that is used to identify the theoretically exact size, profile, orientation, or location of a feature or datum target. Basic dimensions are the basis from which permissible variations are established by tolerances on other dimensions in notes or in feature control frames. A basic dimension is described on a drawing by placing a thin-line box around the dimension. Metric and inch basic dimension numerals follow the same rules applied to other dimension numerals. (Refer to Chapter 11.) Metric basic dimensions contain only the number of decimal places needed to display the intended control. For inch basic dimensions, the basic dimension value is expressed in the same number of decimal places as the related tolerance. (See Figure 14.27.)

GEOMETRIC TOLERANCES

Geometric tolerances are divided into five types: form, profile, orientation, runout, and location. These tolerances are subdi-

vided into thirteen characteristics plus modifying terms, all of which are discussed in this section.

Form Tolerance

A tolerance of form is commonly applied to individual features or elements of single features and is not related to datums. The amount of given form variation must fall within the specified size tolerance zone.

Straightness

The straightness symbol is detailed in Figure 14.28. Perfect straightness exists when a surface element or the axis of a part is a straight line. A straightness tolerance allows for a specified amount of variation from a straight line. The straightness feature control frame may be attached to the surface of the object with a leader or combined with the diameter dimension of the part. When straightness is connected to the surface of the feature with a leader, surface straightness is implied as shown in Figure 14.29. When the straightness feature control frame is combined with the diameter dimension of the part, then axis straightness is specified. (See Figure 14.30.) The tolerance zone shape is cylindrical for this application. While a straightness tolerance is common on cylindrical parts, straightness may also be applied to the surface of noncylindrical parts. When specifying center plane feature straightness, a center plane is implied and the diameter tolerance zone descriptor is omitted. The straight-

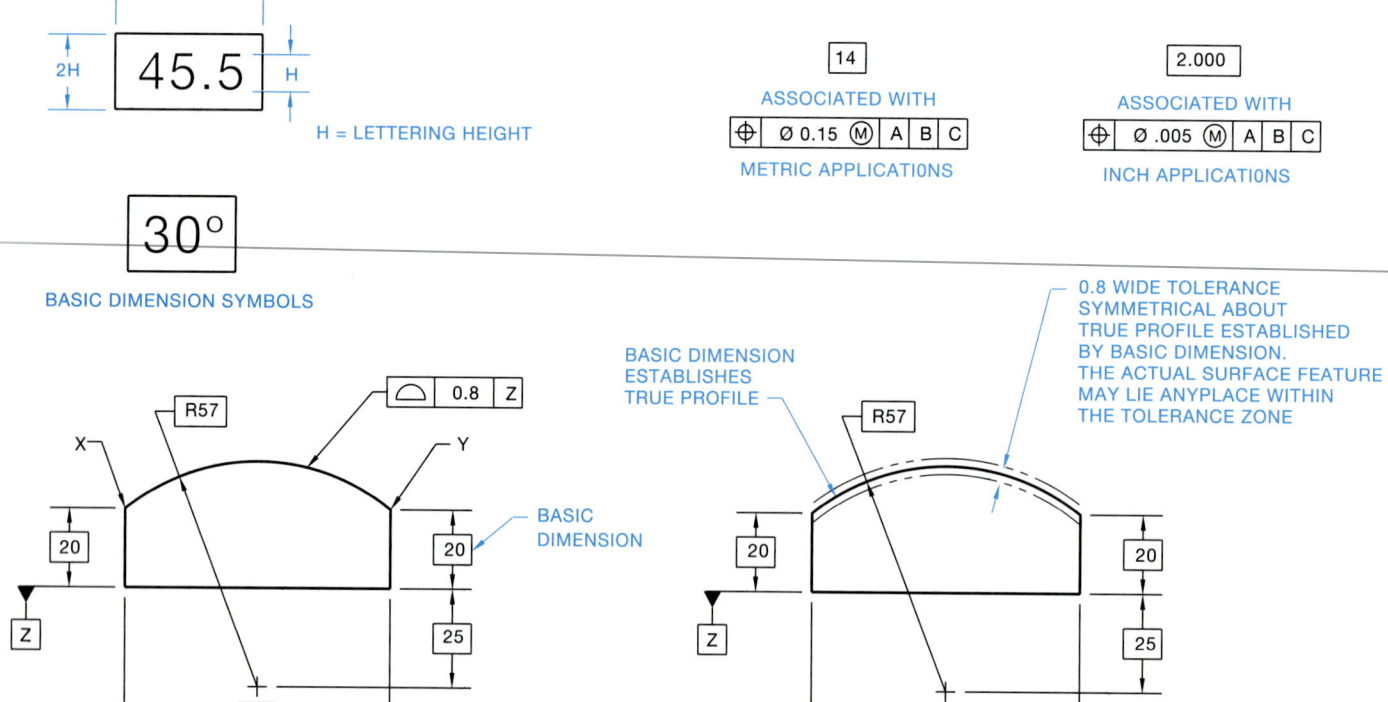

FIGURE 14.27 ■ Basic dimensions.

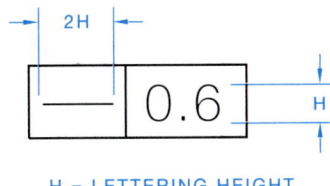

H = LETTERING HEIGHT

FIGURE 14.28 ■ Straightness feature control frame.

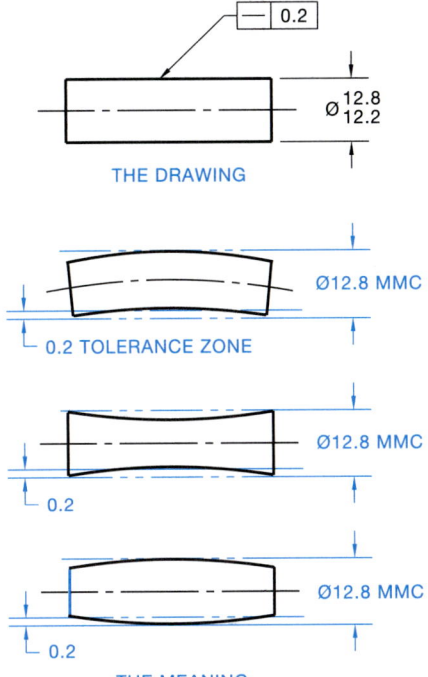

FIGURE 14.29 ■ Surface straightness.

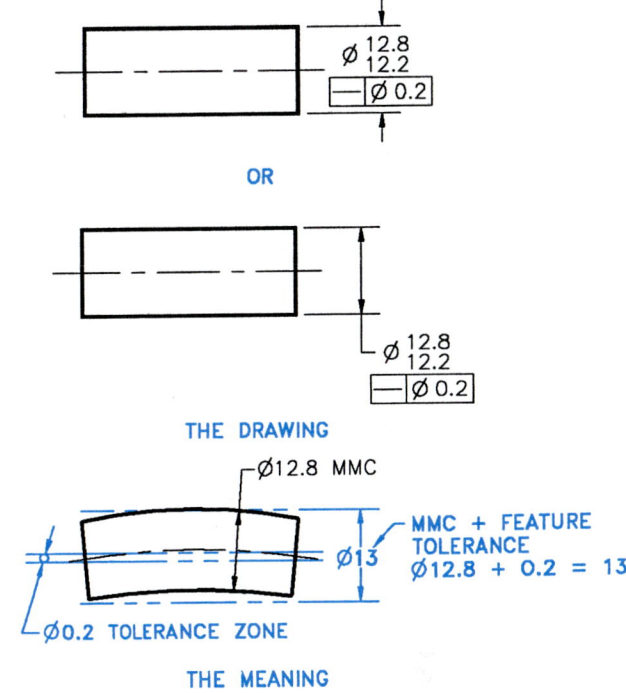

FIGURE 14.30 ■ Axis straightness.

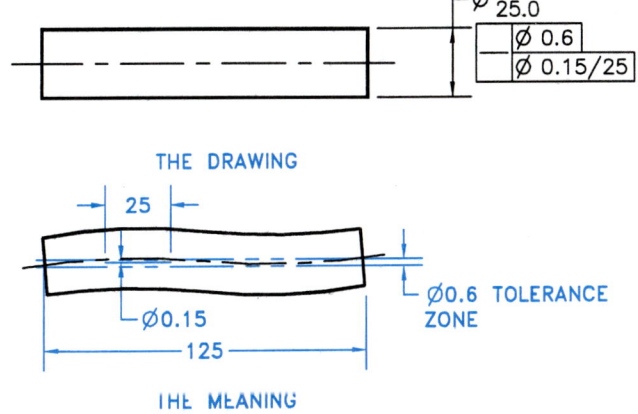

FIGURE 14.31 ■ Unit straightness.

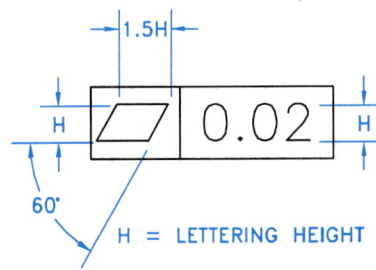

H = LETTERING HEIGHT

FIGURE 14.32 ■ Flatness feature control frame.

ness geometric tolerance must always be less than the size tolerance, thereby refining the size tolerance.

Unit straightness is a situation where a specified tolerance is given per unit of length, for example 25-mm units, and a greater amount of tolerance is provided over the total length of the part. (See Figure 14.31.) This is intended to control excessive waviness in a long part.

Flatness

The flatness tolerance symbol is detailed in relationship to the feature control frame in Figure 14.32. A surface is considered flat when all of the surface elements lie in one plane. A flatness geometric tolerance callout allows for a specified amount of surface variation from a flat plane. The flatness tolerance zone establishes two parallel planes. The actual surface of the object may not extend beyond the boundary of the size tolerance zone and, when associated with the size dimension, the flatness tolerance must be smaller than the size tolerance. The flatness feature control frame may be connected to the edge view of the surface with a leader or with an extension line. (See Figure 14.33.)

Unit flatness may be specified when it is desirable to control the flatness of surface units. The unit size may be 25 × 25 mm or 1.00 × 1.00 in. This is used to prevent abrupt surface varia-

tion. A unit flatness callout can be presented with or without a separate tolerance for the total area. Unit flatness, just as unit straightness, often has a total tolerance in order to avoid a situation where the unit tolerance gets out of control. The size of

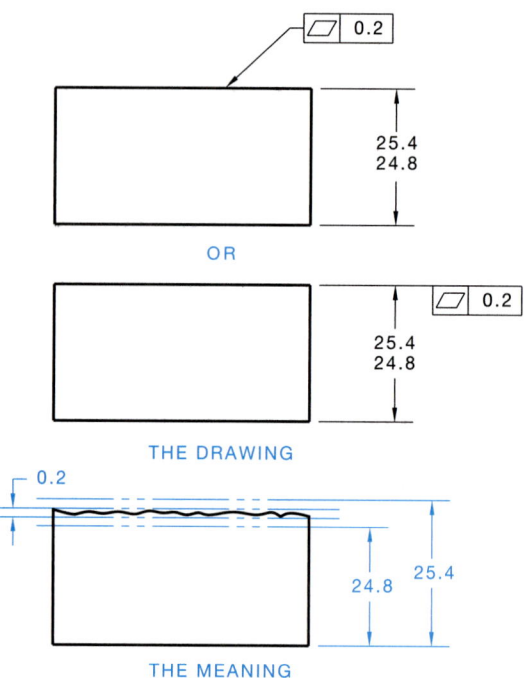

FIGURE 14.33 ■ Flatness representation.

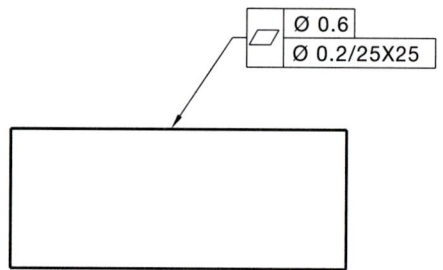

FIGURE 14.34 ■ Unit flatness.

the unit area is given after the unit tolerance specification. The feature control frame is doubled in height and the total area flatness is given first as shown in Figure 14.34.

Circularity

The circularity geometric characteristic symbol is detailed in a feature control frame in Figure 14.35. Circularity can be applied to cylindrical, conical, or spherical shapes. Circularity exists when all of the elements of a circle are the same distance from the center. Circularity is a cross-sectional evaluation of the feature to determine if the circular surface lies between a tolerance zone that is made up of two concentric circles. The cross-sectional toler-

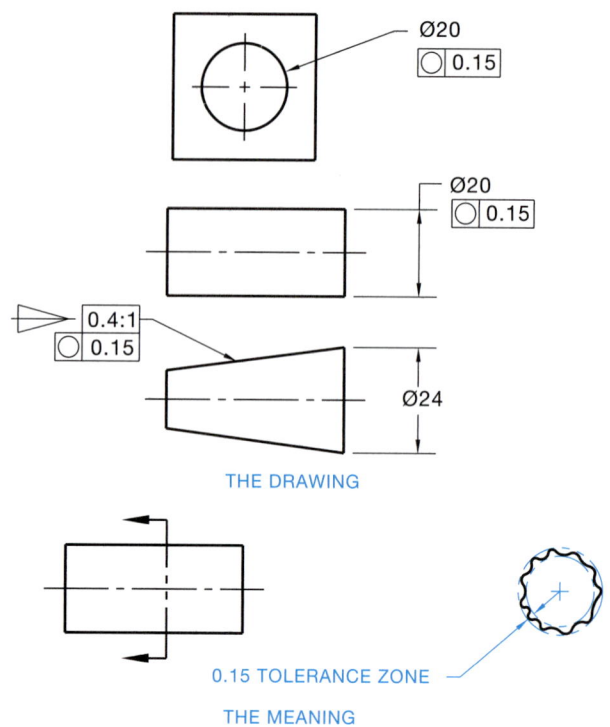

FIGURE 14.36 ■ Circularity representation.

ance zone is established perpendicular to the axis of the feature. The term circular or line element is used to refer to a cross-sectional or single-line tolerance zone as opposed to a blanket or entire surface tolerance zone. The circularity geometric tolerance zone is a radius dimension. The circularity feature control frame may be connected to the part with a leader in the circular or rectangular view as shown in Figure 14.36. The circularity tolerance must always be less than the size tolerance except for parts subject to free state variations. Free state variation is a term used to describe distortion of a part after removal of forces applied during manufacturing, such as a rubber gasket or sheet metal. The part may have to meet the tolerance specifications while in free state, or it may be necessary to hold features in a simulated mating part to verify dimensions. The free state symbol shown in Figure 14.37 is placed in the feature control frame after the geometric tolerance and material condition symbol, if any.

Cylindricity

The cylindricity geometric characteristic symbol is detailed in Figure 14.38. Cylindricity is similar to circularity because both have a radius tolerance zone. The difference is that circularity is

FIGURE 14.35 ■ Circularity feature control frame.

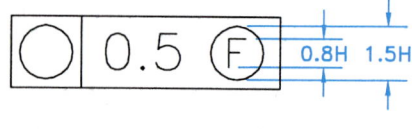

FIGURE 14.37 ■ Using the free state symbol.

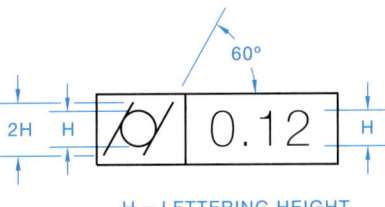

FIGURE 14.38 ■ Cylindricity feature control frame.

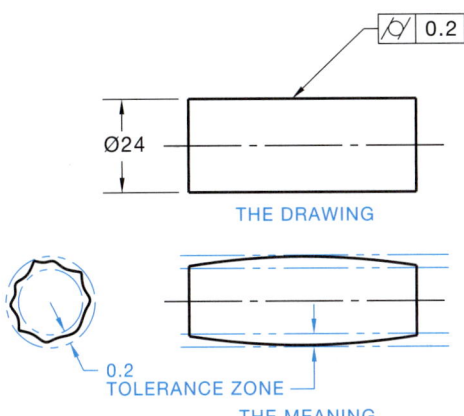

FIGURE 14.39 ■ Cylindricity representation.

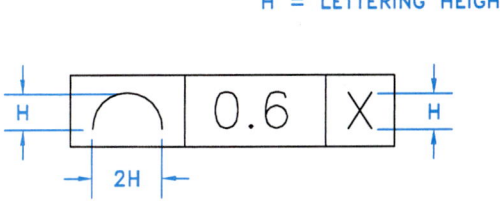

PROFILE OF A LINE

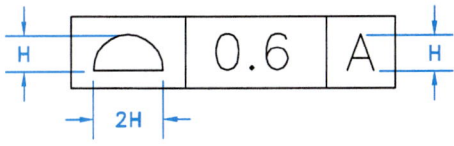

PROFILE OF A SURFACE

FIGURE 14.40 ■ Profile feature control frames.

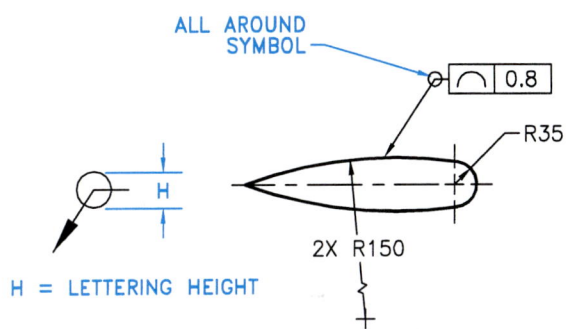

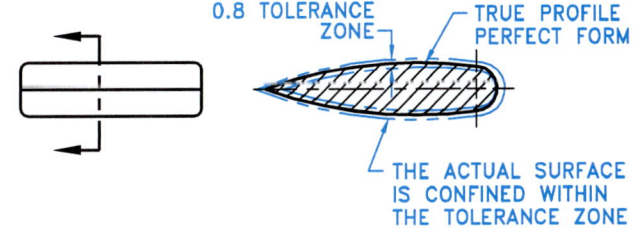

FIGURE 14.41 ■ Profile of a line all around, and the all around symbol.

a cross-sectional tolerance which results in a feature that must lie between two concentric circles, while cylindricity is a blanket tolerance that results in a feature lying between two concentric cylinders. The cylindricity feature control frame may be connected with a leader to either the circular or rectangular view. (See Figure 14.39.) The cylindricity tolerance must be less than the size tolerance.

Profile Tolerance

Profile is used to control form or a combination of size, form, and orientation. Profile callouts are commonly used on arcs, curves, or irregular-shaped features but can also be applied to plane surfaces. The shape and size of the profile may be defined with basic dimensions or tolerance dimensions. When tolerance dimensions are used to establish profile, the profile tolerance zone must be within the size tolerance. The profile feature control frame is connected to the longitudinal view with a leader line. The two types of profile are profile of a line and profile of a surface. (See Figure 14.40.)

The profile of a line is a cross-sectional or single-line tolerance that extends through the specified feature. The profile of a line may be controlled in relation to datums or without datums. The profile of a line may be all around the part by using the all around symbol on the leader, as seen in Figure 14.41, or between two specified points on the part by using the between symbol and identifying the points, as shown in Figure 14.42.

The profile of a surface is a blanket tolerance zone that affects all the surface elements of a feature equally. Profile of a surface is generally referenced to one or more datums so proper orientation of the profile boundary can be maintained. Profile of a surface may be applied all around a part as in Figure 14.43 or between two specified points as shown in Figure 14.44.

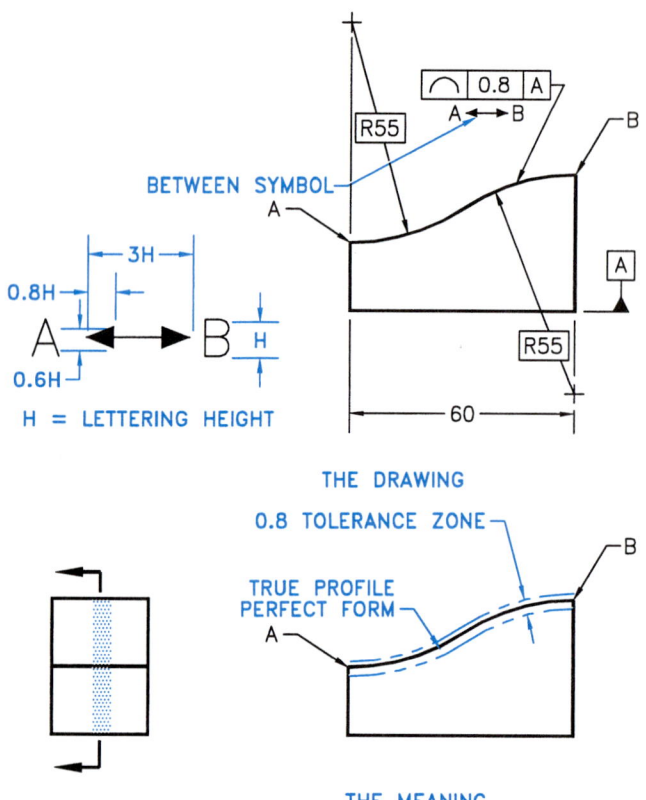

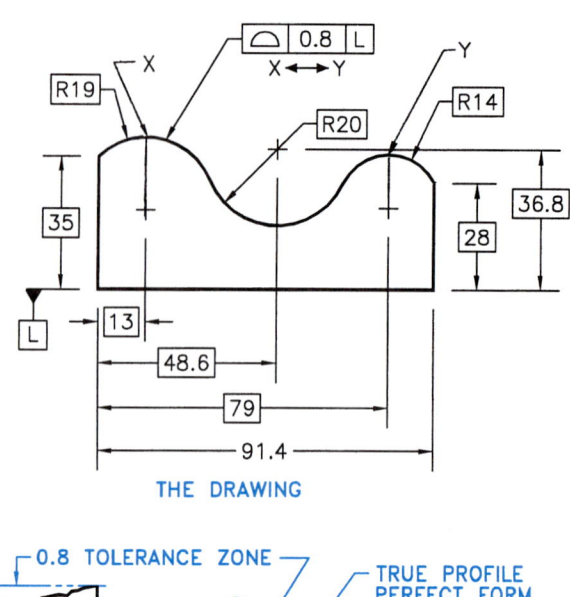

THE DRAWING

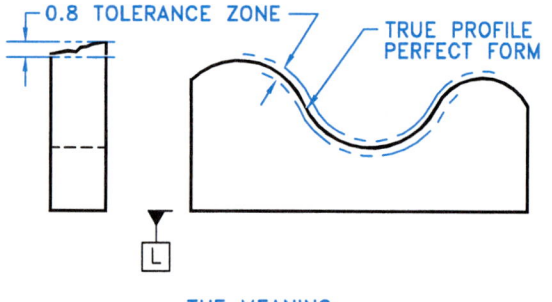

FIGURE 14.42 ■ Profile of a line between two given points, and the between symbol.

THE MEANING

FIGURE 14.44 ■ Profile of a surface between two given points.

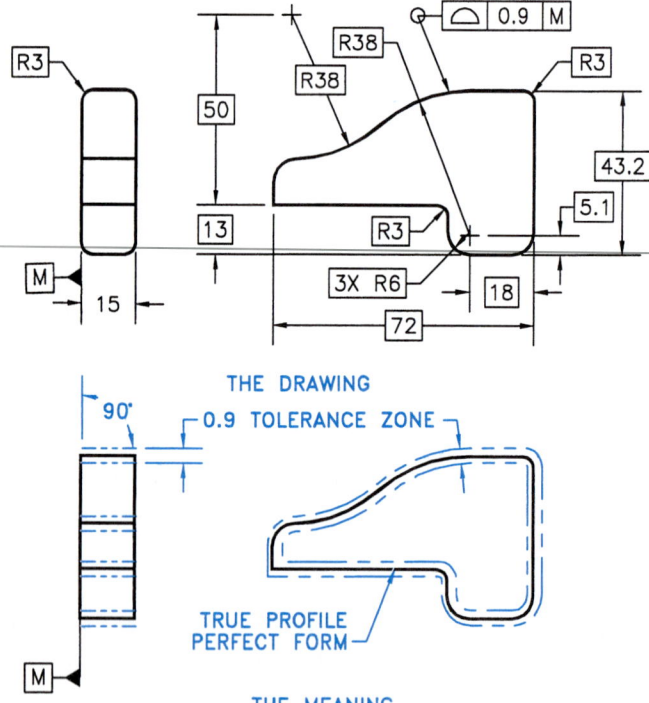

FIGURE 14.43 ■ Profile of a surface all around.

The profile of a line or the profile of a surface is a bilateral tolerance zone when the leader line points to the surface of the part without any additional symbology as was shown in Figures 14.40 through 14.44. A **bilateral profile tolerance** means that the tolerance zone is split equally on each side of the specified perfect form. Either type of profile callout may also have a unilateral tolerance zone specified. A **unilateral tolerance** zone places the entire zone on only one side of the true profile or perfect form. This is accomplished on a drawing with a phantom line placed parallel to the surface where the leader arrowhead touches the part. The phantom line is placed any clear distance from the part surface and is either inside or outside depending on the specified direction of the unilateral tolerance from the true profile. The actual feature is confined between the true profile and the given tolerance reference. (See Figure 14.45.)

The profile of coplanar surfaces also may be specified by placing a phantom line between the surfaces in the view where they appear as edges. The number of coplanar surfaces is identified below the feature control frame with a note, such as 2 SURFACES. The profile feature control frame is then connected to the phantom line with a leader. (See Figure 14.46.) Profile may also be used to control the angle of an inclined surface in relationship to a datum. (See Figure 14.47.)

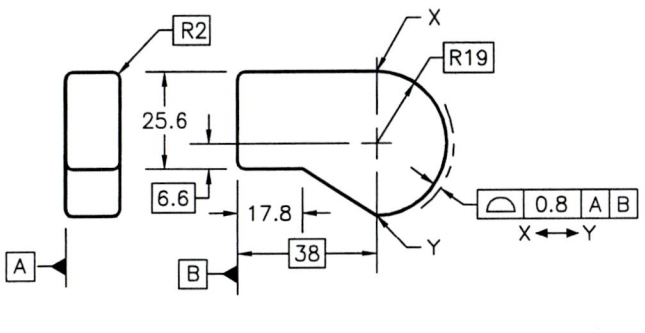

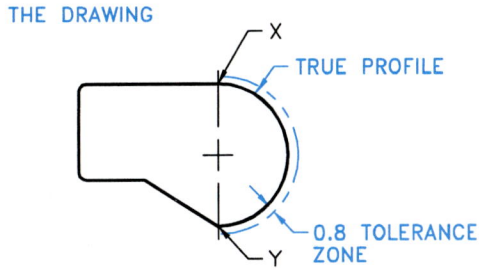

THE DRAWING

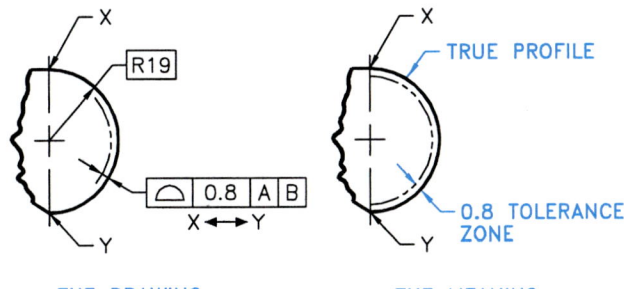

FIGURE 14.45 ■ A unilateral profile representation.

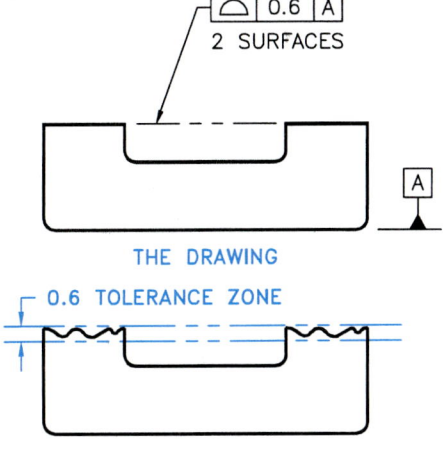

FIGURE 14.46 ■ Profile of coplanar surfaces.

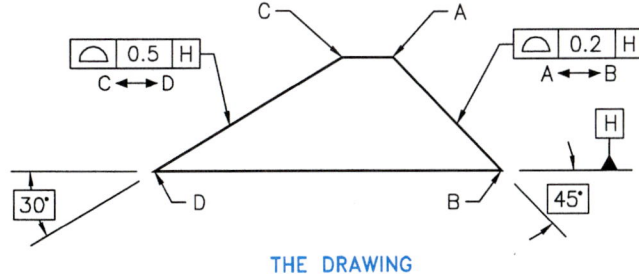

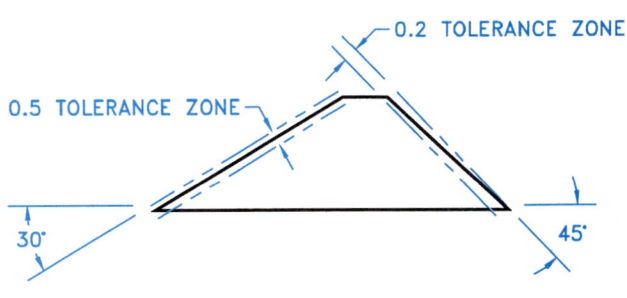

FIGURE 14.47 ■ Profile of an inclined surface.

Orientation Tolerance

Orientation tolerances refer to a specific group of geometric characteristics that establish a relationship between the features of an object. The tolerances that orient one feature to another are parallelism, perpendicularity, angularity and, in some applications, profile. Orientation tolerances require that one or more datums be used to establish the relationship between features. Parallelism, perpendicularity, and angularity also control flatness or cylindricity, depending on the shape of the feature being controlled. The geometric tolerance must fall within the size tolerance of the part.

Parallelism

The parallelism geometric characteristic symbol is detailed in Figure 14.48. A parallelism tolerance zone requires that the actual feature is between two parallel planes or lines that are parallel to a datum. The parallelism feature control frame can be attached to the feature with a leader or on an extension line of the surface. (See Figure 14.49.) Unless otherwise specified, a parallelism tolerance zone is a total tolerance that covers the entire surface. If a single-line element is to be specified rather than the surface, then the note EACH ELEMENT, or EACH RADIAL ELEMENT for an arc shaped surface, must be added below the feature control frame. (See Figure 14.50.) This technique also applies to perpendicularity and angularity.

In certain instances, parallelism also can be applied to a cylindrical feature. The tolerance zone is a cylindrical shape that is parallel to a datum axis reference. The feature control frame that relates to an axis specification is attached to the diameter

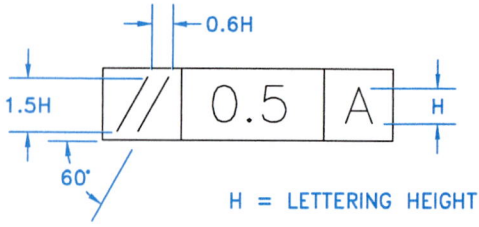

FIGURE 14.48 ■ Parallelism feature control frame.

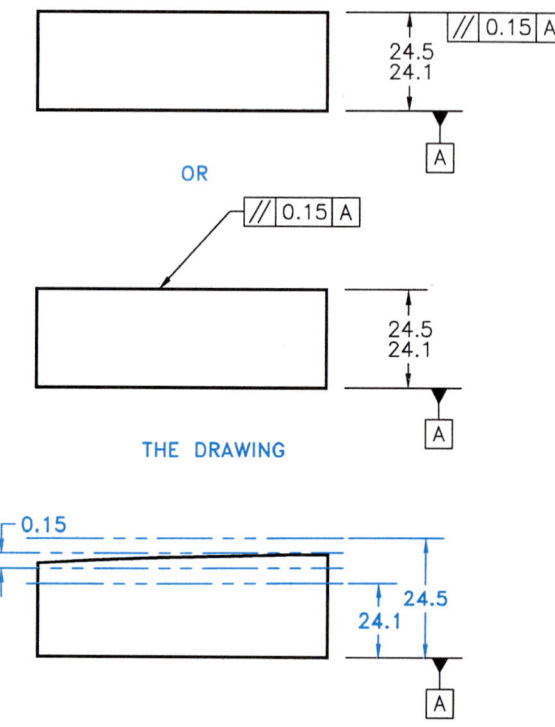

FIGURE 14.49 ■ Parallelism representation.

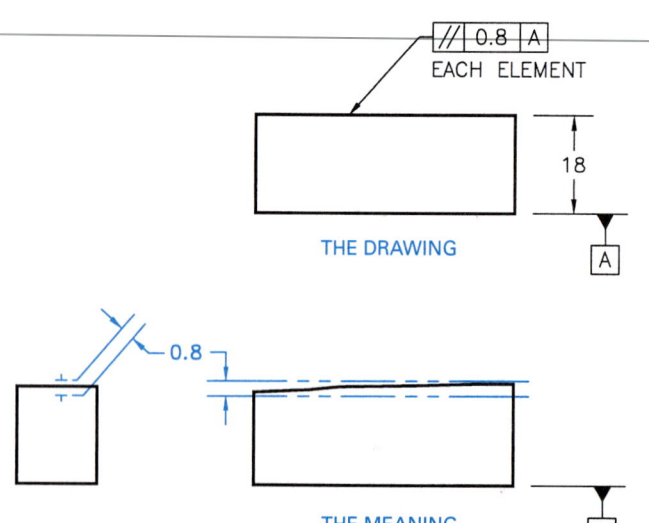

FIGURE 14.50 ■ Single-line element parallelism.

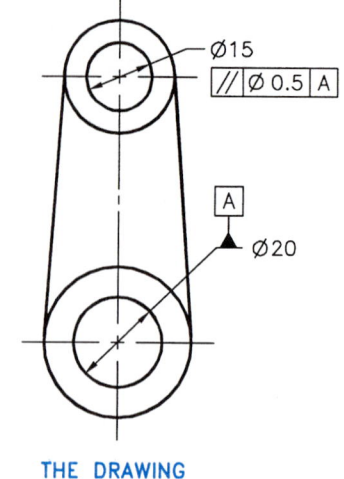

THE DRAWING

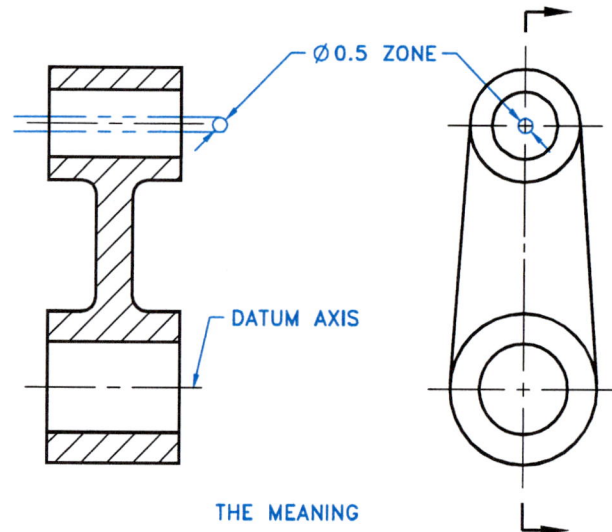

THE MEANING

FIGURE 14.51 ■ Axis parallelism.

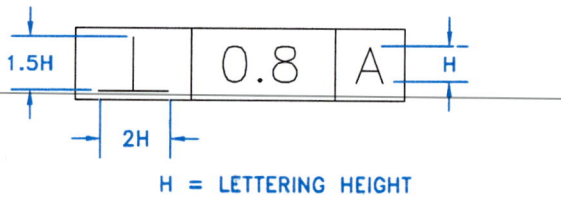

FIGURE 14.52 ■ Perpendicularity feature control frame.

dimension, and a diameter zone descriptor should precede the geometric tolerance. (See Figure 14.51.)

Perpendicularity

The perpendicularity geometric characteristic symbol is detailed in Figure 14.52. A **perpendicularity geometric tolerance** requires that a given feature be located between two parallel planes or lines, or within a cylindrical tolerance zone that is a basic 90° to a datum. The perpendicularity feature control frame may be connected to the feature surface with a leader or an extension line, or attached to the diameter dimension for axis perpendicularity. (See Figure 14.53.)

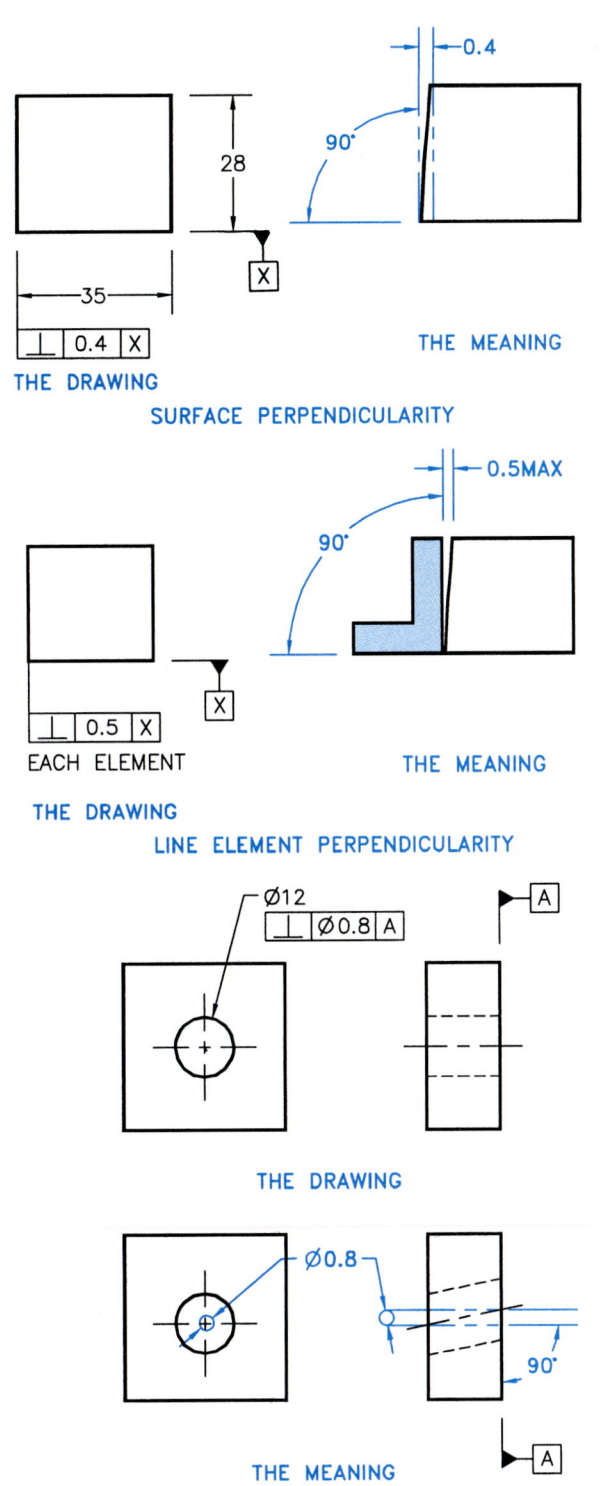

FIGURE 14.53 ▪ Perpendicularity representations.

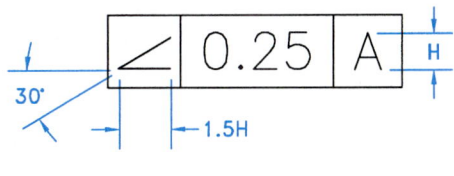

FIGURE 14.54 ▪ Angularity feature control frame.

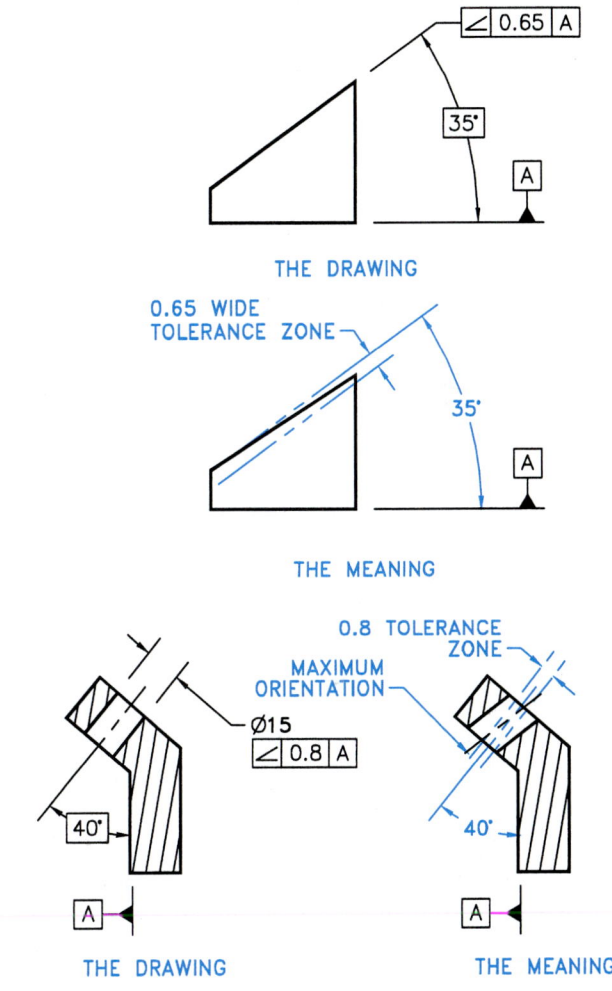

FIGURE 14.55 ▪ Angularity representations.

Angularity

The angularity geometric characteristic symbol is represented in Figure 14.54. An **angularity tolerance zone** places a given feature between two parallel planes that are at a specified basic angular dimension from a datum. The basic angle from the datum can be any amount except 90°. The angularity feature control frame may be connected to the surface with a leader or from an extension line, or attached to the diameter dimension for axis angularity. (See Figure 14.55.)

Runout Tolerance

Runout is used to control the relationship of radial features to a datum axis and features that are 90° to a datum axis. These types of features include cylindrical, tapered, and curved shapes, as well as plane surfaces that are at right angles to a datum axis. A runout tolerance is determined when a dial indicator is placed on the surface to be inspected and the part is rotated 360°. The **full indicator movement (FIM)** on the dial indicator must not exceed the

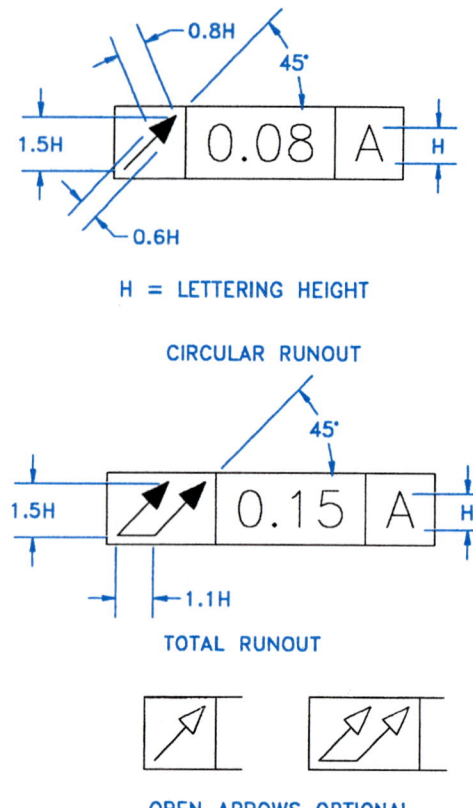

H = LETTERING HEIGHT

CIRCULAR RUNOUT

TOTAL RUNOUT

OPEN ARROWS OPTIONAL

FIGURE 14.56 ■ Runout feature control frames.

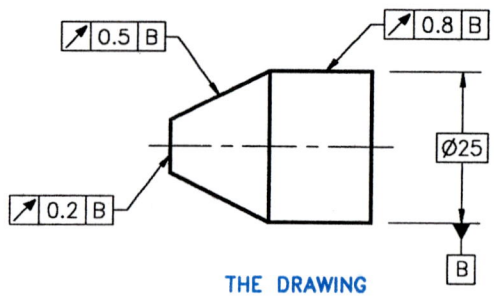

THE DRAWING

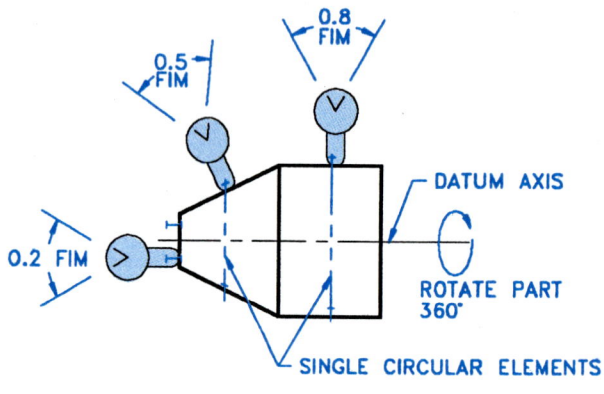

THE MEANING

FIGURE 14.57 ■ Circular runout.

amount of specified tolerance. The runout feature control frame is attached to the required surface with a leader. The runout tolerance applies to the length of the intended surface or until there is a break, or change, in shape or diameter of the surface. There are two types of runout: circular and total. (See Figure 14.56.)

Circular Runout

Circular runout provides control of single circular elements of a surface. Circular runout controls circularity and the relationship of the common axes (coaxial) of parts. This tolerance, established by the FIM of a dial indicator, is placed at one location on the feature as the part is rotated 360°. (See Figure 14.57.)

Total Runout

Total runout provides composite control of all surface elements. The tolerance is applied simultaneously to all circular and profile measuring positions as the part is rotated 360°. (See Figure 14.58.) Where applied to surfaces constructed around a datum axis, total runout is used to control cumulative variations of circularity, straightness, coaxiality, angularity, taper, and profile of a surface. Where applied to surfaces constructed at right angles to a datum axis, total runout controls cumulative variations of perpendicularity to detect wobble and flatness, and to detect concavity or convexity.

A portion of a surface can have a specified runout tolerance if it is not desired to control the entire surface. This is done by

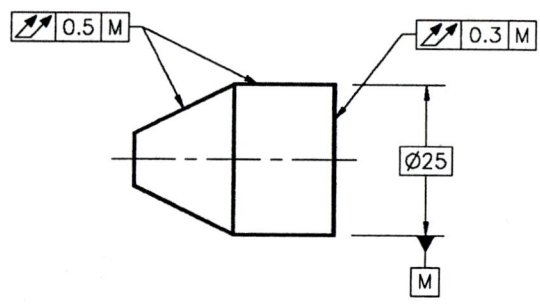

THE DRAWING

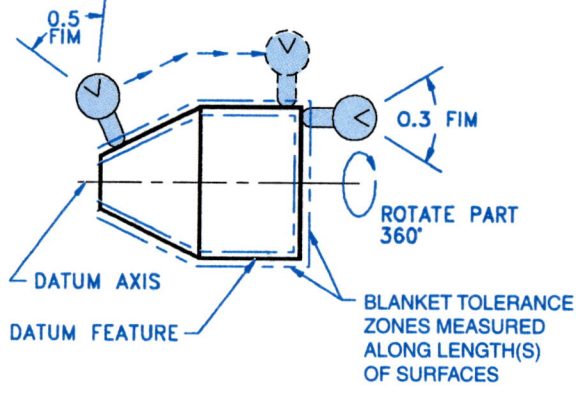

THE MEANING

FIGURE 14.58 ■ Total runout.

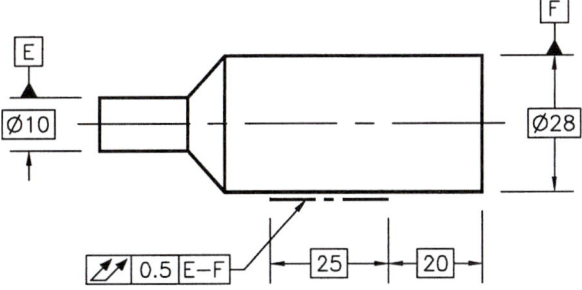

FIGURE 14.59 ■ Partial surface runout.

placing a chain line in the linear view adjacent to the desired location. The chain line is located with basic dimensions as shown in Figure 14.59. When a part has compound datum features (where more than one datum feature is used to establish a common datum), then the combined datum features are shown separated by a dash in the feature control frame. The datums are of equal importance. (See Figure 14.59.)

MATERIAL CONDITION SYMBOLS

Material condition symbols are used in conjunction with the feature tolerance or datum reference in the feature control frame. The material condition symbols are required to establish the relationship between the size or location of the feature and the geometric tolerance. The use of different material condition symbols alters the effect of this relationship. The material condition modifying elements are maximum material condition, MMC; regardless of feature size, RFS; and least material condition, LMC. There is no symbol for RFS because it is assumed for all geometric tolerances and datum references unless another material condition symbol is specified. The standard material condition symbols are shown in Figure 14.60.

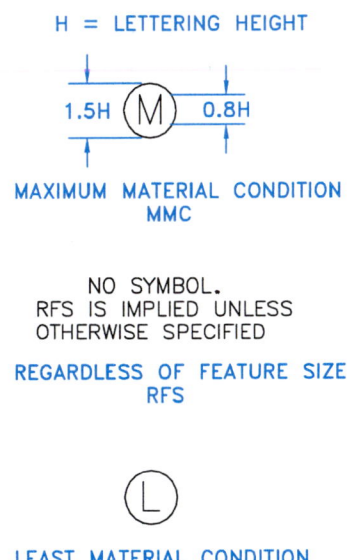

FIGURE 14.60 ■ Material condition symbols.

Perfect Form Envelope

The form of a feature is controlled by the size tolerance limits. The envelope, or boundary, of these size limits is established at MMC. Remember from the discussion in Chapter 11 that MMC is the largest limit for an external feature and the smallest limit for an internal feature. The key is most material. The true geometric form of the feature is at MMC. This is known as the **perfect form boundary**. This is also referred to as the **perfect form envelope**. If the part feature is produced at MMC, it is considered to be at perfect form. When it is desired to permit a surface or surfaces of a feature to exceed the boundary of perfect form at MMC, a note such as PERFECT FORM AT MMC NOT REQUIRED is specified, exempting the pertinent size dimension. If a feature is produced at LMC, the opposite of MMC, the form is allowed to vary within the geometric tolerance zone or to the extent of the MMC envelope.

Regardless of Feature Size (RFS)

Regardless of feature size (RFS) means that the geometric tolerance applies at any produced size. The tolerance remains as the specified value regardless of the actual size of the feature. Regardless of feature size is implied (assumed) for all geometric characteristics and datums unless otherwise specified.

Effect of RFS on Surface Straightness

RFS is implied for the straightness geometric characteristic. When a surface straightness tolerance is used by connecting a leader from the feature control frame to the surface of the part, then the geometric tolerance remains the same regardless of the feature size, and the actual size may not exceed the perfect form envelope at MMC. An acceptable part may be produced between the given size tolerance, and any straightness irregularity may not be greater than the specified geometric tolerance. Perfect form is required at MMC. (See Figure 14.61.)

Effect of RFS on Axis Straightness

When the straightness tolerance is applied to the axis of the feature by a relationship with the diameter dimension, RFS is implied unless otherwise specified, and the actual feature size plus the geometric tolerance may exceed the MMC perfect form envelope. (See Figure 14.62.)

Datum Feature RFS

Datum features that are influenced by size variations, such as diameters and widths, are also subject to variations in form. RFS is implied unless otherwise specified. When a datum feature has a size dimension and a geometric form tolerance, the size of the simulated datum is the MMC size limit. This rule applies except for axis straightness where the envelope is allowed to exceed MMC. Figure 14.63 shows the effect of RFS on the primary datum feature with axis and center plane datums. When the datum

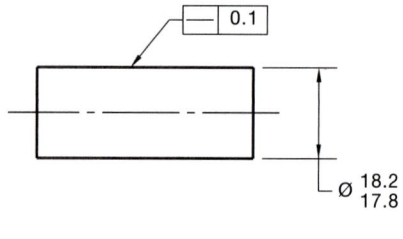

THE DRAWING

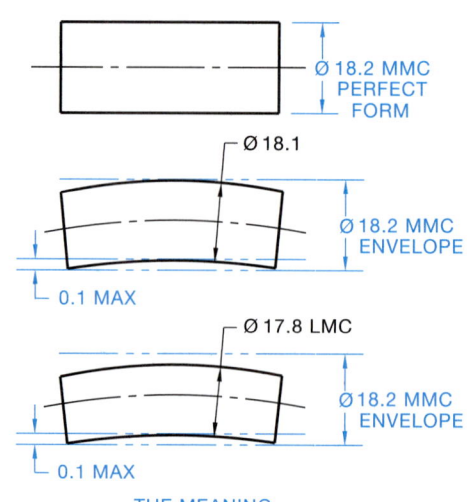

ACTUAL SIZE	GEOMETRIC TOLERANCE
18.2 MMC	0–PERFECT FORM
18.1	0.1
18.0	0.1
17.9	0.1
17.8 LMC	0.1

THE MEANING

FIGURE 14.61 ■ Effect of RFS, regardless of feature size, on surface straightness. Perfect form is required at MMC.

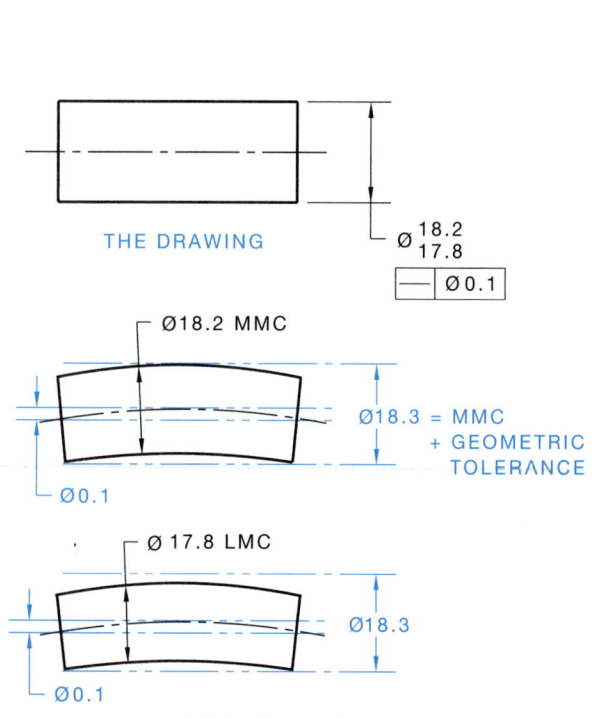

THE DRAWING

THE MEANING

ACTUAL SIZE	GEOMETRIC TOLERANCE
18.2 MMC	0.1
18.1	0.1
18.0	0.1
17.9	0.1
17.8 LMC	0.1

FIGURE 14.62 ■ Effect of RFS on axis straightness.

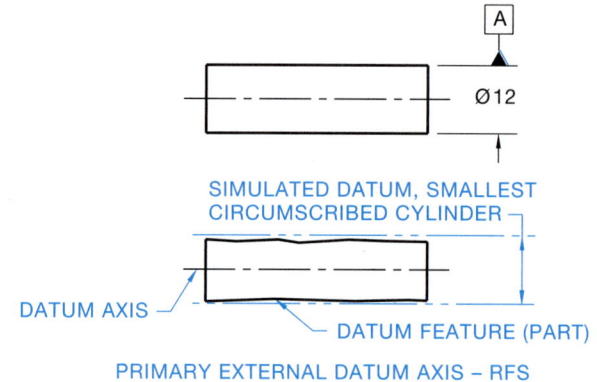

PRIMARY EXTERNAL DATUM AXIS – RFS

PRIMARY INTERNAL DATUM AXIS – RFS

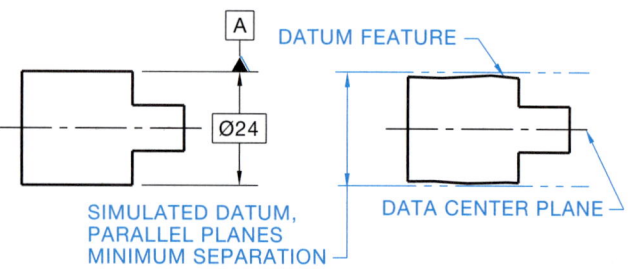

PRIMARY EXTERNAL DATUM CENTER PLANE – RFS

FIGURE 14.63 ■ Effect of RFS on the primary datum feature with axis and center plane datums.

GEOMETRIC TOLERANCING

The implementation of geometric tolerancing and dimensioning into a mechanical drafting CADD program is quite practical. The GT symbology makes this application a bonus to the mechanical drafting system.

Some dimensioning and tolerancing guidelines for use in conjunction with CADD/CAM are outlined, in part, from ASME Y14.5M as follows:

1. Major features of the part should be used to establish the basic coordinate system but are not necessarily defined as datums.

2. Subcoordinated systems that are related to the major coordinates are used to locate and orient features on a part.

3. Define part features in relation to three mutually perpendicular reference planes and along features that are parallel to the motion of CAM equipment.

4. Establish datums related to the function of the part and relate datum features in order of precedence as a basis for CAM usage.

5. Completely and accurately dimension geometric shapes. Regular geometric shapes may be defined by mathematical formulas, although a profile feature that is defined with mathematical formulas should not have coordinate dimensions unless required for inspection or reference.

6. Coordinate or tabular dimensions should be used to identify approximate dimensions on an arbitrary profile.

7. Use the same type of coordinate dimensioning system on the entire drawing.

8. Continuity of profile is necessary for CADD. Clearly define contour changes at the change or point of tangency. Define at least four points along an irregular profile.

9. Circular hole patterns may be defined with polar coordinate dimensioning.

10. When possible, dimension angles in degrees and decimal parts of degrees; for example, 45° 30' = 45.5°.

11. Base dimensions at the mean of a tolerance because the numerical control (NC) programmer will normally split a tolerance and work to the mean. Establish dimensions without limits that conform to the NC machine capabilities and part function where possible. Bilateral profile tolerances are also recommended for the same reason.

12. Geometric tolerancing is necessary to control specific geometric form and location, but is not required if general tolerancing provides sufficient form or location control.

The standard ASME Y14.5.1, *Mathematical Definition of Y14.5,* is the mathematical model of all specifications found in ASME Y14.5M. This allows for convenient digitizing of a part so all gaging is in a CADD system.

features are secondary or tertiary, then the axis or center plane also shall have an angular relationship to the primary datum. (See Figure 14.64.)

Maximum Material Condition (MMC)

The use of MMC in conjunction with the geometric tolerance in a feature control frame means that the given tolerance applies at the MMC produced size. As the feature dimension departs from MMC, the geometric tolerance is allowed to increase equal to the change. The maximum amount of change is at the LMC produced size. MMC must be specified for any geometric characteristic where its application is desired.

Effect of MMC on Form Tolerance

When MMC is specified in conjunction with an axis straightness callout, the MMC feature size envelope is exceeded by the given geometric tolerance. The given geometric tolerance is held at the MMC produced size, then as the actual produced size departs from MMC, the geometric tolerance is allowed to increase equal to the change from MMC to a maximum amount of departure at LMC. (See Figure 14.65.) The same situation also may be demonstrated with a perpendicularity tolerance where a datum axis is perpendicular to a given datum, as in Figure 14.66.

Least Material Condition (LMC)

The application of LMC is not as prevalent as MMC. LMC may be used when it is desirable to control minimum wall thickness. This is discussed later with positional geometric tolerance at LMC. When an LMC material condition symbol is used in conjunction with a geometric tolerance, the specified tolerance applies at the LMC produced size. LMC controls are refinement of size controls. As the actual produced size deviates from LMC toward MMC, the tolerance is allowed an increase equal to the amount of change from LMC. (See Figure 14.67.)

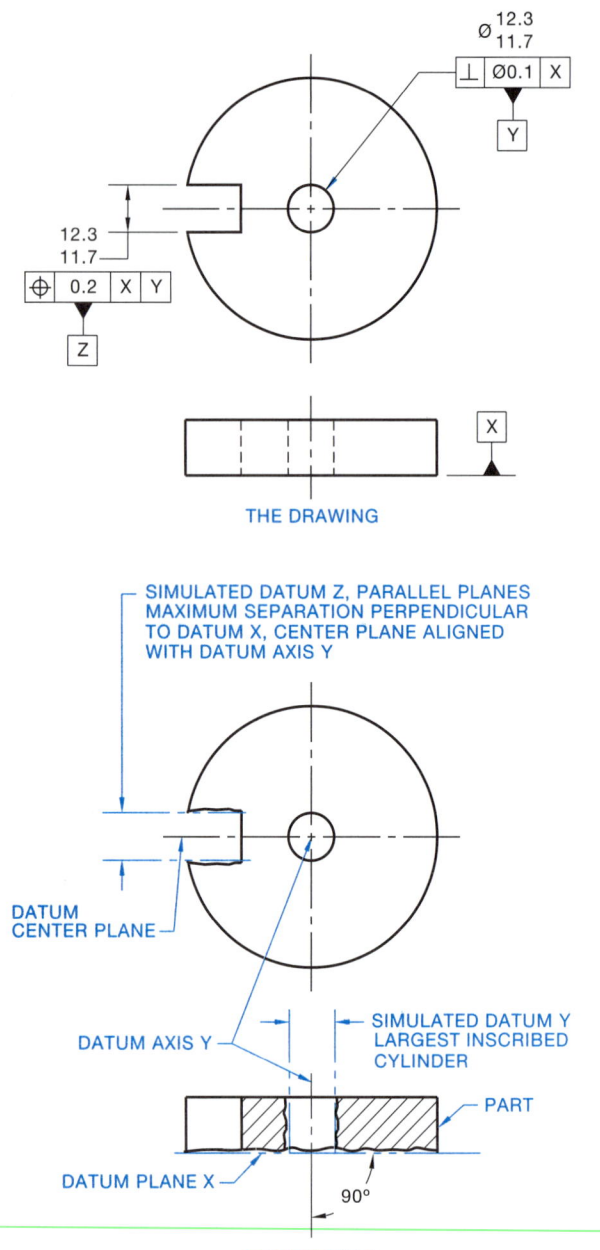

THE DRAWING

SIMULATED DATUM Z, PARALLEL PLANES
MAXIMUM SEPARATION PERPENDICULAR
TO DATUM X, CENTER PLANE ALIGNED
WITH DATUM AXIS Y

DATUM
CENTER PLANE

DATUM AXIS Y

SIMULATED DATUM Y
LARGEST INSCRIBED
CYLINDER

PART

DATUM PLANE X

90°

THE MEANING

FIGURE 14.64 ■ The secondary or tertiary datum relationship to the primary datum.

LOCATION TOLERANCE

Location tolerances include concentricity, symmetry, and position. **True position** is the theoretically exact location of the axis or center plane of a feature. The true position is located using basic dimensions. The locational tolerance specifies the amount that the axis of the feature is allowed to deviate from true position. All geometric tolerances imply RFS. MMC or LMC must be specified if desired. Reference to true position dimensions must be provided as basic dimensions at each required location on the drawing, or a general note may be used to specify that UNTOLERANCED DIMENSIONS LOCATING TRUE POSI-

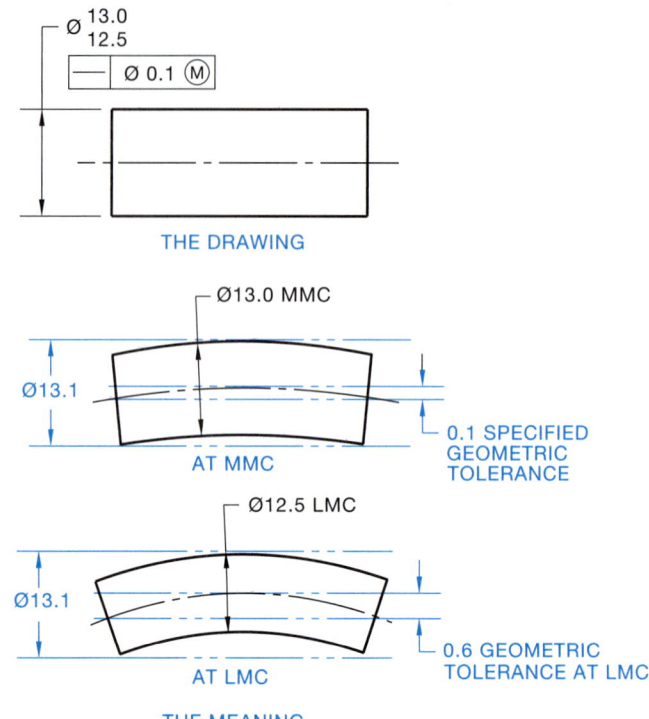

THE DRAWING

Ø13.0 MMC

Ø13.1

AT MMC

0.1 SPECIFIED
GEOMETRIC
TOLERANCE

Ø12.5 LMC

Ø13.1

AT LMC

0.6 GEOMETRIC
TOLERANCE AT LMC

THE MEANING

ACTUAL SIZE	GEOMETRIC TOLERANCE
13.0 MMC	0.1
12.9	0.2
12.8	0.3
12.7	0.4
12.6	0.5
12.5 LMC	0.6

FIGURE 14.65 ■ Effect of maximum material condition, MMC, on form tolerance, axis straightness.

TION ARE BASIC. Basic dimensions are theoretically perfect, so that datum or chain dimensioning may be used equally to locate true position, because there is no tolerance buildup. The location feature control frame is generally added to the note or dimension of the related feature.

Concentricity

The concentricity geometric characteristic symbol is detailed in Figure 14.68. **Concentricity** is the relationship of the axes of cylindrical shapes. Perfect concentricity exists when the axes of two or more cylindrical features are in perfect alignment. (See Figure 14.69.) In a concentricity geometric tolerance, the axis of the feature establishes the axis of the cylindrical tolerance zone in which median points of all correspondingly located elements being controlled must lie. Concentricity is an axis-to-median point control and not an axis-to-axis control. (See Figure 14.70.) The geometric tolerance and related datums imply RFS. It is difficult to control the axis relationship specified by concentricity, so runout is often used. Where the balance of a shaft is critical, concentricity may be used.

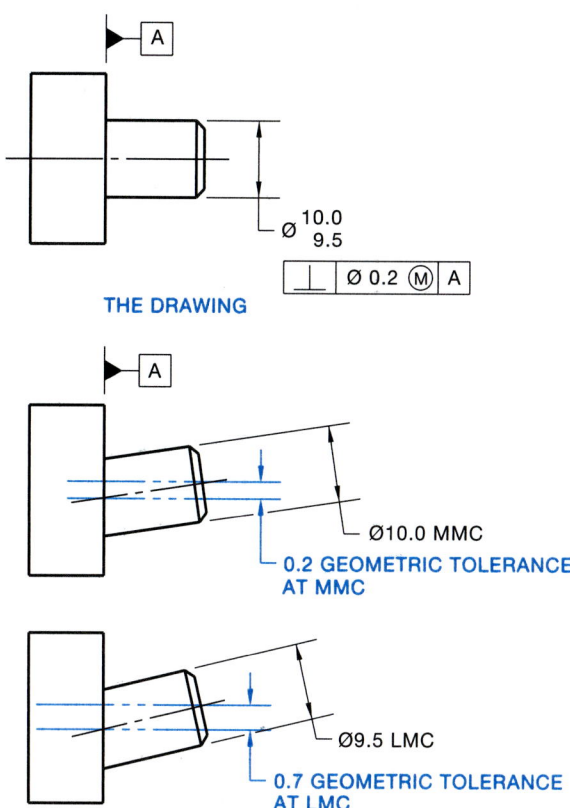

THE DRAWING

Ø 10.0
9.5

THE MEANING

Ø10.0 MMC
0.2 GEOMETRIC TOLERANCE AT MMC

Ø9.5 LMC
0.7 GEOMETRIC TOLERANCE AT LMC

ACTUAL SIZE	GEOMETRIC TOLERANCE
10.0 MMC	0.2
9.9	0.3
9.8	0.4
9.7	0.5
9.6	0.6
9.5 LMC	0.7

FIGURE 14.66 ■ Effect of MMC on form tolerance, axis perpendicularity.

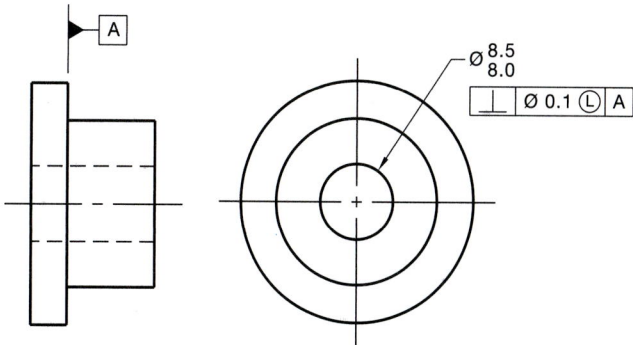

Ø 8.5
8.0

ACTUAL SIZE	GEOMETRIC TOLERANCE
8.0 MMC	0.6
8.1	0.5
8.2	0.4
8.3	0.3
8.4	0.2
8.5 LMC	0.1

FIGURE 14.67 ■ Application of least material condition, LMC.

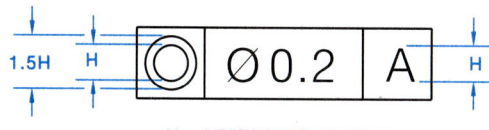

1.5H H Ø 0.2 A H

H = LETTERING HEIGHT

FIGURE 14.68 ■ Concentricity feature control frame.

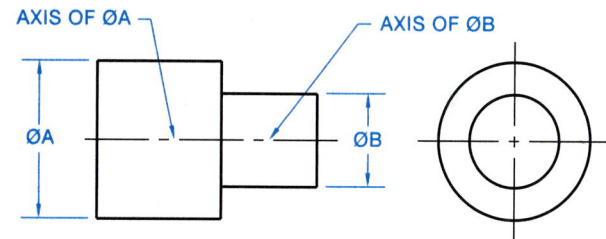

AXIS OF ØA   AXIS OF ØB

ØA   ØB

FIGURE 14.69 ■ Perfect concentricity when both axis coincide.

Position Tolerance of Symmetrical Features

Position tolerancing can be used to locate the center plane of one or more features relative to a datum center plane. Figure 14.71 shows optional feature control frames where the position geometric tolerance and related datum reference are controlled on an RFS or MMC basis. When the position tolerance is at MMC, the related datum feature can be RFS, MMC, or LMC, depending on design requirements. Using MMC causes the geometric tolerance zone to change with the size of datum B, between 26.5-26.1. If the position tolerance is held RFS, then the related datum reference is RFS. RFS is implied unless otherwise specified.

Symmetry

Symmetry is the center plane relationship between two or more features. Perfect symmetry exists when the center plane of two

or more features is in alignment. The geometric characteristic symbol used to specify symmetry is detailed in the feature control frame shown in Figure 14.72. The **symmetry geometric tolerance zone** is a zone in which the median points of opposite symmetrical surfaces align with the datum center plane. The symmetry geometric tolerance and related datum reference are applied only on an RFS basis. (See Figure 14.73.) A zero position tolerance at MMC is used when it is required to control the symmetrical relationship of features within their size limits. An explanation of zero position tolerance at MMC follows.

Position

The position geometric characteristic symbol is detailed in Figure 14.74. RFS is implied for the position geometric tolerance and related datum reference unless otherwise specified. MMC or LMC

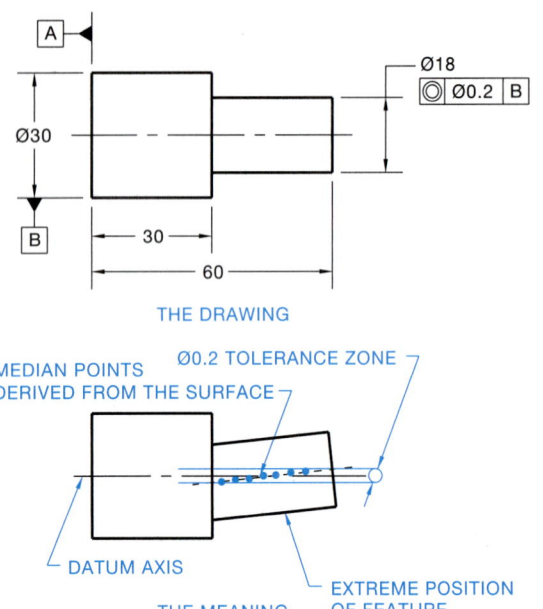

FIGURE 14.70 ■ Concentricity represented.

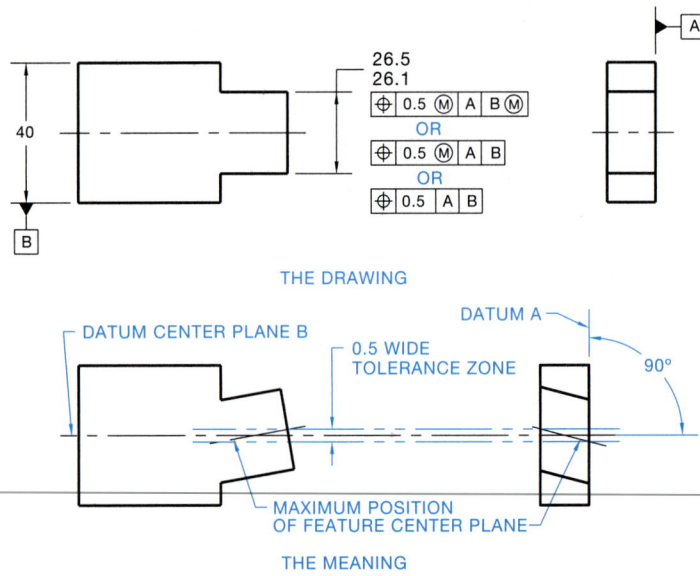

FIGURE 14.71 ■ Position geometric characteristic specifying symmetry.

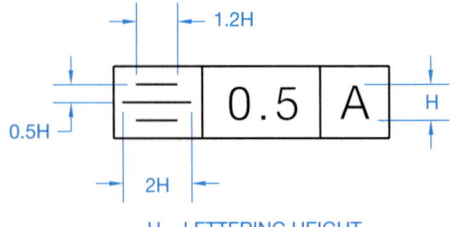

H = LETTERING HEIGHT

FIGURE 14.72 ■ The symmetry symbol detailed in a feature control frame.

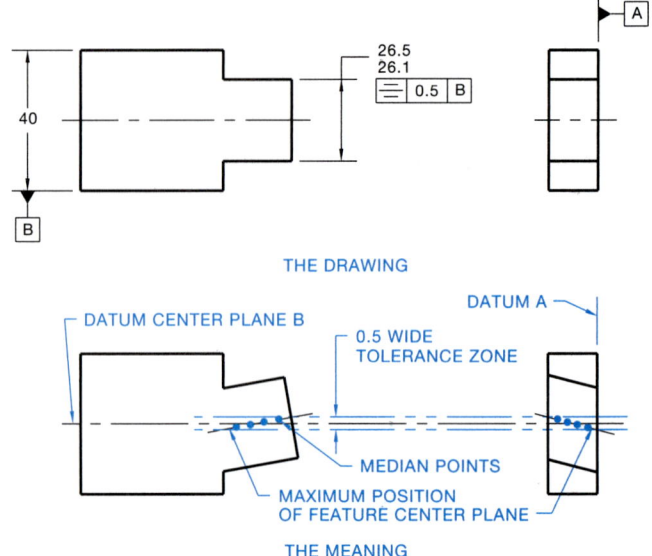

FIGURE 14.73 ■ Symmetry application.

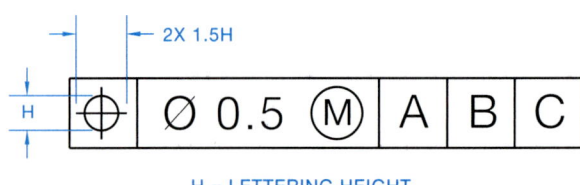

H = LETTERING HEIGHT

FIGURE 14.74 ■ Position feature control frame.

is applied to the position tolerance and datum reference as needed for design. The maximum material condition application is common, although some cases require the use of regardless of feature size or least material condition. The feature control frame is applied to the note or size dimension of the feature. A **positional tolerance zone** defines a zone within which the center, axis, or center plane of a feature of size is permitted to vary from the true, theoretically exact, position. Basic dimensions establish the true position from specified datum features and between interrelated features. A *positional* tolerance is indicated by the position symbol, a tolerance, and appropriate datum references placed in a feature control frame. The location of each feature is given by basic dimensions. Dimensions locating true position must be excluded from the drawing's general tolerances by applying the basic dimension symbol to each basic dimension or by specifying on the drawing or drawing reference the general note UNTOLERANCED DIMENSIONS LOCATING TRUE POSITION ARE BASIC.

Positional Tolerance at Maximum Material Condition (MMC)

Positional tolerance at MMC means that the given tolerance is held at the MMC produced size. Then, as the feature size departs from MMC, the position tolerance increases equal to the amount of change from MMC to the maximum change at LMC. Positional tolerance at MMC can be defined by the feature axis or surface. The datum references commonly establish true position perpendicular to the primary datum and the coordinate location dimensions to the secondary and tertiary datums. The position tolerance zone is a cylinder equal in diameter to the given position tolerance. The cylindrical tolerance zone extends through the thickness of the part unless otherwise specified. The actual centerline of the feature can be anywhere within the cylindrical tolerance zone. (See Figure 14.75.) Another explanation of position may be related to the surface of a hole. The hole surface cannot be inside a cylindrical tolerance zone established by the MMC diameter of the hole less the position tolerance. (See Figure 14.76.)

The previously discussed application of positional tolerancing for cylindrical features, such as for holes and bosses, also applies to non-cylindrical features such as slots, tabs, and elongated

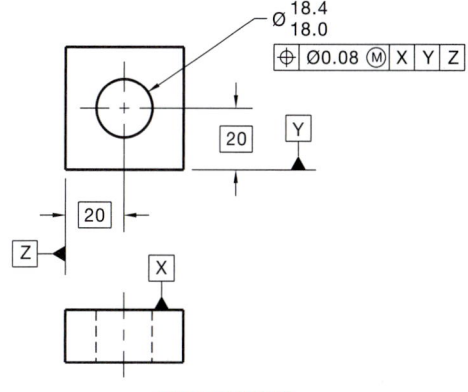

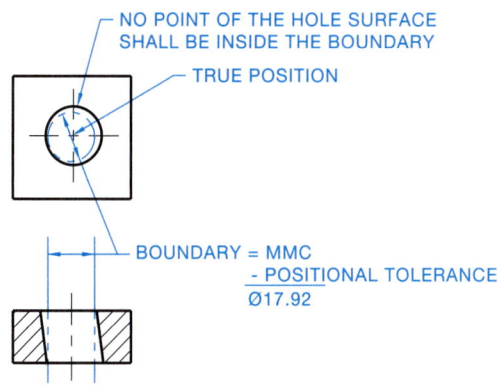

FIGURE 14.76 ■ Position Boundary = MMC − Position Tolerance.

holes. When applied to non-cylindrical features of size, the positional tolerance is used to locate the center plane established by parallel surfaces of the feature. The positional tolerance value is the distance between the two planes. Examples of this application are provided later in this chapter.

Zero Positional Tolerance at MMC

When the positional tolerance is associated with the MMC symbol, the tolerance is allowed to exceed the specified amount when the actual feature size departs from MMC. When the actual sizes of features are manufactured very close to MMC, it is critical that the feature axis or surface not exceed the boundaries discussed previously. When parts are rejected due to this situation, it is possible to increase the acceptability of mating parts by reducing the MMC size of the feature to minimum allowance with the mating part (virtual condition) and providing a zero positional tolerance at MMC. The positional tolerance is dependent on the feature size. When zero positional tolerance is used, no positional tolerance is allowed when the part is produced at MMC. True position is required at MMC. As the actual size departs from MMC, the positional tolerance increases equal to the amount of change to the maximum tolerance at LMC. (See Figure 14.77.)

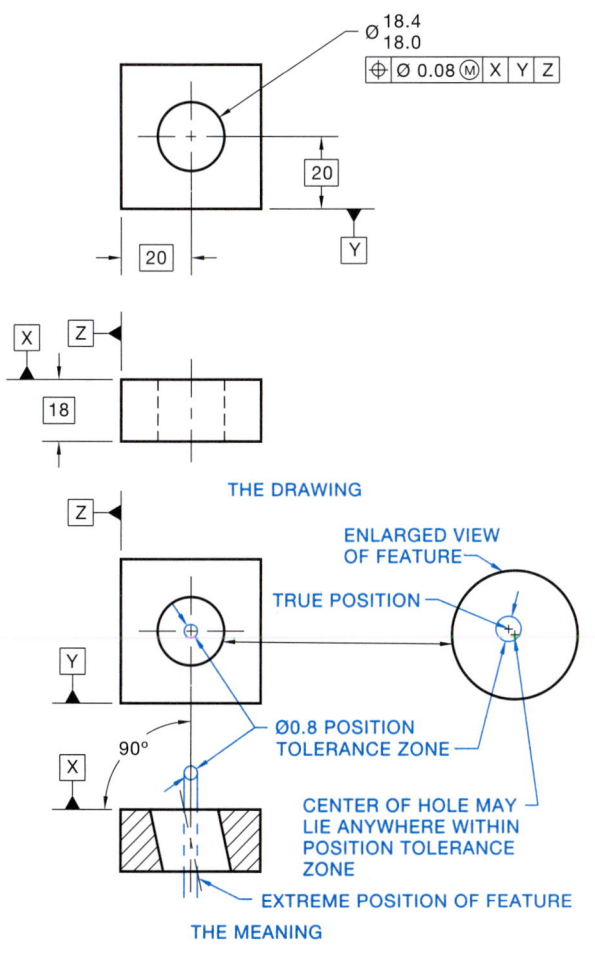

ACTUAL SIZE	POSITIONAL TOLERANCE
18.0 MMC	0.08
18.1	0.18
18.2	0.28
18.3	0.38
18.4 LMC	0.48

FIGURE 14.75 ■ Positional tolerance at maximum material condition, MMC.

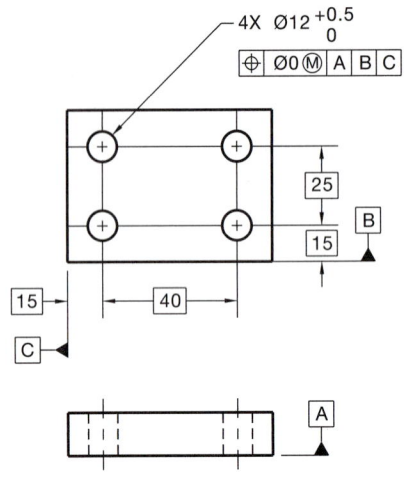

ACTUAL SIZE	POSITIONAL TOLERANCE
12.0 MMC ZERO ALLOWANCE 12mm FASTENER AT MMC	0
12.1	0.1
12.2	0.2
12.3	0.3
12.4	0.4
12.5	0.5

FIGURE 14.77 ■ Zero positional tolerance at MMC.

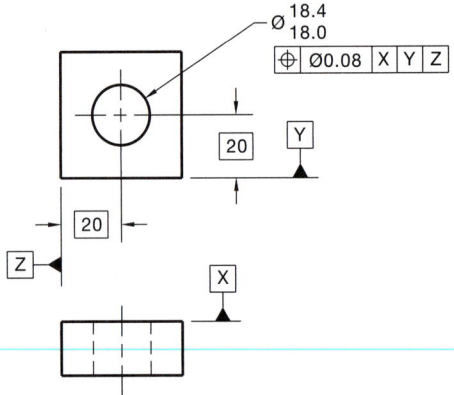

ACTUAL SIZE	POSITIONAL TOLERANCE
18.0 MMC	0.08
18.1	0.08
18.2	0.08
18.3	0.08
18.4 LMC	0.08

FIGURE 14.78 ■ Positional tolerance at RFS.

Positional Tolerance at RFS

RFS may be applied to the positional tolerance when it is desirable to maintain the given tolerance at any produced size. This application results in close positional control. (See Figure 14.78.) No additional tolerance is permitted based on size of feature.

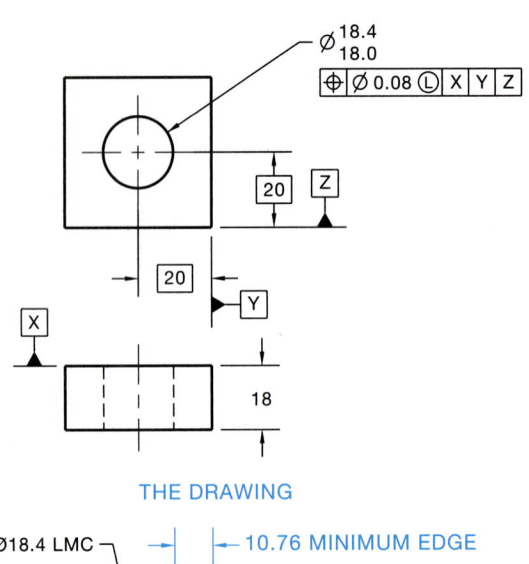

THE DRAWING

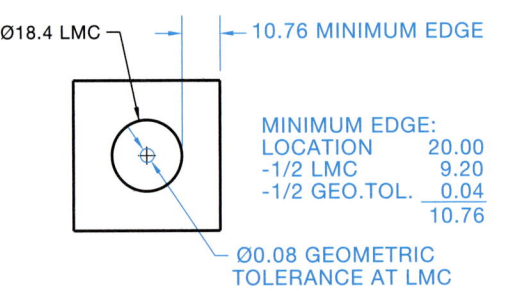

THE MEANING

ACTUAL SIZE	POSITIONAL TOLERANCE
18.0 MMC	0.48
18.1	0.38
18.2	0.28
18.3	0.18
18.4 LMC	0.08

FIGURE 14.79 ■ Positional tolerance at LMC.

Positional Tolerance at LMC

Positional tolerance at LMC is used to control the relationship of the feature surface and the true position at largest hole size. The function of the LMC specification is generally for the control of minimum-edge distances. When LMC is used, there is no requirement of perfect form at LMC. A feature produced at the LMC limit of size is allowed to vary from true form to the maximum at the MMC perfect form boundary. As the actual size departs from LMC toward MMC, the positional tolerance zone is allowed to increase equal to the amount of change. The maximum positional tolerance is at the MMC-produced size. (See Figure 14.79.)

DATUM PRECEDENCE AND MATERIAL CONDITION

The effect of material condition on the datum and related feature can be altered by changing the datum precedence and the

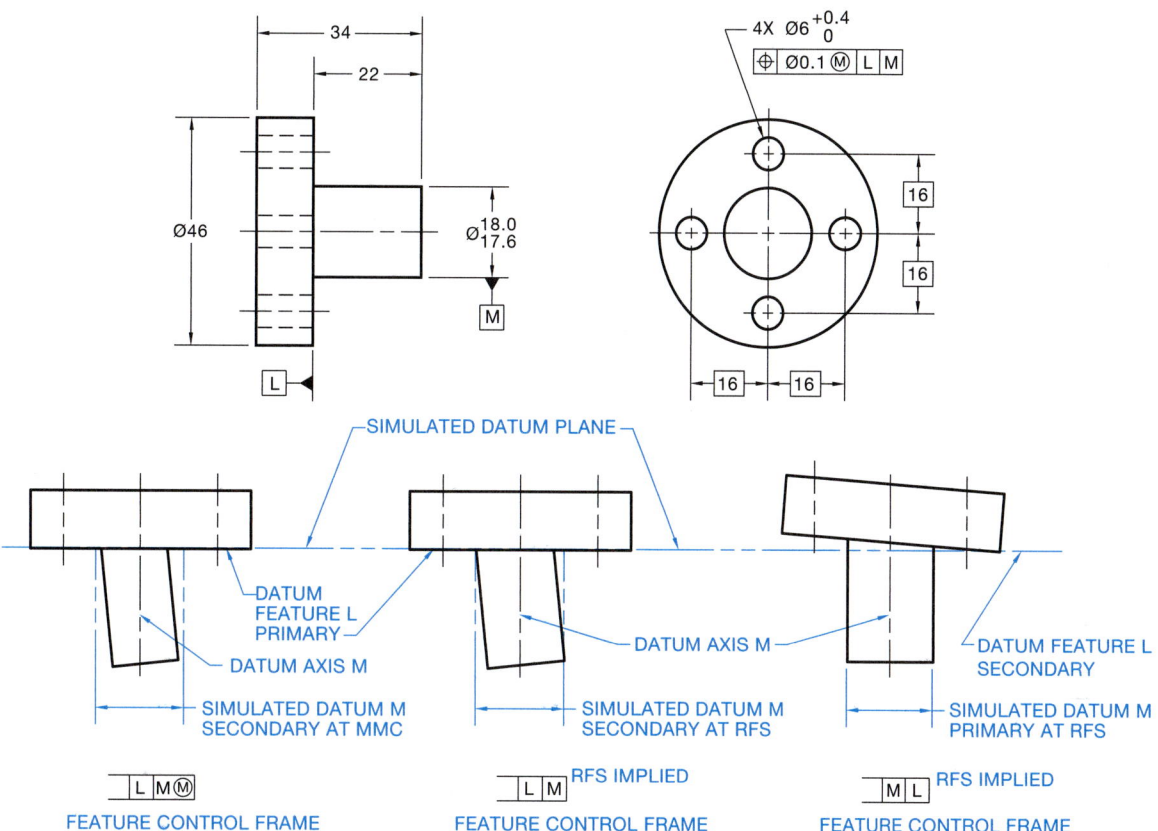

FIGURE 14.80 ■ The effect of altering datum precedence and material condition.

applied material condition symbol. The **datum precedence** is established by the order of placement in the feature control frame. The first datum listed is the primary datum; following datums are secondary and tertiary. Figure 14.80 shows the effect of altering datum precedence and material condition.

POSITION OF MULTIPLE FEATURES

The location of multiple features is handled in a manner similar to the location of a single feature. The true positions of the features are located with basic dimensions using rectangular or polar coordinates. The features are then identified by quantity, size, and position. (See Figure 14.81.) When two or more separate patterns are referenced to the same datums and with the same datum precedence, then the patterns are functionally the same. Verification of location and size is performed together. If this situation occurs and the interrelationship between patterns is not desired, then the specific note SEP REQT, meaning separate requirement, shall be placed beneath each affected feature control frame.

COMPOSITE POSITIONAL TOLERANCING

In some situations, it is permissible to allow the location of individual features in a pattern to differ from the tolerance related to

the items as a group. When this is done, the group of features is given a positional tolerance that is greater in diameter than the zone specified for the individual features. The tolerance zone of the individual features must fall within the group zone. The positional tolerance of the individual elements controls the perpendicularity of the features. An individual tolerance zone can extend partly beyond the group zone only if the feature axis does not fall outside the confines of both zones. When such is the case, the feature control frame is expanded in height and divided into two parts. The upper part, known as the **pattern locating control**, specifies the larger positional tolerance for the pattern of features as a group. The lower entry, commonly called the **feature relating control**, specifies the smaller positional tolerance for the individual features within the pattern. The pattern locating control is located first with basic dimensions. The feature relating control is established at the actual position of the feature center. Only the primary datum is represented in the feature relating control. (See Figure 14.82.) This is called **composite positional tolerancing**.

TWO SINGLE-SEGMENT FEATURE CONTROL FRAMES

The composite position tolerance is specified by a feature control frame doubled in height with one position symbol shown in

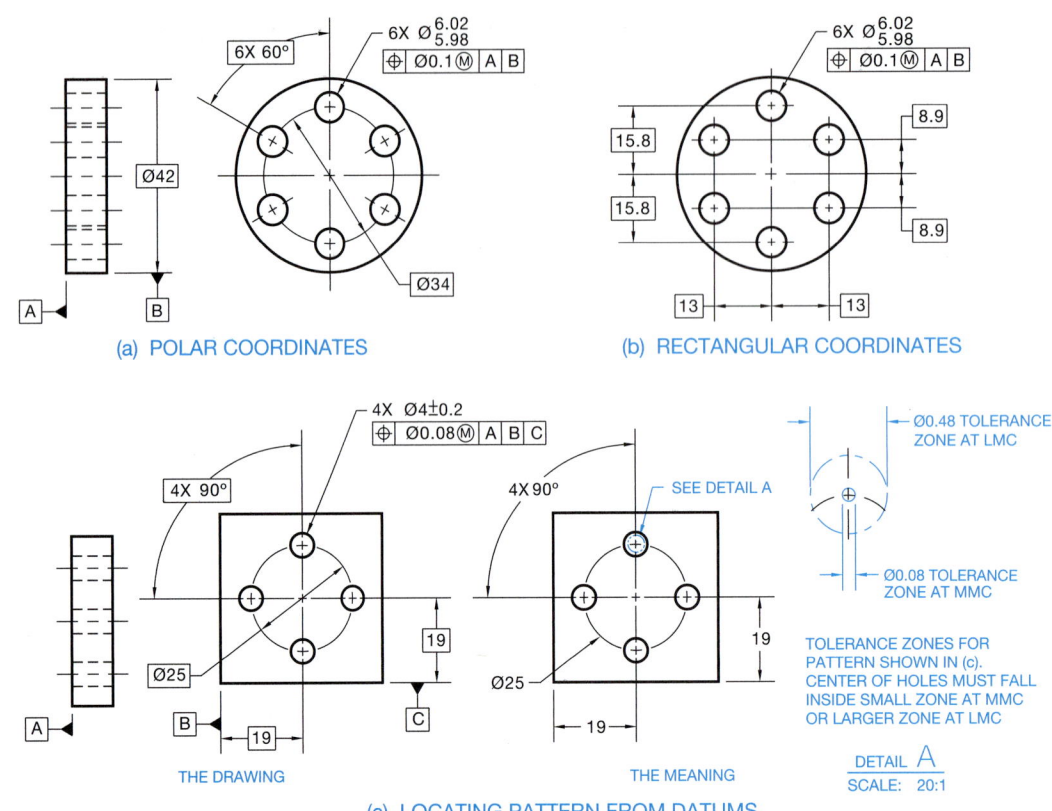

(a) POLAR COORDINATES

(b) RECTANGULAR COORDINATES

THE DRAWING

THE MEANING

TOLERANCE ZONES FOR PATTERN SHOWN IN (c). CENTER OF HOLES MUST FALL INSIDE SMALL ZONE AT MMC OR LARGER ZONE AT LMC

DETAIL A
SCALE: 20:1

(c) LOCATING PATTERN FROM DATUMS

FIGURE 14.81 ■ Position of multiple features.

the first compartment and only a single datum reference for orientation given for the feature relating control. The **two single-segment feature control frame** is similar, except there are two position symbols, each displayed in a separate compartment, and a two-datum reference in the lower feature control frame. (See Figure 14.83.) The top feature control frame is the pattern locating control and works as previously discussed. The lower feature control frame is the feature relating control in which two datums control the orientation and alignment with the pattern locating control. This type of position tolerance provides a tighter relationship of the holes within the pattern than the composite position tolerance.

POSITIONAL TOLERANCE OF TABS

Positional tolerancing of tabs can be accomplished by identifying related datums, dimensioning the relationship between the tabs, and providing the number of units followed by the size and feature control frame. (See Figure 14.84.)

POSITIONAL TOLERANCE OF SLOTTED HOLES

The positional tolerance of slotted holes may be accomplished by locating the slot centers with basic dimensions and providing a feature control frame to both the length and width of the slotted holes. (See Figure 14.85.) The word BOUNDARY is placed below each feature control frame. This means that each slotted feature is controlled by a theoretical boundary of identical shape that is located at true position. The size of each slot must remain within the size limits and no portion of the surface may enter the theoretical boundary, which is the size of the boundary calculated by MMC (each dimension, length and width) − Position Tolerance. Normally, there is a greater position tolerance for the length than the width. When the position tolerance is the same for the length and the width, the feature control frame is separated from the size dimensions and connected to one of the slots with a leader.

POSITIONAL TOLERANCE OF THREADED FEATURES AND GEARS

The orientation or positional tolerance of screw threads applies to the datum axis. The datum feature is established by a pitch diameter cylinder. If a different datum feature is required, the note should be lettered below the feature control frame, for example, MINOR DIA or MAJOR DIA. Similarly, the intended datum feature for gears or splines should be identified below the feature control frame. The options include MAJOR DIA, PITCH DIA, or MINOR DIA.

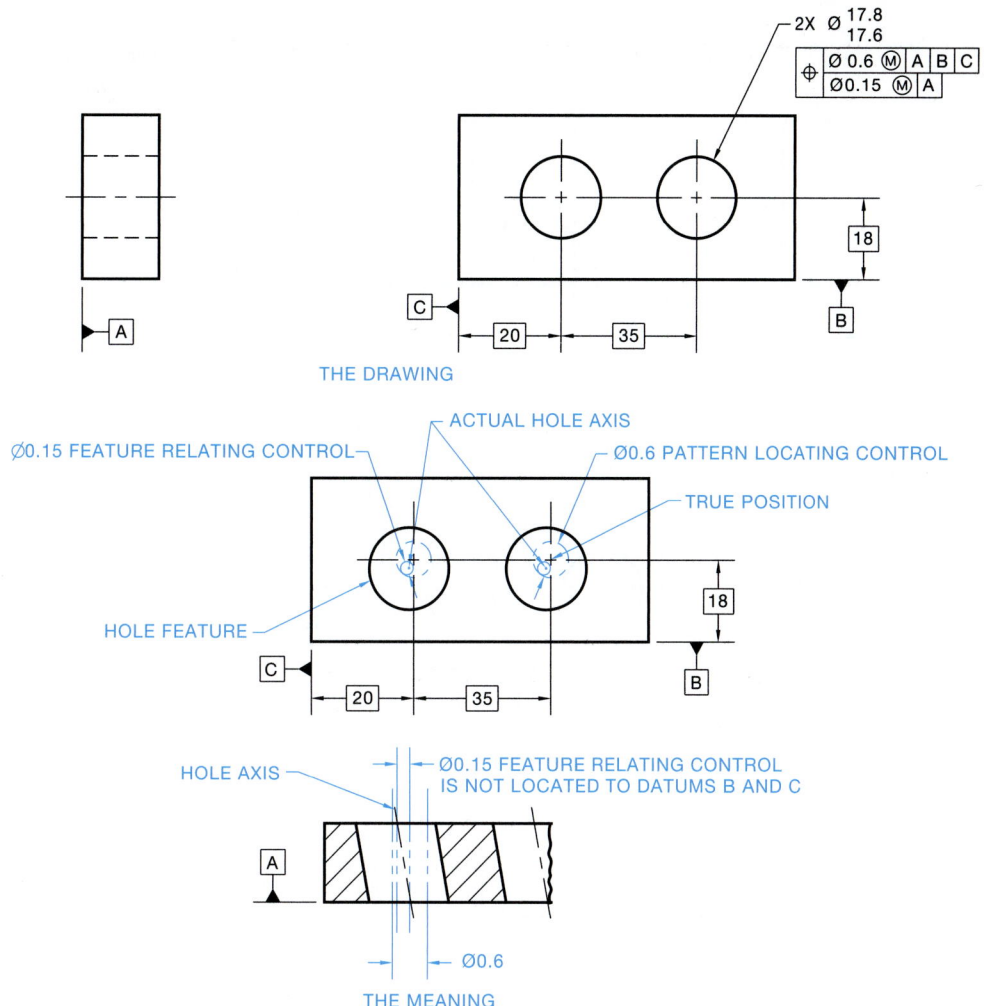

FIGURE 14.82 ■ Composite positional tolerance.

POSITIONAL TOLERANCE OF COUNTERBORED HOLES

Counterbored holes are a type of coaxial feature. Coaxial means that one or more features have the same axis. There are three different ways to provide positional tolerancing to counterbored holes, as shown in Figure 14.86.

POSITIONAL TOLERANCING OF COAXIAL HOLES

Coaxial holes that are aligned through different features may be controlled with positional tolerancing using a feature relating zone for each hole and a pattern locating zone for the holes as a group. Some possible options are shown in Figure 14.87, page 456.

POSITIONAL TOLERANCING OF NONPARALLEL HOLES

Positional tolerancing may be applied to holes that are not parallel to each other. The axis cannot be perpendicular to a surface, as shown in Figure 14.88, page 456. The feature is located with a base longitudinal and angular dimension.

PROJECTED TOLERANCE ZONE

The standard application of a positional tolerance implies that the cylindrical zone extends through the thickness of the part or feature. In some situations where there is the possibility of interference with mating parts, the tolerance zone could be extended or projected away from the primary datum controlling the axis of the related feature. The amount of projection is equal to the thickness

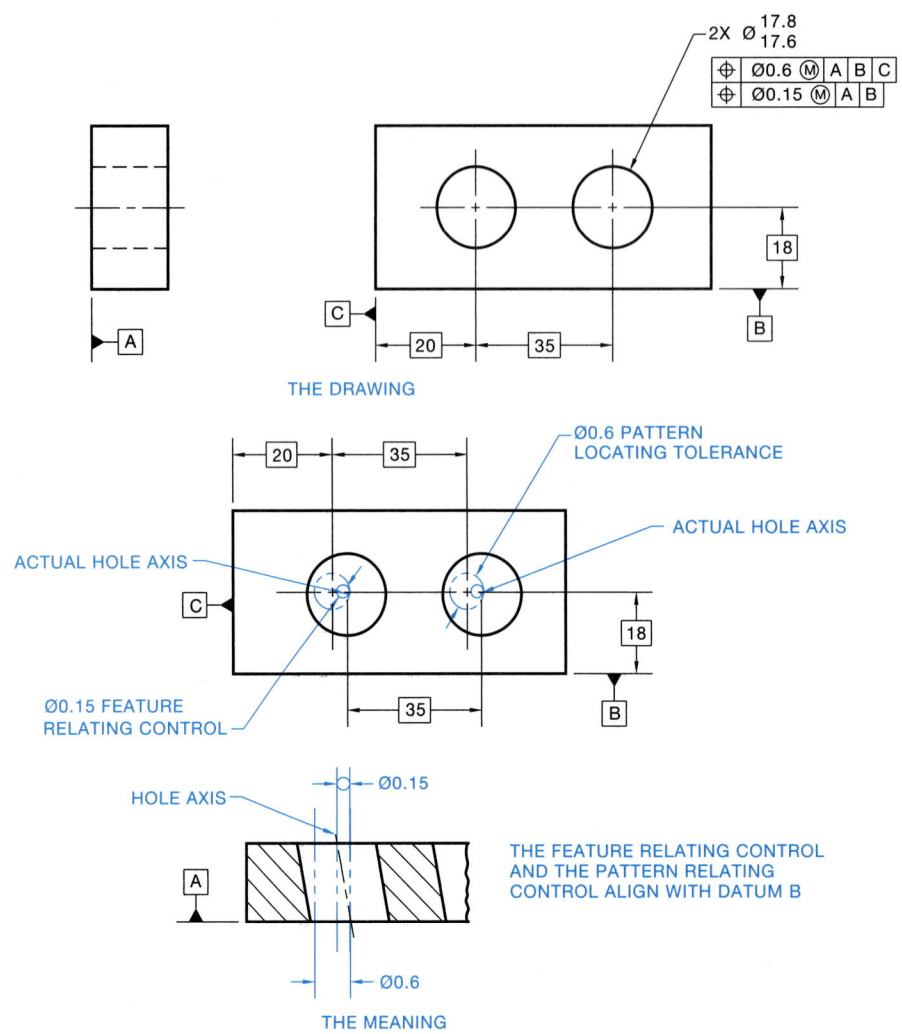

FIGURE 14.83 ■ Two single-segment feature control frame.

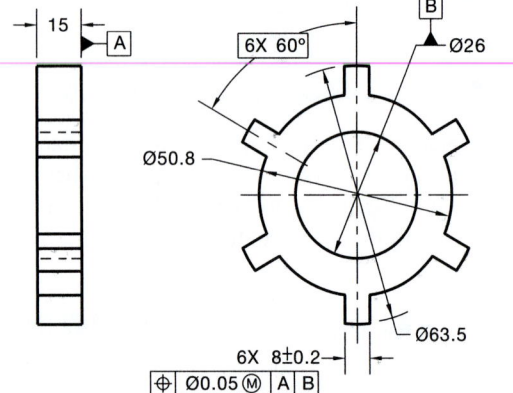

FIGURE 14.84 ■ Positional tolerance of tabs, similar practice for slots.

of the mating part. This application works because the axis of the threaded or press fit feature is limited by the exact location and angle of the thread or hole within which it is assembled. This type of application is especially useful when the mating features are

screws, pins, or studs. These types of conditions are referred to as **fixed fasteners** because the fastener is fixed in the mating part and there is no clearance allowance. The projected tolerance zone can be handled in one of two ways using the projected tolerance zone symbol. (See Figure 14.89.) When calculating the positional tolerance zones that are applied to the parts of a fixed fastener, the following formula may be used: MMC Hole − MMC Fastener (nominal thread size) ÷ 2 = Positional Tolerance Zone of Each Part. In some situations, it is desirable to provide more tolerance to one part than another, for example, 60% to the threaded part and 40% to the unthreaded part.

When parts are assembled with fasteners, such as bolts and nuts or rivets, and where all the parts have clearance holes to accommodate the fasteners, the application is referred to as a **floating fastener**. Floating fasteners require that the fastening device be secured on each side of the part, such as with a bolt and nut. With a fixed fastener, one of the parts is a fastening device. Greater tolerance flexibility with floating fasteners is due to the fastener clearance at each part. When calculating the positional tolerance zone for floating

(*Continued on page 458*)

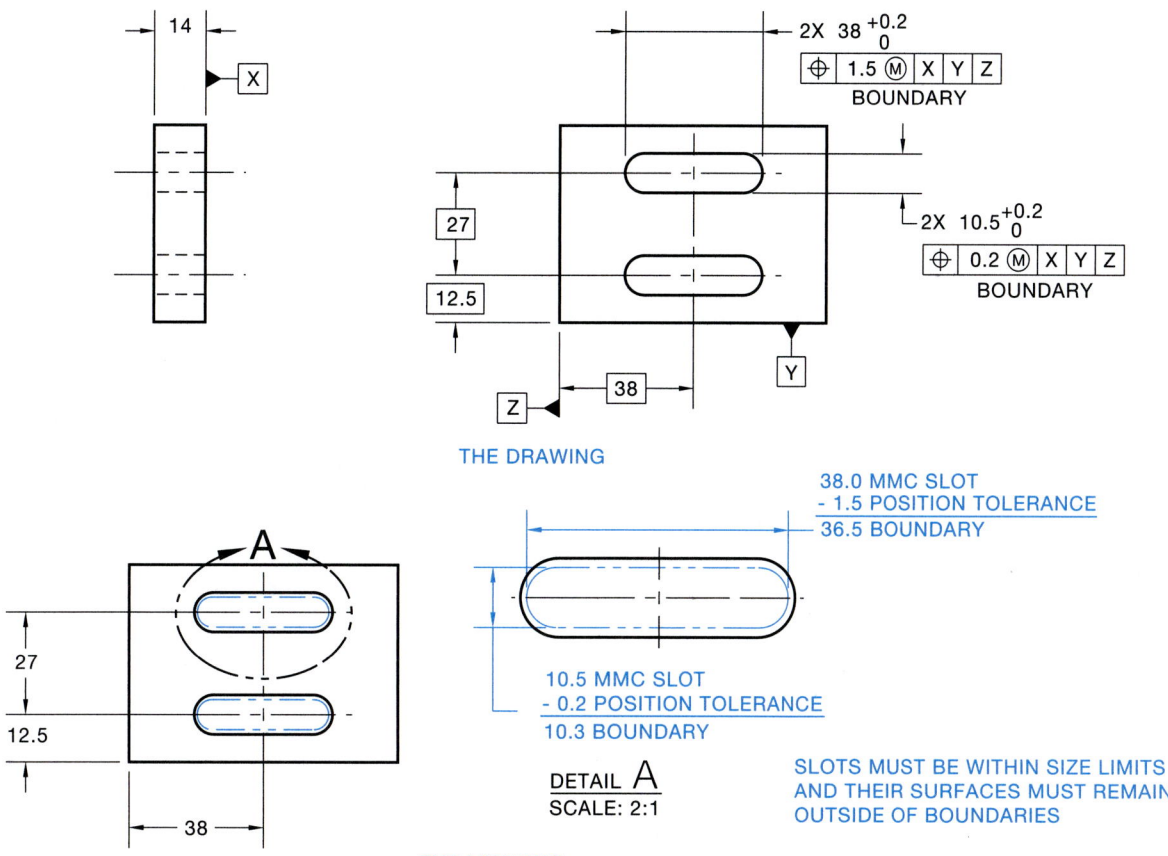

THE DRAWING

DETAIL A
SCALE: 2:1

38.0 MMC SLOT
- 1.5 POSITION TOLERANCE
36.5 BOUNDARY

10.5 MMC SLOT
- 0.2 POSITION TOLERANCE
10.3 BOUNDARY

SLOTS MUST BE WITHIN SIZE LIMITS
AND THEIR SURFACES MUST REMAIN
OUTSIDE OF BOUNDARIES

THE MEANING

FIGURE 14.85 ■ Positional tolerance for slotted holes.

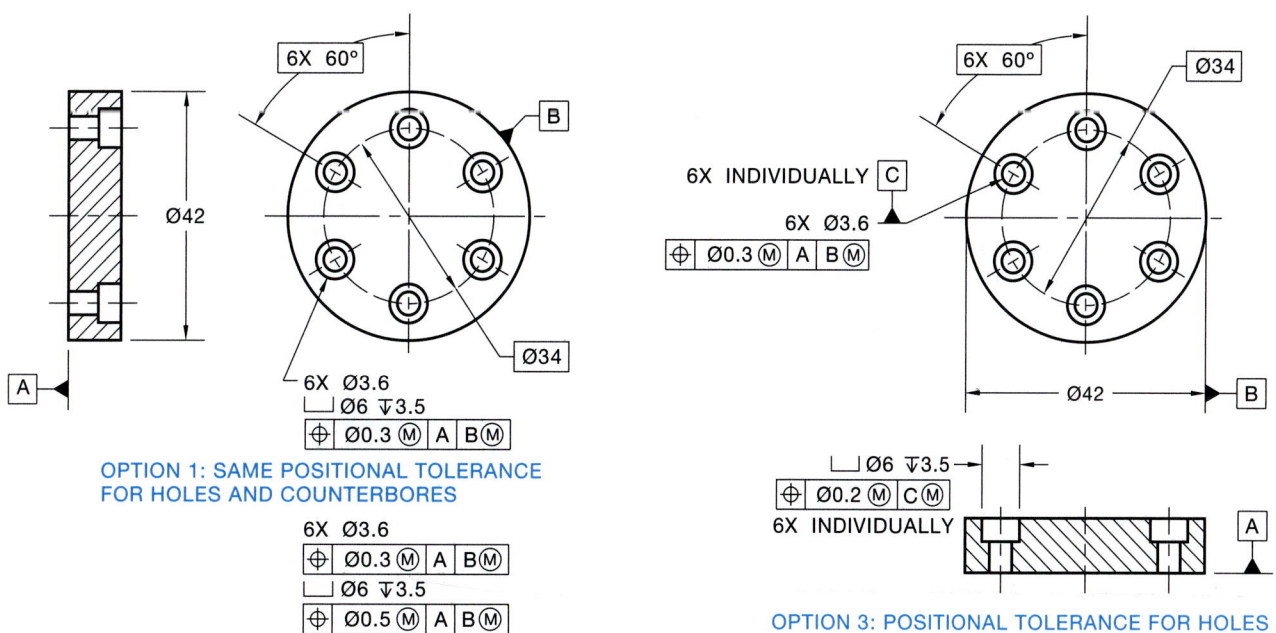

OPTION 1: SAME POSITIONAL TOLERANCE
FOR HOLES AND COUNTERBORES

OPTION 3: POSITIONAL TOLERANCE FOR HOLES

FIGURE 14.86 ■ Positional tolerance of counterbored holes.

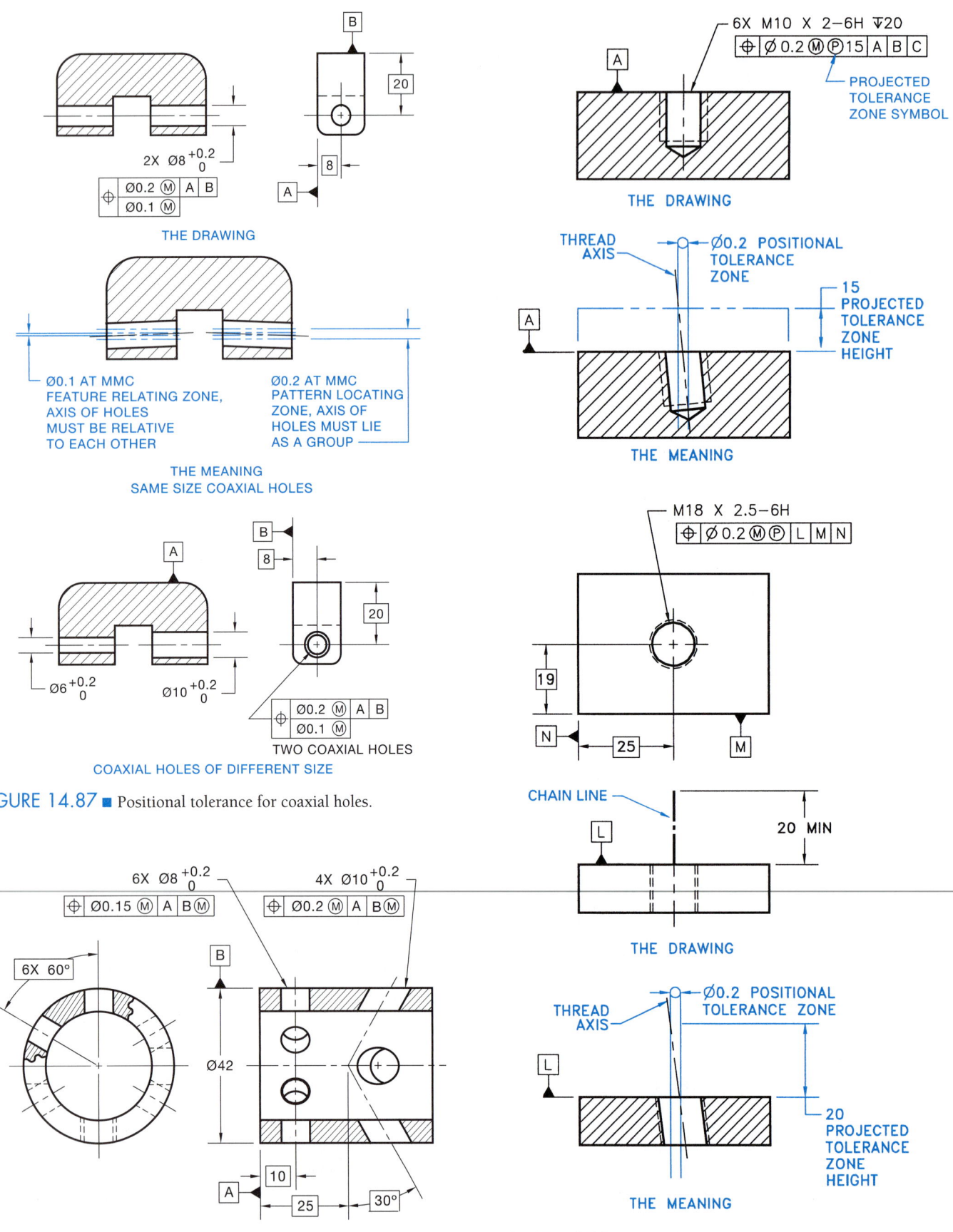

FIGURE 14.87 ■ Positional tolerance for coaxial holes.

FIGURE 14.88 ■ Positional tolerancing of nonparallel holes.

FIGURE 14.89 ■ Projected tolerance zone.

GEOMETRIC DIMENSIONING AND TOLERANCING WITH AUTOCAD

AutoCAD provides the ability to add GD & T symbols to your drawings. The feature control frame and related GD & T symbols may be created using the TOLERANCE and LEADER commands. These commands are used to access the Geometric Tolerance dialog box, which contains the feature control frame symbols. (See Figure 14.90.)

The Geometric Tolerance dialog box is divided into compartments that relate to the compartments found in the feature control frame. These compartments allow you to create the desired feature control frame by entering geometric characteristic symbols, diameter symbols, geometric tolerance values, material condition symbols, and datum references as needed. Pick the Sym box to open the Symbol dialog box shown in Figure 14.91. Pick the desired geometric characteristic symbol from the Symbol dialog box. The selected symbol is then displayed in the Geometric Tolerance dialog box. (See Figure 14.91.) If a diameter symbol is desired, pick the first box in the Tolerance compartment of the Geometric Tolerance dialog box. (See Figure 14.90.) Now, type the desired geometric tolerance in the Tolerance text box as shown in Figure 14.90. Pick the third box in the Tolerance area if a material condition symbol is desired. This displays the Material Condition dialog box shown in Figure 14.92. Select the desired symbol to have it displayed in the Geometric Tolerance dialog box. (See Figure 14.90.) If datum references are applied to the feature control frame,

type the desired datum reference letters in the Datum 1, Datum 2, and Datum 3 text boxes. Pick the box to the right of each datum reference text box if you want to open the Material Condition text box again to add desired material condition symbols with the datum reference.

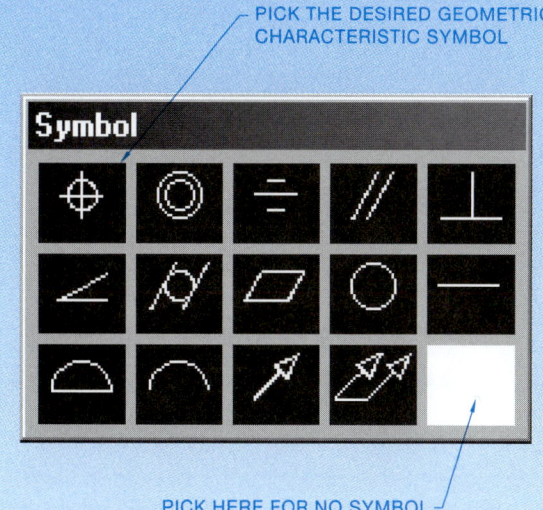

PICK THE DESIRED GEOMETRIC CHARACTERISTIC SYMBOL

PICK HERE FOR NO SYMBOL

FIGURE 14.91 ■ The AutoCAD Symbol dialog box. Pick the desired symbol or pick the blank box for no symbol.

PICK TO ACCESS THE SYMBOL DIALOG BOX

PICK FOR THE DIAMETER SYMBOL

TYPE GEOMETRIC TOLERANCE

PICK TO ACCESS THE MATERIAL CONDITION DIALOG BOX

TYPE DATUM REFERENCE

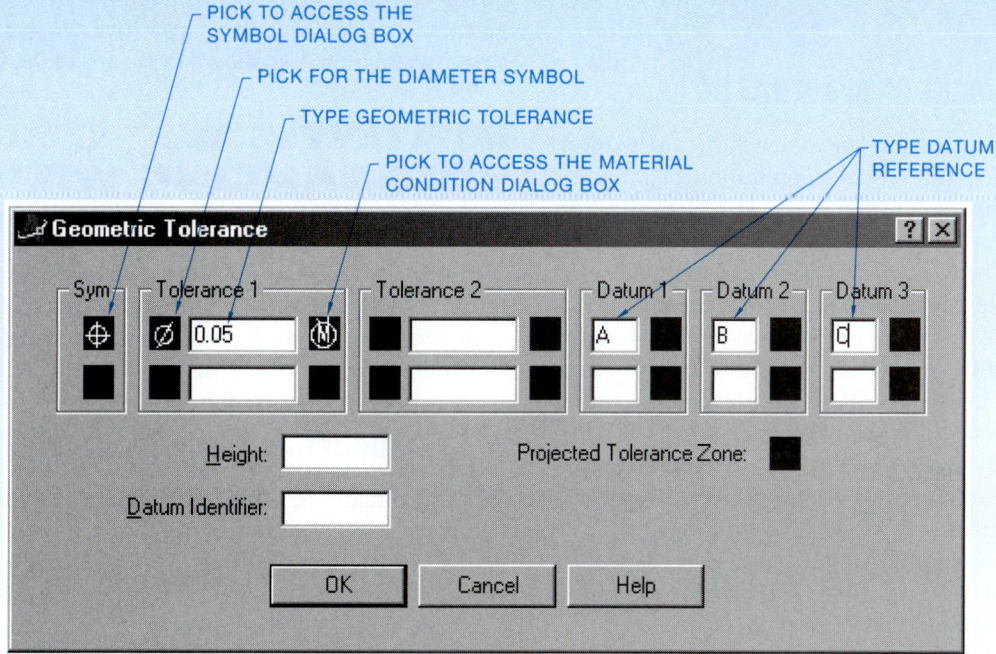

FIGURE 14.90 ■ The AutoCAD Geometric Tolerance dialog box.

(Continued)

Double feature control frames also can be created for applications such as unit straightness, unit flatness, composite profile tolerance, composite positional tolerance, or coaxial positional tolerance. The projected tolerance zone symbol also can be added to the feature control frame by picking the Projected Tolerance Zone box shown in Figure 14.90. Add the desired height of the projected tolerance zone by typing the value in the Height text box.

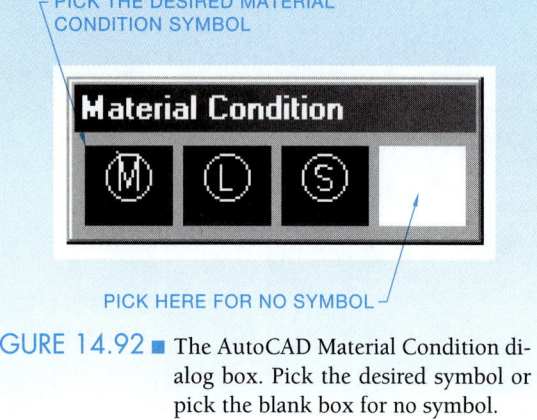

PICK THE DESIRED MATERIAL CONDITION SYMBOL

Material Condition

Ⓜ Ⓛ Ⓢ

PICK HERE FOR NO SYMBOL

FIGURE 14.92 ■ The AutoCAD Material Condition dialog box. Pick the desired symbol or pick the blank box for no symbol.

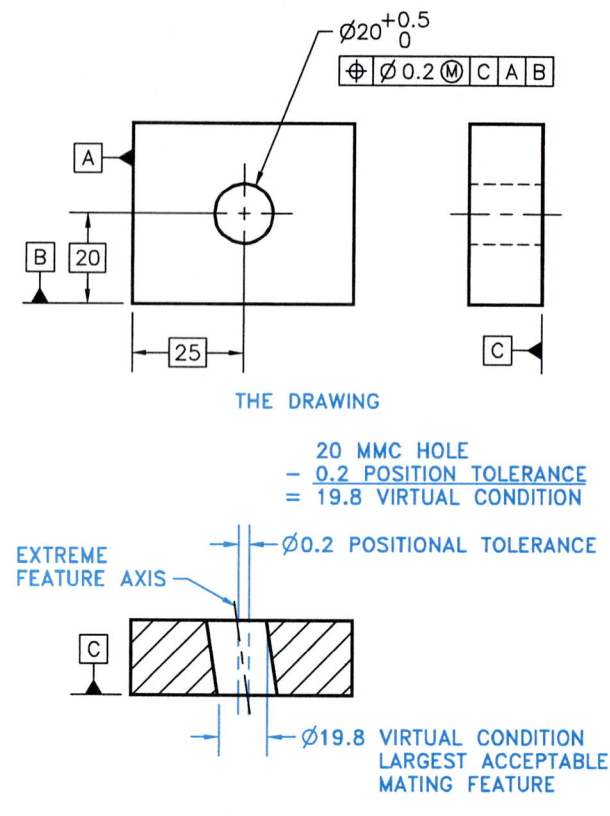

THE DRAWING

20 MMC HOLE
− 0.2 POSITION TOLERANCE
= 19.8 VIRTUAL CONDITION

Ø0.2 POSITIONAL TOLERANCE

EXTREME FEATURE AXIS

Ø19.8 VIRTUAL CONDITION LARGEST ACCEPTABLE MATING FEATURE

THE MEANING

FIGURE 14.93 ■ Virtual Condition of Hole = MMC − Positional Tolerance.

fastener parts, the following formula is used: MMC Hole − MMC Fastener (nominal thread size) = Positional Tolerance Zone of Each Part.

VIRTUAL CONDITION

The tolerances of a feature that relate to size, form, orientation, and location, including the possible application of MMC or RFS, are determined by the function of the part. Consideration must be given to the collective effect of these factors in determining the clearance between mating parts and in establishing gage feature sizes. The boundary created by the combined effects of size MMC, and the geometric tolerance is known as the virtual condition. Virtual condition is the sole condition where a feature size may be outside of MMC. Controlling the clearance between mating parts is critical to the design process. When features are dimensioned using a combination of size and geometric tolerances, the resulting effects of the specifications should be considered to ensure that parts will always fit together.

When a positional tolerance is applied to an internal feature, the Virtual Condition = MMC Hole − Positional Tolerance. This calculation determines the maximum feature size that should be allowed to fit within the hole. (See Figure 14.93.)

When perpendicularity is applied to an external diameter, such as a pin, the Virtual Condition = MMC Feature + Perpendicularity Geometric Tolerance. The virtual condition deter-

mines the smallest acceptable mating feature that fits over the given part while maintaining a positive connection between the surfaces at datum A as shown in Figure 14.94.

STATISTICAL TOLERANCING WITH GEOMETRIC CONTROLS

Methods of tolerancing for statistical process control (SPC) were introduced in Chapter 11. Statistical tolerances are also used with geometric controls by placing the statistical tolerancing symbol in the feature control frame as shown in Figure 14.95.

COMBINATION CONTROLS

In some situations, compatible geometric characteristics are combined in one feature control frame or separate frames associated with the same surface. This is normally done when the combined effect of two different geometric characteristics and tolerance zones is desired. The profile tolerance is used to illustrate the combination of geometric characteristics. Profile and parallelism can be combined to control the profile of a surface and the parallelism of each element to a datum. Profile and runout may be combined to control the line elements within the

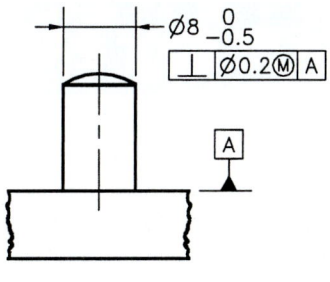

THE DRAWING

8 MMC FEATURE + 0.2 GEOMETRIC TOLERANCE
= 8.2 VIRTUAL CONDITION

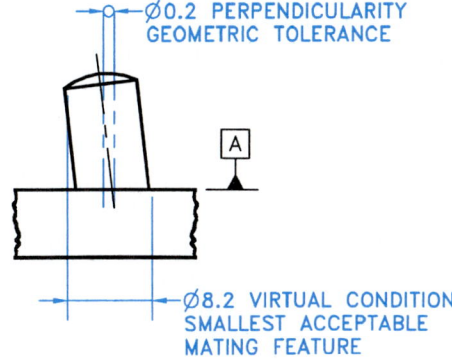

THE MEANING

FIGURE 14.94 ■ Virtual Condition Perpendicular Pin = MMC Feature + Geometric Tolerance.

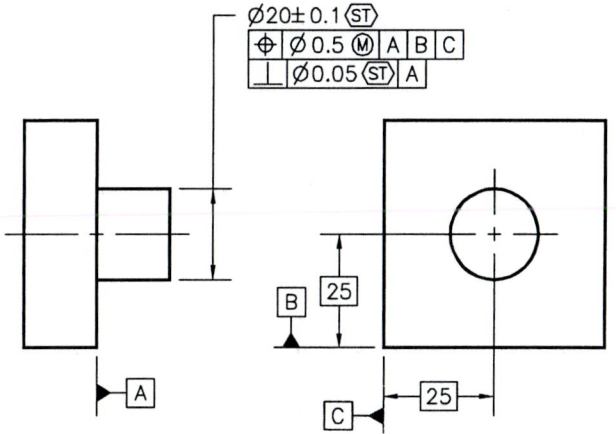

FIGURE 14.95 ■ Statistical tolerancing with geometric controls.

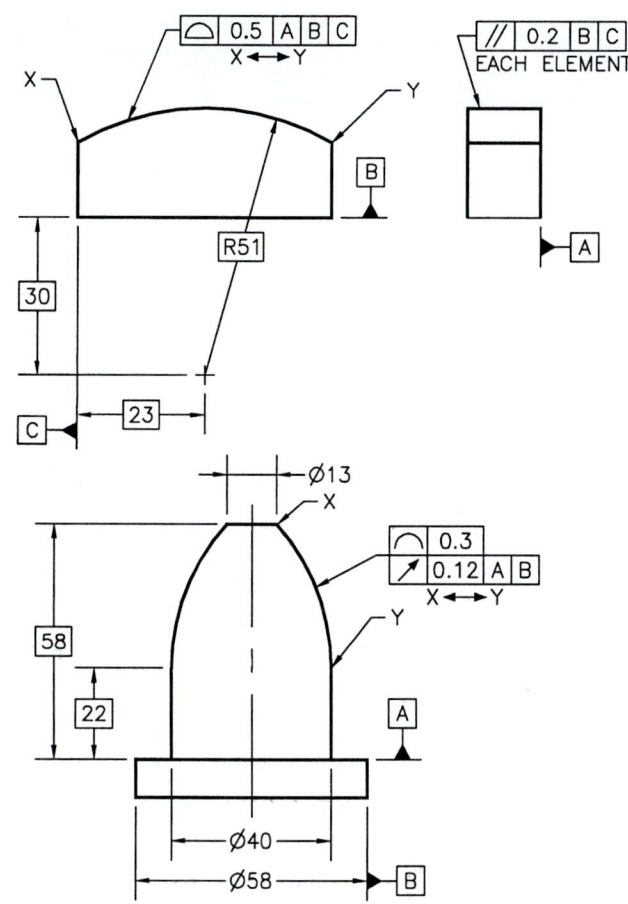

FIGURE 14.96 ■ Combination controls (profile and runout).

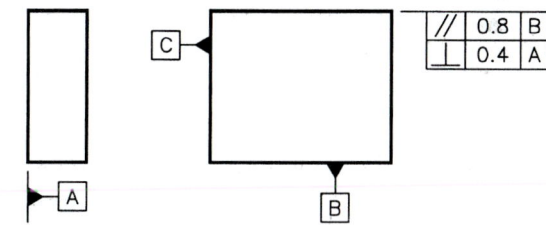

FIGURE 14.97 ■ Combination control (parallelism and perpendicularity).

profile specification and circular elements within the runout tolerance as shown in Figure 14.96. The combined control of parallelism and perpendicularity can be represented on the same feature as shown in Figure 14.97. Other combined geometric characteristics may be used when the callouts are compatible and the design function of the part requires such specific controls.

WEBSITE RESEARCH

The following websites can provide you additional information for research or further study into topics covered in this chapter:

http://www.asme.org—Find information and publications related to the American Society of Mechanical Engineers, including ASME Y14.5M, *Dimensioning and Tolerancing*.

http://www.ansi.org—The American National Standards Institute. Information about national and international drafting standards, including ASME Y14.5M, *Dimensioning and Tolerancing*.

(Continued on page 462)

PROFESSIONAL PERSPECTIVE

*Written By Michael A. Courtier, Supervisor/Instructor
CAD/CAM/CAE Employee Training and Development,
Boeing Aerospace*

In working with engineers and drafters for years on the subject of geometric tolerancing, I find that questions keep popping up in several key areas—datums; limits of size; geometric tolerances as a refinement of size; the tolerance zones (2-D or 3-D, cylindrical or total wide, etc.); and the apparent overlap of controls (circularity, cylindricity, circular runout, total runout, position). It is in these "weak" spots that you should dig a little deeper and fill in your knowledge.

New and old users alike tend to miss the key rules of thumb in selecting datums, and most textbooks on the subject do not address this area.

■ Make sure you select datums that are real, physical surfaces of the part. I have seen many poor drawings with datums on centerlines, at the centers of spheres, or even at the center of gravity.

■ Be sure to select large enough datum surfaces. It is hard to chuck up on a cylinder 1/10" long, or get three noncolinear points of contact on a plane of similar dimensions.

■ Choose datums that are accessible. If manufacturing or quality assurance has to build a special jig to reach the datum, extra costs are incurred for the ultimate part.

■ Given the above constraints, select functional surfaces and corresponding features of mating parts where possible.

Remember, datums are the *basis* for subsequent dimensions and tolerances, so they should be selected carefully. You might also be well advised to assign a little form, orientation, or runout control to the datum surface to ensure an accurate basis.

In the area of limits of size, there are three possible interpretations of a size dimension. Every class I have taught has been divided on what the proper extent of a size dimension's control should be—the overall size (or geometric form), the size at various cross-sectional checks, or both. The correct answer, per definition in the standard, is *BOTH!*

The words refinement of size are a source of general confusion to new users of geometric tolerancing. Let us say a part is dimensioned on the drawing as .500 ± .005" thick, with parallelism control to .003" on the upper surface. If a production part has a convex upper surface with low points at .504, then the high point can only be .505, *not* .507. In other words, only .001" of the .003" of parallelism control can be used since parallelism is a *refinement of size within the overall size limits.*

Another one of the often misunderstood aspects of geometric tolerancing is the tolerance zone. Many engineers, technicians, drafters, machinists, and inspectors misinterpret the *shape* of the tolerance zone. I recommend that all new users make a chart showing the tolerance zones for each type of tolerance and a sampling of its various applications. This chart should include—at a minimum—the 2-D zones between two lines or concentric circles (such as straightness, circularity) versus the 3-D zones between parallel planes (such as flatness, perpendicularity); the cylindrical versus the total wide zones (for example, straightness to an axis versus surface straightness, or position of a hole versus position of a slot); the cross-sectional zones versus the total surface zones (such as circularity versus cylindricity); and many, many others. If you can properly visualize the tolerance zone, you are well on your way to the proper interpretation of the geometric tolerance.

One more thing to watch out for is the *apparent overlap of controls.* While it appears that there is overlap, there are subtle yet distinct differences between all the geometric tolerance controls. Consider cylindrical shapes, for instance. You can use a wide variety of controls for them, including straightness, circularity, cylindricity, concentricity, circular runout, total runout, and position, to name a few. Remember that the first three are form controls, so no datum relationship is controlled. Straightness can control the axis or the surface merely by placing the feature control frame in a different location on the drawing. Straightness is a longitudinal control, whereas circularity is a circular element control (both are 2-D); cylindricity is a 3-D control of the surface relative to itself (no datum). The last four controls are with respect to a datum (or datums), but they have their differences also. Concentricity and the runouts are always RFS, whereas positional tolerancing can be LMC or MMC and get bonus tolerances. Runout is a rotational consideration, while positional tolerancing does not imply rotation.

In summary, ASME Y14.5M is a very powerful tool for engineering drawings. Digging deep into the subtleties will separate the amateur from the professional, who knows it will save time and money in every discipline from design to manufacturing to quality assurance testing. Used properly, the engineering design team can convey more information about the overall design to downstream areas or subcontracting concerns that may have no knowledge of the final assembly. The proper assessment of the symbology requires no interpreter for the reader who is well versed in this international sign language called *geometric tolerancing.* Figure 14.98, is an example of an actual industry drawing. In general, the part is fairly basic, but the geometric tolerancing is very specific and leaves no doubt about the interpretation.

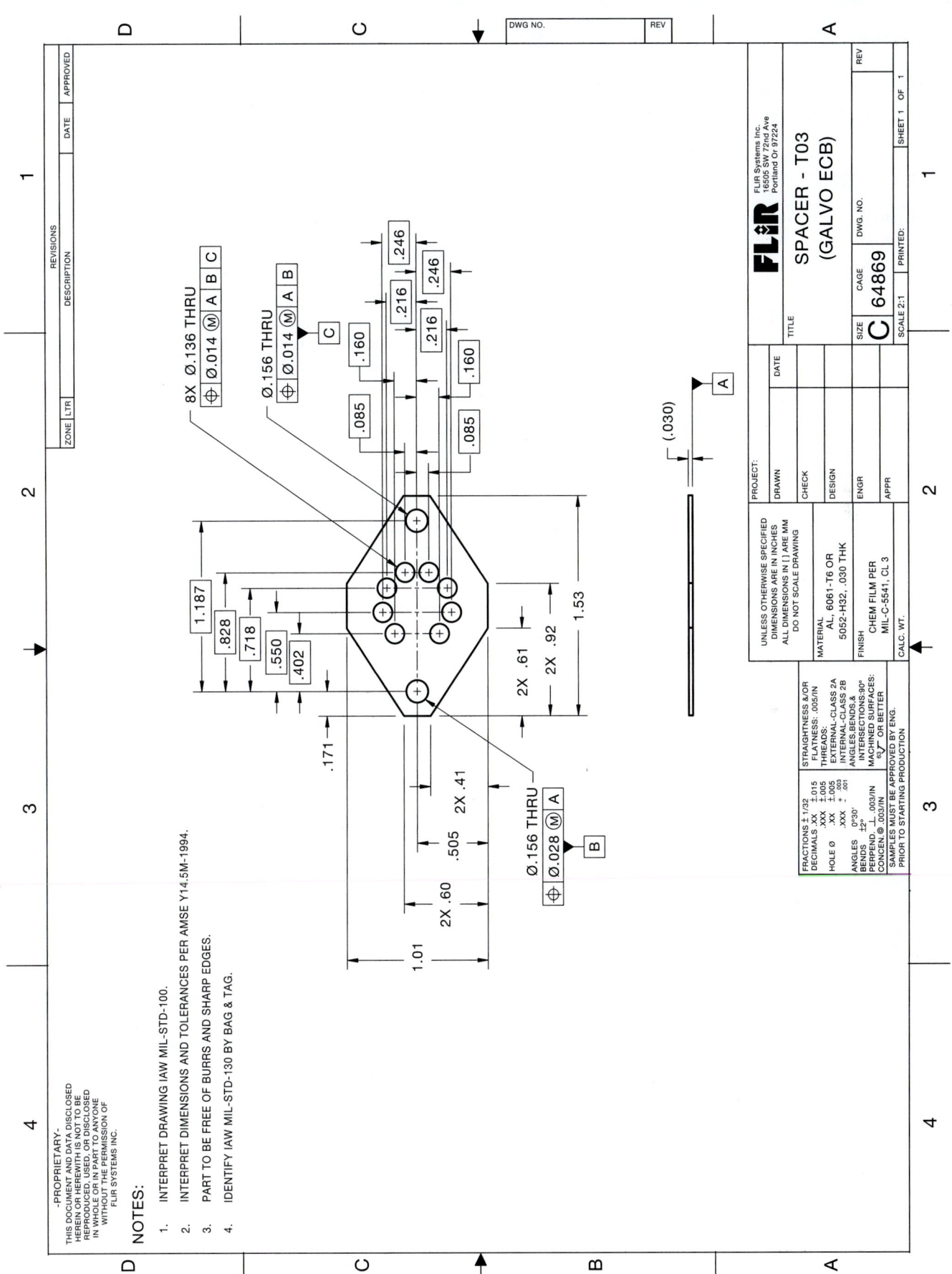

FIGURE 14.98 ■ An actual industry drawing of a part with geometric tolerancing applied. *Courtesy Flir Systems, Inc.*

MATH APPLICATIONS

ROUNDING NUMBERS

Calculators and computers usually indicate more precision in their answers than is warranted by the original data. In addition to the information on rounding in the CD Chapter 3, Engineering Drawing and Design Math Applications, here is the mathematical convention for rounding calculations:

When Adding or Subtracting:
1. Round each figure to the coarsest (least "precise") piece of data.
2. Do the arithmetic.

For example, to add this group 4.39 + 7.9 + 6.42, first round each number to the nearest tenth (because of the 7.9) giving 4.4 + 7.9 + 6.4. Then, add for the answer of **18.7.**

When Multiplying or Dividing:
1. Do the arithmetic.
2. Round the answer to the least number of significant digits in the original data.

Digits are significant when they give information other than holding a decimal point in place; 13.09 has four significant digits, but 13,000,000 has only two.

For example, to calculate $13.78 \times 6.1 \times 4.385$, first multiply, giving 368.59433. Because the 6.1 has only two significant digits, round the answer to 370, which has only two significant figures (the zero of 370 being merely a place-keeper).

When working with complicated formulas or when running computer programs, do not round off until the final answer is obtained. Constants in formulas are not considered data. When using a geometrical formula with π in it, it is best to push the calculator's π button instead of manually entering 3.14.

http://www.adda.org—American Design Drafting Association, *Drafting Reference Guide.*

http://www.industrialpress.com—Information about the *Machinery's Handbook.* This is a valuable resource for manufacturing standards, sizes, tolerances, fits, materials, and anything else you can think of for design and drafting.

http://www.industrialpress.com—Online trigonometry tables.

http://www.industrialpress.com—Prime numbers and factoring information.

http://www.g-w.com—Search Products by Title, Textbook, Geometric Dimensioning and Tolerancing. *Geometric Dimensioning and Tolerancing Basic Fundamentals,* by David A. Madsen. Comprehensive basic text workbook covering geometric dimensioning and tolerancing. Complete with tests, print reading exercises, and drafting problems.

http://amazon.com—Search Books, Geometric Dimensioning and Tolerancing, find *Geometric Dimensioning and Tolerancing* by David A. Madsen.

CHAPTER 14
Geometric Dimensioning and Tolerancing Test

 Access the CD found with this textbook to view the Chapter 14 Test. Confirm the preferred submittal method with your instructor.

CHAPTER 14

Geometric Dimensioning and Tolerancing Problems

DIRECTIONS

1. From the selected problems, determine which views and dimensions should be used to completely detail the part. Use simplified representation unless otherwise specified in the instructions or by your instructor.
2. Make a multiview sketch, to proper proportions, including dimensions and notes.
3. Using the sketch as a guide, draw an original multiview drawing on an adequately sized drawing sheet. Add all necessary dimensions and notes using unidirectional dimensioning. Use manual or computer-aided drafting as required by your course guidelines.
4. Include the following general notes at the lower-left corner of the sheet, .5 in. each way from the corner border lines:

1. INTERPRET DIMENSIONS AND TOLERANCES PER ASME Y14.5M—1994.
2. REMOVE ALL BURRS AND SHARP EDGES.

NOTES

Additional general notes may be required depending on the specifications of each individual assignment. The following should be part of your title block unless otherwise specified by your instructor.

UNSPECIFIED TOLERANCES

DECIMALS	mm	IN.
X	±2.5	±.1
XX	±0.25	±.01
XXX	±0.125	±.005
ANGULAR ±30'		
FINISH	3.2μm	125μIN.

PROBLEM 14.1   Geometric tolerancing (metric)

Part Name: Flow Pin

Material: Bronze

Finish: Finish All Over 0.20μm.

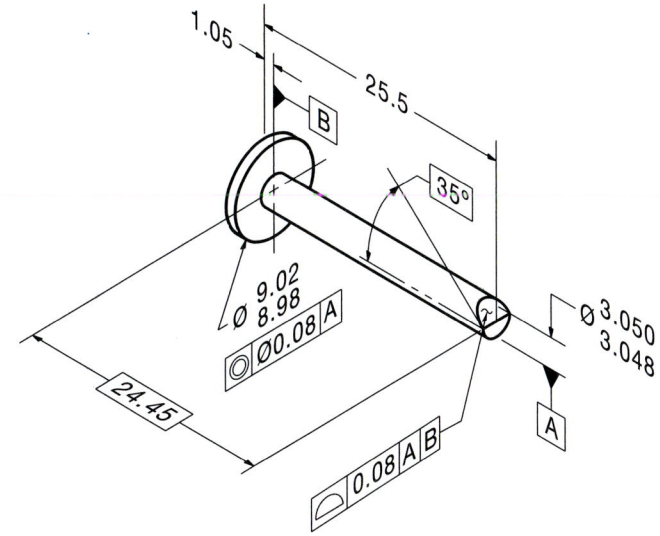

PROBLEM 14.2   Geometric tolerancing (metric)

Part Name: LN2 Test Pump Lock Nut

Material: AMS 5732.

Additional General Notes:

1. ⚠3: Mark per AS478 Class D with 1193125 and applicable dash number.
2. Finish All Over 1.6μm.

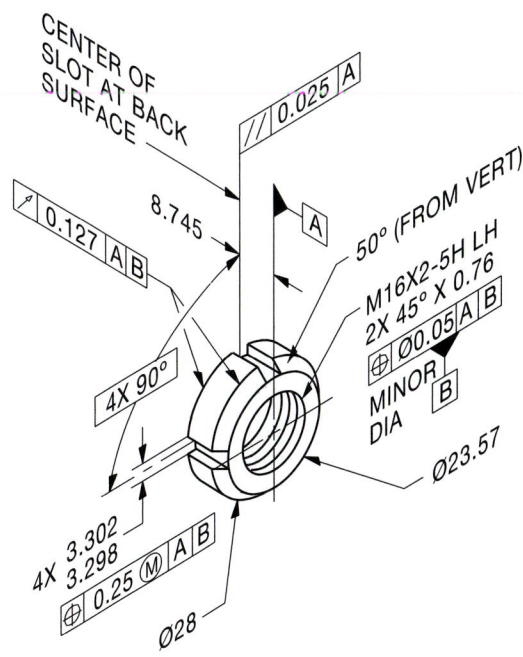

PROBLEM 14.3 **(in.)**

Part Name: Half Coupling

Material: ∅1.250 6061-T6 Aluminum

SPECIFIC INSTRUCTIONS:

Provide MMC material condition after position tolerance except for RFS at threads.

Problem based on original art courtesy TEMCO.

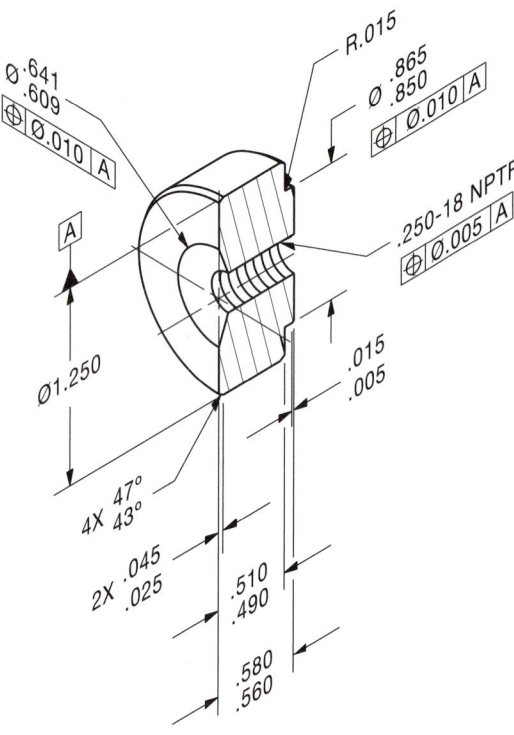

PROBLEM 14.5 **(in.)**

Part Name: Half Coupling

Material: ∅1.625 6061-T6511

Problem based on original art courtesy TEMCO.

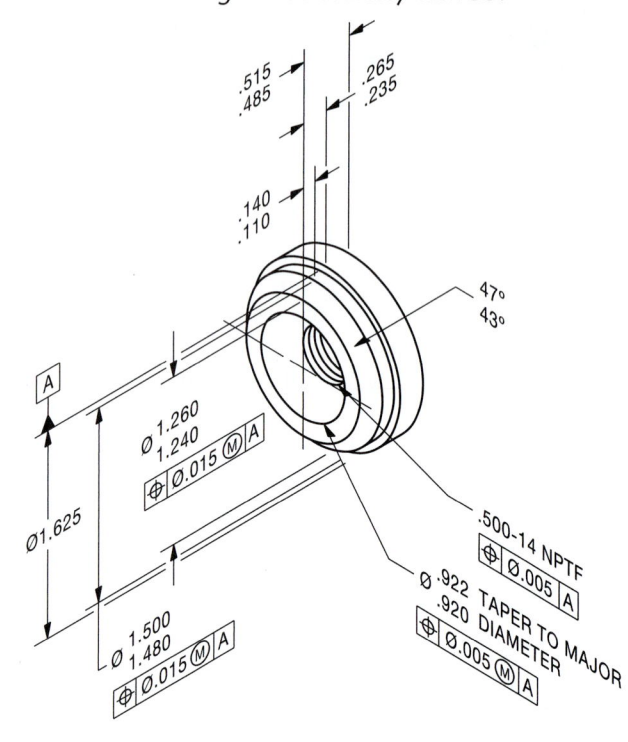

PROBLEM 14.4 **(in.)**

Part Name: Coupling

Material: AISI 1010, Killed

SPECIFIC INSTRUCTIONS:

Provide MMC material condition after position tolerance except for RFS at threads.

Problem based on original art courtesy TEMCO.

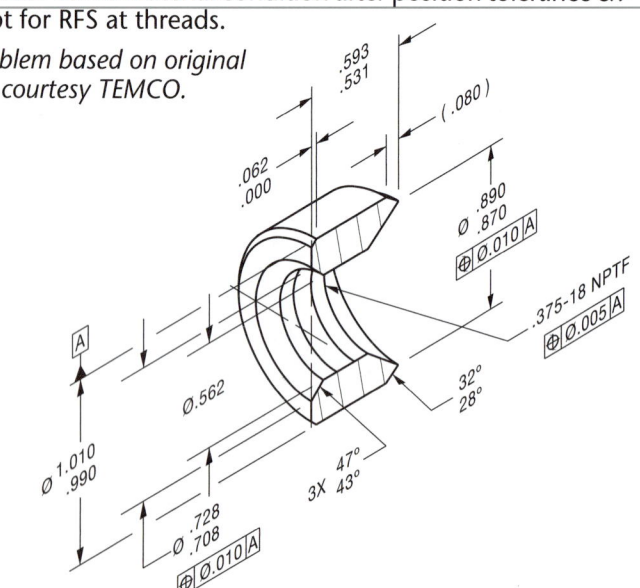

PROBLEM 14.6 **(metric)**

Part Name: Spline Plate

Material: SAE 3135

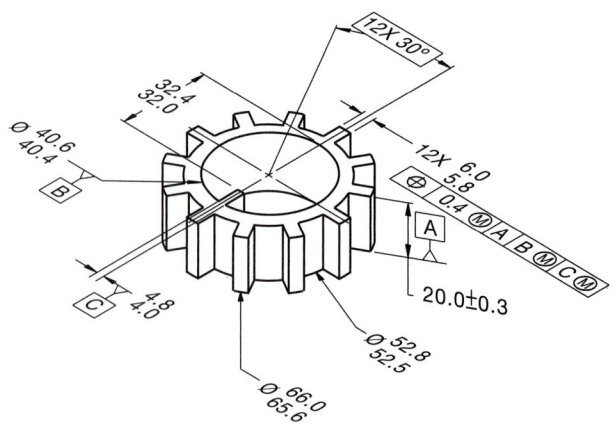

PROBLEM 14.7 **(in.)**

Part Name: Nut

Material: No. 10 Bronze

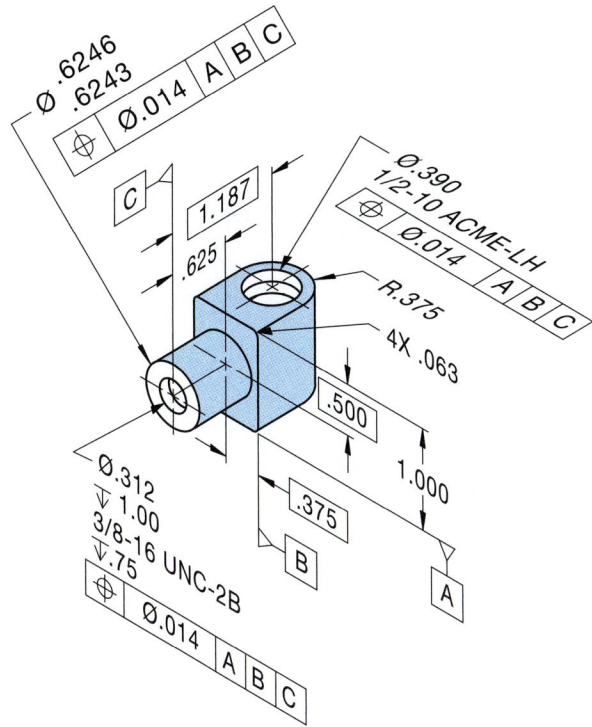

PROBLEM 14.9 **(in.)**

Part Name: Thrust Washer

Material: SAE 5150

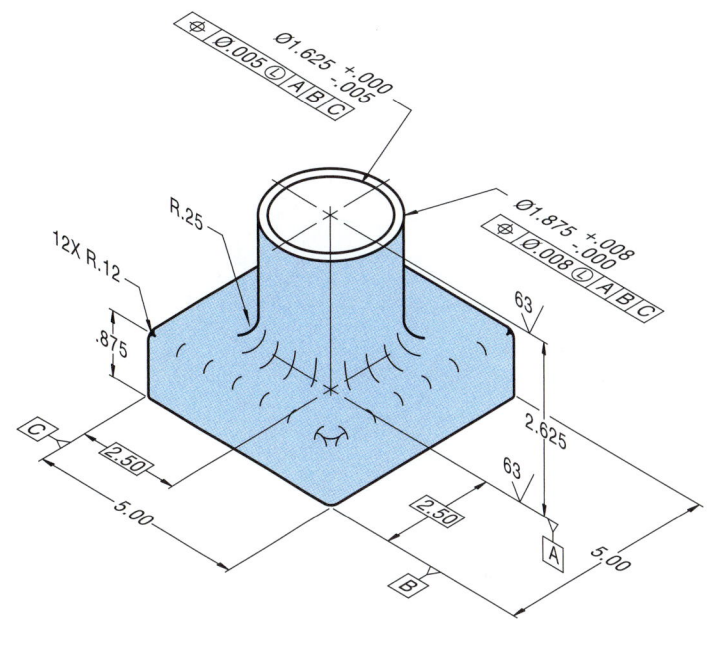

PROBLEM 14.8 **(metric)**

Part Name: Coupling Bracket

Material: SAE 4310 Steel

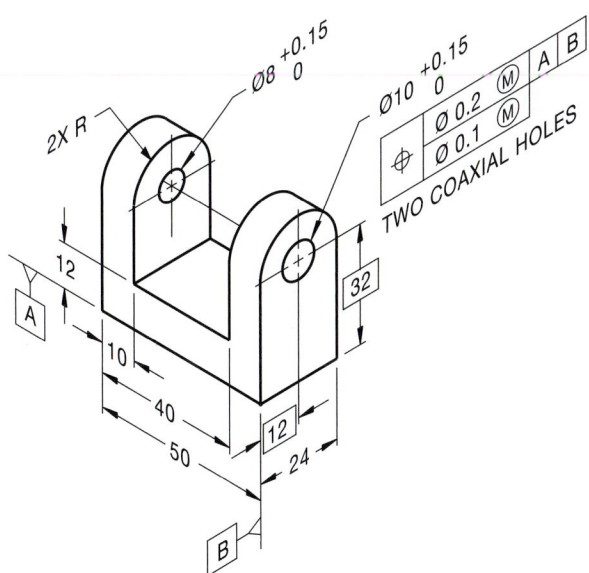

PROBLEM 14.10 **(metric)**

Part Name: Spacer

Material: SAE 4310

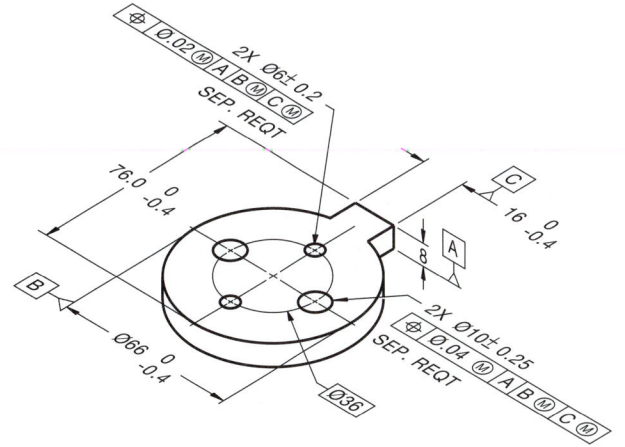

PROBLEM 14.11   (metric)
Part Name: Bearing Support
Material: SAE 1040

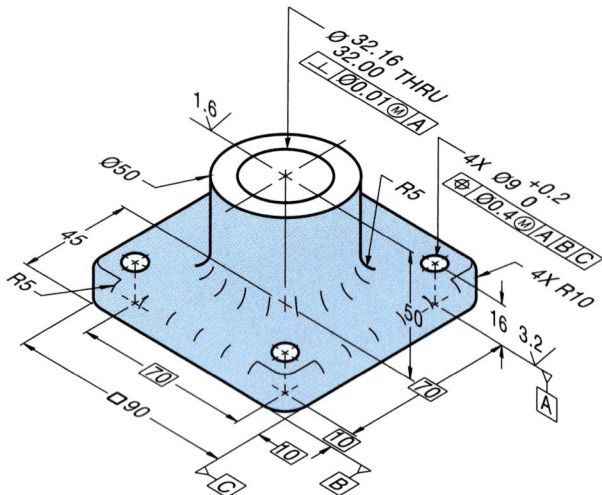

PROBLEM 14.13   (in.)
Part Name: Cover Plate
Material: Phosphor Bronze

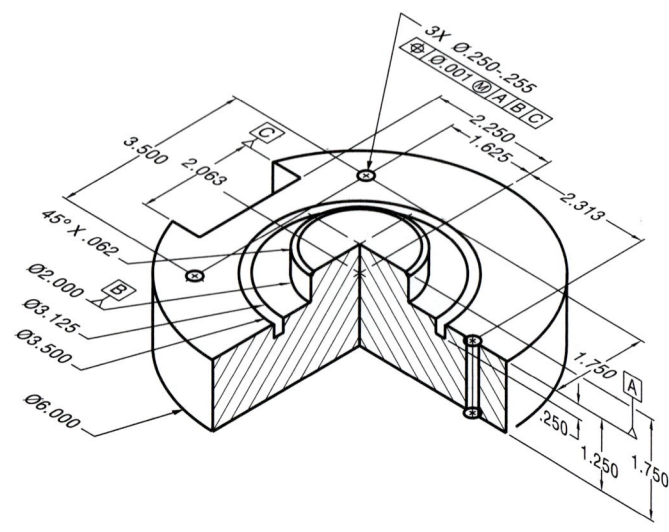

PROBLEM 14.12   (metric)
Part Name: Lock Nut
Material: SAE 3130

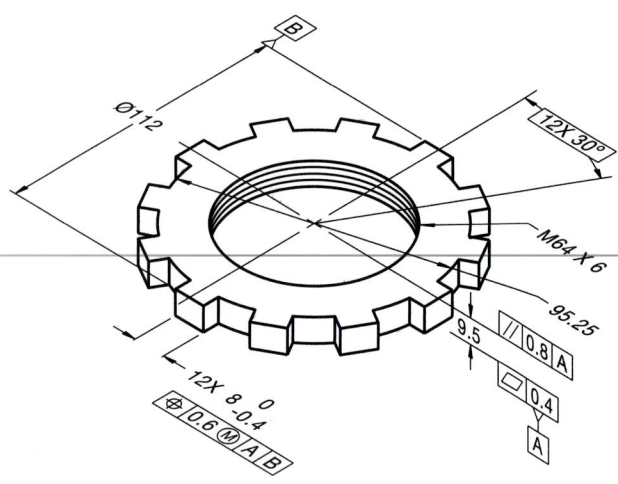

PROBLEM 14.14   (in.)
Part Name: Angle Support Mounting
Material: SAE 3110

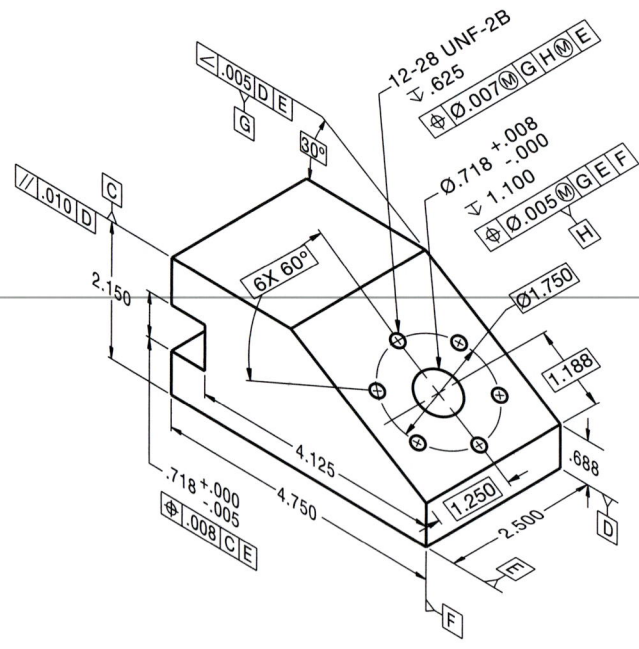

PROBLEM 14.15 **(metric)**

Part Name: Hub

Material: SAE 3310

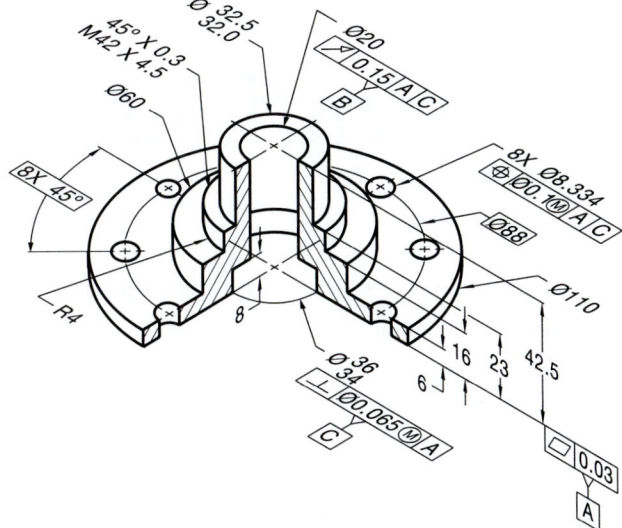

PROBLEMS 14.16 through 14.23: Access the CD found with this textbook and open the problem of your choice, or as assigned by your instructor. Solve the problems using the instructions provided with this chapter or on the CD, unless otherwise specified by your instructor.

DESIGN PROBLEM 14.24: Access the CD found with this textbook and open this design problem, unless otherwise specified by your instructor. Solve the problem using the instructions provided on the CD, unless otherwise specified by your instructor.

MATH PROBLEMS

PROBLEMS 14.25 through 14.34: Access the CD found with this textbook and open the math problem of your choice, or as assigned by your instructor. Solve the problem or problems using the instructions provided.

Pictorial Drawings and Technical Illustrations

LEARNING OBJECTIVES

After completing this chapter, you will:

- Draw three-dimensional objects using 3-D coordinates.
- Construct objects using isometric, diametric, or trimetric methods.
- Construct objects using oblique drawing methods.
- Draw objects using one-, two-, or three-point perspective.
- Apply a variety of shading techniques to pictorial drawings.
- Given an orthographic engineering sketch of a part or assembly, draw it in pictorial form using proper line contrasts and shading techniques.

THE ENGINEERING DESIGN APPLICATION

The design and manufacturing department has proposed a new part using 2-D orthographic sketches. The sketches are extremely crude, and it is your task to construct an isometric drawing that can be used for visualization purposes and to construct a prototype. Your first task is to create a 3-D drawing from the sketch provided. Check with the designers to verify dimensions and sizes. The engineering sketch shown in Figure 15.1a is used to draw an isometric view of the object.

After you confirm the accuracy of the 2-D sketch, begin construction of an isometric view. Use the following steps to construct the part.

1. Choose the view of the object that best shows most of the features of the part. Orient this view facing to the left or right.
2. Use the centerline layout method to locate the axis lines of the circular features.
3. Lay out additional thicknesses and features using the coordinate, or box-in, method.
4. Use proper ellipses to draw circles and arcs.
5. Use different widths for line contrast.
6. Apply shading as required.

The completed object is shown in Figure 15.1b.

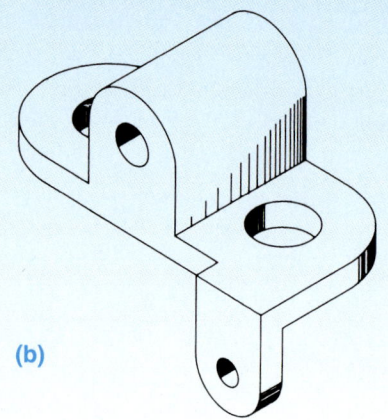

(a)   (b)

FIGURE 15.1 ■ Using an engineering sketch to create an isometric drawing. Dimension values in this figure are in inches.

ASME/ANSI This chapter is developed in accordance with the ASME/ANSI standard for pictorial drawings, ASME Y14.4M, *Pictorial Drawing*. This standard establishes pictorial drawing-related definitions and illustrates the different kinds of three-dimensional (3-D) representation practices commonly used on engineering drawings.

INTRODUCTION TO PICTORIAL DRAWING AND TECHNICAL ILLUSTRATION

The term pictorial is related to pictures, and pictures are representations by painting, drawing, or photograph. The intent of pictorial drawings is for the image to look realistic. Pictorial drawing is an ancient form of graphic communication and, with the 3-D CADD applications discussed in Chapter 4, is one of the most modern forms of graphic communication and is becoming commonly used. Pictorial drawings are often used to accompany two-dimensional (2-D) orthographic multiview drawings to provide a realistic 3-D view to help improve visualization. Figure 15.2 shows an example of a 2-D orthographic drawing with a 3-D pictorial view in the upper-right corner for visualization purposes. Pictorial drawings are useful in mechanical design, manufacturing production, architecture, construction, assembly instructions, service and repair manuals, and sales brochures. Pictorial drawings are used to clarify basic and complicated engineering designs when it can be difficult to interpret 2-D multiview drawings. Pictorial drawings also help designers and engineers work out spatial problems such as clearances and interferences. Pictorial drawing is used, as shown in Chapter 18, to present an exploded assembly drawing with a correlated parts list. There are a variety of pictorial drawing styles, which are used for different purposes and explained throughout this chapter. Some of the pictorial drawing styles are commonly used while others are used occasionally. Two-dimensional drawings can easily be generated from a 3-D model. As designers and engineers continue to move to 3-D modeling software, most 2-D production drawings will be created from the 3-D model.

Many of the techniques provided in this chapter are applied commonly with manual drafting skills, although they can be used with 3-D CADD programs. A review of Chapter 4, 3-D CADD, Animation, and Virtual Reality will give you a full insight into the power of 3-D modeling and its applications.

Solid Models in Pictorial Drawing

Solid models are realistic 3-D pictorial models and allow you to analyze exterior and interior object characteristics. Designers and engineers use solid models to perform interference and collision checks, mass calculations, simulations, and generate machining code for part manufacturing. Parametric solid modeling CADD programs involve the idea of developing solid models that contain parameters, which are controls, limits, and checks allowing you to easily and effectively make changes and updates. This means that when you describe the size, shape, and location of model geometry using specific parameters, you can easily modify those specifications to explore alternative design options. The parametric concept builds intelligence into the model, because parametric modeling occurs as a result of the software programs' ability to store and manage model information. This information includes knowledge of every model characteristic, such as calculations, sketches, features, dimensions, geometric parameters, when each piece of the model was created, and all other model history and properties. Figure 15.3 shows a 3-D solid model of an electric guitar assembly.

PICTORIAL DRAWINGS

Most products are made from multiview drawings that allow you to view an object with the line of sight perpendicular to the surface you are looking at. The one major shortcoming of this form of drawing is the lack of depth. A single view of the object providing a more realistic representation is often helpful for visual relation. This realistic single view is achieved with pictorial drawing.

The most common forms of pictorial drawing used in engineering drawing are isometric and oblique. These two basic forms of pictorial drawing are easy to create, if you can visualize objects in orthographic projection and three dimensions.

Isometric drawing belongs to a family of pictorial representation known as axonometric projection. Two other similar forms of drawing are in this group. Dimetric projection involves the use of two different scales. The simplest is isometric, which uses a single scale for all axes. Trimetric projection is the most involved of the three and uses three different scales for measurement. Figure 15.4 illustrates the differences in scale between isometric, diametric, and trimetric drawings.

The terms drawing and projection should be clarified. A projection is an exact representation of an object projected onto a plane from a specific position. Your line of sight to the various points on the object passes through a projection plane. The representation on a drawing sheet of the points of the object on the projection plane becomes the drawing. Exact projections of objects are time-consuming to make and often involve the use of odd angles and scales. Therefore, most drafters and illustrators work with axonometric drawing techniques rather than true projection techniques. Creating an axonometric drawing involves the use of approximate scales and angles that are close enough to the projection scales and angles to be acceptable.

The most realistic type of pictorial illustration is perspective drawing. The use of vanishing points in the projection of these drawings gives them the depth and distortion that the human eye perceives. Each of these types of pictorial drawings are examined in this chapter with step-by-step construction methods. In addition, you will see how pictorial drawings can be drawn with a computer-aided drafting system.

Technical Illustration

Pictorial drawing is a term that is often used interchangeably with technical illustration. But pictorial drawing includes only

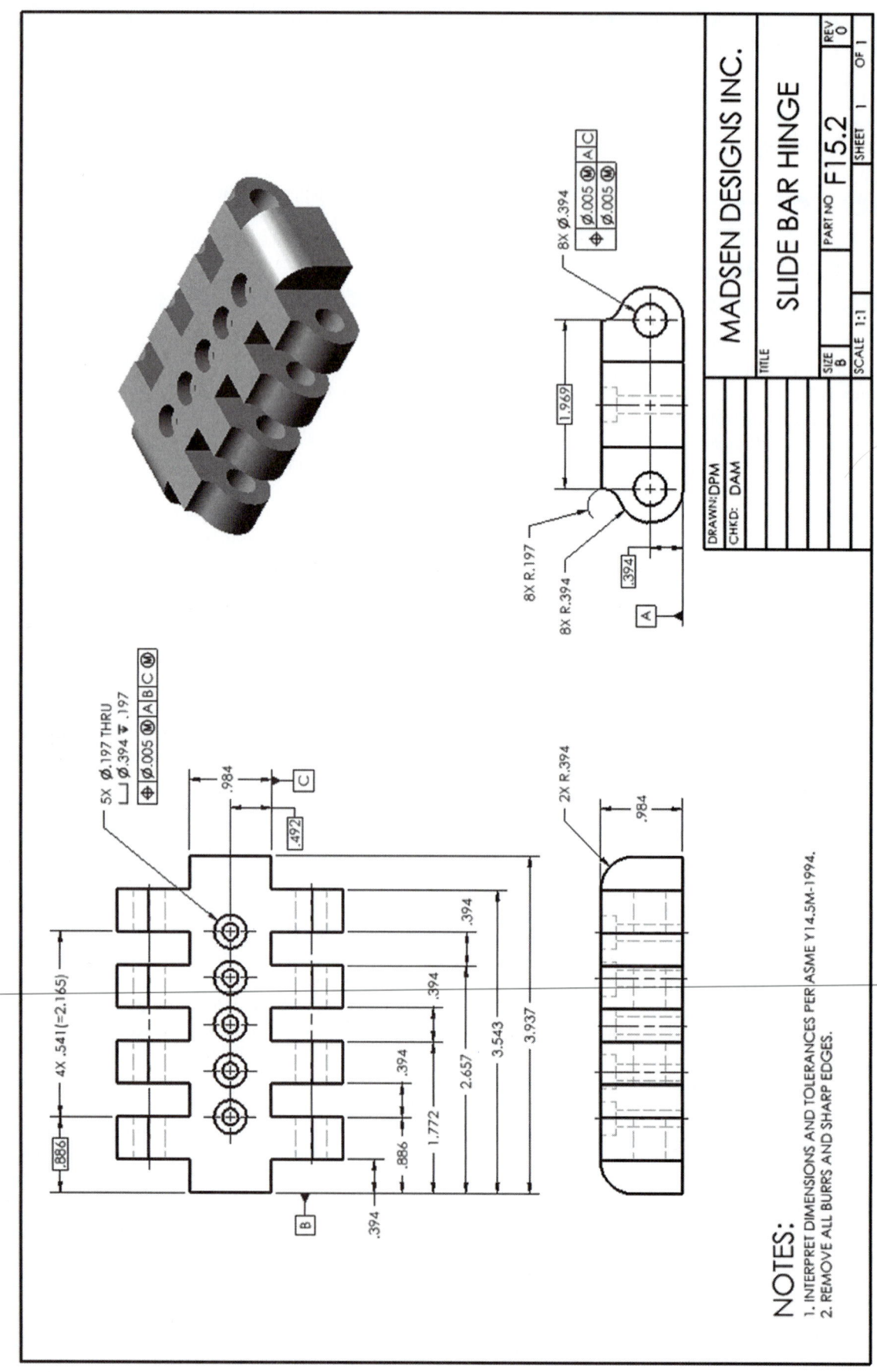

FIGURE 15.2 ■ A 2-D multiview drawing of a part with a pictorial drawing in the upper-right corner to aid in visualization. Dimension values in this figure are in inches.

FIGURE 15.3 ■ A 3-D solid model of an electric guitar assembly. *Courtesy Ethan Collins.*

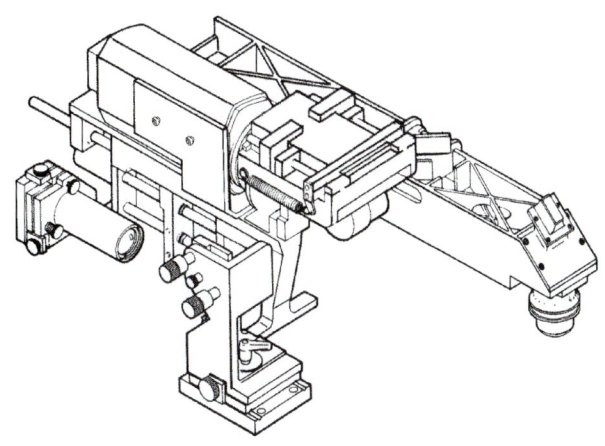

FIGURE 15.5 ■ Pictorial drawing. *Courtesy Industrial Illustrators, Inc.*

line drawings done in one of several three-dimensional methods, whereas technical illustration involves the use of a variety of artistic and graphic arts skills and a wide range of media in addition to pictorial drawing techniques. Figure 15.5 is an example of a pictorial drawing, and Figure 15.6 shows a technical illustration. Pictorial drawings are most often the basis for technical illustrations.

Uses of Pictorial Drawings

Pictorial drawings are excellent aids in the design process because they allow designers and engineers to view the objects at various stages of development. Pictorial drawings are used in instruction manuals, parts catalogs, advertising literature, technical reports, presentations, and as aids in the assembly and construction of products.

FIGURE 15.6 ■ Cutaway technical illustration. *Courtesy Industrial Illustrators, Inc.*

ISOMETRIC PROJECTIONS AND DRAWINGS

The word isometric means equal (iso) measure (metric). The three principal planes and edges make equal angles with the plane of projection. Creating isometric sketches was covered in Chapter 6. An isometric projection is achieved by first revolving

the object, in this case a 1-in cube, 45° in a multiview drawing as shown in Figure 15.7a, then tilting the object forward until the diagonal line AE is perpendicular to the projection plane, as seen in the side view of Figure 15.7b. This creates an angle of 35° 16' between the vertical axis AD and the plane of projection. When viewed in the isometric or front view, this axis appears

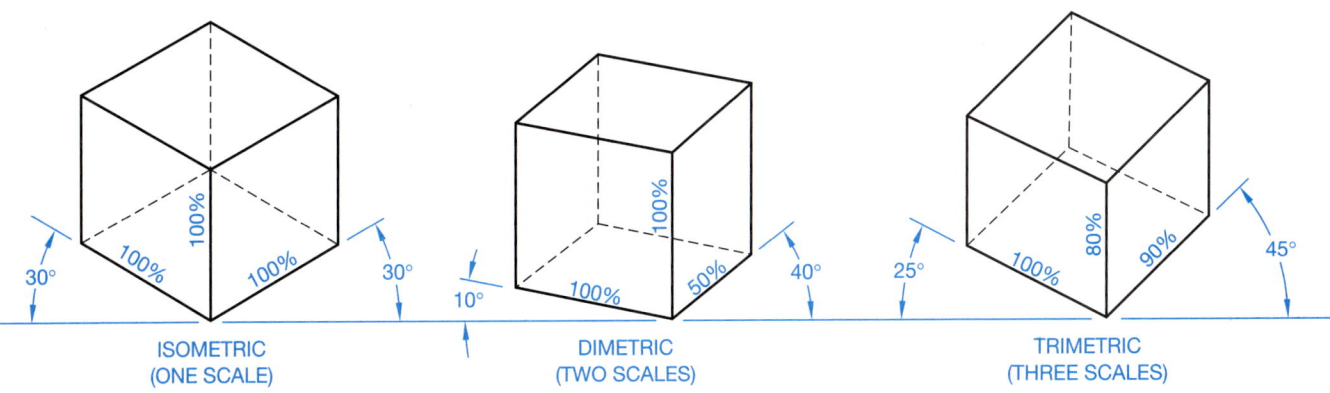

FIGURE 15.4 ■ Three types of axonometric projections.

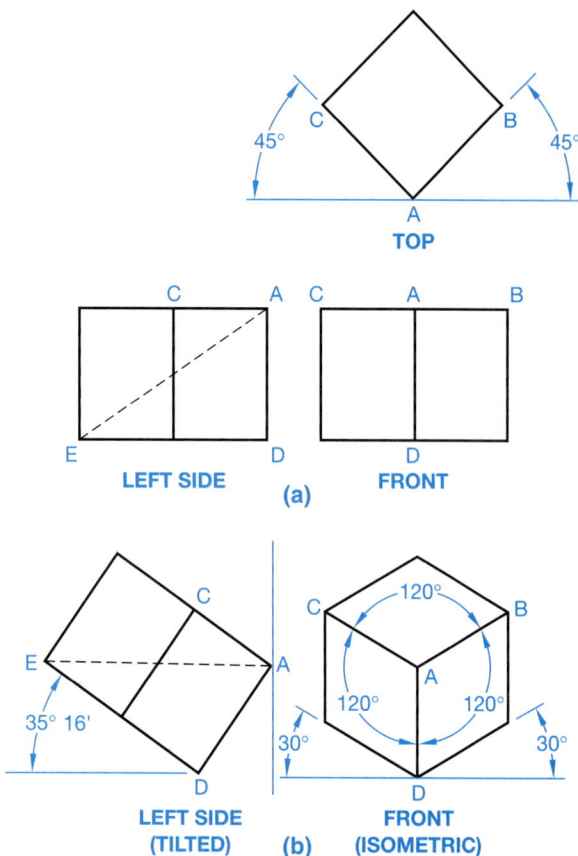

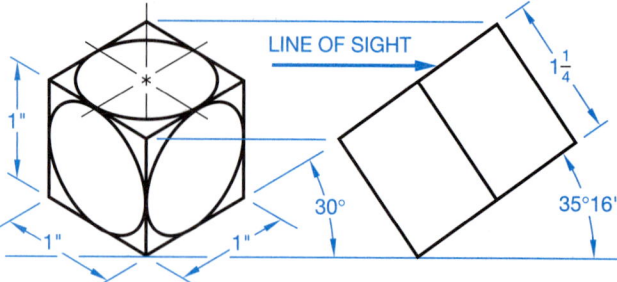

ISOMETRIC DRAWING
BASED ON TRUE MEASUREMENT IN ISOMETRIC VIEW
(PREFERRED METHOD)

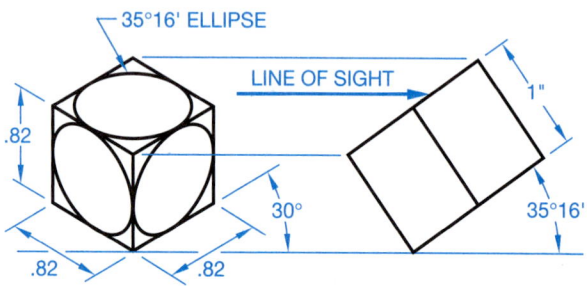

ISOMETRIC PROJECTION
BASED ON TRUE MEASUREMENT IN ORTHOGRAPHIC
(SELDOM USED)

FIGURE 15.9 ■ The differences between isometric drawing and isometric projection.

FIGURE 15.7 ■ Construction of an isometric projection.

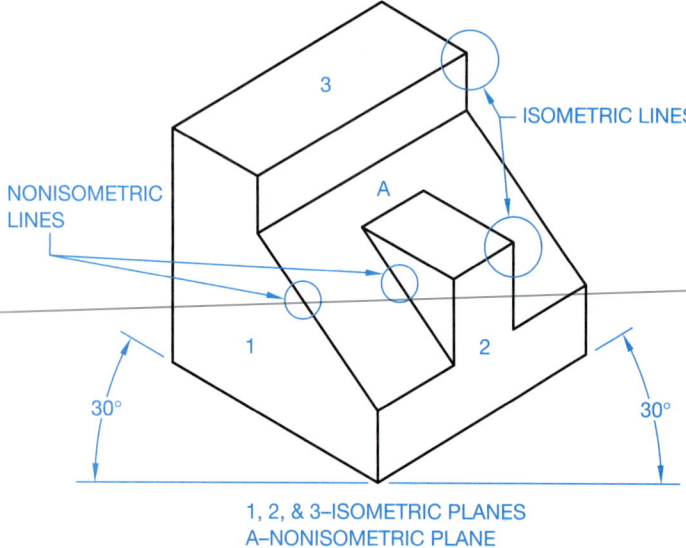

FIGURE 15.8 ■ Isometric and nonisometric planes.

vertical. The remaining two principal axes, AB and AC, are at 30° to a horizontal line.

The three principal axes are called **isometric lines**, and any line parallel to them is also an isometric line. These lines can all be measured. Any lines not parallel to these three axes are **nonisometric lines** and cannot be measured. The angles between each of these three isometric axes are 120°. The three planes be-

tween the isometric axes and any plane parallel to them are called isometric planes. (See Figure 15.8.)

Isometric Scale

An **isometric drawing** is done using a regular scale. This is most common in industry because it does not involve the creation of special scales. The only difference between an isometric drawing and an isometric projection is the size. The drawing appears slightly larger than the projection. Figure 15.9 illustrates the differences between the isometric drawing and the projection.

The **isometric projection** is a true representation of an object rotated and tilted in the manner just described. An isometric projection must be drawn using an isometric scale. An isometric scale is created by first laying a regular scale at 45° and projecting the increments of that scale vertically down to a blank scale drawn at an angle of 30°. The resulting isometric scale is seen in Figure 15.10. A 1-in. measurement on the regular scale now measures .816 in. on the isometric scale.

TYPES OF ISOMETRIC DRAWINGS

Isometric drawing is a form of pictorial drawing in which the receding axes are drawn at 30° from the horizontal, as shown in Figure 15.7. There are three basic forms of isometric drawing: these are known as regular, reverse, and long-axis isometric.

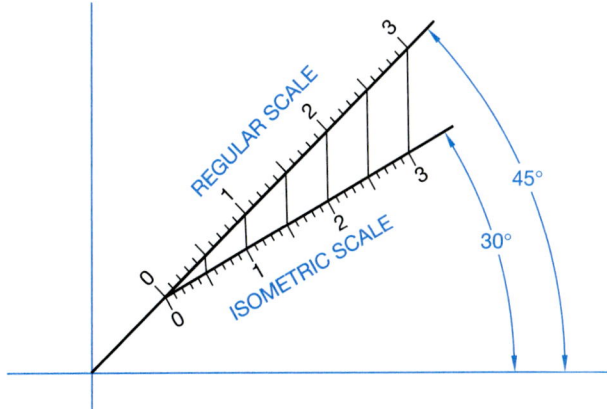

FIGURE 15.10 ■ Projection of regular scale to isometric scale.

You should choose the view that gives the most realistic presentation of the object. For example, if the object is normally seen from below, then the reverse isometric is the proper form to use.

Regular Isometric

The top of an object can be seen in the regular isometric form of drawing. An example is shown in Figure 15.11a. This is the most common form of isometric drawing, and when using it, you can choose to view the object from either side.

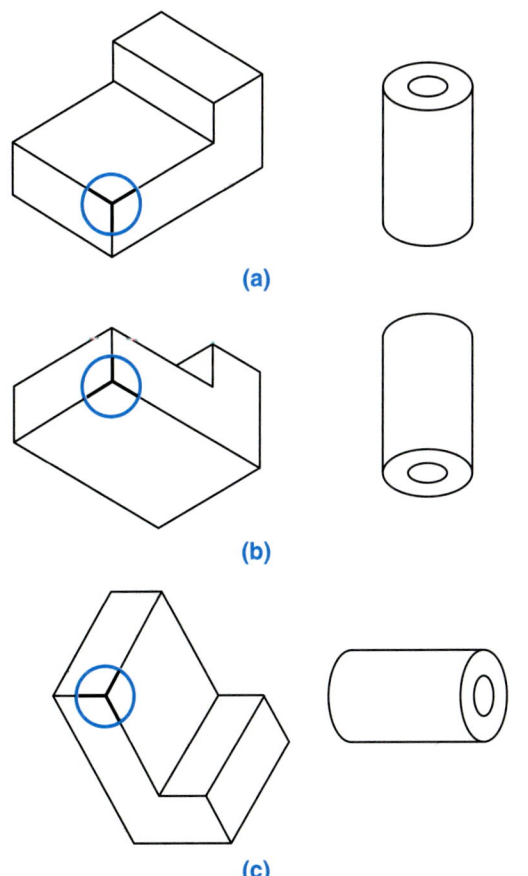

FIGURE 15.11 ■ Isometric axis variations: (a) regular; (b) reverse; (c) long-axis.

Reverse Isometric

The only difference between reverse and regular isometric is that you can view the bottom of the part instead of the top. The 30° axis lines are drawn downward from the horizontal line instead of upward. Figure 15.11b shows an example of reverse isometric.

Long-Axis Isometric

The long-axis isometric drawing is normally used for objects that are long, such as shafts. Figure 15.11c shows an example of the long-axis form.

ISOMETRIC CONSTRUCTION TECHNIQUES

Just as objects differ in their geometric makeup, so do the construction methods used to draw the object. Different techniques exist to assist you in constructing the various shapes.

Box, or Coordinate, Method

The most common form of isometric construction is the box, or coordinate, method and is used on objects that have angular or radial features. The orthographic views of the object to be drawn are shown in Figure 15.12a, and explained as follows:

First, an isometric box the size of the overall dimensions of the object (X, Y, and Z) must be drawn. (See Figure 15.12b.) Then the measurements of the features of the object can be transferred to the isometric box. To locate points D and C, just measure dimension U from points F and G. Point E is located at the lower left corner of the box. Draw a line from E to D. Next, locate point H by measuring up distance W from the bottom right corner of the box. Draw a light construction line at a 30° angle toward the front vertical axis. Now, draw a line from A parallel to line ED until it intersects the line from H. This intersection is point B. The location of point B can also be found by measuring the horizontal dimension T from point A, then the vertical dimension W.

It may be necessary to draw construction lines on the orthographic views and transfer measurements directly from these views to the isometric view. This method works well with irregularly shaped objects such as the one shown in Figure 15.13.

Centerline Layout Method

The centerline method begins with the skeleton of the object, the centerlines. This technique is used on objects with many circles and arcs. The use of this method is seen in Figure 15.14a through d. The center points of all of the circles and arcs should be located first. Begin with a good reference point from which you can work, such as the bottom of the object in Figure 15.14a. Center points A and B are first established, then vertical axis

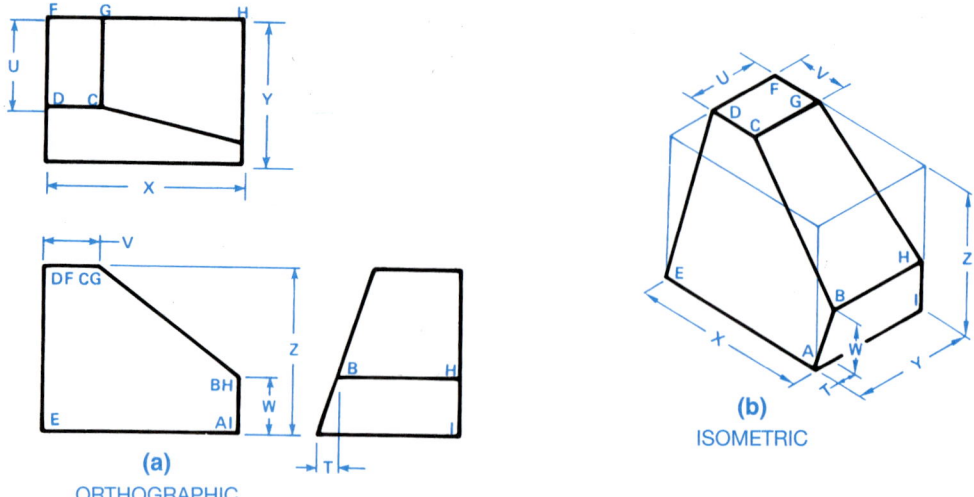

FIGURE 15.12 ■ Box method of isometric construction: (a) orthographic view; (b) isometric view.

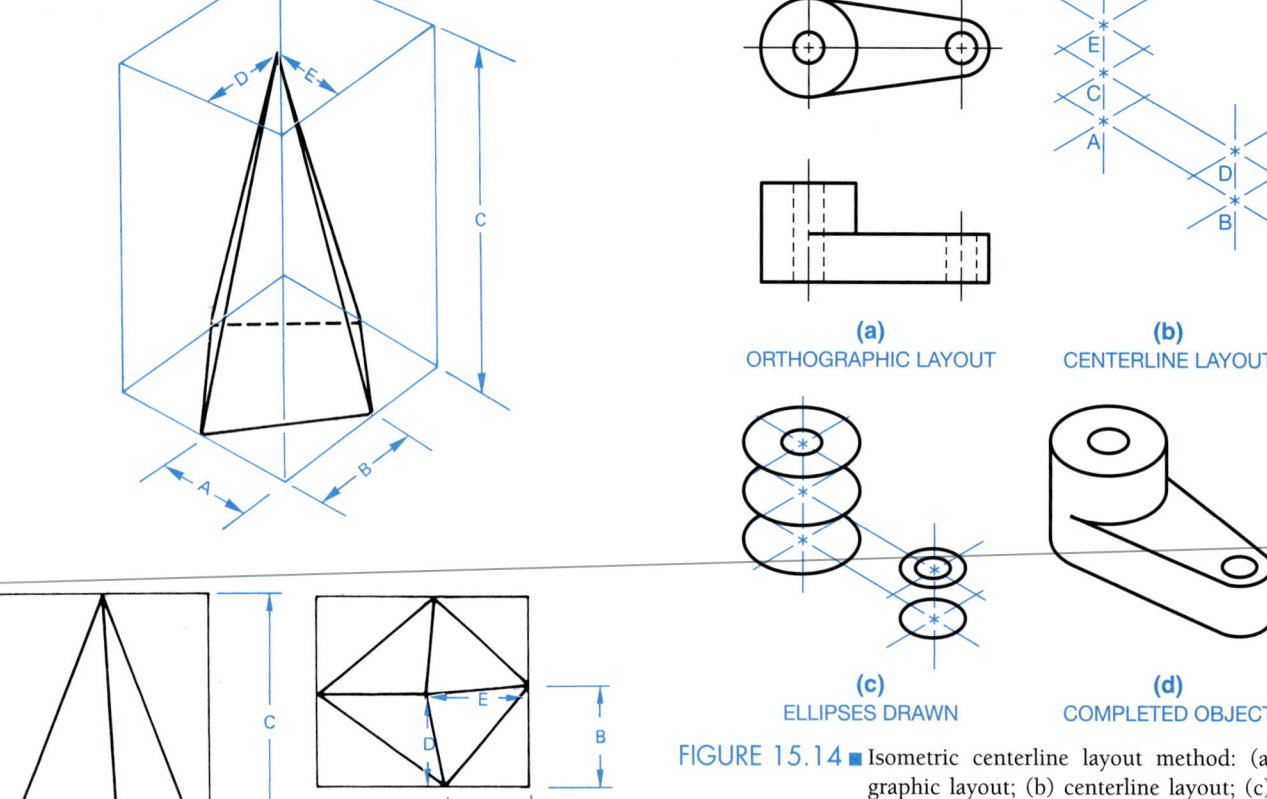

FIGURE 15.13 ■ Isometric box method for an irregular object.

FIGURE 15.14 ■ Isometric centerline layout method: (a) orthographic layout; (b) centerline layout; (c) ellipses drawn; (d) completed object.

Circle and Arc Construction

lines are drawn up from these. Now the vertical locations of center points C, D, and E can be measured as seen in Figure 15.14b. Then determine the various sizes of the ellipses and draw them at their proper locations as shown in Figure 15.14c. The finished drawing is shown in Figure 15.14d.

After the center points of circles and arcs are located, you need to select the correct size of circle from an isometric ellipse template. The isometric ellipse template is only useful for drawing ellipses on isometric surfaces. Each ellipse on the template has several tick marks. A description of these tick marks is shown in Figure 15.15. An isometric ellipse is measured on the marks

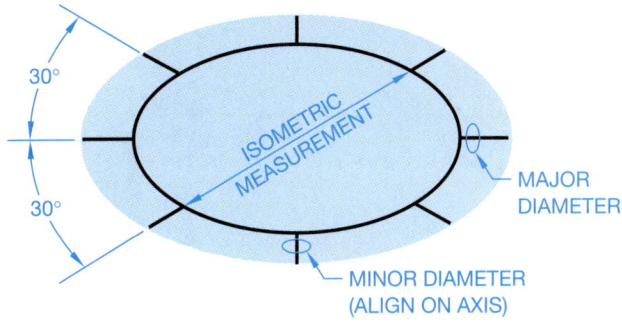

FIGURE 15.15 ■ Meaning of isometric ellipse template markings.

at 30° angles from the large (horizontal) diameter of the ellipse. When aligning an isometric ellipse on the centerlines of your layout, always align the tick marks on the **minor diameter** of the ellipse with the **axis** of the feature, as shown in Figure 15.16. When you do this, two other sets of tick marks on the template align with the ellipse centerlines.

Arc locations are found in the same way as circles, and the use of the isometric ellipse template is also the same, except just a portion of the circle is drawn. First, find the center point of the arc, then the tangent points. (See Figure 15.17.) Draw the axis

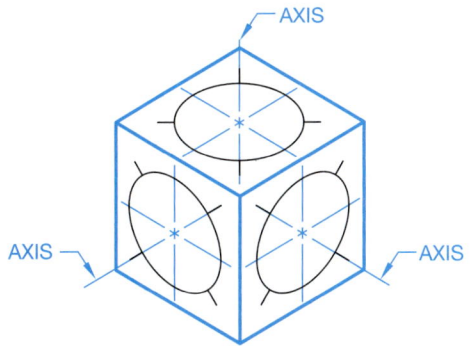

FIGURE 15.16 ■ Align minor diameter of isometric ellipse template on axis of hole.

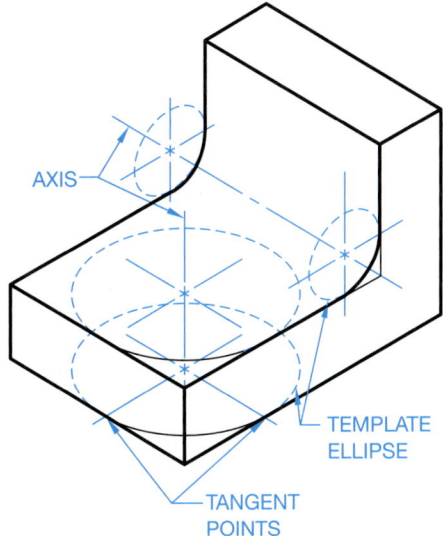

FIGURE 15.17 ■ Isometric arc layout using ellipse template.

line of the arc lightly and again align the minor diameter tick marks with the axis line. Now, draw only the portion of the ellipse that is needed for the arc.

Constructing isometric arcs is easy with CADD software such as AutoCAD by using the isocircle option of the ELLIPSE command. The isogrid must be set for the isocircle option to show. The isometric crosshairs can be toggled while the ellipse is attached to the crosshairs, and you can see the ellipse in the left-, top-, and right-side orientations.

Drawing Fillets and Rounds

The key to drawing isometric circles and arcs is getting the ellipse aligned properly. Study the object in Figure 15.18. Determine the axis of the circular feature, then align the minor diameter tick marks of the template with the axis of the fillet or round.

Establishing Intersections

A common type of intersection is that of a cylindrical hole passing through an oblique plane. To establish this, draw an isometric ellipse at the top of the boxed surface of the object. (See Figure 15.19.) Then draw a series of construction lines parallel to one side of the box that pass through the ellipse. Project each of these lines down the sides of the box and across the oblique surface. Each line forms a trapezoid or parallelogram depending on the shape of the object. The points at which each line intersects the ellipse, such as A and B, should be projected straight down to the opposite side of the parallelogram. These new points, A_1 and B_1, are now points on the perimeter of the hole in the oblique surface.

Another common form of intersection occurs when two cylinders meet. This construction is similar to the one just described. In Figure 15.20, the center point location of the branch cylinder is determined, and the ellipse is drawn. Next, draw as many construction lines passing through the ellipse as needed to produce a smooth curve on the intersection. Project these lines to the end of the main cylinder, down to the edge, then back along the cylinder. Now, project points on the ellipse down to

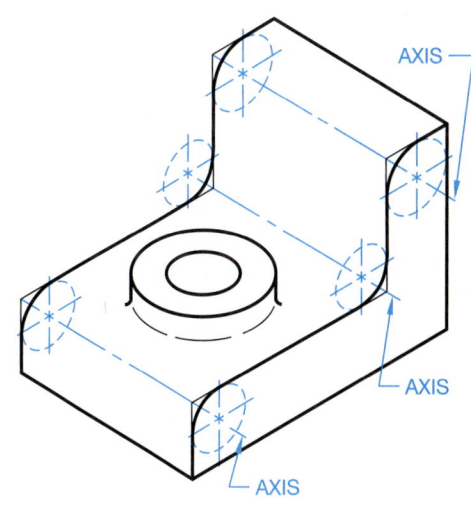

FIGURE 15.18 ■ Isometric fillet and round layout.

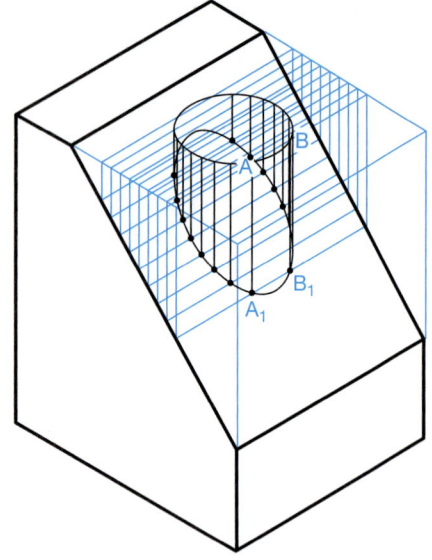

FIGURE 15.19 ■ Construction of ellipse on isometric oblique plane.

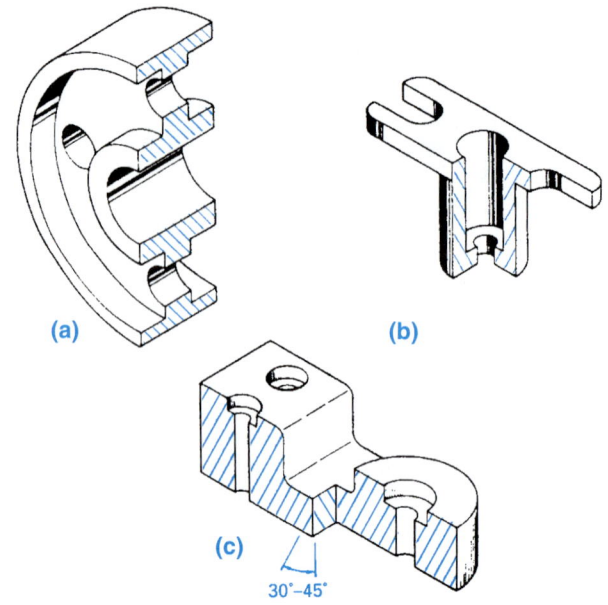

FIGURE 15.21 ■ Isometric sections: (a) full; (b) half; (c) offset.

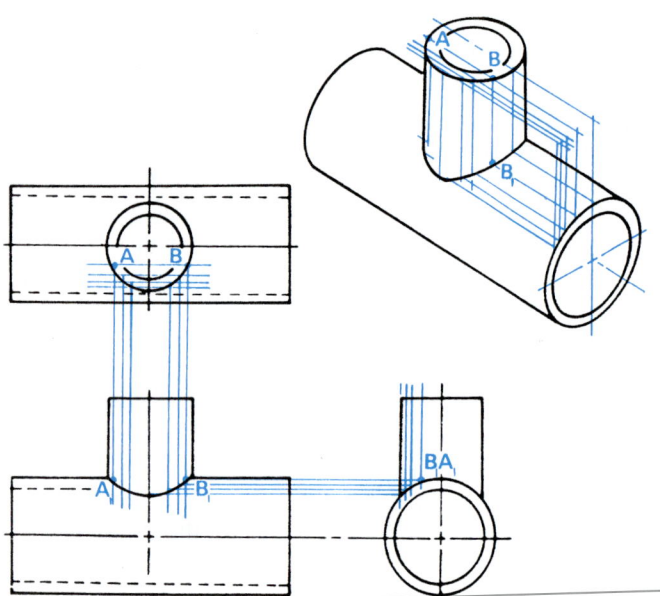

FIGURE 15.20 ■ Isometric construction of two intersecting cylinders.

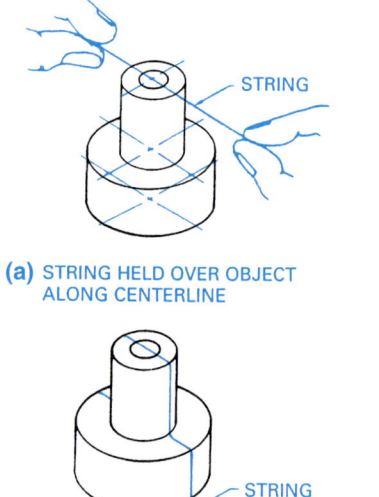

(a) STRING HELD OVER OBJECT ALONG CENTERLINE

(b) STRING FALLS ALONG CENTERLINES

(c) COMPLETED FULL SECTION

FIGURE 15.22 ■ Visualize a string dropped along the cutting plane to create an isometric full section.

the corresponding construction line on the cylinder, A to A_1 and B to B_1, for example. Then connect the points using an irregular curve, or polyline in AutoCAD, to produce the intersection.

Drawing Isometric Sections

Full and half sections are common in technical illustration and should be drawn along the isometric axes. The section lines in full sections should all be drawn in the same direction, while those on a half section should appear in opposite directions. See the section lines illustrated in Figure 15.21a and b. The section lines in offset sections should change directions with each jog in the part, as seen in Figure 15.21c.

There is no preferred way to draw an isometric section. One technique that you might try is to imagine the cutting-plane line is a long string, and you are stretching it out just above the axis along which you plan to cut. If you let the string fall, it comes to rest along the axis where the cut is to be made, as shown in Figure 15.22.

Drawing Isometric Threads

Chapter 12, Fasteners and Springs, covered the drafting of 2-D screw threads. A review of this chapter is recommended as needed. While there are three different types of 2-D thread representations discussed in Chapter 12, the detailed thread representation is commonly used when creating an isometric drawing.

Screw threads can be drawn in isometric by first measuring equal spaces along the shaft or hole to be threaded. Then, using the same size ellipse as the diameter of the shaft or hole, draw a series of parallel ellipses. (See Figure 15.23.) These ellipses represent the crests of the threads. This method achieves a simple isometric representation of threads.

A more detailed thread appearance can be achieved by giving each ellipse rounded ends instead of butting the ellipse into the straight sides of the shaft. (See Figure 15.24.) When using this technique, begin drawing the threads from the head of the shaft. With manual drawing, be sure to lay out guidelines for the shaft diameter so the ends of the ellipse do not fall short of the edge of the shaft or go beyond it, as seen in Figure 15.25. This can create an image known as wandering threads.

Drawing threads, like those shown in Figure 15.24, is quick using software such as AutoCAD. To do this, draw the uppermost thread using an arc or ellipse and edit as needed by trimming or breaking the ellipse so it fits properly. Then use a command such as ARRAY, and create a rectangular array of the required number of threads.

Drawing Isometric Spheres

A sphere drawn isometrically is a true circle. Figure 15.26 shows the construction of a sphere using three isometric ellipses drawn to represent the perpendicular axes of the circle. If you need to draw a sphere, remember to choose a circle that is 1 1/4

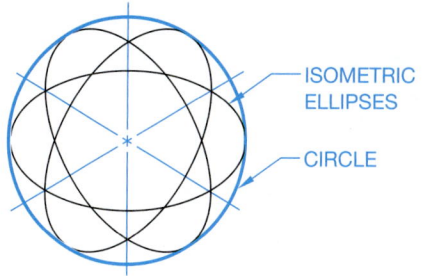

FIGURE 15.26 ■ Isometric sphere construction.

times larger than the actual sphere, because isometric drawings are 1 1/4 times larger than the actual representation.

If you need to draw spheres that have been cut in some manner, refer to Figure 15.27. This figure shows a half sphere, three-quarter sphere, and a sphere with a flat side.

Isometric Dimensioning

It is not common for isometric drawings to be dimensioned; however, some isometric piping drawings rely heavily on dimensioning to convey their message. Industrial pipe drawings are covered in Chapter 22. When dimensioning, one technique that can be used is to letter vertical strokes parallel to extension lines, as shown in Figure 15.28. This gives the appearance of the dimension lying in the plane of the extension lines. A second technique is to draw the heel of the arrowhead parallel with the extension line. This further emphasizes the plane where the dimension is lying. (See Figure 15.28a.) Examples of unidirectional, aligned, and one-plane (horizontal) isometric dimensioning are shown in Figure 15.28b. Many of the problems at the end of chapters throughout this book are dimensioned isometric drawings.

DIMETRIC PICTORIAL REPRESENTATION

Dimetric projection is similar to isometric projection, but instead of all three axes forming equal angles with the plane of

AXIS —

MARK THREAD SPACING WITH TICKS AND CENTERLINES ON AXIS.

FIGURE 15.23 ■ Isometric thread representation.

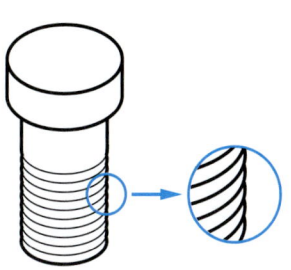

FIGURE 15.24 ■ Detailed isometric thread representation.

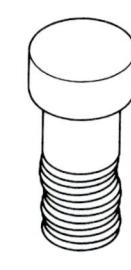

FIGURE 15.25 ■ Off-center ellipses produce "wandering" threads.

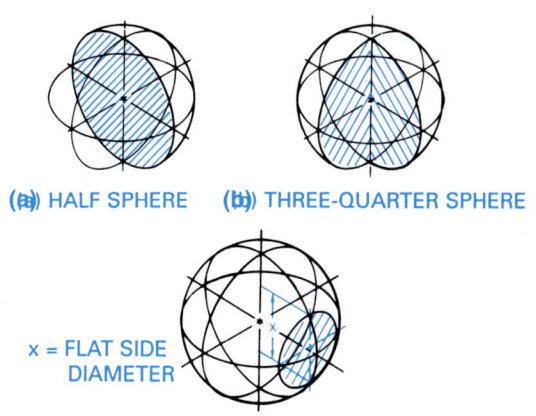

(a) HALF SPHERE   (b) THREE-QUARTER SPHERE

x = FLAT SIDE DIAMETER

(c) SPHERE WITH FLAT SIDE

FIGURE 15.27 ■ Portions of isometric spheres; (a) half sphere; (b) three-quarter sphere; (c) sphere with flat side.

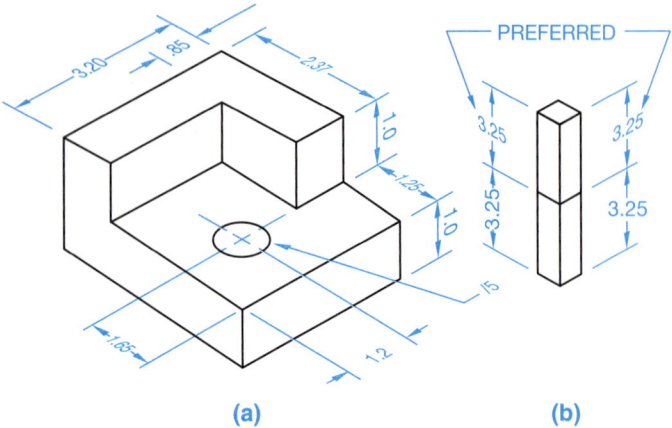

(a)   **(b)**

FIGURE 15.28 ■ Isometric dimensioning: (a) dimensions parallel to extension lines; (b) various styles of dimensioning. Dimension values in this figure are in inches.

projection, only two axes form equal angles. These can be greater than 90° and less than 180° but cannot have an angle of 120°, because that is the isometric angle. The third axis can have an angle less or greater than the two equal axes, depending on the angles chosen. The two equal angles are said to be *fore-shortened* because they are not measured at full scale, and they are foreshortened equally. Since the third axis is projected at a different angle, it is foreshortened at a different scale. Full-size and foreshortened approximate scales are used on the diametric axes. Some common approximate dimetric angles and scales are shown in Figure 15.29.

TRIMETRIC PICTORIAL REPRESENTATION

The term *trimetric* refers to three measurements. It is a type of pictorial drawing in which none of the three principal axes makes an equal angle with the plane of projection. Since all three angles are unequal, the scales used to measure on the three axes also are unequal. Trimetric projection provides an infinite number of projections.

Trimetric projection is similar to dimetric projection. Figure 15.30 shows some common trimetric angles for the width and depth axes and the scales to use on each of the three axes. The size of angle ellipse to use on each principal plane also is indicated.

EXPLODED PICTORIAL DRAWING

Complicated parts and mechanisms are often illustrated as an exploded assembly in order to show the relationship of the parts in the most realistic manner. (See Figure 15.31a.)

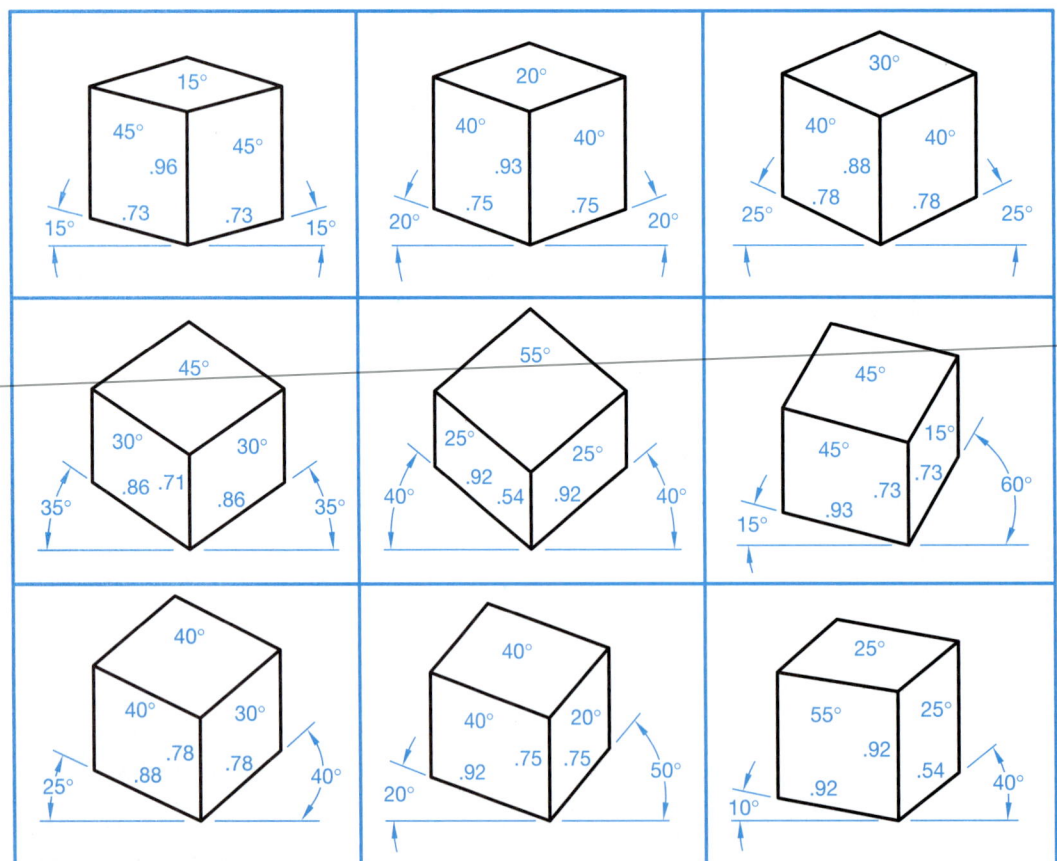

FIGURE 15.29 ■ Common approximate diametric angles, scales, and ellipse angles.

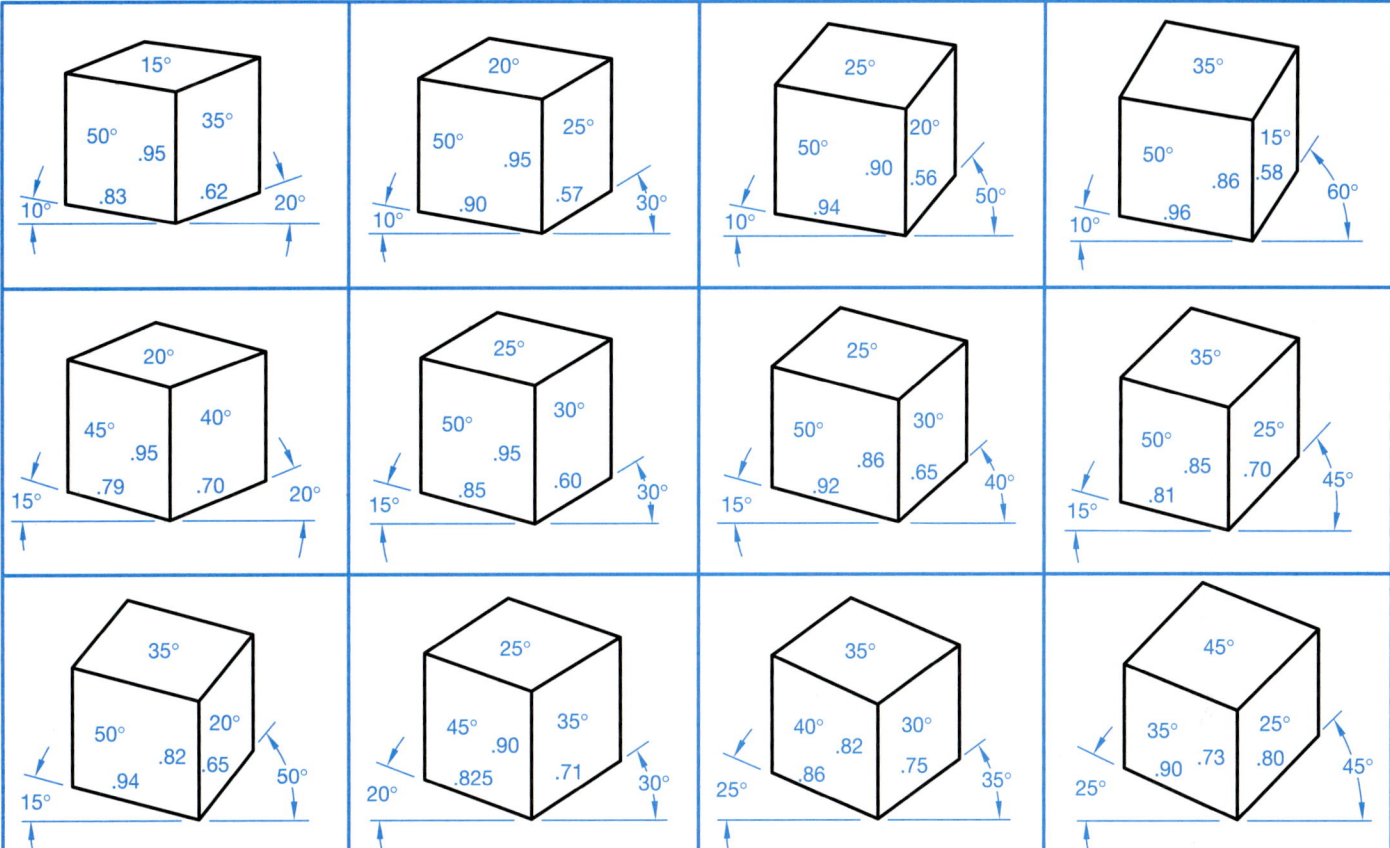

FIGURE 15.30 ■ Common approximate trimetric angles, scales, and ellipse angles.

An **exploded assembly** is a collection of parts, each drawn in the same axonometric method. Any of the pictorial drawing methods mentioned in this chapter can be used to create exploded assemblies. The most important aspect of this type of drawing is to select the viewing direction that illustrates as much of the assembly as possible.

Centerlines are used in exploded views to show the connection between part and subassembly axis. Thin solid extension lines can be used to connect between non-cylindrical features, parts, and subassemblies. These lines are often called **connection lines**, or **trails**. These lines aid the eye in following a part to its position in the assembly. Connection lines should avoid crossing other lines and parts. Lines should always be drawn parallel to the axis lines of the drawing regardless of the drawing method selected. The neatest presentation is achieved by leaving gaps where the line intersects the part to which it applies and the mating feature. This is shown in Figure 15.31b. Some school or company standards prefer to draw parts as close together as clearly possible without the use of connection lines.

When drawing an exploded assembly with CADD, it is not necessary to draw all the parts in their final positions. Draw each part as needed, placing construction lines for the connection lines. When all parts are completed, move them around as needed to construct the final layout.

OBLIQUE DRAWING

Oblique drawing is a form of pictorial drawing in which the plane of projection is parallel to the front surface of the object. The lines of sight are at an angle to the plane of projection and are parallel to each other. This allows you to see three faces of the object. The front face, and any surface parallel to it, is shown in true size and shape, while the other two faces are at an angle. Oblique drawing is useful if one face of an object needs to be shown flat. Choose the orthographic face with the most curves/circles because these are more difficult to draw on other surfaces at an angle.

There are three methods of oblique drawing: cavalier, cabinet, and general. (See Figures 15.32 through 15.34.)

Cavalier Oblique

The **cavalier oblique** projection is one in which the receding lines are drawn true size, or full scale. This form of oblique drawing is usually drawn at an angle of 45° from horizontal, which approximates a viewing angle of 45°. Because of the scale used on the receding axis, it is not a good idea to draw long objects with the long axis perpendicular to the front face, or projection plane. Objects that have a depth that is smaller than the

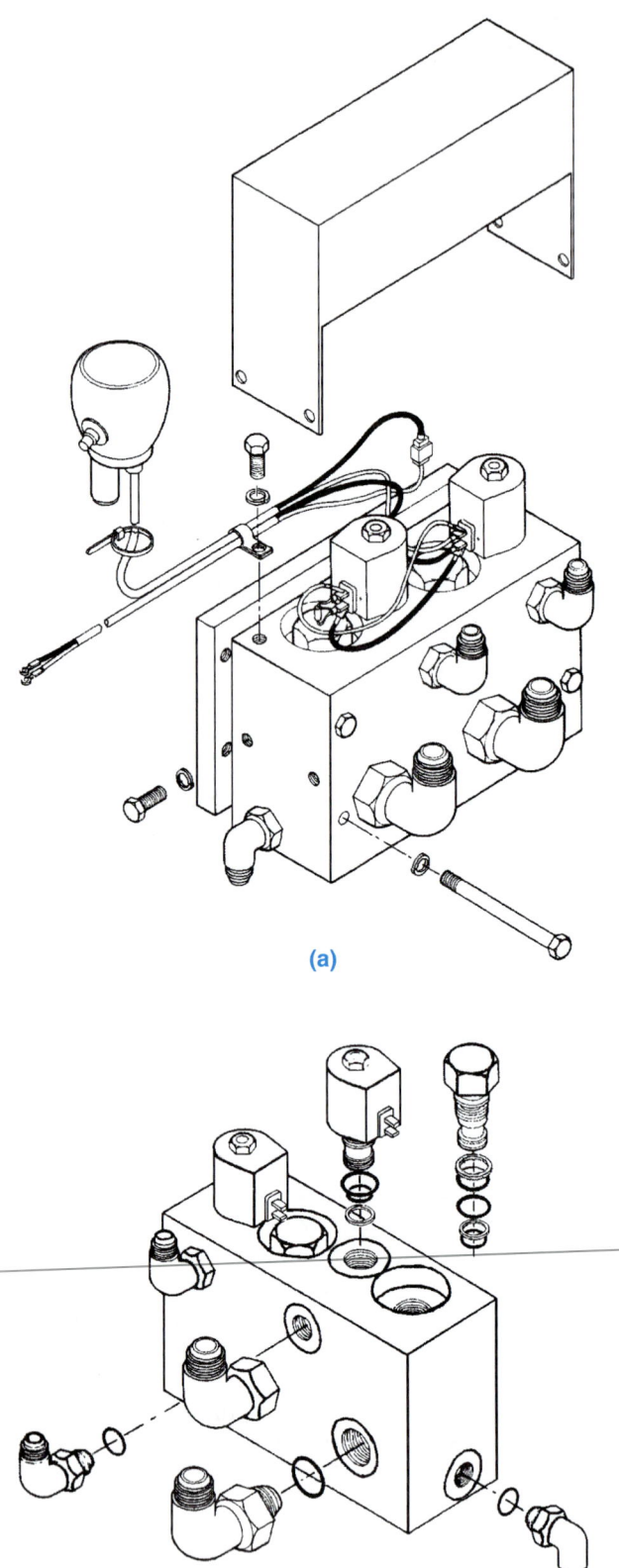

(a)

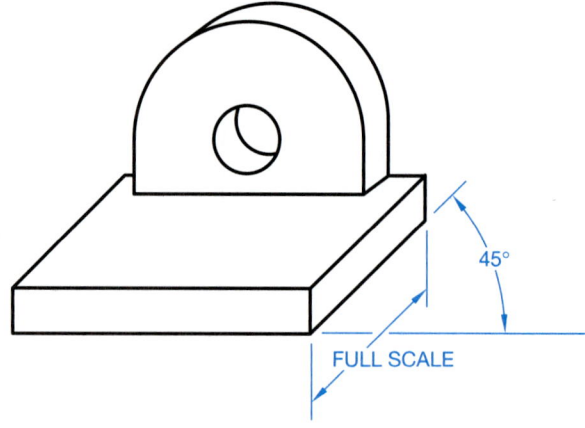

FIGURE 15.32 ■ Cavalier oblique.

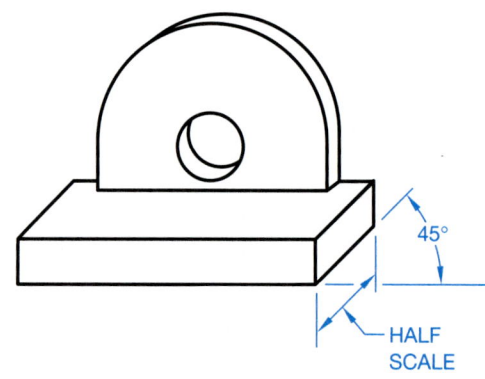

FIGURE 15.33 ■ Cabinet oblique.

(b)

FIGURE 15.31 ■ Exploded assemblies. *Courtesy Industrial Illustrators, Inc.*

width can be drawn in the cavalier form with clarity. Figure 15.32 shows an object drawn in cavalier oblique.

Cabinet Oblique

A **cabinet oblique** drawing is also drawn with a receding angle of 45°, but the scale along the receding axis is half size. Objects having a greater depth than width can be drawn in this form with clarity. Cabinet makers often manually draw their cabinet designs using this type of oblique drawing. A cabinet drawing is shown in Figure 15.33.

General Oblique

The **general oblique** drawing is normally drawn at an angle other than 45°, and the scale on the receding axis is also different from those used in cavalier and cabinet. The most common angles for a general oblique drawing are 30°, 45°, and 60°, but any angle can be used. Any scale from half to full size can be used. A general oblique drawing is shown in Figure 15.34.

PERSPECTIVE DRAWING

Perspective drawing is the most realistic form of pictorial illustration. Perspective drawings give the representation of objects

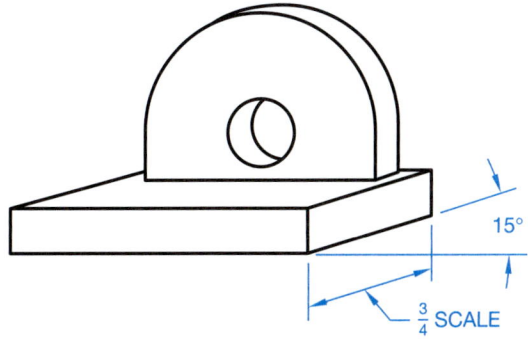

FIGURE 15.34 ■ General oblique.

appearing smaller the farther away they are until they vanish at a point on the horizon. Lines drawn parallel to the principal orthographic planes appear to converge on a vanishing point in perspective drawing.

Three types of perspective drawing techniques take their names from the number of vanishing points used in each. **One-point**, or **parallel**, **perspective** has one vanishing point and is used most often when drawing interiors of rooms. **Two-point**, or **angular**, **perspective** is the most popular, and is used to illustrate exteriors of houses, small buildings, civil engineering projects, and, occasionally, machine parts. The third type, **three-point perspective**, has three vanishing points and is used to illustrate objects having great vertical measurements, such as tall buildings. It is a lengthy process to draw in three-point perspective; therefore, it is not used as often as two-point perspective.

The following discussions cover the construction of perspective drawings using manual drawing methods or CADD 2-D drawing methods. Keep in mind that once an object is drawn in 3-D using some CADD systems, it can be automatically displayed in a perspective format.

General Perspective Drawing Concepts

Two principal components of a perspective drawing are the eye of the person viewing the object, and the location of the person in relation to the object. The eye level of the observer is the **horizon line (HL)**. This line is established in the elevation (front) view. (See Figure 15.35.) The position of the observer in relation to the object is the **station point (SP)** and is established in relation to the plan (top) view. The location of the station point determines how close the observer is to the object and the angle at which the observer is viewing the object. The **ground line (GL)** is the line on which the object rests. The **picture plane (PP)**, or plane of projection, is the surface, or drawing sheet, on which the object is projected. The picture plane can be situated anywhere between the observer and the object. The picture plane can also be located beyond the object. The observer's lines of sight, or visual projection lines, determine what shows on the picture plane and where. Finally, the three vanishing points discussed in the following sections are referred to as **vanishing point right (VPR)**, **vanishing point left (VPL)**, and **vanishing point vertical (VPV)**.

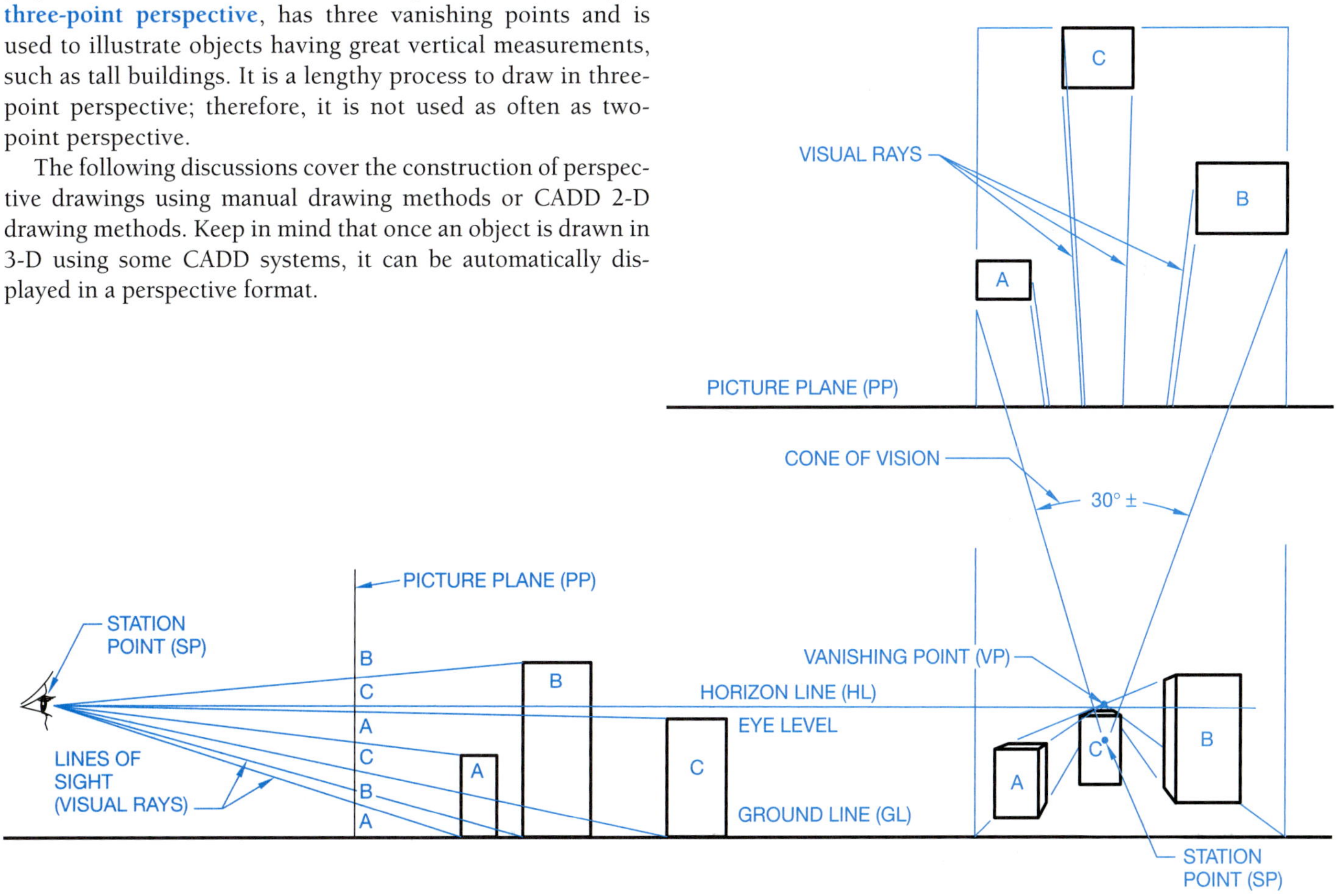

FIGURE 15.35 ■ Principal components of perspective drawings.

ONE-POINT PERSPECTIVE

One-point perspective, also known as parallel perspective, has only one vanishing point. The plan view is oriented so the front surface of the object is parallel to the picture plane. The elevation view is situated below and to the right or left of the plan and rests on the ground line. The following steps allow you to construct a one-point perspective. Refer to Figure 15.36 as you read the instructions.

STEP 1 Use construction lines for all work except the final object, which is thick object lines or shading techniques. Locate the station point between the picture plane and the elevation view of the object. The station point can be anywhere, depending on the part of the object you want to view. The visual projection lines from the station point to the extreme corners of the object should form an included angle of approximately 30° to provide the most realistic perspective.

STEP 2 Determine the eye level of the observer in the elevation view. If you want to look over the object to see the top surface, the horizon line should be drawn above the elevation view. The horizon line can be drawn at any eye level. Constructing the horizon line at 5'6" above the ground line provides an eye level at approximately the average height of a person.

STEP 3 The vanishing point is located on the horizon line directly in line with the station point.

STEP 4 Project all points touching the picture plane, in the plan view, to the corresponding points in the elevation view.

STEP 5 Draw visual projection lines from the station point to the rear corners of the object in the plan view.

STEP 6 The points where the visual projection lines intersect the picture plane are projected vertically down to the perspective view. When drawing complex objects, work with only a portion of these points at one time to avoid confusion.

STEP 7 Project points A, B, and C toward the vanishing point to intersect the corresponding projectors from the picture plane. Now, draw the horizontal line B'C'. Any line that is parallel to the PP in the plan view must be parallel to the GL in the perspective view.

STEP 8 Project the height of the object from the elevation view to points D and E on the two true-height lines (THL). True-height lines are projected from points touching the PP.

STEP 9 Project the height at points D and E back toward the vanishing point to intersect the projectors from the rear portion of the object in the plan view.

STEP 10 Complete the object by connecting the ends of the sloping portion and projecting the base of the sloped feature toward the vanishing point.

TWO-POINT PERSPECTIVE

The two-point perspective method is also called angular perspective because it is turned so two of its principal planes are at an angle to the picture plane. This is the most popular form of perspective drawing. Two vanishing points allow parallel lines on the two principal planes to be projected in two directions, giving another dimension to the depth of the perspective.

The same one-point perspective object is used in this example. The object has been positioned at an angle in the plan view, with one corner touching the PP. The elevation view, SP, HL, and GL have all been established. Refer to Figure 15.37 as you read these instructions:

STEP 1 Draw a line from the station point, parallel to each side of the object in the plan view, to intersect the PP.

STEP 2 Project the points on the PP down to the HL. This establishes vanishing point right (VPR) and vanishing point left (VPL) on the horizon line.

STEP 3 Project point A on the PP down to the GL. This becomes the true-height line AB in the perspective view.

STEP 4 Begin blocking in the object by projecting points A and B to the two vanishing points. This establishes the two angular, or perspective, sides of the object.

STEP 5 Project visual projection lines from the SP to the extreme corners of the object in the plan view.

STEP 6 Project the intersection of the visual projection lines with the PP down to intersect with the projectors from points A and B to the two vanishing points. This blocks in the two sides of the object.

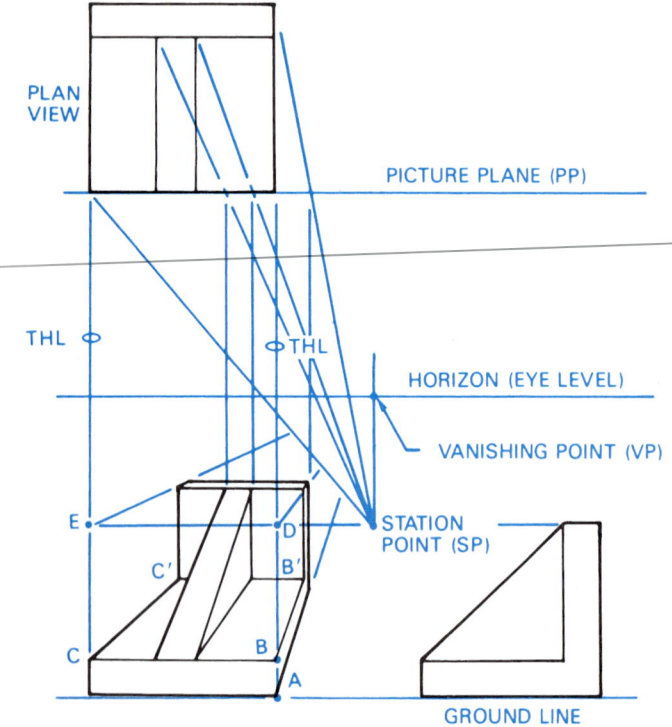

FIGURE 15.36 ■ Constructing a one-point perspective drawing.

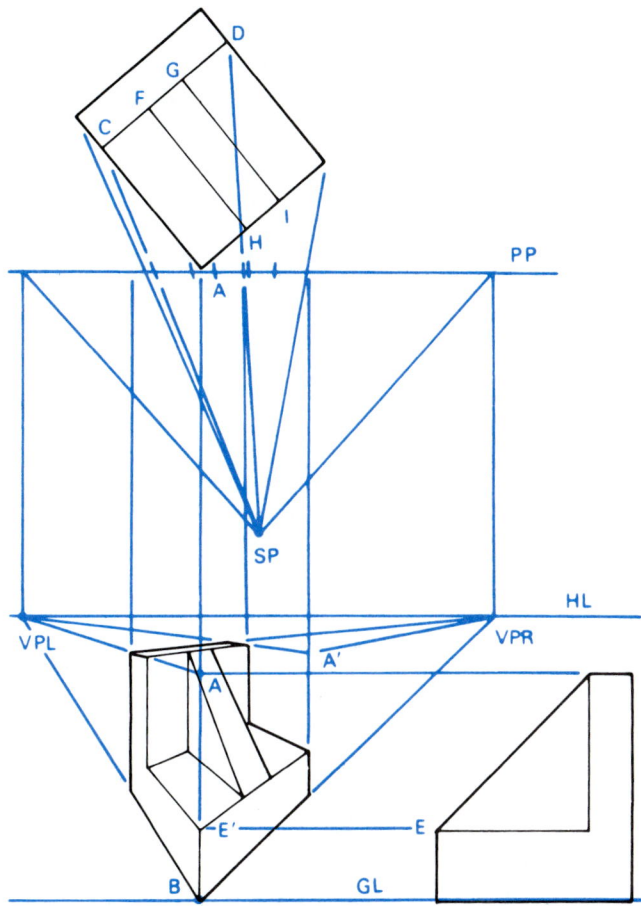

FIGURE 15.37 ■ Constructing a two-point perspective drawing.

STEP 7 Draw lines from the SP to points C and D. Where these lines intersect the PP, project vertical lines down into the perspective view. The height of the object at points A and A' can now be projected toward the VPL to intersect with the projectors from points C and D.

STEP 8 The height at E can be projected to E' on the THL. Project E' toward both vanishing points to create the basic shape of the object.

STEP 9 Project points F, G, H, and I to the station point. Where these lines intersect the PP, drop vertical projectors down to the perspective view. These lines will intersect corresponding points on the perspective drawing. Connect the points as shown in Figure 15.37 to complete the perspective view.

THREE-POINT PERSPECTIVE

Three vanishing points are used in **three-point perspective** drawings. These drawings require more time to construct than two-point perspectives and often occupy a considerable area on the drawing sheet. Three-point perspective drawings are used when certain effects are needed for visual simulation of an object such as a tall building. The method of constructing a three-point perspective that requires the least amount of drawing space is used in the following example. Refer to Figure 15.38 as you read these instructions:

STEP 1 Draw an equilateral triangle to occupy as much of the sheet as possible. Label the three corners as vanishing points VPR, VPL, and VPV.

STEP 2 Construct perpendicular bisectors of each of the three sides. Their intersection at the center of the triangle should be labeled SP (station point). (See Figure 15.38a.)

STEP 3 Draw a horizontal line through the SP. Measure and mark the length of the object to the left of the SP. Measure and mark the width of the object to the right of the SP. This allows you to draw an object in perspective as if you turned the plan view 45° to the pictureplane line. (See Figure 15.38b.)

STEP 4 Draw a line parallel to line VPV, VPR. This becomes the height measuring line. Measure the height of the object from the SP down to the left along this line, as shown in Figure 15.38b. Numbers have been placed at intervals along the measuring lines. These indicate measurements of features on the object.

STEP 5 Project lines from points 1 and 2 to the left of the SP to VPR. Next, draw a line from point 1 at the right of the SP to VPL. Draw lines from points 1 and 2 on the height measuring line VPR. (See Figure 15.38c.) You have now blocked in the overall measurements of the object. The following points have been located: lower front corner (A), upper-right corner (B), upper far corner (C) and upper-left corner (D). Notice in Figure 15.38c that the station point (SP) is the upper-front corner of a box enclosing the object.

STEP 6 Project points B and D to VPV, then project A to VPR and VPL. The intersections of these lines are points E and F, the lower-left and right corners, respectively. Point G is the height of the front corner of the object as shown in Figure 15.38c. Keep in mind that any surface of the object parallel to one of the principal planes of the orthographic view must project to one of the vanishing points as you draw the remainder of the object features. The completed object is shown in Figure 15.38d.

CIRCLES AND CURVES IN PERSPECTIVE

In most cases, circles in perspective appear as ellipses. But if a surface of an object is parallel to the picture plane, any circle on that surface appears as a circle. (See Figure 15.39.) Circles located in planes at an angle to the picture plane appear as ellipses and can be drawn by a method of intersecting lines projected from the elevation and plan view, which is referred to as the coordinate method.

The object having the circles is drawn first in plan and elevation, and all of the necessary lines and points for the

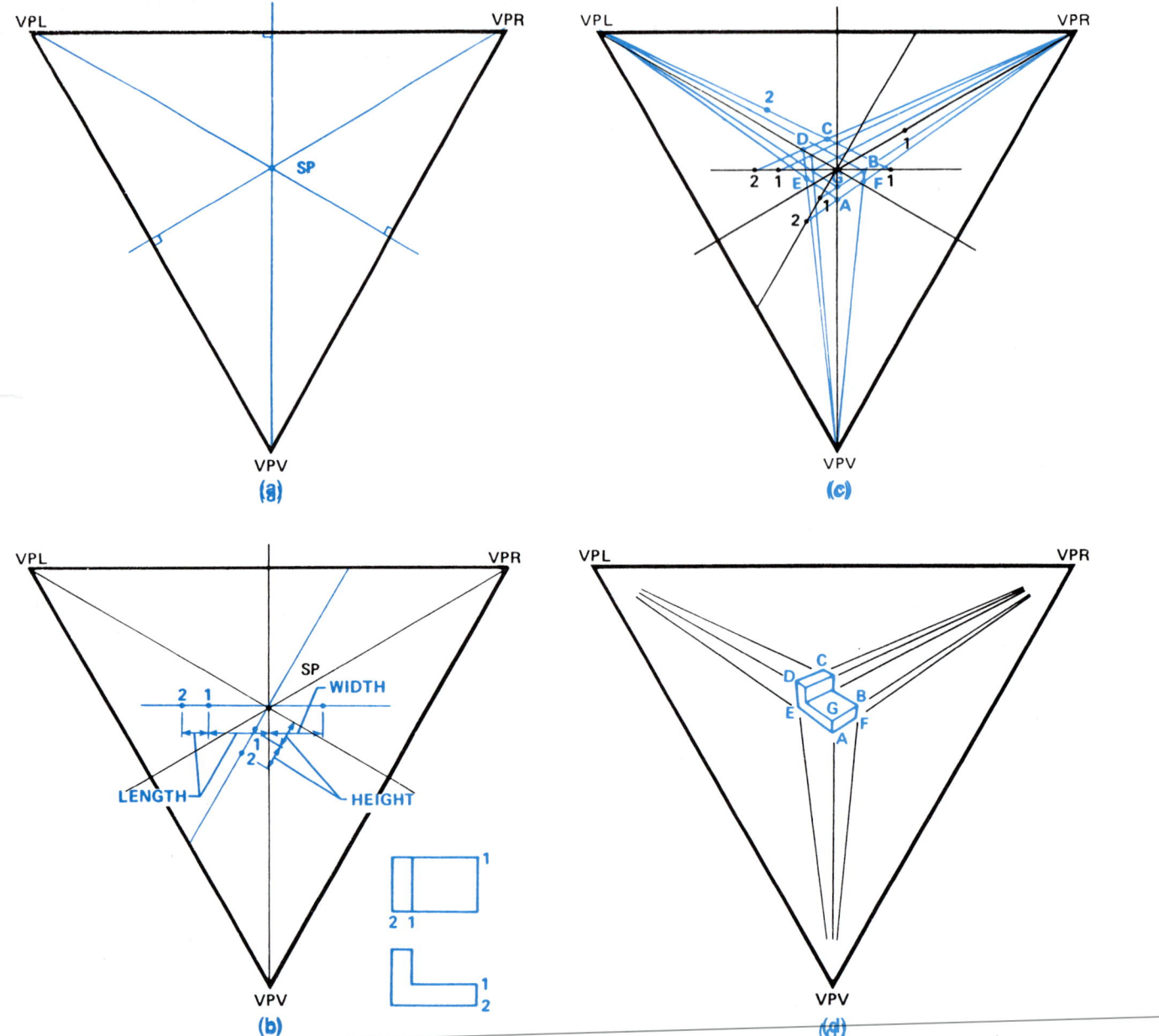

FIGURE 15.38 ■ Constructing a three-point perspective drawing: (a) establishing vanishing points and the station point; (b) establishing the length, width, and height of the object; (c) establishing the outline of the object; (d) the completed object.

perspective drawing are determined and placed on the drawing. (See Figure 15.40.) Next, divide the circle in the elevation view into a convenient number of pie-shaped sections. Project the intersections of these section lines with the circle to the top and side of the object. The points along the side of the object can then be projected onto the perspective view. Now, transfer the distances formed by the intersection of the top of the object and the lines projected from the pie-shaped sections in the elevation view to the plan view. Next, draw visual projection lines from the SP to the points in the plan

view. Where the visual projection lines intersect the picture-plane line, project the points straight to the perspective view. These projectors intersect the ones drawn from the elevation view. Connect these intersection points with an irregular curve or an ellipse template.

The same method can be used to draw irregular curves. Establish a grid, or coordinates, on the curve in the elevation view and place the same divisions on a plan view. Project these two sets of coordinates to the perspective view and then use an irregular curve to connect the points. (See Figure 15.41.)

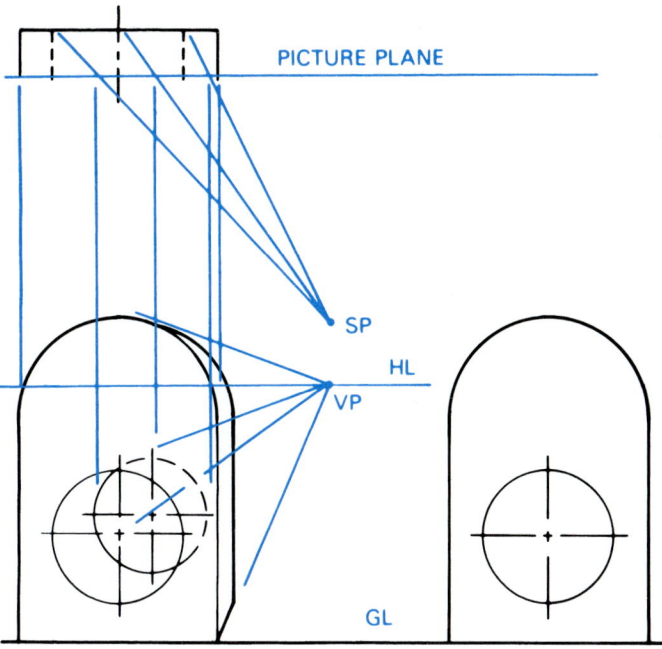

FIGURE 15.39 ■ Circles in surfaces parallel to the picture plane.

BASIC SHADING TECHNIQUES

A variety of shading techniques can be used to enhance your pictorial drawings. The methods discussed in the following text are basic line and block shading. Illustrators, graphic artists, and in-dustrial designers often use tools including pencil, paper, markers, and computers with specialized hardware and software. Traditional pencil and paper with color markers are commonly used in a creative and artistic manner for many pictorial drawings. Airbrush tools also are used for shading applications. The illustrator uses a light touch for layout and harder pressure for solid lines and forms when working on the electronic tablet. A computer stylus and tablet simulate the feeling of pencil and paper, providing a direct relationship between the illustrator's brain and the drawing. The illustrator starts by drawing shapes, then renders as needed, including highlights for depth and appearance. Modifying the computer-generated design sketch is easy and efficient, especially with the aid of commands such as undo, which can be used as often as needed to remove unwanted work. The computer and its peripheral pressure sensitivity tablet are also used by today's illustrator as shown in Figure 15.42.

Line Shading

Most pictorial drawings are created to illustrate shape description and to show the relationship of parts in an assembly. The objects shown in Figure 15.43 illustrate a basic form of shading, called **line-contrast shading**. Vertical lines opposite the light source and bottom edges of the part are drawn with the thick lines. Some illustrators outline the entire object in a heavy line to make it stand out as shown in Figure 15.43b.

Straight-line shading is a series of thin straight lines that can be varied to achieve any desired shading. They can be used on flat and curved surfaces. Notice in Figure 15.44 the two ways that straight lines can be used on curved surfaces.

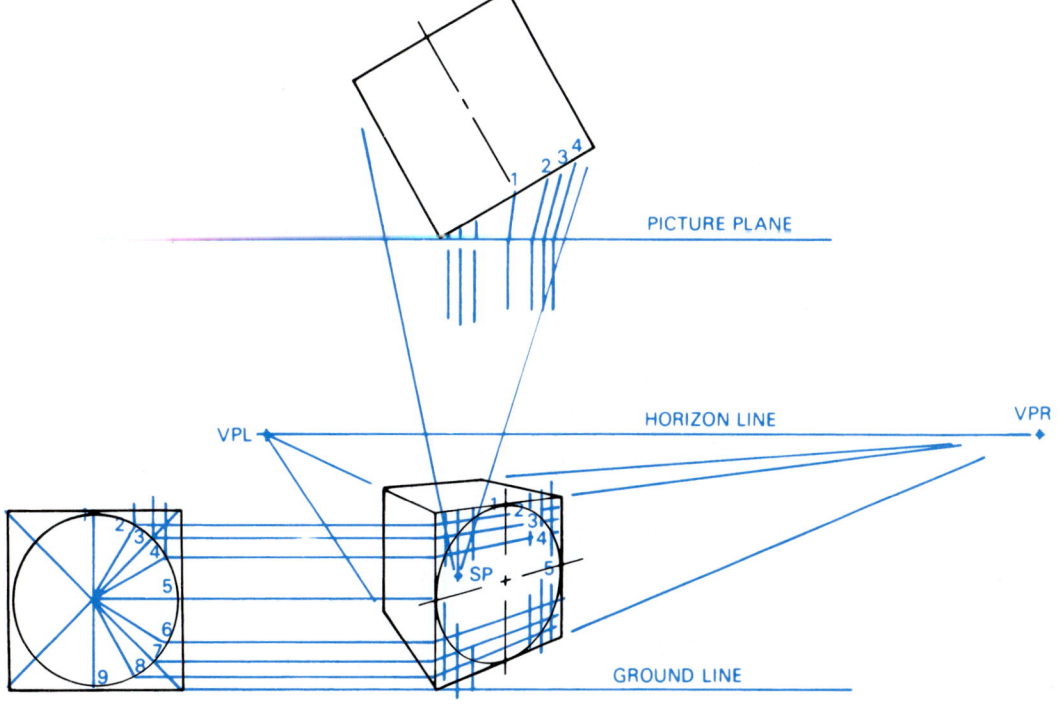

FIGURE 15.40 ■ Circles plotted in perspective by the coordinate method.

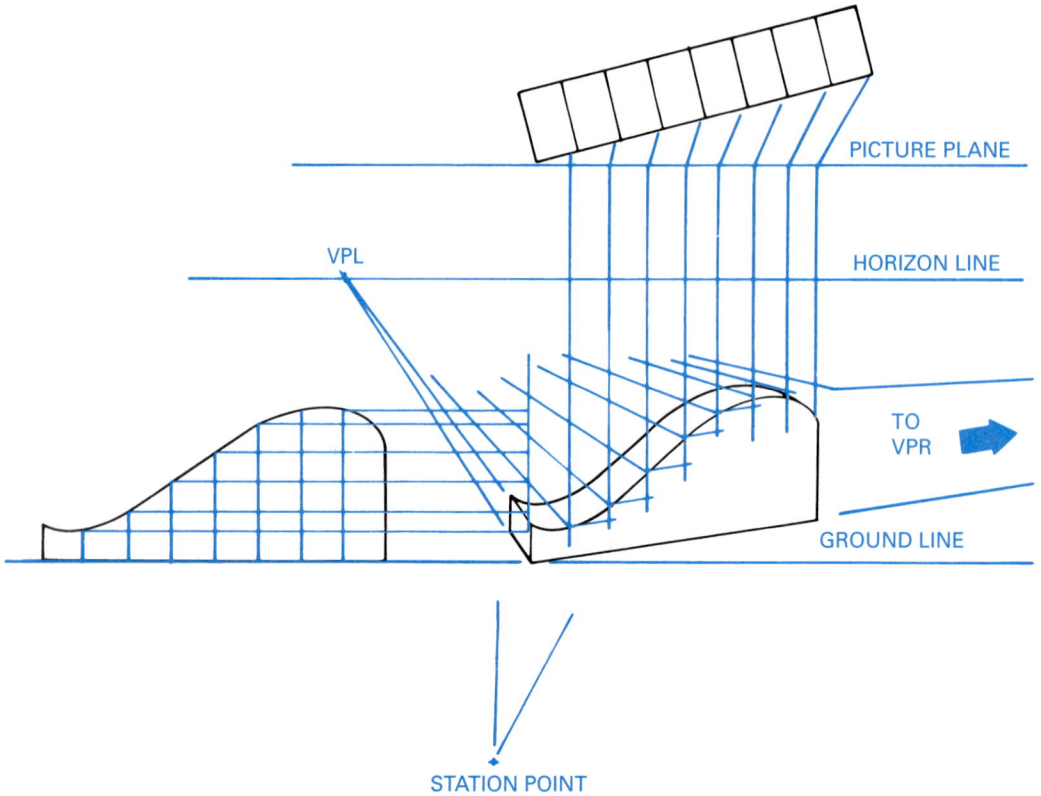

PICTURE PLANE

HORIZON LINE

VPL

TO
VPR

GROUND LINE

STATION POINT

FIGURE 15.41 ■ Irregular curve plotted in perspective by the coordinate method.

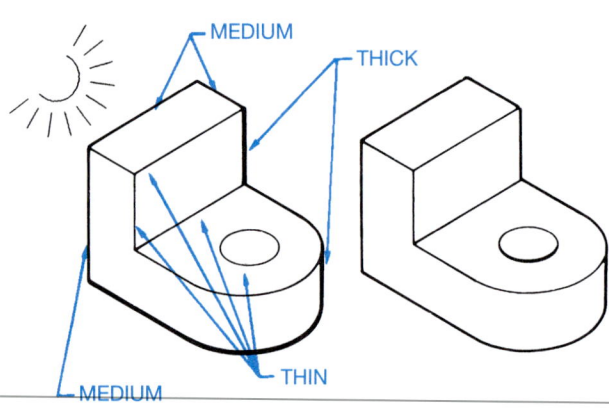

MEDIUM

THICK

THIN

MEDIUM

(a) LINE CONTRAST   **(b)** OUTLINE

FIGURE 15.43 ■ Line-contrast shading.

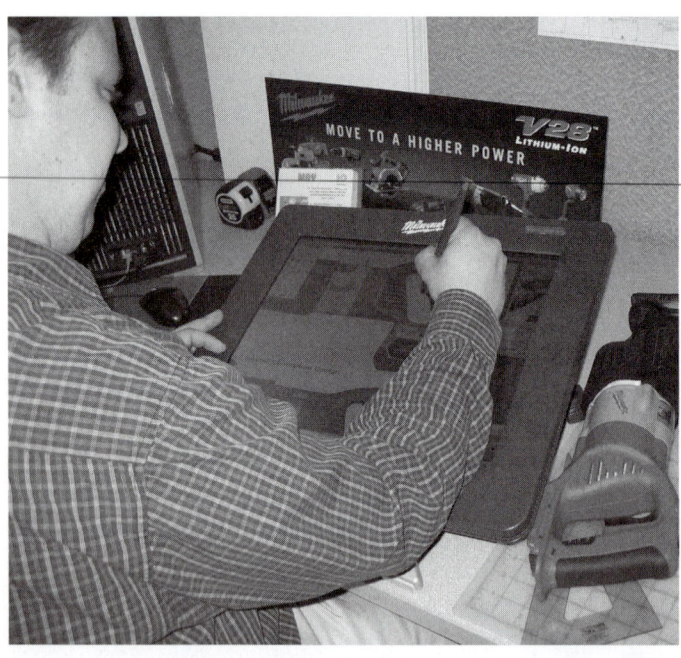

FIGURE 15.42 ■ The computer and its peripheral pressure sensitivity tablet are also used by today's illustrator. *Courtesy Milwaukee Electric Tool Corporation.*

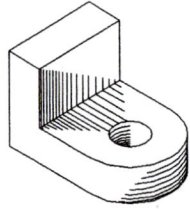

FIGURE 15.44 ■ Straight-line shading.

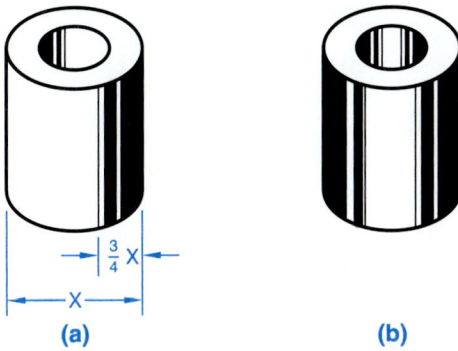

FIGURE 15.45 ■ Block shading.

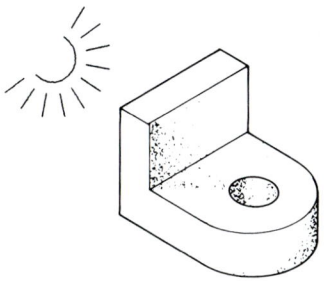

FIGURE 15.46 ■ Stipple shading.

Additional emphasis can be given to curved surfaces with the use of **block shading**. Figure 15.45 shows how block shading on one or both sides of a curved surface can produce a highlight effect. One to three lines of shading are normally used to achieve the block effect. The total amount of block shading should be approximately one-third the width of the feature.

Another type of shading is called **stipple shading**, which is dots. (See Figure 15.46.) The closer the dots are, the darker the shading looks. Although stippling takes longer than line shading, the results can be pleasing.

Representing Fillets and Rounds in Pictorial Drawings

Fillets and rounds can be represented in three ways, as shown in Figure 15.47. Figure 15.47a uses three lines to indicate the two outside edges or tangent points of the radius and the centerline of the radius. Notice the occasional gaps in the outside line, giving a contour effect.

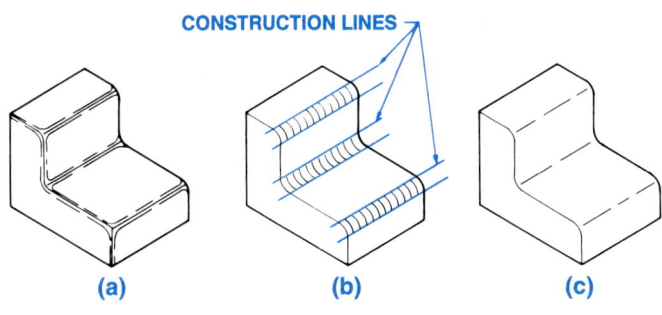

FIGURE 15.47 ■ Depicting fillets and rounds.

With the technique in Figure 15.47b, the same size elliptical arc is moved along the axis of the radius and repeated at regular intervals. It is important to draw construction lines first for the outside edges of the fillet or round. This ensures that all of the elliptical arcs are aligned. Draw these elliptical arcs with a thin pencil point. Using a CADD system, you can copy or array the arcs and you do not need to draw the construction lines.

The example shown in Figure 15.47c is the simplest and requires the least amount of time to draw. Using this method, a broken line is drawn along the axis of the radius to indicate the curved surface.

PICTORIAL DRAWING LAYOUT TECHNIQUES

The best way to begin any pictorial drawing, especially if you are working from an orthographic drawing, is to make a quick freehand sketch. This allows you to see the part in a 3-D form before you try to create a formal drawing.

Objects that have rectangular or angular shapes should be drawn by first laying out the overall sizes of the part by boxing in the shape. Then begin to cut away from the box using measurements from the part or drawing. That is called the **coordinate**, or **box method**. Use construction lines to assist in the layout, whether a manual or a CADD drawing. If the object has circular features, use the following layout techniques:

STEP 1   Lay out the centerlines and axis lines of the circular features. (See Figure 15.48a.)

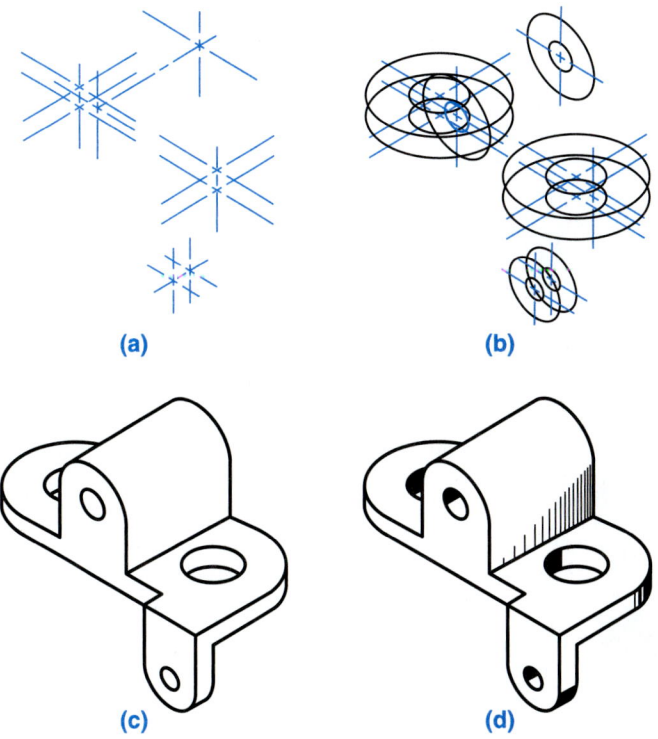

FIGURE 15.48 ■ (a) Lay out centerlines and axis lines of circular features; (b) draw circular features using ellipses; (c) join ellipses with straight lines; (d) complete the drawing with shading.

3-D CAPABILITIES

Many 3-D CADD techniques were given throughout this chapter where they apply to specific applications.

Computer-generated pictorial line drawings have given drafters, designers, and engineers the ability to draw an object once and then create as many different displays of that object as needed. Three-dimensional wire-form capabilities, seen in Figure 15.49, are available with most commercial CADD systems. They allow the operator to view the object as if it is constructed of wire. All edges can be seen at the same time. This is somewhat limiting because complex parts soon become a maze of lines. The surface form shown in Figure 15.50, is a step closer to reality because it shows only the surfaces that would actually be seen.

The greatest realism in pictorial presentational can be achieved by CADD systems that possess color solids modeling capabilities. These systems allow you to draw the object, shade or color its different parts or surfaces, and rotate it in any direction desired. You can also cut into the object at any location to view internal features and then rotate it to achieve the best view. This screen display can then be made into a hardcopy or a video format.

Chapter 4, 3-D CADD, Animation, and Virtual Reality, provides detailed coverage of the various types of 3-D CADD modeling options. A review of Chapter 4 is suggested to give you a comparison and the difference between 3-D CADD applications and the techniques explained in this chapter.

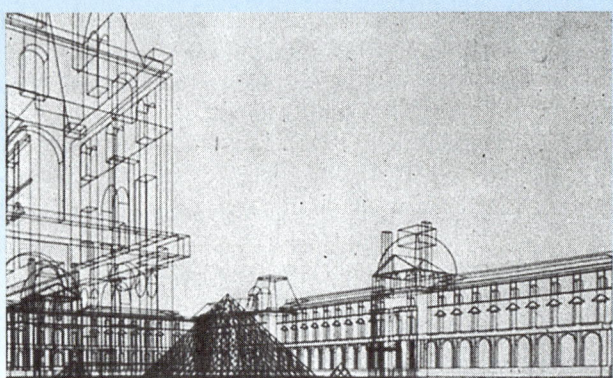

FIGURE 15.49 ■ The Louvre art museum in Paris, France, drawn as a wire form. *Courtesy Computervision Corporation.*

FIGURE 15.50 ■ Another view of the Louvre, drawn as a solid model. *Courtesy Computervision Corporation.*

An isometric drawing is constructed using the actual measurements of the part. But because isometric drawing does not take into account foreshortening, the object appears larger in the drawing than it is in reality. To compensate for this, isometric projection can be used to show the object in its foreshortened appearance. If a part measures one inch on a side, a true isometric projection creates an isometric line measuring .816 in.

In order, to construct an isometric view of an object that is a true isometric projection, it is necessary to multiply the actual measurements by .816 to get the drawing measurements. The following measurements are converted as follows:

1.45 in. × .816 = **1.183 in.**

3.67 in. × .816 = **2.995 in.**

.89 in. × .816 = **.726 in.**

STEP 2 Draw the circular features using ellipses with the centerlines as a guide. (See Figure 15.48b.)

STEP 3 Join ellipse contours with straight lines. (See Figure 15.48c.)

STEP 4 Add additional features such as shading, notes, and item tags. If you are drawing manually, trace the object lines in pencil or ink and add shading as required. Any notes or item tags can be added last. If you are working with a CADD drawing, you can edit the drawing so only the object lines show. If you created a construction layer for your CADD drawing, turn off this layer when the drawing is completed. The finished drawing is shown in Figure 15.48d.

PROFESSIONAL PERSPECTIVE

Pictorial drawing requires a good ability to visualize objects in three-dimensional form. Companies that produce parts catalogs, instruction manuals, and presentation drawings require the services of a technical illustrator or someone skilled in pictorial drawing. A part of your professional portfolio should be 8" × 10" photo reductions or laser prints of your best pictorial drawings.

The field of technical illustration is often more limited than engineering and drafting but is available for people with talent.

The area of industrial design, discussed in Chapter 27, is a great place for you to demonstrate your creative abilities and your illustrative talents. Always keep examples of a variety of pictorial drawing types in your portfolio. Your portfolio should be professional looking and show examples of the different types of designs and drawings you have created. Manila pocket folders that fit into the binder are excellent for holding folded prints. Remember that often a small freelance job can lead to bigger jobs and maybe even to owning your own company.

CHAPTER 15 — *Pictorial Drawings and Technical Illustrations Test*

Access the CD found with this textbook to view the Chapter 15 Test. Confirm the preferred submittal method with your instructor.

CHAPTER 15 — *Pictorial Drawings and Technical Illustrations Problems*

DIRECTIONS

Choose the best axis to show as many features of the object as possible in your axonometric or oblique drawing. Problem assignments are presented as engineering designs or sketches and may not match proper ASME standards.

For axonometric problems—Problems numbered 15.1 through 15.10—draw isometric, dimetric, or trimetric as assigned. For oblique problems—Problems numbered 15.11 through 15.17—draw cavalier, cabinet, or general oblique as assigned. Remember that circular features are best shown in the front plane of the oblique view.

For perspective problems—Problems numbered 15.18 through 15.25—make a one-, two-, or three-point perspective drawing as assigned except for Problem 15.18, which should be done as a one-point perspective view. All objects can be turned at any desired angle on the picture-plane line for viewing from the station point, except Problem 15.18, which should be drawn in the direction indicated.

1. Make a freehand sketch of the object to assist in visualization and layout of axonometric and oblique problems.
2. For axonometric and oblique problems, select a scale to fit the drawing comfortably on an A- or B-size drawing sheet. Use a C- or D-size drawing sheet for drawing an initial layout of the perspective problems on sketch paper or butcher paper.
3. Dimension axonometric or oblique problems only if assigned by your instructor. Do not place dimensions on a perspective view.
4. Perspective objects without dimensions can be measured directly on the given problem and scaled up as indicated or assigned.
5. Trace the final perspective view in pencil or ink on vellum or polyester film, unless otherwise specified by your instructor.
6. Make your drawings using a CADD system if appropriate with course guidelines.

PROBLEM 15.1   **Axonometric projection (in.)**

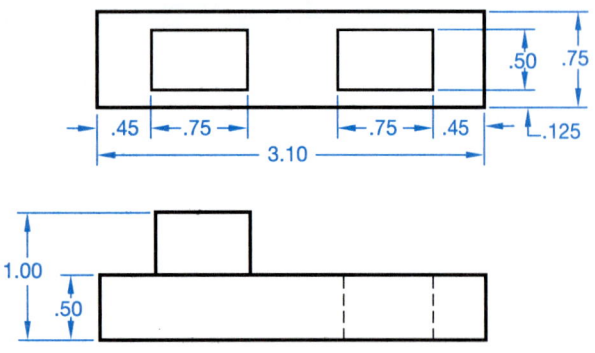

PROBLEM 15.4   **Axonometric projection (metric)**

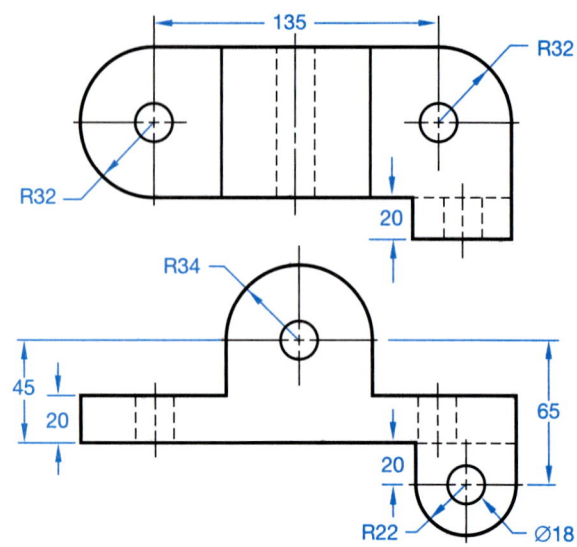

PROBLEM 15.2   **Axonometric projection (in.)**

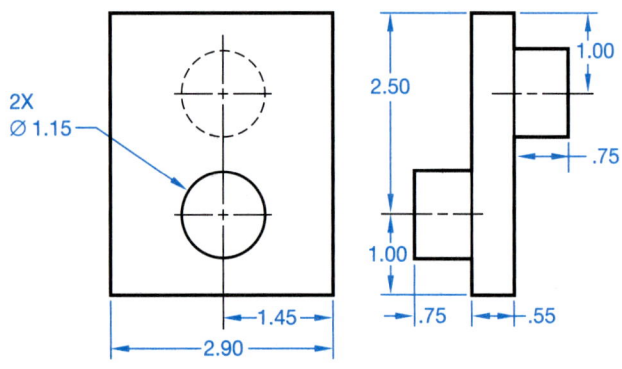

PROBLEM 15.5   **Axonometric projection (in.)**

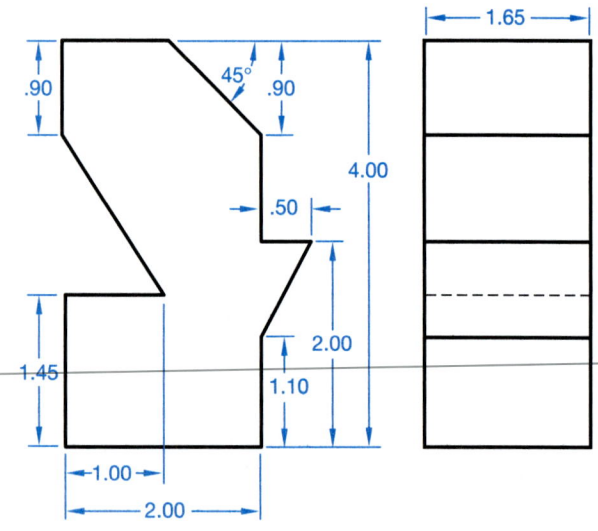

PROBLEM 15.3   **Axonometric projection (in.)**

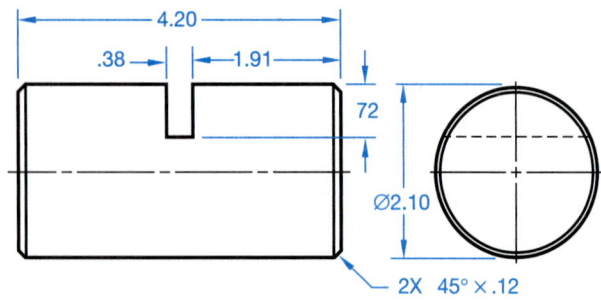

 PROBLEMS 15.6 through 15.10: Access the CD found with this textbook and open the problem of your choice, or as assigned by your instructor. Solve the problems using the instructions provided with this chapter or on the CD, unless otherwise specified by your instructor.

PROBLEM 15.11   Oblique projection (in.)

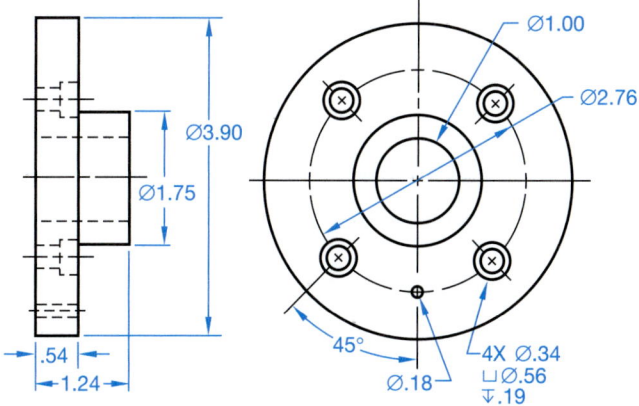

PROBLEM 15.12   Oblique projection (in.)

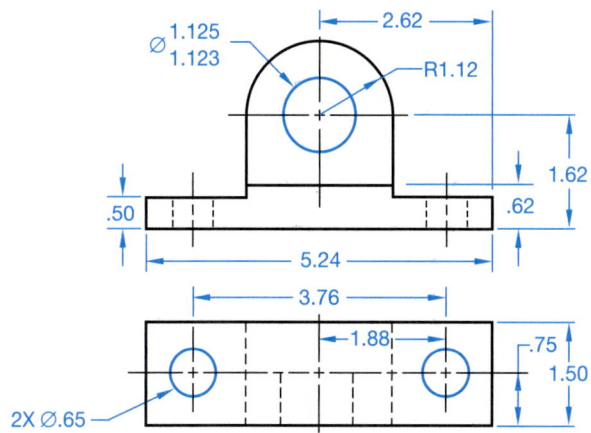

PROBLEM 15.13   Oblique projection (in.)

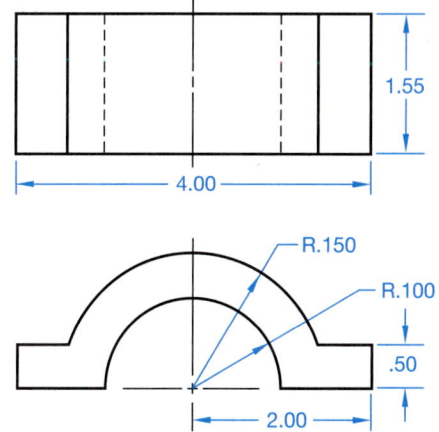

PROBLEMS 15.14 through 15.17: Access the CD found with this textbook and open the problem of your choice, or as assigned by your instructor. Solve the problems using the instructions provided with this chapter or on the CD, unless otherwise specified by your instructor.

PROBLEM 15.18   Perspective

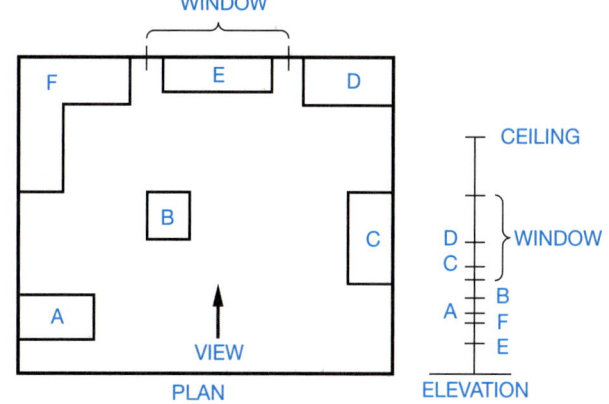

PROBLEM 15.19   Perspective

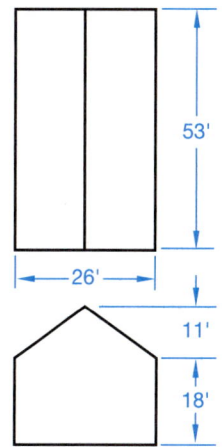

PROBLEMS 15.20 through 15.25: Access the CD found with this textbook and open the problem of your choice, or as assigned by your instructor. Solve the problems using the instructions provided with this chapter or on the CD, unless otherwise specified by your instructor.

MATH PROBLEMS

PROBLEMS 15.26 through 15.30: Access the CD found with this textbook and open the math problem of your choice, or as assigned by your instructor. Solve the problem or problems using the instructions provided.

Descriptive Geometry

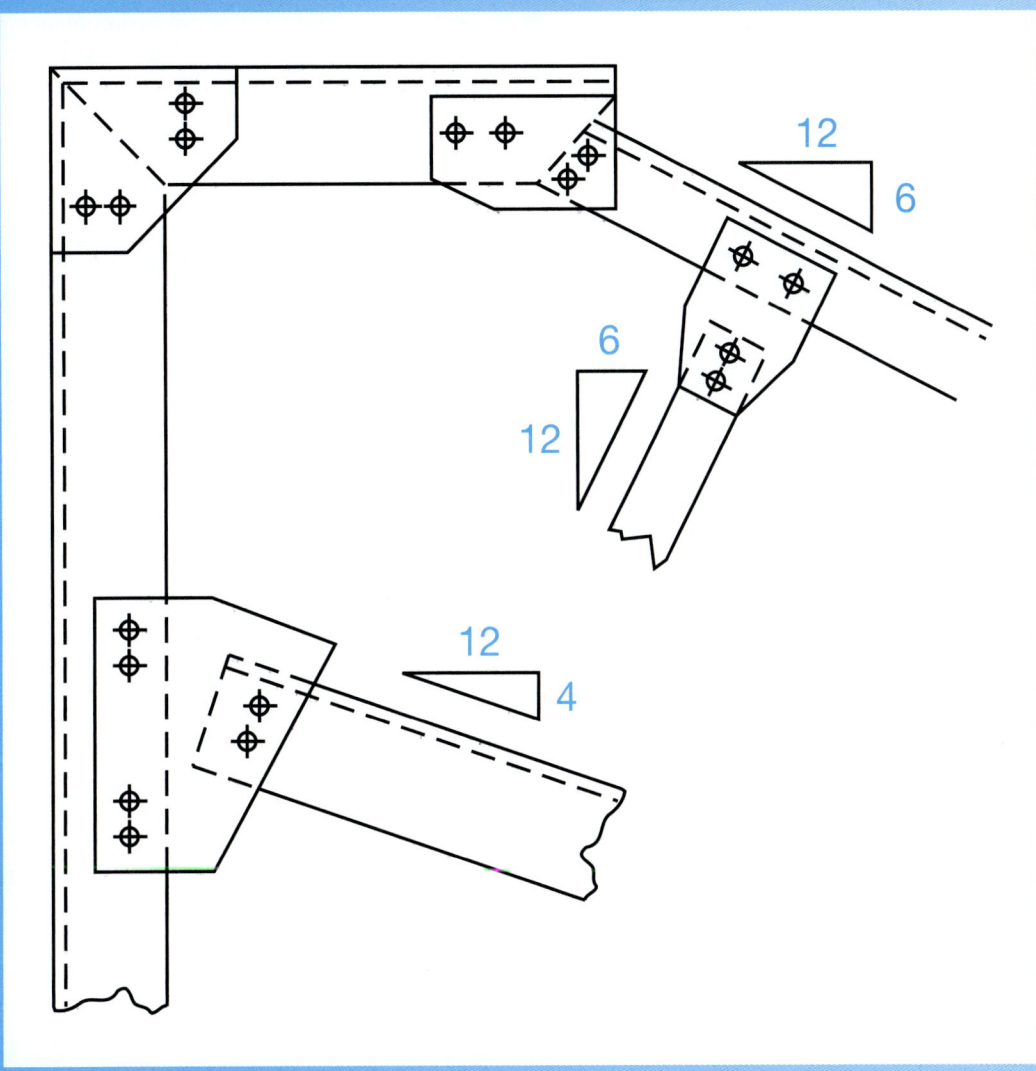

Descriptive Geometry I

LEARNING OBJECTIVES

After completing this chapter, you will be able to:

- Project a point onto various projection planes when given two adjacent views of the point.

- Project a line onto various projection planes; project a line onto a projection plane showing the line in *true length*; and project the true-length line onto a projection plane where it

appears as a point when given two adjacent views of the line.

- Project a plane onto various projection planes; project a plane onto a projection plane that shows the plane as an *edge view*, and project the edge view of the plane onto a projection plane that shows the plane in *true shape* when given two adjacent views of the plane.

THE ENGINEERING DESIGN APPLICATION

Descriptive geometry principles are valuable for determining true shapes of planes, angles between two lines, two planes, or a line and a plane, and for locating the intersection between two planes, a cone and a plane, or two cylinders. Problems are solved *graphically*, by projecting points onto selected adjacent projection planes in an imaginary projection system.

In this chapter, an imaginary projection plane system is used based on the conventional third-angle projection system used in positioning standard multiview drawings. Most of the problems can be solved using just two basic concepts. The first concept is correctly projecting a *point* onto an imaginary projection plane, and the second is knowing when and how to create a fold line (either *parallel* or *perpendicular* to a line) to get the desired results.

In addition to the practical applications of descriptive geometry, these concepts can provide valuable training in visualizing how lines and planes are orientated in space. For those who cannot visualize well, the system of solving descriptive geometry problems used in this chapter works well by memorizing and/or following the step-by-step cookbook approach shown for each example.

There are usually several acceptable ways to solve descriptive geometry problems. An attempt has been made in this chapter to include at least one, and in some cases two, understandable solutions for each problem.

Whether solving descriptive geometry problems by using a straightedge and a triangle on paper or by using CADD, the most important first step in solving the problem is to *define the problem*. The quickest and usually the easiest way to do this is to develop a freehand sketch of the given problem, being sure to include all of the known information.

Secondly, sketch the correct process of your solution, but do not be too concerned with accuracy at this point. When you believe you have sketched a correct process, then draw out the solution, as accurately as possible, using a straightedge and a triangle, or with CADD where, if the process is correct, the solution will be perfect. The CADD system also allows you to work in layers and colors for a clearer picture of the problem and solutions, although if you are using a straightedge and triangle, you can obtain the same effect if different colored pencil leads are used for different lines when solving the problems.

THE DESCRIPTIVE GEOMETRY PROJECTION LAYOUT FORMAT

The way descriptive geometry problems are laid out in this chapter is based on the third-angle projection system used on all regular drawings. An example of the imaginary glass box system drawn about an object is shown in Figure 16.1. The unfolded multiprojection system is shown in a flat layout projection of

that same object in Figure 16.2 where all the projected images of the six standard sides of the object are shown projected onto the six standard views.

In Figure 16.3, the image of a real point located in the space of the imaginary glass box is shown projected onto the three visible sides of the glass box. The real point floating within the glass box is represented with a star and identified with an uppercase letter. Uppercase letter *A* is used in this example, but any letter can be used. The real point is shown projected with projectors

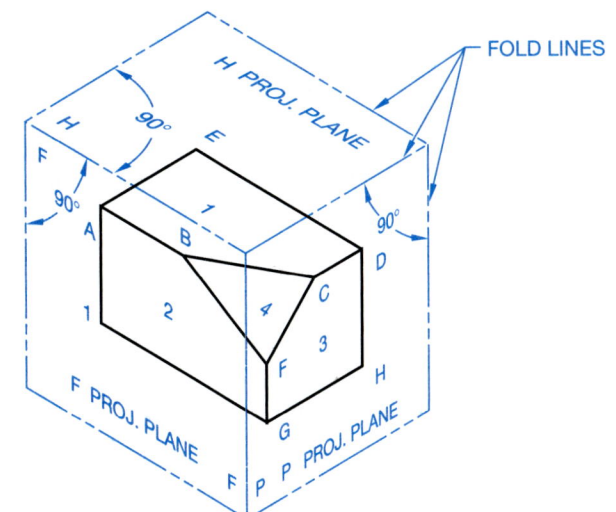

FIGURE 16.1 ■ 3-D box layout projection.

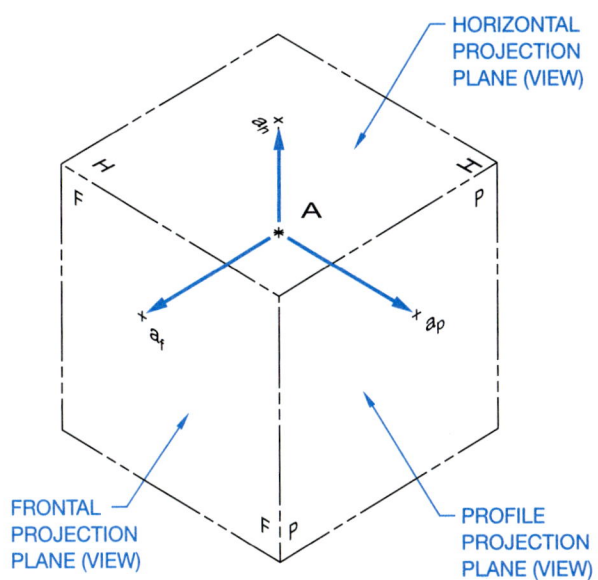

FIGURE 16.3 ■ Image of real-point projection.

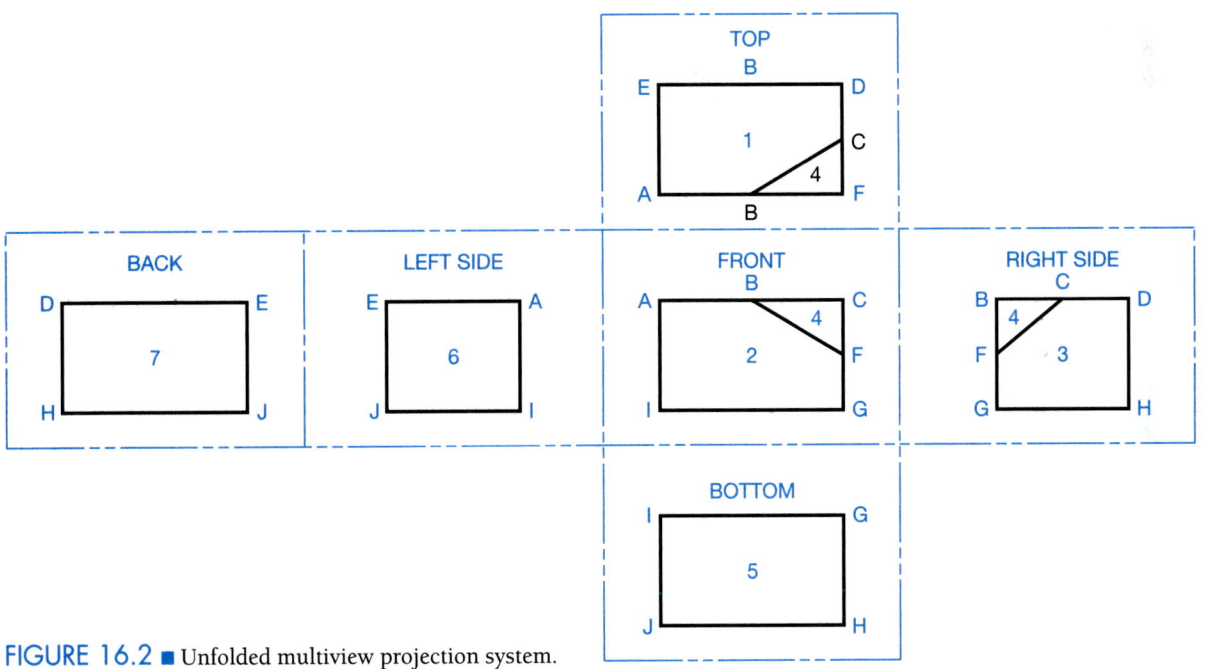

FIGURE 16.2 ■ Unfolded multiview projection system.

radiating **perpendicular (90°)** from the real point onto the three sides of the imaginary glass box. The real point is used to solve any of the problems. Where the real point is projected onto the frontal projection plane of the imaginary glass box, a **cross** is located. This cross represents the image of the real point projected onto the frontal plane and is identified as a_f. The lowercase letter a corresponds to the real point, and the lowercase subscript letter f corresponds to the frontal projection plane. On the profile projection plane, the cross representing the projected point is identified as a_p, and on the horizontal projection plane, the cross representing the projected point is identified as a_h.

Thick phantom lines called **fold lines** represent the edges of the imaginary glass box. Two uppercase letters corresponding to

the two adjacent (next to) projection planes identify each fold line. In this example, F/H identifies the fold line between the frontal and horizontal projection planes, F/P identifies the fold line between the frontal and profile projection planes, and H/P identifies the fold line between the horizontal and profile projection planes.

Figure 16.4 shows the images of the same point A shown in Figure 16.3, but in a multiview flat layout system. The point is identified in the same manner as in Figure 16.3. The projection planes are now called **views** just as they are in the standard third-angle projection system. Each adjacent view is shown separated by the same fold line as seen in Figure 16.3 with the same identification uppercase letters. The other fold lines shown,

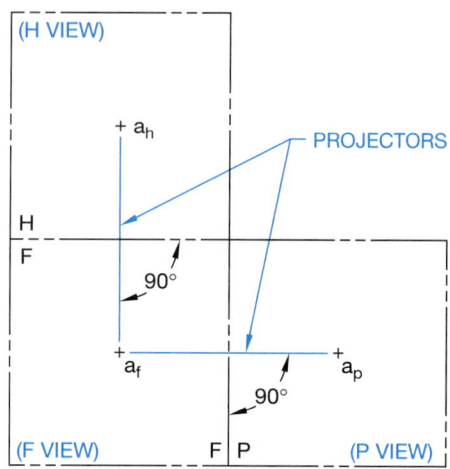

FIGURE 16.4 ■ Flat layout of one-point projection.

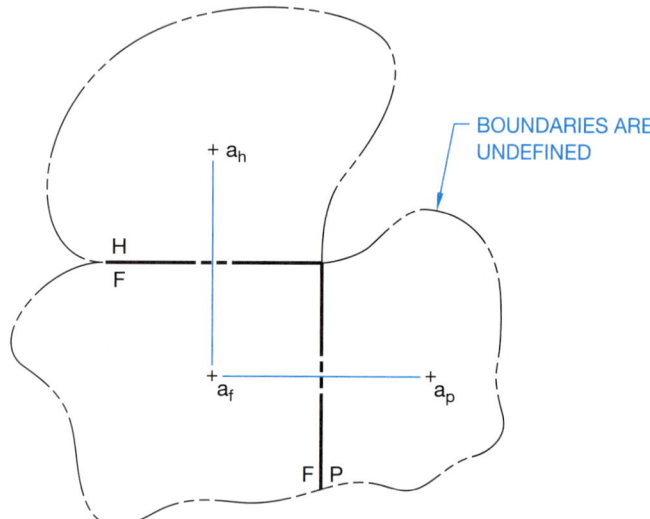

FIGURE 16.5 ■ Flat layout with unidentified boundaries omitted.

which do not have uppercase identification letters, merely represent the other imaginary boundaries of each view. These fold lines are not identified because there was no adjacent view given in Figure 16.3. The lines of projection (projectors) from point A are also included. Notice that the projectors cross each fold line at a 90°, or right angle. This is important to remember when you begin projecting points.

Figure 16.5 shows the transition from the complete multiview flat layout system to using the minimal information layout system that will be used in this chapter. This is exactly the same multiview flat layout system as shown in Figure 16.4, but the unidentified fold lines are now omitted, because they represented arbitrary boundaries that are not used.

Rule 1

All given problems in descriptive geometry must include **two adjacent views** of the given points, lines, and/or planes. Two adjacent views are needed to solve the problems because one view contains only two of the three dimensions needed to solve descriptive geometry problems. In Figure 16.6, notice that the front view includes the up, down, right, and left dimensions only, while the adjacent top and profile views contain the additional forward and back (depth) dimensions needed to complete a three-dimensional system.

Rule 2

When a projector crosses over a fold line, it must always cross *perpendicular* to that fold line. This represents a change of 90° in direction from one view to another as shown in Figure 16.7.

PROJECTION OF A POINT

Knowing how to project an image of a real point in the flat projection system is fundamental to solving most of the problems in applied descriptive geometry. This concept is simple but must be understood completely before attempting to solve the descriptive geometry problems.

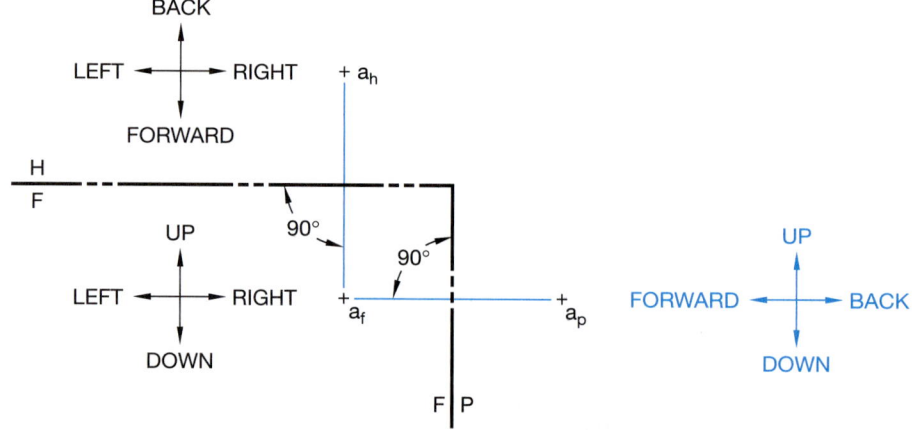

FIGURE 16.6 ■ Example showing that each view contains only four different directions.

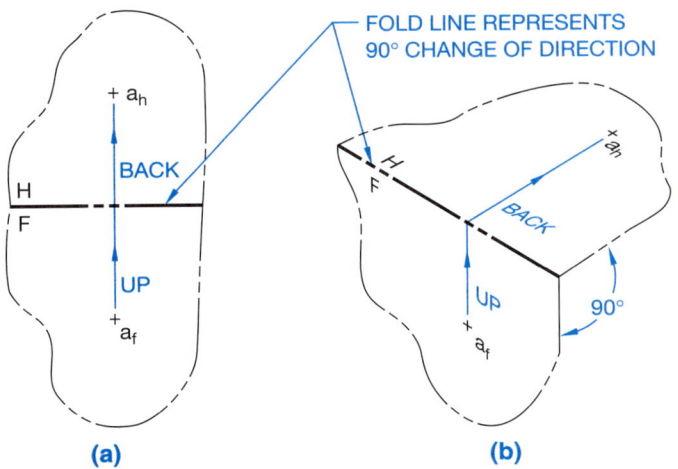

(a)   **(b)**

FIGURE 16.7 ■ (a) Showing fold line changing direction in flat layout. (b) Showing fold line changing directions in 3-D layout.

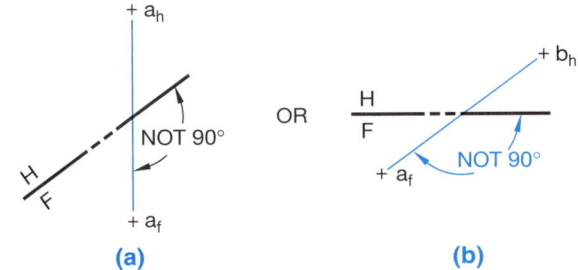

(a)   **(b)**

FIGURE 16.9 ■ (a) Fold line placed incorrectly. (b) Two different points assumed to be the same.

Sample Problem

Project a given point correctly onto a new view. Things that must be given to solve this problem:

1. Two adjacent views of the point.

2. A fold line located correctly between the two given adjacent views.

3. Another fold line that represents one edge of the new view onto which the point is to be projected.

As shown in Figure 16.8, point A is located in the front view with a cross and is identified by the letters a_f. In the horizontal view, point A is also shown with a cross, but is identified with the letters a_h. These two views are adjacent as indicated by the fold line F/H located between them. The combined dimensions shown in both views supply the needed three dimensions to solve the problem. When a projector is drawn connecting both views of point A, it crosses the fold line F/H at a 90° angle. This verifies that point A is correctly located in both adjacent views. If a projector connected to both views of point A (a_f to a_h) did not cross the fold line at 90°, the fold line could be located at a wrong angle between the points, as shown in Figure 16.9a. Or the problem could be that the drawing really shows one view of two different points, as shown in Figure 16.9b. If either condition exists, not enough in-

formation has been given to solve the problem. Finally, another fold line F/P representing one known edge of the P view is given. (See Figure 16.8 again.) This represents the edge of the new view (view P) onto which point A is to be projected.

Solving the Problem

Given the problem shown in Figure 16.10:

STEP 1   Draw a projector (projector 1) from point a_f to point a_h. (See Figure 16.11.) Projector 1 must cross the fold line F/H at a 90° angle, making it perpendicular to the fold line F/H.

STEP 2   Draw a new projector (projector 2) from point a_f perpendicular to the fold line F/P and extend projector 2 into the P view at any given distance at this time. Point A lies somewhere along projector 2 in the P view. (See Figure 16.12.) Next, draw in a little angle bracket close to where projector 2 and the fold line F/P intersect just to reinforce the idea that a projector must always be

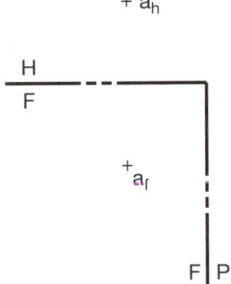

FIGURE 16.10 ■ A given problem.

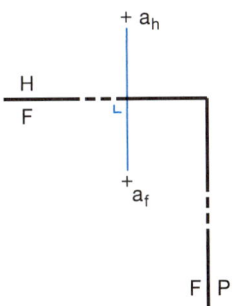

FIGURE 16.8 ■ Correctly given problem to project a point.

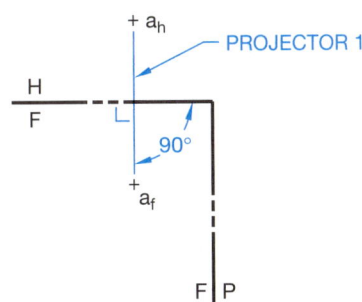

FIGURE 16.11 ■ Drawing the first projector.

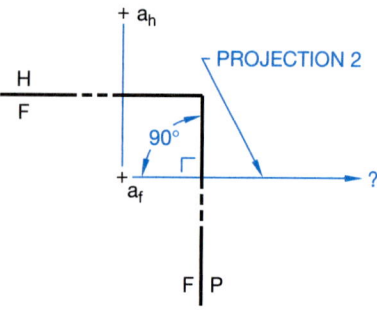

FIGURE 16.12 ■ Drawing the second projector.

drawn at a 90° angle to a fold line and that it was done in this case.

STEP 3 Measure the perpendicular distance from point a_h to the fold line *H/F* in the *H* view. Identify this distance with a bracket and the number 1. Transfer this distance 1 along projector 2, measuring from the fold line *F/P.* Identify this distance with a similar bracket and the same number, 1. Do this just to reinforce the idea that this distance 1 must be the same distance as the other distance 1 and that it was correctly identified. Make a cross representing point *A* on projector 2 at this distance. Point *A* now has been correctly projected onto the *P* view. Identify the point (cross) as a_p. (See Figure 16.13.)

Note: This problem represents only *one point*, point *A*, projected onto three different views, *not three points* projected onto three different views.

To project a point into an auxiliary view, the same process is used. The only difference is that the fold line representing the

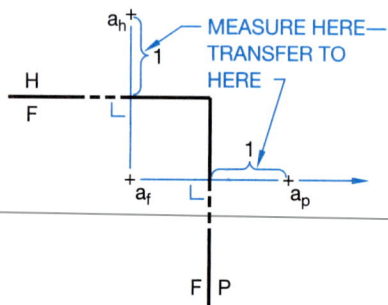

FIGURE 16.13 ■ Transferring dimension 1.

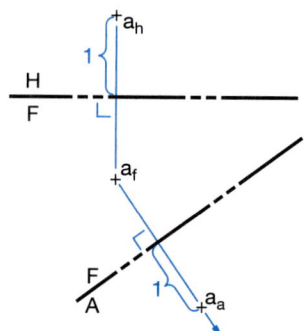

FIGURE 16.14 ■ Locating a point in an auxiliary view.

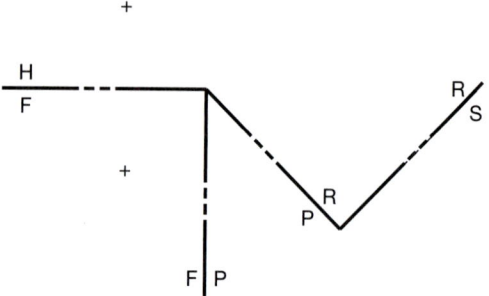

FIGURE 16.15 ■ A given multiview problem.

edge of the given auxiliary view is at a different angle than that in the standard views. (See Figure 16.14.)

To project a point through successive (more than one) auxiliary views, you need to repeat the three-step process for each given auxiliary view. The only "tricky" part of this process is to understand *where* you get the dimension from to locate the point in each additional auxiliary view. The problem given in Figure 16.15 is similar to the problem given in Figure 16.10, but two additional fold lines, *P/R* and *R/S*, have been added. This problem requires you to find point *A* in the auxiliary view *R* and in the auxiliary view *S*.

STEP 1 *Project point A onto the P view.* To solve this problem, you must first project point *A* onto the *P* view just as was done for the problem in Figure 16.10.

STEP 2 *Project point A onto the R view.* To project point *A* onto the *R* view, imagine that only views *F*, *P*, and the edge (fold line *P/R*) of view *R* exist as shown in Figure 16.16. Now project point *A* onto the *R* view using the same three steps as used before.

STEP 2a Draw a third projector, projector 3, from point a_p perpendicular to the fold line *P/R* and extend projector 3 into the *R* view at any given distance at this time. Point *A* will lie somewhere along projector 3 in the *R* view. (See Figure 16.16 again.)

Next, draw in a little angle bracket close to where projector 3 and the fold line *P/R* intersect, again just to reinforce the idea that a projector must always be drawn at a 90° angle to the fold line and that it was done in this case.

STEP 2b Locate the correct placement of point *A* in the *R* view. When determining the correct distance for point *A* in

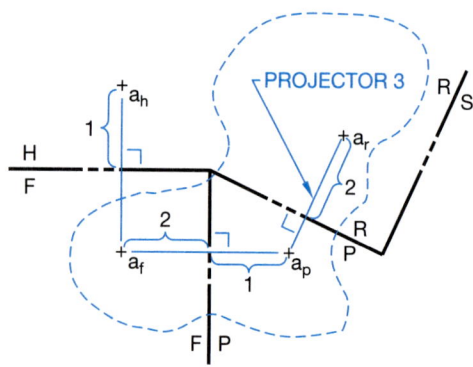

FIGURE 16.16 ■ Transferring distances 1 and 2 to the correct views.

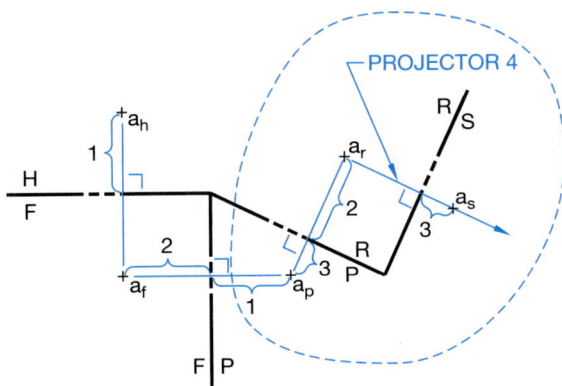

FIGURE 16.17 ■ Transferring distance 3 to the S view.

the *R* view, you need to be careful where to find this correct distance. Notice it will be from the *F/P* fold line to point *a_f* skipping the *P* view. Identify this distance with a bracket and the number 2.

Transfer this distance 2 measuring from the fold line *P/R* along projector 3. Identify this distance with a bracket and the same number 2 just to confirm that the correct distance was used.

Identify the point *A* with a cross and the letters *a_r*.

STEP 3 *Project point A onto the S view.* To project point *A* onto the *S* view, imagine that only views *P, R,* and the edge (fold line *R/S*) of view *S* exist as shown in Figure 16.17. Notice that each part of the problem is being solved using groups of three elements, the *two known views and the one unknown view.* Repeat the process to project point *A* onto the *S* view the same way as was done for the *R* view.

STEP 3a Draw a fourth projector, projector 4, from point *a_r* *perpendicular* to the fold line *R/S* and extend projector 4 into the *R* view at any given distance at this time. Point *A* will lie somewhere along projector 4 in the *S* view. (See Figure 16.17 again).

Next, draw in a little angle bracket close to where projector 4 and the fold line *R/S* intersect, again just to reinforce the idea that a projector must always be drawn at a 90° angle to the fold line and that it was done in this case.

STEP 3b Locate the correct placement of point *A* in the *S* view. When determining the correct distance for point *A* in the *S* view, you need to be careful where to find this correct distance. Notice it is from the *P/R* fold line to point *a_p* skipping the *R* view. Identify this distance with a bracket and the number 3.

Transfer this distance 3 measuring from the fold line *R/S* along projector 4. Identify this distance with a bracket and the same number 3 just to confirm that the correct distance was used.

Identify the point *A* with a cross and the letter *a_s*.

WORKING WITH LINES

The following are definitions related to lines and working with lines:

- **Straight line:** The shortest distance between two points. (See Figure 16.18.)
- **True-length line (TL):** The actual measured length of a line. (See Figure 16.19.)
- **Foreshortened line:** A line that *appears* shorter than it really is. (See Figure 16.20.)
- **Parallel lines:** Lines that are equal distance from each other throughout their lengths. (See Figure 16.21.) Parallel lines will appear parallel in all views with only *two exceptions.* You

FIGURE 16.18 ■ Two points representing a straight line.

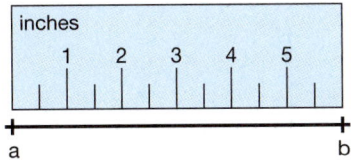

FIGURE 16.19 ■ True length represented by true measurement.

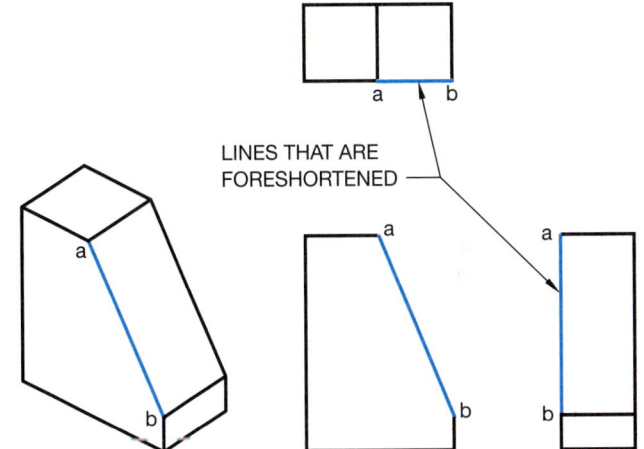

FIGURE 16.20 ■ A foreshortened line represented by line *A-B.*

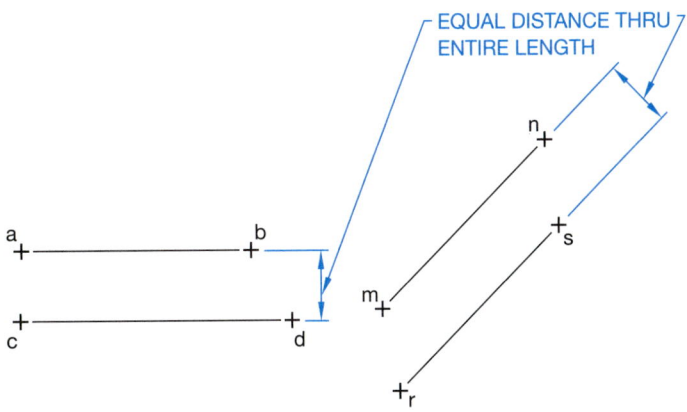

FIGURE 16.21 ■ Parallel lines.

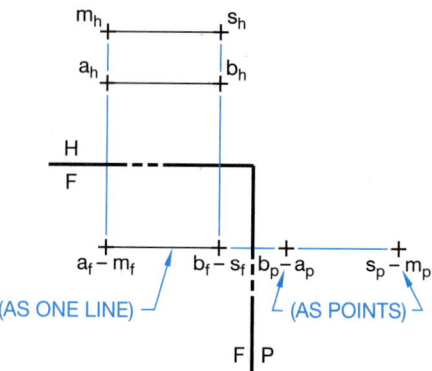

FIGURE 16.22 ■ Parallel lines viewed as one line in the first view and as points in the second view.

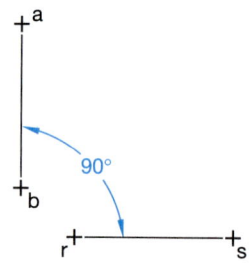

FIGURE 16.23 ■ Perpendicular lines.

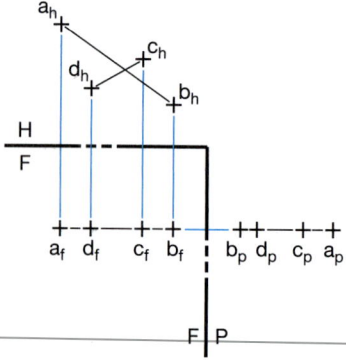

FIGURE 16.24 ■ Intersecting lines.

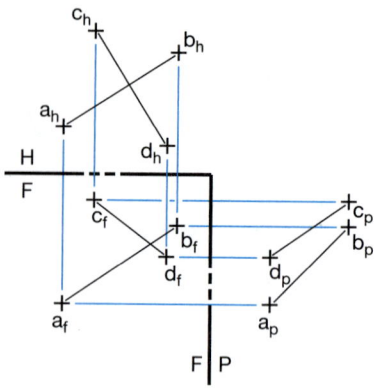

FIGURE 16.25 ■ Skewed lines.

PROJECTION OF A LINE

Projecting a line in the flat projection systems is done in exactly the same way as projecting two points in space. Remember, a straight line is defined as the shortest distance between *two points*.

Sample Problem

Project line *A-B* correctly onto the *P* view. (See Figure 16.26.) Conditions that must be given to solve this problem:

1. Two adjacent views of the line.

2. A fold line located correctly between the two given adjacent views.

3. Another fold line that represents one edge of the new view onto which the line is to be projected.

STEP 1   Draw a projector (projector 1) from point a_f to point a_h and another projector (projector 2) from point b_f to point b_h. (See Figure 16.27.) Projectors 1 and 2 must cross the fold line *F/H* at a 90° angle, making them perpendicular to the fold line *F/H*. Next, draw little angle brackets close to where projectors 1 and 2 and the fold line *F/H* intersect just to reinforce the idea that a projec-

are not able to see that the lines are parallel when one line is located behind the other; they appear as *one line*. And lines do not appear parallel where both lines are viewed at the ends. Here they appear as *points*. (See Figure 16.22.)

■ **Perpendicular lines**: Lines that are positioned at a 90° angle to each other. (See Figure 16.23.)

■ **Intersecting lines**: Lines that actually intersect each other. They will appear to intersect in all views except where they appear as one line. (See Figure 16.24.)

■ **Skew lines**: Lines that are nonparallel and nonintersecting. (See Figure 16.25.)

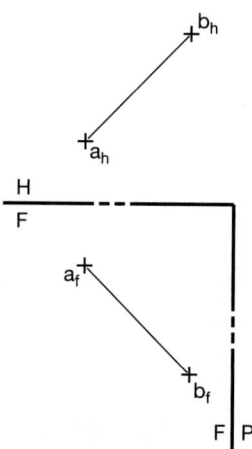

FIGURE 16.26 ■ Setup to project a line in space.

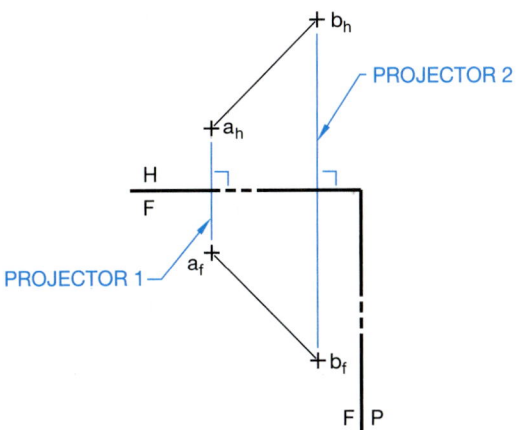

FIGURE 16.27 ■ Confirming views are correct by connecting projectors.

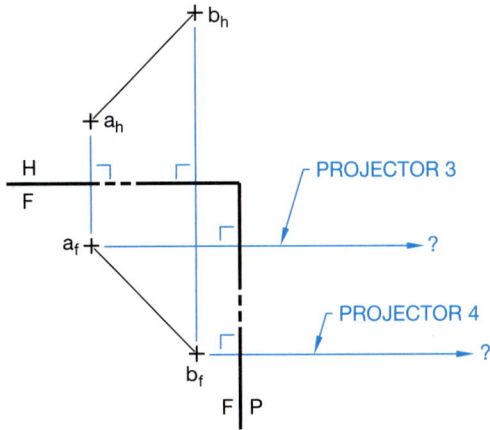

FIGURE 16.28 ■ Locating projectors correctly into new view.

tor must always be drawn at a 90° angle to a fold line and that it was done in this case. (See Figure 16.27 again.)

STEP 2   Draw a new projector (projector 3) from point a_f *perpendicular* to the fold line F/P and extend projector 3 into the P view at any given distance at this time. Point A lies somewhere along projector 3 in the P view. (See Figure 16.28.)

Draw another new projector (projector 4) from point b_f perpendicular to the fold line F/P and extend projector 4 into the P view at any given distance at this time. Point B will lie somewhere along projector 4 in the P view. (See Figure 16.28 again.)

Next, draw little angle brackets close to where projectors 3 and 4 and the fold line F/P intersect just to reinforce the idea that projectors must always be drawn at a 90° angle to a fold line and that it was done in this case. (See Figure 16.28 again.)

STEP 3   Measure the perpendicular distance from point a_h to the fold line H/F in the H view. Identify this distance with a bracket and the number 1. Transfer this distance 1 along

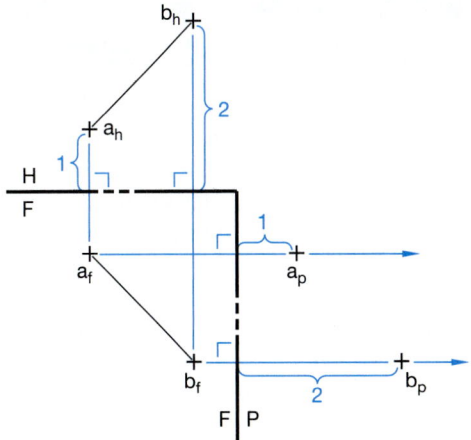

FIGURE 16.29 ■ Transferring distances 1 and 2 correctly.

projector 3 measuring *from* the fold line F/P. Identify this distance with a similar bracket and the same number 1. Do this just to reinforce the idea that this distance 1 must be the same distance as the other distance 1 and that it was correctly identified. Make a cross representing point A on projector 3 at this distance. Point A now has been correctly projected onto the P view. Identify the point (cross) as a_p (See Figure 16.29.)

Next, measure the perpendicular distance from point b_h to the fold line H/F in the H view. Identify this distance with a bracket and the number 2. Transfer this distance 2 along projector 4 measuring *from* the fold line F/P. Identify this distance with a similar bracket and the same number 2. Do this just to reinforce the idea that this distance 2 must be the same distance as the other distance 2 and that it was correctly identified. Make a cross representing point B on projector 4 at this distance. Point B now has been correctly projected onto the P view. Identify the point (cross) as b_p. (See Figure 16.29 again.)

Connect the new points a_p and b_p with a line. This represents the line a_p-b_p. (See Figure 16.30.)

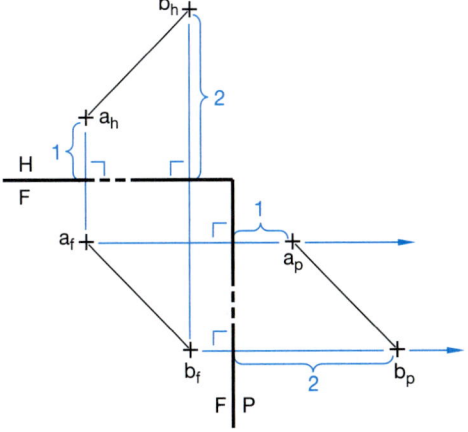

FIGURE 16.30 ■ Showing problem solved.

PROJECTION OF A LINE TO FIND TRUE LENGTH

Knowing how to project a line in space has little value in itself. But knowing how to project the line in a certain manner enables you to find the line in true length (TL) and where the true-length line will appear as a point. These are valuable concepts to know when trying to determine the true distance between lines or the true distance between a line and a plane; when finding the true shape of a plane or the true angle between planes; and when working with other advanced concepts.

There are two basic ways to find the true length of a line that appears foreshortened in the two given views of a problem. The first is the fold-line method, and the second is the revolution method. Both methods are presented because each method of finding the true length of a line has certain advantages over the other when used in solving certain advanced descriptive geometry problems, although either method, fold-line or revolution, can be used equally as well to find the true length of a line.

Fold-Line Method of Finding True Length of a Line

The fold-line method used to find the true length of a line represents the concept of the viewer moving around the line in such a way that both ends of the line are the same distance from the viewer. (See Figure 16.31.)

Sample Problem

Find the true length of line *C-D* by the fold-line method. (See Figure 16.32.) Information that must be given to solve this problem:

1. Two adjacent views of the line.

2. A fold line located correctly between the two given adjacent views.

STEP 1 *Draw a new fold line parallel to the given line C-D in either view.* A new fold line *F/R* as shown in Figure 16.33a or a new fold line *H/A* as shown in Figure 16.33b can be drawn parallel to line c_f-d_f or line c_h-d_h on either side. However, drawing the new fold line *F/R* or *H/A* between line c_f-d_f or line c_h-d_h and the existing fold line *H/F* would make solving the problem much more complicated and confusing, so this is *not* recommended. (See Figure 16.34a and b.)

Also, the distance the fold line *F/R* or *H/A* is positioned *parallel* from line c_f-d_f or line c_h-d_h is not critical and does not change the solution. (See Figure 16.35a.) The fold line *F/R* or *H/A* could actually be drawn on top

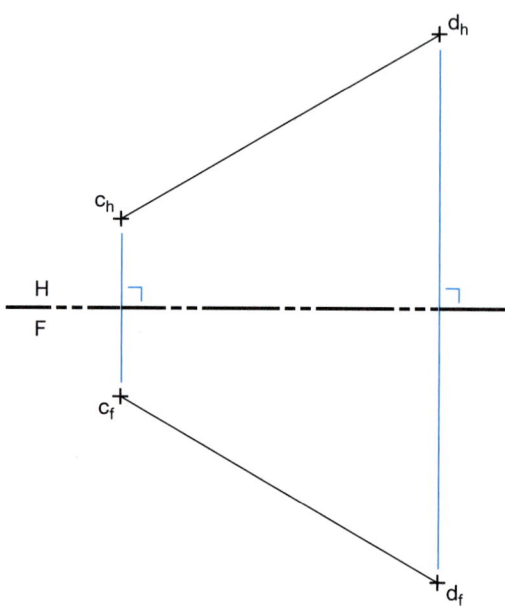

FIGURE 16.32 ■ Setup to find a line in true length by the fold-line method.

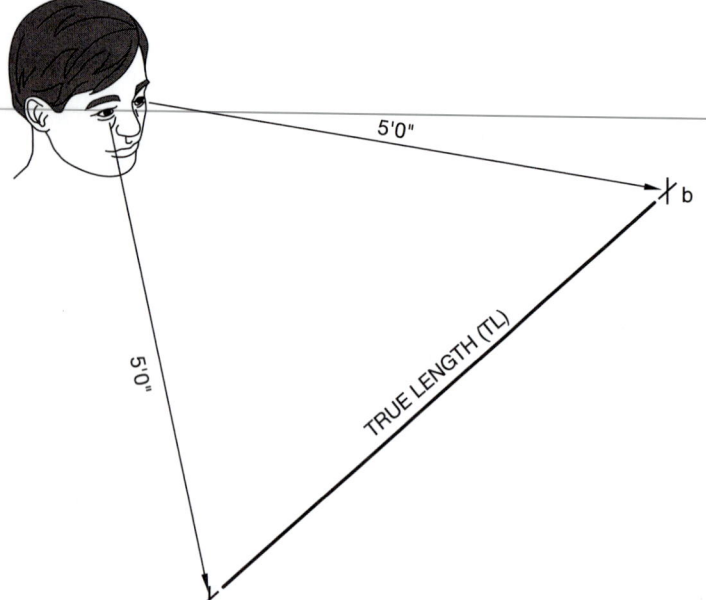

FIGURE 16.31 ■ A line viewed in true length.

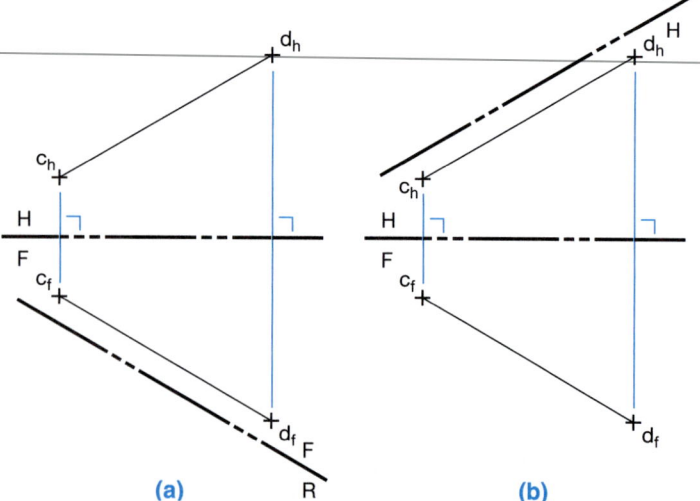

FIGURE 16.33 ■ (a) Possible placement of a new fold line next to the front view. (b) Possible placement of a new fold line next to the top view.

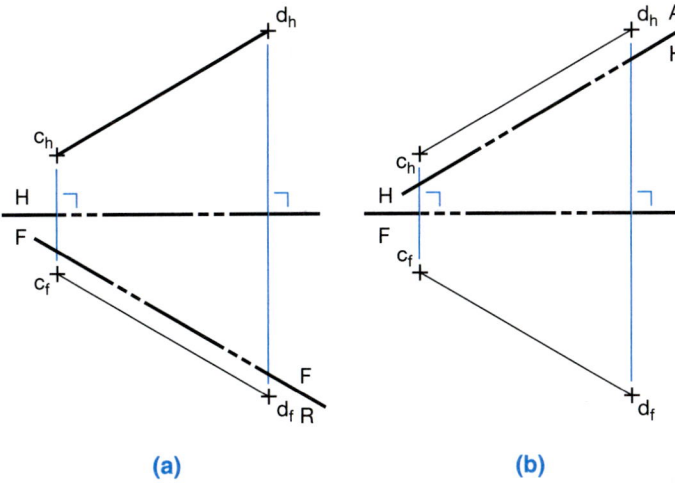

(a)　　　　　　　　　　　　**(b)**

FIGURE 16.34 ■ (a) New fold-line placement next to front view that is *not* recommended. (b) New fold-line placement next to top view that is *not* recommended.

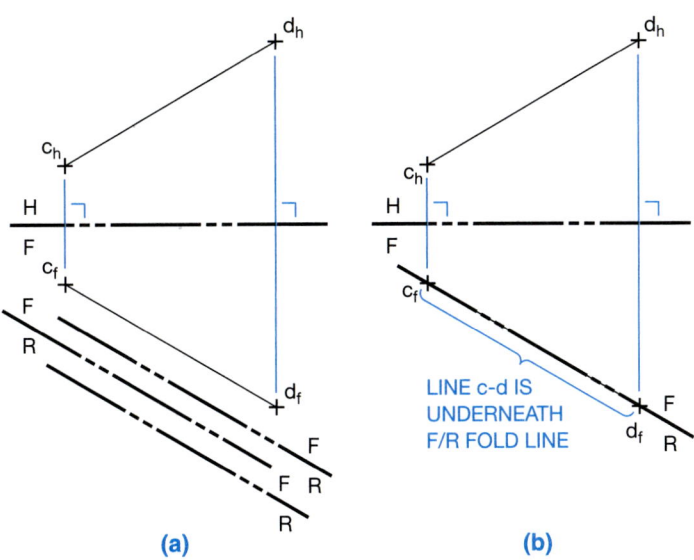

(a)　　　　　　　　　　　　**(b)**

FIGURE 16.35 ■ (a) Other good possible locations for the new fold line F/R. (b) Fold line could work on top of the given line, but this is *not recommended*.

of line c_f-d_f or line c_h-d_h, but again it is **not recommended**. (See Figure 16.35b.)

STEP 2　*Project the line onto the new view.* Project the line c_f-d_f onto view R or project the line c_h-d_h onto view A just the same way as was previously explained in projecting a line onto a new view. The line c_r-d_r or c_a-d_a is the line C-D shown in true length. (See Figure 16.36a and b.)

Hint: Finding the true length of line C-D in both the A and the R views could be used to double check that you have found the correct true length of line C-D. Line c_r-d_r must be the *same* length as line c_a-d_a. If this is not the case, one line or both lines are projected incorrectly, and you have not really found the true length of line C-D.

The trick to solving this problem by the fold-line method is to remember to create a new fold line *parallel* to the given line.

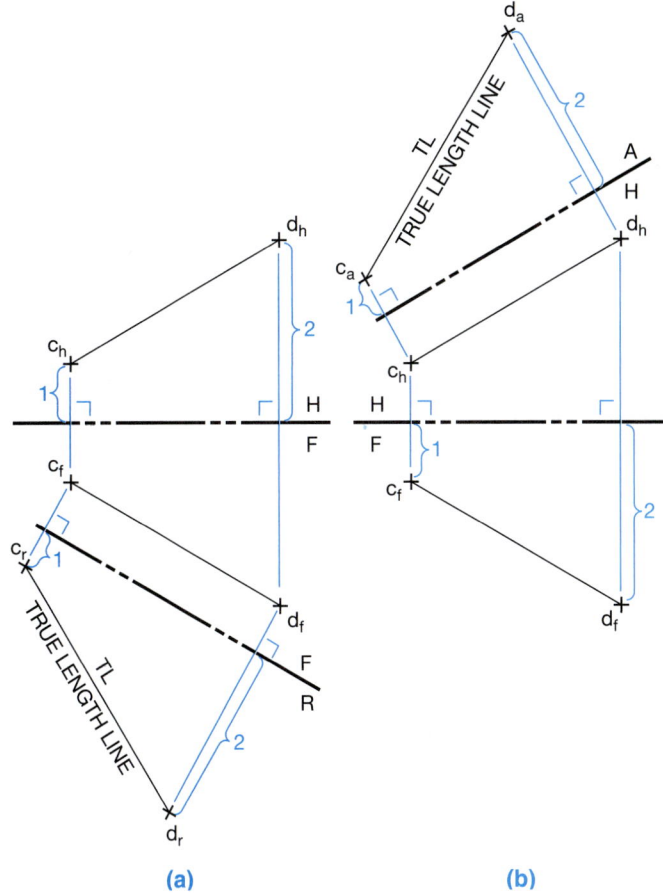

(a)　　　　　　　　　　　　**(b)**

FIGURE 16.36 ■ (a) Creating true length from the front view. (b) Creating true length from the top view.

Then project the line (end points) just as you learned previously when projecting points or a line onto a new view.

Revolution Method of Finding True Length of a Line

The revolution method of finding the true length of a line represents the concepts of moving or **revolving the line** in space in such a way that both ends of the line are the same distance from the viewer. (See Figure 16.37.) An advantage of using this method over the fold-line method is that you can find the true length of a line in the two given views, thus eliminating the need to create a new view. This also gives more accuracy in the solution when using the board method.

Sample Problem

Find the true length of line C-D by the revolution method. (See Figure 16.38.) Information that must be given to solve this problem:

1. Two adjacent views of the line C-D.

2. The fold line F/H located correctly between the two given adjacent views.

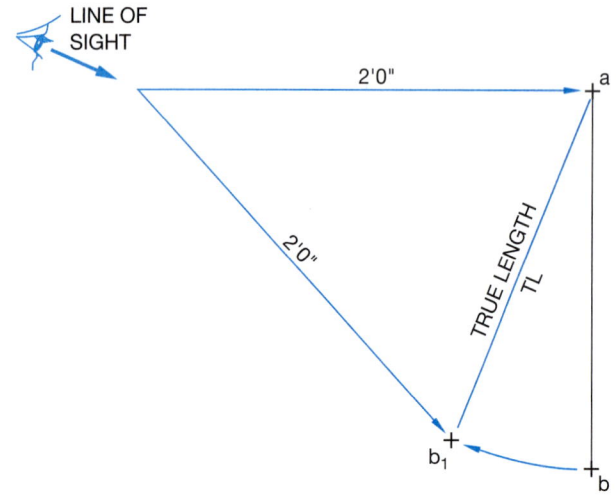

FIGURE 16.37 ■ Revolved line to see in true length.

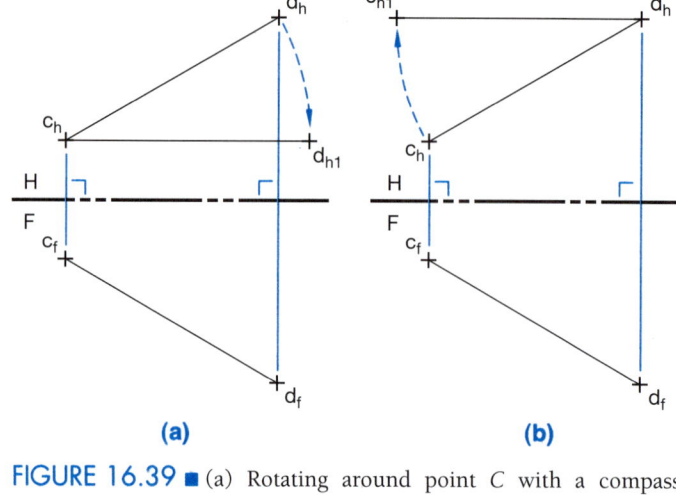

FIGURE 16.39 ■ (a) Rotating around point C with a compass. (b) Rotating around point D with a compass.

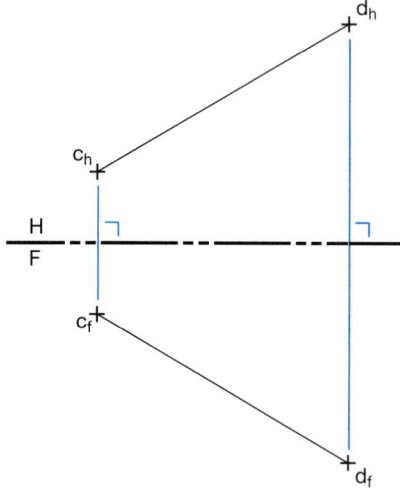

FIGURE 16.38 ■ Setup problem to find true length.

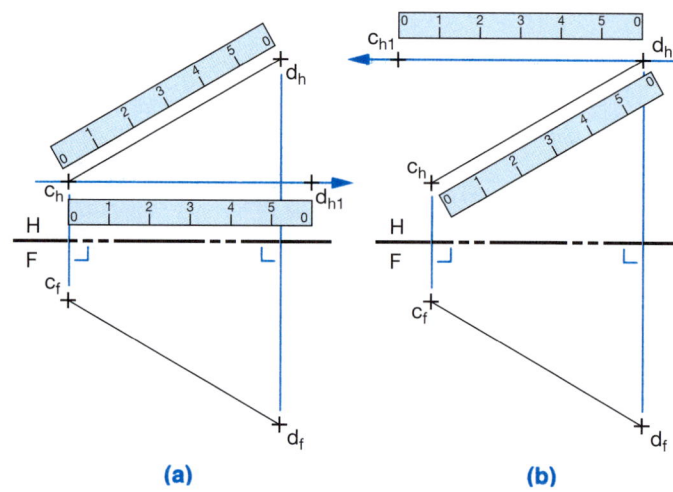

FIGURE 16.40 ■ (a) Rotating without using a compass around point C, Step 1. (b) Rotating D, Step 1 around point without using a compass.

STEP 1   *Move point d_h around point c_h or point c_h around point d_h in the H view to find the true length in the F view.* To rotate point d_h around point c_h graphically, place a compass needle end on point c_h and the compass lead point on point d_h. Draw an arc from point d_h until it is the same distance from the fold line H/F as point c_h. Make a cross at this place and label the cross as d_{h1}. Then connect point c_h to point d_{h1} with a straight line. (See Figure 16.39a.) To rotate point d_h around point c_h, use the same process, but put the compass point on point c_h and proceed just as in the previous example. (See Figure 16.39b.)

If you do not have a compass, you can rotate point d_h by doing the following. (See Figure 16.40.) Draw a light construction line parallel to fold line H/F through point c_h. Measure the length of line c_h-d_h and transfer this length from point c_h along the construction line. As in the previous example, identify the other

end of this length with a cross and as d_{h1}. Then connect point c_h to point d_{h1} with a straight line.

STEP 2   *Project point d_{h1} or point c_{h1} onto the F view.* Draw a projector from point d_{h1} perpendicular to and on through the H/F fold line and beyond point d_f in the F view. (See Figure 16.41a.) To project point c_{h1}, see Figure 16.41b.

STEP 3   *Project point d_f across the projector of d_{h1} or project point c_f across the projector of c_{h1}.* Draw a projector from point d_f parallel to fold line H/F crossing the d_{h1} projector in the F view. Make a cross at this intersection point and label the cross d_{f1}. (See Figure 16.42a.) If projecting point c_f, see Figure 16.42b.

STEP 4   *Create the true length of line C-D.* Connect point c_f to point d_{h1} with a straight line. Line c_f-d_{h1} is the true length of line C-D. Identify the true-length line c_f-d_{h1} with the initials TL. (See Figure 16.43a.) If point c_h was rotated,

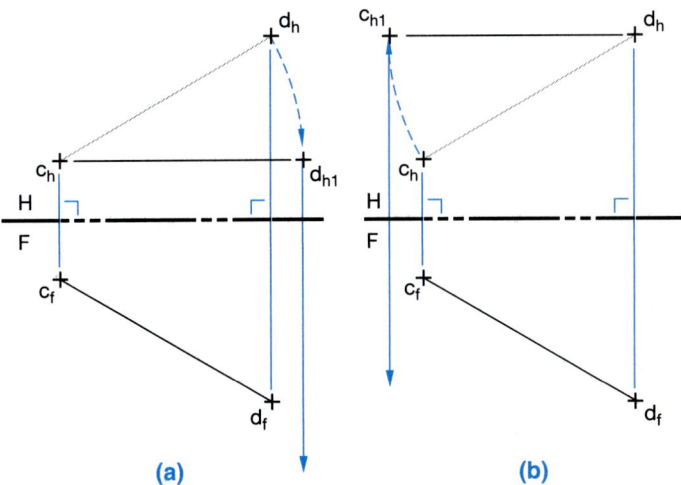

FIGURE 16.41 ■ (a) Rotating around point *C*, Step 2, without using a compass. (b) Rotating around point *D*, Step 2, without using a compass.

connect point d_f to point c_{f1} and identify this true-length line d_f-c_{f1} with the initials TL. This is also the true length of line *C-D*. (See Figure 16.43b.)

Note: The true length of the line could just as well have been found in the *H* view if the revolution process was started in the *F* view.

PROJECTION OF A TRUE-LENGTH LINE TO FIND WHERE IT APPEARS AS A POINT

Knowing how to find where a true-length line appears as a point is necessary to being able to find where an oblique plane appears as an edge view. When a true-length line is viewed at the end, it appears as a *point*. (See Figure 16.44.)

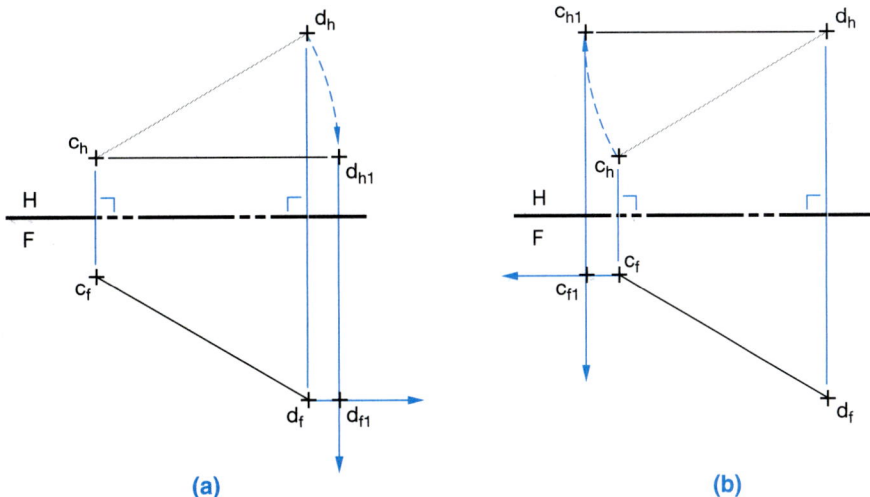

FIGURE 16.42 ■ (a) Rotating around point *C*, Step 3, without using a compass. (b) Rotating around point *D*, Step 3, without using a compass.

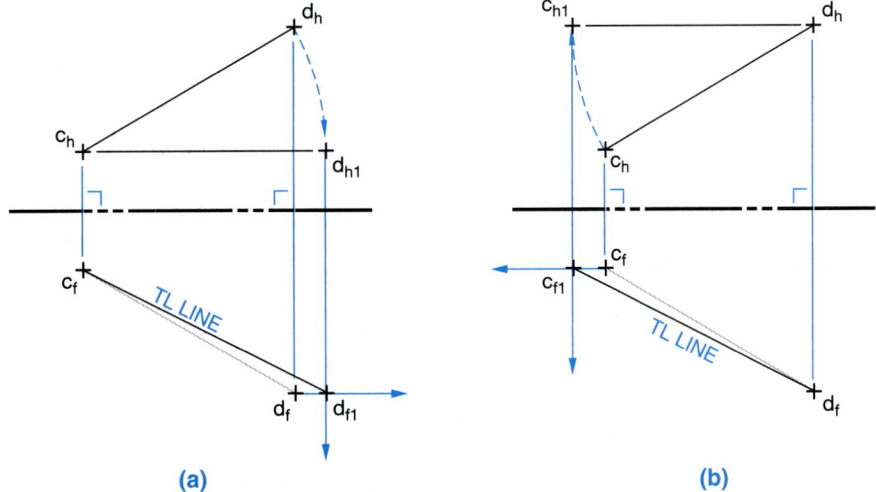

FIGURE 16.43 ■ (a) Rotating around point *C*, Step 4, without using a compass. (b) Rotating around point *D*, Step 4, without using a compass.

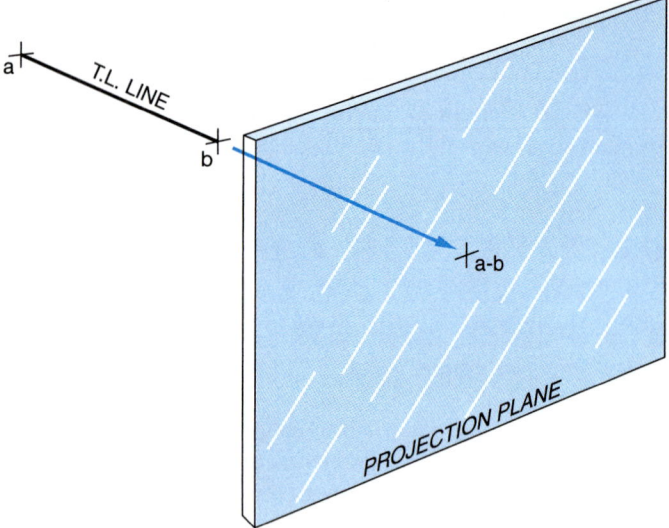

FIGURE 16.44 ■ Viewing a true-length line where it appears as a point.

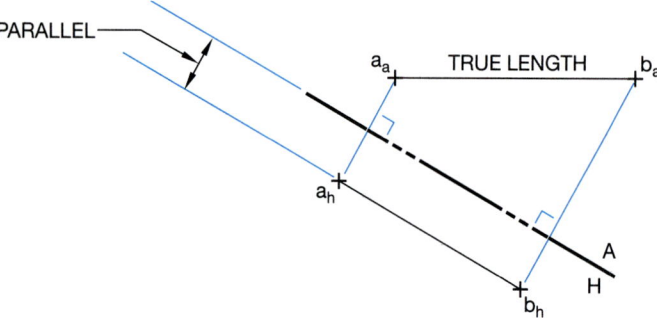

FIGURE 16.45 ■ Problem setup to find true-length line as a point.

Sample Problem

Find where true-length line *A-B* appears as a point. (See Figure 16.45.) Conditions that must be given or must be found first to solve this problem:

1. Two adjacent views of line *A-B* where line *A-B* is in true length in at least in one of the views. *Hint:* To verify if line a_a-b_a is in true length in the *A* view, line a_h-b_h in the *H* view must be *parallel* to the fold line *H/A*. (See Figure 16.45.)

2. The fold line *H/A* must be located correctly between the two given adjacent views.

STEP 1 *Draw a new fold line perpendicular to the true length of line A-B.* Extend the true-length line a_a-b_a with a light construction line. At a short convenient distance, draw a new fold line *A/R* perpendicular to this light construction line. Next, draw in a little angle bracket close to where the light construction line and the fold line *A/R* intersect just to reinforce the idea that the new fold line must always be drawn at a 90° angle to the true-length line. (See Figure 16.46.) This *A/R* fold line can be drawn at any

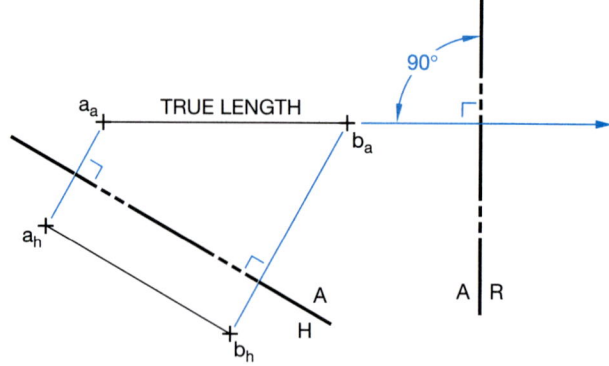

FIGURE 16.46 ■ Locating new fold line.

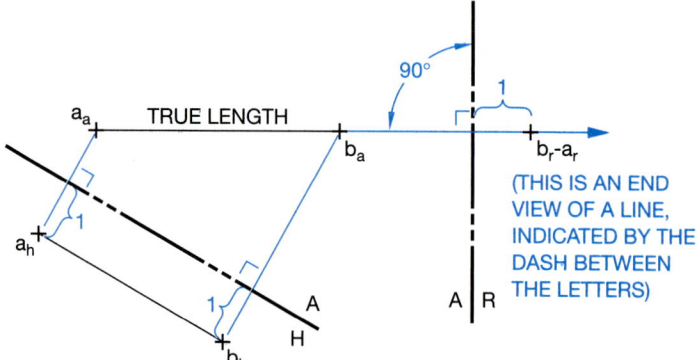

FIGURE 16.47 ■ Projecting line *A-B* onto the *R* view.

perpendicular distance from the true-length line a_a-b_a. It could be drawn through point b_a or through the true-length line a_a-b_a itself, but this is *not* recommended because it is a very confusing way to solve the problem.

STEP 2 *Project line A-B onto the R view.* Project line a_a-b_a onto the *R* view just the same way it was explained earlier in projecting a line in space. Include all the distance brackets, and projectors as before. When you do, notice that point a_a projects on top of point b_a in the *R* view creating what looks like one point. Identify this one point as a_r-b_r so as to remember that this is a line appearing as a point. The dash is included to further clarify that this is the end of the line *A-B*, not just a point. (See Figure 16.47.)

WORKING WITH PLANES

Without question, problems involving planes and plane surfaces are some of the most common found in industry. The basic principles involving planes are applicable in most industrial fields. A **plane** is a flat surface that is not curved or warped, and in which a straight line lies. Figure 16.48 shows four different ways a plane may be represented in graphic form: by three points not in a straight line, by one straight line and one point not on the line, by two intersecting straight lines, and by two parallel straight lines.

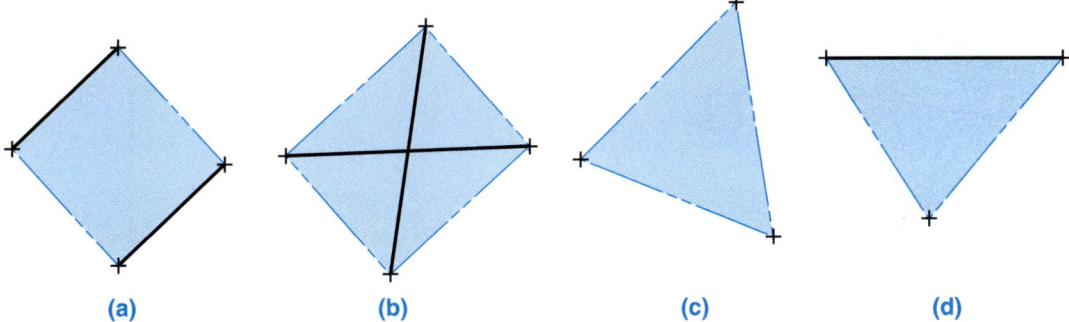

FIGURE 16.48 ■ Different ways of representing planes: (a) two parallel straight lines; (b) two intersecting straight lines; (c) three points not in a straight line; (d) one straight line and one point not on the line.

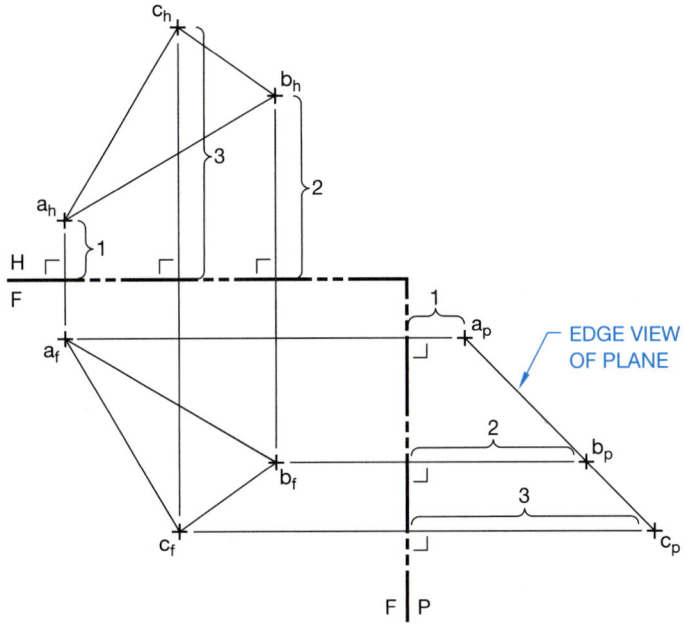

FIGURE 16.49 ■ An inclined plane.

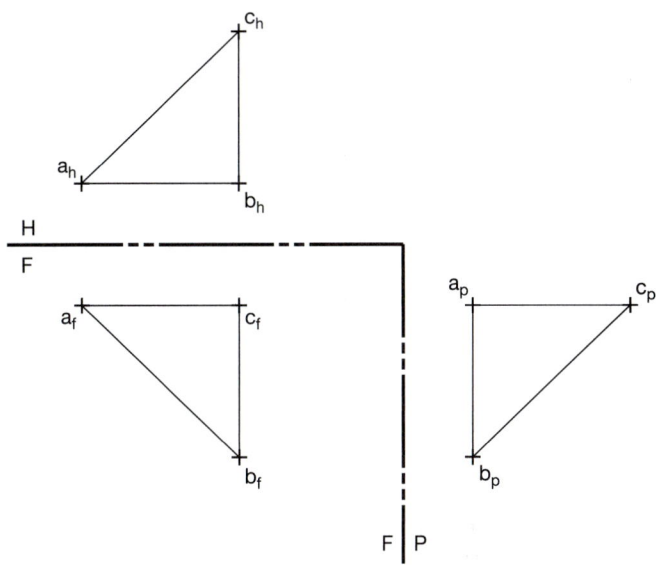

FIGURE 16.50 ■ An oblique plane.

Note: Planes can be any shape such as rectangular, circular, or irregular, but their true shapes are found in exactly the same way—by using at least three points on the plane that are not in a straight line to define the plane.

Inclined Plane

When a plane appears as a line (edge) in at least one of the three standard views of third-angle projection and foreshortened in the other views, it is considered an **inclined plane**. The true shape of the inclined plane can be found in a first auxiliary view. (See Figure 16.49.)

Oblique Plane

When a plane does not appear as a line (edge) in any of the standard views of third-angle projection, it is considered an **oblique plane**, and the true shape of the plane can be found only in a secondary auxiliary view. (See Figure 16.50.)

True Shape of a Plane

The **true shape (TS)** is the actual or real *size* and *shape* of the plane. All measurements and angles can be measured on this view of the plane. In Figure 16.51, plane *A-B-C* is parallel to the horizontal view and appears in **true size and shape** on this view.

PROJECTION OF A PLANE

Knowing how to project a plane in space has little value in itself other than it is a part of the process in finding the true shape of a plane. Using the concept that a plane may be represented by three points not in a straight line that are connected with straight lines, project the three points (plane) onto a new view in exactly the same manner as one point was projected, but do it three times. The projected plane may not necessarily be in true shape. (See Figure 16.52.)

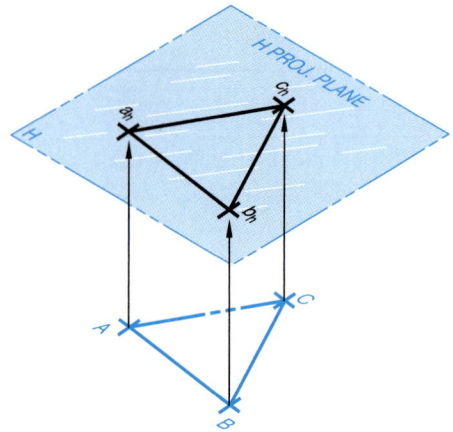

FIGURE 16.51 ■ The true shape of a horizontal plane.

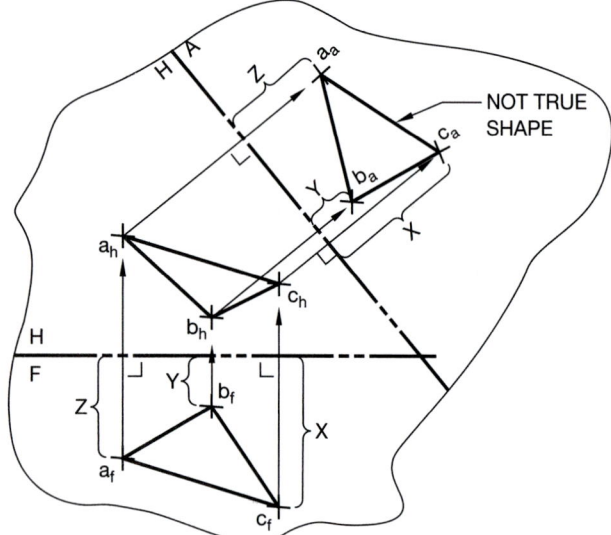

FIGURE 16.52 ■ Projecting a plane onto an auxiliary view.

Finding the True Shape of an Inclined Plane

An inclined plane appears as an edge view (line) in at least one of the standard views of third-angle projection. If in a given problem one of the two given views of the plane appears as a line, it can be considered an inclined plane. (See Figure 16.53.) The true shape of the inclined plane can be found by either the fold-line or the revolution method.

Sample Problem

Find the true shape of the inclined plane *A-B-C* by the fold-line method. (See Figure 16.53 again.) Things that must be given to solve this problem:

1. Two adjacent views of the plane where one view of the plane appears as a line.

2. A fold line located correctly between the two adjacent views.

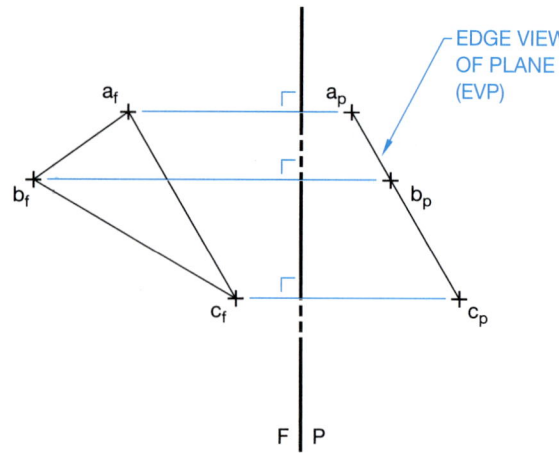

FIGURE 16.53 ■ Inclined plane setup.

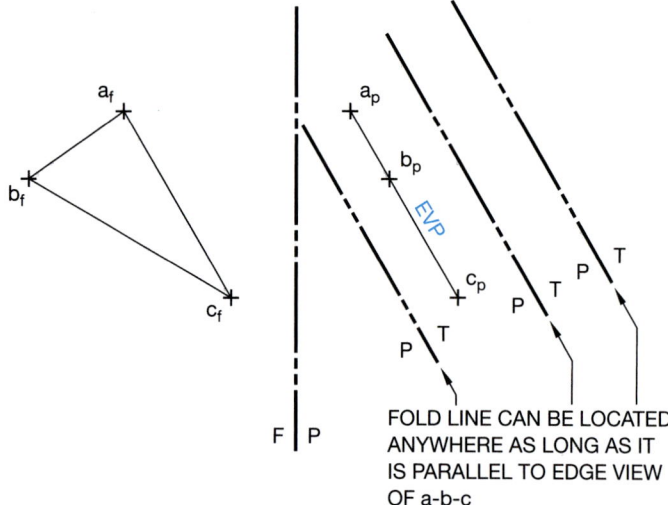

FIGURE 16.54 ■ Possible locations of a new fold line.

STEP 1 *Draw a new fold line parallel to the edge view of plane A-B-C.* Just as when finding the true length of a line, the new fold line can be placed on either side of the edge view at any space from or directly on top of the edge view as long as it is placed *parallel* to the edge view. (See Figure 16.54.) However, placing the new fold line *P/R* about where it is placed in Figure 16.55 works well.

STEP 2 *Project the edge view of plane A-B-C onto the new view.* Project the edge view of plane a_p-b_p-c_p (a line with three points) onto view *R* just the same way as projecting a line with two points onto view *R*. Plane a_r-b_r-c_r is the true shape of the inclined plane. Put the initials TS somewhere on the true shape of the plane. (See Figure 16.56.)

Sample Problem

Find the true shape of the inclined plane *A-B-C* by the revolution method. (See Figure 16.57.) Things that must be given to solve this problem:

1. Two adjacent views of the plane where one view of the plane appears as a line.

2. A fold line located correctly between the two adjacent views.

STEP 1 *Revolve the edge of the plane where all the points of the plane A-B-C are the same distance from the existing fold line. There are several ways to solve this problem. For this example, points b_p and c_p are revolved around point*

a_p where they are the same distance from the fold line F/P. Or, in other words, the edge view of plane A-B-C is revolved *parallel* to the existing F/P fold line.

To rotate points b_p and c_p around point a_p graphically, place a compass needle end on point a_p and the compass lead point on point b_p. Draw an arc from point b_p until it is the same distance from the fold line F/P as point a_p. Make a cross at this place and label the cross as b_{p1}. Again, place a compass needle end on point a_p and the compass lead point on point c_p. Draw an arc from point c_p until it is the same distance from the fold line F/P as point a_p. Make a cross at this place and label the cross as c_{p1}. Then connect point a_p to points b_{p1} and c_{p1} with a straight line. (See Figure 16.58.)

If you do not have a compass, you can rotate the points b_p and c_p by the following way. (See Figure 16.59.) Draw a light construction line *parallel* to fold line F/P through point a_p. Measure the length of line a_p-b_p and transfer this length from point a_p along the light construction line. Identify the other end of this length with a cross and as b_{p1}. Then measure the length of line a_p-c_p

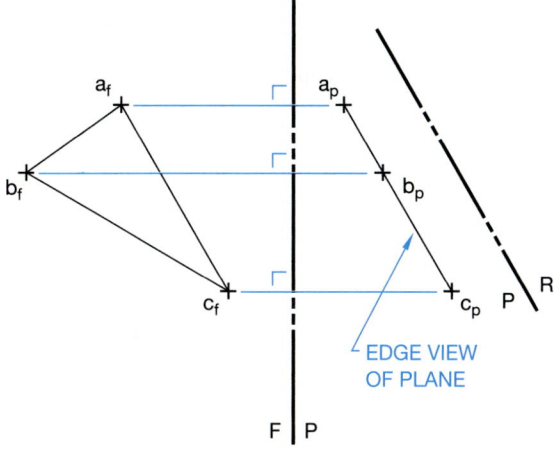

FIGURE 16.55 ■ Locating new fold line for this problem.

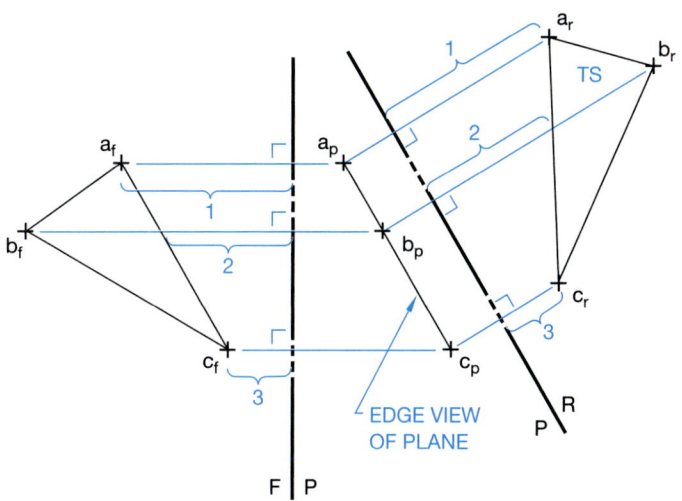

FIGURE 16.56 ■ Creating true shape of plane A-B-C.

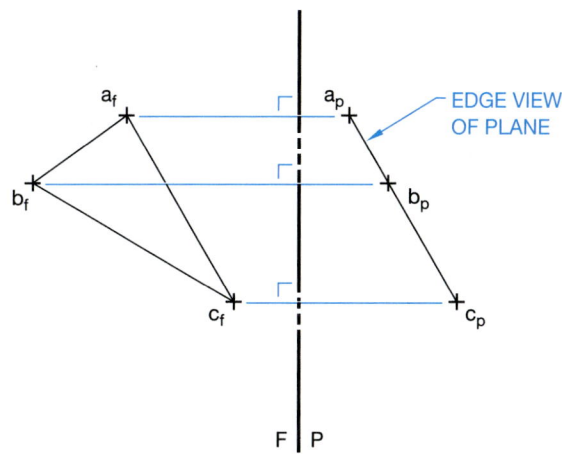

FIGURE 16.57 ■ Problem setup.

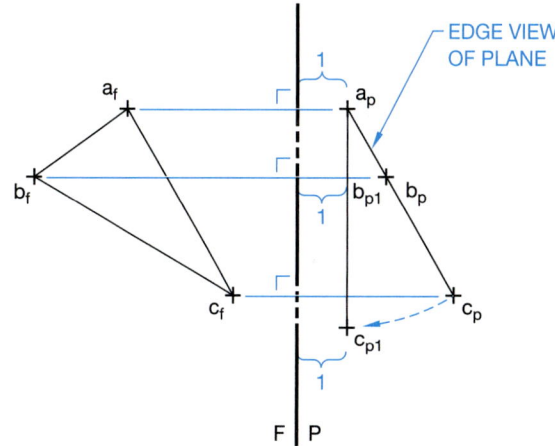

FIGURE 16.58 ■ Step 1, finding true shape of an inclined plane using a compass.

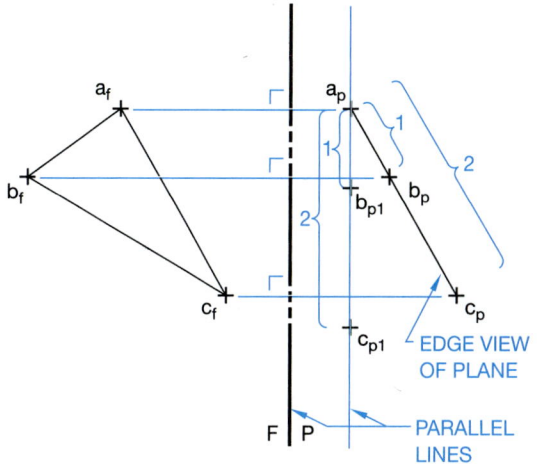

FIGURE 16.59 ■ Step 1, finding true shape of an inclined plane without using a compass.

and transfer this length from point a_p along the light construction line. Identify the other end of this length with a cross and as c_{p1}. Connect points a_p, b_{p1}, and c_{p1} with a straight line.

STEP 2 Project points b_{p1} and c_{p1} onto the F view. Draw a projector from point b_{p1} *perpendicular* to and on through the F/P fold line and beyond point b_f in the F view. Then draw a projector from point c_{p1} perpendicular to and on through the F/P fold line and beyond point c_f in the F view. (See Figure 16.60.)

STEP 3 Project point b_f across the projector of b_{p1} and project point c_f across the projector of c_{p1}. Draw a projector from point b_f *parallel* to fold line F/P crossing the b_{p1} projector in the F view. Make a cross at this intersection point and label the cross b_{f1}. Next draw a projector from point c_f *parallel* to fold line F/P crossing the c_{p1} projector in the F view. Make a cross at this intersection point and label the cross c_{f1} (See Figure 16.61.)

STEP 4 Create the true shape of plane A-B-C. Connect point a_f to points b_{f1} and c_{f1} with straight lines. Label this plane with the initials T.S. to identify this plane as the true shape. (See Figure 16.62.)

Hint: The size of the true shape of a plane can never appear smaller than any other view of the plane. It can appear the same size, in which case the other view is also a true shape. If when you think you have found the true shape of the plane it appears smaller in some other view, that indicates you have made some error and you have not really found the true shape.

Finding the True Shape of an Oblique Plane

An oblique plane does not appear as an edge view (line) in either of the two given views of a problem, nor does it appear as an edge view (line) in any of the standard views of third-angle projection. To find the true shape of an oblique plane, the plane first must be found in a view where it appears as an edge view (line).

Sample Problem

Find the true shape of the oblique plane A-B-C. (See Figure 16.63.) Things that must be given to solve this problem:

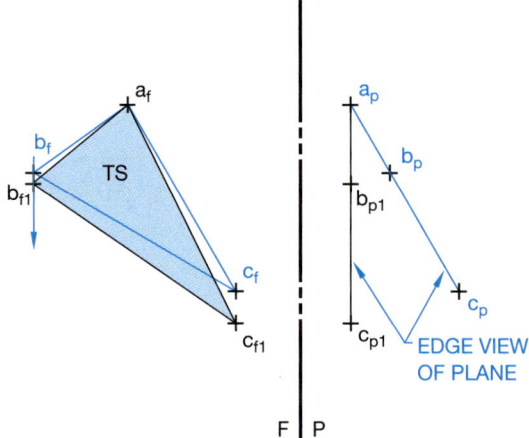

FIGURE 16.62 ■ Step 4 to find true shape of an inclined plane.

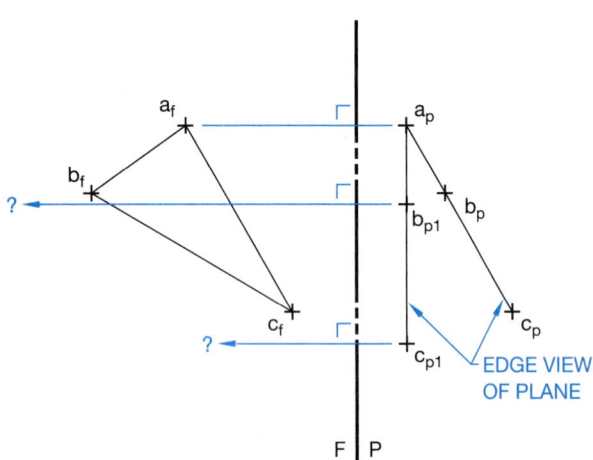

FIGURE 16.60 ■ Step 2 to find true shape of an inclined plane.

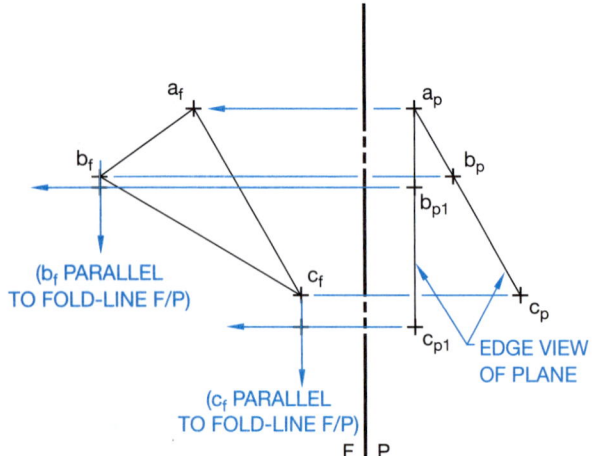

FIGURE 16.61 ■ Step 3 to find true shape of an inclined plane.

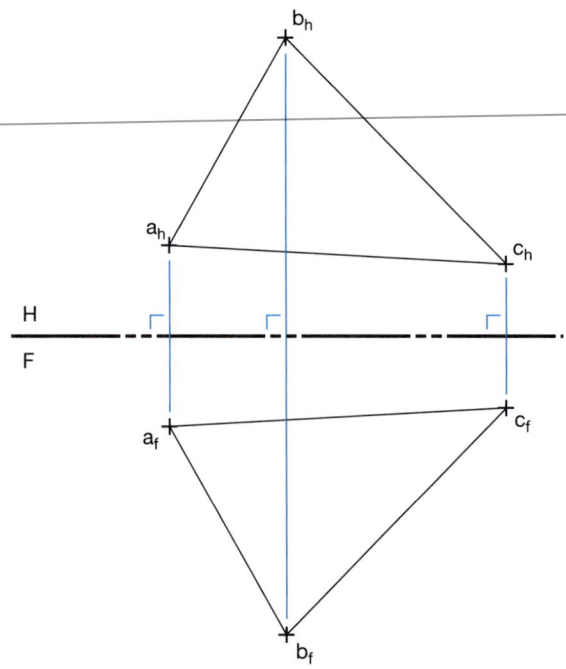

FIGURE 16.63 ■ Problem setup for an oblique plane.

1. Two adjacent views of the plane where neither view of the plane appears as a line.

2. A fold line located correctly between the two adjacent views.

STEP 1 *Find a view where the oblique plane A-B-C appears as an edge.* To find the true shape of an oblique plane, you must first find an edge view of the oblique plane. The concept used in this explanation is based on the idea that when you find a view that shows the end view of any true-length line in the plane (a point), you also see the oblique plane as an edge view (line). There are at least two good ways to do this.

First Method:

STEP 1a *Find one of the edges of the oblique plane A-B-C in true length.* (See Figure 16.64.)

(a) Draw a new fold line *parallel* to any side of the oblique plane in either view. In this case, side b_f-c_f was chosen at random. The new fold line *F/X* was drawn parallel to side b_f-c_f.

(b) Project points a_f, b_f, and c_f onto view *X* just as individual points have been projected in the past. Make sure to include all measurement brackets and perpendicular bracket reminders. Identify

each of the new point locations with a cross and the correct letters and subscript letters.

(c) Connect each of the points a_x, b_x, and c_x with straight lines to form another view of the oblique plane *A-B-C* where the side b_x-c_x is in true length. (See Figure 16.64 again.)

STEP 1b *Find the true-length line b_x-c_x in a view where it appears as a point.* (See Figure 16.65.)

(a) Draw a new fold line *X/Y* perpendicular to the true-length line b_x-c_x of the oblique plane a_x-b_x-c_x.

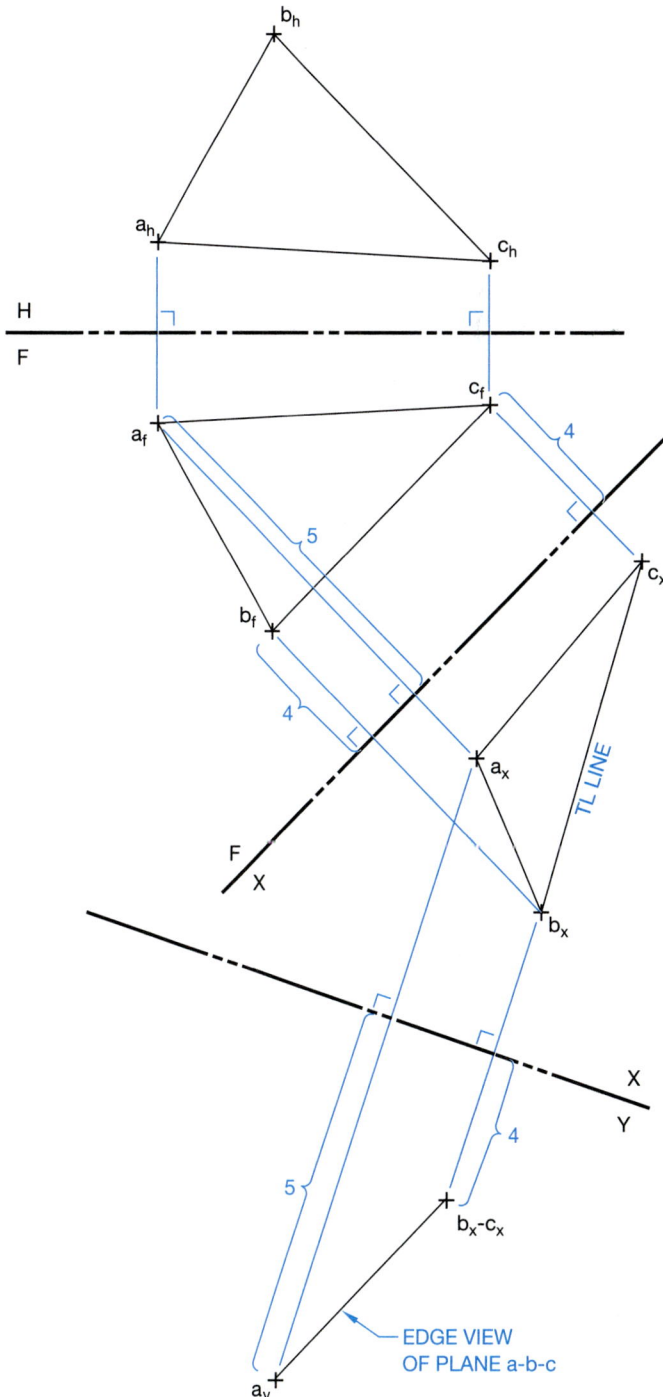

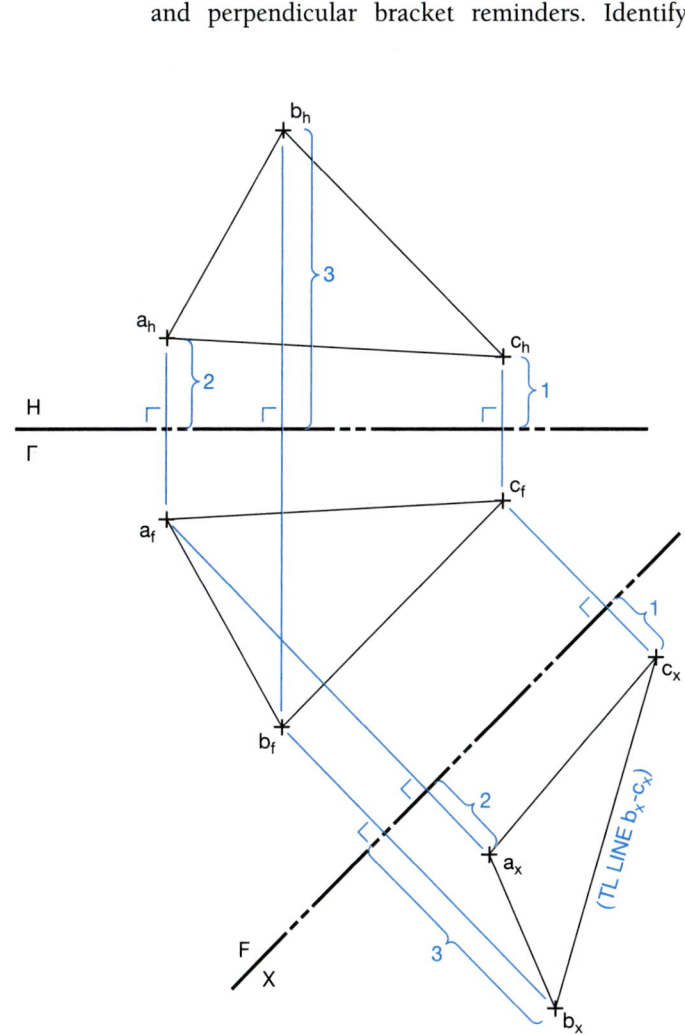

FIGURE 16.64 ■ Step 1a to find true shape of an oblique plane.

FIGURE 16.65 ■ Step 1b to find true shape of an oblique plane.

(b) Project a_x, b_x, and c_x onto view Y just as individual points have been projected in the past. Make sure to include all measurement brackets and perpendicular bracket reminders. Identify each of the new point locations with a cross and the correct letters and subscript letters. Notice that when points c_x and b_x are projected onto the Y view, they are superimposed on top of each other, creating what looks like one point, b_y-c_y.

(c) Connect point a_y to b_y-c_y with a straight line to form the edge view of the oblique plane A-B-C. (See Figure 16.65 again.)

Second Method:

STEP 1a *Create a true-length line in one of the original views of the oblique plane.* (See Figure 16.66.)

(a) Draw a new line on either of the two existing views of the oblique plane A-B-C *parallel* to the fold line H/F. In this case, a new line c_f-x_f is drawn parallel to the fold line H/F in the F view using the existing point c_f as one end and making only one new point. The new line could have been drawn anywhere on the plane as long as it was drawn parallel to the fold line.

(b) Project the new point x_f onto the H view. Draw a projector from point x_f perpendicular to the fold line F/H and on through until it crosses the a_h-b_h line. Identify this point of intersection in the H view with a cross and the letters x_h.

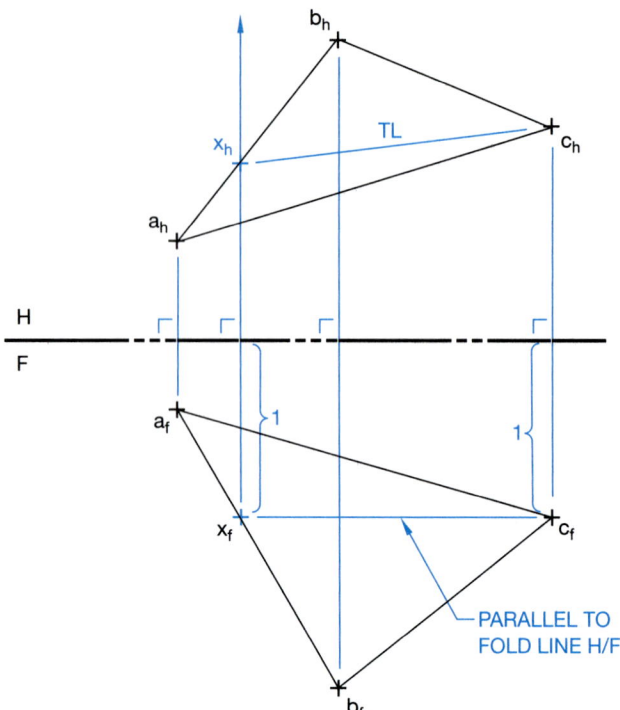

FIGURE 16.66 ■ Another method of finding edge view of an oblique plane, Step 1a.

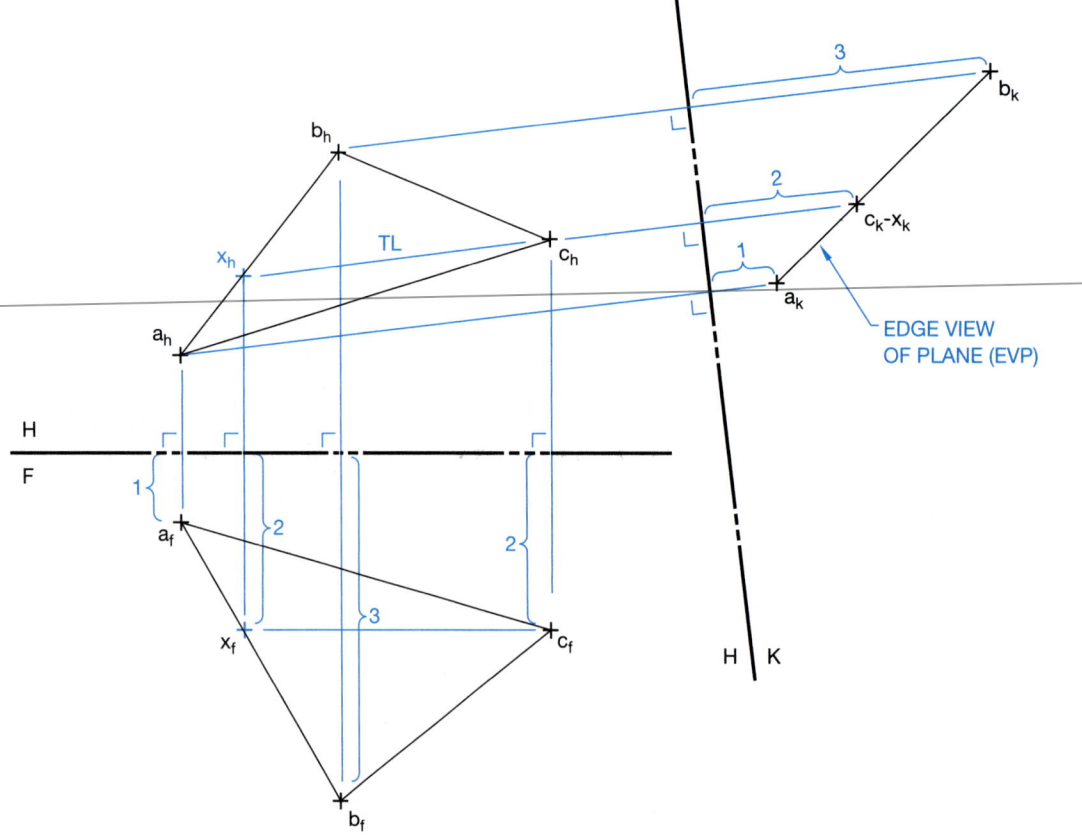

FIGURE 16.67 ■ Step 1b of finding edge view with the second method.

(c) Connect point c_h to x_h to create the true-length line c_h-x_h. (See Figure 16.66.)

STEP 1b *Find the true-length line c_h-x_h in a view where it appears as a point.* (See Figure 16.67.)

    (a) Draw a new fold line *H/K* perpendicular to the true-length line c_h-x_h of the oblique plane a_h-b_h-c_h.

    (b) Project a_h, b_h, c_h, and x_h onto view *K* just as individual points have been projected in the past. Make sure to include all measurement brackets and perpendicular bracket reminders. Identify each of the new point locations with a cross and the correct letters and subscript letters. Notice that when points c_h and x_h are projected onto the *K* view, they are superimposed on top of each other, creating what looks like one point, c_k-x_k.

    (c) Connect points a_k and b_k to c_k-x_k with a straight line to form the edge view of the oblique plane *A-B-C*. (See Figure 16.67 again.)

Hint: Look for a true-length line already in the problem, and you will be able to skip step 1 in either method. (See Figure 16.68.)

STEP 2 To finish solving the problem, use the same method as that used in finding the true shape of an inclined plane as described earlier in this chapter. You can use either the fold-line or the revolution method to solve the problem. (See Figures 16.69 and 16.70 for a quick review.)

WEBSITE RESEARCH

The following websites can provide you additional information for research or further study into topics covered in this chapter:

http://www.asme.org—Find information and publications related to the American Society of Mechanical Engineers.

http://www.ansi.org—The American National Standards Institute. Information about national and international drafting standards.

http://www.adda.org—American Design Drafting Association. Information related to the drafting and design profession.

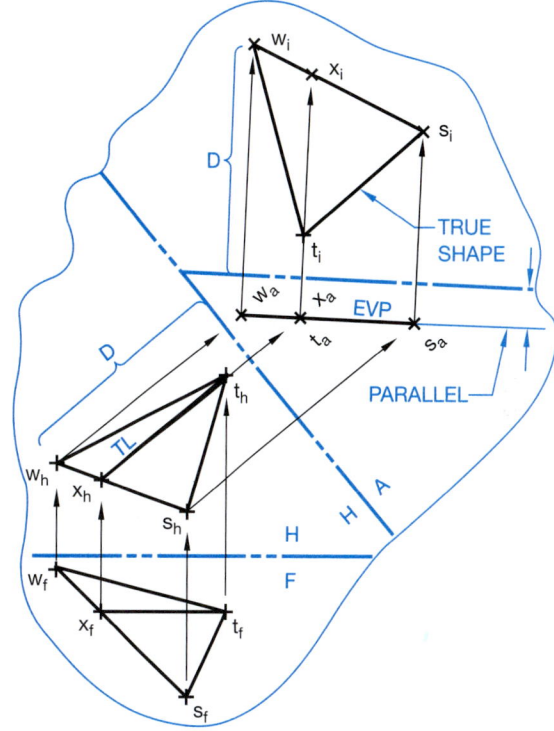

FIGURE 16.69 ■ Finding the true shape of a plane by the fold-line method.

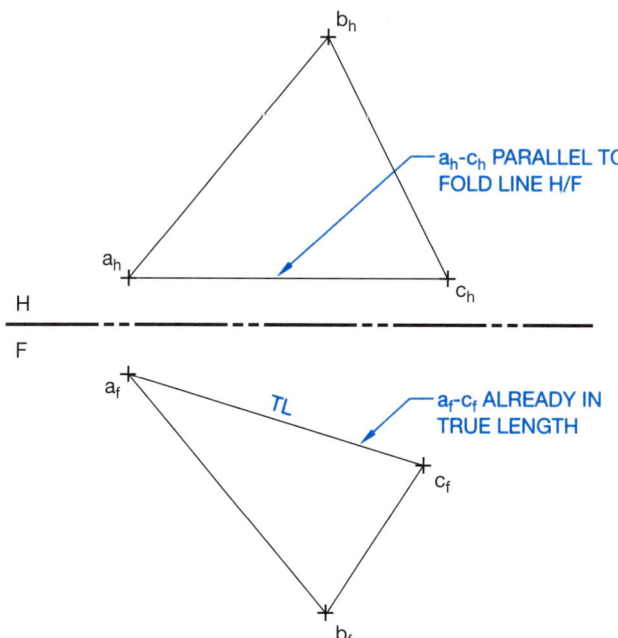

FIGURE 16.68 ■ An oblique plane with an existing true-length line a_f-c_f.

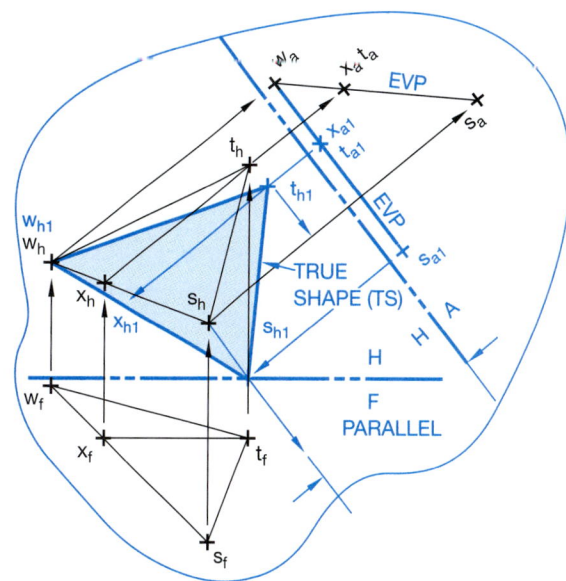

FIGURE 16.70 ■ Finding the true shape of a plane by the revolution method.

PROFESSIONAL PERSPECTIVE

Finding the correct solutions to descriptive geometry problems is mainly dependent on three things: *defining your problem,* using a problem-solving *method,* and *accuracy* in constructing the solution. There are different techniques that can be used to solve each problem graphically, and the challenge is to select the best method. Extreme accuracy is also required when drawing and constructing the various lines on the board. For people who tend to visualize objects easily, descriptive geometry problems are not too difficult solve. Others who do not visualize easily will need to follow the step-by-step cookbook procedure explained in this chapter and memorize specific procedures to obtain the correct answers. Either way, solving descriptive geometry problems can and should be fun and intriguing.

MATH APPLICATIONS

It is important to be able to convert from one type of scale measurement to another when setting up descriptive geometry problems in order to have accuracy in your answers. Otherwise, the process for solution may be correct, but the answer will not be. Some examples of conversions follow.

1 in. = 2.54 cm (25.4 mm) 1 ft. = 30.48 cm
1 cm = .394 in. = 3/8 + in. 1 m = 100 cm = 39.37 in.

You may also need to calculate areas of planes in descriptive geometry problems. The area formulas are included for review. (See Figures 16.71 through 16.76.)

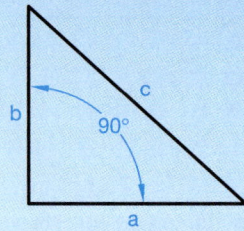

FIGURE 16.71 ■ Area of a right triangle = $ab/2$.

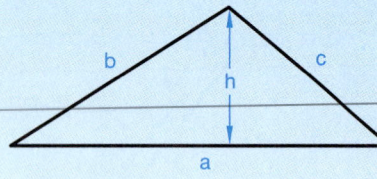

FIGURE 16.72 ■ Area of a general triangle = $ah/2$.

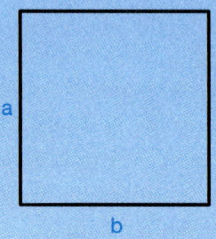

FIGURE 16.73 ■ Area of a square = a^2.

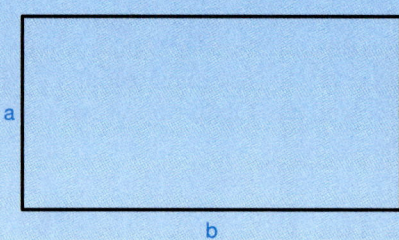

FIGURE 16.74 ■ Area of a rectangular feature = ab.

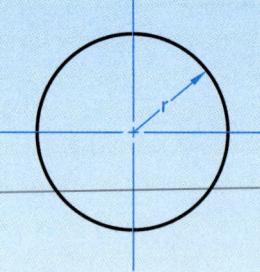

FIGURE 16.75 ■ Area of a circle = πr^2.

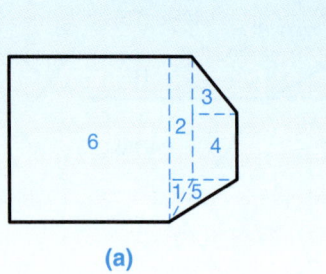

(a)

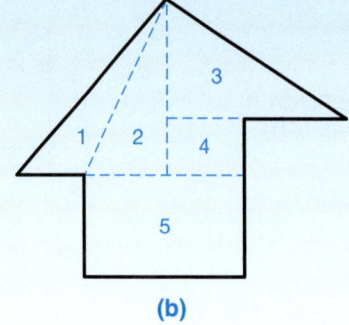

(b)

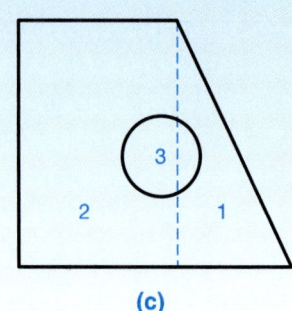

(c)

FIGURE 16.76 ■ Calculating areas of odd-shaped planes: (a) area separated into rectangles and planes; (b) area separated into triangles and rectangles; (c) area separated into a rectangle and triangle minus area of a circle.

CHAPTER 16 *Descriptive Geometry I Test*

Access the CD found with this textbook to view the Chapter 16 Test. Confirm the preferred submittal method with your instructor.

CHAPTER 16 *Descriptive Geometry I Problems*

PROBLEMS 16.1 through 16.25: Access the CD found with this textbook and open the problem of your choice, or as assigned by your instructor. Solve the problems using the instructions provided on the CD, unless otherwise specified by your instructor.

DIRECTIONS

For Problems 16.26 through 16.32, determine and draw the required multiviews and auxiliary views used to set up the problems to establish the true shapes of the surfaces. Do *not* include the given dimensions on your drawing solution; they are only included to help you to create the needed views for each problem.

PROBLEM 16.26 Find the true shape of surface *A* by the fold-line or revolution method.

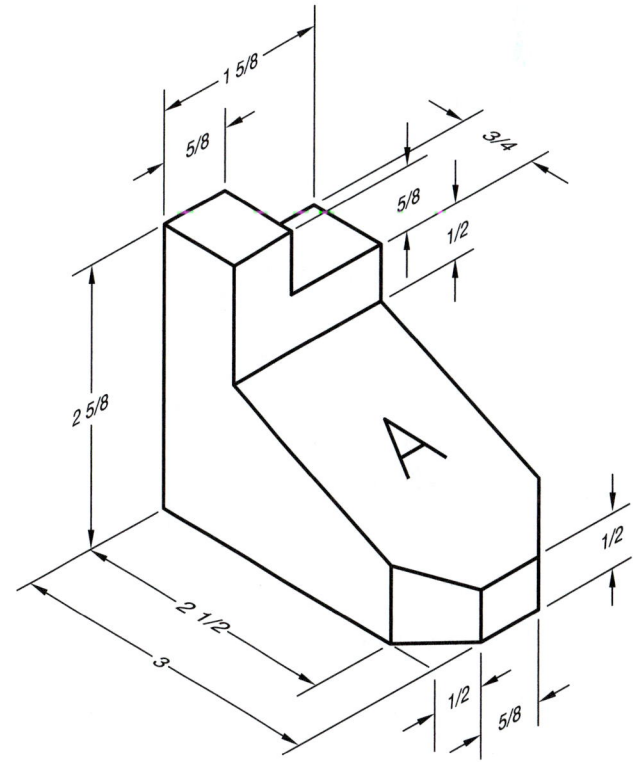

PROBLEM 16.27   Find the true shape of surface *B* by the fold-line or revolution method.

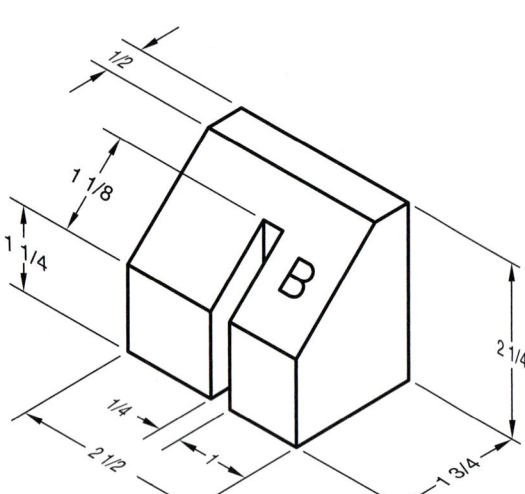

PROBLEM 16.29   Find the true shape of surface *D*.

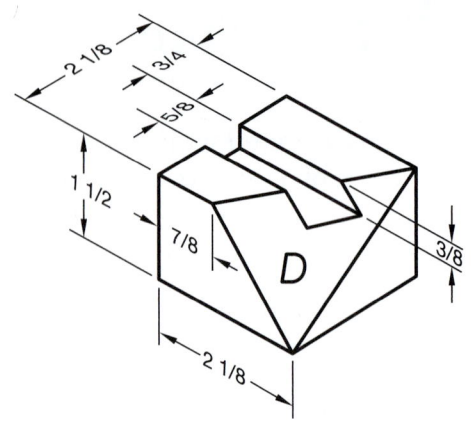

PROBLEM 16.30   True shape of a plane. Find the true shapes of the surfaces *A, B, C, D, E,* and *F* of the concrete pier block.

PROBLEM 16.28   Find the true shapes of surfaces *A* and *B* by the fold-line method.

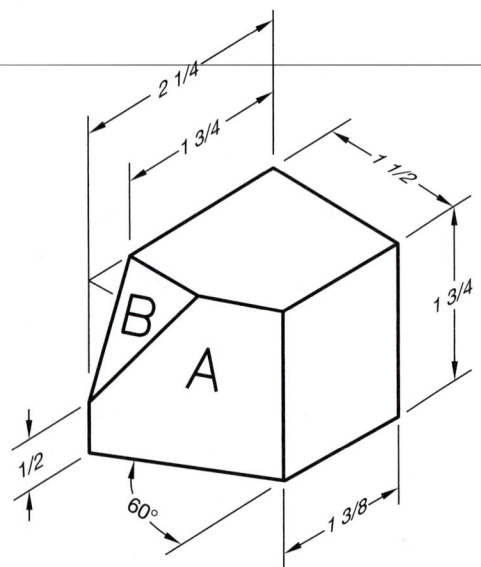

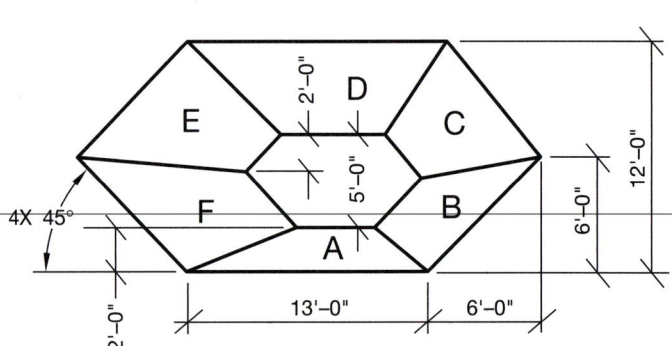

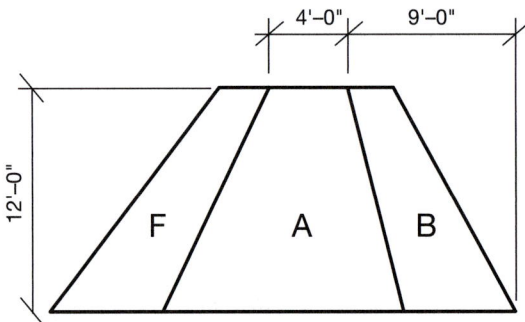

PROBLEM 16.31   From the given engineer's sketch, draw the top, front, and appropriate projection planes of the sloped surface. Use the fold-line method to create the true shape of the sloped surface.

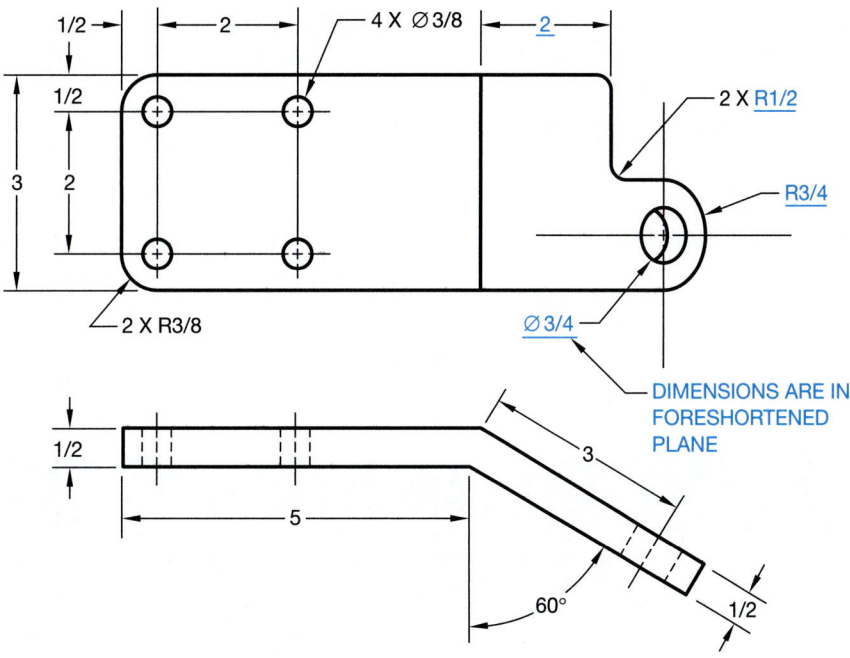

PROBLEM 16.32   From the given engineer's sketch, draw the top, front, and appropriate projection plane of the sloped surface. Use the fold-line method to create the true shape of the sloped surface.

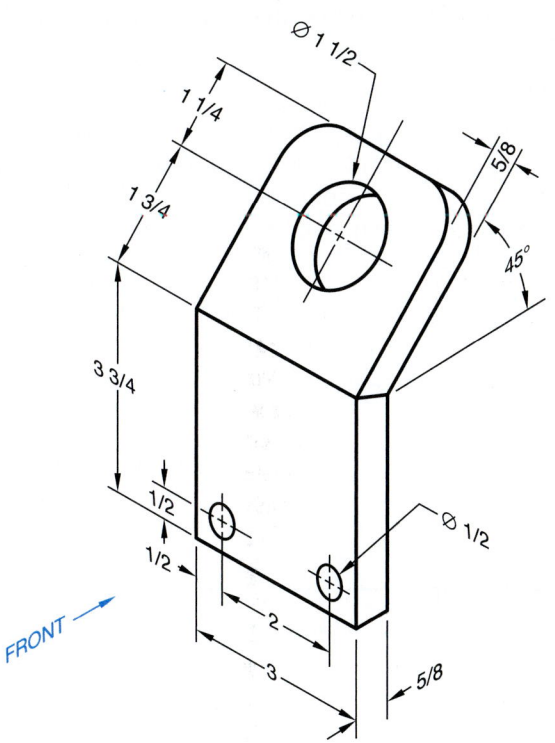

MATH PROBLEMS

 PROBLEMS 16.33 through 16.44: Access the CD found with this textbook and open the math problem of your choice, or as assigned by your instructor. Solve the problem or problems using the instructions provided.

Descriptive Geometry II

LEARNING OBJECTIVES

After completing this chapter, you will:

- Determine bearing, slope, and percent of grade.
- Establish the true angle between a line and a plane.
- Find the true distance between a plane and a point.
- Find the true distance between two skew lines.
- Find the piercing point of a line and a plane.
- Determine the visibility of lines.
- Find the true angle between intersecting planes.
- Graphically solve vector force problems.

THE ENGINEERING DESIGN APPLICATION

A sample descriptive geometry problem solved by the design process follows.

1. *Identify and define the problem.* Given an isometric and partial detail drawing of a building with a chip blower and its discharge pipe location, determine the location and the shape of the roof opening shown in Figure 17.1. Also, the angle between the pipe and the roof must be determined to complete the design of the pipe support bracket. Finally, determine the slope of the discharge pipe. The partial plane view shows the location of a chip blower in a 40' wide × 100' long metal building. The building has 16' high walls and a 4/12 slope on the roof. The blower point 1 is 4' above the floor and the discharge point 2 is 28' above the floor.

2. *Research and generate solutions.* To see how much can be understood from the problem statement, take it word by word. "Determine the location of the roof opening" means there will be a pipe intersecting or going through

the roof. This is the concept of a line intersecting a plane. "Determine the shape of the roof opening" explains itself. The true shape of the opening is asked for here. You are finding the true shape of a plane to determine the angle between the pipe and the roof. The concept of the angle between a line and a plane is used to determine the slope of the discharge pipe. Therefore, the problem requires you to find:

a. The intersection of a line and a plane.

b. The true shape of a plane.

c. The angle between a line and a plane.

d. The slope of a line.

3. *Evaluate the possible solutions.* Draw the top, front, and side projection planes or, as in architectural drawing, the top, front elevation, and side elevation of the given problem. This will help in the evaluation if there is doubt about which part of the information is useful.

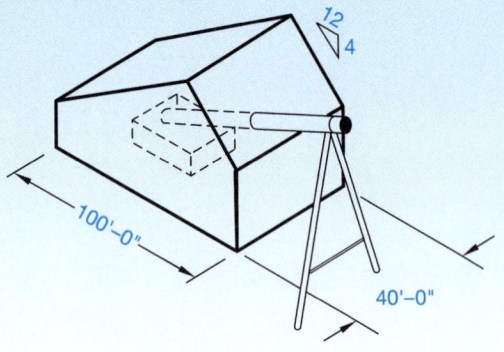

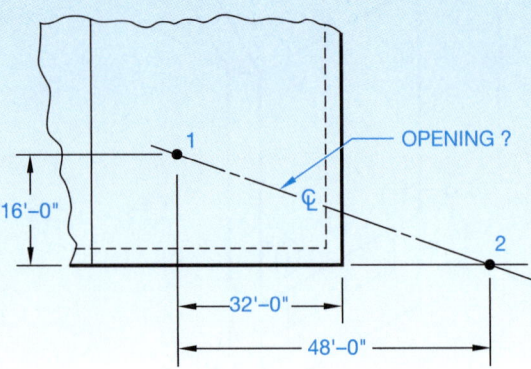

FIGURE 17.1 ■ Problem layout with partial detail.

THE ENGINEERING DESIGN APPLICATION (continued)

Only one solution for each part has been determined, so this step is limited in its usefulness.

4. *Use the best solution.* Decide which views or projection planes will give the necessary information. The front, top, and side views do not give the true lengths, shapes, or angles needed. Auxiliary views must be found and constructed, as in Figure 17.2. The roof plane must be seen as a true shape in order to see the true shape of the hole in it. You must see the edge of the roof plane and the true length of the pipe in the same view to determine where they intersect and the angle between them. To find the slope of the discharge pipe, you must see the true length of the centerline of the pipe in an elevation view.

5. *Evaluate the best solution.* Check the answers by other methods if possible. Does everything in the answer make sense?

6. *Finalize the solution.* Present the solution to your employer or the instructor with the completed notations.

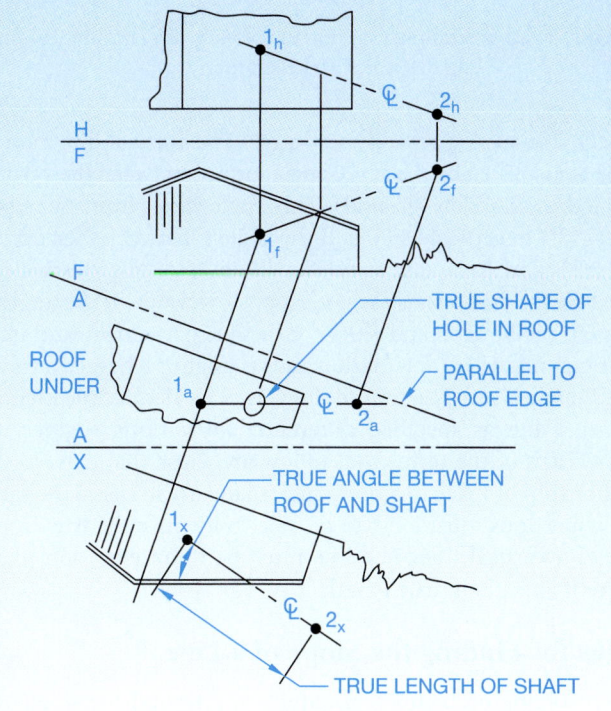

FIGURE 17.2 ■ Fold-line auxiliary view solution.

DIRECTIONS OF LINES AND PLANES

The procedure to locate lines and planes in space was discussed in Chapter 16. To solve many engineering problems, the direction of lines and planes must also be known. The direction of lines and planes is identified in space by a variety of ways, depending on their uses.

Horizontal Directions

Bearings are generally used for map directions and can only be determined in a horizontal view or a top view. Knowing the bearing angle of a line or plane is especially useful in mining and civil engineering. The direction of north is always understood to be toward the top of a drawing layout unless otherwise indicated by a north-pointing arrow pointing in another direction. South is always opposite north, while west is to the left and east is to the right, forming the word *WE* as shown in Figure 17.3. A bearing angle is always 90° or less and is identified either from the north or south. Lines pointing directly north are identified as due north. Also, lines pointing directly east, west, or south are identified as due east, due west, and due south, respectively. The bearings of several angles are shown in Figure 17.4. Notice that the bearing is first identified by a north or south direction indicator. The acute angle formed from the vertical direction and east or west follows next. Lastly, the corresponding side of the acute angle direction east or west is noted. Examples include: N 60° E, N 75° W, S 30° E, and S 45° W.

Bearing Is Found in the Horizontal View

The bearing of a line can only be found in the horizontal view. The line does *not* have to be in true length to represent the bearing.

There is really no such thing as the bearing of a plane; but in mining applications, it is common practice to locate a stratum

FIGURE 17.3 ■ Compass directions.

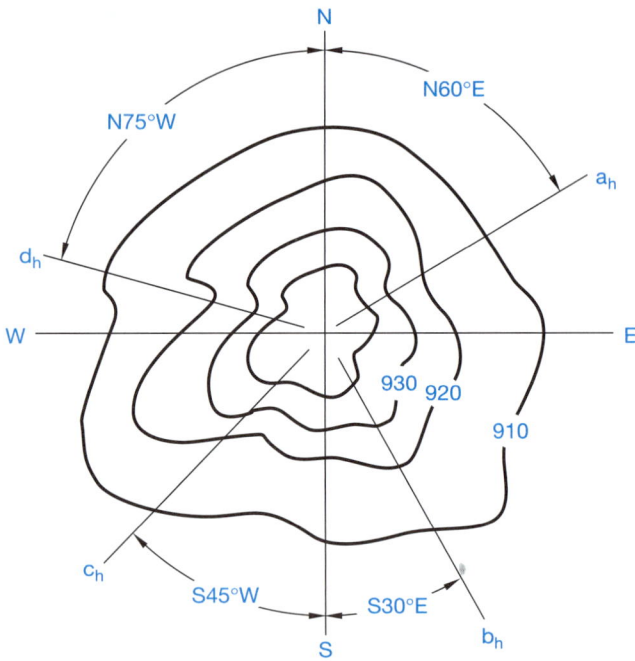

FIGURE 17.4 ■ Bearing angles of lines on a contour map.

by using the bearing of a horizontal line in the plane of a stratum. **Stratums** in their simplest form are horizontal layers of rock. Within a specific area, it is reasonable to assume that a stratum is uniform in thickness and that it lies between two parallel planes. For the discussion in this text, these conditions are assumed. The bearing of a horizontal line in a stratum is called the **strike**. To find the strike of plane M-N-O in Figure 17.5, you can use the same technique found in Chapter 16 by locating a horizontal line in the front view and finding that same line in the horizontal view where it appears as true length.

STEP 1 Draw horizontal line m_f-s_f in the F view.

STEP 2 Find line M-S in the H view. Line m_h-s_h shown in true length is not important, because this is where the bearing of the line is shown whether it is in true length or not.

STEP 3 Measure the angle of line m_h-s_h from the north or south direction. This bearing N 65° E of line M-N as seen in the horizontal view is the strike of the plane.

This discussion of the application of descriptive geometry to mining problems is brief for the purposes of this chapter. It will be necessary to study in more detail the concepts of contour lines, saddles, ridges, ravines, stratas, veins, faults, outcrops, cuts, and fills to adequately utilize descriptive geometry for civil engineering applications. Many of the calculations to determine cross sections for highways, and the designs and drawings from many civil engineering projects, are currently being made using computers.

Vertical Directions

The **vertical direction** of a line can only be found when the following two conditions exist: *the line must be in an elevation view*

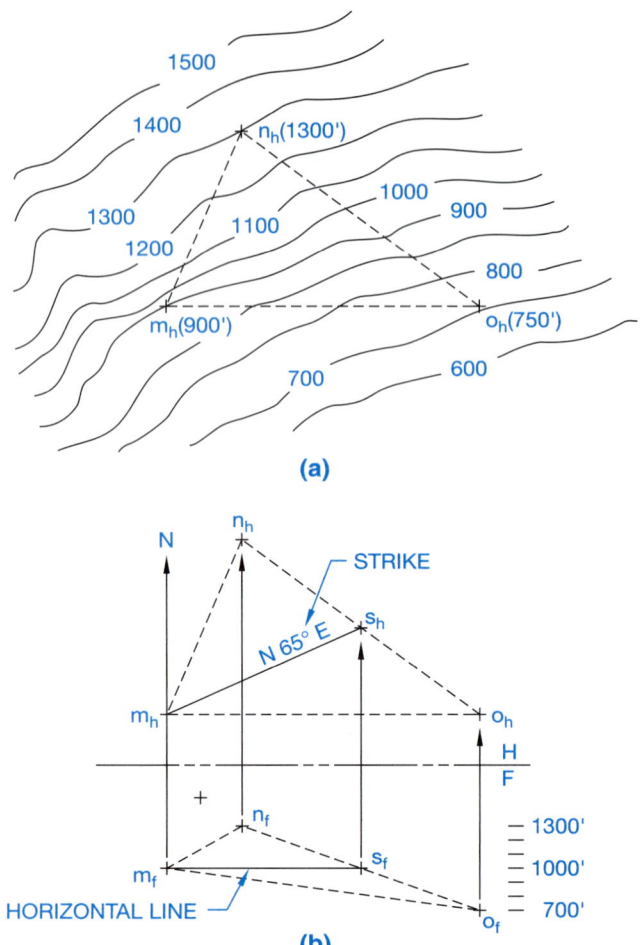

FIGURE 17.5 ■ Stratums: (a) horizontal view of a stratum; (b) finding the strike of the stratum.

and *the line must appear in true length*. The vertical direction of a line is identified by the acute angle formed between the vertical line and the horizon (ground). This angle varies from true horizontal (0°) to vertical (+90°). If the angle is between the horizon and +90°, or the line goes upward from its origin, the angle is called **inclination** and is assigned a positive value. If the angle is between the horizon and −90°, or the line goes downward from its origin, it is called **declination** and is assigned a negative value. (See Figure 17.6.) The terminology used to identify the inclination of a line is specified differently for various engineering fields. Each of the terms that follow are those that express the relationship of a line to a horizontal plane. In all cases, the same two conditions must exist to be able to identify the true direction of a vertical line. The line must be in an elevation plane where it appears in true length.

Rules for Finding the Slope of a Line

When solving problems that require you to find the slope of a line, always start from the horizontal view. For example, when using the fold-line technique, place the fold line parallel to the line where it exits on the horizontal view. When using the revolution method, revolve the line where it exists in the horizon-

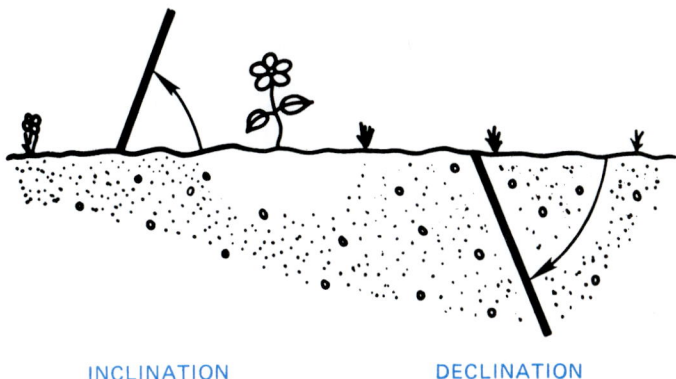

FIGURE 17.6 ■ Inclination and declination.

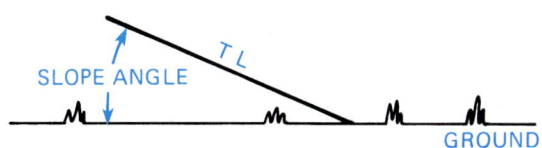

FIGURE 17.7 ■ Slope angle.

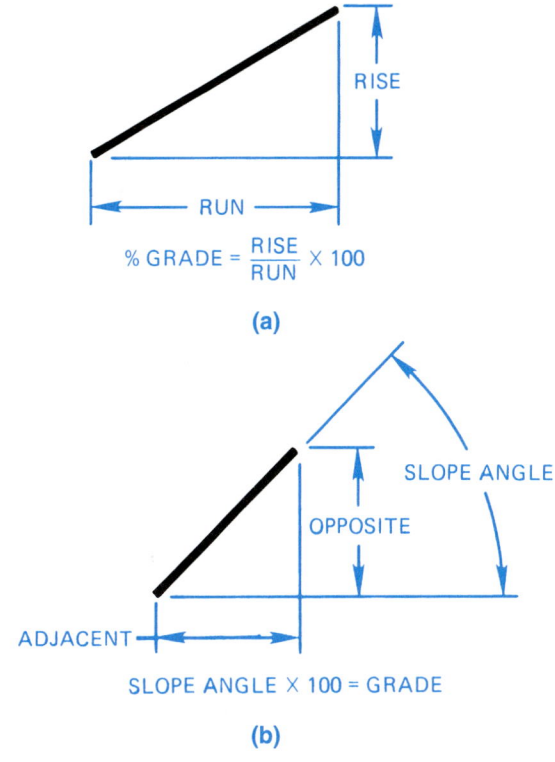

$$\% \text{ GRADE} = \frac{\text{RISE}}{\text{RUN}} \times 100$$

(a)

$$\text{SLOPE ANGLE} \times 100 = \text{GRADE}$$

(b)

FIGURE 17.8 ■ Calculating the grade of a line: (a) formula one; (b) formula two.

tal view parallel to the existing fold line. Either method results in showing the line in true length in an elevation view. This is the only place you can measure the true slope of a line.

The **slope angle** or **slope** of a line is the angle in degrees that the line makes with the horizontal (ground or level) plane. The slope angle can be expressed by the following formula:

Slope angle (s) is the angle where tangent (arctan s) $= \dfrac{\text{Rise}}{\text{Run}}$

Again, the true slope of a line can only be seen in an elevation view where the line appears in true length. (See Figure 17.7.)

The **grade** of a line is expressed in percent form and is found by using the following formula:

Percent grade $= \dfrac{\text{Rise}}{\text{Run}} \times 100$

This is illustrated in Figure 17.8a. Observe also that the grade is the tangent of the slope angle multiplied by 100. From trigonometry, you know that the tangent of an angle is defined as dividing the length of the opposite side of a triangle of the included angle by the adjacent side of that angle as seen in Figure 17.8b. One of the most common ways of measuring run and rise is to use an engineer's scale, which has divisions in multiples of 10, or set the computer scale in CADD to the appropriate scale. Percent of grade is simply another way to express slope; therefore, the conditions for finding the percent of grade of a line are the same as for finding the true slope of a line. (See Figure 17.9.) The term *grade* is usually used when referring to a line that has an angle with less than a 45° inclination. Grade is usually associated with civil engineering but not always. Slope is usually applied to a line that has an angle with greater than 45° inclination, such as the slope of a roof on a house. Slope is used more with architectural engineering but not always.

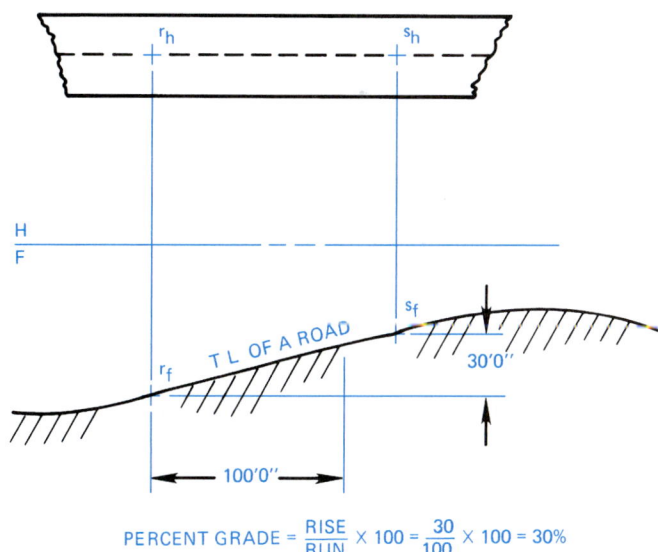

$$\text{PERCENT GRADE} = \frac{\text{RISE}}{\text{RUN}} \times 100 = \frac{30}{100} \times 100 = 30\%$$

FIGURE 17.9 ■ Grade of a road.

In architectural engineering, the slope is also referred to as the **pitch**. When describing the angle the roof plane has with the horizon, it is expressed as the ratio of the vertical rise to 12' of span. The graphical method of showing pitch on a drawing is illustrated in Figure 17.10.

In structural engineering, slope is usually referred to as the **bevel** of a line. Figure 17.11 shows the bevel of several beams.

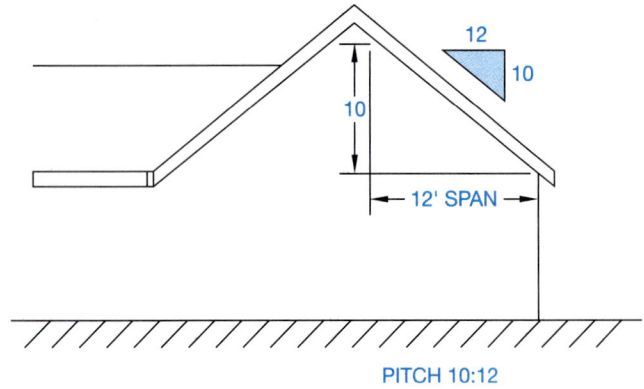

FIGURE 17.10 ■ Pitch of a roof.

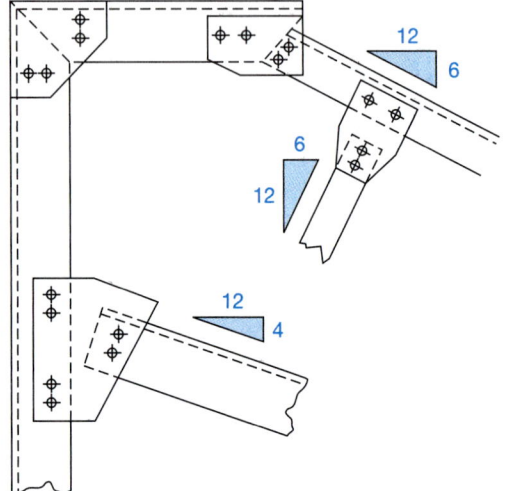

FIGURE 17.11 ■ Bevel of a beam.

In civil engineering, the terms *grade* and *slope* are frequently used as shown in Figure 17.12a. In addition, **batter** is applied when dealing with the slope of concrete footings as shown in Figure 17.12b. **Dip** is used when finding the slope of a stratum (plane). The dip is the slope of the strike. Figure 17.13 shows how to find the dip of a stratum.

In review, *slope, grade, pitch, bevel, batter,* and *dip* are all representations of the angle between a true-length line and the horizon. Finding this angle graphically is done with the same procedure in each situation.

TRUE DISTANCE BETWEEN A POINT AND A LINE

To find the true distance between a point and a line, you must create a view that shows the line and the point appearing as *two points.* This occurs when you are viewing the end of the line and the point on one view. As with most problems, there are several ways to solve this problem. Figure 17.14 shows how to find this distance by the fold-line method.

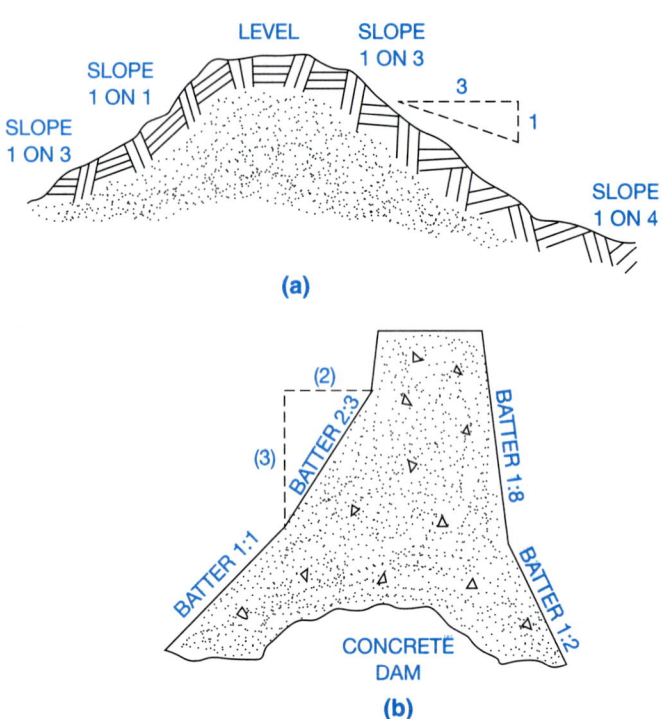

FIGURE 17.12 ■ Terms in civil engineering: (a) slopes of earth dam; (b) batter of concrete dam.

Finding the Distance between Two Points with the Fold-Line Method

The first step in the solution of this type of problem is always to place a fold line parallel to the given line. The line is in true length when it is projected onto the new view. The second step is to place another fold line perpendicular to the true-length line. The line will then appear as a point when it is projected onto the next new view. After projecting the point onto the two new views, it is just a matter of measuring the distance between two points.

STEP 1 Find the true length of line *R-S* and the location of point X on the A view. (See Figure 17.14.)

STEP 2 Find the end view of the true-length line *R-S* and the location of point X on the I view. The distance measured between point X and the end view of line *R-S* on the I view is the true distance between point X and line *R-S*.

TRUE DISTANCE BETWEEN TWO SKEW LINES

Skew lines are two nonparallel, nonintersecting lines. An example of two skew lines is shown in Figure 17.15. The true distance between two skew lines is the perpendicular between them. To be able to determine the true distance between the two skew lines, you must create a view where one line appears as a point. (*Note:* While the one line appears as a point, the other

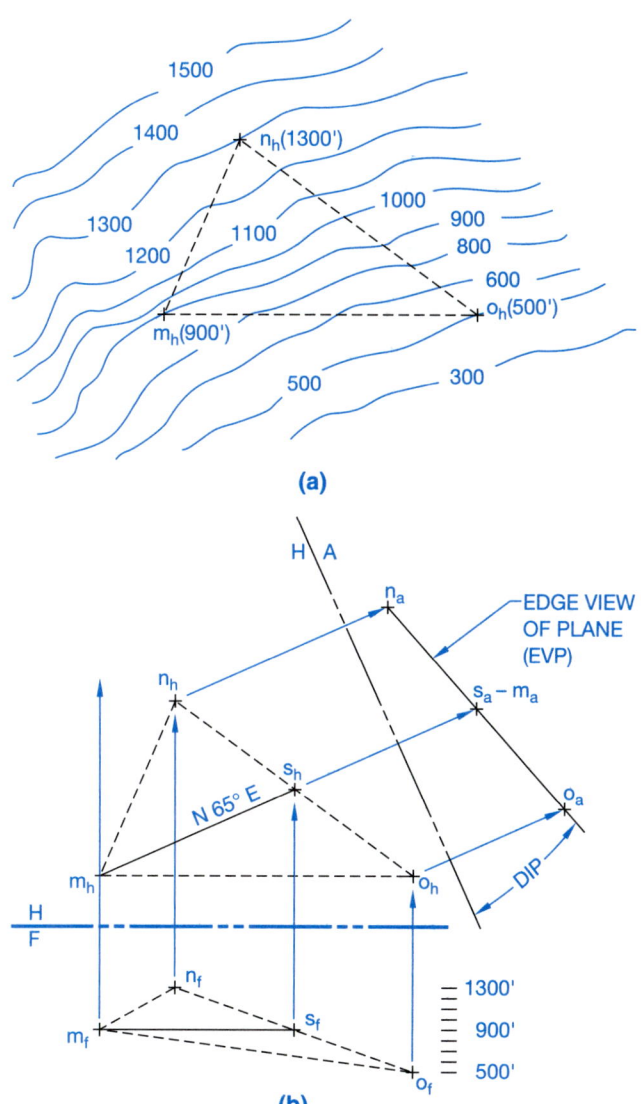

(a)

(b)

FIGURE 17.13 ■ Fold-line method of finding the dip of a stratum. EVP represents the edge view of the plane.

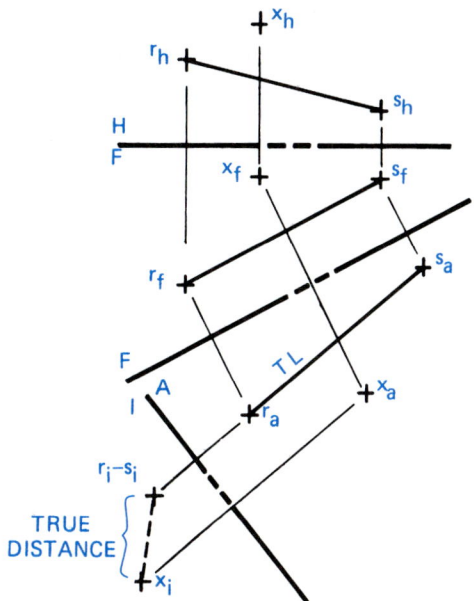

FIGURE 17.14 ■ True distance between a line and a point. TL represents true length.

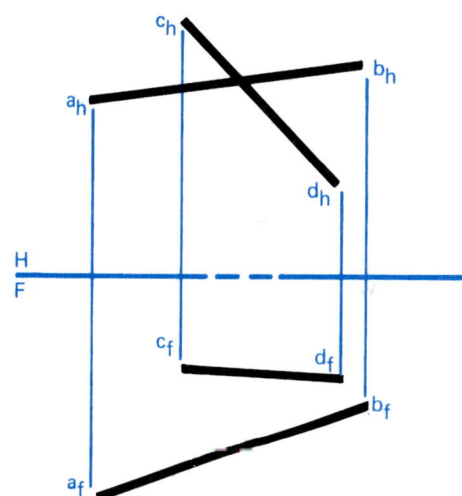

FIGURE 17.15 ■ Skew lines.

line will not necessarily appear in true length.) The perpendicular distance between the line appearing as a point and the other line is the true distance between both lines. (See Figure 17.16.)

STEP 1 Find the true length of one of the lines. In this example, line *R-S* was selected to find the true length. The true length of line *R-S* and the location of line *M-N* has been found on the A view. (See Figure 17.16a.)

STEP 2 Create the view where line *R-S* appears as a point. Line *R-S* appears as a point in the I view. Line *M-N* is also carried onto this view. (See Figure 17.16b.)

STEP 3 On the I view, the true distance can be measured between line *R-S* and line *M-N* by constructing a perpendicular line from line *R-S* (appearing as a point) and line *M-N*. (See Figure 17.16c.)

Sometimes a problem may require you to adjust the distance between two skew lines or make sure there is enough clearance

between the two lines. An application for determining the true distance or clearance determination might be for the location of a valve in a piping system or the clearance between power lines and a nearby metal structure in an electrical power substation. When you have found the true distance between the two skew lines as in step 3 in the previous problem, adjust the point (line *R-S*) in or out along the perpendicular distance from line *M-N*, as shown in Figure 17.16d. You then can relocate the new position of line *R-S* in all the previous views. If line *R-S* cannot be moved in the original problem, you need to solve the problem by starting with line *M-N* rather than line *R-S* as in this example.

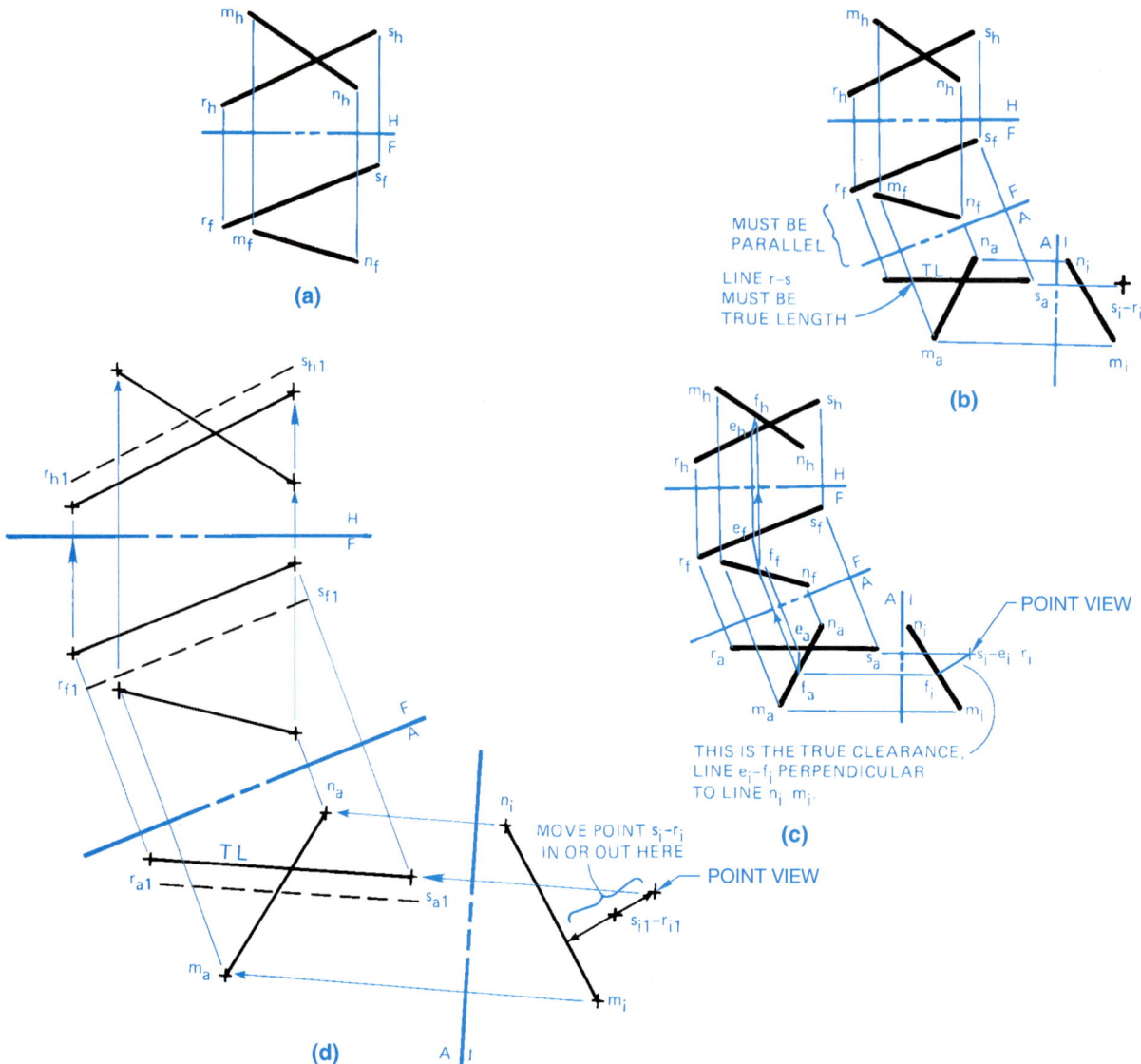

FIGURE 17.16 ■ Finding the true distance between skew lines: (a) step 1; (b) step 2; (c) step 3; (d) adjusting the true distance.

INTERSECTION OF A LINE AND A PLANE

The location and anchoring of guy wires; the location of holes for shafts, cables, pipes, and wires; and the angles that are created by these features are important considerations for the designer in structural and mining problems.

Finding the Piercing Point of a Line and a Plane

To find the point of intersection, the *piercing point*, of a line and a plane, you must create a view where the *plane appears as an edge*. You can then project this apparent point of intersection back into the original views.

Use the following steps if you have a problem that requires finding the intersection of the roof and the circular shaft as shown in Figure 17.17a (piercing point problem).

STEP 1 Find a view that shows the roof plane as an edge. Figure 17.17b shows that the P view contains the roof plane as an edge. When the centerline of the shaft is projected onto the P view, the apparent point of intersection, w_p, of the centerline of the shaft and the roof plane can be seen.

STEP 2 Project this apparent point of intersection, w_p, back onto the previous views. As Figure 17.17c shows, a projector is projected from the A view back to the F view intersecting the centerline of the shaft. This is the actual point of intersection, point w_f of the roof plane shown

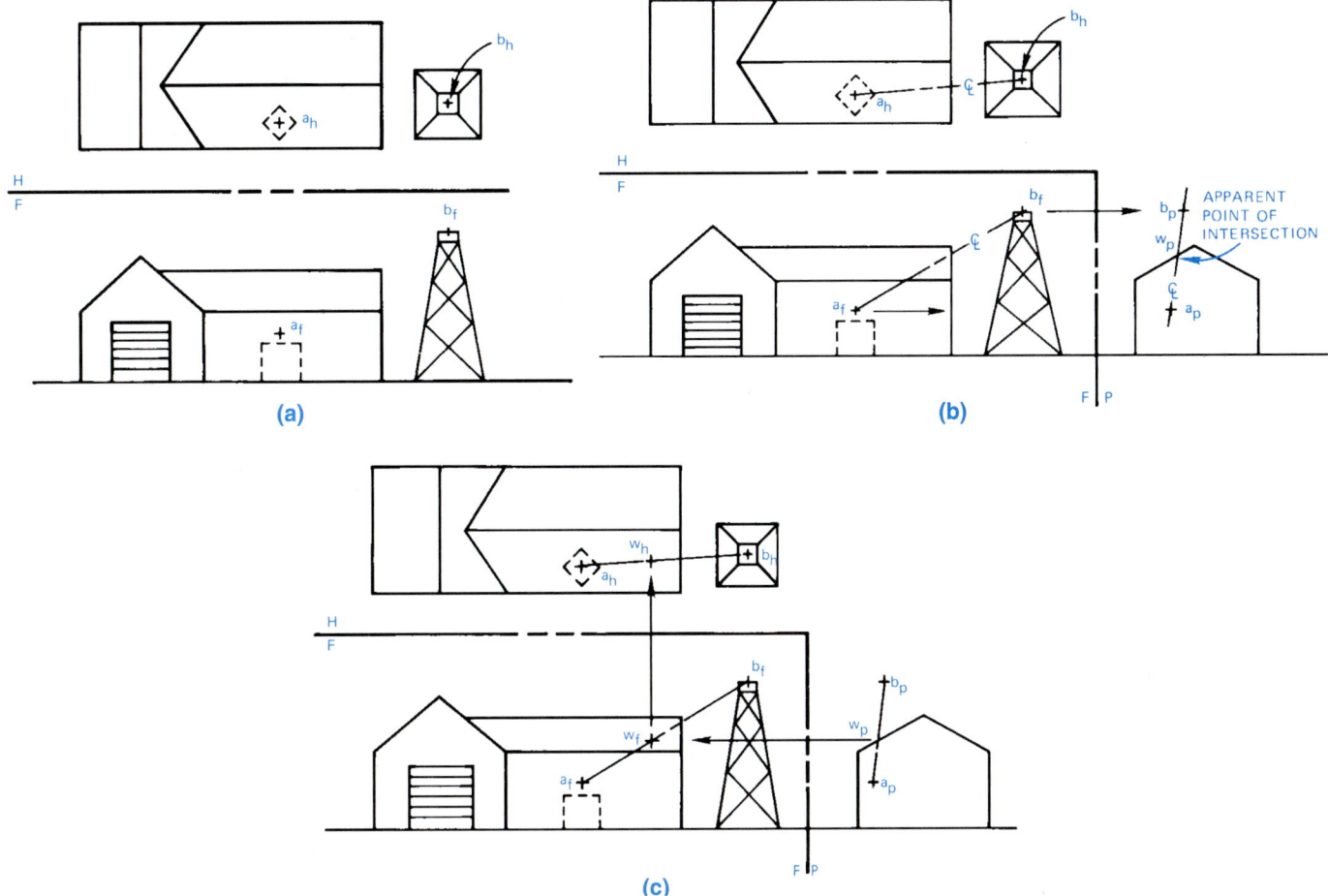

FIGURE 17.17 ■ Piercing point problem: (a) problem setup; (b) step 1, connecting two points and finding the apparent point of intersection; (c) step 2, finding the actual piercing point.

on the F view. You can also show the proper location of point W in the H view by the normal procedure of locating a point in space.

VISIBILITY

When two lines are shown in space, sometimes it is difficult to know which line is on top of which or if they intersect. Problems dealing with the intersection of planes require you to know which line is in front of the other in order to correctly solve and complete the problem as you will see later in this chapter. For example, look at Figure 17.18a. Looking at the F view, you cannot tell if lines A-B and R-S intersect or if line A-B is in front of or behind line R-S. There is a simple way to determine which situation exists.

STEP 1 On the F view shown in Figure 17.18b, project a projector from the apparent point of intersection of lines A-B and R-S perpendicular to fold line F/H. Continue projecting until it crosses one or both of the lines on the H view.

STEP 2 If the projector passes through the apparent point of intersection on the H view, these are intersecting lines and

are the real point of intersection. But this is *not* the case in this example. The projector crosses line R-S in the H view first and verifies that line R-S is closest to you in the F view.

STEP 3 You can apply the same procedure when determining the visibility of lines R-S and A-B on the H view. Figure 17.18c shows a projector being projected from the apparent point of intersection on the H view perpendicular to the H/F fold line. The projector is then continued on until it crosses one or both of lines A-B and R-S in the F view. The projector crosses line A-B in the F view first and verifies that line A-B is closest to you in the H view.

INTERSECTION OF PLANES

Structural sheets, houses, sheet metal forms, and many other features have intersecting surfaces. Determining the lines of intersection is essential in completing each view of a drawing and eventually determining the shape and size of each surface for the construction of each part in the fabrication shop. If two planes are not parallel, they will intersect and form a straight

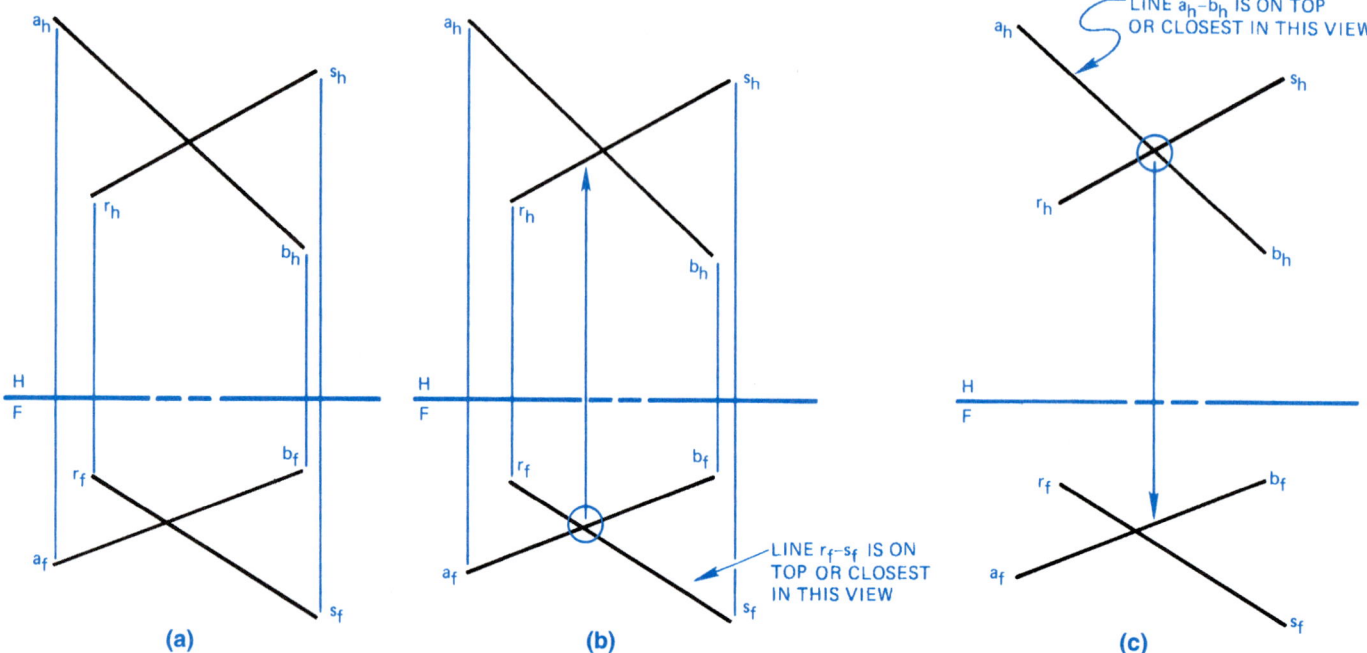

FIGURE 17.18 ■ Visibility: (a) two lines in space; (b) finding visibility of lines in the F view; (c) finding visibility of lines in the H view.

line that is common to both planes. Only two points that lay on both planes need to be found to fix the position of this line of intersection that is common to both planes. To find the line of intersection of two planes, a view must be created that shows both planes where one appears as an edge view. Figure 17.19a shows two views in which the two planes appear to intersect.

STEP 1 Create a projection plane where one of the planes appears as an edge view. Figure 17.19b shows finding plane A-B-C as an edge view in the A view. You already know how to do this from Chapter 16.

STEP 2 Figure 17.19c shows the common points J and K of intersection of the two planes. These points can now be projected back onto the F and H views and completed as the line of intersection.

STEP 3 Identify the visibility of the intersecting lines of each plane on the F and H views by the process previously explained. You will need to do this in order to show which lines are visible and which lines are hidden on the two planes. Figure 17.19d shows this process completed.

DIHEDRAL ANGLES

Working with structural sheets or various sheet metal projects requires knowing the exact angle between surfaces or planes. The true angle between two planes, or dihedral angle, is found in a view in which both of the intersecting planes appear as an edge. (See Figures 17.20 and 17.21.) Figure 17.22a shows the top and front views of a sheet metal funnel on the H and F views. The problem here is to determine the dihedral angle between the front side and the right side of the funnel.

Steps in Finding the True Dihedral Angle

To find the true dihedral angle between two planes, you need to find the end view of the line that is created by the intersection of the two planes. That first step is to place a fold line parallel to this line of intersection. Then, place a fold line perpendicular to this true-length line.

STEP 1 Create a view in which you show the true length of the line of intersection of the two planes for which you are trying to determine the dihedral angle. In Figure 17.22b, the true length of line *D-T*, the line of intersection, has been found in the I view. You already know how to find the true length of a line from Chapter 16. Then, project only the points that identify the rest of the two intersecting planes onto the I view as shown.

STEP 2 Create a view that shows the intersecting line *D-T* appearing as a point. In Figure 17.22c, line *D-T* has been found on the A view. It appears as a point. Notice also that when you project all the other points of the two planes onto the A view, they form two intersecting straight lines. The angle formed between these two planes is the true dihedral angle. This angle can be measured with a protractor or calculated on the computer.

AN INTRODUCTION TO VECTOR GEOMETRY

Vector analysis is a branch of mathematics that includes the manipulation of vectors. Vector geometry is included in this

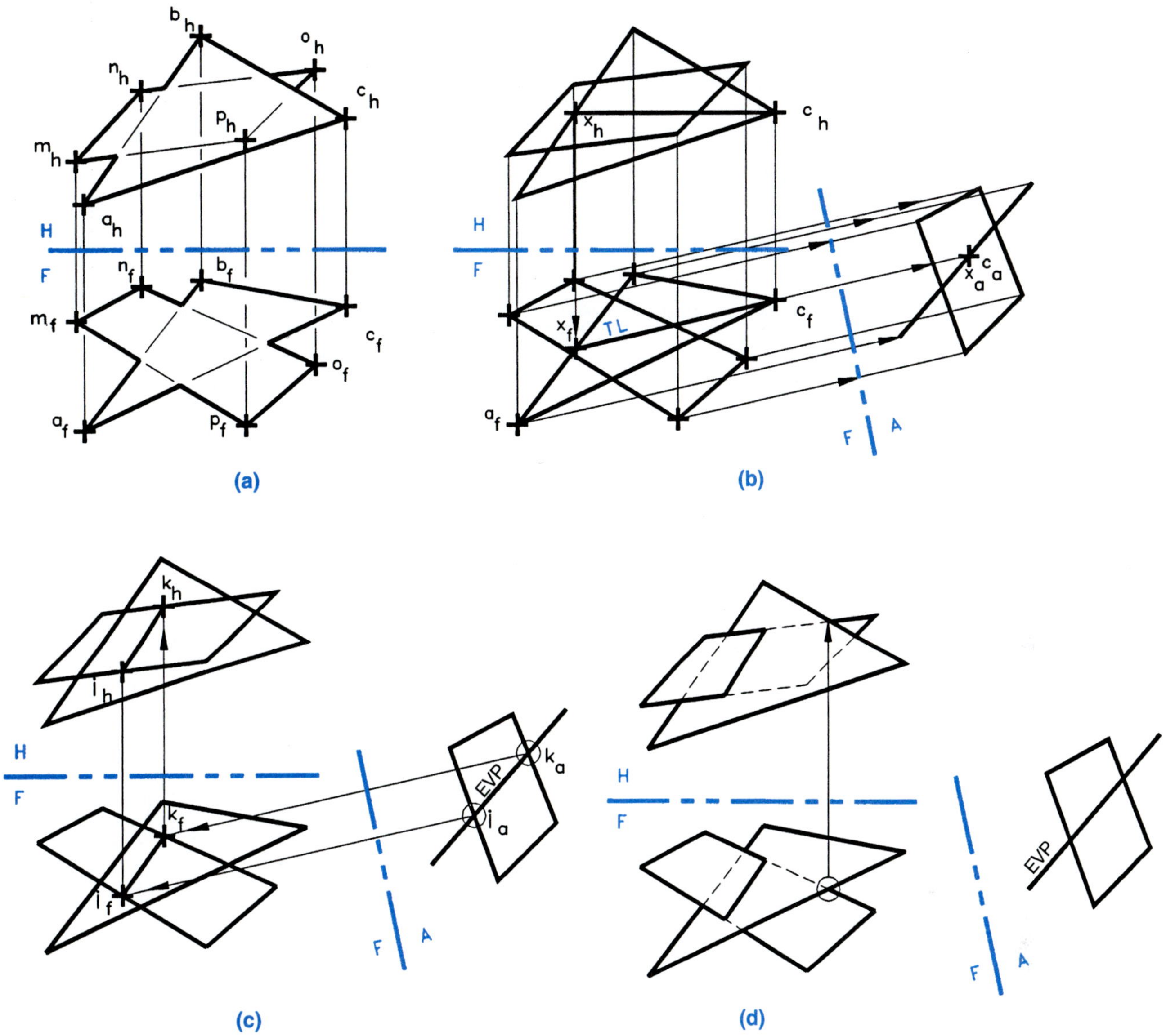

(a)

(b)

(c)

(d)

FIGURE 17.19 ■ Two planes that appear to intersect: (a) problem setup; (b) step 1, finding the edge view of a plane; (c) step 2, finding the line of intersection of two planes; (d) step 3, determining visibility and completing the visible and hidden lines of two planes. EVP represents the edge view of the plane.

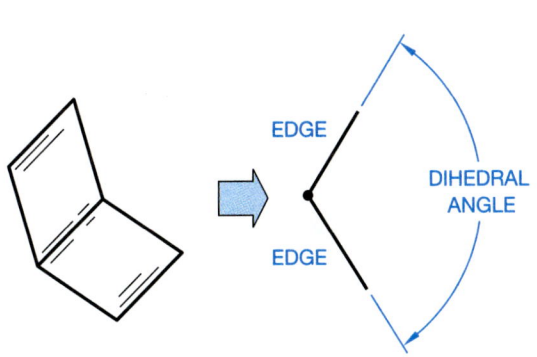

FIGURE 17.20 ■ Pictorial representation of a dihedral angle.

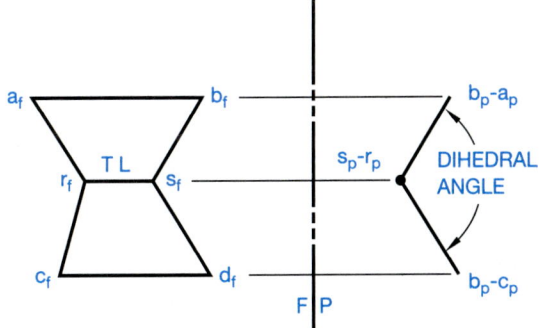

FIGURE 17.21 ■ Descriptive geometry flat layout showing the dihedral angle of two planes.

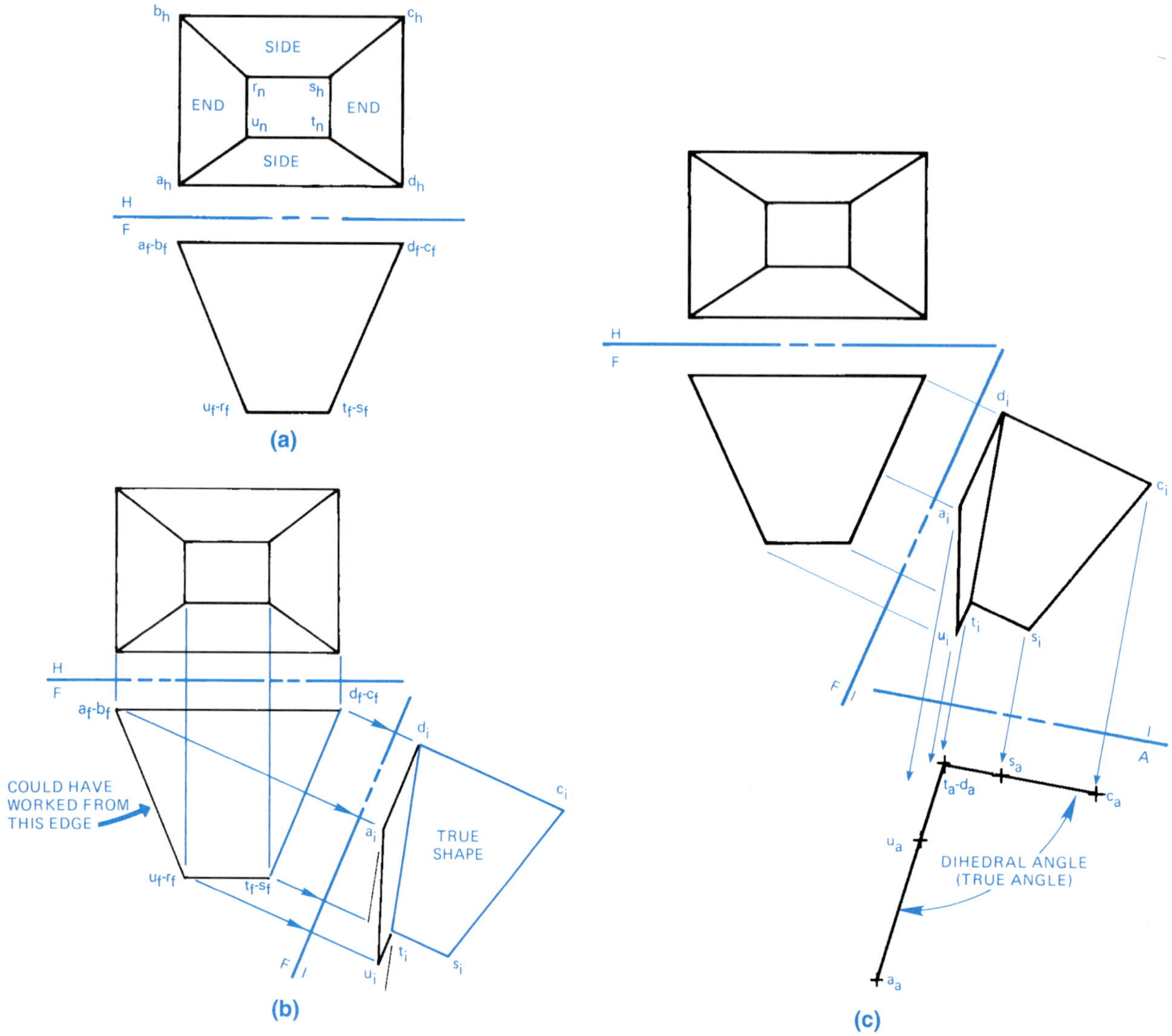

FIGURE 17.22 ■ Dihedral angle problem: (a) two views of a sheet metal funnel; (b) step 1, finding the true length of the line of intersection; (c) step 2, finding the end view of a true-length line and creating the dihedral angle of two planes.

chapter because vectors are a vital part of many of the engineering sciences. Graphic solutions are usually sufficiently accurate for engineering analysis of forces and other directional quantities in structural design, mechanics, and other physical sciences. Vectors can be manipulated in some very complex ways, but the discussion here is limited to vector addition and its practical purposes. Since vectors are segments of straight lines, the principles of descriptive geometry are directly applicable to the graphic solution of problems involving vectors.

A **vector quantity** is one that has **magnitude** and **direction**, such as displacement, acceleration, momentum, force, velocity, and torque. The speed of a space shuttle, for example, describes only the magnitude of its velocity, which is a physical or **scalar** quality. The velocity of that space shuttle, on the other hand, is a vector quality that describes its speed and its direction of motion in space. A vector quantity is usually represented by a straight line with an attached arrowhead pointed in the same direction in which the quantity is acting. The length of the vector

is drawn proportionally to the magnitude of that quantity. An example of a **vector diagram** representation of a force of 10 1bs. directed N 45° E is shown in Figure 17.23. To solve problems using vector geometry, it is essential to know the difference between a vector diagram and a **space diagram**. Figure 17.24 shows the space diagram of the 10-lb vector. Notice that the space diagram

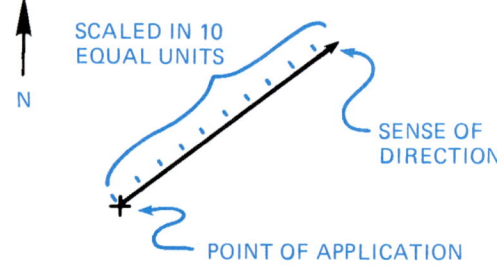

FIGURE 17.23 ■ Vector representation/vector diagram.

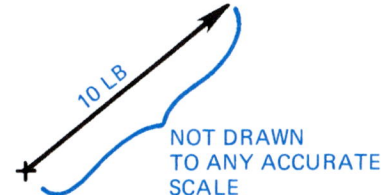

FIGURE 17.24 ■ Space diagram.

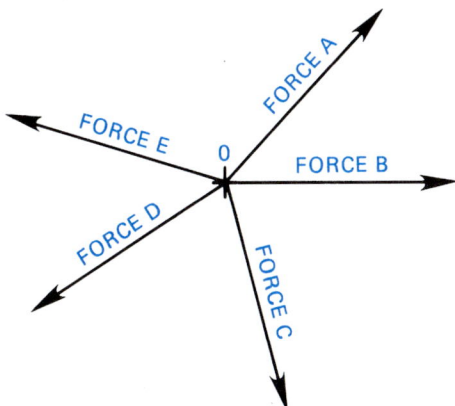

FIGURE 17.25 ■ Space diagram of concurrent forces.

is *not* drawn to scale. It shows sense and the accurate direction. When two or more vectors act on an object, it is called a **vector** or **force system**. The discussion in this chapter is limited to forces that are **concurrent**, because they are the most common. In concurrent forces, vectors pass through a common point, as seen in Figure 17.25. Forces with a line of action lying in one plane are called **coplanar** forces. Figure 17.26 shows five coplanar vectors pictorially and with their corresponding vector diagram. When those forces lie in more than one plane but still meet at one point, they are identified as **noncoplanar**, as shown pictorially and in their corresponding vector diagram in Figure 17.27,

Addition of Concurrent Coplanar Vectors

To determine the net effect of a vector system, vector quantities must be combined. The graphic addition of concurrent coplanar vectors requires you to draw a vector diagram locating the vectors **head-to-tail** in their correct direction. The sum, or the space that is left open, is called the **resultant**. The resultant has the same effect as the total of the individual vectors added together. In Figure 17.28a, the two vectors M and N shown in the space diagram are to be added. The magnitude of vector M is 40 lb and vector N is 60 lb. Remember that the space diagram shows the **correct direction**, so in the vector diagram, Figure 17.28b, the direction of the vectors is duplicated and the vector length is scaled to the same units. After placing each vector head-to-tail, in any order, the space left is connected with a straight line. This line is called the resultant and the direction arrow is in the same direction as the added vectors. Measuring the length of the resultant to scale gives a length of 80 lb. If the direction of the resultant is reversed as shown in Figure 17.28c, it is called the **equilibrant**. An equilibrant is a force that balances all other forces through a point. An equilibrant is the same as the resultant of zero. If you have just two vectors to be added, the **parallelogram method** can be used. This method requires more construction, in that the sides of the parallelogram are drawn parallel to the given vectors. Then the diagonal of the parallelogram is drawn, which is the resultant of the two vectors, as shown in Figure 17.28d. When you have three or more vectors, similar methods can be used. In Figure 17.29a, the vectors are drawn to scale, using the space diagram. In Figure 17.29b, the correct direction is obtained as previously explained. The closing side of the vector polygon is the **resultant**. As before, if the equilibrant is required, the direction of the arrowhead of the resultant is reversed, as shown in Figure 17.29c.

Resolution of Concurrent Coplanar Vectors

Two or more vectors can be added to form a single resultant vector. By reversing that process, a single vector can be resolved

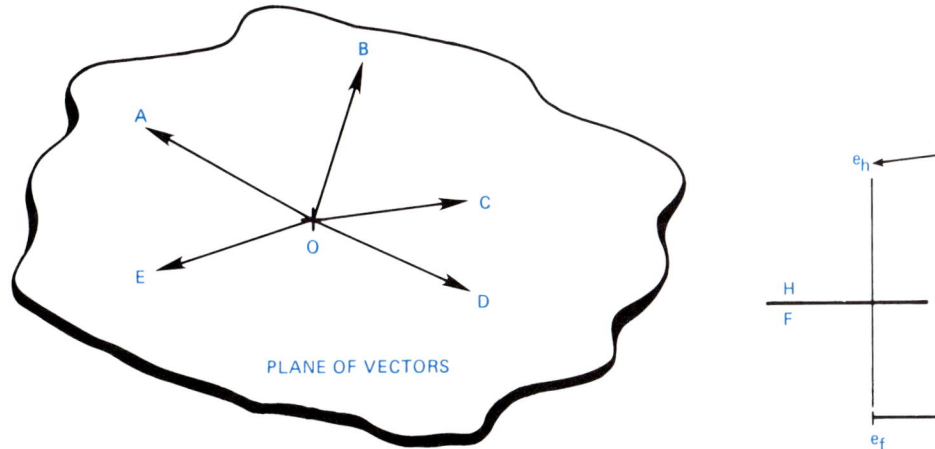
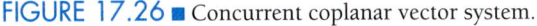

FIGURE 17.26 ■ Concurrent coplanar vector system.

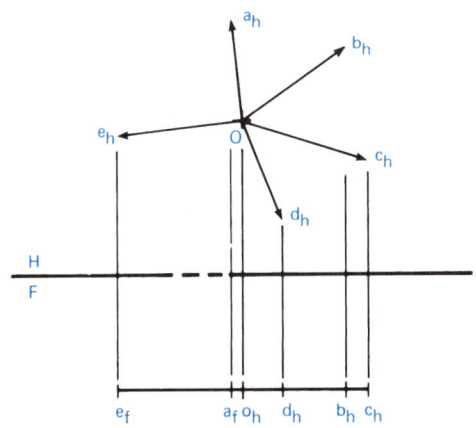

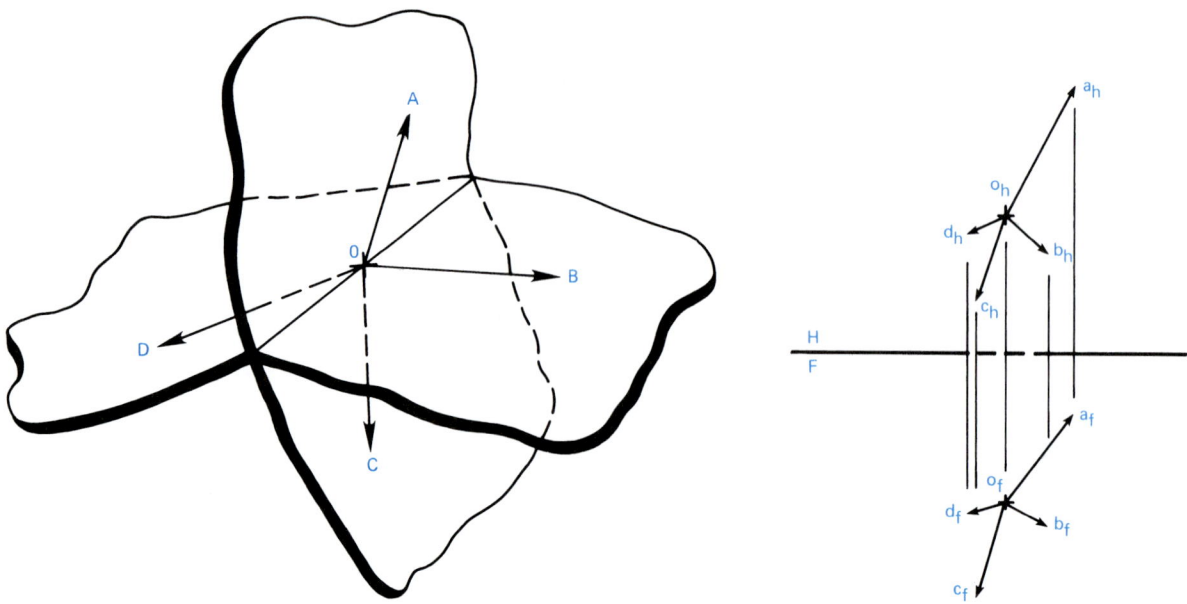

FIGURE 17.27 ■ Concurrent noncoplanar vector system.

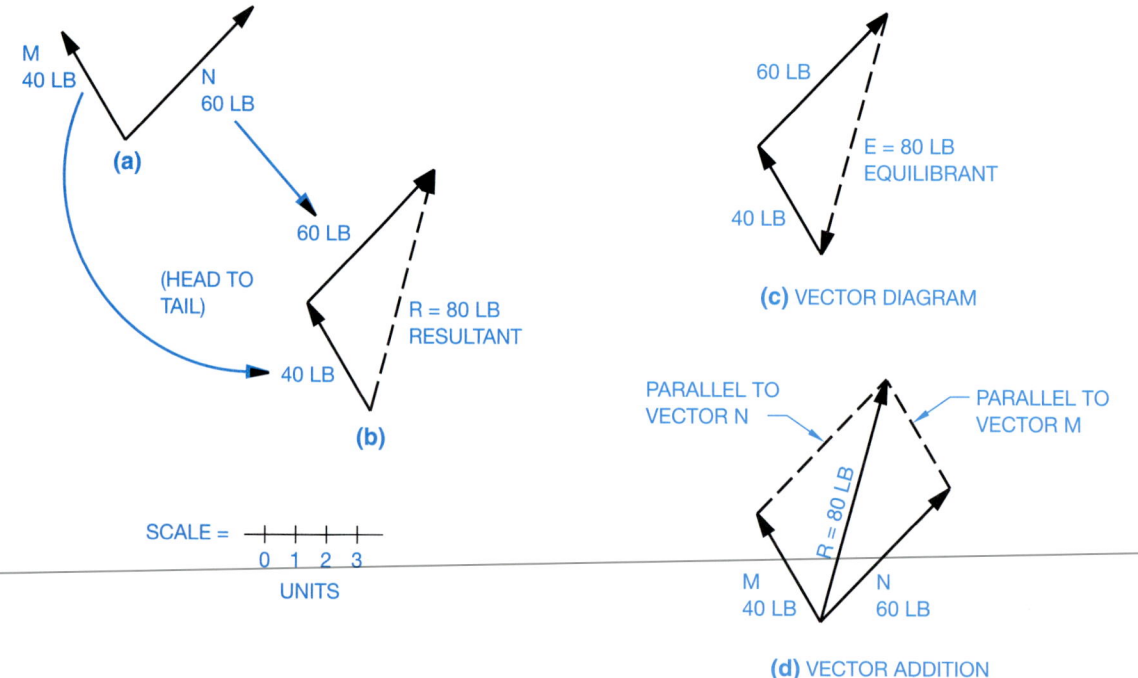

FIGURE 17.28 ■ Addition of concurrent coplanar vectors: (a) space diagram; (b) vector diagram showing the resultant; (c) vector diagram showing the equilibrant; (d) vector addition by parallelogram method.

into two or more component vectors. In the case of coplanar vectors, only two can be unknown in magnitude or direction. Most commonly, a vector must be resolved into two other vectors having specified directions. In Figure 17.30a, a given vector T is to be resolved into two component vectors having direction of OR and OS. The magnitude of each needs to be found. By using the parallelogram method for the addition of two vectors, the solution is obtained in reverse. Vector R is moved parallel until it crosses the arrow end of the given vector.

(See Figure 17.30b.) The S vector is extended through the tail end of the given vector. Where the two moved vectors cross will determine the lengths of each. The direction of each of the two vectors is opposite that of the given vector.

Addition of Concurrent Noncoplanar Vectors

The addition of noncoplanar vectors is the same as for coplanar vectors, except that three dimensions are used now. Remember

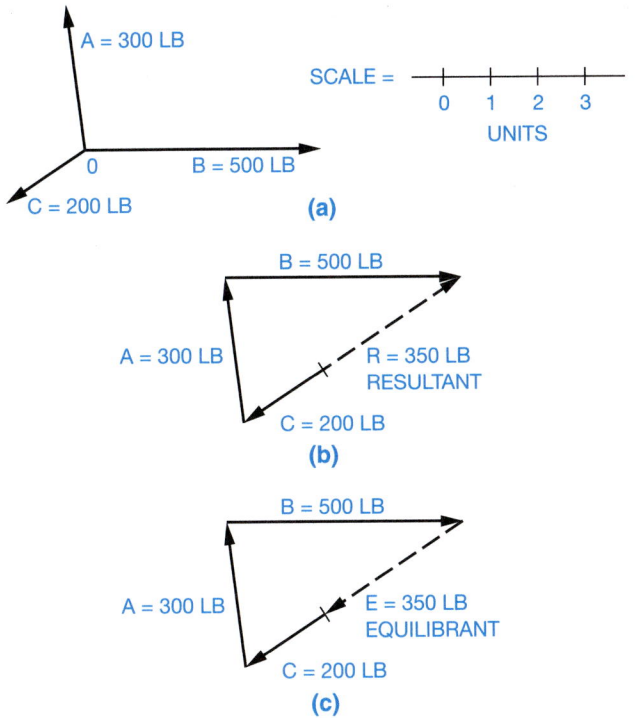

FIGURE 17.29 ■ Addition of concurrent coplanar vectors: (a) space diagram; (b) vector diagram showing the resultant; (c) vector diagram showing the equilibrant.

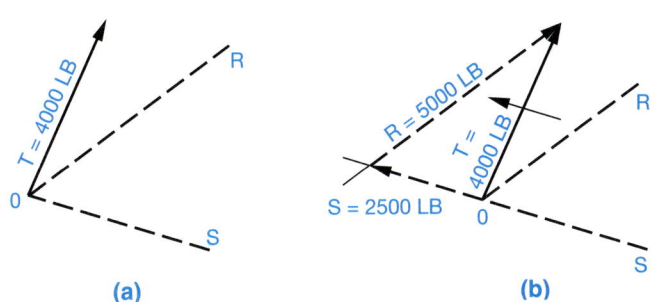

FIGURE 17.30 ■ Resolution of a single vector into two components: (a) space diagram; (b) vector diagram.

that noncoplanar means *in more than one plane*, so the principles of descriptive geometry are needed to solve these problems. Figure 17.31a shows the space diagram on the F and H views for the given vector quantities. To solve this problem, you must create a projection plane that shows the true length of the resultant force.

STEP 1 Set up the space diagram on a frontal view and the adjacent horizontal view as shown in Figure 17.31a.

STEP 2 Find the resultant vector on the H and F views just as you would for coplanar vectors with one major exception: You will need to find the true lengths of those vectors, which are *not* shown in true length on either the H view or the F view. To do this you can use either the fold-

line or the revolution method. Then project the true lengths back onto the H and F views correctly. In Figure 17.31b, the S vector is already shown in true length on the F view. Therefore, you can scale the S vector on the F view. The T vector is shown already in true length on the H projection plane. Therefore, you can scale the T vector on the H view. However, the R vector is not shown in true length on either the H view or the F view. To find the true length of the R vector, use the same procedure as you did when you had to find the slope of a line when given only the bearing of that line as explained in Chapter 16. In this case, line s_h-r_h was established with an arbitrary length after the vector diagram was completed and placed in the correct direction on the H view. Then, the corresponding vector diagram was completed in the F view. Vector R was placed in the correct direction as shown in the space diagram on the F view with the corresponding length as projected from the H view. The true length of line s_f-r_f was found in the H view. Notice that the line needed to be lengthened to equal the scaled 30 lbs. Line s_h-r_{h2} now represents the true length of vector R on the H view. This length is then projected into the vector diagrams as shown in Figure 17.31b.

STEP 3 Find the true length of the resultant force, the space left in the vector diagrams. The resultant force is *not* shown in true length on either the H or F views, so the fold-line method was used this time to find the true length of the resultant force on the I view. The resultant force is measured at 15 lbs. as shown on the I view. (See Figure 17.31b.)

FORCES IN EQUILIBRIUM

Equilibrium is a condition in which the series of forces acting upon a body or structure equals zero and remains at rest. Graphically, the vector polygon closes, with vectors in a continuous head to tail arrangement. (See Figure 17.32.) When a body is known to be in equilibrium under the action of certain known and unknown forces, these graphic conditions can be used to evaluate the unknown forces. However, the number of unknown forces is strictly limited. Coplanar vector systems can have up to three unknown forces. Vector diagrams can more easily be constructed if you isolate the structure or part to which the known forces are applied. This isolation means free from all adjacent bodies and is referred to as **free-body**. Construction of a free-body diagram should be the first step in the analysis of every equilibrium problem.

Finding the Equilibrium of Two Unknowns in a Coplanar System

The following steps may be used to find the equilibrium of two unknowns in a coplanar system when you are given a weight that is supported by two ropes as shown in Figure 17.33a.

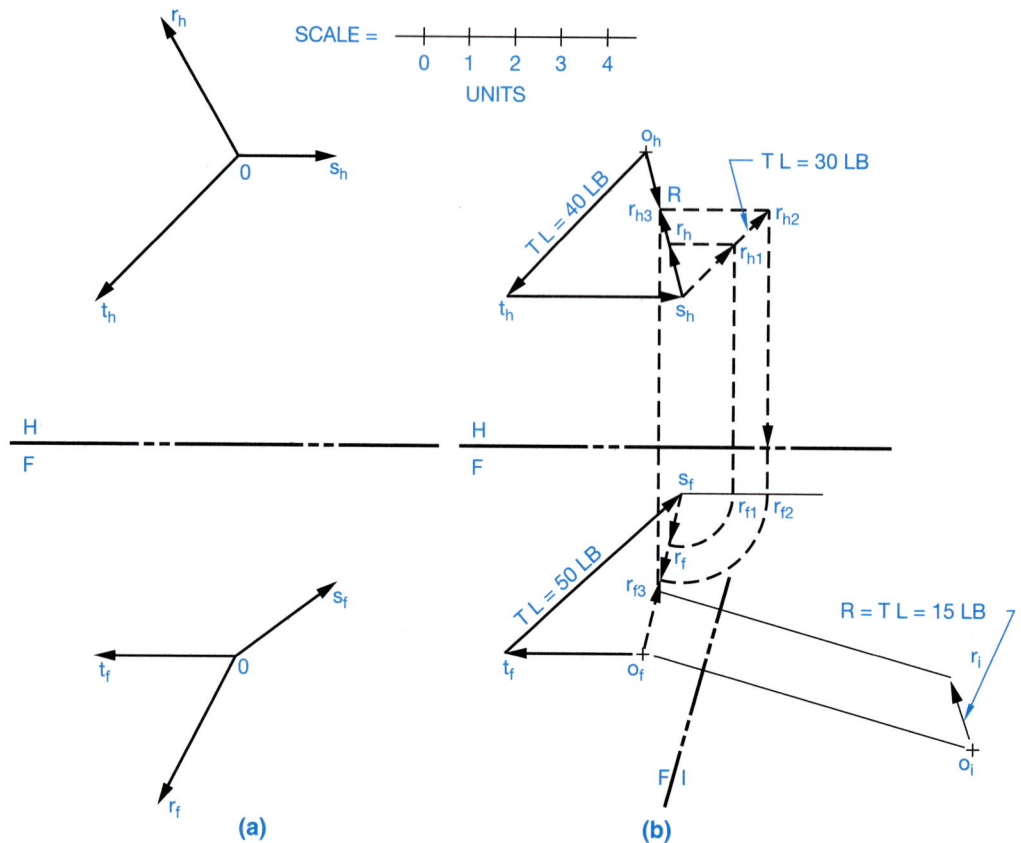

FIGURE 17.31 ■ Addition of concurrent noncoplanar vectors: (a) space diagram; (b) vector diagram.

STEP 1 Imagine the ropes are cut, making the knot an isolated free-body, as shown in Figure 17.33b. Draw a free-body diagram with the tension forces in the free-body diagram being represented with force vectors pointing away from the knot.

STEP 2 Make a vector diagram. Start with the known vector force of 100 lbs. by drawing it to scale with the proper direction. The direction of forces A and B are known but not their magnitudes. It was previously explained that for forces to be in equilibrium, the vector diagram must close. As shown in Figure 17.33c, the vectors A and B are drawn parallel to their corresponding vectors in the free-body diagram shown in Figure 17.33b. When the vectors are drawn properly, point L is established in the vector diagram.

STEP 3 Scale the magnitudes of vectors A and B. In this example, both vector A and vector B measure to be 60 lbs.

A similar type of problem occurs when two directions are unknown. In this situation, you are given the space diagram in Figure 17.34a, which shows a weight of 10 lbs. supported by two cables that pass over pulleys, which, in turn, support two other weights of 6 and 8 lbs.

STEP 1 Draw the space diagram. The forces acting in the cables are known, but the direction of the vectors is not. As

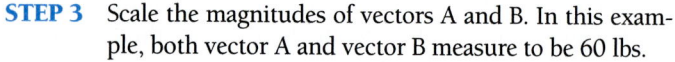

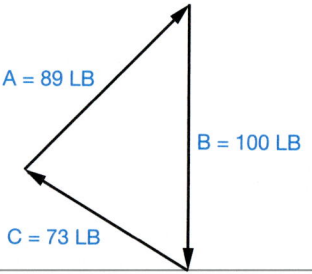

VECTOR DIAGRAM

FIGURE 17.32 ■ Vector diagram showing forces in equilibrium.

shown in Figure 17.34b, start with the vector with the force and direction known. The vector that represents 10 lbs. is drawn in the down position. As shown, draw the other two vector forces at any two different directions as shown in Figure 17.34b. The wavy lines indicate that the real directions are unknown.

STEP 2 Draw the free-body diagram. This is solved by using the method of creating a triangle when three sides of the triangle are known. This concept was covered in Chapter 8 on geometric constructions. Start with the known force of 10 lbs. Draw it in its correct direction

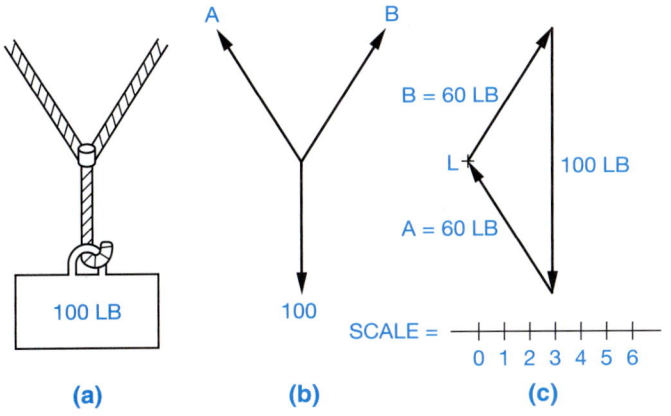

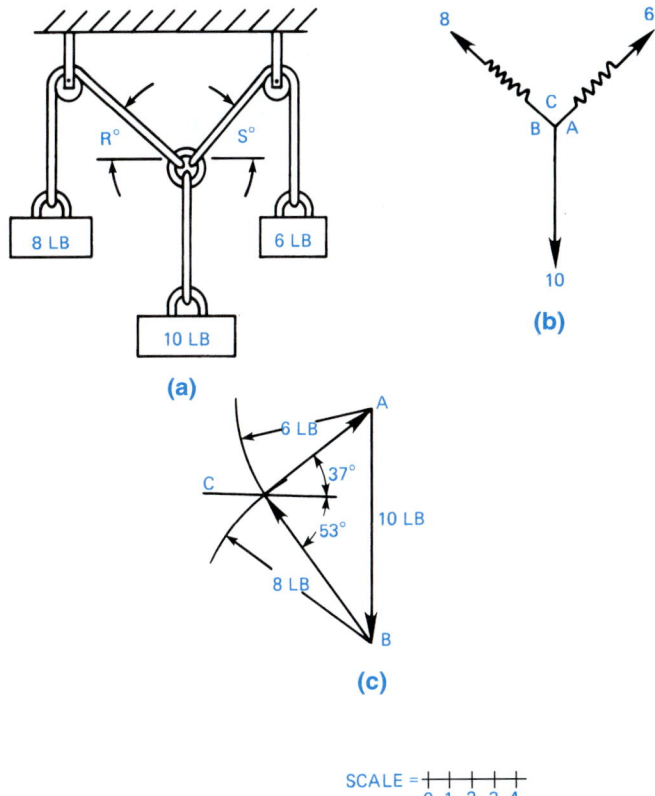

FIGURE 17.33 ■ Forces in equilibrium with two magnitudes unknown: (a) space diagram; (b) free-body diagram; (c) vector diagram.

and magnitude. Next, with a radius equal to 6 lbs., an arc is drawn with A as its center. (See Figure 17.34c). With a radius equal to 8 lbs., another arc is drawn with B as its center. These arcs intersect at C, completing the triangle.

STEP 3 Determine the directions of the 6-lb. and 8-lb. vectors. The directions are indicated by the angles shown and are determined by placing the arrowheads in a head-to-tail manner in the diagram. (See Figure 17.34c.)

FIGURE 17.34 ■ Forces in equilibrium with two directions unknown: (a) space diagram; (b) free-body diagram; (c) vector diagram.

WEBSITE RESEARCH

The following websites can provide you additional information for research or further study into topics covered in this chapter:

http://www.asme.org—Find information and publications related to the American Society of Mechanical Engineers.

http://www.ansi.org—The American National Standards Institute. Information about national and international drafting standards.

http://www.adda.org—American Design Drafting Association. Information related to the drafting and design profession.

PROFESSIONAL PERSPECTIVE

In this chapter, many new concepts have been introduced. But as you may have noticed, the techniques used to solve each of the unknown angles and locations of the intersections are the same as those used in Chapter 16 with, for the most part, only the terminology changed. Again, accuracy in the construction and drawing of lines is essential and natural with CADD. In most cases, the problem situations are more complex; so it is essential to understand how to break each one down into various simple elements. To solve vector problems, the correct terminology must be learned for the different types of vectors and diagrams. This will be especially true if there has been no previous experience with the related physics. Problem solving, while difficult for many, can be made easier if a logical system is used, such as the one used at the beginning of this chapter. This system is basically the one most engineers or anyone can best use to solve any type of problem.

CHAPTER 17

Descriptive Geometry II Test

 Access the CD found with this textbook to view the Chapter 17 Test. Confirm the preferred submittal method with your instructor.

CHAPTER 17

Descriptive Geometry II Problems

DIRECTIONS

Carefully study the problems before you begin working. Solve Problems 17.1 through 17.19 on graph paper or with CADD, showing and identifying all projectors, brackets with distances used, and point identifications in the required view(s). If using graph paper, use a separate sheet of graph paper for each problem and the graph lines included with each problem for proper setup. Use a scale of at least three squares to one to obtain accuracy in the problem solutions.

PROBLEM 17.1 Given the adjacent horizontal and front views showing the cable R-S, find its bearing, slope, and grade.

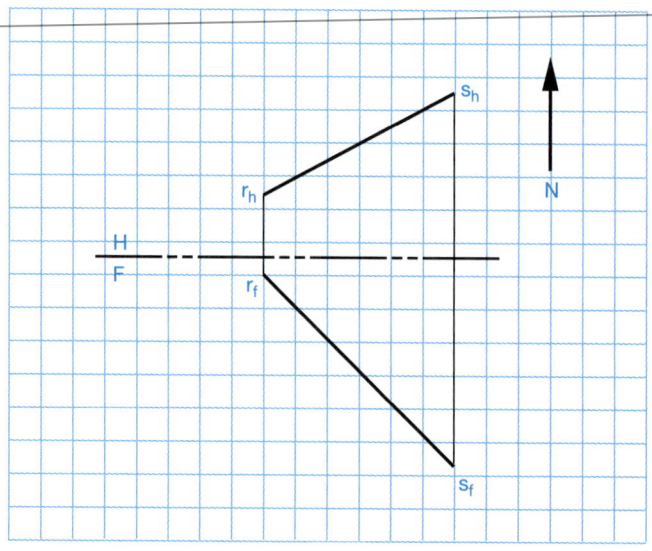

PROBLEM 17.2   True length and slope. A crane is mounted on the bow of a boat as shown. Find the true length and the slope of the cable A-B and the support structure B-C.

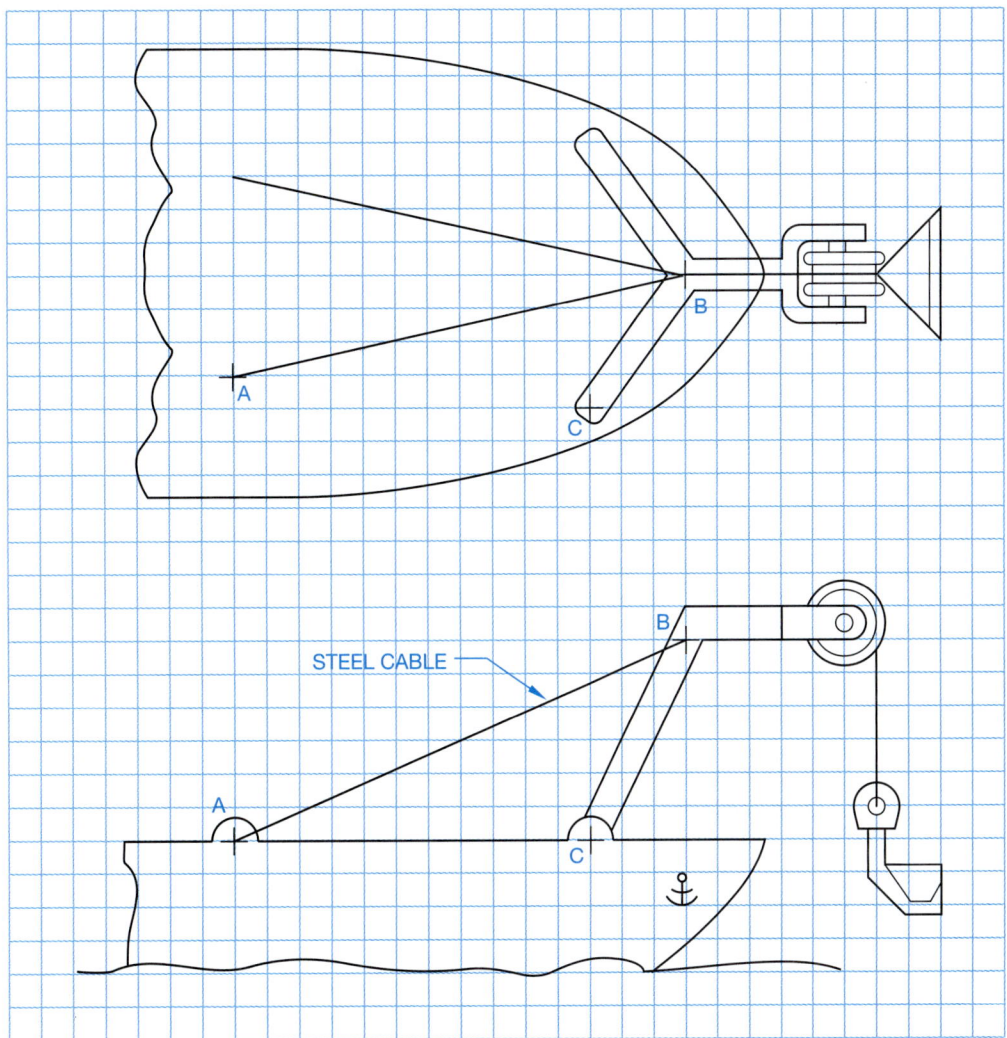

STEEL CABLE

PROBLEM 17.3   Bearing, slope, and true length. From the following given data, draw separate adjacent horizontal and vertical views that contain the following lines:

Line	Bearing	Slope	True Length
B-C	N 22° W	+30°	3 in.
R-S	S 75° E	−10°	4 in.
M-N	N 60° E	+15°	5 in.

PROBLEM 17.4   Slope and grade. In this problem, two sections of pipe are connected together. Using only the centerlines of each pipe, draw the necessary views to answer the following questions. Leave all construction lines on the drawing. Identify on the drawing where the answers appear:

a. Identify and calculate the true length of pipe A-B.
b. Identify and calculate the slope in degrees of pipe B-C.
c. Identify and calculate the grade in percent of pipe A-B.
d. Identify and calculate the true angle between the pipes.

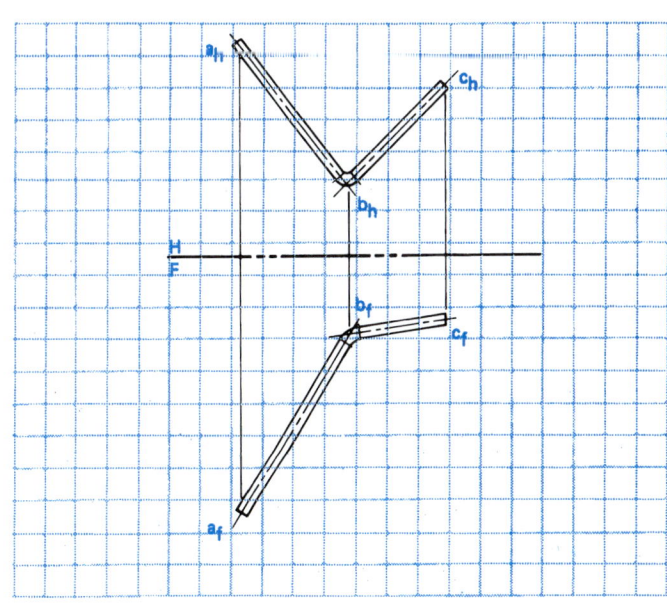

PROBLEM 17.5 True length, true angle. A designer's sketch, below, shows the adjacent front and horizontal views of the roof plane A-B-C-D and the location where the antenna is to be mounted on the roof plane. The top end of the antenna is to project 8' above the ridge of the house.

a. Identify and calculate the true length of the antenna in feet and inches.

b. Find the place where the bottom of the antenna touches the roof in the front view.

c. Create a view that will show the true angle made by the roof plane and then calculate the true angle in degrees.

d. Show, in the horizontal and front views, three equally spaced guy wires anchored *on the roof plane*. Two guy wires are attached 5' from the base of the antenna along the roof plane, and the third guy wire is attached at the point shown in the horizontal view. All three guy wires are anchored to the antenna at a distance of 6' above the ridge.

e. Identify and calculate the total length of guy wire used to stabilize the antenna to the nearest foot.

 PROBLEMS 17.6 through 17.18: Access the CD found with this textbook and open the problem or problems of your choice, or as assigned by your instructor. Solve the problem or problems using the instructions provided.

PROBLEM 17.19 Given the following information on sloping lines, determine their slope, slope angle, and percent of grade.

1. Line A-B: run = 50 feet; rise = 34 feet

2. Line C-D: run = 100 feet; rise = 25 feet

3. Line D-E: run = 148 yards; rise = 100 yards

4. Line F-G: run = 65 inches; rise = 0 inches

5. Line H-J: run = 100 meters; rise = 75 meters

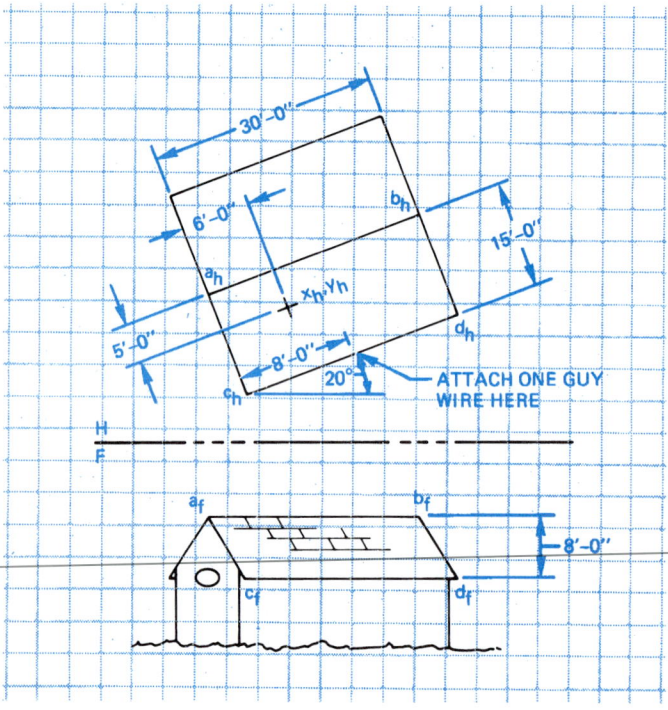

Working Drawings

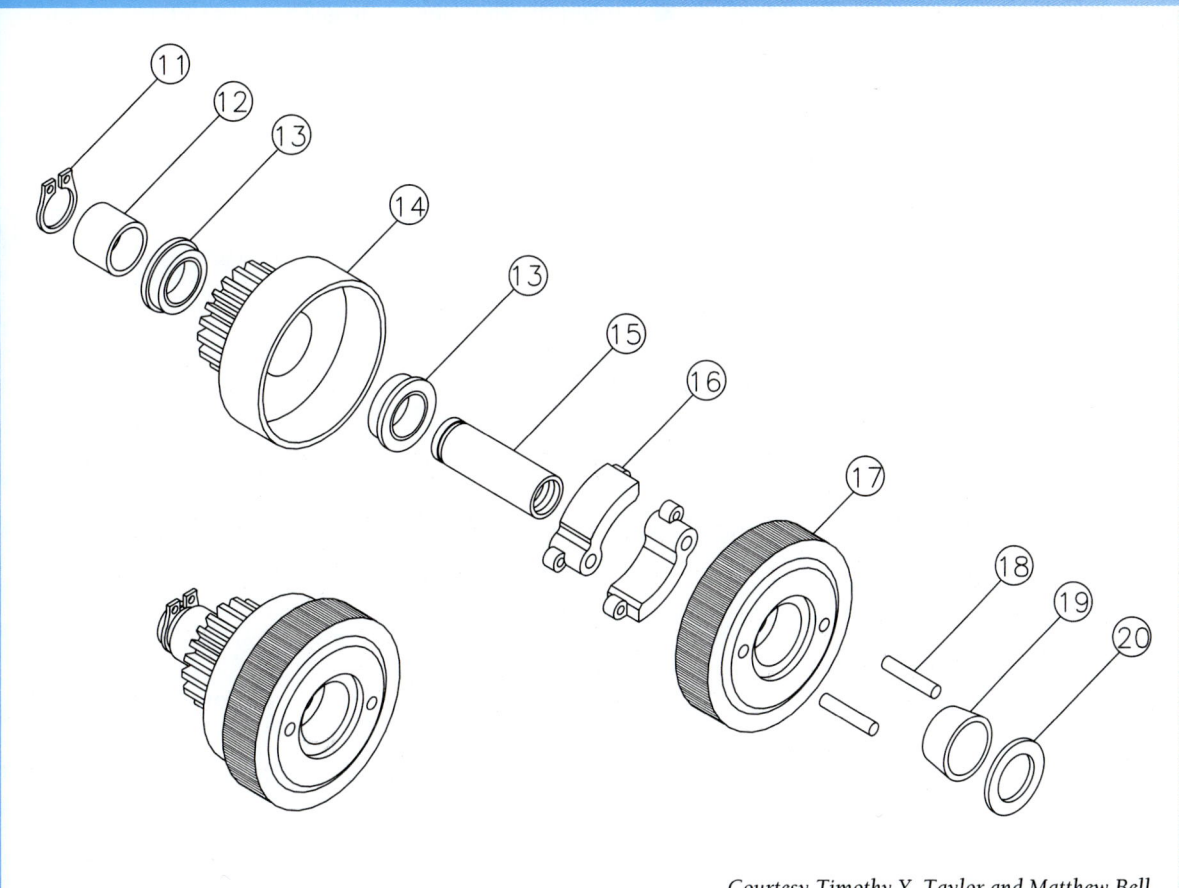

Courtesy Timothy Y. Taylor and Matthew Bell

Working Drawings

LEARNING OBJECTIVES

After completing this chapter, you will:

- Draw complete sets of working drawings, including details, assemblies, and parts lists.

- Prepare written specifications of purchase parts for the parts list.

- Properly group information on the assembly drawing with identification numbering systems.

- Explain the engineering change process and prepare engineering changes.

THE ENGINEERING DESIGN APPLICATION

The engineering department has been asked to develop a new product line consisting of an adjustment knob replacement kit. As the engineering drafter, you have been supplied with appropriate engineering sketches and asked to develop a complete drawing package.

In the new product development process, several departments within your company need specific information.

The manufacturing department needs the dimensional data as well as material and finish specifications. Since one of the items in the kit is a purchase part, the purchasing department must be provided with the necessary information. Packaging information is required to get the product ready for shipping, and the sales department needs some presentation drawings to show to potential clients. Additionally, customer assembly drawings are to be included in an instruction sheet to be supplied with the kit.

It is up to you to develop a complete set of working drawings. Figures 18.1 through 18.6 show all of the drawings and information used to complete this project.

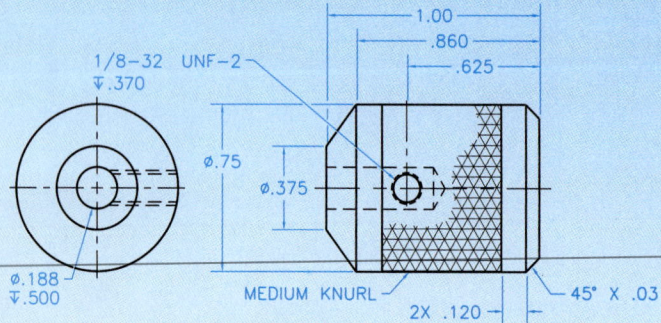

FIGURE 18.1 ■ Detail drawings provide all of the necessary manufacturing information. Dimension values in this figure are in inches.

ITEM	QTY	PART #	DESCRIPTION	MATERIAL
2	1	RAK–002	1/8–32UNF SET SCREW	ACME PART #15–125–32A
1	1	RAK–001	ADJUSTMENT KNOB	A360

FIGURE 18.3 ■ Purchasing information is included in the bill of materials or parts list.

REV	DESCRIPTION	DATE	BY
B	ADDED KNURL	3/7	RJ
A	Ø.375 WAS .438	2/9	RJ

FIGURE 18.2 ■ Engineering changes must be recorded. Dimension values in this figure are in inches.

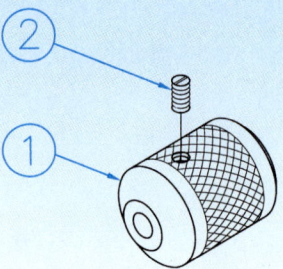

FIGURE 18.4 ■ Isometric assembly drawings show preassembly data and packaging information.

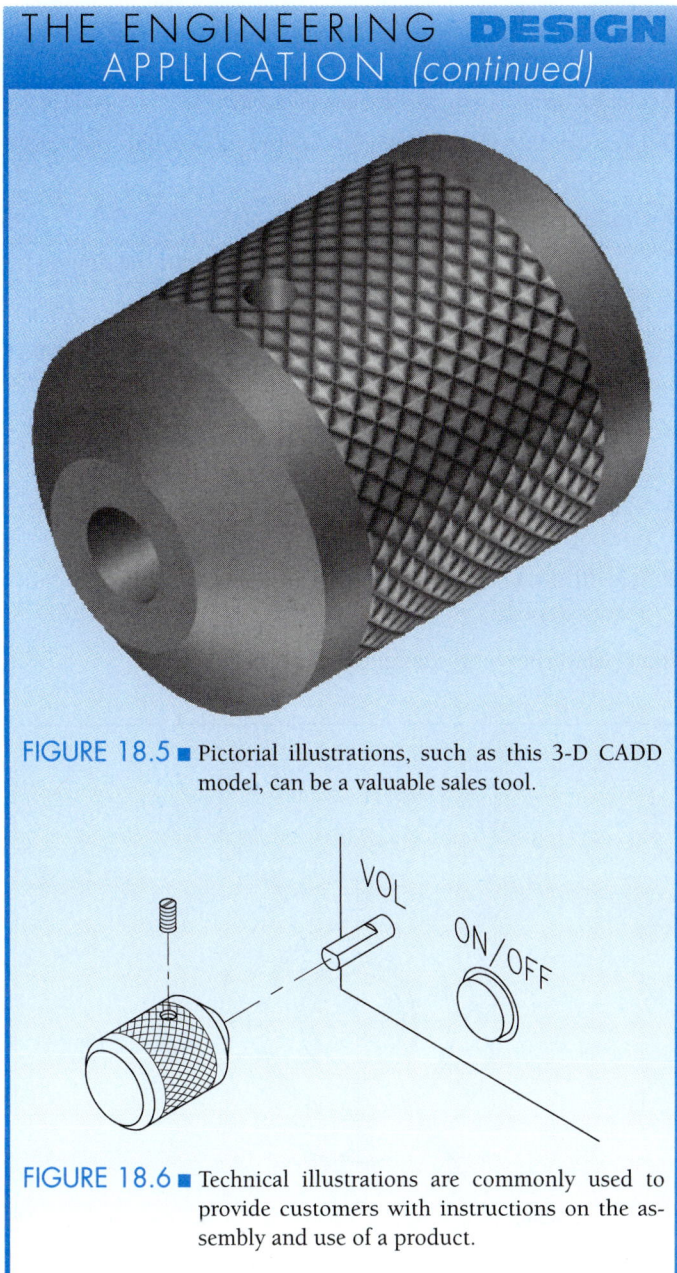

FIGURE 18.5 ■ Pictorial illustrations, such as this 3-D CADD model, can be a valuable sales tool.

FIGURE 18.6 ■ Technical illustrations are commonly used to provide customers with instructions on the assembly and use of a product.

INTRODUCTION TO SECTION FIVE

Everything you have learned up to now has prepared you for the advanced and specialty applications that continue with this section and throughout the rest of this textbook. Section Five is titled Working Drawings because this is where you begin to learn how to create a complete set of drawings for a product. You will be able to apply your previous knowledge and skills as you learn more about the design and drafting industry.

Section Five, Working Drawings, has the following chapters:

Chapter	Title
18	Working Drawings
19	Mechanisms, Linkages, Cams, Gears, and Bearings
20	Belt and Chain Drives
21	Welding Processes and Representations

Creating a set of working drawings is covered first in Chapter 18. A set of working drawings, as you will discover, includes all the detail drawings, assembly drawings, and a parts list for the manufacture and assembly of a product.

Section Five's chapters are organized in the selected manner so you learn about the principles and options that make up a set of working drawings before you continue with features that can be found in the assembly of some products, such as:

■ Linkages, cams, gears, and bearings in Chapter 19.

■ Belt and chain drives in Chapter 20.

■ Assembly of welded parts in Chapter 21.

Most of the Chapter 18 problems can be drawn without studying the other chapters in this section. However, several of the given problems are advanced and challenging, and may best be solved after you study all of Section Five's chapters. If you choose, or if you are assigned to solve these problems in Chapter 18, you should study Chapters 19, 20, and 21 as you encounter parts in your problems that relate to those chapters. This is the kind of challenge that you can face in the real-world of engineering drafting. Oftentimes, you will have to go ahead on your own, or seek additional instruction when you encounter new and varied obstacles.

INTRODUCTION TO WORKING DRAWINGS

Most of the drawings that have been shown as examples or assigned as problems in this text are called detail drawings. Detail drawings are the kind of drawings that most entry-level mechanical drafters prepare. When a product is designed and drawings are made for manufacturing, each part of the product must have a drawing. These drawings of individual parts are referred to as detail drawings. Component parts are assembled to create a final product, and the drawing that shows how the parts go together is called the assembly drawing. Associated with the assembly drawing and coordinated to the detail drawings is the parts list. When the detail drawings, assembly drawing, and parts list are combined, they are referred to as a complete set of working drawings. Working drawings, then, are a set of drawings that supply all the information necessary to manufacture any given product. A set of working drawings includes all the information and instructions needed for the purchase or construction of parts and the assembly of those parts into a product.

DETAIL DRAWINGS

Detail drawings, used by workers in manufacturing, are drawings of each part contained in the assembly of a product. The only parts, that may not have to be drawn are standard parts. Standard parts, also known as purchase parts, are items that can be purchased from an outside supplier more economically than they can be manufactured. Examples of standard or purchased parts are common bolts, screws, pins, keys, and any other product that can be purchased from a vendor. Standard parts do not have to be drawn because a written description clearly identifies the part as shown in Figure 18.7. Detail drawings contain some or all of the following items:

■ Necessary multiviews.

■ Dimensional information.

1/2 - 13UNC-2 X 1.5 LG, SOCKET HEAD CAP SCREW

FIGURE 18.7 ■ Written description of standard or purchase part.

■ Identity of the part, project name, and part number.

■ General notes and specific manufacturing information.

■ The material of which the part is made.

■ The assembly that the part fits (could be keyed to the part number).

■ Number of parts required per assembly.

■ Name(s) of person(s) who worked on or with the drawing.

■ Engineering changes and related information.

In general, detail drawings have information that is classified into three groups:

■ Shape description, which shows or describes the shape of the part.

■ Size description, which shows the size and location of features on the part.

■ Specifications regarding items such as material, finish, and heat treatment.

Figure 18.8 shows an example of a detail drawing.

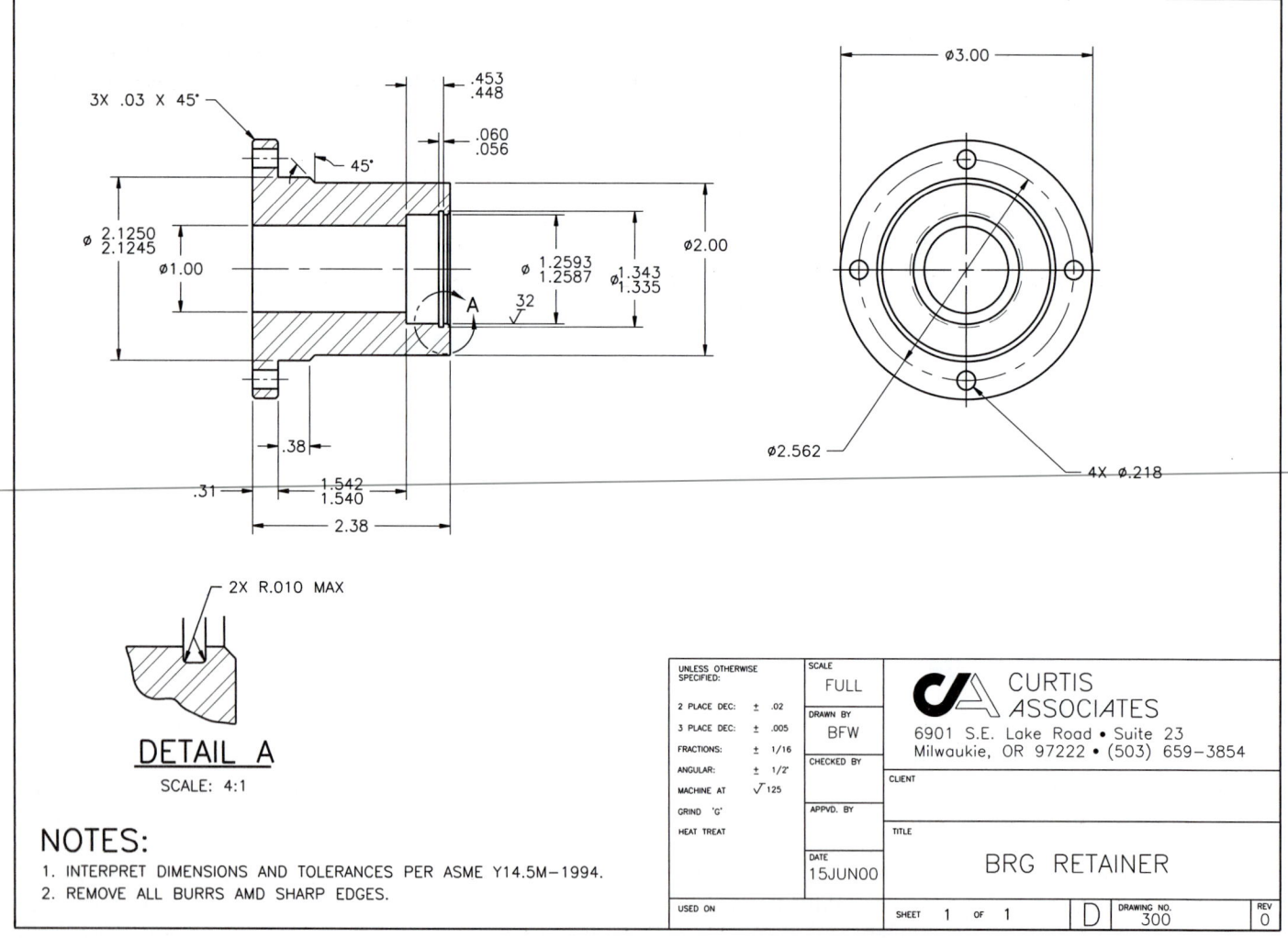

FIGURE 18.8 ■ A monodetail drawing has one part per sheet. Dimension values in this figure are in inches. *Courtesy Curtis Associates.*

Monodetail and Multidetail Drawings

Detail drawings can be prepared with one part per sheet, referred to as a monodetail drawing, as in Figure 18.8 or with several parts grouped on one sheet, which is called a multidetail drawing. The method of presentation depends on the choice of the individual company. Most of the drawing assignments in this text have been done with one detail drawing per sheet, which is a common industry practice. Some companies, however, draw many details per sheet. The advantage of monodetail drawings is that each part stands alone so the drawing of the part can be distributed to manufacturing without several other parts attached. Drawing sheet sizes vary depending on the part size, scale used and information presented. This procedure requires drawings to be filed with numbers that allow the parts to be located in relation to the assembly. The advantage of multidetail drawings per sheet is one of economics. The drafter is able to draw several parts on one sheet depending on the size of the parts, the scale used, and the information associated with the parts. If there are six parts on one sheet, then the number of title blocks that must be completed is reduced by five. The company can use one standard sheet size and can encourage drafters to place as many parts as possible on one sheet. When this practice is used, there are a group of sheets with parts detailed for one assembly. The sheet numbers correlate the sheets to the assembly, and each part is keyed to the assembly. The sheets are given page numbers identifying the page number and total number of pages in the set. For example, if there are three pages in a set, then the first page is identified as 1 of 3, the second as 2 of 3, and the third as 3 of 3. Figure 18.9 shows an example of a multidetail drawing with several detail drawings on one sheet. Some companies use both methods at different times, depending on the purpose of the drawings and the type of product. For example, it is more common for the parts of a weldment to be drawn grouped on sheets as opposed to one per sheet because the parts may be fabricated at one location in the shop.

Sheet Layout

The way in which a detail drawing is laid out normally depends on company practice. Some companies want the drawing to be crowded on as small of a sheet as possible. However, a drawing is easy to read when it is set up on a sheet that provides enough room for views and dimensions without crowding. For example, Figure 18.8 displays two views of a part with dimensions spaced apart far enough to make the drawing clear and easy to read. It is a good idea to provide enough clear space on the drawing to do future revisions without difficulty. An area of the sheet needs to be kept free to provide for the general notes. This area is often in the lower left corner, or above the title block when using ASME standards, or in the upper-left corner when using Military standards. An area should also be left clear of drawing content to provide for engineering change documentation. Preparing engineering changes is covered later in this chapter. The standard location for engineering change information is the upper-right corner of the drawing. It is normally recommended that the space between the revision block and the title block be left clear. Some companies place their revision column above or to the left of the title block, although the ASME standard places this in the upper-right corner.

Advanced sheet layout planning is critical when doing manual drafting, because poor space planning can cause a need for redrawing when engineering changes are done. Proper planning for sheet layout is also important when using CADD, but the sheet size can be quickly and easily changed at any time if needed to provide additional space.

Steps in Making a Detail Drawing

Review Chapter 9, covering the layout of multiviews, Chapter 10 for auxiliary views, and Chapter 11, which explains in detail how to lay out a dimensioned drawing. Review Chapter 12 for fasteners, and Chapter 13 for sectioning techniques.

Detail Drawing Manufacturing Information

Detail drawings are drawn to suit the needs of the manufacturing processes. A detail drawing has all of the information necessary to completely manufacture the part—for example, casting and machining information on one drawing. In some situations, a completely dimensioned machining drawing can be sent to the pattern or die maker. The pattern or die is then made to allow for extra material where machined surfaces are specified. When company standards require, two detail drawings are prepared for each part. One detail gives views and dimensions that are necessary only for the casting or forging process. Another detail is drawn that does not give the previous casting or forging dimensions but provides only the dimensions needed to perform the machining operations on the part. Examples of these drawings are given in Chapter 11.

ASSEMBLY DRAWINGS

Most products are composed of several parts. A drawing showing how all of the parts fit together is called an assembly drawing. Assembly drawings differ in the amount of information provided, and this decision often depends on the nature or complexity of the product. Assembly drawings are generally multiview drawings. Your goal in the preparation of assembly drawings is to use as few views as possible to completely describe how each part goes together. In many cases a single front view does the job. (See Figure 18.10.) Full sections are commonly associated with assembly drawings because the full section exposes the assembly of most or all of the internal features, as shown in Figure 18.11, page 544. If one section or view is not enough to show how the parts fit together, then a number of views or sections can be necessary. In some situations a front view or group of views with broken-out sections is the best method of showing the external features while exposing some of the internal features. (See Figure 18.12, page 544.) You must make the assembly drawing clear enough for the assembly department to put

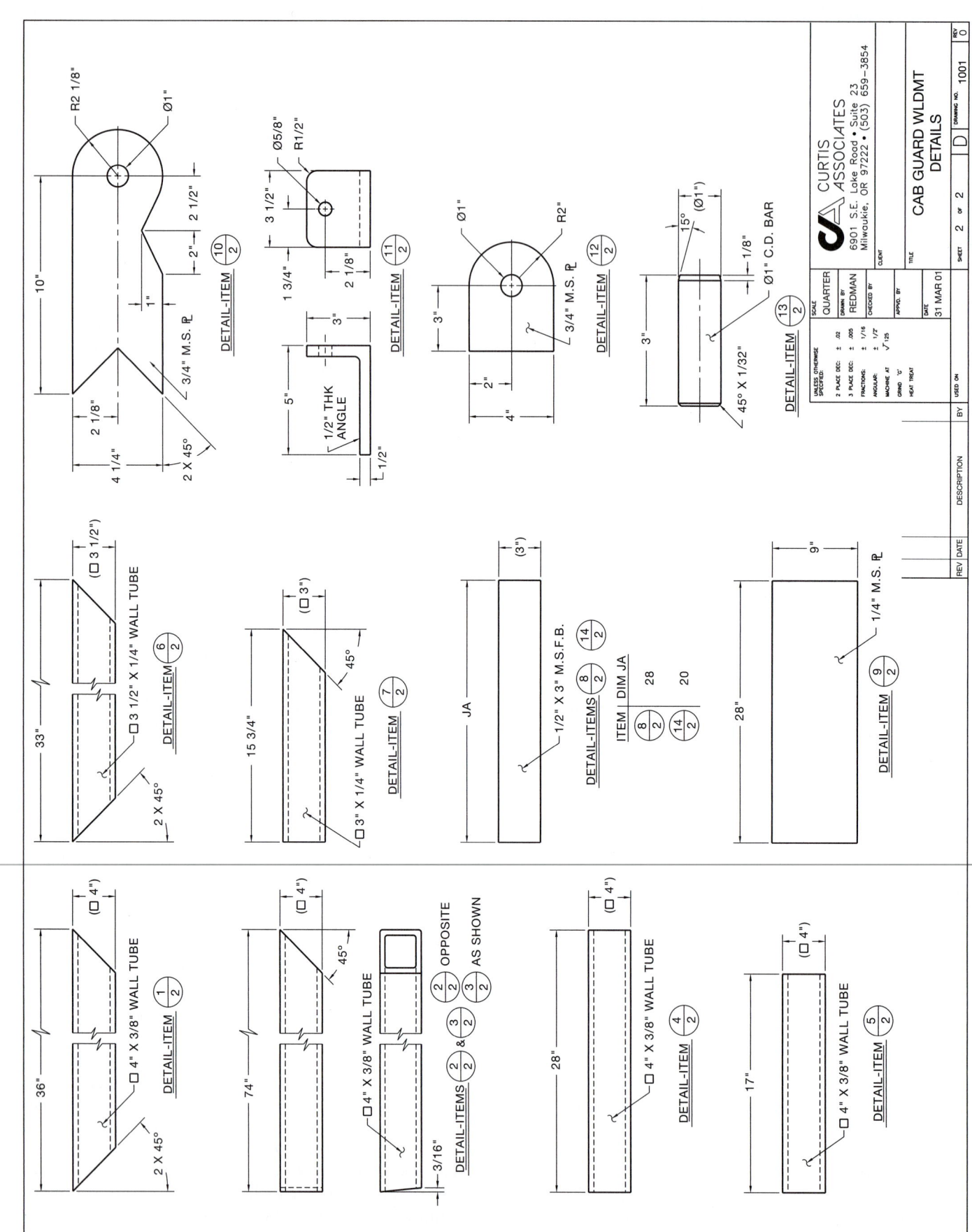

FIGURE 18.9 ■ A multidetail drawing has several detail drawings on one sheet. Dimension values in this figure are in inches. *Courtesy Curtis Associates.*

DETAIL DRAWINGS

CADD systems can be used to produce monodetail and multidetail drawings depending on the purpose or company standard. A single drawing file is typically used to create each monodetail drawing. For example, when using AutoCAD, the process of producing a monodetail drawing involves the following steps:

1. Create a new drawing file based on an existing template.

2. Draw dimensioned views at full scale in Model Space.

3. Access a Layout, also known as Paper Space, to display and plot the detail drawing at a specific scale on a selected sheet size, with a border, title block, and general notes.

These steps are repeated until all the detail drawings have been created in individual drawing files. The typical process of producing a multidetail drawing is similar, except that Model Space contains the dimensioned views of several parts.

An alternative technique for developing monodetail drawings is to create multiple detailed drawings in Model Space, using a single file, as previously described for creating a multidetail drawing. However, instead of using a Layout to display all the detail drawings on a single sheet, use separate Layouts to view each detail drawing. This method is most effective for producing monodetail drawings of small, standard parts, providing a single drawing file that can be used as a parts library, or when company standards require the development of a multidetail drawing and individual monodetail drawings of the same parts. Multidetail drawings can also be drawn by referencing content from multiple files. For example, the XREF command can be used to insert monodetail drawings from several files into a single drawing without significantly increasing file size.

---

the product together. Other elements of assembly drawings that make them different from detail drawings is that they usually contain few or no hidden lines or dimensions. Hidden lines should be avoided on assembly drawings unless absolutely necessary for clarity. The common practice is to draw an exterior view to clarify outside features and a sectional view to expose

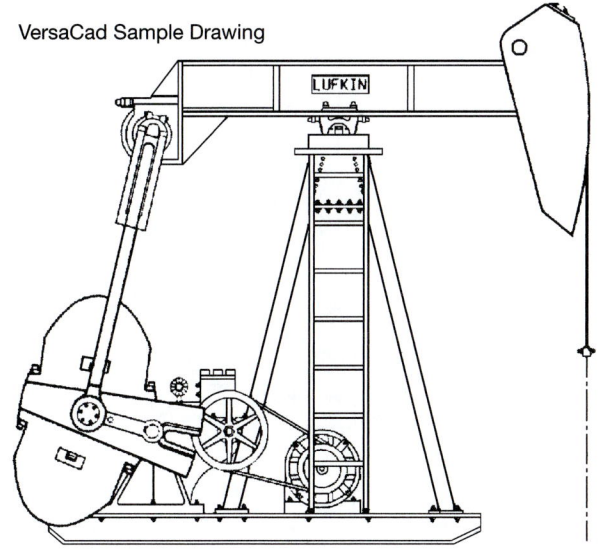

VersaCad Sample Drawing

LUFKIN

FIGURE 18.10 ■ Assembly drawing with single front view. *Courtesy T & W Systems, Inc.*

DETAIL DRAWINGS

As with traditional CADD systems, 3-D CADD programs that combine 2-D drawing capabilities allow you to create monodetail or multidetail drawings. However, the process of producing detail drawings is significantly different, because the drawing geometry, and often the dimensions, are associated with an existing model file. Developing a monodetail drawing based on a 3-D model typically involves the following steps:

1. Create a new drawing file, based on a template. The drawing file usually contains a sheet that represents a specific paper size, and contains a border, title block, and general notes.

2. Place and scale drawing views associated with a single part model.

3. Add dimensions and other specifications as needed.

Then you can add additional monodetail drawings to the same drawing file by inserting additional sheets. Follow the same steps to create a multidetail drawing, but place several detail drawings on the same sheet by referencing multiple part models.

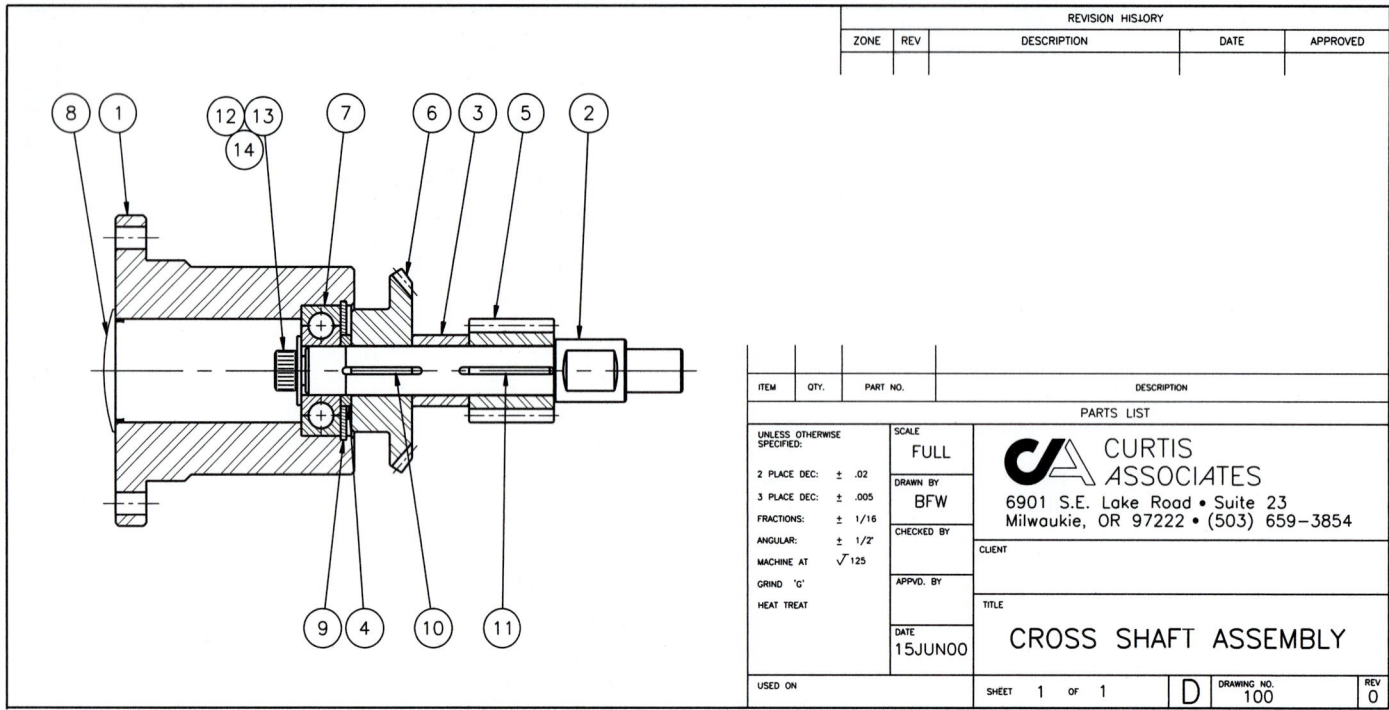

FIGURE 18.11 ■ Assembly in full section. *Courtesy Curtis Associates.*

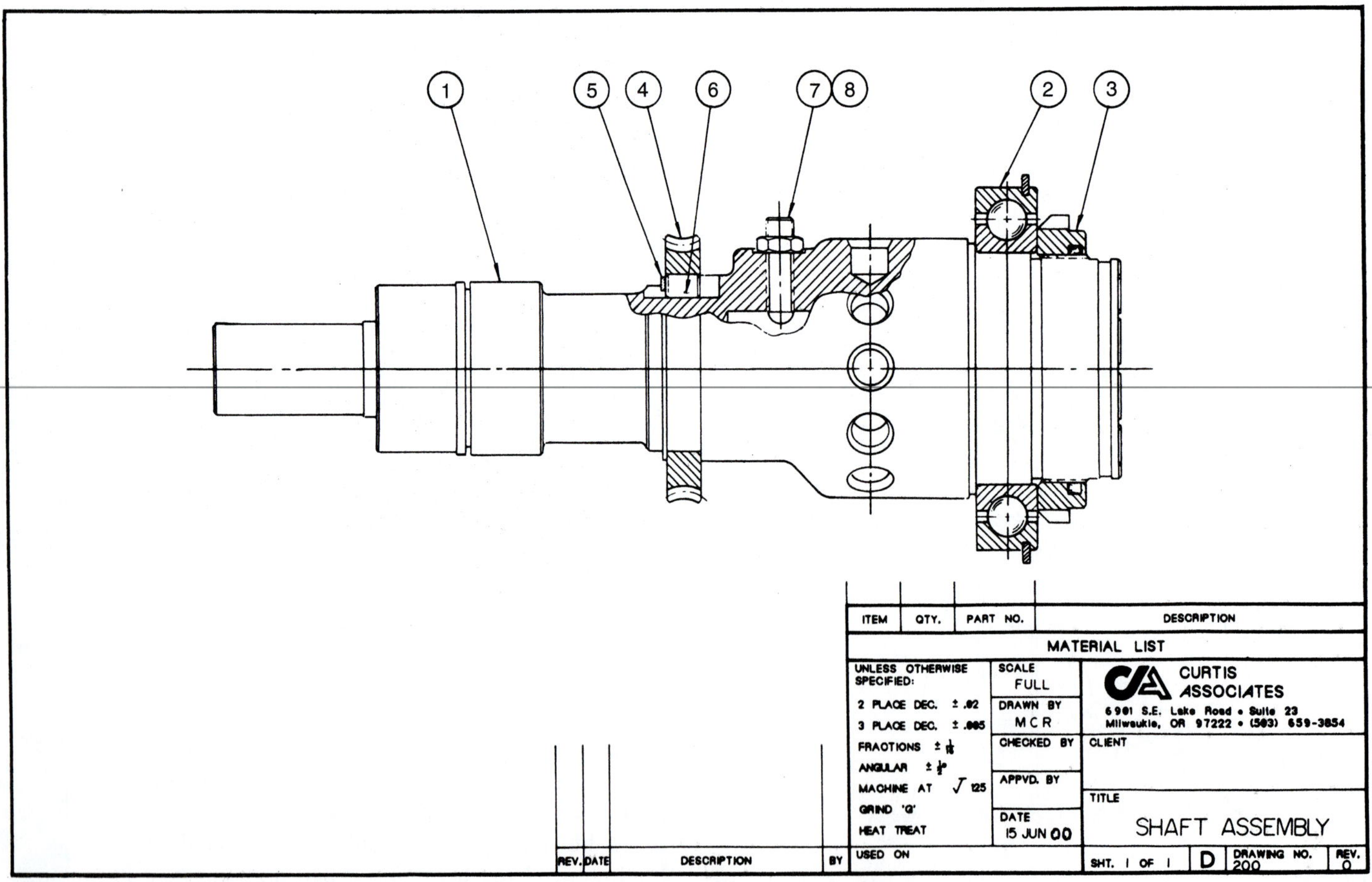

FIGURE 18.12 ■ Assembly with broken-out sections. *Courtesy Curtis Associates.*

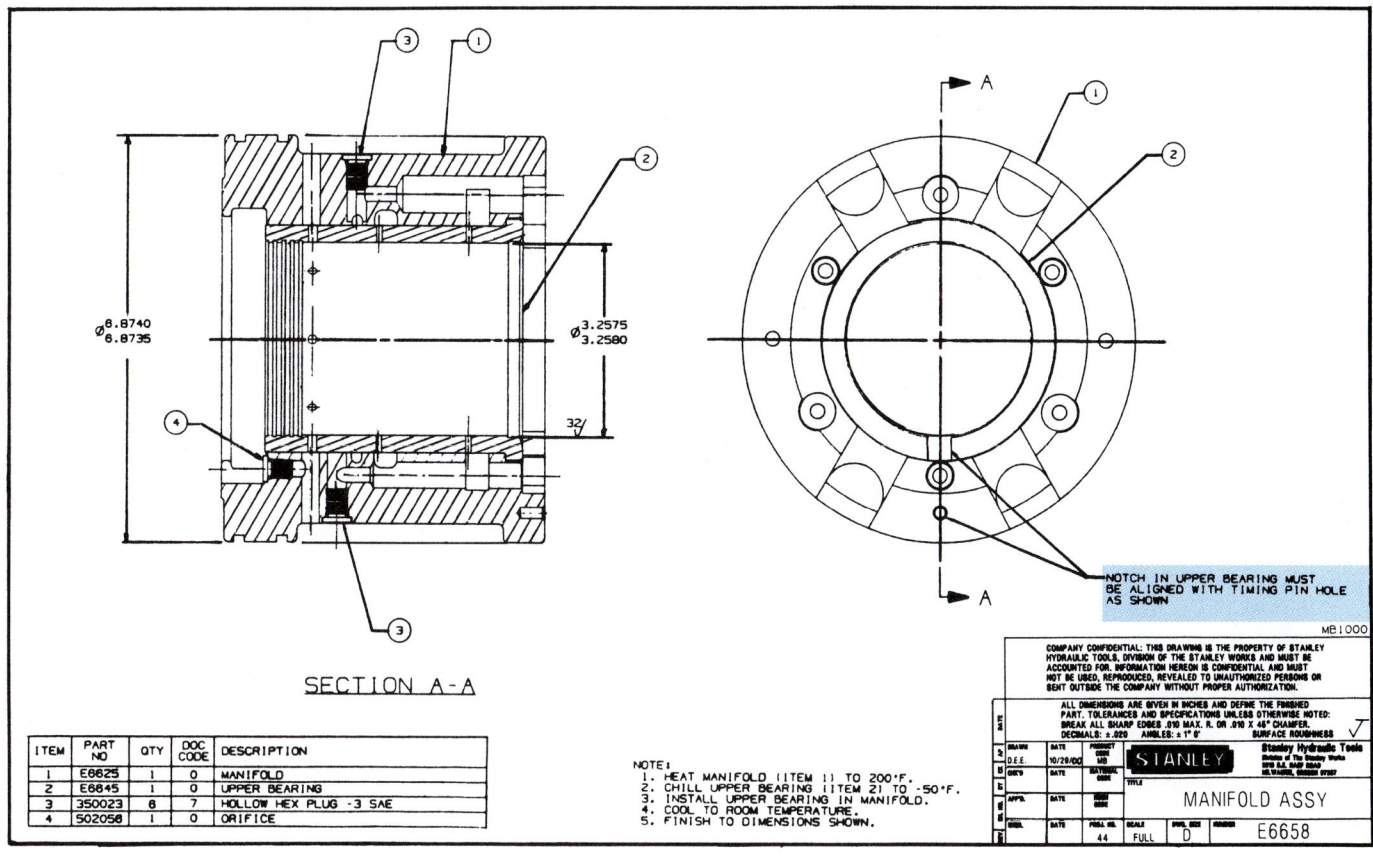

SECTION A-A

ITEM	PART NO	QTY	DOC CODE	DESCRIPTION
1	E6625	1	0	MANIFOLD
2	E6645	1	0	UPPER BEARING
3	350023	8	7	HOLLOW HEX PLUG -3 SAE
4	502056	1	0	ORIFICE

NOTE:
1. HEAT MANIFOLD (ITEM 1) TO 200°F.
2. CHILL UPPER BEARING (ITEM 2) TO -50°F.
3. INSTALL UPPER BEARING IN MANIFOLD.
4. COOL TO ROOM TEMPERATURE.
5. FINISH TO DIMENSIONS SHOWN.

MANIFOLD ASSY — E6658

FIGURE 18.13 ■ A process note on an assembly drawing. Dimension values in this figure are in inches. *Courtesy Stanley Hydraulic Tools Division of The Stanley Works.*

interior features. Dimensions serve no purpose on an assembly drawing unless the dimensions are used to show the assembly relationship of one part to another. These **assembly dimensions** are only necessary when a certain distance between parts must exist before proper assembly can take place. Machining processes and other specifications are generally not given on an assembly drawing unless a machining operation must take place after two or more parts are assembled. Other assembly notes can include bolt tightening specifications, assembly welds, and cleaning, painting, or decal placement that must take place after assembly. Figure 18.13 shows a process note applied to an assembly drawing. Some company standards or drawing presentations prefer that assemblies be drawn with sectioned parts shown without section lines, although this is not a common practice. Typically, sectioned assemblies do not show section lines on objects such as fasteners, pins, keys, and shafts. Figure 18.14 shows a full-section assembly with parts left unsectioned.

Assembly drawings can contain some or all of the following information:

■ One or more views. Auxiliary views can be used as needed.

■ Sections necessary to show internal features, function, and assembly.

■ Any enlarged views necessary to show adequate detail.

■ Arrangement of parts.

■ Overall size and specific dimensions necessary for assembly.

■ Manufacturing processes necessary for or during assembly.

■ Reference or item numbers that key the assembly to a parts list and to the details.

■ Parts list or bill of materials.

TYPES OF ASSEMBLY DRAWINGS

There are several different types of assembly drawings used in industry:

■ Layout, or design, assembly.

■ General assembly.

■ Working-drawing or detail assembly.

■ Erection assembly.

■ Subassembly.

■ Pictorial assembly.

Layout Assembly

Engineers and designers can prepare a design layout in the form of a sketch or as an informal drawing. These engineering design

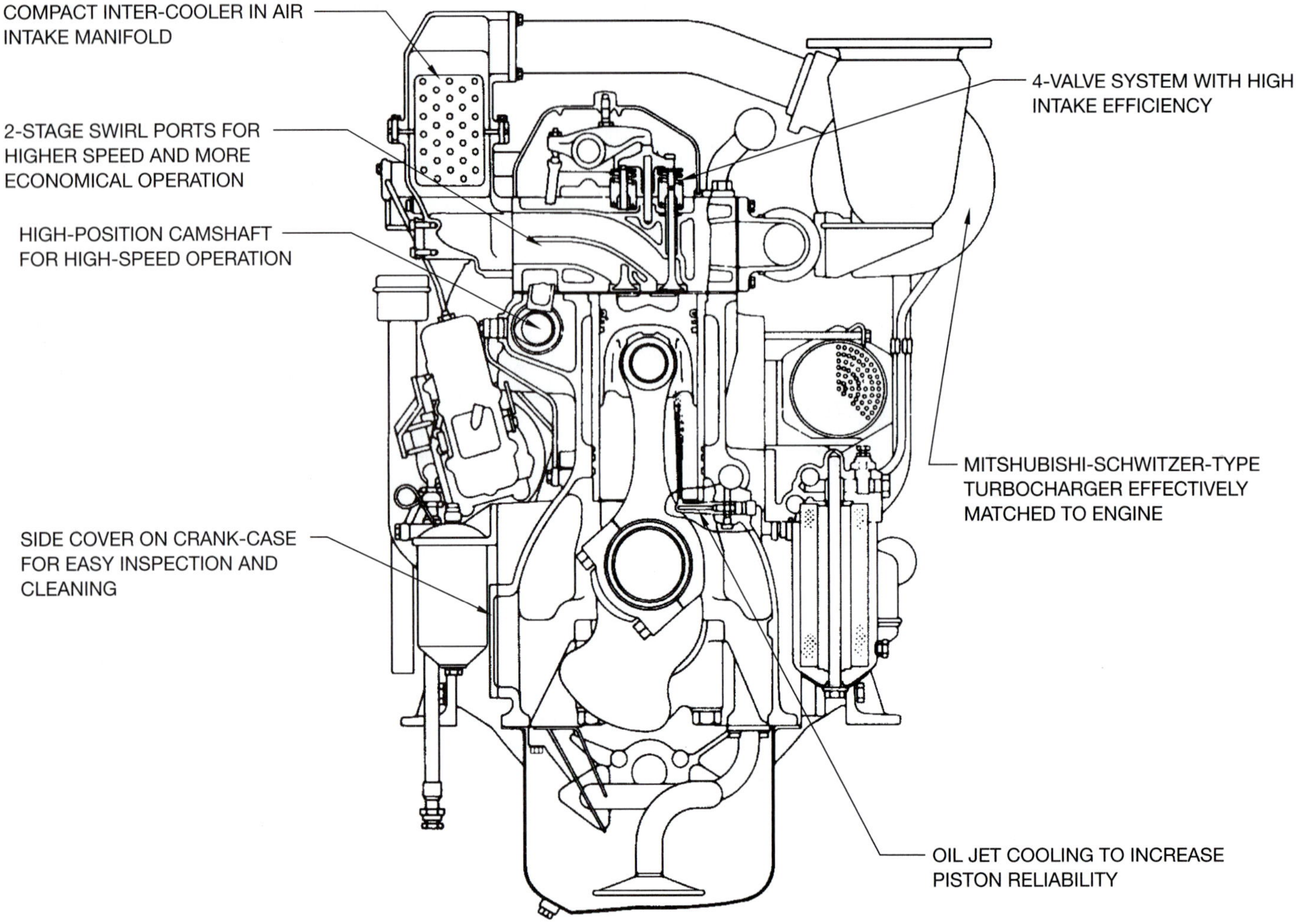

COMPACT INTER-COOLER IN AIR
INTAKE MANIFOLD

2-STAGE SWIRL PORTS FOR
HIGHER SPEED AND MORE
ECONOMICAL OPERATION

HIGH-POSITION CAMSHAFT
FOR HIGH-SPEED OPERATION

SIDE COVER ON CRANK-CASE
FOR EASY INSPECTION AND
CLEANING

4-VALVE SYSTEM WITH HIGH
INTAKE EFFICIENCY

MITSHUBISHI-SCHWITZER-TYPE
TURBOCHARGER EFFECTIVELY
MATCHED TO ENGINE

OIL JET COOLING TO INCREASE
PISTON RELIABILITY

FIGURE 18.14 ■ Full-section assembly with section lines omitted. *Courtesy Mitsubishi Heavy Industries America, Inc.*

drawings are used to establish the relationship of parts in a product assembly. From the layout the engineer prepares sketches or informal detail drawings for prototype construction. This research and development (R & D) is the first step in the process of taking a design from an idea to a manufactured product. Layout, or design, assemblies may take any form depending on the drafting ability of the engineer, the time frame for product implementation, the complexity of the product, or company procedures. In many companies, the engineers work with drafters who help prepare formal drawings from engineering sketches or informal drawings. The R & D department is one of the most exciting places for a drafter to work. Figure 18.15 shows a basic layout assembly of a product in the development stage. The limits of operation are shown using phantom lines.

General Assembly

General assemblies are the most common types of assemblies that are used in a complete set of working drawings. A set of working drawings contains three parts: detail drawings, an assembly drawing, and a parts list. The general assembly contains the features pre-

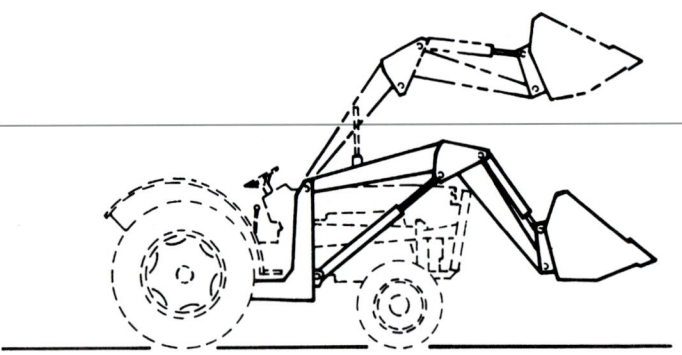

FIGURE 18.15 ■ Layout assembly. The limits of operation are shown with phantom lines.

viously discussed in this chapter, including multiviews, auxiliary views, detail views, and section view selection as needed for the specific product. Each part is identified with a balloon containing a number keyed to the parts list. The parts list identifies every part in the assembly. A typical general assembly is shown in Figures 18.13 and 18.23, page 552. The additional features found on a general assembly are discussed throughout the rest of this chapter.

Working-Drawing, or Detail Assembly

When a drawing is created where details of parts are combined on the same sheet with an assembly of those parts, a detail assembly is the result. While this practice is not as common as general assemblies, it is a practice at some companies. The use of working-drawing assemblies can be a company standard, or this technique can be used in a specific situation even when it is not considered a normal procedure at a particular company.

The detail assembly can be used when the end result dictates that the details and assembly be combined on as few sheets as possible. An example can be a product with few parts that is produced only once for a specific purpose. (See Figure 18.16.) Another example of a detail assembly is a product shown with its component parts assembled in a manufacturer's catalog or on its website, as shown in Figure 18.17.

Erection Assembly

Erection assemblies usually differ from general assemblies in that dimensions and fabrication specifications are commonly included. Typically associated with products that are made of structural steel, or cabinetry, erection assemblies are used for both fabrication and assembly. Figure 18.18 shows an erection assembly with multiviews, fabrication dimensions, and an isometric drawing that also helps display how the parts fit together.

Subassembly

The complete assembly of a product can be made up of several component assemblies. These individual unit assemblies are called subassemblies. A complete set of working drawings can have several subassemblies, each with its own detail drawings, and the general assembly. The general assembly of an automobile, for example, includes the subassemblies of the drive components, the engine components, and the steering column, just to name a few. An assembly, such as an engine, can have other subassemblies, such as carburetor or the generator. Figure 18.19, page 550, shows a subassembly with a parts list.

Pictorial Assembly

Pictorial assemblies are used to display a pictorial rather than multiview representation of the product, which can be used in other types of assembly drawings. Pictorial assemblies can be made from photographs, artistic renderings, or CADD models.

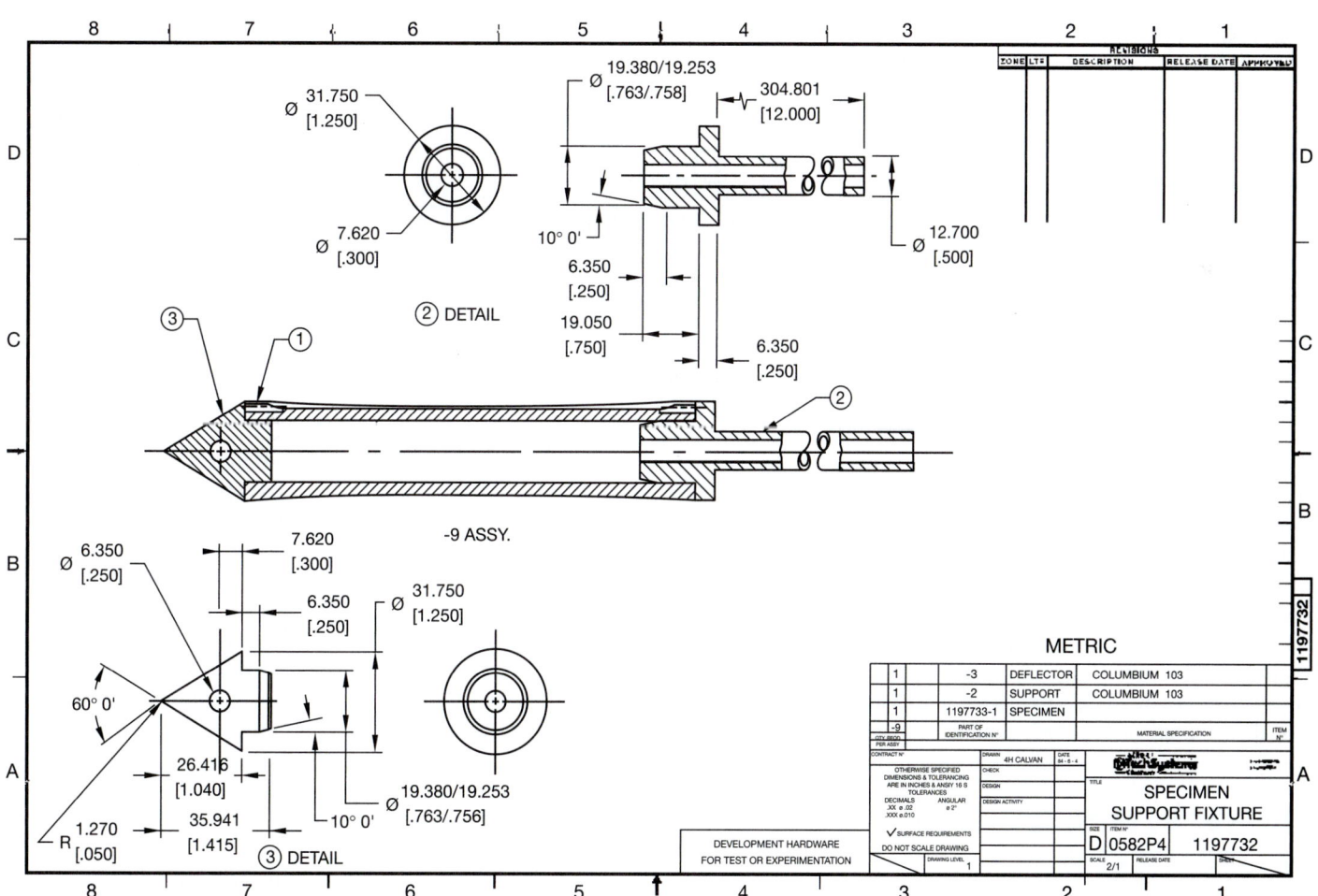

FIGURE 18.16 ■ Working-drawing, or detail assembly. *Courtesy Aerojet Propulsion Division.*

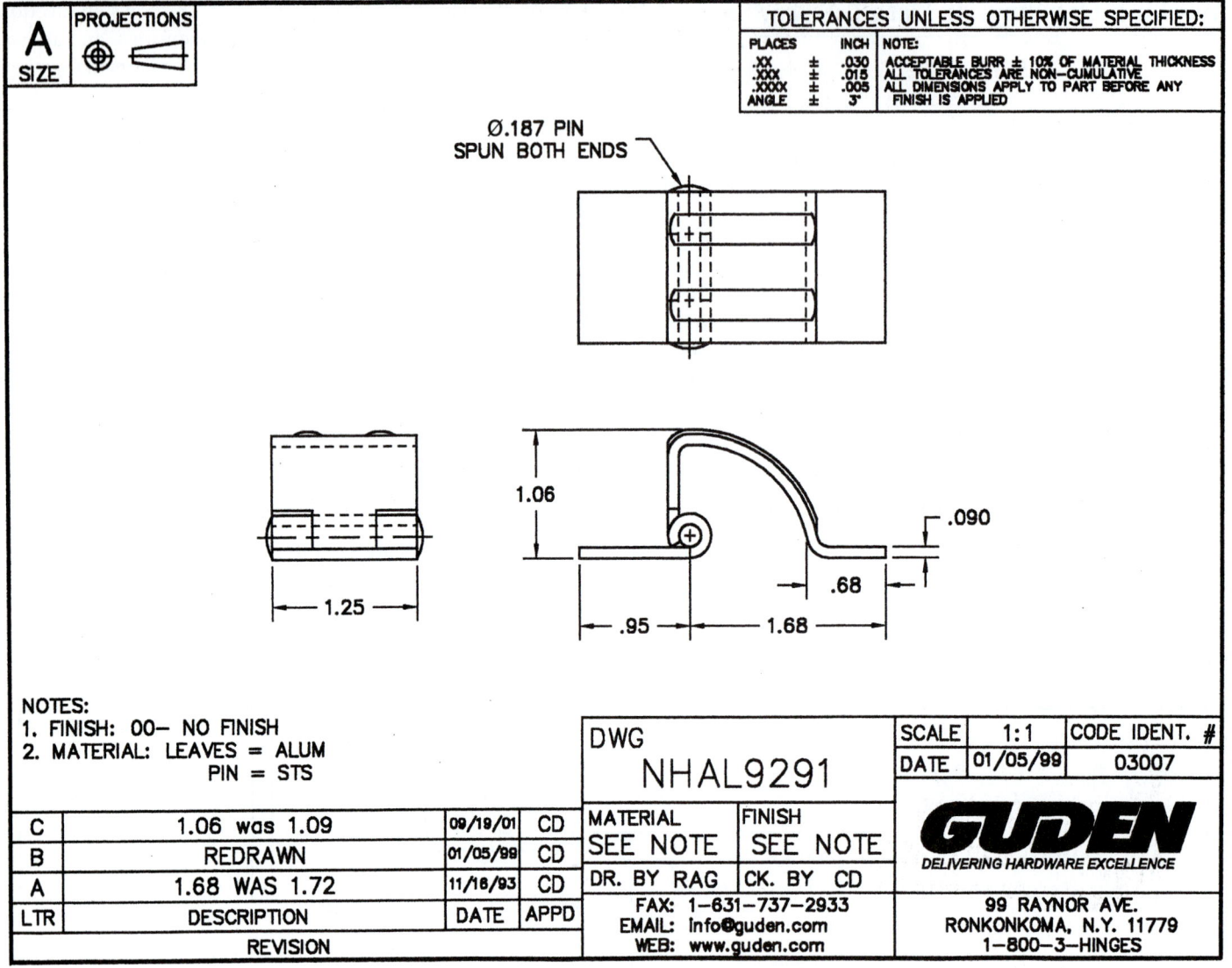

FIGURE 18.17 ■ An example of a detail assembly is this product shown with its component parts assembled in a manufacturer's catalog or on its website. Dimension values in this figure are in inches. *Courtesy H.A. Gluden Co. Inc.*

The pictorial assembly can be as simple as the isometric drawing in Figure 18.18, which was used to more clearly assist workers in the assembly of the product. Pictorial assemblies are commonly used in product catalogs or brochures. The purpose of these pictorial representations can be for sales promotion, customer self-assembly, or maintenance procedures. (See Figure 18.20.) Pictorial assemblies can also take the form of exploded technical illustrations, also commonly known as illustrated parts breakdowns. These exploded multiview or isometric pictorials are used in vendors' catalogs and instruction manuals, for maintenance or assembly. (See Figure 18.21.)

IDENTIFICATION NUMBERS

Identification or item numbers are used to key the parts from the assembly drawing to the parts list. Identification numbers are generally placed in balloons. Balloons are circles that are connected to the related part with a leader line. Several of the assembly drawings in this chapter show examples of identification numbers and balloons. Numbers in balloons are common, although some companies prefer to use identification letters. Balloons are drawn between .375-1 in. (10-25 mm) in diameter depending on the size of the drawing and the amount of information that must be placed in the balloon. All balloons on the drawing are the same size. The leaders connecting the balloons to the parts are thin lines that can be presented in any one of several formats depending on company standards. Figure 18.22 shows the common methods used to connect balloon leaders. Notice that the leaders can terminate with arrowheads or dots. Whichever method of connecting balloons is used, the same technique should be used throughout the entire drawing.

The item numbers in balloons should be grouped so they are in an easy-to-read pattern. This is referred to as information

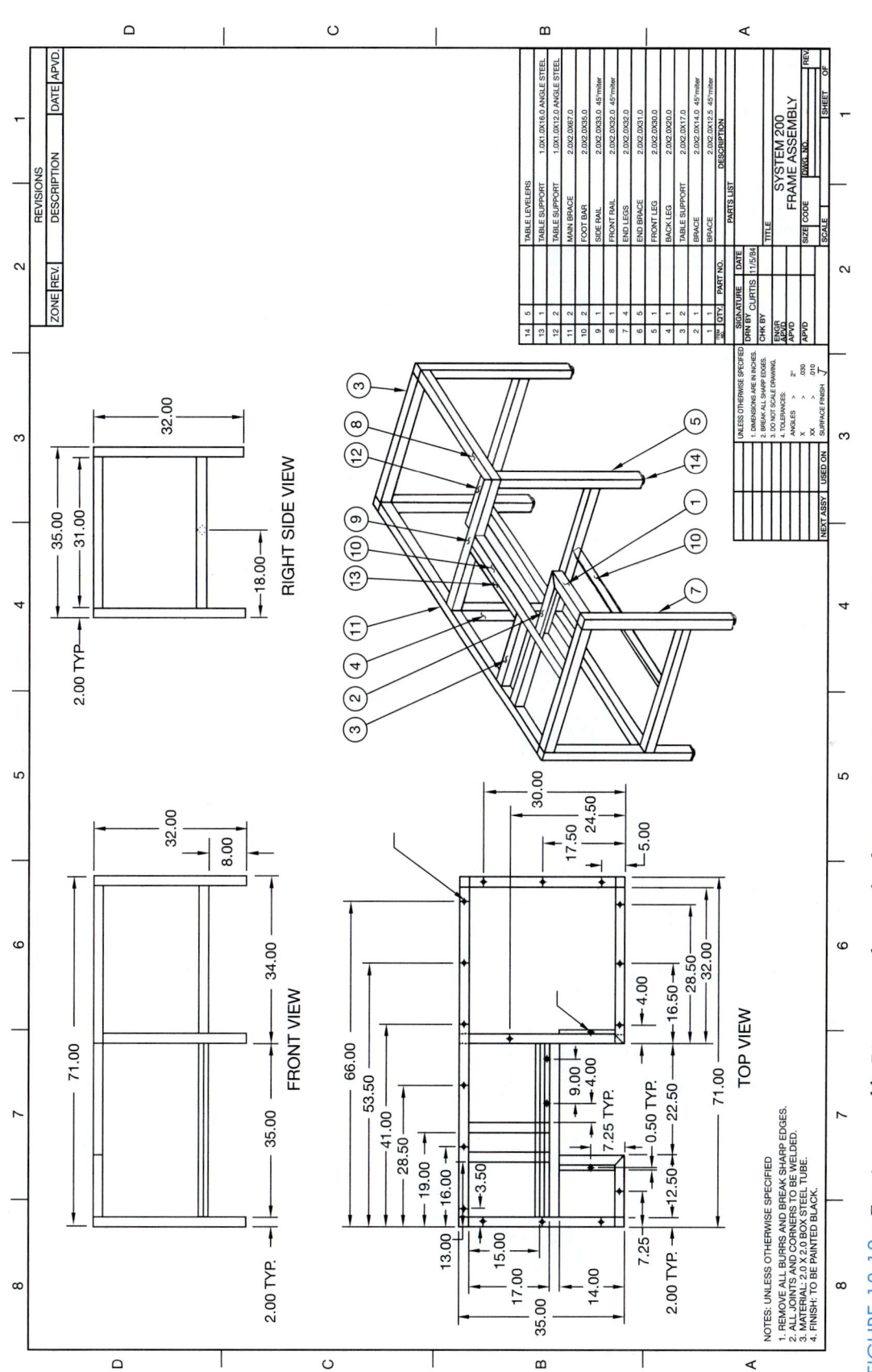

FIGURE 18.18 ■ Erection assembly. Dimension values in this figure are in inches. *Courtesy EFT Systems.*

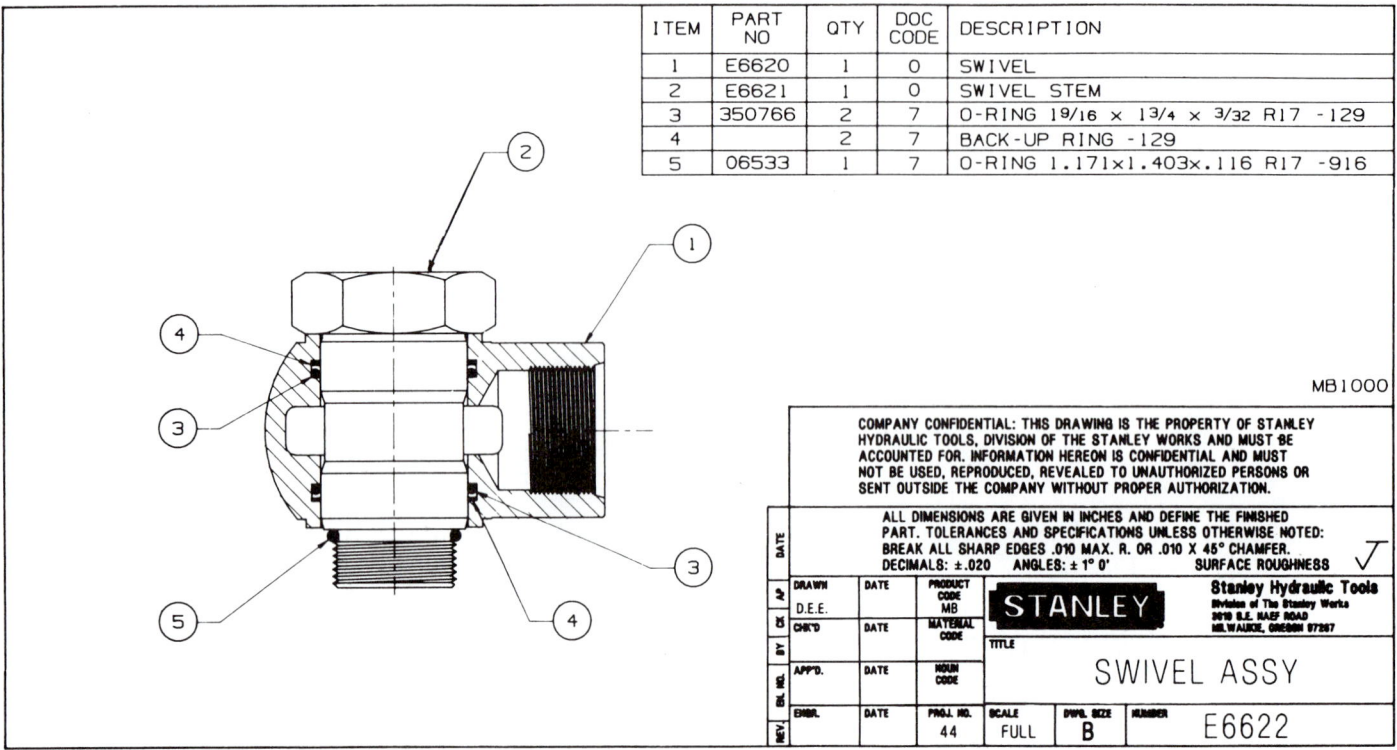

ITEM	PART NO	QTY	DOC CODE	DESCRIPTION
1	E6620	1	0	SWIVEL
2	E6621	1	0	SWIVEL STEM
3	350766	2	7	O-RING 19/16 × 13/4 × 3/32 R17 -129
4		2	7	BACK-UP RING -129
5	06533	1	7	O-RING 1.171×1.403×.116 R17 -916

MB1000

COMPANY CONFIDENTIAL: THIS DRAWING IS THE PROPERTY OF STANLEY HYDRAULIC TOOLS, DIVISION OF THE STANLEY WORKS AND MUST BE ACCOUNTED FOR. INFORMATION HEREON IS CONFIDENTIAL AND MUST NOT BE USED, REPRODUCED, REVEALED TO UNAUTHORIZED PERSONS OR SENT OUTSIDE THE COMPANY WITHOUT PROPER AUTHORIZATION.

ALL DIMENSIONS ARE GIVEN IN INCHES AND DEFINE THE FINISHED PART. TOLERANCES AND SPECIFICATIONS UNLESS OTHERWISE NOTED: BREAK ALL SHARP EDGES .010 MAX. R. OR .010 × 45° CHAMFER. DECIMALS: ±.020   ANGLES: ± 1° 0'   SURFACE ROUGHNESS

STANLEY

Stanley Hydraulic Tools
Division of The Stanley Works
3810 S.E. NAEF ROAD
MILWAUKIE, OREGON 97267

TITLE

SWIVEL ASSY

PROJ. NO. 44   SCALE FULL   DWG. SIZE B   NUMBER E6622

FIGURE 18.19 ■ Subassembly with parts list. *Courtesy Stanley Hydraulic Tools Division of The Stanley Works.*

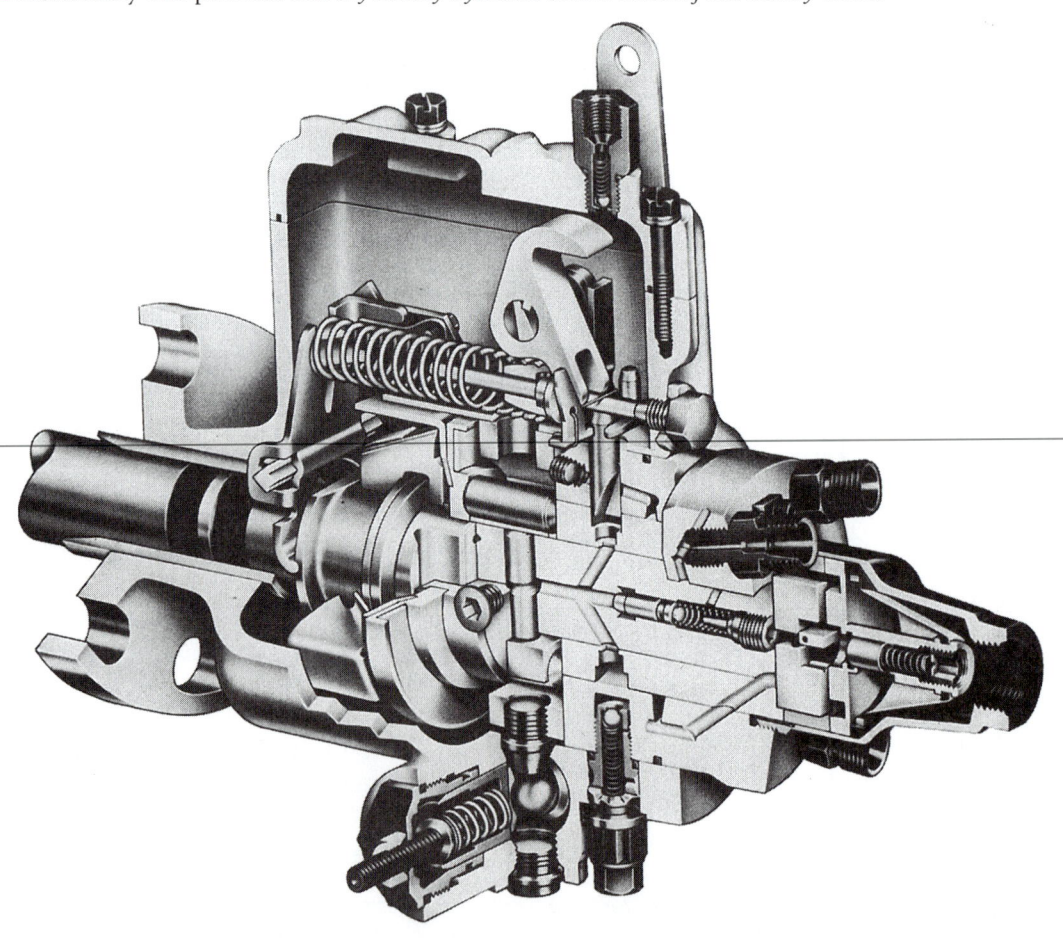

FIGURE 18.20 ■ Pictorial assembly. *Courtesy Stanadyne Diesel Systems.*

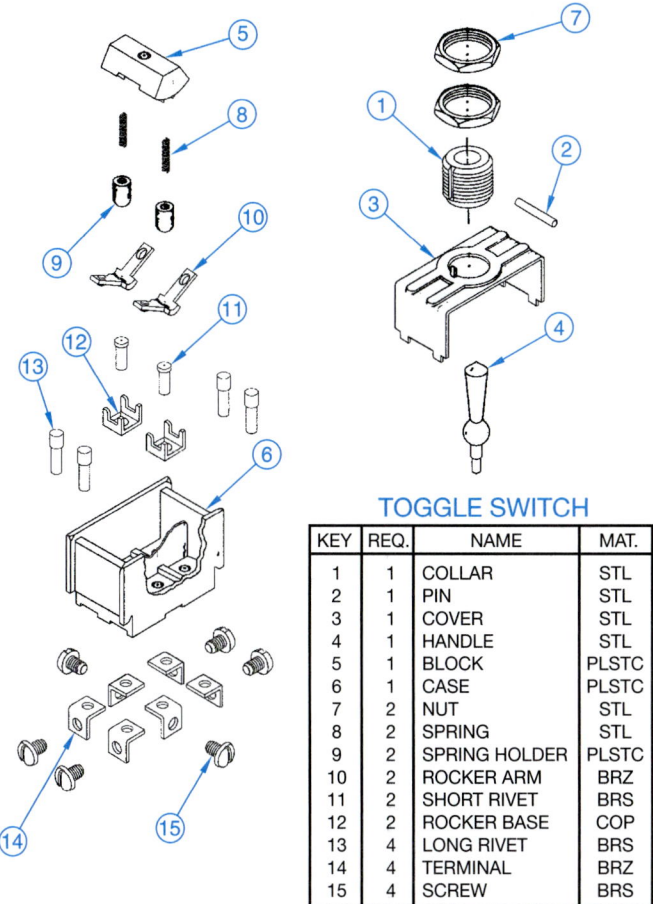

TOGGLE SWITCH

KEY	REQ.	NAME	MAT.
1	1	COLLAR	STL
2	1	PIN	STL
3	1	COVER	STL
4	1	HANDLE	STL
5	1	BLOCK	PLSTC
6	1	CASE	PLSTC
7	2	NUT	STL
8	2	SPRING	STL
9	2	SPRING HOLDER	PLSTC
10	2	ROCKER ARM	BRZ
11	2	SHORT RIVET	BRS
12	2	ROCKER BASE	COP
13	4	LONG RIVET	BRS
14	4	TERMINAL	BRZ
15	4	SCREW	BRS

FIGURE 18.21 ■ Exploded isometric assembly.

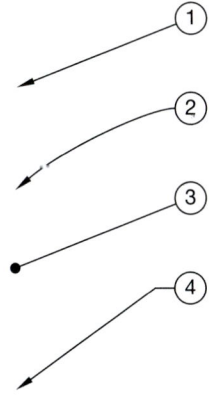

BALLOON OPTIONS

FIGURE 18.22 ■ Balloons and styles of leaders.

grouping. Good information grouping occurs when balloons are in alignment in a pattern that is easy to follow, as opposed to scattered about the drawing. (See Figure 18.23.) In some situations, when particular part groups are so closely related that individual identification is difficult, the identification balloons can be grouped next to each other. For example, a cap screw, lock washer, and flat washer can require that the balloons be placed in a cluster or side by side, as shown with items 12, 13, and 14 in Figure 18.11.

The balloons key the detail drawings to the assembly and parts list, and can also key the assembly drawing and parts list to the page on which the detail drawing is found. (See Figure 18.24.) Figure 18.25 is an assembly drawing and parts list that is located on page 1 of a two-page set of working drawings. Notice how the balloons key the parts from the assembly and parts list to the detail drawings found on page 2. The page 2 details are located in Figure 18.9 (see page 542).

PARTS LISTS

The parts list, also known as a bill of materials, is usually combined with the assembly drawing yet remains one of the individual components of a complete set of working drawings. The information that is associated with the parts list generally includes:

■ Item number—from balloons.

■ Quantity—the number of that particular part needed for this assembly.

■ Part or drawing number, which is a reference back to the detail drawing.

■ Description, which is usually a part name or complete description of a purchase part or stock specification including sizes or dimensions.

■ Material identification—the material used to make the part.

■ Information about vendors for purchase parts.

Parts lists may or may not contain all of the previous information, depending on company standards. The first four elements listed are the most common items. When all six elements are provided, the parts list may also be called a list of materials or bill of materials.

When drawn on the assembly drawing, the parts list may be located above the title block, in the upper right or left corner, or in a convenient location on the drawing field. The location depends on company standards, although the position over the title block is most common. The information on the parts list is usually presented with the first item number followed by consecutive item numbers. When the parts list is so extensive that the columns fill the page, a new group of columns is added next to the first. If additional parts are added to the assembly, space on the parts list is available. This is the reason that parts list data is provided from the bottom of the sheet upward or the top of the sheet downward.

Some companies prepare parts lists on a computer so information can be retrieved and edited easily. When the set of working drawings is prepared on a CADD system, then the parts list can be parametrically associated with the details and assembly, so changes to any are automatically updated on all.

The parts list is not always placed on the assembly drawing. Some companies prefer to prepare parts lists on separate sheets,

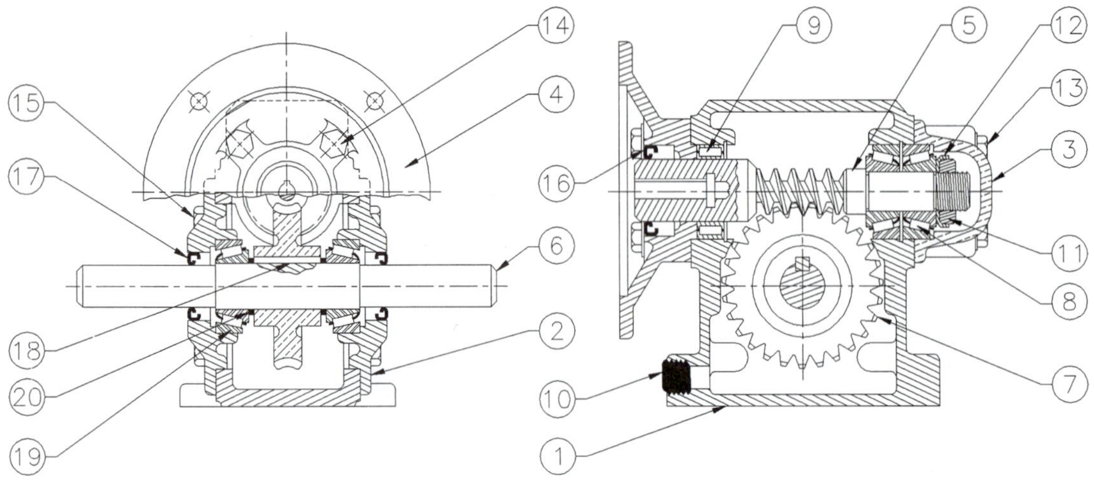

KEY	QTY.	NAME	PARTS DESCRIPTION	PART NO.
20	2	SLOW SPEED SPACER	TIMKEN TW-506	5DT 1020
19	2	SNGL. ROW TAP. ROLLER BEARING	KOYO 32005J	5DT1019
18	1	SLOW SPEED KEYWAY	.1875 X .245 X 1.450	5DT1018
17	2	SLOW SPEED OIL SEAL	PARKER 2-020	5DT1017
16	1	HIGH SPEED OIL SEAL	PARKER 2-028	5DT1016
15	8	MACHINE SCREW	.375-16UNC-2A X .625 HEX HEAD	5DT1015
14	4	MACHINE SCREW	.375-16UNC-2A X 1.875 HEX HEAD	5DT1014
13	4	MACHINE SCREW	.375-16UNC-2A X 2.250 HEX HEAD	5DT1013
12	1	HIGH SPEED LOCKWASHER	TIMKEN TW-105	5DT1012
11	1	HEX NUT	.875-16 UN-2B	5DT1011
10	1	TAPER PLUG	.500-18NPT PLUG	5DT1010
9	1	SNGL. ROW CYL. ROLLER BEARING	KOYO CRL11	5DT1009
8	1	DBL. ROW TAPERED ROLLER BEARING	KOYO46T30305DJ./29.5	5DT1008
7	1	WORM GEAR	BRONZE	10T1005
6	1	SLOW SPEED SHAFT		10T1006
5	1	HIGH SPEED SHAFT		10T1005
4	1	MOTOR ADAPTOR		10T1004
3	1	BEARING CAP		10T1003
2	2	RETAINING PLATE		10T1002
1	1	HOUSING		10T1001

Stennfeld Engineering
Vancouver, Washington

UNLESS OTHERWISE SPECIFIED
INCHES
AND TOLERANCES FOR:
1 PLACE DIMS.: ± .1
2 PLACE DIMS.: ± .01
3 PLACE DIMS.: ± .005
ANGULAR: ± .30°
FRACTIONAL: ± 1/32
FINISH: 125 μ in.

DR: MHS   SCALE: FULL   DATE: 18 MAR 96   APPD:
MATERIAL: VARIES
NAME: WORM GEAR RED.
PART NO. D6DT1000   REV: 0
FIRST USED ON: SIMILAR TO:

THIRD ANGLE PROJECTION

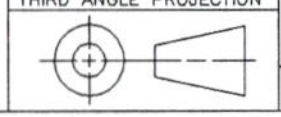

NOTES:
1. INTERPRET DIMENSIONS AND TOLERANCES PER ANSI Y14.5M—1994.

FIGURE 18.23 ■ Assembly drawing and parts list. *Courtesy Mark Stennfeld.*

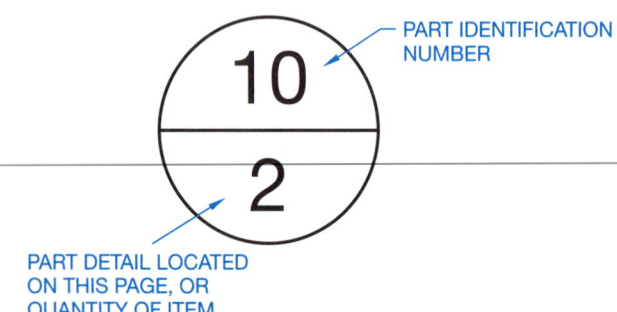

PART IDENTIFICATION NUMBER

PART DETAIL LOCATED ON THIS PAGE, OR QUANTITY OF ITEM

FIGURE 18.24 ■ Balloon with page identification.

which allows for convenient filing. This method also allows for the parts list to be computer-generated separately from the drawings. Separate parts lists are usually prepared on a computer so information can be edited conveniently. Another option is to have parts lists typed on a standard parts list form (see Figure 18.26).

PURCHASE PARTS

Purchase or standard parts, as previously mentioned, are parts that are manufactured and available for purchase from an outside vendor. It is generally more economical for a company to buy such items than to make them. Parts that are available from suppliers can be found in the *Machinery's Handbook*, *Fastener's Handbook*, or vendors' catalogs. Purchase parts do not require a detail drawing, because a written description completely describes the part. For this reason, the purchase parts found in a given assembly must be described clearly and completely in the parts list. Some companies have a purchase parts book or computer directory that is used to record all purchase parts used in their product line. The standard parts book gives a reference number for each part, which is placed on the parts list for convenient identification.

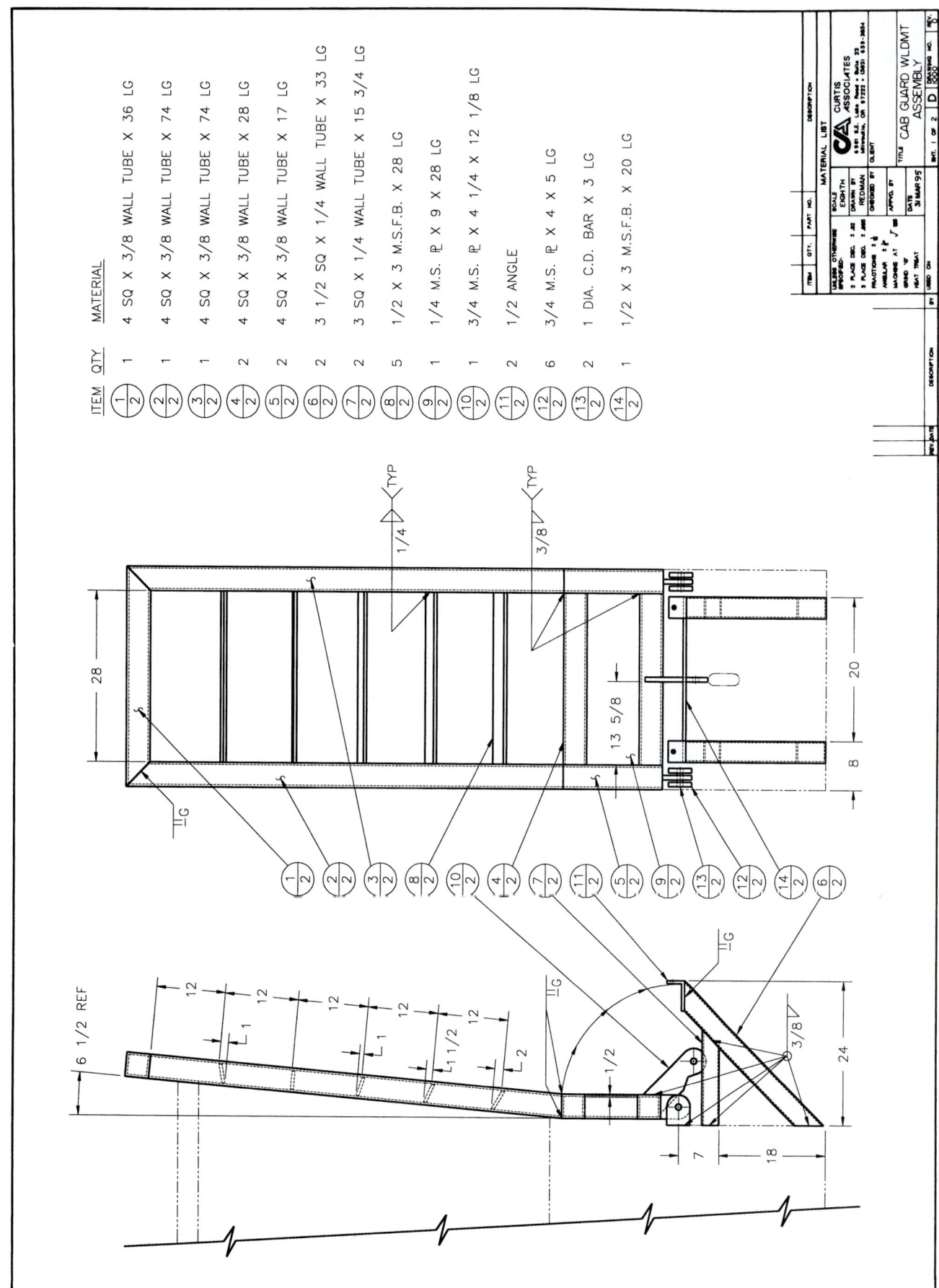

FIGURE 18.25 ■ Assembly and parts list with page identification balloons. Dimension values in this figure are in inches. *Courtesy Curtis Associates.*

ASSEMBLY DRAWINGS

Manual drafters can often spend hours or even days drawing assemblies, depending on the complexity of the product. This amount of time can be necessary because each part has to be redrawn in relationship to the entire assembly. One of the major benefits to using CADD is the ability to reuse existing drawing content. This practice significantly reduces the amount of time spent drawing assemblies and increases accuracy. When using AutoCAD, for example, assemblies are prepared quickly and easily using the following techniques:

1. Create all the detail drawings first. Be sure to use a separate layer for visible object lines, which will be used to represent each part in the assembly drawing.

2. Create a new drawing file based on an existing template, which will be used to generate the assembly.

3. Option 1:
   ■ Insert the detail drawing files into the assembly drawing file as blocks.
   ■ Explode the blocks.
   ■ Erase all items except the principal view of each detail drawing.

Option 2:
   ■ Open each detail drawing and create a block of the principal view.
   ■ Insert the block of each principal detail drawing view into the assembly drawing file.

Option 3:
   ■ Copy the principal drawing view from each detail drawing and paste the view into the assembly drawing view.

4. If needed, move each detail view into place to form an assembly. Use specific coordinate values or Object Snaps to accurately position the drawing views.

5. After all the detail views are assembled, use editing commands such as ERASE and TRIM to clean up the assembly.

6. Add balloons, notes, and dimensions as needed.

7. Add the parts list. Use an existing parts list that can be easily modified, or create a new parts list using the TABLE command.

This method saves a lot of time because you do not have to start over again, drawing each part in the assembly.

ASSEMBLY DRAWINGS

Part models are assembled in the 3-D modeling environment to form an assembly model. Assembly models are excellent for observing the interaction of assembly components and for creating assembly presentations. In addition, assembly models can be used to produce assembly drawings. Three-dimensional modeling software that combines 2-D drawing capabilities usually allows you to produce several types of assembly drawings, depending on the purpose and company standard, including weldment drawings complete with weld symbols and representations. Weld symbols are covered in Chapter 21. Creating an assembly drawing with 3-D modeling software is quick and easy, because all the information needed to produce the drawing is already available in the model. Developing an assembly drawing based on a 3-D model typically involves the following steps:

1. Create a new drawing file, based on an existing template. The drawing file usually contains a sheet that represents a specific paper size, and contains a border, title block, and general notes.

2. Place and scale drawing views associated with an assembly model.

3. Add a parts list, identification numbers, and other dimensions and specifications as needed. Due to the parametric nature of 3-D modeling software, all the information displayed in the parts list is associated with the assembly model, which is associated with each of the assembly components, and as a result, the detail drawings. When you create a parts list, all the required data is automatically added, and all you have to do is select a location for the parts list. For example, if four of the same items are displayed in an assembly, a quantity of four is automatically established in the quantity area for the item in the parts list. Additional information, including the part number, description, and material, is referenced from each component file, eliminating the need to type information in the parts list. Then, when you place balloons, the balloon numbers or letters, are correlated with the parts list and are automatically defined.

ASSEMBLY __Cross Shaft Assembly_____   USED ON _____

NUMBER OF UNITS _____   DATE _____

HAVE	NEED	P/Ø NO. W/Ø NO.	DET. NO.	PART NO.	DWG.	QTY.	PART NAME	DESCRIPTION	VENDOR
			1		B	1	Bearing Retainer	Ø 3" C.D. Bar	
			2		B	1	Cross Shaft	Ø 3/4" C.D. Bar	
			3		A	1	Spacer	Ø 3/4 O.D × 11GA Wall Tube × .738 Thick	
			4		A	1	Spacer	Ø 3/4 O.D × 11GA Wall Tube × .125 Thick	

FIGURE 18.26 ■ Parts list separate from assembly drawing. *Courtesy Curtis Associates.*

ENGINEERING CHANGES

ASME The standards controlling the use of engineering changes are specified in ASME Y14.35M.

Engineering change documents are used to initiate and record changes to products in the manufacturing industry. Changes to engineering drawings can be requested from any branch of a company that deals directly with production and distribution of the product. For example, the engineering department can implement product changes as research and development results show a need for upgrading. Manufacturing requests change as problems arise in product fabrication or assembly. The sales staff can also initiate change proposals that stem from customer complaints. Engineering changes are treated in the same way as original drawings; they are initiated, approved, sent to the drafting department, and filed when completed. It is a good idea to leave the area below the revision block clear for future changes.

Engineering Change Request (ECR)

Before a drawing can be changed by the drafting department, an engineering change request (ECR) is needed. The engineering change request is also commonly called an engineering change order (ECO). The ECR is the document used to initiate a change in a part or assembly. The ECR is usually attached to a print of the part affected. The print and the ECR show by sketches and written descriptions what changes are to be made. The ECR also contains a number that becomes the reference record of the change to be made. Figure 18.27 shows a sample ECR form. Generally when a drafter receives an ECR, the following procedure is used:

STEP 1 Get the drawing original, sepia, or computer file.

STEP 2 Make the requested change using the same media as that used for the original, unless a CADD redraw is required.

STEP 3 Fill out the engineering documents using freehand lettering or a word processor.

STEP 4 Distribute a changed copy of the drawing and the completed engineering change forms for release approval.

STEP 5 Refile the drawing original.

Engineering Change Notice (ECN)

Records of changes are kept so reference can be made between the existing and proposed product. To make sure records of these changes are kept, special notations are made on the drawing, and engineering change records are kept. These records are commonly known as engineering change notices (ECN).

When a change is to be made, an engineering change request initiates the change to the drafting department. The drafter then alters the original drawing or computer file to reflect the change request. When the drawing of a part is changed, a revision letter is placed next to the change. For example, the first change is lettered *A*, the second change is *B*, and so on. The letters *1, O, S, X,* and *Z* are not used, because they can be mistaken for numbers or other symbols. When all of the available letters *A* through *Y* have been used, double letters such as *AA* and *AB*, or *BA* and *BB*, are used next. A circle drawn around the revision letter helps the identification stand out clearly from the other drawing numerals. Figure 18.28 shows a part as it exists and also after a change has been made.

When the drafter chooses to make the change by not altering the drawing of the part but only changing the dimension, that dimension is labeled not-to-scale. The method of making the change is the decision of the drafting department based on the extent of the change and the time required to make it. The not-to-scale symbol can be used to save time. Figure 18.29 shows a change made to a part with the new dimension identified as not-to-scale with a thick straight line placed under the new dimension. With CADD, it is generally easy and preferred to make the change to scale and avoid using the not-to-scale method.

After the part has been changed on the drawing and the proper revision letter is placed next to the change, the drafter records the change in the ECN column of the drawing. The location of the ECN column varies with different companies. The ECN identification column can be found next to the title block or in a corner of the drawing.

ASME The ASME document, *Decimal Inch Drawing Sheet Sizes and Format*, ASME Y14.1, recommends the ECN column be placed in the upper right corner of drawing. (See Chapter 2.)

Aerojet Liquid Rocket Company
ENGINEERING CHANGE REQUEST & ANALYSIS

DATE	ECRANG	PAGE

ORIGINATOR NAME		DEPT	EXT	DATE	DOCUMENT NEED DATE	PROPOSED EFFECTIVITY

PART DOCUMENT NO	CURRENT REV TR	PART DOCUMENT NAME

USED ON NEXT ASSY NO	PROGRAMS AFFECTED	PROGRAMS AFFECTED		
		QTSS ITEM	YES ——— NO ———	
		REQUAL REQD	YES ——— NO ———	
		REVISE QTSS FORM	YES ——— NO ———	

DESCRIPTION OF CHANGE

JUSTIFICATION OF CHANGE

PROJECT ENGINEER SIGNATURE	DEPT	EXT	DATE	CCB REP SIGNATURE	DATE

DESIGN ENGINEERING TECHNICAL EVALUATION

DESIGN ENGINEER SIGNATURE	DEPT	EXT	DATE	CAUSING CONT OR WORK ORDER NO	

ANY AFFECT ON	YES	NO		YES	NO		YES	NO
1 PERFORMANCE			5 WEIGHT			9 OPERATIONAL COMPUTER PROGRAMS		
2 INTERCHANGEABILITY			6 COST SCHEDULE			10 RETROFIT		
3 RELIABILITY			7 OTHER END ITEMS			11 END ITEM IDENT		
4 INTERFACE			8 SAFETY EMI			12 VENDOR CHANGE CRITICAL ITEMS ONLY		

CCB DECISION	SIGNATURES	DEPT	CON CUR	DIS SENT
	QUALITY ASSURANCE			
	MANUFACTURING			
	ENGINEERING			
	PRODUCT SUPPORT			
	MATERIAL			
	TEST OPERATION			

CCB CHAIRMAN SIGNATURE	DATE	CLASS I ☐ CLASS II ☐
CUSTOMER SIGNATURE	DATE	EFFECTIVITY

FIGURE 18.27 ■ Sample engineering change request (ECR) form. *Courtesy Aerojet Propulsion Division.*

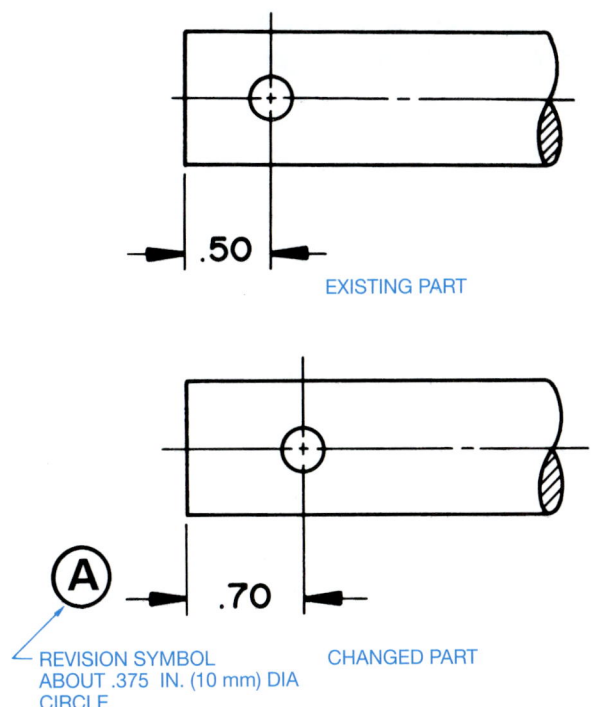

FIGURE 18.28 ■ An existing part and the same part after a change has been made. Dimension values in this figure are in inches.

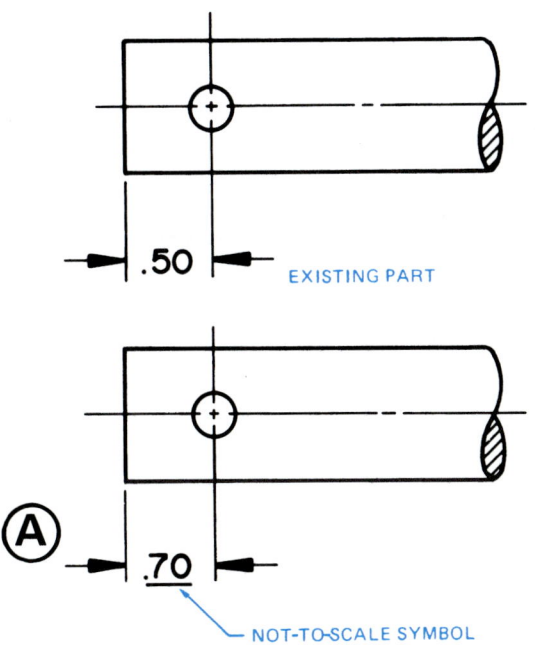

FIGURE 18.29 ■ An existing part and the same part changed using the not-to-scale symbol. Dimension values in this figure are in inches.

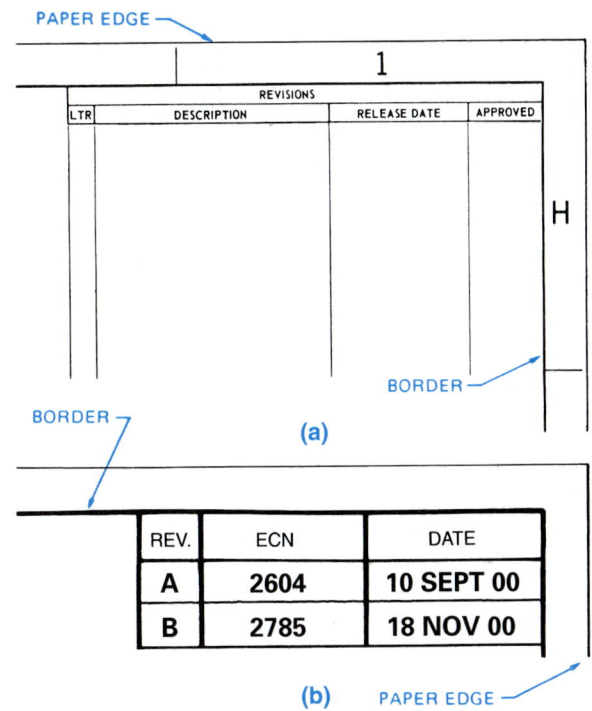

FIGURE 18.30 ■ (a) Expanded revision columns. *Courtesy Aerojet Propulsion Division.* (b) Condensed revision columns.

The drafter records the revision number as the consecutive number of times the drawing has been revised, ECN number (usually given on the ECR), and the date of the change. Some companies also have a column for a brief description of the change (as recommended by ASME) and an approval. Figure 18.30a shows an expanded and condensed ECN column format. Notice that changes are added in alphabetic order from top to bottom.

The condensed ECN column in Figure 18.30b shows the drawing has been changed twice. The first change was initiated by ECN number 2604 on September 10, 2000, and the second by ECN 2785 on November 18, 2000. The number of times a drawing has been changed is identified by providing the letter of the current change in the title block, as shown in Figure 18.31. The drafter updates this letter to reflect each change.

When the drafter has made all of the drawing changes as specified on the ECR, then an engineering change notice (ECN) is filled out. The ECN completes the process and is filed for future reference. Usually the ECN completely describes the part as it existed before the change and the change that was made as presented on the ECR. Figure 18.32 shows a typical ECN form. With a changed drawing and a filed ECN, anyone can verify what the part was before the change and the reason for the change. The ECN number is a reference for reviewing the change. This process allows the history of a product to be maintained.

The general engineering change elements, terminology, and techniques are consistent among companies, although the actual format of engineering change documents can be considerably different. Some formats are very simple while others are much more detailed. One of the first tasks of an entry-level

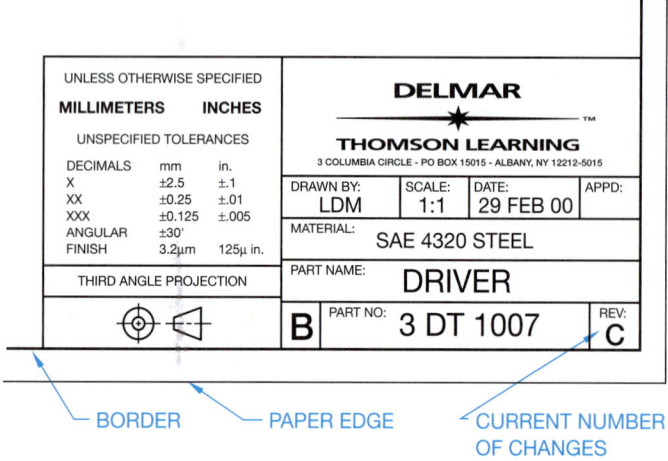

FIGURE 18.31 ■ Title block displaying the current number of times the drawing has been changed.

drafter is to become familiar with the specific method of preparing engineering changes.

Engineering Change Redraw

The drawing revision process previously discussed provides for each drawing change to be documented with a drawing revision containing a balloon with a letter inside at the change and correlated to the same revision letter in the revision block where a description of the change is provided. After a few months or years, depending on the product, the revisions can get so numerous that the drawing is cluttered with revision balloons and revision block content. When the clutter gets extensive, the company can make a decision to do a redraw. Redraw means that all the revision identification is removed and the drawing continues at the current state as a clean drawing without any changes. The redraw process is easy with CADD, and all previous engineering change documentation is kept for future reference.

DRAWING FROM A PROTOTYPE

Occasionally the engineering drafter has to take an existing product or structure and draw a set of working drawings. This may be a prototype. A prototype is a model or original design that has not been released for production. The prototype is often used for testing and performance evaluations. In most cases, there are prototype drawings from which the prototype was built, but during the modification process, changes may have been made. Your job is to complete the production drawings, which are established by measuring the prototype, redrawing the prototype drawings with changes, or a combination of these methods. These revised drawings are often referred to as as-built drawings. In this type of situation, you need to be well versed in the use of measuring equipment such as calipers, micrometers, and surface gages. You need to consult with the engineer, the testing department, and the shop personnel who made any modifications to the prototype. With this research you can fully understand how the new drawings should be created.

You need to be careful to make the drawings accurate and take into account the manufacturing capabilities of your company. Then the set of working drawings that you complete is checked by the checker and the engineer before the product is released for production. Once the product is released for production, changes are often costly because tooling and patterns may have to be altered. These types of changes are usually submitted as engineering change requests. They must be approved by the engineering and manufacturing departments.

ANALYSIS OF A SET OF WORKING DRAWINGS
The Design Sketches

The design of a product generally begins in the research and development (R & D) department of a company. The engineer or

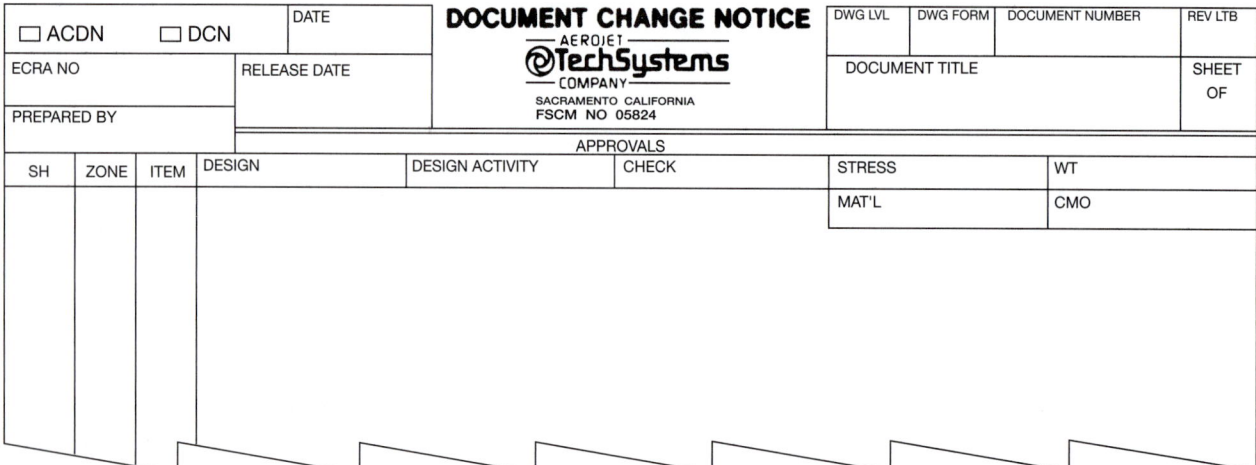

FIGURE 18.32 ■ Sample engineering change notice. *Courtesy Aerojet Propulsion Division.*

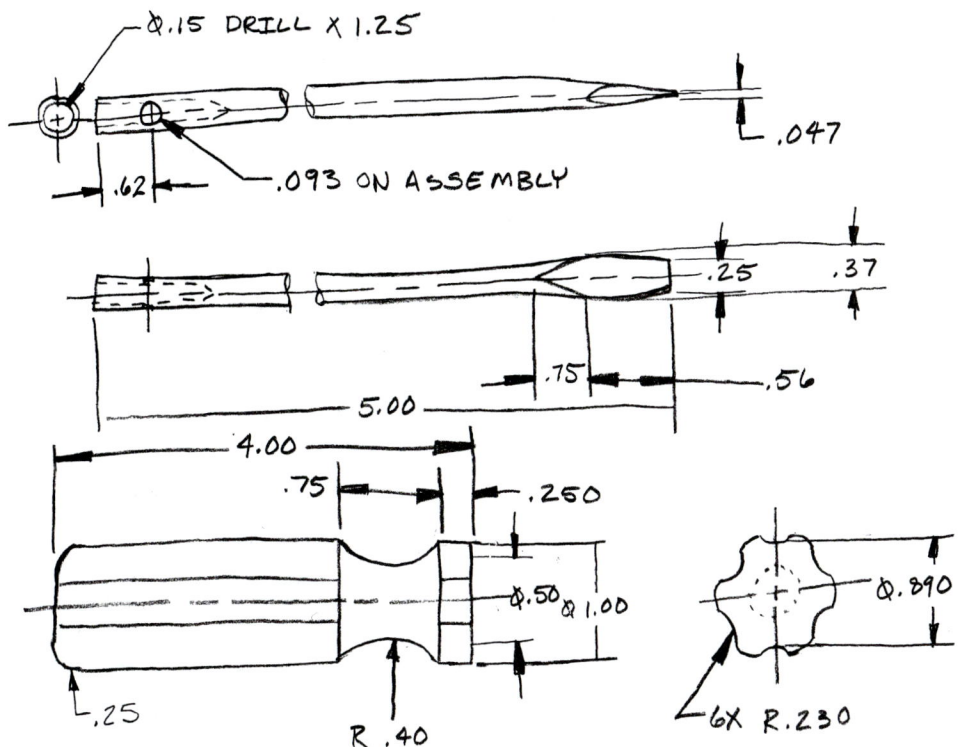

FIGURE 18.33 ■ A very rough engineer's sketch can contain errors. Engineering sketches may not be as good as the one shown here. The engineering drafter must convert the sketch to proper ASME, MIL, or ISO standards as appropriate. Dimension values in this figure are in inches.

designer may prepare engineering sketches or layout drawings. Figure 18.33 shows the engineer's sketches for a screwdriver to be manufactured by the company. Notice in this example that the engineer's sketch is rough. This is common. It is your responsibility, as a professional drafter, to prepare the final drawings in accordance with ASME/ANSI, MIL, or the appropriate standards adopted by the company.

The Detail Drawings

The engineer's sketches go to the drafting department for the preparation of formal drawings. Depending on the complexity of the project, you and the engineer may work together, or there may be a team of engineers and drafters working on various components of the project. Your first job is to complete the detail drawings as shown in Figure 18.34 and Figure 18.35.

The Assembly Drawing and Parts List

As discussed earlier in this chapter, the assembly drawing is used to show the assembly department how the parts fit together and if there are any special instructions required for the assembly process. When CADD is used, you often freeze the dimensions and isolate the individual parts, which are brought together in a separate drawing. Using CADD allows you to scale the parts into one common size and then move them together in the assembled position. Assembly notes and dimensions, if any,

are added. Each part receives a balloon and a correlating parts list is created as shown for the screwdriver design in Figure 18.36. As you look at Figure 18.36, the assembly drawing has two process notes. The note on the left requires the HANDLE hole to be filled with epoxy prior to assembly. This helps secure the DRIVER in the HANDLE. The note on the right is a process note that drills through the HANDLE matching the existing hole in the DRIVER. The PIN is then pressed into the hole, providing additional connection between the HANDLE and the DRIVER. When the complete set of working drawings has been drawn and approved, the product is released to production.

The Product Catalog

The power of 3-D CADD applications makes it possible for companies to display their products using pictorial assemblies or exploded pictorial assemblies. Figure 18.37 shows a pictorial assembly of the screwdriver that was designed in the previous discussion.

Making Engineering Changes

In review, when a product is in production a change can be requested from any department including manufacturing, engineering, or sales. These changes can be based on new ways to manufacture, redesign, customer feedback, or a variety of issues. Before a drawing can be changed by the drafting department, it

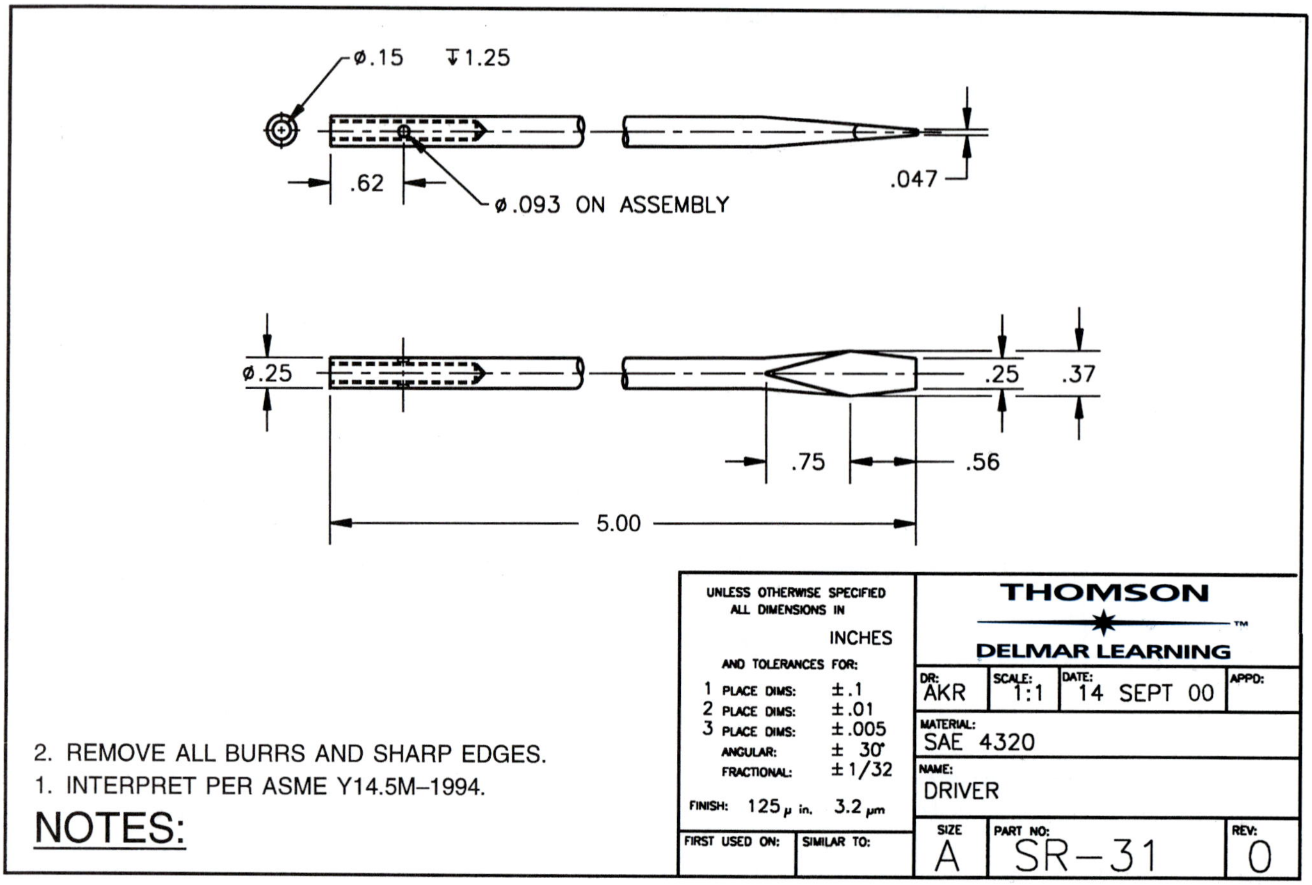

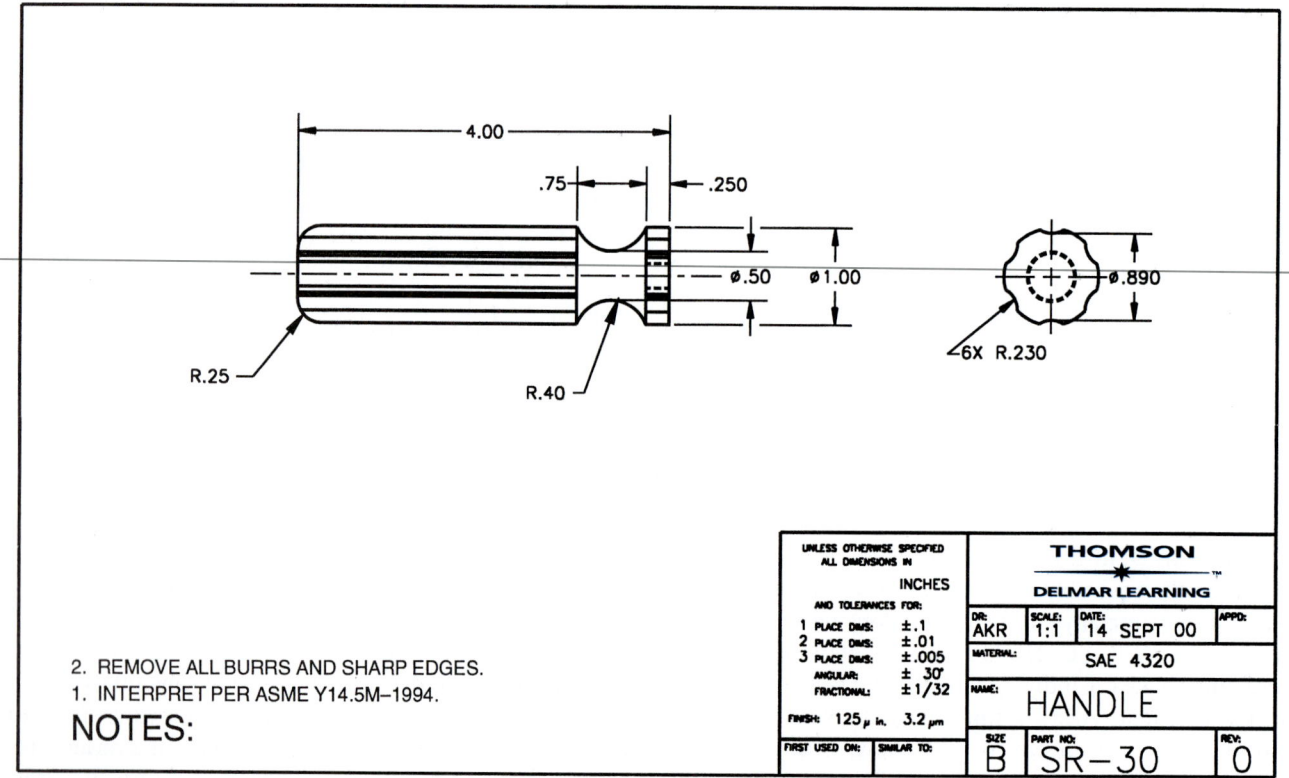

FIGURE 18.34 ■ The DRIVER detail drawing based on the engineer's sketch before an ECR is issued. Notice the "0" in the lower-right corner. This indicates an original unchanged drawing. Dimension values in this figure are in inches.

FIGURE 18.35 ■ Detail drawing of the HANDLE created from the engineer's sketch. Dimension values in this figure are in inches.

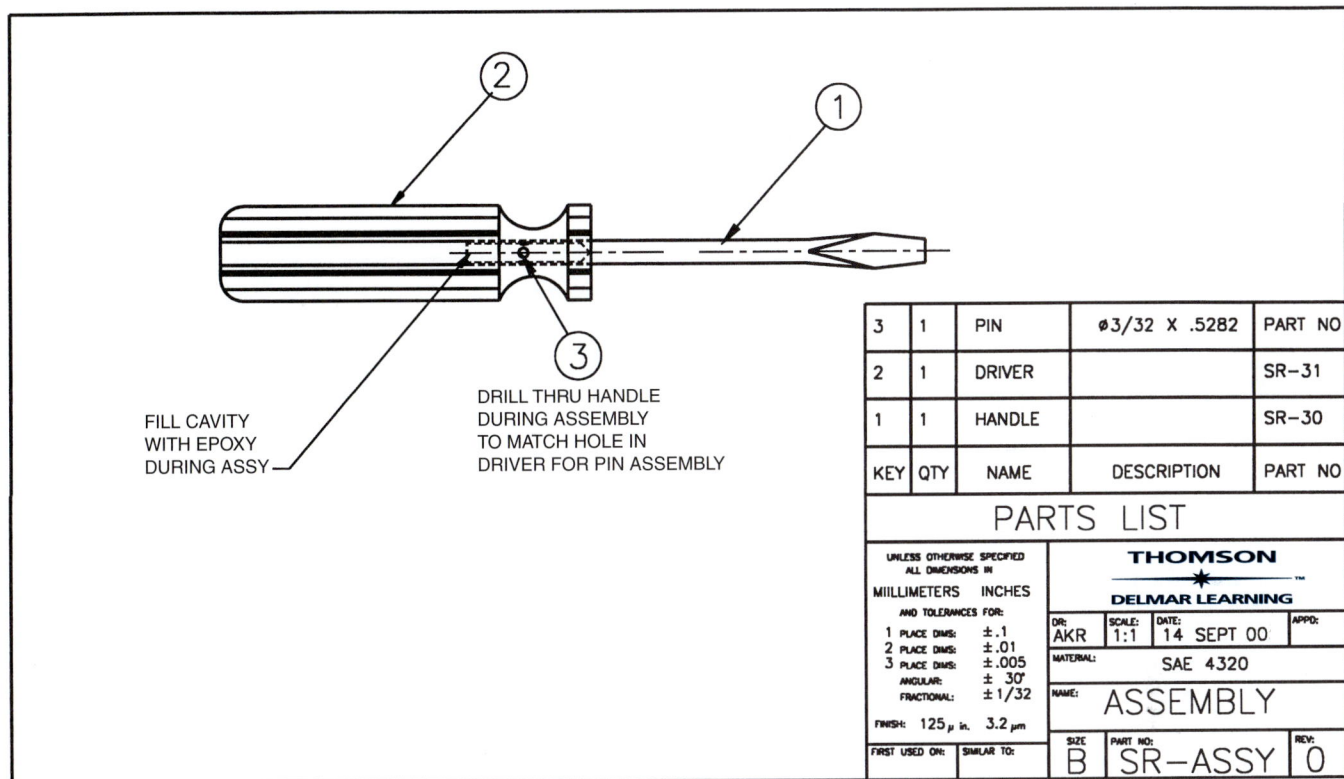

FILL CAVITY
WITH EPOXY
DURING ASSY

DRILL THRU HANDLE
DURING ASSEMBLY
TO MATCH HOLE IN
DRIVER FOR PIN ASSEMBLY

3	1	PIN	⌀3/32 X .5282	PART NO
2	1	DRIVER		SR–31
1	1	HANDLE		SR–30
KEY	QTY	NAME	DESCRIPTION	PART NO

PARTS LIST

UNLESS OTHERWISE SPECIFIED
ALL DIMENSIONS IN
MIILLIMETERS   INCHES
AND TOLERANCES FOR:
1 PLACE DIMS: ±.1
2 PLACE DIMS: ±.01
3 PLACE DIMS: ±.005
ANGULAR: ± 30'
FRACTIONAL: ±1/32
FINISH: 125μ in.   3.2 μm
FIRST USED ON: SIMILAR TO:

THOMSON
DELMAR LEARNING

DR: AKR   SCALE: 1:1   DATE: 14 SEPT 00   APPD:
MATERIAL: SAE 4320
NAME: ASSEMBLY

SIZE B   PART NO: SR–ASSY   REV: 0

FIGURE 18.36 ■ Assembly drawing for the screwdriver.

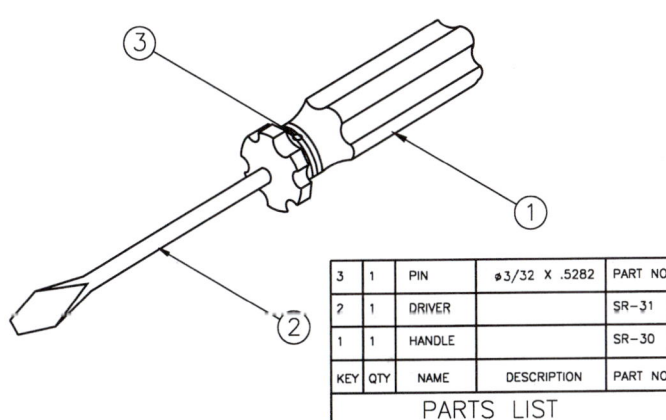

3	1	PIN	⌀3/32 X .5282	PART NO
2	1	DRIVER		SR–31
1	1	HANDLE		SR–30
KEY	QTY	NAME	DESCRIPTION	PART NO

PARTS LIST

FIGURE 18.37 ■ Pictorial assembly drawing of the screwdriver for the company catalog.

usually requires an engineering change request (ECR). The ECR is also referred to as an engineering change order (ECO). The ECR is the document that is used to initiate a change in a part or assembly. The ECR is often attached to a redlined print of the part or drawing affected. The print and the ECR normally contain a change number that is used later by the drafter to properly document the change. Every company has an ECR form that looks different. The ECR form shown in Figure 18.38 displays a change requested for the DRIVER shown in Figure 18.34.

Companies use various methods to change drawings depending on the complexity of the change, the company stan-

dards, and the time allowed. If dimensions are changed and the feature is not drawn to its proper scale, the not-to-scale symbol is used. Changes made to CADD drawings are usually easier to accurately change and scale than those created using manual drawing. When CADD is used, the company often desires to have the changed drawing accurately reflect the product.

After a change is made on the drawing, a balloon is placed next to the drawing revision and a letter identifying the change is placed in the balloon. Balloons are usually 3/8 to 1/2 inch (8 to 12 mm) in diameter depending on the drawing size. Some companies use a square, a hexagon, or another symbol for change balloons. The letter in the balloon designates the change. An A is used for the first change, a B for the second change, and so on. Some companies use R1, R2, and R3. However, letters are common to differentiate from balloon numbers on assembly drawings. Next, you record the change in the revision block of the drawing. The revision block is usually in the upper-right corner in accordance with ASME standards. The current change is also recorded in the title block as "A," "B," or "C" as appropriate. An original drawing generally has a "O" in the title block indicating no changes have been made. Figure 18.39 shows the DRIVER in Figure 18.34 changed in accordance with the ECR in Figure 18.38.

The next procedure in the change process is for you to fill out the engineering change notice (ECN). The change documents are a formal part of the drafting process and must be professionally prepared, recorded, and filed for future reference. The ECN should be hand-lettered, typed, or electronically prepared if the

Engineering Change Request

Request to be completed
Through ECN# **1171**

Engineering:

Engineer

Date: 20 JUL 01

Manufacturing:

Date:

Sales:

Date:

Description of Request:

Refer to part number SR–31 (Driver)

Change 5" length to 6"

Change Hole location dimension from .62 to .69

Reason for Request:

Longer tool dimension will make part more stable when locked in handle.

Hole location was design error.

Castings & Forgings Affected? (yes or no) NO	Disposition of Production Stock:	Use in production	☐
Approved: Date: 20 JUL 01	Drawing by: Student	Scrap ☒ Transfer to service stock	☐

FIGURE 18.38 ■ The engineering change request (ECR). Dimension values in this figure are in inches.

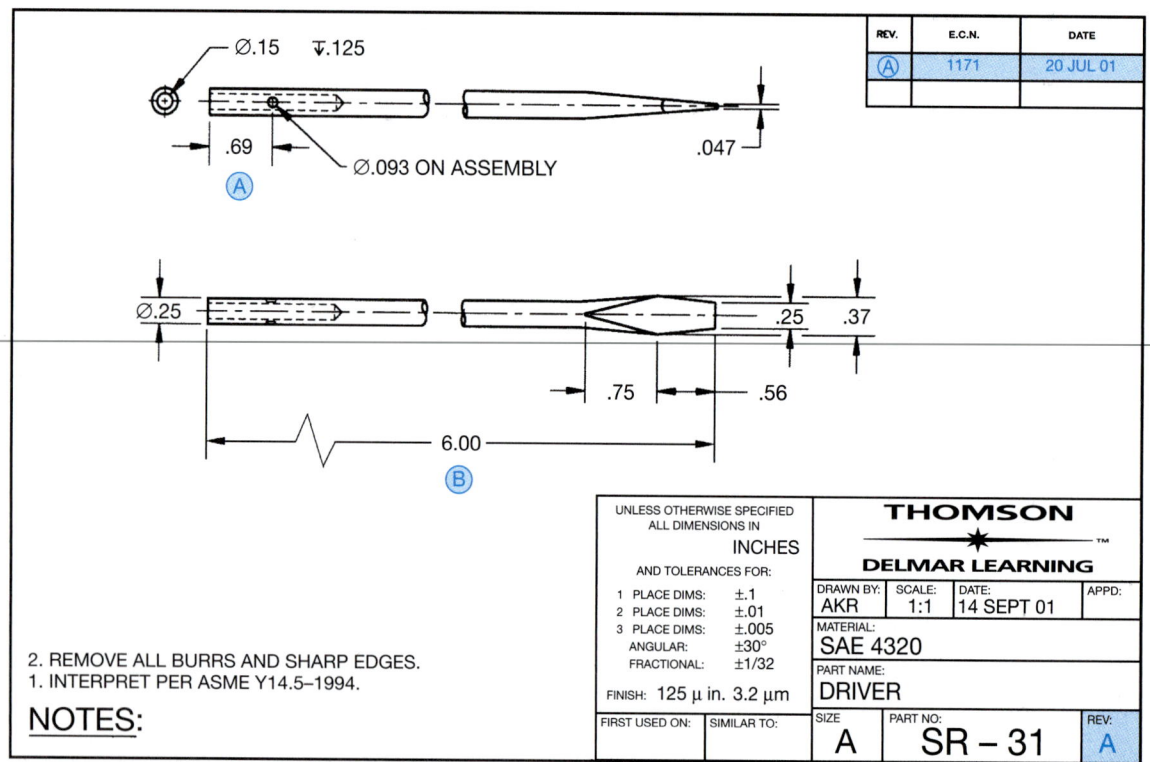

REV.	E.C.N.	DATE
Ⓐ	1171	20 JUL 01

Ø.15   ⊽.125

.69

Ø.093 ON ASSEMBLY

Ⓐ

.047

Ø.25

.25   .37

.75   .56

6.00

Ⓑ

2. REMOVE ALL BURRS AND SHARP EDGES.
1. INTERPRET PER ASME Y14.5–1994.

NOTES:

UNLESS OTHERWISE SPECIFIED
ALL DIMENSIONS IN
INCHES
AND TOLERANCES FOR:

1 PLACE DIMS:	±.1
2 PLACE DIMS:	±.01
3 PLACE DIMS:	±.005
ANGULAR:	±30°
FRACTIONAL:	±1/32

FINISH: 125 μ in. 3.2 μm

THOMSON
™
DELMAR LEARNING

DRAWN BY:	SCALE:	DATE:	APPD:
AKR	1:1	14 SEPT 01	

MATERIAL:
SAE 4320

PART NAME:
DRIVER

FIRST USED ON:	SIMILAR TO:	SIZE **A**	PART NO: **SR – 31**	REV: **A**

FIGURE 18.39 ■ Changes to the DRIVER based on the ECR shown in Figure 18.38. Notice the balloons next to each change, the identification of the change in the upper-right corner, and the current change "A" in the lower-right corner. Dimension values in this figure are in inches.

Engineering Change Notice

ECN NO. 1171 ①

Disposition of production in stock:
A = Alter or rework U = Use in production
T = Transfer to service stock S = Scrap

Qty.	Drawing Size Part No.	R/N	Description	Change	Other Usage in Production	D/S
01	1	A SR–31	A	DRIVER	6" was 5"		S
02	②	③	④	⑤	.69 Hole Location Dimension		⑦
03					was .62 ⑥		
04							
05							
06							
07							
08							
09							
10							
11							
12							
13							
14							
15							
16							
17							
18							

Reason: ⑧
Line 1–This dimension needed to be longer to help make the part more stable when locked in handle.
Lines 2–3 This was a design error.

Castings & forgings affected? ☐ Yes ☒ No ⑨	Design engineer: DAM ⑩	Supervisor approval: DPM ⑪	Release date: 25 NOV 01 ⑬	Page ⑫ 1/1

FIGURE 18.40 ■ The completed engineering change notice (ECN) form. The circled numbers refer to the steps for completing the form given in the text. The ECN form must be professionally typed or hand-lettered.

document is available in a word processor or CADD system. The ECN form used in companies and the process for filling out the forms vary. The ECN form shown in Figure 18.40 has been created from a combination of formats found in industry. The large numbers on the ECN refer to the changes requested for the DRIVER in Figure 18.34 and the ECR in Figure 18.38. The following steps are used when filling out this ECN form.

STEP 1 Put the ECN number in at the top right side of page, on this example. Get this number from the engineering change request. In this case, it is 1171.

STEP 2 List quantity of the part affected. On this drawing, a detail-only part can be affected. *Note:* Quantity = 1.

STEP 3 List drawing size, such as A, B, C, or D and the part number. This is an A size, and the part number is SR-31.

STEP 4 List revision. In this case, it is the first change. A new drawing has N or O here. Put the current revision letter such as A, B, C, or D here. Some companies use numbers. This is the first change on this drawing so an A is used.

STEP 5 Place the title for the drawing or part here (DRIVER).

STEP 6 Identify the changes you made. Describe the changes made in short terms.

STEP 7 List what is to be done with parts already made under the old design. Use a symbol such as A, U, T, or S to show this. The definitions of these letters are conveniently located on the ECR form. (See Figure 18.38.)

STEP 8 Identify reasons for the change in brief but complete statements so in the future someone can find out why the change was made.

STEP 9 To help the machinist, list if casting or forgings are affected.

STEP 10 You need to initial your name or that of the project engineer so if questions arise about the changes, those asking know who to query.

STEP 11 Your supervisor should approve the changes you have made.

STEP 12 Pages are listed as 1/1 for a one-page change document, or 1/3, if three pages are used to complete one change. If there are three pages, the pages are numbered 1/3, 2/3, 3/3, or 1 of 3, 2 of 3, 3 of 3.

STEP 13 The date you made the change must be recorded.

The part or product is released again for continued manufacturing after the engineering change documents have been properly prepared and approved.

WORKING DRAWING SHEETS

A set of working drawings usually contains several sheets. A simple product may only require an assembly drawing and one or two detail drawings, while a complex product can include separate assembly drawings, subassembly drawings, and hundreds of detail drawings. All the sheets in a set of working drawings must be labeled correctly and linked with the other sheets in the set. This process has traditionally involved manually correlating each sheet with the set. However, the advent of parametric CADD software has significantly improved or eliminated this sometimes time-consuming requirement.

A single drawing file can be used to produce a set of working drawings, because you can add multiple drawing sheets to a single file. Drawing sheets are used in the same way that pieces of paper are used to document a set of working drawings. Often an entire design can be documented using one drawing file and multiple drawing sheets. The advantage to using this system is that all the drawings are contained in a single file and are automatically linked together. For example, the order of sheets in the drawing file can be used to define the sheet number in the title block. If there are four sheets in the drawing, the first sheet is automatically labeled 1 of 4 in the title block, while the last sheet is labeled 4 of 4 in the title block. If you want to modify the sheet arrangement, you can drag a sheet before or after another sheet, and the sheet numbers in the title block automatically update according to the new configuration.

DISTANCE BETWEEN TANGENT CIRCLES

Problem: It is desired to find dimension z, the distance between the centers of the two smaller circles of the illustration shown in Figure 18.41.

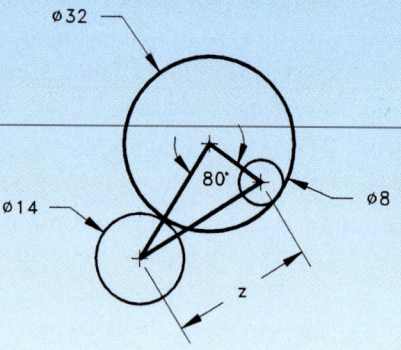

FIGURE 18.41 ■ Tangent circles.

Solution: It is necessary to visualize the triangle connecting the centers of the circles and make use of the known radii to calculate two of the sides. The side of the triangle connecting the 32" and 14" diameter circle centers must be the sum of their radii, 16 + 7 = 23", and the side connecting the 32" and 8" diameter circle centers must be the difference of their radii, 16 − 4 = 12". (See Figure 18.42.)

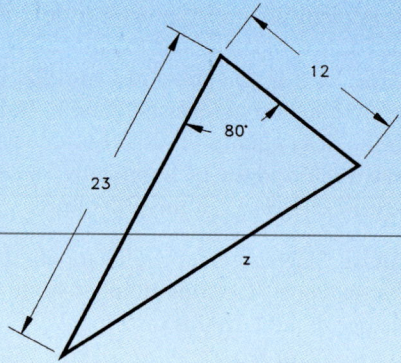

FIGURE 18.42 ■ Oblique triangles for the tangent circles.

Solution: Since the triangle is not a right triangle, you cannot use a single trig function, and since you do not know a side *and* the angle opposite the side, you cannot use the Law of Sines. Instead, you must use the Law of Cosines (See CD Chapter 3):

$$z^2 = (23)^2 + (12)^2 - 2(23)(12)\cos 80°$$
$$z^2 = 529 + 144 - 2(23)(12)(.1736)$$
$$z^2 = 529 + 144 - 95.854$$
$$z^2 = 577.146$$
$$z = \sqrt{577.146} = 24.024 \text{ or } 24.0"$$

WEBSITE RESEARCH

The following websites can assist you in doing additional research on subjects such as manufacturing materials, manufacturing practices, tooling, and related areas:

http://www.asme.org—American Society of Mechanical Engineers.

http://www.industrialpress.com—*Machinery's Handbook*. Manufacturing materials, standard features, and processes.

PROFESSIONAL PERSPECTIVE

Many entry-level drafters are often involved in preparing detail drawings or making changes to detail drawings. When a drafter has gained valuable experience with drafting practices, standards, and company products, there is usually an opportunity to advance to a design drafter position. Drafting jobs at these levels can be exciting, as the drafters work closely with engineers to create new and updated product designs. An individual drafter in a small company may be teamed with an engineer to implement designs. In larger companies, a team of drafters, designers, and engineers may work together to design new products. These are the types of situations where a drafter may have the opportunity to prepare some or all of the drawings for a complete set of working drawings. Generally, in these research and development departments of companies, the preliminary product drawings are used to build prototypes of the designs for testing. After sufficient tests have been performed on the product, the drawings are revised and released for production. The new product becomes a reality.

CHAPTER 18 Working Drawings Test

Access the CD found with this textbook to view the Chapter 18 Test. Confirm the preferred submittal method with your instructor.

CHAPTER 18 Working Drawings Problems

DIRECTIONS

1. From the selected engineering sketches or layouts, prepare a complete set of working drawings, including details assembly and parts list. Determine which views and dimensions should be used to completely detail each part. Also, determine the views, parts list, dimensions, and notes, if any, for the assembly drawing. Use ANSI/ASME standards. Use manual drafting or CADD as required by your course guidelines and objectives.

2. The complete set of working drawings should be prepared with one detail drawing per sheet using multiview projection and with the assembly drawing and parts list on one sheet unless otherwise specified. All purchase (standard) parts will be completely identified in the parts list. Using the sketches as a guide, draw original multiview drawings on adequately sized sheets. Add all necessary dimensions and notes using unidirectional dimensioning. Use computer-aided drafting if required by your course guidelines. Many of the problems are designed to be manufactured as projects in the manufacturing (machine) technology lab.

3. Include the following general notes at the lower-left corner of the sheet .5 in. (13 mm) each way from the corner border lines for each detail drawing: (*Note:* Number 2 does not apply to the assembly drawings).

Notes:
1. INTERPRET DIMENSIONS AND TOLERANCES PER ASME Y14.5M—1994.
2. REMOVE ALL BURRS AND SHARP EDGES.

DIRECTIONS (Continued)

Additional general notes may be required depending on the specifications of each individual assignment. The following should be part of your title block unless otherwise specified by your instructor.

UNSPECIFIED TOLERANCES

DECIMALS	mm	IN.
X	±2.5	±.1
XX	±0.25	±.01
XXX	±0.125	±.005
ANGULAR ± 30'		
FINISH	3.2 μm	125μIN.

4. Most of the Chapter 18 problems can be drawn without studying the other chapters in this section. However, several of the given problems are advanced and challenging, and may best be solved after you study all of Section Five's chapters. If you choose, or if you are assigned to solve these problems, you should study Chapters 19, 20, and 21 as you encounter parts in your problems that relate to those chapters. This is the kind of challenge that you can face in the real world of engineering drafting. Oftentimes you will have to go ahead on your own, or seek additional instruction when you encounter new and varied obstacles. The following are the types of features that will require you to study additional content beyond this chapter:

■ Linkages, cams, gears, and bearings in Chapter 19.

■ Belt and chain drives in Chapter 20.

■ Assembly of welded parts in Chapter 21.

Special Note: Some problems in this chapter contain errors, missing information, or slight inaccuracies. This is intentional and is meant to encourage you to apply appropriate problem-solving methods, engineering, and drafting standards in order to solve the problems. This is meant to force you to think about each part and how parts fit together in the assembly. As in "real-world" projects, the engineering problem should be considered as a basis for your preliminary layouts. Always question inaccuracies in project designs and consult with the proper standards and other sources. In some cases, an error might be the source of engineering changes provided by the instructor; however, this is determined by your specific course objectives. Other situations may require that corrections be made during the development of the original design drawings. This is not intended as a source of frustration but is considered to be part of the engineering drafter's daily responsibility in project development.

PROBLEM 18.1 **Working-drawing assembly (metric)**

Assembly Name: Plumb Bob

SPECIFIC INSTRUCTIONS:

Prepare a working-drawing assembly that has a detail drawing of each part, an assembly drawing, and a parts list on one sheet.

PARTS LIST

ITEM	QTY	NAME	MATERIAL
1	1	PLUMB BOB	BRONZE
2	1	CAP	BRONZE

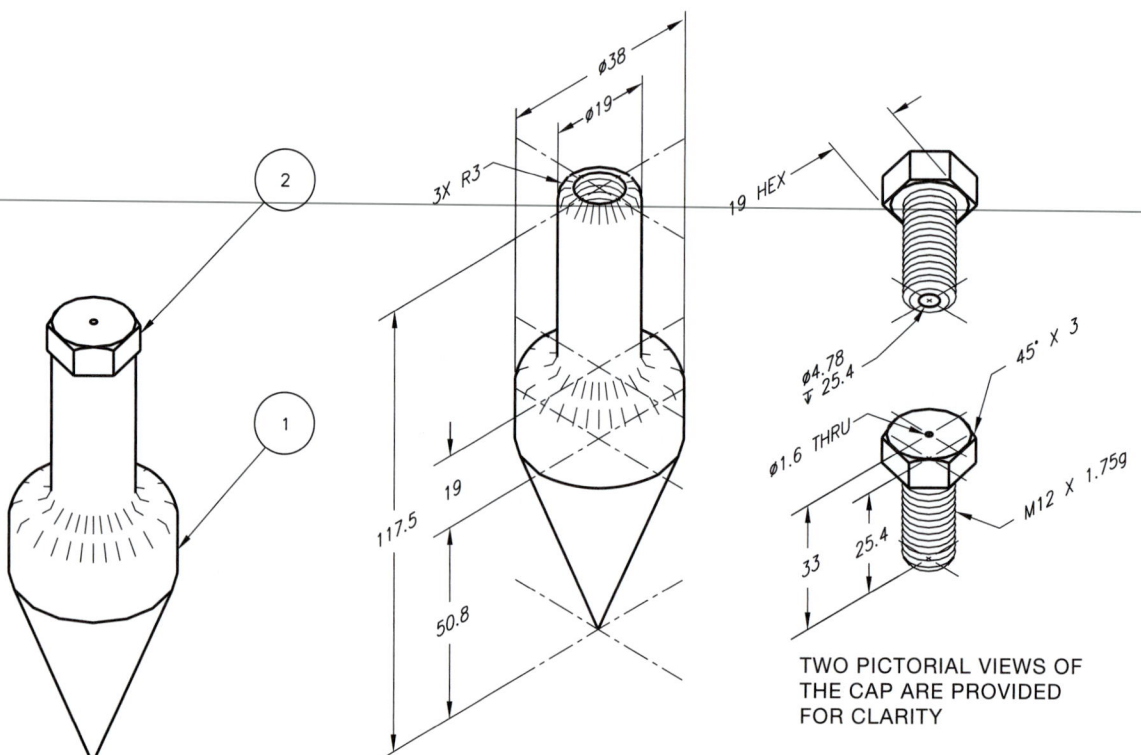

TWO PICTORIAL VIEWS OF
THE CAP ARE PROVIDED
FOR CLARITY

PROBLEM 18.2   Working drawing (in.)

Assembly Name: Hammer

SPECIFIC INSTRUCTIONS:

Prepare a detail drawing for the hammer head and two optional hammer handles on one sheet. Make the assembly drawing and parts list on another sheet.

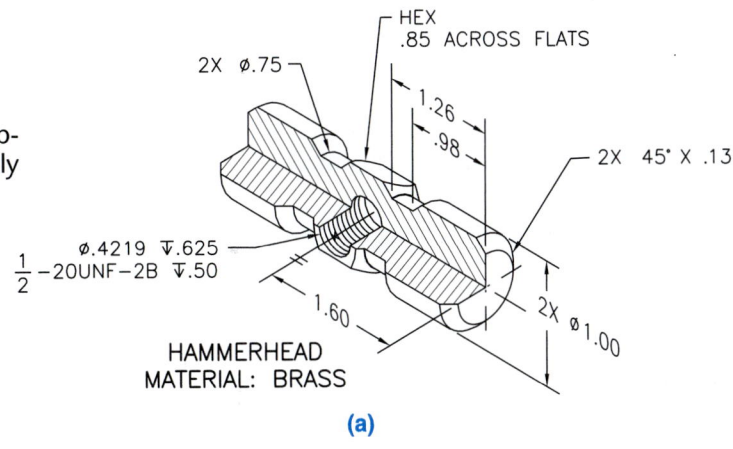

HAMMERHEAD
MATERIAL: BRASS

(a)

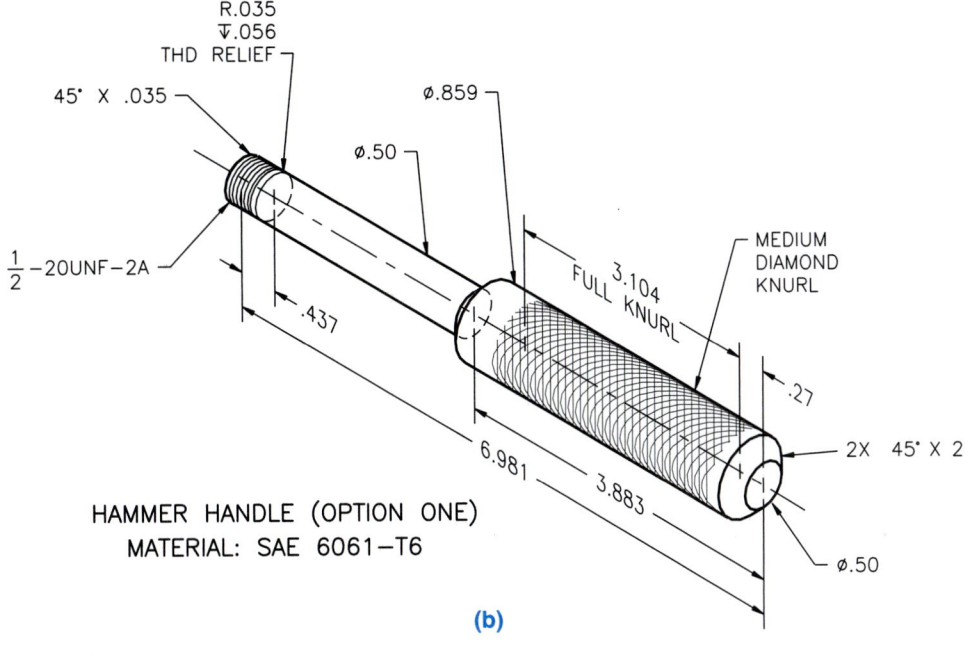

HAMMER HANDLE (OPTION ONE)
MATERIAL: SAE 6061–T6

(b)

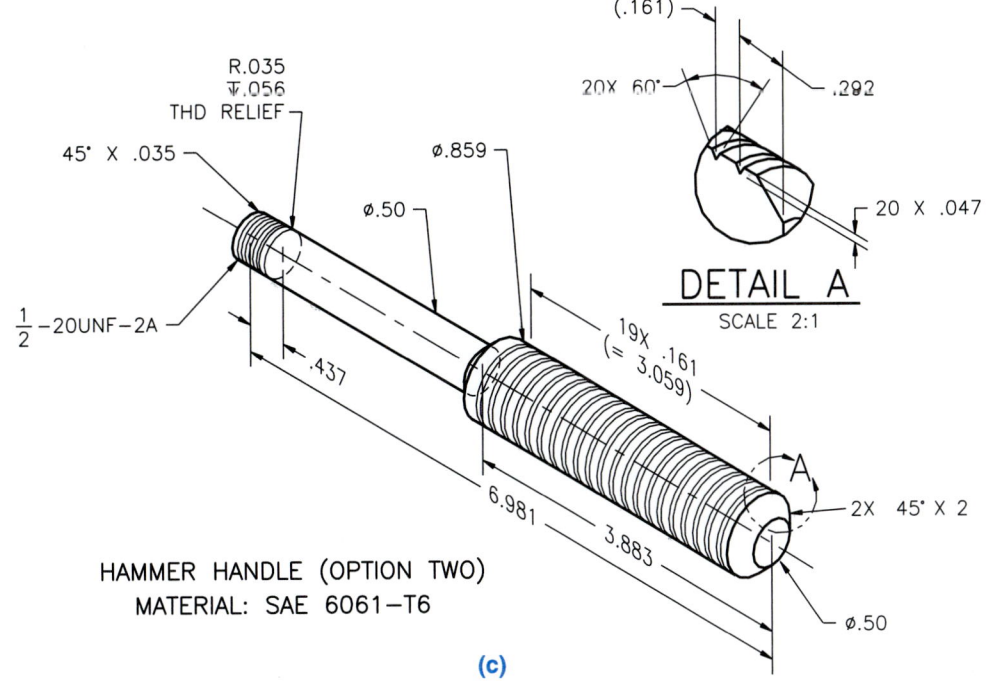

HAMMER HANDLE (OPTION TWO)
MATERIAL: SAE 6061–T6

DETAIL A
SCALE 2:1

(c)

PROBLEM 18.3 **Working drawing (in.)**

Assembly Name: Key Holder

SPECIFIC INSTRUCTIONS:

Prepare a detail drawing for each part, an assembly drawing, and a parts list on one sheet, unless otherwise specified by your instructor.

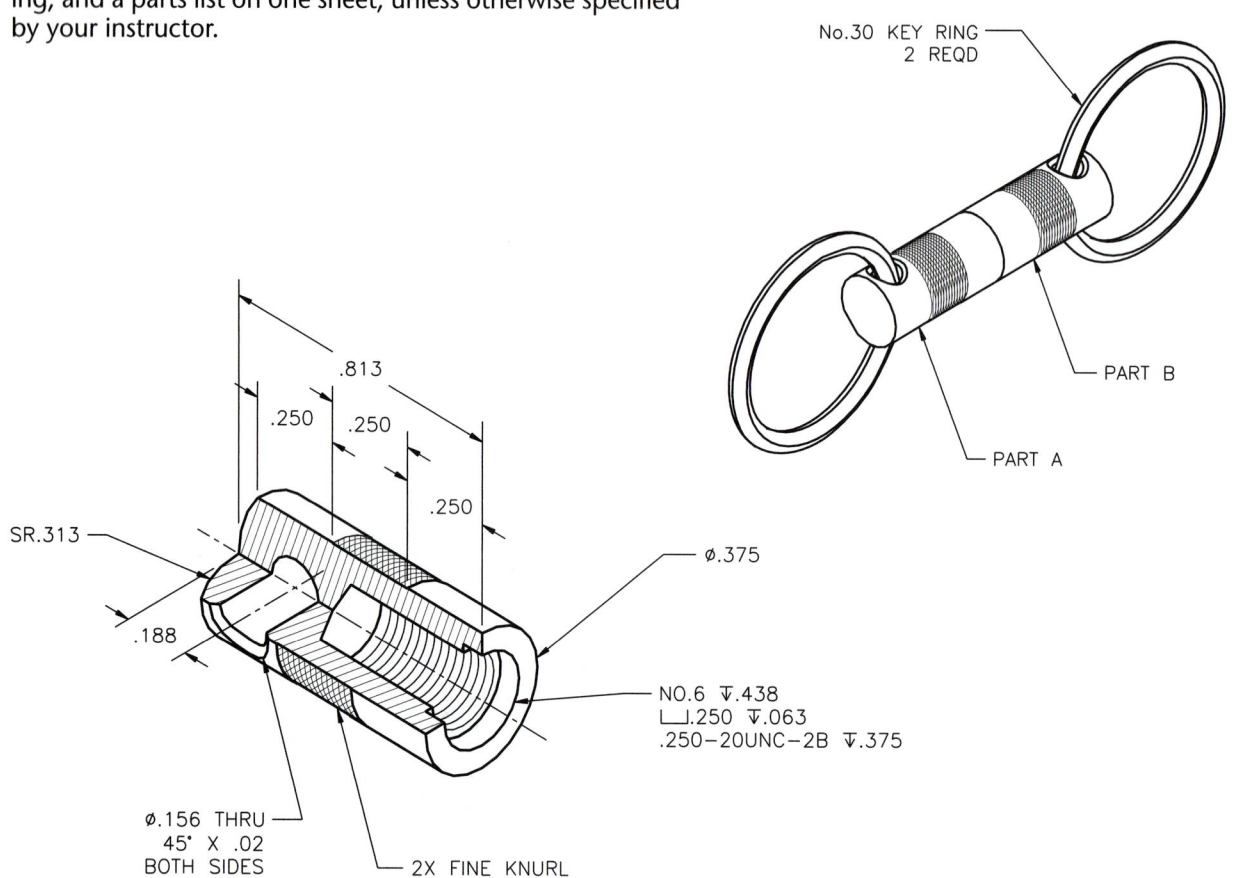

No.30 KEY RING
2 REQD

PART B

PART A

SR.313

.813

.250

.250

.250

Ø.375

.188

NO.6 ∇.438
⌴.250 ∇.063
.250–20UNC–2B ∇.375

Ø.156 THRU
45° X .02
BOTH SIDES

2X FINE KNURL

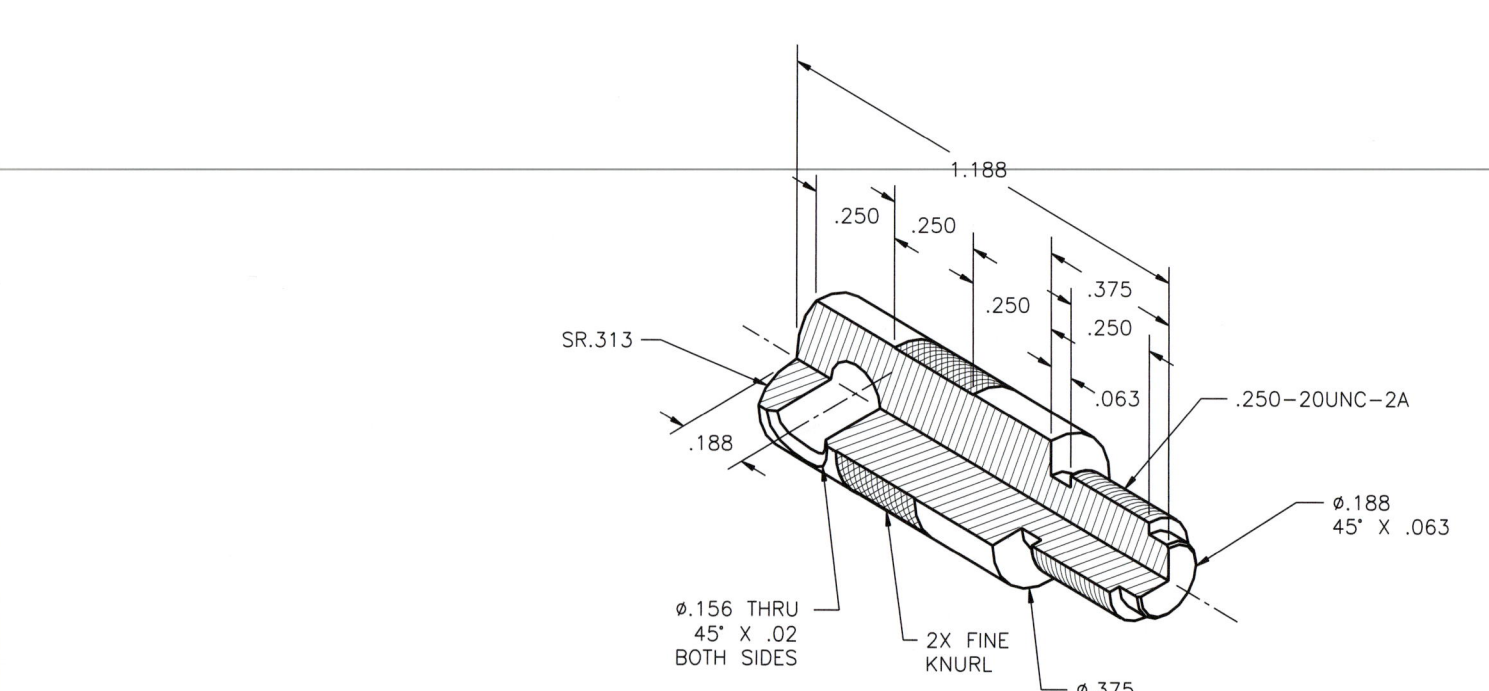

1.188

.250

.250

.250

.375

.250

.063

.250–20UNC–2A

SR.313

.188

Ø.188
45° X .063

Ø.156 THRU
45° X .02
BOTH SIDES

2X FINE
KNURL

Ø.375

PROBLEM 18.4 Working Drawing (in.)

Assembly Name: Lock Nut Wrench

SPECIFIC INSTRUCTIONS:

Prepare a complete set of working drawings with all of the detail drawings on one sheet and the assembly drawing and bill of materials on another sheet. Use multiview projection for view layout.

LOCK NUT WRENCH BILL OF MATERIALS		
ITEM	QTY	DESCRIPTION
1	1	GRIP
2	1	STEEL RIVET, Ø.25 X .625 LONG
3	1	MAIN BODY

(b)

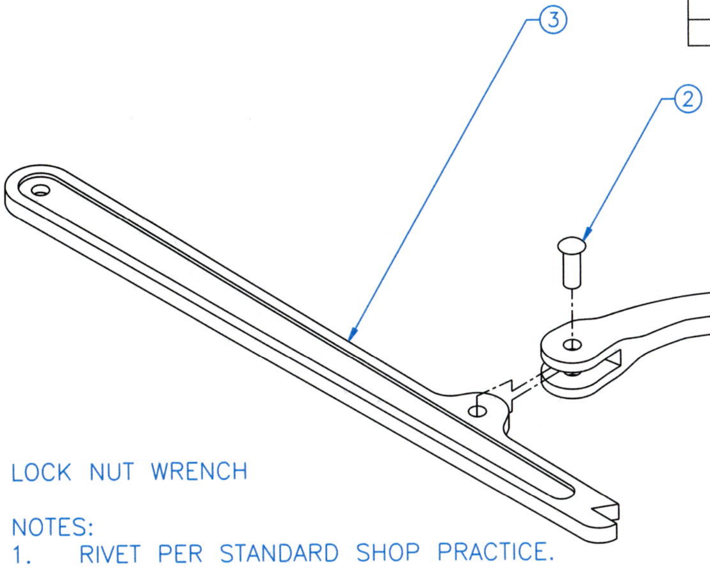

3
2
1
LOCK NUT WRENCH
NOTES:
1. RIVET PER STANDARD SHOP PRACTICE.

(a)

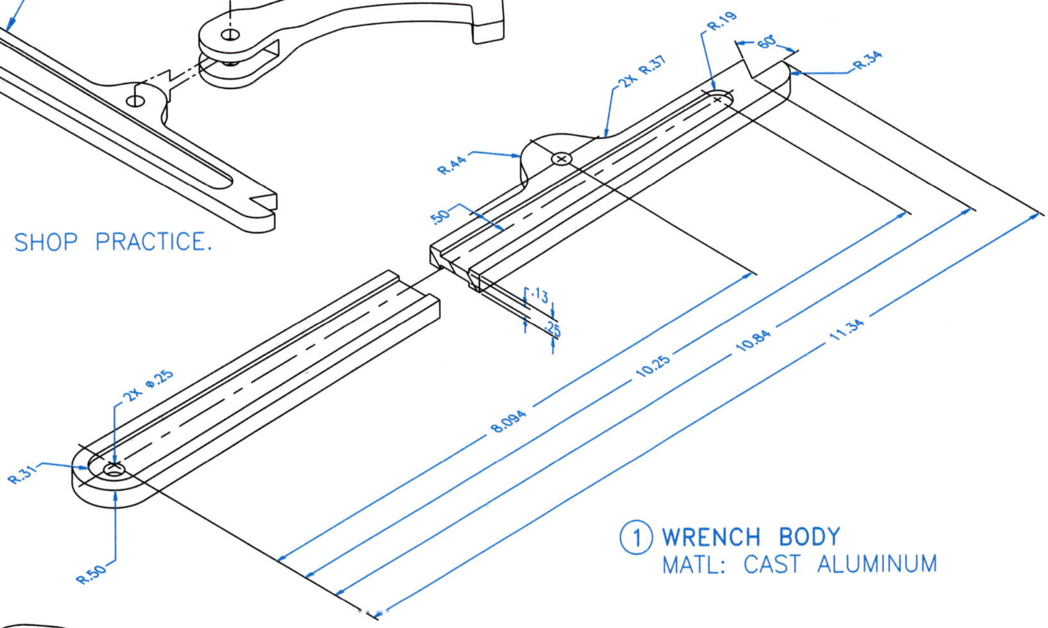

R.19
60°
R.34
2X R.37
R.44
.50
.13
.25
2X ⌀.25
R.31
R.50
8.094
10.25
10.84
11.34
1 WRENCH BODY
MATL: CAST ALUMINUM

(c)

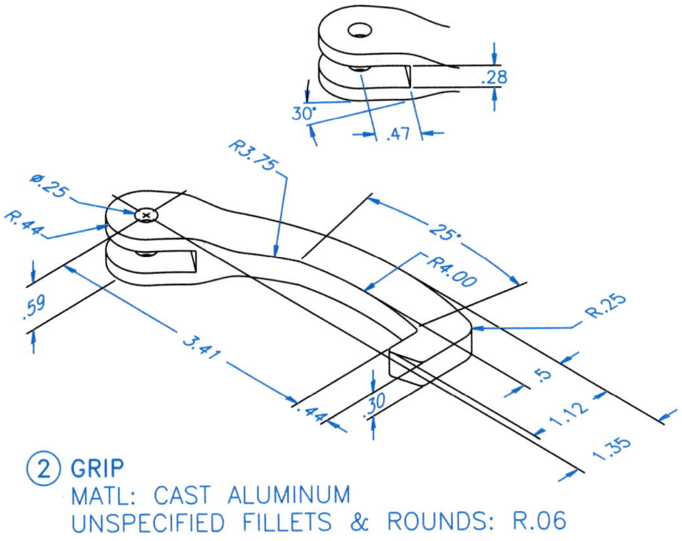

.28
30°
.47
R3.75
⌀.25
R.44
25°
R4.00
.59
R.25
3.41
.5
.30
.44
1.12
1.35
2 GRIP
MATL: CAST ALUMINUM
UNSPECIFIED FILLETS & ROUNDS: R.06

(d)

PROBLEM 18.5 **Working drawing (metric)**

Assembly Name: C-clamp

SPECIFIC INSTRUCTIONS:

Prepare a complete set of working drawings with all of the detail drawings on one sheet and the assembly drawings and parts list on another sheet. Use multiview projection for view layout.

PARTS LIST

ITEM	QTY	NAME	MATERIAL
1	1	PIN	CRMS
2	1	BODY	SAE 4320
3	1	SCREW	SAE 4320
4	1	SWIVEL	SAE 1020

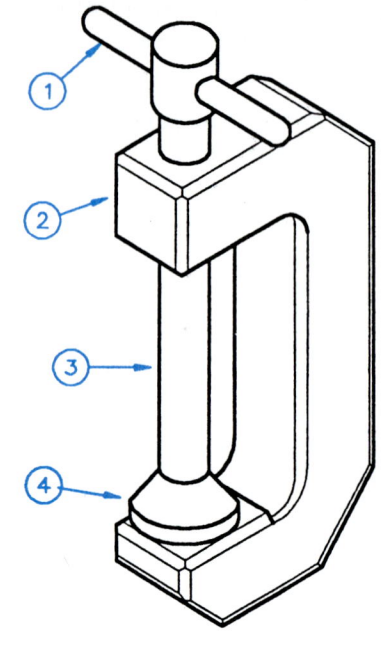

C-CLAMP

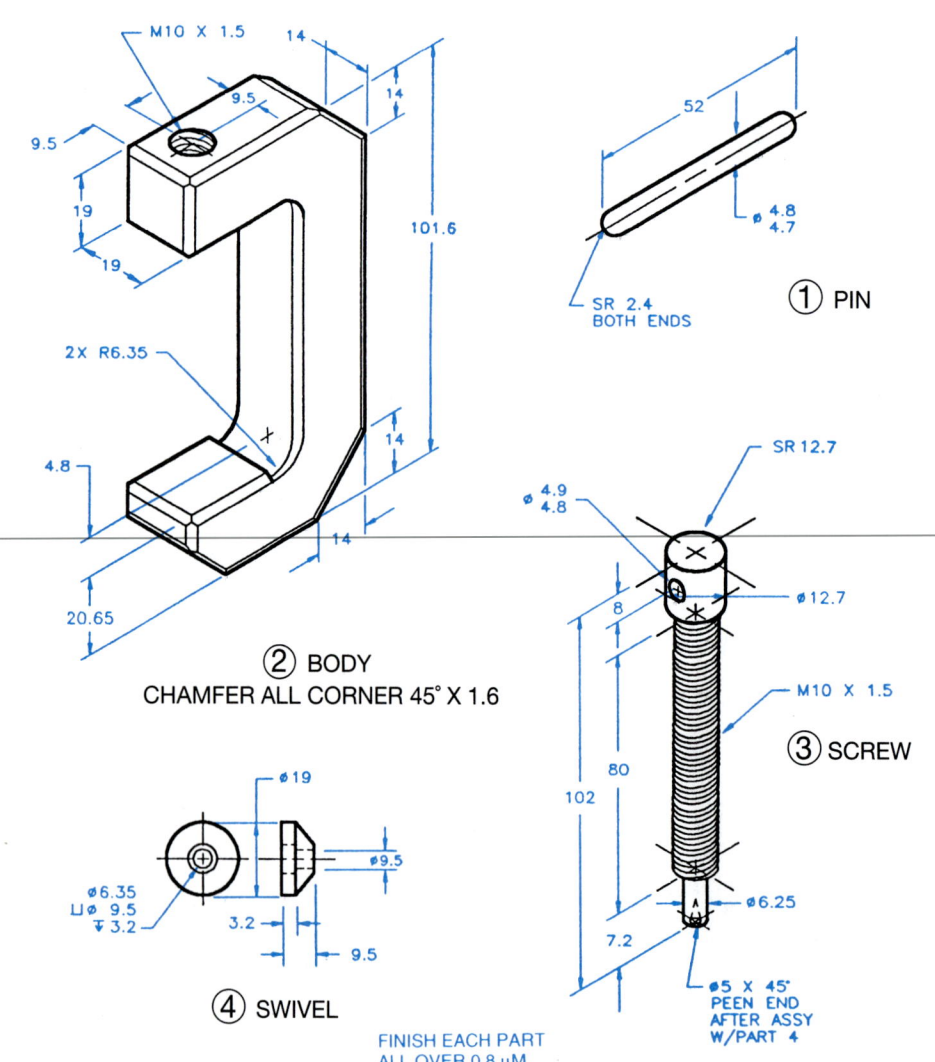

M10 X 1.5

2X R6.35

② BODY
CHAMFER ALL CORNER 45° X 1.6

① PIN

SR 12.7

M10 X 1.5

③ SCREW

ø5 X 45°
PEEN END
AFTER ASSY
W/PART 4

④ SWIVEL

FINISH EACH PART
ALL OVER 0.8 μM

PROBLEM 18.6   Working Drawing (in.)

Assembly Name: Drill Pump

SPECIFIC INSTRUCTIONS:

Prepare a complete set of working drawings with one detail drawing per sheet and the assembly drawing and bill of materials on another sheet. Use multiview projection for view layout.

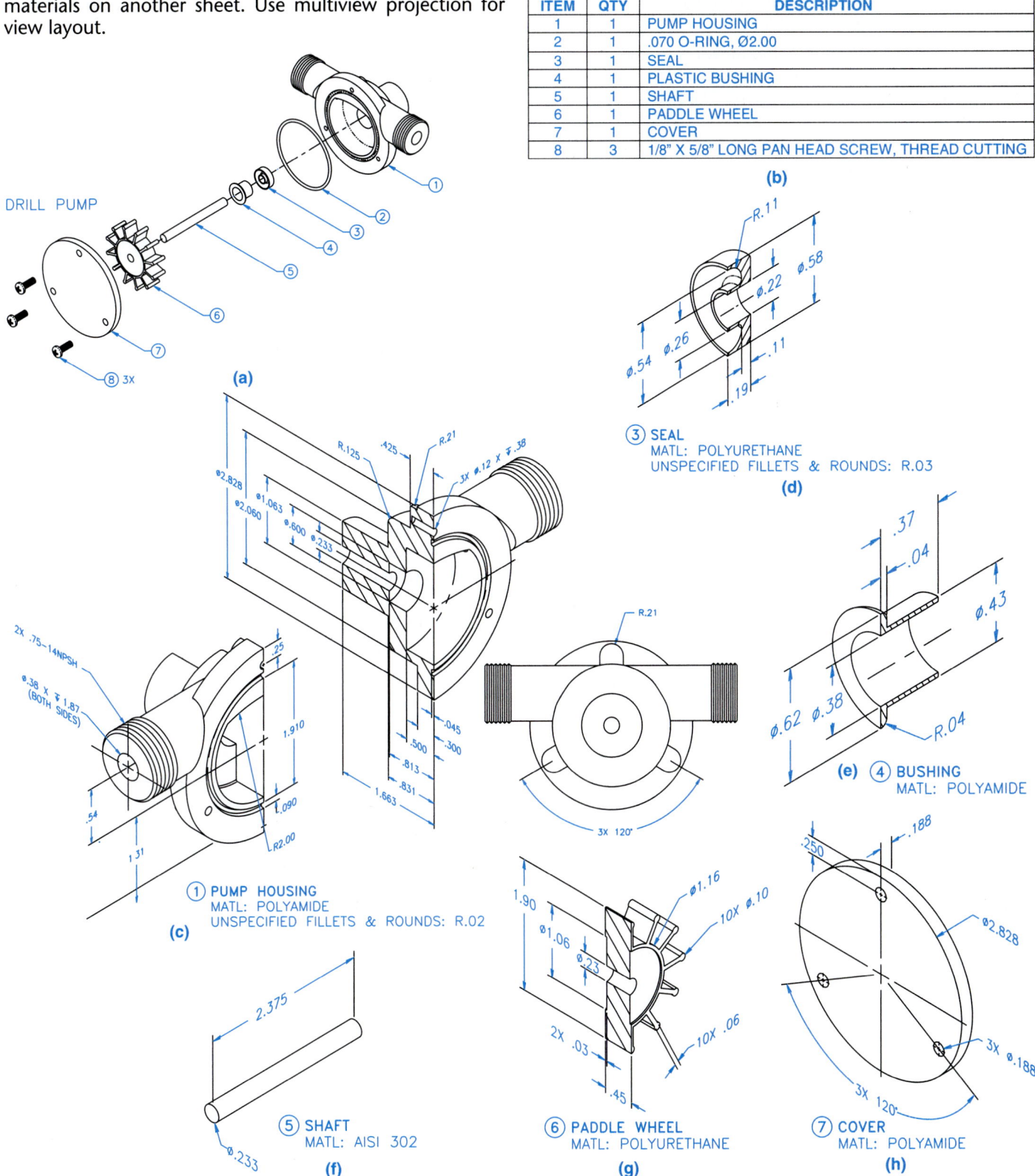

	DRILL PUMP BILL OF MATERIALS	
ITEM	**QTY**	**DESCRIPTION**
1	1	PUMP HOUSING
2	1	.070 O-RING, Ø2.00
3	1	SEAL
4	1	PLASTIC BUSHING
5	1	SHAFT
6	1	PADDLE WHEEL
7	1	COVER
8	3	1/8" X 5/8" LONG PAN HEAD SCREW, THREAD CUTTING

(b)

DRILL PUMP

(a)

③ SEAL
MATL: POLYURETHANE
UNSPECIFIED FILLETS & ROUNDS: R.03
(d)

(e)   ④ BUSHING
MATL: POLYAMIDE

① PUMP HOUSING
MATL: POLYAMIDE
UNSPECIFIED FILLETS & ROUNDS: R.02
(c)

⑤ SHAFT
MATL: AISI 302
(f)

⑥ PADDLE WHEEL
MATL: POLYURETHANE
(g)

⑦ COVER
MATL: POLYAMIDE
(h)

PROBLEM 18.7 **Working drawing (in.)**

Assembly Name: Mill Work Stop

SPECIFIC INSTRUCTIONS:

This problem has basic applications; however, there is a welding symbol involved in the assembly. You should study welding processes and representations covered in Chapter 21 before completing this problem.

Prepare a complete set of working drawings with one detail drawing per sheet and the assembly drawing with parts list on one sheet. Notice: part number 10, TEE STUD, is cut from standard 3/8-16UNC ALL THREAD. Parts 12 and 13, BLANK KNOBS, are purchased as knobs without the threads machined. You will draw the knobs as a "representation" of the purchase part and give the thread note as identified in the parts list. The actual dimensions of the knob to be purchased are not important; however, accurate specifications for the threads to be machined into the purchase part are important.

NOTE: Some design is required throughout the completion of the project.

PARTS LIST

ITEM	QTY	NAME	DESCRIPTION	MATERIAL
1	1	VERTICAL MEMBER	3/4 × 2 × 4	SAE 6061
2	1	BASE MEMBER	1 × 2 × 3.5	SAE 6061
3	2	RIB	1/4 PLATE	SAE 6061
4	1	ARM	3/8 × 1 × 5-1/2	SAE 1018
5	1	CLAMP	1/2 × 1 × 1-5/16	SAE 1018
6	1	TEE PLATE DRILL AND TAP AT CENTER FOR 5/16-18UNC-2 THREAD THRU	5/16 × 1 × 1-1/4	SAE 1018
7	1	STOP ROD	∅5/16 × 6	SAE 303 SS
8	1	WING NUT	1/4-28 × 1/2	
9	1	BUSHING	∅1/2 × .31	SAE 1018
10	1	TEE STUD	3/8-16 × 2 LG ALL THREAD	
11	1	CARRIAGE BOLT	5/16-18UNC-2	
12	21	BLANK KNOB	VLIER HK-3 (DRILL AND TAP FOR 3/8-16UNC-2B)	
13	1	BLANK KNOB	VLIER HK-3 (DRILL AND TAP FOR 5/16-18UNC-2B)	

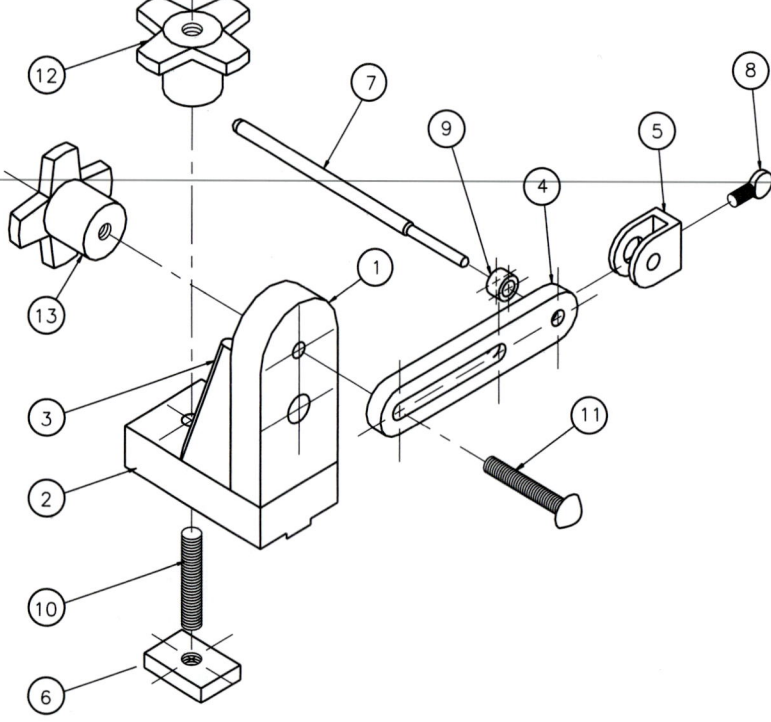

PROBLEM 18.7 (Continued)

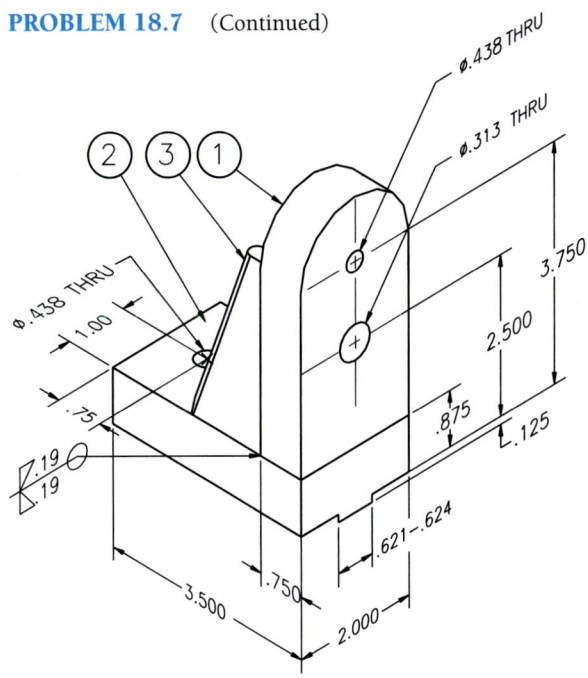

Ø.438 THRU

Ø.313 THRU

3.750

2.500

.875

.125

Ø.438 THRU

1.00

.75

.19
.19

.621–.624

3.500

.750

2.000

DEBUR ALL EDGES .020

BODY ASSEMBLY WELDMENT
FINISH: COLOR ANODIZED

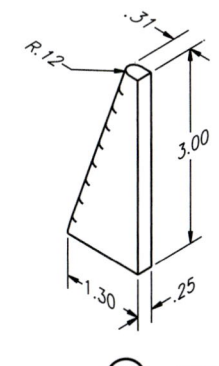

R.12

.31

3.00

1.30   .25

③ **RIB**

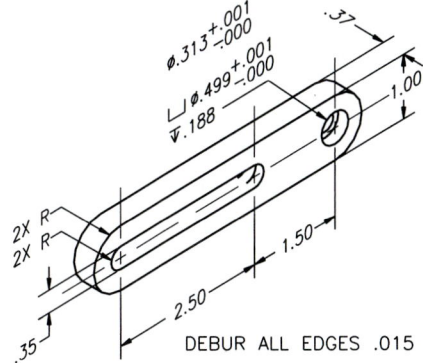

Ø.313 +.001/−.000

⌴ Ø.499 +.001/−.000
▼.188

.37

1.00

2X R
2X R

1.50

2.50

.35

DEBUR ALL EDGES .015

④ **ARM**
FINISH: C HARDEN .005/.010
BLACK OXIDE

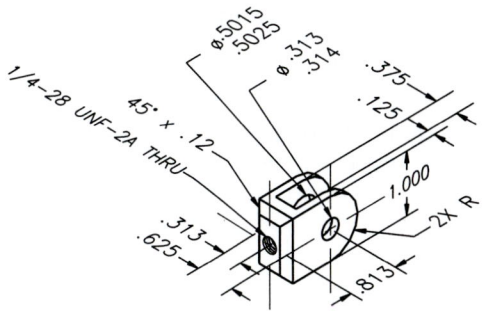

Ø.5015/.5025

Ø.313/.314

.375

.125

45° x .12

1/4−28 UNF−2A THRU

1.000

2X R

.313
.625

.813

DEBUR ALL EDGES .015
EXCEPT Ø.313−.314

⑤ **CLAMP**
FINISH: C HARDEN .005/.010
BLACK OXIDE

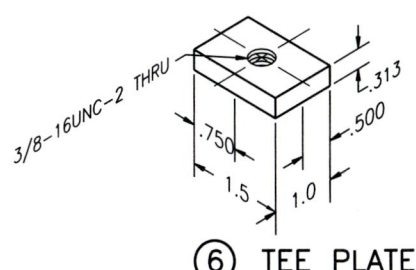

.313

.500

3/8−16UNC−2 THRU

.750

1.5   1.0

⑥ **TEE PLATE**

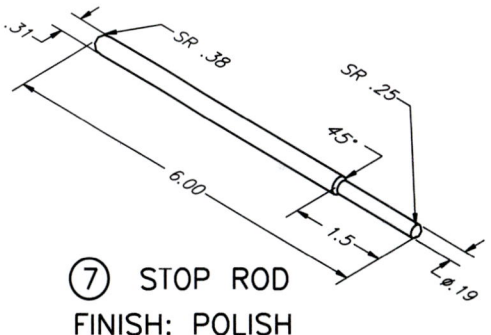

.31

SR .38

SR .25

45°

6.00

1.5

Ø.19

⑦ **STOP ROD**
FINISH: POLISH

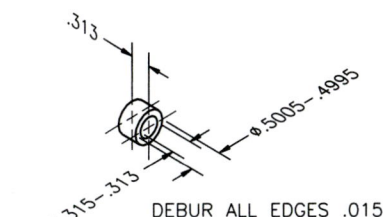

.313

Ø.5005−.4995

Ø.315−.313

DEBUR ALL EDGES .015

⑨ **BUSHING**
FINISH: C HARDEN .005/.010
BLACK OXIDE

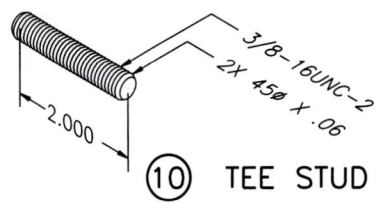

3/8−16UNC−2

2X 45Ø X .06

2.000

⑩ **TEE STUD**

PROBLEM 18.8   Working drawing (in.)

Assembly Name: Fly Tying Vice

SPECIFIC INSTRUCTIONS:

Prepare a complete set of working drawings with one detail drawing per sheet and the assembly and parts list on another sheet. When preparing the assembly drawing, use separate balloons for each part in each view or use only one balloon in the view that most clearly identifies the part. *Problem courtesy of David P. Madsen and John Melloy.*

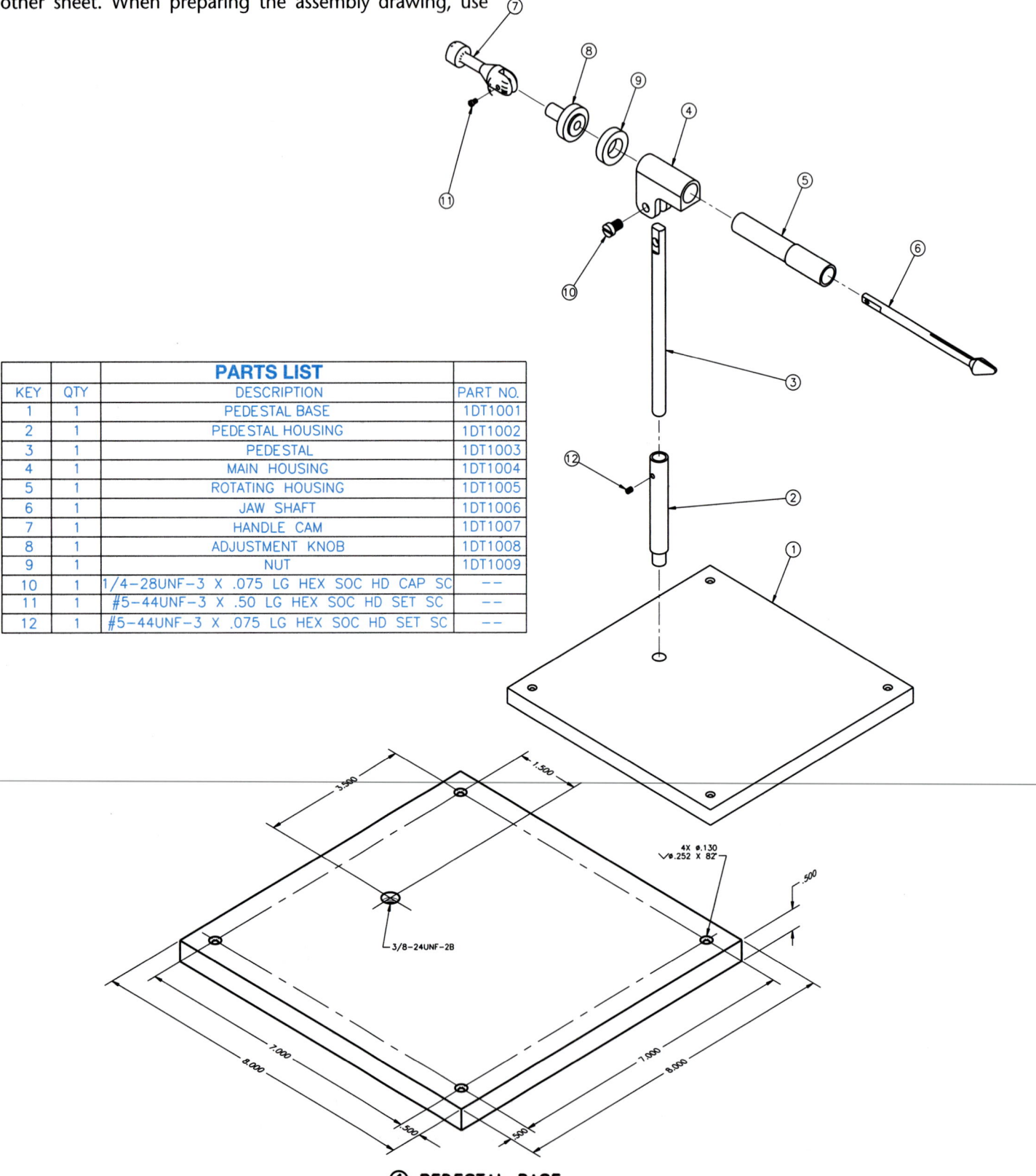

PARTS LIST			
KEY	QTY	DESCRIPTION	PART NO.
1	1	PEDESTAL BASE	1DT1001
2	1	PEDESTAL HOUSING	1DT1002
3	1	PEDESTAL	1DT1003
4	1	MAIN HOUSING	1DT1004
5	1	ROTATING HOUSING	1DT1005
6	1	JAW SHAFT	1DT1006
7	1	HANDLE CAM	1DT1007
8	1	ADJUSTMENT KNOB	1DT1008
9	1	NUT	1DT1009
10	1	1/4−28UNF−3 X .075 LG HEX SOC HD CAP SC	−−
11	1	#5−44UNF−3 X .50 LG HEX SOC HD SET SC	−−
12	1	#5−44UNF−3 X .075 LG HEX SOC HD SET SC	−−

① PEDESTAL BASE

PROBLEM 18.8 (Continued)

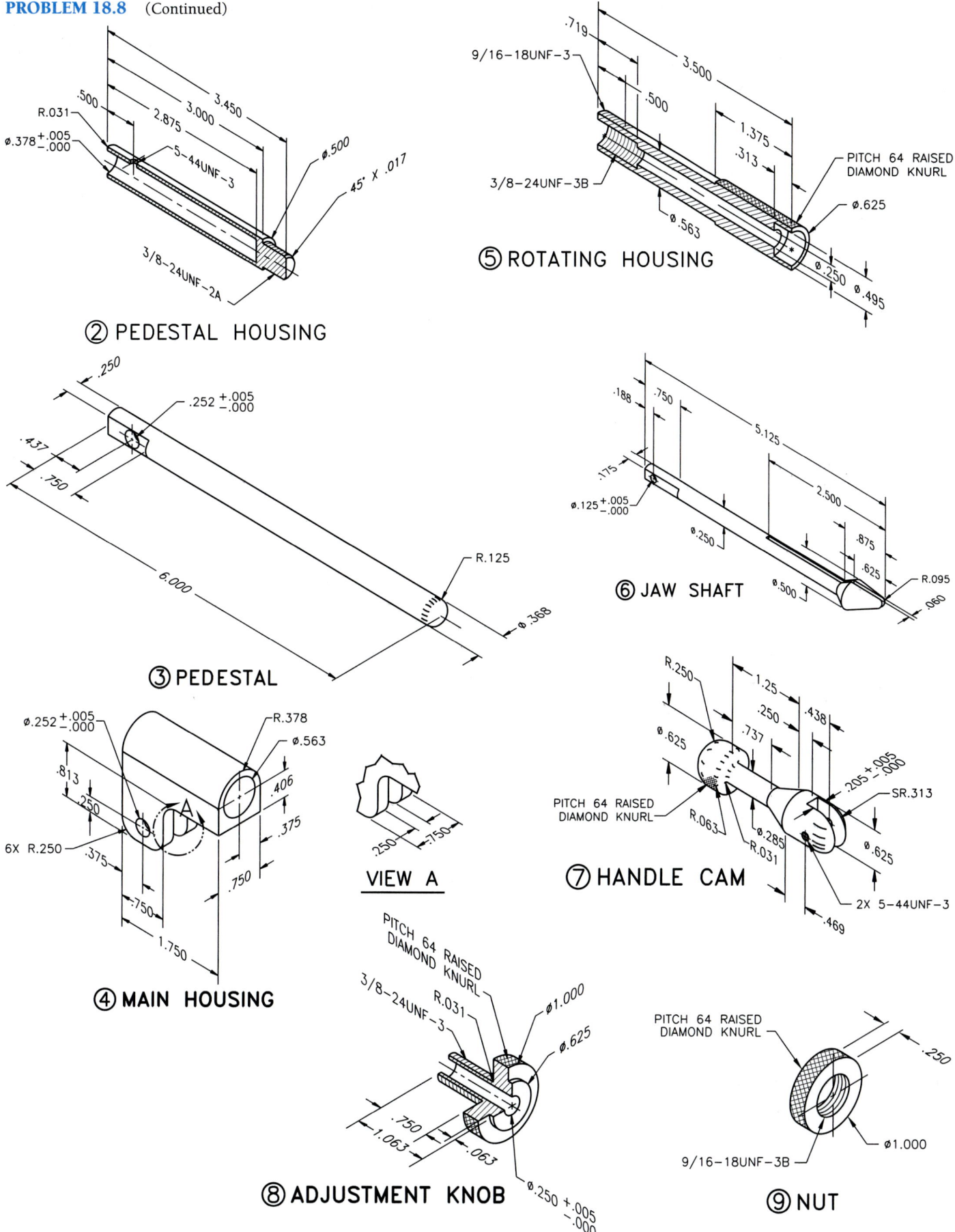

.500
R.031
ø.378 +.005 −.000
3.450
3.000
2.875
5–44UNF–3
ø.500
45° X .017
3/8–24UNF–2A

② PEDESTAL HOUSING

.719
9/16–18UNF–3
3.500
.500
1.375
.313
3/8–24UNF–3B
ø.563
PITCH 64 RAISED DIAMOND KNURL
ø.625
ø.250
ø.495

⑤ ROTATING HOUSING

.250
.252 +.005 −.000
.437
.750
6.000
R.125
ø.368

③ PEDESTAL

.188
.750
.175
5.125
ø.125 +.005 −.000
ø.250
2.500
.875
.625
ø.500
R.095
.060

⑥ JAW SHAFT

ø.252 +.005 −.000
R.378
ø.563
.813
.406
.250
.375
6X R.250
.375
.750
1.750

④ MAIN HOUSING

.250
.750

VIEW A

R.250
1.25
.250
.438
ø.625
.737
205 +.005 −.000
PITCH 64 RAISED DIAMOND KNURL
R.063
R.031
ø.285
SR.313
ø.625
2X 5–44UNF–3
.469

⑦ HANDLE CAM

PITCH 64 RAISED DIAMOND KNURL
3/8–24UNF–3
R.031
ø1.000
ø.625
.750
1.063
.063
ø.250 +.005 −.000

⑧ ADJUSTMENT KNOB

PITCH 64 RAISED DIAMOND KNURL
.250
ø1.000
9/16–18UNF–3B

⑨ NUT

PROBLEM 18.9 **Working drawing (in.)**

Assembly Name: Cork Screw

SPECIFIC INSTRUCTIONS:

Prepare a complete set of working drawings with one detail drawing per sheet and the assembly and parts list on another sheet. When preparing the assembly drawing, use separate balloons for each part in each view or use only one balloon in the view that most clearly identifies the part. Establish tolerances between mating parts. Determine the dimensions for the Pin, Part 6 and create needed detail or purchase part specifications. *Problem courtesy of Kasey Cassell.*

PARTS LIST

ITEM	QTY	NAME	MATERIAL
1	1	BODY	Stainless Steel
2	1	HANDLE	Stainless Steel
3	2	ARM	Stainless Steel
4	2	UPPER PIN	Stainless Steel
5	2	LOWER PIN	Stainless Steel
6	1	PIN	Stainless Steel
7	1	CORK SCREW	Stainless Steel

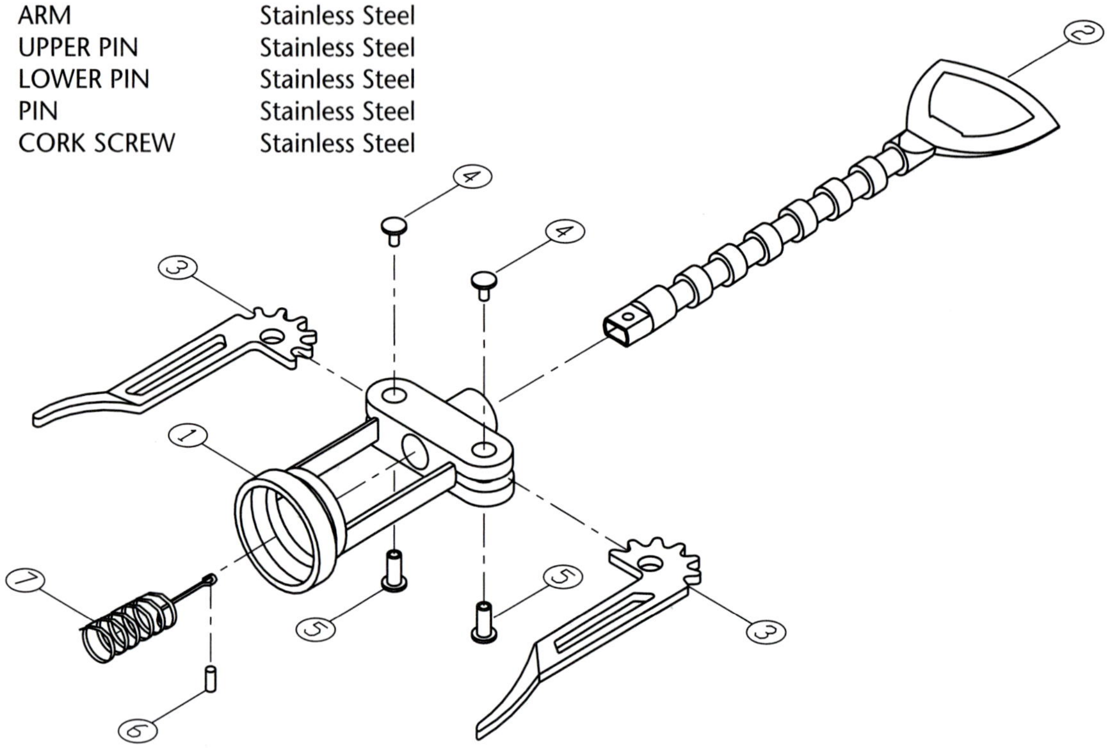

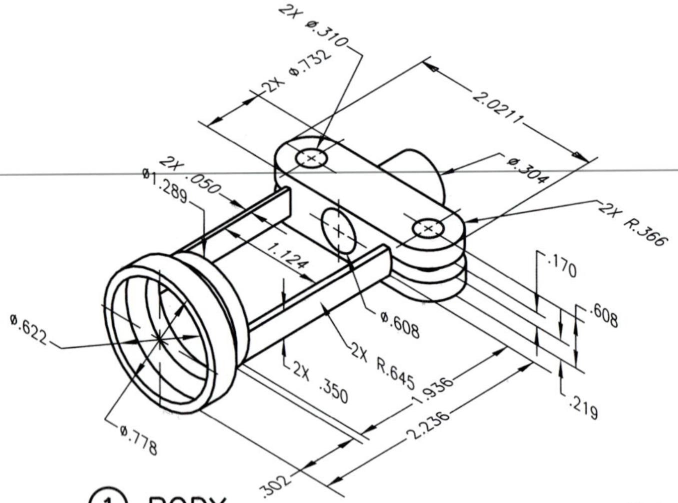

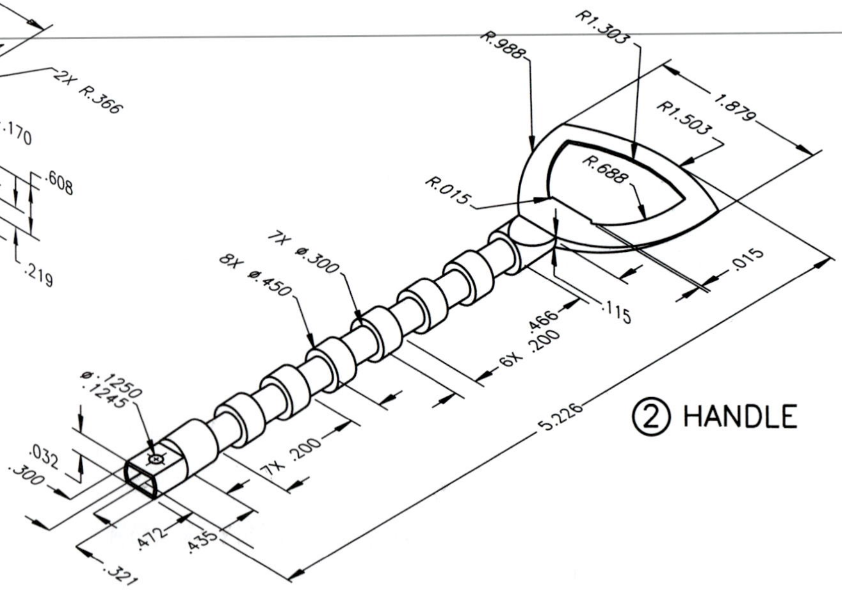

PROBLEM 18.9 (Continued)

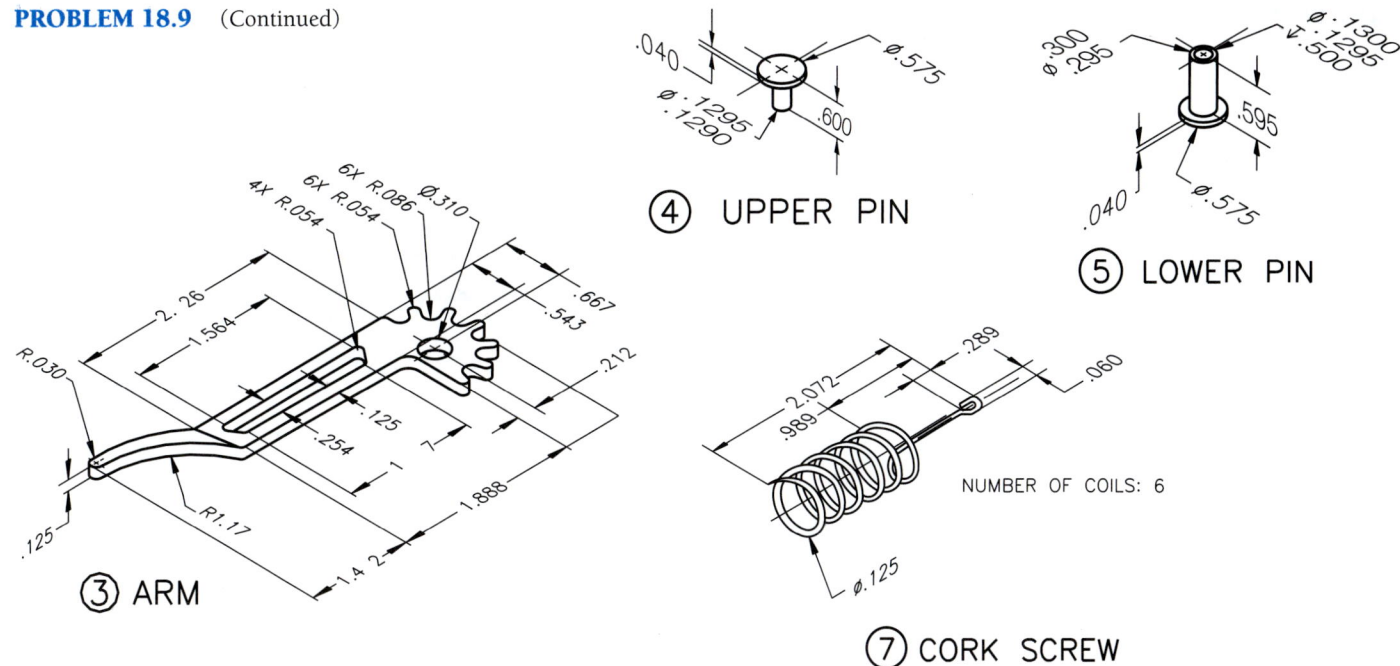

③ ARM

④ UPPER PIN

⑤ LOWER PIN

⑦ CORK SCREW

NUMBER OF COILS: 6

PROBLEM 18.10 Working drawing (in.)

Assembly Name: Ball Valve

SPECIFIC INSTRUCTIONS:

Prepare a complete set of working drawings with one detail drawing per sheet and the assembly and parts list on an-other sheet. When preparing the assembly drawing, use separate balloons for each part in each view or use only one balloon in the view that most clearly identifies the part. This is a prototype project. Errors are likely. Verify dimensions during assembly. Establish tolerances between mating parts. *Problem courtesy of Chris Hurford.*

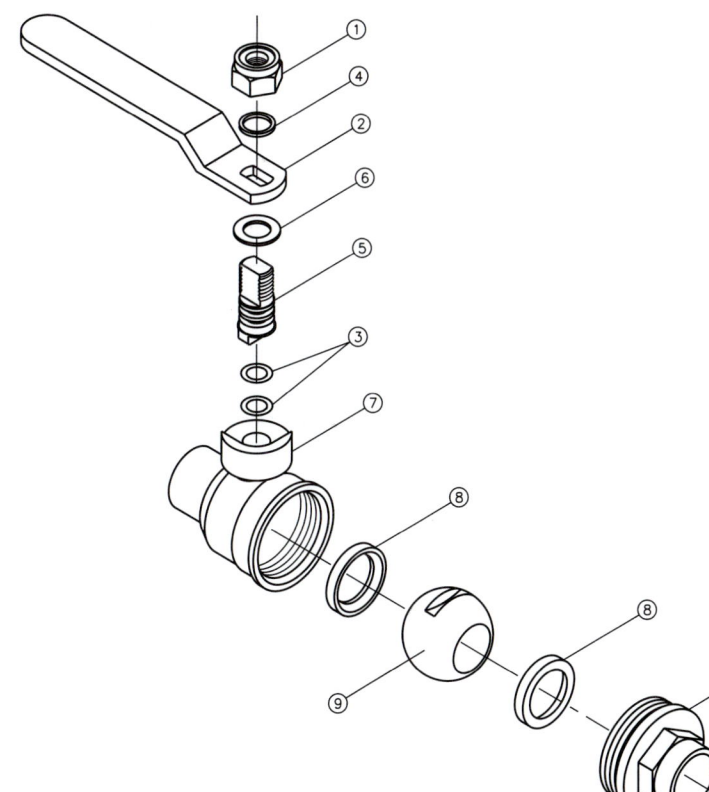

PARTS LIST				
KEY	QTY	NAME	MATERIAL	PART NO.
1	1	NUT	5/16-24UNF-2	5DT1001
2	1	HANDLE	STAINLESS STEEL	1DT1002
3	2	O-RING	PARKER #2-008	5DT1003
4	1	WASHER	PLASTIC	1DT1004
5	1	STEM	STAINLESS STEEL	1DT1005
6	1	WASHER	PLASTIC	1DT1006
7	1	VALVE BODY	BRASS	1DT1007
8	2	WASHER	PLASTIC	1DT1008
9	1	BALL	STAINLESS STEEL	1DT1009
10	1	END CAP	BRASS	1DT1010

PROBLEM 18.10   (Continued)

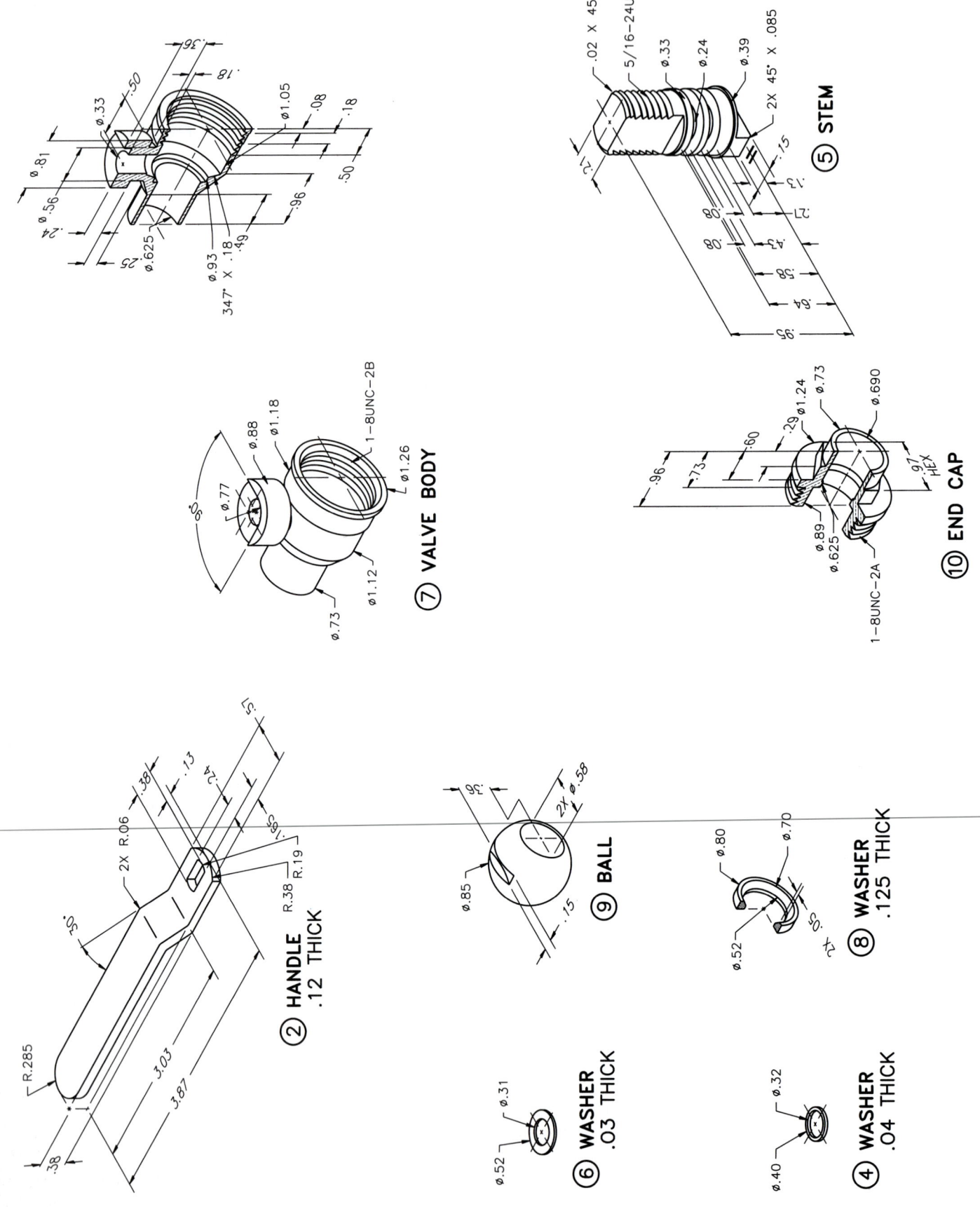

⑤ STEM

⑦ VALVE BODY

⑩ END CAP

② HANDLE
.12 THICK

⑨ BALL

⑧ WASHER
.125 THICK

⑥ WASHER
.03 THICK

④ WASHER
.04 THICK

PROBLEM 18.11 **Working drawing (metric)**

Assembly Name: Tool Holder

SPECIFIC INSTRUCTIONS:

Prepare a complete set of working drawings with one detail drawing per sheet and the assembly and parts list on another sheet. When preparing the assembly drawing, use separate balloons for each part in each view or use only one balloon in the view that most clearly identifies the part.

PARTS LIST

ITEM	QTY	NAME	MATERIAL
1	1	PARTING TOOL, 3/32 in. × 1/2 in. PURCHASE PART	TOOL STEEL
2	1	TOOL HOLDER BODY	06 STEEL
3	1	ADJUSTMENT SCREW	SAE 1035 STEEL
4	1	SHIM	SAE 4320 STEEL
5	1	KNURL NUT	SAE 3130 STEEL
6	1	WASHER	SAE 1060 STEEL
7	1	STUD	SAE 1035 STEEL
8	1	M 10 × 1.5 HEX NUT PURCHASE PART	

(4) SHIM
FAO 3.2μM

(7) STUD
FAO 3.2μM

(3) ADJUSTMENT SCREW
FAO 3.2μM

(5) KNURL NUT
FAO 3.2μM

(6) WASHER
FAO 3.2μM

(2) TOOL HOLDER BODY
FAO 3.2μM UNLESS OTHERWISE SPECIFIED

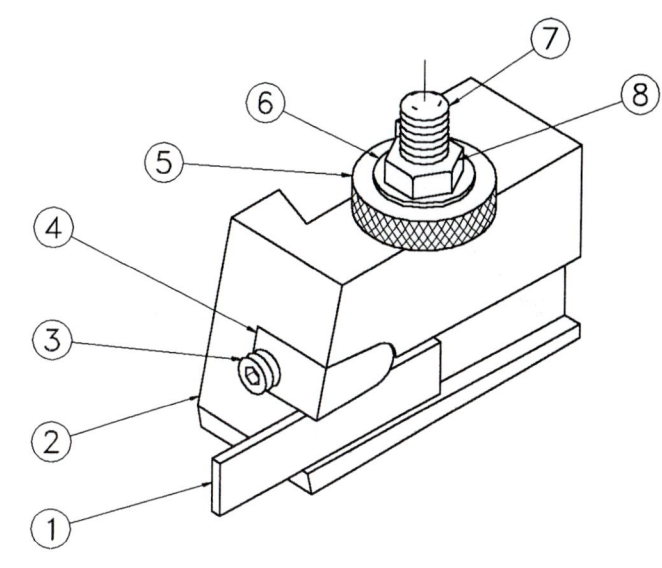

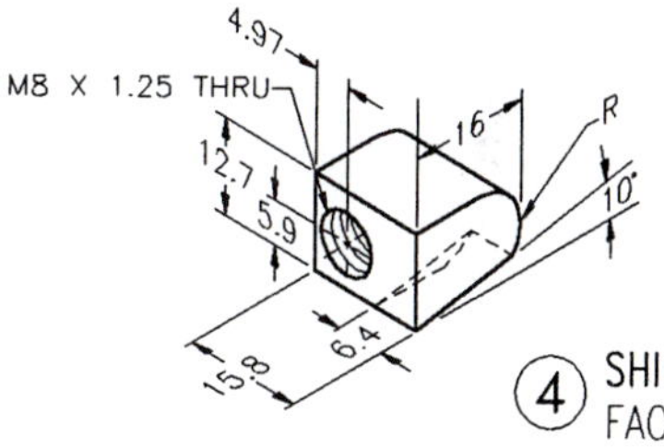

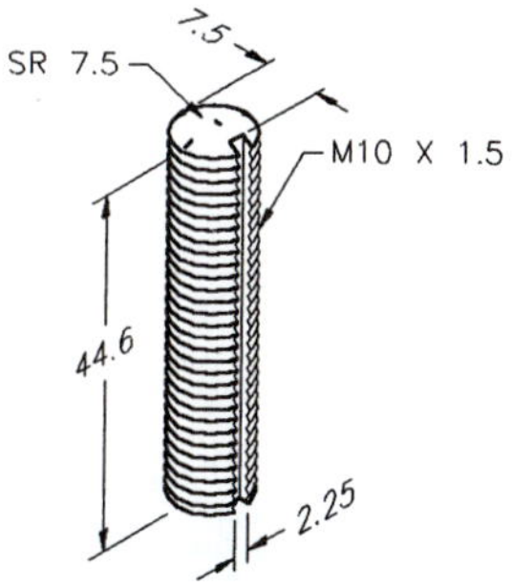

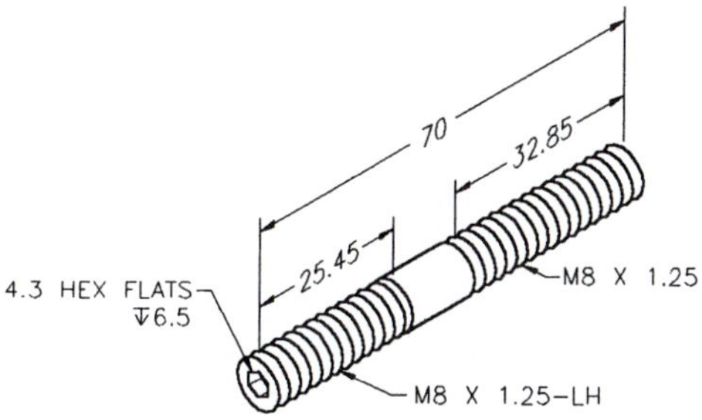

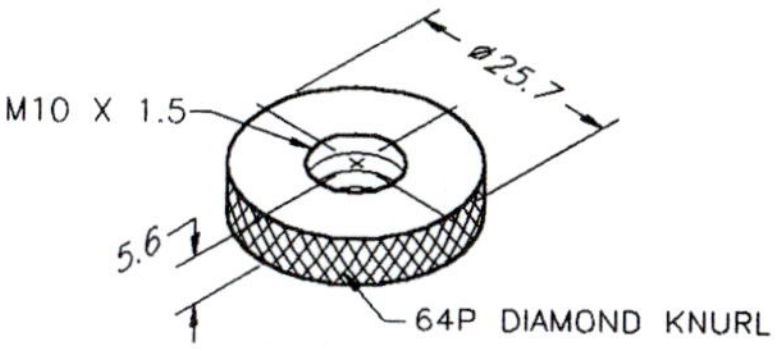

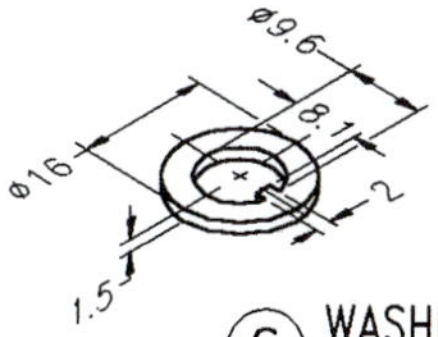

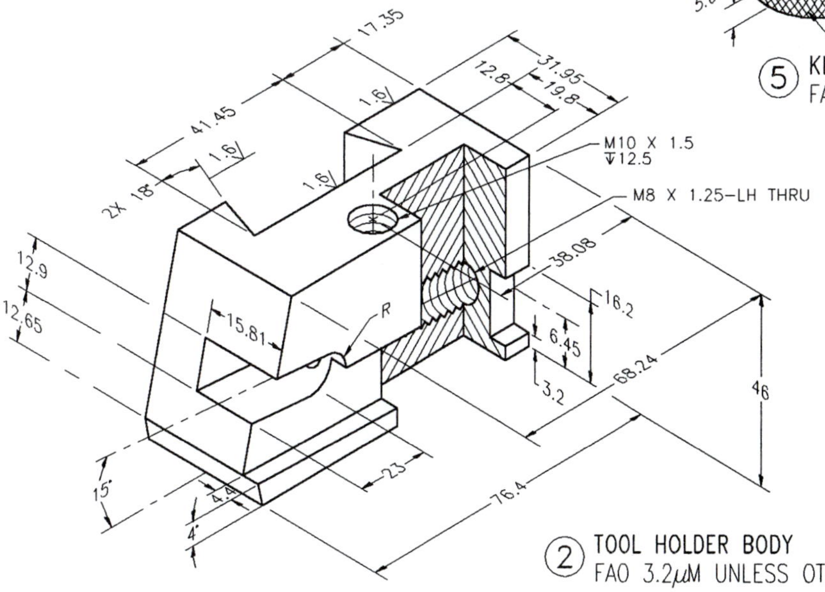

PROBLEM 18.12   Working drawing (in.)

Assembly Name: Adjustable Attachment

SPECIFIC INSTRUCTIONS:

Prepare a complete set of working drawings with detail drawings of individual parts combined on one or more sheets depending on the size of sheet selected. The assembly drawing and parts list will be combined on one sheet.

PARTS LIST

ITEM	QTY	NAME	DESCRIPTION	MATERIAL
1	1	CAP SCREW	.25-20UNC-2 × .75 HEX SOC HEAD	STL
2	1	FRAME		SAE 4340
3	1	ADJUSTING SCREW		SAE1045
4	1	SET SCREW	8-32UNC-2 HEX SOC FLAT POINT	STL
5	1	TAPER PIN	0 × .625	STL
6	1	KNURL KNOB		SAE 1024
7	1	CLAMP SCREW		SAE 2330
8	1	PILOT SCREW		SAE 2330
9	1	ADJUSTING NUT		SAE 3130

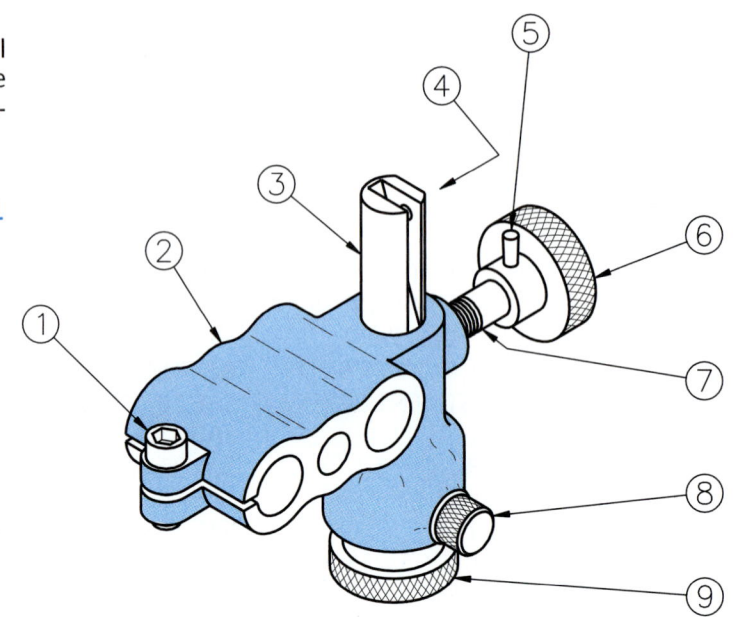

ADJUSTABLE ATTACHMENT

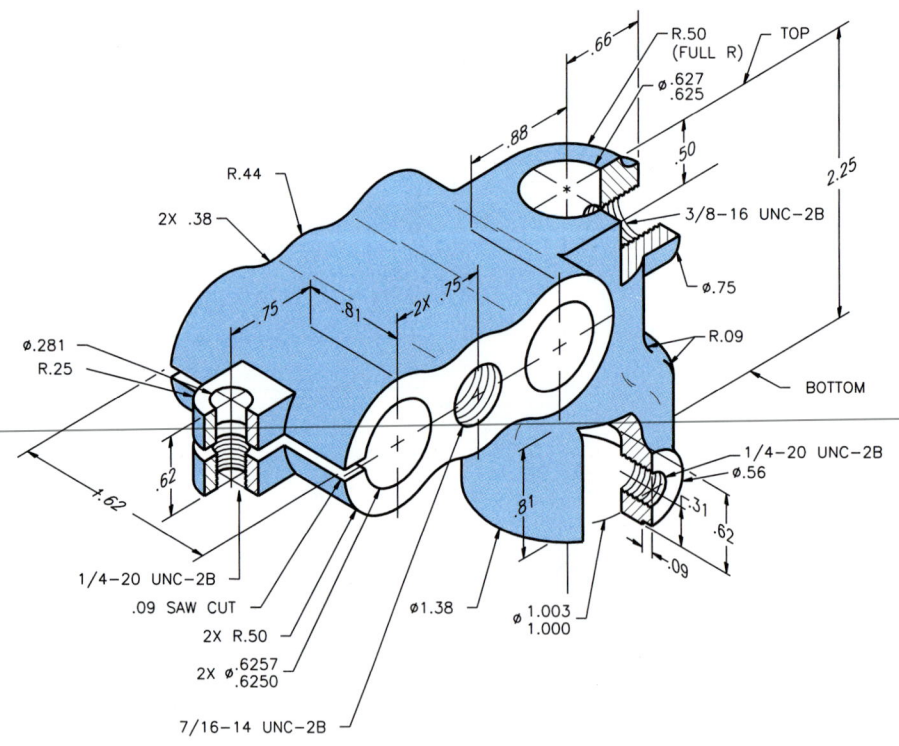

② FRAME

SAE 4340
REMOVE ALL BURRS & SHARP EDGES
FAO
FILLETS AND ROUNDS R.125

PROBLEM 18.12   (Continued)

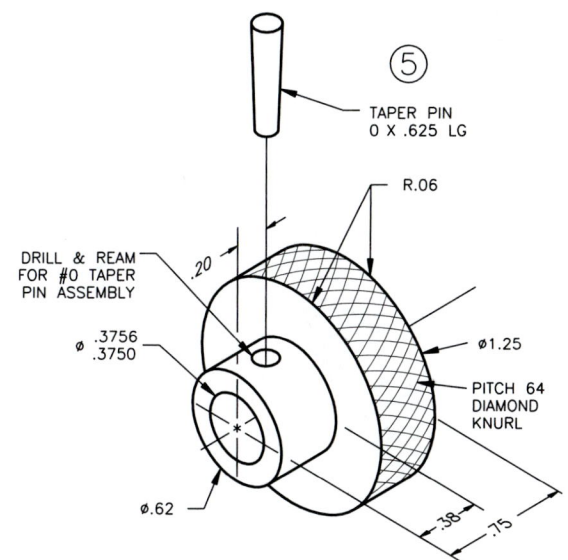

TAPER PIN
0 X .625 LG

DRILL & REAM
FOR #0 TAPER
PIN ASSEMBLY

ø .3756
.3750

ø.62

R.06

.20

ø1.25

PITCH 64
DIAMOND
KNURL

.38

.75

⑤

⑥   **KNURL KNOB**
SAE 1024
REMOVE ALL BURRS & SHARP EDGES
FAO

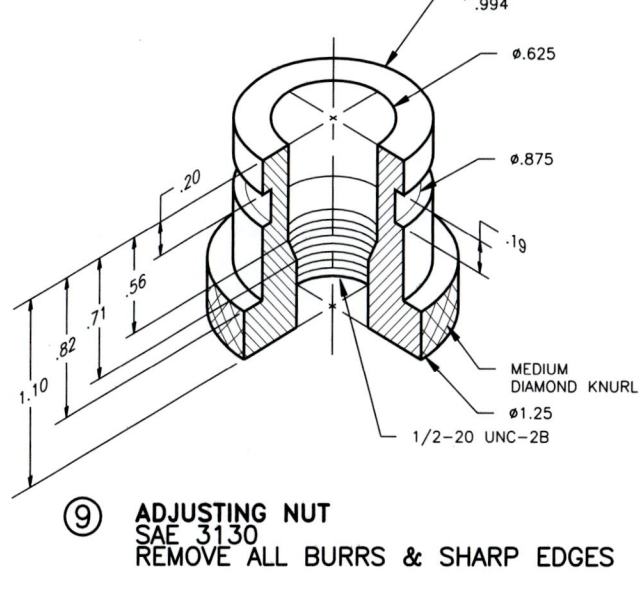

ø .997
.994

ø.625

ø.875

.19

.20

.56

.71

.82

1.10

MEDIUM
DIAMOND KNURL

ø1.25

1/2–20 UNC–2B

⑨   **ADJUSTING NUT**
SAE 3130
REMOVE ALL BURRS & SHARP EDGES

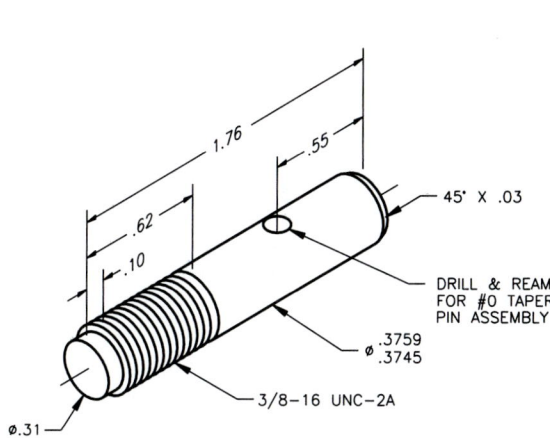

1.76

.55

.62

.10

45° X .03

DRILL & REAM
FOR #0 TAPER
PIN ASSEMBLY

.3759
ø .3745

ø.31

3/8–16 UNC–2A

⑦   **CLAMP SCREW**
SAE 2330
REMOVE ALL BURRS & SHARP EDGES

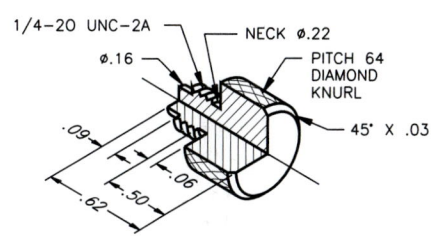

1/4–20 UNC–2A

ø.16

NECK ø.22

PITCH 64
DIAMOND
KNURL

45° X .03

.09

.06

.50

.62

⑧   **PILOT SCREW**
SAE 2330
REMOVE ALL BURRS & SHARP EDGES
FAO

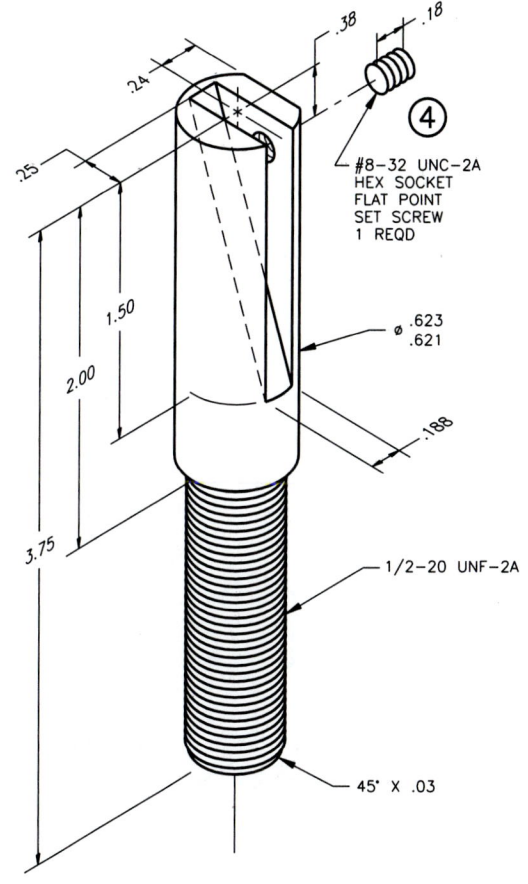

.38

.18

.24

.25

#8–32 UNC–2A
HEX SOCKET
FLAT POINT
SET SCREW
1 REQD

④

1.50

2.00

ø .623
.621

.188

3.75

1/2–20 UNF–2A

45° X .03

③   **ADJUSTING SCREW**
SAE 1045
REMOVE ALL BURRS & SHARP EDGES
FAO

PROBLEM 18.13 Working drawing (in.)

Assembly Name: Precision Vice

PARTS LIST

ITEM	QTY	NAME	DESCRIPTION	MATERIAL
1	1	BASE		SAE 1040
2	1	BALL WASHER		SAE 1040
3	1	NUT		SAE 1040
4	1	CAP SCREW	5/16-24NF × 1 1/2 HEX SOCKET HEAD	STL
5	1	JAW		SAE 1040
6	1	JAW INSERT		SAE 4330
7	2	MACH SCREW	1/4-28NF × 1/2 FLAT HD	STL
8	2	CAP SCREW	1/4-28NF × 3/4 HEX SOCKET HEAD	STL
9	1	JAW INSERT		SAE 4330

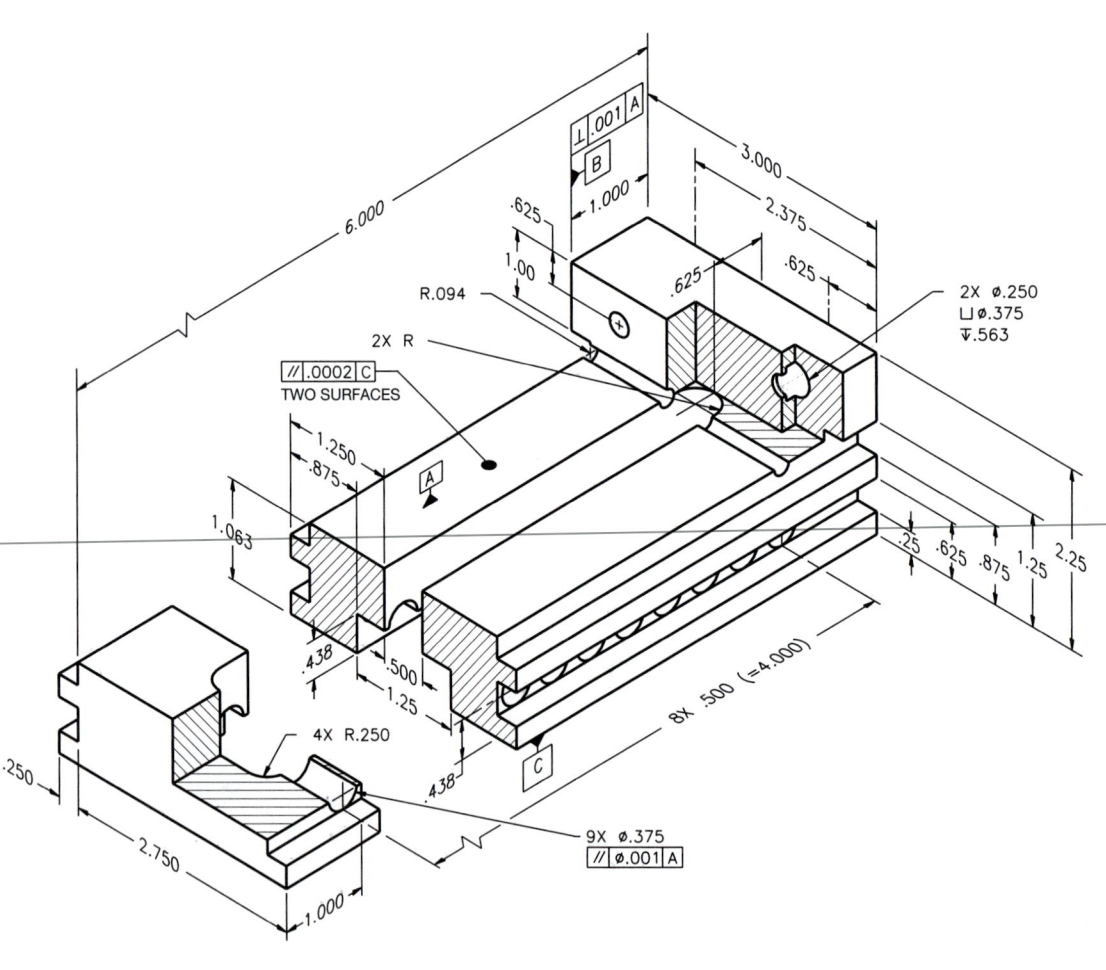

① **BASE**
FAO 32μIN

PROBLEM 18.13 (Continued)

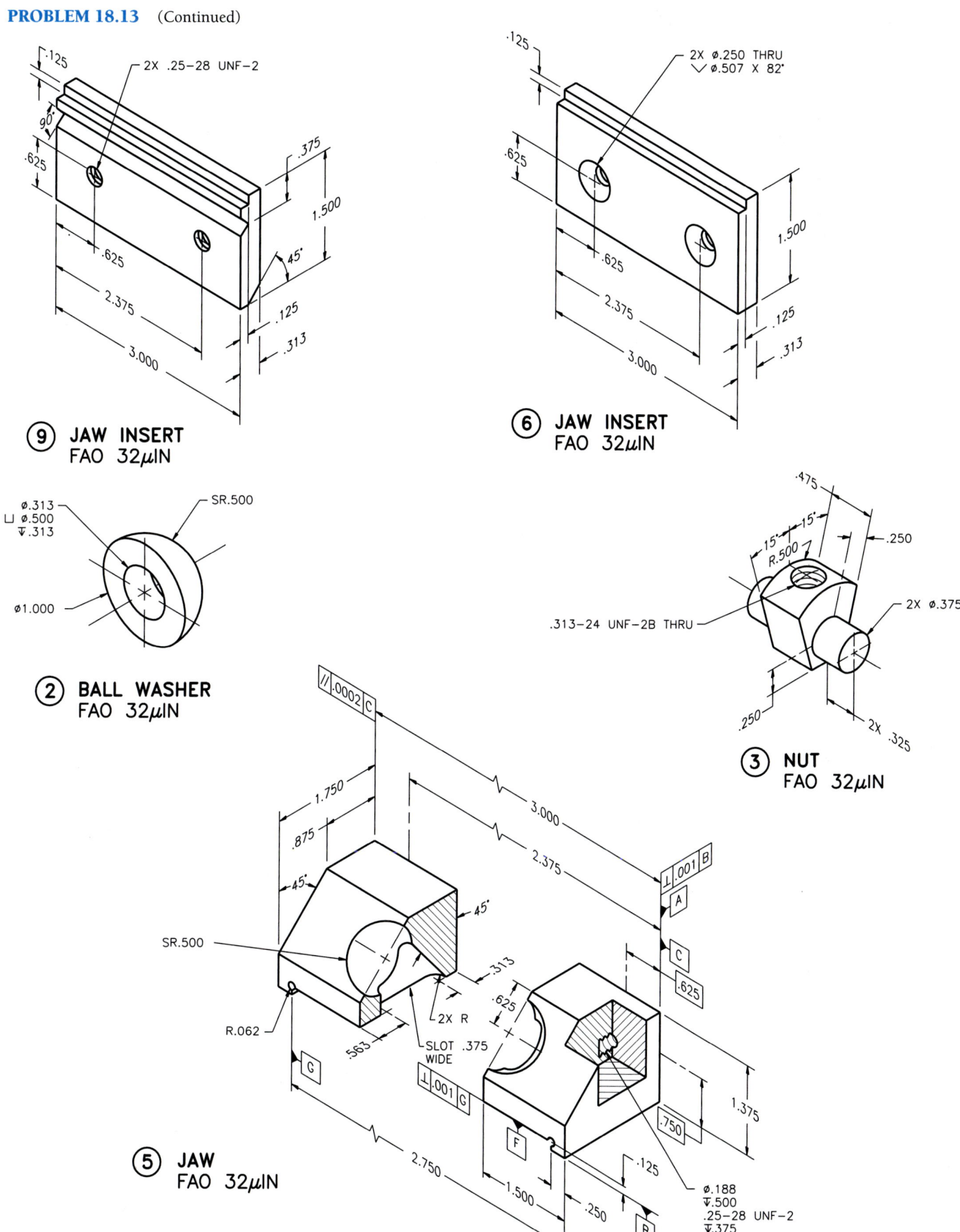

.125

2X .25−28 UNF−2

90°

.625

.375

1.500

.625

45°

2.375

.125

.313

3.000

⑨ **JAW INSERT**
FAO 32μIN

.125

2X ⌀.250 THRU
⌵ ⌀.507 X 82°

.625

1.500

.625

2.375

.125

.313

3.000

⑥ **JAW INSERT**
FAO 32μIN

⌀.313
⌴ ⌀.500
⌵.313

SR.500

⌀1.000

② **BALL WASHER**
FAO 32μIN

.475

15° 15°

R.500

.250

.313−24 UNF−2B THRU

2X ⌀.375

.250

2X .325

③ **NUT**
FAO 32μIN

∥ .0002 C

1.750

.875

45°

3.000

2.375

45°

SR.500

.313

.625

⊥ .001 B

A

C

.625

R.062

2X R

SLOT .375
WIDE

.563

G

⊥ .001 G

F

.750

1.375

⑤ **JAW**
FAO 32μIN

2.750

1.500

.250

.125

B

⌀.188
⌵.500
.25−28 UNF−2
⌵.375

⊕ ⌀.002 Ⓜ A B C

PROBLEM 18.14 Working drawing (in.)

Assembly Name: Machine Vice

SPECIFIC INSTRUCTIONS:

When preparing the assembly drawings, use separate bal-loons for each part in each view, or use only one balloon in the view that most clearly identifies the part.

PARTS LIST

ITEM	QTY	NAME	DESCRIPTION	MATERIAL
1	1	SCREW		SAE 4320
2	1	HANDLE		MS
3	2	CAP		MS
4	2	MACHINE SCREW	(10).190-32UNF-2 × .625 SLOT FIL HD	STL
5	1	SET SCREW	1/4-20UNC-2 × .250 FULL DOG POINT	STL
6	1	MOVABLE JAW		SAE 1020
7	1	MOVABLE JAW PLATE		SAE 4320
8	2	MACHINE SCREW	(10).190-32UNF-2 × .875 SLOT FIL HD	STL
9	1	FIXED JAW PLATE		SAE 4320
10	1	BODY		SAE 4320
11	1	GUIDE		SAE 1020
12	2	MACHINE SCREW	1/4-20UNC-2 × .500 SLOT FIL HD	STL

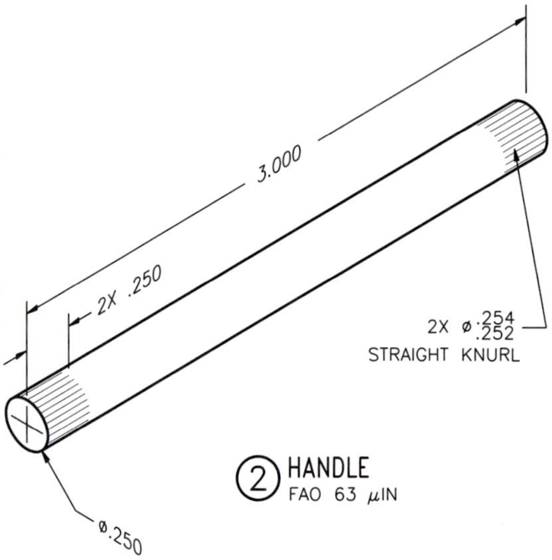

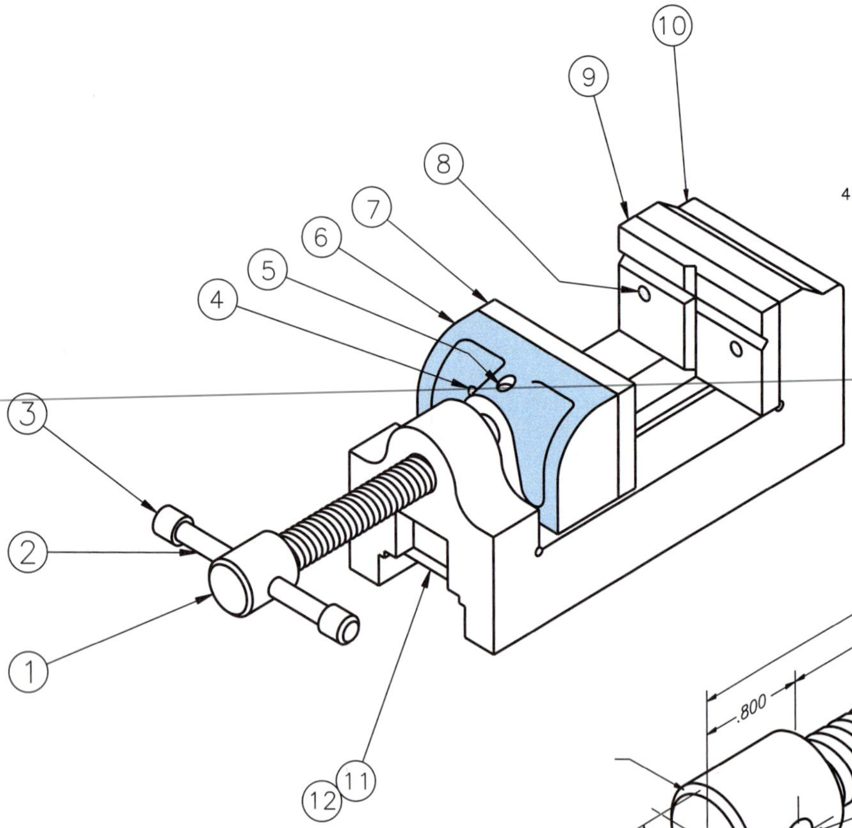

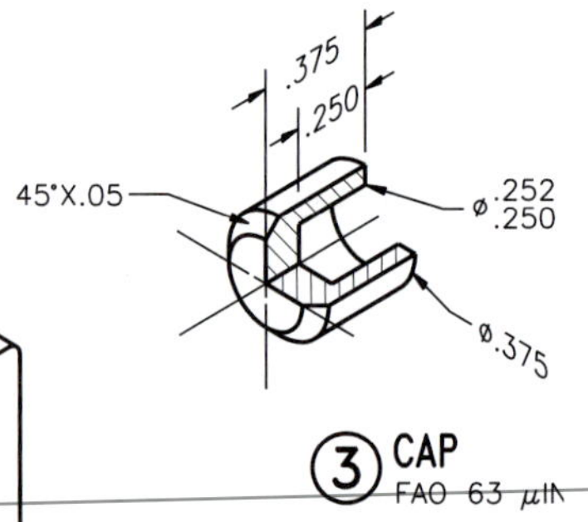

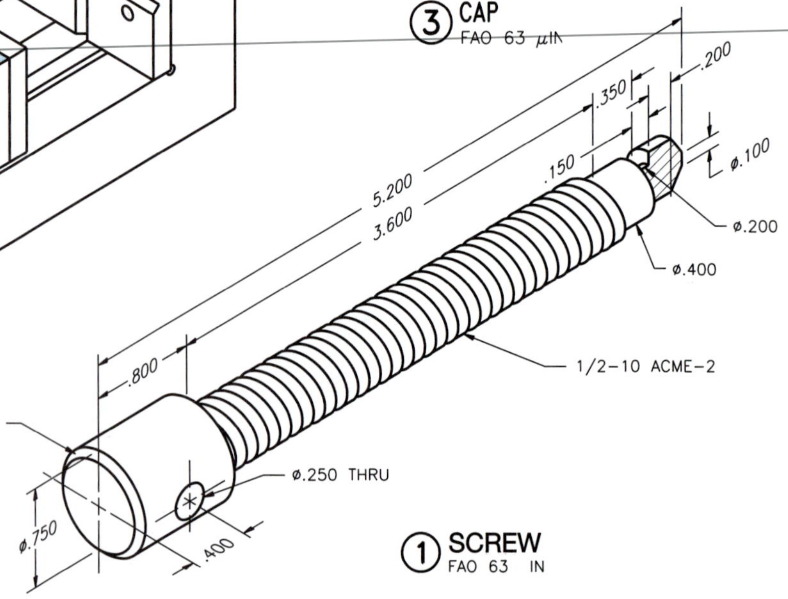

PROBLEM 18.14 (Continued)

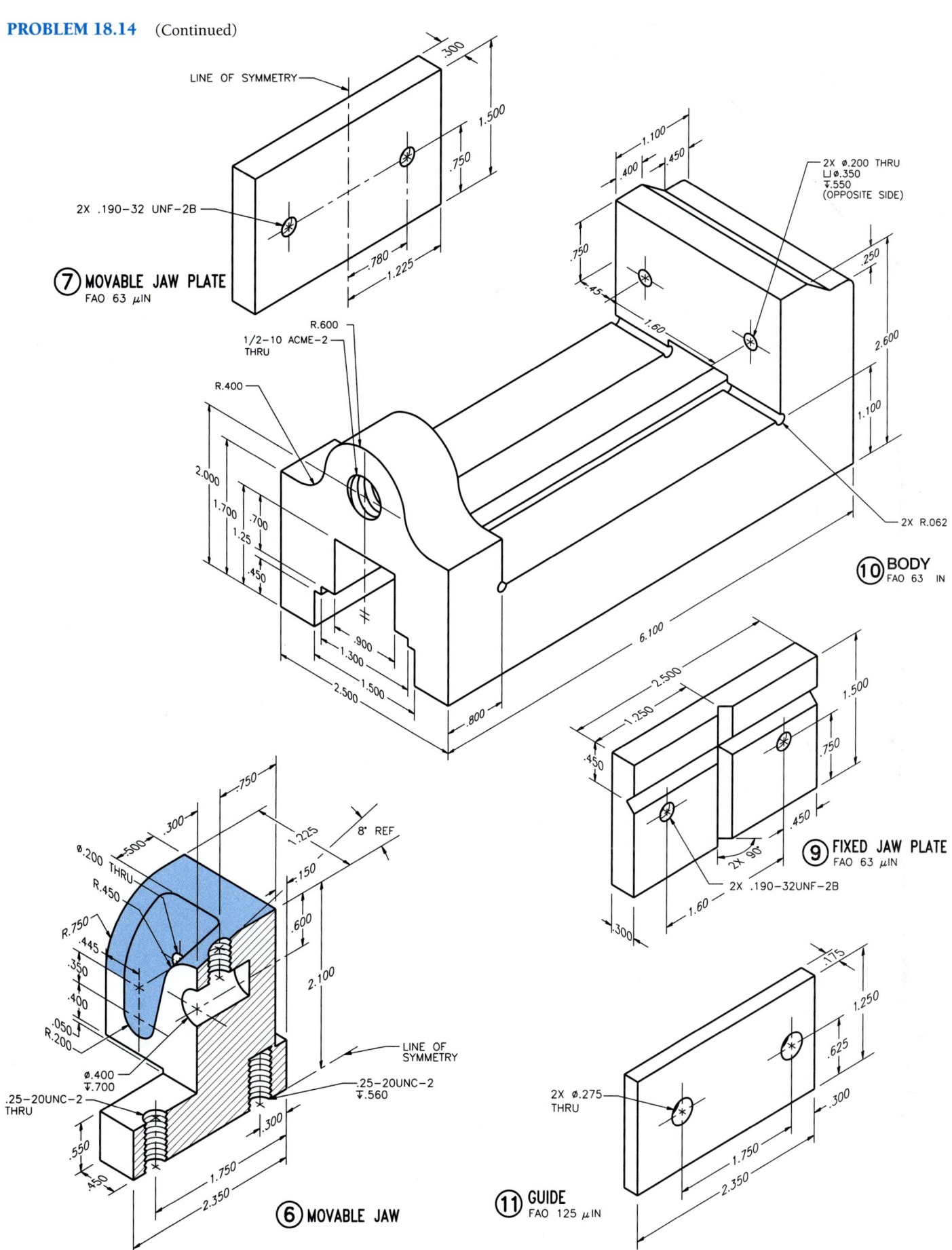

LINE OF SYMMETRY

.300

1.500

.750

2X .190–32 UNF–2B

.780

1.225

⑦ **MOVABLE JAW PLATE**
FAO 63 μIN

1.100

.400 .450

2X ø.200 THRU
⊔ø.350
↧.550
(OPPOSITE SIDE)

.750

.45

.250

2.600

1.60

1.100

2X R.062

⑩ **BODY**
FAO 63 IN

R.600

1/2–10 ACME–2
THRU

R.400

2.000

1.700 .700

1.25

.450

.900

1.300

1.500

2.500

.800

6.100

2.500

1.250

.450

1.500

.750

.450

2X 90

1.60

.300

⑨ **FIXED JAW PLATE**
FAO 63 μIN

2X .190–32UNF–2B

.750

.300

.500

1.225

8° REF

.150

.600

2.100

ø.200 THRU

R.450

R.750

.445

.350

.400

.050

R.200

ø.400
↧.700

.25–20UNC–2
THRU

.550

.450

1.750

2.350

LINE OF
SYMMETRY

.25–20UNC–2
↧.560

.300

⑥ **MOVABLE JAW**

.175

1.250

.625

.300

2X ø.275
THRU

1.750

2.350

⑪ **GUIDE**
FAO 125 μIN

PROBLEM 18.15 Working drawing (in.)

Assembly Name: Arbor Press

SPECIFIC INSTRUCTIONS:

This problem is generally basic, but also contains challenging applications that include gears. You should study gears and related drafting practices covered in Chapter 19 before completing this problem.

PARTS LIST

ITEM	QTY	NAME	DESCRIPTION	MATERIAL
1	1	BASE		SAE 1020
2	1	TABLE PIN		SAE 1020
3	1	TABLE		SAE 1020
4	2	BALL END		SAE 1020
5	1	SLEEVE		SAE 1020
6	1	HANDLE		SAE 1020
7	1	GEAR		SAE 4320
8	1	RACK PAD		SAE 4320

ITEM	QTY	NAME	DESCRIPTION	MATERIAL
9	1	COVER PLATE		SAE 1020
10	4	CAP SCREWS	8-32UNC-2 × .50 HEX SOC	STL
11	1	RACK		SAE 4320
12	1	SCREW		SAE 1040
13	1	COLUMN		SAE 1020
14	1	CAP SCREW	3/8-16UNC-2 × 1.00 HEX SOCKET HEAD	STL

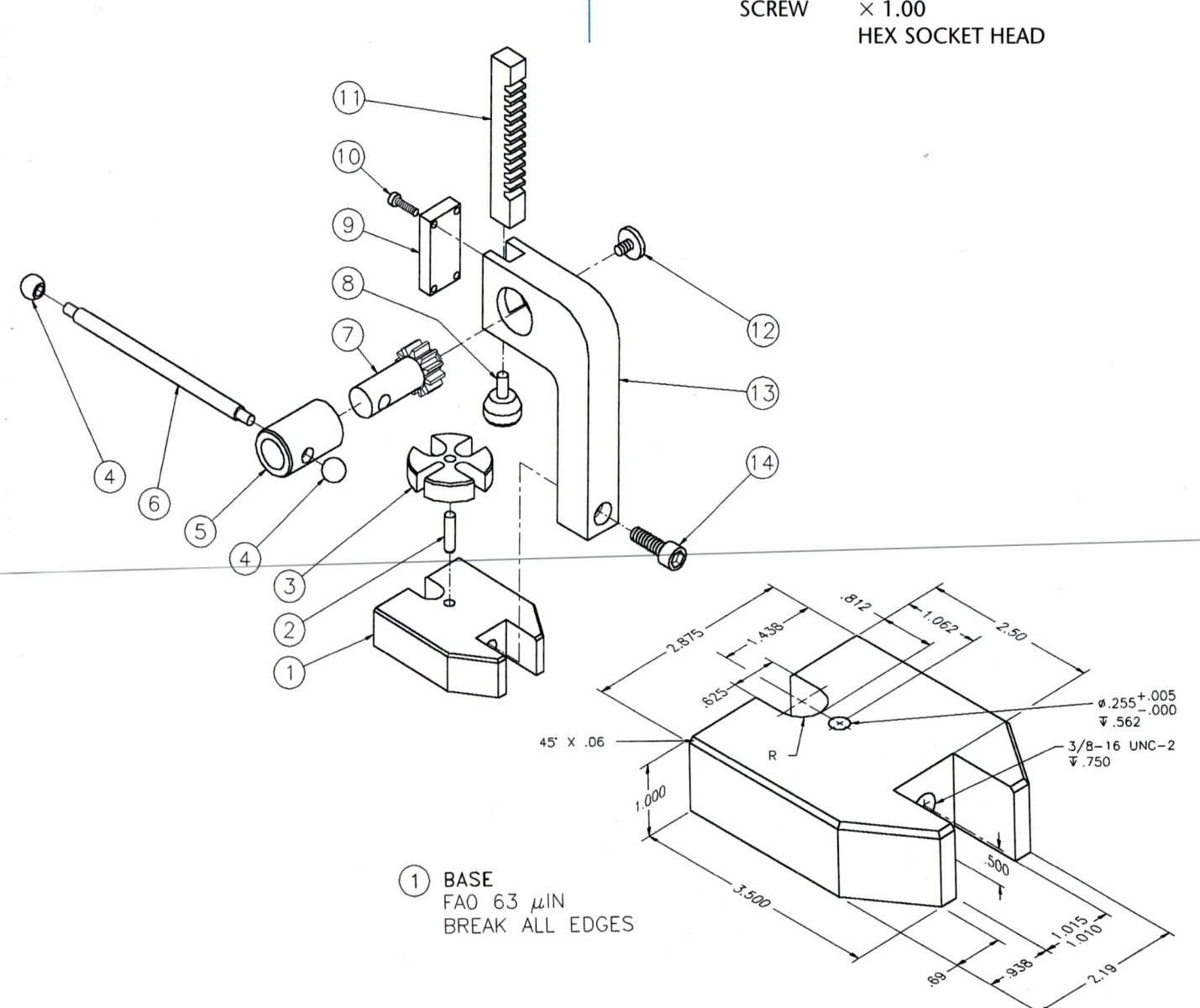

1 BASE
FAO 63 μIN
BREAK ALL EDGES

PROBLEM 18.15 (Continued)

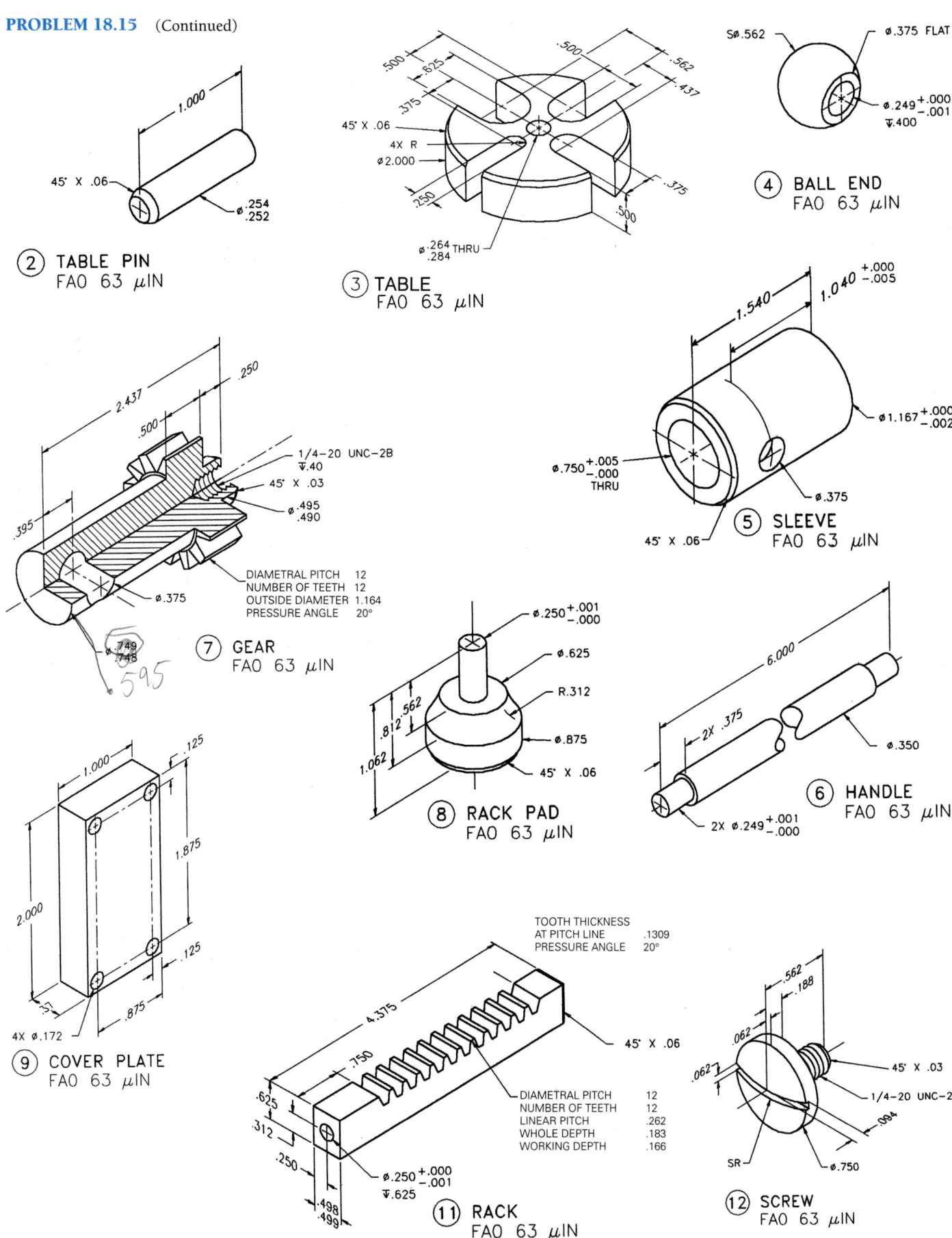

45° X .06

.254
.252

② TABLE PIN
FAO 63 μIN

.500
.625
.375
.500
.562
.437

45° X .06
4X R
ø2.000
.250
.375
.500

ø.264 THRU
.284

③ TABLE
FAO 63 μIN

Sø.562
ø.375 FLAT

ø.249 +.000 −.001
▽.400

④ BALL END
FAO 63 μIN

2.437
.250
.500

1/4−20 UNC−2B
▽.40
45° X .03
ø.495 .490

.395

ø.375

.749 .748
595

DIAMETRAL PITCH 12
NUMBER OF TEETH 12
OUTSIDE DIAMETER 1.164
PRESSURE ANGLE 20°

⑦ GEAR
FAO 63 μIN

1.540
1.040 +.000 −.005

ø.750 +.005 −.000
THRU

ø1.167 +.000 −.002

ø.375

45° X .06

⑤ SLEEVE
FAO 63 μIN

ø.250 +.001 −.000
ø.625
R.312
ø.875
45° X .06

.812 .562
1.062

⑧ RACK PAD
FAO 63 μIN

6.000

2X .375
ø.350

2X ø.249 +.001 −.000

⑥ HANDLE
FAO 63 μIN

1.000
.125

1.875

2.000

.125

.875

4X ø.172

⑨ COVER PLATE
FAO 63 μIN

TOOTH THICKNESS
AT PITCH LINE .1309
PRESSURE ANGLE 20°

4.375
.750
45° X .06

.625
.312
.250

DIAMETRAL PITCH 12
NUMBER OF TEETH 12
LINEAR PITCH .262
WHOLE DEPTH .183
WORKING DEPTH .166

ø.250 +.000 −.001
▽.625

.498 .499

⑪ RACK
FAO 63 μIN

.562
.188
.062
.062

45° X .03
1/4−20 UNC−2
.094

SR
ø.750

⑫ SCREW
FAO 63 μIN

PROBLEM 18.15 (Continued)

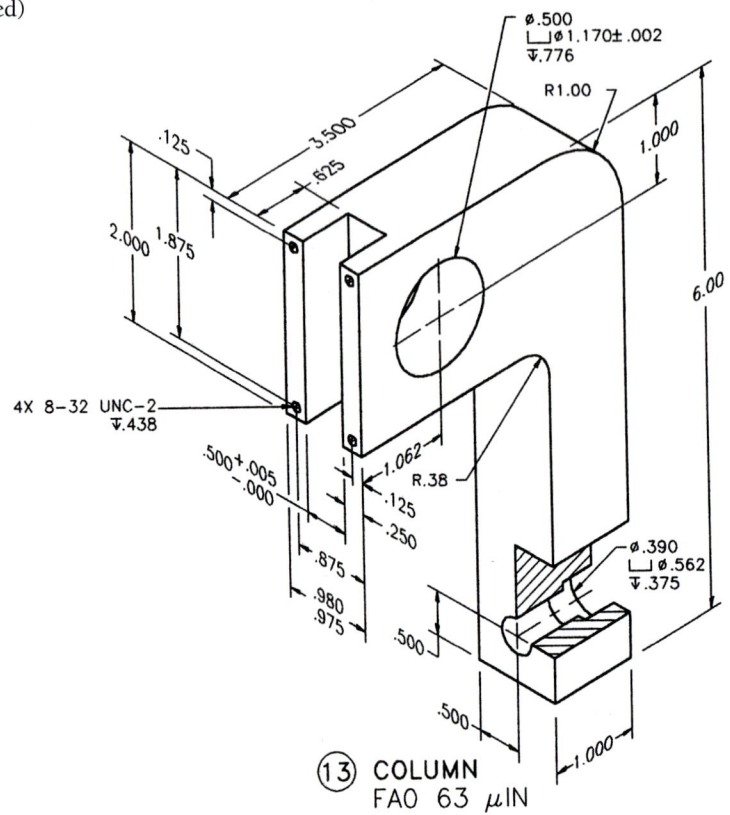

⑬ COLUMN
FAO 63 μIN

PROBLEM 18.16 Working drawing (in.)

Assembly Name: Pen Light

SPECIFIC INSTRUCTIONS:

Prepare a complete set of working drawings with one detail drawing per sheet and the assembly and parts list on an- other sheet. When preparing the assembly drawing, use separate balloons for each part in each view or use only one balloon in the view that most clearly identifies the part. This is an engineering prototype project. Errors are likely. Verify dimensions during assembly. Establish tolerances between mating parts. *Problem courtesy of Kim P. Lockwood.*

PARTS LIST				
KEY	**QTY**	**NAME**	**DESCRIPTION**	**PART NO.**
1	1	MAGNET	MAGNET	5DT1001
2	1	BODY	POLYCARBONATE	1DT1002
3	1	PARKER O-RING	PARKER #2-014	5DT1003
4	1	PRONG	PHOSPHOR BRONZE	1DT1004
5	1	SPRING	PHOSPHOR BRONZE	1DT1005
6	1	WASHER	PHOSPHOR BRONZE	1DT1006
7	1	PLUG	PHOSPHOR BRONZE	1DT1007
8	1	CATALYST	DE OXO TYPE D	5DT1008
9	1	BULB CASING	POLYCARBONATE	1DT1009
10	1	LAMP MODULE	VACUUM METALLIZED	1DT1010
11	1	LAMP	HIGH INTENSITY XENON	5DT1011
12	1	LAMP LENS	POLYCARBONATE	1DT1012
13	1	FIBER OPTIC HOLDER	(PVF) POLYVINYLIDENE FLOURIDE	1DT1013
14	1	FIBER OPTIC	FIBER-OPTIC 3-1/2" × .12"	5DT1014

PROBLEM 18.16 (Continued)

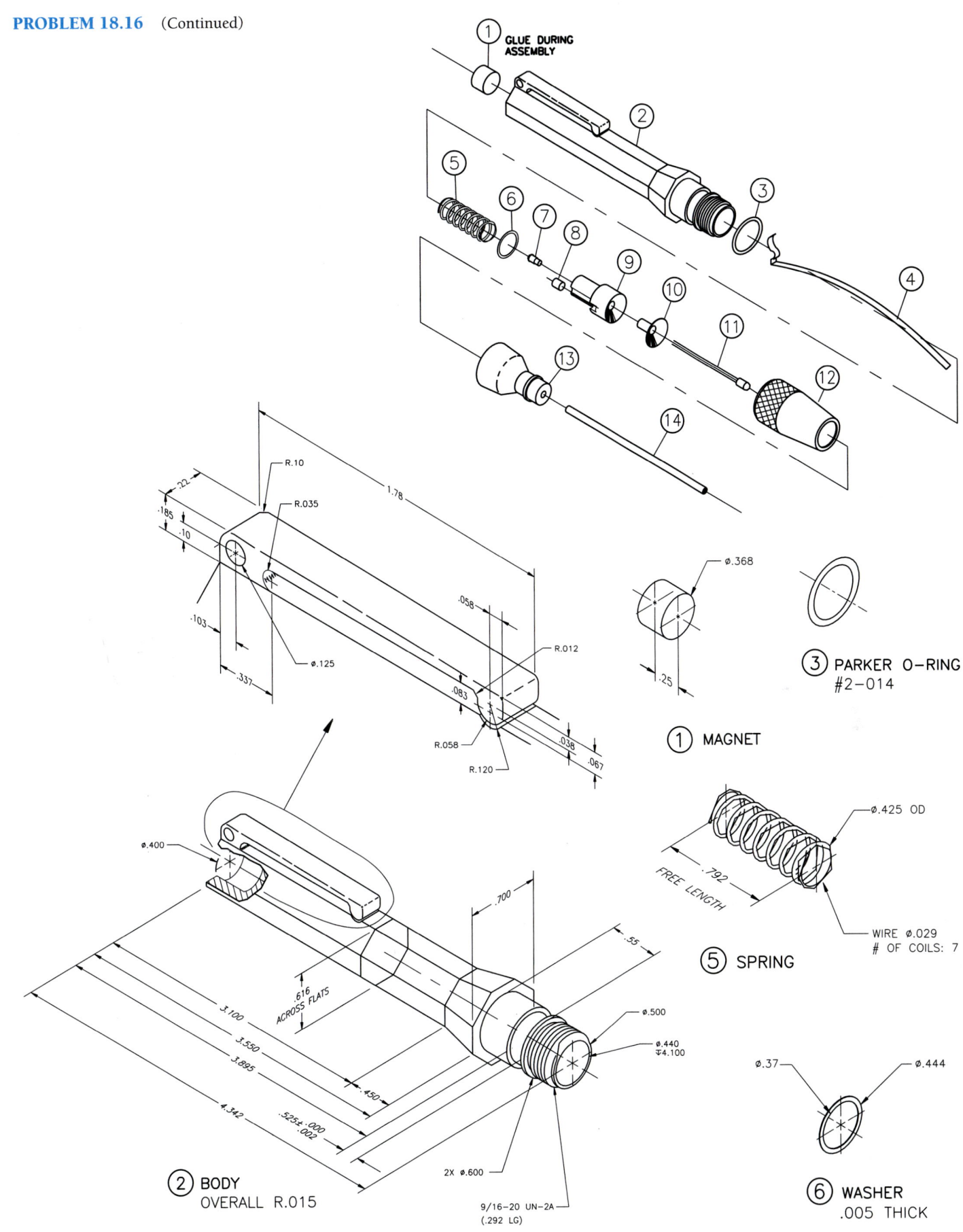

① GLUE DURING ASSEMBLY

③ PARKER O-RING #2-014

① MAGNET
ø.368
.25

R.10
R.035
1.78
.058
R.012
R.058
R.120
.083
.038
.067
ø.125
.337
.103
.10
.185
.22

⑤ SPRING
ø.425 OD
FREE LENGTH .792
WIRE ø.029
OF COILS: 7

② BODY
OVERALL R.015
ø.400
.700
.55
.616 ACROSS FLATS
3.100
3.550
3.895
4.342
.450
.525±.000 .002
2X ø.600
9/16-20 UN-2A
(.292 LG)
ø.500
ø.440 ⌴4.100

⑥ WASHER
.005 THICK
ø.37
ø.444

PROBLEM 18.16 (Continued)

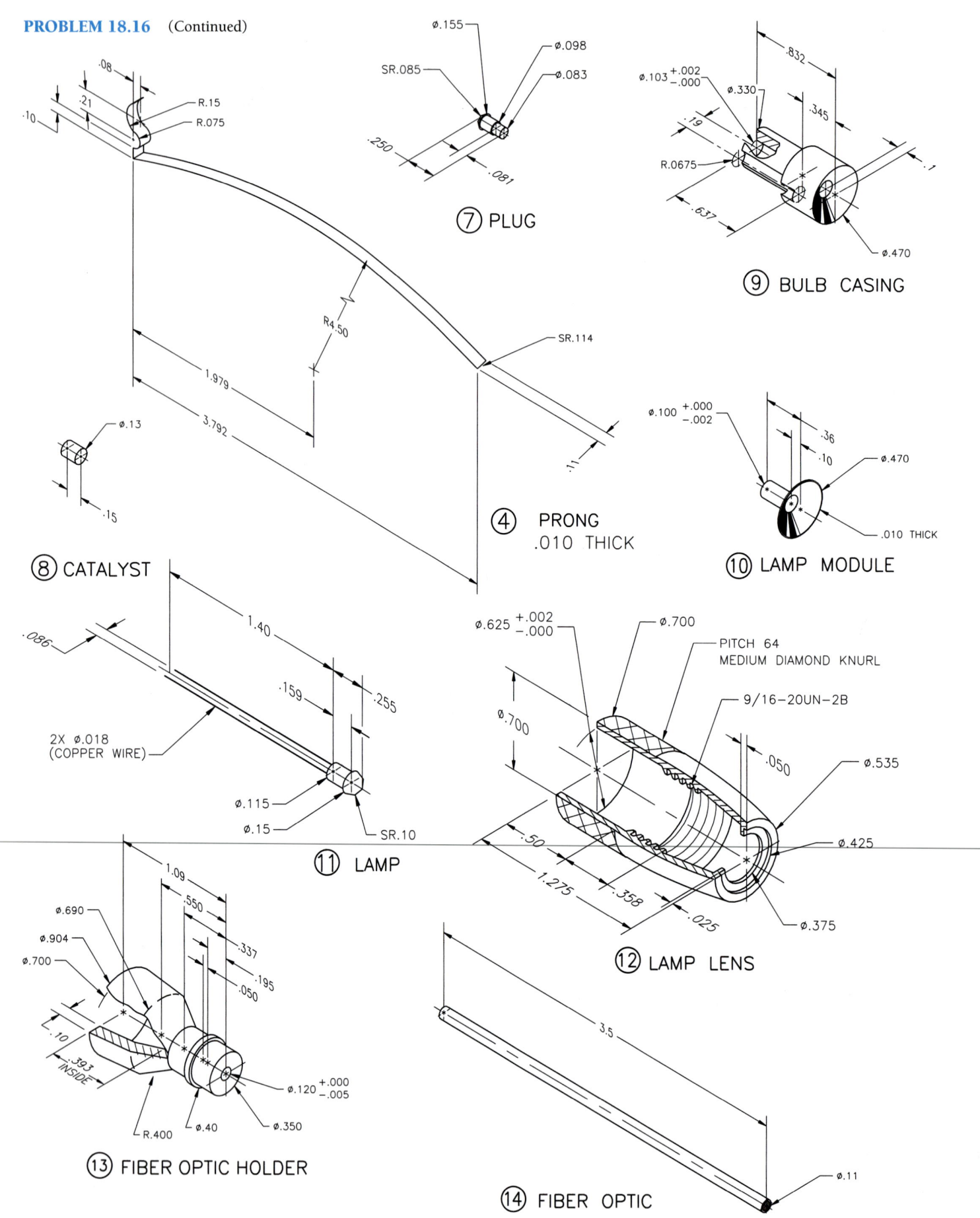

⑦ PLUG

⑨ BULB CASING

④ PRONG
.010 THICK

⑧ CATALYST

⑩ LAMP MODULE

⑪ LAMP

⑫ LAMP LENS

⑬ FIBER OPTIC HOLDER

⑭ FIBER OPTIC

PROBLEM 18.17 **Working drawing (in.)**

Assembly Name: Fluorescent Light Fixture

SPECIFIC INSTRUCTIONS:

Prepare a complete set of working drawings with one detail drawing per sheet and the assembly and parts list on another sheet. When preparing the assembly drawing, use separate balloons for each part in each view or use only one balloon in the view that most clearly identifies the part. This is an engineering prototype project. Errors are likely. Verify dimensions during assembly. Establish tolerances between mating parts. There are several purchase parts in this assembly. Manufacturer research is needed to locate available products. *Problem courtesy of John Melloy.*

ITEM	QTY	NAME	DESCRIPTION
1	1	RACK	
2	1	BASE	
3	1	COVER	
4	2	END MOUNT	
5	1	SWITCH MOUNT	
6	1	OUTLET	SATCO OUTLET 3 WIRE
7	1	SWITCH	RACO PUSH BUTTON
8	2	FLOURESCENT END	EAGLE DOUBLE PIN TUBE
9	1	LIGHTING/VALMONT	89G457 BALLAS 1-20 WATT
10	6	MACHINE SCREW	⌀0.125-24UNC-2A
11	2	HEX NUT	⌀0.125-24UNC-2B

PARTS LIST

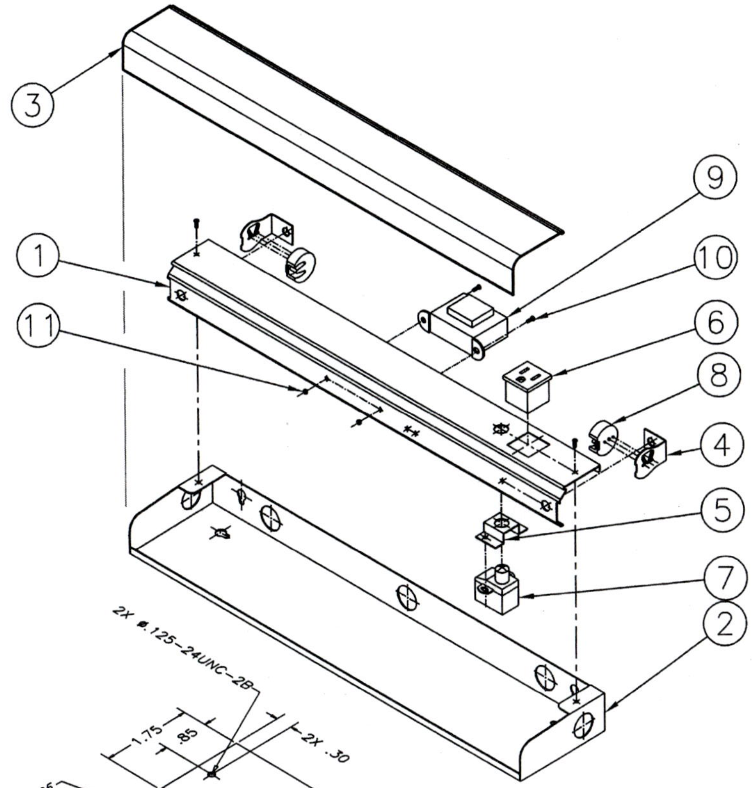

NOTES:
1. ITEM 1, THE RACK HAS TWO RIVETED PIECES ON EACH INTERIOR SIDE TO CONNECT ITEM 4, THE END MOUNTS IN THE ⌀0.40 HOLE.

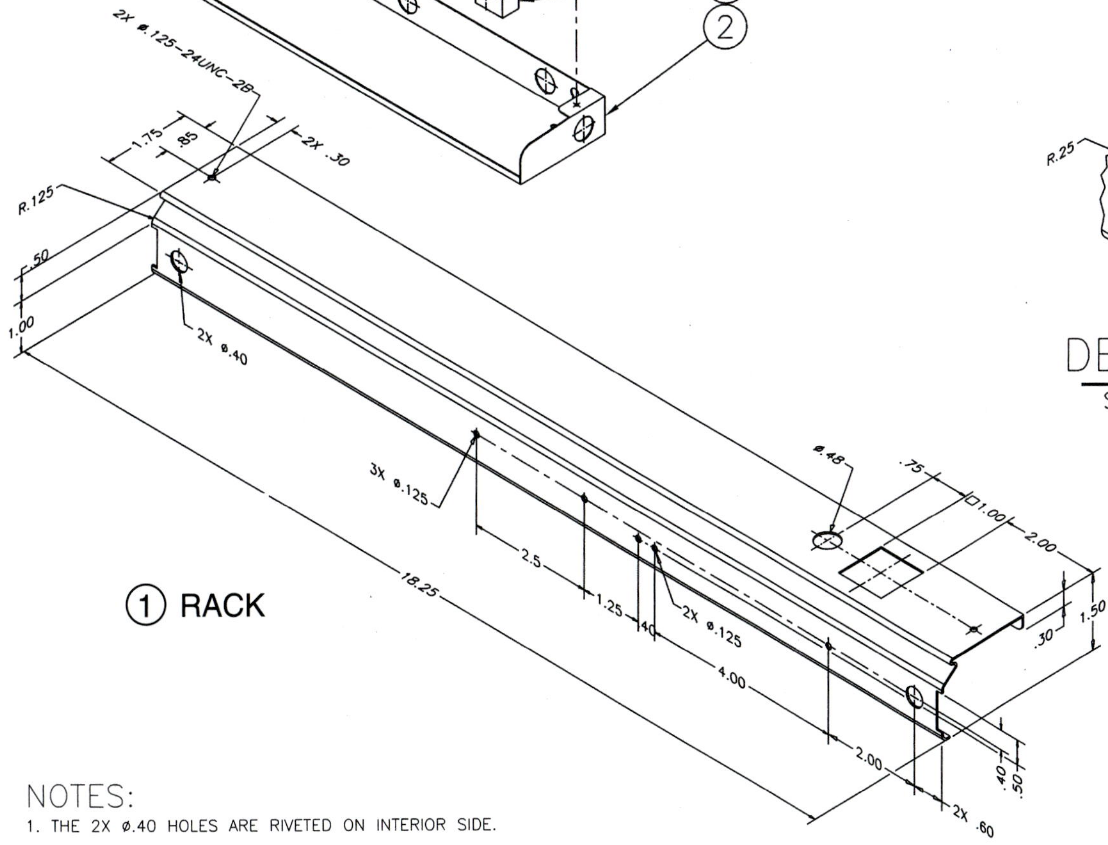

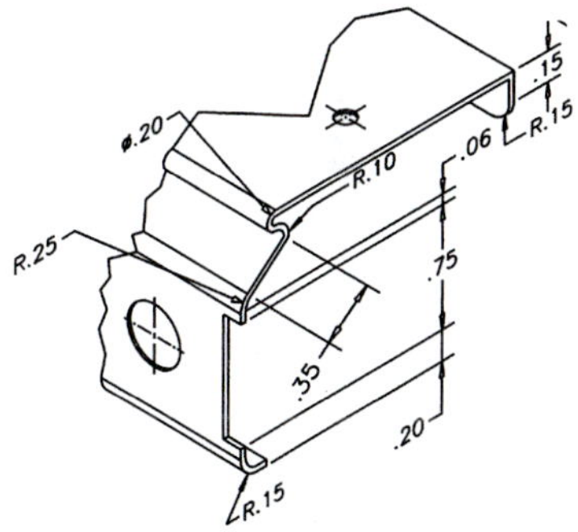

DETAIL A
SCALE: 2:1

① RACK

NOTES:
1. THE 2X ⌀0.40 HOLES ARE RIVETED ON INTERIOR SIDE.

PROBLEM 18.17   (Continued)

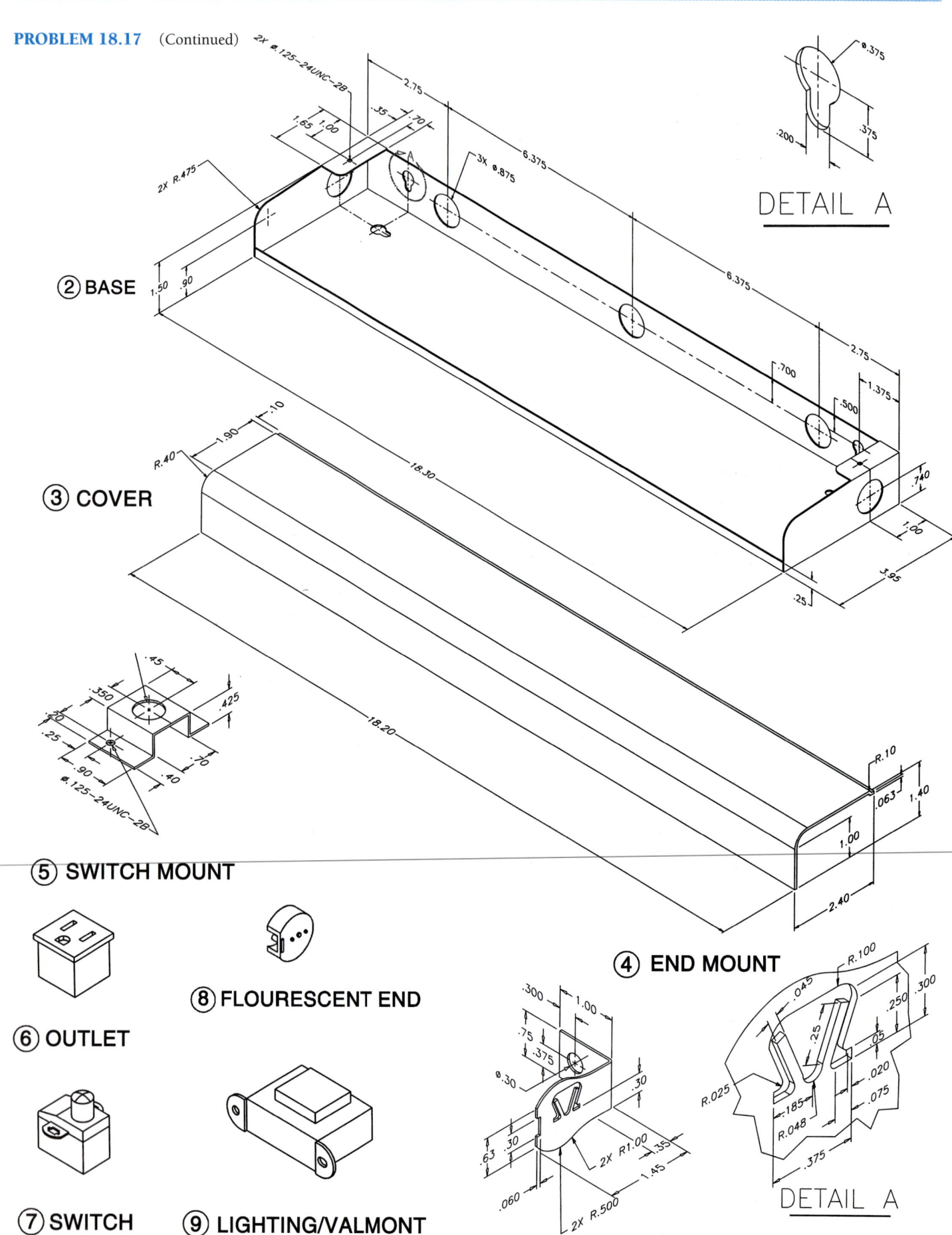

DETAIL A

② BASE

③ COVER

⑤ SWITCH MOUNT

④ END MOUNT

⑥ OUTLET

⑧ FLOURESCENT END

⑦ SWITCH   ⑨ LIGHTING/VALMONT

DETAIL A

PROBLEM 18.18 Working drawing (in.)

Assembly Name: Oil Pump

SPECIFIC INSTRUCTIONS:

This problem is generally basic, but it also contains challenging applications that include gears and seals. You should study gears and related drafting practices covered in Chapter 19 before completing this problem.

Prepare a complete set of working drawings with one detail drawing per sheet and the assembly and parts list on another sheet. When preparing the assembly drawing, use separate balloons for each part in each view or use only one balloon in the view that most clearly identifies the part. This is an engineering prototype project. Errors are likely. Verify dimensions during assembly. Establish tolerances between mating parts. There are several purchase parts in this assembly. Manufacturer research is needed to locate available products. *Problem courtesy of Richard Hertel.*

PARTS LIST

ITEM	DESCRIPTION	QTY
1	PUMP COVER	1
2	1/4 - 20 × 3/4" BUTTON HEAD BOLT	2
3	END COVER	1
4	DRIVE SHAFT	1
5	OIL SEAL	1
6	DRIVE GEAR (RETURN)	1
7	IDLER GEAR (RETURN)	1
8	OIL PUMP BODY	1
9	PLUNGER VALVE	1
10	PLUNGER SPRING	1
11	END CAP	2
12	PSI CONTROL SPRING	1
13	CHECK BALL	1
14	1/4 - 20 × 2" HEX HEAD BOLT	2
15	IDLER GEAR (FEED)	1
16	DRIVE GEAR (FEED)	1
17	1/4 - 20 × 2 1/2" HEX HEAD BOLT	4

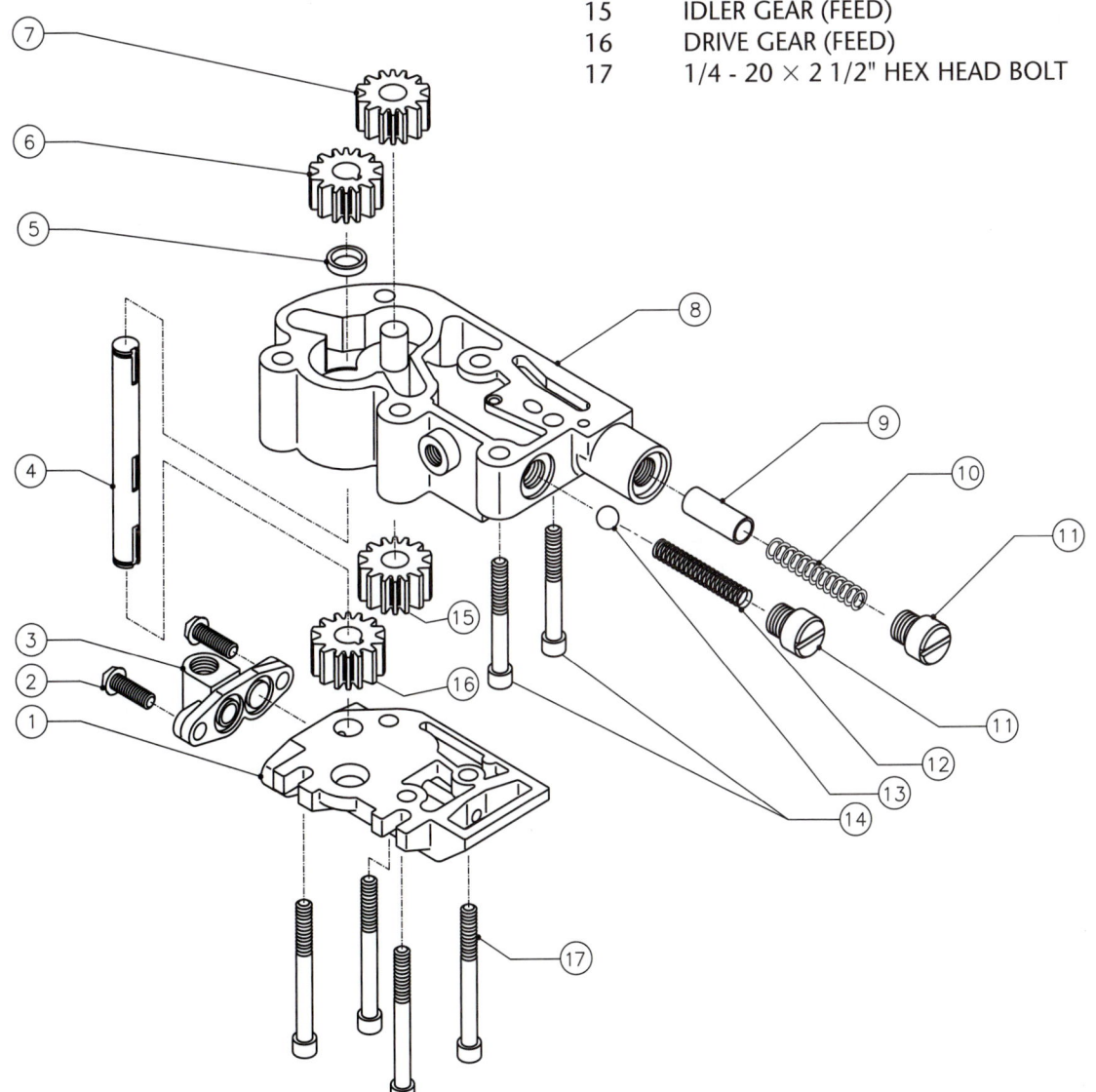

PROBLEM 18.18   (Continued)

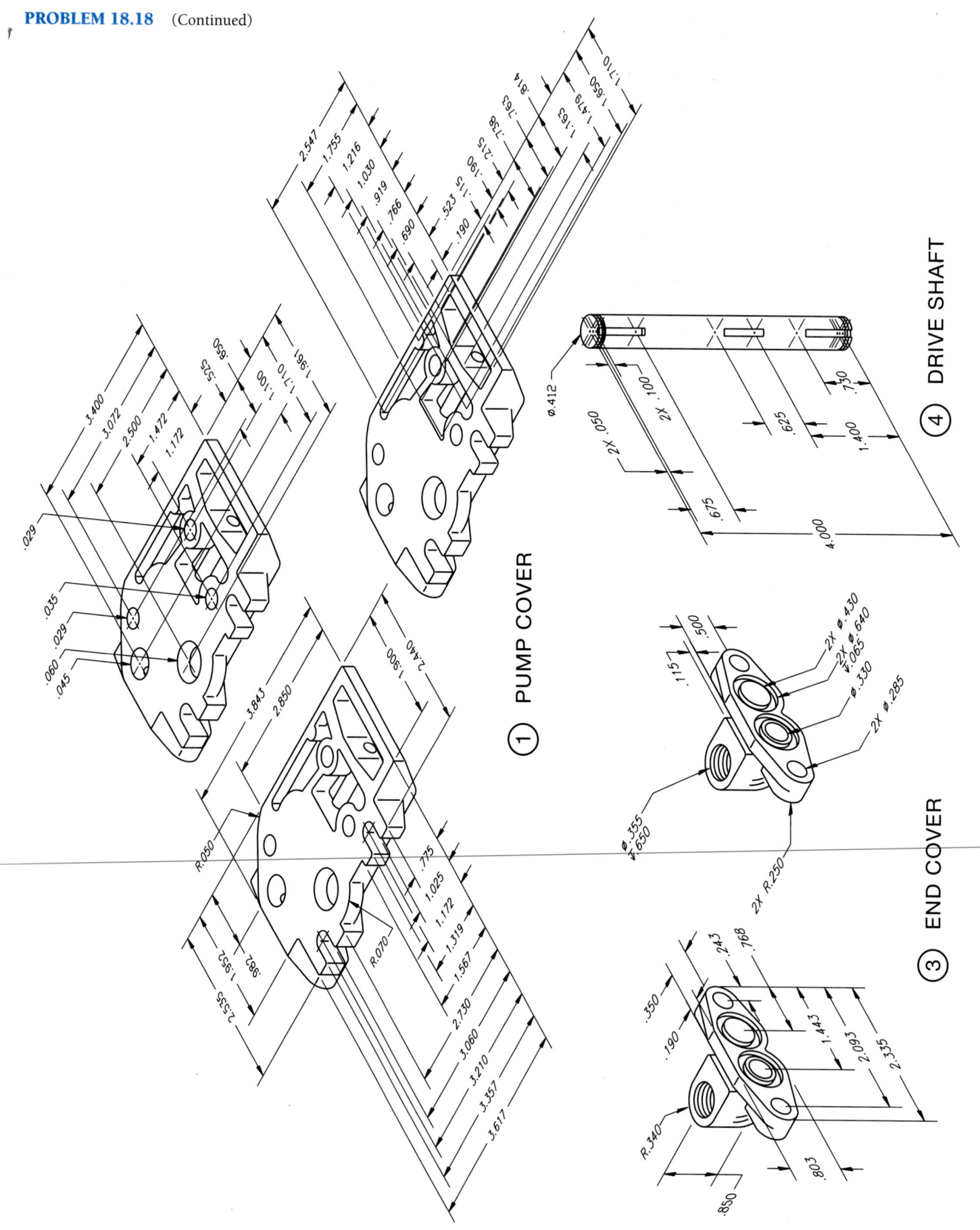

① PUMP COVER

③ END COVER

④ DRIVE SHAFT

PROBLEM 18.18 (Continued)

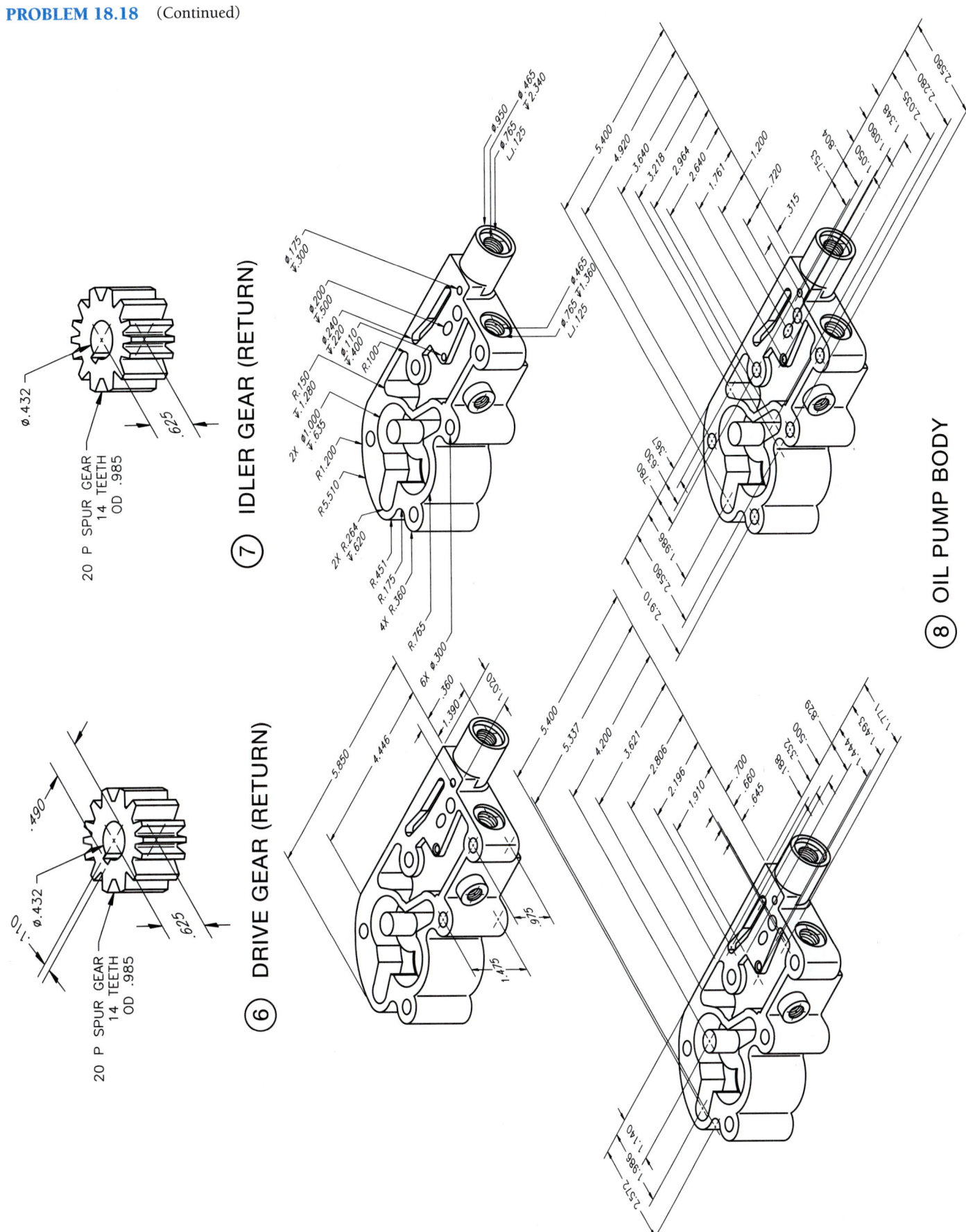

⑥ DRIVE GEAR (RETURN)

⑦ IDLER GEAR (RETURN)

⑧ OIL PUMP BODY

PROBLEM 18.18 (Continued)

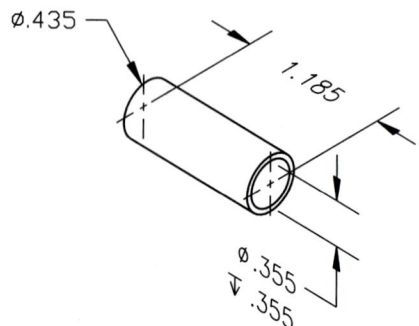

Ø.435
1.185
Ø .355
▽ .355

⑨ PLUNGER VALVE

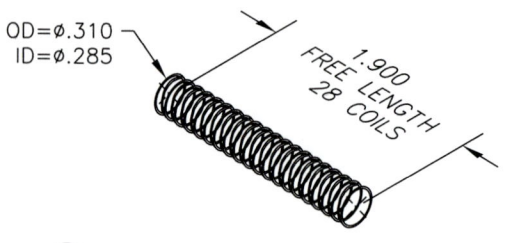

OD=Ø.310
ID=Ø.285
1.900
FREE LENGTH
28 COILS

⑫ PSI CONTROL SPRING

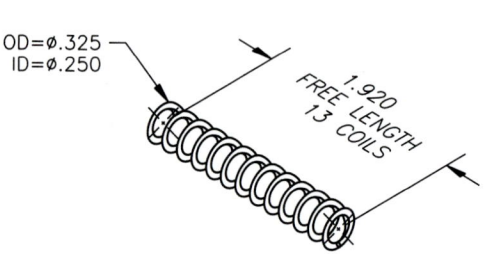

OD=Ø.325
ID=Ø.250
1.920
FREE LENGTH
13 COILS

⑩ PLUNGER SPRING

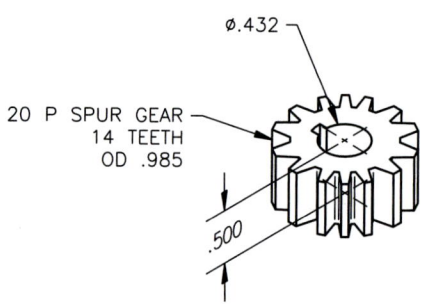

Ø.432
20 P SPUR GEAR
14 TEETH
OD .985
.500

⑮ IDLER GEAR (FEED)

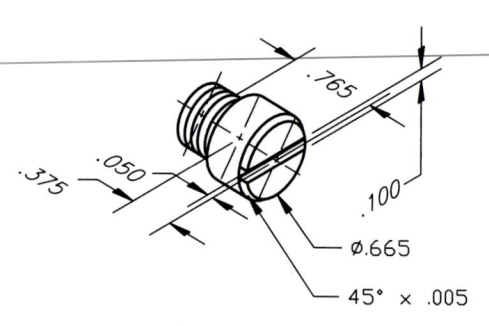

.765
.050
.375
.100
Ø.665
45° × .005

⑪ END CAP

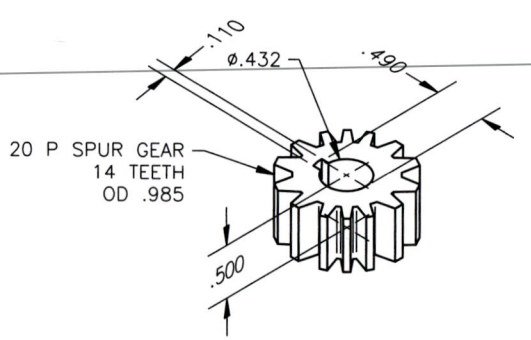

.110
Ø.432
.490
20 P SPUR GEAR
14 TEETH
OD .985
.500

⑯ DRIVE GEAR (FEED)

PROBLEM 18.19 **Working drawing (in.)**

Assembly Name: Landing Gear Retract Assembly

SPECIFIC INSTRUCTIONS:

This problem is generally basic, but it also contains challenging applications that include gears. You should study gears and related drafting practices covered in Chapter 19 before completing this problem.

Prepare a complete set of working drawings with one detail drawing per sheet and the assembly and parts list on another sheet. When preparing the assembly drawing, use separate balloons for each part in each view or use only one balloon in the view that most clearly identifies the part. This is an engineering prototype project. Errors are likely. Verify dimensions during assembly. Establish tolerances between mating parts. There are several purchase parts in this assembly. Manufacturer research is needed to locate available products. *Problem courtesy of Mike McIntyre.*

PARTS LIST

ITEM	QTY	NAME	DESCRIPTION	PART NO.
1	1	AIR CYLINDER	CLIPPERD	CL0002
2	2	E-CLIP	1/8"	EC0018
3	1	BACKPLATE	ALUMINUM	LGR001
4	1	1/4" × 2" BOLT	2" HARD BOLT	BT0002
5	1	SIDE FRAME (LF)	ALUMINUM	LGR003
6	11	CAP SCREW	6-32UNF-2 × 1/4"	CS0012
7	1	SLIDE PIN	STEEL	LGR006

ITEM	QTY	NAME	DESCRIPTION	PART NO.
8	1	CENTER BLOCK	ALUMINUM	LGR005
9	1	45° BEVEL GEAR	MODIFIED	BG0002
10	1	SPACER	ALUMINUM	LGR007
11	1	SIDE FRAME (RT)	ALUMINUM	LGR002
12	1	WASHER	BRASS	WA0014
13	1	PIVOT PIN	HARD STEEL	LGR008
14	1	45° BEVEL GEAR	BROWNING	BG0001

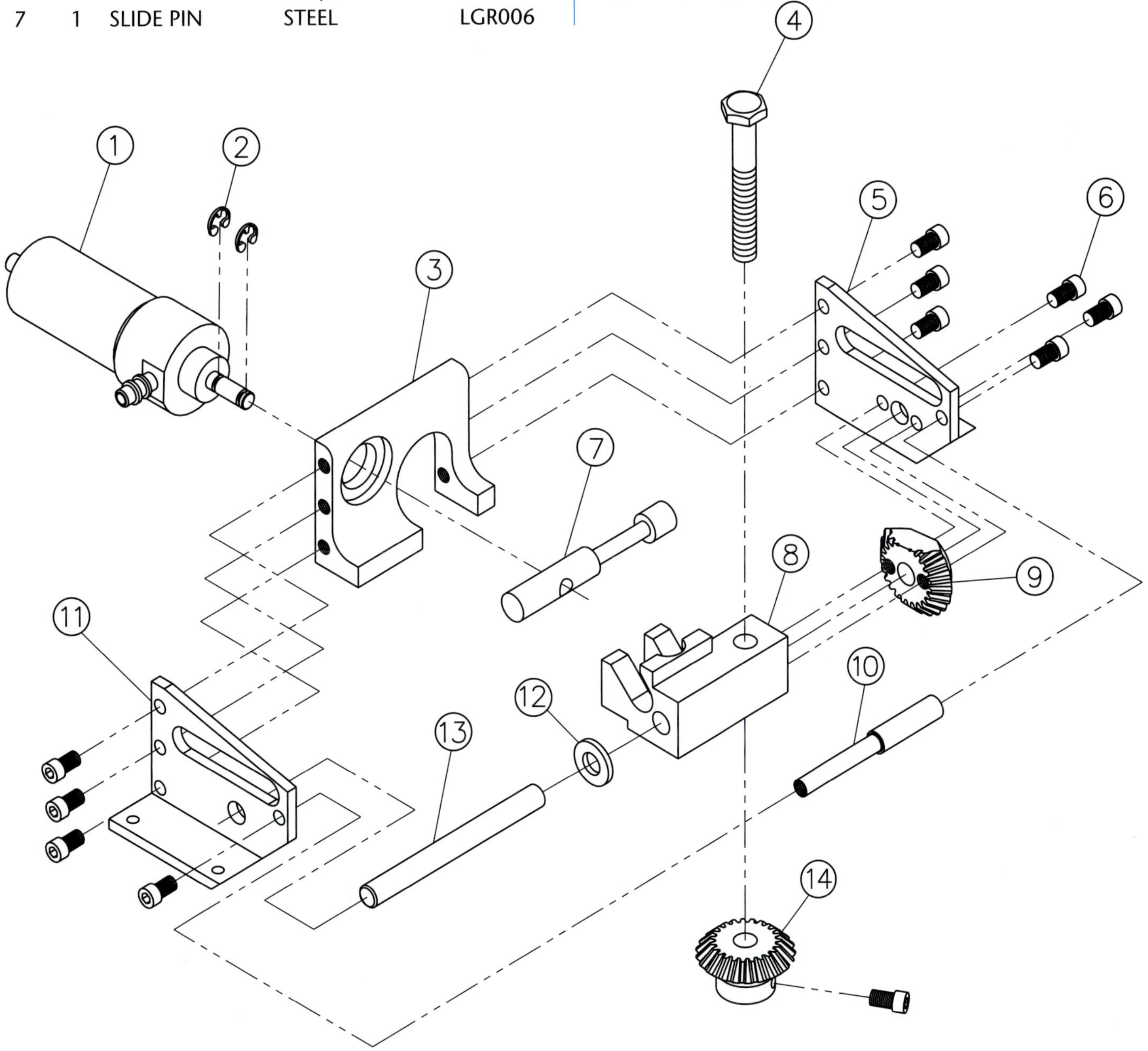

PROBLEM 18.19 (Continued)

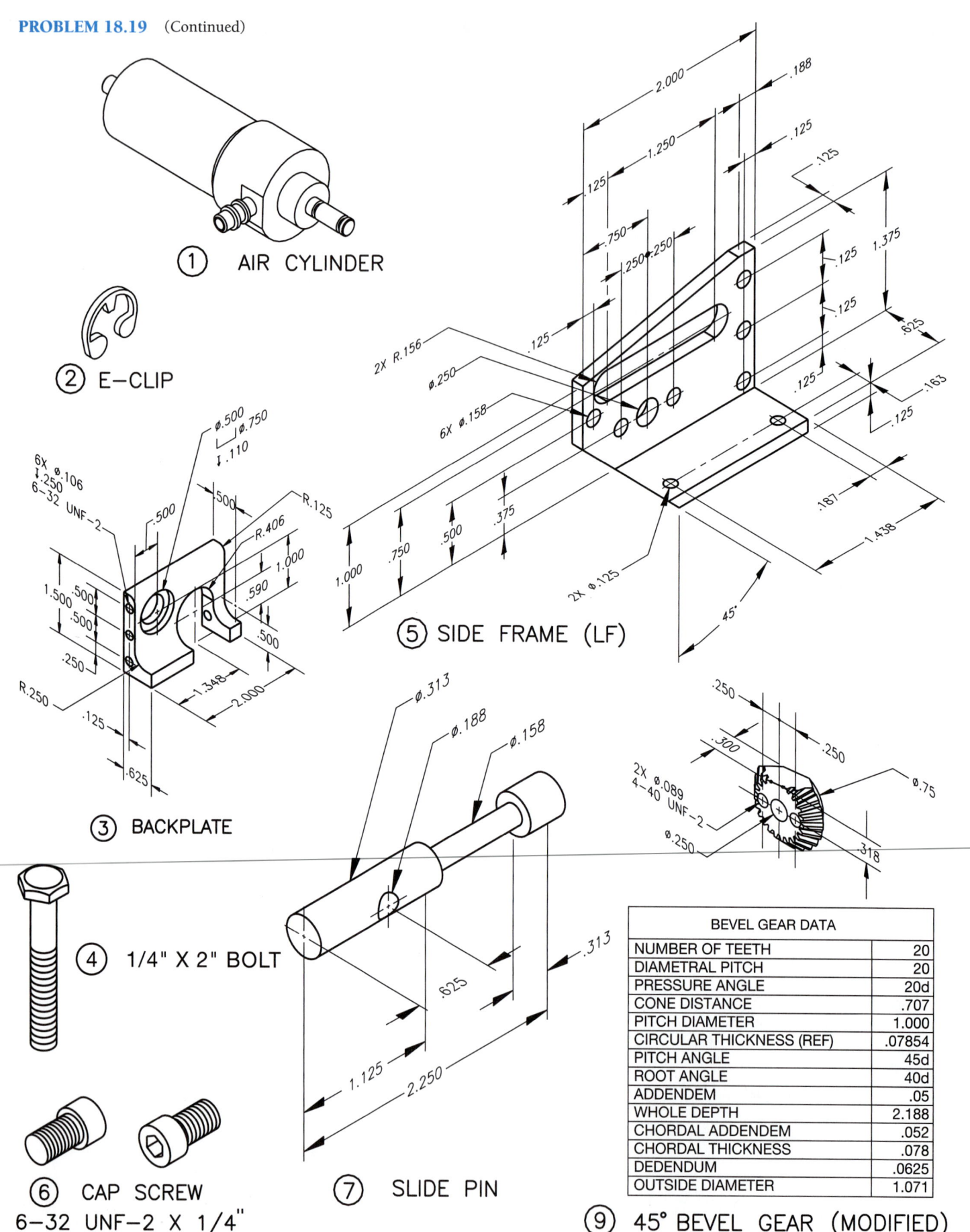

① AIR CYLINDER

② E–CLIP

③ BACKPLATE

6X ø.106
↓.250
6–32 UNF–2

⑤ SIDE FRAME (LF)

④ 1/4" X 2" BOLT

⑥ CAP SCREW
6–32 UNF–2 X 1/4"

⑦ SLIDE PIN

2X ø.089
4–40 UNF–2

⑨ 45° BEVEL GEAR (MODIFIED)

BEVEL GEAR DATA	
NUMBER OF TEETH	20
DIAMETRAL PITCH	20
PRESSURE ANGLE	20d
CONE DISTANCE	.707
PITCH DIAMETER	1.000
CIRCULAR THICKNESS (REF)	.07854
PITCH ANGLE	45d
ROOT ANGLE	40d
ADDENDEM	.05
WHOLE DEPTH	2.188
CHORDAL ADDENDEM	.052
CHORDAL THICKNESS	.078
DEDENDUM	.0625
OUTSIDE DIAMETER	1.071

PROBLEM 18.19   (Continued)

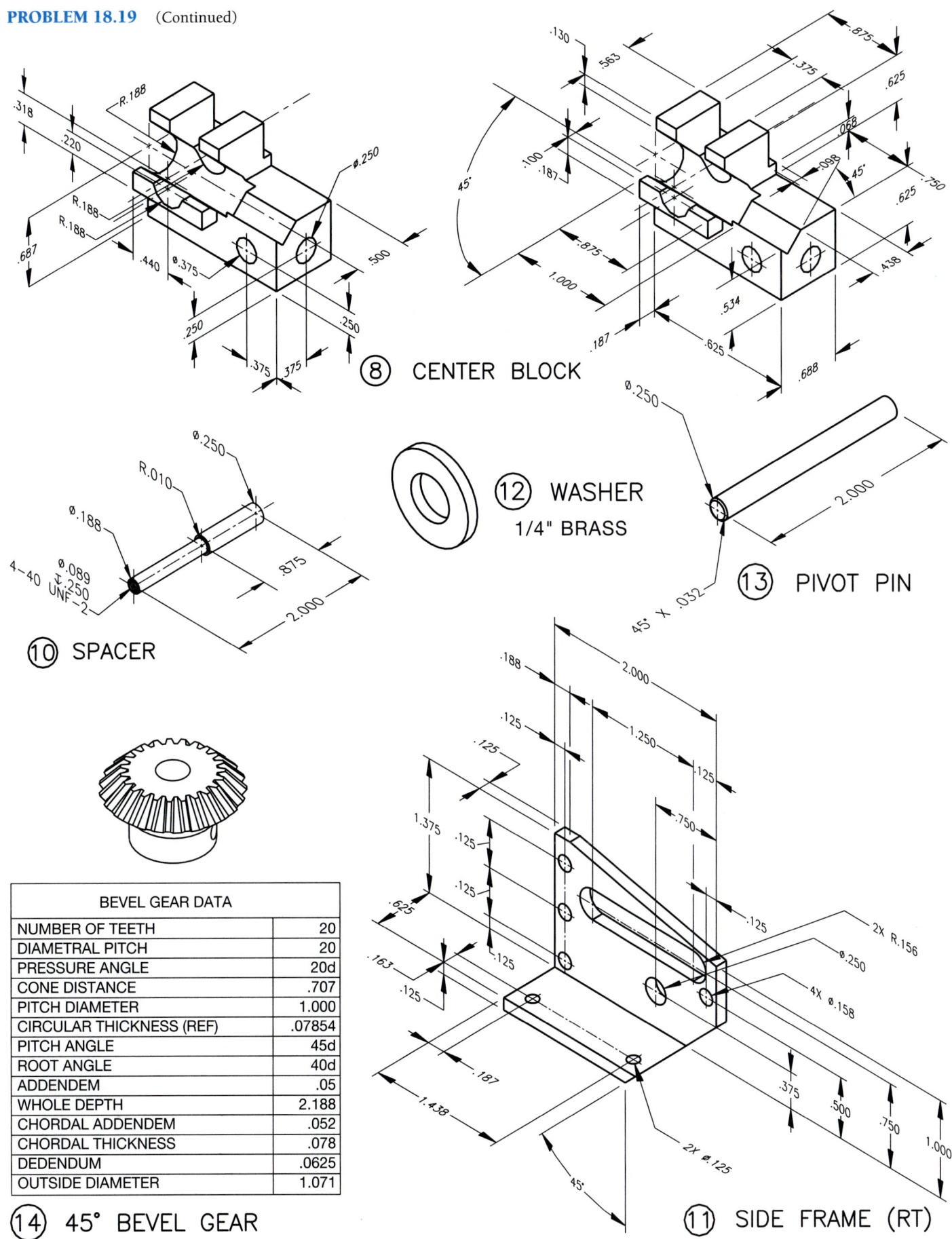

⑧ CENTER BLOCK

⑩ SPACER

⑫ WASHER
1/4" BRASS

⑬ PIVOT PIN

BEVEL GEAR DATA	
NUMBER OF TEETH	20
DIAMETRAL PITCH	20
PRESSURE ANGLE	20d
CONE DISTANCE	.707
PITCH DIAMETER	1.000
CIRCULAR THICKNESS (REF)	.07854
PITCH ANGLE	45d
ROOT ANGLE	40d
ADDENDEM	.05
WHOLE DEPTH	2.188
CHORDAL ADDENDEM	.052
CHORDAL THICKNESS	.078
DEDENDUM	.0625
OUTSIDE DIAMETER	1.071

⑭ 45° BEVEL GEAR

⑪ SIDE FRAME (RT)

PROBLEM 18.20   Working drawing (in.)

Assembly Name: Hydraulic Jack

PARTS LIST

KEY	QTY	NAME	DESCRIPTION	PART NO.	KEY	QTY	NAME	DESCRIPTION	PART NO.
1	1	BASE	JACK	1DT1001	17	1	NUT	10-32 UNF-2B	5DT1017
2*	1	TUBE	CYLINDER	1DT1002	18	1	NUT	.437-14 UNC-2B	5DT1018
3*	1	PISTON	JACK	1DT1003	19	1	CUP	∅1.500 LEATHER	5DT1019
4*	1	SCREW	JACK	1DT1004	20*	1	WASHER	PISTON	1DT1020
5*	1	TUBE	RESERVOIR	1DT1005	21	1	NUT	.437-20 UNF-2B	5DT1021
6*	1	CAP	TOP	1DT1006	22*	1	GUIDE	BRONZE	1DT1022
7	1	PACKING	∅1.500 LEATHER	5DT1007	23*	1	PIN	PIVOT	1DT1023
8*	1	NUT	PACKING	1DT1008	24	1	BALL	∅.250	5DT1024
9	1	SOCKET	PUMP HANDLE	1DT1009	25	1	SPRING	∅.281 × 1.000	5DT1025
10*	1	HANDLE	PUMP	1DT1010				CLOSED ENDS,	
11*	1	PLUNGER	PUMP	1DT1011				6 COILS	
12*	1	PIN	DRIVE	1DT1012	26	1	BALL	∅.3125	5DT1026
13*	1	PIN	STOP	1DT1013	27	1	SPRING	∅.343 × 1.250	5DT1027
14*	1	SUPPORT	PUMP PIVOT	1DT1014				CLOSED ENDS,	
15	1	CUP SEAL	∅.445 NEOPRENE	5DT1015				9 COILS	
16	1	WASHER	ANS TYPE B PLAIN	5DT1016	28*	1	NUT	NEEDLE VALVE	1DT1028
			NO. 10 N SERIES		29*	1	VALVE	NEEDLE	1DT1029
					30*	1	HANDLE	NEEDLE VALVE	1DT1030
					31	5	PLUG	1/8-27 NPT	5DT1031

*Engineering layouts are not provided for these parts. These parts need to be designed to complete this set of working drawings. Possible drawing solutions are provided in the *Solutions Manual,* for *Engineering Drawing and Design,* Fourth Edition.

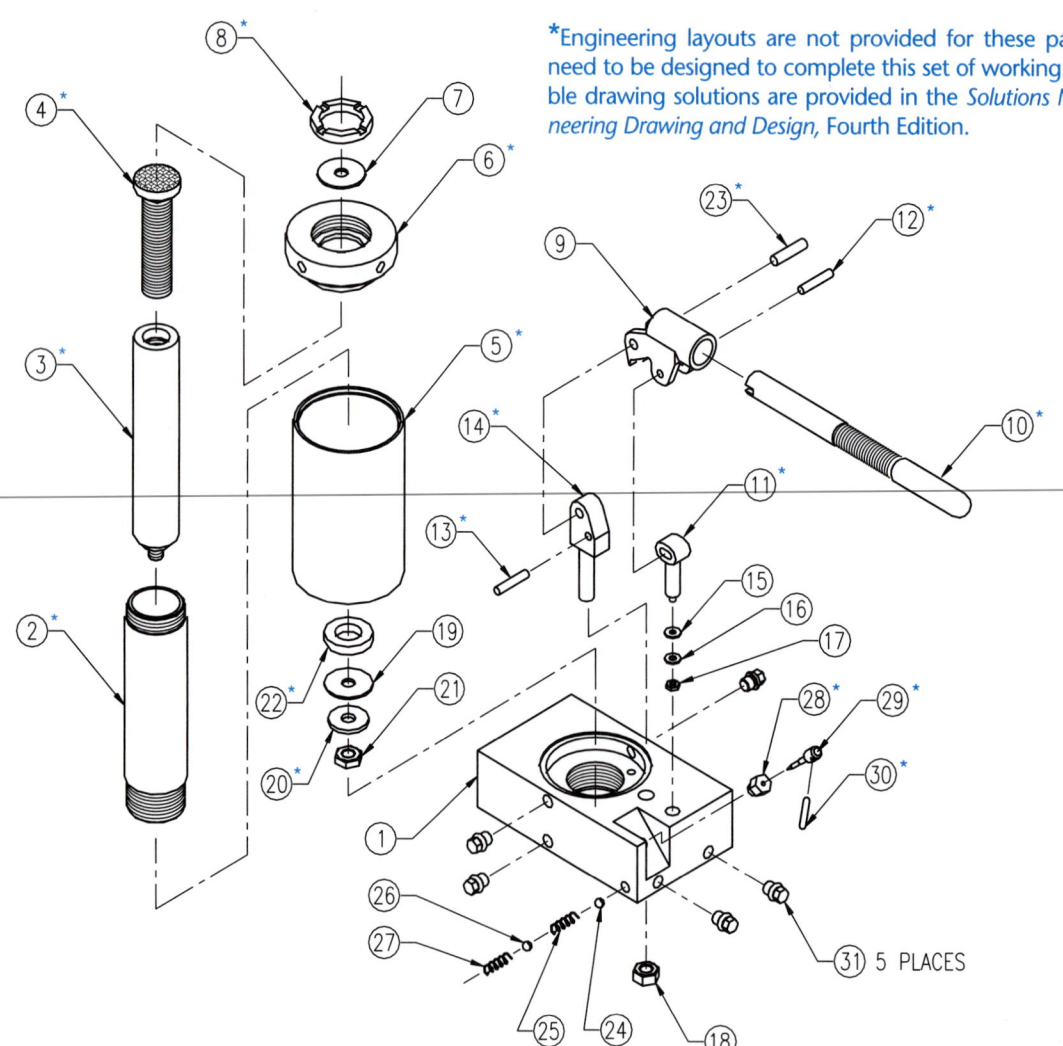

PROBLEM 18.20 (Continued)

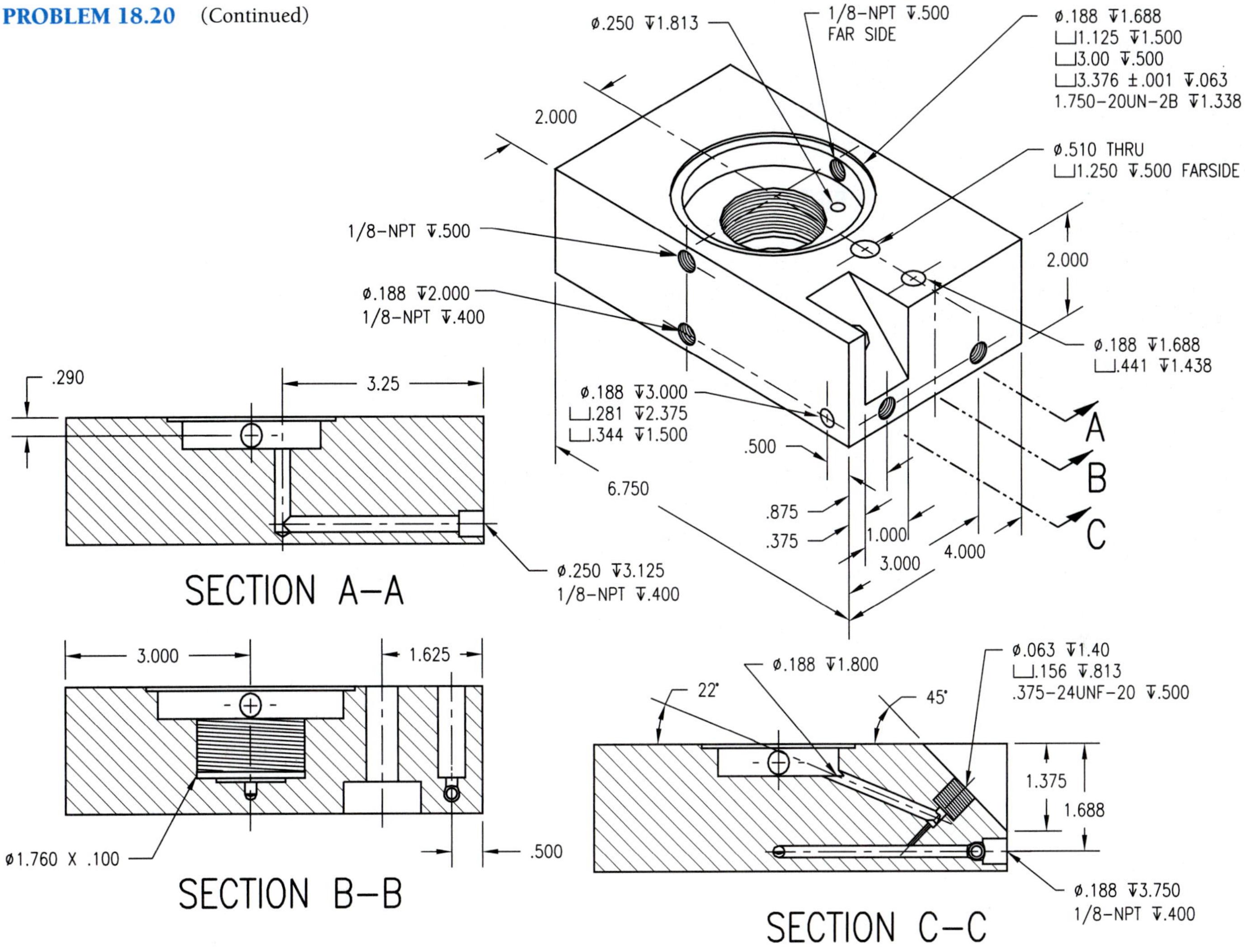

Ø.250 ↧1.813

1/8–NPT ↧.500
FAR SIDE

Ø.188 ↧1.688
⌴1.125 ↧1.500
⌴3.00 ↧.500
⌴3.376 ±.001 ↧.063
1.750–20UN–2B ↧1.338

2.000

Ø.510 THRU
⌴1.250 ↧.500 FARSIDE

1/8–NPT ↧.500

2.000

Ø.188 ↧2.000
1/8–NPT ↧.400

Ø.188 ↧1.688
⌴.441 ↧1.438

Ø.188 ↧3.000
⌴.281 ↧2.375
⌴.344 ↧1.500

.290

3.25

A

B

C

.500

6.750

.875

.375

1.000

4.000

3.000

SECTION A–A

Ø.250 ↧3.125
1/8–NPT ↧.400

3.000

1.625

Ø.188 ↧1.800

Ø.063 ↧1.40
⌴.156 ↧.813
.375–24UNF–20 ↧.500

22°

45°

1.375

1.688

Ø1.760 X .100

.500

SECTION B–B

Ø.188 ↧3.750
1/8–NPT ↧.400

SECTION C–C

①BASE

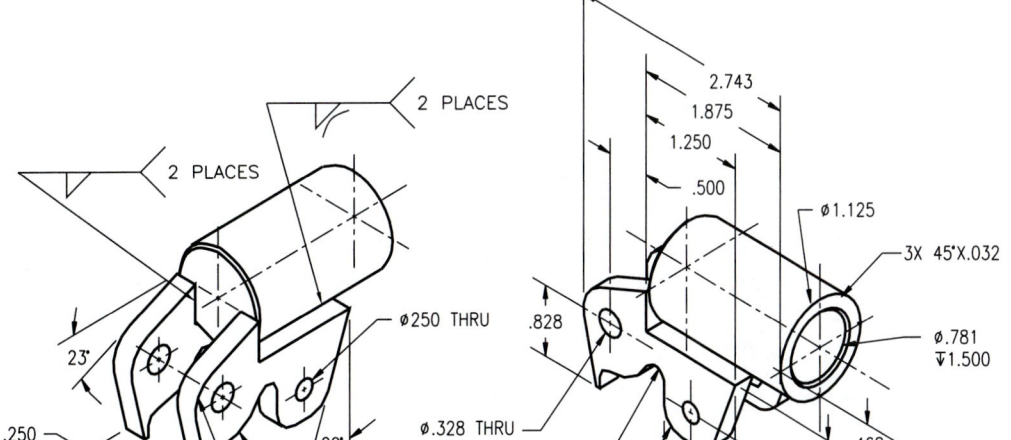

2 PLACES

2 PLACES

2.743

1.875

1.250

.500

Ø1.125

3X 45°X.032

Ø250 THRU

.828

Ø.781
↧1.500

23°

20°

103°

.250

.626

R.562

R.375

Ø.328 THRU

2X R.125

2X R.250

.468

.625

.625

1.00

1.443

⑨SOCKET

PROBLEM 18.21   Working drawing (in.)

Assembly Name: Worm Gear Reducer

This problem is demanding, and it contains challenging applications that include gears and bearings. You should study gears and related drafting practices covered in Chapter 19 before completing this problem.

Advanced project: design changes may be required.

PARTS LIST

ITEM	QTY	NAME	DESCRIPTION
1	1	HOUSING	
2	2	RETAINING PLATE	
3	1	BEARING CAP	
4	1	MOTOR ADAPTER	
5	1	HIGH SPEED SHAFT	
6	1	SLOW SPEED SHAFT	
7	1	WORM GEAR	BRONZE
8	1	DBL. ROW TAPERED ROLLER BEARING	KOYO46T30305DJ/29.5
9	1	SNGL. ROLE CYL. ROLLER BEARING	KOYO CRL11
10	1	TAPER PLUG	500-16NPT PLUG
11	1	HEX NUT	.875-16 UN-28
12	4	HIGH SPEED LOCKWASHER	TIMKEN TW-105
13	4	MACHINE SCREW	.375-16UNC-2A × 1.813 HEX HEAD
14	4	MACHINE SCREW	.375-16UNC-2A × 1.625 HEX HEAD
15	8	MACHINE SCREW	.375-16UNC-2A × .625 HEX HEAD
16	1	HIGH SPEED OIL SEAL	PARKER 2-028
17	2	SLOW SPEED OIL SEAL	PARKER 2-020
18	1	SLOW SPEED KEYWAY	.1875 × .245 × 1.450
19	1	SNGL. ROW TAP ROLLER BEARING	KOYO 32005J
20	2	SLOW SPEED SPACER	TIMKEN TW-506

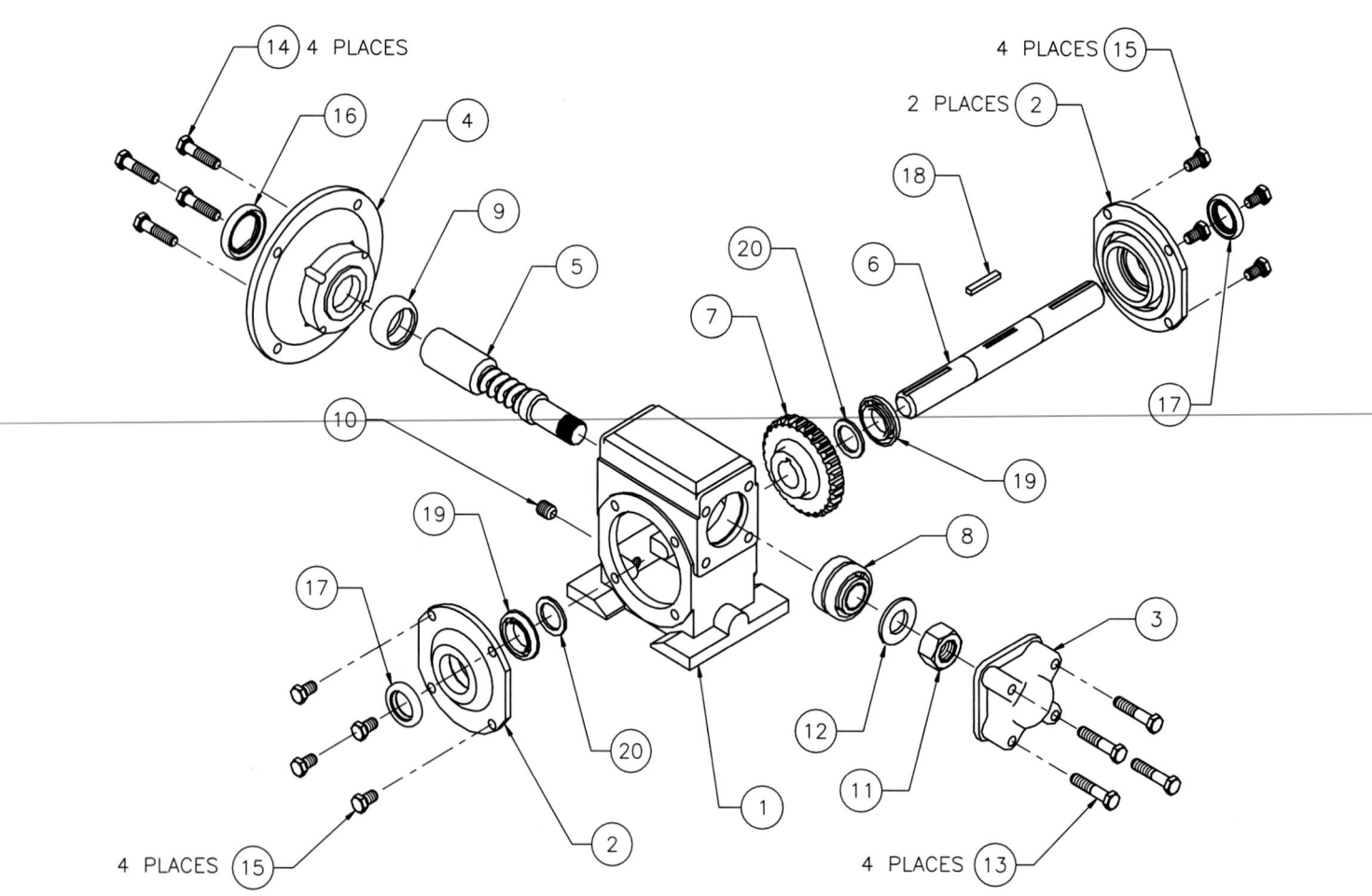

PROBLEM 18.21   (Continued)

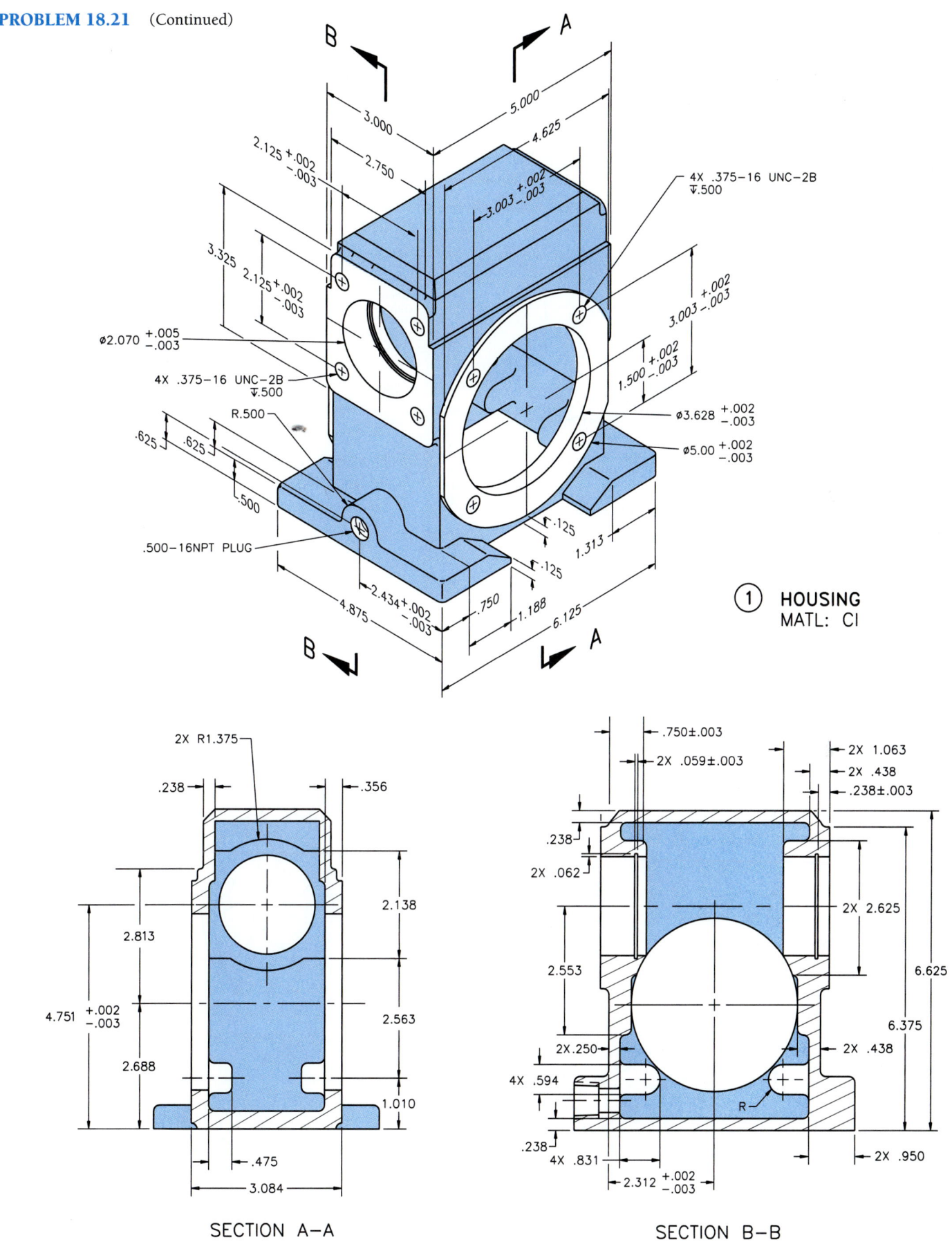

1 HOUSING
MATL: CI

SECTION A–A

SECTION B–B

PROBLEM 18.21   (Continued)

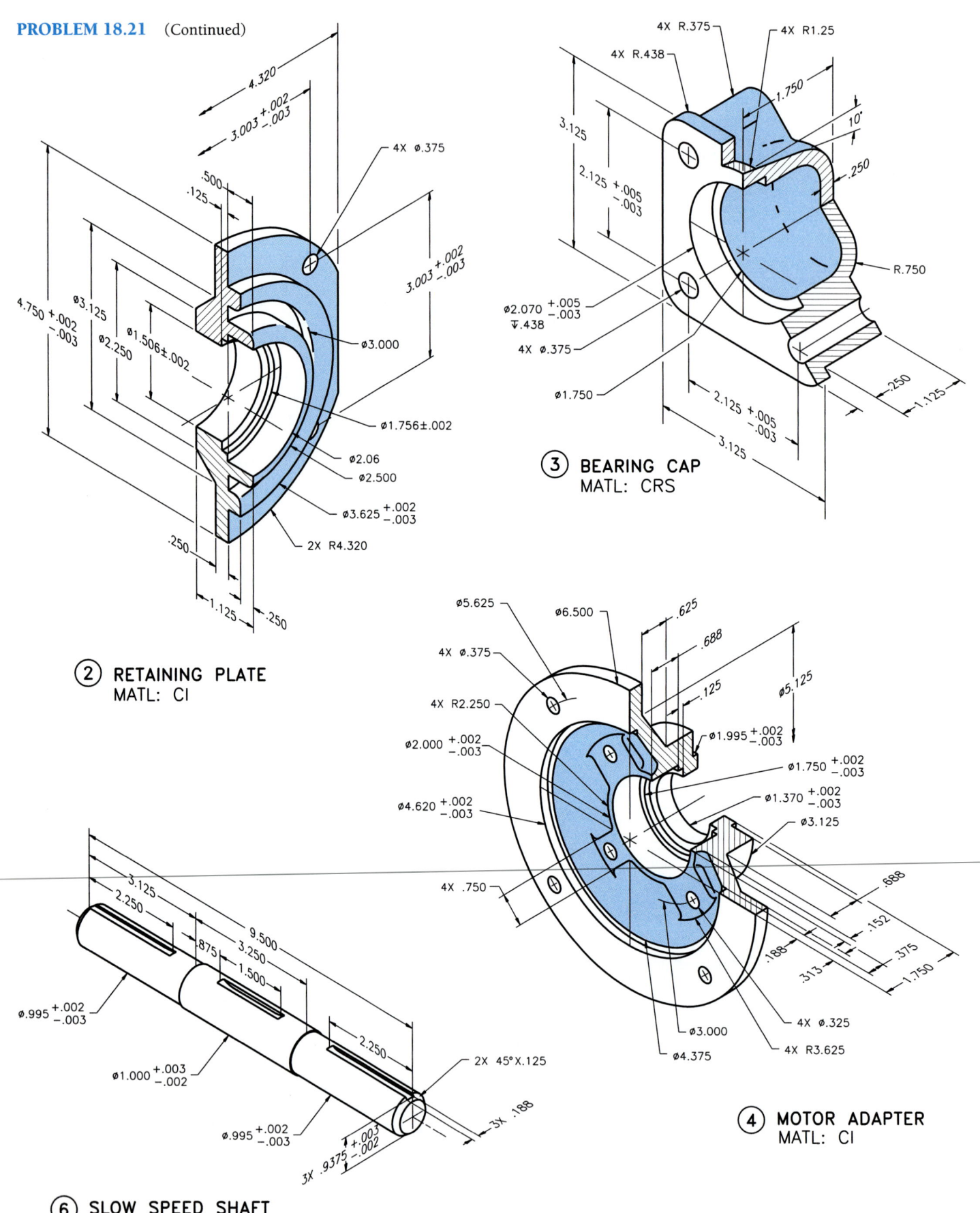

4X ⌀.375

4.320

3.003 +.002 −.003

.500

.125

⌀3.125

4.750 +.002 −.003

⌀1.506±.002

⌀2.250

3.003 +.002 −.003

⌀3.000

⌀1.756±.002

⌀2.06

⌀2.500

⌀3.625 +.002 −.003

.250

2X R4.320

1.125   .250

(2) **RETAINING PLATE**
   MATL: CI

4X R.375   4X R1.25
4X R.438

1.750

3.125

2.125 +.005 −.003

10°

.250

⌀2.070 +.005 −.003
⊽.438

4X ⌀.375

R.750

⌀1.750

2.125 +.005 −.003

3.125

(3) **BEARING CAP**
   MATL: CRS

⌀5.625   ⌀6.500   .625

.688

4X ⌀.375

.125

4X R2.250

⌀5.125

⌀2.000 +.002 −.003

⌀1.995 +.002 −.003

⌀1.750 +.002 −.003

⌀1.370 +.002 −.003

⌀4.620 +.002 −.003

⌀3.125

.688

4X .750

.152

.188

.375

.313

1.750

⌀3.000   4X ⌀.325

⌀4.375   4X R3.625

(4) **MOTOR ADAPTER**
   MATL: CI

3.125

2.250

.875

9.500

3.250

1.500

2.250

⌀.995 +.002 −.003

2X 45°X.125

⌀1.000 +.003 −.002

3X .188

⌀.995 +.002 −.003

3X .9375 +.003 −.002

(6) **SLOW SPEED SHAFT**
   MATL: SAE 4320

PROBLEM 18.21   (Continued)

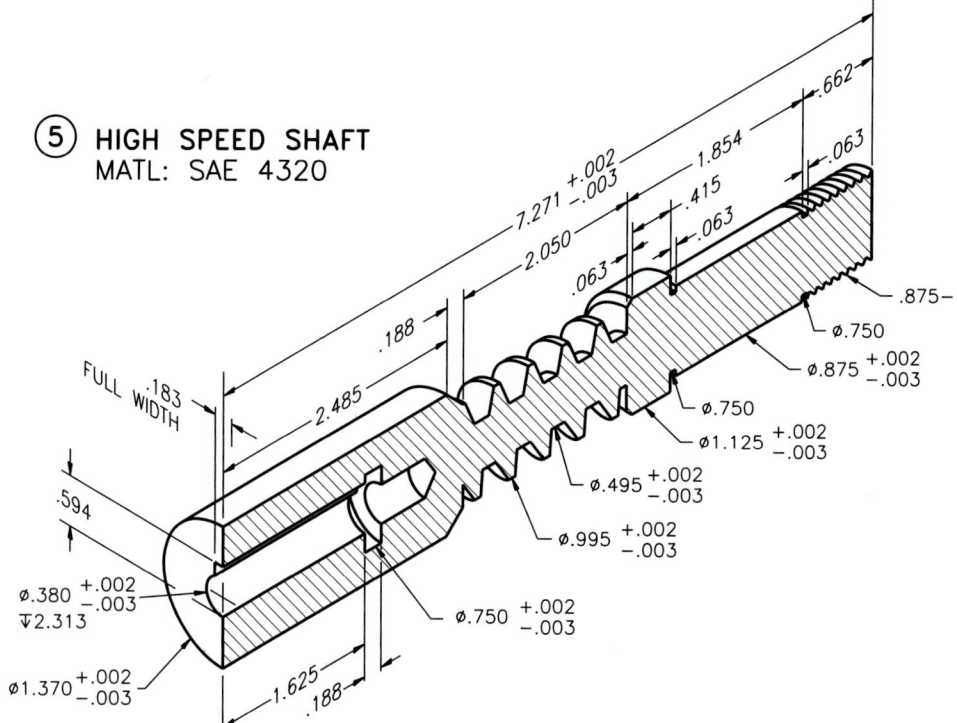

(5) HIGH SPEED SHAFT
MATL: SAE 4320

WORM GEAR DATA	
NUMBER OF THREADS	5
AXIAL PITCH	
PRESSURE ANGLE	20°
PITCH DIAMETER	.750
LEAD RIGHT HAND	.300
LEAD ANGLE	169°
ADDENDUM	.125
WHOLE DEPTH	.250
CHORDAL THICKNESS	.163

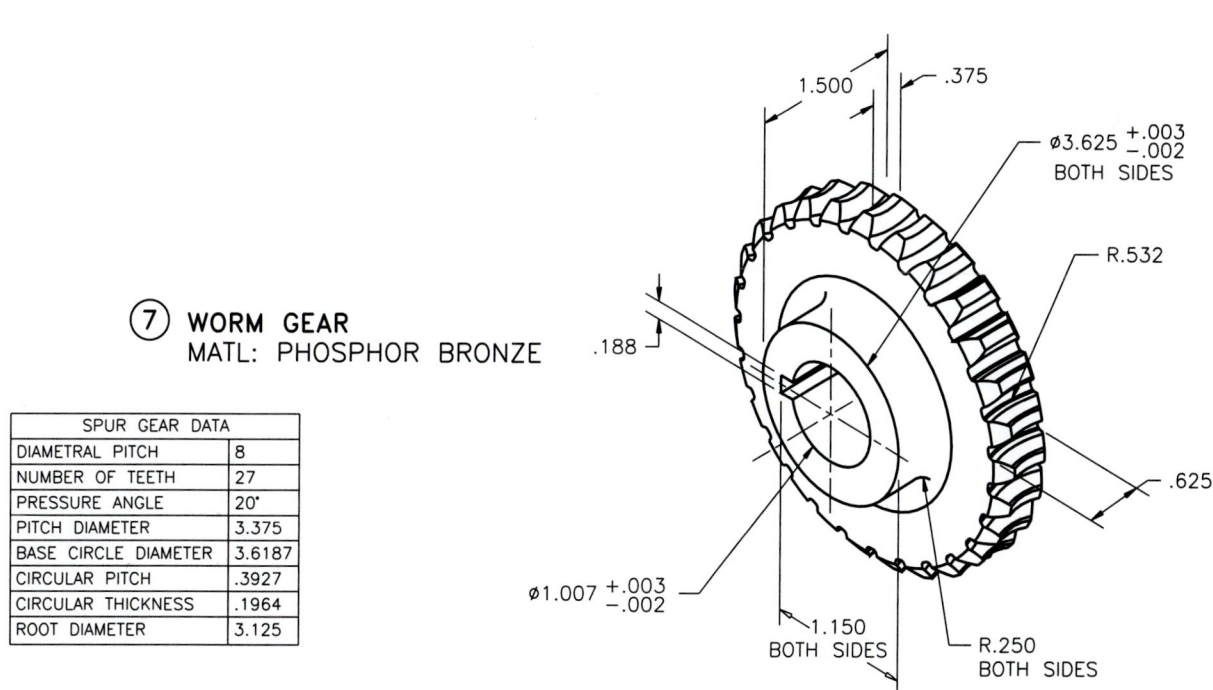

(7) WORM GEAR
MATL: PHOSPHOR BRONZE

SPUR GEAR DATA	
DIAMETRAL PITCH	8
NUMBER OF TEETH	27
PRESSURE ANGLE	20°
PITCH DIAMETER	3.375
BASE CIRCLE DIAMETER	3.6187
CIRCULAR PITCH	.3927
CIRCULAR THICKNESS	.1964
ROOT DIAMETER	3.125

PROBLEM 18.22   Working drawing (in.)

Assembly Name: Table Vice

SPECIFIC INSTRUCTIONS:

Prepare a complete set of working drawings with one detail drawing per sheet and the assembly and parts list on an- other sheet. When preparing the assembly drawing, use separate balloons for each part in each view or use only one balloon in the view that most clearly identifies the part. This is an engineering prototype project. Errors are likely. Verify dimensions during assembly. Establish tolerances between mating parts. *Problem courtesy of Jack Pitcher.*

PARTS LIST

KEY	QTY	DESCRIPTION	KEY	QTY	DESCRIPTION
1	2	CLAMP JAWS	9	1	1/4-28UNF NYLOCK NUT
2	1	1/4 -28UNF X 2" LG HEX HD MACH. SCREW	10	2	3/8 FLAT WASHER
3	1	PIVOT	11	2	ADJUSTING ROD
4	2	SCREW ADJUSTER	12	2	ADJUSTING ROD HANDLE
5	2	1/4-28UNF X 1" LG STANDARD SLOT FLAT COUNTERSUNK HD CAP SCREW	13	2	ADJUSTING HANDLE KNOB
			14	2	8-32UNC X 5/8" LG SLOTTED ROUND HD. MACHINE SCREW
6	1	TABLE	15	2	1/8 SPRING PIN
7	1	BASE	16	2	1/4-28UNF X 1" LG STANDARD SLOT PAN HEAD MACHINE SCREW
8	1	LOCKING HANDLE			

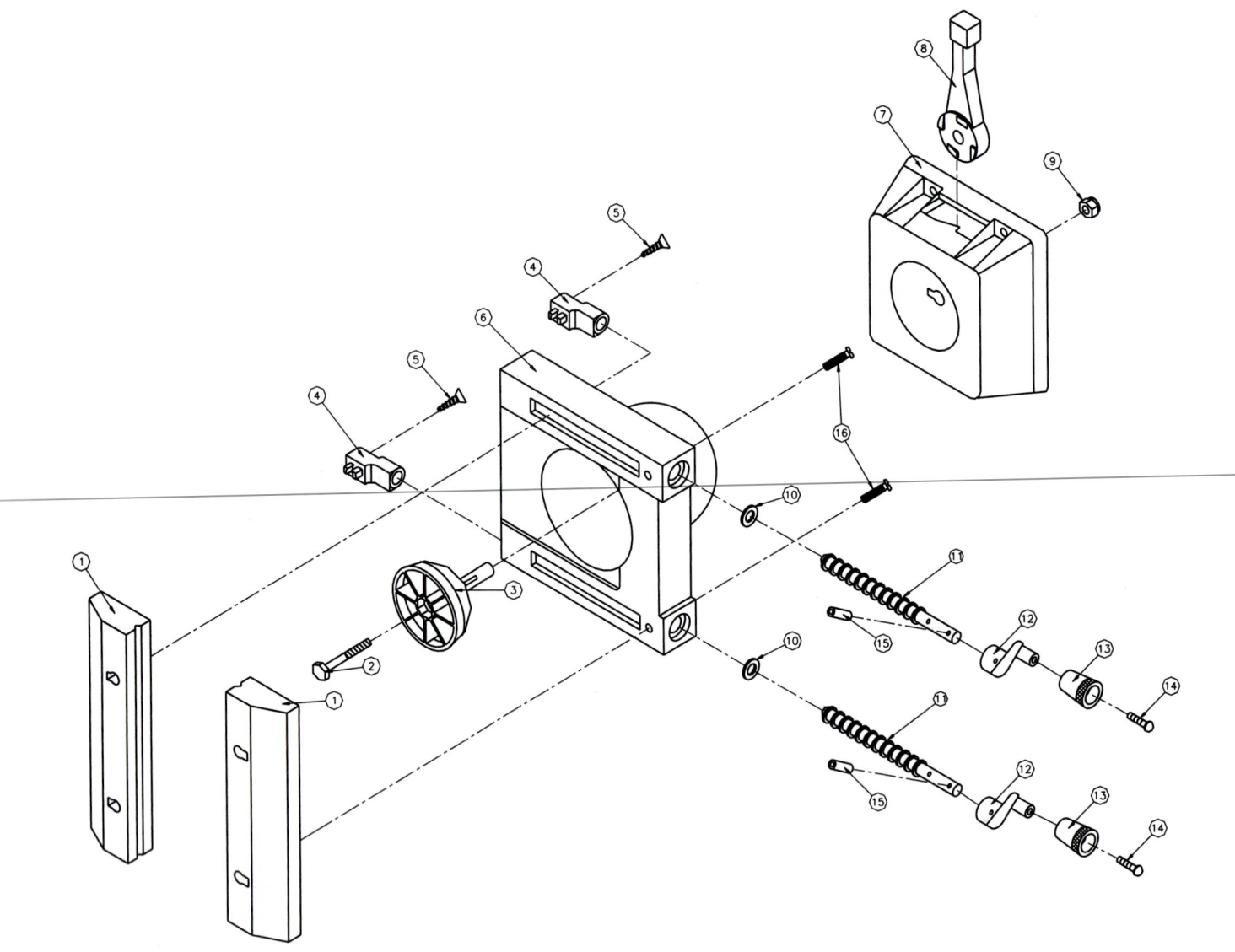

PROBLEM 18.22 (Continued)

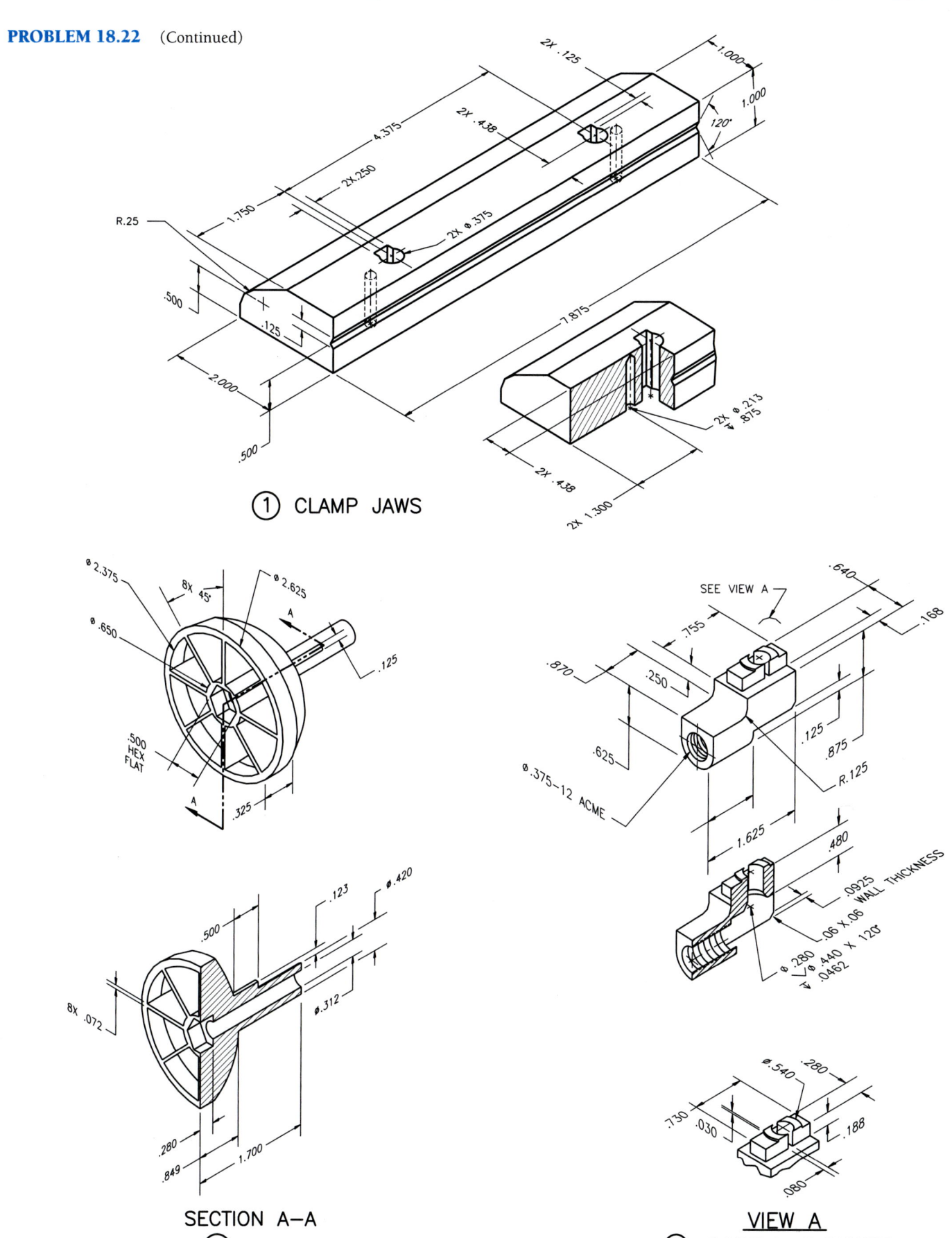

① CLAMP JAWS

SECTION A–A
③ PIVOT

VIEW A
④ SCREW ADJUSTER

PROBLEM 18.22   (Continued)

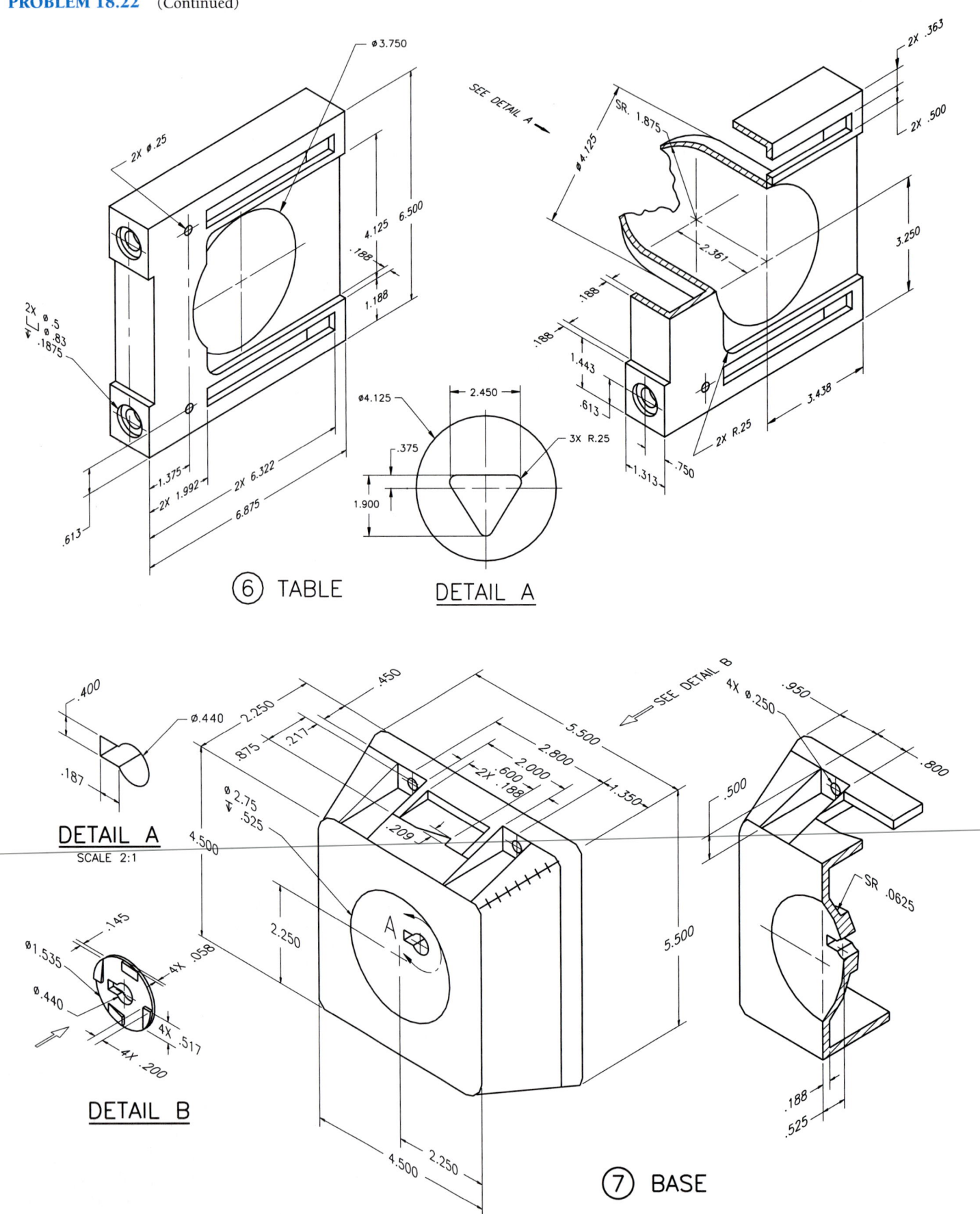

⑥ TABLE

DETAIL A

DETAIL A
SCALE 2:1

DETAIL B

⑦ BASE

PROBLEM 18.22   (Continued)

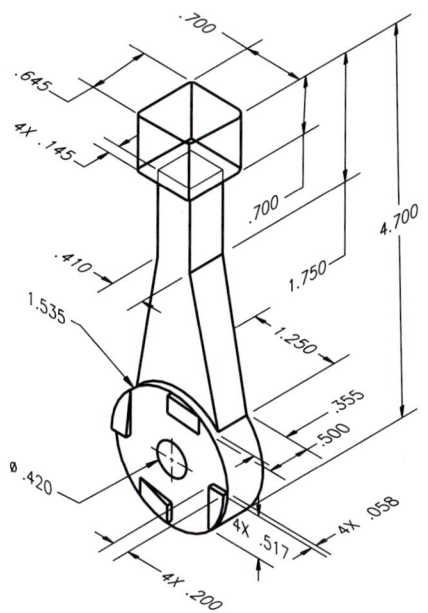

⑧   LOCKING   HANDLE

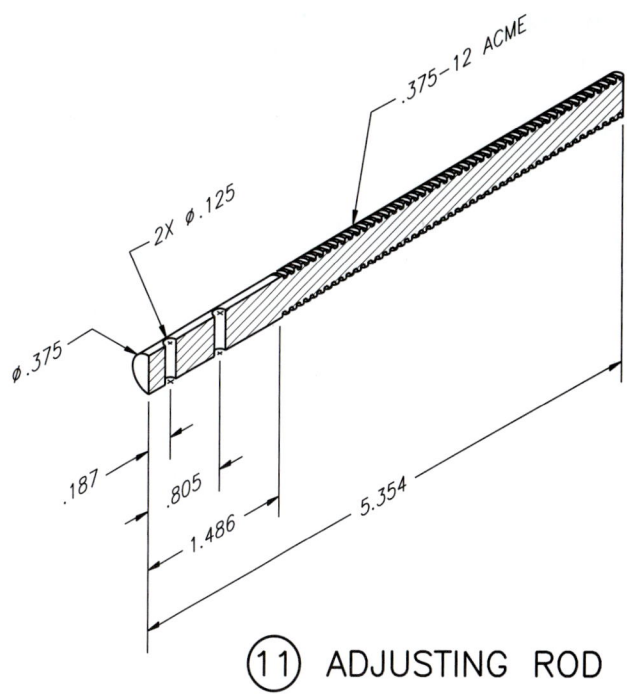

⑪   ADJUSTING   ROD

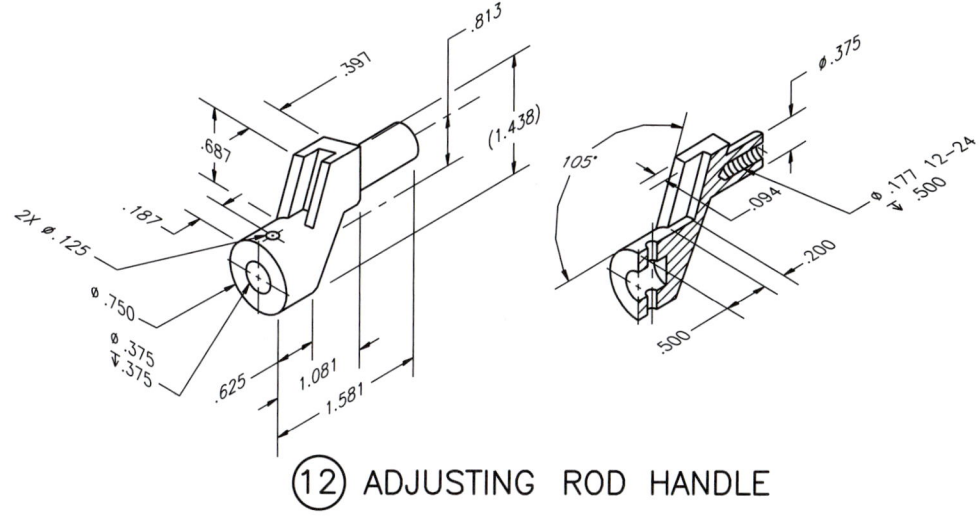

⑫   ADJUSTING   ROD   HANDLE

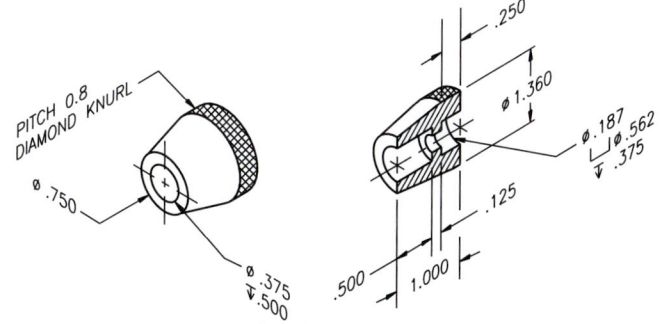

⑬   ADJUSTING   HANDLE   KNOB

ADVANCED TEAM DESIGN PROBLEMS

PROBLEMS 18.23 and 18.24: Access the CD found with this textbook and open the problem of your choice, or as assigned by your instructor. Solve the problem or problems using the instructions provided.

MATH PROBLEMS

PROBLEMS 18.26 through 18.31: Access the CD found with this textbook and open the math problem of your choice, or as assigned by your instructor. Solve the problem or problems using the instructions provided.

PROBLEM 18.25 **Engineering Changes**

Your instructor will use your complete set of working drawings to prepare a series of engineering changes.

Mechanisms: *Linkages, Cams, Gears, and Bearings*

LEARNING OBJECTIVES

After completing this chapter, you will:

- Draw linkage diagrams.
- Create cam displacement diagrams.
- Design cam profile drawings from previously drawn cam displacement diagrams.
- Make detail gear drawings using simplified representations and gear data charts.
- Establish unknown data for gear trains.
- Calculate bearing information from specifications.
- Design a complete gear reducer from engineering data and sketches.

THE ENGINEERING DESIGN APPLICATION

Your latest assignment is to produce a drawing of an in-line follower plate cam profile for the manufacturing department. The engineer has provided the specifications in a cam displacement diagram. (See Figure 19.1.) Your manufacturing department is using a computer-aided manufacturing (CAM) system and needs a CADD drawing from which the required tooling paths can be established.

Using a CADD system, the process is fairly easy and completely accurate. Using appropriate techniques, you find and draw the base circle and the prime circle. Next, you create an array of lines extending outward from the center of the circle at 30° increments. (See Figure 19.2.) Finally, use offsets of the base circle as indicated by the engineer's diagram, to find all the specified control points. Using a spline curve that intersects each of these points, you create a profile drawing to generate tooling paths for the CAM system. (See Figure 19.3.)

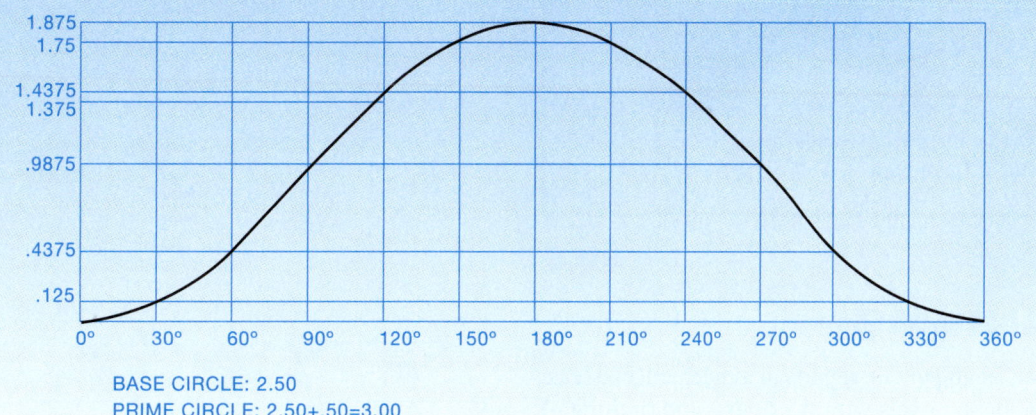

BASE CIRCLE: 2.50
PRIME CIRCLE: 2.50+.50=3.00

FIGURE 19.1 ■ Engineer's cam displacement diagram.

(Continued)

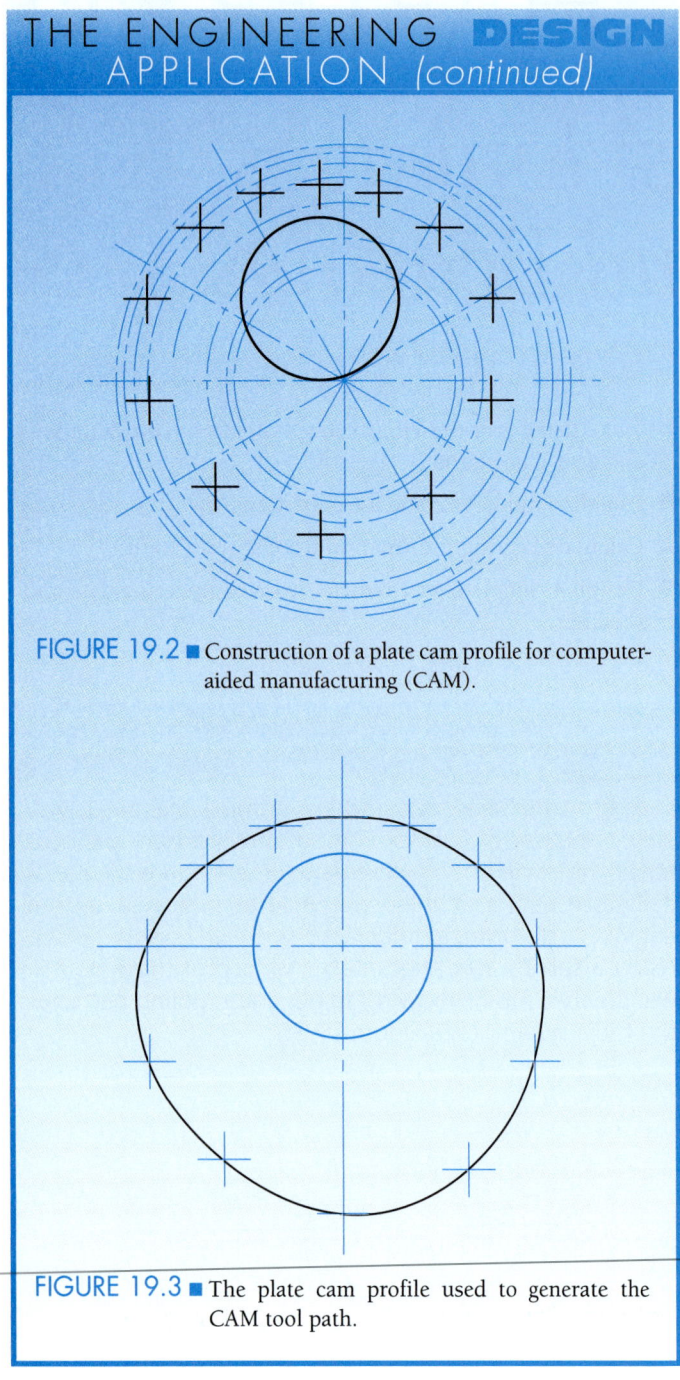

FIGURE 19.2 ■ Construction of a plate cam profile for computer-aided manufacturing (CAM).

FIGURE 19.3 ■ The plate cam profile used to generate the CAM tool path.

MECHANISMS

A mechanism is an arrangement of parts in a mechanical device or machine. This chapter deals with the design and drafting of elements of a mechanism, including linkages, cams, gears, and bearings. The study of mechanisms is part of the physical sciences known as mechanics. Mechanics includes statics and dynamics. Statics is the study of physics dealing with nonmoving objects acting as weight. Dynamics is the branch of physics that studies the motion of objects and the effects of the forces that cause motion. Dynamics is divided into two categories, kinetics and kinematics. Kinetics is an element of physics that deals

with the effects of forces that cause motion in mechanisms. The linkages, cams, and gears discussed in this chapter relate to the branch of physics known as kinematics. Kinematics is the study of mechanisms without reference to the forces that cause the movement.

Mechanisms in Our Daily Lives

Mechanisms play an important role in our daily activities. Every modern convenience, from the toaster used to make part of your breakfast to the automobile that you drive to work, is made up of one or more mechanisms. The car, for example, is a complex combination of mechanisms that includes all of the types of mechanisms to be discussed in this chapter.

LINKAGES

The elements of any mechanism are referred to as links. Links or linkages may be defined as any rigid element of the mechanism. In actual practice all of the mechanism components are links, including levers, bars, sliders, cams, or gears. The frame of the device is even considered a fixed link and is an important part of the mechanism. The first part of this chapter separates the types of linkages that deal primarily with motion caused by levers, rockers, cranks, and sliders.

LINKAGE SYMBOLS

One of the benefits of designing linkage mechanisms is that the drawings are in the form of schematic or single-line representations. After the complete design is created using these schematic drawings, the designer can go to work creating the actual components in relation to the schematic design. Schematic drawings have only a few basic components as shown in Figure 19.4. The

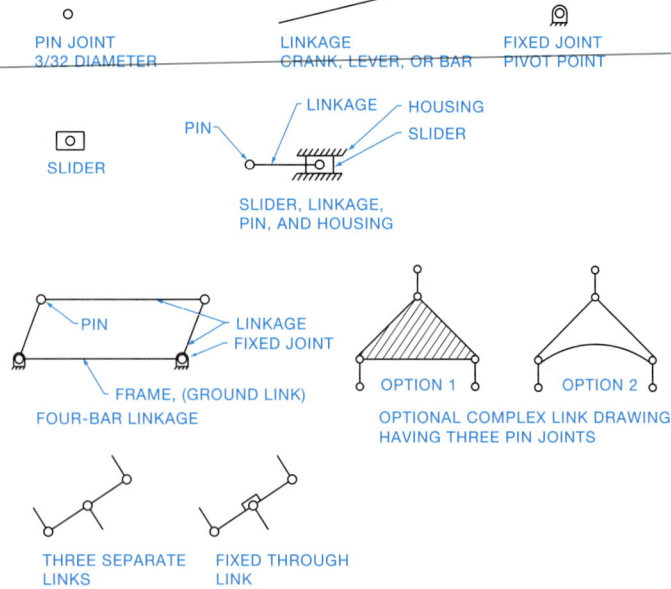

FIGURE 19.4 ■ Linkage diagram symbols.

symbols may be drawn proportional to the examples. The actual scale depends on the size of the drawing. Typical sizes are shown for most applications in this chapter. You can see here how easy it would be to develop a CADD symbols library for use when drawing these mechanisms.

TYPES OF LINKAGES

There are many different combinations of linkage mechanisms. These devices are broken into a few basic elements. The illustrations in this discussion show you the linkage mechanism in a simple pictorial drawing of the actual device and in the schematic representation.

Crank Mechanism

A crank is a link that makes a complete revolution around a fixed point. When working with the crank, keep in mind that the crank link is a fixed distance equal to the radius of the movement as shown in Figure 19.5.

Lever, or Rocker, Mechanism

A lever, or rocker, is a link that moves back and forth or oscillates through a given angle as illustrated in Figure 19.6.

Rocker Arm Mechanism

A rocker arm is different from the rocker previously discussed because it has a pivot point near the center and oscillates through a given angle as shown in Figure 19.7. Notice the symbol for a fixed link. This symbol is used to indicate that the link between points A and B remains rigid.

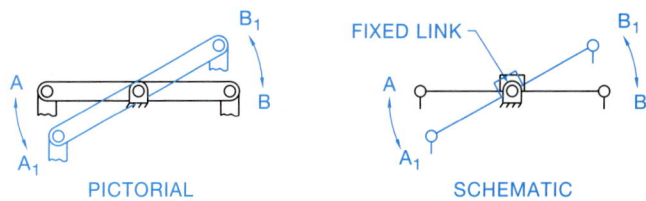

FIGURE 19.7 ■ Rocker arm and fixed link symbol.

Bell Crank Mechanism

Another form of a rocker arm is the bell crank. The bell crank is a more complex form of the linkage mechanism. The bell crank drawing shown in Figure 19.8 is a three-joint mechanism where the distances between points A, B, and C are fixed. Other bell crank designs may contain more than three pivot points depending on the design requirements. Notice that the schematic representation in Figure 19.8 shows two alternatives. Consult with your instructor or employer on the technique to use.

Four-Bar Linkage

The most commonly used linkage mechanism is called a four-bar linkage. There are many alternate designs of the four-bar linkage, but the basic form has four links, one of which is the ground link or machine frame. One of the rotating links is called the driver or crank and the other is called the follower or rocker. The link connected between the crank and rocker is called the connecting rod or coupler. The two pivoted links both rotate through 360°, or one rotates while the other oscillates, or they both oscillate, depending on the lengths and arrangement of the links. (See Figure 19.9.)

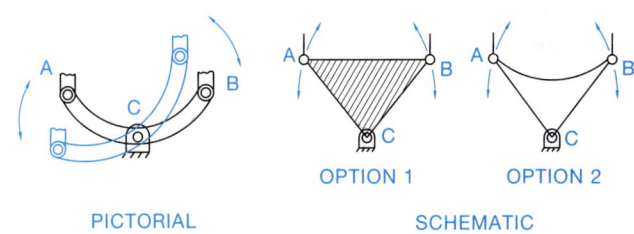

FIGURE 19.8 ■ Bell crank mechanism.

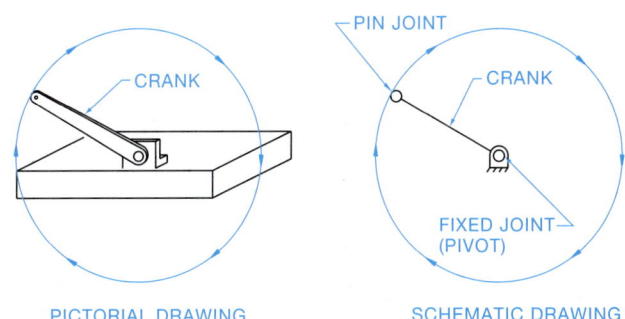

FIGURE 19.5 ■ Crank mechanism.

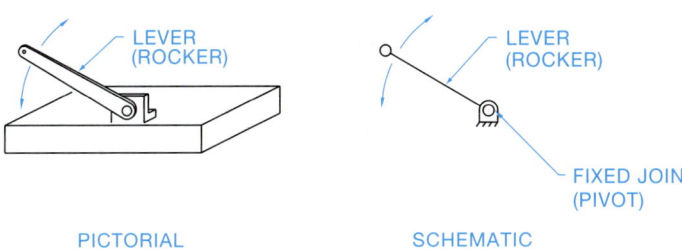

FIGURE 19.6 ■ Lever or rocker mechanism.

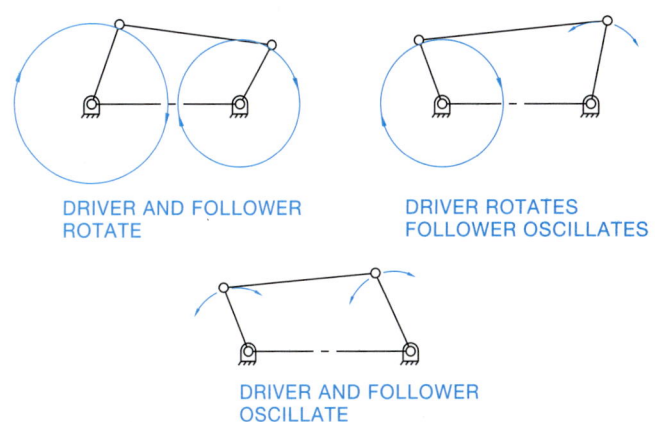

FIGURE 19.9 ■ Four-bar linkage movements.

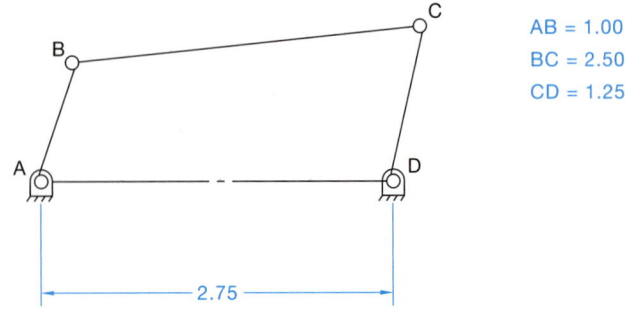

AB = 1.00
BC = 2.50
CD = 1.25

2.75

FIGURE 19.10 ■ Four-bar linkage example. Dimension values are in inches.

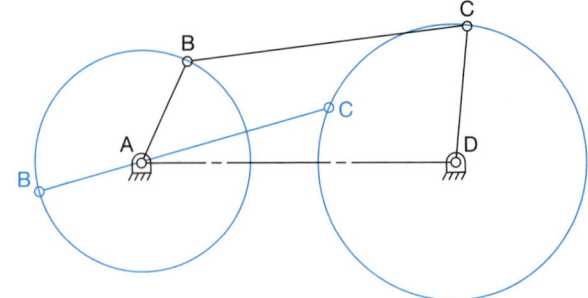

FIGURE 19.12 ■ Step 3, four-bar linkage solution, extreme left position.

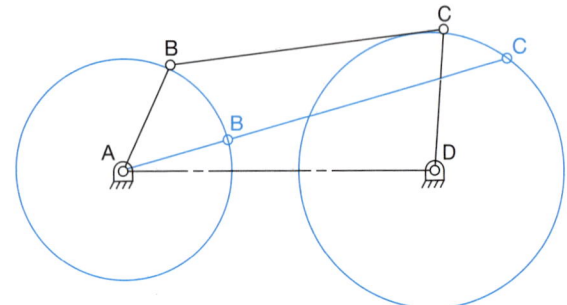

FIGURE 19.11 ■ Steps 1 and 2, four-bar linkage solution, extreme right position.

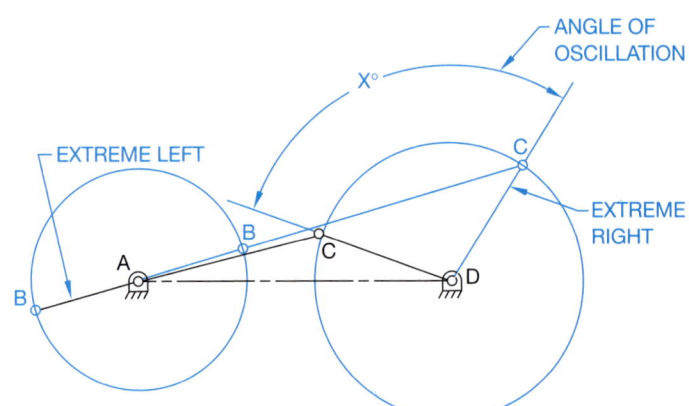

FIGURE 19.13 ■ Step 4, four-bar linkage solution, angle of oscillation.

Normally when determining the function of a four-bar linkage, the designer must show the extreme right and extreme left positions. If the crank rotates and the rocker oscillates, then the angle of oscillation can be established with this technique. For example, determine the angle of oscillation for rocker link CD in the four-bar linkage shown in Figure 19.10. If you are using manual drafting, it is best to use a sharp pencil and make careful measurements. Colored pencils are often helpful in showing the different positions. If you are using CADD to draw the diagrams, then the different positions can be shown on different layers and in different colors for a clear analysis of the operation. Follow these steps:

STEP 1 AND STEP 2 Draw a circle through point B with the center at A and a circle through C with the center at D. Determine the extreme right position of link CD by adding the length of links AB and BC (1.00 + 2.50 = 3.50). This puts AB and BC in a straight line as shown in Figure 19.11.

STEP 3 Establish the extreme left position of link CD by subtracting the length of link AB from BC (2.50 − 1.00 = 1.50). This places link AB and BC in a straight line to the left as shown in Figure 19.12.

STEP 4 Determine the angle between the extreme right and extreme left positions of CD as shown in Figure 19.13.

Slider Crank Mechanism

A **slider crank** is a linkage mechanism that is commonly used in machines such as engine pistons, pumps, and clamping devices where a straight line motion is required. Figure 19.14 shows the

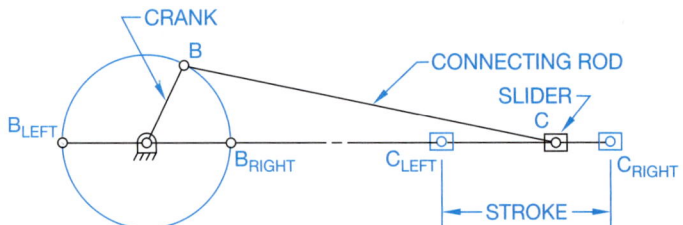

FIGURE 19.14 ■ Slider mechanism.

type of slider design used to move a piston back and forth in a straight line. Notice that the distance the slider travels from the extreme left to the extreme right position is called the **stroke**. When working with the design of an engine, it is important to establish the piston stroke and diameter. The distance the piston travels is the stroke. A combination of the stroke and piston diameter determines the **piston displacement**. Therefore, a four-cylinder, 2000-cc engine has a displacement of 500 cc per piston cylinder. You can graphically show the stroke of a piston if you have an engine with a crank length of 1.25 in. (32 mm) and a connecting rod length of 3.25 in. (82 mm). Determine the length of the piston stroke as follows:

STEP 1 AND STEP 2 Draw an arc with point A as its center and AB = 1.25 in. (32 mm) as the radius. Show the connecting rod attached to the piston as link BC. Add AB to BC (1.25 + 3.25 = 4.50) (32 + 82 = 114 mm). Measure this dis-

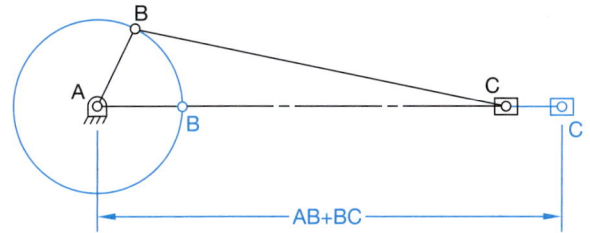

FIGURE 19.15 ■ Steps 1 and 2, slider mechanism, extreme right position.

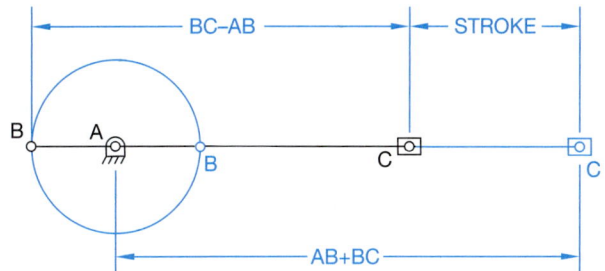

FIGURE 19.16 ■ Step 3, slider mechanism, extreme left position.

tance from point A along the centerline through A and C. This establishes the extreme right position as shown in Figure 19.15.

STEP 3 Subtract AB from BC ($3.25 - 1.25 = 2.00$)($82 - 32 = 50$ mm). Measure from point A along the centerline through A and C. This establishes the extreme left position of point C. Measure the distance between the extreme left and right positions to determine the stroke as shown in Figure 19.16.

Combination Four-Bar Linkage and Slider Mechanism

The design of linkage mechanisms is only limited by the imagination of the designer. There can be any variety of combinations. In the previous examples, the extreme right and left positions were established to determine the function of the mechanism. However, in many situations the designer must establish many positions to completely analyze the movement. One such example is shown in Figure 19.17. This mechanism is a combination four-bar linkage and slider. The objective is to determine the path of point P as the link AB rotates 360°. In order to effectively solve a problem of this type, it is necessary to plot the path of point P by moving link AB a minimum of 30° increments. This is best accomplished if you use a different color or line type for each position. It can become confusing; therefore, careful labeling of each position is important.

STEP 1 Draw a circle through point B with A as the center. Point B must remain on this circle through each movement. Draw an arc through point C with D as the center. Point C must remain on this arc through each movement of the mechanism.

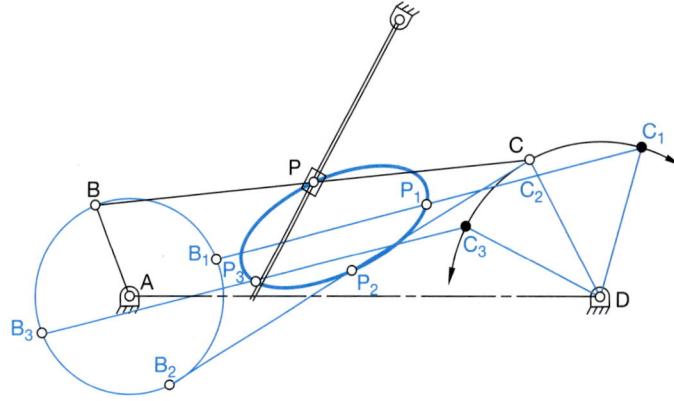

FIGURE 19.17 ■ Solution to four-bar linkage/slider mechanism example.

STEP 2 While 30° increments are recommended, this example uses 90° increments for convenience and clarity. Move point B clockwise to B_1. (See Figure 19.17.) From B_1 draw an arc with a radius equal to BC until it intersects arc CD, establishing point C_1. Draw link B_1C_1. Measure distance BP and transfer it to link B_1C_1, thus establishing point P_1.

STEP 3 Follow the same procedure for positions B_2 and B_3. This determines the path of coupler point P at positions P_2 and P_3.

CAM DESIGN

A **cam** is a rotating mechanism that is used to convert rotary motion into a corresponding straight motion. The timing involved in the rotary motion is often the main design element of the cam. For example, a cam can be designed to make a follower rise a given amount in a given degree of rotation, then remain constant for an additional period of rotation, and finally fall back to the beginning in the last degree of rotation. The total movement of the cam follower happens in one 360° rotation of the cam. This movement is referred to as the **displacement**. Cams are generally in the shape of irregular plates, grooved plates, or grooved cylinders. The basic components of the cam mechanism are shown in Figure 19.19.

Cam Types

There are basically three different types of cams: the plate cam, face cam, and drum cam. (See Figure 19.20.) The plate cam is the most commonly used type of cam.

Cam Followers

There are several types of **cam followers**. The type used depends on the application. The most common type of follower is the roller follower. The **roller follower** works well at high speeds, reduces friction and heat, and keeps wear to a minimum. The arrangement of the follower in relation to the cam shaft may differ depending

MECHANISM DESIGN

Computer-aided design and drafting is a natural in mechanism design and drafting. Mechanism design drawings are created using symbols and techniques that are easily adapted to CADD systems. You will see later that linkage mechanisms, for example, are often designed by displaying the mechanism in several different positions. Using CADD to do this allows you to place each different position of movement on a separate layer and in a different color. Any one or more layers can be turned on or off at your convenience to evaluate the function of the mechanism. In addition, the CADD system is much faster and more accurate than manual design and drafting methods.

There are computer programs available to make the design and drafting of mechanisms easy. For example, there are programs that can be used to simulate the movement of linkage mechanisms. The illustrations in this chapter were created using a CADD system.

Using CADD in linkage design allows you to quickly and accurately establish a series of positions for the mechanism. After the first position is determined, you can easily use commands such as COPY and ROTATE to place the linkages in alternate positions. You can even set up a continuous slide show with each position of the mechanism as a screen display in the slide show. The term **slide show** refers to screen displays that are sequenced in a specific order and for a given period of time. If you allow the slide

show to run fairly fast, the viewer gets a good understanding and display of the mechanism in action. This type of a CADD display can be established in 2-D or 3-D as shown in Figure 19.18.

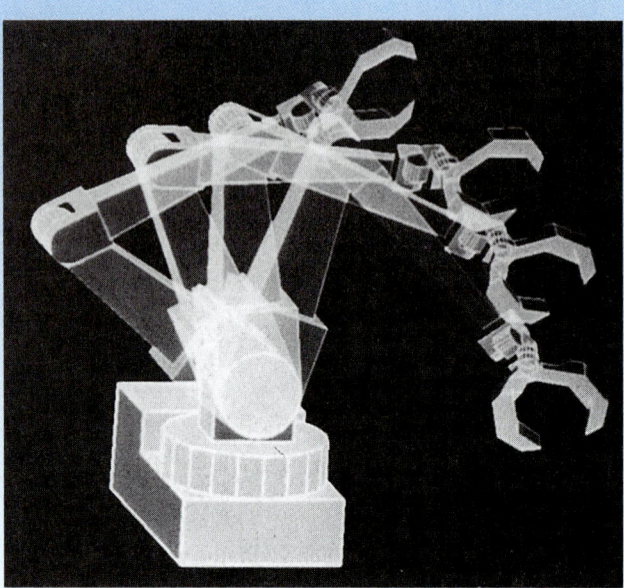

FIGURE 19.18 ■ Linkage mechanism designed with 3-D CADD showing the movement at several positions. *Courtesy Spectragraphics.*

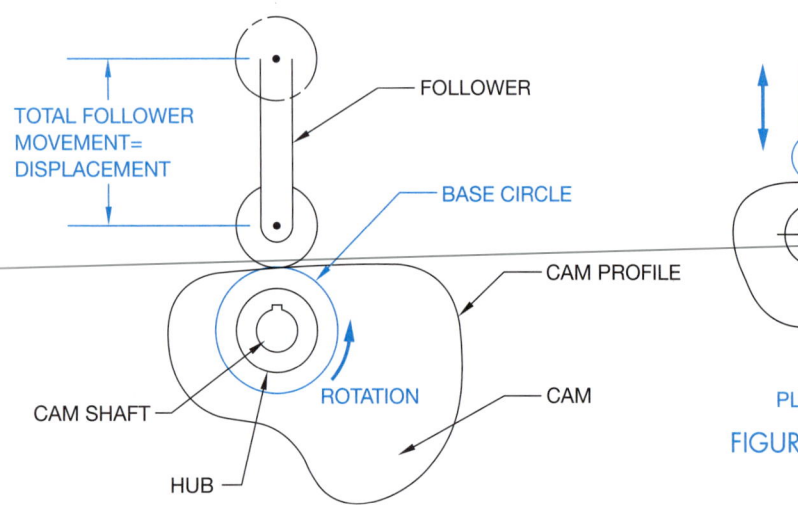

FIGURE 19.19 ■ Elements of a cam mechanism.

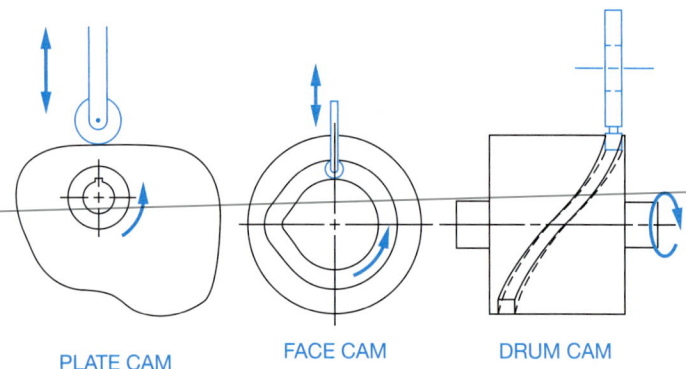

FIGURE 19.20 ■ Types of cam mechanisms.

on the application. The roller followers shown in Figure 19.21 include: the in-line follower where the axis of the follower is in line with the cam shaft; the offset roller follower; and the pivoted follower. The pivoted follower requires spring tension to keep the follower in contact with the cam profile.

Another type of cam follower is the **knife-edged follower** shown in Figure 19.22a. This follower is used for only low-

speed and low-force applications. The knife-edged follower has a low resistance to wear but is very responsive and may be effectively used in situations that require abrupt changes in the cam profile.

The **flat-faced follower** shown in Figure 19.22b is used in situations where the cam profile has a steep rise or fall. Designers often offset the axis of the follower. This practice causes the follower to rotate while in operation. This rotating action allows the follower surface to wear evenly and last longer.

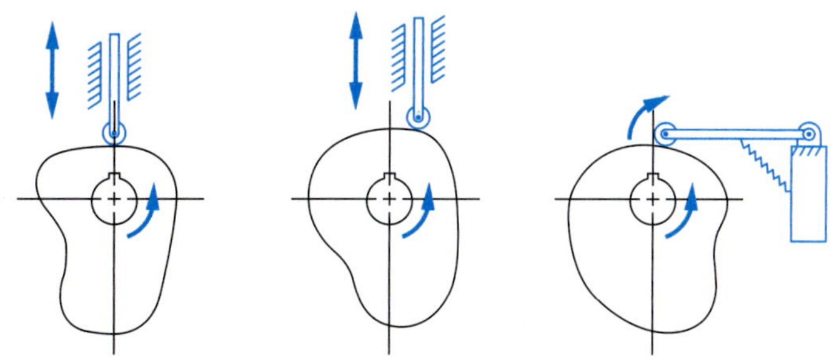

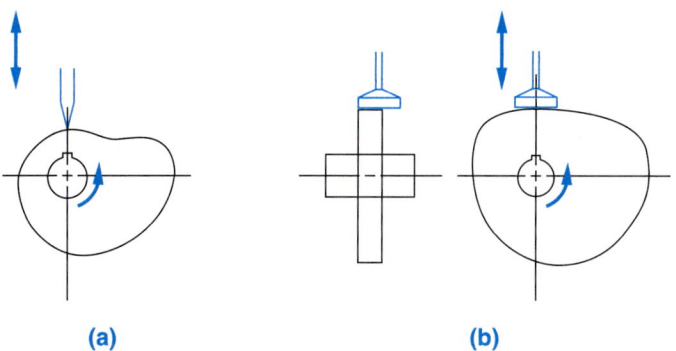

IN-LINE FOLLOWER   OFFSET FOLLOWER   PIVOTED FOLLOWER

FIGURE 19.21 ■ Types of cam roller followers.

(a)   **(b)**

FIGURE 19.22 ■ (a) Knife-edged cam follower. (b) Flat-faced cam follower.

CAM DISPLACEMENT DIAGRAMS

Cams are generally designed to achieve some type or sequence of a timing cycle in the movement of the follower. There are several predetermined types of motion from which cams are designed. These forms of motion can be used alone, in combination, or custom designed to suit specific applications. The following discussion shows you how to set up a cam **displacement diagram** given a specific type of cam motion. The cam displacement diagram is similar to a graph representing the cam profile in a flat pattern of one complete 360° revolution of the cam. The terms associated with the displacement diagram include **cycle**, **period**, **rise**, **fall**, **dwell**, and **displacement**. A complete cam cycle has taken place when the cam rotates 360°. A period of the cam cycle is a segment of follower operation such as rise, dwell, or fall. Rise exists when the cam is rotating and the follower is moving upward. Fall is when the follower is moving downward. Dwell exists when the follower is constant, not moving either up or down. A dwell is shown in the displacement diagram as a horizontal line for a given increment of degrees. When developing the cam displacement diagram, the height of the diagram is drawn to scale and is equal to the total follower displacement. (See Figure 19.23.) The horizontal scale is equal to one cam revolution or 360°. The horizontal scale may be drawn without scale. Some engineering drafters prefer to make this scale equal to the circumference of the base circle, or any convenient length. The base circle is an imaginary circle with its center at the center of the cam shaft and its radius tangent to the cam follower at zero position. The hori-

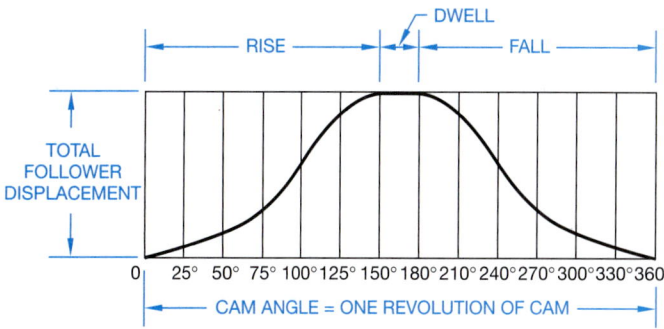

FIGURE 19.23 ■ Cam displacement diagram.

zontal scale is then divided into increments of degrees. Each rise and fall is divided into six increments. So, if the follower rises 150°, then each increment is 150°/6 = 25°. If the cam falls between 180° and 360°, this represents a total fall of 180°. The fall increments are 180°/6 = 30° as shown in Figure 19.23.

Simple Harmonic Motion

Simple harmonic motion can be used for high-speed applications if the rise and fall are equal at 180°. Moderate speeds are recommended if the rise and fall are unequal or if there is a dwell in the cycle. This application causes the follower to jump if the speeds are too high.

Draw a cam displacement diagram using simple harmonic motion when the total displacement is 2.00 in. (50 mm) and the cam follower rises the length of the total displacement in 180° and falls back to 0° in 180°. Use the following procedure to set up the displacement diagram:

STEP 1   Draw a rectangle equal in height (vertical scale) to the total displacement of 2.00 in. (50 mm) and equal in length (horizontal scale) to 360°. The horizontal scale should have 6–30° increments for the rise from 0° to 180° and 6–30° increments for the fall from 180° to 360°. The horizontal scale can be any convenient length. Draw a thin vertical line from each horizontal increment as shown in Figure 19.24.

STEP 2   Draw a half circle at one end of the displacement diagram equal in diameter to the rise of the cam. Divide the half circle into six equal parts as shown in Figure 19.25.

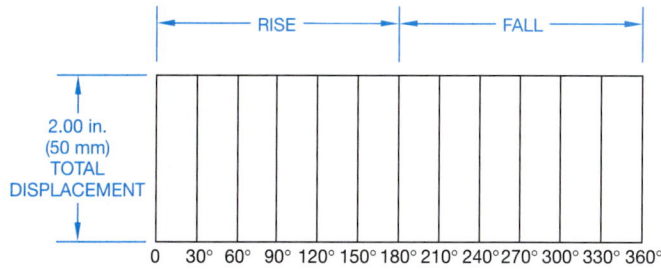

FIGURE 19.24 ■ Step 1, layout for simple harmonic motion cam displacement diagram.

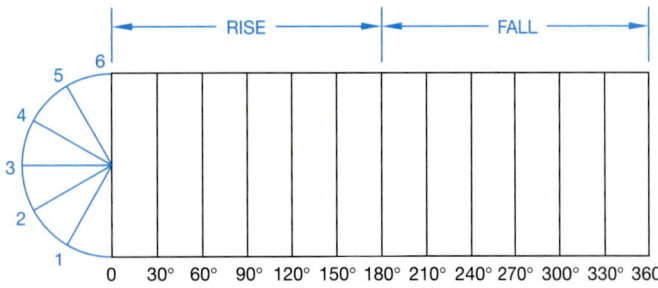

FIGURE 19.25 ■ Step 2, layout for simple harmonic motion cam displacement diagram.

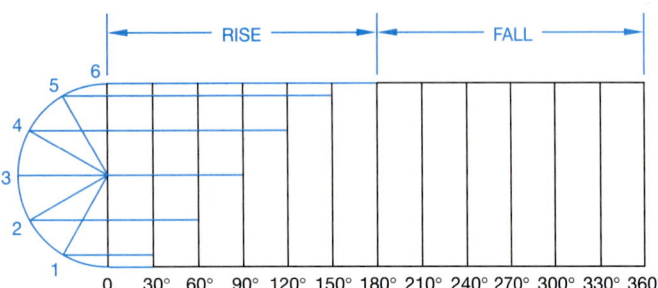

FIGURE 19.26 ■ Steps 3, 4, and 5, layout for simple harmonic motion cam displacement diagram.

STEP 3 The cam follower begins its rise at 0°. The rise continues by projecting point 1 on the half circle over to the first (30°) increment on the horizontal scale. Continue this process for points 2, 3, 4, 5, and 6 on the half circle, each intersecting the next increment on the horizontal scale.

STEP 4 Notice the pattern of points created in the preceding step. Use your irregular curve to carefully connect these points. If you are using CADD, use your curve-fitting command to draw the cam profile.

STEP 5 Develop the fall profile by projecting the points from the half circle in the reverse order discussed in Step 3. (See Figure 19.26.)

Constant Velocity Motion

Constant velocity motion, also known as straight-line motion, is used for the feed control of some machine tools when it is required for the follower to rise and fall at a uniform rate. Constant velocity motion is only used at slow speeds because of the abrupt change at the beginning and end of the motion period. The displacement diagram is easy to draw. All you have to do is draw a straight line from the beginning of the rise or fall to the end as shown in Figure 19.27.

Modified Constant Velocity Motion

Modified constant velocity motion was designed to help reduce the abrupt change at the beginning and end of the motion period. This type of motion can be adjusted to accomplish specific

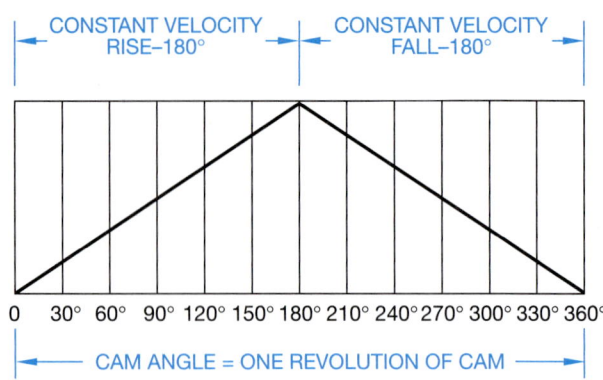

FIGURE 19.27 ■ Cam displacement diagram for constant velocity motion.

results by altering the degree of modification. This is done by placing a curve at the beginning and end of the rise and fall. The radius of this arc depends on the amount of smoothing required, but the radius normally ranges from one-third to full displacement. If the motion were modified to one-third the displacement, then the cam displacement diagram is drawn as shown in Figure 19.28.

Uniform Accelerated Motion

Uniform accelerated motion is designed to reduce the abrupt change at the beginning and end of a period. It is recommended for moderate speeds, especially when associated with a dwell. The advantage of this motion is its use when constant acceleration for the first half of the rise and constant deceleration for the second half of the rise are required.

Use the following technique to draw a uniform accelerated motion displacement diagram where the follower rises a total of 2.00 in. (50 mm) in 180° and falls back to 0° in 180°:

STEP 1 AND STEP 2 Set up the displacement diagram with the height equal to 2.00 in. total rise and the horizontal scale divided into 30° increments. Keep in mind that the horizontal scale is divided into 30° increments if the rise and fall is in 180° each. If the rise, for example, were 120°, then the increments would be 120°/6 = 20° each. Set up a scale with 18 equal divisions at one end of the dis-

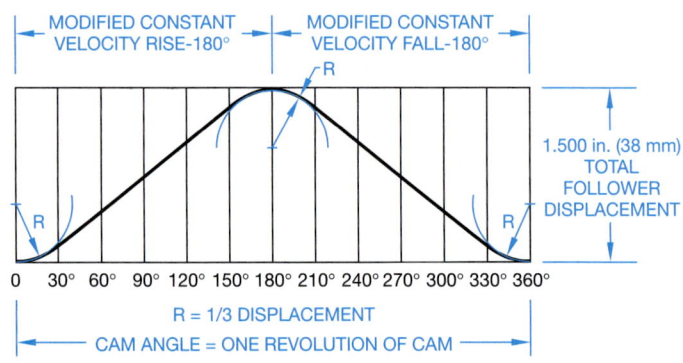

FIGURE 19.28 ■ Cam displacement diagram for modified constant velocity motion.

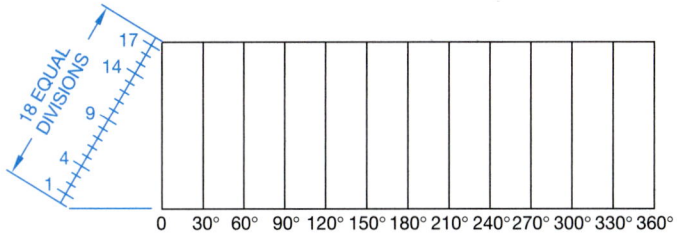

FIGURE 19.29 ■ Steps 1 and 2, layout for uniform accelerated motion cam displacement diagram.

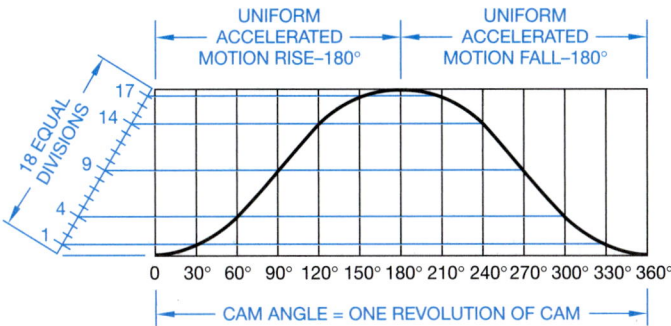

FIGURE 19.30 ■ Step 3, layout for uniform accelerated motion cam displacement diagram.

placement diagram and mark off the first, fourth, ninth, fourteenth and seventeenth divisions as shown in Figure 19.29.

STEP 3 Establish the rise by projecting from the first division on the scale to the 30° increment on the diagram, then continue with the fourth division to 60°, and so on until each division is used. Continue this same procedure in reverse order to establish the profile of the fall. Connect all of the points to complete the displacement diagram as shown in Figure 19.30.

Cycloidal Motion

Cycloidal motion is the most popular cam profile development for smooth-running cams at high speeds. The term **cycloidal** comes from the word **cycloid**. A cycloid is a curved line generated by a point on the circumference of a circle as the circle rolls along a straight line. The cycloidal cam motion is developed in this same manner, and the result is the smoothest possible cam profile. Cycloidal motion is a little more complex to set up than other types of motion. Use the following procedure to develop a cam displacement diagram for cycloidal motion with a total rise of 2.500 in. (64 mm) in 180°:

STEP 1 Begin the displacement diagram with a total rise of 2.500 in. (64 mm) in 180°. Only half of the diagram is shown for this example.

STEP 2 Draw a circle tangent to and centered on the total displacement at one end of the diagram. This circle must have a circumference equal to the displacement. Calculate the diameter using the formula $D = C/\pi$. In this

case $D = 2.500/3.1414 = .7958$ (64/3.1414 = 20.37 mm). Round off to .80 (20 mm) for manual drafting or use as is for a CADD application. Then, beginning at the point where the circle is tangent to the diagram, divide the circle into six equal parts and number them as shown in Figure 19.31.

STEP 3 Draw a vertical line through the center of the circle equal in length to the displacement. Divide this line into six equal parts as shown in Figure 19.32.

STEP 4 Use a radius equal to the radius of the circle to draw arcs from the divisions on the vertical line as shown in Figure 19.33. These arcs should intersect the dashed lines

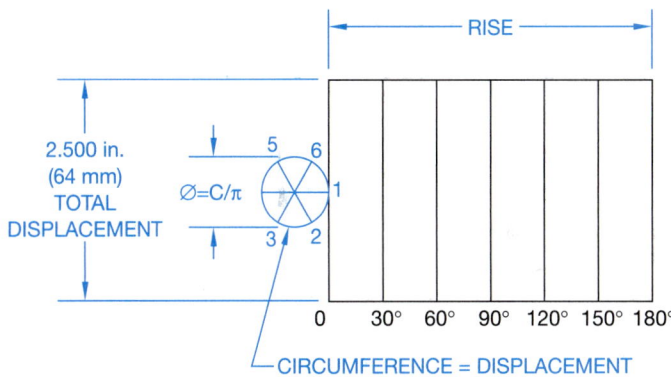

FIGURE 19.31 ■ Steps 1 and 2, layout for cycloidal motion cam displacement diagram.

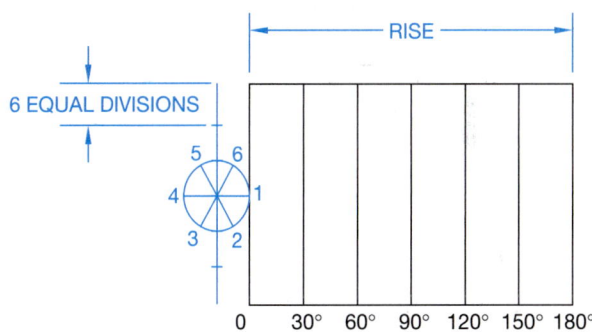

FIGURE 19.32 ■ Step 3, layout for cycloidal motion cam displacement diagram.

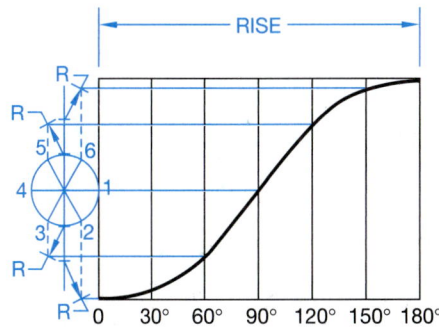

FIGURE 19.33 ■ Step 4, layout for cycloidal motion cam displacement diagram.

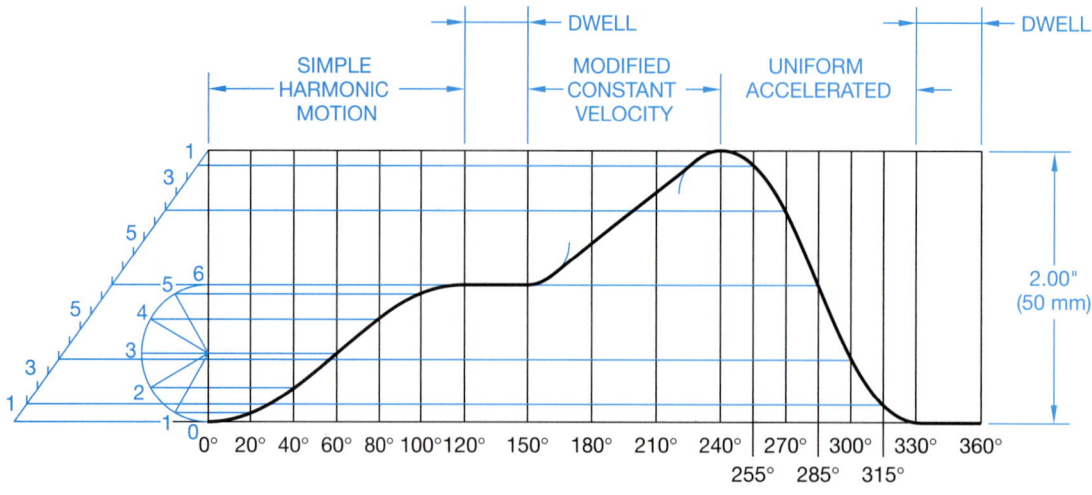

FIGURE 19.34 ■ The development of a cam displacement diagram with different cam motions.

drawn from points 2, 3, 5, and 6 on the circle. Where the dashed lines and the arcs intersect, draw horizontal lines into the displacement diagram, intersecting the appropriate increment from the horizontal scale. Connect the points of intersection with a smooth curve as shown in Figure 19.33.

Developing a Cam Displacement Diagram with Different Cam Motions

Most cam profiles are not as simple as the preceding examples. Many designs require more than one type of cam motion and can also incorporate dwell. Construct a cam displacement diagram from the following information:

■ Total displacement equals 2.00 in. (50 mm)

■ Rise 1.00 in. (25 mm) simple harmonic motion in 120°.

■ Dwell for 30°.

■ Rise 1.00 in. (25 mm) modified constant velocity motion in 90°.

■ Fall 2.00 in. (50 mm) uniform accelerated motion in 90°.

■ Dwell for 30° through the balance of the cycle.

Look at Figure 19.34 as you review the method of development for each type of cam motion displayed.

CONSTRUCTION OF AN IN-LINE FOLLOWER PLATE CAM PROFILE

Each of the cam displacement diagrams can be used to construct the related cam profiles. The **cam profile** is the actual contour of the cam. In operation, the cam follower is stationary and the cam rotates on the cam shaft. In cam profile construction, however, the cam is drawn in one position and the cam follower is moved to a series of positions around the cam in relationship to the cam

displacement diagram. The following technique is used to draw the cam profile for the cam displacement diagram given in Figure 19.35 and for a cam with a 2.50 in. (64 mm) base circle, a .75 in. (19 mm) cam follower, and counterclockwise rotation:

STEP 1   Refer to Figure 19.35. Use construction lines for all preliminary work. Draw the cam follower in place near the

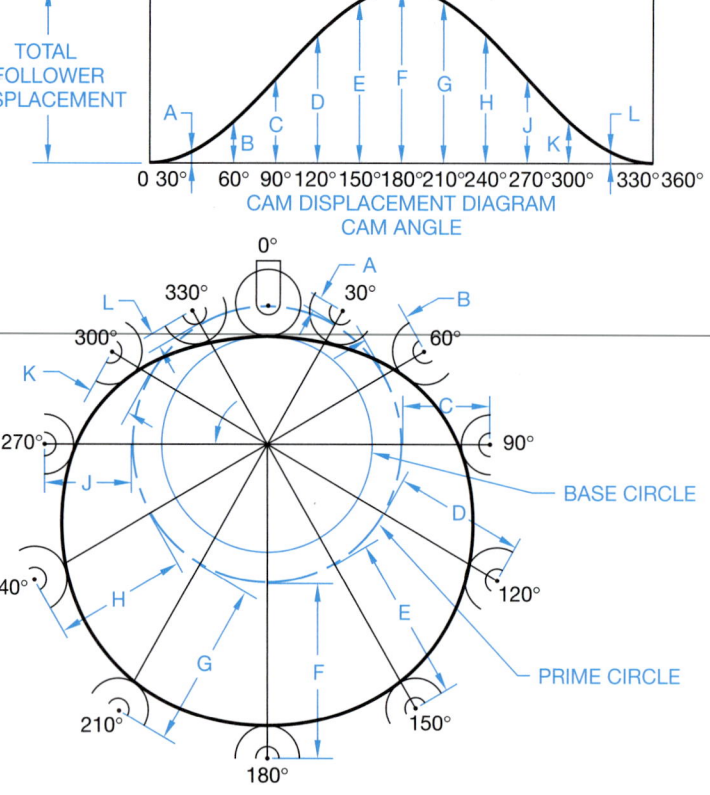

FIGURE 19.35 ■ Construction of an in-line follower plate cam profile.

top of the sheet with phantom lines. Draw the base circle of ∅2.50 in. (64 mm). The base circle is tangent to the follower at the 0° position. Draw the prime circle 2.50 + .75 = ∅3.25 in. (64 + 19 = ∅83 mm). The prime circle passes through the center of the follower at the 0° position. The cam displacement diagram is placed on Figure 19.35 for easy reference. In actual practice, the displacement diagram is next to the cam profile drawing or on a CADD layer to be inserted on the screen for reference.

STEP 2 Begin working in a direction opposite from the rotation of the cam. Since this cam rotates counterclockwise, work clockwise. Starting at 0° draw an angle equal to each of the horizontal scale angles on the displacement diagram. In this case, the angles are 30° increments. This is not true with all cam displacement diagrams. Some have varying increments.

STEP 3 Notice the measurements labeled A through L on the displacement diagram in Figure 19.35. Begin by transferring the distance A along the 30° element (in the profile construction drawing) by measuring from the prime circle along this line. This establishes the center of the cam follower at this position. Do the same with each of the measurements B through L located on each corresponding increment.

STEP 4 Lightly draw the cam followers in position with centers located at each of the points found in step 3 (See Figure 19.35.)

STEP 5 Draw the cam profile by connecting a smooth curve tangent to the cam followers at each position. This is one place where CADD drafting plays an important role regarding accuracy and speed.

PREPARING THE FORMAL PLATE CAM DRAWING

You are ready to prepare the formal plate cam drawing after you have constructed the plate cam profile using the previously described techniques. The previous drawing can be used because all lines are drawn as construction lines or on a CADD layer that can be turned off when the formal drawing is complete. The information needed on the plate cam drawing includes:

■ Cam profile.

■ Hub dimensions, including cam shaft, outside diameter, width; keyway dimensions.

■ Roller follower placed in one convenient location, such as 60°, using phantom lines.

■ The drawing is set up as a chart drawing where A° equals the angle of the follower at each position, and R equals the radius from the center of the cam shaft to the center of the follower at each position.

■ A chart giving the values of the angles A and the radii R at each follower position.

■ Side view showing the cam plate thickness and set screw location with thread specification, if used.

■ Tolerances, unless otherwise specified.

Establish all measurements for dimensions A and R at each of the follower positions. This is done graphically by measuring from the profile construction, or mathematically using trigonometry. If a CADD system is used, the measurements are taken directly from the layout. Figure 19.36 shows a formal plate cam drawing.

CONSTRUCTION OF AN OFFSET FOLLOWER PLATE CAM PROFILE

When an offset cam follower is used, the method of construction is a little more complex than the technique used for the in-line follower. For this example, the follower is offset .75 in. (19 mm) as shown in Figure 19.37. Prepare the cam profile drawing as follows:

STEP 1 Refer to Figure 19.37. Use construction lines or a construction CADD layer for all preliminary work. Draw the cam follower in place near the top of the sheet with phantom lines. Draw the base circle ∅2.50 in. (64 mm). The base circle is tangent to the follower at 0° position. Draw the prime circle: (2.50 + .75 = ∅3.25 in.) (64 + 19 = ∅83 mm). The prime circle passes through the center of the follower at 0° position. Draw the offset circle ∅1.50 in. (38 mm). The offset circle is drawn with a radius equal to the follower offset distance. The cam displacement diagram is placed on Figure 19.37 for easy reference. In actual practice, the displacement diagram is next to the cam profile drawing or on a CADD layer to be inserted on the screen for reference.

STEP 2 Begin working in a direction opposite from the rotation of the cam. Since this cam rotates counterclockwise, work clockwise. Starting at 0° on the offset circle, draw an angle equal to each of the horizontal scale angles on the displacement diagram. In this case the angles are 30° increments. This is not true with all cam displacement diagrams, because some have varying increments.

STEP 3 Notice where each of the angle increments drawn in step 2 intersects the offset circle. At each of these points, draw another line tangent to the offset circle. Make these lines long enough to pass beyond the reference circle.

STEP 4 Notice the measurements labeled A through L on the displacement diagram in Figure 19.37. Begin by transferring the distance A along the line tangent to the 30° element by measuring from the prime circle along this line. This establishes the center of the cam follower at this position. Do the same with each of the measurements B through L located on the tangent line from each corresponding increment.

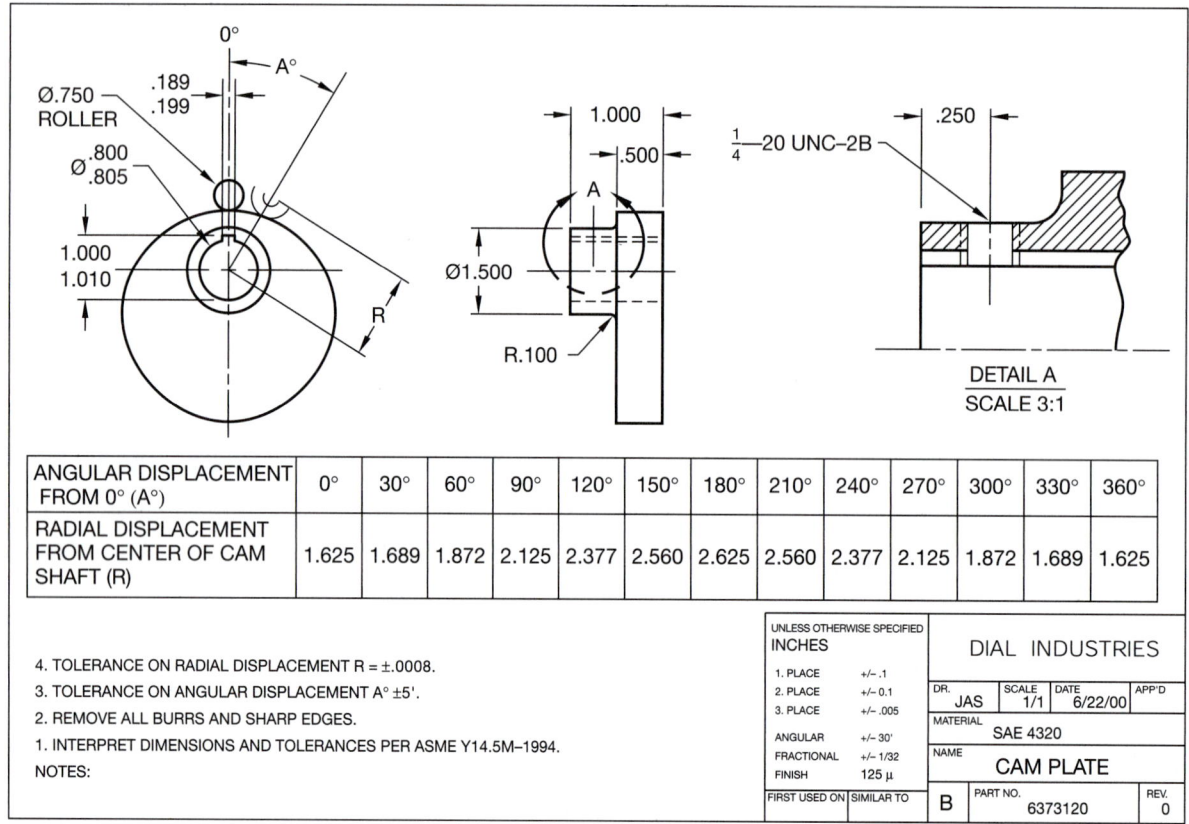

FIGURE 19.36 ■ Formal plate cam drawing. Dimension values in this figure are in inches. *Courtesy Dial Industries.*

ANGULAR DISPLACEMENT FROM 0° (A°)	0°	30°	60°	90°	120°	150°	180°	210°	240°	270°	300°	330°	360°
RADIAL DISPLACEMENT FROM CENTER OF CAM SHAFT (R)	1.625	1.689	1.872	2.125	2.377	2.560	2.625	2.560	2.377	2.125	1.872	1.689	1.625

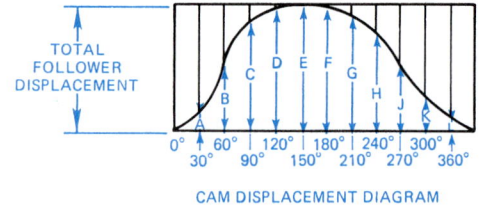

CAM DISPLACEMENT DIAGRAM

STEP 5 Lightly draw the cam followers in position with centers located at each of the points found in Step 4. (Refer to Figure 19.37.)

STEP 6 Draw the cam profile by connecting a smooth curve tangent to the cam followers at each position. This is one place where CADD drafting plays an important role regarding accuracy and speed.

DRUM CAM DRAWING

Drum cams are used when it is necessary for the follower to move in a path parallel to the axis of the cam. The drum cam is a cylinder with a groove machined in the surface the shape of the cam profile. The cam follower moves along the path of the groove as the drum is rotated. The displacement diagram for a drum cam is actually the pattern of the drum surface as if it were rolled out flat. The height of the displacement diagram is equal to the height of the drum. The length of the displacement diagram is equal to the circumference of the drum. Refer to the drum cam drawing in Figure 19.38 as you follow the construction steps:

STEP 1 Draw the top view showing the diameter of the drum, the cam shaft and keyway, and the roller follower in place at 0°. Draw the front view as shown in Figure 19.38. Draw the outline of the cam displacement diagram equal to the height and circumference of the drum.

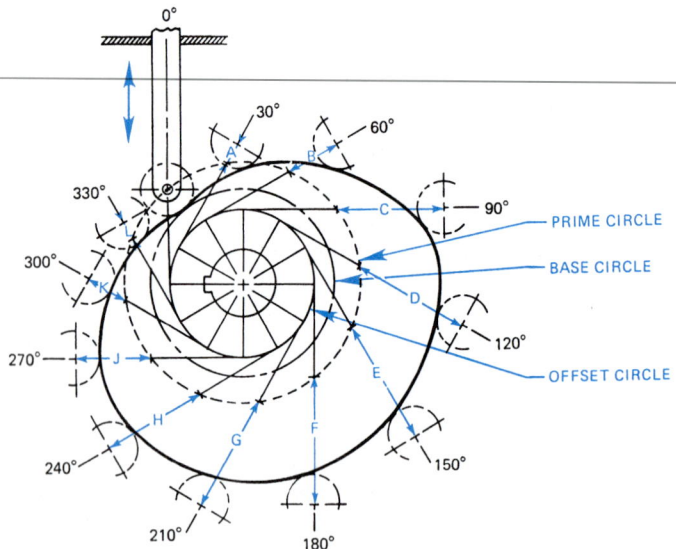

FIGURE 19.37 ■ Construction of an offset follower plate cam profile.

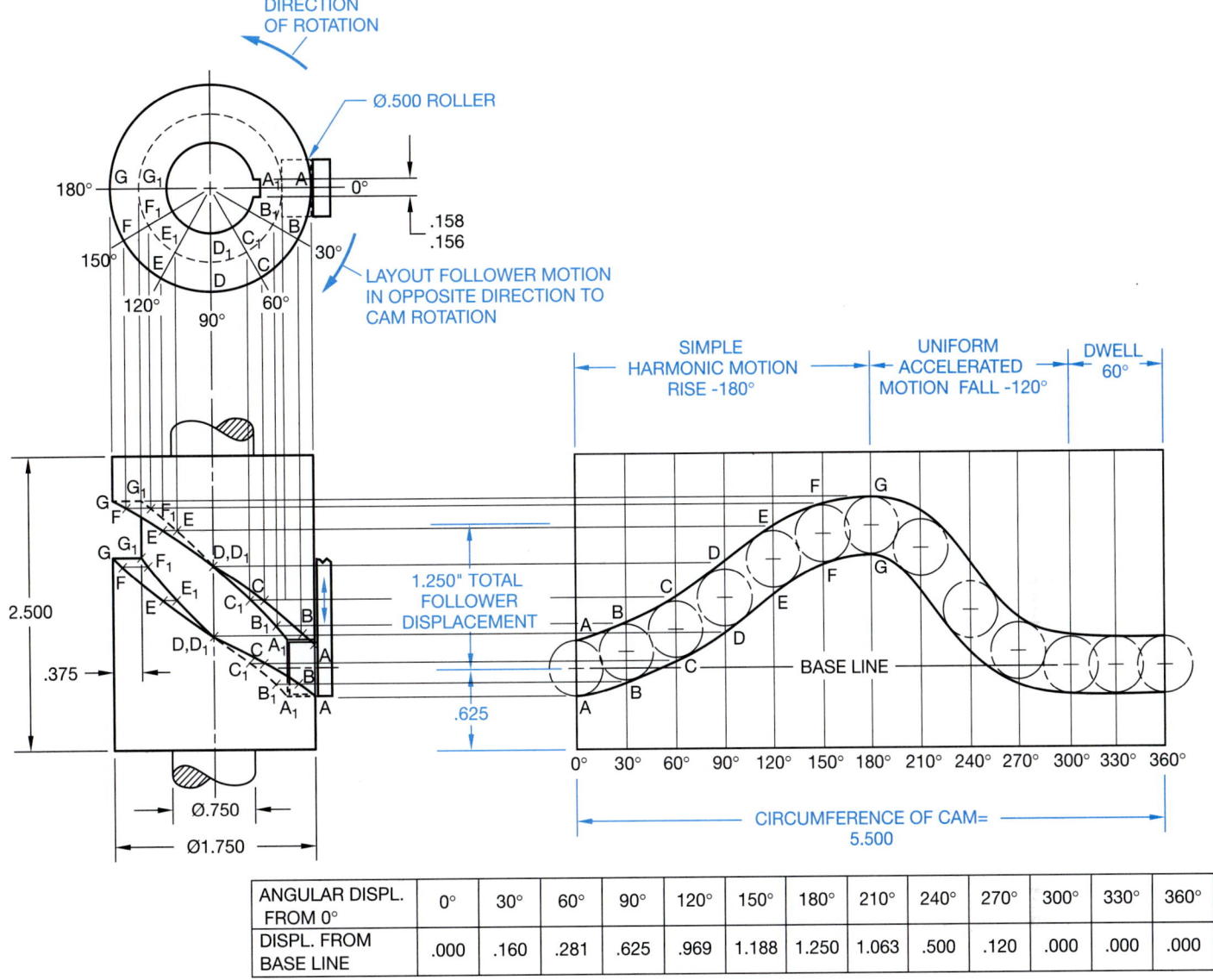

FIGURE 19.38 ■ Construction of a drum cam drawing. Dimension values in this figure are in inches.

ANGULAR DISPL. FROM 0°	0°	30°	60°	90°	120°	150°	180°	210°	240°	270°	300°	330°	360°
DISPL. FROM BASE LINE	.000	.160	.281	.625	.969	1.188	1.250	1.063	.500	.120	.000	.000	.000

2. TOLERANCE ON ANGULAR DISPLACEMENT ± .5°.
1. TOLERANCE ON DISPLACEMENT FROM BASELINE ± .0008.
NOTES:

STEP 2 Draw the roller follower on the displacement diagram at each angular interval.

STEP 3 Draw curves tangent to the top and bottom of each roller follower position. These curves represent the development of the groove in the surface of the cam.

STEP 4 In the top view, draw radial lines from the cam shaft center equal to the angle increments shown on the displacement diagram. Be sure to lay out the angles in a direction opposite the cam rotation. The points where these lines intersect the depth and the outside circumference of the groove are labeled A, A₁, B, B₁, respectively. Notice that the same corresponding points are labeled on the displacement diagram.

STEP 5 From points A, A₁ on the displacement diagram, project horizontally until each point intersects a vertical line from the same corresponding point in the top view. This establishes the points in the front view along the outer and inner edges of the groove, both top and bottom. Continue this process for each pair of points on the drum cam displacement diagram.

GEAR INTRODUCTION

Gears are toothed wheels used to transmit motion and power from one shaft to another. Gears are rugged and durable and can transmit power with up to 98% efficiency with long service life. Gear design involves a combination of material, strength, and

CAM DESIGN

Designing cams is easy with CADD if a parametric program is used. In a system of this type all you have to do is change the variables and automatically create a new cam or gear. For cam design, the variables are:

- Type of cam motion.
- Follower displacement.
- The specific rise, dwell, and fall configuration.
- Prime circle and base circle diameters.
- In-line or offset follower.
- Hub diameter, hub projection, face width, shaft, and keyway specifications.

The program automatically calculates the rest of the data and draws the cam profile in a detail drawing.

CAM DISPLACEMENT DIAGRAMS AND PROFILES

CADD systems make drawing cam displacement diagrams and cam profiles easy and accurate without the need for mathematical calculations. Constructing the cam displacement diagram and converting the information to the cam profile is the same as with manual drafting, except the accuracy is increased substantially. Drawing the cam profile using irregular curves is difficult and time-consuming. Most CADD systems have a curve-fitting command that makes it possible to automatically draw the cam profile through the points of tangency at the cam follower positions.

CAM MANUFACTURING

Modern cams are manufactured using computer numerical control (CNC) machining. In many instances, the drawing is prepared on a CADD system and transferred to a computer-aided manufacturing (CAM) program for immediate coding for the CNC machine. Some CAM programs allow the designer to create the cam profile and transfer the data to the CNC machine tool without ever generating a drawing. Refer to Chapter 5 on Manufacturing Processes for more information.

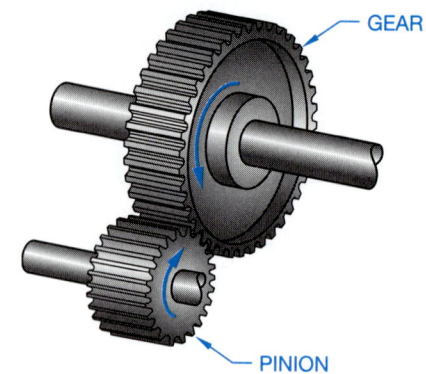

FIGURE 19.39 ■ The gear and pinion.

wear characteristics. Most gears are made of cast iron or steel, but brass and bronze alloys and plastic are also used for some applications. Gear selection and design are often done through vendors' catalogs or the use of standard formulas. A **gear train** exists when two or more gears are in combination for the purpose of transmitting power. Generally, two gears in mesh are used to increase or reduce speed, or change the direction of motion from one shaft to another. When two gears are in mesh, the larger is called the **gear** and the smaller is called the **pinion**. (See Figure 19.39.)

AGMA/ANSI Gear selection generally follows the guidelines of the American Gear Manufacturers Association (AGMA) or the American National Standards Institute (ANSI).

It is important for you to fully understand gear terminology and formulas. However, many drafters do not draw gears, because gears are commonly supplied as purchase parts. When this happens, you may be required to make gear selections for specific applications or draw gears on assembly drawings. Details of gears are often drawn using simplified techniques as described in this chapter. In many situations, the gears are drawn as they actually exist for display on assembly drawings, or in catalogs. When this is necessary, CADD can make the job easy and increase productivity. If manual drafting is used, a gear tooth template is needed. Gear data is readily available in the *Machinery's Handbook*.

GEAR STRUCTURE

Gears are made in a variety of structures depending on the design requirements, but there is some basic terminology that can typically be associated with gear structure. The elements of the gear structure shown in Figure 19.40 include the **outside diameter**, **face**, **hub**, **bore**, and **keyway**.

Types of Hubs

The **hubs** are the lugs or shoulders projecting from one or both faces of some gears. These can be referred to as A, B, or C hubs. An A hub is also called a flush hub, because there is no projection from the gear face. B hubs have a projection on one side of the gear, while C hubs have projections on both sides of the gear face. (See Figure 19.41.)

Keyways, Keys, and Set Screws

Gears are usually held on the shaft with a key, keyway, and set screw. Refer to Chapter 5 for information on the manufacture of keyways, Chapter 11 for proper dimensioning, and Chapter 12 for screw thread specifications. One or more set screws are usually used to keep the key secure in the keyway. (See Figure 19.42.)

SPLINES

Splines are teeth cut in a shaft and a gear or pulley bore and are used to prevent the gear or pulley from spinning on the shaft.

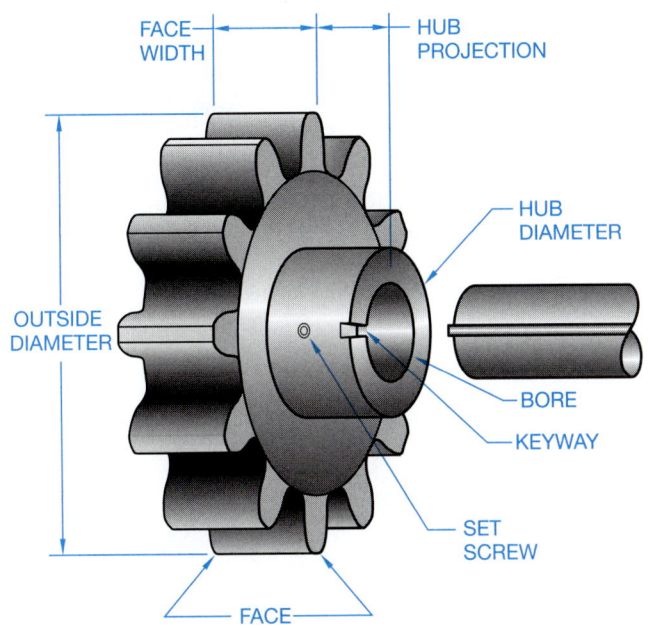

FIGURE 19.40 ▪ Elements of the gear structure.

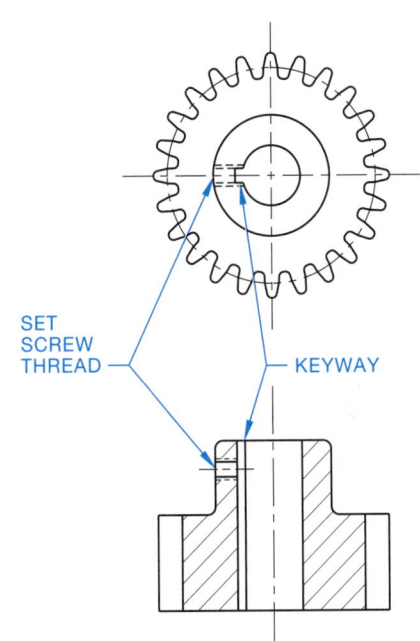

FIGURE 19.42 ▪ Gear, keyways, and set screw.

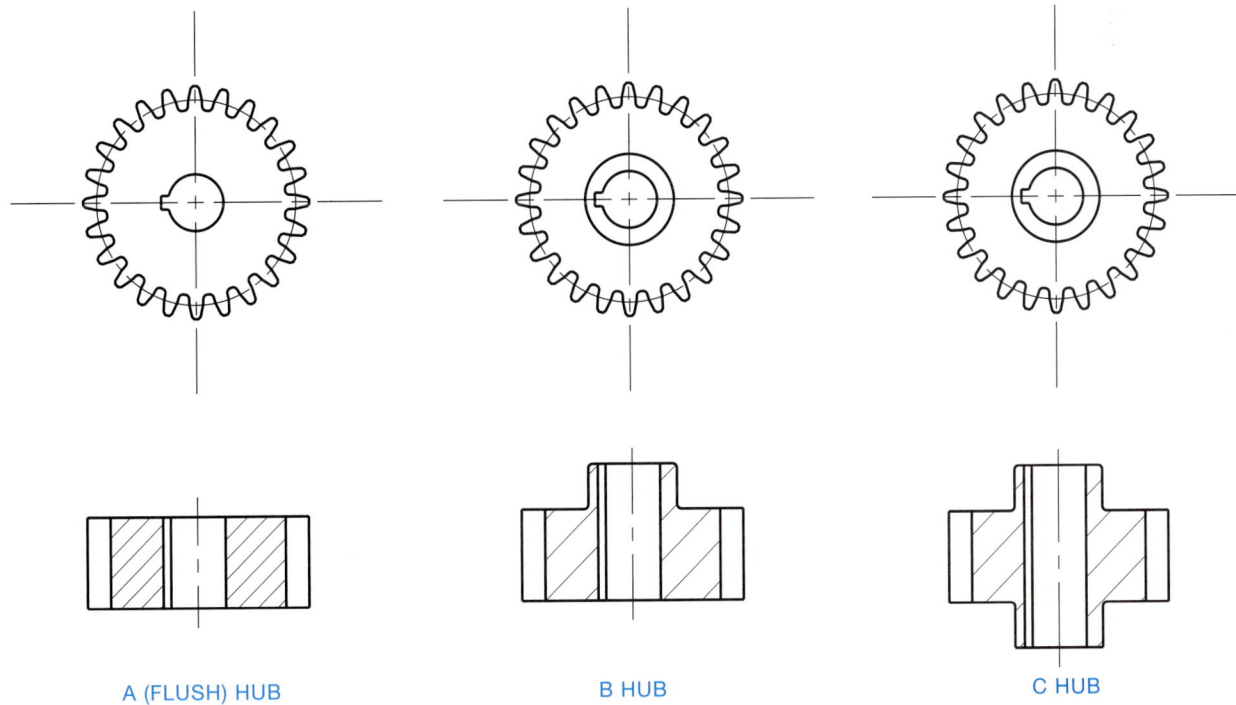

A (FLUSH) HUB   B HUB   C HUB

FIGURE 19.41 ▪ Types of gear hubs.

Splines are often used when it is necessary for the gear or pulley to easily slide on the shaft. Splines can also be nonsliding and in all cases are stronger than keyways and keys. The standardization of splines is established by the Society of Automotive Engineers (SAE) so any two parts with the same spline specifications fit together. The following is an example of an SAE spline specification:

SAE 2 1/2-10   B SPLINE
(A)   (B)   (C) (D)

The note components are described as follows:
   (A) Society of Automotive Engineers.
   (B) The outside diameter of the spline.
   (C) The number of teeth.
   (D) A = a fixed nonsliding spline.
       B = spline slides under no-load conditions.
       C = spline slides under load conditions.

Involute Spline

Another spline standard is the involute spline. The teeth on the involute spline are similar to the curved teeth found on spur gears. The spline teeth generally have a shorter whole depth than standard spur gears, and the pressure angle is normally 30°.

GEAR TYPES

The most common and simplest form of gear is the spur gear. Bevel gears and worm gears are also used. Gear types are designed based on one or more of the following elements:

■ The relationship of the shafts: either parallel, intersecting, nonintersecting shafts, or rack and pinion.

■ Manufacturing cost.

■ Ease of maintenance in service.

■ Smooth and quiet operation.

■ Load-carrying ability.

■ Speed reduction capabilities.

■ Space requirements.

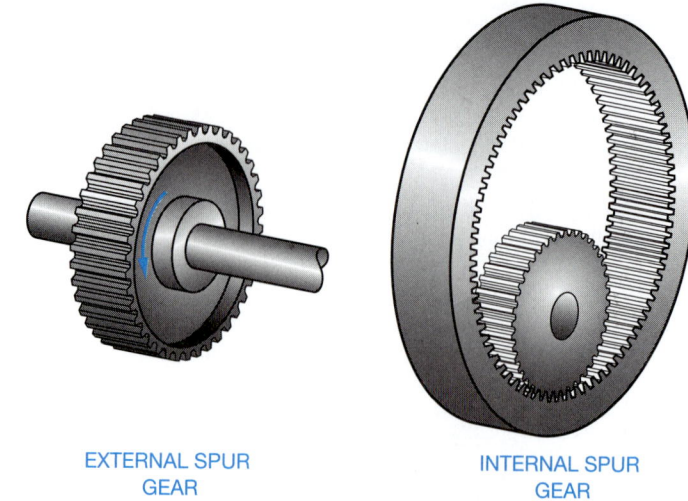

EXTERNAL SPUR GEAR   INTERNAL SPUR GEAR

FIGURE 19.43 ■ Spur gears.

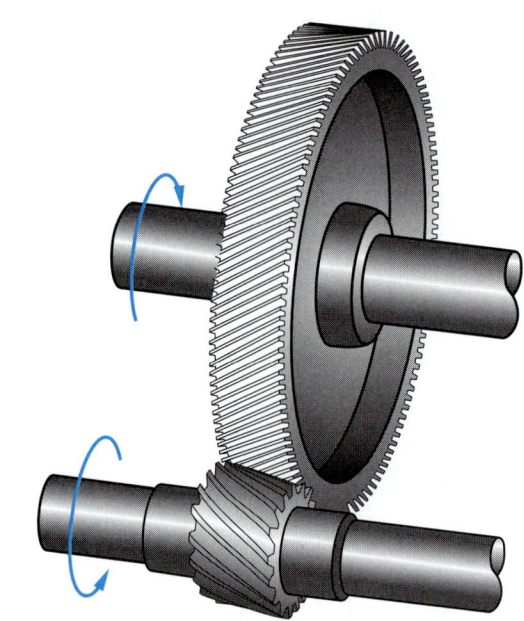

FIGURE 19.44 ■ Helical gear.

Parallel Shafting Gears

Many different types of mating gears are designed with parallel shafts. These include spur and helical gears.

Spur Gears

Spur gears are designed to transmit motion and power between parallel shafts. There are two basic types of spur gears: external and internal spur gears. (See Figure 19.43.) When two or more spur gears are cut on a single shaft, they are referred to as cluster gears. External spur gears are designed with the teeth of the gear on the outside of a cylinder. External spur gears are the most common type of gear used in manufacturing. Internal spur gears have the teeth on the inside of the cylindrical gear. The advantages of spur gears over other types are their low manufac-

turing cost, simple design, and ease of maintenance. The disadvantages include less load capacity and higher noise levels than other types.

Helical Gears

Helical gears have their teeth cut at an angle, allowing more than one tooth to be in contact. (See Figure 19.44.) Helical gears carry more load than equivalent-sized spur gears and operate more quietly and smoothly. The disadvantage of helical gears is that they develop end thrust. End thrust is a lateral force exerted on the end of the gear shaft. Thrust bearings are required to reduce the effect of this end thrust. Double helical gears are designed to eliminate the end thrust and provide long life under heavy loads. However, they are more difficult and

FIGURE 19.45 ■ Herringbone gear.

costly to manufacture. The herringbone gear shown in Figure 19.45 is a double helical gear without space between the two opposing sets of teeth.

Intersecting Shafting Gears

Intersecting shafting gears allow for the change in direction of motion from the gear to the pinion. Different types of intersecting shafting gears include bevel and face gears.

Bevel Gears

Bevel gears are conical in shape, allowing the shafts of the gear and pinion to intersect at 90° or any desired angle. The teeth on the bevel gear have the same shape as the teeth on spur gears except they taper toward the apex of the cone. Bevel gears provide for a speed change between the gear and pinion. (See Figure 19.46.) Miter gears are the same as bevel gears, except both the gear and pinion are the same size and are used when shafts must intersect at 90° without speed reduction. Spiral bevel gears have the teeth cut at an angle, which provides the same advantages as helical gears over spur gears.

Face Gears

The face gear is a combination of bevel gear and spur pinion, or bevel gear and helical pinion. This combination is used when the mounting accuracy is not as critical as with bevel gears. The load-carrying capabilities of face gears is not as good as that of bevel gears.

Nonintersecting Shafting Gears

Gears with shafts that are at right angles but not intersecting are referred to as nonintersecting shafts. Gears that fall into this category are crossed helical, hypoid, and worm gears.

Crossed Helical Gears

Also known as right angle helical gears or spiral gears, crossed helical gears provide for nonintersecting right angle shafts with low load-carrying capabilities. (See Figure 19.47.)

Hypoid Gears

Hypoid gears have the same design as bevel gears except the gear shaft axes are offset and do not intersect. (See Figure 19.48.) The gear and pinion are often designed with bearings mounted on both sides for improved rigidity over standard bevel gears. Hypoid gears are very smooth, strong, and quiet in operation.

Worm Gears

A worm and worm gear are shown in Figure 19.49. This type of gear is commonly used when a large speed reduction is required in a small space. The worm may be driven in either direction. When the gear is not in operation, the worm automatically locks in place. This is a particular advantage when it is important for the gears to have no movement of free travel when the equipment is shut off.

FIGURE 19.47 ■ Crossed helical gears. *Courtesy Browning Mfg. Division of Emerson Electric Co.*

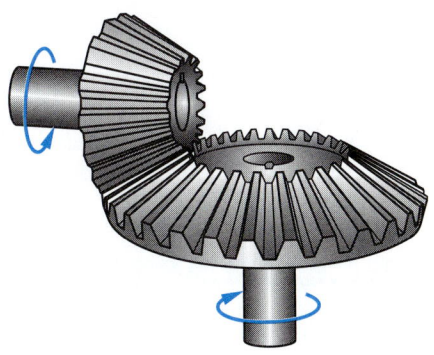

FIGURE 19.46 ■ Bevel gears.

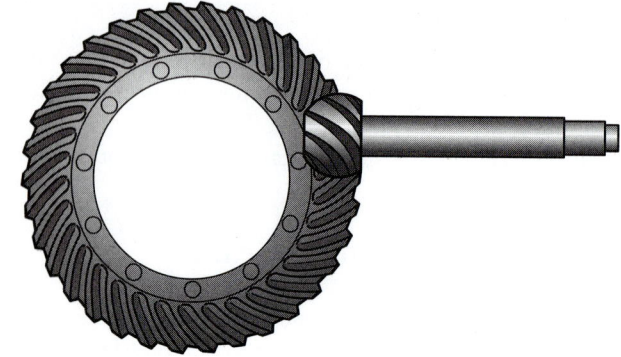

FIGURE 19.48 ■ Hypoid gears.

FIGURE 19.49 ■ Worm and worm gear.

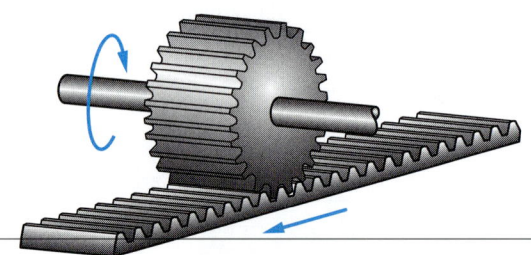

FIGURE 19.50 ■ Rack and pinion.

Rack and Pinion

A **rack and pinion** is a spur pinion operating on a flat straight bar rack. (See Figure 19.50.) The rack and pinion is used to convert rotary motion into straight-line motion.

SPUR GEAR DESIGN

Spur gear teeth are straight and parallel to the gear shaft axis. The tooth profile is designed to transmit power at a constant rate and with a minimum of vibration and noise. To achieve these requirements, an **involute curve** is used to establish the

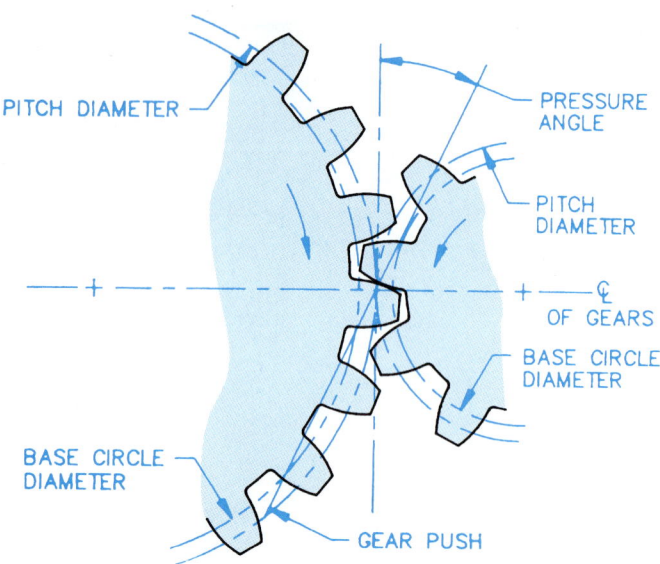

FIGURE 19.51 ■ The spur gear pressure angle and related terminology.

gear tooth profile. An **involute curve** is a spiral curve generated by a point on a chord as it unwinds from the circle. The contour of a gear tooth, based on the involute curve, is determined by a base circle, the diameter of which is controlled by a pressure angle. The **pressure angle** is the direction of push transmitted from a tooth on one gear to a tooth on the mating gear or pinion. (See Figure 19.51.) Two standard pressure angles, 14.5° and 20°, are used in spur gear design. The most commonly used pressure angle is 20° because it provides a stronger tooth for quieter running and heavier load-carrying characteristics. One of the basic rules of spur gear design is to have no fewer than 13 teeth on the running gear and 26 teeth on the mating gear.

Standard terminology and formulas control the drawing requirements for spur gear design and specifications. Figure 19.52 shows a pictorial representation of the spur gear teeth with the components labeled. As an engineering drafter, it is important that you become familiar with the terminology and associated mathematical formulas used to calculate values.

Diametral Pitch

The **diametral pitch** refers to the tooth size and has become the standard for tooth size specifications. As you look at Figure 19.53, notice how the tooth size increases as the diametral pitch decreases. One of the most elementary rules of gear tooth design is that mating teeth must have the same diametral pitch.

GEAR ACCURACY

A system has been established by the American Gear Manufacturers Association (AGMA) for the classification of gears based on the accuracy of the maximum tooth-to-tooth and total composite tolerances allowed. This number is called the **AGMA quality number**. The AGMA quality numbers and corresponding maximum tolerances are established by diametral pitch and pitch diameter. The AGMA quality numbers are listed in *AGMA Gear Handbook, 2000-A88*. The higher the AGMA quality num-

ber, the tighter the tolerance. For example, an AGMA Q6 allows approximately .004 total composite error, Q10 is less than .001, and Q13 is less than .0004. The *AGMA Gear Handbook* also displays a list of gear applications and the quality number suggested for each application.

Term	Description	Formula
Pitch Diameter (D)	The diameter of an imaginary pitch circle on which a gear tooth is designed. Pitch circles of two spur gears are tangent.	$D = N/P$
Diametral Pitch (P)	A ratio equal to the number of teeth on a gear per inch of pitch diameter.	$P = N/D$
Number of Teeth (N)	The number of teeth on a gear.	$N = D \times P$
Circular Pitch (p)	The distance from a point on one tooth to the corresponding point on the adjacent tooth, measured on the pitch circle. Circular pitch is the same as linear pitch on a rack. See Figure 19.58.	$p = 3.1416 \times D / N$ $p = 3.1416/P$
Center Distance (C)	The distance between the axis of two mating gears.	C = sum of pitch DIA/2
Addendum (a)	The radial distance from the pitch circle to the top of the tooth.	$a = 1/P$
Dedendum (b)	The radial distance from the pitch to the bottom of the tooth. (This formula is for 20° teeth only.)	$b = 1.250/P$
Whole Depth (h_t)	The full height of the tooth. It is equal to the sum of the addendum and the dedendum.	$h_t = a + b$ $h_t = 2.250/P$
Working Depth (h_k)	The distance that a tooth occupies in the mating space. A distance equal to two times the addendum.	$h_k = 2a$ $h_k = 2.000/P$
Clearance (c)	The radial distance between the top of a tooth and the bottom of the mating tooth space. It is also the difference between the addendum and dedendum.	$c = b - a$ $c = .250/P$
Outside Diameter (D_o)	The overall diameter of the gear. It is equal to the pitch diameter plus two addendums.	$D_o = D + 2a$
Root Diameter (D_r)	The diameter of a circle coinciding with the bottom of the tooth spaces.	$D_r = D - 2b$ $t = 1.5708/P$
Circular Thickness (t)	The length of an arc between the two sides of a gear tooth measured on the pitch circle.	
Chordal Thickness (t_c)	The straight-line thickness of a gear tooth measured on the pitch circle. Chordal thickness is also referred to as tooth thickness.	$t_c = D \sin (90°/N)$
Chordal Addendum (a_c)	The height from the top of the tooth to the line of the chordal thickness.	$a_c = a + t^2/4D$
Pressure Angle (∅)	The angle of direction of pressure between contacting teeth. It determines the size of the base circle and the shape of the involute teeth.	
Base Circle Diameter (D_B)	The diameter of a circle from which the involute tooth form is generated.	$D_B = D \cos ∅$

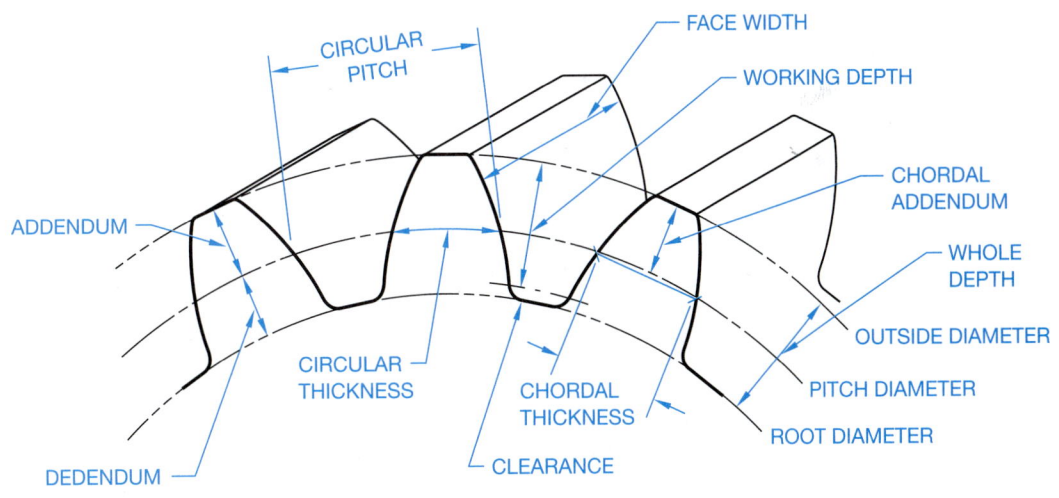

FIGURE 19.52 ■ Gear terminology formulas.

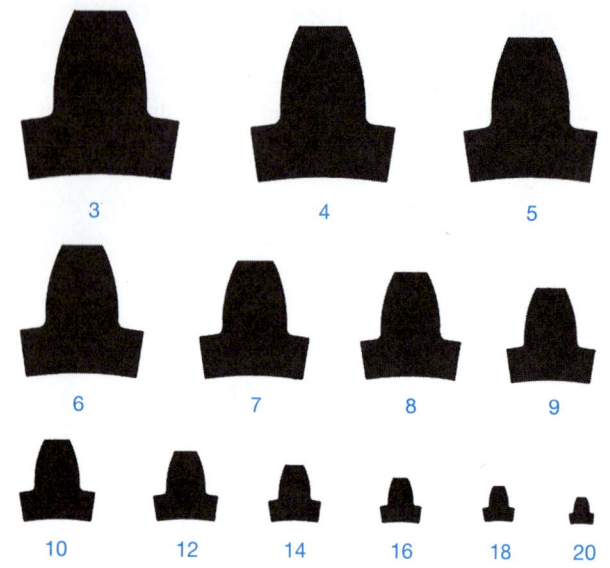

FIGURE 19.53 ■ Diametral pitch.

DRAWING SPECIFICATIONS AND TOLERANCES

The final step in designing a gear is the presentation of a drawing displaying the dimensioned gear using multiviews and gear data charts. While companies do not always provide the complete gear data charts on their drawings, errors can happen when complete data is not provided. Gear data formulas and sample gear data charts are shown with examples in this chapter. However, the AGMA recommended information for spur and helical gears is also in Appendix AA.

DESIGNING AND DRAWING SPUR GEARS

ANSI The drafting standard governing gear drawings is the American National Standards document ANSI Y14.7.1, *Gear Drawing Standards—Part I.*

Because gear teeth are complex and time-consuming to draw, simplified representations can be used to make the practice easier. (See Figure 19.54.) The simplified method shows the outside diameter and the root diameters as phantom lines and the pitch diameter as a centerline in the circular view. In addition, a side view is often required to show width dimensions and related features. (See Figure 19.54a.) If the gear construction has webs, spokes, or other items that require further clarification, then a full section is normally used. (See Figure 19.54b.) Notice in the cross section that the gear tooth is left unsectioned and the pitch diameter is shown as a centerline.

When cluster gears are drawn, the circular view can show both sets of gear tooth representations in simplified form, or two circular views can be drawn. (See Figure 19.55.) When

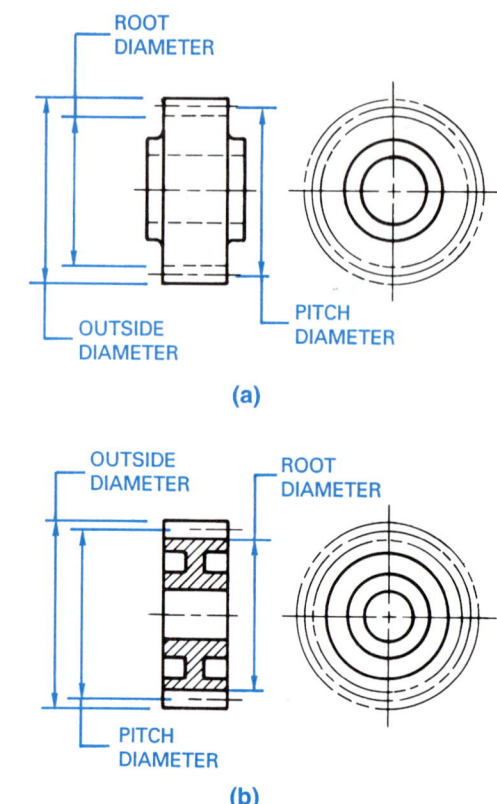

FIGURE 19.54 ■ Typical spur gear drawings using simplified gear teeth representation.

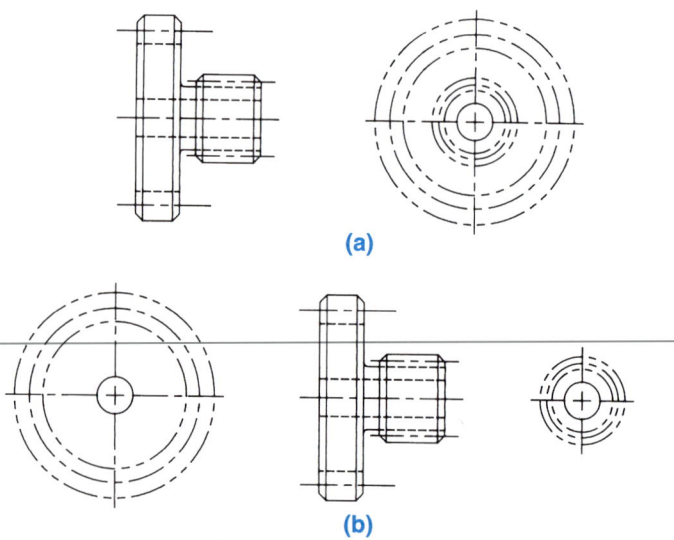

FIGURE 19.55 ■ Cluster gear drawings.

cluster gears are more complex than those shown here, multiple views and removed sections can be required.

One or more teeth can be drawn for specific applications. For example, when a tooth must be in alignment with another feature of the gear, the tooth can be drawn as shown in Figure 19.56.

Gear drawings typically have a chart showing the manufacturing information associated with the teeth and related part detail dimensions placed on the specific views.

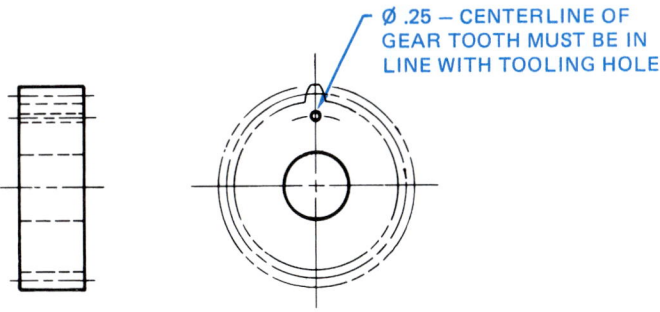

FIGURE 19.56 ■ Showing the relationship of one gear tooth to another feature on the gear. Dimension values in this figure are in inches.

DESIGNING SPUR GEAR TRAINS

A gear train is an arrangement of two or more gears connecting driving and driven parts of a machine. Gear reducers and transmissions are examples of gear trains. The function of a gear train is to:

■ Transmit motion between shafts.

■ Decrease or increase the speed between shafts.

■ Change the direction of motion.

It is important for you to understand the relationship between two mating gears in order to design gear trains. When gears are designed, the end result is often a specific gear ratio. Any two gears in mesh have a gear ratio. The gear ratio is expressed as a proportion, such as 2:1 or 4:1, between two similar values. The gear ratio between two gears is the relationship between the following characteristics:

■ Number of teeth.

■ Pitch diameters.

■ Revolutions per minute (rpm).

If you have gear A (pinion) mating with gear B, as shown in Figure 19.57, the gear ratio is calculated by dividing the number of teeth or the pitch diameter values of the smaller gear into the larger gear as follows:

$$\frac{\text{Number Teeth}_{\text{Gear B}}}{\text{Number Teeth}_{\text{Gear A}}} = \text{Gear Ratio}$$

$$\frac{\text{Pitch Diameter}_{\text{Gear B}}}{\text{Pitch Diameter}_{\text{Gear A}}} = \text{Gear Ratio}$$

The gear ratio can also be calculated by dividing the RPM values of the larger gear into the smaller gear like this:

$$\frac{\text{rpm}_{\text{Gear A}}}{\text{rpm}_{\text{Gear B}}} = \text{GearRatio}$$

Now, calculate the gear ratio for the two mating gears in Figure 19.57 if gear A has 18 teeth, 6-in. pitch diameter, and operates at 1200 rpm, and gear B has 54 teeth, 18-in. pitch diameter, and operates at 400 rpm:

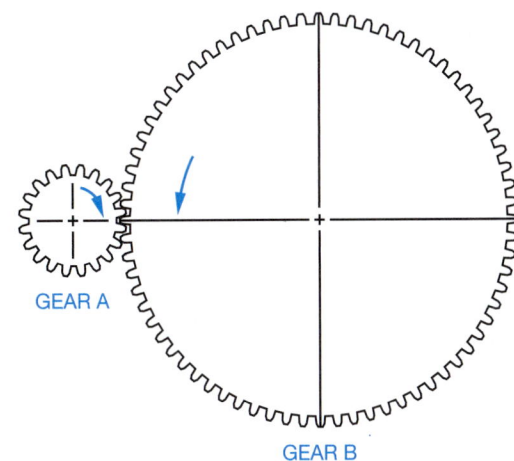

GEAR A

GEAR B

	(D) PITCH DIAMETER	(N) NUMBER OF TEETH	(P) DIAMETRAL PITCH	RPM	DIRECTION
GEAR A	6	18	3	1200	C.WISE
GEAR B	18	54	3	400	C.C.WISE

FIGURE 19.57 ■ Calculating gear data. Dimension values in this figure are in inches.

$$
\begin{aligned}
\text{Number of Teeth} &= 54/18 &= 3:1 \\
\text{Pitch Diameter} &= 18/6 &= 3:1 \\
\text{rpm} &= 1200/400 &= 3:1
\end{aligned}
$$

You can solve for unknown values in the gear train if you know the gear ratio you want to achieve, the number of teeth and pitch diameter of one gear, and the input speed. For example, gear A has 18 teeth, a pitch diameter of 6 in., and an input speed of 1200 rpm, and the ratio between gear A and gear B is 3:1. In order to keep this information well organized, it is recommended that you set up a chart similar to the one shown in Figure 19.57. The unknown values are shown in color for your reference. Determine the number of teeth for gear B as follows:

$$\text{Teeth}_{\text{Gear A}} \times \text{Gear Ratio} = 18 \times 3 = \text{Teeth}_{\text{Gear B}} = 54$$

Or, if you know that gear B has 54 teeth and the gear ratio is 3:1, then:

$$\frac{\text{Teeth}_{\text{Gear B}}}{\text{Gear Ratio}} = \frac{54}{3} = \text{Teeth}_{\text{Gear A}} = 18$$

Determine the rpm of gear B:

$$\frac{\text{rpm}_{\text{Gear A}}}{\text{Gear Ratio}} = \frac{1200}{3} = \text{rpm}_{\text{Gear B}} = 400$$

Determine the pitch diameter of gear B:

$$\text{Pitch Diameter}_{\text{Gear A}} \times \text{Gear Ratio} = 6 \text{ in.} \times 3 = \text{Pitch Diameter}$$
$$\text{Gear B} = 18 \text{ in.}$$

In some situations it is necessary for you to refer to the formulas given in Figure 19.52 to determine some unknown values. In this case it is necessary to calculate the diametral pitch using the formula P = N/D, where P = diametral pitch, N = number of teeth, and D = pitch diameter:

$$P_{\text{Gear A}} = \frac{N}{D} = \frac{18}{6} = 3$$

Because the diametral pitch is the tooth size and the teeth for mating gears must be the same size, the diametral pitch of gear B is also 3.

Keep in mind that the preceding example presents only one set of design criteria. Other situations can be different. Always solve for unknown values based on information that you have and work with the standard formulas presented in this chapter. Following are some important points to keep in mind as you work with the design of gear trains:

- The rpm of the larger gear is always slower than the rpm of the smaller gear.

- Mating gears always turn in opposite directions.

- Gears on the same shaft (cluster gears) always turn in the same direction and at the same speed (rpm).

- Mating gears have the same size teeth (diametral pitch).

- The gear ratio between mating gears is a ratio between the number of teeth, the pitch diameters, and the rpms.

- The distance between the shafts of mating gears is equal to 1/2 $D_{\text{Gear A}}$ + 1/2 $D_{\text{Gear B}}$. This distance is 3 + 9 = 12 in. between shafts in Figure 19.57.

DESIGNING AND DRAWING THE RACK AND PINION

A **rack** is a straight bar with spur teeth used to convert rotary motion to reciprocating motion. Figure 19.58 shows a spur gear **pinion** mating with a rack. Notice the circular dimensions of the pinion become linear dimensions on the rack (Linear Pitch = Circular Pitch).

When preparing a simplified detailed drawing of the rack, the front view is normally shown with the profile of the first and last tooth drawn. Phantom lines are drawn to represent the top and root and a centerline is used for the pitch line. (See Figure 19.59.)

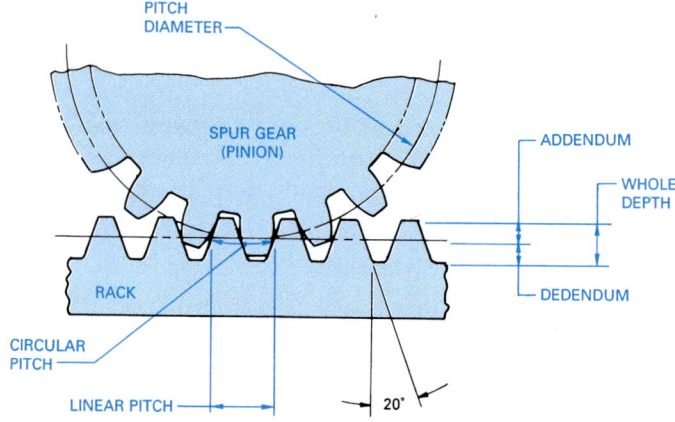

FIGURE 19.58 ■ Rack and pinion terminology.

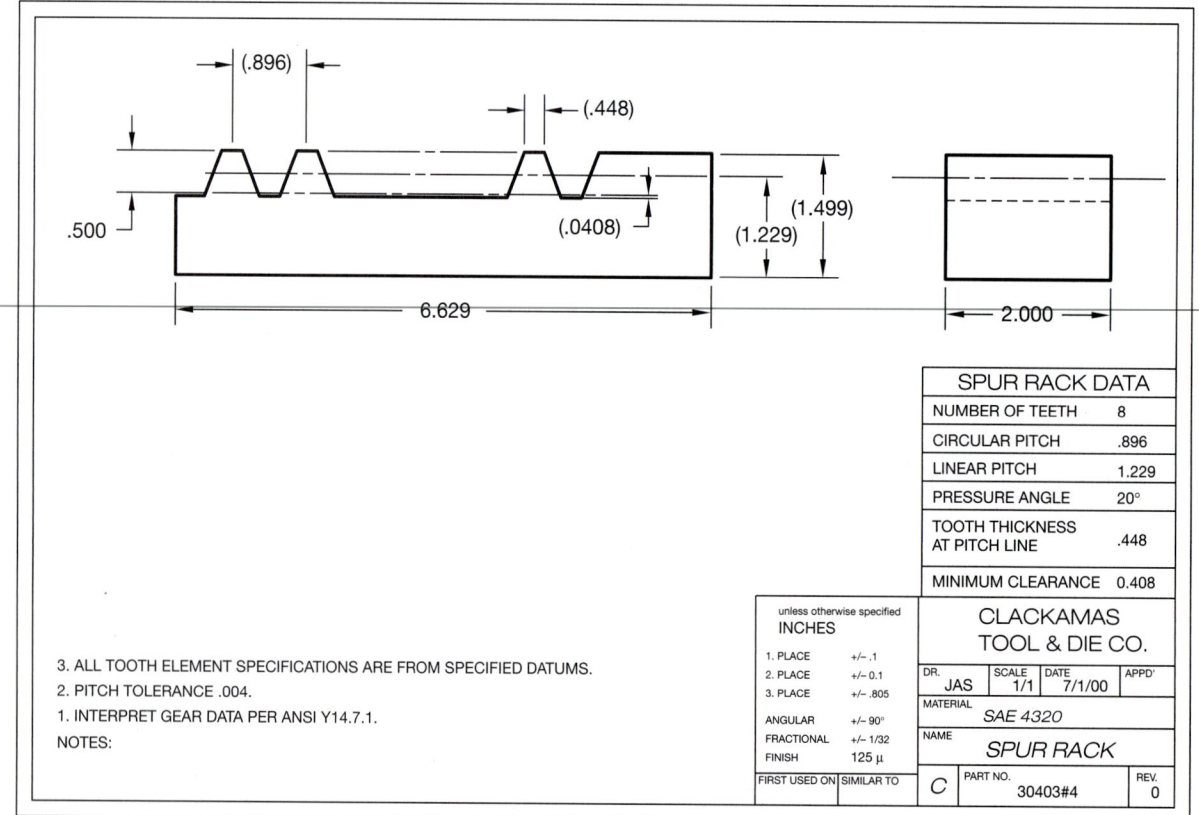

FIGURE 19.59 ■ A detailed drawing of a rack. Dimension values in this figure are in inches.

Related tooth and length dimensions are also placed on the front view, and a depth dimension is placed on the side view. A chart is then placed in the field of the drawing to identify specific gear-cutting information. A rack detail drawing can also have all teeth displayed, especially when CADD is used.

DESIGNING AND DRAWING BEVEL GEARS

Bevel gears are usually designed to transmit power between intersecting shafts at 90°, although they can be designed for any angle. Some gear design terminology and formulas relate specifically to the construction of bevel gears. These formulas and a drawing of a bevel gear and pinion are shown in Figure 19.60. Most of the gear terms discussed for spur gears apply to bevel gears. In many cases information must be calculated using the formulas provided in the spur gear discussion shown in Figure 19.52.

The drawing for a bevel gear is similar to the drawing for a spur gear, but in many cases only one view is needed. This view is often a full section. Another view can be used to show the dimensions for a keyway. As with other gear drawings, a chart is

Term	Description	Formula
Pitch Diameter (D)	The diameter of the base of the pitch cone.	$D = N/P$
Pitch Cone	In Figure 19.60, the pitch cone is identified as XYZ.	$\tan \varnothing_a = D_a/D_b$
Pitch Angle (Δ)	The angle between an element of a pitch cone and its axis. Pitch angles of mating gears depend on their relative diameters (gear a and gear b).	$\tan \varnothing_b = D_a/D_b$
Cone Distance (A)	Slant height of pitch cone.	$A = D/2 \sin \varnothing$
Addendum Angle (δ)	The angle subtended by the addendum. It is the same for mating gears.	$\tan \delta = a/A$
Dedendum Angle (Ω)	The angle subtended by the dedendum. It is the same for mating gears.	$\tan \Omega = b/A$
Face Angle (∅₀)	The angle between the top of the teeth and the gear axis.	$\varnothing_o = \varnothing - 8$
Root Angle (∅ᵣ)	The angle between the bottom of the tooth space and the gear axis.	$\varnothing_r = \varnothing - \Omega$
Outside Diameter (D₀)	The diameter of the outside circle of gear.	$D_o = D + 2a \, (\cos\varnothing)$
Crown Height (X)	The distance between the cone apex and the outer tip of the gear teeth.	$X = .5(D_o)/\tan\varnothing_o$
Crown Backing (Y)	The distance from the rear of the hub to the outer tip of the gear tooth, measured parallel to the axis of the gear.	
Face Width	A distance that should not exceed one-third of the cone distance (A).	

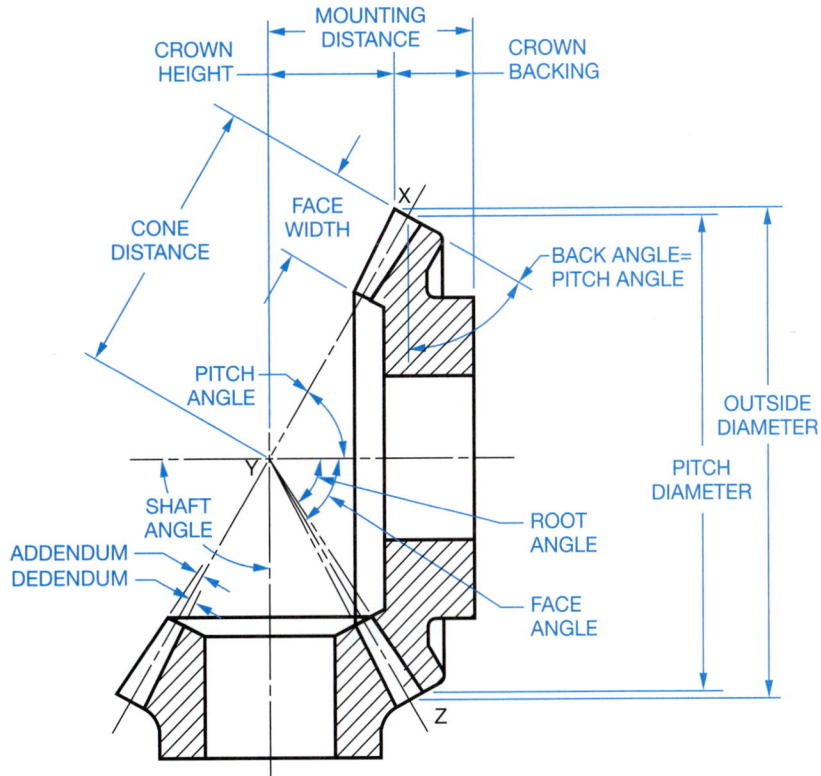

FIGURE 19.60 ■ Bevel gear terminology and formulas.

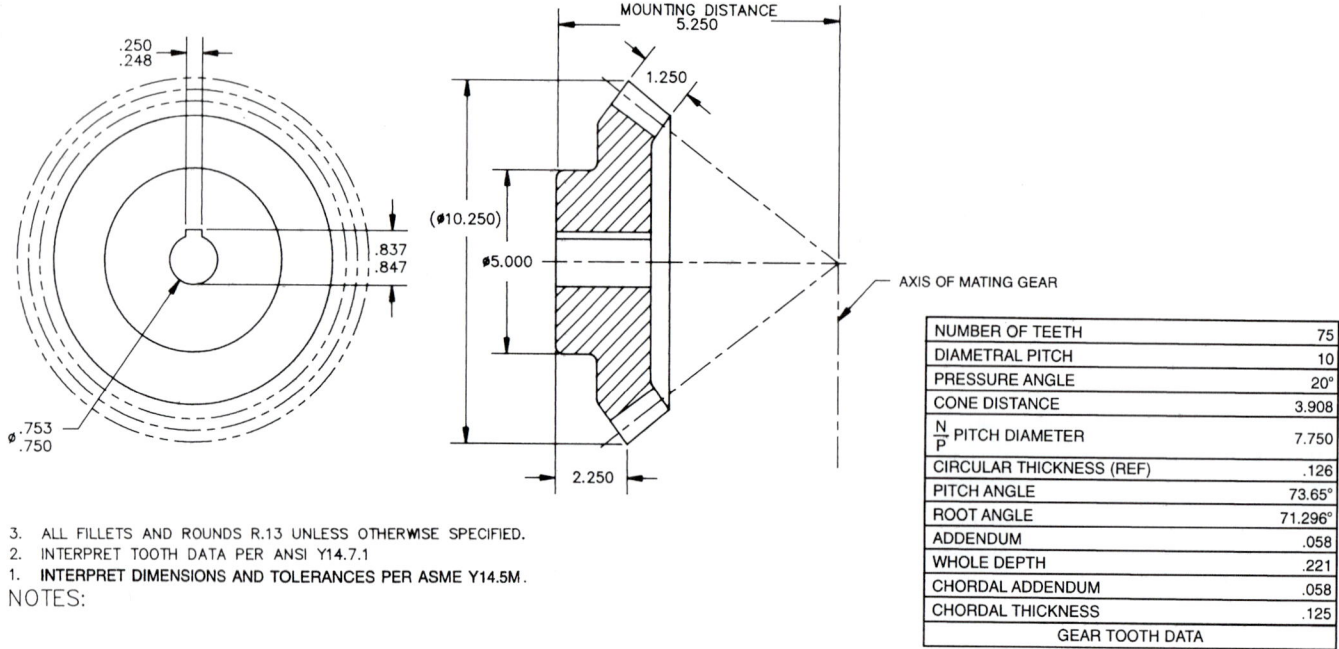

NUMBER OF TEETH	75
DIAMETRAL PITCH	10
PRESSURE ANGLE	20°
CONE DISTANCE	3.908
$\frac{N}{P}$ PITCH DIAMETER	7.750
CIRCULAR THICKNESS (REF)	.126
PITCH ANGLE	73.65°
ROOT ANGLE	71.296°
ADDENDUM	.058
WHOLE DEPTH	.221
CHORDAL ADDENDUM	.058
CHORDAL THICKNESS	.125
GEAR TOOTH DATA	

NOTES:
3. ALL FILLETS AND ROUNDS R.13 UNLESS OTHERWISE SPECIFIED.
2. INTERPRET TOOTH DATA PER ANSI Y14.7.1
1. INTERPRET DIMENSIONS AND TOLERANCES PER ASME Y14.5M.

FIGURE 19.61 ■ A detailed drawing of a bevel gear. Dimension values in this figure are in inches.

used to specify gear-cutting data. (See Figure 19.61.) Notice in this example that a side view is used to specify the bore and keyway.

DESIGNING AND DRAWING WORM GEARS

Worm gears are used to transmit power between nonintersecting shafts, and they are used for large speed reductions in a small space as compared to other types of gears. Worm gears are also strong, move in either direction, and lock in place when the machine is not in operation. A single lead worm advances one pitch with every revolution. A double lead worm advances two pitches with each revolution. As with bevel gears, the worm gear and worm have specific terminology and formulas that ap-

ply to their design. The representative worm gear and worm technology and design formulas are shown in Figure 19.62.

When drawing the worm gear, the same techniques are used as discussed earlier. Figure 19.63 shows a worm gear drawing using a half-section method. A side view is provided to dimension the bore and keyway, and a chart is given for gear-cutting data.

A detail drawing of the worm is prepared using two views. The front view shows the first gear tooth on each end with phantom lines between for a simplified representation. The side view is the same as a spur gear drawing with the keyway specifications. The gear-cutting chart is then placed on the field of the drawing or over the title block. Figure 19.64 shows the worm drawing using CADD to provide a detailed representation of the worm teeth in the front view rather than phantom lines as in the simplified representation.

Term	Description	Formula
Pitch Diameter (worm) (Dw)		$D_w = 2C - D_g$
Pitch Diameter (gear) (D_g)		$D_g = 2C - D_w$
Pitch (P)	The distance from one tooth to the corresponding point on the next tooth measured parallel to the worm axis. It is equal to the circular pitch on the worm gear.	$P = L/T$
Lead (L)	The distance the thread advances axially in one revolution of the worm.	$L = D_g/R$
		$L = P \times T$
Threads (T)	Number of threads or starts on worm.	$T = L/P$
Gear Teeth (N)	Number of teeth on worm gear.	$N = \pi D_g/P$
Ratio (R)	Divide number of gear teeth by the number of worm threads.	$R = N/T$
Addendum (a)	For single and double threads.	$a = .318P$
Whole Depth (WD)	For single and double threads.	$WD = .686P$

NOTE: REFER TO THE FORMULAS ON PAGE 629 FOR ADDITIONAL CALCULATIONS.

FIGURE 19.62 ■ Worm and worm gear terminology and formulas.

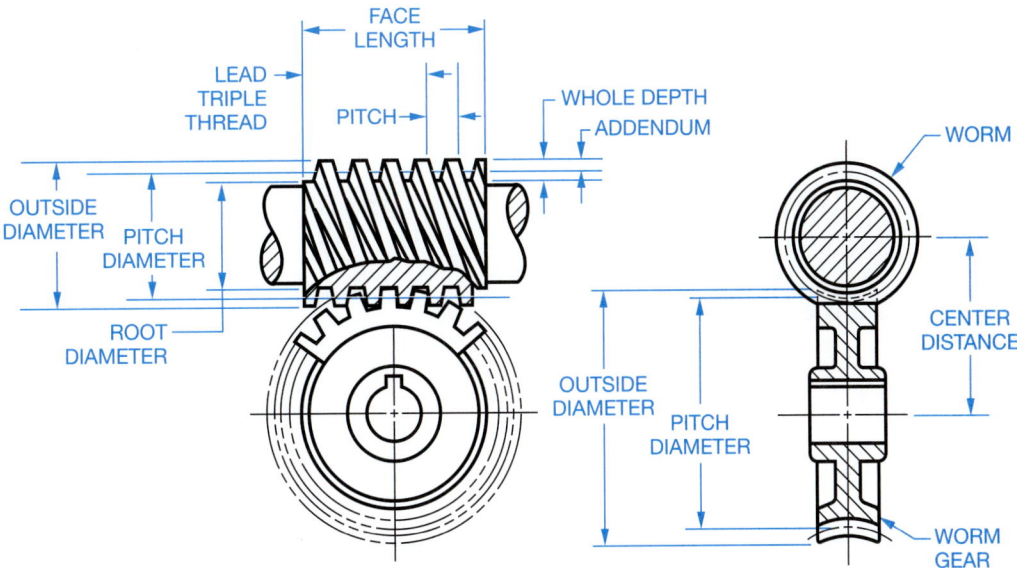

FIGURE 19.62 ■ (Continued)

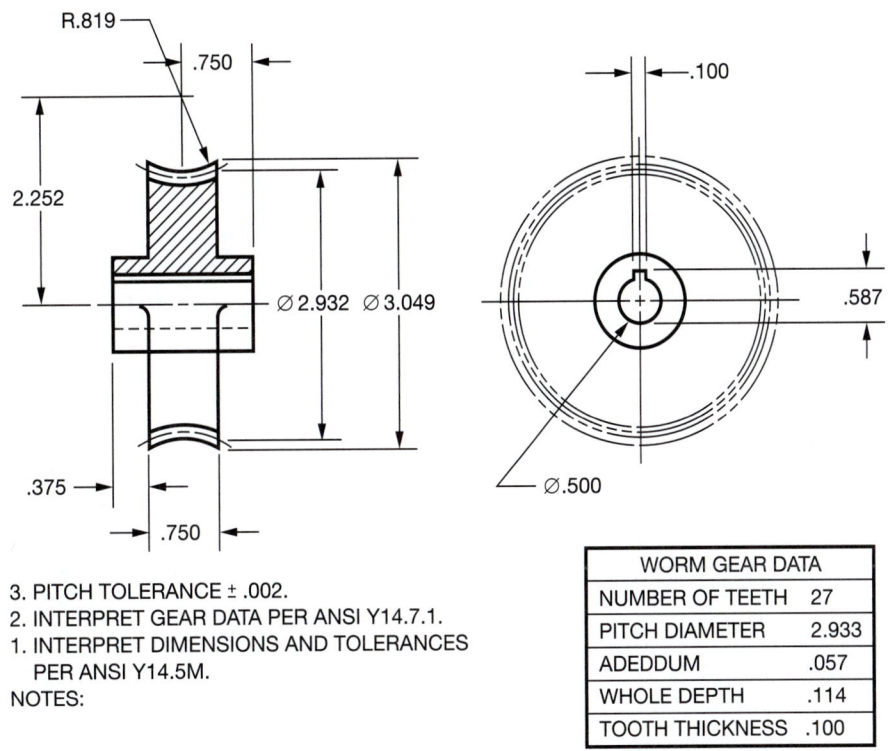

3. PITCH TOLERANCE ± .002.
2. INTERPRET GEAR DATA PER ANSI Y14.7.1.
1. INTERPRET DIMENSIONS AND TOLERANCES
   PER ANSI Y14.5M.
NOTES:

WORM GEAR DATA	
NUMBER OF TEETH	27
PITCH DIAMETER	2.933
ADEDDUM	.057
WHOLE DEPTH	.114
TOOTH THICKNESS	.100

FIGURE 19.63 ■ A detailed drawing of a worm gear. Dimension values in this figure are
in inches.

PLASTIC GEARS

The following is taken in part from *Plastic Gearing* by William McKinlay and Samuel D. Pierson, published by ABA/PGT Inc., Manchester, CT. Gears can be molded of many engineering plastics in various grades and in filled varieties. **Filled plastics** are those in which a material has been added to improve the mechanical properties. The additives normally used in gear plastics

are glass, polytetrafluoroethylene (PTFE), silicones, and molybdenum disulphide. Glass fiber reinforcement can double the tensile strength and reduce the thermal expansion. Carbon fiber is often used to increase strength. Silicones, PTFE, and molybdenum disulphide are used to act as built-in lubricants and provide increased wear resistance. Plastic gears are designed in the same manner as gears made from other materials. However, the physical characteristics of plastics make it necessary to follow

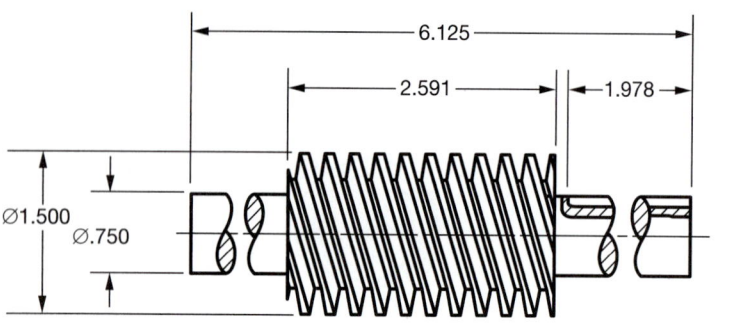

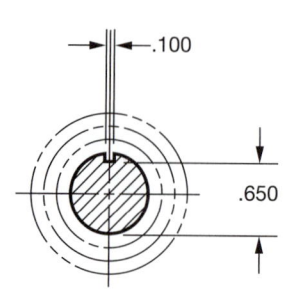

WORM GEAR DATA	
PITCH DIAMETER	1.251
LEAD RIGHT OR LEFT	RIGHT
CENTER DISTANCE	1.858
WORKING DEPTH	.218
CLEARANCE	.0312
PRESSURE ANGLE	20°
ADDENDUM	.125
WHOLE DEPTH	.25
CHORDAL THICKNESS	.125

2. PITCH TOLERANCE ± .002.
1. INTERPRET GEAR DATA PER ANSI Y14.7.1.
NOTES:

FIGURE 19.64 ■ A detailed drawing of a worm. Dimension values in this figure are in inches.

CADD APPLICATIONS

GEAR DRAWINGS AND ANALYSIS

There are custom CADD programs that allow the gear teeth to be drawn automatically. To do this, the computer prompts for variables such as:

■ pitch

■ diameter

■ diametral pitch

■ pressure angle

■ number of teeth

The program then automatically notifies you if the information is accurate, provides all additional data, and creates a detail drawing of the gear. CADD was used to draw the teeth in the gear detail drawing displayed in Figure 19.65. You can easily draw gears displayed in detailed representation or use the simplified technique to save regeneration and plotting time.

Some software programs do more than assist in the design and drafting process. For example, objects such as gear teeth can be subjected to simulated tests and stress analysis on the computer screen as shown in Figure 19.66.

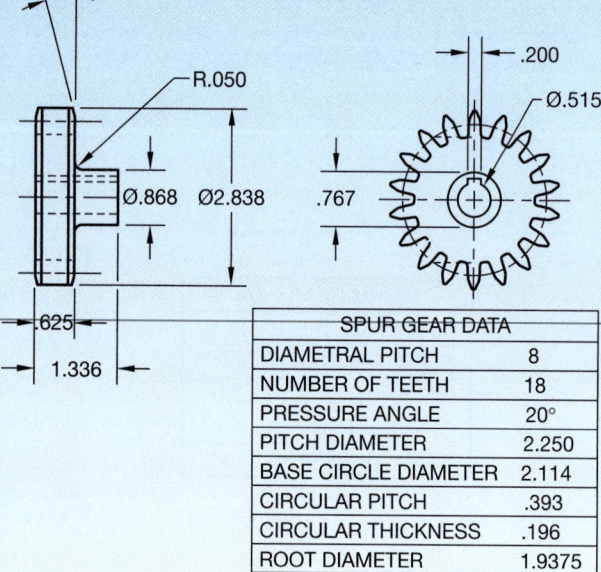

SPUR GEAR DATA	
DIAMETRAL PITCH	8
NUMBER OF TEETH	18
PRESSURE ANGLE	20°
PITCH DIAMETER	2.250
BASE CIRCLE DIAMETER	2.114
CIRCULAR PITCH	.393
CIRCULAR THICKNESS	.196
ROOT DIAMETER	1.9375

5. PROFILE TOLERANCE .003
4. PITCH TOLERANCE .003
3. ALL TOOTH ELEMENT SPECIFICATIONS ARE FROM DATUM A

FIGURE 19.65 ■ Gear detailed drawing using CADD to automatically draw teeth and gear data chart from given specifications. Dimension values in this figure are in inches.

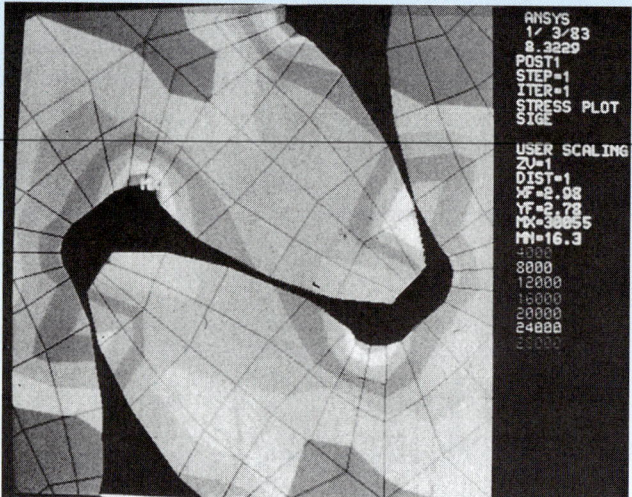

FIGURE 19.66 ■ Objects such as these gear teeth can be subjected to simulated tests and stress analysis on the computer screen. *Courtesy Swanson Analysis Systems, Inc.*

gear design practices more closely than when designing gears that are machined from metals.

Advantages of Molded Plastic Gears

Gears molded of plastic are replacing stamped and cut metal gears in a wide range of mechanisms. Designers are turning to molded plastic gears for one or more of the following reasons:

- Reduced cost.
- Increased efficiency.
- Self-lubrication.
- Increased tooth strength with nonstandard pressure angles.
- Reduced weight.
- Corrosion resistance.
- Less noise.
- Available in colors.

Disadvantages of Molded Plastic Gears

Plastic gearing has the following limitations when compared with metal gearing:

- Lower strength.
- Greater thermal expansion and contraction.
- Limited heat resistance.
- Size change with moisture absorption.

Accuracy of Molded Plastic Gears

Technology permits a very high degree of precision. In general, tooth-to-tooth composite tolerances can be economically held to .0005 or less for fine pitch gears. Total composite tolerance varies depending on configuration, evenness of product cross section, and the selection of the molding material.

BEARINGS

Bearings are mechanical devices used to reduce friction between two surfaces. They are divided into two large groups known as plain and rolling element bearings. Bearings are designed to accommodate either rotational or linear motion. Rotational bearings are used for radial loads, and linear bearings are designed for thrust loads. Radial loads are loads that are distributed around the shaft. Thrust loads are lateral. Thrust loads apply force to the end of the shaft. Figure 19.67 shows the relationship between rotational and linear motion.

Plain Bearings

Plain bearings are often referred to as sleeve, journal bearings, or bushings. Their operation is based on a sliding action between mating parts. A clearance fit between the inside diameter of the bearing and the shaft is critical to ensure proper opera-

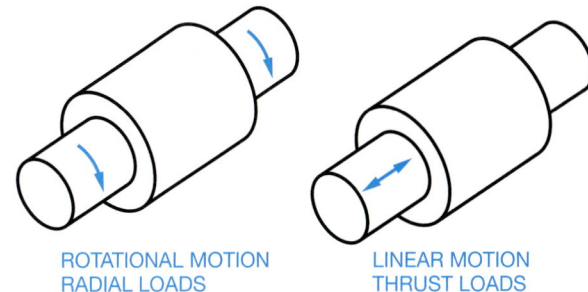

ROTATIONAL MOTION RADIAL LOADS   LINEAR MOTION THRUST LOADS

FIGURE 19.67 ■ Radial and thrust loads.

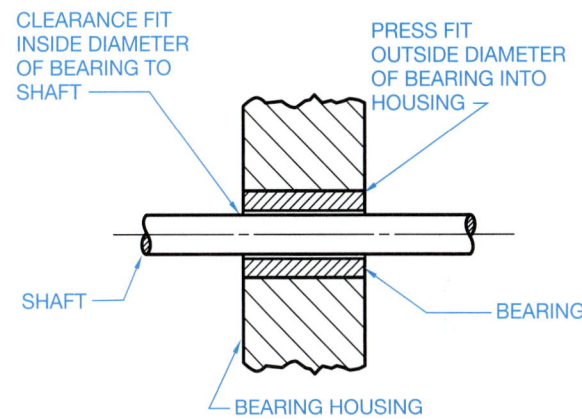

CLEARANCE FIT INSIDE DIAMETER OF BEARING TO SHAFT

PRESS FIT OUTSIDE DIAMETER OF BEARING INTO HOUSING

SHAFT

BEARING

BEARING HOUSING

FIGURE 19.68 ■ Plain bearing terminology and fits.

tion. Refer to fits between mating parts in Chapter 11 for more information. The bearing has an interference fit between the outside of the bearing and the housing or mounting device as shown in Figure 19.68.

The material from which plain bearings are made is important. Most plain bearings are made from bronze or phosphor bronze. Bronze bearings are normally lubricated, while phosphor bronze bearings are commonly impregnated with oil and require no additional lubrication. Phosphor bronze is an excellent choice when antifriction qualities are important and where resistance to wear and scuffing are needed.

Rolling Element Bearings

Ball and roller bearings are the two classes of rolling element bearings. Ball bearings are the most commonly used rolling element bearings. In most cases, ball bearings have higher speed and lower load capabilities than roller bearings. Even so, ball bearings are manufactured for most uses. Ball bearings are constructed with two grooved rings, a set of balls placed radially around the rings, and a separator that keeps the balls spaced apart and aligned as shown in Figure 19.69.

Single-row ball bearings are designed primarily for radial loads, but they can accept some thrust loads. Double-row ball bearings may be used where shaft alignment is important. Angular contact ball bearings support a heavy thrust load and a moderate radial load. Thrust bearings are designed for use in

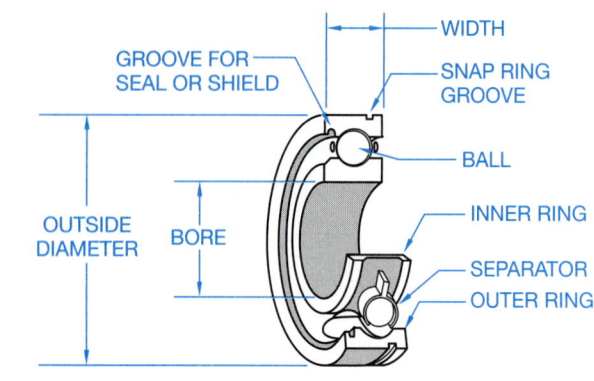

FIGURE 19.69 ■ Ball bearing components.

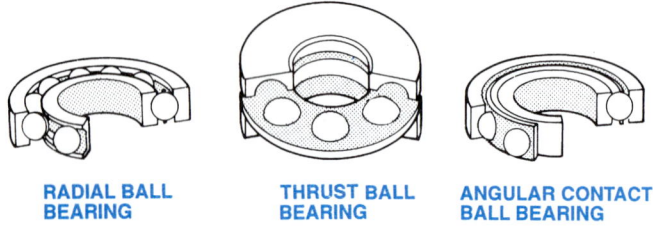

FIGURE 19.70 ■ Typical ball bearings.

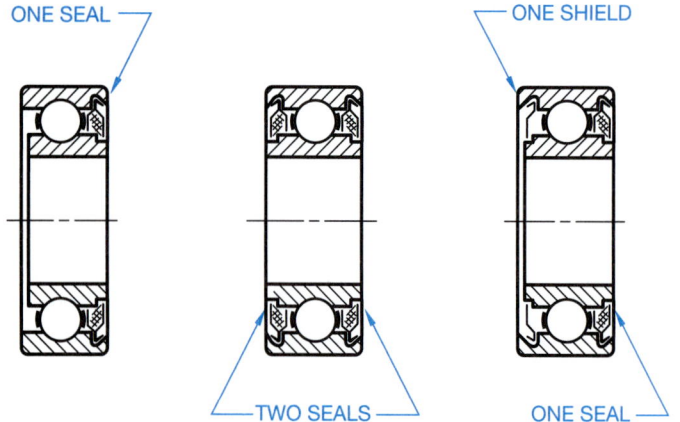

FIGURE 19.71 ■ Bearing seals and shields.

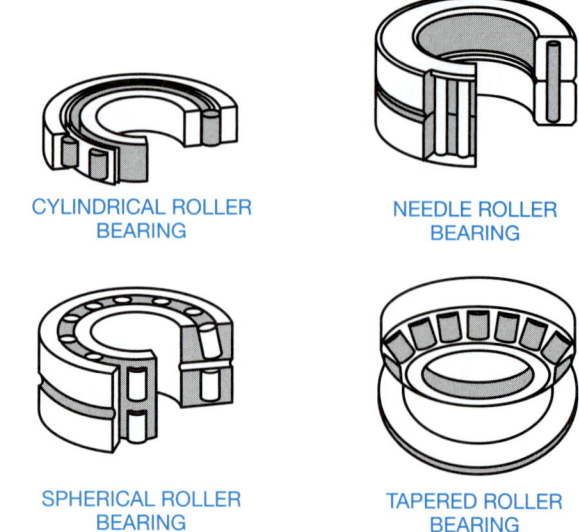

FIGURE 19.72 ■ Typical roller bearings.

small rollers and are designed for the highest load-carrying capacity of all rolling element bearings with shaft sizes under 10 in. **Tapered roller bearings** are used in gear reducers, steering mechanisms, and machine tool spindles. **Spherical roller bearings** offer the best combination of high load capacity, tolerance to shock, and alignment, and are used on conveyors, transmissions, and heavy machinery. Some common roller bearings are displayed in Figure 19.72.

DRAWING BEARING SYMBOLS

Only engineering drafters who work for a bearing manufacturer normally make detailed drawings of roller or ball bearings. In most other industries, bearings are not drawn because they are purchase parts. When bearings are drawn as a representation on assembly drawings or product catalogs, they are displayed using symbols with specific detail required by your company or school.

When CADD is used, the bearing symbols can be drawn and saved in a symbols library for immediate use at any time. This use of CADD helps increase productivity and accuracy over other drafting techniques. Figure 19.73 shows bearing symbols used for a variety of applications.

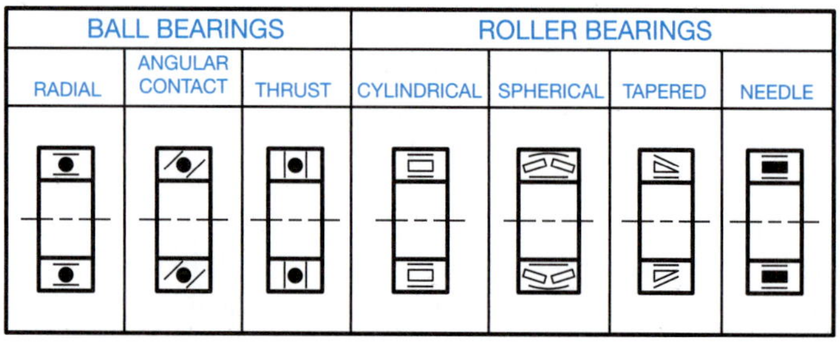

thrust load situations only. When both thrust and radial loads are necessary, both radial and thrust ball bearings are used together. Some typical ball bearings are shown in Figure 19.70.

Ball bearings are available with shields and seals. A **shield** is a metal plate on one or both sides of the bearing. The shields act to keep the bearing clean and retain the lubricant. A **sealed bearing** has seals made of rubber, felt, or plastic placed on the outer and inner rings of the bearing. The sealed bearings are filled with special lubricant by the manufacturer. They require little or no maintenance in service. Figure 19.71 shows the shields and seals used on ball bearings.

Roller bearings are more effective than ball bearings for heavy loads. **Cylindrical roller bearings** have a high radial capacity and assist in shaft alignment. **Needle roller bearings** have

BALL BEARINGS			ROLLER BEARINGS			
RADIAL	ANGULAR CONTACT	THRUST	CYLINDRICAL	SPHERICAL	TAPERED	NEEDLE

FIGURE 19.73 ■ Bearing symbols drawn using CADD.

BEARING CODES

Bearing manufacturers use similar coding systems for the identification and ordering of different bearing products. The bearing codes generally contain the following type of information:

■ Material.

■ Bearing type.

■ Bore size.

■ Lubricant.

■ Type of seals or shields.

A sample bearing numbering system is shown in Figure 19.74.

BEARING SELECTION

A variety of bearing types are available from manufacturers. Bearing design differs depending on use requirements. For example, suppliers have light, medium, and heavy bearings available. Bearings have specially designed outer and inner rings. Bearings are available open without seals or shields, or with one or two shields or seals. Light bearings are generally designed to accommodate a wide range of applications involving light to medium loads combined with relatively high speeds. Medium bearings have heavier construction than light bearings and provide a greater radial and thrust capacity. They are also able to withstand greater shock than light bearings. Heavy bearings are often designed for special service where extra heavy shock loads are required. Bearings are also designed to accommodate radial loads, thrust loads, or a combination of loading requirements.

Bearing Bore, Outside Diameters, and Width

Bearings are dimensioned in relation to the bore diameter, outside diameter, and width. These dimensions are shown in Figure 19.75. After the loading requirements have been established, the bearing is selected in relationship to the shaft size. For example, if an approximate $\varnothing$1.5-inch shaft size is required for a medium service bearing, then a vendor's catalog chart, similar to the one shown in Figure 19.76, is used to select the bearing.

Referring to the chart shown in Figure 19.76, notice the first column is the vendor's bearing number, followed by the bore size (B). To select a bearing for an approximate 1.5-inch shaft, go to the chart and pick the bore diameter of 1.5748, which is close to 1.5. This is the 308K bearing. The tolerance for this bore is specified in the chart as 1.5748 + .0000 and − .0005. Therefore, the limits dimension of the bore in this example is 1.5748 − 1.5743. The outside diameter is 3.5433 + .0000 − .0006 (3.5433 − 3.5427). The width of this bearing is .906 + .000 − .005 (.906 − .901). The fillet radius is the maximum shaft or housing fillet radius in which the bearing corners clear. The fillet radius for the 308K bearing is R.059. Notice the dimensions are also given in millimeters.

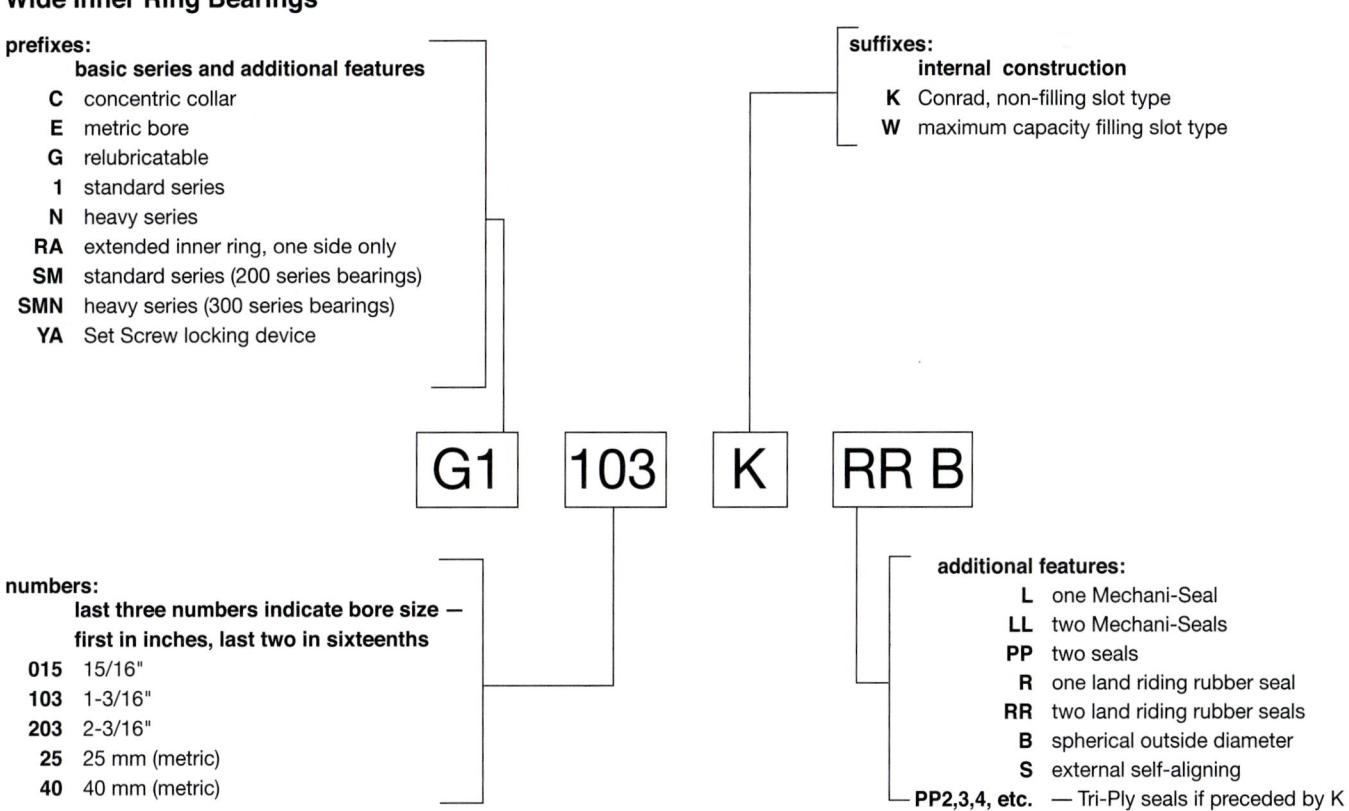

Wide Inner Ring Bearings

prefixes:
basic series and additional features

C concentric collar
E metric bore
G relubricatable
1 standard series
N heavy series
RA extended inner ring, one side only
SM standard series (200 series bearings)
SMN heavy series (300 series bearings)
YA Set Screw locking device

suffixes:
internal construction

K Conrad, non-filling slot type
W maximum capacity filling slot type

G1 | 103 | K | RR B

numbers:
last three numbers indicate bore size —
first in inches, last two in sixteenths

015 15/16"
103 1-3/16"
203 2-3/16"
25 25 mm (metric)
40 40 mm (metric)

additional features:

L one Mechani-Seal
LL two Mechani-Seals
PP two seals
R one land riding rubber seal
RR two land riding rubber seals
B spherical outside diameter
S external self-aligning
PP2,3,4, etc. — Tri-Ply seals if preceded by K

FIGURE 19.74 ■ A sample bearing numbering system. *Courtesy The Torrington Company.*

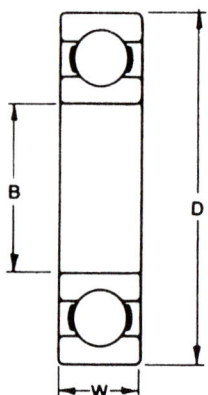

FIGURE 19.75 ■ Bearing dimensions. *Courtesy The Torrington Company.*

Shaft and Housing Fits

Shaft and housing fits are important, because tight fits can cause failure of the balls, rollers, lubricant, or overheating. Loose fits can cause slippage of the bearing in the housing, resulting in overheating, vibration, or excessive wear.

Shaft Fits

In general, for precision bearings, it is recommended that the shaft diameter and tolerance be the same as the bearing bore diameter and tolerance. The shaft diameter used with the 308K bearing is dimensioned ⌀1.5748 − 1.5743.

Housing Fits

In most applications with rotating shafts, the outer ring is stationary and should be mounted with a push of the hand or light

tapping. In general, the minimum housing diameter is .0001 larger than the maximum bearing outside diameter and the maximum housing diameter is .0003 larger than the minimum housing diameter. With this in mind, the housing diameter for the 308K bearing is 3.5433 + .0001 = 3.5434 and 3.5434 + .0003 = 3.5437. The housing diameter limits are 3.5437 − 3.5434.

The Shaft Shoulder and Housing Shoulder Dimensions

Next, you should size the shaft shoulder and housing shoulder diameters. The shaft shoulder and housing shoulder diameter dimensions are represented in Figure 19.77 as S and H. The shoulders should be large enough to rest flat on the face of the bearing and small enough to allow bearing removal. Refer to the chart in Figure 19.78 to determine the shaft shoulder and housing shoulder diameters for the 308K bearing selected in the preceding discussion. Find the basic bearing number 308 and determine the limits of the shaft shoulder and the housing shoulder. The shaft shoulder diameter is 2.00 − 1.93 and the housing shoulder diameter is 3.19 − 3.06. Now you are ready to detail the bearing location on the housing and shaft drawings. A partial detailed drawing of the shaft and housing for the 308K bearing is shown in Figure 19.79.

Surface Finish of Shaft and Housing

The recommended surface finish for precision bearing applications is 32 microinches (0.80 micrometer) for the shaft finish on shafts under 2 inches (50 mm) in diameter. For shafts over 2 inches in diameter, a 63 microinch (1.6 micrometer) finish is suggested. The housing diameter can have a 125 microinch (3.2 micrometer) finish for all applications.

DIMENSIONS — TOLERANCES

Bearing Number	Bore B		tolerance +.0000" +.000 mm to minus		Outside Diameter D		tolerance +.0000" +.000 mm to minus		Width W +.000", −.005" +.00 mm, − 13 mm		Fillet Radius[1]		WL		Static Load Rating C_O		Extended Dynamic Load Rating C_E	
	in.	mm	in.	mm	in.	mm	in.	mm	in.	mm	.in	mm	lbs.	kg	lbs.	N	lbs.	N
300K	.3937	10	.0003	.008	1.3780	35	.0005	.013	.433	11	.024	.6	.12	.054	850	3750	2000	9000
301K	.4724	12	.0003	.008	1.4567	37	.0005	.013	.472	12	.039	1.0	.14	.064	850	3750	2080	9150
302K	.5906	15	.0003	.008	1.6535	42	.0005	.013	.512	13	.039	1.0	.18	.082	1270	5600	2900	13200
303K	.6693	17	.0003	.008	1.8504	47	.0005	.013	.551	14	.039	1.0	.24	.109	1460	6550	3350	15000
304K	.7874	20	.0004	.010	2.0472	52	.0005	.013	.591	15	.039	1.0	.31	.141	1760	7800	4000	17600
305K	.9843	25	.0004	.010	2.4409	62	.0005	.013	.669	17	.039	1.0	.52	.236	2750	12200	5850	26000
306K	1.1811	30	.0004	.010	2.8346	72	.0005	.013	.748	19	.039	1.0	.78	.354	3550	15600	7500	33500
307K	1.3780	35	.0005	.013	3.1496	80	.0005	.013	.827	21	.059	1.5	1.04	.472	4500	20000	9150	40500
308K	1.5748	40	.0005	.013	3.5433	90	.0006	.015	.906	23	.059	1.5	1.42	.644	5600	24500	11000	49000
309K	1.7717	45	.0005	.013	3.9370	100	.0006	.015	.984	25	.059	1.5	1.90	.862	6700	3000	13200	58500
310K	1.9685	50	.0005	.013	4.3307	110	.0006	.015	1.063	27	.079	2.0	2.48	1.125	8000	35500	15300	68000
311K	2.1654	55	.0006	.015	4.7244	120	.0006	.015	1.142	29	.079	2.0	3.14	1.424	9500	41500	18000	80000
312K	2.3622	60	.0006	.015	5.1181	130	.0008	.020	1.220	31	.079	2.0	3.89	1.765	10800	48000	20400	90000

[1]Maximum shaft or housing fillet radius which bearing corners will clear.

FIGURE 19.76 ■ Bearing selection chart. *Courtesy The Torrington Company.*

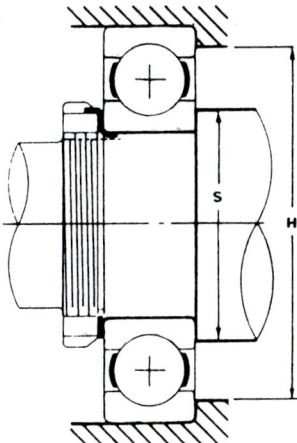

FIGURE 19.77 ■ Shaft shoulder and housing shoulder dimensions. *Courtesy The Torrington Company.*

Bearing Lubrication

It is necessary to maintain a film of lubrication between the bearing surfaces. The factors to consider when selecting lubrication requirements include the:

■ Type of operation, such as continuous or intermittent.

■ Service speed in rpm (revolutions per minute).

■ Bearing load, such as light, medium, or heavy.

Bearings can also be overlubricated, which can cause increased operating temperatures and early failure. Selection of the proper lubrication for the application should be determined by the manufacturer recommendations. The ability of the lubricant is due, in part, to viscosity. Viscosity is the internal friction of a fluid, which makes it resist a tendency to flow. Fluids with low viscosity flow more freely than those with high viscosity. The chart in Figure 19.80 shows the selection of oil viscosity based on temperature ranges and speed factors.

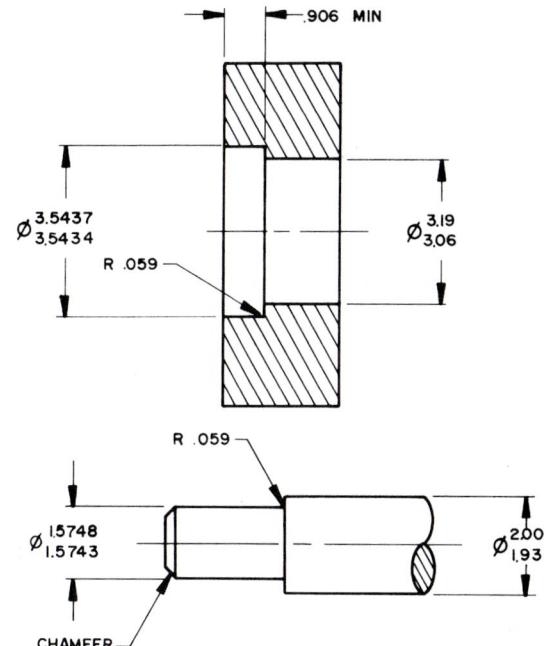

FIGURE 19.79 ■ A partial detail drawing of the shaft and housing for the 308K bearing. Dimension values in this figure are in inches.

Oil Viscosities and Temperature Ranges for Ball Bearing Lubrication

Maximum Temperature Range Degrees F	Optimum Temperature Range Degrees F	Speed Factor, S_i (inner race bore diameter (inches) × RPM)	
		Under 1000	Over 1000
		Viscosity	
–40 to +100	–40 to –10	80 to 90 SSU (at 100 deg. F)	70 to 80 SSU (at 100 deg. F)
–10 to +100	–10 to +30	100 to 115 SSU (at 100 deg. F)	80 to 100 SSU (at 100 deg. F)
+30 to +150	+30 to +150	SAE 20	SAE 10
+30 to +200	+150 to +200	SAE 40	SAE 30
+50 to +300	+200 to +300	SAE 70	SAE 60

FIGURE 19.80 ■ Selection of oil viscosity based on temperature ranges and speed factors.

Extra-Light • 9100 Series								L	ight • 200, 7200WN Series								M	edium • 300, 7300WN Series							
Basic Bearing Number	shaft, S		houlder Diameters				housing, H		Basic Bearing Number	shaft, S		houlder Diameters				housing, H		Basic Bearing Number	shaft, S		houlder Diameters				housing, H
	max.		min.		max.		min.			max.		min.		max.		min.			max.		min.		max.		min.
	in.	mm	in.	mm	in.	mm	in.	mm		in.	mm	in.	mm	in.	mm	in.	mm		in.	mm	in.	mm	in.	mm	in.	mm
9100	.52	13.2	.47	11.9	.95	24.1	.91	23.1	200	.56	14.2	.50	12.7	.98	24.9	.97	24.6	300	.59	15.0	.50	12.7	1.18	30.0	1.15	29.2
9101	.71	18.0	.55	14.0	1.02	25.9	.97	24.6	201	.64	16.3	.58	14.7	1.06	26.9	1.05	26.7	301	.69	17.5	.63	16.0	1.22	31.0	1.21	30.7
9102	.75	19.0	.67	17.0	1.18	30.0	1.13	28.7	202	.75	19.0	.69	17.5	1.18	30.0	1.15	29.2	302	.81	20.6	.75	19.0	1.42	36.1	1.40	35.6
9103	.81	20.6	.75	19.0	1.30	33.0	1.25	31.8	203	.84	21.3	.77	19.6	1.34	34.0	1.31	33.3	303	.91	23.1	.83	21.1	1.61	40.9	1.60	40.6
9104	.98	24.9	.89	22.6	1.46	37.1	1.41	35.8	204	1.00	25.4	.94	23.9	1.61	40.9	1.58	40.1	304	1.06	26.9	.94	23.9	1.77	45.0	1.75	44.4
9105	1.18	30.0	1.08	27.4	1.65	41.9	1.60	40.6	205	1.22	31.0	1.14	29.0	1.81	46.0	1.78	45.2	305	1.31	33.3	1.14	29.0	2.17	55.1	2.09	53.1
9106	1.38	35.1	1.34	34.0	1.93	49.0	1.88	47.8	206	1.47	37.3	1.34	34.0	2.21	56.1	2.16	54.9	306	1.56	39.6	1.34	34.0	2.56	65.0	2.44	62.0
9107	1.63	41.4	1.53	38.9	2.21	56.1	2.15	54.6	207	1.72	43.7	1.53	38.9	2.56	65.0	2.47	62.7	307	1.78	45.2	1.69	42.9	2.80	71.1	2.72	69.1
9108	1.81	46.0	1.73	43.9	2.44	62.0	2.39	60.7	208	1.94	49.3	1.73	43.9	2.87	72.9	2.78	70.6	308	2.00	50.8	1.93	49.0	3.19	81.0	3.06	77.7
9109	2.03	51.6	1.94	49.3	2.72	69.1	2.67	67.8	209	2.13	54.1	1.94	49.3	3.07	78.0	2.97	75.4	309	2.28	57.9	2.13	54.1	3.58	90.9	3.41	86.6
9110	2.22	56.4	2.13	54.1	2.91	73.9	2.86	72.6	210	2.34	59.4	2.13	54.1	3.27	83.1	3.17	80.5	310	2.50	63.5	2.36	59.9	3.94	100.1	3.75	95.2
9111	2.48	63.0	2.33	59.2	3.27	83.1	3.22	81.8	211	2.54	64.5	2.41	61.2	3.68	93.5	3.56	90.4	311	2.75	69.8	2.56	65.0	4.33	110.0	4.13	104.9
9112	2.67	67.8	2.53	64.3	3.47	88.1	3.42	86.9	212	2.81	71.4	2.67	67.8	3.98	101.1	3.87	98.3	312	2.94	74.7	2.84	72.1	4.65	118.1	4.44	112.8

FIGURE 19.78 ■ Shaft shoulder and housing shoulder dimension selection chart. *Courtesy The Torrington Company.*

Oil Grooving of Bearings

In situations where bearings or bushings do not receive proper lubrication, it can be necessary to provide grooves for the proper flow of lubrication to the bearing surface. The bearing grooves help provide the proper lubricant between the bearing surfaces and maintain adequate cooling. There are several methods of designing paths for the lubrication to the bearing surfaces, as shown in Figure 19.81.

Sealing Methods

Machine designs normally include means for stopping leakage and keeping out dirt and other contaminants when lubricants are involved in the machine operation. This is accomplished using static or dynamic sealing devices. Static sealing refers to stationary devices that are held in place and stop leakage by applied pressure. Static seals such as gaskets do not come in contact with the moving parts of the mechanism. Dynamic seals are those that contact the moving parts of the machinery, such as packings.

Gaskets are made from materials that prevent leakage and access of dust contaminants into the machine cavity. Silicone rubber gasket materials are used in applications such as water pumps, engine filter housings, and oil pans. Gasket tapes, ropes, and strips provide good cushioning properties for dampening vibra-

tion, and the adhesive sticks well to most materials. Nonstick gasket materials such as paper, cork, and rubber are available for certain applications. Figure 19.82 shows a typical gasket mounting.

Dynamic seals include packings and seals that fit tightly between the bearing or seal seat and the shaft. The pressure applied by the seal seat or the pressure of the fluid causes the sealing effect. Molded lip packings are available that provide sealing as a result of the pressure generated by the machine fluid. Figure 19.83 shows examples of molded lip packings. Molded ring seals are placed in a groove and provide a positive seal between the shaft and bearing or bushing. Types of molded ring seals include labyrinth, O-ring, lobed ring, and others. Labyrinth which means maze, refers to a seal that is made of a series of spaced strips that are connected to the seal seat, making it difficult for the lubrication to pass. Labyrinth seals are used in heavy machinery where some leakage is permissible. (See Figure 19.84.) The O-ring seal is the most commonly used seal because of its low cost, ease of application, and flexibility. The O-ring can be used for most situations involving rotating or oscillating motion. The O-ring is placed in a groove that is machined in either the shaft or the housing as shown in Figure 19.85. The lobed ring has rounded lobes that provide additional sealing forces over the standard O-ring seal. A typical lobed ring seal is shown in Figure 19.86.

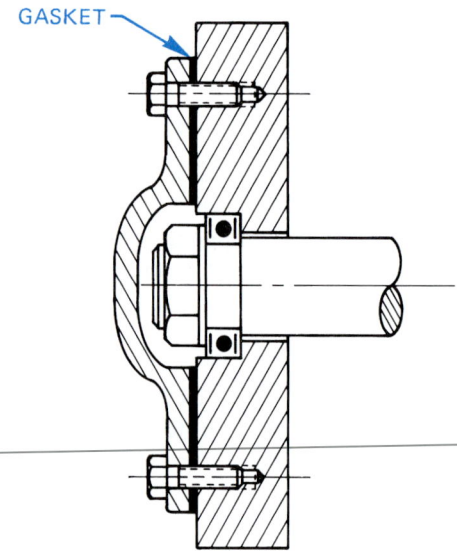

FIGURE 19.82 ■ Typical gasket mounting.

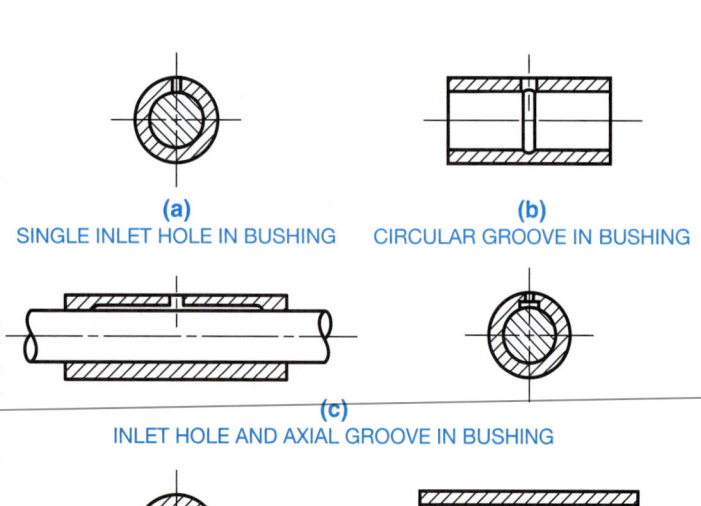

(a)
SINGLE INLET HOLE IN BUSHING

(b)
CIRCULAR GROOVE IN BUSHING

(c)
INLET HOLE AND AXIAL GROOVE IN BUSHING

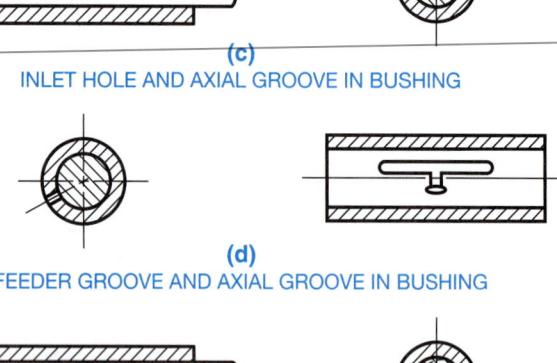

(d)
FEEDER GROOVE AND AXIAL GROOVE IN BUSHING

(e)
FEEDER GROOVE AND STRAIGHT AXIAL GROOVE IN THE SHAFT

FIGURE 19.81 ■ Methods of designing paths for lubrication to bearing surfaces.

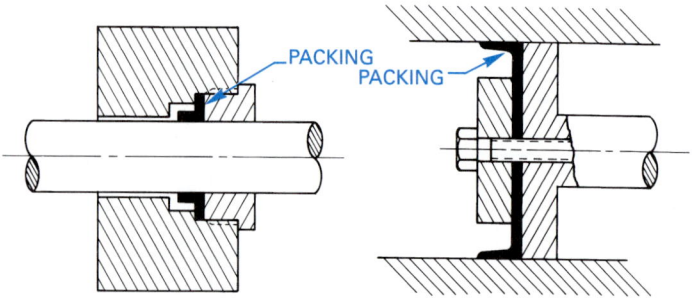

FIGURE 19.83 ■ Molded lip packings.

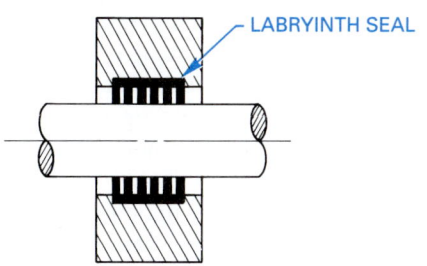

FIGURE 19.84 ■ Labyrinth seal.

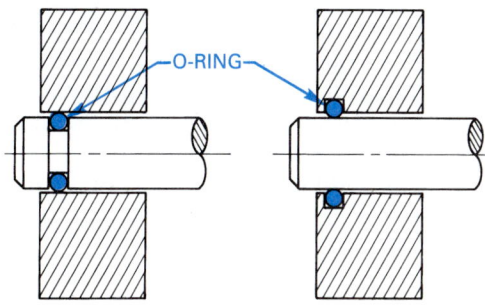

FIGURE 19.85 ■ O-ring seals.

Felt seals and **wool seals** are used where economical cost, lubricant absorption, filtration, low friction, and a polishing action are required. However, the ability to completely seal the machinery is not as positive as with the seals described earlier. (See Figure 19.87.)

Bearing Mountings

There are a number of methods used for holding the bearing in place. Common techniques include a nut and lock washer, a nut and lock nut, or a retaining ring. Other methods can be designed

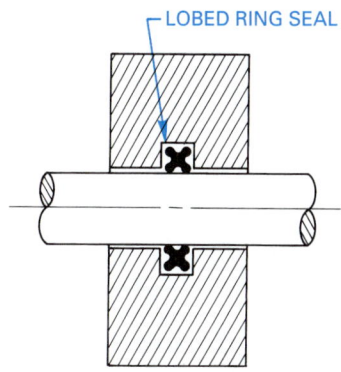

FIGURE 19.86 ■ Lobed ring seal.

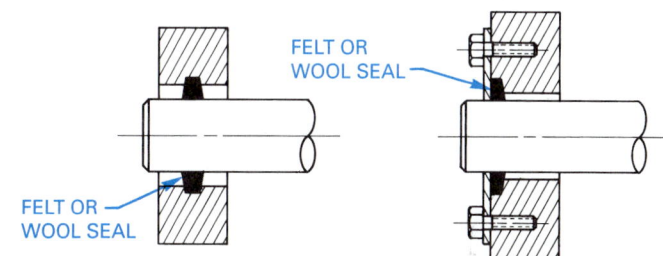

FIGURE 19.87 ■ Felt and wool seals.

to fit the specific application or requirements, such as a shoulder plate. Figure 19.88 shows some examples of mountings.

GEAR AND BEARING ASSEMBLIES

Gear and bearing assemblies show the parts of the complete mechanism as they appear assembled. (See Figure 19.89.) When drawing assemblies, you need to use as few views as

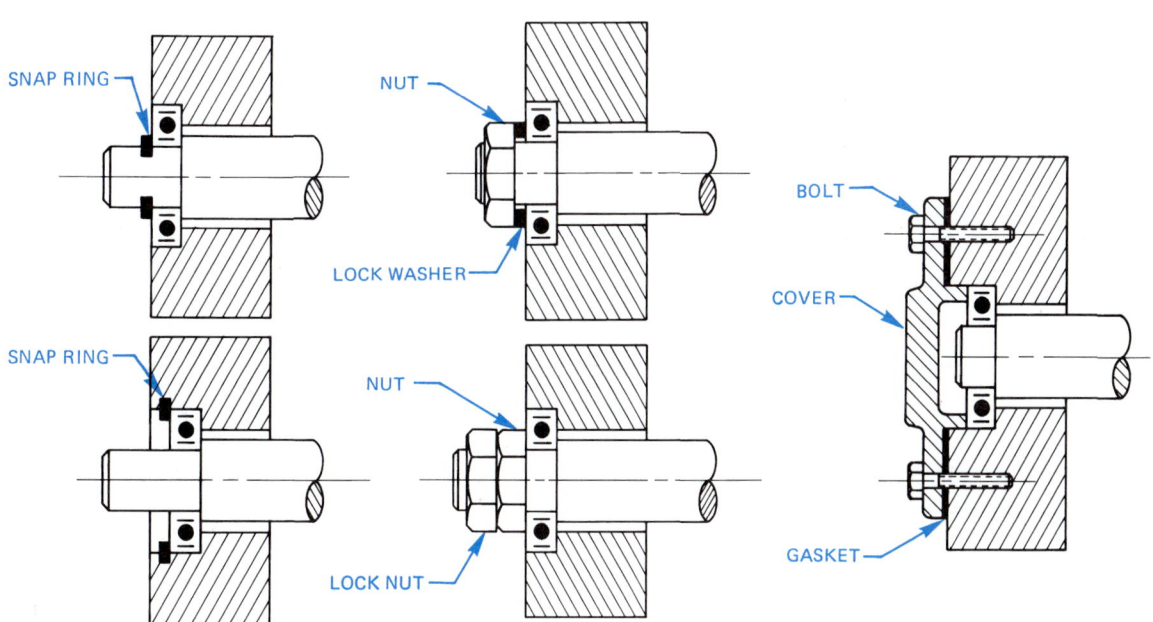

FIGURE 19.88 ■ Typical bearing mountings.

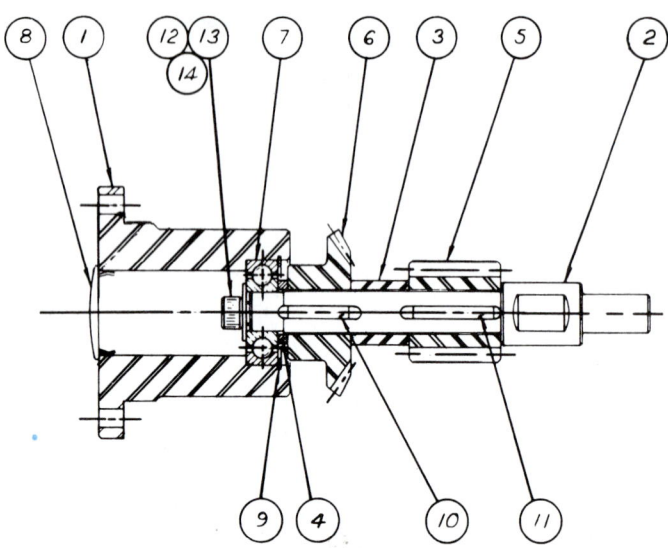

FIGURE 19.89 ■ *Assembly drawing. Courtesy Curtis Associates.*

possible, but enough to adequately display how all the parts fit together. In some situations, all that is needed is a full sectional view that displays all of the internal components. An exterior view such as a front or top view plus a section sometimes works. Dimensions are normally omitted from the assembly unless the dimensions are needed for assembly purposes. For example, when a specific dimension regarding the relationship of one part to another is required to properly assemble the parts, each part is identified with a number in a circle. This circle is referred to as a balloon. The balloons are connected to the part being identified with a leader. The balloons are about one-half inch in diameter. The identification number is .24-inch-high (6 mm) lettering or text. The balloon numbers correlate with a parts list. The parts list is normally placed on the drawing above or adjacent to the title block, or on a separate sheet as shown in Figure 19.90. Assembly includes torque data and lubricant information. Refer to Chapter 18 for complete information covering detail, assembly drawings, and parts lists.

ASSEMBLY <u>Cross Shaft Assembly</u>   USED ON _____

NUMBER OF UNITS _____   DATE _____

HAVE	NEED	P/Ø NO. W/Ø NO.	DET. NO.	PART NO.	DWG.	QTY.	PART NAME	DESCRIPTION	VENDOR
			1		B	1	Bearing Retainer	Ø 3" C.D. Bar	
			2		B	1	Cross Shaft	Ø 3/4" C.D. Bar	
			3		A	1	Spacer	Ø 3/4 O.D. × 11GA Wall Tube × .738 Thick	
			4		A	1	Spacer	Ø 3/4 O.D. × 11GA Wall Tube × .125 Thick	
			5		–	1	Steel Worm 12 D.P. Single Thread	Boston Gear #H 1056 R.H.	
			6		–	1	Bevel Gear 20° P.A. 2" P.D.	Boston Gear #HL 149 Y-G	
			7		–	1	Ball Bearing .4724 Ø Bore	T.R.W. or Equivalent #MRC 201-S22	
			8		–	1	End Plug		
			9		–	1	Snap Ring	Waldes-Truarc #N 5000-125	
			10		–	1	Key Stock	1/8 Sq. × 3/4 Lg.	
			11		–	1	Key Stock	1/8 Sq. × 1 Lg.	
			12		–	1	Socket Head Cap Screw	1/4 UNC × 3/4 Lg.	
			13		–	1	Lockwasher	1/4 Nominal	
			14		–	1	Flat Washer	1/4 Nominal	

FIGURE 19.90 ■ Parts list. *Courtesy Curtis Associates.*

PROFESSIONAL PERSPECTIVE

In every situation regarding linkage, cam, and gear design, there is a need to investigate all of the manufacturing alternatives and provide a solution the customer can afford. Competition is so great in the manufacturing industry that you should evaluate each design to find the best way to produce the product. The best way to understand this concept as an entry-level engineering drafter is to talk to experienced designers, engineers, and machinists. Go to the shop to see how things are done and determine the drawing requirements directly from the people who know. According to one design engineer, "If you don't know how it is going to be made, you're not a good drafter. Know the manufacturing capabilities of each piece of equipment." In addition to your drafting courses, it is a good idea to take some manufacturing technology classes. Math is also an important part of your program. Drafters in this field use a lot of calculations, including geometry and trigonometry. After you have a strong educational background, the experienced engineer says, "Keep an open mind and look at all the alternatives."

For example, you work as an engineering drafter for a foundry. For years the flasks have been handled either by hand or with a lift truck. Both of these methods are time-consuming and dangerous, and the company has some contracts that require founding some castings that are too heavy to handle. So, the engineering department plans the design for a hydraulic

flask handler. Your team is responsible for the handling mechanism that must be designed with this criteria:

■ Use a hydraulic piston with a 6-in. stroke.

■ Handle up to 44-in.-wide flasks.

■ Take up no more than 28 in. in overall height.

Your team begins the problem-solving process and comes up with these steps to solve the problem:

■ Develop a single-line schematic kinematic diagram representing the movement of the mechanism based on the design information.

■ Identify and locate the available materials needed to build the flask handler.

The preliminary design is shown in Figure 19.91. All your team needs to do now is ask the other design teams for input and, after revisions are made, prepare a complete set of working drawings for fabrication.

It is recommended that you review Chapter 5, Manufacturing Materials and Processes, Chapter 9, Multiviews, Chapter 13, Sections and Conventional Revolutions, Chapter 11, Dimensioning and Tolerancing, and CD Chapter 3, Engineering Drawing and Design Math Applications.

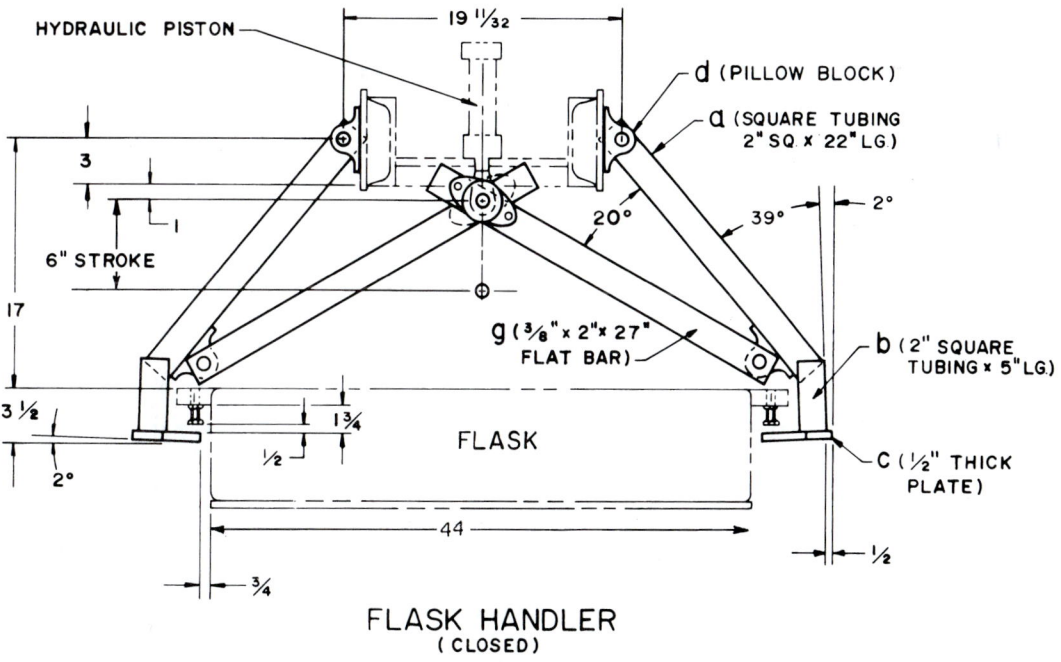

FIGURE 19.91 ■ A preliminary design is created. Dimension values in this figure are in inches.

MATH APPLICATIONS

LENGTH OF A CONNECTING ROD

Problem: Find the length of the connecting rod for the slider mechanism of Figure 19.92.

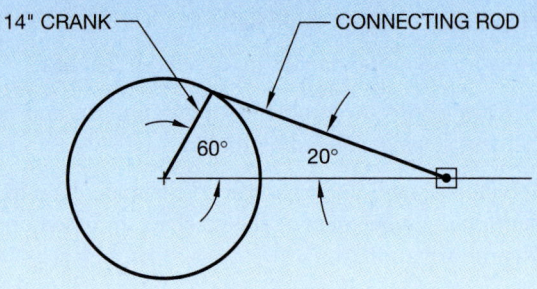

FIGURE 19.92 ■ Slider mechanism.

Solution: Using the Law of Sines (from CD Chapter 3, Engineering Drawing and Design Math Applications.),

$$\frac{a}{\sin A} = \frac{b}{\sin B}$$

$$\frac{14"}{\sin 20°} = \frac{b}{\sin 60°}$$

Solving this proportional equation for b gives b = 35.4".

WEBSITE RESEARCH

The following websites can assist you in doing additional research on subjects such as standards, gears, cams, linkages, bearings, and related areas:

http://www.asme.org—American Society of Mechanical Engineers for ANSI Y14.7.1.

http://www.industrialpress.com—*Machinery's Handbook.* Gear, cam, linkage, and bearing-related information and specifications.

http://www.gearmfg.com—American Gear Manufacturers Association.

http://www.abapgt.com—Plastic gearing technology.

http://www.torrington.com—Bearings.

http://www.thomasregional.com—Gear, cam, and bearing products.

http://www.source4industries.com—Bearings.

CHAPTER 19

Mechanisms: Linkages, Cams, Gears, and Bearings Test

 Access the CD found with this textbook to view the Chapter 19 Test. Confirm the preferred submittal method with your instructor.

CHAPTER 19

Mechanisms: Linkages, Cams, Gears, and Bearings Problems

DIRECTIONS

Read problems carefully before you begin working. Complete each problem on an appropriately sized sheet. Precision work is important for accurate solutions. Increase the drawing scale as needed for clear presentation.

LINKAGE PROBLEMS

PROBLEM 19.1   On the right is a pictorial drawing and a linkage schematic of a vise grip in the closed position. Reproduce the schematic exactly as shown and also show the open position in a second color.

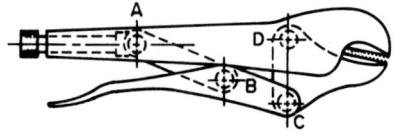
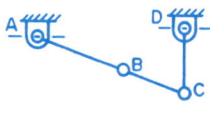

PROBLEM 19.2 Following is a pictorial drawing and a linkage schematic of a toggle clamp in the closed position. Reproduce the schematic exactly as shown and also show the open position in a second color.

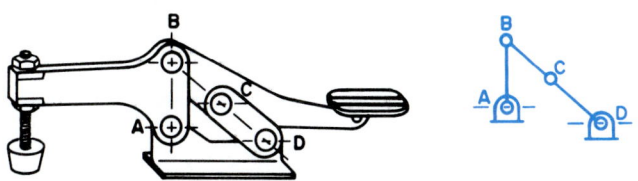

PROBLEM 19.3 Determine the extreme right and left positions of link CD in the figure shown below. Determine and label the angle through which CD oscillates.

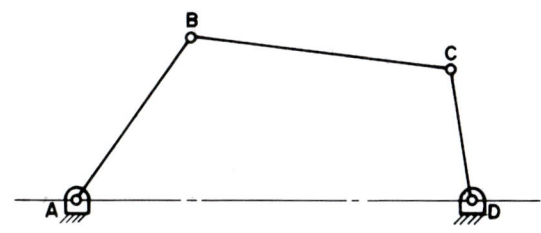

PROBLEM 19.4 Draw the mechanism below in the position shown. Using different colors, draw the mechanism in the extreme right and left positions. Dimension the stroke. (inches)

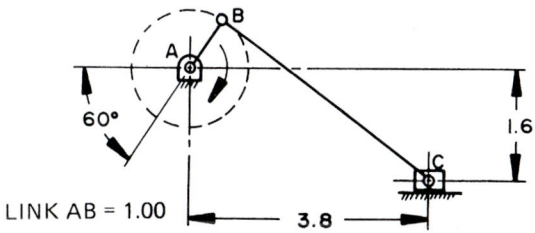

PROBLEM 19.5 Draw the combination bellcrank slider mechanism in the position shown below. Determine the stroke of the slider if A moves to position A¹. Note: Position of features in the sketch may be out of proportion. (inches)

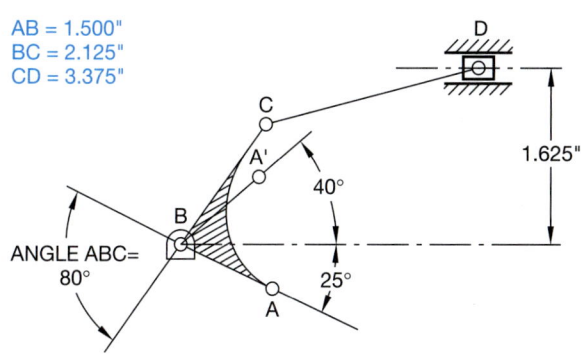

PROBLEM 19.6 The following drawing is a linkage schematic of an oscillating lawn sprinkler. The spray tube, shown in section, is part of link CD. Link AB moves through 360°, while points A and D are stationary. Determine and dimension the angle of oscillation through which the spray moves. (inches)

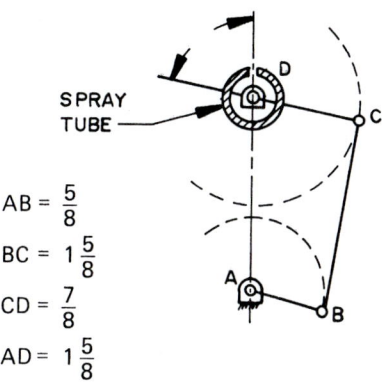

$$AB = \frac{5}{8}$$
$$BC = 1\frac{5}{8}$$
$$CD = \frac{7}{8}$$
$$AD = 1\frac{5}{8}$$

PROBLEM 19.7 Given the windshield wiper mechanism shown below, determine and dimension the angle of oscillation of the wiper blades. The electric motor rotates link ED continuously through 360°. ABC is one link with a 90° angle at B. A tension spring is located in the center of link BG.

AB = FG = ED = 1.25 in.   BC = .875 in.
AF = BG = 16 in.   CD = 7.625 in.

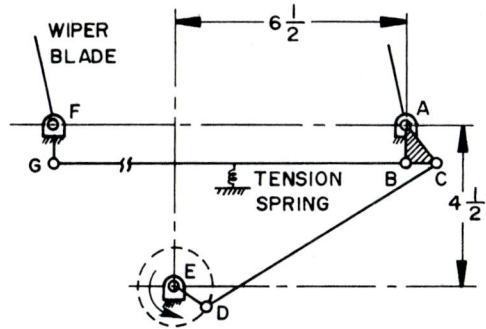

PROBLEM 19.8 Draw the mechanism shown below. Using different colors, show the path of point D in a total of five equally spaced positions, including the extreme right and left positions. (inches)

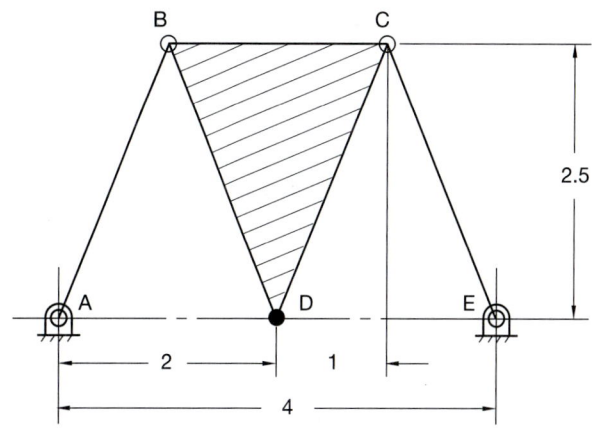

PROBLEM 19.9   Draw the mechanism in the position shown below. Draw the path of point P as linkage AB moves every 30° through a total of 360°. Use a different color and/or line type for each position. (inches)

Dimension the angle through which CD oscillates.
BCP is a through link.
Point P slides on EF.
EF = 6.5 in.

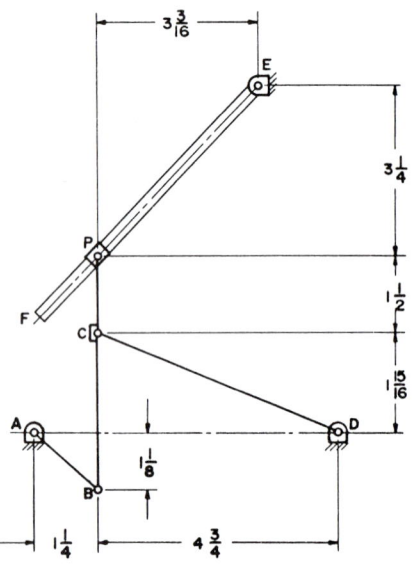

PROBLEM 19.10   Given the mechanism shown below, rotate link BC at 60° intervals clockwise through 360°. Plot and draw the paths of points D, E, and F. Dimension the angle of oscillation of link AF.

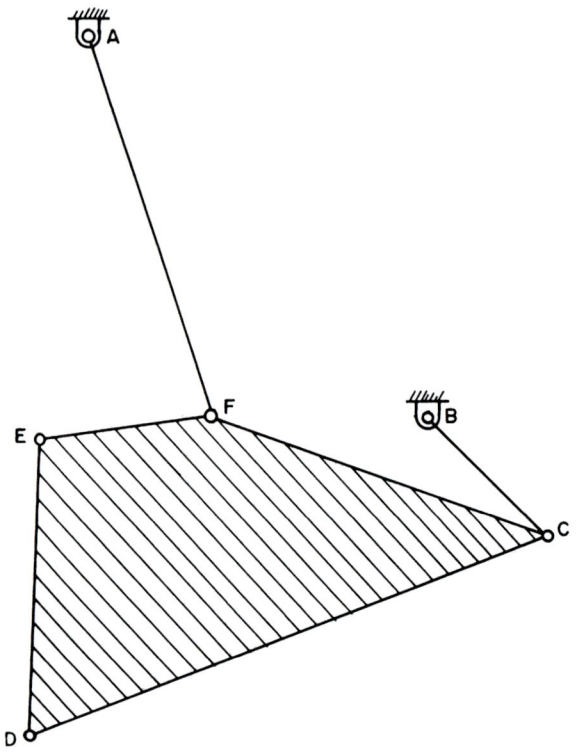

PROBLEM 19.11   Given the assembly drawing of the foundry flask handler shown below, draw a mechanism schematic showing the two extreme positions of movement. The handler is operated by a hydraulic piston with a 6-in. stroke.

a is welded to b.
b is welded to c.
(inches)

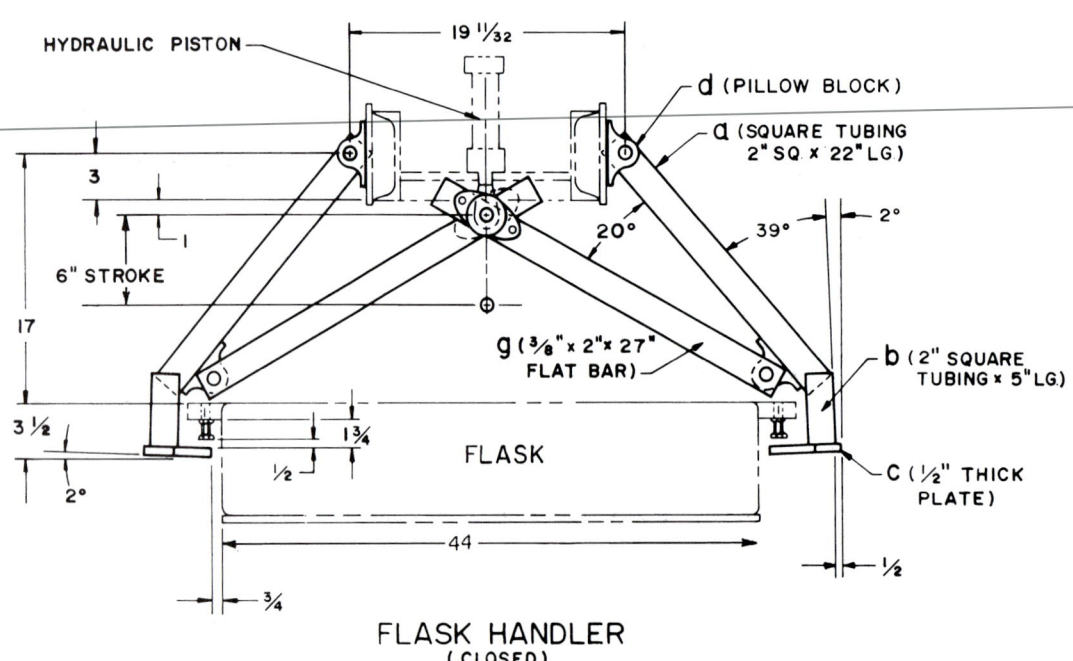

FLASK HANDLER
(CLOSED)

PROBLEM 19.12   Given the pivot hoist shown below, draw and scale exactly as shown. Rotate ADE clockwise so link AE is horizontal. Determine and dimension the extended length of the spring between C and E. Determine and dimension the angle between AE and CE when E is in the new position.

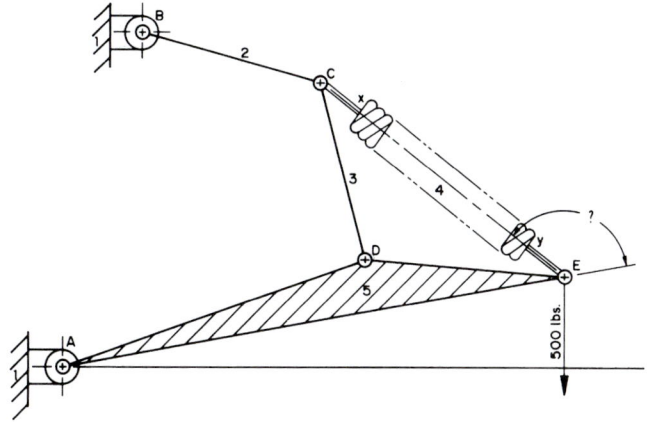

PROBLEM 19.13   Given the figure shown below, determine and dimension the stroke of point E moving in a straight line as link AB rotates 360°. Also dimension the angle of oscillation of link CD. (inches)

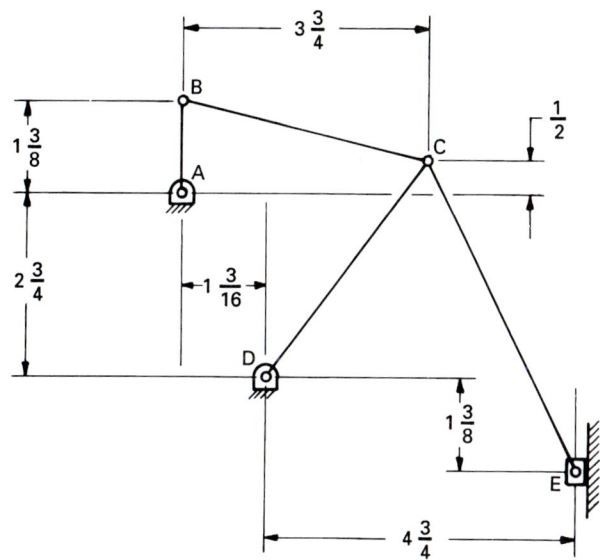

PROBLEM 19.14   Find two examples of linkage mechanisms at home or school. Explain in a short complete statement the function of each mechanism. Using schematic representations, show and dimension the extreme positions of each mechanism.

CAM DISPLACEMENT DIAGRAMS

PROBLEM 19.15   Construct a cam displacement diagram for a cam follower that rises in simple harmonic motion a total of 2.00 in. in 150°, dwells for 30°, falls 2.00 in. simple har-

monic motion in 120°, and dwells for 60°. Draw the horizontal scale 6.00 in.

PROBLEM 19.16   Construct a cam displacement diagram for a cam follower that rises in uniform accelerated motion a total of 2.00 in. in 180°, dwells for 30°, falls 2.00 in. uniform accelerated motion in 120°, and dwells for 30°. Draw the horizontal scale 6.00 in.

PROBLEM 19.17   Construct a cam displacement diagram for a cam follower that rises in modified constant velocity for 3.00 in. in 180°, falls 3.00 in. modified constant velocity motion in 120°, and dwells for 60°. Use a modified constant velocity motion designed with one-third of the displacement.

PROBLEM 19.18   Construct a cam displacement diagram for a cam follower that rises 2.000 in. cycloidal motion in 120°, dwells for 60°, and falls 2.000 in. in cycloidal motion in 180°.

PROBLEM 19.19   Construct a cam displacement diagram for a cam follower that rises in simple harmonic motion a total of 1.250 in. in 90°, dwells for 60°, rises .750 in. in 45° simple harmonic motion, falls 2.00 in. with cycloidal motion in 120°. Draw the horizontal scale 12 in.

PROBLEM 19.20   Construct a cam displacement diagram for a cam follower that rises in modified constant velocity motion (modified to one-third the displacement) for 3.000 in. in 180°, dwells for 30°, and falls 3.000 in. simple harmonic motion in 30°, and dwells to the end of the cycle. Draw the horizontal scale 12 in.

PROBLEM 19.21   Construct a cam displacement diagram for a cam follower that rises 3.500 in. in 90° cycloidal motion, dwells for 45°, falls 2.500 in. cycloidal motion in 135°, dwells for 30°, falls 1.000 in. simple harmonic motion in 30°, and dwells to the end of the cycle. Draw the horizontal scale 12 in.

PROBLEM 19.22   Construct a cam displacement diagram for a cam follower that rises in cycloidal motion for 3.000 in. in 90°, dwells for 30°, falls 1.000 in. in simple harmonic motion in 90°, dwells for 30°, and falls the remaining 2.000 in. in uniform accelerated motion in 120°. Draw the horizontal scale 12 in.

PROBLEM 19.23   Construct a cam displacement diagram for a cam follower that rises in cycloidal motion a total of 2.1875 in. in 150°, dwells for 30°, falls back to the original level in simple harmonic motion in 150°, then dwells through the remainder of the cycle. Use a 12-in. horizontal scale.

PROBLEM 19.24   Construct a cam displacement diagram for a cam follower that rises a .375 in. diameter in-line roller follower 1.500 in. in uniform accelerated motion in 150°, dwells 45°, falls with modified constant velocity (one-third displacement) in 120°, and dwells the remainder of the cycle.

PROBLEM 19.25   Construct a cam displacement diagram for a cam follower that rises 3.500 in. 90° cycloidal motion, dwells for 30°, falls 2.250 in. cycloidal motion in 150°, falls 1.250 in. simple harmonic motion in 60°, and dwells for

30°. Draw the horizontal scale equal in circumference to a 3.000-in. diameter circle.

CAM PROFILE DRAWINGS

PROBLEM 19.26 Use the displacement diagram constructed in Problem 19.15 and the information given in the illustration below to lay out the plate cam profile drawing. The cam rotates counterclockwise. Make a two-view detailed drawing of the cam and dimension it as shown in this chapter. (inches)

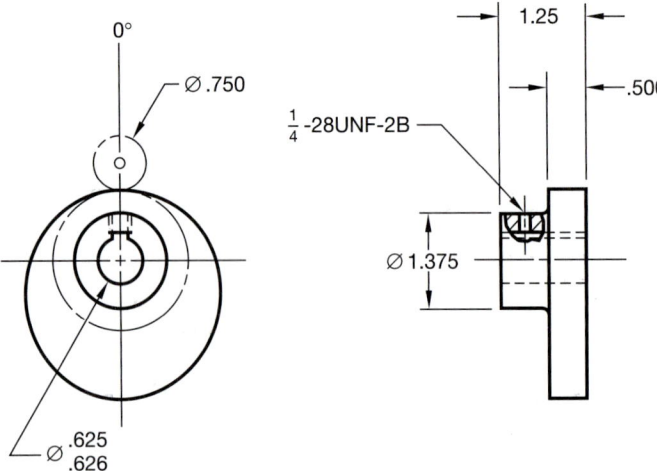

IN-LINE ROLLER FOLLOWER	= ∅.750
BASE CIRCLE	= ∅2.000
KEY SIZE USED	= 1/4 × 1/4 SQ. KEY
PLATE THICKNESS	= .500
HUB THICKNESS	= .750
HUB DIAMETER	= ∅1.375
SHAFT DIAMETER	= ∅.625/.626

PROBLEM 19.27 Use the displacement diagram constructed in Problem 19.18 and the information given in the illustration for Problem 19.26 to lay out the plate cam profile. The cam is rotating clockwise. Make a two-view detailed drawing of the cam, properly toleranced and dimensioned.

PROBLEM 19.28 Make a two-view detailed drawing of a plate cam using the displacement diagram from Problem 19.24. Completely dimension the drawing using the following information:

The cam rotates counterclockwise.
In-line roller follower = ∅.375   Hub projection = .500
Base circle = ∅2.750   Plate thickness = .375
Shaft = ∅1.000   All dimensions in inches.
Hub diameter = ∅1.250

PROBLEM 19.29 Make a two-view detailed drawing of a plate cam using the displacement diagram from Problem 19.20. Completely dimension the drawing using the following information:

The cam rotates counterclockwise.
In-line roller follower = ∅.750   Hub projection = .500
Base circle = ∅2.000   Plate thickness = .500
Shaft = ∅.625   All dimensions in inches.
Hub diameter = ∅1.250

PROBLEM 19.30 Make a two-view detailed drawing of a plate cam using the displacement diagram from Problem 19.18. Completely dimension the drawing using the information shown below: (inches)

The cam rotates clockwise.

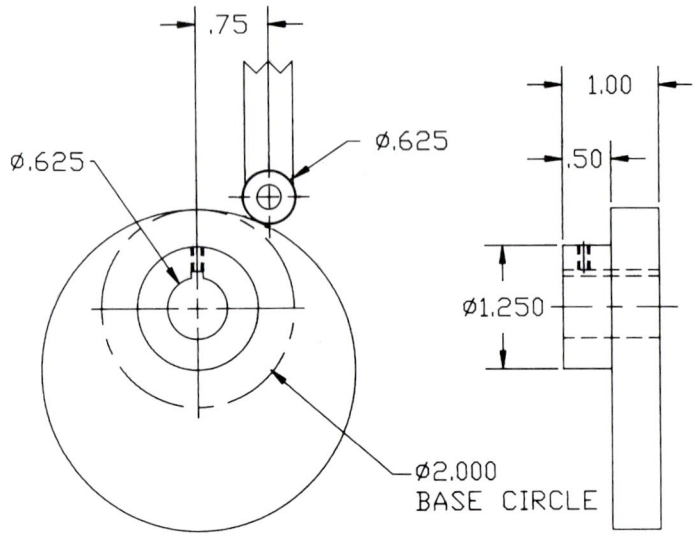

PROBLEM 19.31 Use the displacement diagram constructed in Problem 19.25 and the following illustration to lay out the profile of the groove in the drum cam. The cam rotates clockwise. Make a two-view detailed drawing of the drum cam, with tolerances and dimensions as discussed and shown in this chapter. (inches)

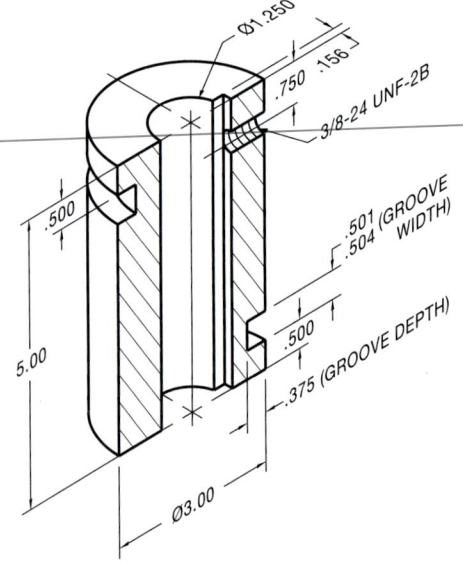

HALF OF OBJECT SHOWN

GEAR PROBLEMS

PROBLEM 19.32 Given the gear train and chart shown, calculate or determine the missing information to complete the chart.

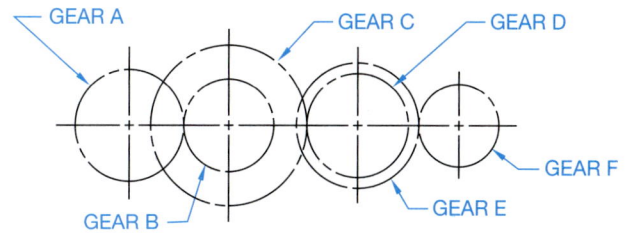

Gear	Diametral Pitch (P)	Number of teeth (N)	Pitch Diameter (D)	RPM	Direction	Center Distance	Gear Ratio
A	4		7.5"	240	Clockwise		
B		18					
C			10.0"	400			
D	5	40					
E	7						
F		14		1500			

PROBLEM 19.33 Given the ten-gear power transmission and chart shown, calculate or determine the missing information to complete the chart.

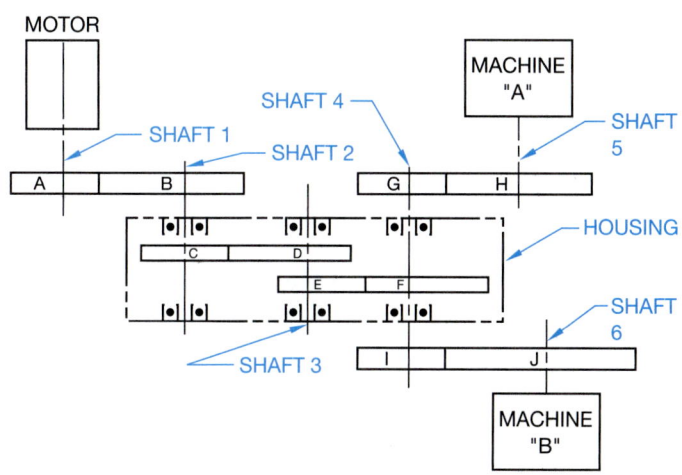

Gear	Pitch Diameter	No. of Teeth	Diametral Pitch	RPM	Ctr. Distance Between Mating Gears	Direction
A	3.00			3600		Counter-clockwise
B			5		4.00	
C		48				
D			12	1080		
E	4.00	40				
F		100				
G	5.00				6.00	
H			6			
I			4			
J		40		108		

PROBLEM 19.34 Given the following information, use ANSI standards to make a detailed drawing of the spur gear shown at right: (inches)

20 teeth	Face width = 2.500 in.
Diametral pitch = 5	Shaft diameter = ⌀1.125
20° pressure angle	Keyway for a .25 in. square key

Place the centerline of the keyway in line with a radial line through the center of one tooth (one tooth profile needed to show alignment). Include the necessary spur gear data in a chart placed over the title block. Use the formulas given in this chapter to solve for unknown values.

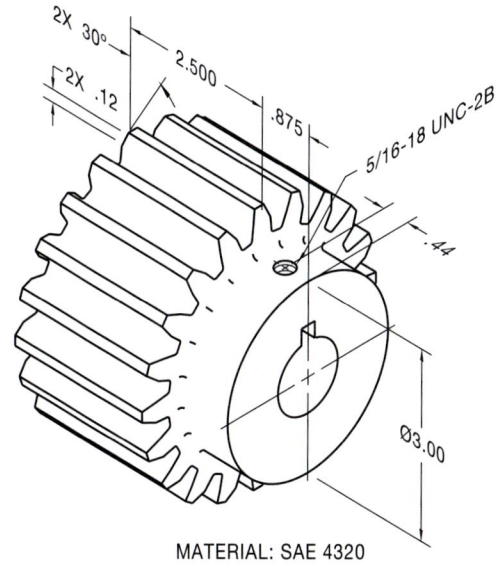

MATERIAL: SAE 4320

PROBLEM 19.35   Use ANSI standards to make a detailed drawing of a rack that mates with the spur gear in Problem 19.34. The overall length is 24 in. (inches)

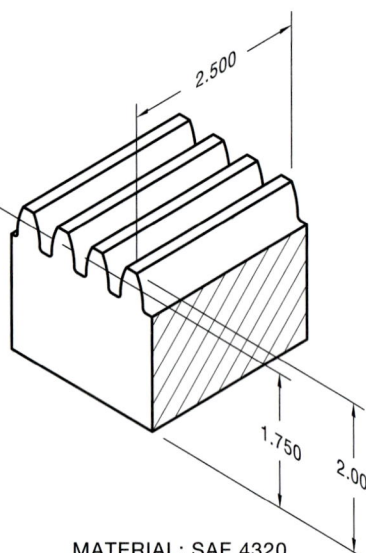

MATERIAL: SAE 4320

PROBLEM 19.36   Use ANSI standards to make a detailed drawing of a straight bevel gear given the following information and the illustration shown below: (inches)

Pitch diameter = ∅8.000 in.   Circular thickness = 4.0939
Pressure angle = 20°   Pitch angle = 65°
32 teeth   Root angle = 62.15°
Diametral pitch = 4   Addendum = .3022
Face width = 1.400   Whole depth = .5493
Shaft diameter = ∅1.125 in.   Cordal addendum = .0496
Use a .250 square key.   Cordal thickness = .7841
Core distance = 4.401

PROBLEM 19.37   Prepare detail drawing of the pinion and gear from the following gear data:

Dimensions for 20° straight bevel gear 90° shaft angle

	Pinion	Gear
Number of teeth	22	75
Diametral pitch	10	10
Face width	1.25	1.25
Pressure angle	20°	20°
Shaft angle	90°	90°
Working depth	0.200	0.200
Whole depth	0.221	0.221
Pitch diameter	2.200	7.500
Pitch angle	16.348°	73.652°
Cone distance	3.908	3.908
Circular pitch	0.314	0.314
Addendum	0.142	0.058
Dedendum	0.077	0.161
Clearance	0.021	0.021
Dedendum angle	1.126°	2.356°
Face angle of blank	18.704°	74.778°
Root angle	15.222°	71.296°
Outside diameter	2.473	7.533
Pitch apex to crown	3.710	1.044
Circular thickness	0.188	0.126
Backlash	0.002	0.002
Chordal thickness	0.186	0.125
Chordal addendum	0.146	0.058
Tooth angle	107.149 min	107.149 min
Limit point width	0.046	0.046
Tool advance	0.002	0.002

4.250
MOUNTING
DISTANCE

2.688

1.563

5/16–13UNC–2B

∅3.750

(8.233)

BEARING PROBLEMS

PROBLEM 19.38   Use the charts shown in this chapter to establish the following medium-service bearing dimensions for an approximate ∅1.25-in. shaft.

Bearing catalog number _____
Bore _____
Outside diameter _____
Width _____
Fillet radius _____
Shaft shoulder diameter _____
Housing shoulder diameter _____
Shaft diameter _____
Housing diameter _____

PROBLEM 19.39   Use the charts shown in this chapter to establish the following medium-service bearing dimensions for an approximate ∅O3.5-in. shaft.

Bearing catalog number _____
Bore _____
Outside diameter _____
Width _____
Fillet radius _____
Shaft shoulder diameter _____
Housing shoulder diameter _____
Shaft diameter _____
Housing diameter _____

PROBLEM 19.40   Use the charts shown in this chapter to establish the following medium-service bearing dimensions for an approximate ∅20-mm shaft.

Bearing catalog number _____
Bore _____
Outside diameter _____
Width _____
Fillet radius _____
Shaft shoulder diameter _____
Housing shoulder diameter _____
Shaft diameter _____
Housing diameter _____

PROBLEM 19.41   Use the charts shown in this chapter to establish the following medium-service bearing dimensions for an approximate ∅60-mm shaft.

Bearing catalog number _____
Bore _____
Outside diameter _____
Width _____
Fillet radius _____
Shaft shoulder diameter _____
Housing shoulder diameter _____
Shaft diameter _____
Housing diameter _____

LINKAGE DESIGN PROBLEMS

 PROBLEMS 19.42 and 19.43: Access the CD found with this textbook and open the problem or problems of your choice, or as assigned by your instructor. Solve the problem or problems using the instructions provided.

GEAR, BEARING SELECTION, AND SHAFT DESIGN PROBLEM

 PROBLEM 19.44: Access the CD found with this textbook and open this problem, unless otherwise specified by your instructor. Solve the problem using the instructions provided.

MATH PROBLEMS

 PROBLEMS 19.45 through 19.50: Access the CD found with this textbook and open the math problem of your choice, or as assigned by your instructor. Solve the problem or problems using the instructions provided.

Belt and Chain Drives

LEARNING OBJECTIVES

After completing this chapter, you will:

■ Design belt drive systems using vendor specifications.

■ Design chain drive systems from vendor specifications.

■ Draw sprocket and chain designs from given engineering layouts.

THE ENGINEERING DESIGN APPLICATION

The problem is to design a two belt sheave for Class D conventional industrial belts. A **sheave** is a grooved wheel in a pulley block. The inside diameter (ID) should be 5.00 in. and the outside diameter (OD) 6.375 in. Provide 1.126 in. between belt centers. Keep the weight down as much as possible; try for .375-in. web thickness. Design for a ∅.875 in. shaft. Refer to the *Machinery's Handbook* for the bore diameter and the recommended keyway for a parallel rectangular key. Your research concludes that you need the following specifications:

■ 38° V in sheave for belts.

■ ∅.874-875 in. bore.

■ .203 X 1.032 in. (measured from bottom of bore) keyway.

■ You decide on a ∅2.00-in. hub for good material strength.

The problem solution is shown in Figure 20.1.

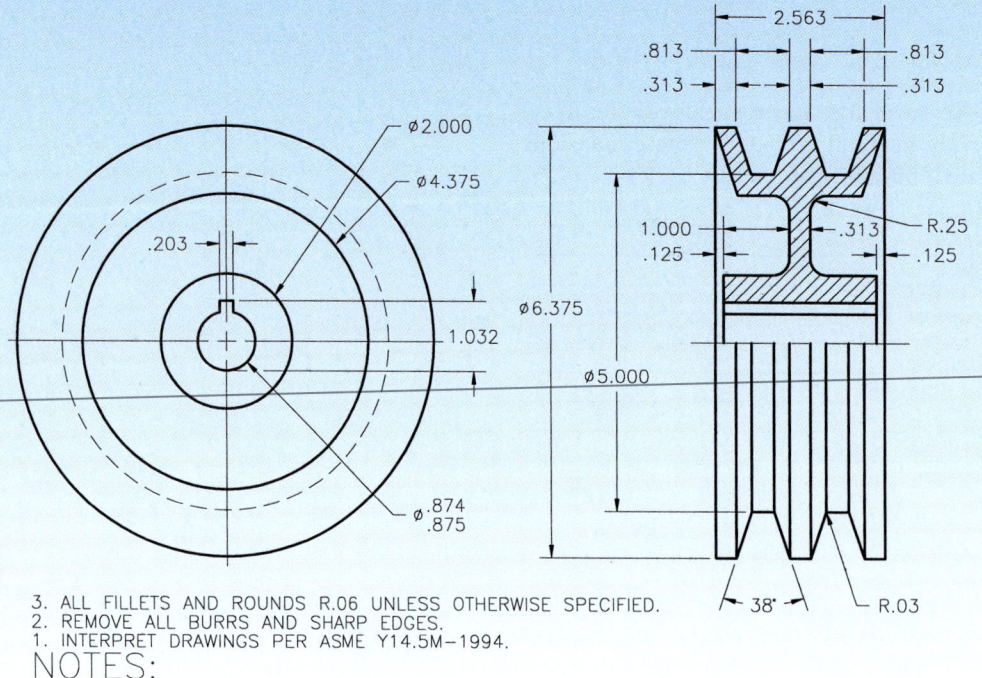

NOTES:
1. INTERPRET DRAWINGS PER ASME Y14.5M-1994.
2. REMOVE ALL BURRS AND SHARP EDGES.
3. ALL FILLETS AND ROUNDS R.06 UNLESS OTHERWISE SPECIFIED.

FIGURE 20.1 ■ Solution to engineering problem. Dimension values in this figure are in inches.

INTRODUCTION

The function of **belt**, **chain**, and **gear drives** is to transmit power between rotating shafts. It is important to evaluate the advantages and disadvantages of each type of mechanism for the proper design and implementation of the final product. Gears were discussed in Chapter 19, because they fall into a class of linkage mechanisms where two or more elements of the linkage system are in direct contact with each other. Gears are also the most rugged and durable of the mechanisms, transmitting motion with little or no slippage. They can satisfy high power demands at efficiencies of up to 98%. The following gives advantages of gear, belt, and chain drives for your reference and future use.

ADVANTAGES OF GEAR DRIVES

Gear drives are commonly used in some applications rather than belt or chain drives for the following reasons:

■ They are more rugged and durable than most belt and chain drives.

■ They are more efficient than belt or chain drives.

■ They are more practical when space limitations require the shortest distance between centers.

■ They have great maximum speed ratios.

■ They work better when high horsepower and load capabilities are important.

■ They are preferable when nonparallel shafting is the design requirement.

■ They can be used where timing and synchronization are required. Chain drives may work effectively in these applications, but belt drives are not recommended because of slippage.

ADVANTAGES OF BELT DRIVES

Belt drives are often the choice of the designer when some of the following characteristics are important:

■ They are up to 95% efficient, especially at high speeds.

■ They are designed to slip when an overload occurs.

■ They resist abrasion.

■ They require no lubrication, because there is no metal-to-metal contact, except at shaft bearings.

■ They are normally smooth running and less noisy.

■ Belt drives, especially those using flat belts, can be used more effectively when very long shaft center distances are required.

■ They can operate effectively at high speed ranges.

■ They have flexible shaft center distances, where gear drives are restricted.

■ They are less expensive than gear or chain drives.

■ They are easy to assemble and install and have flexible tolerances.

■ They absorb shock well.

■ They are easy and inexpensive to maintain.

ADVANTAGES OF CHAIN DRIVES

Chain drive mechanisms have certain design advantages over both gear and belt drives:

■ They have flexible shaft center distances, where gear drives are restricted.

■ They are less expensive than gear drives, in most cases.

■ They have simpler installation and assembly tolerances than gear drives.

■ They provide better shock-absorbing qualities than gears.

■ They have no slippage as compared to belt drives, resulting in more efficient operation.

■ They have lower loads on shaft bearings because tension is not required as with belt drives.

■ They are easy to install.

■ They are not affected by sun, heat, or oil and do not deteriorate with age as do belts.

■ They are more effective at lower speeds than belts.

■ They require little adjustment, while belt drives require frequent adjustment.

BELTS AND BELT DRIVES

Belts are used to transmit power from one shaft to another smoothly, quietly, and inexpensively. Belts are made of continuous construction from materials such as rubberized cord, leather strands, fabric, or, more commonly, reinforced nylon, rayon, steel, or glass fiber. Belts operate on shaft-mounted pulleys and sheaves. Pulleys are wheels constructed with a groove in the circumference to match the shape of the belt and transmit power to the belt. Sheaves are the same as pulleys except that sheaves are generally the drive and pulleys are the driven. Sheaves and pulleys both guide the belt in its operation, so the terms are normally used interchangeably.

BELT TYPES

Different types of belts are designed for individual applications. The type of belt you are probably most familiar with is the V belt, which is used for most automobile and household applications. Other belts, such as flat belts, are used for long center distances and high-speed applications. Positive drive belts fit notched pulleys and are designed for more positive power transmission than other belt designs.

V Belts

V belts are the most commonly used belts. These belts have a wide range of applications and operating conditions. They operate efficiently at speeds between 1500 and 6500 fpm (feet per minute) and a temperature range of -30° to 185°F (-34° to 85°C).

SAE V-belt manufacturers code the belts according to shape to help ensure interchangeability between suppliers. V belts fall into three main categories: automotive, agricultural, and industrial. Automotive belts are manufactured in accordance with six Society of Automotive Engineers (SAE) approved sectional shapes. Agricultural V belts are similar in shape to automotive V belts. They are identified with a letter *H*. Industrial V belts fall into the categories of conventional, narrow, light-duty, and double V cross section.

A comparison of the standard V-belt cross section is shown in Figure 20.2.

The load-carrying capability of a V belt is provided by reinforcing cords in the belt. The reinforcement is made of rayon, nylon, steel, or glass fibers. Figure 20.3 shows the construction of a V belt with steel reinforcing cables. The reinforcing cords are usually embedded in a soft rubber material called a cushion section. The cushion section is surrounded by a hard rubber or

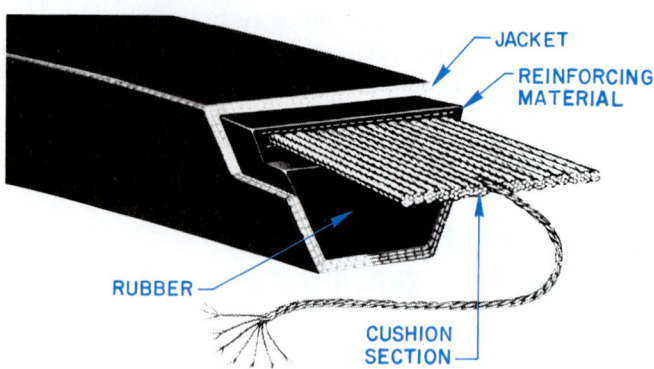

FIGURE 20.3 ■ V belt with steel reinforcing cables. *Courtesy Browning Mfg. Emerson Power Transmission.*

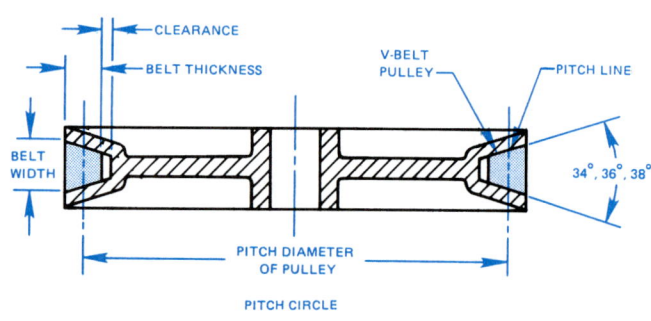

FIGURE 20.4 ■ V belt and pulley (sheave) cross section.

other material and the internal structure is covered with an abrasion-resistant material.

The **pitch line** is the only portion of the belt that does not change length as the belt bends around the pulley. The pitch line is used to determine the pitch diameter of the pulley. The **pitch diameter** is the effective diameter of the pulley, which is used to establish the speed ratio. Most pulleys are made of cast iron or steel. The construction and terminology of a pulley and mating belt are shown in Figure 20.4.

Flat Belts

Flat belts are typically used where high-speed applications are more important than power transmission and when long center distances are necessary. Flat belts are less efficient at moderate speeds, because they tend to slip under load. Flat belts are also used where drives with nonparallel shafts are required, because the belt can be twisted to accommodate the relationship of the shafts.

Positive Drive Belts

Positive drive belts have a notched underside that contacts a pulley with the same design on the circumference. These belts have some of the same advantages as chain and gear drives due

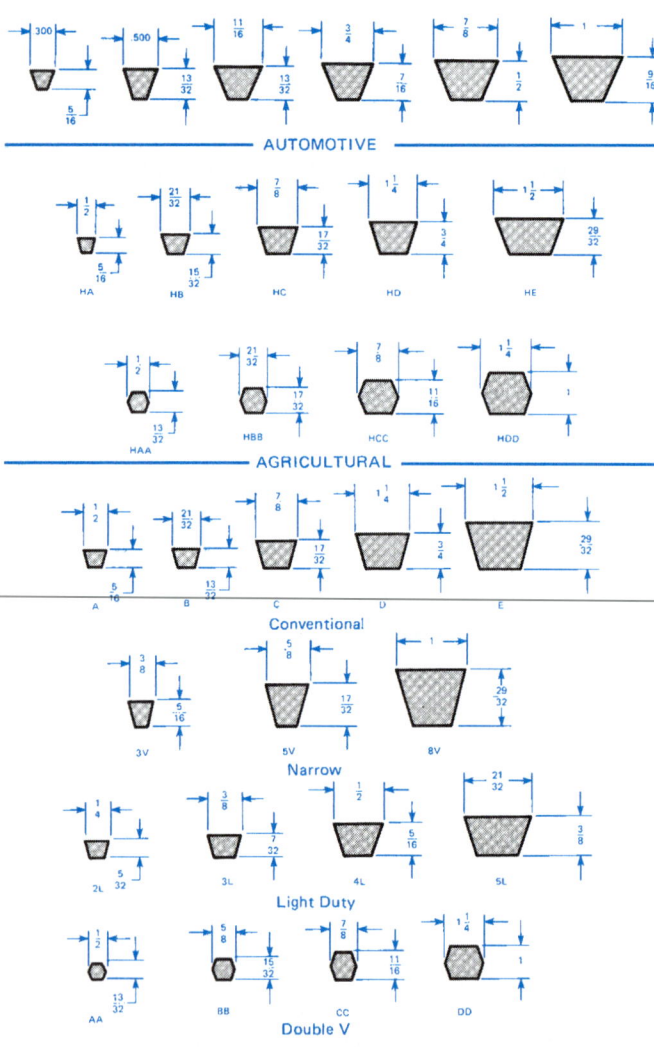

FIGURE 20.2 ■ Standard V-belt types.

FIGURE 20.5 ■ Positive drive belt. *Courtesy Browning Mfg. Emerson Power Transmission.*

to their positive contact between the belt and the pulley. These belts are more suitable for operations requiring high efficiency, timing, or constant velocity. They can also be used to decrease the pulley size and give the same operating performance of a larger sized V-belt pulley. A sample positive drive belt is shown in Figure 20.5.

TYPICAL BELT DRIVE ARRANGEMENTS

Belt drives are normally arranged with or without an idler. The idler is often used to help maintain constant tension on the belt. (See Figure 20.6.)

When belts are in operation, they stretch. Machines that are operated by belt drives must have an idler, an adjustable motor base, or both to provide for belt adjustment. Maintenance of the tension between the shafts is important in controlling the efficiency of the belt drive. Figure 20.7 shows an assembly drawing of an adjustable motor base. There are many different types of adjustable motor bases, including the sliding and tilting types.

In some cases adjustable motor bases are not practical. When this occurs, idler pulleys are used to keep belts tight. A V-belt and flat-belt idler are displayed in Figure 20.8. When idler pulleys are used, their size and location in relation to the belt drive

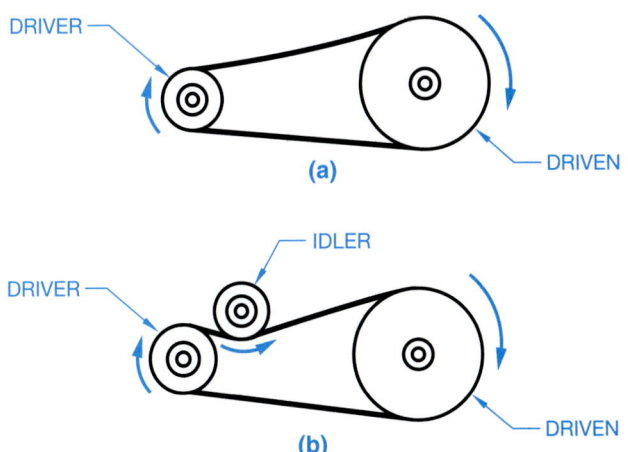

FIGURE 20.6 ■ (a) V-belt drive arrangement. (b) Belt drive with an idler.

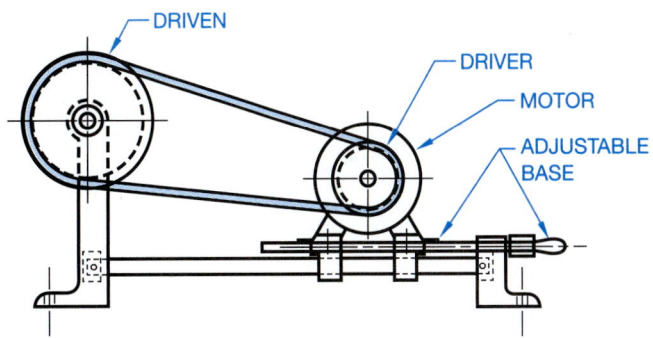

FIGURE 20.7 ■ V-belt drive arrangement with adjustable motor base.

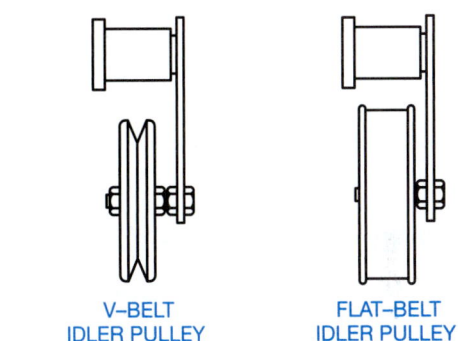

FIGURE 20.8 ■ Idler pulleys. *Courtesy Lovejoy Inc., Downers Grove, IL.*

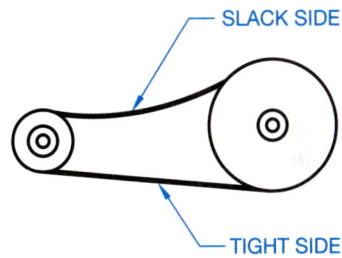

FIGURE 20.9 ■ Belt drive slack side and tight side.

are important. When referring to the placement of the idler pulley, the terms slack side and tight side are used. When the belt is in operation, the slack side is the belt on the top of the drive and the tight side is normally on the bottom of the drive when the driver and driven shafts are in horizontal alignment as shown in Figure 20.9. There are several typical design positions for the location of the idler pulley in relation to the driver pulley. An inside idler pulley is placed near the driver on the slack side of the belt. It should be the same diameter or larger than the driver. An inside idler pulley may also be placed on the tight side near the driven pulley. An outside idler pulley is placed near the driver on the outside of the belt on the slack side. The best design for this application is to have the idler pulley slightly larger in diameter than the driver pulley. When an outside idler pulley is placed on the tight side, it is normally placed near the driven pulley. (See Figure 20.10.)

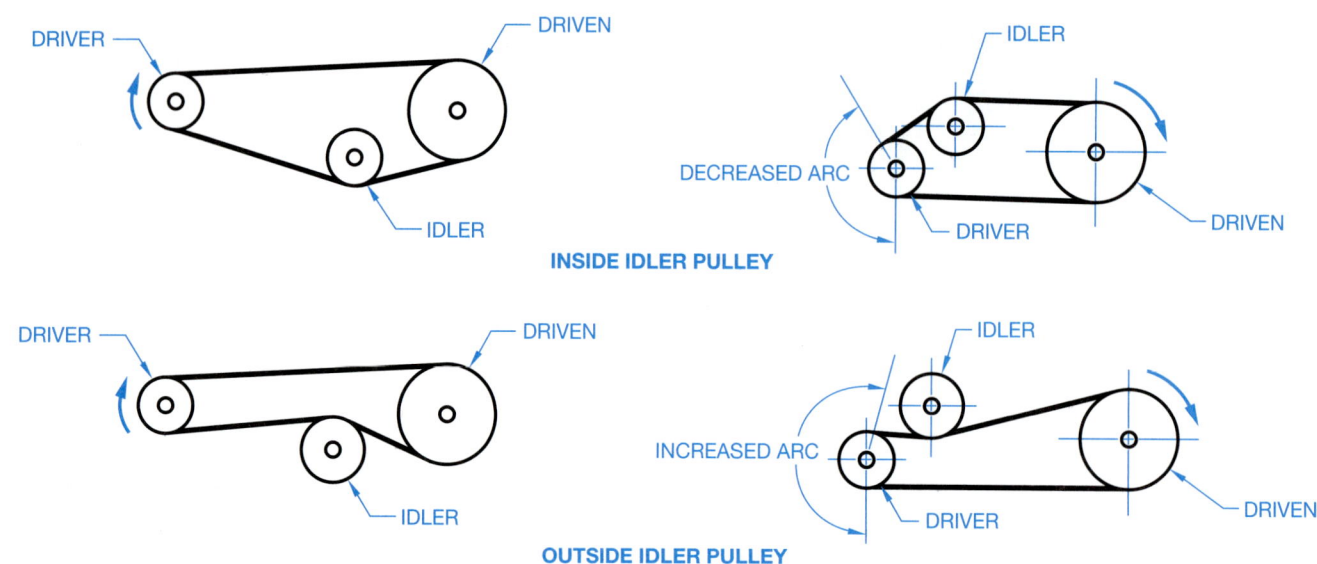

FIGURE 20.10 ■ Idler pulley arrangements.

BELT DRIVE SELECTION

Belts are selected for a specific application based on the function of the machinery. The factors influencing the selection and design of belt drives include:

- Horsepower of driving shaft.
- Speed of driving shaft.
- Service conditions.
- Type of machine.
- Center distance between shafts.
- Belt length.
- Number of belts required.

Designing V-Belt Drives

Belt manufacturers provide formulas and catalogs of the design and selection of belt drives based on the criteria described previously. The following discussion is based on engineering data provided in the Browning Power Transmission catalog, produced by the Browning Manufacturing Division of Emerson Power Transmission Company. In order to select a drive using standard drive selection tables, this sample problem is used, courtesy of Browning Manufacturing Division of Emerson Power Transmission Company.

A drive is required for a 20-HP (horsepower) motor driving a fan 16 hours per day. The motor speed is 3600 rpm and the shaft size is 1.625 in. The fan speed is approximately 2250 rpm and the fan shaft is 1.438 in. The center distance is 38 in. minimum and 41 in. maximum.

STEP 1   Determine the design horsepower. The **design HP** of a machine takes into account the type of service and the efficiency of a particular machine. Design horsepower is calculated using the formula:

$$\text{Design HP} = \text{Rated HP} \times \text{Service Factor}$$

The **rated horsepower** is the horsepower specified on the driver motor. For example, the label on the motor may read 2 HP. The **service factor** is determined by the type of machinery, the number of hours of daily operation, and the type of driven unit. Service factors can be selected from the chart shown in Figure 20.11. Based on the example problem, a fan operating 16 hours per day has a service factor of 1.3. Calculate the design HP:

$$20 \text{ HP} \times 1.3 = 26 \text{ Design HP}$$

STEP 2   Determine the driven speed. To establish the required driven speed, refer to Figure 20.12. Find the "Nominal Driven Speeds and Horsepower per Belt" column. Under the column labeled "3500 RPM Motor," find the "Nominal Driven RPM" column. Select the RPMs equal to or less than the given fan speed of 2250 rpm. Fifteen combinations that provide 2244 rpm for the driven pulley are highlighted in Figure 20.12.

STEP 3   Determine the available selections. Refer again to Figure 20.12 and look under the "HP per Belt" column to determine which sheave combination results in the smallest sheave diameters, the least number of belts, and the lightest belt selection for the sample problem. To begin this process, assume your selection is found in the "Gripnotch" column under HP per Belt. If you use a "Super" belt, another selection would be made. Read down the "Gripnotch" column until you find the largest HP per belt (12.94) without exceeding the Design HP(26) when multiplied by 1, 2, or 3 belts. Your goal is to use the least number of belts. Therefore, the best so-

TYPES OF DRIVEN MACHINES	TYPES OF DRIVING UNITS		
	AC Motors; Normal Torque, Squirrel Cage, Synchronous and Split Phase, DC Motors; Shunt Wound. Multiple Cylinder Internal Combustion Engines.		
	Intermittent Service (3-5 Hours Daily or Seasonal)	Normal Service (8-10 Hours Daily)	Continuous Service (16-24 Hours Daily)
Agitators for Liquids Blowers and Exhausters Centrifugal Pumps and Compressors Fans up to 10 HP Light Duty Conveyors	1.0	1.1	1.2
Belt Conveyors For Sand, Grain, etc. Dough Mixers Fans Over 10 HP Generators Line Shafts Laundry Machinery Machine Tools Punches-Presses-Shears Printing Machinery Positive Displacement Rotary Pumps Revolving and Vibrating Screens Speed Reducers, All Types	1.1	1.2	1.3
Brick Machinery Bucket Elevators Exciters Piston Compressors Conveyors (Drag-Pan-Screw) Hammer Mills Paper Mill Beaters Piston Pumps Positive Displacement Blowers Pulverizers Saw Mill and Woodworking Machinery Textile Machinery	1.2	1.3	1.4
Crushers (Gyratory-Jaw-Roll) Mills (Ball-Rod-Tube) Hoists Rubber Calendars-Extruders-Mills	1.3	1.4	1.5

A minimum Service Factor of 2.0 is suggested for equipment subject to choking. *

* CAUTION—Drives requiring high Overload Services Factors, such as crushing machinery, certain reciprocating compressors, etc. subjected to heavy shock load without suitable fly wheels, may need heavy duty web type sheaves rather than standard arm type. For any such application, consult the Browning Engineers.

FIGURE 20.11 ■ Service factors for belt drives. *Courtesy Browning Mfg. Emerson Power Transmission.*

lution in this case is 2 belts × *12.94 × 25.88. 25.88* is less than the 26-Design HP. These items must be equal to or less than the given data. Of the 15 selections from step 2, the proposed selections are highlighted in Figure 20.12. They are:

Pitch Diameter (PD) Driver = 5.4 in.
PD Driven = 8.4 in.
Belt Section = B
Number of Belts = 1 minimum and 3 maximum (exact number to be determined later)
HP per Belt = 12.94

STEP 4 Determine the belt part number for the approximate center distance and the F factor for the belt. The selected center distance, as stated in the problem, must be between 38 and 41 in. The F factor is a correction factor for the loss of arc of contact. The standard arc contact for a belt to sheave is 180°. When the belt does not contact 180°, there is loss or gain or arc contact. When there is

loss or gain or arc contact, a correction factor must be used to establish the corrected horsepower and number of belts required. Look at Figure 20.13, which is an extension of the chart in Figure 20.12. Find the center distance (CD) in the horizontal column that is closest to the required center distance. If you cannot find an acceptable CD between the given requirements, find a pair of CDs as highlighted in Figure 20.13. In this case, you have located:

Belt B90 with a CD of 35.0 in. and F of 0.99
Belt B103 with a CD of 41.5 and F of 1.02

To establish the exact belt for this application, you must *interpolate*. Interpolate means to make an approximation between existing known values. Interpolate to find the correct belt for the application using this method:

For every inch of belt length difference between belts, there is about 1/2 in. of center distance change. The belt numbers represent a relationship. For example, the B90 belt is 13 in. shorter than the B103 belt (B103 − B90 = 13). For this example, if you subtract 3 in. from a B103 belt, you get a B100 belt. If there is 1/2 in. of CD change for every inch of belt length, then 3 in. of length change equals 1.5 in. of CD change. Therefore, the B100 belt has a CD of 40 in.

Now determine the F factor of the B100 belt by comparing the values of the known belts:

B90   CD = 35 in.   F = .99
B100   CD = 40 in.   F = ?
B103   CD = 41.5 in. F = 1.02

Establish the difference in F factor between the B90 and B103 belts.

1.02 − .99 = .03 (difference in F factor)

Establish the difference in CD between the B90 and B103 belts:

4.1.5 − 35 = 6.5 in. (difference in CD)

Now divide the difference in F factor by the difference in CD to determine the F factor per inch of CD:

.03 divided by 6.5 = .0046
(F factor per inch of CD)

The B100 belt has a 40-in. CD, which is 1.5 in. less than the B103 belt; so, the B100 belt has an F factor of 1.5 × .0046 less than the B103 belt:

1.5 × .0046 = .0069
F 1.02 (B103 belt) − .0069 = 1.01
(rounded from 1.013)
F factor for the B100 belt.

Therefore, the B100 belt has an F factor of 1.01.

STEP 5 Determine the corrected horsepower and number of belts for the drive. The corrected horsepower is an

3500 and 1750 RPM Motors

Ratio	Driver P.D.	Driven P.D.	Belt Section	Min.	Max.	3500 RPM Nominal Driven RPM	Super	Gripnotch	1750 RPM Nominal Driven RPM	Super	Gripnotch	Belt No.	C.D.	F	Belt No.	C.D.	F	Belt No.	C.D.	F
1.54	10.4	16.0	B	1	3	—	—	—	1136	14.97	17.10	B67	13.4	0.87	B74	16.9	0.90	B81	20.5	0.94
1.54	13.0	20.0	C	1	12	—	—	—	1136	27.81	33.05	C85	17.7	0.85	C100	25.3	0.89	C112	31.0	0.91
1.55	3.8	5.9	B	1	3	2258	4.33	8.03	1129	3.32	5.33	B28	7.2	0.73	B45	15.7	0.84	B56	21.3	0.89
1.55	4.0	6.2	A	1	6	2258	5.19	6.35	1129	3.31	4.17	A26	5.5	0.76	A36	10.6	0.84	A46	15.6	0.89
1.55	4.0	6.2	B	1	6	2258	5.00	8.71	1129	3.75	5.76	B28	6.8	0.71	B45	15.3	0.84	B56	20.9	0.87
1.55	9.6	14.9	5V	2	10	—	—	—	1129	27.32	32.36	5V670	14.0	0.85	5V800	20.6	0.89	5V950	28.1	0.93
1.56	3.2	5.0	A	1	6	2244	3.62	4.35	1122	2.32	2.92	A24	6.1	0.78	A35	11.7	0.84	A45	16.7	0.88
1.56	3.2	5.0	B	1	2	2244	2.18	5.87	1122	1.99	4.00	B28	8.4	0.73	B45	16.9	0.84	B56	22.4	0.89
1.56	3.6	5.6	A	1	6	2244	4.43	5.38	1122	2.82	3.56	A24	5.3	0.76	A35	10.9	0.84	A45	15.9	0.88
1.56	3.6	5.6	B	1	6	2244	3.63	7.33	1122	2.88	4.89	B28	7.6	0.73	B45	16.1	0.84	B56	21.6	0.89
1.56	3.9	6.1	B	1	3	2244	4.67	8.38	1122	3.54	5.55	B28	7.0	0.71	B45	15.5	0.84	B56	21.0	0.87
1.56	4.1	6.4	B	1	3	2244	5.33	9.05	1122	3.97	5.97	B28	6.6	0.71	B45	15.1	0.84	B56	20.6	0.87
1.56	4.1	6.4	3V	1	4	2244	7.72	8.76	1122	4.47	4.90	3V280	5.6	0.79	3V335	8.4	0.84	3V400	11.7	0.89
1.56	4.5	7.0	A	1	3	2244	6.07	7.50	1122	3.90	4.92	A29	6.0	0.78	A39	11.0	0.84	A49	16.1	0.90
1.56	4.5	7.0	5V	2	5	2244	—	17.45	1122	—	10.28	5V500	15.9	0.82	5V630	22.4	0.86	5V800	30.9	0.92
1.56	4.8	7.5	A	1	1	2244	6.56	8.15	1122	4.26	5.36	A31	6.3	0.79	A40	10.9	0.85	A49	15.4	0.90
1.56	5.4	8.4	B	1	3	2244	9.06	12.94	1122	6.66	8.65	B34	6.9	0.74	B47	13.5	0.84	B58	19.0	0.88
1.56	5.4	8.4	5V	2	5	2244	—	24.53	1122	—	14.42	5V500	14.1	0.82	5V630	20.6	0.86	5V800	29.1	0.90
1.56	5.5	8.6	A	1	3	2244	7.57	9.53	1122	5.05	6.36	A36	7.4	0.81	A45	12.0	0.87	A54	16.5	0.92
1.56	6.4	10.0	A	1	3	2244	8.58	11.03	1122	6.02	7.58	A42	8.6	0.84	A50	12.6	0.89	A58	16.7	0.93
1.57	2.8	4.4	A	1	2	2229	2.76	3.26	1115	1.80	2.28	A24	6.9	0.78	A35	12.5	0.84	A45	17.5	0.90
1.57	3.0	4.7	A	1	3	2229	3.19	3.81	1115	2.06	2.60	A24	6.5	0.78	A35	12.1	0.84	A45	17.1	0.90
1.57	3.5	5.5	A	1	3	2229	4.23	5.13	1115	2.69	3.40	A24	5.5	0.76	A35	11.0	0.84	A45	16.0	0.88
1.57	3.7	5.8	A	1	3	2229	4.62	5.63	1115	2.94	3.71	A24	5.1	0.75	A35	10.6	0.84	A45	15.7	0.88
1.57	4.2	6.6	A	1	6	2229	5.56	6.83	1115	3.55	4.48	A27	5.5	0.76	A37	10.6	0.84	A47	15.6	0.89
1.58	4.8	7.6	A	1	6	2215	6.56	8.16	1108	4.26	5.36	A31	6.3	0.79	A40	10.8	0.85	A49	15.3	0.90
1.58	5.0	7.9	B	1	1	2215	8.04	11.85	1108	5.86	7.86	B32	6.6	0.73	B46	13.7	0.84	B57	19.2	0.87
1.58	5.2	8.2	A	1	10	2215	7.16	8.97	1108	4.71	5.94	A34	7.0	0.80	A43	11.5	0.86	A52	16.1	0.92
1.58	5.7	9.0	A	1	3	2215	7.83	9.90	1108	5.27	6.64	A37	7.4	0.82	A46	12.0	0.88	A55	16.5	0.93
1.58	6.0	9.5	A	1	1	2215	8.18	10.41	1108	5.60	7.05	A39	7.8	0.82	A47	11.8	0.88	A55	15.9	0.92
1.58	6.6	10.4	B	1	3	2215	11.51	15.74	1108	8.97	10.95	B43	8.8	0.80	B53	13.9	0.86	B63	19.0	0.89
1.58	6.7	10.6	A	1	1	2215	8.85	11.46	1108	6.33	7.98	A44	8.9	0.85	A52	12.9	0.89	A60	16.9	0.94
1.58	6.9	10.9	B	1	1	2215	11.95	16.30	1108	9.52	11.50	B45	9.2	0.81	B55	14.3	0.86	B65	19.3	0.90
1.58	7.2	11.4	C	2	6	—	—	—	1108	12.92	19.41	C51	12.2	0.76	C75	24.2	0.84	C96	34.8	0.89
1.58	7.4	11.7	5V	2	8	—	—	—	1108	18.97	23.39	5V500	9.8	0.79	5V630	16.4	0.85	5V800	24.9	0.90

FIGURE 20.12 ■ Motor speed and drive selection table. *Courtesy Browning Mfg. Emerson Power Transmission.*

3500 and 1750 RPM Motors

Belt No.	C.D.	F	Belt No.	C.D.	F	Belt No.	C.D.	.	Belt No.	C.D.	F	Belt No.	C.D.	F	Belt No.	C.D.	F	Driver P.D.	Driven P.D.	Ratio
5V1120	36.7	0.95	5V1320	46.7	0.98	5V1600	60.7	1.03	5V1900	75.7	1.06	5V2240	92.7	1.08	5V3550	158.2	1.17	9.6	14.9	1.55
A55	21.7	0.95	A65	26.7	0.98	A75	31.7	1.01	A85	36.7	1.05	A95	41.7	1.06	A180	84.2	1.19	3.2	5.0	1.56
B67	27.9	0.93	B78	33.4	0.97	B88	38.4	1.00	B99	43.9	1.02	B136	62.5	1.09	B360	173.7	1.31	3.2	5.0	1.56
A55	20.9	0.95	A65	25.9	0.98	A75	30.9	1.01	A85	35.9	1.04	A95	40.9	1.06	A180	83.4	1.19	3.6	5.6	1.56
B67	27.2	0.93	B78	32.7	0.97	B88	37.7	0.99	B99	43.2	1.02	B136	61.7	1.09	B360	172.9	1.31	3.6	5.6	1.56
B67	26.5	0.93	B78	32.0	0.97	B88	37.0	0.99	B99	42.5	1.01	B136	61.0	1.09	B360	172.3	1.31	3.9	6.1	1.56
B67	26.1	0.93	B78	31.6	0.97	B88	36.6	0.99	B99	42.1	1.01	B136	60.6	1.09	B360	171.9	1.31	4.1	6.4	1.56
3V475	15.5	0.92	3V560	19.7	0.95	3V670	25.2	1.00	3V800	31.7	1.03	3V950	39.2	1.07	3V1400	61.7	1.15	4.1	6.4	1.56
A59	21.1	0.94	A69	26.1	0.99	A79	31.1	1.02	A89	36.1	1.04	A100	41.6	1.07	A180	81.6	1.19	4.5	7.0	1.56
5V1000	40.9	0.95	5V1250	53.5	1.00	5V1600	71.0	1.04	5V2000	91.0	1.08	5V2500	116.0	1.11	5V3550	168.5	1.17	4.5	7.0	1.56
A58	19.9	0.94	A67	24.5	0.97	A76	29.0	1.01	A85	33.5	1.04	A94	38.0	1.05	A180	81.0	1.19	4.8	7.5	1.56
B69	24.5	0.92	B80	30.0	0.97	B90	35.0	0.99	B103	41.5	1.02	B144	62.0	1.11	B360	169.3	1.31	5.4	8.4	1.56
5V1000	39.1	0.95	5V1250	51.6	0.99	5V1600	69.1	1.04	5V2000	89.2	1.08	5V2550	114.2	1.11	5V3550	166.7	1.17	5.4	8.4	1.56
A63	21.0	0.96	A72	25.5	0.98	A81	30.0	1.01	A90	34.5	1.05	A100	39.5	1.07	A180	79.6	1.19	5.5	8.6	1.56
A66	20.7	0.97	A74	24.7	0.99	A82	28.7	1.01	A90	32.7	1.03	A98	36.7	1.07	A180	77.7	1.19	6.4	10.0	1.56
A55	22.5	0.95	A65	27.5	0.98	A75	32.5	1.02	A85	37.5	1.05	A95	42.5	1.06	A180	85.0	1.19	2.8	4.4	1.57
A55	22.1	0.95	A65	27.1	0.98	A75	32.1	1.01	A85	37.1	1.05	A95	42.1	1.06	A180	84.6	1.19	3.0	4.7	1.57
A55	21.1	0.95	A65	26.1	0.98	A75	31.1	1.01	A85	36.1	1.04	A95	41.1	1.06	A180	83.6	1.19	3.5	5.5	1.57
A55	20.7	0.93	A65	25.7	0.98	A75	30.7	1.01	A85	35.7	1.04	A95	40.7	1.05	A180	83.2	1.19	3.7	5.8	1.57
A57	20.6	0.94	A67	25.6	0.99	A77	30.6	1.01	A87	36.6	1.04	A97	40.6	1.07	A180	82.2	1.19	4.2	6.6	1.57
A58	19.9	0.94	A67	24.5	0.97	A76	28.9	1.01	A85	33.4	1.04	A94	37.9	1.05	A180	80.9	1.19	4.8	7.6	1.58
B68	24.7	0.92	B79	30.2	0.97	B89	35.2	0.99	B100	40.7	1.02	B140	60.7	1.10	B360	170.0	1.31	5.0	7.9	1.58
A61	20.6	0.95	A70	25.1	0.98	A79	29.6	1.00	A88	34.1	1.04	A97	38.6	1.07	A180	80.1	1.19	5.2	8.2	1.58
A64	21.0	0.96	A73	25.5	0.98	A82	30.1	1.01	A91	34.6	1.05	A103	40.6	1.07	A180	79.1	1.19	5.7	9.0	1.58
A63	19.9	0.96	A71	23.9	0.98	A79	27.9	1.00	A87	31.9	1.02	A95	35.9	1.05	A180	78.5	1.19	6.0	9.5	1.58
B73	24.0	0.93	B82	29.0	0.94	B92	33.5	0.97	B105	40.0	1.03	B144	59.5	1.10	B360	166.8	1.31	6.6	10.4	1.58
A68	21.0	0.97	A76	25.0	0.99	A84	29.0	1.01	A92	33.0	1.03	A103	38.5	1.05	A180	77.0	1.18	6.7	10.6	1.58
B75	24.3	0.94	B84	28.8	0.96	B94	33.9	0.98	B112	42.9	1.04	B154	63.9	1.11	B360	166.2	1.31	6.9	10.9	1.58
C112	42.8	0.94	C136	54.8	0.98	C162	67.8	1.02	C210	91.8	1.08	C270	120.8	1.14	C420	195.8	1.24	7.2	11.4	1.58
5V1000	34.9	0.93	5V1250	47.4	0.99	5V1600	65.0	1.03	5V2000	85.0	1.07	5V2500	110.0	1.11	5V3550	162.5	1.17	7.4	11.7	1.58

FIGURE 20.13 ■ Nominal center distances, number of belts, arc length factor chart. *Courtesy Browning Mfg. Emerson Power Transmission.*

adjustment of the design horsepower in relation to the F factor. It is calculated with this formula:

F Factor × HP per Belt = Corrected Horsepower

The corrected horsepower for the sample problem is:

1.01 (F) × 12.94 (HP per belt) = 13.06
Corrected HP per Belt

Now determine the number of belts required for the drive using this formula:

Design HP/Corrected HP = Number of Belts

If the number calculated in this formula is one or less, then one belt is required; or two belts for two or less; three belts for three or less; and so on. The number of belts for the sample problem is:

26 (Design HP)/13.06 (Corrected HP) = 1.99
= 2 Belts Required

In conclusion, the design solution to the sample problem is as follows:

26 Design HP
12.94 Design HP per Belt
13.06 Corrected HP per Belt
B100 Belt Number
40 in. Center Distance
2 Belts Required
∅5.4 in. Driver Pitch Diameter (PD)
∅8.4 in. Driven PD
B Belt Section

Determine the Belt Drive Ratio

The belt drive ratio is the relationship between the drive and driven pulleys. The ratio can be determined by dividing the rpm of the driver by the rpm of the driven pulley. If you use the sample problem presented earlier, for example, the motor rpm from Figure 20.12 is 3500 rpm, and the nominal driven rpm from the same chart is 2244 rpm. Therefore, the Ratio = 3500 rpm/2244 rpm = 1.5597 = 1.56. Now, refer back to Figure 20.12 and notice the 1.56 highlighted in the Ratio column at the far left. This 1.56 ratio coincides with the other items selected for the application.

Determine the Belt Velocity

Belt velocity is the speed a belt travels in feet per minute (fpm) for a given application. In some cases, the belt velocity can be important. In many applications, though, it is not a factor. Belt velocity becomes important at very high speeds. For example, cast iron sheaves should not travel at speeds greater than 6500 fpm. Another factor that influences design at high speeds is vibration, which can require balancing of the sheaves. Belt velocity is calculated using this formula:

Belt Velocity fpm = Pitch Diameter (PD)
Sheave × 0.2618 × Sheave Speed rpm
(0.2618 is a constant to be used for all applications)

The belt velocity for the driver sheave of the sample problem is:
∅5.4 (PD Sheave) × 0.2618 × 3500 rpm = 4948 fpm

CHAIN DRIVES

Chain drives, like gear and belt drives, are used to transmit power from one shaft to another. As discussed earlier, chain drives have certain advantages over gear and belt drives. In general, chain drives are less efficient and durable than gear drives, but they offer greater power capacity, more position power transmission, durability, and longer service life than belts, especially at high temperatures.

In chain drive applications, toothed wheels called sprockets mate with a chain to transmit power from one shaft to another. (See Figure 20.14.) A sprocket is a wheel with teeth on its periphery to engage a chain.

CHAIN DRIVE SPROCKETS

Sprockets are designed for various applications. Depending on the shaft mounting requirements, sprockets are available with a hub on one or both sides or without hubs. Sprockets are designed with a solid web in most applications, although weight is reduced by designing recessed webs or spokes.

Sprockets are machined from cast iron, steel, aluminum alloy, powdered metal, or plastic. Split sprockets are used when mounting between bearings requires easy installation. Some applications have sprockets mounted to a removable steel or cast iron hub when frequent replacement is required.

CHAIN CLASSIFICATION AND TYPES

Chains are classified in relation to the accuracy of construction between the sprocket and the chain links. There are three broad categories: precision, nonprecision, and light-duty chains. Precision chains are designed for smooth, free-running operation at

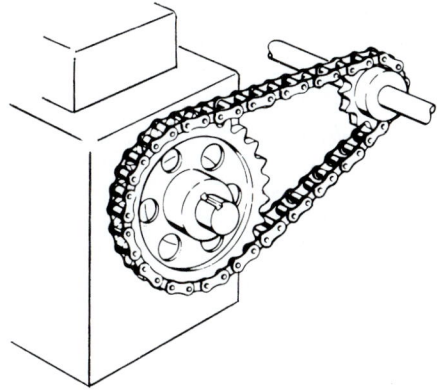

FIGURE 20.14 ■ Chain drive sprockets and chain. *Courtesy Morse Industrial.*

high speed and high power. Common precision chains include the roller, offset sidebar, silent inverted tooth, and double pitch chains. Nonprecision chains do not have as high a degree of precision between the sprocket and chain links. These lower cost chains are generally used for low-speed applications on machinery rated below 50 HP. Nonprecision chains include detachable, pintle, and welded chains. Light-duty chains are designed for application in low-power situations such as equipment control mechanisms for computers and printers or appliance controls.

PRECISION CHAINS

Roller Chain

Roller chain is the most commonly used power transmission chain. Roller chain is rated up to 630 HP for single-chain drives and is referred to as single-strand chain. Roller chains can be added in multiple units, known as multiple-strand chains, to substantially increase the horsepower under which they can operate. Roller chains provide efficient, quiet operation and should be lubricated for the best service life. Figure 20.15 shows the parts of a typical roller chain.

The chain pitch is important when designing a chain drive. The chain pitch is the distance from the center of one pin to the center of the other pin in one link. Figure 20.17 shows a single-strand chain with the pitch dimensioned and a multiple-strand chain. Roller chains are available with pitch lengths ranging from 1/4 to 3 in.

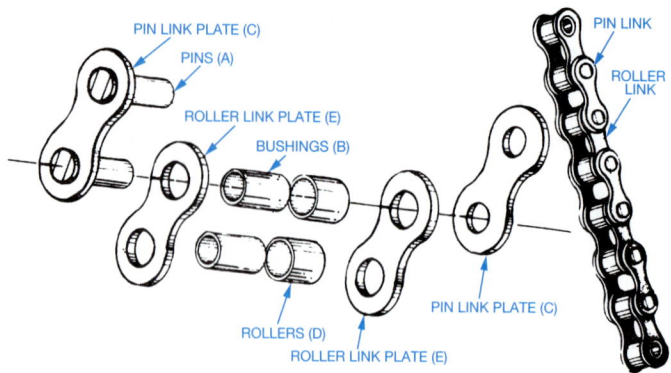

FIGURE 20.15 ■ Roller chain assembly. *Courtesy Rainbow Industrial Products Corporation.*

Double Pitch Roller Chain

Double pitch roller chain is designed mostly for situations with long center distances such as conveyors. While the efficiency of the double pitch chain is as good as the roller chain, it is intended for lighter-duty operation than regular pitch roller chain. (See Figure 20.18.)

Offset Sidebar Roller Chain

The offset sidebar, shown in Figure 20.19, is the least expensive precision chain. It is designed to carry heavier loads than

SPROCKETS

The same simplified representation used in gear CADD drawings can also be used for sprocket CADD drawings, or custom software is available to automatically generate detailed drawings from given data (see Figure 20.16).

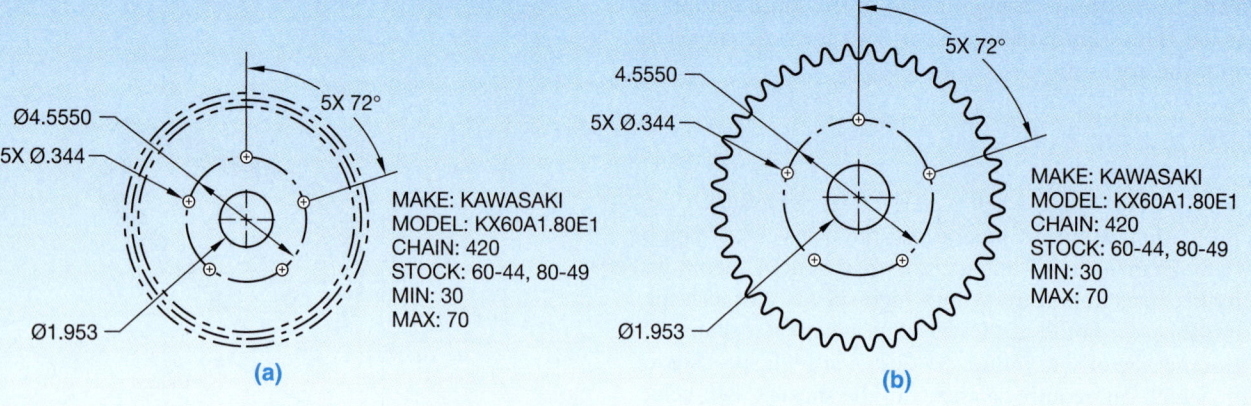

FIGURE 20.16 ■ Sprocket drawing: (a) simplified representation, (b) detailed representation. Dimension values in this figure are in inches.

FIGURE 20.17 ■ Single-strand and multiple-stand roller chain. *Courtesy Morse Industrial.*

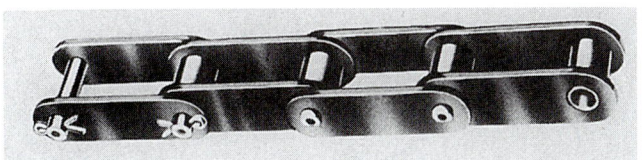

FIGURE 20.18 ■ Double pitch roller chain. *Courtesy Morse Industrial.*

FIGURE 20.19 ■ Offset sidebar roller chain.

non-precision chains. Offset sidebar chains are designed to handle loads up to 425 HP and speeds up to 36 feet per second. One of the advantages of the offset sidebar chain over the other chain types is its open construction, which allows it to withstand dirt and contaminants that might cause other precision chains to bind and wear out rapidly. For this reason and because they are rugged and durable, offset sidebar chains are often used to drive construction machinery.

Inverted Tooth Silent Chain

The most expensive precision chain to manufacture is the **inverted tooth silent chain** shown in Figure 20.20. This chain is used where high speed and smooth, quiet operation are required

FIGURE 20.20 ■ Inverted tooth silent chain. *Courtesy Morse Industrial.*

in rigorous applications. Lubrication is required to keep these chains running trouble free. Applications include machine tools, pumps, and power drive units.

NONPRECISION CHAINS

Detachable Chain

The **detachable chain** shown in Figure 20.21 is the lightest, simplest, and least expensive of all chains. The detachable chain is capable of transmitting power up to 25 HP at low speeds. One common application is farm machinery requiring speeds of up to 350 fpm. Detachable chains do not require lubrication.

Pintle and Welded Steel Chains

The **pintle** and **welded steel chains** shown in Figure 20.22 are a combination design between the detachable and the offset

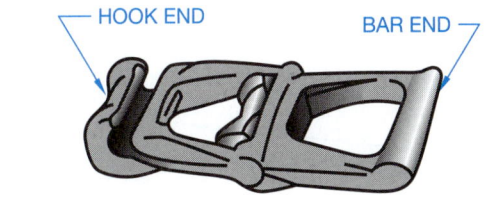

FIGURE 20.21 ■ Detatchable chains.

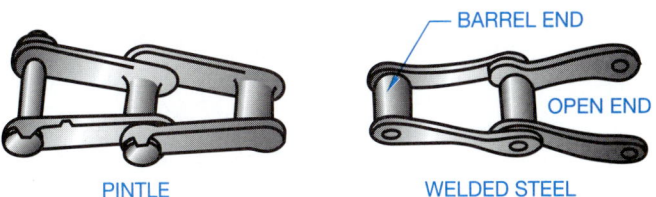

FIGURE 20.22 ■ Pintle and welded steel chains.

sidebar roller chain. They are used for applications similar to the detachable chain where more rigorous service is required for up to 40 HP and 425 fpm. Lubrication is not required, so they may also be used where lubrication is not effective, such as dusty or wet conditions.

LIGHT-DUTY CHAIN

Bead Chain

Bead chains, Figure 20.23, are commonly used for control mechanisms for light-duty, low-power applications. Bead chains are rated from 15 to 75 pounds and are available in standard bead diameters of 3/32, 1/8, 3/16, and 1/4 in.

CHAIN DRIVE ARRANGEMENTS

Chain drives have some of the same characteristics as belt drives in the following ways:

■ Chains elongate when in operation, thus requiring shaft adjustment or link removal to maintain proper tightness.

■ Chains have a tight side and a slack side when in operation.

The arrangement and direction of rotation of the driver sprocket and the driven sprocket are important to the proper function of the chain drive mechanism. For chain drives with long center distances, the slack side should be on the bottom, because upperside slack could cause the chain on the top to meet the bottom as the chain elongates. Figure 20.24 shows the recommended and not recommended arrangements.

On short center drive arrangements, as shown in Figure 20.25, the slack side should be on the bottom because slack on the top could cause the chain to jump out of the sprocket. The relationship of the shaft centers is also important. The shaft centers should be designed in a horizontal position or inclined no more than 45°. When the shafts are inclined up to 45°, the slack side can be on either the top or bottom. (See Figure 20.26.)

Shaft arrangements between 45° and vertical should be avoided. However, when they are required due to the machine design, the chain should be kept tighter than normal applications. This type of arrangement requires that the chain be tightened frequently. Properly designed situations with the shafts inclined greater than 45° provide for the slack to be on the side closest to horizontal. (See Figure 20.27.)

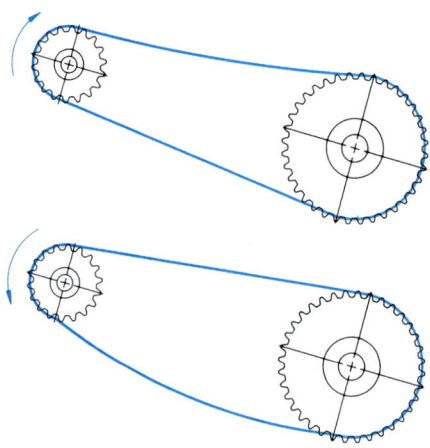

FIGURE 20.25 ■ Chain drives with short center distances should have slack on the lower side.

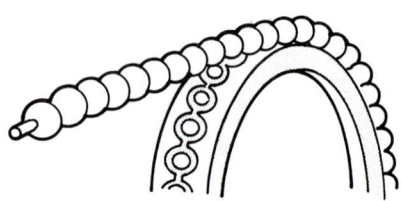

FIGURE 20.23 ■ Bead chain.

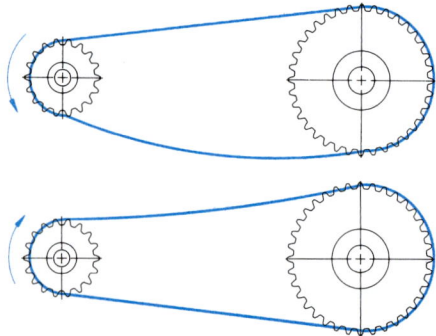

FIGURE 20.24 ■ Chain drives with long center distance should have slack on bottom side.

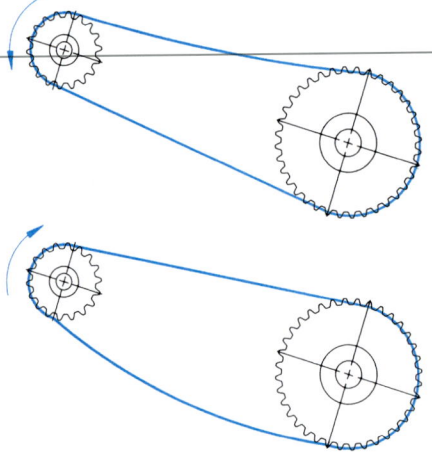

FIGURE 20.26 ■ Chain drive shaft centers should be on a horizontal line or on a line inclined not more than 45°, in which case the slack can be on the upper or lower side.

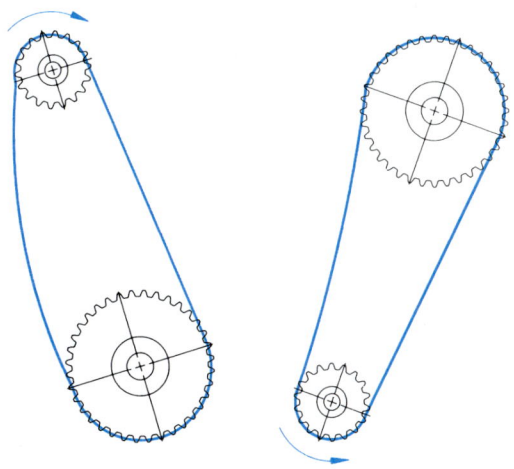

FIGURE 20.27 ■ When the shaft centers are aligned more than 45° from horizontal, the slack side should be in the side closest to horizontal.

ROLLER CHAIN DRIVE SELECTION

Designing roller chain drives includes selecting the sprocket sizes, the chain pitch, the chain length, the distance between shaft centers, and the lubrication requirements. Factors that contribute to selecting the proper chain drive variable are rpm of the sprockets, the **horsepower (HP)** of the motor, and the load conditions.

Chain manufacturers provide formulas and catalogs for the design and selection of chain drives. The following discussion is based on engineering data provided in the Morse Power Transmission Products catalog, provided by the Morse Industrial Products Division, Emerson Power Transmission Corporation.

A chain drive is required for a 2-HP electric motor operating laundry machinery. The drive spocket turns at 1800 rpm, which produces 900 rpm at the driven sprocket. The approximate center distance is 12 in.

STEP 1 Determine the design horsepower. The design horsepower (HP) of a machine takes into account the type of service and the efficiency of the machine. A service factor is applied to the horsepower rating of the drive motor to determine the design horsepower of a chain drive. The service factors are established by the power source, the nature of the load, and the strain or shock on the drive. Three basic operating characteristics are used to establish the service factors. (See Figure 20.28.)

Now, determine the service factor for the given problem. You have a 2-HP electric motor operating laundry machinery. Laundry machinery operates under moderate shock. Refer to the table in Figure 20.29 to determine the service factor. An electric motor operating under moderate shock has a service factor of 1.3.

Establish the design horsepower using the formula:

Motor HP × Service Factor = Design HP

> **Smooth:** Running load is fairly uniform. Starting and peak loads may be somewhat greater than running load, but occur infrequently.
>
> **Moderate Shock:** Running load is variable. Starting and peak loads are considerably greater than running load and occur frequently.
>
> **Heavy Shock:** Starting loads are extremely heavy. Peak loads and overloads occur continuously and are of maximum fluctuation.

FIGURE 20.28 ■ Chain drive operating classifications. *Courtesy Morse Industrial.*

For the sample problem, the design HP is:

2 HP × 1.3 = 2.6 Design HP

STEP 2 Determine the number of teeth of the drive sprocket. Now that you know the design HP (2.6 HP) and the rpm of the drive sprocket (1800 rpm), you can determine the number of teeth of the drive sprocket. The drive sprocket is normally the "Small" sprocket as shown in the left column of the chart in Figure 20.30. Look at Figure 20.30 and find the rpm—Small Sprocket highlighted along the top horizontal column. Move down the 1800 rpm column until you find a design HP equal to or greater than the design HP of 2.6 for the sample problem. Find 2.93 highlighted in Figure 20.30. Now move horizontally to the left from 2.93 to find the number of teeth of the small sprocket in the left column. Notice that 40 teeth is highlighted in Figure 20.30.

STEP 3 Determine the chain pitch. The **chain pitch** is the center distance between pins of one chain link. (See Figure 17.31.) The number of teeth of the small sprocket is taken from the horsepower rating table for a chain with a specific pitch. If you look back at Figure 20.30, you can see the pitch identified in the heading as "No. 25-1/4" pitch." Chains are manufactured with standard size pitches available in increments from 1/4 to 3 in. Chains are also classified by number; for example, a 1/4-in. pitch is No. 25, a 3/8-in. pitch is No. 35, a 1/2-in. pitch is No. 40, and a 5/8-in. pitch is No. 50.

STEP 4 Determine the ratio and the number of teeth for the driven sprocket. The chain drive ratio is the relationship between the drive and driven sprockets. The ratio may be determined by dividing the rpm of the driver by the rpm of the driven sprocket. If you use the sample problem, the driver is 1800 rpm and the driven is 900 rpm. The ratio is 1800 rpm/900 rpm = 2:1. Now, multiply the number of teeth of the small sprocket by the ratio to get the number of teeth for the large sprocket: 40 (teeth small sprocket) × 2 (ratio) = 80 teeth large sprocket.

STEP 5 Determine the center distance in chain pitches. As a general rule, the preferred center distance between shafts is between 30 and 50 chain pitches. The maximum

service factors

Service Factors are selected below for various applications after first determining the prime mover or power source type.

service factor table

Prime Mover	TYPE
Internal Combustion Engine with Hydraulic Coupling or Torque Converter Electric Motor Turbine Hydraulic Motor	A
Internal Combustion Engine with Mechanical Drive	B

APPLICATION	Type of Prime Mover A	Type of Prime Mover B	APPLICATION	Type of Prime Mover A	Type of Prime Mover B	APPLICATION	Type of Prime Mover A	Type of Prime Mover B
AGITATORS (paddle or propeller)			CRUSHING MACHINERY			PAPER INDUSTRY MACHINERY		
Pure liquid	1.1	1.3	Ball mills, crushing rolls, Jaw crushers	1.6	1.8	Agitators, bleachers	1.1	1.3
Liquids—variable density	1.2	1.4	DREDGES			Barker—mechanical	1.6	1.8
BAKER MACHINERY			Conveyors, cable reels	1.4	1.6	Beater, Yankee Dryer	1.3	1.5
Dough Mixer	1.2	—	Jigs & screens	1.6	1.8	Calendars, Dryer & Paper Machines	1.2	1.4
BLOWERS	See Fans		Cutter head drives	Consult Morse		Chippers & winder drums	1.5	1.7
BREWING & DISTILLING EQUIPMENT			Dredge pumps	See Pumps		PRINTING MACHINERY		
Bottling Machinery	1.0	—	FANS & BLOWERS			Embossing & flat bed presses, folders	1.2	—
Brew Kettles, cookers, mash tubs	1.0	—	Centrifugal, propeller, vane	1.3	1.5	Paper cutter, rotary press		
Scale Hopper—Frequent starts	1.2	—	Positive blowers (lobe)	1.5	1.7	& linotype machine	1.1	—
BRICK & CLAY EQUIPMENT			GRAIN MILL MACHINERY			Magazine & newspaper presses	1.5	—
Auger machines, cutting table	1.3	1.5	Sifters, purifiers, separators	1.1	1.3	PUMPS		
Brick machines, dry press,			Grinders and hammer mills	1.2	1.4	Centrifugal, gear, lobe & vane	1.2	1.4
& granulator	1.4	1.6	Roller mills	1.3	1.5	Dredge	1.6	1.8
Mixer, pug mill, & rolls	1.4	1.6	GENERATORS & EXCITERS	1.2	1.4	Pipe line	1.4	1.6
CENTRIFUGES	1.4	1.6	MACHINE TOOLS			Reciprocating		
COMPRESSORS			Grinders, lathes, drill press	1.0	—	3 or more cyl.	1.3	1.5
Centrifugal & rotary (lobe)	1.1	1.3	Boring mills, milling machines	1.1	—	1 or 2 cyl.	1.6	1.8
Reciprocating			MARINE DRIVES	Consult Morse		RUBBER & PLASTICS		
1 or 2 cyl.	1.6	1.8	MILLS			INDUSTRY EQUIPMENT		
3 or more	1.3	1.5	Rotary type:	1.5	1.7	Calendars, rolls, tubers		
CONSTRUCTION EQUIPMENT			Ball, Pebble, Rod, Tube, Roller			Tire-building and Banbury Mills	1.5	1.7
OR OFF-HIGHWAY VEHICLES			Dryers, Kilns, & tumbling barrels	1.6	1.8	Mixers and sheeters	1.6	1.8
Drive line duty, power take-off,			Metal type:			Extruders	1.5	1.7
accessory drives	Consult Morse		Draw bench carriage & main drive	1.5	—	SCREENS		
CONVEYOR			Forming Machines	Consult Morse		Conical & revolving	1.2	1.4
Apron, bucket, pan & elevator	1.4	1.6	MIXERS			Rotary, gravel, stone & vibrating	1.5	1.7
Belt (ore, coal, sand, salt)	1.2	1.4	Concrete	1.6	1.8	STOKERS	1.1	—
Belt—light package, oven	1.0	1.2	Liquid & Semi-liquid	1.1	1.3	TEST STANDS & DYNAMOMETERS	Consult Morse	
Screw & flight (heavy duty)	1.6	1.8	OIL INDUSTRY MACHINERY			TEXTILE INDUSTRY		
CRANES & HOISTS			Compounding Units	1.1	1.3	Spinning frames, twisters, wrappers		
Main hoist—medium duty	1.2	1.4	Pipe line pumps	1.4	1.6	& reels	1.0	—
Main hoist—heavy duty, skip hoist	1.4	1.6	Slush pumps	1.5	1.7	Batchers, calendars & looms	1.1	—
			Draw works	1.8	2.0			
			Chillers, Paraffin filter presses, Kilns	1.5	1.7			

FIGURE 20.29 ■ Chain drive service factors. *Courtesy Morse Industrial.*

recommended spacing between centers is 80 pitches. This is to help ensure that there is clearance between the two sprockets and to allow for a minimum of 120° of chain contact around the small sprocket. (See Figure 20.32.) The center distance between shafts should be no less than the difference between sprocket diameters for ratios greater than 3:1. To determine the center distance in chain pitches, divide the center distance by the chain pitch. For the sample problem, this is:

$$\frac{12 \text{ in. (Center Distance)}}{0.25 \text{ in. (Pitch)}} = 48 \text{ Chain Pitches}$$

STEP 6 Determine the chain length. First, determine the chain length in pitches using the formula:

$$2C + \frac{S}{2} + \frac{K}{C}, \text{ where:}$$

C = Center distances between sprockets in chain pitches.
S = Number of teeth in small sprocket plus number of teeth in large sprocket.
K = Number of teeth in large sprocket minus number of teeth in small sprocket equals D. Look at the table in Figure 20.33, page 668, to find the value K corresponding to D.

For the same problem:

C = 48 (from step 5)
S = 40 + 80 = 120
K = 80 − 40 = 40
D = 40

Look at Figure 20.33 to find K = 40.53.

$$2C + \frac{S}{2} + \frac{K}{C} = 2(48) + \frac{120}{2} + \frac{40.53}{48} =$$
$$96 + 60 + .84 = 156.84$$

No. 25—¼″ pitch   standard single strand roller chain

No. of Teeth Small Spkt.	50	100	300	500	700	900	1200	1500	1800	2100	2500	3000	3500	4000	4500	5000	5500	6000	6500	7000	7500	8000	8500	9000	10000
										Revolutions per Minute—Small Sprocket															
12	0.03	0.06	0.16	0.25	0.34	0.43	0.55	0.68	0.80	0.92	1.07	1.26	1.45	1.57	1.32	1.12	0.97	0.86	0.76	0.68	0.61	0.56	0.51	0.47	0.40
13	0.04	0.06	0.17	0.27	0.37	0.47	0.60	0.74	0.87	1.00	1.17	1.38	1.58	1.77	1.49	1.27	1.10	0.96	0.86	0.77	0.69	0.63	0.57	0.53	0.45
14	0.04	0.07	0.19	0.30	0.40	0.50	0.65	0.80	0.94	1.08	1.27	1.49	1.71	1.93	1.66	1.42	1.23	1.08	0.96	0.86	0.77	0.70	0.64	0.59	0.50
15	0.04	0.07	0.20	0.32	0.43	0.54	0.70	0.86	1.01	1.17	1.36	1.61	1.85	2.08	1.84	1.57	1.36	1.20	1.06	0.95	0.86	0.78	0.71	0.65	0.56
16	0.04	0.08	0.22	0.34	0.47	0.58	0.76	0.92	1.09	1.25	1.46	1.72	1.98	2.23	2.03	1.73	1.50	1.32	1.17	1.05	0.94	0.86	0.78	0.72	0.61
17	0.05	0.08	0.23	0.37	0.50	0.62	0.81	0.99	1.16	1.33	1.56	1.84	2.11	2.38	2.22	1.90	1.64	1.44	1.28	1.14	1.03	0.94	0.86	0.79	0.67
18	0.05	0.09	0.25	0.39	0.53	0.66	0.86	1.05	1.24	1.42	1.66	1.96	2.25	2.53	2.42	2.07	1.79	1.57	1.39	1.25	1.12	1.02	0.93	0.86	0.73
19	0.05	0.09	0.26	0.41	0.56	0.70	0.91	1.11	1.31	1.50	1.76	2.07	2.38	2.69	2.62	2.24	1.94	1.70	1.51	1.35	1.22	1.11	1.01	0.93	0.79
20	0.06	0.10	0.28	0.44	0.59	0.74	0.96	1.75	1.38	1.59	1.86	2.19	2.52	2.84	2.83	2.42	2.10	1.84	1.63	1.46	1.32	1.20	1.09	1.00	0.86
21	0.06	0.11	0.29	0.46	0.62	0.78	1.01	1.24	1.46	1.68	1.96	2.31	2.66	2.99	3.05	2.60	2.26	1.98	1.76	1.57	1.42	1.29	1.17	1.08	0.92
22	0.06	0.11	0.31	0.48	0.66	0.82	1.07	1.30	1.53	1.76	2.06	2.43	2.79	3.15	3.27	2.79	2.42	2.12	1.88	1.69	1.52	1.38	1.26	1.16	0.99
23	0.06	0.12	0.32	0.51	0.69	0.86	1.12	1.37	1.61	1.85	2.16	2.55	2.93	3.30	3.50	2.98	2.59	2.27	2.01	1.80	1.62	1.47	1.35	1.24	1.06
24	0.07	0.13	0.34	0.53	0.72	0.90	1.17	1.43	1.69	1.94	2.27	2.67	3.07	3.46	3.73	3.18	2.76	2.42	2.15	1.92	1.73	1.57	1.44	1.32	1.12
25	0.07	0.13	0.35	0.56	0.75	0.94	1.22	1.50	1.76	2.02	2.37	2.79	3.21	3.61	3.96	3.38	2.93	2.57	2.28	2.04	1.84	1.67	1.53	1.40	1.20
26	0.07	0.14	0.37	0.58	0.79	0.98	1.28	1.56	1.84	2.11	2.47	2.91	3.34	3.77	4.19	3.59	3.11	2.73	2.42	2.17	1.95	1.77	1.62	1.49	1.27
28	0.08	0.15	0.40	0.63	0.85	1.07	1.38	1.69	1.99	2.29	2.68	3.15	3.62	4.09	4.54	4.01	3.47	3.05	2.70	2.42	2.18	1.98	1.81	1.66	1.42
30	0.08	0.16	0.43	0.68	0.92	1.15	1.49	1.82	2.15	2.46	2.88	3.40	3.90	4.40	4.89	4.45	3.85	3.38	3.00	2.68	2.42	2.20	2.01	1.84	1.57
32	0.09	0.17	0.46	0.73	0.98	1.23	1.60	1.95	2.30	2.64	3.09	3.64	4.18	4.72	5.25	4.90	4.25	3.73	3.30	2.96	2.67	2.42	2.21	2.03	1.73
35	0.10	0.19	0.51	0.80	1.08	1.36	1.76	2.15	2.53	2.91	3.41	4.01	4.61	5.20	5.78	5.60	4.86	4.26	3.78	3.38	3.05	2.77	2.53	2.32	1.98
40	0.12	0.22	0.58	0.92	1.25	1.57	2.03	2.48	2.93	3.36	3.93	4.64	5.32	6.00	6.68	6.85	5.93	5.21	4.62	4.13	3.73	3.38	3.09	2.83	2.42
45	0.13	0.25	0.66	1.05	1.42	1.78	2.31	2.82	3.32	3.82	4.47	5.26	6.05	6.82	7.58	8.17	7.08	6.21	5.51	4.93	4.45	4.04	3.69	3.38	2.89
	TYPE A						TYPE B							TYPE C											

TYPE A: Manual or Drip Lubrication (500 fpm max.)
TYPE B: Bath or Disc Lubrication (3500 fpm max.)
TYPE C: Oil Stream Lubrication (Up to max. speed shown).
The limiting RPM for each lubrication type is read from the column to the left of the boundary line shown.
The ratings on this page are in accordance with the standards of the American Chain Association, Copyright 1974.

For Multi. Strand Chain use

No. of Strands	Strand Factor
2	1.7
3	2.5
4	3.3

FIGURE 20.30 ■ Horsepower ratings for single-strand roller chain. *Courtesy Morse Industrial.*

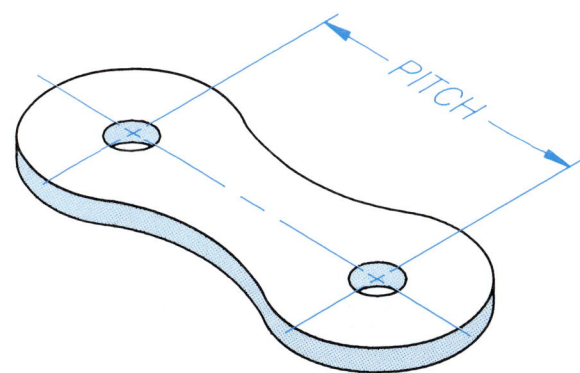

FIGURE 20.31 ■ Chain pitch.

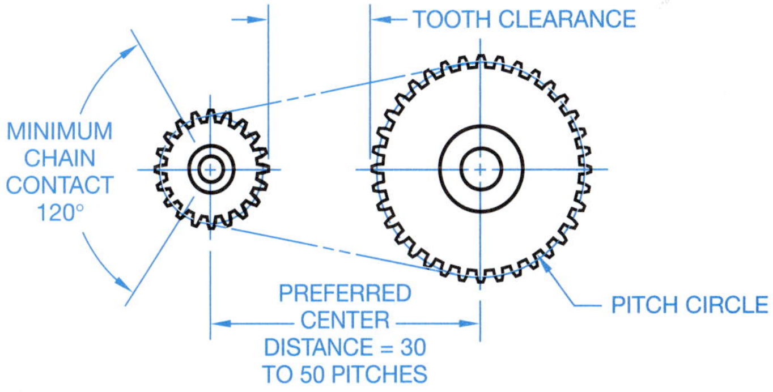

TOOTH CLEARANCE

MINIMUM CHAIN CONTACT 120°

PREFERRED CENTER DISTANCE = 30 TO 50 PITCHES

PITCH CIRCLE

FIGURE 20.32 ■ The center distance between shafts and 120° minimum recommended chain contact.

The chain length in pitches must be the next higher whole number, because you cannot have a partial chain pitch. So, 156.84 = 157 chain length in pitches. Whenever possible, use an even number of chain pitches. Using an odd number of chain pitches requires an offset link, which is not preferred. For this problem, therefore, use 158 chain length in pitches rather than 157. Now,

multiply the chain length in pitches by the chain pitch to find the chain length in inches:

$$158 \times .25 = 39.5 \text{ inches of chain length}$$

STEP 7   Determine chain lubrication. Correct lubrication is important in maintaining long life for chain drives. Lubrication methods are governed by the speed and function

D	K	D	K	D	K	D	K
32	25.94	63	100.54	94	223.82	125	395.79
33	27.58	64	103.75	95	228.61	126	402.14
34	29.28	65	107.02	96	233.44	127	408.55
35	31.03	66	110.34	97	238.33	128	415.01
36	32.83	67	113.71	98	243.27	129	421.52
37	34.68	68	117.13	99	248.26	130	428.08
38	36.58	69	120.60	100	253.30	131	434.69
39	38.53	70	124.12	101	258.39	132	441.36
40	40.53	71	127.69	102	263.54	133	448.07
41	42.58	72	131.31	103	268.73	134	454.83
42	44.68	73	134.99	104	273.97	135	461.64
43	46.84	74	138.71	105	279.27	136	468.51
44	49.04	75	142.48	106	284.61	137	475.42
45	51.29	76	146.31	107	290.01	138	482.39
46	53.60	77	150.18	108	295.45	139	489.41
47	55.95	78	154.11	109	300.95	140	496.47
48	58.36	79	158.09	110	306.50	141	503.59
49	60.82	80	162.11	111	312.09	142	510.76
50	63.33	81	166.19	112	317.74	143	517.98
51	65.88	82	170.32	113	323.44	144	525.25
52	68.49	83	174.50	114	329.19	145	532.57
53	71.15	84	178.73	115	334.99	146	539.94
54	73.86	85	183.01	116	340.84	147	547.36
55	76.62	86	187.34	117	346.75	148	554.83
56	79.44	87	191.73	118	352.70	149	562.36
57	82.30	88	196.16	119	358.70	150	569.93
58	85.21	89	200.64	120	364.76	151	577.56
59	88.17	90	205.18	121	370.86	152	585.23
60	91.19	91	209.76	122	377.02	153	592.96
61	94.25	92	214.40	123	383.22	154	600.73
62	97.37	93	219.08	124	389.48	155	608.56

FIGURE 20.33 ■ Sprocket teeth factors "K." *Courtesy Morse Industrial.*

of the machine. Periodic manual lubrication works well for slow-speed chain drives. To do this, use a brush to apply a medium-consistency mineral oil while a machine is stopped. For moderate-speed chain drives, a drip system is often designed to keep the chain lubricated. For high-speed chain drives, an oil bath or an oil spray is often provided. To determine the type of oil application recommended for the sample problem, refer back to the chart in Figure 20.30. Notice the notes Type A, Type B, and Type C at the bottom of the chart. These refer to the recommended lubrication as follows:

Type A = Manual or drip lubrication.
Type B = Bath or disc lubrication.
Type C = Oil stream lubrication.

Notice the HP rating of 2.93 for the sample problem falls in the Type B lubrication category.

WEBSITE RESEARCH

The following websites can assist you in doing additional research on subjects such as standards, gears, cams, linkages, bearings, and related areas:

http://www.sae.org—Society of Automotive Engineers (SAE).

http://www.industrialpress.com—*Machinery's Handbook.* Gear, cam, linkage, and bearing-related information and specifications.

http://www.phoenix-mfg.com—Phoenix Manufacturing. Pulleys and idlers.

http://www.dxpe.com—Industrial V belts.

http://www.igus.com—IGUS Corporation. Bearings and chains.

http://www.teleflexmorse.com—Teleflex Morse, industrial products.

http://www.emerson-ept.com—Emerson Power Transmission, producer of power transmission drives, components, and bearings.

PROFESSIONAL PERSPECTIVE

This chapter covers the use of vendors' catalogs to find specific belt and chain drives for a given application. As a drafter in any engineering field, you will find that using vendors' catalogs and specifications is a very important part of your job. Most companies have a complete library of catalogs of purchase parts. Purchase parts, often referred to as standard parts, are products that can be purchased already made. Common purchase parts are bolts, nuts, belts, chains, and sprockets. You should become familiar with the types of purchase parts your company uses and how to find them in the vendors' catalogs. Also determine the suppliers that have the best prices, highest quality, and fastest availability.

MATH APPLICATIONS

BELT LENGTH

Problem: Find the length of a driving belt running around two pulleys of radii 14" and 8" if the distance between the centers of the two pulleys is 25", as shown in Figure 20.34.

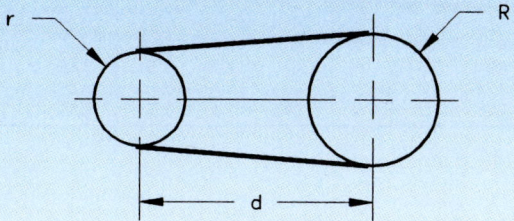

FIGURE 20.34 ■ Drive belt around pulleys.

Solution: The formula for belt length between two pulleys is

$$L = 2(d)(\cos\theta) + \pi(R+r) + 20(R - r)$$
$$\text{where } \theta = \text{Inv sin}\left(\frac{R - r}{d}\right)$$

"Inv sin" stands for "inverse sine" and means "angle whose sine is." For example, Inv sin .7 is equal to 44.4° (or .775 radians) because sin 44.4° = .7. It is important to note that this formula requires angle θ to be in radian measure. The first step in this rather complicated formula to find L is to determine θ. It is easiest to put your calculator in the "radian mode" to work this problem:

$$\theta = \text{Inv sin}\left(\frac{14 - 8}{25}\right) = \text{Inv sin } .24 = .2424 \text{ radians}$$

Then substituting into the formula for L:

$$L = 2(25)(\cos .2424) + \pi(14 + 8) + 2(.2424)(14 - 8)$$
$$L = 2(25)(.9708) + \pi(22) + 2(.2424)(6)$$
$$L = 48.54 + 69.12 + 2.91 = \mathbf{120.57"}$$

CHAPTER 20 *Belt and Chain Drives Test*

Access the CD found with this textbook to view the Chapter 20 Test. Confirm the preferred submittal method with your instructor.

CHAPTER 20 *Belt and Chain Drives Problems*

DIRECTIONS

Solve Problems 20.1 through 20.8 using the information and tables given in this chapter and in Figure 20.35. All motors are AC motors with normal torque, squirrel cage, synchronous, and split phase, or shunt-wound DC motors, or multiple cylinder internal combustion engines. Include the following information with each solution:

■ Service factor.

■ Design horsepower.

■ Driven speed.

■ Pitch diameter driver and driven.

■ Belt section.

■ Number of belts.

■ Belt number and center distance.

■ Corrected horsepower.

■ Belt drive ratio.

■ Belt velocity of driver sheave.

1.50	12.0	18.0	A	1	10	—	—	—	1167	10.57	13.63	A77	15.3	0.96	A81	17.3	0.98	A85	19.4	0.99
1.50	12.0	18.0	C	2	12	—	—	—	1167	25.92	31.20	C75	15.1	0.82	C96	25.7	0.88	C112	33.8	0.92
1.50	13.0	19.5	A	1	3	—	—	—	1167	11.07	14.40	A83	16.3	0.98	A86	17.8	0.99	A89	19.4	0.99
1.51	3.7	5.6	A	1	3	2318	4.61	5.62	1159	2.94	3.71	A24	5.3	0.76	A35	10.8	0.84	A45	15.8	0.88
1.51	3.9	5.9	B	1	3	2318	4.64	8.35	1159	3.53	5.53	B28	7.1	0.73	B45	15.7	0.84	B56	21.2	0.89
1.53	3.4	5.2	B	1	6	2288	2.90	6.59	1144	2.43	4.44	B28	8.1	0.73	B45	16.6	0.84	B56	22.1	0.89
1.53	3.6	5.5	A	1	3	2288	4.42	5.37	1144	2.81	3.55	A24	5.4	0.76	A35	11.0	0.84	A45	16.0	0.88
1.53	3.6	5.5	3V	1	4	2288	6.28	7.30	1144	3.63	4.07	3V250	5.3	0.78	3V300	7.8	0.83	3V355	10.6	0.85
1.53	3.8	5.8	A	1	6	2288	4.81	5.87	1144	3.06	3.86	A24	5.0	0.76	A35	10.6	0.84	A45	15.6	0.88
1.53	3.8	5.8	B	1	6	2288	4.31	8.01	1144	3.31	5.32	B28	7.3	0.73	B45	15.8	0.84	B56	21.3	0.89
1.54	2.4	3.7	A	1	1	2273	1.85	2.12	1136	1.27	1.61	A24	7.8	0.79	A35	13.3	0.86	A45	18.3	0.90
1.54	2.6	4.0	A	1	2	2273	2.31	2.70	1136	1.54	1.94	A24	7.4	0.79	A35	12.9	0.84	A45	18.0	0.90
1.54	3.5	5.4	A	1	3	2273	4.22	5.12	1136	2.69	3.40	A24	5.6	0.76	A35	11.1	0.84	A45	16.1	0.88
1.54	3.7	5.7	A	1	3	2273	4.62	5.62	1136	2.94	3.71	A24	5.2	0.76	A35	10.7	0.84	A45	15.7	0.88
1.54	3.9	6.0	B	1	3	2273	4.66	8.36	1136	3.53	5.54	B28	7.0	0.73	B45	15.6	0.84	B56	21.1	0.89

FIGURE 20.35 ■ Nominal center distances, number of belts, and arc length factor chart.

PART 1

PROBLEM 20.1 A 3-HP, 1750-rpm motor is to operate a furnace blower having a shaft speed of approximately 1115 rpm under normal service. The center distance between the motor and blower shafts is about 16 in.

PROBLEM 20.2 A 3-HP, 1750-rpm motor is used to operate a drill press speed reducer under intermittent service. The spindle speed is about 1136 rpm. The center distance between the motor and spindle shafts is about 20.5 in.

PROBLEM 20.3 A 11/2-HP, 1750-rpm electric motor is used to operate a woodworking band saw with the blade turning at 1144 rpm, intermittent service. The center-to-center distance is about 16 in.

PROBLEM 20.4 A 2-HP electric motor with a shaft speed of 1750 rpm operates a printing machine at normal service. The shaft on the printing machine is to operate at 1167 rpm. The center-to-center distance is about 18 in.

PROBLEM 20.5 A 2-HP electric motor with a shaft speed of 1750 rpm operates a punch machine at continuous service. The shaft on the punch machine is to operate at 1108 rpm. The center-to-center distance is about 17 in.

PROBLEM 20.6 A 1.5-HP motor with a shaft speed of 1750 rpm operates a compressor at normal service. The shaft on the compressor is to operate at 1167 rpm. The center-to-center distance is about 18 in.

PROBLEM 20.7 A 2-HP electric motor with a shaft speed of 1750 rpm operates a printing machine at normal service. The shaft on the printing machine is to operate at 1115 rpm. The center-to-center distance is about 17.5 in.

PROBLEM 20.8 Given the following layout, prepare a detailed drawing of the pulley on appropriately sized sheet or using the CADD system, unless otherwise specified by your instructor. (in.)

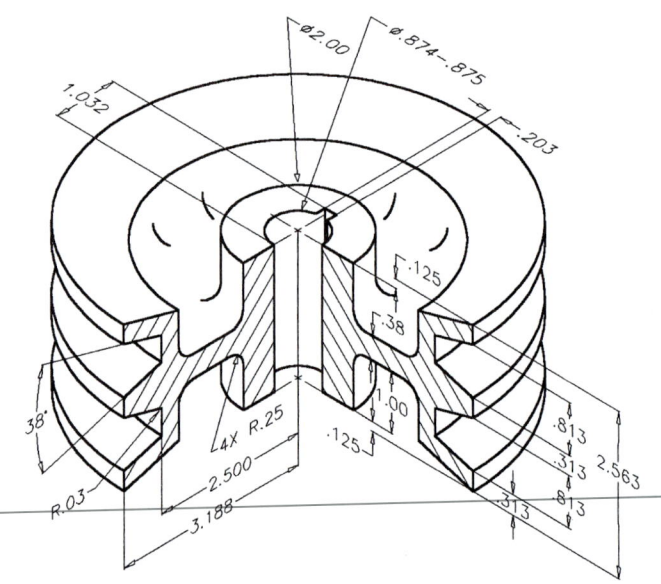

DIRECTIONS

Solve Problems 20.9 through 20.15 using the information and tables given in this chapter. Include the following information with each solution:

■ Service factor.

■ Design horsepower.

■ Number of teeth drive sprocket and driven sprocket.

■ Chain pitch.

■ Ratio.

■ Center distance in chain pitches.

■ Chain length.

■ Lubrication.

PROBLEM 20.9 A chain drive is required for a 1.5-HP electric motor operating a paper machine. The drive sprocket turns 1500 rpm and the driven sprocket turns 600 rpm. The shaft center distance is about 19 in.

PROBLEM 20.10   A chain drive is required for a 1/10-HP electric motor operating a speed reducer for a belt light package conveyor. The drive sprocket turns 50 rpm and the driven sprocket turns 17 rpm. The shaft center distance is about 17 in.

PROBLEM 20.11   A chain drive is required for a 1-HP electric motor operating a centrifuge. The drive sprocket turns 10,000 rpm and the driven sprocket turns 4000 rpm. The shaft center distance is about 18 in.

PROBLEM 20.12   A chain drive is required for a 2-HP electric motor operating a centrifugal pump. The drive sprocket turns 5500 rpm and the driven sprocket turns 1550 rpm. The shaft center distance is about 14 in.

PROBLEM 20.13   A chain drive is required for a 5-HP electric motor operating a flour grinder. The drive sprocket turns 3500 rpm and the driven sprocket turns 1500 rpm. The shaft center distance is about 18 in.

PROBLEM 20.14   A chain drive is required for a 2.5-HP internal combustion tractor engine operating a generator. The drive sprocket turns 3000 rpm and the driven sprocket turns 2000 rpm. The shaft center distance is about 20 in.

PROBLEM 20.15   A chain drive is required for a 2-HP electric motor operating a pure liquid agitator. The drive sprocket turns 2100 rpm and the driven sprocket turns 700 rpm. The shaft center distance is about 20 in.

 PROBLEMS 20.16 through 20.22: Access the CD found with this textbook and open the problem of your choice, or as assigned by your instructor. Solve the problems using the instructions provided with this chapter or on the CD, unless otherwise specified by your instructor.

MATH PROBLEMS

 PROBLEMS 20.23 through 20.28: Access the CD found with this textbook and open the math problem of your choice, or as assigned by your instructor. Solve the problem or problems using the instructions provided.

Welding Processes and Representations

LEARNING OBJECTIVES

After completing this chapter, you will:

- Identify welding processes.

- Draw welding representations and provide proper welding symbols and notes.

- Draw weldments from engineering sketches and actual industrial layouts.

THE ENGINEERING DESIGN APPLICATION

The engineer has just handed you a sketch showing the details for a couple of weldments. The welding details are not shown as standard welding symbols but rather as written instructions and it is up to you to design the appropriate weld and welding symbols for the weldment drawing.

The first weldment sketch shows an assembly of two aluminum bars connected in a lap joint; the instructions are as follows: *Use 3/8" plug weld, 7/16" deep with a 45° included angle. Space the welds at 3" on center, for the length of the material.*

Since aluminum welding requires special considerations, you examine the thickness of the material and decide on a *gas tungsten arc welding* process. Using this information, you design the weld symbol shown in Figure 21.1.

The second weldment sketch shows two low carbon steel plates joined in a T-joint. The instructions are: *Use 1/4" intermittent fillet weld on both sides. The welds should be 2" long with a pitch of 10" and staggered. Field weld at installation site.* The low carbon steel material and T-joint suggest a *shielded metal arc welding* process, and you design an appropriate symbol (Figure 21.2).

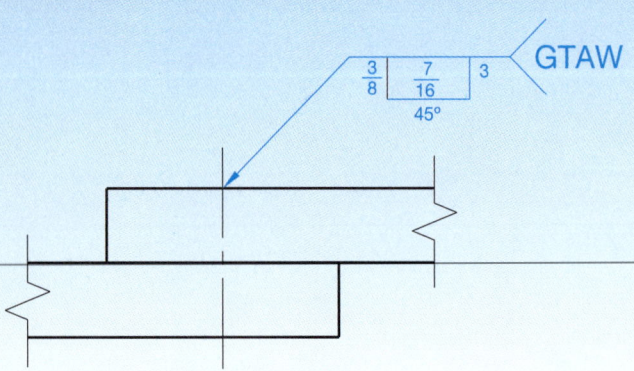

FIGURE 21.1 ■ Plug weld symbol. Dimension values in this figure are in inches.

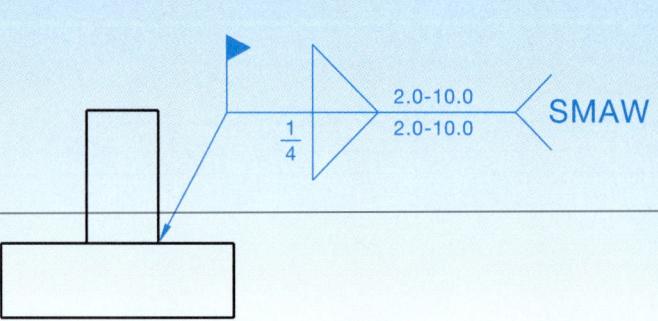

FIGURE 21.2 ■ Fillet weld symbol. Dimension values in this figure are in inches.

Welding is a process of joining two or more pieces of like metals by heating the material to a temperature high enough to cause softening or melting. The location of the weld is where the materials actually combine the grain structure from one piece to the other. The parts that are welded become one, and the properly welded joint is as strong or stronger than the original material. Welding may be performed with or without pressure applied to the materials. Some materials may actually be welded together by pressure alone. Most welding operations, however, are performed by filling a heated joint between pieces with molten metal.

Welding is actually a method of fastening adjacent parts. Welding was not discussed with fasteners in Chapter 12, because the weld is a more permanent fastening application than screw threads or pins, for example. Welding is a common fastening method used in many manufacturing applications and industries from automobile to aircraft manufacturing, and from computers to ship building. Some of the advantages of welding over other fastening methods include better strength, better weight distribution and reduction, a possible decrease in the size of castings or forgings needed in an assembly, and a potential savings of time and manufacturing costs.

WELDING PROCESSES

There are a number of welding processes available for use in industry, as shown in Figure 21.3. The most common welding processes include oxygen gas welding, shielded metal arc welding, gas tungsten arc welding, and gas metal arc welding.

Oxygen Gas Welding

Oxygen gas welding, commonly known as oxyfuel welding or oxyacetylene welding, may also be performed with such fuels

as natural gas, propane, or propylene. Oxyfuel welding is most typically used to fabricate thin materials, such as sheet metal and thin-wall pipe or tubing. Oxyfuel processes are also used for repair work and metal cutting. One advantage of oxyfuel welding is that the equipment and operating costs are less than with other methods. But other welding methods have advanced over the oxyfuel process because they are faster, cleaner, and cause less material distortion. Common oxyfuel welding and cutting equipment is shown in Figure 21.4.

Also associated with oxyfuel applications are soldering, brazing, and braze welding. These methods are more of a bonding process than welding, as the base material remains solid while a filler metal is melted into a joint. Soldering and brazing differ in application temperature. Soldering is done below 840°F (450°C) and brazing above 840°F (450°C). Like alloys may be used depending on their melting temperatures. The filler generally associated with soldering is solder. Solder is an alloy of tin and lead. The filler metal associated with brazing is an alloy of

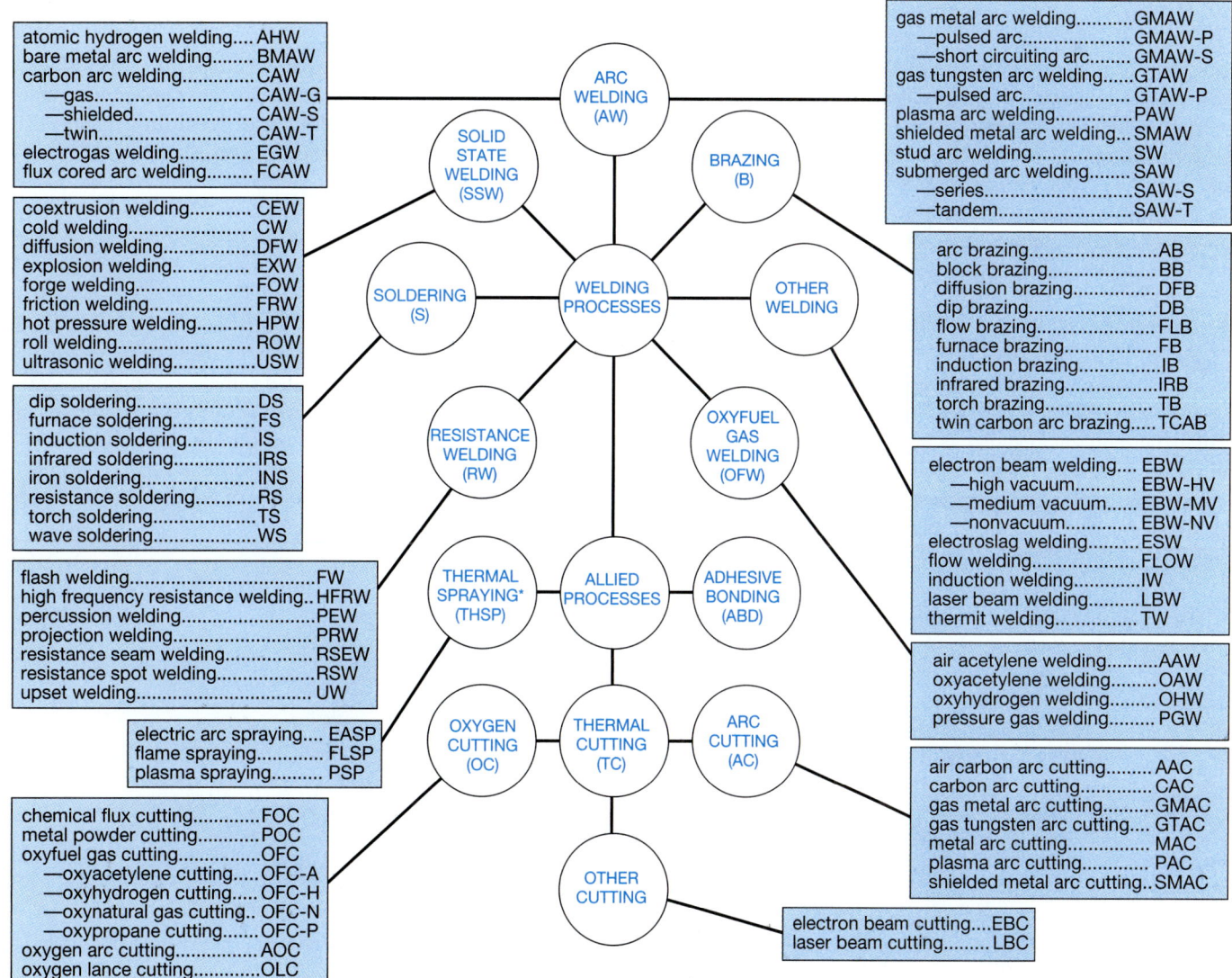

FIGURE 21.3 ■ Master chart of welding and allied processes. *Courtesy American Welding Society.*

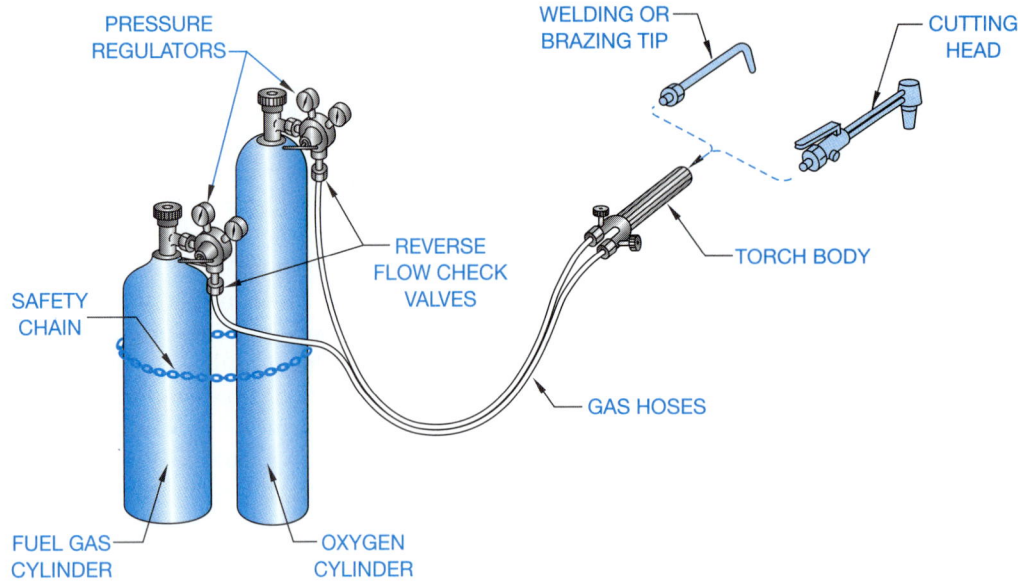

FIGURE 21.4 ■ Oxyfuel welding and cutting equipment.

copper and zinc. **Brazing** is a process of joining two very closely fitting metals by heating the pieces, causing the filler metal to be drawn into the joint by capillary action. **Braze welding** is more of a joint filling process that does not rely on capillary action. Another process that uses an oxyfuel mixture is **flame cutting**. This process uses a high-temperature gas flame to preheat the metal to a kindling temperature, at which time a stream of pure oxygen is injected to cause the cutting action.

Shielded Metal Arc Welding

Shielded metal arc or **stick electrode welding** is the most traditionally used welding method. High-quality welds on a variety of metals and thicknesses can be made rapidly with excellent uniformity. This method uses a flux-covered metal electrode to carry an electrical current forming an arc that melts the work and the electrode. The molten metal from the electrode mixes with the melting base material, forming the weld. Shielded metal arc welding is popular because of low-cost equipment and supplies, flexibility, portability, and versatility. Figure 21.5 shows a shielded metal arc welding setup.

Gas Tungsten Arc Welding

The **gas tungsten arc welding** process is sometimes referred to as **tungsten inert gas welding (TIG)**, or as **Heliarc®**, which is a trademark of the Union Carbide Corporation. Gas tungsten arc welding can be performed on a wider variety of materials than shielded metal arc welding, and produces clean, high-quality welds. This welding process is useful for certain materials and applications. Gas tungsten arc welding is generally limited to thin materials, high-integrity joints, or small parts, because of its slow welding speed and high cost of equipment and materials. (See Figure 21.6.)

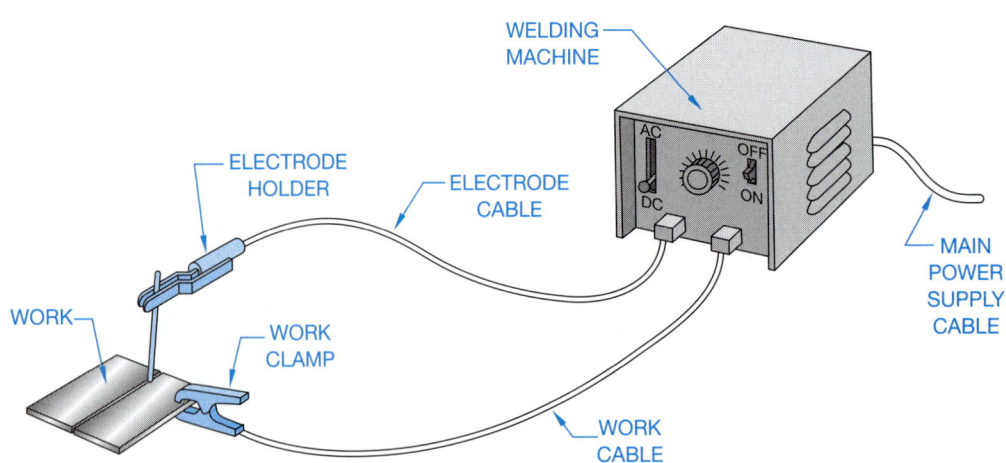

FIGURE 21.5 ■ Shielded metal arc welding equipment.

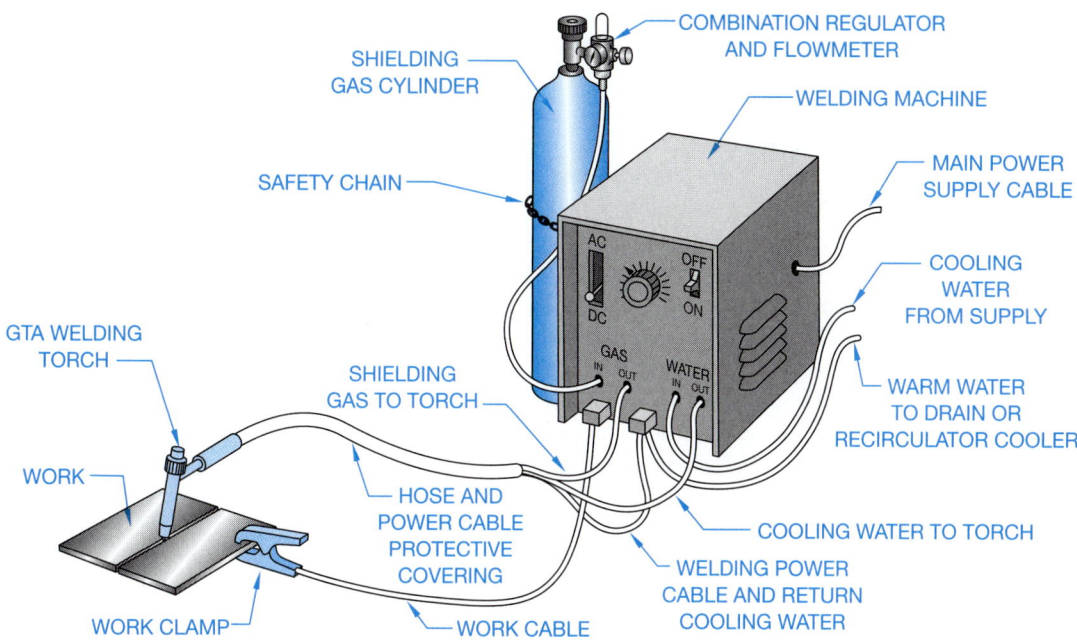

FIGURE 21.6 ■ Gas tungsten arc welding equipment.

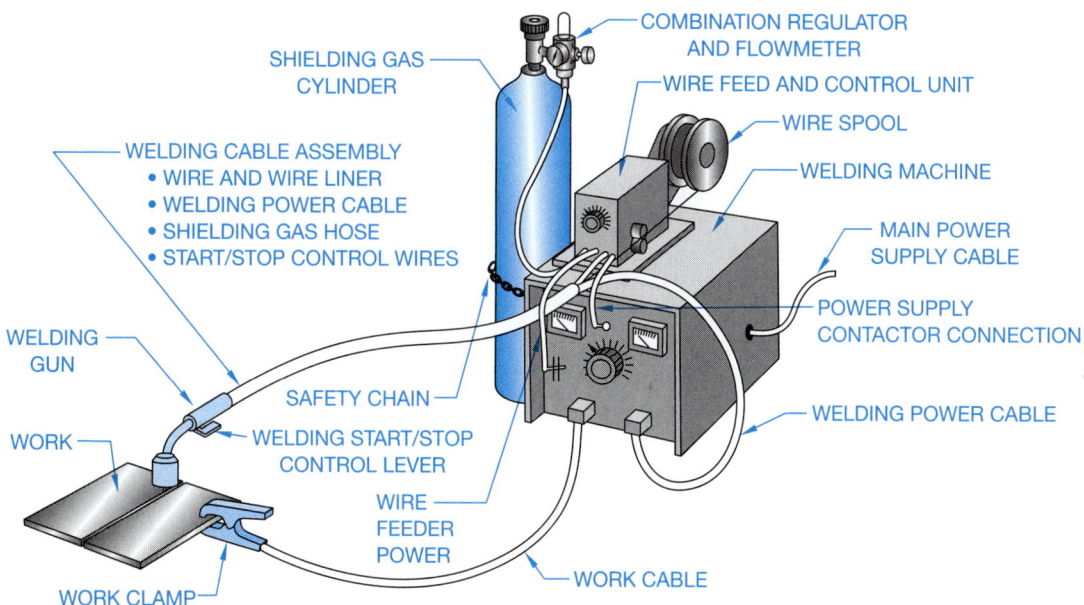

FIGURE 21.7 ■ Gas metal arc welding equipment.

Gas Metal Arc Welding

Another welding process that is extremely fast, economical, and produces a very clean weld is gas metal arc welding. This process may be used to weld thin material or heavy plate. It was originally used previously for welding aluminum using a metal inert gas shield, a process which became referred to as MIG. The present application employs a current-carrying wire that is fed into a joint between pieces to form the weld. This welding process is used in industry with automatic or robotic welding machines to produce rapidly made, high-quality welds in any welding position. While the capital expense of the equipment remains high, the cost is declining due to its popularity. Figure 21.7 shows gas metal arc welding equipment.

ELEMENTS OF WELDING DRAWINGS

ANSI/ASME The standard related to drafting practices covered in this chapter is ANSI/AWS A2.4-93, *Standard Symbols for Welding, Brazing, and Nondestructive Examination.* This American Welding Society (AWS) standard is developed in accordance with rules of the American National Standards Institute (ANSI), and published by the American Welding Society.

Welding drawings are made up of several parts to be welded together. These drawings are usually called **weldments** or **welding assemblies** or **subassemblies**. The welding assembly typically shows the parts together in multiview with all of the fabrication dimensions, types of joints, and weld symbols. **Welding symbols** identify the location of the weld, the welding process, the size and length of the weld, and other **weld information**. The **weld symbol** indicates the type of weld and is part of the welding symbol. The **welding assembly** has a list of materials that generally provides a key to the assembly, the number of each part, part size, and material. Figure 21.8 shows a welding subassembly. When additional clarity of component parts must be given, then detailed drawings of each part are prepared, as shown in Figure 21.9.

Welding and Weld Symbols

AWS The welding symbol represents complete information about the weld. The weld and welding symbols discussed in this chapter are in accordance with the American Welding Society document AWS A2.4-93.

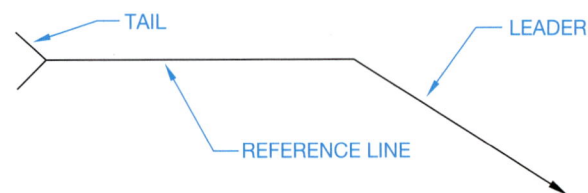

FIGURE 21.10 ■ Welding symbols: reference line, tail, and leader.

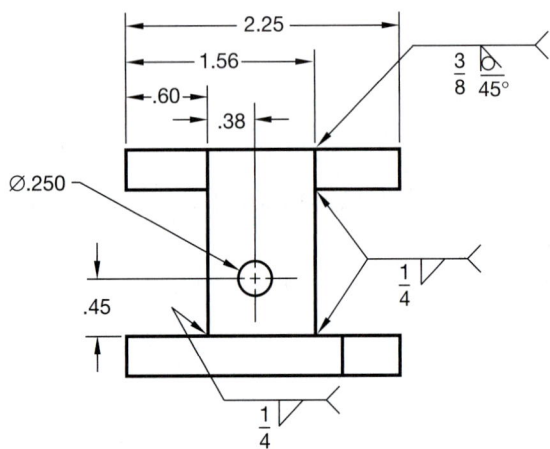

FIGURE 21.8 ■ Welded subassembly.

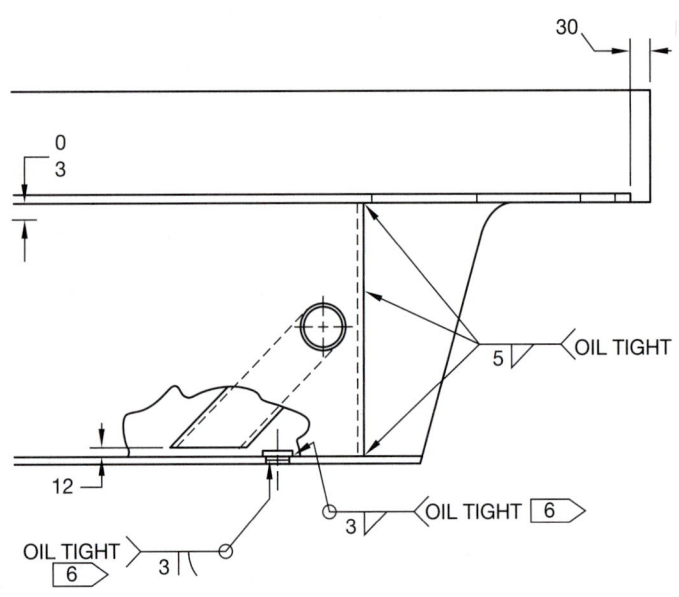

FIGURE 21.11 ■ Welding symbol leader use. *Courtesy Hyster Company.*

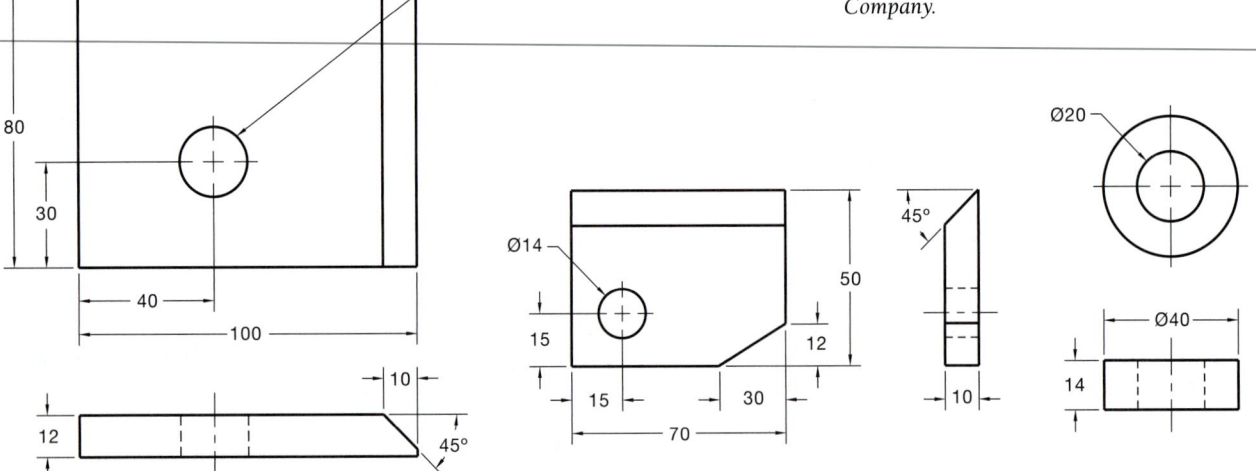

FIGURE 21.9 ■ Drawings of each part of the welded subassembly.

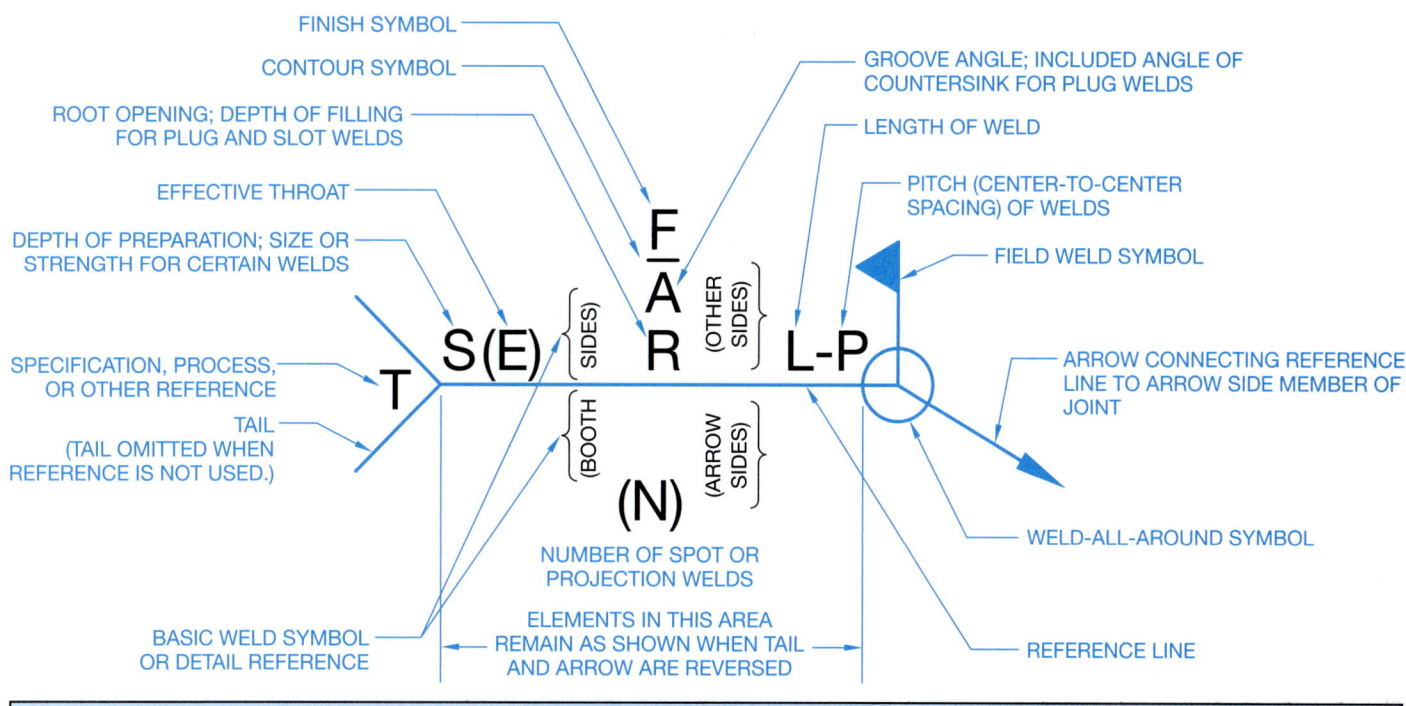

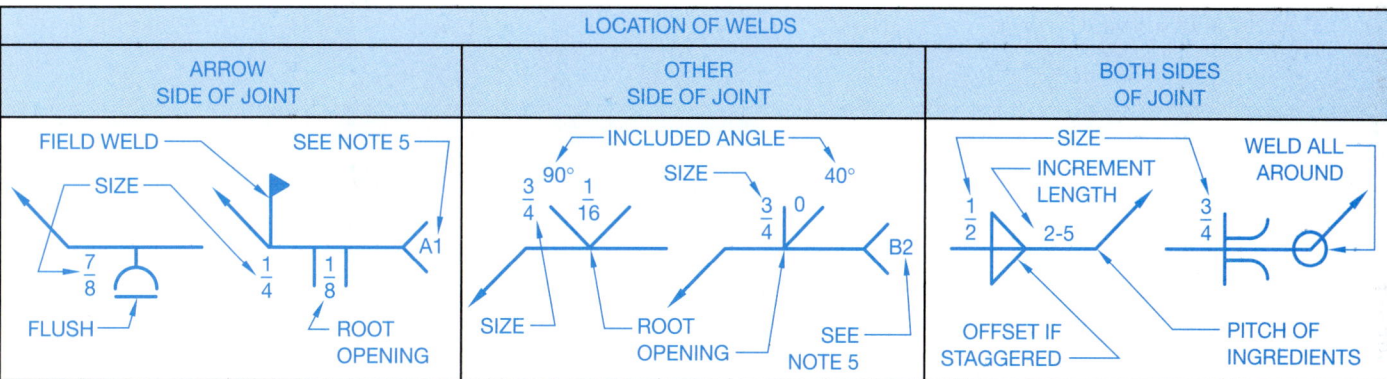

LOCATION OF WELDS		
ARROW SIDE OF JOINT	OTHER SIDE OF JOINT	BOTH SIDES OF JOINT

1. THE SIDE OF THE JOINT TO WHICH THE ARROW POINTS IS THE ARROW SIDE.
2. BOTH-SIDES WELDS OF SAME TYPE ARE OF SAME SIZE UNLESS OTHERWISE SHOWN.
3. SYMBOLS APPLY BETWEEN ABRUPT CHANGES IN DIRECTION OF JOINT OR AS DIMENSIONED (EXCEPT WHERE ALL AROUND SYMBOL IS USED).
4. ALL WELDS ARE CONTINUOUS AND OF USER'S STANDARD PROPORTIONS, UNLESS OTHERWISE SHOWN.
5. TAIL OF ARROW USED FOR SPECIFICATION REFERENCE. (TAIL MAY BE OMITTED WHEN REFERENCE NOT USED.)
6. DIMENSIONS OF WELD SIZES, INCREMENT LENGTHS, AND SPACING IN INCHES.

WELDING SYMBOL TEXT: .12 IN, (3 mm) MINIMUM, OFFSET TEXT .06 (1.5 mm) MINIMUM FROM THE REFERENCE LINE. WELD SYMBOL AND TEXT SHOULD BE PLACED RELATED TO THE REFERENCE LINE AS SHOWN IN THE EXAMPLES THROUGHOUT THIS CHAPTER.

FIGURE 21.12 ■ Standard location of elements of a welding symbol. *Courtesy American Welding Society.*

There are a few basic components of a welding symbol, beginning with the reference line, tail, and leader, which are drawn as thin lines, as shown in Figure 21.10. The reference line is usually the first welding symbol element to be located on the drawing. The reference line leader is drawn using the same rules associated with leaders for notes. The leader can be drawn at any angle, although angles less than 15° or greater than 75° should be avoided. Also, leaders for notes typically run from the shoulder directly to the feature. This practice should be used for welding leaders, although sometimes welding leaders bend to point into difficult-to-reach places, or with more than one leader extending from the same reference line, as shown in Figure 21.11. This practice is not allowed when drawing ASME standard dimensioning leaders.

After the reference line has been established, additional information is placed on the reference line to continue the weld specification. Figure 21.12 shows the standard location of welding symbol elements as related to the reference line, tail, and leader. As previously introduced, the weld symbol indicates the type of weld and is part of the welding symbol. Weld symbols are generally drawn on or near the center of the reference line of the welding symbol. Typical weld symbols are shown in Figure 21.13.

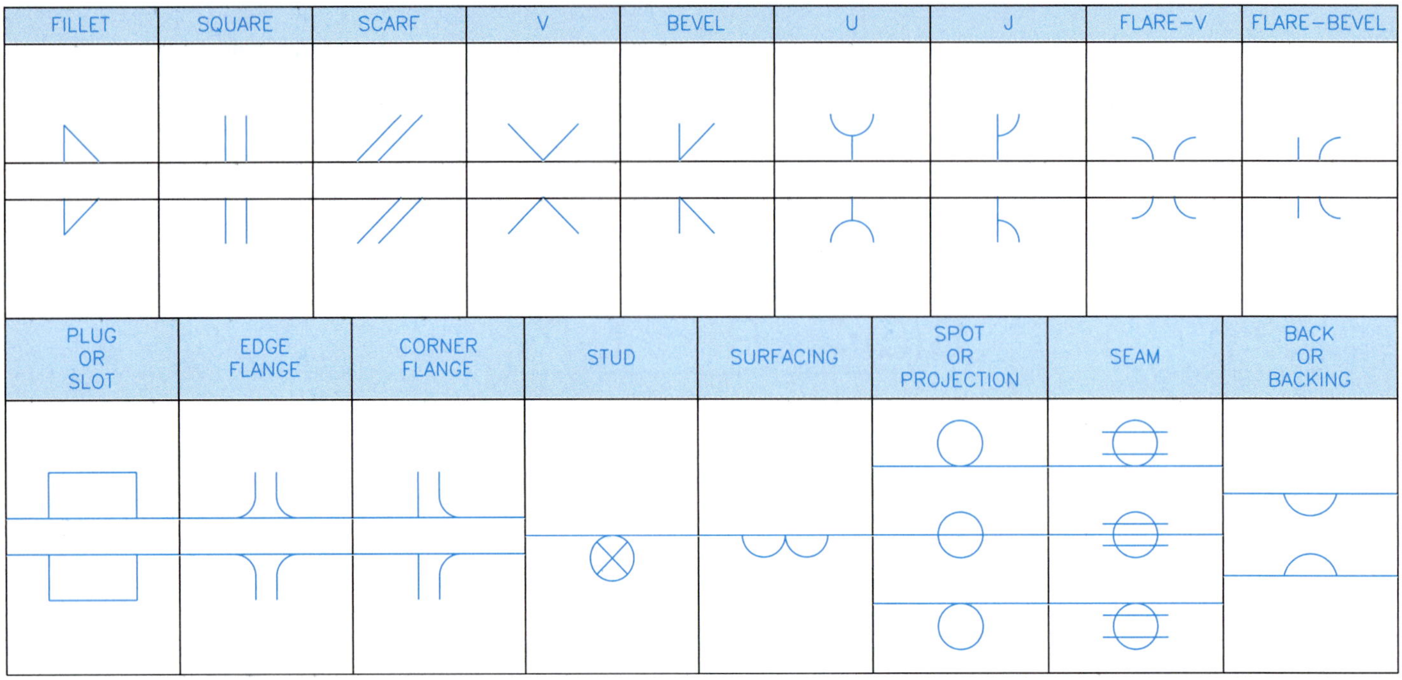

FILLET	SQUARE	SCARF	V	BEVEL	U	J	FLARE–V	FLARE–BEVEL

PLUG OR SLOT	EDGE FLANGE	CORNER FLANGE	STUD	SURFACING	SPOT OR PROJECTION	SEAM	BACK OR BACKING

FIGURE 21.13 ■ Typical weld symbols.

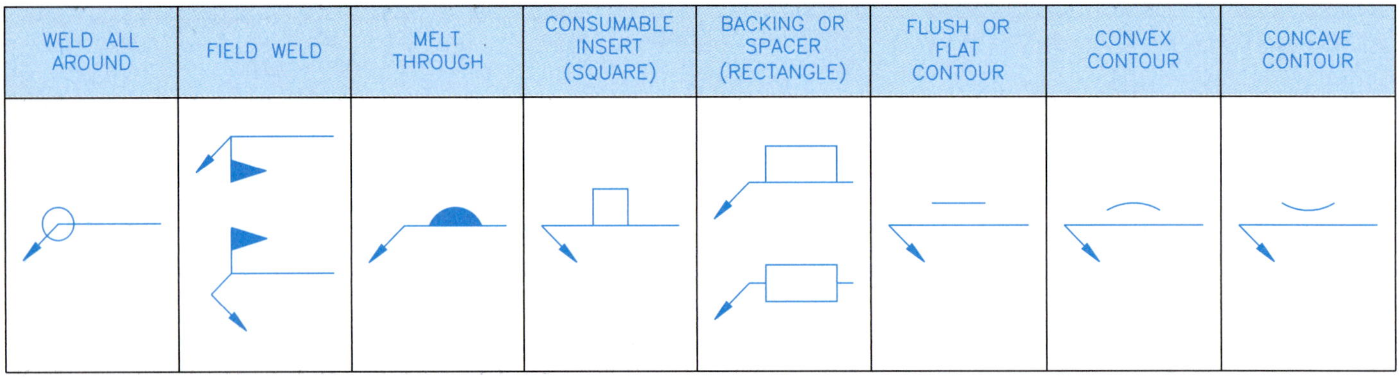

WELD ALL AROUND	FIELD WELD	MELT THROUGH	CONSUMABLE INSERT (SQUARE)	BACKING OR SPACER (RECTANGLE)	FLUSH OR FLAT CONTOUR	CONVEX CONTOUR	CONCAVE CONTOUR

FIGURE 21.14 ■ Supplementary welding symbols.

Supplementary Symbols

Supplementary symbols, shown in Figure 21.14, are used with welding symbols to identify weld features such as how and where the weld is placed.

Welding Symbol Placement

The arrowhead of the welding symbol leader points to the desired weld location in the view that best shows the weld location as object lines; although, drawing a welding symbol arrow to a hidden feature is acceptable, if necessary.

U.S. Customary and Metric Units

Welding symbols and their components are created using the same units as the drawing upon which they are displayed. For example, if the drawing is created using inch units, the welding symbols are drawn in inches. When the drawing is created using metric units, the welding symbols are drawn in millimeters.

Welding and Weld Symbol Sizes

ANSI/AWS A2.4-93 provides inch and metric values in millimeters for use when drawing weld and welding symbols, as shown in Figure 21.15. Use these recommended sizes when creating your manual drawings or for designing a welding CADD symbol library. Welding symbol templates are also available for manual drafting practices.

TYPES OF WELDS

The following information relates types of welds to their correlated weld symbol. The weld symbol is normally the next item you place on the welding symbol. The type of weld is as-

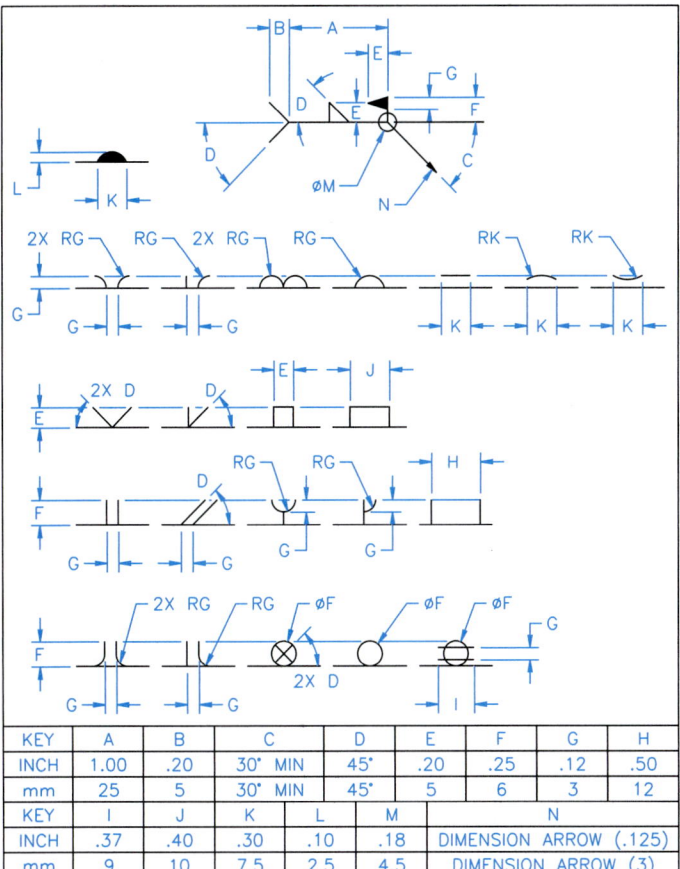

KEY	A	B	C	D	E	F	G	H
INCH	1.00	.20	30° MIN	45°	.20	.25	.12	.50
mm	25	5	30° MIN	45°	5	6	3	12

KEY	I	J	K	L	M	N		
INCH	.37	.40	.30	.10	.18	DIMENSION ARROW (.125)		
mm	9	10	7.5	2.5	4.5	DIMENSION ARROW (3)		

FIGURE 21.15 ■ Specification for drafting welding symbols. The horizontal line segment upon which each of the weld or supplementary symbols rest is the related portion of the reference line. The letters on the drawing match specific inch and metric values in the chart.

sociated with the weld symbol, weld shape, and/or the type of groove to which the weld is applied. Figures 21.13, 21.14, and 21.15 show the information that is associated with the types of welds.

Fillet Weld

A fillet weld is formed in the internal corner of the angle formed by two pieces of metal. The size of the fillet weld is shown on the same side of the reference line as the weld symbol and to the left of the symbol. When both legs of the fillet weld are the same, the size is given once, as shown in Figure 21.16. When the leg lengths are different in size, the vertical dimension is followed by the horizontal dimension. (See Figure 21.17.)

Square Groove Weld

A square groove weld is applied to a butt joint between two pieces of metal. The two pieces of metal are spaced apart a given

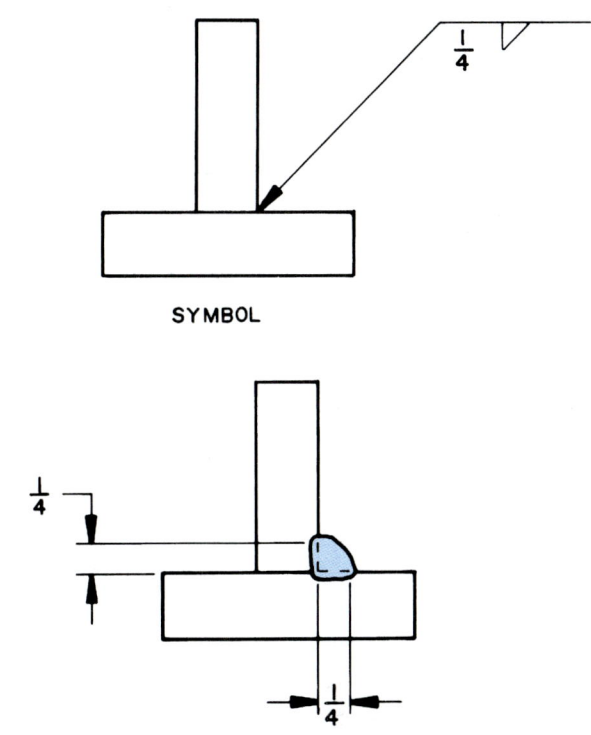

FIGURE 21.16 ■ Fillet weld with equal legs. Dimension values in this figure are in inches.

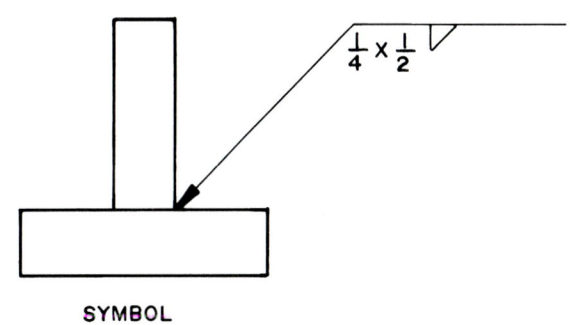

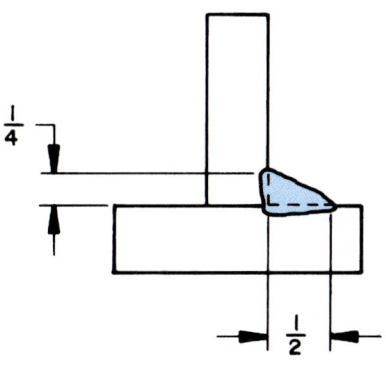

FIGURE 21.17 ■ Fillet weld with different leg lengths. Dimension values in this figure are in inches.

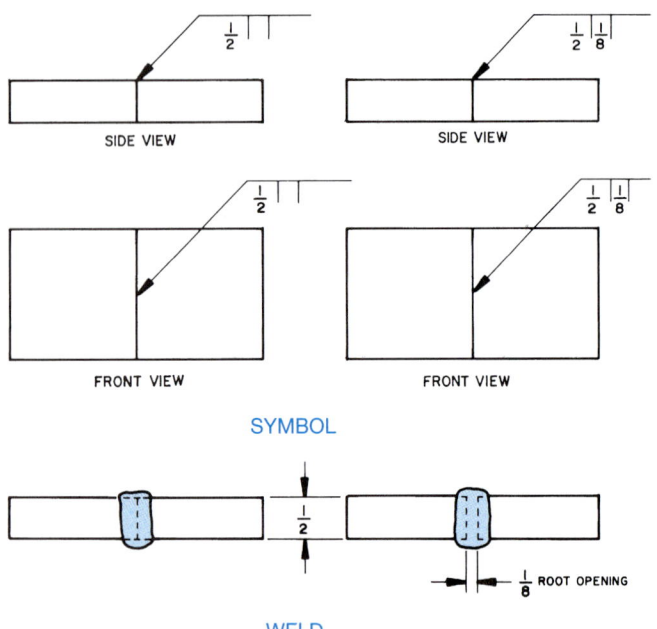

FIGURE 21.18 ■ Square groove weld showing the root opening. Dimension values in this figure are in inches.

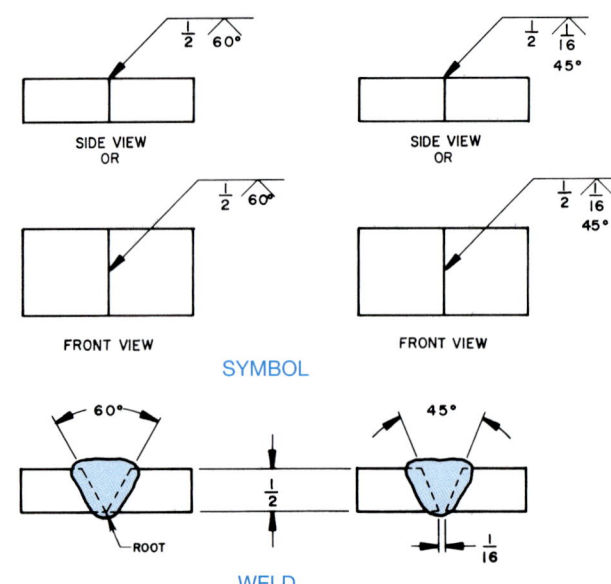

FIGURE 21.19 ■ V groove weld. Dimension values in this figure are in inches.

distance, known as the **root opening**. If the root opening distance is a standard in the company, then this dimension is assumed. If the root opening is not standard, then the specified dimension is given inside of the square groove symbol, as shown in Figure 21.18.

V *Groove Weld*

A **V groove weld** is formed between two adjacent parts when the side of each part is beveled to form a groove between the parts in the shape of a **V**. The included angle of the **V** can be given with or without a root opening, as shown in Figure 21.19.

Bevel *Groove Weld*

The **bevel groove weld** is created when one piece is square and the other piece has a beveled surface. The bevel weld can be given with a bevel angle and a root opening, as shown in Figure 21.20.

U *Groove Weld*

A **U groove weld** is created when the groove between two parts is in the form of a **U**. The angle formed by the sides of the **U** shape, the root, and the weld size are generally given. (See Figure 21.21.)

J *Groove Weld*

The **J groove weld** is necessary when one piece is a square cut and the other piece is in a **J**-shaped groove. The included an-

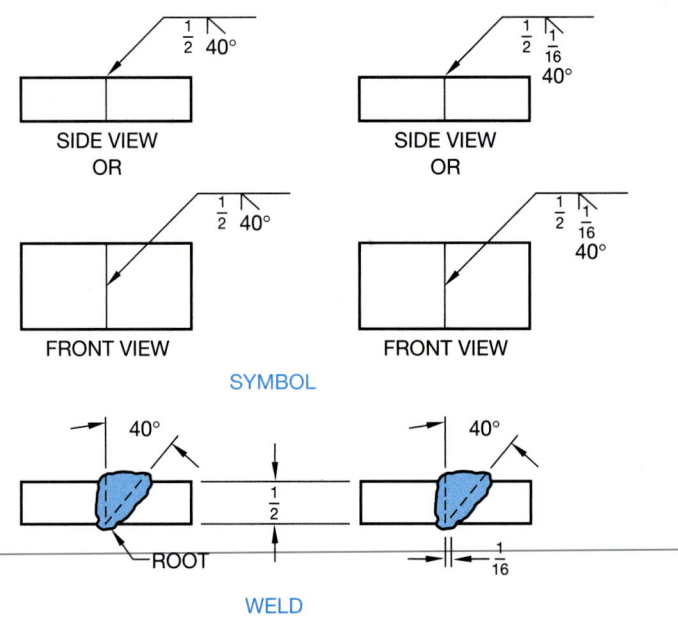

FIGURE 21.20 ■ Bevel groove weld. Dimension values in this figure are in inches.

gle, the root opening, and the weld size are given, as shown in Figure 21.22.

Plug *Weld*

A **plug weld** is made in a hole in one piece of metal that is lapped over another piece of metal. These welds are specified by giving the weld size, angle, depth, and pitch. (See Figure 21.23a.) The same type of weld can be applied to a slot. This is referred to as a **slot weld** as shown in Figure 21.23b.

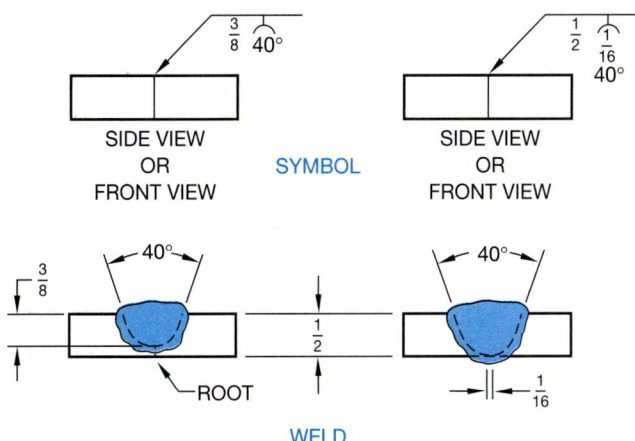

FIGURE 21.21 ■ U groove weld. Dimension values in this figure are in inches.

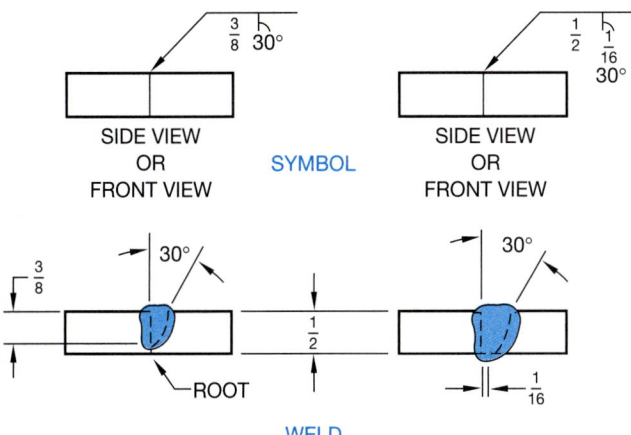

FIGURE 21.22 ■ J groove weld. Dimension values in this figure are in inches.

Flush Contour Weld

Generally the surface contour of a weld is raised above the surface face. If this is undesirable, a flush surface symbol must be applied to the weld symbol. When the **flush contour weld** symbol is applied without any further consideration, then the welder must perform this effect without any finishing. The other option is to specify a flush finish using another process. The letter designating the other process is placed above the flush contour symbol for another side application, or below the flush contour symbol for an arrow-side application. The options include: C = chipping, G = grinding, M = machining, R = rolling, or H = hammering. (See Figure 21.24.)

Spot Weld

Spot welding is a process of resistance welding where the base materials are clamped between two electrodes and a momentary electric current produces the heat for welding at the contact spot. Spot welding is generally associated with welding sheet metal lap seams. The size of the spot weld is given as a diameter to the left of the symbol. The center-to-center pitch is given to the right of the symbol. (See Figure 21.25.) The strength of the spot welds can be given as minimum shear strength in pounds per spot to the left of the symbol, as shown in Figure 21.26a. When a specific number of spot welds is required in a seam or joint, then the quantity is placed above or below the symbol in parentheses, as shown in Figure 21.26b.

Seam Weld

A **seam weld**, another type of resistance weld, is a continuous weld made between or upon overlapping members. The continuous weld can consist of a single weld bead or a series of overlapping spot welds. The dimensions of seam welds, shown on the same side of the reference line as the weld symbol, relate to either size or strength. The weld size width is shown to the left of the symbol in fractional or decimal inches, or millimeters. (See Figure 21.27a.) The weld length, when specified, is provided to the right of the symbol, as shown in Figure 21.27b. The strength of a seam weld is expressed as minimum acceptable shear strength in pounds per linear inch, placed to the left of the weld symbol, as shown in Figure 21.28, page 684.

Flange Welds

Flange welds are used on light-gauge metal joints where the edges to be joined are flanged or flared. Dimensions of flange welds are placed to the left of the weld symbol. Further, the radius and height of the weld above the point of tangency are indicated by showing both the radius and the height separated by a plus (+) symbol. The size of the flange weld is then placed outward of the flange dimensions. (See Figure 21.29, page 684.)

WELD SYMBOL LEADER ARROW RELATED TO WELD LOCATION

Welding symbols are applied to the joint as the basic reference. All joints have an **arrow side** and an **other side**. When fillet and groove welds are used, the welding symbol leader arrows connect the symbol reference line to one side of the joint known as the arrow side. The side opposite the location of the arrow is called the other side. If the weld is to be deposited on the arrow side of the joint, the proper weld symbol is placed *below* the reference line, as shown in Figure 21.30a, page 684. If the weld is to be deposited on the side of the joint opposite the arrow, then the weld symbol is placed *above* the reference line. (See Figure 21.30b, page 684.) When welds are to be deposited on both sides of the joint, then the same weld symbol is shown *above and below* the reference line, as shown in Figure 21.30c and d, page 684.

For plug, spot, seam, or resistance welding symbols, the leader arrow connects the welding symbol reference line to the outer surface of one of the members of the joint at the center line of the desired weld. The member that the arrow points to is

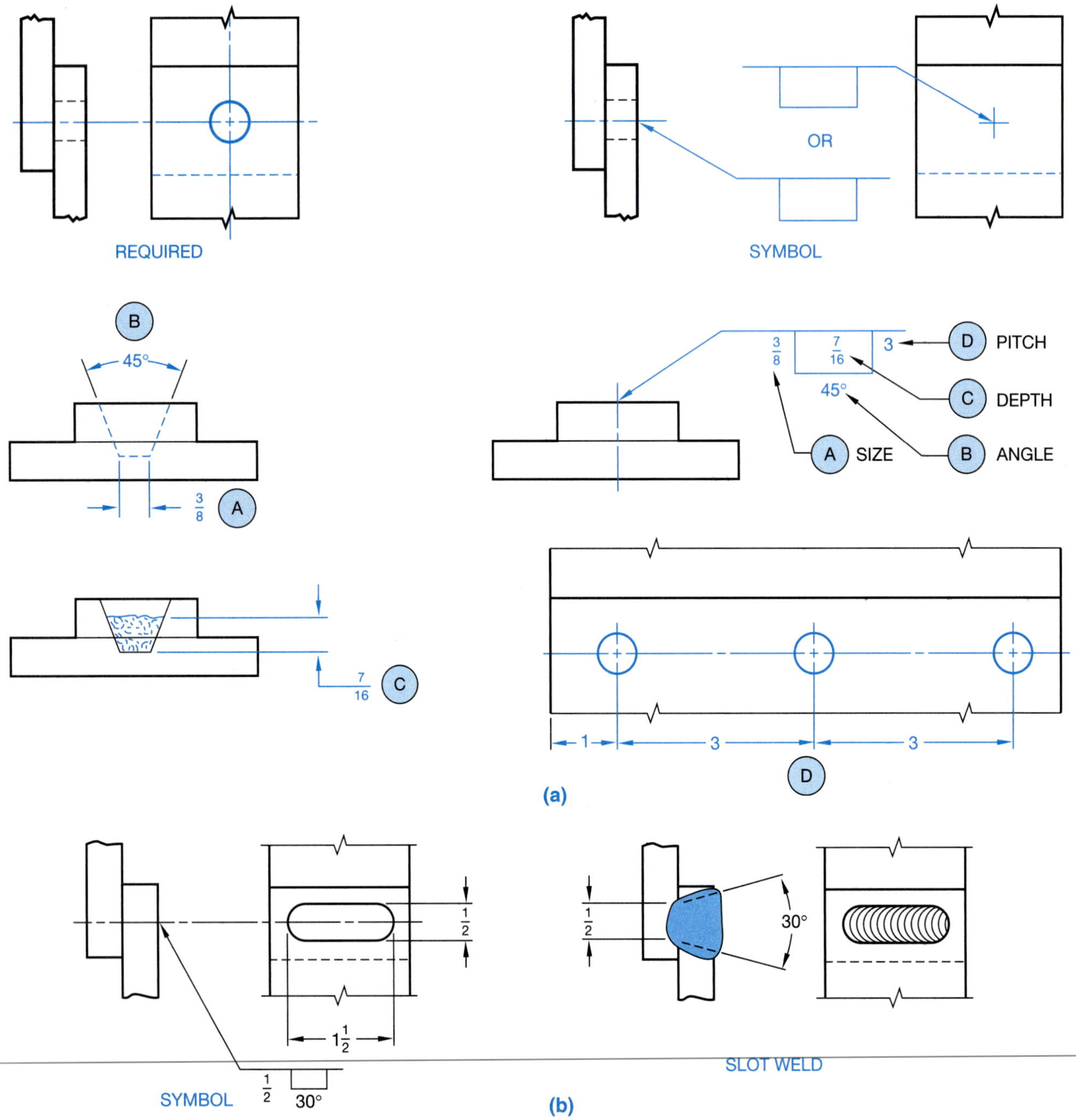

REQUIRED

SYMBOL

OR

B

45°

$\frac{3}{8}$ A

$\frac{7}{16}$ C

$\frac{3}{8}$ $\frac{7}{16}$ 3   D PITCH

45°   C DEPTH

A SIZE   B ANGLE

1   3   3

D

(a)

$\frac{1}{2}$

$1\frac{1}{2}$

$\frac{1}{2}$ 30°

$\frac{1}{2}$ 30°

SYMBOL

SLOT WELD

(b)

FIGURE 21.23 ■ (a) Plug weld. (b) Slot weld. Dimension values in this figure are in inches.

considered the arrow side member. The member opposite of the arrow is considered the other side member.

Metric Reference Line Application Related to Weld Location

Metric welding symbol practices such as those used in the British Standard include a dashed line next to the solid reference line for certain applications. Weld symbols drawn on the solid reference line relate to welds on the arrow side of the item being welded.

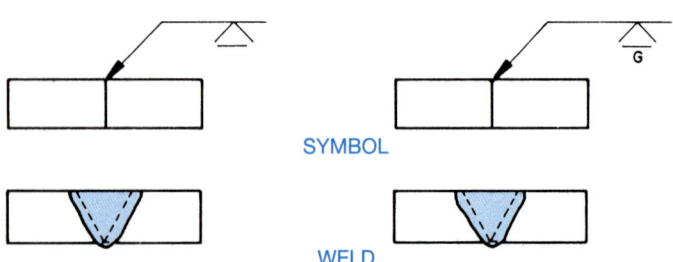

SYMBOL

G

WELD

FIGURE 21.24 ■ Flush contour weld. Dimension values in this figure are in inches.

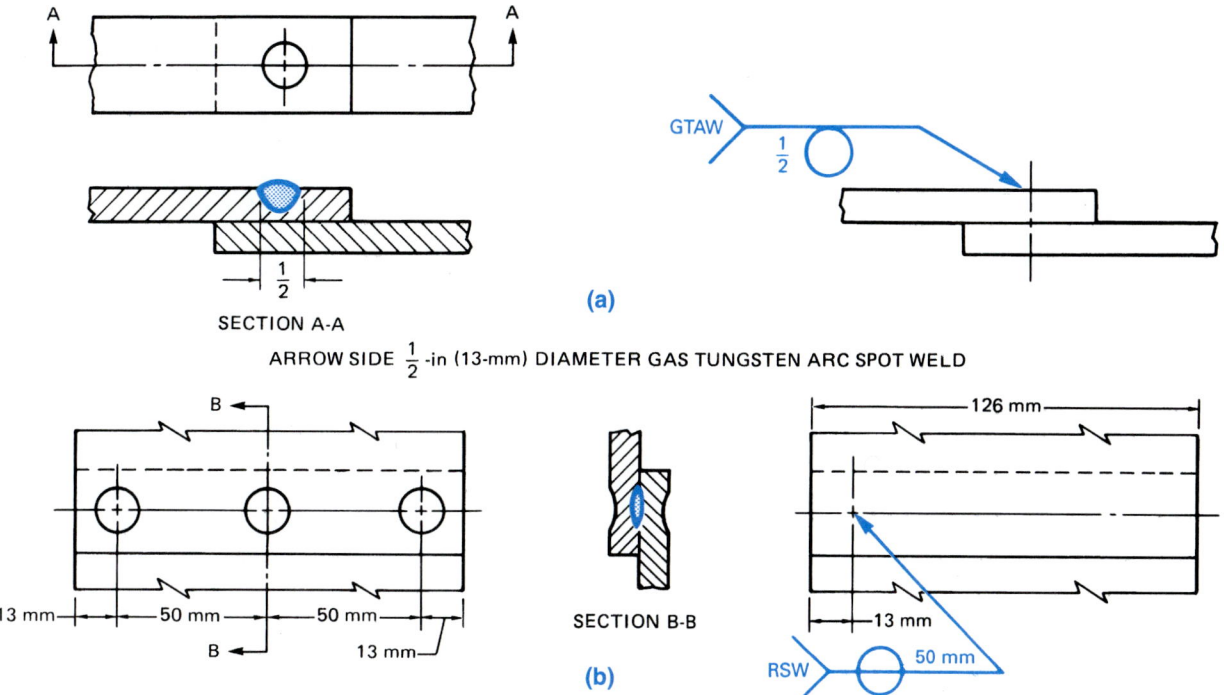

SECTION A-A

ARROW SIDE $\frac{1}{2}$-in (13-mm) DIAMETER GAS TUNGSTEN ARC SPOT WELD

(a)

SECTION B-B

(b)

50-mm PITCH ON A RESISTANCE SPOT WELD

FIGURE 21.25 ■ Designating spot welds.

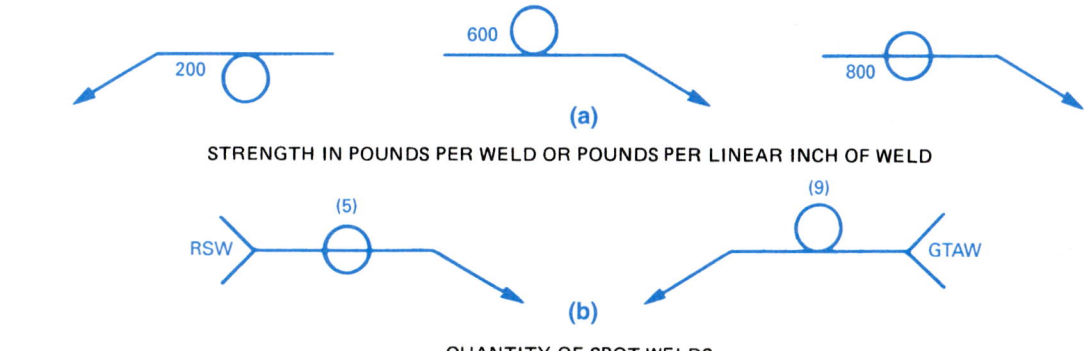

STRENGTH IN POUNDS PER WELD OR POUNDS PER LINEAR INCH OF WELD

(a)

QUANTITY OF SPOT WELDS

(b)

FIGURE 21.26 ■ Designating strength and number of spot welds.

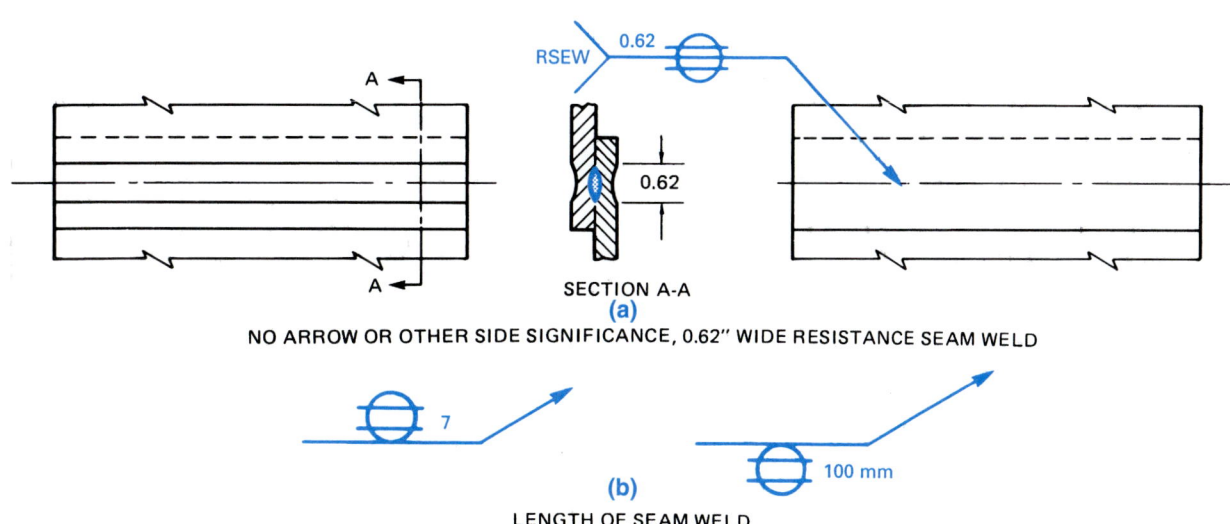

SECTION A-A

(a)

NO ARROW OR OTHER SIDE SIGNIFICANCE, 0.62″ WIDE RESISTANCE SEAM WELD

(b)

LENGTH OF SEAM WELD

FIGURE 21.27 ■ Indicating the length of a seam weld.

A
A
SECTION A-A
200
EBW

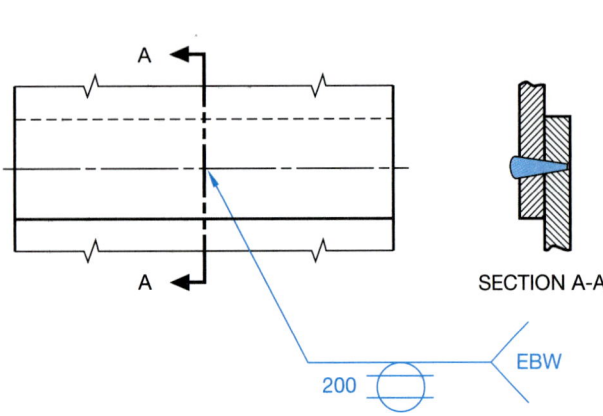

FIGURE 21.28 ■ Indicating the strength of a seam weld.

3/32
1/8
1/16
DESIRED WELD
1/16 + 1/8
3/32
SYMBOL

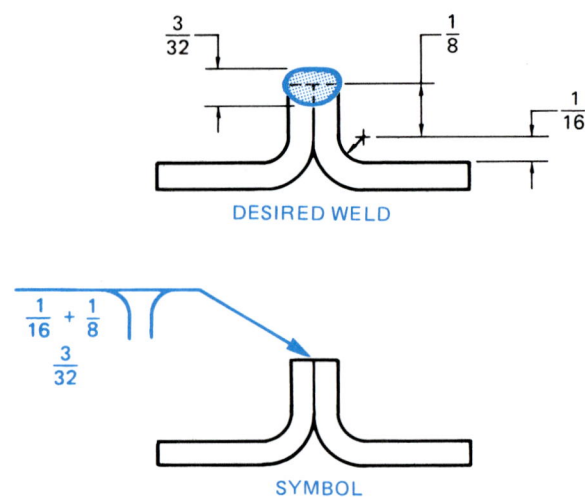

FIGURE 21.29 ■ Applying dimensions to flange welds. Dimension values in this figure are in inches. *Courtesy American Welding Society.*

ARROW SIDE
OTHER SIDE
SYMBOL
(a)
WELD
(b)
BOTH SIDES
BOTH SIDES
FOR TWO JOINTS
SYMBOL
(c)
WELD
(d)

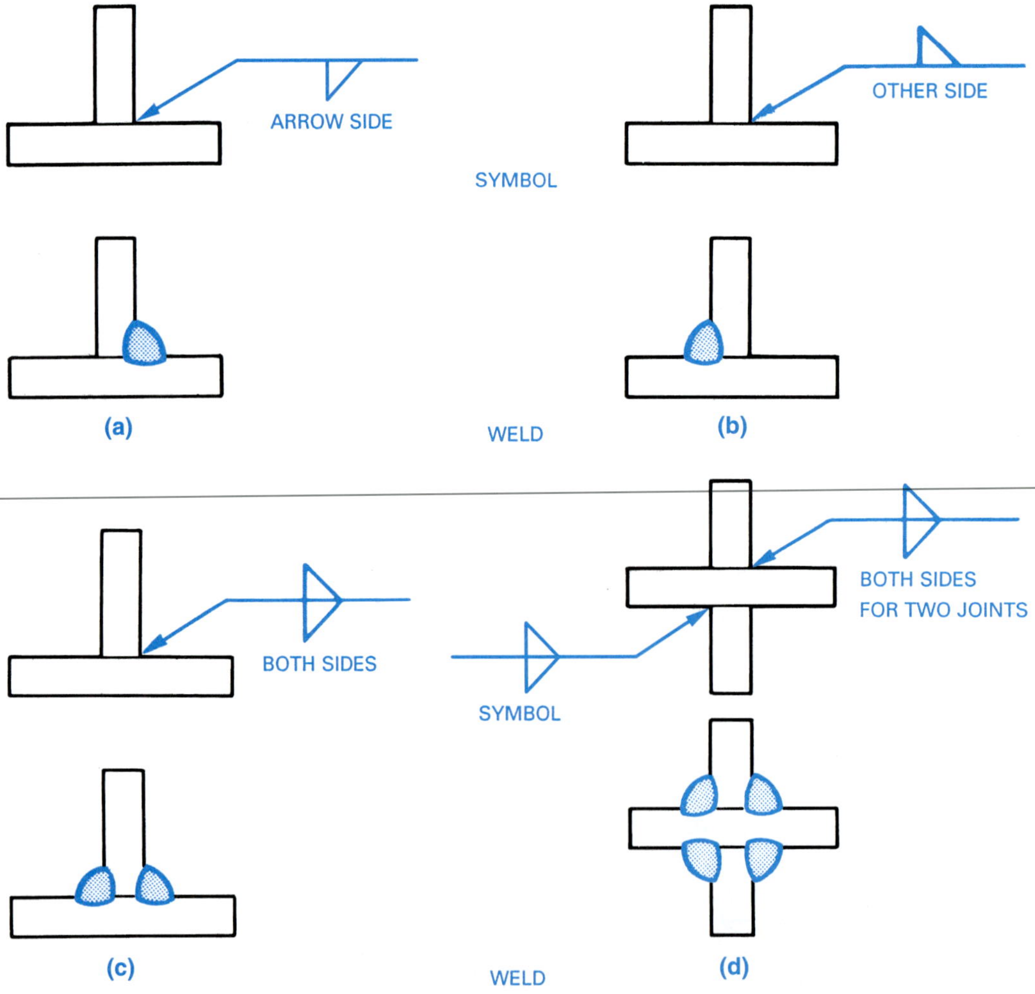

FIGURE 21.30 ■ Designing weld locations. *Courtesy American Welding Society.*

Use caution when comparing this method to the AWS standard, because there are differences that can cause confusion. Weld symbols on the dashed line relate to a weld on the other side of the item. If the dashed line is above the solid line, then the symbol for the arrow side weld is drawn below the solid reference line and the symbol for the far side weld is above the dashed line. Figure 21.31 shows this application. The dashed line is omitted if the welds are symmetrical on both sides of the item being welded.

Drawing a Break in the Welding Symbol Leader

When two parts meet and a bevel, J-groove, or flange is used, you can direct the groove to be placed on only one of the pieces. To do this, draw a jog or bend in the leader line and point the leader to the part where you want the groove placed. The other part has no groove, as shown in Figure 21.32.

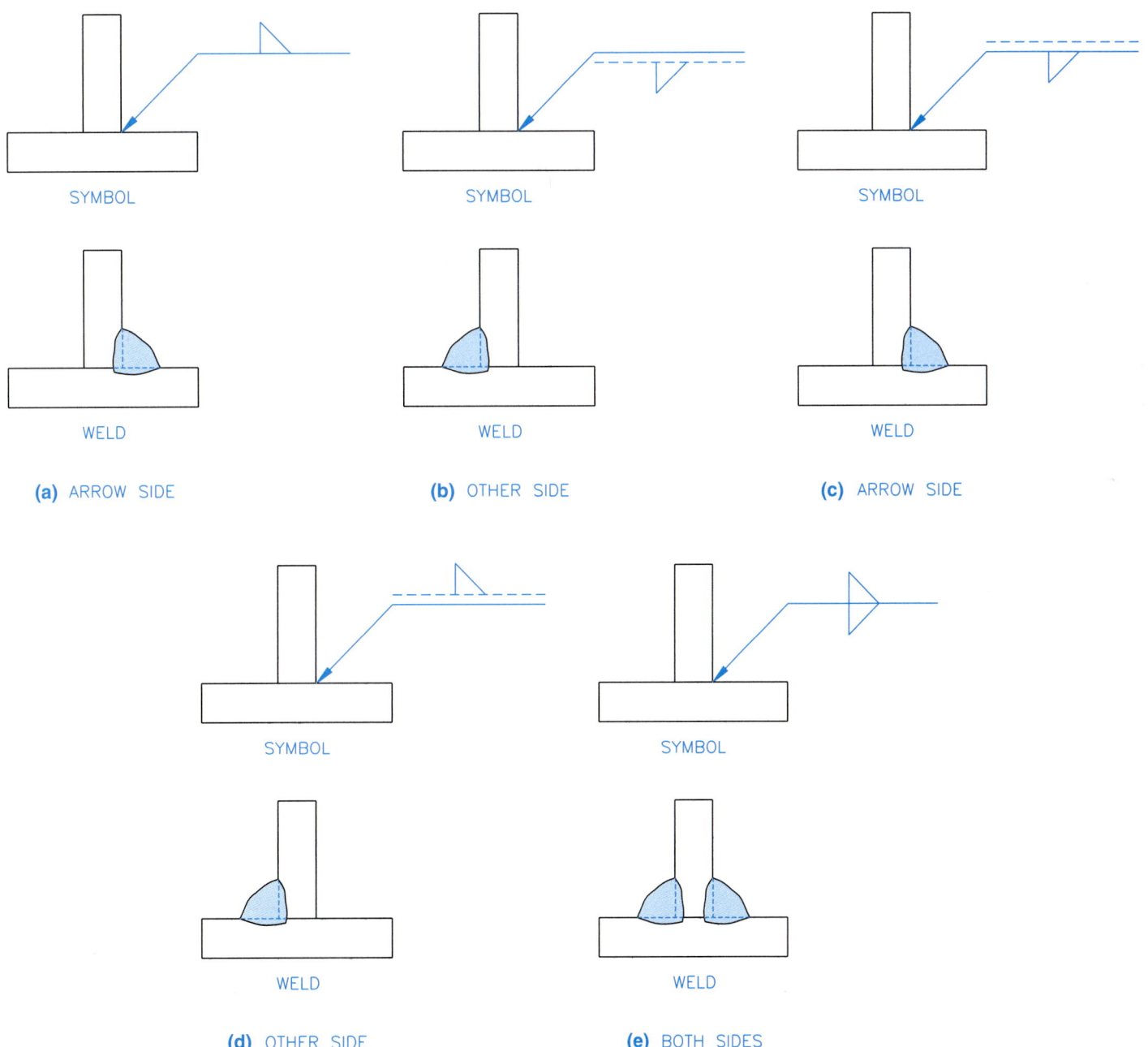

FIGURE 21.31 ■ Metric welding symbol practices such as those used in the British Standard include a dashed line next to the solid reference line. (a) A weld symbol drawn on the solid reference line relates to a weld on the arrow side of the item being welded. (b) A weld symbol on the dashed line relates to a weld symbol on the far side of the item. (c) If the dashed line is above the solid line, then the symbol for the near side weld is drawn below the solid reference line. (d) If the dashed line is above the solid line, then the symbol for the far side weld is above the dashed line. (e) The dashed line is omitted if the welds are symmetrical on both sides of the item being welded.

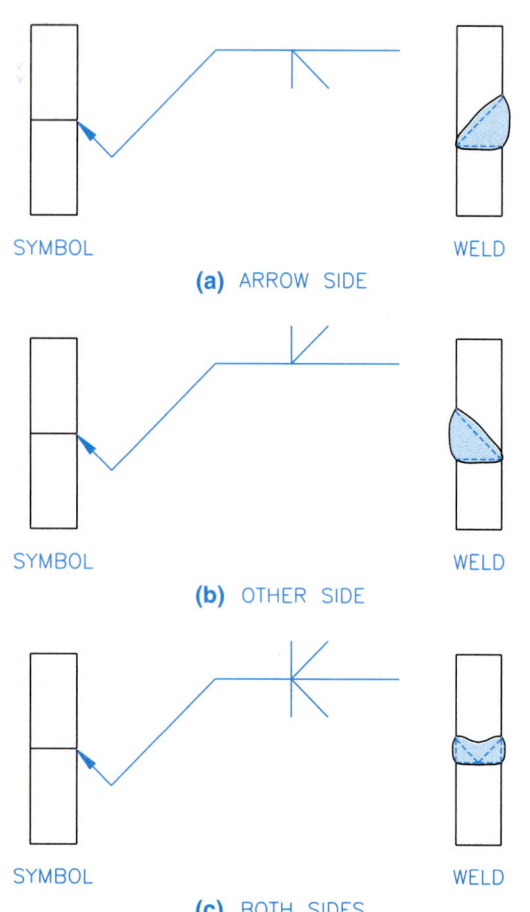

SYMBOL   WELD

(a) ARROW SIDE

SYMBOL   WELD

(b) OTHER SIDE

SYMBOL   WELD

(c) BOTH SIDES

FIGURE 21.32 ■ Using a jog or bend in the leader line to designate the part where you want the groove to be placed.

ADDITIONAL WELD CHARACTERISTICS

Additional weld characteristics can be added to the welding symbol, including field weld, weld penetration, melt through, weld all around, and weld length.

Field Weld

A field weld is a weld that is performed in the field as opposed to in a fabrication shop. The reason for this application may be that the individual components may be easier to transport disassembled or that the mounting procedure may require job site installation. The field weld symbol is a flag attached to the reference line at the leader intersection, as shown in Figures 21.12, 21.14, and 21.15. The field weld symbol can be above or below the reference line, and the flag always points in the direction of the reference line.

Weld Penetration

Unless otherwise specified, a weld penetrates through the thickness of the parts at the joint. The size of the groove weld remains to the left of the weld symbol. Figure 21.33a shows the size of grooved welds with partial penetration. Notice in Figure 21.33b that a weld with partial penetration can specify the depth of the groove followed by the depth of weld penetration in parentheses, with both items placed to the left of the weld symbol.

When a single-groove and symmetrical double-groove welds penetrate completely through the parts being joined, the weld size can be omitted, as shown in Figure 21.34. The depth of penetration of flare-formed groove welds is assumed to extend to the tangent points of the members, as shown in Figure 21.35.

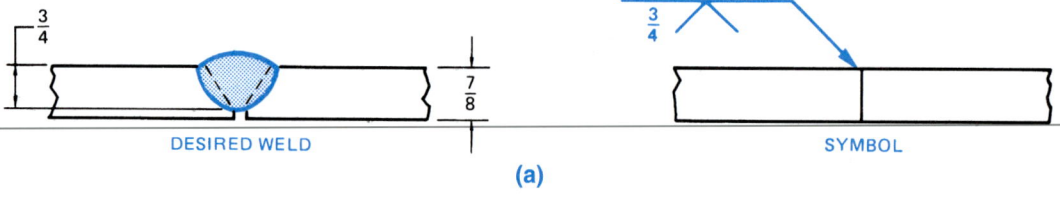

DESIRED WELD   SYMBOL

(a)

DESIGNATING THE SIZE OF GROOVED WELDS WITH PARTIAL PENETRATION

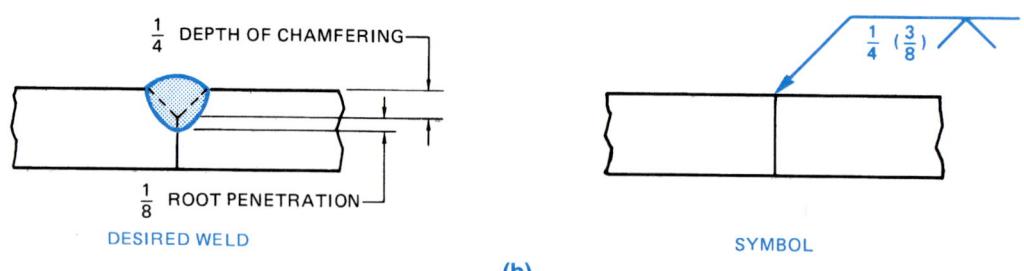

$\frac{1}{4}$ DEPTH OF CHAMFERING

$\frac{1}{8}$ ROOT PENETRATION

DESIRED WELD   SYMBOL

(b)

SHOWING SIZE AND ROOT PENETRATION OF GROOVED WELDS

FIGURE 21.33 ■ (a) Designating the size of grooved welds with partial penetration. (b) Showing size and penetration of grooved welds. Dimension values in this figure are in inches. *Courtesy American Welding Society.*

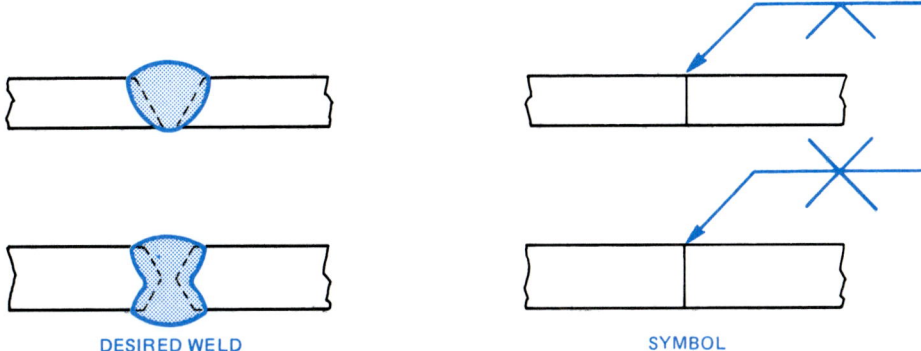

FIGURE 21.34 ■ Designating single- and double-groove welds with complete penetration. *Courtesy American Welding Society.*

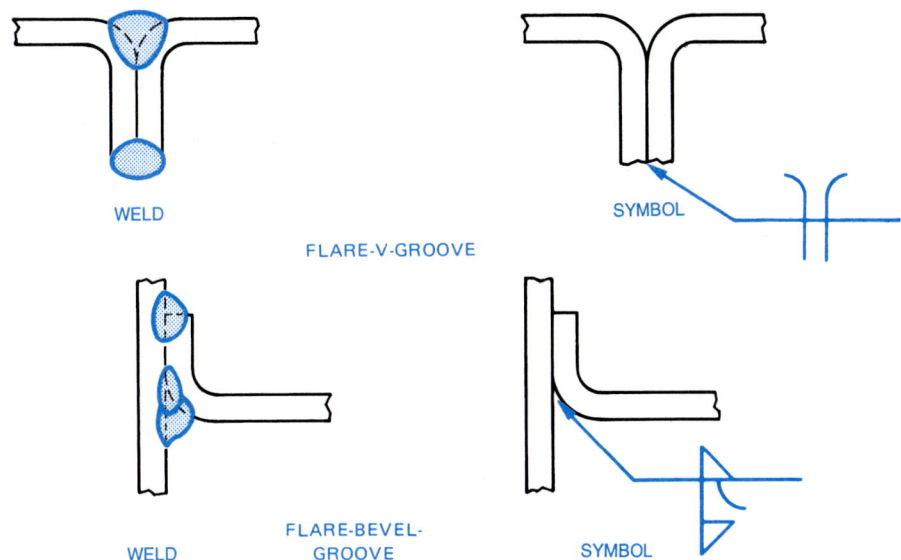

FIGURE 21.35 ■ Designating flare-V and flare-bevel-groove welds. *Courtesy American Welding Society.*

Melt Through

Melt through is a term that refers to the weld melting through the bottom of the weld or the opposite side of where the weld is being applied. When improperly done or when not specified, melt through can be unacceptable. However, melt through is desired when the symbol shown in Figure 21.36 is used. The melt through symbol is placed on the side of the reference line opposite the weld symbol. Melt through is used only when complete joint penetration and visible root reinforcement is required in welds made from one side. The desired melt through distance can be placed to the left of the melt through symbol on the reference line. (See Figure 21.36.)

Weld-All-Around

When a welded connection must be performed all around a feature, the weld-all-around symbol is attached to the reference line at the junction of the leader. (See Figure 21.37.)

Weld Length and Pitch

When a weld is not continuous along the length of a part, then the weld length should be given. In some situations, the weld along the length of a feature is given in lengths spaced a given distance apart. The distance from one point on a weld length to the same corresponding point on the next weld is called the pitch; generally from center to center of welds. The weld length and pitch are shown to the right of the weld symbol, as shown in Figure 21.38. Unless the welding symbol specifies a weld length and pitch, a weld is always continuous.

Welding Process Designation

The tail is added to the welding symbol when it is necessary to designate the welding specification, procedures, or other supplementary information needed to fabricate the weld. (See Figure 21.39.)

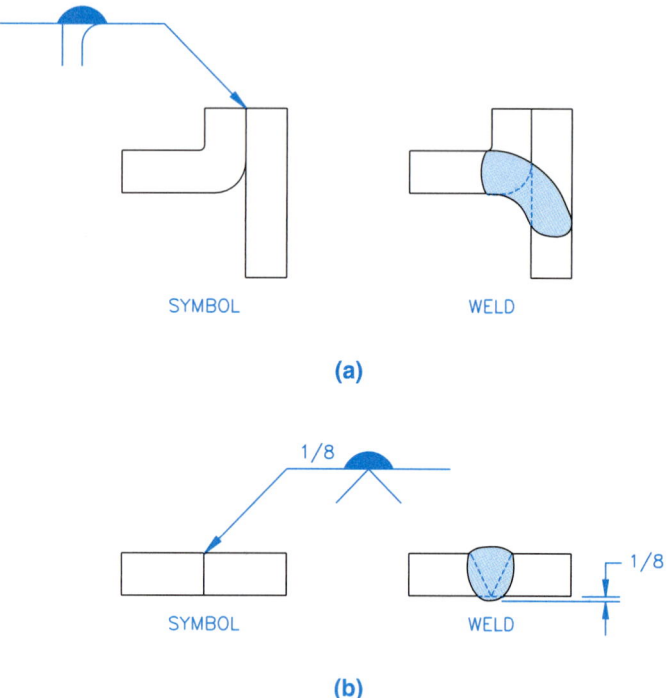

FIGURE 21.36 ■ Application of the melt-through symbol and melt-through penetration dimension represented as 1/8 in. in the lower example.

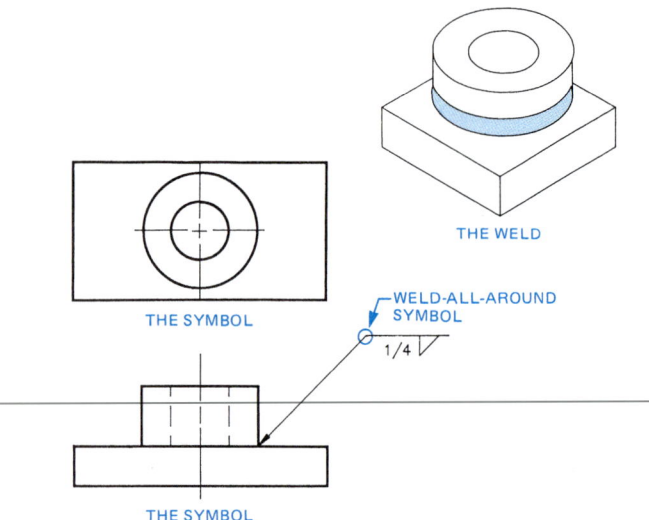

FIGURE 21.37 ■ Weld-all-around. Dimension values in this figure are in inches.

A **weld tolerance** can also be applied to a welding symbol tail, as shown in Figure 21.40, by placing the tolerance value and a note giving a reference to the weld feature where the tolerance is applied.

Weld Joints

The types of weld joints are often closely associated with the types of weld grooves already discussed. The weld grooves can

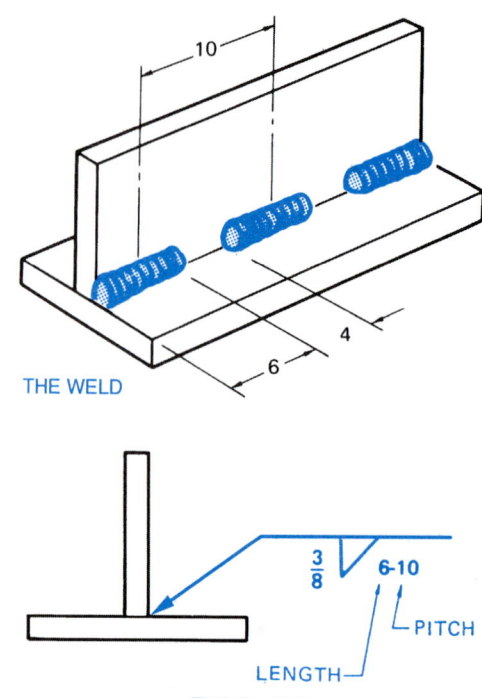

FIGURE 21.38 ■ Intermittent fillet weld. Dimension values in this figure are in inches.

be applied to any of the typical joint types. The weld joints used in most weldments are the butt, lap, tee, outside corner, and edge joints shown in Figure 21.41.

WELDING TESTS

There are two types of welding tests—destructive and nondestructive tests.

Destructive Tests (DT)

Destructive tests (DT) use the application of a specific force on the weld until the weld fails. These types of tests can include the analysis of tensile, compression, bending, torsion, or shear strength. Figure 21.42 shows the relationship of forces that can be applied to a weld. The **continuity** of a weld is when the desired characteristics of the weld exist throughout the weld length. **Discontinuity** or lack of continuity exists when a change in the shape or structure exists. The types of problems that alter the desired weld characteristic can include cracks, bumps, seams or laps, or changes in density. The intent of destructive testing is to determine how much of a discontinuity can exist in a weld before the weld is considered to be flawed. Parts can be periodically selected for destructive testing. The tested weld is unfit for any further use.

Nondestructive Tests (NDS)

Nondestructive tests (NDS) are tests for potential defects in welds that are performed without destroying or damaging the

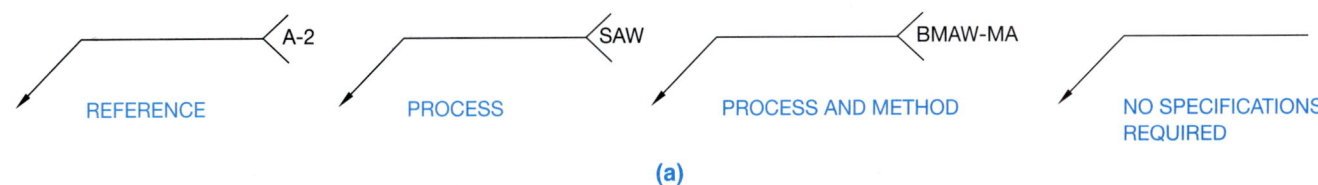

REFERENCE | PROCESS | PROCESS AND METHOD | NO SPECIFICATIONS REQUIRED

(a)

	Welding Process	Letter Designation
Brazing	Torch brazing	TB
	Induction brazing	IB
	Resistance brazing	RB
Flow Welding	Flow welding	FLOW
Induction Welding	Induction welding	IW
Arc Welding	Bare metal arc welding	BMAW
	Submerged arc welding	SAW
	Shielded metal arc welding	SMAW
	Carbon arc welding	CAW
	Oxyhydrogen welding	OHW
Gas Welding	Oxyacetylene welding	OAW

The following suffixes may be added if desired to indicate the method of applying the above processes:

Automatic welding	-AU
Machine welding	-ME
Manual welding	-MA
Semiautomatic welding	-SA

(b)

FIGURE 21.39 ■ (a) Location for weld specifications, process, and other references on weld symbols. (b) Designation of welding processes.

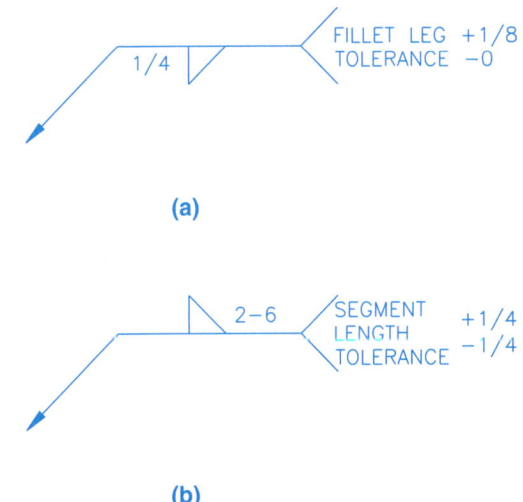

FILLET LEG +1/8
TOLERANCE −0

(a)

SEGMENT LENGTH TOLERANCE +1/4 −1/4

(b)

FIGURE 21.40 ■ A weld tolerance applied to the welding symbol tail. Dimension values in this figure are in inches.

weld or the part. The types of nondestructive tests and the corresponding symbol for each test are shown in Figure 21.43. The testing symbols are used in conjunction with the weld symbol to identify the area to be tested and the type of test to be used. (See Figure 21.44.)

The location of the testing symbol above, below, or placed in a break on the reference line has the same reference to the weld joint as the weld symbol application. Test symbols below the reference line mean arrow side tests. Symbols above the reference line are for other side tests, and a test symbol placed in a break on the line indicates no preference of side to be tested. Test symbols placed on both sides of the reference line require the weld to be tested on both sides of the joint. (See Figure 21.45.)

Two or more different tests can be required on the same section or length of weld. Methods of combining welding test symbols to indicate more than one test procedure are shown in Figure 21.46. The length of the weld to be tested can be shown to the right of the test symbol, or can be provided as a dimension line giving the extent of the test length, as shown in Figure 21.47. The number of tests to be made can be identified in parentheses below the test symbol for arrow side tests or above the symbol for other side tests, as shown in Figure 21.48. The welding symbols and nondestructive testing symbols can be combined as shown in Figure 21.49. The combination symbol is appropriate to help the welder and inspector identify welds requiring special attention. When a radiograph test is needed, a special symbol and the angle of radiation can be specified, as shown in Figure 21.50, page 692.

WELDING SPECIFICATIONS

A welding specification is a detailed statement of the legal requirements for a specific classification or type of product. Products manufactured to code or specification requirements commonly must be inspected and tested to ensure compliance.

There are a number of agencies and organizations that publish welding codes and specifications. The application of a particular

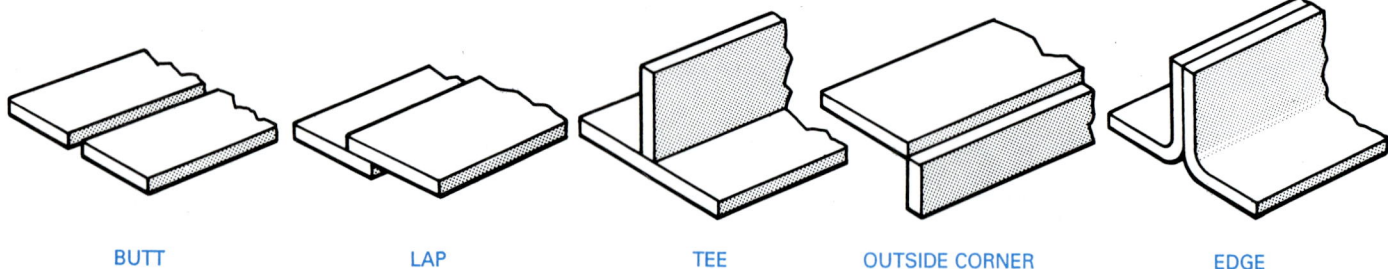

BUTT   LAP   TEE   OUTSIDE CORNER   EDGE

FIGURE 21.41 ■ Types of joints.

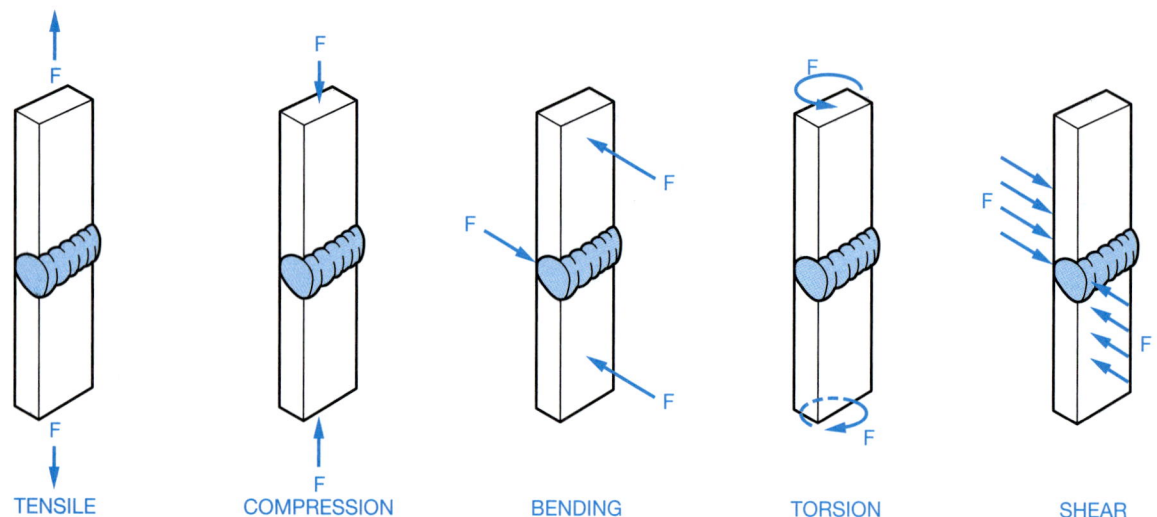

TENSILE   COMPRESSION   BENDING   TORSION   SHEAR

FIGURE 21.42 ■ Forces on a weld.

Type of Nondestructive Test	Symbol
Visual	VT
Penetrant	PT
Dye penetrant	DPT
Fluorescent penetrant	FPT
Magnetic particle	MT
Eddy current	ET
Ultrasonic	UT
Acoustic emission	AET
Leak	LT
Proof	PRT
Radiographic	RT
Neutron radiographic	NRT

FIGURE 21.43 ■ Standard nondestructive testing symbols.

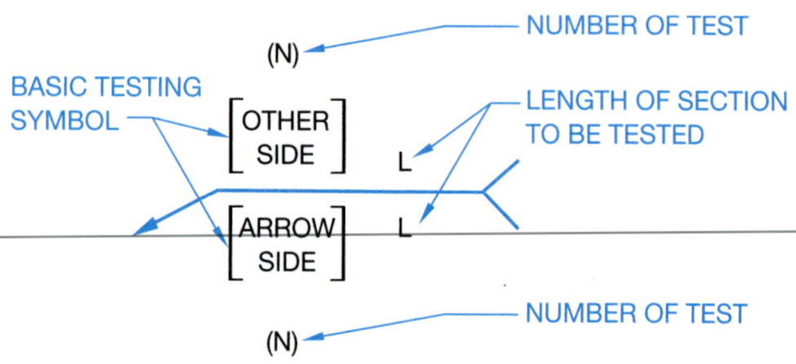

NUMBER OF TEST

BASIC TESTING SYMBOL

OTHER SIDE

LENGTH OF SECTION TO BE TESTED

L

ARROW SIDE   L

(N)

NUMBER OF TEST

FIGURE 21.44 ■ Basic nondestructive testing symbol.

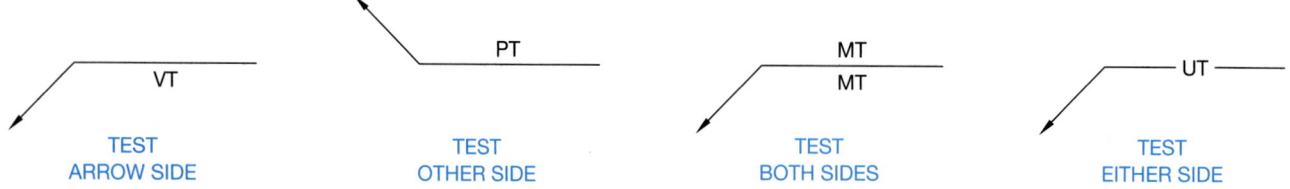

VT   PT   MT / MT   UT

TEST ARROW SIDE   TEST OTHER SIDE   TEST BOTH SIDES   TEST EITHER SIDE

FIGURE 21.45 ■ Testing symbols used to indicate the side to be tested.

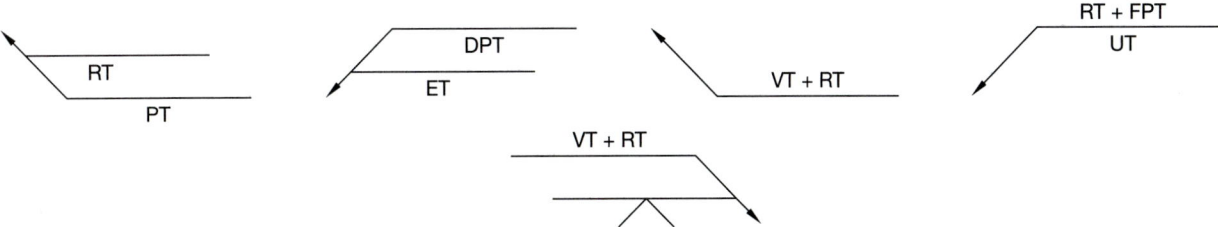

FIGURE 21.46 ■ Methods of combining testing symbols.

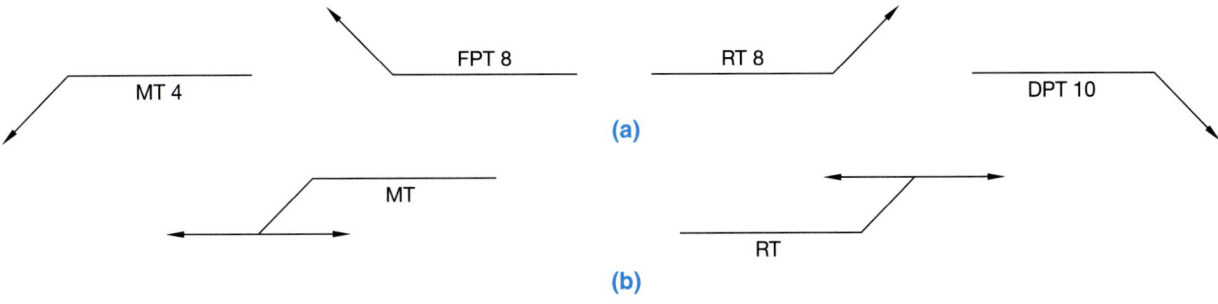

FIGURE 21.47 ■ Two methods of designating the length of weld to be tested.

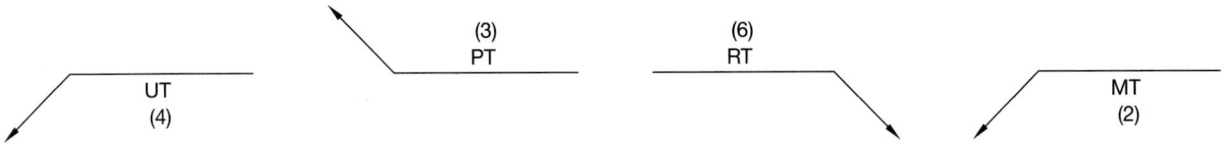

FIGURE 21.48 ■ Method of specifying number of tests to be made.

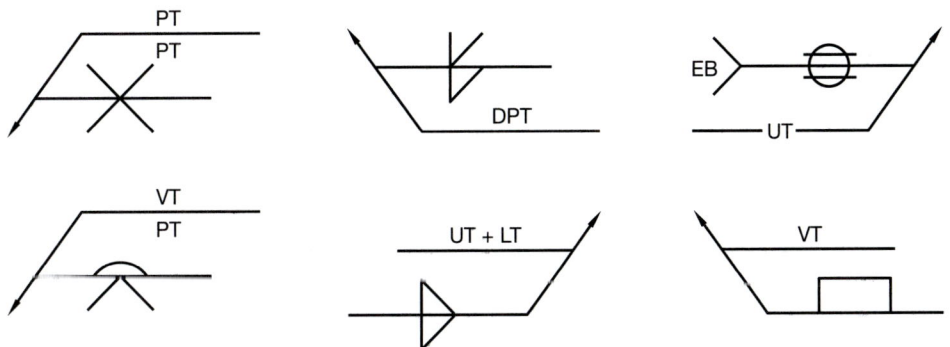

FIGURE 21.49 ■ Combination welding and nondestructive testing symbols.

code or specification to a weldment can be the result of one or more of the following requirements:

- Local, state, or federal government regulations.
- Bonding or insurance company requirements.
- Customer requirements.
- Standard industrial practice.

Commonly used codes include:

- No. 1104, American Petroleum Institute (API). Used for pipeline specifications.

- Section IX, American Society of Mechanical Engineers (ASME). Used to specify welds for pressure vessels.
- D1.1, American Welding Society (AWS). Welding specifications for bridges and buildings.
- AASHT, American Association of State Highway and Transportation Officials.
- AIAA, Aerospace Industries Association of America.
- AISC, American Institute of Steel Construction.
- ANSI, American National Standards Institute.
- AREA, American Railway Engineering Association.

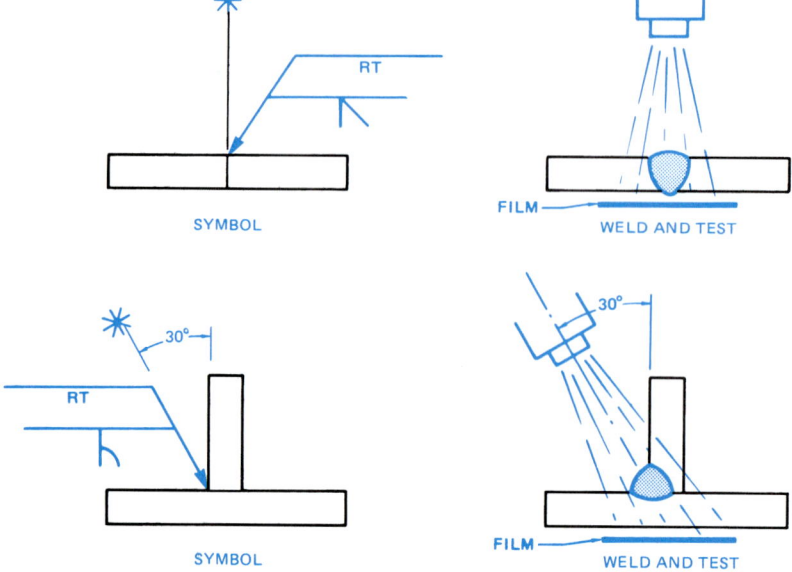

FIGURE 21.50 ■ Combination symbol for welding and radiation location for testing.

- AWWA, American Water Works Association.
- AAR, Association of American Railroads.
- MILSTD, Military Standards, Department of Defense.
- SAE, Society of Automotive Engineers.

Note in the problem assignments that the welding specifications and notes are provided under SPECIFICATIONS and in the general notes.

PREQUALIFIED WELDED JOINTS

AISC/AWS The American Institute of Steel Construction (AISC) and the Structural Welding Code of the American Welding Society (AWS) exempt from tests and qualification most of the common welded joints used in steel construction. These specific common welded joints are referred to as prequalified. Work on prequalified welded joints must be in accordance with the Structural Welding Code. Generally, fillet welds are considered prequalified if they conform to the requirements of the AISC and the AWS code.

WELD DESIGN

In most cases, the weld should be about as wide as the thickness of the metal to be welded. Undersized or oversized welds can result in joint failure. Undersized welds may not have enough area to hold the parts together under loading conditions. Oversized welds could result in a joint that is too stiff. The lack of flexibility in the oversized weld could cause the metal near the joint to become overstressed and result in failure adjacent to the weld

joint. Welds between materials of different thickness should provide for more weld next to the thickest piece. If this design precaution is not taken, the weld may result in good bonding to the thin material and poor bonding to the thick material because the welding heat is more concentrated at the thinner material. Another option would be to reduce the thickness of the thicker material, or to build up the thickness of the thinner material at the weld joint.

Welding design is generally an engineering decision. Weld design can be as basic as previously discussed, or complicated with the use of a variety of engineering formulas. Weld design can be different for each discipline, such as welding of machinery compared to the welding of steel structures. Engineers use a wide variety of different mathematical weld-stress formulas that are determined by factors involved in weld design based on the forces shown in Figure 21.42 and these characteristics:

- Welding process.
- Type of material.
- Weld joint.
- Type of weld, such as fillet or groove weld.
- Material thickness.
- Weld size.
- Bending and torsion properties of the material, weld joint, and weld type. Torsion refers to the twisting forces placed on features.

When weldments are used in the design, you must be able to recognize what the weld means and establish the best location on the drawing for the weld symbol. As entry-level drafters become familiar with company products and the welds used in them, under certain conditions they may be given more design-related tasks.

CADD APPLICATIONS 2-D

WELDING SYMBOLS

Weld information is quickly and accurately added to drawings using CADD. Welding symbols contain shapes such as lines, circles, and arcs, and text is drawn with basic CADD tools and usually created using the dimension format or style, and layer. Specific commands including COPY, MOVE, OFFSET, MIRROR, and TRIM also greatly aid in the process of adding any required welding symbol. Though welding symbols can be drawn using basic drafting commands such as LINE and TEXT, there are specific techniques that quickly and efficiently allow you to produce welding symbols complete with a leader. One example involves using a combination of AutoCAD's BLOCK and LEADER, or QLEADER, commands, for example.

The BLOCK tool allows you to produce a weld symbol that can be stored in the current drawing file or a symbols library. One of the most powerful features of CADD software is the ability to reuse drawing content. This is especially true when designing and drafting welded products that contain multiple welding symbols. Often welding symbols are standard annotations that are used multiple times for several different projects. As a result, creating or purchasing a library of welding symbols can prove very beneficial. Reusing existing welding symbols and using welding symbol libraries can save a significant amount of time and allow designers and drafters to focus on the design and documentation of a product instead of the task of creating individual welding symbols using basic drawing tools. An example of a welding symbol library is shown in Figure 21.51. This particular library contains several commonly used welding symbol blocks that can be easily added to a drawing.

The LEADER command allows you to use predefined dimension style settings to place a leader that references a particular welding symbol block. Once a welding symbol has been created and stored in the form of a block, you can use the block in place of text when you create a leader. This process typically involves picking points to define the length and location of the leader line followed by selecting the desired welding symbol block. Once the base symbol is inserted, you have the option of creating a unique welding symbol by editing the size of the weld bead information. Figure 21.52 shows how a leader may be used in CADD to quickly create a welding symbol by referencing a block.

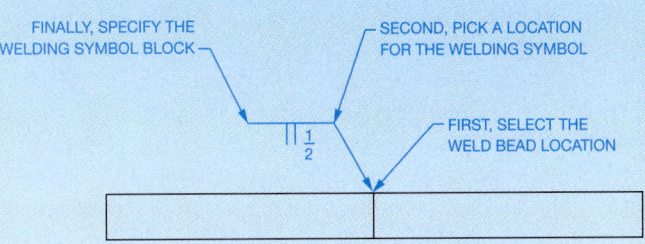

FIGURE 21.52 ■ Creating a welding symbol using a leader and a block. Dimension values in this figure are in inches.

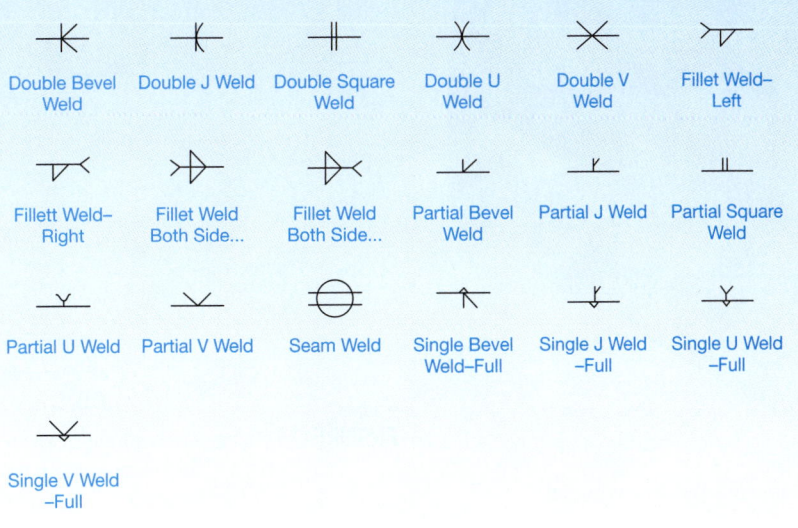

FIGURE 21.51 ■ An example of a welding symbol library.

CADD WELDING MODELS AND DRAWINGS

Some 3-D CADD programs incorporate very realistic welding processes into the modeling environment. One of the primary functions of 3-D CADD software is to model actual manufacturing procedures. The concept of generating a virtual model of a product is also applied to welding, as shown in Figure 21.53. Software is available that allows you to model a complete weldment, from the initial assembly of parts, to pre-welding requirements such as adding a slot, hole, or groove to welding components, and finally post-welding processes such as drilling a hole through welded components.

Typically, when you create a weld bead in the 3-D model, all weld bead information is specified in a dialog box. The result is an accurate representation of a weld according to the required specifications. This information is available in the model for reference and modification and can also be used in a 2-D drawing. As described throughout this book, one of the benefits of using a 3-D CADD program that combines 2-D drawing capabilities is the power to dimension a drawing by referencing existing model information. For example, if you create a fillet weld in a model, the weld specifications are stored in the model,

and it is just a matter of referencing the model weld parameters in the drawing to add a complete welding symbol to the drawing. (See Figure 21.54.) Also, due to the parametric nature of the software, drawing welding symbols automatically update when changes are made to the specifications of the corresponding model.

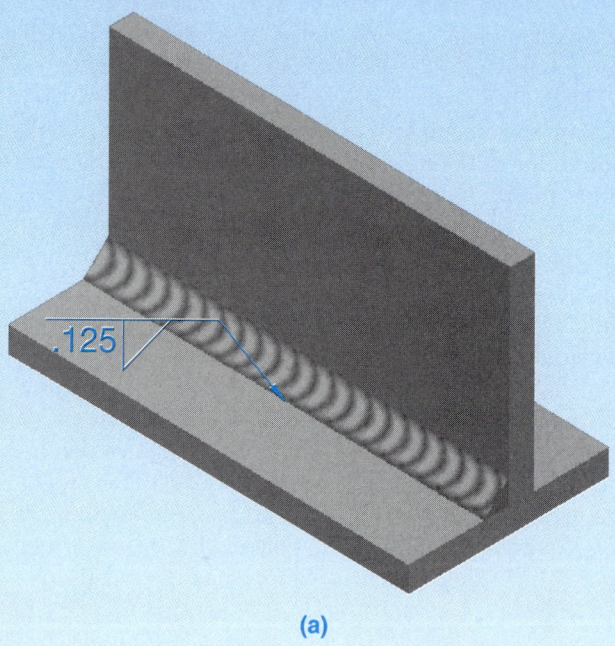

(a)

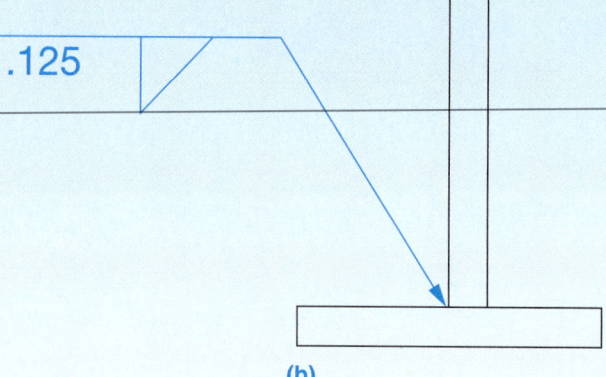

(b)

FIGURE 21.54 ■ An example of (a) a model weldment, and (b) referencing existing welding symbol information in a drawing.

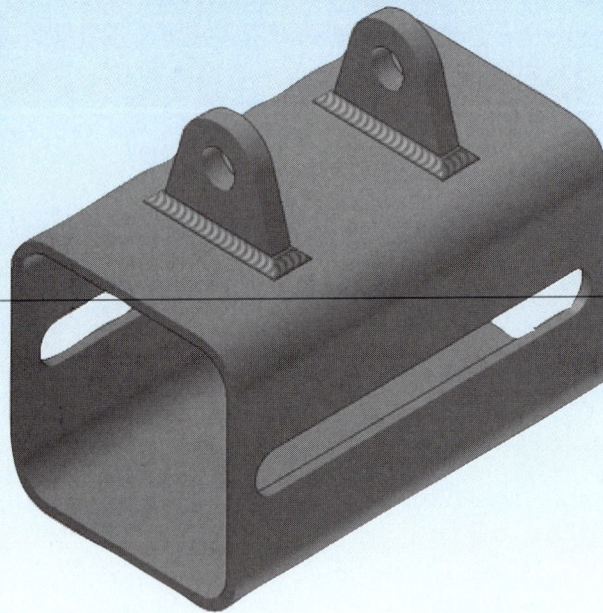

FIGURE 21.53 ■ An example of a weldment model.

PROFESSIONAL PERSPECTIVE

Welding processes are used in many manufacturing situations from heavy equipment manufacturing to electronic chassis fabrication to steel building construction. While the specific applications in these fields are different, the display of welding symbols is similar. If there is a strong chance that you will be employed as an engineering drafter in a field that performs a lot of welding operations, then it is a good idea for you to take a class in welding technology. Several classes may even be necessary for you to gain a good understanding of the welding methods and materials. The welding drafting standards and techniques discussed in this chapter are in accordance with the American Welding Society and the American Institute of Steel Construction. If you work in either mechanical engineering or steel construction and fabrication, you should become familiar with the welding applications demonstrated and discussed in these standards.

As an entry-level drafter, your drafting assignments will probably be engineer's sketches, or prints marked for revision. As you gain some experience, the engineer may explain verbally or in writing what needs to be done. As you gain experience, you will start to design with welding symbols based on the methods used in the past, material thickness, and welding processes. Assume you are working as an engineering drafter with a mechanical engineer and one of the parts for the project you are working on is a weldment. You have a good idea about what to do, but you ask the engineer for input. The engineer says, "Use a fillet weld-all-around on the diameter 41 side and a 4.5 mm deep bevel weld on the diameter 44 side, but be sure

the bevel weld is ground smooth so it doesn't interfere with the mating part." You say, "Okay, now should the process be shielded metal arc welding?" The engineer indicates with a nod, "Yes, you have it!" Now, you go back to work and come up with the drawing shown in Figure 21.55.

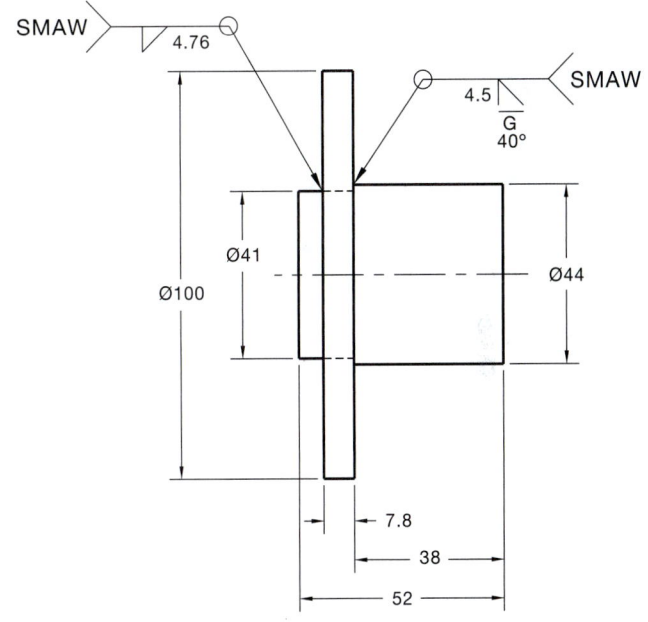

FIGURE 21.55 ■ A welding drawing from engineering information.

WEBSITE RESEARCH

The following website can provide you additional information for research or further study into topics covered in this chapter:

http://www.aws.org—American Welding Society.

http://www.amsi.org—American National Standards Institute.

http://www.aisc.org—American Institute of Steel Construction.

http://www.machinedesign.com—Welding processes nicely listed.

PERCENTAGE OF WELDED PARTS

Problem: Out of a lot of 1600 welded parts, 2% were rejected. How many were accepted?

Solution: This is a problem in percentages. Following percentage problem-solving terminology, the *base* is 1600 and the *rate* is 2 percent. The number of rejected parts is the *part*. To find the part, multiply: 1600 × .02 and get 32. Then the number of accepted parts must be 1600 - 32, or **1568**.

Another way to determine the number of accepted parts is to calculate 98% of the total parts made. If 2% of the parts were rejected, then 98% were accepted (100% − 2% = 98%). Using percentage problem solving again, the base is 1600 and the rate is 98%. To find the number of acceptable parts, multiply: 1600 × .98 = **1568**.

The math problems for this chapter involve finding the base and rate as well as the part.

MATH APPLICATIONS

CHAPTER 21

Welding Processes and Representations Test

 Access the CD found with this textbook to view the Chapter 21 Test. Confirm the preferred submittal method with your instructor.

CHAPTER 21

Welding Processes and Representations Problems

DIRECTIONS

1. Given the engineer's sketches and layouts, draw the required number of views and proper welding symbols for each problem. Completely dimension unless otherwise specified.

2. For Problems 21.1 through 21.7: Given the drawings showing actual welds or drawings and specifications, draw the necessary views and proper welding symbols.

PROBLEM 21.1   Fillet (inches)

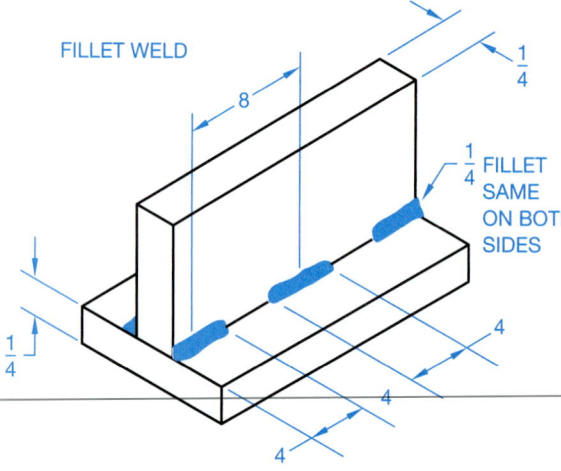

PROBLEM 21.2   Fillet and bevel groove (inches)

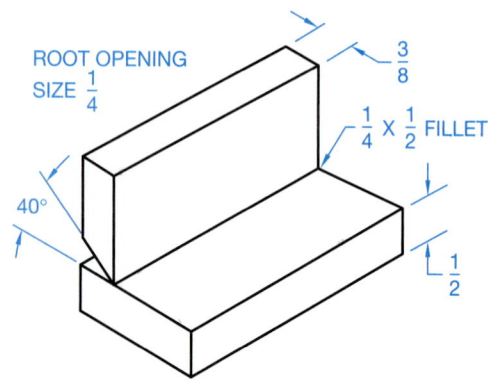

PROBLEM 21.3   V groove (inches)

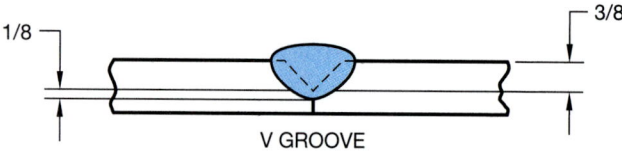

PROBLEM 21.4   Double V groove (inches)

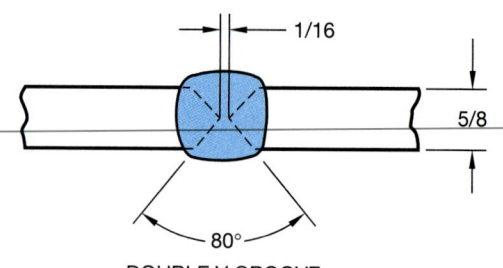

PROBLEM 21.5   Flange (inches)

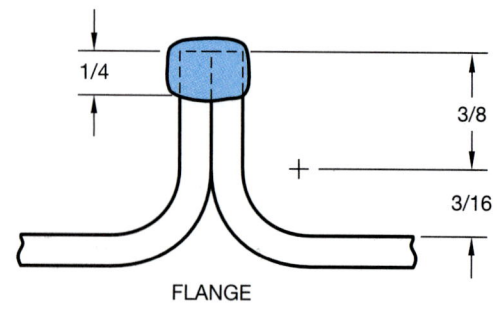

PROBLEM 21.6   J groove weld with radiation test (inches)

PROBLEM 21.7   Resistance spot weld (metric)

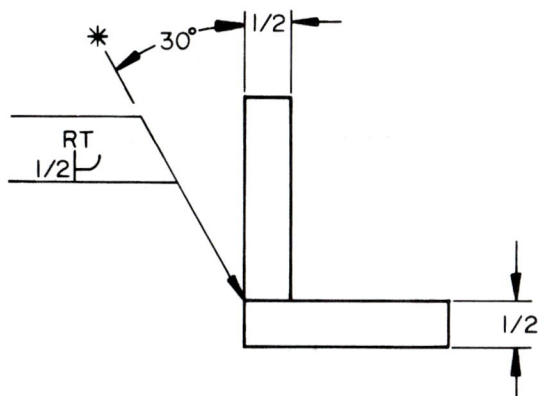

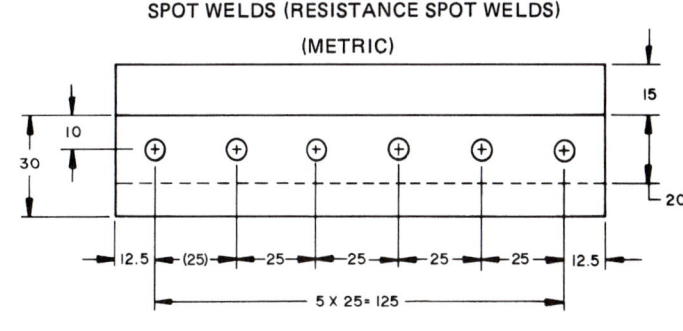

PROBLEM 21.8   (metric)

Part Name: Spring Housing Weldment
Material: MS

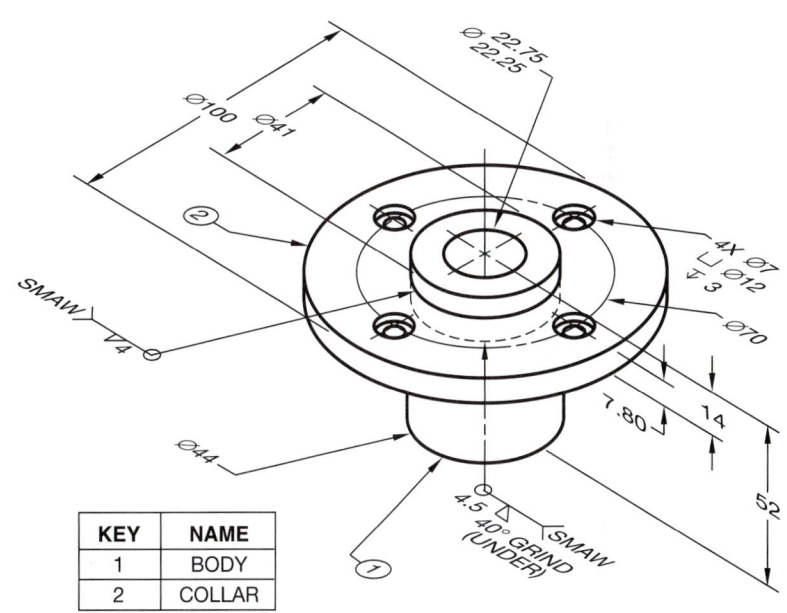

KEY	NAME
1	BODY
2	COLLAR

PROBLEM 21.9   (inches)

Drawing Name: Column Base Plate Detail
Material: MS

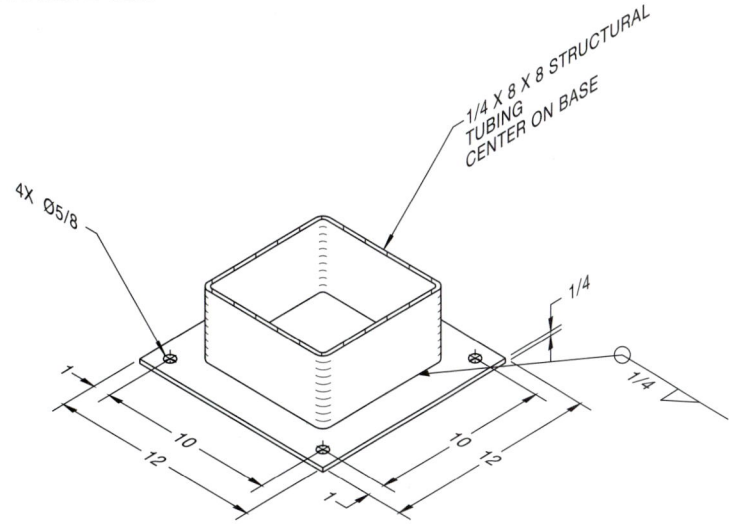

PROBLEM 21.10 (inches)

Drawing Name: Column Base Detail
Material: MS

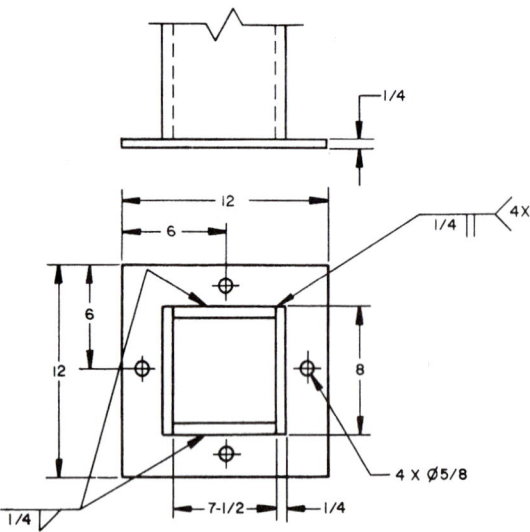

PROBLEM 21.11 (inches)

Drawing Name: Column Intersection Detail
Material: MS

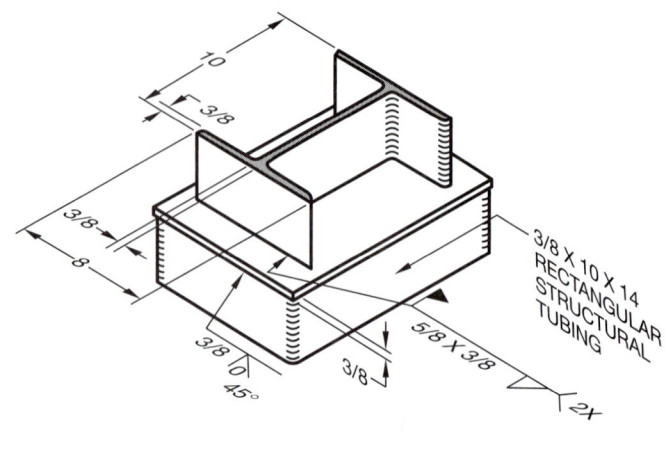

PROBLEM 21.12 (inches)

Drawing Name: Framed Beam Connection

Material: Steel

Dimensions of Members: W10 × 39; depth = 9-7/8", width = 8", web thickness = 5/16", flange thickness = 1/2".

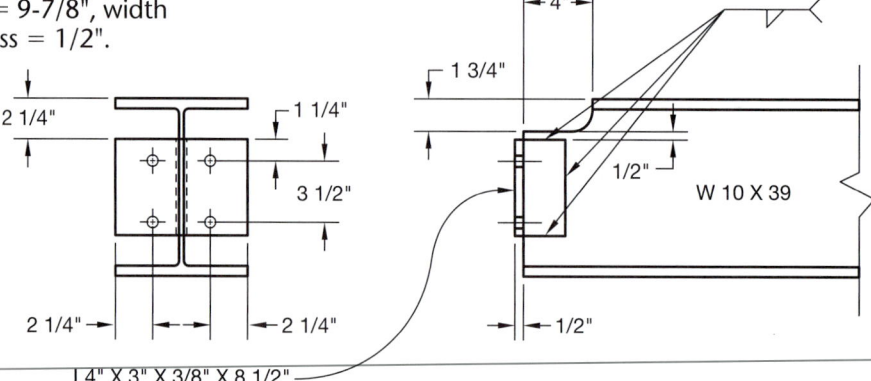

PROBLEM 21.13 (inches)

Drawing Name: Framed Beam Connection

Material: Steel

Dimensions of Members: W12 × 40: depth = 12", width = 8", web thickness = 3/16", flange thickness = 1/2".

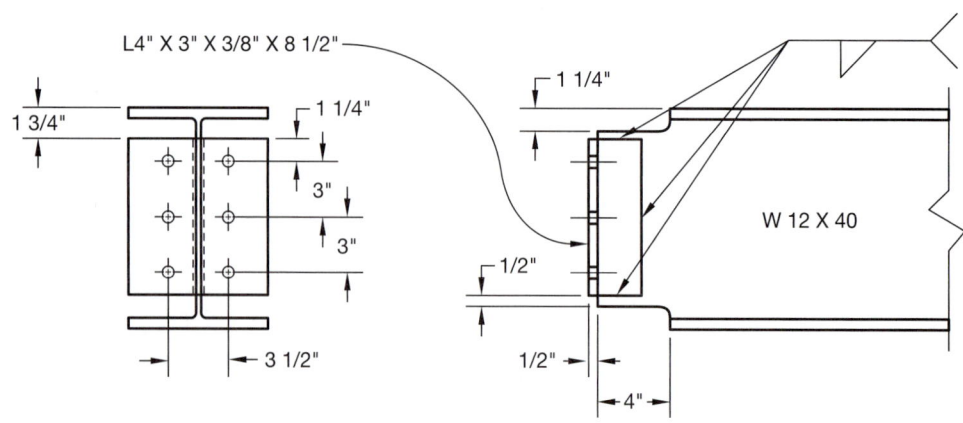

PROBLEM 21.14 **(inches)**

Drawing Name: Motor Support

Material: See Materials List

Dimensions of Members: C4 × 5.4: depth = 4", width = 1-5/8", web thickness = 3/16", flange thickness = 5/16".

Courtesy Production Plastics.

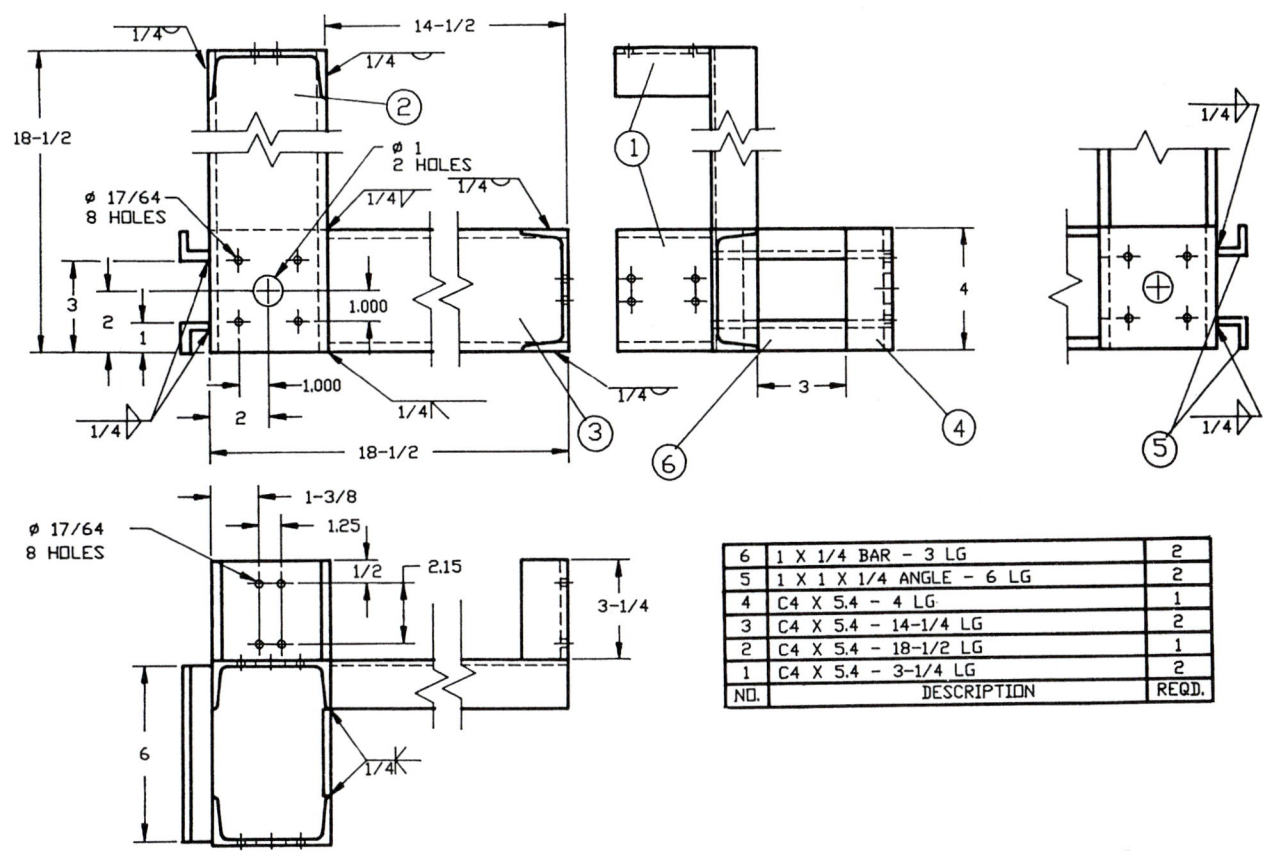

6	1 X 1/4 BAR - 3 LG	2
5	1 X 1 X 1/4 ANGLE - 6 LG	2
4	C4 X 5.4 - 4 LG	1
3	C4 X 5.4 - 14-1/4 LG	2
2	C4 X 5.4 - 18-1/2 LG	1
1	C4 X 5.4 - 3-1/4 LG	2
NO.	DESCRIPTION	REQD.

PROBLEM 21.15 **(inches)**

Drawing Name: Mounting Bracket

Material: See Parts List

Problem based on original art courtesy TEMCO.

2	WELD NUT	OHIO/ SF 3306 (3/8-16)
1	ANGLE	M821W200L2000 (2X2X3/16 A36)
-01	NAME OF PART	P/N DESCRIPTION MATERIAL

PROBLEMS 21.16 through 21.23: Access the CD found with this textbook and open the problem of your choice, or as assigned by your instructor. Solve the problems using the instructions provided with this chapter, or on the CD, unless otherwise specified by your instructor.

MATH PROBLEMS

PROBLEMS 21.24 through 21.33: Access the CD found with this textbook and open the math problem of your choice, or as assigned by your instructor. Solve the problem or problems using the instructions provided.

Specialty Drafting and Design

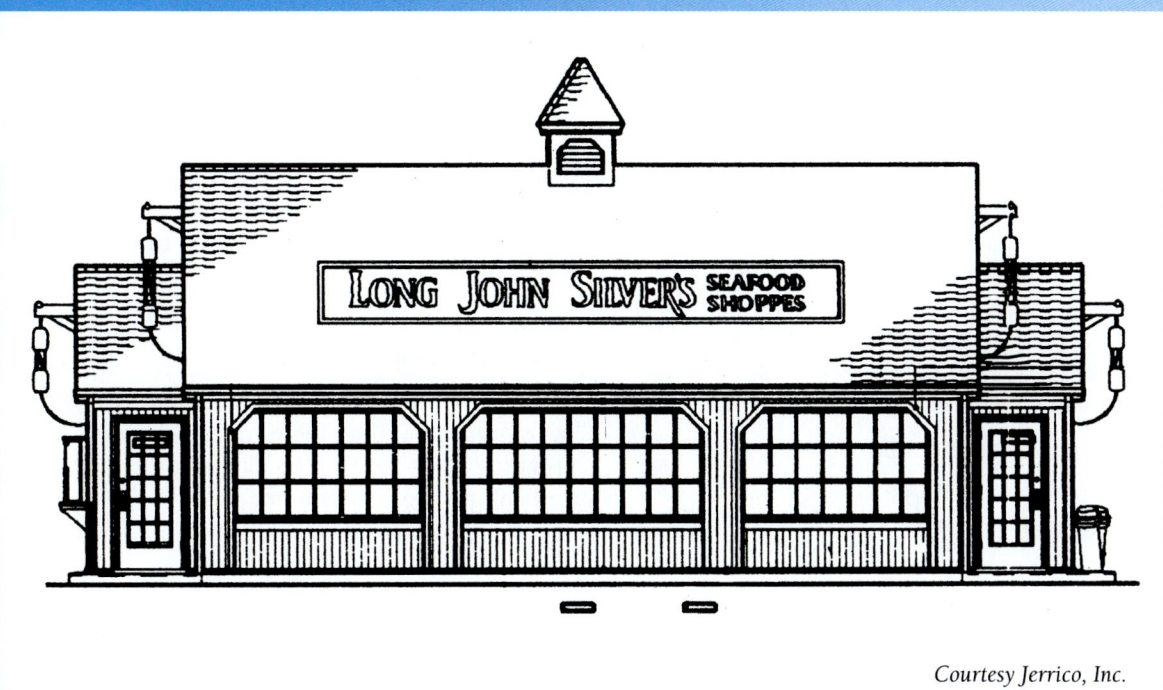

Courtesy Jerrico, Inc.

Industrial Process Piping

LEARNING OBJECTIVES

After completing this chapter, you will:

- Describe the different kinds of pipe and their uses.

- Define the methods of pipe connection and their applications.

- Identify pipe fittings and valves.

- Draw dimensioned, single-line and double-line piping drawings using pipe fittings and valves.

- Construct piping isometric and spool drawings using piping plans and elevations.

- Given an engineering sketch of a piping arrangement, construct an isometric sketch, and calculate linear dimensions and straight lengths of pipe in angular pipe runs.

THE ENGINEERING DESIGN APPLICATION

This problem requires that you use an engineering sketch (see Figure 22.1) to create an isometric freehand sketch then calculate the lengths of straight runs of pipe shown as "1" and "2" in the sketch. If you are not familiar with isometric sketching or drawing, see Chapters 6 and 15 first. The following instructions are used for this problem.

STEP 1 Sketch the pipe run in isometric format.

STEP 2 Calculate the lengths of the hypotenuse of triangles A and B. (See Figure 22.1.)

STEP 3 Use the Pythagorean theorem or trigonometric formulae. Each method is shown in the solution.

STEP 4 Calculate overall lengths of straight runs of pipe (1 and 2) including fittings.

STEP 5 Add the fitting lengths and subtract from the overall length of the pipe run. Use Appendix W to find pipe fitting and flange dimensions. All fittings and flanges are 150# rating.

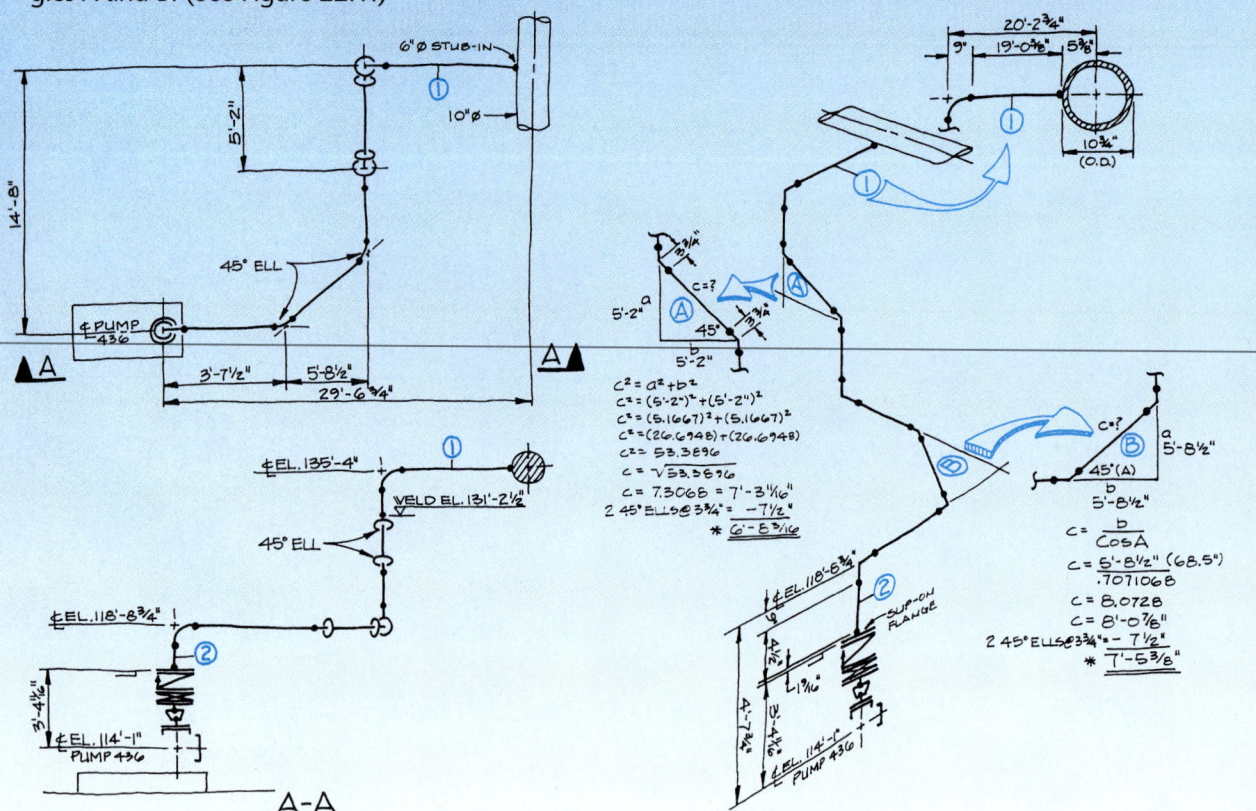

FIGURE 22.1 ■ This engineering sketch is used to construct an isometric sketch and to calculate the lengths of straight pipe in the angular offsets.

The term **piping** can refer to any kind of pipe used in a wide range of applications. The word **plumbing** refers to the small-diameter pipes within residences that carry water, gas, and wastes. Pipes of this kind can be copper, steel, cast iron, and plastic. Large underground pipes that transport water and gas to homes, from public utilities, and collect the waste water that is taken to treatment plants are known as **civil piping** because of their municipal nature. Steel, cast iron, and concrete are the principal materials used for this piping.

Industrial plants involve a process in which raw materials are converted to finished products. Water, air, and steam are used in many industries for processing the raw materials. Piping used to transport fluids between storage tanks and processing equipment is called **process piping**. (See Figure 22.2.)

Large-diameter pipes called **pipelines** carry crude oil, water, petroleum products, gases, coal slurries, and a variety of liquids hundreds of miles. This type of pipe is known as **transportation piping**, as shown in Figure 22.3.

FIGURE 22.2 ■ Process piping transports fluids between storage tanks and processing equipment at this oil refinery. *Courtesy Amoco Corporation.*

FIGURE 22.3 ■ This pipeline is a form of transportation piping. *Courtesy Amoco Corporation.*

WHERE INDUSTRIAL PIPING IS USED

A web of piping lies beneath the ground in your neighborhood. Miles of pipes for water, sewers, storm drains, natural gas, and electrical power provide vital services. (See Figure 22.4.) Oil and gas pipelines provide the compounds that allow you to drive to work and school daily. The petrochemical plants that process the crude oil into a tremendous variety of products employ thousands of miles of piping. (See Figure 22.5.)

The paper that is the page you are reading now was once a tree that was chopped into chips, cooked into a pulpy stew, and pumped in pipes through a series of processes in a paper mill. The food you consume is processed with the aid of pipes filled with water, chemicals, and liquid mixtures. The beverages you drink were moved in pipes through the various stages of production. The electric wires that provide power are resting inside pipes (electrical conduit). The electric power inside the wires is produced at steam and nuclear power plants containing intricate webs of piping. (See Figure 22.6.)

PIPE DRAFTING

Pipe drafting is a specialized field that calls upon the drafter's skill of visualization and the ability to see pipe and fittings in several planes (depth) in an orthographic view. Pipe fittings often must be turned and rotated at angles. Visualization can be confusing for beginning pipe drafters and engineers; therefore, study of pictorial (isometric and 3-D) drafting techniques is important. Pipe can be drawn in two forms, double line and single line, as shown in Figure 22.7. The single-line method usually gives beginners the most visualization problems. This method is discussed later in this chapter.

FIGURE 22.4 ■ Underground piping supplies natural gas for homes and industries. *Courtesy Pacific Gas and Electric.*

FIGURE 22.5 ■ Petroleum refineries employ miles of interwoven piping. *Courtesy Amoco Corporation.*

FIGURE 22.6 ■ Piping is an integral part of steam generation power plants. *Courtesy Washington Public Power Supply System.*

The principal area of learning, other than refining the ability to visualize, is the realm of pipe fittings and joining methods. A major portion of this chapter is spent on these areas. Those who have visualization problems should spend additional time studying and drawing pipe fittings.

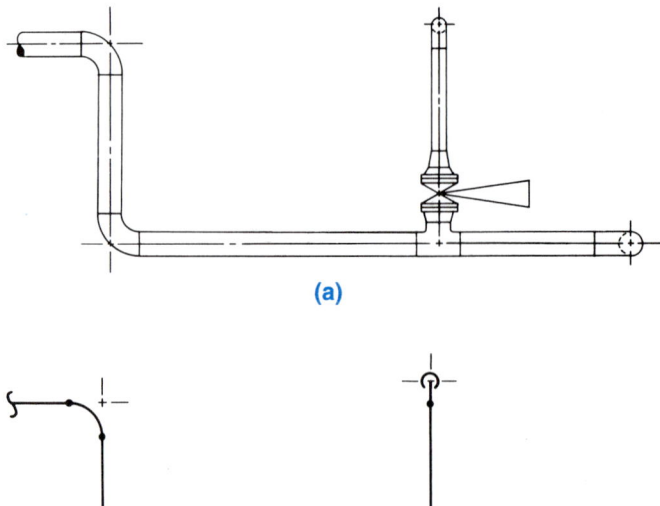

(a)

(b)

FIGURE 22.7 ■ Pipe can be drawn in (a) double-line or (b) single-line representation.

Pipe drafting involves the creation of a variety of drawing types from maps such as site plans, and mechanical drawings such as piping details. One of the easiest types of drawings to construct is the **flow diagram** as shown in Figure 22.8. This is a schematic, nonscale diagram that illustrates the layout and composition of a system using symbols. The **piping drawing** is the most complex. It is a scale drawing that provides plan, elevation, and section views. All equipment, fittings, dimensions, and notes are shown on this type of drawing. Examples of plans and sections are shown in Figure 22.9a and Figure 22.9b. The drafter must use flow diagrams, structural, mechanical, and instrumentation drawings, and vendor catalogs to construct these drawings.

The **piping isometric** is a pictorial drawing that illustrates a pipe run in three-dimensional form. (See Figure 22.10.) Information from this drawing is obtained from the piping drawings. The drafter's ability to view piping runs in three planes is important when working with piping isometrics. Some pipe drawings are done in isometric form because lines can be measured and fits checked. Isometrics are used extensively in chemical process industries.

Subassemblies of pipe and fittings are constructed by pipe fitters and welders who use the piping **spool** drawing. This drawing is usually drawn orthographically and shows all of the pipe and fittings needed to assemble a segment of piping. Spools often are nonscale drawings that show all dimensions needed for assembly. A typical spool drawing is shown in Figure 22.11.

TYPES OF PIPE

When an engineer or designer decides to use a specific type of pipe, the decision is based on a number of considerations. Principal among those are temperature, pressure, and corrosion.

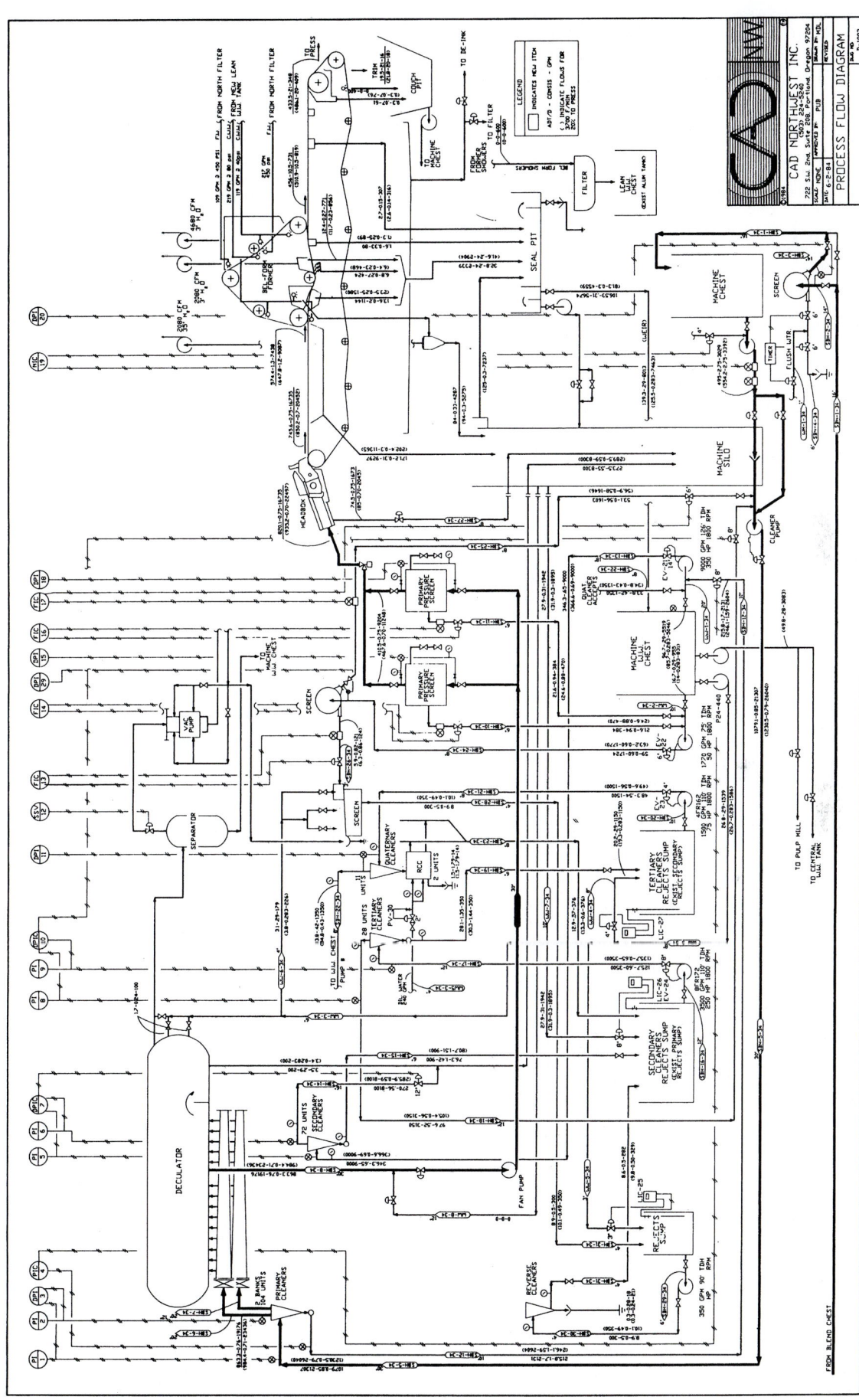

FIGURE 22.8 ■ This piping flow diagram is a nonscale view of a piping system. *Courtesy Autodesk, Inc.*

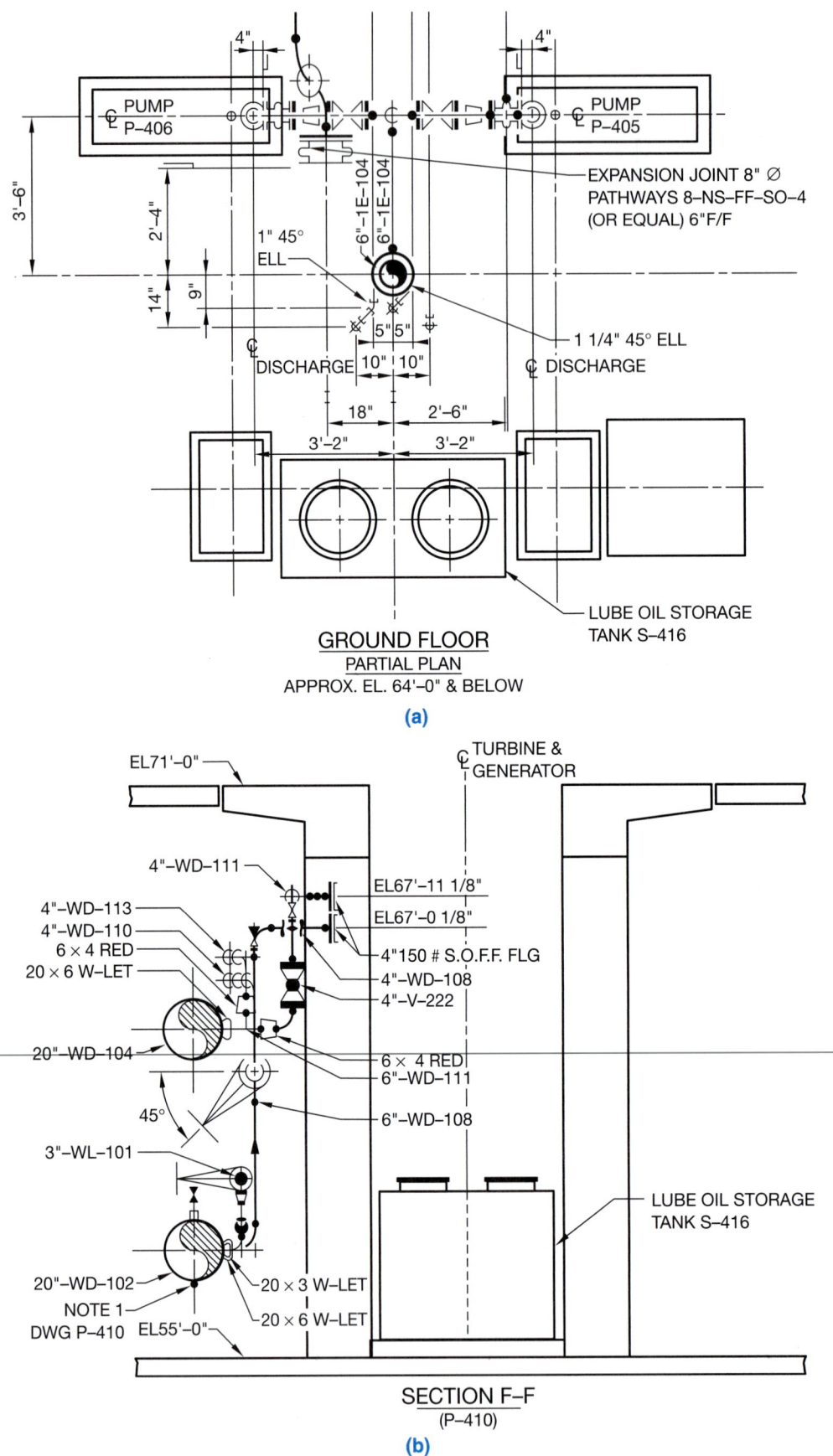

FIGURE 22.9 ■ (a) The piping plan; (b) section containing detailed piping information. *Courtesy Schuchart and Associates.*

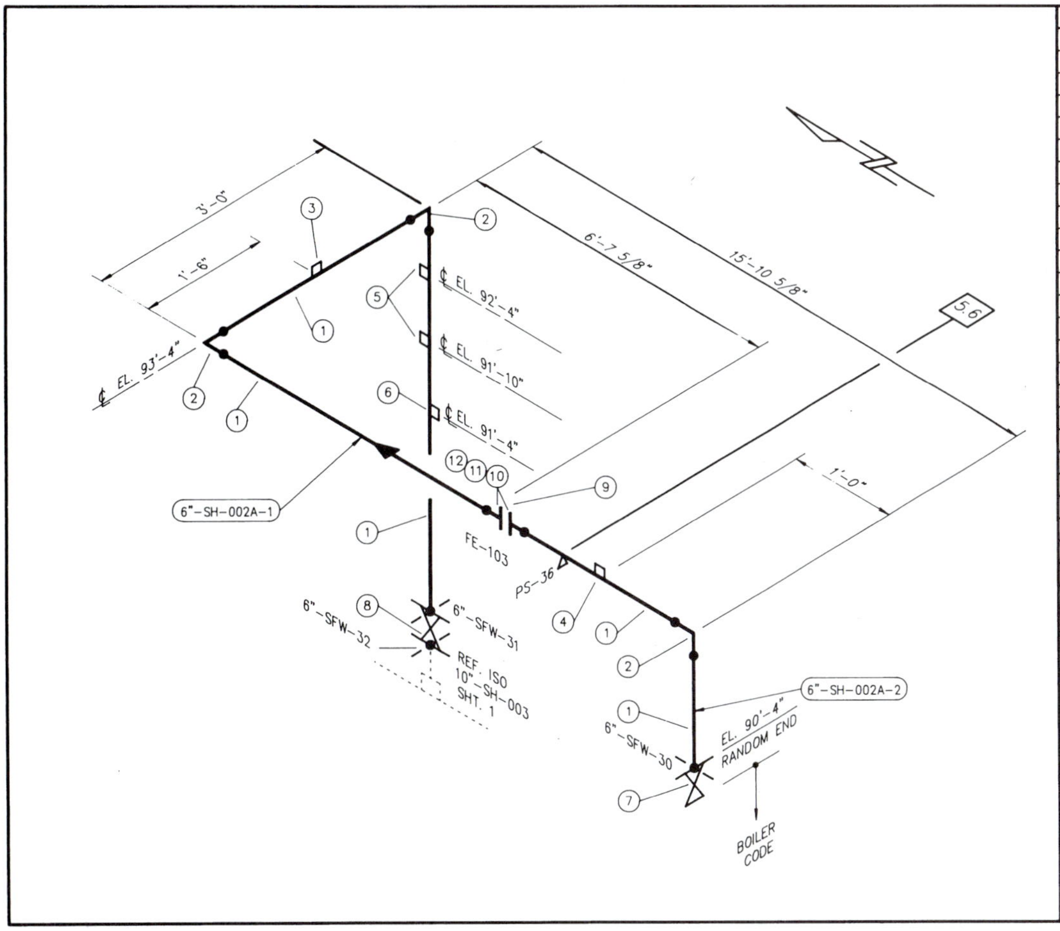

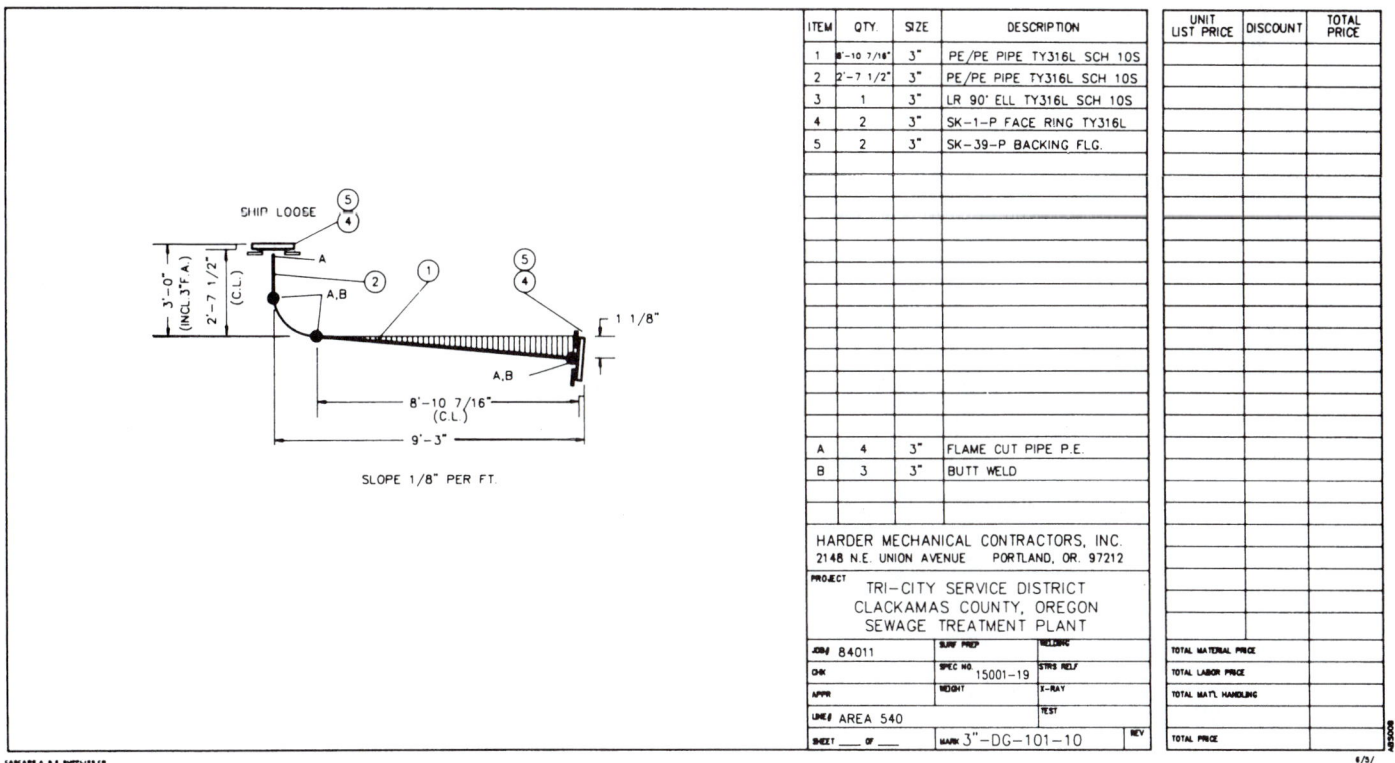

BILL OF MATERIAL

ITEM	QTY	SIZE	DESCRIPTION
1	25'	6"	PIPE SCH80 A106-B SMLS
2	3	6"	LR 90° ELL SCH80 A234
3	1	2"	SOCKOLET 3000# A105
4	1	1"	SOCKOLET 3000# A105
5	2	1"	THREADOLET 3000# A105
6	1	3/4"	THREADOLET 3000# A105
7	1	6"	SH-HV-002
8	1	6"	SH-HV-005
9	1	6"	FE-103
10	2	6"	RF ORIFICE FLG 600# S80 A105
11	2	6"	FLEXITALLIC STYLE C-G
12	12	1"	7" LG. A193 GR-B7 STUDS
			W/A194 SFH GR-2H NUTS

REFERENCE DRAWINGS: 12302- M150, REV.2

HARDER MECHANICAL CONTRACTORS, INC.
2148 N.E. UNION AVE.   PORTLAND, OR 97212

KIEWIT INDUSTRIAL COMPANY
MARION COUNTY
SOLID WASTE TO ENERGY FACILITY
STEAM PIPING

DR.	ELH / CAD	CHK.		APPR.	
DATE					
JOB NO.	85118	DWG. NO.	6"-SH-002A	REV.	0

FIGURE 22.10 ■ The piping isometric shows a pictorial view of a single run of pipe. *Courtesy Harder Mechanical Contractors, Inc.*

ITEM	QTY	SIZE	DESCRIPTION	UNIT LIST PRICE	DISCOUNT	TOTAL PRICE
1	8'-10 7/16"	3"	PE/PE PIPE TY316L SCH 10S			
2	2'-7 1/2"	3"	PE/PE PIPE TY316L SCH 10S			
3	1	3"	LR 90° ELL TY316L SCH 10S			
4	2	3"	SK-1-P FACE RING TY316L			
5	2	3"	SK-39-P BACKING FLG.			
A	4	3"	FLAME CUT PIPE P.E.			
B	3	3"	BUTT WELD			

HARDER MECHANICAL CONTRACTORS, INC.
2148 N.E. UNION AVENUE   PORTLAND, OR. 97212

PROJECT
TRI-CITY SERVICE DISTRICT
CLACKAMAS COUNTY, OREGON
SEWAGE TREATMENT PLANT

JOB#	84011	SURF PREP		WELDING	
CHK		SPEC NO. 15001-19		STRS RELF	
APPR		WEIGHT		X-RAY	
LINE#	AREA 540			TEST	
SHEET ___ OF ___		MARK 3"-DG-101-10		REV	

TOTAL MATERIAL PRICE
TOTAL LABOR PRICE
TOTAL MAT'L HANDLING
TOTAL PRICE

SHIP LOOSE
3'-0" (INCL 3"F.A.)
2'-7 1/2" (C.L.)
8'-10 7/16" (C.L)
9'-3"
1 1/8"
SLOPE 1/8" PER FT.

FIGURE 22.11 ■ A piping spool drawing is a subassembly of pipe and fittings. *Courtesy Harder Mechanical Contractors, Inc.*

Safety and cost also are important factors. And finally, any decision must comply with the project specifications and local and national codes.

ANSI The American National Standards Institute (ANSI) specifications that apply to pipes and fittings are listed in Figure 22.12.

Specification Number	Description
Graphic Standards	
ASA Z32.2.3	Graphical Symbols for Pipe Fittings, Valves, and Piping
ANSI Y32.4	Graphic Symbols for Plumbing Fixtures for Diagrams Used in Architecture and Building Construction
ASA Y32.11	Graphical Symbols for Process Flow Diagrams
Pipe, Steel	
ANSI/ASTM A430–79	Austenitic Steel Forged and Bored Pipe for High-Temperature Service
ANSI/ASTM A524–80	Seamless Carbon Steel Pipe for Process Piping
ANSI B36.19	Stainless Steel Pipe
ANSI B36–10	Welded and Seamless Wrought Steel Pipe
ANSI/ASTM A135–79	Electric-Resistance-Welded Steel Pipe
Pipe and Fittings, Plastic	
ANSI/ASTM D1527–77	Acrylonitrile-Butadiene-Styrene (ABS) Plastic Pipe, Schedules 40 and 80
ANSI/ASTM F443–77	Bell-End Chlorinated Poly(Vinyl Chloride) (CPVC) Pipe
ANSI/ASTM D2446–73	Cellulose Acetate Butyrate (CAB) Plastic Pipe (SDR-PR) and Tubing
ANSI/ASTM D1503–73	Cellulose Acetate Butyrate (CAB) Plastic Pipe, Schedule 40
ANSI/ASTM F441–77	Chlorinated Poly(Vinyl Chloride) (CPVC) Plastic Pipe, Schedule 40, 80, and 120
ANSI/ASTM D2104–74	Polyethylene (PE) Plastic Pipe, Schedule 40
ASTM D2469–76	Socket-Type Acrylonitrile-Butadiene-Styrene (ABS) Plastic Pipe Fittings, Schedule 80
ANSI/ASTM D2466–78	Socket-Type Poly(Vinyl Chloride) (CPVC) Plastic Pipe Fittings, Schedule 40
Pipe, Miscellaneous	
ANSI/ASTM C700–78a	Extra Strength and Standard Strength Clay Pipe and Perforated Clay Pipe
ANSI/ASTM B43–80	Seamless Red Brass Pipe, Standard Sizes
ANSI/ASTM C599–70	Process Glass Pipe and Fittings
ANSI A 40.5	Threaded Cast-Iron Pipe for Drainage, Vent, and Waste Services
ANSI/AWWA C151/A21.51	Ductile-Iron Pipe, Centrifugally Cast, in Metal Molds or Sand-Lined Molds for Water or Other Liquids
Fittings and Flanges	
ANSI B16.26	Cast Copper Alloy Fittings for Flared Copper Tubes
ANSI B16.1	Cast Iron Pipe Flanges and Flanged Fittings, Class 25, 125, 250, and 800
ANSI B16.4	Cast-Iron Threaded Fittings, Class 125 and 250
ANSI B16.9	Factory-Made Wrought Steel Buttwelding Fittings
ANSI B16.42	Fittings, Class 150 and 300, Ductile Iron Pipe Flanges and Flanged Fittings
ANSI/AWWA C207–78	Flanges for Water Works Service, 4 in. through 144 in. Steel
ANSI B16.11	Forged Steel Fittings, Socket-Welding and Threaded
ANSI/AWWA C606–81	Joints, Grooved and Shouldered Type
ANSI B16.3	Malleable-Iron Threaded Fittings, Class 150 and 300
ANSI B16.21	Nonmetallic Flat Gaskets for Pipe Flanges
ANSI B16.5	Pipe Flanges and Flanged Fittings, Steel Nickel Alloy and Other Special Alloys
ANSI B16.39	Pipe Unions (Class 150, 250, and 300), Malleable Iron Threaded
ANSI B16.20	Ring-Joint Gaskets and Grooves for Steel Pipe Flanges
ANSI B16.36	Steel Orifice Flanges, Class 300, 600, 900, 1500, and 2500 (includes supplement ANSI B16.36a)
ANSI B16.22	Wrought Copper and Copper Alloy Solder-Joint Pressure Fittings
ANSI B16.28	Wrought Steel Buttwelding Short Radius Elbows and Returns
Valves	
ANSI B16.34	Valves, Flanged and Butt-Welding End-Steel, Nickel Alloy, and Other Special Alloys
ANSI B16.10	Face-to-Face and End-to-End Dimensions of Ferrous Valves
ANSI/FCI 74-1	Valves Spring Loaded Lift Disc Check
ANSI/AWWA C509–80	Valves, 3 through 12 NPS, for Water and Sewage Systems, Resilient-Seated Gate
ANSI/AWWA C500–80	Valves 3 through 48 NPS, for Water and Sewage Systems, Gate
Pipe Hangers	
ANSI/MSS SP–58	Pipe Hangers and Supports—Materials, Design, and Manufacture

FIGURE 22.12 ■ ANSI specification for pipe, fittings, valves, and pipe hangers.

Cast Iron Pipe

The most common types of pipe used for commercial and industrial applications are steel and cast iron.

AWWA Cast iron is specified by the American Water Works Association (AWWA) for underground water lines. Cast iron pipe is corrosion resistant and has a heavy wall construction. It is good for use under pavement because its long life obviates the need to dig up the pavement due to frequent leaks. Cast iron is used extensively for water, gas, and sewage piping.

Steel Pipe

Carbon steel pipe is the preferred above-ground pipe used in industry today. It is strong, relatively durable, and it can be welded and machined. It is also not as expensive as other materials. In conditions of high temperature it tends to lose strength, so stainless steel and other alloys, that withstand high temperatures, are then used.

Steel pipe is manufactured by several processes. **Forged** pipe is formed to a special **outside diameter** or **O.D.** (usually one inch greater than the finished pipe), then bored to the required **inside diameter (I.D.)**. **Seamless pipe** is created by piercing a solid billet and rolling the resulting cylinder to the required diameter. **Welded pipe** is formed from plate steel. The edges are welded together in a **lap weld** or a **butt weld**. (See Figure 22.13.)

Copper and Copper Alloy Pipe

Copper pipe and copper alloys can be manufactured by the hot piercing and rolling process or by an extrusion method. Copper pipe is corrosion resistant and has good heat transfer properties but has a low melting point and is expensive. It is good for instrument lines and food processing, although stainless steel is used more frequently in these applications. Copper pipe is also used in residential water lines.

Copper tubing is softer and more pliable than rigid copper and brass pipe and is used for steam, air, and oil piping. It is also used extensively for **steam tracing**, the small-diameter pipe that is attached to larger pipes to protect them from freezing or to keep the fluids inside the pipe warm. The outside diameter of copper tubing is the same as its **nominal pipe size (NPS)**. For example, a 3/4" tube has an outside diameter of 3/4".

Plastic Pipe

Thermoplastic pipe was first developed in Germany in 1921, but it was not used to any great extent in the United States until 1951. This first type of plastic pipe was **polyvinyl chloride (PVC)**. PVC piping is now used for acids, salt solutions, alcohols, crude oil, and a variety of highly corrosive chemicals. PVC pipe specifications are given in Appendix Y.

Polyethylene pipe (PE) was introduced into the United States in 1947. Plastic piping can handle temperatures up to 150° F (65.5° C) and is suitable for water and vent piping of corrosive and acidic gases. PE piping is used as conduit for electrical and phone lines, water lines, farm sprinkler systems, salt water disposal, and chemical waste lines. Other forms of plastic piping include **acrylonitrilc-butadienestyrene (ABS)**, which is popular for sewage piping, **cellulose-acetatebutyrate (CAB)**, and **fiberglass-reinforced pipe (FRP)**.

Clay Pipe

Clay pipe is formed under high pressure from fire clays or shales, or a combination of the two. It is dried, then fired at 2100° F (1149° C). Under these extreme temperatures, the clay particles actually fuse or "weld" together to form a solid, durable pipe. Clay pipe is one of the most corrosion-proof pipes available for use in sanitary and industrial sewers and can carry any known chemical waste except hydrofluoric acid.

Glass Pipe

The chemical resistance, transparency, and cleanliness of glass makes it popular for applications in the chemical, food and beverage, and pharmaceutical industries. It can withstand temperatures up to 450° F (232° C).

Wood Pipe

Continuous stave wood pipe is used in the Pacific Northwest because of the abundance of redwood and Douglas fir. Tongue and groove staves are milled to exact radii for the specific diameter of pipe. These staves are fitted together then strapped with wire, flat steel bands, or steel rod threaded on the ends and tightened with bolts. (See Figure 22.14.) The wood used for this pipe is not resistant, especially when water under constant pressure is flowing through it. Wood stave pipe is used almost exclusively for transporting water, and is available in sizes from 10 in. to 16 ft. in diameter.

Steel Tubing

Tubing is small-diameter pipe and is often flexible, thus eliminating the need for many common fittings. Tubing is specified

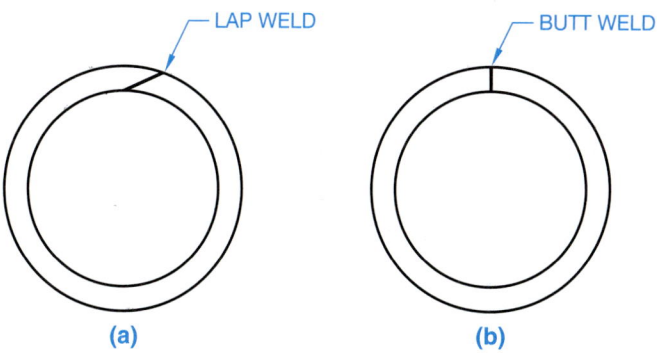

LAP WELD   BUTT WELD

(a)   (b)

FIGURE 22.13 ▪ (a) Lap-welded steel pipe; (b) butt-welded steel pipe.

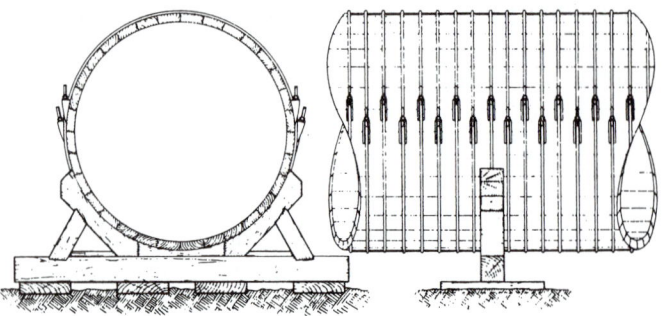

FIGURE 22.14 ■ Front and side views of a wood stave pipe resting in its foundation cradle. *Courtesy National Tank and Pipe Co.*

by its outside diameter and wall thickness. Its uses include external heating applications, boilers, superheaters, and hydraulic lines in the automotive, equipment, and aircraft industries.

Pipe Sizes and Wall Thickness

Pipe is available in sizes from 1/8" to 44" (3–1100 mm) in diameter. Sizes greater than this can be ordered but are not normally stocked by suppliers. The typical range of commonly stocked pipe is from 1/2" to 24" (12–600 mm). These are the sizes most often used in process piping. Sizes from 1/8" to 1/2" (3–12 mm) are used for instrument lines and service piping.

Pipe is specified by its **nominal pipe size (NPS)**. The NPS for pipe 1/8" to 12" (3–300 mm) is the **inside diameter (I.D.)**, and pipe 14" (350 mm) and above uses the **outside diameter (O.D.)** as the NPS. (See Figure 22.15.) If you want to determine the actual inside diameter of a pipe, you should double the wall thickness and subtract that number from the outside diameter.

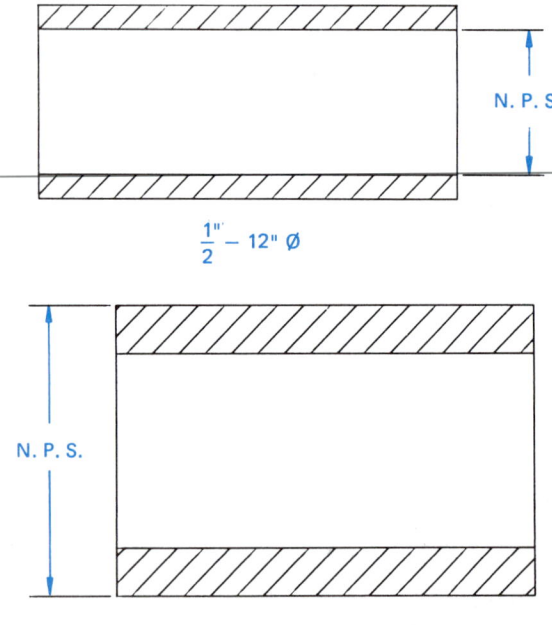

FIGURE 22.15 ■ How to measure nominal pipe size (NPS).

ANSI The wall thickness of pipe varies in relation to the size and weight of the pipe. The ANSI specifications for wall thickness (B36.10M) classify pipe as **schedule (SCH)** numbers. These numbers range from 10 to 160, the higher number representing the thicker wall. The ANSI schedule numbers incorporate the classifications of the ASTM and ASME, which use the designations of standard (STD), extra strong (XS), and double extra strong (XXS). These three classifications are drawn from manufacturers' dimensions. The STD wall thickness compares to SCH 40, XS to SCH 80, and XXS has no comparable schedule number (it is a thicker wall than SCH 160).

PIPE CONNECTION METHODS

Butt-Welded Connection

Butt welding is the most common method of joining pipes used in industry today. Butt welding is used on steel pipe 2 in. (50 mm) diameter and more to create permanent systems.

Butt-welded pipe and fittings provide a uniform wall thickness throughout the system. The smooth inside surface creates gradual direction changes, thus generating little turbulence. One circumferential weld is required to join two pieces of pipe (the number of welds may vary depending on pipe NPS). Figure 22.16 shows a cross section of a butt weld. The weld is strong, leak-proof, and relatively maintenance free. Pipe joined in this manner is self-contained, withstands high temperature and pressure, is easy to insulate, and requires less space for construction and hanging than do other methods.

Pipe and fittings joined by butt welding are prepared with a **beveled end (BE)**. This type of end preparation provides space for the welding operation. (See Figure 22.17.)

Socket-Welded Connection

The **socket-welded** connection shown in Figure 22.18 is used on pipe 2 in. and smaller. It forms a reliable, leak-proof connection. The pipe has **plain end (PE)** preparation and slips into the fitting. One exterior weld is required, thus no weld material protrudes into the pipe. Since the pipe is slipped inside the fitting,

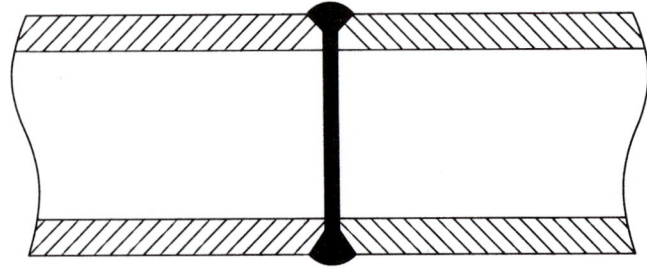

FIGURE 22.16 ■ Cross section of a butt-welded pipe connection.

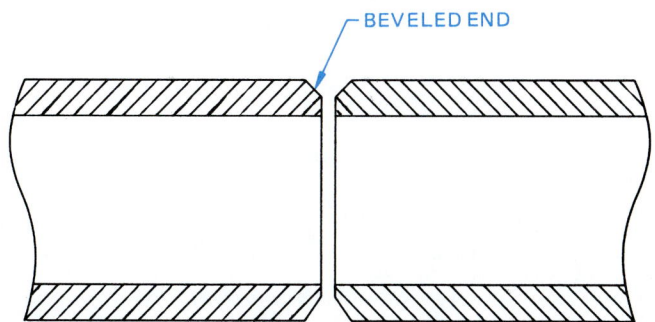

FIGURE 22.17 ■ Beveled-end pipe is used for butt-welded pipe connections.

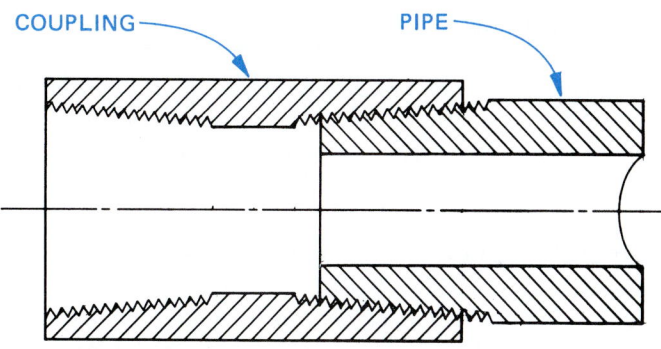

FIGURE 22.19 ■ Cross section of a screwed connection.

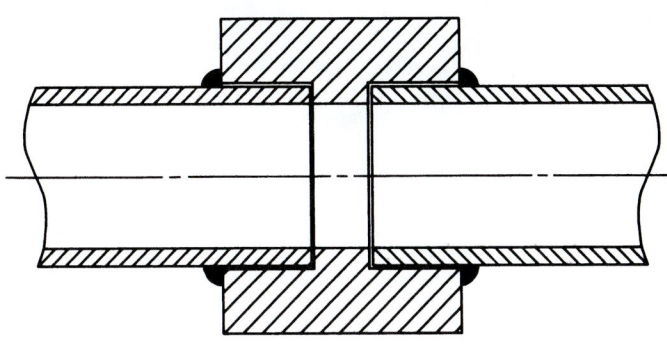

FIGURE 22.18 ■ The socket-welded connection forms a leak-proof joint.

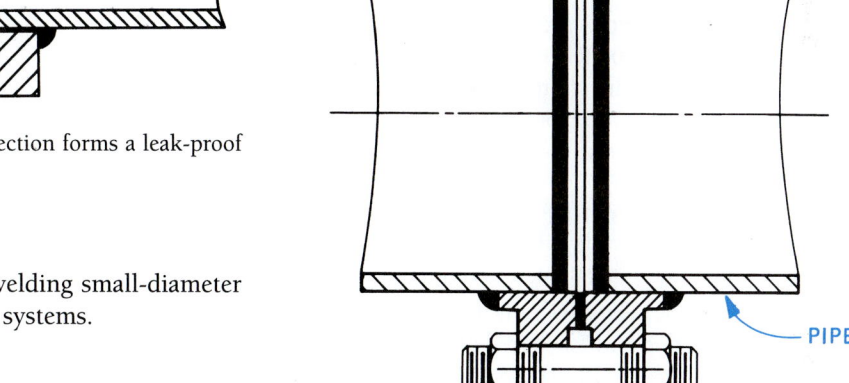

FIGURE 22.20 ■ Cross section of a typical flanged connection.

the connection is self-aligning. Socket welding small-diameter pipe is less expensive than other welded systems.

Screwed Connection

Screwed connections are used on steel, malleable iron, cast iron, and cast brass pipe less than 2 1/2" diameter. A screwed connection is the least leak-proof of the pipe joining methods, and is used where the temperature and pressure are low. The end preparation for screwed pipe is termed threaded and coupled (T&C) because a coupling is usually supplied with a straight length of pipe. American Standard pipe threads are the most common. The tightest fit is achieved by using fittings with straight threads and pipe with tapered threads. Pipe sealing compounds and teflon tape aid in producing a tighter fit. A typical coupling and pipe screwed connection is shown in Figure 22.19.

Flanged Connection

The flanged method of connecting pipes utilizes a fitting called a flange. The flange has an outside diameter greater than the pipe, and contains several bolt holes. Flanges are welded to the ends of pipe and then sections of pipe can be bolted together, as shown in Figure 22.20. Most flanges are forged steel, cast steel, or iron. Flanged pipe is easily assembled and disassembled but is considerably heavier than butt-welded pipe.

Flanged pipe occupies more space and is also more expensive to support or hang.

Two steel flanges bolted together do not form a tight fit. There must be some sort of sealing material placed between the two flange faces. This sealer is called a gasket. The gasket can be a soft or semimetallic material. When the bolts are tightened, the gasket is squeezed until it is pressed into the machining grooves of the flange faces.

Soft gaskets can be cork, rubber, asbestos, or a combination of materials. Semimetallic gaskets contain both metal and soft material. The soft material provides resilience and gives a tight seal while the metal helps to retain the gasket in place against high pressure and temperature. Gaskets are chosen from a wide range of materials that resist deterioration caused by high temperatures and that are not chemically affected by the fluids in the pipe.

Soldered Connection

Copper and brass water tubes are most often joined by soldered connections. Domestic water systems where temperatures and pressures are low are the most common use of this method. Both rigid and soft pipe can be soldered.

Bell (Hub) and Spigot Connection

Underground sewage, water, and gas lines are common applications of the bell and spigot connection method. (See Figure 22.21.) Cast iron pipe is used where pressures and loads are low. There are many variations of the bell and spigot method, but they all use some sort of sealer. The most common is lead and oakum (a fibrous sealer), although cement is used in certain situations.

Mechanical Joint Connection

The mechanical joint is a modification of the bell and spigot connection in which flanges and bolts are used with gaskets, packing rings, or grooved pipe ends providing a seal. Some mechanical joints allow for angular deflection and lateral expansion of the pipe. This type of connection is used in low-pressure applications, and where vibration may be excessive. Figure 22.22 illustrates a mechanical joint.

Solvent Welding Connection

This type of pipe connection is also known as cementing or gluing. It is used on plastic pipe. The solvent is applied to the pipe end, which is then inserted in the "socket" opening of a fitting and twisted. The solvent creates a bonding that acts much the same as a weld. Plastic pipe can also be joined by hot gas welding in which a torch is used to melt the pipe and fitting together.

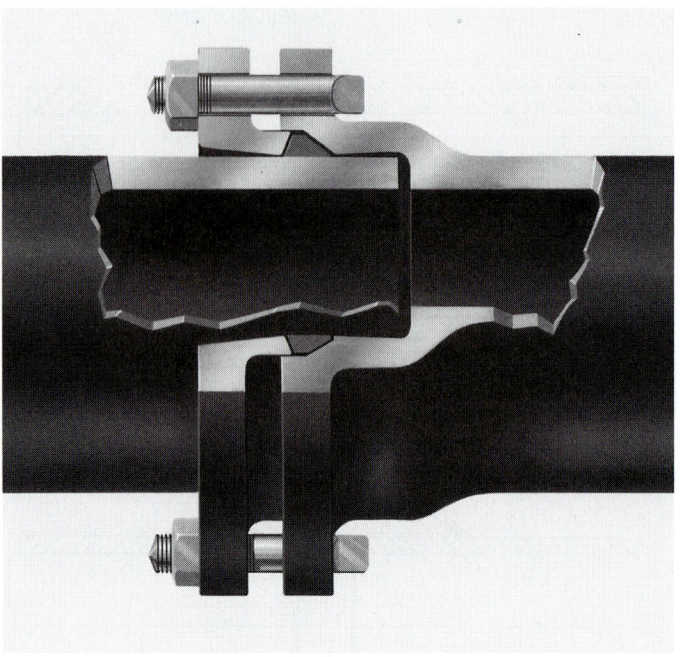

FIGURE 22.22 ■ Typical mechanical joint connection. *Courtesy Griffin Pipe Products Co.*

Flaring Connection

Soft copper pipe and tubing can be joined by a method called flaring. A special tool is clamped on the pipe and the flaring tip is inserted into the end of the pipe. As the tip is rotated, it is forced into the pipe, spreading or expanding the end of the pipe open, or "flaring" it. The end of the pipe resembles an old blunderbuss rifle. A fitting called a swage is then used to connect the flared end to a pipe or a fitting.

PIPE FITTINGS

Pipe fittings allow pipe to change direction and size and provide for branches and connections. Each type of pipe and connection method discussed previously uses the same type of fittings. However, special fittings can be required by the nature of the connection method used. The following is a general discussion of common fittings used in industry today.

Welded Pipe Fittings

Seamless forged steel fittings are prepared with a beveled end to accommodate butt welding. A welding ring (see Figure 22.23a) is placed between pipe and fitting to aid in alignment, provide even spacing, and to prevent weld material from falling into the pipe. (See Figure 22.23b). Since fittings have the same schedule numbers and wall thickness as pipe, a smooth inside surface is produced at the joints when fittings and pipe of the same schedule number are welded.

The most familiar fittings are elbows. These and other welded fittings are shown in Figure 22.24. Standard shapes are

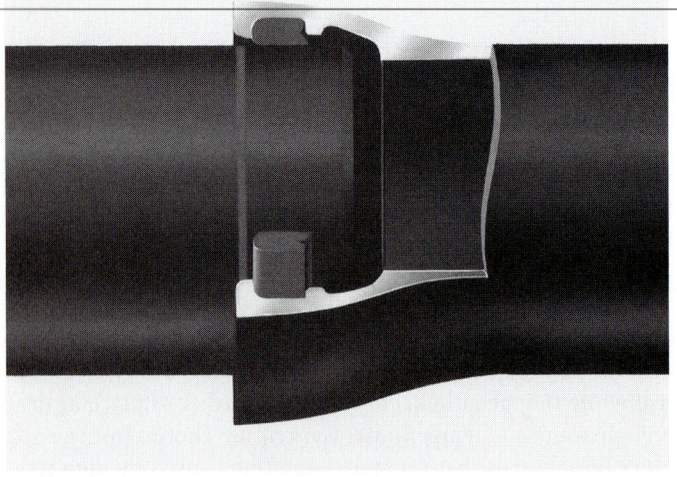

FIGURE 22.21 ■ Bell and spigot connection. *Courtesy Griffin Pipe Products Co.*

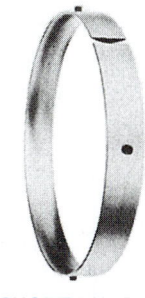

LONG NUBS   SHORT NUBS

(a)

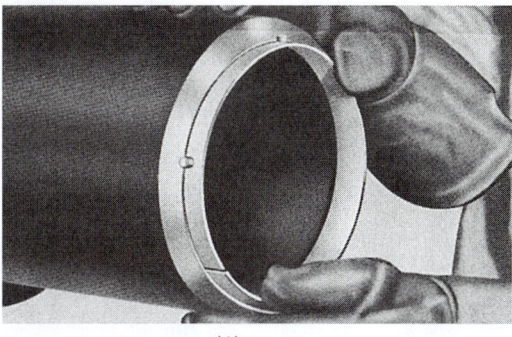

(1)

This photograph shows the ease with which a welder can fit a backing ring into the beveled end of pipe. This is an NPS 6" schedule 80 pipe. Tack welds are not necessary. The spacer nubs melt and become part of the weld metal.

(2)

The adjoining pipe is slipped onto the backing ring until the edge of the bevel touches the spacer nubs. Short pieces of pipe will be self-supported by the close tolerances of the ring diameter.

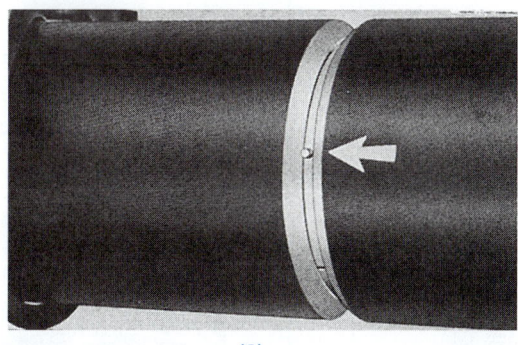

(3)

Note the close fit between the ring and the pipe . . . the minimum gap at the split. The carefully aligned spacer nubs center the backing ring perfectly across the welding groove and provide exact, even spacing between the pipe ends. The metal arrow locates the position of the spacer nubs in X-ray photographs.

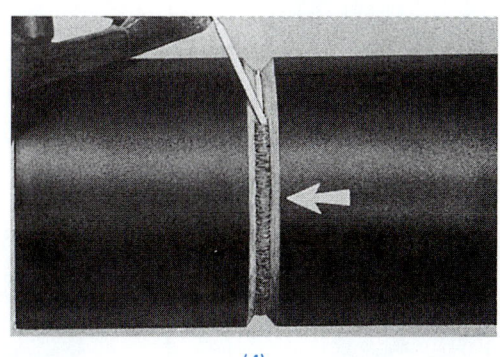

(4)

The weld is started at the bottom of the pipe. As this weld is being made in a fixed horizontal position, the split in the ring is on the bottom of the weld to prevent icicle formation inside the pipe. The ring is designed to allow a gap of only 1/16" at the split, minimizing the operation of filling the gap.

(b)

FIGURE 22.23 ■ (a) Common welding rings with long nubs and short nubs; (b) pipe-joining process using a welding ring. *Courtesy ITT Grinnell Corp.*

the 90° and 45° elbows. The **90° reducing elbow** reduces the pipe size in addition to changing direction. A reversal in direction can be achieved by using a **180° elbow**. Standard elbows can also be cut to any angle required. A **mitered elbow** is produced by cutting and welding straight pipe. The mitered elbow can be composed of one, two, three, or more welded joints. (See Figure 22.25.) It is not used often since it produces considerably more turbulence than standard elbows. An installation of the mitered elbow is shown in Figure 22.26.

The diameter of the pipe can be changed by **reducers**. The **concentric reducer** tapers the pipe equally about the axis centerline of the pipe. The **eccentric reducer** has a flat side that allows one side of the pipe—typically the top or bottom—to remain level.

The **straight tee** is a standard fitting that creates a branch or inlet. The straight tee has the same diameter on all three openings. The **reducing tee** has a smaller opening on the outlet side of the fitting. A branch or inlet can also be created with a **lateral**.

This fitting provides a 45° branch of the **run** pipe. The run is the straight portion of the fitting. The lateral is available in both straight and reducing forms. The **true Y** is similar to the lateral, but produces branches that are at a 90° angle, thus forming a "Y"

90° ELBOW

90° REDUCING ELBOW

45° ELBOW

STRAIGHT TEE

REDUCING TEE

CROSS

CONCENTRIC REDUCER

ECCENTRIC REDUCER

CAP

STRAIGHT LATERAL

REDUCING LATERAL

FIGURE 22.24 ■ Welded seamless fittings. *Courtesy ITT Grinnell Corp.*

with the run pipe. A branching fitting similar to the tee is the **cross**. It has two branches opposite each other. It can be either straight or reducing. It is employed in special situations and tight spaces but is seldom used because of its expense.

There are several small fittings, less expensive than regular fittings, that are used to create branch connections in new in-stallations or on pipe that is already assembled. They are com-monly called **weldolets®** but are manufactured to accept welded, screwed, socket-welded, and brazed pipe. These little fittings can be welded to an elbow (**elbolet®**) at a 45° angle to the run pipe (**latrolet®**), or at a 90° angle to the run pipe (**wel-dolet®**). (See Figure 22.27.)

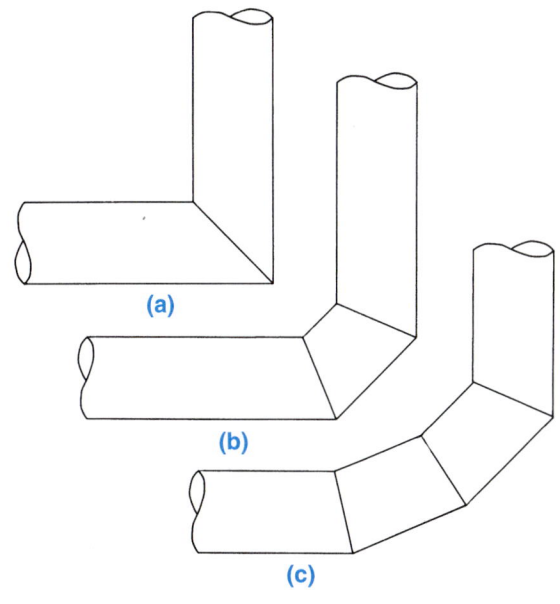

FIGURE 22.25 ■ The mitered elbow is composed of several cut pipe segments: (a) two-piece miter; (b) three-piece miter; and (c) four-piece miter.

FIGURE 22.26 ■ Mitered elbows in a vacuum pump installation at a paper mill. *Courtesy Ingersoll-Rand.*

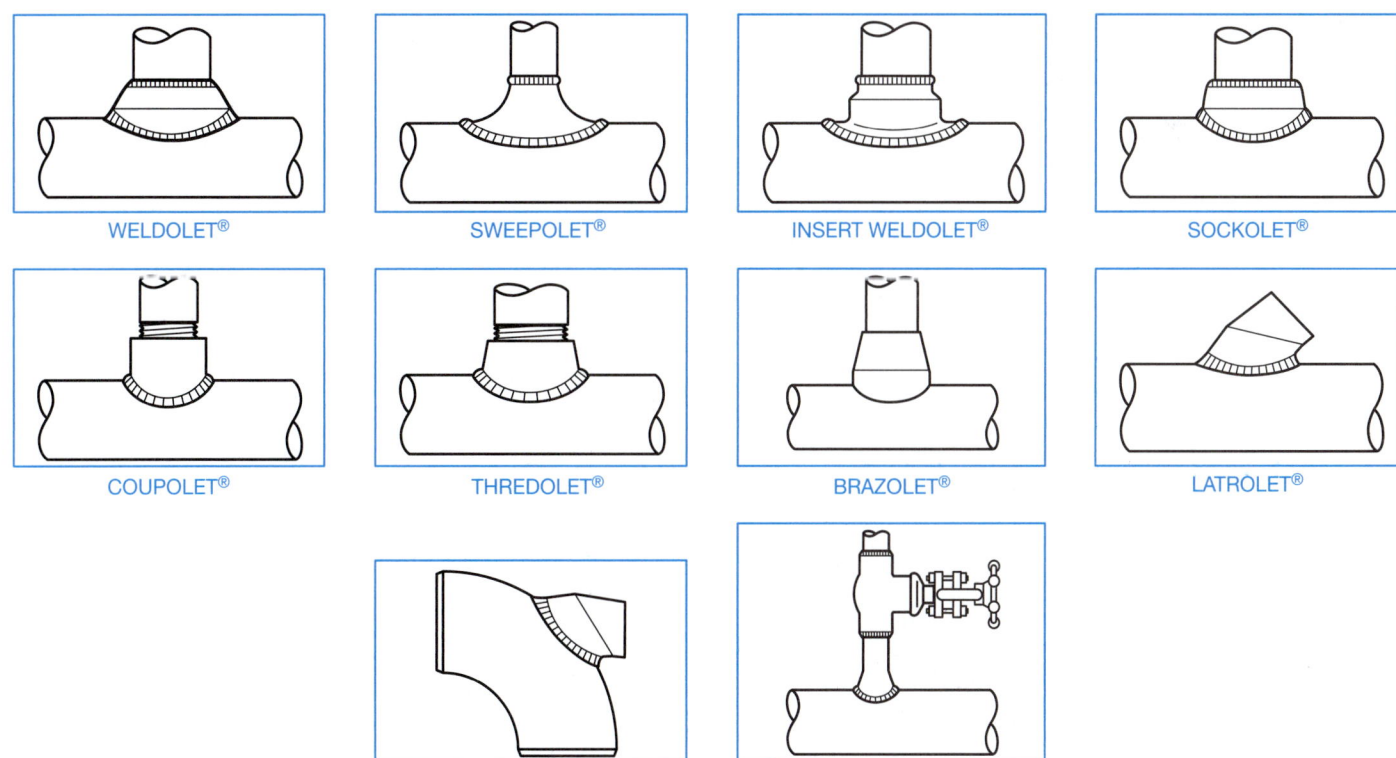

WELDOLET®   SWEEPOLET®   INSERT WELDOLET®   SOCKOLET®

COUPOLET®   THREDOLET®   BRAZOLET®   LATROLET®

ELBOLET®   NIPOLET®

FIGURE 22.27 ■ Branch connections can create inlets or outlets of varying sizes and angles to the main run of pipe and at less expense than reg-ular fittings. *Courtesy Bonney Forge Division, Gulf and Western.*

Dimensions for seamless welded fittings can be found in Appendix W, Table 30.

Screwed Pipe Fittings

All of the welded fittings are available in threaded form. In addition to these, there are several special fittings that are used with screwed pipe. Appendix W, Table 31 provides dimensions for galvanized malleable iron fittings.

- **Union**—Composed of two threaded sleeves and a threaded union ring, this fitting provides a connection point in a straight run of pipe. It allows pipe to be broken apart without tearing down the entire run of pipe. Figure 22.28 illustrates screwed fittings.
- **Coupling**—Threaded at both ends (TBE) with internal threads, it is used to attach two lengths of pipe together.
- **Half coupling**—Threaded at one end (TOE), this fitting is often welded to pipes and used for instrument connections.
- **Street elbow**—This 90° elbow is threaded at one end with internal threads and the other end with external threads. It can be attached directly to a fitting, thus eliminating the need of a short piece of threaded pipe (**nipple**).
- **Bushing**—A reducing fitting used to connect small pipe to larger fittings.
- **Plug**—A fitting with external threads on one end that is used to seal the screwed end of a fitting.

FLANGES

The component that creates a bolted connection point in welded pipe is the flange. It is a circular piece of steel containing a center bore that matches the pipe I.D. to which it is attached. It has several bolt holes evenly spaced around the center bore. Flanges are used most extensively on welded pipe but also are available for most other types of pipe. Figure 22.29 shows several types of flanges. Flange dimensions are given in Appendix W, Table 24.

Slip-On Flange

The slip-on flange can be used only on straight pipe because it is bored to slip over the end of the pipe. Two welds are required to attach this flange to pipe. (See Figure 22.30.) There are two types of slip-on flanges.

ASME When the Type 1 slip-on flange is used, the pipe is set back from the face of the flange. It has two-thirds the strength of a weld-neck flange and is limited to 300 lb service by the ASME Pressure Piping Code. The Type 2 slip-on flange allows the pipe to be welded flush with the face of the flange, which is then machined flat. This type of flange is used on lines having 400 lb pressure and above.

Weld-Neck Flange

The weld-neck flange is forged steel and prepared with a beveled end for butt welding to pipe or fittings. It is always used when a flange must be attached directly to a fitting.

Blind Flange

A pipe can be temporarily sealed with a blind flange It is basically a steel plate with bolt holes in it.

Stub-End or Lap-Joint Flange

The stub-end or lap-joint flange is composed of two parts, the stub end and the flange ring. This flange ring can be carbon steel if expensive pipe such as stainless steel is used. Only the stub end needs to be stainless.

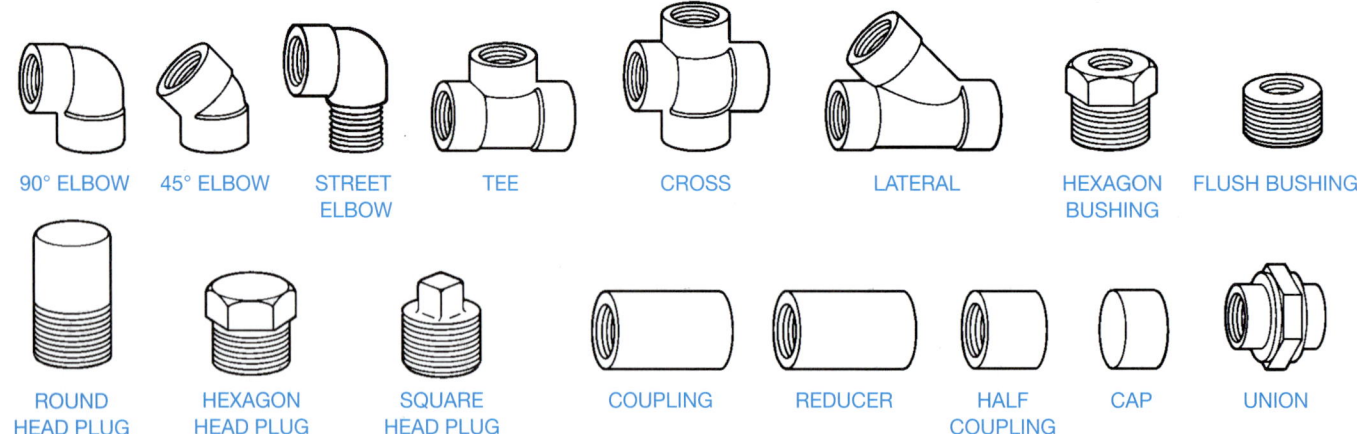

90° ELBOW   45° ELBOW   STREET ELBOW   TEE   CROSS   LATERAL   HEXAGON BUSHING   FLUSH BUSHING

ROUND HEAD PLUG   HEXAGON HEAD PLUG   SQUARE HEAD PLUG   COUPLING   REDUCER   HALF COUPLING   CAP   UNION

FIGURE 22.28 ■ Common screwed fittings. *Courtesy Alaskan Copper Works.*

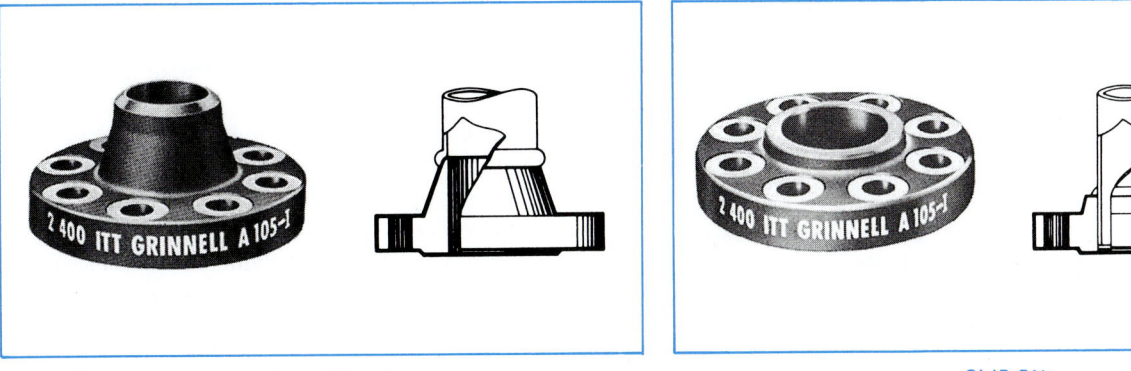

WELDING NECK

SLIP-ON

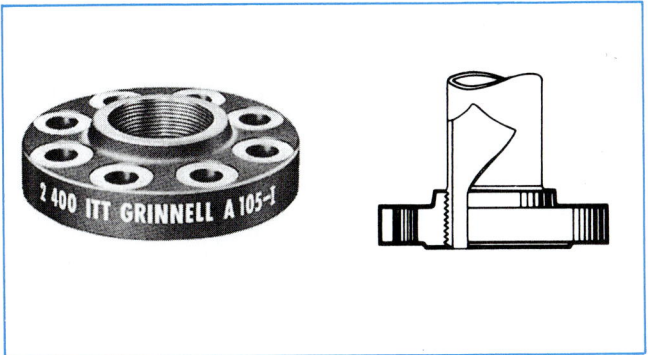

THREADED

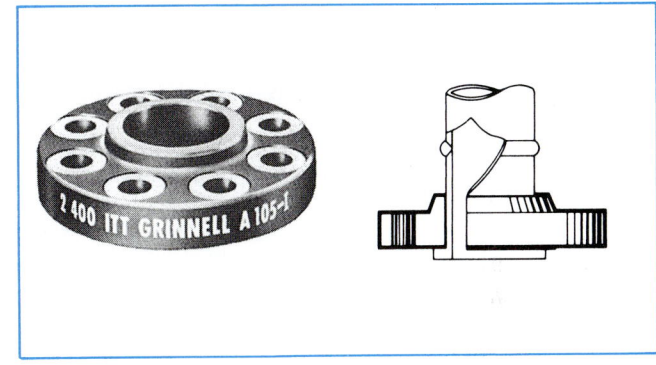

LAP JOINT

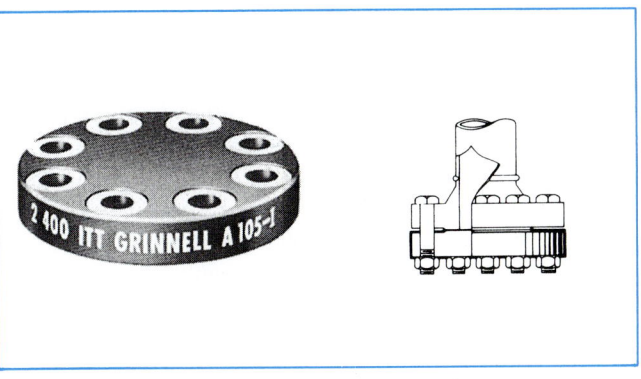

BLIND

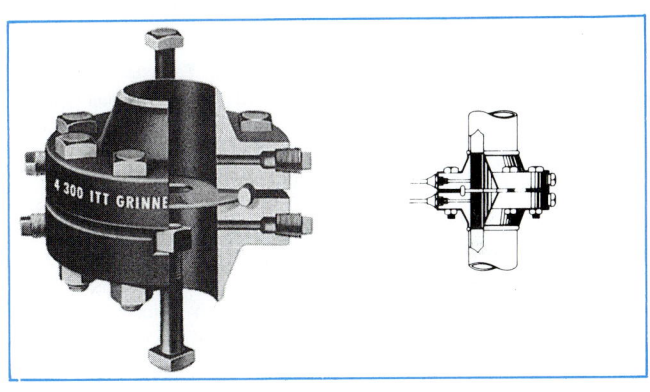

ORIFICE

FIGURE 22.29 ■ Common flanges used on butt-welded steel pipe. *Courtesy ITT Grinnell Corp.*

Reducing and Expander Flange

A change in line size can be achieved with a reducing flange, but it should not be used where increased turbulence would be undesirable. A reduction in line size can also be created with an expander flange. This fitting is a flange/reducer combination and can be used in place of a weld-neck flange and a reducer.

Orifice Flange

The flow rate inside a pipe can be measured using orifice flanges and an orifice plate. The two orifice flanges are drilled and tapped to accommodate tubing and a pressure gauge is used to measure the flow rate. The orifice plate is a flat disc with a small hole drilled in its center. The two flanges are welded to the pipe, the orifice plate and gaskets are placed between them, and the flanges are bolted together. As fluid flows through the hole in the orifice, a pressure differential is created on either side of the plate. This differential can be read on the pressure gauge.

Flange Faces

The facing of a flange is the type of machining that is done to the contact surface of the flange. A common flange used in industry is the raised face (RF) flange. The face of this flange protrudes a certain distance beyond the flange itself. The male and female facings interlock with each other. A recessed face on the female flange accepts both the gasket and the raised face of the male

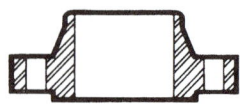

The Raised Face is the most common facing employed with steel flanges; it is 1/16" high for Class 150 and Class 300 flanges and 1/4" high for all other pressure classes. The facing is machine-tool finished with spiral or concentric grooves (approximately 1/64" deep on approximately 1/32" centers) to bite into and hold the gasket. Because both flanges of a pair are identical, no stocking or assembly problems are involved in its use. Raised face flanges generally are installed with soft flat ring composition gaskets. The width of the gasket is usually less than the width of the raised face. Faces for use with metal gaskets preferably are smooth finished.

Male-and-Female Facings are standardized in both large and small types. The female face is 3/16" deep and the male face 1/4" high and both are usually smooth finished since the outer diameter of the female face acts to locate and retain the gasket. The width of the large male and female gasket contact surface, like the raised face, is excessive for use with metal gaskets. The small male and female overcomes this but provides too narrow a gasket surface for screwed flanges assembled with standard weight pipe.

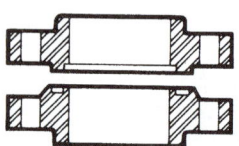

Tongue-and-Groove Facings are also standardized in both large and small types. They differ from male-and-female in that the inside diameters of tongue and groove do not extend to the flange bore, thus retaining the gasket on both its inner and outer diameter; this removes the gasket from corrosive or erosive contact with the line fluid. The small tongue-and-groove construction provides the minimum area of flat gasket it is advisable to use, thus resulting in the minimum bolting load for compressing the gasket and the highest joint efficiency possible with flat gaskets.

Ring Joint Facing is the most expensive standard facing but also the most efficient, partly because the internal pressure acts on the ring to increase the sealing force. Both flanges of a pair are alike, thus reducing the stocking and assembling problem found with both male-and-female and tongue-and-groove joints. Because the surfaces the gasket contacts are below the flange face, the ring joint facing is least likely of all facings to be damaged in handling or erecting. The flat bottom groove is standard.

Flat Faces are a variant of raised faces, sometimes formed by machining off the 1/64" raised face of Class 150 and Class 300 flanges. Their chief use is for mating with Class 125 and Class 250 cast iron valves and fittings. A flat-faced steel flange permits employing a gasket whose outer diameter equals that of the flange or is tangent to the bolt holes. In this manner the danger of cracking the cast iron flange when the bolts are tightened is avoided.

FIGURE 22.30 ■ Cross-section views of common flange faces. *Courtesy ITT Grinnell Corp.*

flange. The tongue-and-groove facing is interlocking. The male tongue-and-groove flange is 1/4" longer than the female flange to accommodate the height of the tongue. The tongue fits into the groove, creating a tighter lock and seal. Figure 22.30 shows cross sections of the flange faces mentioned here.

The ring-type joint (RTJ) flange connection requires two flanges that both have a machined groove in the face. This groove is cut to accept an oval or octagonal ring. The grooves on both flanges fit over the ring, which creates a seal. Examples of the ring joint flange are seen in the cutaway photo of the orifice flange in Figure 22.29, and in the section view in Figure 22.30.

VALVES

Valves are the components in the piping system that control and regulate fluids. Valves provide on/off service in a pipe and are also used to regulate the flow of fluid in the pipe, maintain a constant pressure, prevent dangerous pressure buildup, and prevent backflow in the pipe. A variety of valves are used to perform these tasks.

On/Off Valve

The gate valve is used exclusively to provide on/off service in a pipe. It is the most common valve used in industry today. (See Figure 22.31.) A gate or disc is moved up and down inside the valve manually or automatically. The design of the gate valve is such that fluid can flow through it with a minimum of friction and pressure loss. The valve seat does not interfere with the straight line flow of the fluid. The seat is the material with which the gate makes contact to create a seal. The gate valve is designed specifically for on or off service and infrequent operation. It is unsuitable for throttling or regulating flow. A partially open disc (gate) can cause erosion and wear of the downstream side of the seat and the disc itself.

The sealing mechanism inside a ball valve is just that—a ball. The ball has a hole through it that matches the I.D. of the pipe. (See Figure 22.32.) It is a quick-opening valve, requiring only a one-quarter turn, that is used extensively because of its tight seat. Plastics, nylon, and synthetic rubber are used for the seat material and enable this valve to achieve a tight seal. The ball valve is popular because it has a low profile and low torque requirement for operation. It is also easy and inexpensive to maintain and repair. It is not used in large-diameter lines (greater than 12") because the pressure in the line makes it difficult to open and close.

The plug valve, also known as cock valve, is similar in design to the ball valve and requires only a one-quarter turn to open and close. The opening through the plug can be either rectangular or round. (See Figure 22.33.) When open, there is a pressure drop through the valve but a high flow efficiency due to the contours of the valve. It has low throttling ability and is best used for on/off service. It can achieve a tighter shutoff than a gate valve but is normally used on smaller-diameter lines.

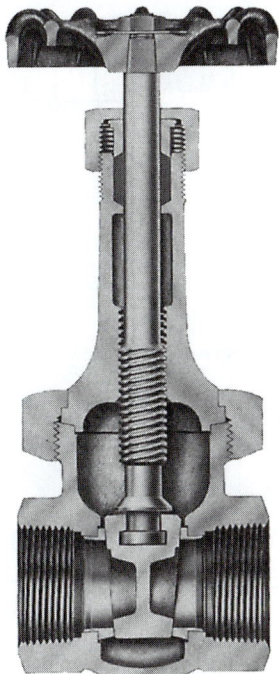

FIGURE 22.31 ■ The gate valve provides on/off service. *Courtesy Crane Co.*

FIGURE 22.32 ■ The ball valve is quick-opening and is used in pipe less than 12" in diameter. *Courtesy The Wm. Powell Co.*

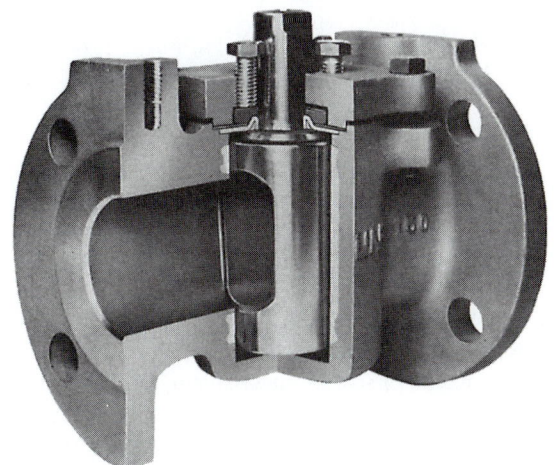

FIGURE 22.33 ■ The plug valve achieves tighter shutoff than a gate valve but is used on small diameter lines. *Courtesy Xomox Corp.*

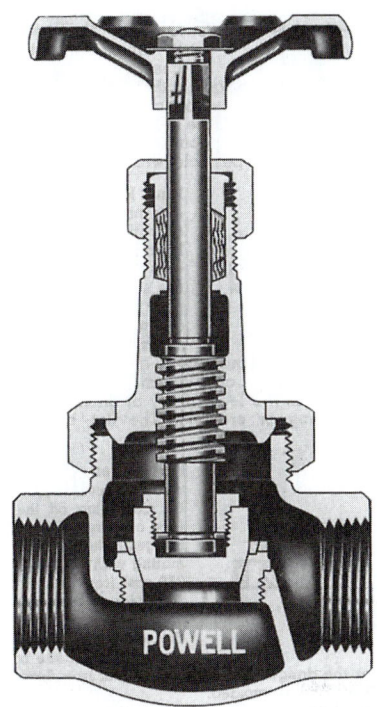

FIGURE 22.34 ■ The globe valve achieves close flow control but creates high flow resistance. *Courtesy The Wm. Powell Co.*

Regulating Valve

The most common type of regulating valve is the globe valve. It is normally used in pipe up to 3" (75 mm) in diameter but can be used in lines with diameters up to 12" (300 mm). Fluid flowing through the globe valve travels in an "S" pattern, which allows the valve to maintain a close control on the flow and to achieve a tight, positive shutoff. This flow pattern is seen in Figure 22.34. The internal design of the globe valve creates a high flow resistance, leading to a significant pressure drop through the valve. Body pockets in the valve are not drained when the flow is stopped. Situations that require frequent valve operation and maintenance are suited for globe valves because the discs and seats are easily replaced.

A special type of globe valve that creates a 90° direction change in the pipe is an angle valve. It is similar in design to a globe valve (see Figure 22.35), and is used in place of a globe valve and a 90° elbow to save money. In high-stress situations, the angle valve is not used.

Low-pressure situations can warrant the use of the butterfly valve. The disc is mounted on a stem that is turned a one-quarter turn to open and close. The entire disc moves inside the valve. (See Figure 22.36.) It is a simple operating mechanism that is excellent for regulating flow. The design creates a minimum pressure drop through the valve. It is light and inexpensive, and it requires a small installation space and is easy to maintain. All-plastic butterfly valves are also available.

The needle valve is so named because the end of the stem is needle shaped. (See Figure 22.37.) The stem has fine threads enabling this valve to be adjusted exactly to achieve accurate

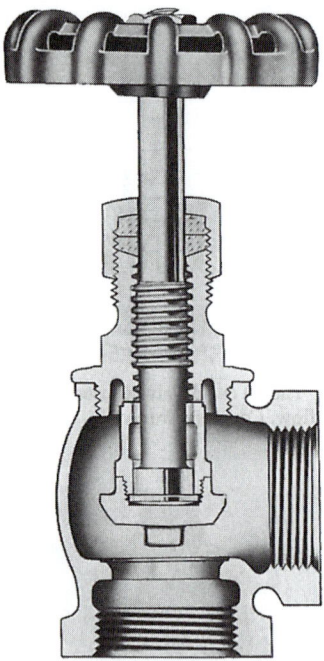

FIGURE 22.35 ■ The angle valve replaces the globe valve and an elbow. *Courtesy Crane Co.*

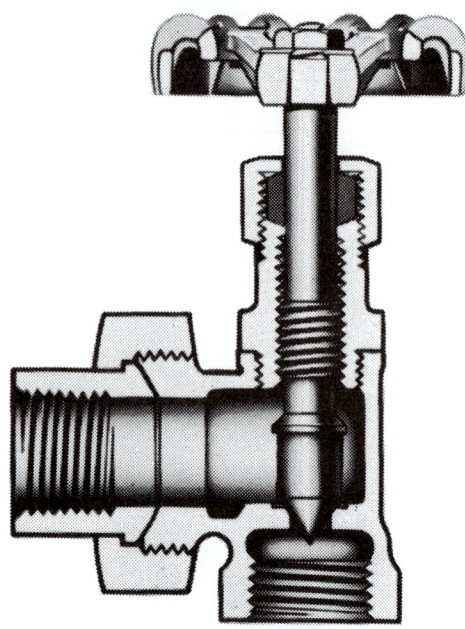

FIGURE 22.37 ■ The needle valve is good for accurate throttling on instrument and meter lines. *Courtesy Crane Co.*

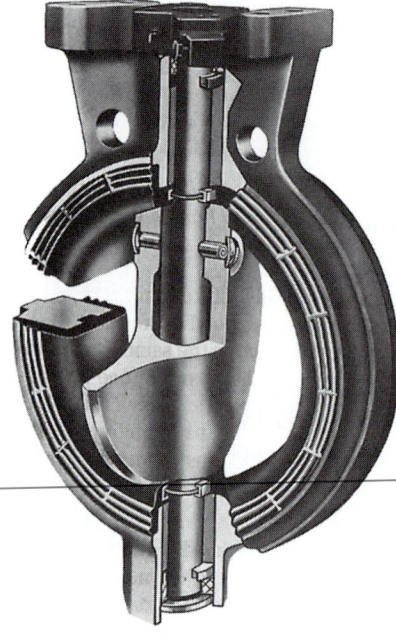

FIGURE 22.36 ■ The butterfly valve is excellent for regulating flow and creates a minimum pressure drop. *Courtesy Crane Co.*

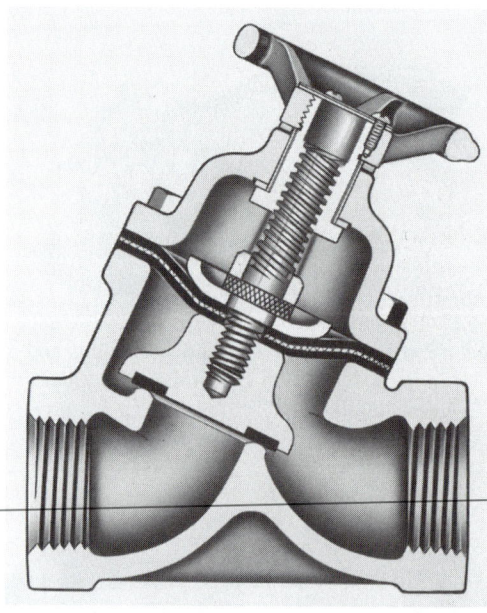

FIGURE 22.38 ■ The diaphragm valve protects the valve mechanism from the fluid. *Courtesy Crane Co.*

throttling. The flow through the needle valve changes direction much like the flow through a globe valve. The needle valve is used for high-temperature and high-pressure service in instrument, gauge, and meter lines.

Some types of service require that the working parts of the valve be sealed from the fluid stream. The protection of the valve parts is important when working with fluids that are corrosive, viscous, fibrous, or contain suspended solids and sludges. The

protection of the **fluid** is most important when dealing with food and beverages. The **diaphragm valve** suits this need. In place of a disc or other type of sealing mechanism, a diaphragm of rubber, neoprene, butyl, silicone, or other flexible material is used. The diaphragm is pushed down by the stem and creates a seal against a seat on the bottom of the valve. (See Figure 22.38.) The diaphragm also serves to protect the working parts of the valve. It has a smooth, streamlined flow, is easy to maintain, achieves positive flow control, and is leak-tight. It is suited for both on/off and regulating services up to 400° F (204° C).

Backflow Valve

Certain situations require that fluids be prevented from flowing backward in the pipe if a power failure or pump breakdown occurs. Check valves prevent backflow by closing when the fluid stops flowing.

The swing check valve as shown in Figure 22.39 is similar in construction to the gate valve and is used with a gate valve. The disc inside the swing check valve operates by gravity or the weight of the disc. It is best used with low-velocity liquids. The lift check valve in Figure 22.40 is similar to the globe valve and is used with a globe valve. A lift check valve operates by gravity and is available in either horizontal or vertical models.

Safety Valve

The high temperatures and pressures within many industrial processes are potentially dangerous and must be controlled and regulated to prevent serious accidents. The responsibility for keeping pressures at or below a given point is the task of safety valves and relief valves.

The pop safety valve actually pops wide open when the pressure in a pipe or piece of equipment reaches a set pressure.

These valves are used for steam, air, and gas lines only—never for liquids. The opening and closing of this valve is instantaneous. (See Figure 22.41.)

The relief valve performs the same function as the pop safety but is used for liquids only. (See Figure 22.42.) It is also set to

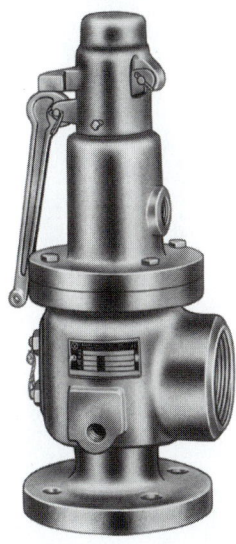

FIGURE 22.41 ■ The pop safety valve provides momentary release of pressure. *Courtesy Kunkle Valve Co., Inc.*

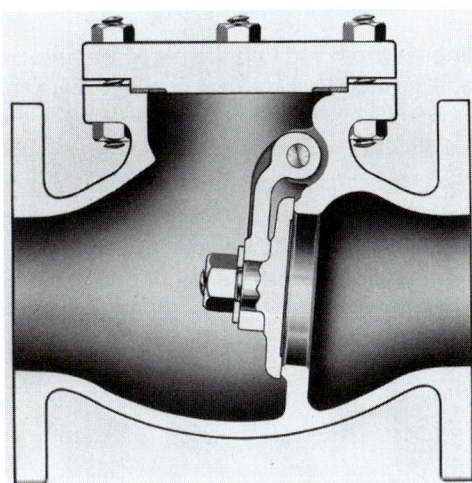

FIGURE 22.39 ■ The disc inside the swing check valve operates by gravity to check backflow. *Courtesy Crane Co.*

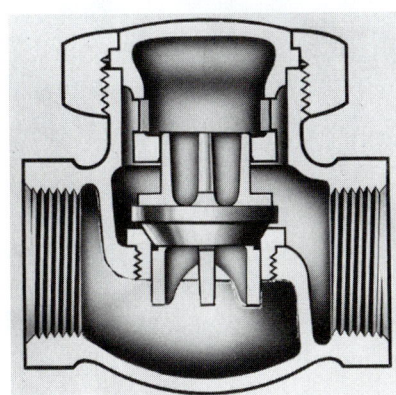

FIGURE 22.40 ■ The lift check valve is used with the globe valve to check backflow. *Courtesy Crane Co.*

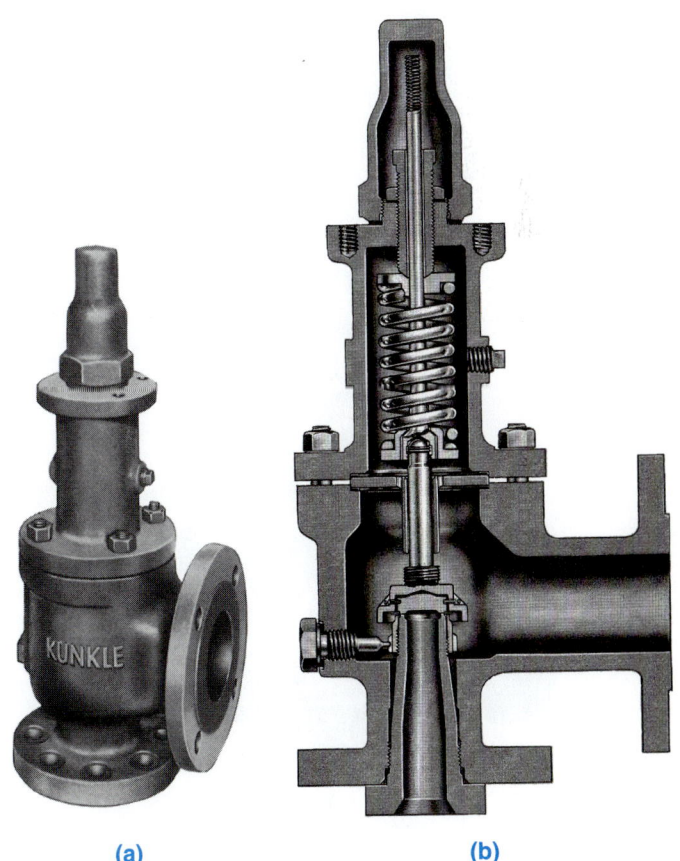

(a)   **(b)**

FIGURE 22.42 ■ (a) Relief valve exterior; (b) cutaway view; provides slow release of fluids and pressure. *Courtesy Kunkle Valve Co., Inc.*

(a)

open at a specific pressure, but it opens and closes slowly in response to changing pressure. This type of valve is found on most home hot water heaters.

Control Valve

The complex processes used in industry often demand instantaneous control and adjustment of flow, pressure, and temperature. That is where control valves are typically used. Most any type of valve can be a control valve. The example in Figure 22.43 shows one of the types used. The identifying component of the control valve is the controller or actuator. This is the mechanism that operates the valve. It can be operated by an electric motor, air cylinder, or hydraulic cylinder.

Pressure Regulating Valve

Processes relying on steam require a constant flow at a steady rate. This task falls on pressure regulators. This mechanism reduces the incoming pressure of the steam to the required service pressure. The regulator maintains the pressure at the specified level and provides a uniform flow. (See Figure 22.44.)

PIPE DRAFTING

Pipe drafting positions are found with consulting engineering companies, large industrial construction firms with engineering offices, and manufacturers of equipment that use piping. The entry-level drafter in these companies can begin by drawing revisions, known as markups, to existing drawings. These can be flow diagrams (nonscale schematic drawings) or piping drawings. As experience is gained, the drafter is assigned more complex drawings that must be constructed from engineering sketches.

(b)

FIGURE 22.43 ■ (a) Cage-guided control valve exterior view; (b) cutaway view of same valve. *Courtesy Dezurik Valve Co.*

FIGURE 22.44 ■ Pressure regulator. *Courtesy Spence Engineering.*

Engineers, designers, and technicians design the piping system, select and size the equipment and pipe, and give this information to the drafter. The drafter's task is to construct the required type of drawing or model, often using vendor catalogs, brochures, and charts, or other drawings containing reference information. Once a working knowledge of the subject is gained, the drafter can be given the responsibility of designing and laying out pipe runs. At this point, the drafter begins to use his or her knowledge of pipe fittings and reference information.

Plans and Sections (Elevations)

The most common and the most complex type of piping drawings are plans and sections that show building outlines, equipment, pipe, and pipe supports. Examples of a plan and several sections taken from that plan are shown in Figure 22.45. Study these drawings carefully. Look at the plan and sections together to try to determine where specific pipe runs are located. A little time spent now aids you in working with this type of drawing in the future.

Equipment shown on piping plan and section drawings is seldom drawn in detail, but instead an outline is used to illustrate clearances and access. Tanks, vessels, heat exchangers, pumps, columns, compressors, boilers, dryers, and reactors are just a few examples of some of the equipment with which a piping drafter may work.

In addition to major pieces of equipment, the drafter can be required to indicate pipe supports and hangers. Note in Figure 22.46, page 728, the extensive use of pipe hangers in the form of pipe clamps and iron rods. Pipe supports also must come from the bottom up, as seen in Figure 22.47, page 729. These are vital to a piping system because they provide stability and anchor points for the pipe. A few common pipe supports and hangers with their application in a sample piping system are shown in Figure 22.48, page 729. If the hanger or support is complex or of a special design, it is often illustrated on a "detail" drawing, as shown in Figure 22.49, page 730. When standard pipe clamps, anchors, hangers, and supports are used, they can be indicated by a symbol or note such as "P.S."

Single-Line Drawings

A time-saving method of drawing pipe is the single-line method. The centerline of the pipe is used to represent the pipe. The single-line technique can be used to better illustrate new pipe that is added to an existing installation or to show small-diameter pipe. Standard pipe symbols are used in single-line drawings, and the pipe and fittings are often drawn as heavy lines to provide contrast with other lines on the drawing.

Double-Line Drawings

The double-line method is the easiest to interpret, because it looks like the actual pipe. But it takes longer to draw unless a CADD program is used. This method is used extensively by equipment manufacturers to illustrate standard installation and in presentation drawings that are seldom revised. (See Figure

22.50, page 731.) The double-line method is also used to represent large diameter pipe, to show existing pipe at an industrial facility, or for small views that are easily visualized as double-line drawings. (See Figure 22.51, page 732.) When new or small pipe is shown in the single-line method on the same drawing, the contrast between the two styles aids in interpretation. The single-line method is seldom used for large-diameter pipe because clearances, interferences, spacing, and distances are not as easily seen. Where these factors are critical, the double-line method is preferred.

When laying out either type of piping drawing, it is best to begin with background information such as building outlines or structural steel column reference points. From these points, the equipment locations can be determined. Pumps often are represented by centerlines. Other large equipment must be drawn. Pipe centerline locations then must be determined. CADD drafters can use pipe centerline location lines of a different color layer or as tick marks on the screen then insert fittings and draw pipe as they go.

PIPING SYMBOLS AND DETAILS

Most piping drawings involve the use of symbols. Symbols need to be drawn only once, saved in a symbol library, and displayed in a menu. When a symbol is needed, it is simply picked from the menu, placed on the drawing, and scaled or rotated as needed. If your CADD system does not have a piping symbol library yet, it will save you time in the future if you begin to create one as soon as possible.

Software companies create standard and specialty symbols that are assembled in libraries. A productive part of the piping drafter's job is the use and placement of these standard symbols and shapes on new drawings.

Many of the symbol programs are "intelligent." That is, they contain pieces of information called attributes or tags that give meaning to the symbol. For example, a valve symbol can contain hidden attributes pertaining to its diameter, pressure rating, material, operating mechanism, specific type of styling, weight, price, and manufacturer. When the valve symbol is used on a drawing, all of the attributes become a part of the drawing's database.

Drafters using CADD systems should keep an easily accessible up-to-date record of all piping details. They are often needed again on the same project or on another project. Time savings are involved by just placing an entire detail on a drawing instead of constructing it again. An individual detail may not be exactly the same as a previous one, but even revising an existing detail is much quicker than redrawing it.

CADD APPLICATIONS 2-D

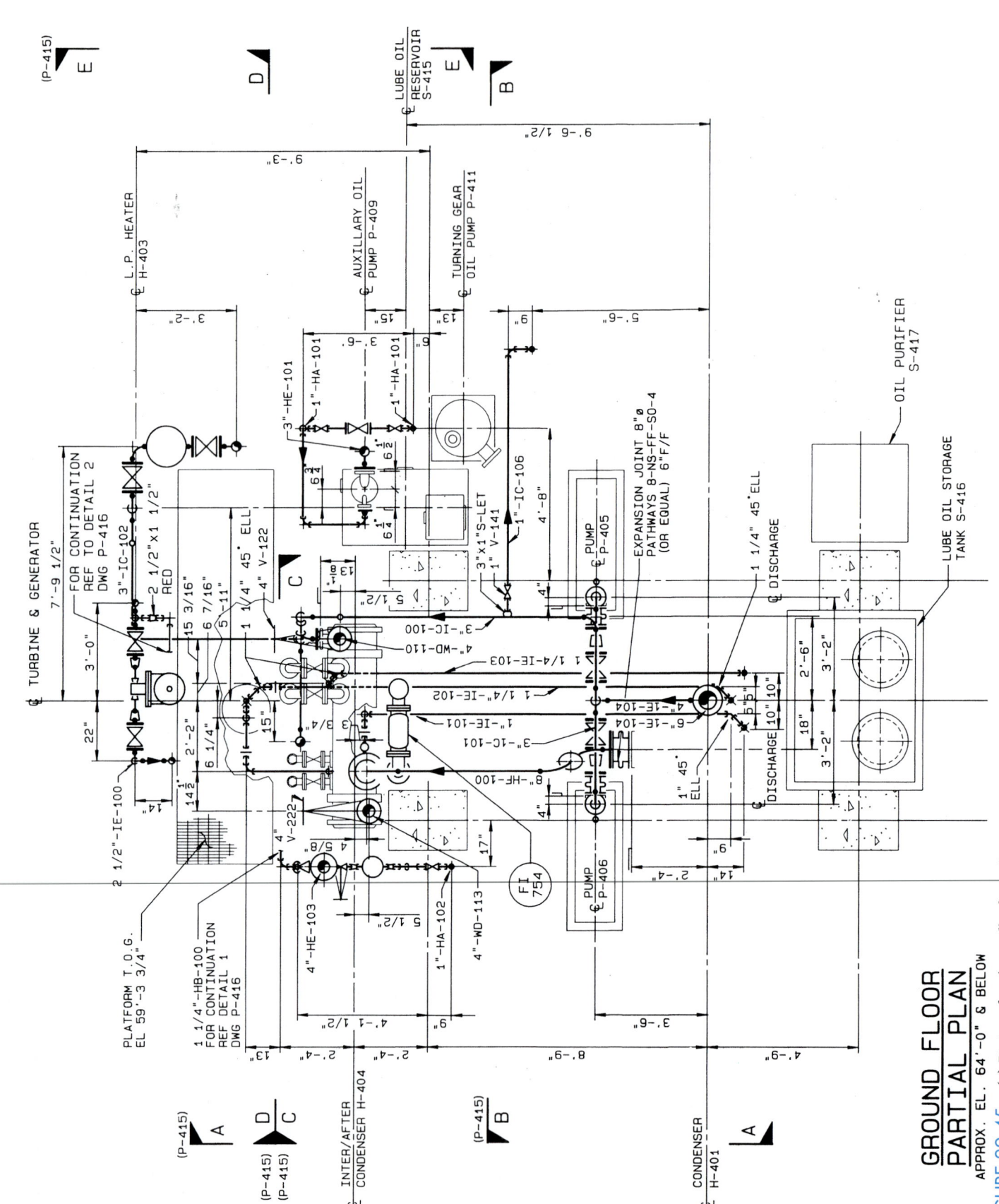

(a)

GROUND FLOOR PARTIAL PLAN

APPROX. EL. 64'-0" & BELOW

FIGURE 22.45 ■ (a) Piping plan (partial) of a turbine generator installation; (b-e) piping sections are taken from plan in (a). *Courtesy Schuchart and Associates.*

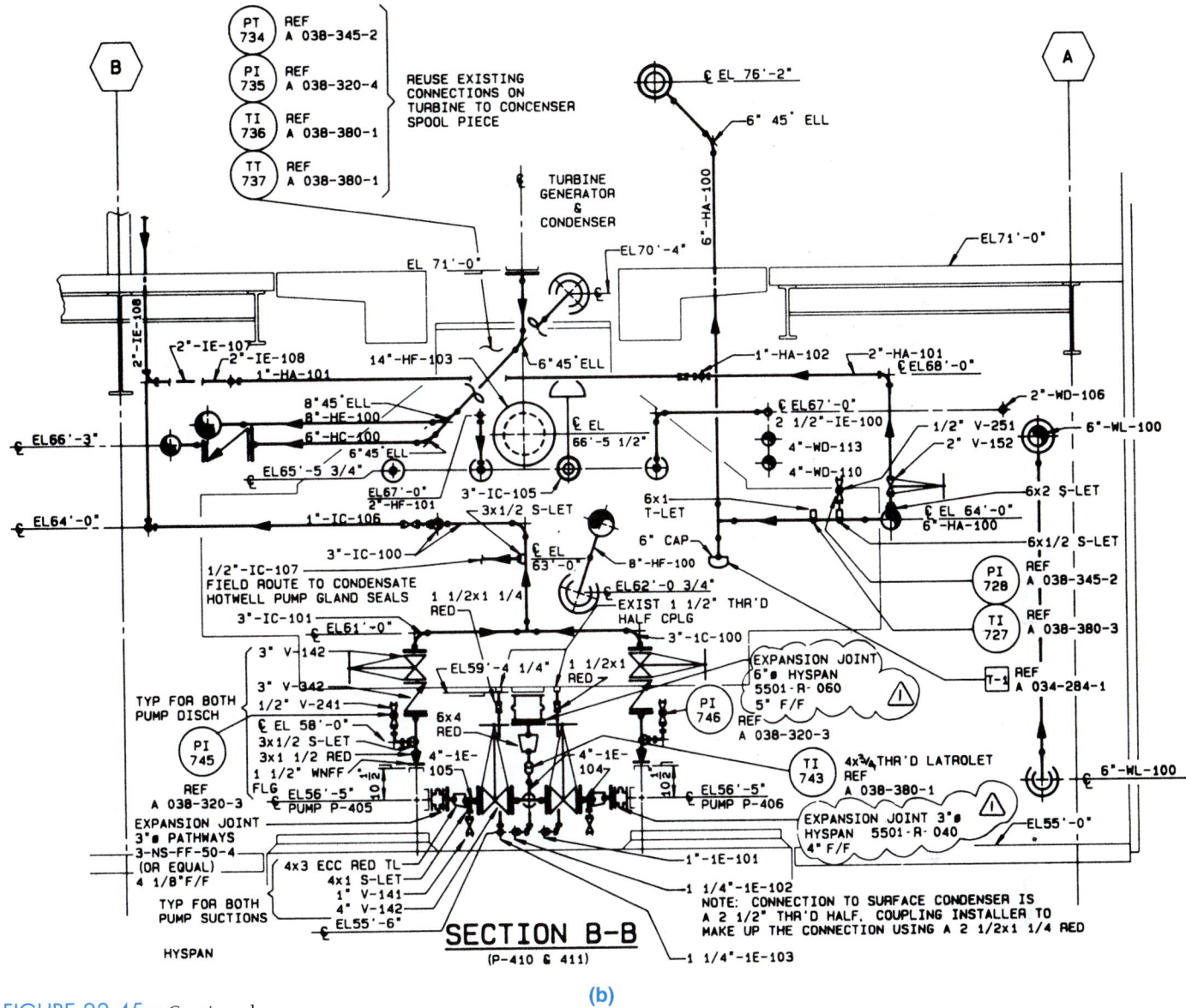

FIGURE 22.45 ▪ Continued.

Fitting and Valve Symbols

A major part of the piping drafters work is drawing fittings and valves. CD Appendix E provides commonly used valve symbols. Manual drafting piping templates are available in a variety of sizes. Some companies may even have their own special templates made for them.

Dimensions and Notes

Piping drawings normally are dimensioned in a style similar to architectural drawings but some companies use the mechanical style with dimension values placed in a break in the dimension lines. The architectural style of dimension lines are unbroken, and values are lettered above horizontal lines and to the left of and reading up on vertical dimension lines. Values are given in feet and inches with a dash between the two. Refer to the plan in Figure 22.45a for examples of piping plan dimensions.

Piping elevations or sections are not normally dimensioned like plan views. Instead of linear dimensions, elevations are used. Occasionally elevations for the bottom of the pipe (B.O.P.) are required. The sections shown in Figure 22.45b–e, pages 725–728, illustrate the use of elevations to call out the centerlines of pipes and the top of concrete (T.O.C.), or top of steel (T.O.S.), for example.

Lay out dimensions before any other written information is added to the drawing. Specified pipe lengths and dimensions to pipe direction changes should be given from reference points such as equipment and structural steel centerlines. Location dimensions of fittings, flanges, and valves (if needed) should be given to center points or flange faces. Pipe fittings and valves that are attached without pipe between them are often left undimensioned.

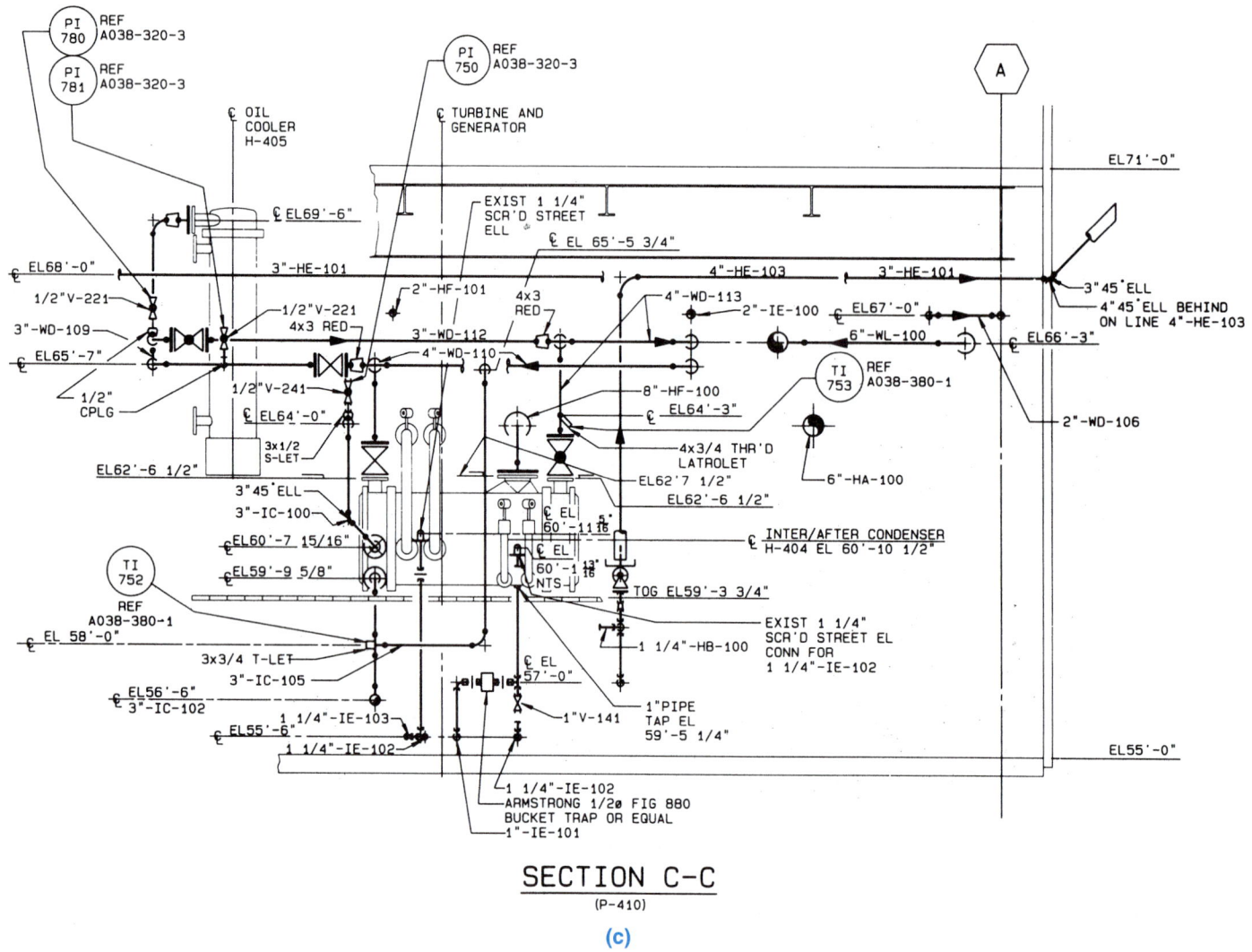

SECTION C-C
(P-410)

(c)

FIGURE 22.45 ■ Continued.

This is termed "fitting-to-fitting," and their location is simply the result of the combined fitting lengths and will not vary.

Dimension lines are important and should not be broken if passing through other lines. Pipes, equipment, or extension lines that must pass through dimension lines should be broken and the dimension line should remain intact.

Once dimension lines have been placed on the drawing, local notes and callouts can be given. These, too, should be placed close to where they apply without crowding other objects or creating a confused layout. Try to group local notes together when possible. This allows those who must interpret the drawing to find similar information without too much searching.

Pipe specification numbers can be written near the pipe and connected to it with a leader line, or the numbers can reside inside a symbol called a line balloon. The line balloon is either rounded or squared on the ends and is 1/4" wide and approximately 1" (6–25 mm) long. The preferred location for the line balloon is inside the pipe, but it may be placed outside the pipe and used with a leader line if space is unavailable. Some examples of pipe specification numbers are shown in Figure 22.52, page 732.

When checking your drawing, be sure all of the following items are provided:

- Pipe length dimensions.
- Dimension locations of all pipe direction changes.
- Locations of all fittings and valves if required.
- Elevation location of all pipe direction changes in section views.
- Size and type of valves.
- Size and type of fittings (if not readily identified).
- Pipe diameter, contents, and identification number.
- Pipe flow arrows.
- Equipment names and numbers.

PIPING DETAILS

A piping detail drawing can be made of any connection, application, or installation that cannot be readily distinguished from

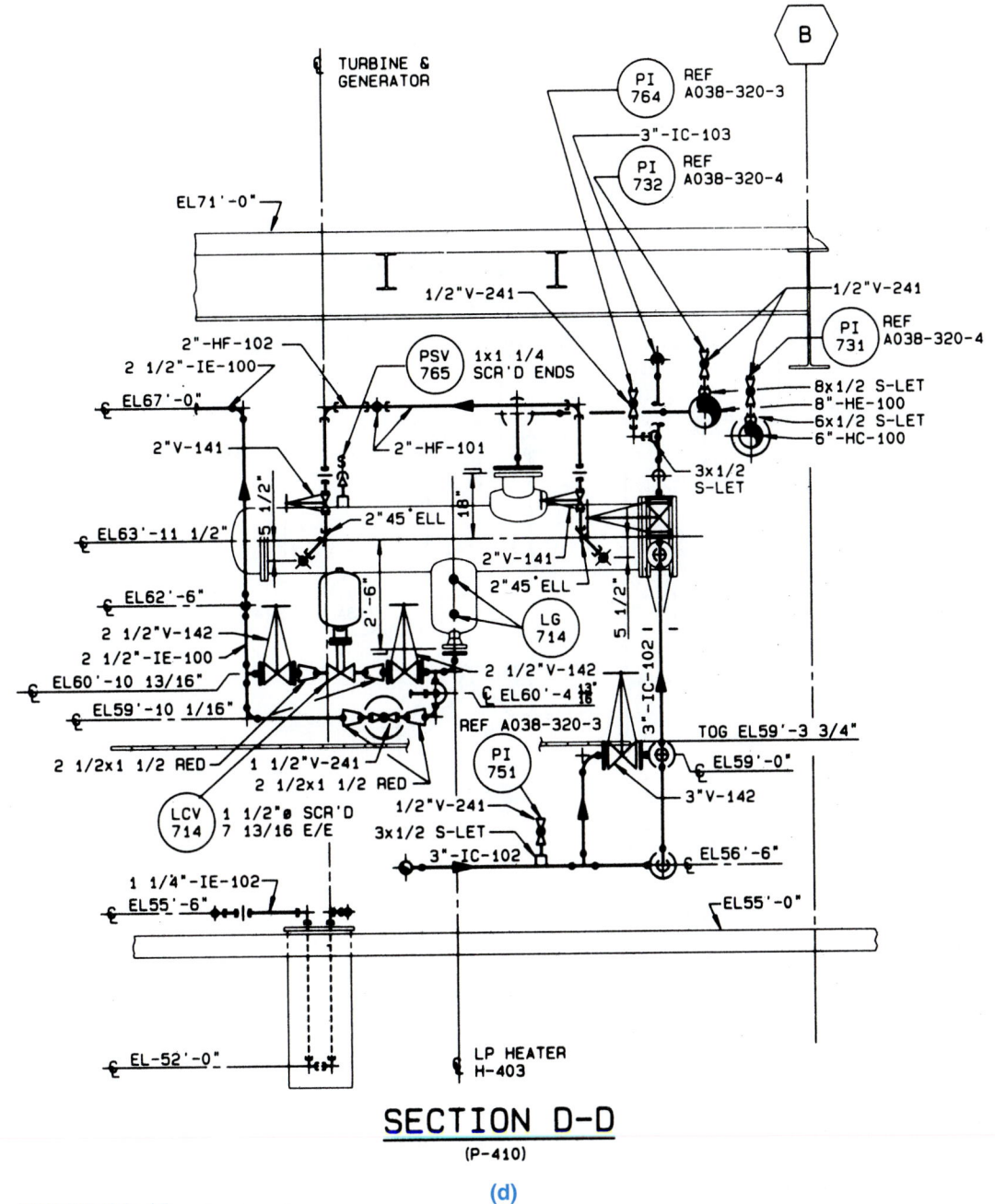

SECTION D-D
(P-410)

(d)

FIGURE 22.45 ■ Continued.

the other piping drawings. Details can often quickly clarify a point that would have taken valuable time to determine. Common piping detail drawings are used for the following:

- Special pipe connections or fittings, or special valve arrangements.
- Small-diameter pipe and fitting assemblies.
- Special pipe support arrangements.
- Tank attachment details.
- Minor structural alterations (concrete, wood, and steel).
- Operating and installation procedures.

The engineers and drafters on the job must determine when piping details are needed and if they are to be drawn orthographically or pictorially. Some common piping details are shown in Figure 22.53, page 733.

Piping Isometric and Spool Drawings

A piping isometric drawing is a pictorial view of a piping system complete with fittings, valves, dimensions, notes, and possibly instrumentation. This single view of the piping system eliminates the need for additional views. It is often done after the plans and elevations are completed.

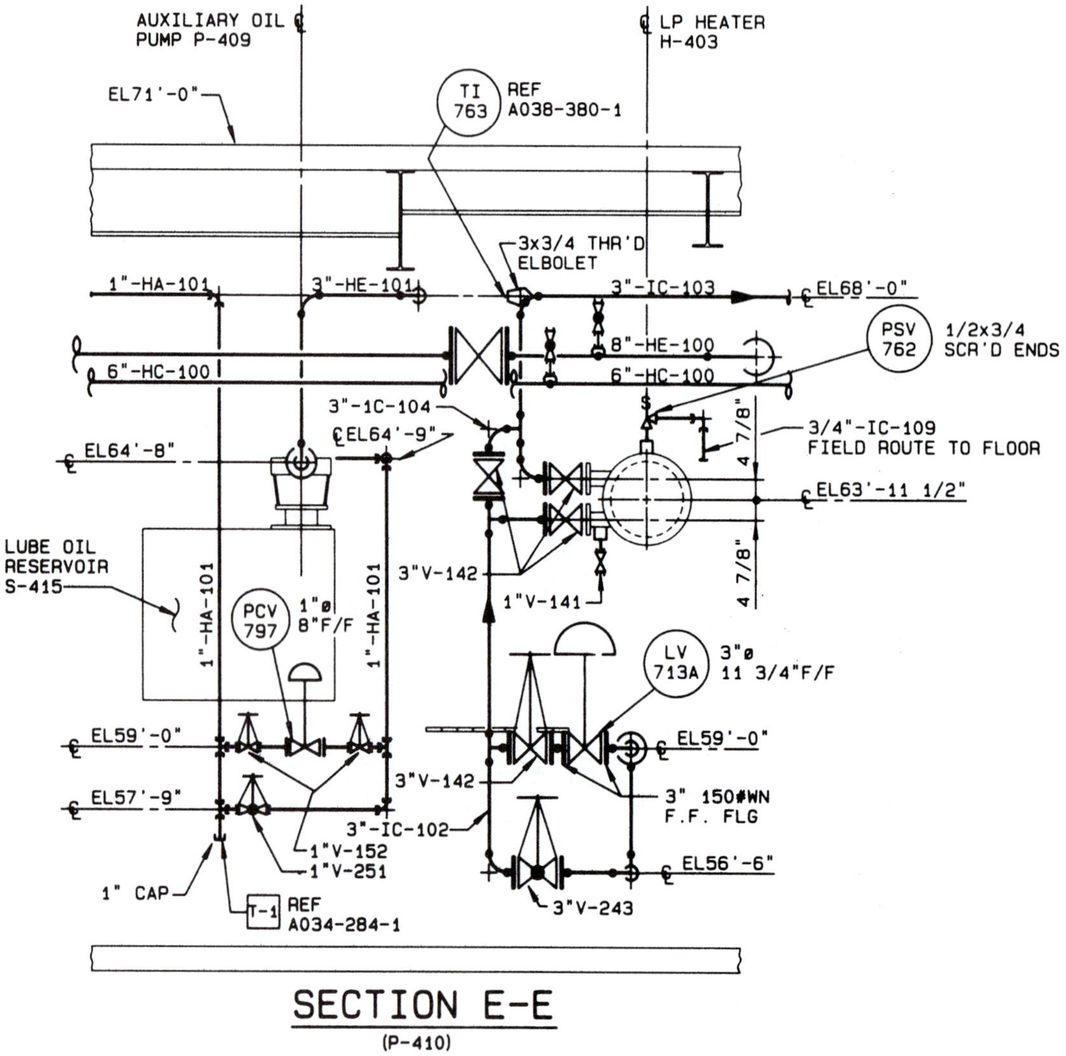

AUXILIARY OIL
PUMP P-409

LP HEATER
H-403

EL71'-0"

TI 763 REF A038-380-1

3x3/4 THR'D ELBOLET

1"-HA-101   3"-HE-101   3"-IC-103   EL68'-0"

PSV 762 1/2x3/4 SCR'D ENDS

8"-HE-100

6"-HC-100   6"-HC-100

3"-IC-104

EL64'-9"

3/4"-IC-109 FIELD ROUTE TO FLOOR

EL64'-8"

EL63'-11 1/2"

LUBE OIL RESERVOIR S-415

3"V-142

PCV 797 1"ø 8"F/F

1"V-141

4 7/8"

4 7/8"

1"-HA-101   1"-HA-101

LV 713A 3"ø 11 3/4"F/F

EL59'-0"   EL59'-0"

3"V-142

3" 150#WN F.F. FLG

EL57'-9"

3"-IC-102

1"V-152

1"V-251

1" CAP

T-1 REF A034-284-1

EL56'-6"

3"V-243

SECTION E-E
(P-410)
(e)

FIGURE 22.45 ■ Concluded.

FIGURE 22.46 ■ Pipe hangers are used to suspend pipe and conduit from the ceiling in this paper mill expansion project. *Courtesy Publisher's Paper Co.*

There are many advantages of a piping isometric drawing. For example, a single run of pipe from beginning to end can be shown in one view and is more easily visualized than corresponding orthographic views. Many companies draw piping isometrics to scale, which makes layout and visualization more accurate. Some firms do *not* draw piping isometrics to scale. If an isometric is not drawn to scale, it eliminates the need for large sheets and is most often drawn on B-size media. Dimensions are given on the isometric view, and straight lengths of pipe are drawn proportionally to indicate long and short pieces. Notice this in the isometric shown in Figure 22.54. (See page 734.)

It is easier to check the drawing for interferences and clearances if the isometric is drawn to scale. A bill of materials is often included on the piping isometric, which allows the pipe fitter, checker, and purchasing agent to cross-check the drawing with a list of the materials required.

Some companies illustrate a piping system or portions of a piping system in a pictorial manner for purposes of orientation, training, assembly, and installation. It is not often practical for large and complex systems to be shown in isometric form if

(Continued on page 731)

FIGURE 22.47 ■ Pipe supports anchored to the ground are used on this air compressor piping at a chemical processing plant. *Courtesy Ingersoll-Rand.*

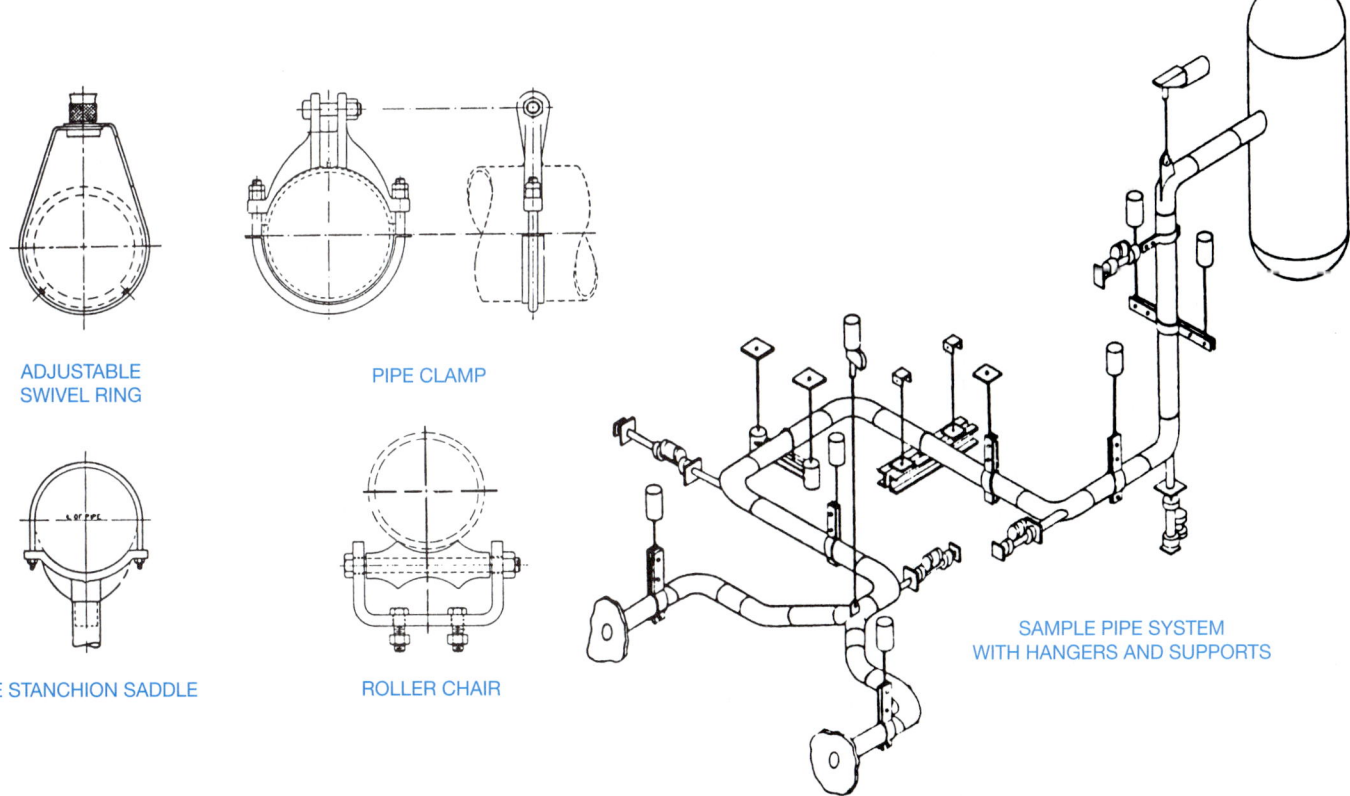

ADJUSTABLE
SWIVEL RING

PIPE CLAMP

PIPE STANCHION SADDLE

ROLLER CHAIR

SAMPLE PIPE SYSTEM
WITH HANGERS AND SUPPORTS

FIGURE 22.48 ■ Examples of common pipe hangers and supports. *Courtesy ITT Grinnell Corp.*

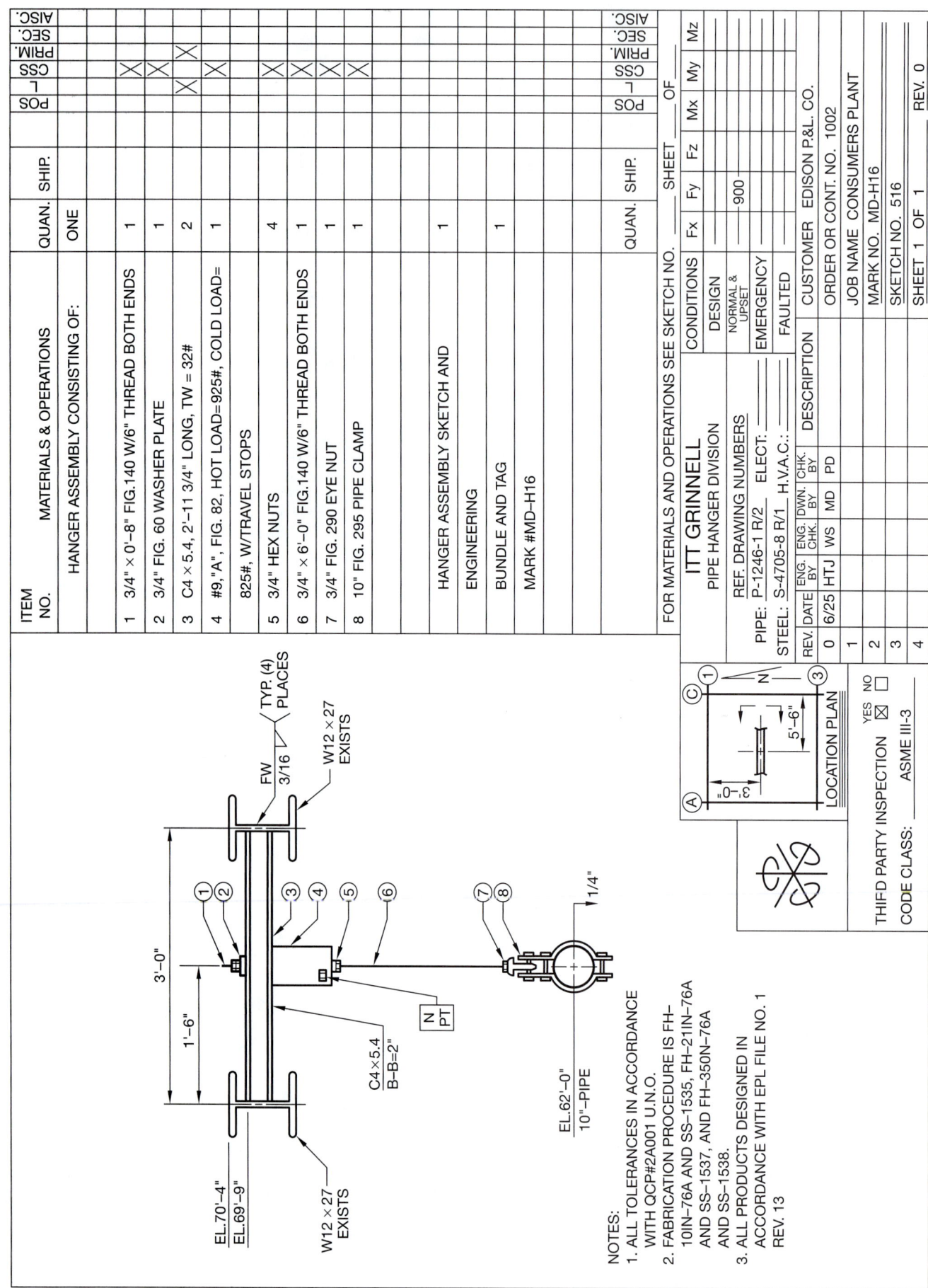

FIGURE 22.49 ■ Detail drawing of pipe hanger installation. *Courtesy ITT Grinnell Corp.*

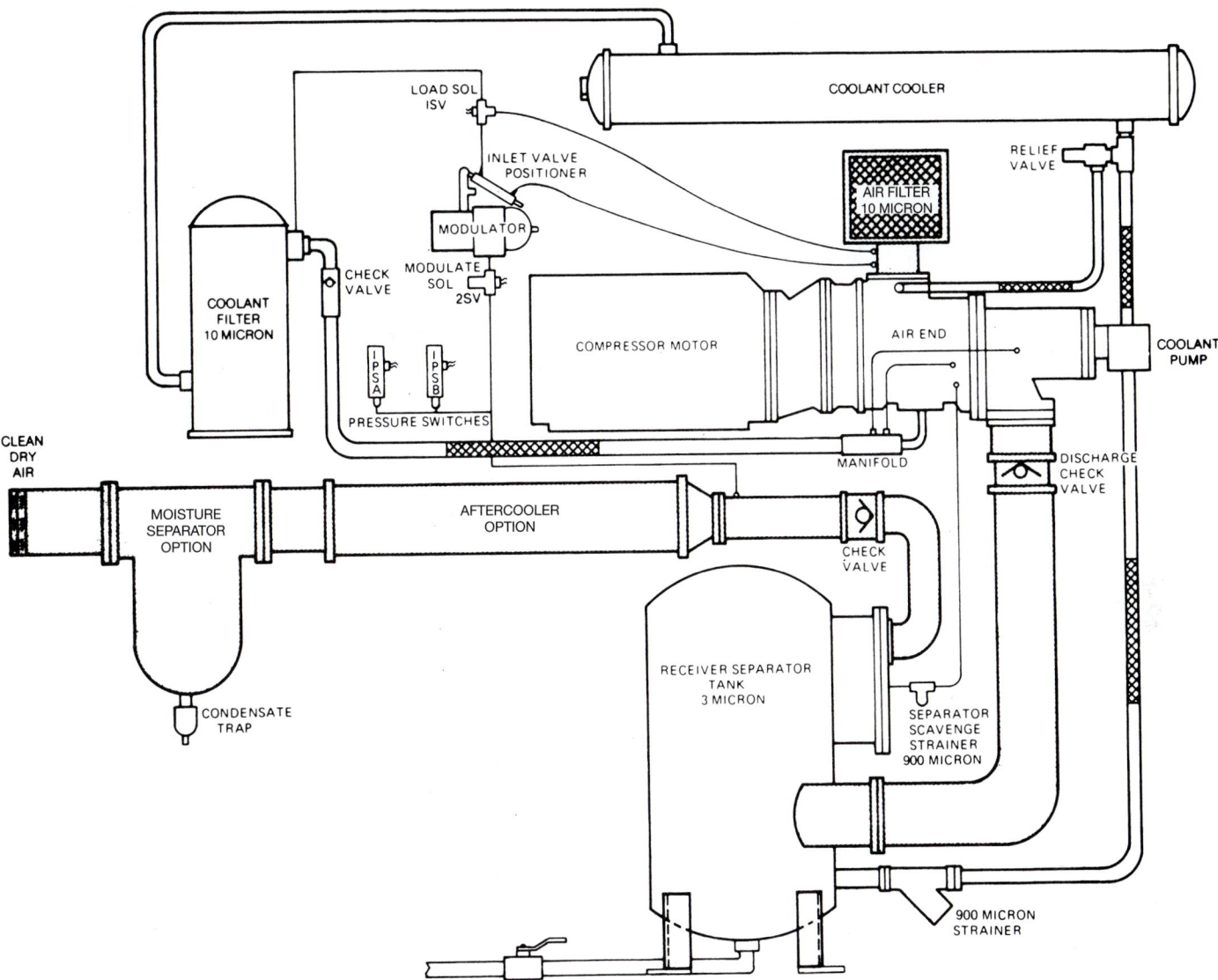

FIGURE 22.50 ■ Manufacturer's double-line piping drawing of a compressed air system. *Courtesy Ingersoll-Rand.*

manual drafting is used. However, CADD systems allow companies to create complete, pictorially modeled systems.

Piping systems are constructed by pipe fitters who must initially weld or thread pipe fittings and pipe together to create pipe spools. (See Figure 22.55, page 735.) The pipe fitters work from assembly instructions called spool drawings, spools, or spool sheets. Spools are also called pipe assemblies. The spools are then transported to the job site and installed. Pipe spools awaiting installation are shown in Figure 22.56, page 735.

Pipe spools are drawn in isometric or orthographic, often on B-size sheet, and may or may not be drawn to scale. (See Figure 22.11, page 707.) Spool sheets are assembly drawings that contain complete dimensions and a bill of materials (B.O.M.) that indicates the exact size and specifications for each fitting. Extra pipe is usually added to pipe lengths to compensate for errors. If the drawing is orthogonal, it only shows the views required to present all of the lengths of straight pipe and fittings. Valves

are never shown on spool drawings because they are installed at the job site when the spools are erected.

Models

The construction of a model can take months, even years, making it impossible to ship to the job site during most of the construction phases. Photos of the model are used at the job site for checking purposes. Drawings of the system must still be generated for use in construction.

One of the greatest advantages of a model is the ability to easily see clearances and interferences. (See Figure 22.57, page 735.) These can be quickly transferred to the drawings, thus eliminating many errors that might occur if only drawings were used. Some companies use a three-dimensional digitizing system to locate pipe, equipment, and structural steel on the model. This information is then fed into the CADD system and used to generate drawings.

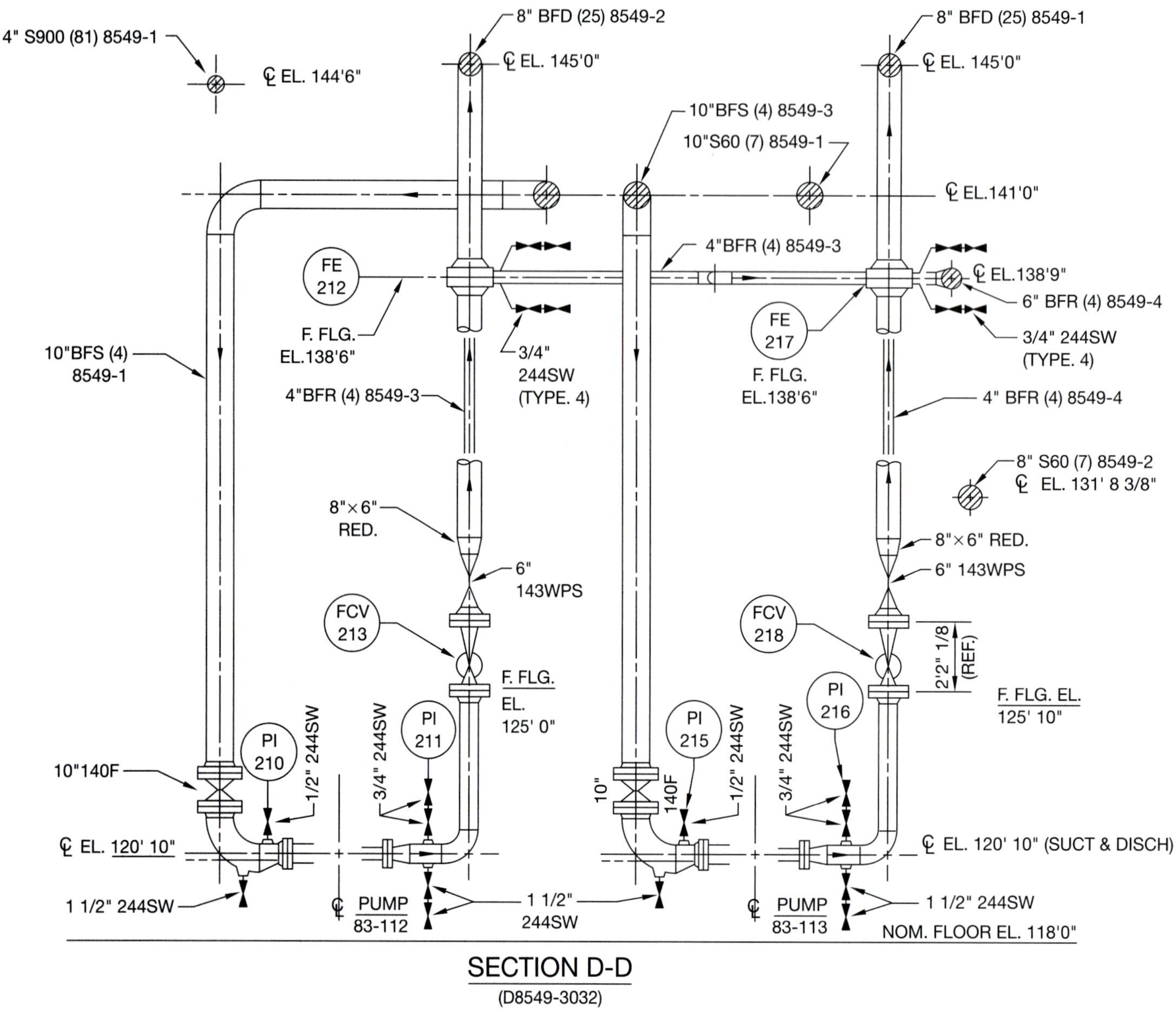

4" S900 (81) 8549-1

Ȼ EL. 144'6"

8" BFD (25) 8549-2

Ȼ EL. 145'0"

8" BFD (25) 8549-1

Ȼ EL. 145'0"

10"BFS (4) 8549-3

10"S60 (7) 8549-1

Ȼ EL.141'0"

FE 212

4"BFR (4) 8549-3

Ȼ EL.138'9"

F. FLG. EL.138'6"

6" BFR (4) 8549-4

FE 217

3/4" 244SW (TYPE. 4)

10"BFS (4) 8549-1

4"BFR (4) 8549-3

3/4" 244SW (TYPE. 4)

F. FLG. EL.138'6"

4" BFR (4) 8549-4

8" S60 (7) 8549-2
Ȼ EL. 131' 8 3/8"

8"×6" RED.

6" 143WPS

8"×6" RED.

6" 143WPS

FCV 213

FCV 218

2'2" 1/8 (REF.)

F. FLG. EL. 125' 0"

F. FLG. EL. 125' 10"

PI 211

PI 215

PI 216

PI 210

1/2" 244SW

3/4" 244SW

1/2" 244SW

3/4" 244SW

10"140F

10" 140F

Ȼ EL. 120' 10"

Ȼ EL. 120' 10" (SUCT & DISCH)

1 1/2" 244SW

Ȼ PUMP 83-112

1 1/2" 244SW

Ȼ PUMP 83-113

1 1/2" 244SW

NOM. FLOOR EL. 118'0"

SECTION D-D
(D8549-3032)

FIGURE 22.51 ■ Sections from a typical double-line piping drawing. *Courtesy Schuchart and Associates.*

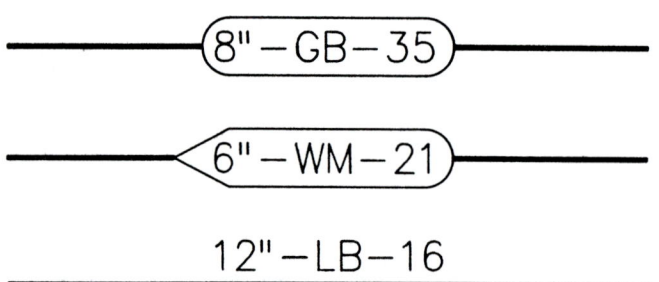

FIGURE 22.52 ■ Methods of indicating pipe line specifications.

Equipment on the model is often shown in the form of block shapes, details seldom being required. Equipment is created from standard, inexpensive materials such as rigid foam, plastic, and wood. Piping and structural steel components are available from model supply vendors.

Upon completion, the model is shipped to the job site where it is used in the final stages of the construction project. It is also used to check the design drawings of the system. Operators, maintenance technicians, and engineers can be trained on the workings of the system using the model. Revisions, alterations, and design changes can be accomplished efficiently using the model to check for clearances, and the addition of new equipment and piping locations.

Piping model kits are available for use in piping and blueprint reading classes. These kits come with all the parts necessary to
(Continued on page 737)

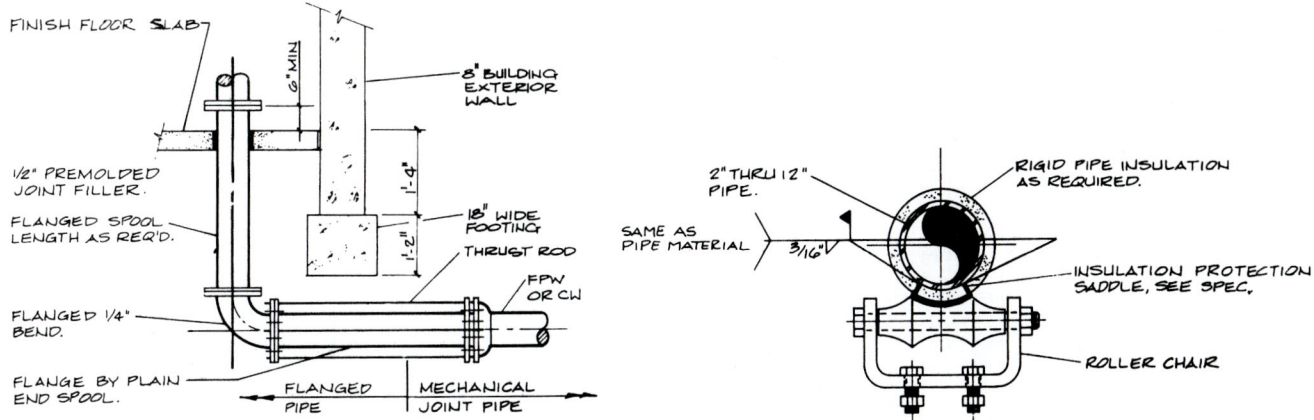

FINISH FLOOR SLAB

6" MIN

8" BUILDING EXTERIOR WALL

1/2" PREMOLDED JOINT FILLER.

FLANGED SPOOL LENGTH AS REQ'D.

1'-4"

1'-2"

18" WIDE FOOTING

THRUST ROD

FPW OR CW

FLANGED 1/4" BEND.

FLANGE BY PLAIN END SPOOL.

FLANGED PIPE

MECHANICAL JOINT PIPE

NOTE:
1. SEE MECHANICAL DRAWINGS FOR LOCATIONS AND SIZES.
2. SEE FOUNDATION DRAWINGS FOR ANY DIFFERENCES IN FOOTING DEPTHS.

⑧ THRUST TIE

(a)

2" THRU 12" PIPE.

RIGID PIPE INSULATION AS REQUIRED.

SAME AS PIPE MATERIAL

3/16" V

INSULATION PROTECTION SADDLE, SEE SPEC.

ROLLER CHAIR

⑤ ROLLERCHAIR PIPE SUPPORT

(c)

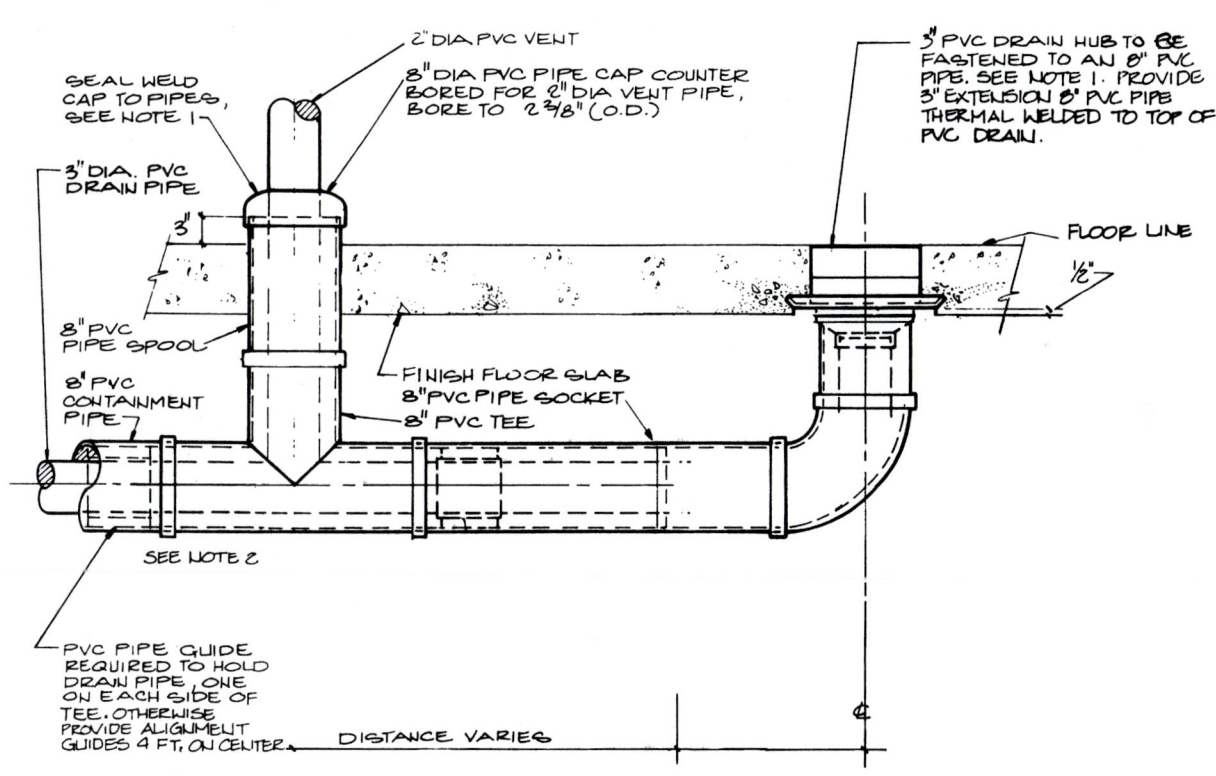

SEAL WELD CAP TO PIPES, SEE NOTE 1

2" DIA PVC VENT

8" DIA PVC PIPE CAP COUNTER BORED FOR 2" DIA VENT PIPE, BORE TO 2 3/8" (O.D.)

3" PVC DRAIN HUB TO BE FASTENED TO AN 8" PVC PIPE. SEE NOTE 1. PROVIDE 3" EXTENSION 8" PVC PIPE THERMAL WELDED TO TOP OF PVC DRAIN.

3" DIA. PVC DRAIN PIPE

3"

8" PVC PIPE SPOOL

8" PVC CONTAINMENT PIPE

FINISH FLOOR SLAB
8" PVC PIPE SOCKET
8" PVC TEE

FLOOR LINE

1/2"

SEE NOTE 2

PVC PIPE GUIDE REQUIRED TO HOLD DRAIN PIPE, ONE ON EACH SIDE OF TEE. OTHERWISE PROVIDE ALIGNMENT GUIDES 4 FT. ON CENTER.

DISTANCE VARIES

₵

NOTES:
1. THERMAL WELD ALL SEAMS WHERE POSSIBLE.
2. TERMINATE 8" CONTAINMENT PIPE 3" INTO TRENCH OR SUMP.
3. SLOPE PIPE AND CONTAINMENT PIPE 1/8"/FT. TOWARD TRENCH OR SUMP.
4. PROVIDE RUNNING TRAP ON 3" PIPE INTO TRENCH OR SUMP.

⑫ **DRAIN PIPING**

(b)

FIGURE 22.53 ■ (a and b) Piping details of buried pipe; (c) pipe hanger. *Courtesy CH₂M Hill—I.D.C.*

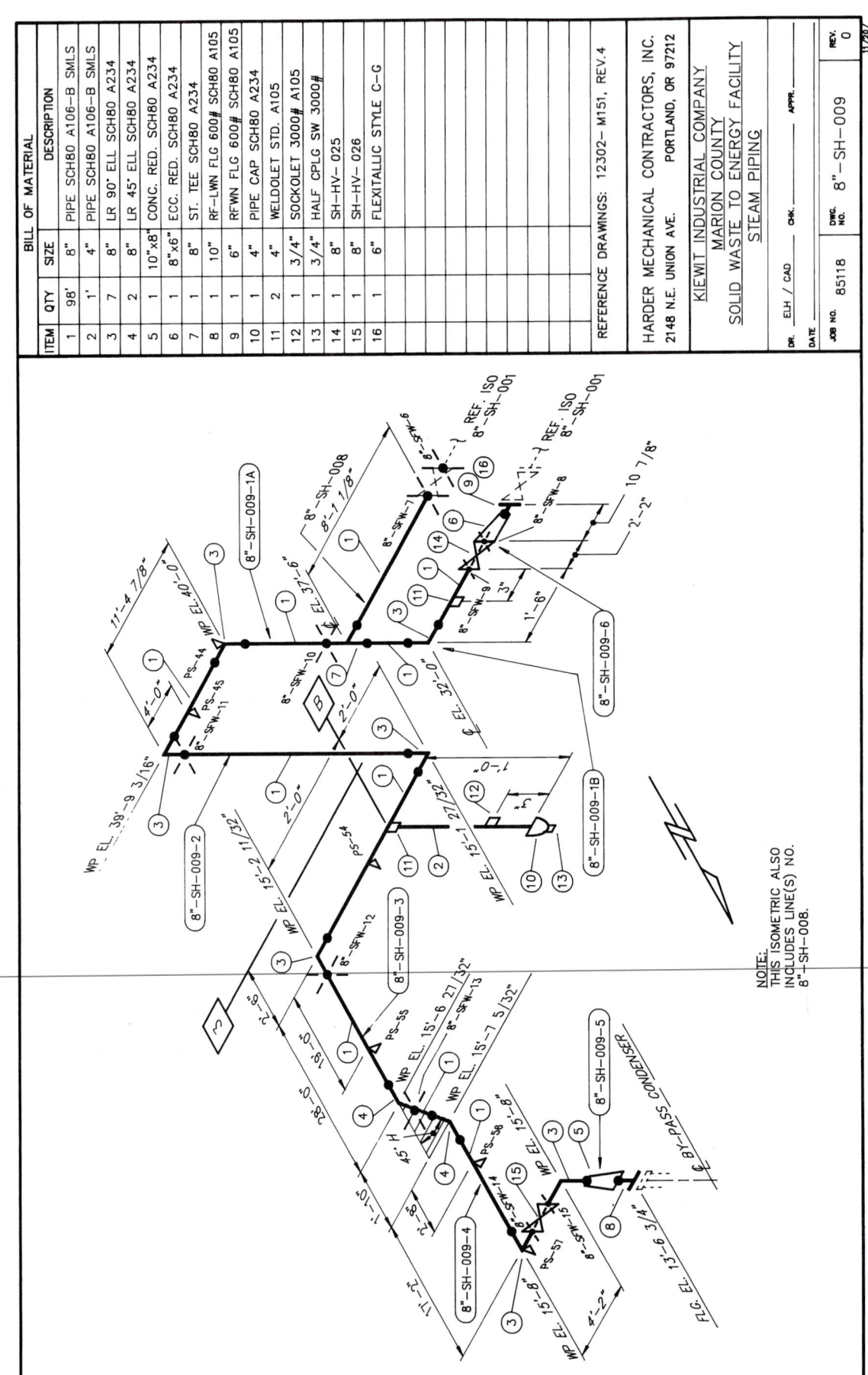

BILL OF MATERIAL

ITEM	QTY	SIZE	DESCRIPTION
1	98'	8"	PIPE SCH80 A106–B SMLS
2	1'	4"	PIPE SCH80 A106–B SMLS
3	7	8"	LR 90° ELL SCH80 A234
4	2	8"	LR 45° ELL SCH80 A234
5	1	10"×8"	CONC. RED. SCH80 A234
6	1	8"×6"	ECC. RED. SCH80 A234
7	1	8"	ST. TEE SCH80 A234
8	1	10"	RF–LWN FLG 600# SCH80 A105
9	1	6"	RFWN FLG 600# SCH80 A105
10	1	4"	PIPE CAP SCH80 A234
11	2	4"	WELDOLET STD. A105
12	1	3/4"	SOCKOLET 3000# A105
13	1	3/4"	HALF CPLG SW 3000#
14	1	8"	SH–HV–025
15	1	8"	SH–HV–026
16	1	6"	FLEXITALLIC STYLE C–G

REFERENCE DRAWINGS: 12302– M151, REV.4

HARDER MECHANICAL CONTRACTORS, INC.
2148 N.E. UNION AVE.   PORTLAND, OR 97212

KIEWIT INDUSTRIAL COMPANY
MARION COUNTY
SOLID WASTE TO ENERGY FACILITY
STEAM PIPING

DR. ELH / CAD	CHK.	APPR.		REV.
DATE	DWG. NO. 8"–SH–009			0
JOB NO. 85118				11/20/

(a)

FIGURE 22.54 ■ (a) Piping isometric with nonscale proportional dimensioning. *Courtesy Harder Mechanical Contractors, Inc.*

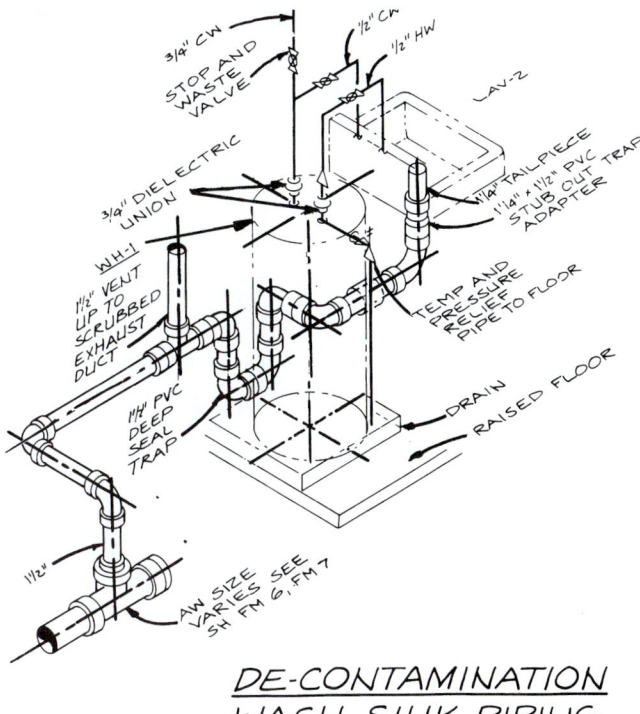

DE-CONTAMINATION
WASH SINK PIPING
DETAIL ③

(b)

FIGURE 22.54 ■ Continued. (b) Nonscale double-line piping
isometric. *Courtesy CH₂M Hill—I.D.C.*

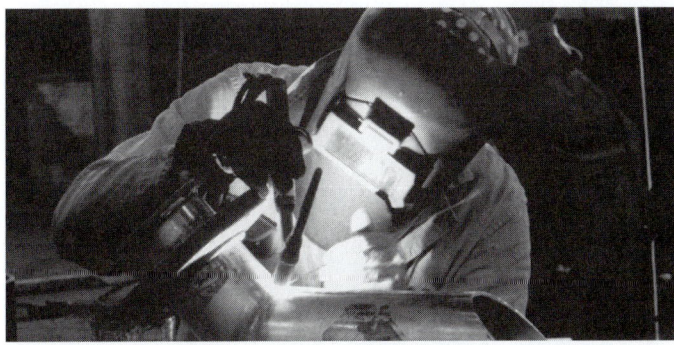

FIGURE 22.55 ■ Pipe fitter assembling a piping spool. *Courtesy
Alaskan Copper Works.*

FIGURE 22.56 ■ Assembled pipe spools awaiting installation. *Courtesy Alaskan Copper Works.*

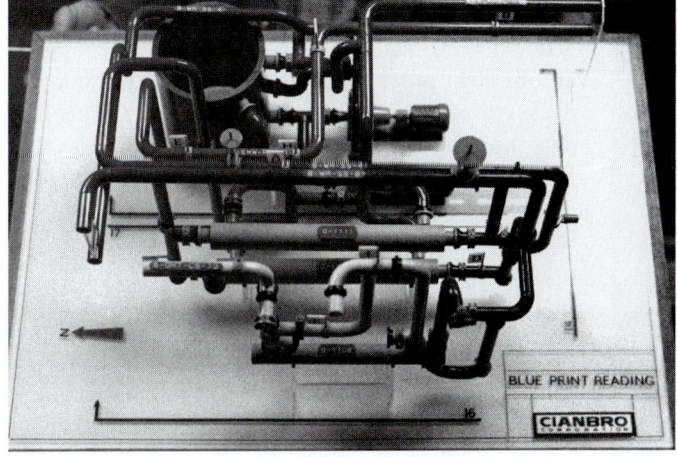

FIGURE 22.57 ■ Piping models are good visualization tools that allow designers to easily check clearances. Models
are also excellent teaching aids. *Courtesy Engineering Model Associates, Inc.*

THREE-DIMENSIONAL SYSTEM MODELS

In addition to increasing productivity in generating drawings and materials lists, some CADD systems allow you to create a 3-D, full-color model of the system. This model can be shown in wire or solid form, colored and shaded in any fashion, and rotated to suit the needs of the user. (See Figure 22.58.) Hard-copy prints of any aspect of the piping system can be generated for use in the design or construction of the system.

A CADD model can be used to create a scale model and/or orthographic and pictorial drawings for the project. This is a major step in the effort to generate a single data base for an entire construction project that is accessible to all engineers, technicians, drafters, and clients. Once all the design information is in the system, any type of data in the form of plotted drawings, reports, tables, material lists, estimates, bills, and a variety of screen displays can be requested. As changes are made to the design, it is reflected in all displays, hard-copy prints, and plotted drawings.

Future product and industrial piping design and construction may be handled completely by a computer system. This central system containing an extensive database including building or site sizes and construction codes, are able to automatically route and lay out pipe, valves, fittings, and equipment. The more routine tasks are performed by the computer, the more time is available for the overall design of the system by the engineer, technician, or

FIGURE 22.58 ■ This 3-D model of a piping system has been shaded and rendered to achieve a realistic presentation. *Courtesy International Software Systems, Inc.*

design drafter. A knowledge of CADD and a continuing interest in computer and software advances helps the piping engineer or drafter/designer to learn and change as the technology advances.

3-D DIGITIZING

Three-dimensional digitizers allow you to pick points on an object in order to input 3-D model data into CADD packages such as AutoCAD and CADKEY. This form of data input is called reverse engineering because the real part is used to construct a model. (See Figure 22.59.)

The 3-D digitizer has an articulated arm that can be moved and rotated in order to pick any point on an existing object or model. The data is transmitted as 3-D points to the CADD software through a cable attached to a serial port on the computer. This process is useful for collecting CADD data of an existing object in order to create a computer model. The computer model can then be revised in order to construct a new part.

This CADD input is also useful for digitizing an existing scale model of an object, building, or industrial site. The digitized data from the model can then be used to create construction drawings and 3-D computer models and renderings.

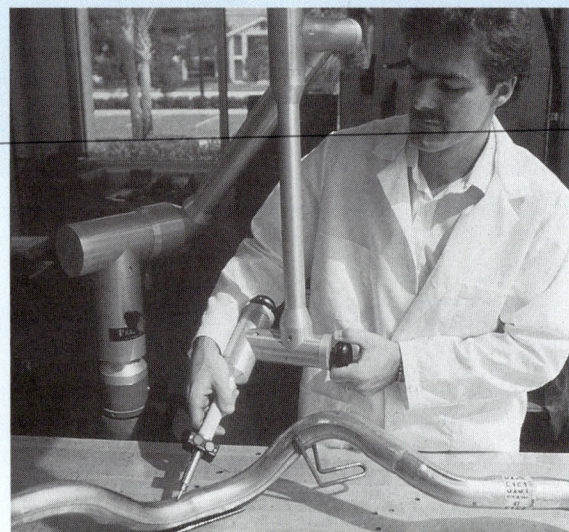

FIGURE 22.59 ■ The dimensions of an existing pipe can be input to the CADD software using a 3-D digitizer. *Courtesy FARO Technologies, Inc.*

construct a small piping and equipment arrangement. These models aid in visualization and enable the student to better construct piping drawings.

DRAWING REVISIONS

Drawing revisions are common in the architectural, structural, piping, and construction industries. Revisions can be caused for a number of reasons; for example, changes requested by the owner, job site corrections, correcting errors, or code changes. Changes are done in a formal manner by submitting an addendum to the contract, which is a written notification of the change or changes and is accompanied by a drawing that represents the change.

Revision Clouds

A revision cloud is placed around the area that is changed. The revision cloud is a cloudlike circle around the change as shown in Figure 22.60. CADD programs that are commonly used for architectural and structural drafting have commands that allow you to easily draw the revision cloud. There is also a triangle with a revision number inside that is placed next to the revision cloud or along the revision cloud line as shown in Figure 22.60. The triangle is commonly called a delta. The number is then correlated to a revision note placed somewhere on the drawing, or in the title block as shown in Figure 22.61. Each company has a desired location for revision notes, although common places are in the corners of the drawing, in a revision block or table, or in the title block. This practice is not as clearly defined as in ANSI/ASME standard drawings. The note is used on the drawing to explain the change. If a reference is given in the title block, then detailed information about the revision is normally provided in the revision document that is filed with the project information. The revision document is typically filled out and filed for reference. Changes can cause increased costs in the project.

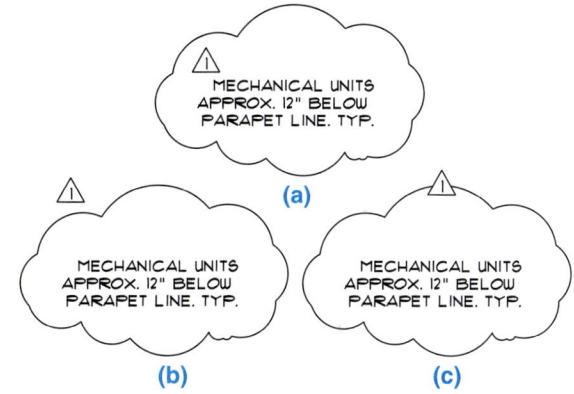

FIGURE 22.60 ■ Placement of the delta reference with the revision cloud: (a) delta inside of the revision cloud; (b) delta outside of the revision cloud; (c) delta inserted in the revision cloud line.

FIGURE 22.61 ■ Revision reference in the title block. The specific information about the revision is found in the job file. *Portion of title block courtesy Ankrom Moisan Architects.*

It is easy to draw revision clouds with CADD. AutoCAD, for example, has a REVCLOUD command that allows you to specify the revision cloud arc length and draw a revision cloud around any desired area. The command works by picking a start point and then moving the cursor in the desired direction to create the revision cloud, as shown in Figure 22.62a. Create the revision cloud in a pattern around the desired area while moving the cursor back toward the start point. Press the pick button when the cursor is back at the start point to complete the revision cloud as shown in Figure 22.62b.

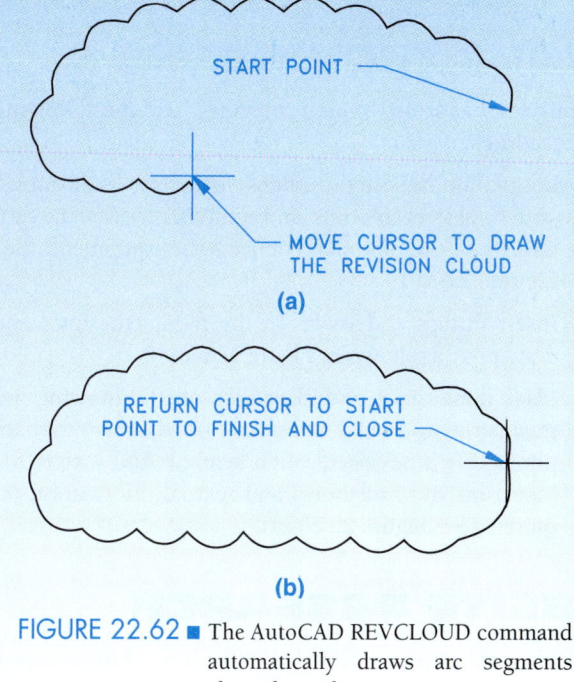

FIGURE 22.62 ■ The AutoCAD REVCLOUD command automatically draws arc segments along the path.

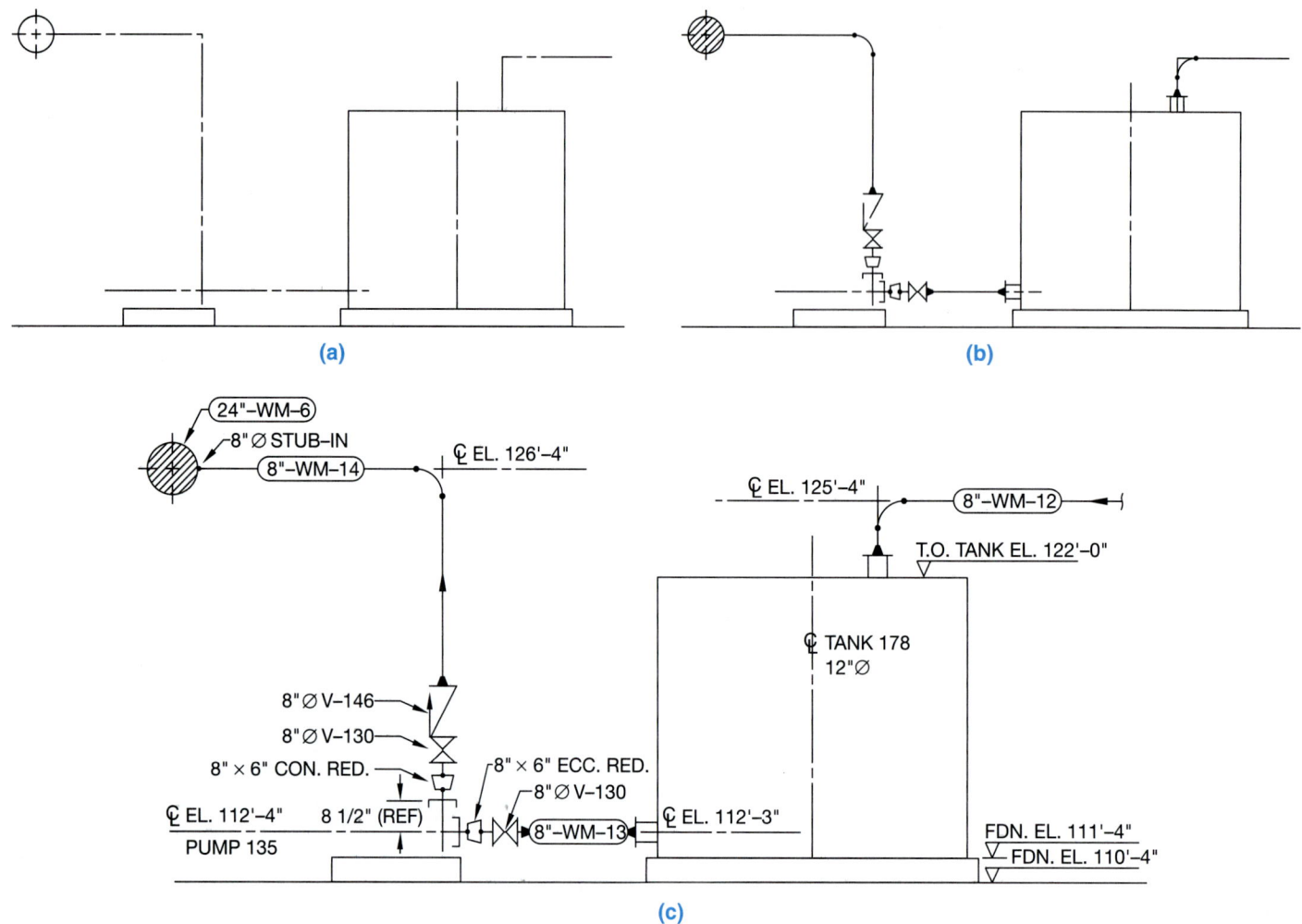

FIGURE 22.63 ■ (a) Equipment outlines and pipe centerlines are located; (b) pipe, fittings, and valves are added; (c) dimensions, elevations, and text are added to complete the drawing.

LAYOUT TECHNIQUES

When constructing scaled piping drawings, use the following layout procedures:

STEP 1 Locate all building outlines, concrete foundations, structural steel columns, and equipment. Draw the centerlines of the pipe that connects the equipment. (See Figure 22.63a.)

STEP 2 Insert fittings and valves in the pipe centerlines and draw pipe, as shown in Figure 22.63b.

STEP 3 Place dimensions and elevations on the drawing, remembering to keep dimensions close to where they apply. Locate pipe specification symbols and text in the pipe runs. Add all notes and textual information required. (See Figure 22.63c.)

WEBSITE RESEARCH

The following websites can provide you additional information for research or further study into topics covered in this chapter:

http://www.asme.org—Find information and publications related to the American Society of Mechanical Engineers.

http://www.ansi.org—The American National Standards Institute. Information about national and international drafting standards.

http://www.adda.org—American Design Drafting Association. Information related to the drafting and design profession.

http://www.americanbackflow.com—American Backflow Specialties.

http://www.craneco.com—Crane Company. Industrial valves and pumps.

http://www.powellvalves.com—Powell Valves. Industrial valves.

http://www.kunklevalve.com—Kunkle Valves. Industrial valves.

http://www.spenceengineering.com—Spence Engineering Company. Piping accessories.

PROFESSIONAL PERSPECTIVE

The field of process piping design and drafting can provide an excellent opportunity for a person to gain entry into engineering. Many consulting engineering companies pay for the education of employees who want to upgrade their skills or work toward an engineering degree. This is often a good method of becoming a designer, technician, or engineer, because you can get valuable job training while working toward a degree. This job training may not be possible if you are going to school full time.

A good piping drafter, designer, or engineer is one who is aware of the actual job site requirements and problems. These "field" situations are often considerably different from the layout that is designed in the office. Therefore, if you are planning to work in the industrial piping profession, do your best to work on projects in which you can gain field experience. Field work is especially important when adding new equipment and pipe in an existing facility. There are many stories of inexperienced piping designers who have created a design in the office and then gone to the field to find a new 4" pipe routed directly through an existing 12" pipe, exactly according to the plan!

USING THE PYTHAGOREAN THEOREM

Industrial pipe design often requires the use of fittings such as 45° elbows to route pipe past obstructions. Exact dimensions of all lengths of pipe and fittings must be calculated before the pipe can be drawn and assembled. When 45° elbows are used, the piping designer must use the **Pythagorean theorem** ($a^2 + b^2 = c^2$) to solve for one side of the triangle.

The designer has laid out a run of pipe shown in Figure 22.64. In order to find the true length, or *travel*, of the angled run of pipe, a 45° triangle is applied to the pipe. The length of one side of the triangle is determined by finding the difference between the two elevations given, 2'–9". The two adjacent sides each have a length of 2'–9". Solve for the angled side of the triangle, or **hypotenuse**. (See Figure 22.65.)

The Pythagorean theorem is stated as follows: The sum of the squares of the two sides is equal to the square of the hypotenuse or:

$$a^2 + b^2 = c^2$$

Therefore, the square root of the hypotenuse equals the travel of the angled run of pipe.

This problem is solved as follows:

$$a^2 + b^2 = c^2$$
$$(2'-9")^2 + (2'-9")^2 = c^2$$
$$(33)^2 + (33)^2 = c^2$$
$$1089 + 1089 = c^2$$
$$2178 = c^2$$
$$\sqrt{2178} = c$$
$$c = 46.669"$$
$$c = 3'-10\ 11/16"$$

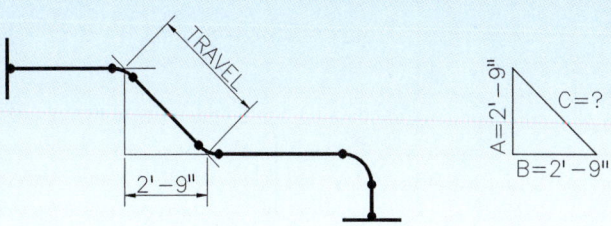

FIGURE 22.64 ■ Find the true length, or travel, of the angled run of pipe.

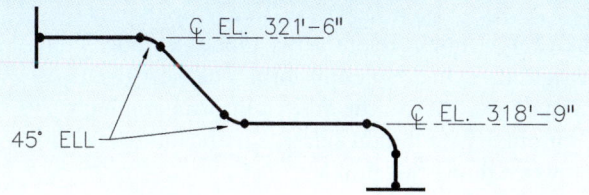

FIGURE 22.65 ■ Solve for the travel (angled) side of the triangle.

MATH APPLICATIONS

CHAPTER 22 *Industrial Process Piping Test*

Access the CD found with this textbook to view the Chapter 22 Test. Confirm the preferred submittal method with your instructor.

CHAPTER

22 *Industrial Process Piping Problems*

DIRECTIONS

1. Read problems carefully before you begin working. Your instructor will assign one or more of the following problems. Complete each problem on an appropriately sized drawing sheet or use the size indicated in the specific instructions.
2. Refer to Appendixes W and X for fitting and valve dimensions. Consult with your instructor to determine if specific vendor catalogs should be used to obtain dimensions.
3. Construct your drawings using a CADD system, if indicated in course guidelines.

PROBLEM 22.1 Draw the following fittings in double-line representation. Use 8" diameter NPS and draw at a scale of 1/2" = 1'–0".

■ 90° elbow

■ 45° elbow

■ Straight tee

■ Concentric reducer

■ Eccentric reducer

Draw a front view and each of the four orthographic views (top, bottom, left, and right sides).

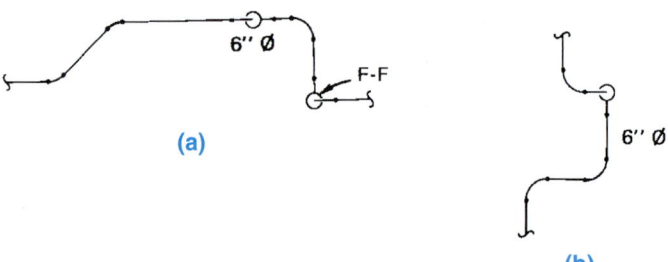

PROBLEM 22.2 Draw the fittings listed in Problem 22.1 in single-line representation. Use the same NPS and the same scale.

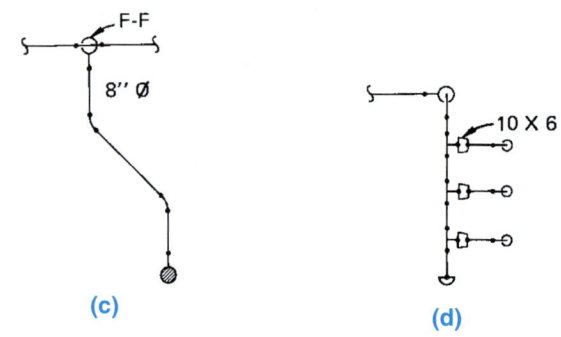

PROBLEM 22.3 Draw each of the five pipe assemblies shown on the right (a–e) in double-line form. The view given is the front view. Draw each of the other four orthographic views (top, bottom, left, and right sides). Use the following information when doing this problem:

■ Draw at a scale of 3/8" = 1'–0".

■ Use pipe diameters indicated.

■ Draw the fittings to scale but straight lengths of pipe can be drawn proportional to the sketch.

■ Pipe that appears to be extending away from the viewer should be drawn "fitting-to-fitting" if indicated with "F-F" on the sketch. This means that there is no straight pipe in the run that is going away from the viewer. Runs that do contain pipe are not shown with "F-F."

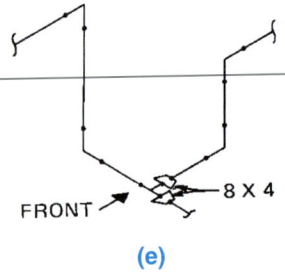

PROBLEM 22.4   Draw the assemblies shown in Problem 22.3 in single-line form. Follow the instructions given in Problem 22.3.

PROBLEM 22.5   Redraw the Ground Floor Partial Plan shown in Figure 22.45a, page 724, on C-size media. Use the following information:

■ Draw at a scale of 1/2" = 1'-0".

■ Pipe more than 3" in diameter should be drawn in double-line form.

■ Use line balloons to indicate pipe specifications.

■ Lettering should be 1/8" high.

Use Figure 22.45b-e, pages 725–728, as references for this drawing.

PROBLEM 22.6   Redraw section E-E (Figure 22.45e) in double-line form at a scale of 1/2" = 1'-0". Use standard dimensions for fittings and valves given in the appendices.

PROBLEM 22.7   Draw a piping detail of the suction piping of pump P-405 in section B-B, Figure 22.45.

PROBLEM 22.8   Redraw section B-B (Figure 22.45b) in single-line form at a scale of 1/2" = 1'-0".

PROBLEM 22.9   Draw the sections shown in Figure 22.45 in single-line form. Use a scale of 3/8" = 1'-0" on C-size media.

PROBLEM 22.10   Redraw section E-E, Figure 22.45e, at a scale of 1/2" = 1'-0". Pipe that is greater than 3" in diameter, draw as a single-line. Pipe less than 3" diameter, draw as double-line.

PROBLEM 22.11   Draw isometric views of the five pipe assemblies shown in Problem 22.3. Draw in single-line form.

PROBLEM 22.12   Draw a piping isometric of lines 1"–HA–101 in section E-E (Figure 22.45e). Use B-size media.

PROBLEM 22.13   Draw a plan and elevation of 8"–SH–009 (Figure 22.54a) in single-line form.

PROBLEM 22.14   Draw a plan and elevation of 8"–SH–009 (Figure 22.54a) in double-line form.

PROBLEM 22.15   Draw the spools required for the isometric 4"–SH–004 shown below. Draw one per sheet of B-size media. Include a bill of materials for each spool. The field weld connection point between the two largest spools is indicated by an "X" just above El. 36'–6". *Courtesy Harder Mechanical Contractors, Inc.*

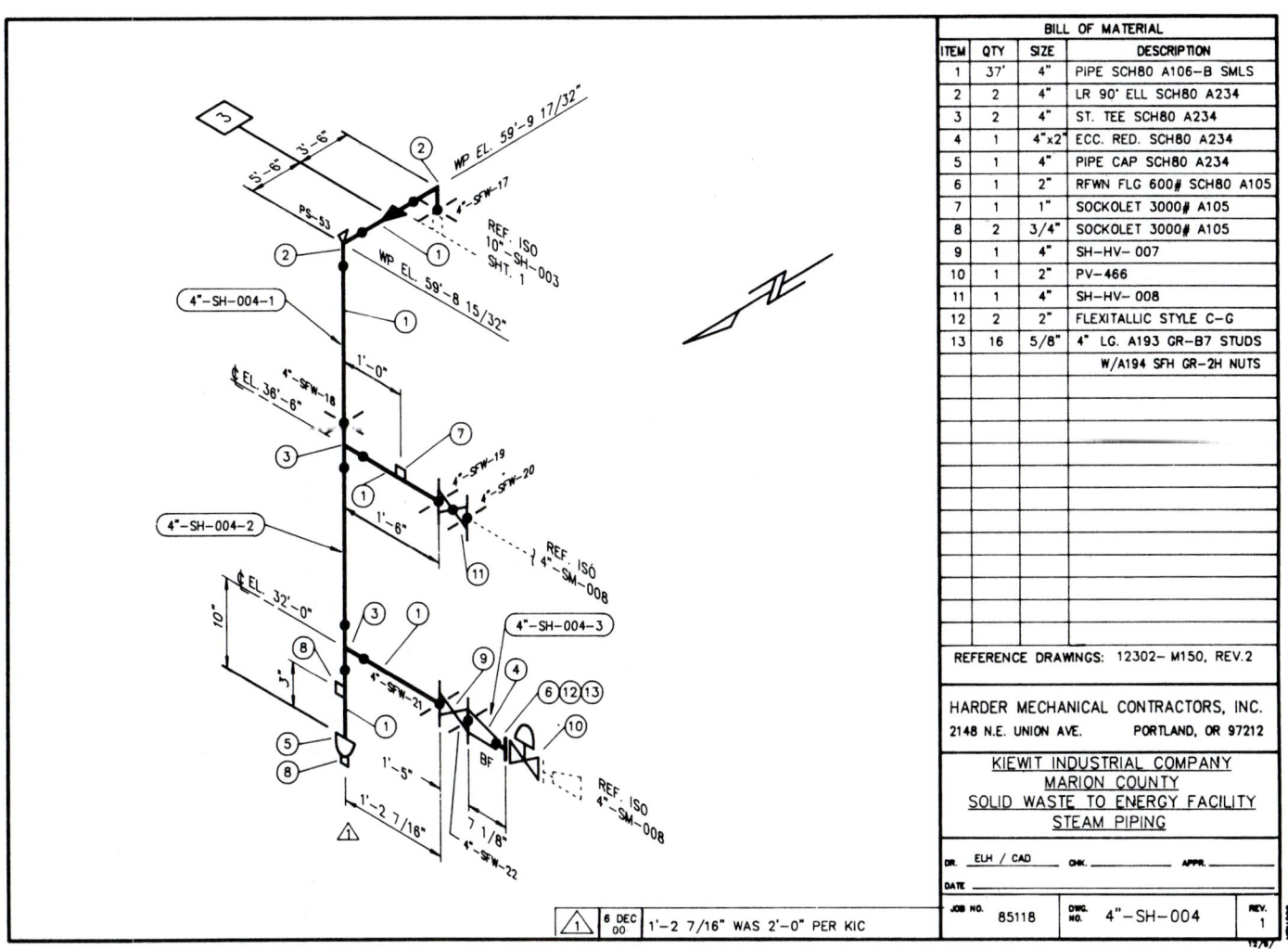

BILL OF MATERIAL

ITEM	QTY	SIZE	DESCRIPTION
1	37'	4"	PIPE SCH80 A106–B SMLS
2	2	4"	LR 90° ELL SCH80 A234
3	2	4"	ST. TEE SCH80 A234
4	1	4"x2"	ECC. RED. SCH80 A234
5	1	4"	PIPE CAP SCH80 A234
6	1	2"	RFWN FLG 600# SCH80 A105
7	1	1"	SOCKOLET 3000# A105
8	2	3/4"	SOCKOLET 3000# A105
9	1	4"	SH–HV– 007
10	1	2"	PV–466
11	1	4"	SH–HV– 008
12	2	2"	FLEXITALLIC STYLE C–G
13	16	5/8"	4" LG. A193 GR–B7 STUDS
			W/A194 SFH GR–2H NUTS

REFERENCE DRAWINGS: 12302– M150, REV.2

HARDER MECHANICAL CONTRACTORS, INC.
2148 N.E. UNION AVE.   PORTLAND, OR 97212

KIEWIT INDUSTRIAL COMPANY
MARION COUNTY
SOLID WASTE TO ENERGY FACILITY
STEAM PIPING

DR. ELH / CAD   CHK. _____   APPR. _____

DATE _____

| JOB NO. 85118 | DWG. NO. 4"–SH–004 | REV. 1 |

| △1 | 6 DEC 00 | 1'-2 7/16" WAS 2'-0" PER KIC |

PROBLEMS 22.16 through 22.22: Access the CD found with this textbook and open the problem of your choice, or as assigned by your instructor. Solve the problems using the instructions provided with this chapter, or on the CD, unless otherwise specified by your instructor.

MATH PROBLEMS

 PROBLEMS 22.23 through 22.25: Access the CD found with this textbook and open the math problem of your choice, or as assigned by your instructor. Solve the problem or problems using the instructions provided.

Structural Drafting

LEARNING OBJECTIVES

After completing this chapter, you will:

■ Identify, describe, and draw various components of the following commercial construction methods: concrete, concrete block, wood, heavy timber, laminated beam, and steel.

■ Prepare a complete set of structural drawings.

■ Draw and document drawing revisions.

■ Draw commercial structural drawings from engineering sketches.

THE ENGINEERING DESIGN APPLICATION

In many situations the drafter prepares formal drawings from engineering calculations and sketches. Registered structural engineers, architects, or designers with experience and training may prepare structural calculations and design. The entry-level engineering drafter may not fully understand the calculations but the drafter can interpret the results of the calculations because the engineer or designer normally highlights the solution to be placed on the drawing by clearly underlining or placing a box around it, as shown in Figure 23.1. When engineering calculations result in a design sketch, the drafter's job is to convert the sketch to a formal drawing, as shown in Figure 23.2b.

The quality and completeness of the engineering sketch often depends on the experience of the drafter. If this is the drafter's first job, the engineer may have to take a little more time providing detailed information until the drafter gains experience. Once the drafter has had some experience, the engineering sketches can be less complete. For example, if the engineering sketch in Figure 23.2a had omitted the note: 3/16" PLY SHIM EA SIDE GLUE AND NAIL W/4-ROWS 8–10d COMMON, an experienced drafter would realize that shims are required when two beams of different thicknesses are joined. In some situations, an experienced drafter can work directly from the engineering calculations without sketches. In all situations, it is very important for the drafter to maintain a current set of vendor's catalogs. These catalogs give the detailed information necessary for reproducing and labeling the drawings. The kinds of information found in a vendor's cat-

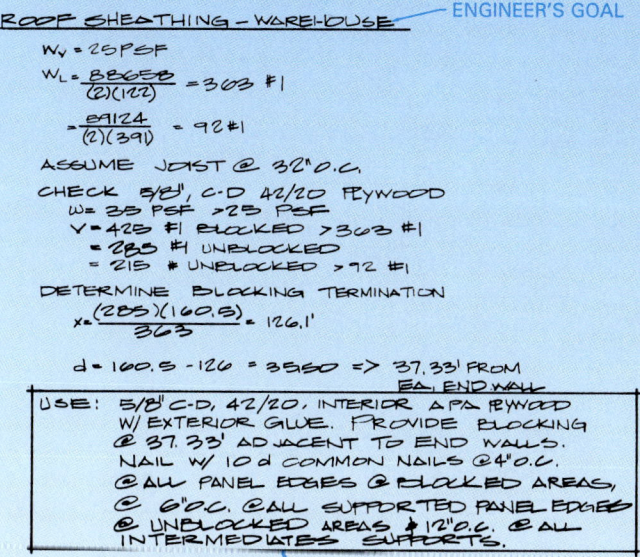

FIGURE 23.1 ■ Engineering calculations typically contain a problem to be solved, mathematical solutions, and specifications to be placed on the drawing. The drawing note or information is placed in a box or otherwise highlighted.

alog that will enable the drafter to complete the drawing in Figure 23.2b with the dimensions and specifications for the MST 27 straps, CC 5-1/4 COLUMN CAP, and L50 or A35 framing anchor.

(Continued)

THE ENGINEERING DESIGN APPLICATION (continued)

FIGURE 23.2 ■ (a) A sample of engineer's calculations to determine the loading criteria at a beam connection over a column. (b) The drafter would draw the detail to convey the information of the engineer's calculation and sketch, using proper drafting techniques.

STRUCTURAL ENGINEERING

A structural engineer works with architects and building designers to engineer the structural components of a building. Structural engineering is generally associated with commercial steel and concrete buildings, and in many situations, it is also used for the structural design of residential buildings. The structural engineer works with civil engineers to design bridges and other structures related to road and highway construction. Mechanical engineers also can be involved in the structural project by designing machinery supports and foundations. There is a wide variety of projects where structural drafting may be involved. Structural drafting techniques are generally the same as mechanical drafting, although a combination of mechanical and architectural methods are used.

Structural engineering drawings and detail drawings show in a condensed form the final results of designing. Drawings, general notes, schedules, and specifications serve as instructions to the contractor. The drawings must be complete and have sufficient detail so no misinterpretation can be made. Structural drawings are usually independent of architectural drawings and other drawings, such as plumbing or heating, ventilating, and air conditioning (HVAC), or, when necessary, structural drawings are clearly cross-referenced to architectural drawings.

LINE WORK

The lines used in structural drafting are generally the same as those used in mechanical drafting. There are a few exceptions. For example, the object line can be drawn thicker than normal when a shape or feature requires extra emphasis. Object lines also may be thinner than the standard when used on a small-scale drawing. (See Figure 23.3.) Dimension lines can be drawn with a space provided for the numeral as in mechanical drafting, or the numeral can be placed above the dimension line as in architectural drafting. The dimension line can be capped on the end with arrowheads, slashes, or dots. (See Figure 23.4.) The architectural style of dimensioning is most commonly used in structural drafting. Refer to Chapter 11 for additional dimensioning practices, terminology, and examples. Cutting-plane lines and symbols are provided one of several ways, as shown in Figure 23.5.

A common way to create the cutting-plane symbol is to draw a circle with a vertical or horizontal line through the middle, depending on the symbol orientation. The line can be at an angle if the symbol is oriented at an angle through the building or view. The top half of the circle has a letter identifying the section, and the bottom half has a number, or letter and number combination, identifying the sheet where the sectional view is located. A solid-filled triangle is generally drawn attached to the circle as shown in Figure 23.5. The sides of the triangle can be tangent to the circle or slightly past the tangency as in the given example. The point of the triangle represents the direction of sight for the cutting plane. The cutting-plane symbol can be duplicated on each side of the building or view, or an arrow representing the continued cutting plane can be placed across the

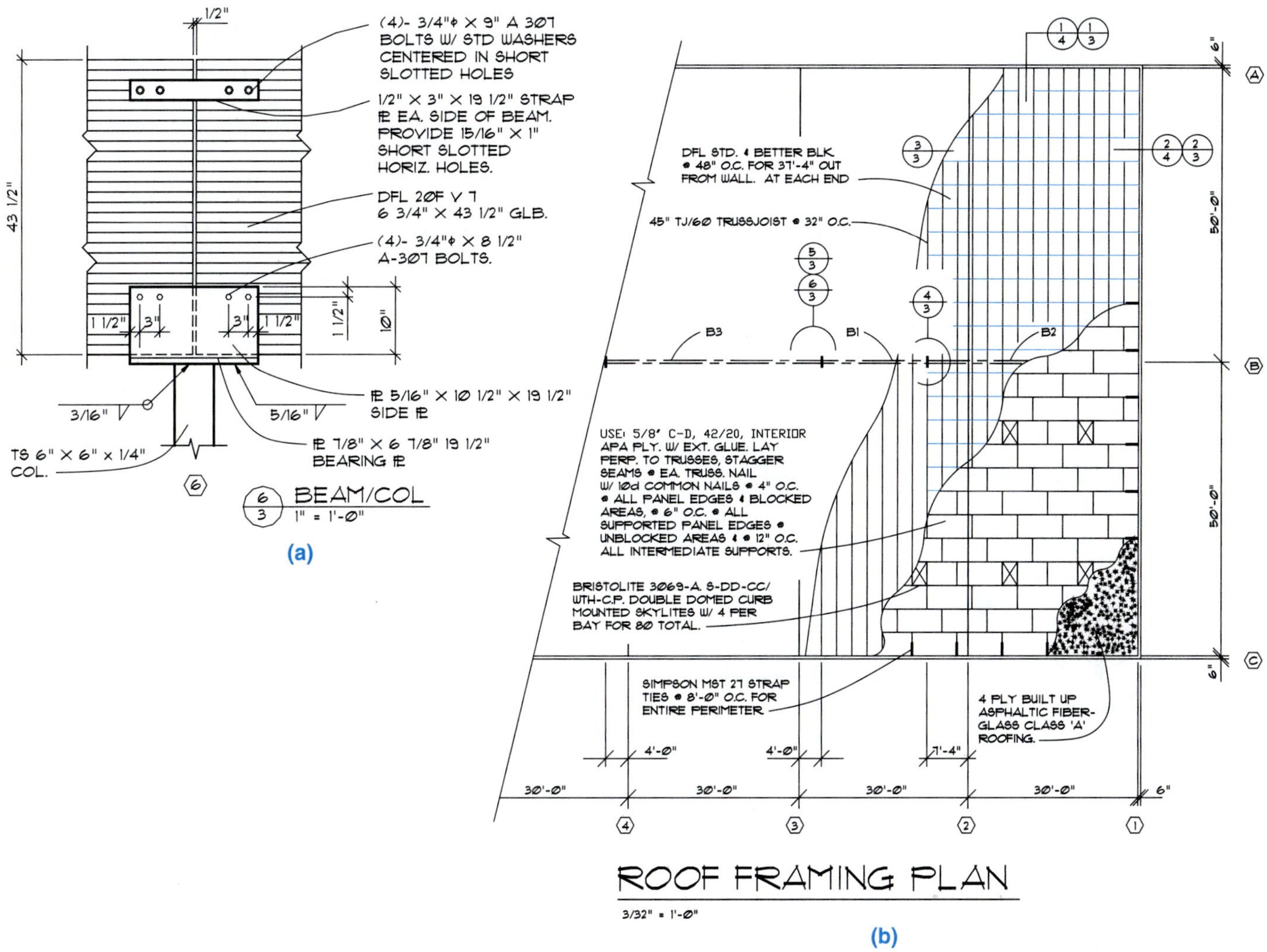

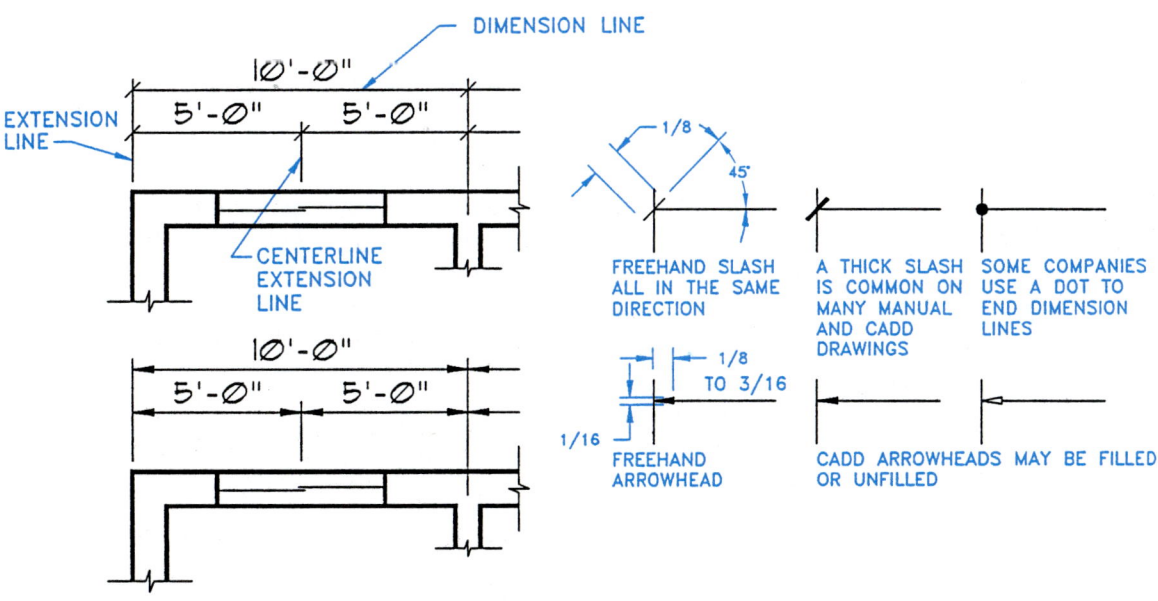

FIGURE 23.3 ■ (a) Thick object lines for emphasis used on a beam/column detail; (b) thin object lines on small-scale drawing used on a panelized roof framing system.

FIGURE 23.4 ■ Dimension line examples.

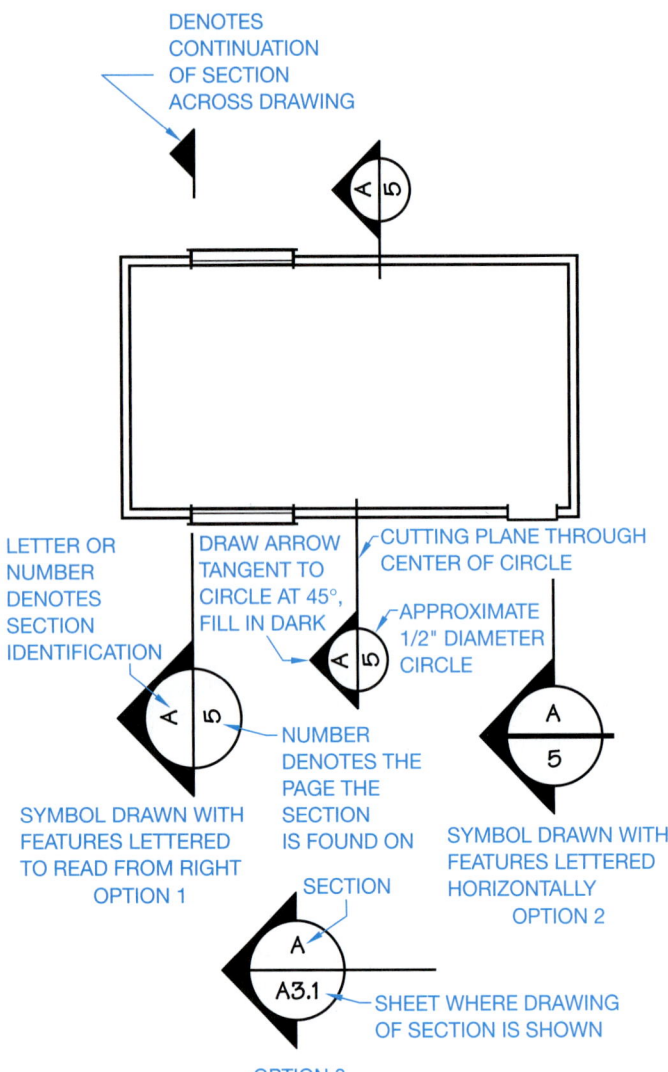

FIGURE 23.5 ■ Cutting-plane line symbols.

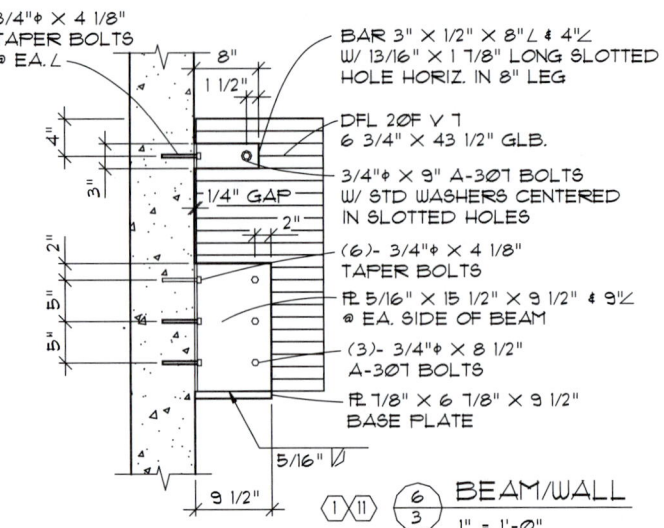

FIGURE 23.6 ■ Common lettering used in structural drafting used on a beam construction detail.

building or view. While cutting-plane representations are similar throughout this chapter, there are some differences depending on company preference. However, the method used should be consistent throughout each set of structural drawings. Review Chapter 13 for detailed information about sectioning practices.

LETTERING

The quality of lettering in structural drafting is equally as important as it is in other drafting fields. There is a great deal of lettering on structural drawings. A well-developed, legible style of lettering makes the job easier and adds professional quality. Lettering is similar to the Gothic style used in mechanical drafting; however, the structural drafter has more freedom of style with an architectural lettering style commonly used. The best structural lettering is simple, easy to read, and quick to produce. Structural drafters sometimes prefer slanted letters. Lettering height on drawings is typically 1/8" to 5/32" (3–4 mm) for all lettering except titles, which are 3/16" to 1/4" (5–6 mm) in height. If drawings are to be micro-

filmed, 5/32" lettering is used. Figure 23.6 shows lettering used in structural drafting. Other lettering rules are presented in Chapter 7.

COORDINATION OF WORKING DRAWINGS

The structural drawings are a portion of a complete set of working drawings. A set of drawings for a commercial project can have over 100 pages. Due to the complex nature of the drawings and the number of disciplines involved, the set of drawings is generally divided into several major groups. The major groups are generally architectural, structural, mechanical, plumbing, electrical, and fixture drawings. The architect normally prepares the architectural drawings. The architect then coordinates with consulting engineering firms that prepare the drawings for their specific disciplines. The structural engineering firm generally prepares the related structural drawings.

Numbering the Pages

A page numbering system needs to be used because of the large number of pages in a set of structural drawings. When a set of drawings has a number of pages that are easy to manage, it is common to see pages numbered in consecutive order such as 1, 2, 3, 4. If a sheet has subpages, then the sheet number might be followed by letters alphabetically, such as 4A, 4B, 4C, 4D. The *Architects Handbook of Professional Practice* recommends a decimal page numbering system. This numbering system is established from the following sample general categories:

Architectural

T1 Title sheet, site demolition and survey

A1 Site plan and details

A2 Grading plan and details

A3   First floor plan and details

A4   Second floor plan, sections and details

A5   Enlarged plans, interior elevations and details

A6   Exterior elevations and details

A7   Building and wall sections and details

A8   Roof plan and details

A9   Details

A10   Reflected ceiling plan and details

Civil

U1   Site utilities

U2   Erosion control plans and details

U3   Public utility plan

U4   Utility details

Landscape

L1   Irrigation system plan and legend

L2   Irrigation system details and notes

L3   Planting plan, details and notes

Structural

S1   Foundation plan

S2   Foundation details

S3   Second floor framing plan

S4   Roof framing plan

S5   Details

Mechanical

M1   First floor plumbing plan, legends

M2   Second floor plumbing plan, notes

M3   Details and schedules

M4   First floor HVAC plans and legends

M5   Second floor HVAC plan

M6   Roof mounted HVAC equipment plan

M7   HVAC details and schedules

Electrical

E1   Notes, legend, riser

E2   First floor lighting plan

E3   Second floor lighting plan

E4   First floor power plan

E5   Second floor power plan

E6   Roof mounted equipment power plan

E7   Details and schedules

E8   First floor communications plan and legend

E9   Second floor communications plan

Fixtures

F1   Fixture plan and schedules

F2   Details

The general groups and elements within the groups can differ depending on the company practice and the building being designed. Each category within a general group can have additional pages. These pages are numbered with the sequential decimals of .1, .2, .3. So, the architectural drawings might have pages such as:

A1.1, A1.2, A1.3

A2.1, A2.2, A2.3, A2.4

The structural drawings can have a series of sheets that are numbered such as:

S1.1, S1.2

S2.1, S2.2, S2.3

S3.1, S3.2, S3.3, S3.4, S3.5

The decimal sheet numbering system can broken down even further by adding .01, .02, .03 to the existing numbers as needed. For example, additional pages in the series of S3.1 are numbered S3.1.01, S3.1.02, S3.1.03.

Coordinating the Details and Sections

Now that a page numbering system has been established, you need to coordinate the elements of drawings between pages. Figure 23.5 shows some examples of how cutting-plane line symbols are correlated to the drawing. Construction details and sections are commonly labeled in a similar manner. The details might be numbered in consecutive order and correlated to the page where the detail is found, as shown in Figure 23.7. Sections are commonly labeled with letters in alphabetical order, but some companies use numbers. Details are generally labeled with numbers. The detail identification symbol is similar to the cutting-plane symbol, except the solid-filled triangle is omitted. Compare the cutting-plane symbols in Figure 23.5 with the detail identification symbols in Figure 23.7.

Laying Out Details and Sections

Detail and section drawings can be placed within the drawing where they relate, if space is available. For example, the foundation details and sections can be on the same sheet as the foundation if they can be placed there clearly and while maintaining a drawing that is easy to read and well organized. If there is not enough space on the same sheet, then the details and sections can be placed on other sheets. Details should be grouped together and organized from left to right and top to bottom in an aligned, orderly manner. If the details are numbered, they should be organized in numerical order. The sections also should be grouped together and organized from left to right and top to bottom. If the sections are labeled with letters, they should be placed in alphabetical order.

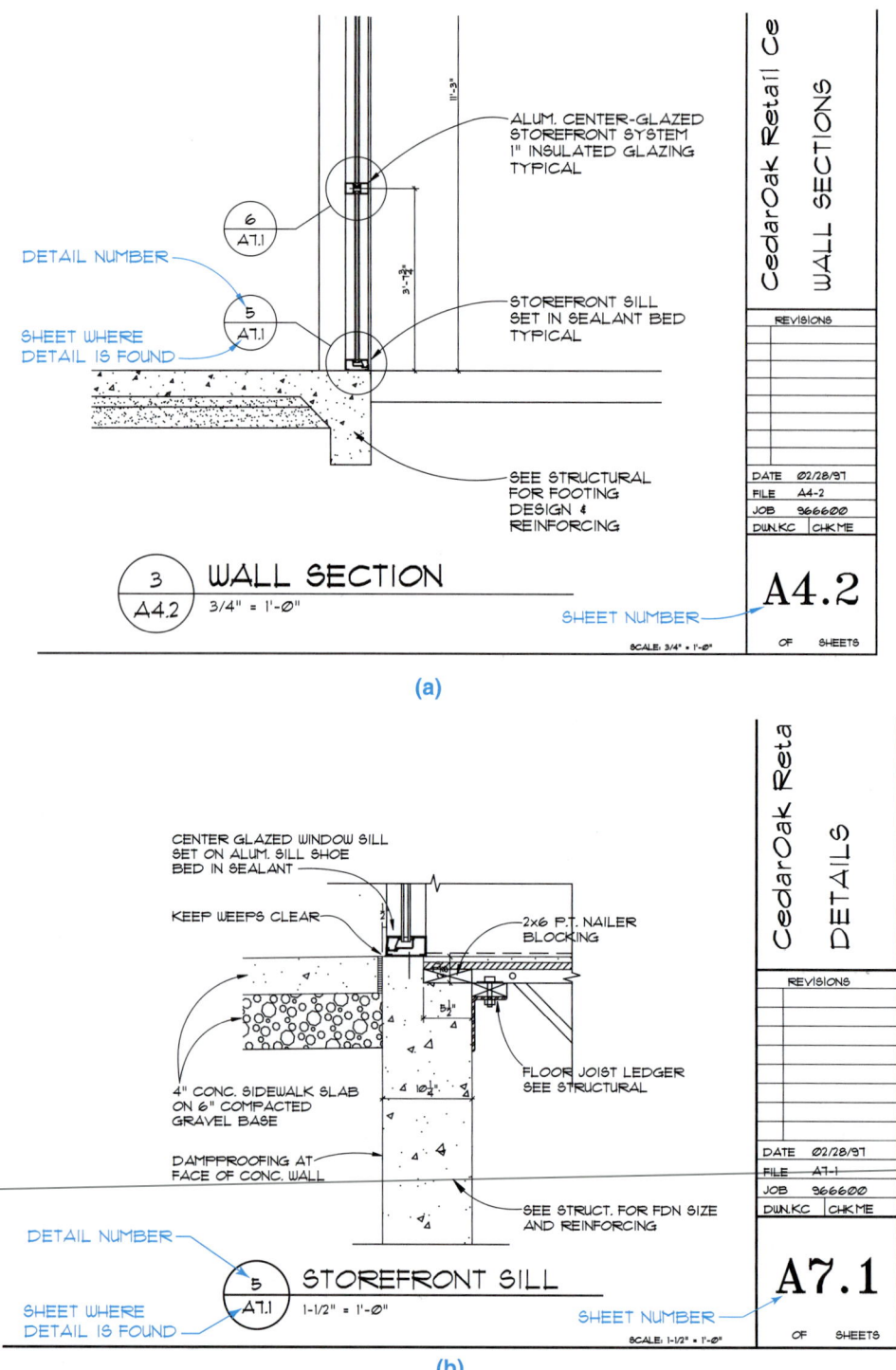

FIGURE 23.7 ■ Coordinating the details and sections to the sheet number. (a) Detail 5 is found on sheet number A7.1; (b) Sheet number A7.1 displays detail number 5. *Courtesy Ankrom Moisan Architects.*

STRUCTURAL DRAFTING RELATED TO CONSTRUCTION SYSTEMS

Different types of construction methods relate directly to the materials to be used, the area of the country where the construction takes place, the type of structure to be built, and even the office practices of the architect or engineer. The structural drafter should have a knowledge of construction materials and techniques. This chapter provides an introduction to construction techniques and materials. Additional resources should be used for reference, as each construction method discussed has volumes of both general and vendor information available. Another valuable way to learn about construction is to visit job sites to talk to builders and see firsthand how things are done.

CONCRETE CONSTRUCTION

Concrete is a mixture of Portland cement, sand, gravel, and water. This mixture is poured into forms that are built of wood or other materials to contain the mix in the desired shape until it is hard. Concrete is a fundamental material used for building foundations. The foundation is the system used to support the building loads and is usually made up of walls, footings, and piers. The term foundation is used in many areas to refer to the footing. The footing is the lowest member of the foundation system used to spread the loads of the structure across supporting soil. Concrete is also used in commercial applications for wall and floor systems. Residential buildings use concrete foundations with or without steel reinforcing, while commercial buildings usually use steel-reinforced concrete, depending on the structural requirements.

Earth is generally excavated prior to building a concrete structure. Excavation refers to removing earth for construction purposes. Before a concrete structure can be built, earth usually needs to be excavated down to firm, undisturbed supporting soil. Detailed engineering specifications are often placed on the drawing identifying the amount of bearing pressure required. The bearing pressure is normally the number of pounds per square foot of pressure the soil is engineered to support. A concrete material symbol can be used when drawing foundations in section view as shown in Figures 23.7 and 23.8. Concrete slabs are also normally found on the foundation drawings. A slab is a concrete floor system, typically made of poured concrete at ground level. Concrete slabs also normally need to be constructed on firm, undisturbed soil, with engineering specifications provided. In many cases, compacted gravel is specified below the concrete slab. The gravel provides for a level or accurate slope surface upon which to pour the concrete slab, and provides for compaction as needed. A concrete material symbol can be used when drawing slabs, and a gravel material symbol can be used in section view as shown in Figure 23.7. A drawing note representing under-slab gravel might read:

4" MIN. 3/4 MINUS COMPACTED GRAVEL OVER FIRM UNDISTURBED SOIL

Concrete can be either poured in place at the job site, formed at the job site and lifted into place, or formed off-site and delivered ready to be erected into place.

Concrete alone has excellent compression qualities. Steel added to concrete improves the tension properties of the material. Concrete poured around steel bars placed in the forms is known as reinforced concrete. The steel bars are referred to as rebar. Steel is the best choice for reinforcing concrete, as its coefficient of thermal expansion is almost the same as cured concrete. The resulting structure has concrete to resist the compressive stress, and steel to resist the tensile stress which is caused by the loads acting on the structure.

Steel reinforcing is available in a number of sizes. Steel reinforcing is deformed steel bars. Deformed reinforcing bars have raised ridges to hold better in concrete. (See Figure 23.8.)

Deformed steel bars have surface projections that increase the adhesion between the concrete and steel. (See Figure 23.8.) Steel re-

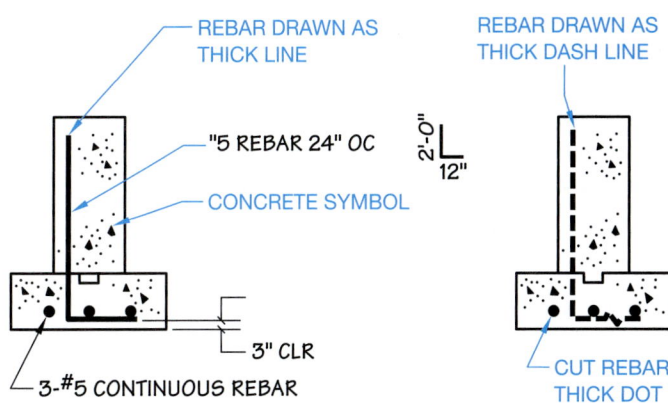

FIGURE 23.8 ■ Representation of reinforcing bars, *rebar*.

inforcing bars are sized by number, starting at #3, which is 3/8" in diameter, and increasing in size at approximately 1/8" intervals to #18, which is 2-1/4" in diameter. No. 4 rebar is 1/2" in diameter.

Another steel concrete reinforcing method is welded wire reinforcement (WWR). Welded wire reinforcement is steel wires spaced a specified distance apart in a square grid, and the wires are welded together as shown in Figure 23.9. Welded wire reinforcement is specified by giving the spacing of the wire grid, such as 6" × 6", which means 6" OC each way, followed by the wire type and size of the wires in the spacing. Wire spacing can be square or rectangular. The wire type is given as plain or deformed. Plain wire is smooth and is designated with a W, and

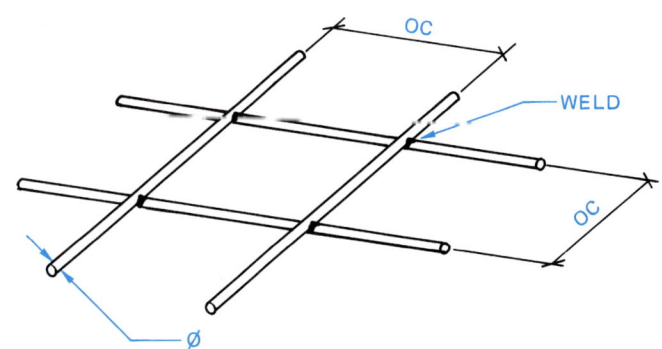

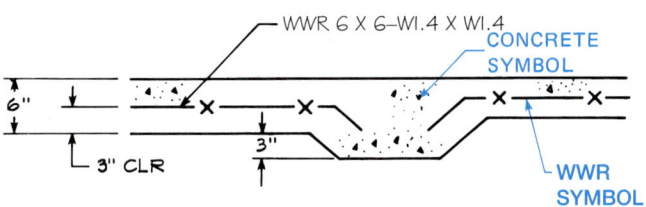

FIGURE 23.9 ■ Welded wire reinforcement representation.

deformed is designated with a D. Deformed steel reinforcing was described earlier. A few of the available wire sizes are W1.4, D1.4, W2.9, D2.9, W11, D11, W15, D15, W20, D20, W45, and D45. Wires are available in many sizes from W1.4 (3/16" dia.) to W45 (3/4" dia.). Most wire sizes are available as either plain or deformed. The wire size numbers relate to the wire area. The wire area is determined by dividing the wire designation by 100, for example, W1.4 is .014 sq. in. $(1.4 \div 100 = .014)$, and W45 is .45 sq. in. The welded wire reinforcement grid can have equally spaced wires and the same wire size, or the spacing and size can be different each way. Sample welded wire reinforcement callouts are written:

WWR 6 × 6-W15 × W15

or

WWR 4 × 18-D20 × W8

Metric welded wire reinforcement specifications have an M preceding the W or D and the wire area is given in square millimeters. The metric equivalent for a W2.9-inch wire is MW19, for example. The wire spacing is given in millimeters. A metric welded wire reinforcement callout is:

WWR 102 × 102-MW20 × MW20

Welded wire reinforcement is purchased in sheets. Sheet sizes can vary, but are normally 8' × 15', 8' × 20', and some can go to 12' widths and 40' lengths, depending on shipping and handling equipment to lift and move bundles of sheets. Sheet sizes can also be controlled by state requirements, where restrictions may exist for greater than standard loads of 8' × 40' and 40,000 lbs. Appendix T provides a list of inch and metric welded wire reinforcement sizes.

ASME/ANSI The American Society of Testing Materials (ASTM) document A615, Volume 01.04 provides standards related to steel reinforcing bars. ASTM A496 and A497 provide standards related to welded wire reinforcement. You can also consult the *Manual of Standard Practices*, published by the Concrete Reinforcing Steel Institute (CRSI), for complete information and examples related to steel concrete reinforcing. Also refer to the Wire Reinforcement Institute (WRI) documents *TECH FACTS, Excellence Set in Concrete* for detailed information about welded wire reinforcement. Appendix S, found in this textbook, gives concrete reinforcing bar specifications, and Appendix T provides standard welded wire reinforcement specifications for inch and metric applications.

Poured-in-Place Concrete

Commercial and residential applications for poured-in-place concrete are similar except that the size of the casting (resulting concrete structure) and the amount of reinforcing are generally more extensive in commercial construction. In addition to the foundation and on-grade (ground) floor systems, concrete is often used for walls, columns, and floors above ground. Lateral soil pressure acting on concrete structures tends to bend the wall inward, thus placing the soil side of the wall in compression and the interior side of the wall in tension. Steel reinforcing is used to increase the concrete's ability to withstand this tensile stress as shown in Figure 23.10. Steel-reinforced walls and columns are constructed by setting steel reinforcing in place and then surrounding it with wooden forms to contain the concrete. Once the concrete has been poured and allowed to set (harden), the forms are removed. As a drafter you will be required to draw details showing the sizes of the feature to be constructed and steel placement within the structure. This typically consists of drawing the vertical steel and the horizontal ties. Ties are wrapped around vertical steel in a column or placed horizontally in a wall or slab to help keep the structure from separating, and they keep the rebar in place

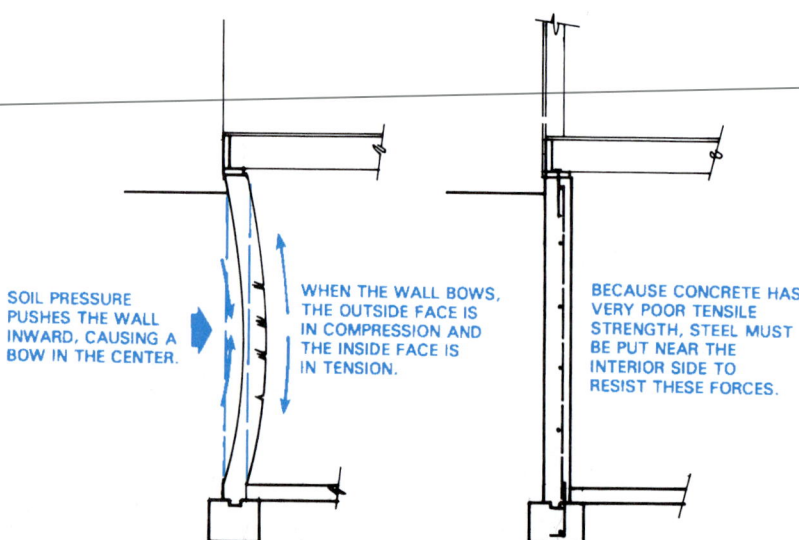

SOIL PRESSURE PUSHES THE WALL INWARD, CAUSING A BOW IN THE CENTER.

WHEN THE WALL BOWS, THE OUTSIDE FACE IS IN COMPRESSION AND THE INSIDE FACE IS IN TENSION.

BECAUSE CONCRETE HAS VERY POOR TENSILE STRENGTH, STEEL MUST BE PUT NEAR THE INTERIOR SIDE TO RESIST THESE FORCES.

FIGURE 23.10 ■ Stresses created in a concrete wall from horizontal forces. The wall serves as a beam spanning between each floor, and the soil is the supported load.

while the concrete is being poured into the forms. Figure 23.11 shows two examples of column reinforcing. The drawings required to detail the construction of a rectangular concrete column are shown in Figure 23.12.

Concrete is also used on commercial projects to build above-ground floor systems. The floor slab can be supported by either a steel deck or self-supported. The steel deck system is typically used on structures constructed with a steel frame. (See Figure 23.13.) Two of the most common poured-in-place concrete floor systems are the ribbed and waffle floor methods, as shown in Figure 23.14. The ribbed system is used in many office buildings. The ribs serve as floor joists to support the slab but are actually part of the slab. Spacing of the ribs varies depending on the span and the amount and size of reinforcing. The span is the horizontal distance between two supporting

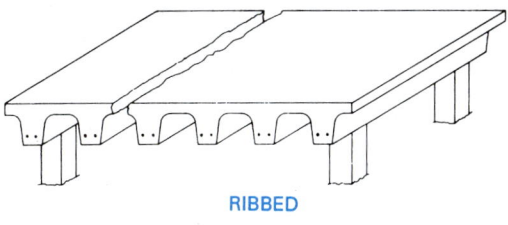

RIBBED

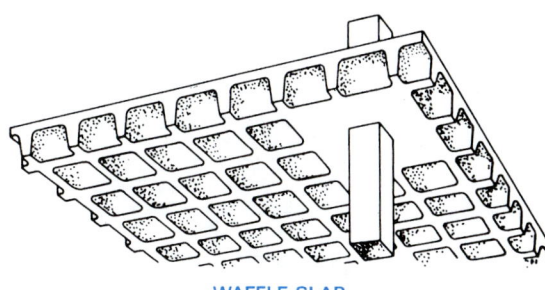

WAFFLE-SLAB

FIGURE 23.14 ■ Two common concrete floor systems.

members. The waffle system is used to provide added support for the floor slab and is typically used in the floor systems of parking garages.

Precast Concrete

Precast concrete construction consists of forming the concrete component off-site at a fabrication plant and transporting it to the construction site. Figure 23.15 shows a precast beam being lifted into place. Drawings for precast components must show how precast members are to be constructed and methods of transporting and lifting the member into place. Precast members often have an exposed metal flange so the member can be connected to other parts of the structure. Common details used for wall connections are shown in Figure 23.16.

Many concrete structures are precast and prestressed. Concrete is prestressed by placing steel cables, wires, or bars held in tension between the concrete forms while the concrete is poured

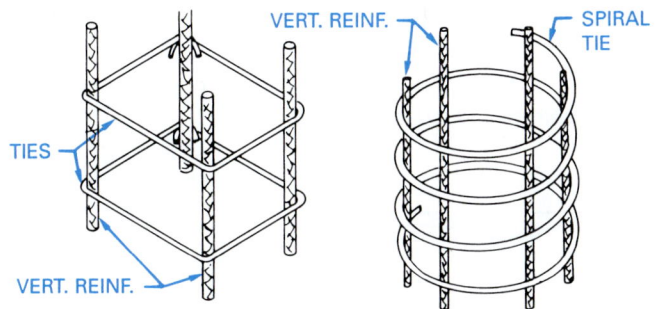

FIGURE 23.11 ■ Examples of column reinforcing: (a) square column ties; (b) round column spiral tie.

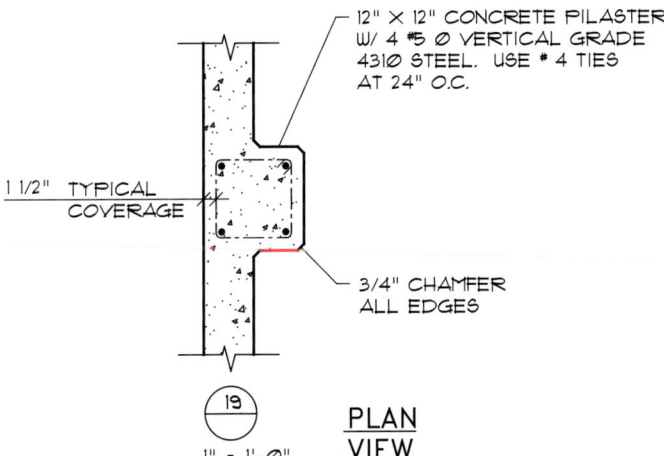

12" X 12" CONCRETE PILASTER W/ 4 #5 Ø VERTICAL GRADE 4310 STEEL. USE # 4 TIES AT 24" O.C.

1 1/2" TYPICAL COVERAGE

3/4" CHAMFER ALL EDGES

19

PLAN VIEW

1" = 1'-Ø"

FIGURE 23.12 ■ Typical reinforcing for a rectangular concrete column.

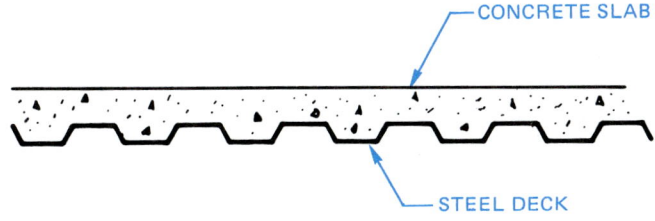

CONCRETE SLAB

STEEL DECK

FIGURE 23.13 ■ Representation of a steel deck system.

FIGURE 23.15 ■ Precast concrete beams and panels are often formed off-site, delivered to the job site, and then set into place with a crane.

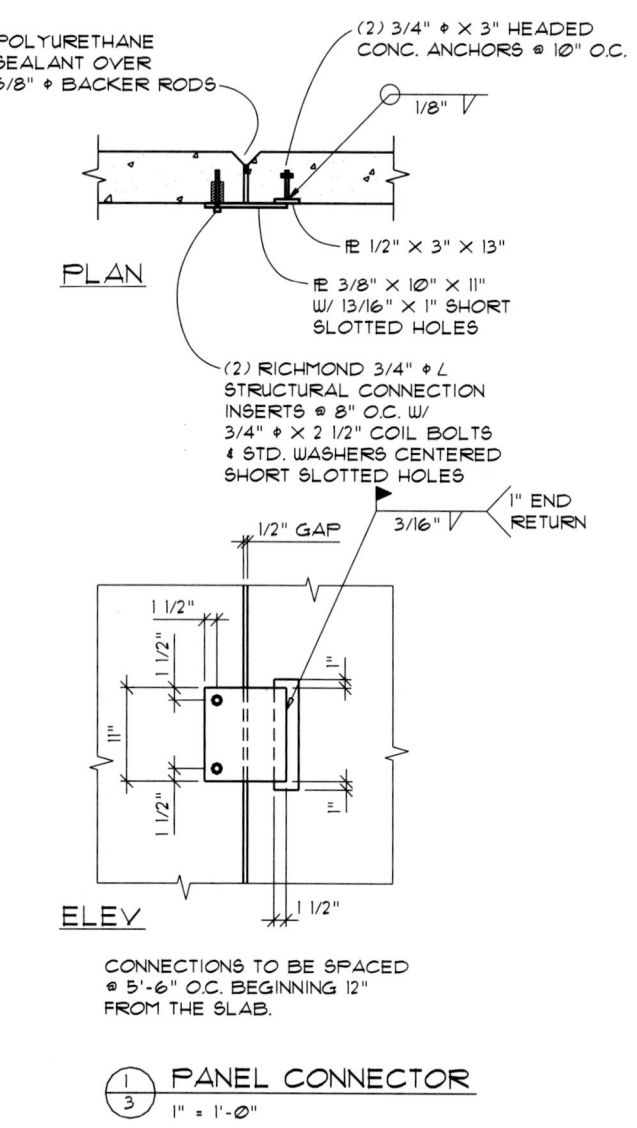

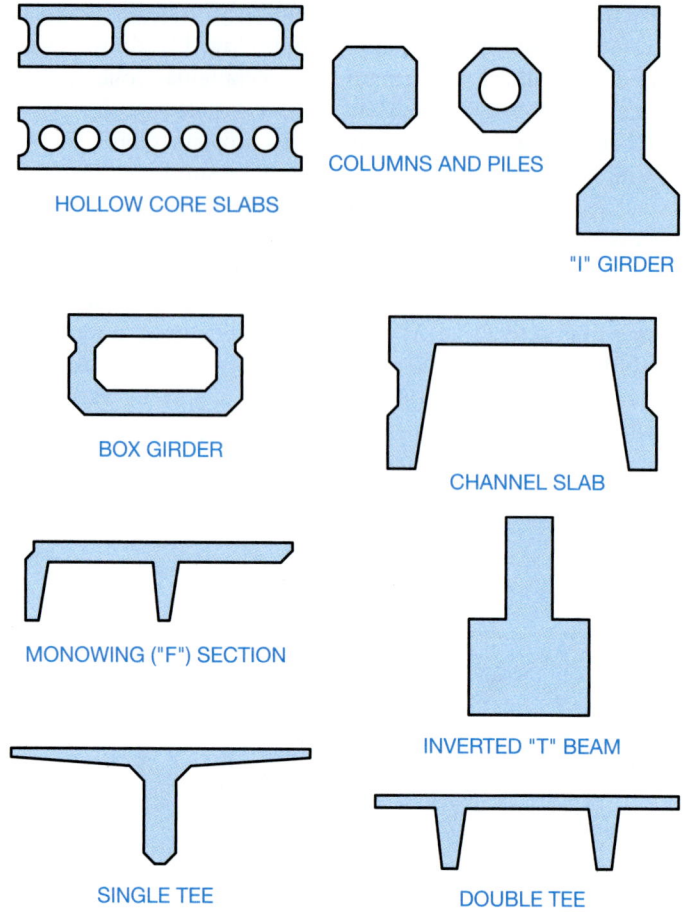

FIGURE 23.16 ■ Typical panel connection detail.

FIGURE 23.17 ■ Common prestressed concrete shapes.

around them. Once the concrete has hardened and the forms are removed, the cables act like big springs. As the cables attempt to regain their original shape, compression pressure is created within the concrete. The compressive stresses built into the concrete member help prevent cracking and deflection. Prestressed concrete members are generally reduced in size in comparison with the same design features of a standard precast concrete member. Prestressed concrete components are commonly used for the structural beams of buildings and bridges. Common prestressed concrete shapes are shown in Figure 23.17.

Tilt-Up Precast Concrete

Tilt-up construction is a precast concrete method using formed wall panels that are lifted or "tilted" into place. Panels may be formed and poured either at or off the job site. Forms for a wall are constructed in a horizontal position and the required steel placed in the form. Concrete is then poured around the steel and allowed to harden. The panel is lifted into place once it has

reached its desired hardening and design strength. When using this type of construction, the drafter usually draws a plan view to specify the panel locations, as shown in Figure 23.18. The location and size of steel placement and openings are also important, as shown in the panel elevation in Figure 23.19.

Figure 23.20 shows a precast concrete panel drawing. A precast concrete beam drawing is displayed in Figure 23.21 and a precast concrete slab drawing is shown in Figure 23.22, page 756. Construction details are commonly used in prestressed concrete construction. Figure 23.23, page 757, shows a precast concrete panel installed on a poured concrete wall and footing.

Standard Structural Callouts for Concrete Reinforcing

When specifying **anchor bolts (AB)** on a drawing, give the quantity, diameter, type, length, spacing on center (OC), and projection (PROJ) of the thread out of the concrete. The quantity can be displayed followed by a dash or in parentheses. The W is the abbreviation for **with** and can also be abbreviated W/. For example:

(12)-3/4" ∅ × 12" STD AB 24" OC W/3" PROJ.

When specifying rebar on a drawing, give the quantity (if required), bar size, length (if required), spacing in inches on

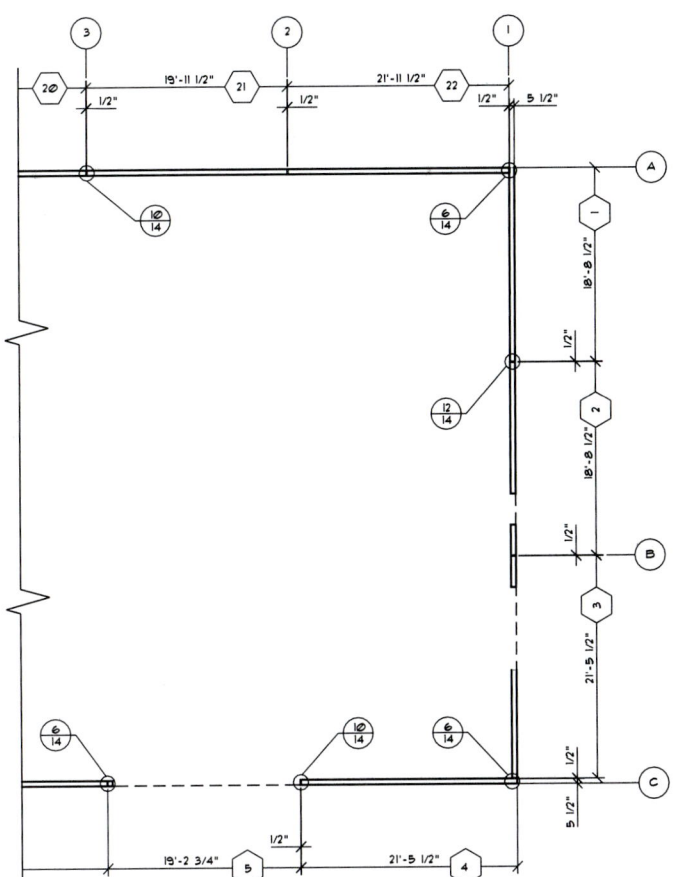

FIGURE 23.18 ■ Tilt-up panel plan.

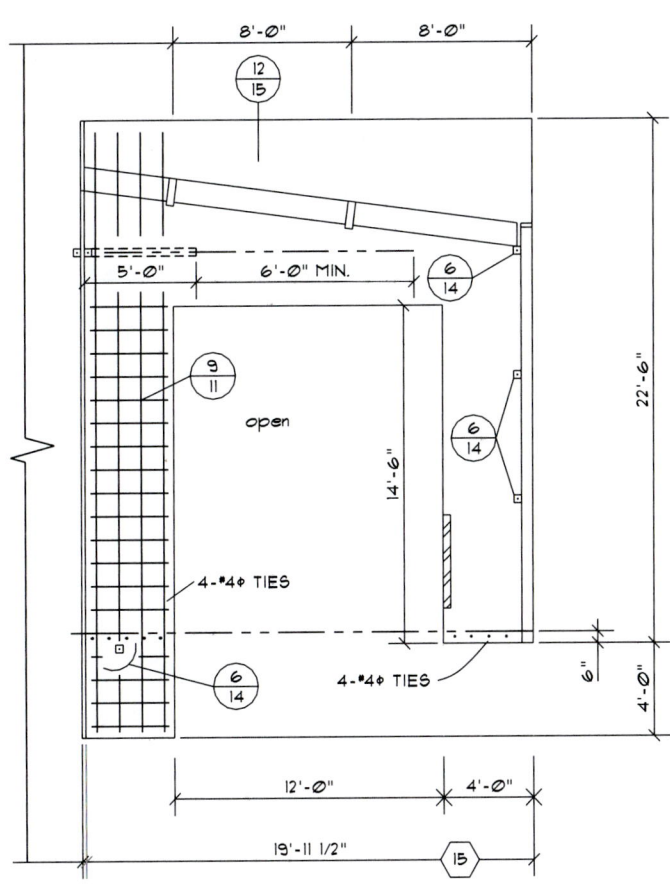

FIGURE 23.19 ■ Steel and opening locations specified in precast panel elevation.

center, horizontal or vertical, and bend information (if required). The (if required) note means that this specific information is given only if needed. These specifications are not needed if general space requirements control the application. Deformed steel rebar is assumed unless otherwise specified. For example:

> #4 @ 12" O.C.
> (25)#8 @ 24" O.C.HORIZ AND 16" O.C.VERT
> #6 @ 24" O.C.EA WAY (OR EW)
> #5 @ 16" O.C. × 12" O.C.

Notice the AT (@) symbol is used in the previous note. This symbol, or the word AT, can be used to separate the material specification from the designated spacing. The word AT or the @ symbol are also sometimes omitted, which is an acceptable option depending on company or personal preference.

Dimensions of bends are to be provided in feet and inches without the (') and (") marks given, for example:

> #5 AT 16" OC VERT W/ 90° × 12 BEND

or the bend diagram can be drawn, as shown in Figure 23.24, page 757.

When specifying welded wire reinforcement on a drawing, give the designation WWR, the wire spacing in inches OC, and the wire size. For example:

> WWR 6 × 6-W1.4 × W1.4

Reinforcing steel rebar lap splices may be shown on the drawing by giving the length of the lap and a lap location dimension. For example.

> #5 REBAR 16" O.C.W/ 24" MIN SPLICE AT FOOTING

Welded wire reinforcement lap splices need not be shown as a specific note on the drawing; however, a general note should clarify the amount of allowable splice overlap, in inches, at the cross wires.

Clear distances should be given from the surface of the concrete to the rebar. This dimension is assumed to be to the edge of the rebar or clarified by the abbreviation CLR, as in 3" CLR. If this dimension is designated to the centerline of the rebar, then OC must be specified.

When structural members are embedded in concrete, the size of holes for rebar to pass through should be specified with the structural member callout. For example:

> #5 AT 16" OC W/ 13/16 ∅ HOLES

Recommended hole sizes for rebar passing through steel or timber are as follows:

Bar #	Hole ∅	Bar #	Hole ∅
4	11/16	8	1 1/4
5	13/16	9	1 3/8
6	1	10	1 9/16
7	1 1/8		

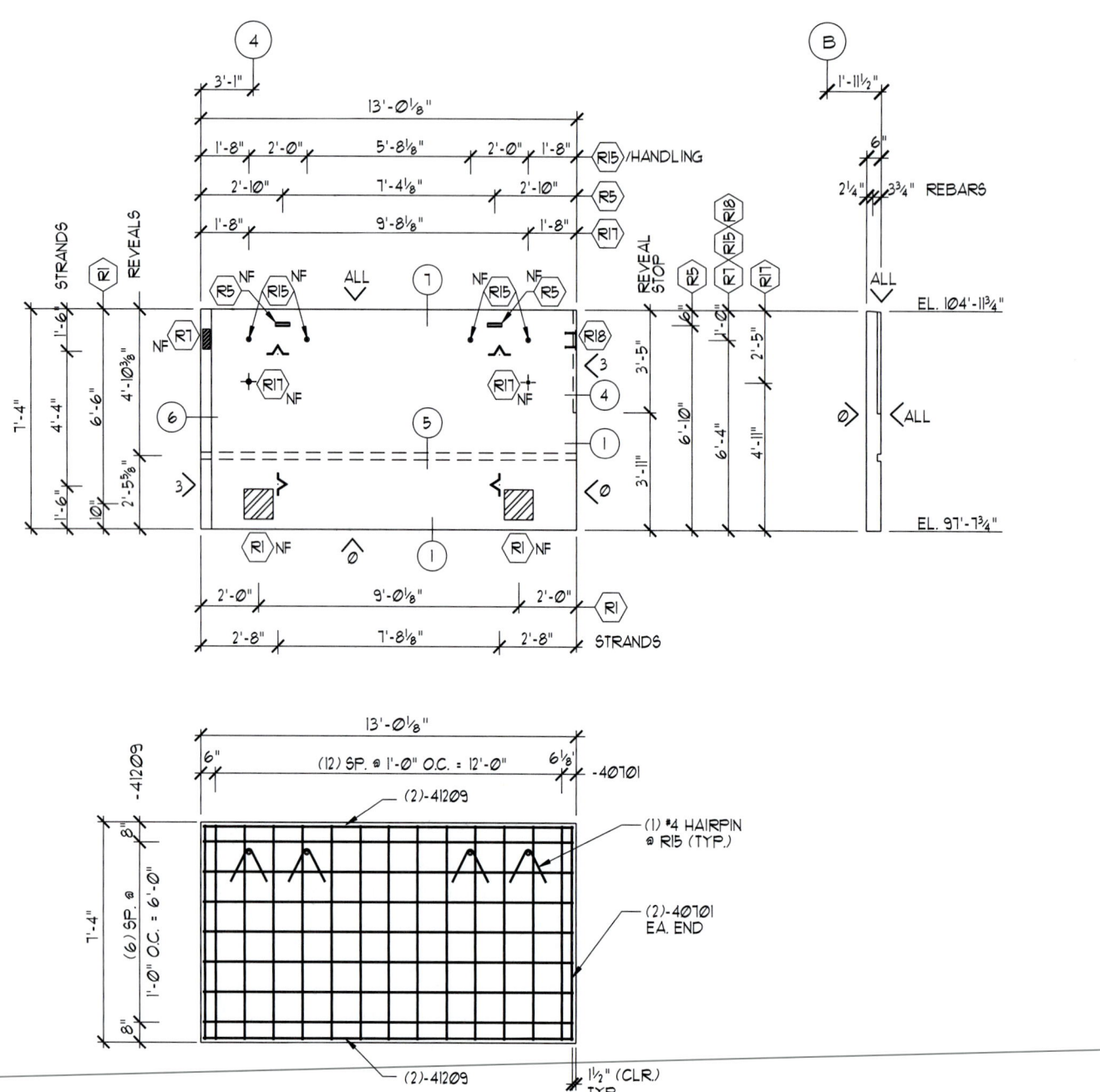

FIGURE 23.20 ■ Precast concrete panel drawing. *Courtesy Morse Bros. Inc.*

When rebar must be driven through timber, a tighter hole tolerance is recommended. For example:

#8 REBAR × 6'-0" @ 12" OC W/1-1/8" ∅ HOLES AT TIMBER

This application for the given rebar is as follows:

Bar #	Hole ∅	Bar #	Hole ∅
4	9/16	8	1 1/8
5	11/16	9	1 1/4
6	7/8	10	1 17/16
7	1		

Slab thickness, concrete wall, or concrete beam and column cross-sectional dimensions should be given in inches or feet and inches. A complete slab-on-grade callout can read as follows:

4" CONC SLAB ON FIRM UNDISTURBED SOIL OR 4" SAND FILL

or

6" THICK 3000 PSI CONC SLAB WITH WWF 6 × 6—
W2.9 × W2.9 3" CLR AT 4" MIN 3/4" MINUS FILL
COMPACTED TO 5% OF SOIL DENSITY

Concrete footing thickness should be given in inches or feet and inches. The footing width can be given followed by the thickness all in one note. Metric values are in millimeters.

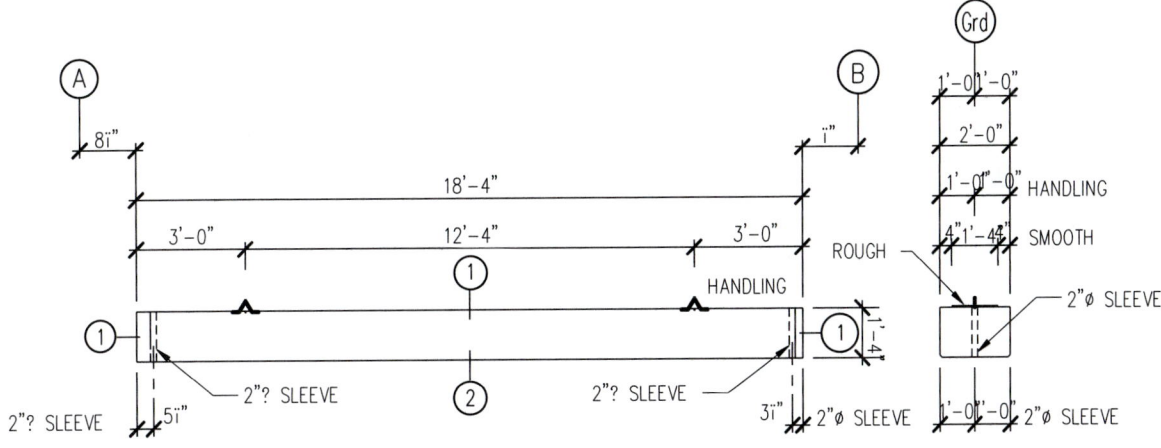

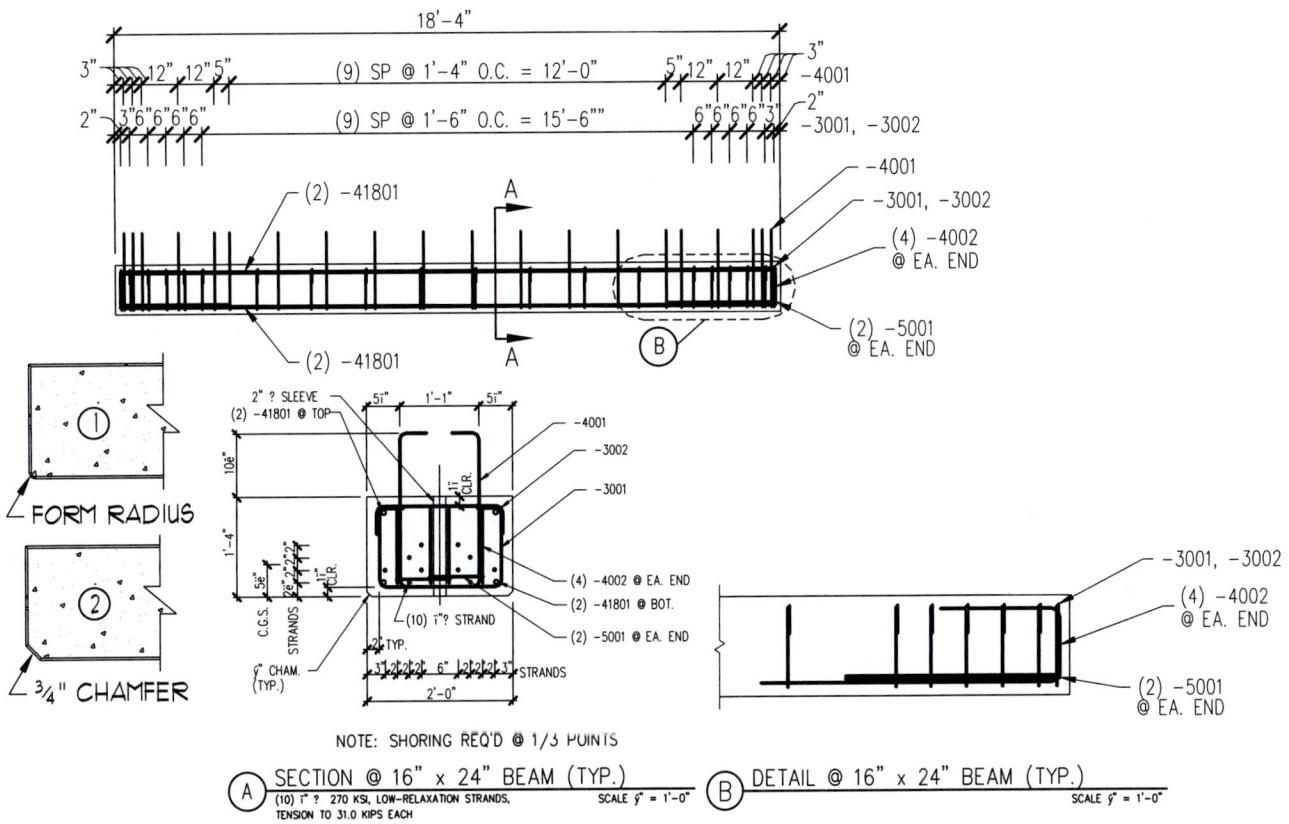

FIGURE 23.21 ■ Precast concrete beam drawing. *Courtesy Morse Bros. Inc.*

Concrete Structural Engineering Drawings

Structural engineering drawings show general information that is required for sales, marketing, engineering, or erection purposes. Concrete structures are drawn in plan (top) view, elevation (side) view, or sectional views at a scale that is dependent on the size of the structure, the amount of detail to be shown, or the size of paper used. Dimensions and notes are used to completely describe the construction characteristics. Concrete material symbols are used where appropriate. Rebar is shown as a very thick line or a thick dashed line in longitudinal view, or as a round dot where the bar appears cut.

Figure 23.25, page 758, illustrates a concrete foundation and related concrete and concrete block details as a typical example of the type of concrete drawings created by the structural drafter.

CONCRETE BLOCK CONSTRUCTION

Concrete block construction is often used for residential foundations but is also used in some above-ground construction. In commercial applications, concrete blocks are used to form the wall systems for many types of buildings. Concrete blocks provide a durable construction material and are relatively inexpensive to

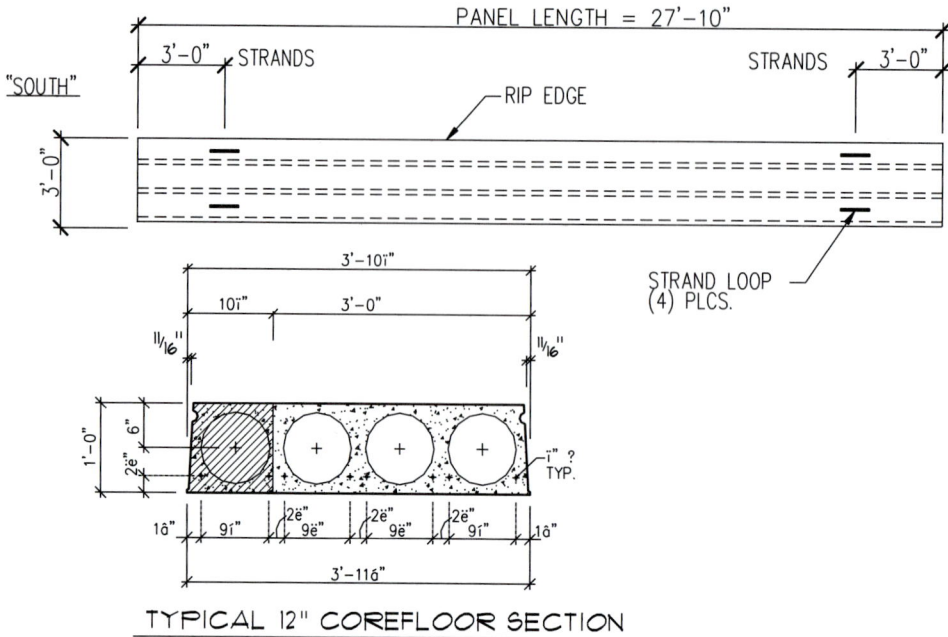

FIGURE 23.22 ■ Precast concrete slab drawing. *Courtesy Morse Bros. Inc.*

install and maintain. In residential and light commercial applications, foam-filled blocks provide excellent insulating characteristics and are often used in desert climates. Blocks are commonly manufactured in nominal size **modules** of 8 × 8 × 16 in., 4 × 8 × 16 in., or 6 × 8 × 16 in. The metric conversion of a nominal 8 × 8 × 16 in. concrete masonry unit is 200 × 200 × 400 mm. The actual size of the block is smaller than the nominal size so **mortar (grout)** joints can be included in the final size. Mortar is a combination of cement, sand, and water used to bond masonry units together. Although the building designer determines the size of the structure, it is important that the drafter be aware of the modular principles of concrete block construction. Wall lengths, opening locations, and wall and opening heights must be based on the modular size of the block being used. Failure to maintain the modular layout can result in a tremendous increase in labor costs to cut and lay the blocks.

The drafter's responsibility when working with concrete block structures is also to detail steel reinforcing patterns. Concrete blocks are often reinforced with a wire mesh at every other course of blocks. Where the risk of seismic activity must be considered, concrete block structures are often required to have reinforcing steel placed within the wall to help tie the blocks together. The steel is placed in a block that has a channel or cell running through it. This cell is then filled with grout or concrete to form a **bond beam** within the wall. The bond beam solidifies and ties the block structure together. A typical bond beam concrete block structure is detailed in Figure 23.26. Figure 23.27, page 760, shows typical concrete block sizes and reinforcing methods. Openings for windows and doors require steel reinforcing for both poured-in-place and concrete block construction. (See Figure 23.28, page 761.)

When the concrete blocks are required to support a load from a beam, a **pilaster** is often placed in the wall to help transfer the beam loads down the wall to the footing. Pilasters are also used to provide vertical support to the wall when the wall is required to span long distances. Examples of pilasters are shown in Figure 23.29, page 761. A structural detail is shown in Figure 23.30, page 762.

WOOD CONSTRUCTION

Wood frame construction is typical in residential construction and is also used in commercial construction applications, especially for multifamily dwellings and office buildings. Other commercial uses include partition framing, upper-level floor framing, and roof framing. Residential and commercial wood wall construction is essentially the same, with the main difference often being the type of covering used. In commercial applications, wood walls can require special finishes to meet fire protection requirements, as shown in Figure 23.31, page 762.

Joists, trusses, and panelized systems are the most commonly used wood roof framing systems. These systems allow for the members to be placed 16, 24 or 32 in. (400–800 mm) OC. Figure 23.3b, page 745, shows a truss roof system drawing. Panelized roof systems use beams placed 20 to 30 ft. (6000–9000 mm) apart with smaller beams called **purlins** placed between the main beams on 8-ft. (2400 mm) centers. Joists that are 2 or 3 in. (75 mm) wide are placed between the purlins at 24 in. (600 mm) OC. The roof is then covered with plywood sheathing. The dimensional values given in this discussion are examples. Actual dimensions must be determined by structural engineering

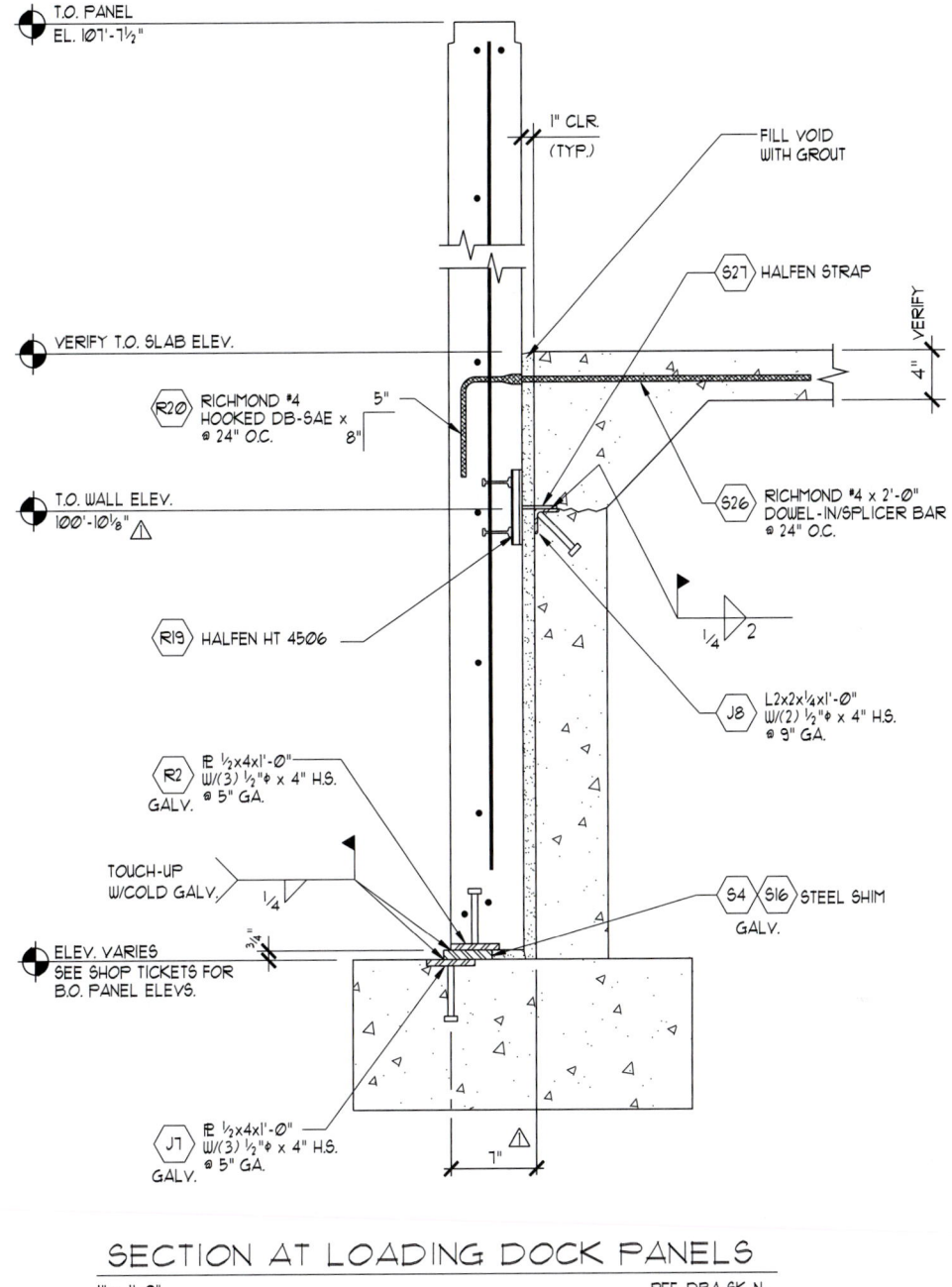

SECTION AT LOADING DOCK PANELS

1" = 1'-0"   REF. DBA SK-N

FIGURE 23.23 ■ Precast concrete panel, and concrete wall, footing, and slab detail drawing. *Courtesy Morse Bros. Inc.*

FIGURE 23.24 ■ Rebar bend diagram.

based on the building design. Figure 23.32, page 762, shows an example of how a panelized roof system is constructed. A roof framing plan for a panelized roof system is shown in Figures 23.3b, page 745, and 23.33, page 763.

When wood frame construction joins concrete or concrete block construction, a rigid connection between the two systems is required. Several methods of connecting wood to concrete block are shown in Figure 23.34 (see page 764).

Heavy Timber Construction

Large wood members are sometimes used for the structural frame of a building. This method of construction is used for appearance, structural purposes, and when material is available to suit the application. Heavy timbers have structural advantages

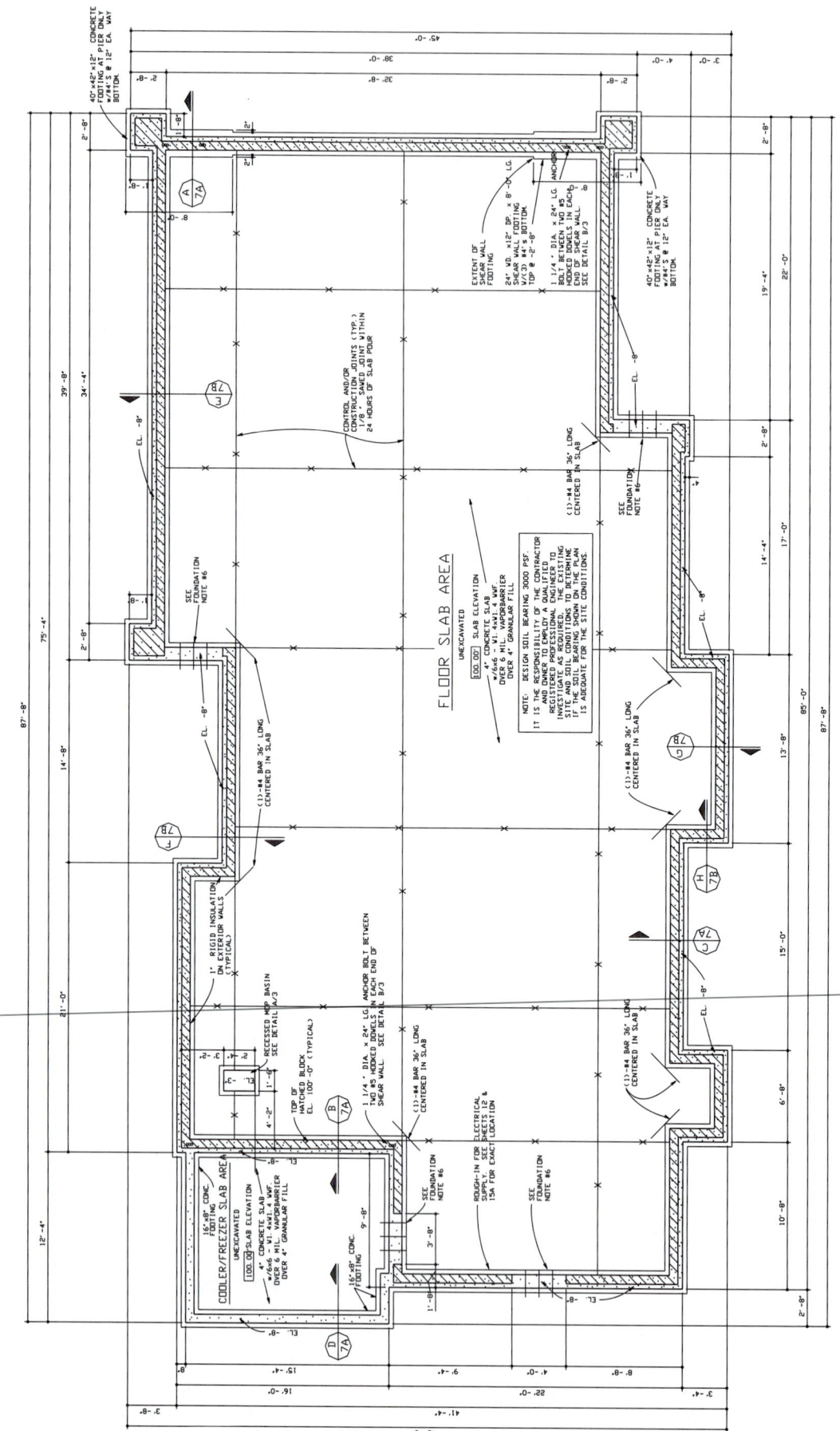

FIGURE 23.25 ■ Concrete foundation and related concrete and concrete block details.

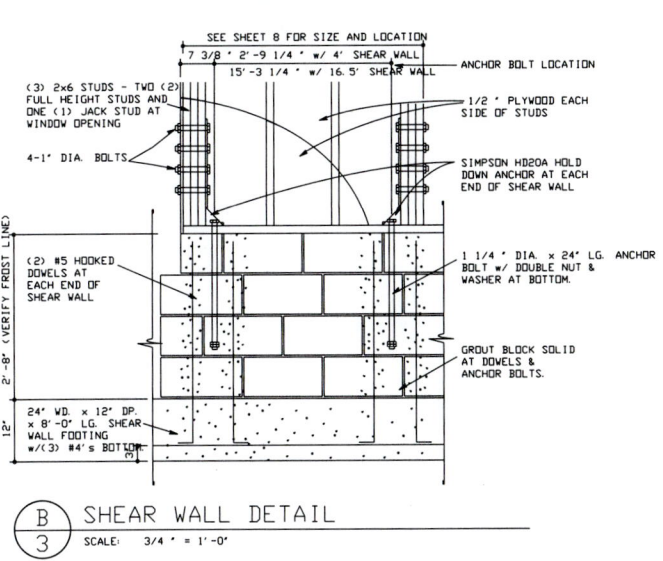

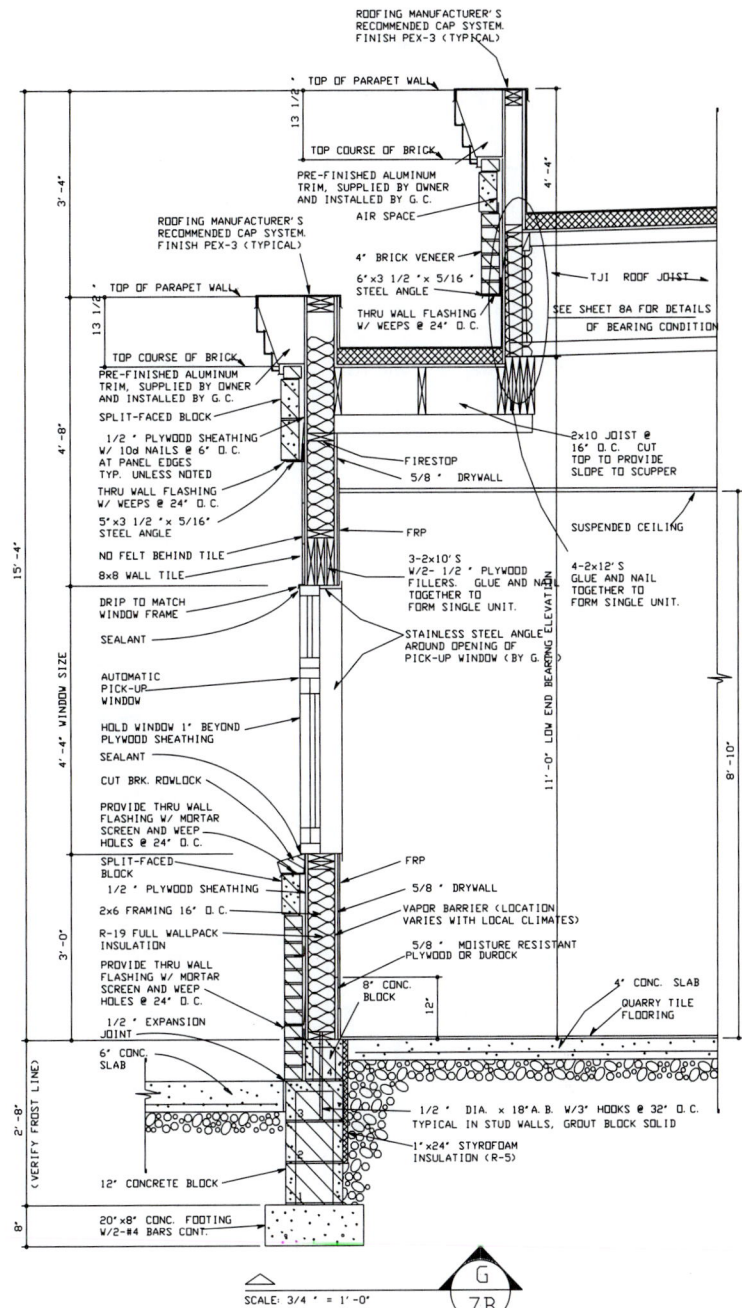

FIGURE 23.25 ■ Continued.

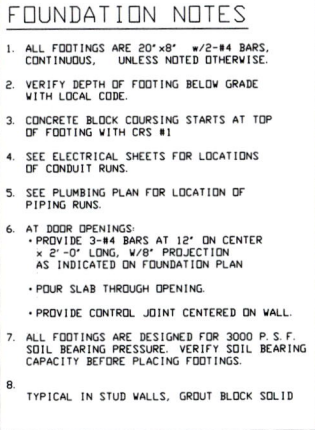

FOUNDATION NOTES

1. ALL FOOTINGS ARE 20"x8" w/2-#4 BARS, CONTINUOUS, UNLESS NOTED OTHERWISE.

2. VERIFY DEPTH OF FOOTING BELOW GRADE WITH LOCAL CODE.

3. CONCRETE BLOCK COURSING STARTS AT TOP OF FOOTING WITH CRS #1

4. SEE ELECTRICAL SHEETS FOR LOCATIONS OF CONDUIT RUNS.

5. SEE PLUMBING PLAN FOR LOCATION OF PIPING RUNS.

6. AT DOOR OPENINGS:
   • PROVIDE 3-#4 BARS AT 12" ON CENTER x 2'-0" LONG, W/8" PROJECTION AS INDICATED ON FOUNDATION PLAN

   • POUR SLAB THROUGH OPENING.

   • PROVIDE CONTROL JOINT CENTERED ON WALL.

7. ALL FOOTINGS ARE DESIGNED FOR 3000 P.S.F. SOIL BEARING PRESSURE. VERIFY SOIL BEARING CAPACITY BEFORE PLACING FOOTINGS.

8. TYPICAL IN STUD WALLS, GROUT BLOCK SOLID

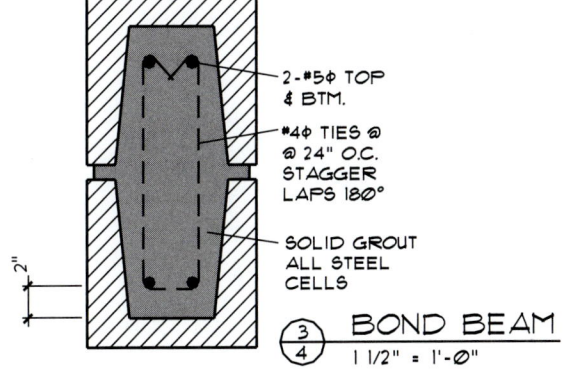

FIGURE 23.26 ■ Concrete block bond beam reinforcing detail.

for short spans and excellent fire retardant qualities. In a fire, heavy wood members will char on the exposed surfaces while maintaining structural integrity long after a steel beam of equal size has failed. A heavy timber roof framing plan is shown in Figure 23.33, page 763.

Laminated Beam Construction

It is difficult and often impossible to produce large-size wood timbers in long lengths, especially when lumber is currently sawn from smaller and smaller logs. When long-span wood beams are required, the solution is laminated (lam) beams, as shown in Figure 23.35, page 765. Notice the beams labeled

(Continued on page 763)

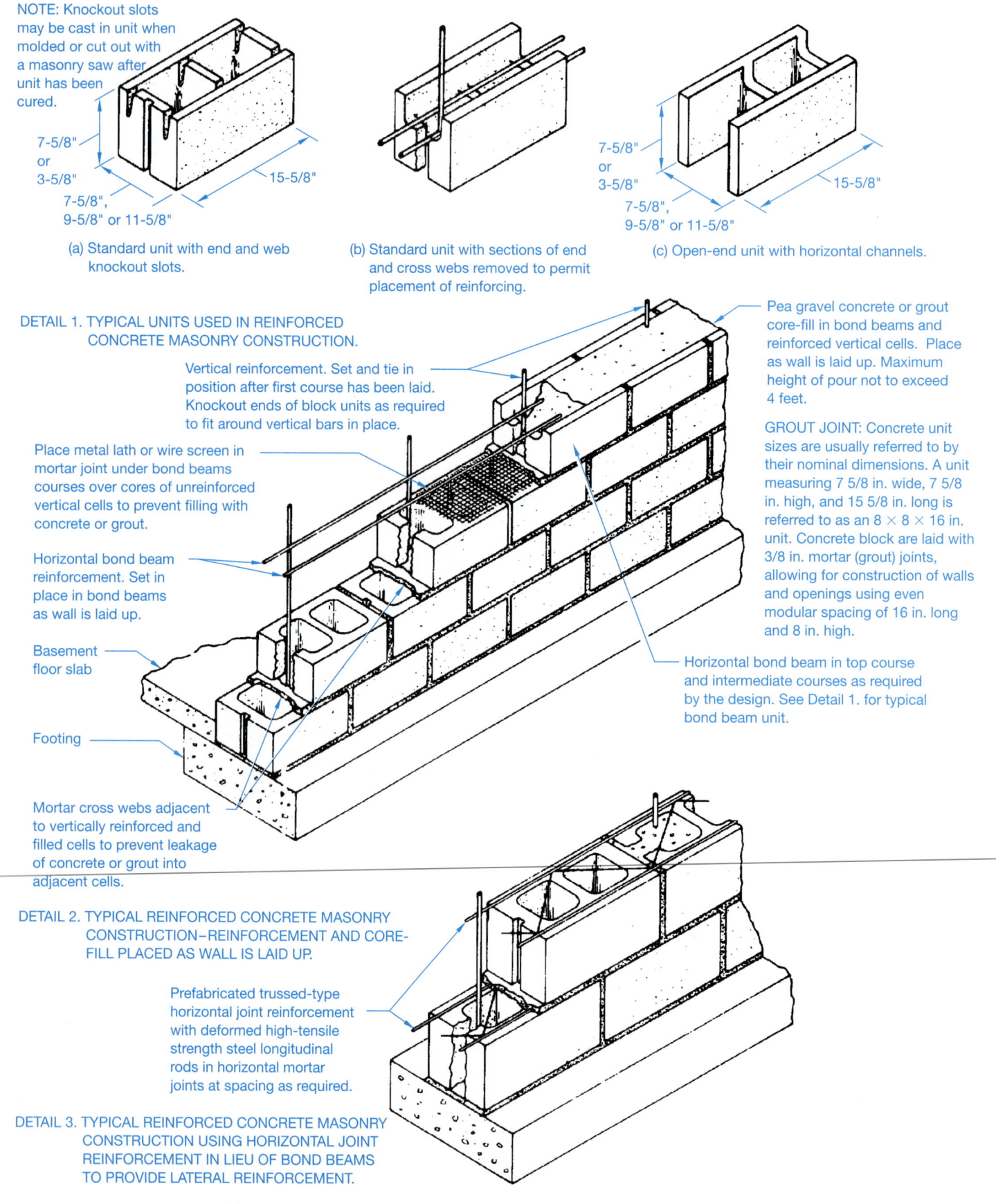

NOTE: Knockout slots may be cast in unit when molded or cut out with a masonry saw after unit has been cured.

7-5/8" or 3-5/8"

15-5/8"

7-5/8", 9-5/8" or 11-5/8"

(a) Standard unit with end and web knockout slots.

(b) Standard unit with sections of end and cross webs removed to permit placement of reinforcing.

7-5/8" or 3-5/8"

15-5/8"

7-5/8", 9-5/8" or 11-5/8"

(c) Open-end unit with horizontal channels.

DETAIL 1. TYPICAL UNITS USED IN REINFORCED CONCRETE MASONRY CONSTRUCTION.

Vertical reinforcement. Set and tie in position after first course has been laid. Knockout ends of block units as required to fit around vertical bars in place.

Place metal lath or wire screen in mortar joint under bond beams courses over cores of unreinforced vertical cells to prevent filling with concrete or grout.

Horizontal bond beam reinforcement. Set in place in bond beams as wall is laid up.

Basement floor slab

Footing

Mortar cross webs adjacent to vertically reinforced and filled cells to prevent leakage of concrete or grout into adjacent cells.

Pea gravel concrete or grout core-fill in bond beams and reinforced vertical cells. Place as wall is laid up. Maximum height of pour not to exceed 4 feet.

GROUT JOINT: Concrete unit sizes are usually referred to by their nominal dimensions. A unit measuring 7 5/8 in. wide, 7 5/8 in. high, and 15 5/8 in. long is referred to as an 8 × 8 × 16 in. unit. Concrete block are laid with 3/8 in. mortar (grout) joints, allowing for construction of walls and openings using even modular spacing of 16 in. long and 8 in. high.

Horizontal bond beam in top course and intermediate courses as required by the design. See Detail 1. for typical bond beam unit.

DETAIL 2. TYPICAL REINFORCED CONCRETE MASONRY CONSTRUCTION–REINFORCEMENT AND CORE-FILL PLACED AS WALL IS LAID UP.

Prefabricated trussed-type horizontal joint reinforcement with deformed high-tensile strength steel longitudinal rods in horizontal mortar joints at spacing as required.

DETAIL 3. TYPICAL REINFORCED CONCRETE MASONRY CONSTRUCTION USING HORIZONTAL JOINT REINFORCEMENT IN LIEU OF BOND BEAMS TO PROVIDE LATERAL REINFORCEMENT.

FIGURE 23.27 ■ Suggested construction details for reinforced concrete masonry foundation walls. *Courtesy National Concrete Masonry Association.*

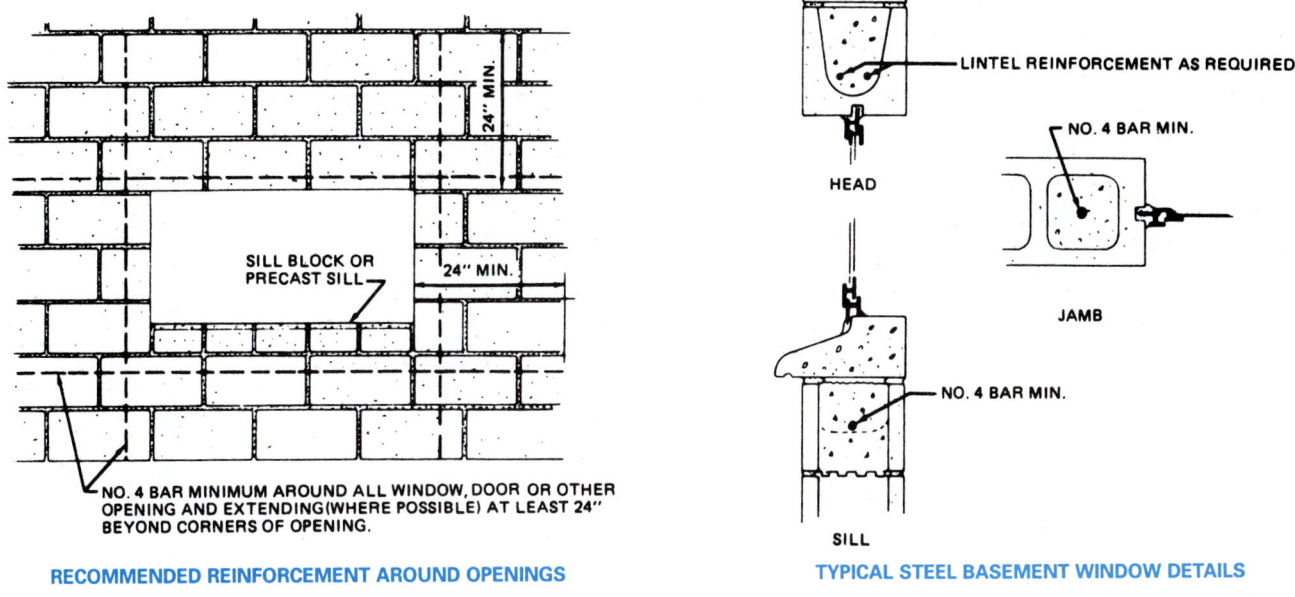

24" MIN.
SILL BLOCK OR
PRECAST SILL
24" MIN.
NO. 4 BAR MINIMUM AROUND ALL WINDOW, DOOR OR OTHER
OPENING AND EXTENDING(WHERE POSSIBLE) AT LEAST 24"
BEYOND CORNERS OF OPENING.
RECOMMENDED REINFORCEMENT AROUND OPENINGS
LINTEL REINFORCEMENT AS REQUIRED
HEAD
NO. 4 BAR MIN.
JAMB
NO. 4 BAR MIN.
SILL
TYPICAL STEEL BASEMENT WINDOW DETAILS

FIGURE 23.28 ■ Typical reinforcing and window detail at opening in reinforced concrete masonry. *Courtesy National Concrete.*

8 × 8 × 16
Alternate
courses
HOLLOW UNIT
4 × 8 × 16 solid units laid flat
4 × 8 × 16 solid units
set on end
Alternate
courses
SOLID UNIT
Special pilaster unit
Hollow cores filled
with concrete or
mortar
Alternate
courses
FILLED CELL HOLLOW UNIT
12 × 8 × 16
Alternate
courses
HOLLOW UNIT
4 × 8 × 16 solid
units set on end
4 × 8 × 16 solid
units laid flat
Alternate
courses
SOLID UNIT
Special pilaster unit with
hollow core reinforced
Wood sash
jamb unit
REINFORCED FILLED CELL
HOLLOW UNIT

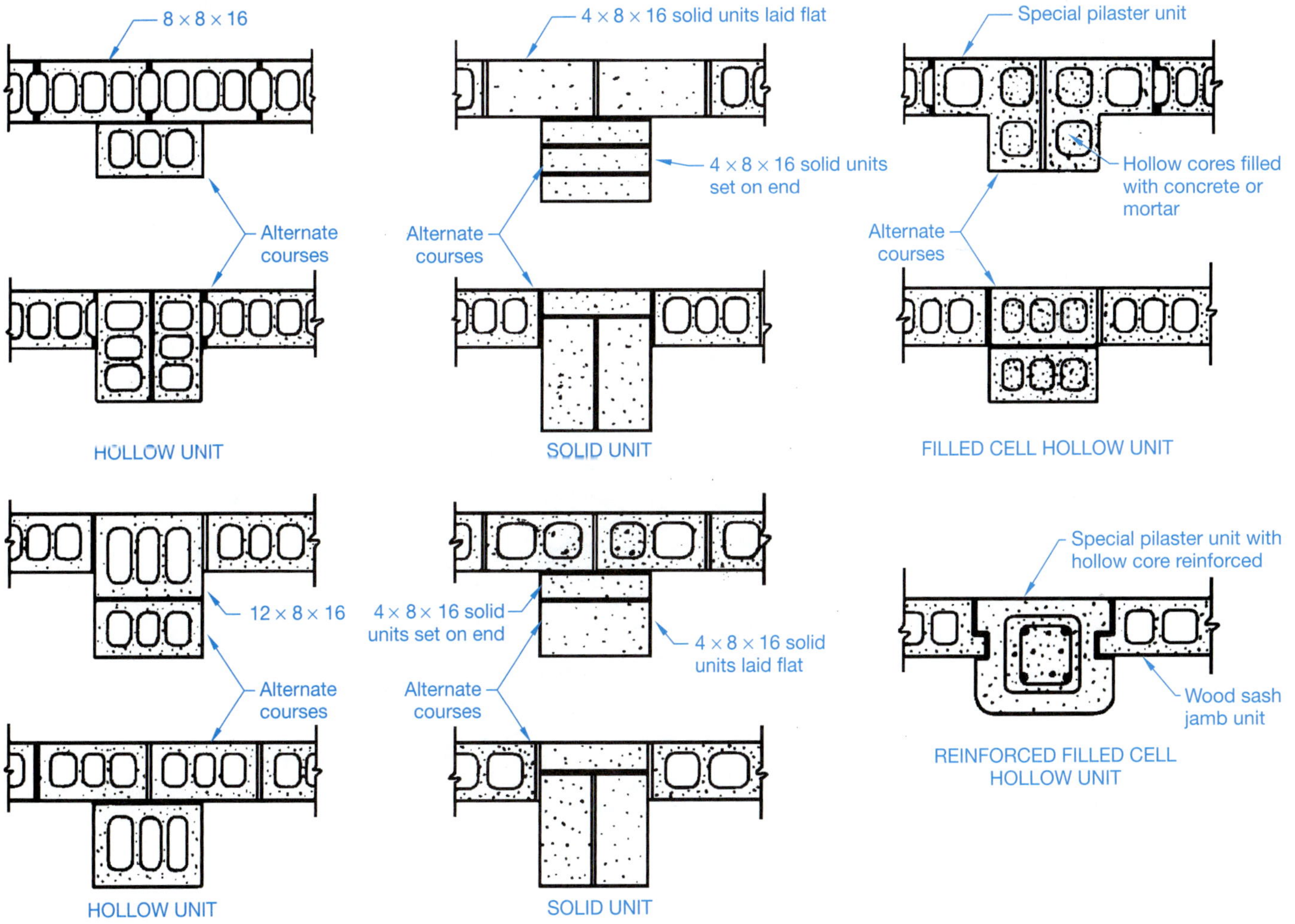

FIGURE 23.29 ■ Typical concrete masonry pilaster designs. *Courtesy National Concrete Masonry Association.*

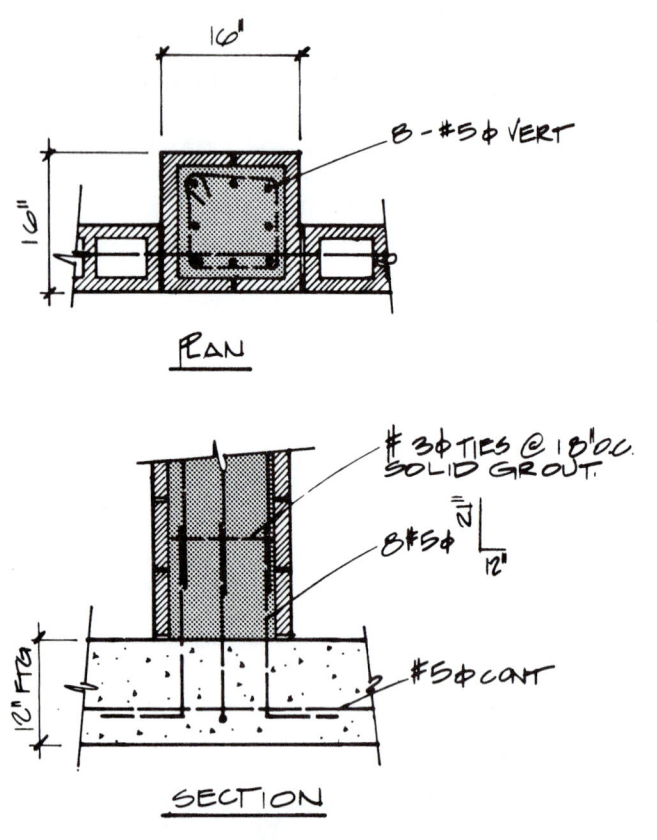

PLAN

B - #5 ∅ VERT

SECTION

#3 ∅ TIES @ 18" O.C.
SOLID GROUT.

8 #5 ∅

#5 ∅ CONT

FIGURE 23.30 ■ Reinforced pilaster detail.

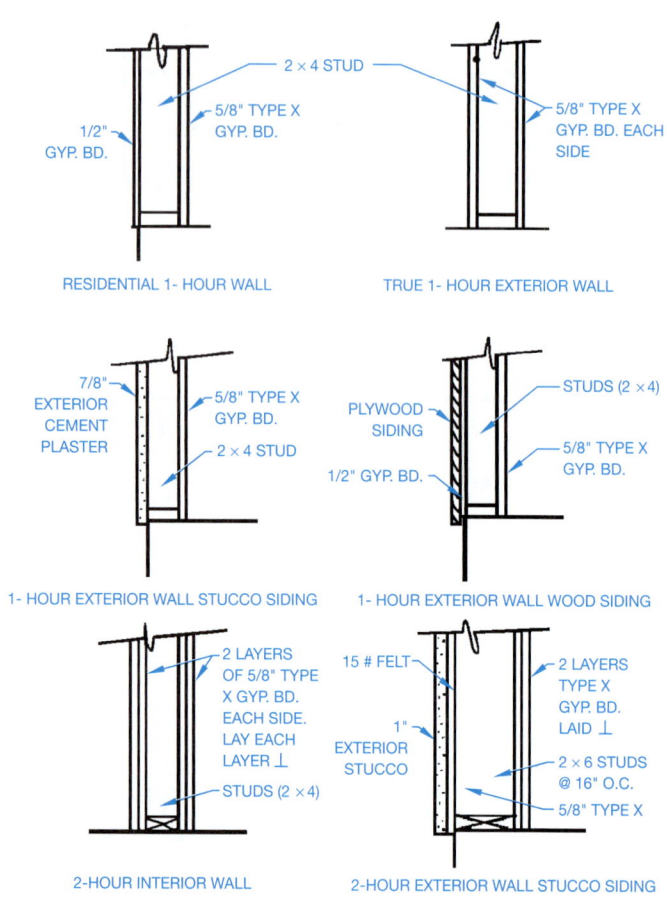

2 × 4 STUD

1/2" GYP. BD.

5/8" TYPE X GYP. BD.

5/8" TYPE X GYP. BD. EACH SIDE

RESIDENTIAL 1- HOUR WALL

TRUE 1- HOUR EXTERIOR WALL

7/8" EXTERIOR CEMENT PLASTER

5/8" TYPE X GYP. BD.

2 × 4 STUD

PLYWOOD SIDING

1/2" GYP. BD.

STUDS (2 × 4)

5/8" TYPE X GYP. BD.

1- HOUR EXTERIOR WALL STUCCO SIDING

1- HOUR EXTERIOR WALL WOOD SIDING

2 LAYERS OF 5/8" TYPE X GYP. BD. EACH SIDE. LAY EACH LAYER ⊥

STUDS (2 × 4)

15 # FELT

1" EXTERIOR STUCCO

2 LAYERS TYPE X GYP. BD. LAID ⊥

2 × 6 STUDS @ 16" O.C.

5/8" TYPE X

2-HOUR INTERIOR WALL

2-HOUR EXTERIOR WALL STUCCO SIDING

FIGURE 23.31 ■ Separation walls often require special treatment to achieve the needed fire rating for certain types of construction.

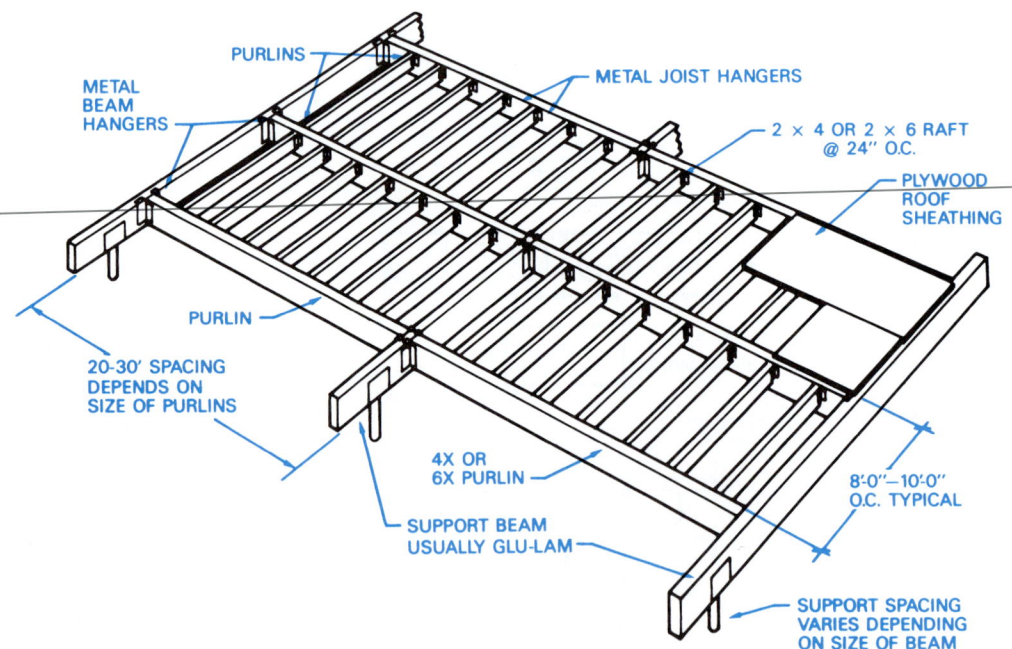

PURLINS

METAL BEAM HANGERS

METAL JOIST HANGERS

2 × 4 OR 2 × 6 RAFT @ 24" O.C.

PLYWOOD ROOF SHEATHING

PURLIN

20-30' SPACING DEPENDS ON SIZE OF PURLINS

4X OR 6X PURLIN

8'-0"–10'-0" O.C. TYPICAL

SUPPORT BEAM USUALLY GLU-LAM

SUPPORT SPACING VARIES DEPENDING ON SIZE OF BEAM

FIGURE 23.32 ■ A panelized roof system is often used to provide roofing for a large area with limited supports.

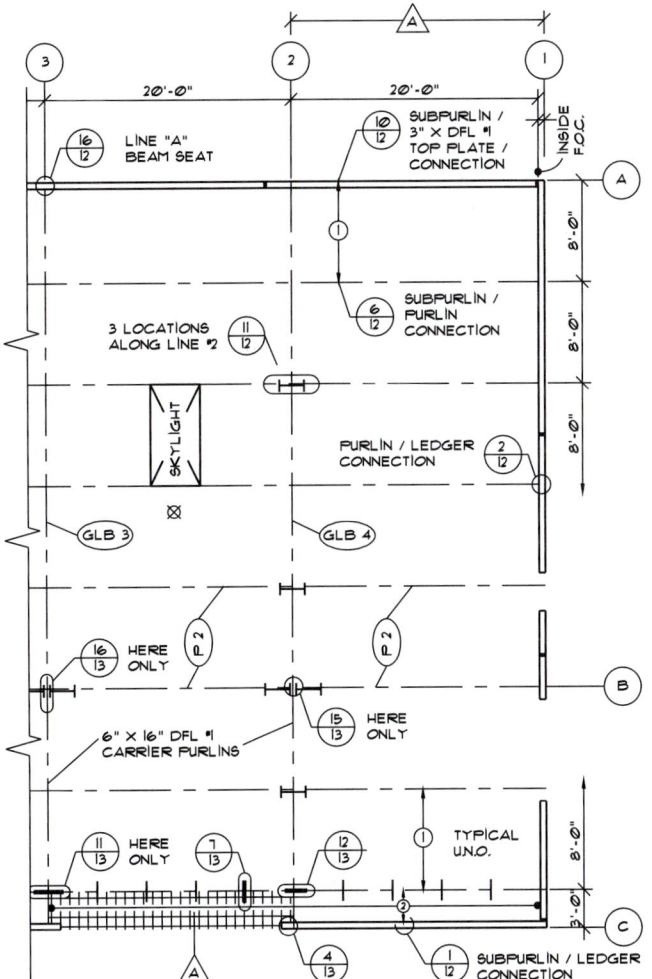

FIGURE 23.33 ■ A panelized roof and heavy timber framing plan. *Courtesy Structureform Masters, Inc.*

drafter's major responsibility when working with either heavy timber construction or laminated beams is in the drawing of connection details. Beam to beam, beam to column, and column to support are among the most common details drawn by drafters. Common types of manufactured connectors used to connect timbers are shown in Figure 23.38, page 766. The drafter is required to draw the fabrication details for a connector when the size of the beam does not match existing connectors, as shown in Figure 23.3a. (See page 745.)

Wood members can be attached to concrete in several ways. Two of the most common methods are by the use of a pocket or seat, as shown in Figure 23.39, page 766, or a metal connector.

Engineered Wood Products

Engineered wood products are a variety of products that have been designed to replace conventional lumber and provide many advantages, such as reducing the industry dependence on natural lumber. Common types of engineered wood products include I-joists, laminated veneer lumber (LVL), and related products such as rim board.

I-Joists

I-joists are generally made of softwood veneers such as fir or pine that are bonded together or solid wood to make the top and bottom flanges. The web or core is then made of composite wood. The name comes from the I shape that is shown in Figure 23.40, page 767.

I-joists provide stronger, more stable performance; are lighter weight, easier to handle, and faster to install; and make more efficient use of valuable wood resources than conventional lumber. They also provide longer lengths and prestamped knockouts for wire or plumbing runs. I-joists are drawn just like other joists in the plan view using solid lines or centerlines. The shape of the I-joist showing the top and bottom flanges and the core is displayed when drawing in sectional view.

Laminated Veneer Lumber

Laminated veneer lumber (LVL) is an engineered structural member that is manufactured by bonding wood veneers with an exterior adhesive. LVL has almost no shrinking, checking, twisting, or splitting; it is excellent for floor and roof framing supports or as headers for doors, windows, and garage doors and columns. LVL has no camber, so LVL products provide flatter, quieter floors. This lumber is available in lengths that are longer than conventional lumber, so builders can avoid on-site splicing and multiple nailing and save on labor costs. Common sizes range from 1-3/4" to 7" (40–180 mm) wide, with depths from 5 1/2" to 18" (140–460 mm) and lengths of up to 66' (20 mm). Figure 23.41, page 767, shows an example of LVL sizes. Figure 23.42, page 767, shows a comparison of physical properties of glu-laminated lumber, laminated veneer lumber, and solid lumber. You can see the physical advantage of the LVL product. The physical properties vary depending on the type of material and the manufacturer. LVL members are drawn like other beams, using a

GLB 1, GLB 2, GLB 3, and GLB 4 in Figure 23.33. These are **glu-lam beams (GLB)**. The intermediate members labeled P1 and P2 are purlins, as previously discussed. Laminated beams are manufactured from smaller, equally sized members glued together to form a larger beam. Laminated beams are used to support heavy loads or accommodate long spans and are also used when the wood appearance is important. The common types of laminated beams are the **single span straight**, **Tudor arch**, and **three-hinged arch beams**. (See Figure 23.36.)

The single span beam is commonly referred to as glu-lam and noted on a drawing with the abbreviation GLB for glu-lam beam. A glu-lam beam can be used to replace a much larger wood timber due to increased structural qualities. Figure 23.37, compares the strength of common wood members with laminated beams. In some applications, laminated beams have a **camber** or curve built into the beam. The camber is designed into the beam to help resist the loads to be carried.

The Tudor and three-hinged arch members are a post-and-beam system combined into one member. These beams are specified on plans in a method similar to other beams. The

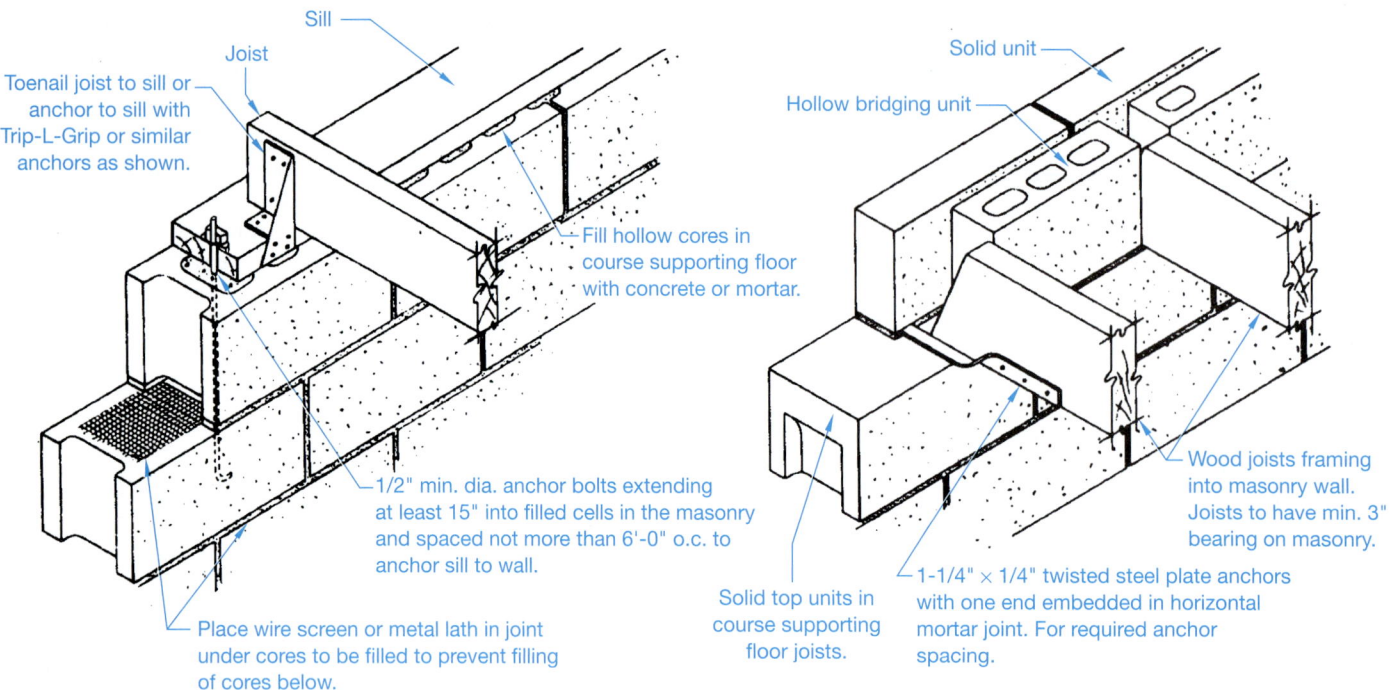

Toenail joist to sill or anchor to sill with Trip-L-Grip or similar anchors as shown.

Sill

Joist

Fill hollow cores in course supporting floor with concrete or mortar.

1/2" min. dia. anchor bolts extending at least 15" into filled cells in the masonry and spaced not more than 6'-0" o.c. to anchor sill to wall.

Place wire screen or metal lath in joint under cores to be filled to prevent filling of cores below.

Solid unit

Hollow bridging unit

Wood joists framing into masonry wall. Joists to have min. 3" bearing on masonry.

1-1/4" × 1/4" twisted steel plate anchors with one end embedded in horizontal mortar joint. For required anchor spacing.

Solid top units in course supporting floor joists.

ANCHORAGE OF WOOD JOISTS TO FOUNDATION IN WOOD FRAME CONSTRUCTION

ANCHORAGE OF WOOD JOISTS TO FOUNDATION IN MASONRY CONSTRUCTION

SUGGESTED METHODS OF ANCHORING WOOD JOISTS BEARING ON CONCRETE MASONRY FOUNDATION WALLS.

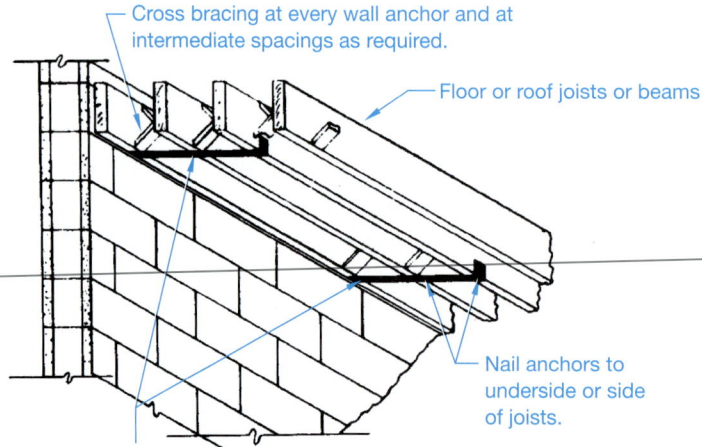

Cross bracing at every wall anchor and at intermediate spacings as required.

Floor or roof joists or beams

Nail anchors to underside or side of joists.

Wall anchors at required intervals. Anchors should have split end embedded in mortar joint or end bent down into block core and core filled with mortar. Length of anchor should be sufficient to engage at least three joists.

TYPICAL DETAILS FOR ANCHORAGE OF CONCRETE MASONRY WALLS TO PARALLEL WOOD JOISTS OR BEAMS.

FIGURE 23.34 ■ Metal angles and straps are typically used to ensure a rigid connection between the wall and floor or roof system. *Courtesy National Concrete Masonry Association.*

FIGURE 23.35 ■ Timbers are often used in commercial construction because of their beauty and structural qualities. The structural system of St. Philip's Episcopal Church is formed by glu-laminated Southern pine arches and beams under a sweeping canopy of Southern pine roof decking. *Courtesy Southern Forest Products Association.*

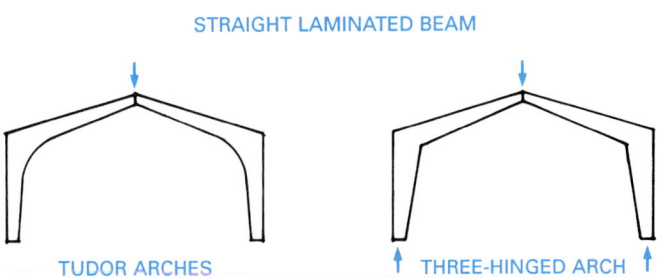

FIGURE 23.36 ■ Common laminated beam shapes.

thick solid line or a thick centerline symbol. The dimensional values given in this discussion are examples. Actual dimensions must be determined by structural engineering based on the building design.

Rim Board

As part of a complete engineered wood construction system, the rim board is used to cap the ends of the I-joist construction just like the rim joist used in conventional framing systems. The advantages of engineered rim joists is that they exactly match I-joist depths, require no special backing for siding products, and are quick and easy to install. Rim boards have almost no shrinking, checking, twisting, or splitting. Rim boards cannot be used as headers or beams.

Plywood Lumber Beams

Plywood lumber beams are also an engineered wood product that are made to precise specifications in size and nailing requirements for specific applications. Plywood lumber beams are also known as box beams. You can see their boxlike construction in Figure 23.43.

Stressed Skin Panels

Stressed skin panels use solid lumber or engineered lumber stringers and headers with a plywood skin on the top and bottom as shown in Figure 23.44. These panels can be used for walls, floors, and roof systems. They allow the builder to quickly erect a building because they cover a large area. They are commonly used in remote areas where they can be shipped in and quickly installed.

Standard Wood Structural Callouts

When specifying the callouts for structural wood sizes on a drawing, give the nominal (rough, before planning) cross-sectional dimensions for sawn lumber and timbers. For example:

$$4 \times 12, 6 \times 14$$

Specis and Commercial Grade	Extreme Fiber Bending F_b	Horizontal Shear F_v	Compression Perpendicular to Grain $F_c\perp$	Modulus of Elasticity E
DFL #1	1350	85	385	1,600,600
22FV4				
DF/DF	2200	165	385	1,700,000
Hem-Fir #1	1050	70	245	1,300,000
22F E2				
HF/HF	2200	155	245	1,400,000
SPF	900	65	265	1,300,000
22F-E-1				
SP/SP	2200	200	385	1,400,00

FIGURE 23.37 ■ Comparative values of common framing lumber with laminated beams of equal material. Values based on the *Uniform Building Code.*

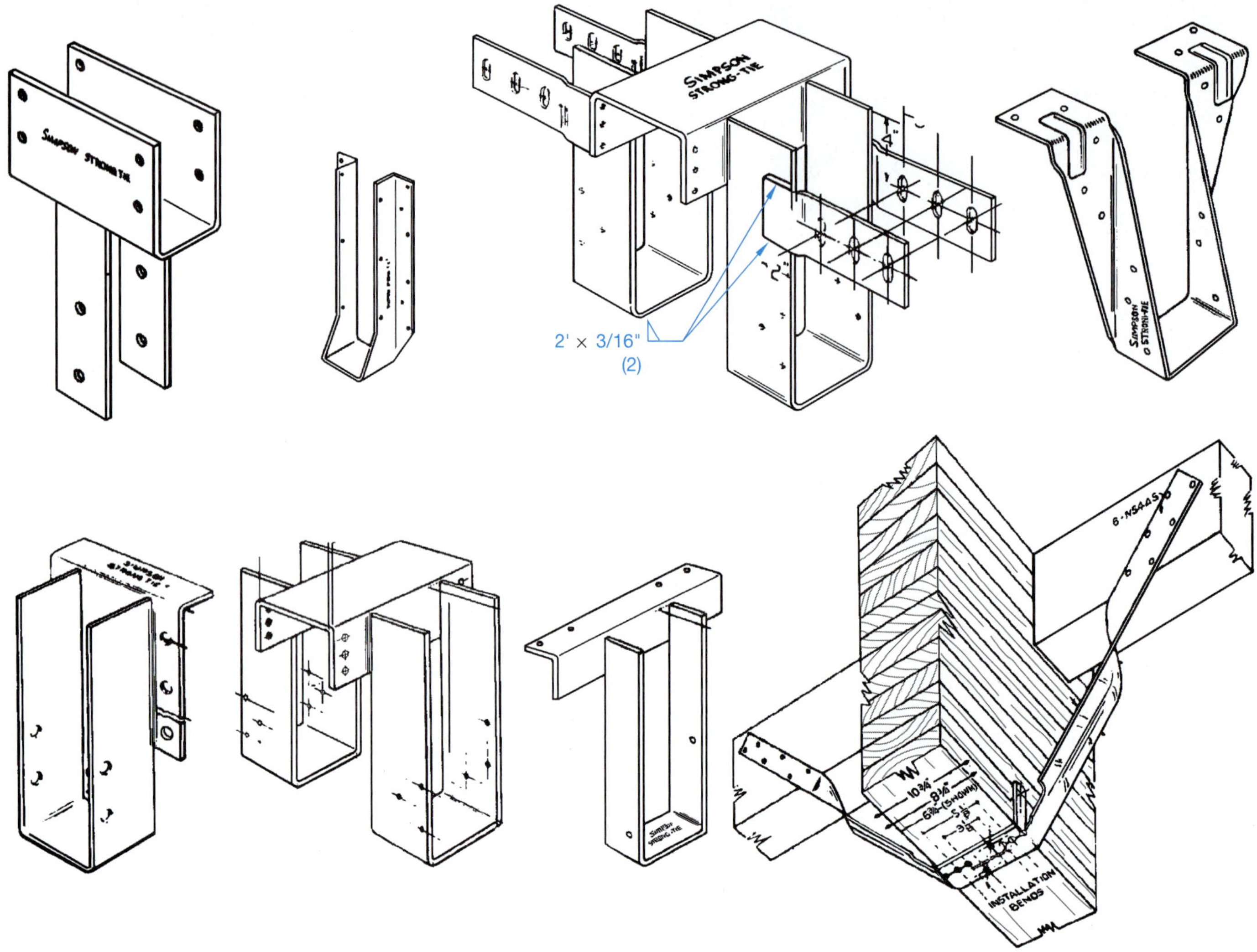

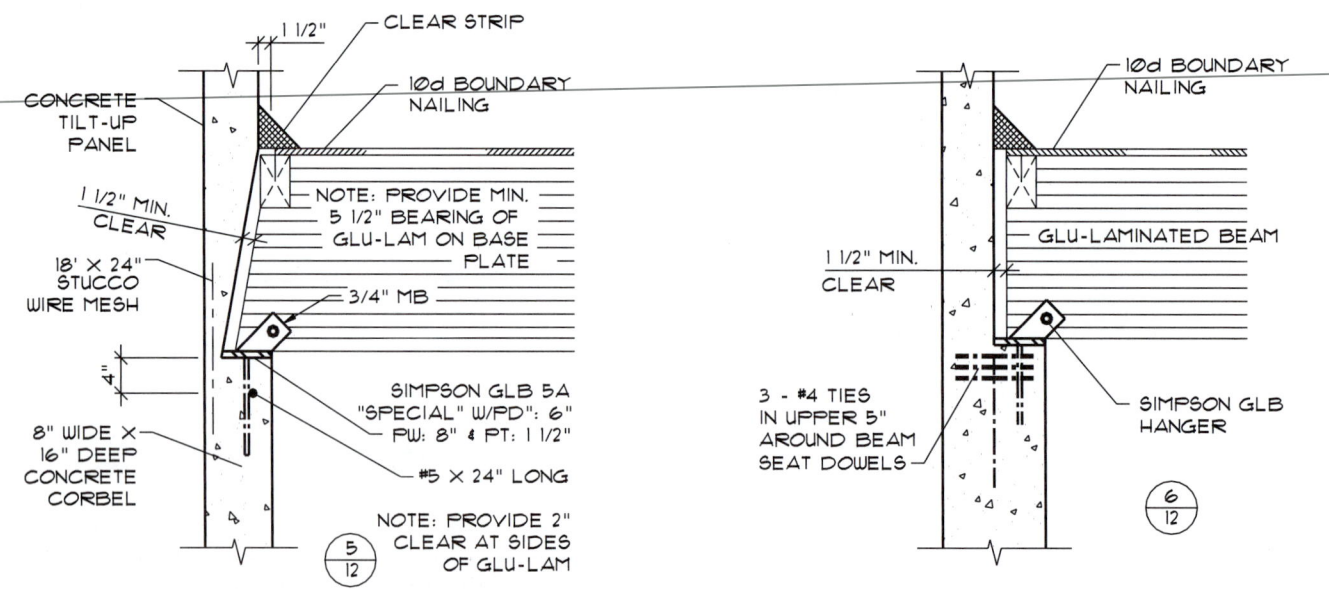

SIMPSON STRONG-TIE
2' × 3/16"
(2)
8-N54AS
INSTALLATION BENDS

FIGURE 23.38 ■ Common manufactured metal beam connectors. *Courtesy Simpson Strong-Tie Company, Inc.*

1 1/2"
CLEAR STRIP
10d BOUNDARY NAILING
CONCRETE TILT-UP PANEL
NOTE: PROVIDE MIN. 5 1/2" BEARING OF GLU-LAM ON BASE PLATE
1 1/2" MIN. CLEAR
18' X 24" STUCCO WIRE MESH
3/4" MB
4"
SIMPSON GLB 5A "SPECIAL" W/PD": 6" PW: 8" & PT: 1 1/2"
8" WIDE X 16" DEEP CONCRETE CORBEL
#5 X 24" LONG
NOTE: PROVIDE 2" CLEAR AT SIDES OF GLU-LAM
5/12
10d BOUNDARY NAILING
GLU-LAMINATED BEAM
1 1/2" MIN. CLEAR
3 - #4 TIES IN UPPER 5" AROUND BEAM SEAT DOWELS
SIMPSON GLB HANGER
6/12

FIGURE 23.39 ■ Wood beams are often designed to rest on a ledge or pocket. *Courtesy Structureform Masters, Inc.*

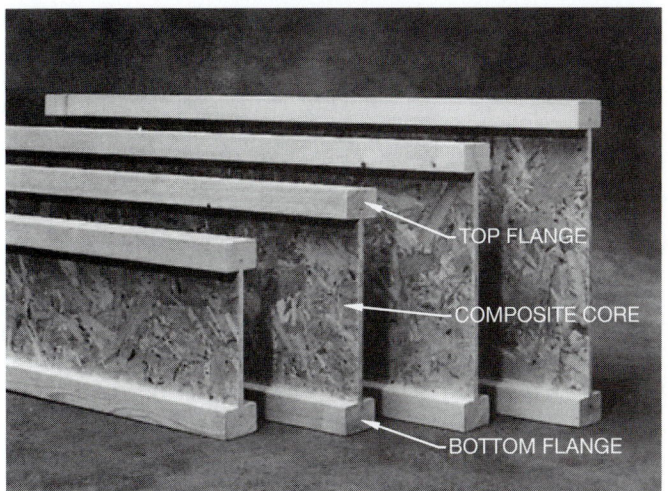

FIGURE 23.40 ■ The I-joist has top and bottom flanges connected by a composite wood core. *Courtesy Louisiana-Pacific.*

TOP FLANGE

COMPOSITE CORE

BOTTOM FLANGE

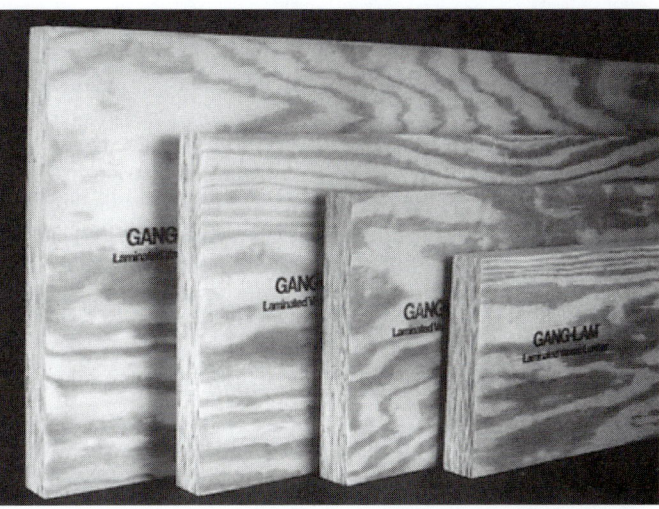

FIGURE 23.41 ■ Laminated veneer lumber (LVL) is manufactured by bonding layers of wood veneer with an exterior adhesive. *Courtesy Louisiana-Pacific.*

Property	LVL[a]	Glued Laminated[b]	Solid Lumber[c]
Extreme fiber in bending (F_b) psi	3000	1600–2400	760–2710
Horizontal shear (H_v) psi	290	140–650	95–120
Compression perpendicular to grain ($F_{c\perp}$) psi	3180	375–650	565–660
Compression parallel with grain (F_c) psi	—	1350–2300	625–1100
Tension parallel with grain (F_t) psi	2300	1050–1450	450–1200
Modulus of elasticity (E) million psi	2.0	1.2–1.8	1.3–1.9

[a]Limited to one size member
[b]Data from one manufacturer
[c]Data for one species of lumber

FIGURE 23.42 ■ Properties of laminated veneer lumber, glu-laminated members, and solid lumber.

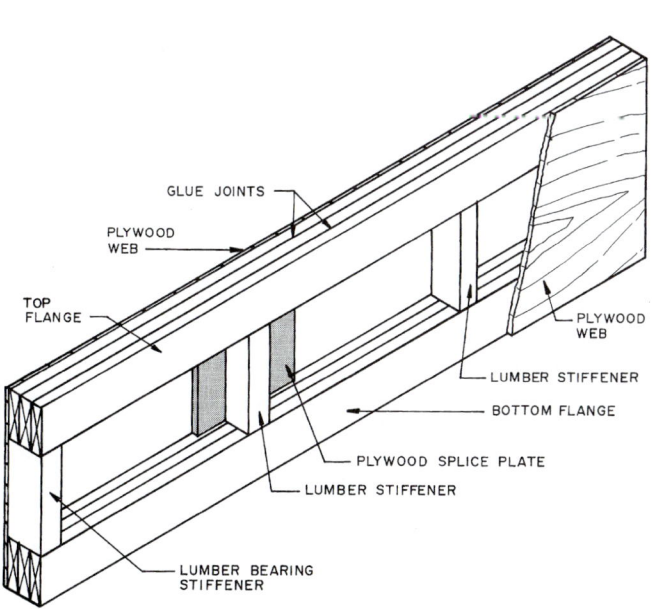

GLUE JOINTS

PLYWOOD WEB

TOP FLANGE

PLYWOOD WEB

LUMBER STIFFENER

BOTTOM FLANGE

PLYWOOD SPLICE PLATE

LUMBER STIFFENER

LUMBER BEARING STIFFENER

FIGURE 23.43 ■ A typical plywood-lumber beam (box-beam) with plywood webs and solid wood flanges and stiffeners.

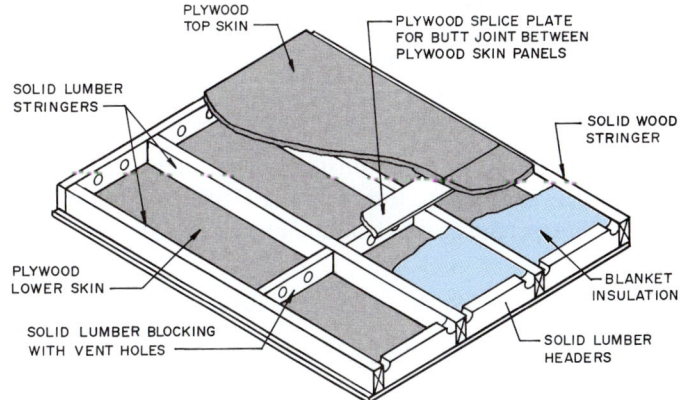

PLYWOOD TOP SKIN

PLYWOOD SPLICE PLATE FOR BUTT JOINT BETWEEN PLYWOOD SKIN PANELS

SOLID LUMBER STRINGERS

SOLID WOOD STRINGER

PLYWOOD LOWER SKIN

BLANKET INSULATION

SOLID LUMBER BLOCKING WITH VENT HOLES

SOLID LUMBER HEADERS

FIGURE 23.44 ■ A stress skin panel.

The nominal size of a wood construction member is before milling. The actual finished size is different. For example, the nominal size before milling is provided in the previous drawing specification. The milling process makes the finished member 3 1/2" × 11 1/2" and 5 1/2" × 13 1/2", respectively. Metric values for lumber are specified in millimeters and are rounded to common metric standards. For example, a 4" × 12" wood member is specified as 100 × 300 mm.

The net (actual) cross-sectional dimension is given for glu-lams. The net dimensions of a glu-lam are followed by the abbreviation GLB and engineering specification. For example:

$$6\text{-}1/8 \times 14 \text{ GLB f2400}$$
(f2400 = actual units of stress in fiber bending)

If the net cross-sectional dimensions for lumber are required, they can be given in parentheses after the nominal dimensions. If lengths (LG) are required for wood members, they should be given in feet and inches or millimeters. For example:

$$6\text{-}3/4 \times 18 \text{ GLB} \times 24'\text{-}0'' \text{ LG}$$

Glu-lam beams can also be designated with a manufacturing number such as:

$$6\text{-}3/4 \times 18 \text{ 22FV4}$$

The 6-3/4 × 18 gives the width and depth, and the 22FV4 designation represents the following values:

22F This is the fiber bending, f or F_b as given in previous examples. Different wood species have different fiber bending values.

V4 This is a designation that varies depending on the wood species and the application for the beam. This specification might be V1, V2, V3, or E1, E2, E3, for example. The options include the species of laminations outside and inside of the beam. This also relates to the application for the beam, such as down loading, up loading, and cantilever. Down loading refers to a beam that is designed to support weight from above. Up loading refers to a beam that can accept forces that push up on the beam, such as wind loads or the forces of weight applied to an adjacent cantilever. Some beams are designed to accept both down loading and up loading. Cantilever beams are designed to project beyond their supporting wall or column. Refer to building codes for glu-lam beams or manufacturers' catalogs for specific information about these glu-lam beam designations.

The specifications for laminated veneer lumber (LVL) vary slightly between manufacturers, but the elements of the note are similar. For example, an LVL beam or header might be specified like this:

$$1\ 3/4'' \times 14'' \text{ BCI VERSALAM — D}$$

The 1 3/4" × 14" is the width and depth, and the other elements are described as follows:

BCI This is the manufacturer, in this case Boise Cascade Corporation.

VERSALAM This is a specific LVL product of the Boise Cascade Corporation called Versa-Lam®.

D This identifies the wood species, Douglas fir in this case.

The specifications for I-joists can also vary between manufacturers, but there are similar components, such as:

$$14'' \text{ BCI 90XL @ 16'' OC.}$$

The following identifies each of the elements of this I-joist note:

14" This is the depth of the I-joist.

BCI This is the manufacturer, in this case Boise Cascade Corporation.

90XL This identifies the strength characteristics based on a specific BCI I-joist.

@16" OC. This is the spacing on center of the I-joists.

When wood is used for floor joists (FJ) or ceiling joists (CJ), rafters or trusses, or stud walls, the member size is followed by the OC spacing in inches up to 24" (600 mm) OC and feet and inches for over 24" OC. For example, abbreviations are shown in parentheses:

$$2 \times 8 \text{ CEILING JOISTS (CJ) 16'' OC,}$$
$$4 \times 14 @ 4'\text{-}0'' \text{ OC}$$
$$2 \times 6 \text{ STUDS 24'' OC,}$$
$$2 \times 12 \text{ FLOOR JOISTS (FJ) 12'' OC}$$

All lumber and timber on a drawing should be dimensioned to the centerlines unless dimensioning to the face of the member is otherwise required by the engineer or architect. The exception to this rule is dimensioning to the top of a beam or other structural timber.

When specifying plywood on a drawing, give the thickness, group, face veneer grades, identification index (if required), and glue type for each different panel used. Also include nailing, blocking, and edge spacing requirements if specific applications are required. Dimension to the face of the panel when surface location dimensions are required. Plywood is wood composed of three or more layers of thin veneer sheets, with the grain of each sheet placed at 90° to each other and bonded together with glue. Plywood sheets are typically 4' × 8' (1200 × 2400 mm) and come in a variety of thicknesses for different structural applications. The following are plywood callout examples:

1. 1/2" CDX 32/16 PLYWOOD SHEATHING.

2. 1/2" CDX SHTG W/10d NAILS 3" OC @ SEAMS AND 8" OC FIELD. (Note: *SEAMS* = edges or splices; *FIELD* = within the sheet at supports.)

3. 1/2" GROUP 1, CC, 24/0 EXTERIOR APA PLYWOOD W/8d NAILS 4" OC @ EDGES AND @ 6" OC FIELD.

4. 5/8" GROUP 2, UNDERLAYMENT CC-PTS EXTERIOR APA PLYWOOD W/8d RINGSHANK NAILS @ 3" OC EDGES AND 6" OC FIELD, BLOCK ALL PANEL EDGES PERPENDICULAR TO SUPPORTS. (Note: *CC* = CC grade outside veneer, *PTS* = plugged and touch sanded.)

When lumber decking is used, the lumber size and specification is followed by nailing information, if required. For example:

$$3 \times 8 \text{ T\&G RANDOM CONTROLLED DECKING W/20d}$$
TOE-NAIL EACH SUPPORT AND 30d RINGSHANK FACE NAIL
EACH SUPPORT.

Random controlled means various lengths, usually 4'-0" modules placed so there are not two adjacent splices in the same support. T&G refers to tongue and groove.

STEEL CONSTRUCTION

Steel construction can be divided into three categories: steel studs, prefabricated steel structures, and steel-framed structures.

Steel Studs

Prefabricated steel studs are used in many types of commercial structures. Steel studs offer lightweight, noncombustible, corrosion-resistant framing for interior partitions and load-bearing exterior walls up to four stories high. Steel members are available for use as studs or joists. Members are designed for rapid assembly and are predrilled for electrical and plumbing conduits. The standard 24" (600 mm) spacing reduces the number of studs required by about one-third when compared with wood studs spaced 16" (400 mm) OC. Steel stud widths range from 3-5/8" to 10" (90–250 mm) but can be manufactured in any width. The material used to make studs ranges from 12- to 20-gage steel depending on the design loads to be supported. Steel studs are mounted in a channel track at the top and bottom of the wall or partition. This channel serves to tie the studs together at the ends. Horizontal bridging is often placed through the predrilled holes in the studs and then welded to the studs to serve as fire blocking within walls and as mid supports. The components of steel stud framing are shown in Figure 23.45.

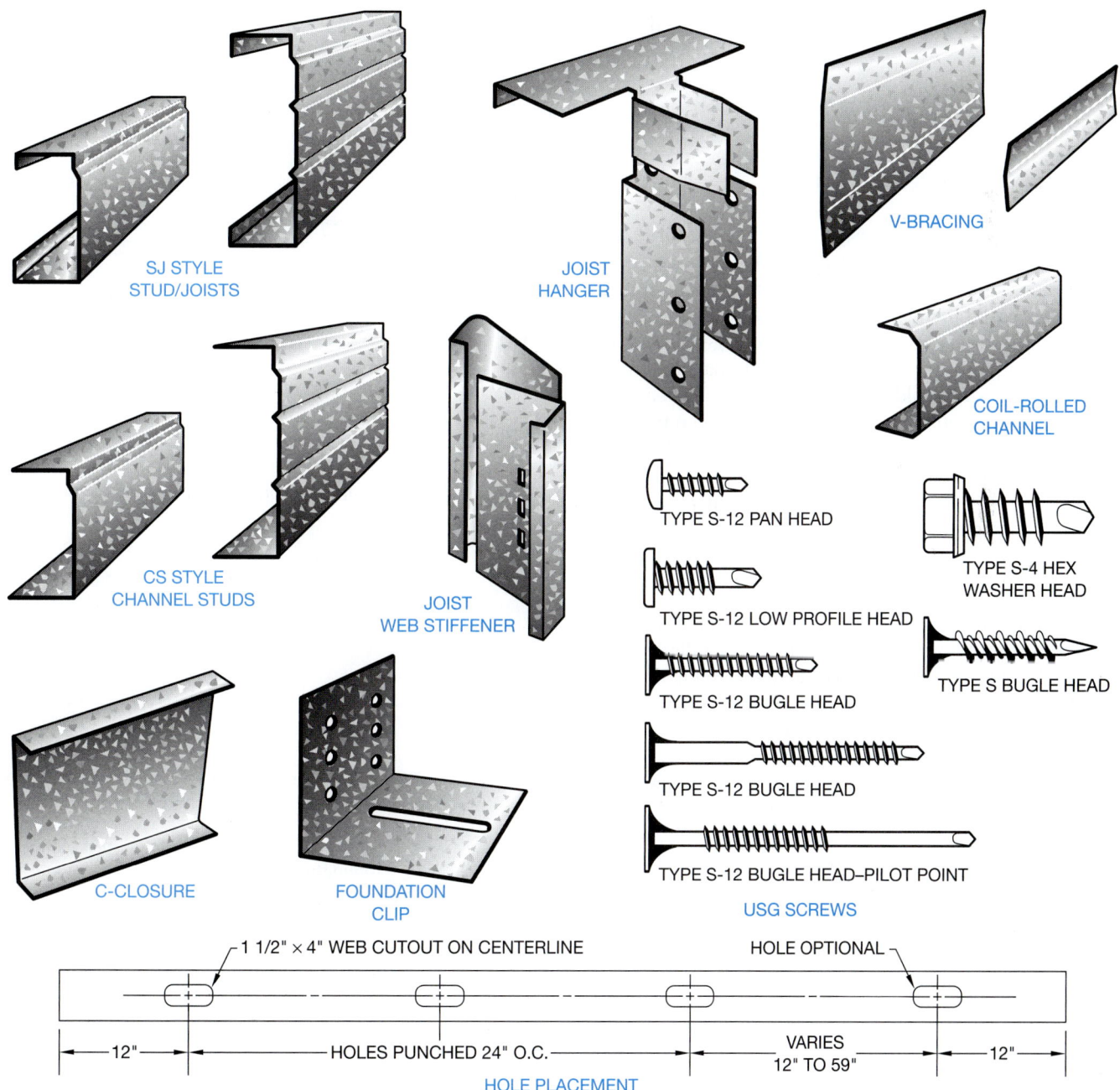

FIGURE 23.45 ■ Common components of steel stud construction. *Courtesy United States Gypsum Company.*

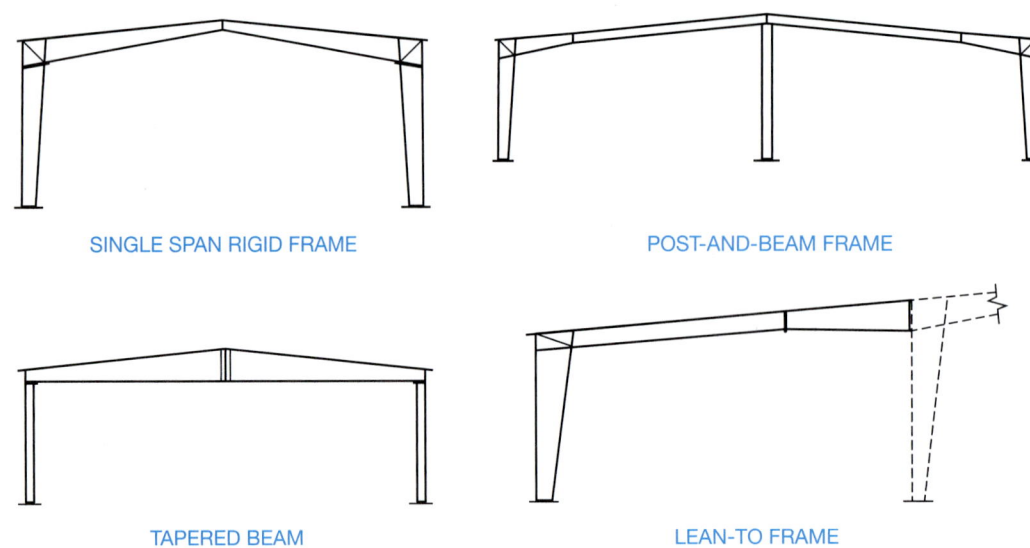

SINGLE SPAN RIGID FRAME

POST-AND-BEAM FRAME

TAPERED BEAM

LEAN-TO FRAME

FIGURE 23.46 ■ Common prefabricated structural systems.

Prefabricated Steel Structures

Prefabricated or metal buildings, as they are often called, have become a common type of construction for commercial and agricultural structures in many parts of the country. Drafters who are involved in the preparation of drawings for premanufactured structures may work in a structural engineering office or for a building manufacturer. Standardized premanufactured steel buildings are often sold as modular units with given spans, wall heights, and lengths in 12' or 20' (3600–6000 mm) increments. Most manufacturers provide a wide variety of design options for custom applications that may be required by the client. One advantage of these structures is faster erection time as compared with other construction methods.

The Prefabricate Steel Structural System

The prefabricate steel structural system is made up of the frame that supports the walls and roof. There are several different types of structural systems commonly used, as shown in Figure 23.46. The wall system is horizontal girts attached to the vertical structure and metal wall sheets attached to the girts. The roof system is horizontal purlins attached to the structure and metal sheets attached to the purlins. (See Figure 23.47.) Steel wall and roof sheets are available from many vendors in a variety of patterns and can be purchased plain, galvanized, or prepainted. A sample pattern design is shown in Figure 23.48.

Steel-Framed Structures

Steel-framed buildings require structural engineering and shop drawings. As a drafter in an engineering or architectural firm, you may be drafting engineering drawings similar to the one shown in Figure 23.49.

AISC Drafters in this field must become familiar with the *Manual of Steel Construction* published by the American

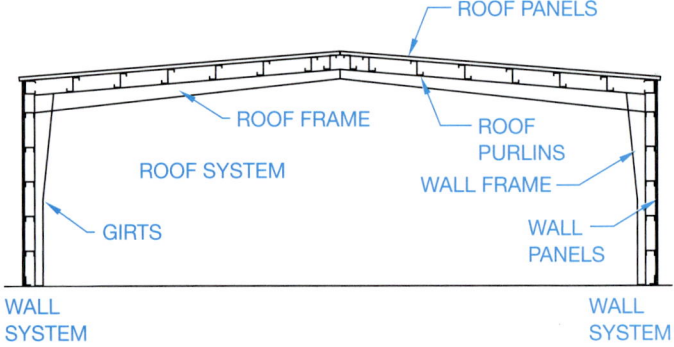

FIGURE 23.47 ■ Components of the prefabricated structural system.

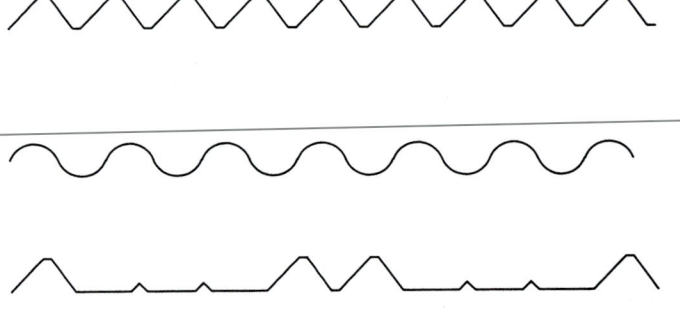

FIGURE 23.48 ■ Typical cross sections of sheet metal siding and roofing material; many pattern shapes and finish colors are available.

Institute of Steel Construction, Inc. (AISC). In addition to the local and national building codes, this manual is one of your prime references, as it helps you determine dimensions and properties of common steel shapes. Another manual that provides information on dimensions for detailing and properties for design work related to steel structural materials is *Structural Steel Shapes*, published by the U.S. Steel Corporation.

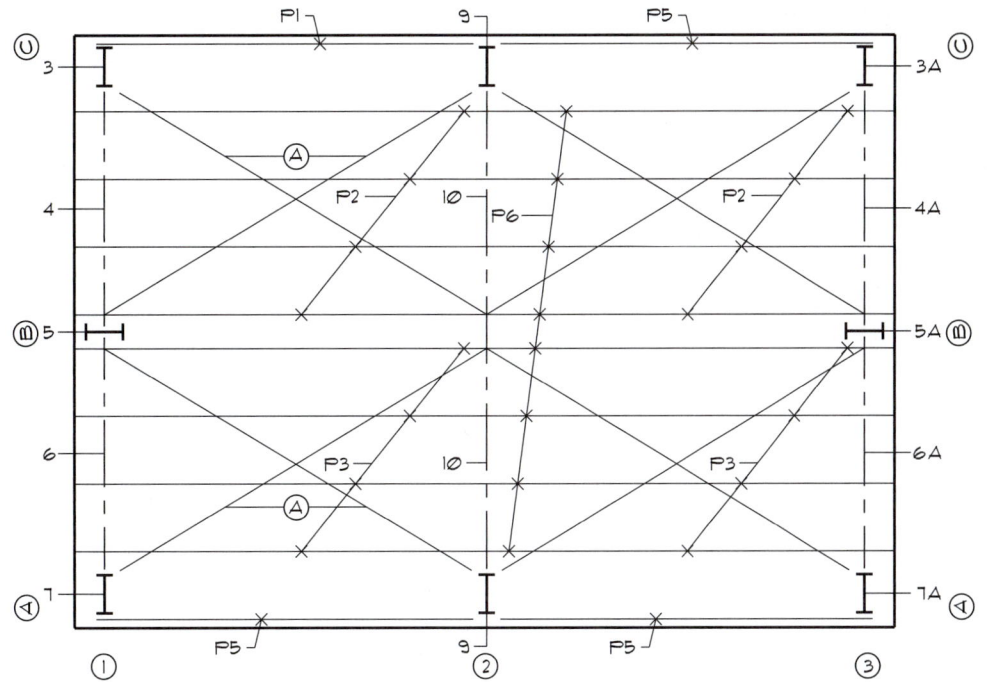

FIGURE 23.49 ■ Structural steel engineering drawing. *Courtesy Pacific Building Systems.*

Common Structural Steel Materials

Structural steels are commonly identified as plates, bars, or shape configurations. **Plates** are flat pieces of steel of various thickness used at the intersection of different members and for the fabrication of custom connectors. Figure 23.50 shows an example of a steel connector that uses top, side, and bottom plates. Plates are typically specified on a drawing by giving the thickness, width, and length in that order. The symbol ℞ is often used to specify plate material. For example:

℞ 1/4 × 6 × 10 (with or without inch marks)

Bars are the smallest of structural steel products and are manufactured in round, square, rectangular, flat, or hexagonal cross sections. Bars are often used as supports or braces for other steel parts or connectors. Flat bars are usually specified on a drawing by giving the width, thickness, and length, in that order. For example:

BAR 3 × 1/2 X 1'–6"

Structural steel is also available in several different manufactured shapes, as shown in Figure 23.51. When specifying a steel shape on a drawing, the shape identification letter is followed by the member's depth, the "by" sign (×), and the weight in number of pounds per linear foot. For example:

W 12 × 22 or C 6 × 10.5

as shown in Figure 23.52. In the AISC *Manual of Steel Construction,* specific information regarding dimensions for detailing and dimensioning is clearly provided along with typical connection details. The representative pages for the W 12 × 22 wide flange and the C 6 × 10.5 channel from Figure 23.52 are

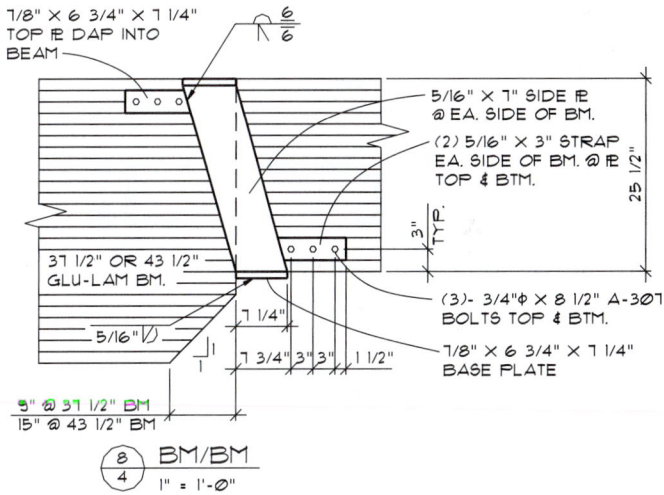

FIGURE 23.50 ■ Steel plates used to fabricate a beam connector.

shown in Figure 23.53. The W, S, and M shapes all have an I-shaped cross section and are often referred to as "I" beams. The three shapes differ in the width of their **flanges**. In addition to varied flange widths, the S-shape flanges are tapered, making them stronger than equivalent sized "W" beams and suitable for train rail or monorail beams. The W shape is commonly used for columns. All can be used for horizontal or vertical members.

Angles are structural steel components that have an L shape. The legs of the angle can be either equal or unequal in length but are usually equal in thickness. Channels have a squared C cross-sectional area and are designated with the letters C or MC.

Structural tees are produced from W, S, and M steel shapes. Common designations include WT, ST, and MT.

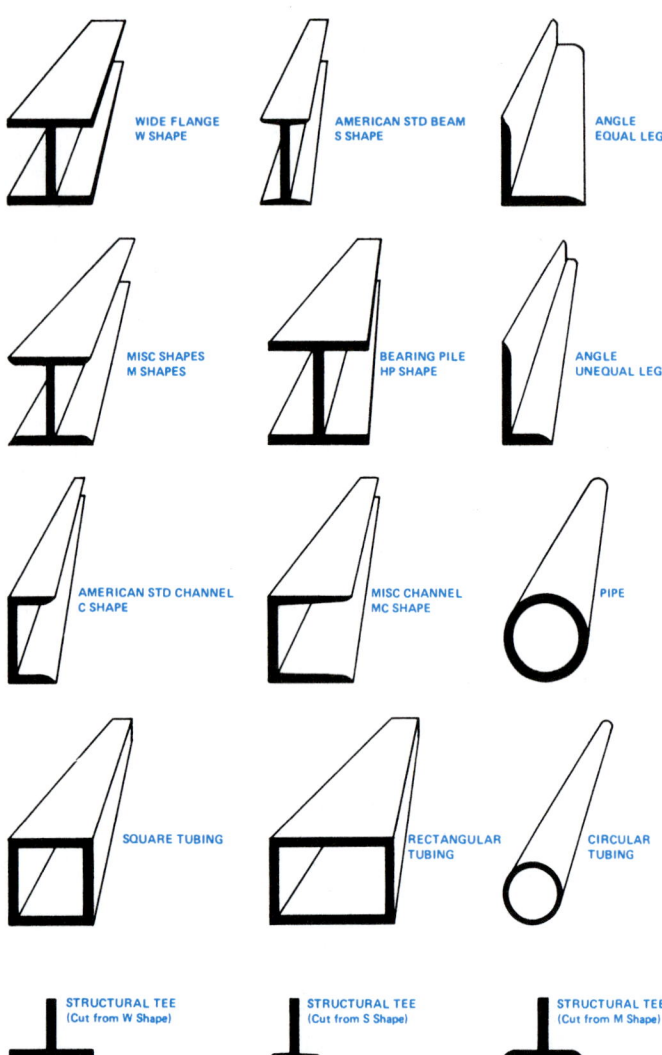

FIGURE 23.51 ■ Standard structural steel shapes.

Structural tubing is manufactured in square, rectangular, and round cross-sectional configurations. These members are used as columns to support loads from other members. Tubes are also commonly used for beams and truss members. Tubes are specified by the size of the outer wall followed by the thickness of the wall.

Steel pipe is also commonly used for columns and bracing. Available steel pipe strengths are standard, extra strong, and double-extra strong. The wall thickness increases with each type.

A variety of templates are available for structural drafting to assist in drawing steel shapes. Many CADD programs are available also to increase structural drafting productivity.

Structural Steel Callouts

Many structural steel materials are specified by shape designation, flange width, and weight in pounds per linear foot. For example:

W 24 × 120

WIDE FLANGE SHAPES

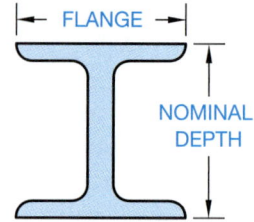

THE DESIGNATION FOR A WIDE FLANGE LOOKS LIKE THIS:

W 12 × 22

WHERE: **W** IS THE SHAPE.
12 IS THE NOMINAL DEPTH.
22 IS THE NUMBER OF POUNDS PER LINEAL FOOT.

CHANNELS

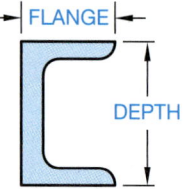

THE DESIGNATION FOR A CHANNEL SHAPE LOOKS LIKE THIS:

C 6 × 10.5

WHERE: **C** IS THE SHAPE.
6 IS THE DEPTH.
10.5 IS THE NUMBER OF POUNDS PER LINEAL FOOT.

FIGURE 23.52 ■ Dimensional elements of the wide flange and channel shapes.

The structural steel shapes that fall into this category are as follows:

W—wide flange shapes

S—American standard beams

M—miscellaneous beam and column shapes

C—American standard channels

MC—miscellaneous channel shapes

WT—structural tees cut from W shapes

ST—structural tees cut from S shapes

MT—structural tees cut from M shapes

T—structural tees

Z—zee shapes

HP—steel H piling

The specifications for square and rectangular structural steel tubing, and sample structural steel shape designations, are provided in Appendices U and V.

Structural materials that are specified by shape designation, type, diameter or outside dimension, and wall thickness are in the following table. (*Note:* If length dimensions are required, they are given at the end of the callout in feet and inches except for plates, which should be given only in inches.)

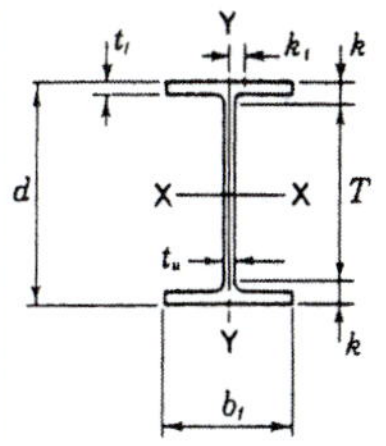

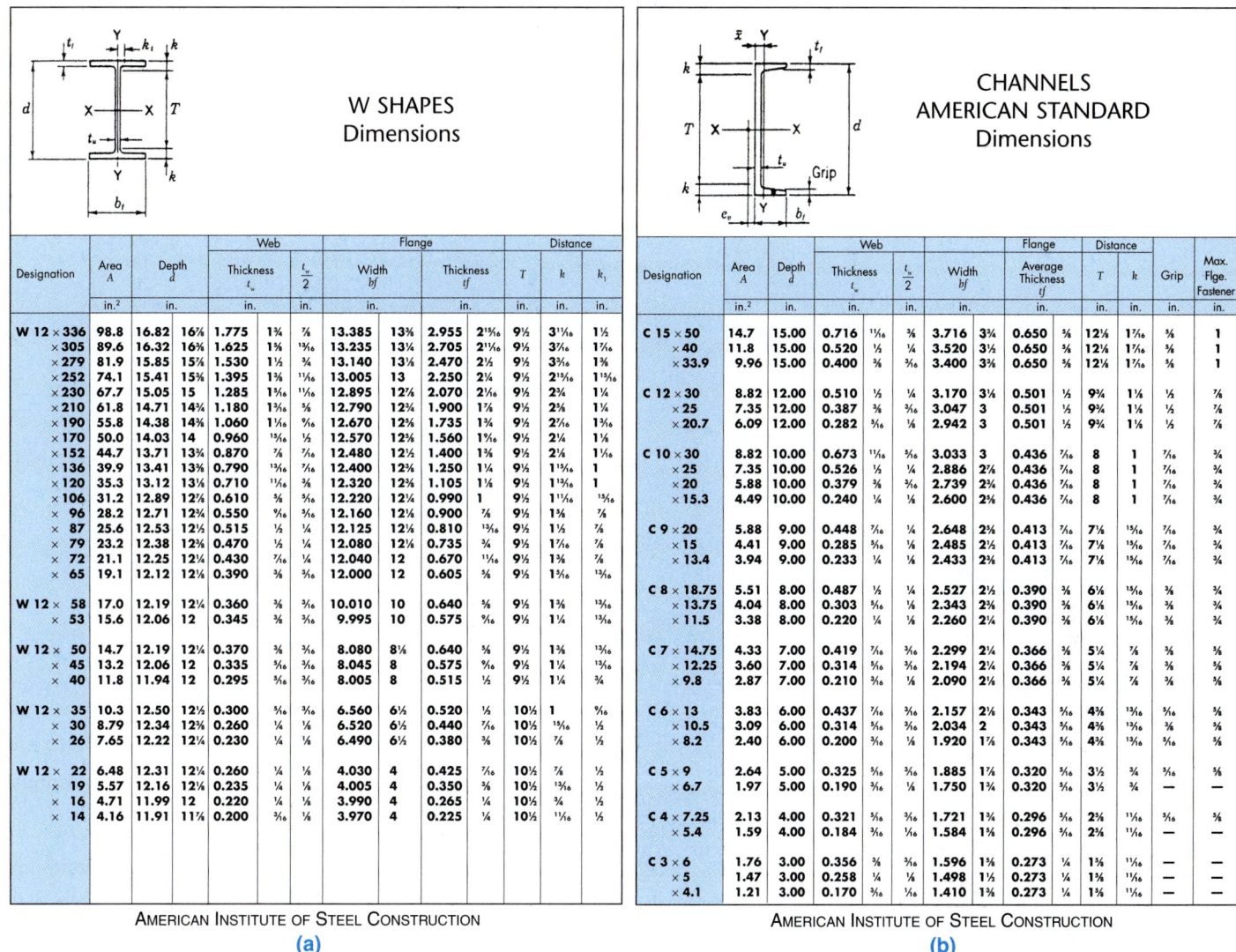

W SHAPES
Dimensions

Designation	Area A	Depth d		Web Thickness t_w		Web $\frac{t_w}{2}$	Flange Width bf		Flange Thickness tf		Distance T	Distance k	Distance k_1
	in.²	in.		in.		in.	in.		in.		in.	in.	in.
W 12 × 336	98.8	16.82	16 7/8	1.775	1 3/4	7/8	13.385	13 3/8	2.955	2 15/16	9 1/2	3 11/16	1 1/2
× 305	89.6	16.32	16 3/8	1.625	1 5/8	13/16	13.235	13 1/4	2.705	2 11/16	9 1/2	3 7/16	1 7/16
× 279	81.9	15.85	15 7/8	1.530	1 1/2	3/4	13.140	13 1/8	2.470	2 1/2	9 1/2	3 3/16	1 3/8
× 252	74.1	15.41	15 3/8	1.395	1 3/8	11/16	13.005	13	2.250	2 1/4	9 1/2	2 15/16	1 15/16
× 230	67.7	15.05	15	1.285	1 5/16	11/16	12.895	12 7/8	2.070	2 1/16	9 1/2	2 3/4	1 1/4
× 210	61.8	14.71	14 3/4	1.180	1 3/16	5/8	12.790	12 3/4	1.900	1 7/8	9 1/2	2 5/8	1 1/4
× 190	55.8	14.38	14 3/8	1.060	1 1/16	9/16	12.670	12 5/8	1.735	1 3/4	9 1/2	2 7/16	1 3/16
× 170	50.0	14.03	14	0.960	15/16	1/2	12.570	12 5/8	1.560	1 9/16	9 1/2	2 1/4	1 1/8
× 152	44.7	13.71	13 3/4	0.870	7/8	7/16	12.480	12 1/2	1.400	1 3/8	9 1/2	2 1/8	1 1/16
× 136	39.9	13.41	13 3/8	0.790	13/16	7/16	12.400	12 3/8	1.250	1 1/4	9 1/2	1 15/16	1
× 120	35.3	13.12	13 1/8	0.710	11/16	3/8	12.320	12 3/8	1.105	1 1/8	9 1/2	1 13/16	1
× 106	31.2	12.89	12 7/8	0.610	5/8	5/16	12.220	12 1/4	0.990	1	9 1/2	1 11/16	15/16
× 96	28.2	12.71	12 3/4	0.550	9/16	5/16	12.160	12 1/8	0.900	7/8	9 1/2	1 5/8	7/8
× 87	25.6	12.53	12 1/2	0.515	1/2	1/4	12.125	12 1/8	0.810	13/16	9 1/2	1 1/2	7/8
× 79	23.2	12.38	12 3/8	0.470	1/2	1/4	12.080	12 1/8	0.735	3/4	9 1/2	1 7/16	7/8
× 72	21.1	12.25	12 1/4	0.430	7/16	1/4	12.040	12	0.670	11/16	9 1/2	1 3/8	7/8
× 65	19.1	12.12	12 1/8	0.390	3/8	3/16	12.000	12	0.605	5/8	9 1/2	1 5/16	13/16
W 12 × 58	17.0	12.19	12 1/4	0.360	3/8	3/16	10.010	10	0.640	5/8	9 1/2	1 3/8	13/16
× 53	15.6	12.06	12	0.345	3/8	3/16	9.995	10	0.575	9/16	9 1/2	1 1/4	13/16
W 12 × 50	14.7	12.19	12 1/4	0.370	3/8	3/16	8.080	8 1/8	0.640	5/8	9 1/2	1 3/8	13/16
× 45	13.2	12.06	12	0.335	5/16	3/16	8.045	8	0.575	9/16	9 1/2	1 1/4	13/16
× 40	11.8	11.94	12	0.295	5/16	3/16	8.005	8	0.515	1/2	9 1/2	1 1/4	3/4
W 12 × 35	10.3	12.50	12 1/2	0.300	5/16	3/16	6.560	6 1/2	0.520	1/2	10 1/2	1	9/16
× 30	8.79	12.34	12 3/8	0.260	1/4	1/8	6.520	6 1/2	0.440	7/16	10 1/2	15/16	1/2
× 26	7.65	12.22	12 1/4	0.230	1/4	1/8	6.490	6 1/2	0.380	3/8	10 1/2	7/8	1/2
W 12 × 22	6.48	12.31	12 1/4	0.260	1/4	1/8	4.030	4	0.425	7/16	10 1/2	7/8	1/2
× 19	5.57	12.16	12 1/8	0.235	1/4	1/8	4.005	4	0.350	3/8	10 1/2	13/16	1/2
× 16	4.71	11.99	12	0.220	1/4	1/8	3.990	4	0.265	1/4	10 1/2	3/4	1/2
× 14	4.16	11.91	11 7/8	0.200	3/16	1/8	3.970	4	0.225	1/4	10 1/2	11/16	1/2

American Institute of Steel Construction

(a)

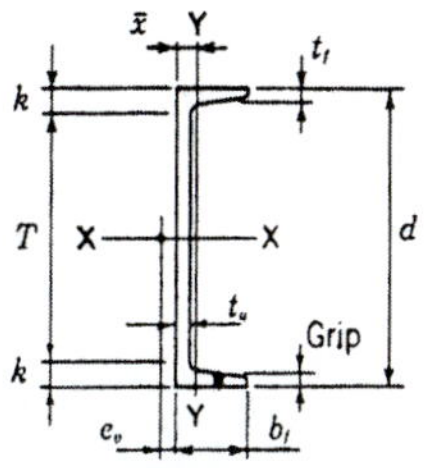

CHANNELS
AMERICAN STANDARD
Dimensions

Designation	Area A	Depth d	Web Thickness t_w		Web $\frac{t_w}{2}$	Width bf		Flange Average Thickness tf		Distance T	Distance k	Grip	Max. Flge. Fastener
	in.²	in.	in.		in.	in.		in.		in.	in.	in.	in.
C 15 × 50	14.7	15.00	0.716	11/16	3/8	3.716	3 3/4	0.650	5/8	12 1/8	1 7/16	5/8	1
× 40	11.8	15.00	0.520	1/2	1/4	3.520	3 1/2	0.650	5/8	12 1/8	1 7/16	5/8	1
× 33.9	9.96	15.00	0.400	3/8	3/16	3.400	3 3/8	0.650	5/8	12 1/8	1 7/16	5/8	1
C 12 × 30	8.82	12.00	0.510	1/2	1/4	3.170	3 1/8	0.501	1/2	9 3/4	1 1/8	1/2	7/8
× 25	7.35	12.00	0.387	3/8	3/16	3.047	3	0.501	1/2	9 3/4	1 1/8	1/2	7/8
× 20.7	6.09	12.00	0.282	5/16	1/8	2.942	3	0.501	1/2	9 3/4	1 1/8	1/2	7/8
C 10 × 30	8.82	10.00	0.673	11/16	5/16	3.033	3	0.436	7/16	8	1	7/16	3/4
× 25	7.35	10.00	0.526	1/2	1/4	2.886	2 7/8	0.436	7/16	8	1	7/16	3/4
× 20	5.88	10.00	0.379	3/8	3/16	2.739	2 3/4	0.436	7/16	8	1	7/16	3/4
× 15.3	4.49	10.00	0.240	1/4	1/8	2.600	2 5/8	0.436	7/16	8	1	7/16	3/4
C 9 × 20	5.88	9.00	0.448	7/16	1/4	2.648	2 5/8	0.413	7/16	7 1/8	15/16	7/16	3/4
× 15	4.41	9.00	0.285	5/16	1/8	2.485	2 1/2	0.413	7/16	7 1/8	15/16	7/16	3/4
× 13.4	3.94	9.00	0.233	1/4	1/8	2.433	2 3/8	0.413	7/16	7 1/8	15/16	7/16	3/4
C 8 × 18.75	5.51	8.00	0.487	1/2	1/4	2.527	2 1/2	0.390	3/8	6 1/8	15/16	3/8	3/4
× 13.75	4.04	8.00	0.303	5/16	1/8	2.343	2 3/8	0.390	3/8	6 1/8	15/16	3/8	3/4
× 11.5	3.38	8.00	0.220	1/4	1/8	2.260	2 1/4	0.390	3/8	6 1/8	15/16	3/8	3/4
C 7 × 14.75	4.33	7.00	0.419	7/16	3/16	2.299	2 1/4	0.366	3/8	5 1/4	7/8	3/8	5/8
× 12.25	3.60	7.00	0.314	5/16	3/16	2.194	2 1/4	0.366	3/8	5 1/4	7/8	3/8	5/8
× 9.8	2.87	7.00	0.210	3/16	1/8	2.090	2 1/8	0.366	3/8	5 1/4	7/8	3/8	5/8
C 6 × 13	3.83	6.00	0.437	7/16	3/16	2.157	2 1/8	0.343	5/16	4 3/8	13/16	5/16	5/8
× 10.5	3.09	6.00	0.314	5/16	3/16	2.034	2	0.343	5/16	4 3/8	13/16	3/8	5/8
× 8.2	2.40	6.00	0.200	3/16	1/8	1.920	1 7/8	0.343	5/16	4 3/8	13/16	5/16	5/8
C 5 × 9	2.64	5.00	0.325	5/16	3/16	1.885	1 7/8	0.320	5/16	3 1/2	3/4	5/16	5/8
× 6.7	1.97	5.00	0.190	3/16	1/8	1.750	1 3/4	0.320	5/16	3 1/2	3/4	—	—
C 4 × 7.25	2.13	4.00	0.321	5/16	3/16	1.721	1 3/4	0.296	5/16	2 5/8	11/16	5/16	5/8
× 5.4	1.59	4.00	0.184	3/16	1/16	1.584	1 5/8	0.296	5/16	2 5/8	11/16	—	—
C 3 × 6	1.76	3.00	0.356	3/8	3/16	1.596	1 5/8	0.273	1/4	1 5/8	11/16	—	—
× 5	1.47	3.00	0.258	1/4	1/8	1.498	1 1/2	0.273	1/4	1 5/8	11/16	—	—
× 4.1	1.21	3.00	0.170	3/16	1/16	1.410	1 3/8	0.273	1/4	1 5/8	11/16	—	—

American Institute of Steel Construction

(b)

FIGURE 23.53 ■ Dimensional information for (a) W12×22, and (b) C6×10.5. *Courtesy American Institute for Steel Construction, Manual of Steel Construction.* Additional samples of structural steel dimensions are found in Appendices U and V.

Designation	Shape	Example	Meaning
PL	plate	PL 1/2 × 6 × 8	THK × WIDTH × LENGTH
BAR	square bar	BAR 1-1/4 □*	1-1/4" WIDTH × THICKNESS
	round bar	BAR 1-1/4∅	1-1/4" DIAMETER
	flat bar	BAR 2 × 3/8	WIDTH × THICKNESS
PIPE	pipe	PIPE 3∅STD	OD [OUTSIDE DIA] SPEC
TS	structural tubing		
	square	TS 4 × 4 × .250	OUTWIDTH × OUTHEIGHT × WALL THICKNESS
	rectangular	TS 6 × 3 × .375	SAME
	round	TS 4 OD × .188	OD × WALL THICKNESS
L	angle		
	unequal leg	L3 × 2 × 1/4	LEG × LEG × THICKNESS
	equal leg	L3 × 3 × 1/2	LEG × LEG × THICKNESS

*Some companies use ⊠ for the square symbol.

When plate material is bent, the minimum bend radius should be given with the plate callout, and the length of the bend legs are dimensioned on the drawing. For example:

3/8 × 10 W/MIN BEND R 5/8"

Location dimensions for structural steel components should be as shown in Figure 23.54. Drafters in architectural or structural engineering firms typically draw structural drawings as shown in Figure 23.55.

Shop Drawings

Shop drawings are used to break each individual component of a structural engineering drawing down into fabrication parts. Shop drawings are also referred to as **fabrication drawings** and are generally drawn by the drafter in the fabrication company. Many large companies do both structural and shop drawings. Depending on the complexity of the structure, the structural

WIDE FLANGES AND
OTHER I SHAPES

TO CENTERLINE IN
BOTH DIRECTIONS

CHANNELS

CENTERLINE ALONG X–X AXIS AND BACK
FACE OF WEB ALONG Y–Y AXIS

Note: An exception for wide flanges and channels is when specifying top of beam.

ANGLES

OUTSIDE FACE OF LEGS

PLATES

TO THE CENTERLINE IN BOTH DIRECTIONS
WHEN DIMENSIONING IN THE PLAN VIEW OF
THE PLATE AND TO THE FACE OF THE PLATE
WHEN DIMENSIONING IN SECTION OR PROFILE.

PLAN

PROFILE

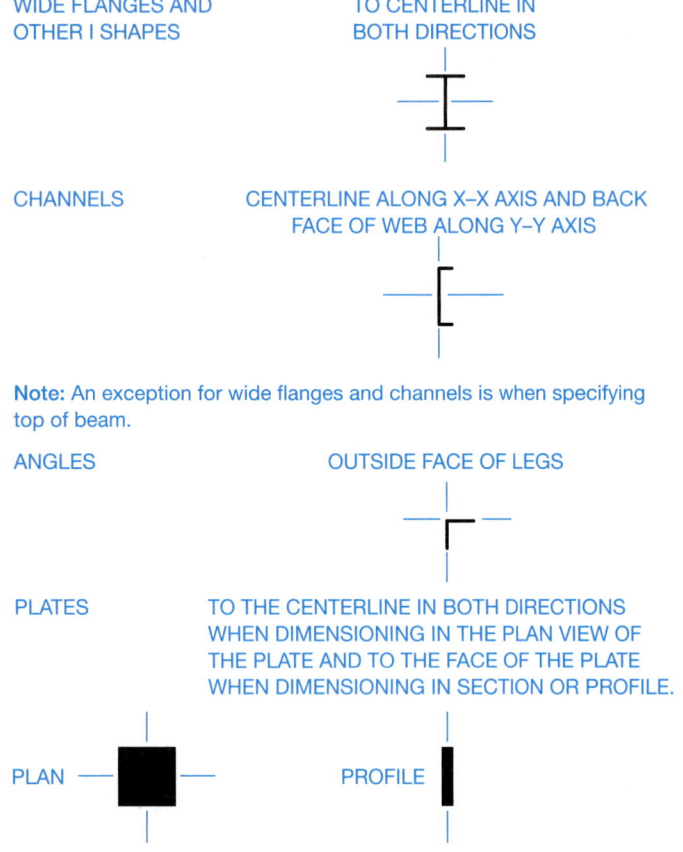

FIGURE 23.54 ■ Location dimensions for structural components.

and shop drawings may be combined on the same drawing. An example of a shop drawing is shown in Figure 23.56.

COMMON CONNECTION METHODS

Bolted Connections

ASTM Bolts are used for many connections in lumber and steel construction. A bolt specification includes the diameter, length, and strength of the bolt. Washers or plates are specified so the bolt head will not pull through the hole made for the bolt. Bolt strength is classified in accordance with the American Society for Testing Materials (ASTM) specifications. Refer to Chapter 12, Fasteners and Springs, for more information.

Standard Structural Bolt Callouts

When specifying bolts on a drawing, give the quantity, diameter, bolt type if special, length in inches, and ASTM specification. If special washer and nut requirements are present, this should also be specified in the bolt callout. Hexagon head bolts and hexagon nuts are assumed unless either is specified differently in the callout. Examples of bolt callouts are:

2-3/4"⌀ BOLTS ASTM A503

4-1/2"⌀ × 10" BOLTS

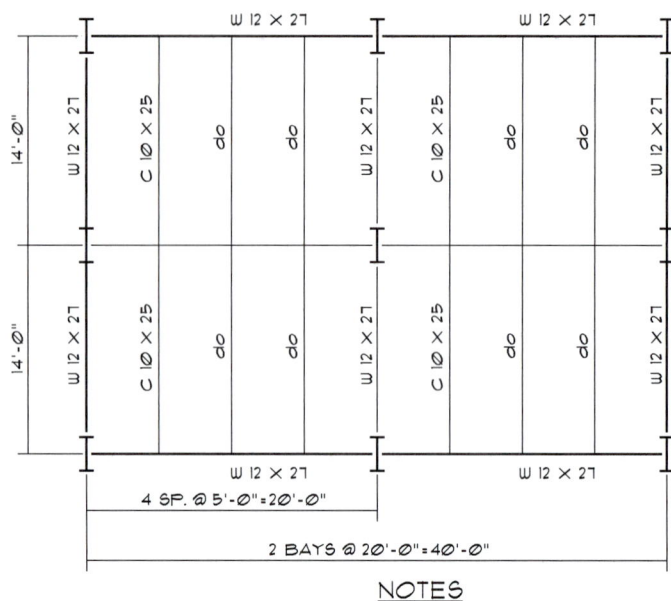

NOTES
ALL BEAMS FLUSH TOP
TOP OF STEEL ELEVATION 113'-0"

FIGURE 23.55 ■ Structural steel engineering drawing.

2-5/8"⌀ × 6" CARRIAGE BOLTS

6-3/4"⌀ BOLTS W/MALLEABLE IRON WASHERS AND HEAVY HEX NUTS

4-1/2"⌀ GALVANIZED BOLTS

Give the hole diameter when holes for standard bolts must be specified on the drawing. In general, holes should be 1/16" (1.5 mm) larger in diameter than the specified bolt for standard steel-to-steel, wood-to-wood, or wood-to-steel construction up to 1" (25 mm) in diameter, and 1/8" (3 mm) larger for hole sizes over 1" in diameter, than the specified bolt for standard steel-to-concrete or wood-to-concrete applications unless otherwise specified by the engineer. For example:

2-3/4"⌀ BOLTS FIELD DRILL 13/16"⌀ HOLES.

Refer to Chapter 12 for bolt and fastener terminology and types and refer to Appendixes J through P for additional fastener specifications.

The following are recommended standard hole diameters in inches for given bolt sizes:

Bolt ⌀	Standard Hole ⌀	Concrete Hole ⌀	Oversize Hole ⌀
1/2	9/16	5/8	11/16
5/8	11/16	3/4	13/16
3/4	13/16	7/8	15/16
7/8	15/16	1	1 1/16
1	1 1/16	1 1/8	1 1/4
1 1/8	1 1/4	1 1/4	1 7/16
1 1/4	1 3/8	1 3/8	1 9/16
1 3/8	1 1/2	1 1/2	1 11/16
1 1/2	1 5/8	1 5/8	1 13/16

When specifying lag bolts on a drawing, the lead or tap hole diameter should be given with the bolt specification. For example:

2-3/4"⌀ × 8 LAG BOLTS W/7/16"⌀ LEAD HOLES.

Floor Plan

The floor plan is generally designed and drawn by the architect. This floor plan drawing is then used as the layout for the other associated drawings from consulting engineers—for example, the mechanical (HVAC), plumbing, electrical, and structural engineers. In some situations and depending on the structure, the structural engineering firm draws the floor plan as part of the set of drawings. (See Figure 23.58.)

Foundation Plan and Details

The purpose of the foundation plan is to show the supporting system for the walls, floor, and roof. The foundation may consist of continuous perimeter footings and walls that support the exterior and main bearing walls of the building. The footings in these locations are rectangular in shape and are continuous concrete supports centered at exterior and interior bearing walls. (See Figure 23.59.) There are also pedestal footings that support the concentrated loads of particular elements of the structure. Interior support for the columns that support upper floor and roof loads are provided by pedestal footings, as shown in Figure 23.60, page 780. Anchor bolts and other metal connectors are shown and located on the foundation plan, as shown in Figure 23.61, page 780. A partial typical foundation plan is shown in Figure 23.62, page 780.

The foundation details are drawn to provide information on the concrete foundation at the perimeter walls, and the foundation and pedestals at the center support columns. Possible details include retaining walls, typical exterior foundation and rebar schedule, typical interior bearing wall, and typical pedestal and rebar schedule. Detail drawings are keyed to the plan view using detail markers. Detail markers are usually drawn as a circle of about 3/8" to 1/2" in diameter on the plan view and a coordinating circle of about 3/4" in diameter under the associated detail. Each detail marker is divided in half. The top half contains the detail number and the bottom identifies the page number on which the detail is drawn. Notice the detail markers associated with the example drawings in this chapter. The rebar schedules are charts placed adjacent to the detail that key to the drawing information about the rebar used, as shown in Figure 23.60.

Concrete Slab Plan and Details

The concrete slab plan is drawn to outline the concrete that will become the floor(s). Items often found in slab plans include floor slabs, slab reinforcing, expansion joints, pedestal footings, metal connectors, anchor bolts, and any foundation cuts for doors or other openings. The openings and other items are located and labeled on the plan, as shown in Figure 23.63, page 780. In situations where tilt-up construction is used, the opening locations are dimensioned on the elevations.

The concrete slab details are drawn to provide information of the intersections of the concrete slabs. These details include interior and perimeter slab joints as in Figure 23.64, page 781, and other slab details. The drawings include such information as slab thicknesses, reinforcing sizes and locations, and slab elevations (heights), as shown in Figure 23.65, page 781.

Roof Framing Plan and Details

The purpose of the roof framing plan is to show the major structural components in plan view that occur at the roof level.

STRUCTURAL DRAFTING

The procedure for preparing structural drawings works especially well with CADD. When a floor plan is drawn, this drawing is used as a base layer upon which each of the other plan views are drawn. For example, the floor plan can be drawn with black lines and labeled as the FLOOR PLAN layer, then the layer is changed and named FOUNDATION PLAN and the color is changed to red. All of the drawing information on the red layer is for the foundation plan. This process of changing layer names and colors continues until all of the drawings are complete. One advantage is that each layer can be turned on or off at your command. It is the same as tracing a drawing in manual drafting. Time is saved, and the drawings are all extremely accurate. When the drawings are to be plotted, each drawing can be converted to an individual file, or the layer to be plotted is turned on while the others are off. Any combination of plots can be reproduced, with select layers turned on.

There is structural CADD software that quickly and accurately draws structural steel shapes to exact specifications in plan, section, or elevation views. These structural steel packages also provide for beam details, integrating cutouts, framing angles, hole groups, welded or bolted connections, and complete dimensioning. An advantage is the capability of such programs to perform tedious standards specification sizing and calculations automatically. You do not have to spend time dimensioning hole patterns, because the program draws bolt or rivet hole patterns to your specifications, automatically displaying them on center dimensions, hole diameters, and flange thicknesses. When framing angles are used, the package automatically draws, sizes, positions, dimensions, and notes the angle specifications along with bolt holes and dimensions. These CADD structural packages are often available with a variety of template menus and symbol libraries. Detail views and symbology are often available from manufacturers of structural systems and components.

CADD APPLICATIONS

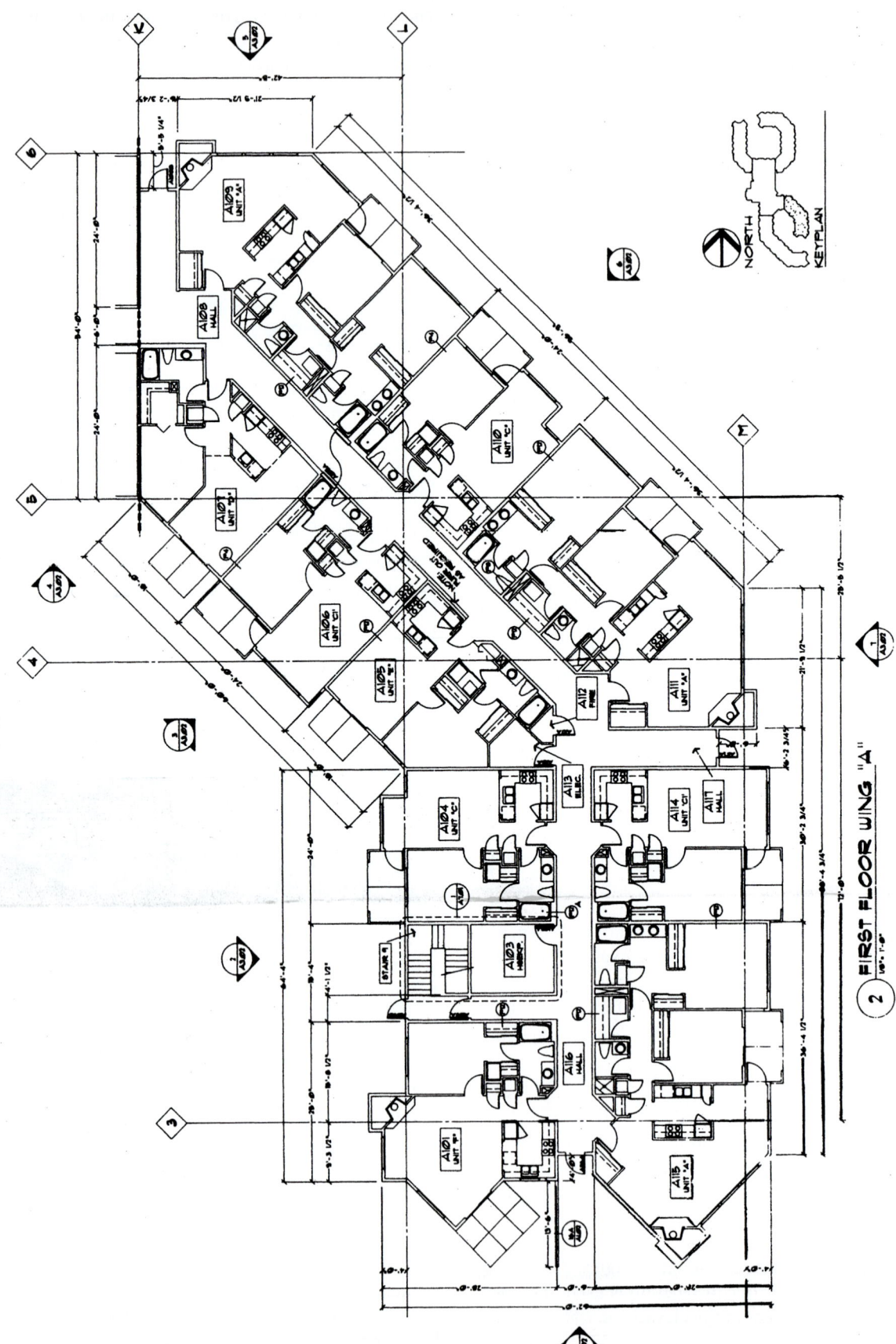

FIGURE 23.58 ■ CADD floor plan for a set of construction drawings. *Courtesy Soderstrom Architects, PC.*

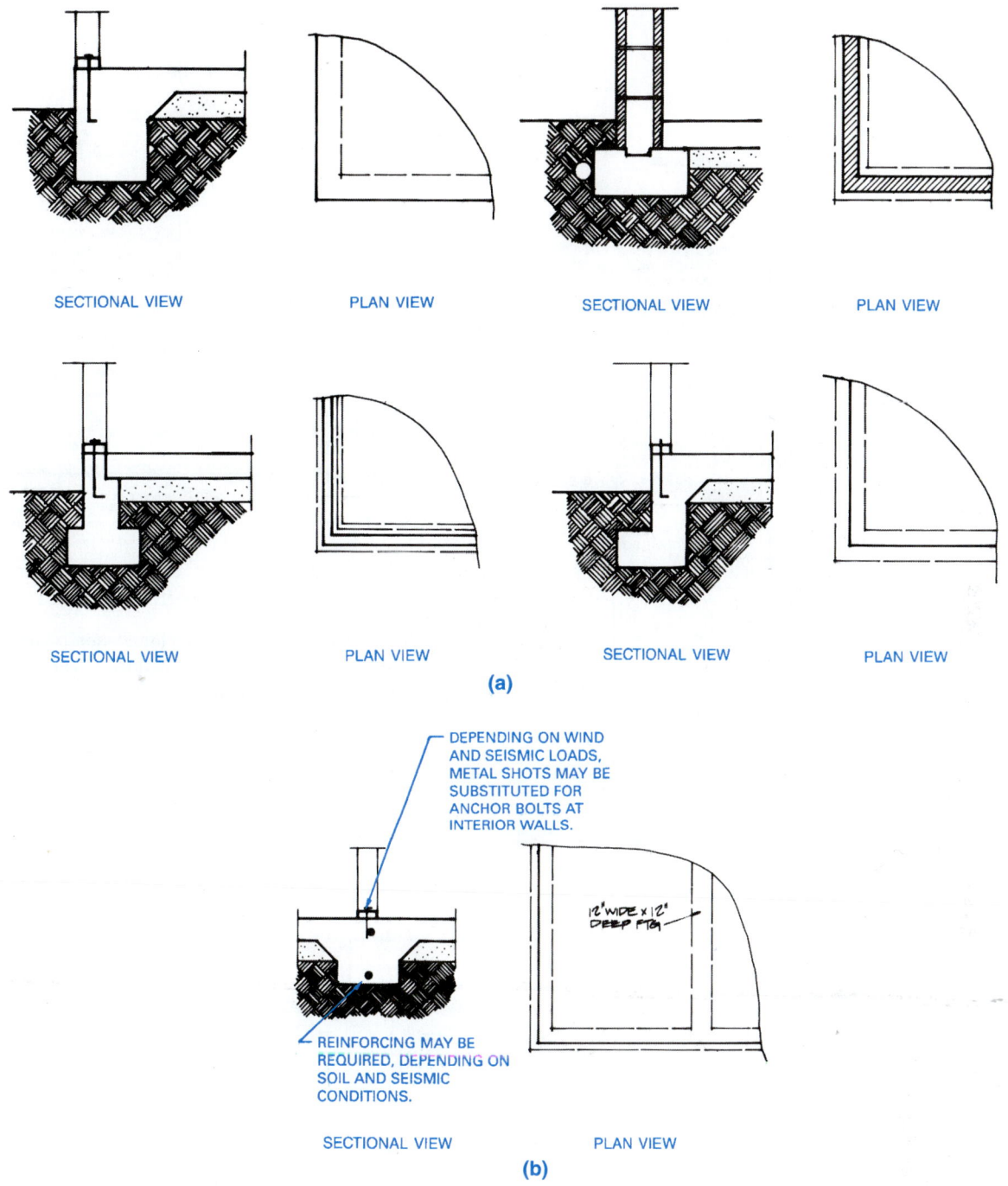

FIGURE 23.59 ■ (a) Common foundation and slab intersections; (b) footing and floor intersections at interior load-bearing wall.

A roof framing plan using truss framing is shown in Figure 23.3b, page 745. A panelized roof framing plan is shown in Figures 23.3b, page 745, and 23.33, page 763.

Roof framing details are required to show the construction methods used at various member intersections in the building. Details may include the following intersections: wall to beam, beam to column, beam splices and connections, truss details, bottom chord bracing plan and details, purlin clips, cantilever locations, and roof drains. (See Figure 23.66, page 782.)

Notice the elevation symbol shown in Figure 23.66e. This symbol is commonly used on structural drawings to give the el-

evation of locations from a known zero elevation. The zero elevation might be at the first floor or other good reference point such as the top of a foundation wall. These elevation symbols are used together with standard dimensioning practice as needed.

Roof Drainage Plan

The roof drainage plan can be part of a set of structural drawings for some buildings, although it may be considered part of the plumbing or piping drawings depending on the particular company's use and interpretation. The purpose of this drawing

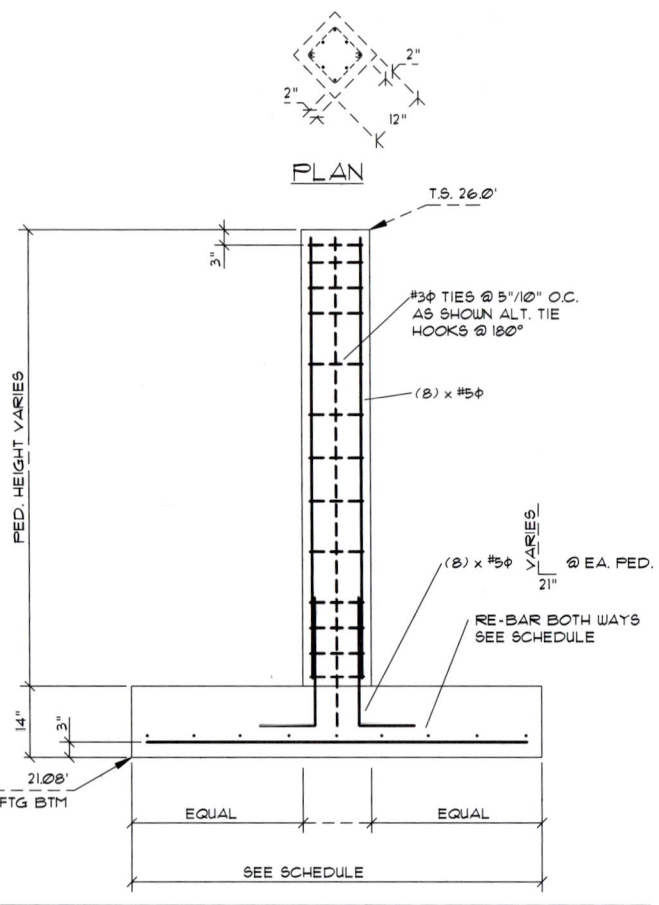

PLAN

#3φ TIES @ 5"/10" O.C.
AS SHOWN ALT. TIE
HOOKS @ 180°

(8) × #5φ

(8) × #5φ @ EA. PED.

RE-BAR BOTH WAYS
SEE SCHEDULE

T.S. 26.0'

PED. HEIGHT VARIES

VARIES

21"

3"

14"
3"

21.08'
FTG BTM

EQUAL   EQUAL

SEE SCHEDULE

FTG	PED HEIGHT	SIZE	REINFORCING STEEL
B-2	7.50	7'-0" ⊡	#6φ @10" O.C.
B-3	6.75	6'-6" ⊡	#6φ @12" O.C.
B-4	5.91	6'-6" ⊡	#6φ @12" O.C.
B-5	5.00	7'-0" ⊡	#6φ @10" O.C.

FIGURE 23.60 ■ Pedestal footing detail.

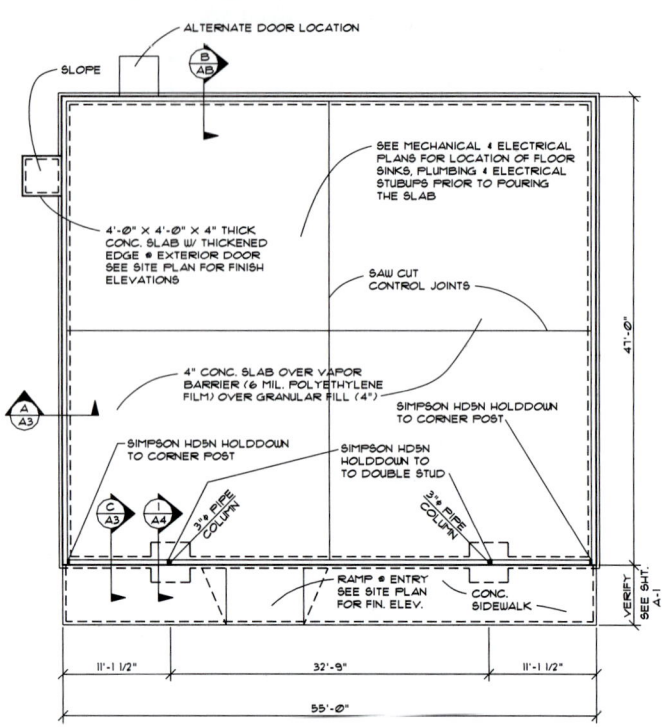

FIGURE 23.62 ■ A concrete slab foundation is commonly used for commercial structures. *Courtesy The Southland Corporation.*

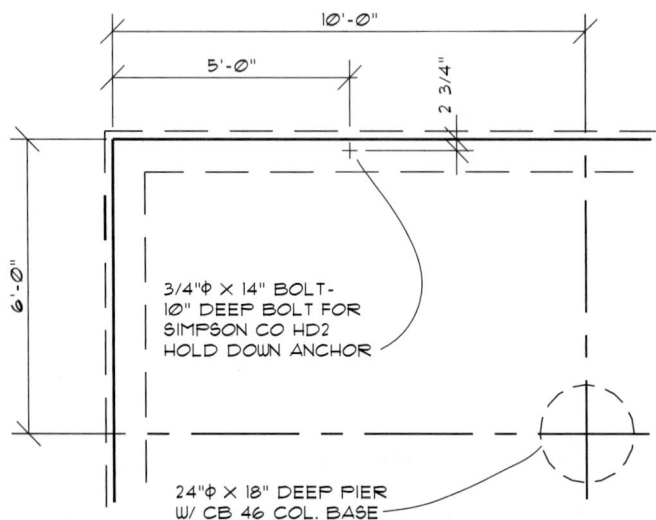

10'-0"

5'-0"

2 3/4"

6'-0"

3/4"φ × 14" BOLT-
10" DEEP BOLT FOR
SIMPSON CO HD2
HOLD DOWN ANCHOR

24"φ × 18" DEEP PIER
W/ CB 46 COL. BASE

FIGURE 23.61 ■ Foundation plan showing location of metal connectors.

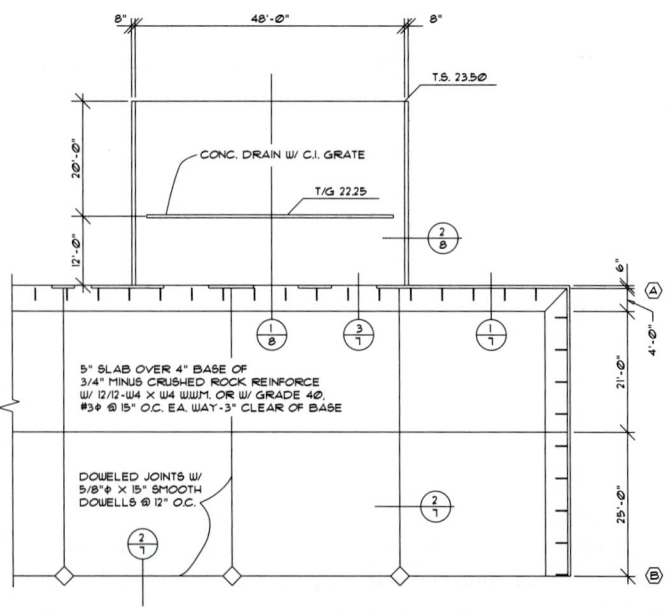

FIGURE 23.63 ■ A slab-on-grade plan shows the size and location of all concrete pours plus reinforcing specifications. *Courtesy Structureform Masters, Inc.*

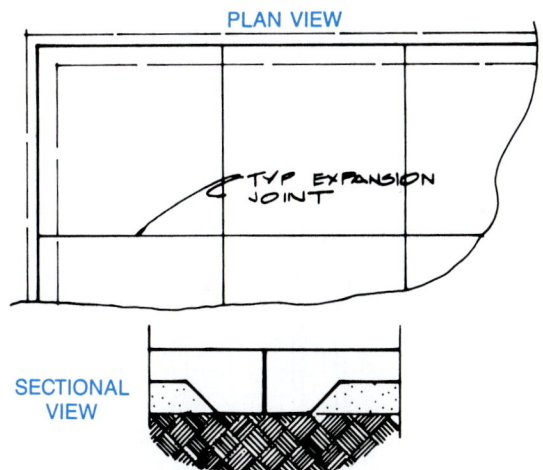

FIGURE 23.64 ■ Control joints are often placed in large slabs to resist cracking and allow the construction crews manageable areas to pour.

is to show the elevations of the roof and provide for adequate water drainage generally associated with low slope and flat roofs. (See Figure 23.67.) Some of the terminology associated with roof drainage plans includes: roof drain (RD)—a screened opening to allow for drainage; overflow drain (OD)—a backup in case the roof drains fail; down spout (DS)—usually a vertical pipe used to transport water from the roof; and scupper or gutter—a water collector usually on the outside of a wall at the roof level to funnel water from the roof drains to the downspouts.

Building Sections

The building section is used to show the relationship between the plans and details previously drawn. This drawing is considered a general arrangement or construction reference, as it is often drawn at a small scale. While some detailed information is provided with regard to building elevations (heights) and general dimensions, the overall section is not intended to provide explanation of building materials. The details more clearly serve this purpose, as they are drawn at a larger scale. Some less complex buildings, however, may show a great deal of construction information on the overall section and use fewer details. These overall sections, commonly called typical cross sections, show the general arrangement of the construction and often have detail correlated to them. (See Figure 23.68.) In some situations, partial sections may be useful in describing portions of the construction that may not be effectively handled with the building section and may be larger areas than normally identified with a detailed section, as shown in Figure 23.69, page 784.

Exterior Elevations

The exterior elevations are drawings that show the external appearance of the building. An elevation is drawn at each side of the building to show the relationship of the building to the final

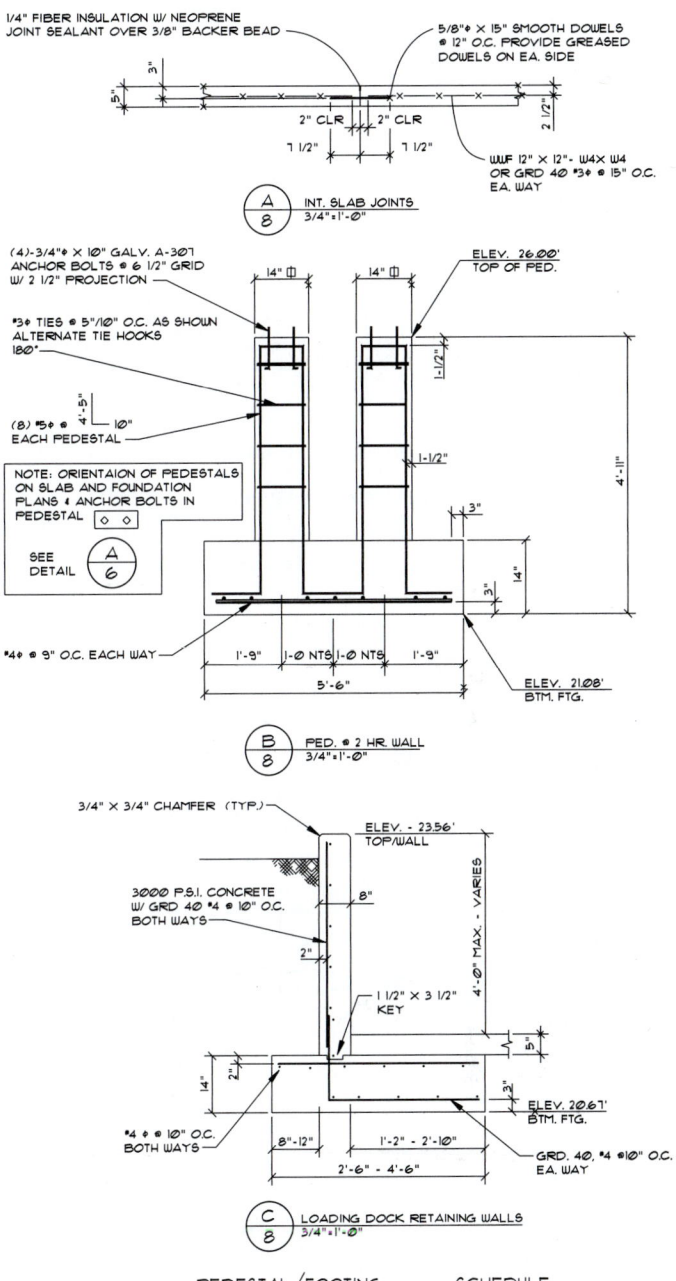

FOOTING	PEDESTAL HEIGHT	SIZE	REINFORCING STEEL
B-2	7.50	7'-0" □	#6φ @10" O.C.
B-3	6.75	6'-6" □	#6φ @12" O.C.
B-4	5.91	6'-6" □	#6φ @12" O.C.
B-5	5.00	7'-0" □	#6φ @10" O.C.
B-6	4.50	6'-0" □	#5φ @10" O.C.
B-7	4.00	7'-0" □	#6φ @10" O.C.

FIGURE 23.65 ■ Typical slab and footing details.

grade, location of openings, wall heights, roof slopes, exterior building materials, and other exterior features. A front elevation is generally the main entry view and is drawn at a 1/4" = 1'-0" scale, depending on the size of the structure. Other elevations may be drawn at a smaller scale. The elevation's scale depends upon the size of the building, the amount of detail shown, and

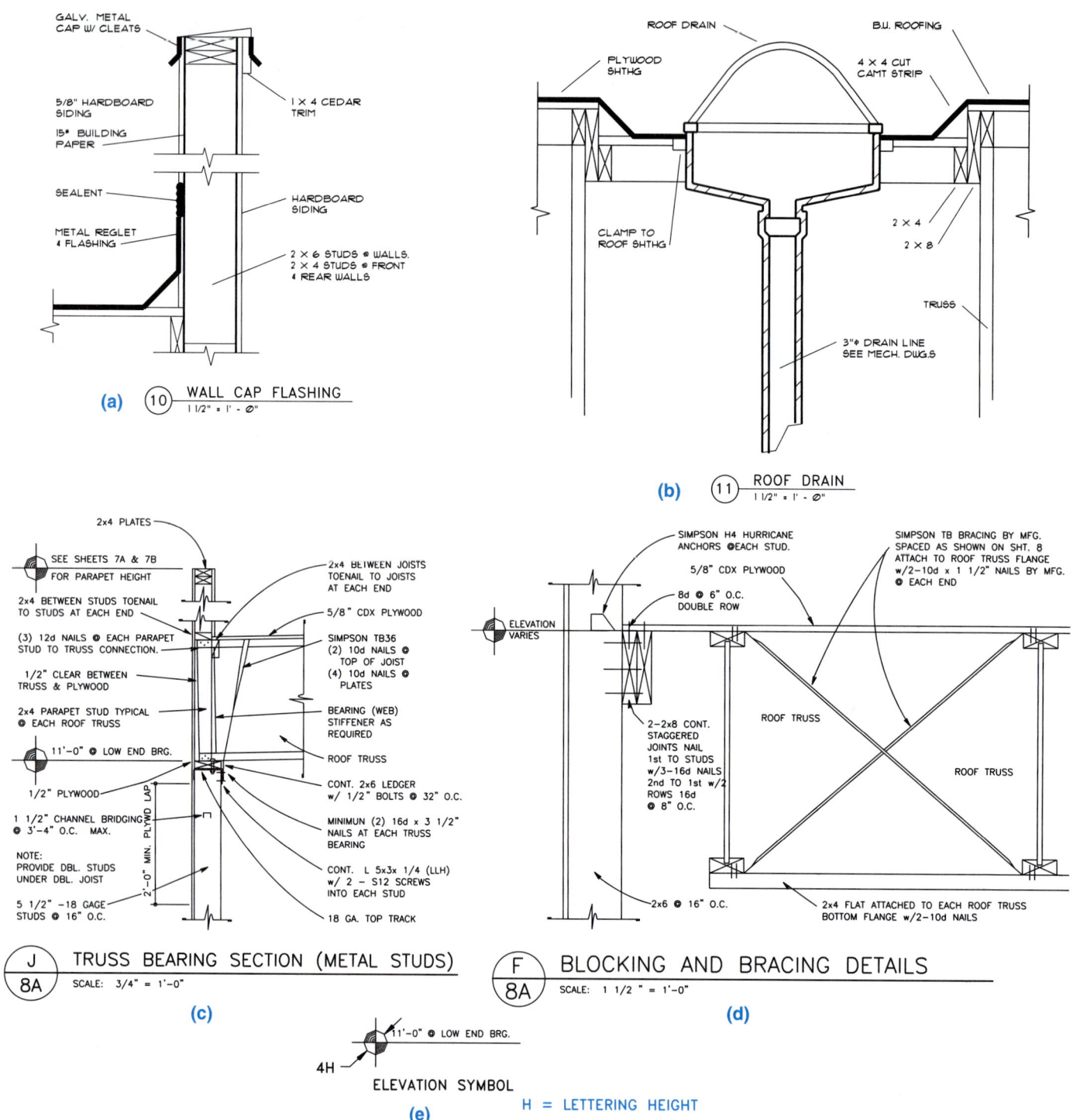

FIGURE 23.66 ■ Roof construction details. (a and b) *Courtesy The Southland Corporation.* (c and d) *Courtesy Wendy's International Inc.* (e) Elevation symbol is used to designate the elevation of a specific location. The elevation is above a known zero elevation on the structure.

the sheet size used. Many companies prefer to draw all elevations at the same scale showing an equal amount of detail in all views. An elevation may be omitted if it is the same as another. When this happens, the elevation may be labeled as RIGHT AND LEFT ELEVATION. Elevations are also often labeled by compass orientation, such as SOUTH ELEVATION. The elevations may be

drawn showing a great deal of detail, as shown in Figure 23.70. Often, elevations for commercial buildings may be drawn at a small scale representing very little detail, as shown in Figure 23.71, page 786. In many situations the drafter may be required to draw interior elevations or details such as the interior finish elevation detail for Farrell's, as shown in Figure 23.72, page 786.

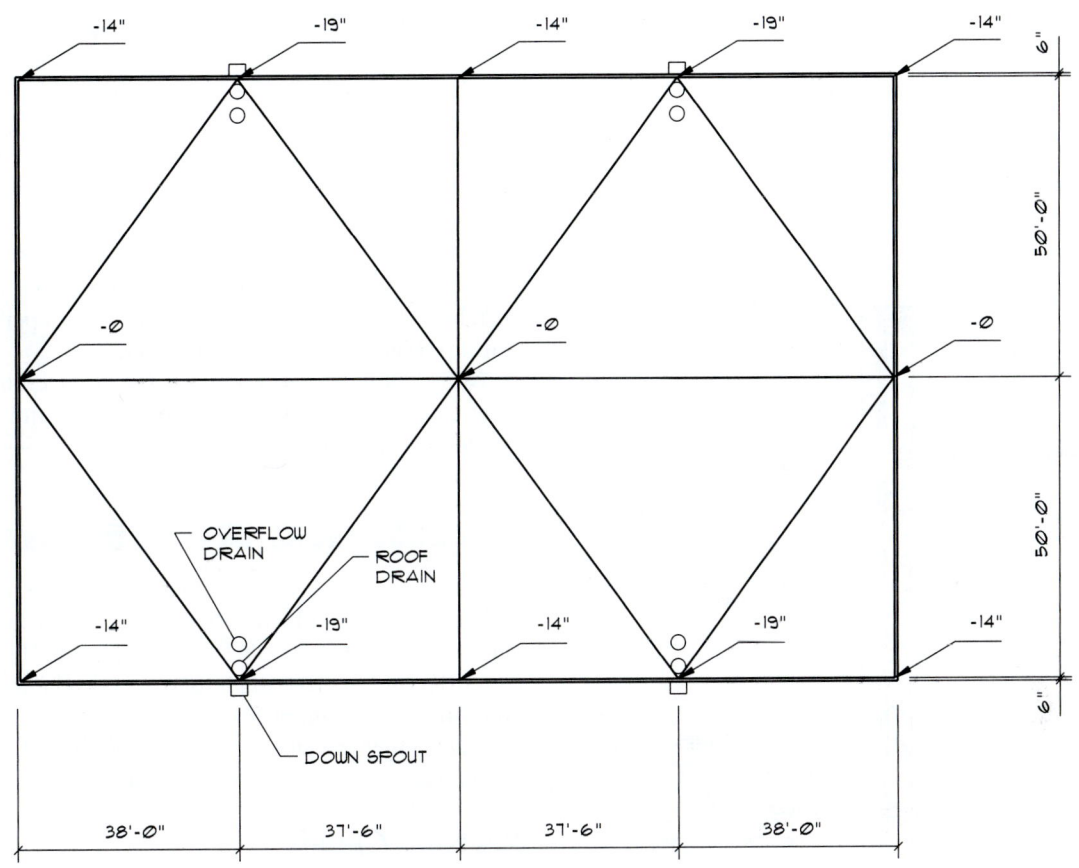

ROOF DRAINAGE PLAN

SCALE: 1/16" ====== 1'-0"

GENERAL NOTES:

1. ROOF AND OVERFLOW DRAINS SHALL BE LARGE GENERAL PURPOSE TYPE W/
   NON-FERROUS DOMES AND 4"φ OUTLETS
2. OVERFLOW DRAINS SHALL BE SET W/ INLET 2" ABOVE ROOF DRAIN INLET AND
   SHALL BE CONNECTED TO DRAINS LINES INDEPENDENT FROM ROOF DRAINS.
3. USE A 4"H X 7W SCUPPER W/5" NOMINAL (3 3/4 X 5) RECTANGULAR CORRUGATED
   DOWNSPOUT.
   PROVIDE A 6" X 9" CONDUCTOR HEAD @ TOP OF DOWNSPOUT.

FIGURE 23.67 ■ Roof drainage plan.

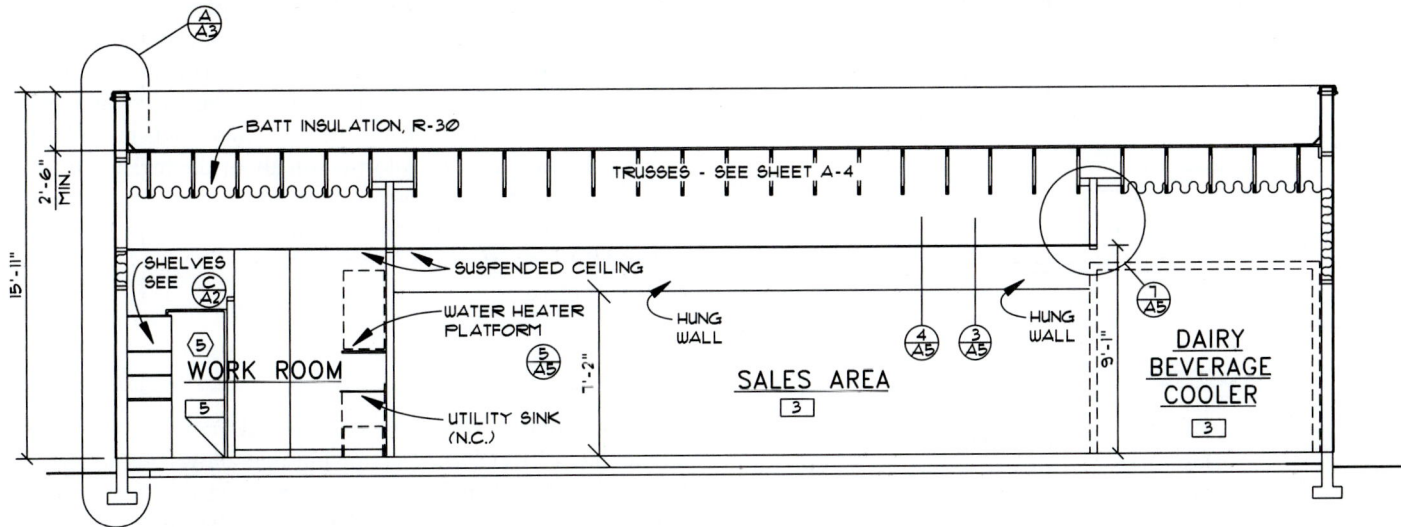

FIGURE 23.68 ■ Typical building sections for commercial construction are often drawn at a small scale to show major types of construction with specific information shown in details. *Courtesy The Southland Corporation.*

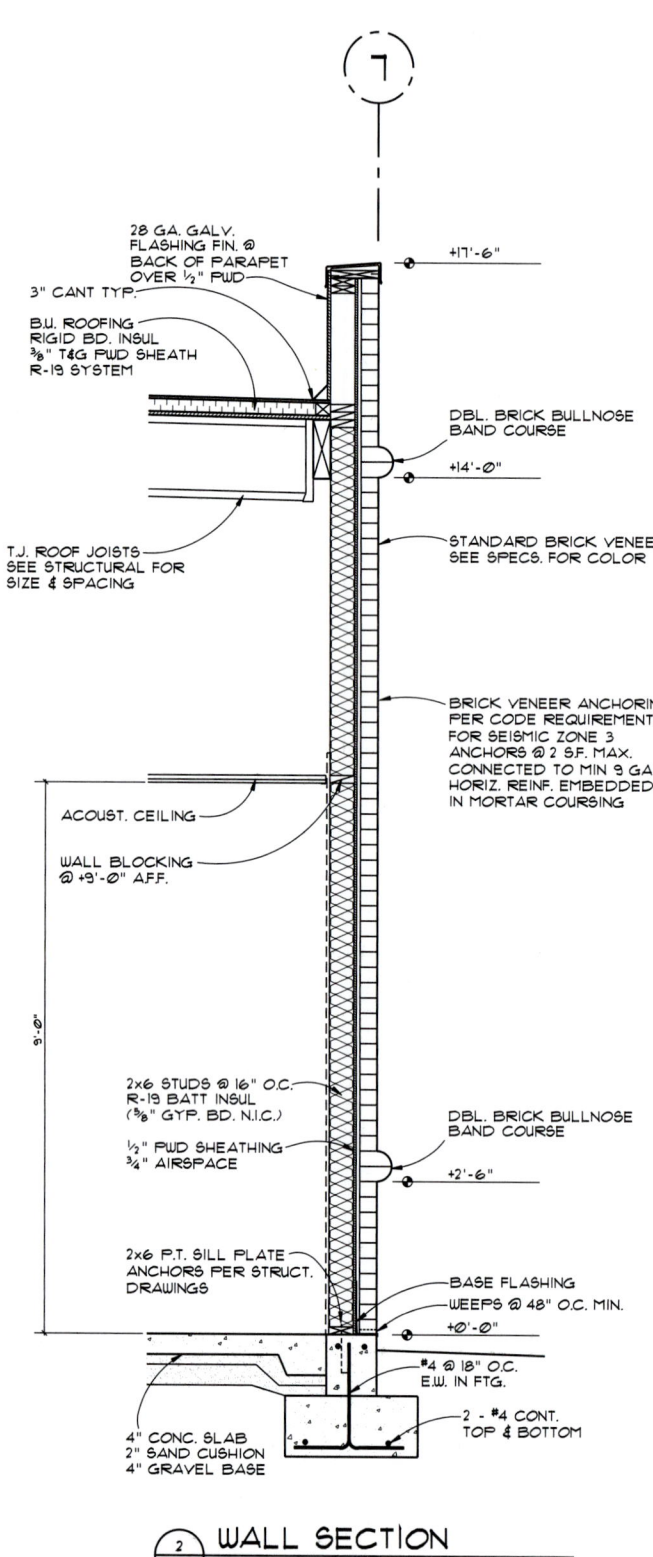

28 GA. GALV.
FLASHING FIN. @
BACK OF PARAPET
OVER ½" PWD

3" CANT TYP.

B.U. ROOFING
RIGID BD. INSUL
⅜" T&G PWD SHEATH
R-19 SYSTEM

T.J. ROOF JOISTS
SEE STRUCTURAL FOR
SIZE & SPACING

ACOUST. CEILING

WALL BLOCKING
@ +9'-0" A.F.F.

2x6 STUDS @ 16" O.C.
R-19 BATT INSUL
(⅜" GYP. BD. N.I.C.)

½" PWD SHEATHING
¾" AIRSPACE

2x6 P.T. SILL PLATE
ANCHORS PER STRUCT.
DRAWINGS

4" CONC. SLAB
2" SAND CUSHION
4" GRAVEL BASE

+17'-6"

DBL. BRICK BULLNOSE
BAND COURSE

+14'-0"

STANDARD BRICK VENEER
SEE SPECS. FOR COLOR

BRICK VENEER ANCHORING
PER CODE REQUIREMENTS
FOR SEISMIC ZONE 3
ANCHORS @ 2 S.F. MAX.
CONNECTED TO MIN 9 GA.
HORIZ. REINF. EMBEDDED
IN MORTAR COURSING

DBL. BRICK BULLNOSE
BAND COURSE

+2'-6"

BASE FLASHING
WEEPS @ 48" O.C. MIN.
+0'-0"

#4 @ 18" O.C.
E.W. IN FTG.

2 - #4 CONT.
TOP & BOTTOM

9'-0"

2
A42   WALL SECTION   3/4" = 1'-0"

FIGURE 23.69 ■ Partial sections are used to clarify construction information through various portions of the structure. *Courtesy Ankrom Moisan Architects.*

Panel Plan, Elevations, and Wall Details

The panel plan is used in tilt-up construction to show the location of the panels, as shown in Figure 23.18, page 753.

Panel elevations are used when the exterior elevations do not clearly show information about items located on or in the walls. Similar to exterior elevations, the panel elevations will show door locations and reinforcing within walls and around openings. Dimensions associated with panel elevations provide both horizontal and vertical dimensions for openings and other features. (See Figure 23.19, page 753.)

Wall details are used to show the connection points of the concrete panels used in tilt-up construction, and connection details at the walls for other types of structures. (See Figure 23.73.)

DRAWING REVISIONS

Drawing revisions are common in the architectural, structural, and construction industry. Revisions can be caused for a number of reasons; for example, changes requested by the owner, job site corrections, correcting errors, or code changes. Changes are done in a formal manner by submitting an addendum to the contract, which is a written notification of the change or changes and is accompanied by a drawing that represents the change.

Revision Clouds

A revision cloud is placed around the area that is changed. The revision cloud is a cloudlike circle around the change as shown in Figure 23.74. CADD programs that are commonly used for architectural and structural drafting have commands that allow you to easily draw the revision cloud. There is also a triangle with a revision number inside that is placed next to the revision cloud or along the revision cloud line as shown in Figure 23.75, page 788. The triangle is commonly called a delta. The number is then correlated to a revision note placed somewhere on the drawing, or in the title block as shown in Figure 23.76, page 788. Each company has a desired location for revision notes, although common places are in the corners of the drawing, in a revision block or table, or in the title block. This practice is not as clearly defined as in ANSI/ASME standard drawings. The note is used to explain the change. If a reference is given in the title block, then detailed information about the revision is normally provided in the revision document that is filed with the project information. The revision document is typically filled out and filed for reference, because changes can cause increased costs in a project.

CADD Application

It is easy to draw revision clouds with CADD. AutoCAD, for example, has a REVCLOUD command that allows you to specify the revision cloud arc length and draw a revision cloud around any desired area. The command works by picking a start point and then moving the cursor in the desired direction to create the

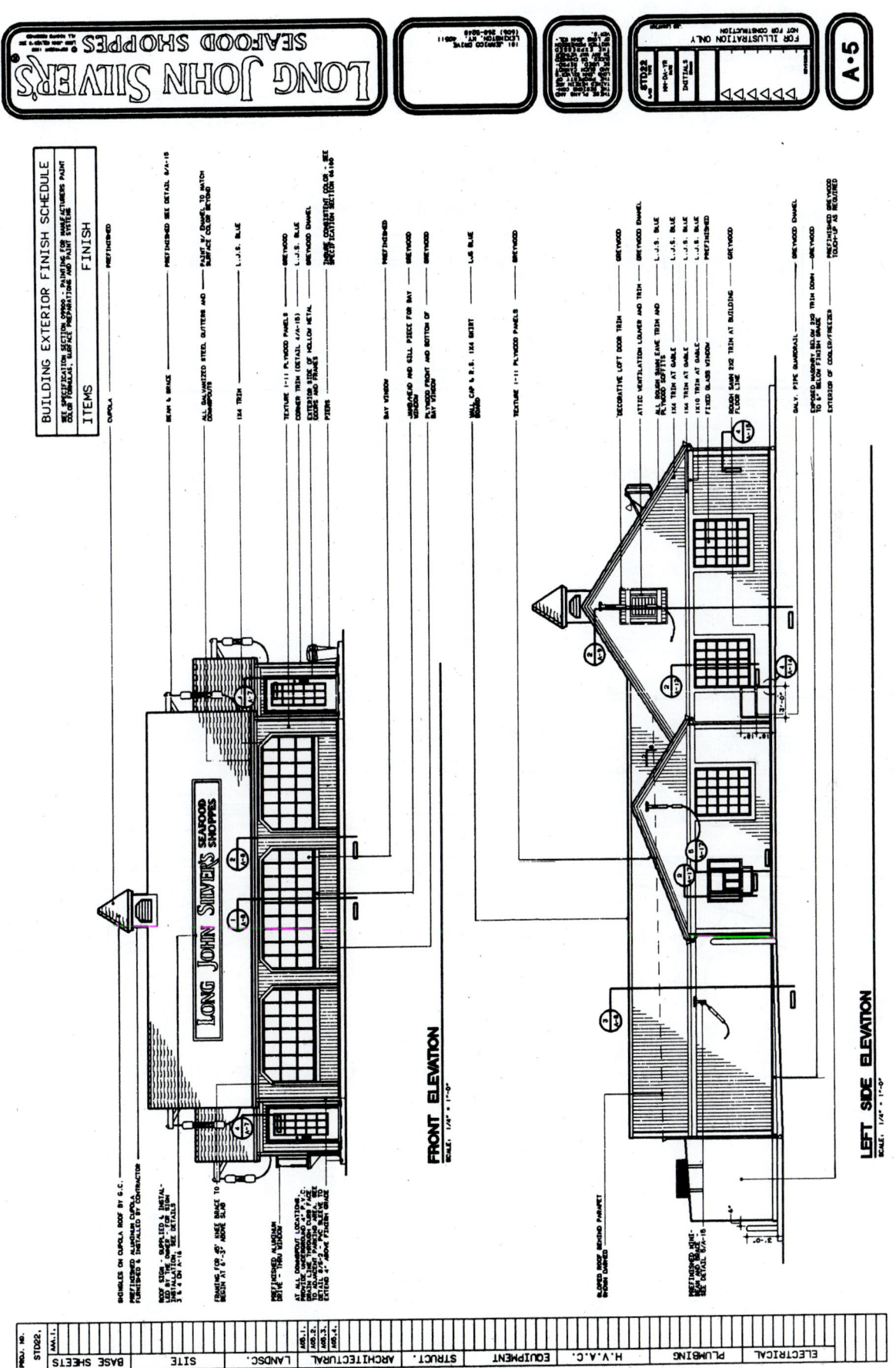

FIGURE 23.70 ■ CADD elevations of Long John Silver's Seafood Shoppes. *Courtesy Jerrico, Inc.*

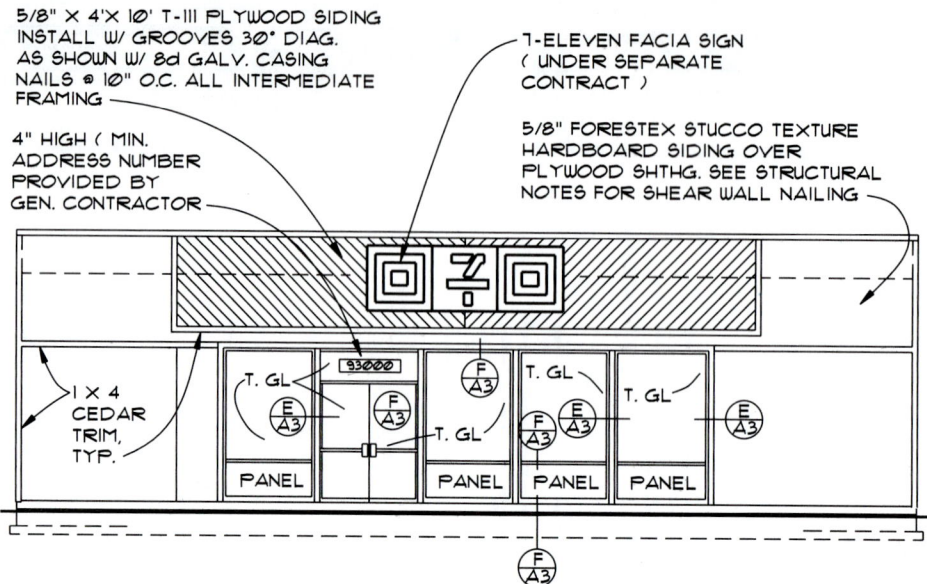

5/8" X 4'X 10' T-111 PLYWOOD SIDING INSTALL W/ GROOVES 30° DIAG. AS SHOWN W/ 8d GALV. CASING NAILS @ 10" O.C. ALL INTERMEDIATE FRAMING
7-ELEVEN FACIA SIGN (UNDER SEPARATE CONTRACT)
4" HIGH (MIN. ADDRESS NUMBER PROVIDED BY GEN. CONTRACTOR
5/8" FORESTEX STUCCO TEXTURE HARDBOARD SIDING OVER PLYWOOD SHTHG. SEE STRUCTURAL NOTES FOR SHEAR WALL NAILING
1 X 4 CEDAR TRIM, TYP.
T. GL
PANEL
93000
E A3
F A3

FIGURE 23.71 ■ Elevations for commercial structures, such as this 7-11 store, often require little detail and may be drawn at a small scale. *Courtesy The Southland Corporation.*

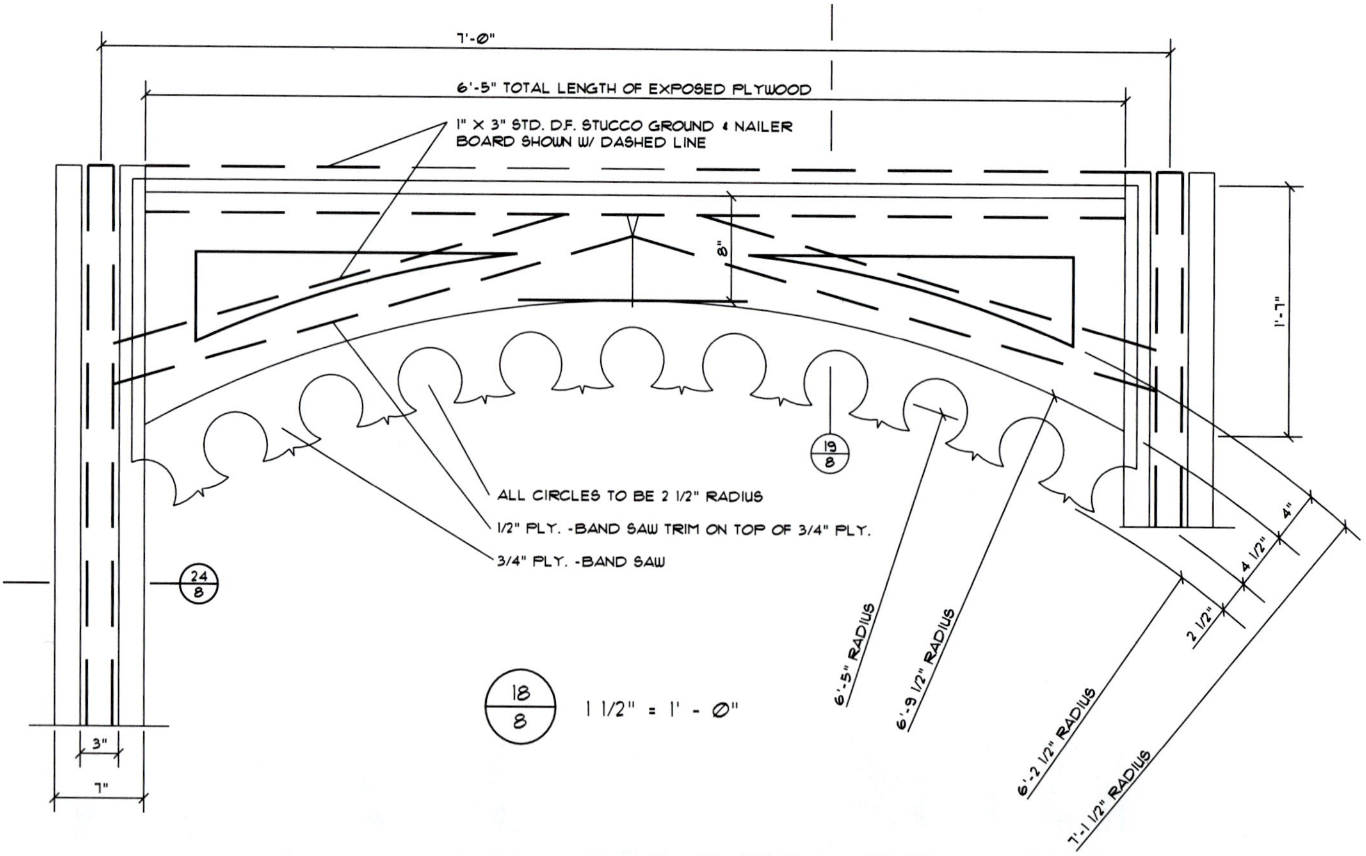

7'-0"
6'-5" TOTAL LENGTH OF EXPOSED PLYWOOD
1" X 3" STD. D.F. STUCCO GROUND & NAILER BOARD SHOWN W/ DASHED LINE
8"
1'-7"
ALL CIRCLES TO BE 2 1/2" RADIUS
1/2" PLY. -BAND SAW TRIM ON TOP OF 3/4" PLY.
3/4" PLY. -BAND SAW
19 8
24 8
18 8
1 1/2" = 1' - 0"
6'-5" RADIUS
6'-9 1/2" RADIUS
6'-2 1/2" RADIUS
7'-1 1/2" RADIUS
4"
4 1/2"
2 1/2"
3"
7"

FIGURE 23.72 ■ In addition to structural details, the drafter may be required to draw interior elevations and details such as this finish detail for Farrell's Restaurant. *Courtesy Structureform Masters, Inc.*

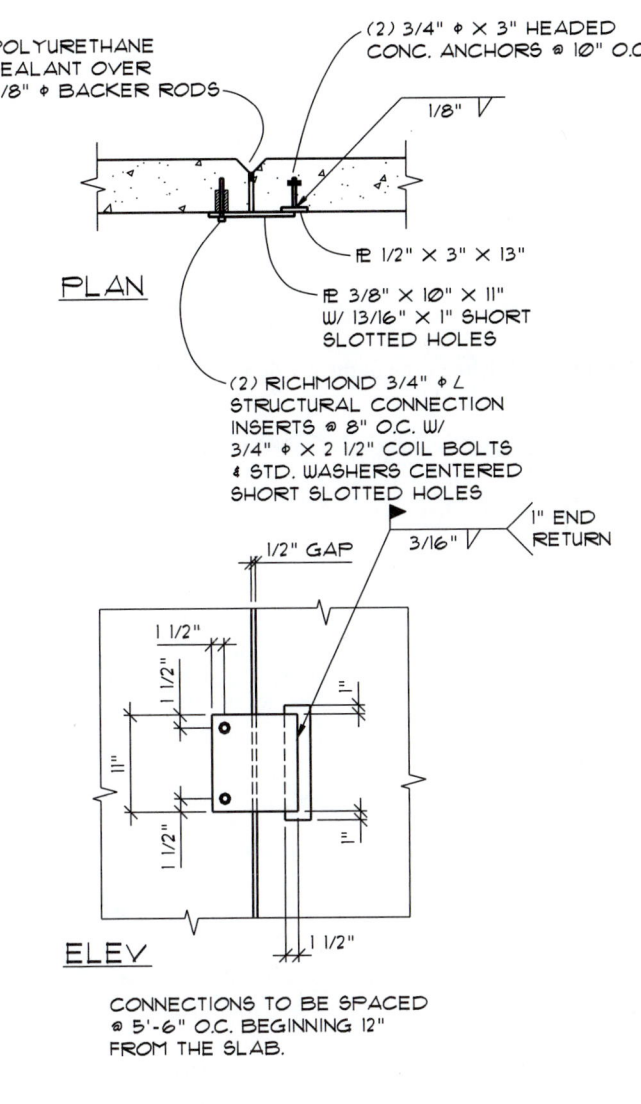

PLAN

POLYURETHANE
SEALANT OVER
5/8" ∅ BACKER RODS

(2) 3/4" ∅ X 3" HEADED
CONC. ANCHORS @ 10" O.C.

1/8"

⊕ 1/2" X 3" X 13"

⊕ 3/8" X 10" X 11"
W/ 13/16" X 1" SHORT
SLOTTED HOLES

(2) RICHMOND 3/4" ∅ L
STRUCTURAL CONNECTION
INSERTS @ 8" O.C. W/
3/4" ∅ X 2 1/2" COIL BOLTS
& STD. WASHERS CENTERED
SHORT SLOTTED HOLES

1/2" GAP

3/16"

1" END
RETURN

1 1/2"

1 1/2"

1"

1"

1"

1 1/2"

1 1/2"

ELEV

CONNECTIONS TO BE SPACED
@ 5'-6" O.C. BEGINNING 12"
FROM THE SLAB.

$\frac{1}{3}$ PANEL CONNECTOR
1" = 1'-0"

FIGURE 23.73 ■ A typical wall connection detail.

revision cloud, as shown in Figure 23.77a. Create the revision cloud in a pattern around the desired area while moving the cursor back toward the start point. Press the pick button when the cursor is back at the start point to complete the revision cloud as shown in Figure 23.77b.

CONSTRUCTION SPECIFICATIONS

The Construction Specifications Institute (CSI) and the Construction Specifications Canada (CSC) is a group of professionals devoted to the standardization of specification documents used in the Architectural, Engineering, and Construction (AEC) fields. This group has created a system of organization for architects, engineers, and contractors called the MasterFormat™. According to information on the CSI Website (*www.csinet.org*), the MasterFormat: Numbers and Titles is a master list of numbers and subject titles for organizing information about construction work results, requirements, products, and activities into groups and subgroups. MasterFormat: Numbers and Titles facilitate standard filing and retrieval format throughout the construction industry. MasterFormat: Numbers and Titles are suitable for use in project manuals, organization of cost data, reference keynotes on drawings, filing product information and technical data, identification of drawing objects, and presentation of construction market data. The system is divided into 49 major groups and subgroups called divisions. Each division is made up of levels. Level one is the major division title, levels two, three, and four define an increasingly detailed area of work results to be specified. The divisions and levels of the Master-Format establish a six-digit numbering system, shown in Figure 23.78. The first two digits designate the division number. The second pair of numbers identifies level two, and the third pair of numbers refers to level three. Level four numbers are preceded by a dot and are only used when additional detail is required. Several Division Numbers have been reserved to provide space for future expansion by the CSI. Look at the CSI website for more information and to see the complete list of divisions and levels.

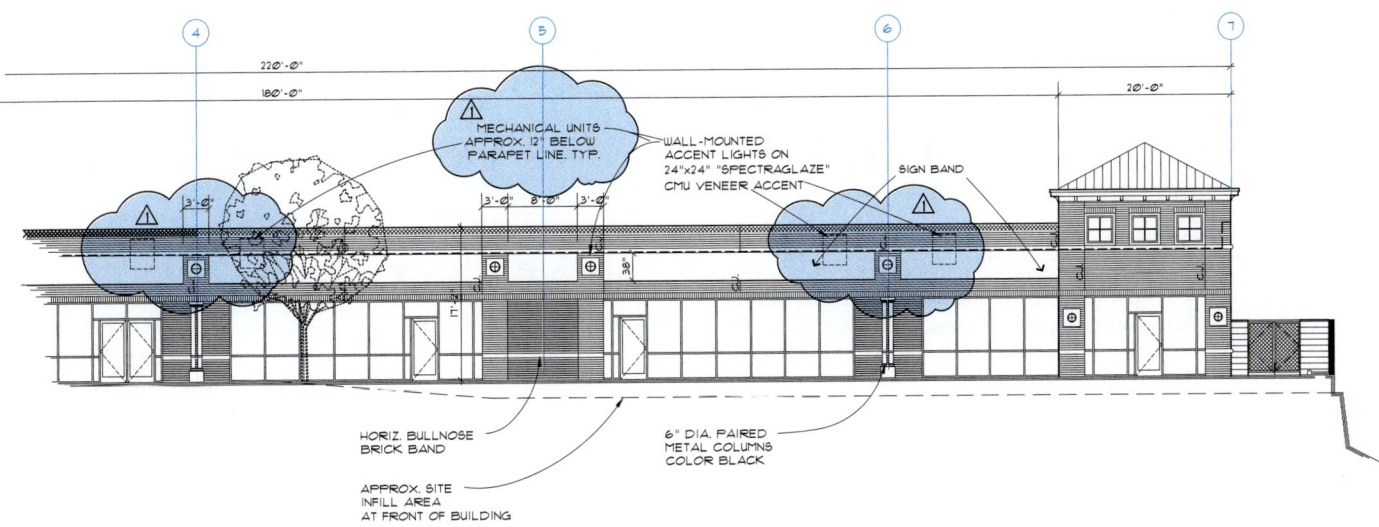

FIGURE 23.74 ■ A typical revision cloud and delta reference. *Portion of drawing courtesy Ankrom Moisan Architects.*

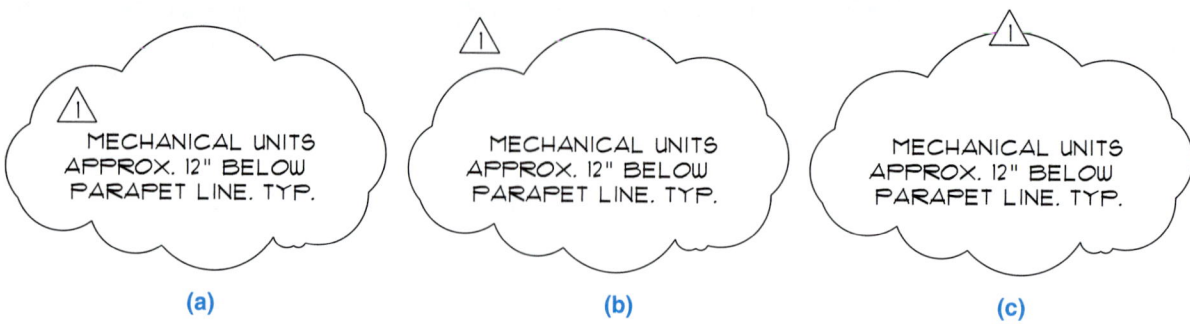

(a)   **(b)**   **(c)**

FIGURE 23.75 ■ Placement of the delta reference with the revision cloud: (a) delta inside of the revision cloud; (b) delta outside of the revision cloud; (c) delta inserted in the revision cloud line.

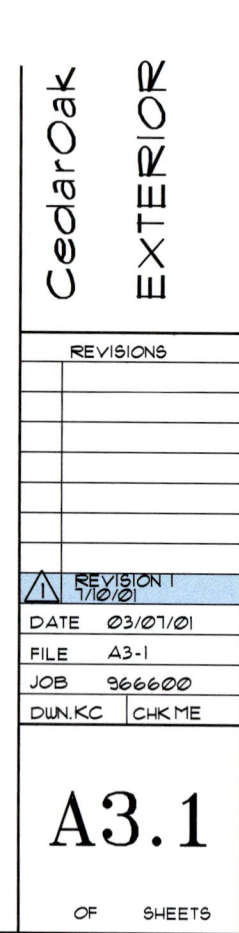

FIGURE 23.76 ■ Revision reference in the title block. The specific information about the revision is found in the job file. *Portion of title block courtesy Ankrom Moisan Architects.*

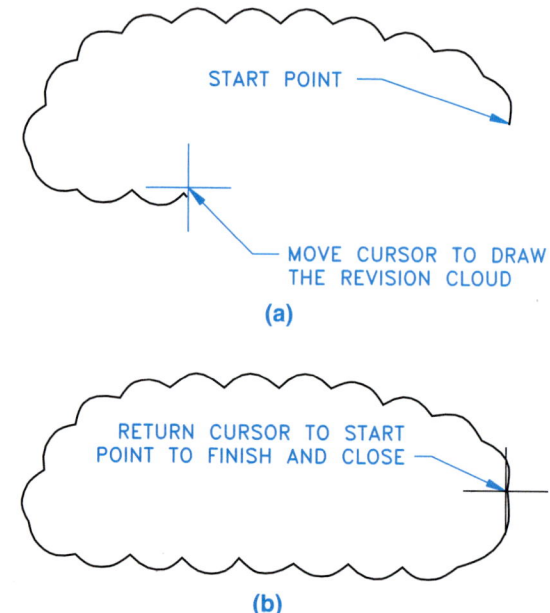

FIGURE 23.77 ■ The AutoCAD REVCLOUD command automatically draws arc segments along the path.

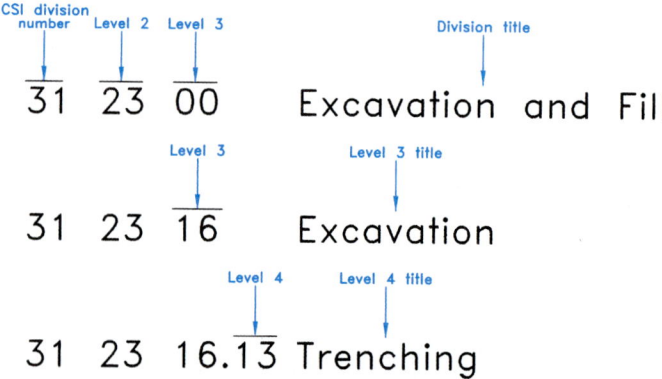

FIGURE 23.78 ■ The divisions and levels of the MasterFormat establish this six-digit numbering system. The first two digits designate the division number. The second pair of numbers identifies level two, and the third pair of numbers refers to level three. Level four numbers are preceded by a dot and are only used when additional detail is required. Several Division Numbers have been reserved to provide space for future expansion by the CSI.

CSI MasterFormat Divisions	
Division Number	Division Title

Procurement and Contracting Requirements Group

00	Procurement and Contracting Requirements

Specifications Group

General Requirements Subgroup

01	General Requirements

Facility Requirements Subgroup

02	Existing Conditions
03	Concrete
04	Masonry
05	Metals
06	Wood, Plastics, and Composites
07	Thermal and Moisture Protection
08	Openings
09	Finishes
10	Specialties
11	Equipment
12	Furnishings
13	Special Construction
14	Conveying Equipment
15–19	Reserved

Facilities Services Subgroup

20	Reserved
21	Fire Suppression
22	Plumbing
23	Heating, Ventilation, and Air Conditioning
24	Reserved
25	Integrated Automation
26	Electrical
27	Communications
28	Electronic Safety and Security
29	Reserved

Site and Infrastructure Subgroup

30	Reserved
31	Earthwork
32	Exterior Improvements
33	Utilities
34	Transportation
35	Waterway and Marine Construction
36–39	Reserved

Process Equipment Subgroup

40	Process Integration
41	Material Processing and Handling Equipment
42	Process Heating, Cooling, and Drying Equipment
43	Process Gas and Liquid Handling, Purification, and Storage Equipment
44	Pollution Control Equipment
45	Industry-Specific Manufacturing Equipment
46–47	Reserved
48	Electrical Power Generation
49	Reserved

BASIC DRAWING LAYOUT STEPS

The following gives you some basic steps that you can use when laying out a set of architectural/structural drawings. Not all elements of a set of drawings are demonstrated, but the steps used in the examples can be applied to any of the components in the set of working drawings.

Laying Out the Plan Views

The structural plan views are normally the drawings that begin the complete set. These include the types of drawings already discussed, such as the floor plan, foundation plan, and roof framing plan. After the plan views have been drawn, the sections and details may be properly correlated.

The Floor Plan, Step 1

Begin by laying out the floor plan at a scale that will provide the clearest representation on the sheet size to be used. Normally a scale range of 1/8" = 1'-0" to 1/4" = 1'-0" (1:100–1:50 metric) is used. Draw the outline of the building, leaving adequate space for dimensions and notes, as shown in Figure 23.79.

The Floor Plan, Step 2

Draw all internal walls, partitions, posts, and fixtures. Draw all dimensions. Structural drafting normally uses aligned dimensioning with the dimension numeral placed above the dimension line, and slashes placed where the dimension and extension lines meet, as shown in Figure 23.80. Finally, notes, symbols, and titles are added to complete the drawing.

The Foundation Plan, Step 1

After the floor plan has been drawn, turn off or freeze the layers that are not used to lay out the foundation, such as partitions

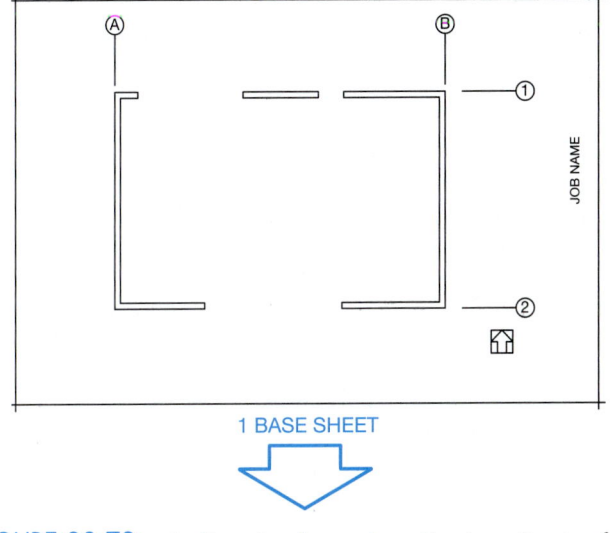

FIGURE 23.79 ■ Outline the floor plan. *Courtesy Structureform Masters, Inc.*

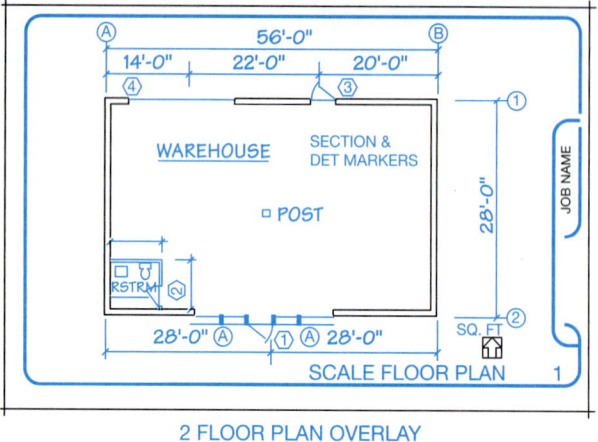

2 FLOOR PLAN OVERLAY

FIGURE 23.80 ■ Draw all interior features, dimensions, notes, and title to complete the floor plan. *Courtesy Structureform Masters, Inc.*

and fixtures. Leave the perimeter wall, bearing wall, and post layers turned on. The floor plan makes an excellent guide for the other plans because it is the base for all other construction.

The Foundation Plan, Step 2

Begin by drawing the foundation walls and footings. Next, add all dimensions, notes, and symbols. Finally, letter the title and scale, as shown in Figure 23.81.

The Roof Framing Plan

The layout of the roof framing plan works in much the same way as the foundation plan. Again, use the floor plan as a basis for the roof plan drawing. This gives you all of the layout features to quickly begin the new drawing. First, place all beams

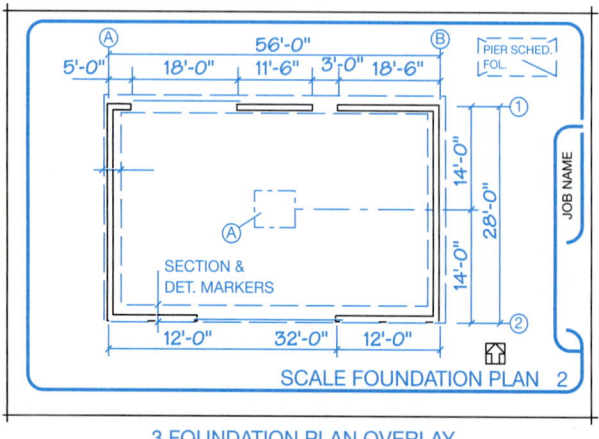

3 FOUNDATION PLAN OVERLAY

FIGURE 23.81 ■ Lay out the foundation with the floor plan as a guide. *Courtesy Structureform Masters, Inc.*

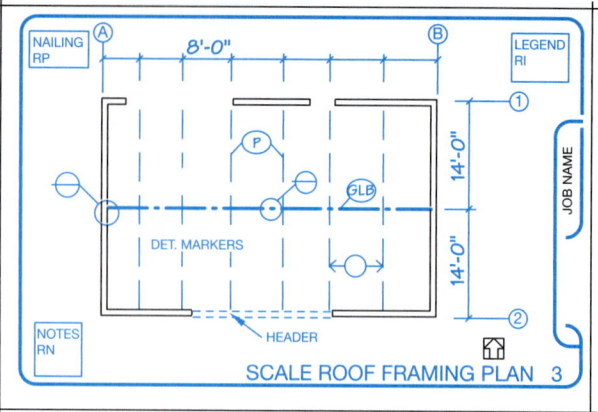

4 ROOF FRAMING PLAN OVERLAY

FIGURE 23.82 ■ Lay out the roof framing plan with the floor plan as a guide. *Courtesy Structureform Masters, Inc.*

and purlins. Second, completely dimension the drawing and add all necessary notes and symbols. Finally, add all schedules and titles, as shown in Figure 23.82.

The Details, Step 1

The detail drawings can be placed with the plan views or on separate sheets. In either case, the details and sections must be clearly correlated to the plans. The scale of the detail drawings depends on the size and complexity of the structure to be drawn. Select a scale that clearly shows all construction details without oversizing the drawing, thus wasting space and drawing time. Common scales for detail drawings range from 1/2" = 1'–0" to 3" = 1'–0" (1:20–1:5 metric). All the details can be the same scale, or the scale can change depending on the complexity of the drawing. In many cases it is necessary to arrange many details on one sheet or in an area on a sheet. Do this by laying out the details, beginning at the top left corner of the working area. Proceed with additional details placed in rows from left to right. The engineering sketch in Figure 23.2a, page 744, is used as a guide for the following layout. Start the detail drawing by blocking out the major components of the structure. As you can see in the engineer's sketch, the beam on the left is 24" high and the beam on the right is 16" (400 mm) high. Thus the maximum beam height is 24" (600 mm). If you add another 12" (300 mm) for the post and 12 more inches for the notes, at a scale of 1" = 1'–0" (1:10 metric) you need a total drawing height of 48" (1200 mm), or an actual height of 4" (100 mm) at the selected scale. The width of the detail is up to you. All you need is enough space to show all of the construction members. So, a width of about 4" should work. (See Figure 23.83.)

The Details, Step 2

Proceed by drawing in all of the construction members to scale. Refer to a vendor's catalog as shown in Figure 23.38, page 766, to determine the actual sizes of the prefabricated connectors. Lay out the detail, as shown in Figure 23.84.

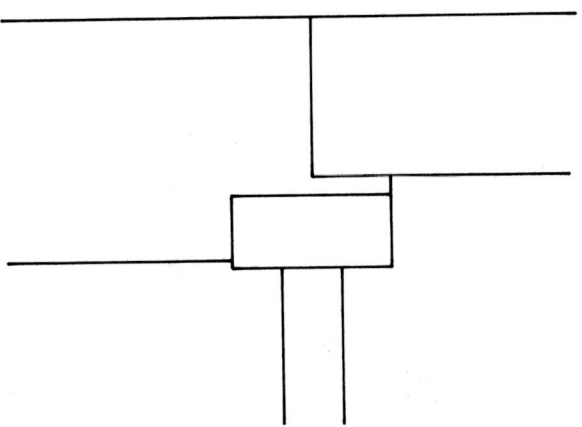

FIGURE 23.83 ■ Step 1—Block out components of the detail.

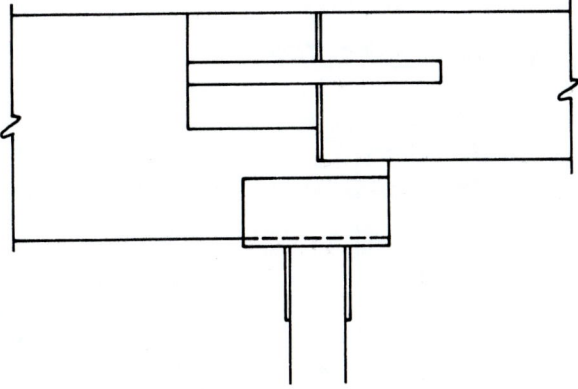

FIGURE 23.84 ■ Step 2—Draw all construction members.

The Details, Step 3

Place all dimensions and notes on the drawing, followed by the title and scale, or the detail identification symbol and scale, as shown in Figure 23.2b, page 744.

PICTORIAL DRAWINGS

Pictorial drawings such as isometrics are sometimes used in the set of architectural/structural drawings when it is determined to be necessary to represent something more clearly than in a two-dimensional drawing. Pictorial drawings are not required, but they are done when there is a need to aid in the visualization of specific construction applications. Figure 23.85 shows the use of pictorial drawings. Pictorial drawings are covered in detail in Chapter 15 of this text.

WEBSITE RESEARCH

The following websites can assist you in doing additional research on subjects such as manufacturing materials, manufacturing practices, tooling, and related areas:

http://www.aia.org—American Institute of Architects.

http://www.aluminum.org—Aluminum Association.

http://www.asme.org—American Society of Mechanical Engineers.

http://www.aist.org—Association for Iron and Steel Technology.

http://www.industrialpress.com—*Machinery's Handbook*. Materials and processes.

http://www.worldsteel.org—International Iron and Steel Institute.

http://www.csinet.org—Construction Specifications Institute.

http://www.aisc.org—American Institute of Steel Construction.

http://www.capterra.com—Construction Documents and Management Software.

http://www.csc-dcc.ca—Construction Specifications Canada.

http://www.hud.org—U.S. Department of Housing and Urban Development.

http://www.nahb.org—National Association of Home Builders.

http://www.crsi.org—Concrete Reinforcing Steel Institute.

http://www.wirereinforcementinstitute.org—Wire Reinforcement Institute.

http://www.amazon.com—Search author, Madsen, for *Architectural Drafting and Design*, *Architectural Drafting Using AutoCAD*, and *Architectural Desktop and Its Applications* textbook titles.

http://www.g-w.com—Publisher of the textbooks *Architectural Drafting Using AutoCAD*, and *Architectural Desktop and Its Applications*, by David A. Madsen and Ron Palma.

http://www.delmar.com—Publisher of the textbooks *Architectural Drafting and Design* by Jefferis and Madsen, and *Print Reading for Architecture and Construction Technology*, by Madsen and Jefferis.

http://www.bcewp.com—Boise Cascade Engineered Wood Products Division.

http://www.prnewswire.com—Willamette Industries.

http://www.awc.org—American Wood Council.

http://www.aci-int.org—American Concrete Institute.

http://www.tilt-up.org—The Tilt-Up Concrete Association.

http://www.concrete.com—The industry portal for concrete services and information.

http://www.precast.org—National Precast Concrete.

http://www.steel-sci.org—The Steel Construction Institute.

http://www.strongtie.com—Simpson Strong Tie information and catalog.

http://www.pci.org—Precast/Prestressed Concrete Institute.

http://www.cpci.ca—Canadian Precast/Prestressed Concrete Institute.

http://www.buildingteam.com—Building and construction industry links.

http://www.arcat.com—Building materials information, specifications, and CAD details.

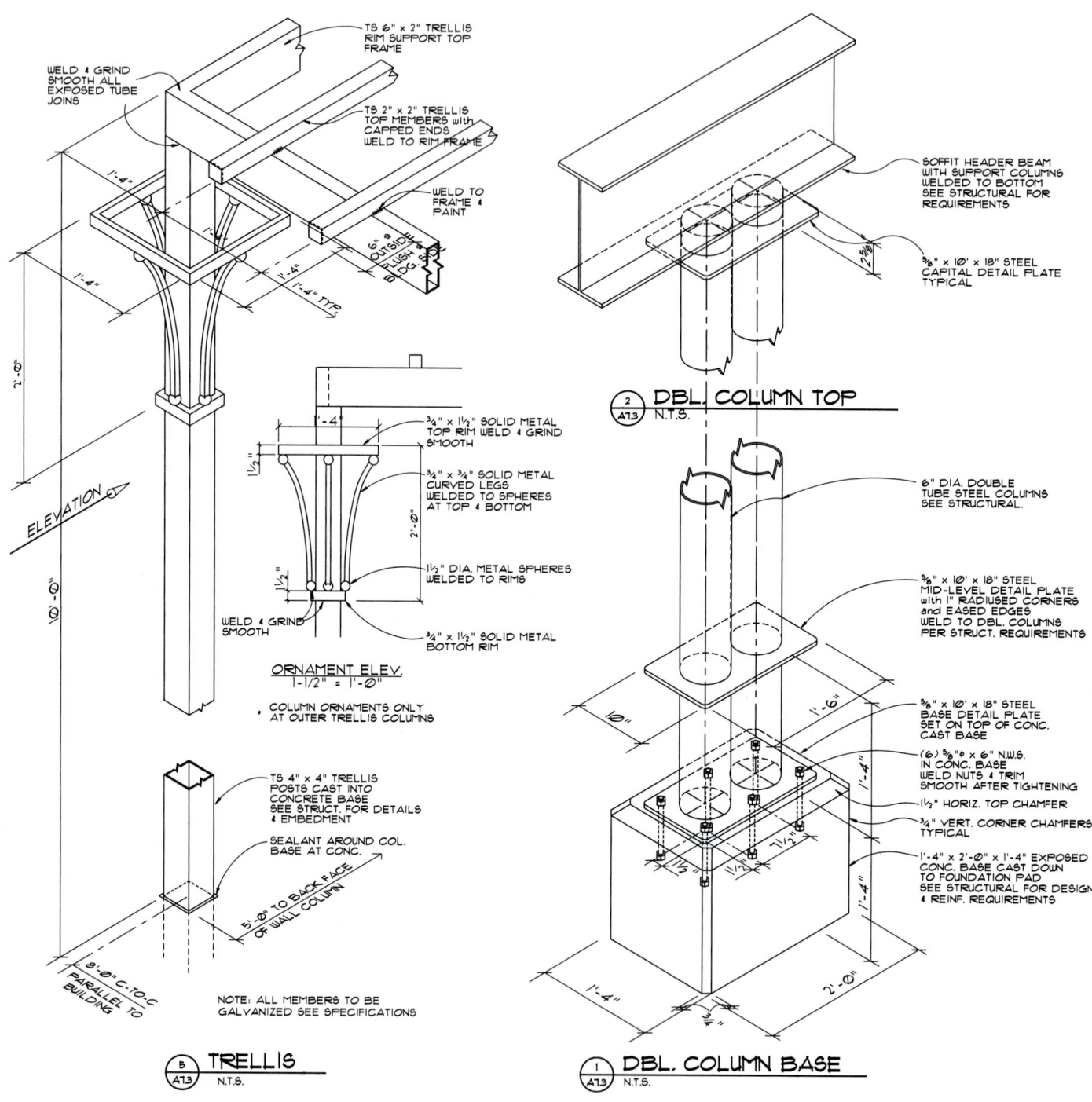

FIGURE 23.85 ■ Pictorial detail drawings. *Courtesy Ankrom Moisan Architects.*

Structural engineering software programs are available that integrate structural analysis, design, and drafting. A 3-D model is made during the design and drafting process as shown in Figure 23.86. The model can be updated at any time from new design information. The model can be used to automatically create framing plans and elevations as shown in Figure 23.87. The software package also creates construction details for a variety of materials. The construction details can be drawn at any scale and stored for later use. Parametric design of welding symbols and specifications allows the designer to change a variable and automatically update the entire drawing. Detail models as shown in Figure 23.88 can be converted to traditional 2-D views, complete with dimensions and specifications.

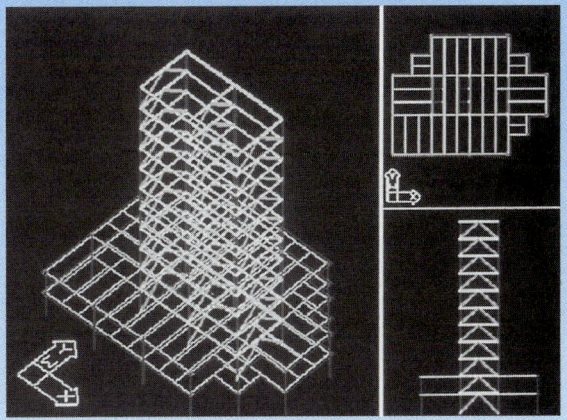

FIGURE 23.87 ■ Creating framing plans and elevations with the 3-D model. *Courtesy Computers and Structures, Inc.*

FIGURE 23.86 ■ A 3-D model is made during the CADD design and drafting process. *Courtesy Softdesk.*

FIGURE 23.88 ■ A 3-D CADD model of a structural detail. *Courtesy Softdesk.*

PROFESSIONAL PERSPECTIVE

The drafting procedure for the preparation of structural drawings is often complicated by the amount of information to be contained in these drawings. Structural drawings are normally done on large sheets with as much information as possible placed on each sheet. The first task is to determine the sheet size. This may already be decided as a standard within the company: for example, 22 × 34. The factors that should be considered when selecting sheet size or when drawing on predetermined sheet sizes are:

■ The size of the plan.

■ The number of dimensions and notes needed.

■ The scale used.

■ The number of details to be placed on the sheet.

■ The amount of free space required for possible future revisions.

All of these issues must be considered, because you do not want to end up with some sheets that are overcrowded and others that have little information. Use CADD layers such as FL1, FL2, FDN, ROOF, ELEV, and NOTES.

DESIGNING AN OBSERVATION PLATFORM

Problem: Suppose a building code states that the rise between stair treads cannot exceed 8" for a 12" tread. Find the minimum distance for dimension L and a corresponding angle A in Figure 23.89 of an observation platform.

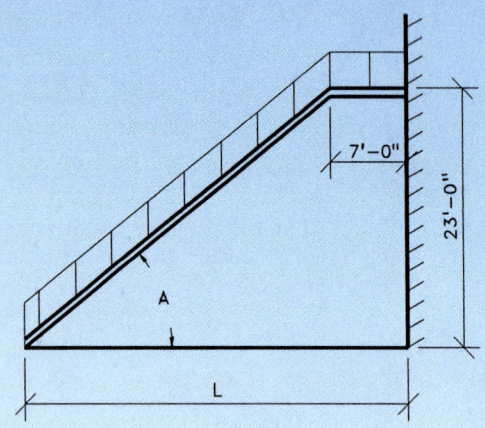

FIGURE 23.89 ■ Observation platform.

Solution: Geometrical figures are *similar* when they have the same shape. They are *congruent* when they have the same shape *and* size. When two triangles are similar, proportional equations can be written. In this design problem, you have similar triangles: the large staircase and the smaller single tread. (See Figure 23.90.)

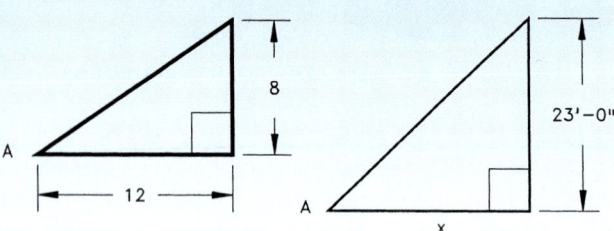

FIGURE 23.90 ■ Two similar triangles.

Set up a proportion

$$\frac{23}{x} = \frac{8}{12}$$

which gives

$$x = 34.5'$$

The dimension L is an additional 7'.

$$L = 34.5 + 7 = \textbf{41.5'} \text{ or } \textbf{41'-6"}$$

Also, being right triangles, angle A can be found by the application of one trig function:

$$A = \text{Inv tan } \frac{8}{12} = 33.7°$$

It is often necessary for the drafter to calculate the weight and cubic yard of concrete for bills of materials, cost estimates, and construction purposes. The following formulas can be used to make these calculations:

(Length × Width × Height) × 150 = Total weight
(Length × Width × Height) ÷ 27 = Cubic yard

Problem: Given a precast concrete panel with the dimensions 18'–4" long, 1'–4" wide, and 2'–0" high, calculate the weight in pounds, and the volume in yards.

Solution:

18'-4" × 1'-4" × 2'-0" = 48.88889 cubic feet
48.88889 × 150 = 7333.3333 lb
48.88889 ÷ 27 = 1.81 cubic yards

If there are holes and/or cutouts, it is necessary to calculate the combined volume and weight of these and subtract them from the total.

CHAPTER

23 *Structural Drafting Test*

 Access the CD found with this textbook to view the Chapter 23 Test. Confirm the preferred submittal method with your instructor.

CHAPTER 23

Structural Drafting Problems

DIRECTIONS

1. Read all related instructions before you begin working.

2. Use manual or computer-aided drafting as required by your course guidelines. Use an architectural lettering style or architectural CADD font style unless otherwise specified by your instructor.

3. Use the engineering sketches to prepare a complete set of drawings for the given problems.

4. Do all drawings on 22" × 34" or 24" × 36" (AI and AO metric) sheets unless otherwise specified by your instructor.

5. Use proper sectioning and detailing techniques as correlated to the engineering sketches. Some recommended scales are provided. You may increase or decrease the scale depending on the available space and the complexity of the section or detail. Sections and details should be drawn at a minimum scale of 3/8" = 1'–0" (1:20 metric). Judgment should be used if a section or detail requires a lot of information; a larger scale ranging from 3/4" to 1-1/2" = 1'–0" (1:10 metric) should then be used.

6. It is recommended that you evaluate the entire set of sketches for each problem before beginning to draw. Information needed to draw one problem may be found on the sketch or data for another. The problems are real drafting situations, so you will interpret rough engineering sketches and prepare formal drawings using the best drafting techniques and standards that you have learned.

7. It is suggested that drawing sheets be organized with as much information as possible without crowding or reducing clarity. If additional drawing area is required, it would be better to add another sheet than to overcrowd the drawing.

8. Some problems in this chapter may contain errors, missing information, or slight inaccuracies. This is intentional and is meant to encourage you to apply appropriate problem-solving methods and engineering, and drafting standards in order to solve the problems. This is meant to force you to think about each part and how parts fit together in the structure. As in "real-world" projects, the engineering problem should be considered as a basis for your preliminary layouts. Always question inaccuracies in project designs and consult with the proper standards and other sources. In some cases, an error might be the source of engineering changes provided by your instructor; however, this is determined by your specific course objectives. Other situations may require that corrections be made during the development of the original design drawings. This is not intended as a source of frustration, but it is considered to be part of the engineering drafter's daily responsibility in project development.

9. After completion of assigned drawings and complete sets of working drawings, your instructor can make drawing changes for you to complete using the drawing revision practices discussed in this chapter. This option depends on your course objectives and your instructor's preference.

PROBLEM 23.1 **Exterior pole light, sign footing, curb details, and sidewalk and paving details**

Draw the following details on one sheet unless otherwise specified by your instructor. *Drawings courtesy of Wendy's International, Inc.*

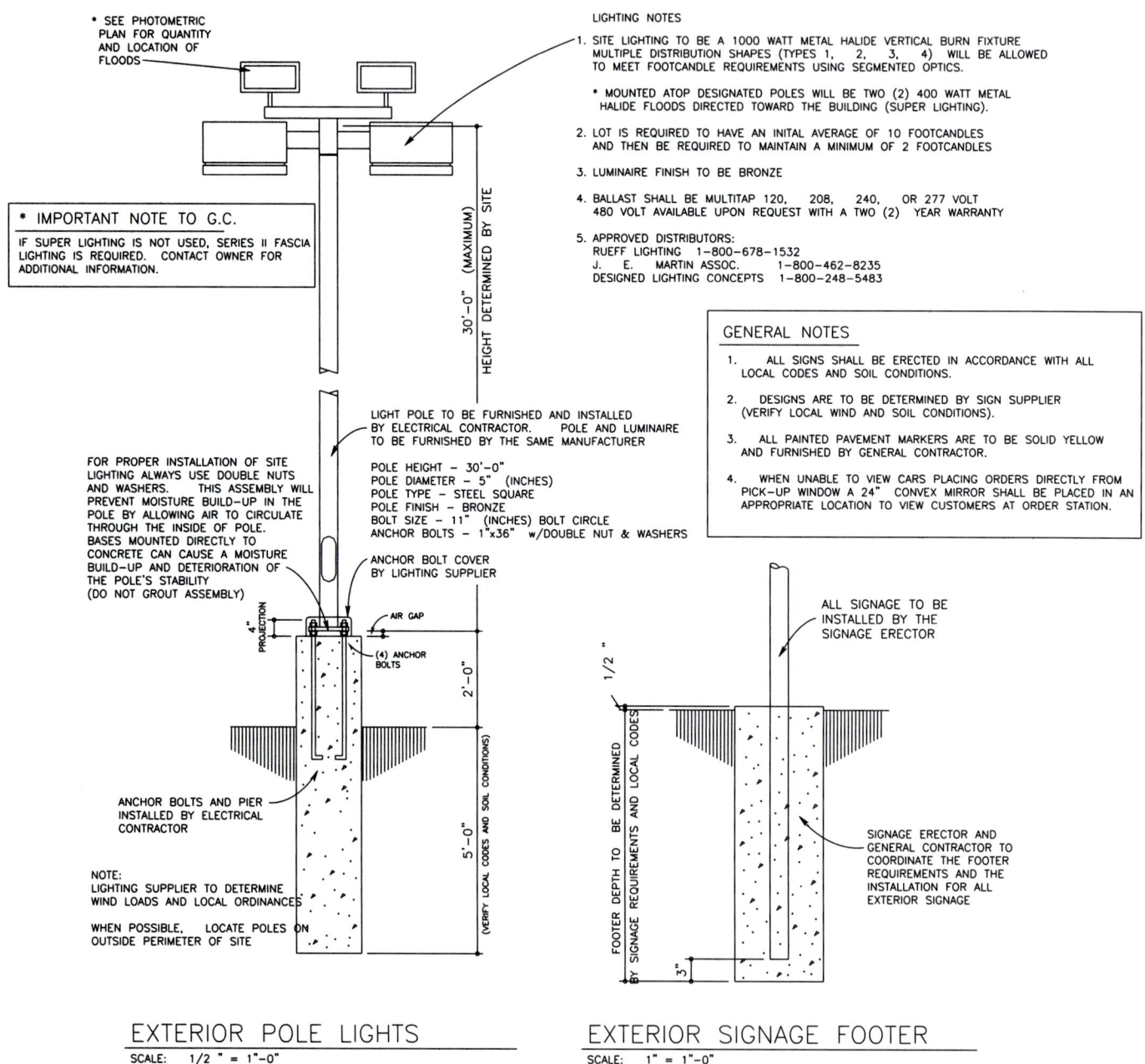

* SEE PHOTOMETRIC PLAN FOR QUANTITY AND LOCATION OF FLOODS

LIGHTING NOTES

1. SITE LIGHTING TO BE A 1000 WATT METAL HALIDE VERTICAL BURN FIXTURE MULTIPLE DISTRIBUTION SHAPES (TYPES 1, 2, 3, 4) WILL BE ALLOWED TO MEET FOOTCANDLE REQUIREMENTS USING SEGMENTED OPTICS.

   * MOUNTED ATOP DESIGNATED POLES WILL BE TWO (2) 400 WATT METAL HALIDE FLOODS DIRECTED TOWARD THE BUILDING (SUPER LIGHTING).

2. LOT IS REQUIRED TO HAVE AN INITAL AVERAGE OF 10 FOOTCANDLES AND THEN BE REQUIRED TO MAINTAIN A MINIMUM OF 2 FOOTCANDLES

3. LUMINAIRE FINISH TO BE BRONZE

4. BALLAST SHALL BE MULTITAP 120, 208, 240, OR 277 VOLT 480 VOLT AVAILABLE UPON REQUEST WITH A TWO (2) YEAR WARRANTY

5. APPROVED DISTRIBUTORS:
   RUEFF LIGHTING 1-800-678-1532
   J. E. MARTIN ASSOC. 1-800-462-8235
   DESIGNED LIGHTING CONCEPTS 1-800-248-5483

* IMPORTANT NOTE TO G.C.

IF SUPER LIGHTING IS NOT USED, SERIES II FASCIA LIGHTING IS REQUIRED. CONTACT OWNER FOR ADDITIONAL INFORMATION.

30'-0" (MAXIMUM) HEIGHT DETERMINED BY SITE

GENERAL NOTES

1. ALL SIGNS SHALL BE ERECTED IN ACCORDANCE WITH ALL LOCAL CODES AND SOIL CONDITIONS.

2. DESIGNS ARE TO BE DETERMINED BY SIGN SUPPLIER (VERIFY LOCAL WIND AND SOIL CONDITIONS).

3. ALL PAINTED PAVEMENT MARKERS ARE TO BE SOLID YELLOW AND FURNISHED BY GENERAL CONTRACTOR.

4. WHEN UNABLE TO VIEW CARS PLACING ORDERS DIRECTLY FROM PICK-UP WINDOW A 24" CONVEX MIRROR SHALL BE PLACED IN AN APPROPRIATE LOCATION TO VIEW CUSTOMERS AT ORDER STATION.

LIGHT POLE TO BE FURNISHED AND INSTALLED BY ELECTRICAL CONTRACTOR. POLE AND LUMINAIRE TO BE FURNISHED BY THE SAME MANUFACTURER

POLE HEIGHT - 30'-0"
POLE DIAMETER - 5" (INCHES)
POLE TYPE - STEEL SQUARE
POLE FINISH - BRONZE
BOLT SIZE - 11" (INCHES) BOLT CIRCLE
ANCHOR BOLTS - 1"x36" w/DOUBLE NUT & WASHERS

FOR PROPER INSTALLATION OF SITE LIGHTING ALWAYS USE DOUBLE NUTS AND WASHERS. THIS ASSEMBLY WILL PREVENT MOISTURE BUILD-UP IN THE POLE BY ALLOWING AIR TO CIRCULATE THROUGH THE INSIDE OF POLE. BASES MOUNTED DIRECTLY TO CONCRETE CAN CAUSE A MOISTURE BUILD-UP AND DETERIORATION OF THE POLE'S STABILITY (DO NOT GROUT ASSEMBLY)

ANCHOR BOLT COVER BY LIGHTING SUPPLIER

ALL SIGNAGE TO BE INSTALLED BY THE SIGNAGE ERECTOR

AIR GAP

4" PROJECTION

(4) ANCHOR BOLTS

2'-0"

1/2 "

ANCHOR BOLTS AND PIER INSTALLED BY ELECTRICAL CONTRACTOR

5'-0" (VERIFY LOCAL CODES AND SOIL CONDITIONS)

FOOTER DEPTH TO BE DETERMINED BY SIGNAGE REQUIREMENTS AND LOCAL CODES

SIGNAGE ERECTOR AND GENERAL CONTRACTOR TO COORDINATE THE FOOTER REQUIREMENTS AND THE INSTALLATION FOR ALL EXTERIOR SIGNAGE

NOTE: LIGHTING SUPPLIER TO DETERMINE WIND LOADS AND LOCAL ORDINANCES

WHEN POSSIBLE, LOCATE POLES ON OUTSIDE PERIMETER OF SITE

3"

EXTERIOR POLE LIGHTS
SCALE: 1/2 " = 1"-0"

EXTERIOR SIGNAGE FOOTER
SCALE: 1" = 1"-0"

PROBLEM 23.1 (Continued)

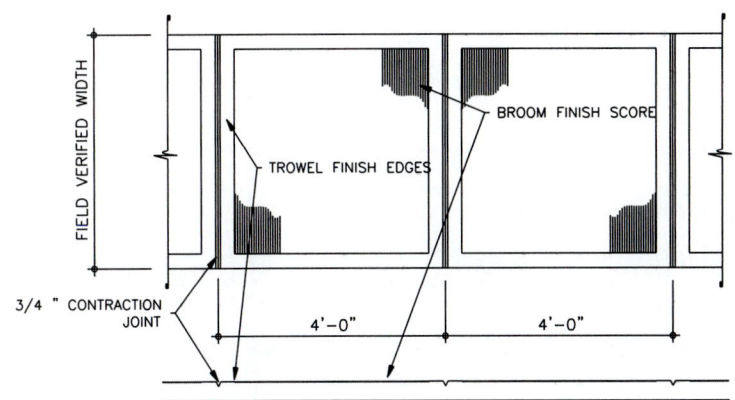

SIDEWALK FINISH DETAIL
NO SCALE

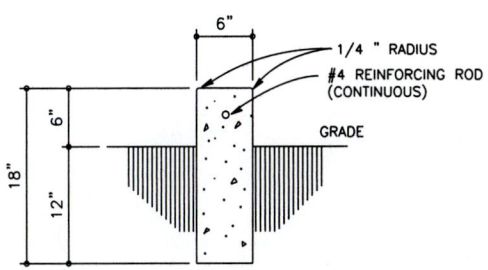

NOTE: ALL CONCRETE CURBS TO HAVE EXPANSION JOINTS
OR SAW CUTS NOT MORE THAN 20'-0" APART

CONCRETE CURB DETAIL
SCALE: 1" = 1"-0"

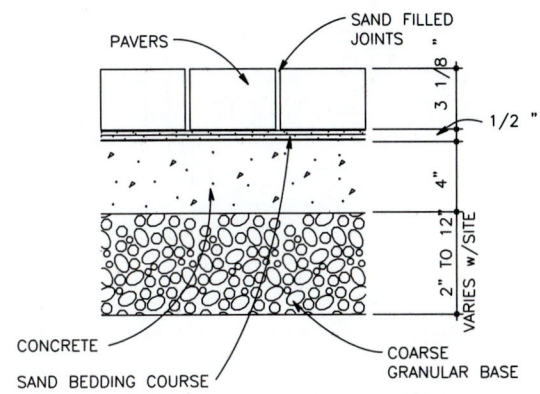

PAVER DETAIL
NO SCALE:

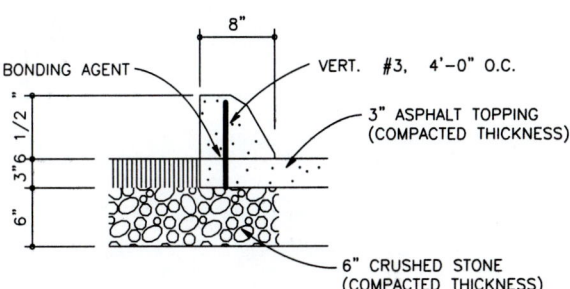

EXTRUDED CURB DETAIL
SCALE: 1" = 1"-0"

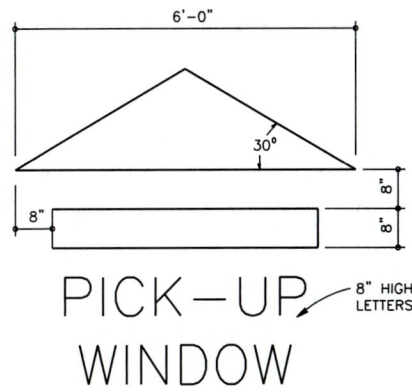

PAVEMENT MARKER DETAILS
SCALE: 1/2 " = 1'-0"

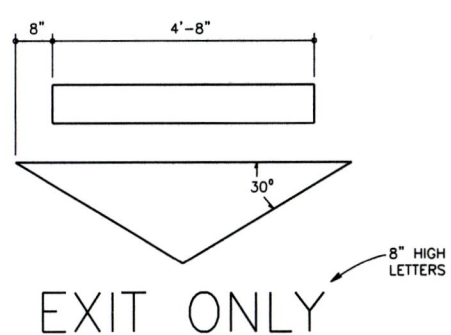

PAVEMENT MARKER DETAILS
SCALE: 1/2 " = 1'-0"

PROBLEM 23.2   Trash enclosure drawings

Given the following layouts, draw the trash enclosure plan, elevation, sections, and details on one sheet, unless otherwise specified by your instructor.

Drawings courtesy of Wendy's International, Inc.

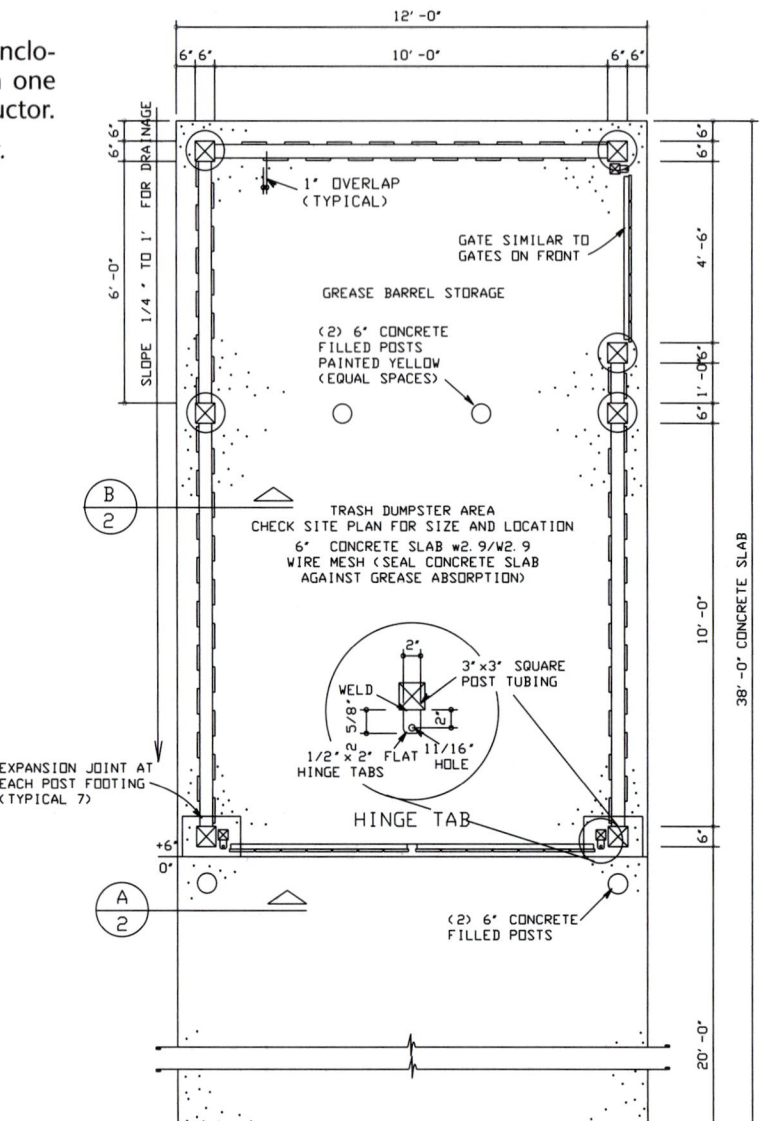

STANDARD TRASH ENCLOSURE PLAN
SCALE: 3/8 " = 1'-0"

HINGE DETAIL

TYPICAL FRONT ELEVATION
SCALE: 3/8 " = 1'-0"

PROBLEM 23.2 (Continued)

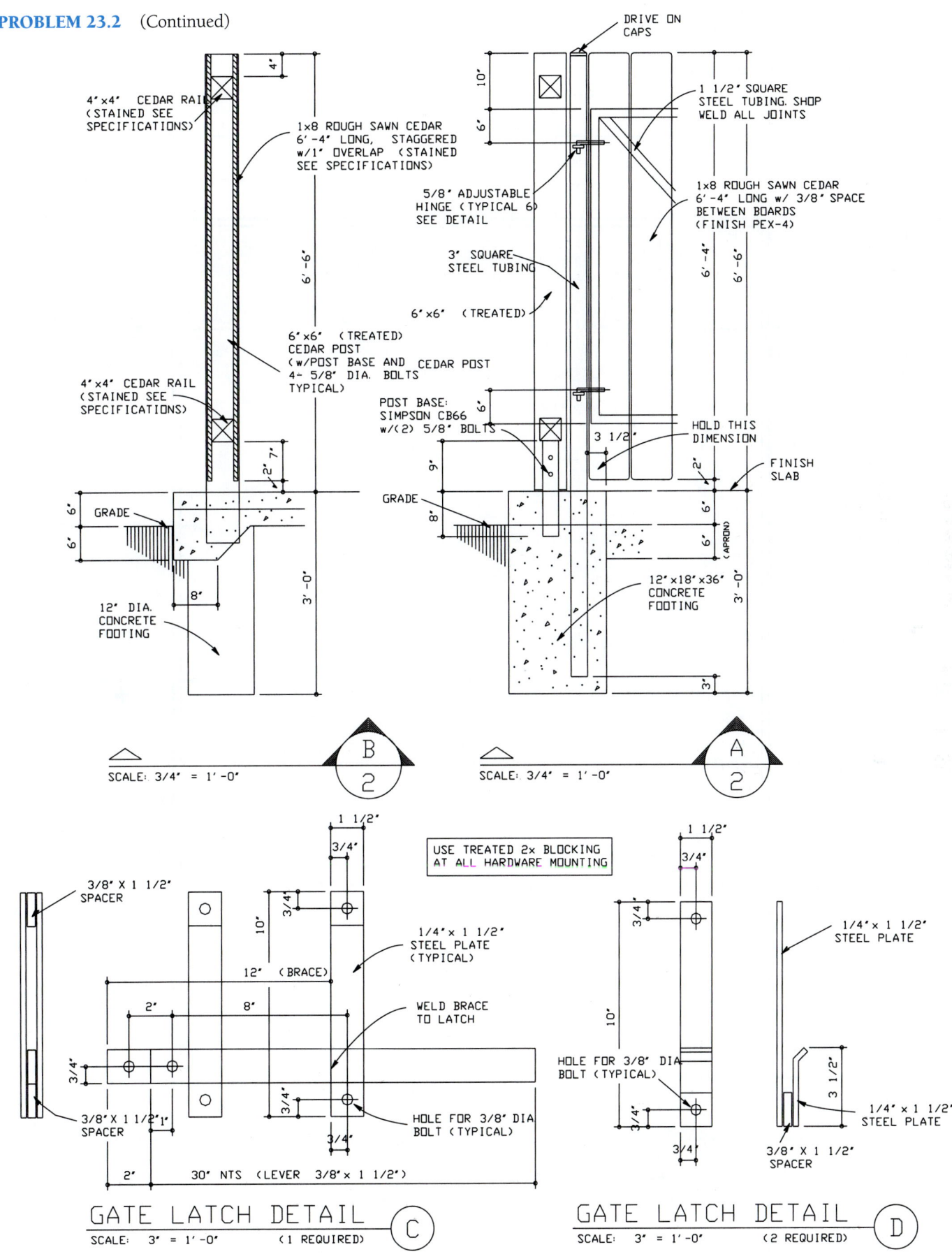

4'x4' CEDAR RAIL (STAINED SEE SPECIFICATIONS)

1x8 ROUGH SAWN CEDAR 6'-4" LONG, STAGGERED w/1" OVERLAP (STAINED SEE SPECIFICATIONS)

5/8" ADJUSTABLE HINGE (TYPICAL 6) SEE DETAIL

3" SQUARE STEEL TUBING

6"x6" (TREATED)

6"x6" (TREATED) CEDAR POST (w/POST BASE AND 4- 5/8" DIA. BOLTS TYPICAL)

CEDAR POST

POST BASE: SIMPSON CB66 w/(2) 5/8" BOLTS

4'x4' CEDAR RAIL (STAINED SEE SPECIFICATIONS)

GRADE

12" DIA. CONCRETE FOOTING

DRIVE ON CAPS

1 1/2" SQUARE STEEL TUBING. SHOP WELD ALL JOINTS

1x8 ROUGH SAWN CEDAR 6'-4" LONG w/ 3/8" SPACE BETWEEN BOARDS (FINISH PEX-4)

HOLD THIS DIMENSION

FINISH SLAB

GRADE

12"x18"x36" CONCRETE FOOTING

B/2   SCALE: 3/4" = 1'-0"

A/2   SCALE: 3/4" = 1'-0"

3/8" X 1 1/2" SPACER

USE TREATED 2x BLOCKING AT ALL HARDWARE MOUNTING

1/4" x 1 1/2" STEEL PLATE (TYPICAL)

WELD BRACE TO LATCH

12" (BRACE)

3/8" X 1 1/2" SPACER

HOLE FOR 3/8" DIA BOLT (TYPICAL)

30" NTS (LEVER 3/8" x 1 1/2")

GATE LATCH DETAIL   C
SCALE: 3" = 1'-0"   (1 REQUIRED)

1/4" x 1 1/2" STEEL PLATE

HOLE FOR 3/8" DIA BOLT (TYPICAL)

1/4" x 1 1/2" STEEL PLATE

3/8" X 1 1/2" SPACER

GATE LATCH DETAIL   D
SCALE: 3" = 1'-0"   (2 REQUIRED)

Problems 23.3 through 23.19 are the drawings for a 2400-square-foot storage building.

PROBLEM 23.3 Storage building floor plan

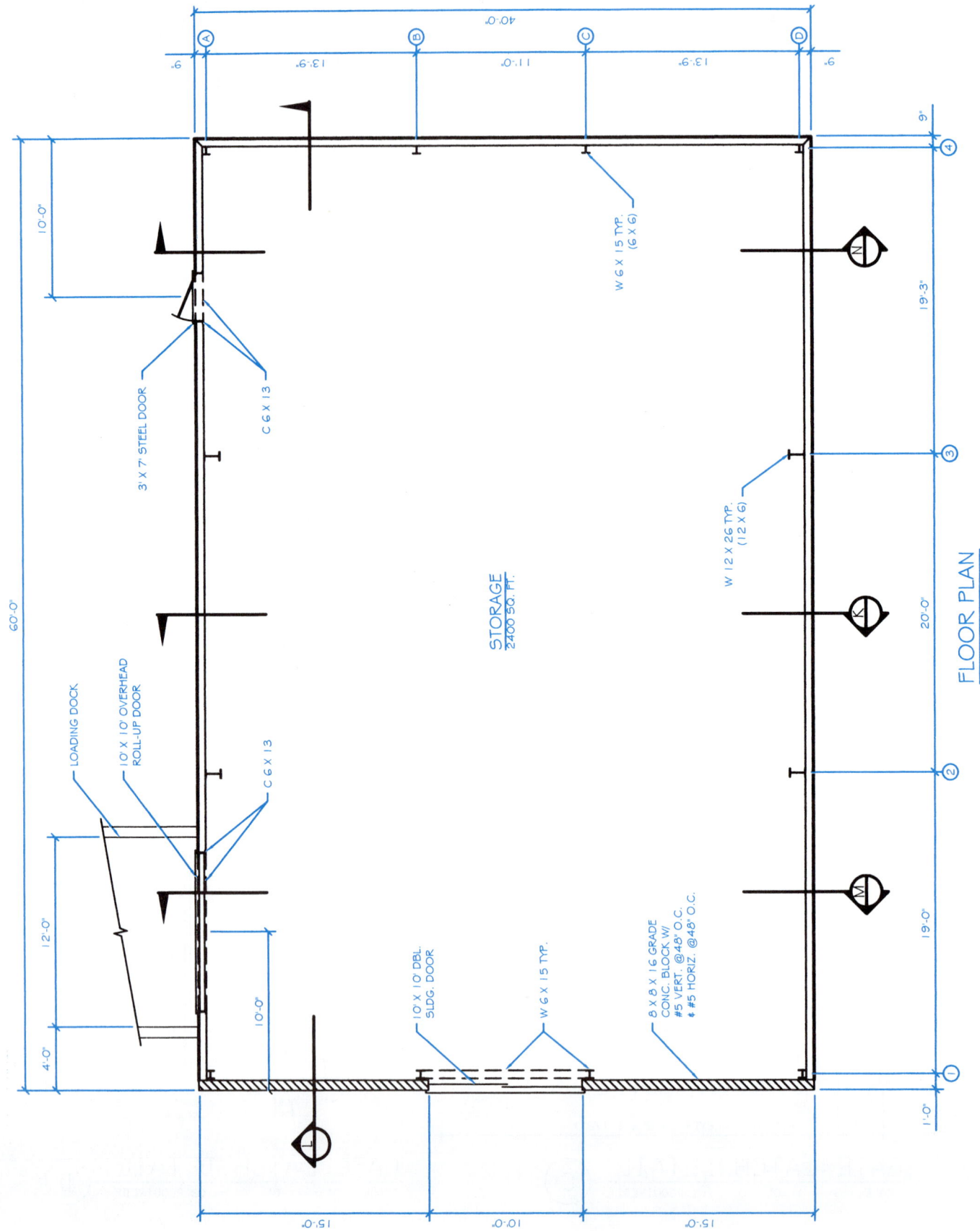

PROBLEM 23.4 Foundation plan

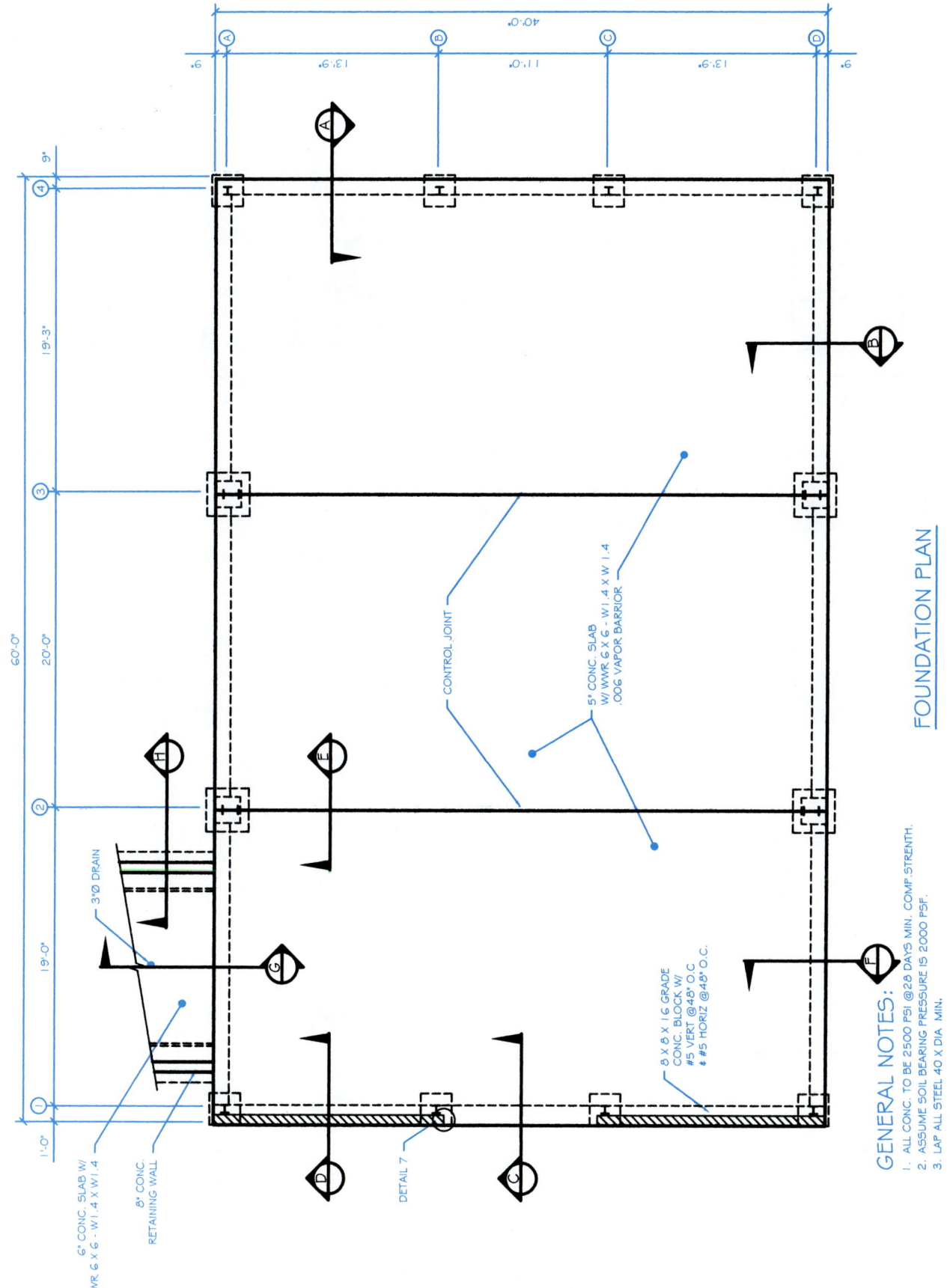

FOUNDATION PLAN

CONTROL JOINT

5" CONC. SLAB
W/ WWR 6 X 6 - W1.4 X W1.4
.006 VAPOR BARRIOR

8 X 8 X 16 GRADE
CONC. BLOCK W/
#5 VERT @48" O.C.
& #5 HORIZ @48" O.C.

3"∅ DRAIN

6" CONC. SLAB W/
WWR 6 X 6 - W1.4 X W1.4

8" CONC.
RETAINING WALL

DETAIL 7

GENERAL NOTES:
1. ALL CONC. TO BE 2500 PSI @28 DAYS MIN. COMP. STRENTH.
2. ASSUME SOIL BEARING PRESSURE IS 2000 PSF.
3. LAP ALL STEEL 40 X DIA MIN.

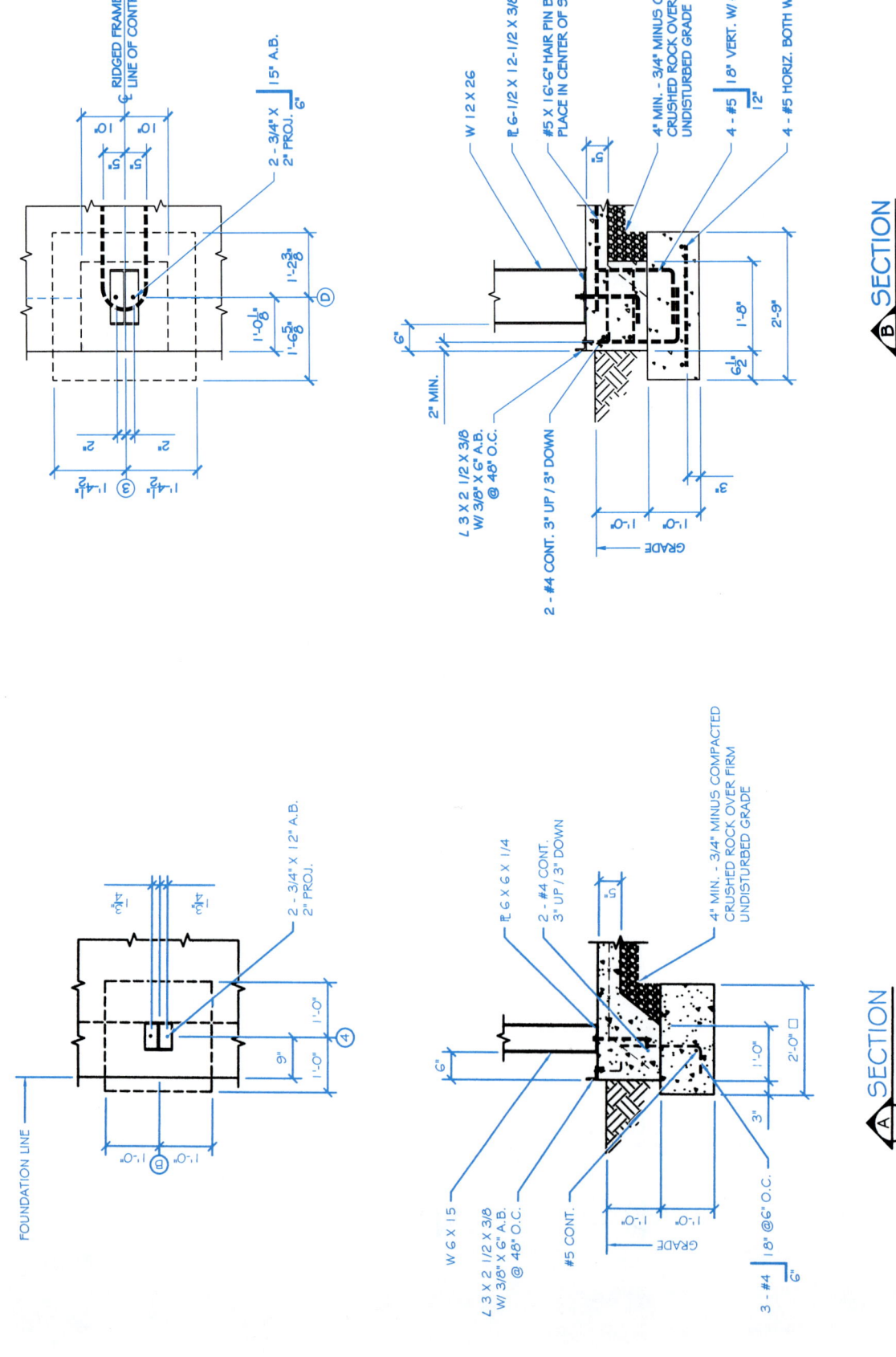

PROBLEM 23.6 Section B, Footing section

PROBLEM 23.5 Section A, Footing section

PROBLEM 23.7 Section C, Footing and concrete block section; Section D, Footing section; Detail 7, Bond beam detail

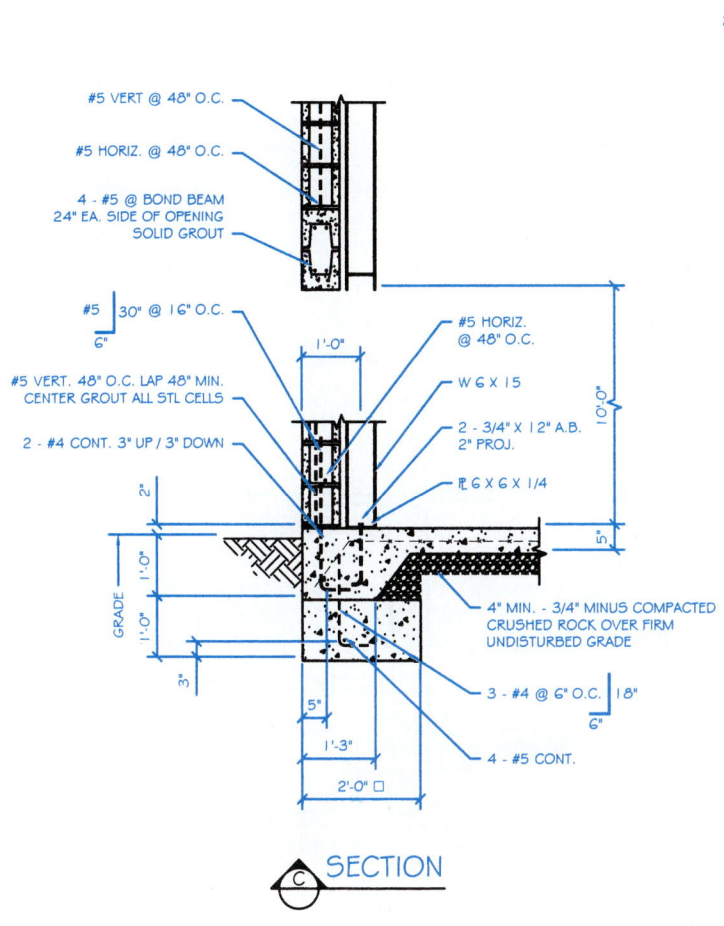

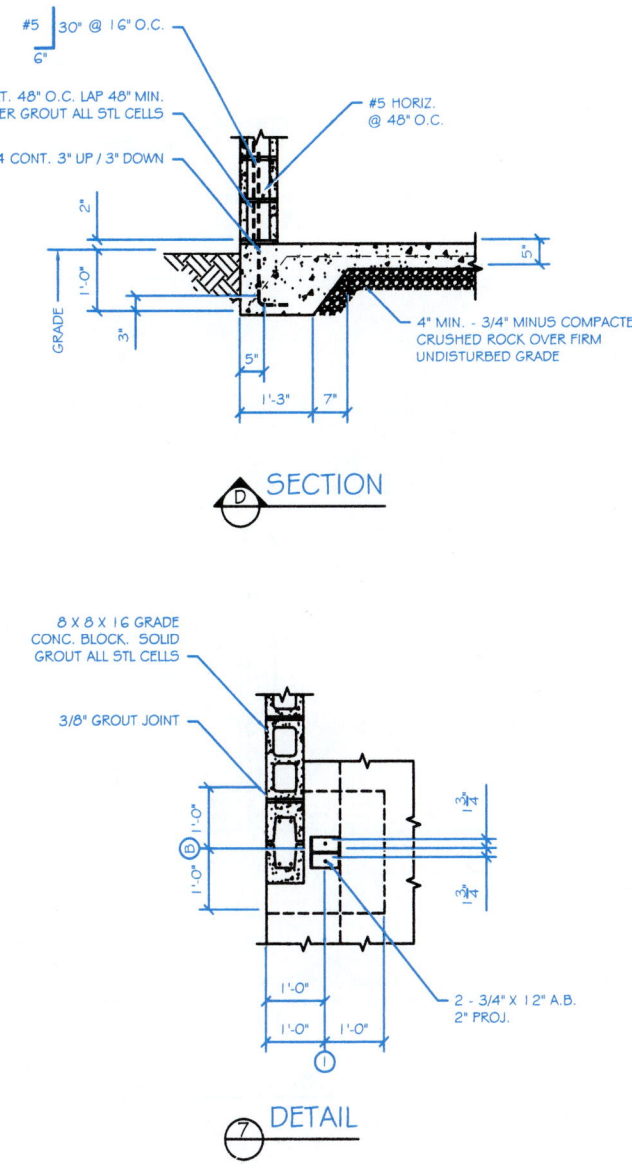

PROBLEM 23.8 Section E, Slab footing section; Section F, Slab joint section

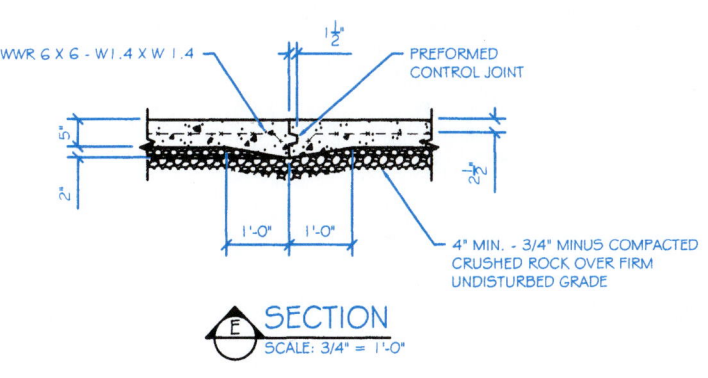

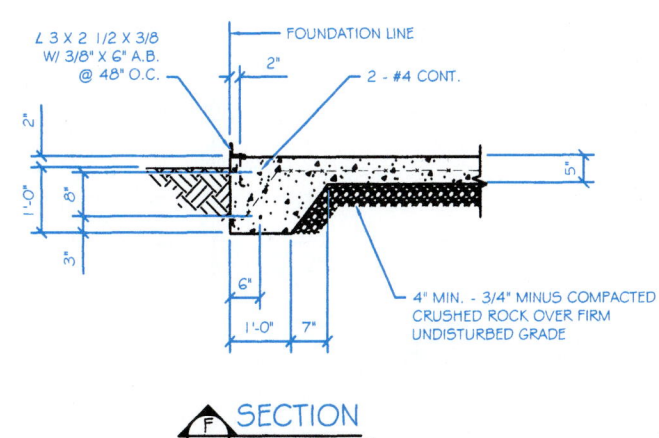

PROBLEM 23.9 Section G, Loading dock ramp section

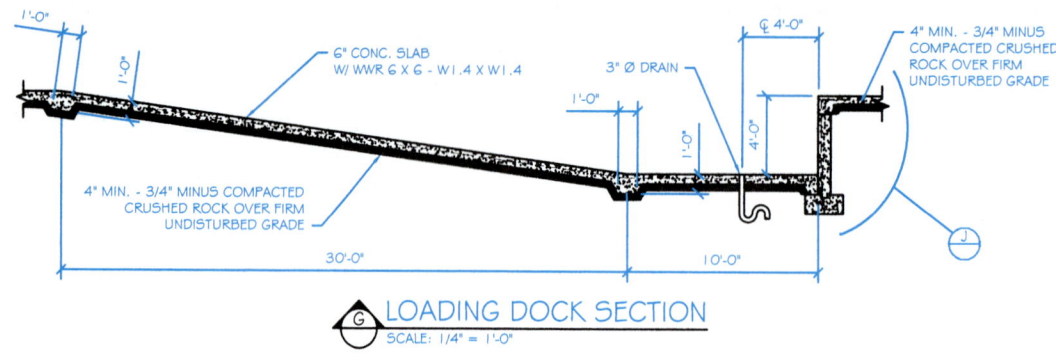

6" CONC. SLAB
W/ WWR 6 X 6 - W1.4 X W1.4

3" Ø DRAIN

₵ 4'-0"

4" MIN. - 3/4" MINUS
COMPACTED CRUSHED
ROCK OVER FIRM
UNDISTURBED GRADE

4" MIN. - 3/4" MINUS COMPACTED
CRUSHED ROCK OVER FIRM
UNDISTURBED GRADE

30'-0"   10'-0"

LOADING DOCK SECTION
G
SCALE: 1/4" = 1'-0"

PROBLEM 23.10 Sections H and J, Loading dock wall sections

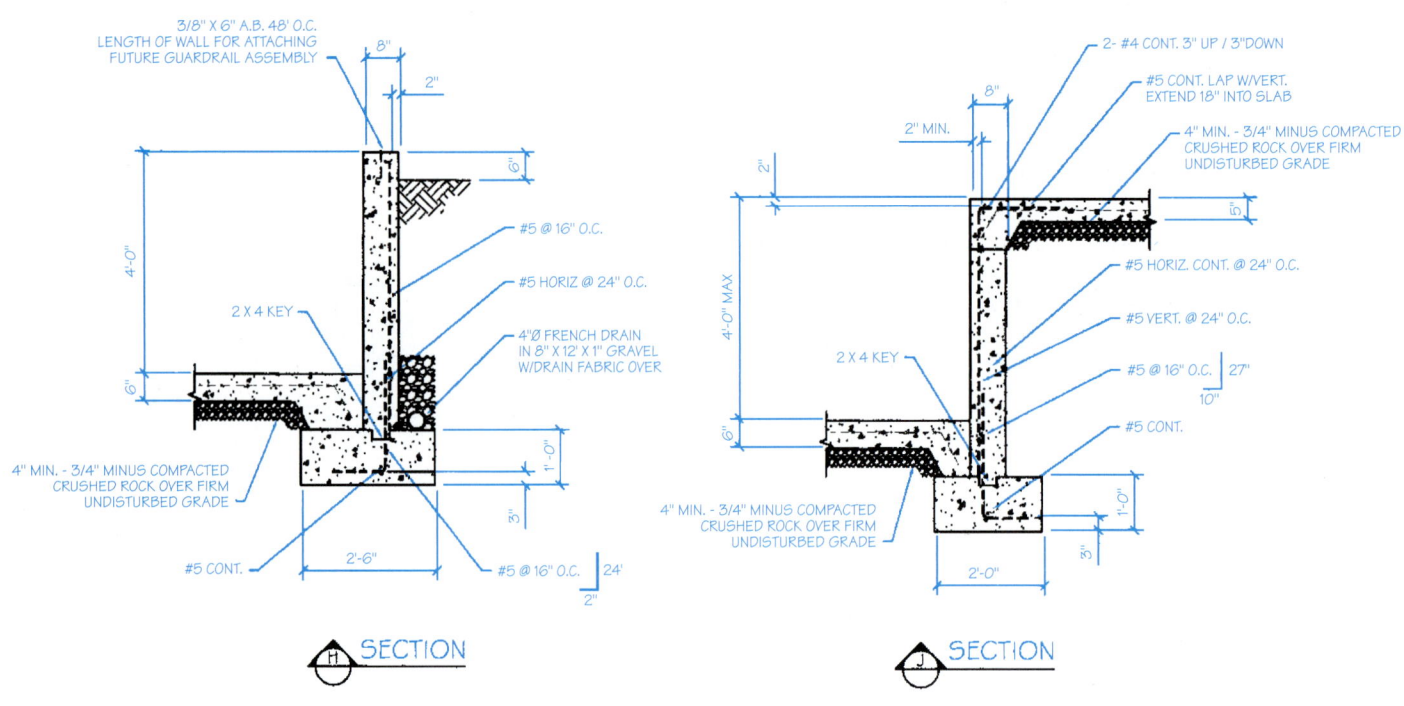

3/8" X 6" A.B. 48' O.C.
LENGTH OF WALL FOR ATTACHING
FUTURE GUARDRAIL ASSEMBLY

#5 @ 16" O.C.

#5 HORIZ @ 24" O.C.

2 X 4 KEY

4"Ø FRENCH DRAIN
IN 8" X 12' X 1" GRAVEL
W/DRAIN FABRIC OVER

4" MIN. - 3/4" MINUS COMPACTED
CRUSHED ROCK OVER FIRM
UNDISTURBED GRADE

#5 CONT.   2'-6"   #5 @ 16" O.C.

SECTION
H

2- #4 CONT. 3" UP / 3"DOWN

#5 CONT. LAP W/VERT.
EXTEND 18" INTO SLAB

4" MIN. - 3/4" MINUS COMPACTED
CRUSHED ROCK OVER FIRM
UNDISTURBED GRADE

#5 HORIZ. CONT. @ 24" O.C.

#5 VERT. @ 24" O.C.

#5 @ 16" O.C.

#5 CONT.

2 X 4 KEY

4" MIN. - 3/4" MINUS COMPACTED
CRUSHED ROCK OVER FIRM
UNDISTURBED GRADE

2'-0"

SECTION
J

PROBLEM 23.11 Roof framing plan

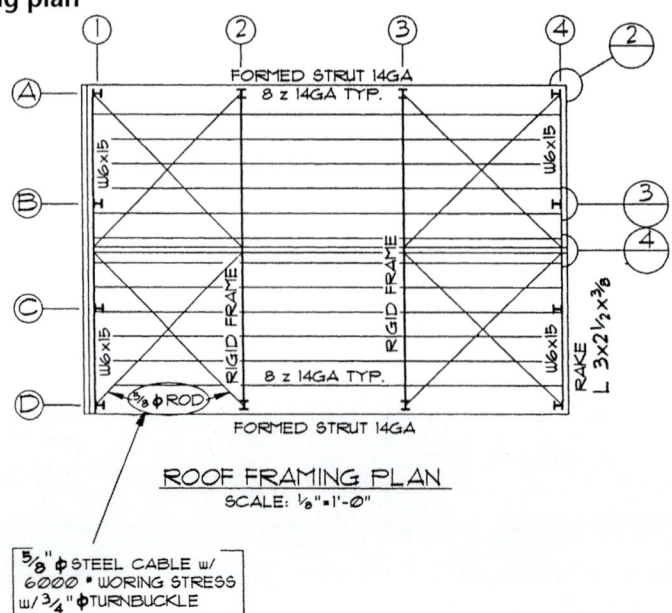

FORMED STRUT 14GA
8 z 14GA TYP.

RIGID FRAME

RIGID FRAME

8 z 14GA TYP.

FORMED STRUT 14GA

⅝ Ø ROD

ROOF FRAMING PLAN
SCALE: ⅛"=1'-0"

⅝" Ø STEEL CABLE w/
6000 # WORING STRESS
w/ ¾" Ø TURNBUCKLE

PROBLEM 23.12 Section K, Typical building section with detail

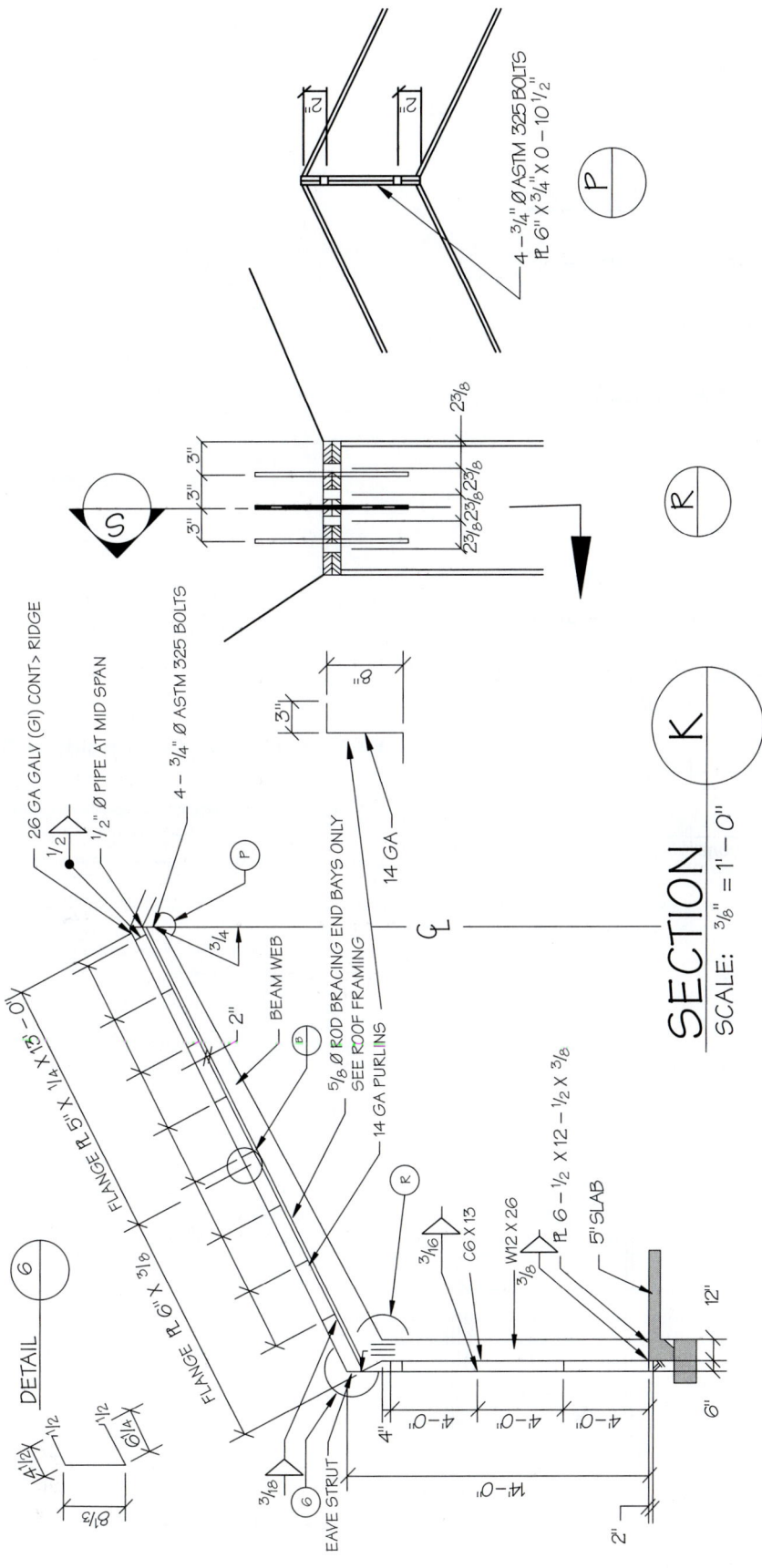

PROBLEM 23.13   Flange detail and connection section

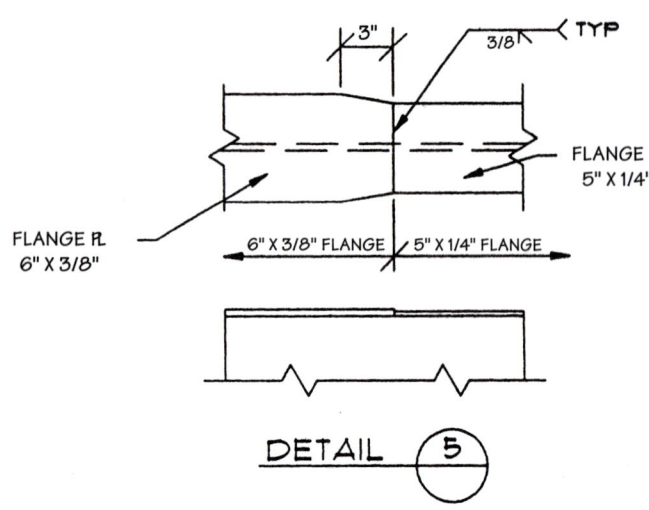

DETAIL ⑤

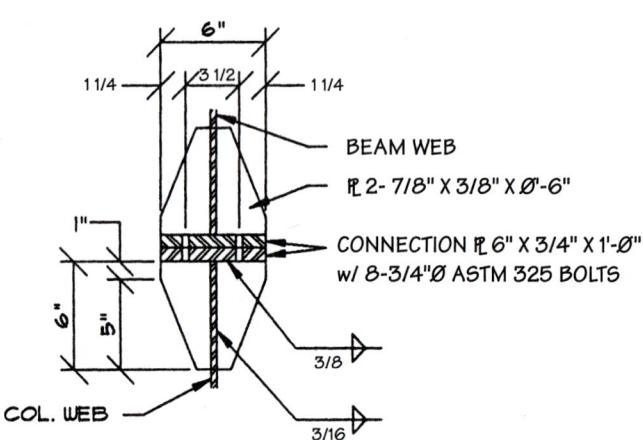

SECTION ⑤

PROBLEM 23.14   End wall section

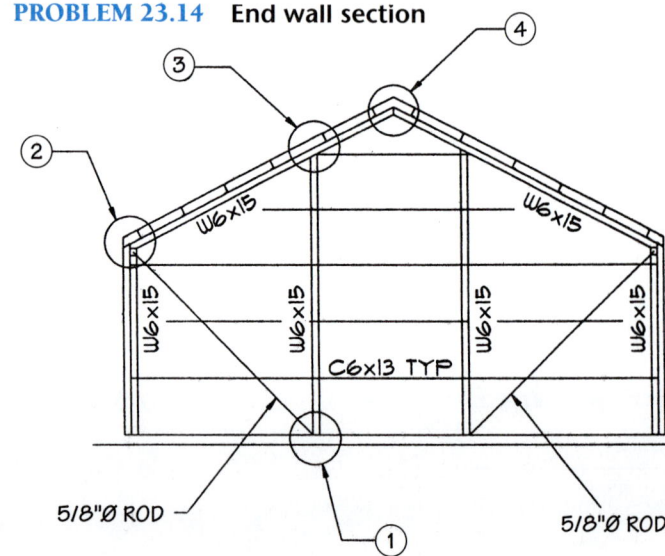

PROBLEM 23.15   Fire wall section

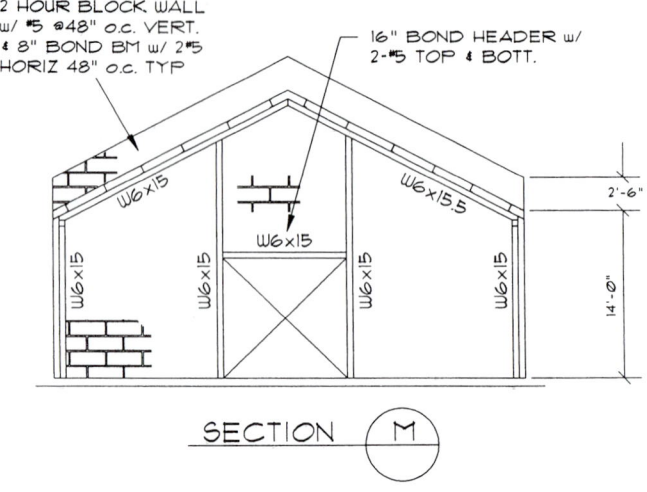

SECTION Ⓜ

NOTE: TYP BLOCK WALL CONN TO BMS &
COLS USE 3/4"Ø ANCHOR BOLTS @ 4'-0" o.c.

PROBLEM 23.16   Section side wall framing

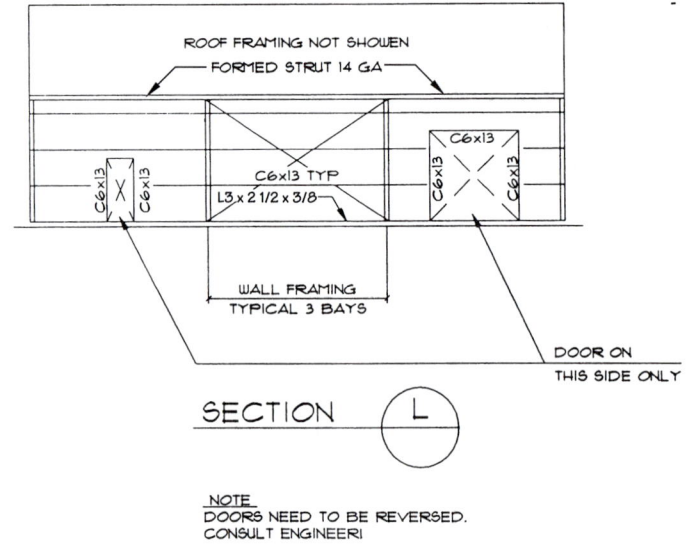

SECTION Ⓛ

NOTE
DOORS NEED TO BE REVERSED.
CONSULT ENGINEER!

PROBLEM 23.17 Column base and cap connection details

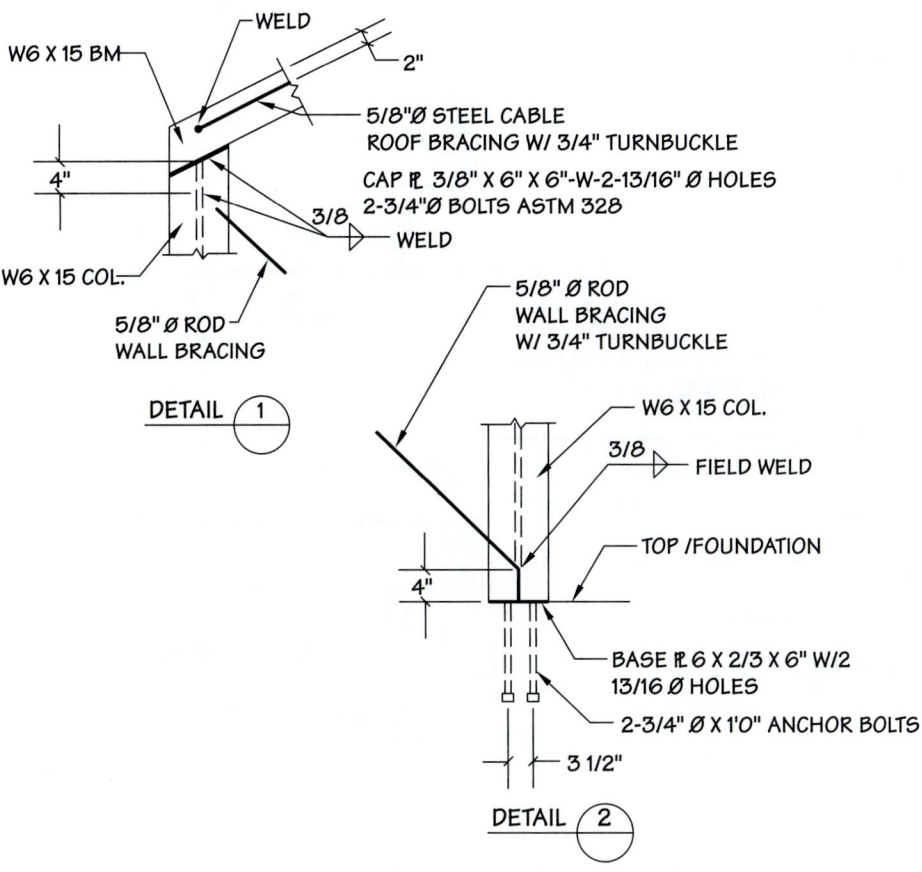

WELD

W6 X 15 BM

2"

5/8"Ø STEEL CABLE
ROOF BRACING W/ 3/4" TURNBUCKLE

CAP ℞ 3/8" X 6" X 6"-W-2-13/16" Ø HOLES
2-3/4"Ø BOLTS ASTM 328

4"

3/8 ▷ WELD

W6 X 15 COL.

5/8" Ø ROD
WALL BRACING

DETAIL ①

5/8" Ø ROD
WALL BRACING
W/ 3/4" TURNBUCKLE

W6 X 15 COL.

3/8 ▷ FIELD WELD

TOP /FOUNDATION

4"

BASE ℞ 6 X 2/3 X 6" W/2
13/16 Ø HOLES

2-3/4" Ø X 1'0" ANCHOR BOLTS

3 1/2"

DETAIL ②

PROBLEM 23.18 **Column cap and ridge connection details**

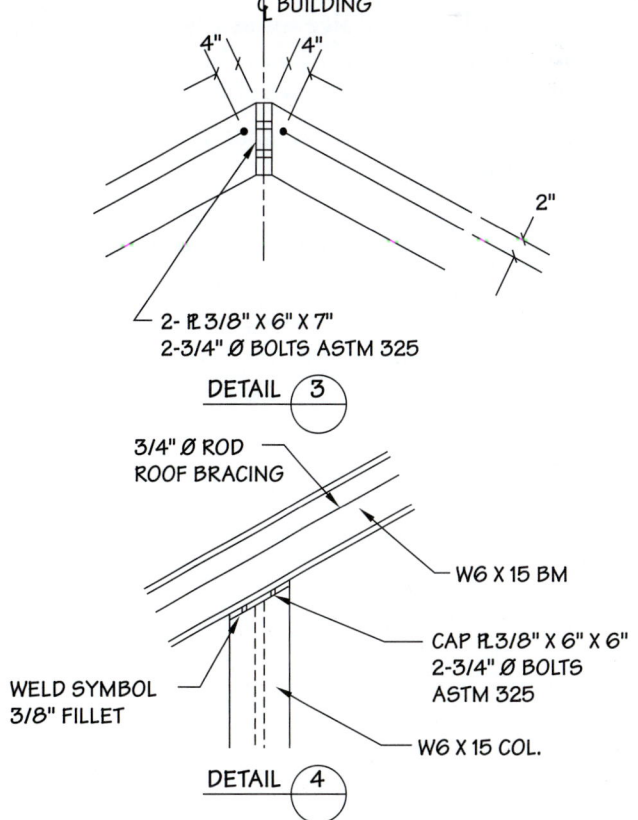

₵ BUILDING

4"   4"

2"

2- ℞ 3/8" X 6" X 7"
2-3/4" Ø BOLTS ASTM 325

DETAIL ③

3/4" Ø ROD
ROOF BRACING

W6 X 15 BM

CAP ℞ 3/8" X 6" X 6"
2-3/4" Ø BOLTS
ASTM 325

WELD SYMBOL
3/8" FILLET

W6 X 15 COL.

DETAIL ④

PROBLEM 23.19 **Elevations**

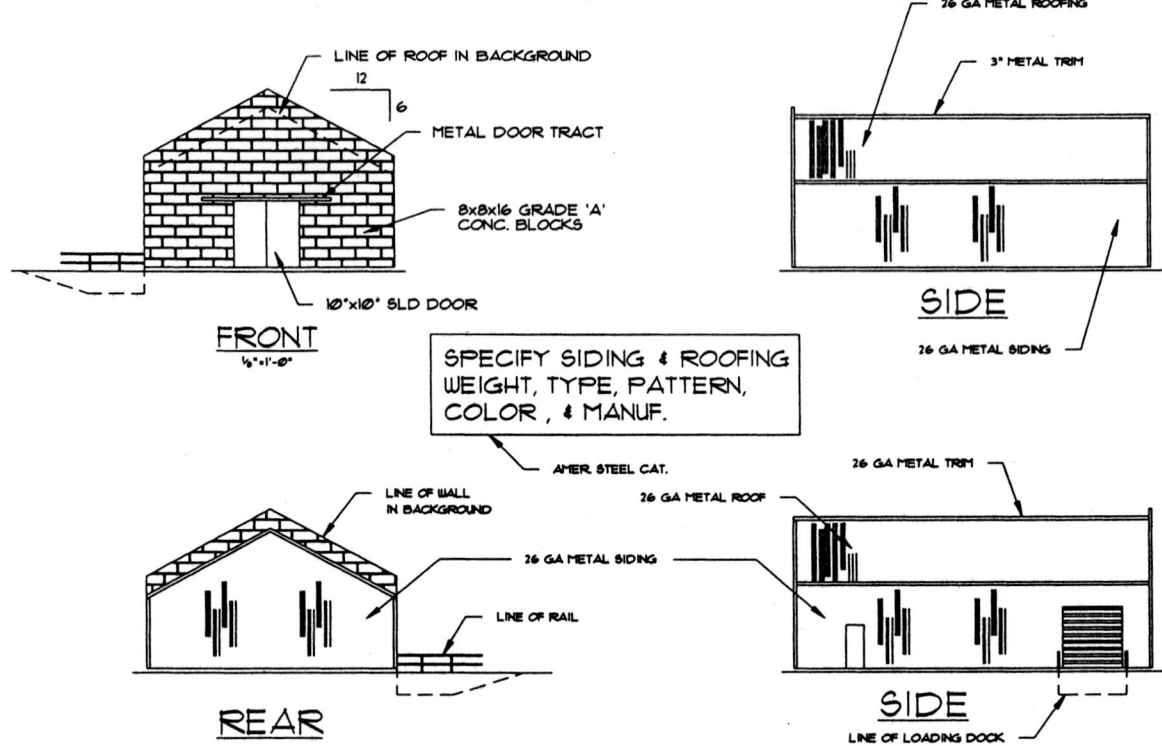

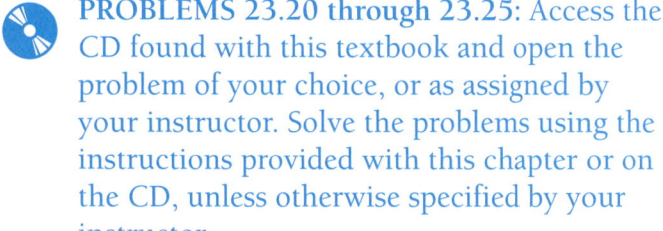

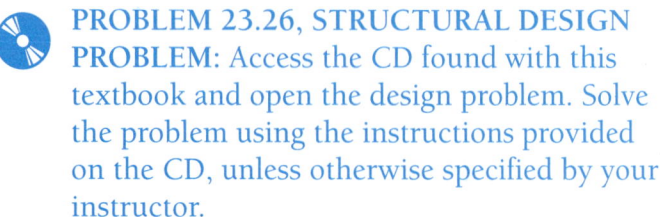

![disc icon] **PROBLEMS 23.20 through 23.25:** Access the CD found with this textbook and open the problem of your choice, or as assigned by your instructor. Solve the problems using the instructions provided with this chapter or on the CD, unless otherwise specified by your instructor.

![disc icon] **PROBLEM 23.26, STRUCTURAL DESIGN PROBLEM:** Access the CD found with this textbook and open the design problem. Solve the problem using the instructions provided on the CD, unless otherwise specified by your instructor.

MATH PROBLEMS

![disc icon] **PROBLEMS 23.27 through 23.34:** Access the CD found with this textbook and open the math problem of your choice, or as assigned by your instructor. Solve the problem or problems using the instructions provided.

CHAPTER 24

Civil Drafting

LEARNING OBJECTIVES

After completing this chapter, you will:

- Calculate unknown values of bearings, slopes, and curve dimensions.

- Draw transit lines for roadway layouts.
- Reduce survey information from field notes.
- Prepare site plans and profiles for cut and fill work.
- Draw plans and profiles from survey information.

THE ENGINEERING DESIGN APPLICATION

A new roadway, Valerian Lane, is going to be built from Erika Street to Quincey Avenue. (See Figure 24.1.) The information in the table in Figure 24.2 is given to you from the designer and survey department.

How would you go about executing this project? First, finish the required mathematical calculations for the chart, then lay out Valerian Lane. To complete the unknown portions of the chart and draw the new roadway, you must understand the concepts of survey and civil drafting. This represents typical work done by a civil drafter and is covered in this chapter.

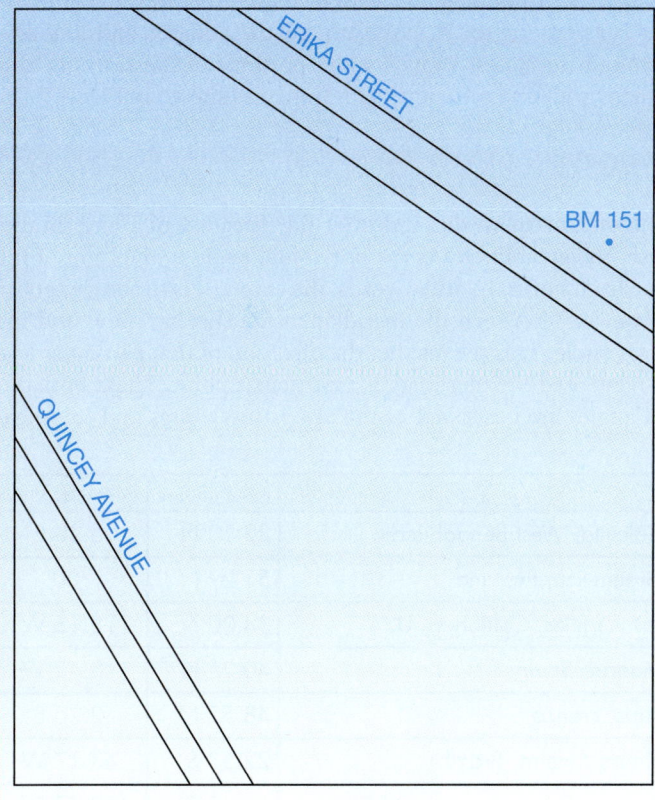

FIGURE 24.1 ■ Example of a roadway layout.

STATION	TRANSIT LINE FOR VALERIAN LANE FROM ERIKA STREET (BEARING N56°16'W) TO QUINCEY AVENUE (BEARING N31°24'W)
0 + 00 Begin Project	245.78'N67°37'W from BM 151 to centerline of Erika Street. Then 132.00'33°44'W to P.C.
STA 1 + 32	P.C. to the left Curve Data R = 200' Δ = 47°23' L=? D=?
STA ?	P.R.C. Curve Data R= 170' Δ = 72°15' L=? D=?
STA ?	P.T.
STA ? End Project	S58°36'W from STA ? to centerline of Quincey Avenue

FIGURE 24.2 ■ Survey information.

INTRODUCTION TO SURVEY: DIRECTION

Think of the earth as having imaginary lines that partition it into sections for the purpose of exactly pinpointing the area to be described. These lines divide the earth horizontally and vertically into lines of latitude (sometimes referred to as parallels) and lines of longitude (sometimes referred to as meridians). These two sets of grid lines are called the graticule of the earth. Parallels and meridians are shown in Figure 24.3.

Lines of latitude do not intersect. They are approximately the same distance apart (about 69 statute miles). The longest line of latitude is around the earth at the equator and the shortest lines are around the North and South Poles. These parallels are numbered from 0 degrees at the equator to 90 degrees at each pole as shown in Figure 24.4. When describing a particular point on the earth, a numbered latitude coordinate is given with N or S to describe north or south of the equator.

Lines of longitude are all about the same length and meet at both Poles. They divide the earth into 360 degrees. Each line of longitude is numbered as shown in Figure 24.5. The zero line of

longitude is the prime meridian and the other lines, including those between the ones shown here, are local meridians. The most generally accepted prime meridian runs through Greenwich, England, although some countries in a spirit of nationalism have defined the prime meridian (0 degrees longitude) as going through their major city. Note that while the length of the lines of longitude remains constant, the area between them decreases the closer they get to the Poles.

Each latitude degree and longitude degree is divided even further into 60 equal sections called minutes, and each minute is divided into 60 equal sections called seconds. The symbol for minute is ('), and the symbol for second is ("). Therefore, a latitude of 29 degrees, 15 minutes, 47 seconds to the north of the equator would be shown as 29° 15'47"N. A longitude of 124 degrees, 12 minutes, 53 seconds to the east of the prime meridian (at Greenwich, England) is shown as 124°12'53"E. Figure 24.6 shows some cities and their latitude and longitude coordinates.

Direction

The direction around the earth from a given point in relation to the latitude and longitude coordinates describes a great circle. There are an infinite number of great circles. The great circle in Figure 24.7 has a starting point in northern Africa. From this one point, which has coordinates of its own, describe the direction the great circle goes. To do this, use the true azimuth or bearing.

In order to understand both of these processes, imagine the earth as having an X axis through the equator and a Y axis through the Poles. Draw smaller portions of the earth on this square grid or plane coordinate system as shown in Figure 24.8.

True Azimuth

The measurement that indicates the direction of a line (in this case, the great circle) is the horizontal angle to the right of the line from north. In other words, the azimuth is the angle (going clockwise) between the meridian at the starting point and the great circle. This means that the direction of that particular line is relative to the meridian at a particular point. These horizontal angles are described as the azimuth readings and can mea-

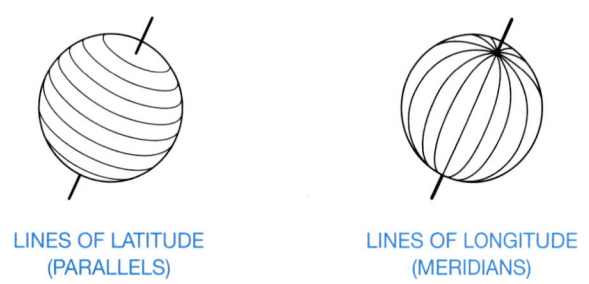

LINES OF LATITUDE
(PARALLELS)

LINES OF LONGITUDE
(MERIDIANS)

FIGURE 24.3 ■ Latitude and longitude.

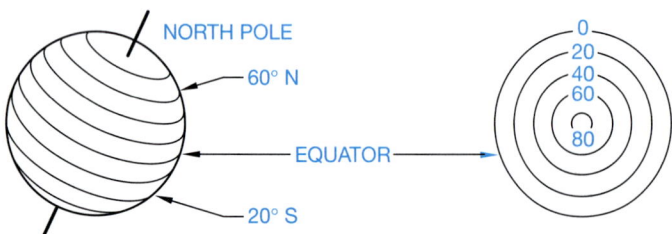

FIGURE 24.4 ■ Latitude coordinates.

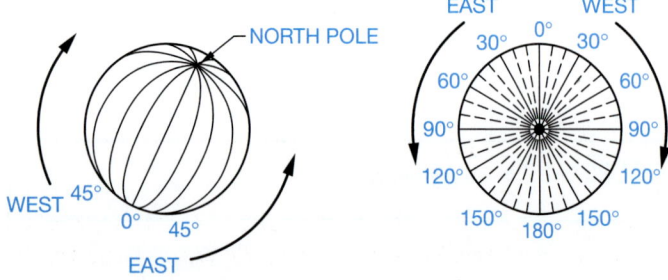

FIGURE 24.5 ■ Longitude coordinates.

City	Latitude	Longitude
Calcutta, West Bengal, India	22.30 N	88.20 E
Greenwich, England	51.29 N	00
Los Angeles, California, USA	34.00 N	118.15 W
Madrid, Spain	40.25 N	3.43 W
Paris, France	48.52 N	2.20 E
Rio de Janeiro, Brazil	22.53 S	43.17 W
Sydney, Queensland, Australia	16.40 S	139.45 E
Tokyo, Japan	35.40 N	139.45 E
Vancouver, British Columbia, Canada	49.13 N	123.06 W

FIGURE 24.6 ■ Cities and their coordinates.

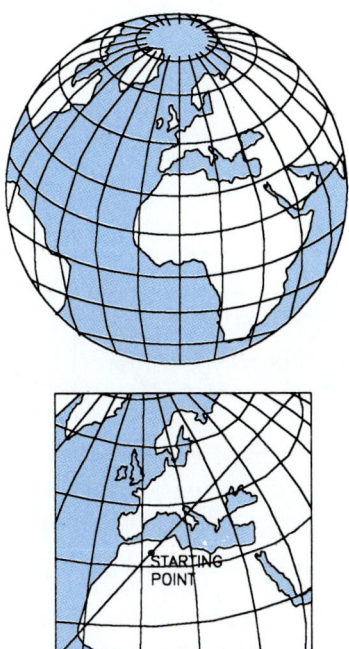

FIGURE 24.7 ■ Great circle.

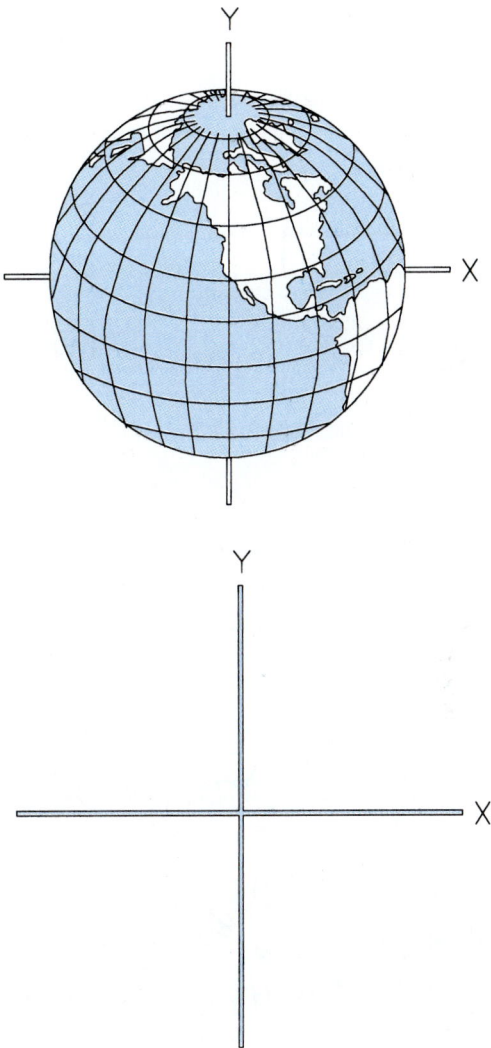

FIGURE 24.8 ■ Coordinates.

sure from 0° to 360°. In Figure 24.9, the azimuth of line AB is 35° and the azimuth of line AC is 110°. All azimuth readings are from a meridian, yet depending on which meridian system is used, a different type azimuth is described:

True azimuth is measured from a true meridian.

Magnetic azimuth is measured from a magnetic meridian.

Grid azimuth is measured from a central meridian on a grid system.

Assumed azimuth is measured from an arbitrary meridian.

Bearings

The **bearing** of a line is basically the same as the azimuth, except it only has values up to 90°, while the azimuth goes through all four quadrants of the grid up to 360°. A bearing is also described by the north or south meridian and by the east or west parallel. For example, a line in the northwest quadrant (using the X and Y coordinates) that has a horizontal angle of 45° counterclockwise from the north meridian would be described as having an azimuth reading of 315°, while its bearing would be N45° W (360° − 45° = 315°). A line 75° to the east from the south meridian has an azimuth of 150° and a bearing of S75°E. Note that a bearing of S75°E has the same angle as a bearing of N75°W. (See Figure 24.10.)

Magnetic Azimuth

The bearings explained so far have been gauged from a true azimuth, or from true north or south. The magnetic north and south are not at the actual North and South Poles where the meridians meet, and where the earth spins on its axis. Figure 24.11 shows the position of the true poles and the magnetic poles. The needle

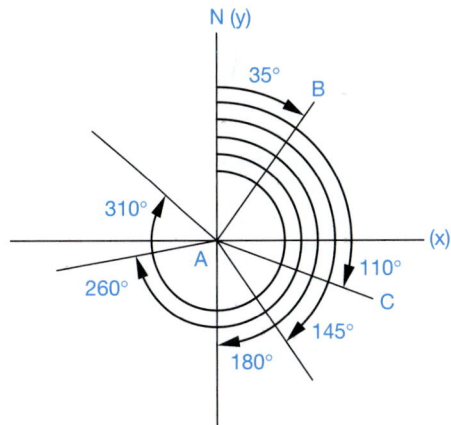

FIGURE 24.9 ■ Azimuth angles.

of the compass points to these magnetic poles. The degree of difference on the compass between magnetic azimuth and true azimuth is called **magnetic declination**. To adjust for this difference, isogonic lines have been established that show how

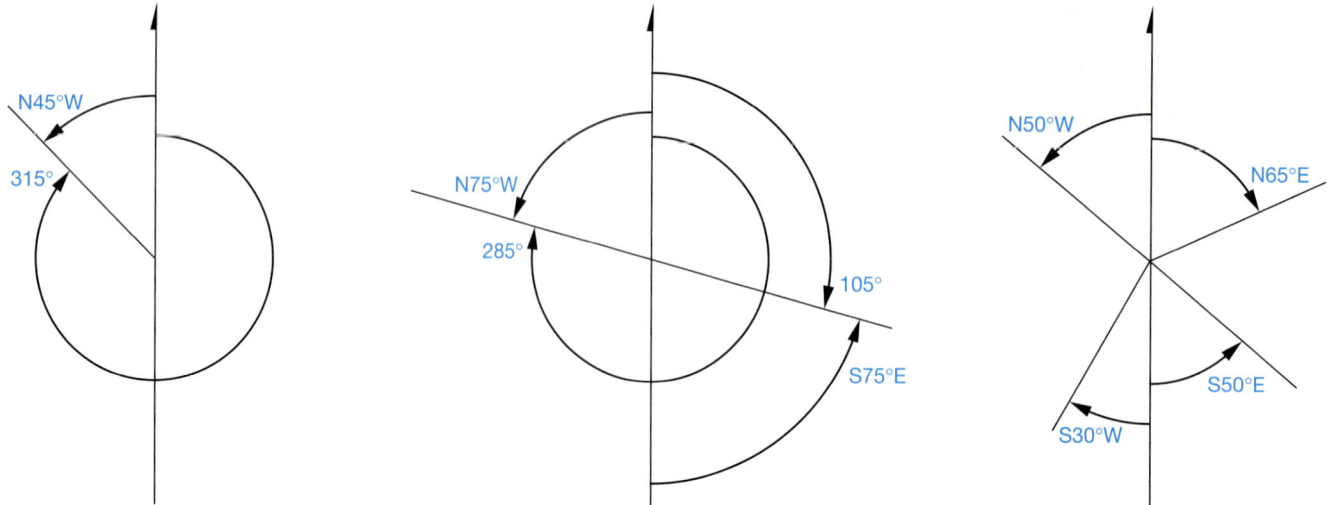

FIGURE 24.10 ■ Bearing examples.

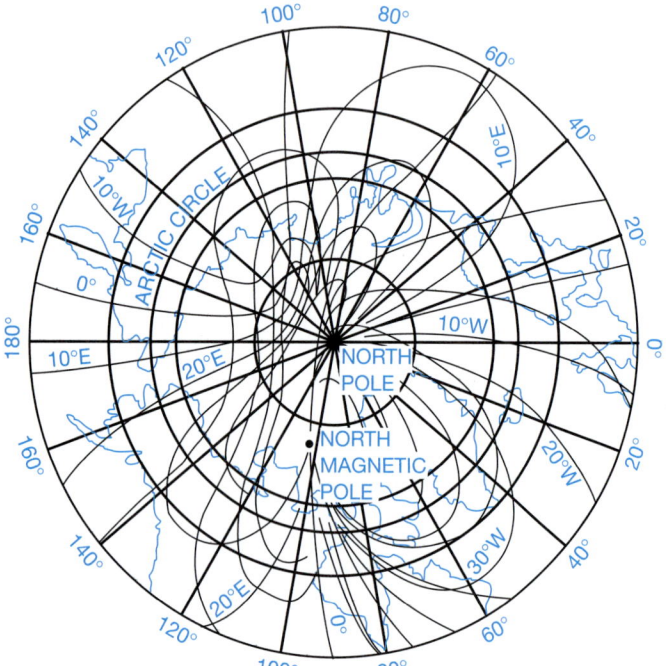

FIGURE 24.11 ■ Azimuthal equidistant projection.

many degrees to the east or west the magnetic pole is from the actual pole. The isogonic lines must be reestablished periodically because the position of the magnetic pole is constantly changing. Therefore, a map showing an arrow for a magnetic pole also gives the year in which the magnetic declination was calculated.

An example of an isogonic chart showing isogonic lines or magnetic declination for the United States is shown in Figure 24.12. This chart, which is based on the U.S. Geological Survey of 1983, shows Chicago, Illinois, as being on the line of zero declination, also known as the agonic line. This means that the magnetic North Pole is directly in line with the actual North Pole and the compass reading is the actual North Pole reading, or true azimuth. Los Angeles, California, is approximately 14.3°E of zero declination. This means that when the compass reads 0° it is pointing toward magnetic north, which is 14.3° east of true north. Therefore, the north arrow on any map around Los Angeles appears as in Figure 24.13.

Grid Azimuth

The grid azimuth is measured from the central meridian in a rectangular grid system. This will be discussed later. It is common in maps to find three north arrows to define true azimuth, magnetic azimuth, and grid azimuth, as shown in Figure 24.14.

SURVEYING

Civil engineering drafting is involved in displaying a portion of the earth. Now that you know about positioning and direction, you can proceed to an understanding of surveying.

Traverses

A traverse is a series of lines that each have a known length and are connected by known angles. Each traverse line is a course, and each point where courses intersect is a traverse station or station point. When traversing, you start with a point of beginning (POB) and proceeds to utilize one of several traverse types.

Open Traverse

An open traverse does not close upon itself, meaning that the courses do not return to the POB. It is used for route surveys when mapping linear features such as highways or power lines. An example is shown in Figure 24.15.

Closed Traverse

In a closed traverse (a) the courses return to the POB, or (b) the courses close on a point different from the POB, but a point that has a known position, called a control point. Figure 24.16 shows both a loop traverse and a connecting traverse. The connecting traverse has known positions at both the POB and at the

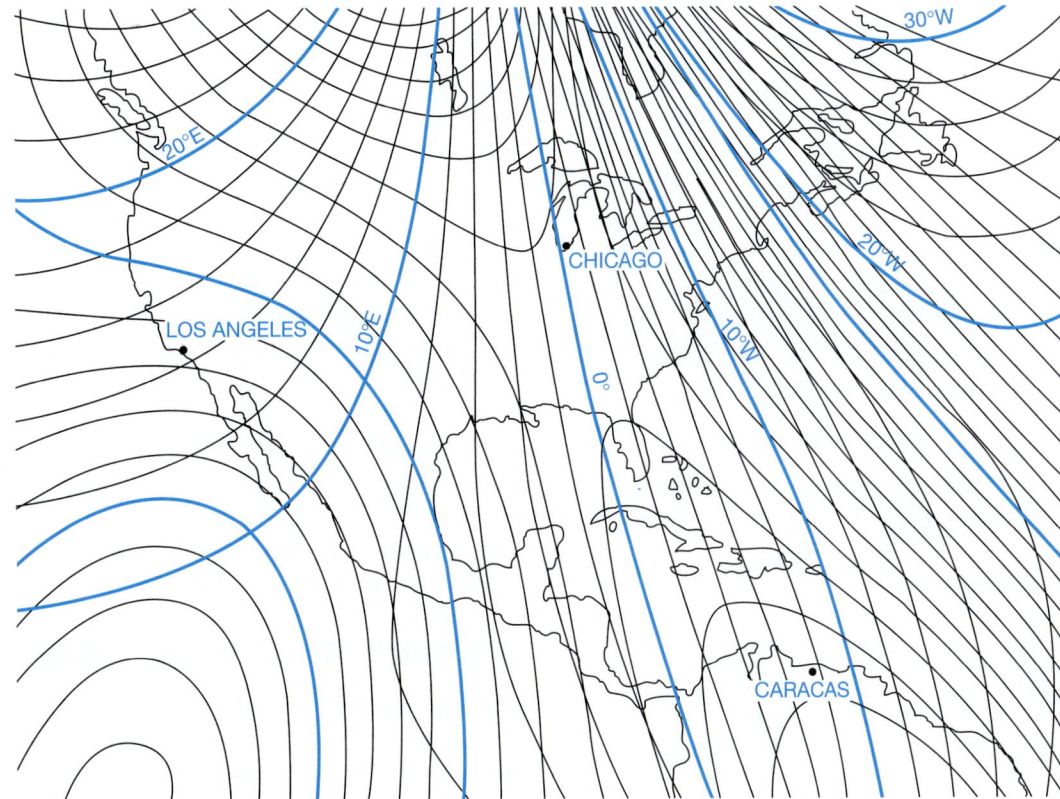

FIGURE 24.12 ■ Isogonic chart.

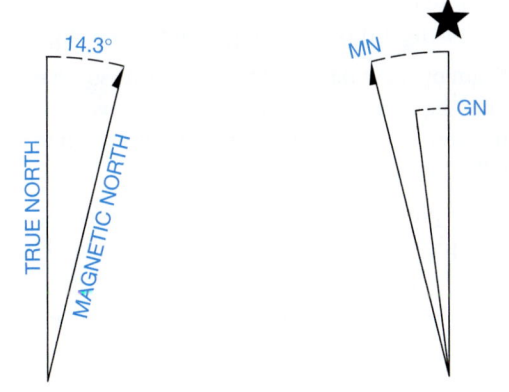

FIGURE 24.13 ■ Approximate mean declination, 1983.

FIGURE 24.14 ■ Universal Traverse Mercator (UTM) grid and 1983 magnetic north declination at center of sheet.

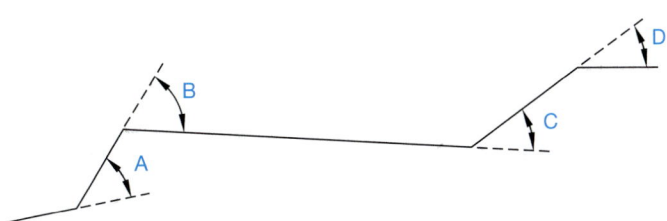

FIGURE 24.15 ■ Open traverse.

end of the survey. They are similar in that they each can be checked for accuracy. These traverses are the ones used exclusively for land and construction surveys.

Direct-Angle Traverse

A direct-angle traverse is used primarily with a loop traverse. The interior angles are measured as the traverse proceeds either clockwise or counterclockwise.

Compass-Bearing Traverse

In a compass-bearing traverse, the bearings of the courses are read directly from the compass. (See Figure 24.17.) When using a compass, you must be aware of the magnetic declination at that particular point.

Deflection-Angle Traverse

In a deflection-angle traverse, each course veers to the right or left from each station measured. Figure 24.18 shows that an *R* or *L* (right or left) must always be shown with the angle. This method is used most often when doing route surveys for features such as highways and railroads.

Azimuth Traverse

An azimuth traverse uses the azimuth angle to show the direction of the next course. The azimuth reference line is usually the north-south line and can either be the magnetic azimuth or the

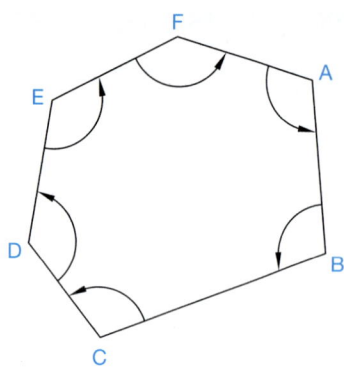

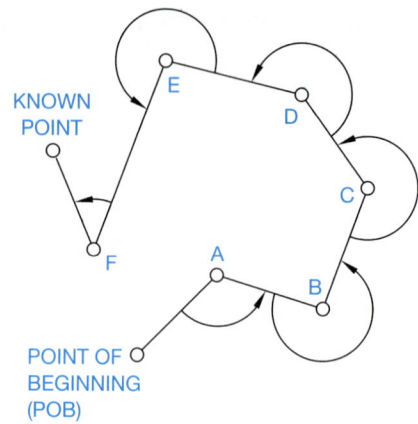

FIGURE 24.16 ■ Closed traverses.

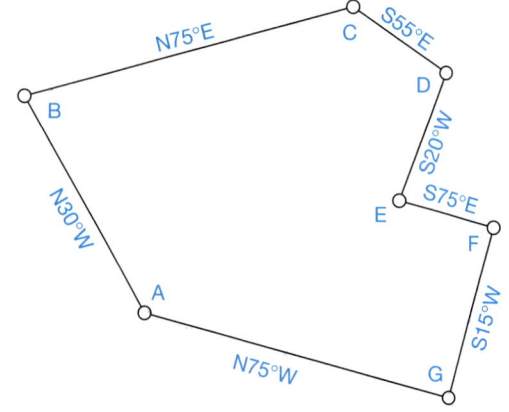

FIGURE 24.17 ■ Compass-bearing traverse.

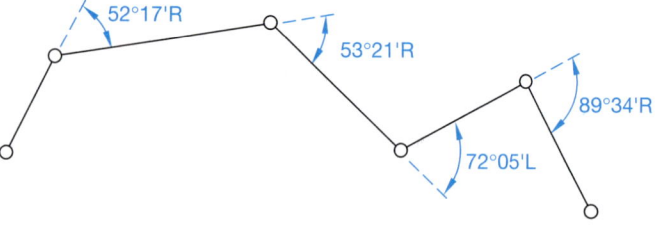

FIGURE 24.18 ■ Deflection-angle traverse.

true azimuth. Remember that the azimuth is always measured clockwise from north. (See Figure 24.19.)

PLOTTING TRAVERSES

The term plotting, as used here, refers to the layout or drafting of a traverse by establishing the endpoints of bearings enclosing the traverse. Plotting used in this sense is not to be confused with plotting a drawing, such as making a hard copy, when using a CADD system. The term course, as used in this discussion, refers to each element or line of the traverse. As with other types of drafting, you should do construction work with a light blue pencil or very lightly drawn construction lines. Follow these steps when plotting a traverse using manual drafting:

STEP 1 Begin by drawing a light vertical north-south line on the drawing surface. Mark the POB wherever you want on the vertical line, then draw another light line horizontally through the POB.

STEP 2 When plotting a compass-bearing traverse, the next step is to place the center of a protractor on the POB. If the bearing of the course from POB is, for example, N25°E, then make a mark on the drawing surface at the point shown in Figure 24.20.

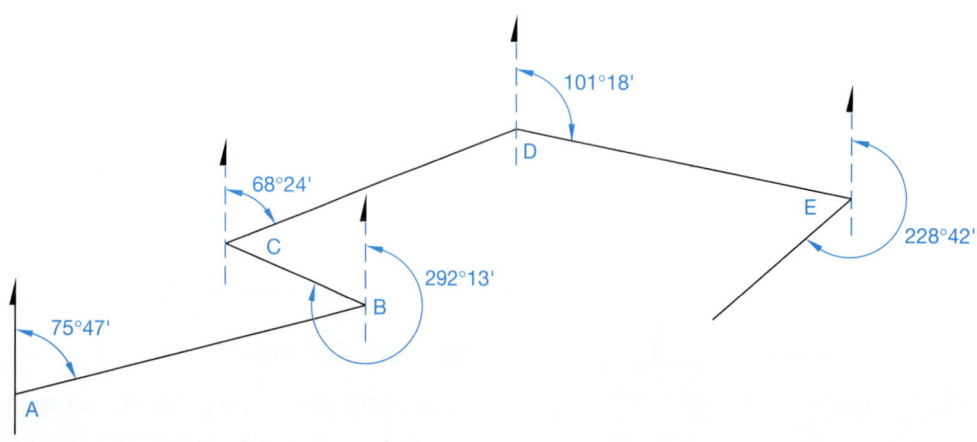

FIGURE 24.19 ■ Azimuth traverse.

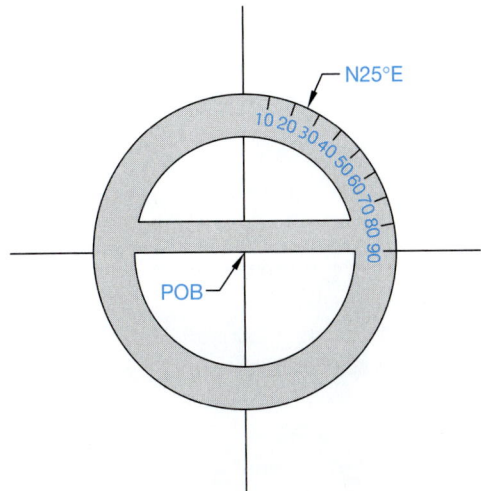

FIGURE 24.20 ■ Using a protractor.

STEP 3 Next, remove the protractor and draw a line through the POB to the N25°E mark. This is the bearing line, which is one element of the course. The other necessary component of the course is the distance. If that distance is 314 feet, then measure from the POB 314 feet along the bearing line using a scale that has been previously determined (for example, 1" = 100'). Make a mark at that point, as shown in Figure 24.21.

STEP 4 Draw another north-south line through that point and plot the next compass-bearing traverse.

STEP 5 Continue this process until you have constructed each line of the traverse.

STEP 6 Draw all final traverse lines as thick object lines, unless otherwise specified by your company or school standards.

There is a major point to keep in mind when plotting traverses. It was mentioned earlier that a bearing of N25°E and a bearing of S25°W have the identical angle. However, when actually plotting the distance on the course, you must plot in the direction that is given. Otherwise, the 314 feet in Figure 24.21 is measured to the southwest from the POB instead of to the northeast.

When plotting traverses with a CADD system, the same concepts are used. Depending on the particular system you are us-

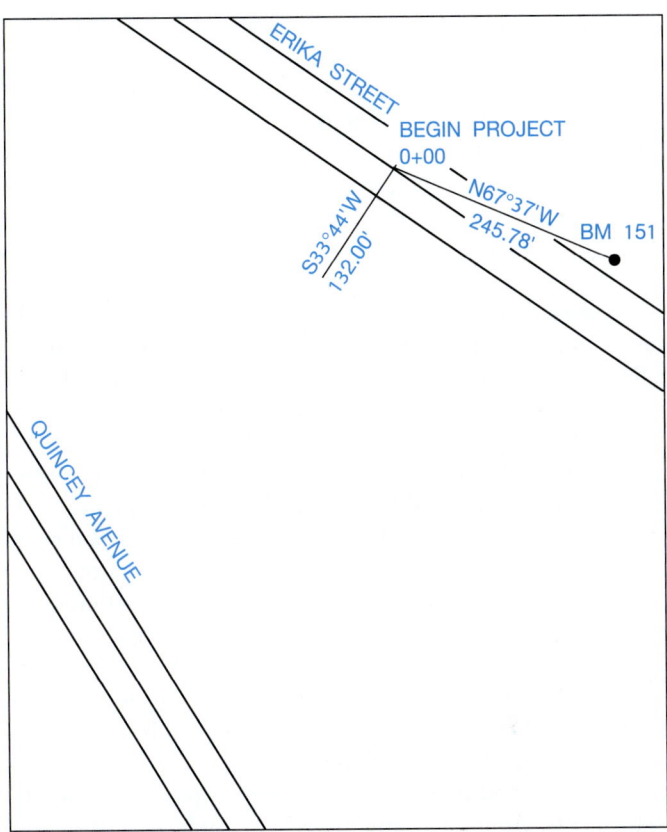

FIGURE 24.22 ■ Finding the point of beginning (POB).

ing, the layout of the traverse follows certain steps. For example, a line can begin at the POB and then be given information relating to the length and bearing, azimuth, or other data. The system then draws the line automatically.

The direct-angle traverse is drawn by plotting the measured interior angles and by using the lengths given for each course.

A deflection-angle traverse is drawn similarly to the direct-angle traverse except that the length of each course is measured and plotted. Then the deflection angle to the right or the left is measured from that point.

The azimuth traverse is plotted similarly to the compass-bearing traverse except that the complete counterclockwise angle is measured from north. Measure each course from information given, make another north-south line at that point, then plot the next course. (See Figure 24.19.)

In the example at the beginning of the chapter, the POB of the project is a specific length and distance from BM 151. **BM** refers to **bench mark**, which is discussed later. Now you know how to find the point of beginning and the first distance of Valerian Lane. (See Figure 24.22.)

DISTANCE AND ELEVATION

So far, you have learned how to establish the direction of a line by using the different azimuths and the bearing of the line. Now you need to know how to calculate the distance and the elevation in order to complete your basic knowledge of making accurate maps.

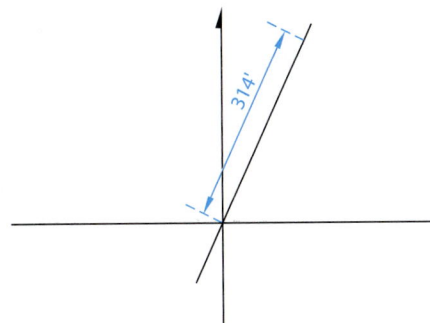

FIGURE 24.21 ■ Measuring length.

Gunter, the mathematician who invented terms such as co-sine and cotangent, also invented a standard of land measurement called Gunter's chain, and is simply referred to as chain. A chain is 66 feet long, or 1/80th of a mile. It was used extensively in the days before using steel tape became common. Each chain is divided into 100 links, each of which is 7.92 feet long. Because of the widespread and long-term use of Gunter's chain, terminology such as breaking chain, rear chainman, and head chainman, all of which are described shortly, is still used.

Distance in modern-day usage is measured several ways. The most common is the stadia technique. With this method, a rod, called a Philadelphia rod, is used along with a level, which is an instrument with crosshairs used to read measurements on the rod at a level line of sight. The Philadelphia rod is 7 feet long, can be extended 12 or 13 feet long, or up to 45 feet long in some models, and is graduated to hundredths of a foot. Figure 24.23 shows a Philadelphia rod. The level is placed on a tripod a certain distance away from the rod. When you look through the level at the rod, the top stadia crosshair and the bottom stadia crosshair encompass an amount that is then multiplied by 100 to equal the distance. (See Figure 24.24.) In this example, the crosshairs encompass the distance between 10 and 11 feet. When multiplied by 100, the distance to the rod is 100 feet.

This type of surveying technique uses mechanical and optical principles and uses mechanical/optical principals (MOD) instruments. Another technique is Electronic Distance Measuring (EDM), which uses electronic principles. The EDM system is unique in that it uses a reflected beam to gauge distances. (See Figure 24.25.) Examples of theodolites and transits, which are instruments of the EDM system, are shown in Figure 24.26, page 818. Theodolites and transits are used instead of compasses because they are more accurate. The magnetic declination is taken into account by an adjustment that can be made on the instrument.

The Philadelphia rod and level or transit also can be used to find the elevation of an unknown point. The method is as follows:

STEP 1 A known elevation is located, and the rod is placed at that point.

STEP 2 The transit (or level) is placed between the rod and the unknown elevation and then leveled.

STEP 3 A reading is taken back to the rod. The rod location is called the backsight. When this height is added to the known elevation, calculate the height of the instrument (HI).

STEP 4 The rod is then placed on the unknown elevation, called the foresight and a reading is taken forward to the rod from the transit. What is read on the rod is then subtracted from the HI to find the unknown elevation.

To find unknown elevations that are a great distance away from the known elevation, several turning points are required. Turning points are locations where the rod is placed. (See Figure 24.27.)

When distances are first measured using Gunter's chains the method for finding elevations is similar to what was just described. The difference is that instead of using an instrument,

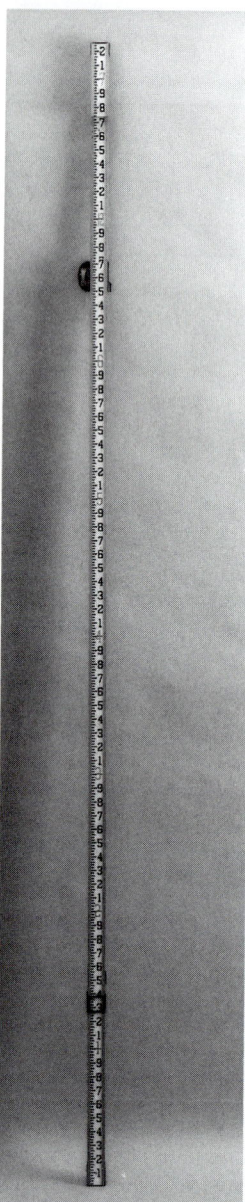

FIGURE 24.23 ■ Philadelphia rod. *Courtesy Chicago Steel Tape.*

such as a transit, between the rod readings, the chain is held level between the rods. Then the elevations are read, and the chain is then brought to the next level to be read again. This is known as breaking chain, referred to earlier. The person holding the chain at the backsight is the rear chainman, and the person holding the chain at the foresight is the head chainman. This process is called leveling. (See Figure 24.28.)

A known elevation can be read from a bench mark or monument, which is considered a permanent, fixed object on which the elevation is marked. Often a bench mark is a brass disk about 3 inches in diameter that can have "U.S. GEOLOGICAL SURVEY" engraved on it, along with the elevation above sea level. Another way to describe turning points is temporary bench marks or stations, often abbreviated as STA.

When the surveyor begins a survey, the stations, or turning points, are shown in different ways. The distances are sectioned

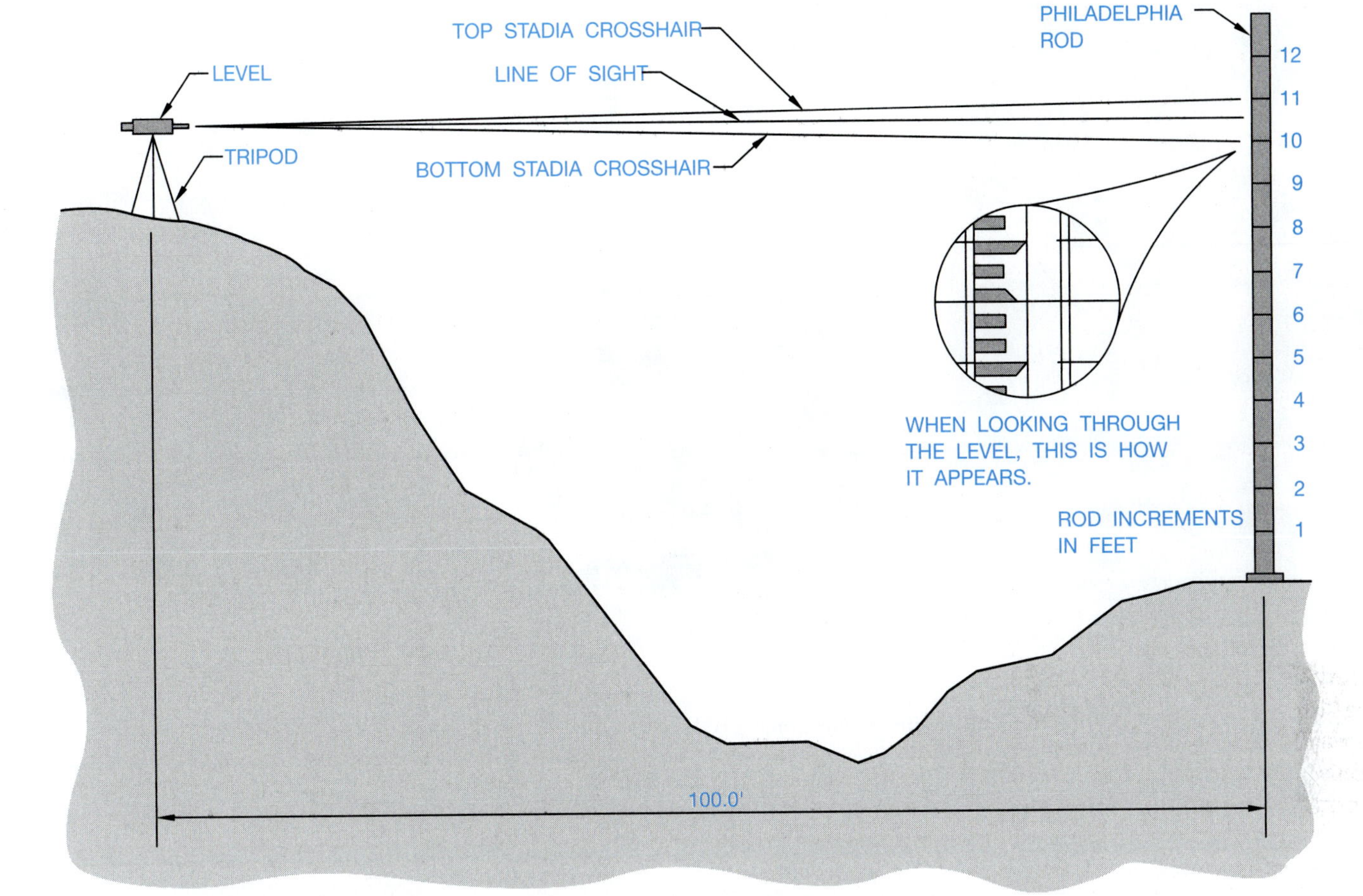

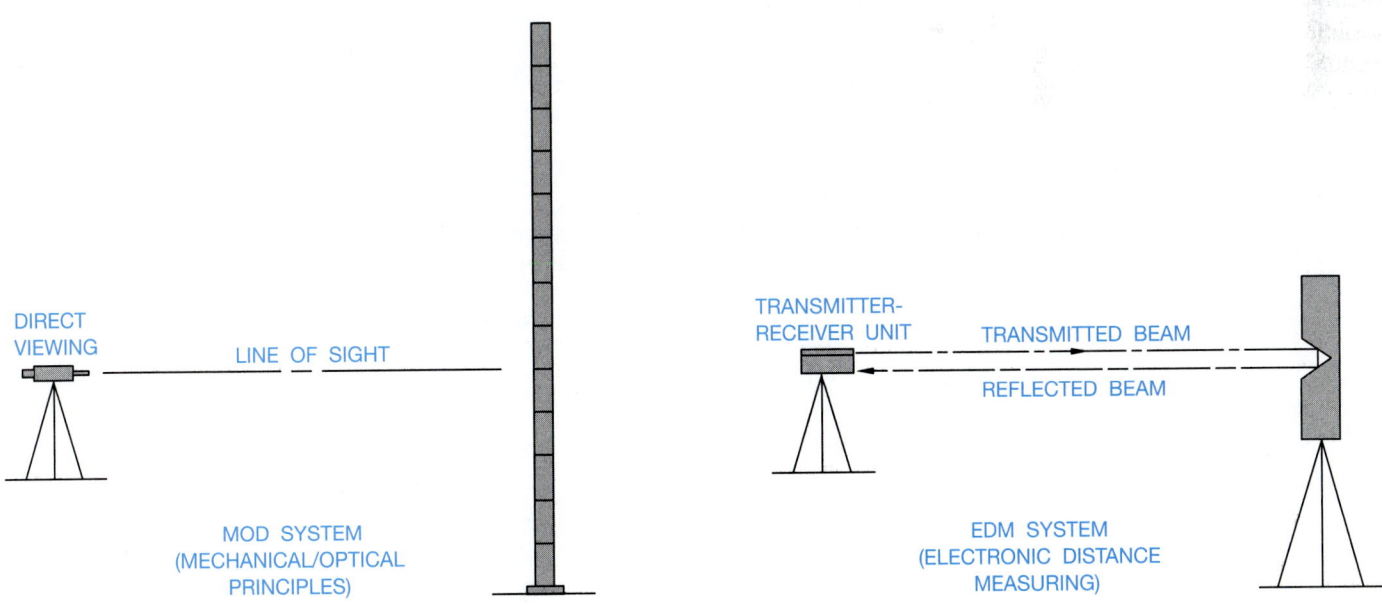

FIGURE 24.24 ■ Stadia technique.

FIGURE 24.25 ■ MOD and EDM systems.

into 100-foot intervals, with the 100s separated from the 10s by a plus (+) sign. For example, the POB is shown as 0 + 00. If the next station is at 57 feet, it is shown as 0 + 57. If the following station is at 244 feet, it is shown as 2 + 44. The beginning of a survey (station 0 + 00) is always positioned according to a

known bench mark so exact locations and elevations can be calculated. A station can also be shown as a turning point. An example would be TP1, meaning turning point #1. BM 35, meaning bench mark #35, can be one of the stations. In Figure 24.29, the surveyor has given these notes to the drafter. These

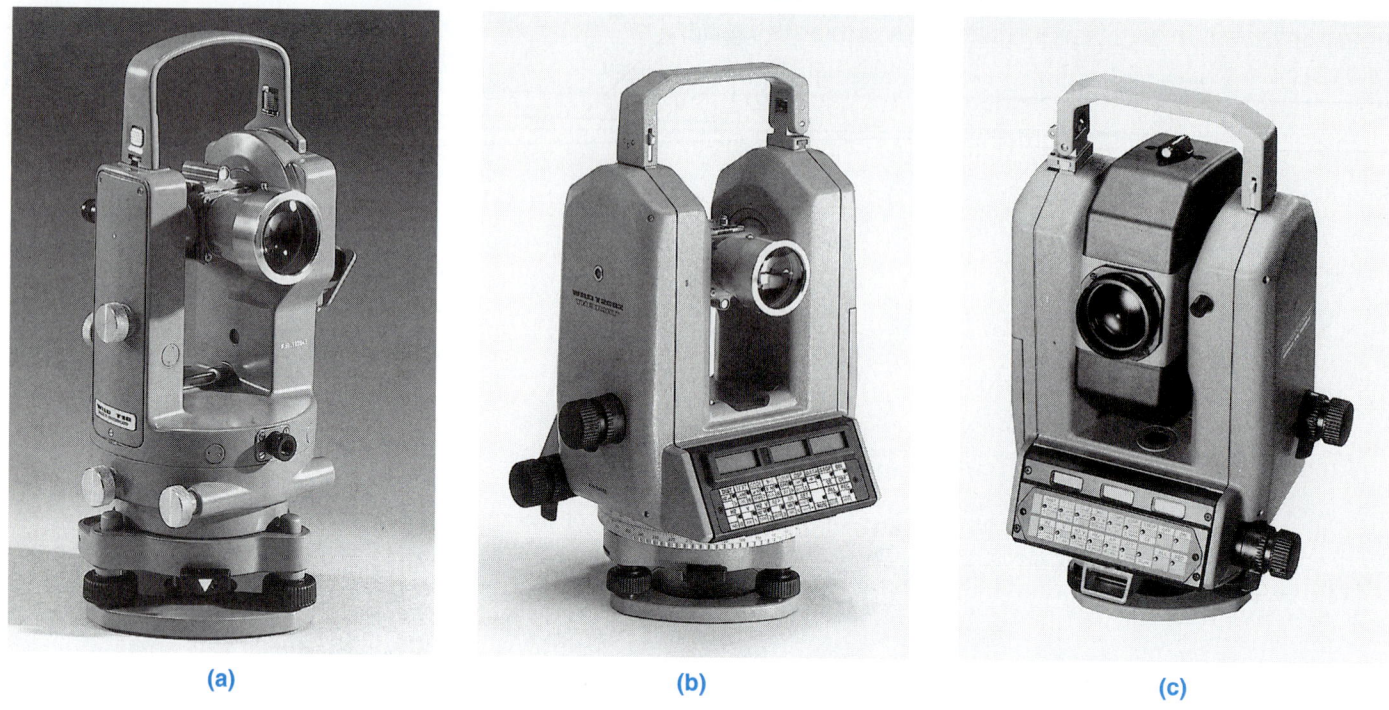

FIGURE 24.26 ■ (a) Transit (optical) theodolite; (b) electronic theodolite; (c) total station theodolite. *Courtesy Leica, Inc.*

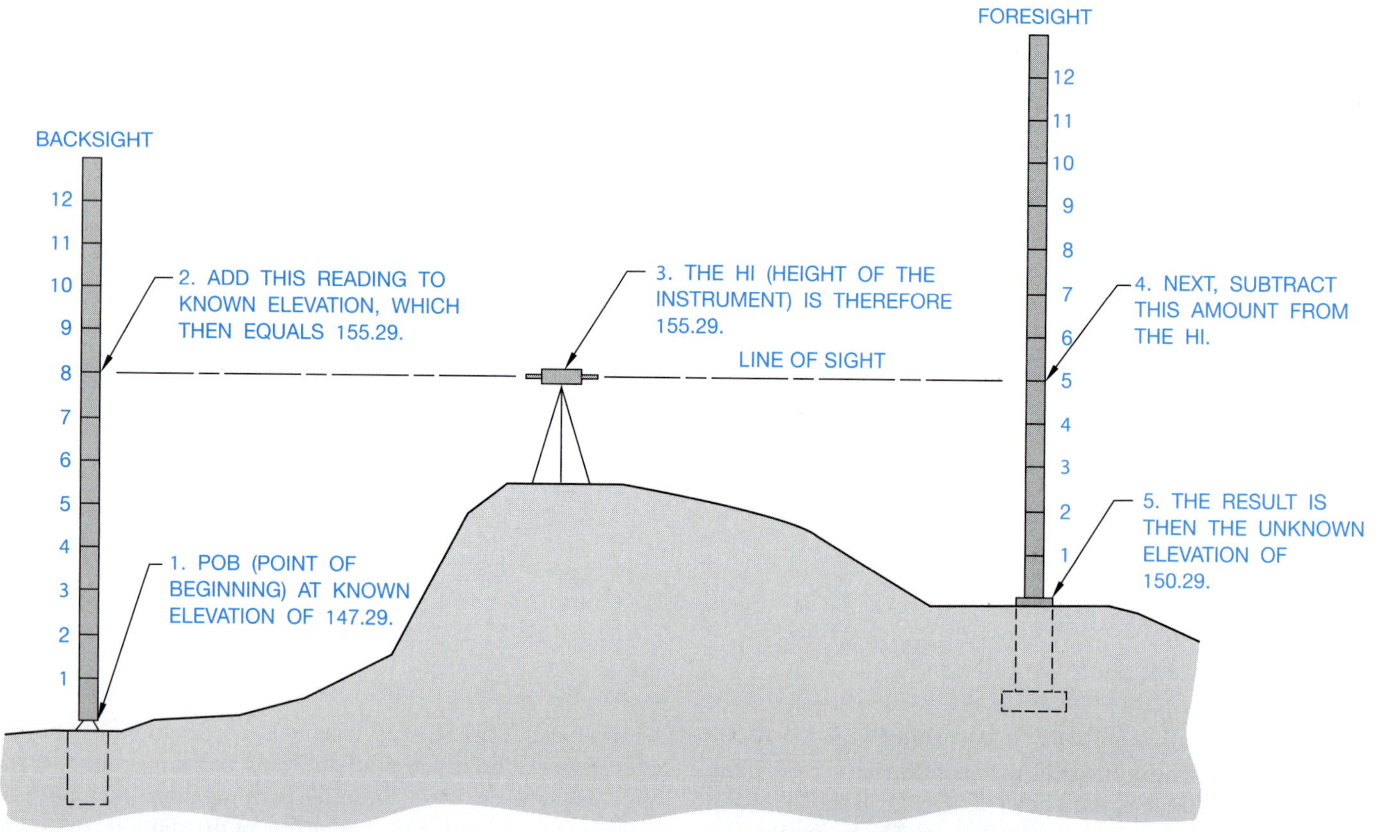

BACKSIGHT

FORESIGHT

12
11
10
9
8
7
6
5
4
3
2
1

12
11
10
9
8
7
6
5
4
3
2
1

2. ADD THIS READING TO KNOWN ELEVATION, WHICH THEN EQUALS 155.29.

3. THE HI (HEIGHT OF THE INSTRUMENT) IS THEREFORE 155.29.

LINE OF SIGHT

4. NEXT, SUBTRACT THIS AMOUNT FROM THE HI.

1. POB (POINT OF BEGINNING) AT KNOWN ELEVATION OF 147.29.

5. THE RESULT IS THEN THE UNKNOWN ELEVATION OF 150.29.

FIGURE 24.27 ■ Elevation reading.

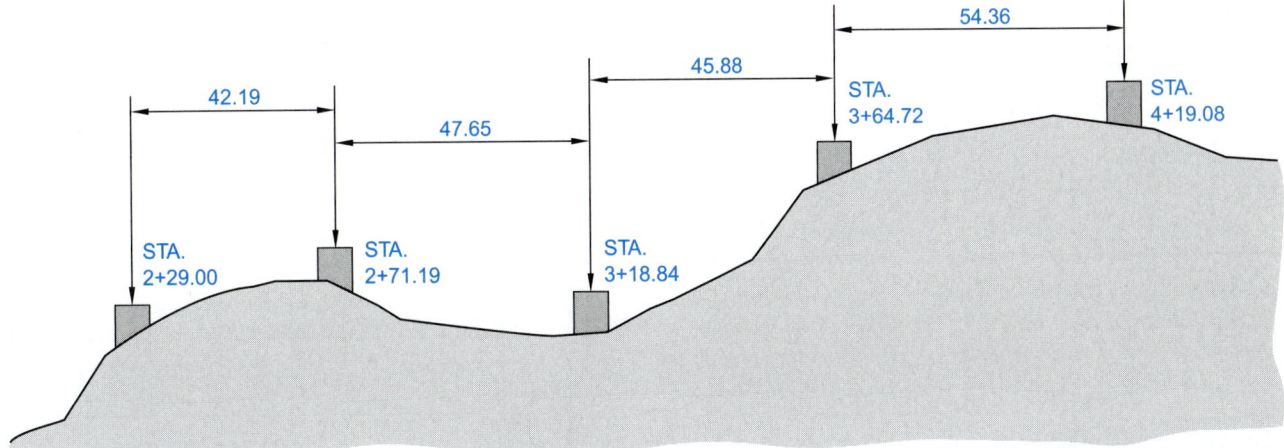

FIGURE 24.28 ■ Leveling.

Station	B.S. (+)	H.I.	F.S. (–)	Elevation
BM 34	7.25		11.92	124.13
0 + 75	5.87		10.29	
1 + 50	5.57		12.62	
TP1	7.04		3.44	
2 + 25	11.70		6.54	
BM 35 116.75				

FIGURE 24.29 ■ Leveling field notes from the surveyor.

Station	B.S. (+)	H.I.	F.S. (–)	Elevation
BM 34	7.25	131.38	11.92	124.13
0 + 75	5.87	125.33	10.29	119.46
1 + 50	5.57	120.61	12.62	115.04
TP1	7.04	115.03	3.44	107.99
2 + 25	11.70	123.29	6.54	111.59
BM 35 116.75				116.75

FIGURE 24.30 ■ Leveling field notes completed by a drafter.

surveyor notes are called field notes. The drafter then calculates the rest of the information, as shown in Figure 24.30. BM 35 can be double-checked to see if the calculation actually adds up to the posted elevation.

PROPERTY DESCRIPTIONS IN CIVIL ENGINEERING

Civil engineering drafting, in general, can be used in a wide variety of applications: highway and traffic design, water and sewage design, park planning, subdivision planning, single-family residence planning, and design, to name a few. When describing parcels or plots of land, there are three primary types of legal descriptions used. One method of describing land parcels is the metes and bounds system, and the other is the rectangular system, which is based on the public land system. The third method used to describe land is the lot and block system, which can be a combination of the other systems, and is used to describe lots within a subdivision. These terms are defined later.

A parcel of land, such as where your home is located, is often referred to as a plot, lot, or site. This parcel typically has its boundary lines established by a legal description containing very specific information about location, property line lengths, and bearings using one of the methods used to describe land, or a combination of methods.

Metes and Bounds System

When the United States was first formed, the method used for describing the boundaries of properties was by measuring the length (metes) and bearings (bounds) of the property. The metes and bounds system is the primary form of legal description for nineteen states, including the original thirteen states. These states are not part of the public land system, where the rectangular system of surveys is used. Metes and bounds are occasionally used in public land system states as shown in the following example. (See Figure 24.31.)

Beginning at the Northwest corner of that certain tract of land conveyed to James H. Wilcox et ux by Deed recorded in Film Volume 95, Page 1488, Deed Records of Yamhill County, Oregon, said point is in the center of County Road No. 87, thence along the center of said County Road No. 87, N57°45'W 16.49 feet; thence N69°33'W 213.18 feet; thence leaving said County Road No. 87, S22°29'W 400.00 feet more or less to a point on the North bank of the Yamhill River; thence Easterly along said North bank of the Yamhill River to the Southwest corner of the aforesaid James H. Wilcox et ux tract of land; thence along the West line thereof N22°29'E 360.00 feet more or less to the point of beginning.

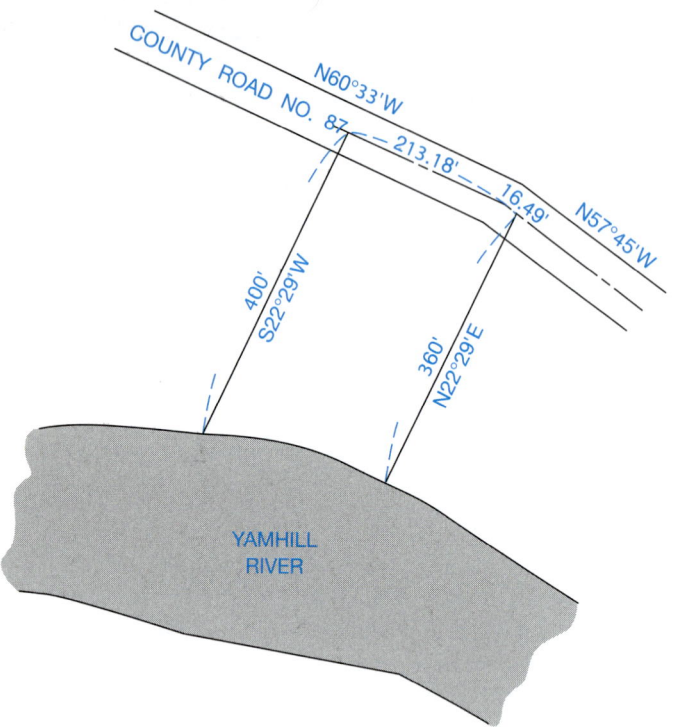

FIGURE 24.31 ■ Metes and bounds tract of land.

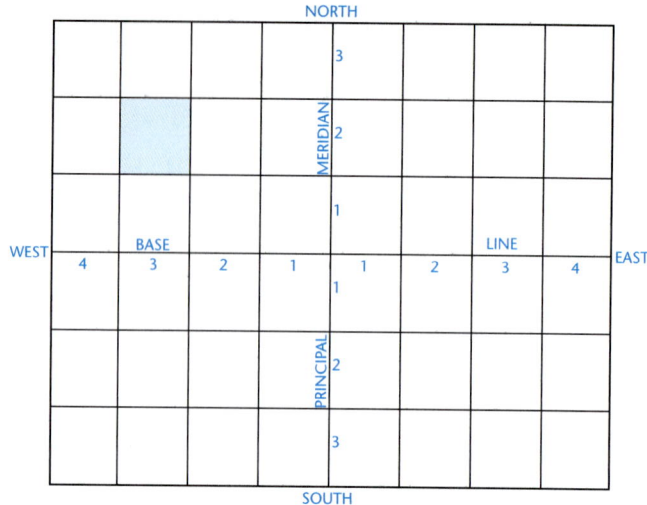

FIGURE 24.32 ■ Baseline and meridian.

6	5	4	3	2	1
7	8	9	10	11	12
18	17	16	15	14	13
19	20	21	22	23	24
30	29	28	27	26	25
31	32	33	34	35	36

FIGURE 24.33 ■ Township divided into sections.

Rectangular System

In the late 1700s, when the remaining land to the Pacific Ocean was opened up to homesteading, a more complete and universal method of determining land ownership was needed. The government, through the General Land Office, now known as the Bureau of Land Management, measured large areas in these public lands and provided a rectangular system of surveys from which ownership can be determined. The public-land states are Alabama, Alaska, Arizona, Arkansas, California, Colorado, Florida, Idaho, Illinois, Indiana, Iowa, Kansas, Louisiana, Michigan, Minnesota, Mississippi, Missouri, Montana, Nebraska, Nevada, New Mexico, North Dakota, Ohio, Oklahoma, Oregon, South Dakota, Utah, Washington, Wisconsin, and Wyoming, in addition to 75 million acres of Texas bought by the federal government. Each large area of public land was a single great survey and had a basic reference point where one specific meridian of longitude and one specific parallel of latitude crossed. The meridians in each of these public-land surveys were named separately with names such as the first principal meridian (which is the western boundary of the state of Ohio), Michigan meridan, Choctaw meridian and the Black Hills meridian. The parallels were all named baseline. There are thirty-one sets of great surveys in the contiguous United States and three in Alaska.

With the baseline and principal meridian as reference points, each survey is arranged into rows of 6-mile-square blocks, each having an area of 36 square miles called townships. From the baseline, each row of townships can be located by referring it to the number and direction from the baseline. For example, a township just to the north of the baseline is identified as Township No. 1 North or abbreviated T1N, and a township in a row

of townships three rows south of the baseline is Township No. 3 South or T3S. The townships are also arranged according to columns called ranges, and each column of townships can be located from the principal meridian by referring to how far east or west it is from the meridian. For example, a township in a column of townships four columns to the east of the principal meridian is labeled Range No. 4 East, or R4E. The marked township in Figure 24.32 is T2N, R3W.

Each of the township's thirty-six squares is called a section, with the sections numbered as shown in Figure 24.33. Each section is approximately one square mile and can be divided into quarter-sections, which are sometimes referred to as quarters. An example is the Southwest Quarter of Section 19, which can also be written SW 1/4 of Sec. 19. The quarter-sections can be divided further into quarter-quarter-sections, as seen in Figure 24.34. Notice that the description of the particular piece of land begins with the smallest portion and proceeds through the largest portion.

Lot and Block System

The lot and block legal description system can be established from either the metes and bounds or the rectangular system. Generally when a portion of land is subdivided into individual building sites,

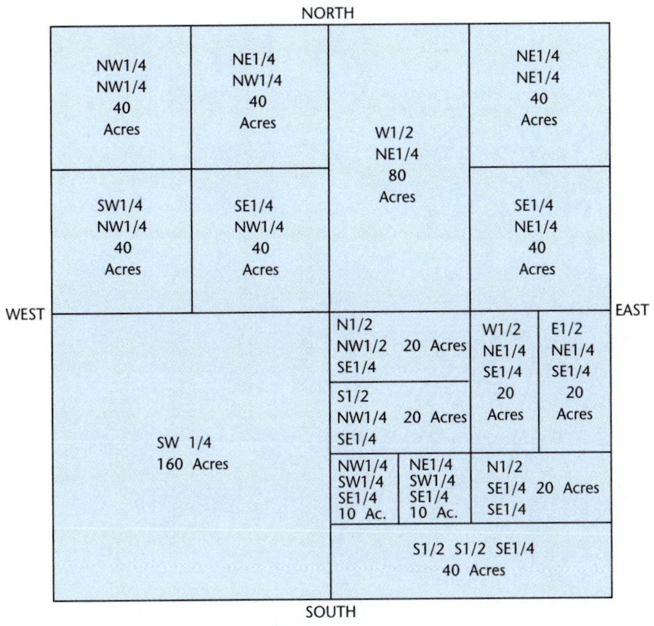

FIGURE 24.34 ■ Section divided into parcels.

the subdivision is established as a legal plot and recorded as such in the local county records. The subdivision is given a name and broken into blocks of lots. A subdivision can have several blocks, each divided into a series of lots. Each lot can be 50 × 100', for example, depending on the zoning requirements of the specific area. Figure 24.33 shows an example of a typical lot and block system. A typical lot and block legal description might read: LOT 14, BLOCK 12, LINCOLN PARK NO. 3, CITY OF SALEM, STATE. This lot is the shaded area in Figure 24.35.

BEGINNING A CIVIL ENGINEERING DRAFTING PROJECT

An engineering map is a large-scale map showing the information necessary for planning an engineering project such as a highway layout or utility layout. The information can come from an engineering survey, an aerial survey, or other reliable source, and includes locations, dimensions, and elevations of buildings, roads, water, and utility lines. This type of map can also include contour lines in addition to showing the boundaries of plats. A plat is a tract of land, such as a subdivision, showing building lots. (See Figure 24.36.) Notice also the section, township, and

FIGURE 24.35 ■ Part of a lot and block subdivision. *Courtesy GLADS Program.*

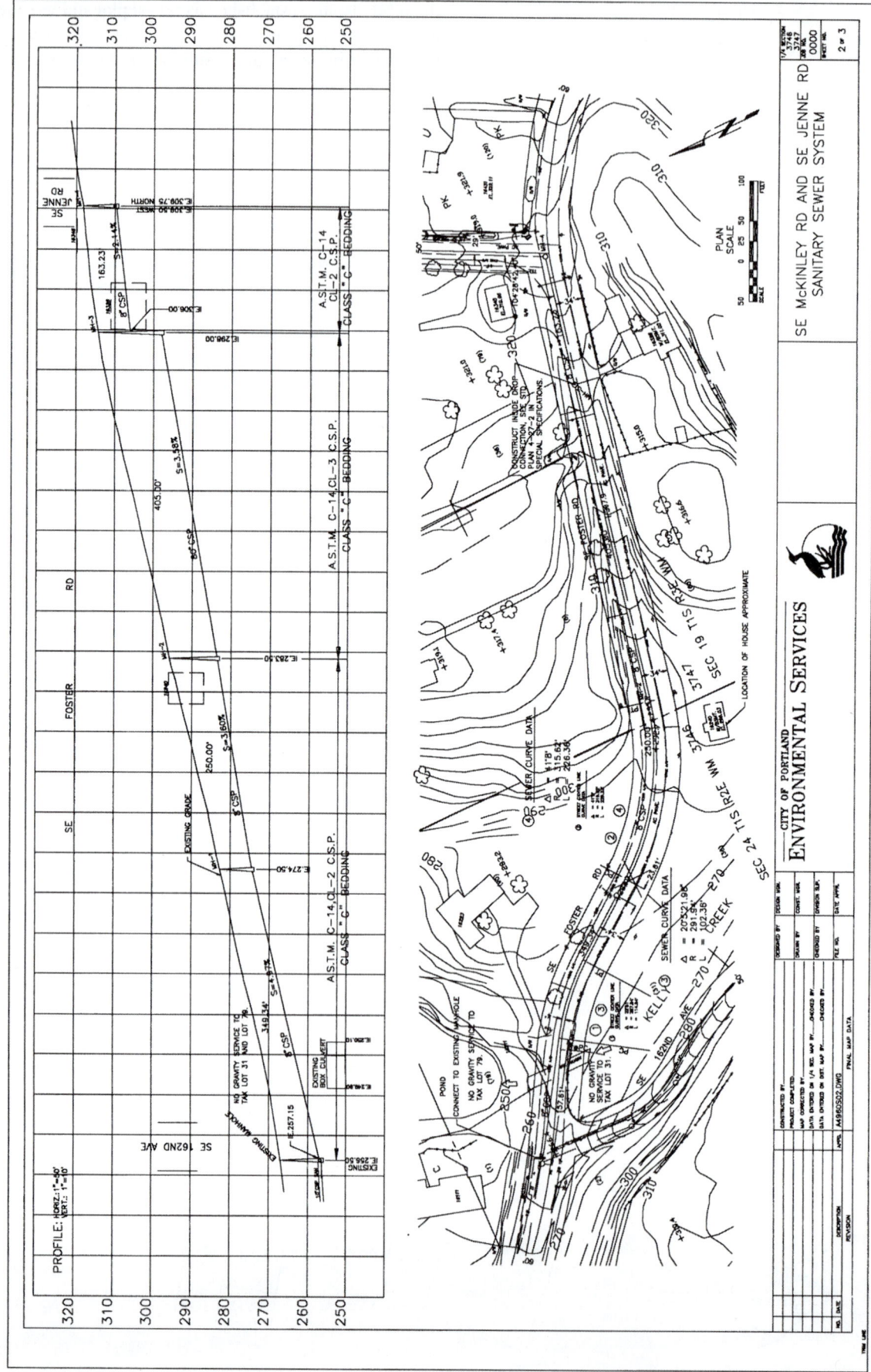

FIGURE 24.36 ■ Engineering map. *Printed by permission of City of Portland.*

range notations. WM is an abbreviation for Willamette meridian, which is the name of the meridian in this great land survey.

Contour Lines

One of the most important symbols in civil engineering drafting is the contour line. Contours are lines joining points of the
same elevation. When looking at a map with contours, you are
able to understand the topography of the land. Figure 24.37
shows the relationship between actual elevations and how the
contour lines appear on the map. Each horizontal plane represents an additional 10 feet in elevation from the horizontal
plane below it. The contour interval in this case is 10 or 10 feet.

Figure 24.38 shows the contour lines around Mount St. Helens in Washington state. The contour lines are spaced more
closely together on steep slopes than they are on gentle slopes.
They also point upstream when depicting creeks.

Some of the contours are drawn with a heavier line than others. These are index contours and can occur every fourth or fifth
contour, depending on the map. Some of the index contours
have their elevations labeled on them. Intermediate contours
are the lighter lines shown between the index contours, and they
are not usually labeled with elevations. In the Mount St. Helens
map, find the index contour at 8,000 feet of elevation. The next

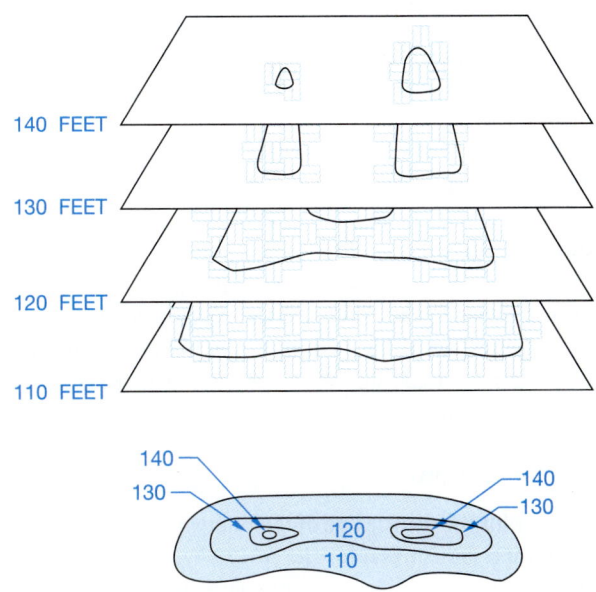

FIGURE 24.37 ■ Contours.

downhill index contour that is labeled is at 7,600 feet. This
means the index contour between these two is at 7,800 feet, and
each intermediate contour is at 40-feet intervals. There are four

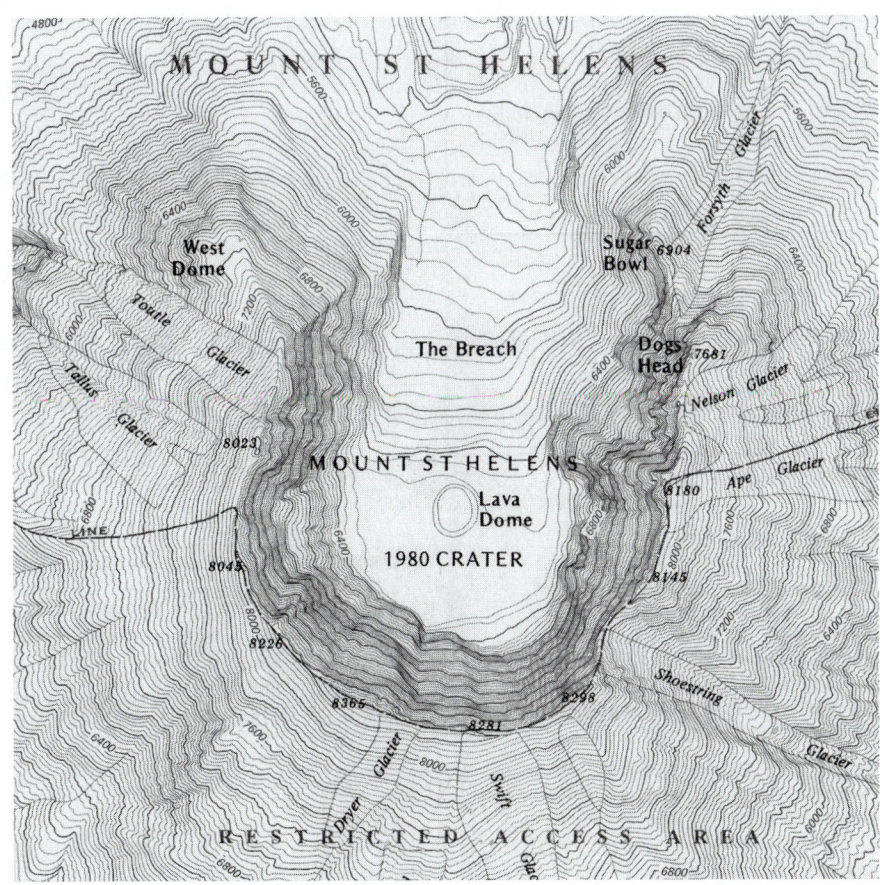

FIGURE 24.38 ■ Contour intervals of Mount St. Helens. *Printed by permission of U.S. Geological Survey.*

intermediate contours between index contours. The contour interval is 40 feet, and this interval is stated on the map. There is an additional discussion of constructing contour lines later in this chapter.

Highway Layout

For a highway layout, you first draw a plan view and then a profile of the proposed highway from the information given in the survey notes. The plan view shows existing features such as roads, buildings, and trees in addition to the centerline (transit line) of the proposed highway. The profile is a section view through the plan at a specific location.

Plan View Layout

In order to plot the transit line of the highway, use curve data information like that shown for the sewer layout in Figure 24.36. The following is basic terminology related to this application:

Transit line is any line of a survey established by a transit or other surveying instruments.

Curve data is any measurements or features used to create the road layout. The following terms are used in curve data:

Point of curve (PC) is the station at which the curve begins.

Radius (R) means the radius of that particular curve. The center point of the curve can be found by measuring the radius distance perpendicularly from the transit line at the PC. For example, to find the center of a curve with a radius of 265 feet in which the transit lines are known, draw two lines parallel to the transit lines on the inside of the curve. Their intersection is the center of the curve. (See Figure 24.39.)

Point of tangency (PT) denotes the end of the curve.

Delta angle (Δ) is that inclusive angle measurement of the curve from the PC to the PT.

Curve length (L) is the actual length of the curve from the PC to the PT.

Degree of curve (D) is the angle within the delta angle that has a chord of 100 feet along the curve. A chord connects two points on the curve with a straight line.

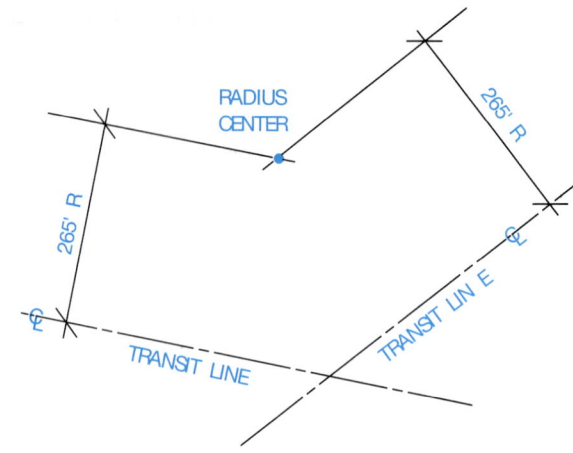

FIGURE 24.39 ■ Locating radius center of highway curve.

Point of reverse curve (PRC) is the point where two curves meet and curve in different directions.

The simplified layout in Figure 24.40 shows the relationship of the transit lines with the curve data.

Station points are given at all significant points along the transit line such as the PC, PRC, PT, and other features such as utility crossings and railroad crossings.

Fill in some of the information in the example at the beginning of the chapter. The mathematical equation for the length of the curve is

$$L = (\pi R\Delta) \div (180)$$

Therefore, for the first curve in Figure 24.42, the length is:

$$[(3.1416)(200)(47°23')] + 180, \text{ or } 165.40'$$

One way to calculate the degree of curve is to imagine the 100-foot chord is divided in half. Then, imagine a triangle is made from the radius length (200 feet), the 50-foot length, and the default length to the center of the curve. (See Figure 24.41.) The sine of the angle at the center of the curve is the opposite side (50 feet) divided by the hypotenuse (200 feet). This number is 0.25. The inverse of the sine gives you the angle, which is 14.477° or 14°28'39". Twice this amount, which is the value for the full 100-foot chord, is rounded to 28°57', or the degree of curve.

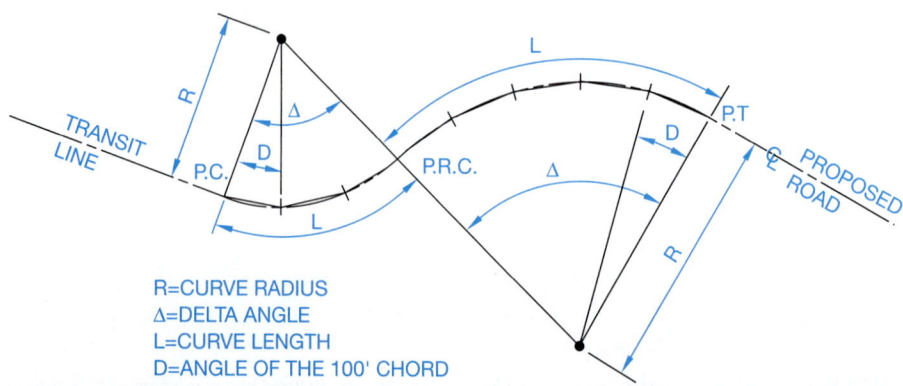

R=CURVE RADIUS
Δ=DELTA ANGLE
L=CURVE LENGTH
D=ANGLE OF THE 100' CHORD

FIGURE 24.40 ■ A simplified highway layout.

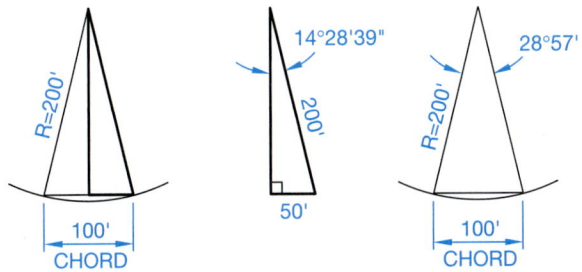

FIGURE 24.41 ■ Finding delta for Figure 24.42.

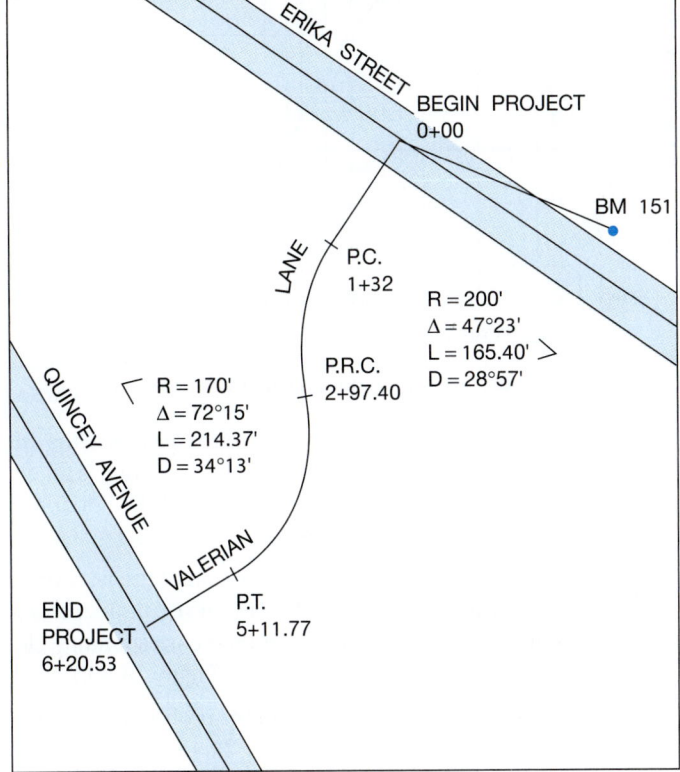

FIGURE 24.42 ■ Plotting curves.

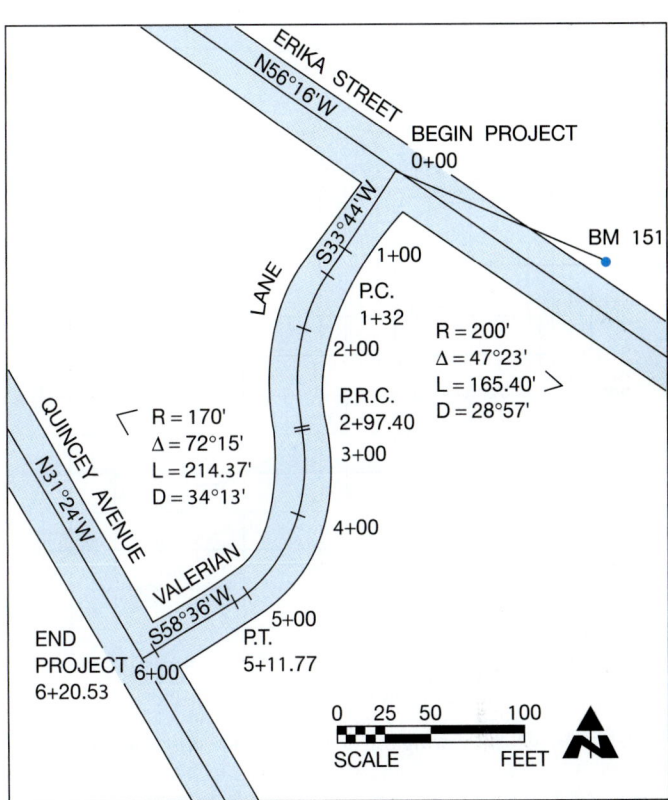

FIGURE 24.43 ■ The completed drawing of Valerian Lane. This completes the layout problem from page 809.

Since you know the length of the first curve in Figure 24.42 is 165.40', you know that the PRC stationing is 1 + 65.40 added to 1 + 32.00, equals 2 + 97.40. Other information for Figure 24.42 can now be calculated and drawn. (See Figure 24.43.)

You can also receive survey notes that establish a **point of intersection (PI)** of a curve instead of a PC or PT. A point of intersection is where two transit lines converge if a curve is not taken into account. Figure 24.39 shows an example of this application.

After the curve center is found, lines are drawn from the center perpendicularly to the transit lines. Where the perpendicular lines intersect the transit lines is the PC and the PT of the curve, and the angle that these two lines include is the delta angle (Δ).

After plotting the transit line of the proposed highway using either method, the right-of-ways can then be plotted by measuring from the transit line. In the example in Figure 24.43, plot the right-of-way 20 feet on each side of the transit line. The term **right-of-way** refers to an easement or deeded strip of land for construction and maintenance of features

such as roads, utilities, or other uses. The right-of-way includes the surface, underground, and overhead space at the designated strip of land. An **easement**, in this application, is a legal right of access over land owned by another for the purpose of access, egress, utilities, or other designated uses. Add station marks at every 1 + 00 station, a north arrow, and a graphic scale. (See Figure 24.43.)

Profile Layout

A person's profile is what the person's face or body looks like from the side. A **profile** for a map is what that map looks like in elevation if a vertical slice is cut through the land. Profiles are established from the contour information of the map and information from the survey. Figure 24.44 shows a portion of a map with contours and a horizontal line through the map. The horizontal line is referred to as the **cutting line** or **cutting plane line** and is where the section is made through the land. The cutting line can be placed other than horizontal. Follow these steps to draw a profile.

STEP 1 Wherever the cutting line crosses a contour line, a perpendicular construction line is drawn to create the profile.

STEP 2 A mark is made at the point where the construction line intersects with the corresponding vertical scale line. The vertical scale, constructed along the left side of the profile in Figure 24.44, is divided into parts with each division representing an elevation found in the plan layout. To establish a complete profile, the scale generally starts

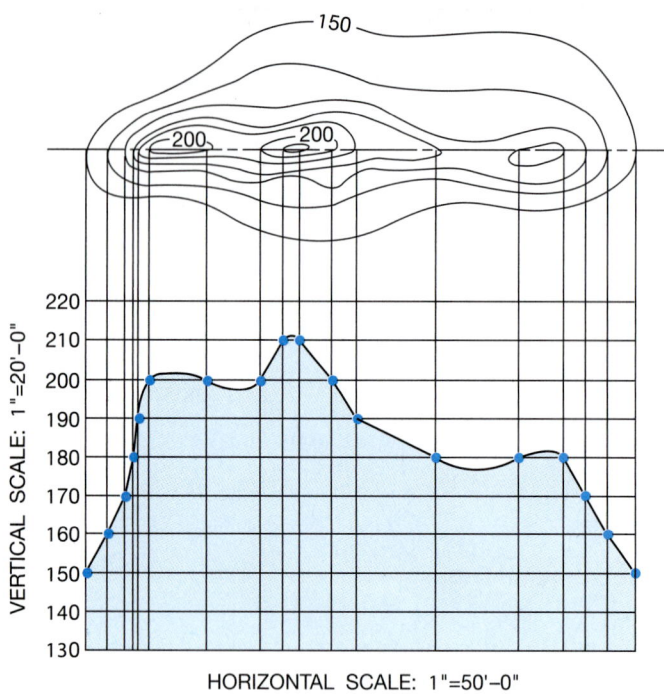

FIGURE 24.44 ■ Drawing the horizontal and vertical scales.

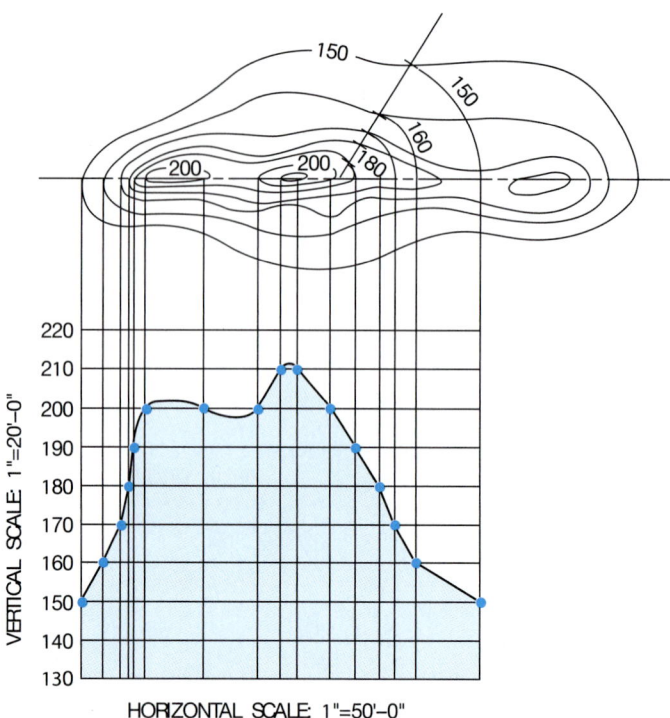

FIGURE 24.45 ■ Drawing the horizontal and vertical scales.

with an elevation less than the lowest elevation in the plan and ends with an elevation greater than found in the plan. You can see this as elevations 130 and 220 in Figure 24.44. The rest of the elevations are spaced between these extremes.

STEP 3 The marks are then connected. The resulting profile shown at the base of the figure has the same horizontal scale as the map, with an exaggerated vertical scale. An exaggerated vertical scale is common in civil drafting because it helps to visualize the slope of the land, especially in areas that are relatively flat. Horizontal scales are typically 1" = 50' or 1" = 100' and vertical scales are commonly 1" = 5' or 1" = 10'.

STEP 4 Connect the points to complete the profile as shown in Figure 24.44. Figure 24.45 shows the profile drawn if the cutting line is placed through the plan at an angle.

In this example, the cutting line is horizontal from left to right until it angles up to the right just past the second 200' elevation. The part of the cutting line that is horizontal is used to construct a profile as previously discussed.

Follow these steps to create the profile with an angled cutting line. Remember the construction lines need to be perpendicular to the cutting line.

STEP 1 Where the horizontal part of the cutting line crosses a contour line, a perpendicular construction line is drawn to create the profile as previously discussed.

STEP 2 The angled cutting line creates a vertex where it joins the horizontal cutting line. The profile construction, at the angled cutting line, is done by drawing arcs with

their centers at the vertex and their radii where the angled cutting line crosses contour lines. Each arc is drawn from these points to the continuation of the horizontal line, as shown in Figure 24.45.

STEP 3 Where the arcs meet the continuation of the horizontal cutting line, project each down into the profile, establishing points where they meet their corresponding elevation from the vertical scale. (See Figure 24.45.)

STEP 4 Connect the elevation points in the profile view to complete the profile line. (See Figure 24.45.)

Plan and profile drawings are also used when drawing proposed utility services.

Follow these steps for this profile application.

STEP 1 The plan of the particular utility line is plotted and drawn showing right-of-way lines, other services, and manholes, with their corresponding elevations and station points.

STEP 2 Draw a profile of the service onto the appropriate vertical scale, usually above the plan on the drawing surface. When drawing a sewer line, for example, if the elevation at the top of MH (manhole) #1 is 76.4 feet at STA 2 + 32, then the manhole is plotted on the profile at 76.4 vertical feet. If possible, this manhole is plotted directly above the 2 + 32 station point. In cases in which the utility line makes turns and angles, this is not possible.

STEP 3 Complete the plan and profile. An example of a plan and profile for a storm sewer layout is shown in Figure 24.46.

Invert elevation (IE) is the lowest elevation of the inside of a sewer pipe at a particular station. The IE of a pipe is shown on

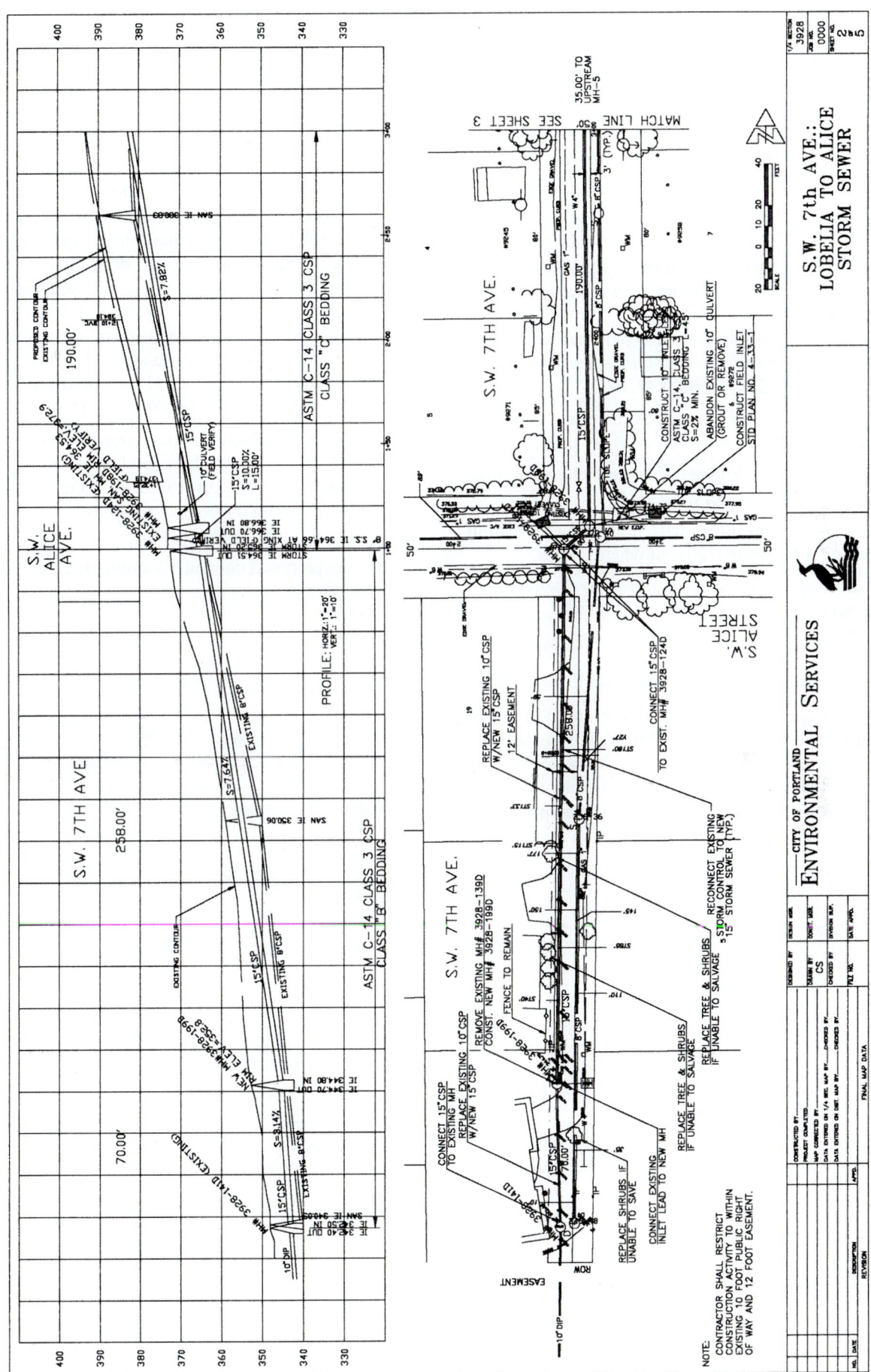

FIGURE 24.46 ■ Storm sewer plan project sheet. *Printed by permission of City of Portland.*

the profile; it is information necessary to determine the "grade slope" of the pipe. The grade slope is the percentage number given to show the amount of slope of the pipe (or road) and is shown on the profile. For example, the grade slope shown in Figure 24.46 is 7.64%. Grade slope is the amount of elevation gain (or loss) divided by the distance. So, the elevation difference is 19.71 feet (364.51 − 344.80), and the distance is 258.00 feet. The slope is 0.0764 or 7.64%.

Vertical Curves on the Profile

A vertical curve is one that is shown on the profile when the road travels over a hill or down a valley, then up again. To draft a vertical curve, you are given the elevation points of the road from engineer or survey people, and these points are plotted on the profile grid. The following are abbreviations used when plotting vertical curves: (See Figure 24.47.)

BVC (begin vertical curve) is used at the point where the curve begins.

EVC (end vertical curve) is used at the point where the curve ends and the road has an even slope again.

VC (vertical curve) is the horizontal length of the curve from the point where it begins to the point where it ends.

PI (point of intersection) is the point where the two road grades intersect.

Cut and Fill Drawing

When designing a road, highway, or building site, portions of earth must sometimes be removed (cut) from hillsides that are too steep and added (filled) to valleys and low spots. The amount of cut and fill needed can be shown on the profile, the plan, or sometimes on both. The following examines profile cut and fill in relation to planning a roadway, then in relation to a building site.

Roadway Cut and Fill

Follow these steps when drawing a cut and fill for a roadway.

STEP 1 The first step in drafting the cut and fill of a roadway is to plot the roadway on the original contour plan.

STEP 2 The necessary cut and fill ratio of the road is placed off the edge of the drawing surface in line with the road. This ratio is given to you by the designer or engineer and is based on information such as soil conditions. The cut and fill ratio is the relationship of run to rise (run being the horizontal length, rise being the vertical height), and is called the angle of repose. The angle of repose is the design slope of the area on either side of the roadway. In Figure 24.48 the angle of repose is 2:1, and the cut and fill ratio is set up so the designed roadway has an elevation of 180 feet. The cut occurs as the elevation increases, and the fill occur as the elevation decreases.

STEP 3 The desired cut and fill lines from the ratio are projected onto the drawing of the road and parallel to the road. Marks are made at the intersection of the new contour intervals and the original contours.

STEP 4 The lines connecting the marks show the boundaries of the proposed cut and fill around the road. (See Figure

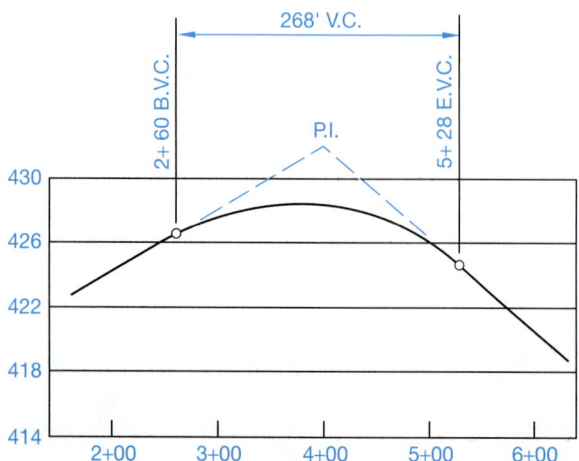

FIGURE 24.47 ■ Layout of vertical curve.

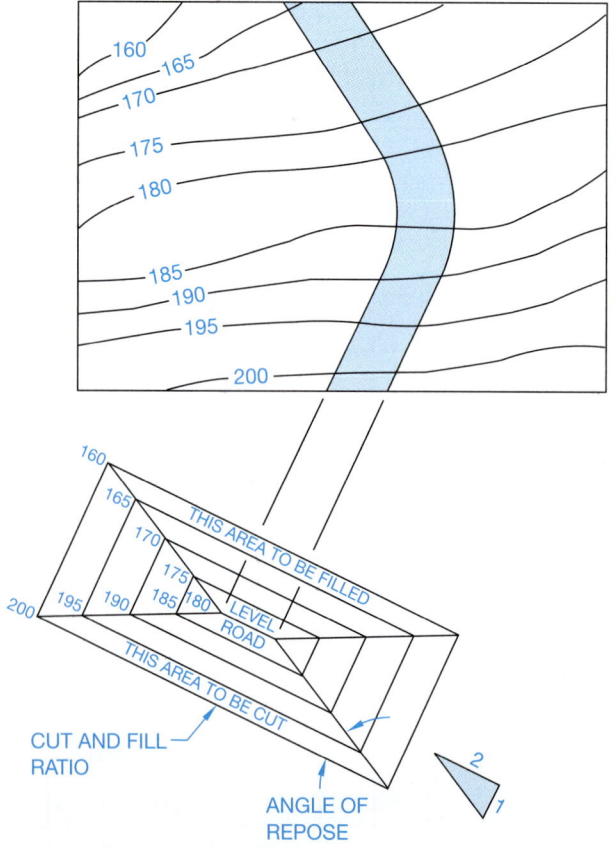

FIGURE 24.48 ■ Highway layout cut and fill.

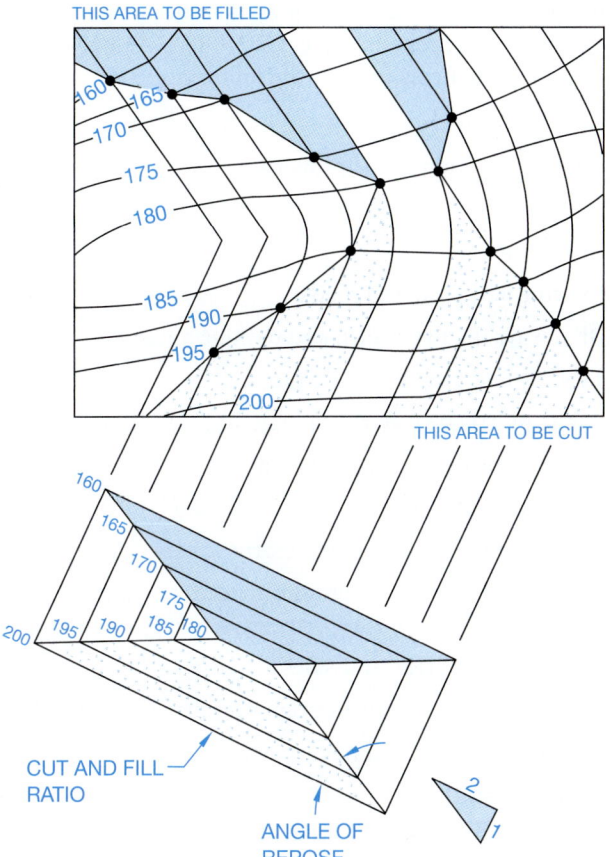

FIGURE 24.49 ■ Highway layout cut and fill.

24.49.) The building site cut and fill construction method is discussed later in this chapter.

INTRODUCTION TO SITE PLANS

A site plan, also known as a plot plan, is a map of a piece of land that can be used for any number of purposes. Site plans can show a proposed construction site for a specific property. Sites can show topography with contour lines, or the numerical value of land elevations can be given at certain locations. Site plans are also used to show how a construction site is excavated and are known as grading plans. While site plans can be drawn to serve any number of required functions, they all have similar characteristics, which include showing the following:

■ A legal description of the property based on a survey.

■ Property line bearings and directions.

■ North direction.

■ Roads and easements.

■ Utilities.

■ Elevations.

■ Map scale.

Topography

Topography is a physical description of land surface showing its variation in elevation, known as relief, and locating other features. Surface relief can be shown with graphic symbols that use shading methods to highlight the character of land, or the differences in elevations can be shown with contour lines. Plot plans that require surface relief identification generally use contour lines. These lines connect points of equal elevation and help show the general lay of the land.

A good way to visualize the meaning of contour lines is to look at a lake or ocean shoreline. When the water is high during the winter or at high tide, a high-water line establishes a contour at that level. As the water recedes during the summer or at low tide, a new lower-level line is obtained. This new line represents another contour. The high-water line goes all around the lake at one level and the low-water line goes all around the lake at another level. These two lines represent contours, which are lines of equal elevation. The vertical distance between contour lines is known as contour interval. When the contour lines are far apart, the contour interval shows relatively flat or gently sloping land. When the contour lines are close together, the contour interval shows land that is much steeper. Drawn contour lines are broken periodically, and the numerical value of the contour elevation above sea level is inserted. Figure 24.50 shows sample contour lines. Figure 24.51 shows a graphic example of land relief in pictorial form and contour lines of the same area.

A site plan with contour lines is shown in Figure 24.52. Site plans do not always require contour lines showing topography. Verify the requirements with the local building codes. In most instances, the only contour-related information required is property corner elevations, street elevation at a driveway, and the elevation of the finished floor levels of the structure. Additionally, slope can be identified and labeled with an arrow.

Elements of a Site Plan

Site plan requirements vary from one local jurisdiction to the next, although there are elements of site plans that are similar around the country. Guidelines for site plans can be obtained from the local building official or building permit department.

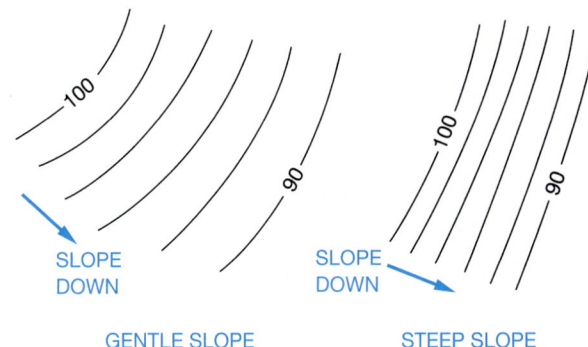

FIGURE 24.50 ■ Contour lines showing both gentle and steep slopes.

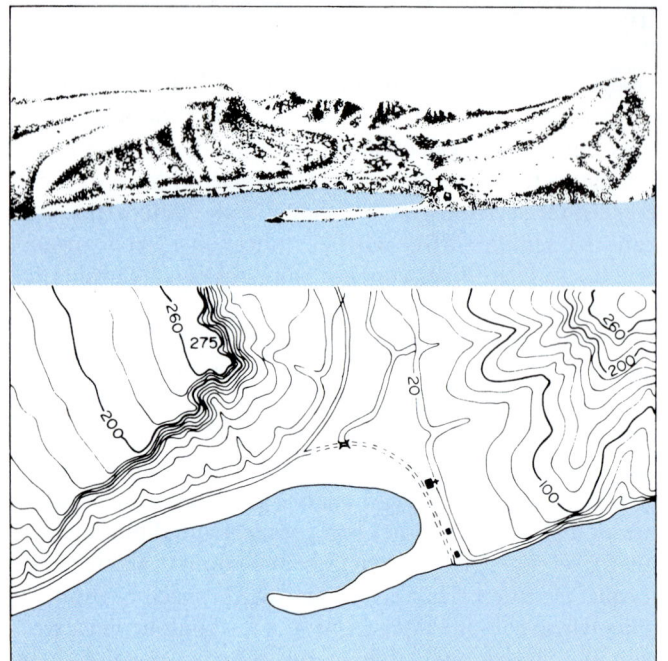

FIGURE 24.51 ■ Land relief pictorial above and contour lines of the same area below. *Courtesy U.S. Department of Interior, Geological Survey.*

Some agencies, for example, require that the site plan be drawn on specific size paper such as 8-1/2" × 14" (A4 or A3 metric). Typical site plan items include the following:

■ Site plan scale.

■ Legal description of the property.

■ Property line bearings and dimensions.

■ North direction.

■ Existing and proposed roads.

■ Driveways, patios, walks, and parking areas.

■ Existing and proposed structures.

■ Public or private water supply.

■ Public or private sewage disposal.

■ Location of utilities.

■ Rain and footing drains, and storm sewers or drainage.

■ Topography including contour lines or land elevations at lot corners, street centerline, driveways, and floor elevations.

■ Setbacks, front, rear, and sides.

■ Specific items on adjacent properties may be required.

■ Existing and proposed trees may be required.

Figure 24.53 shows a site plan layout that is used as an example at a local building codes department. Figure 24.54 shows a basic site plan for a proposed residential addition.

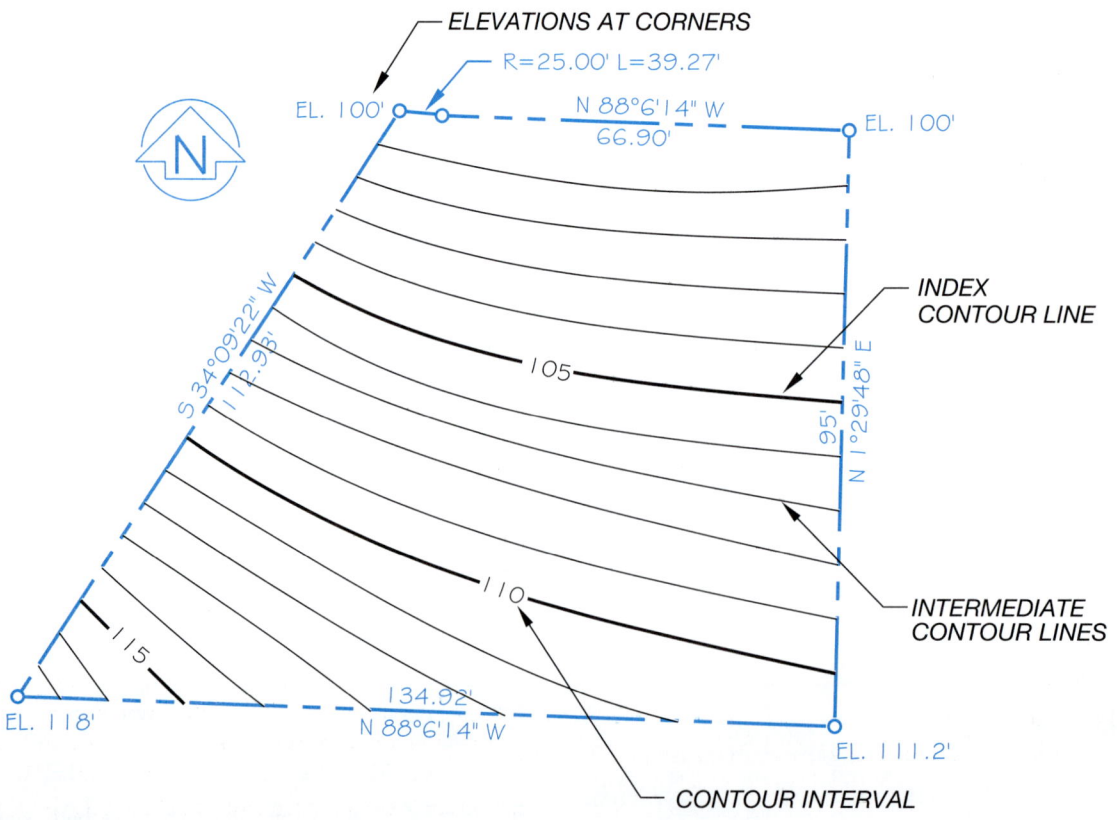

FIGURE 24.52 ■ Site plan with contour lines.

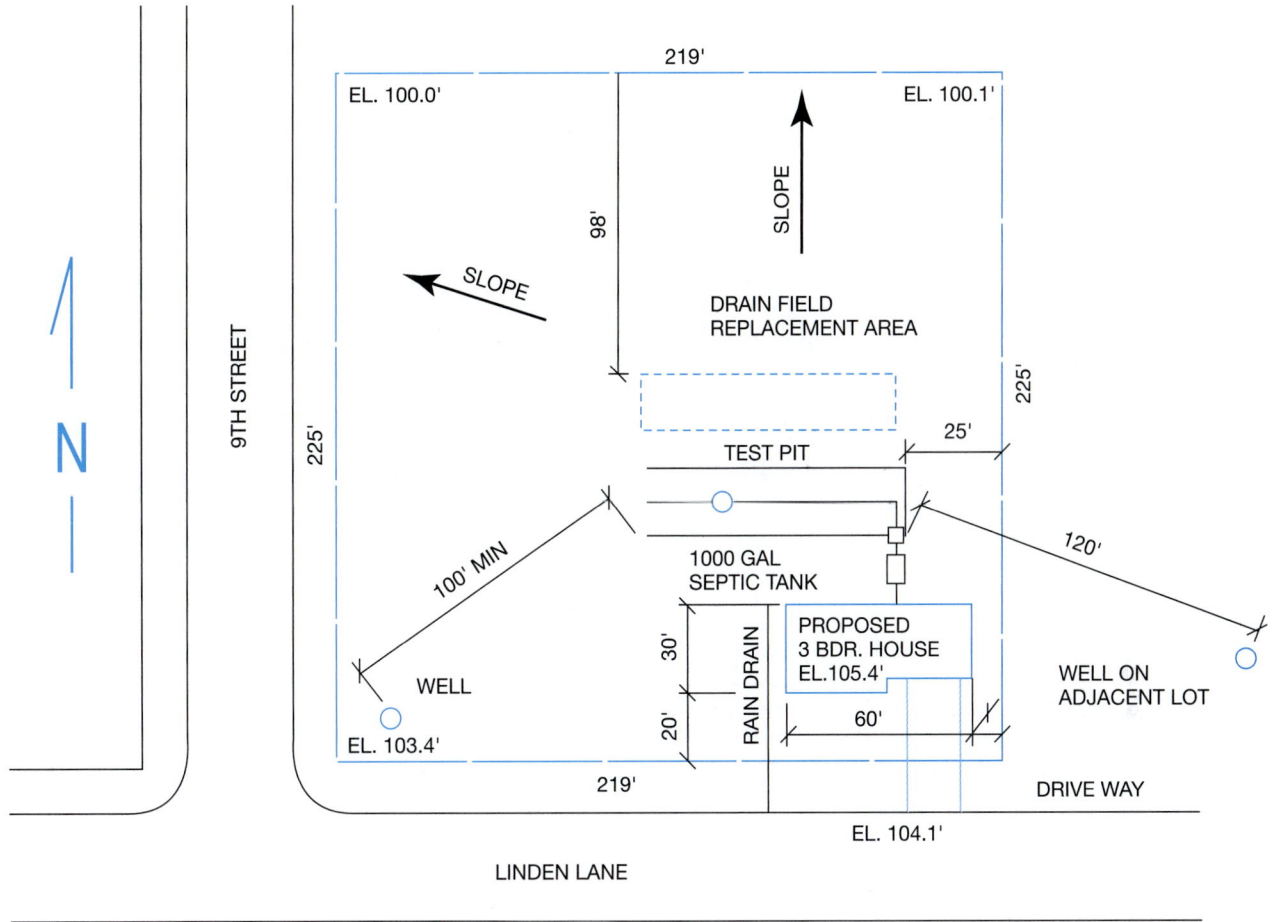

PLAN
SCALE 1" = 60'

FIGURE 24.53 ■ Recommended typical site plan layout.

Site Analysis Plan

Local requirements for subdivisions should be confirmed as there are a variety of procedures depending on local guidelines and zoning rules. In areas where zoning and building permit applications require a design review, a site analysis plan may be required. The site analysis should provide the basis for the proper design relationship of the proposed development to the site and to adjacent properties. The degree of detail of the site analysis is generally appropriate to the scale of the proposed project. A site analysis plan, as shown in Figure 24.55 often includes the following:

■ A vicinity map showing the location of the property in relationship to adjacent properties, roads, and utilities.

■ Site features such as existing structures and plants on the property and adjacent property.

■ The scale.

■ North direction.

■ Property boundaries.

■ Slope shown by contour lines, cross sections, or both.

■ Plan legend.

■ Traffic patterns.

■ Solar site information, if solar application is intended.

■ Pedestrian patterns.

Planned Unit Development

A creative and flexible approach to land development is a planned unit development. Planned unit developments may include such uses as residential areas, recreational areas, open spaces, schools, libraries, churches, or convenient shopping facilities. Developers involved in these projects must pay particular attention to the impact on local existing developments. Generally, the plats (site plans) for these developments must include all of the same information shown on a subdivision plat, plus:

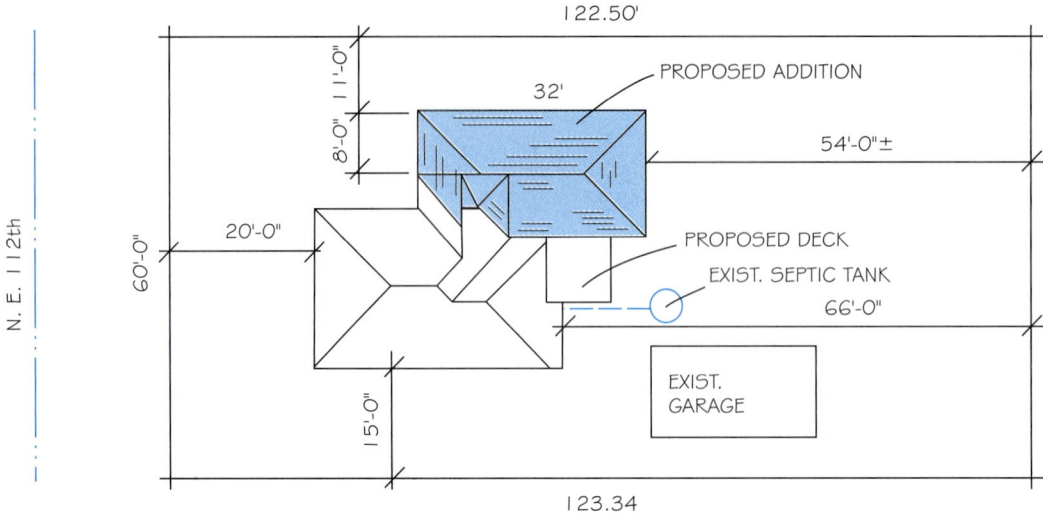

PLOT PLAN

1" ═══════ 20'-0"

LEGAL:

LOT #3
BLOCK F1
VIEW RIDGE
MULTNOMAH COUNTY

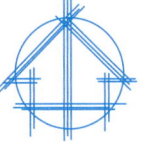

FIGURE 24.54 ■ Sample site plan showing existing home and proposed addition.

■ A detailed vicinity map, as shown in Figure 24.56.

■ Land use summary.

■ Symbol legend.

■ Special spaces such as recreational and open spaces, or other unique characteristics.

Figure 24.57 shows a typical planned unit development plan. These plans, as in any proposed plat plan, can require changes before the final drawings are approved for development.

There are several specific applications for a site or plat plan. The applied purpose of each will be different, although the characteristics of each type of site plan can be similar. Local districts have guidelines for the type of site plan required. Be sure to evaluate local guidelines before preparing a site plan for a specific purpose. Prepare the site plan in strict accordance with the requirements in order to receive acceptance.

INTRODUCTION TO GRADING PLANS

Grading plans are construction drawings that generally show existing and proposed topography. The outline of the structure can be shown with elevations at each building corner and the elevation given for each floor level. Figure 24.58 shows a detailed grading plan for a residential construction site. Notice that the legend identifies symbols for existing and finished contour lines. This particular grading plan provides retaining walls and graded slopes to accommodate a fairly level construction site from the front of the structure to the extent of the rear yard. The finished slope represents an embankment that establishes the relationship of the proposed contour to the existing contour. This grading plan also shows a proposed irrigation and landscaping layout.

Grading plan requirements can differ from one location to the next. Some grading plans also show a cross section through the site at specified intervals or locations to evaluate the contour more fully. This cross section, as previously discussed, is called a profile. A profile through the grading plan in Figure 24.58 is shown in Figure 24.59, page 836. Drawing a grading plan and site profile is explained later in this chapter.

Subdivision Plans

Local requirements for subdivisions should be confirmed because procedures vary, depending on local guidelines and zoning rules.

Some areas have guidelines for minor subdivisions with few sites that differ from major subdivisions with many sites. The term plat, used in the following discussion, refers to a map, plan, or layout of a subdivision indicating the location and boundaries

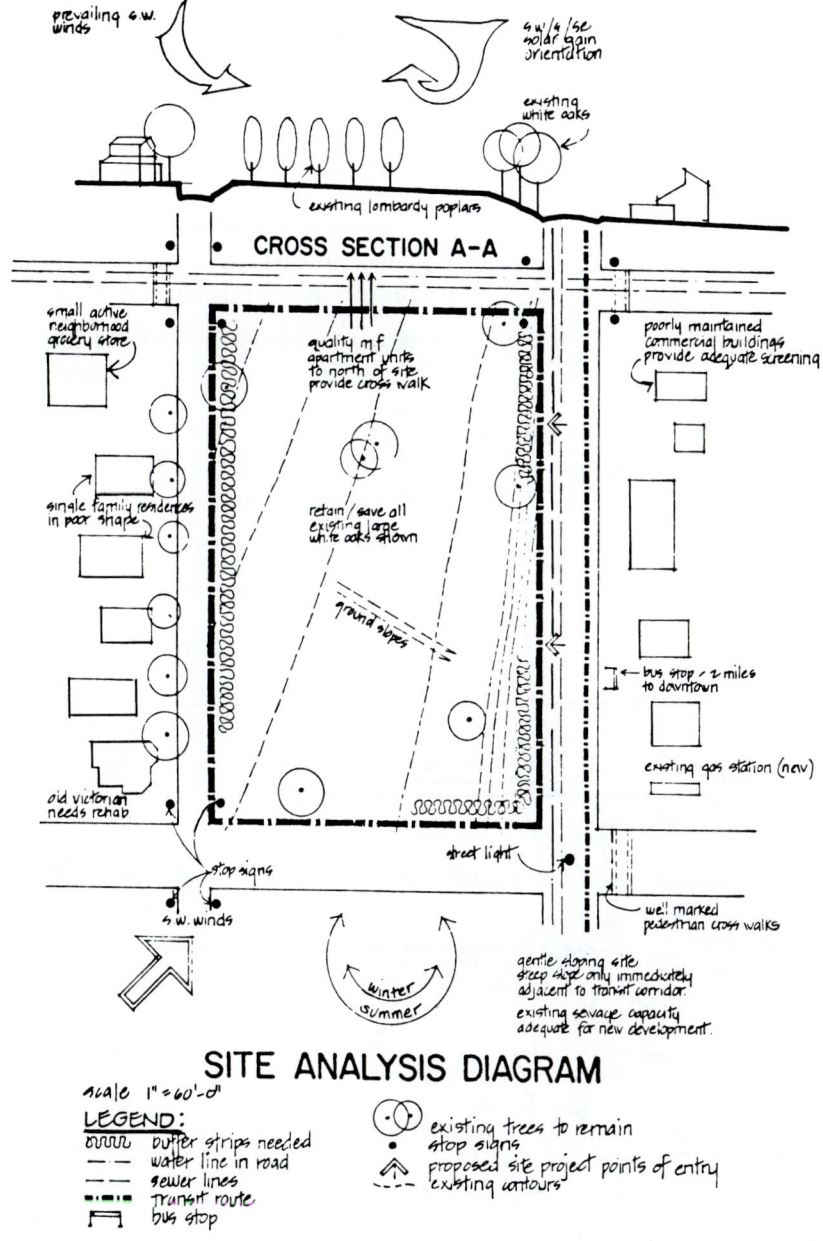

FIGURE 24.55 ■ Site analysis plan. *Courtesy Planning Department, Clackamas County, OR.*

of individual properties, typically called lots. The term subdivision is defined as a tract of land divided into residential lots. The plat required for a minor subdivision can include the following:

■ Legal description.

■ Name, address, and telephone number of applicant.

■ Parcel layout with dimensions.

■ Direction of north.

■ All existing roads and road widths.

■ Number identification of parcels, such as Parcel 1, Parcel 2.

■ Location of well or proposed well, or name of water district.

■ Type of sewage disposal (septic tank or public sanitary sewers). Name of sewer district.

■ Zoning designation.

■ Size of parcel(s) in square feet or acres.

■ Slope of ground (arrows pointing downslope).

■ Setbacks of all existing buildings, septic tanks, and drainfields from new property lines.

■ All utility and drainage easements.

■ Any natural drainage channels. Indicate direction of flow and whether drainage is seasonal or year-round.

PACIFIC OCEAN
BEACH
ENTRANCE
EASTWOOD
COVE
N
HWY 101
CARMEL 15 MILES

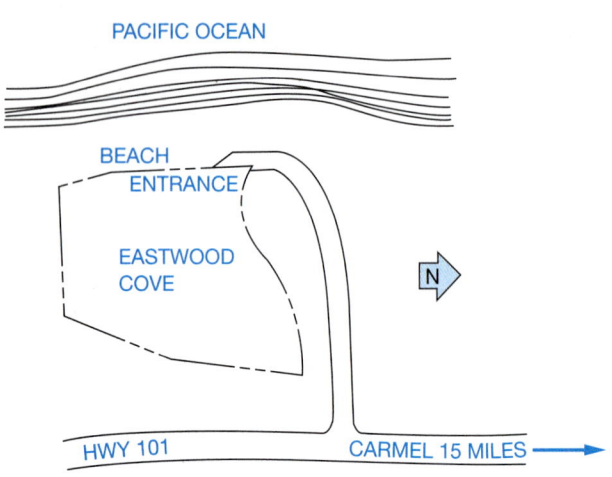

FIGURE 24.56 ■ Vicinity map.

- Map scale.
- Date.
- Building permit application number, if any.

Figure 24.60 shows a typical subdivision of land with three proposed parcels. The local planning agency determines the measurements for a minor or major subdivision.

A major subdivision can require more detailed plats than a minor subdivision. Some of the items that can be included on the plat or in a separate document are the following:

- Name, address, and telephone number of the property owner, applicant, and engineer or surveyor.
- Source of water.
- Method of sewage disposal.
- Existing zoning.
- Proposed utilities.
- Calculations justifying the proposed density.

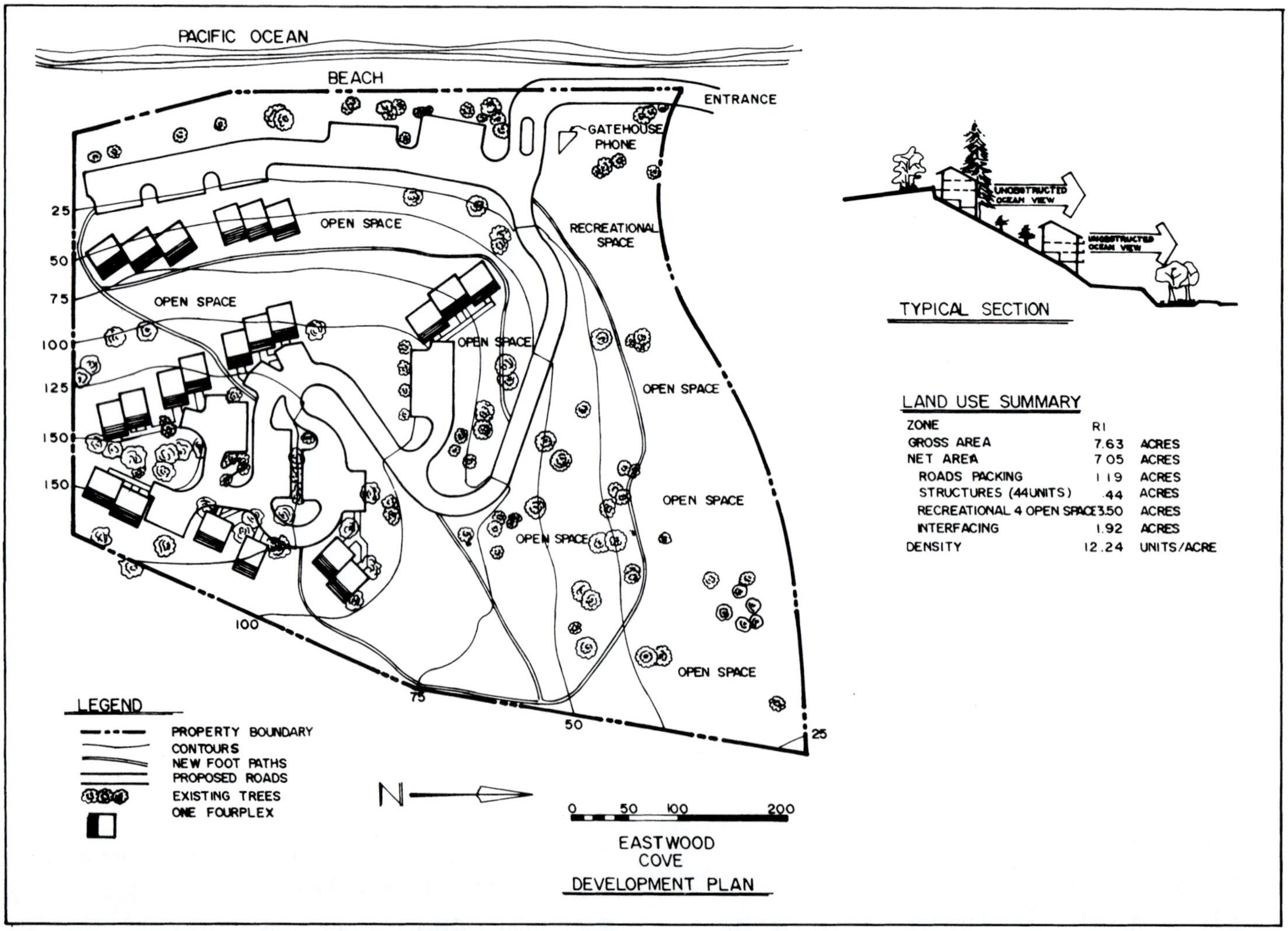

PACIFIC OCEAN
BEACH
ENTRANCE
GATEHOUSE
PHONE
RECREATIONAL SPACE
OPEN SPACE
25
50
75
100
125
150
150
100
75
50
25
TYPICAL SECTION
LAND USE SUMMARY
ZONE R1
GROSS AREA 7.63 ACRES
NET AREA 7.05 ACRES
ROADS PACKING 1.19 ACRES
STRUCTURES (44UNITS) .44 ACRES
RECREATIONAL 4 OPEN SPACE 3.50 ACRES
INTERFACING 1.92 ACRES
DENSITY 12.24 UNITS/ACRE
LEGEND
PROPERTY BOUNDARY
CONTOURS
NEW FOOT PATHS
PROPOSED ROADS
EXISTING TREES
ONE FOURPLEX
N
0 50 100 200
EASTWOOD
COVE
DEVELOPMENT PLAN

FIGURE 24.57 ■ Planned unit development plan.

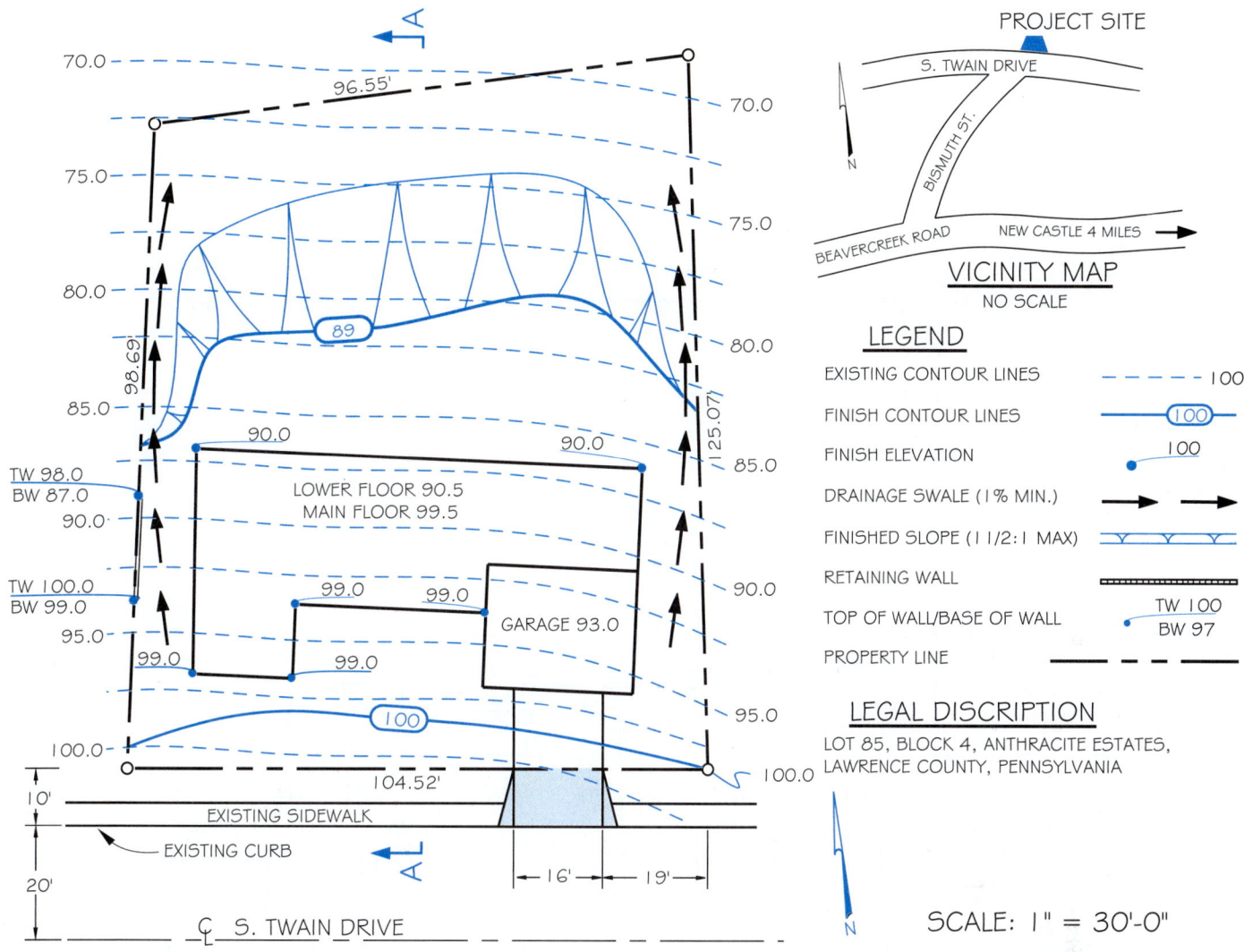

FIGURE 24.58 ■ Grading plan.

- Name of the major partitions or subdivision.
- Date the drawing was made.
- Legal description.
- North arrow.
- Vicinity sketch showing location of the subdivision.
- Identification of each lot or parcel and block by number.
- Gross acreage of property being subdivided or partitioned.
- Dimensions and acreage of each lot or parcel.
- Streets abutting the plat, including name, direction of drainage, and approximate grade.
- Streets proposed, including names, approximate grades, and radius of curves.
- Legal access to subdivision or partition other than public road.

- Contour lines at 2' interval for slopes of 10 percent or less, 5' interval if slopes exceed 10 percent.
- Drainage channels, including width, depth, and direction of flow.
- Locations of existing and proposed easements.
- Location of all existing structures, driveways, and pedestrian walkways.
- All areas to be offered for public dedication.
- Contiguous property under the same ownership, if any.
- Boundaries of restricted areas, if any.
- Significant vegetative areas such as major wooded areas or specimen trees.

Figure 24.61, page 838, shows an example of a small major subdivision plat.

(a)

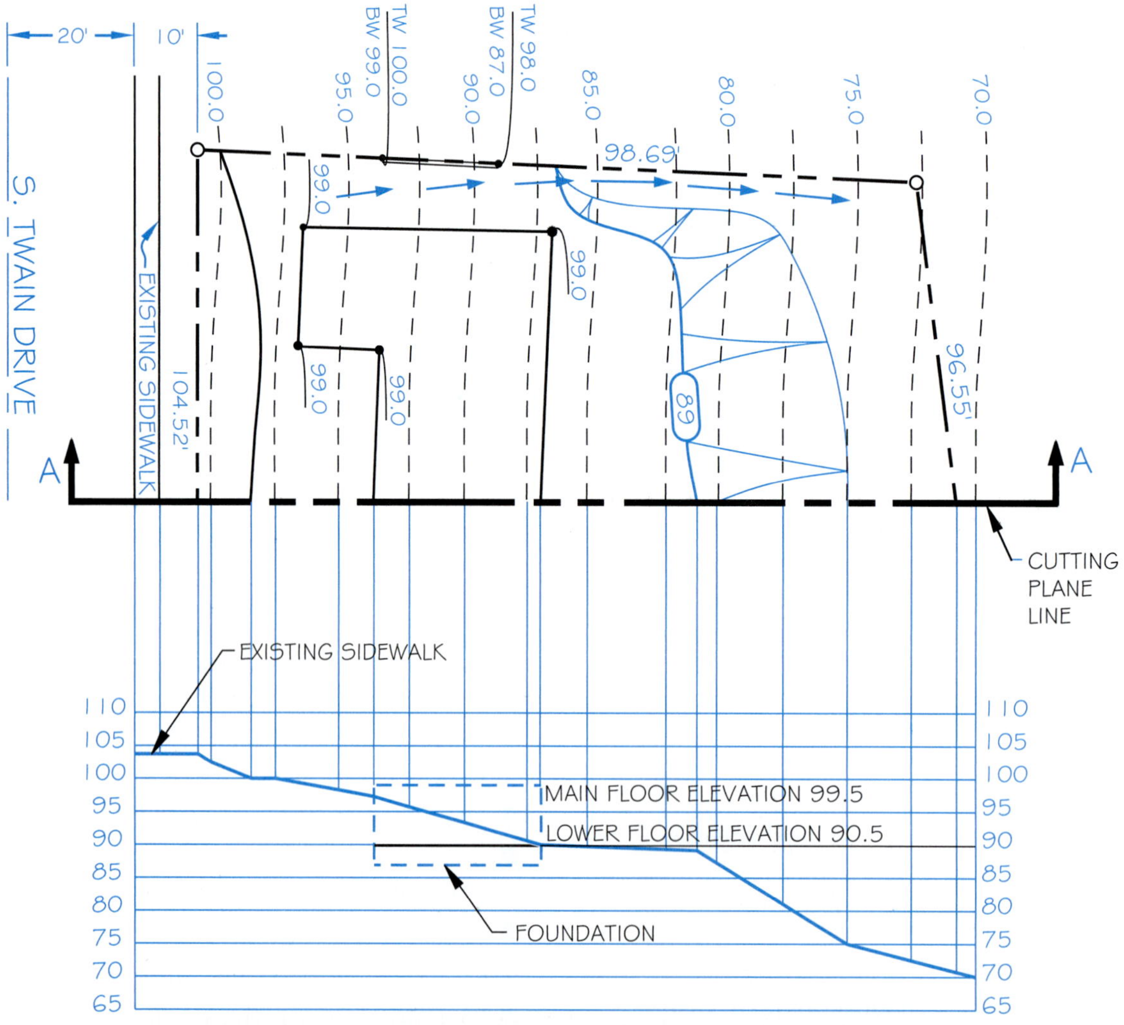

(b)

FIGURE 24.59 ■ Constructing a profile from the grading plan in Figure 24.58. (a) Establish a horizontal grid with a range of elevations found in the grading plan. (b) Project where each elevation crosses the cutting plane line perpendicular to the corresponding elevation at the horizontal grid.

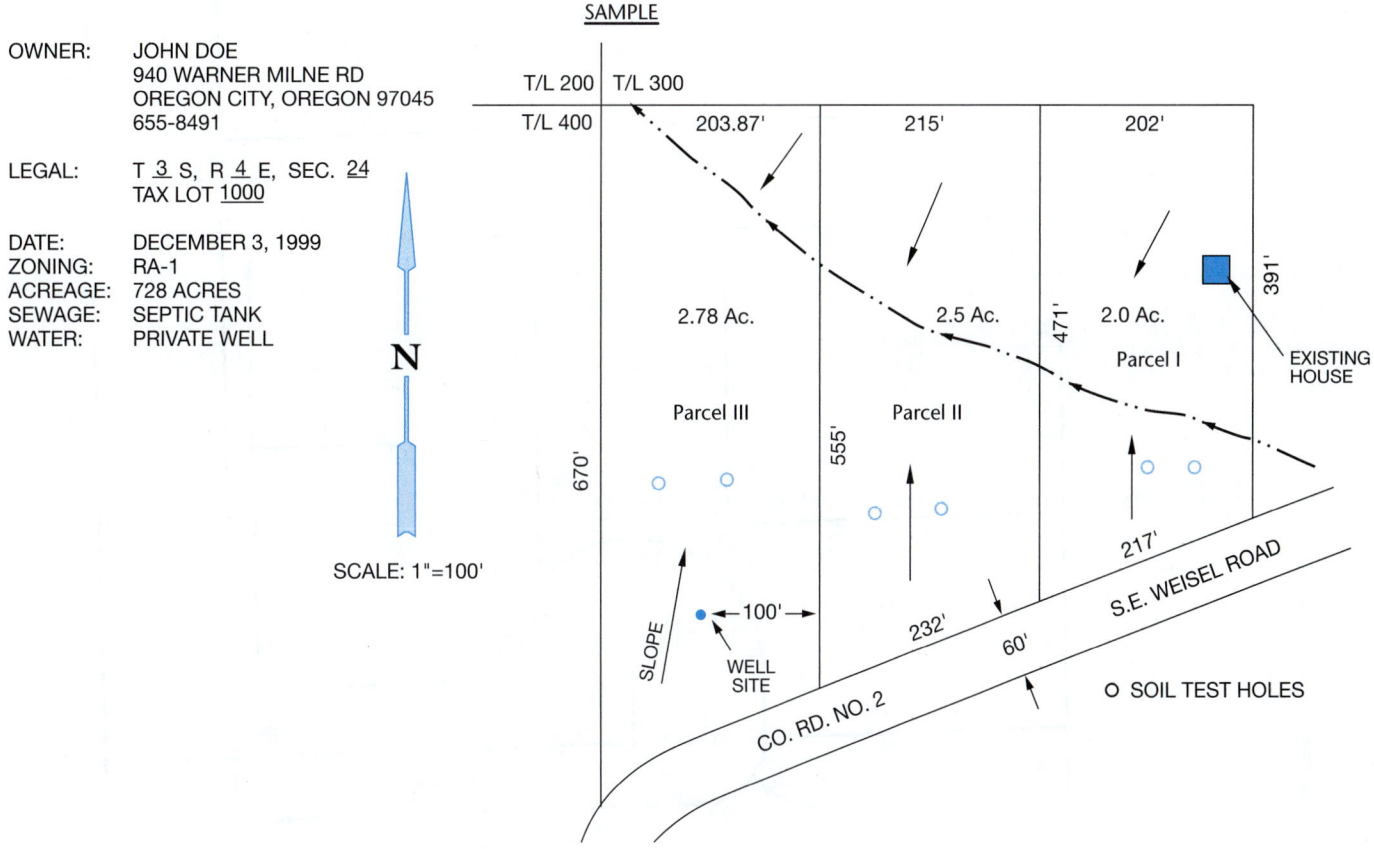

OWNER: JOHN DOE
         940 WARNER MILNE RD
         OREGON CITY, OREGON 97045
         655-8491

LEGAL: T 3 S, R 4 E, SEC. 24
         TAX LOT 1000

DATE: DECEMBER 3, 1999
ZONING: RA-1
ACREAGE: 728 ACRES
SEWAGE: SEPTIC TANK
WATER: PRIVATE WELL

N

SCALE: 1"=100'

SAMPLE

T/L 200 | T/L 300
T/L 400

203.87'   215'   202'

2.78 Ac.   2.5 Ac.   2.0 Ac.

Parcel III   Parcel II   Parcel I

WELL SITE   232'   60'   217'

CO. RD. NO. 2   S.E. WEISEL ROAD

O SOIL TEST HOLES

EXISTING HOUSE

SLOPE

FIGURE 24.60 ■ Subdivision with three proposed lots. *Courtesy Planning Department, Clackamas County, OR.*

METRICS IN SITE PLANNING

The recommended metric values used in the design and drafting of site plans based on surveying, excavating, paving, and concrete construction are as follows:

	Quantity	Unit and Symbol
Surveying	Length	meter (m) and kilometer (km)
	Area	square meter (m^2), hectare (ha), and square kilometer (km^2)
	Plane angle	degree (°), minute ('), second ("), and percent (%)
Excavating	Length	millimeter (mm) and meter (m)
	Volume	cubic meter (m^3)
Paving	Length	millimeter (mm) and meter (m)
	Volume	cubic meter (m^3)
Concrete	Length	millimeter (mm) and meter (m)
	Area	square meter (m^2)
	Volume	cubic meter (m^3)

INTRODUCTION TO SITE PLAN LAYOUT

Site plans can be drawn on media ranging in size from 8 1/2 × 11" up to 34 × 44" (A4 to A0 metric) depending on the purpose of the plan and the guidelines of the local government agency or lending institution requiring the plan. Many local jurisdictions recommend that site plans be drawn on a sheet 8 1/2 × 14" (210 × 360 mm).

Before you begin the site plan layout, there is some important information that you need. This information can often be found in the legal documents for the property, the surveyor's map, the local assessor's office, or the local zoning department. Figure 24.62 is a plat from a surveyor's map that can be used as a guide to prepare the site plan. The scale of the surveyor's plat can vary, although in this case it is 1" = 200' (1:1000 metric). The site plan to be drawn can have a scale ranging from 1" = 10' (1:50 metric) to 1" = 200'. The factors that influence the scale include the following:

■ Sheet size.

■ Plot size.

■ Amount of information required.

■ Amount of detail required.

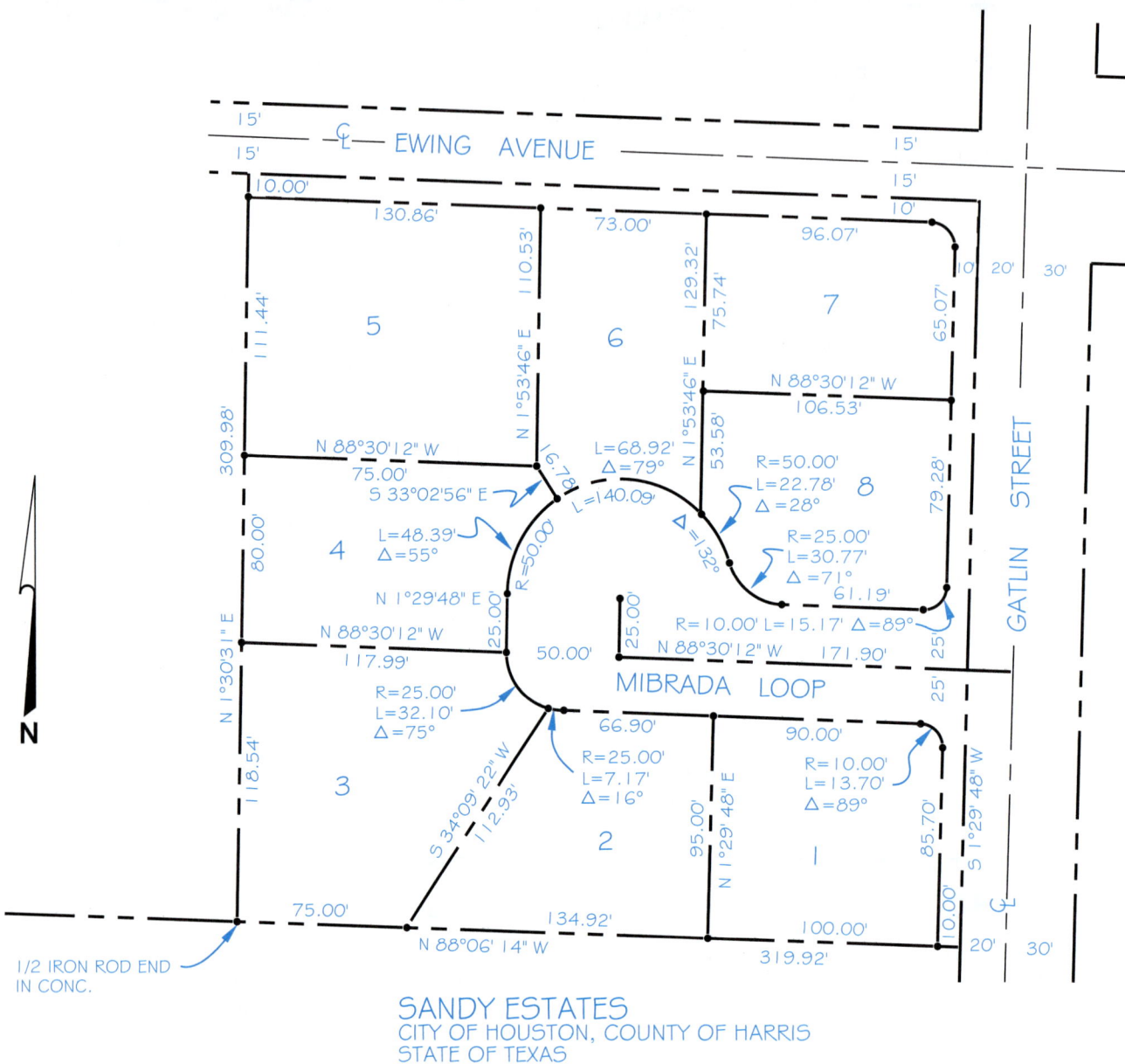

FIGURE 24.61 ■ Small major subdivision plat.

Additional information that should be determined before the site plan can be completed usually includes the following:

■ Legal description.

■ North direction.

■ All existing roads, utilities, water, sewage disposal, drainage, and slope of land.

■ Zoning information, including front, rear, and side yard setbacks.

■ Size of proposed structures.

■ Elevations at property corners, driveway at street, or contour elevations.

SITE DESIGN CONSIDERATIONS

Several issues that can affect the quality of the construction project and adjacent properties should be considered during site design. Some of these factors are outlined as follows:

■ Provide a minimum driveway slope of 1/4" per foot. The maximum slope depends on surface conditions and local requirements.

■ Provide a minimum lawn slope of 1/4" per foot if possible.

■ Single-car driveways should be a minimum of 10' (3 m) wide, and double-car driveways should be a minimum of

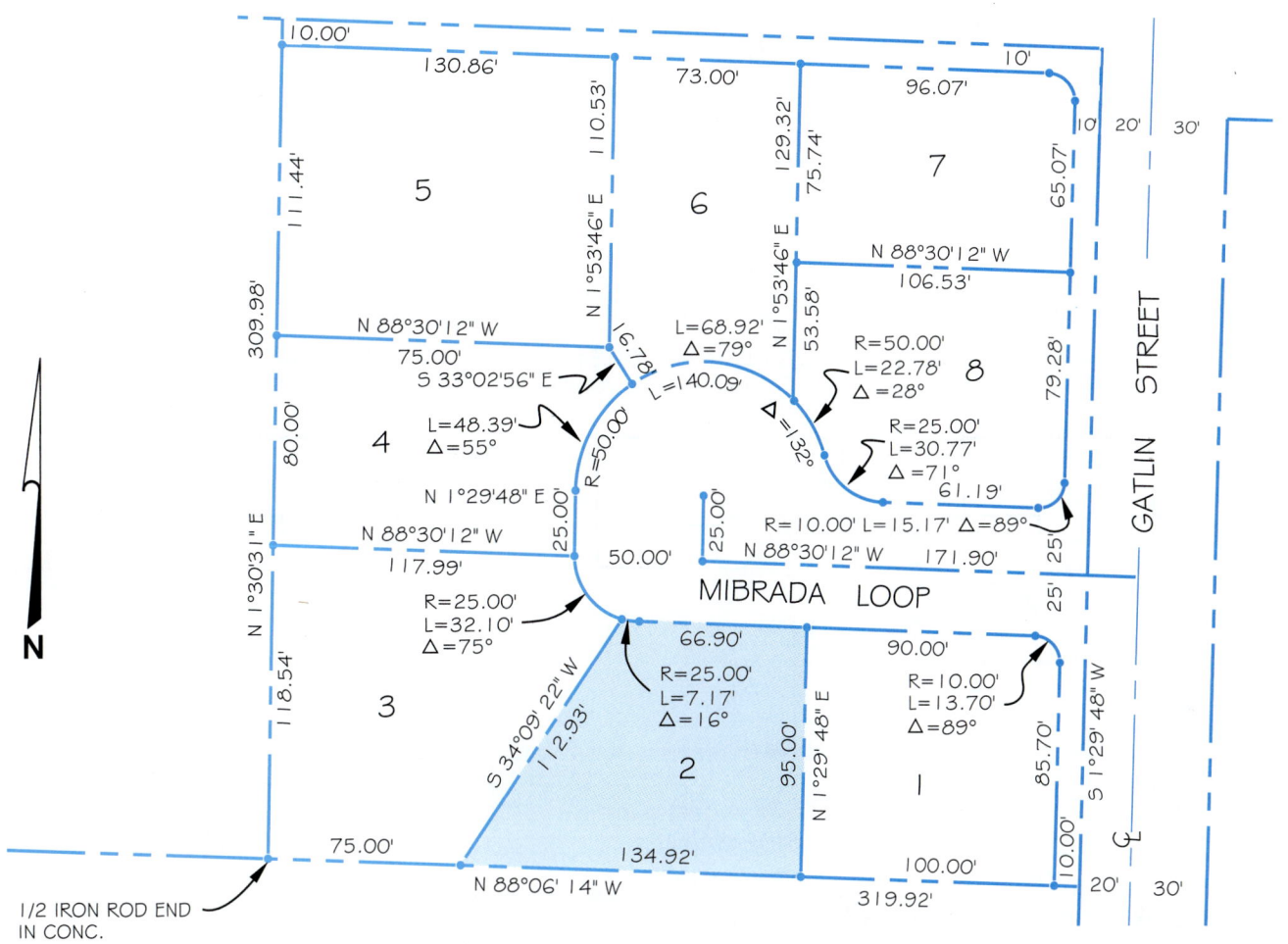

FIGURE 24.62 ■ Plat from a surveyor's map.

18' (5.5 m) wide but can taper to 10' wide at the entrance. Any reduction of driveway width should be centered on the garage door. A turning apron is preferred when space permits. This allows the driver to back into the parking apron and then drive forward into the street, which is safer than backing into the street.

■ The minimum turning radius for a driveway should be 15' (4.5 m). The turning radius for small cars can be less, but more should be considered for trucks. A turning radius of 20' (6 m) is preferred if space permits. Figure 24.63 shows a variety of driveway layouts for you to use as examples. The dimensions are given as commonly recommended minimums for small to standard-sized cars. Additional room should be provided if available.

■ Provide adequate room for the installation of and future access to water, sewer, and electrical utilities.

■ Do not build over established easements. An easement is the right-of-way for access to property and for the purpose of construction and maintenance of utilities.

■ Follow basic grading rules, which include not grading on adjacent property. Do not slope the site so as to cause water drainage onto adjacent property. Slope the site away from the house. Adequate drainage of at least 10' away from the house is recommended.

■ Identify all trees that are to remain on the site after construction.

■ Establish retaining walls where needed to minimize the slope, control erosion, and level portions of the site.

Rural Residential Fire Department Access and Site Planning

Special considerations should be made when designing driveways and turnarounds in rural areas. In urban and suburban areas, fire truck access is normally designed into the subdivision plan and each residence is fairly easy to access for firefighting purposes. Rural locations pose special problems for firefighting because they are often larger pieces of property that frequently

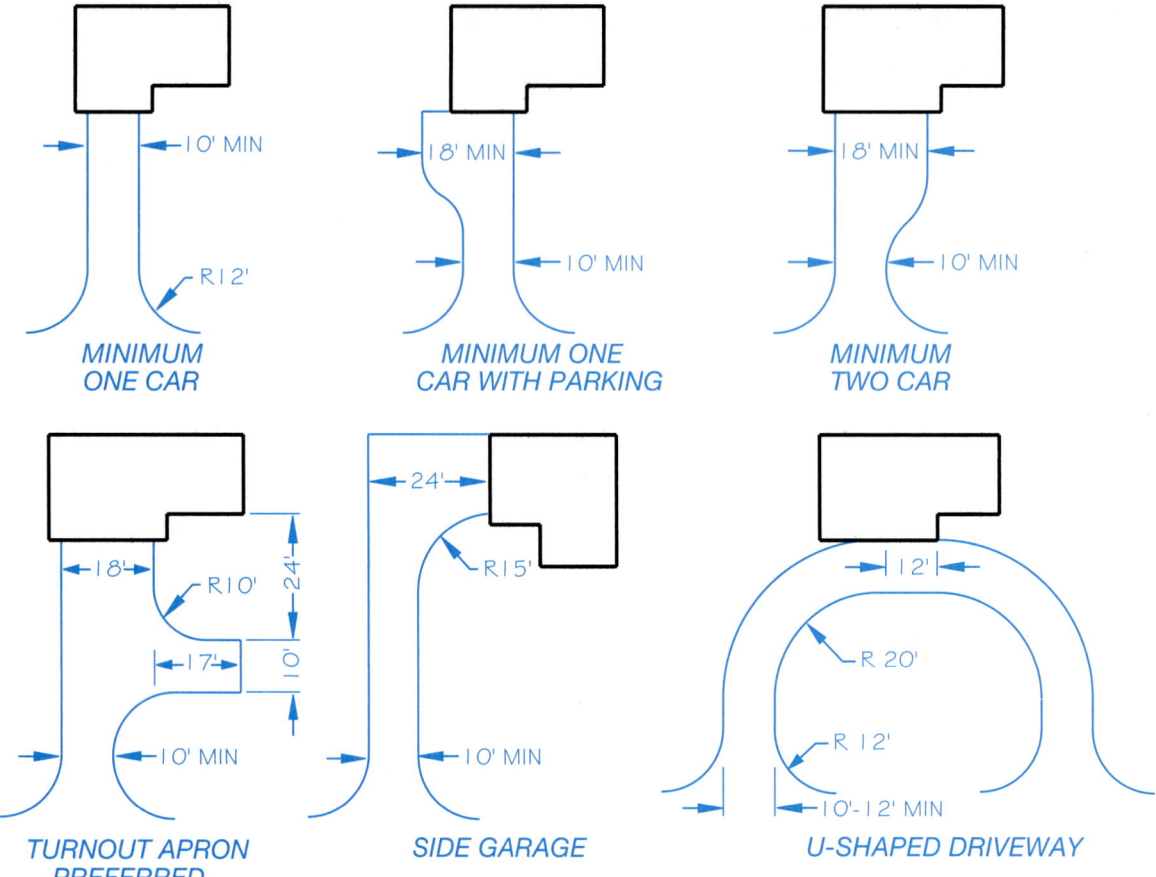

MINIMUM
ONE CAR

MINIMUM ONE
CAR WITH PARKING

MINIMUM
TWO CAR

TURNOUT APRON
PREFERRED

SIDE GARAGE

U-SHAPED DRIVEWAY

FIGURE 24.63 ■ Typical driveway layout options.

have long gravel driveways where access can be limited. Check the regulations for your location. The guidelines governing the site design can differ from one location to the next, but the following are common standards:

■ *Road clearances:* A 15' minimum width all-weather surface driveway must be provided, with an additional 5' width clear of vegetation. The driveway must also be clear of vegetation to a height of 13'−6". All-weather means gravel or paved surface. Figure 24.64 shows several basic driveway design options.

■ *Road load capacities:* The driveway must be engineered for a 12,500-lb wheel load and 50,000 lb gross vehicle load.

■ *Grade:* A 10 percent average minimum road grade is preferred, but up to 15 percent for 200' is acceptable.

■ *Dead ends:* Provide a turnaround if the driveway is longer than 150'.

■ *Turnouts:* Provide a 20' wide by 40' long passage space at the midpoint of every 400' length.

■ *Bridges and culverts:* These features must be designed to support a minimum of 50,000 lb.

■ *Fire Safety zone:* Provide a firebreak at least 30' around all structures. A firebreak requires ground cover no more than

24" in height and all dead vegetation removed. Steep terrain around the structure can require a greater firebreak.

■ *Property identification:* The property address shall be posted on a fire department-approved sign where the driveway meets the main road.

■ *Firefighting water supply:* On-site water supplies, such as a swimming pool, pond, or water storage tank, must be accessible within 15'. A fire sprinkler system designed and installed in the home can be substituted for a water supply.

■ *Roof coverings in wildfire zones:* Wildfire zones are heavily wooded areas. The use of wood roofing material or other combustible materials is restricted in these areas.

LAYING OUT PROPERTY LINES

Earlier in this chapter, you were introduced to the methods for describing properties. Look at Figure 24.62 and notice that the property lines of each lot and the boundaries of the plat are labeled with distances and bearings. An example is the west property line of Lot 2, which is 112.93' for the length and S 34°09'22" W for the bearing. This property line is set up to be drawn as shown in Fig-

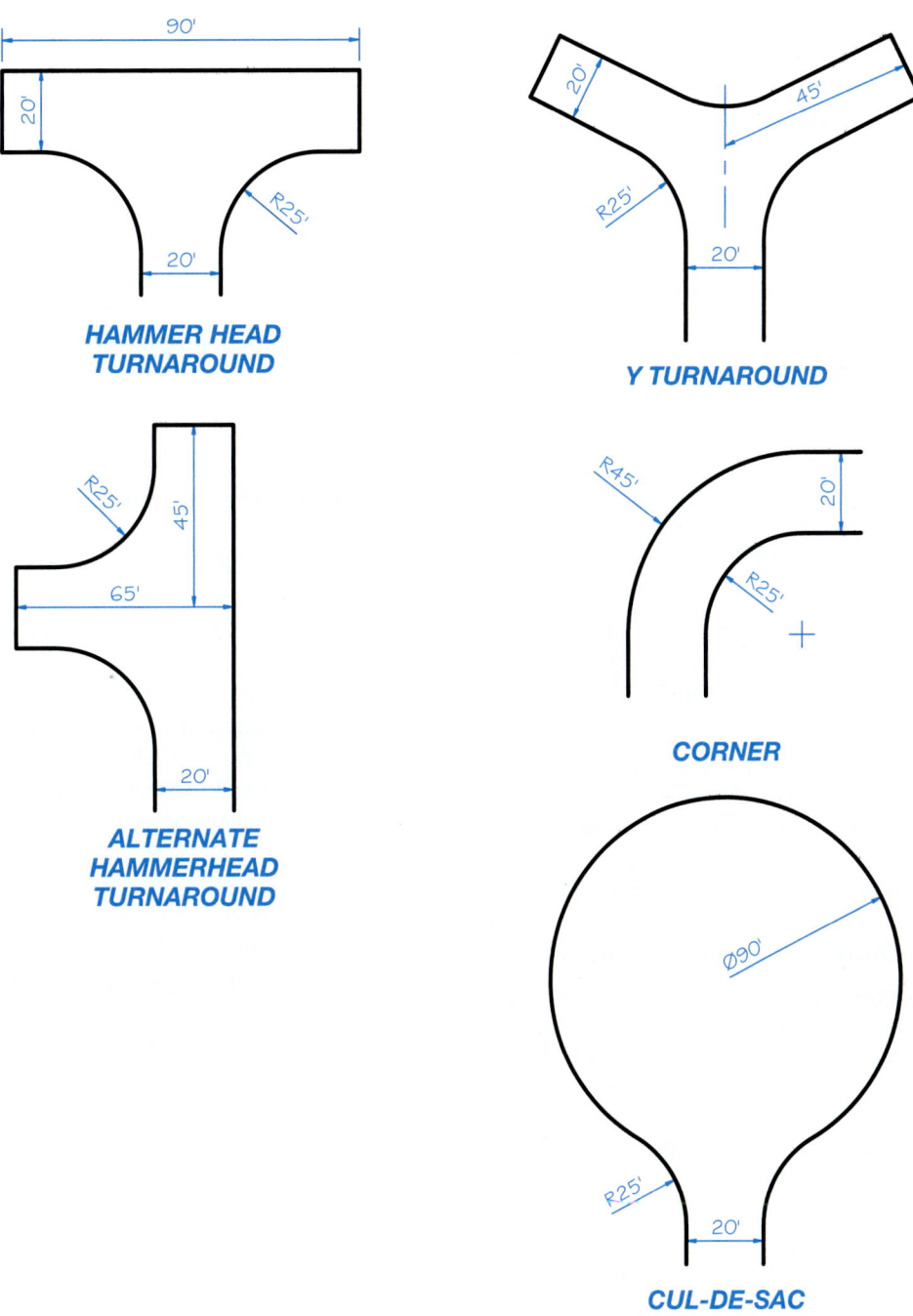

FIGURE 24.64 ■ Driveway turnaround options for rural residential fire department access.

ure 24.65a. After the property lines are drawn, they are labeled with the distance and bearing, as shown in Figure 24.65b.

Many plats have property lines that curve, such as the lines around the cul-de-sac in Figure 24.62. For example, the largest curve is labeled R = 50.00' L = 140.09'. R is the radius of the curve, and L is the length of the curve. There are three sublengths in this curve. L = 48.39' is for Lot 4, L = 68.92' is for Lot 6, and L = 22.78' is for Lot 8. Figure 24.66a shows the setup for drawing this curve using the radius and arc lengths. Plats also typically show a Delta angle for curves, which is represented by the symbol Δ. The Delta angle is the included angle

of the curve. The included angle is the angle formed between the center and the endpoints of the arc, as shown in Figure 24.66b. Figure 24.66c shows the final labeling of the curve.

STEPS IN SITE PLAN LAYOUT

Follow these steps to draw a site plan.

STEP 1   Select the paper size. In this case the size is 8 1/2 × 11" (A4 metric). Evaluate the plot to be drawn. Lot 2 of

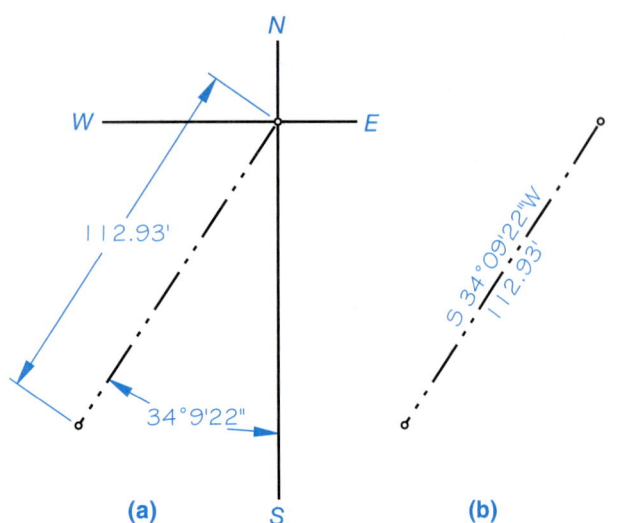

(a)   **(b)**

FIGURE 24.65 ■ Drawing and labeling a property line.

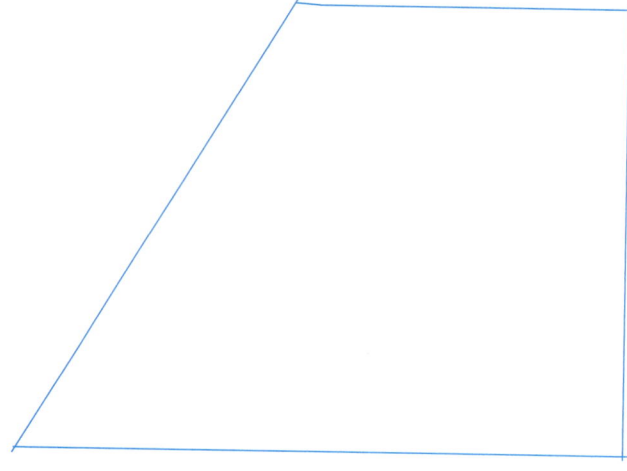

FIGURE 24.67 ■ Step 2: Lay out the plot plan property lines.

Sandy Estates, shown in Figure 24.62, is used. Determine the scale to use by considering how the longest dimension (134.92') fits on the sheet. Always try to leave at least a 1/2" (12 mm) margin around the sheet.

STEP 2 Use the given plat as an example to lay out the proposed site plan. If a plat is not available, then the site plan can be laid out from the legal description by establishing the boundaries using the bearings and dimensions in feet. Lay out the entire site plan using construction lines. If errors are made, the construction lines are very easy to erase, or use CADD layers. See Figure 24.67.

STEP 3 Lay out the proposed structure using construction lines. The proposed structure in this example is 60' long and 36' wide. The house can be drawn on the site plan with or without showing the roof. A common practice is to draw only the outline of the floor plan. In some cases, the roof overhang is considered in the setback. In these situ-

ations, the house outline is drawn as a dashed line under the roof and dimensions are given for the house and the overhang. The front setback is 25', and the east side is 15'. Lay out all roads, driveways, walks, and utilities. Be sure the structure is inside or on the minimum setback requirements. Setbacks are imaginary boundaries beyond which the structure may not be placed. Think of them as property line offsets that are established by local regulations. Minimum setbacks can be confirmed with local zoning regulations. The minimum setbacks for this property are 25' front, 10' sides, and 35' back from the property lines to the house. See Figure 24.68.

STEP 4 Darken all property lines, structures, roads, driveways, walks, and utilities, as shown in Figure 24.69. Some drafters use a thick line or shading for the structure.

STEP 5 Add dimensions and contour lines (if any) or elevations. The property line dimensions are generally placed on the inside of the line in decimal feet, and the bearing is

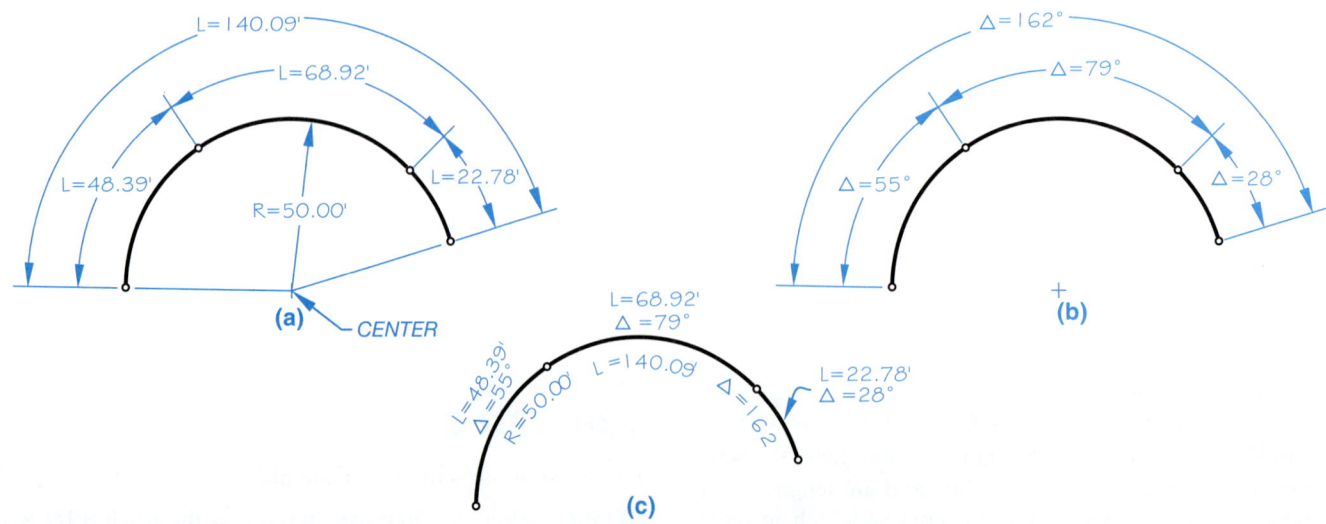

FIGURE 24.66 ■ Drawing and labeling curved property lines.

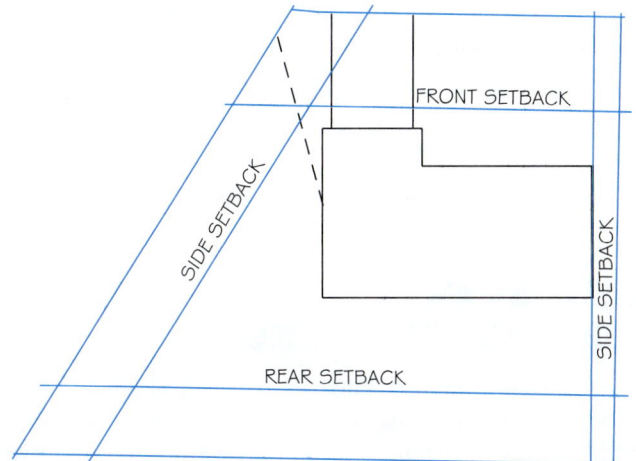

FIGURE 24.68 ■ Step 3: Lay out the structures, roads, driveways, walks, and utilities. Be sure the structure is on or within the minimum required setbacks.

Generally, the plat or legal description provides metes and bounds coordinates for property line boundaries. For example, the south property line for Lot 2 has the coordinates given with the bearing and distance of N 88°06'14"W and 134.92'. CADD systems that are used for either general or architectural applications have a surveyor's units setting. When the CADD system is set to draw lines based on surveyor's units, the south property line of Lot 2 is drawn with this computer prompt: @134.92' < N 88d06' 14"W. Notice that the degree symbol (°) is replaced with the lowercase letter *d* when entering a bearing at the computer prompt.

Most CADD programs have options for drawing arcs that allow you to create the curves in Figure 24.66. The commands typically allow you to draw arcs with a combination of radius, arc length, and included angle. This provides you with the flexibility needed to use the R, L, and Δ information provided on the plat or in the surveyor's notes.

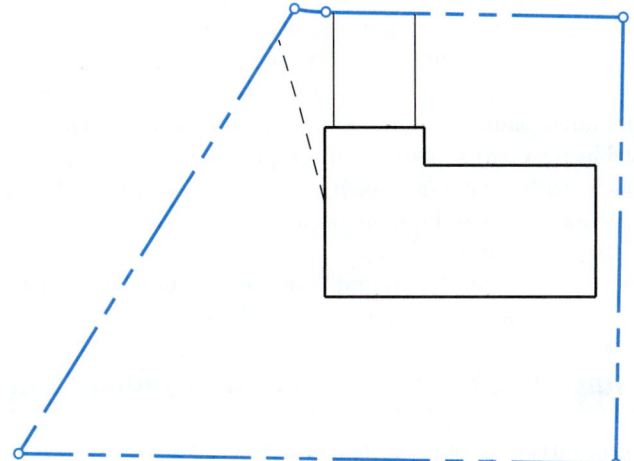

FIGURE 24.69 ■ Step 4: Darken boundary lines, structures, roads, driveways, walks, and utilities.

placed on the outside of the line. The dimensions locating and giving the size of the structure are commonly in feet and inches or in decimal feet. Try to keep the amount of extension and dimension lines to a minimum on the site plan. One way to do this is to dimension directly to the house and place size dimensions inside the house outline. Add all labels, including the road name, property dimensions and bearings (if used), utility names, walks, and driveways, as shown in Figure 24.70.

STEP 6   Complete the site plan by adding the north arrow, the legal description, title, scale, client's name, and other title block information. Figure 24.71 shows the complete site plan.

Site Plan Drawing Checklist

Check off the items in the following list as you work on the basic site plan, to be sure that you have included all of the necessary

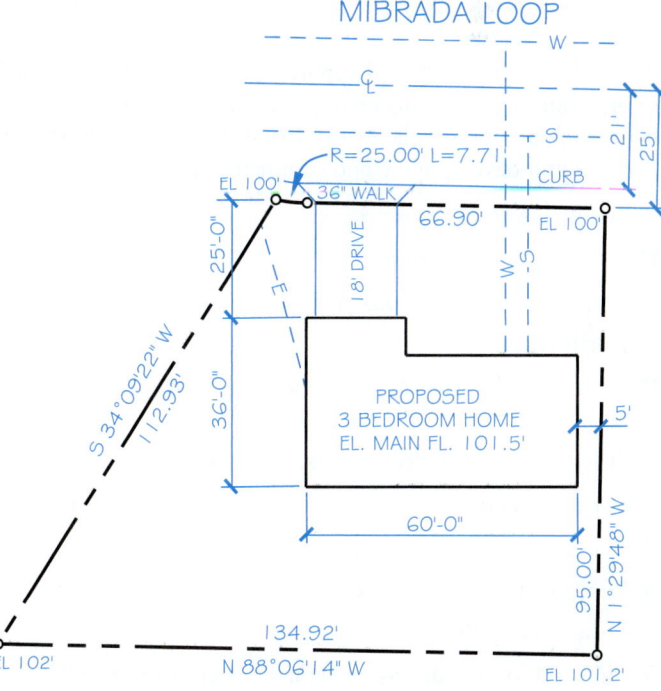

FIGURE 24.70 ■ Step 5: Add dimensions and elevations; then label all roads, driveways, walks, and utilities.

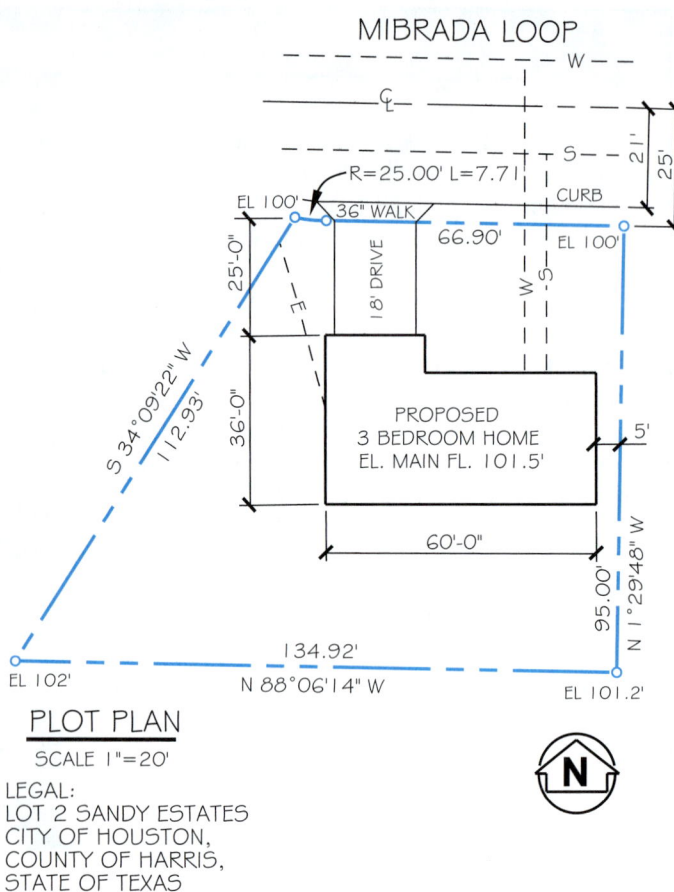

MIBRADA LOOP

R=25.00' L=7.71

EL 100'   36" WALK

66.90'   EL 100'

PROPOSED
3 BEDROOM HOME
EL. MAIN FL. 101.5'

EL 102'   134.92'

N 88°06'14" W   EL 101.2'

PLOT PLAN

SCALE 1"=20'

LEGAL:
LOT 2 SANDY ESTATES
CITY OF HOUSTON,
COUNTY OF HARRIS,
STATE OF TEXAS

FIGURE 24.71 ■ Step 6: Complete the plot plan. Add title, scale, north arrow, legal description, and other necessary information, such as the owner's name if required.

details. Site plans for special applications may require additional information. Refer to this chapter for features found in special plans such as grading plan, subdivision plan, site analysis plan, planned unit development, and commercial plan.

■ Site plan title and scale.

■ Property legal description.

■ Property line dimensions and bearings.

■ North arrow.

■ Existing and proposed roads with the elevation at the center of roads.

■ Driveways, patios, decks, walks, and parking areas.

■ Existing and proposed structures with floor-level elevations.

■ Public or private water supplies.

■ Public or private sewage disposal.

■ Location of utilities.

■ Rain and footing drains and storm sewer or drainage.

■ Topography, including contour lines or elevations at property corners, street centerline, driveways, and floor elevations.

■ Front, side, and rear setbacks dimensioned and in compliance with zoning.

■ Specific items on adjacent properties, if required, such as existing structures, water supply, sewage disposal, trees, or water features.

■ Existing and proposed trees may be required.

DRAWING CONTOUR LINES

Contour lines represent intervals of equal elevation, as explained earlier in this chapter. The following discussion shows you how to lay out contour lines on a construction site. Surveyors can use various methods to establish elevations at points on the ground. This information is recorded in field notes. You then use the field notes to plot the contour lines. This discussion explains the grid survey method. Another common technique is the control point survey, which establishes elevations that are recorded on a map. You then lay out the elevations for the contour lines based on the given elevation points The radial survey is also a common method used for locating property corners, structures, natural features, and elevations points. This system establishes control points using a process called radiation in which measurements are taken from a survey instrument located at a base known as a transit station. From the transit station, a series of angular and distance measurements are established to specific points on the ground. You then establish the property lines, land features, and contours based on these points.

Using the Grid Survey to Draw Contour Lines

A grid survey divides the site into a pattern similar to a checkerboard. Stakes are driven into the ground at each grid intersection. The surveyor then establishes an elevation at each stake and records this information in field notes. The spacing of the stakes depends on the land area and the topography. The stakes may be placed in a grid 10, 20, 50, or 100' apart, for example. An example is shown in Figure 24.72. This grid has lines spaced 20' apart. The vertical lines are labeled with letters, and the horizontal lines are labeled with numbers that are called stations. The station numbers are in two parts: for example, 0 + 20. The first number is hundreds of feet. Zero represents 0 hundreds, or 0'. The first number is followed by a plus (+) sign. This means that you add the first number to the second number to get the actual distance to this station. The second number is in tens of feet. For example, 0' + 20' or 1' + 20' = 120'. The intersection of vertical line B and station 0 + 40 is identified as station B-0 + 40. The field notes record the elevation at each station, as shown in Table 24.1. For example, the elevation at A-0 + 80 is 105'.

Steps in Drawing Contour Lines

Contour lines are drawn from grid survey field notes using the following steps:

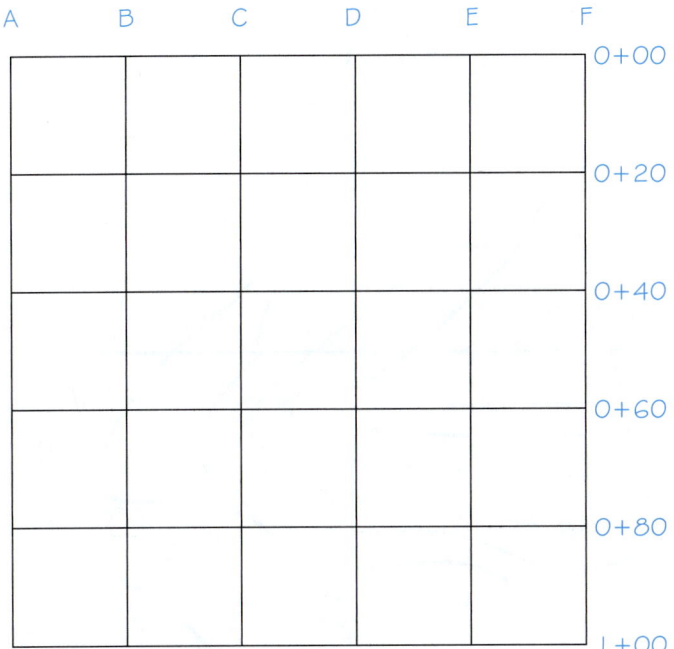

FIGURE 24.72 ■ Draw a grid at a desired scale. Label the vertical lines with letters and the horizontal lines with station numbers, as shown in this example.

TABLE 24.1 GRID SURVEY FIELD NOTES

Station	Elevation	Station	Elevation
A-0 + 00	101	D-0 + 00	95
A-0 + 20	104	D-0 + 20	97
A-0 + 40	108	D-0 + 40	99
A-0 + 60	112	D-0 + 60	105
A-0 + 80	105	D-0 + 80	100
A-1 + 00	102	D-1 + 00	94
B-0 + 00	100	E-0 + 00	94
B-0 + 20	102	E-0 + 20	96
B-0 + 40	105	E-0 + 40	98
B-0 + 60	110	E-0 + 60	100
B-0 + 80	104	E-0 + 80	95
B-1 + 00	100	E-1 + 00	92
C-0 + 00	98	F-0 + 00	92
C-0 + 20	100	F-0 + 20	93
C-0 + 40	102	F-0 + 40	95
C-0 + 60	108	F-0 + 60	98
C-0 + 80	102	F-0 + 80	90
C-1 + 00	95	F-1 + 00	85

STEP 1 Draw a grid at a desired scale similar to Figure 24.72. Use construction lines if you are drafting manually, or establish a GRID or CONSTRUCTION layer if you are using CADD. Label the vertical and horizontal grid lines.

STEP 2 Use the field notes to label the elevation at each grid intersection. The elevations, labeled on the grid in Figure 24.73, are based on the field notes found in Table 24.1.

STEP 3 Determine the desired contour interval and connect the points on the grid that represent contour lines at this interval. The contour interval for the grid in Figure 24.73 is 2'. This means that you will establish contour lines every 2', such as at 92, 94, 96, and 98'.

STEP 4 Establish the contour lines at the desired contour interval by picking points on the grid that represent the desired elevations. If a grid intersection elevation is 100, then this is the exact point for the 100' elevation on the contour line. If the elevation at one grid intersection is 98' and the next is 102', then you estimate the location of 100' between the two points. This establishes another 100' elevation point.

STEP 5 When you have identified the points for all of the 100' elevations, then connect the points to create the contour line. See Figure 24.74. Do this for the elevation at each contour interval.

STEP 6 Darken the contour lines. The intermediate lines are thin. The index contour lines are broken and labeled with the elevation and are generally drawn thicker than intermediate lines, as shown in Figure 24.75.

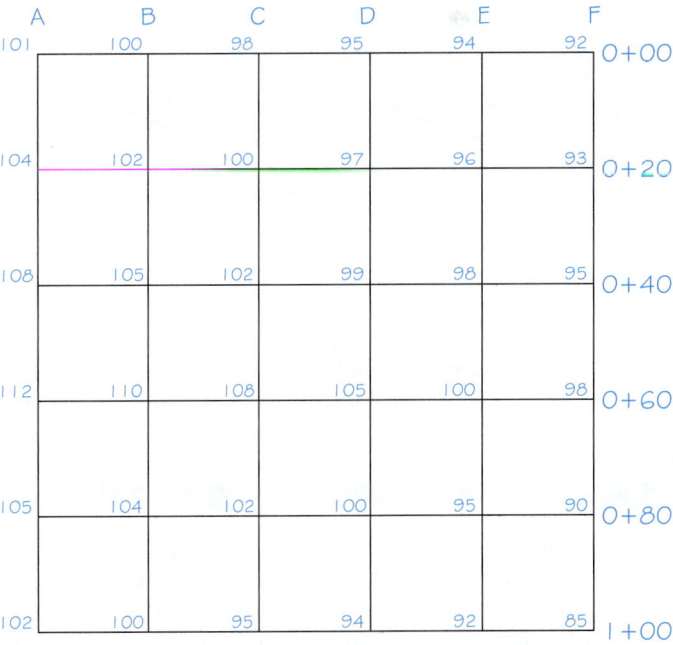

FIGURE 24.73 ■ Use the field notes in Table 24.1 to label the elevation at each grid intersection.

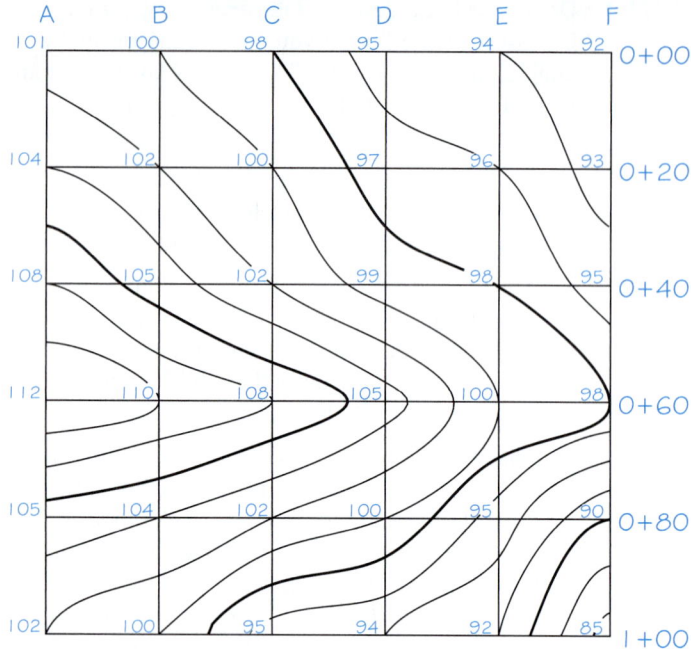

FIGURE 24.74 ■ Connect the points at the elevations for each contour interval.

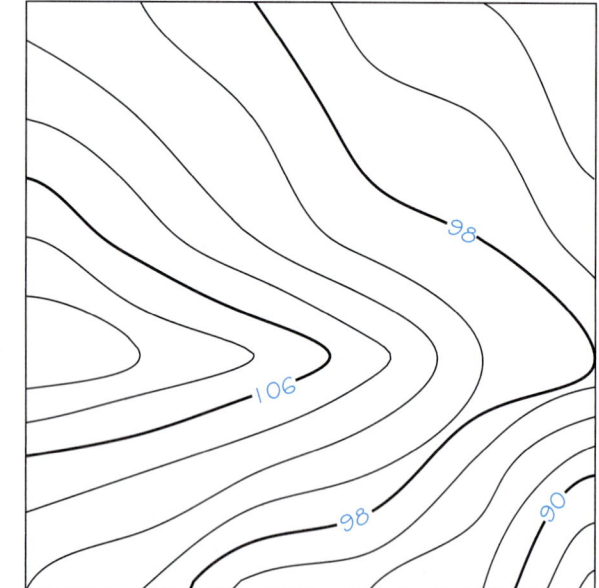

FIGURE 24.75 ■ Darken the contour lines. The intermediate lines are thin, and the index contour lines are broken and labeled with the elevation of the contour line.

DRAWING SITE PROFILES

As described earlier, a profile is a vertical section of the surface of the ground, and/or of underlying earth that is taken along any desired fixed line. The profile of a construction site is usually through the building excavation location, but more than one profile can be drawn as needed. The profile for road construction is normally placed along the centerline. Profiles are drawn

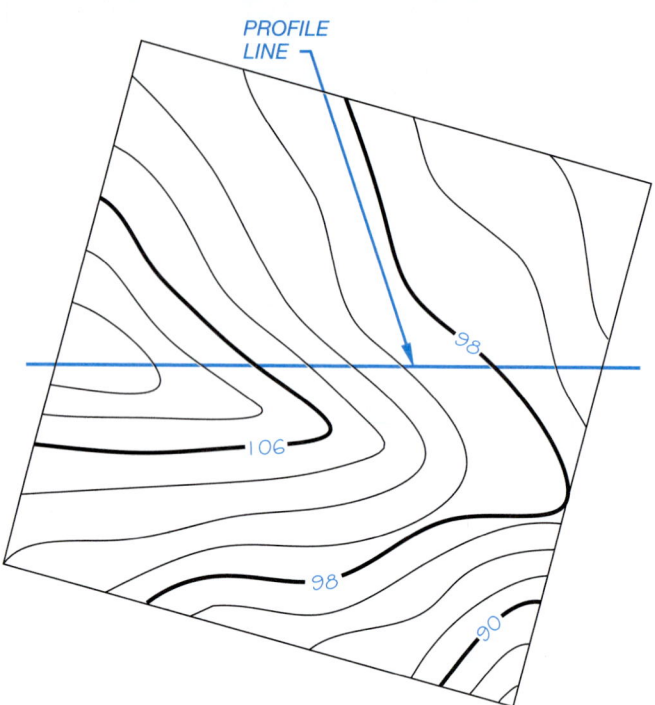

FIGURE 24.76 ■ Draw a straight line on the contour map at the location of the desired profile.

from the contour lines at the section location. The contour map and its related profile are commonly referred to as the **plan and profile**. Profiles have uses such as showing road grades and site excavation. Projecting directly from the desired cut location on the contour map following these steps creates the profile:

STEP 1   Draw a straight line on the contour map at the location of the desired profile. See Figure 24.76.

STEP 2   Set up the profile vertical scale. The horizontal scale is the same as the map, because you are projecting directly from the map. The vertical scale can be the same or it can be exaggerated. Exaggerated vertical scales are used to give a clearer representation of the contour when needed. Establish a vertical scale increment that is above the maximum elevation and one that is below the minimum elevation. Figure 24.77 shows how the vertical scale is set up. Notice that the profile is projected 90° from the profile line on the map.

STEP 3   Project a line from the location where every contour line crosses the profile line on the contour map. See Figure 24.78.

STEP 4   Draw the profile by connecting the points where every two vertical and horizontal lines of the same elevation intersect, as shown in Figure 24.79.

DRAWING THE GRADING PLAN

If you have a specific location on a site where a level excavation must take place for the proposed construction, you can lay out

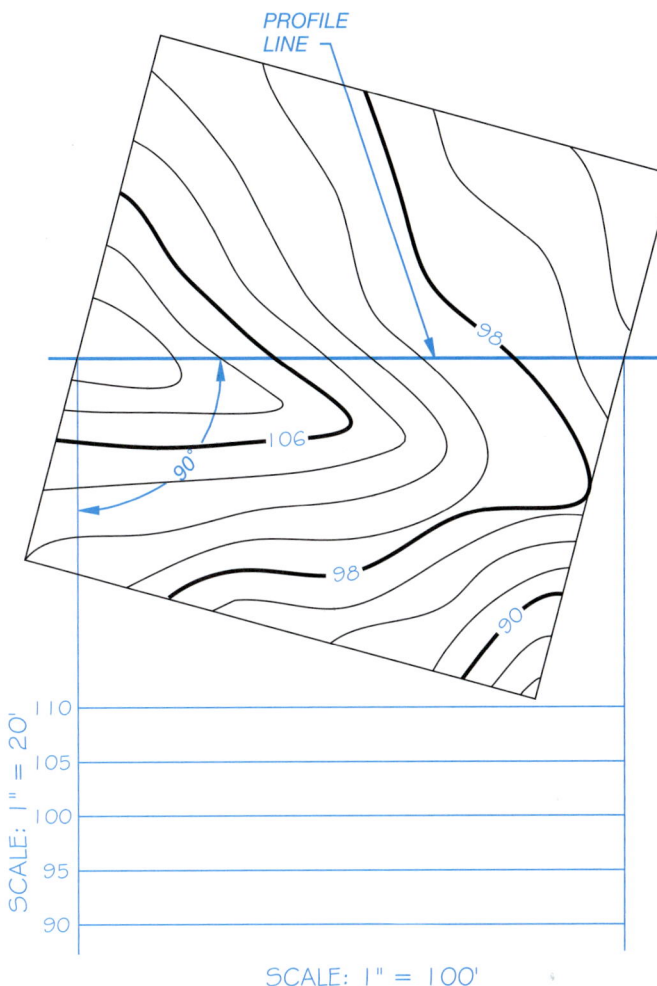

FIGURE 24.77 ■ Project the profile 90° from the start of the profile line. Set up the vertical scale. Label the vertical scale and the elevation of each contour along the vertical scale. Draw a horizontal line at each contour interval. Label the horizontal scale.

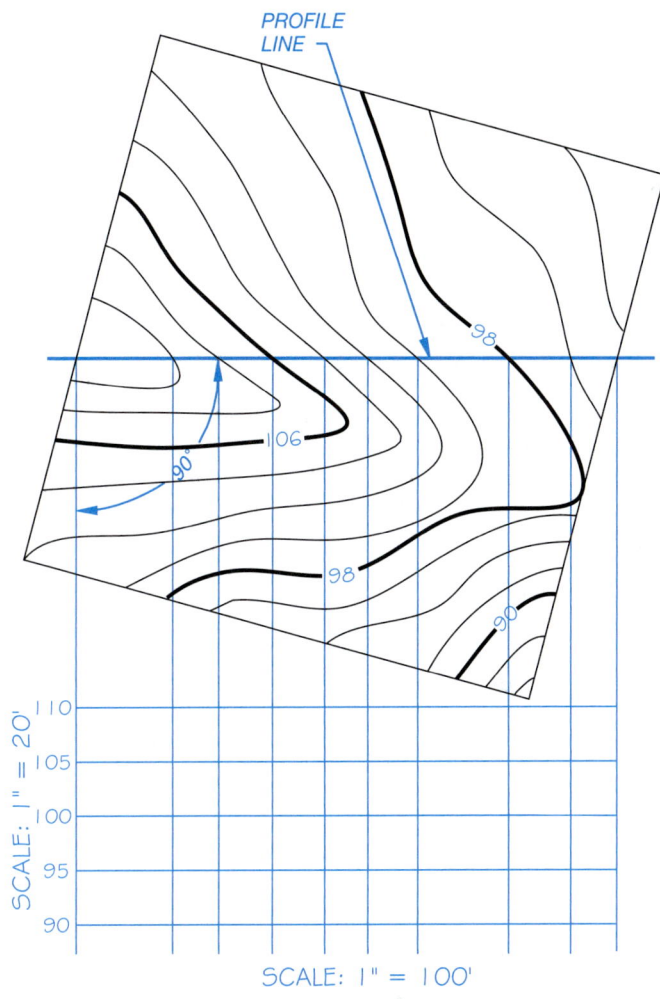

FIGURE 24.78 ■ Project a line 90° from the location where every contour line crosses the profile line.

a grading plan. The **grading plan** shows the elevation of the site after excavation. This plan shows where areas need to be cut and filled. This is referred to as **cut and fill** as previously discussed. Figure 24.80 shows the site plan location for the desired level excavation. The following steps can be used to draw the grading plan for a level construction site:

STEP 1 Determine the **angle of repose**, which is the slopes of cut and fill from the excavation site measured in feet of horizontal run to feet of vertical rise. One unit of rise to one unit of run is specified as 1:1. See Figure 24.81. The actual angle of repose for cuts and fills is normally determined by approved soils engineering or engineering geology reports. The slope of cut surfaces can be no steeper than is safe for the intended use and cannot exceed 1 unit vertical in 2 units horizontal (1:2). Alternative designs may be allowed if soils engineering and/or engineering geology reports state that the site has been

investigated and give an opinion that a cut at a steeper slope will be stable and not create a hazard to property. Fill slopes cannot be constructed on natural slopes steeper than 1 unit vertical in 2 units horizontal (1:2). The ground surface must be prepared to receive fill by removing vegetation, unstable fill material, topsoil, and other unsuitable materials to provide a bond with the new fill. Other requirements include soil engineering where stability, steeper slopes, and heights are issues. Soil engineering can require benching the fill into sound material and specific drainage and construction methods. A **bench** is a fairly level step excavated into the earth material on which fill is placed.

STEP 2 Draw parallel lines around the excavation site, with each line representing the elevation at the cut and fill. If the angle of repose is 1:1 and the contour interval is 2', then these parallel lines are 2' from the level excavation site. The contour interval is determined by multiplying rise by contour interval ($1 \times 2 = 2$ in this example). (See Figure 24.82.)

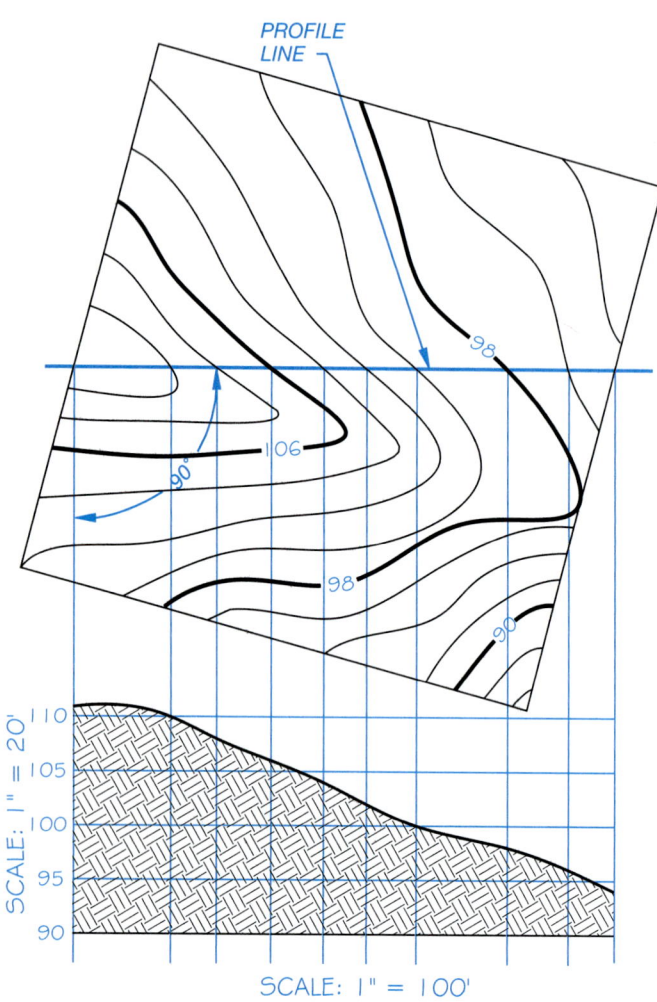

PROFILE LINE

SCALE: 1" = 20'

SCALE: 1" = 100'

FIGURE 24.79 ■ Draw the profile by connecting the points wherever a vertical and a horizontal line of the same elevation intersect.

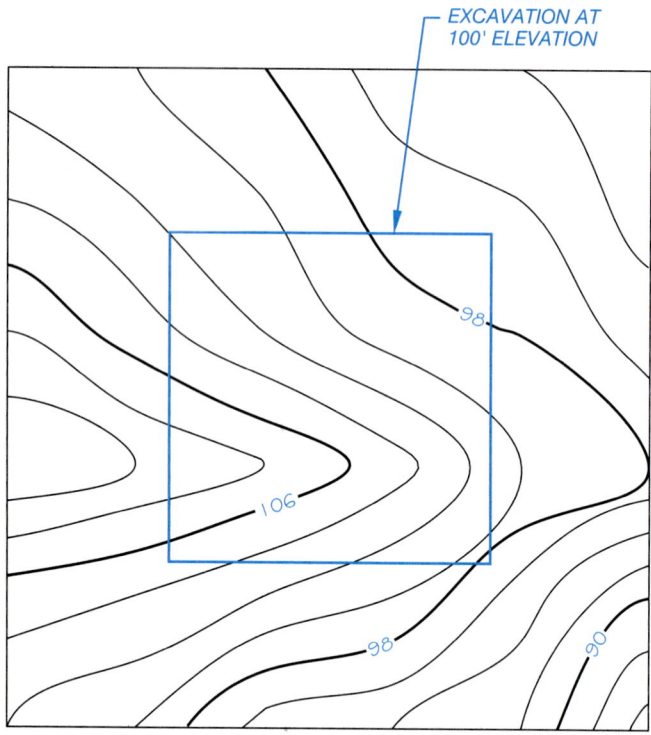

EXCAVATION AT 100' ELEVATION

FIGURE 24.80 ■ Site plan location for the desired level excavation at 100'.

STEP 3   The elevation of the excavation is 100'. Elevations above this are considered cuts, and elevations below this are fills. To establish the cut and fills, mark where the elevations of the parallel lines around the excavation match the corresponding elevations of the contour lines, as shown in Figure 24.83.

STEP 4   Connect the points established in step 3, as shown in Figure 24.84. The cut and fill areas can be shaded or left unshaded.

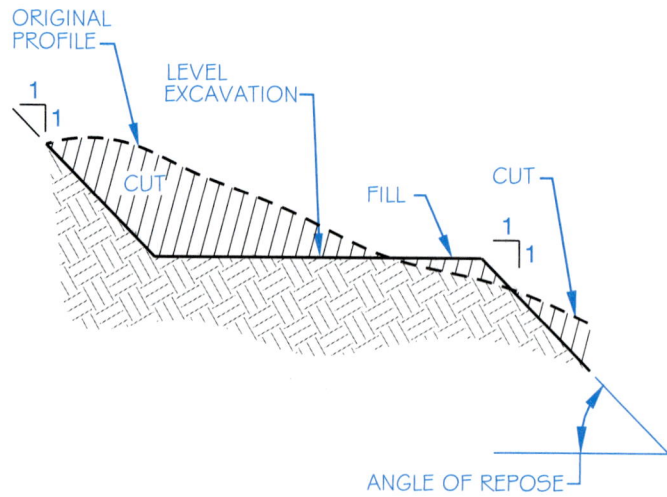

ORIGINAL PROFILE

LEVEL EXCAVATION

CUT

FILL

CUT

ANGLE OF REPOSE

FIGURE 24.81 ■ Angle of repose.

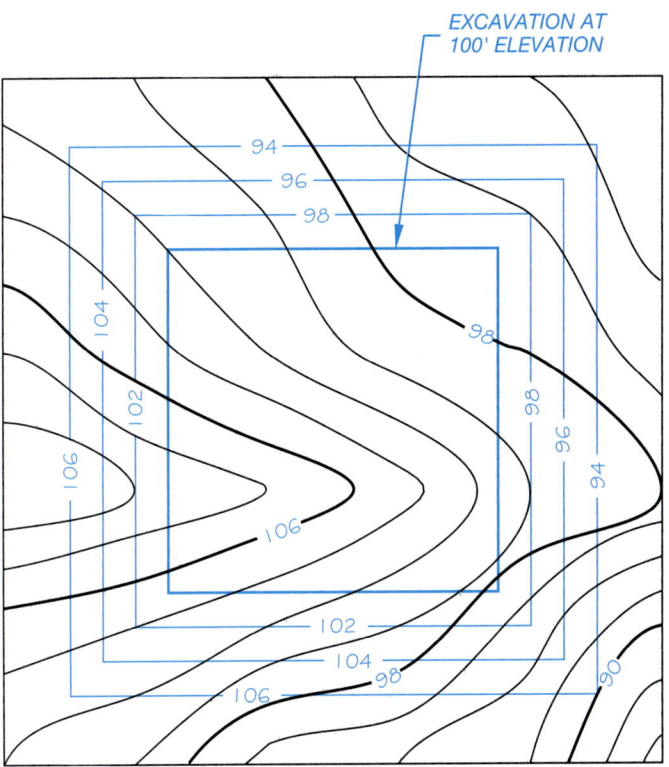

FIGURE 24.82 ■ Draw parallel lines around the excavation site, with each line representing the elevation of the cut and fill.

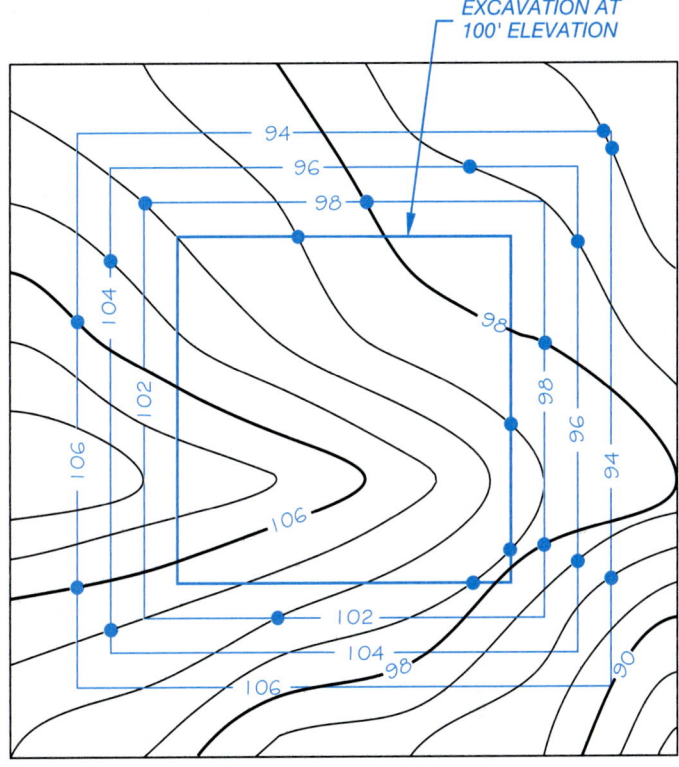

FIGURE 24.83 ■ To establish the cut and fill, mark where the elevations of the parallel lines around the excavation match the corresponding elevations of the contour lines.

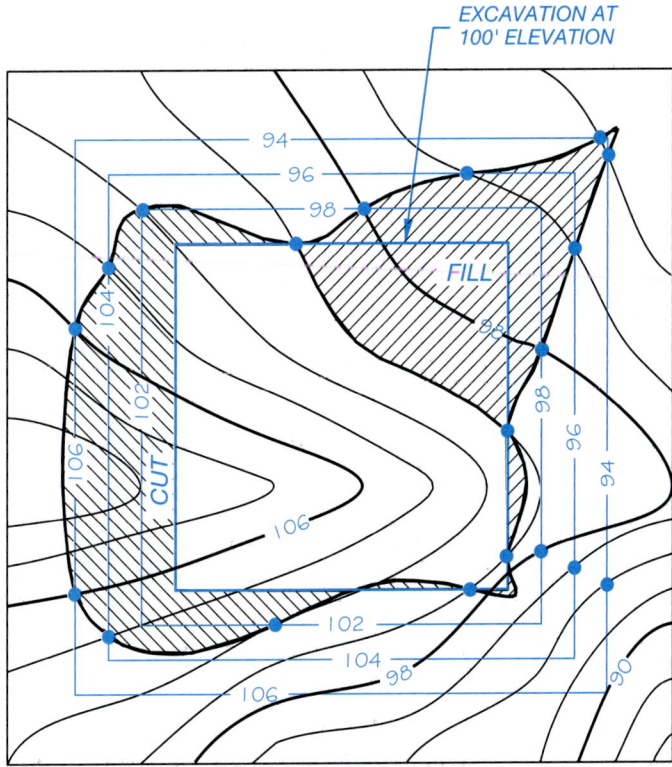

FIGURE 24.84 ■ Complete the cut and fill drawing by connecting the points. The cut and fill areas can be labeled, and they can be shaded or left unshaded.

CADD APPLICATIONS 2-D

USING CADD TO DRAW SITE PLANS

There are CADD software packages that can be customized to assist in drawing site plans. Also available are complete CADD mapping packages that allow you to draw topographic maps, terrain models, grading plans, and land profiles. It all depends on the nature of your business and how much power you need in the CADD mapping program. One of the benefits of CADD over manual drafting is accuracy. For example, you can draw a property boundary line by giving the length and bearing. The computer automatically draws the line, then labels the length and bearing. Continue by entering information from the surveyor's notes to draw the entire property boundary in just a few minutes.

Such features increase the speed and accuracy of drawing site plans. A residential site plan drawn using CADD mapping software is shown in Figure 24.85.

The needs of the commercial site plan are a little different from the residential requirements. The commercial CADD site plan package uses the same features as the residential application and, additionally, has the ability to design street and parking lot layouts. The commercial CADD drafter uses features from symbol libraries, including utility symbols; street, curb, and gutter designs; landscaping; parking lot layouts; titles; and scales. A commercial site plan is shown in Figure 24.86.

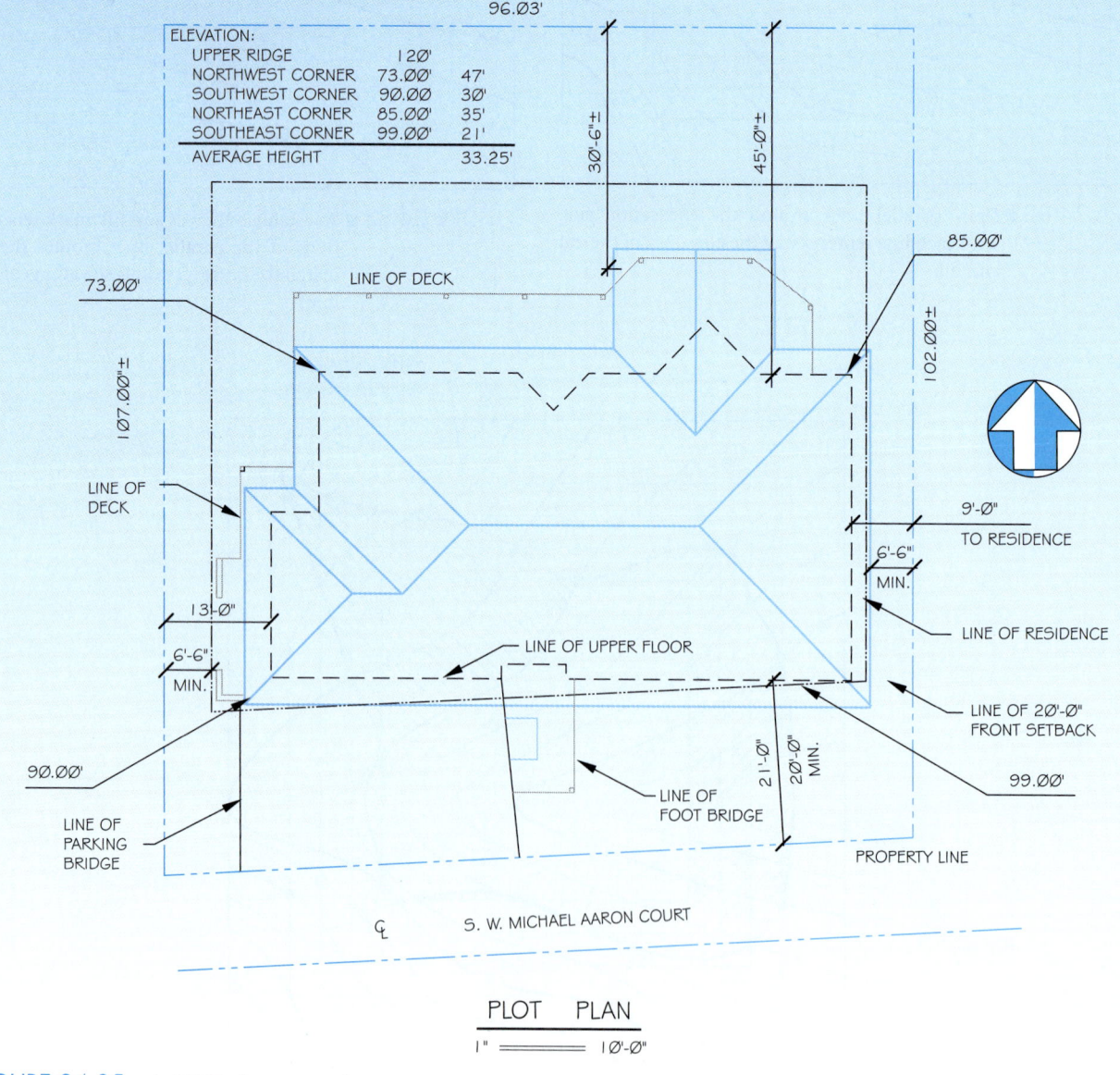

FIGURE 24.85 ■ A CADD-drawn site plan.

CADD APPLICATIONS 2-D

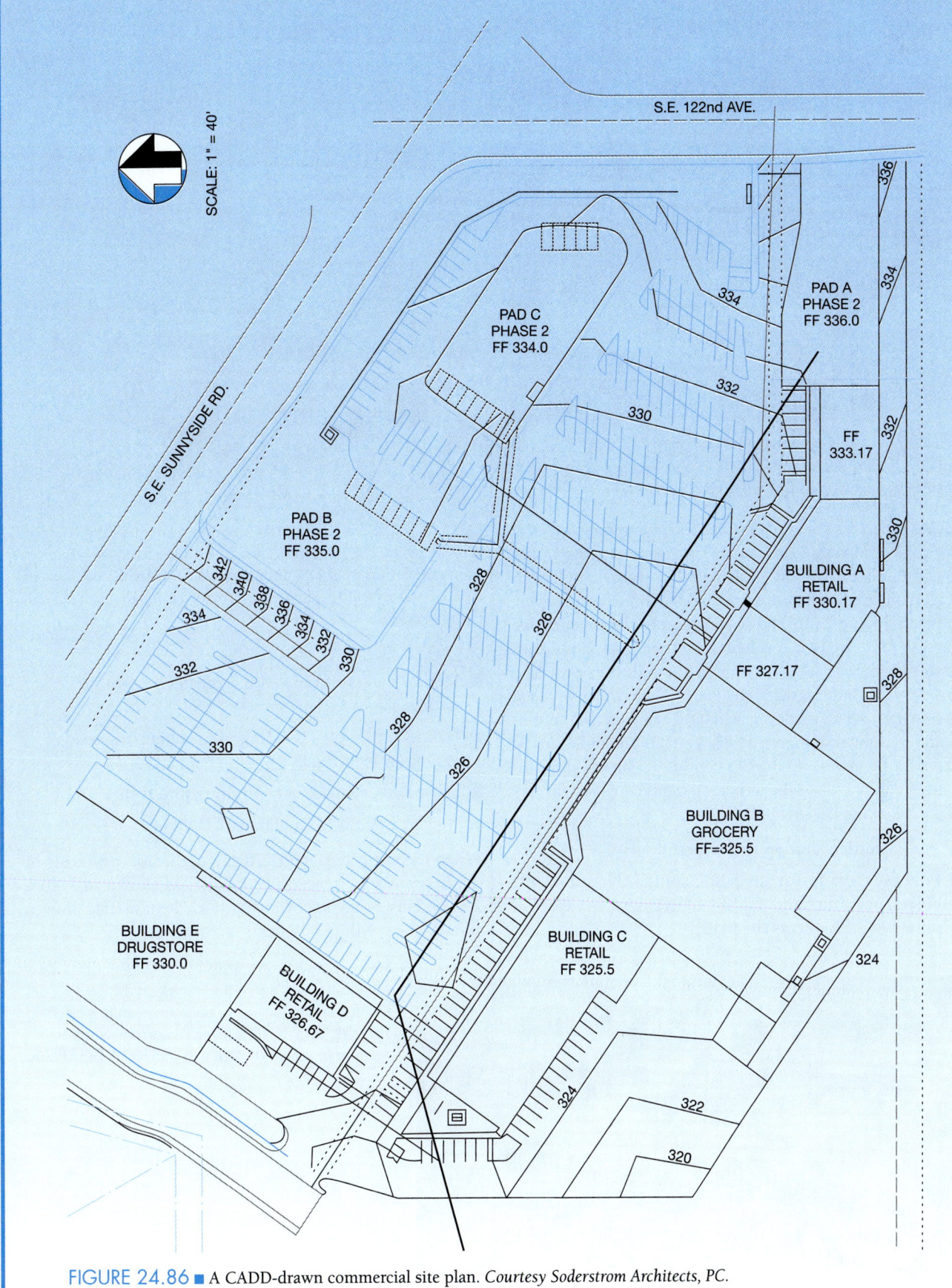

SCALE: 1" = 40'

S.E. 122nd AVE.

S.E. SUNNYSIDE RD.

PAD C
PHASE 2
FF 334.0

PAD A
PHASE 2
FF 336.0

FF
333.17

PAD B
PHASE 2
FF 335.0

BUILDING A
RETAIL
FF 330.17

FF 327.17

BUILDING B
GROCERY
FF=325.5

BUILDING E
DRUGSTORE
FF 330.0

BUILDING D
RETAIL
FF 326.67

BUILDING C
RETAIL
FF 325.5

336
334
332
330
328
326
324
322
320

342
340
338
336
334
332
330

334
332
330

FIGURE 24.86 ■ A CADD-drawn commercial site plan. *Courtesy Soderstrom Architects, PC.*

DEVELOPING A CADD TERRAIN MODEL

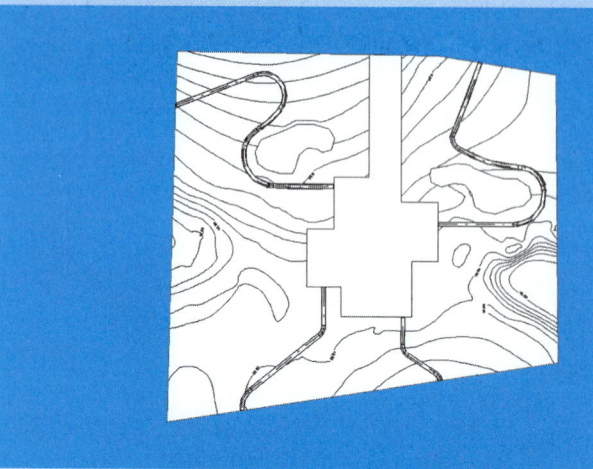

FIGURE 24.87 ■ A topographic site plan to be used for developing a terrain model. *ArchiCAD® images courtesy Graphisoft®, planned by architects Csaba Szakál and Geörgy Péchy.*

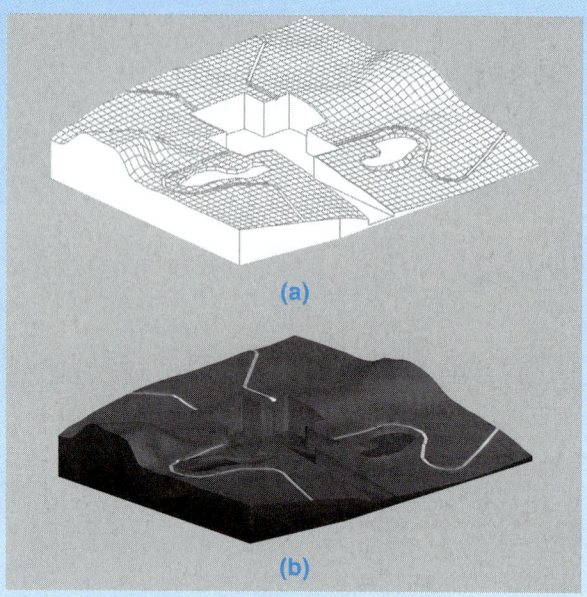

(a)

(b)

FIGURE 24.88 ■ The CADD program recognizes the contour lines from the site plan in Figure 24.87 and creates a 3-D wireframe model (a) or a 3-D terrain model (b). *ArchiCAD® images courtesy Graphisoft®, planned by architects Csaba Szakál and Geörgy Péchy.*

CADD programs are available that allow you to include site work, landscaping, roads, driveways, retaining walls, and construction excavation as part of the project. A terrain model can be created from survey data or from a topographic site plan. Figure 24.87 shows an example of a topographic site plan to be used for developing a terrain model. A terrain model shows the land contour in three dimensions. The CADD program recognizes the contour lines or survey data and creates a 3-D model in wireframe or as a 3-D terrain model, as shown in Figure 24.88. The terrain model can be viewed from any angle to help you fully visualize the contours of the site. Now, the terrain model can be used for any of the following design applications:

■ Define borders and property lines.

■ Find the elevation at any point, contour line, surface object, or feature and modify the elevation to determine how this affects the model.

■ Modify the terrain model by editing the contour shapes.

■ Define and display building excavation sites, roads, and other features.

■ Show and calculate cut and fill requirements.

■ Display the model in plan or 3-D view.

When the site is designed as desired, the 3-D rendering of the home can be placed on the site, as shown in Figure 24.89. This is an excellent way to demonstrate how a client's home will look when it is finished.

FIGURE 24.89 ■ When the site is designed as desired, a 3-D rendering of the house can be placed on the site. *ArchiCAD® images courtesy Graphisoft®, planned by architects Csaba Szakál and Geörgy Péchy.*

CADD LAYERS FOR SITE PLAN DRAWINGS

The American Institute of Architects (AIA) CADD Layer Guidelines establish the heading Civil Engineering and Site Work as the major group for CADD layers related to site plans. The following are some of the recommended CADD layer names for site plan applications:

Layer Name	Description
C-PROP	Property lines and survey benchmarks.
C-TOPO	Proposed contour lines and elevations.
C-BLDG	Proposed building footprints. The term *footprint* is often used to describe the area directly below a structure.
C-PKNG	Parking lots.
C-ROAD	Roads.
C-STRM	Storm drainage, catch basins and manholes, C-COMM Site communications systems, such as telephone.
C-WATR	Domestic water.
C-FIRE	Fire protection hydrants, connections.
C-NGAS	Natural gas systems.
C-SSWR	Sanitary sewer systems.
C-ELEV	Elevations.
C-SECT	Sections.
C-DETL	Details.
C-PSIT	Site plan.
C-PELC	Site electrical systems plan.
C-PUTL	Site utility plan.
C-PGRD	Grading plan.
C-PPAV	Paving plan.
C-P***	Other site, landscape.

Landscape plans also have CADD layer designations, with the major heading Landscape Architecture. Landscape plans generally show the suitable plants for the site, and specify the plants by their proper Latin names and sometimes their common names. The architect or designer often sends the site plan computer drawing file to the landscape architect or designer where the file is used to specify the size, type, and location of the plants. Details for maintaining the plants, such as a water sprinkler system and other care requirements are also provided on the landscape plan. Figure 24.90 shows a CADD landscape plan without detailed plant information given on the drawing. In this case, the plant information was given in a set of written specifications. The following are some of the commonly used layer names:

Layer Name	Description
L-PLNT	Plant and landscape materials.
L-IRRG	Irrigation systems.
L-WALK	Walks and steps.

(Continued)

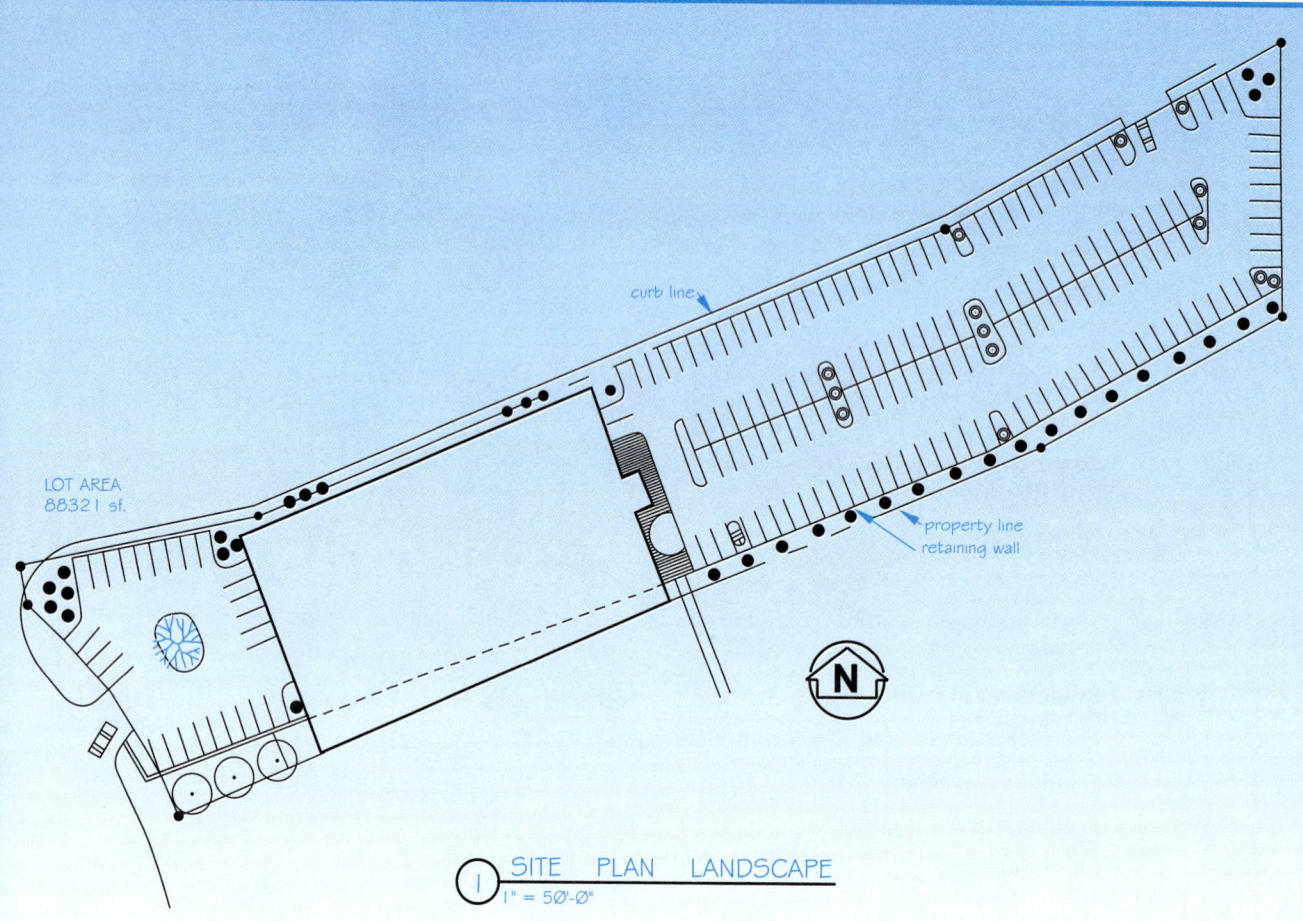

LOT AREA
88321 sf.

curb line

property line
retaining wall

N

SITE PLAN LANDSCAPE
1" = 50'-0"

FIGURE 24.90 ■ The landscape plan shows the location of all trees and plantings. Plant types can be included on the landscape plan or in separate documents. *Courtesy Soderstrom Architects.*

When finding the degree of curve (D) for a roadway or utility line, the use of the Law of Cosines is helpful. To use this law, you must know two sides and the included angle, or know all three sides. In finding the degree of curve, you know the three sides: two sides are simply the radius of the curve, and the third side is the 100-foot chord. The Law of Cosines follows:

$$a^2 = b^2 + c^2 - 2bc(\text{cosine } A)$$

in which A = the angle to be determined

a = the side opposite the A angle
(the 100-foot chord)

b and c = the radius of the curve

The formula can then be rewritten for your purposes.

$$a^2 = 2b^2 - 2b^2(\cos A) \text{ or } \cos A = (2b^2) - a^2) + (2b^2)$$

For example, if the radius of a curve is 145.62 feet, then the formula would appear as follows:

$$\cos A = [2(145.62)^2 - 100^2] + [2(145.62)^2]$$
$$\cos A = 32,410.3688 + 42,410.3688$$
$$\cos A = 0.764208$$

Use the second function on your calculator to obtain the inverse cosine, and the angle is 40°9'48".

The use of the simple formula shown in the discussion with Figure 24.41, page 825, gives the same answer. Double the angle after you find the sine of the angle using the 50-foot side.

PROFESSIONAL PERSPECTIVE

This chapter has touched on the basic portions of civil engineering drafting, with an overview of surveying fundamentals. The civil drafter needs to keep in mind the overall scope of the project when putting it together. The various phases of completing a project require the drafter to work with the different disciplines. It takes a great deal of well-organized thought and planning to synthesize the material and produce a high-quality graphic document. In your work, keep the broad scope of the project in mind.

CHAPTER 24 Civil Drafting Test

 Access the CD found with this textbook to view the Chapter 24 Test. Confirm the preferred submittal method with your instructor.

CHAPTER 24 Civil Drafting Problems

DIRECTIONS

Use polyester film, vellum, or a CADD system for all of the problems that call for a drawing to be created. Determine the sheet size based on the drawing requirements. Metric scales can be substituted if preferred by your supervisor or instructor. Include a north arrow and note the scale where applicable.

PROBLEM 24.1   Draw an open traverse at a scale of 1" = 20' using the following information:

From POB, traverse 34.23' at S61°42'E, then 15.69' at S44°7'E, then 40.89' at N78°48'E, then 39.33' at N56°14'E. Be sure to label each course.

PROBLEM 24.2   Give the azimuth angles as well as the bearings of the lines shown. Fill in the values in the following table.

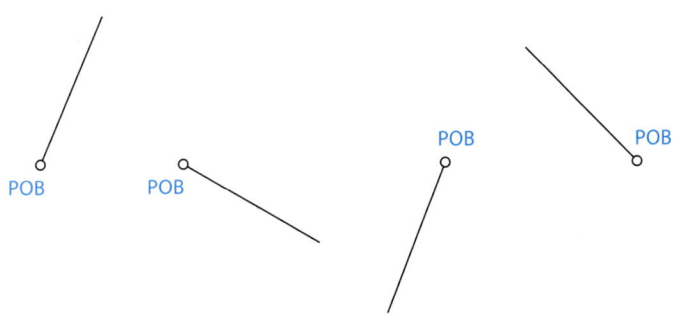

Line	Azimuthal Angle	Bearing
AB		
CD		
EF		
GH		

PROBLEM 24.3   Using either polyester film, vellum, or a CADD system, draw a traverse at a scale of 1" = 20' using the following lengths and bearings. What kind of traverse is this?

21 feet at N40°7'E, 77 feet at S84°9'E, 19 feet at S24°29'E, 48 feet at S39°22'W, 31 feet at S76°9'W, 65 feet at N34°57'W.

o
POB

PROBLEM 24.4   Using either polyester film, vellum, or a CADD system, transfer the information shown on the following figure. Then draw a traverse at a scale of 1" = 20' using the following lengths and azimuth angles. What kind of traverse is this?

23 feet at an angle of 131°40', 11 feet at an angle of 231°8', 36 feet at an angle of 113°24', 49 feet at an angle of 64°39', 27 feet at an angle of 303°50', 36 feet at an angle of 65°33'.

o
POB

o
KNOWN
POINT

PROBLEM 24.5 Using either polyester film, vellum, or a
CADD system, transfer the information shown on the figure at the right. Then draw a deflection angle traverse at a
scale of 1" = 40' using the following lengths and bearings.
Be sure to use proper notation for the deflections. What
type of traverse is this?

48' at S63°46'E, 33' at S37°32'E, 71' at N86°10'E, 23' at S81°4'E,
73' at S73°9'E.

○
POB

○
BENCH
MARK

PROBLEM 24.6 Complete the leveling notes that were gathered by the survey crew. Create a leveling notes table and fill in the HI
and elevations for each station.

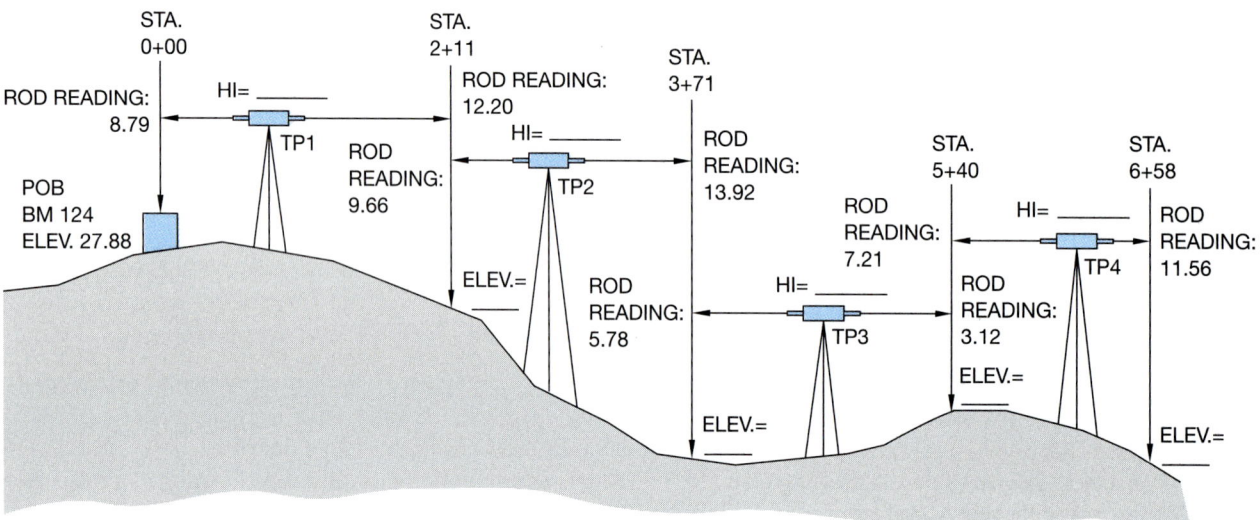

PROBLEM 24.7 Draw the grid as shown on a separate piece
of polyester film, vellum, or on a CADD system. Mark the
following townships as shown:

A Township: T2S, R2W

B Township: T3N, R4E

C Township: T1N, R2E

D Township: T2S, R2E

E Township: TIN, R4W

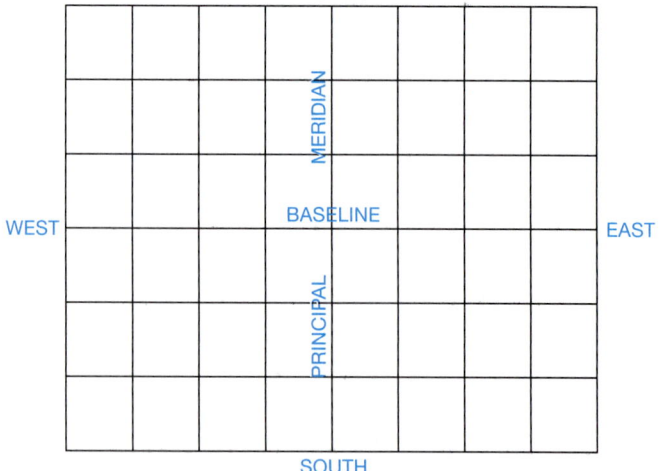

PROBLEM 24.8 Draw the section grid as shown on a separate piece of polyester film, vellum, or on a CADD system. Mark the following parcels of land as shown:

A Parcel: SW 1/4, SE 1/4

B Parcel: W 1/2, NW 1/4

C Parcel: W 1/2, NE 1/4, SW 1/4

D Parcel: NE 1/4, NE 1/4, NW 1/4

E Parcel: E 1/2, SE 1/4, SE 1/4

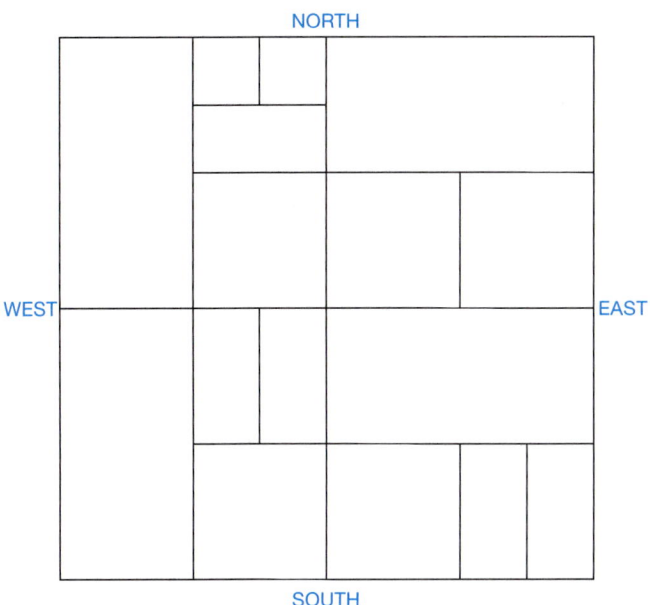

PROBLEM 24.9 Complete the leveling notes that were gathered by the survey crew. Then plot the resulting profile using 1" = 10' for the vertical scale and 1" = 40' for the horizontal scale.

Station	B.S. (+)	H.I.	F.S. (−)	Elevation
BM 12	17.31		7.46	34.12
0 + 30	6.83		4.87	
0 + 55	9.66		7.72	
1 + 05	14.30		10.13	
1 + 45	12.20		13.29	
1 + 85	7.12		15.82	
2 + 15	10.04		12.73	
2 + 40				BM 13

PROBLEM 24.10 Answer the following question using the drawing on page 858.

What is the vertical scale?

What is the horizontal scale?

What is the degree of curve of the sewer?

Where is the property located in relation to the public-land survey?

If the IE at MH-4 is 309.75, and the IE at MH-5 is 313.50, then what is the grade slope?

What is the contour interval?

Printed by permission of City of Portland.

PROBLEM 24.11 Complete the unknown values that were given at the beginning of the chapter. (See Figure 24.2, page 809.)

Station	Transit Line for Valerian Lane From Erika Street (Bearing N56°16'W) to Quincey Avenue (Bearing N31°24'W)
0 + 00 Begin Project	245.78'N67°37'W from BM 151 to centerline of Erika Street. Then 132.00'S33°44'W to P.C.
STA 1 + 32	P.C. to the left Curve Data R = 200' Δ = 47°23' L = —— D = ——
STA ___	P.R.C. Curve Data R = 170' Δ = 72°15' L = —— D = ——
STA ___	P.T.
STA ___ End Project	S58°36'W from STA ___ to centerline of Quincey Avenue

PROBLEM 24.10
(Continued)

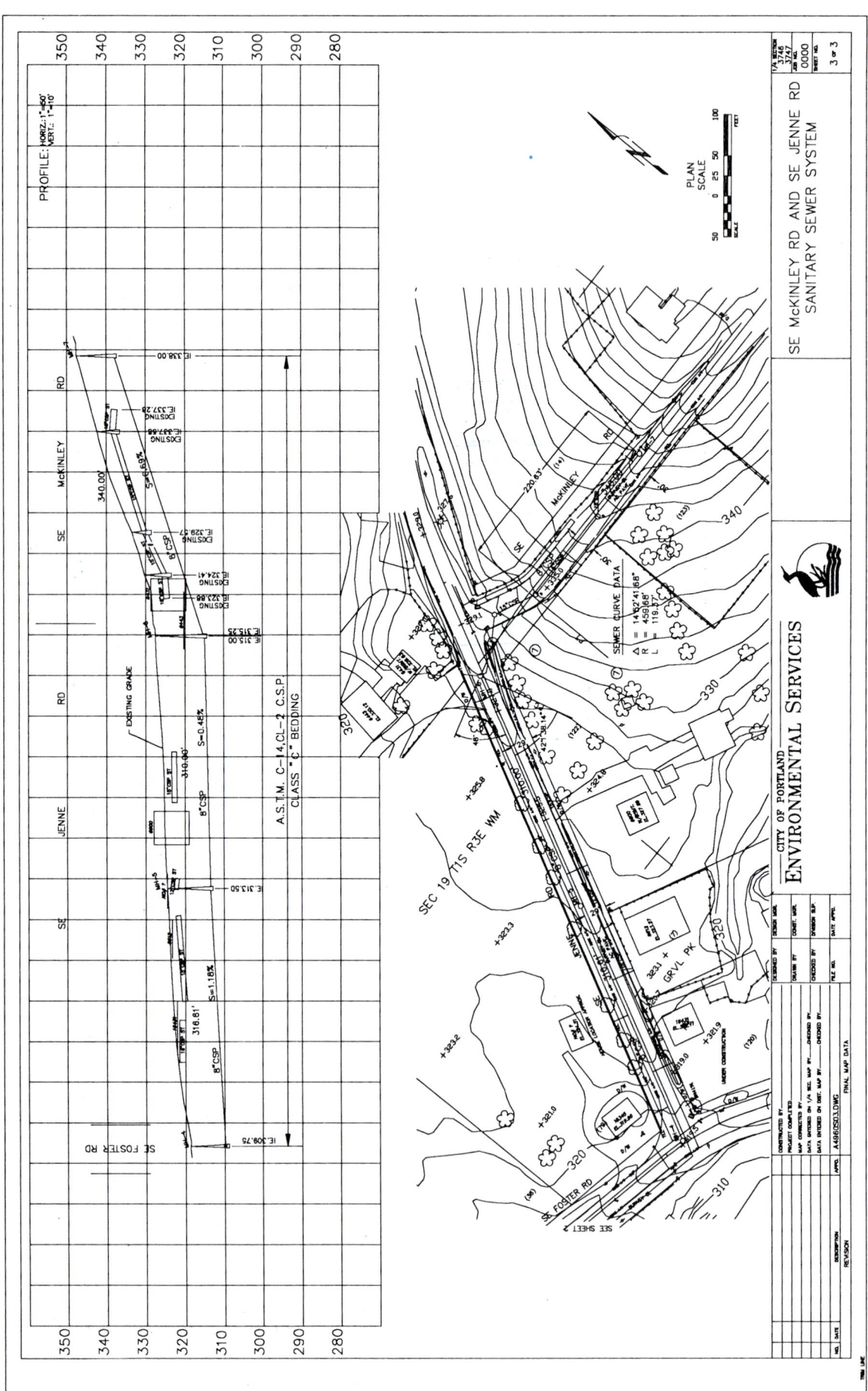

PROBLEMS 24.12 through 24.19: Access the CD found with this textbook and open the problem of your choice, or as assigned by your instructor. Solve the problems using the instructions provided with this chapter or on the CD, unless otherwise specified by your instructor.

DIRECTIONS
Problems 24.20 through 24.26: Draw the plot plans from the given sketches or layouts.

PROBLEM 24.20

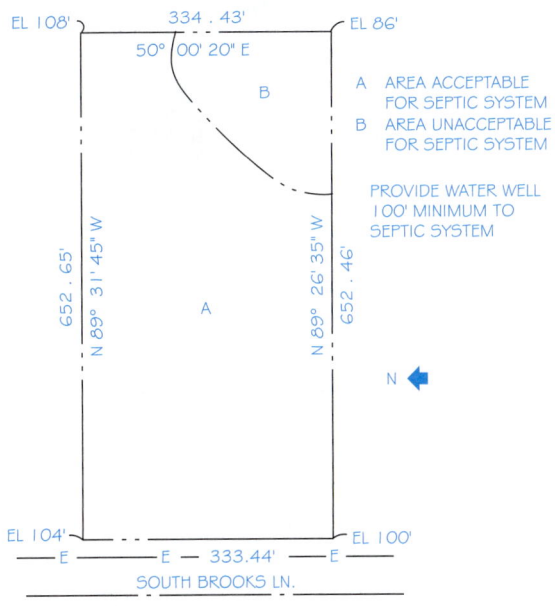

PROBLEM 24.22

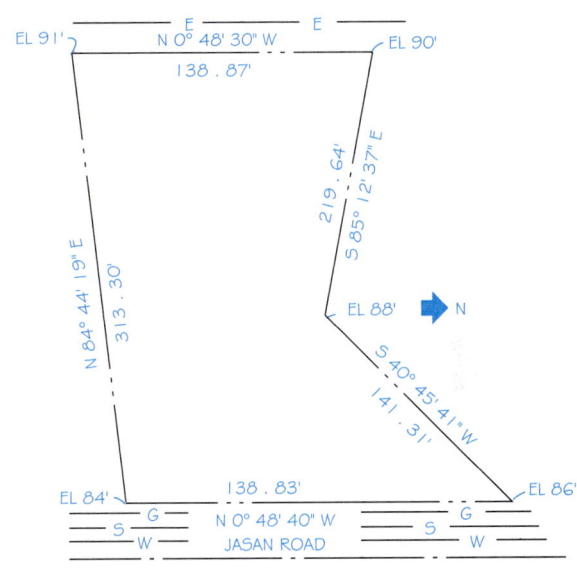

PROBLEM 24.21

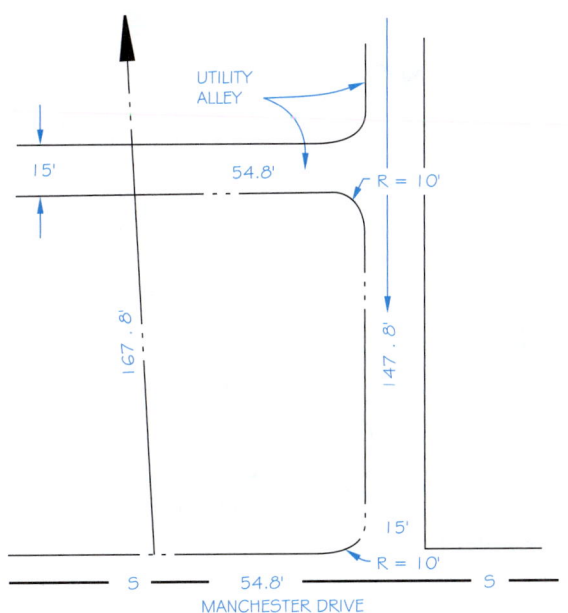

PROBLEM 24.23

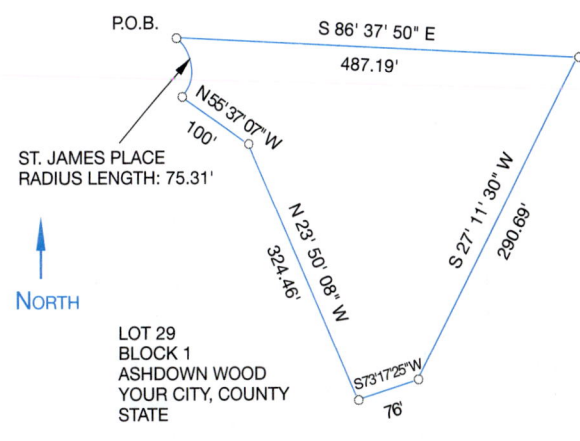

PROBLEM 24.24

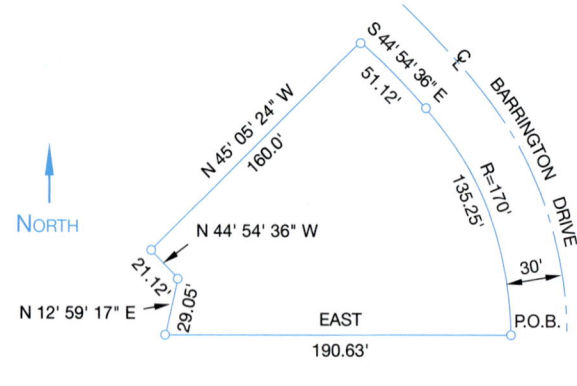

LOT 15
BLOCK 3
BARRINGTON HEIGHTS
YOUR CITY, COUNTY
STATE

PROBLEM 24.25

LOT 25
BLOCK 4
BARRINGTON HEIGHTS
YOUR CITY, COUNTY
STATE

PROBLEM 24.26

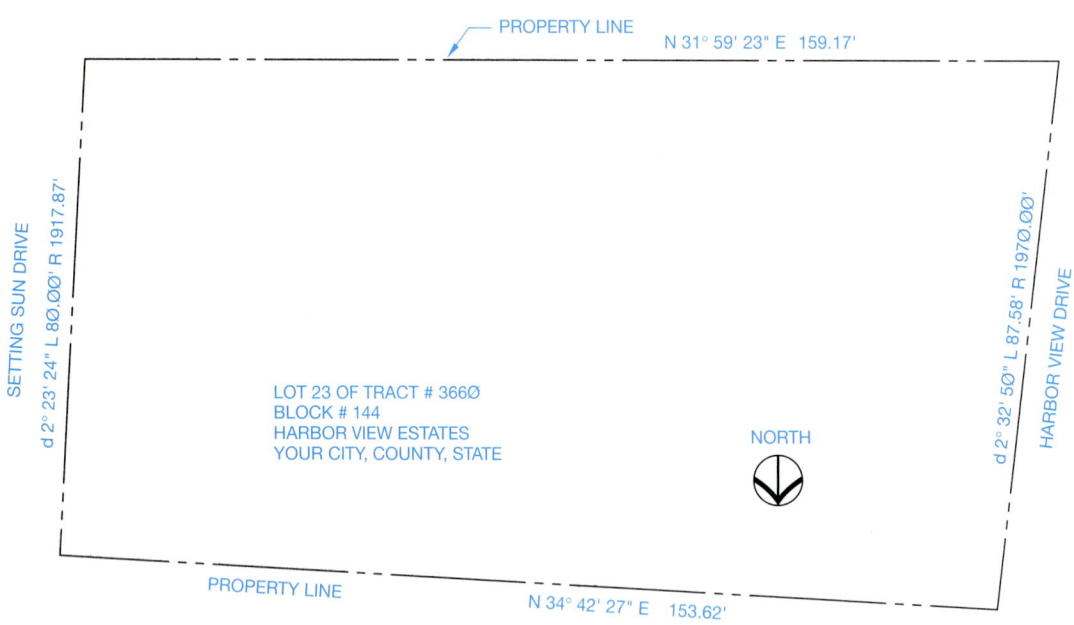

PROPERTY LINE   N 31° 59' 23" E 159.17'

SETTING SUN DRIVE
d 2° 23' 24" L 80.00' R 1917.87'

LOT 23 OF TRACT # 3660
BLOCK # 144
HARBOR VIEW ESTATES
YOUR CITY, COUNTY, STATE

NORTH

HARBOR VIEW DRIVE
d 2° 32' 50" L 87.58' R 1970.00'

PROPERTY LINE   N 34° 42' 27" E 153.62'

PROBLEMS 24.27 and 24.28: Access the CD found with this textbook and open the problem of your choice, or as assigned by your instructor. Solve the problems using the instructions provided with this chapter or on the CD, unless otherwise specified by your instructor.

MATH PROBLEMS

PROBLEMS 24.29 and 24.30: Access the CD found with this textbook and open the math problem of your choice, or as assigned by your instructor. Solve the problem or problems using the instructions provided.

CHAPTER 25

Heating, Ventilating, and Air-Conditioning (HVAC), Pattern Development, and Precision Sheet Metal Drafting

LEARNING OBJECTIVES

After completing this chapter, you will:

- Discuss the purpose and function of HVAC systems.
- Prepare complete HVAC drawings, including plans, schedules, and details.
- Draw sheet metal pattern developments and intersections.
- Calculate and apply bend allowances to sheet metal components.
- Draw and completely dimension precision sheet metal fabrication drawings.
- Use an engineering problem as an example for HVAC and sheet metal drawing solutions.

THE ENGINEERING DESIGN APPLICATION

Your company produces a wide range of sheet metal products for various customers, ranging from HVAC ductwork to precision housing structures for electronic equipment. Your current drawing project is a chassis for an electronic testing device.

With HVAC drawings, it is not usually necessary to provide bend allowances. In this project, however, the final development requires tighter tolerances for all of the components to mount properly within the chassis. Calculating bend allowances allows you to determine the required size of the flat pattern.

While there are a number of methods for determining bend allowance, your company standards reference the tables listed in the *Machinery's Handbook*. By checking the table for the type of metal you are using, plus the bend radius and thickness of material, you are able to determine the appropriate bend allowances.

Applying the bend allowance to each bend in the flat pattern ensures that when the material is formed into its final shape, the desired final dimensions can be achieved. Figure 25.1 shows an example flat pattern layout.

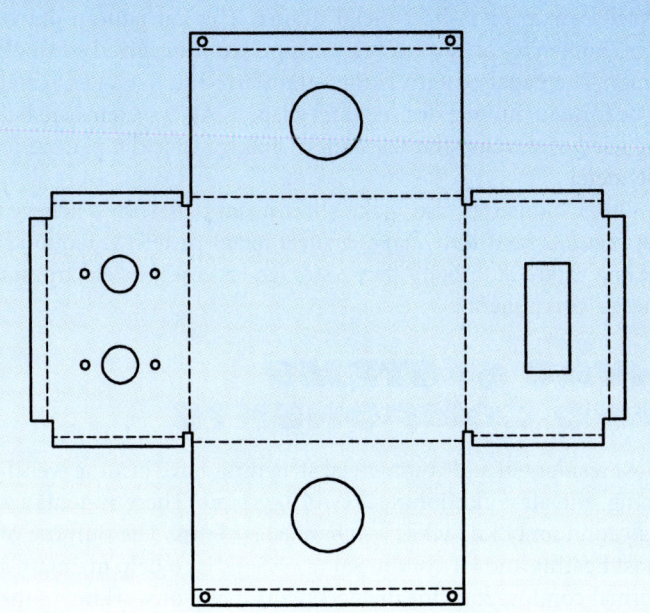

FIGURE 25.1 ■ Sample flat pattern layout for the engineering problem.

INTRODUCTION

This chapter covers three fundamental areas of drafting and design related to sheet metal fabrication. The first content involves the heating, ventilation, and air-conditioning (HVAC) industry for residential and commercial architecture, engineering, and construction. The HVAC area is followed by pattern development, which involves laying out the geometric sheet metal shapes, used in HVAC design, into their true size and shape flat patterns. These flat patterns are then used in sheet metal fabrication shops to make the fittings used in HVAC construction. The last field covered in this chapter also involves design of flat patterns, but this area uses what is called precision sheet metal drafting. Precision sheet metal drafting requires tighter tolerances and bend calculations that are not generally needed in the fabrication of products for the HVAC industry.

INTRODUCTION TO HVAC SYSTEMS

Heating, ventilating, and air-conditioning (HVAC) systems are made up of mechanical equipment such as the furnace, air conditioner or ventilator, and ductwork. The duct is generally sheet metal pipe designed as the passageway for conveying air from the HVAC equipment to and from the source. Types of duct shapes include cylindrical, oval, rectangular, and square. Connectors such as elbows and transition shapes that allow conversion from round to square or rectangular are also used. Any duct shape is possible depending on the application. In most residential applications, standard premanufactured ductwork is used. In residential and commercial structures, the ductwork can be premanufactured or custom-built in a sheet metal fabrication shop. Flat patterns are made when custom sheet metal shapes are required. These patterns are made by a sheet metal layout person or a sheet metal drafter. The flat pattern drawings, known as pattern developments, are transferred to thick paper. The paper pattern is then transferred to flat sheet metal to be formed into the desired duct shape. CADD systems are being used to develop the flat pattern and transfer the pattern to the metal.

Other industries also make sheet metal patterns. Whenever flat metal is bent into shape, a sheet metal pattern is required. Examples are auto body parts, storage facilities, or electronics chassis components.

HVAC SYSTEMS AND COMPONENTS

Most residential and commercial structures have heating, ventilating, and air-conditioning (HVAC) systems. These systems are also commonly known as mechanical systems. The purpose of most heating and air-conditioning systems is to help maintain a normal comfort zone for the occupants living or working in the structure. In other applications, such as meat lockers, the product can be the reason for environmental controls. The function of ventilating systems is to provide air movement or exchange within the structure. Ventilation is required when excess heat,

fumes, moisture, odor, or pollutants must be removed and fresh air replaced. Most residential structures do not use HVAC plans unless the system is complex, or required by the lending or code enforcement agency. The architect of a commercial or public building often consults with an HVAC engineer for the design, drafting, and installation supervision of the HVAC system. This consulting engineer is called a mechanical engineer, named after the mechanical system being designed. The consulting engineer prepares all of the designs and specifications for the project and the drafter either works directly from written specifications or from engineering sketches. In many cases the engineer prepares rough sketches directly on a copy of the floor plans. The HVAC drafter then creates a new drawing using proper lines, symbols, dimensions, and notes. Some companies use an HVAC overlay where the base sheet is the floor plan and another sheet, known as the overlay or layer, is used to draw the HVAC plan.

Central Forced-Air Systems

Central forced-air systems, among the most common systems for climate heating and air-conditioning, circulate the air from the living spaces through or around heating or cooling devices. A thermostat starts the cycle as a fan forces the air into ducts. These ducts connect to openings called diffusers, or air supply registers, which put warm air (WA) or cold air (CA) in the room. The air enters the room and either heats or cools as needed. Air then flows from the room through another opening called a return-air register (RA) and into the return duct. The return duct directs the air from the room over a heating or cooling device, depending on which is needed. If cool air is required, the return air is passed over the surface of a cooling coil. If warm air is required, the return air is passed over either the surface of a combustion chamber (the part of a furnace where fuel is burned) or a heating coil. The conditioned air is picked up again by the fan and the cycle is repeated. Figure 25.2 shows the air cycle in a forced-air system.

Refrigeration

Refrigeration, the most common type of cooling system, is based on the principle that as a liquid changes to vapor it absorbs large amounts of heat. The boiling point of a liquid can be altered by changing the pressure applied to the liquid so when a gas is changed to a liquid, it gives up the heat. The basic parts of a refrigeration system are the cooling coil (evaporator), compressor (air pump), condenser (where vaporized refrigerant is liquefied), and expansion valve. Common refrigerants boil at low temperatures. Figure 25.3 shows a pictorial diagram of the refrigeration cycle.

Hot Water System

In a hot water system, water is heated as it circulates around the combustion chamber of a fuel-fired boiler. The water is then circulated through pipes to radiators or convectors in the rooms. In a one-pipe system, one pipe leaves the boiler and runs through the building and back to the boiler as shown in Figure

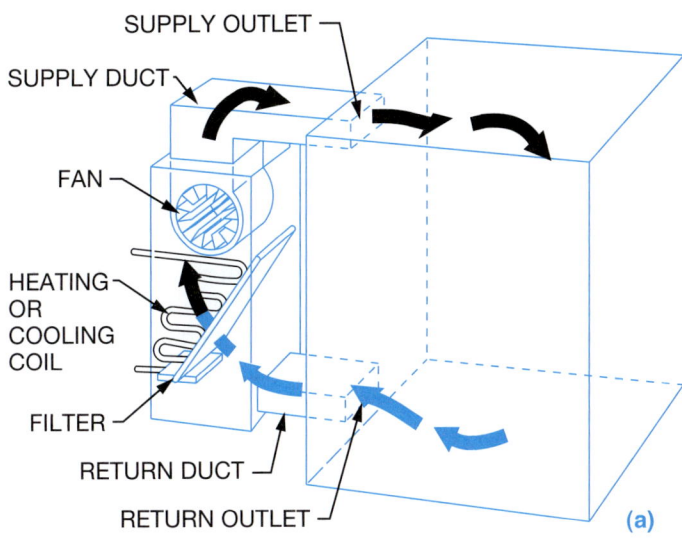

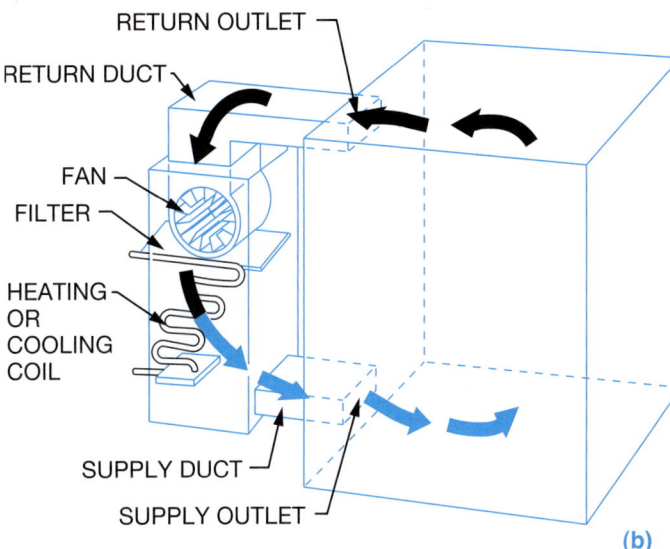

FIGURE 25.2 ■ (a) Down draft forced air system air cycle. (b) Up draft forced air system air cycle.

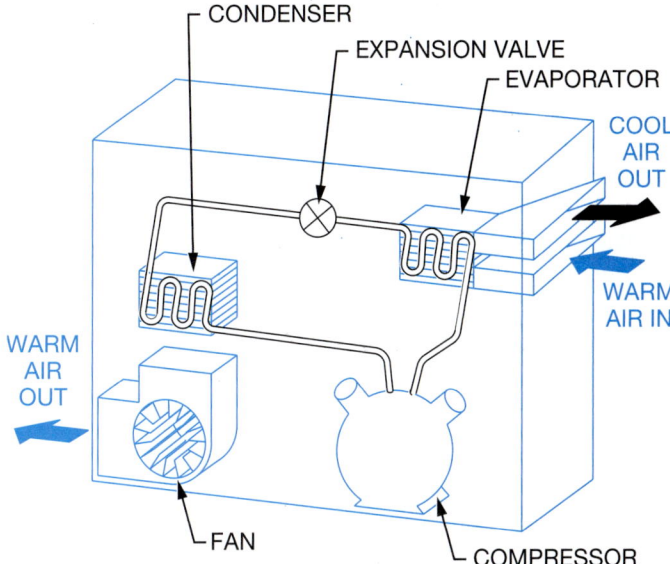

FIGURE 25.3 ■ Pictorial diagram of refrigeration cycle. *From Lang. Principles of Air Conditioning, 3d ed., Delmar Publishers Inc.*

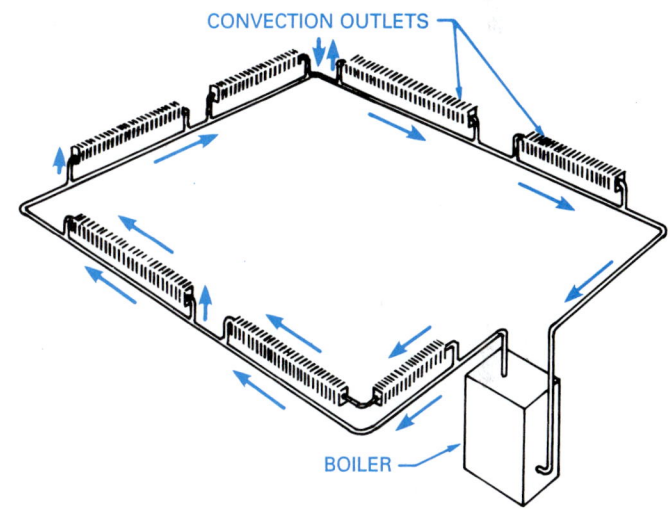

FIGURE 25.4 ■ One-pipe hot water system.

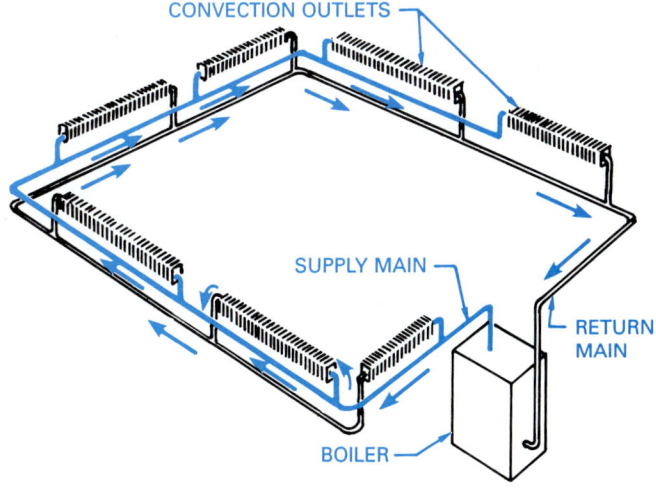

FIGURE 25.5 ■ Two-pipe hot water system.

25.4. In a two-pipe system, one pipe supplies heated water to all the outlets, while the other is a return pipe, which carries the water back to the boiler for reheating as shown in Figure 25.5. Hot water systems use a pump, called a circulator, to move the water through the system. The water is kept at a temperature of 150°–180°F (65.5°-82°C) in the boiler. When heat is needed, a thermostat starts the circulator pump.

Zoned Control System

A zoned system allows for one or more heaters and one thermostat per room. No ductwork is required, and only the heaters in occupied rooms need to be turned on. One of the major differences between a zoned and central system is flexibility. A zoned system allows the occupant to determine how many areas are heated and how much energy is used.

Radiant Heat

Radiant heating and cooling systems function on the basis of providing a comfortable environment by means of controlling surface temperatures and minimizing excessive air movement within the space. Surface-mounted radiant panels provide comfort at a lower thermostat temperature than other systems. Radiant systems vary from oil or gas hot water piping in the ceiling or floor, to electric coils, wiring, or elements at the ceiling.

Heat Pump System

The heat pump is a forced-air central heating and cooling system that operates using a compressor and a circulating liquid gas refrigerant. Heat is extracted from the outside air and pumped inside the structure. The heat pump supplies up to three times as much heat per year for the same amount of electrical consumption. A standard electrical forced-air heating system works best when outside air is above 20°F. In the summer, the cycle is reversed and the unit operates as an air conditioner. In this mode the heat is extracted from the inside air and pumped outside. On the cooling cycle, the heat pump also acts as a dehumidifier. Figure 25.6 shows how a heat pump works.

Ventilation

There are a number of reasons why ventilation of a structure or area is necessary. Residential applications include bath, kitchen, and laundry exhaust fans. In commercial applications, ventilation can be necessary to exhaust fumes, pollutants, or moisture.

Sources of Pollutants

There are a number of sources of pollutants that make it necessary to plan ventilation systems. Moisture in the form of relative humidity can cause structural damage and health problems such as respiratory troubles and microbial growth. Each individual can produce up to one gallon of water vapor per day.

Indoor combustion from such items as gas-fired or wood-burning appliances, or fireplaces, can generate a variety of pollutants, including carbon monoxide and nitrogen oxides.

Humans and pets can transmit diseases through the air by exhaling a variety of bacterial and viral contaminants.

Tobacco smoke can contribute chemical compounds to the air environment. The pollution can affect smokers and nonsmokers.

Formaldehyde in glues used in construction materials such as plywood, particleboard, or even carpet and furniture causes pollution, as do the components of certain insulations. Formaldehyde has been considered a factor in certain diseases, eye irritation, and respiratory problems.

Radon is a naturally occurring radioactive gas that breaks down into compounds that can cause cancer when large quantities are inhaled over a long period of time. Radon can be more apparent in a structure that contains a great deal of concrete or is located in certain geographic areas of the country. Radon can be scientifically monitored at a nominal cost. Barriers can be built that help reduce the concern of radon contamination.

Household products such as aerosols and crafts materials such as glues and paints can contribute a number of toxic pollutants.

Air-to-Air Heat Exchangers

The air-to-air heat exchanger is a heat-recovery ventilation device that pulls stale, polluted, warm air from the living or working space through a duct system and transfers the heat in that air to the fresh cold air being pulled into the structure. Heat exchangers do not produce heat; they only move heat from one airstream to the other. The heat transfers to the fresh airstream in the core of the heat exchanger. The core is usually designed to avoid mixing of the two airstreams to ensure that indoor pollutants are removed. Moisture in the stale air condenses in the core and is drained from the unit. Figure 25.7 shows the basic components and function of an air-to-air heat exchanger. Knowledgeable mechanical engineers, designers, and contrac-

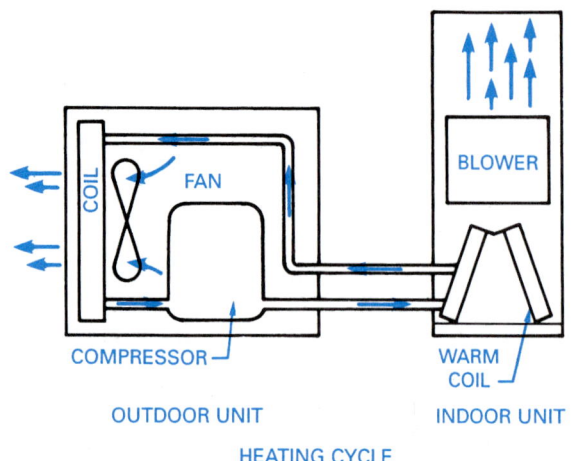

HEATING CYCLE

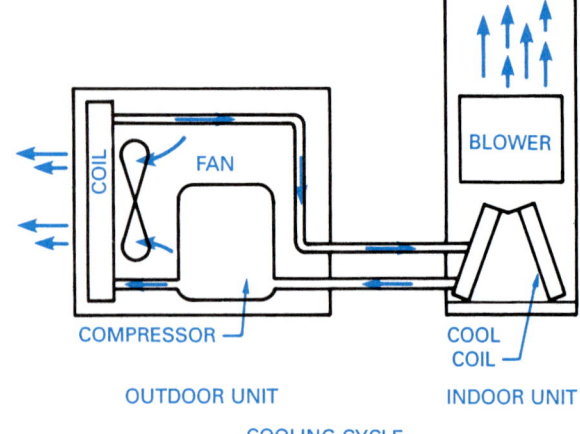

COOLING CYCLE

FIGURE 25.6 ■ Heat pump heating and cooling cycle. *Courtesy Lennox Industries, Inc.*

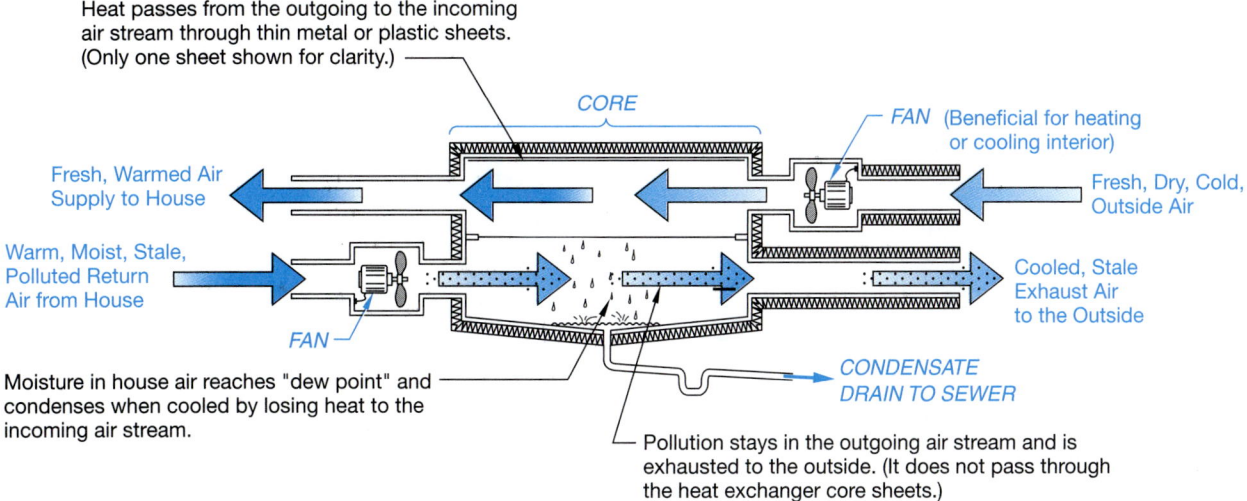

FIGURE 25.7 ■ Components and function of an air-to-air heat exchanger. *Courtesy U.S. Department of Energy.*

tors are able to implement air-to-air heat exchanger technology by the selection of the proper size of a ducted system.

Thermostat

The thermostat is an automatic mechanism for controlling the amount of heating or cooling given by a central or zoned heating or cooling system. The thermostat symbol is shown in Figure 25.8.

The location of the thermostat is an important consideration to the proper functioning of the system. For zoned units, there may be thermostats placed in each room or a central panel placed in a convenient location. For central systems, there may be one or more thermostats, depending on the layout of the system or the number of units required to service the structure. For example, an office complex may have a split system where the structure is divided into two or more zones. Each individual segment of the system has its own thermostat.

Several factors contribute to the effective placement of the thermostat for a central system. A good common location is near the center of the structure and close to a return-air duct. The air entering the return-air duct is usually temperate, thus causing little variation in temperature. This is a location where an average temperature reading can be achieved. There should be no drafts. Avoid locations where sunlight or a heat register could cause an unreliable reading. Avoid placement near an exterior opening or on an outside wall. Avoid placement near stairs or a similar traffic area where significant bouncing or shaking could cause the mechanism to alter the actual reading.

FIGURE 25.8 ■ Thermostat floor plan symbol.

HVAC SYMBOLS

ANSI Standard symbols adopted by ANSI are available in the document *Graphical Symbols for Heating, Ventilating, and Air Conditioning*, ANSI Y32.2.4. Standard symbols are often modified to fit individual needs, with a legend placed on the drawing for interpretation.

There are over a hundred HVAC symbols that can be used in residential and commercial heating plans. Only a few of the possible symbols are typically used in residential HVAC drawings. The symbols are divided into heating, ventilating, and air-conditioning symbols. Figure 25.9 shows some common HVAC symbols. Line symbols are also commonly used in HVAC drafting to represent related piping. These line symbols are commonly solid or dashed lines that are broken periodically with an abbreviation inserted in the break. The abbreviation identifies the pipe application as shown in Figure 25.9. Sheet metal conduit (duct) and air-conditioning templates are available to help speed manual drafting, and CADD menus and symbols libraries may be used to customize drafting practices.

HVAC DRAWINGS

Drawings for the HVAC system show the size and location of all equipment, ductwork, and components with accurate symbols, specifications, notes, and schedules that form the basis of contract requirements for construction. Specifications are documents that accompany the drawings and contain all pertinent written information related to the HVAC system.

When a complete HVAC layout is necessary, drawings can be prepared by the architect, designer, architectural drafter, or the heating contractor. Figure 25.10, page 868, shows a heating plan for a residential structure.

For commercial structures, the HVAC plan can be prepared by a mechanical engineer as a consultant for the architect. The consulting engineer is responsible for the HVAC design and installation. The engineer determines the placement of all equipment

(Continued on page 868)

EQUIPMENT SYMBOLS

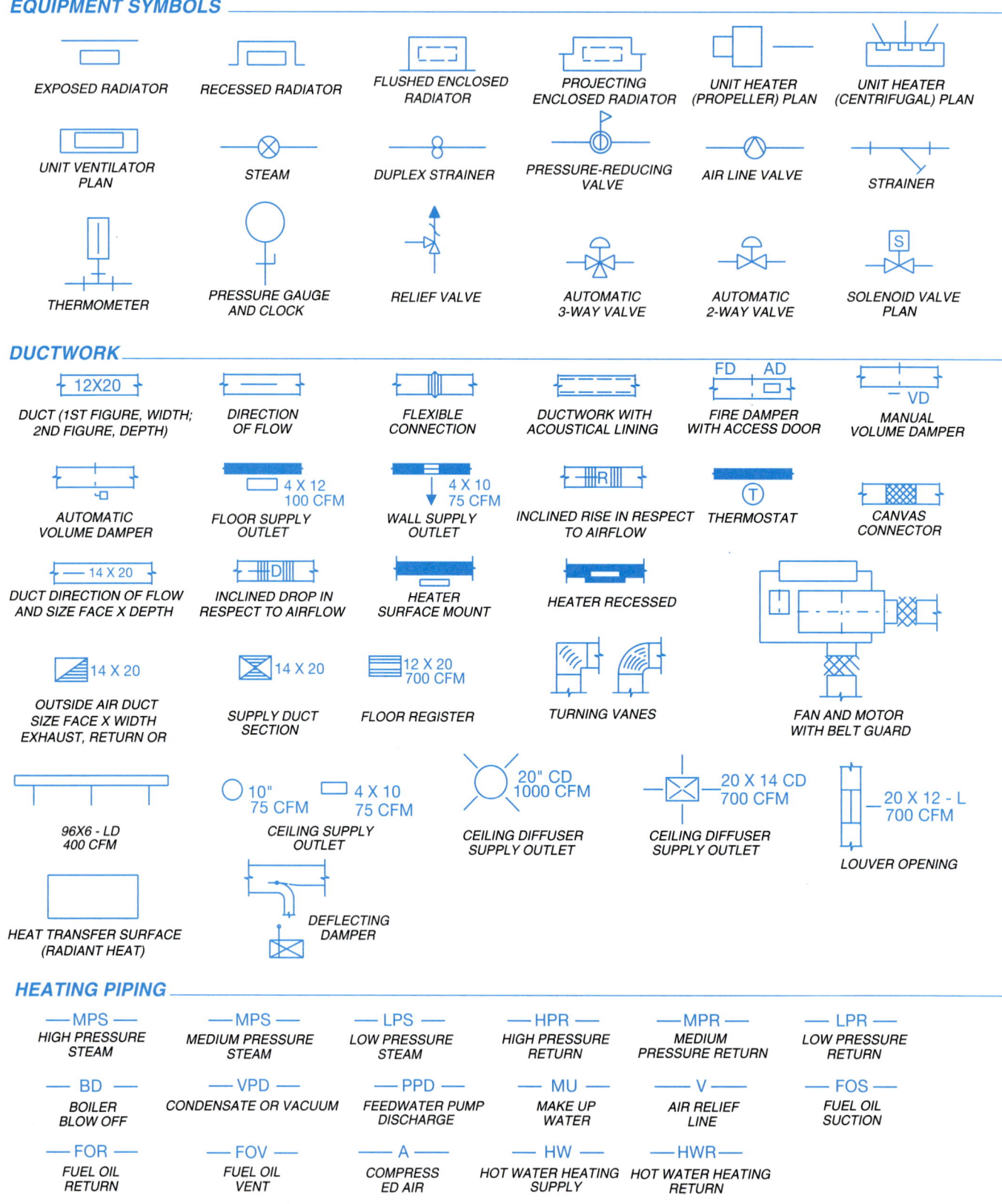

FIGURE 25.9 ■ Common HVAC symbols.

AIR-CONDITIONING PIPING

— RL —	— RD —	— RS —	—CWS—	—CWR—	— CHWS —
REFRIGERANT LIQUID	REFRIGERANT DISCHARGE	REFRIGERANT SUCTION	CONDENSER WATER SUPPLY	CONDENSER WATER RETURN	CHILLED WATER SUPPLY

—CHWR—	— MU —	— H —	— D —
CHILLED WATER RETURN	MAKE UP WATER	HUMIDIFICATION LINE	DRAIN

REFRIGERATION SYMBOLS

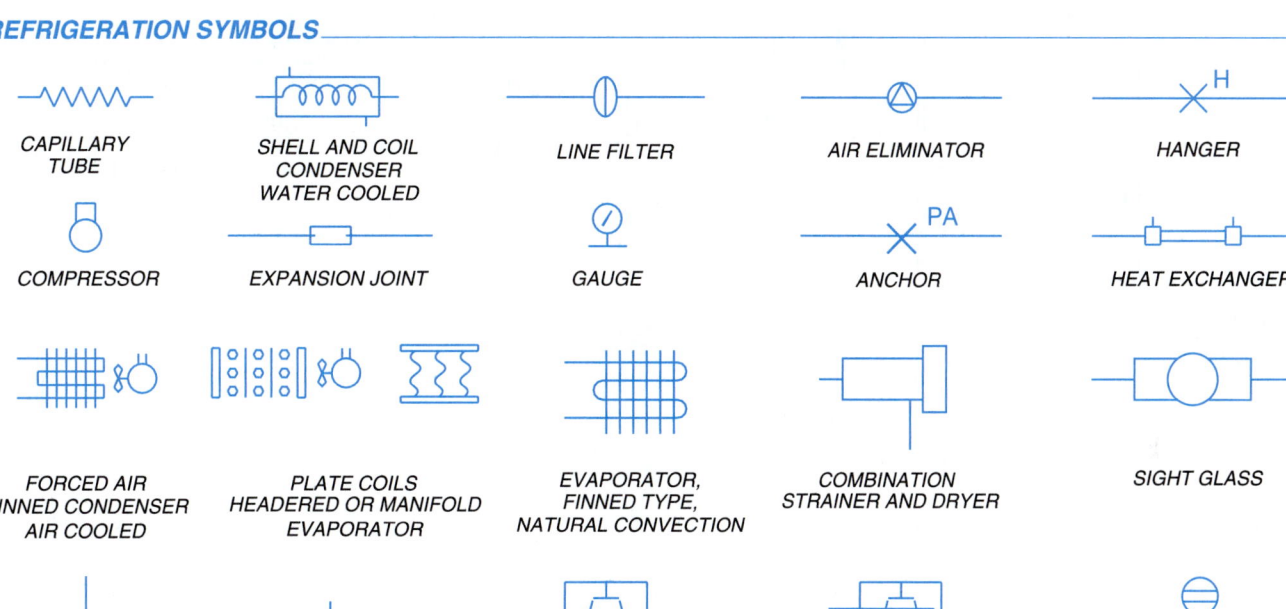
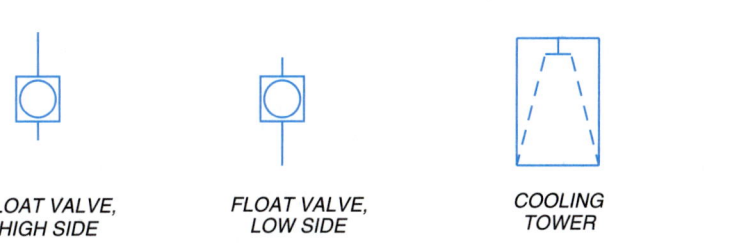
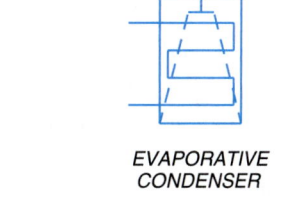
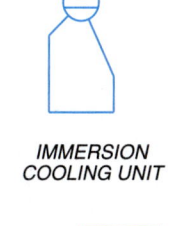

CAPILLARY TUBE	SHELL AND COIL CONDENSER WATER COOLED	LINE FILTER	AIR ELIMINATOR	HANGER
COMPRESSOR	EXPANSION JOINT	GAUGE	ANCHOR	HEAT EXCHANGER

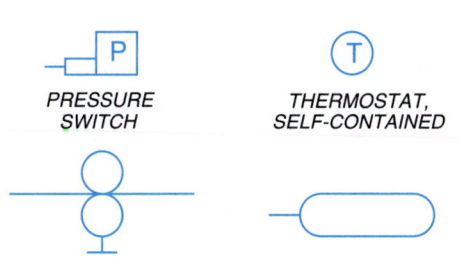
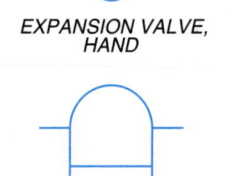
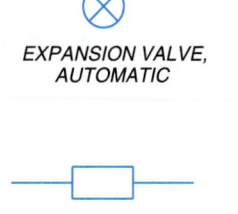

FORCED AIR FINNED CONDENSER AIR COOLED	PLATE COILS HEADERED OR MANIFOLD EVAPORATOR	EVAPORATOR, FINNED TYPE, NATURAL CONVECTION	COMBINATION STRAINER AND DRYER	SIGHT GLASS

FLOAT VALVE, HIGH SIDE	FLOAT VALVE, LOW SIDE	COOLING TOWER	EVAPORATIVE CONDENSER	IMMERSION COOLING UNIT

PRESSURE SWITCH	THERMOSTAT, SELF-CONTAINED	EXPANSION VALVE, HAND	EXPANSION VALVE, AUTOMATIC	EXPANSION VALVE, THERMOSTATIC

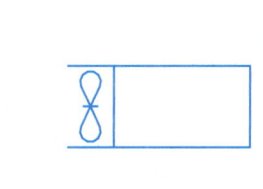
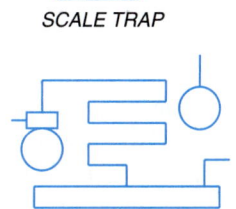
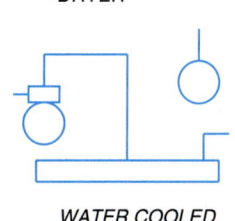

PRESSURE SWITCH	THERMAL BULB	SCALE TRAP	DRYER	FILTER AND STRAINER

EVAPORATOR, FORCED CONVECTION	AIR COOLED CONDENSING UNIT	WATER COOLED CONDENSING UNIT

FIGURE 25.9 ■ Common HVAC symbols. (*Continued*)

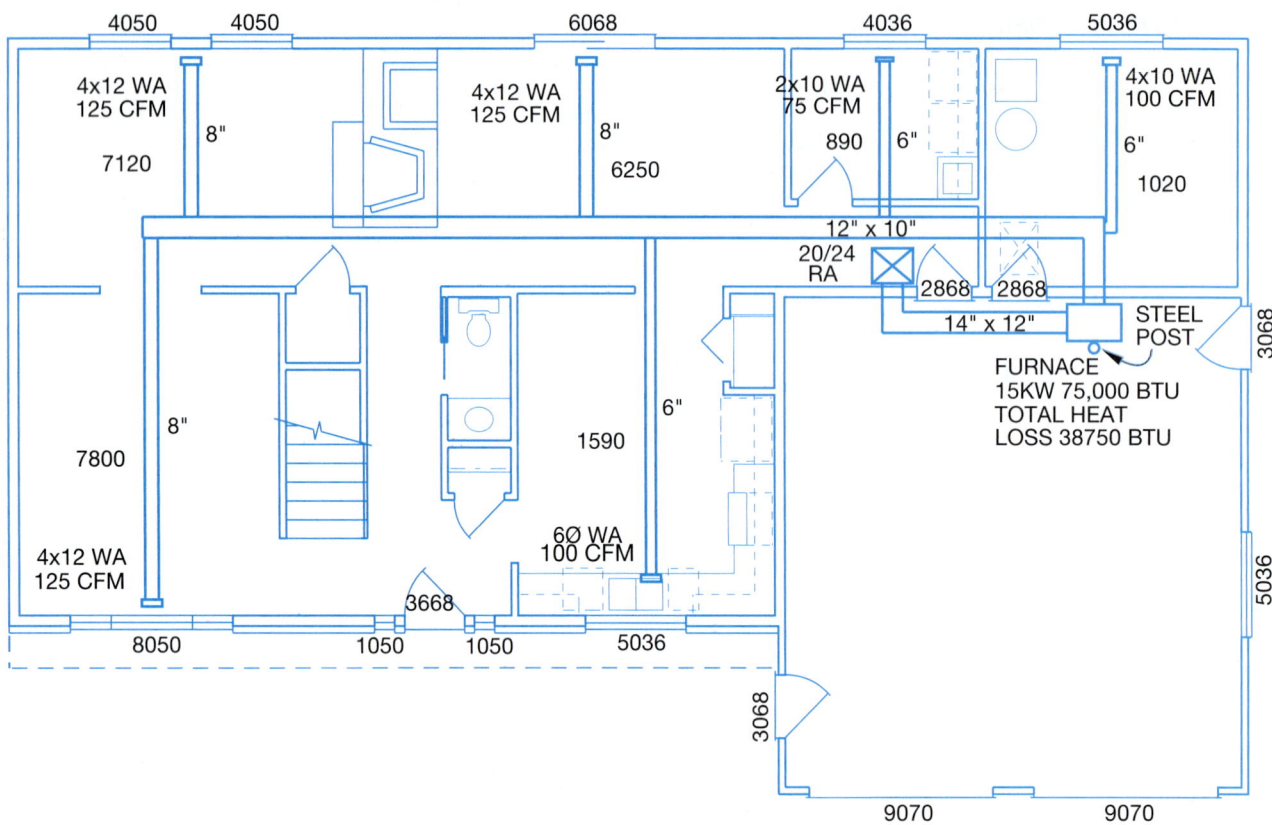

FIGURE 25.10 ■ A detailed forced-air plan for a residential structure.

and the location of all duct runs and components. The engineer also determines all of the specifications for unit and duct size based on calculations of structure volume, exterior surface areas and construction materials, rate of airflow, and pressure. The engineer can prepare single-line sketches or submit data and calculations to a design drafter who prepares design sketches or final drawings. Drafters without design experience work from engineering or design sketches to prepare formal drawings. A single-line engineer's sketch is shown in Figure 25.11. The next step in the HVAC design is for the drafter to convert the rough sketch into a preliminary drawing. This preliminary drawing goes back to the engineer and architect for verification and corrections or changes. The final step in the design process is for the drafter to implement the design changes on the preliminary drawing to establish the final HVAC drawing. The final HVAC drawing is shown in Figure 25.12.

Convert an engineering sketch to a formal drawing using manual or CADD in this manner:

STEP 1 Draw duct runs using thick .03 or .035 in. (0.7 or 0.9 mm) line widths.

STEP 2 Label duct sizes within the duct when appropriate, or use a note with a leader to the duct in other situations.

STEP 3 Duct sizes can be noted as 22 × 12 (560 × 300 mm) or 22/12, where the first number, 22, is the duct width and the second numeral, 12, indicates the duct depth.

STEP 4 Place notes on the drawing to avoid crowding. Aligned techniques may be used where horizontal notes read from the bottom of the sheet and vertical notes read from the right side of the sheet. Make notes clear and concise.

STEP 5 Refer to schedules to get specific drawing information that may not otherwise be available on the sketch.

STEP 6 Label equipment either blocked out or bold to clearly stand out from other information on the drawing

Several examples of duct system elements are shown comparing the engineering sketch and formal drawing in Figure 25.13.

Single- and Double-Line HVAC Plans

HVAC plans are drawn over the outline of the floor plan or as an overlay. The floor plan layout is drawn first using thin lines as a base sheet for the HVAC layout and other overlays. The HVAC plan is then drawn using thick lines and notes for contrast with the floor plan. The HVAC plan shows the placement of equipment and ductwork. The size (in inches) and shape (with symbols, ∅ = round, □ = square or rectangular) of ductwork and system component labeling is placed on the drawing or keyed to schedules. Drawings may be either single-line or double-line, depending on the needs of the client or how much detail must be shown. Single-line drawings are easier and faster

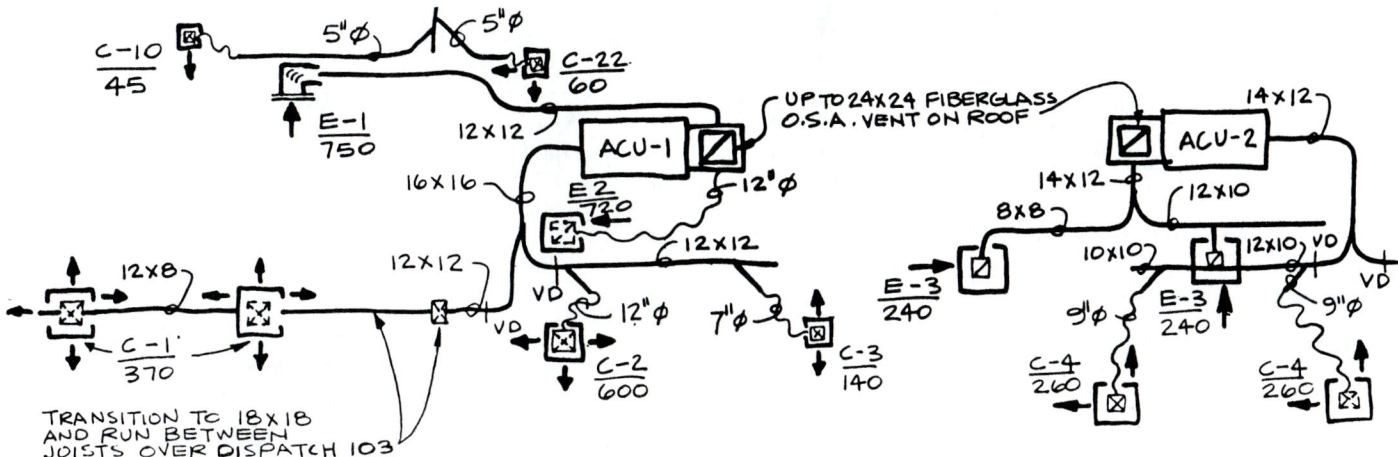

FIGURE 25.11 ■ Single-line HVAC engineer's sketch.

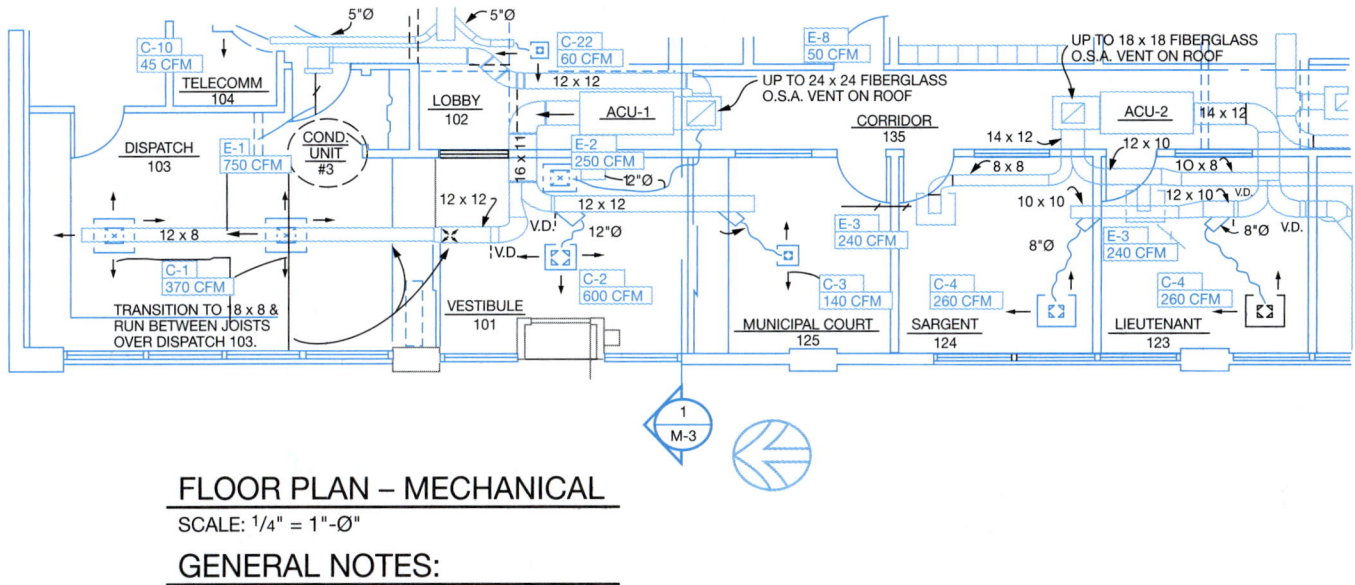

FLOOR PLAN – MECHANICAL

SCALE: 1/4" = 1"-∅"

GENERAL NOTES:

① VERIFY ALL EXISTING CONDITIONS AT SITE.

② SEE ARCHITECTURAL FLOOR PLAN FOR 1-HR RATED AREAS. PROVIDE FIRE DAMPERS AS SCHEDULED AND AS REQ'D BY CODE.

FIGURE 25.12 ■ HVAC plan for the engineer's sketch shown in Figure 24.11. *Courtesy W. Alan Gold Consulting Mechanical Engineer and Robert Evenson Associates AIA Architects.*

to draw. In many situations they are adequate to provide the equipment placement and duct routing as shown in Figure 25.14. Double-line drawings take up more space and are more time-consuming to draw than single-line, but they are often necessary when complex systems require more detail, as shown in Figure 25.12.

HVAC Symbol Specifications

There are a large variety of HVAC symbols that can be placed on a drawing. Symbols and notes are used to show and label floor, ceiling, and wall ducts, diffusers, and grills. Duct runs can be shown in plan view and in section. Ducts and components can be drawn with double-line or single-line representations. The size of double-line ducts is represented by the width of the duct in plan view. The outline of the duct runs is drawn with thick lines so the ducts and related connections and equipment contrast with the rest of the drawing. The sectional view displays both the width and height. Single-line drawings are commonly created with a very thick line representing the duct run and with symbols to show the related connections and equipment. Common symbol specifications for double-line and single-line

SKETCHES

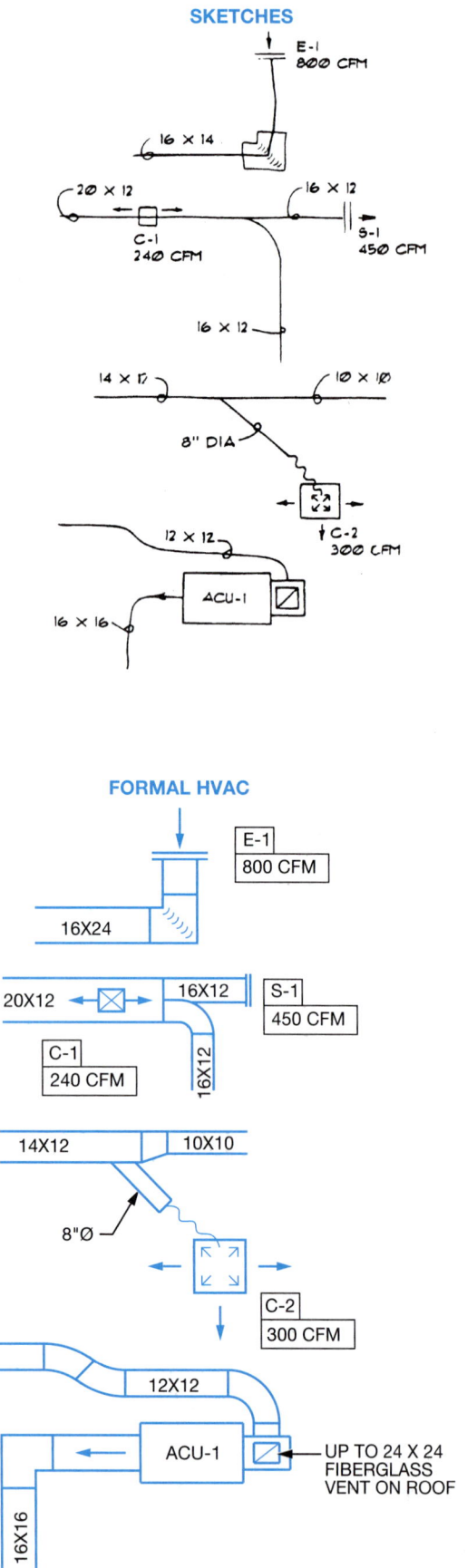

FORMAL HVAC

FIGURE 25.13 ■ Examples showing engineering sketches converted to formal HVAC drawings.

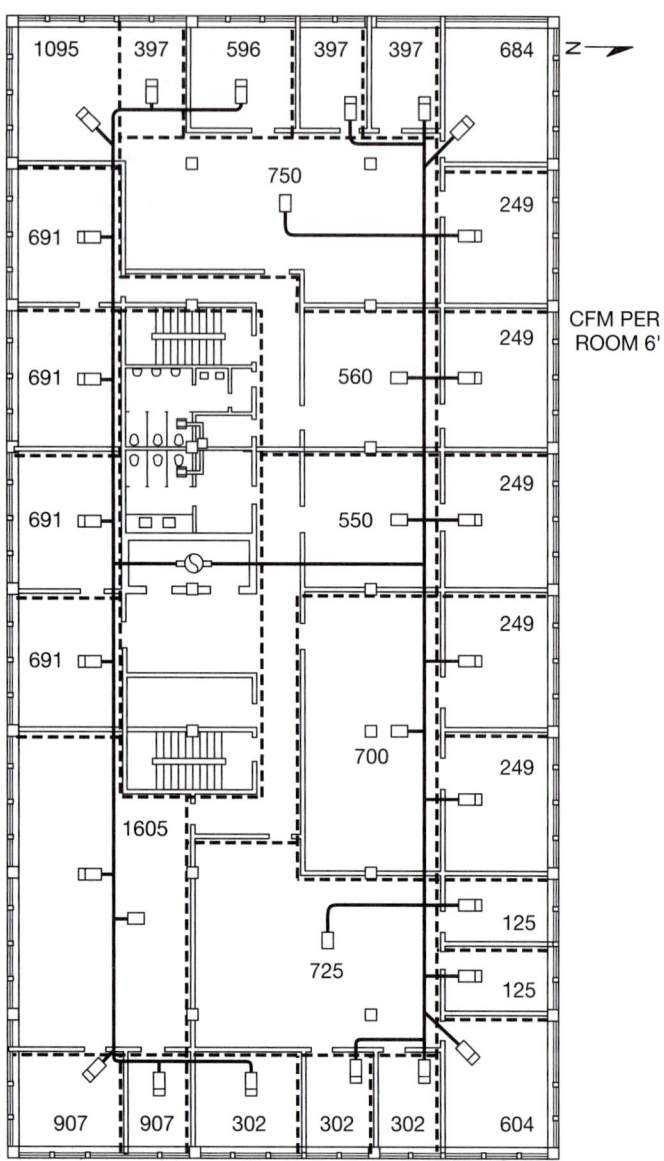

FIGURE 25.14 ■ Single-line ducted HVAC system showing a layout of the proposed trunk and runout ductwork. *Courtesy The Trane Company, La Crosse, WI.*

HVAC drawings are shown in Figure 25.15a. Many companies include an **HVAC legend** on their drawings. This legend also provides information about the standards that are used by the company. Figure 25.15b shows an example of an HVAC legend. This legend is another good place to find information about how to create HVAC drawings.

Written Specifications

HVAC drafting standards and specifications are important whether the drawings are created manually or with CADD. Many companies have a standards manual or instructions that correlate with the way the HVAC drawings are created. The numbered standards instructions in Figure 25.16a, page 873, correlate to the numbers found on the sample drawing in Figure 25.16b, page 874. Follow

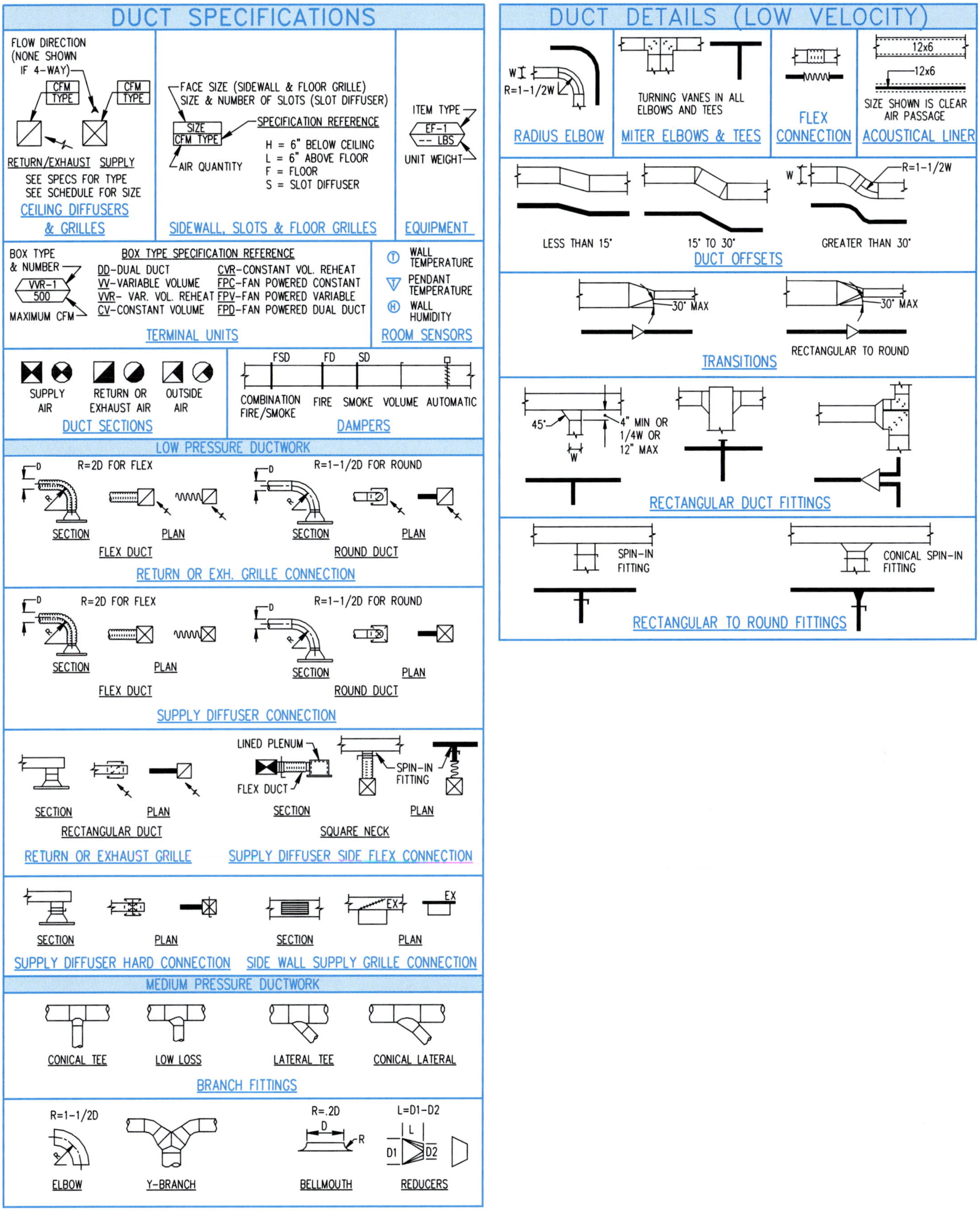

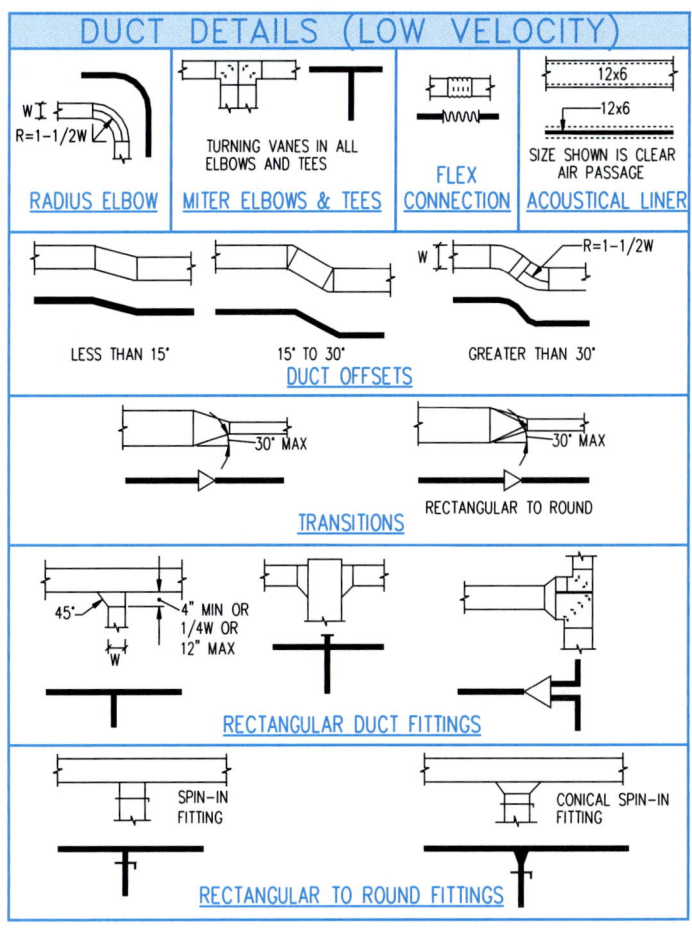

DUCT SPECIFICATIONS

CEILING DIFFUSERS & GRILLES

FLOW DIRECTION (NONE SHOWN IF 4-WAY)

CFM TYPE / CFM TYPE

RETURN/EXHAUST SUPPLY
SEE SPECS FOR TYPE
SEE SCHEDULE FOR SIZE

SIDEWALL, SLOTS & FLOOR GRILLES

FACE SIZE (SIDEWALL & FLOOR GRILLE)
SIZE & NUMBER OF SLOTS (SLOT DIFFUSER)

SIZE / CFM TYPE — SPECIFICATION REFERENCE
AIR QUANTITY

H = 6" BELOW CEILING
L = 6" ABOVE FLOOR
F = FLOOR
S = SLOT DIFFUSER

EQUIPMENT

ITEM TYPE
EF-1
-- LBS
UNIT WEIGHT

TERMINAL UNITS

BOX TYPE & NUMBER
VVR-1
500
MAXIMUM CFM

BOX TYPE SPECIFICATION REFERENCE
DD-DUAL DUCT   CVR-CONSTANT VOL. REHEAT
VV-VARIABLE VOLUME   FPC-FAN POWERED CONSTANT
VVR- VAR. VOL. REHEAT FPV-FAN POWERED VARIABLE
CV-CONSTANT VOLUME   FPD-FAN POWERED DUAL DUCT

ROOM SENSORS

T WALL TEMPERATURE
▽ PENDANT TEMPERATURE
H WALL HUMIDITY

DUCT SECTIONS

SUPPLY AIR   RETURN OR EXHAUST AIR   OUTSIDE AIR

DAMPERS

FSD   FD   SD

COMBINATION FIRE/SMOKE   FIRE   SMOKE   VOLUME   AUTOMATIC

LOW PRESSURE DUCTWORK

RETURN OR EXH. GRILLE CONNECTION

R=2D FOR FLEX
SECTION   PLAN
FLEX DUCT

R=1-1/2D FOR ROUND
SECTION   PLAN
ROUND DUCT

SUPPLY DIFFUSER CONNECTION

R=2D FOR FLEX
SECTION   PLAN
FLEX DUCT

R=1-1/2D FOR ROUND
SECTION   PLAN
ROUND DUCT

RETURN OR EXHAUST GRILLE

SECTION   PLAN
RECTANGULAR DUCT

SUPPLY DIFFUSER SIDE FLEX CONNECTION

LINED PLENUM
FLEX DUCT
SPIN-IN FITTING
SECTION   PLAN
SQUARE NECK

SUPPLY DIFFUSER HARD CONNECTION   SIDE WALL SUPPLY GRILLE CONNECTION

SECTION   PLAN
SECTION   PLAN

MEDIUM PRESSURE DUCTWORK

BRANCH FITTINGS

CONICAL TEE   LOW LOSS   LATERAL TEE   CONICAL LATERAL

R=1-1/2D
ELBOW

Y-BRANCH

R=.2D
D   R
BELLMOUTH

L=D1-D2
L
D1   D2
REDUCERS

DUCT DETAILS (LOW VELOCITY)

RADIUS ELBOW

W
R=1-1/2W

MITER ELBOWS & TEES

TURNING VANES IN ALL ELBOWS AND TEES

FLEX CONNECTION

ACOUSTICAL LINER

12x6
12x6
SIZE SHOWN IS CLEAR AIR PASSAGE

DUCT OFFSETS

LESS THAN 15°   15° TO 30°   GREATER THAN 30°
W
R=1-1/2W

TRANSITIONS

30° MAX   30° MAX
RECTANGULAR TO ROUND

RECTANGULAR DUCT FITTINGS

45°
4" MIN OR 1/4W OR 12" MAX
W

RECTANGULAR TO ROUND FITTINGS

SPIN-IN FITTING   CONICAL SPIN-IN FITTING

(a)

FIGURE 25.15 ■ (a) Common HVAC symbol specifications. *Courtesy PAE Consulting Engineers.* (Continued)

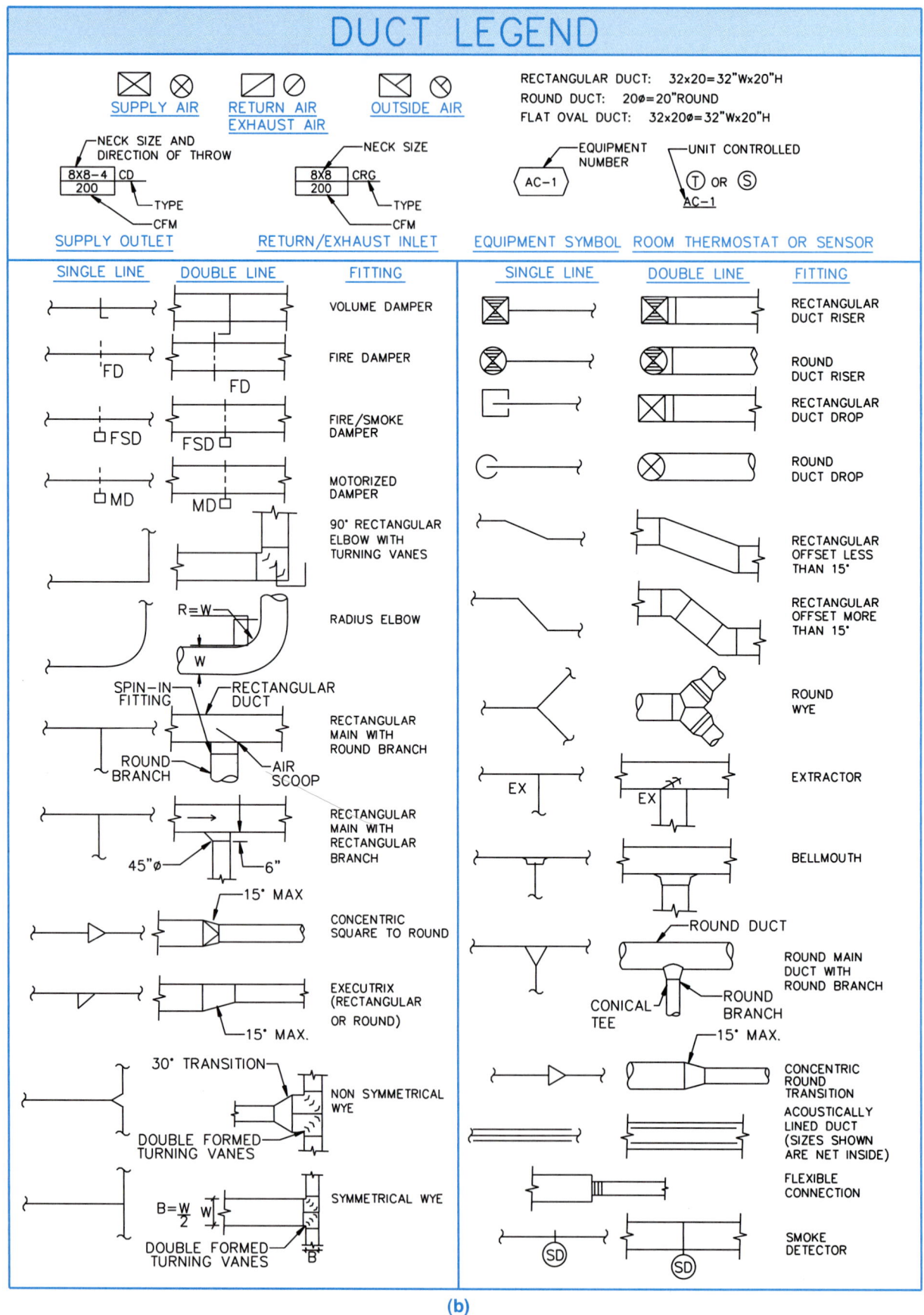

(b)

FIGURE 25.15 ■ *(Continued).* (b) An HVAC legend is commonly found on a drawing to help communicate drafting standards. *Courtesy Interface Engineering.*

INSTRUCTIONS FOR NOTES:

1. DO NOT PUT TEXT ON/UNDER DUCT LINES. MOVE TEXT IF NEW DUCT GOES OVER OLD TEXT.

2. INSERT REVISION CLOUD DELTA IN CLOUD LINE.

   Example

3. IF POSSIBLE STRETCH OR EXTEND LINES INSTEAD OF ADDING SHORT SEGMENTS TO EXISTING LINES.

4. BREAK PIPE LINES FOR ALL UNIONS, VALVES AND OTHER PLUMBING FITTINGS. Example

5. USE THICK LINES FOR WASTE AND EXTRA THICK LINES FOR SINGLE LINE DUCT WORK.

6. HVAC VENT LAYER SHOULD BE USED FOR STEAM OR FUEL OIL VENTING ONLY, USE PLUMBING VENT IN ALL OTHER VENT CASES.

7. DO NOT BREAK LEADER LINES FOR OTHER LEADERS UNLESS INSTRUCTED OTHERWISE BY ENGINEER. TRY TO MOVE THINGS UNTIL ROOM FOR CORRECT LEADER AND NOTES ARE FOUND.

8. TO LABEL TERMINAL UNITS:
   USE LARGE EQUIPMENT BOX TO LABEL ALL TERMINAL UNITS.

   Example   °R-12 / 1000

9. TO LABEL EQUIPMENT:
   USE LARGE EQUIPMENT BOX TO LABEL ALL EQUIPMENT.

   Example   AHU-2 / XX LBS

10. DIFFUSER BOXES:
    USE THE SMALL DIFFUSER BOX FOR RETURN/EXHAUST AND SUPPLY CEILING DIFFUSERS AND GRILLES.

    Example   6x48

    USE THE LARGE DIFFUSER BOX FOR SIDEWALL, SLOTS AND FLOOR GRILLES.   Example   500 / C-1   400 H-12

11. DO NOT PUT DIMENSIONS OR LABELS FOR DUCTS OR OTHER EQUIPMENT IN A DUCT THAT IS NOT THE ONE BEING LABELED.

12. READ AND USE THE MECHANICAL LEGENDS FOR EXAMPLES AND OTHER INFORMATION NOT SHOWN ON THIS SHEET.

13. WHEN LABELING A DUCT UP OR DOWN, WITH AN ARROW, THE FIRST NUMBER OF THE LABEL IS THE HORIZONTAL DIMENSION AND THE SECOND IS THE VERTICAL DIMENSION REGARDLESS OF WHICH SIDE THE ARROW POINTS.   Example   40x24 / 40x24

DRAWING INSTRUCTIONS

1. TERMINAL UNITS (°R'S & °S) GO ON DUCT LAYER.

2. IF LABEL IS °R (VARIABLE VOLUME REHEAT) THEN USE TERMINAL UNIT W/REHEAT COIL.

3. LEADERS ON MULTI-LINE TEXT COME FROM THE UPPER LEFT OR THE LOWER RIGHT OF THE TEXT BLOCK

4. LEFT JUSTIFY ALL MULTI-LINE NOTES.

5. BREAKS FOR HIDDEN LINES ON DUCT & DIFFUSERS OR BREAKS IN PIPE LINES THAT CROSS NEED GAPS IN THE LINES. USE:
   'SB' BREAK COMMAND FOR BREAKING LINES IN DIFFUSERS AND SPECIAL CASES ONLY.
   'B' BREAK COMMAND FOR BREAKING LINES OVER PIPING AND DOUBLE LINE DUCT LINES.
   'BB' BREAK COMMAND FOR BREAKING A LINE OVER SINGLE LINE DUCT AND WASTE PIPE LINES.

6. TYPICAL OF #'s ARE WRITTEN (TYP. #), NOT (TYP #) OR (TYP. OF #).

7. THERMOSTATS – USE CIRCLE T, THERMDOT & ARC – 3 FROM DOUBLE DUCT MENU TO KEEP ON CORRECT LAYERS.

8. USE STRAIGHT SPIN IN FITTINGS FOR ROUND DUCT OFF OF SQUARE DUCT (LOW PRESSURE DUCT ONLY).

9. USE CONICAL FITTINGS FOR ROUND TAKE-OFFS FROM ROUND DUCT (MEDIUM PRESSURE ONLY).

10. USE TICKS AND NOT ARROWS FOR ANY DIMENSIONING ON DRAWINGS.

11. ALL LETTERED ('SYMBOLS') HEX NOTES ARE TO BE PRIOR (LEFT SIDE) TO WRITTEN NOTES DISREGARDING WHICH SIDE LEADER COMES OFF. ALL 'NUMBER' HEX NOTES ARE TO BE AFTER (RIGHT SIDE) WRITTEN NOTES DISREGARDING WHICH SIDE LEADER COMES OFF.

12. ALL SUPPLY AND RETURN BRANCHES WILL HAVE DAMPERS. UNLESS OTHERWISE NOTED BY ENGINEER.

13. USE LOOP LEADERS TO LABEL SINGLE LINE PIPE AND ARROWS TO LABEL DROPS, RISERS AND VALVES ETC.

14. USE THE ARC ARROW (<–C–) TO LABEL RELOCATE EXISTING AND PUT HEX WITH R IN LINE. MOVE DIFFUSER LABEL TO LABEL NEW LOCATION AND PUT HEX R WITH LABEL DIFFUSER IN NEW LOCATION IS ON NEW, NOT EXISTING LAYER.

15. DEMO LAYER FOR HVAC & DOUBLE LINE PIPE USES AN L3 LINE (MAGENTA) AND HIDDENA LINETYPE. EACH ENTITY TYPE SHOULD HAVE ITS OWN DEMO LAYER (DUCT, DIFFUSERS, DOUBLE LINE PIPES ETC.).

16. TO DEMO SINGLE LINE PIPE USE "DEMO X's" ROUTINE FROM THE LISP DROP-DOWN MENU. EACH ROUTINE CREATES A LAYER AND USES CORRECT COLOR FOR X's. (HVAC PIPING TYPES SUCH AS CHWS, HTWS, CDS ETC. CAN BE FOUND IN HVAC PIPING LEGEND.)

17. BREAK DUCT (OR PIPE) BETWEEN CAP AND DUCT (OR PIPE).

18. PIPES TO °R'S DO NOT HAVE DROPS UNLESS REQUESTED BY ENGINEER.

19. USE SQUARE DUCT BREAK FROM HVAC DROP-DOWN MENU AND PUT BREAK ON DUCT LAYER OF TYPE BEING BROKEN.

20. INTERNAL SOUND INSULATION IS ON THE DUCT LAYER OFFSET 3" USING THE HIDDENA LINETYPE AND COLOR 3 (GREEN).

(a)

FIGURE 25.16 ■ The (a) numbered standards instructions used for creating standard HVAC symbols and drafting applications correlate to the circled numbers on (b) the sample drawing. *Courtesy PAE Consulting Engineers.* (Continued)

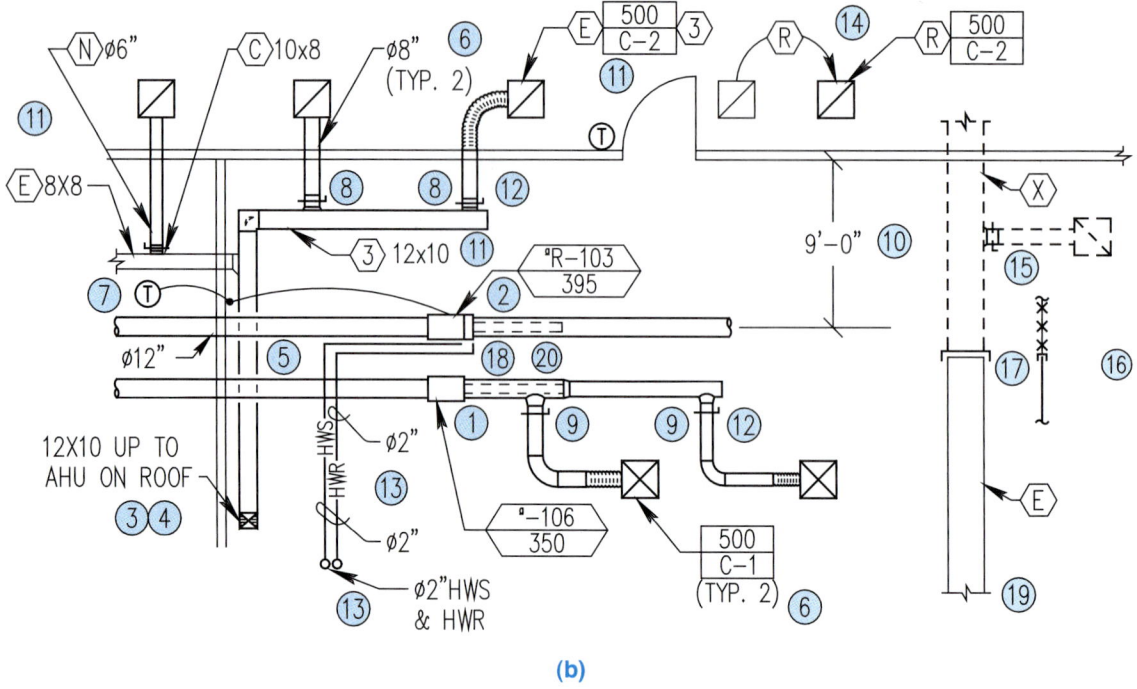

(b)

FIGURE 25.16 ■ *(Continued).* (b) The sample drawing. *Courtesy PAE Consulting Engineers.*

these instructions as you create your own HVAC drawings. This is a common example of HVAC **specifications**. Keep in mind that other slight variations are found in industry.

Detail Drawings

Detail drawings are used to clarify specific features of the HVAC plan. Single- and double-line drawings are intended to establish the general arrangement of the system; they do not always provide enough information to fabricate specific components. When further clarification of features is required, detail drawings are made. A detail drawing is an enlarged view(s) of equipment, equipment installations, duct components, or any feature that is not defined on the plan. Detail drawings may be scaled or unscaled and provide adequate views and dimensions for sheet metal shops to prepare fabrication patterns, as shown in Figure 25.17.

Section Drawings

Sections or **sectional views** are used to show and describe the interior portions of an object or structure that would otherwise be difficult to visualize. Section drawings may be used to provide a clear representation of construction details or a profile of the HVAC plan as taken through one or more locations in the building. There are two basic types of section drawings used in HVAC. One method is used to show the construction of the HVAC system in relationship to the structure. In this case, the building is sectioned and the duct system is shown unsectioned.

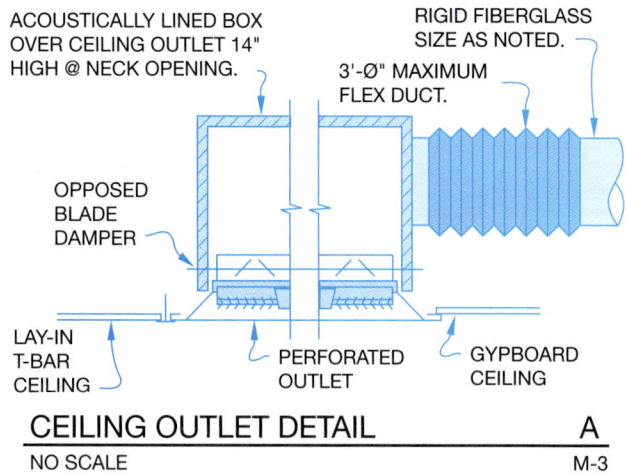

CEILING OUTLET DETAIL   **A**
NO SCALE   M-3

FIGURE 25.17 ■ Sample detail drawing. *Courtesy W. Alan Gold Consulting Mechanical Engineer and Robert Evenson Associates AIA Architects.*

This drawing provides a profile of the HVAC system. There may be one or more sections taken through the structure, depending on the complexity of the project. The building structure can be drawn using thin lines as shown in Figure 25.18. Figure 25.18 is a section through the HVAC plan shown in Figure 25.12. The other sectioning method is used to show detail of equipment or to show how parts of an assembly fit together. (See Figure 25.19.)

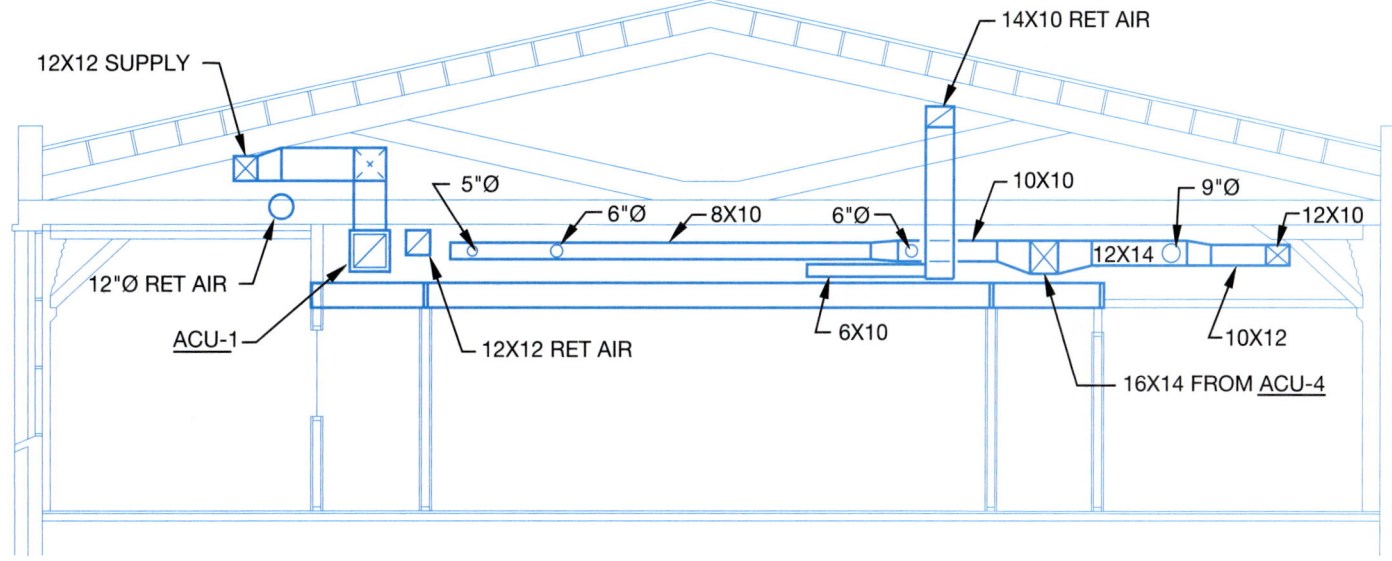

FIGURE 25.18 ■ Section drawing. *Courtesy W. Alan Gold Consulting Mechanical Engineer and Robert Evenson Associates AIA Architects.*

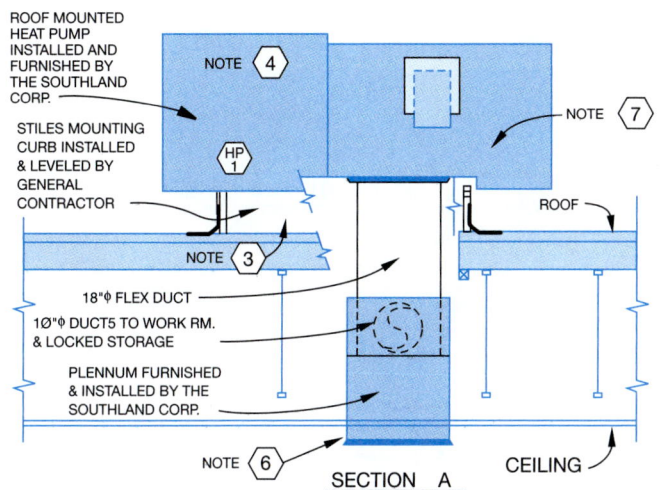

FIGURE 25.19 ■ Detailed section showing HVAC equipment installation. *Courtesy The Southland Corporation.*

Schedules

Numbered symbols are used on the HVAC plan to key specific items to charts known as **schedules**. These schedules are used to describe items such as ceiling outlets, supply and exhaust grills, hardware, and equipment. Schedules are charts of materials or products that include size, description, quantity used, capacity, location, vendor's specification, and any other information needed to construct or finish the system. Schedules aid the drawing by keeping it clear of unnecessary notes. They are generally placed in any convenient area of the drawing field or on a separate sheet. Items on the plan can be keyed to schedules by using a letter and number combination such as C-1 for CEILING OUTLET NO. 1, E-1 for EXHAUST GRILL NO. 1, or ACU-1 for

EQUIPMENT UNIT NO. 1. The exhaust grill schedule keyed to the HVAC plan in Figure 25.12 can be set up as a chart as shown in Figure 25.20.

Pictorial Drawings

Pictorial drawings can be isometric or oblique as shown in Figure 25.21. Isometric, oblique, and perspective techniques are discussed in Chapter 15. Pictorial drawings are usually not drawn to scale. They are used in HVAC for a number of applications, such as assisting in visualization of the duct system, and when the plan and sectional views are not adequate to show difficult duct routing. Manually drawn pictorials are time-consuming and generally not used unless necessary. CADD models are valuable as you will see in the following CADD Applications on pages 878–881.

DRAWING REVISIONS

Drawing revisions are common on HVAC projects. Revisions can be caused for a number of reasons, for example, changes requested by the owner, job site corrections, correcting errors, or code changes. Changes are done in a formal manner by submitting an **addendum** to the contract, which is a written notification of the change that is accompanied by a drawing representing the change.

Revision Clouds

A revision cloud is placed around the area that is changed. The **revision cloud** is a cloudlike circle around the change as shown in Figure 25.22. CADD programs that are commonly used for architectural and structural drafting have commands that allow

EXHAUST GRILL SCHEDULE

SYMBOL	SIZE	CFM	LOCATION	FIRE DPR.	KEY OP. OPPBLD	KEY OP. EXTR	NO DPR.	TYPE	REMARKS
				DAMPER TYPE					
E–1	24x12	750	HIGH WALL	X	X			4	
E–2	18x18	720	CEILING	X	X			2	
E–3	10x10	240	CEILING			X		1	24x24 PANEL
E–4	10x10	350	CEILING			X		1	
E–5	12x12	280	CEILING			X		1	↓
E–6	10x10	500	CEILING			X		1	
E–7	6x6	350				X		2	
E–8	12x6	50				X		3	
E–9	12x6	200	↓			X		3	↓
E–10	12x8	150				X		3	
E–11	10x10	290				X		3	
E–12	9x4	160				X		1	24x24 PANEL
E–13	9x4	75	HIGH WALL	X	X			4	

TYPE 1: KRUGER 1190 SERIES STEEL PERFORATED FRAME 23 FOR LAY-IN TILE

TYPE 2: KRUGER 1190 SERIES STEEL PERFORATED FRAME 22 FOR SURFACE MOUNT

TYPE 3: KRUGER EGC-5: 1/2"x1/2"x1/2" ALUMINUM GRID.

TYPE 4: KRUGER S80H: 35° HORIZ. BLADES 3/4" O.C.

FIGURE 25.20 ■ Exhaust grill schedule.

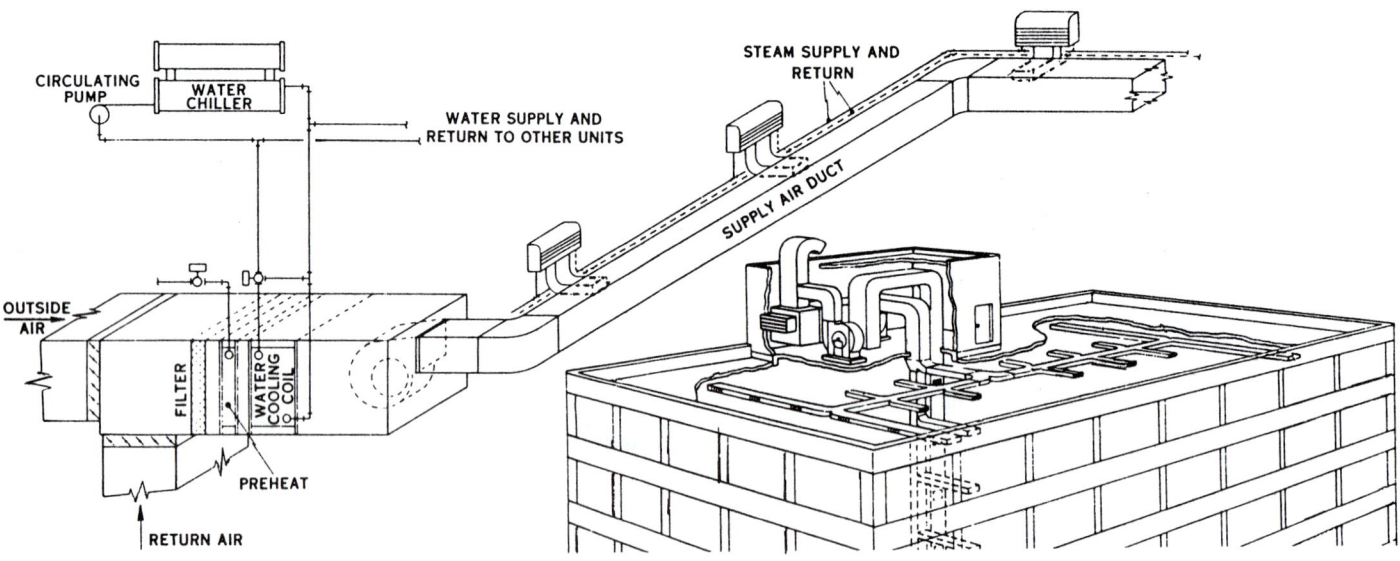

FIGURE 25.21 ■ Pictorial drawings. *Courtesy The Trane Company, La Crosse, WI.*

you to easily draw the revision cloud. There is also a triangle with a revision number inside that is placed next to the revision cloud or along the revision cloud line as shown in Figure 25.23. The triangle is commonly called a delta. The number is then correlated to a revision note placed somewhere on the drawing, or in the title block as shown in Figure 25.24. Each company has a desired location for revision notes, although common places are in the corners of the drawing, in a revision block or table, or in the title block. This practice is not as clearly defined on architectural drawings as in ANSI/ASME standard drawings. The note is used to explain the change. If a reference is given in the title block, then detailed information about the revision is normally provided in the revision document that is filed with the project information. The revision document is typically

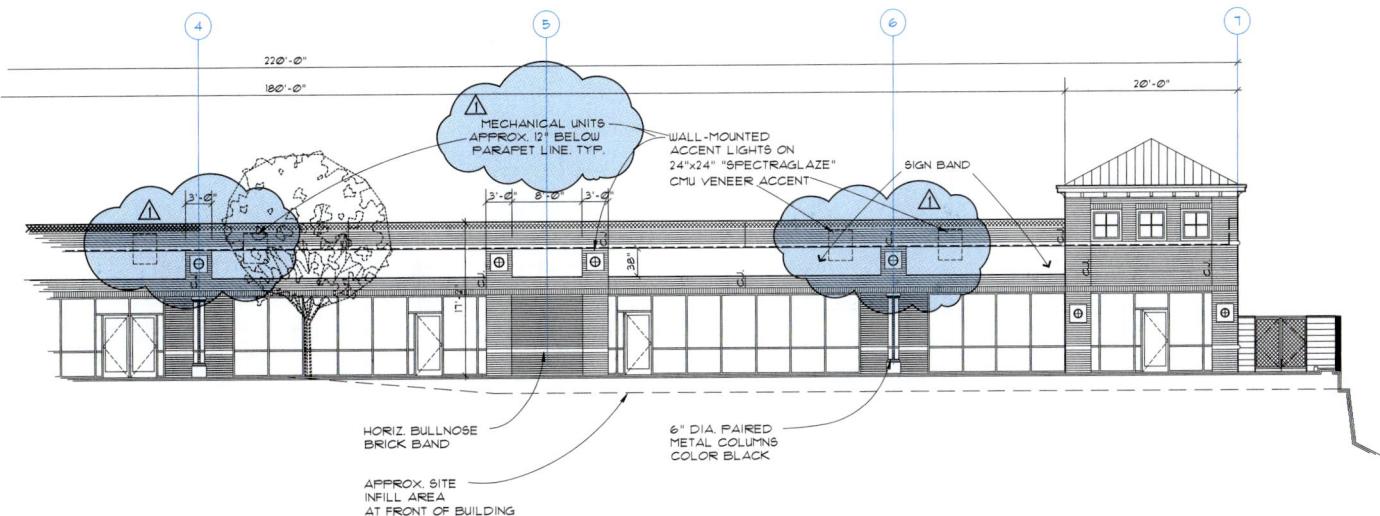

FIGURE 25.22 ■ A typical revision cloud and delta reference. *Portion of drawing courtesy Ankrom Moisan Architects.*

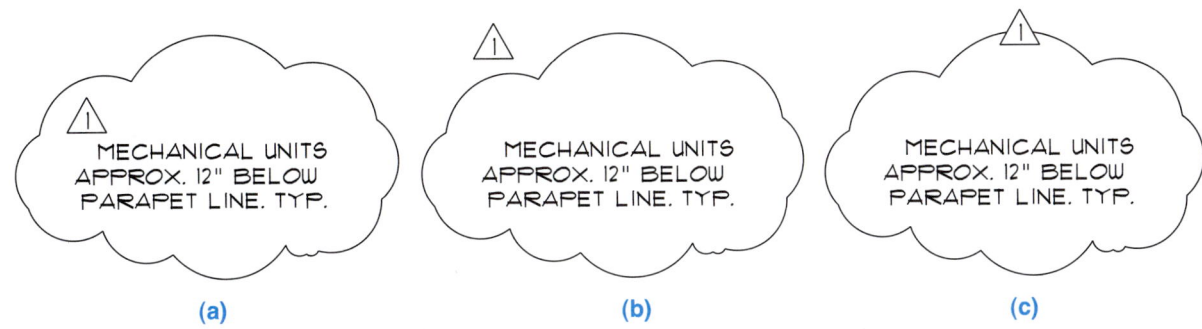

(a)   (b)   (c)

FIGURE 25.23 ■ Placement of the delta reference with the revision cloud: (a) delta inside of the revision cloud; (b) delta outside of the revision cloud; (c) delta inserted in the revision cloud line.

filled out and filed for reference. Changes can cause increased costs in the project.

It is easy to draw revision clouds with CADD. AutoCAD, for example, has a REVCLOUD command that allows you to specify the revision cloud arc length and draw a revision cloud around any desired area. The command works by picking a start point and then moving the cursor in the desired direction to create the revision cloud, as shown in Figure 25.25a. Create the revision cloud in a pattern around the desired area while moving the cursor back toward the start point. Press the pick button when the cursor is back at the start point to complete the revision cloud as shown in Figure 25.25b.

SHEET METAL DRAFTING AND DESIGN

Sheet metal drafting is used in any industry where flat material is used for fabrication into desired shapes. One of the most common applications is the HVAC industry, although sheet metal shapes are common in the automotive, electronics, and other related industries. Sheet metal drafting is also called pattern development.

PATTERN DEVELOPMENT

The principle of pattern development is based on laying out geometric shapes in true size and shape flat patterns. The fundamental concepts involved in making patterns for basic geometric shapes can be used in the development of any pattern. In most situations a front and top or bottom view is drawn to help establish true-length lines and true size shapes. *The key to pattern development is any line or element used in the development must be in true length.* Use construction lines for all preliminary layout work so errors may be easily erased. Use a construction layer when working with CADD. Finding true length lines and true size and shape of surfaces was discussed in detail in Chapter 16, Descriptive Geometry I. It is recommended that you review this chapter before continuing, if you need additional instruction.

Stretch-Out Line

A stretch-out line is typically the beginning line upon which measurements are made and the pattern development is established.

(Continued on page 881)

FIGURE 25.24 ■ Revision reference in the title block. The specific information about the revision is found in the job file. *Portion of title block courtesy Ankrom Moisan Architects.*

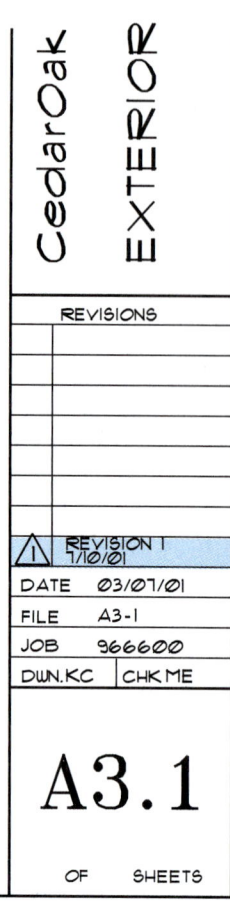

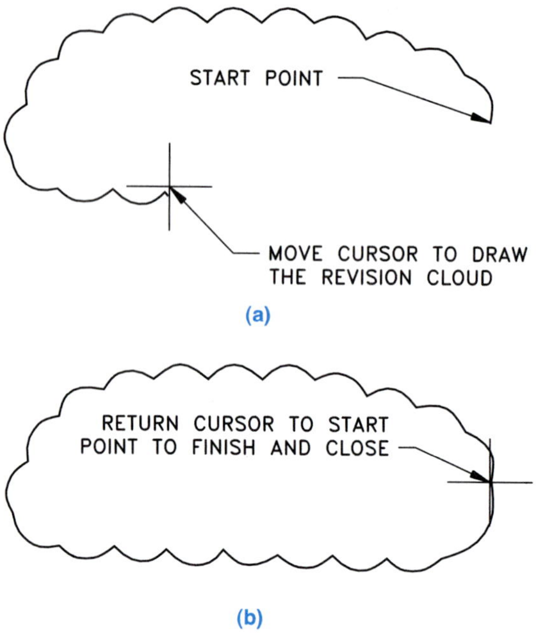

CedarOak
EXTERIOR

REVISIONS

⚠ REVISION 1
7/10/01

DATE 03/07/01
FILE A3-1
JOB 966600
DWN.KC CHK ME

A3.1

SCALE: 1" = 8'-0"   OF   SHEETS

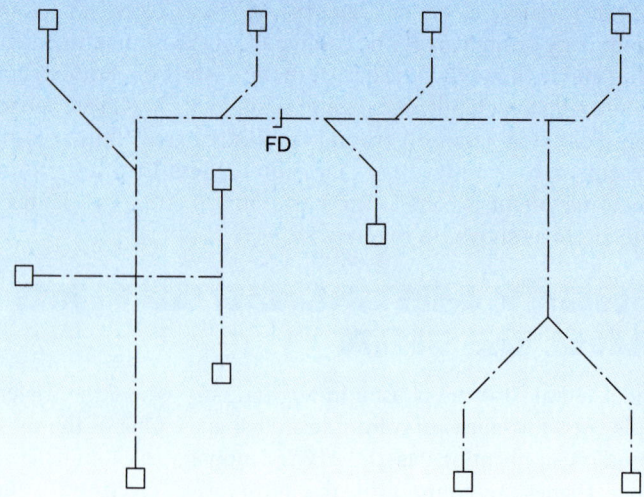

FIGURE 25.25 ■ The AutoCAD REVCLOUD command automatically draws arc segments along the path of your cursor movement.

HVAC PLANS

HVAC CADD software is available that allows you to place duct fittings and then automatically size ducts in accordance with common mechanical equipment suppliers' specifications. The CADD drafter typically uses the floor plan as a reference layer and develops the HVAC plan as a separate layer using the following steps:

STEP 1 The drawing begins with the preliminary layout drawn as the duct centerlines. (See Figure 25.26.)

STEP 2 Select supply and return registers from the template menu symbols library and add the symbols to the end of the centerlines where appropriate. (See Figure 25.27.)

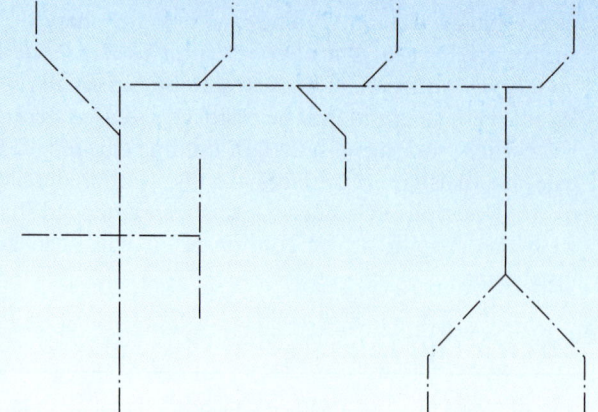

FIGURE 25.26 ■ The drawing begins with the preliminary layout drawn as duct centerlines.

FIGURE 25.27 ■ Select supply and return registers from a menu.

CADD APPLICATIONS 2-D

CADD APPLICATIONS 2-D

STEP 3 The program then automatically identifies and records the lengths of individual duct runs, and tags each run. Fittings are located and identified by the type of intersection. (See Figure 25.28.) While all of this drawing information is added to the layout, the computer automatically gathers design information into a file for duct sizing based on a specific mechanical manufacturer's specifications that you select.

STEP 4 After the fitting location and sizes are determined, the program transforms each fitting into accurate double-line symbols according the ANSI Y32.24 standard (See Figure 25.29.)

STEP 5 When the fittings are in place, the program calculates and draws the connecting ducts, adding couplings automatically at the maximum duct

lengths. If a transition is needed in a duct run, the program recommends the location and all you have to do is pick a transition fitting from the menu library. See Figure 25.30 for the complete HVAC layout.

An added advantage to using HVAC CADD software is that the program automatically records information, while you draw, to generate a complete bill of materials. The systems that offer you the greatest flexibility and productivity are designed as a parametric package. This type of program allows you to set the design parameters that you want and then the computer automatically draws and details according to these settings. As you draw, information such as the type of fitting, CFM, and gauge is placed with each fitting. A partial HVAC plan is shown in Figure 25.31.

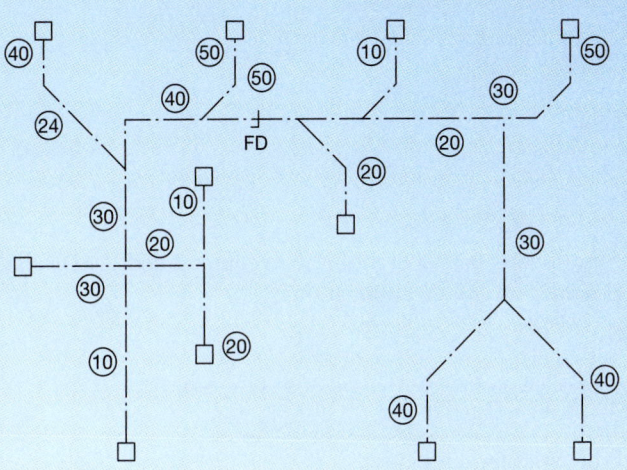

FIGURE 25.28 ■ Fittings are located and identified by the type of intersection.

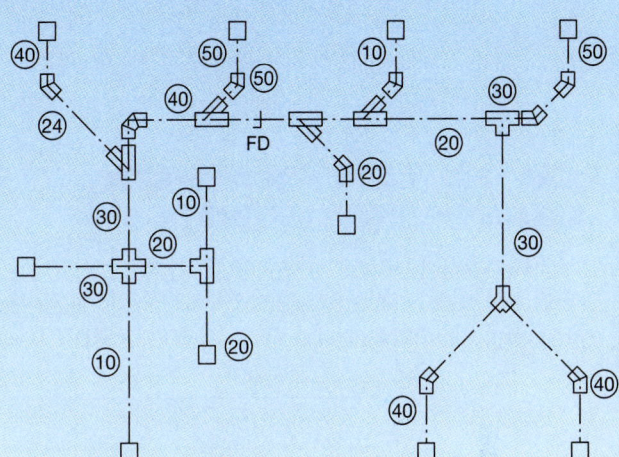

FIGURE 25.29 ■ HVAC symbols are drawn as accurate double-line symbols exactly to ANSI Y32.2.4 standards.

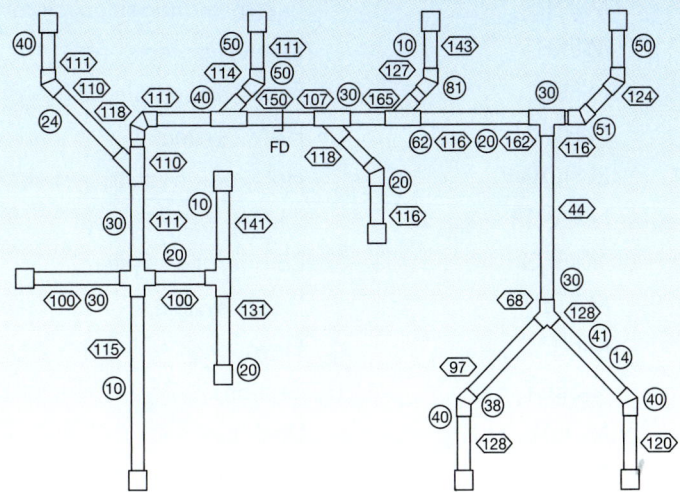

FIGURE 25.30 ■ The finished HVAC plan.

(Continued)

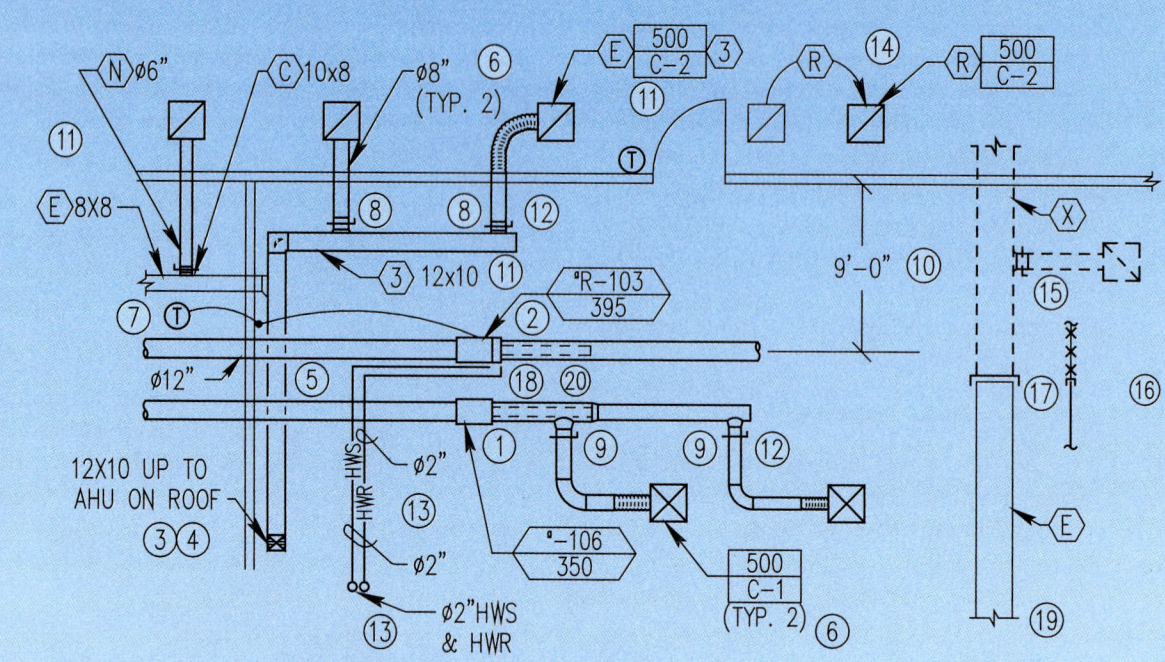

FIGURE 25.31 ■ A partial HVAC plan created using CADD.

CADD LAYERS FOR HVAC DRAWINGS

The American Institute of Architects (AIA) CADD Layer Guidelines establish the heading "Mechanical" as the major group identification for HVAC-related CADD layers.

Following are some of the recommended CADD layer names for HVAC applications:

Layer Name	Description
M-CHIM	Prefabricated chimneys
M-CMPA	Compressed air systems
M-CONT	Controls and instrumentation
M-DUST	Dust and fume collection systems
M-ENER	Energy management systems
M-EXHS	Exhaust systems
M-FUEL	Fuel system piping
M-HVAC	HVAC system
M-HOTW	Hot water heating system
M-CWTR	Chilled water system
M-REFG	Refrigeration system
M-STEM	Steam system
M-ELEV	Elevations
M-SECT	Sections
M-DETL	Details
M-SCHD	Schedules and title block sheets

HVAC MODELS

Three-dimensional CADD software provides designers and drafters with the ability to produce a pictorial representation, or model, of an HVAC system. Figure 25.32 shows an example of a 3-D model complete with HVAC duct routing. An HVAC model offers several advantages over a 2-D HVAC plan. Probably the most obvious benefit of an HVAC model is the ability to visualize and communicate an HVAC system design by three-dimensionally viewing a model from any angle or orientation. A 3-D HVAC layout also greatly aids in the design of an HVAC

system. For example, some CADD programs automatically analyze the HVAC layout for obstacles where an error in design can result in a duct that does not have a clear path. This type of feature allows a 3-D model of an entire HVAC system to be designed and tested before the actual structure is ever built. In addition, due to the parametric nature of many CADD applications, when changes are made to the HVAC model, the changes are corrected on all drawings, schedules, and lists of materials at the same time.

FIGURE 25.32 ■ CADD-generated model of HVAC duct routing. *Courtesy Computervision Corporation.*

Carefully observe how the stretch-out lines are established for each of the following developments, as this is the first process in making a layout. Specific developments that do not begin with a stretch-out line are identified. The instruction provided for the following pattern development is broken down into the most basic procedures. Individual shortcuts can be taken after adequate experience has been gained. It is extremely important to maintain a high degree of accuracy. Scale drawings and transfer dimensions very carefully. Another important consideration is to accurately label the elements of the views with numbers and/or letters and then transfer these labels to the pattern. This procedure may seem unnecessary on simple developments, but on complex patterns it is absolutely necessary. It is recommended that beginners label all developments as suggested in the given procedures.

Seams and Hems

A seam is the line formed when two or more edges come together. When sheet metal parts are bent and formed into the

desired shape, a seam results where the ends of the pattern come together. The fastening method depends on the kind and thickness of material, on the fabrication processes available, and on the end use of the part. Sheet metal components that must hold gases or liquid or that are pressurized can require soldering or welding. Other applications can use mechanical seams, which hold the parts together by pressure lapped metal, metal clips, pop rivets, or other fasteners. Some of the most common seams used in the sheet metal fabrication industry are shown in Figure 25.33. Extra material can be required on the pattern to allow for seaming. For the purpose of discussion, in the following procedures and problems, either a single- or double-lap seam is used. For a single-lap seam, add the given amount to one side of the pattern. For double-lap seams, add the given amount to both sides of the pattern. The corner of a seam can be cut off at an angle that is usually 45° if it interferes with adjacent parts during fastening.

A hem provides extra material on the pattern for strength and connection at the seams. Hemmed edges are necessary

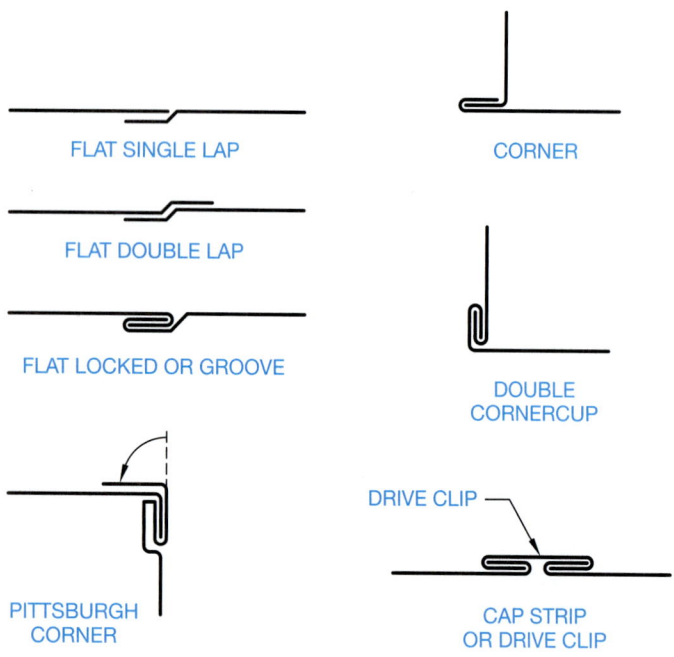

FLAT SINGLE LAP

FLAT DOUBLE LAP

FLAT LOCKED OR GROOVE

CORNER

DOUBLE CORNERCUP

PITTSBURGH CORNER

DRIVE CLIP

CAP STRIP OR DRIVE CLIP

FIGURE 25.33 ■ Common seams.

when an exposed edge of a pattern must be strengthened. When hems are used, extra material must be added to the pattern on the side of the hem. Figure 25.34 shows some common hems.

Developing a Rectangular (Right) or Square Prism

Commonly referred to as a box (open ended in this example), the rectangular or right prism can be developed as follows:

STEP 1 Draw the front and top view, label the corners, and establish the stretch-out line off the base of the front view and perpendicular to the height line. (See Figure 25.35.)

STEP 2 Beginning at 1 in the top view, use dividers to measure the true-length distance from 1 to 2. Transfer this dimension to the stretch-out line, starting at any point near the front view. Continue this process by transferring the distance from 2 to 3, 3 to 4, and 4 to 1 to the stretch-out line. (See Figure 25.36.) You must end at the point you began; point 1 in this example.

STEP 3 From each of the points established in step 2, draw vertical construction lines to meet a horizontal line drawn from the true-length (TL) height in the front view. (See Figure 25.37.)

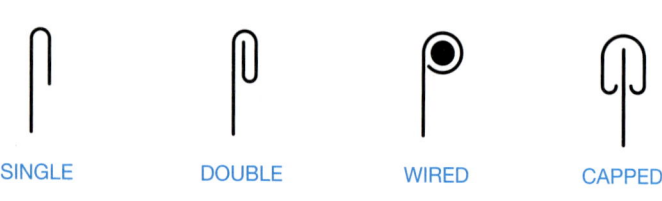

SINGLE   DOUBLE   WIRED   CAPPED

FIGURE 25.34 ■ Common hems.

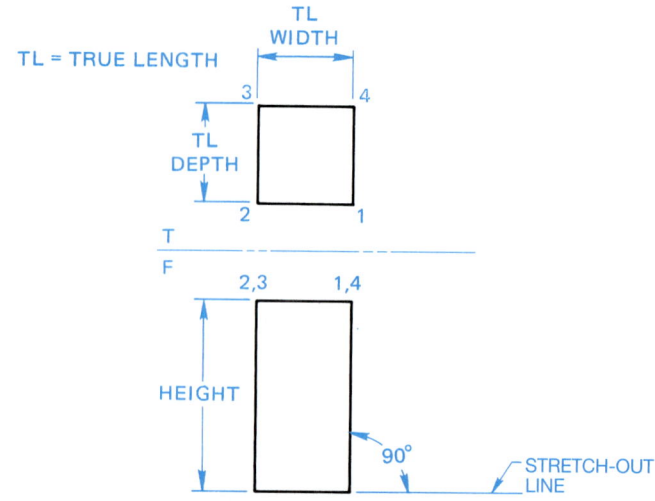

FIGURE 25.35 ■ Step 1—right prism development.

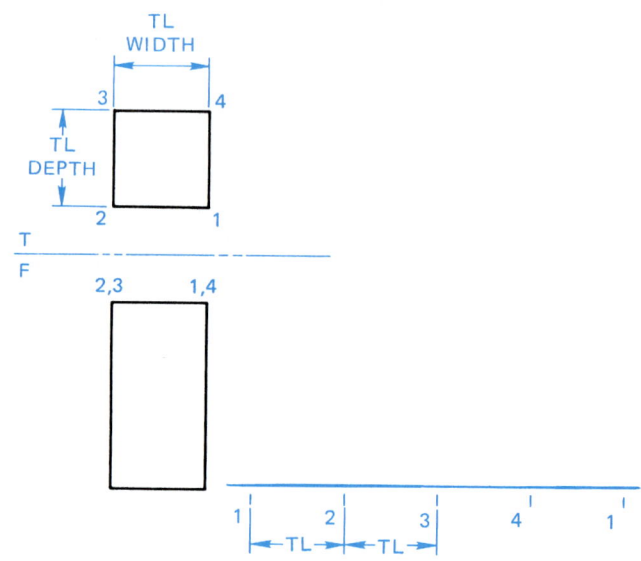

FIGURE 25.36 ■ Step 2—right prism development.

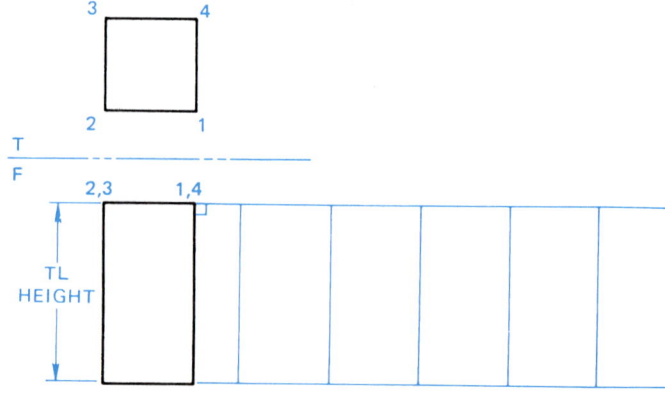

FIGURE 25.37 ■ Step 3—right prism development.

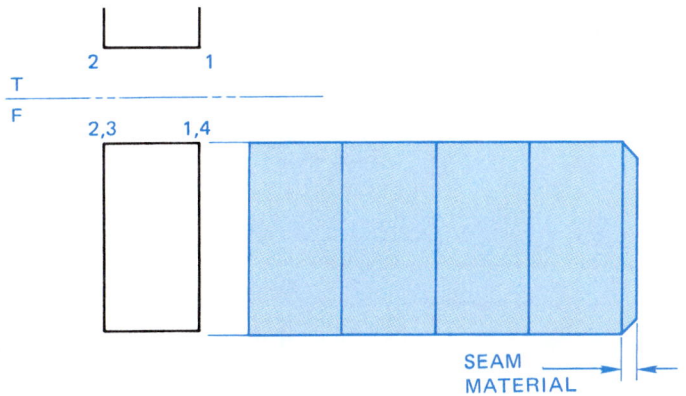

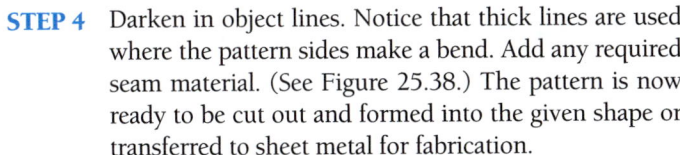

FIGURE 25.38 ■ Step 4—right prism development.

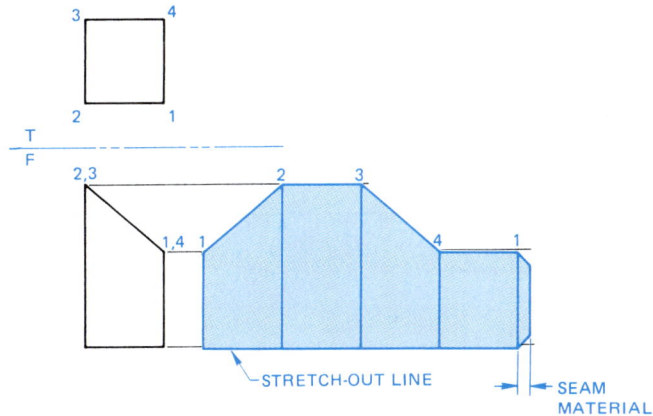

FIGURE 25.40 ■ Step 2—truncated prism development.

STEP 4 Darken in object lines. Notice that thick lines are used where the pattern sides make a bend. Add any required seam material. (See Figure 25.38.) The pattern is now ready to be cut out and formed into the given shape or transferred to sheet metal for fabrication.

Truncated Prism Pattern Development

A sheet metal part such as a prism, pyramid, or cone is considered to be **truncated** if a portion is cut off, generally at an angle.

STEP 1 Proceed as described in steps 1 through 3 for a right prism. Begin with the shortest element as the seam, when possible. The shortest seam is stronger, easier to fabricate, and requires less materials. This is true for all pattern development. (See Figure 25.39.)

STEP 2 Darken all object lines by connecting the ends of the true-height elements to form the outline of the object. Darken in all bend lines and add seam material as shown in Figure 25.40.

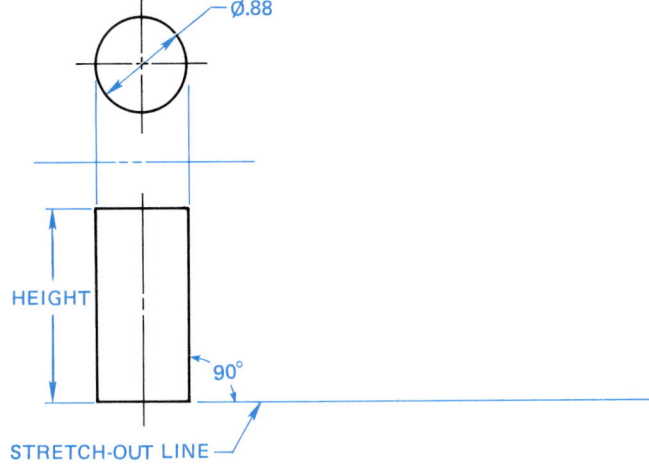

FIGURE 25.41 ■ Step 1—right cylinder development.

Developing a Regular (Right) Cylinder

One of the most common shapes is the cylinder. The following procedure can be used for the development of any right cylindrical shape:

STEP 1 Draw the top and front views. Establish the stretch-out line perpendicular to the side of the front view. The stretch-out line can be placed anywhere next to the front view, but it must be perpendicular to the side. (See Figure 25.41.)

STEP 2 Establish the length of the stretch-out line equal to the circumference of the circle with the formula $C = \pi D$. The diameter is .88 in., so $C = 3.14(\pi) \times .88 = 2.76$ in. Now, from the ends of the stretch-out line, draw perpendicular lines that meet a line projected from the true height in the front view as shown in Figure 25.42.

STEP 3 Darken the outline of the pattern and add seam material as shown in Figure 25.43, which shows a double-lap seam.

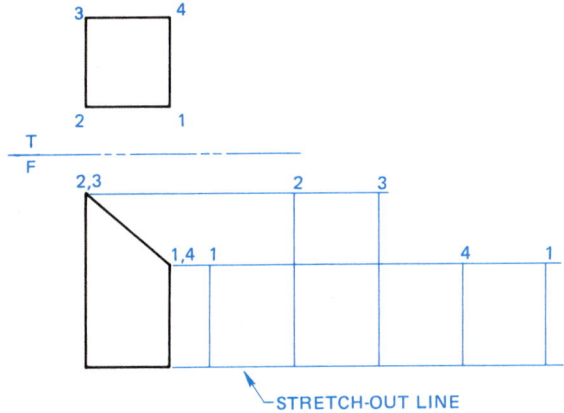

FIGURE 25.39 ■ Step 1—truncated prism development.

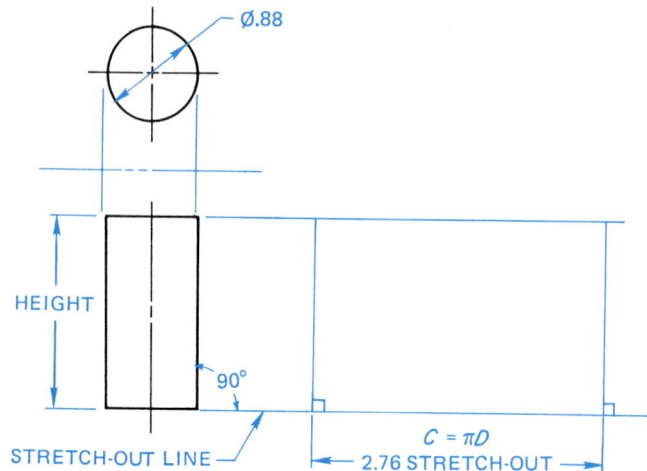

FIGURE 25.42 ■ Step 2—right cylinder development.

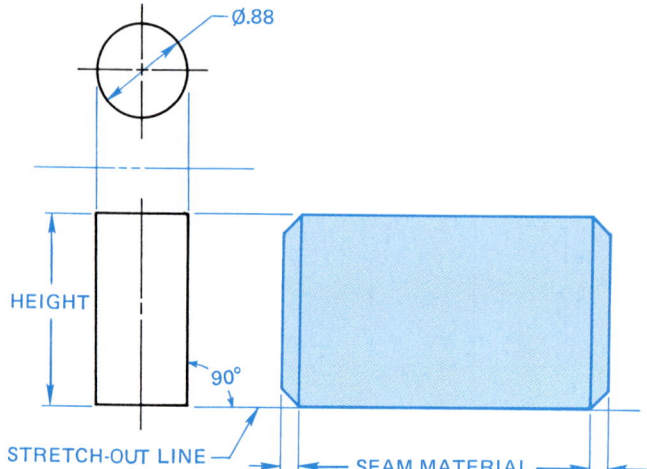

FIGURE 25.43 ■ Step 3—right cylinder development.

Truncated Cylinder Pattern Development

Truncated means to cut a geometric shape or object off at an angle with a plane. Truncated cylinders have many HVAC applications with development procedures similar to the regular cylinder.

STEP 1 Draw the top and front views. Divide the top view into twelve equal parts. More divisions establish better accuracy; less give less accuracy. Twelve divisions have been selected because of ease and effectiveness. Work carefully with a sharp pencil to maintain the best accuracy possible. Number each element in the top view and extend the numbering system into the front view. (See Figure 25.44.)

STEP 2 Establish the stretch-out line perpendicular to the side of the front view and the length with the formula $C = \pi D$ as previously discussed. Divide the stretch-out into twelve equal parts. Refer to dividing a line into equal parts in Chapter 8. Number each part from 1 through

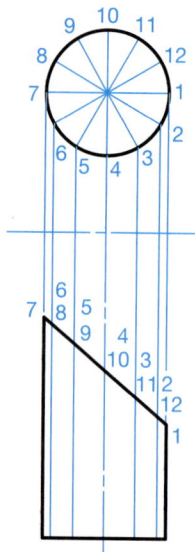

FIGURE 25.44 ■ Step 1—truncated cylinder development.

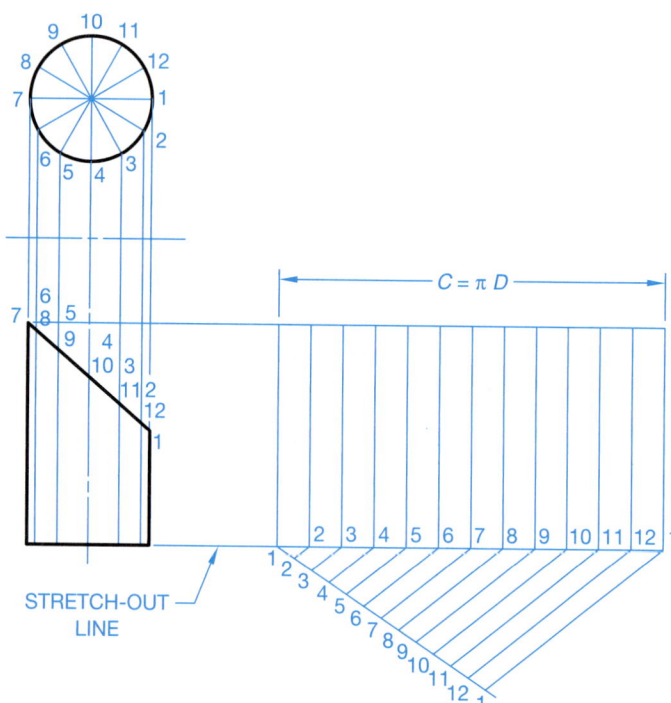

FIGURE 25.45 ■ Step 2—truncated cylinder development.

12, ending at 1 where the seam is located. From each part on the stretch-out line, draw a perpendicular construction line equal to the total height of the cylinder. (See Figure 25.45.)

STEP 3 Project the true height of each line segment in the front view to the correspondingly numbered line in the pattern layout. Where these lines intersect, a pattern of points is established. Use an irregular curve to connect the points forming the curved outline of the truncated cylinder pattern. Darken the outline of the pattern and

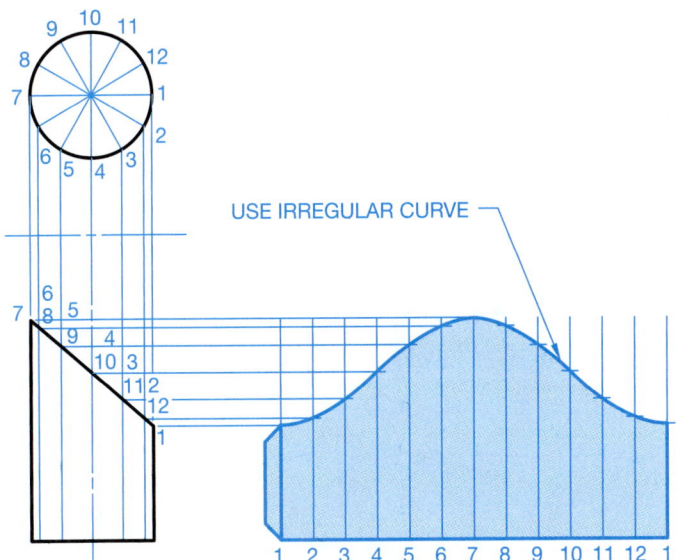

USE IRREGULAR CURVE

FIGURE 25.46 ■ Step 3—truncated cylinder development.

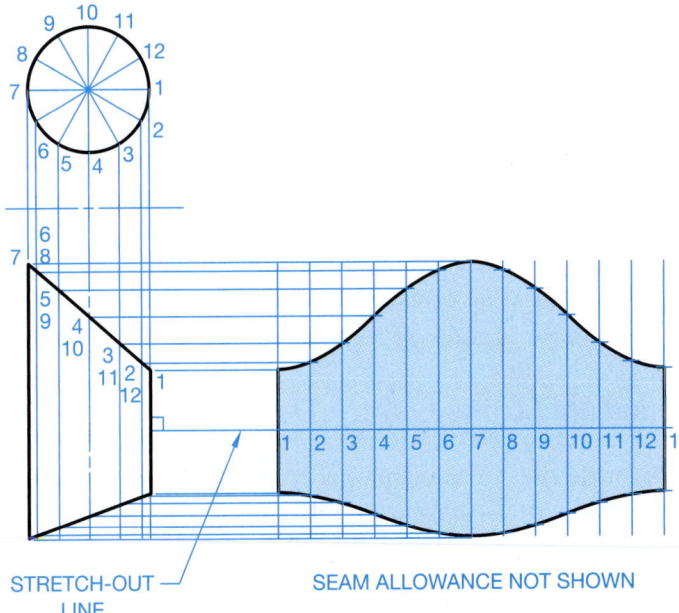

STRETCH-OUT LINE

SEAM ALLOWANCE NOT SHOWN

FIGURE 25.47 ■ Development of cylinder truncated on both ends.

add seams. The division lines can be left as thin lines to represent a curved object. Construction lines should not have to be erased if properly drawn. Turn off or freeze the construction layer when using CADD. (See Figure 25.46.)

When the cylinder is truncated on both ends, the process is the same, with the stretch-out line established perpendicular to the side at any desired location. (See Figure 25.47.)

Developing a Cylindrical Elbow

Cylindrical elbows are used to make turns or corners in ductwork. The turns can be any number of degrees. Standard cylin-

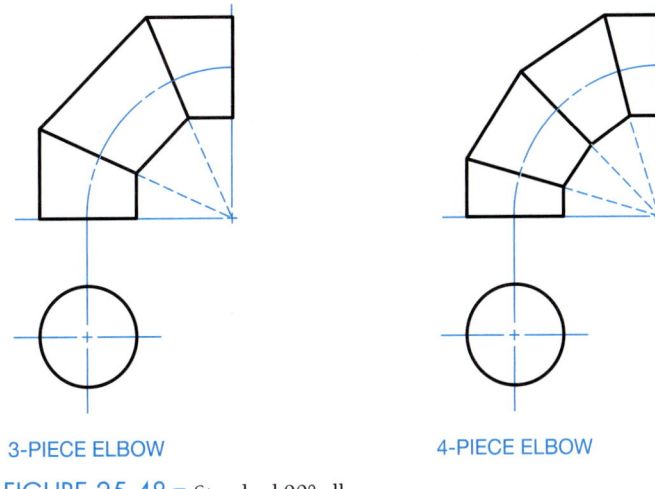

3-PIECE ELBOW   4-PIECE ELBOW

FIGURE 25.48 ■ Standard 90° elbows.

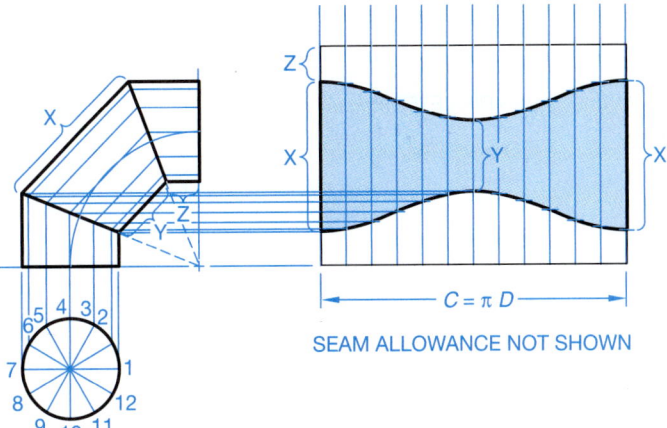

$C = \pi D$

SEAM ALLOWANCE NOT SHOWN

FIGURE 25.49 ■ Development of a three-piece cylindrical elbow.

drical elbows are made of any given number of truncated cylinders. Each piece of the elbow can be developed as demonstrated in the previous discussion. Figure 25.48 shows standard 90° three-piece and four-piece elbows. Each piece of the three-piece elbow is developed as a truncated cylinder, with the patterns developed separately or together by alternating starting elements as shown in Figure 25.49.

Developing a Cone

The size of the cone is established by the base diameter, and the height and can be developed as follows:

STEP 1 Given the height and base diameter, draw the front and top or bottom views. Divide the top or bottom view into twelve equal parts. Number each part and project these to the cone base in the front view. At the base in the front view, extend each point to the vertex of the cone. (See Figure 25.50.)

STEP 2 The stretch-out line for a right circular cone development is a true circular arc. The radius of the arc is taken from the true-length element measured from zero to one

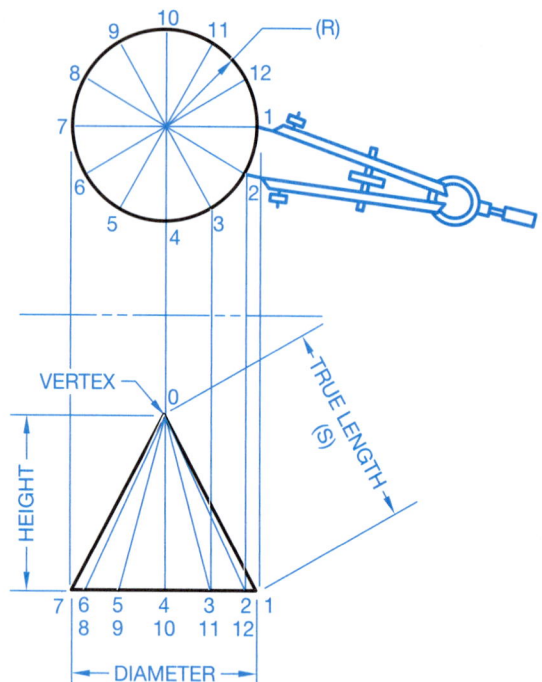

FIGURE 25.50 ■ Step 1—cone development.

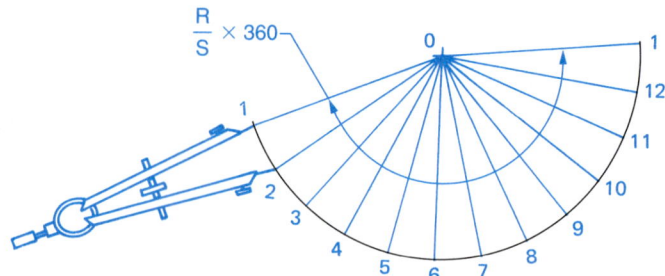

FIGURE 25.52 ■ Step 3—cone development.

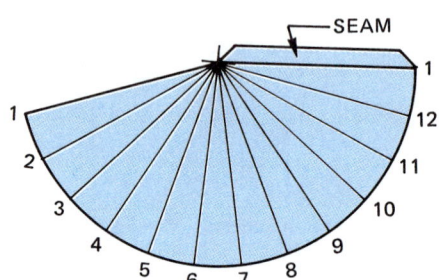

FIGURE 25.53 ■ Step 4—cone development.

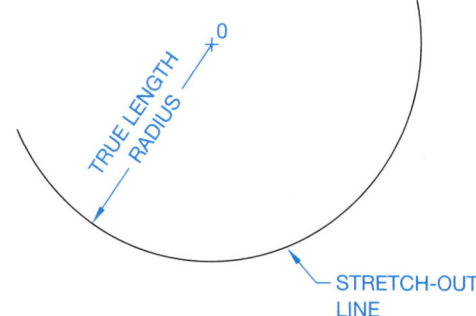

FIGURE 25.51 ■ Step 2—cone development.

in the front view. (See Figure 25.50.) At any convenient location where you have adequate space, draw an arc with the true length radius. (See Figure 25.51.)

STEP 3 Go to the top view and use dividers to establish the increment from point 1 to point 2. Rotate the dividers around the circle, adjusting the dividers until you have accurately established the increment at least half way around the circle. (See Figure 25.50.) Beginning at any point on the stretch-out line, use the established increment to locate twelve equal spaces. Remember, if you begin at 1, you must end at 1. Now, connect each point along the stretch-out line to the vertex 0 with construction lines. (See Figure 25.52.) When using CADD, this procedure can be done quickly and accurately. The AutoCAD MEASURE and QCAL commands can be used to establish the circumference on the stretch-out

line followed by the DIVIDE command to find the twelve equal spaces. Alternately, the angle forming the sides of the cone in the Figure 25.52 pattern layout, can be calculated using the formula: $R/S \times 360$, where R is the radius of the cone base, and S is the true length of the side of the cone as shown in Figure 25.50.

STEP 4 Darken the outline of the cone pattern and add seam material as necessary. (See Figure 25.53.)

Truncated Cone Pattern Development

The procedure for developing a truncated cone begins the same as the regular cone previously described. Always start by developing a pattern for a full regular cone. After step 3, the true length of each line segment from the vertex to the truncated line in the front view must be established. Project each point to the true-length line. Then measure the true length of each element from the vertex along the true-length line as shown in Figure 25.54. Transfer the true length of each individual line to the corresponding line in the development. (See Figure 25.54.) Finally, darken the outline using an irregular curve, where necessary, and add seam material as shown in Figure 25.54.

Developing an Offset Cone

The offset cone has the vertex offset from the center of the base. The procedure is slightly different from the development of a regular cone, because the true length of each line element must be found. Refer to Figure 25.55 as you read the following instructions:

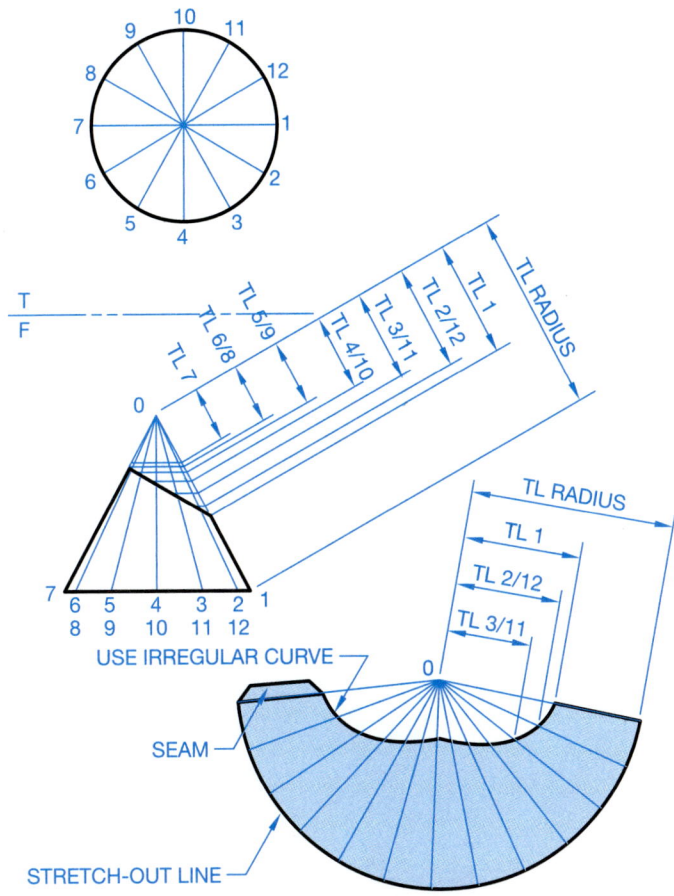

FIGURE 25.54 ■ Truncated cone development.

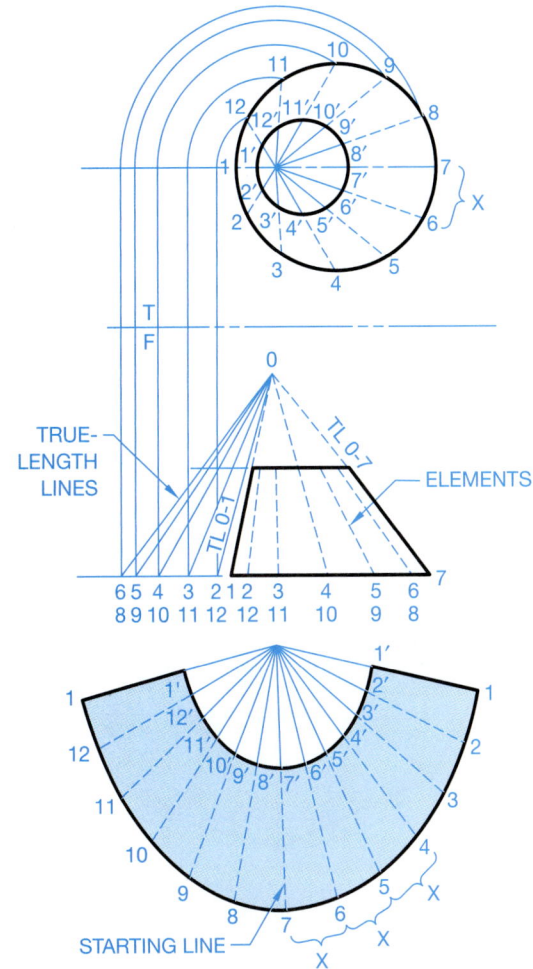

FIGURE 25.55 ■ Offset cone development.

STEP 1 Draw the top and front views. Divide the top view base circle into twelve equal parts. Project each point to the vertex and establish the same elements in the front view.

STEP 2 The true lengths of elements 0–1 and 0–7 are true length in the front view. Establish the true-length diagram for the other elements using the revolution method as shown in color in Figure 25.55.

STEP 3 There is no stretch-out line in this development. Begin the development by laying out the true length of element 0–7 in a convenient location. Take a compass and lay out the true length of 0–6 and 0–8 on each side of 0–7. Set another compass with the radius equal to the distance X in the top view. This measurement is used several times, so keep your compass at this setting. Using the distance X, set the compass point at 7 and draw an arc that intersects the arcs drawn for 0–6 and 0–8. This procedure locates the exact position of elements 0–6 and 0–8. Follow this same method for each of the elements, 0–5 and 0–9, 0–4 and 0–10, 0–3 and 0–11, 0–2 and 0–12, and 0–1 at each end. You now have a series of points along the base that may be connected with an irregular curve.

STEP 4 The true lengths of elements 0–1 and 0–7 are in the front view. Use the true-length diagram to establish the true

lengths of the other elements, 0–2 through 0–12. Transfer these true lengths to the corresponding elements in the development and connect the points with an irregular curve. Darken the outline and add seam material, if required. (See Figure 25.55.)

Developing a Pyramid

Pyramids are developed in a manner similar to a cone as described in the following steps:

STEP 1 Draw the top and front views and establish the true length of the edge of a side or lateral edge. Set the compass with a radius equal to the true length of an edge and draw an arc in a convenient location. This is the stretch-out line. (See Figure 25.56.)

STEP 2 Use dividers or a compass to lay out the true lengths from points 1–2, 2–3, 3–4, and 4–1 in the top view and transfer these distances consecutively to the stretch-out line. Connect points 1, 2, 3, 4, and 1 in the development and also connect these points to the vertex 0. Darken all lines and add seam allowances as shown in Figure 25.57.

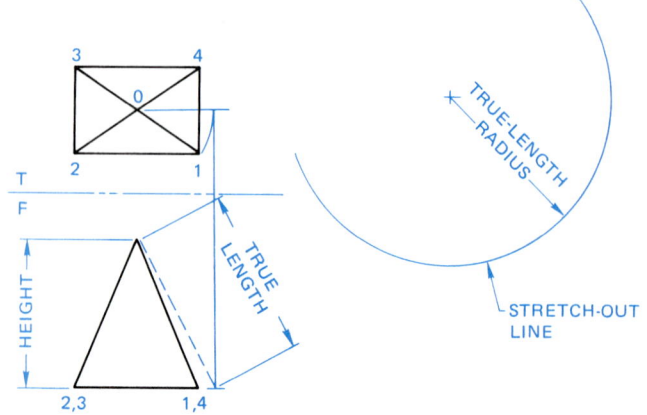

FIGURE 25.56 ■ Right pyramid development.

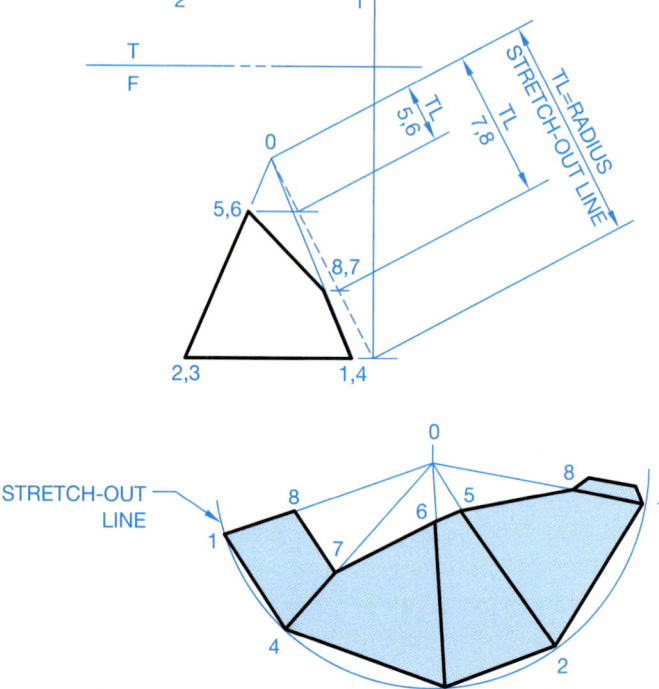

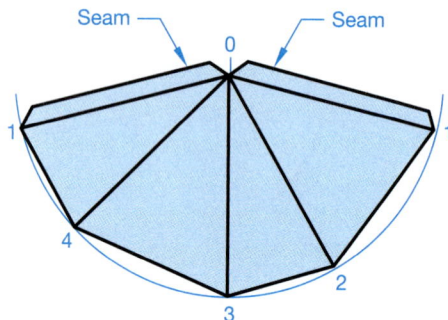

FIGURE 25.57 ■ Pyramid development.

FIGURE 25.58 ■ Truncated pyramid development.

Pattern Development of a Truncated Pyramid

The procedure for the development of a truncated pyramid is the same as for a regular pyramid except that the true length of each element from the vertex to the truncated line must be established and transferred to the pattern as shown in Figure 25.58.

Transition Piece Pattern Development

Also known as a **square to round**, a **transition piece** is a duct component providing a change in shape from square or rectangular to round. Transition pieces can be designed to fit any given situation, but the pattern development technique is always as follows: (*Note: Accuracy and the use of a number and letter system is very important. Do all layout work using construction lines or a CADD construction layer. There may be other techniques that work in some unique situations, but the method described here works with any transition piece configuration.*)

STEP 1 Draw the top and front views. Divide the circle in the top view into twelve equal parts. Connect the points in each quarter of the circle to the adjacent corner of the square. Project the same corresponding system to the front view. Number and letter each point in each view as shown in Figure 25.59. The transition piece shown in Figure 25.59 is made up of a series of triangles; for example, 1,A,D; 1,2,A; and 2,3,A. The development progresses by

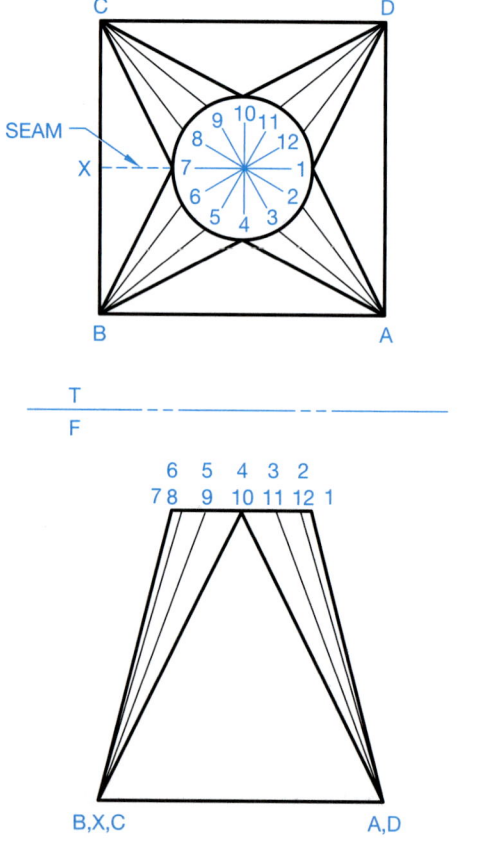

FIGURE 25.59 ■ Step 1—right transition piece view setup.

attaching the true size and shape of each triangle to-gether in order. This technique is known as *triangulation*. There is no stretch-out line.

STEP 2 Each element of the development must be in true length. Establish the true length diagram using revolution as shown in Figure 25.60. Only one set of true-length elements is required because the given problem is a right transition piece that is symmetrical about both axes.

STEP 3 Begin the pattern development on a sheet in an area where there is a lot of space. Start with the true size and shape of triangle 1,A,D. The true lengths of 1–A and 1–D are the same and can be found in the true-length diagram. The true length of A–D is in the top view. (See Figure 25.61.) A review of Chapter 8, Geometric Construction, and Chapters 16 and 17, Descriptive Geometry I and II, is helpful here.

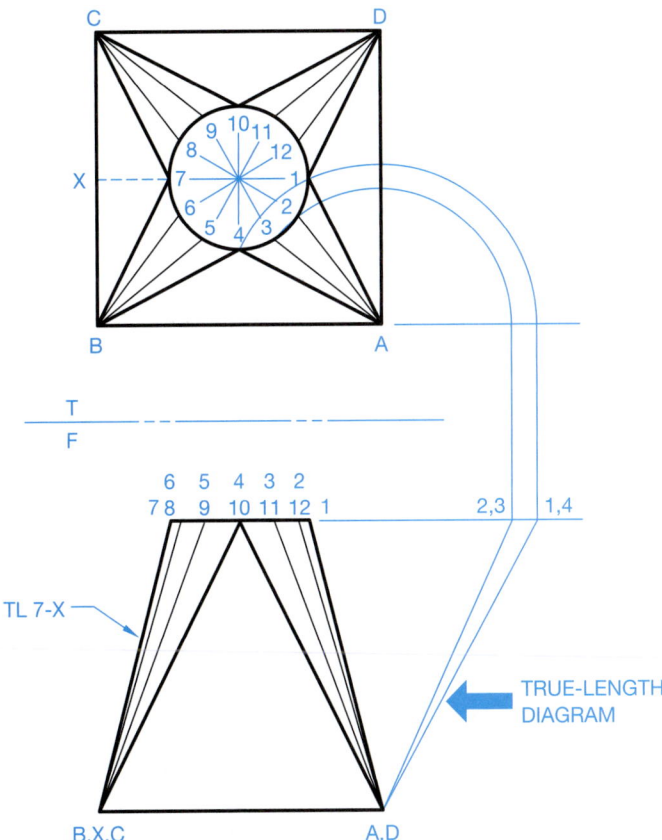

THE FOLLOWING TRUE LENGTHS ARE THE
SAME FOR THIS RIGHT TRANSITION PIECE:

GROUP I	GROUP II
1-A	2-A
4-A	3-A
4-B	5-B
7-B	6-B
7-C	8-C
10-C	9-C
10-D	11-D
1-D	12-D

FIGURE 25.60 ■ Step 2—transition piece development; true-length diagram.

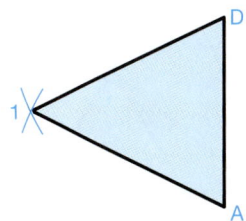

FIGURE 25.61 ■ Step 3—transition piece development; triangulation, true size and shape of starting triangle.

STEP 4 Work both ways from triangle 1,A,D in Figure 25.61 to develop the adjacent triangles. From points A and D, draw arcs equal to A–2 and D–12, respectively. Set a compass for the distance from 1–2. Keep this compass setting, as it is used several times. With this setting, scribe two arcs from point 1 that intersect the previously drawn arcs. These intersections are points 2 and 12. Connect points 2–A and 12–D. Do not connect points 1,2 and 12 yet. (See Figure 25.62.) Do all work with construction lines until complete. *Note: Measuring directly from point 1–2 in the top view establishes a cord of the arc from 1–2. Refer to Chapter 8 as needed. The cord of the arc is not the actual arc length. This inaccuracy is normally acceptable when drawing a transition piece pattern development for an HVAC application, because the fabrication tolerances are generally large enough. Confirm this with your employer or instructor. If additional accuracy is required, calculate the circumference of the circle, divide by 12, and use this value for the 1–2 measurement. This more accurate practice is easy and expected when using CADD.*

STEP 5 Continue the same procedure, working both ways around the transition piece until the entire development is complete. Every triangle must be included and the pattern ends on both sides with the element 7–X, which is true length in the front view. The two final triangles are right triangles, because each is half of the full side that is an equilateral triangle. Remember that every line transferred from the views to the pattern must be true length. It is recommended that the numbering system be used to avoid errors. When all points have been established, the outline is darkened. A series of points form the inside curve. Connect these points with an irregular curve. If any point is out of alignment with the

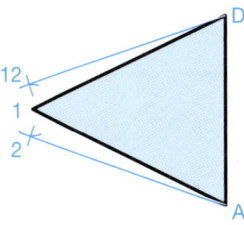

FIGURE 25.62 ■ Step 4—transition piece development; continue the pattern development both ways from the starting triangle.

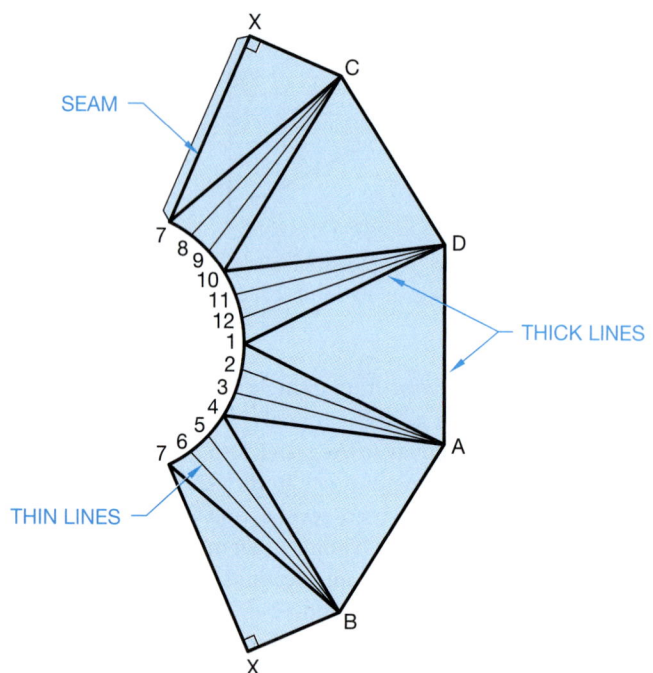

FIGURE 25.63 ■ Step 5—complete the transition piece development.

others, you have made an error at this location. Notice in Figure 25.63 that thick lines in the pattern are either outlines or bend lines, while thin lines form a smooth curve or contour. Cut out your pattern and see for yourself how it fits together.

The right transition piece used in the process previously described is the easiest transition to develop because it is symmetrical. If the round is offset as shown in Figure 25.64, then true-length diagrams are required for both halves.

When the round is offset, as in Figure 25.65, true-length diagrams are necessary for the elements on each quarter of the transition. In these more complex transitions, a numbering system and accurate work are very important. Colored lines can be used to help keep true-length diagrams clearly separate.

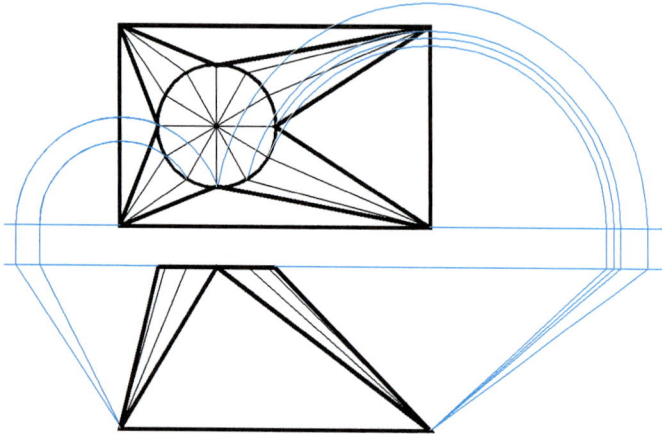

FIGURE 25.64 ■ Setting up a symmetrical offset transition piece; two true-length diagrams are required.

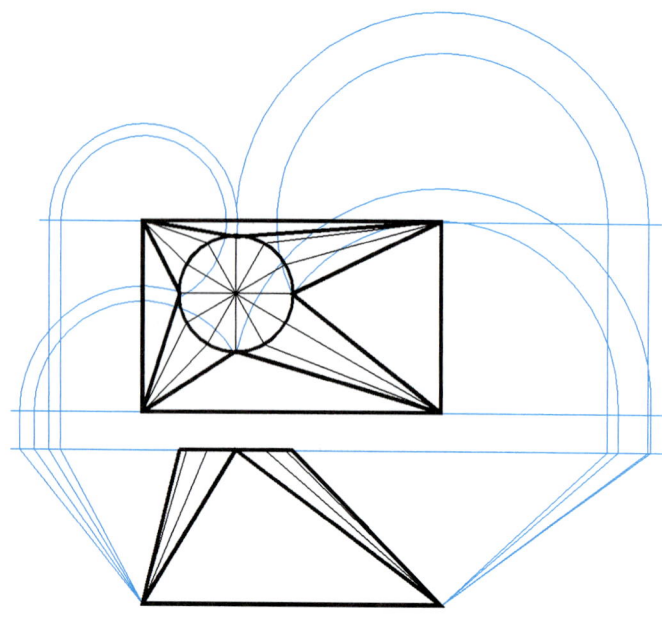

ONLY THE TRUE LENGTHS OF QUADRANT ELEMENTS ARE SHOWN FOR CLARITY

FIGURE 25.65 ■ Setting up an offset transition piece; four true-length diagrams are required.

Curve-to-Curve Triangulation

Triangulation is a technique that is used for the development of transition pieces or any other pattern where a series of triangles is used to form the desired shape. Figure 25.66 shows a part that

FIGURE 25.66 ■ Curve-to-curve transition.

makes a transition from one curved shape to another. In this situation, the adjacent curves must be divided into a series of triangles before the pattern can be developed. The triangles help define one shape in relationship to the other. (See Figure 25.67.) After the views have been drawn, the following procedure is similar to the development of a transition piece:

STEP 1 The part is symmetrical, so one set of true-length lines must be established. (See Figure 25.68.) Be careful when laying out and numbering the true-length lines. Some drafters even use colored lines to keep elements coordinated. A numbering system and accuracy are critical for this type of problem.

STEP 2 After the true-length elements have been established, the development technique is the same as the transition piece. (See Figure 25.69.)

ESTABLISHING INTERSECTIONS

When one geometric shape meets another, the line of intersection between the shapes must be determined before the pattern of each piece can be developed. The key to determining the intersection between parts is setting up views with numbered true-length elements that correlate between views. The points of intersection of line elements at planes can then be projected between views.

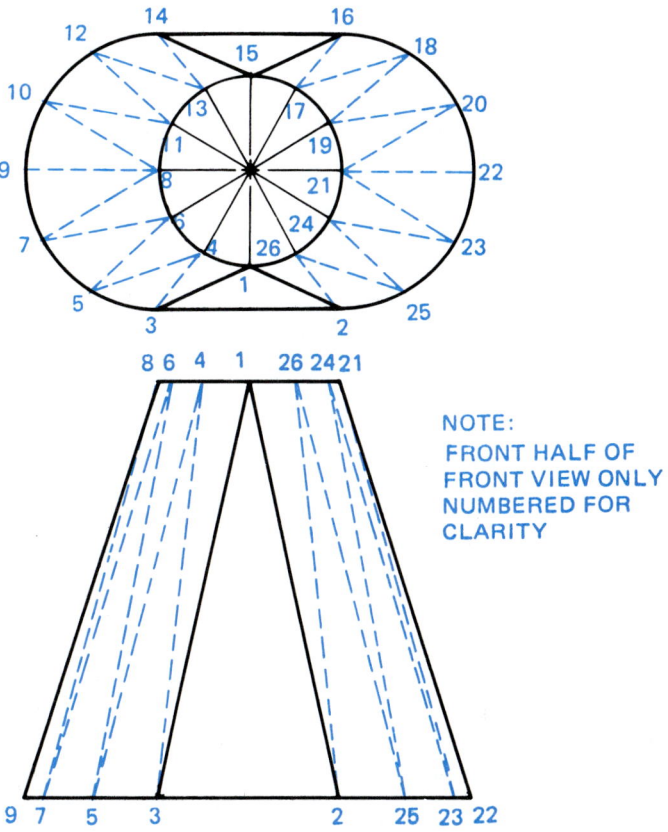

FIGURE 25.67 ■ Setting up the triangulation on a curve-to-curve transition.

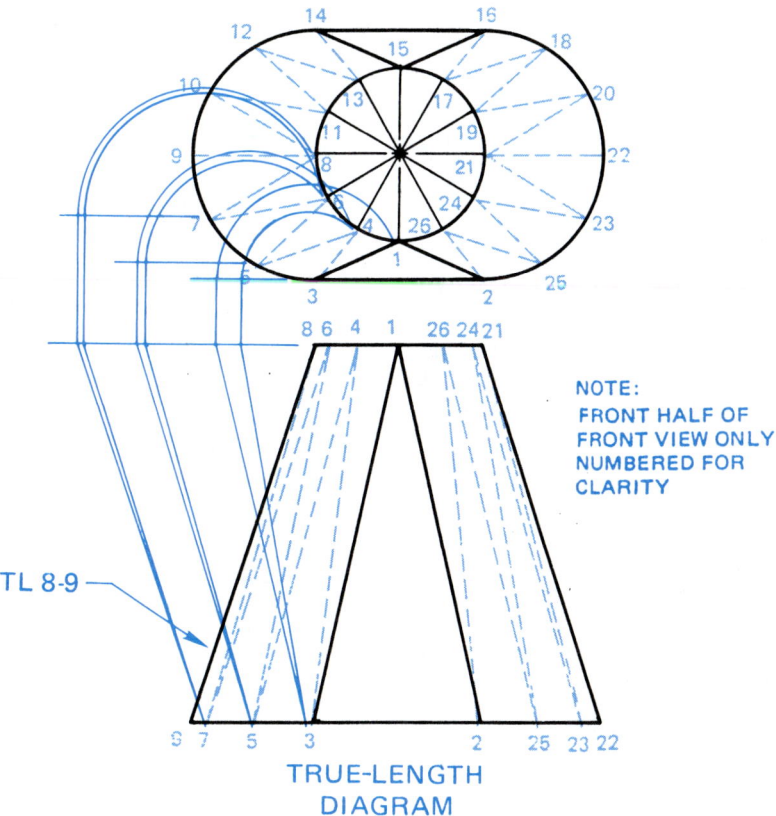

FIGURE 25.68 ■ Curve-to-curve transition, true-length diagram.

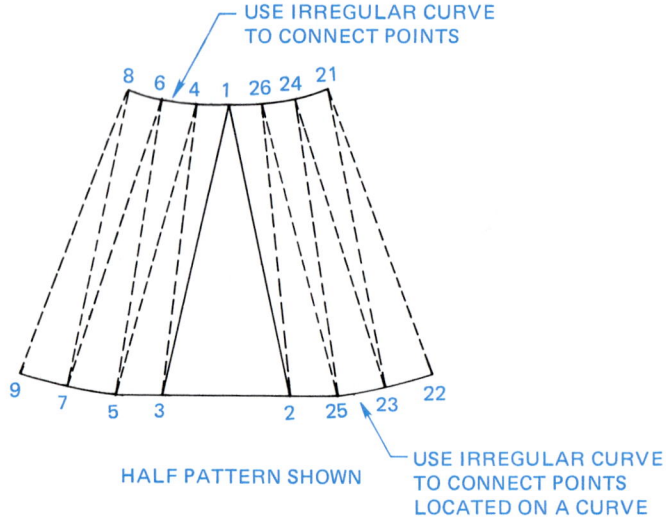

FIGURE 25.69 ■ Curve-to-curve transition development.

Intersecting Prisms

The line of intersection between intersecting prisms can be found by projecting the true lengths of individual line elements between views. The corresponding points of intersection where lines intersect planes establish a series of points that are connected to form the lines of intersection. Several examples are shown in Figure 25.70. After the lines of intersection are determined, the pattern development can be made of the resulting shape.

Intersecting Cylinders

The line of intersection between intersecting cylinders is determined in a manner that is similar to that for intersecting prisms. A series of line elements is established on the intersecting cylinder. Points of intersection of these line elements are then plotted. When the points are connected, the line of intersection is formed. The following procedure provides a typical example:

STEP 1   Establish two adjacent views and show the circular view of the intersecting cylinder in both views. Divide the circular views of the intersecting cylinder into twelve equal parts in each view and number the points correspondingly. Be sure that your numbering system correlates between views as shown in Figure 25.71.

STEP 2   Project point 1 in the front view down until it meets the intersecting circle. Then project to the side view until it meets the same corresponding point 1 projected down in the side view. This is the point of intersection for line element 1. Follow the same procedure for each of the other eleven points. When complete, you have a series of points of intersection as shown in Figure 25.72.

STEP 3   Connect the points of intersection in the side view to establish the line of intersection as shown in Figure 25.73. Now pattern developments of each cylinder can be made as discussed earlier in this chapter and as shown in Figure 25.73.

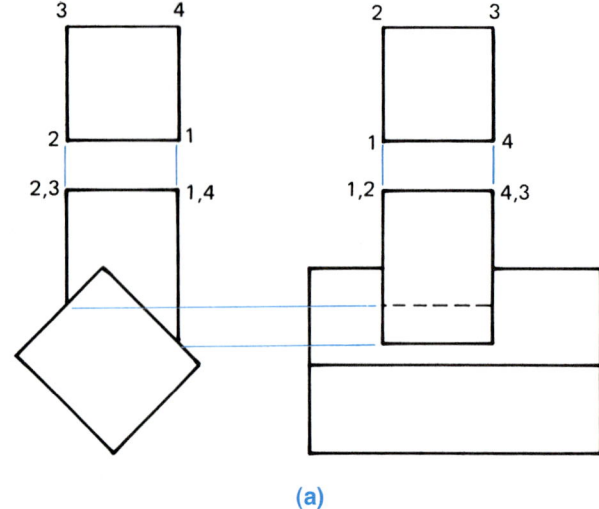

(a)

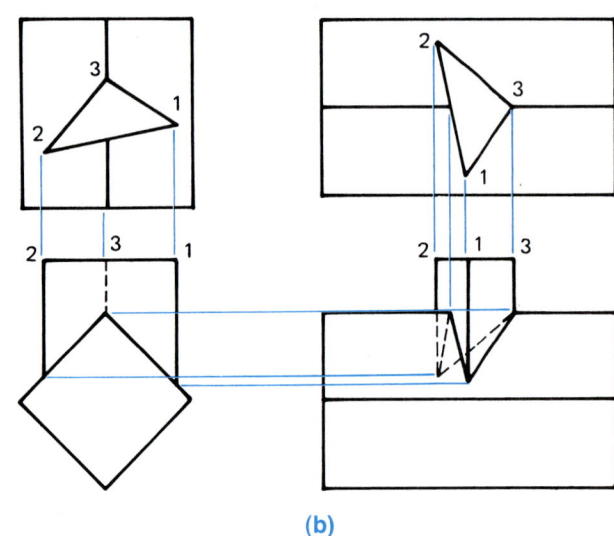

(b)

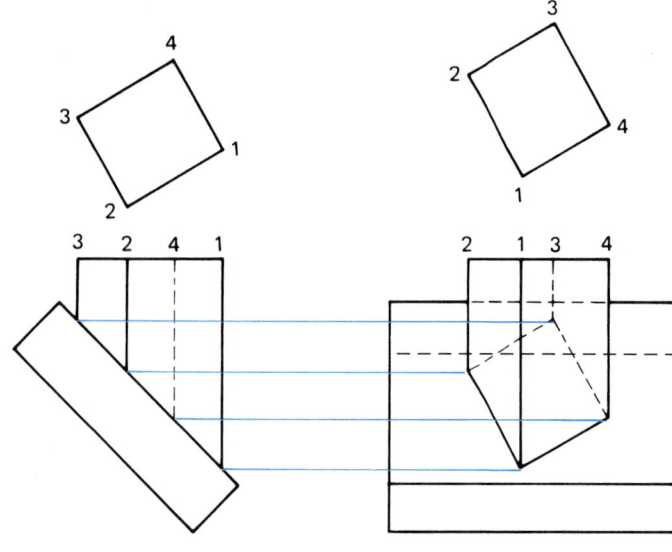

(c)

FIGURE 25.70 ■ Intersecting prisms.

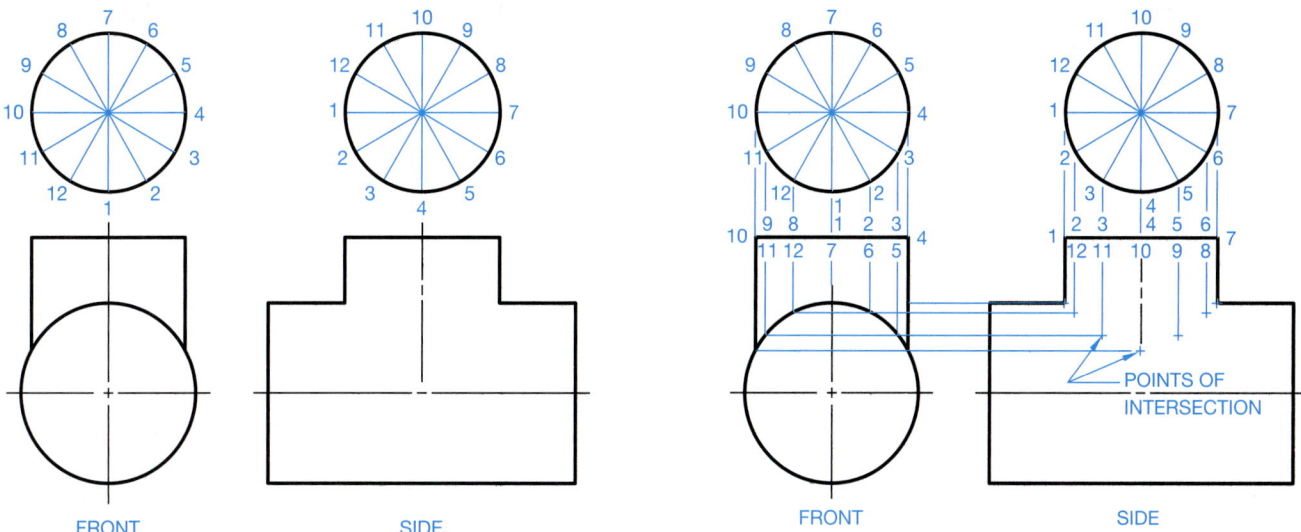

FIGURE 25.71 ■ Setting up intersecting cylinders.

FIGURE 25.72 ■ Establishing points of intersection for intersecting cylinders.

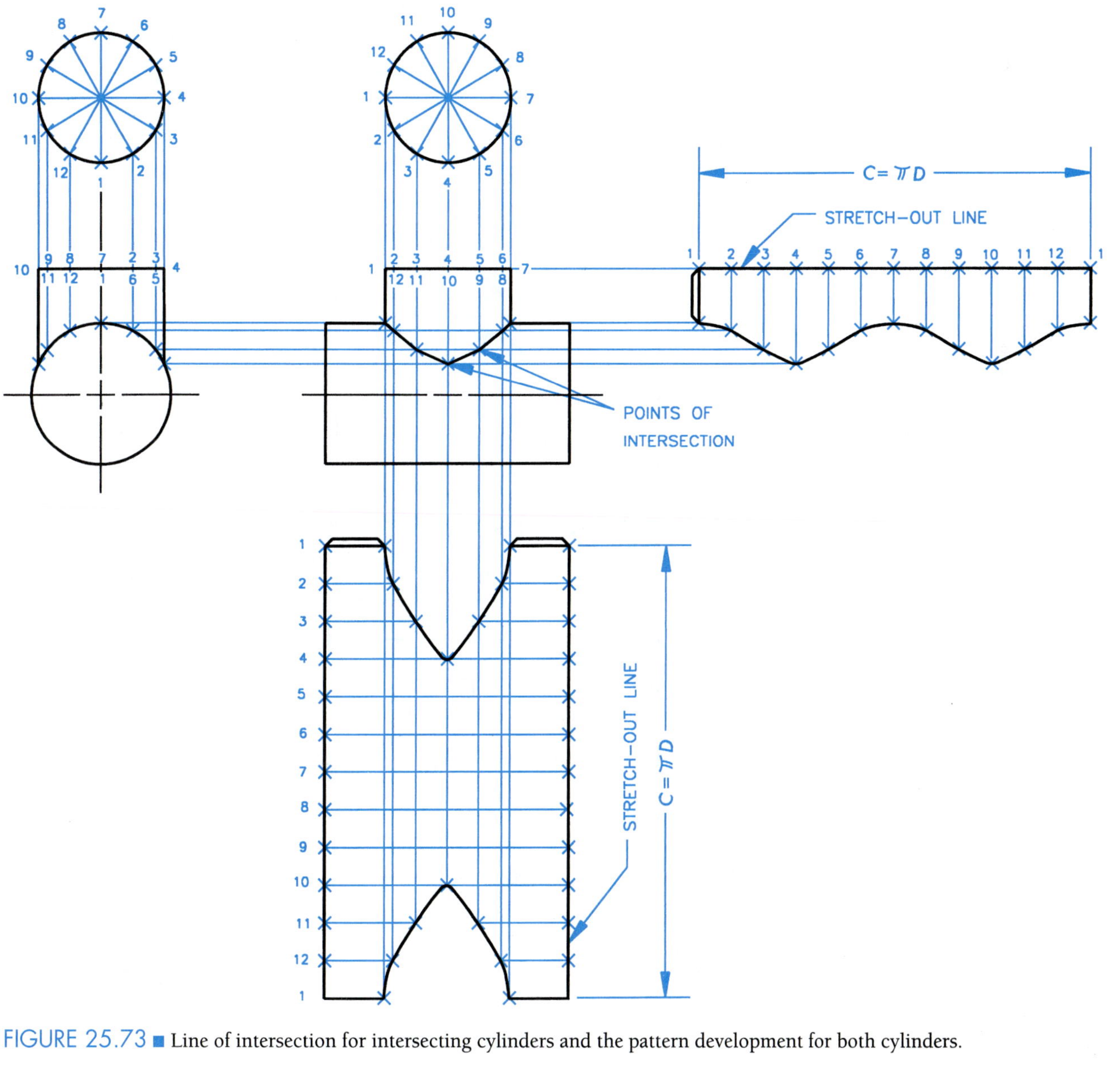

FIGURE 25.73 ■ Line of intersection for intersecting cylinders and the pattern development for both cylinders.

No matter how the intersecting cylinders are arranged, the technique is the same. The circular views are always drawn looking into the intersecting cylinders, even when the cylinders intersect at an angle as in Figure 25.74. When the cylinders are offset, the back half of the intersecting cylinder appears hidden as shown in Figure 25.74. This example also shows how the patterns for the cylinders are drawn.

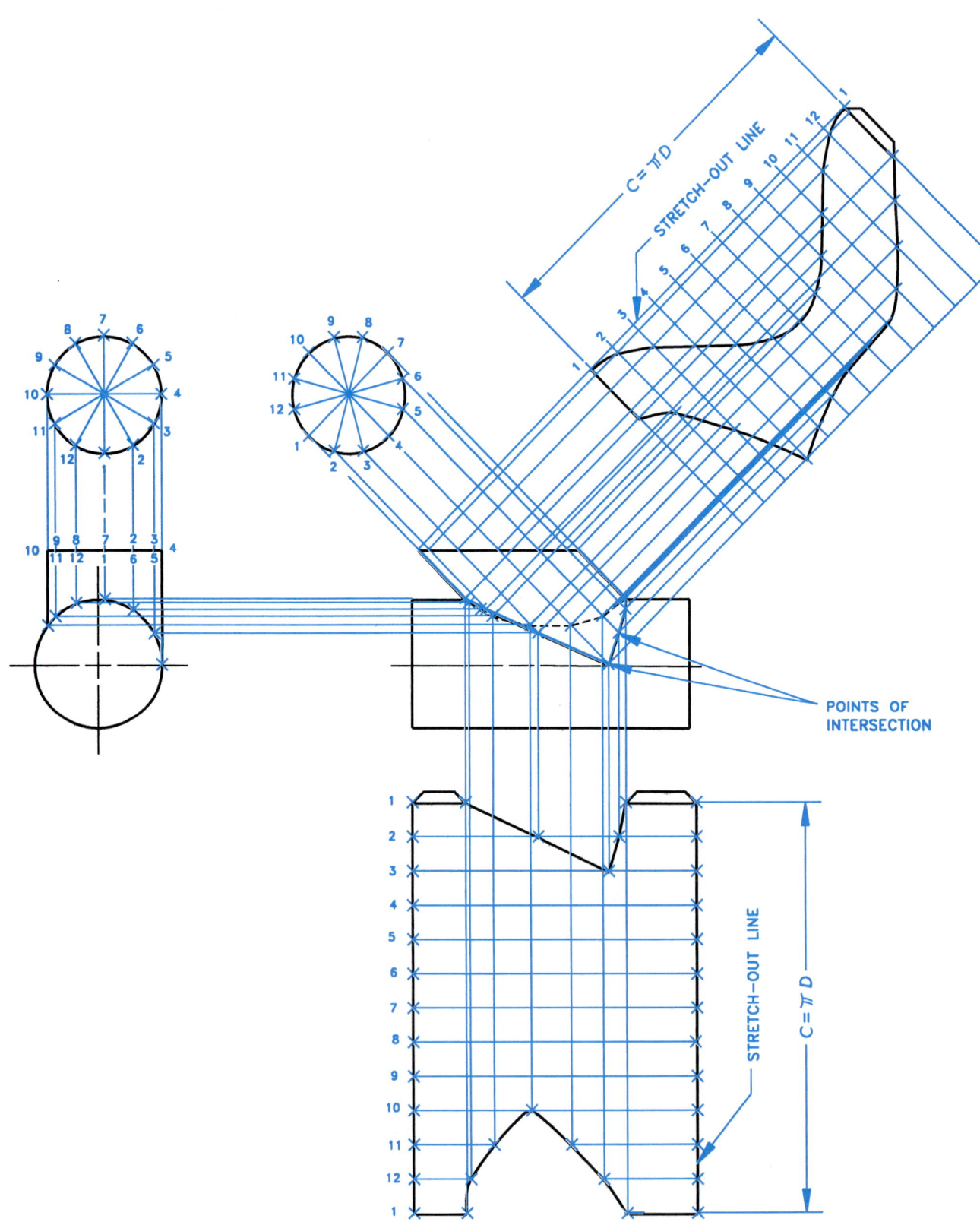

FIGURE 25.74 ■ Offset intersecting cylinders and the pattern development for both cylinders.

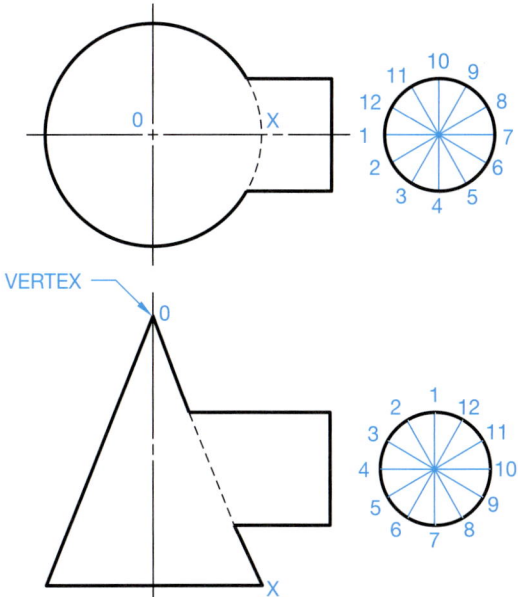

FIGURE 25.75 ■ Setting up the intersection between a cone and a cylinder.

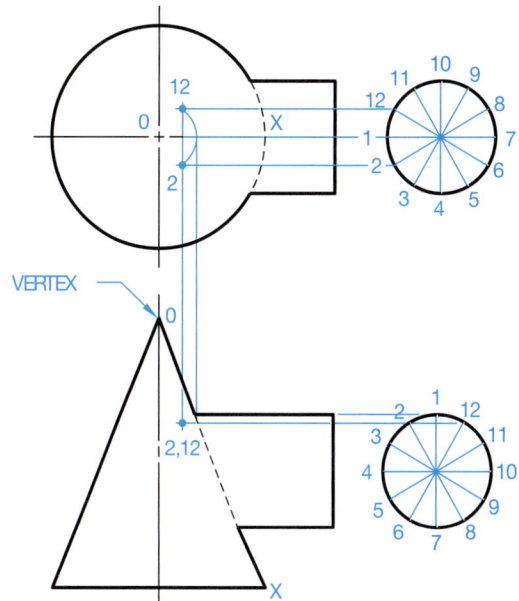

FIGURE 25.76 ■ Determining the points of intersection between a cone and a cylinder.

Cylinder Intersecting Cone

The procedure used to get the line of intersection between a cylinder and cone is a little more complex, but it may be done given any situation if the following steps are used:

STEP 1 Draw the front and top views. Draw the circular view of the intersecting cylinder in both views and divide these circles into twelve equal parts. Number the parts of each circle so the numbering system correlates between views. (See Figure 25.75.)

STEP 2 Beginning with point number 1 in the front view, project this point until it intersects the true-length element of the cone, 0–X. This is the point of intersection of point 1 in the front view. From this point of intersection, project to the top view until the projection meets the same corresponding point projected from the circle in the top view. This is the point of intersection of point 1 in the top view. (See Figure 25.76.)

STEP 3 Now, project point 2 from the front view to the true-length element of the cone, 0–X (points 2 and 12 are on the same projection line if the object is symmetrical). From this point, project point 2 into the top view until intersecting 0–X. Set a compass with a radius from 0 to this point of intersection and draw an arc. Project points 2 and 12 in the top view until they intersect this arc. These are the points of intersection of points 2 and 12 in the top view. Now, project points 2 and 12 from the top view until they intersect the projection line for 2 and 12 in the front view. This is the point of intersection for 2 and 12 in the front view. (See Figure 25.76.)

STEP 4 Continue this process until each point has been located in each view. (See Figure 25.77.) Darken the line of intersection by connecting the points of intersection with

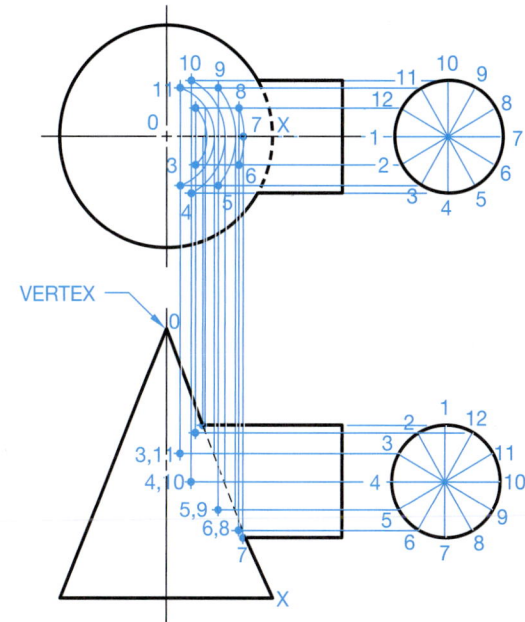

FIGURE 25.77 ■ Continuing the points of intersection between a cone and a cylinder.

an irregular curve. Be sure hidden lines are properly represented. (See Figure 25.78.)

STEP 5 The pattern development for the cone is started by using the true-length side of the cone to draw the stretch-out line arc. (This is the edge view.) Look at Figure 25.78 as you follow these steps: Lay out the circumference of the cone base along the stretch-out line, as you learned earlier in this chapter. Then, establish element 0–X in the middle of the pattern; this line coincides with 0–X in the front and top views. In the front view, project a radius from the vertex and begin where the projection

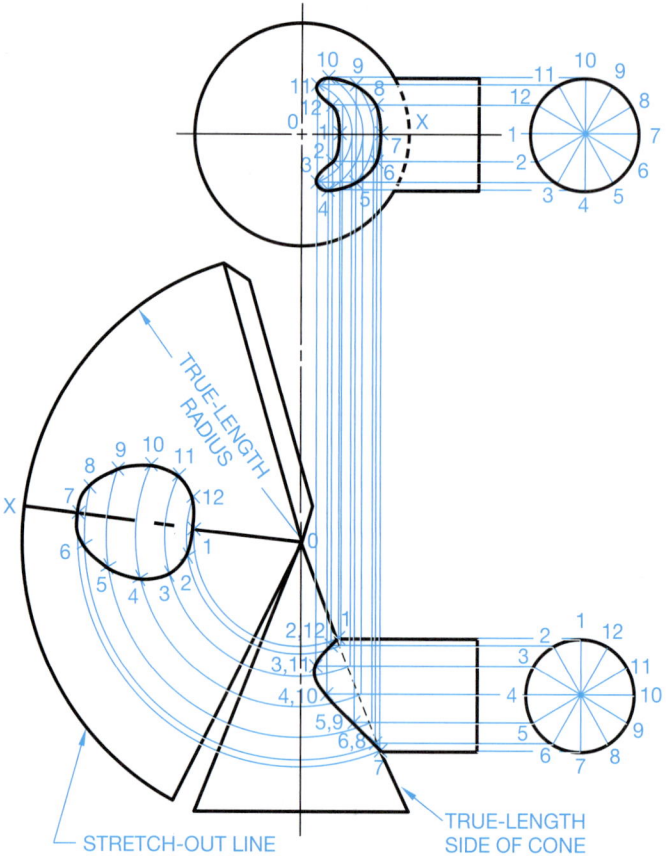

FIGURE 25.78 ■ Connecting the points of intersection to establish the line of intersection between a cone and a cylinder and the pattern development for the cone. Follow previous cylinder development examples for developing the cylinder pattern.

lines from the circle next intersect the true-length line (0–X); the points for elements 1 and 7 are where the arcs from the front view cross line 0–X in the pattern. Repeat this for each intersection, and extend these radii past the 0–X line in the pattern.

In the top view, take measurements from the 0–X line at the point where the arcs passing through points 2 and 12 intersect. Measure from this intersection to point 2. The distance from 0–X to point 2 should be the same as from 0–X to point 12 because this object is symmetrical. Take this measurement and transfer it to the pattern by measuring from the intersection of the 2,12 radius on the 0–X line, and place a point on each side of the 0–X line on this radius. Continue this process for each pair of points (3,11; 4,10; 5,9; and 6,8) until each point has been transferred into the pattern. Connect the points in the pattern to establish the cutout where the cylinder intersects. Make a pattern development for the cylinder

using the same procedure you have used for other cylinder developments.

PRECISION SHEET METAL DRAFTING, MATERIAL BENDING, AND CHASSIS LAYOUT

In precision sheet metal applications such as fabrication for electronics chassis components or sheet metal appliance body parts, the condition of material when bent must be taken into consideration.

Calculating Bend Allowance

Bend allowance is the amount of extra material needed for a bend to compensate for the compression during the bending process. This consideration is often not critical to HVAC sheet metal bending due to the larger tolerances applied to these components. Bend allowance is important when close tolerances must be held or when thick material must be bent or formed into desired shapes. The purpose of a bend allowance calculation is to determine the overall dimension of the flat pattern, so when bent, the desired final dimension is achieved. There are a number of slightly different methods used to calculate bend allowance. There are different formulas in the *Machinery's Handbook, ASME Handbook*, and in most textbooks. Many companies use formulas derived from a proven method or by individual experimentation. The amount of bend allowance depends on:

■ the material thickness;

■ type of material;

■ bending process;

■ degree of bend;

■ bend radius;

■ grain of material;

■ surface condition;

■ amount of lubrication used.

The only exact way to establish bend allowance for a specific application is to experiment with the equipment and material to be used. Some companies have made these tests and have developed charts that give the bend allowance for the type of material, material thickness, and bend radius. Use these charts when available.

When material bends, there is compression on the inside of the bend and stretching on the outside. Somewhere between, there is a **neutral zone** where neither stretching nor compression occurs; this is called the **neutral axis**. The neutral axis is approximately four-tenths of the thickness from inside of the bend, but this depends on the material and other factors. Information related to calculating the bend allowance and length of

CADD APPLICATIONS 2-D

SHEET METAL PATTERN DESIGN

Sheet metal patterns are drawn using CADD techniques that are similar to manual drafting. Many of the manual drafting geometric construction methods described in this chapter are effective for producing any type of sheet metal pattern. However, CADD tools such as Layers, and specific commands including COPY, MOVE, OFFSET, ARRAY, TRIM, and DIVIDE greatly aid in the process of drawing any required sheet metal pattern.

Specialized CADD software is also available that significantly automates the process of sheet metal pattern design and drafting. These programs can be used to generate nearly any sheet metal flat pattern layout needed, including the examples shown in Figure 25.79. Sheet metal pat-terns created using CADD software can be plotted full-size, or the information can be transferred to computer numerical control burners, lasers, punch presses, or arc machines so the pattern can be cut or punched as necessary. The programs can automatically calculate sheet layout for the best material conservation possible. Parametric design of sheet metal fabrication packages allows you to enter specific information about the design; for example, the variables H, Y, X, OX, and D shown on the transition piece in Figure 25.80. After you input the variable information, the program automatically draws the flat pattern on the screen and provides a print file of all the development information for your reference as shown in Figure 25.81.

SAMPLE SHAPES:

PATTERN SHAPES:

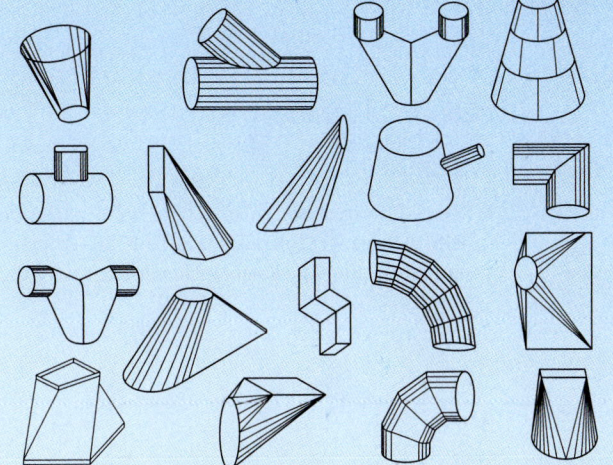

FIGURE 25.79 ■ A variety of sheet metal shapes and patterns can be easily drawn on a CADD system. *Courtesy C.A.M. Systems, Inc.*

USER INPUT SAMPLE:

Easy to operate comprehensive, self-prompting. Simply enter dimensions of desired shape.

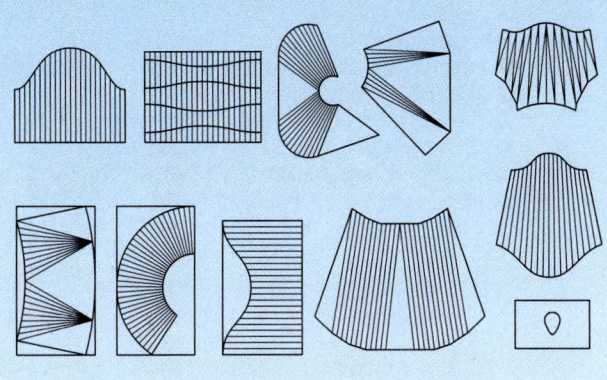

RECTANGLE TO ROUND
ONE OFFSET — NO ANGLES

Rectangle to Length X Direction	X	(12.125)
Rectangle to Length Y Direction	Y	(10.500)
Height	D	(5.375)
Number of Breaks per Corner	H	(12.562)
Select X in Y Offset	B	(5.0)
Offset in X Direction	OX	(12.125)

(Offset may be in X or Y direction from
lower left corner of the rectangle)

FIGURE 25.80 ■ Parametric CADD programs are available to design duct fittings by giving variable information and having the pattern development quickly created. Any changes made to the pattern or the model automatically effects the same change in the other. *Courtesy C.A.M. Systems, Inc.*

(Continued)

CADD APPLICATIONS 2-D

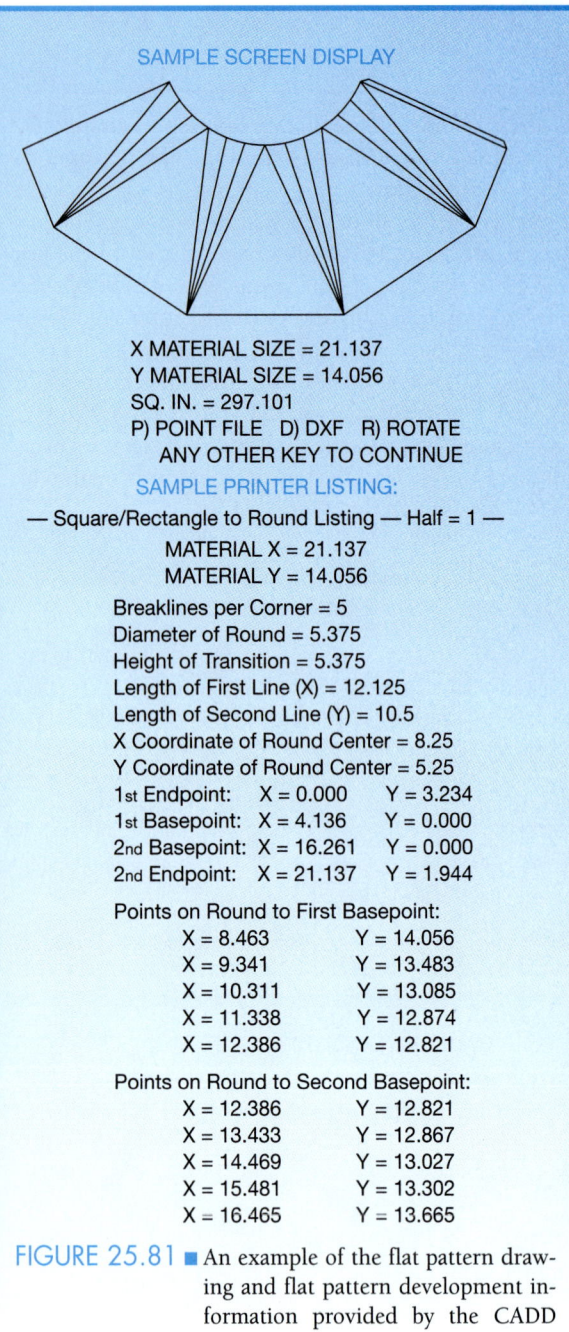

SAMPLE SCREEN DISPLAY

X MATERIAL SIZE = 21.137
Y MATERIAL SIZE = 14.056
SQ. IN. = 297.101
P) POINT FILE   D) DXF   R) ROTATE
ANY OTHER KEY TO CONTINUE
SAMPLE PRINTER LISTING:

— Square/Rectangle to Round Listing — Half = 1 —
MATERIAL X = 21.137
MATERIAL Y = 14.056

Breaklines per Corner = 5
Diameter of Round = 5.375
Height of Transition = 5.375
Length of First Line (X) = 12.125
Length of Second Line (Y) = 10.5
X Coordinate of Round Center = 8.25
Y Coordinate of Round Center = 5.25
1st Endpoint: X = 0.000   Y = 3.234
1st Basepoint: X = 4.136   Y = 0.000
2nd Basepoint: X = 16.261   Y = 0.000
2nd Endpoint: X = 21.137   Y = 1.944

Points on Round to First Basepoint:
X = 8.463   Y = 14.056
X = 9.341   Y = 13.483
X = 10.311   Y = 13.085
X = 11.338   Y = 12.874
X = 12.386   Y = 12.821

Points on Round to Second Basepoint:
X = 12.386   Y = 12.821
X = 13.433   Y = 12.867
X = 14.469   Y = 13.027
X = 15.481   Y = 13.302
X = 16.465   Y = 13.665

FIGURE 25.81 ■ An example of the flat pattern drawing and flat pattern development information provided by the CADD system. *Courtesy C.A.M. Systems, Inc.*

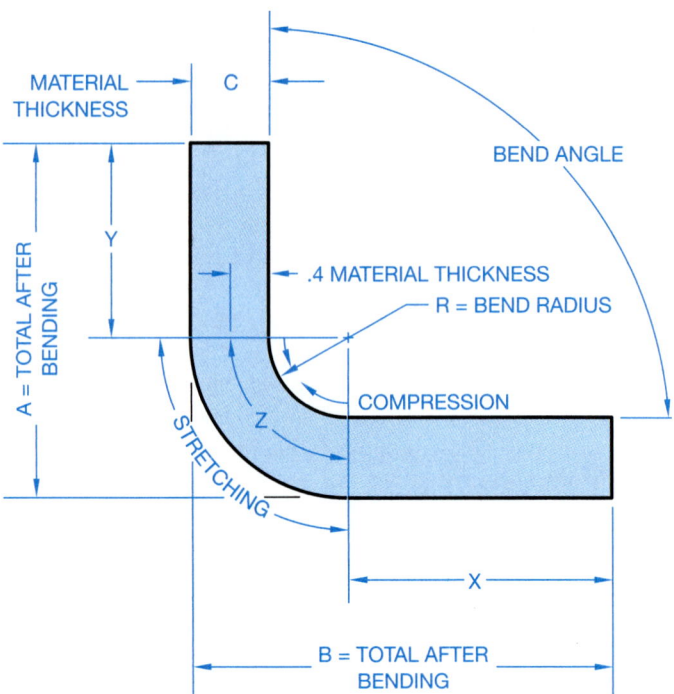

X = HORIZONTAL DIMENSION TO BEND
Y = VERTICAL DIMENSION TO BEND
Z = LENGTH OF NEUTRAL AXIS

FIGURE 25.82 ■ Bend allowance variables. There are a variety of bend allowance formulas available in company standards, the *Machinery's Handbook*, and other sources.

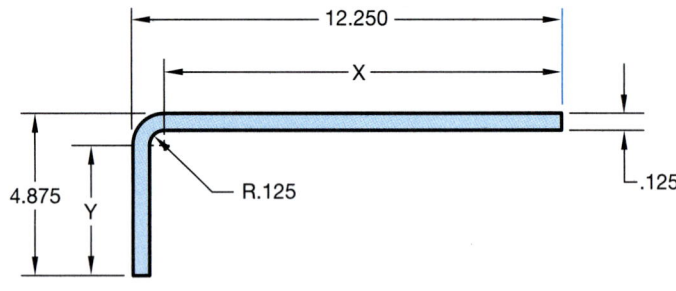

FIGURE 25.83 ■ Sample bend allowance problem.

Given the sheet metal bend shown in Figure 24.83, determine the length of the flat pattern.

$$X = B - R - C = 12.250 - .125 - .125 = 12.000$$
$$Y = A - R - C = 4.875 - .125 - .125 = 4.625$$
$$Z = 2(R + .4C) \times \pi \div 4 = 2(.125 + .4 \times .125)$$
$$\times 3.14 \div 4 = .275$$

Length of Flat Pattern = X + Y + Z = 12.000 + 4.625 + .275 = 16.900.

Bend Relief

A corner with two edges bent in the same direction has internal stresses that can cause a crack at the corner and interference

the flat pattern is shown in Figure 25.82. The following formula is used to calculate the length of the flat pattern (straight stock before bending):

Length of Flat Pattern = X + Y + Z
X = Horizontal Dimension to Bend = B − R − C
Y = Vertical Dimension to Bend = A − R − C
Z = Length of Neutral Axis
C = Material Thickness
R = Bend Radius
$$Z = 2(R + .4C) \times \pi \div 4 \text{ (90° bend)}$$

when bending. When necessary, some material at the corner can be cut away to help relieve this stress. This cutout at the corner is called **bend relief**. (See Figure 25.84.)

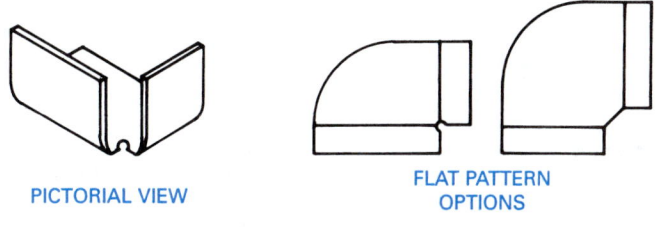

PICTORIAL VIEW

FLAT PATTERN OPTIONS

FIGURE 25.84 ■ Bend relief examples.

CHASSIS AND PRECISION SHEET METAL LAYOUT OPTIONS

Electronic assemblies often require shields, frames, panels, and chassis to be fabricated. Layout drawings are required for the fabrication of these items. Preparing chassis layouts and other precision sheet metal drawings requires close tolerances and bend allowance calculations. The method of dimensioning these parts includes standard unidirectional, arrowless, or tabular dimensioning systems. (Refer to Chapter 11, Dimensioning.)

PRECISION SHEET METAL DRAFTING AND DESIGN

The accuracy and clarity of drawings produced using CADD software is extremely beneficial for precision sheet metal design and drafting applications. CADD applications are available specifically for developing flat pattern layouts and converting existing drawings into flat layouts. These packages can automatically calculate bend allowances and notch locations based on material thickness and type, and bend radius you enter. The added advantage of many CADD systems is the ability to interface with CAM software, which allows information that is generated at the CADD workstation to be sent to the fabrication shop to run a computer numerical control metal break or punch press. A 2-D drawing generated by a CADD precision sheet metal program is shown in Figure 25.85.

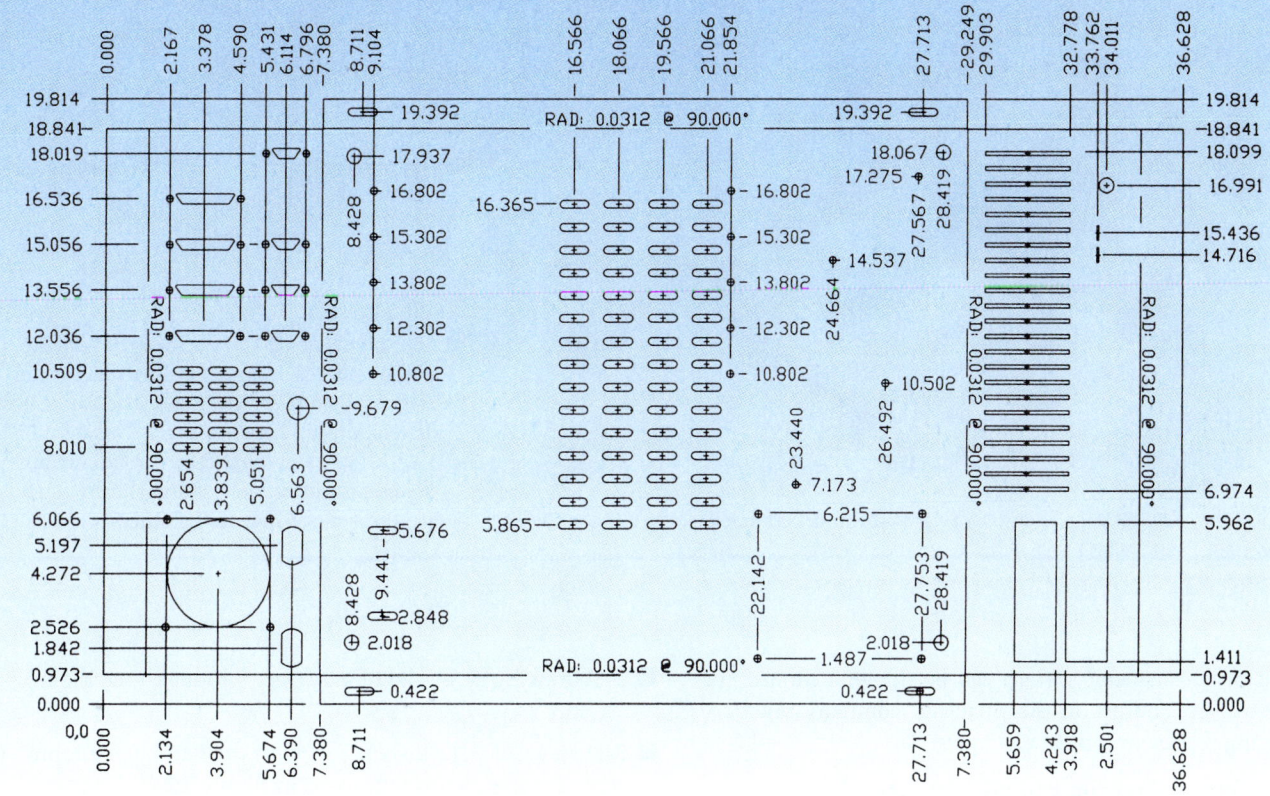

FIGURE 25.85 ■ A CADD-generated precision sheet metal fabrication drawing. *Courtesy C.A.M. Systems, Inc.*

CADD APPLICATIONS 2-D

SHEET METAL MODELING

The process of creating sheet metal parts using 3-D CADD programs is typically very similar to the process of creating any other part model. Usually, you begin with a sketched feature and then add additional features to the part. Most 3-D CADD software packages can be used to develop sheet metal models. However, some programs have specific sheet metal tools available for building sheet metal parts, and precision sheet metal patterns.

Most sheet metal products use a certain metal thickness, type of material, bend radii, and relief size that remains the same throughout fabrication. Some specialized 3-D sheet metal CADD systems accommodate this requirement by allowing you to enter sheet metal parameters before developing the model, so the style characteristics are applied to the part while you develop the design. This technique reproduces real-world sheet metal part creation by forming parts using a specific type of sheet metal and fabrication preferences. After you establish sheet metal parameters, you can use a variety of tools to prepare a sheet metal part, including features such as flanges, hems, holes, bends, and punches. Figure 25.86 shows an example of a 3-D sheet metal part model created using specialized sheet metal tools and options. Once you create a sheet metal part, or anytime during sheet metal model development, you can unfold the entire part and create a sheet metal pattern. Unfolding a sheet metal part is often as easy as picking a single button, depending on the software program. The pattern tool calculates all the features and bends in the part and creates a flat pattern. A sheet metal pattern can exist in the modeling environment, as shown in Figure 25.87, or can be accessed in a drawing and fully detailed. (See Figure 25.88.)

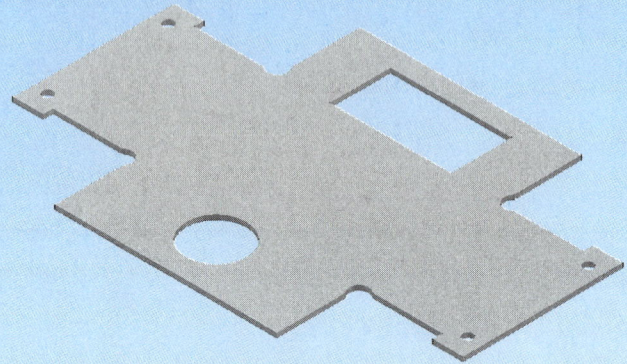

FIGURE 25.87 ■ An example of a 3-D sheet metal pattern model created from the sheet metal model shown in Figure 25.86.

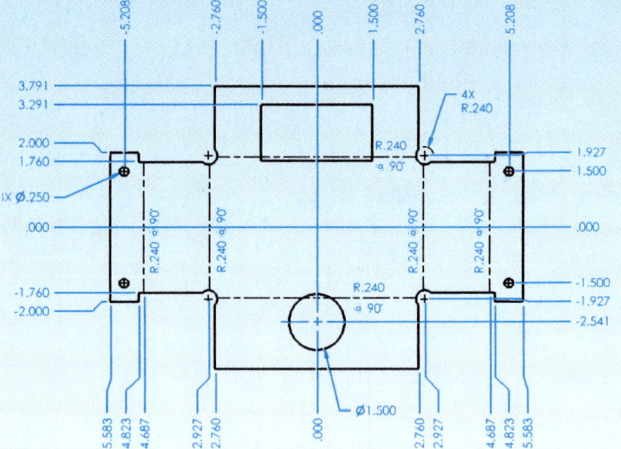

FIGURE 25.88 ■ An example of a 2-D sheet metal pattern created by referencing the existing sheet metal model shown in Figure 25.86. The parametric relationship between the 3-D model, the 3-D pattern, and the 2-D pattern allows any changes made to one to automatically establish the same change in the others.

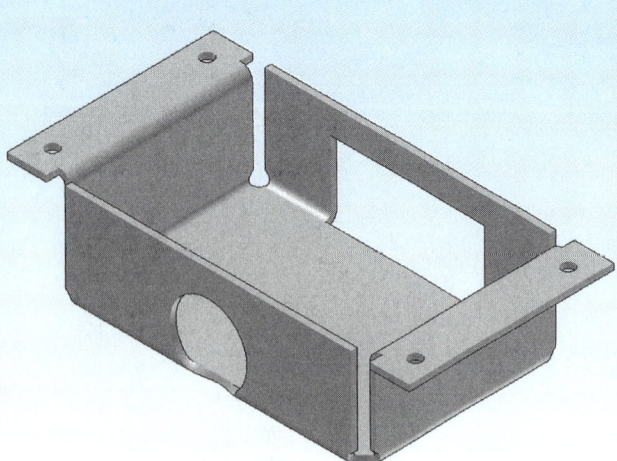

FIGURE 25.86 ■ An example of a 3-D sheet metal model.

A precision sheet metal pattern can be drawn a number of different ways depending on the preferred company practice. The following represent typical examples:

■ A flat pattern as shown in Figures 25.85 and 25.88.

■ The finished product multiviews and flat pattern as shown in Figure 25.89.

■ 3-D drawing or model of a finished product. (See Figure 25.86.)

■ 2-D views of the finished part with the flat pattern represented using phantom lines as in Figure 25.90.

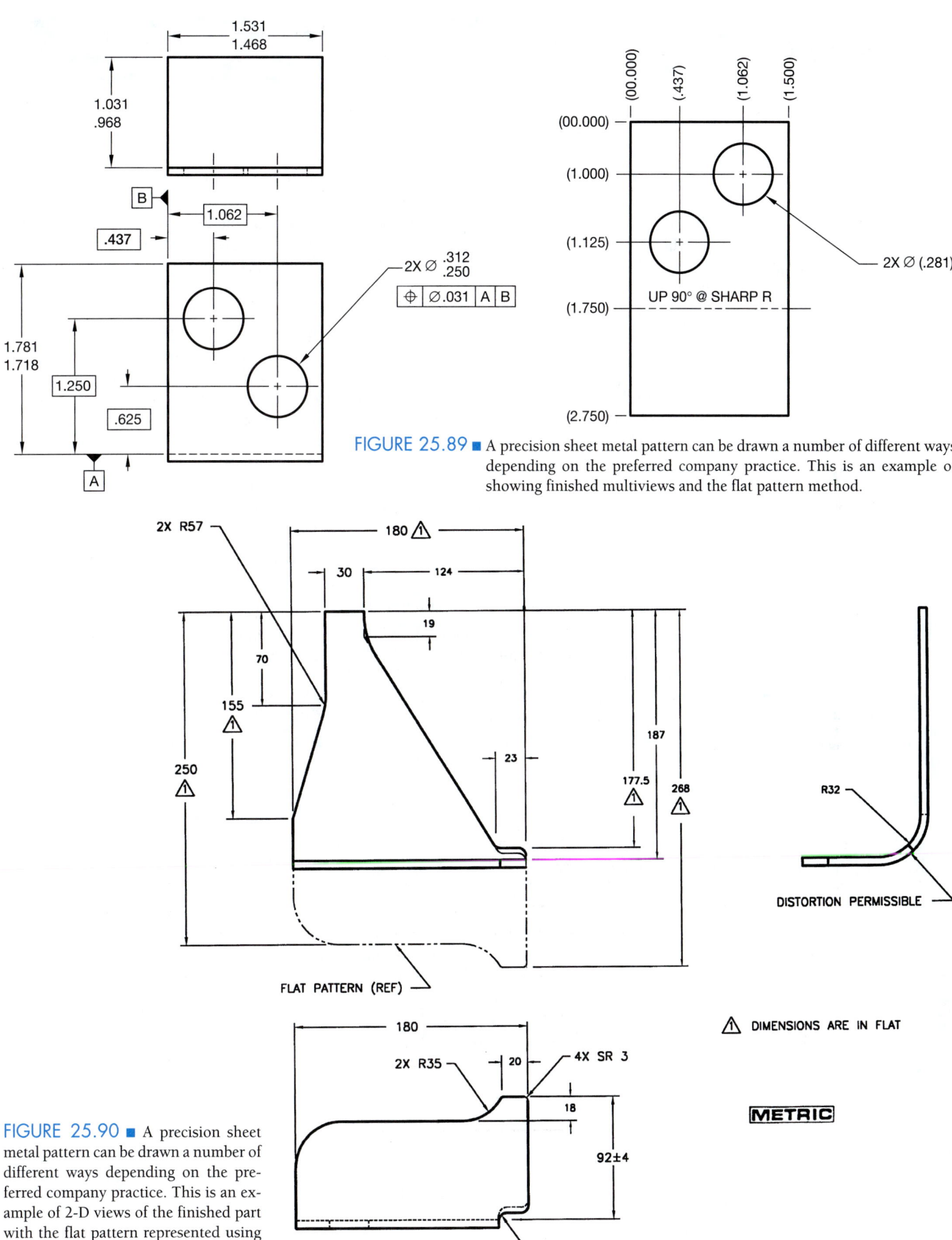

FIGURE 25.89 ■ A precision sheet metal pattern can be drawn a number of different ways depending on the preferred company practice. This is an example of showing finished multiviews and the flat pattern method.

UP 90° @ SHARP R

2X Ø .312 / .250

⊕ Ø.031 A B

2X Ø (.281)

DISTORTION PERMISSIBLE

R32

⚠ DIMENSIONS ARE IN FLAT

METRIC

FLAT PATTERN (REF)

2X R57

2X R35

4X SR 3

2X R6

92±4

FIGURE 25.90 ■ A precision sheet metal pattern can be drawn a number of different ways depending on the preferred company practice. This is an example of 2-D views of the finished part with the flat pattern represented using phantom lines.

PROFESSIONAL PERSPECTIVE

This chapter has covered the engineering drafting that takes place in three different, but closely allied, engineering fields. The first is HVAC drafting, the second is sheet metal fabrication, and the third is precision sheet metal fabrication. The first two, HVAC and sheet metal fabrication, are the most closely allied. If you are a drafter for a mechanical (HVAC) engineer, you are likely to draw HVAC plans, details, and schedules. The drawings from your office are sent to a sheet metal fabricator so the duct components can be built and delivered to the construction site. Your colleague in the fabrication industry takes the drawing from the mechanical engineer's office and converts the duct shapes to flat pattern layouts for fabrication. The third type of drafting deals with sheet metal shapes and parts. Although these components are not used in the HVAC industry, they are used in the precision sheet metal fabrication industry. The difference is that this industry normally deals with closer

tolerances. The types of items made in the precision sheet metal industry are electronics chassis and automotive (cars, trucks, and tractors) sheet metal components. In many cases, HVAC fabrication drawings do not even have dimensions; the pattern itself is used to fabricate the duct shape. On the other hand, the dimensioning and tolerancing for precision sheet metal parts are critical, and arrowless, datum, and tabular dimensioning is often used with close tolerances to achieve an accurate layout.

If you enter the HVAC industry, you should become very familiar with HVAC vendors' catalogs and manuals, and duct design applications engineering information available from mechanical suppliers such as Trane, Apec, and Elite. If you enter the precision sheet metal fabrication industry, you need to have a good understanding of tolerancing, material bending, and precision sheet metal fabrication methods.

MATH APPLICATIONS

SHEET METAL BEND ANGLES

It is required to find bend angle A for a sheet of metal having the cross section shown in Figure 25.91. The solution to this problem requires working with *two* right triangles because thickness of the material must be taken into account. Figure 25.92 is a drawing of the solution. OML stands for outside mold line. The steps that were taken are the following:

1. Find the angle of the large triangle with Inv tan
   $\left(\dfrac{.75}{1.25}\right) = 31.0°$.

2. Find the hypotenuse of both triangles with
   $\sqrt{1.25^2 + 75^2} = 1.4578$.

3. Find the angle of the slender, shaded triangle with Inv
   sin $\left(\dfrac{.125}{1.4578}\right) = 4.9°$

4. Subtract the two angles to find the bend angle.

$$A = 31.0 - 4.9 = \mathbf{26.1°}$$

If the true distance along the bend (OML to OML) is known then the problem is much simpler and requires working with only one right triangle.

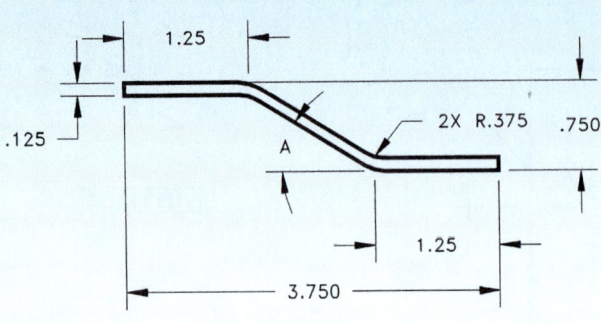

FIGURE 25.91 ■ Sheet metal with two identical bends. Dimension values in this figure are in inches.

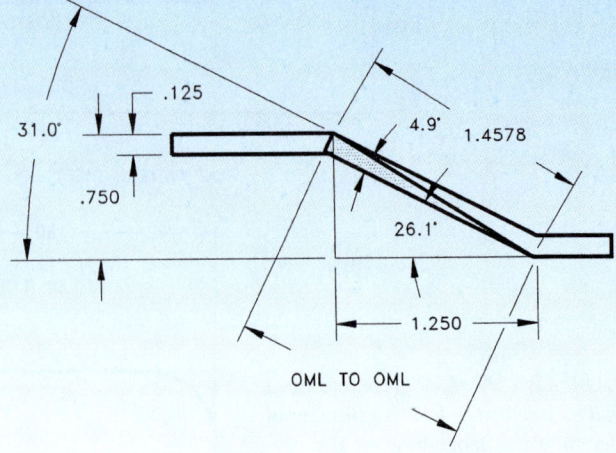

FIGURE 25.92 ■ Triangles drawn to assist in solving the bend angle problem in Figure 25.91. Dimension values in this figure are in inches.

CHAPTER 25

HVAC, Pattern Development, and Precision Sheet Metal Drafting Test

Access the CD found with this textbook to view the Chapter 25 Test. Confirm the preferred submittal method with your instructor.

CHAPTER 25

HVAC, Pattern Development, and Precision Sheet Metal Drafting Problems

DIRECTIONS

1. Read all related instructions before you begin working. Specific information will be provided for each problem.

2. Use manual or computer-aided drafting as required by your course guidelines.

PROBLEM 25.1   Residential HVAC plan

Given: Residential heating engineering sketch of main floor plan and basement. Do the following on appropriately sized sheets (two B-size sheets or one C-size sheet is recommended):

1. Make a formal double-line HVAC floor plan layout at a 1/4" = 1'-0" scale.

2. Approximate the location of undimensioned items such as windows.

3. Use thin lines for the floor plan and use thick lines for the heating equipment and duct runs.

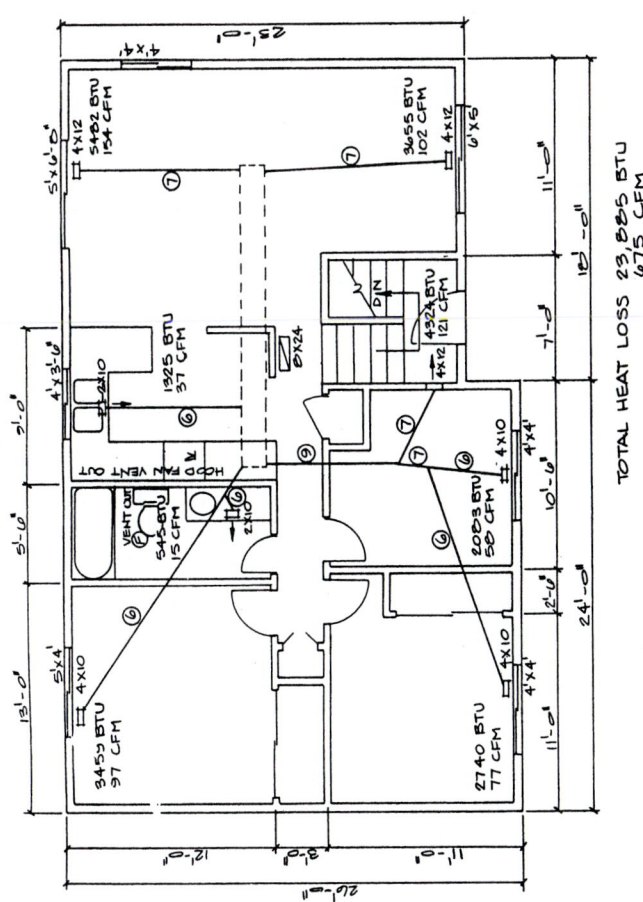

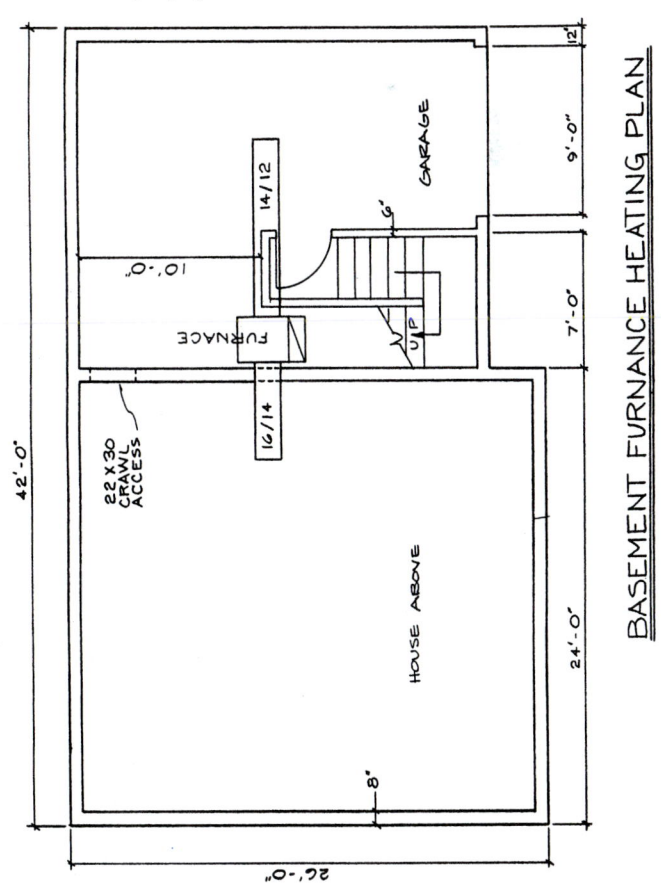

PROBLEM 25.2 Residential air-to-air heat exchanger plan

Given: Residential air-to-air heat exchanger ducting engineering sketch of basement floor plan. Do the following on appropriately sized sheet (one B- or C-size sheet is recommended):

1. Make a formal single-line air-to-air heat exchanger floor plan layout at a 1/4" = 1'−0" scale.

2. Approximate the location of undimensioned items such as doors.
3. Use thin lines for the floor plan and use thick lines for the air-to-air heat exchanger equipment and duct runs.

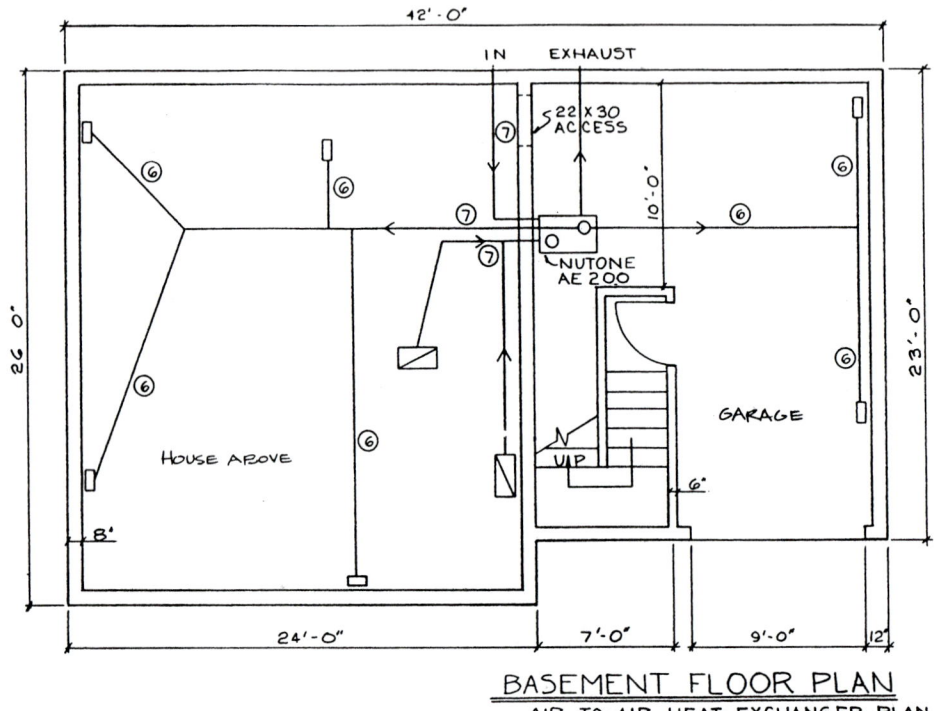

BASEMENT FLOOR PLAN
AIR-TO-AIR HEAT EXCHANGER PLAN

PROBLEM 25.3 Commercial HVAC plan

Given the following:

1. HVAC floor plan engineering layout at approximately 1/16" = 1'−0". The engineer's layout is rough, so round off dimensions to the nearest convenient units at 6" intervals. For example, if the dimension you scale reads 24 ft. 3 in., round off to 24 ft. 0 in. The floor plan will not require dimensioning; therefore, the representation is more important than the specific dimensions.
2. Related schedules.
3. Engineer's sketch for exhaust hood.

SCHEDULES				
CEILING OUTLET SCHEDULE				
Symbol	Size	CFM	Damper Type	Panel Size
C-10	9 × 9	230	Key Operated	12 × 12
C-11	8 × 8	185	Key Operated	12 × 12
C-12	6 × 6	40	Key Operated	12 × 12
C-13	6 × 6	45	Key Operated	12 × 12
C-14	6 × 18	300	Fire Damper	24 × 24

SUPPLY GRILL SCHEDULE				
Symbol	Size	CFM	Location	Damper Type
S-1	20 × 8	450	High Wall	Key Operation
S-2	12 × 12	450	High Wall	External Operation

EXHAUST GRILL SCHEDULE				
Symbol	Size	CFM	Location	Damper Type
E-5	18 × 24	1000	Low Wall	No Damper

ROOF EXHAUST FAN SCHEDULE			
Symbol	Area Served	CFM	Fan Specifications
REF-1	Solvent Tank	900	1/4 HP, 12 in. Nonspark Wheel, 1050 Max. Outlet Velocity

Do the following on appropriately sized sheet (D size is recommended; all required items will fit on one sheet with careful planning):

1. Make a formal double-line HVAC floor plan layout at a 1/4" = 1'−0" scale. (Note: You measured the given engineer's sketch at 1/16" = 1'−0".) Now, convert the established dimensions to a formal drawing at 1/4" = 1'−0". Approximate the location of the HVAC duct runs and

equipment in proportion to the presentation on the sketch. Assume that the single-line sketch is the centerline of the ducts.

2. Prepare correlated schedules in the space available. Set up the schedules in a manner *similar* to the examples in Figure 25.20, page 876, for layout.

3. Make a detail drawing of the exhaust hood either scaled or unscaled. Make the detail large enough to clearly show the features. Refer to Figure 25.17, page 874, for an example of a detail drawing.

4. Approximate the location of doors, windows, and fixtures.
5. Draw the floor plan using thin lines and the HVAC components with thick lines for contrast. Use appropriate CADD layers.

Note: Do not include notes and dimensions for wall thickness, door sizes, and tangent. Top view of exhaust hood detail is drawn as a transition piece, similar to Figure 25.59, page 888.

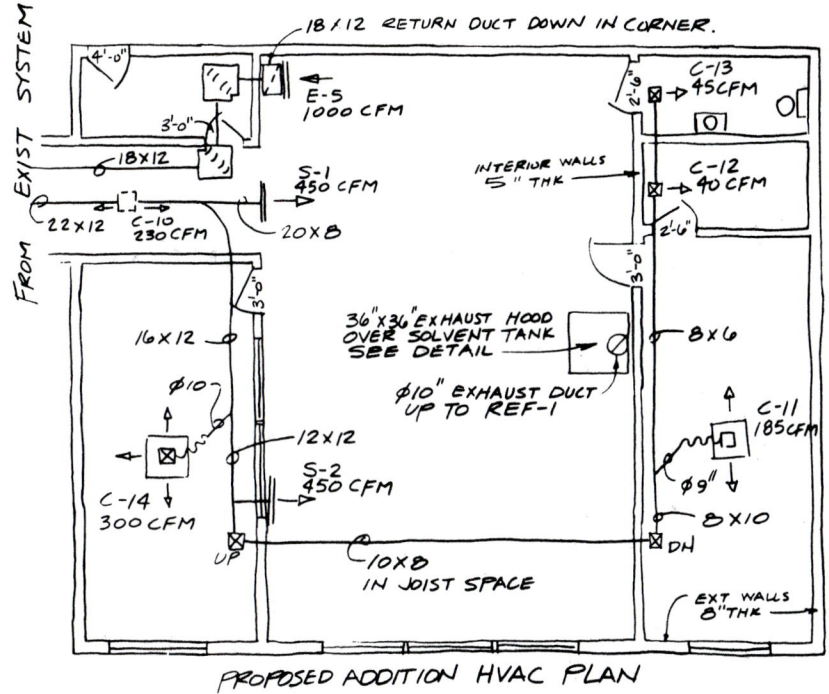

PROPOSED ADDITION HVAC PLAN

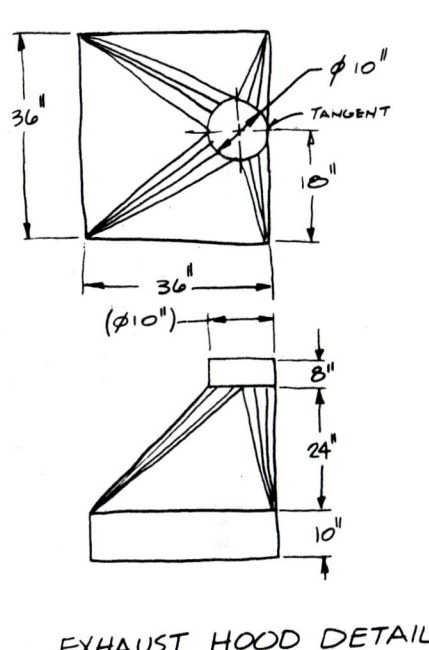

EXHAUST HOOD DETAIL

 PROBLEMS 25.4 and 25.5: Access the CD found in this textbook and open the problem of your choice, or as assigned by your instructor. Solve the problems using the instructions provided with this chapter or on the CD, unless otherwise specified by your instructor.

PROBLEM 25.6   Exhaust duct system (inch)

Given: The engineer's sketch and specifications for an exhaust duct system. The sketch displays the top, front, and partial left side views of an exhaust duct system that could be found in any commercial solid-fuel exhaust. The exhaust pickup is rectangular in shape and the discharge throat is cylindrical. The directional path of the system is often obstructed and closely confined for reasons of design and operation of the system.

Do the following on an appropriately sized sheet or sheets:

1. Make a pattern development for each of the five exhaust duct components: (There will be five individual pattern development drawings.)
    A. Truncated cylinder
    B. Truncated cone
    C. Three-piece elbow
    D. Square-to-round transition piece
    E. Rectangular transitional elbow
2. Use full scale unless otherwise specified by your instructor. A beam compass is necessary for manual drafting or use CADD.
3. Use a 3/8-in. single-lap seam on individual parts and between adjacent parts.
4. Show all layout and construction. Do not erase your construction lines.

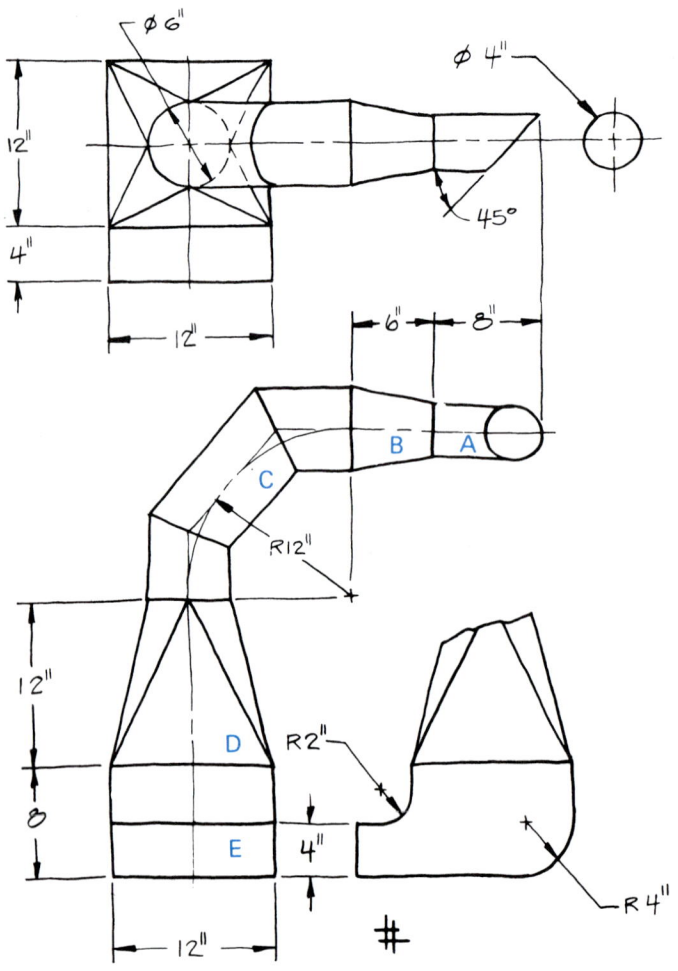

PROBLEM 25.7   Welding booth hood (inch)

Given: The engineer's sketch of a fabrication shop's welding booth hood.

Do the following:

1. Make a pattern development drawing of the pyramid-shaped hood and the shroud base at a scale of 1 1/2" = 1'-0", or full scale for CADD.
2. Provide the cutout for the window to be added later. Be careful to find the true location, and true size and shape of the cutout in the pattern.
3. No seam material allowance is required as the seams will be welded.
4. Show all layout and construction.

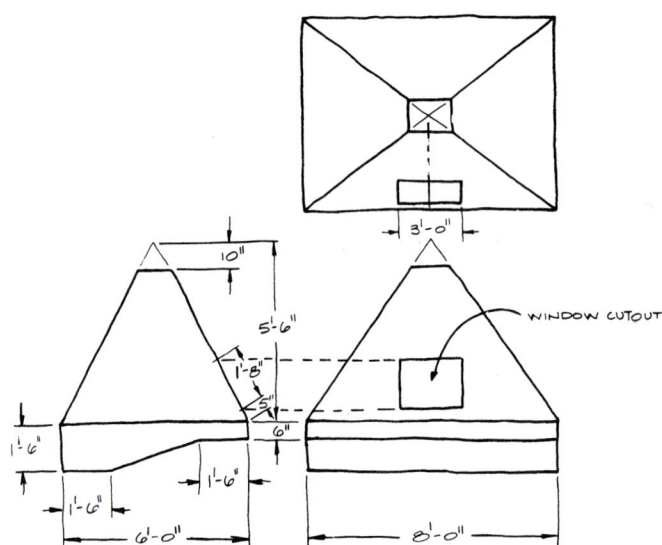

PROBLEM 25.8   Exhaust hood (inch)

Given: The exhaust hood detail from Problem 25.3.

Do the following:

1. Make a pattern development on appropriately sized layout for the transition piece, the top collar, and the base collar.
2. Use a half scale, or full scale in CADD.
3. Provide a 1-in. single-lap seam for each part and between adjacent parts.
4. Show all layout and construction.

PROBLEM 25.9   Chemistry laboratory hood (inch)

Given: The engineer's rough sketch of the chemistry laboratory hood.

Do the following:

1. Make a pattern development drawing of the chemistry laboratory hood, including the top and bottom collars.
2. Use a 3" = 1'−0" scale, or full scale for CADD.
3. No seams required.
4. Show all layout and construction.

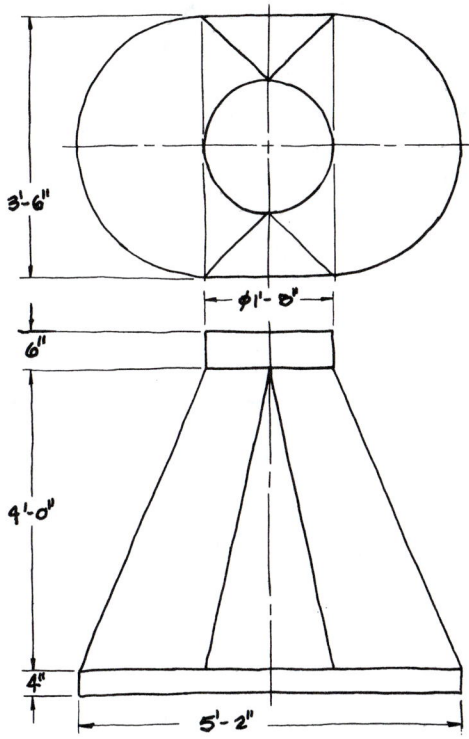

PROBLEM 25.11   Grain hopper (inch)

Given the following:

1. Use a 3/4" = 1'−0" scale, or full scale for CADD.
2. The engineer's sketch of the grain hopper.
   Do the following:

1. Determine the line of intersection between the cylinder and the cone in both views.
2. Make the resulting pattern development of the cone and side intersecting cylinder.
3. No seam material allowance required.
4. Show all layout and construction.

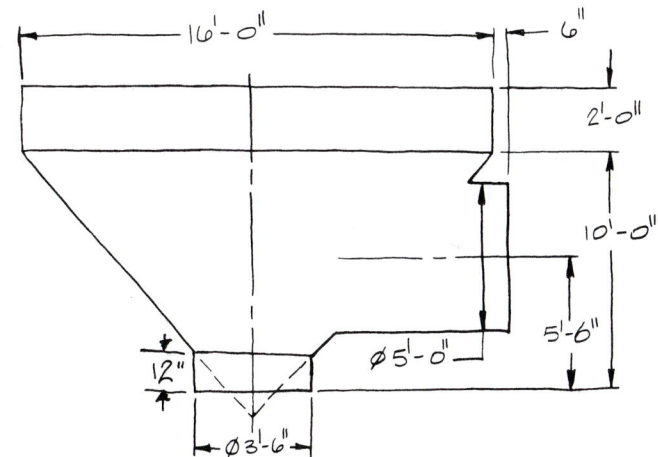

PROBLEM 25.10   Cylindrical duct intersection (inch)

Given: The engineer's computer sketch of intersecting cylindrical ducts.

Do the following:

1. Use a 1" = 1'−0" scale, or full scale for CADD.
2. Find the intersection between the cylindrical ducts.
3. Make a pattern development for each cylinder.
4. Use a 1-in. (scale) double-lap seam.
5. Show all layout and construction.

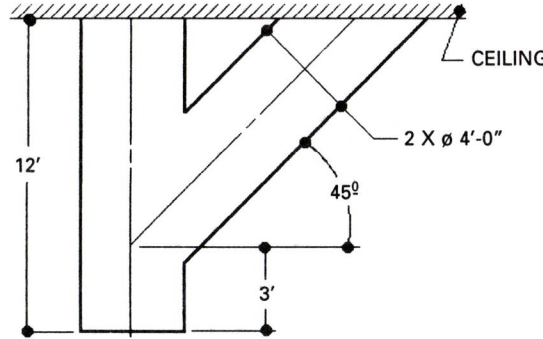

PROBLEM 25.12   Intersection

Given the following drawing, use your scale to double the size shown and draw the object. Determine the line of intersection between the parts.

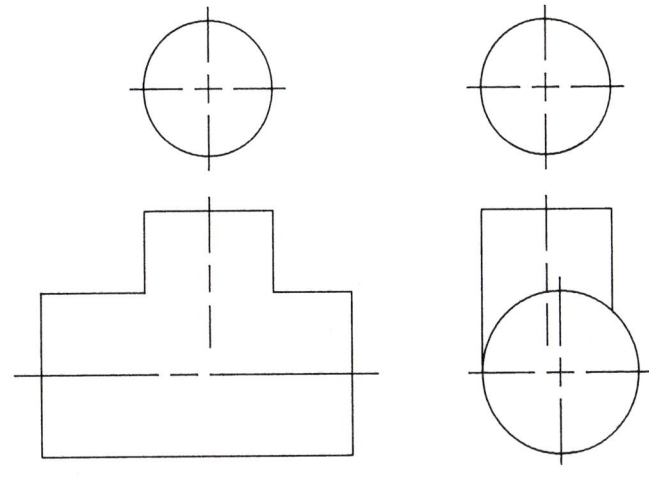

PROBLEM 25.13 Intersection

Given the following drawing, use your scale to double the size shown and draw the object. Determine the line of intersection between the parts.

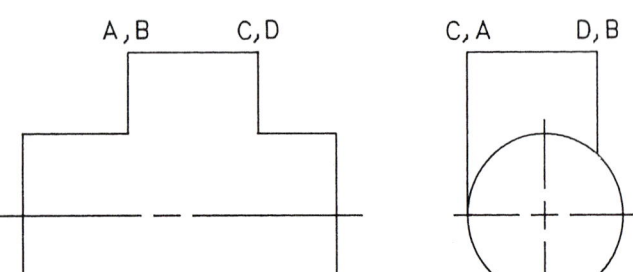

PROBLEM 25.14 Intersection

Given the following drawing, use your scale to double the size shown and draw the object. Complete the right-side view and determine the line of intersection between the parts in both views.

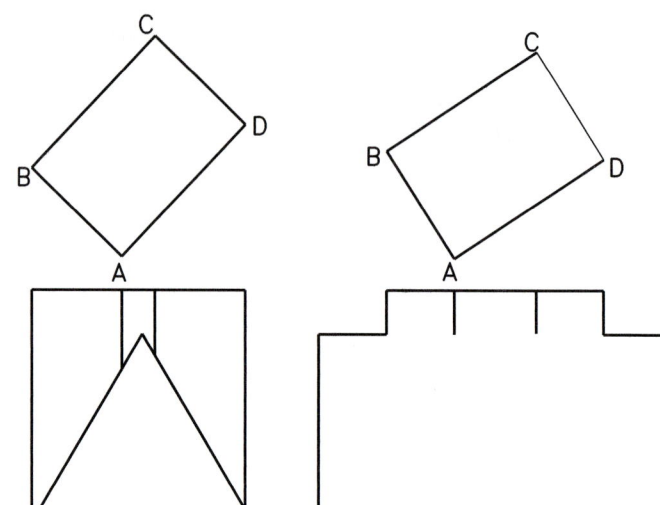

PROBLEM 25.15 Drawing displayed in form view and flat pattern (inch)

Draw the following object as shown using correct ASME standards:

Part Name: Mounting Bracket

Material: .25 THK 5086-H32

Courtesy TEMCO.

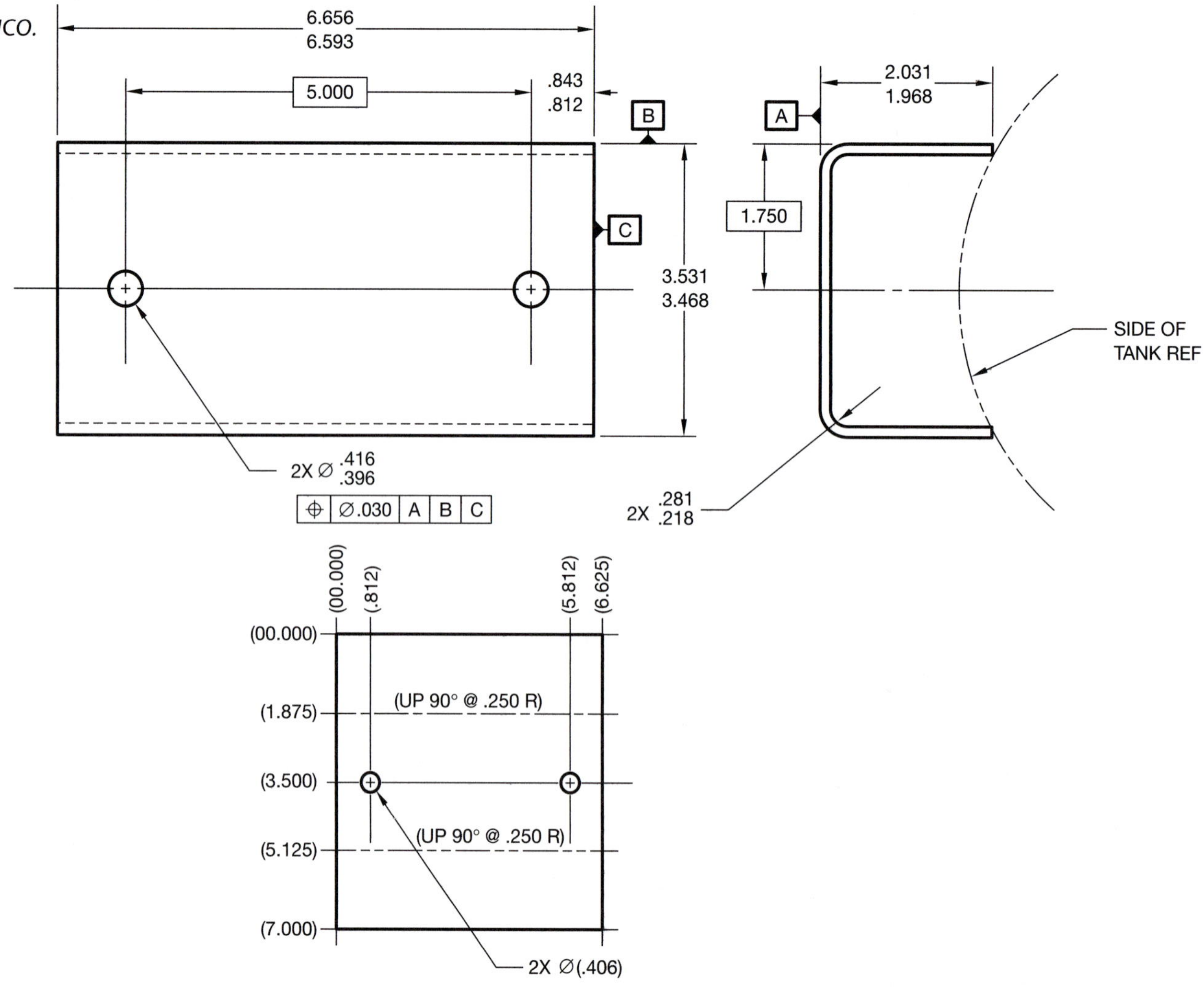

6.656
6.593

5.000

.843
.812

B

C

2.031
1.968

1.750

3.531
3.468

SIDE OF
TANK REF

A

2X Ø .416
.396

⊕ | Ø.030 | A | B | C

2X .281
.218

(00.000)
(.812)
(5.812)
(6.625)

(00.000)

(1.875)

(UP 90° @ .250 R)

(3.500)

(5.125)

(UP 90° @ .250 R)

(7.000)

2X Ø(.406)

FLAT PATTERN
(SCALE .50X)

PROBLEM 25.16 Drawing displayed in form view and flat pattern (inch) *Courtesy TEMCO.*

Draw the following object as shown using correct ASME dimensioning standards:

Part Name: Bracket

Material: 14 GA GALV CRS

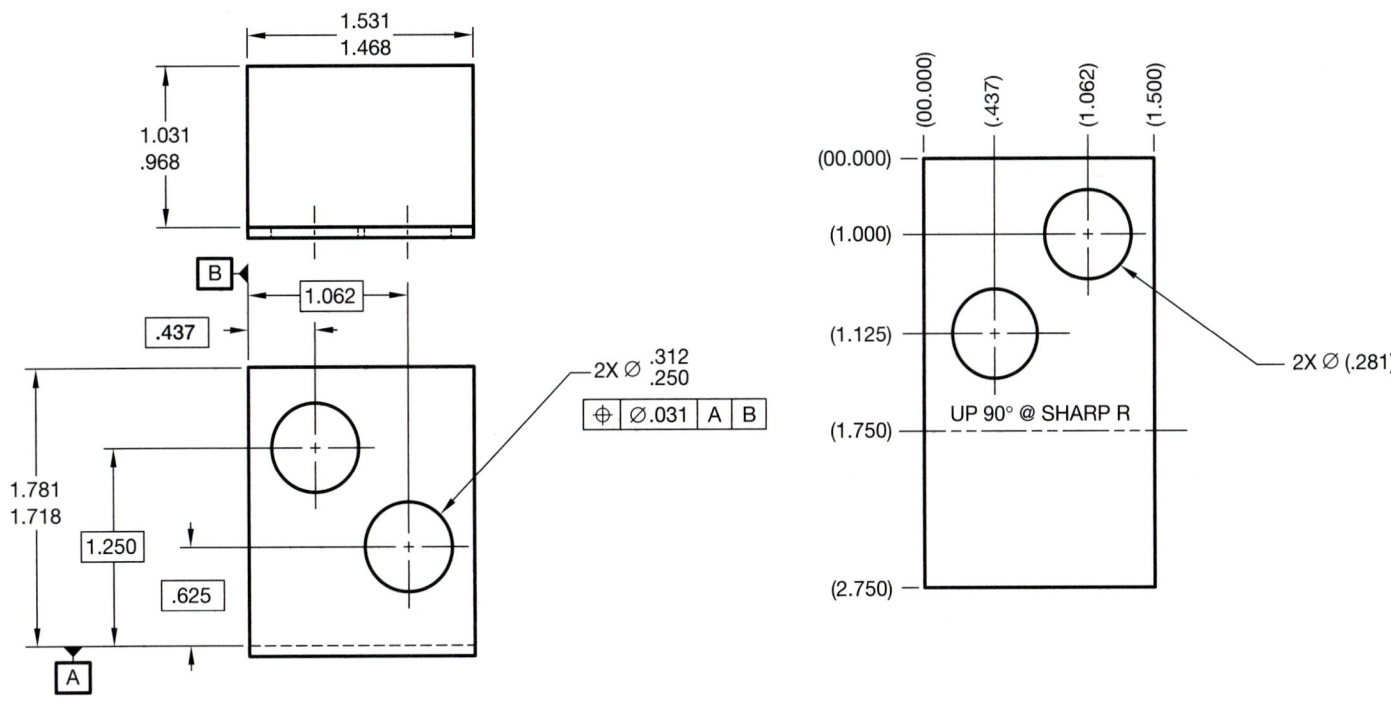

PROBLEM 25.17 Chassis layout (inch)

Part Name: Chassis

Material: Aluminum

Given: The engineer's rough sketch of a computer component chassis.

Do the following using correct ASME standards:

1. Make a flat pattern drawing of the given chassis on properly sized sheet. Full scale is recommended.

2. Use arrowless tabular dimensioning from the given datums.

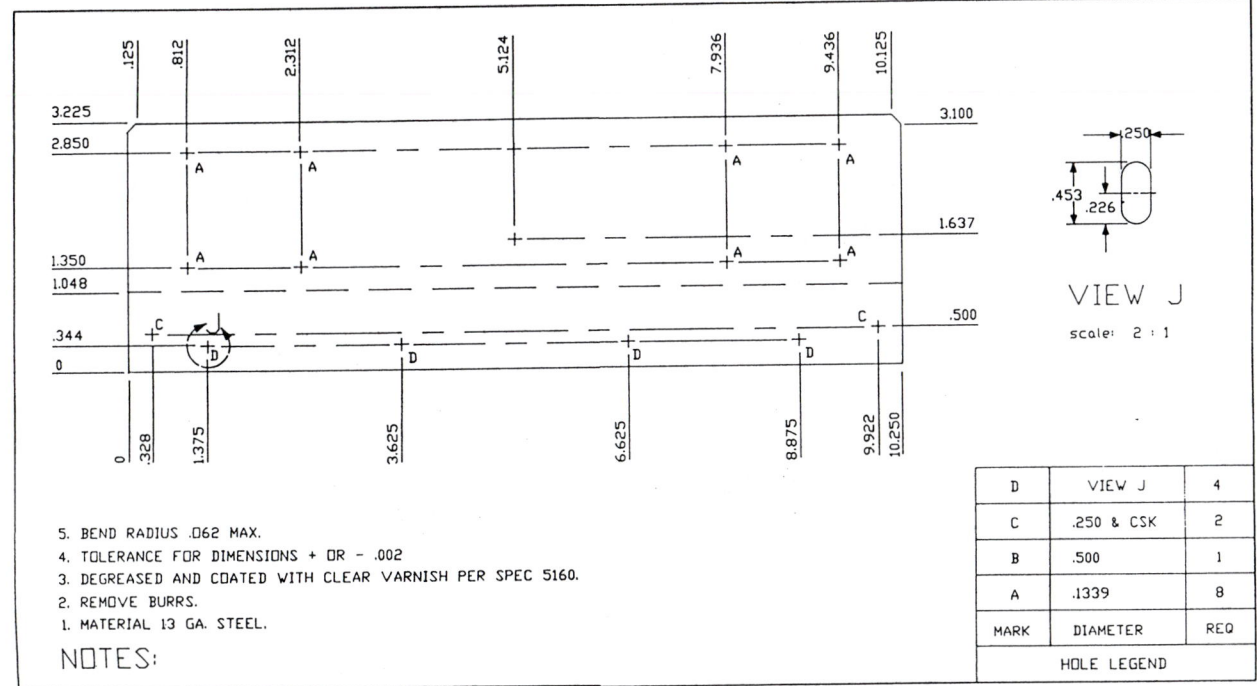

PROBLEM 25.18   Chassis layout

Given: The engineer's rough sketch of a computer component chassis.

Do the following:

1. Make a flat pattern drawing of the given chassis on properly sized sheet. Full scale is recommended. Establish the flat pattern layout dimension by bend allowance calculations. Show all math formulas and calculations on another paper.
2. Use arrowless tabular dimensioning from the given datums.

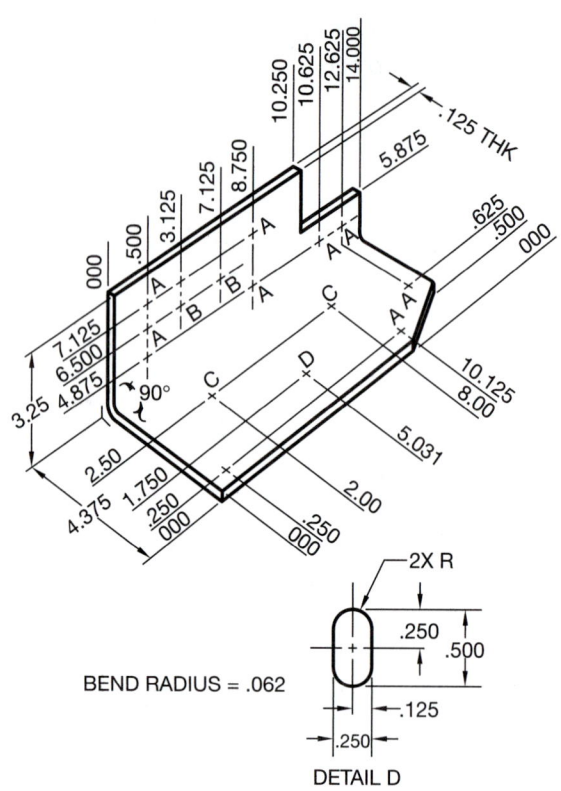

BEND RADIUS = .062

DETAIL D

DIMENSIONING TABLE			
Hole Symbol	Hole Diameter	Depth	Quantity
A	.125	Thru	9
B	.250	Thru	2
C	.469	.030	2
D	See Detail D	Thru	1

3. Standard dimensioning is needed for the total length of flat pattern and a dimension from one datum to the bend line in the flat pattern.

PROBLEM 25.19   Display of part in form view with flat pattern shown as phantom line (inch)

Draw the following object as shown:

Part Name: Plate-formed

Material: HC-112 6 mm THK

Problem based on original art courtesy Hyster Company.

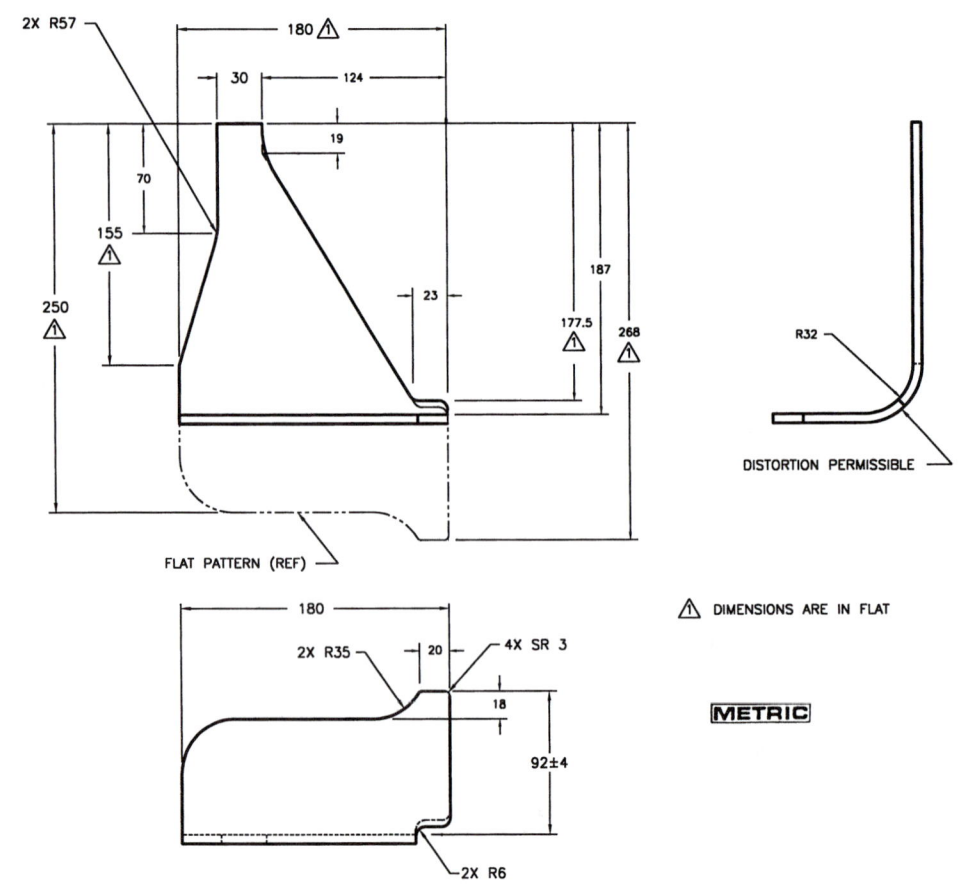

FLAT PATTERN (REF)

DISTORTION PERMISSIBLE

⚠ DIMENSIONS ARE IN FLAT

METRIC

PROBLEM 25.20   Display of part in form view and in flat pattern (inch)

Given the following engineer's layout, draw the flat pattern, and the formed view. Use geometric dimensioning and tolerancing as shown.

Part Name: Mounting Bracket

Material: 11 GA A569

Courtesy TEMCO.

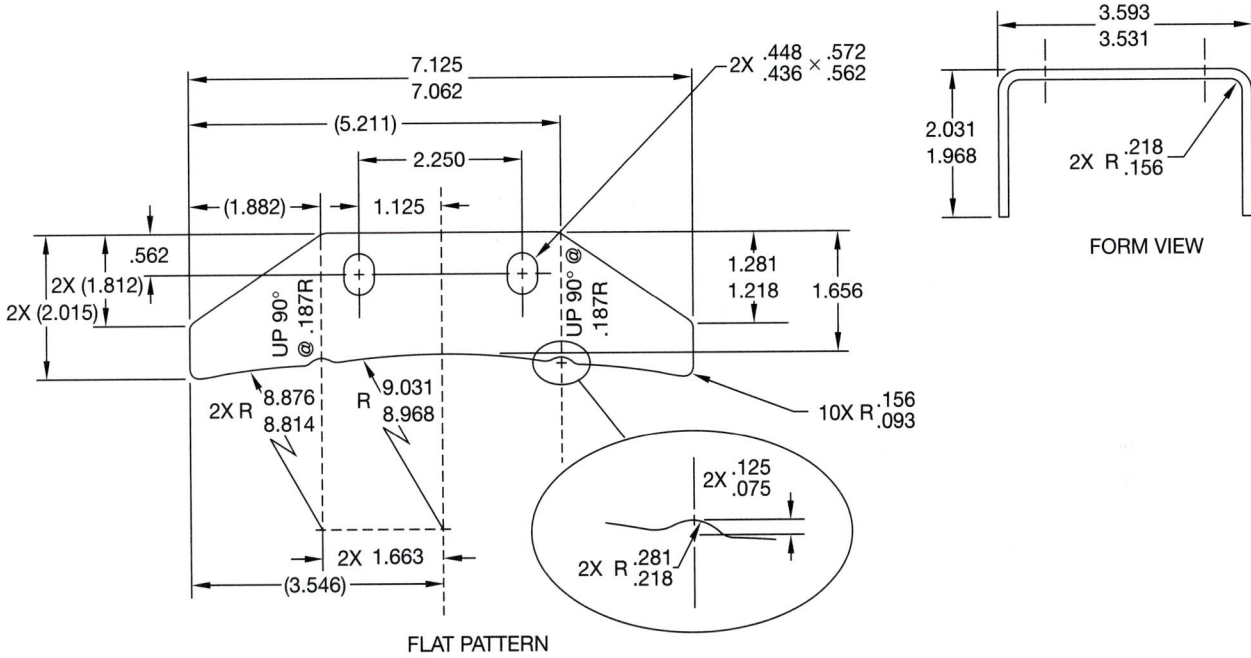

FORM VIEW

FLAT PATTERN

MATH PROBLEMS

PROBLEMS 25.21 through 25.26: Access the CD found with this textbook and open the math problem of your choice, or as assigned by your instructor. Solve the problem or problems using the instructions provided.

Electrical and Electronic Schematic Drafting

LEARNING OBJECTIVES

After completing this chapter, you will:

- Draw the following electrical diagrams: block, elementary, highway, wireless, and cable.

- Draw a cable assembly.

- Draw a set of electrical power system substation plans, including plot plan, bus plan and elevation, grounding layout and details, conduit installation details, electrical equipment layout, power panel detail, electrical floor plan, lighting plan, and security lighting plan.

- Draw an industrial electrical schematic.

- Prepare electrical drawings from engineering sketches.

- Draw electronic block and schematic diagrams.

- Create logic diagrams.

- Prepare a printed circuit board layout.

- Complete a marking and drilling drawing.

- Prepare electronics pictorial drawings.

- Make electronics drawings from given engineering sketches.

THE ENGINEERING DESIGN APPLICATION

In both electronic and electrical drafting, much of your work involves the use of symbols to show the components of your schematic or diagram. Having the proper tools available can greatly simplify your work in this area. If you are using manual drafting techniques, drawing templates are very useful for quick construction of standard symbols. CADD drafting offers additional advantages, however, because once a symbol has been drawn it can be used over and over again very simply. Additionally, CADD symbols can be created that automatically prompt for and place the required notational data in your drawing to properly define the symbol.

In this example, you have been provided a rough sketch of a circuit diagram. (See Figure 26.1.) By utilizing your CADD symbol library and adding the required notation, you are able to create the finished drawing with a minimum of time and effort (Figure 26.2).

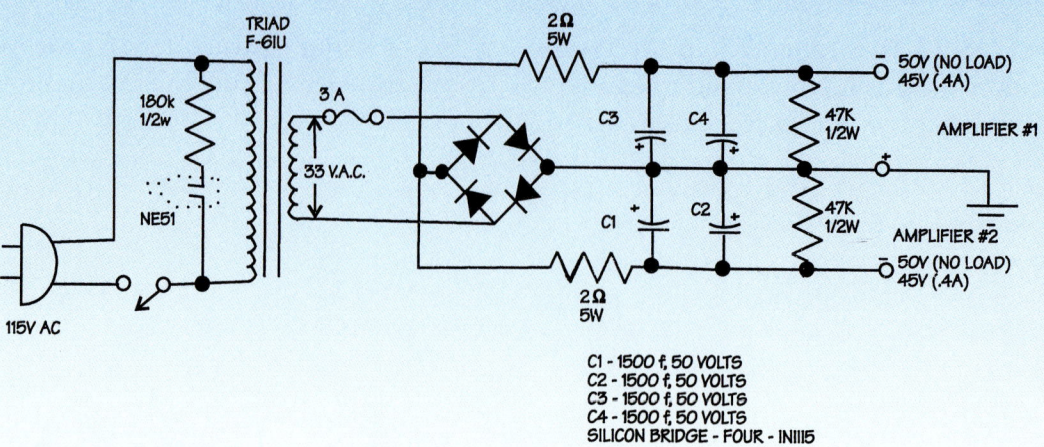

FIGURE 26.1 ■ Engineer's rough computer sketch of a circuit diagram.

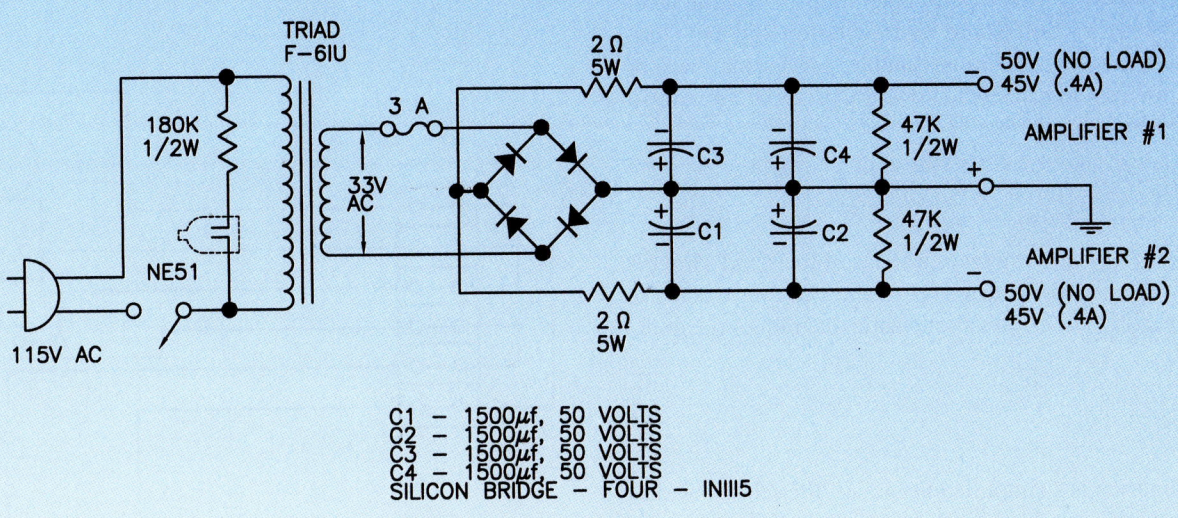

C1 — 1500 μf, 50 VOLTS
C2 — 1500 μf, 50 VOLTS
C3 — 1500 μf, 50 VOLTS
C4 — 1500 μf, 50 VOLTS
SILICON BRIDGE — FOUR — 1N1115

FIGURE 26.2 ■ The finished drawing from the engineer's sketch.

In the fully automated electronics industry, design begins with an engineer's sketch, which goes to an electronics technician who inputs the design into a computer. The computer system automatically evaluates the design and produces a prototype fabrication. The engineer then evaluates the system and makes modifications. After modifications are made and the system is completely tested and approved, the computer layout is sent to the drafter for engineering documentation. The drafter uses the CADD system to check design and drafting standards, evaluate engineering information for documentation accuracy, and correct component values. The drafter adds device tables that relate exactly how each component is attached to the board. The entire process is computerized. The drawing then progresses through the system to the automatic generation of the circuit boards. The drafter works closely with the circuit board computer program to ensure the accuracy of the product. This system may vary from one industry to the next, but as you can see, the electronics drafter is an important link between the engineering department and the final product.

INTRODUCTION TO ELECTRICAL AND ELECTRONIC DRAFTING

Electrical drafting deals with concepts and symbols that relate to high-voltage applications from the production of electricity in power plants through distribution to industry and homes. While there is often a fine line between electrical and electronic drafting, electronic drafting is more oriented toward the design of electronic circuitry for radios, computers, and other low-voltage equipment.

ANSI The key to effective communication on electrical drawings is the use of standardized symbols so that anyone who uses the diagram makes the same interpretation. To ensure proper standardization, these engineering drawings and related documents should be prepared in accordance with the American National Standards Institute publication ANSI Y32.2.

This chapter divides the content into the two areas of electrical drafting and electronic schematic drafting. You can focus your interest on electrical drafting for power transmission, electronic drafting for low-voltage applications, or study both fields for a broad range of knowledge and skills.

FUNDAMENTALS OF ELECTRICAL DIAGRAMS

The purpose of electrical diagrams is to communicate information about the electrical system or circuit in a simple, easy-to-understand format of lines and symbols. Electrical circuits provide the path for electrical flow from the source of electricity through system components and connections and back to the source. Electrical diagrams are generally not drawn to scale. The responsibility of the drafter is to organize the information in a logical, orderly manner without crowding or large variations in spacing layout.

Pictorial Diagram

Pictorial diagrams represent the electrical circuit as a three-dimensional drawing. This type of diagram provides the most realistic and easy-to-understand representation and can commonly be used in sales brochures, catalogs, service manuals, or assembly drawings. Figure 26.3 shows the pictorial drawing of a simple doorbell circuit.

Schematic Diagram

Schematic diagrams are drawn as a series of lines and symbols representing the electrical current path and the components of the circuit. Figure 26.4 shows a schematic diagram of the doorbell circuit.

Block Diagram

The block diagram is a simplified version of the schematic diagram. Simplified symbols exhibit a minimum of detail of the component and generally no connections at individual terminals as shown in Figure 26.5.

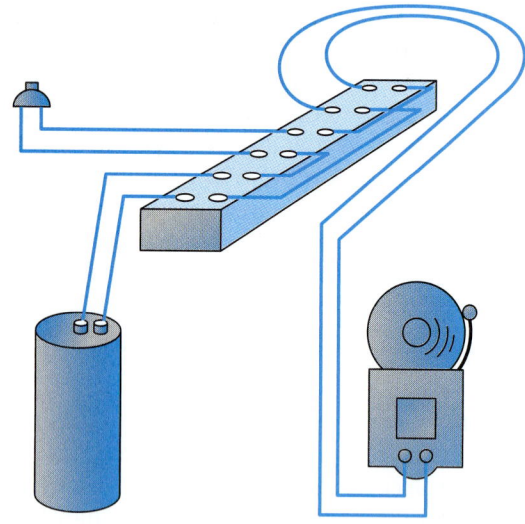

FIGURE 26.3 ■ Pictorial drawing of a simple doorbell circuit.

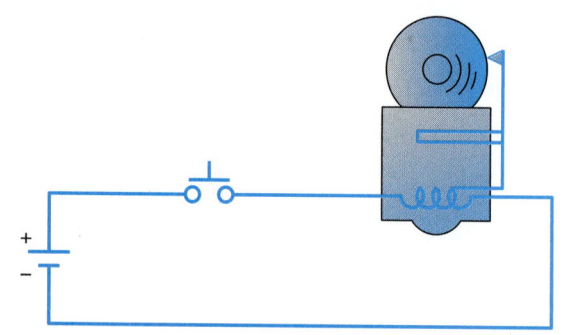

FIGURE 26.4 ■ Schematic diagram of doorbell circuit shown in Figure 26.3.

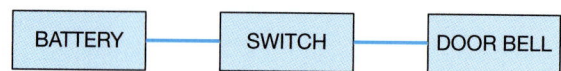

FIGURE 26.5 ■ Block diagram of the doorbell circuit shown in Figure 26.3.

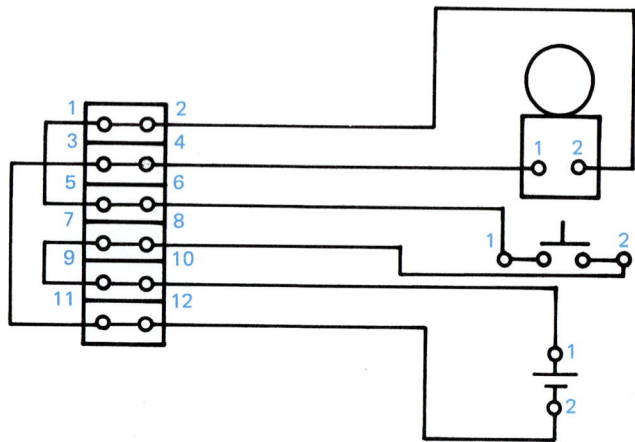

FIGURE 26.6 ■ Wiring diagram of the doorbell circuit shown in Figure 26.3.

Wiring Diagram

The wiring diagram is a type of schematic that shows all of the interconnections of the system components. These diagrams are often referred to as point-to-point interconnecting wiring diagrams. The wiring diagram is much more detailed than a standard schematic diagram because it shows the layout of individual wire runs. Figure 26.6 shows a wiring diagram of the doorbell circuit.

Schematic Wiring Diagram

A schematic wiring diagram combines the simplicity of a schematic diagram and the completeness of a wiring diagram. The complete circuit is drawn as a series of lines and symbols that represent the electrical current path and the components of the circuit, plus the connection terminals are shown in their proper locations along the circuit. (See Figure 26.7.)

Highway Wiring Diagram

Also known as highway diagrams, these drawings are used for fabrication, quality control, and troubleshooting of the wiring of electrical circuits and systems. A highway wiring diagram is a simplified or condensed representation of a point-to-point interconnecting wiring diagram. Highway wiring diagrams can be used when it becomes difficult to draw individual connection lines because of diagram complexity or when it is not necessary to show all of the wires between terminal blocks.

The wiring lines are merged at convenient locations into main trunk lines, called highways, that run horizontally or vertically between component symbols. The lines that run from the

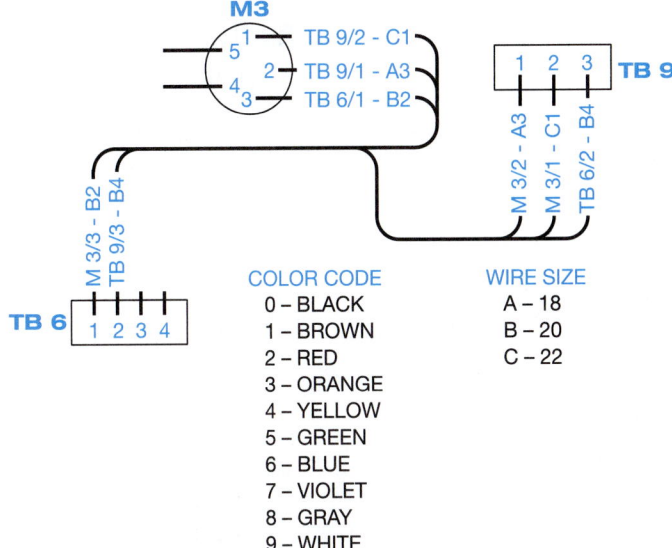

FIGURE 26.9 ■ Highway wiring diagram.

COLOR CODE
0 – BLACK
1 – BROWN
2 – RED
3 – ORANGE
4 – YELLOW
5 – GREEN
6 – BLUE
7 – VIOLET
8 – GRAY
9 – WHITE

WIRE SIZE
A – 18
B – 20
C – 22

FIGURE 26.7 ■ Schematic wiring diagram of the doorbell circuit shown in Figure 26.3.

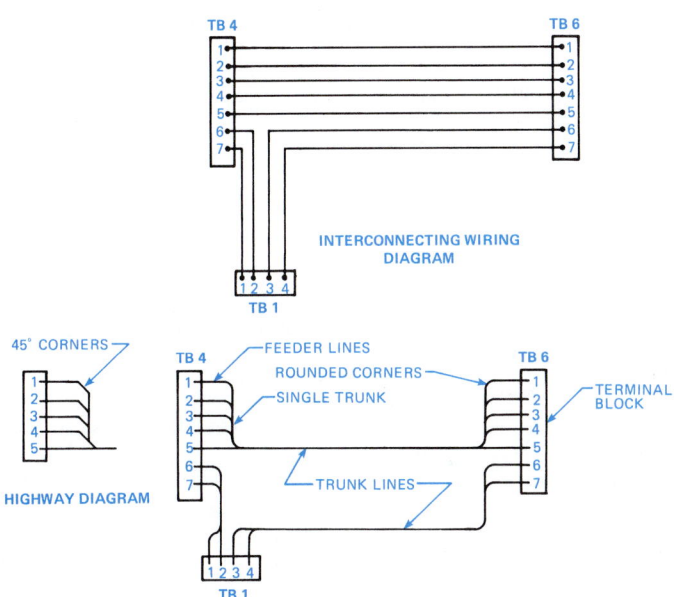

FIGURE 26.8 ■ An interconnecting wiring diagram and the same electrical system converted to a highway diagram.

component to the trunk lines are called feed lines. The feed lines are identified by code letter, number, or a combination letter/number at the point where each line leaves the component. By reducing the number of lines drawn, the wiring diagram becomes easier to draw and interpret. Figure 26.8 shows an interconnecting wiring diagram and the same electricity system converted to a highway diagram.

A complete highway diagram has an identification system that guides the reader through the system, with a code at one terminal to the same corresponding terminal on another component. An identification system recommended by the American Standards Association (ASA) is a code made up of the wire destination, terminal number at the destination, wire size, and

wire-covering color. Look at Figure 26.9 as you interpret the following code:

Component—M3	Component—TB6
TB6/1-B2	M3/3-B2
TB6 = Destination	M3 = Destination
1 = Terminal at Destination	3 = Terminal at Destination
B = Wire Size	B = Wire Size
2 = Color of Wire Covering	2 = Color of Wire Covering

Wireless Diagram

Wireless diagrams are similar to highway diagrams except that interconnecting lines are omitted as shown in Figure 26.10.

Cable Diagram and Assemblies

Cable diagrams are associated with multiconductor systems. A multiconductor is a cable or group of insulated wires put together in one sealed assembly. For example, the trunk line shown in the highway diagram in Figure 26.8 or 26.9 can be a multiconductor. Cable systems are made up of insulated conductor wires, protective outer jacket or some other means of holding the wires together, and connectors at one end or both ends. Cables are used to connect components, equipment assemblies, or systems together. Cable diagrams usually provide circuit destination, conductor size, number of leads, conductor type, and power rating. A cable diagram is shown in Figure 26.11.

Also known as cable harness diagrams, cable assemblies are drawn to scale with dimensions and include a parts list that is coordinated with the drawing by identification balloons as shown in Figure 26.12.

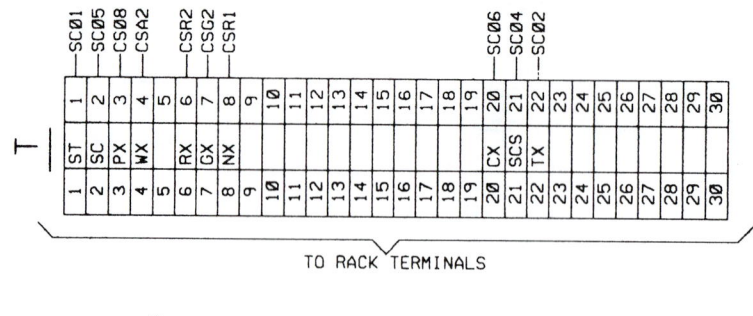

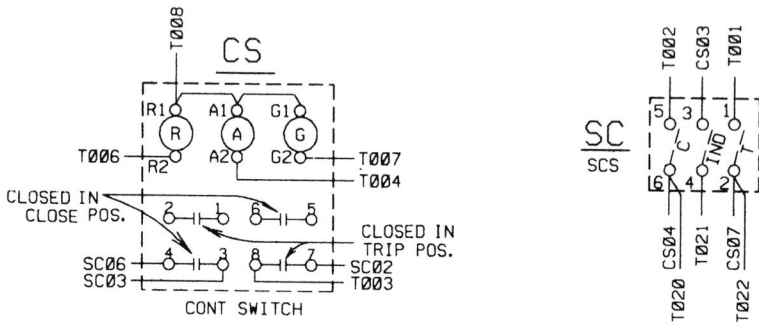

FIGURE 26.10 ■ CADD-generated wireless diagram. *Courtesy Bonneville Power Administration.*

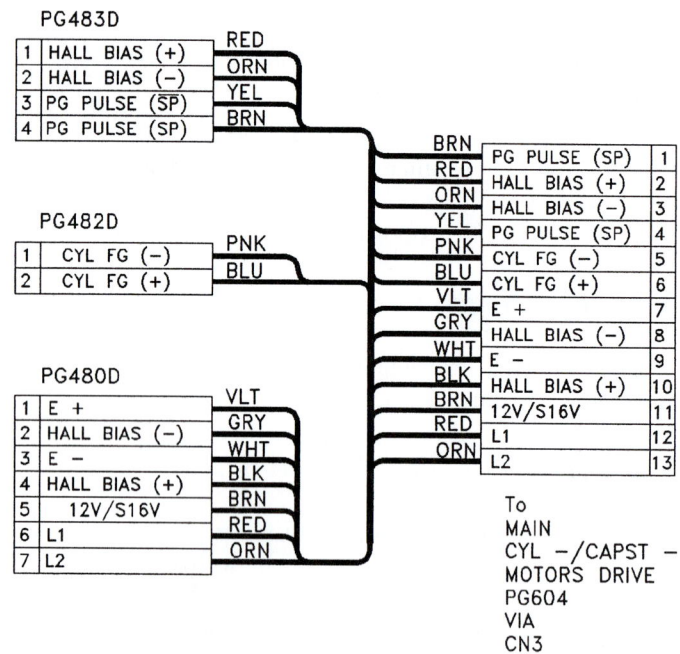

FIGURE 26.11 ■ Cable diagram. *Reproduced by permission of RCA Consumer Electronics.*

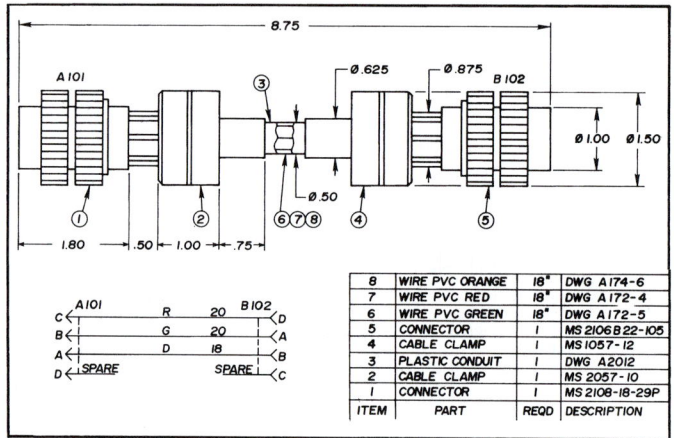

FIGURE 26.12 ■ Cable assembly.

former and sent through high-voltage lines to a switching station where the electricity is retransmitted to various locations. Before the electricity can be used, it goes to a substation where a step-down transformer converts the high voltage to a lower voltage for heavy industry or transmission through lines to a distribution substation for further voltage reduction and distribution to light commercial and residential users.

GENERATION, TRANSMISSION, AND DISTRIBUTION OF ELECTRICITY

Electricity is generated around the world in hydroelectric, coal burning, and nuclear power plants. The electricity created at the generating station is increased in voltage at a step-up trans-

ELECTRIC POWER SUBSTATION DESIGN DRAWINGS

A substation is the part of the electrical transmission system where electricity is switched or transformed from a very high voltage to a conveniently usable form for distribution to homes or businesses. Substation design drawings are an important part of any power supply system.

One-Line Diagrams

The one-line diagram is a simple way for electrical engineers and drafters to communicate the design of an electrical power substation as shown in Figure 26.13.

Elementary Diagrams

Elementary diagrams provide the detail necessary for engineering analysis and operation of the substation equipment by operators or maintenance people.

Elementary diagram, DC (direct current) circuits, shows the direct current circuits that operate the relaying and controls for the substation equipment. Elementary diagram, AC (alternating current) circuits, depicts the circuits that provide information for both the protective relays and the instruments and meters used by people who work in the substation.

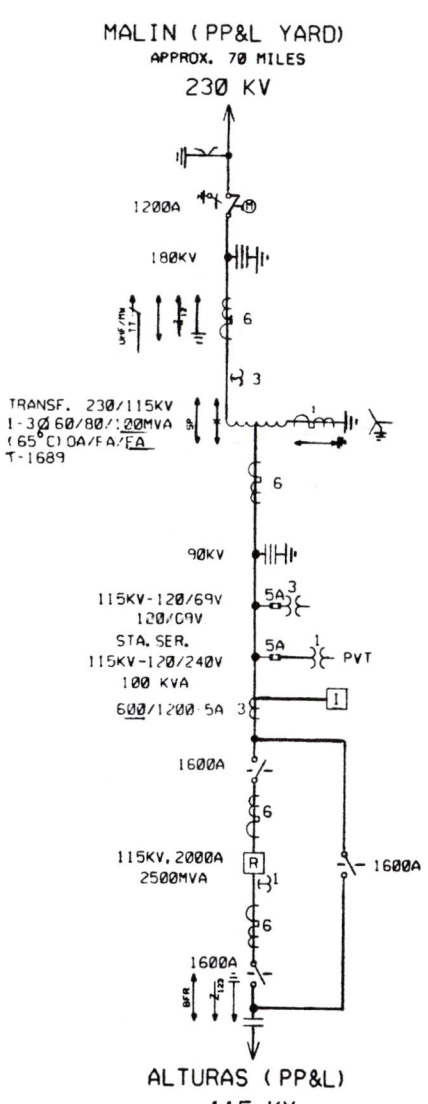

FIGURE 26.13 ■ CADD-generated one-line diagram. *Courtesy Bonneville Power Administration.*

Schematic Diagrams

The schematic diagram is used to show the relationship of the components and equipment located in the substation. The schematic diagram is more detailed than the one-line diagram.

Electric Power Systems Schematic Symbols and Terminology

ANSI Most standards for the preparation of electrical diagrams use symbols in accordance with the ANSI Y32.2 document. Any special variation to symbols should be shown in a legend or described in a general note. Commonly used electrical equipment symbols are shown in Figure 26.14.

Electrical relays are magnetic switching devices.

Nondirectional relays are relays that operate when current flows in either direction. Directional relays operate only when current flows in one direction.

Differential relays provide a switching connection between a circuit with two different voltage values. Pilot relay systems are controlled by communication devices and are designated by the type of communication circuit or the function of the relay system.

Supervisory control relays are used to check, monitor, or control other devices.

Plot Plan Drawing

The entire layout of a substation is shown by a plot-plan drawing. The plot plan is drawn similar to a map, showing the relationship between the elements of the substation in correct orientation to compass direction. A plot plan can also be referred to as a site plan. The scale used for the plot plan generally ranges from 1" = 20' to 1" = 50' (1:200, 1:500 metric). Typical plot-plan symbols are shown in Figure 26.15, page 920. The items that are commonly found on a plot plan include:

■ Property boundaries and developed-yard boundary with chain link fence or block wall.

■ Primary bearing lines. Bearing is direction in relation to the Northwest or Northeast and Southwest or Southeast quadrants of a compass. For example, N 50°30'15"W is a line that is located at an angle of 50°30'15" toward West from North.

■ Service and access roads.

■ Buildings and other nonelectric structures.

■ System electrical components such as switches, fuses, transformers, and racks.

■ Complete dimensioning.

Bus Layout

A bus is an aluminum or copper plate or tubing that carries the electrical current. The bus layout drawing is used by construction crews for the construction of the current-carrying portion

(Continued on page 920)

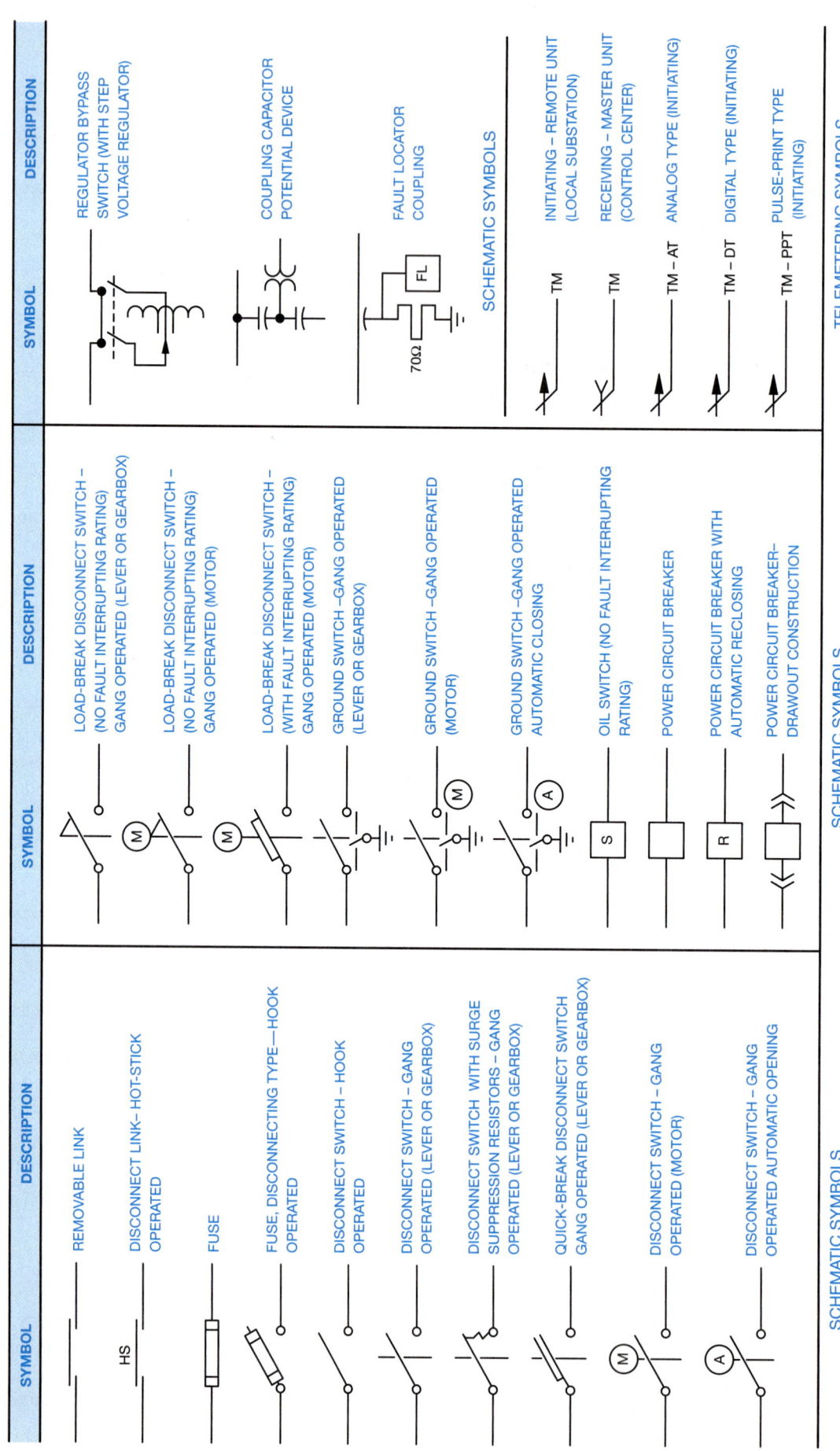

FIGURE 26.14 ■ Electric power systems schematic symbols.

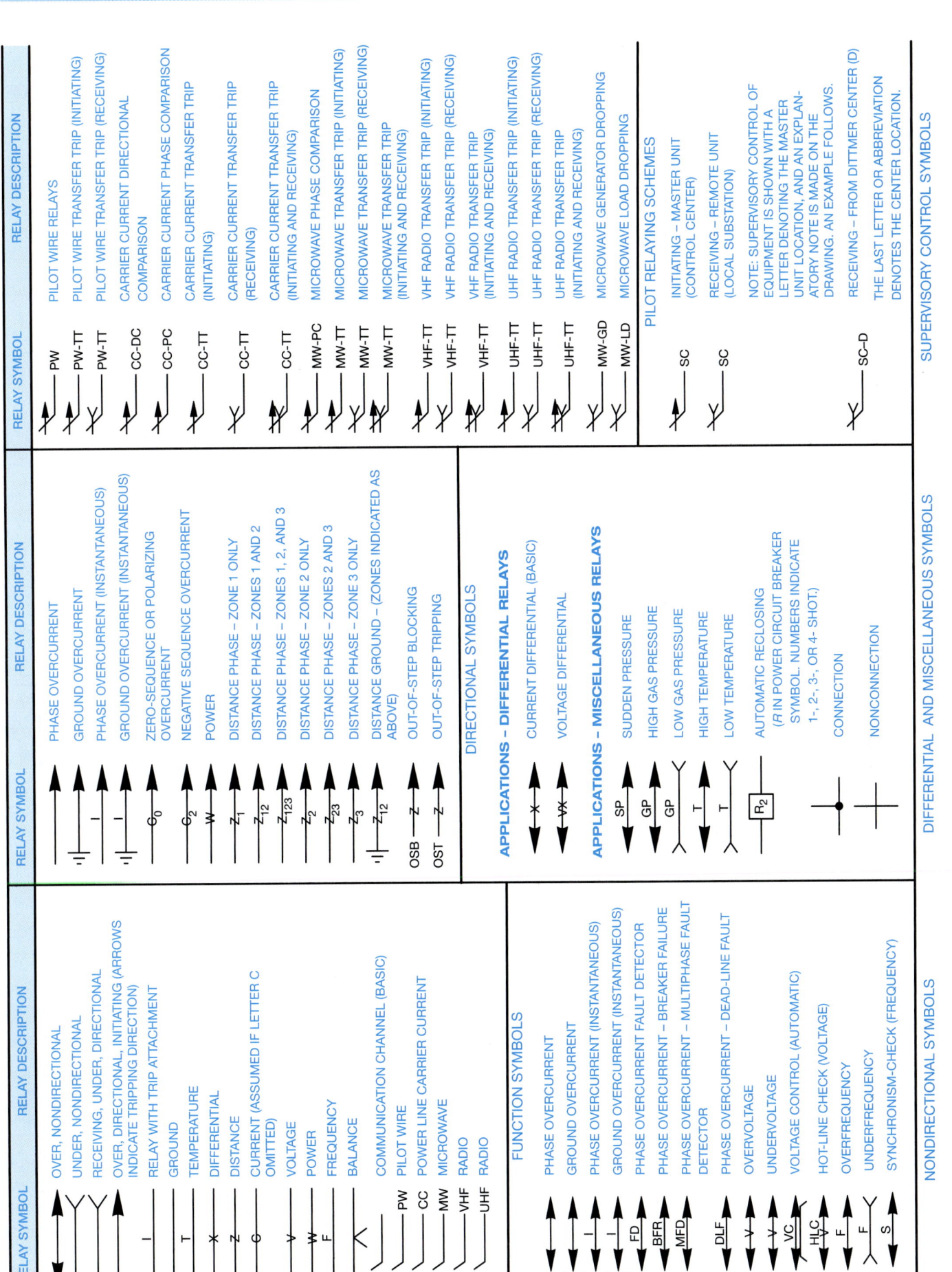

FIGURE 26.14 ■ (*Continued*) Electric power systems schematic symbols.

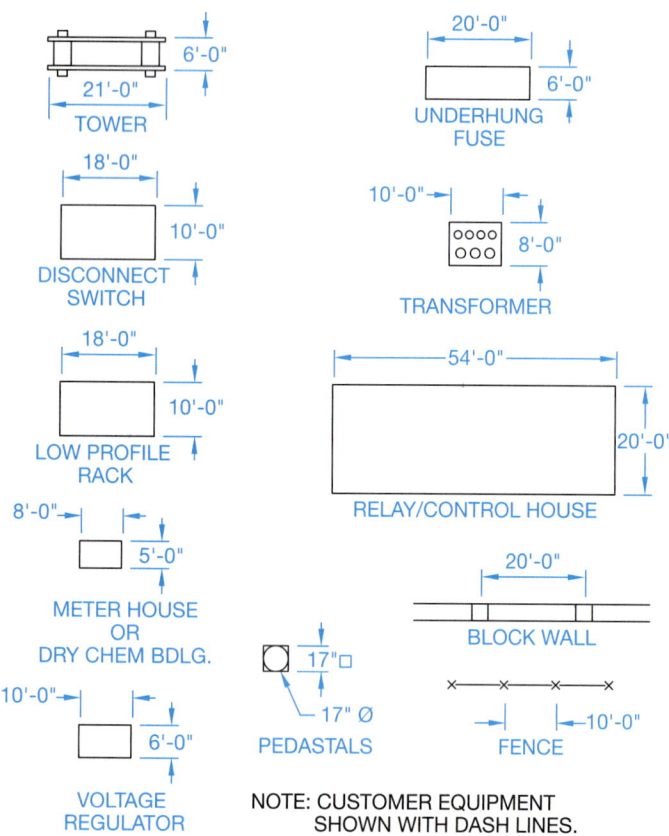

FIGURE 26.15 ■ Plot-plan symbols.

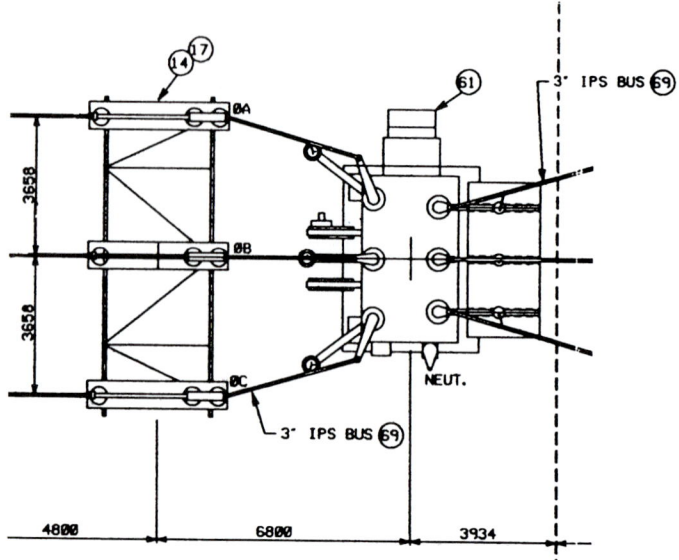

FIGURE 26.16 ■ CADD-generated bus layout plan. *Courtesy Bonneville Power Administration.*

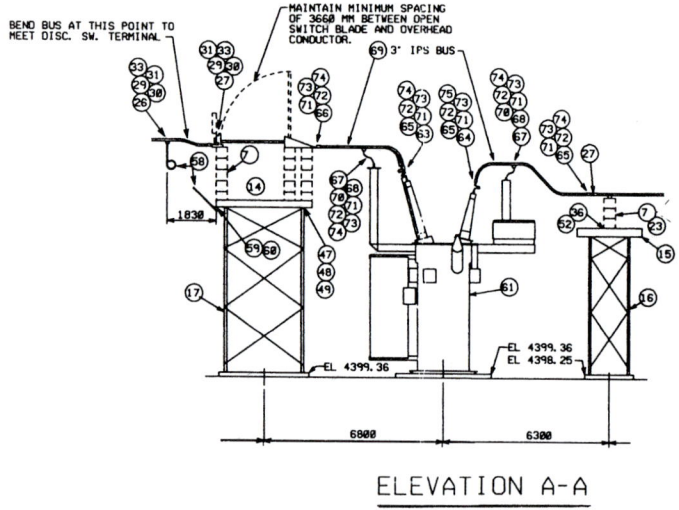

FIGURE 26.17 ■ CADD-generated bus elevations. *Courtesy Bonneville Power Administration.*

of the substation. Bus layout drawings show the system in plan (top) view, as shown in Figure 26.16, and elevation (side) view as shown in Figure 26.17.

The components are identified with numbered balloons, which key each item to a bill of materials. Some common bus layout symbols are shown in Figure 26.18.

Grounding Layout and Details

There is a tremendous hazard in substations due to the possibility of high voltage occurring on pieces of equipment during fault conditions. The voltage at different pieces of equipment varies with the location in the yard and the fault current available. The amount of potential electrical shock depends on the voltages available during a fault condition. A fault condition is a short circuit, which is a zero resistance path for current flow. The voltage difference between the metal parts of the equipment and the ground surface on which a person stands must be maintained at a very low level. This is determined by the amount and location of the ground grid, which is the grounding system. Grounding layouts are often drawn at a scale of 1" = 10'-0" (1:50 metric) as shown in Figure 26.19.

Grounding details provide an enlarged view of how a structure or piece of equipment is grounded. Details are keyed to the grounding layout with letters; for example, "DETAIL H1," as shown in Figure 26.20.

Conduit Installation Layout and Details

Each component in a substation has a specific function and is a different type of electrical equipment. The relay control house is a building that interconnects each piece of equipment using multiple conductor cables. (See Figure 26.21.)

Conduit detail drawings coordinate with the layout by providing construction details and the locations of various fittings, junction boxes, and brackets. Details correlate to the layout with callouts that give the detail identification and the page where the detail is located; for example, DETAIL N/2. These details are either drawn at a scale of 1/2" = 1'-0" (1:20 metric)

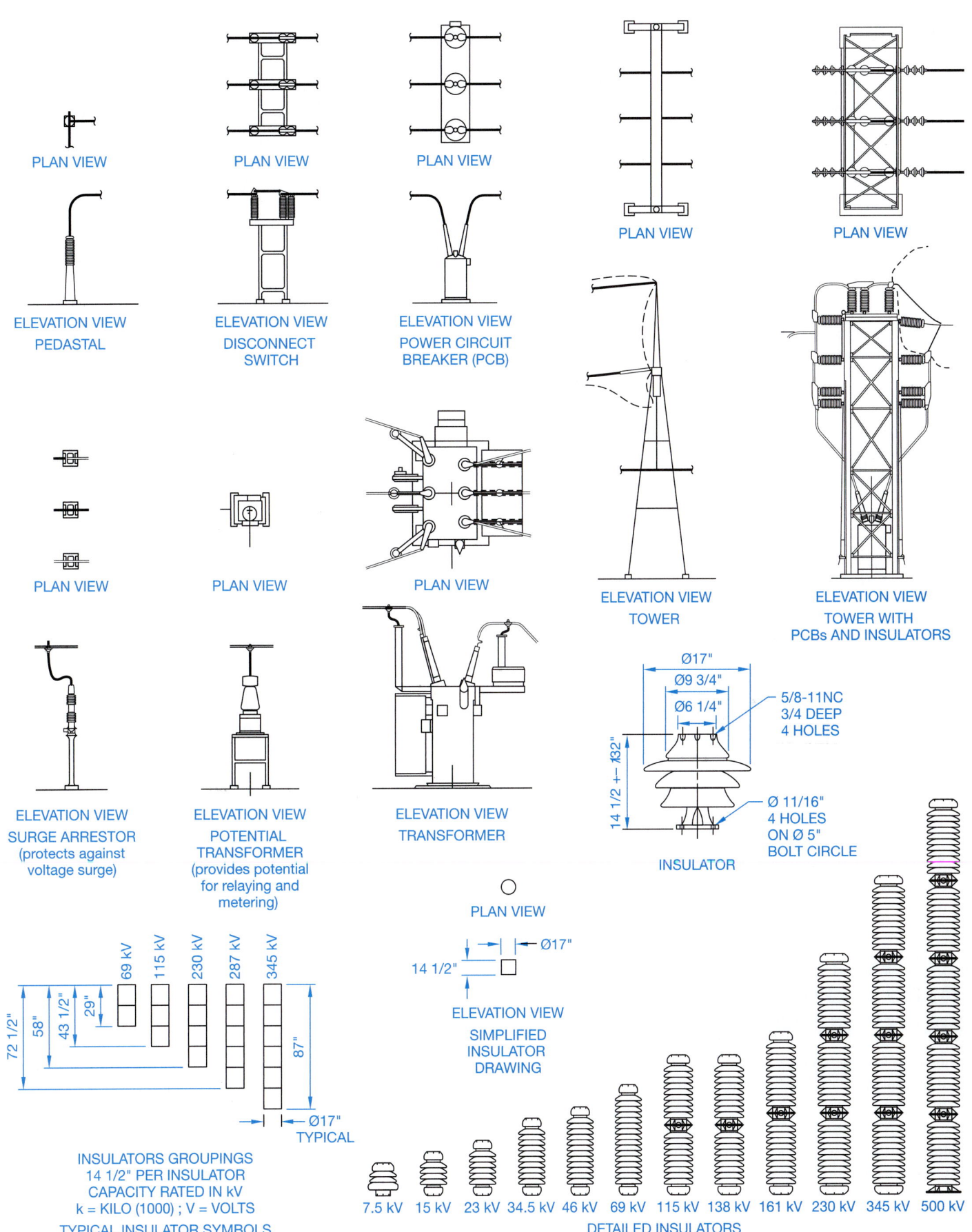

FIGURE 26.18 ■ Common bus layout symbols.

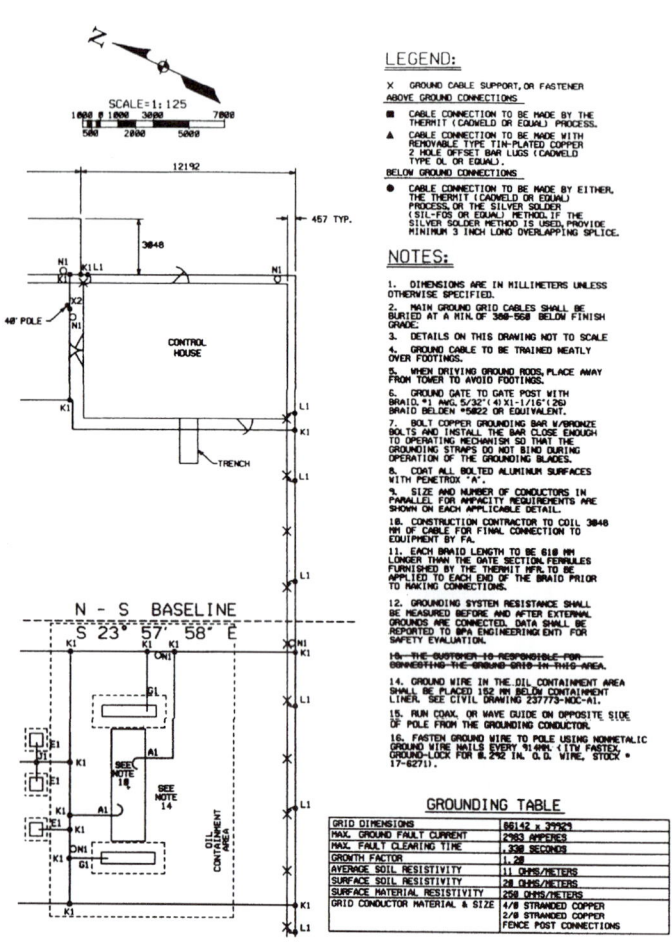

FIGURE 26.19 ■ Grounding layout. *Courtesy Bonneville Power Administration.*

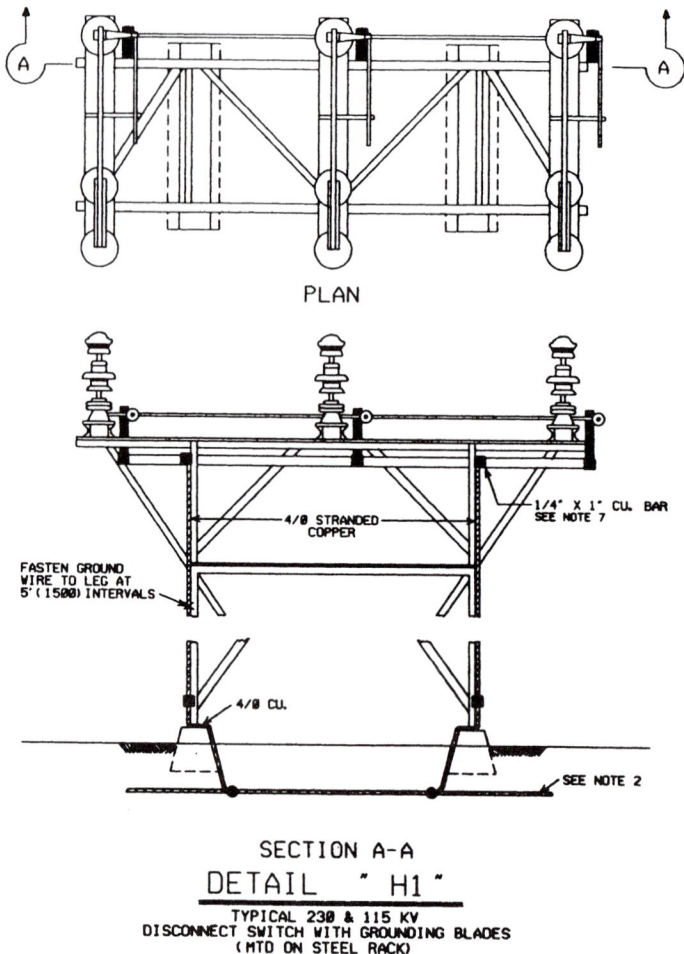

FIGURE 26.20 ■ Grounding details. *Courtesy Bonneville Power Administration.*

or can be drawn without scale. (See Figure 26.22.) Various components in the drawing are keyed by balloons to a bill of materials.

RESIDENTIAL AND COMMERCIAL ELECTRICAL PLANS

The design of the electrical system is an important part of the total livability of a home and the function and safety of a commercial or industrial facility. Careful analysis should be given to the design placement of equipment and furniture and the planned use of each room. Local and national electrical codes provide specifications for installations and layout. Layout planning should play an important role in conjunction with code guidelines. An evaluation of need and code requirements should be closely compared so the electrical layout is not over- or underdesigned.

Architectural Electrical Symbols

Symbols and lines are used to show the electrical layout in a structure. In residential applications the electrical layout is often part of the floor plan. Commercial electrical plans are commonly an overlay of the base floor-plan sheet. Electrical symbols are generally 1/8 in. (3 mm) in height, where applicable. The electrical plan should be drawn in a clear, concise manner so the layout remains uncluttered. All lettering for switches and other notes should be 1/8 in. high, although a 5/32-in. (4 mm) height is used by some companies. Common electrical symbols are shown in Figure 26.23.

Switch symbols are drawn perpendicular to the wall and are placed to read from the right side or bottom of the sheet. The switch relay dashed line should intersect the symbol at right angles to the wall or the relay may begin next to the symbol. Do not mix methods. Verify the preferred procedure with your instructor or employer. (See Figure 26.24.) Figure 26.25 shows several typical electrical installations with switches to light outlets.

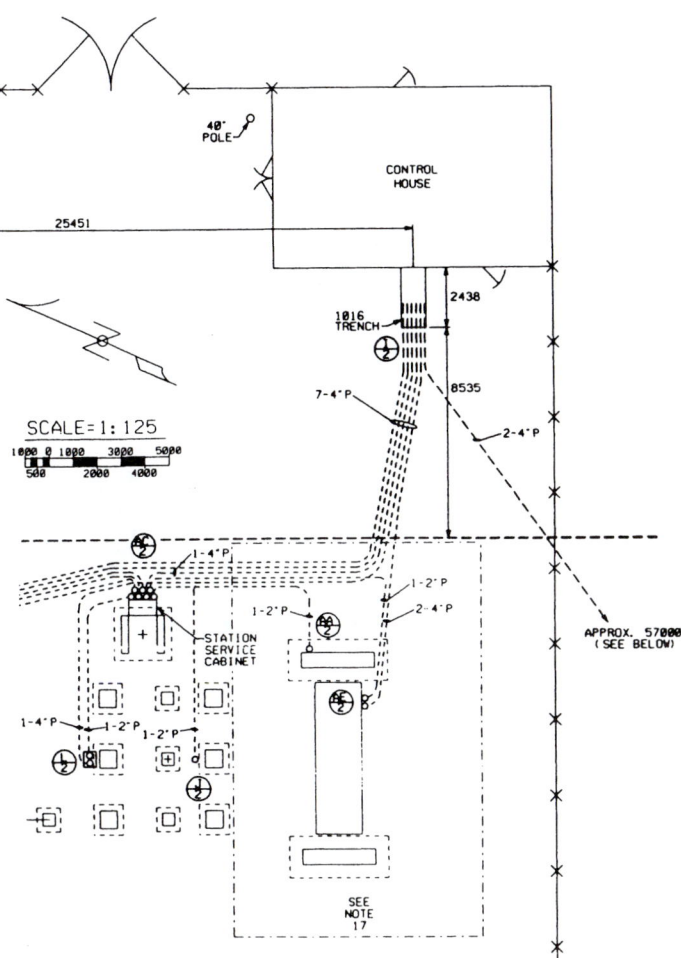

FIGURE 26.21 ■ Conduit installation layout drawing. *Courtesy Bonneville Power Administration.*

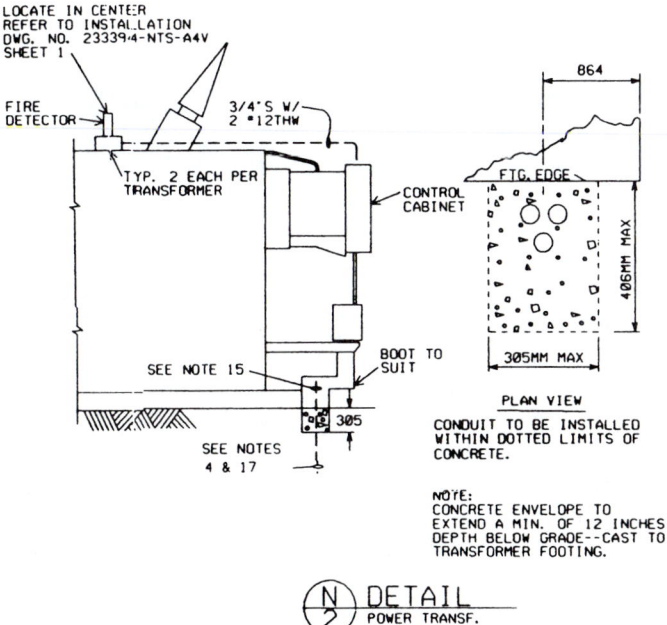

FIGURE 26.22 ■ Conduit detail. *Courtesy Bonneville Power Administration.*

When special characteristics are required, such as a specific size fixture, a location requirement, or any other specification, a local note may be applied next to the electrical symbol to briefly describe the situation. (See Figure 26.26.) Where a specification affects electrical installations on the entire layout, general notes may be used. Some common errors related to the placement and practice of electrical floor-plan layout are shown in Figure 26.27.

Residential Electrical Plan Examples

Residential electrical plans are generally less complex than commercial plans. Figure 26.28 shows some maximum spacing recommendations for installation of wall outlets in a residential structure. A typical residential electrical layout is shown in Figure 26.29, page 926. A typical kitchen layout is shown in Figure 26.30, page 926. Refer to *Architectural Drafting and Design* by Jefferis and Madsen, Delmar Publishers, for complete architectural applications.

Commercial Electrical Plan Examples

Commercial electrical plans often follow much more detailed installation guidelines than residential applications. Electrical plan overlays are commonly used so the only information provided is that of electrical installations.

The electrical circuit switch legs for commercial applications are generally drawn as solid lines rather than dashed, as in residential electrical plans. The electrical circuit lines that continue from an installation to the service distribution panel are terminated next to the fixture and capped with an arrowhead, meaning that the circuit continues to the distribution panel. When multiple arrowheads are shown, this indicates the number of circuits in the electrical run. (See Figure 26.31, page 926.) For many installations where a number of circuit wires are used, the number of wires is indicated by slash marks placed in the circuit run. The number of slash marks equals the number of wires. (See Figure 26.32, page 926.)

There can be more than one commercial electrical overlay, for example, floor-plan lighting, electrical-plan power supplies, or reflected ceiling plan. The floor-plan lighting layout provides the location and identification of lighting fixtures and circuits. It is usually coordinated with a lighting fixture schedule as shown in Figure 26.33, page 926. In some applications a power supply plan is used to show all electrical outlets, junction boxes, and related circuits. (See Figure 26.34, page 926.) The reflected ceiling plan is used to show the layout for the suspended ceiling system as shown in Figure 26.35 (see page 927). Electrical plans for equipment installations may also be needed to supplement the power supply and lighting plans. Figure 26.36 (see page 927) shows the plan for roof installation of equipment. You are also often required to draw schematic diagrams for specific electrical installations as shown in Figure 26.37 (see page 927).

(Continued on page 927)

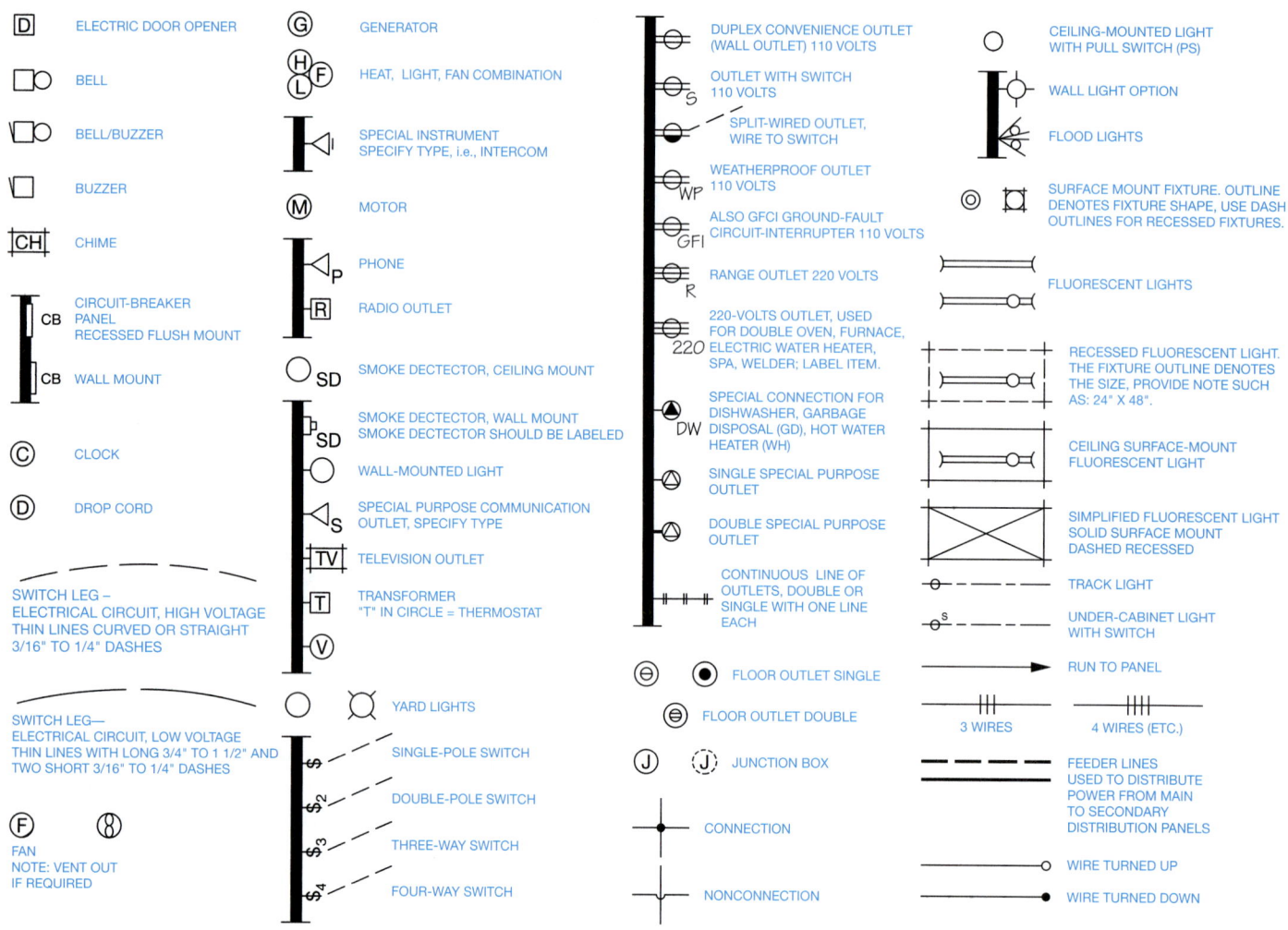

FIGURE 26.23 ■ Common floor-plan electrical symbols.

PROFESSIONAL PERSPECTIVE

The responsibilities of the electrical drafter are varied depending on the electrical needs of the client or employer. This chapter has touched on the basics of electrical drafting as related to electrical power systems, and residential and commercial electrical applications. Engineering drafting related to these fields requires practical experience associated with the specific engineering needs of the company.

In addition to a thorough understanding of company requirements, you must know electrical terminology and know how the components go together. It is important for an engineering drafter in this business to visualize the systems and understand electrical clearance requirements. Schooling in descriptive geometry, communication skills, and problem-solving techniques, along with electrical theory and practices, is essential.

When you go to work in this field, you will find that it is extremely important for you to check your work completely, making sure that electrical connections made are necessary. A *schematic* method of checking your own work is recommended. This method involves running a print of your drawing and then marking the print with a colored pencil to compare it to the original engineer's sketch or job assignment. You will find that drawings develop from one to the next. Beginning with the single-line diagram, each stage of project development becomes more specific. The checking process continues as an evolutionary process. With each stage, checking and accuracy become more critical. There are many tools available in modern electronics design analysis products to help eliminate circuit errors and to ensure correct design.

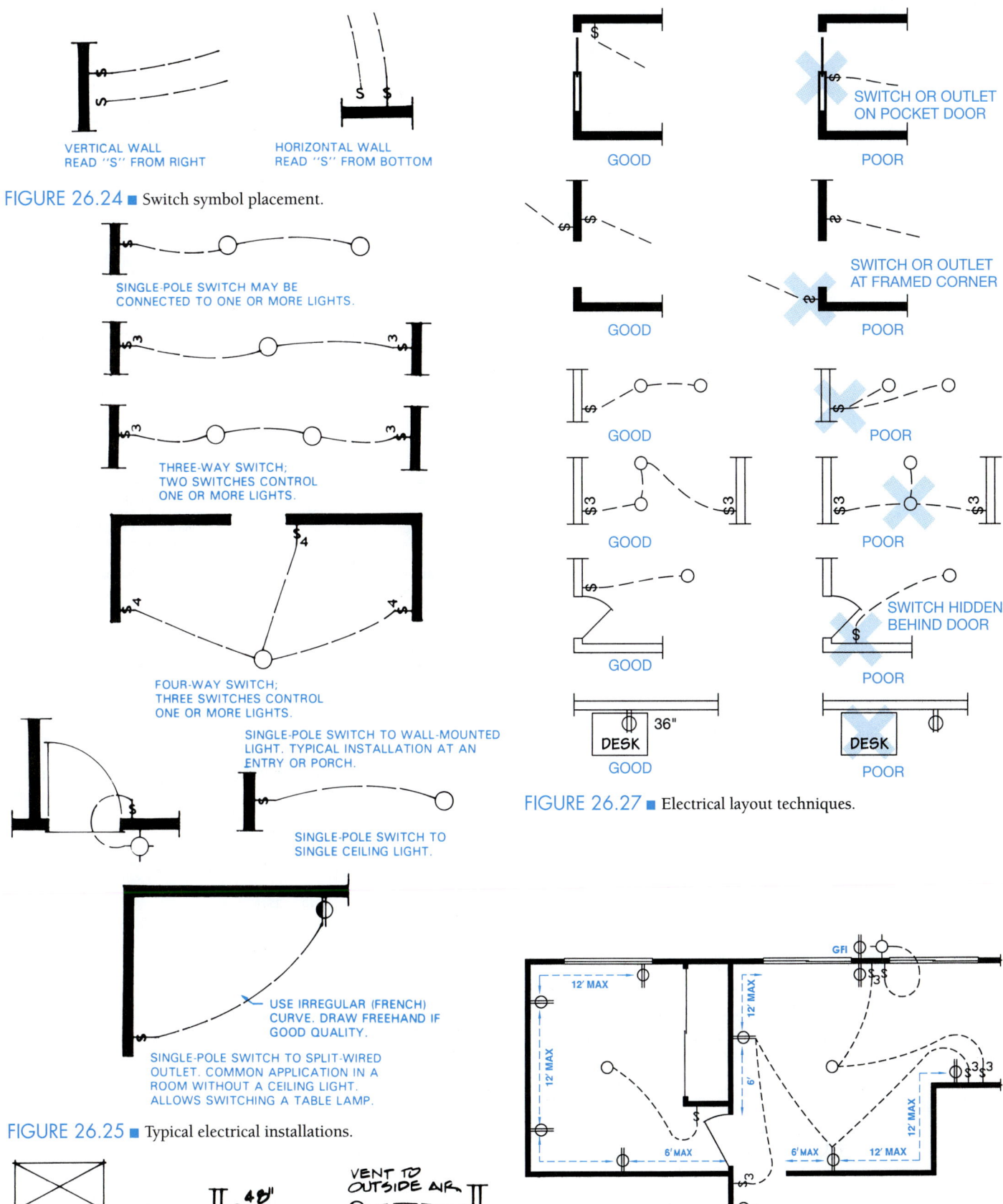

VERTICAL WALL
READ "S" FROM RIGHT

HORIZONTAL WALL
READ "S" FROM BOTTOM

FIGURE 26.24 ■ Switch symbol placement.

SINGLE-POLE SWITCH MAY BE
CONNECTED TO ONE OR MORE LIGHTS.

THREE-WAY SWITCH;
TWO SWITCHES CONTROL
ONE OR MORE LIGHTS.

FOUR-WAY SWITCH;
THREE SWITCHES CONTROL
ONE OR MORE LIGHTS.

SINGLE-POLE SWITCH TO WALL-MOUNTED
LIGHT. TYPICAL INSTALLATION AT AN
ENTRY OR PORCH.

SINGLE-POLE SWITCH TO
SINGLE CEILING LIGHT.

USE IRREGULAR (FRENCH)
CURVE. DRAW FREEHAND IF
GOOD QUALITY.

SINGLE-POLE SWITCH TO SPLIT-WIRED
OUTLET. COMMON APPLICATION IN A
ROOM WITHOUT A CEILING LIGHT.
ALLOWS SWITCHING A TABLE LAMP.

FIGURE 26.25 ■ Typical electrical installations.

24"×48"
FIXTURE SIZE

48"
OUTLET HEIGHT

VENT TO
OUTSIDE AIR
F   TIMED

FIGURE 26.26 ■ Special notes for electrical fixtures.

GOOD   SWITCH OR OUTLET
ON POCKET DOOR
POOR

GOOD   SWITCH OR OUTLET
AT FRAMED CORNER
POOR

GOOD   POOR

GOOD   POOR

GOOD   SWITCH HIDDEN
BEHIND DOOR
POOR

36"
DESK
GOOD

DESK
POOR

FIGURE 26.27 ■ Electrical layout techniques.

GFI

12' MAX
12' MAX
12' MAX
12' MAX
6'
6' MAX   6' MAX   12' MAX

SHORT WALL OVER 24" LONG
REQUIRES ONE OUTLET.

FIGURE 26.28 ■ Maximum spacing requirements.

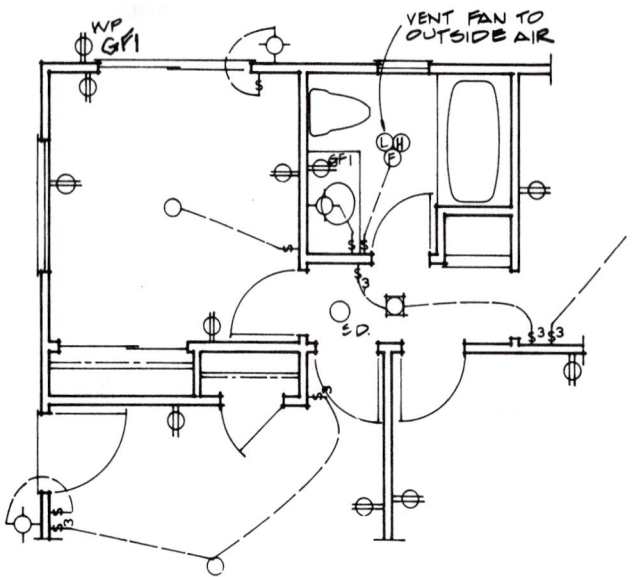

FIGURE 26.29 ■ Typical electrical layout.

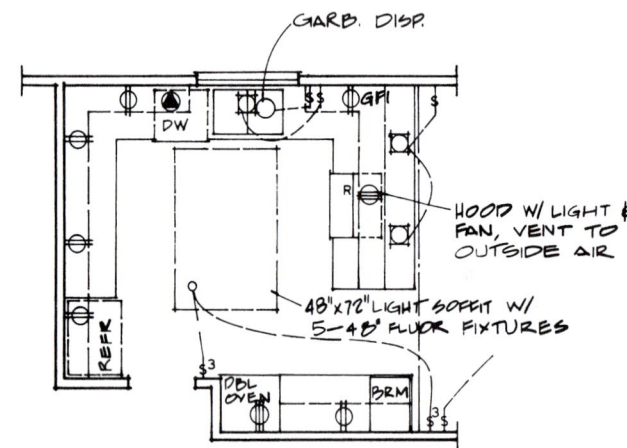

FIGURE 26.30 ■ Kitchen electrical layout.

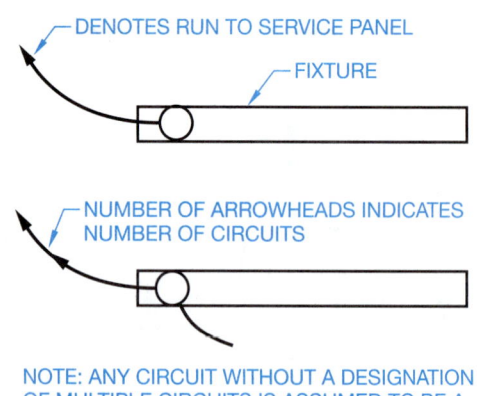

NOTE: ANY CIRCUIT WITHOUT A DESIGNATION OF MULTIPLE CIRCUITS IS ASSUMED TO BE A TWO-WIRE CIRCUIT.

FIGURE 26.31 ■ Electrical circuit designations.

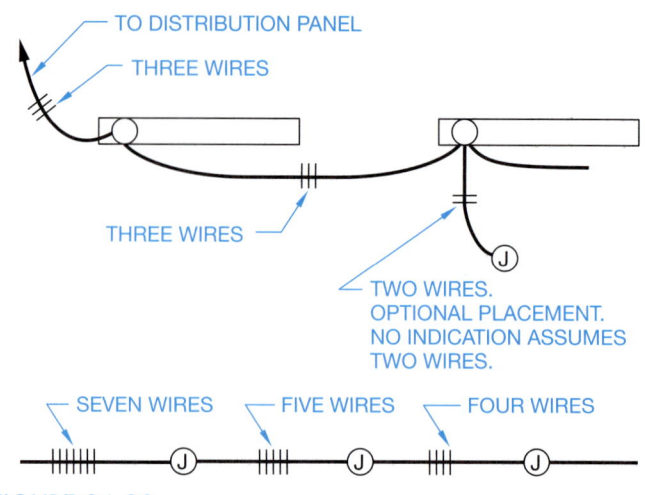

FIGURE 26.32 ■ Number of wires designated in circuit runs.

LIGHTING FIXTURE SCHEDULE

F1 – Surface mounted 8' open strip fluorescent. Lamps: (1) F96T12/LW/WM (75 watt). Manufacturer: Lithonia PUN 196 – 120V

F2 – Surface mounted 8' open strip fluorescent with damp location label and low temperature ballast. Lamps: (1) F96T12/LW/WM (75 watt). Manufacturer: Lithonia PUN 196 – DL – 120V

F3 – Surface mounted 4' open strip fluorescent. Lamps: (2) F48T12/LW/WM (30 watt). Manufacturer: Lithonia PUN 248 – 120V

F4 – Surface ceiling mounted vapor-tight incandescent with cast guard. Lamp: (1) 100W A19 Manufacturer: Steneo QVCXL—11GC

F5 – Surface mounted incandescent with prismatic lexan cylinder and dump location label. Lamp: (1) 100W A19 Manufacturer: Marco QB5NP – SA

F6 – Surface well mounted sodium vapor security flood light. Lexan lens and weather tight. Lamp: 70w Manufacturer: Crousshinds Sc – 711 – 70W HPS

F7 – Recessed ceiling mounted incandescent fan/light combination. Lamp: 10w Manufacturer: Broan Q678

F8 – 16' pole mounted sodium vapor flood area luminaire. Type III distribution flat lens. Bronze finish. Pole to be 16' straight square steel. Coated with paint to match fixture. See detail. Lamp: LU150 – 55 Manufacturer: ELSCO ZCHL – 150 – MPS – 16 – DP – 120 – B2A Alternate: Nu-Art QULT – III – MPS – 150

FIGURE 26.33 ■ Lighting needs are shown using an overlay of the floor plan and a lighting fixture schedule. *Courtesy The Southland Corporation.*

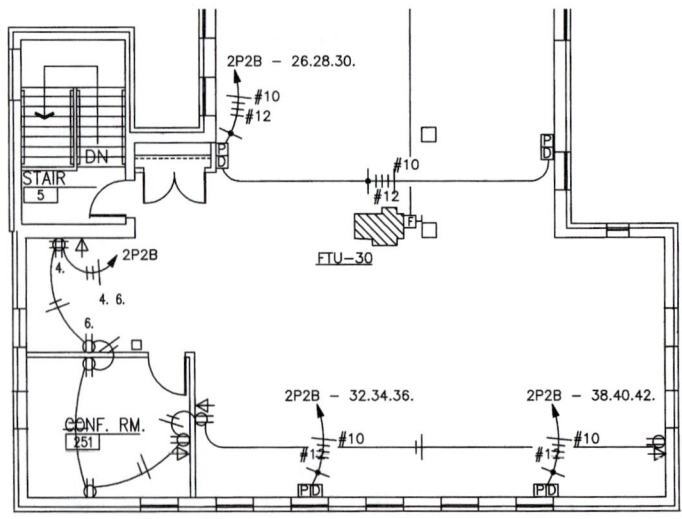

FIGURE 26.34 ■ Power supply plan. *Courtesy System Design Consultants.*

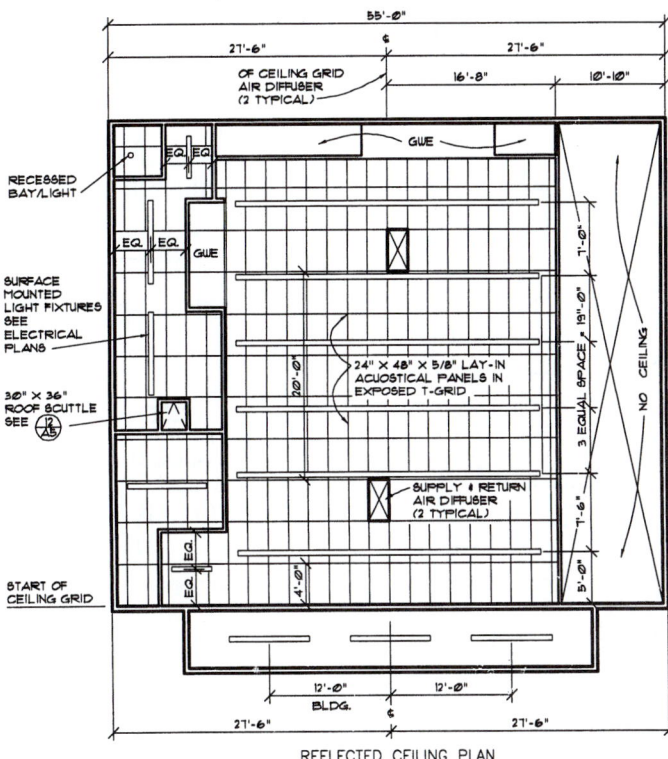

FIGURE 26.35 ■ A reflected ceiling plan to show the suspended ceiling layout. *Courtesy The Southland Corporation.*

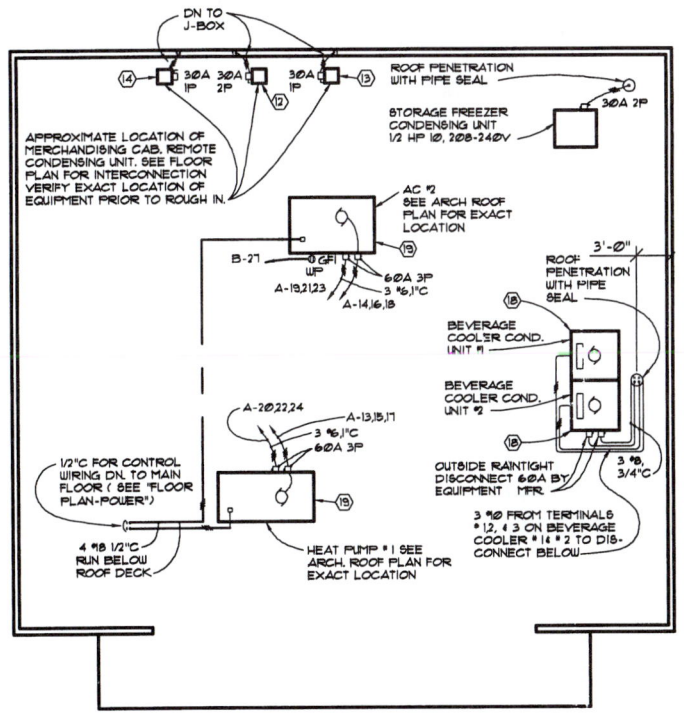

FIGURE 26.36 ■ To supplement the power supply and lighting plan, a plan showing the electrical needs of equipment on the roof is also drawn. *Courtesy The Southland Corporation.*

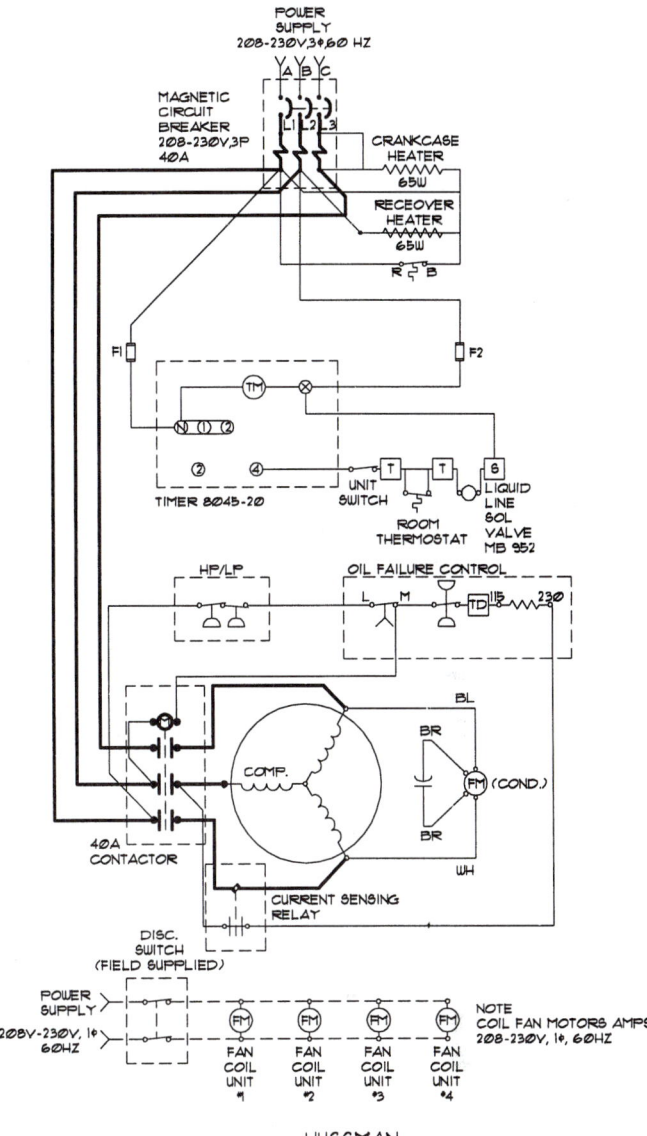

FIGURE 26.37 ■ Schematic wiring diagrams for specific applications. *Courtesy The Southland Corporation.*

ELECTRONIC SCHEMATIC DRAFTING

Electronics is the control of electrons for use in devices that are dependent on low voltage, amperage, and signal paths. Electronic drafters are typically referred to as technicians in the electronics industry. The technician's responsibility is to convert engineering sketches or instructions to formal drawings, or to revise existing drawings. This task dictates that drawings be prepared in a neat, organized manner using proper symbols. It is advisable for technicians to become familiar with the fundamentals of electronic technology, and as experience is gained in the industry, in-depth knowledge of product function is important for advancement into design and engineering.

ELECTRONIC DIAGRAMS

Many of the same types of diagrams used in electrical drafting are used in electronic drafting. There is only a slight difference between the appearance of electrical and electronic drawings. This text separates the two areas due to the specialized difference that exists between electrical power transmission systems and intricate electronic systems.

Block Diagrams

A **block diagram** outlines the path of a signal through a series of "steps or operations." The purpose of these drawings is to provide a quick interpretation of the relationship between the different components in electronic equipment. The steps or operations are shown in "blocks," which summarize a schematic drawing by omitting the details. A square or rectangular block is the usual basic symbol used in block diagrams. The size of the block is determined by the largest title going into the block, such as CONVERTER or AMPLIFIER. The block diagram should start in the upper left corner of the drawing and the components read from left to right, as shown in Figure 26.38. Graphic symbols are often used to demonstrate input and output devices, such as speakers, microphones, or antennas, as shown in Figure 26.39.

You may be given a schematic diagram where the engineer has clearly defined the components to be placed in the block diagram. Designers in this field are able to determine the block diagram elements directly from a given schematic diagram. Figure 26.40 shows a block diagram prepared from a schematic diagram. In some situations, you may prepare a block diagram from a pictorial diagram, as shown in Figure 26.41. Separate or optional components can be represented with a dashed line.

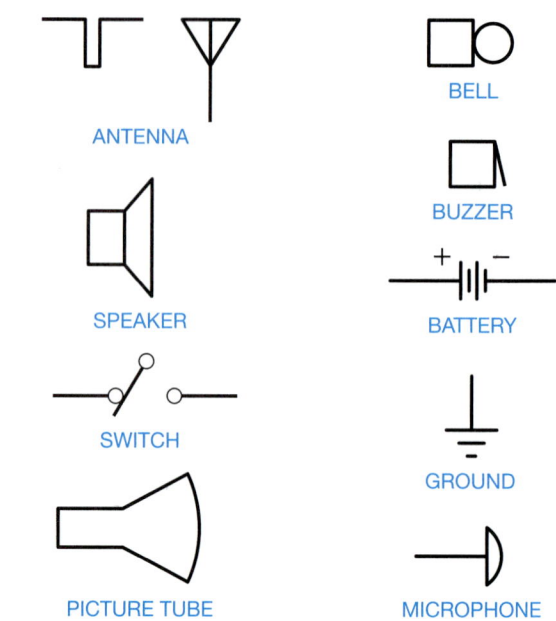

FIGURE 26.39 ■ Graphic symbols for block diagrams.

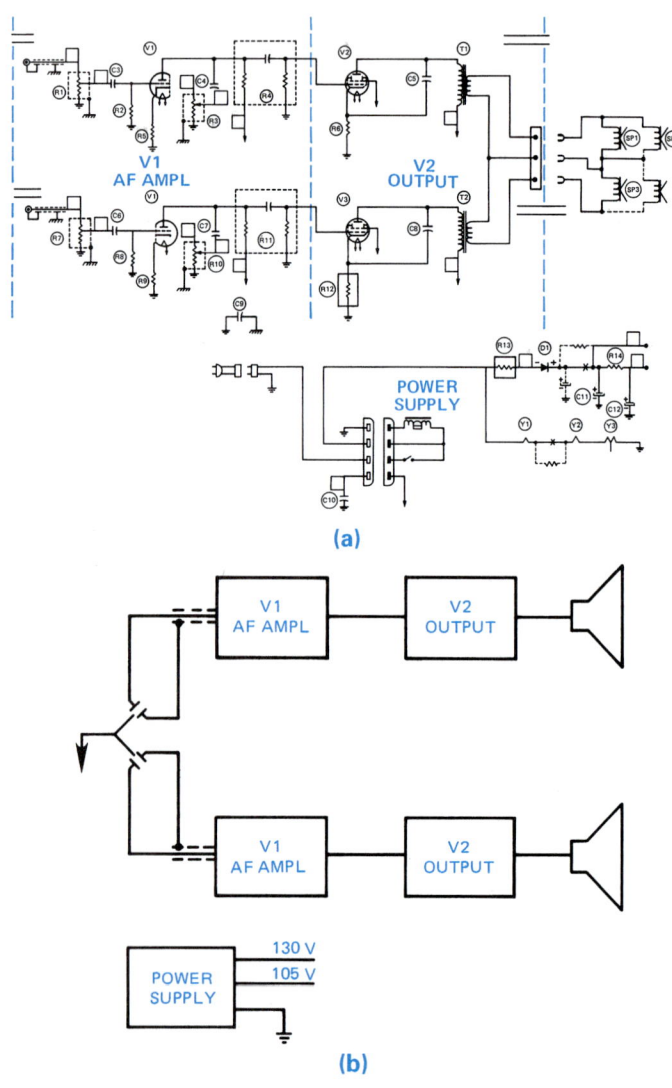

(a)

(b)

FIGURE 26.40 ■ (a) Schematic diagram is converted into (b) a block diagram.

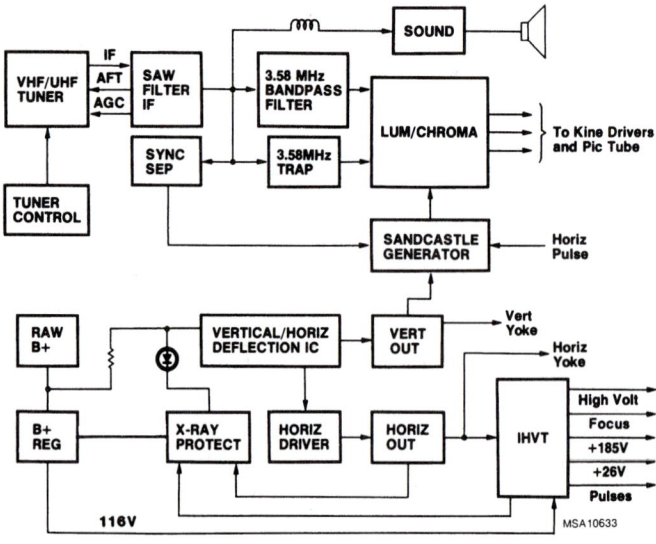

FIGURE 26.38 ■ Block diagram. *Reproduced by permission of RCA Consumer Electronics.*

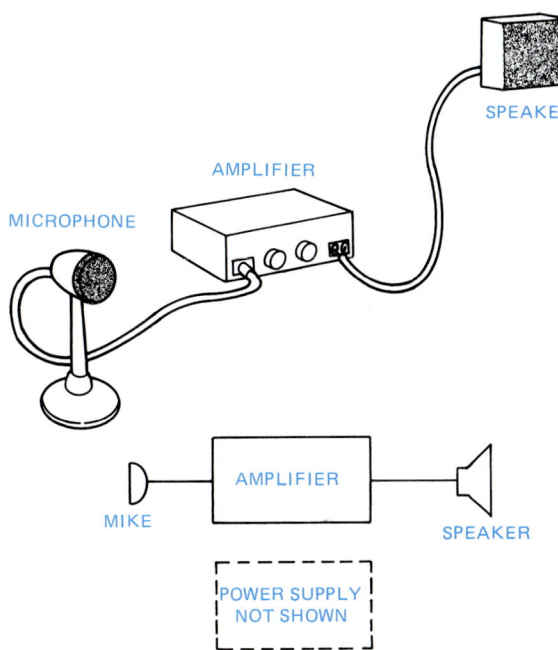

FIGURE 26.41 ■ Block diagram from a given pictorial drawing and optional components shown with dashed lines.

Schematic Diagrams

Preparation of a schematic diagram is typically the first stage of product development. Schematic diagrams provide the basic circuit connection information for electronic products. Components of the electronic system are often drawn in stages that represent the function of the device. Individual components are labeled with "reference designations," values, and supplier identification. Reference designations tie the component directly with the schematic drawing. A reference designation is a letter (or letters) identifying a component, as follows:

Component	Reference Designation
Capacitor	C
Inductor	L
Resistor	R
Diode	D
Transistor	Q
Transformer	T
Switch	S
Semiconductor	CR
Integrated circuit	U

You should know the product well enough to prepare schematics that are uncluttered and easy to read, follow standards, fit the size and shape of the product, and are technically correct. The following discussion provides enough electronic information to help you prepare a basic schematic.

Basic Electronic Symbols

The symbol and a brief definition of the following components are provided only for general information. A full understanding of function requires technical training.

Capacitor—C. A simple capacitor consists of two metal plates with wire connectors separated by an insulator. A capacitor in an electronic circuit opposes a change in voltage and stores electronic charge. (See Figure 26.42.)

Coil or Inductor—L. A coil or inductor is a conductor wound on a form or in a spiral that has inductive properties. Inductance is the property in an electronic circuit that opposes a change in current flow or where energy may be stored in a magnetic field, as in a transformer. (See Figure 26.43.)

Resistor—R. Resistors are components that resist the flow of electricity. They are used in applications where the circuit requires protection by reducing the flow of current. Resistors are available in fixed or variable values. Variable resistors allow the user to adjust the resistance. An example of a variable resistor is the volume control of a radio. The symbol is shown in Figure 26.44.

Semiconductor—Q, D, or CR. A simplified definition of a semiconductor is a device that provides a degree of resistance in an electronic circuit. There are various kinds of semiconductors used in electronics. Specific types of semiconductors are diodes, transistors, and others that are discussed later. Under certain conditions, these devices allow the current to pass through them freely, and under other conditions, they block the flow of current. The simplest of these devices is the diode. (See Figure 26.45.) The symbols picture the basic characteristics of the device. The arrow points in the direction in which conventional current flows. The arrow and bar are present in some form in almost all semiconductor symbols. The bar end of the diode is referred to as the cathode or negative side. The direction the symbol arrow points is important to the drafter and may be the only difference in the symbol for two different components. Figure 26.45 is not a complete

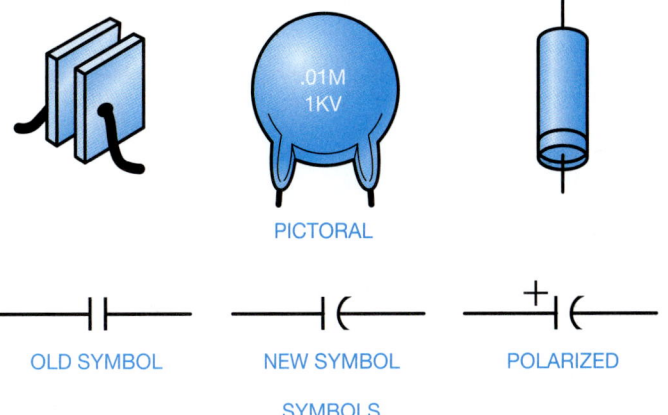

PICTORAL

OLD SYMBOL   NEW SYMBOL   POLARIZED

SYMBOLS

FIGURE 26.42 ■ Capacitor symbols.

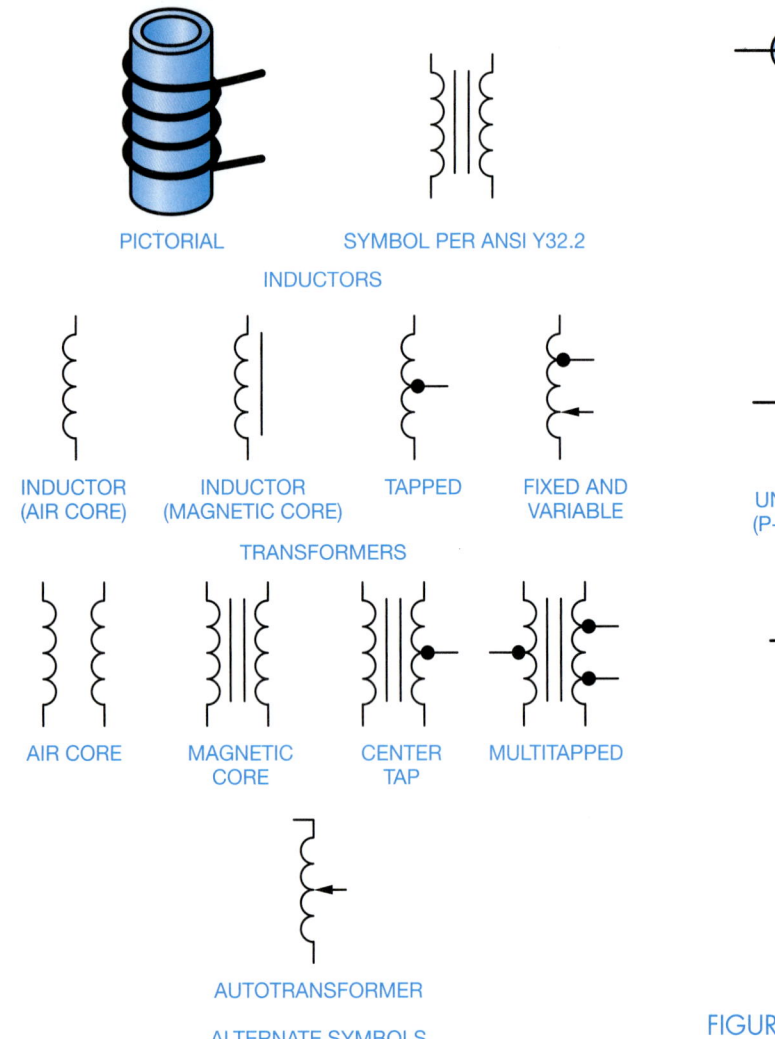

FIGURE 26.43 ■ Coil (inductor) symbols.

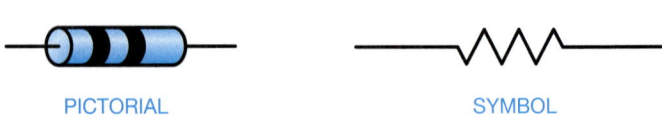

FIGURE 26.44 ■ Resistor symbol.

illustration of semiconductor devices, but it does show how slight changes are used to indicate different components.

Symbol Variations

An arrow through or to a symbol changes the component from a fixed-value device to a variable-value device. (See Figure 26.46.) A screwdriver symbol added to a schematic tells the technician that the component has a variable value by service adjustment, as shown in Figure 26.47.

Another group of symbol variations exists in coils, as shown in Figure 26.48.

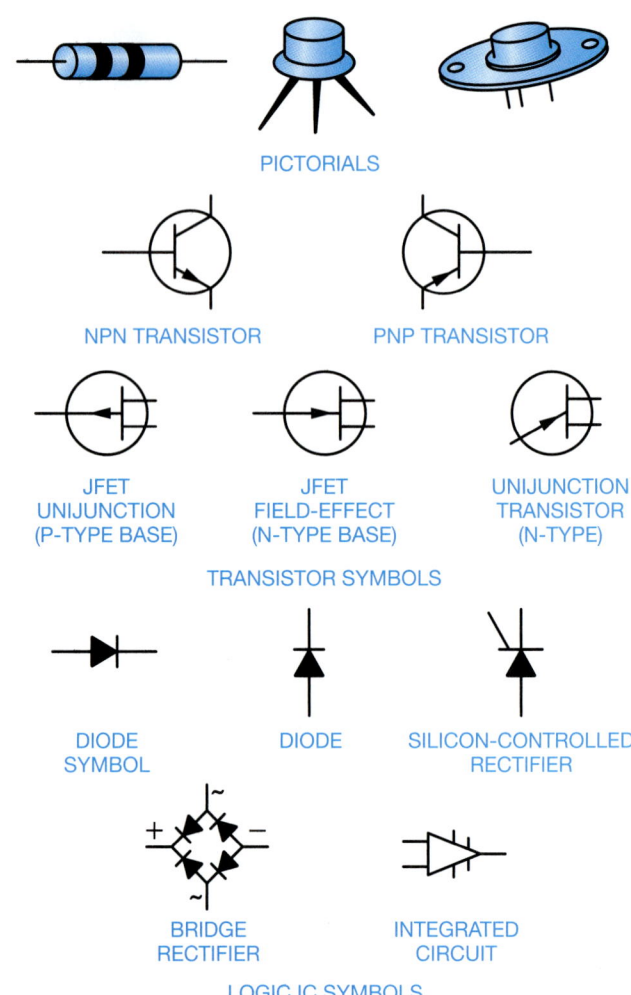

FIGURE 26.45 ■ Semiconductor symbols.

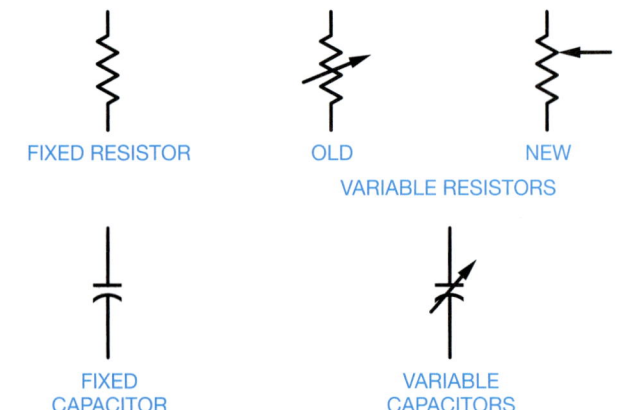

FIGURE 26.46 ■ Fixed and variable component symbols.

Symbol Sizes

The actual size of symbols can vary slightly from one company or computer-aided design and drafting software to another. The uniformity of symbols within a drawing is important. Figure 26.49 shows the recommended sizes of commonly used electronics symbols based on the drawing text height.

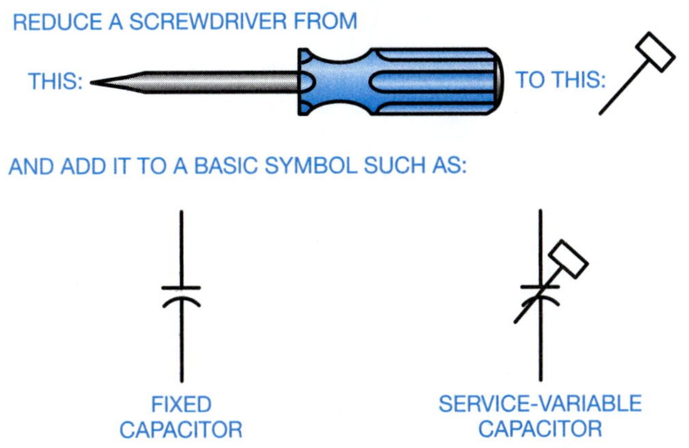

REDUCE A SCREWDRIVER FROM
THIS:
TO THIS:
AND ADD IT TO A BASIC SYMBOL SUCH AS:
FIXED CAPACITOR
SERVICE-VARIABLE CAPACITOR

FIGURE 26.47 ■ Fixed and service-variable component symbols.

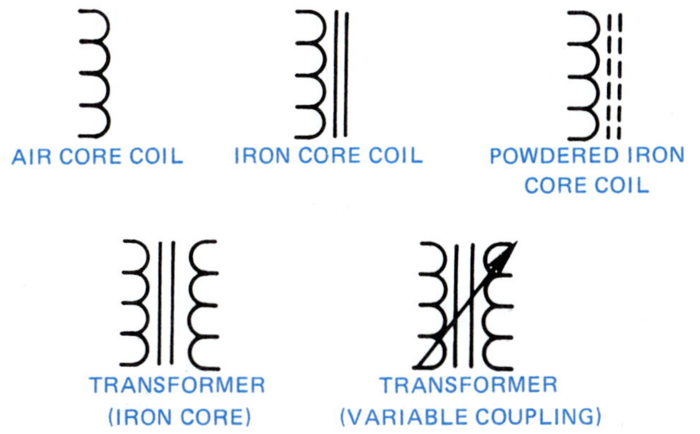

AIR CORE COIL
IRON CORE COIL
POWDERED IRON CORE COIL
TRANSFORMER (IRON CORE)
TRANSFORMER (VARIABLE COUPLING)

FIGURE 26.48 ■ Coil symbol variations.

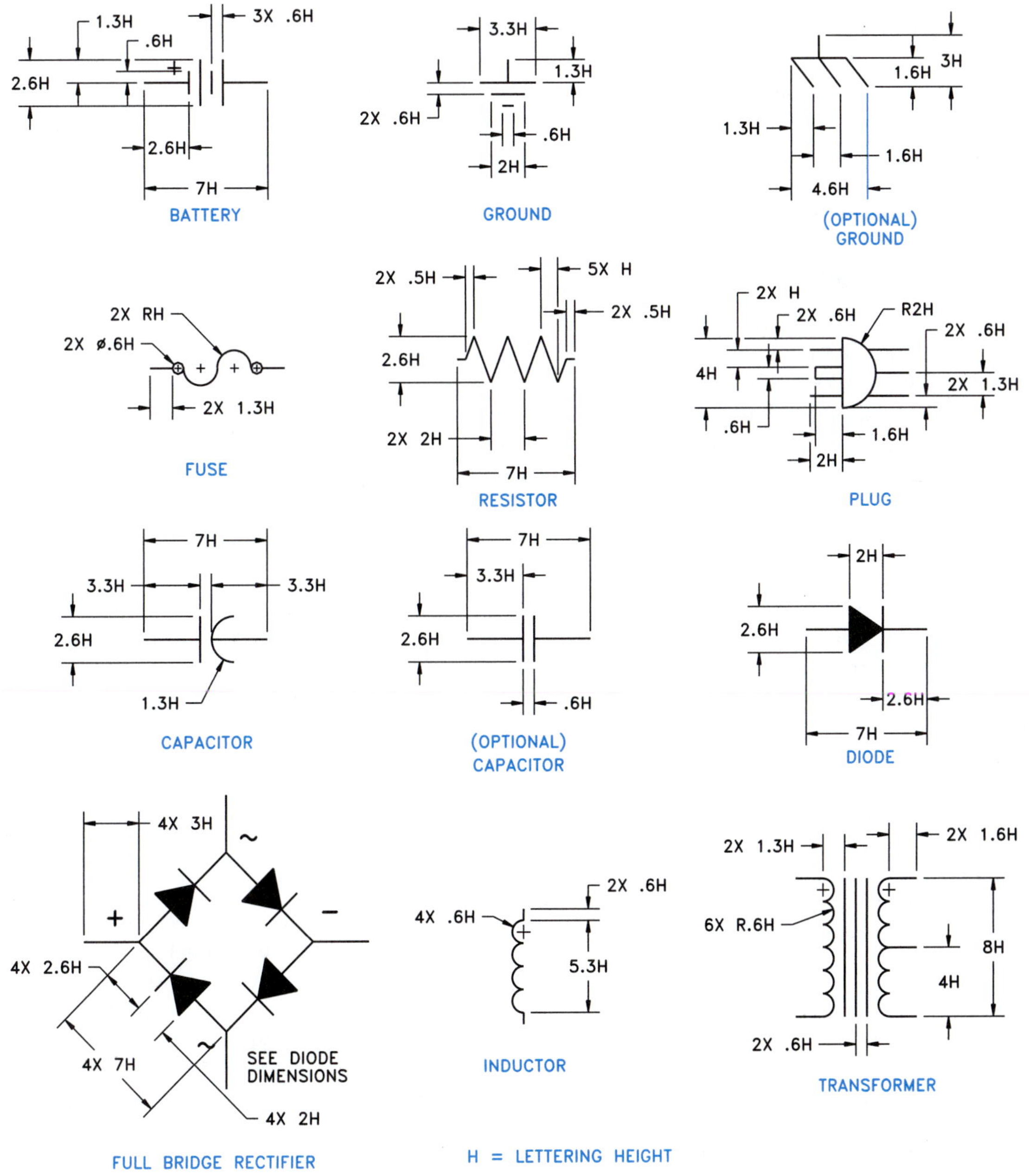

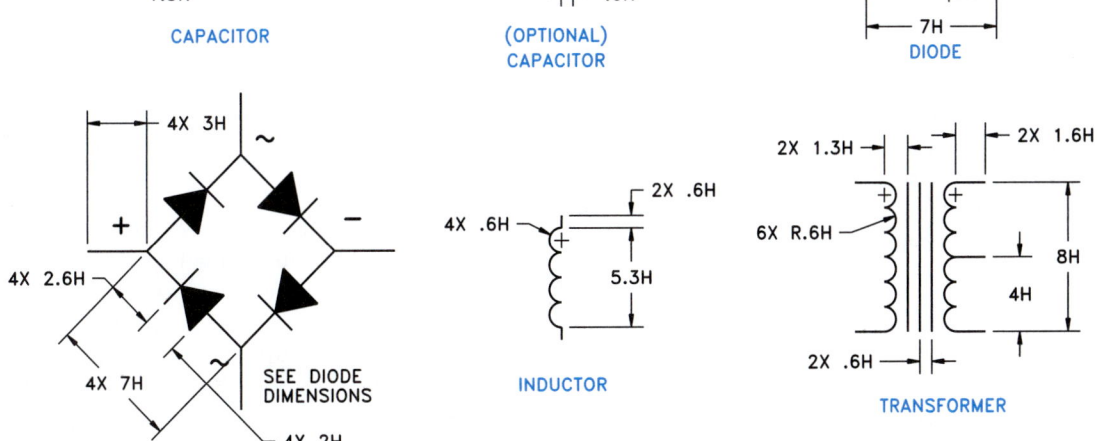

1.3H
3X .6H
.6H
2.6H
2.6H
7H
BATTERY
3.3H
1.3H
2X .6H
.6H
2H
GROUND
3H
1.6H
1.3H
1.6H
4.6H
(OPTIONAL) GROUND
2X RH
2X ø.6H
2X 1.3H
FUSE
2X .5H
5X H
2X .5H
2.6H
2X 2H
7H
RESISTOR
2X H
2X .6H
R2H
2X .6H
4H
2X 1.3H
.6H
1.6H
2H
PLUG
7H
3.3H
3.3H
2.6H
1.3H
CAPACITOR
7H
3.3H
2.6H
.6H
(OPTIONAL) CAPACITOR
2H
2.6H
2.6H
7H
DIODE
4X 3H
4X 2.6H
4X 7H
SEE DIODE DIMENSIONS
4X 2H
FULL BRIDGE RECTIFIER
2X .6H
4X .6H
5.3H
INDUCTOR
2X 1.3H
2X 1.6H
6X R.6H
8H
4H
2X .6H
TRANSFORMER
H = LETTERING HEIGHT

FIGURE 26.49 ■ Recommended sizes of commonly used electronics symbols based on drawing text height.

As an electronic drafter you may be working from rough engineering sketches.

Here are some sample potential communication problems:

■ Know current standards so if the symbol —⊣⊢ is on a sketch, you know it is an old symbol of an electrical application for a capacitor.

■ Question symbols with odd details such as the one in Figure 26.50. Does the line across the bottom of the resistor mean something special, or was the engineer's pencil resting on the paper? In this case it could mean either, but be careful, as it could symbolize a variable resistor with a stop on the adjustment that will not allow it to go all the way to zero. If in doubt, ask the engineer.

Correct and Incorrect Applications for Schematic Diagrams

There are a variety of electronic drafting techniques that are more acceptable than others. While there are some methods that the beginning drafter should learn, there may be some situations when common applications may change due to the given situation or a difference between national and company standards. The examples shown in Figure 26.51 are some examples of recommended correct and incorrect applications.

Active Devices

An active device is an electronic component that contains voltage or current sources, such as a transistor and integrated circuit. Transistors are semiconductor devices in that they are conductors of electricity with resistance to electron flow, and are used to transfer or amplify an electronic signal. The key to the schematic layout is the location of the active devices. When schematics are designed, the active devices are often established first. Follow the engineering sketch as a guide for overall size and layout balance, and begin the schematic diagram by positioning the active devices. Inputs are usually on the left, and outputs on the right. Signal flow is across the device from left to right.

FIGURE 26.50 ■ Confusing symbol sketch.

Bias Circuitry

Bias is the voltage applied to a circuit element to control the mode of operation. Bias may be fixed, forward, or reversed. The basic circuitry that makes any electronic device function is known as the DC biasing. The configuration of this basic circuit is resistors that are connected to an active device. Resistors are components that maintain resistance to the flow of an electric current. The resistors should be placed as close as possible to the active device without crowding or otherwise destroying the balance of the drawing. Resistors are normally placed vertically in the schematic, as shown in Figure 26.52. Transistors may be drawn one of two basic ways, depending on the bias direction by changing the arrow direction, as shown in Figure 26.53. After the active devices have been established on the drawing sheet, the next step is to add the resistors. There are a few common bias arrangements for transistor circuits, as shown in Figure 26.54.

Coupling Circuitry

After the bias circuitry has been established, the next step is generally the layout of the coupling circuitry. This is the point where each schematic becomes unique. The drafter's goal is to provide adequate space for all of the information while keeping the schematic compact and readable.

Making a Schematic Layout Sketch

Now that some fundamental guidelines have been provided for the schematic diagram layout, it is time to make a layout sketch. Making this sketch on graph paper helps in estimating the space required for component labels and values. You often work from an engineer's design sketch. This sketch can be rough, but gives the layout characteristics, as shown in Figure 26.55. The engineering sketch can be difficult to read and interpret. For example, notice in Figure 26.55 the screened area where the lines cross. These crossing lines represent a question that you should answer. Is there a connection at this point or do the lines cross without making a connection? In this case, the resistors are part of the universal bias pattern as previously discussed, so you can conclude that a connection is intended. The standard representations for connections and nonconnections are shown in Figure 26.56. If there is any doubt about how to interpret the engineer's sketch, ask the engineer.

Now, make a layout sketch on graph paper, as shown in Figure 26.57. The blocked-out areas next to the symbols represent intended component labeling.

There is more than one correct way to lay out a schematic. The same schematic with an optional layout is shown in Figure 26.58. The basic difference between this layout and the previous one is that the components attached to the top side of the active devices have been "folded down," as shown in Figure 26.59. The result of this folded-down form is a long, narrow schematic. This layout can be used to an advantage where the height of the schematic is fixed but the length can vary, as in fold-out pages in instruction manuals. This form gives the advantage of providing a clear space above the active device for labeling or circuit notations.

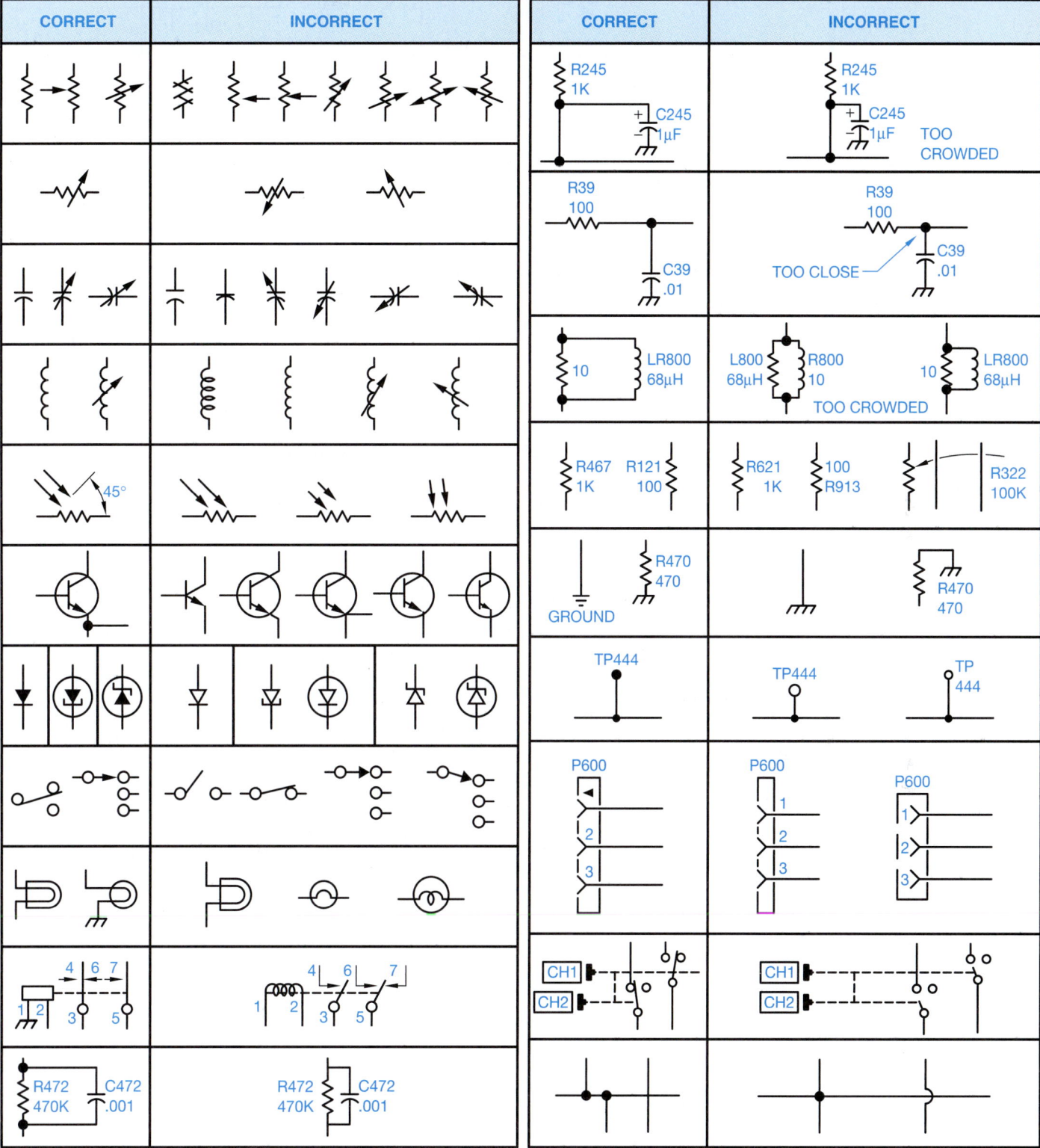

FIGURE 26.51 ■ Schematic do's and don'ts.

Labeling and Circuit Notations

There are two parts to the schematic layout—the symbols and the labeling. Without labeling, the symbols mean nothing more than the component they represent. The designating system has developed around a letter indicating the type of component

followed by a number that gives the sequence of the part in the circuit. For example, the first resistor in the upper left of the schematic is designated R1. The next resistor to the right is R2, and so on across the schematic. The same holds true for capacitors, for example, C1, C2, and C3. Some of the other common component reference designators are: Transistor = Q,

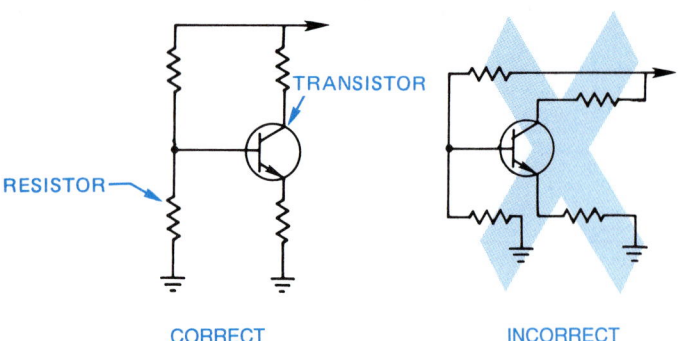

CORRECT   INCORRECT

FIGURE 26.52 ■ Resistors placed in relation to the active device.

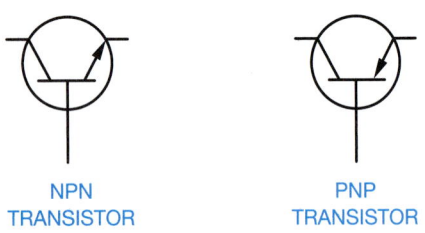

NPN
TRANSISTOR

PNP
TRANSISTOR

FIGURE 26.53 ■ Changing the bias direction of a transistor.

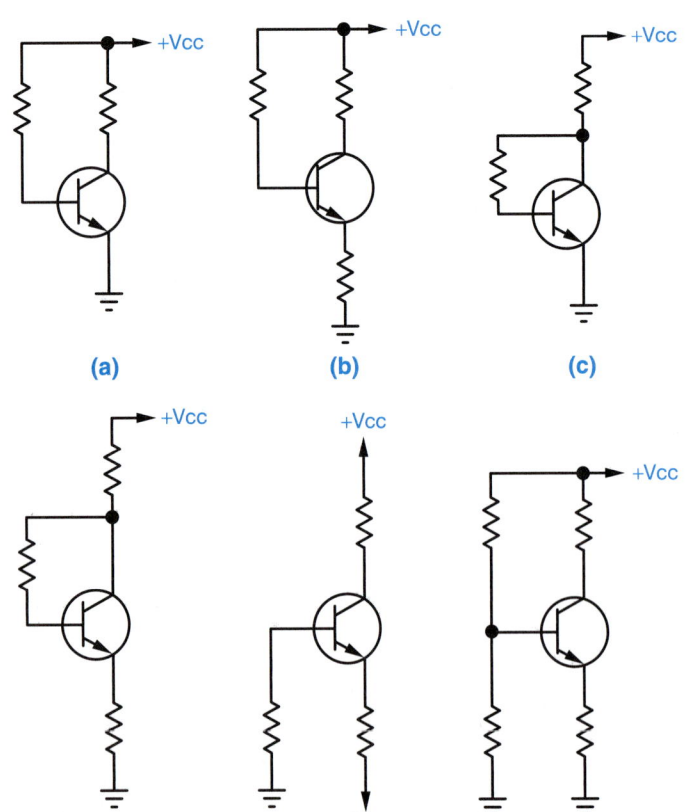

(a)   (b)   (c)

(d)   (e)   (f)

FIGURE 26.54 ■ Common transistor bias circuit arrangements.

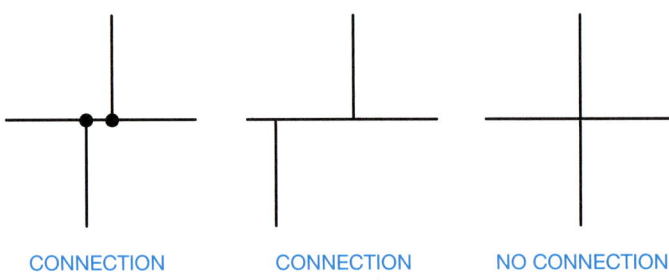

MICROPHONE PREAMPLIFIER

FIGURE 26.55 ■ Engineer's rough sketch.

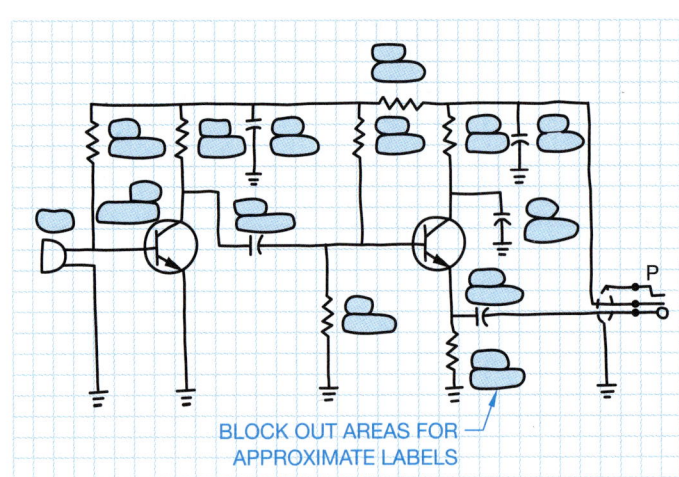

CONNECTION   CONNECTION   NO CONNECTION

FIGURE 26.56 ■ Standard connection and non-connection details.

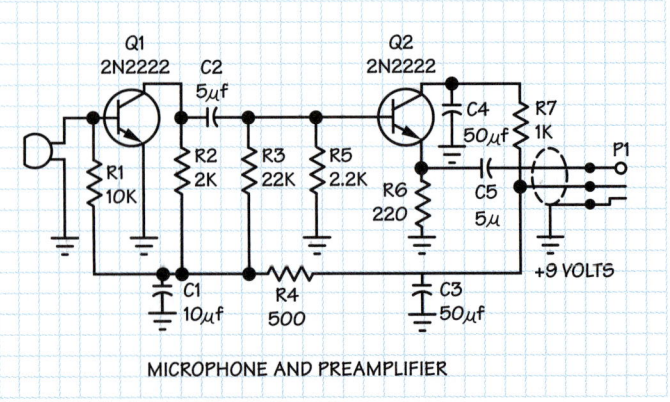

BLOCK OUT AREAS FOR
APPROXIMATE LABELS

FIGURE 26.57 ■ Schematic layout sketch.

MICROPHONE AND PREAMPLIFIER

FIGURE 26.58 ■ Optional layout sketch.

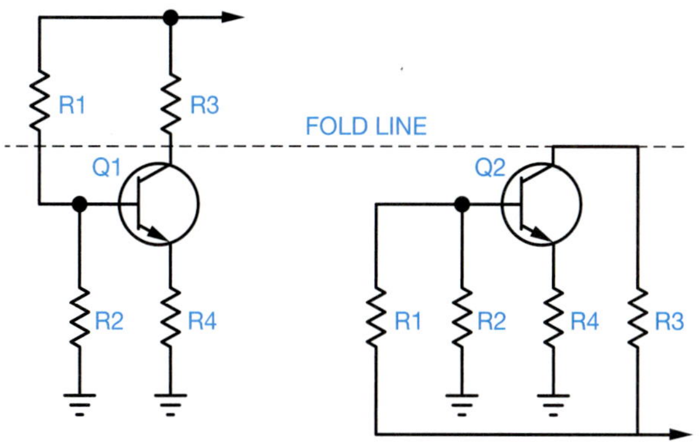

FIGURE 26.59 ■ Folding down the schematic.

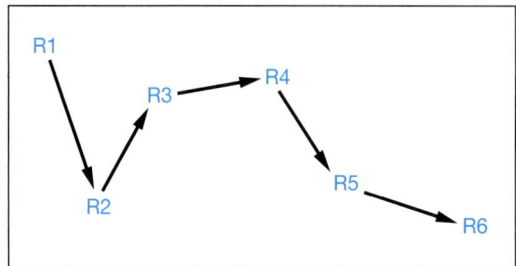

FIGURE 26.60 ■ Left-to-right component numbering sequence.

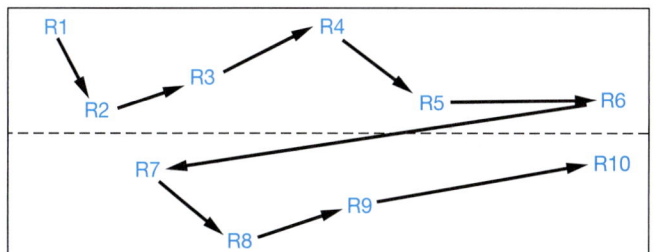

FIGURE 26.61 ■ Component numbering sequence applied in layers.

Inductor = L, Diode = D, Jack = J, and Plug = P. On simple circuits, the number sequence flows from left to right, as shown in Figure 26.60. In more complex schematics there can be two or more levels with each level flowing from left to right and top to bottom, as shown in Figure 26.61. Another method, used in units that contain subassemblies (individual groups of components that make up a complete system) or modular construction (component groups made up of equal components), is to assign the components a set of three- or four-digit numbers. For example, Figure 26.62 shows the vertical sweep module containing numbers from 100 to 199, the horizontal sweep module containing numbers from 200 to 299, the 300 series numbers ranging from 300 to 399, the 400 series numbers ranging from 400 to 499, and the 500 series numbers ranging from 500 to 599. The sequence of number series in schematics may be scrambled, established on the basis of physical layout, or arranged from left to right and top to bottom. This is also sometimes used for multiple sheet schematics. Notice the page numbers labeled in Figure 26.62. The annotation is usually

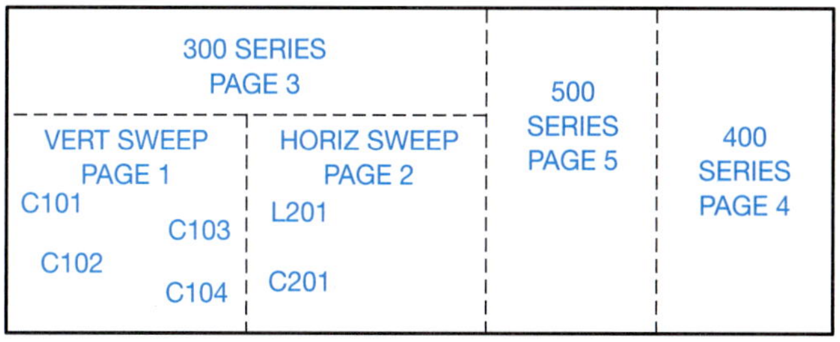

300 SERIES PAGE 3		500 SERIES PAGE 5	400 SERIES PAGE 4
VERT SWEEP PAGE 1	HORIZ SWEEP PAGE 2		
C101			
C103	L201		
C102			
C104	C201		

FIGURE 26.62 ■ Numbering components in groups.

done automatically by a computer program that generates numbers from left to right and top to bottom. The term **annotation** refers to numbers and text.

Units Used for Parts Values

Electronic component values run from extremely small values (0.000000000005) to extremely large values (2,500,000,000,000). In either case, numbers in this form take up too much space on the schematic. The numbers 1500 or 0.0003 are probably acceptable. A zero should precede the decimal point for values less than one. It has become common practice to move the decimal point in groups of three and then modify the numerical name to indicate the number of places the decimal point has been moved. For example, if you have a 2,200-ohm resistor, you can move the decimal point three places to the left, drop the zeros, and change the numerical name to **kilohms** which means 1000 ohms. The result is the designation 2.2 kilohms. The word *kilohms* may be abbreviated "k," thus reducing the notation to 2.2 k for a savings of more drafting space and time. The word *ohms* may be omitted because all resistors are rated in ohms. The following terms are used to rate or size common components:

■ Resistors = **ohms** = the unit of measurement of resistance. Symbol = Ω. Typical ranges are 0.1 ohm to 10m ohm.

■ Capacitor = **farad** = the unit of measure of capacitance. **Capacitance** is the property of an electric circuit to oppose a change in voltage. Symbol = f. Typical range is 10 pf to 10,000 µf.

■ Inductor or coil = **henry** = the unit of measurement of inductance. **Inductance** is the property of a component in an electric circuit that opposes a change in current flow. Symbol = H. Typical range is 1nH to 10H.

The following chart shows the values achieved by moving decimal points in given numbers:

Number	Move Decimal	Add Prefix	Abbreviation
1,000,000,000,000	Left 12 places	1 tera	1 T
1,000,000,000	Left 9 places	1 giga	1 G
1,000,000	Left 6 places	1 mega	1 M
1,000	Left 3 places	1 kilo	1 k
1	No change	1	1
0.001	Right 3 places	1 milli	1 m
0.000001	Right 6 places	1 micro	1 µ
0.000000001	Right 9 places	1 nano	1 n
0.000000000001	Right 12 places	1 pico	1 p

There are some gray areas in this system. For example, there are no firm rules on which conversion for 500,000 is best—0.5 M or 500 k. However, 500 k is normally preferred. The number 1200 can be left as is or reduced to 1.2 k.

Military Identification Systems

Schematic diagrams drawn to meet military specifications provide a more elaborate component identification system. For example, a conventional part identification may be R304. In the military system, this part is labeled something like 14A2A4R304. This means that component R304 is located in subassembly 4 of subassembly 2A of assembly 14A. Military specifications are available covering every element used in the design, construction, and packaging of electronic equipment.

MIL-STD Some of the Military Standard Specifications (MIL-STD) include: MIL-STD-454, Standard General Requirements for Electronic Equipment, and MIL-STD-681, Identification Coding for Hook-up Wire.

Placement of Identification Numbers and Values

The placement of identification numbers is standard. The reference designator and value numbers are typically placed above horizontal components. (See Figure 26.63.) Other arrangements could cause confusion when components are not adequately spaced as shown in Figure 26.64.

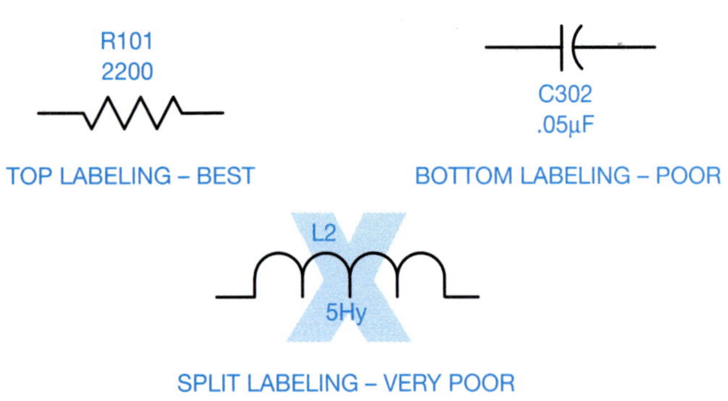

For vertically drawn components the reference designator and the value are commonly placed on the right side of the symbol. (See Figure 26.65.)

The preferred technique in component labeling is placement of part identification at the top of all horizontal symbols and to the right of all vertical symbols, as shown in Figure 26.66.

A final drawing of the schematic used in the sketching layout examples is shown in Figure 26.67.

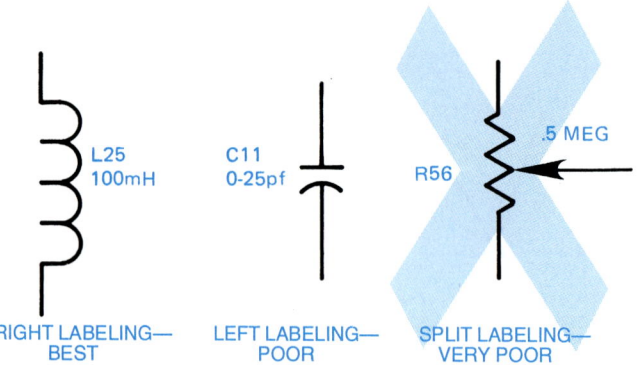

FIGURE 26.65 ■ Vertical component labeling.

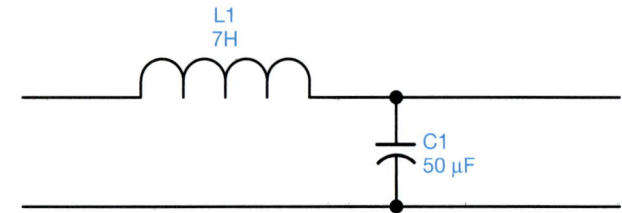

FIGURE 26.66 ■ Preferred labeling at the top of horizontal components and to the right of vertical components.

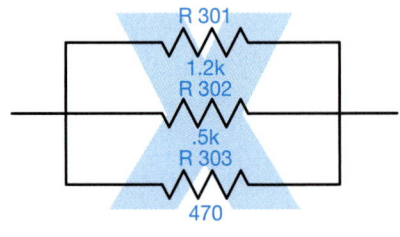

FIGURE 26.63 ■ Horizontal component labeling.

VERY CROWDED AND IMPROPERLY LABELED
(SPLIT LABELING CAUSES CONFUSION)

FIGURE 26.64 ■ An overcrowded schematic using split labeling. Avoid this practice.

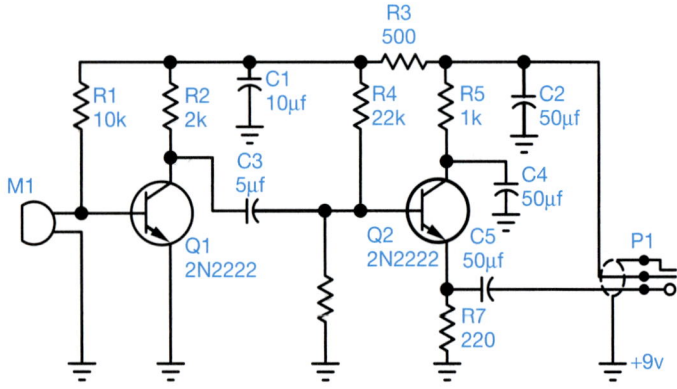

FIGURE 26.67 ■ Final drawing of schematic used in layout sketches (Figures 26.55 through 26.58).

Operational Amplifiers and Integrated Circuit Schematics

An **amplifier (AMP)** is a device that allows an input signal to control power and is capable of an output signal greater than the input signal. **An operational amplifier (OP AMP)** is a high-gain amplifier created from an integrated circuit. An **integrated circuit (IC)** is an electronic circuit that has been fabricated with extremely small components as an inseparable assembly known as a **chip**. You can become involved with integrated circuit schematics by drawing the schematic of the internal circuitry involved, or drawing the schematic for a system in which integrated circuits or operational amplifiers are used.

Internal Integrated Circuit Schematics

The internal circuitry of an integrated circuit is shown in Figure 26.68. The only exception is that in the IC diagram, the transistor symbols are not circled. The diagram is also sometimes enclosed by dashed lines in the form of a box. This box represents the package within which the IC is contained. Points are shown and labeled with **pin numbers**, which are external connectors for attachment to other circuitry. This usually means that all of the components shown in the schematic are made up of one piece of semiconductor material. Thus the definition of integrated circuit—all of the components of the circuit have been put together on one small piece of material. The schematic for an IC can be very simple or so complex that it contains several million components and hundreds of pins.

IC Symbols and Logic Circuits

In **logic circuits**, which are computer-oriented circuits, the schematics become a cross between a flow diagram and a schematic diagram. The internal components of an IC are self-contained for a specific function and may not be altered. Therefore, symbols are used that identify the IC. The shape of the

symbol tells the technician what the device is expected to do. The numbers next to the input and output lines or pins tell the reader how to connect the IC to make it work. The basic format of IC symbols is shown in Figure 26.69. The IC symbol that represents the internal IC shown in Figure 26.68 is shown in Figure 26.70. This symbol is called a NAND gate. The **gate** is the part of the system that makes the electronic circuit operate and permits an output only when a predetermined set of input conditions are met. The code NAND is the predetermined function of the gates. There are also AND, OR, and NOR functions. Simply put, the functions relate to a condition or pattern of conditions whereby an electrical pulse can be received at one pin when the input at another pin or pins is triggered at specific times.

The precise operation of each logic function is required knowledge for electronic technicians and design drafters, but premature for drafting fundamentals and beyond the scope of this study. Your concern at this time is to prepare a well-balanced, accurate drawing from engineering sketches. Notice how the IC schematic diagram shown in Figure 26.71 compares to previous schematics. The difference is in the IC symbols. While the integrated circuits may be complex in nature, the drafting task becomes easier when ICs are used. For example, the IC labeled TAA-500 in Figure 26.72 is an operational amplifier that contains 11 transistors, 5 diodes, 13 resistors, and 1 capacitor.

Logic diagrams are a type of schematic that are used to show the logical sequence of events in an electrical or electronic system. The basic electronic logic symbols and their functions are shown in Figure 26.73. Part of an electronic logic diagram is shown in Figure 26.74. These symbols are currently used but may become obsolete.

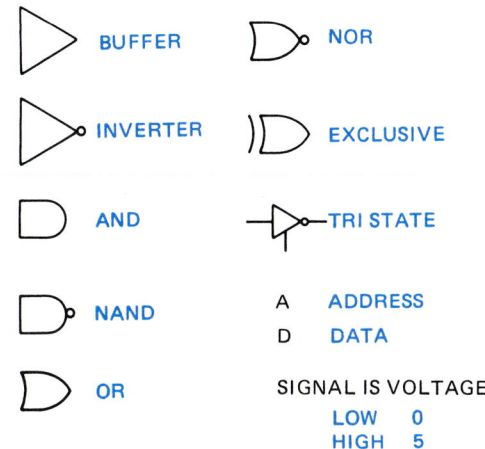

FIGURE 26.69 ■ The basic format of IC symbols.

FIGURE 26.70 ■ The IC symbol of the internal IC schematic shown in Figure 26.68.

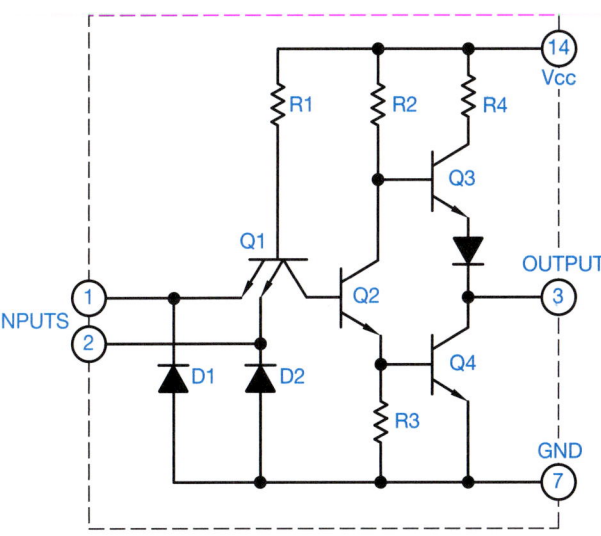

FIGURE 26.68 ■ The internal components of a logic integrated circuit.

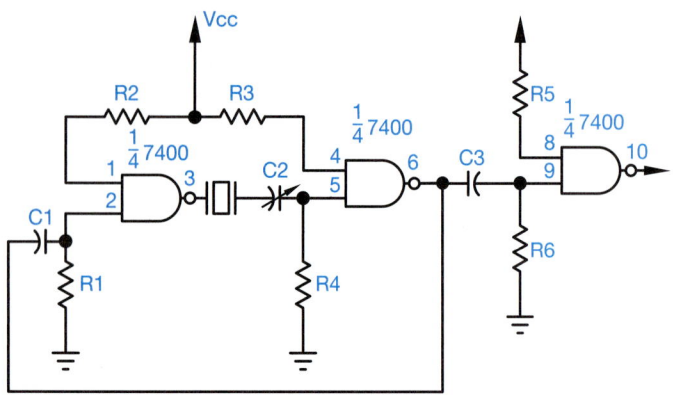

FIGURE 26.71 ■ A schematic diagram with an IC included.

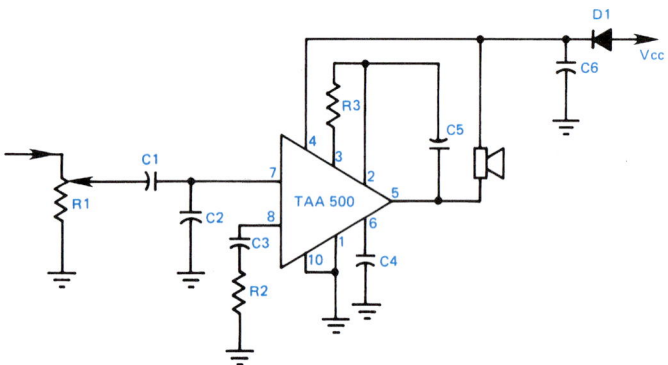

FIGURE 26.72 ■ Amplifier IC.

TITLE	SYMBOL	
AND	AND	ALL INPUTS MUST BE PRESENT FOR OUTPUT.
OR	OR	ANY INPUT WILL ALLOW OUTPUT.
NOT	NOT	AN INPUT WILL RESULT IN NO OUTPUT.
TIME DELAY	TD 0-5	AN INPUT WILL ALLOW AN OUTPUT AFTER A TIME DELAY.
ON OFF	ON OFF	INPUT "A" WILL ALLOW AN OUTPUT EXCEPT WHEN THERE IS AN INPUT AT "B."

FIGURE 26.73 ■ Basic electronic logic symbols are currently being used but are becoming obsolete.

Large-Scale Integration

The next generation beyond the specific symbols, such as the gates, came about because of the rapid advancement of the electronic industry's ability to put more and more circuits on a sin-

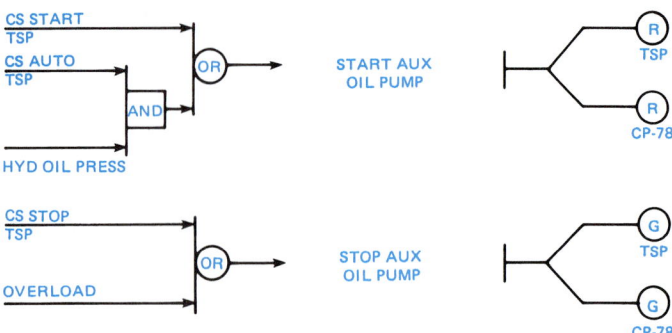

FIGURE 26.74 ■ Part of an electronic logic diagram.

gle, small IC chip. Many calculators contain only one IC with hundreds of gates. The trend in drawing schematics for this large-scale integration (LSI) is simply to show the chip as a box. This type of schematic becomes like a block diagram with the name or function identified within the rectangle for each LSI. The LSI pin terminals are shown connected to the rest of the schematic. An LSI schematic diagram is shown in Figure 26.75. Keep in mind that each LSI block in the schematic may contain hundreds or thousands of components.

Symbol Sizes

The actual size of symbols can vary slightly from one company or computer-aided design software to another. The uniformity of symbols within a drawing is important. Figure 26.76 shows the recommended sizes of commonly used electronics symbols based on the drawing text height.

PRINTED CIRCUIT TECHNOLOGY

Schematic diagrams are designed and drawn to show the location of electronic components using symbols and lines to represent circuit paths or connections. In the production of the actual electronic product, the symbols shown on the schematic become electronic devices and the lines become wires that connect these devices. At one time, wires were soldered to component terminals and used as the circuit path between components. Today, wires are used as the connection cables that provide the source of current between pieces of electronic equipment. As electronic equipment designs have become increasingly smaller, the internal connection between electronic devices must also take up less space, be easier to install, and be extremely accurate. Printed circuit technology is the answer to these needs.

Schematic and Printed Circuit Board Accuracy

Accuracy and close attention to detail in the preparation of the schematic and printed circuit board are absolutely essential. Everything depends on the accuracy of these two items. The schematic and the printed circuit board cannot have a single

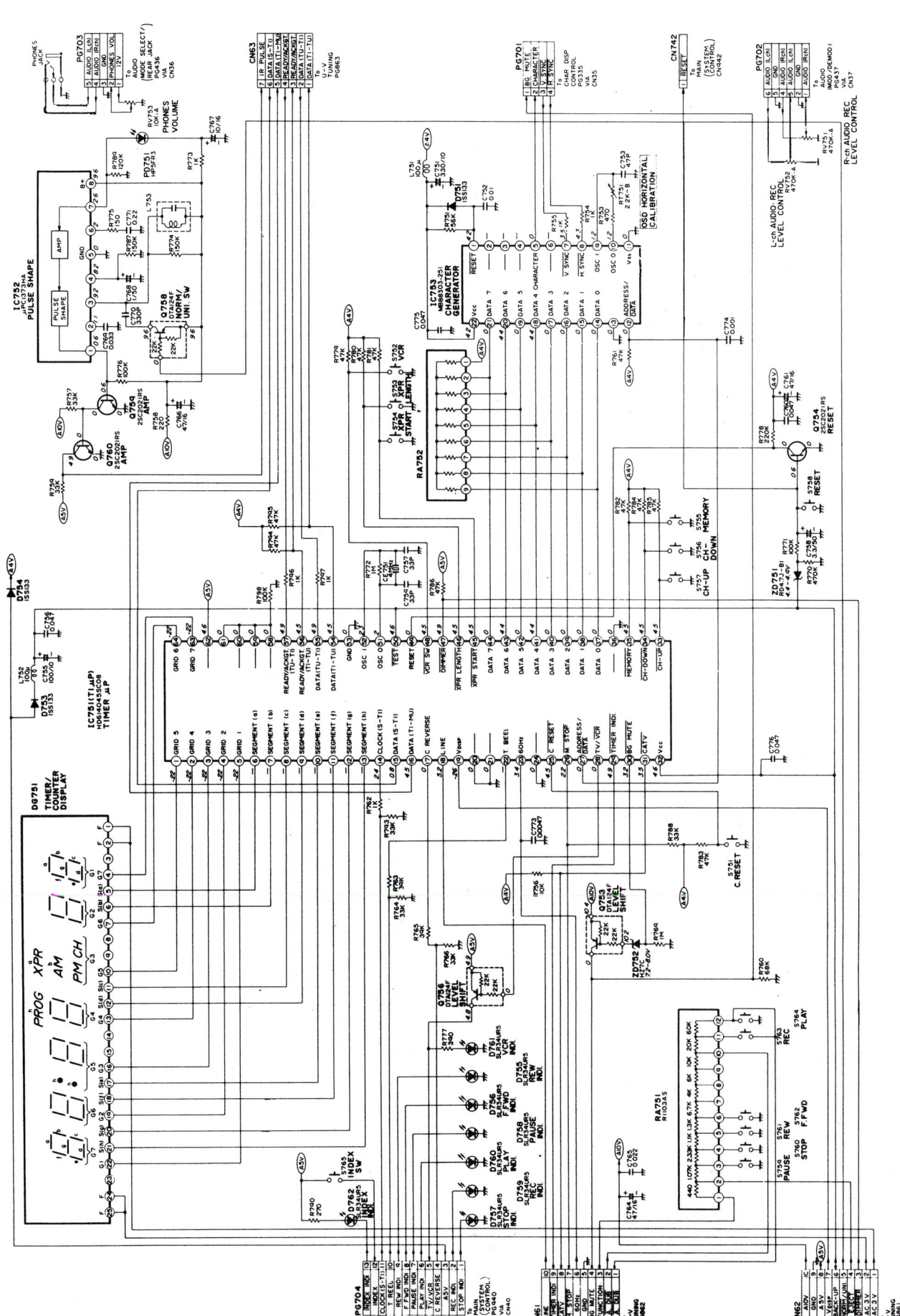

FIGURE 26.75 ■ A large-scale integration (LSI) schematic. *Reproduced by permission of RCA Consumer Electronics.*

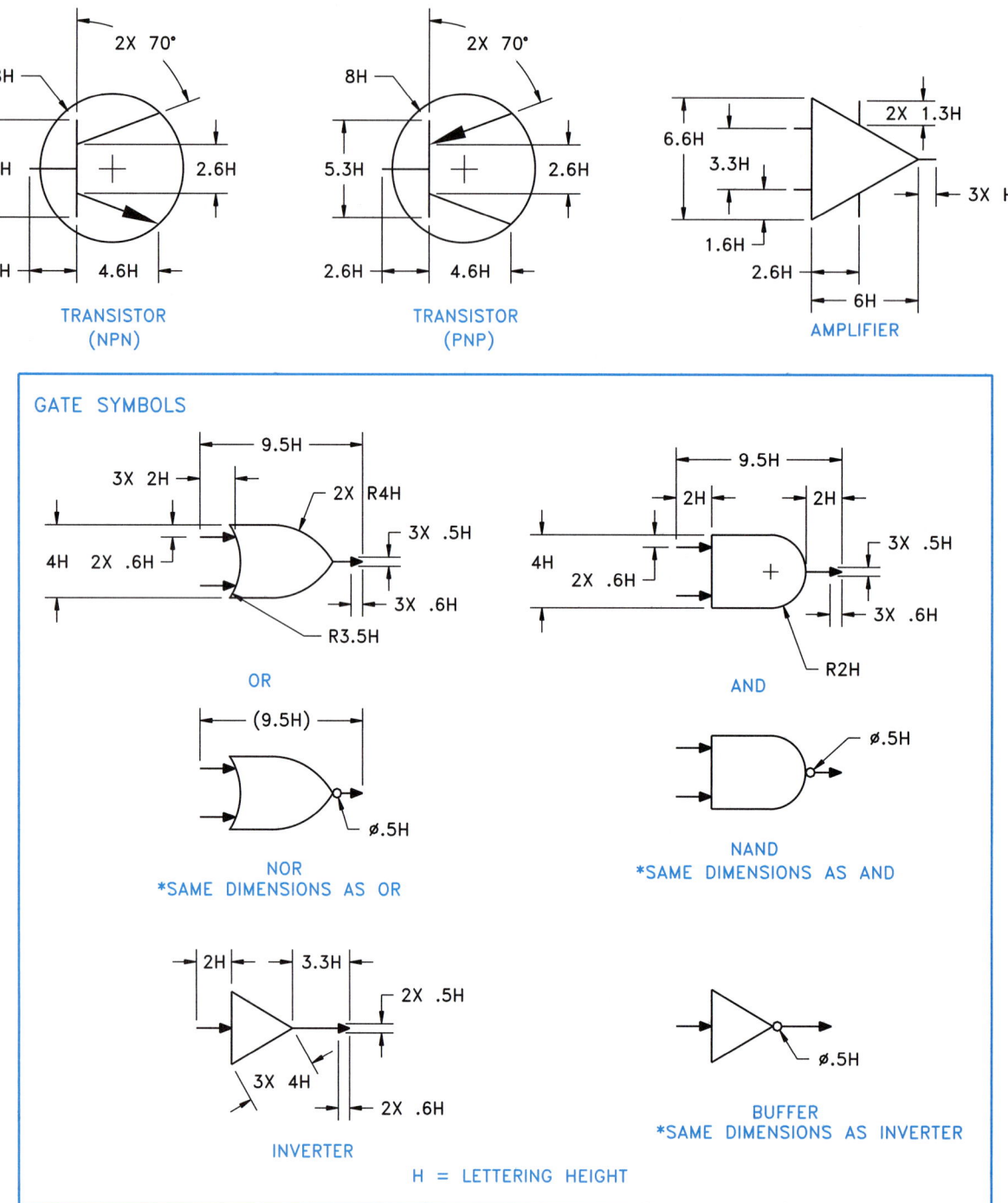

FIGURE 26.76 ■ Recommended sizes of commonly used electronics symbols based on the drawing text height.

mistake, because any mistake affects the master artwork, the drilling drawing, and the assembly and bill of materials.

Printed Circuit Design and Layout

Printed circuits (PC) form the interconnection between electronic devices. The base materials for circuit boards are special paper, plastic, glass, or Teflon. The quality of the printed circuit board (PCB) begins with the quality of the base material. De-

pending on the complexity of the electronics, printed circuits are prepared on one side or both sides of the board. In many applications, there are multiple boards or layers for one piece of electronic equipment.

The printed circuits consist of pads and conductor lines that are made of thin conductive material such as copper on a base sheet or board. The pads or lands are the circuit termination locations where the electronic devices are attached. The pad normally has a location where a hole is to be drilled in the circuit

ROUND
OR
"DONUT"
CUT
ROUND PADS
OVAL
ELLIPTICAL
HEXAGONAL
SQUARE
RECTANGULAR
TEARDROP
FILLETS
TERMINAL AREA (PAD) SHAPES
FLAT PACK
PATTERN
PLATFORM
PATTERN
IC CIRCUIT
PATTERN

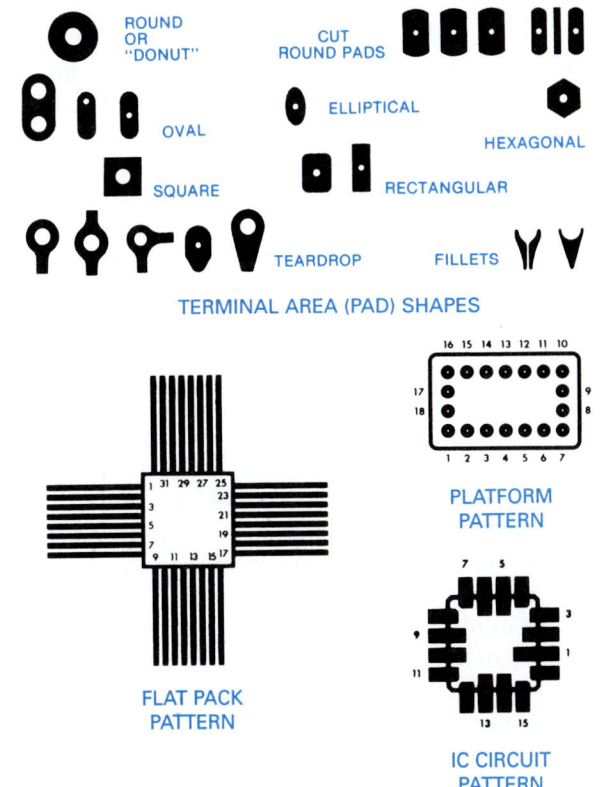

FIGURE 26.77 ■ Mounting pads. *Courtesy Bishop Graphics, Inc.*

RECOMMENDED
NOT RECOMMENDED
AVOID SHARP EXTERNAL ANGLES
WHICH CAN CAUSE FOIL DELAMINATION.
GOOD
GOOD
OK
AVOID ACUTE INTERNAL ANGLES.
ALWAYS USE THE SHORTEST
PRACTICAL CIRCUIT
ROUTING.
MAINTAIN EQUAL SPACING
WHERE CONDUCTORS PASS
BETWEEN TERMINAL AREAS.
AVOID LARGE MULTIPLE HOLE
TERMINAL AREAS, WHICH MAY
CAUSE THERMAL SOLDERING
PROBLEMS AND NON-
SYMMETRICAL SOLDER FILLETS.
MAINTAIN UNIFORM PATTERN
AROUND HOLE TO PRODUCE
SYMMETRICAL SOLDER
FILLETS.
AVOID USING CONDUCTORS
THE SAME SIZE AS TERMINALS,
WHICH WILL CAUSE SOLDER TO
FLOW AWAY FROM TERMINAL.
PLATING BAR SHOULD EXTEND
OUT FROM AND BEYOND BOARD
EDGE TO FACILITATE FABRICATION.

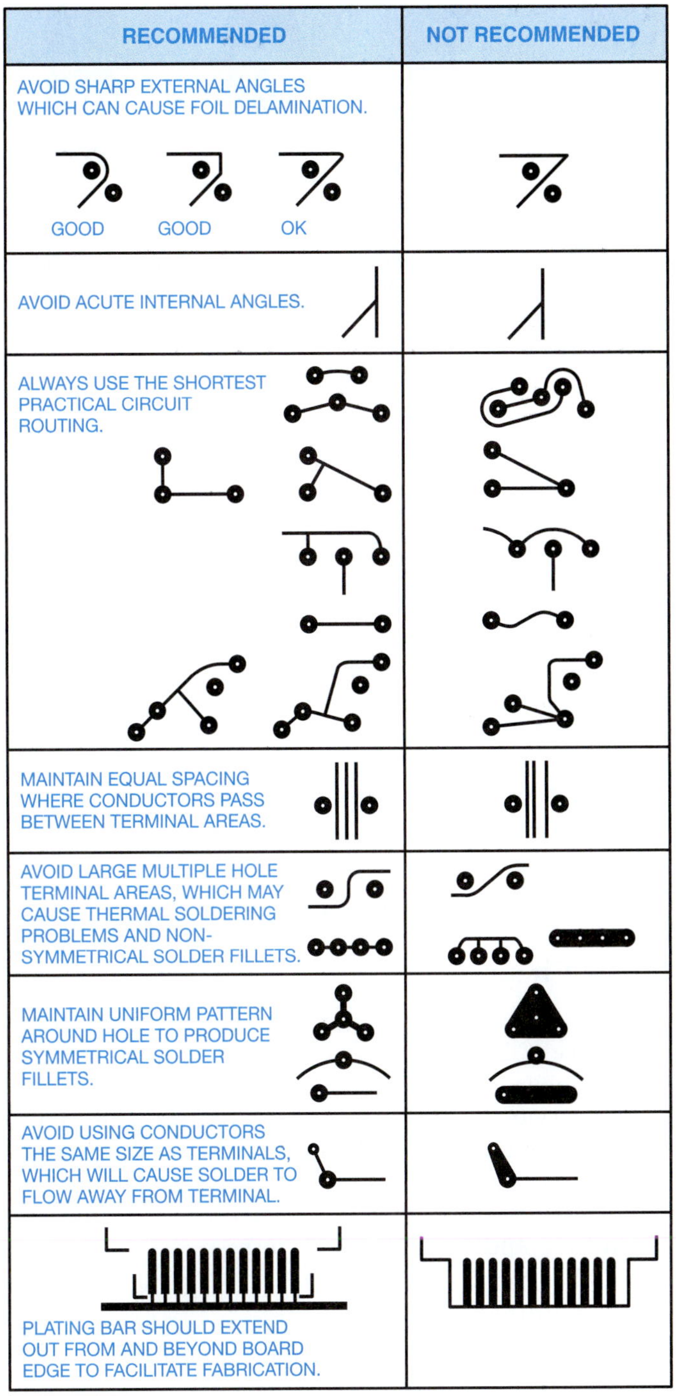

FIGURE 26.78 ■ Artwork pattern configuration, conductor traces, and taping techniques. *Courtesy Bishop Graphics, Inc.*

board for mounting the device. Pads can be placed individually or in patterns depending on the connection characteristics of the device. (See Figure 26.77.) **Conductor traces (lines)** connect the pads to complete the circuit design. Design characteristics of conductors are shown in Figure 26.78. The size depends on the amount of current, the temperature, and the type of board specified. Conductors are .008 in. (0.2 mm) wide with a .007 in. (0.18 mm) space for many applications, but may be designed for applications with a width of .50 in. (12.7 mm) or more for large current-carrying requirements. Factors that influence conductor spacing include product requirements and current specifications. For general uses, conductor spacing should be .007 in. (0.18 mm) minimum.

Printed Circuit Board Artwork

The **printed circuit artwork** is an accurate, scaled, undimensioned drawing used to produce the master pattern from which the actual board is manufactured. The artwork should be prepared from a design layout at an enlarged scale on polyester film using computer-aided design drafting. In addition to all electrical circuitry, the board should contain board edge identification marks; registration, datum, and indexing marks; photo reduction dimensions; scale designation; an identification number; and a revision number. Letters to be etched on the board should be 0.62 in. (1.57 mm) in height (minimum) and .015 in. (0.38 mm) in thickness (minimum).

Master Pattern Artwork

The **master pattern** is a one-to-one scale circuit pattern that is used to produce a printed circuit board. The master pattern artwork is normally prepared at an enlarged scale so that when it is reduced to a 1:1 ratio, its quality is enhanced.

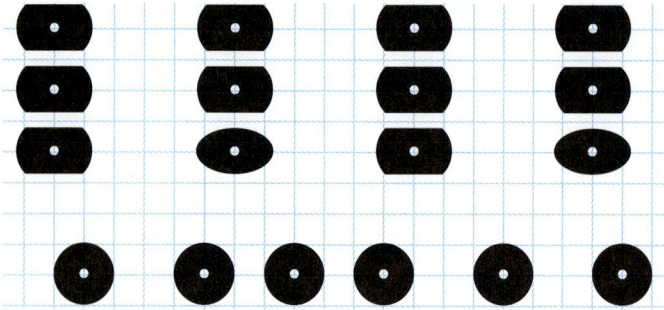

FIGURE 26.79 ■ Grid system. *Courtesy Bishop Graphics, Inc.*

Grid System

The use of a grid system is essential in laying out and preparing the master pattern artwork. The grid system aids in the placement of pads and conductor traces. A **grid** is a network of equally spaced parallel lines running vertically and horizontally on a glass or polyester film sheet, as shown in Figure 26.79. Standard spacing is .100 in. (2.54 mm), .050 in. (1.27 mm), or .025 in. (0.635 mm).

> **MIL-STD** MIL-STD 275D specifies any multiple of .005 in. (0.127 mm).
>
> **IEC** The International Electrotechnical Commission (IEC) specifies 0.5-mm and 0.1-mm increments.

Printed Circuit Scale

The printed circuit board layout should be prepared at an enlarged scale of 2:1 or 4:1 so possible drafting errors or slight flaws are minimized when the layout is reduced to actual size. The accuracy of the PC layout is also critical. Sometimes it is necessary to prepare the PC layout at a 100:1 scale to ensure clarity and accuracy.

Board Size and Number of Layers

Printed circuit boards can be designed in three basic configurations: single layer, multilayer, and multilayer sandwich. **Single-layer boards** contain all printed wiring on one side with the components on the opposite side. **Multilayer boards** have printed circuits on both sides with most of the components on one side and the circuitry on the other. **Multilayer sandwich boards** consist of many thin boards laminated together, with the components on one or both sides of the external layers. As boards increase in complexity, they also cost more. However, the cost of a single multilayer board may be less than several single-layer boards for the same system.

The circuit board should be just large enough to contain the components and interconnections and remain economical to manufacture. One way to estimate the board size is to randomly place scaled component cutouts, allowing enough space for interconnections. Allow extra space for IC interconnections.

Solder Masks

Wave solder has become a common method of attaching components to printed circuit boards. A wave of molten solder is passed over the noncomponent side of the board, making all solder connections. A polymer coating, called a **solder mask**, or **solder resist mask**, is applied to the board, covering all conductors except pads, connector lands, and test points. Solder masks are used to prevent the bridging of solder between pads or conductor traces, to cut down on the amount of solder used, and to reduce the weight of the board.

Conductor Width and Spacing

Careful consideration must be given to conductor trace width and spacing during printed circuit design. Width and spacing that are too small can cause service problems in the circuitry. Width and spacing that are too great waste space and increase costs. Verify minimum width and spacing requirements with product and company specifications. Standard charts and specifications are available; but in general, conductor widths of .050 in. (1.27 mm) or .062 in. (1.57 mm), and minimum spacing between conductors of from .031 in. (0.79 mm) to .050 in. (1.27 mm) are recommended for low-voltage applications.

Component Terminal Holes

A printed circuit board should have a separate mounting hole for each component lead or terminal. Unsupported holes contain no conductive material. To determine the diameter of unsupported holes use the formula:

$$\text{Minimum hole diameter} = \text{Maximum lead diameter} + \text{Minimum drill tolerance}$$

Plated-through holes have conductive material plated on the inside wall to form a conductive connection between layers of the circuit board, as shown in Figure 26.80. Plated-through holes are drilled prior to plating. A general note "ALL HOLES TO BE PLATED THROUGH" should be specified. Usually plating thickness is specified as a minimum with an accepted tolerance of −0 + 100%. For plated holes, the minimum diameter should be greater than the minimum lead size plus .028 in. (0.71 mm).

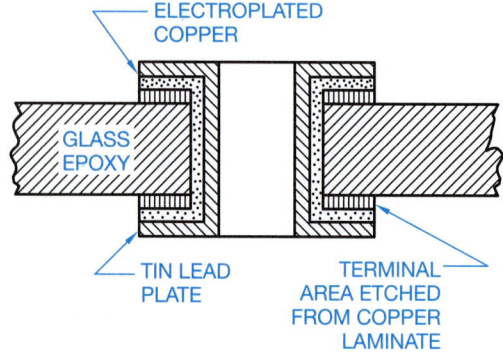

FIGURE 26.80 ■ Cross section of plated-through hole. *Courtesy Bishop Graphics, Inc.*

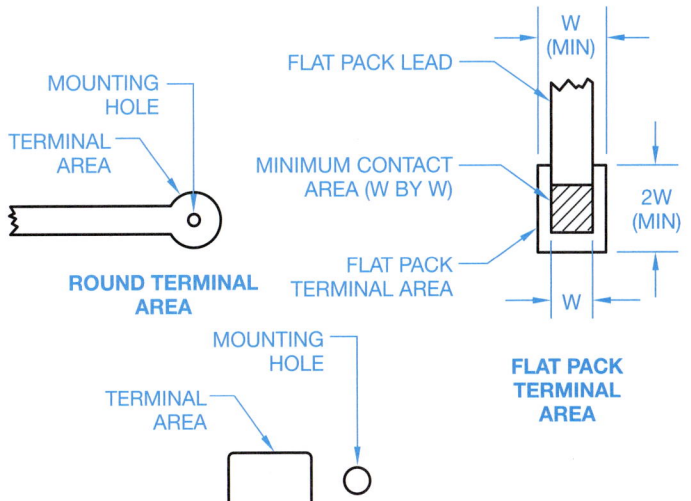

FIGURE 26.81 ■ Terminal areas. *Courtesy Bishop Graphics, Inc.*

A second drilling process is required if the predrilled holes do not take into account the plating thickness. Drilling twice adds cost to the board. The number of different hole sizes should be kept to a minimum to help save manufacturing costs. Plating is specified in ounces (oz) of copper. The base is typically 1/2 oz plating but can be up to 1–2 oz. Additionally, .0014" (0.035 mm) thick is specified as 1 oz (31 grams) of copper, which equals 1 oz per square foot.

Terminal Pads

There should be a separate terminal pad for each component lead or wire attachment. Terminal pads vary with designer preference and component characteristics. (See Figures 26.77 and 26.81.) The **minimum required annular ring** is the smallest part of the circular strip of conductive material surrounding a mounting hole that meets design requirements.

MIL-STD MIL-STD 275D specifies .015 in. (0.38 mm) minimum for unsupported holes, .005 in. (0.13 mm) minimum for plated-through holes on external layers, and .002 in. (0.05 mm) for internal layers of multilayer sandwich boards.

Ground Planes

A **ground plane** is a continuous conductive area used as a common reference point for circuit returns, signal potentials, shielding, or heat sinks. Ground plane patterns should be broken up so the conductive area is equal to about half of the non-conductive area. (See Figure 26.82.) Clearance should be provided between terminal pads and the ground plane, as shown in Figure 26.83.

Printed Circuit Board Layout

Before starting a printed circuit board layout, you should have a schematic or logic diagram, a parts list, design specifications, component sizes, lead-and-trace pattern and spacing, hole and terminal sizes, grid system, scale, board size and number of layers, and know the manufacturing processes to be used.

Silk Screen Artwork

The circuit board artwork can have a separate sheet called **silk screen artwork** containing component outlines and orientation symbols. This pattern is printed on the component side of the board after etching and plating. The reference designations should be placed so they are visible after component assembly. The component outlines should be located in the same position that the actual components occupy after assembly. The silk screen artwork should include registration marks that align with the pattern. (See Figure 26.84.)

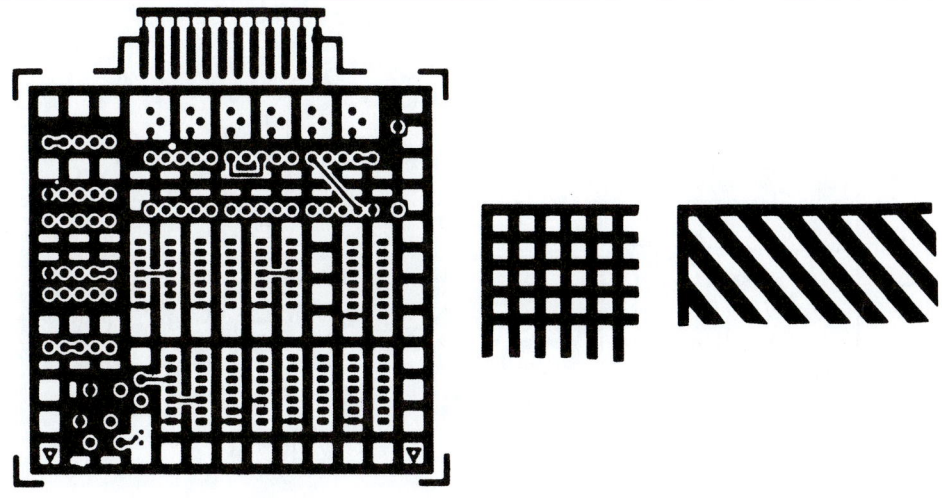

FIGURE 26.82 ■ Ground planes. *Courtesy Bishop Graphics, Inc.*

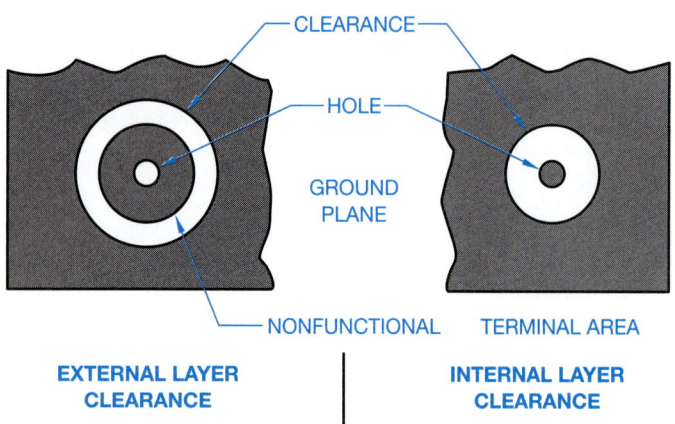

FIGURE 26.83 ■ Layer clearance at ground planes. *Courtesy Bishop Graphics, Inc.*

Drilling Drawings

Drilling drawings are prepared after the master layout is complete. The drilling drawing is used to provide size and location dimensions for component and chassis mounting holes and the final dimensions for trimming the board. Drilling drawings are often set up using datum arrowless tabular dimensioning in accordance with computer numerical control manufacturing methods. (See Figure 26.85.) Refer to the previous discussion

on hole sizes and to Chapter 11 for a more in-depth discussion of dimensioning practices.

Assembly Drawings

The printed circuit board assembly drawing is a complete engineering drawing, including components, assembly and fastening or soldering specifications, and a parts list or bill of materials. The parts list should include the part number, item identification keyed to the assembly, quantity of each part, electronics designations for components, and Federal Code Identification, if required by contract. (See Figure 26.86.)

Printed Circuit Preparation

There are several methods used to prepare the final printed circuit board from a printed circuit layout; however, the two fundamental methods are the additive and the etching processes.

Additive Process

The additive process takes place on a board that is covered with a chemically etched material that will accept copper. The PCB layout design image is then transferred to the board using a silk screen or photo printing process. Copper conductor is then deposited on the image using electroless. Electroless is the de-

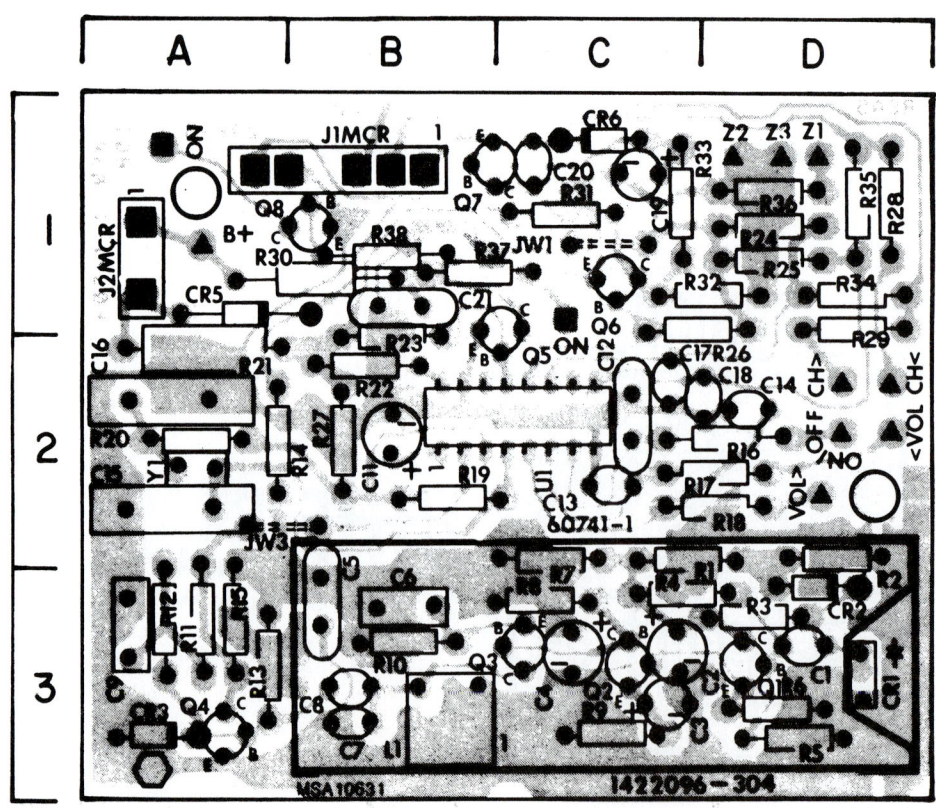

FIGURE 26.84 ■ Silk screen artwork. *Reproduced by permission of RCA Consumer Electronics.*

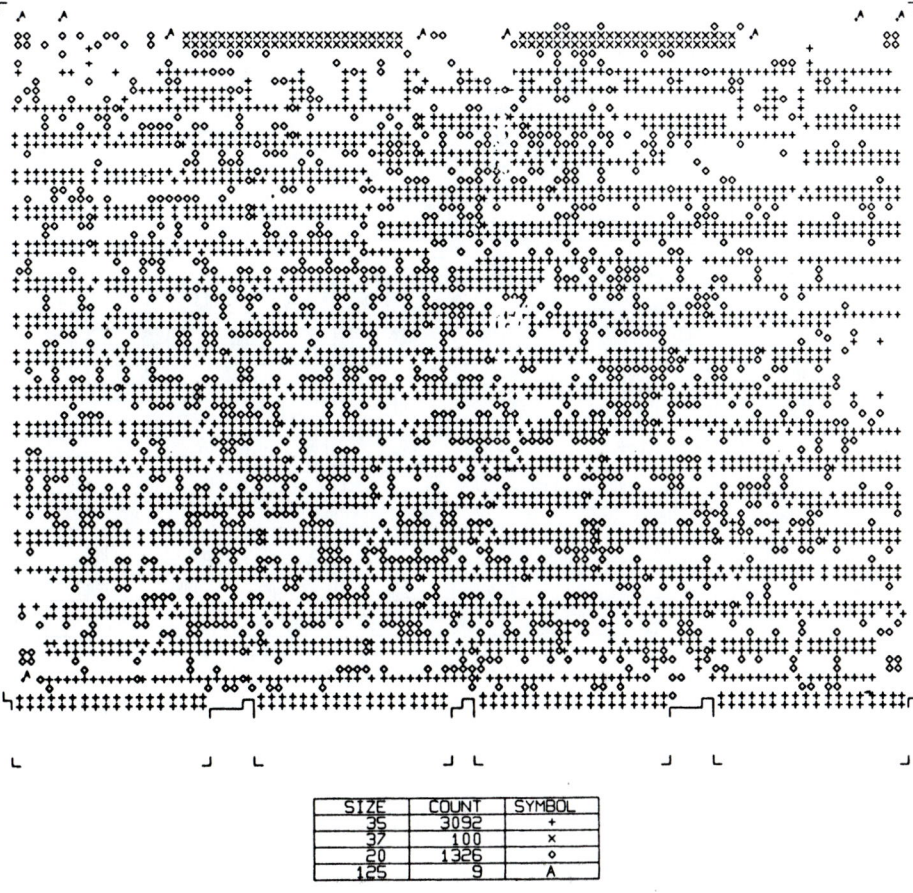

SIZE	COUNT	SYMBOL
35	3092	+
37	100	×
20	1326	○
125	9	A

FIGURE 26.85 ■ Drilling drawing. *Courtesy Floating Point Systems, Inc.*

positing of a metal on another material through the action of an electric current. A solder resist mask is then applied to the copper deposit. This process prevents solder from bridging between closely spaced conductors when components are added to the printed circuit board. The board is then coated with a lacquer to protect the surface from corrosion.

Etched Process

A copper-covered base is used in the etched process. (Other conductive material may also be used.) The printed circuit layout is normally prepared at 2:1 or 4:1 scale and then photographically reduced to a photo negative at full scale. The copper-covered base is cleaned and treated with a light-sensitive emulsion. The base is then exposed to light through the negative. During this process, the emulsion hardens and an acid resist is formed at the printed circuit areas. The board is then placed in a chemical solution that etches away the copper coating in areas that are not protected by the acid resist. The board is then rinsed to stop the etching process and a protective coating is added. The resulting printed circuit is ready for components to be added.

Surface-Mount Technology (SMT)

Surface-mounted devices (SMD) have emerged from the increasing need to get more electronic components into smaller places. Surface-mounted devices commonly take up to less than one-third the space of feed-through printed circuits. Conventional printed circuit board connectors require plated-through holes for each pin or connection point. Surface-mounted devices save the area, and are especially useful when designing very complex boards. The only drilled holes that are required are for chassis mounting or feed-through between layers.

In surface-mount technology, the traditional component lead-through is replaced with a solder paste to hold or glue the components in place on the surface of the printed circuit board. A solder paste template is made that allows for the placement of the paste at required areas. Then the components are automatically or manually placed into the solder paste, although most companies use robotic fabrication. A baking process allows the solder to liquefy and then, upon solidification, the components remain soldered to the printed circuit. Some solderless connections have also been designed for surface-mounted devices.

Surface-mount technology is responsible for new types of electrical components and materials used in electronic circuits; for example, leadless plastic and ceramic IC chip housings. There are some problems, however, with surface-mounted devices; they are so dense that solder connections often become difficult, and testing electrical circuits becomes a problem because testing equipment may have difficulty connecting to very small fine-pitch parts. In most cases, surface-mounted boards are replaced, rather than serviced, when maintenance is required.

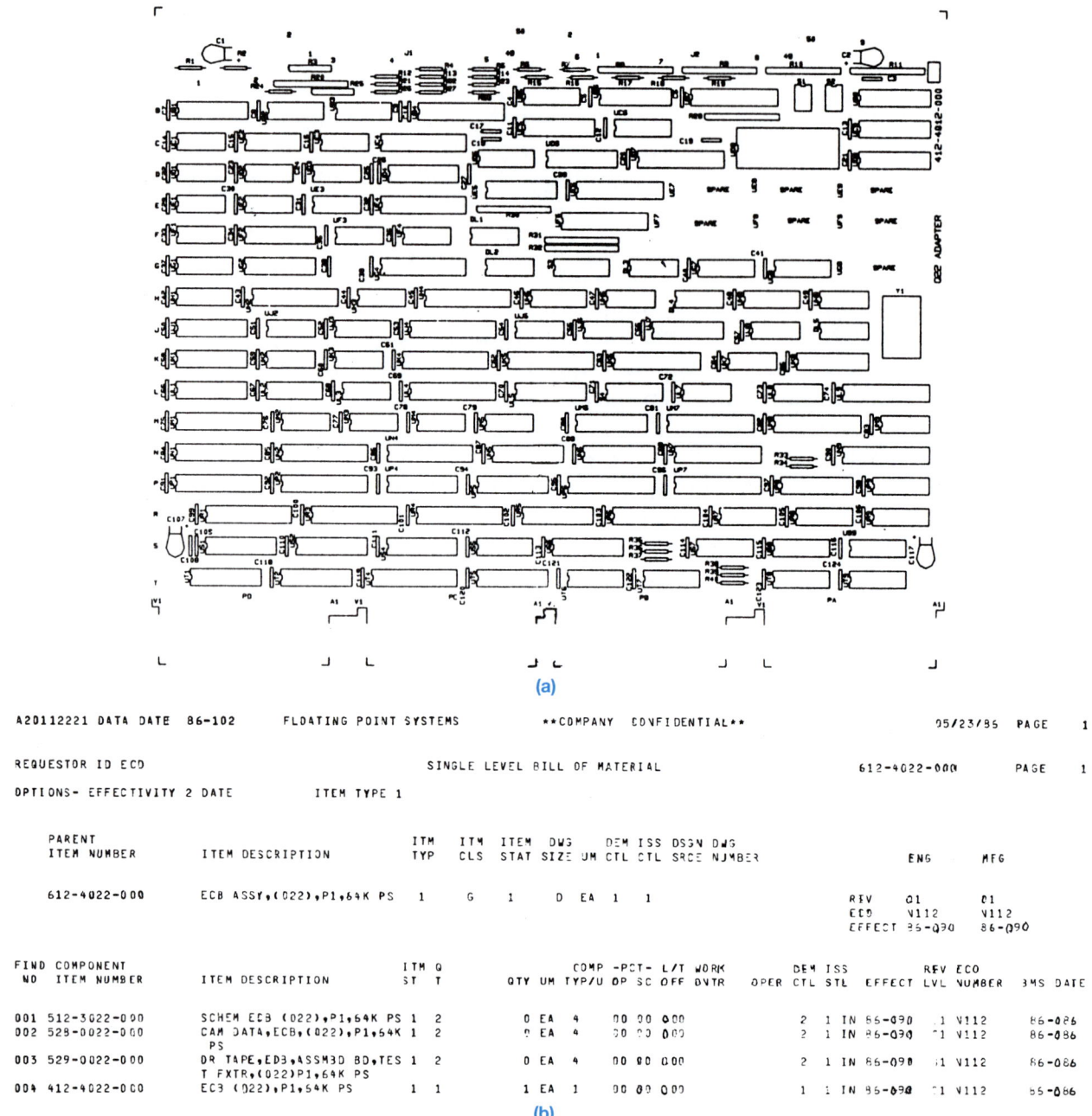

A20112221 DATA DATE 86-102   FLOATING POINT SYSTEMS   **COMPANY CONFIDENTIAL**   05/23/85 PAGE   1

FIGURE 26.86 ■ (a) Assembly drawing. (b) Partial computerized parts list. *Courtesy Floating Point Systems, Inc.*

PICTORIAL DRAWINGS

Three-dimensional drawings are frequently made to show pictorial representations of products for display in vendors' catalogs or instruction manuals. Pictorial assembly drawings are often used to show the physical arrangement of components so components are properly placed by factory workers during assembly. These drawings are also helpful for maintenance as they provide a realistic representation of the product.

One of the most effective uses of photodrafting is found in the electronics field. Photodrafting is where a photo is taken of a complete assembly or product and drafting is used to label individual components. A common method of pictorial representation is the exploded technical illustration often used for parts identification and location, as shown in Figure 26.87.

Another type of pictorial diagram, known as a semipictorial wiring diagram, uses two-dimensional images of components in a diagram form for use in electrical or electronic applications. The idea is to show components as features that are recognizable by laypersons. This type of pictorial schematic is commonly used in automobile operations manuals, as shown in Figure 26.88.

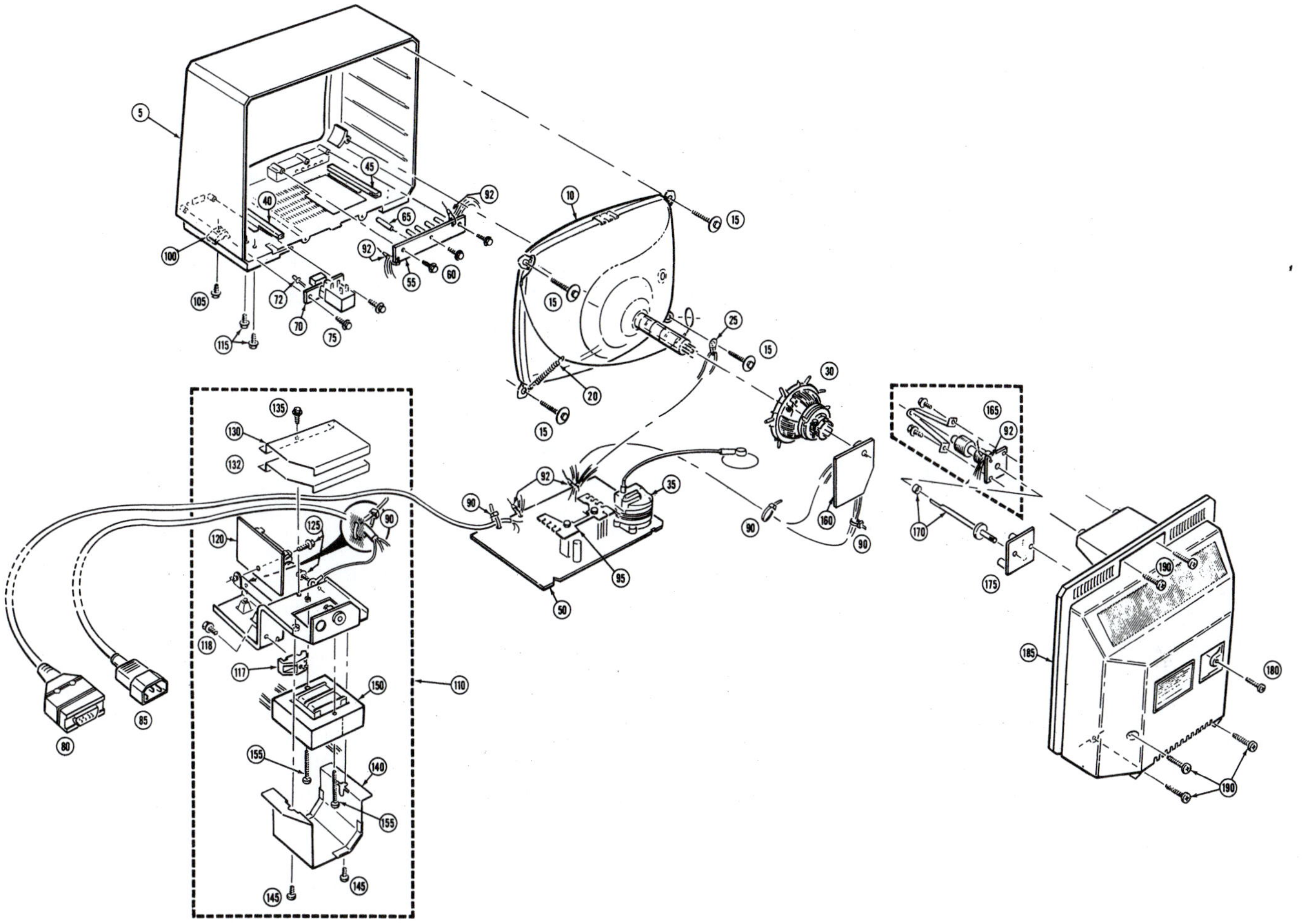

FIGURE 26.87 ■ Parts pictorial, exploded technical illustration. *Reprinted by permission of HEATH COMPANY.*

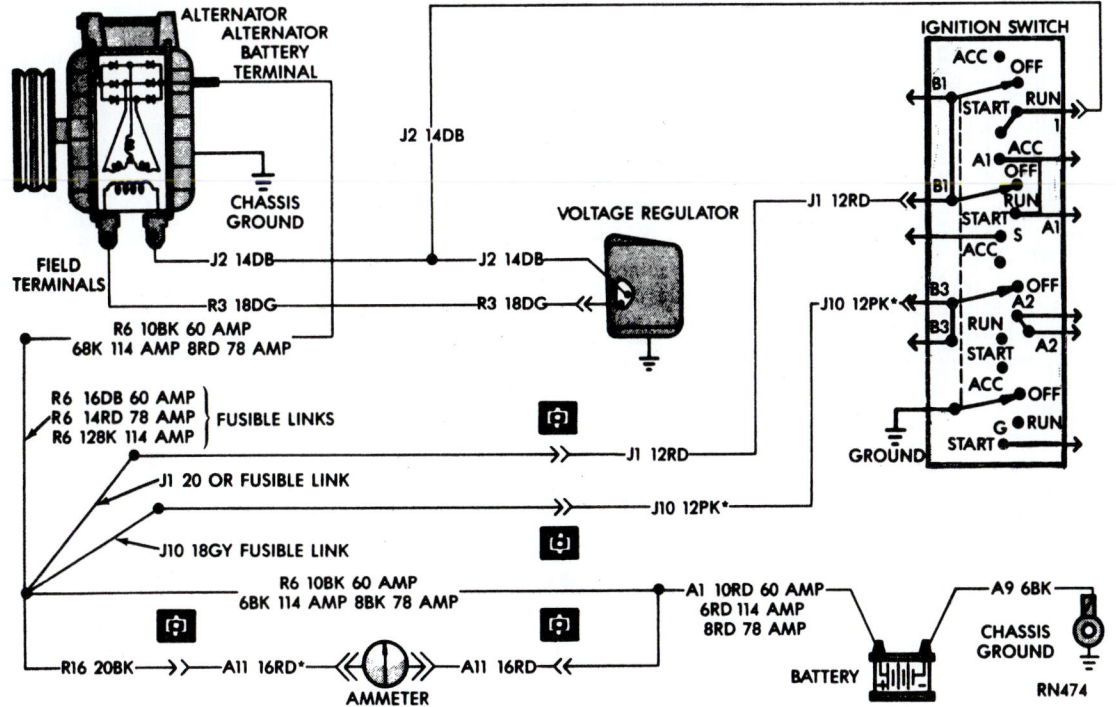

FIGURE 26.88 ■ Semipictorial wiring diagram. *Courtesy Chrysler Corporation.*

CADD APPLICATIONS 2-D

COMPUTER-AIDED ENGINEERING (CAE)

With electronic symbols established in a CADD template or symbols library, the design and layout of block diagrams, schematic diagrams, and other electronics drawings are greatly enhanced. The implementation of computers in this industry has advanced into what is known as computer-aided engineering (CAE), where the CADD function is taken further into the complete design, engineering, and functional analysis of the product. Beginning with the development of a computer-aided schematic design, the CAE program can perform the following functions:

- It stimulates the circuit operation.
- It tests the system for possible problems.
- It can identify and analyze design complexity to help reduce manufacturing costs.
- Thermal characteristics of the circuitry are evaluated to identify possible overheating situations where heat sinks may be needed or mechanical cooling is required.
- Arced or 45° corners, rather than sharp corners, are automatically placed in the conductor traces during the routing process.
- Artwork masters are generated at 1:1 scale for all circuit levels, eliminating costly and error-prone photoreduction, as shown in Figure 26.89.
- The solder resist mask is generated from the same database as the artwork master. This improves the manufacturability of the board and reduces external changes. (See Figure 26.90.)
- The component layout that is created as part of the symbols can be extracted to create a marking drawing silk screen master, as shown in Figure 26.91.

- The drill template can be generated from the same database as the artwork master and is used to determine plating requirements: generate a drill chart with sizes, quantities, and X-Y coordinate location dimensions, and create a drill pattern or provide direct communication to a computer numerical control (CNC) drilling machine to achieve repeatable board-to-board registration.
- A description and reference designator may be placed on each symbol and may be used to simultaneously complete assembly drawings, parts lists, and other documentation.
- A large-format digitizing table may be used to enable the designer to go from rough layout or existing works to right, on-screen design.
- Laser-directed photoplotters provide PCB artwork with accuracy tolerances up to 0.4 mil (0.000010 mm) along with image quality and speed that exceed the highest industry standards.
- Printed circuit conductor routing may be performed automatically with completion on success on most boards of 97−100%. In complex board design, the incomplete part of the trace routing, if any, may be quickly engineered in the manual mode.
- The designer or engineer may stop the routing process at any time to make component or connection changes and restart the automatic routing process at the new location.
- Component locations are designed along with trace routing.

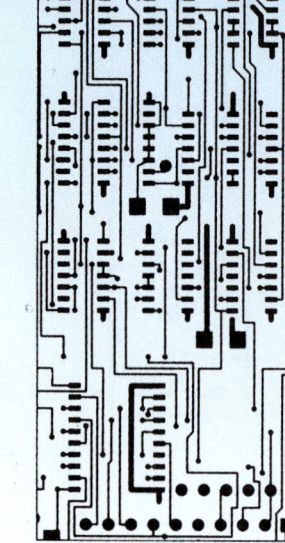

FIGURE 26.89 ■ Full scale CAE artwork eliminating the need to photoreduce. *Courtesy The Gerber Scientific Instrument Company.*

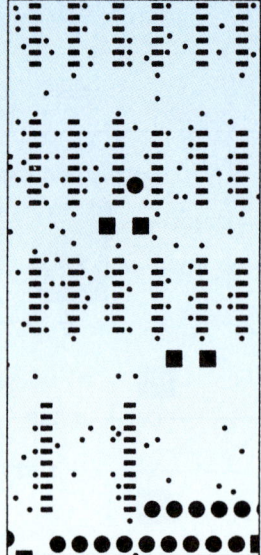

FIGURE 26.90 ■ The solder resist mask is generated from the same database as the artwork master shown in Figure 26.87. *Courtesy The Gerber Scientific Instrument Company.*

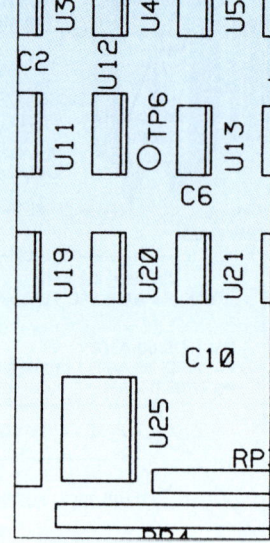

FIGURE 26.91 ■ Marking drawing created from symbols library used to design the artwork master. *Courtesy The Gerber Scientific Instrument Company.*

PROFESSIONAL PERSPECTIVE

As an entry-level drafter with training in electronics schematic drafting, you are able to make engineering drawing changes and prepare drawings from engineering sketches without specific knowledge of how electronic components and the systems function. However, in order to advance in skill level to designer or engineering technician, you need a thorough understanding of company requirements, the ability to communicate electrical and electronic terminology, and the knowledge of how the components go together. It is important for an engineering drafter in this business to visualize the systems and understand electrical clearance requirements. A lot of this can be learned on the job because the engineers are responsible for the design, but as you become experienced, you can assist in identifying engineering problems and troubleshooting systems. Schooling in communication skills and problem-solving techniques, along with electrical and electronic theory and practices, is essential. Helpful college courses in addition to your electronic drafting could include:

■ Computer basics—Windows.

■ Basic AC/DC electronics.

■ Solid-state devices.

■ Digital logic.

■ Basic microprocessors.

■ Electrical physics.

When you go to work in this field, you will find that it is extremely important for you to check your work completely, making sure that electrical connections made are necessary. Nearly all drafting is done on a CADD system, so it is important for you to fully understand the computer operating systems. When you are bringing electronic symbols from a menu library, be sure that you leave adequate space to display the symbol, labeling, and circuit notations. Be sure that you set up the schematic in a logical layout sequence so the components, test points, and other items are identified from left to right and top to bottom. You need to know and understand the numbering systems relating to the schematic and the PCB design. This is all part of the knowledge you will gain with experience in the industry. An engineering drafter at Intel warns, "Keep an open mind on new CADD systems and better ways to do things. The hardest part of the job is to constantly learn new systems, because the industry changes so fast. To sum it up, be flexible."

WORKING WITH POWERS OF TEN

Powers of ten (scientific notation) is used to express the very large and small numbers found in electronics compactly, making multiplying or dividing them much easier. Here are some of the powers of ten.

$$10^4 = 10,000$$
$$10^3 = 1,000$$
$$10^2 = 100$$
$$10^1 = 10$$
$$10^0 = 1$$
$$10^{-1} = .1$$
$$10^{-2} = .01$$
$$10^{-3} = .001$$

A number may be changed to a power of ten form by first writing the digits as a number between 1 and 10, then multiplying it by the appropriate power of ten. Here are some examples to illustrate this:

$$3,400 = 3.4 \times 10^3$$
$$53,000,000 = 5.3 \times 10^7$$
$$.000051 = 5.1 \times 10^{-5}$$
$$.02 = 2 \times 10^{-2}$$

To Multiply: Multiply the numbers out front and add the exponents.

To Divide: Divide the numbers out front and subtract the exponent in the bottom from the exponent on top. Here are examples to illustrate:

$$(2 \times 10^3)(7 \times 10^8) = 14 \times 10^{11} = 1.4 \times 10^{12}$$
$$(3 \times 10^{-6})(2 \times 10^5) = 6 \times 10^{-6+5} = 6 \times 10^{-1}$$
$$(5 \times 10^{-20})(7 \times 10^{-3}) = 35 \times 10^{-23} = 3.5 \times 10^{-22}$$

These results can be obtained on the calculator using the EE, EXP, or EEX buttons, depending on the make of your calculator. For the last example, many calculators would use this sequence of buttons to obtain the final answer: 1.4 EE 15 ÷ 2 EE 5 +/− =. The calculator may display the answer without the "× 10" like this: 7^{19}. However, the answer should never be written down and communicated like this because it would be confused with 7 raised to the 19th power, which is an entirely different number! Always include the "× 10" when writing down a power of 10 number.

MATH APPLICATIONS

CHAPTER 26
Electrical and Electronic Schematic Drafting Test

 Access the CD found with this textbook to view the Chapter 26 Test. Confirm the preferred submittal method with your instructor.

CHAPTER 26
Electrical and Electronic Schematic Drafting Problems

DIRECTIONS

1. Read all instructions before you begin working, unless otherwise specified by your instructor.
2. Use manual or computer-aided drafting as required by your course guidelines.
3. Use the attached engineering layouts to prepare each drawing. Refer to chapter coverage for symbol and component dimensions and also refer to previous drawings for information as you progress to each drawing.
4. Prepare well-balanced, easy-to-read drawings. Most drawings have no scale.
5. Many of the following problems contain symbols that are duplicated multiple times. These electrical problems demonstrate the power of CADD when you create one symbol and have the ability to use it several times on one or more drawings.
6. Additional information is provided for some problems.
7. Estimate unknown dimensions.

PART 1

PROBLEM 26.1 **Switch panel**

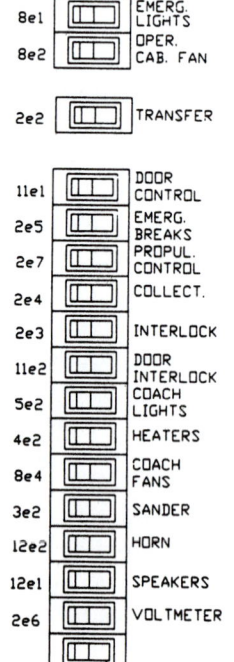

PROBLEM 26.2 **Schematic**

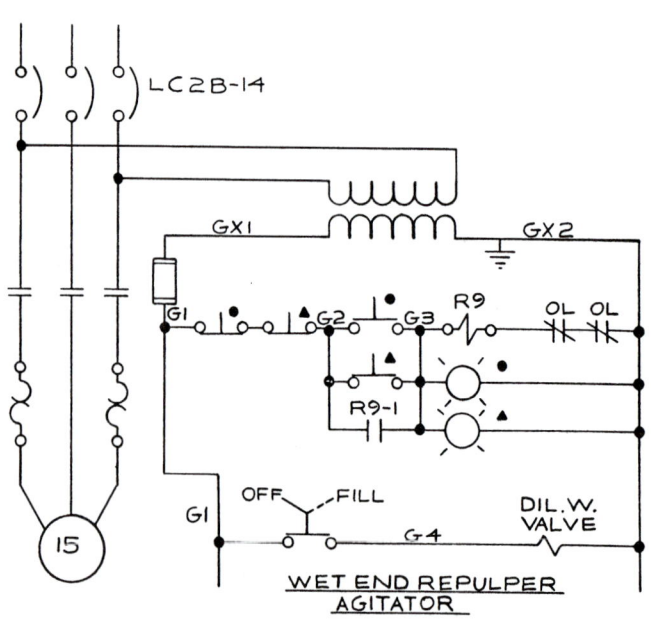

PROBLEM 26.3 Schematic

PROBLEM 26.5 Schematic

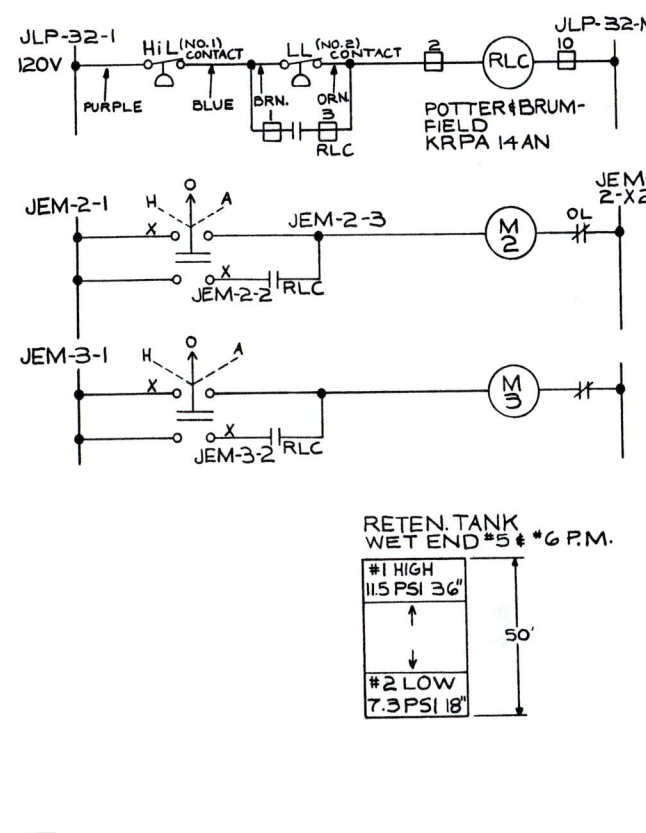

PROBLEM 26.4 Schematic

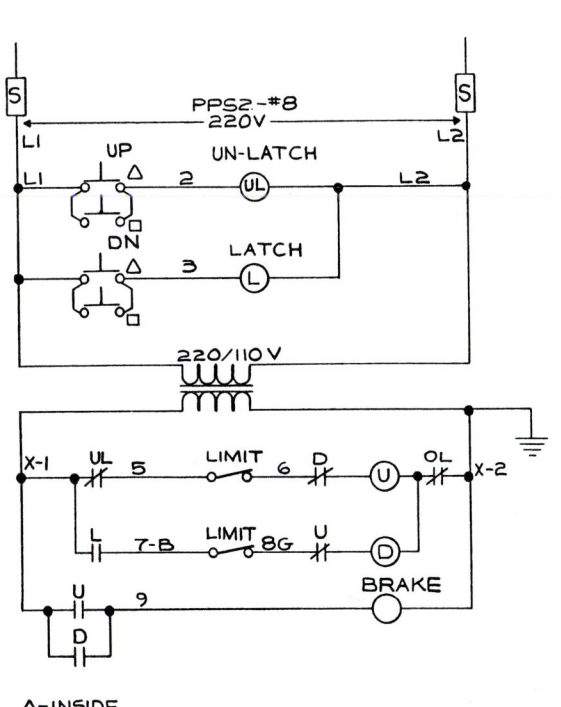

PROBLEM 26.6 Schematic

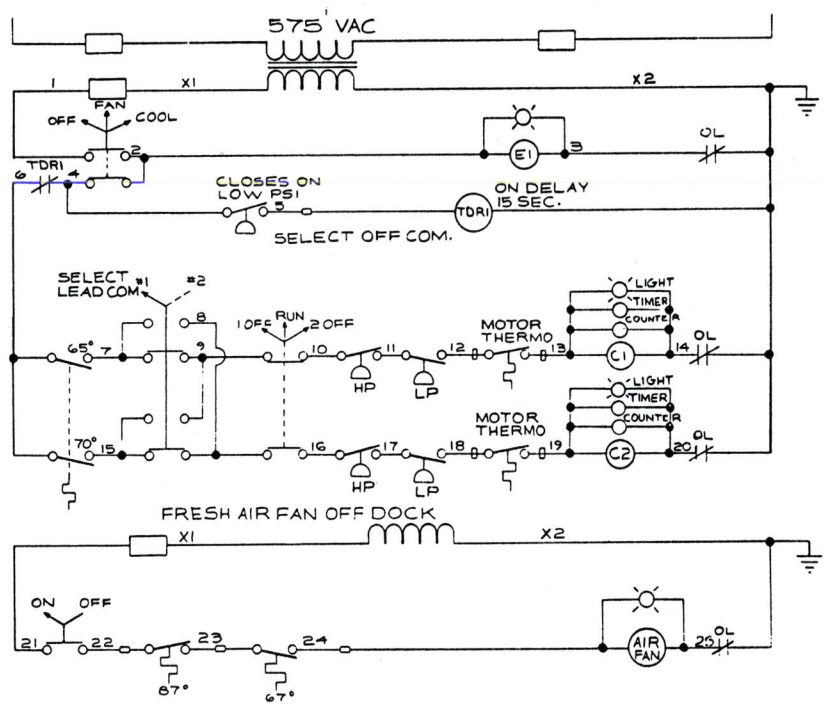

PROBLEM 26.7 Block diagram

Given the following engineering layout of the Potlatch-Pearl Substation, draw the block diagram on 11 × 17 polyester film or vellum unless otherwise specified by your instructor. There is no scale. The layout should neatly fill the sheet without crowding symbols or notes. Make all connection points ⌀3/32 in. Provide the general note: INTERPRET PER ANSI Y32.2. *Courtesy Bonneville Power Administration.*

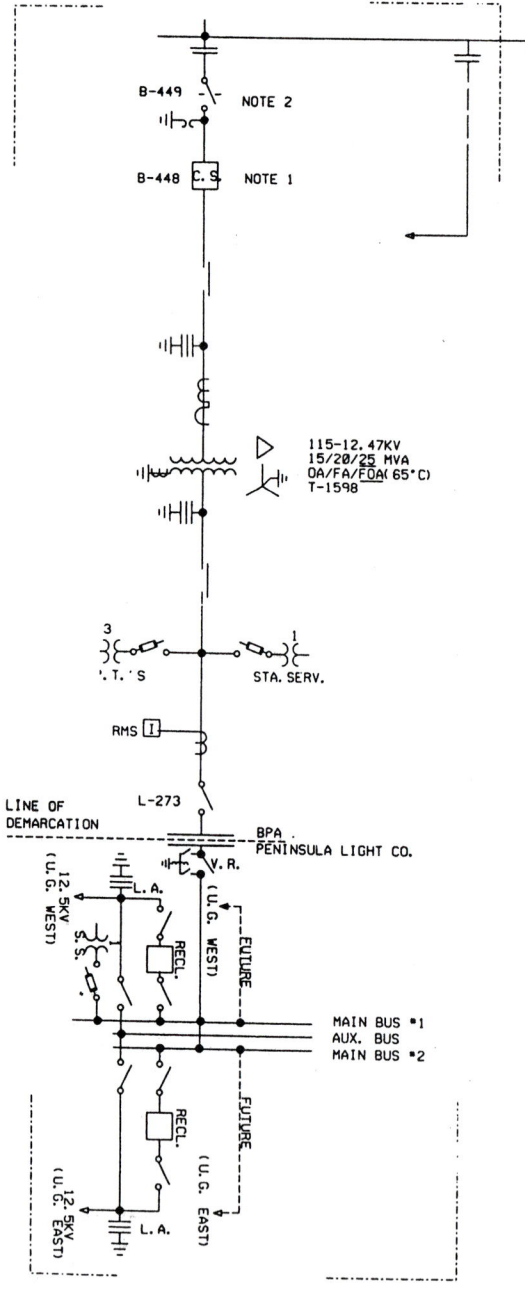

PROBLEM 26.8 Elementary diagram

Metering System.

Courtesy Bonneville Power Administration.

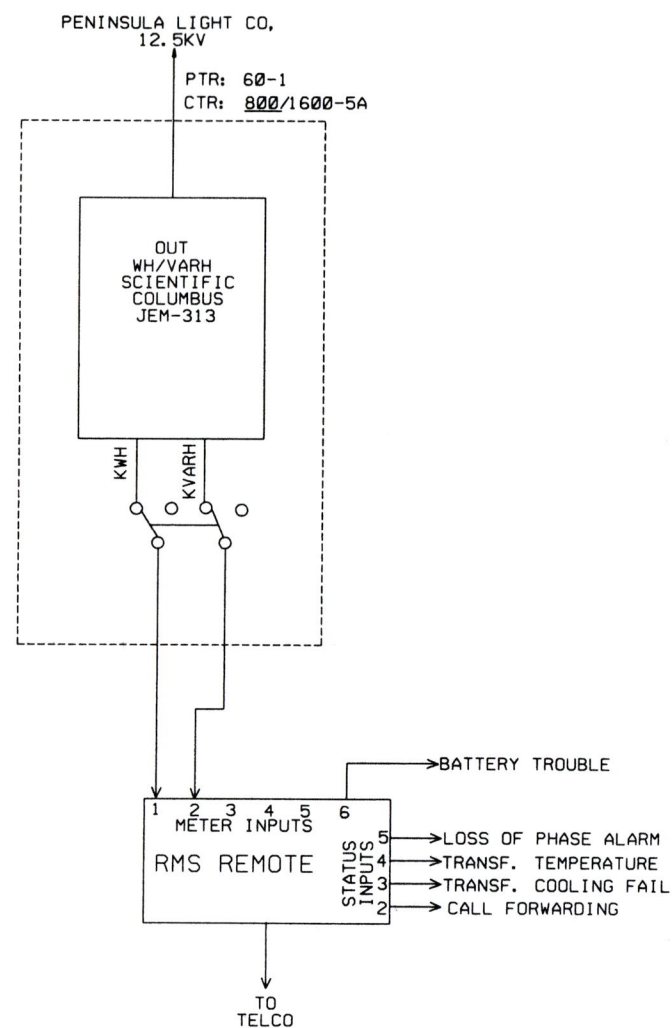

PROBLEM 26.9 Highway diagram

Draw all component outlines and feeder lines .020 in. (0.5 mm) wide and trunk lines .039 in. (1.00 mm) wide.

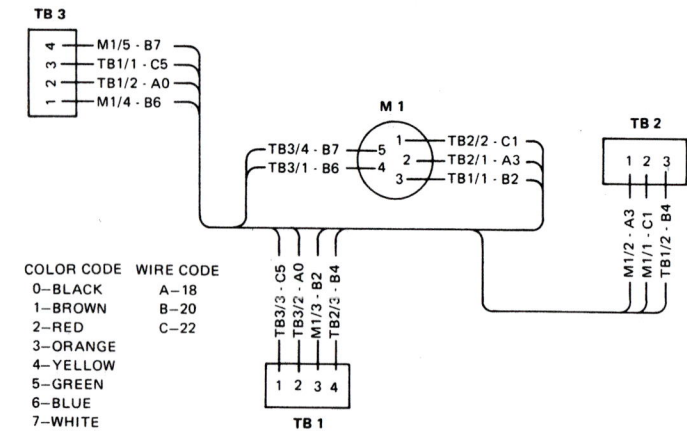

PROBLEM 26.10 **Wireless diagram**

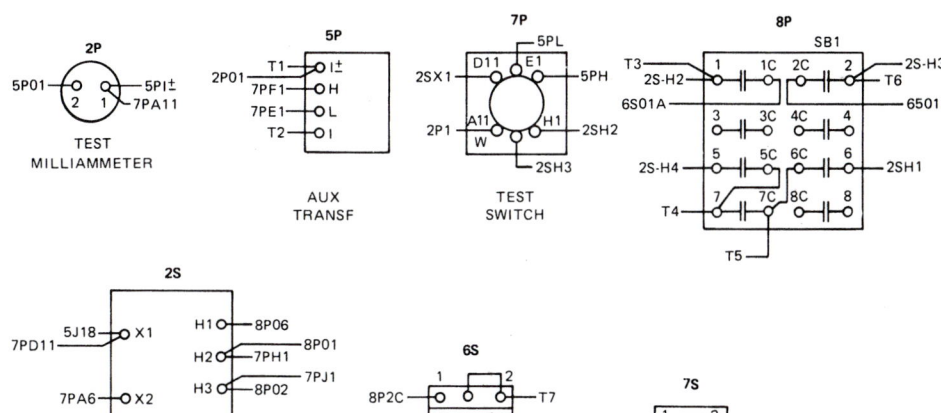

 PROBLEMS 26.11 through 26.15: Access the CD found with this textbook and open the problem of your choice, or as assigned by your instructor. Solve the problems using the instructions provided with this chapter or on the CD, unless otherwise specified by your instructor.

PROBLEM 26.16 **Wiring harness**

Courtesy Flir Systems, Inc.

1. INTERPRET DRAWING IAW MIL-STD-100, CLASSIFICATION PER MIL-T-31000, PARA 3.6.4.

2. INTERPRET DIMENSIONS AND TOLERANCES PER ASME Y14.5M-1994.

3. COVER SOLDER JOINTS USING ITEM 8 (TUBING-SHRINK 3/16").

4. PREP WIRES AND ASSEMBLE CONTACTS PER XXXXXXXX.

5. MARK REFERENCE DESIGNATORS ON HOUSING PER MIL-M-13231.

6. BAG ITEM AND IDENTIFY IAW MIL-STD-130, INCLUDE CURRENT REV LEVEL: 64869ASSYXXXXXXXX REV_.

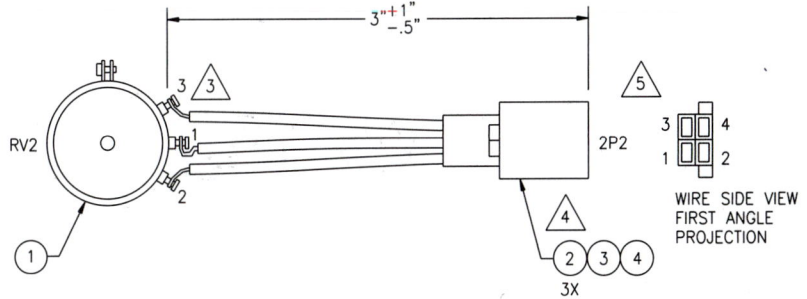

WIRE LIST				
WIRE #	GA/COLOR/ITEM	FROM	TO	SIGNAL
1	22/ORN/6	RV2-1	2P2-2	AZ +12V
2	22/GRN/7	RV2-2	2P2-3	AZ REF
3	22/BLU/5	RV2-3	2P2-4	AZ -12V
UNUSED POSITIONS: J3-D, E, F				

8		TUBING - SHRINK 3/16" BLK	.75"
7		WIRE - #22 STRD, TFE, GRN	4"
6		WIRE - #22 STRD, TFE, ORN	4"
5		WIRE - #22 STRD, TFE, BLU	4"
4		SHROUD - LATCHING DOUBLE ROW 4 POS	1
3		HOUSING - .025 SQ DUAL 4 POS LATCH	1
2		CONTACT - CRIMP PIN .025 SQ 22-26 GA	3
1		RESISTOR - VARI, 10K OHM SERVO MOUNT	1

PROBLEM 26.17   Wiring harness

Courtesy Flir Systems, Inc.

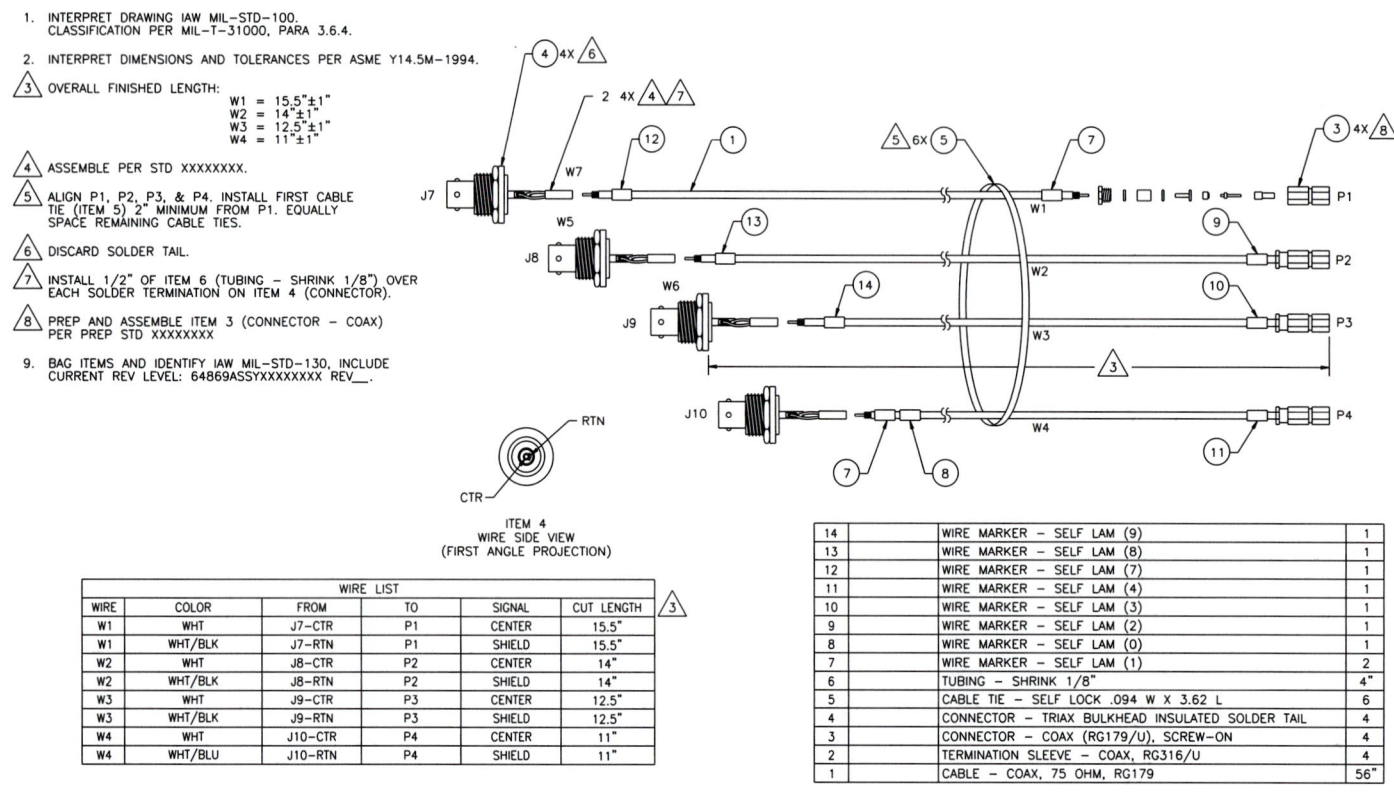

1. INTERPRET DRAWING IAW MIL-STD-100.
   CLASSIFICATION PER MIL-T-31000, PARA 3.6.4.

2. INTERPRET DIMENSIONS AND TOLERANCES PER ASME Y14.5M-1994.

3. OVERALL FINISHED LENGTH:
      W1 = 15.5"±1"
      W2 = 14"±1"
      W3 = 12.5"±1"
      W4 = 11"±1"

4. ASSEMBLE PER STD XXXXXXXX.

5. ALIGN P1, P2, P3, & P4. INSTALL FIRST CABLE
   TIE (ITEM 5) 2" MINIMUM FROM P1. EQUALLY
   SPACE REMAINING CABLE TIES.

6. DISCARD SOLDER TAIL.

7. INSTALL 1/2" OF ITEM 6 (TUBING – SHRINK 1/8") OVER
   EACH SOLDER TERMINATION ON ITEM 4 (CONNECTOR).

8. PREP AND ASSEMBLE ITEM 3 (CONNECTOR – COAX)
   PER PREP STD XXXXXXXX

9. BAG ITEMS AND IDENTIFY IAW MIL-STD-130, INCLUDE
   CURRENT REV LEVEL: 64869ASSYXXXXXXXX REV__.

ITEM 4
WIRE SIDE VIEW
(FIRST ANGLE PROJECTION)

WIRE LIST

WIRE	COLOR	FROM	TO	SIGNAL	CUT LENGTH
W1	WHT	J7-CTR	P1	CENTER	15.5"
W1	WHT/BLK	J7-RTN	P1	SHIELD	15.5"
W2	WHT	J8-CTR	P2	CENTER	14"
W2	WHT/BLK	J8-RTN	P2	SHIELD	14"
W3	WHT	J9-CTR	P3	CENTER	12.5"
W3	WHT/BLK	J9-RTN	P3	SHIELD	12.5"
W4	WHT	J10-CTR	P4	CENTER	11"
W4	WHT/BLU	J10-RTN	P4	SHIELD	11"

14	WIRE MARKER – SELF LAM (9)	1
13	WIRE MARKER – SELF LAM (8)	1
12	WIRE MARKER – SELF LAM (7)	1
11	WIRE MARKER – SELF LAM (4)	1
10	WIRE MARKER – SELF LAM (3)	1
9	WIRE MARKER – SELF LAM (2)	1
8	WIRE MARKER – SELF LAM (0)	1
7	WIRE MARKER – SELF LAM (1)	2
6	TUBING – SHRINK 1/8"	4"
5	CABLE TIE – SELF LOCK .094 W X 3.62 L	6
4	CONNECTOR – TRIAX BULKHEAD INSULATED SOLDER TAIL	4
3	CONNECTOR – COAX (RG179/U), SCREW-ON	4
2	TERMINATION SLEEVE – COAX, RG316/U	4
1	CABLE – COAX, 75 OHM, RG179	56"

PROBLEM 26.18   Photocell wiring diagram

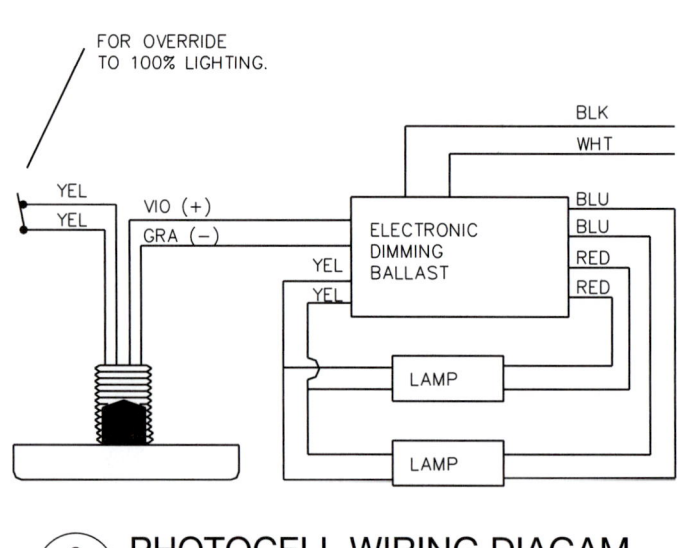

FOR OVERRIDE
TO 100% LIGHTING.

PHOTOCELL WIRING DIAGAM
NO SCALE

PROBLEM 26.19   **Block diagram**

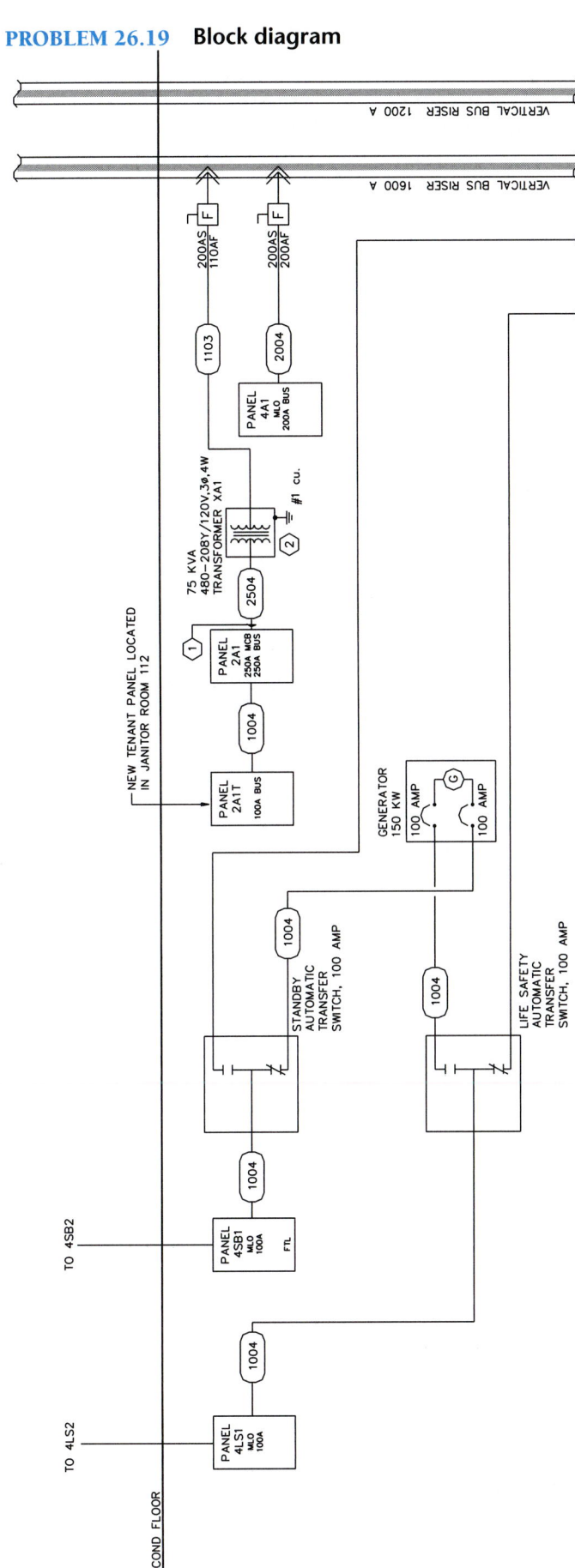

PROBLEM 26.20   **Cable assembly**

Given the cable assembly engineering layout in Figure 26.12, page 916, do the following (approximate dimensions not given):

1. Draw the cable assembly, wiring diagram, and bill of materials.

2. Use ASME standard line widths for cable assembly and 0.5 mm lines for wiring diagram.

3. Use 1/8-in. lettering with 1/4-in. high letters in balloons and titles.

4. Make balloons Ø1/2 in.

5. Letter the following general notes:

1. INTERPRET DIMENSIONING AND TOLERANCING PER ASME Y14.5M—1994.

2. INTERPRET DRAWING PER ANSI Y32.2.

3. DIMENSIONS ARE IN INCHES WITH TOLERANCES: FRACTIONS = ± 1/64, .XX = ±.010, .XXX = ±.005.

4. 32 MICRO IN FINISH ON METAL PARTS.

5. MATERIAL BRONZE.

6. USE SPEC 6712 FOR WIRE END PREPARATION.

PROBLEM 26.21   **Bus layout plan view**

Given the engineering layout shown in Figure 26.16, page 920, for Narrows Substation, draw the bus layout using a 3/16" = 1'–0" (1:50 metric) scale. The following line widths are recommended:

.014 in. (0.35 mm)   Extension, dimension, balloon lines all connections
.020 in. (0.50 mm)   Component and bus lines, and lettering
.024 in. (0.60 mm)   Titles

Draw balloons Ø9/32 in. (7-8 mm). Draw north arrow S45° W. Bus lines are to be drawn from given dimensions and filled in solid except for connections. Dimensions not given are to be estimated.

PROBLEM 26.22   **Bus elevations**

Given the partial engineering layout shown in Figure 26.17, page 920, for Narrows Substation, draw the bus elevation. Follow all other instructions given in Problem 26.21.

PROBLEM 26.23   **Grounding layout**

Given the partial engineering layout shown in Figure 26.19, page 922, of the Narrows Substation, draw the grounding layout at a scale of 1" = 10'–0" (1:100 metric). Include notes and grounding table.

PROBLEM 26.24   **Grounding details**

Given the partial engineering layout shown in Figure 26.20, page 922, of the Narrows Substation, draw the grounding detail.

PROBLEM 26.25   **Conduit installation layout**

Given the partial engineering layout shown in Figure 26.21, page 923, of the Narrows Substation, draw the conduit installation layout. A magnifying glass may be needed to read the engineering layout.

PROBLEM 26.26 Conduit installation details

Given the partial engineering layout of the conduit installation detail shown in Figure 26.22, page 923, for the Narrows Substation, draw the given views.

PROBLEM 26.28 Power panel detail (shown below)

Narrows Substation.

Courtesy Bonneville Power Administration.

PROBLEM 26.27 Residential electrical

Given the typical electrical layouts shown in Figures 26.29 and 26.30, page 926, draw each at a scale of 1/4" = 1'–0" (1:50 metric). Place one next to the other. Estimate dimensions making your drawing about twice the size of the given drawing.

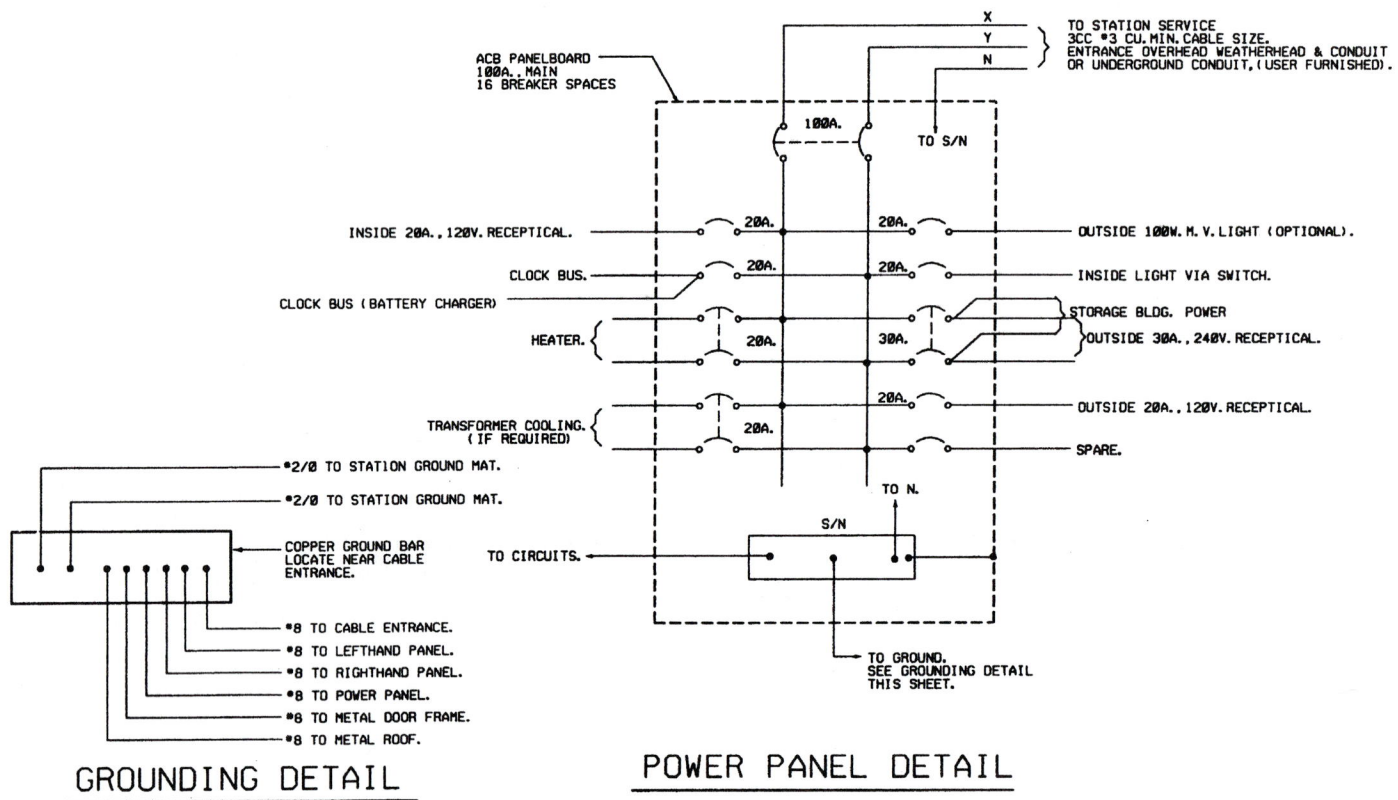

GROUNDING DETAIL

POWER PANEL DETAIL

PROBLEM 26.29 Electrical floor plan

Given the partial engineering upper-floor power layout shown in Figure 26.34, page 926, draw the plan using a 1/4" = 1'–0" (1:50 metric) scale. Estimate dimensions, making your drawing about three times the size of the given drawing.

PROBLEM 26.30 Reflected ceiling plan

Given the engineering layout shown in Figure 26.35, page 927, draw the reflected ceiling plan at a 1/4" = 1'–0" (1:50 metric) scale. Estimate unknown dimensions.

PROBLEM 26.31 Roof plan—electrical

Given the partial engineering layout shown in Figure 26.36, page 927, draw the roof plan—electrical at a 1/4" = 1'–0" (1:50 metric) scale. Estimate unknown dimensions.

PROBLEM 26.32 Commercial schematic wiring diagram

Given the engineering layout shown in Figure 26.37, page 927, draw the beverage cooler condenser unit wiring diagram.

PROBLEM 26.33 **First-floor lighting plan**

Use the given problem layout to create the partial lighting plan shown. Use a 1/4" = 1'–0" (1:50 metric) scale to draw the floor plan. Make the dimensions to your own specifications, but proportional to the given engineering drawing. Use appropriate CADD layers.

Courtesy Interface Engineering.

LUMINAIRE SCHEDULE

TYPE 'A': RECESSED FLUORESCENT PARABOLIC LUMINAIRE. 2' BY 4' STEEL HOUSING. LUMINAIRE PROVIDED BY LANDLORD. THREE 32W T8 LAMPS. NOMINAL INPUT WATTS: 93.

TYPE 'A1': SAME AS TYPE 'A' EXCEPT CONTINUOUS DIMMING BALLAST.

LUMINAIRE SCHEDULE GENERAL NOTES

1 THIS LUMINAIRE SCHEDULE IS NOT COMPLETE WITHOUT A COPY OF THE PROJECT MANUAL CONTAINING THE ELECTRICAL SPECIFICATIONS.

2 T8 FLUORESCENT LAMPS TO BE 3500K WITH A MINIMUM CRI OF 75.

NOTES THIS SHEET

1 2' X 4' PARABOLIC LUMINAIRES ARE TO BE PROVIDED BY LANDLORD

2 TYPE 'B' LUMINAIRES LOCATED BELOW CABINETS.

3 PHOTO CELL TO CONTROL ALL AMBIENT SECTORS AS SHOWN ON DRAWINGS. ROUTE LOW VOLTAGE CONTROL CIRCUIT VIA PHOTOCELL. SEE DIAGRAM 2/E2. VERIFY AMOUNT AND LOCATIONS OF PHOTO CELLS WITH MANUFACTURES REP. PRIOR TO ROUGH-IN. PROVIDE WATTSTOPPER OR EQUAL.

4 CONNECT TO EMERGENCY LIGHTING CIRCUIT IN CORRIDOR.

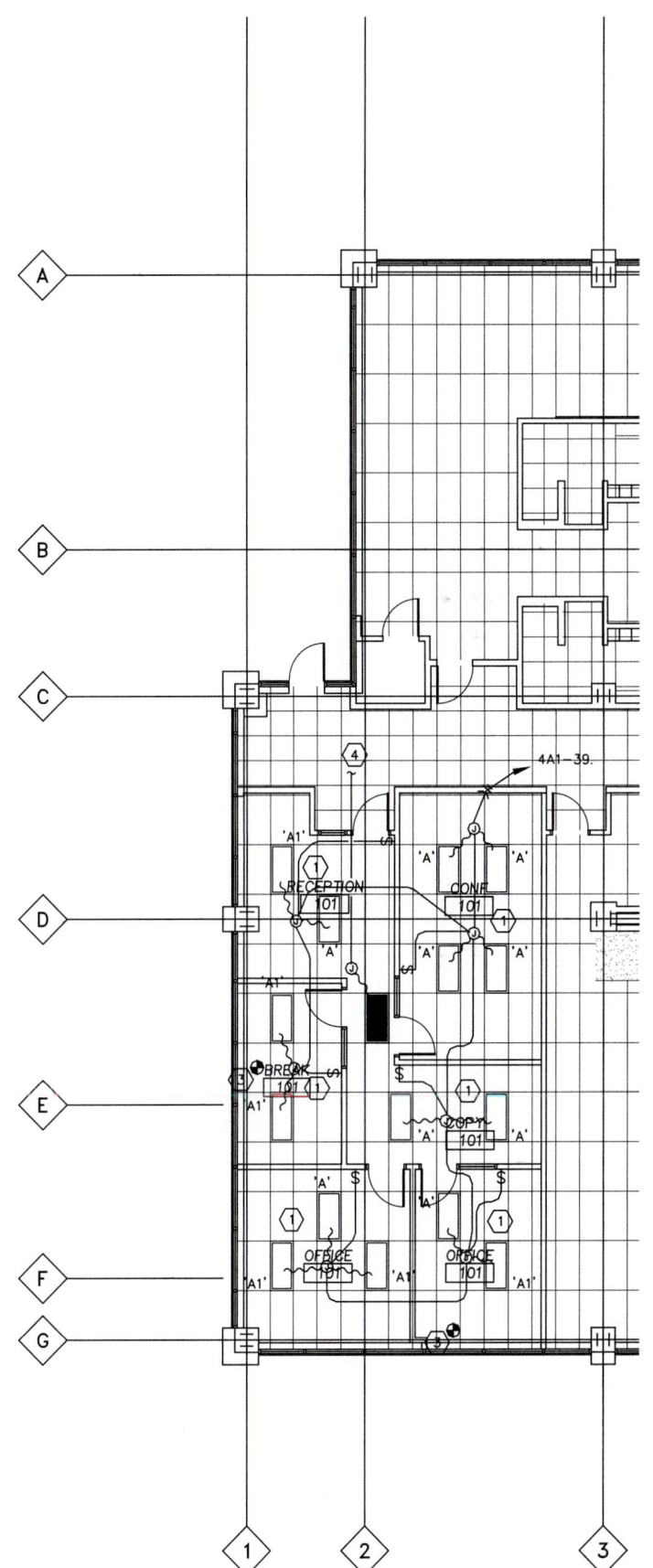

PROBLEM 26.34   First-floor power plan

Use the partial floor plan that you drew in Problem 26.33 as a guide to create the partial power plan shown in the given engineering drawing. Use appropriate CADD layers.
Courtesy Interface Engineering.

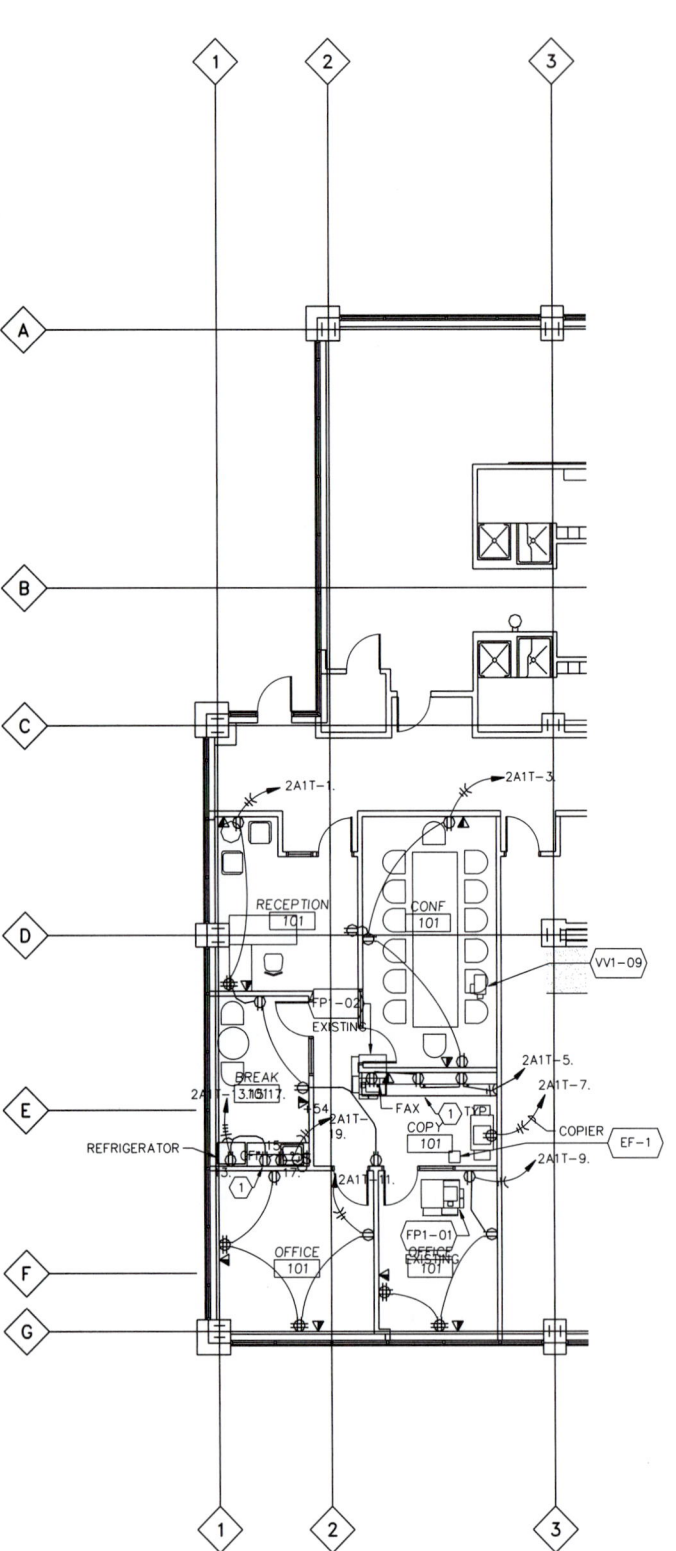

PART 2

DIRECTIONS

1. Follow previous instructions.
2. Use the selected engineering layouts and sketches to prepare each drawing. Keep in mind that engineering sketches may contain slight errors in format and symbol accuracy. Verify proper representation before drawing each symbol.

BLOCK AND SCHEMATIC DIAGRAMS

PROBLEM 26.35   Block diagram

Given the schematic diagram sketch that has been divided into stages, prepare a block diagram using 11 × 17 (A3 metric) size. unless otherwise specified by your instructor. Make all lines .020 in. (0.5 mm) wide. Do all lettering 1/8" (3 mm) high with titles 1/4" (6 mm) high. Sketch does not show connection dots.

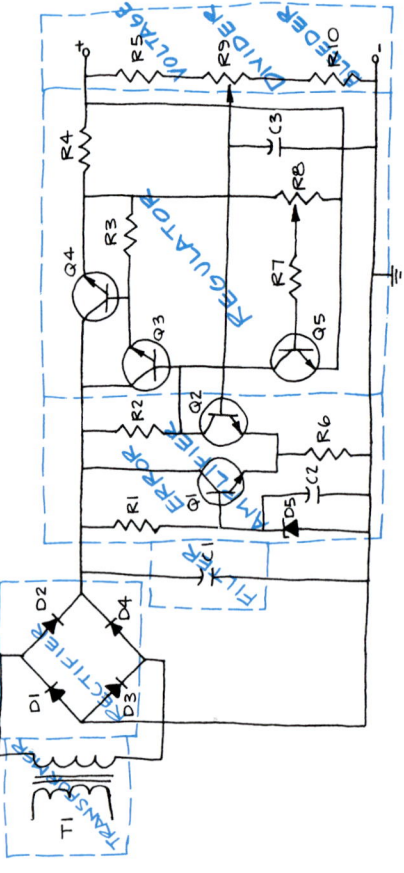

PROBLEM 26.36 Television receiver block diagram

Given the block diagram engineering layout, draw the block diagram using 11 × 17 (A3 metric) size unless otherwise specified by your instructor. Do all lettering 1/8" (3 mm) high with titles 1/4" (6 mm) high. Use .020 in. (0.5 mm) thick lines unless otherwise specified on the engineering sketch. *Courtesy RCA Consumer Electronics.*

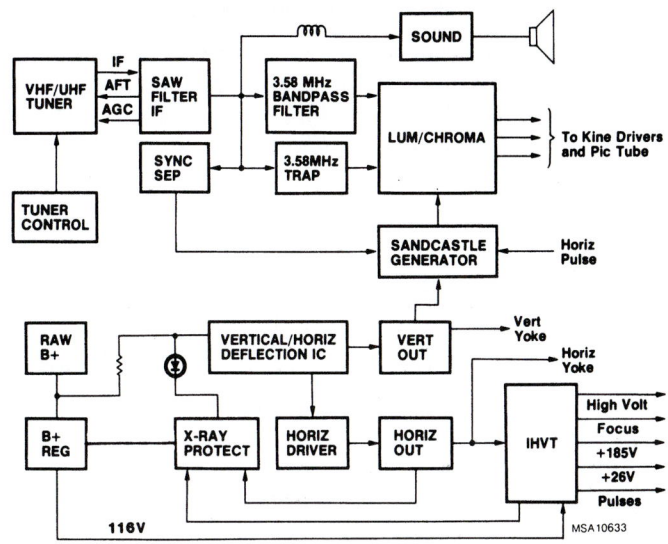

PROBLEM 26.37 Cylinder speed block diagram

Given the block diagram engineering layout, draw the block diagram using 17 × 22 (A2 metric) size, unless otherwise specified by your instructor. Do all lettering 1/8" (3 mm) high with titles 1/4" (6 mm) high. Use .020 in. (0.5 mm) thick lines unless otherwise specified on the engineering sketch. *Courtesy RCA Consumer Electronics.*

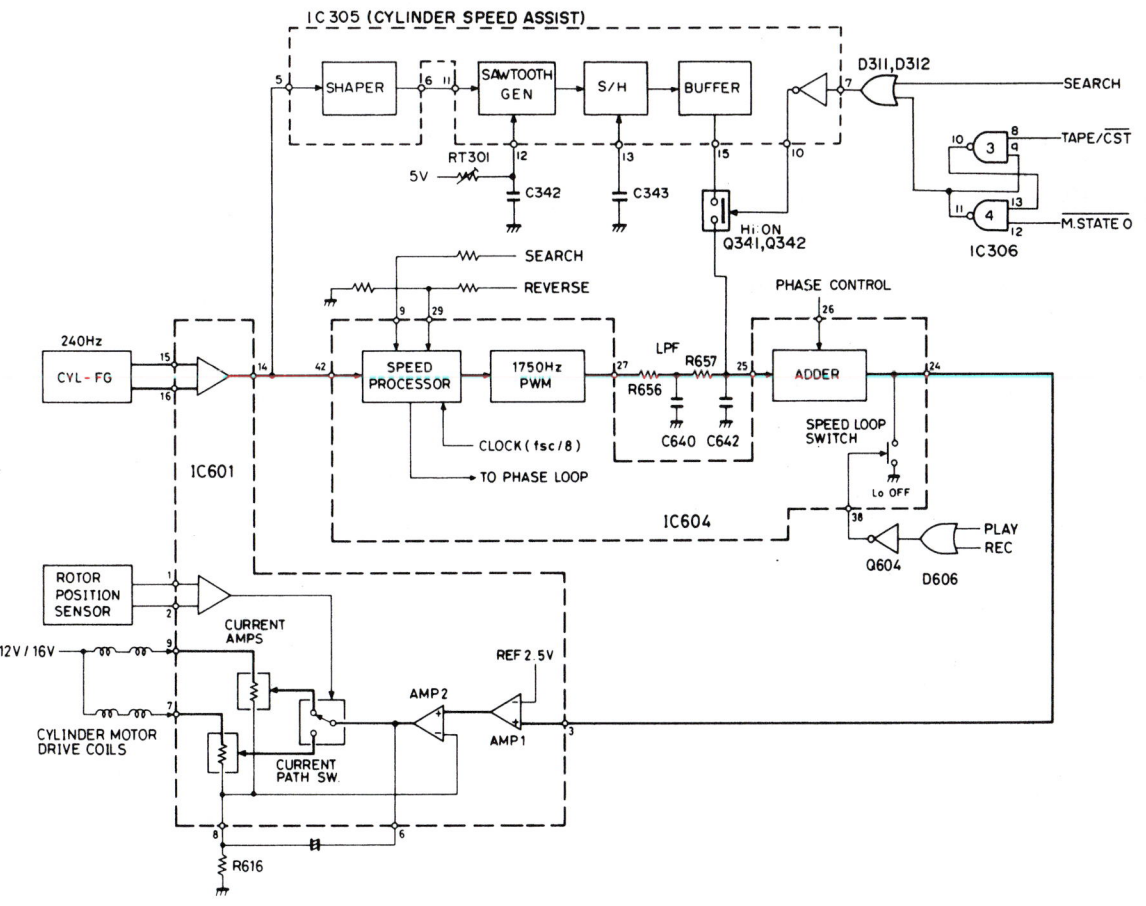

PROBLEM 26.38 Schematic diagram

Given the schematic engineering sketch, make a schematic diagram using 11 × 17 (A3 metric) size. There is no scale, and the drawing must be balanced, uncluttered, and easy to read. Use the following instructions unless otherwise specified by your instructor:

1. Draw the entire schematic with .020 in. (0.5 mm) wide lines.

2. Do all lettering 1/8" (3 mm) high with titles 1/4" (6 mm) high.

3. Follow the left-to-right and top-to-bottom labeling system using coding as described in this chapter: R1, R2, R3 . . ., C1, C2. . . .

4. Label horizontal components above and vertical components to the right.

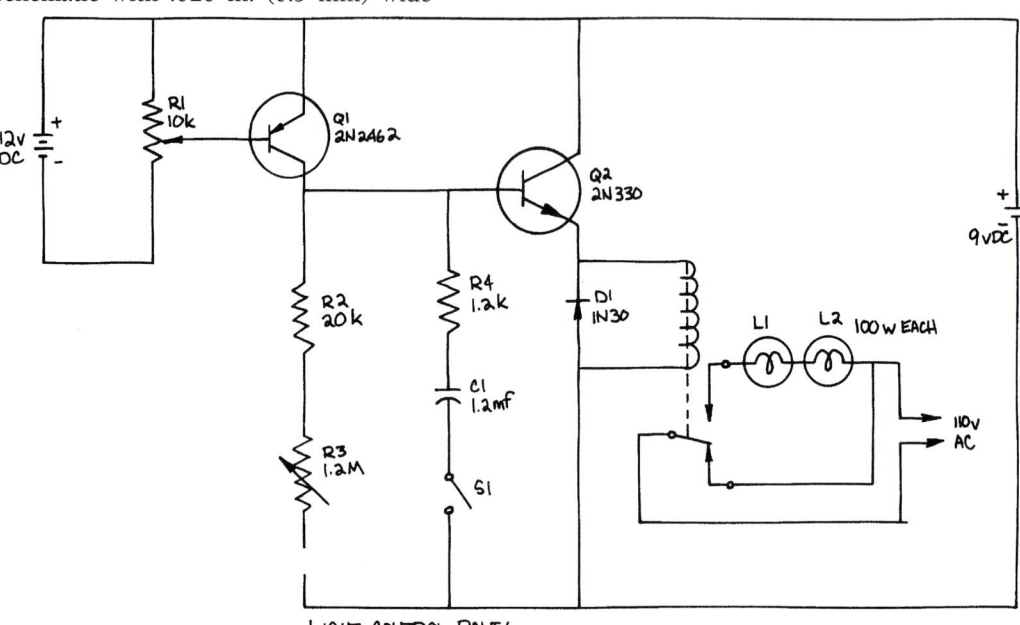

PROBLEM 26.39 Tuner schematic diagram

Given the schematic engineering layout, make a schematic diagram using 11 × 17 (A3 metric) size. There is no scale, and the drawing must be balanced, uncluttered, and easy to read. Use the following instructions unless otherwise specified by your instructor.

1. Draw the entire schematic with .020 in. (0.5 mm) wide lines, and dashed stage lines .028 in. (0.7 mm) wide.

2. Do all lettering 1/8" (3 mm) high with titles 1/4" (6 mm) high.

3. Reference designators are not shown; use the left-to-right and top-to-bottom labeling system, using coding as described in this chapter: R1, R2, R3 . . ., C1, C2. . . .

4. Label horizontal components above and vertical components to the right.

5. Place connection dots as necessary.

Courtesy RCA Consumer Electronics.

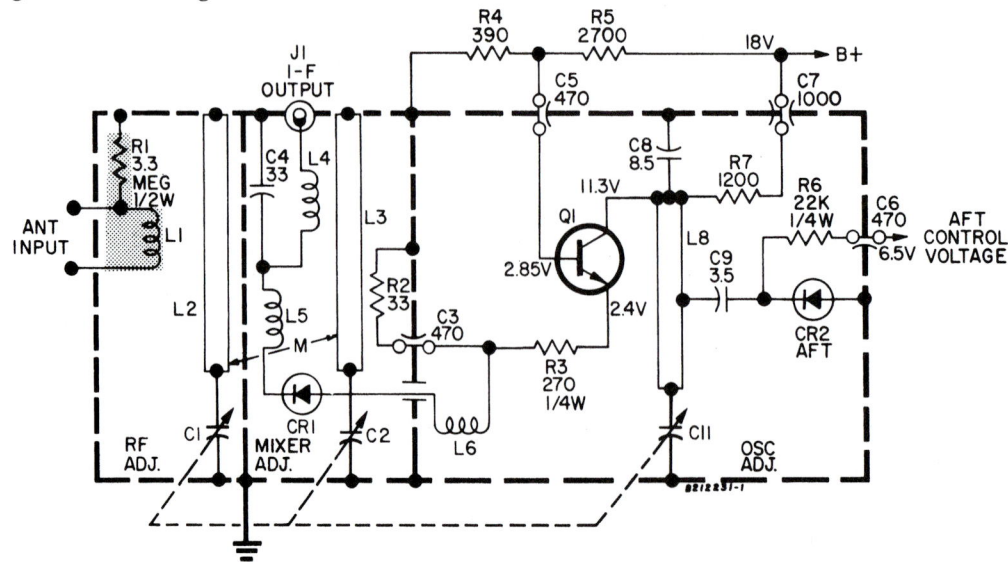

PROBLEM 26.40 **Television receiver tuner schematic diagram**

Given the schematic engineering sketch, make a schematic diagram using 17 × 22 (A2 metric) size. There is no scale, and the drawing must be balanced, uncluttered, and easy to read. Use the following instructions unless otherwise specified by your instructor:

1. Draw the entire schematic with .020 in. (0.5 mm) wide lines, and dashed stage lines .028 in. (0.7 mm) wide.

2. Do all lettering 1/8" (3 mm) high with titles 1/4" (6 mm) high.
3. Reference designators are not shown; use the left-to-right and top-to-bottom labeling system, using coding as described in this chapter: R1, R2, R3 . . ., C1, C2. . . .
4. Label horizontal components above and vertical components to the right.
5. Place connection dots as necessary.

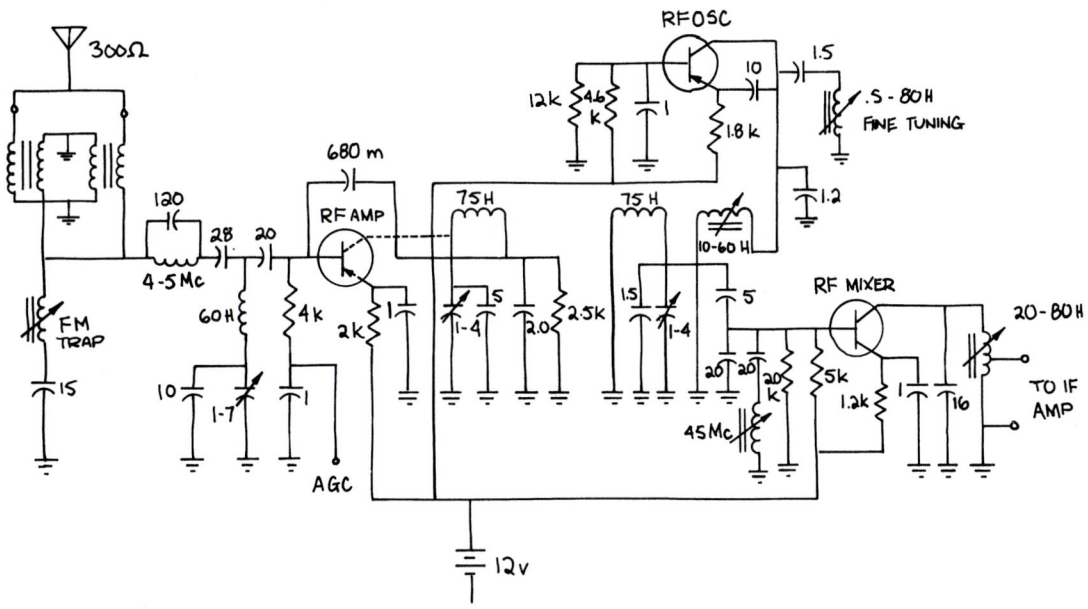

PROBLEM 26.41 **Electronics schematic**

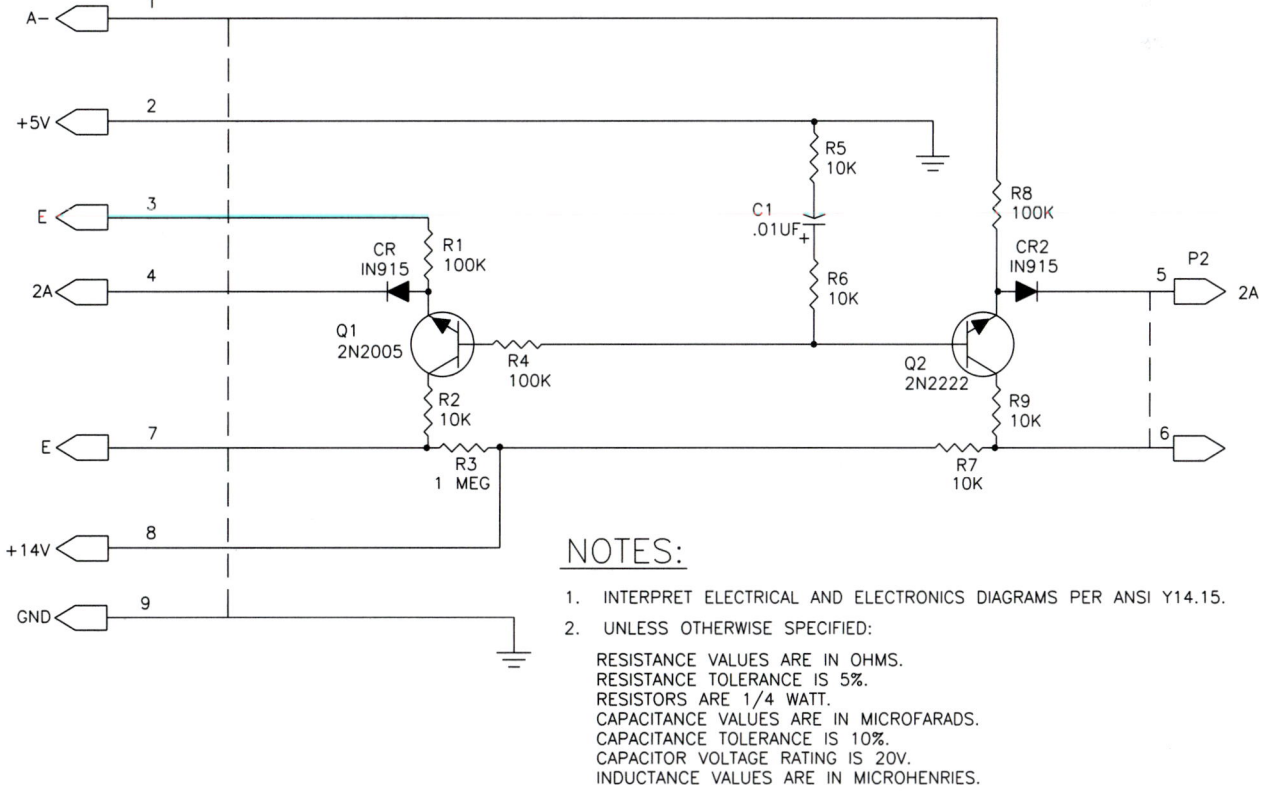

NOTES:

1. INTERPRET ELECTRICAL AND ELECTRONICS DIAGRAMS PER ANSI Y14.15.
2. UNLESS OTHERWISE SPECIFIED:

   RESISTANCE VALUES ARE IN OHMS.
   RESISTANCE TOLERANCE IS 5%.
   RESISTORS ARE 1/4 WATT.
   CAPACITANCE VALUES ARE IN MICROFARADS.
   CAPACITANCE TOLERANCE IS 10%.
   CAPACITOR VOLTAGE RATING IS 20V.
   INDUCTANCE VALUES ARE IN MICROHENRIES.

PROBLEM 26.42 **FM tuner schematic**

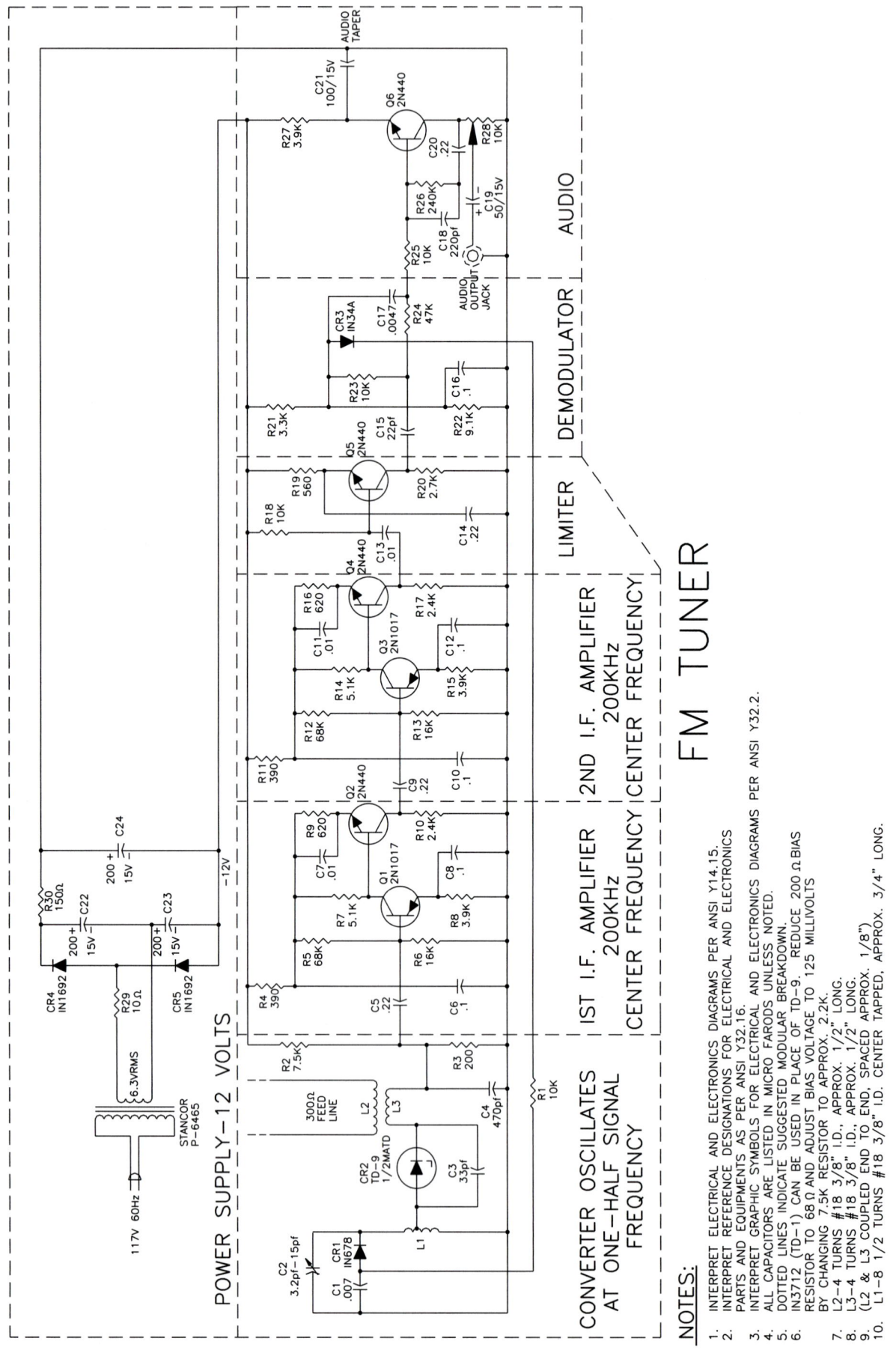

NOTES:

1. INTERPRET ELECTRICAL AND ELECTRONICS DIAGRAMS PER ANSI Y14.15.
2. INTERPRET REFERENCE DESIGNATIONS FOR ELECTRICAL AND ELECTRONICS PARTS AND EQUIPMENTS AS PER ANSI Y32.16.
3. INTERPRET GRAPHIC SYMBOLS FOR ELECTRICAL AND ELECTRONICS DIAGRAMS PER ANSI Y32.2.
4. ALL CAPACITORS ARE LISTED IN MICRO FARODS UNLESS NOTED.
5. DOTTED LINES INDICATE SUGGESTED MODULAR BREAKDOWN.
6. IN3712 (TD-1) CAN BE USED IN PLACE OF TD-9. REDUCE 200 Ω BIAS RESISTOR TO 68 Ω AND ADJUST BIAS VOLTAGE TO 125 MILLIVOLTS BY CHANGING 7.5K RESISTOR TO APPROX. 2.2K.
7. L2-4 TURNS #18 3/8" I.D., APPROX. 1/2" LONG.
8. L3-4 TURNS #18 3/8" I.D., APPROX. 1/2" LONG.
9. (L2 & L3 COUPLED END TO END, SPACED APPROX. 1/8")
10. L1-8 1/2 TURNS #18 3/8" I.D. CENTER TAPPED, APPROX. 3/4" LONG.

PROBLEM 26.43 Logic diagram

Given the IC schematic engineering layout for the logic diagram, make a schematic drawing using 17 × 22 (A2 metric) size. Use the following instructions unless otherwise specified by your instructor.

1. Do all line work and lettering using .020 in. (0.5 mm) width.

2. Do all lettering 1/8" (3 mm) high with titles 1/4" (6 mm) high.

3. All connection points will be ∅3/32" filled in.

4. Provide the general note: INTERPRET PER ANSI Y 32.2.

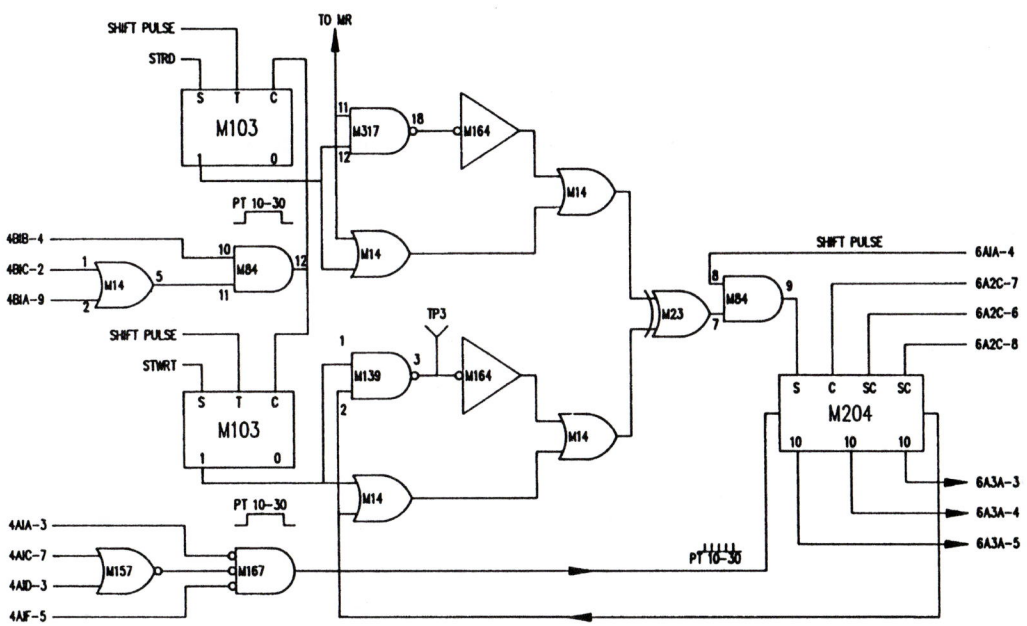

PROBLEM 26.44 Logic diagram

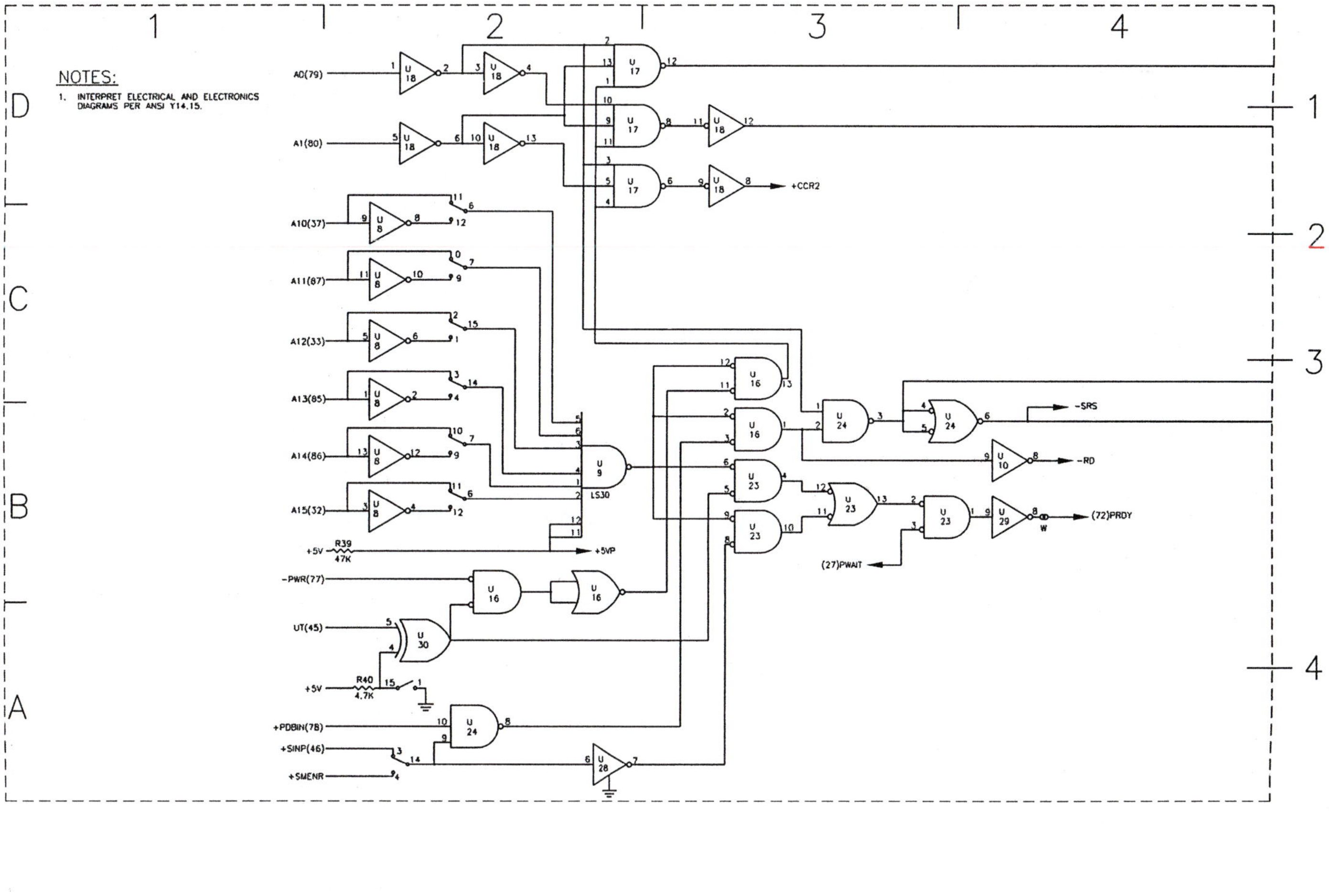

PROBLEM 26.45   Schematic diagram

Given the schematic engineering sketch, make a schematic diagram using 17 × 22 (A2 metric) size. There is no scale, and the drawing must be balanced, uncluttered, and easy to read. Use the following instructions unless otherwise specified by your instructor:

1. Draw the entire schematic with .020 in. (0.5 mm) wide lines, and dashed stage lines .028 in. (0.7 mm) wide.

2. Do all lettering 1/8" (3 mm) high with titles 1/4" (6 mm) high.

3. Reference designators are not shown; use the left-to-right and top-to-bottom labeling system, using coding as described in this chapter: R1, R2, R3 . . ., C1, C2. . . .

4. Label horizontal components above and vertical components to the right.

5. Place connection dots as necessary.

   Notes: UNLESS OTHERWISE SPECIFIED.

1. Reference designators are for reference only and may not appear on part.

2. Resistor values are in DHMS.

3. For PWB/component assembly, see dwg. 100/01.

4. For parts, see SPL 100/02.

5. For printed wiring board, see dwg. 100/03.

   Courtesy RCA Consumer Electronics.

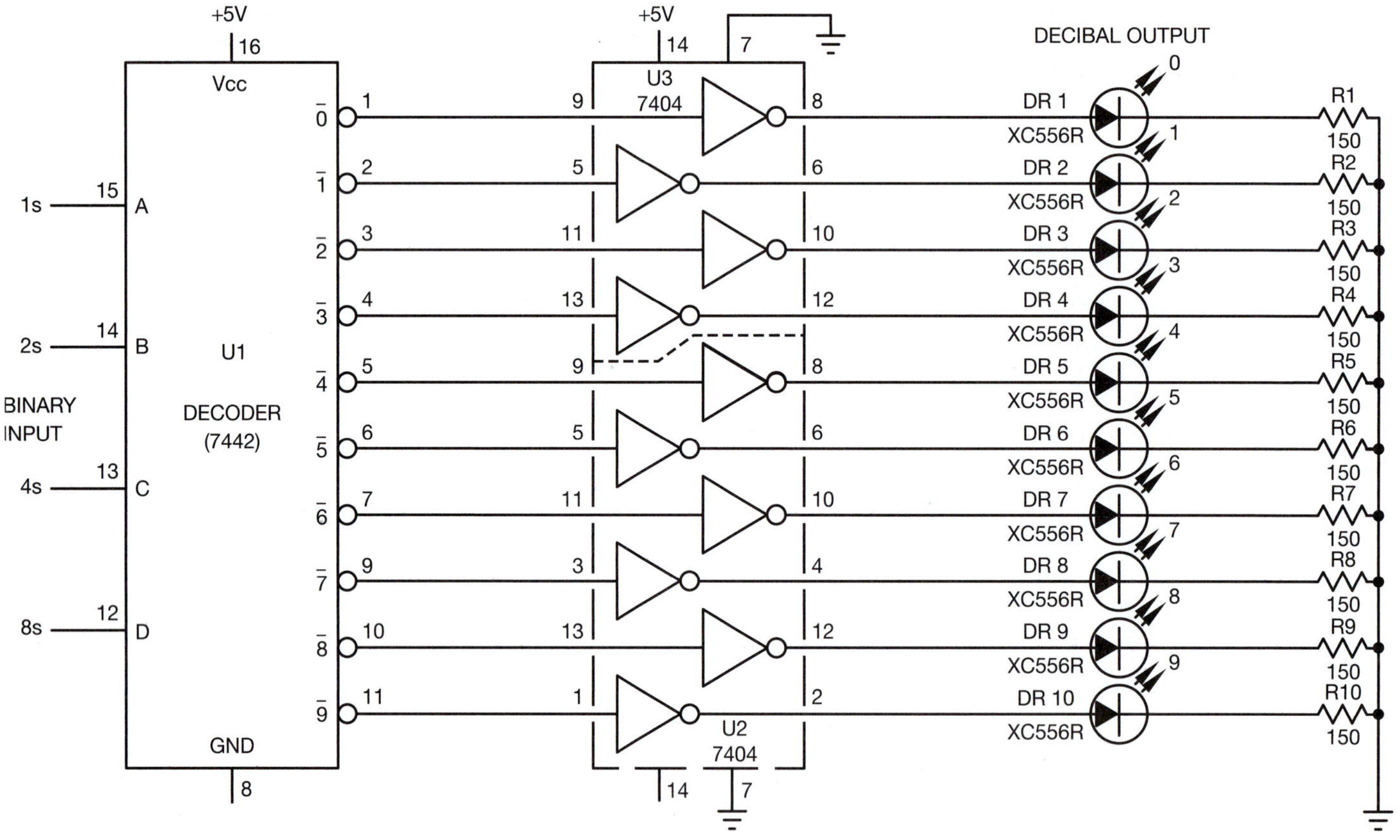

PROBLEMS 26.46 through 26.52: Access the CD found with this textbook and open the problem of your choice, or as assigned by your instructor. Solve the problems using the instructions provided with this chapter or on the CD, unless otherwise specified by your instructor.

MATH PROBLEMS

PROBLEMS 26.53 through 26.62: Access the CD found with this textbook and open the math problem of your choice, or as assigned by your instructor. Solve the problem or problems using the instructions provided.

Engineering Design

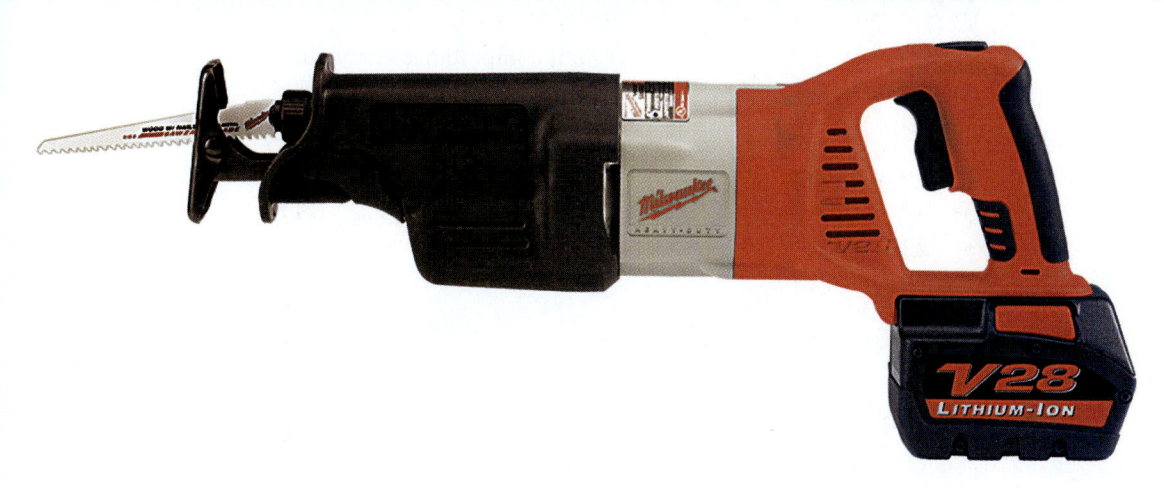

Courtesy Milwaukee Electric Tool Corporation

The Engineering Design Process

LEARNING OBJECTIVES

After completing this chapter, you will:

- Describe various factors that are changing the design process.
- Discuss the steps in a design analysis process.
- Explain the importance of creativity and innovation in the design product.
- Describe the problem-solving steps used in a design process.
- Explain the importance of concurrent engineering and teams in the development of a product.
- Define today's engineering design models.
- Explain the design review process.
- Describe design deliverables.

THE ENGINEERING DESIGN APPLICATION

TRANSITIONS TO A SYSTEMS APPROACH: A HISTORIC POINT OF VIEW

As you begin the study of the engineering design process, it is important to reflect a moment on the past and on how in recent times design systems have changed. Traditionally an engineering process has been a collection of activities that takes one or more kinds of input and creates an output that is of value to a customer. The delivery of the product to the customer or client is the value that the process creates. This is still true, but how the design process works today is significantly different from how it worked in the past. Engineering has used a design process that goes back to Adam Smith's model for production. Mr. Smith's model focused on individual tasks in a process of designing a product.

Traditionally, engineering, testing, manufacturing, marketing, cost analysis, and so on have been components of the design cycle. Each component has functioned independently with communication occurring only when the design was passed on to the next group. Often the larger objective of customer satisfaction and product quality was lost with the individual focus on the task at hand. Revisions and engineering change notices were the rule of the day. Communication from one end of the process to the other was limited to each group's perspective. Cost of production, time to market, and the demand for quality products have driven a change from this linear design engineering process to an integrated approach that brings together a team of people from all aspects of product development. The team includes engineering, marketing, production, and the customer. This system is referred to as the whole systems approach to design and is the primary way design is occurring in engineering firms today. Whole-systems is a process through which the interconnections between systems are actively considered, and solutions are sought that address multiple problems at the same time. Some refer to this process as the search for solution multipliers. (See Figure 27.1.)

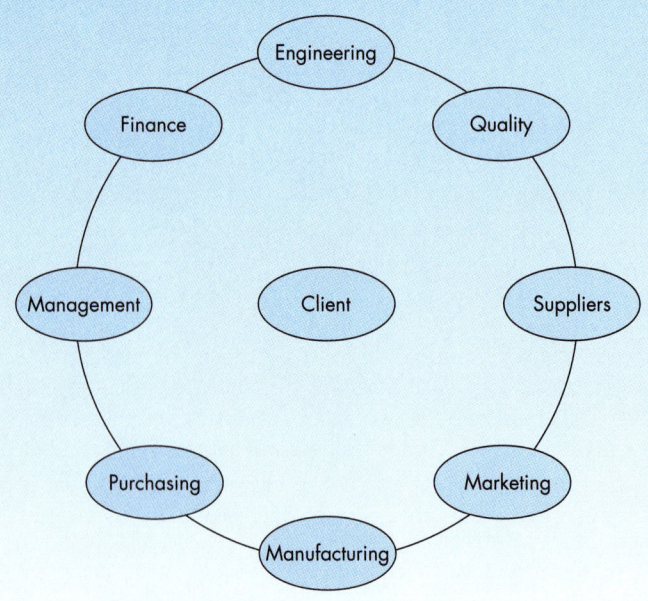

FIGURE 27.1 ■ The design team.

THE ENGINEERING DESIGN APPLICATION (continued)

The design process today is influenced by the need for better, less expensive, and faster-to-market products. Behind these changes in design is global competition, communication technologies that have accelerated the speed and volumes of information available to designers. Also, a compression of time has resulted from the Information Age of Technology, which creates a need to bring designs faster to market. The need and tools are at hand to design a quality product quickly, using all parties in the design process concurrently.

CHANGING BUSINESS MODEL

DESIGNER (YESTERDAY)	KNOWLEDGE WORKER (TODAY) TEAM MEMBER
DETAIL WORK	INTELLECTUAL WORK
FACTOR OF PRODUCTION	KNOWLEDGE PRODUCER
QUESTION NOTHING	QUESTION EVERYTHING
DO AS YOU ARE TOLD	DETERMINE WHAT TO DO
REPETITIVE TASKS	CONTINUOUS CHALLENGES
SEGMENTED WORK	HOLISTIC WORK
DIRECT SUPERVISION	AUTONOMY

FIGURE 27.2 ■ The changing business model.

INTRODUCTION

Today's designs are driven by cost, market trends, and customer service, as well as by quality and technology. Because of these realities, engineers, designers, drafters, and technologists now need to be aware of how their designs fit into the broader function of product development. In turn, they need to understand how their function fits into the entire product development cycle. And finally, this knowledge needs to be carried one step further into the areas of sales and marketing to understand exactly what the company is trying to sell and produce, to whom, and for what purpose. These are the forces behind the change to a whole systems approach to engineering design.

As a future designer or technician in the engineering profession, it is vital to possess abilities in areas such as problem definition, resource/information acquisition, design review, problem solving, technical communication and teamwork skills, business systems, project management, and above all, the ability to see the whole system. The basic premise of a whole systems approach to engineering design is the need for all people in the process to know all the necessary information and actions involved in the design process.

This sharing of knowledge, varying viewpoints, and ideas leads to better, faster, and more efficient problem solving. Accumulated knowledge and the availability of information in the engineering profession are explosive. The ability and tools to gather and archive information are available to the engineering design team more quickly and with less effort than ever before. This surge of information and its availability to everyone have brought a new model for decision making in the design process. The decisions are now made with input from everyone on a design team, not with just a few at the top passing it down the line. A person can use this shared data bank of product information for inventory planning, scheduling, purchasing, labor forecasting, or customer service. The power of all this input is a quality product that is cost-efficient and produced faster than in the past.

Traditionally, designers were trained in design fundamentals, the nature of materials, the capabilities of tools and equipment, and manufacturing processes. These skills are still very important to the development of an engineering designer or drafter. Today these skills are expanded to include knowledge of market trends, safety, data acquisition and management, teaming concepts, packaging, distribution, and storage. This is a full systems approach to design. Today's designer faces a world of complex dependencies and interrelationships that involve federal and state regulations, environmental impacts, consumer perceptions, economic/societal needs, cultural trends, and demographics impacting the design process. These forces have changed the world of engineering design and the fundamental nature of work. The change has moved from a designer assigned to a project to a knowledge-based workforce. (See Figure 27.2.)

TODAY'S ENGINEERING DESIGN MODELS

The following briefly reviews some of the systems approaches used by the engineering world to address the constantly evolving engineering design process.

Concurrent Engineering (CE)

Concurrent engineering (CE) is an integrated approach to design, production, and customer service that emphasizes the advantages of simultaneous, or concurrent, product design by employing individuals from various areas of the business in the up-front concept and design phase, with special emphasis on customers and their needs. The focus is on all aspects of the design simultaneously, which is accomplished by the integration of people, processes, standards, tools, and methods to achieve a quality solution to the design problem.

Life Cycle Engineering

The concept behind life cycle engineering is that the entire life of a product should be evaluated at the beginning of the design process. The optimal design cannot be achieved unless performance, start and sustaining costs, reliability, maintainability, disposability, and market trends are addressed up front. Built into this design process is the consideration for product replacement timelines, withdrawal and disposal of the product from the market, and product replacement.

Integrated Product Development (IPD)

Integrated product development (IPD) is the process by which the product is designed and developed to satisfy all the conditions the product encounters in its product life. Evaluating all the factors that impact the design during the product life at the initial design phase increases the likelihood that unforeseen circumstances will not accumulate and damage the product's longevity in the marketplace.

Knowledge-Based Engineering (KBE)

Knowledge-based engineering (KBE) is the use of computer models to simulate the best-known engineering processes. Typically KBE is used by creating a set of engineering data. This data is comprised of components such as CADD data, manufacturing data, tooling data, and structural information. This data is then used in an integrated manner to develop a more detailed and comprehensive design plan. The inclusion of so much information allows for better analysis and evaluation throughout the process.

Total Quality Management (TQM)

Total Quality Management (TQM) is a philosophy that calls for the integration of all organizational activities to achieve the goal of serving customers. It seeks to achieve this goal by establishing process standards, maximizing production efficiency, implementing quality improvement processes, and employing integrated teams to effectively design customer-driven products. TQM strives to eliminate all non value-added activities in the production process and to achieve satisfaction of all its customers, both internal and external.

AN ENGINEERING DESIGN PROCESS

Previously, you were introduced to the importance of the driving forces behind the engineering design process, such as market trends, customer desires, competitive pricing, quality, and technology. You also reviewed several different engineering design models used in industry. Bringing new or revised products to market quickly is critical in a competitive world. As a result, the design process has evolved into a system that focuses on quality and speed to market. While getting a product to market fast is necessary, the actual length of time it takes to go from an idea to a final manufactured product, on the assembly line, varies with each company, their products, and their marketplace. If the driving force is to make a profit, then the company must be very responsive to customer needs. This process also depends on the engineering design model used. For example, using the concurrent engineering model might reduce time to market by 40 percent, in some cases, over the traditional model.

The following discussion gives you an insight into an engineering design process that is used widely and focuses on how

one company in the manufacturing industry successfully uses the process. The process normally follows a step-by-step approach that can be modified as needed to meet specific objectives. It can take anywhere from one week to 10 years to bring a new product to market. The amount of time required to bring a product to market depends on the complexity of the product, the company, flexibility, maturity of technology, commitment, design, and documentation requirements.

Today's industry involves a cross-functional team approach, where everyone is involved in cooperation as a team, rather than a situation where one person or group does something and then passes the completed portion on to the next person or group. The entire process from a design idea to final production goes much more smoothly if everyone communicates about, and is involved with, the product development process.

The following information introduces you to marketing, the project leader, and a variety of other people who work to help ensure the success of a design from idea through production. The product description/business case is created by marketing. After thorough market research, marketing typically prepares the product description/business case for approval by management. It is important that the new product fit within the goals and long-term strategy of the corporation. The product description/business case is a formal document, which specifies the features, pricing, and costs of a new product. This document communicates the customer needs to engineering and provides guidance for the actions of the engineering team. Marketing gathers input from customers in the target market and also from departments such as purchasing, engineering, and sales. The person who takes charge of the project after marketing releases a product description/business case is the project leader. The project leader coordinates across the functions within the company's different disciplines, such as purchasing, quality control, manufacturing, and marketing as the project moves from design through manufacturing. The project leader does not generally have other departments reporting directly to him or her, so it is important for the project leader to develop good working relationships where cooperation and a common objective are achieved. It is also to the benefit of all groups and people involved to have a quality rapport with the project leader for the good of the project and the company.

You can use most of the concepts covered in the following discussion when designing your own ideas or for the completion of design projects at the end of this chapter. Some of the principles are highly technical in nature and are influenced by specific company activities, products, and objectives. You may not be able to use some of the information in your designs now, but you should be aware of the concepts for future applications when you work in industry.

Much of the following discussion was created from information obtained from Milwaukee Electric Tool Corporation. Paralleling the general design process information is the examination of a case study on a new product design from concept through production. The product used for this design sequence example is the Milwaukee Electric Tool Corporation's V28™ Lithium-

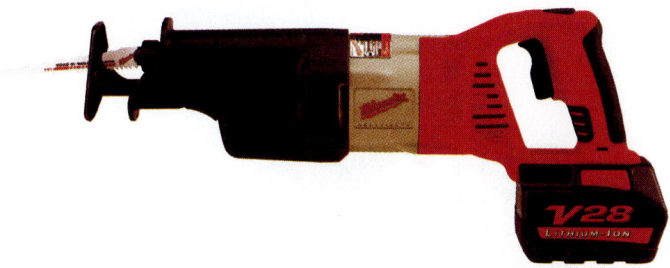

FIGURE 27.3 ■ The V28™ Lithium-Ion Sawzall® reciprocating saw. *Courtesy Milwaukee Electric Tool Corporation.*

Ion Sawzall® reciprocating saw. The V28 Lithium-Ion Sawzall® reciprocating saw is shown in Figure 27.3. In addition to the acknowledgment provided in the Preface, special thanks is given here to David P. Serdynski, Senior Design Engineer, and Troy Thorson, Design Engineer with Milwaukee Electric Tool Corporation, and their colleagues who supported this content.

Project Portfolio Management

A project portfolio is a compilation of all potential new projects that are under consideration for implementation. Project portfolio management is the methodology management uses to select projects for execution and the order in which they are completed. Project portfolio management is also used to determine the number of projects a company puts into action at any one time with the available resources. Several elements go into managing the project portfolio, including input from groups within the company and market research. Project portfolio management allows the company to prioritize new projects, and determine which projects to eliminate or save for implementation in the future. Project portfolio management helps the company identify what items are sensitive to the outcomes of business decisions. Project portfolio management can be supplemented using techniques from operations management, also known as management science, which is mathematics and optimization techniques used to prioritize the project portfolio. Management science uses a mathematical process in which a business or engineering solution is simulated to determine optimum results. The participants determine the independent variables which effectively simulate reality and the output as an optimized prioritization of projects. These mathematical processes are based on quantitative data rather than emotional opinions or political agendas. For example, if you have to evaluate several different products that you want to take to market within the next year, you determine the value or attractiveness of each as inputs that your company feels are correct to evaluate the projects under consideration.

Research Activities

Most of the accomplishments achieved in modern industry are the result of extensive research, planning, and evaluation. The following information describes the two basic types of research, including advanced research and existing product research.

Advanced Research

A competitive edge in modern industry is often established through advanced research. Advanced research is used to determine if there are areas of technology currently not utilized. This unutilized technology can be commonly found in some applications while untapped or currently unattainable in others. Unutilized technology also can be completely new. In either case, new technology is something the company strives to develop. The goal of advanced research is to create a product no one else has or a technology no one else can make work. Advanced research also involves working with specific partners in industry to develop a technology that exceeds anything currently available. The desired results can put the company in a position of being the only one who has this technology available to sell to customers. This also makes very high barriers for other companies to get into the same market.

An excellent example of effective advanced research is the Milwaukee Electric Tool Corporation V28™ Lithium Ion (28-volt) battery technology. Cordless electric power tools have been available for years, but none of them have used the superior technology of the lithium battery. Other industries have used Lithium Ion technology, such as cellular phones and laptop computers, but this technology has not previously been used on power tools. This example shows how advanced research has resulted in the following gains in this technology:

■ Up to twice the run time of 18-volt tools.

■ Delivers up to 40-50% more power.

■ Provides 28 volts of power at the weight of 18-volt batteries.

Figure 27.4 shows the complete line of V28 powered portable electric tools.

Existing Product Research

A common approach to the design of new and updated products is based on existing product research, and is also called bench marking, which is discussed later. Existing product research involves redesign of a current product. To be successful in this process, it is important to know everything about the function of the existing product and evaluate how it is used. It is also critical to look at competitive products and carefully compare their features and operation to the product being redesigned. Redesign can involve use and modification of features available in other industries. Existing product research can also involve modernization of a product with new features such as ergonomic design, shape, and color.

Project Termination

The project team has the responsibility to stop the project at any time if it becomes unprofitable or if market conditions change. A project can get terminated based on new market information, new competitor releases, and changes in company direction or resources. Termination of a proposed project should occur as soon as obstacles mount and belief that future efforts would be a poor use of company resources. A decision to continue or stop

FIGURE 27.4 ■ The complete line of V28™ Lithium Ion powered portable electric tools. *Courtesy Milwaukee Electric Tool Corporation.*

a project is made by a broad cross section of people who provide information to the project team. The project can be canceled if feedback indicates continuing the project is a poor idea. Stopping a proposed project can be a strategic business decision or a cost savings issue.

THE PHASE GATE DESIGN PROCESS

Modern industry uses a variety of processes to help ensure success when creating a new product design. While there is no guarantee of achieving success, it is important to do the required research, survey customers, and make smart business decisions based on all available information and technology. The phase gate design process helps to prevent failures in new product development. Past experiences help a corporation identify the correct system of checks and balances leading to successful projects. The phase gate design process is a practice that makes sure you get the right work done at each phase of design before you are able to go to the next phase. The idea involves an imaginary gate, at the end of each phase, that must be opened

before you can continue. You must complete certain tasks in order to open each gate. The project can be terminated before you pass through the next gate if specific criteria are not met. The costs associated with continuing a project rise after each phase gate. Therefore, it is important to discontinue a project as soon as it becomes unattractive. Figure 27.5 shows a flow chart providing the phases and gates of the phase gate design process. The "YES" indicators, in the Figure 27.5 flowchart, represent paths leading to success. Sometime a YES path allows the process to continue as the result of a deficiency correction in the process. The "NO" indicators mean that there is a problem that needs to be corrected in the previous step. NO paths can lead to corrections, which result in a revised YES path, or they can lead to additional NO paths where more work is required to correct a problem. If deficiencies cannot be corrected, NO paths can lead to canceling the project.

Phase 1: The Concept Phase

The first phase of the design process is the concept phase. The concept phase begins after a marketing team creates the previously

New Product Development Process

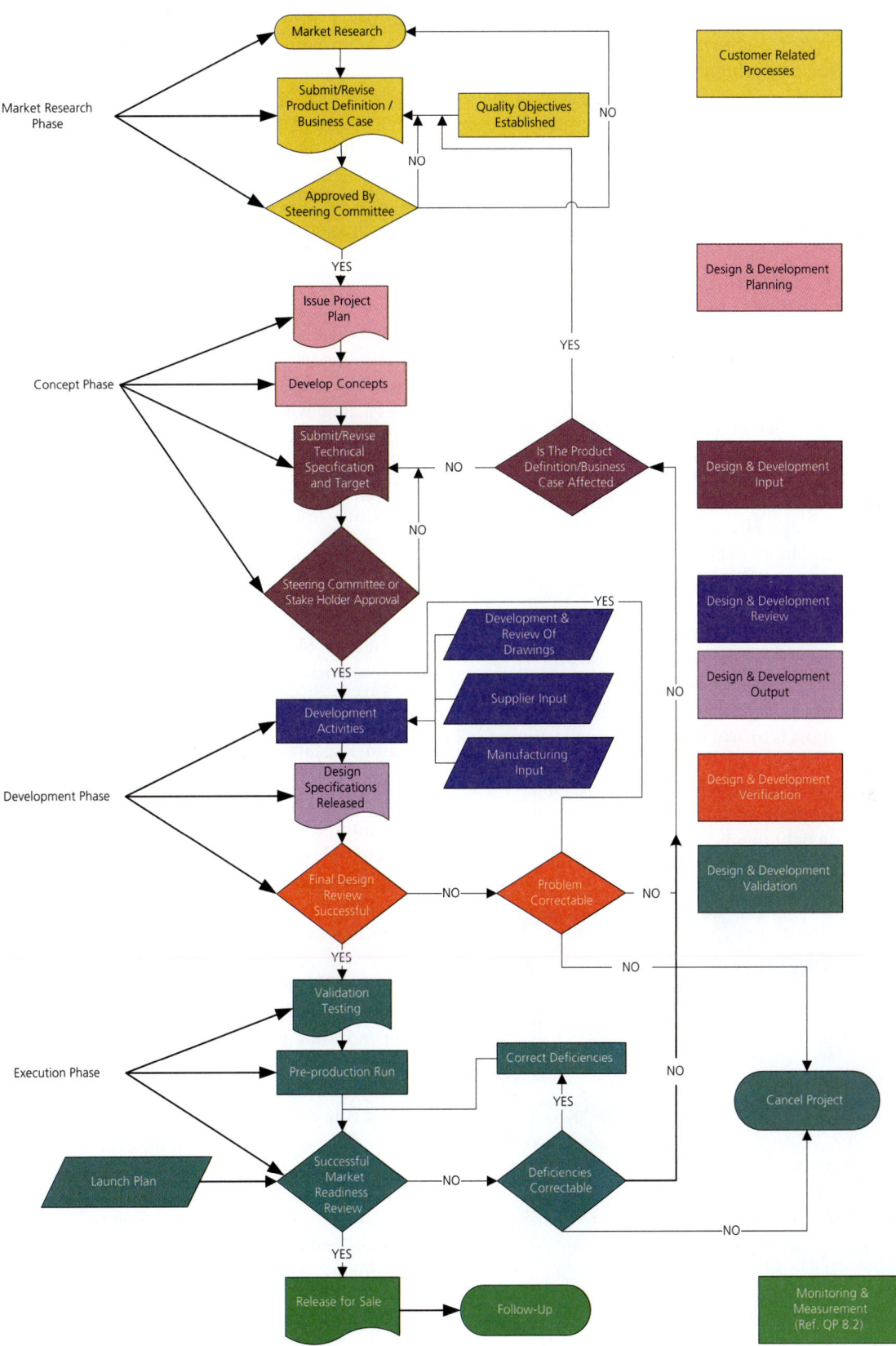

FIGURE 27.5 ■ A flowchart providing the phases and gates of the phase gate design process. *Courtesy Milwaukee Electric Tool Corporation.*

discussed product description/business case. In this phase, multiple concepts are typically explored and evaluated, leading to the selection of the best solution.

Customer satisfaction is critical to the success of all products and is specifically addressed during the concept phase. There are a variety of ways to determine customer satisfaction depending on the specific product, the size of the customer base, and the desired results.

A project plan is also created by the project leader, which issues a schedule communicating the anticipated implementation timeline for the project. The project team is also formed at this time. Each team member is given their assignments and expectations are established. A major benefit of this practice is to ensure all critical tasks are identified and accounted for.

Concept Activities

Industrial Design in the Concept Phase. The concept team is primarily made up of industrial designers, designers, and engineers. Industrial design (ID) is a team of highly skilled and creative people who have extensive product knowledge and experience. The ID team has the flexibility of being extremely creative and their knowledge of engineering and production processes is essential. The ability to be creative and work within engineering, manufacturing, and cost constraints is a key to having a quality industrial design group.

ID tools include pencil, paper, markers, and computers with specialized hardware and software. Traditional pencil and paper with color markers are commonly used in a creative and artistic manner for many initial design concept sketches. Basic drawing tools still allow for the most intuitive and rapid visualization of multiple concepts.

The computer and a peripheral pressure-sensitive tablet are also used by the industrial designer as shown in Figure 27.6. The industrial designer uses a light touch for layout and harder pres-

FIGURE 27.6 ■ The computer and its peripheral pressure-sensitivity tablet are also used by today's industrial designer. *Courtesy Milwaukee Electric Tool Corporation.*

sure for solid lines and forms when working on the electronic tablet. A computer stylus and tablet simulate the feeling of pencil and paper, providing a direct relationship between the designer's brain and the sketch. The designer starts by drawing shapes, then renders as needed, including highlights for depth and appearance. Modifying the computer-generated design sketch is easy, and efficient, especially with the aid of commands such as undo, which can be used as often as needed to remove unwanted work. Different software applications have different commands for this activity. The ID team uses a variety of software applications, but the industrial designers with Milwaukee Electric Tool Corporation prefer Alias® Sketchbook™ Pro (*http://www.autodesk.com*), which provides close to a natural feel and appearance according to Scott Bublitz, Industrial Designer.

The process begins with a meeting held to start the project. Marketing, the project team, and industrial designers attend this meeting. The initial direction is communicated to the industrial designers along with any research that has been done on existing products and the target market. Engineering communicates the overall approach planned for production. The concept design phase begins following this meeting. Concept design also involves further research into the product and observation of similarly existing products in use. This is also the point at which ergonomics are researched for the given product. The consideration of ergonomics is very important, especially when the product is a handheld tool, such as the design example used throughout this discussion. Sketches are then created illustrating various layouts in multiview or pictorial, features, and styling. Usually six to twelve idea renderings are created in the concept design phase. These are then presented to marketing and engineering for their input and assistance in narrowing down the concepts to one or two desired possibilities. This is sometimes as easy as picking a concept that everyone likes, but often involves combining features of multiple concepts to create a hybrid that satisfies the design intent. This is also a good checkpoint for engineering to examine any issues that may be inherent to a specific design or approach. This allows the industrial designer to make needed adjustments or corrections in the process rather than having to clean up a mess later as an afterthought. Next, the industrial designer begins concept refinement, where final renderings and multiple views are created. This is also usually the time when the creation of foam mockups is helpful, although this sometimes happens even earlier. The refined concept or concepts are then presented again to engineering and marketing. Once everyone is in agreement on a concept direction, CADD models are created either by industrial designers or CADD designers. A comparison between the industrial design and the modified CADD design for the V28 Sawzall® reciprocating saw is shown in Figure 27.7.

As discussed earlier, the development of a concept involves a cross-functional team approach, where everyone is involved in cooperation as a team, rather than a situation where one person or group does something and then passes the completed portion on to the next person or group. The entire process, from a design idea to final production, goes much more smoothly if everyone agrees about, and is involved with, what is being done during every step. For the cross-functional team approach to

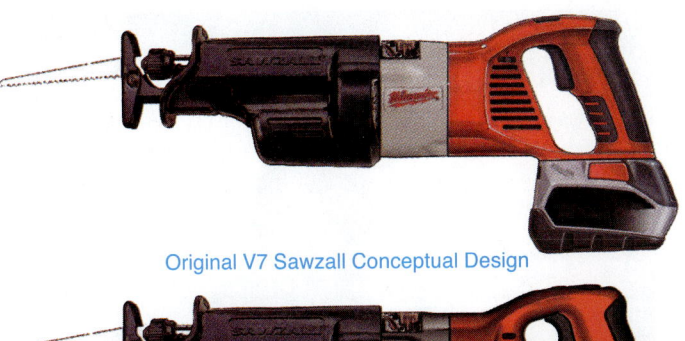

Original V7 Sawzall Conceptual Design

Current, Engineered V7 Sawzall

FIGURE 27.7 ■ A comparison between the industrial design and the modified CADD design for the V28 Sawzall® reciprocating saw. *Courtesy Milwaukee Electric Tool Corporation.*

work, engineering parameters are communicated at the initial meeting, and concepts challenging this are assessed by engineering in process. Industrial designers then work with engineers to solve the problem in a manner that pleases engineering, industrial design, and marketing. In this way, industrial design essentially becomes the cross-functional liaison between what marketing wants and what engineering can accomplish.

Quality Function Deployment (QFD). Information from customers is a key element of the concept phase, and one method of determining the voice of the customer (VOC) is by using a system known as quality function deployment (QFD). QFD is a product and service planning process that starts and ends with input from customers. Customer feedback is the driving force behind the development requirements for a new or revised product or service. QFD is generally added to the process depending on the scope of the project and the need for customer response. A team of marketing and research professionals generally come together to implement the QFD process, but the method can be done by one person, depending on the complexity of the task. The format of the QFD process is designed to remove many of the barriers found in organizations, including emotions, and unsupported perceptions about the way products or services are developed. When the QFD process works as it is intended, the customer knowledge available in the marketing department is coupled with the product knowledge found in the engineering department. Companies deciding to use QFD must listen to what customers want and investigate how to respond to customer needs. QFD is important to use when companies take the creation of new or revised products and services seriously. It is essential to use a process such as QFD for these reasons:

■ It is difficult to be impartial when you do product planning for the company where you are employed.

■ The process is mechanical rather than emotional or political.

■ It helps highlight customer desires.

■ It uses an analytical approach to rank customer wishes.

It is sometimes preferable to hire an outside firm to do market research, in an effort to isolate the company from the research and reduce as much favoritism as possible. Other times, the expertise held within a corporation lend it better to complete this research directly. Proper use of customer research allows the company to seek the truth rather than make the results say what may be preferred. Customer input can be established through one or more of the following means:

■ Mail or phone surveys or questionnaires.

■ Telephone or field interviews with customers.

■ Product clinics at retailers or by customer invitation.

■ Focus groups made up of key customers, retailers, or wholesalers.

■ Listening to customer complaints and recommendations from feedback on your website or through other correspondence.

■ Evaluating warranty service documentation.

QFD can be done before the product description/business case, so it can be used as part of the product description/business case. Projects at the end of this chapter allow you to design products and do your own customer research.

The Design Specification. During the concept phase, technical specifications and performance targets are established. These give guidelines to the project team and help identify what constitutes a successful design. Various factors are considered when developing the design specification, including, but not limited to, competitive product investigation, customer research, expected life or warranty, and certification requirements. QFD can be used as a tool to help translate customer needs into the design specifications. The design specification is a document typically made for each project with requirements such as:

■ Dimensions

■ Weight

■ Materials

■ General shape

■ Power requirements

■ Voltage requirements

■ Amperage specifications

■ Product life and durability requirements

■ Required agency approvals

■ Performance requirements

■ Included and optional accessories

■ General and special features

■ Environmental factors

Gate 1: Passing the Concept Phase Gate

Throughout the concept phase, all of the design possibilities are investigated and a desired product concept is selected. Mockups and cost estimates are made to aid the decision process. After the design team has completed their investigation and arrived at a selected concept, a concept review is held to officially approve the concept and to get a "*go*" decision to move into development. The concept review team is a group of people representing engineering, marketing, quality control, manufacturing, and management.

Phase 2: The Development Phase

When the concept phase is successfully completed, it is time to start the development phase. The development phase is where a full-functioning prototype model is made that operates at the desired quality level. Prototype parts can be machined or created using rapid prototyping. Parts are assembled into the desired product and then tested to determine if the design meets specific metrics. Metrics is a term frequently used in engineering to stand for product requirements such as weight, dimensions, operation performance criteria, comfort, or other physical factors included in the design. For example, a requirement specified in a product description/business case may be for a product to weigh a maximum of five pounds. Using this example, if the product weighs five pounds or less the design is successful based on the weight metric, but if the product is over weight, the design or the specification must be modified. Another example of metrics might be a functional quantity that the product must meet when in use, such as a portable battery-operated automated screwdriver that must place 500 screws per battery charge. During capacity testing, if the tool performs at only 300 screws per battery charge, something needs to be done to improve the performance by designing a more efficient mechanism or a better battery. The testing performed on the functioning prototype can be robotic or human testing to determine if the proposed product meets a number of use cycles and is good enough to continue further development.

Creating a Functioning Prototype and Establishing Purchase Parts.

Part of the development phase includes making prototypes and working models that allow testing and confirmation with marketing, sales, and customers that the product is appropriate. A functioning prototype is built to verify product performance and to finalize part tolerances. Figure 27.8 shows a rapid prototype part for the V28 Sawzall® reciprocating saw.

The design might have to return to the concept phase for reevaluation if it is found that some aspects of the design do not perform as intended, or manufacturing process are estimated to be too costly.

After the functioning prototype has been built and tested, drawings are created for assembly of all components. Parts made by others are called purchase parts. Purchase parts and purchase part drawings are discussed in Chapter 18. Figure 27.9 shows a purchase part drawing. These drawings define the crit-

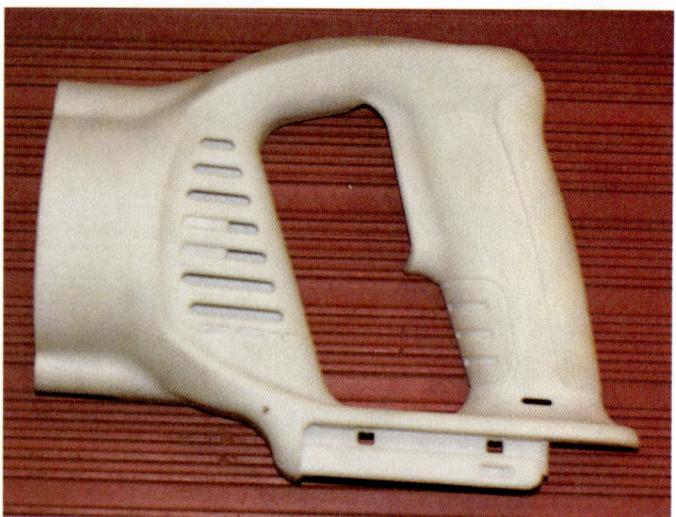

FIGURE 27.8 ■ A rapid prototype part for the V28 Sawzall® reciprocating saw. *Courtesy Milwaukee Electric Tool Corporation.*

ical performance functions, dimensions, and tolerances, and are used by suppliers to provide price quotes. The purchasing department identifies possible suppliers, gets quotes, and then makes the final supplier selections. Parts made in-house go through a similar process in cooperation with manufacturing engineering. In-house refers to any operations conducted inside the company. At this time, the Quality Engineering department generally reviews all drawings to insure the parts can be inspected in production.

Checking Intellectual Properties.

In addition to creating a functioning prototype, other concerns need to be resolved, such as intellectual properties. As taken from the World Intellectual Property Organization website (*http://www.wipo.int*), intellectual property is divided into two categories: Industrial property, and Copyright. Industrial property includes inventions (patents), trademarks, and industrial designs. Copyright includes literary works such as novels, poems, plays, films, and musical works, and artistic works such as drawings, paintings, photographs, sculptures, and architectural designs. Intellectual properties have to be investigated to be sure the design does not infringe upon something that is already on the market or will be on the market soon. A popular form of intellectual property investigation is called patent research. Patent research is done to ensure there is no infringement of an existing patent. It is essential to be sure there is no violation of an existing patent before significant investment is made in tooling and manufacturing equipment. Invention disclosure is the process that establishes an idea with a written dated document securing the design as yours. This is followed by filing a patent application. This process is critical, because competitors are also doing the same types of exploration with their market research, making it possible that they can come up with the same design at approximately the same time. It has been known to happen that the same design is submitted for a patent within days of a previous submittal. Intellectual property and workplace ethics were also discussed in Chapter 1.

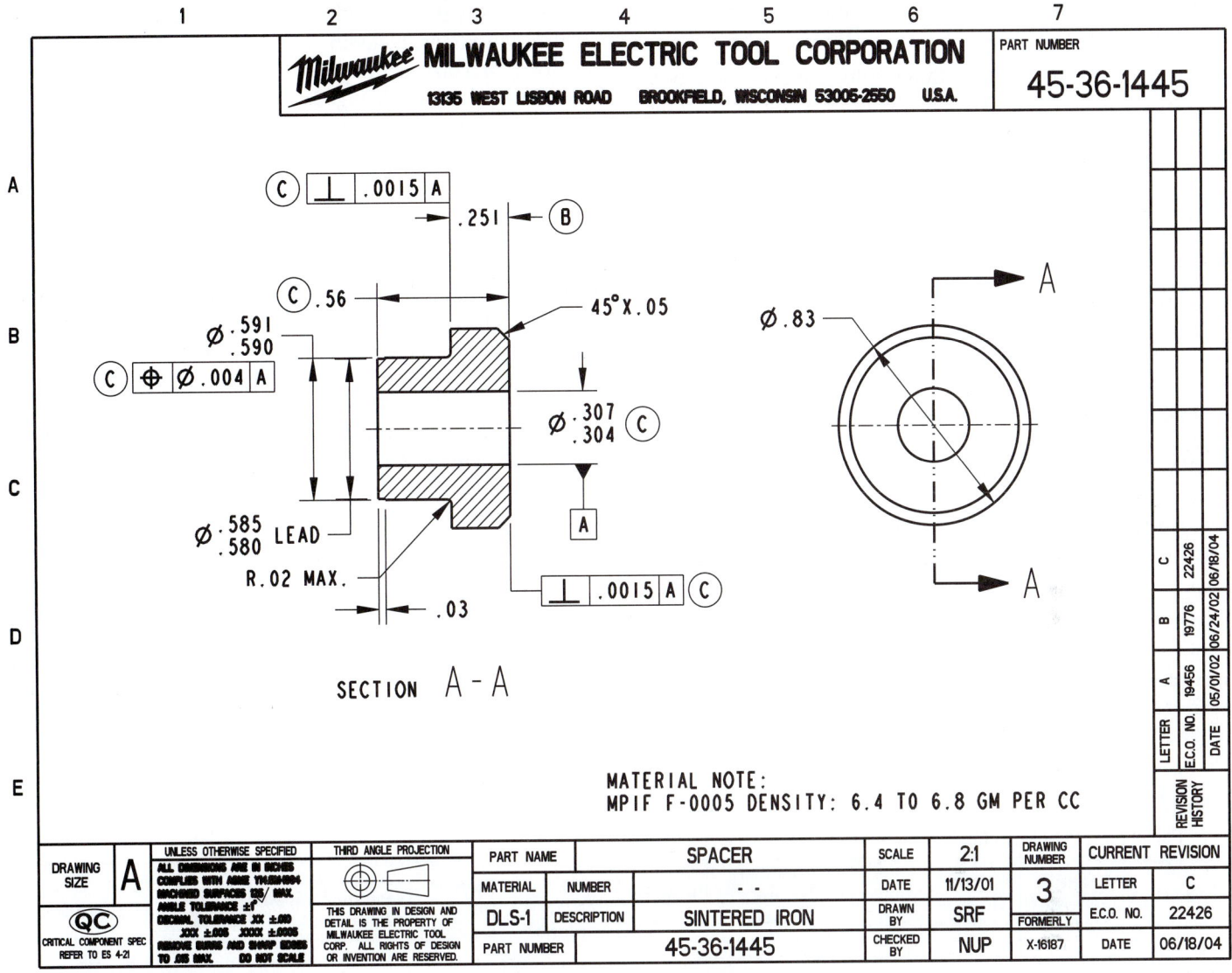

FIGURE 27.9 ■ A purchase part drawing. *Courtesy Milwaukee Electric Tool Corporation.*

The Product Bill of Materials in the Development Phase

A product bill of materials is created to establish the final product cost and to help organize the manufacturing process. Capital that needs to be spent to make parts in production, such as tooling, die, and pattern costs, also needs to be determined at this time. If cost or any other process is unacceptable, the project can be delayed until the situation is improved. For example, if a product is estimated to be $9 overbudget, the project leader may be responsible for reducing the cost before proceeding. Review Chapter 18 for a discussion and examples of parts lists and bills of material.

Failure Mode Effect Analysis (FMEA)

Once the development phase is underway, the design team initiates other necessary activities. It is important for an engineering design to function under all expected conditions and within all tolerances ranges. A tool used to help achieve this is called failure mode effect analysis (FMEA). FMEA is a technique used to determine possible failures and ways to eliminate them. If a failure cannot be eliminated, the design should minimize its impact on the product. The FMEA is also used to prioritize issues. For each potential failure, the following issues would be addressed:

■ The potential effect of the failure. What events will take place if this failure happens?

■ The severity of the failure. Severity can range from loss of life to minor irritation. More severe problems must be addressed first.

■ The probability of occurrence. Failures most likely to occur are addressed first.

■ The probability of not detecting the failure until serious damage is done. An example of this would be fatigue cracks on an airplane wing. A crack detected while the plane is on the ground is less harmful than a rupture while in midflight. Hard-to-detect failures get priority.

■ Potential failure causes and mechanisms. It may take considerable time and testing to identify the main causes.

■ Current controls and procedures. What controls and procedures are currently in place to eliminate and/or mitigate the failure?

■ Recommended actions. What future actions need to be taken?

■ Verification. Is evidence available that the problem has been addressed? Verification can come from laboratory testing, field tests, or the performance of a similar product over time.

Companies commonly identify special product characteristics with an appropriate symbol on an FMEA worksheet. These special characteristics are normally items that affect regulatory compliance, such as features that should be given consumer warnings or special process controls. Figure 27.10 shows the FMEA worksheet for the V28 Sawsall® reciprocating saw.

Computational Dynamics and Finite Element Analysis in the Development Phase

A variety of software products are available for mathematically analyzing features such as gear, cam, and bearing design. Finite element analysis (FEA) software programs allow for mathematical solutions to structural and thermal problems. FEA consists of a computer model of a material or design that is stressed and analyzed for specific results. It is used in new product design and existing product modification. Using FEA, a company is able to verify that a proposed design is able to perform to desired specifications, including tolerances and fits, before manufacturing or construction. Two basic types of FEA are 2-D modeling and 3-D modeling. Two-dimensional modeling allows the analysis to be run on most computers, but generally provides less accurate results.

In comparison, 3-D modeling results in better accuracy, but high-end FEA programs must be run on powerful computers. Many parametric modeling packages have stripped down FEA applications that can be run on most PCs. When using an FEA program, the engineer provides a variety of mathematical functions to make the product or structure perform under desired stress conditions. Figure 27.11 shows a 3-D model being subjected to simulated tests and stress analysis.

Another benefit of FEA is its use in bench marking, which means that you analyze the product where you start, establishing the bench mark. Then, when you go through the redesign or make desired changes, you reanalyze to determine the difference from the bench mark condition to the revised condition.

An example of an FEA software program is TK Solver, provided by Universal Technical Systems, Inc. (*http://www.uts.com*), which is software that allows you to perform the finite element analysis functions previously discussed and also a process referred to as back solving. An example of back solving is when you are designing a bearing, you input data, and the software determines the force it takes to press fit the bearing on a shaft. The inputs for this operation include the shaft size, hole size, and material specifications. In reverse, back solve is when you

Partial Sawzall Design FMEA

Function	Potential Failure Mode	Potential Effect(s) of Failure	s e v	Potential Failure Cause/ Mechanism	Prob of Occur	Current Design Controls	Prob of Not Detecting	Recommended Actions	Action Status	Verification
Cut nail embedded wood	Cut too slow	Customer dissatisfaction	H	Not enough power in motor	L	Use proven motor	L	None		Verified by testing and performance of previous models
Cut nail embedded wood	Cut too slow	Customer dissatisfaction	H	User using wrong cutting speeds	H	SAWDUST helpline, distributor training, and available user training	L	Review customer feedback records to see if current training is effective	Sales reps contacted	Analysis to date: users report improved results after training
Plunge cut	Battery hits workpiece before achieving approach angle	Customer dissatisfaction	H	Not enough space provided by design	L	Design provides sufficient battery clearance	L	None		Annex testing verifies that battery does not interfere with plunge cuts at several different angles

For severity (sev), probability of occurrence, probability of not detecting - low is better, high is worse

FIGURE 27.10 ■ FMEA worksheet for the V28 Sawsall® reciprocating saw. *Courtesy Milwaukee Electric Tool Corporation.*

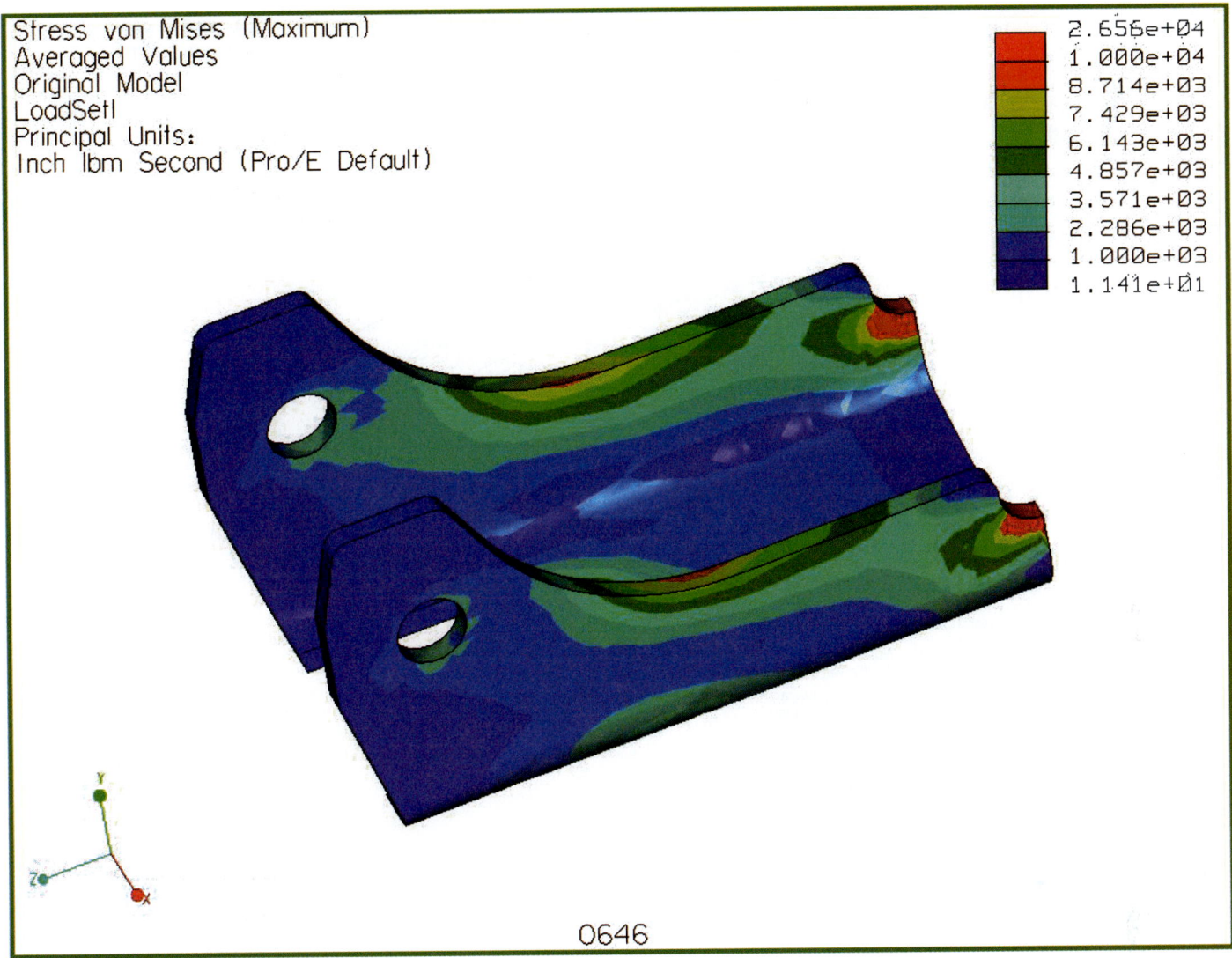

Stress von Mises (Maximum)
Averaged Values
Original Model
LoadSetl
Principal Units:
Inch lbm Second (Pro/E Default)

2.656e+04
1.000e+04
8.714e+03
7.429e+03
6.143e+03
4.857e+03
3.571e+03
2.286e+03
1.000e+03
1.141e+01

0646

FIGURE 27.11 ■ A 3-D model being subjected to simulated tests and stress analysis. *Courtesy Milwaukee Electric Tool Corporation.*

provide the force data and the software calculates the design criteria, such as shaft size, hole size, and material.

Design Documentation

The following discussion provides a typical example of the transition from design documentation to release documentation. The term release, as used here, means that documentation has made the transition from the development phase to the execution phase, or changes in documentation that occur between revisions of the product. The execution phase is covered later in this chapter. While companies can have different part numbering systems, there are similarities. For example, all sketches and drawings in the development stage might be created with prototype part numbers using a P-prefix to designate the part is a prototype. When the product passes through the development stage, prototype part numbers change to release part numbers and the P-prefix changes to represent the final product. Companies use different part numbering systems, but a common format can have three elements, such as 31-50-0019. In this example, the prefix 31 represents the

product code, the 50 relates to the assembly or subassembly, and 0019 is the specific part number. Most companies have a control database for all files, including prototype, release drawings, and revisions. This system can be created to make sure all revisions are backward compatible and to determine which products can be serviced with which parts. Backward compatible means that new parts also function in previously manufactured versions of products. For part and product tracking, everything in the CADD system is controlled by product release level. This means that the CADD system allows the company to design and draw parts and to control changes in releases.

Finished component assembly drawings are created for each product and its components. The set of working drawings can include a 3-D and/or 2-D assembly drawing, sub-assembly drawings, 2-D detail drawings, specifications, wiring drawings, and assembly instructions. The 3-D assembly drawing, service parts list, assembly notes, 3-D wiring diagram, and wiring specifications for the V28 Sawzall® reciprocating saw are shown in Figure 27.12. Refer to Chapter 18 for information on the variety of detail and assembly drawings and for part lists used in industry.

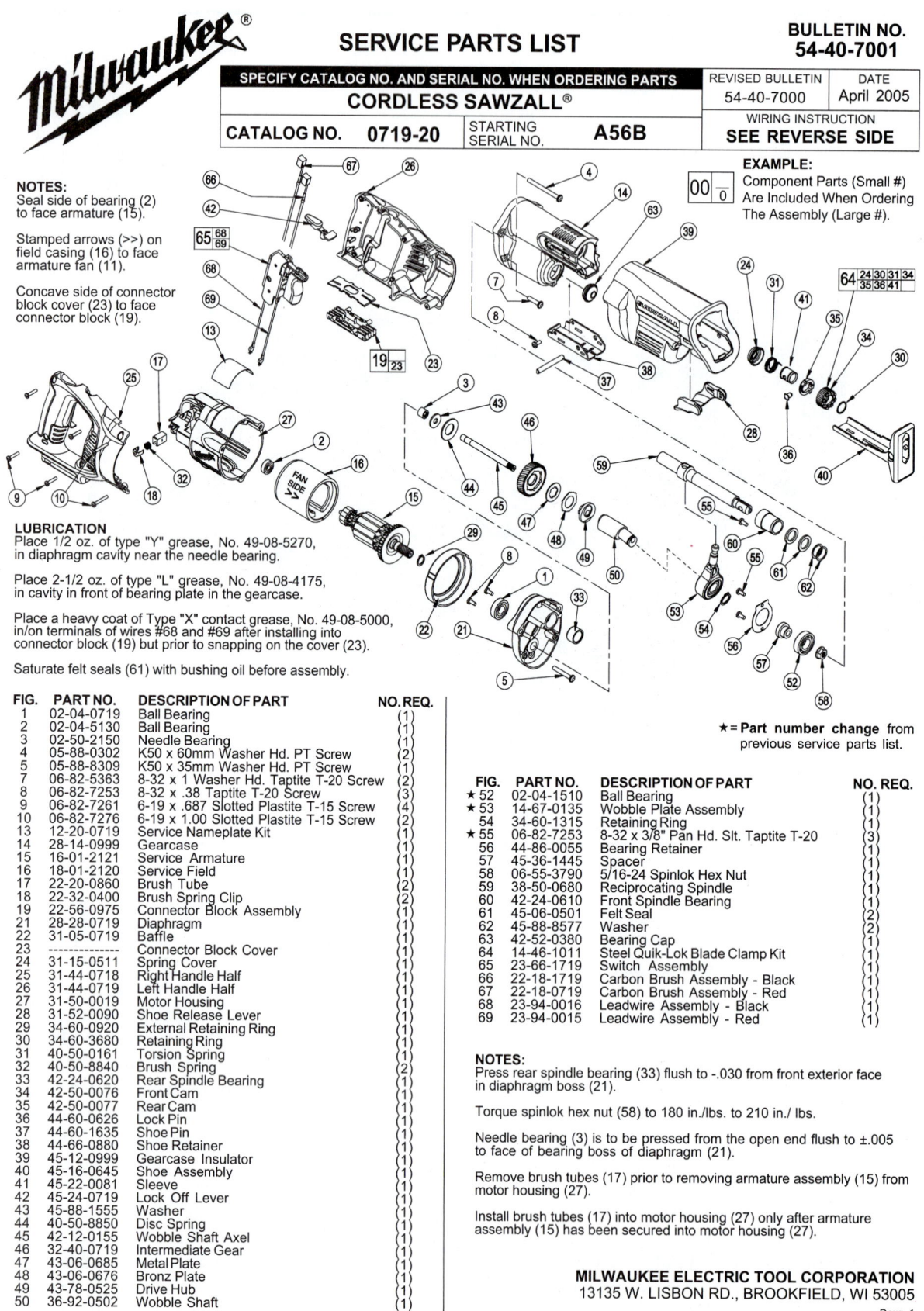

SERVICE PARTS LIST

BULLETIN NO.
54-40-7001

Milwaukee®

SPECIFY CATALOG NO. AND SERIAL NO. WHEN ORDERING PARTS

CORDLESS SAWZALL®

REVISED BULLETIN	DATE
54-40-7000	April 2005

CATALOG NO.	**0719-20**	STARTING SERIAL NO.	**A56B**

WIRING INSTRUCTION
SEE REVERSE SIDE

NOTES:
Seal side of bearing (2) to face armature (15).

Stamped arrows (>>) on field casing (16) to face armature fan (11).

Concave side of connector block cover (23) to face connector block (19).

EXAMPLE:
Component Parts (Small #) Are Included When Ordering The Assembly (Large #).

LUBRICATION
Place 1/2 oz. of type "Y" grease, No. 49-08-5270, in diaphragm cavity near the needle bearing.

Place 2-1/2 oz. of type "L" grease, No. 49-08-4175, in cavity in front of bearing plate in the gearcase.

Place a heavy coat of Type "X" contact grease, No. 49-08-5000, in/on terminals of wires #68 and #69 after installing into connector block (19) but prior to snapping on the cover (23).

Saturate felt seals (61) with bushing oil before assembly.

★ = **Part number change** from previous service parts list.

FIG.	PART NO.	DESCRIPTION OF PART	NO. REQ.
1	02-04-0719	Ball Bearing	(1)
2	02-04-5130	Ball Bearing	(1)
3	02-50-2150	Needle Bearing	(1)
4	05-88-0302	K50 x 60mm Washer Hd. PT Screw	(2)
5	05-88-8309	K50 x 35mm Washer Hd. PT Screw	(1)
7	06-82-5363	8-32 x 1 Washer Hd. Taptite T-20 Screw	(2)
8	06-82-7253	8-32 x .38 Taptite T-20 Screw	(3)
9	06-82-7261	6-19 x .687 Slotted Plastite T-15 Screw	(4)
10	06-82-7276	6-19 x 1.00 Slotted Plastite T-15 Screw	(2)
13	12-20-0719	Service Nameplate Kit	(1)
14	28-14-0999	Gearcase	(1)
15	16-01-2121	Service Armature	(1)
16	18-01-2120	Service Field	(1)
17	22-20-0860	Brush Tube	(2)
18	22-32-0400	Brush Spring Clip	(2)
19	22-56-0975	Connector Block Assembly	(1)
21	28-28-0719	Diaphragm	(1)
22	31-05-0719	Baffle	(1)
23	--------------	Connector Block Cover	(1)
24	31-15-0511	Spring Cover	(1)
25	31-44-0718	Right Handle Half	(1)
26	31-44-0719	Left Handle Half	(1)
27	31-50-0019	Motor Housing	(1)
28	31-52-0090	Shoe Release Lever	(1)
29	34-60-0920	External Retaining Ring	(1)
30	34-60-3680	Retaining Ring	(1)
31	40-50-0161	Torsion Spring	(1)
32	40-50-8840	Brush Spring	(2)
33	42-24-0620	Rear Spindle Bearing	(1)
34	42-50-0076	Front Cam	(1)
35	42-50-0077	Rear Cam	(1)
36	44-60-0626	Lock Pin	(1)
37	44-60-1635	Shoe Pin	(1)
38	44-66-0880	Shoe Retainer	(1)
39	45-12-0999	Gearcase Insulator	(1)
40	45-16-0645	Shoe Assembly	(1)
41	45-22-0081	Sleeve	(1)
42	45-24-0719	Lock Off Lever	(1)
43	45-88-1555	Washer	(1)
44	40-50-8850	Disc Spring	(1)
45	42-12-0155	Wobble Shaft Axel	(1)
46	32-40-0719	Intermediate Gear	(1)
47	43-06-0685	Metal Plate	(1)
48	43-06-0676	Bronz Plate	(1)
49	43-78-0525	Drive Hub	(1)
50	36-92-0502	Wobble Shaft	(1)

FIG.	PART NO.	DESCRIPTION OF PART	NO. REQ.
★ 52	02-04-1510	Ball Bearing	(1)
★ 53	14-67-0135	Wobble Plate Assembly	(1)
54	34-60-1315	Retaining Ring	(1)
★ 55	06-82-7253	8-32 x 3/8" Pan Hd. Slt. Taptite T-20	(3)
56	44-86-0055	Bearing Retainer	(1)
57	45-36-1445	Spacer	(1)
58	45-55-3790	5/16-24 Spinlok Hex Nut	(1)
59	38-50-0680	Reciprocating Spindle	(1)
60	42-24-0610	Front Spindle Bearing	(1)
61	45-06-0501	Felt Seal	(2)
62	45-88-8577	Washer	(2)
63	42-52-0380	Bearing Cap	(1)
64	14-46-1011	Steel Quik-Lok Blade Clamp Kit	(1)
65	23-66-1719	Switch Assembly	(1)
66	22-18-1719	Carbon Brush Assembly - Black	(1)
67	22-18-0719	Carbon Brush Assembly - Red	(1)
68	23-94-0016	Leadwire Assembly - Black	(1)
69	23-94-0015	Leadwire Assembly - Red	(1)

NOTES:
Press rear spindle bearing (33) flush to -.030 from front exterior face in diaphragm boss (21).

Torque spinlok hex nut (58) to 180 in./lbs. to 210 in./ lbs.

Needle bearing (3) is to be pressed from the open end flush to ±.005 to face of bearing boss of diaphragm (21).

Remove brush tubes (17) prior to removing armature assembly (15) from motor housing (27).

Install brush tubes (17) into motor housing (27) only after armature assembly (15) has been secured into motor housing (27).

MILWAUKEE ELECTRIC TOOL CORPORATION
13135 W. LISBON RD., BROOKFIELD, WI 53005
Drwg. 1

FIGURE 27.12 ■ The 3-D assembly drawing, service parts list, assembly notes, 3-D wiring diagram, and wiring specifications for the V28 Sawzall® reciprocating saw. *Courtesy Milwaukee Electric Tool Corporation.*

REMOVING THE STEEL QUIK-LOK® BLADE CLAMP

- Remove external retaining ring (30) and pull front cam (34) off.
- Pull lock pin (36) out and remove remainder of parts and discard.

REASSEMBLY OF THE STEEL QUIK-LOK® BLADE CLAMP

- Coat new lock pin with powdered graphite.
- Hold tool in a vertical position.
- Place spring cover (24) onto spindle.
- Slide torsion spring (31) onto spindle with spring leg on hole side of spindle.
- Slide sleeve (41) onto spindle aligning hole on sleeve with hole in spindle.
- Slide rear cam (35) over sleeve until it bottoms on sleeve shoulder, <u>ensure spring leg inserts into hole in rear cam.</u>
- Rotate rear cam in the direction of the arrows located on spring cover until there is clearance for lock pin (36) to be inserted into sleeve/spindle holes. Insert lock pin.
- Align front cam (34) inner ribs with rear cam outer slots and slide front cam onto sleeve until it bottoms. Retaining ring (30) groove should be completely visible.
- Attach retaining ring by separating coils and inserting end of ring into groove, then wind remainder of ring into groove. Ensure ring is seated in groove.
- Blade clamp should rotate freely. During normal usage, debris may not allow blade clamp to rotate freely. The use of spray lubricant can help free blade clamp. In extreme conditions, follow these instructions to remove, clean and reassemble blade clamp.

WIRING INSTRUCTIONS

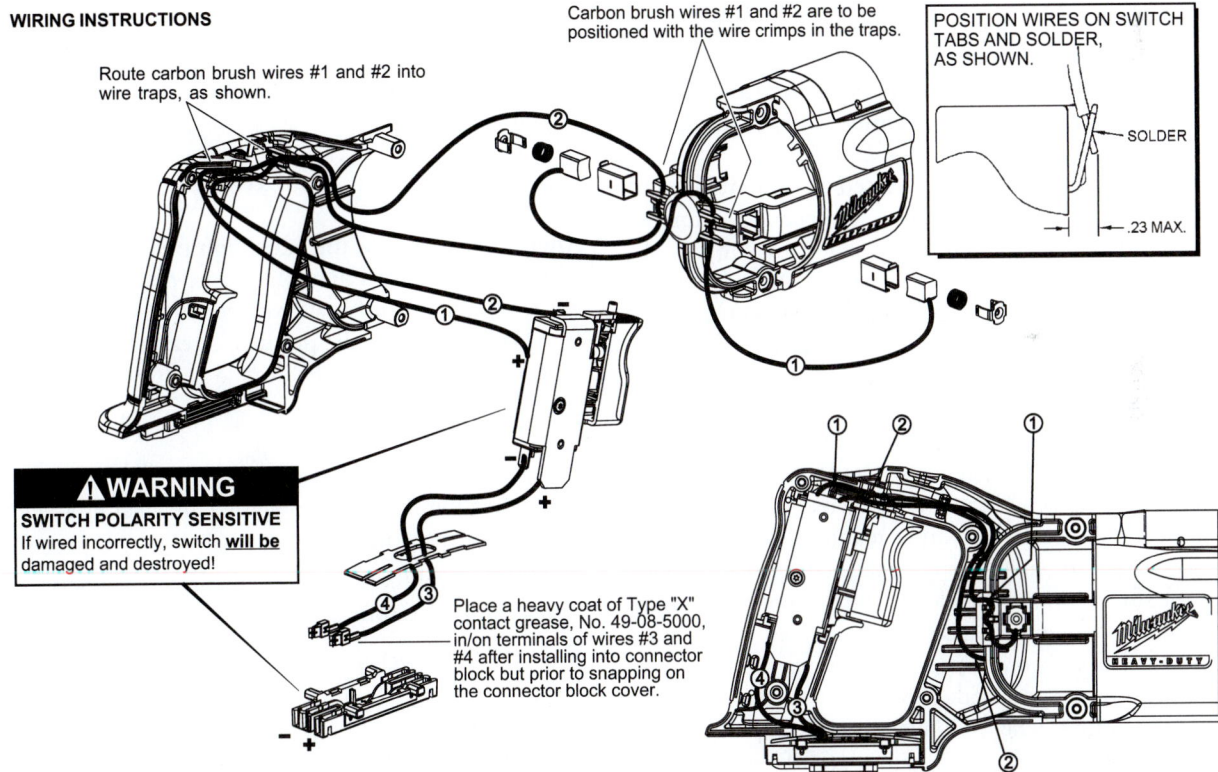

WARNING

SWITCH POLARITY SENSITIVE
If wired incorrectly, switch **will be** damaged and destroyed!

Place a heavy coat of Type "X" contact grease, No. 49-08-5000, in/on terminals of wires #3 and #4 after installing into connector block but prior to snapping on the connector block cover.

WIRING SPECIFICATIONS				
Wire No.	Wire Color	Origin or Gauge	Length	Terminals, Connectors and 1 or 2 End Wire Preparation
1	Red	22-18-0719	-----	Carbon brush assembly.
2	Black	22-18-1719	-----	Carbon brush assembly.
3	Red	23-94-0015	-----	Leadwire assembly.
4	Black	23-94-0016	-----	Leadwire assembly.

BULK LEAD WIRE - BULLETIN NO. 58-01-0003

TERMINAL DESCRIPTION		
Code	Part No.	Qnty.

FIGURE 27.12 ■ The 3-D assembly drawing, service parts list, assembly notes, 3-D wiring diagram, and wiring specifications for the V28 Sawzall® reciprocating saw. *Courtesy Milwaukee Electric Tool Corporation. (Continued)*

Gate 2: Passing the Development Phase Gate

When the development phase is complete as desired, the design review committee evaluates the product development through the current status before the design can pass through the product development phase gate. If the product development is approved, the committee certifies that there is enough confidence in the design to continue. There may not be one-hundred percent certainty, but enough has been done to show that the proposed product is appropriate, and reasonable success has been provided from prototype testing. This testing ensures high confidence that all requirements of the Design Specification have been met. The execution phase starts when the design review committee gives final approval of the development phase.

Phase 3: The Execution Phase

The **execution phase** begins when the company is ready to invest in the final production of the product. This investment can include capital acquisitions for machinery, tooling, patterns, molds, and assembly equipment. This process also includes the coordination of all manufacturing activities, such as setting up production lines and purchasing components needed in preparation for the first **pre-production build**, which is discussed later in this section.

Once engineering drawings are released to the vendors and in-house manufacturing, **schedule tracking** becomes an important project management function. Schedule tracking establishes a system that prioritizes and organizes the work to be done through manufacture, assembly, packaging, and shipping of the product. This scheduling includes defining all tasks, applying resources where needed, assigning personnel, and establishing desired outcomes.

When samplings of the parts come back from the outside vendor, they are carefully inspected to ensure they meet design requirements for elements such as dimensional tolerances, functionality, weight, color, and other required characteristics. The company's quality engineering department generally confirms that the suppliers have met design dimensions by carefully measuring and verifying the parts submitted.

Once all parts have been inspected and are confirmed to be within specifications, a pre-production build is performed. The **pre-production build** is where a limited number of products are manufactured and assembled for the purpose of validating the manufacturing process and final product conformance. Additional testing is done on the pre-production build to determine how the first products off the production line meet the established Design Specification. This allows the company to make needed adjustments before a full production begins. The pre-production build is also used for required regulatory approvals.

Regulatory Approvals

Regulatory approval is part of the execution phase where established institutions, normally not for profit and for consumer protection, test the product with varying levels of intensity depending on their purpose. Examples of tests performed include:

- Impact testing to determine the durability of the product during use or when dropped.
- Flame retardant ability to determine if the product resists fire or self-extinguishes when on fire.
- Dielectric tests to verify insulation systems.
- Vibration testing for operational smoothness and related ergonomics.
- Sound level testing.

The Underwriters Laboratories (UL) is probably the most commonly recognized certification agency. The American National Standards Institute (ANSI) is another.

The Production Sample Inspection

Random samples are commonly taken from the pre-production build for inspection. The quality engineering department typically coordinates this inspection process. Successful sample inspection is required before full-scale production can begin. Problems detected during the sample inspection have to be corrected, and if necessary, more than one pre-production build might be required.

Gate 3: Passing the Market Readiness Gate

After all elements of the execution phase are complete, many companies employ a final phase gate that assesses overall market readiness for the new product. At this phase gate, evidence is provided to verify the product is ready for shipment from all aspects of the company. Factors to be considered include:

- Ship testing.
- Field testing results.
- Verification that the packaging Universal Product Code (UPC) scans correctly.
- All technical literature is available, such as service parts lists and operator manuals.
- Target inventory quantities are in stock.

After a successful final review, all groups sign off and the product goes into full production at this stage. The phase gate design process is completed and the product goes to distributors, suppliers, and stores for sale.

AFTER FULL PRODUCTION

The product development process continues even beyond the time when the newly designed product goes into full production. In the modern manufacturing environment, companies are constantly evaluating the quality of their products and processes using continuous improvement methods. The following briefly describes the procedures that can be used to achieve these objectives for the new product previously discussed.

Three Months After Full Production

Approximately three months after full production start up, a post-mortem review is conducted. The post-mortem review assesses the effectiveness of the product development process and the product development team. Representatives across departments are invited to the post-mortem review meeting to provide a critique of the project's successes and failures. A grading system to determine effectiveness of the process during post-mortem assessment addresses the following questions and issues:

■ Were meetings timely and productive?

■ Was the schedule met?

■ Were there design changes after release?

■ Were there things that had to be fixed on the product and in the process?

■ How successful were sample inspection reports?

■ Was supplier and vendor coordination and delivery successful?

■ Were all quality inspection processes acceptable?

■ Did the packaging system work and were products delivered on time?

■ Were agency approvals achieved without difficulty?

■ How effective was the coordination between teams and departments?

Twelve Months After Full Production

A post-production review is conducted approximately twelve months after the start of production to assess the commercial success of the product. Issues to be considered include:

■ Is the product achieving the desired volume?

■ Is the product producing the desired income?

■ Are field performance reports acceptable?

■ Are the customers satisfied?

■ Have there been service problems that need to be addressed?

■ Are there any unresolved issues with production?

■ Have there been warranty issues that need to be resolved?

■ How good was the marketing research?

The Milwaukee Electric Tool Corporation's V28 Lithium-Ion Sawzall® reciprocating saw and its related V28 products represent a success story for the manufacturing industry. Figure 27.13 shows a marketing piece used to promote the V28 Lithium-Ion Sawzall® reciprocating saw.

Product End of Life

Companies continuously evaluate their products to be sure they maintain the preferred demand and sales required to have a desired profit. Just as with any income-producing business, product manufacturing and sales must constantly assess how well their products are selling and ask serious questions about the validity of keeping a product in production when sales decline. A process to discontinue a product is used when it is determined that the product is no longer needed, is at the end of the product life cycle, or the product is replaced by a newer, improved model.

CREATIVITY AND INNOVATION IN DESIGN

Creativity is very important in the world of engineering and design. Today's competitive market for products is placing a major need on the engineering design process to be more creative and innovative in order to meet the demands of faster, less expensive, and better products. The only way to meet these demands is to do things differently, which is where creativity and innovation enter the picture. By definition, creativity is the ability to produce through imaginative skill, to make or bring into existence something new, to form new associations and see patterns and relationships between diverse information. Creativity's partner is innovation. Innovation can be defined as the process of transforming a creative idea into a tangible product, process, or service. Innovation is about improving the quality of a specific thing and allowing for more and better choices.

Using creativity techniques can lead to more effective problem-solving skills that facilitate productive and satisfying designs. Creativity may be required in the following situations:

■ *New methods:* Problems in industry and construction often must be solved in areas that require original ideas or revisions of existing designs.
Creativity tools allow for new ideas to be generated.

■ *Determination:* A designer must have a lot of determination in order to keep working on the project until the problems are solved.
Creativity provides a way to look at problems from a different angle, making it easier to problem-solve.

■ *Attitude:* The designer needs a positive attitude and must realize the possibilities of modern technology.
Creativity is all about possibilities.

■ *Confidence:* A good designer must have confidence in his or her ability to solve a problem within the marketing and manufacturing or construction requirements.
Creativity contributes to confidence by providing alternative solutions.

■ *Experimentation:* Careful testing and recording of data is a key to successful design. Experimentation, prototyping, testing, and analysis are used in the design process.
Creativity is the oil for experimentation and analysis.

■ *Logic:* A designer may lose ideas and waste valuable time without a logical approach to problem solving. A step-by-step design process works when followed and documented.
Creativity generates the ideas. The designer puts them in logical order.

FIGURE 27.13 ■ A marketing piece used to promote the V28 Lithium-Ion Sawzall® reciprocating saw. *Courtesy Milwaukee Electric Tool Corporation.*

Principles of Creativity

There are basic principles of creativity. These principles may not be natural, but with practice and determination, they can be cultivated and improved upon to increase creativity. People who are looked upon as creative seem to have the ability to see the entire picture and make connections and combinations of seemingly unrelated subjects to produce new original ideas. These are the basic creativity principles:

■ *The ability to see relationships and patterns:* This means taking existing objects and combining them in different ways for new purposes.

■ *The belief that you are creative:* The power of positive thinking.

■ *The ability to look at a problem from a different point of view:* This involves changing the position or role in how a new idea or problem is approached.

■ *Playfulness and humor:* Be a dreamer, brainstorm ideas, and surround yourself with inspiring ideas.

■ *A work environment that is flexible, open, and autonomous:* In an ideal world, an unusual idea is never criticized or declared a failure. It may spark a creative solution.

■ *Imagery and visualization:* Daydream, sketch, and visualize the future.

■ *Subconscious thoughts:* Allow an idea to simmer awhile before making judgments. Ideas keep working in the subconscious mind even while you sleep.

In thinking of ways to become more creative in designs, remember the creative individual asks more "*why*" questions as opposed to "*what*" questions. (See Figure 27.14.)

These are just a few of the ways to foster the creative process. One that works well for designers is to draw a sketch of the problem instead of using words. Many times a visual representation provides a fresh perspective and insights.

1. Look beyond the immediate object or situation—see the Big Picture.
2. Look for potential future consequences.
3. Always ask "Why" questions.
4. Always question and remain curious.
5. Keep an interest in a wide variety of areas—the broad base of knowledge helps make connections to new ideas.
6. Keep a daily journal of ideas, thoughts, and sketches.
7. Lateral thinking: Look at the idea from a number of different directions.
8. Mindmapping: Use free word association and brainstorming. Then organize the ideas into related areas.
9. Brainstorming: Use free association of ideas from a group to generate as many ideas as possible without passing judgment on any idea presented.

FIGURE 27.14 ■ Methods for developing creativity.

FIGURE 27.15 ■ Teamwork in action.

Creativity Works Well in a Group

The world of engineering is moving to self-directed teams and concurrent team groups through which the real advantage for creative problem solving is realized. A group of people is able to bring diverse and varied perspectives to the problem or idea at hand. A group possesses accumulated knowledge and experience that one individual cannot even begin to exhibit. A team of people is in a position to learn from each other and consider various ideas and solutions, which creates a combined effort that leads to creativity. (See Figure 27.15.) The cross-functional team approach was discussed throughout this chapter.

Creativity and Innovation

Creativity and innovation require a diverse, information- and interaction-rich environment. This is brought about with people who have different perspectives working together toward a common goal. This team of people requires accurate, up-to-date information and the proper tools.

Only in this kind of environment—where the organization is continuously learning about its products, services, processes, customers, technologies, industry competitors, and environment—can innovation thrive. Existing tools such as CAD/CAM/CAE are good for routine problem solving but fall short in generating breakthrough concepts. To meet current market demand and customer requirements, creativity must transcend into innovative new ideas and processes.

In addition, remember that failure can be a learning experience and a necessary step in the design process. Sometimes design solutions fail. Failure prompts you to look at other solutions and promotes new ideas that lead to innovations. So when a design fails a test, think of it as an opportunity, not a dead end.

CHANGE AND THE IMPACT ON THE DESIGN PROCESS

The designer naturally embraces change as part of the design process. Every design or idea goes through an evolution before its final outcome. Today the designer and the engineering team are attempting to reduce the amount of change in the design process by taking an integrated approach to the design process at the early stages of the design inception. By considering all aspects in the development of a product, the design team can troubleshoot in advance any potential problems that would result in the change of a design. Solid modeling and rapid product development tools aid in early identification of design flaws and eliminate the need for change down the line.

Within the engineering profession, the ability to manage change is crucial to a project's success. Although there are many variations of the engineering change process, the structure and purpose are fairly standard. The objective of any engineering change process is to incorporate design changes as quickly and accurately as possible with a minimum of disruption and cost. Changes in the engineering design process serve many functions:

- Satisfy customer requests.
- Improve the product.
- Incorporate improvements in production and manufacturing.
- Resolve design problems.
- Integrate new technologies.

The engineering change process requires thoroughly documented entrance and exit criteria to ensure design changes are accomplished accurately and in a timely manner, as well as to ensure the maintainance of a complete history of the changes.

Tracking Documents: ECO/ECR

As introduced in Chapter 18, a tracking document commonly is referred to as an Engineering Change Order (ECO) or an Engineering Change Request (ECR). The style and procedure for ECO or ECR may differ from engineering firm to engineering firm, but the basic process is the same. Standard information is included on most document change forms. (See Figure 27.16.)

The engineering change process necessary to incorporate design change properly can be summarized in three steps:

- Engineering drawings
- Parts list or bill of materials
- Production specifications
- ECO/ECR document—problem description
- Engineering Change Order number
- Status of engineering change
- Effective date for change
- Approval list for reviewing changes

FIGURE 27.16 ■ Documents and information required for an ECO/ECR.

1. Communicating the change.
2. Documenting the change.
3. Tracking the change.

In considering a potential change to a design, remember that change has an impact on the whole system within which the design is developed. A positive change for the designer may be a negative change for someone else who has a role in the design process. This is where a team meeting may be of value, since ideas can be considered by everyone involved.

DESIGN DELIVERABLES

As a design process comes to an end, many different types of documents are created to document the product, process, installation procedures, and training plans. These materials are known in the industry as deliverables. Typical deliverables include such items as engineering and/or construction drawings or CADD files, prototypes, life cycle plans, installation manuals, and instruction guides. Deliverables can also include service/maintenance manuals, manufacturing process specifications, product support equipment, materials, spare parts, customer training plans, and a final project report. (See Figure 27.17.)

THE DESIGN PROCESS RESPONDS TO CHANGES IN ENGINEERING

In examining all the changes in technology and the way in which engineering and product development occurs, the design process has responded to these changes with emerging engineering systems, such as the following:

- *Timing of the design:* Do it early in the design cycle; doing it later costs more.
- *Reduce parts in a design:* This reduces costs throughout the whole assembly process.
- *Standardize parts:* This reduces cost by making parts interchangeable for different designs.
- *Keep the design simple:* The higher the tooling costs, the more complicated the design.
- *Utilize modular designs:* These reduce cost in design time because parts can be used in other designs. Modular designs also reduce assembly time.
- *Design with gravity in mind:* It is easier and more time-efficient to assemble from the top down.
- *Eliminate fasteners:* This saves time and cost in assembly as well as maintenance.
- *Optimize part handling:* Costs go down and quality goes up with minimal proper assembly sequence.
- *Design for easy part mating:* Easier assembly through alignment for insertion speeds up assembly.
- *Provide nesting features in the design:* These help to show where features go.
- *Optimize manufacturing process sequence:* This increases the speed of getting the product to market.
- *Form follows function:* Design for the user.
- *Utilize design generations:* An initial design can be reused in the next generation of the product.

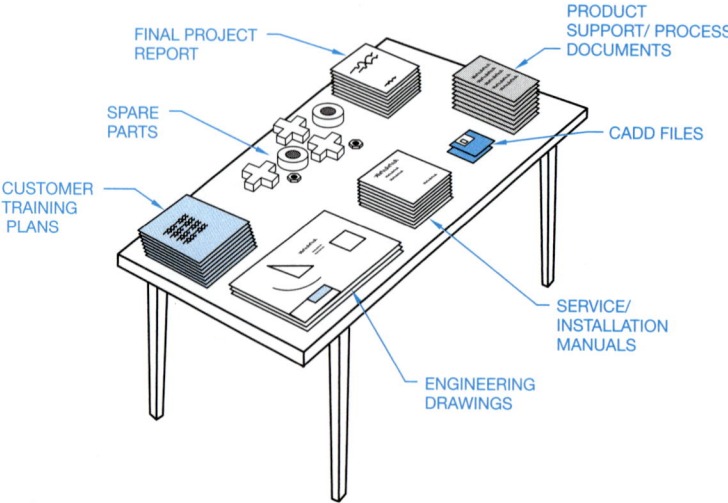

FINAL PROJECT REPORT

PRODUCT SUPPORT/ PROCESS DOCUMENTS

SPARE PARTS

CADD FILES

CUSTOMER TRAINING PLANS

SERVICE/ INSTALLATION MANUALS

ENGINEERING DRAWINGS

FIGURE 27.17 ■ Items delivered to the customer.

PROFESSIONAL PERSPECTIVE

ONLY THE FAST SURVIVE

—*Jim Leonard, Colorado Manufacturing Competitiveness, Denver, Colorado*

Every morning in Africa a gazelle wakes up, knowing it must outrun the fastest lion or it will be killed. At the same time a lion wakes up, knowing it must run faster than the slowest gazelle or it will starve. It doesn't matter whether you're a gazelle or a lion, when the sun comes up you'd better be running. It's the law of the competitive jungle—only the fastest survive.

Source: A gazelle somewhere in Africa

Every product development team member has heard or read the law of the competitive jungle; many have the law stamped on their foreheads. So, why is rapid product development, a.k.a., quick time to market, so important? There are many reasons, but the main one is: *Satisfy customers' needs first and best.*

Design Tools for Engineering Teams: An Integrated Approach describes many tools and practices that product developers use to improve product quality and reduce development time. But the toolbox of processes, practices, and technologies is constantly changing. What is state-of-the-art today will generally be mainstream in two or three years. State-of-the-art tools are rarely visible to anyone other than the most insightful industry observers because best-in-class product developers keep their new tools and practices secret as long as possible. Here is a brief list of practices and technologies that best-in-class teams are using right now.

Customer Focus

The 1990s were characterized by the increased role of "customer focus," particularly in leading-edge manufacturers. That leading role will intensify in the 2000s, as articulated by Jacques Nasser, CEO of Ford, " . . . the real work at hand; getting inside the mind of the consumer to understand what he or she wants, aspires to become, and will be needed long after a purchase has been made." Product development professionals will be employing all kinds of innovative ways to understand customer needs, wants, fantasies, including living with customers, having customers as permanent members of product development teams, drilling down into and analyzing mountains of consumer data collected over the Internet.

E-business

Current applications of the Internet provide a distinct competitive advantage. Although Internet applications are appearing everywhere to improve customer service and reduce cycle times, most of the highest value-added applications are in product development. New, innovative applications will continue to appear. Every one of the practices mentioned in this brief list applies Internet technologies in some way.

Collaborative Engineering

Collaborative engineering goes beyond concurrent engineering. It is the cooperative exchange of resources, information, and ideas among a virtual team focused on an engineering-intensive project and having on overall common creative purpose. [A] virtual team is one whose members are not physically collocated, but connected by distance communication technologies such as videoconferencing and e-mail.

Predictive Engineering, Simulation, and Virtual Prototypes

With mainstream use of 3-D solid modeling, products can now be designed, built, tested to failure, and redesigned all in digital form. The designs can be nearly optimized before time, funds, and effort are spent on hardware. Entire manufacturing plants can be simulated and optimized before the first cubic yard of concrete is poured.

Visual Engineering

3-D solid modeling also allows all team members to see and understand information in the same way. Now the marketing, finance, and other nontechnical members can see product designs without needing the skill to read 2-D engineering drawings. Realistic images of engineered products can be shown more quickly and more affordably than building physical prototypes.

Supply Chain Integration, and Value Chain Integration

Companies and teams now coordinate the series of activities and processes that purchase, design, manufacture, and deliver products and/or services to customers. The Supply Chain or Value Chain is a network of company relationships that support materials, information, and funds flows from a firm's supplier's supplier to its customer's customer.

(Continued)

PROFESSIONAL PERSPECTIVE (Continued)

Design for Everything (DfX) Has New Meaning

DfX used to mean "design for manufacturability, assembly, test, service, and environment." Now DfX is taking on several new functions:

■ *Design for supply chain:* Optimizes use of the distinctive competencies of all supply chain members.

■ *Design for postponement:* Allows for incorporation of distinctive/customizable features into a product until the latest possible production step.

■ *Design for recycle, reuse, rebuild, and disposal:* Design for environment used to mean "consider ease of recycling." In

the future, expensive, high-information-content products will be designed for easy, inexpensive rebuild/refurbish, which eliminates the need for recycle or disposal for at least another life cycle. Less expensive products will be designed for convenient, environmentally acceptable disposal, which may include recycling or components made of biodegradable material.

This list is not comprehensive. Enlightened product development professionals constantly scan the horizon for new practices, processes, and technologies and then intelligently apply them in their own projects. This is the only way to consistently *satisfy customers' needs first and best.*

WEBSITE RESEARCH

The following websites can provide you with information about the engineering design process.

http://www.designnews.com—Trade magazine.

http://www.partnerwerks.com—Information on design teams.

http://www.sme.org—Society of Manufacturing Engineering.

http://www.ame.org—Association for Manufacturing Excellence.

http://www.manufacturingnews.com—Current articles.

http://www.world.std.com—User interface engineering.

http://www.milwaukeetool.com—Milwaukee Electric Tool Corporation.

http://www.wipo.int—World Intellectual Property Organization.

http://www.dimensionprinting.com—3-D printing technology.

http://www.globalspec.com—Technical products and services database.

http://www.machinedesign.com—Design processes listings.

http://www.avdel.textron.com—Fastener technical information.

http://www.engineeringzone.com—Online handbooks.

http://www.engr.usask.ca—FEA shareware.

http://www.tenlinks.com—CAD sites.

http://www.fmeainfocentre.com—FMEA worksheet.

http://www.sv.vt.edu—Introduction to Finite Element Analysis.

CHAPTER
27 *The Engineering Design Process Test*

 Access the CD found with this textbook to view the Chapter 27 Test. Confirm the preferred submittal method with your instructor.

CHAPTER 27

The Engineering Design Process Problems

PROBLEM 27.1 The following topics require research and/or industrial visitations. It is recommended that your research current professional magazines, Internet websites, visit local industries, or interview professionals in the engineering field. Your reports should emphasize:

- The link between manufacturing and engineering
- Current technological advances
- The design process
- Team applications

Select one or more of the following engineering fields or as assigned by your instructor and write a 250 word report for each.

Mechanical Engineering

Structural/Architectural Engineering

Civil Engineering

Electrical/Electronic Engineering

HVAC/Sheet Metal

PROBLEM 27.2 The following topics are changing the way engineering design is currently conducted. Research, observe, visit industry or product vendors, or search the Internet to find ways that the following items are allowing product development to be less expensive, faster, and of higher quality. Your presentation should emphasize the following:

- Process
- Current technological advances
- Team applications
- Problem solving tools
- Design applications for engineering

Select *one or more* of the topics listed below or as assigned by your instructor and prepare an oral presentation for each.

Concurrent engineering

Life cycle engineering

Integrated product development

Knowledge-based engineering

Total quality management

Cross-functional team approach

Project portfolio management

A comparison between advanced research and existing product research

One or more of the phase gate design process stages

Industrial design

Quality function deployment

Design specification

Functioning prototypes

Intellectual properties

Patent research and invention disclosure

Failure mode effect analysis

Computational dynamics and finite element analysis

Design documentation

Pre-production build

Regulatory approval

After full production

The end of product

Creativity and innovation in design and principles of creativity

Change and the impact on the design process

Tracking documents

Design deliverables

DESIGN PROBLEMS

The following problems involve using elements of the engineering design process identified in this chapter to design products from given concepts or for your own design projects. These design projects are best solved using a team approach, but they also can be done individually depending on your course objectives and available participants. When using a team, identify a project leader and the responsibility of each team member, such as industrial designer, CADD designers and drafters, prototype maker, research specialist, marketing, and specification developers. Some of the following steps may not be possible to accomplish due to unavailable facilities, skills, or resources. Confirm this with your course objectives and instructor approval. Using the content of the Engineering Design Process sections found in this chapter, include the following documents as part of the completion for each design:

The Product Description/Business Case. This is a formal document, which specifies the features, pricing, and costs of a new product. This document communicates the customer needs to engineering and is the guidance upon which engineering acts. Marketing gathers input from customers in the target market and also from departments such as purchasing, engineering, and sales.

The Project Plan. This document is created by the project leader, who issues a schedule communicating the anticipated implementation timeline for the project. The project team is also formed at this time. Each team member is given their assignments, and expectations are established. A major benefit of this practice is to ensure all critical tasks are identified and accounted. List the team member names and identify their specific responsibilities and estimated deadlines.

Industrial Design (ID) Team. Select team members who are possibly the most creative. These should be people who have the ability to be creative and work within engineering, manufacturing, and cost constraints. Create sketches illustrating various layouts, features, and styling. Usually six to twelve idea renderings are created in this concept design phase. Present the sketches to marketing and engineering for their input and assistance in narrowing down the concepts to one or two desired possibilities.

Concept Refinement. In this step, the industrial designer begins concept refinement, where final renderings and multiple views are created. This is also usually when mockups are created. The refined concept or concepts are then presented again to engineering and marketing. Once everyone is in agreement on a concept, one or more team members are selected to create CADD models.

Customer Research. Customer input may involve classmates, friends, and family and can be established through one or more of the following means:

- Hardcopy distribution, phone surveys, or questionnaires.
- Telephone or field interviews with customers.
- Listening to and gathering customer recommendations.
- Using a website or other correspondence.

Create the Design Specification. Establish technical specifications and performance targets. These are guidelines to the project team and help identify what constitutes a successful design. The design specification is a document with project requirements such as:

- Dimensions
- Weight
- Materials
- General shape
- Power requirements
- Voltage requirements
- Amperage specifications
- Product life and durability requirements
- Required agency approvals
- Performance requirements
- Included and optional accessories
- General and special features
- Environmental factors

The Development Phase. Construct a fully functioning prototype model. If facilities or skills are not available for the construction of a full-functioning prototype model, then build a model using other materials such as paper, cardboard, or foam.

A Product Bill of Materials. Develop a product bill of materials to the best of your ability and knowledge. This document is used to establish the final product cost and to help organize the manufacturing process. Capital that must be spent to make parts in production, such as tooling, die, and pattern costs also should be determined at this time.

Design Documentation. Prepare a complete set of working drawings for the product. This includes an assembly drawing, parts list or bill of materials, detail drawings, and specifications. Establish a part or drawing numbering system. The assembly drawing can be in a 2-D or 3-D format as determined by your course objectives. Use the content in Chapter 18 for reference.

The Pre-Production Build. If facilities and skills are available, manufacture and assemble at least one of the products. Do additional testing to determine how the first products off the production line meet the established design specifications.

Marketing. Create a marketing plan along with at least one promotional flyer or brochure.

PROBLEM 27.3   Trailer hitch design

Design a standard receiver trailer hitch for a pick-up truck, light truck, or SUV with the following requirements:

- Quick change height adjustment.
- Quick change trailer ball to accommodate at least two ball sizes, such as 2 and 2.5 in. (50–60 mm).
- Corrosion resistant material or finish.

PROBLEM 27.4   Cheese cutter

Design a countertop cheese cutter with the following requirements:

- Compact design to accommodate a typical two-pound (1 kg) cheese block.
- Flat base to rest on any countertop or table.
- Adjustable cutting width.
- Corrosion-resistant material and finish.
- Easy to clean.
- Easy to store.

PROBLEM 27.5   Electronics component door

Design a door covering the controls of an electronics component, such as a television, radio receiver, or DVD player with the following requirements:

- One touch opens.
- Opens slowly when door is released.

PROBLEM 27.6   Ski cycle

Design a track drive system that can replace the rear wheel on a motorcycle, and design a ski to be used to replace the front wheel to allow a light-weight motorcycle to be used on snow.

PROBLEM 27.7   ATV loader

Design hardware that can be mounted in a pick-up truck bed that will allow one person to load and unload an ATV into the truck.

PROBLEM 27.8   Paddle-powered boat

Design a paddle-powered boat with a rudder system.

PROBLEM 27.9   Enclosed wheel lock

Design a lockable device to enclose the wheel of a boat trailer or utility trailer to keep the wheel from being stolen.

PROBLEM 27.10   Manual can crusher

Design a manual can crusher with ejection system. This is used to crush standard metal cans for recycling.

PROBLEM 27.11   Electric-powered can crusher

Redesign Problem 27.10 as an electric-powered can crusher with ejection system.

PROBLEM 27.12   Double-deck trailer

Design a double-deck trailer that can be used to load two ATVs, one above the other.

PROBLEM 27.13   Pop-up target system

Design a pop-up target system that will automatically reset itself for target-shooting enthusiasts.

PROBLEM 27.14   Appliance dolly

Design an appliance dolly, powered or non-powered, that will make it easier to use on stairs than a conventional dolly.

PROBLEM 27.15   Skateboard trucks

Design skateboard trucks to your own specifications in 150, 200, or 250 mm as desired. Incorporate the following features:

- Durability.
- Non-corrosive material.

PROBLEM 27.16   Stock car roll cage

Design a stock car roll cage that enhances driver protection, comfort, and ease of escape.

PROBLEM 27.17   Bicycle helmet

Design a bicycle helmet with the following general characteristics:

- Lightweight and impact resistant material.
- Provide three variable size groups for children, adolescents, and adults.
- Aerodynamic.
- Shock resistance.

PROBLEM 27.18   Bicycle seat

Design a bicycle seat with the following general characteristics:

- Wide for maximum comfort.
- Fits any standard bicycle seat shaft.
- Absorbs shock for a smooth ride.

PROBLEM 27.19   Golf club cart

Design a golf club cart that carries a full set of golf clubs and has the following general features:

- Separates clubs for easy access and security.
- Places for extra balls, tees, scorecards, and other desired compartments.
- Large tires for stability and ease of movement.
- Folding leg to stand cart upright.
- Backpack style and hand carrying features.

PROBLEM 27.20   Collapsible music stand

Design a mobile collapsible music stand that has the following general features:

- Folds into a maximum size of 8.5 × 11 in. (215 × 280 mm).
- Adjustable height from 11 in. (215 mm) to 48 in. (1200 mm).
- Sturdy, lightweight construction.

PROBLEM 27.21   Mobile cappuccino store

Design a mobile cappuccino store with the following general features:

- Mobile building with trailer hookup.
- Approximate dimensions 8 × 20 ft. (2400 × 6000 mm).
- Preparation facilities and storage.
- Service counter and window.
- Refrigeration unit.
- Coverings over wheels and tires, and trailer tongue.
- Modern or traditional exterior architectural style as desired.
- Signage.

PROBLEM 27.22   Convenience store

Design a convenience store with the following general features:

- Maximum 1,500 square feet (140 square meters).
- Cold cases.
- Freezer.
- Service counter.
- Restroom.
- Modern or traditional exterior architectural style as desired.

PROBLEM 27.23 **Office and warehouse combination facility**

Design an office and warehouse combination facility for a small distribution company with the following general features:

■ Office area maximum 1,000 square feet (90 square meters) combined with a restroom and employee/guest lounge.

■ Warehouse area maximum 2,000 square feet (180 square meters).

■ Loading dock with large door or doors.

PROBLEM 27.24 **Creative design**

Design a product of your own creation or from the ideas gathered from a team.

Specialty Drafting and Design on the Companion CD

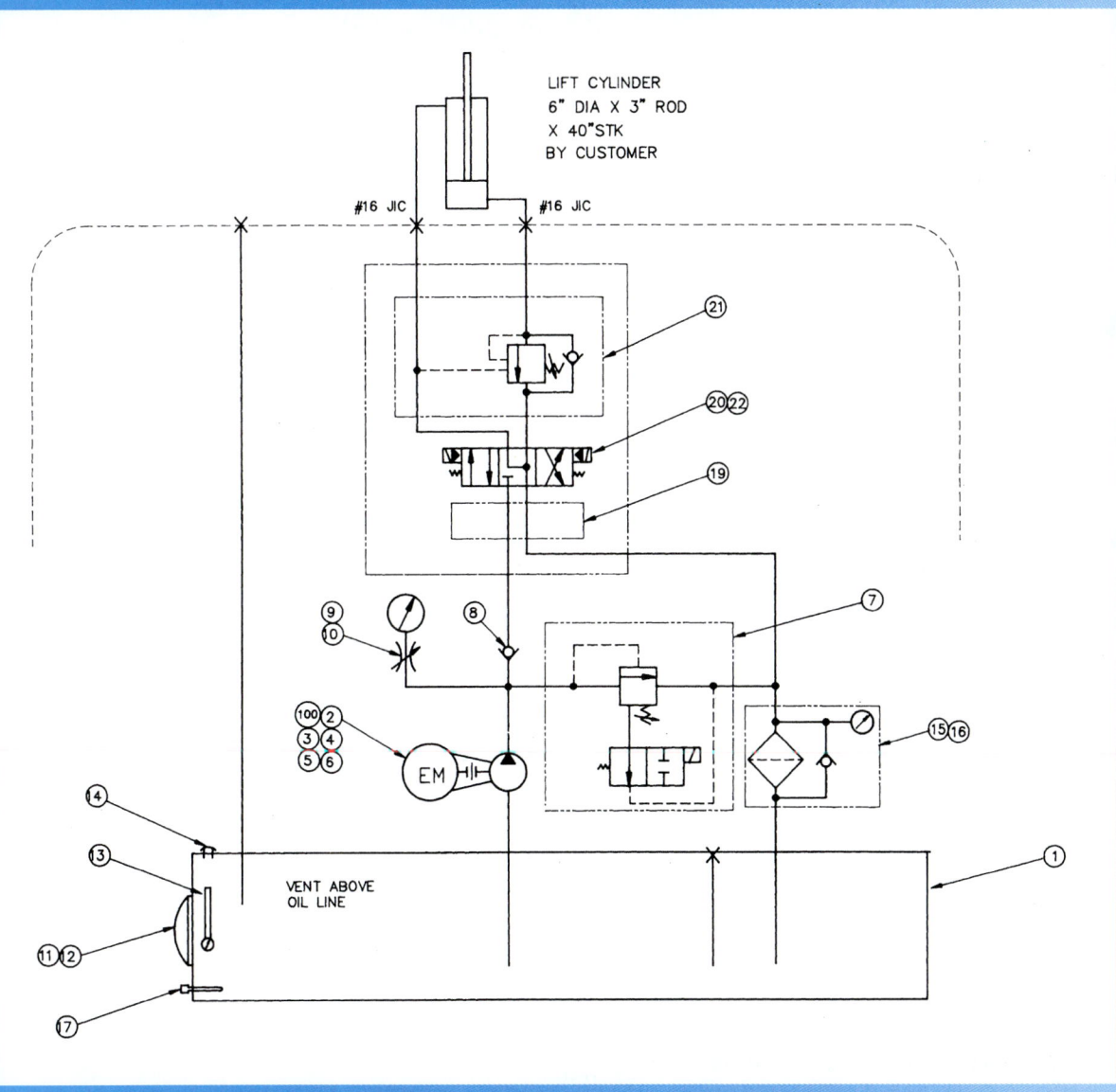

LIFT CYLINDER
6" DIA X 3" ROD
X 40"STK
BY CUSTOMER

#16 JIC   #16 JIC

VENT ABOVE
OIL LINE

There is a CD available at the back of *Engineering Drawing and Design*, 4e. The CD contains a variety of valuable features for you to use as you learn engineering drawing and design. The CD icon, placed throughout this textbook, directs you to features found on the CD.

CD INSTRUCTIONS

Access the CD to view the CD chapters, appendices, chapter tests, and selected chapter problems:

- Place the CD in your CD drive.
- The CD should open (start) automatically.
- If the CD does not start automatically, pick the Start button in the lower left corner of your screen, and select Run . . . , followed by accessing the CD drive on your computer.
- Pick the desired chapter, appendix, test, or problem application button on the left side of the CD window.

CD CONTENTS

Supplemental Chapters

CD Chapter 1: Fluid Power

Complete coverage of fluid power drafting and design applications, including hydraulic, fluid power, and pneumatic terminology, rules, symbols, systems, and diagrams.

CD Chapter 2: Engineering Charts and Graphs

The most comprehensive coverage available on the design and drafting of engineering charts and graphs.

CD Chapter 3: Engineering Drawing and Design Math Applications

Comprehensive math instruction for engineering drafting and related fields. The content parallels the math applications and problems in chapters throughout this textbook. Presented with numerous examples in a manner that is easy to use and understand.

Supplemental Appendices

CD Appendix A: American National Standards of Interest to Designers, Architects, and Drafters

CD Appendix B: ASME Standard Line Types

CD Appendix C: Dimensioning and Tolerancing Symbols

CD Appendix D: Designation of Welding and Allied Processes by Letters

CD Appendix E: Symbols for Pipe Fittings and Valves

CD Appendix F: Computer Terminology and Hardware

CD Appendix G: ASME/ANSI Exercises

Supplemental Chapter Tests and Problems

Chapter 1: Introduction to Engineering Drawing and Design

Test

Chapter 2: Manual Drafting Equipment, Media, and Reproduction Methods

Test

Chapter 3: Computer-Aided Design and Drafting

Test

Chapter 4: 3-D CADD, Animation, and Virtual Reality

Test

Chapter 5: Manufacturing Materials and Processes

Test

Chapter 6: Sketching

Test
Problems

Chapter 7: Lines and Lettering

Test
Problems

Chapter 8: Geometric Construction

Test
Problems

Chapter 9: Multiviews

Test
Problems

Chapter 10: Auxiliary Views

Test
Problems

Chapter 11: Dimensioning and Tolerancing

Test
Problems

APPENDICES

CD APPENDICES

Access the CD accompanying *Engineering Drawing and Design,* 4e for the following appendices:

There is a CD available at the back of *Engineering Drawing and Design,* 4e. The CD contains a variety of valuable features for you to use as you learn engineering drawing and design. The CD icon, placed throughout this textbook, directs you to features found on the CD.

CD INSTRUCTIONS

Access the CD to view the CD chapters, appendices, chapter tests, and selected chapter problems:

- Place the CD in your CD drive.
- The CD should open (start) automatically.
- If the CD does not start automatically, pick the Start button in the lower left corner of your screen, and select Run . . . , followed by accessing the CD drive on your computer.
- Pick the desired chapter, appendix, test, or problem application button on the left side of the CD window.

APPENDIX A
ABBREVIATIONS

The following are standard abbreviations commonly used on working drawings:

Above finished floor . . . . .AFF
Accessory .ACCESS
Accumulate . . . . .ACCUM
Adaption . . . . .ADAPT
Addendum .ADD
Addition . . . . .ADD
Airfoil. . . . . .AF
Air pressure drop . . . . .APD
Alteration . . . . .ALT
Alternate . . . . .ALT
Alternating current . . . . .AC
Alternative .ALT
Altitude . . . . .ALT
Aluminum . . . . .AL
American Design
  Drafting Association . . . . .ADDA
American Iron and
  Steel Institute . . . . .AISI
American Society for
  Testing Materials . . . . .ASTM
American Society of
  Mechanical Engineers . . . . .ASME
American Wire Gage . . . . .AWG
Ampere . . . . .AMP
Apparatus housing plenum . . . . .AHP
Approved . . . . .APPD
Approximate, Approximately . . . .APPROX
Architect, Architectural . . . . .ARCH
Assembly . . . . .ASM
Attach . . . . .ATT
Authorize . . . . .AUTH
Automatic . . . . .AUTO
Auxiliary . . . . .AUX
Average . . . . .AVG
Backdraft damper . . . . .BDD
Backward inclined . . . . .BI
Balance . . . . .BAL
Basement . . . . .BSMT
Battery . . . . .BAT
Bearing . . . . .BRG
Bill of materials . . . . .B/M
Bracket . . . . .BRKT
Brass . . . . .BRS
British thermal unit . . . . .BTU
British thermal units
  per hour .BTUH
Bronze . . . . .BRZ
Brown and Sharpe . . . . .B&S
Building . . . . .BLDG
Bushing . . . . .BUSH

Cadmium . . . . .CAD
Capacity . . . . .CAP
Carbon steel . . . . .CS
Cast iron CI
Cast steel . . . . .CS
Casting . . . . .CST
Ceiling or cooling . . . . .CLG
Center . . . . .CTR
Center of gravity . . . . .CG
Centerline . . . .(℄) or CL
Centigrade C
Centimeter .CM
Chamfer . . . . .CHAM
Change . . . . .CHG
Check . . . . .CHK
Chief . . . . .CH
Chromium .CHR
Circular pitch . . . . .CP
Circumference . . . . .CIRC
Clockwise . . . . .CW
Cold drawn steel . . . . .CDS
Cold rolled steel . . . . .CRS
Company . . . . .CO
Composition . . . . .COMP
Compression, Compressor . . . . . . .COMP
Concentric .CONC
Concrete . . . . .CONC
Condition .COND
Conductor .COND
Conduit . . . . .CDT
Connect, Connection,
  Connector . . . . .CONN
Continue, Continued,
  Continuation . . . . .CONT
Control . . . . .CONT
Copper . . . . .CU
Copyright . . . . .copr., ©, or ®
Corporation . . . . .CORP
Correspond . . . . .CORRES
Corrosion resistant steel . . . . .CRES
Counterbalance . . . . .CBAL
Counterbore . . . . .CBORE
Counterclockwise . . . . .CCW
Counterdrill . . . . .CDRILL
Countersink . . . . .CSK
Counterweight . . . . .CTWT
Cubic . . . . .CU
Cubic centimeter CC
Cubic feet per hour . . . . .CFH
Cubic feet per minute . . . . .CFM
Cubic feet per second . . . . .CFS
Cubic foot .CU FT
Cubic inch .CU IN
Cycle . . . . .CY
Dedendum .DED
Deflection .DEFL

Degree . . . .(°) or DEG
Department . . . . .DEPT
Detail . . . . .DET
Develop . . . . .DEV
Deviation . . . . .DEV
Dew point .DP
Diagonal . . . . .DIAG
Diameter . . . . .DIA, ∅
Diametral pitch . . . . .DP
Dimension .DIM
Direct current . . . . .DC
Direct digital control . . . . .DDC
Distance . . . . .DIST
Division . . . . .DIV
Double width double inlet . . . . .DWDI
Down . . . . .DN
Drawing . . . . .DWG
Each . . . . .EA
Eccentric . . . . .ECC
Effective . . . . .EFF
Efficiency . . . . .EFF
Electric, Electrical . . . . .ELEC
Elevation . . . . .ELEV
Engineer . . . . .ENGR
Engineering . . . . .ENGRG
Entering air temperature . . . . .EAT
Entering dry bulb . . . . .EDB
Entering fluid temperature . . . . .EFT
Entering water temperature . . . . . . .EWT
Entering wet bulb . . . . .EWB
Equal . . . . .EQ
Equipment . . . . .EQUIP
Equivalent . . . . .EQUIV
Estimate . . . . .EST
Etcetera . . . . .ETC
Exhaust . . . . .EXH
Exhaust air .EA
Exhaust air damper . . . . .EAD
Existing . . . . .EXIST
Expansion .EXP
Extension . . . . .EXT
External . . . . .EXT
External static pressure . . . . .ESP
Extractor . . . . .EX
Extrusion . . . . .EXTR
Face velocity . . . . .FV
Fahrenheit .F
Feet, Foot . . . . .FT
Feet per minute . . . . .FPM
Feet per second . . . . .FPS
Figure . . . . .FIG
Fillister head . . . . .FIL HD
Filter . . . . .FILT
Finish . . . . .FIN
Fitting . . . . .FTG
Fixture unit .FU

Flat . . . . .FL	Kilowatt hour .KWH	Of true position . . . . .OTP
Flat head . . . . .FHD	Knock out . . . . .KO	On Center, On center
Flexible . . . . .FLEX	Laboratory . . . . .LAB	distance . . . . .OC
Fluid pressure drop . . . . .FPD	Leaving . . . . .LVG	Operate . . . . .OPER
Foot . . . .(') or FT	Leaving air temp . . . . .LAT	Opposite . . . . .OPP
Foot pounds . . . . .FT LB	Leaving dry bulb . . . . .LDB	Optional . . . . .OPT
Forging . . . . .FORG	Leaving fluid temperature . . . . .LFT	Original . . . . .ORIG
Forward . . . . .FWD	Leaving water	Ounce . . . . .OZ
Forward curved . . . . .FC	temperature . . . . .LWT	Outside air .OSA
Front . . . . .FRT	Leaving wet bulb . . . . .LWB	Outside air damper . . . . .OAD
Future . . . . .FUT	Left hand . . . . .LH	Outside diameter . . . . .OD
Gage (Gauge) . . . . .GA	Length . . . . .L or LG	Outside radius . . . . .OR
Gallon . . . . .GAL	Linear feet . . . . .LF	Oval head . . . . .OV HD
Gallons per hour . . . . .GPH	Lock washer . . . . .LWASH	Overall . . . . .OA
Gallons per minute . . . . .GPM	Longitude, Longitudinal . . . . .LONG	Oxygen . . . . .OXY
Galvanize, Galvanized . . . . .GALV	Lower . . . . .LWR	Package, Packing . . . . .PKG
Gasket . . . . .GSKT	Lubricate . . . . .LUB	Page . . . . .P
Generator . . . . .GEN	Machine .MACH	Part . . . . .PT
Glycol . . . . .GLY	Magnesium . . . . .MG	Parting line .PL
Grain . . . . .GRN	Maintenance . . . . .MAINT	Patent . . . . .PAT
Gravity . . . . .G	Malleable .MAL	Pattern . . . . .PATT
Grind . . . . .GRD	Manufacture, Manufacturer . . . . .MFR	Perpendicular . . . . .PERP
Harden . . . . .HDN	Material .MATL	Phase . . . . .PH
Hardware . . . . .HDW	Maximum . . . . .MAX	Pint . . . . .PT
Head . . . . .HD	Maximum material	Pitch . . . . .P
Heating . . . . .HTG	condition . . . . .MMC	Pitch circle .PC
Heat treat . . . . .HT TR	Mechanical . . . . .MECH	Pitch diameter . . . . .PD
Height . . . . .H or HGT	Medium .MED	Plan view . . . . .PV
Hexagon .HEX	Memorandum . . . . .MEMO	Point on line . . . . .POL
Horizontal . . . . .HORIZ	Mercury . . . . .HG	Polypropylene . . . . .PP
Horsepower . . . . .HP	Mile . . . . .MI	Polyvinyl chloride . . . . .PVC
Hot rolled steel . . . . .HRS	Miles per hour . . . . .MPH	Position .POSN
Hour . . . . .HR	Millimeter . . . . .MM	Positive . . . .(+) or POS
Hydraulic . . . . .HYD	Minimum . . . . .MIN	Pound . . . . .LB
Illustration . . . . .ILLUS	Minute . . . .(') or MIN	Pounds per square inch . . . . .PSI
Inch, Inches . . . .(") or IN.	Miscellaneous . . . . .MISC	Preliminary . . . . .PRELIM
Inch ounce . . . . .IN. OZ	Modify . . . . .MOD	Pressure . . . . .PRESS
Inch pounds . . . . .IN. LB	Molding . . . . .MLDG	Process .PROC
Inclusive . . . . .INCL	Mounted . . . . .MTD	Product, Production . . . . .PROD
Information . . . . .INFO	Mounting . . . . .MTG	PVC coated steel . . . . .PVS
Inside diameter, Inside	National .NATL	Quality .QUAL
dimension . . . . .ID	National Electrical Mfg.	Quantity .QTY
Inside radius . . . . .IR	Association . . . . .NEMA	Quart . . . . .QT
Inspection . . . . .INSP	National Machine Tool	Quarter .QTR
Installation . . . . .INSTL	Builders Association . . . . .NMTBA	Radius . . . . .R (RAD)
Instrument . . . . .INST	Negative . . . .(−) or NEG	Rear view .RV
Insulation . . . . .INSL or INSUL	Nickel . . . . .NI	Rectangular . . . . .RECT
Interchangeable . . . . .INTCHG	No drawing . . . . .ND	Reduction . . . . .RED
Intermediate . . . . .INTER	Nominal . . . . .NOM	Reference .REF
Internal . . . . .INT	Nonstandard . . . . .NONSTD	Regardless of feature size . . . . .RFS
Invert elevation . . . . .IE	Normally closed . . . . .NC	Regular . . . . .REG
Isolator, Isolation . . . . .ISOL	Normally open . . . . .NO	Reinforce . . . . .REINF
Joint . . . . .JT	Not in contract . . . . .NIC	Relative humidity . . . . .RH
Kilometer .KM	Number . . . . .NO	Remove .REM
Kilowatt .KW	Obsolete .OBS	Require . . . . .REQ

Required . . . . .REQD
Resistor . . . . .RES
Return air . . . . .RA
Return air damper . . . . .RAD
Reverse . . . . .REV
Revision . . . . .REV
Revolution . . . . .REV
Revolutions per minute . . . . .RPM
Right hand . . . . .RH
Root diameter .RD
Round . . . . .RD
Round head . . . . .RD HD
Rubber . . . . .RUB
Screw . . . . .SCR
Screw threads:
   American National coarse . . . . .NC
   American National fine . . . . .NF
   American National extra fine . . . . . .NEF
   American National 8 pitch 8N
   American National 12 pitch 12N
   American National 16 pitch 16N
   American Standard
      straight pipe coupling . . . . .NPSC
   American Standard taper pipe . . . . .NPT
   American Standard taper . . . . .NPTF
   Unified screw thread coarse . . . . . .UNC
   Unified screw thread fine . . . . .UNF
   Unified screw thread extra fine . . .UNEF
   Unified screw thread 8 thread 8UN
   Unified screw thread 12 thread . . .12UN
   Unified screw thread 16 thread . . .16UN
   Unified screw thread special . . . . . .UNS
Second . . . .(") or SEC
Section . . . . .SECT
Sensible . . . . .SENS
Serial, Series . . . . .SER
Sheet . . . . .SH
Side view . . . . .SV
Similar . . . . .SIM

Single wall plenum . . . . .SWP
Single width single inlet . . . . .SWSI
Sketch . . . . .SK
Smoke damper . . . . .SD
Society of Automotive Engineers . . . .SAE
Special . . . . .SPL
Specification . . . . .SPEC
Specific gravity . . . . .SP GR
Spherical .SPHER
Spot face . . . . .SF
Spring . . . . .SPR
Square . . . . .SQ
Square foot (feet) . . . . .SF
Square inch(es) . . . . .SQ IN
Stainless steel . . . . .SS
Standard . . . . .STD
Standard cubic feet per minute . . . . .SCFM
Static pressure . . . . .SP
Steel . . . . .STL
Structure, Structural . . . . .STRUCT
Supply air . . . . .SA
Support . . . . .SUPT
Switch . . . . .SW
Symbol . . . . .SYM
Symmetrical . . . . .SYM
Synthetic . . . . .SYN
Tangent . . . . .TAN
Technical . . . . .TECH
Teeth . . . . .T
Temperature . . . . .TEMP
Tensile strength . . . . .TS
Terminal . . . . .TERM
Theoretical . . . . .THEO
Thickness . . . . .THK
Thousand BTU per hour . . . . .MBH
Thread . . . . .THD
Through . . . . .THRU
Tolerance . . . . .TOL
Total static pressure . . . . .TSP

Tracer . . . . .TCR
Trademark . . . . .TM
Transformer . . . . .TRANS
Transmission . . . . .TRANS
Transverse . . . . .TRANSV
True length . . . . .TL
True position .TP
True view . . . . .TV
Turnbuckle . . . . .TRNBKL
Typical . . . . .TYP
Ultimate . . . . .ULT
United States .US
United States of America
   Standards Institute . . . . .ANSI
United States gage . . . . .USG
Universal . . . . .UNIV
Unless otherwise specified . . . . .UOS
Upper . . . . .UPR
Vacuum . . . . .VAC
Velocity . . . . .V or VEL
Vent through roof . . . . .VTR
Versus . . . . .VS
Vertical . . . . .VERT
Volt(s) V
Volume . . . . .VOL
Volume damper . . . . .VD
Water gauge .WG
Water line . . . . .WL
Water pressure drop . . . . .WPD
Watt . . . . .W
Weight . . . . .WT
Width . . . . .W
With . . . . .W/
Without . . . . .W/O
Wood . . . . .WD
Wood screw . . . . .WD SCR
Wrought iron .WI
Yard . . . . .YD
Year . . . . .YR

APPENDIX B
CONVERSION CHARTS

TABLE 1 INCHES TO MILLIMETERS

in.	mm	in.	mm	in.	mm	in.	mm
1	25.4	26	660.4	51	1295.4	76	1930.4
2	50.8	27	685.8	52	1320.8	77	1955.8
3	76.2	28	711.2	53	1346.2	78	1981.2
4	101.6	29	736.6	54	1371.6	79	2006.6
5	127.0	30	762.0	55	1397.0	80	2032.0
6	152.4	31	787.4	56	1422.4	81	2057.4
7	177.8	32	812.8	57	1447.8	82	2082.8
8	203.2	33	838.2	58	1473.2	83	2108.2
9	228.6	34	863.6	59	1498.6	84	2133.6
10	254.0	35	889.0	60	1524.0	85	2159.0
11	279.4	36	914.4	61	1549.4	86	2184.4
12	304.8	37	939.8	62	1574.8	87	2209.8
13	330.2	38	965.2	63	1600.2	88	2235.2
14	355.6	39	990.6	64	1625.6	89	2260.6
15	381.0	40	1016.0	65	1651.0	90	2286.0
16	406.4	41	1041.4	66	1676.4	91	2311.4
17	431.8	42	1066.8	67	1701.8	92	2336.8
18	457.2	43	1092.2	68	1727.2	93	2362.2
19	482.6	44	1117.6	69	1752.6	94	2387.6
20	508.0	45	1143.0	70	1778.0	95	2413.0
21	533.4	46	1168.4	71	1803.4	96	2438.4
22	558.8	47	1193.8	72	1828.8	97	2463.8
23	584.2	48	1219.2	73	1854.2	98	2489.2
24	609.6	49	1244.6	74	1879.6	99	2514.6
25	635.0	50	1270.0	75	1905.0	100	2540.0

The above table is exact on the basis: 1 in. = 25.4 mm

TABLE 2 MILLIMETERS TO INCHES

in.	mm	in.	mm	in.	mm	in.	mm
1	0.039370	26	1.023622	51	2.007874	76	2.992126
2	0.078740	27	1.062992	52	2.047244	77	3.031496
3	0.118110	28	1.102362	53	2.086614	78	3.070866
4	0.157480	29	1.141732	54	2.125984	79	3.110236
5	0.196850	30	1.181102	55	2.165354	80	3.149606
6	0.236220	31	1.220472	56	2.204724	81	3.188976
7	0.275591	32	1.259843	57	2.244094	82	3.228346
8	0.314961	33	1.299213	58	2.283465	83	3.267717
9	0.354331	34	1.338583	59	2.322835	84	3.307087
10	0.393701	35	1.377953	60	2.362205	85	3.346457
11	0.433071	36	1.417323	61	2.401575	86	3.385827
12	0.472441	37	1.456693	62	2.440945	87	3.425197
13	0.511811	38	1.496063	63	2.480315	88	3.464567
14	0.551181	39	1.535433	64	2.519685	89	3.503937
15	0.590551	40	1.574803	65	2.559055	90	3.543307
16	0.629921	41	1.614173	66	2.598425	91	3.582677
17	0.669291	42	1.653543	67	2.637795	92	3.622047
18	0.708661	43	1.692913	68	2.677165	93	3.661417
19	0.748031	44	1.732283	69	2.716535	94	3.700787
20	0.787402	45	1.771654	70	2.755906	95	3.740157
21	0.826772	46	1.811024	71	2.795276	96	3.779528
22	0.866142	47	1.850394	72	2.834646	97	3.818898
23	0.905512	48	1.889764	73	2.874016	98	3.858268
24	0.944882	49	1.929134	74	2.913386	99	3.897638
25	0.984252	50	1.968504	75	2.952756	100	3.937008

The above table is approximate on the basis: 1 in. = 25.4 mm, 1/25.4 = 0.039370078740+

TABLE 3 INCH/METRIC EQUIVALENTS

Fraction		Decimal Equivalent Customary (in.)	Decimal Equivalent Metric (mm)	Fraction		Decimal Equivalent Customary (in.)	Decimal Equivalent Metric (mm)
	1/64	.015625	0.3969		33/64	.515625	13.0969
1/32		.03125	0.7938	17/32		.53125	13.4938
	3/64	.046875	1.1906		35/64	.546875	13.8906
1/16		.0625	1.5875	9/16		.5625	14.2875
	5/64	.078125	1.9844		37/64	.578125	14.6844
3/32		.09375	2.3813	19/32		.59375	15.0813
	7/64	.109375	2.7781		39/64	.609375	15.4781
1/8		.1250	3.1750	5/8		.6250	15.8750
	9/64	.140625	3.5719		41/64	.640625	16.2719
5/32		.15625	3.9688	21/32		.65625	16.6688
	11/64	.171875	4.3656		43/64	.671875	17.0656
3/16		.1875	4.7625	11/16		.6875	17.4625
	13/64	.203125	5.1594		45/64	.703125	17.8594
7/32		.21875	5.5563	23/32		.71875	18.2563
	15/64	.234375	5.9531		47/64	.734375	18.6531
1/4		.250	6.3500	3/4		.750	19.0500
	17/64	.265625	6.7469		49/64	.765625	19.4469
9/32		.28125	7.1438	25/32		.78125	19.8438
	19/64	.296875	7.5406		51/64	.796875	20.2406
5/16		.3125	7.9375	13/16		.8125	20.6375
	21/64	.328125	8.3384		53/64	.828125	21.0344
11/32		.34375	8.7313	27/32		.84375	21.4313
	23/64	.359375	9.1281		55/64	.859375	21.8281
3/8		.3750	9.5250	7/8		.8750	22.2250
	25/64	.390625	9.9219		57/64	.890625	22.6219
13/32		.40625	10.3188	29/32		.90625	23.0188
	27/64	.421875	10.7156		59/64	.921875	23.4156
7/16		.4375	11.1125	15/16		.9375	23.8125
	29/64	.453125	11.5094		61/64	.953125	24.2094
15/32		.46875	11.9063	31/32		.96875	24.6063
	31/64	.484375	12.3031		63/64	.984375	25.0031
1/2		.500	12.7000	1		1.000	25.4000

APPENDIX B CONVERSION CHARTS (Continued)

TABLE 4 INCH/METRIC—CONVERSION

Measures of Length

1 millimeter (mm) = 0.03937 inch
1 centimeter (cm) = 0.39370 inch
1 meter (m) = 39.37008 inches
= 3.2808 feet
= 1.0936 yards
1 kilometer (km) = 0.6214 mile
1 inch = 25.4 millimeters (mm)
= 2.54 centimeters (cm)
1 foot = 304.8 millimeters (mm)
= 0.3048 meter (m)
1 yard = 0.9144 meter (m)
1 mile = 1.609 kilometers (km)

Measures of Area

1 square millimeter = 0.00155 square inch
1 square centimeter = 0.155 square inch
1 square meter = 10.764 square feet
= 1.196 square yards
1 square kilometer = 0.3861 square mile
1 square inch = 645.2 square millimeters
= 6.452 square centimeters
1 square foot = 929 square centimeters
= 0.0929 square meter
1 square yard = 0.836 square meter
1 square mile = 2.5899 square kilometers

Measures of Capacity (Dry)

1 cubic centimeter (cm^3) = 0.061 cubic inch
1 liter = 0.0353 cubic foot
= 61.023 cubic inches
1 cubic meter (m^3) = 35.315 cubic feet
= 1.308 cubic yards
1 cubic inch = 16.38706 cubic centimeters (cm^3)
1 cubic foot = 0.02832 cubic meter (m^3)
= 28.317 liters
1 cubic yard = 0.7646 cubic meter (m^3)

Measures of Capacity (Liquid)

1 liter = 1.0567 U.S. quarts
= 0.2642 U.S. gallon
= 0.2200 Imperial gallon
1 cubic meter (m^3) = 264.2 U.S. gallons
= 219.969 Imperial gallons
1 U.S. quart = 0.946 liter
1 Imperial quart = 1.136 liters
1 U.S. gallon = 3.785 liters
1 Imperial gallon = 4.546 liters

Measures of Weight

1 gram (g) = 15.432 grains
= 0.03215 ounce troy
= 0.03527 ounce avoirdupois
1 kilogram (kg) = 35.274 ounces avoirdupois
= 2.2046 pounds
1000 kilograms (kg) = 1 metric ton (t)
= 1.1023 tons of 2000 pounds
= 0.9842 ton of 2240 pounds
1 ounce avoirdupois = 28.35 grams (g)
1 ounce troy = 31.103 grams (g)
1 pound = 453.6 grams
= 0.4536 kilogram (kg)
1 ton of 2240 pounds = 1016 kilograms (kg)
= 1.016 metric tons
1 grain = 0.0648 gram (g)
1 metric ton = 0.9842 ton of 2240 pounds
= 2204.6 pounds

APPENDIX C
MATHEMATICAL RULES RELATED TO THE CIRCLE

MATHEMATICAL RULES RELATED TO THE CIRCLE			
To Find Circumference—			
Multiply diameter by	3.1416 .... Or divide diameter by		0.3183
To Find Diameter—			
Multiply circumference by	0.3183 .... Or divide circumference by		3.1416
To Find Radius—			
Multiply circumference by	0.15915 .... Or divide circumference by		6.28318
To Find Side of an Inscribed Square—			
Multiply diameter by	0.7071		
Or multiply circumference by	0.2251 .... Or divide circumference by		4.4428
To Find Side of an Equal Square—			
Multiply diameter by	0.8862 .... Or divide diameter by		1.1284
Or multiply circumference by	0.2821 .... Or divide circumference by		3.545
Square—			
A side multiplied by	1.4142	equals diameter of its circumscribing circle.	
A side multiplied by	4.443	equals circumference of its circumscribing circle.	
A side multiplied by	1.128	equals diameter of an equal circle.	
A side multiplied by	3.547	equals circumference of an equal circle.	
To Find the Area of a Circle—			
Multiply circumference by one-quarter of the diameter.			
Or multiply the square of diameter by	0.7854		
Or multiply the square of circumference by	.07958		
Or multiply the square of 1/2 diameter by	3.1416		
To Find the Surface of a Sphere or Globe—			
Multiply the diameter by the circumference.			
Or multiply the square of diameter by	3.1416		
Or multiply four times the square of radius by	3.1416		

APPENDIX D
GENERAL APPLICATIONS OF SAE STEELS

Application	SAE No.	Application	SAE No.
Adapters	1145	Chain pins, transmission	4320
Agricultural steel	1070	" " "	4815
" "	1080	" " "	4820
Aircraft forgings	4140	Chains, transmission	3135
Axles, front or rear	1040	" "	3140
" " "	4140	Clutch disks	1060
Axle shafts	1045	" "	1070
" "	2340	" "	1085
" "	2345	Clutch springs	1060
" "	3135	Coil springs	4063
" "	3140	Cold-headed bolts	4042
" "	3141	Cold-heading steel	30905
" "	4063	Cold-heading wire or rod	rimmed*
" "	4340	" " " "	1035
Ball-bearing races	52100	Cold-rolled steel	1070
Balls for ball bearings	52100	Connecting-rods	1040
Body stock for cars	rimmed*	"	3141
Bolts, anchor	1040	Connecting-rod bolts	3130
Bolts and screws	1035	Corrosion resisting	51710
Bolts, cold-headed	4042	"	30805
Bolts, connecting-rod	3130	Covers, transmission	rimmed*
Bolts, heat-treated	2330	Crankshafts	1045
Bolts, heavy-duty	4815	"	1145
" "	4820	"	3135
Bolts, steering-arm	3130	"	3140
Brake levers	1030	"	3141
"	1040	Crankshafts, Diesel engine	4340
Bumper bars	1085	Cushion, springs	1060
Cams, free-wheeling	4615	Cutlery, stainless	51335
"	4620	Cylinder studs	3130
Camshafts	1020	Deep-drawing steel	rimmed*
"	1040	"	30905
Carburized parts	1020	Differential gears	4023
" "	1022	Disks, clutch	1070
" "	1024	"	1060
" "	1320	Ductile steel	30905
" "	2317	Fan blades	1020
" "	2515	Fatigue resisting	4340
" "	3310	"	4640
" "	3115	Fender stock for cars	rimmed*
" "	3120	Forgings, aircraft	4140
" "	4023	Forgings, carbon steel	1040
" "	4032	" "	1045
" "	1117	Forgings, heat-treated	3240
" "	1118	" "	5140

Application	SAE No.	Application	SAE No.
Forgings, heat-treated	6150	Key stock	1030
Forgings, high-duty	6150	" "	2330
Forgings, small or medium	1035	" "	3130
Forgings, large	1036	Leaf springs	1085
Free-cutting carbon steel	1111	" "	9260
"	1113	Levers, brake	1030
Free-cutting chro.-ni.steel	30615	" "	1040
Free-cutting mang. steel	1132	Levers, gear shift	1030
"	1137	Levers, heat-treated	2330
Gears, carburized	1320	Lock-washers	1060
"	2317	Mower knives	1085
"	3115	Mower sections	1070
"	3120	Music wire	1085
"	3310	Nuts	3130
"	4119	Nuts, heat-treated	2330
"	4125	Oil-pans, automobile	rimmed*
"	4320	Pinions, carburized	3115
"	4615	"	3120
"	4620	"	4320
"	4815	Piston-pins	3115
"	4820	"	3120
Gears, heat-treated	2345	Plow beams	1070
Gears, car and truck	4027	Plow disks	1080
" "	4032	Plow shares	1080
Gears, cyanide-hardening	5140	Propeller shafts	2340
Gears, differential	4023	"	2345
Gears, high duty	4640	"	4140
"	6150	Races, ball-bearing	52100
Gears, oil-hardening	3145	Ring gears	3115
" "	3150	"	3120
" "	4340	"	4119
" "	5150	Rings, snap	1060
Gears, ring	1045	Rivets	rimmed*
"	3115	Rod and wire	killed*
"	3120	Rod, cold-heading	1035
"	4119	Roller bearings	4815
Gears, transmission	3115	Rollers for bearings	52100
"	3120	Screws and bolts	1035
"	4119	Screw stock, Bessemer	1111
Gears, truck and bus	3310	"	1112
"	4320	"	1113
Gear shift levers	1030	Screw stock, open hearth	1115
Harrow disks	1080	Screws, heat-treated	2330
"	1095	Seat springs	1095
Hay-rake teeth	1095	Shafts, axle	1045

Application	SAE No.	Application	SAE No.
Shafts, cyanide-hardening	5140	Steel, cold-heading	30905
Shafts, heavy-duty	4340	Steel, free-cutting carbon	11111
	6150		1113
" "	4615	Steel, free-cutting chro.-ni.	30615
" "	4620	Steel, free-cutting mang.	1132
Shafts, oil-hardening	5150		0000
Shafts, propeller	2340	Steel, minimum distortion	4615
" "	2345	" " "	4620
" "	4140	" " "	4640
Shafts, transmission	4140	Steel, soft ductile	30905
Sheets and strips	rimmed*	Steering arms	4042
Snap rings	1060	Steering-arm bolts	3130
Spline shafts	1045	Steering knuckles	3141
"	1320	Steering-knuckle pins	4815
"	2340	" "	4820
"	2345	Studs	1040
"	3115		1111
"	3120	Studs, cold-headed	4042
"	3135	Studs, cylinder	3130
"	3140	Studs, heat-treated	2330
"	4023	Studs, heavy-duty	4815
Spring clips	1060		4820
Springs, coil	1095	Tacks	rimmed*
" "	4063	Thrust washers	1060
" "	6150	Thrust washers, oil-harden	5150
Springs, clutch	1060	Transmission shafts	4140
Springs, cushion	1060	Tubing	1040
Springs, leaf	1085	Tubing, front axle	4140
" "	1095	Tubing, seamless	1030
" "	4063	Tubing, welded	1020
" "	4068	Universal joints	1145
" "	9260	Valve springs	1060
" "	6150	Washers, lock	1060
Springs, hard-drawn coiled	1066	Welded structures	30705
Springs, oil-hardening	5150	Wire and rod	killed*
Springs, oil-tempered wire	1066	Wire, cold-heading	rimmed*
Springs, seat	1095		1035
Springs, valve	1060	Wire, hard-drawn spring	1045
Spring wire	1045		1055
Spring wire, hard-drawn	1055	Wire, music	1085
Spring wire, oil-tempered	1055	Wire, oil-tempered spring	1055
Stainless irons	51210	Wrist-pins, automobile	1020
	51710	Yokes	1145
Steel, cold-rolled	1070		

*The "rimmed" and "killed" steels listed are in the SAE 1008, 1010, and 1015 group. See general description of these steels.

Reprinted by permission from Oberg, Jones, and Horton, *Machinery's Handbook*, 24th ed. (New York: Industrial Press, Inc., 1992), table 6, pp. 382–84.

APPENDIX E
SURFACE ROUGHNESS PRODUCED BY COMMON PRODUCTION METHODS

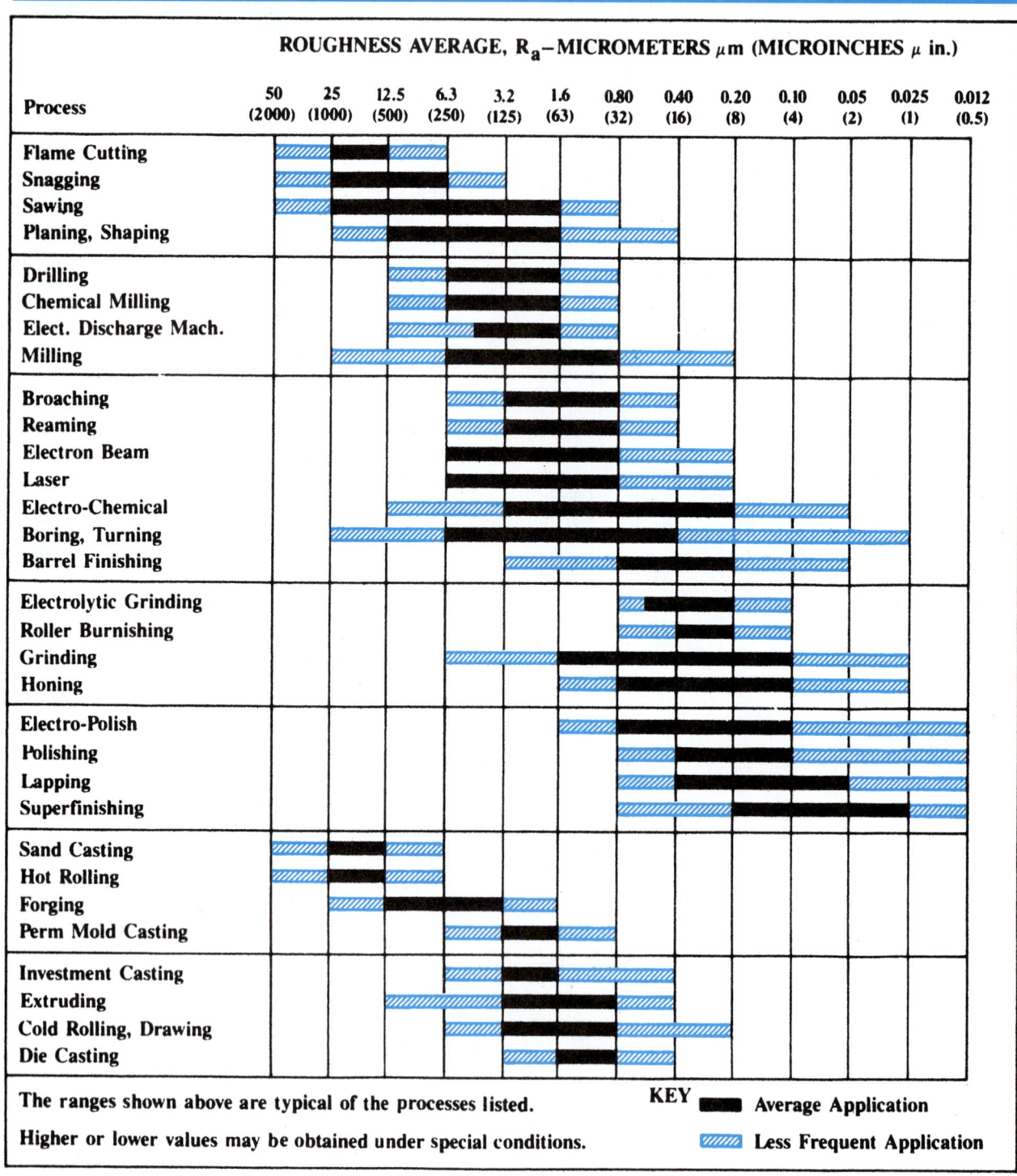

SURFACE ROUGHNESS PRODUCED
BY COMMON PRODUCTION METHODS

ROUGHNESS AVERAGE, R_a—MICROMETERS μm (MICROINCHES μ in.)

The ranges shown above are typical of the processes listed.

Higher or lower values may be obtained under special conditions.

KEY ▬ Average Application

░░ Less Frequent Application

Reprinted by permission from Oberg, Jones, and Horton, *Machinery's Handbook*, 24th ed. (New York: Industrial Press, Inc., 1992), figure 5, p. 672.

APPENDIX F
WIRE GAGES (INCHES)

	WIRE GAGES (INCHES)							
No. of Wire Gage	American Wire or Brown & Sharpe Gage	Steel Wire Gage (U.S.)*	British Standard Wire Gage (Imperial Wire Gage)	Music or Piano Wire Gage	Birmingham or Stub's Iron Wire Gage	Stub's Steel Wire Gage	No. of Wire Gage	Stub's Steel Wire Gage
7/0	. . .	0.4900	0.5000	. . .	. . .	. . .	51	0.066
6/0	0.5800	0.4615	0.4640	0.004	. . .	. . .	52	0.063
5/0	0.5165	0.4305	0.4320	0.005	0.5000	. . .	53	0.058
4/0	0.4600	0.3938	0.4000	0.006	0.4540	. . .	54	0.055
3/0	0.4096	0.3625	0.3720	0.007	0.4250	. . .	55	0.050
2/0	0.3648	0.3310	0.3480	0.008	0.3800	. . .	56	0.045
1/0	0.3249	0.3065	0.3240	0.009	0.3400	. . .	57	0.042
1	0.2893	0.2830	0.3000	0.010	0.3000	0.227	58	0.041
2	0.2576	0.2625	0.2760	0.011	0.2840	0.219	59	0.040
3	0.2294	0.2437	0.2520	0.012	0.2590	0.212	60	0.039
4	0.2043	0.2253	0.2320	0.013	0.2380	0.207	61	0.038
5	0.1819	0.2070	0.2120	0.014	0.2200	0.204	62	0.037
6	0.1620	0.1920	0.1920	0.016	0.2030	0.201	63	0.036
7	0.1443	0.1770	0.1760	0.018	0.1800	0.199	64	0.035
8	0.1285	0.1620	0.1600	0.020	0.1650	0.197	65	0.033
9	0.1144	0.1483	0.1440	0.022	0.1480	0.194	66	0.032
10	0.1019	0.1350	0.1280	0.024	0.1340	0.191	67	0.031
11	0.0907	0.1205	0.1160	0.026	0.1200	0.188	68	0.030
12	0.0808	0.1055	0.1040	0.029	0.1090	0.185	69	0.029
13	0.0720	0.0915	0.0920	0.031	0.0950	0.182	70	0.027
14	0.0641	0.0800	0.0800	0.033	0.0830	0.180	71	0.026
15	0.0571	0.0720	0.0720	0.035	0.0720	0.178	72	0.024
16	0.0508	0.0625	0.0640	0.037	0.0650	0.175	73	0.023
17	0.0453	0.0540	0.0560	0.039	0.0580	0.172	74	0.022
18	0.0403	0.0475	0.0480	0.041	0.0490	0.168	75	0.020
19	0.0359	0.0410	0.0400	0.043	0.0420	0.164	76	0.018
20	0.0320	0.0348	0.0360	0.045	0.0350	0.161	77	0.016
21	0.0285	0.0318	0.0320	0.047	0.0320	0.157	78	0.015
22	0.0253	0.0286	0.0280	0.049	0.0280	0.155	79	0.014
23	0.0226	0.0258	0.0240	0.051	0.0250	0.153	80	0.013
24	0.0201	0.0230	0.0220	0.055	0.0220	0.151	. . .	. . .
25	0.0179	0.0204	0.0200	0.059	0.0200	0.148	. . .	. . .
26	0.0159	0.0181	0.0180	0.063	0.0180	0.146	. . .	. . .
27	0.0142	0.0173	0.0164	0.067	0.0160	0.143	. . .	. . .
28	0.0126	0.0162	0.0149	0.071	0.0140	0.139	. . .	. . .
29	0.0113	0.0150	0.0136	0.075	0.0130	0.134	. . .	. . .
30	0.0100	0.0140	0.0124	0.080	0.0120	0.127	. . .	. . .
31	0.00893	0.0132	0.0116	0.085	0.0100	0.120	. . .	. . .
32	0.00795	0.0128	0.0108	0.090	0.0090	0.115	. . .	. . .
33	0.00708	0.0118	0.0100	0.095	0.0080	0.112	. . .	. . .
34	0.00630	0.0104	0.0092	0.100	0.0070	0.110	. . .	. . .
35	0.00561	0.0095	0.0084	0.106	0.0050	0.108	. . .	. . .
36	0.00500	0.0090	0.0076	0.112	0.0040	0.106	. . .	. . .
37	0.00445	0.0085	0.0068	0.118	. . .	0.103	. . .	. . .
38	0.00396	0.0080	0.0060	0.124	. . .	0.101	. . .	. . .
39	0.00353	0.0075	0.0052	0.130	. . .	0.099	. . .	. . .
40	0.00314	0.0070	0.0048	0.138	. . .	0.097	. . .	. . .
41	0.00280	0.0066	0.0044	0.146	. . .	0.095	. . .	. . .
42	0.00249	0.0062	0.0040	0.154	. . .	0.092	. . .	. . .
43	0.00222	0.0060	0.0036	0.162	. . .	0.088	. . .	. . .
44	0.00198	0.0058	0.0032	0.170	. . .	0.085	. . .	. . .
45	0.00176	0.0055	0.0028	0.180	. . .	0.081	. . .	. . .
46	0.00157	0.0052	0.0024	. . .	. . .	0.079	. . .	. . .
47	0.00140	0.0050	0.0020	. . .	. . .	0.077	. . .	. . .
48	0.00124	0.0048	0.0016	. . .	. . .	0.075	. . .	. . .
49	0.00111	0.0046	0.0012	. . .	. . .	0.072	. . .	. . .
50	0.00099	0.0044	0.0010	. . .	. . .	0.069	. . .	. . .

APPENDIX G
SHEET METAL GAGES (INCHES)

				SHEET METAL GAGES (INCHES)				
No. of Sheet-Metal Gage	Manufacturers' Standard Gage for Steel*	Birmingham Gage (B.G.) for Sheets, Hoops	Galvanized Sheet Gage	Zinc Gage	No. of Sheet-Metal Gage	Manufacturers' Standard Gage for Steel*	Birmingham Gage (B.G.) for Sheets, Hoops	Galvanized Sheet Gage	Zinc Gage
15/0	. . .	1.000	. . .	. . .	20	0.0359	0.0392	0.0396	0.070
14/0	. . .	0.9583	. . .	. . .	21	0.0329	0.0349	0.0366	0.080
13/0	. . .	0.9167	. . .	. . .	22	0.0299	0.03125	0.0336	0.090
12/0	. . .	0.8750	. . .	. . .	23	0.0269	0.02782	0.0306	0.100
11/0	. . .	0.8333	. . .	. . .	24	0.0239	0.02476	0.0276	0.125
10/0	. . .	0.7917	. . .	. . .	25	0.0209	0.02204	0.0247	. . .
9/0	. . .	0.7500	. . .	. . .	26	0.0179	0.01961	0.0217	. . .
8/0	. . .	0.7083	. . .	. . .	27	0.0164	0.01745	0.0202	. . .
7/0	. . .	0.6666	. . .	. . .	28	0.0149	0.01562	0.0187	. . .
6/0	. . .	0.6250	. . .	. . .	29	0.0135	0.01390	0.0172	. . .
5/0	. . .	0.5883	. . .	. . .	30	0.0120	0.01230	0.0157	. . .
4/0	. . .	0.5416	. . .	. . .	31	0.0105	0.01100	0.0142	. . .
3/0	. . .	0.5000	. . .	. . .	32	0.0097	0.00980	0.0134	. . .
2/0	. . .	0.4452	. . .	. . .	33	0.0090	0.00870	. . .	. . .
1/0	. . .	0.3964	. . .	. . .	34	0.0082	0.00770	. . .	. . .
1	. . .	0.3532	. . .	. . .	35	0.0075	0.00690	. . .	. . .
2	. . .	0.3147	. . .	. . .	36	0.0067	0.00610	. . .	. . .
3	0.2391	0.2804	. . .	0.006	37	0.0064	0.00540	. . .	. . .
4	0.2242	0.2500	. . .	0.008	38	0.0060	0.00480	. . .	. . .
5	0.2092	0.2225	. . .	0.010	39	. . .	0.00430	. . .	. . .
6	0.1943	0.1981	. . .	0.012	40	. . .	0.00386	. . .	. . .
7	0.1793	0.1764	. . .	0.014	41	. . .	0.00343	. . .	. . .
8	0.1644	0.1570	0.1681	0.016	42	. . .	0.00306	. . .	. . .
9	0.1495	0.1398	0.1532	0.018	43	. . .	0.00272	. . .	. . .
10	0.1345	0.1250	0.1382	0.020	44	. . .	0.00242	. . .	. . .
11	0.1196	0.1113	0.1233	0.024	45	. . .	0.00215	. . .	. . .
12	0.1046	0.0991	0.1084	0.028	46	. . .	0.00192	. . .	. . .
13	0.0897	0.0882	0.0934	0.032	47	. . .	0.00170	. . .	. . .
14	0.0747	0.0785	0.0785	0.036	48	. . .	0.00152	. . .	. . .
15	0.0673	0.0699	0.0710	0.040	49	. . .	0.00135	. . .	. . .
16	0.0598	0.0625	0.0635	0.045	50	. . .	0.00120	. . .	. . .
17	0.0538	0.0556	0.0575	0.050	51	. . .	0.00107	. . .	. . .
18	0.0478	0.0495	0.0516	0.055	52	. . .	0.00095	. . .	. . .
19	0.0418	0.0440	0.0456	0.060	. . .	. . .	. . .	. . .	. . .

APPENDIX H
SHEET METAL THICKNESSES (MILLIMETERS)

SHEET METAL THICKNESS (MILLIMETERS)					
Preferred Thickness	Second Preference	Third Preference	Preferred Thickness	Second Preference	Third Preference
0.050	. . .	. . .	. . .	. . .	3.6
0.060	. . .	. . .	. . .	3.8	. . .
0.080	. . .	. . .	4.0	. . .	. . .
0.10	. . .	. . .	. . .	4.2	. . .
0.12	. . .	. . .	. . .	4.5	. . .
. . .	0.14	. . .	. . .	4.8	. . .
0.16	. . .	. . .	5.0	. . .	. . .
. . .	0.18	. . .	. . .	5.5	. . .
0.20	. . .	. . .	6.0	. . .	. . .
. . .	0.22	. . .	. . .	. . .	6.5
0.25	. . .	. . .	. . .	7.0	. . .
. . .	0.28	. . .	. . .	. . .	7.5
0.30	. . .	. . .	8.0	. . .	. . .
. . .	0.35	. . .	. . .	9.0	. . .
0.40	. . .	. . .	10	. . .	. . .
. . .	0.45	. . .	. . .	11	. . .
0.50	. . .	. . .	12	. . .	. . .
. . .	0.55	. . .	. . .	14	. . .
0.60	. . .	. . .	16	. . .	. . .
. . .	0.65	. . .	. . .	18	. . .
. . .	0.70	. . .	20	. . .	. . .
. . .	. . .	0.75	. . .	22	. . .
0.80	. . .	. . .	25	. . .	. . .
. . .	. . .	0.85	. . .	28	. . .
. . .	0.90	. . .	30	. . .	. . .
. . .	. . .	0.95	. . .	32	. . .
1.0	. . .	. . .	35	. . .	. . .
. . .	. . .	1.05	. . .	38	. . .
. . .	1.1	. . .	40	. . .	. . .
1.2	. . .	. . .	. . .	45	. . .
. . .	. . .	1.3	50	. . .	. . .
. . .	1.4	. . .	. . .	55	. . .
. . .	. . .	1.5	60	. . .	. . .
1.6	. . .	. . .	. . .	70	. . .
. . .	. . .	1.7	80	. . .	. . .
. . .	1.8	. . .	. . .	90	. . .
. . .	. . .	1.9	100	. . .	. . .
2.0	. . .	. . .	. . .	110	. . .
. . .	. . .	2.1	120	. . .	. . .
. . .	2.2	. . .	. . .	130	. . .
. . .	. . .	2.4	140	. . .	. . .
2.5	. . .	. . .	. . .	150	. . .
. . .	. . .	2.6	160	. . .	. . .
. . .	2.8	. . .	180	. . .	. . .
3.0	. . .	. . .	200	. . .	. . .
. . .	3.2	. . .	250	. . .	. . .
. . .	. . .	3.4	300	. . .	. . .
3.5	. . .	. . .			

APPENDIX I
STANDARD ALLOWANCES, TOLERANCES, AND FITS

TABLE 5 ALLOWANCES AND TOLERANCES

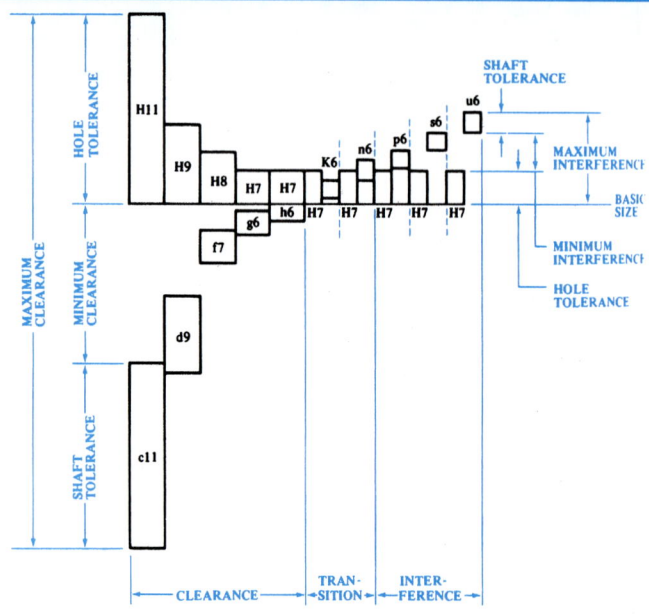

Preferred Hole Basis Fits

Reprinted by permission from Oberg, Jones, and Horton, *Machinery's Handbook*, 24th ed. (New York: Industrial Press, Inc., 1992), figure 2, p. 662.

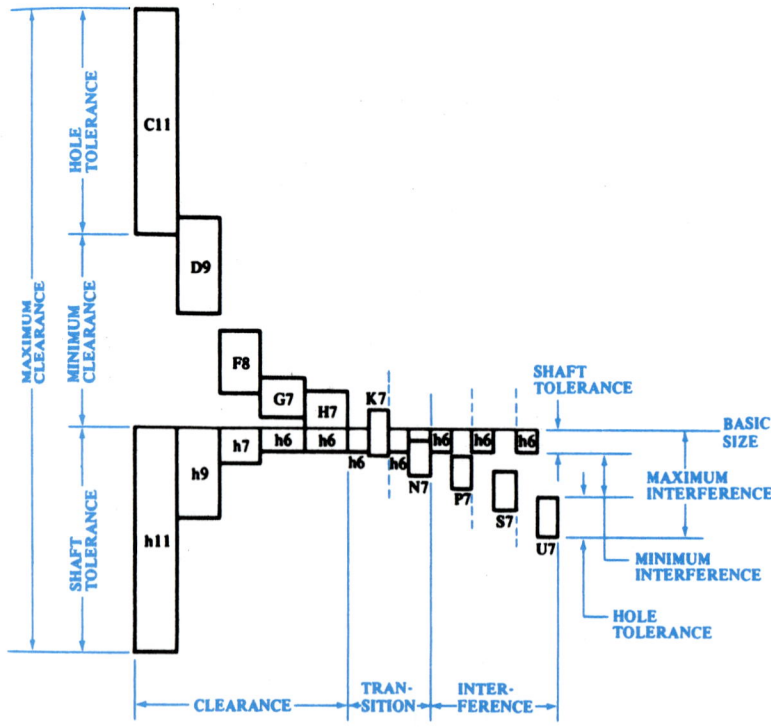

Preferred Shaft Basis Fits

Reprinted by permission from Oberg, Jones, and Horton, *Machinery's Handbook*, 24th ed. (New York: Industrial Press, Inc., 1992), figure 3, p. 623.

APPENDIX I (Continued)

TABLE 5 (CONTINUED)

ISO SYMBOL		DESCRIPTION
Hole Basis	**Shaft Basis**	
H11/c11	C11/h11	*Loose running* fit for wide commercial tolerances or allowances on external members.
H9/d9	D9/h9	*Free running* fit not for use where accuracy is essential, but good for large temperature variations, high running speeds, or heavy journal pressures.
H8/f7	F8/h7	*Close running* fit for running on accurate machines and for accurate location at moderate speeds and journal pressures.
H7/g6	G7/h6	*Sliding* fit not intended to run freely, but to move and turn freely and locate accurately.
H7/h6	H7/h6	*Locational clearance* fit provides snug fit for locating stationary parts; but can be freely assembled and disassembled.
H7/k6	K7/h6	*Locational transition* fit for accurate location, a compromise between clearance and interference.
H7/n6	N7/h6	*Locational transition* fit for more accurate location where greater interference is permissible.
H7/p6[1]	P7/h6	*Locational interference* fit for parts requiring rigidity and alignment with prime accuracy of location but without special bore pressure requirements.
H7/s6	S7/h6	*Medium drive* fit for ordinary steel parts or shrink fits on light sections, the tightest fit usable with cast iron.
H7/u6	U7/h6	*Force* fit suitable for parts which can be highly stressed or for shrink fits where the heavy pressing forces required are impractical.

(Left margin labels: Clearance Fits, Transition Fits, Interference Fits)
(Right margin labels: More Clearance, More Interference)

[1] Transition fit for basic sizes in range from 0 through 3 mm.

Description of Preferred Fits

Reprinted by permission from Oberg, Jones, and Horton, *Machinery's Handbook*, 24th ed. (New York: Industrial Press, Inc., 1992), figure 4, p. 624.

APPENDIX I (Continued)

TABLE 6 AMERICAN NATIONAL STANDARD FITS

RUNNING AND SLIDING FITS

Limits are in thousandths of an inch.

Limits for hole and shaft are applied algebraically to the basic size to obtain the limits of size for the parts.

Data in **boldface** are in accordance with ABC agreements.

Symbols H5, g5, etc., are Hole and Shaft designations used in ABC System (Appendix I).

Nominal Size Range Inches (Over – To)	Class RC 1 Limits of Clearance	Class RC 1 Std Limits Hole H5	Class RC 1 Std Limits Shaft g4	Class RC 2 Limits of Clearance	Class RC 2 Std Limits Hole H6	Class RC 2 Std Limits Shaft g5	Class RC 3 Limits of Clearance	Class RC 3 Std Limits Hole H7	Class RC 3 Std Limits Shaft f6	Class RC 4 Limits of Clearance	Class RC 4 Std Limits Hole H8	Class RC 4 Std Limits Shaft f7
0 – 0.12	0.1 / 0.45	+ 0.2 / 0	– 0.1 / – 0.25	0.1 / 0.55	+ 0.25 / 0	– 0.1 / – 0.3	0.3 / 0.95	+ 0.4 / 0	– 0.3 / – 0.55	0.3 / 1.3	+ 0.6 / 0	– 0.3 / – 0.7
0.12 – 0.24	0.15 / 0.5	+ 0.2 / 0	– 0.15 / – 0.3	0.15 / 0.65	+ 0.3 / 0	– 0.15 / – 0.35	0.4 / 1.12	+ 0.5 / 0	– 0.4 / – 0.7	0.4 / 1.6	+ 0.7 / 0	– 0.4 / – 0.9
0.24 – 0.40	0.2 / 0.6	0.25 / 0	– 0.2 / – 0.35	0.2 / 0.85	+ 0.4 / 0	– 0.2 / – 0.45	0.5 / 1.5	+ 0.6 / 0	– 0.5 / – 0.9	0.5 / 2.0	+ 0.9 / 0	– 0.5 / – 1.1
0.40 – 0.71	0.25 / 0.75	+ 0.3 / 0	– 0.25 / – 0.45	0.25 / 0.95	+ 0.4 / 0	– 0.25 / – 0.55	0.6 / 1.7	+ 0.7 / 0	– 0.6 / – 1.0	0.6 / 2.3	+ 1.0 / 0	– 0.6 / – 1.3
0.71 – 1.19	0.3 / 0.95	+ 0.4 / 0	– 0.3 / – 0.55	0.3 / 1.2	+ 0.5 / 0	– 0.3 / – 0.7	0.8 / 2.1	+ 0.8 / 0	– 0.8 / – 1.3	0.8 / 2.8	+ 1.2 / 0	– 0.8 / – 1.6
1.19 – 1.97	0.4 / 1.1	+ 0.4 / 0	– 0.4 / – 0.7	0.4 / 1.4	+ 0.6 / 0	– 0.4 / – 0.8	1.0 / 2.6	+ 1.0 / 0	– 1.0 / – 1.6	1.0 / 3.6	+ 1.6 / 0	– 1.0 / – 2.0
1.97 – 3.15	0.4 / 1.2	+ 0.5 / 0	– 0.4 / – 0.7	0.4 / 1.6	+ 0.7 / 0	– 0.4 / – 0.9	1.2 / 3.1	+ 1.2 / 0	– 1.2 / – 1.9	1.2 / 4.2	+ 1.8 / 0	– 1.2 / – 2.4
3.15 – 4.73	0.5 / 1.5	+ 0.6 / 0	– 0.5 / – 0.9	0.5 / 2.0	+ 0.9 / 0	– 0.5 / – 1.1	1.4 / 3.7	+ 1.4 / 0	– 1.4 / – 2.3	1.4 / 5.0	+ 2.2 / 0	– 1.4 / – 2.8
4.73 – 7.09	0.6 / 1.8	+ 0.7 / 0	– 0.6 / – 1.1	0.6 / 2.3	+ 1.0 / 0	– 0.6 / – 1.3	1.6 / 4.2	+ 1.6 / 0	– 1.6 / – 2.6	1.6 / 5.7	+ 2.5 / 0	– 1.6 / – 3.2
7.09 – 9.85	0.6 / 2.0	+ 0.8 / 0	– 0.6 / – 1.2	0.6 / 2.6	+ 1.2 / 0	– 0.6 / – 1.4	2.0 / 5.0	+ 1.8 / 0	– 2.0 / – 3.2	2.0 / 6.6	+ 2.8 / 0	– 2.0 / – 3.8
9.85 – 12.41	0.8 / 2.3	+ 0.9 / 0	– 0.8 / – 1.4	0.8 / 2.9	+ 1.2 / 0	– 0.8 / – 1.7	2.5 / 5.7	+ 2.0 / 0	– 2.5 / – 3.7	2.5 / 7.5	+ 3.0 / 0	– 2.5 / – 4.5
12.41 – 15.75	1.0 / 2.7	+ 1.0 / 0	– 1.0 / – 1.7	1.0 / 3.4	+ 1.4 / 0	– 1.0 / – 2.0	3.0 / 6.6	+ / 0	– 3.0 / – 4.4	3.0 / 8.7	+ 3.5 / 0	– 3.0 / – 5.2
15.75 – 19.69	1.2 / 3.0	+ 1.0 / 0	– 1.2 / – 2.0	1.2 / 3.8	+ 1.6 / 0	– 1.2 / – 2.2	4.0 / 8.1	+ 1.6 / 0	– 4.0 / – 5.6	4.0 / 10.5	+ 4.0 / 0	– 4.0 / – 6.5
19.69 – 30.09	1.6 / 3.7	+ 1.2 / 0	– 1.6 / – 2.5	1.6 / 4.8	+ 2.0 / 0	– 1.6 / – 2.8	5.0 / 10.0	+ 3.0 / 0	– 5.0 / – 7.0	5.0 / 13.0	+ 5.0 / 0	– 5.0 / – 8.0
30.09 – 41.49	2.0 / 4.6	+ 1.6 / 0	– 2.0 / – 3.0	2.0 / 6.1	+ 2.5 / 0	– 2.0 / – 3.6	6.0 / 12.5	+ 4.0 / 0	– 6.0 / – 8.5	6.0 / 16.0	+ 6.0 / 0	– 6.0 / –10.0
41.49 – 56.19	2.5 / 5.7	+ 2.0 / 0	– 2.5 / – 3.7	2.5 / 7.5	+ 3.0 / 0	– 2.5 / – 4.5	8.0 / 16.0	+ 5.0 / 0	– 8.0 / –11.0	8.0 / 21.0	+ 8.0 / 0	– 8.0 / –13.0
56.19 – 76.39	3.0 / 7.1	+ 2.5 / 0	– 3.0 / – 4.6	3.0 / 9.5	+ 4.0 / 0	– 3.0 / – 5.5	10.0 / 20.0	+ 6.0 / 0	–10.0 / –14.0	10.0 / 26.0	+10.0 / 0	–10.0 / –16.0
76.39 –100.9	4.0 / 9.0	+ 3.0 / 0	– 4.0 / – 6.0	4.0 / 12.0	+ 5.0 / 0	– 4.0 / – 7.0	12.0 / 25.0	+ 8.0 / 0	–12.0 / –17.0	12.0 / 32.0	+12.0 / 0	–12.0 / –20.0
100.9 –131.9	5.0 / 11.5	+ 4.0 / 0	– 5.0 / – 7.5	5.0 / 15.0	+ 6.0 / 0	– 5.0 / – 9.0	16.0 / 32.0	+10.0 / 0	–16.0 / –22.0	16.0 / 36.0	+16.0 / 0	–16.0 / –26.0
131.9 –171.9	6.0 / 14.0	+ 5.0 / 0	– 6.0 / – 9.0	6.0 / 19.0	+ 8.0 / 0	– 6.0 / –11.0	18.0 / 38.0	+ 8.0 / 0	–18.0 / –26.0	18.0 / 50.0	+20.0 / 0	–18.0 / –30.0
171.9 –200	8.0 / 18.0	+ 6.0 / 0	– 8.0 / –12.0	8.0 / 22.0	+10.0 / 0	– 8.0 / –12.0	22.0 / 48.0	+16.0 / 0	–22.0 / –32.0	22.0 / 63.0	+25.0 / 0	–22.0 / –38.0

APPENDIX I (Continued)

TABLE 6 (CONTINUED)—AMERICAN NATIONAL STANDARD FITS

RUNNING AND SLIDING FITS (continued)

Limits are in thousandths of an inch.

Limits for hole and shaft are applied algebraically to the basic size to obtain the limits of size for the parts.

Data in **boldface** are in accordance with ABC agreements.

Symbols H8, e7, etc., are Hole and Shaft designations used in ABC System (Appendix I).

Class RC 5 Limits of Clearance	Hole H8	Shaft c7	Class RC 6 Limits of Clearance	Hole H9	Shaft e8	Class RC 7 Limits of Clearance	Hole H9	Shaft d8	Class RC 8 Limits of Clearance	Hole H10	Shaft c9	Class RC 9 Limits of Clearance	Hole H11	Shaft	Over	To
0.6 1.6	+0.6 −0	−0.6 −1.0	0.6 2.2	+1.0 0	−0.6 −1.2	1.0 2.6	+1.0 0	−1.0 −1.6	2.5 5.1	+1.6 0	−2.5 −3.5	4.0 8.1	+2.5 0	−4.0 −5.6	0	0.12
0.8 2.0	+0.7 −0	−0.8 −1.3	0.8 2.7	+1.2 0	−0.8 −1.5	1.2 3.1	+1.2 0	−1.2 −1.9	2.8 5.8	+1.8 0	−2.8 −4.0	4.5 9.0	+3.0 0	−4.5 −6.0	0.12	0.24
1.0 2.5	+0.9 −0	−1.0 −1.6	1.0 3.3	+1.4 0	−1.0 −1.9	1.6 3.9	+1.4 0	−1.6 −2.5	3.0 6.6	+2.2 0	−3.0 −4.4	5.0 10.7	+3.5 0	−5.0 −7.2	0.24	0.40
1.2 2.9	+1.0 −0	−1.2 −1.9	1.2 3.8	+1.6 0	−1.2 −2.2	2.0 4.6	+1.6 0	−2.0 −3.0	3.5 7.9	+2.8 0	−3.5 −5.1	6.0 12.8	+4.0 0	−6.0 −8.8	0.40	0.71
1.6 3.6	+1.2 −0	−1.6 −2.4	1.6 4.8	+2.0 0	−1.6 −2.8	2.5 5.7	+2.0 0	−2.5 −3.7	4.5 10.0	+3.5 0	−4.5 −6.5	7.0 15.5	+5.0 0	−7.0 −10.5	0.71	1.19
2.0 4.6	+1.6 −0	−2.0 −3.0	2.0 6.1	+2.5 0	−2.0 −3.6	3.0 7.1	+2.5 0	−3.0 −4.6	5.0 11.5	+4.0 0	−5.0 −7.5	8.0 18.0	+6.0 0	−8.0 −12.0	1.19	1.97
2.5 5.5	+1.8 −0	−2.5 −3.7	2.5 7.3	+3.0 0	−2.5 −4.3	4.0 8.8	+3.0 0	−4.0 −5.8	6.0 13.5	+4.5 0	−6.0 −9.0	9.0 20.5	+7.0 0	−9.0 −13.5	1.97	3.15
3.0 6.6	+2.2 −0	−3.0 −4.4	3.0 8.7	+3.5 0	−3.0 −5.2	5.0 10.7	+3.5 0	−5.0 −7.2	7.0 15.5	+5.0 0	−7.0 −10.5	10.0 24.0	+9.0 0	−10.0 −15.0	3.15	4.73
3.5 7.6	+2.5 −0	−3.5 −5.1	3.5 10.0	+4.0 0	−3.5 −6.0	6.0 12.5	+4.0 0	−6.0 −8.5	8.0 18.0	+6.0 0	−8.0 −12.0	12.0 28.0	+10.0 0	−12.0 −18.0	4.73	7.09
4.0 8.6	+2.8 −0	−4.0 −5.8	4.0 11.3	+4.5 0	−4.0 −6.8	7.0 14.3	+4.5 0	−7.0 −9.8	10.0 21.5	+7.0 0	−10.0 −14.5	15.0 34.0	+12.0 0	−15.0 −22.0	7.09	9.85
5.0 10.0	+3.0 0	−5.0 −7.0	5.0 13.0	+5.0 0	−5.0 −8.0	8.0 16.0	+5.0 0	−8.0 −11.0	12.0 25.0	+8.0 0	−12.0 −17.0	18.0 38.0	+12.0 0	−18.0 −26.0	9.85	12.41
6.0 11.7	+3.5 0	−6.0 −8.2	6.0 15.5	+6.0 0	−6.0 −9.5	10.0 19.5	+6.0 0	−10.0 −13.5	14.0 29.0	+9.0 0	−14.0 −20.0	22.0 45.0	+14.0 0	−22.0 −31.0	12.41	15.75
8.0 14.5	+4.0 0	−8.0 −10.5	8.0 18.0	+6.0 0	−8.0 −12.0	12.0 22.0	+6.0 0	−12.0 −16.0	16.0 32.0	+10.0 0	−16.0 −22.0	25.0 51.0	+16.0 0	−25.0 −35.0	15.75	19.69
10.0 18.0	+5.0 0	−10.0 −13.0	10.0 23.0	+8.0 0	−10.0 −15.0	16.0 29.0	+8.0 0	−16.0 −21.0	20.0 40.0	+12.0 0	−20.0 −28.0	30.0 62.0	+20.0 0	−30.0 −42.0	19.69	30.09
12.0 22.0	+6.0 0	−12.0 −16.0	12.0 28.0	+10.0 0	−12.0 −18.0	20.0 36.0	+10.0 0	−20.0 −26.0	25.0 51.0	+16.0 0	−25.0 −35.0	40.0 81.0	+25.0 0	−40.0 −56.0	30.09	41.49
16.0 29.0	+8.0 0	−16.0 −21.0	16.0 36.0	+12.0 0	−16.0 −24.0	25.0 45.0	+12.0 0	−25.0 −33.0	30.0 62.0	+20.0 0	−30.0 −42.0	50.0 100	+30.0 0	−50.0 −70.0	41.49	56.19
20.0 36.0	+10.0 0	−20.0 −26.0	20.0 46.0	+16.0 0	−20.0 −30.0	30.0 56.0	+16.0 0	−30.0 −40.0	40.0 81.0	+25.0 0	−40.0 −56.0	60.0 125	+40.0 0	−60.0 −85.0	56.19	76.39
25.0 45.0	+12.0 0	−25.0 −33.0	25.0 57.0	+20.0 0	−25.0 −37.0	40.0 72.0	+20.0 0	−40.0 −52.0	50.0 100	+30.0 0	−50.0 −70.0	80.0 160	+50.0 0	−80.0 −110	76.39	100.9
30.0 56.0	+16.0 0	−30.0 −40.0	30.0 71.0	+25.0 0	−30.0 −46.0	50.0 91.0	+25.0 0	−50.0 −66.0	60.0 125	+40.0 0	−60.0 −85.0	100 200	+60.0 0	−100 −140	100.9	131.9
35.0 67.0	+20.0 0	−35.0 −47.0	35.0 85.0	+30.0 0	−35.0 −55.0	60.0 110.0	+30.0 0	−60.0 −80.0	80.0 160	+50.0 0	−80.0 −110	130 260	+80.0 0	−130 −180	131.9	171.9
45.0 86.0	+25.0 0	−45.0 −61.0	45.0 110.0	+40.0 0	−45.0 −70.0	80.0 145.0	+40.0 0	−80.0 −105.0	100 200	+60.0 0	−100 −140	150 310	+100 0	−150 −210	171.9	200

(Reprinted from The American Society of Mechanical Engineers—ANSI B4.1–1967 (R1987).)

APPENDIX I (Continued)

TABLE 6 (CONTINUED)—AMERICAN NATIONAL STANDARD FITS
CLEARANCE LOCATIONAL FITS (LIMITS ARE IN THOUSANDTHS OF AN INCH)

Values shown below are in thousandths of an inch

Nominal Size Range, Inches (Over–To)	Class LC 1 Clearance*	LC 1 Hole H6	LC 1 Shaft h5	Class LC 2 Clearance*	LC 2 Hole H7	LC 2 Shaft h6	Class LC 3 Clearance*	LC 3 Hole H8	LC 3 Shaft h7	Class LC 4 Clearance*	LC 4 Hole H10	LC 4 Shaft h9	Class LC 5 Clearance*	LC 5 Hole H7	LC 5 Shaft g6
0– 0.12	0 / 0.45	+0.25 / 0	0 / −0.2	0 / 0.65	+0.4 / 0	0 / −0.25	0 / 1	+0.6 / 0	0 / −0.4	0 / 2.6	+1.6 / 0	0 / −1.0	0.1 / 0.75	+0.4 / 0	−0.1 / −0.35
0.12– 0.24	0 / 0.5	+0.3 / 0	0 / −0.2	0 / 0.8	+0.5 / 0	0 / −0.3	0 / 1.2	+0.7 / 0	0 / −0.5	0 / 3.0	+1.8 / 0	0 / −1.2	0.15 / 0.95	+0.5 / 0	−0.15 / −0.45
0.24– 0.40	0 / 0.65	+0.4 / 0	0 / −0.25	0 / 1.0	+0.6 / 0	0 / −0.4	0 / 1.5	+0.9 / 0	0 / −0.6	0 / 3.6	+2.2 / 0	0 / −1.4	0.2 / 1.2	+0.6 / 0	−0.2 / −0.6
0.40– 0.71	0 / 0.7	+0.4 / 0	0 / −0.3	0 / 1.1	+0.7 / 0	0 / −0.4	0 / 1.7	+1.0 / 0	0 / −0.7	0 / 4.4	+2.8 / 0	0 / −1.6	0.25 / 1.35	+0.7 / 0	−0.25 / −0.65
0.71– 1.19	0 / 0.9	+0.5 / 0	0 / −0.4	0 / 1.3	+0.8 / 0	0 / −0.5	0 / 2	+1.2 / 0	0 / −0.8	0 / 5.5	+3.5 / 0	0 / −2.0	0.3 / 1.6	+0.8 / 0	−0.3 / −0.8
1.19– 1.97	0 / 1.0	+0.6 / 0	0 / −0.4	0 / 1.6	+1.0 / 0	0 / −0.6	0 / 2.6	+1.6 / 0	0 / −1	0 / 6.5	+4.0 / 0	0 / −2.5	0.4 / 2.0	+1.0 / 0	−0.4 / −1.0
1.97– 3.15	0 / 1.2	+0.7 / 0	0 / −0.5	0 / 1.9	+1.2 / 0	0 / −0.7	0 / 3	+1.8 / 0	0 / −1.2	0 / 7.5	+4.5 / 0	0 / −3	0.4 / 2.3	+1.2 / 0	−0.4 / −1.1
3.15– 4.73	0 / 1.5	+0.9 / 0	0 / −0.6	0 / 2.3	+1.4 / 0	0 / −0.9	0 / 3.6	+2.2 / 0	0 / −1.4	0 / 8.5	+5.0 / 0	0 / −3.5	0.5 / 2.8	+1.4 / 0	−0.5 / −1.4
4.73– 7.09	0 / 1.7	+1.0 / 0	0 / −0.7	0 / 2.6	+1.6 / 0	0 / −1.0	0 / 4.1	+2.5 / 0	0 / −1.6	0 / 10.0	+6.0 / 0	0 / −4	0.6 / 3.2	+1.6 / 0	−0.6 / −1.6
7.09– 9.85	0 / 2.0	+1.2 / 0	0 / −0.8	0 / 3.0	+1.8 / 0	0 / −1.2	0 / 4.6	+2.8 / 0	0 / −1.8	0 / 11.5	+7.0 / 0	0 / −4.5	0.6 / 3.6	+1.8 / 0	−0.6 / −1.8
9.85–12.41	0 / 2.1	+1.2 / 0	0 / −0.9	0 / 3.2	+2.0 / 0	0 / −1.2	0 / 5	+3.0 / 0	0 / −2.0	0 / 13.0	+8.0 / 0	0 / −5	0.7 / 3.9	+2.0 / 0	−0.7 / −1.9
12.41–15.75	0 / 2.4	+1.4 / 0	0 / −1.0	0 / 3.6	+2.2 / 0	0 / −1.4	0 / 5.7	+3.5 / 0	0 / −2.2	0 / 15.0	+9.0 / 0	0 / −6	0.7 / 4.3	+2.2 / 0	−0.7 / −2.1
15.75–19.69	0 / 2.6	+1.6 / 0	0 / −1.0	0 / 4.1	+2.5 / 0	0 / −1.6	0 / 6.5	+4 / 0	0 / −2.5	0 / 16.0	+10.0 / 0	0 / −6	0.8 / 4.9	+2.5 / 0	−0.8 / −2.4

APPENDIX I (Continued)

TABLE 6 (CONTINUED)—AMERICAN NATIONAL STANDARD FITS
CLEARANCE LOCATIONAL FITS (LIMITS ARE IN THOUSANDTHS OF AN INCH)

Values shown below are in thousandths of an inch

Nominal Size Range, Inches Over	To	Class LC 6 Clearance*	Hole H9	Shaft f8	Class LC 7 Clearance*	Hole H10	Shaft e9	Class LC 8 Clearance*	Hole H10	Shaft d9	Class LC 9 Clearance*	Hole H11	Shaft c10	Class LC 10 Clearance*	Hole H12	Shaft	Class LC 11 Clearance*	Hole H13	Shaft
0–	0.12	0.3 / 1.9	+1.0 / 0	−0.3 / −0.9	0.6 / 3.2	+1.6 / 0	−0.6 / −1.6	1.0 / 3.6	+1.6 / 0	−1.0 / −2.0	2.5 / 6.6	+2.5 / 0	−2.5 / −4.1	4 / 12	+4 / 0	−4 / −8	5 / 17	+6 / 0	−5 / −11
0.12–	0.24	0.4 / 2.3	+1.2 / 0	−0.4 / −1.1	0.8 / 3.8	+1.8 / 0	−0.8 / −2.0	1.2 / 4.2	+1.8 / 0	−1.2 / −2.4	2.8 / 7.6	+3.0 / 0	−2.8 / −4.6	4.5 / 14.5	+5 / 0	−4.5 / −9.5	6 / 20	+7 / 0	−6 / −13
0.24–	0.40	0.5 / 2.8	+1.4 / 0	−0.5 / −1.4	1.0 / 4.6	+2.2 / 0	−1.0 / −2.4	1.6 / 5.2	+2.2 / 0	−1.6 / −3.0	3.0 / 8.7	+3.5 / 0	−3.0 / −5.2	5 / 17	+6 / 0	−5 / −11	7 / 25	+9 / 0	−7 / −16
0.40–	0.71	0.6 / 3.2	+1.6 / 0	−0.6 / −1.6	1.2 / 5.6	+2.8 / 0	−1.2 / −2.8	2.0 / 6.4	+2.8 / 0	−2.0 / −3.6	3.5 / 10.3	+4.0 / 0	−3.5 / −6.3	6 / 20	+7 / 0	−6 / −13	8 / 28	+10 / 0	−8 / −18
0.71–	1.19	0.8 / 4.0	+2.0 / 0	−0.8 / −2.0	1.6 / 7.1	+3.5 / 0	−1.6 / −3.6	2.5 / 8.0	+3.5 / 0	−2.5 / −4.5	4.5 / 13.0	+5.0 / 0	−4.5 / −8.0	7 / 23	+8 / 0	−7 / −15	10 / 34	+12 / 0	−10 / −22
1.19–	1.97	1.0 / 5.1	+2.5 / 0	−1.0 / −2.6	2.0 / 8.5	+4.0 / 0	−2.0 / −4.5	3.0 / 9.5	+4.0 / 0	−3.0 / −5.5	5.0 / 15.0	+6 / 0	−5.0 / −9.0	8 / 28	+10 / 0	−8 / −18	12 / 44	+16 / 0	−12 / −28
1.97–	3.15	1.2 / 6.0	+3.0 / 0	−1.0 / −3.0	2.5 / 10.0	+4.5 / 0	−2.5 / −5.5	4.0 / 11.5	+4.5 / 0	−4.0 / −7.0	6.0 / 17.5	+7 / 0	−6.0 / −10.5	10 / 34	+12 / 0	−10 / −22	14 / 50	+18 / 0	−14 / −32
3.15–	4.73	1.4 / 7.1	+3.5 / 0	−1.4 / −3.6	3.0 / 11.5	+5.0 / 0	−3.0 / −6.5	5.0 / 13.5	+5.0 / 0	−5.0 / −8.5	7 / 21	+9 / 0	−7 / −12	11 / 39	+14 / 0	−11 / −25	16 / 60	+22 / 0	−16 / −38
4.73–	7.09	1.6 / 8.1	+4.0 / 0	−1.6 / −4.1	3.5 / 13.5	+6.0 / 0	−3.5 / −7.5	6 / 16	+6 / 0	−6 / −10	8 / 24	+10 / 0	−8 / −14	12 / 44	+16 / 0	−12 / −28	18 / 68	+25 / 0	−18 / −43
7.09–	9.85	2.0 / 9.3	+4.5 / 0	−2.0 / −4.8	4.0 / 15.5	+7.0 / 0	−4.0 / −8.5	7 / 18.5	+7 / 0	−7 / −11.5	10 / 29	+12 / 0	−10 / −17	16 / 52	+18 / 0	−16 / −34	22 / 78	+28 / 0	−22 / −50
9.85–	12.41	2.2 / 10.2	+5.0 / 0	−2.2 / −5.2	4.5 / 17.5	+8.0 / 0	−4.5 / −9.5	7 / 20	+8 / 0	−7 / −12	12 / 32	+12 / 0	−12 / −20	20 / 60	+20 / 0	−20 / −40	28 / 88	+30 / 0	−28 / −58
12.41–	15.75	2.5 / 12.0	+6.0 / 0	−2.5 / −6.0	5.0 / 20.0	+9.0 / 0	−5 / −11	8 / 23	+9 / 0	−8 / −14	14 / 37	+14 / 0	−14 / −23	22 / 66	+22 / 0	−22 / −44	30 / 100	+35 / 0	−30 / −65
15.75–	19.69	2.8 / 12.8	+6.0 / 0	−2.8 / −6.8	5.0 / 21.0	+10.0 / 0	−5 / −11	9 / 25	+10 / 0	−9 / −15	16 / 42	+16 / 0	−16 / −26	25 / 75	+25 / 0	−25 / −50	35 / 115	+40 / 0	−35 / −75

APPENDIX I (Continued)

TABLE 6 (CONTINUED)—AMERICAN NATIONAL STANDARD FITS TRANSITION LOCATIONAL FITS (LIMITS ARE IN THOUSANDTHS OF AN INCH)

Nominal Size Range, Inches	Class LT 1			Class LT 2			Class LT 3			Class LT 4			Class LT 5			Class LT 6		
		Standard Tolerance Limits			Standard Tolerance Limits			Standard Tolerance Limits			Standard Tolerance Limits			Standard Tolerance Limits			Standard Tolerance Limits	
Over To	Fit*	Hole H7	Shaft js6	Fit*	Hole H8	Shaft js7	Fit*	Hole H7	Shaft k6	Fit*	Hole H8	Shaft k7	Fit*	Hole H7	Shaft n6	Fit*	Hole H7	Shaft n7
	Values shown below are in thousandths of an inch																	
0– 0.12	−0.12 +0.52	+0.4 0	+0.12 −0.12	−0.2 +0.8	+0.6 0	+0.2 −0.2							−0.5 +0.15	+0.4 0	+0.5 +0.25	−0.65 +0.15	+0.4 0	+0.65 +0.25
0.12– 0.24	−0.15 +0.65	+0.5 0	+0.15 −0.15	−0.25 +0.95	+0.7 0	+0.25 −0.25							−0.6 +0.2	+0.5 0	+0.6 +0.3	−0.8 +0.2	+0.5 0	+0.8 +0.3
0.24– 0.40	−0.2 +0.8	+0.6 0	+0.2 −0.2	−0.3 +1.2	+0.9 0	+0.3 −0.3	−0.5 +0.5	+0.6 0	+0.5 +0.1	−0.7 +0.8	+0.9 0	+0.7 +0.1	−0.8 +0.2	+0.6 0	+0.8 +0.4	−1.0 +0.2	+0.6 0	+1.0 +0.4
0.40– 0.71	−0.2 +0.9	+0.7 0	+0.2 −0.2	−0.35 +1.35	+1.0 0	+0.35 −0.35	−0.5 +0.6	+0.7 0	+0.5 +0.1	−0.8 +0.9	+1.0 0	+0.8 +0.1	−0.9 +0.2	+0.7 0	+0.9 +0.5	−1.2 +0.2	+0.7 0	+1.2 +0.5
0.71– 1.19	−0.25 +1.05	+0.8 0	+0.25 −0.25	−0.4 +1.6	+1.2 0	+0.4 −0.4	−0.6 +0.7	+0.8 0	+0.6 +0.1	−0.9 +1.1	+1.2 0	+0.9 +0.1	−1.1 +0.2	+0.8 0	+1.1 +0.6	−1.4 +0.2	+0.8 0	+1.4 +0.6
1.19– 1.97	−0.3 +1.3	+1.0 0	+0.3 −0.3	−0.5 +2.1	+1.6 0	+0.5 −0.5	−0.7 +0.9	+1.0 0	+0.7 +0.1	−1.1 +1.5	+1.6 0	+1.1 +0.1	−1.3 +0.3	+1.0 0	+1.3 +0.7	−1.7 +0.3	+1.0 0	+1.7 +0.7
1.97– 3.15	−0.3 +1.5	+1.2 0	+0.3 −0.3	−0.6 +2.4	+1.8 0	+0.6 −0.6	−0.8 +1.1	+1.2 0	+0.8 +0.1	−1.3 +1.7	+1.8 0	+1.3 +0.1	−1.5 +0.4	+1.2 0	+1.5 +0.8	−2.0 +0.4	+1.2 0	+2.0 +0.8
3.15– 4.73	−0.4 +1.8	+1.4 0	+0.4 −0.4	−0.7 +2.9	+2.2 0	+0.7 −0.7	−1.0 +1.3	+1.4 0	+1.0 +0.1	−1.5 +2.1	+2.2 0	+1.5 +0.1	−1.9 +0.4	+1.4 0	+1.9 +1.0	−2.4 +0.4	+1.4 0	+2.4 +1.0
4.73– 7.09	−0.5 +2.1	+1.6 0	+0.5 −0.5	−0.8 +3.3	+2.5 0	+0.8 −0.8	−1.1 +1.5	+1.6 0	+1.1 +0.1	−1.7 +2.4	+2.5 0	+1.7 +0.1	−2.2 +0.4	+1.6 0	+2.2 +1.2	−2.8 +0.4	+1.6 0	+2.8 +1.2
7.09– 9.85	−0.6 +2.4	+1.8 0	+0.6 −0.6	−0.9 +3.7	+2.8 0	+0.9 −0.9	−1.4 +1.6	+1.8 0	+1.4 +0.2	−2.0 +2.6	+2.8 0	+2.0 +0.2	−2.6 +0.4	+1.8 0	+2.6 +1.4	−3.2 +0.4	+1.8 0	+3.2 +1.4
9.85–12.41	−0.6 +2.6	+2.0 0	+0.6 −0.6	−1.0 +4.0	+3.0 0	+1.0 −1.0	−1.4 +1.8	+2.0 0	+1.4 +0.2	−2.2 +2.8	+3.0 0	+2.2 +0.2	−2.6 +0.6	+2.0 0	+2.6 +1.4	−3.4 +0.6	+2.0 0	+3.4 +1.4
12.41–15.75	−0.7 +2.9	+2.2 0	+0.7 −0.7	−1.0 +4.5	+3.5 0	+1.0 −1.0	−1.6 +2.0	+2.2 0	+1.6 +0.2	−2.4 +3.3	+3.5 0	+2.4 +0.2	−3.0 +0.6	+2.2 0	+3.0 +1.6	−3.8 +0.6	+2.2 0	+3.8 +1.6
15.75–19.69	−0.8 +3.3	+2.5 0	+0.8 −0.8	−1.2 +5.2	+4.0 0	+1.2 −1.2	−1.8 +2.3	+2.5 0	+1.8 +0.2	−2.7 +3.8	+4.0 0	+2.7 +0.2	−3.4 +0.7	+2.5 0	+3.4 +1.8	−4.3 +0.7	+2.5 0	+4.3 +1.8

 **APPENDIX I** (Continued)

	Class LN 1			Class LN 2			Class LN 3		
		Standard Limits			Standard Limits			Standard Limits	
Nominal Size Range, Inches	Limits of Interference	Hole H6	Shaft n5	Limits of Interference	Hole H7	Shaft p6	Limits of Interference	Hole H7	Shaft t6
Over To	Values shown below are in thousandths of an inch								
0– 0.12	0 0.45	+0.25 0	+0.45 +0.25	0 0.65	+0.4 0	+0.65 +0.4	0.1 0.75	+0.4 0	+0.75 +0.5
0.12– 0.24	0 0.5	+0.3 0	+0.5 +0.3	0 0.8	+0.5 0	+0.8 +0.5	0.1 0.9	+0.5 0	+0.9 +0.6
0.24– 0.40	0 0.65	+0.4 0	+0.65 +0.4	0 1.0	+0.6 0	+1.0 +0.6	0.2 1.2	+0.6 0	+1.2 +0.8
0.40– 0.71	0 0.8	+0.4 0	+0.8 +0.4	0 1.1	+0.7 0	+1.1 +0.7	0.3 1.4	+0.7 0	+1.4 +1.0
0.71– 1.19	0 1.0	+0.5 0	+1.0 +0.5	0 1.3	+0.8 0	+1.3 +0.8	0.4 1.7	+0.8 0	+1.7 +1.2
1.19– 1.97	0 1.1	+0.6 0	+1.1 +0.6	0 1.6	+1.0 0	+1.6 +1.0	0.4 2.0	+1.0 0	+2.0 +1.4
1.97– 3.15	0.1 1.3	+0.7 0	+1.3 +0.8	0.2 2.1	+1.2 0	+2.1 +1.4	0.4 2.3	+1.2 0	+2.3 +1.6
3.15– 4.73	0.1 1.6	+0.9 0	+1.6 +1.0	0.2 2.5	+1.4 0	+2.5 +1.6	0.6 2.9	+1.4 0	+2.9 +2.0
4.73– 7.09	0.2 1.9	+1.0 0	+1.9 +1.2	0.2 2.8	+1.6 0	+2.8 +1.8	0.9 3.5	+1.6 0	+3.5 +2.5
7.09– 9.85	0.2 2.2	+1.2 0	+2.2 +1.4	0.2 3.2	+1.8 0	+3.2 +2.0	1.2 4.2	+1.8 0	+4.2 +3.0
9.85–12.41	0.2 2.3	+1.2 0	+2.3 +1.4	0.2 3.4	+2.0 0	+3.4 +2.2	1.5 4.7	+2.0 0	+4.7 +3.5
12.41–15.75	0.2 2.6	+1.4 0	+2.6 +1.6	0.3 3.9	+2.2 0	+3.9 +2.5	2.3 5.9	+2.2 0	+5.9 +4.5
15.75–19.69	0.2 2.8	+1.6 0	+2.8 +1.8	0.3 4.4	+2.5 0	+4.4 +2.8	2.5 6.6	+2.5 0	+6.6 +5.0

TABLE 6 (CONTINUED)—AMERICAN NATIONAL STANDARD FITS
INTERFERENCE LOCATIONAL FITS
(LIMITS ARE IN THOUSANDTHS OF AN INCH)

APPENDIX I (Continued)

TABLE 6 (CONTINUED)—AMERICAN NATIONAL STANDARD FIT FORCE AND SHRINK FITS (LIMITS ARE IN THOUSANDTHS OF AN INCH)

Values shown below are in thousandths of an inch

Nominal Size Range, Inches (Over – To)	Class FN 1 Interference*	Class FN 1 Hole H6	Class FN 1 Shaft	Class FN 2 Interference*	Class FN 2 Hole H7	Class FN 2 Shaft s6	Class FN 3 Interference*	Class FN 3 Hole H7	Class FN 3 Shaft t6	Class FN 4 Interference*	Class FN 4 Hole H7	Class FN 4 Shaft u6	Class FN 5 Interference*	Class FN 5 Hole H8	Class FN 5 Shaft x7
0 – 0.12	0.05 / 0.5	+0.25 / 0	+0.5 / +0.3	0.2 / 0.85	+0.4 / 0	+0.85 / +0.6				0.3 / 0.95	+0.4 / 0	+0.95 / +0.7	0.3 / 1.3	+0.6 / 0	+1.3 / +0.9
0.12 – 0.24	0.1 / 0.6	+0.3 / 0	+0.6 / +0.4	0.2 / 1.0	+0.5 / 0	+1.0 / +0.7				0.4 / 1.2	+0.5 / 0	+1.2 / +0.9	0.5 / 1.7	+0.7 / 0	+1.7 / +1.2
0.24 – 0.40	0.1 / 0.75	+0.4 / 0	+0.75 / +0.5	0.4 / 1.4	+0.6 / 0	+1.4 / +1.0				0.6 / 1.6	+0.6 / 0	+1.6 / +1.2	0.5 / 2.0	+0.9 / 0	+2.0 / +1.4
0.40 – 0.56	0.1 / 0.8	+0.4 / 0	+0.8 / +0.5	0.5 / 1.6	+0.7 / 0	+1.6 / +1.2				0.7 / 1.8	+0.7 / 0	+1.8 / +1.4	0.6 / 2.3	+1.0 / 0	+2.3 / +1.6
0.56 – 0.71	0.2 / 0.9	+0.4 / 0	+0.9 / +0.6	0.5 / 1.6	+0.7 / 0	+1.6 / +1.2				0.7 / 1.8	+0.7 / 0	+1.8 / +1.4	0.8 / 2.5	+1.0 / 0	+2.5 / +1.8
0.71 – 0.95	0.2 / 1.1	+0.5 / 0	+1.1 / +0.7	0.6 / 1.9	+0.8 / 0	1.9 / +1.4				0.8 / 2.1	+0.8 / 0	+2.1 / +1.6	1.0 / 3.0	+1.2 / 0	+3.0 / +2.2
0.95 – 1.19	0.3 / 1.2	+0.5 / 0	+1.2 / +0.8	0.6 / 1.9	+0.8 / 0	+1.9 / +1.4	0.8 / 2.1	+0.8 / 0	+2.1 / +1.6	1.0 / 2.3	+0.8 / 0	+2.3 / +1.8	1.3 / 3.3	+1.2 / 0	+3.3 / +2.5
1.19 – 1.58	0.3 / 1.3	+0.6 / 0	+1.3 / +0.9	0.8 / 2.4	+1.0 / 0	+2.4 / +1.8	1.0 / 2.6	+1.0 / 0	+2.6 / +2.0	1.5 / 3.1	+1.0 / 0	+3.1 / +2.5	1.4 / 4.0	+1.6 / 0	+4.0 / +3.0
1.58 – 1.97	0.4 / 1.4	+0.6 / 0	+1.4 / +1.0	0.8 / 2.4	+1.0 / 0	+2.4 / +1.8	1.2 / 2.8	+1.0 / 0	+2.8 / +2.2	1.8 / 3.4	+1.0 / 0	+3.4 / +2.8	2.4 / 5.0	+1.6 / 0	+5.0 / +4.0
1.97 – 2.56	0.6 / 1.8	+0.7 / 0	+1.8 / +1.3	0.8 / 2.7	+1.2 / 0	+2.7 / +2.0	1.3 / 3.2	+1.2 / 0	+3.2 / +2.5	2.3 / 4.2	+1.2 / 0	+4.2 / +3.5	3.2 / 6.2	+1.8 / 0	+6.2 / +5.0
2.56 – 3.15	0.7 / 1.9	+0.7 / 0	+1.9 / +1.4	1.0 / 2.9	+1.2 / 0	+2.9 / +2.2	1.8 / 3.7	+1.2 / 0	+3.7 / +3.0	2.8 / 4.7	+1.2 / 0	+4.7 / +4.0	4.2 / 7.2	+1.8 / 0	+7.2 / +6.0
3.15 – 3.94	0.9 / 2.4	+0.9 / 0	+2.4 / +1.8	1.4 / 3.7	+1.4 / 0	+3.7 / +2.8	2.1 / 4.4	+1.4 / 0	+4.4 / +3.5	3.6 / 5.9	+1.4 / 0	+5.9 / +5.0	4.8 / 8.4	+2.2 / 0	+8.4 / +7.0
3.94 – 4.73	1.1 / 2.6	+0.9 / 0	+2.6 / +2.0	1.6 / 3.9	+1.4 / 0	+3.9 / +3.0	2.6 / 4.9	+1.4 / 0	+4.9 / +4.0	4.6 / 6.9	+1.4 / 0	+6.9 / +6.0	5.8 / 9.4	+2.2 / 0	+9.4 / +8.0

APPENDIX I (Continued)

TABLE 6 (CONTINUED)—AMERICAN NATIONAL STANDARD FIT FORCE AND SHRINK FITS (LIMITS ARE IN THOUSANDTHS OF AN INCH)

Values shown below are in thousandths of an inch

Nominal Size Range, Inches Over–To	Class FN 1 Interference*	Class FN 1 Hole H6	Class FN 1 Shaft	Class FN 2 Interference*	Class FN 2 Hole H7	Class FN 2 Shaft s6	Class FN 3 Interference*	Class FN 3 Hole H7	Class FN 3 Shaft t6	Class FN 4 Interference*	Class FN 4 Hole H7	Class FN 4 Shaft u6	Class FN 5 Interference*	Class FN 5 Hole H8	Class FN 5 Shaft x7
4.73– 5.52	1.2 / 2.9	+1.0 / 0	+2.9 / +2.2	1.9 / 4.5	+1.6 / 0	+4.5 / +3.5	3.4 / 6.0	+1.6 / 0	+6.0 / +5.0	5.4 / 8.0	+1.6 / 0	+8.0 / +7.0	7.5 / 11.6	+2.5 / 0	+11.6 / +10.0
5.52– 6.30	1.5 / 3.2	+1.0 / 0	+3.2 / +2.5	2.4 / 5.0	+1.6 / 0	+5.0 / +4.0	3.4 / 6.0	+1.6 / 0	+6.0 / +5.0	5.4 / 8.0	+1.6 / 0	+8.0 / +7.0	9.5 / 13.6	+2.5 / 0	+13.6 / +12.0
6.30– 7.09	1.8 / 3.5	+1.0 / 0	+3.5 / +2.8	2.9 / 5.5	+1.6 / 0	+5.5 / +4.5	4.4 / 7.0	+1.6 / 0	+7.0 / +6.0	6.4 / 9.0	+1.6 / 0	+9.0 / +8.0	9.5 / 13.6	+2.5 / 0	+13.6 / +12.0
7.09– 7.88	1.8 / 3.8	+1.2 / 0	+3.8 / +3.0	3.2 / 6.2	+1.8 / 0	+6.2 / +5.0	5.2 / 8.2	+1.8 / 0	+8.2 / +7.0	7.2 / 10.2	+1.8 / 0	+10.2 / +9.0	11.2 / 15.8	+2.8 / 0	+15.8 / +14.0
7.88– 8.86	2.3 / 4.3	+1.2 / 0	+4.3 / +3.5	3.2 / 6.2	+1.8 / 0	+6.2 / 5.0	5.2 / 8.2	+1.8 / 0	+8.2 / +7.0	8.2 / 11.2	+1.8 / 0	+11.2 / +10.0	13.2 / 17.8	+2.8 / 0	+17.8 / +16.0
8.86– 9.85	2.3 / 4.3	+1.2 / 0	+4.3 / +3.5	4.2 / 7.2	+1.8 / 0	+7.2 / +6.0	6.2 / 9.2	+1.8 / 0	+9.2 / +8.0	10.2 / 13.2	+1.8 / 0	+13.2 / +12.0	13.2 / 17.8	+2.8 / 0	+17.8 / +16.0
9.85–11.03	2.8 / 4.9	+1.2 / 0	+4.9 / +4.0	4.0 / 7.2	+2.0 / 0	+7.2 / +6.0	7.0 / 10.2	+2.0 / 0	+10.2 / +9.0	10.0 / 13.2	+2.0 / 0	+13.2 / +12.0	15.0 / 20.0	+3.0 / 0	+20.0 / +18.0
11.03–12.41	2.8 / 4.9	+1.2 / 0	+4.9 / +4.0	5.0 / 8.2	+2.0 / 0	+8.2 / +7.0	7.0 / 10.2	+2.0 / 0	+10.2 / +9.0	12.0 / 15.2	+2.0 / 0	+15.2 / +14.0	17.0 / 22.0	+3.0 / 0	+22.0 / +20.0
12.41–13.98	3.1 / 5.5	+1.4 / 0	+5.5 / +4.5	5.8 / 9.4	+2.2 / 0	+9.4 / +8.0	7.8 / 11.4	+2.2 / 0	+11.4 / +10.0	13.8 / 17.4	+2.2 / 0	17.4 / +16.0	18.5 / 24.2	+3.5 / 0	+24.2 / +22.0
13.98–15.75	3.6 / 6.1	+1.4 / 0	+6.1 / +5.0	5.8 / 9.4	+2.2 / 0	+9.4 / +8.0	9.8 / 13.4	+2.2 / 0	+13.4 / +12.0	15.8 / 19.4	+2.2 / 0	+19.4 / +18.0	21.5 / 27.2	+3.5 / 0	+27.2 / +25.0
15.75–17.72	4.4 / 7.0	+1.6 / 0	+7.0 / +6.0	6.5 / 10.6	+2.5 / 0	+10.6 / +9.0	9.5 / 13.6	+2.5 / 0	+13.6 / +12.0	17.5 / 21.6	+2.5 / 0	+21.6 / +20.0	24.0 / 30.5	+4.0 / 0	+30.5 / +28.0
17.72–19.69	4.4 / 7.0	+1.6 / 0	+7.0 / +6.0	7.5 / 11.6	+2.5 / 0	+11.6 / +10.0	11.5 / 15.6	+2.5 / 0	+15.6 / +14.0	19.5 / 23.6	+2.5 / 0	+23.6 / +22.0	26.0 / 32.5	+4.0 / 0	+32.5 / +30.0

AMERICAN NATIONAL STANDARD
PREFERRED METRIC LIMITS AND FITS

ANSI B4.2–1978

TABLE 7 METRIC LIMITS AND FITS

APPENDIX I (Continued)

PREFERRED HOLE BASIS CLEARANCE FITS

Dimensions in mm.

BASIC SIZE		LOOSE RUNNING			FREE RUNNING			CLOSE RUNNING			SLIDING			LOCATIONAL CLEARANCE		
		Hole H11	Shaft c11	Fit	Hole H9	Shaft d9	Fit	Hole H8	Shaft f7	Fit	Hole H7	Shaft g6	Fit	Hole H7	Shaft h6	Fit
1	MAX	1.060	0.940	0.180	1.025	0.980	0.070	1.014	0.994	0.030	1.010	0.998	0.018	1.010	1.000	0.016
	MIN	1.000	0.880	0.060	1.000	0.955	0.020	1.000	0.984	0.006	1.000	0.992	0.002	1.000	0.994	0.000
1.2	MAX	1.260	1.140	0.180	1.225	1.180	0.070	1.214	1.194	0.030	1.210	1.198	0.018	1.210	1.200	0.016
	MIN	1.200	1.080	0.060	1.200	1.155	0.020	1.200	1.184	0.006	1.200	1.192	0.002	1.200	1.194	0.000
1.6	MAX	1.660	1.540	0.180	1.625	1.580	0.070	1.614	1.594	0.030	1.610	1.598	0.018	1.610	1.600	0.016
	MIN	1.600	1.480	0.060	1.600	1.555	0.020	1.600	1.584	0.006	1.600	1.592	0.002	1.600	1.594	0.000
2	MAX	2.060	1.940	0.180	2.025	1.980	0.070	2.014	1.994	0.030	2.010	1.998	0.018	2.010	2.000	0.016
	MIN	2.000	1.880	0.060	2.000	1.955	0.020	2.000	1.984	0.006	2.000	1.992	0.002	2.000	1.994	0.000
2.5	MAX	2.560	2.440	0.180	2.525	2.480	0.070	2.514	2.494	0.030	2.510	2.498	0.018	2.510	2.500	0.016
	MIN	2.500	2.380	0.060	2.500	2.455	0.020	2.500	2.484	0.006	2.500	2.492	0.002	2.500	2.494	0.000
3	MAX	3.060	2.940	0.180	3.025	2.980	0.070	3.014	2.994	0.030	3.010	2.998	0.018	3.010	3.000	0.016
	MIN	3.000	2.880	0.060	3.000	2.955	0.020	3.000	2.984	0.006	3.000	2.992	0.002	3.000	2.994	0.000
4	MAX	4.075	3.930	0.220	4.030	3.970	0.090	4.018	3.990	0.040	4.012	3.996	0.024	4.012	4.000	0.020
	MIN	4.000	3.855	0.070	4.000	3.940	0.030	4.000	3.978	0.010	4.000	3.988	0.004	4.000	3.992	0.000
5	MAX	5.075	4.930	0.220	5.030	4.970	0.090	5.018	4.990	0.040	5.012	4.996	0.024	5.012	5.000	0.020
	MIN	5.000	4.855	0.070	5.000	4.940	0.030	5.000	4.978	0.010	5.000	4.988	0.004	5.000	4.992	0.000
6	MAX	6.075	5.930	0.220	6.030	5.970	0.090	6.018	5.990	0.040	6.012	5.996	0.024	6.012	6.000	0.020
	MIN	6.000	5.855	0.070	6.000	5.940	0.030	6.000	5.978	0.010	6.000	5.988	0.004	6.000	5.992	0.000
8	MAX	8.090	7.920	0.260	8.036	7.960	0.112	8.022	7.987	0.050	8.015	7.995	0.029	8.015	8.000	0.024
	MIN	8.000	7.830	0.080	8.000	7.924	0.040	8.000	7.972	0.013	8.000	7.986	0.005	8.000	7.991	0.000
10	MAX	10.090	9.920	0.260	10.036	9.960	0.112	10.022	9.987	0.050	10.015	9.995	0.029	10.015	10.000	0.024
	MIN	10.000	9.830	0.080	10.000	9.924	0.040	10.000	9.972	0.013	10.000	9.986	0.005	10.000	9.991	0.000
12	MAX	12.110	11.905	0.315	12.043	11.950	0.136	12.027	11.984	0.061	12.018	11.994	0.035	12.018	12.000	0.029
	MIN	12.000	11.795	0.095	12.000	11.907	0.050	12.000	11.966	0.016	12.000	11.983	0.006	12.000	11.989	0.000
16	MAX	16.110	15.905	0.315	16.043	15.950	0.136	16.027	15.984	0.061	16.018	15.994	0.035	16.018	16.000	0.029
	MIN	16.000	15.795	0.095	16.000	15.907	0.050	16.000	15.966	0.016	16.000	15.983	0.006	16.000	15.989	0.000
20	MAX	20.130	19.890	0.370	20.052	19.935	0.169	20.033	19.980	0.074	20.021	19.993	0.041	20.021	20.000	0.034
	MIN	20.000	19.760	0.110	20.000	19.883	0.065	20.000	19.959	0.020	20.000	19.980	0.007	20.000	19.987	0.000
25	MAX	25.130	24.890	0.370	25.052	24.935	0.169	25.033	24.980	0.074	25.021	24.993	0.041	25.021	25.000	0.034
	MIN	25.000	24.760	0.110	25.000	24.883	0.065	25.000	24.959	0.020	25.000	24.980	0.007	25.000	24.987	0.000
30	MAX	30.130	29.890	0.370	30.052	29.935	0.169	30.033	29.980	0.074	30.021	29.993	0.041	30.021	30.000	0.034
	MIN	30.000	29.760	0.110	30.000	29.883	0.065	30.000	29.959	0.020	30.000	29.980	0.007	30.000	29.987	0.000

APPENDIX I (Continued)

TABLE 7 (CONTINUED)—METRIC LIMITS AND FITS

AMERICAN NATIONAL STANDARD PREFERRED METRIC LIMITS AND FITS   ANSI B4.2–1978

Dimensions in mm.

PREFERRED HOLE BASIS CLEARANCE FITS (Continued)

BASIC SIZE		LOOSE RUNNING			FREE RUNNING			CLOSE RUNNING			SLIDING			LOCATIONAL CLEARANCE		
		Hole H11	Shaft c11	Fit	Hole H9	Shaft d9	Fit	Hole H8	Shaft f7	Fit	Hole H7	Shaft g6	Fit	Hole H7	Shaft h6	Fit
40	MAX	40.160	39.880	0.440	40.062	39.920	0.204	40.039	39.975	0.089	40.025	39.991	0.050	40.025	40.000	0.041
	MIN	40.000	39.720	0.120	40.000	39.858	0.080	40.000	39.950	0.025	40.000	39.975	0.009	40.000	39.984	0.000
50	MAX	50.160	49.870	0.450	50.062	49.920	0.204	50.039	49.975	0.089	50.025	49.991	0.050	50.025	50.000	0.041
	MIN	50.000	49.710	0.130	50.000	49.858	0.080	50.000	49.950	0.025	50.000	49.975	0.009	50.000	49.984	0.000
60	MAX	60.190	59.860	0.520	60.074	59.900	0.248	60.046	59.970	0.106	60.030	59.990	0.059	60.030	60.000	0.049
	MIN	60.000	59.670	0.140	60.000	59.826	0.100	60.000	59.940	0.030	60.000	59.971	0.010	60.000	59.981	0.000
80	MAX	80.190	79.850	0.530	80.074	79.900	0.248	80.046	79.970	0.106	80.030	79.990	0.059	80.030	80.000	0.049
	MIN	80.000	79.660	0.150	80.000	79.826	0.100	80.000	79.940	0.030	80.000	79.971	0.010	80.000	75.981	0.000
100	MAX	100.220	99.830	0.610	100.087	99.880	0.294	100.054	99.964	0.125	100.035	99.988	0.069	100.035	100.000	0.057
	MIN	100.000	99.610	0.170	100.000	99.793	0.120	100.000	99.929	0.036	100.000	99.966	0.012	100.000	99.978	0.000
120	MAX	120.220	119.820	0.620	120.087	119.880	0.294	120.054	119.964	0.125	120.035	119.988	0.069	120.035	120.000	0.057
	MIN	120.000	119.600	0.180	120.000	119.793	0.120	120.000	119.929	0.036	120.000	119.966	0.012	120.000	119.978	0.000
160	MAX	160.250	159.790	0.710	160.100	159.855	0.345	160.063	159.957	0.146	160.040	159.986	0.079	160.040	160.000	0.065
	MIN	160.000	159.540	0.210	160.000	159.755	0.145	160.000	159.917	0.043	160.000	159.961	0.014	160.000	159.975	0.000
200	MAX	200.290	199.760	0.820	200.115	199.830	0.400	200.072	199.950	0.168	200.046	199.985	0.090	200.046	200.000	0.075
	MIN	200.000	199.470	0.240	200.000	199.715	0.170	200.000	199.904	0.050	200.000	199.956	0.015	200.000	199.971	0.000
250	MAX	250.290	249.720	0.860	250.115	249.830	0.400	250.072	249.950	0.168	250.046	249.985	0.090	250.046	250.000	0.075
	MIN	250.000	249.430	0.280	250.000	249.715	0.170	250.000	249.904	0.050	250.000	249.956	0.015	250.000	249.971	0.000
300	MAX	300.320	299.670	0.970	300.130	299.810	0.450	300.081	299.944	0.189	300.052	299.983	0.101	300.052	300.000	0.084
	MIN	300.000	299.350	0.330	300.000	299.680	0.190	300.000	299.892	0.056	300.000	299.951	0.017	300.000	299.968	0.000
400	MAX	400.360	399.600	1.120	400.140	399.790	0.490	400.089	399.938	0.208	400.057	399.982	0.111	400.057	400.000	0.093
	MIN	400.000	399.240	0.400	400.000	399.650	0.210	400.000	399.881	0.062	400.000	399.946	0.018	400.000	399.964	0.000
500	MAX	500.400	499.520	1.280	500.155	499.770	0.540	500.097	499.932	0.228	500.063	499.980	0.123	500.063	500.000	0.103
	MIN	500.000	499.120	0.480	500.000	499.615	0.230	500.000	499.869	0.068	500.000	499.940	0.020	500.000	499.960	0.000

Reprinted from The American Society of Mechanical Engineers—ANSI B4.2–1978 (R1994)

AMERICAN NATIONAL STANDARD
PREFERRED METRIC LIMITS AND FITS

ANSI B4.2–1978

Dimensions in mm.

APPENDIX I (Continued)

TABLE 7 (CONTINUED)—METRIC LIMITS AND FITS

PREFERRED SHAFT BASIS CLEARANCE FITS

BASIC SIZE		LOOSE RUNNING Hole C11	LOOSE RUNNING Shaft h11	LOOSE RUNNING Fit	FREE RUNNING Hole D9	FREE RUNNING Shaft h9	FREE RUNNING Fit	CLOSE RUNNING Hole F8	CLOSE RUNNING Shaft h7	CLOSE RUNNING Fit	SLIDING Hole G7	SLIDING Shaft h6	SLIDING Fit	LOCATIONAL CLEARANCE Hole H7	LOCATIONAL CLEARANCE Shaft h6	LOCATIONAL CLEARANCE Fit
1	MAX	1•120	1•000	0•180	1•045	1•000	0•070	1•020	1•000	0•030	1•012	1•000	0•018	1•010	1•000	0•016
	MIN	1•060	0•940	0•060	1•020	0•975	0•020	1•006	0•990	0•006	1•002	0•994	0•002	1•000	0•994	0•000
1•2	MAX	1•320	1•200	0•180	1•245	1•200	0•070	1•220	1•200	0•030	1•212	1•200	0•018	1•210	1•200	0•016
	MIN	1•260	1•140	0•060	1•220	1•175	0•020	1•206	1•190	0•006	1•202	1•194	0•002	1•200	1•194	0•000
1•6	MAX	1•720	1•600	0•180	1•645	1•600	0•070	1•620	1•600	0•030	1•612	1•600	0•018	1•610	1•600	0•016
	MIN	1•660	1•540	0•060	1•620	1•575	0•020	1•606	1•590	0•006	1•602	1•594	0•002	1•600	1•594	0•000
2	MAX	2•120	2•000	0•180	2•045	2•000	0•070	2•020	2•000	0•030	2•012	2•000	0•018	2•010	2•000	0•016
	MIN	2•060	1•940	0•060	2•020	1•975	0•020	2•006	1•990	0•006	2•002	1•994	0•002	2•000	1•994	0•000
2•5	MAX	2•620	2•500	0•180	2•545	2•500	0•070	2•520	2•500	0•030	2•512	2•500	0•018	2•510	2•500	0•016
	MIN	2•560	2•440	0•060	2•520	2•475	0•020	2•506	2•490	0•006	2•502	2•494	0•002	2•500	2•494	0•000
3	MAX	3•120	3•000	0•180	3•045	3•000	0•070	3•020	3•000	0•030	3•012	3•000	0•018	3•010	3•000	0•016
	MIN	3•060	2•940	0•060	3•020	2•975	0•020	3•006	2•990	0•006	3•002	2•994	0•002	3•000	2•994	0•000
4	MAX	4•145	4•000	0•220	4•060	4•000	0•090	4•028	4•000	0•040	4•016	4•000	0•024	4•012	4•000	0•020
	MIN	4•070	3•925	0•070	4•030	3•970	0•030	4•010	3•988	0•010	4•004	3•992	0•004	4•000	3•992	0•000
5	MAX	5•145	5•000	0•220	5•060	5•000	0•090	5•028	5•000	0•040	5•016	5•000	0•024	5•012	5•000	0•020
	MIN	5•070	4•925	0•070	5•030	4•970	0•030	5•010	4•988	0•010	5•004	4•992	0•004	5•000	4•992	0•000
6	MAX	6•145	6•000	0•220	6•060	6•000	0•090	6•028	6•000	0•040	6•016	6•000	0•024	6•012	6•000	0•020
	MIN	6•070	5•925	0•070	6•030	5•970	0•030	6•010	5•988	0•010	6•004	5•992	0•004	6•000	5•992	0•000
8	MAX	8•170	8•000	0•260	8•076	8•000	0•112	8•035	8•000	0•050	8•020	8•000	0•029	8•015	8•000	0•024
	MIN	8•080	7•910	0•080	8•040	7•964	0•040	8•013	7•985	0•013	8•005	7•991	0•005	8•000	7•991	0•000
10	MAX	10•170	10•000	0•260	10•076	10•000	0•112	10•035	10•000	0•050	10•020	10•000	0•029	10•015	10•000	0•024
	MIN	10•080	9•910	0•080	10•040	9•964	0•040	10•013	9•985	0•013	10•005	9•991	0•005	10•000	9•991	0•000
12	MAX	12•205	12•000	0•315	12•093	12•000	0•136	12•043	12•000	0•061	12•024	12•000	0•035	12•018	12•000	0•029
	MIN	12•095	11•890	0•095	12•050	11•957	0•050	12•016	11•982	0•016	12•006	11•989	0•006	12•000	11•989	0•000
16	MAX	16•205	16•000	0•315	16•093	16•000	0•136	16•043	16•000	0•061	16•024	16•000	0•035	16•018	16•000	0•029
	MIN	16•095	15•890	0•095	16•050	15•957	0•050	16•016	15•982	0•016	16•006	15•989	0•006	16•000	15•989	0•000
20	MAX	20•240	20•000	0•370	20•117	20•000	0•169	20•053	20•000	0•074	20•028	20•000	0•041	20•021	20•000	0•034
	MIN	20•110	19•870	0•110	20•065	19•948	0•065	20•020	19•979	0•020	20•007	19•987	0•007	20•000	19•987	0•000
25	MAX	25•240	25•000	0•370	25•117	25•000	0•169	25•053	25•000	0•074	25•028	25•000	0•041	25•021	25•000	0•034
	MIN	25•110	24•870	0•110	25•065	24•948	0•065	25•020	24•979	0•020	25•007	24•987	0•007	25•000	24•987	0•000
30	MAX	30•240	30•000	0•370	30•117	30•000	0•169	30•053	30•000	0•074	30•028	30•000	0•041	30•021	30•000	0•034
	MIN	30•110	29•870	0•110	30•065	29•948	0•065	30•020	29•979	0•020	30•007	29•987	0•007	30•000	29•987	0•000

APPENDIX I (Continued)

TABLE 7 (CONTINUED)—METRIC LIMITS AND FITS

AMERICAN NATIONAL STANDARD PREFERRED METRIC LIMITS AND FITS

ANSI B4.2–1978

Dimensions in mm.

PREFERRED SHAFT BASIS CLEARANCE FITS (Continued)

BASIC SIZE		LOOSE RUNNING Hole C11	LOOSE RUNNING Shaft h11	LOOSE RUNNING Fit	FREE RUNNING Hole D9	FREE RUNNING Shaft h9	FREE RUNNING Fit	CLOSE RUNNING Hole F8	CLOSE RUNNING Shaft h7	CLOSE RUNNING Fit	SLIDING Hole G7	SLIDING Shaft h6	SLIDING Fit	LOCATIONAL CLEARANCE Hole H7	LOCATIONAL CLEARANCE Shaft h6	LOCATIONAL CLEARANCE Fit
40	MAX	40.280	40.000	0.440	40.142	40.000	0.204	40.064	40.000	0.089	40.034	40.000	0.050	40.025	40.000	0.041
	MIN	40.120	39.840	0.120	40.080	39.938	0.080	40.025	39.975	0.025	40.009	39.984	0.009	40.000	39.984	0.000
50	MAX	50.290	50.000	0.450	50.142	50.000	0.204	50.064	50.000	0.089	50.034	50.000	0.050	50.025	50.000	0.041
	MIN	50.130	49.840	0.130	50.080	49.938	0.080	50.025	49.975	0.025	50.009	49.984	0.009	50.000	49.984	0.000
60	MAX	60.330	60.000	0.520	60.174	60.000	0.248	60.076	60.000	0.106	60.040	60.000	0.059	60.030	60.000	0.049
	MIN	60.140	59.810	0.140	60.100	59.926	0.100	60.030	59.970	0.030	60.010	59.981	0.010	60.000	59.981	0.000
80	MAX	80.340	80.000	0.530	80.174	80.000	0.248	80.076	80.000	0.106	80.040	80.000	0.059	80.030	80.000	0.049
	MIN	80.150	79.810	0.150	80.100	79.926	0.100	80.030	79.970	0.030	80.010	79.981	0.010	80.000	79.981	0.000
100	MAX	100.390	100.000	0.610	100.207	100.000	0.294	100.090	100.000	0.125	100.047	100.000	0.069	100.035	100.000	0.057
	MIN	100.170	99.780	0.170	100.120	99.913	0.120	100.036	99.965	0.036	100.012	99.978	0.012	100.000	99.978	0.000
120	MAX	120.400	120.000	0.620	120.207	120.000	0.294	120.090	120.000	0.125	120.047	120.000	0.069	120.035	120.000	0.057
	MIN	120.180	119.780	0.180	120.120	119.913	0.120	120.036	119.965	0.036	120.012	119.978	0.012	120.000	119.978	0.000
160	MAX	160.460	160.000	0.710	160.245	160.000	0.345	160.106	160.000	0.146	160.054	160.000	0.079	160.040	160.000	0.065
	MIN	160.210	159.750	0.210	160.145	159.900	0.145	160.043	159.960	0.043	160.014	159.975	0.014	160.000	159.975	0.000
200	MAX	200.530	200.000	0.820	200.285	200.000	0.400	200.122	200.000	0.168	200.061	200.000	0.090	200.046	200.000	0.075
	MIN	200.240	199.710	0.240	200.170	199.885	0.170	200.050	199.954	0.050	200.015	199.971	0.015	200.000	199.971	0.000
250	MAX	250.570	250.000	0.860	250.285	250.000	0.400	250.122	250.000	0.168	250.061	250.000	0.090	250.046	250.000	0.075
	MIN	250.280	249.710	0.280	250.170	249.885	0.170	250.050	249.954	0.050	250.015	249.971	0.015	250.000	249.971	0.000
300	MAX	300.650	300.000	0.970	300.320	300.000	0.450	300.137	300.000	0.189	300.069	300.000	0.101	300.052	300.000	0.084
	MIN	300.330	299.680	0.330	300.190	299.870	0.190	300.056	299.948	0.056	300.017	299.968	0.017	300.000	299.968	0.000
400	MAX	400.760	400.000	1.120	400.350	400.000	0.490	400.151	400.000	0.208	400.075	400.000	0.111	400.057	400.000	0.093
	MIN	400.400	399.640	0.400	400.210	399.860	0.210	400.062	399.943	0.062	400.018	399.964	0.018	400.000	399.964	0.000
500	MAX	500.880	500.000	1.280	500.385	500.000	0.540	500.165	500.000	0.228	500.083	500.000	0.123	500.063	500.000	0.103
	MIN	500.480	499.600	0.480	500.230	499.845	0.230	500.068	499.937	0.068	500.020	499.960	0.020	500.000	499.960	0.000

Reprinted from The American Society of Mechanical Engineers—ANSI B4.2–1978 (R1994)

APPENDIX I (Continued)

TABLE 8 METRIC TOLERANCE ZONES

Metric Tolerance Zones for Internal (Hole) Dimensions (A14 through A9 and B14 through B9) Dimensions in mm.

BASIC SIZE		A14	A13	A12	A11	A10	A9	B14	B13	B12	B11	B10	B9
OVER 0	TO 3	+0•520 +0•270	+0•410 +0•270	+0•370 +0•270	+0•330 +0•270	+0•310 +0•270	+0•295 +0•270	+0•390 +0•140	+0•280 +0•140	+0•240 +0•140	+0•200 +0•140	+0•180 +0•140	+0•165 +0•140
OVER 3	TO 6	+0•570 +0•270	+0•450 +0•270	+0•390 +0•270	+0•345 +0•270	+0•318 +0•270	+0•300 +0•270	+0•440 +0•140	+0•320 +0•140	+0•260 +0•140	+0•215 +0•140	+0•188 +0•140	+0•170 +0•140
OVER 6	TO 10	+0•640 +0•280	+0•500 +0•280	+0•430 +0•280	+0•370 +0•280	+0•338 +0•280	+0•316 +0•280	+0•510 +0•150	+0•370 +0•150	+0•300 +0•150	+0•240 +0•150	+0•208 +0•150	+0•186 +0•150
OVER 10	TO 14	+0•720 +0•290	+0•560 +0•290	+0•470 +0•290	+0•400 +0•290	+0•360 +0•290	+0•333 +0•290	+0•580 +0•150	+0•420 +0•150	+0•330 +0•150	+0•260 +0•150	+0•220 +0•150	+0•193 +0•150
OVER 14	TO 18	+0•720 +0•290	+0•560 +0•290	+0•470 +0•290	+0•400 +0•290	+0•360 +0•290	+0•333 +0•290	+0•580 +0•150	+0•420 +0•150	+0•330 +0•150	+0•260 +0•150	+0•220 +0•150	+0•193 +0•150
OVER 18	TO 24	+0•820 +0•300	+0•630 +0•300	+0•510 +0•300	+0•430 +0•300	+0•384 +0•300	+0•352 +0•300	+0•680 +0•160	+0•490 +0•160	+0•370 +0•160	+0•290 +0•160	+0•244 +0•160	+0•212 +0•160
OVER 24	TO 30	+0•820 +0•300	+0•630 +0•300	+0•510 +0•300	+0•430 +0•300	+0•384 +0•300	+0•352 +0•300	+0•680 +0•160	+0•490 +0•160	+0•370 +0•160	+0•290 +0•160	+0•244 +0•160	+0•212 +0•160
OVER 30	TO 40	+0•930 +0•310	+0•700 +0•310	+0•560 +0•310	+0•470 +0•310	+0•410 +0•310	+0•372 +0•310	+0•790 +0•170	+0•560 +0•170	+0•420 +0•170	+0•330 +0•170	+0•270 +0•170	+0•232 +0•170
OVER 40	TO 50	+0•940 +0•320	+0•710 +0•320	+0•570 +0•320	+0•480 +0•320	+0•420 +0•320	+0•382 +0•320	+0•800 +0•180	+0•570 +0•180	+0•430 +0•180	+0•340 +0•180	+0•280 +0•180	+0•242 +0•180
OVER 50	TO 65	+1•080 +0•340	+0•800 +0•340	+0•640 +0•340	+0•530 +0•340	+0•460 +0•340	+0•414 +0•340	+0•930 +0•190	+0•650 +0•190	+0•490 +0•190	+0•380 +0•190	+0•310 +0•190	+0•264 +0•190
OVER 65	TO 80	+1•100 +0•360	+0•820 +0•360	+0•660 +0•360	+0•550 +0•360	+0•480 +0•360	+0•434 +0•360	+0•940 +0•200	+0•660 +0•200	+0•500 +0•200	+0•390 +0•200	+0•320 +0•200	+0•274 +0•200
OVER 80	TO 100	+1•250 +0•380	+0•920 +0•380	+0•730 +0•380	+0•600 +0•380	+0•520 +0•380	+0•467 +0•380	+1•090 +0•220	+0•760 +0•220	+0•570 +0•220	+0•440 +0•220	+0•360 +0•220	+0•307 +0•220
OVER 100	TO 120	+1•280 +0•410	+0•950 +0•410	+0•760 +0•410	+0•630 +0•410	+0•550 +0•410	+0•497 +0•410	+1•110 +0•240	+0•780 +0•240	+0•590 +0•240	+0•460 +0•240	+0•380 +0•240	+0•327 +0•240
OVER 120	TO 140	+1•460 +0•460	+1•090 +0•460	+0•860 +0•460	+0•710 +0•460	+0•620 +0•460	+0•560 +0•460	+1•260 +0•260	+0•890 +0•260	+0•660 +0•260	+0•510 +0•260	+0•420 +0•260	+0•360 +0•260
OVER 140	TO 160	+1•520 +0•520	+1•150 +0•520	+0•920 +0•520	+0•770 +0•520	+0•680 +0•520	+0•620 +0•520	+1•280 +0•280	+0•910 +0•280	+0•680 +0•280	+0•530 +0•280	+0•440 +0•280	+0•380 +0•280
OVER 160	TO 180	+1•580 +0•580	+1•210 +0•580	+0•980 +0•580	+0•830 +0•580	+0•740 +0•580	+0•680 +0•580	+1•310 +0•310	+0•940 +0•310	+0•710 +0•310	+0•560 +0•310	+0•470 +0•310	+0•410 +0•310
OVER 180	TO 200	+1•810 +0•660	+1•380 +0•660	+1•120 +0•660	+0•950 +0•660	+0•845 +0•660	+0•775 +0•660	+1•490 +0•340	+1•060 +0•340	+0•800 +0•340	+0•630 +0•340	+0•525 +0•340	+0•455 +0•340
OVER 200	TO 225	+1•890 +0•740	+1•460 +0•740	+1•200 +0•740	+1•030 +0•740	+0•925 +0•740	+0•855 +0•740	+1•530 +0•380	+1•100 +0•380	+0•840 +0•380	+0•670 +0•380	+0•565 +0•380	+0•495 +0•380
OVER 225	TO 250	+1•970 +0•820	+1•540 +0•820	+1•280 +0•820	+1•110 +0•820	+1•005 +0•820	+0•935 +0•820	+1•570 +0•420	+1•140 +0•420	+0•880 +0•420	+0•710 +0•420	+0•605 +0•420	+0•535 +0•420
OVER 250	TO 280	+2•220 +0•920	+1•730 +0•920	+1•440 +0•920	+1•240 +0•920	+1•130 +0•920	+1•050 +0•920	+1•780 +0•480	+1•290 +0•480	+1•000 +0•480	+0•800 +0•480	+0•690 +0•480	+0•610 +0•480
OVER 280	TO 315	+2•350 +1•050	+1•860 +1•050	+1•570 +1•050	+1•370 +1•050	+1•260 +1•050	+1•180 +1•050	+1•840 +0•540	+1•350 +0•540	+1•060 +0•540	+0•860 +0•540	+0•750 +0•540	+0•670 +0•540
OVER 315	TO 355	+2•600 +1•200	+2•090 +1•200	+1•770 +1•200	+1•560 +1•200	+1•430 +1•200	+1•340 +1•200	+2•000 +0•600	+1•490 +0•600	+1•170 +0•600	+0•960 +0•600	+0•830 +0•600	+0•740 +0•600
OVER 355	TO 400	+2•750 +1•350	+2•240 +1•350	+1•920 +1•350	+1•710 +1•350	+1•580 +1•350	+1•490 +1•350	+2•080 +0•680	+1•570 +0•680	+1•250 +0•680	+1•040 +0•680	+0•910 +0•680	+0•820 +0•680
OVER 400	TO 450	+3•050 +1•500	+2•470 +1•500	+2•130 +1•500	+1•900 +1•500	+1•750 +1•500	+1•655 +1•500	+2•310 +0•760	+1•730 +0•760	+1•390 +0•760	+1•160 +0•760	+1•010 +0•760	+0•915 +0•760
OVER 450	TO 500	+3•200 +1•650	+2•620 +1•650	+2•280 +1•650	+2•050 +1•650	+1•900 +1•650	+1•805 +1•650	+2•390 +0•840	+1•810 +0•840	+1•470 +0•840	+1•240 +0•840	+1•090 +0•840	+0•995 +0•840

Reprinted from The American Society of Mechanical Engineers—ANSI B4.2–1978 (R1994).

APPENDIX I (Continued)

TABLE 8 (CONTINUED)—METRIC TOLERANCE ZONES

Tolerance Zones for External (Shaft) Dimensions (a14 through a9 and b14 through b9) Dimensions in mm.

BASIC SIZE		a14	a13	a12	a11	a10	a9	b14	b13	b12	b11	b10	b9
OVER TO	0 3	−0•270 −0•520	−0•270 −0•410	−0•270 −0•370	−0•270 −0•330	−0•270 −0•310	−0•270 −0•295	−0•140 −0•390	−0•140 −0•280	−0•140 −0•240	−0•140 −0•200	−0•140 −0•180	−0•140 −0•165
OVER TO	3 6	−0•270 −0•570	−0•270 −0•450	−0•270 −0•390	−0•270 −0•345	−0•270 −0•318	−0•270 −0•300	−0•140 −0•440	−0•140 −0•320	−0•140 −0•260	−0•140 −0•215	−0•140 −0•188	−0•140 −0•170
OVER TO	6 10	−0•280 −0•640	−0•280 −0•500	−0•280 −0•430	−0•280 −0•370	−0•280 −0•338	−0•280 −0•316	−0•150 −0•510	−0•150 −0•370	−0•150 −0•300	−0•150 −0•240	−0•150 −0•208	−0•150 −0•186
OVER TO	10 14	−0•290 −0•720	−0•290 −0•560	−0•290 −0•470	−0•290 −0•400	−0•290 −0•360	−0•290 −0•333	−0•150 −0•580	−0•150 −0•420	−0•150 −0•330	−0•150 −0•260	−0•150 −0•220	−0•150 −0•193
OVER TO	14 18	−0•290 −0•720	−0•290 −0•560	−0•290 −0•470	−0•290 −0•400	−0•290 −0•360	−0•290 −0•333	−0•150 −0•580	−0•150 −0•420	−0•150 −0•330	−0•150 −0•260	−0•150 −0•220	−0•150 −0•193
OVER TO	18 24	−0•300 −0•820	−0•300 −0•630	−0•300 −0•510	−0•300 −0•430	−0•300 −0•384	−0•300 −0•352	−0•160 −0•680	−0•160 −0•490	−0•160 −0•370	−0•160 −0•290	−0•160 −0•244	−0•160 −0•212
OVER TO	24 30	−0•300 −0•820	−0•300 −0•630	−0•300 −0•510	−0•300 −0•430	−0•300 −0•384	−0•300 −0•352	−0•160 −0•680	−0•160 −0•490	−0•160 −0•370	−0•160 −0•290	−0•160 −0•244	−0•160 −0•212
OVER TO	30 40	−0•310 −0•930	−0•310 −0•700	−0•310 −0•560	−0•310 −0•470	−0•310 −0•410	−0•310 −0•372	−0•170 −0•790	−0•170 −0•560	−0•170 −0•420	−0•170 −0•330	−0•170 −0•270	−0•170 −0•232
OVER TO	40 50	−0•320 −0•940	−0•320 −0•710	−0•320 −0•570	−0•320 −0•480	−0•320 −0•420	−0•320 −0•382	−0•180 −0•800	−0•180 −0•570	−0•180 −0•430	−0•180 −0•340	−0•180 −0•280	−0•180 −0•242
OVER TO	50 65	−0•340 −1•080	−0•340 −0•800	−0•340 −0•640	−0•340 −0•530	−0•340 −0•460	−0•340 −0•414	−0•190 −0•930	−0•190 −0•650	−0•190 −0•490	−0•190 −0•380	−0•190 −0•310	−0•190 −0•264
OVER TO	65 80	−0•360 −1•100	−0•360 −0•820	−0•360 −0•660	−0•360 −0•550	−0•360 −0•480	−0•360 −0•434	−0•200 −0•940	−0•200 −0•660	−0•200 −0•500	−0•200 −0•390	−0•200 −0•320	−0•200 −0•274
OVER TO	80 100	−0•380 −1•250	−0•380 −0•920	−0•380 −0•730	−0•380 −0•600	−0•380 −0•520	−0•380 −0•467	−0•220 −1•090	−0•220 −0•760	−0•220 −0•570	−0•220 −0•440	−0•220 −0•360	−0•220 −0•307
OVER TO	100 120	−0•410 −1•280	−0•410 −0•950	−0•410 −0•760	−0•410 −0•630	−0•410 −0•550	−0•410 −0•497	−0•240 −1•110	−0•240 −0•780	−0•240 −0•590	−0•240 −0•460	−0•240 −0•380	−0•240 −0•327
OVER TO	120 140	−0•460 −1•460	−0•460 −1•090	−0•460 −0•860	−0•460 −0•710	−0•460 −0•620	−0•460 −0•560	−0•260 −1•260	−0•260 −0•890	−0•260 −0•660	−0•260 −0•510	−0•260 −0•420	−0•260 −0•360
OVER TO	140 160	−0•520 −1•520	−0•520 −1•150	−0•520 −0•920	−0•520 −0•770	−0•520 −0•680	−0•520 −0•620	−0•280 −1•280	−0•280 −0•910	−0•280 −0•680	−0•280 −0•530	−0•280 −0•440	−0•280 −0•380
OVER TO	160 180	−0•580 −1•580	−0•580 −1•210	−0•580 −0•980	−0•580 −0•830	−0•580 −0•740	−0•580 −0•680	−0•310 −1•310	−0•310 −0•940	−0•310 −0•710	−0•310 −0•560	−0•310 −0•470	−0•310 −0•410
OVER TO	160 180	−0•580 −1•580	−0•580 −1•210	−0•580 −0•980	−0•580 −0•830	−0•580 −0•740	~0•580 ~0•680	−0•310 −1•310	−0•310 −0•940	−0•310 −0•710	−0•310 −0•560	−0•310 −0•470	−0•310 −0•410
OVER TO	180 200	−0•660 −1•810	−0•660 −1•380	−0•660 −1•120	−0•660 −0•950	−0•660 −0•845	−0•660 −0•775	−0•340 −1•490	−0•340 −1•060	−0•340 −0•800	−0•340 −0•630	−0•340 −0•525	−0•340 −0•455
OVER TO	200 225	−0•740 −1•890	−0•740 −1•460	−0•740 −1•200	−0•740 −1•030	−0•740 −0•925	−0•740 −0•855	−0•380 −1•530	−0•380 −1•100	−0•380 −0•840	−0•380 −0•670	−0•380 −0•565	−0•380 −0•495
OVER TO	225 250	−0•820 −1•970	−0•820 −1•540	−0•820 −1•280	−0•820 −1•110	−0•820 −1•005	−0•820 −0•935	−0•420 −1•570	−0•420 −1•140	−0•420 −0•880	−0•420 −0•710	−0•420 −0•605	−0•420 −0•535
OVER TO	250 280	−0•920 −2•220	−0•920 −1•730	−0•920 −1•440	−0•920 −1•240	−0•920 −1•130	−0•920 −1•050	−0•480 −1•780	−0•480 −1•290	−0•480 −1•000	−0•480 −0•800	−0•480 −0•690	−0•480 −0•610
OVER TO	280 315	−1•050 −2•350	−1•050 −1•860	−1•050 −1•570	−1•050 −1•370	−1•050 −1•260	−1•050 −1•180	−0•540 −1•840	−0•540 −1•350	−0•540 −1•060	−0•540 −0•860	−0•540 −0•750	−0•540 −0•670
OVER TO	315 355	−1•200 −2•600	−1•200 −2•090	−1•200 −1•770	−1•200 −1•560	−1•200 −1•430	−1•200 −1•340	−0•600 −2•000	−0•600 −1•490	−0•600 −1•170	−0•600 −0•960	−0•600 −0•830	−0•600 −0•740
OVER TO	355 400	−1•350 −2•750	−1•350 −2•240	−1•350 −1•920	−1•350 −1•710	−1•350 −1•580	−1•350 −1•490	−0•680 −2•080	−0•680 −1•570	−0•680 −1•250	−0•680 −1•040	−0•680 −0•910	−0•680 −0•820
OVER TO	400 450	−1•500 −3•050	−1•500 −2•470	−1•500 −2•130	−1•500 −1•900	−1•500 −1•750	−1•500 −1•655	−0•760 −2•310	−0•760 −1•730	−0•760 −1•390	−0•760 −1•160	−0•760 −1•010	−0•760 −0•915
OVER TO	450 500	−1•650 −3•200	−1•650 −2•620	−1•650 −2•280	−1•650 −2•050	−1•650 −1•900	−1•650 −1•805	−0•840 +2•390	−0•840 −1•810	−0•840 −1•470	−0•840 −1•240	−0•840 −1•090	−0•840 −0•995

Reprinted from The American Society of Mechanical Engineers—ANSI B4.2–1978 (R1994).

APPENDIX J
UNIFIED SCREW THREAD VARIATIONS

THE FOLLOWING list is a ready reference of available standard series and selected combinations of Unified screw threads. Each thread is given as part of a proper thread note including major diameter, threads per inch, and series identification.

0–80 UNF	7/16–27 UNS	7/8–28 UN	1-3/8–18 UNEF	1-15/16–6 UN	2-7/8–12 UN
1–64 UNC	7/16–28 UNEF	7/8–32 UN	1-3/8–20 UN	1-15/16–8 UN	2-7/8–16 UN
1–72 UNF	7/16–32 UN	15/16–12 UN	1-3/8–28 UN	1-15/16–12 UN	2-7/8–20 UN
2–56 UNC	1/2–12 UNS	15/16–16 UN	1-7/16–6 UN	1-15/16–16 UN	3–4 UNC
2–64 UNF	1/2–13 UNC	15/16–20 UNEF	1-7/16–8 UN	1-15/16–20 UN	3–6 UN
3–48 UNC	1/2–14 UNS	15/16–28 UN	1-7/16–12 UN	2-4 1/2 UNC	3–8 UN
3–56 UNF	1/2–16 UN	15/16–32 UN	1-7/16–16 UN	2–6 UN	3–10 UNS
4–40 UNC	1/2–18 UNS	1–8 UNC	1-7/16–18 UNEF	2–8 UN	3–12 UN
4–48 UNF	1/2–20 UNF	1–10 UNS	1-7/16–20 UN	2–10 UNS	3–14 UNS
5–40 UNC	1/2–24 UNS	1–12 UNF	1-7/16–28 UN	2–12 UN	3–16 UN
5–44 UNF	1/2–27 UNS	1–14 UNS	1-1/2–6 UNC	2–14 UNS	3–18 UNS
6–32 UNC	1/2–28 UNEF	1–16 UN	1-1/2–8 UN	2–16 UN	3–20 UN
6–40 UNF	1/2–32 UN	1–18 UNS	1-1/2–10 UNS	2–18 UNS	3-1/8–6 UN
8–32 UNC	9/16–12 UNC	1–20 UNEF	1-1/2–12 UNF	2–20 UN	3-1/8–8 UN
8–36 UNF	9/16–14 UNS	1–24 UNS	1-1/2–14 UNS	2-1/16–16 UNS	3-1/8–12 UN
10–24 UNC	9/16–16 UN	1–27 UNS	1-1/2–16 UN	2-1/8–6 UN	3-1/8–16 UN
10–28 UNS	9/16–18 UNF	1–28 UN	1-1/2–18 UNEF	2-1/8–8 UN	3-1/4–4 UNC
10–32 UNF	9/16–20 UN	1–32 UN	1-1/2–20 UN	2-1/8–12 UN	3-1/4–6 UN
10–36 UNS	9/16–24 UNEF	1-1/16–8 UN	1-1/2–24 UNS	2-1/8–16 UN	3-1/4–8 UN
10–40 UNS	9/16–27 UNS	1-1/16–12 UN	1-1/2–28 UN	2-1/8–20 UN	3-1/4–10 UNS
10–48 UNS	9/16–28 UN	1-1/16–16 UN	1-9/16–6 UN	2-3/16–16 UNS	3 1/4–12 UN
10–56 UNS	9/16–32 UN	1-1/16–18 UNEF	1-9/16–8 UN	2-1/4–4-1/2 UNC	3 1/4–14 UNS
12–24 UNC	5/8–11 UNC	1-1/16–20 UN	1-9/16–12 UN	2-1/4–6 UN	3 1/4–16 UN
12–28 UNF	5/8–12 UN	1-1/16–28 UN	1-9/16–16 UN	2-1/4–8 UN	3-1/4–18 UNS
12–32 UNEF	5/8–14 UNS	1-1/8–7 UNC	1-9/16–18 UNEF	2-1/4–10 UNS	3-3/8–6 UN
12–36 UNS	5/8–16 UN	1-1/8–8 UN	1-9/16–20 UN	2-1/4–12 UN	3-3/8–8 UN
12–40 UNS	5/8–18 UNF	1-1/8–10 UNS	1-5/8–6 UN	2-1/4–14 UNS	3-3/8–12 UN
12–48 UNS	5/8–20 UN	1-1/8–12 UNF	1-5/8–8 UN	2-1/4–16 UN	3-3/8–16 UN
12–56 UNS	5/8–24 UNEF	1-1/8–14 UNS	1-5/8–10 UNS	2-1/4–18 UNS	3-1/2–4 UNC
1/4–20 UNC	5/8–27 UNS	1-1/8–16 UN	1-5/8–12 UN	2-1/4–20 UN	3-1/2–6 UN
1/4–24 UNS	5/8–28 UN	1-1/8–18 UNEF	1-5/8–14 UNS	2-5/16–16 UNS	3-1/2–8 UN
1/4–27 UNS	5/8–32 UN	1-1/8–20 UN	1-5/8–16 UN	2-3/8–6 UN	3-1/2–10 UNS
1/4–28 UNF	11/16–12 UN	1-1/8–24 UNS	1-5/8–18 UNEF	2-3/8–8 UN	3-1/2–12 UN
1/4–32 UNEF	11/16–16 UN	1-1/8–28 UN	1-5/8–20 UN	2-3/8–12 UN	3-1/2–14 UNS
1/4–36 UNS	11/16–20 UN	1-3/16–8 UN	1-5/8–24 UNS	2-3/8–16 UN	3-1/2–16 UN
1/4–40 UNS	11/16–24 UNEF	1-3/16–12 UN	1-11/16–6 UN	2-3/8–20 UN	3-1/2–18 UNS
1/4–48 UNS	11/16–28 UN	1-3/16–16 UN	1-11/16–8 UN	2-7/16–16 UNS	3-5/8–6 UN
1/4–56 UNS	11/16–32 UN	1-3/16–18 UNEF	1-11/16–12 UN	2-1/2–4 UNC	3-5/8–8 UN
5/16–18 UNC	3/4–10 UNC	1-3/16–20 UN	1-11/16–16 UN	2-1/2–6 UN	3-5/8–12 UN
5/16–20 UN	3/4–12 UN	1-3/16–28 UN	1-11/16–18 UNEF	2-1/2–8 UN	3-5/8–16 UN
5/16–24 UNF	3/4–14 UNS	1-1/4–7 UNC	1-11/16–20 UN	2-1/2–10 UNS	3-3/4–4 UNC
5/16–27 UNS	3/4–16 UNF	1-1/4–8 UN	1-3/4–5 UNC	2-1/2–12 UN	3-3/4–6 UN
5/16–28 UN	3/4–18 UNS	1-1/4–10 UNS	1-3/4–6 UN	2-1/2–14 UNS	3-3/4–8 UN
5/16–32 UNEF	3/4–20 UNEF	1-1/4–12 UNF	1-3/4–8 UN	2-1/2–16 UN	3-3/4–10 UNS
5/16–36 UNS	3/4–24 UNS	1-1/4–14 UNS	1-3/4–10 UNS	1-1/2–18 UNS	3-3/4–12 UN
5/16–40 UNS	3/4–27 UNS	1-1/4–16 UN	1-3/4–12 UN	2-1/2–20 UN	3-3/4–14 UNS
5/16–48 UNS	3/4–28 UN	1-1/4–18 UNEF	1-3/4–14 UNS	2-5/8–6 UN	3-3/4–16 UN
3/8–16 UNC	3/4–32 UN	1-1/4–20 UN	1-3/4–16 UN	2-5/8–8 UN	3-3/4–18 UNS
3/8–18 UNS	13/16–12 UN	1-1/4–24 UNS	1-3/4–18 UNS	2-5.8–12 UN	3-7/8–6 UN
3/8–20 UN	13/16–16 UN	1-1/4–28 UN	1-3/4–20 UN	2-5.8–16 UN	3-7/8–8 UN
3/8–24 UNF	13/16–20 UNEF	1-5/16–8 UN	1-13/16–6 UN	2-5/8–20 UN	3-7/8–12 UN
3/8–27 UNS	13/16–28 UN	1-5/16–12 UN	1-13/16–8 UN	2-3/4–4 UNC	3-7/8–16 UN
3/8–28 UN	13/16–32 UN	1-5/16–16 UN	1-13/16–12 UN	2-3/4–6 UN	4–4 UNC
3/8–32 UNEF	7/8–9 UNC	1-5/16–18 UNEF	1-13/16–20 UN	2-3/4–8 UN	4–6 UN
3/8–36 UNS	7/8–10 UNS	1-5/16–20 UN	1-7/8–6 UN	2-3/4–10 UNS	4–8 UN
3/8–40 UNS	7/8–12 UN	1-5/16–28 UN	1-7/8–8 UN	2-3/4–12 UN	4–10 UNS
.390–27 UNS	7/8–14 UNF	1-3/8–6 UNC	1-7/8–10 UNS	2-3/4–14 UNS	4–12 UN
7/16–14 UNC	7/8–16 UN	1-3/8–8 UN	1-7/8–12 UN	2-3/4–16 UN	4–14 UNS
7/16–16 UN	7/8–18 UNS	1-3/8–10 UNS	1-7/8–14 UNS	2-3/4–18 UNS	4–16 UN
7/16–18 UNS	7/8–20 UNEF	1-3/8–12 UNF	1-7/8–16 UN	2-3/4–20 UN	
7/16–20 UNF	7/8–24 UNS	1-3/8–14 UNS	1-7/8–18 UNS	2-7/8–6 UN	
7/16–24 UNS	7/8–27 UNS	1-3/8–16 UN	1-7/8–20 UN	2-7/8–8 UN	

APPENDIX J (Continued)

TABLE 9 UNIFIED STANDARD SCREW THREAD SERIES

Sizes Primary	Sizes Secondary	Basic Major Diameter	Coarse UNC	Fine UNF	Extra fine UNEF	4UN	6UN	8UN	12UN	16UN	20UN	28UN	32UN	Sizes
			Series with graded pitches			Series with constant pitches								
0		0.0600	–	80	–	–	–	–	–	–	–	–	–	0
	1	0.0730	64	72	–	–	–	–	–	–	–	–	–	1
2		0.0860	56	64	–	–	–	–	–	–	–	–	–	2
	3	0.0990	48	56	–	–	–	–	–	–	–	–	–	3
4		0.1120	40	48	–	–	–	–	–	–	–	–	–	4
5		0.1250	40	44	–	–	–	–	–	–	–	–	–	5
6		0.1380	32	40	–	–	–	–	–	–	–	–	UNC	6
8		0.1640	32	36	–	–	–	–	–	–	–	–	UNC	8
10		0.1900	24	32	–	–	–	–	–	–	–	–	UNF	10
	12	0.2160	24	28	32	–	–	–	–	–	–	UNF	UNEF	12
¼		0.2500	20	28	32	–	–	–	–	–	UNC	UNF	UNEF	¼
⁵⁄₁₆		0.3125	18	24	32	–	–	–	–	–	20	28	UNEF	⁵⁄₁₆
⅜		0.3750	16	24	32	–	–	–	–	UNC	20	28	UNEF	⅜
⁷⁄₁₆		0.4375	14	20	28	–	–	–	–	16	UNF	UNEF	32	⁷⁄₁₆
½		0.5000	13	20	28	–	–	–	–	16	UNF	UNEF	32	½
⁹⁄₁₆		0.5625	12	18	24	–	–	–	UNC	16	20	28	32	⁹⁄₁₆
⅝		0.6250	11	18	24	–	–	–	12	16	20	28	32	⅝
	¹¹⁄₁₆	0.6875	–	–	24	–	–	–	12	16	20	28	32	¹¹⁄₁₆
¾		0.7500	10	16	20	–	–	–	12	UNF	UNEF	28	32	¾
	¹³⁄₁₆	0.8125	–	–	20	–	–	–	12	16	UNEF	28	32	¹³⁄₁₆
⅞		0.8750	9	14	20	–	–	–	12	16	UNEF	28	32	⅞
	¹⁵⁄₁₆	0.9375	–	–	20	–	–	–	12	16	UNEF	28	32	¹⁵⁄₁₆
1		1.0000	8	12	20	–	–	UNC	UNF	16	UNEF	28	32	1
	1¹⁄₁₆	1.0625	–	–	18	–	–	8	12	16	20	28	–	1¹⁄₁₆
1⅛		1.1250	7	12	18	–	–	8	UNF	16	20	28	–	1⅛
	1³⁄₁₆	1.1875	–	–	18	–	–	8	12	16	20	28	–	1³⁄₁₆
1¼		1.2500	7	12	18	–	–	8	UNF	16	20	28	–	1¼
	1⁵⁄₁₆	1.3125	–	–	18	–	–	8	12	16	20	28	–	1⁵⁄₁₆
1⅜		1.3750	6	12	18	–	UNC	8	UNF	16	20	28	–	1⅜
	1⁷⁄₁₆	1.4375	–	–	18	–	6	8	12	16	20	28	–	1⁷⁄₁₆
1½		1.5000	6	12	18	–	UNC	8	UNF	16	20	28	–	1½
	1⁹⁄₁₆	1.5625	–	–	18	–	6	8	12	16	20	–	–	1⁹⁄₁₆
1⅝		1.6250	–	–	18	–	6	8	12	16	20	–	–	1⅝
	1¹¹⁄₁₆	1.6875	–	–	18	–	6	8	12	16	20	–	–	1¹¹⁄₁₆
1¾		1.7500	5	–	–	–	6	8	12	16	20	–	–	1¾
	1¹³⁄₁₆	1.8125	–	–	–	–	6	8	12	16	20	–	–	1¹³⁄₁₆
1⅞		1.8750	–	–	–	–	6	8	12	16	20	–	–	1⅞
	1¹⁵⁄₁₆	1.9375	–	–	–	–	6	8	12	16	20	–	–	1¹⁵⁄₁₆
2		2.0000	4½	–	–	–	6	8	12	16	20	–	–	2
	2⅛	2.1250	–	–	–	–	6	8	12	16	20	–	–	2⅛
2¼		2.2500	4½	–	–	–	6	8	12	16	20	–	–	2¼
	2⅜	2.3750	–	–	–	–	6	8	12	16	20	–	–	2⅜
2½		2.5000	4	–	–	UNC	6	8	12	16	20	–	–	2½
	2⅝	2.6250	–	–	–	4	6	8	12	16	20	–	–	2⅝
2¾		2.7500	4	–	–	UNC	6	8	12	16	20	–	–	2¾
	2⅞	2.8750	–	–	–	4	6	8	12	16	20	–	–	2⅞
3		3.0000	4	–	–	UNC	6	8	12	16	20	–	–	3
	3⅛	3.1250	–	–	–	4	6	8	12	16	–	–	–	3⅛
3¼		3.2500	4	–	–	UNC	6	8	12	16	–	–	–	3¼
	3⅜	3.3750	–	–	–	4	6	8	12	16	–	–	–	3⅜
3½		3.5000	4	–	–	UNC	6	8	12	16	–	–	–	3½
	3⅝	3.6250	–	–	–	4	6	8	12	16	–	–	–	3⅝
3¾		3.7500	4	–	–	UNC	6	8	12	16	–	–	–	3¾
	3⅞	3.8750	–	–	–	4	6	8	12	16	–	–	–	3⅞
4		4.0000	4	–	–	UNC	6	8	12	16	–	–	–	4
	4⅛	4.1250	–	–	–	4	6	8	12	16	–	–	–	4⅛
4¼		4.2500	–	–	–	4	6	8	12	16	–	–	–	4¼
	4⅜	4.3750	–	–	–	4	6	8	12	16	–	–	–	4⅜
4½		4.5000	–	–	–	4	6	8	12	16	–	–	–	4½
	4⅝	4.6250	–	–	–	4	6	8	12	16	–	–	–	4⅝
4¾		4.7500	–	–	–	4	6	8	12	16	–	–	–	4¾
	4⅞	4.8750	–	–	–	4	6	8	12	16	–	–	–	4⅞
5		5.0000	–	–	–	4	6	8	12	16	–	–	–	5
	5⅛	5.1250	–	–	–	4	6	8	12	16	–	–	–	5⅛
5¼		5.2500	–	–	–	4	6	8	12	16	–	–	–	5¼
	5⅜	5.3750	–	–	–	4	6	8	12	16	–	–	–	5⅜
5½		5.5000	–	–	–	4	6	8	12	16	–	–	–	5½
	5⅝	5.6250	–	–	–	4	6	8	12	16	–	–	–	5⅝
5¾		5.7500	–	–	–	4	6	8	12	16	–	–	–	5¾
	5⅞	5.8750	–	–	–	4	6	8	12	16	–	–	–	5⅞
6		6.0000	–	–	–	4	6	8	12	16	–	–	–	6

APPENDIX K
METRIC SCREW THREAD VARIATIONS

THE FOLLOWING list is a ready reference of available standard coarse pitch series ISO metric screw threads. Each thread is given as part of a proper thread note, including metric symbol, major diameter, and thread pitch.

M1 × 0.25	M2.2 × 0.45	M6 × 1	M14 × 2	M30 × 3.5	M52 × 5
M1.1 × 0.25	M2.5 × 0.45	M7 × 1	M16 × 2	M33 × 3.5	M56 × 5.5
M1.2 × 0.25	M3 × 0.5	M8 × 1.25	M18 × 2.5	M36 × 4	M60 × 5.5
M1.4 × 0.3	M3.5 × 0.6	M9 × 1.25	M20 × 2.5	M39 × 4	M64 × 6
M1.6 × 0.35	M4 × 0.7	M10 × 1.5	M22 × 2.5	M42 × 4.5	M68 × 6
M1.8 × 0.35	M4.5 × 0.75	M11 × 1.5	M24 × 3	M45 × 4.5	
M2 × 0.4	M5 × 0.8	M12 × 1.75	M27 × 3	M48 × 5	

APPENDIX L
ASTM AND SAE GRADE MARKINGS FOR STEEL BOLTS AND SCREWS

ASTM AND SAE GRADE MARKINGS FOR STEEL BOLTS AND SCREWS		
Grade Marking	**Specification**	**Material**
NO MARK	SAE—Grade 1	Low or Medium Carbon Steel
	ASTM—A307	Low Carbon Steel
	SAE—Grade 2	Low or Medium Carbon Steel
	SAE—Grade 5	Medium Carbon Steel, Quenched and Tempered
	ASTM—A 449	
	SAE—Grade 5.2	Low Carbon Martensite Steel, Quenched and Tempered
A 325	ASTM—A 325 Type 1	Medium Carbon Steel, Quenched and Tempered Radial dashes optional
A 325	ASTM—A 325 Type 2	Low Carbon Martensite Steel, Quenched and Tempered
A 325	ASTM—A 325 Type 3	Atmospheric Corrosion (Weathering) Steel, Quenched and Tempered
BC	ASTM—A 354 Grade BC	Alloy Steel, Quenched and Tempered
	SAE—Grade 7	Medium Carbon Alloy Steel, Quenched and Tempered, Roll Threaded After Heat Treatment
	SAE—Grade 8	Medium Carbon Alloy Steel, Quenched and Tempered
	ASTM—A 354 Grade BD	Alloy Steel, Quenched and Tempered
	SAE—Grade 8.2	Low Carbon Martensite Steel, Quenched and Tempered
A 490	ASTM—A 490 Type 1	Alloy Steel, Quenched and Tempered
A 490	ASTM—A 490 Type 3	Atmospheric Corrosion (Weathering) Steel, Quenched and Tempered

Reprinted from The American Society of Mechanical Engineers—ANSI B18.2.1–1981 (R1992).

APPENDIX M
CAP SCREW SPECIFICATIONS

TABLE 10 DIMENSIONS OF HEX CAP SCREWS (FINISHED HEX BOLTS)

Nominal Size or Basic Product Dia (18)		E Body Dia (8) Max	E Body Dia (8) Min	F Width Across Flats Basic	F Max	F Min	G Width Across Corners (4) Max	G Min	H Height Basic	H Max	H Min	J Wrenching Height (4) Min	LT Thread Length 6 in. and Shorter Basic	LT Over 6 in. Basic	Y Transition Thread Length (10) Max	Runout of Bearing Surface FIM (5) Max
1/4	0.2500	0.2500	0.2450	7/16	0.438	0.428	0.505	0.488	5/32	0.163	0.150	0.106	0.750	1.000	0.250	0.010
5/16	0.3125	0.3125	0.3065	1/2	0.500	0.489	0.577	0.557	13/64	0.211	0.195	0.140	0.875	1.125	0.278	0.011
3/8	0.3750	0.3750	0.3690	9/16	0.562	0.551	0.650	0.628	15/64	0.243	0.226	0.160	1.000	1.250	0.312	0.012
7/16	0.4375	0.4375	0.4305	5/8	0.625	0.612	0.722	0.698	9/32	0.291	0.272	0.195	1.125	1.375	0.357	0.013
1/2	0.5000	0.5000	0.4930	3/4	0.750	0.736	0.866	0.840	5/16	0.323	0.302	0.215	1.250	1.500	0.385	0.014
9/16	0.5625	0.5625	0.5545	13/16	0.812	0.798	0.938	0.910	23/64	0.371	0.348	0.250	1.375	1.625	0.417	0.015
5/8	0.6250	0.6250	0.6170	15/16	0.938	0.922	1.083	1.051	25/64	0.403	0.378	0.269	1.500	1.750	0.455	0.017
3/4	0.7500	0.7500	0.7410	1 1/8	1.125	1.100	1.299	1.254	15/32	0.483	0.455	0.324	1.750	2.000	0.500	0.020
7/8	0.8750	0.8750	0.8660	1 5/16	1.312	1.285	1.516	1.465	35/64	0.563	0.531	0.378	2.000	2.250	0.556	0.023
1	1.0000	1.0000	0.9900	1 1/2	1.500	1.469	1.732	1.675	39/64	0.627	0.591	0.416	2.250	2.500	0.625	0.026
1 1/8	1.1250	1.1250	1.1140	1 11/16	1.688	1.631	1.949	1.859	11/16	0.718	0.658	0.461	2.500	2.750	0.714	0.029
1 1/4	1.2500	1.2500	1.2390	1 7/8	1.875	1.812	2.165	2.066	25/32	0.813	0.749	0.530	2.750	3.000	0.714	0.033
1 3/8	1.3750	1.3750	1.3630	2 1/16	2.062	1.994	2.382	2.273	27/32	0.878	0.810	0.569	3.000	3.250	0.833	0.036
1 1/2	1.5000	1.5000	1.4880	2 1/4	2.230	2.175	2.598	2.480	1 5/16	0.974	0.902	0.640	3.250	3.500	0.833	0.039
1 3/4	1.7500	1.7500	1.7380	2 5/8	2.625	2.538	3.031	2.893	1 3/32	1.134	1.054	0.748	3.750	4.000	1.000	0.046
2	2.0000	2.0000	1.9880	3	3.000	2.900	3.464	3.306	1 7/32	1.263	1.175	0.825	4.250	4.500	1.111	0.052
2 1/4	2.2500	2.2500	2.2380	3 3/8	3.375	3.262	3.897	3.719	1 3/8	1.423	1.327	0.933	4.750	5.000	1.111	0.059
2 1/2	2.5000	2.5000	2.4880	3 3/4	3.750	3.625	4.330	4.133	1 17/32	1.583	1.479	1.042	5.250	5.500	1.250	0.065
2 3/4	2.7500	2.7500	2.7380	4 1/8	4.125	3.988	4.763	4.546	1 11/16	1.744	1.632	1.151	5.750	6.000	1.250	0.072
3	3.0000	3.0000	2.9880	4 1/2	4.500	4.350	5.196	4.959	1 7/8	1.935	1.815	1.290	6.250	6.500	1.250	0.079

30° +0 −15

APPENDIX M (Continued)

TABLE 11 DIMENSIONS OF HEXAGON AND SPLINE SOCKET HEAD CAP SCREWS (1960 SERIES)

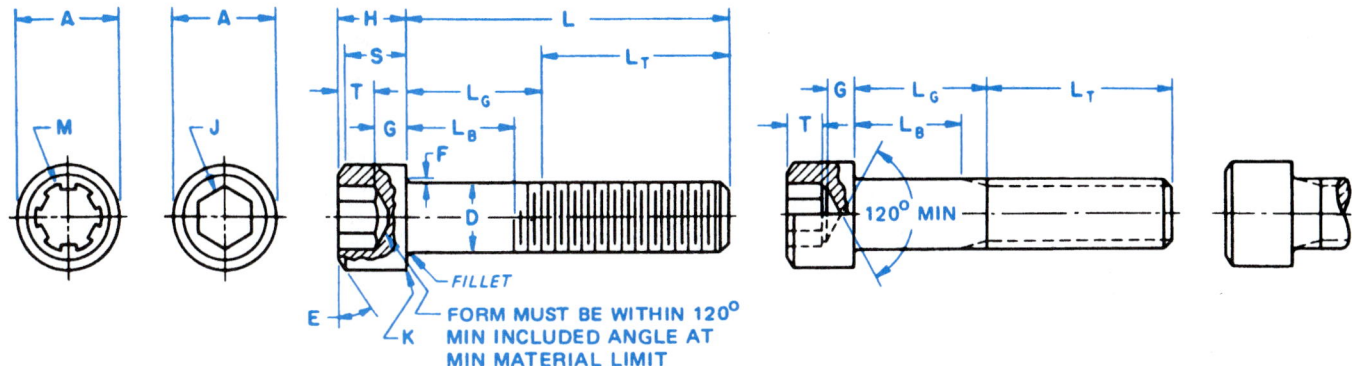

Nominal Size or Basic Screw Diameter		D Body Diameter		A Head Diameter		H Head Height		S Head Side Height	M Spline Socket Size	J Hexagon Socket Size		T Key Engagement	G Wall Thickness	K Chamfer or Radius
		Max	Min	Max	Min	Max	Min	Min	Nom	Nom		Min	Min	Max
0	0.0600	0.0600	0.0568	0.096	0.091	0.060	0.057	0.054	0.060		0.050	0.025	0.020	0.003
1	0.0730	0.0730	0.0695	0.118	0.112	0.073	0.070	0.066	0.072	1/16	0.062	0.031	0.025	0.003
2	0.0860	0.0860	0.0822	0.140	0.134	0.086	0.083	0.077	0.096	5/64	0.078	0.038	0.029	0.003
3	0.0990	0.0990	0.0949	0.161	0.154	0.099	0.095	0.089	0.096	5/64	0.078	0.044	0.034	0.003
4	0.1120	0.1120	0.1075	0.183	0.176	0.112	0.108	0.101	0.111	3/32	0.094	0.051	0.038	0.005
5	0.1250	0.1250	0.1202	0.205	0.198	0.125	0.121	0.112	0.111	3/32	0.094	0.057	0.043	0.005
6	0.1380	0.1380	0.1329	0.226	0.218	0.138	0.134	0.124	0.133	7/64	0.109	0.064	0.047	0.005
8	0.1640	0.1640	0.1585	0.270	0.262	0.164	0.159	0.148	0.168	9/64	0.141	0.077	0.056	0.005
10	0.1900	0.1900	0.1840	0.312	0.303	0.190	0.185	0.171	0.183	5/52	0.156	0.090	0.065	0.005
1/4	0.2500	0.2500	0.2435	0.375	0.365	0.250	0.244	0.225	0.216	3/16	0.188	0.120	0.095	0.008
5/16	0.3125	0.3125	0.3053	0.469	0.457	0.312	0.306	0.281	0.291	1/4	0.250	0.151	0.119	0.008
3/8	0.3750	0.3750	0.3678	0.562	0.550	0.375	0.368	0.337	0.372	5/16	0.312	0.182	0.143	0.008
7/16	0.4375	0.4375	0.4294	0.656	0.642	0.438	0.430	0.394	0.454	3/8	0.375	0.213	0.166	0.010
1/2	0.5000	0.5000	0.4919	0.750	0.735	0.500	0.492	0.450	0.454	3/8	0.375	0.245	0.190	0.010
5/8	0.6250	0.6250	0.6163	0.938	0.921	0.625	0.616	0.562	0.595	1/2	0.500	0.307	0.238	0.010
3/4	0.7500	0.7500	0.7406	1.125	1.107	0.750	0.740	0.675	0.620	5/8	0.625	0.370	0.285	0.010
7/8	0.8750	0.8750	0.8647	1.312	1.293	0.875	0.864	0.787	0.698	3/4	0.750	0.432	0.333	0.015
1	1.0000	1.0000	0.9886	1.500	1.479	1.000	0.988	0.900	0.790	3/4	0.750	0.495	0.380	0.015
1 1/8	1.1250	1.1250	1.1086	1.688	1.665	1.125	1.111	1.012	. . . .	7/8	0.875	0.557	0.428	0.015
1 1/4	1.2500	1.2500	1.2336	1.875	1.852	1.250	1.236	1.125	. . . .	7/8	0.875	0.620	0.475	0.015
1 3/8	1.3750	1.3750	1.3568	2.062	2.038	1.375	1.360	1.237	. . . .	1	1.000	0.682	0.523	0.015
1 1/2	1.5000	1.5000	1.4818	2.250	2.224	1.500	1.485	1.350	. . . .	1	1.000	0.745	0.570	0.015
1 3/4	1.7500	1.7500	1.7295	2.625	2.597	1.750	1.734	1.575	. . . .	1 1/4	1.250	0.870	0.665	0.015
2	2.0000	2.0000	1.9780	3.000	2.970	2.000	1.983	1.800	. . . .	1 1/2	1.500	0.995	0.760	0.015
2 1/4	2.2500	2.2500	2.2280	3.375	3.344	2.250	2.232	2.025	. . . .	1 3/4	1.750	1.120	0.855	0.031
2 1/2	2.5000	2.5000	2.4762	3.750	3.717	2.500	2.481	2.250	. . . .	1 3/4	1.750	1.245	0.950	0.031
2 3/4	2.7500	2.7500	2.7262	4.125	4.090	2.750	2.730	2.475	. . . .	2	2.000	1.370	1.045	0.031
3	3.0000	3.0000	2.9762	4.500	4.464	3.000	2.979	2.700	. . . .	2 1/4	2.250	1.495	1.140	0.031
3 1/4	3.2500	3.2500	3.2262	4.875	4.837	3.250	3.228	2.925	. . . .	2 1/4	2.250	1.620	1.235	0.031
3 1/2	3.5000	3.5000	3.4762	5.250	5.211	3.500	3.478	3.150	. . . .	2 3/4	2.750	1.745	1.330	0.031
3 3/4	3.7500	3.7500	3.7262	5.625	5.584	3.750	3.727	3.375	. . . .	2 3/4	2.750	1.870	1.425	0.031
4	4.0000	4.0000	3.9762	6.000	5.958	4.000	3.976	3.600	. . . .	3	3.000	1.995	1.520	0.031

Reprinted from The American Society of Mechanical Engineers—ANSI/ASME B18.3–1986 (R1993).

APPENDIX M   (Continued)

TABLE 12 DIMENSIONS OF HEXAGON AND SPLINE SOCKET FLAT COUNTERSUNK HEAD CAP SCREWS

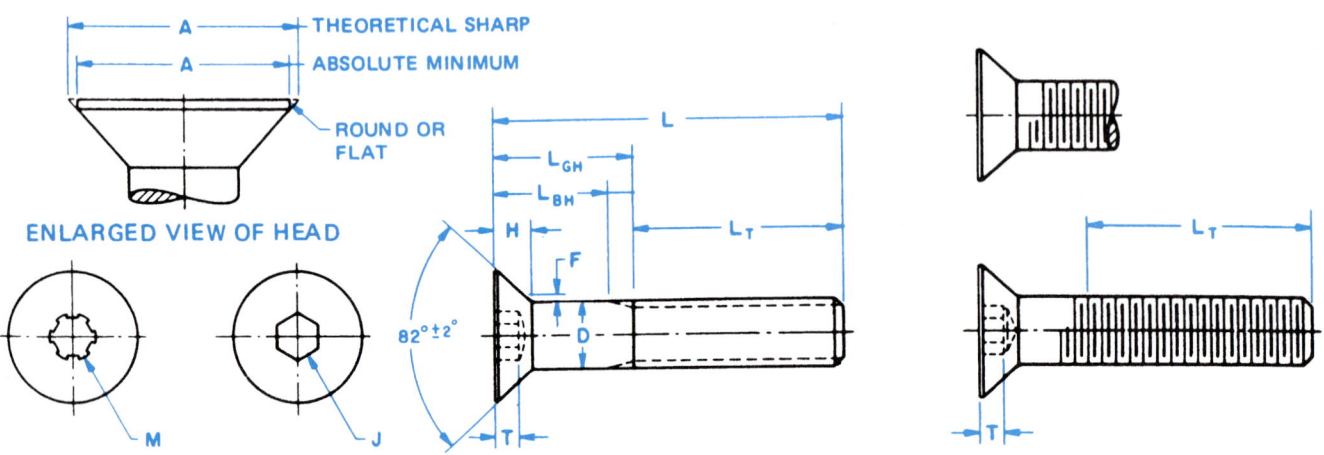

Nominal Size or Basic Screw Diameter		D Body Diameter		A Head Diameter		H Heat Height		M Spline Socket Size	J Hexagon Socket Size		T Key Engagement	F Fillet Extension Above D Max
				Theoretical			Flushness					
		Max	Min	Sharp Max	Abs. Min	Reference	Tolerance		Nom		Min	Max
0	0.0600	0.0600	0.0568	0.138	0.117	0.044	0.006	0.048	0.035		0.025	0.006
1	0.0730	0.0730	0.0695	0.168	0.143	0.054	0.007	0.060	0.050		0.031	0.008
2	0.0860	0.0860	0.0822	0.197	0.168	0.064	0.008	0.060	0.050		0.038	0.010
3	0.0990	0.0990	0.0949	0.226	0.193	0.073	0.010	0.072	1/16	0.062	0.044	0.010
4	0.1120	0.1120	0.1075	0.255	0.218	0.083	0.011	0.072	1/16	0.062	0.055	0.012
5	0.1250	0.1250	0.1202	0.281	0.240	0.090	0.012	0.096	5/64	0.078	0.061	0.014
6	0.1380	0.1380	0.1329	0.307	0.263	0.097	0.013	0.096	5/64	0.078	0.066	0.015
8	0.1640	0.1640	0.1585	0.359	0.311	0.112	0.014	0.111	3/32	0.094	0.076	0.015
10	0.1900	0.1900	0.1840	0.411	0.359	0.127	0.015	0.145	1/8	0.125	0.087	0.015
1/4	0.2500	0.2500	0.2435	0.531	0.480	0.161	0.016	0.183	5/32	0.156	0.111	0.015
5/16	0.3125	0.3125	0.3053	0.656	0.600	0.198	0.017	0.216	3/16	0.188	0.135	0.015
3/8	0.3750	0.3750	0.3678	0.781	0.720	0.234	0.018	0.251	7/32	0.219	0.159	0.015
7/16	0.4375	0.4375	0.4294	0.844	0.781	0.234	0.018	0.291	1/4	0.250	0.159	0.015
1/2	0.5000	0.5000	0.4919	0.938	0.872	0.251	0.018	0.372	5/16	0.312	0.172	0.015
5/8	0.6250	0.6250	0.6163	1.188	1.112	0.324	0.022	0.454	3/8	0.375	0.220	0.015
3/4	0.7500	0.7500	0.7406	1.438	1.355	0.396	0.024	0.454	1/2	0.500	0.220	0.015
7/8	0.8750	0.8750	0.8647	1.688	1.604	0.468	0.025	. . .	9/16	0.562	0.248	0.015
1	1.0000	1.0000	0.9886	1.938	1.841	0.540	0.028	. . .	5/8	0.625	0.297	0.015
1 1/8	1.1250	1.1250	1.1086	2.188	2.079	0.611	0.031	. . .	3/4	0.750	0.325	0.031
1 1/4	1.2500	1.2500	1.2336	2.438	2.316	0.683	0.035	. . .	7/8	0.875	0.358	0.031
1 3/8	1.3750	1.3750	1.3568	2.688	2.553	0.755	0.038	. . .	7/8	0.875	0.402	0.031
1 1/2	1.5000	1.5000	1.4818	2.938	2.791	0.827	0.042	. . .	1	1.000	0.435	0.031

Reprinted from The American Society of Mechanical Engineers—ANSI/ASME B18.3–1986 (R1993).

APPENDIX M (Continued)

TABLE 13 DIMENSIONS OF SLOTTED FLAT COUNTERSUNK HEAD CAP SCREWS

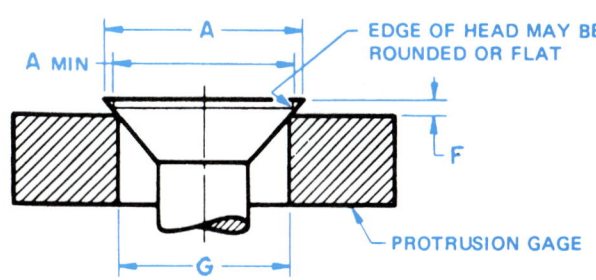

CAP SCREWS

FLAT

Type of Head

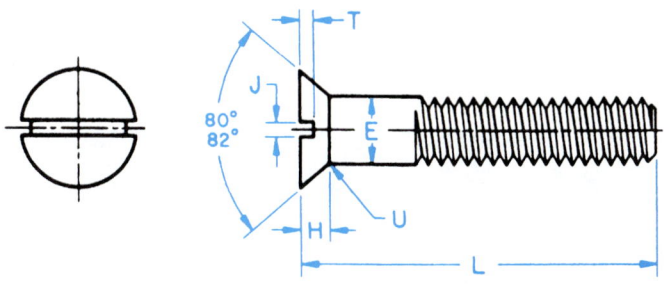

Nominal Size[1] or Basic Screw Diameter		E Body Diameter		A Head Diameter		H[2] Head Height	J Slot Width		T Slot Depth		U Fillet Radius	F[3] Protrusion Above Gaging Diameter		G[3] Gaging Diameter
		Max	Min	Max, Edge Sharp	Min, Edge Rounded or Flat	Ref	Max	Min	Max	Min	Max	Max	Min	
1/4	0.2500	0.2500	0.2450	0.500	0.452	0.140	0.075	0.064	0.068	0.045	0.100	0.046	0.030	0.424
5/16	0.3125	0.3125	0.3070	0.625	0.567	0.177	0.084	0.072	0.086	0.057	0.125	0.053	0.035	0.538
3/8	0.3750	0.3750	0.3690	0.750	0.682	0.210	0.094	0.081	0.103	0.068	0.150	0.060	0.040	0.651
7/16	0.4375	0.4375	0.4310	0.812	0.736	0.210	0.094	0.081	0.103	0.068	0.175	0.065	0.044	0.703
1/2	0.5000	0.5000	0.4930	0.875	0.791	0.210	0.106	0.091	0.103	0.068	0.200	0.071	0.049	0.756
9/16	0.5625	0.5625	0.5550	1.000	0.906	0.244	0.118	0.102	0.120	0.080	0.225	0.078	0.054	0.869
5/8	0.6250	0.6250	0.6170	1.125	1.020	0.281	0.133	0.116	0.137	0.091	0.250	0.085	0.058	0.982
3/4	0.7500	0.7500	0.7420	1.375	1.251	0.352	0.149	0.131	0.171	0.115	0.300	0.099	0.068	1.208
7/8	0.8750	0.8750	0.8660	1.625	1.480	0.423	0.167	0.147	0.206	0.138	0.350	0.113	0.077	1.435
1	1.0000	1.0000	0.9900	1.875	1.711	0.494	0.188	0.166	0.240	0.162	0.400	0.127	0.087	1.661
1 1/8	1.1250	1.1250	1.1140	2.062	1.880	0.529	0.196	0.178	0.257	0.173	0.450	0.141	0.096	1.826
1 1/4	1.2500	1.2500	1.2390	2.312	2.110	0.600	0.211	0.193	0.291	0.197	0.500	0.155	0.105	2.052
1 3/8	1.3750	1.3750	1.3630	2.562	2.340	0.665	0.226	0.208	0.326	0.220	0.550	0.169	0.115	2.279
1 1/2	1.5000	1.5000	1.4880	2.812	2.570	0.742	0.258	0.240	0.360	0.244	0.600	0.183	0.124	2.505

[1] Where specifying nominal size in decimals, zeros preceding decimal and in the fourth decimal place shall be omitted.
[2] Tabulated values determined from formula for maximum H, Appendix III.
[3] No tolerance for gaging diameter is given. If the gaging diameter of the gage used differs from tabulated value, the protrusion will be affected accordingly and the proper protrusion values must be recalculated using the formulas shown in Appendix II.

FOOTNOTES REFER TO ANSI B18.6.2–1972 (R1993).

Reprinted from The American Society of Mechanical Engineers—ANSI B18.6.2–1972 (R1993).

APPENDIX M (Continued)

TABLE 14 DIMENSIONS OF SLOTTED ROUND HEAD CAP SCREWS

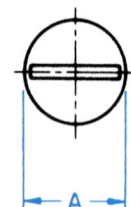

CAP SCREWS
ROUND

Type of Head

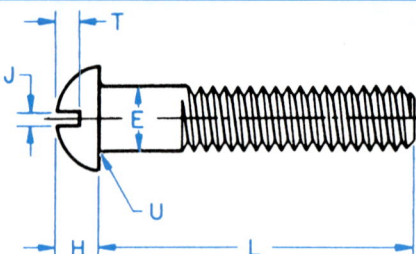

Nominal Size[1] or Basic Screw Diameter	E Body Diameter		A Head Diameter		H Head Height		J Slot Width		T Slot Depth		U Fillet Radius	
	Max	Min	Max	Min	Max	Min	Max	Min	Max	Min	Max	Min
1/4 0.2500	0.2500	0.2450	0.437	0.418	0.191	0.175	0.075	0.064	0.117	0.097	0.031	0.016
5/16 0.3125	0.3125	0.3070	0.562	0.540	0.245	0.226	0.084	0.072	0.151	0.126	0.031	0.016
3/8 0.3750	0.3750	0.3690	0.625	0.603	0.273	0.252	0.094	0.081	0.168	0.138	0.031	0.016
7/16 0.4375	0.4375	0.4310	0.750	0.725	0.328	0.302	0.094	0.081	0.202	0.167	0.047	0.016
1/2 0.5000	0.5000	0.4930	0.812	0.786	0.354	0.327	0.106	0.091	0.218	0.178	0.047	0.016
9/16 0.5625	0.5625	0.5550	0.937	0.909	0.409	0.378	0.118	0.102	0.252	0.207	0.047	0.016
5/8 0.6250	0.6250	0.6170	1.000	0.970	0.437	0.405	0.133	0.116	0.270	0.220	0.062	0.031
3/4 0.7500	0.7500	0.7420	1.250	1.215	0.546	0.507	0.149	0.131	0.338	0.278	0.062	0.031

[1]Where specifying nominal size in decimals, zeros preceding decimal and in the fourth decimal place shall be omitted.
Reprinted from The American Society of Mechanical Engineers—ANSI B18.6.2–1972.

TABLE 15 DIMENSIONS OF SLOTTED FILLISTER HEAD CAP SCREWS

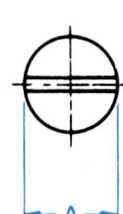

CAP SCREWS
FILLISTER

Type of Head

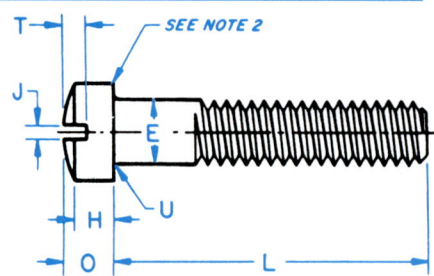

Nominal Size[1] or Basic Screw Diameter	E Body Diameter		A Head Diameter		H Head Side Height		O Total Head Height		J Slot Width		T Slot Depth		U Fillet Radius	
	Max	Min	Max	Min	Max	Min	Max	Min	Max	Min	Max	Min	Max	Min
1/4 0.2500	0.2500	0.2450	0.375	0.363	0.172	0.157	0.216	0.194	0.075	0.064	0.097	0.077	0.031	0.016
5/16 0.3125	0.3125	0.3070	0.437	0.424	0.203	0.186	0.253	0.230	0.084	0.072	0.115	0.090	0.031	0.016
3/8 0.3750	0.3750	0.3690	0.562	0.547	0.250	0.229	0.314	0.284	0.094	0.081	0.142	0.112	0.031	0.016
7/16 0.4375	0.4375	0.4310	0.625	0.608	0.297	0.274	0.368	0.336	0.094	0.081	0.168	0.133	0.047	0.016
1/2 0.5000	0.5000	0.4930	0.750	0.731	0.328	0.301	0.413	0.376	0.106	0.091	0.193	0.153	0.047	0.016
9/16 0.5625	0.5625	0.5550	0.812	0.792	0.375	0.346	0.467	0.427	0.118	0.102	0.213	0.168	0.047	0.016
5/8 0.6250	0.6250	0.6170	0.875	0.853	0.422	0.391	0.521	0.478	0.133	0.116	0.239	0.189	0.062	0.031
3/4 0.7500	0.7500	0.7420	1.000	0.976	0.500	0.466	0.612	0.566	0.149	0.131	0.283	0.223	0.062	0.031
7/8 0.8750	0.8750	0.8660	1.125	1.098	0.594	0.556	0.720	0.668	0.167	0.147	0.334	0.264	0.062	0.031
1 1.0000	1.0000	0.9900	1.312	1.282	0.656	0.612	0.803	0.743	0.188	0.166	0.371	0.291	0.062	0.031

[1]Where specifying nominal size in decimals, zeros preceding decimal and in the fourth decimal place shall be omitted.
[2]A slight rounding of the edges at periphery of head shall be permissible provided the diameter of the bearing circle is equal to no less than 90 percent of the specified minimum head diameter.
Reprinted from The American Society of Mechanical Engineers—ANSI B18.6.2–1972.

APPENDIX N
MACHINE SCREW SPECIFICATIONS

TABLE 16 DIMENSIONS OF SLOTTED FLAT COUNTERSUNK HEAD MACHINE SCREWS

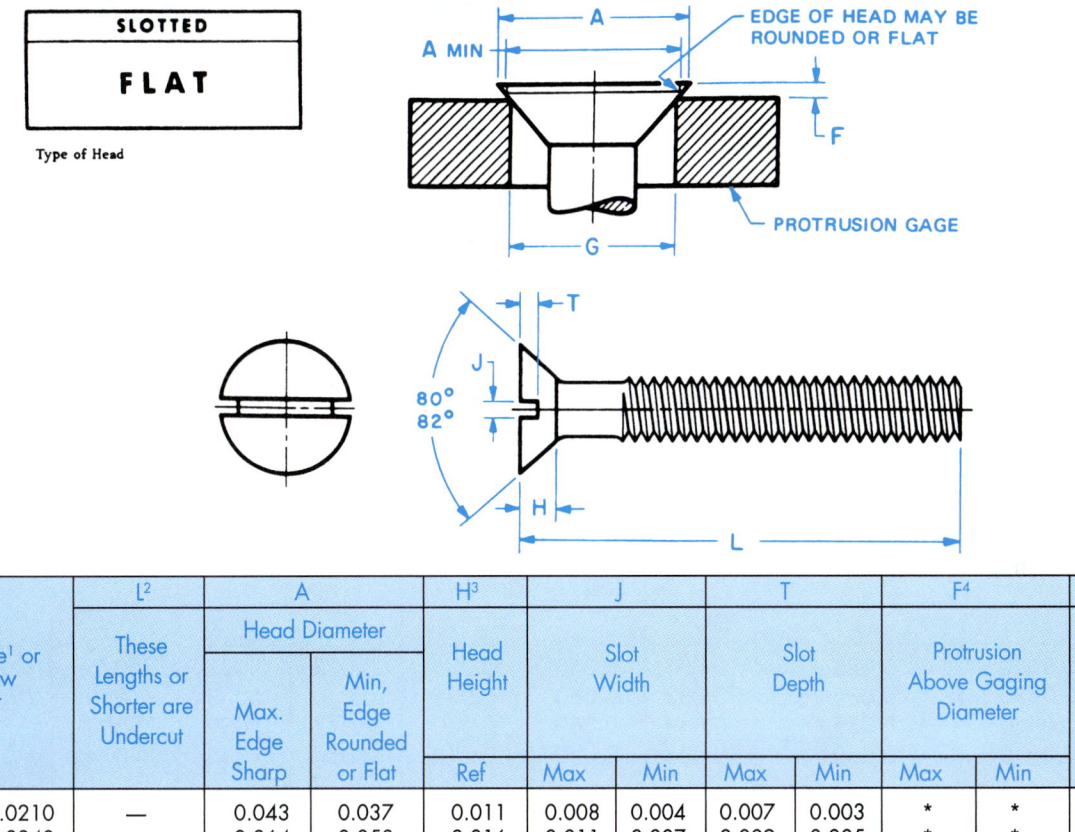

Nominal Size[1] or Basic Screw Diameter		L[2] These Lengths or Shorter are Undercut	A Head Diameter		H[3] Head Height	J Slot Width		T Slot Depth		F[4] Protrusion Above Gaging Diameter		G[4] Gaging Diameter
			Max. Edge Sharp	Min, Edge Rounded or Flat	Ref	Max	Min	Max	Min	Max	Min	
0000	0.0210	—	0.043	0.037	0.011	0.008	0.004	0.007	0.003	*	*	*
000	0.0340	—	0.064	0.058	0.016	0.011	0.007	0.009	0.005	*	*	*
00	0.0470	—	0.093	0.085	0.028	0.017	0.010	0.014	0.009	*	*	*
0	0.0600	1/8	0.119	0.099	0.035	0.023	0.016	0.015	0.010	0.026	0.016	0.078
1	0.0730	1/8	0.146	0.123	0.043	0.026	0.019	0.019	0.012	0.028	0.016	0.101
2	0.0860	1/8	0.172	0.147	0.051	0.031	0.023	0.023	0.015	0.029	0.017	0.124
3	0.0990	1/8	0.199	0.171	0.059	0.035	0.027	0.027	0.017	0.031	0.018	0.148
4	0.1120	3/16	0.225	0.195	0.067	0.039	0.031	0.030	0.020	0.032	0.019	0.172
5	0.1250	3/16	0.252	0.220	0.075	0.043	0.035	0.034	0.022	0.034	0.020	0.196
6	0.1380	3/16	0.279	0.244	0.083	0.048	0.039	0.038	0.024	0.036	0.021	0.220
8	0.1640	1/4	0.332	0.292	0.100	0.054	0.045	0.045	0.029	0.039	0.023	0.267
10	0.1900	5/16	0.385	0.340	0.116	0.060	0.050	0.053	0.034	0.042	0.025	0.313
12	0.2160	3/8	0.438	0.389	0.132	0.067	0.056	0.060	0.039	0.045	0.027	0.362
1/4	0.2500	7/16	0.507	0.452	0.153	0.075	0.064	0.070	0.046	0.050	0.029	0.424
5/16	0.3125	1/2	0.635	0.568	0.191	0.084	0.072	0.088	0.058	0.057	0.034	0.539
3/8	0.3750	9/16	0.762	0.685	0.230	0.094	0.081	0.106	0.070	0.065	0.039	0.653
7/16	0.4375	5/8	0.812	0.723	0.223	0.094	0.081	0.103	0.066	0.073	0.044	0.690
1/2	0.5000	3/4	0.875	0.775	0.223	0.106	0.091	0.103	0.065	0.081	0.049	0.739
9/16	0.5625	—	1.000	0.889	0.260	0.118	0.102	0.120	0.077	0.089	0.053	0.851
5/8	0.6250	—	1.125	1.002	0.298	0.133	0.116	0.137	0.088	0.097	0.058	0.962
3/4	0.7500	—	1.375	1.230	0.372	0.149	0.131	0.171	0.111	0.112	0.067	1.186

[1] Where specifying nominal size in decimals, zeros preceding decimal and in the fourth decimal place shall be omitted.

[2] Screws of these lengths and shorter shall have undercut heads as shown in Table 5.

[3] Tabulated values determined from formula for maximum H, Appendix V.

[4] No tolerance for gaging diameter is given. If the gaging diameter of the gage used differs from tabulated value, the protrusion will be affected accordingly and the proper protrusion values must be recalculated using the formulas shown in Appendix I.

* Not practical to gage.

For additional requirements refer to General Data on Pages 3, 4 and 5.

FOOTNOTES REFER TO ANSI B18.6.3—1972 (R1991).

Reprinted from The American Society of Mechanical Engineers—ANSI B18.6.3–1972 (R1991).

APPENDIX O
SET SCREW SPECIFICATIONS

TABLE 17 DIMENSIONS OF HEXAGON AND SPLINE SOCKET SET SCREWS

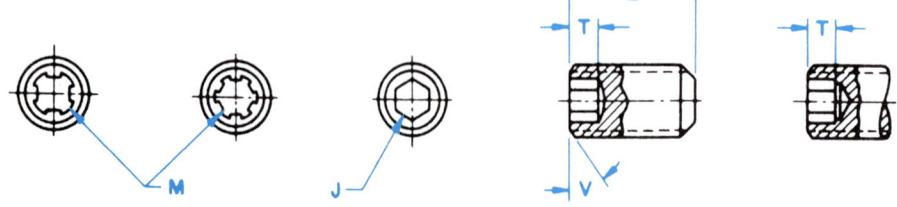

Nominal Size or Basic Screw Diameter		J Hexagon Socket Size		M Spline Socket Size	T Min Key Engagement to Develop Functional Capability of Key		C Cup and Flat Point Diameters		R Oval Point Radius	Y Cone Point Angle 90° ±2° for These Nominal Lengths or Longer; 118° ±2° for Shorter Nominal Lengths
					Hex Socket T_H Min	Spline Socket T_S Min	Max	Min	Basic	
		Nom		Nom						
0	0.0600	0.028		0.033	0.050	0.026	0.033	0.027	0.045	5/64
1	0.0730	0.035		0.033	0.060	0.035	0.040	0.033	0.055	3/32
2	0.0860	0.035		0.048	0.060	0.040	0.047	0.039	0.064	7/64
3	0.0990	0.050		0.048	0.070	0.040	0.054	0.045	0.074	1/8
4	0.1120	0.050		0.060	0.070	0.045	0.061	0.051	0.084	5/32
5	0.1250	1/16	0.062	0.072	0.080	0.055	0.067	0.057	0.094	3/16
6	0.1380	1/16	0.062	0.072	0.080	0.055	0.074	0.064	0.104	3/16
8	0.1640	5/64	0.078	0.096	0.090	0.080	0.087	0.076	0.123	1/4
10	0.1900	3/32	0.094	0.111	0.100	0.080	0.102	0.088	0.142	1/4
1/4	0.2500	1/8	0.125	0.145	0.125	0.125	0.132	0.118	0.188	5/16
5/16	0.3125	5/32	0.156	0.183	0.156	0.156	0.172	0.156	0.234	3/8
3/8	0.3750	3/16	0.188	0.216	0.188	0.188	0.212	0.194	0.281	7/16
7/16	0.4375	7/32	0.219	0.251	0.219	0.219	0.252	0.232	0.328	1/2
1/2	0.5000	1/4	0.250	0.291	0.250	0.250	0.291	0.270	0.375	9/16
5/8	0.6250	5/16	0.312	0.372	0.312	0.312	0.371	0.347	0.469	3/4
3/4	0.7500	3/8	0.375	0.454	0.375	0.375	0.450	0.425	0.562	7/8
7/8	0.8750	1/2	0.500	0.595	0.500	0.500	0.530	0.502	0.656	1
1	1.0000	9/16	0.562	. . .	0.562	. . .	0.609	0.579	0.750	1 1/8
1 1/8	1.1250	9/16	0.562	. . .	0.562	. . .	0.689	0.655	0.844	1 1/4
1 1/4	1.2500	5/8	0.625	. . .	0.625	. . .	0.767	0.733	0.938	1 1/2
1 3/8	1.3750	5/8	0.625	. . .	0.625	. . .	0.848	0.808	1.031	1 5/8
1 1/2	1.5000	3/4	0.750	. . .	0.750	. . .	0.926	0.886	1.125	1 3/4
1 3/4	1.7500	1	1.000	. . .	1.000	. . .	1.086	1.039	1.312	2
2	2.0000	1	1.000	. . .	1.000	. . .	1.244	1.193	1.500	2 1/4

APPENDIX O (Continued)

TABLE 17 (CONTINUED)—DIMENSIONS OF HEXAGON AND SPLINE SOCKET SET SCREWS

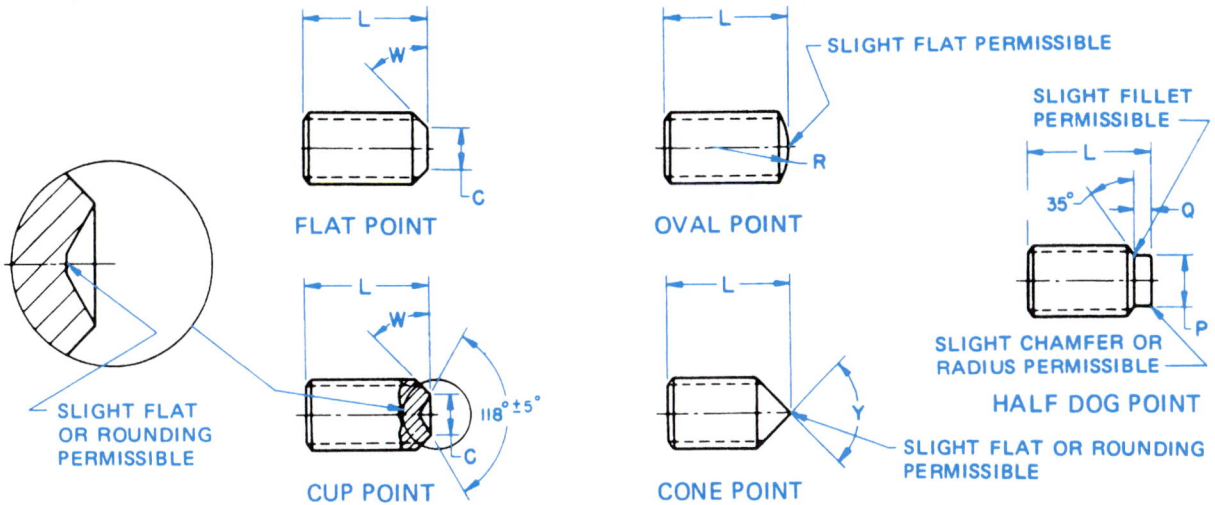

Nominal Size or Basic Screw Diameter		P		Q		B			B₁		
		Half Dog Point				Shortest Optimum Nominal Length To Which Column T_H Applies			Shortest Optimum Nominal Length To Which Column T_S Applies		
		Diameter		Length		Cup and Flat Points	90° Cone and Oval Points	Half Dog Points	Cup and Flat Points	90° Cone and Oval Points	Half Dog Point
		Max	Min	Max	Min						
0	0.0600	0.040	0.037	0.017	0.013	7/64	1/8	7/64	1/16	1/8	7/64
1	0.0730	0.049	0.045	0.021	0.017	1/8	9/64	1/8	3/32	9/64	1/8
2	0.0860	0.057	0.053	0.024	0.020	1/8	9/64	9/64	3/32	9/64	9/64
3	0.0990	0.066	0.062	0.027	0.023	9/64	5/32	5/32	3/32	5/32	5/32
4	0.1120	0.075	0.070	0.030	0.026	9/64	11/64	5/32	3/32	11/64	5/32
5	0.1250	0.083	0.078	0.033	0.027	3/16	3/16	11/64	1/8	3/16	11/64
6	0.1380	0.092	0.087	0.038	0.032	11/64	13/64	3/16	1/8	13/64	3/16
8	0.1640	0.109	0.103	0.043	0.037	3/16	7/32	13/64	3/16	7/32	13/64
10	0.1900	0.127	0.120	0.049	0.041	3/16	1/4	15/64	3/16	1/4	15/64
1/4	0.2500	0.156	0.149	0.067	0.059	1/4	5/16	19/64	1/4	5/16	19/64
5/16	0.3125	0.203	0.195	0.082	0.074	5/16	25/64	23/64	5/16	25/64	23/64
3/8	0.3750	0.250	0.241	0.099	0.089	3/8	7/16	7/16	3/8	7/16	7/16
7/16	0.4375	0.297	0.287	0.114	0.104	7/16	35/64	31/64	7/16	35/64	31/64
1/2	0.5000	0.344	0.334	0.130	0.120	1/2	39/64	35/64	1/2	39/64	35/64
5/8	0.6250	0.469	0.456	0.164	0.148	5/8	49/64	43/64	5/8	49/64	43/64
3/4	0.7500	0.562	0.549	0.196	0.180	3/4	29/32	51/64	3/4	29/32	51/64
7/8	0.8750	0.656	0.642	0.227	0.211	7/8	1 1/8	63/64	7/8	1 1/8	63/64
1	1.0000	0.750	0.734	0.260	0.240	1	1 17/64	1 1/8	. . .	. . .	. . .
1 1/8	1.1250	0.844	0.826	0.291	0.271	1 1/8	1 25/64	1 3/16	. . .	. . .	. . .
1 1/4	1.2500	0.938	0.920	0.323	0.303	1 1/4	1 1/2	1 5/16	. . .	. . .	. . .
1 3/8	1.3750	1.031	1.011	0.354	0.334	1 3/8	1 21/32	1 7/16	. . .	. . .	. . .
1 1/2	1.5000	1.125	1.105	0.385	0.365	1 1/2	1 51/64	1 9/16	. . .	. . .	. . .
1 3/4	1.7500	1.312	1.289	0.448	0.428	1 3/4	2 7/32	1 61/64	. . .	. . .	. . .
2	2.0000	1.500	1.474	0.510	0.490	2	2 25/64	2 5/64	. . .	. . .	. . .

Reprinted from The American Society of Mechanical Engineers—ANSI/ASME B18.3–1986 (R1993).

APPENDIX P
HEX NUT SPECIFICATIONS

TABLE 18 DIMENSIONS OF HEX NUTS AND HEX JAM NUTS

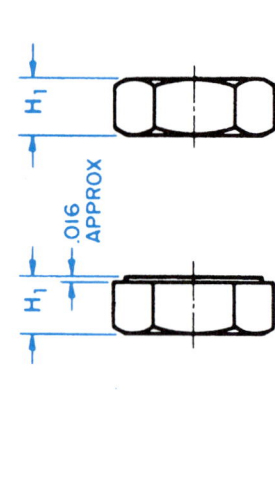

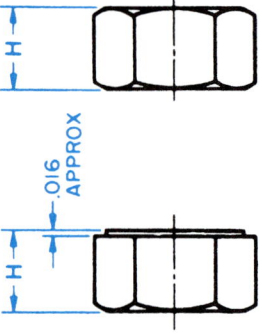

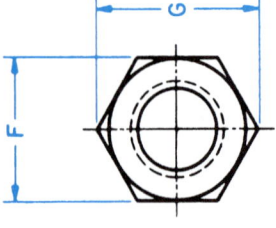

Nominal Size or Basic Major Dia of Thread		F Width Across Flats			G Width Across Corners		H Thickness Hex Nuts			H₁ Thickness Hex Jam Nuts			Runout of Bearing Face, FIR Max		
													Hex Nuts Specified Proof Load		Jam Nuts All Strength Levels
		Basic	Max	Min	Max	Min	Basic	Max	Min	Basic	Max	Min	Up to 150,000 psi	150,000 psi and Greater	
1/4	0.2500	7/16	0.438	0.428	0.505	0.488	7/32	0.226	0.212	5/32	0.163	0.150	0.015	0.010	0.015
5/16	0.3125	1/2	0.500	0.489	0.577	0.557	17/64	0.273	0.258	3/16	0.195	0.180	0.016	0.011	0.016
3/8	0.3750	9/16	0.562	0.551	0.650	0.628	21/64	0.337	0.320	7/32	0.227	0.210	0.017	0.012	0.017
7/16	0.4375	11/16	0.688	0.675	0.794	0.768	3/8	0.385	0.365	1/4	0.260	0.240	0.018	0.013	0.018
1/2	0.5000	3/4	0.750	0.736	0.866	0.840	7/16	0.448	0.427	5/16	0.323	0.302	0.019	0.014	0.019
9/16	0.5625	7/8	0.875	0.861	1.010	0.982	31/64	0.496	0.473	5/16	0.324	0.301	0.020	0.015	0.020
5/8	0.6250	15/16	0.938	0.922	1.083	1.051	35/64	0.559	0.535	3/8	0.387	0.363	0.021	0.016	0.021
3/4	0.7500	1 1/8	1.125	1.088	1.299	1.240	41/64	0.665	0.617	27/64	0.446	0.398	0.023	0.018	0.023
7/8	0.8750	1 5/16	1.312	1.269	1.516	1.447	3/4	0.776	0.724	31/64	0.510	0.458	0.025	0.020	0.025
1	1.0000	1 1/2	1.500	1.450	1.732	1.653	55/64	0.887	0.831	35/64	0.575	0.519	0.027	0.022	0.027
1 1/8	1.1250	1 11/16	1.688	1.631	1.949	1.859	31/32	0.999	0.939	39/64	0.639	0.579	0.030	0.025	0.030
1 1/4	1.2500	1 7/8	1.875	1.812	2.165	2.066	1 1/16	1.094	1.030	23/32	0.751	0.687	0.033	0.028	0.033
1 3/8	1.3750	2 1/16	2.062	1.994	2.382	2.273	1 11/64	1.206	1.138	25/32	0.815	0.747	0.036	0.031	0.036
1 1/2	1.5000	2 1/4	2.250	2.175	2.598	2.480	1 9/32	1.317	1.245	27/32	0.880	0.808	0.039	0.034	0.039

Reprinted from The American Society of Mechanical Engineers—ANSI B18.2.2-1987 (R1993).

APPENDIX Q
KEY AND KEYSEAT SPECIFICATIONS

TABLE 19 WOODRUFF KEY DIMENSIONS

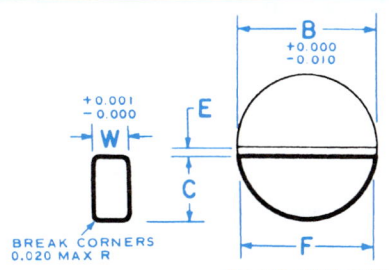

FULL RADIUS TYPE

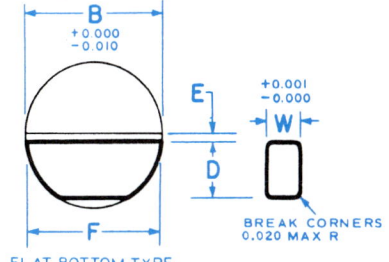

FLAT BOTTOM TYPE

Key No.	Nominal Key Size W × B	Actual Length F +0.000-0.010	Height of Key				Distance Below Center E
			C		D		
			Max	Min	Max	Min	
202	¹⁄₁₆ × ¼	0.248	0.109	0.104	0.109	0.104	¹⁄₆₄
202.5	¹⁄₁₆ × ⁵⁄₁₆	0.311	0.140	0.135	0.140	0.135	¹⁄₆₄
302.5	³⁄₃₂ × ⁵⁄₁₆	0.311	0.140	0.135	0.140	0.135	¹⁄₆₄
203	¹⁄₁₆ × ⅜	0.374	0.172	0.167	0.172	0.167	¹⁄₆₄
303	³⁄₃₂ × ⅜	0.374	0.172	0.167	0.172	0.167	¹⁄₆₄
403	⅛ × ⅜	0.374	0.172	0.167	0.172	0.167	¹⁄₆₄
204	¹⁄₁₆ × ½	0.491	0.203	0.198	0.194	0.188	³⁄₆₄
304	³⁄₃₂ × ½	0.491	0.203	0.198	0.194	0.188	³⁄₆₄
404	⅛ × ½	0.491	0.203	0.198	0.194	0.188	³⁄₆₄
305	³⁄₃₂ × ⅝	0.612	0.250	0.245	0.240	0.234	¹⁄₁₆
405	⅛ × ⅝	0.612	0.250	0.245	0.240	0.234	¹⁄₁₆
505	⁵⁄₃₂ × ⅝	0.612	0.250	0.245	0.240	0.234	¹⁄₁₆
605	³⁄₁₆ × ⅝	0.612	0.250	0.245	0.240	0.234	¹⁄₁₆
406	⅛ × ¾	0.740	0.313	0.308	0.303	0.297	¹⁄₁₆
506	⁵⁄₃₂ × ¾	0.740	0.313	0.308	0.303	0.297	¹⁄₁₆
606	³⁄₁₆ × ¾	0.740	0.313	0.308	0.303	0.297	¹⁄₁₆
806	¼ × ¾	0.740	0.313	0.308	0.303	0.297	¹⁄₁₆
507	⁵⁄₃₂ × ⅞	0.866	0.375	0.370	0.365	0.359	¹⁄₁₆
607	³⁄₁₆ × ⅞	0.866	0.375	0.370	0.365	0.359	¹⁄₁₆
707	⁷⁄₃₂ × ⅞	0.866	0.375	0.370	0.365	0.359	¹⁄₁₆
807	¼ × ⅞	0.866	0.375	0.370	0.365	0.359	¹⁄₁₆
608	³⁄₁₆ × 1	0.992	0.438	0.433	0.428	0.422	¹⁄₁₆
708	⁷⁄₃₂ × 1	0.992	0.438	0.433	0.428	0.422	¹⁄₁₆
808	¼ × 1	0.992	0.438	0.433	0.428	0.422	¹⁄₁₆
1008	⁵⁄₁₆ × 1	0.992	0.438	0.433	0.428	0.422	¹⁄₁₆
1208	⅜ × 1	0.992	0.438	0.433	0.428	0.422	¹⁄₁₆
609	³⁄₁₆ × 1⅛	1.114	0.484	0.479	0.475	0.469	⁵⁄₆₄
709	⁷⁄₃₂ × 1⅛	1.114	0.484	0.479	0.475	0.469	⁵⁄₆₄
809	¼ × 1⅛	1.114	0.484	0.479	0.475	0.469	⁵⁄₆₄
1009	⁵⁄₁₆ × 1⅛	1.114	0.484	0.479	0.475	0.469	⁵⁄₆₄
610	³⁄₁₆ × 1¼	1.240	0.547	0.542	0.537	0.531	⁵⁄₆₄
710	⁷⁄₃₂ × 1¼	1.240	0.547	0.542	0.537	0.531	⁵⁄₆₄
810	¼ × 1¼	1.240	0.547	0.542	0.537	0.531	⁵⁄₆₄
1010	⁵⁄₁₆ × 1¼	1.240	0.547	0.542	0.537	0.531	⁵⁄₆₄
1210	⅜ × 1¼	1.240	0.547	0.542	0.537	0.531	⁵⁄₆₄
811	¼ × 1⅜	1.362	0.594	0.589	0.584	0.578	³⁄₃₂
1011	⁵⁄₁₆ × 1⅜	1.362	0.594	0.589	0.584	0.578	³⁄₃₂
1211	⅜ × 1⅜	1.362	0.594	0.589	0.584	0.578	³⁄₃₂
812	¼ × 1½	1.484	0.641	0.636	0.631	0.625	⁷⁄₆₄
1012	⁵⁄₁₆ × 1½	1.484	0.641	0.636	0.631	0.625	⁷⁄₆₄
1212	⅜ × 1½	1.484	0.641	0.636	0.631	0.625	⁷⁄₆₄

All dimensions given are in inches.

The key numbers indicate nominal key dimensions. The last two digits give the nominal diameter B in eighths of an inch and the digits preceding the last two give the nominal width W in thirty-seconds of an inch.

Example: No. 204 indicates a key ²⁄₃₂ × ⁴⁄₈ or ¹⁄₁₆ × ½.
　　　　　No. 808 indicates a key ⁸⁄₃₂ × ⁸⁄₈ or ¼ × 1.
　　　　　No. 1212 indicates a key ¹²⁄₃₂ × ¹²⁄₈ or ⅜ × 1½.

APPENDIX Q (Continued)

TABLE 19 (CONTINUED)—WOODRUFF KEY DIMENSIONS

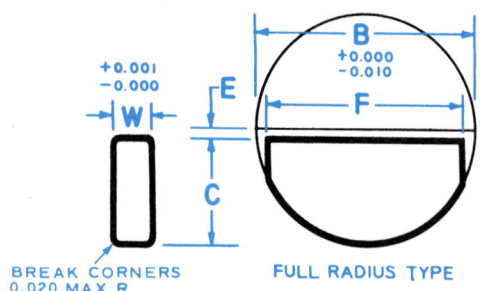

FULL RADIUS TYPE

BREAK CORNERS 0.020 MAX R

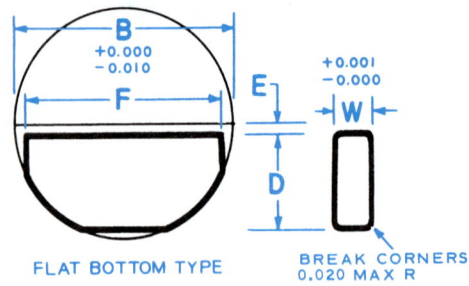

FLAT BOTTOM TYPE

BREAK CORNERS 0.020 MAX R

Key No.	Nominal Key Size W × B	Actual Length F +0.000−0.010	Height of Key				Distance Below Center E
			C		D		
			Max	Min	Max	Min	
617-1	3/16 × 2 1/8	1.380	0.406	0.401	0.396	0.390	21/32
817-1	1/4 × 2 1/8	1.380	0.406	0.401	0.396	0.390	21/32
1017-1	5/16 × 2 1/8	1.380	0.406	0.401	0.396	0.390	21/32
1217-1	3/8 × 2 1/8	1.380	0.406	0.401	0.396	0.390	21/32
617	3/16 × 2 1/8	1.723	0.531	0.526	0.521	0.515	17/32
817	1/4 × 2 1/8	1.723	0.531	0.526	0.521	0.515	17/32
1017	5/16 × 2 1/8	1.723	0.531	0.526	0.521	0.515	17/32
1217	3/8 × 2 1/8	1.723	0.531	0.526	0.521	0.515	17/32
822-1	1/4 × 2 3/4	2.000	0.594	0.589	0.584	0.578	25/32
1022-1	5/16 × 2 3/4	2.000	0.594	0.589	0.584	0.578	25/32
1222-1	3/8 × 2 3/4	2.000	0.594	0.589	0.584	0.578	25/32
1422-1	7/16 × 2 3/4	2.000	0.594	0.589	0.584	0.578	25/32
1622-1	1/2 × 2 3/4	2.000	0.594	0.589	0.584	0.578	25/32
822	1/4 × 2 3/4	2.317	0.750	0.745	0.740	0.734	5/8
1022	5/16 × 2 3/4	2.317	0.750	0.745	0.740	0.734	5/8
1222	3/8 × 2 3/4	2.317	0.750	0.745	0.740	0.734	5/8
1422	7/16 × 2 3/4	2.317	0.750	0.745	0.740	0.734	5/8
1622	1/2 × 2 3/4	2.317	0.750	0.745	0.740	0.734	5/8
1228	3/8 × 3 1/2	2.880	0.938	0.933	0.928	0.922	13/16
1428	7/16 × 3 1/2	2.880	0.938	0.933	0.928	0.922	13/16
1628	1/2 × 3 1/2	2.880	0.938	0.933	0.928	0.922	13/16
1828	9/16 × 3 1/2	2.880	0.938	0.933	0.928	0.922	13/16
2028	5/8 × 3 1/2	2.880	0.938	0.933	0.928	0.922	13/16
2228	11/16 × 3 1/2	2.880	0.938	0.933	0.928	0.922	13/16
2428	3/4 × 3 1/2	2.880	0.938	0.933	0.928	0.922	13/16

All dimensions given are in inches.

The key numbers indicate nominal key dimensions. The last two digits give the nominal diameter B in eighths of an inch and the digits preceding the last two give the nominal width W in thirty-seconds of an inch.

Example:

        No. 617 indicates a key 6/32 × 17/8 or 3/16 × 2 1/8
        No. 822 indicates a key 8/32 × 22/8 or 1/4 × 2 1/4
        No. 1228 indicates a key 12/32 × 28/8 or 3/8 × 3 1/2

The key numbers with the -1 designation, while representing the nominal key size have a shorter length F and due to a greater distance below center E are less in height than the keys of the same number without the -1 designation.

Reprinted from The American Society of Mechanical Engineers—ANSI B17.2–1967 (R1990).

APPENDIX Q (Continued)

TABLE 20 WOODRUFF KEYSEAT DIMENSIONS

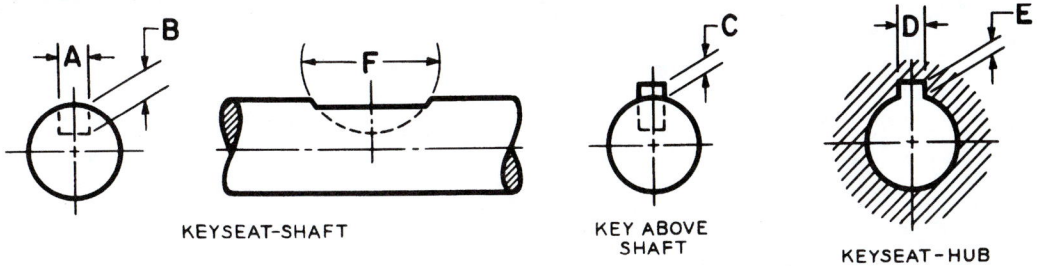

KEYSEAT-SHAFT   KEY ABOVE SHAFT   KEYSEAT-HUB

Key Number	Nominal Size Key	Keyseat – Shaft					Key Above Shaft	Keyseat – Hub	
		Width A*		Depth B	Diameter F		Height C	Width D	Depth E
		Min	Max	+0.005 −0.000	Min	Max	+0.005 −0.005	+0.002 −0.000	+0.005 −0.000
202	1/16 × 1/4	0.0615	0.0630	0.0728	0.250	0.268	0.0312	0.0635	0.0372
202.5	1/16 × 5/16	0.0615	0.0630	0.1038	0.312	0.330	0.0312	0.0635	0.0372
302.5	3/32 × 5/16	0.0928	0.0943	0.0882	0.312	0.330	0.0469	0.0948	0.0529
203	1/16 × 3/8	0.0615	0.0630	0.1358	0.375	0.393	0.0312	0.0635	0.0372
303	3/32 × 3/8	0.0928	0.0943	0.1202	0.375	0.393	0.0469	0.0948	0.0529
403	1/8 × 3/8	0.1240	0.1255	0.1045	0.375	0.393	0.0625	0.1260	0.0685
204	1/16 × 1/2	0.0615	0.0630	0.1668	0.500	0.518	0.0312	0.0635	0.0372
304	3/32 × 1/2	0.0928	0.0943	0.1511	0.500	0.518	0.0469	0.0948	0.0529
404	1/8 × 1/2	0.1240	0.1255	0.1355	0.500	0.518	0.0625	0.1260	0.0685
305	3/32 × 5/8	0.0928	0.0943	0.1981	0.625	0.643	0.0469	0.0948	0.0529
405	1/8 × 5/8	0.1240	0.1255	0.1825	0.625	0.643	0.0625	0.1260	0.0685
505	5/32 × 5/8	0.1553	0.1568	0.1669	0.625	0.643	0.0781	0.1573	0.0841
605	3/16 × 5/8	0.1863	0.1880	0.1513	0.625	0.643	0.0937	0.1885	0.0997
406	1/8 × 3/4	0.1240	0.1255	0.2455	0.750	0.768	0.0625	0.1260	0.0685
506	5/32 × 3/4	0.1553	0.1568	0.2299	0.750	0.768	0.0781	0.1573	0.0841
606	3/16 × 3/4	0.1863	0.1880	0.2143	0.750	0.768	0.0937	0.1885	0.0997
806	1/4 × 3/4	0.2487	0.2505	0.1830	0.750	0.768	0.1250	0.2510	0.1310
507	5/32 × 7/8	0.1553	0.1568	0.2919	0.875	0.895	0.0781	0.1573	0.0841
607	3/16 × 7/8	0.1863	0.1880	0.2763	0.875	0.895	0.0937	0.1885	0.0997
707	7/32 × 7/8	0.2175	0.2193	0.2607	0.875	0.895	0.1093	0.2198	0.1153
807	1/4 × 7/8	0.2487	0.2505	0.2450	0.875	0.895	0.1250	0.2510	0.1310
608	3/16 × 1	0.1863	0.1880	0.3393	1.000	1.020	0.0937	0.1885	0.0997
708	7/32 × 1	0.2175	0.2193	0.3237	1.000	1.020	0.1093	0.2198	0.1153
808	1/4 × 1	0.2487	0.2505	0.3080	1.000	1.020	0.1250	0.2510	0.1310
1008	5/16 × 1	0.3111	0.3130	0.2768	1.000	1.020	0.1562	0.3135	0.1622
1208	3/8 × 1	0.3735	0.3755	0.2455	1.000	1.020	0.1875	0.3760	0.1935
609	3/16 × 1 1/8	0.1863	0.1880	0.3853	1.125	1.145	0.0937	0.1885	0.0997
709	7/32 × 1 1/8	0.2175	0.2193	0.3697	1.125	1.145	0.1093	0.2198	0.1153
809	1/4 × 1 1/8	0.2487	0.2505	0.3540	1.125	1.145	0.1250	0.2510	0.1310
1009	5/16 × 1 1/8	0.3111	0.3130	0.3228	1.125	1.145	0.1562	0.3135	0.1622

APPENDIX Q (Continued)

TABLE 20 (CONTINUED)—WOODRUFF KEYSEAT DIMENSIONS

Key Number	Nominal Size Key	Keyseat – Shaft					Key Above Shaft	Keyseat – Hub	
		Width A*		Depth B	Diameter F		Height C	Width D	Depth E
		Min	Max	+0.005 −0.000	Min	Max	+0.005 −0.005	+0.002 −0.000	+0.005 −0.000
610	3/16 × 1¼	0.1863	0.1880	0.4483	1.250	1.273	0.0937	0.1885	0.0997
710	7/32 × 1¼	0.2175	0.2193	0.4327	1.250	1.273	0.1093	0.2198	0.1153
810	¼ × 1¼	0.2487	0.2505	0.4170	1.250	1.273	0.1250	0.2510	0.1310
1010	5/16 × 1¼	0.3111	0.3130	0.3858	1.250	1.273	0.1562	0.3135	0.1622
1210	3/8 × 1¼	0.3735	0.3755	0.3545	1.250	1.273	0.1875	0.3760	0.1935
811	¼ × 1⅜	0.2487	0.2505	0.4640	1.375	1.398	0.1250	0.2510	0.1310
1011	5/16 × 1⅜	0.3111	0.3130	0.4328	1.375	1.398	0.1562	0.3135	0.1622
1211	3/8 × 1⅜	0.3735	0.3755	0.4015	1.375	1.398	0.1875	0.3760	0.1935
812	¼ × 1½	0.2487	0.2505	0.5110	1.500	1.523	0.1250	0.2510	0.1310
1012	5/16 × 1½	0.3111	0.3130	0.4798	1.500	1.523	0.1562	0.3135	0.1622
1212	3/8 × 1½	0.3735	0.3755	0.4485	1.500	1.523	0.1875	0.3760	0.1935
617-1	3/16 × 2⅛	0.1863	0.1880	0.3073	2.125	2.160	0.0937	0.1885	0.0997
817-1	¼ × 2⅛	0.2487	0.2505	0.2760	2.125	2.160	0.1250	0.2510	0.1310
1017-1	5/16 × 2⅛	0.3111	0.3130	0.2448	2.125	2.160	0.1562	0.3135	0.1622
1217-1	3/8 × 2⅛	0.3735	0.3755	0.2135	2.125	2.160	0.1875	0.3760	0.1935
617	3/16 × 2⅛	0.1863	0.1880	0.4323	2.125	2.160	0.0937	0.1885	0.0997
817	¼ × 2⅛	0.2487	0.2505	0.4010	2.125	2.160	0.1250	0.2510	0.1310
1017	5/16 × 2⅛	0.3111	0.3130	0.3698	2.125	2.160	0.1562	0.3135	0.1622
1217	3/8 × 2⅛	0.3735	0.3755	0.3385	2.125	2.160	0.1875	0.3760	0.1935
822-1	¼ × 2¾	0.2487	0.2505	0.4640	2.750	2.785	0.1250	0.2510	0.1310
1022-1	5/16 × 2¾	0.3111	0.3130	0.4328	2.750	2.785	0.1562	0.3135	0.1622
1222-1	3/8 × 2¾	0.3735	0.3755	0.4015	2.750	2.785	0.1875	0.3760	0.1935
1422-1	7/16 × 2¾	0.4360	0.4380	0.3703	2.750	2.785	0.2187	0.4385	0.2247
1622-1	½ × 2¾	0.4985	0.5005	0.3390	2.750	2.785	0.2500	0.5010	0.2560
822	¼ × 2¾	0.2487	0.2505	0.6200	2.750	2.785	0.1250	0.2510	0.1310
1022	5/16 × 2¾	0.3111	0.3130	0.5888	2.750	2.785	0.1562	0.3135	0.1622
1222	3/8 × 2¾	0.3735	0.3755	0.5575	2.750	2.785	0.1875	0.3760	0.1935
1422	7/16 × 2¾	0.4360	0.4380	0.5263	2.750	2.785	0.2187	0.4385	0.2247
1622	½ × 2¾	0.4985	0.5005	0.4950	2.750	2.785	0.2500	0.5010	0.2560
1228	3/8 × 3½	0.3735	0.3755	0.7455	3.500	3.535	0.1875	0.3760	0.1935
1428	7/16 × 3½	0.4360	0.4380	0.7143	3.500	3.535	0.2187	0.4385	0.2247
1628	½ × 3½	0.4985	0.5005	0.6830	3.500	3.535	0.2500	0.5010	0.2560
1828	9/16 × 3½	0.5610	0.5630	0.6518	3.500	3.535	0.2812	0.5635	0.2872
2028	5/8 × 3½	0.6235	0.6255	0.6205	3.500	3.535	0.3125	0.6260	0.3185
2228	11/16 × 3½	0.6860	0.6880	0.5893	3.500	3.535	0.3437	0.6885	0.3497
2428	¾ × 3½	0.7485	0.7505	0.5580	3.500	3.535	0.3750	0.7510	0.3810

Width A values were set with the maximum keyseat (shaft) width as that figure which will receive a key with the greatest amount of looseness consistent with assuring the key's sticking in the keyseat (shaft). Minimum keyseat width is that figure permitting the largest shaft distortion acceptable when assembling maximum key in minimum keyseat.

Dimensions A, B, C, D are taken at side intersection.

Reprinted from The American Society of Mechanical Engineers—ANSI B17.2–1967 (R1990).

APPENDIX Q (Continued)

TABLE 21 KEY SIZE VERSUS SHAFT DIAMETER

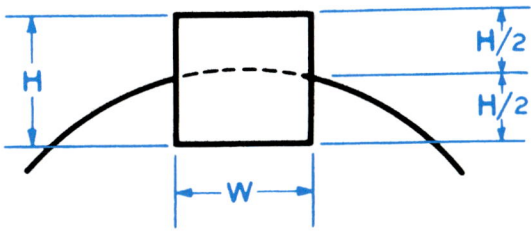

Nominal Shaft Diameter		Nominal Key Size			Nominal Keyseat Depth	
			Height, H		H/2	
Over	To (Incl)	Width, W	Square	Rectangular	Square	Rectangular
5/16	7/16	3/32	3/32		3/64	
7/16	9/16	1/8	1/8	3/32	1/16	3/64
9/16	7/8	3/16	3/16	1/8	3/32	1/16
7/8	1-1/4	1/4	1/4	3/16	1/8	3/32
1-1/4	1-3/8	5/16	5/16	1/4	5/32	1/8
1-3/8	1-3/4	3/8	3/8	1/4	3/16	1/8
1-3/4	2-1/4	1/2	1/2	3/8	1/4	3/16
2-1/4	2-3/4	5/8	5/8	7/16	5/16	7/32
2-3/4	3-1/4	3/4	3/4	1/2	3/8	1/4
3-1/4	3-3/4	7/8	7/8	5/8	7/16	5/16
3-3/4	4-1/2	1	1	3/4	1/2	3/8
4-1/2	5-1/2	1-1/4	1-1/4	7/8	5/8	7/16
5-1/2	6-1/2	1-1/2	1-1/2	1	3/4	1/2
6-1/2	7-1/2	1-3/4	1-3/4	1-1/2*	7/8	3/4
7-1/2	9	2	2	1-1/2	1	3/4
9	11	2-1/2	2-1/2	1-3/4	1-1/4	7/8
11	13	3	3	2	1-1/2	1
13	15	3-1/2	3-1/2	2-1/2	1-3/4	1-1/4
15	18	4		3		1-1/2
18	22	5		3-1/2		1-3/4
22	26	6		4		2
26	30	7		5		2-1/2

*Some key standards show 1-1/4 in. Preferred size is 1-1/2 in. All dimensions given in inches.

Shaded areas:

For a stepped shaft, the size of a key is determined by the diameter of the shaft at the point of location of the key, regardless of the number of different diameters on the shaft.

Square-keys are preferred through 6 1/2-inch diameter shafts and rectangular keys for larger shafts. Sizes and dimensions in unshaded area are preferred.

If special considerations dictate the use of a keyseat in the hub shallower than the preferred nominal depth shown in Table 1, it is recommended that the tabulated preferred nominal standard keyseat be used in the shaft in all cases.

Reprinted from The American Society of Mechanical Engineers—ANSI B17.1–1967 (R1989).

APPENDIX Q (Continued)

TABLE 22 KEY DIMENSIONS AND TOLERANCES

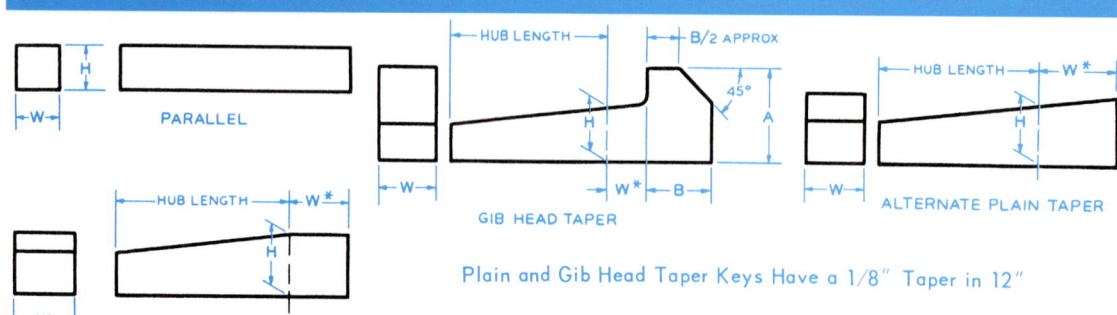

Plain and Gib Head Taper Keys Have a 1/8" Taper in 12"

Key			Nominal Key Size Width, W		Tolerance			
			Over	To (Incl)	Width, W		Height, H	
Parallel	Square	Bar Stock	—	3/4	+0.000	−0.002	+0.000	−0.002
			3/4	1-1/2	+0.000	−0.003	+0.000	−0.003
			1-1/2	2-1/2	+0.000	−0.004	+0.000	−0.004
			2-1/2	3-1/2	+0.000	−0.006	+0.000	−0.006
		Keystock	—	1-1/4	+0.001	−0.000	+0.001	−0.000
			1-1/4	3	+0.002	−0.000	+0.002	−0.000
			3	3-1/2	+0.003	−0.000	+0.003	−0.000
	Rectangular	Bar Stock	—	3/4	+0.000	−0.003	+0.000	−0.003
			3/4	1-1/2	+0.000	−0.004	+0.000	−0.004
			1-1/2	3	+0.000	−0.005	+0.000	−0.005
			3	4	+0.000	−0.006	+0.000	−0.006
			4	6	+0.000	−0.008	+0.000	−0.008
			6	7	+0.000	−0.013	+0.000	−0.013
		Keystock	—	1-1/4	+0.001	−0.000	+0.005	−0.005
			1-1/4	3	+0.002	−0.000	+0.005	−0.005
			3	7	+0.003	−0.000	+0.005	−0.005
Taper	Plain or Gib Head Square or Rectangle		—	1-1/4	+0.001	−0.000	+0.005	−0.000
			1-1/4	3	+0.002	−0.000	+0.005	−0.000
			3	7	+0.003	−0.000	+0.005	−0.000

* For locating position of dimension *H*. Tolerance does not apply. All dimensions given in inches.
Reprinted from The American Society of Mechanical Engineers—ANSI B17.1–1967 (R1989).

TABLE 23 GIB HEAD NOMINAL DIMENSIONS

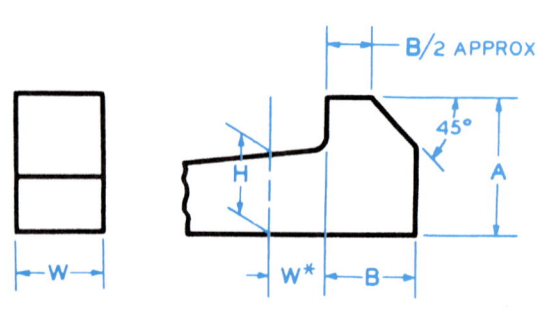

Nominal Key Size Width, W	Square			Rectangular		
	H	A	B	H	A	B
1/8	1/8	1/4	1/4	3/32	3/16	1/8
3/16	3/16	5/16	5/16	1/8	1/4	1/4
1/4	1/4	7/16	3/8	3/16	5/16	5/16
5/16	5/16	1/2	7/16	1/4	7/16	3/8
3/8	3/8	5/8	1/2	1/4	7/16	3/8
1/2	1/2	7/8	5/8	3/8	5/8	1/2
5/8	5/8	1	3/4	7/16	3/4	9/16
3/4	3/4	1-1/4	7/8	1/2	7/8	5/8
7/8	7/8	1-3/8	1	5/8	1	3/4
1	1	1-5/8	1-1/8	3/4	1-1/4	7/8
1-1/4	1-1/4	2	1-7/16	7/8	1-3/8	1
1-1/2	1-1/2	2-3/8	1-3/4	1	1-5/8	1-1/8
1-3/4	1-3/4	2-3/4	2	1-1/2	2-3/8	1-3/4
2	2	3-1/2	2-1/4	1-1/2	2-3/8	1-3/4
2-1/2	2-1/2	4	3	1-3/4	2-3/4	2
3	3	5	3-1/2	2	3-1/2	2-1/4
3-1/2	3-1/2	6	4	2-1/2	4	3

*For locating position of dimension *H*.
For larger sizes the following relationships are suggested as guides for establishing *A* and *B*.

$$A = 1.8 H \qquad B = 1.2 H$$

All dimensions given in inches.
Reprinted from The American Society of Mechanical Engineers—ANSI B17.1–1967 (R1989).

APPENDIX Q (Continued)

TABLE 24 CLASS 2 FIT FOR PARALLEL AND TAPER KEYS

Type of Key	Key Width		Side Fit			Top and Bottom Fit			
	Over	To (Incl)	Width Tolerance		Fit Range*	Depth Tolerance			Fit Range*
			Key	Keyset		Key	Shaft Keyseat	Hub Keyseat	
Parallel Square	—	1-1/4	+0.001 −0.000	+0.002 −0.000	0.002 CL 0.001 INT	+0.001 −0.000	+0.000 −0.015	+0.010 −0.000	0.030 CL 0.004 CL
	1-1/4	3	+0.002 −0.000	+0.002 −0.000	0.002 CL 0.002 INT	+0.002 −0.000	+0.000 −0.015	+0.010 −0.000	0.030 CL 0.003 CL
	3	3-1/2	+0.003 −0.000	+0.002 −0.000	0.002 CL 0.003 INT	+0.003 −0.000	+0.000 −0.015	+0.010 −0.000	0.030 CL 0.002 CL
Parallel Retangular	—	1-1/4	+0.001 −0.000	+0.002 −0.000	0.002 CL 0.001 INT	+0.005 −0.005	+0.000 −0.015	+0.010 −0.000	0.035 CL 0.000 CL
	1-1/4	3	+0.002 −0.000	+0.002 −0.000	0.002 CL 0.002 INT	+0.005 −0.005	+0.000 −0.015	+0.010 −0.000	0.035 CL 0.000 CL
	3	7	+0.003 −0.000	+0.002 −0.000	0.002 CL 0.003 INT	+0.005 −0.005	+0.000 −0.015	+0.010 −0.000	0.035 CL 0.000 CL
Taper	—	1-1/4	+0.001 −0.000	+0.002 −0.000	0.002 CL 0.001 INT	+0.005 −0.000	+0.000 −0.015	+0.010 −0.000	0.005 CL 0.025 INT
	1-1/4	3	+0.002 −0.000	+0.002 −0.000	0.002 CL 0.002 INT	+0.005 −0.000	+0.000 −0.015	+0.010 −0.000	0.005 CL 0.025 INT
	3	Δ	+0.003 −0.000	+0.002 −0.000	0.002 CL 0.003 INT	+0.005 −0.000	+0.000 −0.015	+0.010 −0.000	0.005 CL 0.025 INT

*Limits of variation. CL = Clearance; INT = Interference

Δ To (Incl) 3-1/2 Square and 7 Rectangular key widths.

All dimensions given in inches.

Reprinted from The American Society of Mechanical Engineers—ANSI B17.1–1967 (R1989).

APPENDIX R
TAP DRILL SIZES

TABLE 25 DECIMAL EQUIVALENTS AND TAP DRILL SIZES (LETTER AND NUMBER DRILL SIZES)

Fraction or Drill Size	Decimal Equivalent	Tap Size
Number Size Drills 80	.0135	
1/64 — 79	.0145	
	.0156	
78	.0160	
77	.0180	
76	.0200	
75	.0210	
74	.0225	
73	.0240	
72	.0250	
71	.0260	
70	.0280	
69	.0292	
1/32 — 68	.0310	
	.0312	
67	.0320	
66	.0330	
65	.0350	
64	.0360	
63	.0370	
62	.0380	
61	.0390	
60	.0400	
59	.0410	
58	.0420	
57	.0430	
56	.0465	
3/64	.0469	0–80
55	.0520	
54	.0550	1–56
53	.0595	1–64, 72
1/16	.0625	
52	.0635	
51	.0670	
50	.0700	2–56, 64
49	.0730	
48	.0760	
5/64	.0781	
47	.0785	3–48
46	.0810	
45	.0820	3–56, 4–32
44	.0860	4–36
43	.0890	4–40
42	.0935	4–48
3/32	.0938	
41	.0960	
40	.0980	
39	.0995	
38	.1015	5–40
37	.1040	5–44
36	.1065	6–32
7/64	.1094	
35	.1100	
34	.1110	6–36
33	.1130	6–40
32	.1160	
31	.1200	
1/8	.1250	
30	.1285	
29	.1360	8–32,36
28	.1405	8–40
9/64	.1406	
27	.1440	
26	.1470	
25	.1495	10–24
24	.1520	
23	.1540	
5/32	.1562	
22	.1570	10–30
21	.1590	10–32
20	.1610	
19	.1660	
18	.1695	
11/64	.1719	
17	.1730	
16	.1770	12–24
15	.1800	
14	.1820	12–28
13	.1850	12–32
3/16	.1875	
12	.1890	
11	.1910	
10	.1935	
9	.1960	
8	.1990	
7	.2010	1/4–20
13/64	.2031	
6	.2040	
5	.2055	
4	.2090	
3	.2130	1/4–28
7/32	.2188	
2	.2210	
Letter Size Drills 1	.2280	
A	.2340	
15/64	.2344	
Letter Size Drills B	.2380	
C	.2420	
D	.2460	
1/4 — E	.2500	
F	.2570	5/16–18
G	2610	
17/64	.2656	
H	.2660	
I	.2720	5/16–24
J	.2770	
K	.2810	
9/32	.2812	
L	.2900	
M	.2950	
19/64	.2969	
N	.3020	
5/16	.3125	3/8–16
O	.3160	
P	.3230	
21/64	.3281	
Q	.3320	3/8–24
R	.3390	
11/32	.3438	
S	.3480	
23/64	.3580	
T	.3594	
U	.3680	7/16–14
3/8	.3750	
V	.3770	
W	.3860	
25/64	.3906	7/16–20
X	.3970	
Y	.4040	
13/32	.4062	
Z	.4130	
27/64	.4219	1/2–13
7/16	.4375	
29/64	.4531	1/2–20
15/32	.4688	
31/64	.4844	9/16–12
1/2	.5000	
33/64	.5156	9/16–18
17/32	.5312	5/8–11
35/64	.5469	
9/16	.5625	
37/64	.5781	5/8–18
19/32	.5938	11/16–11
39/64	.6094	
5/8	.6250	11/16–16
41/64	.6406	
21/32	.6562	3/4–10
43/64	.6719	
11/16	.6875	3/4–16
45/64	.7031	
23/32	.7188	
47/64	.7344	
3/4	.7500	
49/64	.7656	7/8–9
25/32	.7812	
51/64	.7969	
13/16	.8125	7/8–14
53/64	.8281	
27/32	.8438	
55/64	.8594	
7/8	.8750	1–8
57/64	.8906	
29/32	.9062	
59/64	.9219	
15/16	.9375	1–12, 14
61/64	.9531	
31/32	.9688	
63/64	.9844	1 1/8–7
1	1.0000	
1 3/64	1.0469	1 1/8–12
1 7/64	1.1094	1 1/4–7
1 1/8	1.1250	
1 11/64	1.1719	1 1/4–12
1 7/32	1.2188	1 3/8–6
1 1/4	1.2500	
1 19/64	1.2969	1 3/8–12
1 11/32	1.3438	1 1/2–6
1 3/8	1.3750	
1 27/64	1.4219	1 1/2–12
1 1/2	1.5000	

Pipe Thread Sizes

Thread	Drill	Thread	Drill
1/8–27	R	1 1/2–11 1/2	1 47/64
1/4–18	7/16	2–11 1/2	2 7/32
3/8–18	37/64	2 1/2–8	2 5/8
1/2–14	23/32	3–8	3 1/4
3/4–14	59/64	3 1/2–8	3 3/4
1–11 1/2	1 5/32	4–8	4 1/4
1 1/4–11 1/2	1 1/2		

Courtesy The L. S. Starrett Company, Athel, Massachusetts.

APPENDIX S
CONCRETE REINFORCING BAR (REBAR) SPECIFICATIONS

REINFORCING BAR SIZES			
US		Metric	
Size	Inch	Size	mm
#3	.375	#10	10
#4	.5	#13	13
#5	.625	#16	16
#6	.75	#19	19
#7	.875	#22	22
#8	1.0	#25	25
#9	1.125	#29	29
#10	1.25	#32	32
#11	1.375	#36	36
#14	1.75	#43	43
#18	2.25	#57	57

REINFORCING BAR GRADE			
Billet Steel			
Inch	Grade	Metric	Grade
#3–#6	40	#10–#19	300
#3–#18	60	#10–#57	420
#6–#18	75	#19–#57	520

SYMBOL FOR TYPE OF STEEL			
Symbol	Type	Grade	ASTM Standard
S	Billet steel	40–75	A615
		300–520	A615M
I	Rail steel	40–60	A616
		300–520	A996M
R	Rail steel	40–60	A616
A	Axle steel	40–60	A617
		300–420	A996M
W	Low alloy	60	A706
		420	A706M

American Society for Testing and Materials (ASTM) requires reinforcing bar identification letters and numbers on the rebar in this order:

- Mill letter or symbol
- Bar size
- Symbol for type of steel
- Grade number and lines

Grade, type, and class are terms used to classify steel products. Different steel grades have specific uses. *Grade* is used to indicate chemical composition. *Type* is used to specify deoxidation practice. *Class* is used to describe some other feature, such as strength level or surface smoothness.

Billet, rail, axle, and low alloy steel are types and classes of steel with different tensile and bending strength, deformation, and chemical composition specifications available from ASTM.

APPENDIX T   COMMON WELDED WIRE REINFORCEMENT SPECIFICATIONS

		CUSTOMARY UNITS			METRIC UNITS (conversions)		
W & D Wire Size* Plain*	W & D Metric Wire Size (Conversion) Plain**	Area (sq. in.)	Diameter (in.)	Nominal Weight (lb./ft.)	Nominal Area (mm²)	Nominal Diameter (mm)	Nominal Mass (kg/m)
W45	MW 290	.45	.757	1.530	290	19.23	2.28
W31	MW 200	.31	.628	1.054	200	15.96	1.57
W20	MW 130	.200	.505	.680	129	12.8	1.01
	MW 122	.189	.490	.643	122	12.4	0.96
W18	MW 116	.180	.479	.612	116	12.2	0.91
	MW 108	.168	.462	.571	108	11.7	0.85
W16	MW 103	.160	.451	.544	103	11.5	0.81
	MW 94	.146	.431	.495	94	10.9	0.74
W14	MW 90	.140	.422	.476	90	10.7	0.71
	MW 79	.122	.394	.414	79	10.0	0.62
W12	MW 77	.120	.391	.408	77	9.9	0.61
W11	MW 71	.110	.374	.374	71	9.5	0.56
W10.5	MW 68	.105	.366	.357	68	9.3	0.53
	MW 67	.103	.363	.351	67	9.2	0.52
W10	MW 65	.100	.357	.340	65	9.1	0.51
W9.5	MW 61	.095	.348	.323	61	8.8	0.48
W9	MW 58	.090	.338	.306	58	8.6	0.45
	MW 56	.086	.331	.292	55.5	8.4	0.43
W8.5	MW 55	.085	.329	.289	54.9	8.4	0.43
W8	MW 52	.080	.319	.272	52	8.1	0.40
W7.5	MW 48	.075	.309	.255	48.4	7.8	0.38
W7	MW 45	.070	.299	.238	45	7.6	0.35
W6.5	MW 42	.065	.288	.221	42	7.3	0.33
	MW 41	.063	.283	.214	41	7.2	0.32
W6	MW 39	.060	.276	.204	39	7.0	0.30
W5.5	MW 36	.055	.265	.187	35.5	6.7	0.28
	MW 35	.054	.263	.184	34.8	6.7	0.27
W5	MW 32	.050	.252	.170	32	6.4	0.25
	MW 30	.047	.244	.158	30	6.2	0.24
	MW 29	.045	.239	.153	29	6.1	0.23
W4	MW 26	.040	.226	.136	26	5.7	0.20
W3.5	MW 23	.035	.211	.119	23	5.4	0.18
W2.9	MW 19	.029	.192	.098	19	4.9	0.15
W2.0	MW 13	.020	.160	.068	13	4.1	0.10
W1.4	MW 9	.014	.135	.048	9	3.4	0.07

Table title: **TABLE 26 WIRE SIZE COMPARISON (WHEN CUSTOMARY UNITS ARE SPECIFIED)**

* For deformed wire, change W to D. ** For deformed wire (metric) change MW to MD.
Provided Courtesy of Wire Reinforcement Institute, Copyright July 2001.

APPENDIX T (Continued)

TABLE 26 (CONTINUED)—WIRE SIZE COMPARISON (WHEN METRIC UNITS ARE SPECIFIED)

Metric Units				Inch-pound Units (conversions)				
Size* (MW = Plain) (mm²)	Area (mm²)	Diameter (mm)	Mass (kg/m)	Nominal Size* (W = Plain) (in²×100)	Area (in²)	Diameter (in)	Weight (lb./ft.)	Gage Guide
MW290	290	19.23	2.28	W45	.450	.757	1.53	
MW200	200	15.96	1.57	W31	.310	.628	1.054	
MW130	130	12.9	1.02	W20.2	.202	.507	.687	
MW120	120	12.4	.941	W18.6	.186	.487	.632	7/0
MW100	100	11.3	.784	W15.5	.155	.444	.527	6/0
MW90	90	10.7	.706	W14.0	.140	.422	.476	5/0
MW80	80	10.1	.627	W12.4	.124	.397	.422	
MW70	70	9.4	.549	W10.9	.109	.373	.371	4/0
MW65	65	9.1	.510	W10.1	.101	.359	.343	3/0
MW60	60	8.7	.470	W9.3	.093	.344	.316	
MW55	55	8.4	.431	W8.5	.085	.329	.289	2/0
MW50	50	8.0	.392	W7.8	.078	.314	.263	
MW45	45	7.6	.353	W7.0	.070	.298	.238	1/0
MW40	40	7.1	.314	W6.2	.062	.283	.214	1
MW35	35	6.7	.274	W5.4	.054	.262	.184	2
MW30	30	6.2	.235	W4.7	.047	.245	.160	3
MW26	26	5.7	.204	W4.0	.040	.226	.136	4
MW25	25	5.6	.196	W3.9	.039	.223	.133	
MW20	20	5.0	.157	W3.1	.031	.199	.105	
MW19	19	4.9	.149	W2.9	.029	.192	.098	6
MW15	15	4.4	.118	W2.3	.023	.171	.078	
MW13	13	4.1	.102	W2.0	.020	.160	.068	8
MW10	10	3.6	.078	W1.6	.016	.143	.054	
MW9	9	3.4	.071	W1.4	.014	.135	.048	10

Note * Wires may be deformed, use prefix MD or D, except where only MW or W is required by building codes (usually less than MW26 or W4). For other available wire sizes, consult other WRI publications or discuss with WWR manufacturers.

APPENDIX T (Continued)

	A[1] (mm²/m)	Metric Styles (MW = Plain wire)[2]	Wt. (kg/m²)	Equivalent US Customary Style	A[1] (in²/ft)	Wt (lbs/CSF)
A[1 & 4]	88.9	102×102—MW9×MW9	1.51	4×4—W1.4×W1.4	.042	31
	127.0	102×102—MW13×MW13	2.15	4×4—W2.0×W2.0	.060	44
	184.2	102×102—MW19×MW19	3.03	4×4—W2.9×W2.9	.087	62
	254.0	102×102—MW26×MW26	4.30	4×4—W4.0×W4.0	.120	88
	59.3	152×152—MW9×MW9	1.03	6×6—W1.4×W1.4	.028	21
	84.7	152×152—MW13×MW13	1.46	6×6—W2.0×W2.0	.040	30
	122.8	152×152—MW19×MW19	2.05	6×6—W2.9×W2.9	.058	42
	169.4	152×152—MW26×MW26	2.83	6×6—W4.0×4.0	.080	58
B[1]	196.9	102×102—MW20×MW20	3.17	4×4—W3.1×W3.1	.093	65
	199.0	152×152—MW30×MW30	3.32	6×6—W4.7×W4.7	.094	68
	199.0	305×305—MW61×MW61	3.47	12×12—W9.4×W9.4	.094	71
	362.0	305×305—MW110×MW110	6.25	12×12—W17.1×W17.1	.171	128
C[1]	342.9	152×152—MW52×MW52	5.66	6×6—W8.1×W8.1	.162	116
	351.4	152×152—MW54×MW54	5.81	6×6—W8.3×W8.3	.166	119
	192.6	305×305—MW59×MW59	8.25	12×12—W9.1×W9.1	.091	69
	351.4	305×305—MW107×MW107	9.72	12×12—W16.6×W16.6	.166	125
D[1]	186.3	152×152—MW28×MW28	3.22	6×6—W4.4×W4.4	.088	63
	338.7	152×152—MW52×MW52	5.61	6×6—W8×W8	.160	115
	186.3	305×305—MW57×MW57	3.22	12×12—W8.8×W8.8	.088	66
	338.7	305×305—MW103×MW103	5.61	12×12—W16×W16	.160	120
E[1]	177.8	152×152—MW27×MW27	3.08	6×6—W4.2×W4.2	.084	60
	317.5	152×152—MW48×MW48	5.52	6×6—W7.5×W7.5	.150	108
	175.7	305×305—MW54×MW54	3.08	12×12—W8.3×W8.3	.083	63
	317.5	305×305—MW97×MW97	5.52	12×12—W15×W15	.150	113

TABLE 27 COMMON STYLES OF METRIC WELDED WIRE REINFORCEMENT (WWR) WITH EQUIVALENT US CUSTOMARY UNITS[3]

[1] Group A—Compares areas of WWR at a minimum f_y = 65,000 psi.
Group B—Compares areas of WWR at a minimum f_y = 70,000 psi
Group C—Compares areas of WWR at a minimum f_y = 72,000 psi ⎫ with areas of #3 or #4 rebar at 12″ o.c.
Group D—Compares areas of WWR at a minimum f_y = 75,000 psi ⎬ at minimum f_y = 60,000 psi
Group E—Compares areas of WWR at a minimum f_y = 80,000 psi ⎭

[2]Wires may also be deformed, use prefix MD or D, except where only MW or W is required by building codes (usually less than a MW26 or W4). Also wire sizes can be specified in 1mm² (metric) or .001 in.² (US Customary) Increments.

[3]For other available styles or wire sizes, consult other WRI publications or discuss with WWR manufacturers.

[4]Styles may be obtained in roll form. Note: It is recommended that rolls be straightened and cut to size before placement.

APPENDIX U

ASTM A500 SQUARE AND RECTANGULAR STRUCTURAL TUBING SPECIFICATIONS

Size	Wall (nominal)	Weight per Foot
2 1/2" x 2 1/2"	0.125	3.90
2 1/2" x 2 1/2"	0.188	5.59
2 1/2" x 2 1/2"	0.250	7.10
3" x 3"	0.125	4.75
3" x 3"	0.188	6.87
3" x 3"	0.250	8.80
3" x 3"	0.313	10.57
3 1/2" x 3 1/2"	0.125	5.60
3 1/2" x 3 1/2"	0.188	8.14
3 1/2" x 3 1/2"	0.250	10.51
3 1/2" x 3 1/2"	0.313	12.69
3 1/2" x 3 1/2"	0.375	14.71
4" x 4"	0.125	6.46
4" x 4"	0.188	9.42
4" x 4"	0.250	12.21
4" x 4"	0.313	14.82
4" x 4"	0.375	17.26
5" x 5"	0.125	8.16
5" x 5"	0.188	11.97
5" x 5"	0.250	15.61
5" x 5"	0.313	19.08
5" x 5"	0.375	22.36
5" x 5"	0.500	28.41
6" x 6"	0.188	14.52
6" x 6"	0.250	19.01
6" x 6"	0.313	23.33
6" x 6"	0.375	27.47
6" x 6"	0.500	35.22
7" x 7"	0.188	17.08
7" x 7"	0.250	22.42
7" x 7"	0.313	27.58
7" x 7"	0.375	32.57
7" x 7"	0.500	42.03
8" x 8"	0.188	19.63
8" x 8"	0.250	25.82
8" x 8"	0.313	31.84
8" x 8"	0.375	37.68
8" x 8"	0.500	48.83
10" x 10"	0.250	32.63
10" x 10"	0.313	40.35
10" x 10"	0.375	47.89
10" x 10"	0.500	62.45

Size	Wall (nominal)	Weight per Foot
3" x 2"	0.125	3.90
3" x 2"	0.188	5.59
3" x 2"	0.250	7.10
3 1/2" x 2 1/2"	0.125	4.75
3 1/2" x 2 1/2"	0.188	6.87
3 1/2" x 2 1/2"	0.250	8.80
4" x 2"	0.125	4.75
4" x 2"	0.188	6.87
4" x 2"	0.250	8.80
4" x 2"	0.313	10.57
4" x 3"	0.125	5.60
4" x 3"	0.188	8.14
4" x 3"	0.250	10.51
4" x 3"	0.313	12.69
4" x 3"	0.375	14.71
5" x 2"	0.125	5.60
5" x 2"	0.188	8.14
5" x 2"	0.250	10.51
5" x 2"	0.313	12.69
5" x 2"	0.375	14.71
5" x 3"	0.125	6.46
5" x 3"	0.188	9.42
5" x 3"	0.250	12.21
5" x 3"	0.313	14.82
5" x 3"	0.375	17.26
5" x 4"	0.125	7.31
5" x 4"	0.188	10.70
5" x 4"	0.250	13.91
5" x 4"	0.313	16.95
5" x 4"	0.375	19.81
6" x 2"	0.125	6.46
6" x 2"	0.188	9.42
6" x 2"	0.250	12.21
6" x 2"	0.313	14.82
6" x 2"	0.375	17.26
6" x 3"	0.125	7.31
6" x 3"	0.188	10.70
6" x 3"	0.250	13.91
6" x 3"	0.313	16.95
6" x 3"	0.375	19.81

Size	Wall (nominal)	Weight per Foot
6" x 4"	0.125	8.16
6" x 4"	0.188	11.97
6" x 4"	0.250	15.61
6" x 4"	0.313	19.08
6" x 4"	0.375	22.36
6" x 4"	0.500	28.41
6" x 5"	0.188	13.25
6" x 5"	0.250	17.31
6" x 5"	0.313	21.20
6" x 5"	0.375	24.92
7" x 3"	0.188	11.97
7" x 3"	0.250	15.61
7" x 3"	0.313	19.08
7" x 3"	0.375	22.36
7" x 4"	0.188	13.25
7" x 4"	0.250	17.31
7" x 4"	0.313	21.20
7" x 4"	0.375	24.92
7" x 5"	0.188	14.52
7" x 5"	0.250	19.01
7" x 5"	0.313	23.33
7" x 5"	0.375	27.47
7" x 5"	0.500	35.22
8" x 2"	0.188	11.97
8" x 2"	0.250	15.61
8" x 3"	0.188	13.25
8" x 3"	0.250	17.31
8" x 3"	0.313	21.20
8" x 3"	0.375	24.92
8" x 4"	0.188	14.52
8" x 4"	0.250	19.01
8" x 4"	0.313	23.33
8" x 4"	0.375	27.47
8" x 4"	0.500	35.22
8" x 6"	0.188	17.08
8" x 6"	0.250	22.42
8" x 6"	0.313	27.58
8" x 6"	0.375	32.57
8" x 6"	0.500	42.03
10" x 2"	0.188	14.52
10" x 2"	0.250	19.01

Size	Wall (nominal)	Weight per Foot
10" x 4"	0.188	17.08
10" x 4"	0.250	22.42
10" x 4"	0.313	27.58
10" x 4"	0.375	32.57
10" x 4"	0.500	42.03
10" x 6"	0.188	19.63
10" x 6"	0.250	25.82
10" x 6"	0.313	31.84
10" x 6"	0.375	37.68
10" x 6"	0.500	48.83
10" x 8"	0.250	29.22
10" x 8"	0.313	36.09
10" x 8"	0.375	42.78
10" x 8"	0.500	55.64
12" x 4"	0.250	25.82
12" x 4"	0.313	31.84
12" x 4"	0.375	37.68
12" x 4"	0.500	48.83
12" x 6"	0.250	29.22
12" x 6"	0.313	36.09
12" x 6"	0.375	42.78
12" x 6"	0.500	55.64
12" x 8"	0.250	32.63
12" x 8"	0.313	40.35
12" x 8"	0.375	47.89
12" x 8"	0.500	62.45

Courtesy Columbia Structural Tubing.

APPENDIX V
STRUCTURAL METAL SHAPE DESIGNATIONS

TABLE 28

W SHAPES — Dimensions

Designation	Area A (In.²)	Depth d (In.)		Web Thickness t_w (In.)		Web t_w/2 (In.)	Flange Width b_f (In.)		Flange Thickness t_f (In.)		Distance T (In.)	Distance k (In.)	Distance k_1 (In.)
W 10×112	32.9	11.36	11 3/8	0.755	3/4	3/8	10.415	10 3/8	1.250	1 1/4	7 5/8	1 7/8	15/16
×100	29.4	11.10	11 1/8	0.680	11/16	3/8	10.340	10 3/8	1.120	1 1/8	7 5/8	1 3/4	7/8
× 88	25.9	10.84	10 7/8	0.605	5/8	5/16	10.265	10 1/4	0.990	1	7 5/8	1 5/8	13/16
× 77	22.6	10.60	10 5/8	0.530	1/2	1/4	10.190	10 1/4	0.870	7/8	7 5/8	1 1/2	13/16
× 68	20.0	10.40	10 3/8	0.470	1/2	1/4	10.130	10 1/8	0.770	3/4	7 5/8	1 3/8	3/4
× 60	17.6	10.22	10 1/4	0.420	7/16	1/4	10.080	10 1/8	0.680	11/16	7 5/8	1 5/16	3/4
× 54	15.8	10.09	10 1/8	0.370	3/8	3/16	10.030	10	0.615	5/8	7 5/8	1 1/4	11/16
× 49	14.4	9.98	10	0.340	5/16	3/16	10.000	10	0.560	9/16	7 5/8	1 3/16	11/16
W 10× 45	13.3	10.10	10 1/8	0.350	3/8	3/16	8.020	8	0.620	5/8	7 5/8	1 1/4	11/16
× 39	11.5	9.92	9 7/8	0.315	5/16	3/16	7.985	8	0.530	1/2	7 5/8	1 1/8	11/16
× 33	9.71	9.73	9 3/4	0.290	5/16	3/16	7.960	8	0.435	7/16	7 5/8	1 1/16	11/16
W 10× 30	8.84	10.47	10 1/2	0.300	5/16	3/16	5.810	5 3/4	0.510	1/2	8 5/8	15/16	1/2
× 26	7.61	10.33	10 3/8	0.260	1/4	1/8	5.770	5 3/4	0.440	7/16	8 5/8	7/8	1/2
× 22	6.49	10.17	10 1/8	0.240	1/4	1/8	5.750	5 3/4	0.360	3/8	8 5/8	3/4	1/2
W 10× 19	5.62	10.24	10 1/4	0.250	1/4	1/8	4.020	4	0.395	3/8	8 5/8	13/16	1/2
× 17	4.99	10.11	10 1/8	0.240	1/4	1/8	4.010	4	0.330	5/16	8 5/8	3/4	1/2
× 15	4.41	9.99	10	0.230	1/4	1/8	4.000	4	0.270	1/4	8 5/8	11/16	7/16
× 12	3.54	9.87	9 7/8	0.190	3/16	1/8	3.960	4	0.210	3/16	8 5/8	5/8	7/16

AMERICAN INSTITUTE OF STEEL CONSTRUCTION

W SHAPES — Dimensions

Designation	Area A (In.²)	Depth d (In.)		Web Thickness t_w (In.)		Web t_w/2 (In.)	Flange Width b_f (In.)		Flange Thickness t_f (In.)		Distance T (In.)	Distance k (In.)	Distance k_1 (In.)
W 8×67	19.7	9.00	9	0.570	9/16	5/16	8.280	8 1/4	0.935	15/16	6 1/8	1 7/16	11/16
×58	17.1	8.75	8 3/4	0.510	1/2	1/4	8.220	8 1/4	0.810	13/16	6 1/8	1 5/16	11/16
×48	14.1	8.50	8 1/2	0.400	3/8	3/16	8.110	8 1/8	0.685	11/16	6 1/8	1 3/16	5/8
×40	11.7	8.25	8 1/4	0.360	3/8	3/16	8.070	8 1/8	0.560	9/16	6 1/8	1 1/16	5/8
×35	10.3	8.12	8 1/8	0.310	5/16	3/16	8.020	8	0.495	1/2	6 1/8	1	9/16
×31	9.13	8.00	8	0.285	5/16	3/16	7.995	8	0.435	7/16	6 1/8	15/16	9/16
W 8×28	8.25	8.06	8	0.285	5/16	3/16	6.535	6 1/2	0.465	7/16	6 1/8	15/16	9/16
×24	7.08	7.93	7 7/8	0.245	1/4	1/8	6.495	6 1/2	0.400	3/8	6 1/8	7/8	9/16
W 8×21	6.16	8.28	8 1/4	0.250	1/4	1/8	5.270	5 1/4	0.400	3/8	6 5/8	13/16	1/2
×18	5.26	8.14	8 1/8	0.230	1/4	1/8	5.250	5 1/4	0.330	5/16	6 5/8	3/4	7/16
W 8×15	4.44	8.11	8 1/8	0.245	1/4	1/8	4.015	4	0.315	5/16	6 5/8	3/4	1/2
×13	3.84	7.99	8	0.230	1/4	1/8	4.000	4	0.255	1/4	6 5/8	11/16	7/16
×10	2.96	7.89	7 7/8	0.170	3/16	1/8	3.940	4	0.205	3/16	6 5/8	5/8	7/16
W 6×25	7.34	6.38	6 3/8	0.320	5/16	3/16	6.080	6 1/8	0.455	7/16	4 3/4	13/16	7/16
×20	5.87	6.20	6 1/4	0.260	1/4	1/8	6.020	6	0.365	3/8	4 3/4	3/4	7/16
×15	4.43	5.99	6	0.230	1/4	1/8	5.990	6	0.260	1/4	4 3/4	5/8	3/8
W 6×16	4.74	6.28	6 1/4	0.260	1/4	1/8	4.030	4	0.405	3/8	4 3/4	3/4	7/16
×12	3.55	6.03	6	0.230	1/4	1/8	4.000	4	0.280	1/4	4 3/4	5/8	3/8
× 9	2.68	5.90	5 7/8	0.170	3/16	1/8	3.940	4	0.215	3/16	4 3/4	9/16	3/8
W 5×19	5.54	5.15	5 1/8	0.270	1/4	1/8	5.030	5	0.430	7/16	3 1/2	13/16	7/16
×16	4.68	5.01	5	0.240	1/4	1/8	5.000	5	0.360	3/8	3 1/2	3/4	7/16
W 4×13	3.83	4.16	4 1/8	0.280	1/4	1/8	4.060	4	0.345	3/8	2 3/4	11/16	7/16

AMERICAN INSTITUTE OF STEEL CONSTRUCTION

APPENDIX V (Continued)

TABLE 28 (CONTINUED)

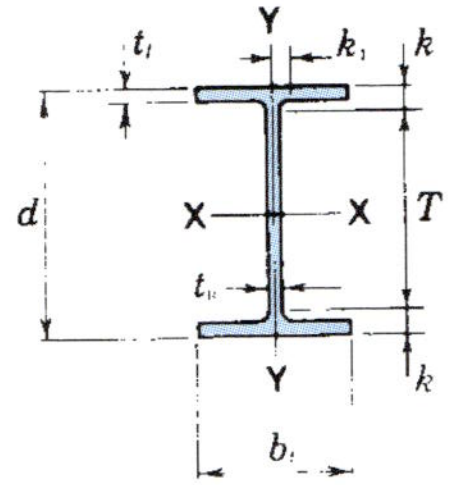

W SHAPES
Dimensions

Designation	Area A	Depth d		Web Thickness t_w		Web $\frac{t_w}{2}$	Flange Width b_f		Flange Thickness t_f		Distance T	Distance k	Distance k_1
	In.²	In.		In.		In.	In.		In.		In.	In.	In.
W 12x336	98.8	16.82	16 7/8	1.775	1 3/4	7/8	13.385	13 3/8	2.955	2 15/16	9 1/2	3 11/16	1 1/2
x305	89.6	16.32	16 3/8	1.625	1 5/8	13/16	13.235	13 1/4	2.705	2 11/16	9 1/2	3 7/16	1 7/16
x279	81.9	15.85	15 7/8	1.530	1 1/2	3/4	13.140	13 1/8	2.470	2 1/2	9 1/2	3 3/16	1 3/8
x252	74.1	15.41	15 3/8	1.395	1 3/8	11/16	13.005	13	2.250	2 1/4	9 1/2	2 15/16	1 5/16
x230	67.7	15.05	15	1.285	1 5/16	11/16	12.895	12 7/8	2.070	2 1/16	9 1/2	2 3/4	1 1/4
x210	61.8	14.71	14 3/4	1.180	1 3/16	5/8	12.790	12 3/4	1.900	1 7/8	9 1/2	2 5/8	1 1/4
x190	55.8	14.38	14 3/8	1.060	1 1/16	9/16	12.670	12 5/8	1.735	1 3/4	9 1/2	2 7/16	1 3/16
x170	50.0	14.03	14	0.960	15/16	1/2	12.570	12 5/8	1.560	1 9/16	9 1/2	2 1/4	1 1/8
x152	44.7	13.71	13 3/4	0.870	7/8	7/16	12.480	12 1/2	1.400	1 3/8	9 1/2	2 1/8	1 1/16
x136	39.9	13.41	13 3/8	0.790	13/16	7/16	12.400	12 3/8	1.250	1 1/4	9 1/2	1 15/16	1
x120	35.3	13.12	13 1/8	0.710	11/16	3/8	12.320	12 3/8	1.105	1 1/8	9 1/2	1 13/16	1
x106	31.2	12.89	12 7/8	0.610	5/8	5/16	12.220	12 1/4	0.990	1	9 1/2	1 11/16	15/16
x 96	28.2	12.71	12 3/4	0.550	9/16	5/16	12.160	12 1/8	0.900	7/8	9 1/2	1 5/8	7/8
x 87	25.6	12.53	12 1/2	0.515	1/2	1/4	12.125	12 1/8	0.810	13/16	9 1/2	1 1/2	7/8
x 79	23.2	12.38	12 3/8	0.470	1/2	1/4	12.080	12 1/8	0.735	3/4	9 1/2	1 7/16	7/8
x 72	21.1	12.25	12 1/4	0.430	7/16	1/4	12.040	12	0.670	11/16	9 1/2	1 3/8	7/8
x 65	19.1	12.12	12 1/8	0.390	3/8	3/16	12.000	12	0.605	5/8	9 1/2	1 5/16	13/16
W 12x 58	17.0	12.19	12 1/4	0.360	3/8	3/16	10.010	10	0.640	5/8	9 1/2	1 3/8	13/16
x 53	15.6	12.06	12	0.345	3/8	3/16	9.995	10	0.575	9/16	9 1/2	1 1/4	13/16
W 12x 50	14.7	12.19	12 1/4	0.370	3/8	3/16	8.080	8 1/8	0.640	5/8	9 1/2	1 3/8	13/16
x 45	13.2	12.06	12	0.335	5/16	3/16	8.045	8	0.575	9/16	9 1/2	1 1/4	13/16
x 40	11.8	11.94	12	0.295	5/16	3/16	8.005	8	0.515	1/2	9 1/2	1 1/4	3/4
W 12x 35	10.3	12.50	12 1/2	0.300	5/16	3/16	6.560	6 1/2	0.520	1/2	10 1/2	1	9/16
x 30	8.79	12.34	12 3/8	0.260	1/4	1/8	6.520	6 1/2	0.440	7/16	10 1/2	15/16	1/2
x 26	7.65	12.22	12 1/4	0.230	1/4	1/8	6.490	6 1/2	0.380	3/8	10 1/2	7/8	1/2
W 12x 22	6.48	12.31	12 1/4	0.260	1/4	1/8	4.030	4	0.425	7/16	10 1/2	7/8	1/2
x 19	5.57	12.16	12 1/8	0.235	1/4	1/8	4.005	4	0.350	3/8	10 1/2	13/16	1/2
x 16	4.71	11.99	12	0.220	1/4	1/8	3.990	4	0.265	1/4	10 1/2	3/4	1/2
x 14	4.16	11.91	11 7/8	0.200	3/16	1/8	3.970	4	0.225	1/4	10 1/2	11/16	1/2

American Institute of Steel Construction

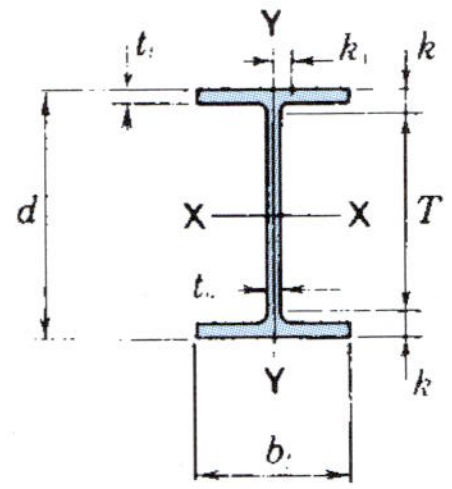

W SHAPES
Dimensions

Designation	Area A	Depth d		Web Thickness t_w		Web $\frac{t_w}{2}$	Flange Width b_f		Flange Thickness t_f		Distance T	Distance k	Distance k_1
	In.²	In.		In.		In.	In.		In.		In.	In.	In.
W 14x132	38.8	14.66	14 5/8	0.645	5/8	5/16	14.725	14 3/4	1.030	1	11 1/4	1 11/16	15/16
x120	35.3	14.48	14 1/2	0.590	9/16	5/16	14.670	14 5/8	0.940	15/16	11 1/4	1 5/8	15/16
x109	32.0	14.32	14 3/8	0.525	1/2	1/4	14.605	14 5/8	0.860	7/8	11 1/4	1 9/16	7/8
x 99	29.1	14.16	14 1/8	0.485	1/2	1/4	14.565	14 5/8	0.780	3/4	11 1/4	1 7/16	7/8
x 90	26.5	14.02	14	0.440	7/16	1/4	14.520	14 1/2	0.710	11/16	11 1/4	1 3/8	7/8
W 14x 82	24.1	14.31	14 1/4	0.510	1/2	1/4	10.130	10 1/8	0.855	7/8	11	1 5/8	1
x 74	21.8	14.17	14 1/8	0.450	7/16	1/4	10.070	10 1/8	0.785	13/16	11	1 9/16	15/16
x 68	20.0	14.04	14	0.415	7/16	1/4	10.035	10	0.720	3/4	11	1 1/2	15/16
x 61	17.9	13.89	13 7/8	0.375	3/8	3/16	9.995	10	0.645	5/8	11	1 7/16	15/16
W 14x 53	15.6	13.92	13 7/8	0.370	3/8	3/16	8.060	8	0.660	11/16	11	1 7/16	15/16
x 48	14.1	13.79	13 3/4	0.340	5/16	3/16	8.030	8	0.595	5/8	11	1 3/8	7/8
x 43	12.6	13.66	13 5/8	0.305	5/16	3/16	7.995	8	0.530	1/2	11	1 5/16	7/8
W 14x 38	11.2	14.10	14 1/8	0.310	5/16	3/16	6.770	6 3/4	0.515	1/2	12	1 1/16	5/8
x 34	10.0	13.98	14	0.285	5/16	3/16	6.745	6 3/4	0.455	7/16	12	1	5/8
x 30	8.85	13.84	13 7/8	0.270	1/4	1/8	6.730	6 3/4	0.385	3/8	12	15/16	5/8
W 14x 26	7.69	13.91	13 7/8	0.255	1/4	1/8	5.025	5	0.420	7/16	12	15/16	9/16
x 22	6.49	13.74	13 3/4	0.230	1/4	1/8	5.000	5	0.335	5/16	12	7/8	9/16

American Institute of Steel Construction

APPENDIX V (Continued)

TABLE 28 (CONTINUED)

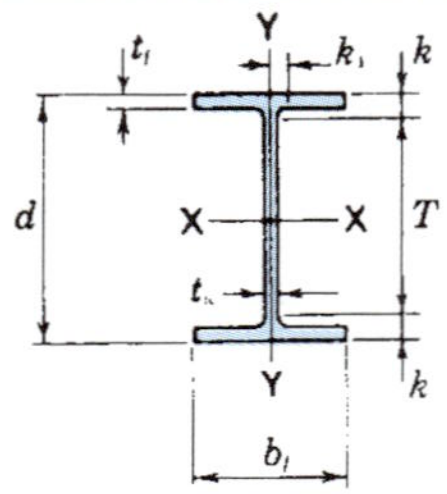

W SHAPES Dimensions

Designation	Area A	Depth d		Web Thickness t_w		$\frac{t_w}{2}$	Flange Width b_f		Flange Thickness t_f		Distance T	Distance k	Distance k_1
	In.^{2}	In.		In.		In.	In.		In.		In.	In.	In.
W 14x730	215.0	22.42	22 3/8	3.070	3 1/16	1 9/16	17.890	17 7/8	4.910	4 15/16	11 1/4	5 9/16	2 3/16
x665	196.0	21.64	21 5/8	2.830	2 13/16	1 7/16	17.650	17 5/8	4.520	4 1/2	11 1/4	5 3/16	2 1/16
x605	178.0	20.92	20 7/8	2.595	2 5/8	1 5/16	17.415	17 3/8	4.160	4 3/16	11 1/4	4 13/16	1 15/16
x550	162.0	20.24	20 1/4	2.380	2 3/8	1 3/16	17.200	17 1/4	3.820	3 13/16	11 1/4	4 1/2	1 13/16
x500	147.0	19.60	19 5/8	2.190	2 3/16	1 1/8	17.010	17	3.500	3 1/2	11 1/4	4 3/16	1 3/4
x455	134.0	19.02	19	2.015	2	1	16.835	16 7/8	3.210	3 3/16	11 1/4	3 7/8	1 5/8
W 14x426	125.0	18.67	18 5/8	1.875	1 7/8	15/16	16.695	16 3/4	3.035	3 1/16	11 1/4	3 11/16	1 9/16
x398	117.0	18.29	18 1/4	1.770	1 3/4	7/8	16.590	16 5/8	2.845	2 7/8	11 1/4	3 1/2	1 1/2
x370	109.0	17.92	17 7/8	1.655	1 5/8	13/16	16.475	16 1/2	2.660	2 11/16	11 1/4	3 5/16	1 7/16
x342	101.0	17.54	17 1/2	1.540	1 9/16	13/16	16.360	16 3/8	2.470	2 1/2	11 1/4	3 1/8	1 3/8
x311	91.4	17.12	17 1/8	1.410	1 7/16	3/4	16.230	16 1/4	2.260	2 1/4	11 1/4	2 15/16	1 5/16
x283	83.3	16.74	16 3/4	1.290	1 5/16	11/16	16.110	16 1/8	2.070	2 1/16	11 1/4	2 3/4	1 1/4
x257	75.6	16.38	16 3/8	1.175	1 3/16	5/8	15.995	16	1.890	1 7/8	11 1/4	2 9/16	1 3/16
x233	68.5	16.04	16	1.070	1 1/16	9/16	15.890	15 7/8	1.720	1 3/4	11 1/4	2 3/8	1 3/16
x211	62.0	15.72	15 3/4	0.980	1	1/2	15.800	15 3/4	1.560	1 9/16	11 1/4	2 1/4	1 1/8
x193	56.8	15.48	15 1/2	0.890	7/8	7/16	15.710	15 3/4	1.440	1 7/16	11 1/4	2 1/8	1 1/16
x176	51.8	15.22	15 1/4	0.830	13/16	7/16	15.650	15 5/8	1.310	1 5/16	11 1/4	2	1 1/16
x159	46.7	14.98	15	0.745	3/4	3/8	15.565	15 5/8	1.190	1 3/16	11 1/4	1 7/8	1
x145	42.7	14.78	14 3/4	0.680	11/16	3/8	15.500	15 1/2	1.090	1 1/16	11 1/4	1 3/4	1

American Institute of Steel Construction

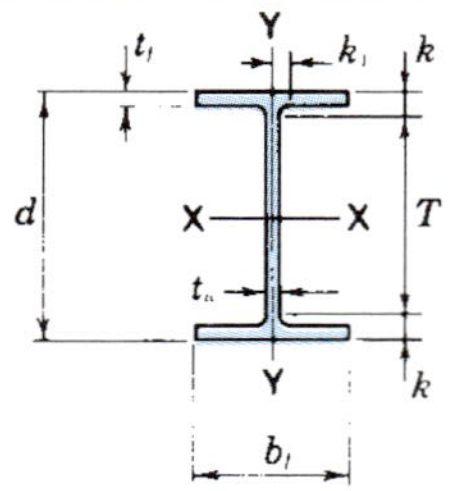

W SHAPES Dimensions

Designation	Area A	Depth d		Web Thickness t_w		$\frac{t_w}{2}$	Flange Width b_f		Flange Thickness t_f		Distance T	Distance k	Distance k_1
	In.^{2}	In.		In.		In.	In.		In.		In.	In.	In.
W 18x119	35.1	18.97	19	0.655	5/8	5/16	11.265	11 1/4	1.060	1 1/16	15 1/2	1 3/4	15/16
x106	31.1	18.73	18 3/4	0.590	9/16	5/16	11.200	11 1/4	0.940	15/16	15 1/2	1 5/8	15/16
x 97	28.5	18.59	18 5/8	0.535	9/16	5/16	11.145	11 1/8	0.870	7/8	15 1/2	1 9/16	7/8
x 86	25.3	18.39	18 3/8	0.480	1/2	1/4	11.090	11 1/8	0.770	3/4	15 1/2	1 7/16	7/8
x 76	22.3	18.21	18 1/4	0.425	7/16	1/4	11.035	11	0.680	11/16	15 1/2	1 3/8	13/16
W 18x 71	20.8	18.47	18 1/2	0.495	1/2	1/4	7.635	7 5/8	0.810	13/16	15 1/2	1 1/2	7/8
x 65	19.1	18.35	18 3/8	0.450	7/16	1/4	7.590	7 5/8	0.750	3/4	15 1/2	1 7/16	7/8
x 60	17.6	18.24	18 1/4	0.415	7/16	1/4	7.555	7 1/2	0.695	11/16	15 1/2	1 3/8	13/16
x 55	16.2	18.11	18 1/8	0.390	3/8	3/16	7.530	7 1/2	0.630	5/8	15 1/2	1 5/16	13/16
x 50	14.7	17.99	18	0.355	3/8	3/16	7.495	7 1/2	0.570	9/16	15 1/2	1 1/4	13/16
W 18x 46	13.5	18.06	18	0.360	3/8	3/16	6.060	6	0.605	5/8	15 1/2	1 1/4	13/16
x 40	11.8	17.90	17 7/8	0.315	5/16	3/16	6.015	6	0.525	1/2	15 1/2	1 3/16	13/16
x 35	10.3	17.70	17 3/4	0.300	5/16	3/16	6.000	6	0.425	7/16	15 1/2	1 1/8	3/4
W 16x100	29.4	16.97	17	0.585	9/16	5/16	10.425	10 3/8	0.985	1	13 5/8	1 11/16	15/16
x 89	26.2	16.75	16 3/4	0.525	1/2	1/4	10.365	10 3/8	0.875	7/8	13 5/8	1 9/16	7/8
x 77	22.6	16.52	16 1/2	0.455	7/16	1/4	10.295	10 1/4	0.760	3/4	13 5/8	1 7/16	7/8
x 67	19.7	16.33	16 3/8	0.395	3/8	3/16	10.235	10 1/4	0.665	11/16	13 5/8	1 3/8	13/16
W 16x 57	16.8	16.43	16 3/8	0.430	7/16	1/4	7.120	7 1/8	0.715	11/16	13 5/8	1 3/8	7/8
x 50	14.7	16.26	16 1/4	0.380	3/8	3/16	7.070	7 1/8	0.630	5/8	13 5/8	1 5/16	13/16
x 45	13.3	16.13	16 1/8	0.345	3/8	3/16	7.035	7	0.565	9/16	13 5/8	1 1/4	13/16
x 40	11.8	16.01	16	0.305	5/16	3/16	6.995	7	0.505	1/2	13 5/8	1 3/16	13/16
x 36	10.6	15.86	15 7/8	0.295	5/16	3/16	6.985	7	0.430	7/16	13 5/8	1 1/8	3/4
W 16x 31	9.12	15.88	15 7/8	0.275	1/4	1/8	5.525	5 1/2	0.440	7/16	13 5/8	1 1/8	3/4
x 26	7.68	15.69	15 3/4	0.250	1/4	1/8	5.500	5 1/2	0.345	3/8	13 5/8	1 1/16	3/4

American Institute of Steel Construction

APPENDIX V (Continued)

TABLE 28 (CONTINUED)

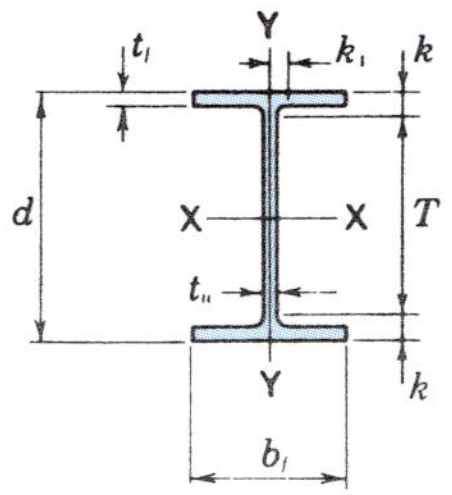

W SHAPES Dimensions

Designation	Area A	Depth d		Web Thickness t_w		Web $\frac{t_w}{2}$	Flange Width b_f		Flange Thickness t_f		Distance T	Distance k	Distance k_1
	In.²	In.		In.		In.	In.		In.		In.	In.	In.
W 36x300	88.3	36.74	36 3/4	0.945	15/16	1/2	16.655	16 5/8	1.680	1 11/16	31 1/8	2 13/16	1 1/2
x280	82.4	36.52	36 1/2	0.885	7/8	7/16	16.595	16 5/8	1.570	1 9/16	31 1/8	2 11/16	1 1/2
x260	76.5	36.26	36 1/4	0.840	13/16	7/16	16.550	16 1/2	1.440	1 7/16	31 1/8	2 9/16	1 1/2
x245	72.1	36.08	36 1/8	0.800	13/16	7/16	16.510	16 1/2	1.350	1 3/8	31 1/8	2 1/2	1 7/16
x230	67.6	35.90	35 7/8	0.760	3/4	3/8	16.470	16 1/2	1.260	1 1/4	31 1/8	2 3/8	1 7/16
W 36x210	61.8	36.69	36 3/4	0.830	13/16	7/16	12.180	12 1/8	1.360	1 3/8	32 1/8	2 5/16	1 1/4
x194	57.0	36.49	36 1/2	0.765	3/4	3/8	12.115	12 1/8	1.260	1 1/4	32 1/8	2 3/16	1 3/16
x182	53.6	36.33	36 3/8	0.725	3/4	3/8	12.075	12 1/8	1.180	1 3/16	32 1/8	2 1/8	1 3/16
x170	50.0	36.17	36 1/8	0.680	11/16	3/8	12.030	12	1.100	1 1/8	32 1/8	2	1 3/16
x160	47.0	36.01	36	0.650	5/8	5/16	12.000	12	1.020	1	32 1/8	1 15/16	1 1/8
x150	44.2	35.85	35 7/8	0.625	5/8	5/16	11.975	12	0.940	15/16	32 1/8	1 7/8	1 1/8
x135	39.7	35.55	35 1/2	0.600	5/8	5/16	11.950	12	0.790	13/16	32 1/8	1 11/16	1 1/8
W 33x241	70.9	34.18	34 1/8	0.830	13/16	7/16	15.860	15 7/8	1.400	1 3/8	29 3/4	2 3/16	1 3/16
x221	65.0	33.93	33 7/8	0.775	3/4	3/8	15.805	15 3/4	1.275	1 1/4	29 3/4	2 1/16	1 3/16
x201	59.1	33.68	33 5/8	0.715	11/16	3/8	15.745	15 3/4	1.150	1 1/8	29 3/4	1 15/16	1 1/8
W 33x152	44.7	33.49	33 1/2	0.635	5/8	5/16	11.565	11 5/8	1.055	1 1/16	29 3/4	1 7/8	1 1/8
x141	41.6	33.30	33 1/4	0.605	5/8	5/16	11.535	11 1/2	0.960	15/16	29 3/4	1 3/4	1 1/16
x130	38.3	33.09	33 1/8	0.580	9/16	5/16	11.510	11 1/2	0.855	7/8	29 3/4	1 11/16	1 1/16
x118	34.7	32.86	32 7/8	0.550	9/16	5/16	11.480	11 1/2	0.740	3/4	29 3/4	1 9/16	1 1/16
W 30x211	62.0	30.94	31	0.775	3/4	3/8	15.105	15 1/8	1.315	1 5/16	26 3/4	2 1/8	1 1/8
x191	56.1	30.68	30 5/8	0.710	11/16	3/8	15.040	15	1.185	1 3/16	26 3/4	1 15/16	1 1/16
x173	50.8	30.44	30 1/2	0.655	5/8	5/16	14.985	15	1.065	1 1/16	26 3/4	1 7/8	1 1/16
W 30x132	38.9	30.31	30 1/4	0.615	5/8	5/16	10.545	10 1/2	1.000	1	26 3/4	1 3/4	1 1/16
x124	36.5	30.17	30 1/8	0.585	9/16	5/16	10.515	10 1/2	0.930	15/16	26 3/4	1 11/16	1
x116	34.2	30.01	30	0.565	9/16	5/16	10.495	10 1/2	0.850	7/8	26 3/4	1 5/8	1
x108	31.7	29.83	29 7/8	0.545	9/16	5/16	10.475	10 1/2	0.760	3/4	26 3/4	1 9/16	1
x 99	29.1	29.65	29 5/8	0.520	1/2	1/4	10.450	10 1/2	0.670	11/16	26 3/4	1 7/16	1

American Institute of Steel Construction

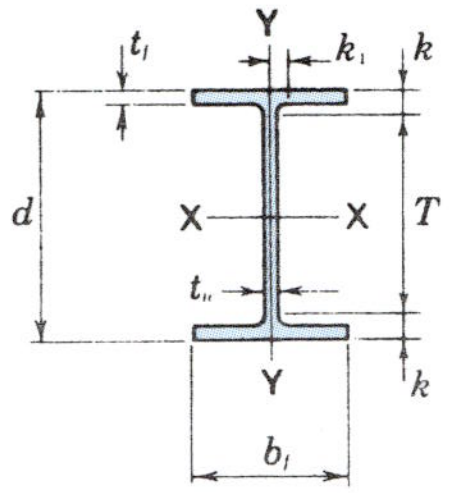

W SHAPES Dimensions

Designation	Area A	Depth d		Web Thickness t_w		Web $\frac{t_w}{2}$	Flange Width b_f		Flange Thickness t_f		Distance T	Distance k	Distance k_1
	In.²	In.		In.		In.	In.		In.		In.	In.	In.
W 27x178	52.3	27.81	27 3/4	0.725	3/4	3/8	14.085	14 1/8	1.190	1 3/16	24	1 7/8	1 1/16
x161	47.4	27.59	27 5/8	0.660	11/16	3/8	14.020	14	1.080	1 1/16	24	1 13/16	1
x146	42.9	27.38	27 3/8	0.605	5/8	5/16	13.965	14	0.975	1	24	1 11/16	1
W 27x114	33.5	27.29	27 1/4	0.570	9/16	5/16	10.070	10 1/8	0.930	15/16	24	1 5/8	15/16
x102	30.0	27.09	27 1/8	0.515	1/2	1/4	10.015	10	0.830	13/16	24	1 9/16	15/16
x 94	27.7	26.92	26 7/8	0.490	1/2	1/4	9.990	10	0.745	3/4	24	1 7/16	15/16
x 84	24.8	26.71	26 3/4	0.460	7/16	1/4	9.960	10	0.640	5/8	24	1 3/8	15/16
W 24x162	47.7	25.00	25	0.705	11/16	3/8	12.955	13	1.220	1 1/4	21	2	1 1/16
x146	43.0	24.74	24 3/4	0.650	5/8	5/16	12.900	12 7/8	1.090	1 1/16	21	1 7/8	1 1/16
x131	38.5	24.48	24 1/2	0.605	5/8	5/16	12.855	12 7/8	0.960	15/16	21	1 3/4	1 1/16
x117	34.4	24.26	24 1/4	0.550	9/16	5/16	12.800	12 3/4	0.850	7/8	21	1 5/8	1
x104	30.6	24.06	24	0.500	1/2	1/4	12.750	12 3/4	0.750	3/4	21	1 1/2	1
W 24x 94	27.7	24.31	24 1/4	0.515	1/2	1/4	9.065	9 1/8	0.875	7/8	21	1 5/8	1
x 84	24.7	24.10	24 1/8	0.470	1/2	1/4	9.020	9	0.770	3/4	21	1 9/16	15/16
x 76	22.4	23.92	23 7/8	0.440	7/16	1/4	8.990	9	0.680	11/16	21	1 7/16	15/16
x 68	20.1	23.73	23 3/4	0.415	7/16	1/4	8.965	9	0.585	9/16	21	1 3/8	15/16
W 24x 62	18.2	23.74	23 3/4	0.430	7/16	1/4	7.040	7	0.590	9/16	21	1 3/8	15/16
x 55	16.2	23.57	23 5/8	0.395	3/8	3/16	7.005	7	0.505	1/2	21	1 5/16	15/16
W 21x147	43.2	22.06	22	0.720	3/4	3/8	12.510	12 1/2	1.150	1 1/8	18 1/4	1 7/8	1 1/16
x132	38.8	21.83	21 7/8	0.650	5/8	5/16	12.440	12 1/2	1.035	1 1/16	18 1/4	1 13/16	1
x122	35.9	21.68	21 5/8	0.600	5/8	5/16	12.390	12 3/8	0.960	15/16	18 1/4	1 11/16	1
x111	32.7	21.51	21 1/2	0.550	9/16	5/16	12.340	12 3/8	0.875	7/8	18 1/4	1 5/8	15/16
x101	29.8	21.36	21 3/8	0.500	1/2	1/4	12.290	12 1/4	0.800	13/16	18 1/4	1 9/16	15/16
W 21x 93	27.3	21.62	21 5/8	0.580	9/16	5/16	8.420	8 3/8	0.930	15/16	18 1/4	1 11/16	1
x 83	24.3	21.43	21 3/8	0.515	1/2	1/4	8.355	8 3/8	0.835	13/16	18 1/4	1 9/16	15/16
x 73	21.5	21.24	21 1/4	0.455	7/16	1/4	8.295	8 1/4	0.740	3/4	18 1/4	1 1/2	15/16
x 68	20.0	21.13	21 1/8	0.430	7/16	1/4	8.270	8 1/4	0.685	11/16	18 1/4	1 7/16	7/8
x 62	18.3	20.99	21	0.400	3/8	3/16	8.240	8 1/4	0.615	5/8	18 1/4	1 3/8	7/8
W 21x 57	16.7	21.06	21	0.405	3/8	3/16	6.555	6 1/2	0.650	5/8	18 1/4	1 3/8	7/8
x 50	14.7	20.83	20 7/8	0.380	3/8	3/16	6.530	6 1/2	0.535	9/16	18 1/4	1 5/16	7/8
x 44	13.0	20.66	20 5/8	0.350	3/8	3/16	6.500	6 1/2	0.450	7/16	18 1/4	1 3/16	7/8

American Institute of Steel Construction

APPENDIX V (Continued)

TABLE 28 (CONTINUED)

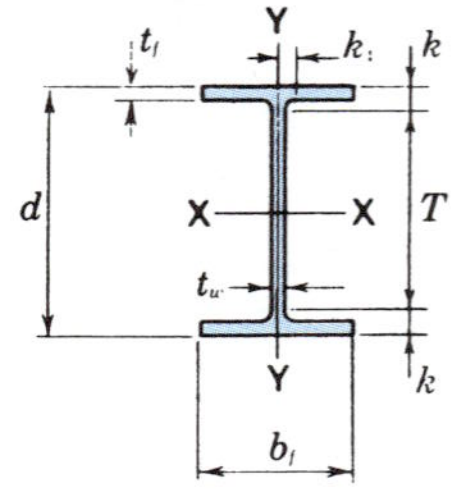

M SHAPES
Dimensions

Designation	Area A	Depth d		Web Thickness t_w		Web $\frac{t_w}{2}$	Flange Width b_f		Flange Thickness t_f		Distance T	Distance k	Grip	Max. Flge. Fas-ten-er
	In.^{2}	In.		In.		In.	In.		In.		In.	In.	In.	In.
M 14x18	5.10	14.00	14	0.215	3/16	1/8	4.000	4	0.270	1/4	12 3/4	5/8	1/4	3/4
M 12x11.8	3.47	12.00	12	0.177	3/16	1/8	3.065	3 1/8	0.225	1/4	10 7/8	9/16	1/4	—
M 10x9	2.65	10.00	10	0.157	3/16	1/8	2.690	2 3/4	0.206	3/16	8 7/8	9/16	3/16	—
M 8x6.5	1.92	8.00	8	0.135	1/8	1/16	2.281	2 1/4	0.189	3/16	7	1/2	3/16	—
M 6x20	5.89	6.00	6	0.250	1/4	1/8	5.938	6	0.379	3/8	4 1/4	7/8	3/8	7/8
M 6x4.4	1.29	6.00	6	0.114	1/8	1/16	1.844	1 7/8	0.171	3/16	5 1/8	7/16	3/16	—
M 5x18.9	5.55	5.00	5	0.316	5/16	3/16	5.003	5	0.416	7/16	3 1/4	7/8	7/16	7/8
M 4x13	3.81	4.00	4	0.254	1/4	1/8	3.940	4	0.371	3/8	2 3/8	13/16	3/8	3/4

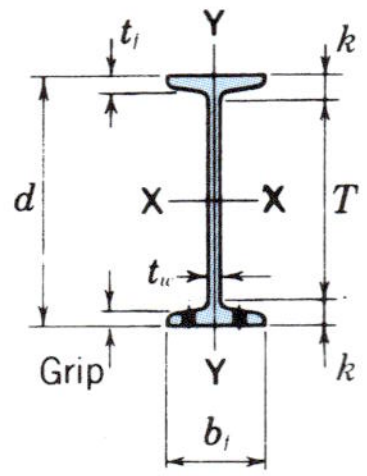

S SHAPES
Dimensions

Designation	Area A	Depth d		Web Thickness t_w		Web $\frac{t_w}{2}$	Flange Width b_f		Flange Thickness t_f		Distance T	Distance k	Grip	Max. Flge. Fas-ten-er
	In.^{2}	In.		In.		In.	In.		In.		In.	In.	In.	In.
S 24x121	35.6	24.50	24 1/2	0.800	13/16	7/16	8.050	8	1.090	1 1/16	20 1/2	2	1 1/8	1
x106	31.2	24.50	24 1/2	0.620	5/8	5/16	7.870	7 7/8	1.090	1 1/16	20 1/2	2	1 1/8	1
S 24x100	29.3	24.00	24	0.745	3/4	3/8	7.245	7 1/4	0.870	7/8	20 1/2	1 3/4	7/8	1
x90	26.5	24.00	24	0.625	5/8	5/16	7.125	7 1/8	0.870	7/8	20 1/2	1 3/4	7/8	1
x80	23.5	24.00	24	0.500	1/2	1/4	7.000	7	0.870	7/8	20 1/2	1 3/4	7/8	1
S 20x96	28.2	20.30	20 1/4	0.800	13/16	7/16	7.200	7 1/4	0.920	15/16	16 3/4	1 3/4	15/16	1
x86	25.3	20.30	20 1/4	0.660	11/16	3/8	7.060	7	0.920	15/16	16 3/4	1 3/4	15/16	1
S 20x75	22.0	20.00	20	0.635	5/8	5/16	6.385	6 3/8	0.795	13/16	16 3/4	1 5/8	13/16	7/8
x66	19.4	20.00	20	0.505	1/2	1/4	6.255	6 1/4	0.795	13/16	16 3/4	1 5/8	13/16	7/8
S 18x70	20.6	18.00	18	0.711	11/16	3/8	6.251	6 1/4	0.691	11/16	15	1 1/2	11/16	7/8
x54.7	16.1	18.00	18	0.461	7/16	1/4	6.001	6	0.691	11/16	15	1 1/2	11/16	7/8
S 15x50	14.7	15.00	15	0.550	9/16	5/16	5.640	5 5/8	0.622	5/8	12 1/4	1 3/8	9/16	3/4
x42.9	12.6	15.00	15	0.411	7/16	1/4	5.501	5 1/2	0.622	5/8	12 1/4	1 3/8	9/16	3/4
S 12x50	14.7	12.00	12	0.687	11/16	3/8	5.477	5 1/2	0.659	11/16	9 1/8	1 7/16	11/16	3/4
x40.8	12.0	12.00	12	0.462	7/16	1/4	5.252	5 1/4	0.659	11/16	9 1/8	1 7/16	5/8	3/4
S 12x35	10.3	12.00	12	0.428	7/16	1/4	5.078	5 1/8	0.544	9/16	9 5/8	1 3/16	1/2	3/4
x31.8	9.35	12.00	12	0.350	3/8	3/16	5.000	5	0.544	9/16	9 5/8	1 3/16	1/2	3/4
S 10x35	10.3	10.00	10	0.594	5/8	5/16	4.944	5	0.491	1/2	7 3/4	1 1/8	1/2	3/4
x25.4	7.46	10.00	10	0.311	5/16	3/16	4.661	4 5/8	0.491	1/2	7 3/4	1 1/8	1/2	3/4
S 8x23	6.77	8.00	8	0.441	7/16	1/4	4.171	4 1/8	0.426	7/16	6	1	7/16	3/4
x18.4	5.41	8.00	8	0.271	1/4	1/8	4.001	4	0.426	7/16	6	1	7/16	3/4
S 7x20	5.88	7.00	7	0.450	7/16	1/4	3.860	3 7/8	0.392	3/8	5 1/8	15/16	3/8	5/8
x15.3	4.50	7.00	7	0.252	1/4	1/8	3.662	3 5/8	0.392	3/8	5 1/8	15/16	3/8	5/8
S 6x17.25	5.07	6.00	6	0.465	7/16	1/4	3.565	3 5/8	0.359	3/8	4 1/4	7/8	3/8	5/8
x12.5	3.67	6.00	6	0.232	1/4	1/8	3.332	3 3/8	0.359	3/8	4 1/4	7/8	3/8	—
S 5x14.75	4.34	5.00	5	0.494	1/2	1/4	3.284	3 1/4	0.326	5/16	3 3/8	13/16	5/16	—
x10	2.94	5.00	5	0.214	3/16	1/8	3.004	3	0.326	5/16	3 3/8	13/16	5/16	—
S 4x9.5	2.79	4.00	4	0.326	5/16	3/16	2.796	2 3/4	0.293	5/16	2 1/2	3/4	5/16	—
x7.7	2.26	4.00	4	0.193	3/16	1/8	2.663	2 5/8	0.293	5/16	2 1/2	3/4	5/16	—
S 3x7.5	2.21	3.00	3	0.349	3/8	3/16	2.509	2 1/2	0.260	1/4	1 5/8	11/16	1/4	—
x5.7	1.67	3.00	3	0.170	3/16	1/8	2.330	2 3/8	0.260	1/4	1 5/8	11/16	1/4	—

APPENDIX V (Continued)

TABLE 28 (CONTINUED)

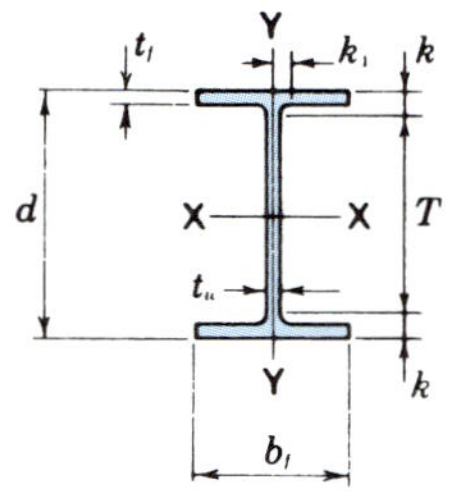

HP SHAPES
Dimensions

Designation	Area A	Depth d		Web: Thickness t_w		Web: $\frac{t_w}{2}$	Flange: Width b_f		Flange: Thickness t_f		Distance: T	Distance: k	Distance: k_1
	In.^{2}	In.		In.		In.	In.		In.		In.	In.	In.
HP 14x117	34.4	14.21	14 1/4	0.805	13/16	7/16	14.885	14 7/8	0.805	13/16	11 1/4	1 1/2	1 1/16
x102	30.0	14.01	14	0.705	11/16	3/8	14.785	14 3/4	0.705	11/16	11 1/4	1 3/8	1
x 89	26.1	13.83	13 7/8	0.615	5/8	5/16	14.695	14 3/4	0.615	5/8	11 1/4	1 5/16	15/16
x 73	21.4	13.61	13 5/8	0.505	1/2	1/4	14.585	14 5/8	0.505	1/2	11 1/4	1 3/16	7/8
HP 13x100	29.4	13.15	13 1/8	0.765	3/4	3/8	13.205	13 1/4	0.765	3/4	10 1/4	1 7/16	1
x 87	25.5	12.95	13	0.665	11/16	3/8	13.105	13 1/8	0.665	11/16	10 1/4	1 3/8	15/16
x 73	21.6	12.75	12 3/4	0.565	9/16	5/16	13.005	13	0.565	9/16	10 1/4	1 1/4	15/16
x 60	17.5	12.54	12 1/2	0.460	7/16	1/4	12.900	12 7/8	0.460	7/16	10 1/4	1 1/8	7/8
HP 12x 84	24.6	12.28	12 1/4	0.685	11/16	3/8	12.295	12 1/4	0.685	11/16	9 1/2	1 3/8	1
x 74	21.8	12.13	12 1/8	0.605	5/8	5/16	12.215	12 1/4	0.610	5/8	9 1/2	1 5/16	15/16
x 63	18.4	11.94	12	0.515	1/2	1/4	12.125	12 1/8	0.515	1/2	9 1/2	1 1/4	7/8
x 53	15.5	11.78	11 3/4	0.435	7/16	1/4	12.045	12	0.435	7/16	9 1/2	1 1/8	7/8
HP 10x 57	16.8	9.99	10	0.565	9/16	5/16	10.225	10 1/4	0.565	9/16	7 5/8	1 3/16	13/16
x 42	12.4	9.70	9 3/4	0.415	7/16	1/4	10.075	10 1/8	0.420	7/16	7 5/8	1 1/16	3/4
HP 8x 36	10.6	8.02	8	0.445	7/16	1/4	8.155	8 1/8	0.445	7/16	6 1/8	15/16	5/8

American Institute of Steel Construction

TABLE 29

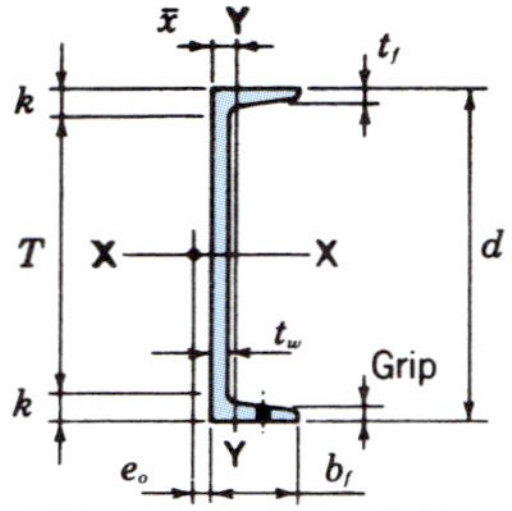

CHANNELS
AMERICAN STANDARD
Dimensions

Designation	Area A	Depth d	Web: Thickness t_w		Web: $\frac{t_w}{2}$	Flange: Width b_f		Flange: Average thickness t_f		Distance: T	Distance: k	Grip	Max. Flge. Fas-ten-er
	In.^{2}	In.	In.		In.	In.		In.		In.	In.	In.	In.
C 15x50	14.7	15.00	0.716	11/16	3/8	3.716	3 3/4	0.650	5/8	12 1/8	1 7/16	5/8	1
x40	11.8	15.00	0.520	1/2	1/4	3.520	3 1/2	0.650	5/8	12 1/8	1 7/16	5/8	1
x33.9	9.96	15.00	0.400	3/8	3/16	3.400	3 3/8	0.650	5/8	12 1/8	1 7/16	5/8	1
C 12x30	8.82	12.00	0.510	1/2	1/4	3.170	3 1/8	0.501	1/2	9 3/4	1 1/8	1/2	7/8
x25	7.35	12.00	0.387	3/8	3/16	3.047	3	0.501	1/2	9 3/4	1 1/8	1/2	7/8
x20.7	6.09	12.00	0.282	5/16	1/8	2.942	3	0.501	1/2	9 3/4	1 1/8	1/2	7/8
C 10x30	8.82	10.00	0.673	11/16	5/16	3.033	3	0.436	7/16	8	1	7/16	3/4
x25	7.35	10.00	0.526	1/2	1/4	2.886	2 7/8	0.436	7/16	8	1	7/16	3/4
x20	5.88	10.00	0.379	3/8	3/16	2.739	2 3/4	0.436	7/16	8	1	7/16	3/4
x15.3	4.49	10.00	0.240	1/4	1/8	2.600	2 5/8	0.436	7/16	8	1	7/16	3/4
C 9x20	5.88	9.00	0.448	7/16	1/4	2.648	2 5/8	0.413	7/16	7 1/8	15/16	7/16	3/4
x15	4.41	9.00	0.285	5/16	1/8	2.485	2 1/2	0.413	7/16	7 1/8	15/16	7/16	3/4
x13.4	3.94	9.00	0.233	1/4	1/8	2.433	2 3/8	0.413	7/16	7 1/8	15/16	7/16	3/4
C 8x18.75	5.51	8.00	0.487	1/2	1/4	2.527	2 1/2	0.390	3/8	6 1/8	15/16	3/8	3/4
x13.75	4.04	8.00	0.303	5/16	1/8	2.343	2 3/8	0.390	3/8	6 1/8	15/16	3/8	3/4
x11.5	3.38	8.00	0.220	1/4	1/8	2.260	2 1/4	0.390	3/8	6 1/8	15/16	3/8	3/4
C 7x14.75	4.33	7.00	0.419	7/16	3/16	2.299	2 1/4	0.366	3/8	5 1/4	7/8	3/8	5/8
x12.25	3.60	7.00	0.314	5/16	3/16	2.194	2 1/4	0.366	3/8	5 1/4	7/8	3/8	5/8
x 9.8	2.87	7.00	0.210	3/16	1/8	2.090	2 1/8	0.366	3/8	5 1/4	7/8	3/8	5/8
C 6x13	3.83	6.00	0.437	7/16	3/16	2.157	2 1/8	0.343	5/16	4 3/8	13/16	5/16	5/8
x10.5	3.09	6.00	0.314	5/16	3/16	2.034	2	0.343	5/16	4 3/8	13/16	3/8	5/8
x 8.2	2.40	6.00	0.200	3/16	1/8	1.920	1 7/8	0.343	5/16	4 3/8	13/16	5/16	5/8
C 5x 9	2.64	5.00	0.325	5/16	3/16	1.885	1 7/8	0.320	5/16	3 1/2	3/4	5/16	5/8
x 6.7	1.97	5.00	0.190	3/16	1/8	1.750	1 3/4	0.320	5/16	3 1/2	3/4	—	—
C 4x 7.25	2.13	4.00	0.321	5/16	3/16	1.721	1 3/4	0.296	5/16	2 5/8	11/16	5/16	5/8
x 5.4	1.59	4.00	0.184	3/16	1/16	1.584	1 5/8	0.296	5/16	2 5/8	11/16	—	—
C 3x 6	1.76	3.00	0.356	3/8	3/16	1.596	1 5/8	0.273	1/4	1 5/8	11/16	—	—
x 5	1.47	3.00	0.258	1/4	1/8	1.498	1 1/2	0.273	1/4	1 5/8	11/16	—	—
x 4.1	1.21	3.00	0.170	3/16	1/16	1.410	1 3/8	0.273	1/4	1 5/8	11/16	—	—

American Institute of Steel Construction

APPENDIX V (Continued)

TABLE 29 (CONTINUED)

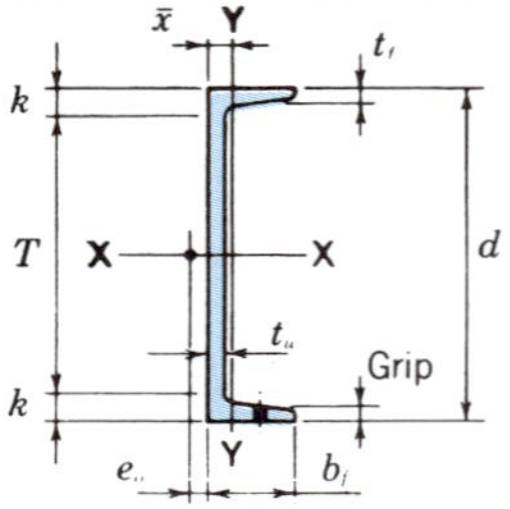

CHANNELS MISCELLANEOUS Dimensions

Designation	Area A	Depth d	Web Thickness t_w		$\frac{t_w}{2}$	Flange Width b_f		Flange Average thickness t_f		Distance T	Distance k	Grip	Max. Flge. Fas-ten-er
	In.^{2}	In.	In.		In.	In.		In.		In.	In.	In.	In.
MC 9x25.4	7.47	9.00	0.450	7/16	1/4	3.500	3 1/2	0.550	9/16	6 5/8	1 3/16	9/16	7/8
x23.9	7.02	9.00	0.400	3/8	3/16	3.450	3 1/2	0.550	9/16	6 5/8	1 3/16	9/16	7/8
MC 8x22.8	6.70	8.00	0.427	7/16	3/16	3.502	3 1/2	0.525	1/2	5 5/8	1 3/16	1/2	7/8
x21.4	6.28	8.00	0.375	3/8	3/16	3.450	3 1/2	0.525	1/2	5 5/8	1 3/16	1/2	7/8
MC 8x20	5.88	8.00	0.400	3/8	3/16	3.025	3	0.500	1/2	5 3/4	1 1/8	1/2	7/8
x18.7	5.50	8.00	0.353	3/8	3/16	2.978	3	0.500	1/2	5 3/4	1 1/8	1/2	7/8
MC 8x 8.5	2.50	8.00	0.179	3/16	1/16	1.874	1 7/8	0.311	5/16	6 1/2	3/4	5/16	5/8
MC 7x22.7	6.67	7.00	0.503	1/2	1/4	3.603	3 5/8	0.500	1/2	4 3/4	1 1/8	1/2	7/8
x19.1	5.61	7.00	0.352	3/8	3/16	3.452	3 1/2	0.500	1/2	4 3/4	1 1/8	1/2	7/8
MC 7x17.6	5.17	7.00	0.375	3/8	3/16	3.000	3	0.475	1/2	4 7/8	1 1/16	1/2	3/4
MC 6x18	5.29	6.00	0.379	3/8	3/16	3.504	3 1/2	0.475	1/2	3 7/8	1 1/16	1/2	7/8
x15.3	4.50	6.00	0.340	5/16	3/16	3.500	3 1/2	0.385	3/8	4 1/4	7/8	3/8	7/8
MC 6x16.3	4.79	6.00	0.375	3/8	3/16	3.000	3	0.475	1/2	3 7/8	1 1/16	1/2	3/4
x15.1	4.44	6.00	0.316	5/16	3/16	2.941	3	0.475	1/2	3 7/8	1 1/16	1/2	3/4
MC 6x12	3.53	6.00	0.310	5/16	1/8	2.497	2 1/2	0.375	3/8	4 3/8	13/16	3/8	5/8

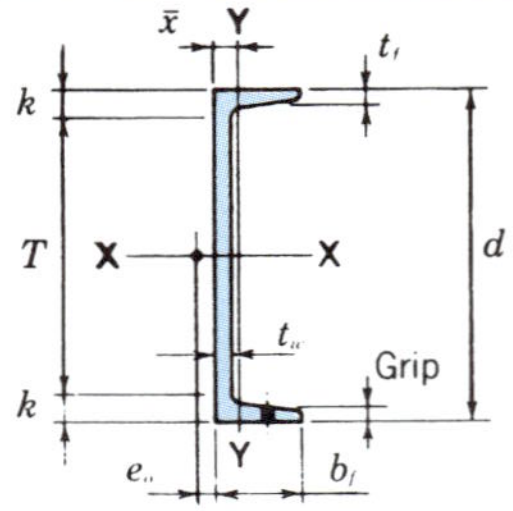

CHANNELS MISCELLANEOUS Dimensions

Designation	Area A	Depth d	Web Thickness t_w		$\frac{t_w}{2}$	Flange Width b_f		Flange Average thickness t_f		Distance T	Distance k	Grip	Max. Flge. Fas-ten-er
	In.^{2}	In.	In.		In.	In.		In.		In.	In.	In.	In.
MC 18x58	17.1	18.00	0.700	11/16	3/8	4.200	4 1/4	0.625	5/8	15 1/4	1 3/8	5/8	1
x51.9	15.3	18.00	0.600	5/8	5/16	4.100	4 1/8	0.625	5/8	15 1/4	1 3/8	5/8	1
x45.8	13.5	18.00	0.500	1/2	1/4	4.000	4	0.625	5/8	15 1/4	1 3/8	5/8	1
x42.7	12.6	18.00	0.450	7/16	1/4	3.950	4	0.625	5/8	15 1/4	1 3/8	5/8	1
MC 13x50	14.7	13.00	0.787	13/16	3/8	4.412	4 3/8	0.610	5/8	10 1/4	1 3/8	5/8	1
x40	11.8	13.00	0.560	9/16	1/4	4.185	4 1/8	0.610	5/8	10 1/4	1 3/8	9/16	1
x35	10.3	13.00	0.447	7/16	1/4	4.072	4 1/8	0.610	5/8	10 1/4	1 3/8	9/16	1
x31.8	9.35	13.00	0.375	3/8	3/16	4.000	4	0.610	5/8	10 1/4	1 3/8	9/16	1
MC 12x50	14.7	12.00	0.835	13/16	7/16	4.135	4 1/8	0.700	11/16	9 3/8	1 5/16	11/16	1
x45	13.2	12.00	0.712	11/16	3/8	4.012	4	0.700	11/16	9 3/8	1 5/16	11/16	1
x40	11.8	12.00	0.590	9/16	5/16	3.890	3 7/8	0.700	11/16	9 3/8	1 5/16	11/16	1
x35	10.3	12.00	0.467	7/16	1/4	3.767	3 3/4	0.700	11/16	9 3/8	1 5/16	11/16	1
MC 12x37	10.9	12.00	0.600	5/8	5/16	3.600	3 5/8	0.600	5/8	9 3/8	1 5/16	5/8	7/8
x32.9	9.67	12.00	0.500	1/2	1/4	3.500	3 1/2	0.600	5/8	9 3/8	1 5/16	9/16	7/8
x30.9	9.07	12.00	0.450	7/16	1/4	3.450	3 1/2	0.600	5/8	9 3/8	1 5/16	9/16	7/8
MC 12x10.6	3.10	12.00	0.190	3/16	1/8	1.500	1 1/2	0.309	5/16	10 5/8	11/16	—	—
MC 10x41.1	12.1	10.00	0.796	13/16	3/8	4.321	4 3/8	0.575	9/16	7 1/2	1 1/4	9/16	7/8
x33.6	9.87	10.00	0.575	9/16	5/16	4.100	4 1/8	0.575	9/16	7 1/2	1 1/4	9/16	7/8
x28.5	8.37	10.00	0.425	7/16	3/16	3.950	4	0.575	9/16	7 1/2	1 1/4	9/16	7/8
MC 10x28.3	8.32	10.00	0.477	1/2	1/4	3.502	3 1/2	0.575	9/16	7 1/2	1 1/4	9/16	7/8
x25.3	7.43	10.00	0.425	7/16	3/16	3.550	3 1/2	0.500	1/2	7 3/4	1 1/8	1/2	7/8
x24.9	7.32	10.00	0.377	3/8	3/16	3.402	3 3/8	0.575	9/16	7 1/2	1 1/4	9/16	7/8
x21.9	6.43	10.00	0.325	5/16	3/16	3.450	3 1/2	0.500	1/2	7 3/4	1 1/8	1/2	7/8
MC 10x 8.4	2.46	10.00	0.170	3/16	1/16	1.500	1 1/2	0.280	1/4	8 5/8	11/16	—	—
MC 10x 6.5	1.91	10.00	0.152	1/8	1/16	1.127	1 1/8	0.202	3/16	9 1/8	7/16	—	—

APPENDIX W
CORROSION-RESISTANT PIPE FITTINGS

TABLE 30 WELDING FITTINGS—DIMENSIONS

90° LONG RAD. WeldELL | 90° REDUCING L.R. WeldELL | 45° LONG RAD. WeldELL | 180° LONG RADIUS WeldELL | 90° SHORT RAD. WeldELL | 180° SHORT RAD. WeldELL | CAP | LAP JOINT | STUB ENDS

Nom. Pipe Size	Pipe O.D.	WeldELL A	B	D	K	V	CAPS E	G O.D. of Lap	Type A F (Length) ANSI Std.	Type A I (Length) MSS Std.	Type A Corner Radius	Type B,C▲ I (Length) MSS Std.	Type B,C▲ Corner Radius	Nom. Pipe Size
1/2	.840	1 1/2	5/8	–	1 7/8	–	1	1 3/8	3	2	1/8	2	1/32	1/2
3/4	1.050	1 1/8	–	–	1 11/16	–	1	1 11/16	3	2	1/8	2	1/32	3/4
1	1.315	1 1/2	7/8	1	2 3/16	1 5/8	1 1/2	2	4	2	1/8	2	1/32	1
1 1/4	1.660	1 7/8	1	1 1/4	2 3/4	2 1/16	1 1/2	2 1/2	4	2	3/16	2	1/32	1 1/4
1 1/2	1.900	2 1/4	1 1/8	1 1/2	3 1/4	2 7/16	1 1/2	2 7/8	4	2	1/4	2	1/32	1 1/2
2	2.375	3	1 3/8	2	4 3/16	3 3/16	1 1/2	3 5/8	6	2 1/2	5/16	2 1/2	1/32	2
2 1/2	2.875	3 3/4	1 3/4	2 1/2	5 3/16	3 15/16	1 1/2	4 1/8	6	2 1/2	5/16	2 1/2	1/32	2 1/2
3	3.500	4 1/2	2	3	6 1/4	4 3/4	2	5	6	2 1/2	3/8	2 1/2	1/32	3
3 1/2	4.000	5 1/4	2 1/4	3 1/2	7 1/4	5 1/2	2 1/2	5 1/2	6	3	3/8	3	1/32	3 1/2
4	4.500	6	2 1/2	4	8 1/4	6 1/4	2 1/2	6 3/16	6	3	7/16	3	1/32	4
5	5.563	7 1/2	3 1/8	5	10 5/16	7 3/4	3	7 5/16	8	3	7/16	3	1/16	5
6	6.625	9	3 3/4	6	12 5/16	9 5/16	3 1/2	8 1/2	8	3 1/2	1/2	3 1/2	1/16	6
8	8.625	12	5	8	16 5/16	12 5/16	4	10 5/8	8	4	1/2	4	1/16	8
10	10.750	15	6 1/4	10	20 3/8	15 3/8	5	12 3/4	10	5	1/2	5	1/16	10
12	12.750	18	7 1/2	12	24 3/8	18 3/8	6	15	10	6	1/2	6	1/16	12
14	14.000	21	8 3/4	14	28	21	6 1/2	16 1/4	12	–	1/2	–	–	14
16	16.000	24	10	16	32	24	7	18 1/2	12	–	1/2	–	–	16
18	18.000	27	11 1/4	18	36	27	8	21	12	–	1/2	–	–	18
20	20.000	30	12 1/2	20	40	30	9	23	12	–	1/2	–	–	20
24	24.000	36	15	24	48	26	10 1/2	27 1/2	12	–	1/2	–	–	24
30	30.000	45	18 1/2	30	60	45	10 1/2	–	–	–	1/2	–	–	30

STRAIGHT TEE | REDUCING TEE | CONCENTRIC REDUCER | ECCENTRIC REDUCER

Nom. Pipe Size	Outlet	C	M	H	Nom. Pipe Size	Outlet	C	M	H	Nom. Pipe Size	Outlet	C	M	H	Nom. Pipe Size	Outlet	C	M	H
3/4	3/4	1 1/8	...	1 1/8		3 1/2	3 3/4	...	...	10	10	8 1/2	...	...	20	20	15	...	20
	1/2	1 1/8	...	1 1/2	3 1/2	3	3 3/4	3 5/8	4		8	8 1/2	8	7		18	15	14 1/2	20
1	1	1 1/2	...	...		2 1/2	3 3/4	3 3/4	4		6	8 1/2	7 5/8	7		16	15	14	20
	3/4	1 1/2	1 1/2	2		2	3 3/4	3 1/4	4		5	8 1/2	7 1/2	7		14	15	14	20
	1/2	1 1/2	1 1/2	2		1 1/2	3 3/4	3 1/8	4		4	8 1/2	7 1/4	7		12	15	13 5/8	20
1 1/4	1 1/4	1 7/8	...	...		4	4 1/8	...	...	12	12	10	...	...		10	15	13 1/8	20
	1	1 7/8	1 7/8	2		3 1/2	4 1/8	...	4		10	10	9 1/2	8		8	15	12 3/4	20
	3/4	1 7/8	1 7/8	2	4	3	4 1/8	3 7/8	4		8	10	9	8	24	24	17	...	...
	1/2	1 7/8	1 7/8	2		2 1/2	4 1/8	3 3/4	4		6	10	8 5/8	8		20	17	17	20
1 1/2	1 1/2	2 1/4	...	...		2	4 1/8	3 1/2	4		5	10	8 1/2	8		18	17	16 1/2	20
	1 1/4	2 1/4	2 1/4	2 1/2		1 1/2	4 1/8	3 3/8	4	14	14	11	...	...		16	17	16	20
	1	2 1/4	2 1/4	2 1/2		5	4 7/8	...	...		12	11	10 5/8	13		14	17	16	20
	3/4	2 1/4	2 1/4	2 1/2		4	4 7/8	4 5/8	5		10	11	10 1/8	13		12	17	15 5/8	20
	1/2	2 1/4	2 1/4	2 1/2	5	3 1/2	4 7/8	4 1/2	5		8	11	9 3/4	13		10	17	15 1/8	20
2	2	2 1/2	...	...		3	4 7/8	4 3/8	5		6	11	9 3/8	13	30	30	22	...	...
	1 1/2	2 1/2	2 3/8	3		2 1/2	4 7/8	4 1/4	5	16	16	12	...	...		24	22	21	24
	1 1/4	2 1/2	2 1/4	3		2	4 7/8	4 1/8	5		14	12	12	14		20	22	20	24
	1	2 1/2	2	3		6	5 5/8	...	...		12	12	11 5/8	14		18	22	19 1/2	24
	3/4	2 1/2	1 3/4	3		5	5 5/8	5 3/8	5 1/2		10	12	11 1/8	14		16	22	19	24
2 1/2	2 1/2	3	...	...		4	5 5/8	5 1/8	5 1/2		8	12	10 3/4	14		14	22	19	24
	2	3	2 3/4	3 1/2	6	3 1/2	5 5/8	5	5 1/2		6	12	10 3/8	14	36	36	26 1/2	...	...
	1 1/2	3	2 5/8	3 1/2		3	5 5/8	4 7/8	5 1/2	18	18	13 1/2	...	...		30	26 1/2	25	24
	1 1/4	3	2 1/2	3 1/2		2 1/2	5 5/8	4 3/4	5 1/2		16	13 1/2	13	15		24	26 1/2	24	24
	1	3	2 1/4	3 1/2		8	7	...	...		14	13 1/2	13	15		20	26 1/2	23	24
3	3	3 3/8	...	...		6	7	6 5/8	6		12	13 1/2	12 5/8	15		18	26 1/2	22 1/2	24
	2 1/2	3 3/8	3 1/4	3 1/2	8	5	7	6 3/8	6		10	13 1/2	12 1/8	15		16	26 1/2	22	24
	2	3 3/8	3	3 1/2		4	7	6 1/8	6		8	13 1/2	11 3/4	15	42	42	30	...	...
	1 1/2	3 3/8	2 7/8	3 1/2		3 1/2	7	6	6							36	30	28	24
	1 1/4	3 3/8	2 3/4	3 1/2												30	30	28	24
																24	30	26	24
																20	30	26	24

NOTES:
Fittings having wall thicknesses of 0.065" are furnished with square ends unless otherwise specified.

STUB ENDS:
Type A for use with Lap Joint Flanges.
Types B and C for use with Slip-On-Flanges.

Types A and B are furnished with square inside corner, gasket surfaces machined and minimum lap thickness equal to nominal wall of barrel.

▲ Dimensions of Type C are same as tabulated for Type B in 5S and 10S thicknesses. Type C is not available in 40S thicknesses. Type C has no fixed corner radius, lap face is not machined. Type C is only available in MSS length. Type B, usually purchased in the MSS length is also available in the ANSI length in Schedules 10S and 40S.

APPENDIX W (Continued)

TABLE 30 (CONTINUED)

FORGED FLANGES – DIMENSIONS

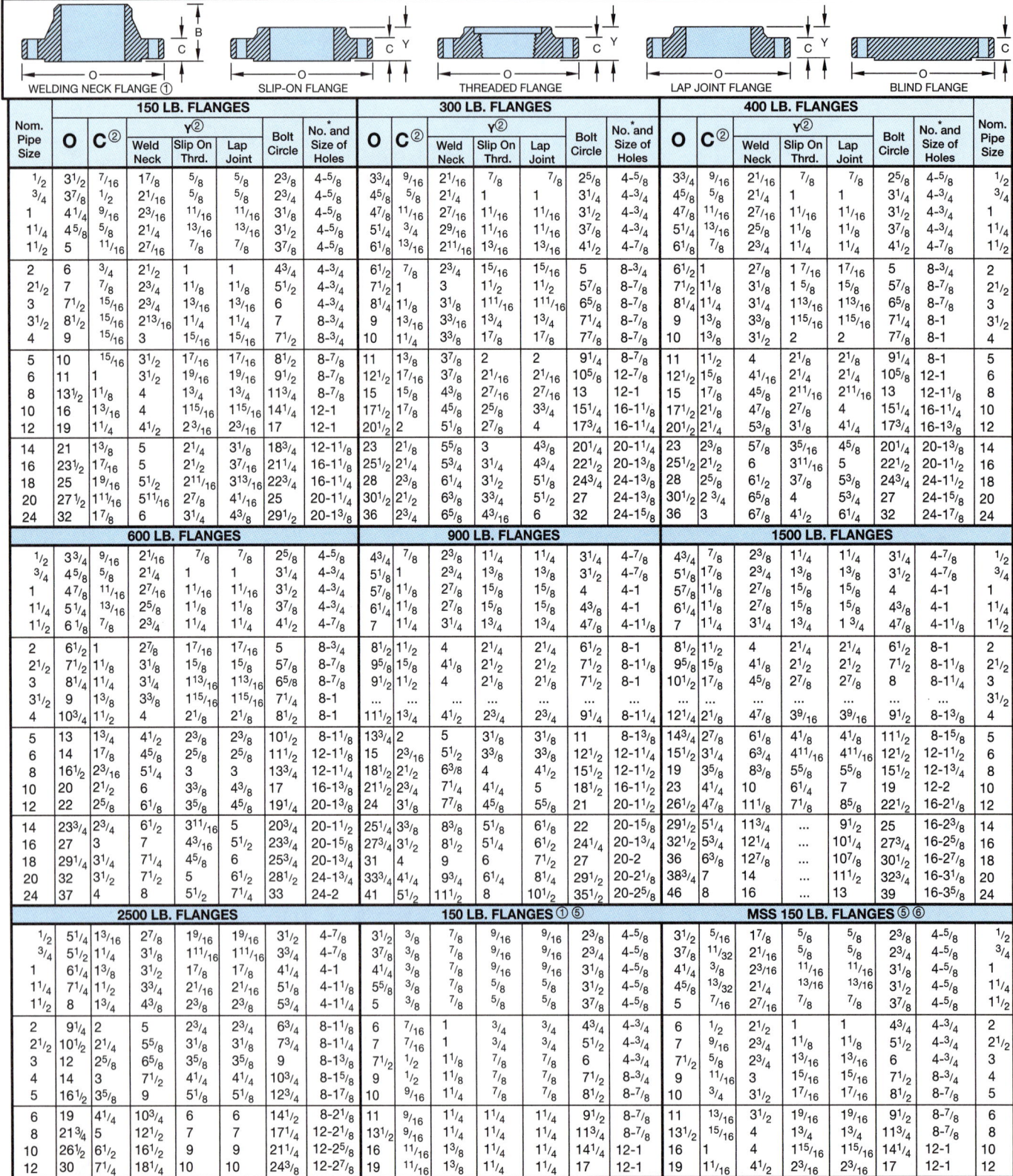

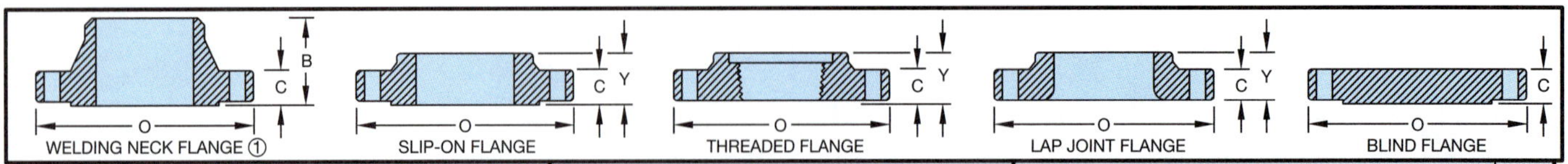
B
C
O
WELDING NECK FLANGE ①
C
Y
O
SLIP-ON FLANGE
C
Y
O
THREADED FLANGE
C
Y
O
LAP JOINT FLANGE
C
O
BLIND FLANGE

150 LB. FLANGES

Nom. Pipe Size	O	C②	Y② Weld Neck	Y② Slip On Thrd.	Y② Lap Joint	Bolt Circle	No.* and Size of Holes
1/2	3 1/2	7/16	1 7/8	5/8	5/8	2 3/8	4-5/8
3/4	3 7/8	1/2	2 1/16	5/8	5/8	2 3/4	4-5/8
1	4 1/4	9/16	2 3/16	11/16	11/16	3 1/8	4-5/8
1 1/4	4 5/8	5/8	2 1/4	13/16	13/16	3 1/2	4-5/8
1 1/2	5	11/16	2 7/16	7/8	7/8	3 7/8	4-5/8
2	6	3/4	2 1/2	1	1	4 3/4	4-3/4
2 1/2	7	7/8	2 3/4	1 1/8	1 1/8	5 1/2	4-3/4
3	7 1/2	15/16	2 3/4	1 3/16	1 3/16	6	4-3/4
3 1/2	8 1/2	15/16	2 13/16	1 1/4	1 1/4	7	8-3/4
4	9	15/16	3	1 5/16	1 5/16	7 1/2	8-3/4
5	10	15/16	3 1/2	1 7/16	1 7/16	8 1/2	8-7/8
6	11	1	3 1/2	1 9/16	1 9/16	9 1/2	8-7/8
8	13 1/2	1 1/8	4	1 3/4	1 3/4	11 3/4	8-7/8
10	16	1 3/16	4	1 15/16	1 15/16	14 1/4	12-1
12	19	1 1/4	4 1/2	2 3/16	2 3/16	17	12-1
14	21	1 3/8	5	2 1/4	3 1/8	18 3/4	12-1 1/8
16	23 1/2	1 7/16	5	2 1/2	3 7/16	21 1/4	16-1 1/8
18	25	1 9/16	5 1/2	2 11/16	3 13/16	22 3/4	16-1 1/4
20	27 1/2	1 11/16	5 11/16	2 7/8	4 1/16	25	20-1 1/4
24	32	1 7/8	6	3 1/4	4 3/8	29 1/2	20-1 3/8

300 LB. FLANGES

Nom. Pipe Size	O	C②	Y② Weld Neck	Y② Slip On Thrd.	Y② Lap Joint	Bolt Circle	No.* and Size of Holes
1/2	3 3/4	9/16	2 1/16	7/8	7/8	2 5/8	4-5/8
3/4	4 5/8	5/8	2 1/4	1	1	3 1/4	4-3/4
1	4 7/8	11/16	2 7/16	1 1/16	1 1/16	3 1/2	4-3/4
1 1/4	5 1/4	3/4	2 9/16	1 1/16	1 1/16	3 7/8	4-3/4
1 1/2	6 1/8	13/16	2 11/16	1 3/16	1 3/16	4 1/2	4-7/8
2	6 1/2	7/8	2 3/4	1 5/16	1 5/16	5	8-3/4
2 1/2	7 1/2	1	3	1 1/2	1 1/2	5 7/8	8-7/8
3	8 1/4	1 1/8	3 1/8	1 11/16	1 11/16	6 5/8	8-7/8
3 1/2	9	1 3/16	3 3/16	1 3/4	1 3/4	7 1/4	8-7/8
4	10	1 1/4	3 3/8	1 7/8	1 7/8	7 7/8	8-7/8
5	11	1 3/8	3 7/8	2	2	9 1/4	8-7/8
6	12 1/2	1 7/16	3 7/8	2 1/16	2 1/16	10 5/8	12-7/8
8	15	1 5/8	4 3/8	2 7/16	2 7/16	13	12-1
10	17 1/2	1 7/8	4 5/8	2 5/8	3 3/4	15 1/4	16-1 1/8
12	20 1/2	2	5 1/8	2 7/8	4	17 3/4	16-1 1/4
14	23	2 1/8	5 5/8	3	4 3/8	20 1/4	20-1 1/8
16	25 1/2	2 1/4	5 3/4	3 1/4	4 3/4	22 1/2	20-1 3/8
18	28	2 3/8	6 1/4	3 1/2	5 1/8	24 3/4	24-1 3/8
20	30 1/2	2 1/2	6 3/8	3 3/4	5 1/2	27	24-1 3/8
24	36	2 3/4	6 5/8	4 3/16	6	32	24-1 5/8

400 LB. FLANGES

Nom. Pipe Size	O	C②	Y② Weld Neck	Y② Slip On Thrd.	Y② Lap Joint	Bolt Circle	No.* and Size of Holes
1/2	3 3/4	9/16	2 1/16	7/8	7/8	2 5/8	4-5/8
3/4	4 5/8	5/8	2 1/4	1	1	3 1/4	4-3/4
1	4 7/8	11/16	2 7/16	1 1/16	1 1/16	3 1/2	4-3/4
1 1/4	5 1/4	13/16	2 5/8	1 1/8	1 1/8	3 7/8	4-3/4
1 1/2	6 1/8	7/8	2 3/4	1 1/4	1 1/4	4 1/2	4-7/8
2	6 1/2	1	2 7/8	1 7/16	1 7/16	5	8-3/4
2 1/2	7 1/2	1 1/8	3 1/8	1 5/8	1 5/8	5 7/8	8-7/8
3	8 1/4	1 1/4	3 1/4	1 13/16	1 13/16	6 5/8	8-7/8
3 1/2	9	1 3/8	3 3/8	1 15/16	1 15/16	7 1/4	8-1
4	10	1 3/8	3 1/2	2	2	7 7/8	8-1
5	11	1 1/2	4	2 1/8	2 1/8	9 1/4	8-1
6	12 1/2	1 5/8	4 1/16	2 1/4	2 1/4	10 5/8	12-1
8	15	1 7/8	4 5/8	2 11/16	2 11/16	13	12-1 1/8
10	17 1/2	2 1/8	4 7/8	2 7/8	4	15 1/4	16-1 1/8
12	20 1/2	2 1/4	5 3/8	3 1/8	4 1/4	17 3/4	16-1 3/8
14	23	2 3/8	5 7/8	3 5/16	4 5/8	20 1/4	20-1 3/8
16	25 1/2	2 1/2	6	3 11/16	5	22 1/2	20-1 1/2
18	28	2 5/8	6 1/2	3 7/8	5 3/8	24 3/4	24-1 1/2
20	30 1/2	2 3/4	6 5/8	4	5 3/4	27	24-1 5/8
24	36	3	6 7/8	4 1/2	6 1/4	32	24-1 7/8

600 LB. FLANGES

Nom. Pipe Size	O	C②	Y② Weld Neck	Y② Slip On Thrd.	Y② Lap Joint	Bolt Circle	No.* and Size of Holes
1/2	3 3/4	9/16	2 1/16	7/8	7/8	2 5/8	4-5/8
3/4	4 5/8	5/8	2 1/4	1	1	3 1/4	4-3/4
1	4 7/8	11/16	2 7/16	1 1/16	1 1/16	3 1/2	4-3/4
1 1/4	5 1/4	13/16	2 5/8	1 1/8	1 1/8	3 7/8	4-3/4
1 1/2	6 1/8	7/8	2 3/4	1 1/4	1 1/4	4 1/2	4-7/8
2	6 1/2	1	2 7/8	1 7/16	1 7/16	5	8-3/4
2 1/2	7 1/2	1 1/8	3 1/8	1 5/8	1 5/8	5 7/8	8-7/8
3	8 1/4	1 1/4	3 1/4	1 13/16	1 13/16	6 5/8	8-7/8
3 1/2	9	1 3/8	3 3/8	1 15/16	1 15/16	7 1/4	8-1
4	10 3/4	1 1/2	4	2 1/8	2 1/8	8 1/2	8-1
5	13	1 3/4	4 1/2	2 3/8	2 3/8	10 1/2	8-1 1/8
6	14	1 7/8	4 5/8	2 5/8	2 5/8	11 1/2	12-1 1/8
8	16 1/2	2 3/16	5 1/4	3	3	13 3/4	12-1 1/4
10	20	2 1/2	6	3 3/8	4 3/8	17	16-1 3/8
12	22	2 5/8	6 1/8	3 5/8	4 5/8	19 1/4	20-1 3/8
14	23 3/4	2 3/4	6 1/2	3 11/16	5	20 3/4	20-1 1/2
16	27	3	7	4 3/16	5 1/2	23 3/4	20-1 5/8
18	29 1/4	3 1/4	7 1/4	4 5/8	6	25 3/4	20-1 3/4
20	32	3 1/2	7 1/2	5	6 1/2	28 1/2	24-1 3/4
24	37	4	8	5 1/2	7 1/4	33	24-2

900 LB. FLANGES

Nom. Pipe Size	O	C②	Y② Weld Neck	Y② Slip On Thrd.	Y② Lap Joint	Bolt Circle	No.* and Size of Holes
1/2	4 3/4	7/8	2 3/8	1 1/4	1 1/4	3 1/4	4-7/8
3/4	5 1/8	1	2 3/4	1 3/8	1 3/8	3 1/2	4-7/8
1	5 7/8	1 1/8	2 7/8	1 5/8	1 5/8	4	4-1
1 1/4	6 1/4	1 1/8	2 7/8	1 5/8	1 5/8	4 3/8	4-1
1 1/2	7	1 1/4	3 1/4	1 3/4	1 3/4	4 7/8	4-1 1/8
2	8 1/2	1 1/2	4	2 1/4	2 1/4	6 1/2	8-1
2 1/2	9 5/8	1 5/8	4 1/8	2 1/2	2 1/2	7 1/2	8-1 1/8
3	9 1/2	1 1/2	4	2 1/8	2 1/8	7 1/2	8-1
3 1/2	...	...	...	...	...	...	...
4	11 1/2	1 3/4	4 1/2	2 3/4	2 3/4	9 1/4	8-1 1/4
5	13 3/4	2	5	3 1/8	3 1/8	11	8-1 3/8
6	15	2 3/16	5 1/2	3 3/8	3 3/8	12 1/2	12-1 1/4
8	18 1/2	2 1/2	6 3/8	4	4 1/2	15 1/2	12-1 1/2
10	21 1/2	2 3/4	7 1/4	4 1/4	5	18 1/2	16-1 1/2
12	24	3 1/8	7 7/8	4 5/8	5 5/8	21	20-1 1/2
14	25 1/4	3 3/8	8 3/8	5 1/8	6 1/8	22	20-1 5/8
16	27 3/4	3 1/2	8 1/2	5 1/4	6 1/2	24 1/4	20-1 3/4
18	31	4	9	6	7 1/2	27	20-2
20	33 3/4	4 1/4	9 3/4	6 1/4	8 1/4	29 1/2	20-2 1/8
24	41	5 1/2	11 1/2	8	10 1/2	35 1/2	20-2 5/8

1500 LB. FLANGES

Nom. Pipe Size	O	C②	Y② Weld Neck	Y② Slip On Thrd.	Y② Lap Joint	Bolt Circle	No.* and Size of Holes
1/2	4 3/4	7/8	2 3/8	1 1/4	1 1/4	3 1/4	4-7/8
3/4	5 1/8	1	2 3/4	1 3/8	1 3/8	3 1/2	4-7/8
1	5 7/8	1 1/8	2 7/8	1 5/8	1 5/8	4	4-1
1 1/4	6 1/4	1 1/8	2 7/8	1 5/8	1 5/8	4 3/8	4-1
1 1/2	7	1 1/4	3 1/4	1 3/4	1 3/4	4 7/8	4-1 1/8
2	8 1/2	1 1/2	4	2 1/4	2 1/4	6 1/2	8-1
2 1/2	9 5/8	1 5/8	4 1/8	2 1/2	2 1/2	7 1/2	8-1 1/8
3	10 1/2	1 7/8	4 5/8	2 7/8	2 7/8	8	8-1 1/4
3 1/2	...	...	...	...	...	...	...
4	12 1/4	2 1/8	4 7/8	3 9/16	3 9/16	9 1/2	8-1 3/8
5	14 3/4	2 7/8	6 1/8	4 1/8	4 1/8	11 1/2	8-1 5/8
6	15 1/2	3 1/4	6 3/4	4 11/16	4 11/16	12 1/2	12-1 1/2
8	19	3 5/8	8 3/8	5 5/8	5 5/8	15 1/2	12-1 3/4
10	23	4 1/4	10	6 1/4	7	19	12-2
12	26 1/2	4 7/8	11 1/8	7 1/8	8 5/8	22 1/2	16-2 1/8
14	29 1/2	5 1/4	11 3/4	...	9 1/2	25	16-2 3/8
16	32 1/2	5 3/4	12 1/4	...	10 1/4	27 3/4	16-2 5/8
18	36	6 3/8	12 7/8	...	10 7/8	30 1/2	16-2 7/8
20	38 3/4	7	14	...	11 1/2	32 3/4	16-3 1/8
24	46	8	16	...	13	39	16-3 5/8

2500 LB. FLANGES

Nom. Pipe Size	O	C②	Y② Weld Neck	Y② Slip On Thrd.	Y② Lap Joint	Bolt Circle	No.* and Size of Holes
1/2	5 1/4	1 3/16	2 7/8	1 9/16	1 9/16	3 1/2	4-7/8
3/4	5 1/2	1 1/4	3 1/8	1 11/16	1 11/16	3 3/4	4-7/8
1	6 1/4	1 3/8	3 1/2	1 7/8	1 7/8	4 1/4	4-1
1 1/4	7 1/4	1 1/2	3 3/4	2 1/16	2 1/16	5 1/8	4-1 1/8
1 1/2	8	1 3/4	4 3/8	2 3/8	2 3/8	5 3/4	4-1 1/4
2	9 1/4	2	5	2 3/4	2 3/4	6 3/4	8-1 1/8
2 1/2	10 1/2	2 1/4	5 5/8	3 1/8	3 1/8	7 3/4	8-1 1/4
3	12	2 5/8	6 5/8	3 5/8	3 5/8	9	8-1 3/8
4	14	3	7 1/2	4 1/4	4 1/4	10 3/4	8-1 5/8
5	16 1/2	3 5/8	9	5 1/8	5 1/8	12 3/4	8-1 7/8
6	19	4 1/4	10 3/4	6	6	14 1/2	8-2 1/8
8	21 3/4	5	12 1/2	7	7	17 1/4	12-2 1/8
10	26 1/2	6 1/2	16 1/2	9	9	21 1/4	12-2 5/8
12	30	7 1/4	18 1/4	10	10	24 3/8	12-2 7/8

150 LB. FLANGES ① ⑤

Nom. Pipe Size	O	C②	Y② Weld Neck	Y② Slip On Thrd.	Y② Lap Joint	Bolt Circle	No.* and Size of Holes
1/2	3 1/2	3/8	7/8	9/16	9/16	2 3/8	4-5/8
3/4	3 7/8	3/8	7/8	9/16	9/16	2 3/4	4-5/8
1	4 1/4	3/8	7/8	9/16	9/16	3 1/8	4-5/8
1 1/4	5 5/8	3/8	7/8	5/8	5/8	3 1/2	4-5/8
1 1/2	5	3/8	7/8	5/8	5/8	3 7/8	4-5/8
2	6	7/16	1	3/4	3/4	4 3/4	4-3/4
2 1/2	7	7/16	1	3/4	3/4	5 1/2	4-3/4
3	7 1/2	1/2	1 1/8	7/8	7/8	6	4-3/4
4	9	1/2	1 1/8	7/8	7/8	7 1/2	8-3/4
5	10	9/16	1 1/4	7/8	7/8	8 1/2	8-7/8
6	11	9/16	1 1/4	1 1/4	1 1/4	9 1/2	8-7/8
8	13 1/2	9/16	1 1/4	1 1/4	1 1/4	11 3/4	8-7/8
10	16	11/16	1 3/8	1 1/4	1 1/4	14 1/4	12-1
12	19	11/16	1 3/8	1 1/4	1 1/4	17	12-1

MSS 150 LB. FLANGES ⑤ ⑥

Nom. Pipe Size	O	C②	Y② Weld Neck	Y② Slip On Thrd.	Y② Lap Joint	Bolt Circle	No.* and Size of Holes
1/2	3 1/2	5/16	1 7/8	5/8	5/8	2 3/8	4-5/8
3/4	3 7/8	11/32	2 1/16	5/8	5/8	2 3/4	4-5/8
1	4 1/4	3/8	2 3/16	11/16	11/16	3 1/8	4-5/8
1 1/4	4 5/8	13/32	2 1/4	13/16	13/16	3 1/2	4-5/8
1 1/2	5	7/16	2 7/16	7/8	7/8	3 7/8	4-5/8
2	6	1/2	2 1/2	1	1	4 3/4	4-3/4
2 1/2	7	9/16	2 3/4	1 1/8	1 1/8	5 1/2	4-3/4
3	7 1/2	5/8	2 3/4	1 3/16	1 3/16	6	4-3/4
4	9	11/16	3	1 5/16	1 5/16	7 1/2	8-3/4
5	10	3/4	3 1/2	1 7/16	1 7/16	8 1/2	8-7/8
6	11	13/16	3 1/2	1 9/16	1 9/16	9 1/2	8-7/8
8	13 1/2	15/16	4	1 3/4	1 3/4	11 3/4	8-7/8
10	16	1	4	1 15/16	1 15/16	14 1/4	12-1
12	19	1 1/16	4 1/2	2 3/16	2 3/16	17	12-1

NOTES:
1. Always specify bore when ordering.
2. Includes 1/16" raised face in 150 lb. and 300 lb. Standard. Does not include 1/4" raised face in 400 lb. and heavier standards.
3. Other types, sizes, and facings on application.
4. For low pressure service up to 150 PSI at 500°F, 225 PSI at 150°F.
5. Drilling and OD match ANSI B16.5 150 lb. steel flange standard, MSS SP-42 150 lb. Corrosion resistant value standards and ANSI B16. 1 125 lb. cast iron flange standard. This class flange has flat face.
6. Thicknesses conform to MSS Standards.

*Bolt holes are 1/8" larger than recommended bolt.

APPENDIX W (Continued)

TABLE 31 THREADED FITTINGS AND THREADED COUPLINGS, REDUCERS, AND CAPS

Threaded Fittings—Class 2000, 3000 and 6000
DIMENSIONS (Inches)

	NOMINAL PIPE SIZE	1/8	1/4	3/8	1/2	3/4	1	1 1/4	1 1/2	2	2 1/2	3	4
Class 2000	A		7/8	31/32	1 1/8	1 5/16	1 1/2	1 3/4	2	2 3/8	3	3 3/8	4 3/16
	B		29/32	1 1/16	1 5/16	1 9/16	1 27/32	2 7/32	2 1/2	3 1/32	3 11/16	4 5/16	5 3/4
	F		3/4	3/4	1	1 1/8	1 1/4	1 5/16	1 3/8	1 11/16	2 1/16	2 1/2	3 1/8
Class 3000	A	7/8	31/32	1 1/8	1 5/16	1 1/2	1 3/4	2	2 3/8	2 1/2	3 1/4	3 3/4	4 1/2
	B	29/32	1 1/16	1 5/16	1 9/16	1 27/32	2 7/32	2 1/2	3 1/32	3 11/32	4	4 3/4	6
	F	3/4	3/4	1	1 1/8	1 1/4	1 5/16	1 3/8	1 11/16	1 3/4		2 1/2	3 1/8
	H	1 1/4	1 1/4	1 1/2	1 5/8	1 7/8	2 1/4	2 5/8	2 15/16	3 5/16			
	J	7/8	7/8	1	1 1/8	1 3/8	1 3/4	2	2 1/8	2 1/2			
	K		1 7/8	2 1/8	2 9/16	3	3 1/2	3 15/16	4 1/4	5			
	L		2 11/16	3	3 9/16	4 1/8	4 13/16	5 3/8	6 7/16	6 5/8			
Class 6000	A	31/32	1 1/8	1 5/16	1 1/2	1 3/4	2	2 3/8	2 1/2	3 1/4	3 3/4	4 3/16	4 1/2
	B	1 1/16	1 5/16	1 9/16	1 27/32	2 7/32	2 1/2	3 1/32	3 11/32	4	4 3/4	5 3/4	6
	F	3/4	1	1 1/8	1 1/4	1 5/16	1 3/8	1 11/16	1 3/4	2 1/2		3 1/8	
	H	1 1/4	1 1/2	1 5/8	1 7/8	2 1/4	2 5/8	2 15/16	3 5/16				
	J	7/8	1	1 1/8	1 3/8	1 3/4	2	2 1/8	2 1/2				
	K			2 9/16	3	3 1/2	3 15/16	4 3/4	5				
	L			3 9/16	4 1/8	4 13/16	5 3/8	6 7/16	6 5/8				

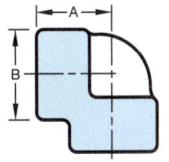 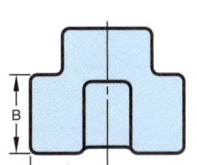 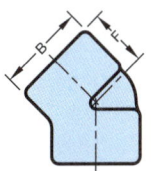 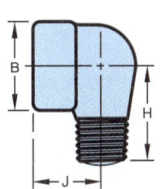 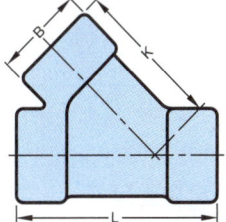 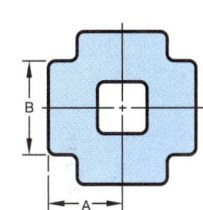

Threaded Couplings, Reducers and Caps—Class 3000 and 6000
DIMENSIONS (Inches)

	NOMINAL PIPE SIZE	1/8	1/4	3/8	1/2	3/4	1	1 1/4	1 1/2	2	2 1/2	3	4
Class 3000	A	1 1/4	1 3/8	1 1/2	1 7/8	2	2 3/8	2 5/8	3 1/8	3 3/8	3 5/8	4 1/4	4 3/4
	B	3/4	3/4	7/8	1 1/8	1 3/8	1 3/4	2 1/4	2 1/2	3	3 5/8	4 1/4	5 1/2
	C	5/8	11/16	3/4	15/16	1	1 3/16	1 5/16	1 9/16	1 11/16	1 13/16	2 1/8	2 3/8
	D	15/16	1	1	1 1/4	1 7/16	1 5/8	1 3/4	1 3/4	1 7/8	2 3/8	2 9/16	2 11/16
Class 6000	A	1 1/4	1 3/8	1 1/2	1 7/8	2	2 3/8	2 5/8	3 1/8	3 3/8	3 5/8	4 1/4	4 3/4
	B	7/8	1	1 1/4	1 1/2	1 3/4	2 1/4	2 1/2	3	3 5/8	4 1/4	5	6 1/4
	C	5/8	11/16	3/4	15/16	1	1 3/16	1 5/16	1 9/16	1 11/16	1 13/16	2 1/8	2 3/8
	D	1	1 1/16	1 1/16	1 5/16	1 1/2	1 11/16	1 13/16	1 7/8	2	2 1/2	2 11/16	2 15/16

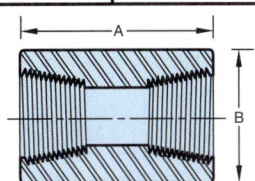 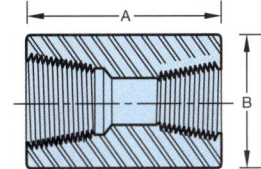 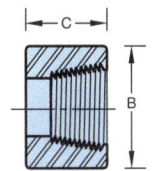 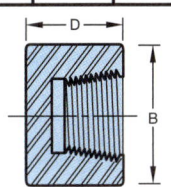

Courtesy Bonney Forge.

APPENDIX X
VALVE SPECIFICATIONS

CRANE

No. 438

GATE VALVES
CLASS 125
¼" to 3"

**Non-Rising Stem
Screwed Bonnet
Solid Wedge Disc**

No. 438, Threaded

RATINGS

Temp. F.	Psi Non-Shock
–20 to 150°	200
200	185
250	170
300	155
350	140
406	125
450	120

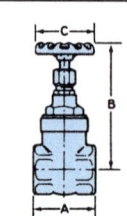
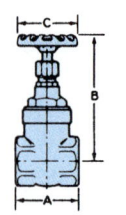

Weights and Dimensions

Valve N.P.S.	Weight—Pounds	Dimensions—Inches		
		A	B	C
¼	.6	1.64	3.44	1.75
⅜	.6	1.64	3.44	1.75
½	1.0	1.90	3.75	2.06
¾	1.4	2.14	4.38	2.75
1	2.1	2.47	4.88	2.75
1¼	3.1	3.08	5.63	3.06
1½	4.2	3.11	6.44	3.63
2	6.2	3.39	7.50	4.50
2½	11.3	4.25	9.06	5.00
3	16.0	4.59	9.69	5.00

"B" dimension is with valve open

CRANE

No. 1

No. 2

GLOBE AND ANGLE VALVES
CLASS 125
⅛" to 3"

Screwed Bonnet

**Globe
No. 1, Threaded**

**Angle
No. 2, Threaded**

RATINGS

Temp. F.	Psi Non-Shock
–20 to 150°	200
200	185
250	170
300	155
350	140
406	125
450	120

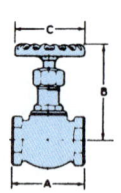

Weights and Dimensions

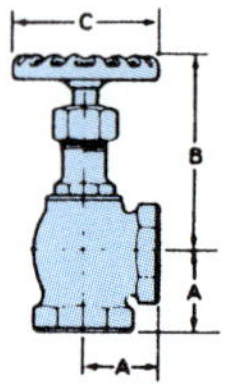

Valve N.P.S.	Weight—Pounds		Dimensions—Inches			
	No. 1	No. 2	A No. 1	A No. 2	B	C
⅛	.3	.3	1.44	.75	2.75	1.75
¼	.4	.4	1.63	.81	3.00	1.75
⅜	.5	.4	1.88	.94	3.25	2.03
½	1.0	.8	2.19	1.13	3.75	2.75
¾	1.4	1.4	2.69	1.38	4.25	2.75
1	2.2	2.0	3.19	1.63	5.00	3.00
1¼	3.3	3.0	3.69	1.81	5.50	3.72
1½	4.7	4.5	4.19	2.06	6.25	4.50
2	7.6	7.6	5.13	2.50	7.50	5.00
2½	11.6	—	6.06	—	7.50	5.00
3	19.5	—	7.06	—	9.25	6.00

"B" dimension is with valve open

APPENDIX X (Continued)

CRANE

No. 366E

LIFT CHECK VALVES CLASS 300 ¼″ to 3″

No. 366E Threaded

RATINGS Non-Shock

Temp. F.	Psi ¼″-2″	Psi 2½″-3″
–20-150°	1000	600
200	920	560
250	830	525
300	740	490
350	650	450
400	560	410
450	480	375
500	390	340
550	300	300

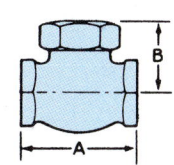
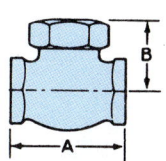

Weights and Dimensions

Valve N.P.S.	Weight—Pounds	Dimensions—Inches A	Dimensions—Inches B
¼	0.4	1.82	1.00
⅜	0.6	2.00	1.12
½	0.9	2.50	1.38
¾	1.5	2.94	1.88
1	2.6	3.50	2.00
1¼	4.2	4.06	2.38
1½	5.4	4.62	2.62
2	10.8	5.75	3.25
2½	15.6	6.88	3.88
3	24.0	8.00	4.50

CRANE

WEDGE GATE VALVES CLASS 125 2″ to 48″

Non-Rising Stem No. 461

Bronze Trim No. 460, Threaded No. 461, Flanged

All Iron No. 473, Flanged

RATINGS

Temp. F.	Psi, Non-Shock 2 to 12″	14 to 24″	30 to 48″
–20 to 150°	200	150	150
200	190	135	115
225	180	130	100
250	175	125	85
275	170	120	65
300	165	110	50
325	155	105	
350	150	100	
375	145	*	
400	140	*	
425	130	*	
450	125	*	

*Use Crane 150-pound steel valves.

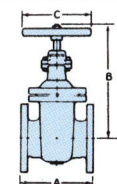
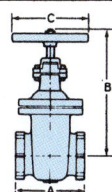

Weights and Dimensions

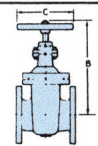

Valve N.P.S.	Weight—Pounds 460	461	473	Threaded A	Flanged A	Flanged B	Flanged C
2	25	30	30	5.38	7.00	11.31	8.0
2½	31	40	40	6.62	7.50	12.40	8.0
3	44	56	56	7.00	8.00	13.25	8.0
4	71	90	90	8.00	9.00	16.31	10.0
5	—	126	—	—	10.00	18.00	10.0
6	—	152	152	—	10.50	20.69	12.0
8	—	260	260	—	11.50	24.12	14.0

Valve N.P.S.	Weight—Pounds 461	Dimensions—Inches A	B	C
10	475	13.00	33.00	20.0
12	680	14.00	36.50	20.0
14	850	15.00	40.50	20.0
16	1280	16.00	47.25	22.0
18	1480	17.00	50.25	24.0
20	1840	18.00	54.75	24.0
24	2860	20.00	65.00	30.0
30	On Request	24.00	76.00	36.0
36	On Request	28.00	85.00	42.0
42	On Request	33.00	106.00	42.0
48	On Request	36.00	111.25	42.0

APPENDIX X (Continued)

No. 353

CRANE

GLOBE AND ANGLE VALVES CLASS 125 2″ to 10″

Outside Screw & Yoke Bronze Trim

Globe No. 351, Flanged

Angle No. 353, Flanged

RATINGS

Temp. F.	Psi, Non-Shock	Temp. F.	Psi Non-Shock
–20 to 150°	200	325	155
200	190	350	150
225	180	375	145
250	175	400	140
275	170	425	130
300	165	450	125

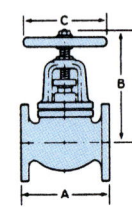

Weights and Dimensions

Valve N.P.S.	Weight—Pounds		Dimensions—Inches				
	351	353	351	353	351	353	C
			A	A	B	B	
2	34	32	8.00	4.00	11.12	11.00	8.00
2½	40	38	8.50	4.25	11.50	11.50	8.00
3	57	54	9.50	4.75	13.25	12.75	9.00
4	95	88	11.50	5.75	15.50	15.00	10.00
5	126	—	13.00	—	17.50	—	10.00
6	176	158	14.00	7.00	19.50	19.50	12.00
8	344	—	19.50	—	25.00	—	16.00
10	570	—	24.50	—	30.50	—	18.00

"B" dimension is with valve open.

CRANE

SWING CHECK VALVES CLASS 125 2″ to 24″

Bolted Cap

Bronze Trim No. 372, Threaded No. 373, Flanged

All Iron No. 373½, Flanged

No. 373

RATINGS

Temp. F.	Psi, Non-Shock	
	Sizes 2″-12″	Sizes 14″-24″
–20 to 150°	200	150
200	190	135
225	180	130
250	175	125
275	170	120
300	165	110
325	155	105
350	150	100
375	145	—
400	140	—
425	130	—
450	125	—

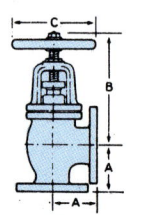

Weights and Dimensions

Valve N.P.S.	Weight—Pounds			Dimensions—Inches		
	372	373	373½	Threaded A	Flanged A	B
2	18	25	25	6.12	8.00	4.50
2½	22	34	34	7.25	8.50	5.38
3	29	44	44	8.00	9.50	5.88
4	54	75	75	9.25	11.50	6.62
5	—	103	—	—	13.00	7.75
6	—	127	127	—	14.00	8.25
8	—	230	230	—	19.50	10.25
10	—	510	510	—	24.50	11.60
12	—	695	695	—	27.50	13.60
14	—	875	—	—	31.00	15.75
16	—	1410	—	—	34.00	17.00
18	—	1540	—	—	38.50	17.40
20	—	1940	—	—	38.50	19.50
24	—	3000	—	—	51.00	20.50

Courtesy Crane Co.

APPENDIX X (Continued)

TABLE 32 WROUGHT STEEL PIPE[a] AND TAPER PIPE THREADS[b]—AMERICAN NATIONAL STANDARD

All dimensions are in inches except those in last two columns.

Nominal Pipe Size	D Outside Diameter of Pipe	Threads per Inch	L_1[c] Normal Engagement by Hand Between External and Internal Threads	L_2[c] Length of Effective Thread	Nominal Wall Thickness										Length of Pipe, Feet, per Square Foot External Surface[f]	Length of Standard Weight Pipe, Feet, Containing 1 cu. ft.[f]
					Sched. 10	Sched. 20[d]	Sched. 30[d]	Sched. 40[d]	Sched. 60	Sched. 80[e]	Sched. 100	Sched. 120	Sched. 140	Sched. 160		
1/8	.405	27.0	.1615	.2639	…	…	…	.068	…	.095	…	…	…	…	9.431	2,533.800
1/4	.540	18	.2278	.4018	…	…	…	.088	…	.119	…	…	…	…	7.073	1,383.8
3/8	.675	18	.240	.4078	…	…	…	.091	…	.126	…	…	…	…	5.658	754.36
1/2	.840	14	.320	.5337	…	…	…	.109	…	.147	…	…	…	.188	4.547	473.91
3/4	1.050	14	.339	.5457	…	…	…	.113	…	.154	…	…	…	.219	3.637	270.03
1	1.315	11.5	.400	.6828	…	…	…	.133	…	.179	…	…	…	.250	2.904	166.62
1 1/4	1.660	11.5	.420	.7068	…	…	…	.140	…	.191	…	…	…	.250	2.301	96.275
1 1/2	1.900	11.5	.420	.7235	…	…	…	.145	…	.200	…	…	…	.281	2.010	70.733
2	2.375	11.5	.436	.7565	…	…	…	.154	…	.218	…	…	…	.344	1.608	42.913
2 1/2	2.875	8	.682	1.1375	…	…	…	.203	…	.276	…	…	…	.375	1.328	30.077
3	3.500	8	.766	1.2000	…	…	…	.216	…	.300	…	…	…	.438	1.091	19.479
3 1/2	4.000	8	.821	1.2500	…	…	…	.226	…	.318	…	…	…	…	.954	14.565
4	4.500	8	.844	1.3000	…	…	…	.237	…	.337	…	.438	…	.531	.848	11.312
5	5.563	8	.937	1.4063	…	…	…	.258	…	.375	…	.500	…	.625	.686	7.199
6	6.625	8	.958	1.5125	…	…	…	.280	…	.432	…	.562	…	.719	.576	4.984
8	8.625	8	1.063	1.7125	…	.250	.277	.322	.406	.500	.594	.719	.812	.906	.443	2.878
10	10.750	8	1.210	1.9250	…	.250	.307	.365	.500	.594	.719	.844	1.000	1.125	.355	1.826
12	12.750	8	1.360	2.1250	…	.250	.330	.406	.562	.688	.844	1.000	1.125	1.312	.299	1.273
14 OD	14.000	8	1.562	2.2500	.250	.312	.375	.438	.594	.750	.938	1.094	1.250	1.406	.273	1.065
16 OD	16.000	8	1.812	2.4500	.250	.312	.375	.500	.656	.844	1.031	1.219	1.438	1.594	.239	.815
18 OD	18.000	8	2.000	2.6500	.250	.312	.438	.562	.750	.938	1.156	1.375	1.562	1.781	.212	.644
20 OD	20.000	8	2.125	2.8500	.250	.375	.500	.594	.812	1.031	1.281	1.500	1.750	1.969	.191	.518
24 OD	24.000	8	2.375	3.5000	.250	.375	.562	.688	.969	1.219	1.531	1.812	2.062	2.344	.159	.358

[a] ANSI/ASME B36.10M–1995.
[b] ANSI/ASME B1.20.1–1983 (R1992).
[c] Refer to §13.22 and Fig. 13.20.
[d] Boldface figures correspond to "standard" pipe.
[e] Boldface figures correspond to "extra strong" pipe
[f] Calculated values for Schedule 40 pipe.

APPENDIX X (Continued)

TABLE 33 CAST IRON PIPE SCREWED FITTINGS,[a] 125 LB—AMERICAN NATIONAL STANDARD

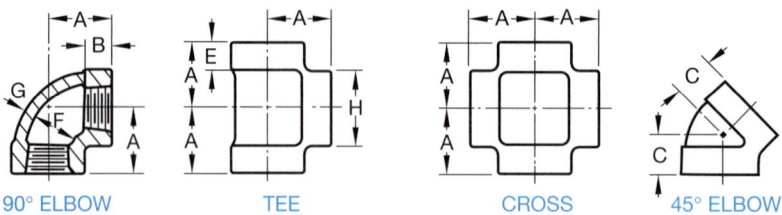

90° ELBOW   TEE   CROSS   45° ELBOW

DIMENSIONS OF 90° AND 45° ELBOWS, TEES, AND CROSSES (STRAIGHT SIZES)

All dimensions given in inches.
Fittings that have right- and left-hand threads shall have four or more ribs or the letter L cast on the band at end with left-hand thread.

Nominal Pipe Size	Center to End, Elbows, Tees, and Crosses A	Center to End, 45° Elbows C	Length of Thread, Min. B	Width of Band, Min. E	Inside Diameter of Fitting F		Metal Thickness G	Diameter of Band, Min. H
					Max.	Min.		
1/4	.81	.73	.32	.38	.58	.54	.11	.93
3/8	.95	.80	.36	.44	.72	.67	.12	1.12
1/2	1.12	.88	.43	.50	.90	.84	.13	1.34
3/4	1.31	.98	.50	.56	1.11	1.05	.15	1.63
1	1.50	1.12	.58	.62	1.38	1.31	.17	1.95
1 1/4	1.75	1.29	.67	.69	1.73	1.66	.18	2.39
1 1/2	1.94	1.43	.70	.75	1.97	1.90	.20	2.68
2	2.25	1.68	.75	.84	2.44	2.37	.22	3.28
2 1/2	2.70	1.95	.92	.94	2.97	2.87	.24	3.86
3	3.08	2.17	.98	1.00	3.60	3.50	.26	4.62
3 1/2	3.42	2.39	1.03	1.06	4.10	4.00	.28	5.20
4	3.79	2.61	1.08	1.12	4.60	4.50	.31	5.79
5	4.50	3.05	1.18	1.18	5.66	5.56	.38	7.05
6	5.13	3.46	1.28	1.28	6.72	6.62	.43	8.28
8	6.56	4.28	1.47	1.47	8.72	8.62	.55	10.63
10	8.08[b]	5.16	1.68	1.68	10.85	10.75	.69	13.12
12	9.50[b]	5.97	1.88	1.88	12.85	12.75	.80	15.47

[a] From ANSI/ASME B16.4–1992.
[b] This applies to elbows and tees only.

APPENDIX Y
PVC PIPE DIMENSIONS IN INCHES

PVC PIPE DIMENSIONS IN INCHES					
Schedule 40 Dimensions					
Nominal Pipe Size	O.D.	Average I.D.	Minimum Wall Thickness	Nominal Weight/Ft.	Maximum Water Pressure PSI
1/8″	0.405	0.249	0.068	0.051	810
1/4″	0.540	0.344	0.088	0.086	780
3/8″	0.675	0.473	0.091	0.115	620
1/2″	0.840	0.602	0.109	0.170	600
3/4″	1.050	0.804	0.113	0.226	480
1″	1.315	1.029	0.133	0.333	450
1-1/4″	1.660	1.360	0.140	0.450	370
1-1/2″	1.900	1.590	0.145	0.537	330
2″	2.375	2.047	0.154	0.720	280
2-1/2″	2.875	2.445	0.203	1.136	300
3″	3.500	3.042	0.216	1.488	260
3-1/2″	4.000	3.521	0.226	1.789	240
4″	4.500	3.998	0.237	2.118	220
5″	5.563	5.016	0.258	2.874	190
6″	6.625	6.031	0.280	3.733	180
8″	8.625	7.942	0.322	5.619	160
10″	10.750	9.976	0.365	7.966	140
12″	12.750	11.889	0.406	10.534	130
14″	14.000	13.073	0.437	12.462	130
16″	16.000	14.940	0.500	16.286	130
18″	18.000	16.809	0.562	20.587	130
20″	20.000	18.743	0.593	24.183	120
24″	24.000	22.544	0.687	33.652	120

Courtesy Harvel Plastics, Inc. and ASTM.

APPENDIX Y (Continued)

PVC PIPE DIMENSIONS IN INCHES (CONTINUED)					
Schedule 80 Dimensions					
Nominal Pipe Size	O.D.	Average I.D.	Minimum Wall Thickness	Nominal Weight/Ft.	Maximum Water Pressure PSI
1/8"	.405	.195	0.095	0.063	1230
1/4"	.540	.282	0.119	0.105	1130
3/8"	.675	.403	0.126	0.146	920
1/2"	.840	.526	0.147	0.213	850
3/4"	1.050	.722	0.154	0.289	690
1"	1.315	.936	0.179	0.424	630
1-1/4"	1.660	1.255	0.191	0.586	520
1-1/2"	1.900	1.476	0.200	0.711	470
2"	2.375	1.913	0.218	0.984	400
2-1/2"	2.875	2.290	0.276	1.500	420
3"	3.500	2.864	0.300	2.010	370
3-1/2"	4.000	3.326	0.318	2.452	350
4"	4.500	3.786	0.337	2.938	320
5"	5.563	4.768	0.375	4.078	290
6"	6.625	5.709	0.432	5.610	280
8"	8.625	7.565	0.500	8.522	250
10"	10.750	9.493	0.593	12.635	230
12"	12.750	11.294	0.687	17.384	230
14"	14.000	12.410	0.750	20.852	220
16"	16.000	14.213	0.843	26.810	220
18"	18.000	16.014	0.937	33.544	220
20"	20.000	17.814	1.031	41.047	220
24"	24.000	21.418	1.218	58.233	210

APPENDIX Z
RECTANGULAR AND ROUND HVAC DUCT SIZES

This chart provides rectangular and round duct sizes for air flows between 100–50,0000 cfm.

CFM (Cubic Feet per Minute)	Equivalent Rectangular Duct Sizes (Inches)	Equivalent Diameter Round Duct Sizes (Inches)	CFM (Cubic Feet per Minute)	Equivalent Rectangular Duct Sizes (Inches)	Equivalent Diameter Round Duct Sizes (Inches)
100	3 × 4	4	6000	14 × 20 15 × 18	19
200	3 × 7 4 × 5	5	7000	12 × 26 16 × 20	20
300	4 × 7 5 × 6	6	8000	12 × 30 14 × 25	21
400	4 × 9 5 × 7 6 × 6	7	9000	12 × 34 15 × 25	22
500	6 × 7	8	10000	12 × 36 16 × 25 20 × 20	23
750	5 × 12 6 × 10 7 × 8	9	12500	12 × 45 16 × 30 20 × 24	24
1000	7 × 10 8 × 9	10	15000	16 × 36 18 × 30 23 × 25	26
1250	8 × 10 9 × 9	10	17500	16 × 40 20 × 32 25 × 25	28
1500	8 × 12 10 × 10	12	20000	16 × 45 20 × 35 25 × 28	30
1750	8 × 14 9 × 12 10 × 11	12	25000	16 × 55 20 × 43 25 × 38	32
2000	8 × 15 10 × 12	12	30000	20 × 50 30 × 32	34
2500	10 × 14 12 × 12	14	35000	20 × 55 30 × 35	36
3000	12 × 14	14	40000	25 × 48 30 × 40	38
3500	12 × 15	15	45000	25 × 25 32 × 40	40
4000	10 × 22 14 × 15	16	50000	32 × 45 35 × 40	42
4500	12 × 19 14 × 16	17			
5000	10 × 25 12 × 20 15 × 16	17			

APPENDIX AA
SPUR AND HELICAL GEAR DATA

	Spur Gear Data	Suggested Number of Decimal Places
BASIC SPECIFICATIONS	NUMBER OF TEETH	
	DIAMETRAL PITCH	XX.XXXX
	PRESSURE ANGLE	XX°
	STANDARD PITCH DIAMETER	X.XXXX
	TOOTH FORM	
	ADDENDUM	.XXXX
	WHOLE DEPTH	.XXXX
	CALC. CIR. TOOTH THICKNESS ON STD. PITCH CIRCLE	.XXXX MAX. .XXXX MIN.
MANUFACTURING AND INSPECTION	GEAR TESTING RADIUS	X.XXXX MAX. X.XXXX MIN.
	AGMA QUALITY NUMBER	
	MAX. TOTAL COMPOSITE TOLERANCE	.XXXX
	MAX. TOOTH-TO-TOOTH COMPOSITE TOLERANCE	.XXXX
	MASTER GEAR SPECIFICATIONS	
	TESTING PRESSURE (OUNCES)	XX
	DIAMETER OF MEASURING PIN	.XXXX
	MEASUREMENT OVER TWO PINS (FOR SETUP ONLY)	X.XXXX MAX. X.XXXX MIN.
	OUTSIDE DIAMETER	+.000 X.XXX − .00X
	MAX. ROOT DIAMETER	X.XXX
ENGINEERING REFERENCES	MATING GEAR PART NUMBER	
	NUMBER OF TEETH IN MATING GEAR	
	OPERATING CENTER DISTANCE	X.XXXX MAX. X.XXXX MIN.

	Helical Data	Suggested Number of Decimal Places
BASIC SPECIFICATIONS	NUMBER OF TEETH	
	DIAMETRAL PITCH	XX.XXXX
	NORMAL DIAMETRAL PITCH	XX.XXXX
	NORMAL PRESSURE ANGLE	XX°
	HELIX ANGLE	XX.XXXX°
	HAND OF HELIX	L.H. OR R.H.
	STANDARD PITCH DIAMETER	X.XXXX
	TOOTH FORM	
	ADDENDUM	.XXXX
	WHOLE DEPTH	.XXXX
	CALC. NORMAL CIR. TOOTH THICKNESS ON STD. PITCH CIRCLE	.XXXX MAX. .XXXX MIN.
MANUFACTURING AND INSPECTION	MANUFACTURING AND INSPECTION	
	GEAR TESTING RADIUS	X.XXXX MAX. X.XXXX MIN.
	AGMA QUALITY NUMBER	
	MAX. TOTAL COMPOSITE TOLERANCE	.XXXX
	MAX. TOOTH-TO-TOOTH COMPOSITE TOLERANCE	.XXXX
	MASTER GEAR SPECIFICATIONS	
	TESTING PRESSURE (OUNCES)	XX
	DIAMETER OF MEASURING PIN	.XXXX
	MEASUREMENT OVER TWO PINS (FOR SETUP ONLY)	X.XXXX MAX. X.XXXX MIN.
	LEAD	
	OUTSIDE DIAMETER	+ .000 X.XXX − .00X
	MAX. ROOT DIAMETER	X.XXXX
ENGINEERING REFERENCES	ENGINEERING REFERENCES	
	MATING GEAR PART NUMBER	
	NUMBER OF TEETH IN MATING GEAR	
	OPERATING CENTER DISTANCE	X.XXXX MAX. X.XXXX MIN.

APPENDIX BB
CADD DRAWING SHEET SIZES, SETTINGS, AND SCALE FACTORS

CADD DRAWING SHEET SIZES, SETTINGS, AND SCALE FACTORS

	Prototype Drawing Sheet Parameters		
Drawing Scale	D-size (34″ × 22″) Drawing Limits	C-size (22″ × 17″) Drawing Limits	B-size (17″ × 11″) Drawing Limits
1″ = 1″	34,22	22,17	17,11
1/2″ = 1″	68,44	44,34	34,22
1/4″ = 1″	136,88	88,68	68,44
1/8″ = 1″	272,176	176,136	136,88
1″ = 1′–0″	408,264	264,204	204,132
3/4″ = 1′–0″	544,352	352,272	272,176
1/2″ = 1′–0″	816,528	528,408	408,264
3/8″ = 1′–0″	1088,704	704,544	544,352
1/4″ = 1′–0″	1632,1056	1056,816	816,528
3/16″ = 1′–0″	2176,1408	1408,1088	1088,704
1/8″ = 1′–0″	3264,2112	2112,1632	1632,1056
3/32″ = 1′–0″	4352,2816	2816,2176	2176,1408
1/16″ = 1′–0″	6528,4224	4224,3264	3264,2112

Prototype Drawing Scale Parameters			
Drawing Scale	Dimension Scale (DIMSCALE)	Linetype Scale (LTSCALE)	Inversion Scale of Border & Parts List Blocks
1″ = 1″	1	.5	1 = 1
1/2″ = 1″	2	1	1 = 2
1/4″ = 1″	4	2	1 = 4
1/8″ = 1″	8	4	1 = 8
1″ = 1′–0″	12	6	1 = 12
3/4″ = 1′–0″	16	8	1 = 16
1/2″ = 1′–0″	24	12	1 = 24
3/8″ = 1′–0″	32	16	1 = 32
1/4″ = 1′–0″	48	24	1 = 48
3/16″ = 1′–0″	64	32	1 = 64
1/8″ = 1′–0″	96	48	1 = 96
3/32″ = 1′–0″	128	64	1 = 128
1/16″ = 1′–0″	192	96	1 = 192

APPENDIX BB (Continued)

CADD DRAWING SHEET SIZES, SETTINGS, AND SCALE FACTORS

Architectural Sheet Size and Settings

Paper Size (in)	Approx. Drawing Area	Scale	Actual Sheet Limits	Approx. Drawing Limits	Text Height 1/8″	Text Height 1/4″	Scale Factor	Ltscale
A 12 × 9	10 × 7.5	1″ = 1′−0″ 1/2″ = 1′−0″ 1/4″ = 1′−0″ 1/8″ = 1′−0″	12′ × 9′ 24′ × 18′ 48′ × 36′ 96′ × 72′	10′ × 7.5′ 20′ × 15′ 40′ × 30′ 80′ × 60′	1.5 3.0 6.0 12.0	3.0 6.0 12.0 24.0	12 24 48 96	6 12 24 48
B 18 × 12	16 × 11	1″ = 1′−0″ 1/2″ = 1′−0″ 1/4″ = 1′−0″ 1/8″ = 1′−0″	18′ × 12′ 36′ × 24′ 72′ × 48′ 144′ × 96′	16′ × 11′ 32′ × 20′ 64′ × 40′ 128′ × 80′				
C 24 × 18	22 × 16	1″ = 1′−0″ 1/2″ = 1′−0″ 1/4″ = 1′−0″ 1/8″ = 1′−0″	24′ × 18′ 48′ × 36′ 96′ × 72′ 192′ × 144′	22′ × 16′ 44′ × 32′ 88′ × 64′ 176′ × 28′				
D 36 × 24	34 × 22	1″ = 1′−0″ 1/2″ = 1′−0″ 1/4″ = 1′−0″ 1/8″ = 1′−0″	36′ × 24′ 72′ × 48′ 144′ × 96′ 288′ × 192′	34′ × 22′ 68′ × 44′ 136′ × 88′ 272′ × 176′				
E 48 × 36	46 × 34	1″ = 1′−0″ 1/2″ = 1′−0″ 1/4″ = 1′−0″ 1/8″ = 1′−0″	48′ × 36′ 96′ × 72′ 192′ × 144′ 384′ × 288′	46′ × 34′ 92′ × 68′ 184′ × 136′ 368′ × 272′				

Mechanical Sheet Size and Settings

Paper Size (in)	Approx. Drawing Area	Scale	Actual Sheet Limits	Approx. Drawing Limits	Text Height 1/8″	Text Height 1/4″	Scale Factor	Ltscale
A 11 × 8.5	9 × 7	2″ = 1″ 3/4″ = 1″ 1/2″ = 1″ 1/4″ = 1″	5.5 × 4.25 14.67 × 11.33 22 × 17 44 × 34	4.5″ × 3.5″ 12″ × 9.33″ 18″ × 14″ 36″ × 28″	.0625 .167 .25 .5	.125 .33 .5 1.0	.5 1.33 2 4	.25 .67 1 2
B 17 × 11	15 × 10	2″ = 1″ 3/4″ = 1″ 1/2″ = 1″ 1/4″ = 1″	8.5 × 5.5 22.67 × 14.67 34 × 22 68 × 44	7.5″ × 5″ 20″ × 13.33″ 30″ × 20″ 60″ × 40″				
C 22 × 17	20 × 15	2″ = 1″ 3/4″ = 1″ 1/2″ = 1″ 1/4″ = 1″	11 × 8.5 29.33 × 14.67 44 × 34 88 × 68	10″ × 7.5″ 26.67″ × 20″ 40″ × 30″ 80″ × 60″				
D 34 × 22	32 × 20	2″ = 1″ 3/4″ = 1″ 1/2″ = 1″ 1/4″ = 1″	17 × 11 45.33 × 29.33 68 × 44 136 × 88	16″ × 10″ 42.67″ × 26.67″ 64″ × 40″ 128″ × 80″				
E 44 × 34	42 × 32	2″ = 1″ 3/4″ = 1″ 1/2″ = 1″ 1/4″ = 1″	22 × 17 58.67 × 45.33 88 × 68 176 × 136	21″ × 16″ 56″ × 42.67″ 84″ × 64″ 168″ × 128″				

APPENDIX BB (Continued)

CADD DRAWING SHEET SIZES, SETTINGS, AND SCALE FACTORS

Civil Sheet Size and Settings

Paper Size (in)	Approx. Drawing Area	Scale	Actual Sheet Limits	Approx. Drawing Limits	Text Height 1/8"	1/4"	Scale Factor	Ltscale
A 11 × 8.5	9 × 7	1″ = 10′ 1″ = 20′ 1″ = 30′ 1″ = 50′	110′ × 85′ 220′ × 170′ 330′ × 255′ 550′ × 425″	90′ × 70′ 180′ × 140′ 270′ × 210′ 450′ × 350′	15 30 45 75	30 60 90 150	120 240 360 600	60 120 180 300
B 17 × 11	15 × 10	1″ = 10′ 1″ = 20′ 1″ = 30′ 1″ = 50′	170′ × 110′ 340′ × 220′ 510′ × 330′ 850′ × 550′	150′ × 100′ 300′ × 200′ 450′ × 300′ 750′ × 500′				
C 22 × 17	20 × 15	1″ = 10′ 1″ = 20′ 1″ = 30′ 1″ = 50′	220′ × 170′ 440′ × 340′ 660′ × 510′ 1100′ × 850′	200′ × 150′ 400′ × 300′ 600′ × 450′ 1000′ × 750′				
D 34 × 22	32 × 20	1″ = 10′ 1″ = 20′ 1″ = 30′ 1″ = 50′	340′ × 220′ 680′ × 440′ 1020′ × 660′ 1700′ × 1100′	320′ × 200′ 640′ × 400′ 960′ × 600′ 1600′ × 1000′				
E 44 × 34	42 × 32	1″ = 10′ 1″ = 20′ 1″ = 30′ 1″ = 50′	440′ × 340′ 880′ × 680′ 1320′ × 1020′ 2200′ × 1700′	420′ × 320′ 840′ × 640′ 1260′ × 960′ 2100′ × 1600′				

GLOSSARY

Acme　A thread system used especially for feed mechanisms.

Adaptive Parts (3-D CADD)　In assemblies, modify automatically if another part changes.

Addendum (Spur Gear)　The radial distance from the pitch circle to the top of the tooth.

Addendum Angle (Bevel Gear)　The angle subtended by the addendum.

Aligned Section　The cutting plane is staggered to pass through offset features of an object.

Alignment Charts　Designed to graphically solve mathematical equation values using three or more scaled lines.

Allowance　The tightest possible fit between two mating parts.

Alloys　A mixture of two or more metals.

Amplifier (AMP)　A device that allows an input signal to control power; capable of having an output signal greater than the input signal.

Angle of Repose　The run-to-rise ratio of highway cut and fill.

Angular Dimension Line　A dimension line drawn as an arc with the center of the arc from the vertex of the angle.

Animation　The process of making drawings or models move and change according to a sequence of predefined images.

Annealing　Under certain heating and cooling conditions and techniques, steel is softened using this method.

Apparent Intersection　This is a condition where lines or planes *look* like they may be intersecting, but in reality they may not be intersecting.

Arrowheads　Used to cap the dimension line and leader line ends.

Assembly　A grouping of one or more design components. Components can include part models and subassemblies.

Assembly Files　Used to create assemblies and subassemblies, and to reference multiple part and subassembly files.

Attribute (CADD)　See text.

Auxiliary View　A view that is required when a surface is not parallel to one of the principal planes of projection; the auxiliary projection plane is parallel to the inclined surface so that the surface may be viewed in its true size and shape.

Axis　The centerline of a cylindrical feature.

Azimuth　The clockwise measurement of the angle of a line, measured from the north or its reference meridian.

Backsight　In surveying, the rod reading behind the level toward the point of beginning.

Ball Bearing　A friction-reducer where balls roll in two grooved rings.

Base Circle (CAM)　The smallest circle tangent to the CAM follower at the bottom of displacement.

Base Circle Diameter (Spur Gear)　The diameter of a circle from which the involute tooth is generated.

Basic Dimension　A numerical value used to describe the theoretically exact size, profile, orientation, or location of a feature or datum target. It is the basis from which permissible variations are established by tolerances on other dimensions, in notes, or in feature control frames.

Bearing (Civil)　The measurement of the angle of a line, measured from either the north or the south meridian, whichever is nearer.

Bearing (Mechanical)　A mechanical device that reduces friction between two surfaces.

Bearing Angle　The bearing angle of a line is always 90 degrees or less and is identified either from the north or the south.

Bearing of a Line　The angular relationship of the horizontal projection of the line relative to the compass, expressed in degrees.

Bearing Seal　A rubber, felt, or plastic seal on the outer and inner ring of a bearing. Generally, it is filled with a special lubricant by the manufacturer.

Bearing Shield　A metal plate on one or both sides of the bearing; serves to retain the lubricant and keep the bearing clean.

Bell and Spigot　A pipe connection in which one end of a piece of pipe has a bell-shaped opening and the other end is tapered or notched to fit into the bell.

Bellcrank　A link pivoted near the center, that oscillates through an angle.

Bench Mark (Civil)　The name for a known point with a known elevation that is part of the geodetic control system. Another name for bench mark is *monument*.

Bench Marking (Engineering Design)   Analysis of the product where you start, establishing the bench mark. Then, when you go through the re-design or make desired changes, you re-analyze to determine the difference from the bench mark condition to the revised condition.

Bend Allowance   The amount of extra material needed for a bend to compensate for compression during the bending process.

Bend Relief   Cutting away material at a corner to help relieve stress.

Bevel   The term used to denote the slope of beams, as in structural engineering.

Bevel Gear   Used to transmit power between intersecting shafts; takes the shape of a frustum of a cone.

Bias   The voltage applied to a circuit element to control the mode of operation.

Bilateral Tolerance   A tolerance in which variation is permitted in both directions from the specified dimension.

Bill of Materials (BOM)   *See* parts list.

Bit   Binary digit.

Bolt Circle   Holes located in a circular pattern.

Bore   To enlarge a hole with a single pointed machine tool in a lathe, drill press, or boring mill.

Boss   A cylindrical projection on the surface of a casting or forging.

Bottom-Up Assembly   Method involves inserting existing parts and sub-assemblies into an assembly file, followed by constraining, or mating, the component.

Bow's Notation   A system of notation used to label a vector system. A letter is given to the space on each side of the vector, and each vector is then identified by the two letters on either side of it, read in a clockwise direction.

Broken Out Section   A portion of a part is broken away to clarify an interior feature; there is no associated cutting-plane line.

Bus   An aluminum or copper plate or tubing that carries the electrical current.

Bushing   A replaceable lining or sleeve used as a bearing surface.

Butt Weld   A form of pipe manufacture in which the seam of the pipe is a welded flat-faced joint. Also, a form of welding in which two pieces of material are "butted" against each other and welded.

Cabinet Oblique Drawing   A form of oblique drawing in which the receding lines are drawn at half scale, and usually at a 45-degree angle from horizontal.

CAD   Computer-aided design, also referred to as computer-aided drafting.

CAD/CAM   Computer-aided design/computer-aided manufacturing.

CADD   Computer-aided design and drafting.

CAE   Computer-aided engineering.

CAGE Code   A five number code assigned by the Defense Logistic Service Center (DLSC) to all Department of Defense contractors. CAGE stands for Commercial and Government Entity. The CAGE Information report also contains contact information for the vendor.

Cam   A machine part used to convert constant rotary motion into timed irregular motion.

Cam Motion   The base point from which to begin cam design. There are four basic types of motion: simple harmonic, constant velocity, uniform accelerated, and cycloidal.

Capacitor   An electronic component that opposes a change in voltage and storage of electronic energy.

Caburization   A process where carbon is introduced into the metal by heating to a specified temperature range while in contact with a solid, liquid, or gas material consisting of carbon.

Cartesian Coordinate System   A measurement system based on rectangular grids to measure width, height, and depth (X, Y, and Z).

Casting   An object or part produced by pouring molten metal into a mold.

Catalog Features   A copy of an existing feature or set of features you create and then save in a library to be used in other models. Also known as library features, they are similar to placed features, because you can place a catalog feature onto an existing feature or part.

Cavalier Oblique Drawing   A form of oblique drawing in which the receding lines are drawn true size, or full scale. Usually drawn at an angle of 45 degrees from horizontal.

Center   Located in the exact middle of the circle.

Central Processing Unit (CPU)   The processor and main memory chips in a computer. Specifically, the CPU is just the processor, but generally it refers to the computer.

Chain Dimensioning   Also known as point-to-point dimensioning when dimensions are established from one point to the next.

Chamfer   A slight surface angle used to relieve a sharp corner.

Chordal Addendum (Spur Gear)   The height from the top of the tooth to the line of the chordal thickness.

Chordal Thickness (Spur Gear)   The straight line thickness of a gear tooth on the pitch circle.

CIM   Computer-integrated manufacturing that combines CADD, CAM, and CAE into a controlled system.

Circular Pattern (CADD)   Copies and organizes a feature around an imaginary circle a designated number of times and places each feature a specified distance from the other.

Circular Pitch (Spur Gear)   The distance from a point on one tooth to the corresponding point on the adjacent tooth, measured on the pitch circle.

Circular Thickness (Spur Gear)   The length of an arc between the two sides of a gear tooth on the pitch circle.

Circumference   The distance around the circle, on the circle edge.

Circumscribed Polygon   Measured from the polygon flats.

Clearance (Spur Gear)   The radial distance between the top of a tooth and the bottom of the mating tooth space.

Coded Section Lines   Used if you want to represent specific material section line symbols in the sectional view.

Coil or Inductor   A conductor wound on a form or in a spiral; contains inductance.

Cold Rolled Steel (CRS)   The additional cold forming of steel after initial hot rolling; cleans up hot formed steel.

Command   A specific instruction issued to the computer by the operator. The computer performs a function or task in response to a command.

Component Used to describe the individual parts and subassemblies used to create an assembly.

Compressive Pushing toward the point of currency, as in forces that are compressed.

Computer Animations Made by defining, or recording, a series of still images in various positions of incremental movement that, when played back, no longer appear individually as static images, but appear to be unbroken motion.

Computer Numerical Control (CNC) Also known as numerical control (NC); control of a process or machine by encoded commands that are commonly prepared by a computer.

Concentric Two or more circles sharing the same center.

Concurrent Forces Forces acting on a common point.

Cone Distance (Bevel Gear) The slant height of the pitch cone.

Constraints (3-D CADD) Same as relations, describing the relationship of geometric shapes. For example, a constraint or relation can define if the length of one line is equal to another, if the center of a small circle is concentric to a larger circle, or if two lines are perpendicular.

Construction Lines Very lightly drawn, nonreproducing lines used for the layout of a drawing.

Container (CADD) Any source of drawing content or other information accessible by a disk, folder, file, or URL.

Contour Interval The distance in elevation between contour lines.

Contour Line Denotes a series of connected points at a particular elevation.

Coplanar Forces Forces that all lie in the same plane.

Counterbore To cylindrically enlarge a hole; generally to allow the head of a screw or bolt to be recessed below the surface of an object.

Counterdrill A machined hole that looks similar to a countersink/counterbore combination.

Countersink Used to recess the tapered head of a fastener below the surface of an object.

Course (Civil) Refers to each element or line of the traverse.

Crank A link, usually a rod or bar, that makes a complete revolution about a fixed point.

Cross-Functional Team Approach (Engineering Design) Where everyone is involved in cooperation as a team, rather than a situation where one person or group does something and then passes the completed portion on to the next person or group.

Crown Backing (Bevel Gear) The distance between the cone apex and the outer tip of the gear teeth.

CRT Cathode ray tube.

Cursor A small rectangle, underline, or set of crosshairs that indicates present location on a video display screen. Also, a handheld input device used in conjunction with a digitizer.

Curve Data (Civil) Any measurements or features used to create the road layout. The following terms are used in curve data:

curve radius (R), curve length (L), and included angle of the curve (Δ).

Curve Length The distance from one endpoint of a curve to the other endpoint along the circumference.

Datum A theoretically exact point, axis, or plane derived from the true geometric counterpart of a specified datum feature. The origin from which the location or geometric characteristics of features of a part are established.

Datum Dimensioning A dimensioning system where each dimension originates from a common surface, plane, or axis.

Declination A line that goes downward from its origin; assigned negative values.

Dedendum (Spur Gear) The radial distance from the pitch circle to the bottom of the tooth.

Dedendum Angle (Bevel Gear) The angle subtended by the dedendum.

Default An action taken by computer software unless the operator specifies differently.

Delta Angle (Δ) The included angle of a curve.

Delta Note A specific note placed with general notes and keyed to the drawing with a delta symbol (Δ).

Derived Components Similar to catalog features, but more complicated and can contain a complete model consisting of several features, or even multiple parts.

Design Specifications During the concept phase, technical specifications and performance targets are established. These give guidelines to the project team and help identify what constitutes a successful design. Examples of these activities include competitor testing, performance goals and life goals, certification requirements, voice of the customer, and design input from the product description/business case. These technical specifications are then communicated in a document called the Design Specification.

Detail Assembly A product shown with its component parts assembled in a manufacturer's catalog or on their website.

Detail Drawing A drawing of an individual part that contains all of the views, dimensions, and specifications necessary to manufacture the part.

Diameter The distance across the circle, from one side to another, through the center. Diameter is identified using the diameter symbol: Ø.

Diametral Pitch A ratio equal to the number of teeth on a gear per inch of pitch diameter.

Diazo A printing process that produces blue, black, or brown lines on various media (other resultant colors are also produced with certain special products). The print process is a combination of exposing an original in contact with a sensitized material exposed to an ultraviolet light, and then running the exposed material through an ammonia chamber to activate the remaining sensitized image to form the desired print. This is a fast and economical method of making prints, once commonly used in drafting.

Digitize The act of locating points and selecting commands using an input device (puck or stylus, used with a digitizer tablet).

Digitizer An electronically sensitized flat board or tablet that serves as a drawing surface for the input of graphics data. Images can be drawn or traced, and commands and symbols can be selected from a menu attached to the digitizer.

Dihedral Angle The angle that is formed by two intersecting planes.

Dimension Lines Indicate the length of the dimension. They are thin lines capped on the ends with arrowheads and broken along the length, providing a space for the dimension numeral. The gap between the dimension line and the dimension numeral varies but is commonly .06 in. (1.5 mm). This standard varies with architectural drafting.

Dimension Text   Text displaying dimension value, normally .12 in. (3 mm) high centered in the space provided in the dimension line.

Dimetric Drawing   A pictorial drawing in which two axes form equal angles with the plane of projection. These can be greater than 90 but less than 180 and cannot have an angle of 120 degrees. The third axis may have an angle less or greater than the two equal axes.

Dip   The slope of a stratum.

Direct Numerical Control (DNC)   The CAD/CAM system is electronically connected to the machine tool. This electronic connection is called networking. This direct link is referred to as direct numerical control (DNC) and requires no additional media, such as paper, floppy disks, CDs, or tape to transfer information from engineering to manufacturing.

Displacement Diagram   A graph; the curve on the diagram is a graph of the path of the cam follower. In the case of a drum cam displacement diagram, the diagram is actually the developed cylindrical surface of the cam.

Documentation   Instruction manuals, guides, and tutorials provided with any computer hardware/software system.

Dowel Pin   A cylindrical fastener used to retain parts in a fixed position or to keep parts aligned.

Draft   The taper on the surface of a pattern for castings or the die for forgings, designed to help facilitate removal of the pattern from the mold or the part from the die. Draft is often 7–10 degrees but depends on the material and the process.

Drag and Drop   A computer term referring to an activity where you perform operations by moving an icon of an object with the mouse into another file, window, or icon. For example, files can be copied or moved, which is applied by dragging (picking on the desired item) from one location and dropping (release the pick button) to another location.

Drawing Area Cursor   Often displayed as a crosshair but can be shown as a wand or point, depending on the CADD program or applications.

Drawing Template   Stores standard CADD drawing settings and objects. All settings and content of a template file are included in a new drawing.

Drilling Drawing   Used to provide size and location dimensions for trimming the printed circuit board.

Drum Cam   A drum, or cylindrical, cam is a cylinder with a groove in its surface. As the cam rotates, the follower moves through the groove, producing a reciprocating motion parallel to the axis of the camshaft.

Drum Plotter   A graphics pen plotter that can accommodate continuous feed paper, or in which sheet paper or film is attached to a sheet of flexible material mounted to a drum. The pen moves in one direction and the drum in the other.

Duct   Sheet metal, plastic, or other material pipe designed as the passageway for conveying air from the HVAC equipment to the source.

Ductility   The ability to be stretched, drawn, or hammered without breaking.

Dumb Solids (3-D CADD)   Also known as basic solids are created using basic solid modeling methods and contain very little, if any, information about specific parameters, dimensions, constraints, part history, or features. Only fundamental volume information, such as length, width, and height, are stored in dumb solids.

Easement (Civil)   A legal right of access over land owned by another for the purpose of access, egress, utilities, or other designated uses.

Eccentric Circles   Not having the same center.

Electrical Relays   Magnetic switching devices.

Electro-Discharge Machining (EDM)   A process where material to be machined and an electrode are submerged in a fluid that does not conduct electricity, forming a barrier between the part and the electrode. A high-current, short-duration electrical charge is then used to remove material.

Electroless   The depositing of metal on another material through the action of an electric current.

Electron Beam (EB)   Generated by a heated tungsten filament used to cut or machine very accurate features in a part.

Element   Any line, group of lines, shape, or group of shapes and text that is so defined by the computer operator.

Elementary Diagrams   Diagrams that provide the detail necessary for engineering analysis and operation or maintenance of substation equipment.

Engineering Change Documents   Documents used to initiate and implement a change to a production drawing; engineering change request (ECR) and engineering change notice (ECN) are examples.

Entity   *See* element. Also referred to as object.

Equilibrant   A vector that is equal in magnitude to the resultant and has the opposite direction and sense.

Equilibrium   When a vector system has a resultant of zero, the system is said to be in equilibrium.

Existing Product Research   A common approach to the design of new and updated products is based on existing product research. Existing product research involves redesign of a current product.

Exploded Assembly   A pictorial assembly showing all parts removed from each other and aligned along axis lines.

Extension Lines   Thin lines used to establish the extent of a dimension, starting with a small offset of .06 in. (1.5 mm) from the object and extend .12 in. (3 mm) past the last dimension line.

Face Angle (Bevel Angle)   The angle between the top of the teeth and the gear axis.

Failure Mode Effect Analysis (FMEA)   A technique used to determine possible failures and ways to eliminate them.

Fault Condition   A short circuit that is a zero resistance path for electrical current flow.

Fillet   A curve formed at the interior intersection between two or more surfaces.

Finite Element Analysis (FEA)   Software programs that allow for mathematical solutions to structural and thermal problems.

Fixture   A device for holding work in a machine tool.

Flange   A thin rim around a part.

Flat-Bed Plotter   A pen plotter where the drawing surface (bed) is oriented horizontally, and paper or film is attached to the surface by a vacuum or an electrostatic charge. The pen moves in both the x and y directions.

Floppy Disk   A thin, circular, magnetic storage medium encased in a cover. It comes in 8", 5 1/4", and 3 1/2" sizes.

Flowcharts   Used to show organizational structure, steps, or progression in a process or system.

Flow Diagram   A chart-type drawing that illustrates the organization of a system in a symbolic format.

Fold Lines   The reference line of intersection between two reference planes in orthographic projection.

Follower   The cam follower is a reciprocating device whose motion is produced by contact with the cam surface.

Font   A specific type face, such as Helvetica or Gothic.

Foreshortened Line   A line that appears shorter than its actual length, because it is at an angle to the line of sight.

Four-Bar Linkage   The most commonly used linkage mechanism. It contains four links: a fixed link called the ground link, a pivoting link called a driver, another pivoted link called a follower, and a link between the driver and follower, called a coupler.

Free-Body Diagram   A diagram that isolates and studies a part of the system of forces in an entire structure.

Full Section   A section view where the cutting plane extends completely through the object.

Fully-Constrained (3-D CADD)   Ensures a model is parametric and the design intentions are not violated.

Function Keys   Extra keys on an alphanumeric keyboard that can be utilized in a computer program to represent different commands and functions. The active commands for function keys may change several times in a program.

Fused Deposition Modeling (FDM)   A type of rapid prototyping (RP) or rapid manufacturing (RM) technology commonly used within engineering design. The FDM technology is marketed exclusively by Stratasys Inc. *Also see* rapid prototyping.

Gate   The part of an electronic system that makes the electronic circuit operate; permits an output only when a predetermined set of input conditions are met.

Gear   A cylinder or cone with teeth on its contact surface; used to transmit motion and power from one shaft to another.

Gear Ratio   Any two mating gears have a relationship to each other called a gear ratio. This relationship is the same between any of the following: RPMs, number of teeth, and pitch diameters of the gears.

Gear Train   Formed when two or more gears are in contact.

General Assembly   Contains features including multiviews, auxiliary views, detail views, and section view selection as needed for the specific product. Each part is identified with a balloon containing a number keyed to the parts list. The parts list identifies every part in the assembly.

Grade of a Line   A way to describe the inclination of a line in relation to the horizontal plane. The percent grade is the vertical rise divided by the horizontal run multiplied by 100.

Grade Slope   The percentage given to show amount of slope.

Graphical Kinematic Analysis   The process of drawing a particular mechanism in several phases of a full cycle to determine various characteristics of the mechanism.

Graphics Tablet   *See* digitizer.

Great Circle   One of an infinite number of circles from any point on the earth that is described by longitude and latitude.

Gunter's Chain   A 66-foot-long chain that Edmund Gunter invented; it is made up of 100 links. It is used in surveying.

Half Section   Used typically for symmetrical objects; the cutting-plane line actually cuts through one quarter of the part. The sectional view shows half of the interior and half of the exterior at the same time.

Hard Copy   A paper copy.

Hardware   The physical computer equipment.

HI   Height of instrument. In surveying, the calculation of the level, which is one factor needed to determine elevation.

Highway Diagram   A simplified or condensed representation of a point-to-point, interconnecting writing diagram for an electrical circuit.

Hone   A method of finishing a hole or other surface to a desired close tolerance and fine surface finish using an abrasive.

Hybrid CADD Modeling   Software programs incorporating the functionality of wireframe, surface, and basic and parametric solid modeling. Hybrid modeling systems are often described as separate CADD applications.

Hypertext   A computer-based text retrieval system that allows you to access specific locations in web pages or other electronic documents by clicking on links within the web pages or documents.

Inclination   A line that goes upward from its origin; assigned positive values.

Included Angle   The angle formed between the center and endpoints of an arc.

Inductance   The property in an electronic circuit that opposes a change in current flow or where energy may be stored in a magnetic field, as in a transformer.

In-House   Refers to any operations conducted in a company's own facility instead of being outsourced.

Industrial Design (ID) Team   A team of highly skilled and creative people who have extensive product knowledge and experience. The ID team has the flexibility of being extremely creative and their knowledge of engineering and production processes is essential.

Inscribed Polygon   Measured from the polygon corners.

Integrated Circuit (IC)   All of the components in a schematic are made up of one piece of semiconductor material.

Intellectual Properties   As taken from the World Intellectual Property Organization Website (*http://www.wipo.int*), intellectual property is divided into two categories: Industrial property and Copyright. Industrial property includes inventions (patents), trademarks, and industrial designs. Copyright includes literary and artistic works such as novels, poems, plays, films, and musical works, and artistic works such as drawings, paintings, photographs and sculptures and architectural designs.

Intelligence   The parametric concept is also referred to as intelligence because parametric modeling occurs as a result of the software programs ability to store and manage model information. This information includes knowledge of every model characteristic such as calculations, sketches, features, dimensions, geometric parameters, when each piece of the model was created, and all other model history and properties.

Intersecting Lines   When lines are intersecting, the point of intersection is a point that lies on both lines.

Isogonic Chart   A chart showing isogonic lines.

Isogonic Lines   Shows how many degrees to the east or west the magnetic north or south pole is from the true North or South Pole.

Isometric Drawing   A form of pictorial drawing in which all three drawing axes form equal angles (120 degrees) with the plane of projection.

Jig   A device used for guiding a machine tool in the machining of a part or feature.

Joint   The connection point between two links.

Joystick   A graphics input device composed of a lever mounted in a small box that allows the user to control the movement of the cursor on the video display screen.

Kerf   A groove created by the cut of a saw.

Kinematics   The study of motion without regard to the forces causing the motion.

Lap Weld   A form of pipe manufacture in which the seam of the pipe is an angular "lap."

Large-Scale Integration (LSI)   More circuits on a single small IC chip.

Laser (Light Amplification by Stimulated Emission of Radiation)   A device that amplifies focused light waves and concentrates them in a narrow, very intense beam.

Latitude   The parallels around the earth that do not intersect. The equator is the longest line of latitude.

Lay   Describes the basic direction or configuration of the predominant surface pattern in a surface finish.

Layer   An individual aspect of a CADD drawing that makes a complete drawing when combined.

Layout (CADD)   Can contain various views of the model, a title block, border, and other annotations such as general notes. In addition, the layout includes page setup information such as paper size and margins and plotter configuration data.

Layouts (CADD)   Created using the Layout tab at the bottom of the drawing area, in an environment called paper space.

Lead (Thread)   The distance that the thread advances axially in one revolution of the worm or thread.

Leader Line   A thin line used to connect a specific note to a feature on the drawing. The leader line can be at any angle between 15°–75° with 45° preferred. There is a short, horizontal shoulder between .12 and .24 in. (3–6 mm) centered where it meets the note.

Leveling   The process of determining elevation using backsight and foresight.

Lever   A link that moves back and forth through an angle; also known as a rocker.

Light Pen   A video display screen input device. It is a light-sensitive stylus connected to the terminal by a wire; enables the user to draw or select menu options directly on the screen.

Line of Sight   An imaginary straight line from the eye of the observer to a point on the object being observed. All lines of sight for a particular view are assumed to be parallel and are perpendicular to the projection plane involved.

Logic Diagrams   A type of schematic that is used to show the logical sequence in an electronic system.

Longitude   The meridians of the earth, which run from the North Pole to the South Pole. The lines of longitude are basically the same length.

Magnetic Declination   The degree difference between magnetic azimuth and true azimuth.

Malleable   The ability to be hammered or pressed into shape without breaking.

Management Science   *See* operations management.

Master Pattern   A one-to-one scale circuit pattern that is used to produce a printed circuit board.

Mates (3-D CADD)   Used in the assembly environment; they link individual components together to create the assembly.

Maxwell Diagram   A combination vector diagram used to analyze the forces acting in a truss.

Mechanical Engineer (HVAC)   This consulting engineer is called a mechanical engineer, named after the mechanical (HVAC) system being designed.

Mechanical Engineer (Manufacturing)   An engineer who designs manufactured products.

Mechanical Joint   A pipe connection that is a modification of the "bell and spigot" in which flanges and bolts are used with gaskets, packing rings, or grooved pipe ends providing a seal.

Mechanism   A combination of two or more machine members that work together to perform a specific motion.

Metes and Bounds   The system of describing portions of land by using lengths and boundaries.

Mirrored Feature (CADD)   Mirrors an existing feature across a plane.

Model (CADD)   Composed of various objects such as lines, circles, and text that make up the bulk of the drawing.

Model Space (CADD)   Model space can be thought of as an environment where you draw and design. The model is drawn in the AutoCAD Model tab found at the bottom of the drawing area and is created in model space.

Modeling Failure (3-D CADD)   The result of dimensions or geometric controls that are impossible to apply to the model. Over-constrained and modeling failure situations must be resolved before you can create a feature and fully parametric model. A sketch or model also cannot contain enough parameters.

Monodetail Drawing   A drawing of a single part on one sheet.

Mouse   A handheld input device connected to the terminal by a wire. It is moved across a flat surface to control the movement of the cursor on the screen. It rolls on a small ball that sends directional signals to the computer and may have one or more buttons that serve as function keys. Wireless models are available.

Multidetail Drawing   A drawing of several parts on one sheet.

Multiview Projection   The views of an object as projected upon two or more picture planes in orthographic projection.

Neck   A groove around a cylindrical part.

Nominal Size   The designation of the size for a commerical product.

Nomographs   A graphic representation of the relationship between two or more variables of a mathematical equation.

Nondestructive Testing (NDS)   Tests for potential defects in welds; they do not destroy or damage the weld or the part.

Normal Plane   A plane surface that is parallel to any of the primary projection planes.

Normalizing   A process of heating steel to a specific temperature and then allowing the material to cool slowly by air, bringing the steel to a normal state.

Numerical Control (NC)   A system of controlling a machine tool by means of numeric codes that direct the commands for the machine movements; computer numerical control (CNC) is a computer command control of the machine movement.

Oblique Drawing   A form of pictorial drawing in which the plane of projection is parallel to the front surface of the object and the receding angle is normally 45°.

Oblique Line   A straight line that is not parallel to any of the six principal planes.

Oblique Plane   Inclined to all of the principal projection planes.

Offset Section   The cutting plane is offset through staggered interior features of an object to show those features in section as if they were in the same plane.

Operational Amplifier (OPAMP)   A high-gain amplifier created from an integrated circuit.

Operations Management   Also known as management science, which is mathematics and optimization techniques used to prioritize the project portfolio. Management science or operations management use a mathematical process in which a business or engineering solution is simulated to determine optimum results.

Outside Diameter (Spur Gear)   The overall diameter of the gear; equal to the pitch diameter plus two times the addendum.

Over-Constrained Sketch (3-D CADD)   A sketch that has too many dimensions or geometric controls and may not allow the software to recognize or resolve the sketch geometry.

Pads   Or lands; the circuit-termination locations where the electronic devices are attached to the printed circuit board.

Paper Space (CADD)   The space you use to lay out the drawing or model to be plotted. Paper space can be thought of as an imaginary piece of paper on which model space information is added and scaled so that the drawing can be plotted correctly on a specific paper size.

Parameter-Driven Assembly (3-D CADD)   Allows changes made to individual parts to be reproduced automatically as changes are made in the assembly and assembly drawing.

Parameters (3-D CADD)   Controls, limits, and checks that allow you to easily and effectively make changes and updates. This way, when you describe the size, shape, and location of model geometry using specific parameters, you can easily modify those specifications to explore alternative design options.

Parametric Solid Modeling (3-D CADD)   Involves the basic idea of developing solid models that contain parameters (controls), limits, and checks that allow you to easily and effectively make changes and updates. This way, when you describe the size, shape, and location of model geometry using specific parameters, you can easily modify those specifications to explore alternative design options.

Parcel of Land (Civil)   Area of land often referred to as a plot, lot, or site. This parcel typically has its boundary lines established by a legal description containing very specific information about location, property line lengths, and bearings using one of the methods used to describe land, or a combination of methods.

Part Files (3-D CADD)   Used to create single parts and parts that are used in assemblies.

Parts List   Identifies every part in the assembly.

Pattern Development   Based on laying out geometric forms in true size and shape flat patterns.

Pattern Features (3-D CADD)   Developed using an existing feature or group of features such as extrusions or holes, in order to make a pattern of the feature or features.

Pentagon   A five-sided polygon. Each side of a regular pentagon is equal and its interior angles are 108°.

Perspective Drawing   A form of pictorial drawing in which vanishing points are used to provide the depth and distortion that is seen with the human eye. Perspective drawings can be drawn using one, two, and three vanishing points.

Phase Gate Design Process   The idea involves an imaginary gate, at the end of each phase, that must be opened before you can continue. You must complete certain tasks in order to open each gate.

Philadelphia Rod   A long pole with numbers and graduated sections that is used in surveying to determine elevation or distance.

Photodrafting   A combination of a photograph(s) with line work and lettering on a drawing.

Pictorial   Related to pictures, and pictures are representations by painting, drawing, or photograph. The intent of pictorial drawings is for the image to look realistic.

Pictorial Drawing   A form of drawing that shows an object's depth. Three sides of the object can be seen in one view.

Pie Charts   Used for presentation purposes where portions of a circle represent quantity.

Piercing Point   A point where a particular line intersects a plane.

Pinion Gear   When two gears are mating, the pinion gear is the smaller, usually the driving gear.

Pitch   A distance of uniform measure determined at a point on one unit to the same corresponding point on the next unit; used in threads, springs, and other machine parts.

Pitch (Worm)   The distance from one tooth to the corresponding point on the next tooth measured parallel to the worm axis; equal to the circular pitch on the worm gear.

Pitch Angle (Bevel Gear)   The angle between an element of a pitch cone and its axis.

Pitch Diameter (Bevel Gear)   The diameter of the base of the pitch cone.

Pitch Diameter (Spur Gear)   The diameter of an imaginary pitch circle on which a gear tooth is designed. Pitch circles of two spur gears are tangent.

Placed Features (3-D CADD)   Features placed on to an existing feature without using a sketch. Placed features are also known by a few other names including built-in, added, or automated features. Shells, fillets, chamfers, threads, and face drafts are considered placed features.

Plain Bearing   Based on a sliding action between the mating parts; also called sleeve or journal bearings.

Plane   A surface that is not curved or warped. It is a surface in which any two points may be connected by a straight line, and the straight line will always lie completely within the surface.

Plat (Civil)   A tract of land showing building lots.

Plate Cam   A cam in the shape of a plate or disk. The motion of the follower is in a plane pendicular to the axis of the camshaft.

Plotting (CADD)   Creating a hard copy of a drawing using a plotter.

Plotting (Civil)   Refers to the layout or drafting of a traverse by establishing the endpoints of bearings enclosing the traverse. Plotting used in this sense is not to be confused with plotting a drawing, such as making a hard copy when using a CADD system.

POB   Point of beginning. Any point that has been determined to be the beginning of a survey.

Polar Charts   Designed by establishing polar coordinate scales where points are determined by an angle and distance from a center or pole.

Polar Coordinates (CADD)   A point located using the distance from a fixed point at a given angle and can be an absolute polar coordinate measured from the origin, or relative polar coordinates measured from the previous point. When preceded by the @ symbol, a polar coordinate is measured from the previous point. If the @ symbol is not included, the coordinate is located relative to the origin. When entering a polar coordinate in AutoCAD, the < symbol separates the two values and establishes that a polar or angular increment follows.

Polyester Drafting Film   A high-quality drafting material with excellent reproduction, durability, and dimensional stability; also known by the trade name Mylar®.

Polygons   Enclosed figures such as triangles, squares, rectangles, parallelograms, and hexagons.

Post-Mortem Review   A process used to assess the effectiveness of the project, project management, product development process, and the project team.

Post-Production Review   Conducted no more than twelve months after the start of production. This is when the company looks at the cost and profit margin and determines if the product does what was expected in the marketplace.

Product Bill of Materials   Created to determine the cost of the complete product. The product bill of materials is used to get quotes through buyers or directly from suppliers and to estimate the cost of manufactured parts within the company or from subcontractors who make the parts for the product.

Pressure Angle   The direction of pressure between contacting gear teeth. It determines the size of the base circle and the shape of the involute spur gear tooth, commonly 20°.

Prime Circle (CAM)   A circle with a radius equal to the sum of the base circle radius and the roller follower radius.

Prime Meridian   The line of longitude that is given the 0 degree designation and from which all other lines of longitude are measured.

Printed Circuits (PC)   Electronic circuits printed on a board that form the interconnection between electronic devices.

Printer   A device that receives data from the computer and converts it into alphanumeric or graphic printed images.

Profile   Shows what a section of land (or utility pipe, etc.) looks like in elevation.

Profile Line   A profile line is one that is parallel to the profile projection plane; its projection appears in true length in the profile view.

Project Portfolio   A compilation of all potential new projects that are under consideration for implementation.

Project Portfolio Management   The methodology management uses to select projects for execution and the order in which they are completed.

Projection Line   A projection line is a straight line at 90° to the fold line, which connects the projection of a point in a view to the projection of the same point in the adjacent view.

Projection Plane   A projection plane is an imaginary surface on which the view of the object is projected and drawn. This surface is imagined to exist between the object and the observer.

Public Land System   The land that was divided by a rectangular system of surveys, in which the main subdivisions are townships and sections.

Quality Function Deployment (QFD)   Also known as voice of the customer (VOC), QFD is a product and service planning process that starts and ends with input from customers. Customer feedback is the driving force behind the development requirements for a new or revised product or service.

Quench   To cool suddenly by plunging into water, oil, or other liquid.

Rack   Basically a straight bar with teeth on it. Theoretically, it is a spur gear with an infinite pitch diameter.

Radial Motion   Exists when the path of the motion forms a circle, the diameter of which is perpendicular to the center of the shaft; also known as rotational motion.

Radius   Identified using an uppercase R, is the distance from the center of the circle to the circumference. A circle's radius is always half of the diameter.

Rapid Prototyping (RP)   Turning 3-D models in the computer into actual, physical 3-D objects. Machines are available that build prototypes from various materials such as paper and liquid polymer. Increasingly, such machines are used to build actual parts. *See* stereolithography.

Ratio Scales   Special scales that are referred to as logarithmic and semi-logarithmic.

Ream   To enlarge a hole slightly with a machine tool called a reamer to produce greater accuracy.

Rectangular Pattern (CADD)   Copies and organizes a feature into a designated number of rows and columns and places the features a specified distance apart from each other.

Rectangular System of Surveys   A system of describing the land that is part of a public land survey. Each one of these public land surveys uses townships, sections, quarter-sections, etc. to describe a particular piece of land.

Rectilinear Charts Charts that are set up on a horizontal and vertical grid where the vertical axis identifies the quantities or values related to the horizontal values; also known as line charts.

Reference Features (3-D CADD) Also commonly known as work features or reference geometry, are used to direct the location and arrangement of other features. Reference features are used primarily for construction purposes in order to develop additional sketches and features in areas where no geometry is available.

Reference Planes (3-D CADD) Used to create construction planes, reference planes are used to create construction lines, or axes, and reference points are used to create construction points anywhere on a feature or in three-dimensional space.

Relief A slight groove between perpendicular surfaces to provide clearance between the surfaces for machining.

Removed Section A sectional view taken from the location of the section cutting plane and placed in any convenient location of the drawing, generally labeled in relation to the cutting plane.

Resistors Components that contain resistance to the flow of electric current.

Revolution An alternate method for solving descriptive geometry problems in which the observer remains stationary and the object is rotated to obtain various views.

Revolved Section A sectional view established by revolving 90 degrees, in place, into a plane perpendicular to the line of sight, generally used to show the cross section of a part or feature that has consistent shape throughout the length.

Rib A thin metal section between parts to reinforce while reducing weight in a part.

Right Angle An angle of 90 degrees.

Right-of-Way (Civil) Refers to an easement or deeded strip of land for construction and maintenance of features such as roads, utilities, or other uses. The right-of-way includes the surface, underground, and overhead space at the designated strip of land.

Rocker A link that moves back and forth through an angle; also known as a lever.

Roller Bearings A bearing composed of two grooved rings and a set of rollers. The rollers are the friction-reducing element.

Root Diameter (Spur Gear) The diameter of a circle coinciding with the bottom of the tooth spaces.

Round Two or more exterior surfaces rounded at their intersection.

RPM Revolutions per minute.

Rubber An elastic hydrocarbon polymer which occurs as a milky emulsion, known as latex, in the sap of a number of plants but which can also be produced synthetically.

Runouts Characteristics of intersecting features, determined by locating the line of intersection between the mating parts.

Schematic Diagrams Drawn as a series of lines and symbols that represent the electrical current path and the components of the circuit. Provides the basic circuit connection information for electronic products.

Semiconductors Devices that provide a degree of resistance in an electronic circuit; types include diodes and transistors.

Sketching Freehand drawing without the aid of drafting equipment.

Sketches (3-D CADD) The first step in the creation of a model. Sketches provide the profile, or foundation and guide, used for developing models containing sketched features.

Sketched Features (3-D CADD) Products of sketches. Every parametric solid modeling system contains several sketch feature tools. Sketched features are typically created as extrusions, revolutions, sweeps, lofts, coils, and other shapes depending on the design, sketch geometry, and available CADD software tools.

Skew Lines Lines that are neither parallel nor intersecting.

Slider A link that moves back and forth in a straight line.

Slope Angle The angle in degrees that the line makes with the horizontal plane.

Socket Weld A form of pipe connection in which a plain-end pipe is slipped into a larger opening or "socket" of a fitting. One exterior weld is required; thus, no weld material protrudes into the pipe.

Software Computer programs stored on magnetic tape or disk that enable a computer to perform specific functions to accomplish a task.

Solder An alloy of tin and lead.

Solder Mask A polymer coating to prevent the bridging of solder between pads or conductor traces on a printed circuit board.

Solid Imaging *See* stereolithography.

Solid Models (3-D CADD) Contain information about object edges, the intersection of those edges and surfaces, and also include data about object volume and mass.

Solid Modeling A design and engineering process in which a 3-D model of the actual part is created on the screen as a solid part showing no hidden features.

Solid Primitives (3-D CADD) Objects such as boxes, cones, spheres, and cylinders that are combined, subtracted, and edited to produce a final model.

Space Diagram A drawing of a vector system showing the correct direction and sense but not drawn to scale.

Specifications Any written information or instructions included on the drawing or with a set of drawings, giving all necessary information not shown in the drawing field. Specifications include such items as quality requirements; manufacturer name, type, part number, and details for purchase parts or material applications and finishes; and instructions defining the manner in which work is to be done. When possible, specifications are commonly included with the general notes.

Spline One of a series of keyways cut around a shaft and mating hole; generally used to transfer power from a shaft to a hub while allowing a sliding action between the parts.

Spur Gear The simplest, most common type of gear used for transmitting motion between parallel shafts. Its teeth are straight and parallel to the shaft axis.

Stadia Technique of measuring distance using a Philadelphia rod and level.

Station Point In surveying, a fixed point from which measurements are made.

Stereolithography A layer manufacturing technology in which the layers are formed by using a laser to cure the surface of a bath of photosensitive polymer resin in the desired shape. A CAD model is used to guide a laser through a polymer solution. The laser hardens the solution within the boundary of the part, one thin layer at a time, to produce a real, three-dimensional prototype, model, or part.

Stretch-Out Line   Typically, the beginning line upon which measurements are made and the pattern development is established.

Subassembly   An assembly that is added to another assembly.

Subassembly (3-D CADD)   Groups of components that are often standard product items and can be used several times in various assemblies.

Subdivision (Civil)   Tract of land divided into residential lots.

Surface Charts   Designed to show values represented by the extent of a shaded area; also known as area charts.

Surface Finish   Refers to the roughness, waviness, lay, and flaws of a machine surface.

Surface Models (3-D CADD)   Contain information about object edges and the intersection of those edges but go a step further to include information about object surfaces.

Surface Mount Technology (SMT)   The traditional component lead through is replaced with a solder paste to hold the electronic components in place on the surface of the printed circuit board and take up to less than one-third of the space of conventional PC boards.

Tangent   A straight or curved line that intersects a circle or arc at one point only; is always 90° relative to the center.

Taper   A conical shape on a shaft or hole, or the slope of a plane surface.

Tempering   A process of reheating normalized or hardened steel to a specified temperature, followed by cooling at a predetermined rate to achieve certain hardening characteristics.

Tensile Forces   Forces that pull away.

Tensile Strength   Ability to be stretched.

Text (CADD)   Lettering is called text, or attributes, when using CADD. Adding text to a CADD drawing is one of the most efficient and effective tasks associated with computer-aided drafting, because the usually tedious task of freehand lettering is not involved. The process of placing text varies depending on the CADD system, the tools and commands used to generate the text, and the purpose of the text. However, typically adding text is just a matter of defining and selecting the desired text style, picking the text location, and typing characters on the keyboard.

Thermoplastic   Plastic material may be heated and formed by pressure. Upon reheating, the shape can be changed.

Thermoset   Plastics are formed into permanent shape by heat and pressure and may not be altered after curing.

Thermostat   An automatic mechanism for controlling the amount of heating or cooling given by a central or zoned heating or cooling system.

Threaded Rod   The external thread can be machined on a fastener, such as the hexagon head bolt, or on a shaft without a head. This is referred to as a threaded rod.

3-D Printing   *See* stereolithography.

Tilt-Up Construction   Method using formed wall panels that are lifted or tilted into place.

Tolerance   The total permissible variation in a size or location dimension.

Tool Palettes (CADD)   Similar to floating toolbars and contain often used block symbols and patterns that can be used in a drawing. Toolbars contain various buttons that activate commands. Tool palettes are customizable with blocks and patterns to suit your own specific needs.

Top-Down Assembly   Technique involves building individual components in an assembly file without inserting existing components. This option is commonly referred to as creating components in-place.

Torsion   Refers to the twisting forces placed on features.

Trackball   An input device consisting of a smooth ball mounted in a small box. A portion of the ball protrudes above the tope of the box and is rotated with the hand to move the cursor on the screen.

Transistors   Semiconductor devices in that they are conductors of electricity with resistance to electron flow applied and are used to transfer or amplify an electronic signal.

Transit Line (Civil)   Any line of a survey established by a transit or other surveying instruments.

Transition Piece   A duct component that provides a change from square or rectangular to round; also known as a square to round.

Translational Motion   Linear motion.

Traverse   In surveying, a series of lines with directions and lengths that are connected at station points.

Triangulation   A technique used to lay out the true size and shape of a triangle with the true lengths of the sides; used in pattern development on objects such as the transition piece.

Trilinear Chart   Designed in the shape of an equilateral triangle; used to show the interrelationship between three variables on a three-dimensional diagram.

Trimetric Drawing   A type of pictorial drawing in which all three of the principal axes do not make equal angles with the plane of projection.

True Azimuth   The azimuth measured from the actual North or South Pole.

True Length or True Size and Shape   When the line of sight is perpendicular to a line, surface, or feature.

True Position   The theoretically exact location of a feature established by basic dimensions.

Turning Point   In surveying, this is each point at which the Philadelphia rod is placed and measured.

Ultrasonic Machining   A process where a high-frequency mechanical vibration is maintained in a tool designed to a desired shape; also known as impact grinding.

Under-Constrained Sketch or Model (3-D CADD)   Has elements that are unclear, can be changed or moved, and remain undefined and unstable.

Undercut   Sometimes a neck is machined where the external thread meets the fastener head or the larger diameter shaft. This neck is referred to as an undercut. The undercut is used to eliminate the possibility of a machine radius at the head and to allow for a tight fit between the thread and the head when assembled.

Uniform Resource Locator (URL)   An Internet address (for example, *http://www.delmarlearning.com*), usually consisting of the access protocol (*http*), the domain name (*www.delmarlearning.com*), and optionally the path to a file or resource found on the server.

Unilateral Tolerance   A tolerance in which variation is permitted in only one direction from the specified dimension.

UNLESS OTHERWISE SPECIFIED   Often placed with notes to remind the reader that the given specification generally applies but can be modified by other information provided on the drawing or in other documents.

Upset A forging metal used to form a head or enlarged end on a shaft by pressure or hammering between dies.

USB Flash Drives Compact and easy-to-use devices that are similar in use to a computer hard drive. USB flash drives slip into your pocket or on a keychain for ultimate portable storage, providing complete freedom and mobility. USB flash drives currently can hold up to two gigabytes of data, or about three times the content of a standard compact disc. If you share a computer, USB flash drives are a great way to store personal information and can be used instead of the computer's hard drive.

Valve Any mechanism, such as a gate, ball, flapper, or diaphragm, used to regulate the flow of fluids through a pipe.

Vector Analysis A branch of mathematics that includes the manipulation of vectors.

Vector Diagram A drawing of the vector system in which the vectors are drawn with the correct magnitude and sense, and to scale.

Vector Quantity A quantity that requires both magnitude and direction for its complete description.

Vellum A drafting paper with translucent properties.

View Tools View tools allow you to look at a drawing in a variety of ways, but do not actually modify the size, location, or shape of drawing objects. Do not confuse zoom or any other view tool with the commands used to edit a drawing.

Viewing-Plane Line Represents the location of where a view is established.

Visualization The process of recreating a three-dimensional image of an object in a person's mind.

Web *See* rib.

Weld Symbol Indicates the type of weld and is part of the welding symbol.

Whole Depth (Spur Gear) The full height of the tooth. It is equal to the sum of the addendum and the dedendum.

Wire Form A three-dimensional form in which all edges and features show as lines thus appearing to be constructed of wire.

Wireframe Models The most basic 3-D CADD models that contain only information about object edges and the intersection of edges.

Wireless Diagram Similar to highway diagrams except that interconnecting lines are omitted. The interconnection of terminals is provided by coding.

Wiring Diagram A type of schematic that shows all of the interconnections of the system components, also referred to as a point-to-point interconnecting wiring diagram.

Working Depth (Spur Gear) The distance that a tooth occupies in the mating space. It is equal to two times the addendum.

Working Drawings A complete set of working drawings containing the detail drawings, subassemblies, assembly, parts list or bill of materials, and written specifications related to a product.

Worm Gears Used to transmit power between nonintersecting shafts. The worm is like a screw and has teeth similar to the teeth on a rack. The teeth on the worm gear are similar to the spur gear teeth but they are curved to form the teeth on the worm.

Zero Declination Places on the earth where the compass points exactly toward the true North or South Pole.

Zip Drive A removable storage device that securely stores all your computer data magnetically. Zip drives and Zip disks are durable, portable, easy-to-use, and extremely efficient. One 250MB Zip disk can store the same amount of data as 173 floppy disks. Zip disks are available in 100MB, 250MB, and 750MB formats.

Zoning A system of numbers along the top and bottom and letters along the left and right margins of a drawing used for ease of reading and locating items.

INDEX

Page numbers in italics indicate figures.